Marine Species
and
Their Distribution in China's Seas

Marine Species and Their Distribution in China's Seas

Edited
by
Huang Zongguo

Translated
by
Junda Lin

Sponsored
by
Smithsonian Institution

Krieger Publishing Company
Malabar, Florida
2001

English Edition 2001

Originally published in Chinese under the title:
中 国 海 洋 生 物 种 类 与 分 布
Copyright © China Ocean Press, Beijing, 1994
All rights reserved.

Printed and Published by
Krieger Publishing Company
Krieger Drive
Malabar, Florida 32950

Copyright © 2001 by Smithsonian Institution

Originally published in China under the title *Marine Species and Their Distributions in China's Seas* Copyright © China Ocean Press, Beijing 1994 and Copyright © Smithsonian Institution, All Rights Reserved. No copyright protection may be available in the United States for any portion of the Work prepared by an employee of the United States as part of his or her official duties. Publisher acknowledges that parts of the Work are a work of the United States as defined by the Copyright Act of 1976, and therefore not subject to copyright protection.

> **FROM A DECLARATION OF PRINCIPLES JOINTLY ADOPTED BY A COMMITTEE OF THE AMERICAN BAR ASSOCIATION AND A COMMITTEE OF PUBLISHERS:**
>
> This publication is designed to provide accurate and authoritative information in regard to the subject matter covered. It is sold with the understanding that the publisher is not engaged in rendering legal, accounting, or other professional service. If legal advice or other expert assistance is required, the services of a competent professional person should be sought.

Library of Congress Cataloging-In Publication Data

Marine species and their distribution in China's Seas / edited by Zongguo Huang;
translated by Junda Lin. -- English ed.
   p. cm.
  Includes bibliographical references and index (p.).
  ISBN 1-57524-103-X (alk. paper)
  1. Marine organisms--China. I. Huang, Zongguo.

QH181 .M38 2001
578.77'457--dc21   2001020379

10 9 8 7 6 5 4 3 2

Contents

Translator's Preface 1
Introduction to the English Translation 2
Foreword to the Original Chinese Edition 15
Introduction to the Original Chinese Edition 16
Contributors 17
Editorial Notes, Revised from the Original Chinese Edition 18

MONERA

Bacteria 21
Actinobacteria 23
Cyanobacteria [Cyanophyta] 24
Chloroxybacteria 27

PROTISTA

Bacillariophyta 28
Chrysophyta 49
Cryptophyta 50
Xanthophyta 50
Pyrrophyta 50
Ciliophora 56
Sarcomastigophora 61
    Radiolaria 61
    Foraminifera 69

FUNGI

Yeast 129
Other Fungi 129
    North of Xiamen 129
    South of Xiamen 130
Mycophycophyta [Lichens] 132

PLANTAE

    ALGAE
Rhodophyta 133
Phaeophyta 140
Chlorophyta 144
    TRACHEOPHYTA
Pteridophyta 150
Gymnospermae 150

Angiospermae 150
    Dicotyledoneae...150
    Monocotyledoneae 159

ANIMALIA

Porifera [Spongia]..164
Coelenterata [Cnidaria] 166
    Hydrozoa..166
        Hydroidomedusae..166
    Scyphomedusae 175
    Anthozoa..179
        Zoanthidea...179
        Actiniaria 179
        Ceriantharia 180
        Scleractinia 181
        Antipatharia..185
        Helioporacea...186
        Stolonifera 186
        Alcyonacea 187
        Pennatulacea 190
        Telestacea 190
Ctenophora 190
Platyhelminthes...191
    Turbellaria 191
    Trematoda..192
    Cestoidea 206
Nemertinea..207
Kinorhyncha..208
Nematoda..209
    Parasitic 209
    Non-parasitic 210
Acanthocephala 213
Rotifera 214
Priapulida 214
Annelida..214
    Polychaeta 214
    Oligochaeta 233
    Hirudinea...235
Sipuncula...235
Echiura 237
Mollusca..237
    Polyplacophora [Amphineura]...237
    Bivalvia [Lamellibranchia] 239
    Scaphopoda 261
    Gastropoda 261
        Prosobranchia...261
        Opisthobranchia 284
        Pulmonata 294
    Cephalopoda..294
Arthropoda 297
    Merostomata..297
    Pycnogonida 297
    Arachnoidea 297

- Insecta 299
- Crustacea 300
  - Branchiopoda 300
    - Anostraca 300
    - Cladocera 300
  - Ostracoda 301
  - Copepoda 306
  - Cirripedia 318
  - Malacostraca 323
    - Mysidacea 323
    - Cumacea 324
    - Tanaidacea 325
    - Isopoda 325
    - Amphipoda 328
      - Gammaridea 328
      - Hyperiidea 330
      - Caprellidea 334
      - Ingolfiellidea 334
    - Euphausiacea 334
    - Decapoda 335
      - Dendrobranchiata 335
      - Pleocyamata 338
        - Stenopodidea 338
        - Caridea 338
        - Astacidea 345
        - Thalassinidea 345
        - Palinuridea 346
        - Anomura 348
        - Brachyura 353
    - Stomatopoda 368
- Bryozoa [Ectoprocta] 369
- Entoprocta 383
- Brachiopoda 384
- Phoronida 384
- Chaetognatha 385
- Echinodermata 386
  - Crinoidea 386
  - Holothuroidea 387
  - Asteroidea 389
  - Echinoidea 392
  - Ophiuroidea 395
- Hemichordata [Stomochordata, Branchiotrema] 399
- Urochordata 400
  - Appendiculata [Copelata] 400
  - Thaliacea 400
  - Ascidiacea 402
- Chordata 403
  - Cephalochordata 403
  - Vertebrata 404
    - Cyclostomata [Marsipobranchii] 404
    - Chondrichthyes 404
    - Osteichthyes 409
    - Amphibia 463
    - Reptilia 463

        Testudiformes 463
        Serpentiformes 464
    Aves 464
    Mammalia 468
        Cetacea 468
        Pinnipedia 469
        Sirenia 469
Index 471

Translator's Preface

When I first heard of the book with a comprehensive list of marine species and their distributions in China from Dr. Joe S. Y. Lee in Hong Kong in the summer of 1994, shortly after the book's publication, I determined that I would translate it into English and have it published. With the help of Dr. Anson "Tuck" Hines and Dr. Greg Ruiz, we finally obtained funding—from the Smithsonian Institution's Antherton Seidell Fund—for the translation and publication.

Revisions were made throughout the book. Hirudinea was added by Prof. Huang Zongguo. Bryozoa and Entoprocta (by Prof. Liu Xixing), and Echinoderm (by Prof. Liao Yulin) listings were completely revised. Information presented on most of the other taxa was also revised to various extents by Professors Cai Bingji, Cai Ruxing, Cai Yaoguo, Chen Changsheng, Chen Huilian, Chen Ruixiang, Cheng Zhaodi, Dai Aiyun, Dai Yanyu, Fu Zhaoxian, Huang Zongguo, Li Chupu, Li Fenglu, Li Rongguan, Lian Guangshan, Lin Jinghong, Lin Jinmei, Lin Mao, Lin Na, Lin Shuangdan, Liu Ruiyu, Pei Zunan, Ren Xianqiu, Shen Jiwei, Song Weibo, Sun Ruiping, Sun Shichun, Tan Zhiyuan, Wang Zhenrui, Xu Fengsan, Xu Zhenzu, Yang Dejian, Yang Qingliang, Yang Siliang, Zhang Raoting, Zhang Zhinan, Zheng Chengxing, Zheng Shouyi, Zhong Zhenru, Zhou Jinming, and Zou Renlin. For the parasitic species, the common (English) and scientific names of the hosts were added, largely based on J. S. Nelson's book, *Fishes of the World*, 3rd Edition, 1994, John Wiley & Sons, Inc. (600 pp.). Most references cited come from original Chinese articles that contained English titles and abstracts. The original translation of the titles was used. For names of locations and authors, the pinyin system was used.

To help readers who are not familiar with China's geography, the distribution range for each species was generally arranged from north to south and from east to west, with the province or major water body provided in parentheses for each location. Several maps were produced for this translated edition to further aid in finding localities for the species.

This project would not have been accomplished without the assistance from many people. Prof. Huang Zongguo spent numerous hours correcting and revising the book, promptly answering my many questions, providing references, and coordinating the revisions of and corrections to various taxa. Mr. Liu Wenhua supplied most of the translated reference citations. Prof. Lin Rongcheng and Prof. Yao Jianhua also assisted in providing the references and facilitating communications. Dr. Tuck Hines administered the Seidell grant. The late Dr. Kerry Clark reviewed various taxa. Mr. Jian Yan and Dr. Richard Turner provided much needed assistance in translating the Preface and Introduction. Prof. Zheng Shouyi translated the geological times for Foraminifera. My father, Prof. Lin Heng, translated the locations and author names using the pinyin system. Dr. Yi-Ching "Eric" Li scanned in the Index, thereby saving many hours of typing. Dr. Weihe Guan assisted in finding locations in the maps. Dr. Jon Shenker provided the common names of fish hosts not found in J. S. Nelson's book. Dr. Xu Runlin provided references on the major marine water bodies of China. Florida Institute of Technology provided me a sabbatical leave in the fall semester of 1998 that allowed me to complete the translation. Mr. Donald Krieger and staff at Krieger Publishing Company have been very patient and helpful.

Last, but certainly not least, I thank my family members, Fanghua, Kurt, and Kyle, for their endurance, support, and love. They are surely as glad as I am that the project has finally been completed.

Junda Lin, PhD
Florida Institute of Technology
Melbourne, Florida USA

Introduction to the English Translation

At a time of unprecedented assault on marine biodiversity, we lack good species inventories for many parts of the ocean. This book provides an amazing compilation of more than 20,000 species along 5000 km and 20 degrees of latitude (20°N to 40°N) of the China coast, plus incomplete surveys on islands farther south to nearly 10°N in the South China Sea. First published in Chinese in 1994, the goal for our translation is to make this important inventory more accessible to scientists and marine conservationists in the western world. We also hope that the inventory will serve as a source of data on biodiversity and biogeography of marine species, and that greater accessibility of this inventory will stimulate both further analysis of the major patterns and additional research on marine ecosystems of China. This inventory of marine species reflects the rich biodiversity of the transition zone between the tropical Indo-West Pacific and the temperate waters of the Western Pacific. It also must reflect the historical strengths and biases in the taxonomy and research interests of the region. Clearly, this can stimulate much needed collaboration among eastern and western systematists, biogeographers and ecologists.

Biodiversity can be complicated to measure (Pielou 1975). However, on large geographic scales, biodiversity can be considered as species richness, or simply the number of species, and this inventory includes over 20,000 species. At a higher taxonomic level, the number of families can be a good indicator of biodiversity, and 1938 families are in this book. Considering the major kingdoms, animals contribute most of the diversity in this inventory for both the number of families (73%) and number of species (67%), and protistans making up 15% of the families and 26% of the species, while plants comprise only about 9% of the families and 6% of the species (see Figure 1).

Within kingdoms, patterns of species richness and family diversity in this book are complex (see Figures 2–6). Certain apparent contradictions to generally accepted taxonomic patterns of diversity may indicate emphases in the systematic research of the region. Among animals, for example, arthropods are generally considered to include much greater diversity in both species and families than chordates. However, in this inventory, chordates include more families (330) and species (3263) than arthropods (265 and 2926 respectively) (see Figure 2). This apparent contradiction may reflect greater emphasis on fish diversity as a food resource in China. For an inventory of marine species, the book includes a surprising diversity of vascular plants (Spermatophyta), which comprise 33% of the species and 49% of the families of plants, including many species that are not really marine (see Figure 3). Among protistans, the Sarcomastigophora clearly dominate the diversity, with 69% of the families and 65% of the species (see Figure 4). For Fungi (see Figure 5) and Monera (see Figure 6), as groups with the most problematic taxonomy, the recognized diversity is generally much lower than the other kingdoms.

It is useful to consider the species richness known for China compared to other regions. Such comparisons provide a benchmark for assessing the intensity of study of the region's biodiversity. Nearly 750 species of marine algae are included in this book, which is less than the approximately 850 species of the eastern tropical and subtropical coasts of the Americas and Caribbean region (Taylor 1960; M. Littler and D. Littler, personal communication). Substantially greater numbers of species are listed in this inventory for major groups of marine invertebrates in China than several other regions that are considered to be well studied by western scientists (see Table 1). However, these numbers should be adjusted proportionately among regions for length of shoreline or area. The inventory of species in this book shows reasonable concordance with other inventories for several well-studied groups, such as echinoderms and bivalve molluscs. Bernard et al. (1993) list about 1140 species of bivalve molluscs living in China waters, compared to 804 species in this book. Liao and Clark (1995) include 457 species of echinoderms for southern China, compared to 523 species in this book. The level of knowledge for these well-studied groups has grown from early British surveys in the late 1800s and more recent extensive collections (e.g., Bell 1894 cited in Liao and Clark 1995; Liao 1975, 1978, 1980, 1983 cited in Liao and Clark 1995). These groups also illustrate the value of collaboration between Eastern and Western scientists, especially British and Chinese systematists (e.g., Bernard et al. 1993; Liao and Clark 1995).

China's biodiversity stems from the complex biogeography of the region, which extends from temperate waters to subtropical zones with physical geography that encompasses four main seas: Bo Sea (Bohai), Yellow Sea (Huang Hai), East

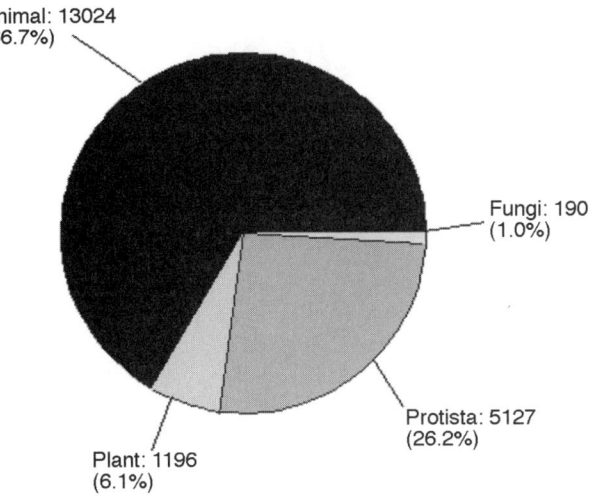
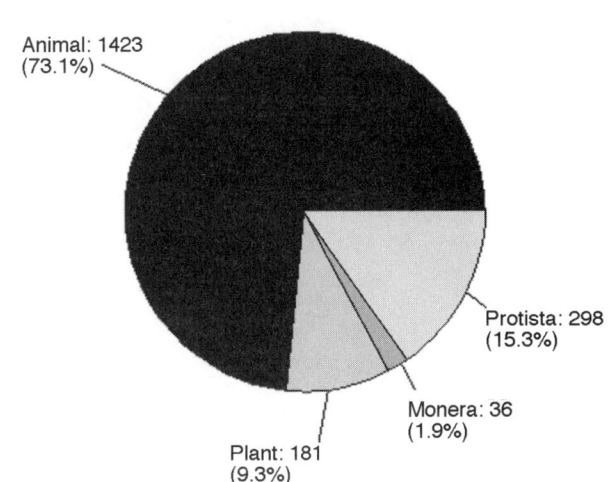

Figure 1. Species and Families per Kingdom.

China Sea (Dong Hai), and South China Sea (Nan Hai) (see Figure 7). An important aspect of this book is the many annotations on species distribution and ranges, and we provide maps that show the location of most of the coastal locations listed in these notes (see Figures 8A and 8B). The four seas differ greatly in size and depth from the small shallow enclosed Bo Sea to the large deep South China Sea (see Table 2) (Chen et al. 1992).

Seasonal oscillation of the monsoon cycle interacts with the major current systems to influence the distribution of organisms along the coast (see Figures 9A and 9B show summer and winter current patterns). Offshore, the Kuroshio Current provides a northward flow of warm tropical water that sends an arm inshore of Taiwan and in the north divides around Japan. Inshore, the coastal current under the effects of the Northern and Southern Monsoons switches direction of flow from the south in summer and the north in winter, interacting with the Taiwan Warm Current and South China Sea Current. The Yellow Sea is separated from the Sea of Japan in the north by the Korean Peninsula, which limits the flow of currents from the south. The Yellow Sea is affected by the Taiwan Warm Current flowing from the south along the coast

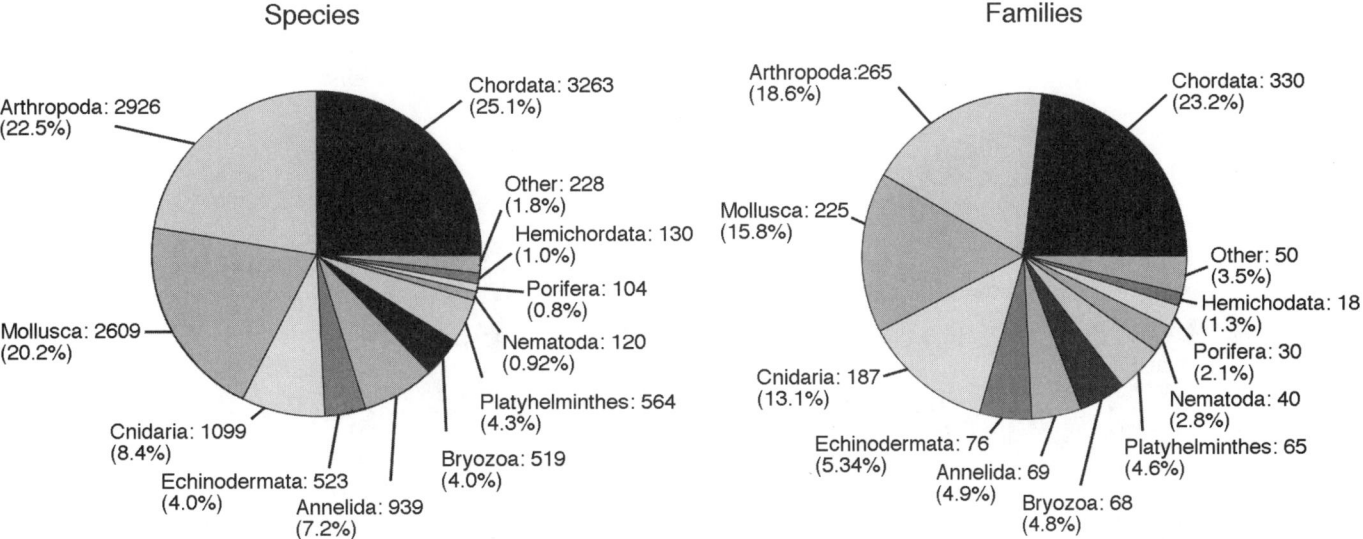

Figure 2. Animal Kingdom.

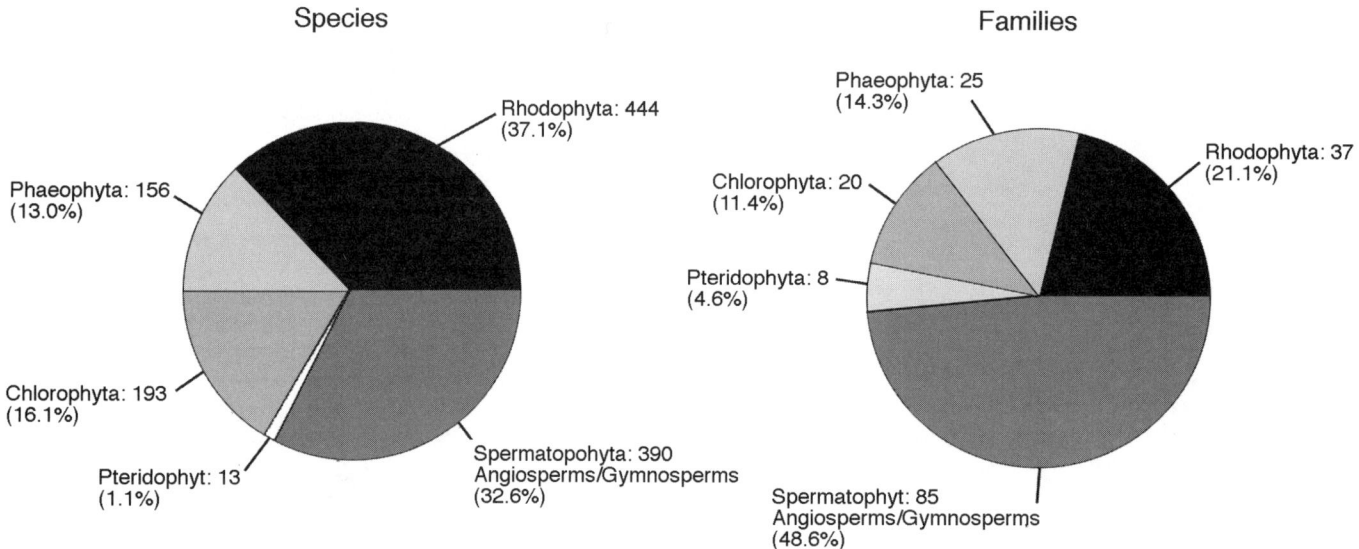

Figure 3. Plant Kingdom.

between the mainland and Taiwan (Formosa), and by the Yellow Sea Warm Current, a branch of the warm Tsushima Current that flows from the Kuroshio Current between Korea and Japan.

The seasonal changes in current patterns regulate variation in temperature and salinity along the continental shelf (see Figures 10A and 10B showing summer and winter sea surface conditions). In the northeast, the shallow embayment forming the Bo Sea has surface temperatures that fluctuate seasonally from 0°C in winter to 28°C in winter. In the south, the Beibuwan (Gulf of Tongking), delimited by Vietnam, Guangdong Province and Hainan Island, is primarily influenced by a gyre of warm tropical waters, and sea surface temperatures fluctuate seasonally much more narrowly from 29°C in summer to 21°C in winter.

Two to three biogeographic provinces are recognized within this complex system of currents and geographic barriers, depending on the researchers and taxonomic group.

Biogeographers have broadly considered two major biogeographic regions for the area: the tropical Indo-Pacific re-

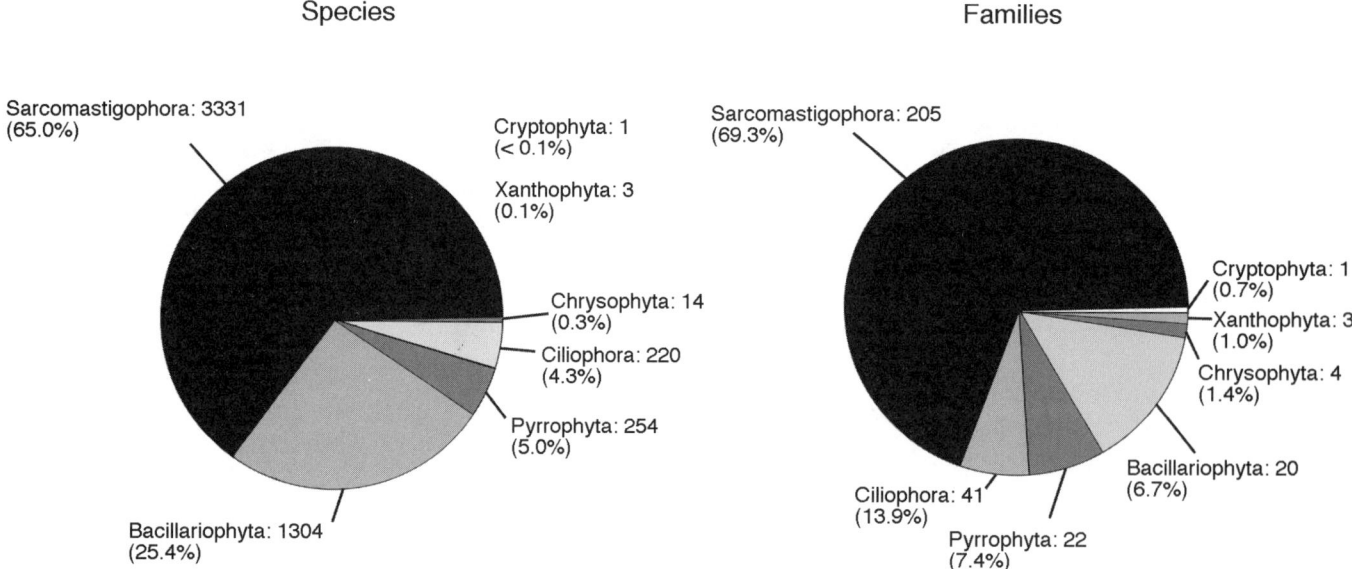

Figure 4. Protista Kingdom.

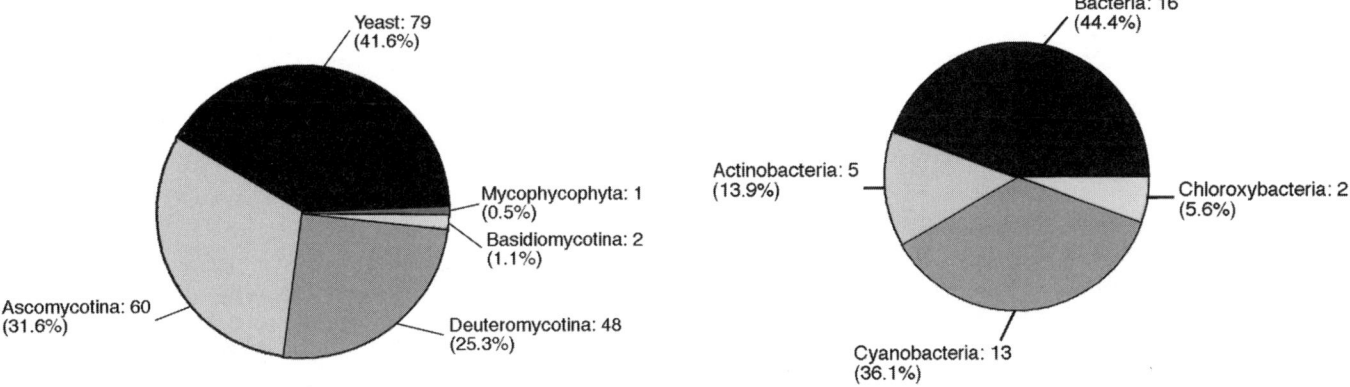

Figure 5. Number of Species per Fungi Phylum.

Figure 6. Number of Families per Monera Phylum.

Table 1. Number of Species by Geographic Region

Taxon	China	Northeast Pacific	Central California	Northwest Europe	Mid-Atlantic
Polychaeta	873	836	275	270	321
Copepoda	513	605	n/a	n/a	n/a
Cumacea	16	44	6	20	23
Amphipoda	272	518	164	210	149
Decapoda	1383	297	92	158	113
Mollusca	2609	1470	399	325	489
Bryozoa	519	276	85	104	125
Echinodermata	523	267	41	75	95
Totals (minus copepods)	6289	3865	1169	1245	1365
Totals	6802	4470	1169	1245	1365

n/a, not available.
References: China, this book; Northeast Pacific, Austin 1985; Central California, Smith and Carlton 1975; Northwest Europe, Hayward and Ryland 1990; Mid-Atlantic, Gosner 1971.

Table 2. Geographic Features of China Seas

	Area (km^2)	Average depth (m)	Maximum depth (m)	Boundaries
Bo Sea (Bohai)	77,000	18	80	Line between the southern tip of eastern Liaoning Peninsula (southwest of Dalian) and Penglai (Shandong)
Yellow Sea (Huang Hai)	380,000	44	140	Line between mouth of Changjian (Yangtze River) and Jizhou Island (Cheju Do), Korea
East China Sea (Dong Hai)	752,000	349	2,719	Line between Dongshan at southern end of Fujian and southern tip of Taiwan, including Taiwan Strait
South China Sea (Han Hai)	3,500,000	1,212	5,559	Includes Beibuwan (Gulf of Tongking), Dongsha (Pratas Islands), Xisha (Paracel Islands), Zhongsha (Macclefield Bank), and Nansha (Spratly Islands)

After Chen et al. 1992.

Figure 7. Overview Map.

Figure 8A. North China Coast.

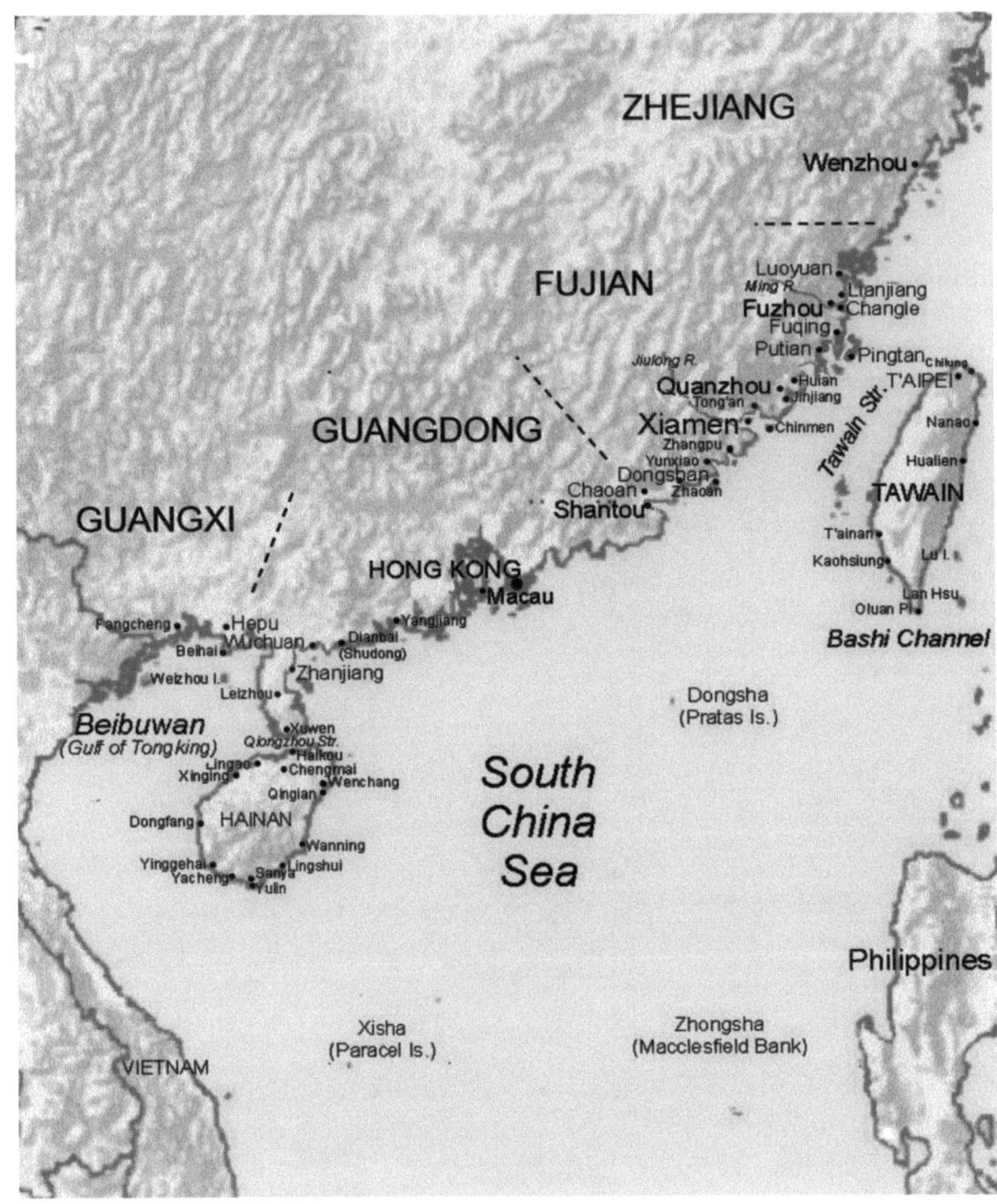

Figure 8B. South China Coast.

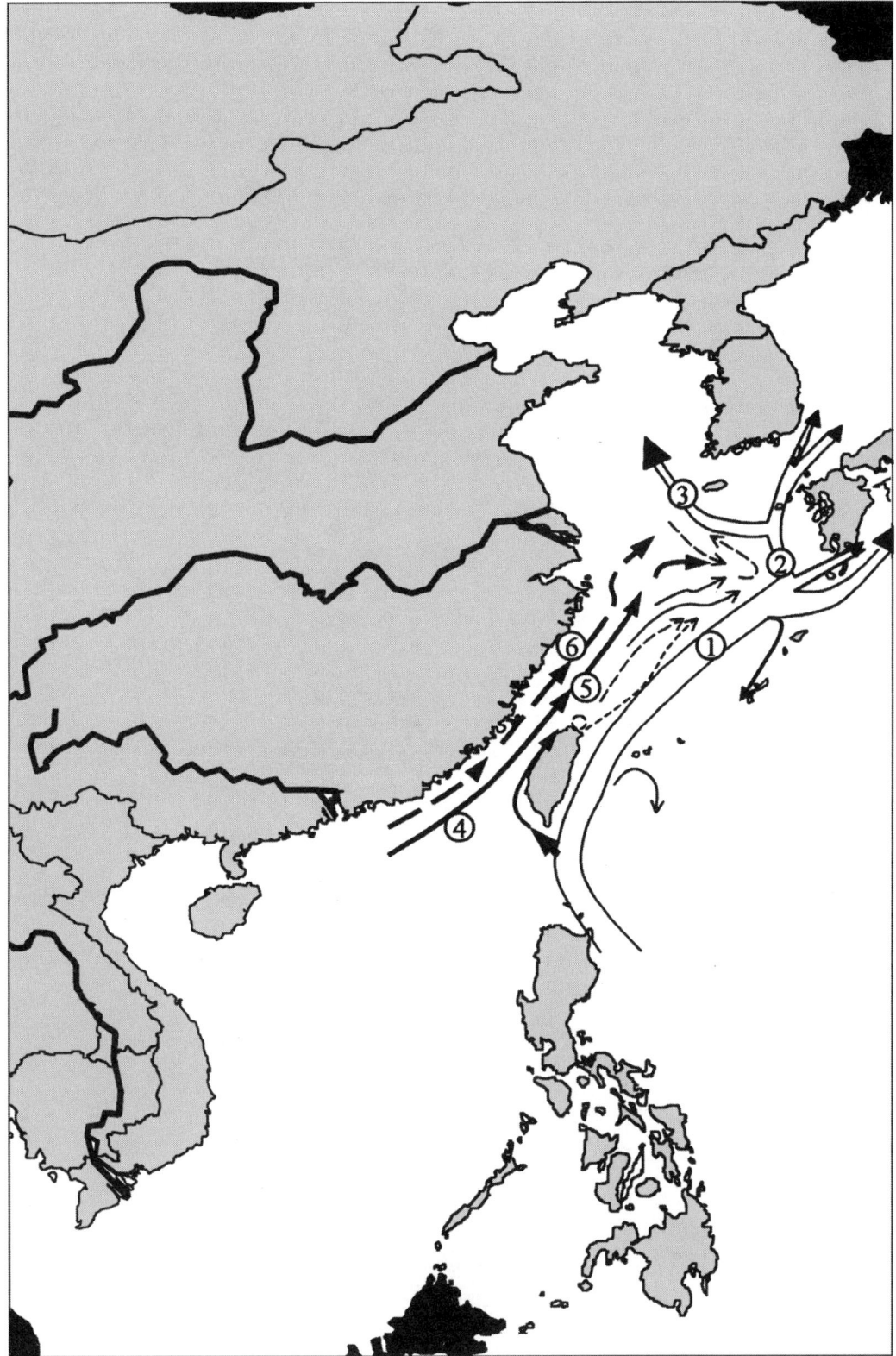

Figure 9A. Current system in the China Seas during summer. 1. Kuroshio Current. 2. Tsushima Current. 3. Yellow Sea (Huang Hai) Warm Current. 4. South China Sea (Nan Hai) Warm Current. 5. Taiwan Warm Current. 6. Coastal Current. (After Liao and Clark 1995.)

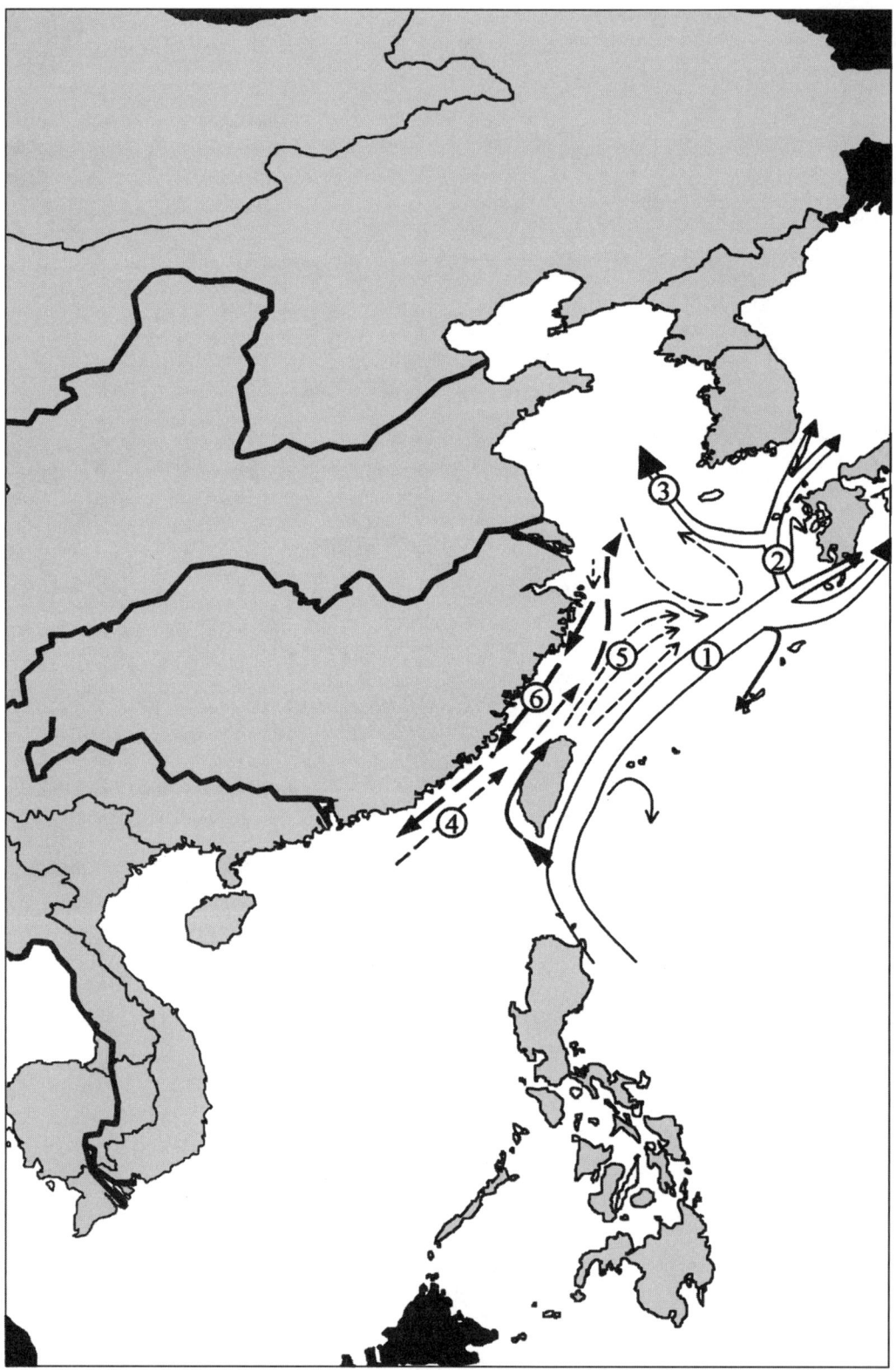

Figure 9B. Current system in the China Seas during winter. 1. Kuroshio Current. 2. Tsushima Current. 3. Yellow Sea (Huang Hai) Warm Current. 4. South China Sea (Nan Hai) Warm Current. 5. Taiwan Warm Current. 6. Coastal Current. (After Liao and Clark 1995.)

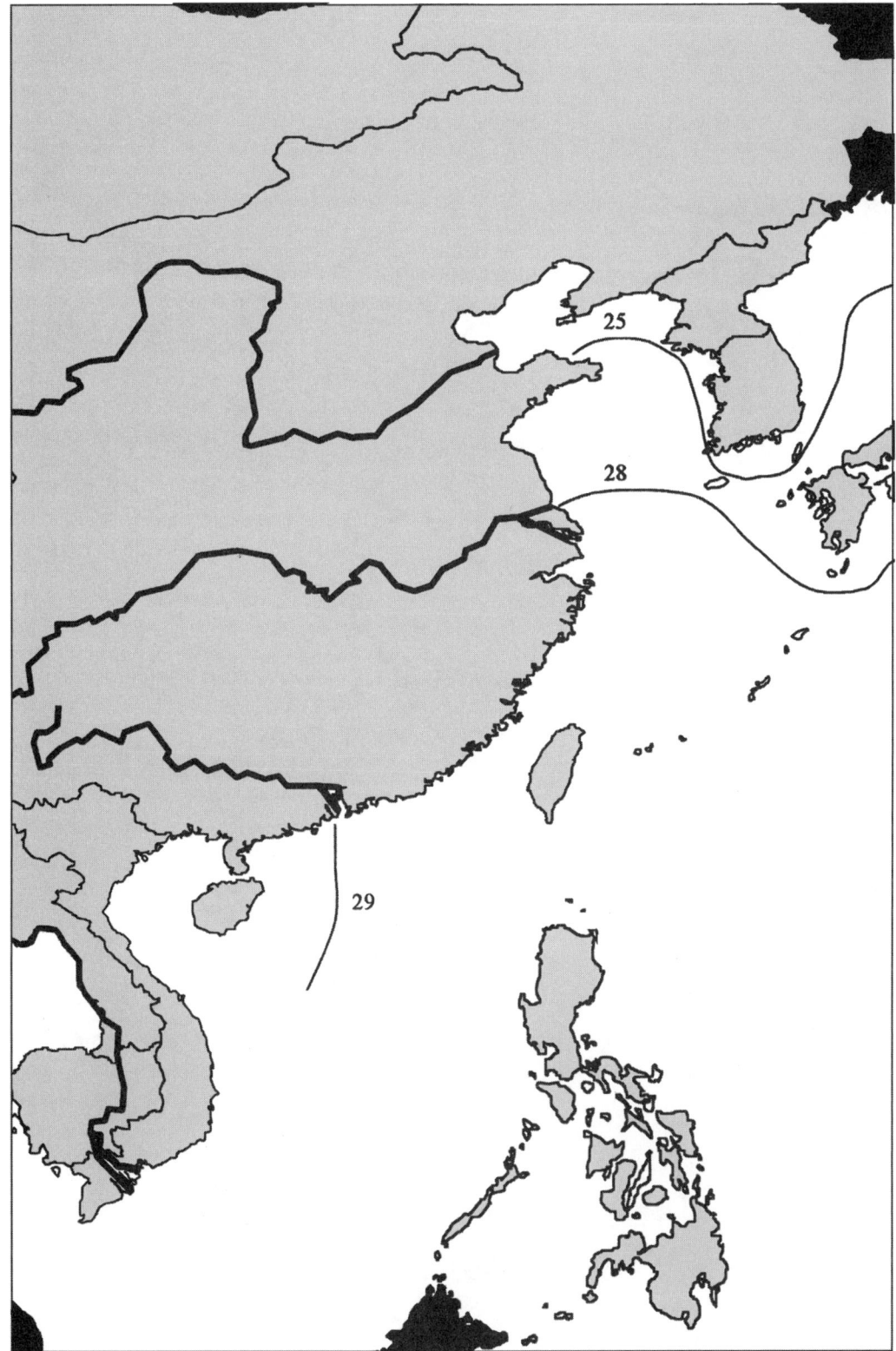

Figure 10A. Summer sea surface temperatures along China coast. Isotherms are shown in degrees Celsius. (After Bernard et al. 1993.)

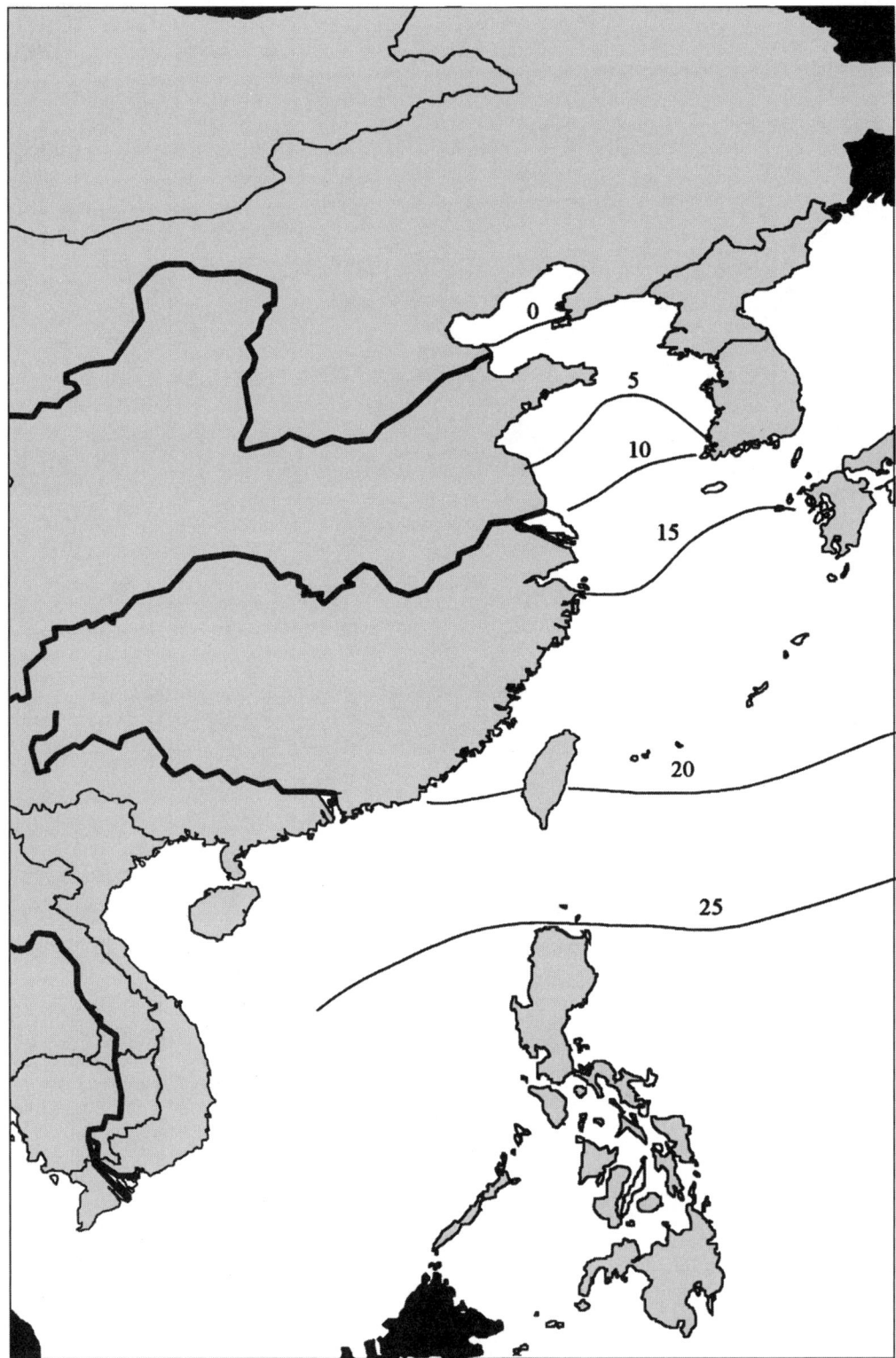

Figure 10B. Winter sea surface temperatures along China coast. Isotherms are shown in degrees Celsius. (After Bernard et al. 1993.)

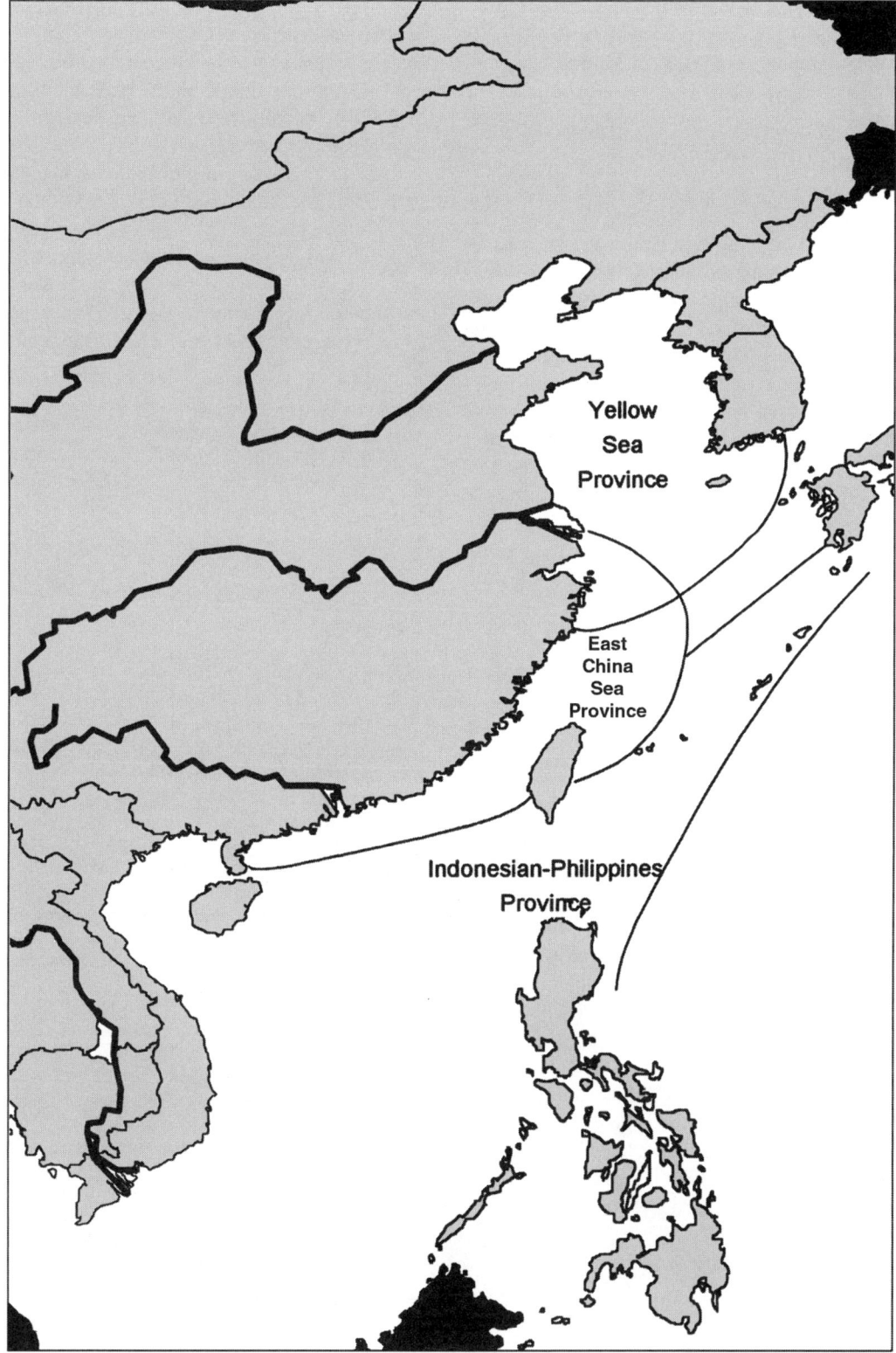

Figure 11. Marine biogeographic provinces off the China coast. (After Bernard et al. 1993.)

gion and a temperate Western Pacific-East Asian region (Ekman 1953; Briggs 1974). However, Bernard et al. (1993) recognize three biogeographic provinces for marine bivalve molluscs of China (see Figure 11):

(1) The northern Yellow Sea Province is contained by the Korean Peninsula and extends southward to the Pearl River to approximately 23°N, and the biota of the Bo Sea (Bohai) is a subset of the Yellow Sea Province;

(2) Along the southern coast, the East China Sea Province overlaps extensively in the north with the Yellow Sea Province along southern Jiangsu and northern Zhejiang provinces, and it extends to the south to Hainan Island and the Leizhou Peninsula (the East China Sea Province includes the northern three-quarters of Taiwan);

(3) The Indonesian-Philippines Province is an extension of the Indo-Pacific biogeographic region and extends from the Indonesian Peninsula and Vietnam to form an offshore arc of the South China Sea including the Paracel Islands and extending from Hainan to the southern end of Taiwan and continuing to the Nansei-shoto (Ryukyus) Islands and Kyushu, Japan.

All are under the strong influence of the northerly flowing Kuroshio Current, with coral and mangrove communities present on many of the islands and shorelines. These three biogeographic provinces are roughly equivalent to the Far East Subregion, the Sino-Japanese Subregion, and the Indo-Malayan Region, respectively (Liao and Clark 1995).

Overlap of species is a good indicator of biogeographic similarity to other regions. For 457 species of echinoderms in southern China, for example, Liao and Clark (1995) showed that, while there are about 20% of species endemic to southern China or the Sino-Japanese Region, all are of southern origin and 60% of the species are shared with the Philippines, 66% with the East Indies, 47% with Northern Australia, 35 % with the Bay of Bengal, 33% with Southern Japan, and 25% with the Ryukyu Islands. Conversely, of 1027 species of Indo-West Pacific echinoderms, 298 (29%) extend to southern China (Liao and Clark 1995). The proportion of species shared with the Indo-West Pacific is greater (76-84% overlap) in echinoderm species found in the China portion of the Kuroshio Current (Liao and Clark 1995), as would be expected for the region defined as the same biogeographic province.

The biodiversity and biogeography embedded in this book represent a summary of current knowledge of marine ecosystems in China. The species lists and notes provide only a starting point for further research, however. The information should be regarded as data deserving collaborative analysis to determine patterns and to test hypotheses for marine ecology and conservation. Chinese marine biologists are anxious to share their knowledge and to develop new partnerships with Western scientists.

Translation and printing of this book were funded by a grant from the Atherton Seidell Fund of the Smithsonian Institution. We thank Don Krieger of Krieger Publishing for his patience and assistance. We also thank Rob Andrews of the Smithsonian Environmental Research Center for technical assistance in producing the figures. This work comprises contribution numbers 511 and 512 of the Smithsonian Marine Station.

Anson H. Hines and Gregory M. Ruiz
Smithsonian Environmental Research Center
Edgewater, Maryland USA

Junda Lin
Florida Institute of Technology
Melbourne, Florida USA

LITERATURE CITED

Austin, W.C. 1985. *An annotated checklist of marine invertebrates of the cold temperate Northeast Pacific. Cowichan Bay, British Columbia.* Khoyatan Marine Laboratory, Vols. 1–3, 682 p.

Bernard, F.R., Y.-Y. Cai, and B. Morton. 1993. *Catalogue of the living marine bivalve molluscs of China.* Hong Kong University Press, 146 p.

Briggs, J.C. 1974. *Marine zoogeography.* McGraw-Hill, New York, 475 p.

Chen, Z.-Y., F.-Q. Wang, X.-H. Fang, D.-X. Zhang, T.-H. Zhou, E.-Q. Zhu, and T.-X. Lin. 1992. *Concepts of marine sciences.* Qingdao Ocean University Press. Qingdao, China (in Chinese), 257 p.

Ekman, S. 1953. *Zoogeography of the sea.* Sidgwick and Jackson, London, 404 p.

Gosner, K.L. 1971. *Guide to identification of marine and estuarine invertebrates. Cape Hatteras to the Bay of Fundy.* John Wiley & Sons, New York, 693 p.

Hayward, P.J. and J.S. Ryland (eds.). 1990. *The marine fauna of the British Isles and North West Europe.* Vols. 1 and 2. Clarendon Press, Oxford, 996 p.

Liao, Y., and A.M. Clark. 1995. *The echinoderms of southern China.* Science Press, Beijing, New York, 614 p.

Pielou, E.C. 1975. *Ecological diversity.* John Wiley & Sons, New York, 165 p.

Smith, R.I. and J.T. Carlton. 1975. *Light's manual: Intertidal invertebrates of the central California coast.* Third Edition. University of California Press, Berkeley, California, 716 p.

Taylor, W.R. 1960. *Marine Algae of the eastern tropical and subtropical coasts of the Americas.* University of Michigan Press, Ann Arbor. 870 p.

Foreword To The Original Chinese Edition

Among the natural sciences, life science is one of the most complex fields, with numerous mysteries awaiting exploration. Today, molecular biology, biotechnology, and other high-tech branches have already attracted great attention. Taxonomy is the foundation of all the life science branches. Early taxonomy relied upon morphology and embryology. In recent years, molecular biology, biochemistry (isozymes, RNA, DNA, etc.) have been introduced into taxonomy. Electron microscopy has enhanced the understanding of microstructure, enables the classification of small and microbial organisms, and makes taxonomy closer to natural classification and systematic evolution.

China has many accurate records of marine organism classification since ancient times. For example, the recording of the oyster and *Caloglossa leprieurii*, etc., in Li Shizhen's *Compendium of Moteria Medica* is still used today. But the modern classification regime, i.e., universal binomial nomenclature, was adopted in China only after the May 4th movement in 1919. The classification of marine organisms has about 70 years of history in China. In the 1920s, it consisted of sporadic research, mainly conducted by foreign scientists. In the 1930s, Chinese scientists began their own research of marine biology and made impressive achievements. The major research site was Xiamen, although some activities were seen in Beiping (Beijing) and Qingdao. Overall progress was slow. In the 1950s, there was some degree of collection and research in Qingdao and Xiamen; the early 1960s and late 1970s was the most active period. A series of surveys and collections made by the Marine Research Institute, Academic Sinica, and many other institutions during the 1958 nationwide marine survey have provided specimens for taxonomic study. Taxonomy is an international study. The recent open and reform policies have provided the opportunities for Chinese researchers to interact with the outside world.

The longer the history of taxonomic research is, the more new species or new records will be found. In the meantime, there will be different names for the same species. To answer the question of how many marine species are in China still requires long-term research. Moreover, the problem of how many marine species in China have been recorded has not been precisely resolved. This book gives relatively accurate answer to the question of how many species have been recorded. The book, with listings of the recorded species in China's seas in the last 70 years, is very timely and appropriate. It is not only a significant event in the history of China's marine biological research, but also a contribution to global species diversity.

Comprehensive, systematic, in-depth, and meticulous are the characteristics of this project. Not just a compilation of species names, the book also includes the Chinese common name, scientific (Latin) name, author(s), synonym(s), horizontal and vertical distribution, host(s) of parasites, as well as reference(s) for each species. This book has similar functions of a Latin-Chinese marine species dictionary, but can hardly be called such. Its depth is not nearly close to that of voluminous annals, but, as a single volume, it does give readers a broad overview of the field.

One Chinese proverb runs, "As in the Yangtze River, the waves behind drive on those before, so each new generation excels the last one." I am very pleased to see middle-aged and young researchers become more and more mature in their fields, which means there is no lack of successors to carry out the marine taxonomic research. Ten years ago, comrade Huang Zongguo organized 88 experts to compile and publish the Dictionary of Marine Biology. Afterward, he organized an even larger team of experts (132 from 40 institutions) to research, compile, and edit this book. These twin books are of high quality and will certainly play important and profoundly influential roles in marine biological research, teaching, and resource utilization.

Zeng Chengkui (C. K. Tseng)
Member of National Academy of Sciences
Honored President, Chinese
Oceanography and Limnology Society
Academia Sinica
Member of The Third World
Academy of Sciences
March 1, 1993

Introduction To The Original Chinese Edition

The unique environment and natural history of the earth produced life. Evolution of life has created a diverse and colorful biological world. Biodiversity and the value of its components in ecology, genetics, society, economics, science, education, culture, recreation, and aesthetics is not only the foundation of human existence, but also the strong supporting force for its continuous development.

Marine organisms are the major components of global biodiversity. The marine biota has greater diversity than the terrestrial one in terms of higher taxonomic groups: among the 23 phyla found in the marine environment, 13 are endemic; only one of the 11 phyla found on land is endemic; and none of the 14 phyla found in freshwater habitats is endemic. This illustrates the importance of marine biodiversity. Marine environments are totally different natural environments from those of the terrestrial: seawater not only provides a good habitat for marine organisms, but also a "refuge" for some "escaping" terrestrial species during periods of great changes in the geological and biological histories, which allows certain species to avoid extinction. Therefore, studying and protecting marine biodiversity has special significance.

But, in recent history, because of man-made and natural causes or their interactions, marine and terrestrial biodiversity is seriously threatened. Some endangered species are continuously going extinct. It has been reported that the average global extinction rate has reached one species per day, an alarming speed. The cruel facts have forced upon us a historically urgent task: to act without any delay and adopt realistic and effective measures to protect global biodiversity. For this purpose, the 1992 International Environment and Development Conference in Brazil sponsored by the United Nations adopted the "United Nations Convention on Biological Diversity." The "Convention" emphasizes that every country has responsibilities to protect its biodiversity, and can make use of its biological resources in a sustained manner.

China's seas, located in the western Pacific Ocean, have unique geological development processes and geographic advantages. With regard to the formation and development of biodiversity, there are the following characteristics: 1. China's seas have a certain degree of isolation from the oceanic water. This restricts biological exchange and has resulted in a number of endemic species; 2. China's seas cover a large latitudinal expanse (more than 38°). They include tropical, subtropical, as well as temperate regions, providing habitats for species adapted to different temperature regimes. The distributional boundary between Pacific tropical and temperate marine species is located in the South China Sea; 3. The Chinese coastline is curving and long, with complex and diverse geomorphology. There is a large volume of freshwater inflow, carrying with it large quantities of nutrients into the seas. These nutrients enrich the coastal water, providing favorable environments for growth and reproduction of the organisms and formation of fishing grounds; 4. There are certain areas of highly productive marine ecosystems such as coral reefs, mangrove forests, wetland swamps, and upwelling areas; and 5. There are large numbers of marine islands in China, with different environmental types and conditions. These provide another favorable condition for high biodiversity. A great variety of marine species occur in the large expanse of China's seas with its complex habitats, including warm-water species and temperate species. Many of them are endemic species. It has been reported that there are about 20,000 species recorded in China's seas, therefore, China's seas are one of the most representative marine regions in global marine biodiversity. Enhancing the protection and management of Chinese marine biodiversity not only has regional importance, but also global significance. However, in recent decades, because of human disturbance and destruction or natural changes, the biodiversity in China's seas, sharing the "fates" of many other regions, has steadily decreased. If we do not take action to enhance protection, it will impact the ecological equilibrium and biological resource richness, and subsequently the sustainable human development.

In order to proceed with protection and management of Chinese marine biodiversity, the "Chinese Marine Biodiversity Protection Action Plan Committee" was established in 1991 under the leadership of the State Oceanic Administration. It has organized the experts from different institutions to formulate the "action plan," based on extensive surveys and research and the status, problems, and needs of Chinese marine biodiversity. In the meantime, we have asked professor Huang Zongguo of the Third Institute of Oceanography, State Oceanic Administration, to be responsible for editing the book, *Marine Species and Their Distributions in China's Seas*. The objectives were to comprehensively and systematically collect and compile recorded species and their distributions in China's seas. These would provide the foundation for marine biodiversity research, protection, and management in our country. One hundred thirty-two researchers from 40 institutions have participated in the project, compiling and analyzing the research conducted in China's seas in the last 70 years (1922-1992). What needs to be emphasized here is that some senior experts, with many years of experience and the spirit of going all-out and putting all their "eggs" in one basket, have incorporated their years of research achievements into this project, thus greatly enhancing the representation and authority of the book. We are deeply grateful to all those who have participated in and supported the project.

Because this is the first time for this kind of undertaking and it was completed in a short period of time, there are inevitable omissions and inaccuracies in the book, despite the great efforts the authors have made to avoid omission or repetition of the species. Also, there will be new discoveries of biological species in future marine surveys and research. Therefore, we plan to publish a revised version based on the additions, revisions and correction, to this book.

Lu Shouben
Director and Senior Engineer
Department of Ocean Management and Monitoring
State Oceanic Administration
February 26, 1993

Contributors

Editor:

Huang Zongguo

Contributing Editors:

Zeng Chengui (C. K. Tseng), Zheng Zhong (C. Cheng), Jin Dexiang (T. G. Chin), Liu Luiyu, Wu Baoling, Shao Kwang Tsao, Chen Qingchao, Brian Morton

Contributors:

Ding Meili, Ma Xiutong (Ma Siutung), Wang Peilie, Wang Wei, Wang Yongliang, Wang Fuzhen, Wang Jianjun, Wang Shaowu, Wang Zhenrui, Yin Weiping, Qiu Shuyan (S. Y. Chiu), Qiu Jianwen, Shi Jihua, Shen Jiwei, Ren Xianqiu, Liu Heng, Liu Shicheng, Liu Jianhua, Liu Xixing, Qi Zhongyan, Guan Liping (L. P. Vrijmoed), Xu Zhenzhu (Hsu Chen-Tsu), Jiang Jinxiang, Zhuang Qiqian, Sun Shichun, Sun Ruiping, Sun Daoyan, Lian Guangshan, Yan Songkai, Wang Song, Wang Puqin, Li Shaoqing, Li Fuxue, Li Rongguan, Li Chuanyan, Li Jinhe, Li Fenglan, Li Fenglu, Li Mingfeng, Li Chengye (S. Y. Lee), Li Chupu, Yang Hequan, Yang Siliang, Yang Dejian, Yang Jiaju, Yang Qingliang, Wu Qiquan, He Minghai, Zou Renlin, Song Weibou, Zhang Shuijin, Zhang Jinbiao (Chin-piao Chang), Zhang Zhinan, Zhang Peishan (Catherine Cheung Pui Shan), Zhang Fuyun, Zhang Chaochang, Zhang Junfu (C. F. Chang), Zhang Raoting, Zhang Songling, Chen Changshen, Chen Huilian, Chen Ruixiang, Lin Mao, Lin Shuangdan, Lin Jinmei, Lin Guangyu, Lin Jinghong, Lin Na, Lin Yuhui, Lin Fushen, Zhou Hong, Zhou Jinming, Zhou Meiying, Zhou Zongchen, Zhou Qiulin, Zheng Shouyi, Zheng Zuoxin (Cheng Tsohsin), Zheng Zhicheng, Zheng Chengxing, Zhao Dadong, Hu Chengye, Hu Qinbo, Zhong Zhenru, Hou Shumin, Hong Junchao, Hong Huixin, Fei Lian, Ni Chunzhi, Ni Dashu, Xu Bin, Xu Huizhou, Xu Zhimin, Xu Fengsan, Xu Huaishu, Tang Zhaohe, Tang Zhongzhang, Tang Zhichan, Tang Chongti, Tang Senming, Huang Zongguo, Huang Xiuming, Huang Meijun, Lu Shouben, Lian Guanyao, Lai Jinyang (Lai Kinyang), Jian Jialun, Dong Zhengzhi, Dong Jinhai, Fu Zhaoxian, Cheng Zhaodi, Zeng Guoshou, Cai Erxi, Cai Bingji, Cai Ruxing, Cai Yaoguo, Pei Zunan, Liao Yulin, Tang Zhiyuan, Pan Haihong, Xue Deqing, Dai Aiyun, Dai Yanyu, Wei Chongde, Ilse Bartsch

Note: All Chinese names listed on this page are in the traditional order, family name first.

Editorial Notes
Revised From The Original Chinese Editon

1. This book lists the species collected from China's seas during the period 1922 to 1992.

2. All living species recorded from such waters by Chinese and foreign scientists and some foraminiferan species from the Cenozoic era are included. As well as oceanic species, some estuarine and freshwater species, intertidal insects, the vascular plants of the high intertidal, and birds (both oceanic, coastal, and migratory species) are included.

3. Each entry includes the species name, authority, synonyms or incorrect name (if applicable), distribution, host (for parasitic species), and reference(s).

   Examples:

   A. *Laminaria japonica* Aresch [*L. ochotensis*]: Bohai, Yellow Sea, East China Sea. Subtidal. (17,...)

   B. *Nicolla epinepheli* (Wang): Pingtan. Host: Red-spotted grouper. (41)

4. All sections are followed by a reference list and a list of the responsible editors. The reference lists comprise taxonomic papers by Chinese and foreign scientists, some taxonomy-related ecological and biological works and unpublished manuscripts by members of the editorial board. For sections about which specialized volumes have been published, only references published subsequent to the specialized volumes are included.

5. The main text is sectioned according to the five-kingdom classification system. (The Phylum Protozoa has been abolished; the Cyanophyta is grouped under the Monera). Taxa lower than phylum are adopted following the advice of members of the editorial board.

6. An alphabetical index follows the main text. This includes all species' names, their synonyms and incorrect names mentioned in the main text, and all other taxa in Latin (all capitalized).

7. All taxa can be found either by the taxonomic hierarchy adopted in the main text or by using the index. The index will allow reference to the Chinese name, the correct species name and synonyms, authority, distribution (horizontal and vertical, hosts), the relevant literature, and the higher taxa to which the organism belongs.

Zongguo Huang
Third Institute of Oceanology, SOA,
Xiamen 361005, China

Taxonomic Listing

MONERA

BACTERIA

THIOBACILLALES
THIOBACILLACAE

Thiobacillus denitrificans Beijerinck: Qingdao (Shandong) shoreline. (16, 30)

T. thiooxidans Waksman & Jofe: Qingdao (Shandong) shoreline. (14)

T. thioparus Beijerinck: Qingdao (Shandong) shoreline. (16)

DESULFOVIBRIONALES
DESULFOVIBRIONACEAE

Desulfovibrio desulfuricans (Beijerinck) Kluyver & Van Niel: Qingdao (Shandong) shoreline. (16)

PSEUDOMONADALES
PSEUDOMONADACEAE

Pseudomonas: Bohai; Yellow Sea; Dalian Bight (Liaoning); Jiaozhou Bay and Lingshan Island (Shandong); Jiangsu; mouth of Changjiang (=Yangtze River); Xiamen (Fujian); central South China Sea; Nansha (=Spratly Islands); skin, digestive track, and cavity fluid of Stichopodidae (sea cucumber) species along the Qingdao and Weihai (Shandong) shoreline. (1, 2, 3, 12, 14, 15, 21, 23, 26, 27)

Xanthomonas: Sea water, skin and cavity fluid of Stichopodidae (sea cucumber) species along the northern Yellow Sea; Qingdao and Lingshan Island (Shandong) shoreline; Xiamen harbor (Fujian); Nansha (=Spratly Islands). (12, 21, 22, 27)

AZOTOBACTERACEAE

Azotobactera halophilum v. *marinus* Cheng: Qingdao (Shandong) near shore waters. (17)

Genera with undeterminated taxonomic position

Alcaligenes: Bohai; Yellow Sea; Jiaozhou Bay (Shandong); East China Sea; mouth of Changjiang (=Yangtze River); Jiangsu; Xiamen harbor (Fujian); central South China Sea; Nansha (=Spratly Islands). (1, 3, 12, 14)

Flavobacterium: Bohai; Yellow Sea; Jiaozhou Bay (Shandong); East China Sea; mouth of Changjiang (=Yangtze River); Jiangsu; Xiamen harbor (Fujian); central South China Sea; Nansha (=Spratly Islands). (1, 3, 12, 14)

ACETOBACTERACEAE

Brevibacterium: Nansha (=Spratly Islands). (12)
Chromobacterium: Nansha (=Spratly Islands). (12)
Gluconobacter: Xiamen harbor (Fujian); central South China Sea. (21, 22)
Zymomonas: Nansha (=Spratly Islands). (12)

NEISSERIALES
NEISSERIACEAE

Acinetobacter: Bohai; Yellow Sea; East China Sea; mouth of Changjiang (=Yangtze River); central South China Sea; Nansha (=Spratly Islands); Rushan, Weihai, Jiaozhou Bay and Lingshan Island (Shandong); Xiamen harbor (Fujian). (1, 2, 12, 14, 20, 21, 27)

Neisseria: Coasts of Lingshan Island, Weihai and Qingdao (Shandong). Skin, digestive tract, cavity fluid, and habitat of Stichopodidae (sea cucumber) species. (14)

ENTEROBACTERALES
ENTEROBACTER

Enterobacter: Bohai; Yellow Sea; East China Sea. (26, 27)
E. coli Castellani & Chalmers: China's seas. (7)
Proteus vulgaris Hauser: China's seas. (7)

VIBRIONACEAE

Aeromonas: Shrimp farms around Qingdao (Shandong); Yellow Sea; East China Sea; Jiangsu; Nansha (=Spratly Islands). (12, 14, 26, 27)

Lucibacterium harveyi (Johnson et al): Yellow Sea; mouth of Changjiang (=Yangtze River); South China Sea. (9, 10, 12, 28)

Photobacterium: Northern Yellow Sea; Jiaozhou Bay (Shandong); East China Sea; Jiangsu; Nansha (=Spratly Islands). (12, 26, 27)

P. fiseheri Beijerinck: Qingdao (Shandong); Yellow Sea; Lianyungang (Jiangsu); East China Sea. Skin of *Sphyraena* (barracuda). (6, 10, 11)

P. leiognathi: Yellow Sea; Qingdao (Shandong); Lianyungang (Jiangsu); South China Sea. (10, 11, 20)

P. phosphsreum (Cohu) Ford: Yellow Sea; East China Sea. (20, 29)

Plesiomonas: Xiamen (Fujian); South China Sea. (12, 21, 22)

Vibrio alginolyticus Sakazaki et al: Bohai; Yellow Sea; Zhejiang; Xiamen (Fujian); Daya Bay (Guangdong); Nansha (=Spratly Islands). Pathogen of black gill disease in cultured shrimps. (13, 19, 22)

V. anguillarum Bergeman: Yellow Sea; East China Sea; Fujian. Pathogen of red leg disease in cultured shrimps. (3, 8)

V. anguillarum I: Nansha (=Spratly Islands). (13)

V. campbelli Baumann, Barmann, Bang & Woolkalis: Yellow Sea; Hong Kong; Nansha (=Spratly Islands). (7, 13, 32, 35)

V. cholerae Pacini: Hong Kong; Nansha (=Spratly Islands). (13, 34)

V. cholerae non-01: Zhejiang. Pathogen of blind disease in cultured shrimps. (24)

V. costicola: Nansha (=Spratly Islands). (13)

V. damsela: East China Sea; Fujian. (27, 28)

V. fischeri (Beijerinck) Lehmann & Neumann: Yellow Sea; East China Sea; Hong Kong; Nansha (=Spratly Islands). (6, 10, 13, 32)

V. fluvialis Lee, Shread, Furniss & Bryant: East China Sea; Fujian; Daya Bay (Guangdong). (22)

V. fluvialis I: Nansha (=Spratly Islands). (13)

V. harveyi Baumann, Baumann, Bang & Woolkalis: Yellow Sea; East China Sea; South China Sea. (10, 11, 12, 32, 33)

V. marinus: Nansha (=Spratly Islands). (13)

V. metschnikovii: Nansha (=Spratly Islands). (13)

V. migripulchritudo Baumann Bang & Woolkalis: Qingdao (Shandong). (4)

V. mimicus: Xiamen (Fujian); Daya Bay (Guangdong). (22)

V. natriegens: Nansha (=Spratly Islands). (13)

V. nereis: Nansha (=Spratly Islands). (13)

V. orientalis Yang: Yellow Sea; East China Sea. (10, 36)

V. parahaemolyticus Fujino et al: Qingdao (Shandong); Zhejiang; Xiamen (Fujian); Daya Bay (Guangdong); Nansha (=Spratly Islands). Pathogen of white muscle disease in cultured shrimps. (13, 22, 33, 36)

V. pelagius I: Zhejiang; Nansha (=Spratly Islands). Pathogen of black gill disease in cultured shrimps. (13, 18)

V. splendidus biotype I: Yellow Sea; East China Sea. (20)

V. vulunicus: Xiamen harbor (Fujian); Daya Bay (Guangdong); Nansha (=Spratly Islands). (13, 22)

BACILLALES
BACILLACEAE

Bacillus: Bohai; Yellow Sea; East China Sea; South China Sea. (2, 12, 14, 15, 21, 22, 26, 27, 32)

B. brovis Migula: Jiaozhou Bay (Shandong). (32)

B. fastidiosus den Dooren de Jong: Jiaozhou Bay (Shandong) (32)

B. firmus Bredemann & Werner: Jiaozhou Bay (Shandong) (32)

B. sphaerieus Meyer & Neide: Jiaozhou Bay (Shandong). (32)

CORYNEBACTERIACES
CORYNEBACTERIACEAE

Arthrobacter: Bohai; Yellow Sea [Weihai, Qingdao and Lingshan Island (Shandong)]; Xiamen harbor (Fujian); Nansha (=Spratly Islands). (2, 12, 14, 21, 22)

Corynebacterium: Bohai; Qingdao (Shandong); Yellow Sea; East China Sea [mouth of Changjiang (=Yangtze River)]; Xiamen harbor (Fujian); Nansha (=Spratly Islands). (2, 12, 14, 21, 22, 26)

Microbacterium: Xiamen harbor (Fujian); central South China Sea. (21, 22)

LACTOBACILLACEAE

Lactobacillus: Bohai; Yellow Sea; East China Sea; Jiangsu; Nansha (=Spratly Islands). (2, 12, 26)

CYTOPHAGALES
CYTOPHAGACEAE

Caulobacter: Qingdao (Shandong) [skin and digestive track of Stichopodidae (sea cucumber) species]; East China Sea. (14)

Cytophaga: Northern Yellow Sea; Nansha (=Spratly Islands). (12, 27)

Flexibacter: Nansha (=Spratly Islands). (12)

MICROCOCCALES
MICROCOCCACEAE

Micrococcus: Bohai; Yellow Sea; East China Sea; South China Sea. (12, 15, 21, 22, 26, 27, 31, 32)

M. candidus Cohn: Around Qingdao (Shandong). (31, 32)

M. cimnabareus: Around Qingdao (Shandong). (31, 32)

M. cinnabarens Flugge-strain A: (12, 15, 21, 22, 26, 27, 31, 32)

M. cinnabarens Flugge-strain B: (12, 15, 21, 22, 26, 27, 31, 32)

M. cinnabarens Flugge-strain C: (12, 15, 21, 22, 26, 27, 31, 32)

M. cinnabarens Flugge-strain D: (12, 15, 21, 22, 26, 27, 31, 32)

M. cinnabarens Flugge-strain E: Around Qingdao (Shandong). (31, 32)

M. citreus v. *migula*: Around Qingdao (Shandong). (31, 32)

M. ridleyi: Around Qingdao (Shandong). (31, 32)

M. ridleyi corbert-strain A: Around Qingdao (Shandong) (31, 32)

M. ridleyi corbert v. *marinus*: Around Qingdao (Shandong) (31, 32)

M. sulfureus Zimmerman-strain A: (12, 15, 21, 22, 26, 27, 31, 32)

M. sulfureus Zimmerman-strain B: Around Qingdao (Shandong). (31, 32)

Stphylococcus: Bohai; Yellow Sea; East China Sea; South China Sea. (3, 5, 12, 26, 27)

STREPTOCOCCACEAE

Streptococcus: Jiangsu; Daya Bay (Guangdong). (3, 23)

PEPTOCOCCACEAE

Planococcus: Nansha (=Spratly Islands). (12)
Sarcina: Jiaozhou Bay (Shandong). (32)

REFERENCES*

(1). Ding Meili et al., 1979. Distribution of oil-degradation bacteria in Jiaozhou Bay. Weishengwuxue Tongbao, 6(6):11–14.
(2). Wang Wenxing et al., 1980. Distribution and biochemical characteristics of oil-degradation bacteria from the water and sediments of continental shelf in the East China Sea. Marine Research 13(2):77–81.
(3). Wang Wenxing et al., 1983. Studies on opportunistic pathogens and bacterial flora in the shrimp-rearing water environment. Journal of Oceanography of Huanghai & Bohai Seas, 1(2):68–79.
(4). Sheng Ju et al., 1991. Effect of marine bacterial mucus on the attachment to *Argopecten irradians*. In: Abstract Collections of the First Seminar on Marine and Freshwater Microorganism, p:11.
(5). Shi Junxian et al., 1984. Ecological distribution of heterotrophic bacteria in seawater and sediments of the Changjiang Estuary. Marine Science Bulletin, 3(6):59–63.
(6). Wang Baohua et al., 1985. One strain of *Vibrio fischeri* isolated from the body surface of fish in East China Sea. J. East China Normal University, 2:108–112.
(7). Xu Bin et al., 1992. *Proteus vulgaris*: A new pathogen in *Penaeus orientalis*. Journal of Fisheries of China, 16(2):130–136.
(8). Xu Bin et al., Studies of pathogens and the pathogenic mechanism. (manuscript)
(9). Liu Tiancai et al., 1989. Studies on pathogenicity and biological characteristics of "red leg disease" bacteria isolated from *Penaeus penicillus*. Fujian Fisheris, 2:1–3.
(10). Shen Jianwei et al., 1987. The isolation and identification of marine luminous bacteria from the waters of the Huanghai Sea. Oceanologia & Limnologia Sinica, 18(4):333–339.
(11). Shen Jianwei et al., 1988. The isolation and identification of marine luminous bacteria from the waters of the South China Sea. Oceanologia & Limnologia Sinica, 19(1):76–80.
(12). Shen Heqin et al., 1991. Ecological distribution of heterotrophic bacteria in the waters around the Nansha Islands. Contributions on the study of Marine Biology in the Nansha Islands and Neighbouring Waters (II). China Ocean Press, 1–17.
(13). Shen Heqin et al., 1991. Ecological distribution of heterotrophic vibrio in Nansha Islands waters. Contributions on the study of Marine Biology in the Nansha Islands and Neighbouring Waters (II). China Ocean Press, 18–33.
(14). Chen Ma et al., 1982. Ecological distribution of heterotrophic bacteria in continental shelf of the East China Sea. Studia Marina Sinica, 19:1–10.
(15). Chen Shiyang et al., 1987. The investigation of sea pollution and heterotrophic bacterial flora in Rushan Bay. Journal of Shandong College of Oceanology, 17(4):86–94.
(16). Xu Huaisu et al., 1989. Steel corrosion by marine bacteria. Journal of Ocean University of Qingdao, 19(1):104–111.
(17). Chen Siyang, 1961. Studies on marine azotobacteria, I. Isolation of aerobic autotrophic azotobacteria and determination of azoto-ability. Journal of Shandong College of Oceanology, 1:31–41.
(18). Chen Gongli et al., 1991. The first determination of vibrio from *Penaeus penicillus*. Journal of Public Health of China, 10(2):9–13.
(19). Zhuang Tiecheng, 1988. A preliminary study on determination of pathogenic vibrio from bivalve-culture ground. Journal of Environmental Pollution and Prevention, 10(5): 13–17.
(20). Yang Yikang et al., 1984. Characteristics of marine luminous bacteria from the waters of the East China Sea and Huanghai Sea and description of a new species. Oceanologia & Limnologia Sinica, 15(3): 258–264.
(21). Zhou Zongchen et al., 1989. Ecological distribution of heterotrophic bacteria in central waters of the South China Sea. Marine Science Bulletin, 8(3):57–64.
(22). Ni Chunzi et al.,1983. Studies on marine oil-degradation bacteria, I. Ecological distribution of oil-degradation bacteria in Xiamen Harbour. Acta Oceanologica Sinica, 5(5):637–644.
(23). Ni Chunzi et al., 1990. Vibrionic ecology in waters around Nuclear Power Station in Daya Bay. Collections of Papers on Marine Ecology in the Daya Bay (II). China Ocean Press, 442–450.
(24). Zheng Guoxing et al., 1986. Identification and pathogenicity of *Vibrio cholerae* (Non-01) isolated from diseased penaeid shrmp. Journal of Fisheries of China, 10(2):195–203.
(25). Zheng Guoxing et al., 1990. *Vibrio anguillarum* as a cause of disease in *Penaeus orientalis* Kishnouye. Journal of Fisheries of China, 14(1):1–17.
(26). Lin Wenna et al., 1984. Antifouling study of fixed nets—analysis of adhesive bacteria. Marine Science Bulletin, 3(6):68–71.
(27). Lin Fenggao et al., 1989. Study on the ecology of marine bacteria in the experimental dump site of northern Huanghai Sea for nontoxic waters, I. Marine Environmental Science, 8(3):10–13.
(28). Huang Weizhen et al., 1990. A study on control of *Penaeus penicillatus* "red leg disease". Fujian Fisheries, 4:1–14.
(29). Cao Yunhui et al., 1990. Distribution and composition of luminous bacteria in Changjiang River estuary. Acta Oceanologica Sinica, 4(1):89–93.
(30). Hsueh, T. Y. and K.Y. Sun, 1959. Studies on the isolation and cultivation of *Thiobacilli* from marine mud. Oceanologia & Limnologia Sinica, 2(2):75–80.
(31). Hsueh, T. Y. and K.Y. Sun, 1960. Studies on marine micrococci from Kiaochow Bay. Oceanologia & Limnologia Sinica, 3(1):1–11.
(32). Qian Zhenru et al., 1985. Study on death mechanism of domestic sewage bacteria in seawater, I. Bactericidal effect of marine bacteria in Jiaozhou Bay on bacteria in sewage. Acta Scientiae Circumstantiae, 5(1):64–69.
(33). Chen, S. Y. et al., 1989. The habitat of *Vibrio parahaemolyticus* in the seawater and seafoods of coastal area around Qingdao district. J. Micro-Ecology China. 1(2):60–67.
(34). Huai-Shu Xu et al., 1988. Occurrence and distribution of *Vibrio* spp. in fish and shellfish in subtropical water of Hong Kong. Programme and Abstracts of GIAM VIII (Eighth International Conference on Global Impacts of Applied Microbiology) and INCABB (International Conference on Applied Biology and Biotechnology). p.101, MBB-1-2
(35). Kwong-Yu Chen et al., 1986. Occurrence and Distribution of Holophilic *Vibrio* in Subtropical Control Water of Hong Kong. Appl. Environ. Microbiol., 52(6):1407–1411.
(36). Yang, Y. et al., 1983. Characterization of marine luminous bacteria off the coast of China and description of *Vibrio orientalis* sp. Nov. Curr. Microbiol. 8:95–100.

*: (1)–(32) in Chinese
Compiled by Ni Chunzhi and Xu Huaishu; Edited by Ding Meili

ACTINOBACTERIA

ACTINOMYCETALES
STREPTOMYCETACEAE

Roseoporus group
Streptomyces xiahaiensis Zhou: Xiamen (Fujian). Mangrove roots, 5–35 m from shore. Depth 10–15 cm, muddy substrate. (3)

Cinereus group
Streptomyces diastatochromogenes No. 79 Sakagami & Yamabayashi: Longhai (Fujian). Mangrove roots, 5–35 m from shore. Depth 10–15 cm, muddy substrate. (6)
Flavus group
Streptomyces albidoflavus v. *marine* Zheng & Zhou: Xiamen (Fujian). Mangrove roots, 5–35 m from shore. Depth 10–15 cm, muddy substrate. (7)
Aureus group: Southern Fujian (intertidal), mouth of Jiulong River (Fujian) (mangrove roots, 5–35 m from shore, depth 10–15 cm, muddy substrate). (4, 5)
Griseofuscus group
Streptomyces rutgersensis gulangyuensis Zhang, Yan & Su: Xiamen (Fujian). 200 m from shore, depth 23 cm, muddy substrate. (1, 3)
Griseobroviolaceus group: Intertidal area along southern Fujian and mangrove roots in Xiamen (Fujian), 5–35 m from shore. Depth 10–15 cm, muddy substrate. (4, 5)
Hygroscopicus group: Intertidal area along southern Fujian and mangrove roots in the mouth of Jiulong River (Fujian), 5–35 m from shore. Depth 10–15 cm, muddy substrate. (4, 5)
Lavendulae group:
Streptomyces lavendulae Waksman & Curtis: Fujian. Deep sea mud. (8)
Cyaneus group: Intertidal area along southern Fujian and mangrove roots in the mouth of Jiulong River (Fujian), 5–35 m from shore. Depth 10–15 cm, muddy substrate. (4, 5)
Viridis group: Intertidal area along southern Fujian and mangrove roots in the mouth of Jiulong River (Fujian), 5–35 m from shore. Depth 10–15 cm, muddy substrate. (4, 5)
Albosporus group: Intertidal area along southern Fujian and mangrove roots in the mouth of Jiulong River (Fujian), 5–35 m from shore. Depth 10–15 cm, muddy substrate. (4, 5)
Glaucus group: Intertidal area along southern Fujian and mangrove roots in the mouth of Jiulong River (Fujian), 5–35 m from shore. Depth 10–15 cm, muddy substrate. (4, 5)

NOCARDIACEAE

Nocardia: Mangrove roots in the mouth of Jiulong River (Fujian), 5–35 m from shore. Depth 10–15 cm, muddy substrate. (5)

MICROMONOSPORACEAE

Micromorospora: Mangrove roots in the mouth of Jiulong River (Fujian), 5–35 m from shore. Depth 10–15 cm, muddy substrate. (5)

MICROPOLYSPORACEAE

Micropolyspora: Mangrove roots in the mouth of Jiulong River (Fujian), 5–35 m from shore. Depth 10–15 cm, muddy substrate. (5)
Microtetraspora viridis xiamensis Su: Xiamen harbor (Fujian). Water depth 14 m, muddy substrate. (2)

ACTINOPLANACEAE

Actinoplanes: Mangrove roots in the mouth of Jiulong River (Fujian), 5–35 m from shore. Depth 10–15 cm, muddy substrate. (5)
Streptosporangium: Mangrove roots in the mouth of Jiulong River (Fujian), 5–35 m from shore. Depth 10–15 cm, muddy substrate. (5)

REFERENCES*

(1). Su Guocheng, Huang Weizhen and Liu Tiancai, 1989. Studies on marine actinomycetes, I. Identification of *Streptomyces rutyersensis*. Acta Microbiologica Sinica, 29(3):225–227.
(2). Su Guocheng et al., 1989. Studies on marine actinomycetes, II. Identification of a new subspecies of *Microtetra spora viridis*. Acta Microbiologica Sinica, 29(6):460–463.
(3). Zhou Meiying, Zheng Zhicheng and Yao Bingxin, 1988. Identification and antibacterial activity in a strain halo-tolerant *Streptomyces*. Journal of Xiamen University (Natural Sciences), 27(5):567–571.
(4). Zhou Meiying et al., 1987. Studies on the distribution and antibacterial activity of *Streptomyces* in intertidal zone of Minnan area. Acta Oceanologica Sinica, 11(2):218–225.
(5). Zhen Zhicheng et al., 1989. Composition and biological activity of *Actinomycetes* in mangrove. Journal of Xiamen University (Natural Sciences), 28(3):306–310.
(6). Zheng Zhicheng and Zhou Meiying. Studies on halo-tolerant *Streptomyces diastatochromogenes* and its produced antibiotic. (manuscript)
(7). Zheng Zhicheng and Zhou Meiying. Studies on a new variant species of *Streptomyces albidoflavus* and its produced anti-fungi antibiotic. (manuscript)
(8). Lin Wenliang et al., 1986. Studies on *Streptomyces* C-141 and its metabolite. Antibiotic, 11(6):512–513.

*: all in Chinese
Compiled by Zheng Zhicheng and Zhou Meiying

CYANOBACTERIA [CYANOPHYTA]

CHROOCOCCALES
CHROOCOCCACEAE

Anacystis elabens (Breb.) S. & G.: Fujian. (8, 12)
Aphanothee saxicola Naeg.: Hainan. (18)
Chroococcus limneticus Lemm. v. *subsalsus* Lemm.: Xisha (=Paracel Islands). Mid- to low-intertidal. (4)
C. membraninus (Meneghini) Naeg: Xisha (=Paracel Islands). Mid-intertidal. (4)
C. minor (Kuetz.) Naeg.: Xisha (=Paracel Islands). Mid- to low-intertidal. (4)

C. turgidus (Kuetz.) Naeg.: Widely distributed along China's shoreline. (7, 11a, 14, 18)
C. varius A. Braun: Xisha (=Paracel Islands). Low-intertidal. (4)
Chroothece littorinae Tseng & Hua: Hainan. (18)
Gomphosphaeria aponina Kuetz.: Common species along China's shoreline. (18, 21)
Merismopedia glauca (Ehrenb.) Naeg.: Qingdao (Shandong). (11b, 14, 18)
M. warmingiana Lagerh: Xiamen harbor (Fujian). Planktonic. (11c)
Microcystis aeruginosa Kuetz.: Guangdong; Xisha (=Paracel Islands). Mid- to low- intertidal. (4)
M. ichthyoblabe Kuetz.: Guangdong. (18)
Synechocystis aquatilis Sauva: Xisha (=Paracel Islands). (6)
S. pevalekii Ercegovic: Shandong. (18)

ENTOPHYSALIDACEAE

Entophysalis granulosa Kuetz.: Common species along China's shoreline. (7, 11a, 14, 18)
Hormathonema enilithicum Ercegovic: Xisha (=Paracel Islands). (4)

TUBIELLACEAE

Chorlogloea lutea S. & G.: Fujian. (11c)
Johannesbaptistia pellucida (Dickie) Tayor & Drouet: Hainan; Xisha (=Paracel Islands). (4, 5, 18)

CHAMAESIPHONALES
DERMOCARPACEAE

Dermocarpa fucicola Saunders: Yellow Sea. (7)
D. prasina (Rein.) Born. & Flah.: Yellow Sea. (7)
D. solitaria Collins & Hervey: Guangdong; Xisha (=Paracel Islands). (6, 18)
D. sphaerica S. & C.: Xiamen (Fujian); Hainan; Xisha (=Paracel Islands). (17, 18)
D. violacea Crouan: Yellow Sea. (7)

PLEUROCAPSALES
PLEUROCAPSACEAE

Dalmatella buaensis Erceg: Xisha (=Paracel Islands). (6)
Pleurocapsa minuta Geitler: Dalian (Liaoning); Qingdao and Yantai (Shandong). (7, 14)
Xenococcus acervatus S. & G.: Shandong. (18)
X. chaetomorphae S. & G.: Shandong. (18)
X. cladopharae (Tilden) S. & G.: Yellow Sea. (7)

HYELLACEAE

Hyella caespitosa Born. & Flah.: Common species along China's shoreline. (6, 7, 18)
H. simplex Chu & Hua: Guangdong; Xisha (=Paracel Islands). (18)

OSCILLATORIALES
OSCILLATORIACEAE

Arthrospira maxima Setch. & Gardn.: Fujian. (11c)
Borzia xishaensis Hua: Guangdong; Xisha (=Paracel Islands). (18)
Hydrocoleus cantharidosmus (Mont.) Gom: Pingtan (Fujian); South China Sea. (2, 11a, 18)
H. glutinossum Gom.: Fujian. (11c)
H. lyngbyaceus Kuetz.: Qingdao (Shandong). (7)
Lyngbya adherens Setchell. & Gardner: Xisha (=Paracel Islands). (6)
L. aestuarii Liebm: Hong Kong; Hainan; Xisha (=Paracel Islands). (18, 20)
L. baculum Gom.: Xisha (=Paracel Islands). (3)
L. confervoides C. Ag.: Qingdao and Yantai (Shandong); Xiamen (Fujian); Hong Kong; Xisha (=Paracel Islands). (1, 7, 14, 18, 20)
L. epiphytica Hieron: Widely distributed along China's shoreline. (1, 7, 14, 18)
L. infixa Fremy: Xisha (=Paracel Islands). (2)
L. kuetzingii Schmidle: Xisha (=Paracel Islands). (2)
L. lulea (Ag.) Gom.: Hong Kong. (21)
L. majuscula Harvey: Widely distributed along China's shoreline; Xisha (=Paracel Islands). (1, 2, 18)
L. nordgaardii Wille: Qingdao (Shandong); Xisha (=Paracel Islands). (3)
L. semiplena (C. Ag.) J. Ag.: Widely distributed along China's shoreline; Xisha (=Paracel Islands). (10, 18)
L. willei S. & G.: Qingdao (Shandong). (14)
Microcoleus chthonoplastes Thur.: Common species along China's shoreline. (1, 18)
M. confluens S. & G.: Dalian (Liaoning); Weihai, Qingdao, and Yantai (Shandong). (7, 14)
M. majuscula Tseng & Hua: Guangdong; Hainan. (18)
M. tenerrimus Gom.: Yellow Sea; South China Sea. (18)
M. vaginatus (Vauch.) Gom.: Fujian. (10)
Oscillatoria amphibia C. Ag.: Qingdao (Shandong). (14)
O. articulata Gardn.: Qingdao (Shandong). (7, 14)
O. bonnemaisonii Crouan: Common species along China's shoreline. (3, 18)
O. brevis Kuetz.: Xisha (=Paracel Islands). (3)
O. chalybea Mertens: Jinyin Island (Xisha=Paracel Islands). Mid- to low-intertidal. (3, 16, 21)
O. corallinae Kuetz. Gom.: Jinyin Island (Xisha=Paracel Islands). Mid-intertidal. (3)
O. formosa Bory: East China Sea. (18)

O. laetevirens (Crouan) Gom.: Qingdao and Yantai (Shandong); Xiamen (Fujian); Chenhang Island (Xisha=Paracel Islands). (3, 7, 18)

O. limosa C. Ag.: Xisha (=Paracel Islands). (1)

O. nigro-viridis Thwaites: Common species along China's shoreline. (3, 18)

O. sancta Kuetz.: Hainan; Xisha (=Paracel Islands). (18)

O. subbrevis Schmidle: Shandong; Xisha (=Paracel Islands). (18)

O. subuliformis Kuetz.: Yongxing Island (Xisha=Paracel Islands). Low intertidal. (4)

Phormidium corium (Ag.) Gom.: Yellow Sea; Xisha (=Paracel Islands). (6, 7, 12, 13)

P. crosbyanum Tilden: Hainan, Xisha (=Paracel Islands). (12, 18)

P. fragile Gom.: Yellow Sea; Hainan; Xisha (=Paracel Islands). (6, 7, 12, 18)

P. naveanum Grunow v. *marina* Tseng & Hua: Hainan; Xisha (=Paracel Islands). (12, 18)

P. penicillatum Gom.: Xisha (=Paracel Islands). Low intertidal. (2)

P. submenbranaceum (Ard. & Straff.) Gom.: Hainan. (18)

P. tenue (Menegh.) Gom.: Along the shoreline from Bohai to South China Sea; Xisha (=Paracel Islands). (1, 18)

Schizothrix lacustris A. Braun: Xisha (=Paracel Islands). Low-intertidal. (2)

Sirocoleus kurzii (Zeller) Gom.: Guangdong; Xisha (=Paracel Islands). (6, 18)

Spirulina labyrinthiformis (Menegh.) Gom.: Xisha (=Paracel Islands). (1)

S. major Kuetz.: Xiamen (Fujian). (1)

S. subsalsa Oerst.: Widely distributed along China's shoreline. (1, 11b, 14, 18)

S. subtillissima Kuetz.: Qingdao (Shandong). (7, 14)

S. tenerrima Kuetz.: Xisha (=Paracel Islands). (3)

S. versicolor Cohn: Qingdao (Shandong). (9)

Symploca caespitosa Tseng & Hua: Guangdong; Hainan. (18)

S. hydnoides Kuetz.: Qingdao (Shandong). (14)

S. muscorum (Ag.) Gom.: Hainan. (18)

Trichodesmium erythraeum Ehr.: Bohai; Yellow Sea; East China Sea; Taiwan Strait; South China Sea. Red tide species. (11a, 15, 18)

T. hildebrandtii (Gomont) J. De Toni: East China Sea; Taiwan Strait; South China Sea. Red tide species. (11c, 12)

T. maceii Li, Bian, Lewin, Cheng & Pan: Qingdao (Shandong). (9)

T. thiebautii Gom. [*T. contortum*]: East China Sea; Taiwan Strait; South China Sea. Red tide species. (6, 12, 18)

NOSTOCALES
NOSTOCACEAE

Anabaena variabilis Kuetz.: Qingdao (Shandong); widely distributed along South China Sea's shoreline; Xisha (=Paracel Islands). (5, 7, 18)

Nodularia harveyana Thuret: Xisha (=Paracel Islands). (6)

N. hawaiiensis Tilden: Hainan, Xisha (=Paracel Islands). (6, 18)

Richelia intracellularis J. Schmidt: East China Sea; Taiwan Strait; South China Sea. (5, 18)

RIVULARIACEAE

Calothrix aeruginea (Kuetz.) Thur.: Hainan. (18)

C. confervicola (Roth) Ag.: Widely distributed along China's shoreline; Xisha (=Paracel Islands). (6, 7, 18)

C. contarenii (Zanard.) Born. & Flah: Common species in South China Sea; Xisha (=Paracel Islands). (6, 18)

C. crustacea Thur.: Widely distributed along China's shoreline. (7, 11c, 18)

C. fusco-violacea Crouan: Xisha (=Paracel Islands). (18)

C. parasitica (Chauv.) Thur.: Dalian (Liaoning); Weihai and Qingdao (Shandong). (7, 14)

C. pilosa Harv.: Hainan. (18)

C. robusta S. & G.: Fujian. (11c)

C. scopulorum (W. & M.) Ag.: Yellow Sea; Fujian; Xisha (=Paracel Islands). (1, 5, 7)

C. vivipara Harv.: Fujian. (11c)

Dichothrix bornetiana Howe: Xisha (=Paracel Islands). (5)

D. fucicola (Kuetz.) Born. & Flah.: Xisha (=Paracel Islands). (5, 18)

D. olivacea (Hooker) Born. & Flah.: Xisha (=Paracel Islands). (18)

D. penicillata Zanard.: Xisha (=Paracel Islands). (5)

D. seriata S. & G.: Fujian. (11c)

D. zhongjianensis: Xisha (=Paracel Islands). (5)

Gardnerula fasciculata Tseng & Hua: Guangdong; Xisha (=Paracel Islands). (18)

G. tenuissima Tseng & Hua: Guangdong; Xisha (=Paracel Islands). (18)

G. xishaensis Tseng & Hua: Guangdong; Xisha (=Paracel Islands). (18)

Isactis plana Thur.: Bohai; Yellow Sea; Hainan, Xisha (=Paracel Islands). (7, 18)

Kyrtuthrix maculans (Gom.) Umezaki: Common species in South China Sea. (18)

Rivularia atra Roth: Yellow Sea; Fujian. (7, 17, 18)

R. blasolettiana Menegh: Fujian. (11c)

SCYTONEMATACEAE

Scytonema javanicum (Kuetz.) Born.: Hainan. (18)

S. polycystum Bornet. & Flah.: Xisha (=Paracel Islands). (18)

S. rivulare Borzi.: Xisha (=Paracel Islands). (18)

Tolypothrix subsalsa Tseng & Hua: Guangdong; Hainan. (18)

MICROCHAETACEAE

Camptyolnemopsis major Tseng & Hua: Leizhou (Guangdong). (18, 22)

Microchaete aeruginea Balters: Xisha (=Paracel Islands). (6, 18)

M. grisea Thur.: Dalian (Liaoning); common species in South China Sea. (5, 17, 18)

M. tapahiensis Setchell: Xisha (=Paracel Islands). (5)

M. vitiensis (Askenasy) J. de Teni: Xisha (=Paracel Islands). (5)

STIGONEMATALES
NOSTOCHOPSIDACEAE

Amphithrix janthina (Montagne) Born. & Flah.: Qingdao (Shandong). (7)

Brachytrichia maculans Gom.: Hong Kong. (19)

B. quoyi (Ag.) Born. & Flah.: Yellow Sea; Taiwan; Hainan; widely distributed in South China Sea. (18)

Homoeothrix rubra (Cr.) Fremy: Qingdao (Shandong). (7)

Mastigocoleus testarum Lagerh.: Widely distributed along China's shoreline; Xisha (=Paracel Islands). (6, 7, 18)

Oncobyrsa adriatica Hauck: Shidao (Shandong). (19)

PARENCHYMORPHATACEAE

Parenchymorpha xishanica Tseng & Hua: Xisha (=Paracel Islands). (23)

REFERENCES*

(1). Hua Maosen, 1978. Studies on some marine blue-green algae from Xisha Islands (I). Studia Marina Sinica, 12:59–66.

(2). Hua Maosen, 1981. A new marine blue-green algae of the genus *Borzia* from Xisha Islands. Oceanol. Limnol. Sinica, 12(3):265–269.

(3). Hua Maosen, 1983. Studies on some marine blue-green algae from Xisha Islands (II). Studia Marina Sinica, 20:55–67.

(4). Hua Maosen and Zeng Chengkui (C. K. Tseng), 1985. Studies on some marine blue- green algae of Xisha Islands (III). Studia Marina Sinia, 24:1–9.

(5). Hua Maosen and Zeng Chengkui (C. K. Tseng), 1985. Studies on some marine blue- green algae of Xisha Islands (IV). Studia Marina Sinica, 24:11–26.

(6). Hua Maosen and Zeng Chengkui (C. K. Tseng), 1985. Studies on some marine blue- green algae of Xisha Islands (V). Studia Marina Sinica, 24:27–50.

(7). Chu Haolan, 1959. Studies on the microscopic marine algae from Yellow Sea. Journal of Nanjing University, 2:1–22.

(8). Chu Haolan and Liu Xuexian, 1980. The marine Cladophoraceae of Xisha Islands. Studia Marina Sinica, 17:11–20.

(9). Li Mingren et al., 1984. Laboratory cultivation of a new species of planktonic blue-green algae, *Trichodesmium maccii*, from the Qingdao Coast. Journal of Shandong College of Oceanology, 14(1):93–110.

(10). Cheng Changsheng and Zhang Jinrong, 1991. A study on marine algae in the intertidal zone of Fujian Province, I. The species and their distribution. Journal of Xiamen Fisheries College, 13(2):26–35.

(11a). Chen Yaqu, 1982. Analysis of the red tide caused by *Trichodesmium* in East China Sea in the year 1972. Journal of Fisheries, China, 6(2):181–189.

(11b). Yang Qingliang, 1988. Comprehensive survey report on marine environment of Xiamen Harbour, II. Phytoplankton. Journal of Oceanography in Taiwan Strait, 7(1):1–9.

(11c). Chou Zhenying and Chen Zhuohua, 1983. A list of marine algae from Fujian coast. Journal of Oceanography in Taiwan Strait, 2(1):91–102.

(12). Guo Yujie and Yang Zheyu, 1982. Study on phytoplankton ecology of the continental shelf of the East China Sea in the summer of 1976. Studia Marina Sinica, 19:11–32.

(13). Tseng Chengkuei and Zheng Bolin, 1954. Studies on the marine algae of Qingdao. Acta. Bot. Sinica, 3(1):105–123.

(14). Tseng Chengkuei and C. F. Chang, 1964. A critical review of the records of the benthic marine algae as reported from the western Yellow Sea coast. Studia Marina Sinica, 6:1–26.

(15). Fan, K. C. et al., 1975. Studies on the marine algae of Xisha Islands, China, II. New species and new records of the Family Nemalionaceae (Rhodophyta). Acta Phytotax. Sinica, 13(2):71–77.

(16). Tseng, C. K. 1936. On marine algae new to China. Bull. Fan. Mem. Inst. Biol. (Bet), 7(5):169–196.

(17). Tseng, C. K. 1936. Notes on the marine algae from Amoy. Amoy Mar. Biol. Bull., 1(1): 1–86.

(18). Tseng, C. K. 1983. Common Seaweeds of China. China Science Press.

(19). Tseng, C. K., Chang C. F., Xia Enchan and Xua Bangmei, 1980. Studies on some marine red algae from Hong Kong. The marine flora and fauna of Hong Kong.

(20). Hodgkiss, I. J. and K. Y. Lee, 1981. Checklist and keys to the marine algae of Hong Kong. Memoirs of the Hong Kong Natural History Society, No.16.

(21). Hodgkiss, I. J. and K. Y. Lee, 1983. Hong Kong seaweeds. An Urban Council Publication.

(22). Chang, C. F. and Xia Bangmei, 1988. On two new *Gracilaria* (Gigartinales, Rhodophyta) from South China. In: Abbott, I. A. ed. Taxonomy of economic seaweeds: with reference to some Pacific and Caribbean species. Vol. 2, California Sea Grant College Program, La Jolla, California, USA, pp:131–136.

(23). Chang, C. F. and Xia Bangmei, 1988. Studies on two new *Gracilaria* from south China and summary of *Gracilaria* species in China. In: Abbot, I. A. ed. Taxonomy of economic seaweeds: with reference to some Pacific and Caribbean species. Vol. 2:1–3, California Sea Grant College Program, La Jolla, California, USA.

*: (1)–(15) in Chinese

Compiled by Zhang Junfu (C. F. Chang), Xu Zhimin, Chen Changsheng, Zhang Shuijin, Jiang Jialun, and Yin Weiping; Edited by Zheng Chengkui (C. K. Tseng), Zhang Junfu (C. F. Chang), and Zhang Shuijin

CHLOROXYBACTERIA

PROCHLOROPHYCEAE
PROCHLORALES
PROCHLORACEAE

Prochloron sp.: Sanya (Hainan); Xisha (=Paracel Islands). On ascidians, corals, and algae; around low intertidal. (1)

REFERENCES

(1). Tseng, C. K. 1983. Common seaweeds of China. Science Press, Beijing. p:246, pl. 123, Figs., 1–2.

Compiled by Huang Zongguo; Edied by Zhang Junfu (C. F. Chang) and Zeng Chengkui (C. K. Tseng)

PROTISTA

BACILLARIOPHYTA

CENTRICEAE
COSCINODISCALES
COSCINODISCACEAE

Bacteriastrum comosum Pavillard: From Bohai to South China Sea; southeastern Taiwan. (1, 2, 3, 30a)

B. comosum v. *hispida* (Castr.) Schroeder: From Yellow Sea to South China Sea. (2, 28, 29)

B. delicatulum Cleve: From Bohai to South China Sea; southeastern Taiwan. (2, 12, 30a)

B. elongatum Cleve: From Yellow Sea to South China Sea. (2, 28)

B. elongatum v. *diversum* Ikari: East China Sea; Taiwan Strait. (2, 15)

B. hyalinum Lauder: From Bohai to Okinawa Trough; South China Sea; southeastern Taiwan; Taiwan continental shelf. (2, 30a, 31)

B. hyalinum v. *princeps* (Castr.) Ikari: From Yellow Sea to South China Sea. (1, 2, 3)

B. mediterraneum Pavillard: East China Sea; South China Sea. (1, 2, 3)

B. minus Karsten: Taiwan Strait. (7, 30)

B. seticulosa Grunow: Taiwan continental shelf. (31)

B. varians Lauder: From Bohai to South China Sea; southeastern Taiwan. (1, 2, 3).

Cerataulina bergonii Peragollo: From Bohai to South China Sea; southeastern Taiwan. (5, 12, 13, 27, 29, 30a)

C. compacta Ostenfeld & A. Schmidt: From Yellow Sea to South China Sea. (2, 28)

C. pelagica (Cleve) Hendey [*Syringidium daemon* Greville]: From Bohai to South China Sea. (1, 2, 3)

C. zhongshaensis Kuo: South China Sea. (1, 2, 3)

Chrysanthemodiscus floriatus Mann: From Yellow Sea to South China Sea. (2, 23a, 28)

Corethron hystrix Hansen: From Bohai to South China Sea; southeastern Taiwan. (1, 2, 3, 30a)

C. pelagicum Brun: Taiwan Strait. (29)

Coscinodiscus africanus Janish: Okinawa Trough. (1, 2, 3)

C. agapetos Rattray: East China Sea. (1, 2, 3)

C. angstii Gran: South China Sea. (12, 16)

C. anguste-lineatus A. Schmidt: Bohai; Yellow Sea; East China Sea; Okinawa Trough; Taiwan Strait; South China Sea. (2, 12)

C. apiculatus Ehrenberg: South China Sea. (1, 2, 3)

C. apiculatus v. *ampigus* Grunow: South China Sea. (14)

C. argus Ehrenberg: Bohai; Yellow Sea; East China Sea; Taiwan Strait; South China Sea. (2, 4, 12)

C. asteromphalus Ehrenberh: Bohai; Yellow Sea; Taiwan Strait; South China Sea. (2, 31)

C. asteromphalus v. *pulchra* Grunow: South China Sea. (14)

C. asteromphalus v. *subbuliens* (Joerg.) Cleve-Euler: Bohai; Yellow Sea; East China Sea; South China Sea. (4, 14)

C. bathyomphalus v. *hispidus* Cheng & Chin: Taiwan Strait. (1, 2, 3)

C. bipartitus Rattray: Bohai; Yellow Sea; East China Sea; Taiwan Strait; South China Sea. (2, 12, 27)

C. blandus A. Schmidt: Bohai; Yellow Sea; East China Sea; Taiwan Strait; South China Sea. (1, 2, 3)

C. centralis Ehrenberg: Bohai; Yellow Sea; East China Sea; Okinawa Trough; Taiwan Strait; Taiwan continental shelf; South China Sea. (2, 31)

C. cinctus Kuetzing: South China Sea. (1, 2, 3)

C. concinnus W. Smith: Bohai; Yellow Sea; East China Sea; Taiwan Strait; South China Sea. (1, 2, 3)

C. confusus Rattray: Okinawa Trough. (1, 2, 3)

C. crenulatus Grunow: East China Sea; Okinawa Trough. (1, 2, 3)

C. curvatulus Grunow: Bohai; Yellow Sea; East China Sea; Taiwan Strait; Taiwan continental shelf; South China Sea. (1, 12, 31)

C. curvatulus v. *minor* (Ehr.) Grunow: Bohai; Yellow Sea; East China Sea; Taiwan Strait. (1, 2, 3)

C. debilis Grove: Bohai; Yellow Sea; South China Sea. (12, 14)

C. decrescens Grunow: Bohai; Yellow Sea; East China Sea; Okinawa Trough; Taiwan Strait; South China Sea. (1, 2, 3)

C. decrescens v. *valida* Grunow: East China Sea; Okinawa Trough; Taiwan Strait; South China Sea. (1, 2, 3)

C. deformatus Mann: East China Sea; Taiwan Strait; South China Sea. (1, 29)

C. denarius A. Schmidt: Bohai; Yellow Sea; East China Sea; Taiwan Strait; South China Sea. (1, 2, 3)

C. denarius v. *sinensis* Meister: South China Sea. (1, 2, 3)

C. diversus Grunow: East China Sea. (1, 2, 3)

C. divisus Grunow: Bohai; Yellow Sea; East China Sea; Taiwan Strait; South China Sea. (1, 2, 3)

C. excentricus Ehrenberg: Bohai; Yellow Sea; East China Sea; Okinawa Trough; Taiwan Strait; South China Sea. (1, 2, 3)

C. excentricus v. *fasciculata* Hustedt: Taiwan Strait. (1, 2, 3)

C. fimbriatus Ehrenberg: East China Sea; Okinawa Trough; Taiwan Strait; South China Sea. (1, 2, 3)

C. gigas Ehrenberg: Yellow Sea; East China Sea; Taiwan Strait; South China Sea. (2, 28)

C. gigas v. *preaetexta* (Janisch) Hustedt: East China Sea; Taiwan Strait; South China Sea. (1, 2, 3)

C. granii Gough: Bohai; Yellow Sea; East China Sea; Taiwan Strait; South China Sea. (1, 2, 3)

C. hainanensis Kuo: South China Sea. (1, 2, 3)

C. hexagonus Cheng & Chin: Taiwan Strait; South China Sea. (1, 2, 3)

C. inclusus Rattray: East China Sea; Okinawa Trough. (1, 2, 3)

C. janischii A. Schmidt: Bohai; Yellow Sea; East China Sea; Taiwan Strait; South China Sea. (1, 2, 3)

C. jonesianus (Grev.) Ostenfeld: Bohai; Yellow Sea; East China Sea; Taiwan Strait; South China Sea. (1, 2, 3)

C. jonesianus v. *commutata* (Grun.) Hustedt: Bohai; Yellow Sea; East China Sea; Taiwan continental shelf; Taiwan Strait; South China Sea. (2, 31)

C. kuetzingii A. Schmidt: Bohai; Yellow Sea; East China Sea; Taiwan continental shelf; Taiwan Strait; South China Sea. (2, 30)

C. latimarginatus Kuo: South China Sea. (14)

C. lentiginosus Janisch: Bohai; Yellow Sea. (1, 2, 3)

C. lineatus Ehrenberg: Bohai; Yellow Sea; East China Sea; Okinawa Trough; Taiwan continental shelf; Taiwan Strait; South China Sea. (2, 31)

C. marginato-lineatus A. Schmidt: East China Sea; Taiwan Strait; South China Sea. (2, 4)

C. marginatus Ehrenberg: Bohai; Yellow Sea; East China Sea; Okinawa Trough; Taiwan continental shelf; Taiwan Strait; South China Sea. (2, 23b, 31)

C. minor Ehrenberg: East China Sea; Okinawa Trough; Taiwan Strait; South China Sea. (1, 2, 3)

C. minutifasciculatus Kuo: South China Sea. (1, 2, 3)

C. nitidus Gregory: Bohai; Yellow Sea; Lanyu (Taiwan); Taiwan Strait; South China Sea. (2, 30, 30b)

C. nobilis Grunow: Bohai; Yellow Sea; East China Sea; Taiwan Strait; South China Sea. (2, 12)

C. nodulifer A. Schmidt: Bohai; Yellow Sea; East China Sea; Okinawa Trough; Lanyu (Taiwan); Taiwan continental shelf; Taiwan Strait; South China Sea. (2, 31)

C. obscurus A. Schmidt: Bohai; Yellow Sea; East China Sea; Taiwan Strait; South China Sea. (1, 2, 3)

C. oculatus (Fauv.) Petit: Bohai; Yellow Sea; East China Sea; Taiwan continental shelf; Taiwan Strait; South China Sea. (2, 31)

C. oculus-iridis Ehrenberg: Bohai; Yellow Sea; East China Sea; Okinawa Trough; Taiwan Strait; South China Sea. (1, 2, 3)

C. oculus-iridis v. *borealis* (Bail.) Cleve: Bohai; Yellow Sea. (4)

C. paradoxus Cheng & Chin: Taiwan Strait. (1, 2, 3)

C. perforatus Ehrenberg: Bohai; Yellow Sea; East China Sea; Taiwan continental shelf. (4, 12, 31)

C. perforatus v. *cellulosa* Grunow: Bohai; Yellow Sea; East China Sea; South China Sea. (4, 12, 14)

C. perforatus v. *pavillard* (Forti) Hustedt: Bohai; Yellow Sea; East China Sea; South China Sea. (4, 12, 14, 15)

C. planithecus Kuo: South China Sea. (1, 2, 3)

C. punctulatus Gregory: Taiwan Strait. (1, 2, 3)

C. radiatus Ehrenberg: Bohai; Yellow Sea; East China Sea; Okinawa Trough; Taiwan continental shelf; Taiwan Strait; South China Sea. (2, 31)

C. reniformis Castracane: Okinawa Trough; Taiwan Strait. (2, 29)

C. rothii (Ehr.) Grunow: Bohai; Yellow Sea; East China Sea; Taiwan Strait; South China Sea. (2, 4, 12)

C. rothii v. *grandiuscula* Rattray: Taiwan Strait. (1, 2, 3)

C. rothii v. *normani* (Greg.) Grunow: Taiwan. (31)

C. scitulus Mann: Okinawa Trough; South China Sea. (1, 2, 3)

C. shantouensis Kuo: South China Sea. (1, 2, 3)

C. spiculatus Ehrenberg: South China Sea. (1, 2, 3)

C. spinosus Chin: Bohai; Yellow Sea; East China Sea; Taiwan Strait; South China Sea. (1, 2, 3)

C. stellaris Roper: Taiwan Strait; South China Sea. (2, 12)

C. stellaris v. *symbolophora* (Grun.) Jörgensen: Taiwan Strait. (1, 2, 3)

C. subaulacodiscoidalis Rattray: East China Sea; Taiwan Strait. (1, 2, 3)

C. sub-bulliens Jörgensen: East China Sea; Taiwan Strait. (1, 2, 3)

C. subconcavus Grunow: East China Sea; Taiwan Strait; South China Sea. (1, 2, 3)

C. subconcavus v. *tenuior* Rattray: East China Sea; Okinawa Trough. (1, 2, 3)

C. subtilis Ehrenberg: Bohai; Yellow Sea; East China Sea; Okinawa Trough; Taiwan continental shelf; Taiwan Strait; South China Sea. (2, 31)

C. subtilis v. *minorus* Kuo: South China Sea (14).

C. suspectus Janisch: South China Sea. (1, 2, 3)

C. temperei Brun: East China Sea; Taiwan Strait. (1, 2, 3)

C. tenuithecus Kuo: South China Sea. (1, 2, 3)

C. thorii Pavillard: Bohai; Yellow Sea; East China Sea; Taiwan Strait; South China Sea. (1, 2, 3)

C. turgidus Rattray: Bohai; Yellow Sea; East China Sea; South China Sea. (1, 2, 3)

C. wailesii Gran & Angst: Bohai; Yellow Sea; East China Sea; Taiwan Strait; South China Sea. (1, 2, 3)

C. wittianus Pantocsek: East China Sea; Okinawa Trough; Taiwan Strait; South China Sea. (1, 2, 3)

Cyclotella caspia Grunow: Taiwan Strait; South China Sea. (6, 23a)

C. comta v. *oligactis* (Ehr.) Grunow: East China Sea. (1, 2, 3)

C. frigida Cleve-Euler: Bohai; Yellow Sea. (12)

C. ladogensis Cleve-Euler: Bohai; Yellow Sea. (12)

C. meneghiniana Kuetzing: Bohai; Yellow Sea; Taiwan Strait; Lanyu (Taiwan); South China Sea. (1, 2, 3, 30b)

C. stelligera (Cl. & Grun.) Van Heurck: South China Sea. (24)

C. striata (Kuetz.) Grunow: From Bohai to South China Sea; Lanyu (Taiwan). (1, 2, 3, 30b)

C. striata v. *ambigua* Cheng, Gao & Chin: Taiwan Strait. (20)

C. striata v. *baltica* Grunow: From Bohai to Taiwan Strait. (1, 2, 3)

C. striata v. *intermedia* Grunow: Taiwan Strait. (1, 2, 3)

C. striata v. *mucronulata* Chi & Chang: South China Sea. (6)

C. stylorum Brightwell: Bohai; Yellow Sea; East China Sea; Taiwan Strait; Taiwan continental shelf; South China Sea. (2, 31)

Cymatodiscus planetophorus (Meister) Hendey: East China Sea; Taiwan Strait; South China Sea. (1, 2, 3)

Cymatotheca minima Voigt: Taiwan Strait. (23a)

C. weissflogii (Grun.) Hendey: Taiwan Strait. (23a)

C. weissflogii v. *densestriata* Voigt: Bohai; Yellow Sea; East China Sea; South China. (1, 2, 3)

Dactyliosolen mediterraneus (Perag.) Peragallo: From Bohai to South China Sea. (1, 2, 3)

Endictya oceanica Ehrenberg: East China Sea; Taiwan Strait; South China Sea. (1, 2, 3, 23b)

Ethmodiscus gazellae (Janisch) Hustedt: Okinawa Trough; Taiwan Strait; South China Sea. (1, 2, 3)

Gossleriella tropica Schuett: From Yellow Sea to South China Sea; eastern Taiwan. (2, 3, 28)

Guinardia blavyana Peragallo: South China Sea. (1, 2, 3)

G. flaccida (Castr.) Peragallo: From Bohai to South China Sea. (2, 12)

Hyalodiscus ambiguus (Grun.) Tempere & Peragallo: Taiwan Strait; South China Sea (1, 2, 3)

H. radiatus (O'Meara) Grunow: East China Sea; Taiwan Strait; South China Sea. (1, 2, 3)

H. subtilis Bailey: From Bohai to South China Sea; Lanyu (Taiwan). (1, 2, 3, 30b)

Lauderia annulata Cleve: South China Sea. (1, 2, 3)

L. borealis Gran: From Bohai to South China Sea; southeastern Taiwan. (1, 2, 3, 30a)

Leptocylindrus adriaticus Schroeder: Bohai; Yellow Sea. (12)

L. danicus Cleve: From Bohai to South China Sea; southeastern Taiwan. (1, 2, 3, 30a)

Liradiscus reniformis Chin & Cheng: Okinawa Trough. (1, 2, 3)

Melosira architecturalis Brun: Bohai, Yellow Sea. (1, 2, 3)

M. arctica Ehrenberg: Taiwan Strait. (1, 2, 3)

M. granulata (Ehr.) Ralfs: East China Sea; South China Sea. (6, 12)

M. islandica O. Mueller: East China Sea; South China Sea. (1, 11)

M. juergensi Agardh: Taiwan Strait; South China Sea. (1, 2, 3)

M. moniliformis (Muell.) Agardh: Bohai; Yellow Sea; East China Sea; Taiwan Strait; South China Sea. (1, 2, 3)

M. nummuloides (Dillw.) Agardh: East China Sea; Taiwan Strait; South China Sea. (1, 2, 3)

M. sol (Ehr.) Kuetzing: East China Sea continental shelf. Sediment surface. (1, 2, 3)

M. sulcata (Ehr.) Kuetzing [*Paralia sulcata* (Ehr.) Cleve]: From Bohai to South China Sea; Taiwan continental shelf. (1, 2, 3)

M. sulcata f. *radiata* (Grun.) Peragallo & Peragallo: Bohai; Yellow Sea. (1, 2, 3)

Microsolenia simplex Takano: Taiwan Strait. (23a)

Minidiscus chilensis Rivera: Taiwan Strait. (19)

M. comicus Takano: Taiwan Strait. (19)

M. ocellatus Guo, Cheng & Chin: Taiwan Strait. (20)

M. spinulosus Guo, Cheng, & Chin: Taiwan Strait. (20)

M. subtilis Gao, Cheng & Chin: Taiwan Strait. (19)

M. triocutatus (F. J. R. Taylor) Hasle: Taiwan Strait. (19)

Planktoniella sol (Wallich) Schuett: From Bohai to South China Sea. (1, 2, 3)

Podosira argus Grunow: East China Sea. (1, 2, 3)

P. granulata Liu: East China Sea. (1, 2, 3)

P. hormoides (Mont.) Kuetzing: East China Sea; Taiwan Strait; South China Sea. (1, 2, 3, 30c)

P. maxima (Kuetz.) Grunow: South China Sea. (1, 2, 3, 23b)

P. stelliger (Bailey) Mann [*Hyalodiscus stelliger* Bailey]: From Bohai to South China Sea; Okinawa Trough; Taiwan continental shelf. (1, 31)

Pyxidicula mediterranea Grunow: Okinawa Trough. (1, 2, 3)

P. weyprechtii Grunow: East China Sea; Taiwan Strait. (1, 2, 3)

Rocella marina Cheng & Gao: Taiwan Strait. (22)

Schroederella delicatula (Perag.) Pavillard: From Yellow Sea to South China Sea. (2, 28)

S. delicatula f. *schroederi* (Bergon) Sournia [*S. schroederi* (Berg.) Lebour]: East China Sea; Taiwan Strait; South China Sea. (2, 4, 29)

Skeletonema costatum (Grev.) Cleve: From Bohai to South China Sea; southeastern Taiwan. (1, 2, 3, 30a)

S. munzelli Guillard, Carpenter & Reimann: Taiwan Strait. (22, 23)

S. potamos (Weber) Hasle: Taiwan Strait. (23)

S. tropicum Cleve: Taiwan Strait. (23)

Stephanodiscus astraea v. *minutula* (Kuetz.) Grunow: Taiwan Strait. (1, 2, 3)

Stephanopyxis aculeata (Ehr.) Grunow: South China Sea. (1, 2, 3)

S. nipponica Gran & Yendo: East China Sea; Taiwan Strait; South China Sea. (15, 16, 30)

S. palmeriana (Grev.) Grunow: Bohai; Yellow Sea. (1, 2, 3)

S. turris (Grev. & Arndt) Ralfs: Bohai; Yellow Sea. (1, 2, 3)

S. turris v. *polaris* Grunow: Taiwan continental shelf. (31)

Thalassiosira antiqua (Grun.) Cleve-Euler: Okinawa Trough; South China Sea. (1, 2, 3)

T. baltica (Grun.) Ostenfeld: Bohai; Yellow Sea; Taiwan Strait. (12, 30)

T. binata G. Fryxell: Taiwan Strait. (23a)

T. binata v. *bibinata* Gao & Cheng: Taiwan Strait. (21)

T. binata v. *miner* Gao & Cheng: Taiwan Strait. (21)

T. condensata (Cleve) Lebour: Bohai; Yellow Sea; East China Sea; southeastern Taiwan; South China Sea. (4, 12, 15, 16, 30a)

T. decipiens (Grun.) E. Joergensen: From Bohai to East China Sea; Okinawa Trough; Taiwan Strait. (3, 4, 12, 15, 30)

T. eccentrica (Ehr.) Cleve: Taiwan Strait; Taiwan continental shelf; South China Sea. (3, 31)

T. gravida Cleve: Yellow Sea; East China Sea; Taiwan Strait; South China Sea. (3, 28, 30)

T. hyalina (Grun.) Gran: Bohai; Yellow Sea. (4, 12, 25)

T. hydra Gombos: Taiwan Strait. (1, 2, 3)

T. laevissima Gao & Cheng: Taiwan Strait. (21)

T. lineata Jouse: Okinawa Trough. (1, 2, 3)

T. nordenskioldii Cleve: From Bohai to Taiwan Strait. (2, 4, 12, 27, 28)

T. oestrupii (Ostenfeld) Proschkina-Lavernko: East China Sea; Okinawa Trough; South China Sea. (1, 2, 3)

T. pacifica Gran & Angst.: From Bohai to South China Sea. (2, 12, 23b, 23e)

T. punctigera Hasle: Taiwan Strait. (8)

T. rotula Meunier: From Bohai to South China Sea; southeastern Taiwan. (2, 12, 30a)

T. scrotiformis sp. nov Shen: Yellow Sea. (10)

T. simonsenii Hasle & Fryxell: Taiwan Strait. (31)

T. subtilis (Ostenf.) Gran.: From Bohai to South China Sea. (1, 2, 3)

T. symmetrica Fryxell & Hasle: Taiwan Strait; South China Sea. (1, 2, 3)

T. weissflogii (Grun.) Fryxell & Hasle: Bohai; Yellow Sea; Taiwan Strait. (1, 2, 3)

Xanthiopyxis microspinosa v. *elliptica* Liu: East China Sea; Taiwan Strait. (1, 2, 3)

EUPODISCACEAE

Actinocyclus alienus Grunow: Bohai; Yellow Sea; East China Sea. (1, 2, 3)

A. appendiculatus (Grun.) Rattray: Taiwan Strait; South China Sea. (1, 2, 3)

A. australis Grunow: Taiwan Strait; South China Sea. (1, 2, 3)

A. curvatulus Janisch: East China Sea; Lanyu (Taiwan). (1, 2, 3, 30b)

A. ehrenbergii Ralf: Bohai; Yellow Sea; East China Sea; Okinawa Trough; Lanyu (Taiwan); Taiwan continental shelf; Taiwan Strait; South China Sea. (2, 31)

A. ehrenbergii v. *crassa* (W. Smith) Hustedt: Bohai; Yellow Sea; East China Sea; Okinawa Trough; Taiwan continental shelf; Taiwan Strait; South China Sea. (2, 31)

A. ehrenbergii v. *ralfsii* (W. Smith) Hustedt [*A. ralfsii*]: Bohai; Yellow Sea; East China Sea; Taiwan Strait; South China Sea. (1, 2, 3)

A. ehrenbergii v. *tenella* (Breb.) Hustedt: Bohai; Yellow Sea; East China Sea; Taiwan Strait; South China Sea. (1, 2, 3)

A. ellipticus Grunow: East China Sea; Taiwan Strait. (1, 2, 3)

A. ellipticus f. *lanceolata* Volbe: Bohai; Yellow Sea; East China Sea; Taiwan Strait. (1, 2, 3)

A. fasciculatus Castracane: South China Sea. (1, 2, 3)

A. ingens Rattray: South China Sea. (1, 2, 3)

A. normani (Greg.) Hustedt: Okinawa Trough; South China Sea. (1, 2, 3)

A. roperi (Breb.) Grunow: East China Sea; Taiwan Strait; South China Sea. (2, 12, 23b)

A. subtilis (Greg.) Ralfs: South China Sea. (1, 2, 3)

A. varicus Mann [*Stictocyclus varicus* Mann]: South China Sea. (1, 2, 3)

Aulacodiscus affinis Grunow: South China Sea. (1, 2, 3, 23b)

A. argus (Ehr.) A. Schmidt: Taiwan; South China Sea. (23b, 31)

A. margaritaceus Ralfs.: Taiwan continental shelf; Taiwan Strait. (1, 2, 3, 23b)

A. reticulata Pantocsek: Taiwan continental shelf. (31)

Auliscus caelatus Bailey: East China Sea; Okinawa Trough; South China Sea. (1, 2, 3, 23b)

A. incertus A. Schmidt: East China Sea; Taiwan Strait. (1, 2, 3)

A. punctatus Bailey: Taiwan Strait; South China Sea. (1, 2, 3, 23b)

A. sculptus (W. Smith) Ralfs: East China Sea; Taiwan Strait; South China Sea. (1, 2, 3, 23b)

Eupodiscus radiatus Bailey: South China Sea. (1, 2, 3)

Hemidiscus cuneiformis Wallich: Yellow Sea; East China Sea; Okinawa Trough; Taiwan Strait; South China Sea. (2, 31)

H. cuneiformis v. *orbicularis* (Castr.) Hustedt: East China Sea; Okinawa Trough; Taiwan continental shelf. (2, 31)

H. cuneiformis v. *recta* (Castr.) Hustedt: Okinawa Trough. (1, 2, 3)

H. cuneiformis v. *ventricosa* (Castr.) Hustedt: Okinawa Trough; Taiwan Strait. (2, 30)

H. hardmannianus (Grev.) Mann: Bohai; Yellow Sea; East China Sea; Taiwan Strait; South China Sea. (2, 12, 28)

H. ovalis Lohman: Taiwan Strait; South China Sea. (1, 2, 3)

Roperia excentrica Chin & Cheng: East China Sea; Okinawa Trough; South China Sea. (1, 2, 3)

R. latiovala Chen & Qian: Yellow Sea; East China Sea. (9, 28)

R. tesselata (Roper) Grunow: Bohai; Yellow Sea; East China Sea; Taiwan Strait; South China Sea. (2, 12)

ACTINODISCACEAE

Actinoptychus annulatus (Wallich) Grunow: Bohai; Yellow Sea; East China Sea; Taiwan Strait; South China Sea. (1, 2, 3, 23b)

A. annulatus v. *minor* Grunow: East China Sea; Taiwan Strait. (1, 2, 3)

A. australis (Grun.) Andrews: Bohai; Yellow Sea; Taiwan Strait. (1, 2, 3)

A. hexagonus Grunow: South China Sea. (1, 2, 3)

A. marylandicus Andrews: East China Sea; South China Sea. (1, 2, 3)

A. notabilis Grunow: Taiwan continental shelf. (31)

A. pericavatus Brun: East China Sea. (1, 2, 3)

A. splendens (Shadb.) Ralfs: Bohai; Yellow Sea; East China Sea; Okinawa Trough; Taiwan continental shelf; Taiwan Strait; South China Sea. (2, 31)

A. stella v. *thunii* A. Schmidt: Taiwan Strait; South China Sea. (2, 29)

A. subangulatus A. Schmidt: East China Sea. (1, 2, 3)

A. triacriformis Cheng & Chin: East China Sea; Okinawa Trough; Taiwan Strait. (1, 2, 3)

A. trilingulatus (Brightw.) Ralfs: Bohai; Yellow Sea; East China Sea; Taiwan Strait; South China Sea. (2, 12, 23b, 29)

A. undulatus (Bailey) Ralfs: Bohai; Yellow Sea; East China Sea; Okinawa Trough; Taiwan continental shelf; Taiwan Strait; South China Sea. (2, 31)

A. vulgaris Schumann: Taiwan Strait. (1, 2, 3, 23b)

Arachnoidiscus ehrenbergii Bailey: Bohai; Yellow Sea; East China Sea; Taiwan Strait; South China Sea. (1, 2, 3)

A. ehrenbergii v. *monteryana* A. Schmidt: Taiwan Strait. (1, 2, 3)

A. ornarus Ehrenberg: East China Sea; Taiwan Strait; South China Sea. (1, 2, 3)

Asterolampra marylandica Ehrenberg: Yellow Sea; East China Sea; Okinawa Trough; Taiwan Strait; Taiwan continental shelf; South China Sea. (2, 28, 29, 30, 30a, 30b)

A. vanheurckii Brun: Yellow Sea; East China Sea; Taiwan Strait; South China Sea. (2, 7, 28)

Asteromphalus arachne (Breb.) Ralfs: East China Sea; Okinawa Trough; South China Sea. (2, 11)

A. brookei Bailey: East China Sea. (1, 2, 3)

A. cleveanus Grunow: Bohai; Yellow Sea; East China Sea; Taiwan continental shelf; Taiwan Strait; South China Sea. (2, 4, 31)

A. elegans Greville: East China Sea; Okinawa Trough; Taiwan Strait; South China Sea. (2, 7, 12, 23e)

A. flabellatus (Breb.) Greville: Bohai; Yellow Sea; East China Sea; Okinawa Trough; Taiwan Strait; South China Sea. (1, 2, 3)

A. heptactis (Breb.) Ralfs: Okinawa Trough; Taiwan Strait; southeastern Taiwan; South China Sea. (2, 23b, 23c, 29, 30, 30a)

A. robustus Castracane: East China Sea; Taiwan Strait. (1, 2, 3)

A. roperianus (Grev.) Ralfs: Taiwan Strait. (7)

Mastogonia crux Ehrenberg: Taiwan Strait. (1, 2, 3)

M. heptagona Ehrenberg: Bohai; Yellow Sea. (12)

M. ocella Chin & Cheng: Okinawa Trough. (1, 2, 3)

Stictodiscus argus A. Schmidt: East China Sea. (1, 2, 3)

S. buryanus Greville: East China Sea. (1, 2, 3)

S. buryanus f. *subtriangularis* Truan & Witt: East China Sea. (1, 2, 3)

S. californicus Greville: East China Sea; South China Sea. (1, 2, 3)

S. californicus v. *nitida* Grove & Sturt [*S. nitidus* (Grove & Sturt) A. Schmidt]: Taiwan continental shelf. (31)

S. johnsonianus Greville: East China Sea; Okinawa Trough. (1, 2, 3)

S. varians Castracane: South China Sea. (1, 2, 3)

BIDDULPHIALES
BIDDULPHIACEAE

Attheya zachariasi Brun: East China Sea. (1, 2, 3)

Bellerochea malleus (Brightw.) Van Heurck: Bohai; Yellow Sea; East China Sea; Taiwan Strait; South China Sea. (1, 2, 3)

Biddulphia aurita (Lyngb.) Brebisson & Godey: Bohai; Yellow Sea; East China Sea; Taiwan Strait; South China Sea. (1, 2, 3)

B. dubia (Brightw.) Cleve: Bohai; Yellow Sea; East China Sea; Taiwan Strait; South China Sea. (2, 12)

B. granulata Roper: Bohai; Yellow Sea; East China Sea; Taiwan Strait. (2, 12, 27)

B. gruendleri A. Schmidt: Bohai; Yellow Sea; East China Sea; Taiwan Strait; South China Sea. (1, 2, 3)

B. heteroceros Grunow: Bohai; Yellow Sea; East China Sea; Taiwan Strait; South China Sea. (2, 7)

B. laevis Ehrenberg: Bohai; Yellow Sea; Taiwan Strait; South China Sea. (2, 12)

B. longicruris Greville: Yellow Sea; East China Sea; Taiwan Strait; South China Sea. (2, 25)

B. mobiliensis (Bailey) Grunow: Bohai; Yellow Sea; East China Sea; Taiwan Strait; southeastern Taiwan; South China Sea. (1, 2, 3, 30a)

B. obtusa: (Kuetz.) Ralfs: Bohai; Yellow Sea; East China Sea; Taiwan Strait; South China Sea. (1, 2, 3)

B. plana A. Schmidt: Taiwan continental shelf. (31)

B. pulchella Gray: Bohai; Yellow Sea; East China Sea; Lanyu (Taiwan); Taiwan continental shelf; Taiwan Strait; South China Sea. (2, 30b, 31)

B. regia (Schultze) Ostenfeld: Bohai; Yellow Sea; East China Sea; Taiwan Strait; South China Sea. (1, 2, 3)

B. reticulata Roper: Bohai; Yellow Sea; East China Sea; Taiwan Strait; South China Sea. (1, 2, 3)

B. retiformis Mann: Taiwan Strait. (1, 2, 3)

B. rhombus (Ehr.) W. Smith: Bohai; Yellow Sea; East China Sea; Taiwan Strait; South China Sea. (1, 2, 3)

B. rhombus v. *trigona* (Cleve) Hustedt: Taiwan Strait. (1, 2, 3)

B. roperiana Greville: East China Sea. (12)

B. schroederiana Schussnig: South China Sea. (1, 2, 3)

B. seticulosa Grunow: Taiwan. (31)

B. sinensis Greville: Bohai; Yellow Sea; East China Sea; Taiwan Strait; South China Sea. (1, 2, 3)

B. tridens Ehrenberg [*B. tuomeyi* (Bailey) Roper]: Bohai; Yellow Sea; East China Sea; Taiwan Strait; South China Sea. (1, 2, 3)

Cerataulus smithii Ralfs: East China Sea. (1, 2, 3)

C. turgidus (Ehr.) Ehrenberg: East China Sea; Taiwan Strait; South China Sea. (2, 12)

Climacodium biconcavum Cleve: Bohai; Yellow Sea; East China Sea; Taiwan Strait; southeastern Taiwan; South China Sea. (1, 2, 3, 30a)

C. frauenfeldianum Grunow: Yellow Sea; East China Sea; Taiwan Strait; southeastern Taiwan; South China Sea. (2, 27, 30a)

Ditylum brightwelli (West) Grunow: Bohai; Yellow Sea; East China Sea; Taiwan Strait; South China Sea. (1, 2, 3)

D. sol Grunow: Bohai; Yellow Sea; East China Sea; Taiwan Strait; southeastern Taiwan; South China Sea. (1, 2, 3, 30a)

Eucampia cornuta (Cleve) Grunow: Yellow Sea; East China Sea; Taiwan Strait; southeastern Taiwan; South China Sea. (2, 28, 30a)

E. zoodiacus Ehrenberg: Bohai; Yellow Sea; East China Sea; Taiwan Strait; southeastern Taiwan; South China Sea. (1, 2, 3, 30a)

Hemiaulus chinensis Greville: Bohai; Yellow Sea; East China Sea; Taiwan Strait; South China Sea. (1, 2, 3)

H. hauckii Grunow: Bohai; Yellow Sea; East China Sea; Taiwan Strait; southeastern Taiwan; South China Sea. (1, 2, 3, 30a)

H. indicus Karsten: Bohai; Yellow Sea; East China Sea; Taiwan Strait; South China Sea. (2, 29)

H. membranaceus Cleve: Bohai; Yellow Sea; East China Sea; Taiwan Strait; South China Sea. (1, 2, 3)

H. sinensis Grunow: Bohai; Yellow Sea; East China Sea; Taiwan Strait; southeastern Taiwan; South China Sea. (2, 7, 30a)

Hydrosera triquetra Wallich: Taiwan continental shelf. (31)

H. whampoensis Schwarz [*Triceratium whampoense* Schwarz]: East China Sea; Taiwan Strait; South China Sea. (6, 12)

Isthmia japonica (Cast.) Sournia: South China Sea. (1, 2, 3)

I. minima Harvey & Bailey: South China Sea. (1, 2, 3)

I. nervosa Kuetzing: East China Sea; South China Sea. (1, 2, 3, 14, 23b)

Leudugeria janischii (Grun.) Van Heurck: Taiwan Strait; South China Sea. (1, 2, 3)

Lithodesmium undulatus Ehrenberg: East China Sea; Taiwan Strait; South China Sea. (12, 30)

L. variabile Takano: Taiwan Strait. (23a)

Streptothece indica Karsten: Yellow Sea; East China Sea; Taiwan Strait; South China Sea. (1, 2, 3)

S. thamesis Shrubsole: Bohai; Yellow Sea; East China Sea; Taiwan Strait; South China Sea. (1, 2, 3)

Triceratium affine Grunow: Bohai; Yellow Sea; East China Sea; Taiwan Strait; South China Sea. (2, 12, 29)

T. americanum Ralfs: South China Sea. (16)

T. antedeluvianum (Ehr.) Grunow: East China Sea; Taiwan Strait; South China Sea. (2, 12)

T. arcticum v. *japonica* Grunow: South China Sea. (1, 2, 3)

T. balearicum f. *biquadrata* (Janisch) Hustedt: Taiwan Strait. (1, 2, 3)

T. broeckii Leuduger-Fortmorel: Taiwan Strait. (1, 2, 3)

T. campechianum Grunow: East China Sea. (1, 2, 3)

T. cultum A. Schmidt: East China Sea. (1, 2, 3)

T. cuspidatum Janisch: East China Sea; Okinawa Trough; South China Sea. (1, 2, 3)

T. dubium Brightwell: Okinawa Trough; Taiwan Strait; Lanyu (Taiwan); South China Sea. (1, 2, 3, 30b)

T. favus Ehrenberg: Bohai; Yellow Sea; East China Sea; Taiwan Strait; Taiwan continental shelf; South China Sea. (2, 31)

T. favus f. *quadrata* Grunow: South China Sea. (1, 2, 3, 23b)

T. formosum Brightwell: Bohai; Yellow Sea; Taiwan Strait; Lanyu (Taiwan); Taiwan continental shelf; South China Sea. (2, 4, 30b, 31)

T. formosum f. *pentagonale* (A. S.) Hustedt: South China Sea. (1, 2, 3)

T. formosum f. *quadrangularis* (Grev.) Hustedt: South China Sea. (1, 2, 3, 23b)

T. gallapagense Cleve: East China Sea. (1, 2, 3)

T. gibbosum Bail: South China Sea. (16)

T. japonicum A. Schmidt: Taiwan Strait. (1, 2, 3)

T. junctum A. Schmidt: Taiwan Strait. (1, 2, 3)

T. orbiculatum Shadbolt: South China Sea. (1, 2, 3)

T. pelagicum (Schroeder) Sournia: South China Sea. (16)

T. pellucida (Castr.) Kuo, Ye & Zhou: South China Sea. (1, 2, 3)

T. pentacrinus (Ehr.) Wallich: Bohai; Yellow Sea; East China Sea; Taiwan Strait; South China Sea. (2, 12, 23b)

T. pentacrinus f. *quadrata* Hustedt: Taiwan Strait. (1, 2, 3)

T. perpendiculare Lin & Chin: East China Sea; Taiwan Strait. (2, 12)

T. reticulum Ehrenberg: East China Sea; Taiwan Strait; Taiwan continental shelf; South China Sea. (1, 2, 3, 23b, 31)

T. rostratum Petit: East China Sea. (1, 2, 3)

T. scitulum Brightwell: Taiwan Strait; South China Sea. (1, 2, 3)

T. shadboltianum Greville: Taiwan Strait; South China Sea. (1, 2, 3)

T. zonulatum Greville: South China Sea. (1, 2, 3, 23b)

Trinacria regina v. *tetragona* Grunow: Taiwan Strait. (1, 2, 3)

CHAETOCERACEAE

Chaetoceros abnormis Proschkina-Lavrenko: Bohai; Yellow Sea; East China Sea; Taiwan Strait; South China Sea. (2, 12, 27)

C. aequatoriale Cleve: Yellow Sea; East China Sea; Taiwan Strait; South China Sea. (2, 28, 29)

C. affinis Lauder: Bohai; Yellow Sea; East China Sea; Taiwan Strait; southeastern Taiwan; South China Sea. (1, 2, 3, 30a)

C. affinis v. *circinalis* (Meunier) Hustedt: Bohai; Yellow Sea; East China Sea; Taiwan Strait; South China Sea. (2, 12)

C. affinis v. *willei* (Gran) Hustedt: Bohai; Yellow Sea; East China Sea; South China Sea. (1, 2, 3)

C. anastomosans Grunow: Bohai; Yellow Sea; East China Sea; Taiwan Strait; South China Sea. (2, 7)

C. atlanticus Cleve: Yellow Sea; East China Sea; Taiwan Strait; southeastern Taiwan; South China Sea. (2, 16, 28, 30a)

C. atlanticus v. *neapolitana* (Schroeder) Hustedt: Yellow Sea; East China Sea; Taiwan Strait; South China Sea. (2, 16, 28)

C. atlanticus v. *skeleton* (Schuett) Hustedt: Yellow Sea; East China Sea; Taiwan Strait; South China Sea. (2, 28)

C. aurivillii Cleve [*C. seychellarum* Karsten]: Yellow Sea; East China Sea; Taiwan Strait; eastern Taiwan; South China Sea. (2, 15, 28, 29)

C. bacteriastroides Karsten: East China Sea; South China Sea. (2, 15)

C. borealis Bailey: Yellow Sea; East China Sea; Taiwan Strait; South China Sea. (2, 12, 28)

C. brevis Schuett: Bohai; Yellow Sea; East China Sea; Taiwan Strait; southeastern Taiwan; South China Sea. (1, 2, 3, 11, 30a)

C. buceros Karsten: Taiwan Strait; South China Sea. (2, 7)

C. castracanei Karsten: Bohai; Yellow Sea; East China Sea; Taiwan Strait; South China Sea. (2, 12, 29)

C. cinctus Gran: Bohai; Yellow Sea; East China Sea; Taiwan Strait; South China Sea. (2, 12)

C. coarctatus Lauder: Bohai; Yellow Sea; East China Sea; Taiwan Strait; South China Sea. (1, 2, 3)

C. compressus Lauder: Bohai; Yellow Sea; East China Sea; Taiwan Strait; southeastern Taiwan; South China Sea. (1, 2, 3, 30a)

C. constrictus Gran: Bohai; Yellow Sea; East China Sea; Taiwan Strait. (1, 2, 3)

C. convolutus Castracane: Bohai; Yellow Sea; East China Sea; Taiwan Strait; South China Sea. (2, 4, 30)

C. costatus Pavillard: Bohai; Yellow Sea; East China Sea; Taiwan Strait; South China Sea. (2, 4, 12)

C. crinitus Schuett: Bohai; Yellow Sea; East China Sea; Taiwan Strait; South China Sea. (2, 7)

C. curvisetus Cleve: Bohai; Yellow Sea; East China Sea; Taiwan Strait; southeastern Taiwan; South China Sea. (1, 2, 3, 30a)

C. dadayi Pavillard: Yellow Sea; East China Sea; South China Sea. (2, 15, 28)

C. danicus Cleve: Bohai; Yellow Sea; East China Sea; Taiwan Strait; South China Sea. (2, 12)

C. debilis Cleve: Bohai; Yellow Sea; East China Sea; Taiwan Strait; South China Sea. (1, 2, 3)

C. decipiens Cleve: Bohai; Yellow Sea; East China Sea; Taiwan Strait; South China Sea. (1, 2, 3)

C. decipiens f. *singularis* Gran: Yellow Sea; East China Sea; Taiwan Strait; South China Sea. (2, 4, 27)

C. densus (Cleve) Cleve: Bohai; Yellow Sea; East China Sea; Taiwan Strait; South China Sea. (1, 2, 3)

C. denticulatus Lauder: Bohai; Yellow Sea; East China Sea; Taiwan Strait; South China Sea. (2, 4, 12, 25, 27)

C. denticulatus v. *angusta* Hustedt: East China Sea; South China Sea. (1, 2, 3)

C. diadema (Ehr.) Grum: South China Sea. (32)

C. dichaeta Ehrenberg: South China Sea. (5)

C. didymus Ehrenberg: Bohai; Yellow Sea; East China Sea; Okinawa Trough; Taiwan Strait; South China Sea. (1, 2, 3)

C. didymus v. *anglica* (Grun.) Gran: Bohai; Yellow Sea; Taiwan Strait; South China Sea. (2, 29, 30)

C. didymus f. *protubernas* (Lauder) Gran & Yendo: Yellow Sea; East China Sea; Taiwan Strait; South China Sea. (2, 28)

C. dipyrenops Meunier: Bohai; Yellow Sea; East China Sea. (4, 15)

C. distans Cleve: Bohai; Yellow Sea; East China Sea; Taiwan Strait; southeastern Taiwan; South China Sea. (1, 2, 3, 30a)

C. diversus Cleve: Bohai; Yellow Sea; East China Sea; Taiwan Strait; South China Sea. (1, 2, 3)

C. eibenii Grunow: Bohai; Yellow Sea; East China Sea; Taiwan Strait; South China Sea. (1, 2, 3)

C. filformis Meunier: East China Sea; South China Sea. (4)

C. hirundinellus Qian: Yellow Sea; East China Sea; South China Sea. (2, 28)

C. holsaticus Schuett: Bohai; Yellow Sea; East China Sea. (2, 12)

C. indicum Karsten: East China Sea; Taiwan Strait; South China Sea. (1, 2, 3)

C. knipowitschi Henckel: Bohai; Yellow Sea. (4)

C. laciniosus Schuett: Bohai; Yellow Sea; East China Sea; Taiwan Strait; South China Sea. (1, 2, 3)

C. laevis Leuduger-Fortmorel: Yellow Sea; East China Sea; Taiwan Strait; southeastern Taiwan; South China Sea. (2, 28, 30a)

C. lauderi Ralfs: Bohai; Yellow Sea; East China Sea; Taiwan Strait; South China Sea. (2, 12, 27)

C. lorenzianus Grunow: Bohai; Yellow Sea; East China Sea; Taiwan Strait; southeastern Taiwan; South China Sea. (1, 2, 3, 30a)

C. lorenzianus f. *subsalinus* Proschkina-Lavrenko: Bohai; Yellow Sea. (12)

C. messanensis Castracane: Yellow Sea; East China Sea; Taiwan Strait; southeastern Taiwan; South China Sea. (2, 28, 30a)

C. mitra (Bailey) Cleve: Bohai; Yellow Sea. (1, 2, 3)

C. muelleri Lemmermann: Bohai; Yellow Sea; East China Sea; South China Sea. (2, 12, 27)

C. nipponica Ikari: Bohai; Yellow Sea; East China Sea; Taiwan Strait; South China Sea. (2, 29)

C. paradox Cleve: Bohai; Yellow Sea; East China Sea; Taiwan Strait; South China Sea. (1, 2, 3)

C. pelagicus Cleve: Bohai; Yellow Sea; East China Sea; Taiwan Strait; South China Sea. (2, 12)

C. pendulum Karsten: East China Sea; Taiwan Strait; South China Sea. (2, 7)

C. perpusillus Cleve: South China Sea. (5)

C. peruvianus Brightwell: Bohai; Yellow Sea; Taiwan Strait; South China Sea. (1, 2, 3)

C. peruvianus f. *robusta* (Cleve) Husted: Yellow Sea; East China Sea; South China Sea. (4, 13, 28)

C. pseudoaurivilli Ikari: Yellow Sea; East China Sea. (28)

C. pseudocurvisetus Mangin: Bohai; Yellow Sea; East China Sea; Taiwan Strait; South China Sea. (1, 2, 3)

C. pseudodichaeta Ikari: Taiwan Strait. (7)

C. radians Schuett: Bohai; Yellow Sea; East China Sea; Taiwan Strait; South China Sea. (1, 2, 3)

C. radicans Schuett: Bohai; Yellow Sea; East China Sea; Taiwan Strait; South China Sea. (2, 12, 28)

C. rostratus Lauder: Bohai; Yellow Sea; East China Sea; Taiwan Strait; South China Sea. (2, 12, 28, 29)

C. rostratus f. *glandazii* (Mang.) Taylor: Yellow Sea; East China Sea. (28)

C. seiracanthus Gran: Bohai; Yellow Sea; East China Sea; South China Sea. (2, 4)

C. setoensis Ikari: Yellow Sea; East China Sea; Taiwan Strait. (2, 28, 30)

C. siamense Ostenfeld: Bohai; Yellow Sea; East China Sea; Taiwan Strait; South China Sea. (1, 2, 3)

C. similis Cleve: Bohai; Yellow Sea. (1, 2, 3)

C. socialis Lauder: Bohai; Yellow Sea; East China Sea; Taiwan Strait; South China Sea. (2, 4, 12, 27)

C. subsecundus (Grun.) Hustedt: Bohai; Yellow Sea; East China Sea; Taiwan Strait. (1, 2, 3)

C. subtilis Cleve: Bohai; Yellow Sea; East China Sea; Taiwan Strait; South China Sea. (2, 12, 27)

C. teres Cleve: Bohai; Yellow Sea; East China Sea; Taiwan Strait. (1, 2, 3)

C. tetrastichon Cleve: Yellow Sea; East China Sea; South China Sea. (2, 15, 28)

C. tortissimus Gran: Bohai; Yellow Sea; East China Sea; Taiwan Strait; southeastern Taiwan; South China Sea. (2, 4, 30a)
C. vanheurcki Gran: Bohai; Yellow Sea; East China Sea; Taiwan Strait; South China Sea. (1, 2, 3)
C. weissflogii Schuett: East China Sea; Taiwan Strait; South China Sea. (2, 4, 12)
C. xishaensis Kuo: Taiwan Strait; South China Sea. (2, 29)

RUTILARIACEAE

Eunotogramma debile Grunow: Taiwan Strait; South China Sea. (1, 2, 3)
E. frauenfeldii Grunow: Taiwan Strait; South China Sea. (1, 2, 3)
E. laevis (Leve) Grunow: Taiwan Strait; South China Sea. (1, 2, 3)
E. rostratum Hustedt: Taiwan Strait; South China Sea. (1, 2, 3)

RHIZOSOLENIALES
RHIZOSOLENIACEAE

Rhizosolenia acuminata (Perag.) Peragallo & Peragallo: Yellow Sea; East China Sea; Taiwan Strait; southeastern Taiwan; South China Sea. (2, 28, 30a)
R. alata Brightwell: Bohai; Yellow Sea; East China Sea; Taiwan Strait; southeastern Taiwan; South China Sea. (1, 2, 3, 30a)
R. alata f. *curvirostris* Gran: East China Sea. (1, 2, 3)
R. alata f. *gracillima* (Cleve) Grunow: Bohai; Yellow Sea; East China Sea; Taiwan Strait; South China Sea. (1, 2, 3)
R. alata f. *indica* (Perag.) Hustedt: Bohai; Yellow Sea; East China Sea; Taiwan Strait; South China Sea. (1, 2, 3)
R. alata f. *inermis* (Castr.) Hustedt: East China Sea. (18)
R. bergonii Peragallo: Yellow Sea; East China Sea; Taiwan Strait; southeastern Taiwan; South China Sea. (1, 2, 3, 30a)
R. calcar-avis Schultze: Bohai; Yellow Sea; East China Sea; Taiwan Strait; southeastern Taiwan; South China Sea. (1, 2, 3, 30a)
R. castracanei Peragallo: Yellow Sea; East China Sea; Taiwan Strait; southeastern Taiwan; South China Sea. (2, 28, 30a)
R. castracanei v. *rhomboidea* Subrahmanyan: East China Sea. (1, 2, 3)
R. cleivei Ostenfeld: Yellow Sea; East China Sea; Taiwan Strait; South China Sea. (2, 28, 29)
R. crassispina Schroeder: Yellow Sea; East China Sea; Taiwan Strait; South China Sea. (1, 2, 3)
R. cylindrus Cleve: Yellow Sea; East China Sea; Taiwan Strait; southeastern Taiwan; South China Sea. (2, 28, 30a)
R. delicatula Cleve: Bohai; Yellow Sea; East China Sea; Taiwan Strait; South China Sea. (1, 2, 3)
R. fragilissima Bergon: Bohai; Yellow Sea; East China Sea; Taiwan Strait; South China Sea. (1, 2, 3)
R. hebetata Bailey: Bohai; Yellow Sea. (1, 2, 3)
R. hebetata f. *semispina* (Hansen) Gran: Bohai; Yellow Sea; Taiwan Strait; South China Sea. (1, 2, 3)
R. hyalina Ostenfeld & A. Schmidt: East China Sea; Taiwan Strait; South China Sea. (4, 11, 12, 15, 29a)
R. imbricata Brightwell: Bohai; Yellow Sea; East China Sea; Taiwan Strait; South China Sea. (1, 2, 3)
R. imbricata v. *shrubsolei* (Cleve) Van Heurck: East China Sea; Taiwan Strait; South China Sea. (1, 2, 3)
R. robusta Norman & Ralfs: Bohai; Yellow Sea; East China Sea; Taiwan Strait; South China Sea. (2, 12, 27)
R. setigera Brightwell: Bohai; Yellow Sea; East China Sea; Taiwan Strait; southeastern Taiwan; South China Sea. (2, 30, 30a)
R. sinensis Qian: Yellow Sea; South China Sea. (26, 28)
R. stolterforthii Peragallo: Bohai; Yellow Sea; East China Sea; Taiwan Strait; southeastern Taiwan; South China Sea. (1, 2, 3, 30a)
R. styliformis Brightwell: Bohai; Yellow Sea; East China Sea; Taiwan Strait; southeastern Taiwan; South China Sea. (1, 2, 3, 30a)
R. styliformis v. *latissima* Brightwell: East China Sea; Taiwan Strait; South China Sea. (2, 7)
R. styliformis v. *longispina* Hustedt: Bohai; Yellow Sea; East China Sea; Taiwan Strait; South China Sea. (2, 12, 29)

PENNATAE
NAVICULALES
NAVICULACEAE

Amphipleura micans (Lyn.) Cleve: South China Sea. (24)
A. rutilans (Trentep.) Cleve: Taiwan Strait; South China Sea. (2, 23d, 24)
A. rutilans v. *antarctica* (Har.) Cleve: South China Sea. (24)
Amphiprora alata (Ehr.) Kuetzing: Bohai; Yellow Sea; East China Sea; Taiwan Strait; South China Sea. (1, 2, 3)
A. paludosa v. *hyalina* (Eulenst.) Cleve: Bohai; Yellow Sea; South China Sea. (1, 2, 3)
A. venusta Greville: South China Sea. (1, 2, 3)
Anomoeoneis costata (Kuetz.) Hustedt: East China Sea; Taiwan Strait. (1, 2, 3)
Caloneis aemula (A. Schmidt) Cleve: Taiwan Strait; South China Sea. (1, 2, 3)
C. amphisbaena v. *fuscata* (Schum.) Cleve: Taiwan Strait. (1, 2, 3)
C. bacillaris (Greg.) Cleve: Taiwan Strait. (1, 2, 3)
C. brevis (Greg.) Cleve: East China Sea; Taiwan Strait; South China Sea. (1, 2, 3)
C. brevis v. *distoma* (Grun.) Cleve: Taiwan Strait. (1, 2, 3)
C. brevis v. *vexans* (Grun.) Cleve: Taiwan Strait; South China Sea. (1, 2, 3)
C. castracanei v. *petitiana* (Gru.) Cleve: South China Sea. (1, 2, 3)
C. consimilis (A. Schmidt) Cleve: Taiwan Strait; South China Sea. (1, 2, 3)
C. eccentrica (Grun.) Boyer: South China Sea. (1, 2, 3)
C. elongata (Grun.) Boyer: Bohai; Yellow Sea; East China Sea; Taiwan Strait; South China Sea. (1, 2, 3)
C. elongata v. *constricta* Cheng & Chin: Taiwan Strait. (1, 2, 3)

C. formosa (Greg.) Cleve: Bohai; Yellow Sea; East China Sea; Taiwan Strait; South China Sea. (1, 2, 3)

C. frater Cleve: South China Sea. (1, 2, 3)

C. galapagensis v. *japonica* Cleve: South China Sea. (1, 2, 3)

C. janischiana (Rab.) Boyer: South China Sea. (1, 2, 3)

C. liber (W. Smith) Cleve: Bohai; Yellow Sea; Taiwan Strait; South China Sea. (1, 2, 3, 30c)

C. linearis (Grun.) Boyer: Bohai; Yellow Sea; Taiwan Strait; South China Sea. (1, 2, 3)

C. ophiocephala (Cleve & Grove) Cleve: East China Sea; Taiwan continental shelf. (3, 31)

C. oregonica (Ehr.) Patrick: Taiwan Strait. (1, 2, 3)

C. permagna (Bailey) Cleve: Bohai; Yellow Sea; East China Sea; Taiwan Strait. (1, 2, 3)

C. platycephala Cheng & Chin: Taiwan Strait. (1, 2, 3)

C. robusta (Grun.) Cleve: Taiwan Strait. (1, 2, 3)

C. wittii Grunow: South China Sea. (1, 2, 3)

Cistula lorenziana (Grun.) Cleve: East China Sea; Taiwan Strait. (1, 2, 3)

Cymatoneis sulcata (Grev.) Cleve: South China Sea. (1, 2, 3)

Dictyoneis marginata (Lewis) Cleve: Taiwan continental shelf; Taiwan Strait. (23b, 31)

D. thumii Cleve: South China Sea. (1, 2, 3)

Diploneis adonis v. *ganymedes* Cleve: Okinawa Trough. (1, 2, 3)

D. advena (A. Schmidt) Cleve: South China Sea. (1, 2, 3)

D. aestiva (Donkin) Cleve: Bohai; Yellow Sea. (1, 2, 3)

D. beyrichiana (A. Schmidt) Amossé: Taiwan Strait; South China Sea. (1, 2, 3, 23b)

D. bomboides (A. Schmidt) Cleve: Taiwan Strait; South China Sea; Taiwan. (2, 31)

D. bomboides v. *media* (Grun.) Hustedt: South China Sea. (1, 2, 3)

D. bombus Ehrenberg: Bohai; Yellow Sea; East China Sea; Okinawa Trough; Taiwan continental shelf; Taiwan Strait; Lanyu (Taiwan); South China Sea. (2, 30b, 31)

D. bombus v. *bombiformis* (Cleve) Hustedt: South China Sea. (1, 2, 3)

D. borealis (Grun.) Cleve: Taiwan Strait. (1, 2, 3)

D. campylodiscus Grunow: East China Sea; Okinawa Trough. (1, 2, 3)

D. chersonensis (Grun.) Cleve: Bohai; Yellow Sea; East China Sea; Taiwan Strait; South China Sea. (2, 31)

D. chinensis Cleve: South China Sea. (1, 2, 3)

D. coffaeiformis (A. Schmidt) Cleve: East China Sea; Taiwan Strait. (1, 2, 3)

D. crabro (Ehr.) Ehrenberg: Bohai; Yellow Sea; East China Sea; Okinawa Trough; Taiwan continental shelf; Taiwan Strait; Lanyu (Taiwan); South China Sea. (2, 30b, 31)

D. crabro v. *excavata* Hustedt: East China Sea; South China Sea. (1, 2, 3)

D. crabro v. *omeari* (Grun.) Cleve: Taiwan Strait. (1, 2, 3)

D. crabro f. *pandura* (Bréb.) Cleve: Bohai; Yellow Sea; East China Sea; Taiwan Strait; South China Sea. (1, 2, 3)

D. crabro v. *subelliptica* Cleve: Okinawa Trough; South China Sea. (1, 2, 3)

D. crabro v. *suspecta* (A. Schmidt) Hustedt: Bohai; Yellow Sea; East China Sea; Okinawa Trough; Taiwan Strait; South China Sea. (1, 2, 3)

D. dalmatica (Grun.) Cleve: East China Sea; Okinawa Trough; Taiwan continental shelf. (2, 31)

D. debyi (Pantocs.) Cleve: Taiwan Strait. (1, 2, 3)

D. decipiens Cleve-Euler: South China Sea. (1, 2, 3)

D. decipiens v. *parallela* Cleve-Euler: East China Sea; Taiwan Strait; South China Sea. (1, 2, 3)

D. diplosticta (Grun.) Hustedt: Taiwan continental shelf. (31)

D. eudoxia (A. Schmidt) Joergensen: East China Sea. (1, 2, 3)

D. eugenia (A. Schmidt) Boyer: East China Sea. (1, 2, 3)

D. fusca (Greg.) Cleve: Bohai; Yellow Sea; East China Sea; Taiwan Strait; Taiwan continental shelf; South China Sea. (1, 2, 3)

D. fusca v. *pelagi* (A. Schmidt) Cleve: East China Sea; Okinawa Trough; Taiwan Strait; South China Sea. (1, 2, 3)

D. gemmata v. *pristiophora* (Janisch) Cleve: East China Sea; Okinawa Trough; South China Sea. (1, 2, 3)

D. gorjanovici v. *major* Chin & Lin: Taiwan Strait. (1, 2, 3)

D. graeffii (Grun.) Cleve: East China Sea; Okinawa Trough; South China Sea. (1, 2, 3)

D. gruendleri (A. Schmidt) Cleve: Bohai; Yellow Sea; East China Sea; Taiwan Strait; South China Sea. (1, 2, 3)

D. guinardinana (Brun) Cleve: East China Sea. (1, 2, 3)

D. hospes A. Schmidt: Taiwan Strait. (1, 2, 3)

D. incurvata (Greg.) Cleve: East China Sea; Okinawa Trough; Taiwan Strait; South China Sea. (1, 2, 3)

D. inscripta Cleve: South China Sea. (1, 2, 3)

D. interrupta (Kuetz.) Cleve: East China Sea; Taiwan Strait; South China Sea. (1, 2, 3)

D. interrupta v. *caffra* Giffen: East China Sea. (1, 2, 3)

D. lineata (Donkin) Cleve: Bohai; Yellow Sea; South China Sea. (1, 2, 3)

D. littoralis (Donkin) Cleve: Taiwan Strait; South China Sea. (1, 2, 3)

D. mediterranea (Grun.) Cleve: Okinawa Trough. (1, 2, 3)

D. nitescens (Gregory) Cleve: Bohai; Yellow Sea; East China Sea; Okinawa Trough; Taiwan continental shelf; Taiwan Strait; South China Sea. (2, 31)

D. notabilis (Grev.) Cleve: Taiwan Strait; South China Sea. (1, 2, 3)

D. novaesselandiae (A. Schmidt) Hustedt: East China Sea; Taiwan Strait; South China Sea. (1, 2, 3)

D. papula (A. Schmidt) Cleve: Taiwan Strait; South China Sea. (1, 2, 3)

D. parca (A. Schmidt) Boyer: East China Sea. (1, 2, 3)

D. puella (Schumam) Cleve: South China Sea. (24)

D. schmidtii Cleve: Bohai; Yellow Sea; East China Sea; Taiwan Strait; South China Sea. (1, 2, 3)

D. sejuncta (A. Schmidt) Joergensen: East China Sea. (1, 2, 3)

D. serratula (Grun.) Hustedt: Bohai; Yellow Sea; East China Sea; Okinawa Trough; Taiwan Strait; South China Sea. (1, 2, 3)

D. smithii (Brég.) Cleve: Bohai; Yellow Sea; East China Sea; Okinawa Trough; Taiwan Strait; South China Sea. (1, 2, 3)
D. smithii v. *constricta* Heiden: East China Sea. (1, 2, 3)
D. smithii v. *dilatata* (M. Per.) Terry: Bohai; Yellow Sea; East China Sea; Taiwan Strait; South China Sea. (1, 2, 3)
D. smithii v. *rhombica* Mereschkowsky: East China Sea; South China Sea. (1, 2, 3)
D. smithii v. *rombica* Mereschkowsky: Taiwan continental shelf. (31)
D. splendida (Greg.) Cleve: Bohai; Yellow Sea; Taiwan Strait; South China Sea. (2, 12)
D. splendida v. *puella* A. Schmidt: East China Sea; Taiwan Strait; South China Sea. (1, 2, 3)
D. subcincta (A. Schmidt) Cleve: Okinawa Trough. (1, 2, 3)
D. suborbicularis (Greg.) Cleve: Bohai; Yellow Sea; East China Sea; Taiwan Strait; Taiwan continental shelf; South China Sea. (2, 31)
D. vacillans (A. Schmidt) Cleve: East China Sea; South China Sea. (1, 2, 3)
D. vetula (A. Schmidt) Cleve: South China Sea. (1, 2, 3)
D. weissflogii (A. Schmidt) Cleve: Bohai; Yellow Sea; Taiwan Strait; Taiwan continental shelf; South China Sea. (2, 31)
D. zanzibarica (Grun.) Hustedt: South China Sea. (1, 2, 3)
Frustulia interposita (Lewis) De Toni: Taiwan Srait; South China Sea. (1, 2, 3)
F. interposita v. *chinensis* Skvortzow: Taiwan Strait. (1, 2, 3)
F. interposita v. *dispar* Liu & Chin: Taiwan Strait. (1, 2, 3)
F. lewisiana (Grev.) De Toni: East China Sea; Taiwan Strait; South China Sea. (1, 2, 3)
Gyrosigma acuminatum (Kuetz.) Rabenhorst: Bohai; Yellow Sea; East China Sea; Taiwan Strait; South China Sea. (1, 2, 3)
G. acuminatum v. *gallica* (Grun.) Cleve: Taiwan Strait. (1, 2, 3)
G. balticum (Ehr.) Rabenhorst: Bohai; Yellow Sea; East China Sea; Taiwan Strait; South China Sea. (1, 2, 3)
G. balticum v. *brevius* Chin & Liu: Taiwan Strait; South China Sea. (1, 2, 3)
G. balticum v. *sinensis* (Ehr.) Cleve: Bohai; Yellow Sea; East China Sea; Taiwan Strait; South China Sea. (1, 2, 3)
G. balticum v. *sinicum* Chin & Liu: Taiwan Strait; South China Sea. (1, 2, 3)
G. compactum (Grev.) Cleve: South China Sea. (1, 2, 3)
G. distortum (W. Smith) Griffith & Henfrey: Taiwan Strait. (1, 2, 3)
G. fasciola v. *arcuata* (Donkin) Cleve: Bohai; Yellow Sea; East China Sea; Taiwan Strait. (1, 2, 3)
G. fasciola v. *sulcata* (Grun.) Cleve: Taiwan Strait; South China Sea. (1, 2, 3)
G. fasciola v. *tenuirostris* (Grun.) Cleve: East China Sea; Taiwan Strait. (2, 22)
G. grovei (Peraf.) Cleve: Taiwan Strait. (1, 2, 3)
G. macrum (W. Smith) Griffith & Henfrey: Taiwan Strait. (1, 2, 3)
G. nodiferum West: Taiwan Strait. (1, 2, 3)
G. nodiferum v. *latum* Chin & Liu: Taiwan Strait; South China Sea. (1, 2, 3)
G. obliquum (Grun.) Boyer: Taiwan Strait. (1, 2, 3)
G. rectum (Donkin) Cleve: Taiwan Strait. (1, 2, 3)
G. sciotense (Sull. & Wormley) Cleve: Taiwan Strait; South China Sea. (1, 2, 3)
G. spencerii (W. Smith) Griffith & Henfrey: Bohai; Yellow Sea; East China Sea; Taiwan Strait; South China Sea. (1, 2, 3)
G. strigilis (W. Smith) Griffith & Henfrey: Bohai; Yellow Sea; Taiwan Strait. (1, 2, 3)
G. strigilis v. *excentriraphe* Chin & Liu: Bohai; Yellow Sea; Taiwan Strait. (1, 2, 3)
G. tenuissimum (W. Smith) Griffith & Henfrey: Taiwan Strait. (1, 2, 3)
G. terryanum (Perag.) Cleve: Taiwan Strait. (1, 2, 3)
G. wansbeckii (Donkin) Cleve: Taiwan Strait; South China Sea. (1, 2, 3)
G. wormleyi (Sull.) Boyer: East China Sea; Taiwan Strait. (1, 2, 3)
Mastogloia achnanthioides Mann: South China Sea. (1, 2, 3)
M. achnanthioides v. *elliptica* Husted: South China Sea. (1, 2, 3)
M. acustiuscula Grunow: South China Sea. (1, 2, 3)
M. acutiuscula v. *elliptica* Hustedt: Taiwan Strait; Lanyu (Taiwan). (1, 2, 3, 30b)
M. acutiuscula v. *vairaensis* Ricard: South China Sea. (1, 2, 3)
M. adriatica v. *linearis* Voigt: South China Sea. (23c)
M. affirmata (Leud.-Fort.) Cleve: South China Sea. (1, 2, 3)
M. amoyensis Voigt: Taiwan Strait; South China Sea. (1, 2, 3)
M. angulata Lewis: South China Sea. (1, 2, 3)
M. apiculata W. Smith: Bohai; Yellow Sea; East China Sea; Taiwan Strait; South China Sea. (1, 2, 3)
M. aspera Voigt: South China Sea. (1, 2, 3)
M. aspera f. *lanceolata* Ricard: South China Sea. (1, 2, 3)
M. asperula Grunow: South China Sea. (1, 2, 3)
M. asperuloides Hustedt: East China Sea; South China Sea. (1, 2, 3)
M. bahamensis Cleve: South China Sea. (1, 2, 3)
M. baldjikiana Grunow: South China Sea. (1, 2, 3)
M. bellatula Voigt: South China Sea. (1, 2, 3)
M. biapiculata Hustedt: South China Sea. (1, 2, 3)
M. binotata (Grun.) Cleve: Bohai; Yellow Sea; Taiwan Strait; South China Sea. (1, 2, 3)
M. binotata f. *sparsipunctata* Voigt: South China Sea. (1, 2, 3)
M. bourrellvana Ricard: South China Sea. (1, 2, 3)
M. brauni Grunow: Taiwan Strait; South China Sea. (1, 2, 3)
M. brauni v. *constricta* Liu & Chin: Taiwan Strait. (1, 2, 3)
M. brauni f. *elongata* Voigt: Taiwan Strait; South China Sea. (1, 2, 3)
M. citroides Ricard: South China Sea. (1, 2, 3)
M. citrus (Cleve) De Toni: Bohai; Yellow Sea; East China Sea; Taiwan Strait; South China Sea. (1, 2, 3)
M. cocconeiformis Grunow: South China Sea. (1, 2, 3)
M. composita Voigt: South China Sea. (1, 2, 3)
M. corallum Paddock & Kemp: South China Sea. (32)

M. corsicana Grunow: Taiwan Strait; Lanyu (Taiwan); South China Sea. (1, 2, 3, 30b)
M. cribrosa Grunow: South China Sea. (1, 2, 3)
M. cruciala (Grun.) Cleve: Taiwan Strait; Lanyu (Taiwan); South China Sea. (1, 2, 3, 30b)
M. cruciata v. *elliptica* Voigt: South China Sea. (1, 2, 3)
M. cucurbita Voigt: South China Sea. (1, 2, 3)
M. cyclops Voigt: Lanyu (Taiwan); South China Sea. (1, 2, 3, 30b)
M. decipiens Hustedt: South China Sea. (1, 2, 3)
M. decussata Grunow: Taiwan Strait; South China Sea. (1, 2, 3)
M. densestriata Hustedt: South China Sea. (1, 2, 3)
M. depressa Hustedt: South China Sea. (23c, 32)
M. dicephala Voigt: South China Sea. (1, 2, 3)
M. dissimilis Hustedt: Bohai; Yellow Sea. (1, 2, 3)
M. dubitabilis Meister: Bohai; Yellow Sea; East China Sea; Taiwan Strait; South China Sea. (1, 2, 3)
M. elegantula Hustedt: South China Sea. (1, 2, 3)
M. elliptica v. *dansei* (Thwaites) Cleve: Bohai; Yellow Sea; Taiwan Strait. (1, 2, 3)
M. emarginata Hustedt: East China Sea; South China Sea. (1, 2, 3)
M. erythraea Grunow: South China Sea. (1, 2, 3)
M. erythraea v. *biocellata* Grunow: South China Sea. (1, 2, 3)
M. exilis Hustedt: South China Sea. (1, 2, 3)
M. fallax Cleve: Taiwan Strait; South China Sea. (1, 2, 3)
M. fascistriata Liu & Chin: Taiwan Strait. (1, 2, 3)
M. fimbriata (Brightw.) Cleve: Bohai; Yellow Sea; Taiwan Strait; South China Sea. (1, 2, 3)
M. graciloides Hustedt: South China Sea. (32)
M. grana Ricard: South China Sea. (33)
M. grevillei W. Smith: Fujian coast. (3)
M. grunowi (Grunow) A. Smidt: East China Sea. (1, 2, 3)
M. hainanensis Voigt: South China Sea. (1, 2, 3)
M. horvathiana Grunow: South China Sea. (1, 2, 3)
M. hustedtii Meister: East China Sea; South China Sea. (1, 2, 3)
M. imitatrix Mann: Bohai; Yellow Sea; Taiwan Strait. (1, 2, 3)
M. inaequalis Cleve: Taiwan Strait; South China Sea. (1, 2, 3)
M. indonesiana Voigt: South China Sea. (1, 2, 3)
M. intrita Voigt: South China Sea. (1, 2, 3)
M. jaoi Voigt: South China Sea. (1, 2, 3)
M. jelinecki Grunow: South China Sea. (1, 2, 3)
M. jelinecki v. *extensa* Voigt: South China Sea. (1, 2, 3)
M. jelineckiana Grunow: Taiwan continental shelf. (31)
M. labuensis Cleve: South China Sea. (1, 2, 3)
M. lacrimata Voigt: East China Sea. (1, 2, 3)
M. lanceolata Thwaites: South China Sea. (1, 2, 3)
M. lata Hustedt: South China Sea. (1, 2, 3)
M. latecostata Hustedt: South China Sea. (1, 2, 3)
M. lemniscata Leuduger-Fortmorel: Taiwan Strait; Taiwan continental shelf; South China Sea. (2, 31)
M. lentiformis Voigt: South China Sea. (1, 2, 3)
M. levis Voigt: South China Sea. (1, 2, 3)
M. liaotungensis Voigt: Bohai; Yellow Sea. (1, 2, 3)
M. lineata Cleve & Grove: South China Sea. (1, 2, 3)

M. lunula Voigt: South China Sea. (1, 2, 3)
M. macdonaldi Greville: South China Sea. (1, 2, 3)
M. mammosa Voigt: South China Sea. (1, 2, 3)
M. manokwariensis Cholnoky: Taiwan Strait; South China Sea. (1, 2, 3)
M. mauritiana Brun: East China Sea; South China Sea. (1, 2, 3)
M. mauritiana v. *capitata* Voigt: South China Sea. (1, 2, 3)
M. mediterranea Hustedt: South China Sea. (1, 2, 3)
M. mediterranea v. *elliptica* Voigt: Bohai; Yellow Sea. (1, 2, 3)
M. minutissima Voigt: South China Sea. (1, 2, 3)
M. muralis Voigt: South China Sea. (1, 2, 3)
M. nebulosa Voigt: South China Sea. (1, 2, 3)
M. neorugosa Voigt: Taiwan Strait; South China Sea. (1, 2, 3)
M. nuiensis Ricard: South China Sea. (1, 2, 3)
M. obesa Cleve: South China Sea. (1, 2, 3)
M. occulta Voigt: South China Sea. (1, 2, 3)
M. omissa Voigt: South China Sea. (1, 2, 3)
M. ovalis A. Schmidt: South China Sea. (1, 2, 3)
M. ovata Grunow: South China Sea. (1, 2, 3)
M. ovulum Hustedt: South China Sea. (1, 2, 3)
M. ovum paschale (A. Schmidt) Mann: Taiwan Starit; South China Sea. (1, 2, 3, 23b)
M. paracelsiana Voigt: South China Sea. (1, 2, 3)
M. paradoxa Grunow: South China Sea. (1, 2, 3)
M. peracuta Janisch: East China Sea. (1, 2, 3)
M. peragalli Cleve: South China Sea. (1, 2, 3)
M. pisciculus Cleve: Bohai; Yellow Sea; Taiwan Strait; South China Sea. (1, 2, 3)
M. pseudexilis Voigt: South China Sea. (1, 2, 3)
M. pseudolatericia Voigt: South China Sea. (1, 2, 3)
M. pseudomauritiana Voigt: South China Sea. (1, 2, 3)
M. pulchella Cleve: South China Sea. (1, 2, 3)
M. pumila (Cleve & Moeller) Cleve: Bohai; Yellow Sea; Taiwan Strait. (1, 2, 3)
M. pumila v. *papuarum* Cholnoky: South China Sea. (1, 2, 3)
M. pumila v. *rennellensis* Foged: South China Sea. (32)
M. punctatissima (Greville) Ricard: South China Sea. (1, 2, 3)
M. punctifera Brun: South China Sea. (1, 2, 3)
M. qionzhouensis Liu: South China Sea. (32)
M. quinquecostata Grunow: East China Sea; Taiwan Strait; South China Sea. (1, 2, 3)
M. rimosa Cleve: South China Sea. (3)
M. robusta Hustedt: Taiwan Strait; South China Sea. (1, 2, 3)
M. rostrate (Wallich) Hustedt [*Stigmophora rostrata*]: Taiwan Strait; South China Sea. (11, 29a, 30)
M. savensis Jurily: Taiwan Strait; South China Sea. (1, 2, 3)
M. schmidtii Heidenet & Kolbe: South China Sea. (1, 2, 3)
M. serians Voigt: South China Sea. (33)
M. serrata Voigt: East China Sea. (1, 2, 3)
M. seychellensis Grunow: South China Sea. (1, 2, 3)
M. similis Hustedt: South China Sea. (1, 2, 3)
M. simplex Klaus-Kemp: South China Sea. (23c)
M. singaporensis Voigt: South China Sea. (1, 2, 3)
M. smithii Thwaites & W. Smith: Bohai; Yellow Sea; East China Sea; Taiwan Strait; South China Sea. (1, 2, 3, 23b)

M. smithii v. *excentrica* Liu & Chin: Lanyu (Taiwan); South China Sea. (1, 2, 3, 30b)
M. splendida (Greg.) Cleve & Moller: South China Sea. (1, 2, 3)
M. subaspera Hustedt: South China Sea. (1, 2, 3)
M. sublatericia Hustedt: South China Sea. (1, 2, 3)
M. sulcata Cleve: South China Sea. (1, 2, 3)
M. tenuis Hustedt: South China Sea. (1, 2, 3)
M. tenuissima Hustedt: South China Sea. (1, 2, 3)
M. testudinea Voigt: East China Sea; South China Sea. (1, 2, 3)
M. umbilicata Voigt: South China Sea. (1, 2, 3)
M. undulata Grunow: South China Sea. (1, 2, 3)
M. varians Hustedt: South China Sea; Lanyu (Taiwan). (1, 2, 3, 30b)
M. viperina Voigt: South China Sea. (1, 2, 3)
M. vulnerata Voigt: South China Sea. (1, 2, 3)
M. woodiana F. J. R. Taylor: South China Sea. (11)
M. xishaensis Liu: South China Sea. (31)
Navicula abrupta (Greg.) Donkin: Taiwan Strait; South China Sea. (1, 2, 3)
N. alpha Cleve: Taiwan Strait; South China Sea. (1, 2, 3, 30c)
N. approximata Greville: South China Sea; Taiwan. (3, 31)
N. approximata v. *niceaensis* (Perag.) Hendey: East China Sea. (1, 2, 3)
N. arabica Grunow: South China Sea. (1, 2, 3)
N. asymmetrica Cleve: Taiwan Strait. (1, 2, 3)
N. australica (A. S.) Cleve: East China Sea; Okinawa Trough; Taiwan Strait; South China Sea. (1, 2, 3, 23b)
N. biformis (Grun.) Mann: South China Sea. (1, 2, 3)
N. bolleana (Grun.) Cleve: South China Sea. (1, 2, 3)
N. brasiliensis Grunow: Taiwan Strait; South China Sea. (1, 2, 3)
N. bruchii Grunow: South China Sea. (1, 2, 3)
N. caeca Mann: South China Sea. (1, 2, 3)
N. cancellata Donkin: Bohai; Yellow Sea; East China Sea; Taiwan Strait; South China Sea. (1, 2, 3)
N. cancellata v. *apiculata* (Greg.) Peragallo & Peragallo: East China Sea. (1, 2, 3)
N. cancellata v. *retusa* (Bréb.) Cleve: South China Sea. (1, 2, 3)
N. carinifera Grunow: Bohai; Yellow Sea; Taiwan Strait; South China Sea. (1, 2, 3)
N. cincta (Ehr.) Ralfs: Bohai; Yellow Sea; East China Sea; Taiwan Strait. (1, 2, 3)
N. circumsecta (Grun.) Grunow: East China Sea; Okinawa Trough; Taiwan Strait. (1, 2, 3, 23b)
N. clavata Gregory: East China Sea; Okinawa Trough; Taiwan continental shelf; Taiwan Strait; South China Sea. (2, 31)
N. clavata v. *indica* (Grev.) Cleve: East China Sea; Okinawa Trough. (1, 2, 3)
N. climacospheniae Booth: South China Sea. (23a)
N. cluthensis Gregory: South China Sea. (1, 2, 3)
N. complanata Grunow: South China Sea. (1, 2, 3)
N. consors A. Schmidt: South China Sea. (1, 2, 3)
N. corymbosa (Ag.) Cleve: East China Sea; Taiwan Strait; South China Sea. (1, 2, 3)
N. crucicula (W. Smith) Dokin: East China Sea; Taiwan Strait (1, 2, 3)
N. crucicula v. *orientalis* Skvortzow: Bohai; Yellow Sea. (1, 2, 3)
N. cruciculoides Brockmann: East China Sea. (1, 2, 3)
N. cryptocephaloides Hustedt: South China Sea. (1, 2, 3)
N. cryptotenella Lange-Bertalot: South China Sea. (23a)
N. cuspidata Kuetzing: Taiwan Strait; South China Sea. (1, 2, 3)
N. delta Cleve: South China Sea. (1, 2, 3)
N. digito-radiata (Greg.) Ralfs: Taiwan Strait; Lanyu (Taiwan); South China Sea. (1, 2, 3, 30b)
N. directa (W. Smith) Ralfs: Bohai; Yellow Sea; East China Sea; Taiwan Strait; Taiwan continental shelf; South China Sea. (2, 31)
N. directa v. *javanica* Cleve: Taiwan Strait; South China Sea. (1, 2, 3)
N. directa v. *remota* Grunow: East China Sea; Taiwan Strait; Lanyu (Taiwan); South China Sea. (1, 2, 3, 30b)
N. distans (W. Smith) Ralfs: Taiwan Strait; South China Sea. (1, 2, 3)
N. epsilon Cleve: East China Sea; South China Sea. (1, 2, 3)
N. eta Cleve: Taiwan Strait. (1, 2, 3)
N. eymei Coste & Ricard: South China Sea. (1, 2, 3)
N. forcipata Greville: Bohai; Yellow Sea; Okinawa Trough; Taiwan Strait; South China Sea. (1, 2, 3)
N. forcipata v. *densestriata* A. Schmidt: Taiwan Strait; South China Sea. (1, 2, 3)
N. fortis (Greg.) Ralfs: Taiwan Strait. (1, 2, 3)
N. fujianensis Chin & Cheng: Taiwan Strait. (1, 2, 3)
N. genuflexa Kuetzig: Taiwan Strait. (1, 2, 3)
N. glacialis (Cleve) Grunow: Taiwan Strait. (1, 2, 3)
N. glacialis v. *neglecta* (Thwaites) Grunow: East China Sea. (1, 2, 3)
N. granulata Bailey: Bohai; Yellow Sea; East China Sea; Taiwan Strait; South China Sea. (1, 2, 3)
N. H-album Cleve: Taiwan Strait; South China Sea. (1, 2, 3)
N. hamulifera Grunow: South China Sea. (24)
N. hennedyi W. Smith: East China Sea; Okinawa Trough; Taiwan continental shelf; Taiwan Strait; South China Sea. (2, 31)
N. hennedyi f. *california* (Grev.) Hustedt: East China Sea. (1, 2, 3)
N. hennedyi v. *nebulosa* (Greg.) Cleve: Bohai; Yellow Sea; East China Sea; Taiwan Strait. (1, 2, 3)
N. hetero-punctata Chin & Cheng: Taiwan Strait. (1, 2, 3)
N. hochstetteri Grunow: South China Sea. (1, 2, 3)
N. howeana Hagelstein: South China Sea. (1, 2, 3)
N. humerosa Brébisson: Taiwan Strait; South China Sea. (1, 2, 3)
N. humerosa v. *constricta* Cleve: Taiwan Strait; South China Sea. (1, 2, 3)
N. humerosa v. *minor* Heicen: Taiwan Strait; South China Sea. (1, 2, 3)
N. impressa Grunow: Taiwan Strait; South China Sea. (1, 2, 3)
N. inhalata A. Schmidt: East China Sea; Taiwan continental shelf. (3, 31)
N. inserata Hustedt: Taiwan Strait. (1, 2, 3)
N. integra (W. Smith) Ralfs: Taiwan Strait. (1, 2, 3)

N. integra v. *maculata* Chin & Cheng: Taiwan Strait. (1, 2, 3)
N. jamalinensis Cleve: South China Sea. (1, 2, 3)
N. jejuna A. Schmidt: Taiwan Strait; South China Sea. (1, 2, 3)
N. latissima Gregory: South China Sea. (1, 2, 3)
N. liaotungiensis Skvortzow: Bohai; Yellow Sea. (1, 2, 3)
N. longa (Greg.) Ralfs: Bohai; Yellow Sea; East China Sea; Okinawa Trough; Taiwan continental shelf; Taiwan Strait; South China Sea. (1, 2, 3)
N. lorenzii (Grun.) Hustedt: Taiwan Strait. (1, 2, 3)
N. luxuriosa Greville: East China Sea; Taiwan Strait; South China Sea. (1, 2, 3)
N. lyra Ehrenberg: Bohai; Yellow Sea; East China Sea; Okinawa Trough; Taiwan continental shelf; Taiwan Strait; South China Sea. (2, 31)
N. lyra v. *dilatata* A. Schmidt: Bohai; Yellow Sea; East China Sea; Taiwan Strait; South China Sea. (1, 2, 3)
N. lyra v. *elliptica* A. Schmidt: East China Sea; South China Sea. (1, 2, 3)
N. lyra v. *insignis* A. Schmidt: East China Sea; Taiwan Strait; South China Sea. (1, 2, 3)
N. lyra v. *recta* Greville: Bohai; Yellow Sea; East China Sea; Taiwan Strait; South China Sea. (1, 2, 3)
N. lyra v. *signata* A. Schmidt: Taiwan Strait. (1, 2, 3)
N. lyra v. *subtypica* A. Schmidt: East China Sea; South China Sea. (1, 2, 3)
N. lyroides Hendey: Bohai; Yellow Sea; East China Sea; Taiwan Strait; Taiwan continental shelf; Lanyu (Taiwan). (2, 30b, 31)
N. maculata (Bailey) Edwards: Bohai; Yellow Sea; Taiwan Strait; South China Sea. (1, 2, 3)
N. marina Ralfs: Bohai; Yellow Sea; East China Sea; Taiwan Strait; Taiwan continental shelf; South China Sea. (2, 31)
N. membranacea Cleve: Bohai; Yellow Sea; Taiwan Strait; southeastern Taiwan; South China Sea. (1, 2, 3, 30a)
N. mollis (W. Smith) Cleve: Bohai; Yellow Sea; East China Sea; Taiwan Strait; South China Sea. (1, 2, 3)
N. monilifera Cleve: Taiwan Strait; South China Sea; Taiwan continental shelf; Lanyu (Taiwan). (2, 30b, 31)
N. mutica Kuetzing: Taiwan Strait. (1, 2, 3)
N. my Cleve: South China Sea. (1, 2, 3)
N. nitescens (Greg.) Ralfs: Taiwan Strait. (1, 2, 3)
N. northumbrica Donkin: East China Sea; South China Sea. (1, 2, 3)
N. nummularia Greville: Taiwan Strait; South China Sea. (1, 2, 3)
N. orthoneoides Hustedt: South China Sea. (1, 2, 3)
N. pantocsekiana De Toni: Bohai; Yellow Sea; Taiwan Strait; South China Sea. (1, 2, 3)
N. parva (Ehr.) Ralfs: Taiwan Strait. (1, 2, 3)
N. pavillardi Hustedt: East China Sea; Taiwan Strait; South China Sea. (2, 12)
N. penna A. Schmidt: South China Sea. (1, 2, 3)
N. perplexoides Hustedt: Taiwan continental shelf. (31)
N. perrhombus Hustedt: Taiwan Strait. (1, 2, 3)
N. pi Cleve: South China Sea. (1, 2, 3)
N. pinna Chin & Cheng: Taiwan Strait; South China Sea. (1, 2, 30)
N. plicatula Grunow: South China Sea. (1, 2, 3)
N. praetexta Ehrenberg: Bohai; Yellow Sea; East China Sea; Taiwan Strait; Okinawa Trough; South China Sea. (1, 2, 3)
N. punctulata W. Smith: Okinawa Trough. (1, 2, 3)
N. pupula v. *elliptica* Hustedt: East China Sea; Taiwan Strait; South China Sea. (1, 2, 3)
N. pygmaea Kuetzing: East China Sea; South China Sea. (1, 2, 3)
N. quincunx Cleve: South China Sea. (1, 2, 3)
N. raeana (Castr.) De Toni: Taiwan Strait; South China Sea. (1, 2, 3)
N. ramosissima (Ag.) Cleve: Bohai; Yellow Sea; Taiwan Strait; South China Sea. (1, 2, 3)
N. restituta A. Schmidt: South China Sea. (1, 2, 3)
N. rhaphoneis (Ehr.) Grunow: South China Sea. (1, 2, 3)
N. rho Cleve: South China Sea. (1, 2, 3)
N. rhynchocephala Kuetzing: East China Sea; Taiwan Strait; South China Sea. (2, 12)
N. robertisiana Greville: Taiwan continental shelf. (31)
N. satura A. Schmidt: Taiwan Strait. (1, 2, 3)
N. schaarschmidtii Pantocsek: South China Sea. (1, 2, 3)
N. scintillans A. Schmidt: East China Sea; Taiwan Strait. (1, 2, 3)
N. scopulorum Brébisson: Taiwan Strait; South China Sea. (1, 2, 3)
N. scutelloides W. Smith: South China Sea. (1, 2, 3)
N. scutiformis Grunow: Taiwan Strait. (1, 2, 3)
N. semistauros Mann: South China Sea. (1, 2, 3)
N. sibayiensis Archibald: South China Sea. (1, 2, 3)
N. spectabilis Gregory: Bohai; Yellow Sea; East China Sea; Okinawa Trough; Taiwan continental shelf; Taiwan Strait; South China Sea. (2, 31)
N. spectabilis v. *excavata* (Grev.) Cleve: Okinawa Trough. (1, 2, 3)
N. stercus muscarum Cleve: South China Sea. (1, 2, 3)
N. subcarinata (Grun.) Hendey: East China Sea; South China Sea. (1, 2, 3)
N. superimposita A. Schmidt: South China Sea. (1, 2, 3)
N. takoradiensis Hendey: Taiwan Strait. (1, 2, 3)
N. toulaae Pantocsek: Bohai; Yellow Sea; Taiwan Strait. (1, 2, 3)
N. transfuga Grunow: South China Sea. (1, 2, 3)
N. tuscula v. *cuneata* Cleve-Euler: Bohai; Yellow Sea; East China Sea; Taiwan Strait. (1, 2, 3)
N. viridula v. *slesvicensis* (Grun.) Grunow: East China Sea; Taiwan Strait. (1, 2, 3)
N. yarrensis Grunow: South China Sea. (1, 2, 3)
N. zostereti Grunow: Bohai; Yellow Sea; Taiwan Strait; South China Sea. (1, 2, 3)
Neidium amphirhynchus (Ehr.) Pfizer: Taiwan Strait. (1, 2, 3)
N. iridis v. *amphigomphus* (Ehr.) Van Heurck: Taiwan Strait. (1, 2, 3)

Oestrupia musca (Greg.) Hustedt: Taiwan Strait; Taiwan continental shelf; South China Sea. (3, 31)
Pinnularia bistriata (Leud.-Fortm.) Cleve: South China Sea. (1, 2, 3)
P. major (Kuetz.) Cleve: Taiwan Strait. (1, 2, 3)
P. trevelyana (Donk.) Rebenhorst: South China Sea. (1, 2, 3)
Pleurosigma acutum Norman & Ralfs: East China Sea; South China Sea. (1, 2, 3, 12)
P. acutum v. *latum* Chin & Liu: East China Sea; Taiwan Strait. (1, 2, 3, 12)
P. aestuarii (Bréb.) W. Smith: East China Sea; Taiwan Strait; South China Sea. (1, 2, 3)
P. affine Grunow: Bohai; Yellow Sea; East China Sea; Taiwan Strait; South China Sea. (2, 5)
P. angulatum (Quekett) W. Smith: Bohai; Yellow Sea; East China Sea; Taiwan Strait. (1, 2, 3)
P. angulatum v. *falcatum* Liu & Chin: East China Sea; Taiwan Strait. (1, 2, 3)
P. angulatum v. *quadratum* (W. Smith) Van Heurck: Bohai; Yellow Sea; East China Sea; Taiwan Strait; South China Sea. (2, 12)
P. decorum W. Smith: East China Sea; Taiwan Strait; South China Sea. (1, 2, 3)
P. delicatulum W. Smith: Bohai; Yellow Sea; Taiwan Strait. (1, 2, 3)
P. diminutum Grunow: Bohai; Yellow Sea. (1, 2, 3)
P. diverse-striatum Heister: Bohai; Yellow Sea; Taiwan Strait. (1, 2, 3)
P. elongatum W. Smith: Bohai; Yellow Sea; East China Sea; Taiwan Strait; South China Sea. (1, 2, 3)
P. elongatum v. *sinica* Skvortzow: Bohai; Yellow Sea; East China Sea; Taiwan Strait. (1, 2, 3)
P. falx Mann: Bohai; Yellow Sea; East China Sea; Okinawa Trough; Taiwan Strait; South China Sea. (1, 2, 3)
P. finmarchicum Grunow: Taiwan Strait; South China Sea. (1, 2, 3)
P. formosum W. Smith: Bohai; Yellow Sea; East China Sea; Taiwan Strait; South China Sea. (1, 2, 3)
P. intermedium W. Smith: Bohai; Yellow Sea; Taiwan Strait; South China Sea. (1, 2, 3)
P. intermedium v. *dongshanense* Chin & Liu: Taiwan Strait; South China Sea. (1, 2, 3)
P. intermedium v. *nubecula* (W. Smith) Van Heurck: South China Sea. (1, 2, 3)
P. longum v. *inflata* (Perag.) Cleve: Taiwan Strait. (1, 2, 3)
P. major Liu & Chin: Taiwan Strait; South China Sea. (1, 2, 3)
P. marinum Donkin: Taiwan Strait. (30)
P. minutum Grunow: Bohai; Yellow Sea; Taiwan Strait. (1, 2, 3)
P. naviculaceum Brébisson: East China Sea; Okinawa Trough; Taiwan Strait; South China Sea. (1, 2, 3)
P. naviculaceum f. *minuta* Cleve: East China Sea; Taiwan Strait; Okinawa Trough; South China Sea. (1, 2, 3)
P. normanii Ralfs [*P. affine* Grunow]: Bohai; Yellow Sea; East China Sea; Taiwan Strait; South China Sea. (1, 2, 3, 5)
P. normanii v. *fossilis* (Grun.) Cleve: East China Sea; Okinawa Trough; Taiwan Strait; South China Sea. (1, 2, 3)

P. obtusum Mann: South China Sea. (1, 2, 3)
P. pelagicum (Perag.) Cleve: Bohai; Yellow Sea; East China Sea; Taiwan Strait; South China Sea. (1, 2, 3)
P. rhombeum (Grun.) Peragallo: Taiwan Strait; South China Sea. (1, 2, 3)
P. rigidium W. Smith: East China Sea; Taiwan Strait; South China Sea. (1, 2, 3)
P. rostratum Husedt: Taiwan Strait. (1, 2, 3)
P. salinarum Grunow: Bohai; Yellow Sea. (1, 2, 3)
P. speciosum W. Smith: Bohai; Yellow Sea; East China Sea; Taiwan Strait; South China Sea. (1, 2, 3)
P. strigosum W. Smith [*P. angulatum* v. *strigosa*]: Taiwan Strait; Taiwan continental shelf; South China Sea. (2, 31)
P. tahitianum: East China Sea; Taiwan Strait; South China Sea. (1, 2, 3)
Raphidivergens bacilliformis Chin & Cheng: Taiwan Strait. (1, 2, 3)
Rossia elliptica Voigt: South China Sea. (1, 2, 3)
Scoliopleura tumida (Bréb.) Rabenhorst: Taiwan Strait. (1, 2, 3)
Stauroneis amphioxys v. *obtum* Hendey: South China Sea. (1, 2, 3)
S. constricta Ehrenberg: Bohai; Yellow Sea; Taiwan Strait; South China Sea. (1, 2, 3)
S. pellucida v. *orientalis* Skvortzow: Bohai; Yellow Sea. (1, 2, 3)
S. phoenicenteron (Nitz.) Ehrenberg: East China Sea; Taiwan Strait; South China Sea. (1, 2, 3)
Trachyneis antillarum (Cleve & Grun.) Cleve: Bohai; Yellow Sea; East China Sea; Okinawa Trough; Taiwan continental shelf; Taiwan Strait; South China Sea. (1, 31)
T. aspera (Ehr.) Ehrenberg: Bohai; Yellow Sea; East China Sea; Okinawa Trough; Taiwan continental shelf; Taiwan Strait; southeastern Taiwan; South China Sea. (2, 30a, 31)
T. aspera v. *angusta* Cleve: Bohai; Yellow Sea; East China Sea; Okinawa Trough; Taiwan; Taiwan Strait; South China Sea. (2, 31)
T. aspera v. *contermina* (A. Schmidt) Cleve: East China Sea; Taiwan Strait; Okinawa Trough. (1, 2, 3)
T. aspera v. *oblonga* (Bailey) Cleve: Bohai; Yellow Sea; East China Sea; Taiwan Strait; South China Sea. (1, 2, 3)
T. aspera v. *orientalis* Skvortzow: Bohai; Yellow Sea. (1, 2, 3)
T. aspera v. *perobliqua* Cleve: Taiwan Strait; South China Sea. (1, 2, 3)
T. aspera v. *producta* Chin & Cheng: Bohai; Yellow Sea; East China Sea; Taiwan Strait. (1, 12)
T. aspera v. *pulchella* (A. Schmidt) Cleve: Bohai; Yellow Sea; East China Sea; Taiwan Strait; South China Sea. (1, 2, 3)
T. aspera v. *residua* (A. Schmidt) Cleve: Bohai; Yellow Sea; Taiwan Strait; South China Sea. (1, 2, 3)
T. aspera v. *unilatera* Chin & Cheng: Bohai; Yellow Sea; East China Sea; Taiwan Strait; South China Sea. (1, 2, 3)
T. aspera v. *vulgaris* Cleve: Bohai; Yellow Sea; East China Sea; Taiwan Strait; South China Sea. (1, 2, 3)
T. brunii (Cleve & Brun) Cleve: East China Sea; Taiwan Strait. (1, 2, 3)
T. clepsydra (Donkin) Cleve: East China Sea; Taiwan Strait; Taiwan continental shelf; South China Sea. (2, 31)

T. debyi (Leuduger-Fortm.) Cleve: East China Sea; Taiwan Strait; Okinawa Trough; South China Sea. (1, 2, 3)

T. formosa Meister: East China Sea; South China Sea. (1, 2, 3)

T. johnsoniana (Grev.) Cleve: Taiwan Strait; South China Sea. (1, 2, 3)

T. minor Chin & Cheng: Taiwan Strait. (1, 2, 3)

T. olivaeformis Chin & Cheng: Bohai; Yellow Sea; East China Sea; Taiwan Strait; South China Sea. (1, 2, 3)

T. velata (A. Schmith) Cleve: Bohai; Yellow Sea; East China Sea; Taiwan Strait; South China Sea. (1, 2, 3)

T. velata v. *oblonga* Chin & Cheng: Bohai; Yellow Sea; Taiwan Strait; South China Sea. (1, 2, 3)

T. velatoides Ricard: Taiwan Strait; South China Sea. (1, 2, 3)

Tropidoneis chinensis Cleve: South China Sea. (1, 2, 3)

T. constricta Li, Cheng & Chin: Taiwan Strait. (8)

T. gibberula (Grun.) Cleve: South China Sea. (1, 2, 3)

T. lepidoptera (Greg.) Cleve: Bohai; Yellow Sea; Taiwan Strait. (1, 2, 3)

T. longa (Cleve) Cleve: Taiwan Strait. (1, 2, 3)

T. maxima (Greg.) Cleve: East China Sea; Taiwan Strait; South China Sea. (1, 2, 3)

T. maxima v. *sinensis* Skvortzow: Bohai; Yellow Sea. (1, 2, 3)

T. pusilla (Greg.) Cleve: Taiwan Strait; South China Sea. (1, 2, 3)

CYMBELLACEAE

Amphora acuta Gregory: South China Sea. (1, 2, 3)

A. angusta Gregory: Bohai; Yellow Sea; East China Sea; Taiwan Strait; South China Sea. (1, 2, 3)

A. angusta v. *chinensis* Skvortzow: Bohai; Yellow Sea; Taiwan Strait. (1, 2, 3)

A. angusta v. *diducta* (A. Schmidt) Cleve: Bohai; Yellow Sea; East China Sea; Taiwan Strait; South China Sea. (1, 2, 3)

A. angusta v. *eulensteinii* (Grun.) Cleve: Taiwan Strait; South China Sea. (1, 2, 3)

A. arenaria Donkin: Taiwan Strait; South China Sea. (1, 2, 3)

A. arenicola (Grun.) Cleve: East China Sea; South China Sea. (1, 2, 3)

A. arenicola v. *major* Cleve: South China Sea. (1, 2, 3)

A. arenicola v. *subaequalis* Cleve: South China Sea. (1, 2, 3)

A. bigibba Grunow: Bohai; Yellow Sea; East China Sea; Taiwan Strait; South China Sea. (1, 2, 3)

A. coffeaeformis (Ag.) Kuetzing: Bohai; Yellow Sea; East China Sea; Taiwan Strait; South China Sea. (1, 2, 3)

A. coffeaeformis v. *acutiuscula* (Kuetz.) Hustedt: East China Sea; Taiwan Strait; Lanyu (Taiwan). (1, 2, 3, 30b)

A. commutata Grunow: Bohai; Yellow Sea; Taiwan Strait. (1, 2, 3)

A. costata W. Smith: Bohai; Yellow Sea; East China Sea; Taiwan Strait; Taiwan continental shelf; South China Sea. (2, 31)

A. costata v. *inflata* (Grun.) Peragallo & Peragallo: Taiwan Strait; South China Sea. (1, 2, 3)

A. crassa Gregory: East China Sea; Taiwan Strait; South China Sea. (1, 2, 3)

A. crassa v. *canpechiana* Grunow: Taiwan continental shelf. (31)

A. crassa v. *interrupta* Lin & Chin: Taiwan Strait. (1, 2, 3)

A. crassa v. *punctata* Grunow: Taiwan continental shelf. (31)

A. cymbelloides Grunow: East China Sea; South China Sea. (2)

A. cymbifera Gregory: Taiwan continental shelf. (31)

A. decussata Grunow: Bohai; Yellow Sea; Taiwan Strait; South China Sea. (1, 2, 3)

A. egregia Ehrenberg: Okinawa Trough; South China Sea. (1, 2, 3)

A. eunotia Cleve: South China Sea. (1, 2, 3)

A. eunotia v. *gigantea* (Grun.) Cleve: South China Sea. (1, 2, 3)

A. exigua Gregory: Bohai; Yellow Sea; Taiwan Strait; South China Sea. (1, 2, 3)

A. exsecta Grunow: South China Sea. (1, 2, 3)

A. gigantea v. *fusca* (A. Schmidt) Cleve: South China Sea. (1, 2, 3)

A. graeffii (Grun.) Cleve: South China Sea. (1, 2, 3)

A. grevilleana Gregory: South China Sea. (1, 2, 3)

A. grovei Cleve: South China Sea. (1, 2, 3)

A. helenensis Giffen: South China Sea. (24)

A. hyalina Kuetzing: East China Sea. (1, 2, 3)

A. javanica A. Schmidt: Taiwan Strait; South China Sea. (1, 2, 3, 30c)

A. laevis Gregory: South China Sea. (1, 2, 3)

A. libyca Ehrenberg: Taiwan. (31)

A. lineaolata v. *chinensis* (A. Schmidt) Cleve: South China Sea. (1, 2, 3)

A. lunula Cleve: South China Sea. (1, 2, 3)

A. lyrata Gregory: South China Sea. (1, 2, 3)

A. macilenta Gregory: South China Sea. (1, 2, 3)

A. macilenta v. *ergadensis* (Greg.) Cleve: East China Sea. (1, 2, 3)

A. maria Hanna & Grant: Taiwan continental shelf. (31)

A. marina Van Heurck: Bohai; Yellow Sea; South China Sea. (1, 2, 3)

A. mexicana A. Schmidt: South China Sea. (1, 2, 3)

A. micrometra Giffen: South China Sea. (1, 2, 3)

A. obtusa Gregory: South China Sea. (1, 2, 3)

A. obtusa v. *oceanica* (Cast.) Cleve: South China Sea. (1, 2, 3)

A. ocellata Donkin: South China Sea. (1, 2, 3)

A. oculus A. Schmidt: South China Sea; Taiwan continental shelf. (3, 31)

A. ostrearia Brébisson: Taiwan Strait; South China Sea. (1, 2, 3)

A. ostrearia v. *vitrea* (Cleve) Cleve: Taiwan Strait. (1, 2, 3)

A. ovalis Kuetzing: East China Sea; Taiwan Strait; South China Sea. (1, 2, 3)

A. ovalis v. *pediculus* (Kuetz.) Van Heurck: East China Sea; Taiwan Strait; South China Sea. (1, 2, 3)

A. pediculus (Kuetz.) Grunow: South China Sea. (33)

A. proteus Gregory: Bohai; Yellow Sea; East China Sea; Taiwan Strait; Taiwan continental shelf; South China Sea. (2, 7, 31)

A. proteus v. *oculata* Peragallo & Peragallo: East China Sea; Taiwan Strait; South China Sea. (1, 2, 3)

A. rhombica Kittom: Bohai; Yellow Sea; Taiwan Strait; South China Sea. (1, 2, 3)

A. rhombica v. *sinica* Skvortzow: Bohai; Yellow Sea; Taiwan Strait. (1, 2, 3)

A. robusta Gregory: South China Sea. (1, 2, 3)

A. schmidtii Grunow: South China Sea. (1, 2, 3)

A. spectabilis Gregory: South China Sea; Taiwan. (3, 31)

A. staurophora (Cast.) Cleve: South China Sea. (1, 2, 3)

A. terroris Ehrenberg: Bohai; Yellow Sea; East China Sea; Taiwan Strait; South China Sea. (1, 2, 3)

A. turgida Gregory: South China Sea. (1, 2, 3)

GOMPHONEMACEAE

Gomphonema salinarum (Pant.) Cleve: South China Sea. (1, 2, 3)

AURICULACEAE

Auricula complexa (Greg.) Cleve: Taiwan Strait. (1, 2, 3)

A. insecta (Grun.) Cleve: South China Sea. (1, 2, 3)

A. intermedia (Lewis) Cleve: South China Sea. (1, 2, 3)

A. minuta Cleve: South China Sea. (1, 2, 3)

DIATOMALES
DIATOMACEAE

Arcocellulus cornucer Hasle, Von Stosch & Syverten: Taiwan Strait. (22, 23a)

A. mammifer Hasle, Von Stosch & Syverten: Taiwan Strait. (22)

Asterionella formosa Hass: East China Sea. (12)

A. glacialis Castracane: Taiwan Strait. (23a)

A. japonica Cleve: Bohai; Yellow Sea; East China Sea; Taiwan Strait; South China Sea. (1, 2, 3)

A. kariana Grunow: Bohai; Yellow Sea; East China Sea; Taiwan Strait. (2, 12, 27)

A. notata (Grun.) Grunow: Yellow Sea; East China Sea; Taiwan Strait; South China Sea. (2, 27)

Campylosira cymbelliformis (A. Schmidt) Grunow: East China Sea. (4)

Climacosphenia elongata Bailey: Lanyu (Taiwan); South China Sea. (30b, 33)

C. moniligera Ehrenberg: East China Sea; Taiwan Strait; Lanyu (Taiwan); South China Sea. (1, 2, 3, 30b)

Cymatosira gibberula Cheng & Gao: Taiwan Strait. (22, 23a)

C. lorenziana Grunow: East China Sea; Taiwan Strait. (2, 12)

Delphineis augustata (Panto.) Andrews: South China Sea. (1, 2, 3)

Diatoma hyalinum Kueting [*Fragilaria hyalina*]: Taiwan Strait; South China Sea. (16, 30)

Dimeregramma fulvum (Greg.) Ralfs: East China Sea; South China Sea. (1, 2, 3)

D. fusiformis Huang, Cheng & Chin: Taiwan Strait; South China Sea. (24)

D. opulens Mann: East China Sea. (1, 2, 3)

Fragilaria aurivillii Cleve: South China Sea. (1, 2, 3)

F. brevistriata Grunow: Taiwan Strait; South China Sea. (1, 2, 3)

F. crotonensis Kitton: Taiwan Strait. (1, 2, 3)

F. cylindrus Grunow: Taiwan Strait; South China Sea. (16)

F. heidenii Oestrup: Taiwan Strait. (8)

F. hyalina (Kuetz.) Grunow: South China Sea. (1, 2, 3)

F. longissima v. *protenta* Lin, Cheng & Chin: Taiwan Strait. (8)

F. oceanica Cleve: Taiwan Strait. (8)

F. striatula Lyngbye: Bohai; Yellow Sea; East China Sea; Taiwan Strait; South China Sea. (2, 4)

Gephyria media Arnott: East China Sea. (1, 2, 3)

Grammatophora angulosa Ehrenber: Taiwan Strait. (1, 2, 3)

G. fundata Mann: East China Sea. (1, 2, 3)

G. fundata v. *spinosa* Chin & Cheng: Okinawa Trough. (1, 2, 3)

G. hamulifera Kuetzing: East China Sea; Taiwan Strait; Lanyu (Taiwan); South China Sea. (1, 2, 3, 30b)

G. marina (Lyngb.) Kuetzing: Bohai; Yellow Sea; East China Sea; Taiwan Strait; South China Sea. (1, 2, 3)

G. oceanica Ehrenberg: East China Sea; Lanyu (Taiwan); South China Sea. (1, 2, 3, 30b)

G. oceanica v. *macilenta* (W. Smith) Grunow: Taiwan Strait; South China Sea. (1, 2, 3)

G. serpentina (Ralfs) Ehrenberg: East China Sea; South China Sea. (1, 2, 3)

G. undulata Ehrenberg: Bohai; Yellow Sea; East China Sea; Taiwan Strait; Lanyu (Taiwan); South China Sea. (1, 2, 3, 30b)

G. undulata v. *japonica* Grunow: Bohai; Yellow Sea. (1, 2, 3)

Licmophora abbreviata Agardh: Bohai; Yellow Sea; East China Sea; Taiwan Strait; South China Sea. (1, 2, 3)

L. california Grunow: Bohai; Yellow Sea; East China Sea; Taiwan Strait; South China Sea. (2, 12)

L. ehrenbergii (Kuetz.) Grunow: South China Sea. (1, 2, 3)

L. ehrenbergii v. *ovata* (W. Smith) Van Heurck: Taiwan Strait; South China Sea. (1, 2, 3)

L. flabellata Agardh: Taiwan Strait; South China Sea. (1, 2, 3)

L. gracilis (Ehr.) Grunow: South China Sea. (1, 2, 3)

L. gracilis v. *elongata* (Kuetz.) De Toni: South China Sea. (1, 2, 3)

L. paradoxa (Lyngb.) Agardh: Taiwan Strait; South China Sea. (1, 2, 3)

L. tenuis (Kuetz.) Grunow: Bohai; Yellow Sea; Taiwan Strait; South China Sea. (1, 2, 3)

Minutocellus polymorphus (Harger. & Guill.) Hasle, Von Stosch & Syverten: Taiwan Strait. (22, 23a)

Opephora gemmata (Grun.) Hustedt: Okinawa Trough. (1, 2, 3)

O. martyi Heribaud: Taiwan Strait; South China Sea. (1, 2, 3)

O. pacifica (Grun.) Petit: Taiwan Strait. (1, 2, 3)

Perissonoë cruciata (Janisch & Rabenhorst) Andrews & Stoelzel [*Rhaphoneis amphiceros* v. *tetragona* Grunow]: East China Sea; Taiwan Strait; South China Sea. (1, 2, 3)

Plagiogramma antillarum Cleve: South China Sea. (1, 2, 3)

P. atomus Greville: South China Sea. (1, 2, 3)
P. pulchellum Greville: Bohai; Yellow Sea; Taiwan Strait; South China Sea. (1, 2, 3)
P. reimeri Liu: South China Sea. (33)
P. staurophorum (Greg.) Heiberg: Taiwan Strait; South China Sea. (1, 2, 3)
P. vanheurckii Grunow: East China Sea; South China Sea. (1, 2, 3)
Rhabdonema adriaticum Kuetzing: Bohai; Yellow Sea; East China Sea; Taiwan Strait; Lanyu (Taiwan); South China Sea. (1, 2, 3, 30b)
R. arcuatum (Ag.) Kuetzing: Taiwan Strait; South China Sea. (2, 16)
R. mirificum W. Smith: South China Sea. (1, 2, 3)
R. punctatum (Harvey & Bailey) Stodder: South China Sea. (1, 2, 3)
R. sutum Mann: East China Sea; South China Sea. (1, 2, 3)
Rhaphoneis amphiceros (Ehr.) Ehrenberg: Bohai; Yellow Sea; East China Sea; Okinawa Trough; Taiwan Strait; South China Sea. (1, 2, 3)
R. atlantica Andrews: South China Sea. (33)
R. belgica (Grun.) Grunow: Bohai; Yellow Sea; East China Sea; Taiwan Strait; South China Sea. (1, 2, 3)
R. belgica v. *densestriata* Chen, Cheng & Chin: South China Sea. (24)
R. castracanei Grunow: South China Sea. (1, 2, 3)
R. elliptica Ehrenberg: Taiwan Strait. (1, 2, 3)
R. lancettula Grunow: South China Sea. (24)
R. paralis Hanna: South China Sea. (24)
R. rhomoides Hendey: Taiwan continental shelf; Lanyu (Taiwan). (30b, 31)
R. surirella (Ehr.) Grunow: East China Sea; Okinawa Trough; Taiwan Srait. (1, 2, 3)
R. surirella v. *australis* (Petit) Grunow: Bohai; Yellow Sea; East China Sea; Okinawa Trough; Taiwan Strait; South China Sea. (1, 2, 3)
Striatella delicatula (Kuetz.) Grunow: Taiwan Strait. (1, 2, 3)
S. interrupta (Ehr.) Heiberg: South China Sea. (1, 2, 3)
S. nanhainica Guo, Zhou & Ye: South China Sea. (1, 2, 3)
S. unipunctata (Lyngb.) Agardh: Yellow Sea; East China Sea; South China Sea. (2, 28)
Synedra barbatula Kuetzing: South China Sea. (24)
S. crystallina (Ag.) Kuetzing: Bohai; Yellow Sea; Taiwan Strait; South China Sea. (1, 2, 3)
S. formosa Hantzsch & Rabenhorst: Taiwan Strait; South China Sea. (1, 2, 3)
S. fulgens (Grev.) W. Smith: Bohai; Yellow Sea; Taiwan Strait; South China Sea. (1, 2, 3)
S. gaillonii (Bory) Ehrenberg: Bohai; Yellow Sea; Taiwan Strait; Taiwan continental shelf; South China Sea. (2, 31)
S. hennedyana Gregory: Taiwan Strait; South China Sea. (1, 2, 3)
S. investiens W. Smith: South China Sea. (1, 2, 3)
S. laevigata Grunow [*Hyalosynedra laevigata* (Grun.) Williams & Round]: Lanyu (Taiwan). (30b)
S. pulcherrima Hantzsch & Rabenhorst: South China Sea. (1, 2, 3)

S. robusta Ralfs: Taiwan Strait; South China Sea. (1, 2, 3)
S. robusta v. *sinica* Skvortzow: Taiwan Strait. (1, 2, 3)
S. rostrata (Hantzsch) Hustedt: South China Sea. (17)
S. tabulata (Ag.) Kuetzing: Bohai; Yellow Sea; East China Sea; Okinawa Trough; Taiwan Strait; South China Sea. (1, 2, 3)
S. tabulata v. *acuminata* (Grunow) Hustedt: South China Sea. (1, 2, 3)
S. tabulata v. *fasciculata* (Kuetz.) Hustedt: South China Sea. (1, 2, 3)
S. tabulata v. *parva* (Kuetz.) Hustedt: Bohai; Yellow Sea; Taiwan Strait; South China Sea. (1, 2, 3)
S. ulna (Nitz.) Ehrenberg: Bohai; Yellow Sea; East China Sea; Taiwan Strait. (3, 12)
S. undulata Bailey: Taiwan Strait; South China Sea. (1, 2, 3)
S. undulata v. *curvata* Heiden & Kolbe: Lanyu (Taiwan). (30b)
Synedrosphenia gomphonema (Janisch & Rabenh.) Hustedt: South China Sea. (1, 2, 3)
Tabellaria flocculosa v. *asterionelloides* Grunow: South China Sea. (33)
T. nodosa Ehrenberg: South China Sea. (24)
Tetracyclus javanicus Hustedt: South China Sea. (1, 2, 3)
Thalassionema bacillaris (Heiden) Kolbe: South China Sea. (11)
T. nitzschioides (Grun.) Van Heurck: Bohai; Yellow Sea; East China Sea; Okinawa Trough; Taiwan Strait; southeastern Taiwan; South China Sea. (1, 2, 3, 30a)
T. nitzschioides v. *parva* Heiden: South China Sea. (11)
Thalassiothrix delicatula Cupp: East China Sea; Taiwan Strait; South China Sea. (4, 7, 15, 16)
T. frauenfeldii (Grun.) Grunow: Bohai; Yellow Sea; East China Sea; Taiwan Strait; southeastern Taiwan; South China Sea. (1, 2, 3, 30a)
T. gibberula Hasle: South China Sea. (11)
T. longissima Cleve & Grunow: Bohai; Yellow Sea; East China Sea; Taiwan Strait; Taiwan continental shelf; southeastern Taiwan; South China Sea. (2, 30a, 31)
T. mediterranea Pavillard: East China Sea; Taiwan Strait; South China Sea. (4, 7)
T. mediterranea v. *pacifica* Cupp: Taiwan shallow water. (5)
T. sinensis Meister: South China Sea. (1, 2, 3)
T. vanhoeffenii Heiden: East China Sea; Taiwan Strait; South China Sea. (11, 18, 29a)

ACHNANTHALES
COCCONEIACEAE

Anorthoneis eurystoma Cleve: Taiwan Strait. (33)
A. excentrica (Donkin) Grunow: East China Sea; Taiwan Strait; South China Sea. (1, 2, 3)
Campyloneis grevillei (W. Smith) Grunow: Okinawa Trough; Taiwan Strait; Lanyu (Taiwan); South China Sea. (2, 23b, 30b, 31)
Cocconeis britannica Naegeli: East China Sea. (1, 2, 3)
C. costata Gregory: Taiwan Strait. (1, 2, 3)
C. diminuta Pantocsek: South China Sea. (1, 2, 3)

C. dirupta Gregory: Bohai; Yellow Sea; East China Sea; Taiwan Strait; South China Sea. (1, 2, 3)
C. dirupta v. *africana* A. Schmidt: South China Sea. (1, 2, 3)
C. dirupta v. *flexella* (Jan. & Rab.) Grunow: South China Sea. (1, 2, 3)
C. diruptoides Hustedt: Taiwan Strait. (1, 2, 3)
C. disculoides Hustedt: Taiwan continental shelf. (31)
C. distans Gregory: East China Sea; Okinawa Trough; Taiwan continental shelf; South China Sea. (2, 31)
C. distans v. *bahusiensis* Cleve-Euler: Taiwan Strait. (1, 2, 3)
C. fasciolata (Ehr.) Brown: East China Sea; South China Sea. (1, 2, 3)
C. fluminensis (Grun.) Peragallo H. & M.: Taiwan Strait. (1, 2, 3)
C. heteroidea Hantzsch: East China Sea; Taiwan Strait; Lanyu (Taiwan); South China Sea. (1, 2, 3, 30b)
C. heteroidea v. *curvirotunda* (Tempere & Brun) Cleve: East China Sea; Taiwan Strait; South China Sea. (1, 2, 3)
C. lyra A. Schmidt: South China Sea. (1, 2, 3)
C. molesta Kuetzing: Taiwan Strait. (1, 2, 3)
C. notata Petit: East China Sea. (1, 2, 3)
C. pediculus Ehrenberg: Taiwan Strait; South China Sea. (3, 31)
C. pellucida Hantzsch: East China Sea; Okinawa Trough; Taiwan Strait; Lanyu (Taiwan). (1, 2, 3, 30b)
C. pellucida v. *minor* Grunow: Taiwan Strait; South China Sea. (1, 2, 3)
C. pelta v. *sinica* Skvortzow: East China Sea. (1, 2, 3)
C. placentula Ehrenberg: East China Sea; South China Sea. (1, 2, 3)
C. placentula v. *euglypta* (Ehr.) Cleve: Bohai; Yellow Sea; East China Sea; Taiwan Strait; South China Sea. (1, 2, 3, 23b)
C. placentula v. *lineata* (Ehr.) Cleve: East China Sea; Taiwan Strait; Lanyu (Taiwan). (1, 2, 3, 30b)
C. problematica Van Landingham: East China Sea. (1, 2, 3)
C. pseudomarginata Gregory: Bohai; Yellow Sea; East China Sea; Taiwan Strait; Taiwan continental shelf; Lanyu (Taiwan); South China Sea. (2, 31)
C. pseudomarginata v. *formosa* Skvortzow: Bohai; Yellow Sea. (1, 2, 3)
C. pseudomarginata v. *intermedia* Grunow: Bohai; Yellow Sea; Taiwan Strait; South China Sea. (1, 2, 3)
C. scutellum Ehrenberg: Bohai; Yellow Sea; East China Sea; Okinawa Trough; Taiwan Strait; South China Sea. (1, 2, 3)
C. scutellum v. *japonica* (A. Schmidt) Skvortzow: Bohai; Yellow Sea. (1, 2, 3)
C. scutellum v. *minutissima* Grunow: East China Sea; Taiwan Strait. (1, 2, 3)
C. scutellum v. *parva* (Grun.) Cleve: Bohai; Yellow Sea; East China Sea; Taiwan Strait. (1, 2, 3)
C. scutellum v. *stauroneiformis* Rabenhorst: East China Sea; Taiwan Strait. (1, 2, 3)
C. sublittoralis Hendey: East China Sea; Taiwan Strait. (1, 2, 3)
C. subtilis A. Schmidt: East China Sea. (1, 2, 3)
C. tenuistriata Lin & Chin: Taiwan Strait. (1, 2, 3)
C. versicolor Brun: South China Sea. (1, 2, 3)
C. vitrea Brun: East China Sea. (1, 2, 3)

ACHNANTHACEAE

Achnanthes bengalensis Grunow: Taiwan Strait. (1, 2, 3)
A. brevipes Agardh: Bohai; Yellow Sea; East China Sea; Taiwan Strait; South China Sea. (1, 2, 3)
A. brevipes v. *angustata* (Grev.) Cleve [*A. angustata* Greville]: Bohai; Yellow Sea; East China Sea; Taiwan Strait; Taiwan continental shelf; Lanyu (Taiwan); South China Sea. (2, 30b, 31)
A. brevipes v. *intermedia* (Kuetz.) Cleve: Bohai; Yellow Sea; East China Sea; Taiwan Strait; South China Sea. (1, 2, 3)
A. brevipes v. *leudugeri* (Temp. & Brun) Cleve: Bohai; Yellow Sea; Taiwan Strait; South China Sea. (1, 2, 3)
A. brevipes v. *parvula* (Kuetz.) Cleve: Bohai; Yellow Sea; East China Sea; Taiwan Strait; South China Sea. (1, 2, 3)
A. citronella (Mann) Hustedt: Taiwan Strait; South China Sea. (1, 2, 3)
A. clevei Grunow: South China Sea. (1, 2, 3)
A. coarctata (Bréb.) Grunow: Taiwan Strait. (33)
A. conspicus v. *brevistrista* Hustedt: South China Sea. (33)
A. cotteriensis Foged: East China Sea. (33)
A. curvirostrum Brun: South China Sea. (1, 2, 3)
A. danica (Floegel) Grunow: South China Sea. (1, 2, 3)
A. delicatula (Kuetz.) Grunow: Taiwan Strait. (1, 2, 3)
A. dispar v. *angulata* Hustedt: East China Sea. (1, 2, 3)
A. hauckiana Grunow: Taiwan Strait; South China Sea. (1, 2, 3, 23d)
A. javanica Grunow: Bohai; Yellow Sea; East China Sea; Taiwan Strait; Lanyu (Taiwan); South China Sea. (1, 2, 3, 23d, 30b)
A. javanica v. *subconstricta* Meister: Bohai; Yellow Sea; East China Sea; Taiwan Strait; South China Sea. (1, 2, 3)
A. javanica v. *tenuistauros* (Mann) Meister: Bohai; Yellow Sea; Taiwan Strait; South China Sea. (1, 2, 3)
A. kryophila Petersen: South China Sea. (33)
A. kuwaitensis Hendey: South China Sea. (33)
A. longipes Agardh: Bohai; Yellow Sea; East China Sea; Taiwan Strait; Lanyu (Taiwan); South China Sea. (1, 2, 3)
A. microcephala (Kuetz.) Cleve: Taiwan Strait. (1, 2, 3)
A. minutissima Kuetzing: South China Sea. (1, 2, 3)
A. oblongella Ostrup: South China Sea. (33)
A. orientalis Petit: Bohai; Yellow Sea; Taiwan Strait; South China Sea. (1, 2, 3)
A. ploenensis (Ploenesis) Hustedt: East China Sea. (1, 2, 3)
A. radiata Du & Cheng: Taiwan Strait. (1, 2, 3)
Rhoicosphenia curvata (Kuetz.) Grunow: East China Sea; Taiwan Strait. (1, 2, 3)

EUNOTIALES
EUNOTIACEAE

Eunotia eruca Ehrenberg: South China Sea. (1, 2, 3)
Pseudoeunotia doliolus (Wallich) Grunow: East China Sea; Okinawa Trough; South China Sea. (2, 11)

PHAEODACTYLALES
PHAEODACTYLACEAE

Phaeodactylum tricornutum Bohlin: Bohai; Yellow Sea; Taiwan Strait. (1, 2, 3)

SURIRELLALES
EPITHEMIACEAE

Denticula subtilis Grunow: East China Sea; Taiwan Strait; South China Sea. (1, 2, 3)
Rhopalodia gibberula (Ehr.) O. Mueller: Bohai; Yellow Sea; Taiwan Strait; South China Sea. (1, 2, 3)
R. gibberula v. *vanheurck* O. Mueller: South China Sea. (1, 2, 3)
R. musculus (Kuetz.) O. Mueller: Taiwan Strait; Lanyu (Taiwan); South China Sea. (1, 2, 3, 30b)
R. musculus v. *constricta* (Bréb) Peragallo & Peragallo: Taiwan Strait; South China Sea. (1, 2, 3)
R. uncinata O. Mueller: East China Sea. (1, 2, 3)

NITZSCHIACEAE

Bacillaria paradoxa Gmelin [*Nitzschia paradoxa* (Gmel.) Grunow]: Bohai; Yellow Sea; East China Sea; Taiwan Strait; South China Sea. (1, 2, 3)
Cylindrotheca closterium (Ehr.) Reimann & Lewin: Bohai; Yellow Sea; East China Sea; Taiwan Strait. (1, 2, 3)
C. closterium v. *californica* (Mereschk.) Reimann & Lewin: Taiwan Strait. (1, 2, 3)
C. gracilis (Bréb.) Grunow: Taiwan Strait. (1, 2, 3)
Gomphonitzschia chinensis Skvortzow: Taiwan Strait. (1, 2, 3)
Hantzschia marina (Donkin) Grunow: East China Sea; Taiwan Strait; South China Sea. (1, 2, 3)
H. virgata (Roper) Grunow: Taiwan Strait; South China Sea. (1, 2, 3)
H. virgata v. *gracilis* Hustedt: South China Sea. (1, 2, 3)
Nitzschia acuminata (W. Smith) Grunow: Taiwan Strait; South China Sea. (1, 2, 3)
N. aequatorialis Heiden: Taiwan. (31)
N. amphibiodes v. *chenghaensis* Liu: South China Sea. (33)
N. angularis W. Smith: Bohai; Yellow Sea; Taiwan Strait; South China Sea. (1, 2, 3)
N. angularis v. *affinis* (Grun.) Grunow: East China Sea. (1, 2, 3)
N. angustata Grunow: East China Sea; South China Sea. (1, 2, 3)
N. antillarum (Cleve) Meister: East China Sea; Okinawa Trough; Taiwan Strait. (1, 2, 3)
N. apiculata v. *liaotungiensis* Skvortzow: Bohai; Yellow Sea. (1, 2, 3)
N. bicapitata Cleve: East China Sea. (18)
N. brevissima Grunow: East China Sea; Taiwan Strait. (1, 2, 3)
N. capuluspalae Simonsen: East China Sea. (1, 2, 3)
N. coarctata Grunow: Taiwan Strait; South China Sea. (1, 2, 3)
N. cocconeiformis Grunow: Bohai; Yellow Sea; East China Sea; Taiwan Strait; South China Sea. (1, 2, 3)
N. compressa (Bailey) Boyer: East China Sea. (1, 2, 3)
N. constricta (Greg.) Grunow: East China Sea; Taiwan Strait; South China Sea. (1, 2, 3)
N. corpulenta Hendey: East China Sea; Taiwan Strait. (1, 2, 3)
N. cursoria (Donkin) Grunow: Taiwan Strait. (1, 2, 3)
N. delicatissima Cleve [*Pseudonitzschia delicatissima* (Cleve) Heiden]: Bohai; Yellow Sea; East China Sea; Taiwan Strait; South China Sea. (2, 4, 28)
N. denticula Grunow: East China Sea; Taiwan Strait. (1, 2, 3)
N. didyma Liu & Chin: Bohai; Yellow Sea; East China Sea; Taiwan Strait; South China Sea. (1, 2, 3)
N. dissipata (Kuetz.) Grunow: Taiwan Strait; Taiwan. (31, 23a)
N. distans Gregory: East China Sea; Lanyu (Taiwan); South China Sea. (1, 2, 3, 30b)
N. distantoides Hustedt: South China Sea. (1, 2, 3)
N. epithemoides Grunow: Taiwan Strait. (1, 2, 3)
N. fasciculata (Grun.) Grunow: East China Sea; Taiwan Strait; South China Sea. (1, 2, 3)
N. filiformis (W. Smith) Van Heurck: Taiwan Strait. (1, 2, 3)
N. fluminensis Grunow: Bohai; Yellow Sea; Taiwan Strait; Taiwan continental shelf; South China Sea. (2, 31)
N. frigida Grunow: Taiwan Strait. (1, 2, 3)
N. frustulum (Kuetz.) Grunow: East China Sea; Taiwan Strait; South China Sea. (1, 2, 3)
N. granulata Grunow: Taiwan Strait; South China Sea. (1, 2, 3)
N. habirshawii H. L. Smith: Bohai; Yellow Sea. (1, 2, 3)
N. hungarica Grunow: Bohai; Yellow Sea; East China Sea; Taiwan Strait; South China Sea. (1, 2, 3)
N. hybrida Grunow: Taiwan Strait; South China Sea. (1, 2, 3)
N. insignis Gregory: East China Sea; Taiwan Strait. (1, 2, 3)
N. insignis v. *lanceolata* Hustedt: Taiwan Strait. (1, 2, 3)
N. jelineckii Grunow: South China Sea. (1, 2, 3)
N. kolaczeckii Grunow: East China Sea; South China Sea. (11, 18)
N. lanceola (Grun.) Grunow: Taiwan Strait. (1, 2, 3)
N. lanceolata W. Smith: East China Sea; Taiwan Strait; South China Sea. (1, 2, 3)
N. lanceolata v. *chinensis* Skvirtzow: Taiwan Strait. (1, 2, 3)
N. lanceolata v. *incrustoms* Grunow: Taiwan Strait. (1, 2, 3)
N. lanceolata v. *minor* Grunow: East China Sea; Taiwan Strait; South China Sea. (1, 2, 3)
N. linkei Hustedt: South China Sea. (1, 2, 3)
N. littoralis Grunow: Bohai; Yellow Sea; Taiwan Strait; Taiwan continental shelf. (3, 31)
N. longissima (Bréb.) Grunow: Bohai; Yellow Sea; East China Sea; Taiwan Strait; southeastern Taiwan; South China Sea. (1, 2, 3, 30a)
N. longissima v. *chinensis* Grunow: South China Sea. (1, 2, 3)
N. longissima v. *costata* A. Schmidt: South China Sea. (17)
N. longissima v. *reversa* Grunow: Bohai; Yellow Sea; East China Sea; Taiwan Strait; South China Sea. (2, 12, 27)
N. lorenziana Grunow: Bohai; Yellow Sea; East China Sea; Taiwan Strait; South China Sea. (2, 12)
N. lorenziana v. *densestriata* (Perag. & Perag.) Hustedt: Taiwan Strait. (1, 2, 3)

N. macilenta Gregory: East China Sea. (1, 2, 3)

N. macilenta f. *abbreviata* Grunow: Taiwan Strait. (1, 2, 3)

N. majuscula v. *lineata* Liu & Chin: East China Sea; Taiwan Strait; South China Sea. (1, 2, 3)

N. marginulata Grunow: Taiwan continental shelf. (31)

N. marginulata v. *didyma* Grunow: Taiwan Strait; South China Sea. (1, 2, 3)

N. marginulata v. *subconstricta* Grunow: Bohai; Yellow Sea; East China Sea; Taiwan Strait; South China Sea. (1, 2, 3)

N. marina Grunow: East China Sea; Okinawa Trough; Taiwan continental shelf; Taiwan Strait; South China Sea. (2, 11, 29a, 30, 31)

N. maxima Grunow: Taiwan Strait. (1, 2, 3)

N. navicularis (Bréb.) Grunow: Taiwan Strait. (1, 2, 3)

N. nelsonii Hanna & Grant: Taiwan continental shelf. (31)

N. obtusa W. Smith: Bohai; Yellow Sea; East China Sea; Taiwan Strait; South China Sea. (1, 2, 3)

N. obtusa v. *scalpelliformis* Grunow: Bohai; Yellow Sea; East China Sea; Taiwan Strait; South China Sea. (1, 2, 3)

N. ossiformis (Taylor) Simonsen: South China Sea. (11)

N. palea v. *minuta* (Bleisch) Grunow: East China Sea. (1, 2, 3)

N. paleacea Grunow: South China Sea. (1, 2, 3)

N. panduriformis Gregory: Bohai; Yellow Sea; East China Sea; Okinawa Trough; Taiwan continental shelf; Taiwan Strait; South China Sea. (2, 31)

N. panduriformis v. *elegans* (Lagerst.) Cleve-Euler: Okinawa Trough; South China Sea. (1, 2, 3)

N. panduriformis v. *minor* Grunow: Bohai; Yellow Sea; East China Sea; Taiwan Strait; South China Sea. (1, 2, 3)

N. parvula W. Smith: Taiwan Strait. (1, 2, 3)

N. petitiana Grunow: Taiwan Strait; South China Sea. (1, 2, 3)

N. plana W. Smith: Taiwan continental shelf. (31)

N. pulcherrima Grunow: East China Sea. (1, 2, 3)

N. punctata (W. Smith) Grunow: Bohai; Yellow Sea; East China Sea; Taiwan Strait; South China Sea. (1, 2, 3)

N. punctata v. *coarctata* Grunow: Bohai; Yellow Sea; East China Sea; Taiwan Strait; South China Sea. (2, 3, 17)

N. punctata v. *elongata* (Grun.) Grunow: Taiwan Strait. (1, 2, 3)

N. pungens Grunow [*Pseudonitzschia pungens* (Grun.) Hasle]: Bohai; Yellow Sea; East China Sea; Taiwan Strait; South China Sea. (1, 2, 3, 23b)

N. pungens v. *atlantica* Cleve: Bohai; Yellow Sea; East China Sea. (2, 28)

N. scalaris (Ehr.) W. Smith: East China Sea; Taiwan Strait. (1, 2, 3)

N. seriata Cleve [*Pseudonitzschia seriata* (Cleve) H. Peragallo]: Bohai; Yellow Sea; East China Sea; Taiwan Strait; southeastern Taiwan; South China Sea. (2, 11, 30a)

N. sigma (Kuetz.) W. Smith: Bohai; Yellow Sea; East China Sea; Taiwan Strait; Taiwan continental shelf; South China Sea. (2, 31)

N. sigma v. *intercedens* Grunow: Bohai; Yellow Sea; East China Sea; Taiwan Strait. (2, 31)

N. sigma v. *rigida* (Kuetz.) Grunow: Taiwan Strait; South China Sea. (1, 2, 3)

N. sigma v. *sigmatella* Grunow: East China Sea; Taiwan Strait; South China Sea. (1, 2, 3)

N. sigmoidea (Nitz.) W. Smith: East China Sea; Taiwan Strait; Taiwan continental shelf; South China Sea. (2, 31)

N. sinensis Liu: East China Sea; Taiwan Strait. (1, 2, 3)

N. socialis Gregory: Bohai; Yellow Sea; Taiwan Strait. (1, 2, 3)

N. spathulata Brébisson: Bohai; Yellow Sea; Taiwan Strait. (1, 30)

N. spathulata v. *hyalina* (Greg.) Grunow: East China Sea; Taiwan Strait. (1, 2, 3)

N. spectabilis (Ehr.) Ralfs: South China Sea. (16)

N. subsalsa Cholnoky: East China Sea. (1, 2, 3)

N. subtilis Grunow: Bohai; Yellow Sea; East China Sea; Taiwan Strait; South China Sea. (1, 2, 3)

N. tryblionella Hantzsch: Taiwan Strait. (1, 2, 3)

N. tryblionella v. *subsalina* f. *subconstricta* Hustedt: Bohai; Yellow Sea; Taiwan Strait. (1, 2, 3)

N. tryblionella v. *victoriae* (Grun.) Grunow: Taiwan Strait. (1, 2, 3)

N. ventricosa Kitton: Taiwan Strait; South China Sea. (1, 2, 3)

N. vermicularis (Kuetz.) Hantzsch: South China Sea; Taiwan continental shelf. (31)

N. vidovichii (Grun.) Grunow: East China Sea; South China Sea. (1, 2, 3)

N. vitraea Norman: Taiwan Strait; South China Sea. (1, 30)

Nitzschiella incurva (Grun.) Peragallo: Bohai; Yellow Sea. (1, 2, 3)

N. longissima (Bréb.) Rabenhorst: Bohai; Yellow Sea; Taiwan Strait. (1, 2, 3)

Pseudonitzschia delicatissima (Cleve) Heiden: Bohai; Yellow Sea; East China Sea; Taiwan Strait; South China Sea. (2, 12)

P. sicula (Castr.) Peragallo & Peragallo: Okinawa Trough. (1, 2, 3)

P. sicula v. *bicuneata* (Grun.) Hasle: Bohai; Yellow Sea; Taiwan Strait; South China Sea. (1, 2, 3)

P. sicula v. *migrans* (Cleve) Hasle: Taiwan Strait. (1, 2, 3)

SURIRELLACEAE

Campylodiscus adriaticus Grunow: Okinawa Trough; Taiwan Strait. (1, 2, 3)

C. australis Grunow: Taiwan continental shelf. (31)

C. biangulatus Greville: East China Sea; Taiwan Strait; Taiwan continental shelf; South China Sea. (2, 31)

C. birostratus Deby: Taiwan Strait; Taiwan continental shelf; South China Sea. (2, 31)

C. brightwellii Grunow: East China Sea; Taiwan Strait; southeastern Taiwan; South China Sea. (1, 2, 3, 30a)

C. daemelianus Grunow: East China Sea; Taiwan Strait. (1, 2, 3, 23b)

C. daemelianus v. *comminuta* A. Schmidt: Okinawa Trough; Taiwan Strait. (1, 2, 3, 23b)

C. decorus Brébisson: East China Sea; Taiwan Strait; Taiwan continental shelf; South China Sea. (2, 31)

C. ecclesianus Greville: Taiwan Strait. (1, 2, 3)

C. echeneïs Ehrenberg: Okinawa Trough. (1, 2, 3)
C. heufleri Grunow: East China Sea. (1, 2, 3)
C. hodgsonii W. Smith: Taiwan Strait. (1, 2, 3)
C. horologium Williamson: East China Sea. (1, 2, 3)
C. incertus A. Schmidt: Bohai; Yellow Sea; Taiwan Strait; South China Sea. (1, 2, 3)
C. intermedius Grunow: Bohai; Yellow Sea; South China Sea. (1, 2, 3)
C. kittonianus Greville: Okinawa Trough. (1, 2, 3)
C. latus Shadbolt: Taiwan Strait; South China Sea. (1, 2, 3)
C. ralfsii W. Smith: Taiwan Strait; Lanyu (Taiwan); South China Sea. (1, 2, 3, 30b)
C. triumphans S. Schmidt: East China Sea; South China Sea. (1, 2, 3)
C. wallichianus Greville: Taiwan Strait; South China Sea. (1, 2, 3)
C. wallichianus v. *normanicus* (Grev.) De Toni: Taiwan Strait. (1, 2, 3)
Podocystis adriatica Kuetzing: South China Sea. (1, 2, 3)
P. spathulata (Shabolt) Van Heurck: Bohai; Yellow Sea; Taiwan Strait; Taiwan continental shelf; Lanyu (Taiwan); South China Sea. (2, 30b, 31)
Stenopterobia intermedia (Lewis) Van Heurck: Taiwan Strait. (1, 2, 3)
Surirella apiae Witt: East China Sea. (1, 2, 3)
S. arabica Grunow: Bohai; Yellow Sea; East China Sea; Taiwan Strait; South China Sea. (1, 2, 3)
S. armoricana Peragallo & Peragallo: Taiwan Strait; Lanyu (Taiwan). (2, 30, 30b)
S. biseriata Brébisson: South China Sea. (12)
S. campechiana Hustedt: East China Sea. (1, 2, 3)
S. collare A. Schmidt: South China Sea. (1, 2, 3)
S. elegans v. *norvegica* (Eulenst.) Brun: Taiwan Strait. (1, 2, 3)
S. eximia Greville: Taiwan Strait. (30)
S. fastuosa Ehrenberg: Bohai; Yellow Sea; East China Sea; Taiwan Strait; Lanyu (Taiwan); South China Sea. (1, 2, 3, 30b)
S. fastuosa v. *cuneata* (A. S.) Witt: Bohai; Yellow Sea; East China Sea; Taiwan Strait; Taiwan continental shelf; South China Sea. (2, 31)
S. fastuosa v. *plusieura* Petit: East China Sea. (1, 2, 3)
S. fastuosa v. *recens* (A. S.) Cleve: Bohai; Yellow Sea; East China Sea; Okinawa Trough; South China Sea. (1, 2, 3)
S. fastuosa v. *spinlifera* (A. S.) Witt: Taiwan continental shelf. (31)
S. fastuosa v. *suborbicularis* (Grun.) Peragallo & Peragallo: Bohai; Yellow Sea; South China Sea. (1, 2, 3)
S. fluminensis Grunow: Bohai; Yellow Sea; East China Sea; Okinawa Trough; Taiwan continental shelf; Taiwan Strait; South China Sea. (2, 31)
S. gemma Ehrenberg: Bohai; Yellow Sea; East China Sea; Taiwan Strait; South China Sea. (1, 2, 3)
S. gravis Mann: East China Sea. (1, 2, 3)
S. japonica Ehrenberg: East China Sea. (1, 2, 3)
S. kurzii Grunow: East China Sea; Taiwan Strait. (1, 2, 3)
S. lata W. Smith: East China Sea; Taiwan Strait. (1, 2, 3, 23b)
S. lata v. *robusta* Witt: East China Sea. (1, 2, 3)
S. liaotungiensis Skvortzow: Bohai; Yellow Sea. (1, 2, 3)
S. liaotungiensis v. *minuta* Skvortzow: Bohai; Yellow Sea. (1, 2, 3)
S. mexicana A. Schmidt: East China Sea. (1, 2, 3)
S. nervata (Grun.) Mereschkowsky [*Plagiodiscus nervatus* Grunow]: Taiwan Strait; South China Sea. (1, 2, 3)
S. palmeriana Greville: South China Sea. (1, 2, 3)
S. schleinitzii Tanisch: East China Sea. (1, 2, 3)
S. schmidtii Witt: East China Sea. (1, 2, 3)
S. seychellarum Hustedt: South China Sea. (1, 2, 3)
S. seychellarum v. *biseriata* Hustedt: South China Sea. (1, 2, 3)
S. significans Mann: East China Sea. (1, 2, 3)
S. voigtii Skvortzow: Bohai; Yellow Sea; Taiwan Strait; South China Sea. (1, 2, 3)
Tryblioptychus cocooneiformis (Cleve) Hendey: Bohai; Yellow Sea; East China Sea; Taiwan Strait; Taiwan continental shelf; South China Sea. (2, 31)
T. hainanensis Voigt: South China Sea. (1, 2, 3)

Taxonomic Position Undeterminated

Van Heurchella admirabilis Pantocsek: Okinawa Trough. (1, 2, 3)

REFERENCES*

(1). Chin, T. G., Cheng Zhaodi, Lin Junmin and Lu Shicheng, 1982. Marine benthic diatoms from China, Vol. I. China Ocean Press, 323pp.
(2). Chin, T. G., 1981. The geographical distribution of the marine diatoms in China. Nova Hedwigia, 35:763–792
(3). Chin, T. G., Cheng Zhaodi, Lu Shichen and Ma Junxiang, 1992. Marine benthic diatoms from China, Vol II. China Ocean Press, 437pp.
(4). The Committee of Science and Technology, the Oceanographic Group, 1977. Studies on offshore plankton in China Seas. Reports on Comprehensive Survey of China Seas, Vol. 8, Chapter 10, pp:1–159.
(5). South China Sea Institute of Oceanology, Acdemia Sinica, 1987. Report on comprehensive survey and research about Zengmu Ansha-the southern boundary of China. China Science Press, Part 3:115–131.
(6). Qi Yuzao and Zhang Zian, 1977. Studies on the taxonomy of diatoms by scanning electronic microscope. Journal of Zhongshan University, 15(2):113–119.
(7). Feng Jifang, 1991. Composition and distribution of species on phytoplankton in Minnan-Taiwan Bank Fishing Ground. Ecosystem studies in the upwelling regions of Minnan-Taiwan Bank Fishing Ground, China Science Press, pp:388–406.
(8). Li Yaqin, Cheng Zhaodi and Chin T.G., 1991. Notes on some new species and new records of diatoms from Xiamen Harbour. Acta Oceanologica Sinica, 13(5):682–685, pl.1.
(9). Chen Guowei and Qian Shuben, 1984. A new species of the genus *Roperia* (*Roperia latiovala* sp. nov.). Journal of Shandong College of Oceanology, 14(3):77–80.
(10). Chen Guowei and Qian Shuben, 1984. A new species of the genus *Thalassiosira* (*Thalassiosira scrotiformis* sp. nov.). Journal of Shandong College of Oceanology, 14(4):74–78.

(11). State Oceanic Administration, 1988. The achievement in comprehensive investigation on environmental resources in the central part of the South China Sea. China Ocean Press, Chapter 12, pp:223– 227.
(12). State Oceanic Administration et al., 1992. Comprehensive survey on the resources in the coastal area and the shore of China Seas. China Ocean Press, pp:97–110.
(13). Yu Jianluan, Zhang Ziyun and Cheng Zhaodi, 1983. Distribution of diatoms in continental shelf fo the East China Sea. Acta Oceanologica Sinica, 5(4):519–525.
(14). Guo Y. J., 1981. Studies on the planktonic *Coscinodiscus* (diatoms) of the South China Sea. Studia Marina Sinica, 18: 149–175, pls:I–IV.
(15). Guo Y. J. and Yang Z.Y., 1982. The ecological studies on the phytoplankton over the continental shelf of the East China Sea in the summer of 1976. Studia Marina Sinica, 19:11, 2, 32.
(16). Guo Y. J. and Zhou H. Q., 1985. The planktonic diatom flora around the Zhongsha and Xisha Islands. Studia Marina Sinica, 24: 87–97.
(17). Guo Y. J. and H. Q. Zhou.1985. Taxonomy for pennatae (Bacillariophyta) of the Xisha Islands (I). Studia Marina Sinica, 24:99–118.
(18). Xu Zhimin, Jiang Jinlun and Lu Douding, 1990. Standing crop and composition of phytoplankton in Kuroshio and adjacent water in the spring of 1986. Essays on the investigation of Kuroshio (I), China Ocean Press, pp:215–228.
(19). Gao Yahui, Cheng Zhaodi and Chin T. G., 1992. *Minidiscus*, a new recorded nanodiatom genus for China. Acta Phytotaxonomica Sinica, 30(3):273–276.
(20). Gao Yahui, Cheng Zhaodi and Chin T.G., 1992. Two new species of *Mindiscs* and a new variety of *Cyclotella*. Journal of Xiamen University (Natural Science), 31(1):74–77.
(21). Gao Yahui and Cheng Zhaodi, 1992. A new species and two new varieties of *Thalassiosira*. Journal of Xiamen University (Natural Science), 31(3):291–294.
(22). Cheng Zhaodi and Gao Yahui, 1993. Nanodiatoms from Xiamen Harbour. Acta Phytotaxonomica Sinica, 31(2):197–200, pl.1 and 2.
(23). Cheng Zhaodi and Liu Shicheng, 1992. A taxonomy study on *Skeletonema greville* from Xiamen Harbour. Journal of Xiamen University (Natural Science), 31(3):295–297.
(23a). Cheng Zhaodi, Gao Yahui and Liu Shicheng, 1993. Nanodiatoms from Fujian Coast. China Ocean Press, pp:1–91, pls.:1, 2, 34.
(23b). Cheng Zhaodi, Gao Yahui and Mick Dickman, 1996. Colour plates of the diatoms. China Ocean Press.
(23c). Lou Shicheng and Cheng Zhaodi, 1996. New records in China for *Mastogloia thwaites*. Journal of Xiamen University (Natural Science), 35(2):283–287.
(23d). Fan Hangqing, Cheng Zhaodi, Liu Shicheng and Gao Yahui, 1993. Species of benthic diatoms in Mangrove Habitats of Guangxi Province, China. Journal of the Guangxi Province, China. Journal of the Guangxi Academy of Sciences, 9(2):37–42.
(23e). Lang Dongzhao, Cheng Zhaodi and Liu Shicheng, 1995. Diatoms in late quanternary sediment core from South China Sea. China Ocean Press, 103pp.
(24). Huang Bangqin, Chen Ping, Cheng Zhaodi and Jin Dexiang, 1989. Notes on new species and new records of attaching diatoms from Daya Bay, Guangdong, China. Journal of Oceanography in Taiwan Strait, 8(2):140–143.
(25). Qian Shuben, Chen Kouwei and Tang Tingyao, 1981. Studies on the phytoplankton from the offshore waters of Rongcheng, Shandong Province. Journal of Shandong College of Oceanology , 11(3):52– 70.
(26). Qian Shuben, 1981. A new diatom of the genus *Rhizosolenia* (*Rhizosolenia sinensis* sp. nov.). Journal of Shandong College of Oceanology, 11(4):53–57.
(27). Qian Shuben, Wang Xiaoqing and Chen Gouwei, 1983. The phytoplankton of the Jiaozhou Bay. Journal of Shandong College of Oceanology,13(1):39–56.
(28). Qian Shuben, Chen Gouwei, 1986. Comprehensive survey and research report on the waters adjacent to the Changjiang River estuary and Chejudo Island, Chapter 5 (section 2): the ecological studies on the phytoplankton. Journal of Shandong College of Oceanology, 6(2): 26–55.
(29). Fujian Institute of Oceanology, 1988. A comprehensive oceanographic survey of the central and northern part of Taiwan Strait (VII): Composition and distribution on the plankton. China Science Press, pp: 259–269.
(29a). Yang Qingliang, 1996. Species composition and distribution of the planktonic diatoms in the western Taiwan Strait. Acta Oceanologica Sinica, 15(3):409–421.
(30). Huang, R., 1988. The influence of hydrography on the distribution of phytoplankton in the southern Taiwan Strait. Estuarine, Coastal and Shelf Science 26:643–656.
(30a). Huang, R., 1993. Phytoplankton distribution I the South China Sea and Kuroshio- flowing region of Taiwan. Acta Oceanographic Taiwanica, 31:73–82.
(30b). Huang, R., 1979. Marine diatoms of Langyu Island, Taiwan. Acta Oceanographica Taiwainca, Science Reports of the National Taiwan University, 10:190–200.
(30c). Huang, R., 1984. Marine diatoms of Chinmen Island. Acta Oceanographica Taiwanica, 15:181–200.
(31). Huang, R., 1990. Diatoms in some surface sediments of the Taiwan Continental Shelf. Nova Hedwigia 50(1–2):213–231.
(32). Liu, Shicheng. 1993a. Marine diatoms of the Xisha Islands, South China Sea. I. *Mastogloia* Thw. Ex. Wm.: Species of the group Sulcatae. Proceedings of the International conference on the Marine Biology of Hong Kong and the South China Sea, pp:705–728.
(33). Liu, Shicheng. 1993b. Marine diatoms of the Xisha Islands, South China Sea. II. Three new species of diatoms (Bacilleriophyceae). Proceedings of of the International conference on the Marine Biology of Hong Kong and the South China Sea, pp:729–734.

*: (1)-(29a) in Chinese
Compiled by Yang Qingliang, Cheng Zhaodi, Liu Shicheng; Edited by Jin Dexiang (T. G. Chin) and Cheng Zhaodi

CHRYSOPHYTA

CHRYSOPHYCEAE
SILICOFLAGELLATALES
DICTYOCHACEAE

Dictyocha fibula Ehrenberg: Bohai; Yellow Sea; East China Sea; northern Taiwan; Taiwan Strait; South China Sea. (1–7)
D. fibula v. *aculeata* Lemmermann: Northern Taiwan. (9)
D. fibula v. *messanensis* (Haeck.) Lemmermann: Northern Taiwan. (9)
D. fibula v. *pentagona* Schulz: Northern Taiwan. (9)
D. fibula v. *stapedia* Haech.: South China Sea. (4)
D. pseudofibula v. *complex* Tsumura: Northern Taiwan. (9)
Distephanus octonarius v. *octonarius* Glezer: Northern Taiwan. (9)
D. octonarius v. *polyactis* (Ehr.) Glezer: Northern Taiwan. (9)
D. speculum (Ehr.) Haeckel: East China Sea; Taiwan Strait; South China Sea. (2, 3, 7)
D. speculum v. *octonarium* (Ehr.) Joergensen: Taiwan Strait; South China Sea. (2, 7)
D. speculum v. *septenarius* (Ehr.) Joergensen: Northern Taiwan. (9)

CHRYSOMONODALES
MALLOMONODACEAE

Mallomonas annulata Harris: Taiwan Strait. (8)
M. liturata Nicholis: Taiwan Strait. (8)

SYNURACEAE

Synura petersenii Korshihov: Taiwan Strait. (8)

REFERENCES*

(1). South China Sea Institute of Oceanology, Academia Sinica, 1987. Report on comprehensive survey and research about Zengmu Ansha- the southern boundary of China. China Science Press, pp:115–131.
(2). State Oceanic Administration, 1988. The achievement in comprehensive investigation on environmental resources in the central part of South China Sea. Chapter 12, Plankton. China Ocean Press, pp:223–227.
(3). State Oceanic Administration et al., 1992. Comprehensive survey on the resources in the coastal area and the shore of China Seas, China Ocean Press, pp:97–110.
(4). Lin Yongshui and Lin Qiuyan, 1991. Distribution characteristics of phytoplankton in the waters around Nansha Islets. In: Collection of research papers on marine rganisms in the waters around and near Nansha Islets (II). China Ocean Press, 86pp.
(5). Guo Yujie and Yang Zeyu, 1982. The ecological studies on the phytoplankton over the continental shelf of East China Sea in the summer of 1976. Studia Marina Sinica, 19: 11–32.
(6). Qian Shuben and Chen Gouwei, 1986. The ecological studies on the phytoplankton. In: Reports on comprehensive survey and research about the waters adjacent to Changjiang River estuary and Chejudo Island (Chapter 5, Section 2). Journal of Shandong College of Oceanology, 6(2):26–55.
(7). Fujian Institute of Oceanology, 1988. A comprehensive oceanographic survey of the central and northern Taiwan Strait (VII). China Science Press, pp:259–269.
(8). Cheng Zhaodi and Gao Yahui, 1992. Some species of Chrysophyta from Xiamen Harbour. (manuscript)
(9). Huang, R., 1979. Study on the silicoflagellates along the northern coast of Taiwan. Acta Oceanographica Taiwanica, 9:119–125.

*: (1)–(8) in Chinese
Compiled by Yang Qingliang; Edited by Cheng Zhaodi

CRYPTOPHYTA

CRYPTOPHYCEAE
CRYPTOMONADALES
CRYPTOMONADACEAE

Cryptomonas baltica (Karsten) Butcher. [*Rhodomonas baltica*]: Dalian Bight (Liaoning). (1)

REFERENCE

(1). Xu Chengyuan, 1982. Surveys on the causal organisms of the red tide in Dalian Bight. Acta Fishery Sinica, 6(2):173–180. (in Chinese)

Compiled by Yang Qingliang and Cheng Zhaodi; Edited by Jin Dexiang (T. G. Chin)

XANTHOPHYTA

XANTHOPHYCEAE
MISCHOCOCCALES [HETEROCOCCALES]
HALOSPHAERACEAE

Halosphaera virdis Schmitz: Bohai; Yellow Sea; East China Sea; South China Sea. Planktonic. (1, 3–6)

HETEROCHLORIDALES

Heterosigma akashiwo (Hada): Dalian Bight (Liaoning). (2)

VAUCHERIALES
VAUCHERIACEAE

Pseudodictomosiphon constricta Yam: Guangdong; Hainan. Muddy high intertidal. (7)

REFERENCES*

(1). The Committee of Science and Technology, the Oceanographic Group, 1977. Studies on the plankton in the near-shore waters of China. In: Comprehensive investigation report on China Seas (Vol. 8, Chapter 10). pp:1–159.
(2). Guo, Y. J. 1994. Studies on *Heterosigma akashiwo* (Hada) Hada in the Dalian Bight, Liaoning, China. Oceanologia & Limnogia Sinica, 25(2):211–215.
(3). State Oceanic Administration, 1988. Chapter 12: Phytoplankton. In: The achievement in omprehensive investigation on environmental resources in the central part of South China Sea. China Ocean Press, pp:223–227.
(4). State Oceanic Administration et al., 1992. Comprehensive survey on the resources in the coastal area and the shore of China Seas, pp:97–110.
(5). Guo Yujie and Yang Zeyu, 1982. The ecological studies on the phytoplankton over the continental shelf of East China Sea in the summer of 1976. Studia Marina Sinica, 19:11–32.
(6). Qian Shuben and Chen Gouwei, 1986. The ecological studies on the phytoplankton. In: Reports on comprehensive survey and research about the waters adjacent to Changjiang River estuary and Chejudo Island (Chapter 5, section 2). Journal of Shandong College of Oceanology, 6(2):26–55.
(7). Tseng, C. K., 1983. Common seaweeds of China. China Science Press, 316pp.

*: (1)–(6) in Chinese
Compiled by Yang Qingliang and Cheng Zhaodi; Edited by Jin Dexiang (T. G. Chin)

PYRROPHYTA

DESMOPHYCEAE
PROROCENTRALAS
PROROCENTRACEAE

Exuviaella marina Ciekowski: East China Sea. (10, 28)
Prorocentrum cordatum (Ost.) Dodge: Nansha (=Spratly Islands). (7a)

P. lenticulatum (Matz.) Taylor: Nansha (=Spratly Islands). (7a)
P. magnum (Gaarder) Dodge: Nansha (=Spratly Islands). (7a)
P. micans Ehrenberg: East China Sea; South China Sea. Widely distributed species. (4, 8, 9, 10, 11, 27)
P. oblongum (Pavill.) Taylor: Nansha (=Spratly Islands). (7a)
P. triestinum Schiller: East China Sea. Neritic low salinity species. (4)

DINOPHYCEAE
GYMNODINIALES
GYMNODINIACEAE

Gymnodinium aeruginosum Stein: East China Sea. (14)
G. coeruleum Dogiel: East China Sea. (14)
G. microadriaticum [*Zooxanthellae*]: Hong Kong; Xisha (=Paracel Islands). Coral zooxanthellae. (17)

POLYKRIKACEAE

Polykrikos schwarzi Butschli: East China Sea. (24)

NOCTILUCACEAE

Noctiluca scintillans (Macartney) Kofoid & Swezy: Red tide species, widely distributed in neritic low salinity water along China's shoreline. (2, 4, 8–11, 14, 25, 27, 28)

KOFOIDINIACEAE

Kofoidinium splendens Cach. & Cash: East China Sea; South China Sea. Warm-water species. (9, 11)
K. velelloides Pavillard: East China Sea. species. (11)

DINOPHYSIALES
AMPHISOLENIACEAE

Amphisolenia asymmetrica Kofoid: East China Sea; South China Sea. Warm-water species. (2, 9)
A. bidentata Schröder: East China Sea; South China Sea. Widely distributed in high temperature and high salinity water. (1, 2, 4, 5, 7, 8, 9, 10, 11, 14)
A. extensa Kofoid: South China Sea. Warm-water species. (9)
A. globifera Stein: East China Sea; South China Sea. Warm-water species. (2, 9)
A. palaeatheroides Kofoid: South China Sea. Warm-water species. (5)
A. rectangulata Kofoid: South China Sea. Warm-water species. (5)
A. schauinslandi Lammermann: East China Sea; South China Sea. Warm-water species. (2, 8, 9, 11)
A. schroederi Kofoid: Nansha (=Spratly Islands). (7a)
A. spinulosa Kofoid: South China Sea. Warm-water species. (9, 11)

A. thrinax Schutt: East China Sea; South China Sea. Warm-water species. (2, 5, 7, 9, 11, 12)
Triposolenia bicornis Kofoid: South China Sea. Warm-water species. (9, 11)
T. intermedia Kofoid & Skogs: South China Sea. Warm-water species. (7)

DINOPHYSIACEAE

Dinophysis acuminata Clap. & Lach: Bohai. (27)
D. caudata Saville-Kent: Yellow Sea; East China Sea; South China Sea. Widely distributed species. (2, 7, 8, 9, 10, 11, 12, 14, 25, 28)
D. caudata v. *abbreviata* Jörgensen: East China Sea. (1)
D. fortii Pavillard: Bohai. (27)
D. hastata Stein: East China Sea. (2)
D. miles Cleve: East China Sea; South China Sea. Warm-water species. (2, 9, 11, 12, 14)
D. schutii Murray & Whitting: South China Sea. (26)
Histioneis biremis Stein: East China Sea; Nansha (=Spratly Islands). (7a)
H. depressa Schiller: Nansha (=Spratly Islands). (2, 7a)
H. hisppoperoides Kof. & Mich.: Nansha (=Spratly Islands). (7a)
H. hyalina Kof. & Mich.: Nansha (=Spratly Islands). (7a)
H. pulchra Kofoid: East China Sea. (2)
Ornithocercus magnificus Stein: East China Sea; South China Sea. Warm-water species.
O. quadratus Schutt: East China Sea; South China Sea. Warm-water species. (2, 4, 8, 9, 11, 12)
O. splendidus Stein: East China Sea; South China Sea. Warm-water species. (1, 2, 4, 7, 9, 11, 12, 14)
O. steinii Schutt: East China Sea; South China Sea. Warm-water species. (3, 4, 9, 11, 26)
O. thurnii (Schmidt) Kofoid & Skogsberg: East China Sea; South China Sea. Warm water species. (1, 2, 4, 8, 9, 11, 12, 24)
Phalacroma cuneus Schutt [*Dinophysis cuneus* (Schütt) Abe]: East China Sea; South China Sea. Warm-water species. (2, 3, 9, 11, 12)
P. doryphorum Stein: East China Sea. (2)
P. favus Kofoid & Mthener: East China Sea. (2)
P. mitra Schutt: East China Sea. (2)
P. ovum Schutt: East China Sea. (3)

PERIDINIALES
CERATIACEAE

Ceratium arietinum Cleve [*C. bucephalum* v. *haterocamptum*]: East China Sea; South China Sea. Warm-water species. (3, 8, 9, 11)
C. arietium v. *gracilentum* (Long.) Sournia: Nansha (=Spratly Islands). (7a)
C. axiale Kofoid: East China Sea. Warm-water species. (2)

C. azoricum Cleve: East China Sea; South China Sea. Warm-water species. (2, 8, 11, 26)

C. belone Cleve: East China Sea; South China Sea. Warm-water species. (2, 4, 8, 9, 11, 15)

C. bigelowii Kofoid: East China Sea; South China Sea. Warm-water species. (8, 9, 11, 15)

C. breve (Ost. & Schmidt) Schröder: Yellow Sea; East China Sea; South China Sea. Widely distributed Warm-water species. (2, 8–12, 14, 15, 18, 27)

C. breve v. *parallelum* (Schmidt) Jörgensen: East China Sea; South China Sea. Warm water species. (2, 11, 12, 15, 18)

C. candelabrum (Ehr.) Stein: Yellow Sea; East China Sea; South China Sea. Warm water species. (2, 8, 9, 11, 15, 18)

C. candelabrum v. *depressum* (Pouchet) Jörgensen: East China Sea; South China Sea. (14, 18)

C. carriense Gourret: East China Sea; South China Sea. Warm-water species. (1, 4, 8, 9, 11, 14, 18)

C. carriense f. *volans* (Cleve): East China Sea; South China Sea. Warm-water species. (2, 7, 12, 14, 15, 18, 26)

C. cephalatum (Lemm.) Jörgensen: East China Sea; South China Sea. Warm-water species. (4, 9, 12, 15, 26)

C. contortum (Gourret) Cleve [*C. longinum*]: East China Sea; South China Sea. Warm water species. (1, 2, 4, 8, 9, 11, 12, 15, 18)

C. contortum v. *robustum* (Karsten) Sournia: East China Sea. Warm-water species. (9)

C. contortum v. *saltans* (Schröder) Jörgensen: South China Sea. Warm-water species. (9)

C. contortum f. *subcontortum* (Schröder) St. Nielsen: South China Sea. Warm-water species. (9)

C. contrarium (Gourret) Pavillard [*C. inflexum*]: Yellow Sea; East China Sea; South China Sea. Warm-water species. (3, 8, 9, 11, 26)

C. contrarium v. *claviceps* Schröder: South China Sea. Warm-water species. (9, 11)

C. declinatum Karsten: Yellow Sea; East China Sea; South China Sea. Warm-water species. (2, 8, 9, 11, 18, 26)

C. deflexum (Kofoid) Jörgensen: Yellow Sea; East China Sea; South China Sea. Warm water species. (1, 2, 7, 8, 9, 11, 12, 14, 15, 18, 26)

C. dens Ostenfeld & Schmidt: East China Sea; South China Sea. Warm-water species. (2, 8, 9, 11, 14, 18)

C. digitatum Schutt: Xisha (=Paracel Islands). (6)

C. digitatum v. *rotundatum* Jörgensen: Xisha (=Paracel Islands). (6)

C. ehrenbergii Kofoid: South China Sea. Warm-water species. (18)

C. euarcuatum Jörgensen [*C. arcuatum*]: East China Sea; South China Sea. Warm-water species. (2, 3, 8, 11, 26)

C. extensum (Gourret) Cleve [*C. biceps*]: East China Sea; South China Sea. Warm-water species. (1, 2, 4, 8, 9, 11, 12, 18, 25)

C. extensum f. *strictum* St. Nielsen: East China Sea; South China Sea. Warm-water species. (8, 9, 11, 14)

C. falcatiforme Jörgensen: East China Sea; South China Sea. Warm-water species. (9, 11)

C. falcatum (Kofoid) Jörgensen: East China Sea; South China Sea. Warm-water species. (2, 11, 12, 15)

C. furca (Ehr.) Claparede & Lachmann: Bohai; Yellow Sea; East China Sea; South China Sea. Widely distributed species. (1, 2, 8–12, 15, 18, 25, 26, 27)

C. furca v. *berghii* (Jörgensen) Schiller: East China Sea; South China Sea. (14, 18)

C. furca v. *eugrammum* (Ehr.) Schiller: East China Sea; South China Sea. Warm-water species. (10, 11, 18)

C. fusus (Ehr.) Dujardin: Bohai; Yellow Sea; East China Sea; South China Sea. Widely distributed species. (1, 2, 4, 8–11, 18, 25, 28)

C. fusus v. *schuttii* Lemm: Bohai; East China Sea. (2, 28)

C. fusus v. *seta* (Ehr.) Jörgensen: East China Sea; South China Sea. Warm-water species. (2, 8, 9, 12, 14, 15, 18, 27)

C. geniculatum (Lemm.) Cleve: East China Sea; South China Sea. Warm-water species. (8, 9, 12, 15)

C. gibberum Gourret: Yellow Sea; East China Sea; South China Sea. Warm-water species. (2, 4, 8, 9, 11, 14)

C. gibberum v. *dispar* (Pouchet) Sournia: Bohai; Yellow Sea; East China Sea; South China Sea. Warm-water species. (2, 9–12, 14, 15, 18, 28)

C. gravidum Gourret: East China Sea; South China Sea. Warm-water species. (1, 2, 4, 9, 11, 12, 15, 26)

C. gravidum v. *angustum* Jörgensen: East China Sea; South China Sea. Warm-water species. (8, 9, 11, 26)

C. hexacanthum Gourret: East China Sea; South China Sea. Warm-water species. (4, 8, 9, 11, 14)

C. hexacanthum v. *contortum* Lemm.: East China Sea; South China Sea. Warm-water species. (9, 11, 12)

C. hexacanthum f. *spirale* (Kofoid) Schiller: East China Sea; South China Sea. Warm water species. (2, 7, 8, 9, 11, 12, 15)

C. horridum (Cleve) Gran [*C. intermedium*]: Bohai; Yellow Sea; East China Sea; South China Sea. Widely distributed species. (2, 3, 8, 9, 11, 12, 25, 27, 28)

C. horridum v. *claviger* (Kofoid) Graham & Bronikovsky: East China Sea; South China Sea. Warm-water species. (2, 3, 9, 11)

C. horridum v. *denticulatum* Jörgensen: East China Sea. Warm-water species. (11)

C. horridum v. *inclinatum* Kofoid: East China Sea; South China Sea. Warm-water species. (9, 11)

C. horridum v. *patentissimum* (Ost. & Schmidt) Taylor: East China Sea; South China Sea. Warm-water species. (8, 9, 11)

C. humile Jörgensen: East China Sea. Warm-water species. (2, 3, 8, 11)

C. incisum (Karsten) Jörgensen: East China sea; South China Sea. Warm-water species. (2, 4, 9, 11, 12, 15)

C. inflatum (Kofoid) Jörgensen [*C. pennatum*]: Yellow Sea; East China Sea; South China Sea. Warm-water species. (2, 8, 9, 10, 14, 18)

C. karstenii Pavillard [*C. schrankii*]: East China Sea; South China Sea. Warm-water species. (2, 3, 8, 9, 11, 14, 26)

C. kofoidii Jörgensen: Southern Yellow Sea; East China Sea; South China Sea. Warm water species. (8, 9, 10, 11, 12, 17, 26)

C. limulus Gourret: East China Sea; South China Sea. Warm-water species. (2, 6, 26)

C. lineatum (Ehr.) Cleve: Bohai; Yellow Sea; East China Sea. (9, 11, 24, 28)

C. longipes (Bailey) Gran: Bohai; Yellow Sea; East China Sea. (1, 11, 28)

C. longirostrum Gourret: East China Sea; South China Sea. Warm-water species. (8, 9, 11, 14, 15)

C. longissimum (Schröder) Kofoid: East China Sea; South China Sea. Warm-water species. (1, 2, 8, 9, 11, 15)

C. lunula Schimper: Yellow Sea; East China Sea; South China Sea. Warm-water species. (2, 8–12, 14, 15, 18, 26, 27)

C. lunula f. *brachyceros* Jörgensen: South China Sea. Warm-water species. (9)

C. macroceros (Ehr.) Cleve: Bohai; Yellow Sea; East China Sea; South China Sea. Widely distributed species. (1, 2, 4, 8–12, 14, 25, 28)

C. macroceros v. *gallicum* (Kofoid) Jörgensen: Yellow Sea; East China Sea; South China Sea. Warm-water species. (2, 4, 8–11, 15, 18)

C. massiliense (Gourret) Karsten: Southern Yellow Sea; East China Sea; South China Sea. Warm-water species. (1, 2, 4, 8–12, 15, 18, 26, 28)

C. massiliense v. *armatum* (Karsten) Jörgensen: East China Sea. Warm-water species. (11, 14)

C. minutum Jörgensen: East China Sea. Warm-water species. (11, 14)

C. molle Kofoid: Bohai; Yellow Sea; East China Sea; South China Sea. Widely distributed species. (8, 9, 11, 12, 28)

C. paradoxides Cleve: East China Sea. Warm-water species. (12)

C. pentagonum Gourret: East China Sea; South China Sea. Widely distributed species. (2, 4, 8, 9, 11, 12, 14, 26)

C. pentagonum v. *longisetum* (Ostenf. & Schmidt) Jörgensen: East China Sea; South China Sea. (2, 15)

C. platycornia Von Dady: East China Sea; South China Sea. Warm-water species. (2, 4, 8, 9, 11, 12, 14, 15)

C. praelongum (Lemm.) Kofoid: East China Sea; South China Sea. Warm-water species. (1, 2, 4, 9, 15)

C. pulchellum B. Schröder: Bohai; Yellow Sea; East China Sea; South China Sea. (1, 8–11, 17, 26, 27)

C. pulchellum f. *semipulchellum* Jörgensen: East China Sea; South China Sea. Warm water species. (2, 11, 12, 14, 18)

C. ranipes Cleve: East China Sea; South China Sea. Warm-water species. (2, 4, 8, 9, 11, 18)

C. ranipes v. *palmatum* (Schröder) Jörgensen: East China Sea; South China Sea. Warm water species. (2, 7, 8, 9, 12, 15)

C. reflexum Cleve: East China Sea; South China Sea. Warm-water species. (3, 9, 11, 12)

C. schmidti Jörgensen: Yellow Sea; East China Sea; South China Sea. Warm-water species. (8, 9, 11, 18, 27)

C. schroeteri B. Schröder: East China Sea; South China Sea. Warm-water species. (11, 15)

C. setaceum Jörgensen: East China Sea; South China Sea. Warm-water species. (9, 11, 26)

C. symmetricum Pavillard [*C. gracile*]: East China Sea; South China Sea. Warm-water species. (2, 8, 9, 11, 12, 26)

C. symmetricum v. *coarctatum* (Pavill.) Graham & Bron: East China Sea; South China Sea. Warm-water species. (2, 9, 11, 15)

C. symmetricum v. *orthoceros* (Jörgensen) Graham & Bron: East China Sea. Warm water species. (11)

C. tenue (Ostenfeld & Schmidt) Jörgensen: East China Sea; South China Sea. Warm water species. (8, 9, 11, 12, 18)

C. teres Kofoid: East China Sea; South China Sea. Warm-water species. (2, 8, 9, 11, 12, 18)

C. trichoceros (Ehr.) Kofoid: Yellow Sea; East China Sea; South China Sea. Warm water species. (1, 2, 4, 7–9, 11, 12, 14, 15, 18, 27)

C. tripos (O. F. Müller) Nitzsch.: Bohai; Yellow Sea; East China Sea; South China Sea. Widely distributed species. (1, 2, 4, 8-12, 14, 18, 25, 27)

C. tripos v. *atlanticum* (Ost.) Paulsen: Yellow Sea; East China Sea; South China Sea. Widely distributed species. (2, 8, 11, 12, 14, 18, 25)

C. tripos f. *balticum* Schütt: East China Sea. (12)

C. tripos v. *subsalsum* Ostenfeld: Bohai; Yellow Sea. (11, 28)

C. vultur Cleve: East China Sea; South China Sea. Warm-water species. (2, 4, 11, 14, 15)

C. vultur f. *angustum* Jörgensen: East China Sea; South China Sea. Widely distributed Warm-water species. (8, 9, 11)

C. vultur v. *japonicus* (Schröder) Jörgensen: East China Sea; South China Sea. Warm water species. (2, 9, 11)

C. vultur f. *recuvun* Jörgensen: East China Sea; South China Sea. Warm-water species. (2, 9, 11, 15)

C. vultur v. *sumatranum* (Karsten) St. Nielsen [*C. sumatranum*]: East China Sea; South China Sea. Widely distributed Warm-water species. (2, 8, 9, 11, 12, 14, 15, 18, 26)

CERATOCORYACEAE

Ceratocorys bipes (Cleve) Kofoid: South China Sea. (5)

C. gourretii Paulsen: South China Sea. Warm-water species. (9)

C. horrida Stein: Yellow Sea; East China Sea; South China Sea. Warm-water species. (2, 4, 5, 8, 9, 11, 12, 14)

C. magna Kofoid: East China Sea. Warm-water species. (8, 11)

C. reticulata Graham: East China Sea; South China Sea. (5, 9, 11)

CLADOPYXIDACEAE

Cladopyxis brachiolata Stein: South China Sea. (9)

OSTREOPSIACEAE

Ostreopsis siamensis Schmidt: Xisha (=Paracel Islands). (6)

GONYAULACEAE

Gonyaulax ceratocoroides (Murr. & Whitt) Kofoid: East China Sea. (12)
G. diegensis Kofoid: Bohai. (27)
G. digitale (Pouchet) Kofoid: Bohai; East China Sea. (10, 11, 27)
G. fusiformis Graham: Xisha (=Paracel Islands). (6)
G. mitra (Schütt) Kofoid: Xisha (=Paracel Islands). (6)
G. pacifica Kofoid: East China Sea. (2)
G. polyedra Stein: East China Sea; South China Sea. (2, 8, 9, 10, 12)
G. polygramma Stein: East China Sea; South China Sea. (9, 10, 12, 26)
G. spinifera (Claparede & Lachmann) Diesing: Bohai. (27)
Spiraulax jollifei (Murray & Whitting) Kofoid: South China Sea. (9)

HETERAULACACEAE

Gambierdiscus toxicus Rukuro & Yasuwo: Xisha (=Paracel Islands). (6)
Heteraulacus polyedricus (Pouch.) Drugg & Loeblich: East China Sea. (9, 10, 11, 27)

HETERODINIACEAE

Heterodinium blackmanii (Murr. & Whitt) Kofoid: East China Sea; South China Sea. Warm-water species. (2, 9)
H. gesticulatum Kofoid: East China Sea. Warm-water species. (12)
H. globosum Kofoid: South China Sea. (27)
H. praetextum Kofoid: South China Sea. Warm-water species. (9)
H. rigdenae Kofoid: East China Sea. Warm-water species. (3)
H. whittingae Kofoid: South China Sea. Warm-water species. (9)

OXYTOXACEAE

Oxytoxum challengeroides Kofoid: South China Sea. Warm-water species. (25)
O. sceptrum (Stein) Schröder: East China Sea; South China Sea. Warm-water species. (25)
O. scolopax Stein: East China Sea; South China Sea. Warm-water species. (2, 9)

PERIDINIACEAE

Diplopeltopsis minor (Paulsen) Pavillard [*Zygabikodinium lenticulatum*]: Bohai. (27)
Diplopsalis asymmetrica (Mangin) Lind. [*Peridiniopsis asymmetrica*]: South China Sea. (22)
D. excentrica Nie [*Peridiniopsis excentrica*]: South China Sea. (22)
D. hainanensis Nie [*Peridiniopsis hainanensis*]: South China Sea. (22)
D. lenticula Bergh: East China Sea; South China Sea. (10, 22, 26)
D. lenticula v. *lebourii* Nie: South China Sea. (22)
D. pingi Nie [*Peridiniopsis pingi*]: South China Sea. (22)
Peridiniopsis rotunda LeBour: Bohai. (27)
Peridinium abei Paulsen: Yellow Sea; East China Sea. (10, 16)
P. achromaticum Levander: South China Sea. (19)
P. angustum P. Dangeard [*P. wiesneri*]: South China Sea. (19)
P. asymmetricum Karsten: Yellow Sea; East China Sea. Warm-water species. (11, 16)
P. biconicum P. Dangeard: South China Sea. (19)
P. brochii Kofoid & Swezy: East China Sea; South China Sea. Warm-water species. (10, 11, 19)
P. claudicans Paulsen: Yellow Sea; East China Sea; South China Sea. Widely distributed species. (4, 8, 12, 16, 26)
P. compressum (Abe) Nie: South China Sea. (19)
P. conicoides Paulsen: East China Sea. (24)
P. conicum (Gran) Ostenfeld & Schmidt: Bohai; Yellow Sea; East China Sea. Widely distributed species. (1, 2, 10, 11, 24, 26, 27)
P. crassipes Kofoid: Bohai; Yellow Sea; East China Sea. Widely distributed species. (8, 10, 11, 12, 16, 25, 28)
P. deficiens Meunier: Yellow Sea. (16)
P. depressum Bailey: Yellow Sea; East China Sea; South China Sea. Widely distributed species. (1, 2, 4, 8–12, 16, 20, 25)
P. diabolus Cleve [*P. globosum*]: East China Sea. (3, 8, 11)
P. divergens Ehrenberg: Bohai; Yellow Sea; East China Sea; South China Sea. Widely distributed species. (2, 8, 10, 11, 27, 28)
P. elegans Cleve: Yellow Sea; East China Sea; South China Sea. Warm-water species. (2, 3, 8, 9, 11, 12)
P. excentricum Paulsen: Yellow Sea. (16)
P. fatulipes Kofoid: East China Sea. (12)
P. gatunense Nygaard: South China Sea. (19)
P. globulus Stein [*P. sphaericum*]: Bohai; East China Sea; South China Sea. (3, 10, 12, 20, 28)
P. grande Kofoid: Bohai; Yellow Sea; East China Sea; South China Sea. (2, 9, 10, 11, 12, 28)
P. granii Ostenfeld & Paulsen: East China Sea. (10)
P. humile Schiller: Yellow Sea. (16)
P. latispinum Mangin: East China Sea; South China Sea. (9, 10, 20)

P. latissimum Kofoid: Yellow Sea; East China Sea; South China Sea. (8, 10, 11, 16, 20)
P. leonis Pavillard: Bohai; Yellow Sea; East China Sea. (10, 12, 16, 28)
P. longipes Karsten: East China Sea. (2)
P. marielebourae Paulsen: Yellow Sea. (16)
P. matsenaueri Gaarder: Yellow Sea. (16)
P. minutum Kofoid: Yellow Sea; East China Sea; South China Sea. (10, 16, 19)
P. murray Kofoid: East China Sea; South China Sea. Warm-water species. (2, 3, 8, 9, 11)
P. nux Schiller: Yellow Sea. (16)
P. oceanicum Vanhöffen: Yellow Sea; East China Sea; South China Sea. (2, 3, 8–12, 26, 27)
P. ovum Schiller [*P. rectum*]: Yellow Sea; East China Sea. (2, 10, 16).
P. pallidum Ostenfeld: Yellow Sea; East China Sea. Widely distributed species. (10, 11, 16)
P. parainerme Nie & Wang: Yellow Sea. (16)
P. paralletum Broth: East China Sea. (1, 3)
P. parapentagonum Wang: Bohai. (27)
P. pellucidum (Bergh) Schütt: Bohai; Yellow Sea; East China Sea. (10, 16, 28)
P. pentagonum Gran: Bohai; Yellow Sea; East China Sea; South China Sea. (1, 2, 5, 11, 21)
P. punctulatum Paulsen: Yellow Sea; East China Sea. (12, 16)
P. pyriforme Paulsen: South China Sea. (9)
P. remotum Karsten: South China Sea. (9)
P. solidicorne Mangin: Yellow Sea; East China Sea. (10, 12, 16)
P. steinii Jörgensen [*P. michaelis*]: East China Sea; South China Sea. Widely distributed species. (2, 8, 10, 11, 12, 20, 27)
P. subinerme Paulsen: Yellow Sea; East China Sea. Widely distributed species. (10, 11, 16)
P. subpyriforme P. Dangeard [*P. globulus* v. *quarerense*]: Yellow Sea; East China Sea. (10, 11, 16)
P. thorianum Paulsen: Yellow Sea; East China Sea. (16, 19)
P. tsingtaoensis Nie & Wang: Yellow Sea. (16)
P. tubum Schiller [*P. sinaicum*]: South China Sea. (9)
P. tumidum Okamura: East China Sea; South China Sea. Warm-water species. (2, 3, 9, 11)
P. vanustum Matzenauer: Yellow Sea; East China Sea; South China Sea. (12, 16, 20)

PODOLAMPADACEAE

Blepharocysta denticulata Nie: South China Sea. (18)
B. splendor-maris Ehrenberg: South China Sea. (18)
Podolampas bipes Stein: East China Sea; South China Sea. Warm-water species. (2, 4, 9, 11, 12, 22, 23, 26)
P. bipes v. *reticulata* (Kofoid) Statnov.: East China Sea. (2)
P. elegans Schütt: South China Sea. Warm-water species. (9)
P. palmipes Stein: East China Sea; South China Sea. Warm-water species. (2, 10, 11, 22, 23)
P. spinifera Okamura: East China Sea; South China Sea. Warm-water species. (2, 9, 26)

GONIODOMACEAE

Goniodoma polyedricum (Pouchet) Jörgensen: East China Sea; South China Sea. (12, 21)
G. sphaericum Murr. & Whitt: South China Sea. (20)

PYROCYSTACEAE

Dissodinium lunula (Schütt) Pascher: East China Sea; South China Sea. Warm-water species. (2, 8, 9, 11, 12, 14)
Pyrocystis acuta Kofoid: East China Sea. Warm-water species. (4, 14)
P. fusiformis Murray: Yellow Sea; East China Sea; South China Sea. Warm-water species. (4, 8, 9, 12, 13, 14)
P. fusiformis f. *bicornia* Kofoid: East China Sea; South China Sea. Warm-water species. (2, 8, 9, 11, 13)
P. gerbautii Pavillard [*Dissodinium gerbautii*]: Yellow Sea; East China Sea; South China Sea. Warm-water species. (2, 10, 11, 22)
P. hamulus Cleve: East China Sea; South China Sea. Warm-water species
P. hamulus v. *inaeaqualis* Schröder: East China Sea; South China Sea. Warm-water species. (2, 4, 8, 9, 11, 13)
P. hamulus v. *semicircuralis* Schröder: East China Sea; South China Sea. Warm-water species. (2, 4, 9, 11, 12, 14)
P. lanceolata Schröder: East China Sea. Warm-water species. (3, 4, 11, 12)
P. obtusa Pavillard: Yellow Sea; East China Sea. Warm-water species. (11)
P. pseudonoctiluca Murray: South China Sea. Warm-water species. (9)
P. rhomboides Matzenauer: South China Sea. Warm-water species. (9)
P. robusta Kofoid: East China Sea; South China Sea. Warm-water species. (2, 4, 8, 9, 13, 14, 23).

PYROPHACACEAE

Pyrophacus horologicum v. *steinii* Schiller: Bohai; Yellow Sea; East China Sea; South China Sea. (2, 8–12, 14, 28)

REFERENCES*

(1). Mao Xinghua and Li Ruixiang, 1984. Distribution and the ecological characteristics of planktonic dinoflagellates in the continental shelf

of the northern East China Sea. Acta Oceanologica Sinica, 6(5): 672–677.
(2). Feng Jifang, 1991. Species composition and number distribution of plankton in Minnan-Taiwan Bank Fishing Ground. In: Minna-Taiwan Bank Fishing Ground Upwelling Ecosystem Study. China Science Press, pp:388–406.
(3). Li Ruixiang and Mao Xinghua, 1985. Dinoflagellates in the continental shelf of the East China Sea. East China Sea Marine Science, 3(1): 41–45.
(4). Lu Douding, 1991. Distribution characteristics of Kuroshio indicative dinoflagellates in the East China Sea. Essays on the investigation of Kuroshio (III), China Ocean Press, pp:287–296.
(5). Chen Guowei, 1981. Studies on the dinoflagellata in adjacent waters of the Xisha Islands, I. On the thecal morphology of Ceratocorys; II. On the thecal morphology of Amphisolenia. Oceanologia & Limnologia Sinica, 12(1):91–99.
(6). Chen Guowei, 1989. Studies on the dinoflagellata in adjacent waters of the Xisha Islands, III. Some rare tropical ocean species. Oceanologia & Limnologia Sinica, 20(3):230–237.
(7). Lin Yongshui, 1985. Marine organisms in the South China Sea, 2. Plankton. Research Report on Comprehensive Survey of the South China Sea (II). China Science Press, pp:332–378.
(7a). Lin Yongshui and Lin Qiuyan, 1991. Distribution characteristics of plankton in the waters around Nasha Islets. Collection of research papers on marine organisms in the waters around and near Nansha Islets (II). China Ocean Press, pp:66–87
(8). Lin Jinmei, 1988. Distribution of planktonic Pyrophyta in the western waters of Taiwan Strait. Journal of Oceanography in Taiwan Strait, 7(2):163–172.
(9). Lin Minmei, 1988. Ecological studies on planktonic Pyrophyta in central waters of the South China Sea. Marine Science Bulletin, 7(3): 47–53.
(10). Lin Jinmei, 1990. Distribution of planktonic Pyrophyta in the waters around Xiamen Island. Acta Ecologica Sinica, 10(2):139–144.
(11). Lin Jinmei. Distribution of planktonic dinoflagellates in the East China Sea. (manuscript)
(12). Qian Shuben and Chen Guowei, 1986. The ecological studies on phytoplankton. Comprehensive survey and research report on the water areas adjacent to Changjiang River Estuary and Chejudo Island (2). Journal of Shandong College of Oceanology, 16(2):26–55.
(13). Guo Yujie and Zhou Hanqiu, 1979. Taxonomic studies on the Pyrocystis of the adjacent region of Zhongsha and Xisha Islands, Guangdong Province, China. Studia Marina Sinica, 15:47–55.
(14). Guo Y. J. and Yang Z.Y, 1982. The ecological studies on the phytoplankton over the continental shelf of East China Sea in the summer of 1976. Studia Marina Sinica, 19:11–32.
(15). Guo Yujie, Ye Jiasong and Zhou Hanqiu, 1983. On some *Ceratium* in the waters around Xisha Islands and the Zhongsha Islands, Guangdong Province, China. Studia Marina Sinica, 20:69–108.
(16). Ni Dashu et al., 1964. Studies on *Peridinium* in Kiaozhou Bay. Collections of paper abstract from acdemic annual meeting in 1963 of the Chinese Society of Oceanography and Limnology. China Science Press, pp:124–126.
(17). Morton B. and J. Morton, 1983. The sea shore ecology of Hong Kong, Hong Kong University Press. 350pp.
(18). Nie, D., 1936. Dinoflagellata of the Hainan Region I. *Ceratium*. Contr. Biol. Lab. Sc. Soc. China, 12(3):29–73.
(19). Nie, D., 1939a, Dinoflagellata of the Hainan Region II. On the thecal morphology of *Blepharocysta*, with a description of a new species. Contr. Biol. Lab. Sc. Soc. China 13(1–6):23–43.
(20). Nie, D., 1939b, On the thecal morphology of *Peridinium*, with special reference to the ventral area. Science, 23(10):584–600.
(21). Nie, D. and Wang, C. C., 1942. Dinoflagellata of the Hainan Region V. on the thecal morphology of the genus *Goniodoma*, with description of species of the Region. Sinensia 13(1–6):61–68.
(22). Nie, D., 1942. Dinoflagellata of the Hainan Region IV. on the thecal morphology of *Podolampas* with description of species. Sinensia, 13(1–6):53–58.
(23). Nie, D., 1943. Dinoflagellata of the Hainan Region VI. on the genus *Diplopsalis*, Sinensia, 14(1–6):1–21.
(24). Nie, D., 1943. Dinoflagellata of the Hainan Region VII. on the thecal morphology of *Ornithocercus thurni* (Schmidt) Kofoid & Skogsberg. Sinensia, 14(1–6):23–28.
(25). Sproston, N. G., 1949. A preliminary survey of the plankton in the Cushan Region, with a review of the relevant literature. Sinensia, 20(1–6):58–126.
(26). Tu, H. K. and Chiang, Y. M., 1972. Dinoflagellata collected from northeastern part of the South China Sea. Acta Oceanographic Taiwanica Science Reports of the National Taiwan University, 2:134–146.
(27). Wang, C. C., and Nie, D., 1932. A survey of the marine protozoa of Amoy. Contr. Biol. Lab. Sc. Soc. China, 8(9):285–385.
(28). Wang, C. C., 1936. Dinoflagellata of the Gulf of Bohai. Sinensisa, 7(2): 128–171.

*: (1)–(16) in Chinese
Compiled by Lin Jinmei; Edited by Ni Dashu (D. Nie)

CILIOPHORA

KINETOFRAGMINOPHORA
KARYORELICTIDA
TRACHELOCERCIDAE

Trachelocerca phoenicopterus Cohn: Xiamen harbor (Fujian). (28)

HARTORIDA
ENCHELYIDAE

Chaenera vorax Quennerstedt: Qingdao (Shandong). (25)
Choenea limicola Levander-Lauterborn: Xiamen harbor (Fujian). (28)
Lacrymaria marina Kahl: Qingdao (Shandong). (25)

DIDINIIDAE

Mesodinium pujlex (C. & L.): Xiamen harbor (Fujian). (28)

PSEUDOTRACHELOCERCIDAE

Pseudotrachelocerca trepida (Kahl): Sheyang (Jiangsu). (18)

PROSTOMATIDA
PLACUIDAE

Placus livida (Wang & Nie) [*Thoracophrya luciae*]: Xiamen harbor (Fujian). (28)

PLEUROSTOMATIDA
AMPHILEPTIDAE

Amphileptus gutta Cohn: Xiamen harbor (Fujian). (28)
A. litonotiformis Song: Sheyang (Jiangsu). (20)
Litonotus fasciola Ehrenberg: Xiamen harbor (Fujian). (28)
L. paracygnus Song: Qingdao (Shandong). (12)
L. yinae Song: Qingdao (Shandong). (12)
Loxophyllum rostratum Cohn: Qingdao (Shandong). (13)
L. setigerum Quennerstedt: Xiamen harbor (Fujian). (28)

CYRTOPHORIDA
DYSTERIIDAE

Dysteria navicula Kahl: Xiamen harbor (Fujian). (28)

CHILODONELLIDAE

Pseudochilodonopsis marina Song: Qingdao (Shandong). (19)

OLIGOHYMENOPHORA
HYMENOSTOMATIDA
GLAUCOMIDAE

Glaucoma hyalina Wang & Nie: Xiamen harbor (Fujian). (28)

SCUTICOCILIATIDA
URONEMATIDA

Uronema marina Dujardin: Xiamen harbor (Fujian). (28)

CRYPTOCHILIDAE

Biggaria bermuderse (Biggar & Wenrich): Xiamen. (3)
B. echinometris (Biggar & Wenrich): Dongshan (Fujian); Shanwei and Aotou (Guangdong); Hainan. (3)
B. polynucleatum (Nie): Xiamen (Fujian); Shanwei and Aotou (Guangdong); Hainan. (3)
Cryptochilidium echini (Maupas): Shanwei and Aotou (Guangdong); Sanya (Hainan). (3)
C. sigmoides Yagiu: Yantai (Shandong). (14)

ENTORHIPIDIIDAE

Entorhipidium triangularis Poljansky: Yantai and Qingdao (Shandong); Shengsi (Zhejiang); Dongshan (Fujian). (3)

ENTODISCIDAE

Entodiscus borealis (Hentschel): Yantai and Qingdao (Shandong); Shengsi (Zhejiang). (3)

E. fukuii (Uyemura): Qingdao (Shandong); Shengsi (Zhejiang). (3)
E. indomitus Madsen: Qingdao (Shandong). (3)
E. minor (Ygaiu) Yantai and Qingdao (Shandong); Shengsi (Zhejiang). (3)

PHILASTERIDAE

Homalogastra binucleata Song: Qingdao (Shandong). (11)
Paranophrys carcini spiralis Zhou et al.: Laizhou (Shandong). (14)
P. carnivora Czapik & Wilbert: Qingdao (Shandong). (10)
P. elongata Biggar: Dongshan (Fujian); Shanwei (Guangdong); Hainan. (3)
P. sarcophaga (Cohn): Xiamen harbor (Fujian). (28)
Parauronema virginianum Thompson: Weifang (Shandong). (10)

CYCLIDIDAE

Cyclidium amoyensis Wang & Nie: Xiamen harbor (Fujian). (28)
C. citrullus Cohn: Laizhou and Rizhao (Shandong). (10)
C. glaucoma Müller: Xiamen harbor (Fujian). (28)
C. ozakii Yagiu: Yantai and Qingdao (Shandong); Shanwei (Guangdong); Sanya (Hainan). (3)

COHNILEMBIDAE

Cohnilembus pusillus Quennerstedt: Xiamen harbor (Fujian). (28)
C. reesi Kahl: Laizhou (Shandong). (10)

PLEURONEMATIDAE

Pleuronema coronata v. *marina* (Duj.-Mob.): Xiamen harbor (Fujian). (28)

PERITRICHA
VORTICELLIDAE

Vorticella chydroridicola Sramek-Husek: Bohai; Yellow Sea. (4)
V. cylindrica Dons: Bohai; Yellow Sea. (5)
V. fornicata Dons: Bohai; Yellow Sea. (4)
V. hamata Ehrenberg: Bohai; Yellow Sea. (5)
V. marina Greef: Bohai; Yellow Sea; Xiamen harbor (Fujian). (5, 28)
V. nebulifera Müller: Bohai; Yellow Sea. (5)
V. patellina Müller: Bohai; Yellow Sea; Xiamen harbor (Fujian). (5, 28)
V. pulchella Sommer: Bohai; Yellow Sea. (5)

V. striata Dujardin: Bohai; Yellow Sea. (5)
V. utriculus Stokes: Bohai; Yellow Sea. (5)

ZOOTHAMNIDAE

Intranstylum asellicola Kahl: Liaoning; Shandong; Jiangsu. (8)
I. intermedium Song: Mouping (Shandong). (4, 8)
Myoschiston simile Song: Liaoning; Tianjin; Qingdao (Shandong); Jiangsu. (4, 6)
Pseudocarchesium aselli (Engel.): Shandong; Jiangsu. (6)
Zoothamnium affine Stein: Shandong coast. (6)
Z. alternans C. & L.: Xiamen harbor (Fujian). (28)
Z. commune Kahl: Shandong coast. (6)
Z. cupiferum Song: Wendeng (Shandong). (4, 6)
Z. duplicatum Kahl: Bohai; Hebei; Shandong. (6)
Z. gammari Dietz: Bohai; Yellow Sea. (4)
Z. hadzii Stiller: Bohai; Yellow Sea. (4)
Z. intermedium Precht: Wendeng (Shandong). (6)
Z. maximum Song: Shandong; Jiangsu. (4, 6)
Z. niveum Ehrenberg: Xiamen harbor (Fujian). (28)
Z. paraentzii Song: Bohai; Yellow Sea. (6)
Z. paragammari Song: Wendeng (Shandong). (6)
Z. penaei Song: Bohai; Yellow Sea
Z. rigidum Precht: Rushan (Shandong). (6)
Z. sinensis Song: Tanggu (Tianjin); Wendeng and Qingdao (Shandong). (6)
Z. thiophilum Stiller: Wendeng, Rongcheng, and Rushan (Shandong). (6)

EPISTYLIDAE

Cothurnia acuta Wang & Nie: Xiamen harbor (Fujian). (28)
C. calix Kahl: Shandong coast. (8)
C. ceramicola Kahl: Rongcheng and Wendeng (Shandong). (8)
C. maritima Ehrenberg: Xiamen harbor (Fujian). (28)
Epistylis acuminata Song: Hebei coast. (4, 8)
E. aselli Stiller: Bohai; Yellow Sea. (8)
E. carcini Presht: Shandong coast. (8)
E. elongata Stokes: Wendeng (Shandong). (8)
E. harpacticola Kahl: Liaoning; Tangu (Tianjin); Shandong. (8)
E. uyemurai Song: Rongcheng and Wendeng (Shandong). (4)
Rhabdostyla scyphoides Song: Wendeng (Shandong) coast. (4, 8)
Vaginicola crystallina marina Song: Wendeng (Shandong). (8)

POLYHYMENOPHORA
HETEROTRICHIDA
LICNOPHORIDAE

Licnophora hippocampi Meng & Yu: Rizhao (Shandong). (15)

FOLLICULINIDAE

Folliculina ampulla Müller: Xiamen harbor (Fujian). (28)

CONDYLOSTOMATIDAE

Condylostoma magnum (Spiegel): Kenlin (Shandong). (26)
C. patens (Dujardin): Xiamen harbor (Fujian). (28)

SPIROSTOMIDAE

Blepharisma minima Lepsi: Xiamen harbor (Fujian). (28)
Protocruzia tuzeti Villeneuve-Brachon: Sheyang (Jiangsu). (26)

CLIMACOSTOMIDAE

Fabrea salina Henneguy: Sheyang (Jiangsu). (26)

METOPIDAE

Metopus circumlabeus Biggar & Wenrich: Xiamen harbor (Fujian); Guangdong; Hainan. (3)
M. ellipsoidis Tchang: Sanya (Hainan). (3)
M. phyllopharius Tchang: Sanya (Hainan). (3)

OLIGOTRICHIDA
STROBILIDIIDAE

Strobilidium clavellinae Buddenbrock: Qingdao (Shandong). (25)
S. paraglobosum Song & Packroff: Qingdao (Shandong). (25)
S. raplum Yagiu: Yantai and Qingdao (Shandong); Shengsi (Zhejiang); Xiamen and Dongshan (Fujian); Shanwei (Guangdong). (3)
S. styliferum Levander: Qingdao (Shandong). (25)
S. typicum (Lankester): Xiamen harbor (Fujian). (28)

TINTINNIDIDAE

Leprotintinnus nordqvisti (Brandt): Yellow Sea; East China Sea. (1, 28)

CODONELLIDAE

Tintinnopsis acuminata (Daday): Jiaozhou Bay (Shandong). (1)
T. amoyensis Wang & Nie: East China Sea; South China Sea. (16, 17)
T. angusta Meunier: Jiaozhou Bay (Shandong). (1)
T. aperta Brandt: Bohai. (29)
T. beroidea Stein: Bohai; Yellow Sea; South China Sea. (1, 17, 29)

T. brasiliensis Kofoid & Campbell: Jiaozhou Bay (Shandong). (1)
T. brevicollis Hada: Yellow Sea. (1)
T. chinglanensis Nie & Cheng: Yellow Sea; South China Sea. (1, 17)
T. cochleata (Brandt): Yellow Sea. (1)
T. compressa (Daddy): Yellow Sea. (1)
T. dadayi Kofoid: East China Sea. (16)
T. digita Nie & Cheng: Yellow Sea; South China Sea. (1, 17)
T. directa Hada: Yellow Sea; South China Sea. (1, 17)
T. gracilis Kofoid & Campbell: Bohai; Yellow Sea; East China Sea. (1, 7–29)
T. hemispiralis Yin: Jiaozhou Bay (Shandong). (2)
T. inflata Nie: Yellow Sea; East China Sea. (1, 16)
T. japonica Hada: Jiaozhou Bay (Shandong). (1)
T. karajacensis Brandt: Yellow Sea; East China Sea; South China Sea. (1, 16, 17)
T. kiaochowensis Ying: Jiaozhou Bay (Shandong). (2)
T. lohmanni (Jörgensen): Bohai; Yellow Sea. (1, 29)
T. loricata (Brandt): East China Sea. (28)
T. major Meunier: East China Sea; South China Sea. (16, 17)
T. mayeri Daday: Jiaozhou Bay (Shandong). (1)
T. minima Wang & Nie: Xiamen harbor (Fujian). (28)
T. nana Lohmann: South China Sea. (17)
T. nitida Brandt: East China Sea. (17)
T. nucula (Fol.): South China Sea. (17)
T. orientalis Kofoid & Campbell: East China Sea. (16)
T. pallida (Brandt): Jiaozhou Bay (Shandong). (1)
T. parva Merkle: East China Sea. (16)
T. radix (Imhof): Bohai; Yellow Sea; East China Sea; South China Sea. (1, 16, 17, 27, 29)
T. rapa Meunier: Yellow Sea. (1)
T. rotundata (Jörgensen): East China Sea. (28)
T. schotti (Brandt): South China Sea. (17)
T. spiralis Kofoid & Campbell: East China Sea. (28)
T. tentaculata Nie & Cheng: South China Sea. (17)
T. tocantinensis Kofoid & Campbell: Bohai; Yellow Sea; East China Sea; South China Sea. (1, 17, 29)
T. tsingtaoensis Yin: Jiaozhou Bay (Shandong). (2)
T. tubulosa Levander: East China Sea; South China Sea. (17, 27)
T. turgida Kofoid & Campbell: East China Sea; South China Sea. (16, 17)

CODONELLOPSIDAE

Codonellopsis hainanensis Nie & Cheng: South China Sea. (17)
C. mobilis Wang: Bohai; Yellow Sea. (1, 29)
C. ostenfeldi (Schmidt): East China Sea; South China Sea. (17, 28)
C. pehaiensis Wang: Bohai. (29)
C. rodunda Wang & Nie: East China Sea. (28)
Stenosemella epunctata Wang: Bohai. (29)
S. nivalis (Meunier): South China Sea. (17)

S. pacifica Kofoid & Campbell: Bohai. (29)
S. parvicollis (Marshall): Yellow Sea; East China Sea. (1, 27)
S. ventricosa Clap. & Lachm: East China Sea. (27)
Wangiella dicollaria Nie: Xiamen harbor (Fujian). (16)

COXLIELLIDAE

Coxliella annulata (Daday): East China Sea; South China Sea. (16, 17)
Helicostomella longa (Brandt): Yellow Sea; South China Sea. (1, 17)

CYTTAROCYLIDAE

Favella amoyensis Wang & Nie: East China Sea. (28)
F. arcuata (Brandt): Bohai. (29)
F. azorica (Cleve): South China Sea. (17)
F. companula (Schmidt): Yellow Sea; South China Sea. (1, 17)
F. cylindrica Wang: Bohai. (29)
F. ehrenbergi (Claparede & Lachmann): South China Sea. (17)
F. franciscana Kofoid & Campbell: South China Sea. (17)
F. hainanensis Nie & Cheng: South China Sea. (17)
F. hainanensis v. *parva* Nie & Cheng: South China Sea. (17)
F. panamensis Kofoid & Campbell: Yellow Sea; East China Sea. (1, 28)
F. shintsuenensis Nie & Cheng: South China Sea. (17)
F. undulata Wang & Nie: East China Sea. (28)
Parafavella elongata Wang: Bohai. (29)

PTYCHOCYLIDAE

Epiplocylis calyx v. *lodiosa* (Kofoid & Campbell): South China Sea. (17)
E. constricta Kofoid & Campbell: South China Sea. (17)
E. undella v. *blanda* Jörgensen: South China Sea. (17)

PETALOTRICHIDAE

Metacylis jörgensenii (Cleve): South China Sea. (17)
M. oviformis Nie & Cheng: South China Sea. (17)
M. sanyahensis Nie & Cheng: Yellow Sea; South China Sea. (1, 17)
Petalotricha aperta (Marshall): South China Sea. (17)

RHABDONELLIDAE

Protorhabdonella curta (Cleve): South China Sea. (17)
P. simplex (Cleve): South China Sea. (17)
Rhabdonella amor (Cleve): South China Sea. (17)
R. conica Kofoid & Campbell: South China Sea. (17)
R. elegans Jörgensen: South China Sea. (17)
R. sanyahensis Nie & Cheng: South China Sea. (17)

UNDELLIDAE

Proplectella globosa (Brandt): South China Sea. (17)
P. ovata (Jörgensen): South China Sea. (17)

TINTINNIDAE

Amphorella brandti (Jörgensen): Yellow Sea; South China Sea. (1, 17)
Amphorellopsis acuta (Schmidt): Bohai; Yellow Sea; East China Sea; South China Sea. (1, 16, 17, 29)
A. pentagona Nie: East China Sea. (16)
Dadayiella ganymedes (Entz.): South China Sea. (17)
Eutintinus tenuis Kofoid & Campbell: Yellow Sea. (1)
Tintinnus fraknoii Daday: South China Sea. (17)
T. lusus-undae Entz: South China Sea. (17)
T. lusus-undae v. *exigua* (Hada): East China Sea; South China Sea. (17, 29)
T. striatus Nie & Cheng: South China Sea. (17)
T. tenuis Kofoid & Campbell: East China Sea. (16)
T. tubulosa Ostenf: East China Sea. (27)

HYPOTRICHIDAE
UROSTYLIDAE

Holosticha diademata (Rees): Qingdao (Shandong). (26)
H. kessleri (Wrzesniowsky): Xiamen harbor (Fujian). (28)
H. lacazei Maupas: Qingdao (Shandong). (25)
H. rubra (Ehrenberg): Xiamen harbor (Fujian). (28)
H. simplicis Wang & Nie: Xiamen harbor (Fujian). (28)
Strongylidium maritimum Wang & Nie: Xiamen harbor (Fujian). (28)
Urostyla limboonkengi Wang & Nie: Xiamen harbor (Fujian). (28)

OXYTRICHIDAE

Balladyna parvula Kowalewsky: Xiamen harbor (Fujian). (28)
Gonostomum pediculiforme (Cohn): Xiamen harbor (Fujian). (28)
Oxytricha ferruginea Stein: Xiamen harbor (Fujian). (28)
O. saltans (Cohn): Qingdao (Shandong). (22)

EUPLOTIDAE

Diophrys appendiculata (Ehrenberg): Qingdao (Shandong). (25)
D. oligothrix Borrow: Qingdao (Shandong). (24)
D. scutum (Dujardin): Qingdao (Shandong). (23)
Euplotes charon Ehrenberg: Xiamen harbor (Fujian). (28)
E. charonopsis Song & Packroff: Qingdao (Shandong). (25)
E. harpa Stein: Xiamen harbor (Fujian). (28)
E. rariseta Curds et al: Qingdao (Shandong). (25)
E. vanus (Müller): Qingdao (Shandong). (25)
Uronychia bivalvorum Fenchel: Qingdao (Shandong). (26)
U. transfuga (Müller): Shandong (including Qingdao); Xiamen (Fujian). (26, 28)
U. uncinata Kahl: Qingdao, Kenli, Laizhou, Rizhao, and Weifang (Shandong). (26)

ASPIDISCIDAE

Aspidisca leptaspos Fresenius: Qingdao (Shandong). (26)
A. lynceus (Müller): Xiamen harbor (Fujian). (28)
A. magna Kahl: Qingdao (Shandong). (26)
A. turrita (Ehrenberg): Xiamen harbor (Fujian). (28)
Onychaspis hexeris (Quennerstedt): Xiamen harbor (Fujian). (28)
O. polystyla (Stein): Xiamen harbor (Fujian). (28)

REFERENCES*

(1). Yen, K. T. 1953. A preliminary investigation of arenaceous shell ciliates from Kiaochou Bay. Journal of Shandong College of Oceanology, 2:36–56.
(2). Yen, K. T. 1956. On three new species of arenaceous shell ciliates in Kiaochou Bay. Journal of Shandong College of Oceanology, 4: 64–69.
(3). Tchang Tso-run, 1963. Studies on the intertidal ciliates from Echinoderms on the coast of China. Oceanologia & Limnologia Sinica, 5(3): 215–229.
(4). Song, Weibo, 1986. Descriptions of seven species of Pertrichs on *Penaeus orientalis* (Pertrichida: Zoothamnidae, Episthlidae). Acta Zootaxonomica Sinica, 11(3):225–235.
(5). Song, Weibo 1991. Contribution to the commensal ciliates on *Penaeus orientalis*, I. Ciliophora, Pertrichida). Journal of Ocean University of Qingdao, 21(3):119–128.
(6). Song, Weibo 1991. Contribution to the commensal ciliates on *Penaeus orientalis*, II. Ciliophora, Pertrichida). Journal of Ocean University of Qingdao, 21(4):45–55.
(7). Song, Weibo 1991. Description of a new species of commensal ciliate on *Penaeus orientalis* (Ciliphora, Pertrichida). Zoologica Research, 12:355–359.
(8). Song, Weibo 1992. Contribution to the commensal ciliates on *Penaeus orientalis*, III. Ciliophora, Pertrichida). Journal of Ocean University of Qingdao, 22(2):107–117.
(9). Song, Weibo 1992. A new marine ciliates, *Zoothamnium penaei* sp. nov. (Ciliophora, Pertrichida). Oceanologia & Limnologia Sinica, 23(1):90–95.
(10). Song, Weibo 1993. A brief revision of desease-causing ciliates (Protozoa, Ciliophora) from marine culture water bodies. Marine Sciences, 4:41–47.
(11). Song, Weibo, 1993. Studies on the morphology of *Homalogastra binucleata* nov. sp. and the contribution to the genus diagnosis (Ciliophora, Scuticociliatida). Oceanologia & Limnologia Sinica, 24(2):143–150.
(12). Song, Weibo 1993. A new marine species, *Litonotus paracygnus* nov. sp. and studies on the morphology. J. Zool., in press.
(13). Song, Weibo 1993. Studies on the morphology and systematic status of *Loxophyllum rostratum* COHN, 1866 (Ciliophora, Pleurostomatida). Journal of Oceanography of Yellow and Bohai Sea, 11(1): 44–49.

(14). Zhou Li et al., 1991. Description of the subspecies, *Paranophrys carcini* Spiralis nov., a parasitic ciliates of artificial over-wintering prawn, *Penaeus orientalis* Kishononye (Ciliophora: Scuticociliatida: Philasteridae). Journal of Ocean University of Qingdao, 21(2):90–98.

(15). Meng Qingxian and Yu Kaikang, 1985. A new species of ciliata, *Licnophora hippocampi* sp. nov., from the sea horse, *Hippocampus trimaculatus* Leach, with considerations of its control in the host. Acta Zoologica Sinica, 31(1):65–69.

(16). Nie, D., 1934. Notes on Tintinnoinea from the bay of Amoy. M. B. A. C. Third Annual Report, pp:71–80.

(17). Nie, D., P. S. Cheng, 1947. Tintinnoinea of the Hainan region. Cont. biol. lab. sc. soc. China, Vol. 46, No. 3.

(18). Song, W. 1990. Infraciliature and silverline systems of the ciliate *Pseudotrachelocerca trepida* nov. comb., nov. gen. and establishment of a new family, the Pseudotrachelocercidae nov. fam. Europ. J. Protistol. 26:160–166.

(19). Song, W. 1991a. Morphologies and infraciliature von *Pseudotrachelodonopsis marina* nov. spec. Zoll. Jb. Syst. 118:79–86.

(20). Song, W. 1991b. A new marine ciliate, *Amphileptus litonotuformis* nov. spec. from Sheyang, Jiangsu Province. J. Mar. Limnol. Sinica 17:247–253.

(21). Song, W. 1991. *Litonotus yinae* n. sp. (Protozoa, Ciliophora), a periphytic and commensal species from Qingdao. Ophelia 34: 181–189.

(22). Song, W., Shin, M. K. and Won, K., 1991. Morphology and infraciliature of the marine ciliate *Oxytricha saltans* (Cohn, 1866) Kahl. 1932 (Protozoa, Ciliophora, Hypotrichida). Korean J. Syst. Zool. 7:233–240.

(23). Song, W. and Packroff, G., 1992. Beitrag zur Morphogenese des marinen Ciliaten *Diophrys scutum* (Dujardin, 1841) (Ciliophora, Hypotrichida). Zool. Jb. Anat. 122:1–11.

(24). Song, W. 1993. Morphogenetic studies during the ontogeny of a Marine Hypotrich, *Diophrys oligothrix* Borror, 1965 (Protozoa, Ciliaophora), with particular attention to its division features compared with other congeners. Europ. J. Protistol. 29, in press.

(25). Song, W. and Packroff, G., 1993. Beitrag zu marinen Ciliatenfauna der Küsten in Shandong, China (Ciliaphora, Protozoa). Acta Protozool.

(26). Song, W., Packroff, G. and Wilbert, N., 1993. Taxonomic studies on some free living ciliates from the sea beach near Qingdao (Protozoa, Ciliaphora). Arch. Protistenkd. 140, in press.

(27). Sproston, N. G., 1949. A preliminary survey of the plankton of the Chusan region, with a review of the relevant literature. Sinensia, Vol. 20, No.1–6, 114–126.

(28). Wang, C. C. and Nie, D., 1932. A survey of the marine protozoa of Amoy. Contr. biol. lab. sc. soc. China, Vol. 8, No. 9, 285–385.

(29). Wang, C. C. 1936. Notes on Tintinnoinea from the Gulf of PE-HAI. Sinensia, Vol. 7, No. 3, 353–370.

*: (1)–(15) in Chinese
Compiled by Lin Jinmei; Edited by Song Weibou

SARCOMASTIGOPHORA

RADIOLARIA
ACANTHARIA
HOLACANTHA
ACANTHOCHIASMIDAE

Acanthochiasma ruboscens Krohn: Northern South China Sea. (17)

ACANTHOPLEGMIDAE

Acanthocolla cruciata (Haeckel): Western East China Sea; Xisha (=Paracel Islands). (16, 17)
A. solidissima Popofsky: Xisha (=Paracel Islands). (16)

SYMPHYACANTHA
ASTROLITHIDAE

Acantholithium dicopum (Haeckel): Xisha (=Paracel Islands). (16)
Astrolithium bulbiferum Haeckel: Northern South China Sea. (17)
Heliolithium aureum Schwiakoff: Northern South China Sea. (17)

AMPHILITHIDAE

Amphilithium clavarium Haeckel: Northern South China Sea; Xisha (=Paracel Islands). (16, 17)
A. concretum (Haeckel): Western East China Sea; Xisha (=Paracel Islands). (6, 16)

PSEUDOLITHIDAE

Amphibelone anomala (Haeckel): Western and northern South China Sea; Xisha (=Paracel Islands). (5, 16, 17)
A. hydrotomica (Haeckel): Western East China Sea; northern South China Sea; Xisha (=Paracel Islands). (5, 16, 17)
Haliommatidium müller Haeckel: Northern South China Sea; Xisha (=Paracel Islands). (16, 17)
Pseudolithium compressum Haeckel: Xisha (=Paracel Islands). (16)

GIGARTACONIDAE

Gigartacon denticulatus (Haeckel): Western East China Sea. (6)
G. fragilis Haeckel: Northern South China Sea. (17)
Heteracon biformis (Popofsky): Northern South China Sea. (17)

ARTHACANTHA
ACANTHOMETRIDAE

Acanthometra pellucida J. Müller: Western East China Sea; northern South China Sea; Xisha (=Paracel Islands). (6, 16, 17)
Amphilonche elongata (J. Müller): Western East China Sea. (6)

DORATASPIDAE

Coleaspis cronata Haeckel: Xisha (=Paracel Islands). (16)

Dictyaspis furcata (Haeckel): Western East China Sea; northern South China Sea; Xisha (=Paracel Islands). (6, 16, 17)

Dorataspis choanopora Tchang: Western East China Sea; northern South China Sea. (6, 17)

D. gladiata Haeckel: Northern South China Sea; Xisha (=Paracel Islands). (16, 17)

D. loricata Haeckel: Western East China Sea; northern South China Sea; Xisha (=Paracel Islands). (6, 16, 17)

D. microspora Haeckel: Western East China Sea; northern South China Sea; Xisha (=Paracel Islands). (6, 16, 17)

D. micropora v. *collosa* (Popofsky): Western East China Sea; northern South China Sea; Xisha (=Paracel Islands). (6, 16, 17)

Hexaconus ciliatus Haeckel: Western East China Sea; northern South China Sea; Xisha (=Paracel Islands). (6, 16, 17)

Hystrichaspis dorsata Haeckel: Northern South China Sea; Xisha (=Paracel Islands). (16, 17)

H. fruticata Haeckel: Western East China Sea. (6)

Lychnaspis giltschi Haeckel: Western East China Sea; northern South China Sea; Xisha (=Paracel Islands). (6, 16, 17)

L. longissima Haeckel: Western East China Sea; northern South China Sea; Xisha (=Paracel Islands). (6, 16, 17)

L. polyancistra (Haeckel): Western East China Sea; northern South China Sea; Xisha (=Paracel Islands). (6, 16, 17)

L. serrata Haeckel: Western East China Sea; northern South China Sea. (6, 17)

Pleuraspis costata (J. Müller): Western East China Sea; northern South China Sea. (6, 17)

P. sarmentosa Tchang & Tan: Western East China Sea; northern South China Sea; Xisha (=Paracel Islands). (6, 16, 17)

Stauraspis echinoides (Haeckel): Western East China Sea; northern South China Sea. (6, 17)

DIPLOCONIDAE

Diploconus fasces Haeckel: Western East China Sea; northern South China Sea; Xisha (=Paracel Islands). (6, 16, 17)

D. nitidus Popofsky: Western East China Sea; northern South China Sea; Xisha (=Paracel Islands). (6, 16, 17)

PHRACTOPELTIDAE

Phractopelta cruciata (Haeckel): Western East China Sea; northern South China Sea. (7, 17)

PHYLLOSTAURIDAE

Acanthostaurus nordgaardi Jörgensen: Northern South China Sea. (17)

A. prupurascens Haeckel: Xisha (=Paracel Islands). (16)

Amphistaurus tetrapterus (Haeckel): Western East China Sea. (6)

Lonchostaurus rhombicus Haeckel: Western East China Sea. (6)

Phyllostaurus cuspidatus (Haeckel): Western East China Sea; northern South China Sea; Xisha (=Paracel Islands). (6, 16, 17)

P. siculus (Haeckel): Xisha (=Paracel Islands). (16)

Zygostaurus amphitectus Haeckel: Northern South China Sea; Xisha (=Paracel Islands). (16, 17)

STAURACANTHIDAE

Xiphacantha alata (J. Müller): Western East China Sea; northern South China Sea; Xisha (=Paracel Islands). (6, 16, 17)

DICTYACANTHIDAE

Dictyacantha tetragonopa (Haeckel): Northern South China Sea. (17)

SPUMELLARIA
COLLOIDEA
COLLOZIDAE

Collozoum contortum Haeckel: Hainan. (19)

C. inerme (J. Müller): Hainan; western South China Sea. (13, 14)

C. ovatum Haeckel: Hainan. (19)

BELOIDEA
SPHAEROZOIDAE

Belonozoum italicum (Haeckel): Western East China Sea. (13)

Rhaphidozoum acuferum (J. Müller): Western East China Sea. (13)

Sphaerozoum fuscum Meyen: Western East China Sea. (13)

S. punctatum (Meyen): Hainan. (19)

S. cf. *strigulosum* Breckner: Xisha (=Paracel Islands). (10)

S. verticillum Haeckel: Hainan. (19)

SPHAEROIDEA
LIOSPHAERIDAE

Carposphaera globosa Clark & Campbell: East China Sea continental shelf. (15)

Cenosphaera corota Haeckel: Xisha (=Paracel Islands). (8)

C. melifica Haeckel: Xisha (=Paracel Islands). (8)

Plegmosphaera cf. *entodictyon* Haeckel: Xisha (=Paracel Islands). (18)

P. leptoplegma Haeckel: Western East China Sea. (13)

P. ovata Su: Xisha (=Paracel Islands). (10)

Sphaeropyle mespilus Dreyer: Xisha (=Paracel Islands). (8)

Thecosphaera grecoi Vinassa de Regny: East China Sea continental shelf; Xisha (=Paracel Islands). (8)

COLLOSPHAERIDAE

Acrosphaera collina Haeckel: East China Sea continental shelf; Xisha (=Paracel Islands). (8, 15)

A. spinosa (Haeckel): East China Sea continental shelf; Okinawa Trough; Taiwan Starit; Hainan. (2, 12, 15, 19)
Buccinosphaera invaginata Haeckel: Taiwan Strait; Guangdong; Hainan. (12, 19, 21)
Choenicosphaera flammabunda Haeckel: East China Sea continental shelf; Taiwan Strait; South China Sea; Xisha (=Paracel Islands). (8, 12, 15, 21)
Collosphaera armata Brandt: East China Sea continental shelf. (15)
C. brattstroimi BjØklund and Goll: Xisha (=Paracel Islands). (18)
C. elliptica Chen & Tan: South China Sea. (1)
C. gigantopora Chen: East China Sea. (4, 20)
C. huxleyi Müller: Western East China Sea; East China Sea continental shelf; Hainan. (13, 15, 19)
C. macropora Popofsky: East China Sea continental shelf; Taiwan Strait. (12, 15)
C. planca Su: Xisha (=Paracel Islands). (10)
Disolenia quadrata (Ehrenberg): South China Sea. (21)
Odonthosphaera cyrtodon Haeckel: Hainan. (19)
Otosphaera auriculata (Ehrenberg): East China Sea continental shelf. (15)
Siphonosphaera ardys Su: Xisha (=Paracel Islands). (9)
S. donghaiense Chen: East China Sea. (23)
S. marginata Haeckel: East China Sea continental shelf. (15)
S. martensi Brandt: East China Sea continental shelf. (15)
S. pericyclis Chen: South China Sea. (23)
S. polypora Chen: East China Sea; South China Sea. (23)
S. polysiphonia Haeckel: Okinawa Trough; Taiwan Strait; South China Sea. (2, 3, 12, 21)
S. socialis Haeckel: Xisha (=Paracel Islands). (8)
S. tenera Brandt: East China Sea continental shelf. (15)
Solenosphaera pandora Haeckel: Western East China Sea; Taiwan Strait; Hainan. (12, 13, 19)
S. zanguebarica (Ehrenberg): Western East China Sea; Taiwan Strait; Hainan. (12, 13, 19)
Tribonosphaera centripetalis Haeckel: East China Sea continental shelf. (15)
Xanthiosphaera lappacea Haeckel: Okinawa Trough; Taiwan Strait; Xisha (=Paracel Islands). (2, 8, 19)

STYLOSPHAERIDAE

Amphisphaera palliatum (Haeckel): Xisha (=Paracel Islands). (8)
Amphistylus xishaensis Sun: Xisha (=Paracel Islands). (8)
Carpocanthum lubricum Chen & Tan: South China Sea. (1)
Saturnalis circularis Haeckel: Central South China Sea. (3)
Xiphosphaera gaea Haeckel: Xisha (=Paracel Islands). (18)
X. tesseractis Dreyer: Xisha (=Paracel Islands). (18)

CUBOSPHAERIDAE

Hexacontium anaximandrii (Haeckel): Xisha (=Paracel Islands). (8)

H. aff. *axotrias* Haeckel: East China Sea continental shelf; Xisha (=Paracel Islands). (8, 15)
H. castanetum Tan: Xisha (=Paracel Islands). (18)
H. enthacanthum Jörgensen: Xisha (=Paracel Islands). (18)
H. favosum (Haeckel): Xisha (=Paracel Islands). (8)
H. hostile Cleve: Xisha (=Paracel Islands). (8)
H. mellarium Tan & Su: East China Sea continental shelf; Taiwan Strait; Xisha (=Paracel Islands). (12, 15, 18)
H. pachydermum Jörgensen: Xisha (=Paracel Islands). (18)
H. quadratum Tan: Xisha (=Paracel Islands). (18)
H. retrospiculum Chen: South China Sea. (23)
H. sarmentum Su: Xisha (=Paracel Islands). (18)
H. senticetum Tan & Su: East China Sea continental shelf. (15)
H. typanum Tan: Xisha (=Paracel Islands). (18)
Hexastylus aristarchi Haeckel: Taiwan Strait. (12)
H. dimensivas Haeckel: East China Sea continental shelf; Taiwan Strait; Xisha (=Paracel Islands). (8, 15)
H. heracliti Haeckel: Xisha (=Paracel Islands). (8)
H. phaenaxonius Haeckel: East China Sea continental shelf; Xisha (=Paracel Islands). (8, 15)
H. philosophica Haeckel: East China Sea continental shelf. (15)
H. cf. *pythagoraea* Haeckel: East China Sea continental shelf. (15)
H. cf. *triaxonius* Haeckel: East China Sea continental shelf; Taiwan Strait. (12, 15)
Hexacromyum elegans Haeckel: Xisha (=Paracel Islands). (18)
Hexadendron bipinnatum Haeckel: Xisha (=Paracel Islands). (18)
Stylacontarium octatigum Ten & Su: East China Sea continental shelf. (15)

ASTROSPHAERIDAE

Acanthosphaera aff. *barbati* Campbell & Clark: East China Sea continental shelf. (15)
A. capillaris Haeckel: Xisha (=Paracel Islands). (8)
A. dodecastyla Mast: Xisha (=Paracel Islands). (18)
A. insignis (Hertwig): East China Sea continental shelf; Xisha (=Paracel Islands). (8, 15)
A. nanhaiensis Sun: Xisha (=Paracel Islands). (8)
Actinomma arcadophorum Haeckel: East China Sea continental shelf; Okinawa Trough; Taiwan Strait; Xisha (=Paracel Islands). (2, 8, 12, 16)
A. bareale Cleve: Xisha (=Paracel Islands). (8)
A. capillaceum Haeckel: Xisha (=Paracel Islands). (18)
A. eriosperma Tan: Xisha (=Paracel Islands). (18)
A. leptodermum Jörgensen: Okinawa Trough; Xisha (=Paracel Islands). (8)
A. medianum Nigrine: East China Sea continental shelf; Okinawa Trough. (2, 15)
A. mediterranensis Hollande & Enjumet: Xisha (=Paracel Islands). (8)
A. multispinula Su: Xisha (=Paracel Islands). (9)
A. popofskii (Petroshevsdaya): Xisha (=Paracel Islands). (8)

A. saccoi Carnevale: East China Sea continental shelf. (15)
Arachnosphaera myriacantha Haeckel: Taiwan Strait. (12)
Astrophaera hexagonalis Haeckel: Western East China Sea; Taiwan Strait. (6, 12)
Centrocubus cladostylus Haeckel: Western East China Sea; Taiwan Starit; Xisha (=Paracel Islands). (8, 12, 13)
Cladococcus indicus Mast: Xisha (=Paracel Islands). (18)
C. scoparius Haeckel: Western East China Sea; Taiwan Strait. (6, 12)
C. sp. cf. *stalactites* Haeckel: East China Sea continental shelf; Xisha (=Paracel Islands). (8, 15)
Cromyechinus antarctica (Dreyer): Xisha (=Paracel Islands). (18)
Diplosphaera spinosa Hertwing: Xisha (=Paracel Islands). (18)
Drymosphaera dendrophora Haeckel: Western East China Sea; Taiwan Strait. (6, 12)
Echinimma frugifera Su: Xisha (=Paracel Islands). (9)
E. polyacantha Chen: East China Sea; South China Sea. (23)
Elaphococcus cervicornis Haeckel: Xisha (=Paracel Islands). (18)
Haliomma erinaceum Haeckel: Xisha (=Paracel Islands). (18)
H. macrodoras Haeckel: Taiwan Strait; Xisha (=Paracel Islands). (8, 12)
Heliaster hexagonium Hollande and Enjumet: Xisha (=Paracel Islands). (18)
Heliosphaera octacantha Mast: Xisha (=Paracel Islands). (18)
Heterosoma heptacanthum Mast: Xisha (=Paracel Islands). (18)
Octodendron hamuliferum Hollande & Enjumet: Xisha (=Paracel Islands). (18)
O. nidum Tan & Tchang: Western East China Sea. (13)
O. spathillatum Haeckel: Taiwan Strait. (12)
Pityomma drymodes Haeckel: East China Sea continental shelf. (15)
Rhizoplegma lychnosphaera Haeckel: Xisha (=Paracel Islands). (18)
Rhizosphaera serrata Haeckel: Western East China Sea. (13)
Rhizospongus arachnoideus Mast: Xisha (=Paracel Islands). (18)
Spongiomma spinatum Chen & Tan: South China Sea. (1)
Spongodendron macrodoras Hollande & Enjumet: Xisha (=Paracel Islands). (18)
Spongopila gracilis Mast: Xisha (=Paracel Islands). (18)
S. verticillata Haeckel: Xisha (=Paracel Islands). (18)
Spongosphaera streptacantha Haeckel: Western East China Sea; Taiwan Strait. (12, 13)
Tetrapentalon elegans Hollande and Enjumet: Xisha (=Paracel Islands). (18)
Tetrasphaera spongiosa Popofsky: Xisha (=Paracel Islands). (18)

PRUNOIDEA
ELLIPSIDAE

Cenellipsis ellipticum Chen: East China Sea; South China Sea. (23)

DRUPPULIDAE

Druppatractus irregularia Popofsky: East China Sea continental shelf. (15)
Druppula aspera Tan: Xisha (=Paracel Islands). (18)
Lithatractus nucleolus Tan & Su: East China Sea continental shelf. (15)
Stylatractus neptunus Haeckel: Xisha (=Paracel Islands). (8)
Xiphatractus trachyphloius Chen & Tan: South China Sea. (1)
X. xiphydrion Chen: East China Sea; South China Sea. (23)

SPONGURIDAE

Spongocore polyacantha Popofsky: East China Sea continental shelf. (15)
S. puella Haeckel: Xisha (=Paracel Islands). (8)
Spongoliva ellipsoides Popofsky: Taiwan Strait. (12)

CYPHINIDAE

Cypassis irregularis Nigrine: Xisha (=Paracel Islands). (18)
Cyphocolpus virginis Haeckel: Western East China Sea; Taiwan Strait. (12, 13)

PANARTIDAE

Panartus tetrathalamum Haeckel: East China Sea; Taiwan Strait; Xisha (=Paracel Islands). (2, 3, 8, 12, 13)

DISCOIDEA
PHACODISCIDAE

Amphibrachium sponguroides Haeckel: East China Sea continental shelf. (15)
Amphirhopalum ypsilon Haeckel: East China Sea continental shelf; Okinawa Trough; Taiwan Strait. (2, 12, 15)
Cyclastrum trifastifiatum Tan & Tchang: Western East China Sea; Taiwan Strait. (12, 13)
Euchitonia aequipondata Popofsky: Western East China Sea; Taiwan Strait. (12, 13)
E. elegans (Ehrenberg): East China Sea; Okinawa Trough; Taiwan Starit; Xisha (=Paracel Islands). (2, 4, 8, 13)
E. furcata Ehrenberg: Okinawa Trough; Taiwan Strait; Xisha (=Paracel Islands). (2, 8, 12)
E. triangulum (Ehrenberg): Xisha (=Paracel Islands). (8)
E. cf. *triangulum* (Ehrenberg): Western East China Sea. (13)
Heliodiscus asteriscus Haeckel: East China Sea; Taiwan Strait; South China Sea. (2, 12, 16, 21)
H. echiniscus Haeckel: East China Sea continental shelf; Taiwan Strait; South China Sea; Xisha (=Paracel Islands). (8, 12, 15, 21)
H. phacodiscus (Haeckel): Western East China Sea; East China Sea continental shelf; Taiwan Strait; Xisha (=Paracel Islands). (12, 13, 15)

Hymeniastrum euclidis Haeckel: Xisha (=Paracel Islands); South China Sea. (3, 8, 9)
Plectodiscus circularia (Clark & Campbell): Xisha (=Paracel Islands). (8)
Porodiscus ellipticus (Stöhr): East China Sea continental shelf; Xisha (=Paracel Islands). (8, 13)
Rhopalastrum cf. *hexaceros* Haeckel: Xisha (=Paracel Islands). (18)
Sethodiscus lenticula Haeckel: Xisha (=Paracel Islands). (18)
S. macrococcus Haeckel: Xisha (=Paracel Islands). (18)
Spongohagiastrum digitatum Tan & Tchang: Western East China Sea. (13)
Stylochlamydium aequale (Stöhr): Western East China Sea. (13)
S. asteriscus Haeckel: Xisha (=Paracel Islands); Xisha (=Paracel Islands). (8, 18)
S. venustum (Bailey): East China Sea continental shelf; Xisha (=Paracel Islands). (8, 15)
Stylodictya arachnia (Müller): Western East China Sea; East China Sea continental shelf; Taiwan Strait. (12, 13, 15)
S. dujardinii Haeckel: East China Sea; Taiwan Strait; Xisha (=Paracel Islands). (8, 12, 13, 15)
S. cf. *gracilis* Ehrenberg: East China Sea continental shelf; Taiwan Strait. (12, 13)
S. lasiacantha Tan & Tchang: Western East China Sea. (13)
S. multispina Haeckel: Western East China Sea. (13)
S. polygonia Popofsky: East China Sea continental shelf; Taiwan Strait. (12, 15)
S. validispina Haeckel: Western East China Sea; Taiwan Strait; Xisha (=Paracel Islands). (8, 12, 13)
S. sp. cf. *validispina* Jörgensen: Xisha (=Paracel Islands). (18)
Triastrum aurivillii Cleve: Western East China Sea; Taiwan Strait. (12, 13)

PYLODISCIDAE

Hexapyle spinulosa Chen & Tan: South China Sea. (1)
Pylodiscus echinatus Tan & Su: East China Sea continental shelf; Taiwan Strait; Xisha (=Paracel Islands). (8, 12, 15)
Pylolena armata Haeckel: Taiwan Strait. (12)

SPONGODISCIDAE

Dictyocoryne profunda Ehrenberg: Western East China Sea; Taiwan Strait; Xisha (=Paracel Islands). (8, 12, 13)
D. tricupiformis Chen & Tan: South China Sea. (1)
D. trimaculatum Tan & Tchang: Western East China Sea; Taiwan Strait. (12, 13)
D. truncatum (Ehrenberg): Xisha (=Paracel Islands). (8)
Spongaster pentas Riedel & Sanfillippo: East China Sea continental shelf. (15)
S. tetras Ehrenberg: Western East China Sea; Okinawa Trough; Taiwan Strait; South China Sea; Xisha (=Paracel Islands). (2, 8, 12, 13)
Spongobrachium froudum Tan & Su: East China Sea continental shelf. (15)
S. pentagrama Chen: South China Sea. (20)
S. toxon Tan & Su: East China Sea continental shelf. (15)
Spongodiscus asiaensis Chen & Tan: South China Sea. (1)
S. biconcavus Haeckel: Western East China Sea; Taiwan Strait; Xisha (=Paracel Islands). (8, 12, 13)
S. craticulatus (Stöhr): Western East China Sea. (13)
S. flos Tan & Su: East China Sea continental shelf. (15)
Spongotrochus glacialis Popofsky: Western East China Sea; Taiwan Strait. (12, 13)
S. multispinus Haeckel: Taiwan Strait. (12)

LARCOIDEA
DARCOPYLIDAE

Larcopyle bütschlii Dreyer: East China Sea continental shelf; Taiwan Strait; Xisha (=Paracel Islands). (8, 12, 15)

PYLONIDAE

Monozonium pachystylum Popofsky: East China Sea continental shelf; Taiwan Strait. (12, 15)
Octopyle circinata Tan & Chen: South China Sea. (22)
O. clypeata Tan & Chen: South China Sea. (22)
O. fruticosa Tan & Chen: South China Sea. (22)
O. hexagona Tan & Chen: East China Sea; South China Sea. (22)
O. octospinosa Tan & Tchang: Western East China Sea; East China Sea continental shelf; Taiwan Strait. (12, 13, 15)
O. polystyle Chen: South China Sea. (23)
O. stenozona Haeckel: South China Sea; Xisha (=Paracel Islands). (8, 22)
Pylonium claviflorum Tan & Chen: South China Sea. (22)
P. scitulum Tan & Chen: South China Sea. (22)
P. scutatulum Chen & Tan: South China Sea. (1)
Tetrapyle circularis Haeckel: Western East China Sea; Taiwan Strait; South China Sea. (12, 13, 22)
T. nephropyle Haeckel: Western East China Sea; Taiwan Strait. (12, 13)
T. octacantha Müller: East China Sea continental shelf; Taiwan Strait; South China Sea; Xisha (=Paracel Islands). (8, 12, 15, 22)
T. pachyderma Tan & Chen: South China Sea. (22)
T. quadriloba (Ehrenberg): East China Sea; South China Sea. (15, 22)
T. rotundospinosa Tan: Xisha (=Paracel Islands). (18)
Tetrapylonium pyrum Tan & Chen: South China Sea. (22)
T. strobilinum Tan & Chen: South China Sea. (22)

THOLONIIDAE

Amphitholonium pylonium Su: Xisha (=Paracel Islands). (9)
A. trasversarum Chen: East China Sea; South China Sea. (23)
Amphitholus cf. *acanthometra* Haeckel: East China Sea continental shelf. (15)

Cubotholonium polystylum Chen: East China Sea; South China Sea. (23)

Cubotholus regularia Haeckel: Xisha (=Paracel Islands). (18)

LITHELIDAE

Larcospira minor Jörgensen: Xisha (=Paracel Islands). (18)

L. quadrangula Haeckel: East China Sea; Taiwan Strait; South China Sea. (2, 12, 13)

Lithelius amphistylis Chen: East China Sea; South China Sea. (23)

L. cf. *alveolina* Haeckel: East China Sea continental shelf. (15)

L. nerites Tan & Su: East China Sea continental shelf; Taiwan Strait. (12, 15)

L. solaria Haeckel: Western East China Sea; Taiwan Strait. (12, 13)

L. spiralis Haeckel: East China Sea continental shelf; Taiwan Strait. (12, 15)

L. xanthiformis Tan & Su: East China Sea continental shelf; Taiwan Strait; Xisha (=Paracel Islands). (8, 12, 15)

Pylospira octopyle Haeckel: Western East China Sea; Taiwan Strait. (12, 13)

Spirema haliomma (Ehrenberg): East China Sea continental shelf; Taiwan Strait. (12, 15)

STREBLONIDAE

Streblacantha circumtexta (Jörgensen): East China Sea continental shelf; Taiwan Strait; Xisha (=Paracle Islands). (8, 15)

PHORTICIDAE

Phorticium polycladum Tan & Tchang: Western East China Sea; Taiwan Strait. (12, 13)

P. pylonium Haeckel: Western East China Sea; Taiwan Strait; South China Sea. (12, 13, 22)

SOREUMIDAE

Soreuma gibbulosum Tan & Tchang: Western East China Sea. (13)

NASSELLARIA
PLECTOIDEA
PLECTANIDAE

Rhizoplecta trithyris Freguelli: Western East China Sea; East China Sea continental shelf; Taiwan Strait. (5, 13, 15)

STEPHOIDEA
STEPHONIDAE

Zygocircus longispininus Tan & Tchang: Western East China Sea. (13)

SEMANTIDIDAE

Neosematis distephanus Popofsky: Western East China Sea. (13)

CORONIDAE

Acanthodesmia viniculata (Müller): South China Sea. (21)

Eucoronis challengeri Haeckel: Western East China Sea; East China Sea continental shelf; Taiwan Strait; Xisha (=Paracel Islands). (8, 12, 13, 15)

E. nephrospyris Haeckel: Western East China Sea; Taiwan Strait. (12, 13)

ZYGOSPYRIDAE

Dorcadospyris pentagona (Ehrenberg): South China Sea. (21)

Liriospyris pulmoformis Tan & Su: Xisha (=Paracel Islands). (14)

L. rotunda Tan & Su: Xisha (=Paracel Islands). (14)

Tristylospyris palmipes Haeckel: East China Sea continental shelf. (15)

THOLOSPYRIDAE

Lophospyris pentagona (Ehrenberg): Xisha (=Paracel Islands). (8)

Tholospyris fenestrata Haeckel: East China Sea continental shelf; Taiwan Strait. (12, 15)

T. cf. *scaphipes* (Haeckel): East China Sea continental shelf. (15)

T. tripodiscus Haeckel: East China Sea continental shelf; central South China Sea; Xisha (=Paracel Islands). (3, 8, 15)

PHORMOSPYRIDAE

Desmospyris stabilis (Goll): South China Sea. (3)

ANDROSPYRIDAE

Amphispyris bria Tan & Su: East China Sea continental shelf; Taiwan Strait. (12, 15)

A. costata Haeckel: Taiwan Strait. (12)

A. reticulata (Ehrenberg): Taiwan Strait; Xisha (=Paracel Islands). (8, 12)

A. sinensis Sun: Xisha (=Paracel Islands). (8)

Nephrospyris docris Renz: South China Sea. (3)

N. paradictyum Haeckel: Taiwan Strait. (12)
N. renilla Haeckel: South China Sea. (3)

BOTRYODEA
CANNOBOTRYIDAE

Centrobotrys thermophila Petrushevskaya: South China Sea. (3)

GLYCOBOTRYDIDAE

Xiphobotrys clavata Tchang & Tan: East China Sea. (6)
X. passerina Tchang & Tan: East China Sea. (6)

LITHOBOTRYIDAE

Acrobotrissa cribosa Popofsky: East China Sea continental shelf; South China Sea. (3, 15)
Botryopyle setosa Cleve: Western East China Sea. (15)

PYLOBOTRYDIDAE

Botryocyrtis scutum (Harting): Taiwan Strait; Okinawa Trough; South China Sea. (2, 3, 12)

CYRTOIDEA
TRIPOCALPIDAE

Archipera dipleura Tan & Tchang: Western East China Sea; East China Sea continental shelf. (13, 15)
A. triclavigera Tan & Tchang: Western East China Sea. (13)
Archipilium eburneforme Chen & Tan: South China Sea. (1)
Pteroscenium pinnatum Haeckel: East China Sea continental shelf. (15)
Peridium ramosum Su: Xisha (=Paracel Islands). (9)
P. spinipes Haeckel: Taiwan Strait. (12)

PHAENOCALPIDAE

Bathropyramis bicornuta Chen & Tan: South China Sea. (1)
Cinclopyramis infundibulum Haeckel: Xisha (=Paracel Islands). (8)

CYRTOCALPIDAE

Cornutella profunda Ehrenberg: Central South China Sea. (3)

TRIPOCYRTIDAE

Clathrocanium coarctatum Ehrenberg: South China Sea. (21)
C. diadema Haeckel: East China Sea continental shelf; Taiwan Strait. (12, 15)
Dictyophimus arabicus (Ehrenberg): East China Sea continental shelf. (15)

Eucecryphalus cervus (Ehrenberg): South China Sea. (21)
E. sestrodiscus (Haeckel): Xisha (=Paracel Islands). (8)
Holotholus histricosa Jörgensen: Western East China Sea; East China Sea continental shelf. (13, 15)
Lamprodiscus quadricuspis Haeckel: South China Sea. (3)
Lampromitra amabilis Su: Xisha (=Paracel Islands). (9)
L. cachoni Petrushevskaya: Xisha (=Paracel Islands). (8)
L. coronata Haeckel: East China Sea continental shelf. (15)
L. quadricuspis Haeckel: Xisha (=Paracel Islands). (8)
L. sinuosa Popofsky: Western East China Sea; East China Sea continental shelf. (13, 15)
Lithomelissa buetschlii Haeckel: Western East China Sea. (13)
L. campanulaeformis Campbell & Clark: East China Sea continental shelf. (15)
L. monoceras Popofsky: Western East China Sea; Taiwan Strait. (12, 15)
L. spinosissima Tan & Tchang: Western East China Sea; Taiwan Strait. (12, 13)
L. thoracites Haeckel: East China Sea continental shelf; Taiwan Strait. (12, 15)
Lithopera bacca Ehrenberg: Western East China Sea; Xisha (=Paracel Islands). (8, 13)
Lychnodictyum challengeri Haeckel: Western East China Sea; East China Sea continental shelf. (13, 15)

ANTHOCYRTIDAE

Acanthocorys castanoides Tan & Tchang: Western East China Sea. (13)
A. umbelifera Haeckel: Western East China Sea. (13)
A. variabilis Popofsky: Western East China Sea; Taiwan Strait. (12, 13)
Anthocyrtidium ophirense (Ehrenberg): East China Sea continental shelf; Okinawa Trough; Taiwan Strait; South China Sea; Xisha (=Paracel Islands). (2, 8, 12, 21)
A. zanguebaricus (Ehrenberg): East China Sea continental shelf. (15)
Carpocanium diadema Haeckel: East China Sea. (2)
C. ensigerum Tan & Su: East China Sea continental shelf. (15)
Sethophormis pentalactis Haeckel: East China Sea continental shelf. (15)
Sethopyramis quadrata Haeckel: East China Sea continental shelf. (15)

CARPOCANIIDAE

Carpocanopsis obovata Tan & Su: East China Sea continental shelf. (15)

SETHOCYRTIDAE

Sethoconus myxobrachia Strelkow & Reschetnjak: Western East China Sea. (13)

Sethocorys odysseus Haeckel: East China Sea continental shelf. (15)

PODOCYRTIDAE

Corocalyptra cervus (Ehrenberg): Taiwan Strait. (12)
Dictyoceras prismaticum Tan & Tchang: Western East China Sea; East China Sea continental shelf. (13, 15)
D. virchowii Haeckel: Western East China Sea; Taiwan Strait. (12, 13)
Lithopilium macroceras Popofsky: Taiwan Strait. (12)
Pterocanium korotnevi (Dogiell): Okinawa Trough. (2)
P. pretextum (Ehrenberg): Western East China Sea; East China Sea continental shelf; Okinawa Trough. (2, 13, 15)
P. trilobum (Haeckel): Western East China Sea; East China Sea continental shelf; Okinawa Trough; Taiwan Strait; South China Sea; Xisha (=Paracel Islands). (2, 8, 12, 13)
Pterocorys campanula Haeckel: East China Sea continental shelf; western East China Sea; Taiwan Strait. (13, 15)
P. zancleus (Müller): Xisha (=Paracel Islands). (8)
Pteropilium clausum Su: Xisha (=Paracel Islands). (9)
Theopilium cranoides Haeckel: East China Sea continental shelf. (15)
T. cucullatum Tan & Su: East China Sea continental shelf. (15)
T. galeatum Tan & Tchang: Western East China Sea. (13)
T. germinis Tan & Su: East China Sea continental shelf. (15)
T. tricostatum Haeckel: Western East China Sea; East China Sea continental shelf; Taiwan Strait. (13, 15)

PHORMOCYRTIDAE

Clathrocyclas alcmenae Haeckel: East China Sea continental shelf; Xisha (=Paracel Islands). (8, 15)
C. cf. *coscinodiscus* Haeckel: East China Sea continental shelf. (15)
C. danaës Haeckel: East China Sea continental shelf. (15)
Diplocyclas bicorona Haeckel: East China Sea continental shelf. (15)
Lamprocyclas maritalis Haeckel: East China Sea continental shelf; Okinawa Trough; Taiwan Strait; central South China Sea; Xisha (=Paracel Islands). (2, 3, 8, 12, 15)
L. maritalis polypora Nigrini: Okinawa Trough. (2)
Lamprocyrtis neoheteroporas Kling: Xisha (=Paracel Islands). (8)
Theocapsa democriti Haeckel: Western East China Sea; East China Sea continental shelf. (13, 15)
Theoconus hertwigii (Haeckel): Taiwan Strait. (12)
Theocorythium trachelium trachelium Haeckel: East China Sea continental shelf; Okinawa Trough; Taiwan Strait; South China Sea; Xisha (=Paracel Islands). (2, 8, 12, 15, 21)
Theophormis callipilium Haeckel: Taiwan Strait. (12)

PODOCAMPIDAE

Artropilium fusiforme Tan & Tchang: Western East China Sea; East China Sea continental shelf. (13, 15)
A. sitularius Tan & Tchang: Western East China Sea. (13)
Stichopilium bicorne Haeckel: South China Sea. (3)
S. campanulatum Haeckel: Western East China Sea; East China Sea continental shelf; Taiwan Strait. (12, 13, 15)
S. obliqum Tan & Su: East China Sea continental shelf. (15)
S. rapaeformis Popofsky: Western East China Sea; East China Sea continental shelf; Taiwan Strait. (12, 13, 15)
S. thoracopterum (Haeckel): Western East China Sea; Taiwan Strait. (12, 13)

LITHOCAMPIDAE

Botryostrobus aquilonaris: Xisha (=Paracel Islands). (8)
Eucyrtidium acuminatum Ehrenberg: Xisha (=Paracel Islands). (8)
E. dictyopodium (Haeckel): Xisha (=Paracel Islands). (8)
E. hexagonatum Haeckel: East China Sea; Okinawa Trough; South China Sea. (3, 4, 21)
E. urceolatum Chen & Tan: South China Sea. (1)
Lithomitra lineata (Ehrenberg): East China Sea continental shelf. (15)
Siphocampe corbula (Harting): East China Sea continental shelf; Okinawa Trough. (2, 6)
Spirocyrtis scalaris Haeckel: Taiwan Strait; South China Sea; Xisha (=Paracel Islands). (3, 8, 12)
S. submerospira Chen: East China Sea; South China Sea. (23)

ARTOSTROBIIDAE

Artostrobium auritum (Ehrenberg): Taiwan Strait; Xisha (=Paracel Islands). (8, 12)

PHAEODARIA
PHAEOCYSTINA
ASTRACANTHIDAE

Astracantha paradoxa Haeckel: Western East China Sea. (13)

PHAEOSPHAERIA
AULOSPHAERIDAE

Aularia ternaria Haeckel: Xisha (=Paracel Islands). (11)

PHAEOCALPIA
CASTANELLIDAE

Castanidium elegans Schmidt: Xisha (=Paracel Islands). (11)
C. moseleyi Haeckel: Xisha (=Paracel Islands). (11)
C. variabile Borgert: Xisha (=Paracel Islands). (11)

CIRCOPORIDAE

Circoporus oxyacanthus Borgert: Xisha (=Paracel Islands). (11)

POROSPATHIDAE

Porospathis holostoma (Cleve): Xisha (=Paracel Islands). (11)

PHAEOGROMIA
CHALLENGERIDAE

Challengeron dioden Haeckel: Xisha (=Paracel Islands). (11)
C. willemoesii Haeckel: Xisha (=Paracel Islands). (11)
Euphysetta elegans Borgert: Xisha (=Paracel Islands). (11)
Gazelletta hexanema Haeckel: Western East China Sea. (13)
Medusetta ansata Borgert: Xisha (=Paracel Islands). (11)
M. inflata Borget: Western East China Sea; Taiwan Strait; Xisha (=Paracel Islands). (11, 13, 14)
Protocystis nautiloides Borget: Xisha (=Paracel Islands). (11)
P. xiphodon (Haeckel): Xisha (=Paracel Islands). (11)

PHAEOCONCHIA
CONCHARIIDAE

Conchoceras caudatum Haeckel: Xisha (=Paracel Islands). (11)

REFERENCES*

(1). Chen Muhong and Tan Zhiyuan, 1989. Description of a new genus and 12 new species of radiolaria in sediments from the South China Sea. Tropic Oceanology, 8(1):1–9, pls. I–II
(2). Chen Wenbin, 1986. Radiolaria layer in Okinawa Trough and its implications in paleo-climate. Oceanologica Sinica, 8(2):191–196.
(3). Chen Wenbin, 1987. Radiolaria in surface sediments of the South China Sea. East China Sea Marine Science, 5(1–2): 60–76.
(4). Chen Wenbin and Wang Baoshui, 1982. A preliminary study on Radiolaria in surface sediments of the East China Sea. Marine Geology Study, 2(2):59–67, pl. I
(5). Tchang Tso-Run, Tan Zhiyuan, 1964. Studies on the radiolaria of the East China Sea I. Acantharia. Studia Marina Sinica, 6:33–78.
(6). Tchang Tso-Run, Tan Zhiyuan, 1964. Sur la nouvelle preuve pour certifer rhizoplecta trithyris frenguelli qui est surement une espece de radiolaria. Studia Marina Sinica, 6:79–81.
(7). Tchang Tso-Run and Tan Zhiyuan, 1965. A new genus and two new species of Glycobotrydidae (Radiolaria) in the East China Sea. Studia Marina Sinica, 7: 15–19.
(8). Hao Yichun et al., 1989. Xisha Islands' Quarternary microbiol community and its geological significance. China Geology University Press.
(9). Su Xinghui, 1982. Description of 11 new species of radiolaria from the Xisha Islands, Guangdong Province, China. Oceanologia & Limnologia Sinica, 13(3):275–284, pls. I–II.
(10). Su Xinghui and Tan Zhiyuan, 1985. A preliminary investigation of Colonial Radiolarias of the Xisha Islands, Guangdong Province, China. Studia Marina Sinica, 24:125–132, pl. I.
(11). Su Xinghui and Tan Zhiyuan, 1985. A preliminary report on the Phaeodarian Radiolaria from the Xisha Islands, Guangdong Province, China. Studia Marina Sinica, 24:135–152.
(12). Fujian Oceanography Institute, 1988. A comprehensive oceanographic survey of the central and northern part of the Taiwan Strait, 353–405, pls. IV–VII, China Science Press.
(13). Tan Zhiyuan and Tchang Tso-Run, 1976. Studies on the radiolaria of the East China Sea II. Spumellaria, Nassellaria, Phacoclaria, Sticholonchea. Studia Marina Sinica, 11:217–313, pls. I–III.
(14). Tan Zhiyuan and Su Xinghui, 1981. Two new species of *Liriospyris* (Radiolaria: Trissocyelidae) from the Xisha Islands, China, with a discussion on their skeletal structures. Acta Zootaxonomica Sinica, 6(4):337–346, pls. I–III
(15). Tan Zhiyuan and Su Xinghui, 1982. Radiolarias from sediments on the continental shelf of the East China Sea, China. Studia Marina Sinica, 19:129–236
(16). Tan Zhiyuan and Su Xinghui, 1983. Acantharia from the reef flats of the Xisha Islands, Guangdong Province, China. Studia Marina Sinica, 20:141–152, pls. I–IV.
(17). Tan Zhiyuan and Su Xinghui, 1985. The Acantharia of the northern South China Sea. Studia Marina Sinica , 25:103–122, pls. I–II
(18). Tan Zhiyuan, 1993. The spumellarian radiolaria of the Xisha Islands. Studia Marina Sinica, 34:181–226, pls. I–IX
(19) Stalekov, A. A. and B. B. Lasibolako, 1962. Spumellaria (Radiolaria) in South China Sea (around southern tip of Hainan). Studia Marina Sinica, 1:121–139.
(20). Chen, Wenbin, 1987. Some new species of radiolaria from surface sediments of the East China Sea and the South China Sea. Chinese Journal of Oceanography and Limnology pp:222–227, pls. I–II.
(21). Hsin, Yi Ling, 1972. Polycystine radiolaria from surface sediments of the South China Sea and the adjacent of Taiwan. Acta Oceanographica Taiwanic, No. 2, pp:159–178, figs. 1–2, pls. 1–2
(22). Tan Zhiyuan and Chen, Muhong, 1990. Some revisions of Pylonidae. Chinese Journal of Oceanology and Limnology, Vol. 8, No. 2, pp:109–125, pls. I–II.

*: (1)–(19) in Chinese
Compiled by Tan Zhiyuan

FORAMINIFERA

RHIZOPODA
GRANULORETICULOEA
FORAMINIFERIDA
ALLOGROMIINA
ALLOGROMIIDAE

1. *Argillotuba argillacea* (Earland): South China Sea. (30)
2. *Nodellum membranaceum* (Brady): South China Sea. (30)

TEXTULARIINA
ASTRORHIZACEA
ASTRORHIZIDAE

3. *Astrorhiza arenaria* Carpenter: East China Sea; Nansha (=Spratly Islands). (28, 30)
4. *A. granulosa* (Brady): East China Sea. (28, 30)
5. *A. limnicola* Sandahl: East China Sea; Nansha (=Spratly Islands). (28, 30)

6. *Astrorhizoides cornutus* (Brady): East China Sea; South China Sea. (28, 30)
7. *Pelosina cylindrica* Brady: East China Sea; South China Sea. (28, 30)

BATHYSIPONIDAE

8. *Bathysiphon albus* Hofker: South China Sea. (30)
9. *B. arenacea* Cushman [184]: East China Sea; South China Sea. (28, 30)
10. *B. capillare* de Folin: East China Sea. (30)
11. *B. echinatus* de Folin: East China Sea. (28, 30)
12. *B. eocenicus* Cushman & Hanna: Taiwan (Lower Oligocene). (44)
13. *B. filiformis* M. Sars: East China Sea; South China Sea. (28, 30)
14. *B. flavidus* de Folin: East China Sea; South China Sea (28, 30)
15. *B. folini* Gooday: South China Sea. (30)
16. *B. giganteus* Cushman: East China Sea. (30)
17. *B. hystrix* Gooday: South China Sea. (30)
18. *B. macilentus* Zheng: East China Sea. (28, 30)
19. *B. major* de Folin: East China Sea; South China Sea. (28, 30)
20. *B. rufescens* Cushman: South China Sea. (30)
21. *B. rufus* de Folin: East China Sea; South China Sea. (28, 30, 66)
22. *Rhabdamminella cylindrica* (Brady): East China Sea; South China Sea. (28, 30)

RHABDAMMINIDAE

23. *Marsipella cylindrica* Brady [22]: East China Sea; South China Sea. (28, 30)
24. *M. dextrospiralis* Chapman & Parr: East China Sea; South China Sea. (28, 30)
25. *M. elongata* Norman: South China Sea. (30)
26. *M. gigantea* Cushman: East China Sea; South China Sea. (28, 30)
27. *Oculosiphon linearis* (Brady): South China Sea. (30)
28. *Rhabdammina abyssorum* M. Sars: East China Sea. (28, 30)
29. *R. cornuta* (Brady) [6]: East China Sea; South China Sea. (28, 30)
30. *R. discreta* Brady: Nansha (=Spratly Islands). (66)
31. *R. eocenica* Cushman & Hanna: East China Sea continental shelf (Eocene). (3)
32. *R. fusiformis* Rhumbler: East China Sea. (28, 30)
33. *R. neglecta* Gooday: South China Sea. (30)
34. *R. scabra* Höglund: East China Sea; South China Sea. (28, 30)
35. *R. triangularis* (Earland): East China Sea. (28, 30)

36. *Rhizammina algaeformis* Brady: East China Sea; South China Sea. (28, 30)
37. *R. indivisa* Brady: East China Sea; South China Sea. (28, 30)
38. *Dendrophrya ramosa* Cushman: East China Sea; South China Sea. (28, 30)
39. *Dendronina arborescens* Heron-Allen & Earland: East China Sea. (28, 30)

DRYORHIZOPSIDAE

40. *Sagenina divaricans* Cushman: East China Sea. (28, 30)

PSAMMOSPHAERIDAE

41. *Psammofax consociata* Rhumbler: South China Sea. (30)
42. *Psammosphaera fusca* Schulze: Yellow Sea; East China Sea; South China Sea. (28, 30)
43. *P. parva* Flint: East China Sea; South China Sea. (28, 30)
44. *P. rustica* Heron-Allen & Earland [117]: East China Sea. (28, 30)

SACCAMMINIDAE

45. *Lagenammina asymmetrica* Stschdrina: Northern Yellow Sea. (28, 65)
46. *L. atlantica* Cushman: Bohai; southern Yellow Sea; East China Sea; South China Sea. (28, 30, 31)
47. *L. difflugiformis* Brady: Bohai; Yellow Sea; East China Sea. (28, 30, 75)
48. *L. longicollis* Zheng: Bohai; northern Yellow Sea. (28, 30)
49. *L. magna* Zheng: East China Sea. (28, 30)
50. *L. pseudodifflugiformis* Stschedrina: Yellow Sea. (30, 65)
51. *L. rhombiformis* Zheng: East China Sea. (28, 30)
52. *L. testacea* (Flint): East China Sea; South China Sea. (28, 30, 66)
53. *Proteonella atlantica* Cushman [46]: Bohai; southern Yellow Sea; East China Sea; South China Sea. (28, 30, 31)
54. *P. asymmetrica* Stschedrina [45]: Northern Yellow Sea. (28, 65)
55. *P. magna* Zheng [49]: East China Sea. (28, 30)
56. *P. pseudodifflugiformis* Stschedrina [50]: Yellow Sea. (30, 65)
57. *P. rhombiformis* Zheng [51]: East China Sea. (28, 30)
58. *P. testacea* (Flint) [52]: East China Sea; South China Sea. (28, 30, 66)
59. *Saccammina anglica* Cushman: East China Sea. (28, 30)
60. *S. atlantica* (Cushman): Yellow Sea; Taiwan Strait. (64)
61. *S. huanghaiensis* Zheng: Southern Yellow Sea. (30)
62. *S. minuta* Zheng: Northern Yellow Sea. (30)
63. *S. sphaerica* Brady: Southern Yellow Sea; East China Sea; South China Sea. (28, 30, 66)
64. *Technitella arenacea* Zheng: East China Sea. (28, 30)
65. *T. legumen* Norman: South China Sea. (30)
66. *Astrammina rara* Rhumbler: South China Sea. (30)

67. *Orbulinelloides agglutinans* Saidova: Nansha (=Spratly Islands). (30)
68. *Thurammina basispiculata* Zheng: East China Sea. (28, 30)
69. *T. papillata* Brady: South China Sea. (30, 66)
70. *T. papyracea* Cushman: East China Sea; South China Sea. (28, 30)

HEMISPHAERAMMINIDAE

71. *Hemisphaerammina bradyi* Loeblich & Tappan: Southern Yellow Sea. (65)

HIPPOCREPINACEA
HIPPOCREPINIDAE

72. *Hyperammina bradyi* Stschedrina: South China Sea. (30)
73. *H. clavellata* Zheng: East China Sea. (28, 30)
74. *H. clavigera* Heron-Allen & Earland: South China Sea. (30)
75. *H. cylindrica* Parr: East China Sea; South China Sea. (28, 30)
76. *H. distorta* Cushman: South China Sea. (30)
77. *H. elongata* Brady: East China Sea; South China Sea. (28, 30)
78. *H. friabilis* Brady: East China Sea. (28, 30)
79. *H. laevigata* Wright: East China Sea; South China Sea. (28, 30)
80. *H. maxima* Cushman: East China Sea; South China Sea. (28, 30)
81. *H. novaezealandiae* Heron-Allen & Earland: East China Sea; Nansha (=Spratly Islands). (28, 30)
82. *Saccorhiza ramosa* (Brady): East China Sea; South China Sea. (28, 30)
83. *Jaculella acuta* Brady: East China Sea. (28, 30)
84. *J. obtusa* Brady: East China Sea. (28, 30)

AMMODISCACEA
AMMODISCIDAE

85. *Ammodiscoides turbinatus* Cushman: Nansha (=Spratly Islands). (66)
86. *Ammodiscus ambiguus* Wang, He, & Lu: East China Sea continental shelf (Paleocene). (3)
87. *A. anthosatus* Guliov: East China Sea continental shelf (Paleocene). (3)
88. *A. argenteus* Zheng: Southern Yellow Sea; East China Sea. (28, 30)
89. *A. evolutus* Zheng: East China Sea. (28, 30)
90. *A. gullmarensis* Höglund: East China Sea; Nansha (=Spratly Islands). (28, 30)
91. *A. hoeglundi* (Uchio): East China Sea; South China Sea. (28, 30)
92. *A. incertus* (d'Orbigny): Xisha (=Paracel Islands); Zhongsha (=Macclesfield Bank). (48)
93. *A. intermedius* Höglund: East China Sea; South China Sea. (28, 30, 74)
94. *A. pacificus* Cushman & Valentine: South China Sea. (30)
95. *A. planus* Höglund: South China Sea. (30)
96. *A. tenuis* Brady: Nansha (=Spratly Islands). (66)
97. *Arenoturrispirillina catinus* Höglund: South China Sea. (30)
98. *Ammolagena clavata* (Jones & Parker): East China Sea; South China Sea. (28, 30)
99. *Tolypammina vagans* (Brady): East China Sea; South China Sea. (28, 30)
100. *Glomospira charoides* (Jones & Parker) [103]: East China Sea; South China Sea. (28, 30, 66)
101. *G. glomerata* Höglund: East China Sea; Nansha (=Spratly Islands). (28, 30)
102. *G. gordialis* (Jones & Parker): Yellow Sea; East China Sea; South China Sea. (28, 30, 48, 66)
103. *Repmanina charoides* (Jones & Parker): East China Sea; South China Sea. (28, 30, 74)

RZEHAKINACEA
RZEHAKINIDAE

104. *Miliammina earlandi* Loeblich & Tappan: Shenzhen Bay (Guangdong); Lingshui (Hainan). (7)
105. *M. fusca* (Brady): Bohai; East China Sea. (19, 20, 21, 37)
106. *M. obliqua* Heron-Allen & Earland: South China Sea; Jiangsu (Quaternary). (9, 30)
107. *Rzehakina epigona* (Rzehak): East China Sea continental shelf (Paleocene). (3)
108. *Silicosigmoilina calcareoarenacea* Stschedrina: Yellow Sea. (65)
109. *S. californica* Cushman & Church: East China Sea (Paleocene). (3)
110. *S. elegantissima* Serova: East China Sea. (28, 30)
111. *Spirosigmoilinella collaris* Wang, He & Lu: East China Sea continental shelf (Miocene). (3)
112. *S. compressa* Matsunaga: East China Sea continental shelf (Miocene). (3)
113. *S. digitata* Wang, He, & Lu: East China Sea continental shelf (Miocene). (3)
114. *S discoformis* Wang, He, & Lu: East China Sea continental shelf (Miocene). (3)
115. *S. irregularis* Wang, He & Lu: East China Sea continental shelf (Miocene). (3)

HORMOSINACEA
ASCHEMONELLIDAE

116. *Aschemonella scabra* Brady: East China Sea; South China Sea. (28, 30)

TELAMMINIDAE

117. *Aggerostramen rusticum* (Heron-Allen & Earland): East China Sea. (29, 30)

HORMOSINIDAE

118. *Hormosinella distans* (Brady): East China Sea; South China Sea. (28, 30, 66, 74)
119. *Nodulina dentaliniformis* (Brady): Southern Yellow Sea; East China Sea. (28, 30, 66, 74)
120. *Reophax aduncus* Brady [181]: East China Sea. (28, 30)
121. *R. advenus* Cushman: Southern Yellow Sea; East China Sea; Nansha (=Spratly Islands). (28, 30)
122. *R. agglutinatus* Cushman: Southern Yellow Sea; East China Sea; South China Sea. (28, 30)
123. *R. apiculatus* Zheng: East China Sea. (28, 30)
124. *R. arayaensis* Bermudez & Seiglie: South China Sea. (30)
125. *R. armatus* Goës: South China Sea. (30)
126. *R. atlanticus* (Cushman): Nansha (=Spratly Islands). (66)
127. *R. barwonensis* Collins: East China Sea. (28, 30)
128. *R. bermudezi* Hofker: Nansha (=Spratly Islands). (28, 30)
129. *R. bilocularis* Flint: Bohai; Yellow Sea; East China Sea; South China Sea; Nansha (=Spratly Islands). (28, 30, 75)
130. *R. bradyi* Brönnimann & Whittaker: Southern Yellow Sea; Nansha (=Spratly Islands). (28, 30)
131. *R. brevis* Parr: East China Sea; South China Sea. (28, 30).
132. *R. capitatus* Zheng: East China Sea; South China Sea. (28, 30)
133. *R. caribensis* Seiglie & Bermudez: Nansha (=Spratly Islands). (30)
134. *R. catenulatus* Cushman: East China Sea; South China Sea. (28, 30)
135. *R. communis* Lacroix: Bohai; northern Yellow Sea. (30, 75)
136. *R. curtus* Cushman: Bohai; Yellow Sea; East China Sea; South China Sea. (19, 28, 30)
137. *R. davepopei* Smith: East China Sea. (28, 30)
138. *R. davisi* Parr: East China Sea; South China Sea. (28, 30, 66)
139. *R. dentaliniformis* Brady: Southern Yellow Sea; East China Sea. (28, 30)
140. *R. depressus* Natland: Southern Yellow Sea; East China Sea; South China Sea. (28, 30)
141. *R. distans* Brady [118]: East China Sea; South China Sea. (28, 30, 66, 75)
142. *R. donhaiensis* L. Zheng & Zhang: East China Sea. (19)
143. *R. enormis* Hada: South China Sea. (30)
144. *R. excentricus* Cushman: Bohai; South China Sea; Kashi Basin (Xinjiang) (Lower Tertiary). (28, 30, 38)
145. *R. eximius* Zheng: East China Sea. (28, 30)
146. *R. fusiformis* (Williamson): South China Sea. (30)
147. *R. guttifer* Brady: East China Sea; South China Sea. (28, 30)
148. *R. hempsteadensis* Harris & Jobe: East China Sea. (28, 30)
149. *R. hispidulus* Cushman: East China Sea. (28, 30)
150. *R. insectus* Goës: East China Sea. (28, 30)
151. *R. irregularis* Parker: East China Sea. (28, 30)
152. *R. littoralis* Lacroix: Bohai; Xisha (= Paracel Islands). (26, 30)
153. *R. longicollaris* Zheng: Yellow Sea; East China Sea. (28, 30)
154. *R. micaceous* Earland: Northern Yellow Sea. (30, 75)
155. *R. miculatus* Zheng: East China Sea. (28, 30)
156. *R. minimus* Zheng: East China Sea. (28, 30)
157. *R. moniliforme* Sidall: East China Sea; South China Sea. (28, 30)
158. *R. nodulosus* (Brady) [189]: South China Sea. (30, 66, 74)
159. *R. obscuratus* Zheng: East China Sea. (28, 30)
160. *R. orientalis* Zheng: Southern Yellow Sea; East China sea; South China Sea. (28, 30)
161. *R. pauciloculatus* (Rhumbler): Southern Yellow Sea; East China Sea; South China Sea. (28, 30)
162. *R. paucus* Hada: Southern Yellow Sea; East China Sea; South China Sea. (28, 30)
163. *R. pilulifer* Brady: East China Sea; South China Sea. (28, 30)
164. *R. pisiformis* Zheng: East China Sea; South China Sea. (28, 30)
165. *R. pseudobacillaris* Cushman: South China Sea. (28, 30, 74)
166. *R. pseudodistans* Cushman: Nansha (=Spratly Islands). (30)
167. *R. pulchrus* Zheng: East China Sea; South China Sea. (28, 30)
168. *R. pyrifera* Rhumbler: Nansha (=Spratly Islands). (30)
169. *R. regularis* Höglund: Yellow Sea; East China Sea; Taiwan Strait; South China Sea. (2, 28, 30, 50)
170. *R. rostratus* Höglund: Bohai; northern Yellow Sea; East China Sea. (28, 30)
171. *R. scorpiurus* Montfort: Bohai; Yellow Sea; East China Sea; South China Sea. (8, 19, 30, 48, 66, 75)
172. *R. spiculifer* Brady: East China Sea. (28, 30)
173. *R. subcapitatus* Zheng: Southern Yellow Sea. (30)
174. *R. subdentaliniformis* Parr: East China Sea; South China Sea. (28, 30)
175. *R. subfusiformis* Earland: Bohai; Yellow Sea; East China Sea; South China Sea. (28, 30)
176. *R. tappuensis* Asano: Bohai. (30, 69)
177. *R. tenuis* Parr: East China Sea; Nansha (=Spratly Islands). (28, 30)
178. *R. torquiformis* Zheng: East China Sea. (28, 30)
179. *R. turbo* Goës: Nansha (=Spratly Islands). (66)
180. *R. tubulus* Zheng: East China Sea; South China Sea. (28, 30)
181. *Subreophax aduncus* Brady: East China Sea. (28, 30)
182. *Cuneata arctica* (Brady): Bohai; northern Yellow Sea. (28, 30, 75)
183. *Sulcophax palustris* Warren: Okinawa Trouh. (45)

184. *Archimerismus arenaceus* (Cushman): East China Sea; South China Sea. (28, 30)
185. *Hormosina globulifera* (Brady): South China Sea. (30, 66)
186. *H. monile* Brady: East China Sea; Nansha (=Spratly Islands). (28, 30, 66)
187. *H. normanni* Brady: South China Sea. (30, 62)
188. *H. spiculifera* Hofker: East China Sea; South China Sea. (28, 30)
189. *Pseudonodosinella nodulosus* (Brady): South China Sea. (30, 66, 74)
190. *Reophanus gracilis* (Earland): South China Sea. (30)
191. *R. oviculus* (Brady): South China Sea. (30)
192. *Nodosinum gaussicum* (Rhumbler): South China Sea. (30)
193. *N. nodulosus* (Brady) [189]: East China Sea; South China Sea. (28, 30, 74)

LITUOLACEA
HAPLOPHRAGMOIDIDAE

194. *Buzasina ringens* (Brady): East China Sea; South China Sea. (28, 30, 48, 66)
195. *Cribrostomoides anomalinoides* (Rhumbler): East China Sea. (28, 30)
196. *C. bradyi* Cushman: Okinawa Trough (Quaternary). (45)
197. *C. crassimargo* (Norman) [234]: East China Sea. (28, 30)
198. *C. nitidum* (Goës): East China Sea; South China Sea. (28, 30, 66)
199. *C. pseudocanariensis* Zheng: Bohai; Yellow Sea; East China Sea; South China Sea. (28, 30, 71)
200. *C. ringens* (Brady) [194]: East China Sea; South China Sea. (28, 30, 48, 66)
201. *C. robusta* (Cushman & McCulloch): Bohai; Yellow Sea. (30)
202. *C. scitulum* (Brady): East China Sea. (28, 30)
203. *C. soldani* (Earland): East China Sea; South China Sea. (28, 30)
204. *C. spiculotesta* Zheng: Xisha (=Paracel Islands); Zhongsha (=Mcclesfield Bank). (26, 27)
205. *C. subglobosum* (G. O. Sars): East China Sea; South China Sea. (28, 30, 66)
206. *C. tenuis* (Cushman): East China Sea. (28, 30)
207. *C. turgimentum* Zheng: East China Sea. (28, 30)
208. *C. weddellensis* (Earland): East China Sea; South China Sea. (28, 30)
209. *C. wiesneri* (Parr) [355]: Yellow Sea; East China Sea; South China Sea. (28, 30)
210. *C. yeni* Chang: Taiwan (Oligocene). (44)
211. *Haplophragmoides applanata* Wang, Min, & Bian: Southern Yellow Sea. (13)
212. *H. australensis* Albani: East China Sea. (28, 30)
213. *H. bradyi* (Robertson): East China Sea; Nansha (=Spratly Islands). (19, 66)
214. *H. breviculus* Krasheninnikov: East China Sea continental shelf (Eocene). (3)
215. *H. canariensis* (d'Orbigny): Yellow Sea; East China Sea; Okinawa Trough; Jiangsu (Quaternary). (9, 28, 30, 31, 36, 44, 64)
216. *H. carinatum* Cushman & Renz: East China Sea continental shelf (Oligocene). (3)
217. *H. chilenum* Todd & Kniker: East China Sea continental shelf (Eocene). (3)
218. *H. dibollensis* Cushman & Applin: East China Sea continental shelf (Eocene). (3)
219. *H. emaciatum* (Brady): Northern Yellow Sea. (30)
220. *H. grandiformis* Cushman: East China Sea; South China Sea. (28, 30, 67)
221. *H. hetha* Berry: East China Sea continental shelf (Eocene). (3)
222. *H. lingfengensis* Wang, He, & Lu: East China Sea continental shelf (Paleocene). (3)
223. *H. membranaceum* Höglund: Northern Yellow Sea. (28, 30)
224. *H. minimum* Zheng: East China Sea. (28, 30)
225. *H. neobradyi* Uchio: East China Sea. (28, 30)
226. *H. praecarinatum* Wang, He, & Lu: East China Sea continental shelf (Eocene). (3)
227. *H. rugosa* Cushman & Waters: East China Sea continental shelf (Eocene). (3)
228. *H. shikiyamaensis* Asano & Murata: East China Sea continental shelf (Eocene). (3)
229. *H. shenzhenensis* Sun: Shenzhen Bay (Guangdong). (7)
230. *H. sphaeriloculum* Cushman: East China Sea; South China Sea. (28, 30, 41)
231. *H. taiwanensis* Nakamura: Northern Taiwan. (44)
232. *H. trullisatum* (Brady): East China Sea; South China Sea. (28, 30, 66)
233. *Labrospira columbiensis* (Cushman): South China Sea. (30)
234. *L. crassimargo* (Norman): East China Sea. (28, 30)
235. *L. evolutum* Cushman & McCulloch: Nansha (=Spratley Islands). (30)
236. *L. kosterensis* Höglund: South China Sea. (30)

DISCAMMINIDAE

237. *Ammoscalaria agrestis* (Cushman & Applin): Bohai; northern and southern Yellow Sea. (30)
238. *A. fontinense* (Terquem): South China Sea. (30, 74)
239. *A. pseudospiralis* (Williamson): Bohai; Yellow Sea; East China Sea; South China Sea. (28, 30, 66, 74, 75)
240. *A. tenuimargo* (Brady): East China Sea; South China Sea. (28, 30, 66)
241. *Discammina compressa* (Goës): East China Sea; South China Sea. (28, 30)

242. *Glaphyrammina americanus* (Cushman): Southern Yellow Sea; East China Sea. (30)

SPHAERAMMINIDAE

243. *Canepaia brasiliensis* Boltovskoy: South China Sea. (30)

LITUOTUBIDAE

244. *Lituotuba lituiformis* (Brady): East China Sea; South China Sea. (28, 30, 66)
245. *Trochamminoides coronatum* (Brady): East China Sea; South China Sea. (28, 30)
246. *T. proteus* (Karrer): East China Sea; South China Sea. (28, 30)

LITUOLIDAE

247. *Ammobaculites agglutinans* (d'Orbigny): Yellow Sea; Taiwan Strait; South China Sea. (28, 30, 50, 66, 67)
248. *A. akabiraensis* Asano: East China Sea continental shelf (Middle Eocene). (3)
249. *A. calcareus* (Brady): East China Sea; Nansha (=Spratly Islands). (28, 30, 66)
250. *A. catenulatus* Cushman & McCulloch: Southern Yellow Sea; East China Sea. (28, 30, 75)
251. *A. crassaformis* Zheng: East China Sea; South China Sea. (28, 30)
252. *A. cyclindricus* Cushman: Taiwan (Upper Tertiary). (44)
253. *A. exiguus* Cushman & Brönnimann: Southern Yellow Sea; South China Sea. (30, 65, 74)
254. *A. filiformis* Earland: Bohai. (30)
255. *A. foliaceus* (Brady) [267]: Hong Kong. (62)
256. *A. formosensis* Nakamura: Bohai; Yellow Sea; East China Sea; South China Sea; Taiwan (Tertiary). (5, 7, 11, 30, 44, 74, 75)
257. *A. hayasakai* Ishizaki: Taiwan (Upper Tertiary). (44)
258. *A. huanghaiensis* P. Wang: Bohai; Yellow Sea; East China Sea. (13, 21, 28, 31).
259. *A. josephi* Acosta: East China Sea. (28, 30)
260. *A. reophaciformis* Cushman [606]: Xuwen (Guangdong) (Pliocene). (33)
261. *A. robusta* Zheng: South China Sea. (30)
262. *Ammomarginulina ensis* Weisner: Northern Yellow Sea. (30)
263. *A. tenerissima* Stschedrina: Bohai; Yellow Sea. (30, 65, 75)
264. *Ammotium minutum* Zheng: South China Sea. (30)
265. *A. palustre* Warren: Xisha (=Paracel Islands). (26)
266. *A. stenostomum* Wang, Min, & Bian: Southern Yellow Sea. (13)
267. *Eratidus foliaceus* (Brady): Hong Kong. (62)
268. *Lituola hispida* Zheng: East China Sea; South China Sea. (28, 30, 66)

PLACOPSILINIDAE

269. *Placopsilina bradyi* Cushman & McCulloch: East China Sea; South China Sea. (28, 30)

HAPLOPHRAGMIACEA
AMMOSPHAEROIDINIDAE

270. *Adercotryma glomerata* (Brady): South China Sea. (30, 74)
271. *Ammosphaeroidina sphaeroidiniformis* (Brady): East China Sea; Xisha (=Paracel Islands). (25, 28, 30)
272. *Cystammina pauciloculata* (Brady): East China Sea; South China Sea. (28, 30, 66)
273. *C. spiculifera* Zheng: East China Sea. (28, 30)
274. *Saccaminoides subcarpathicus* Wang, He, & Lu: East China Sea continental shelf (Oligocene). (3)
275. *Recurvoides contortus* Earland: Southern Yellow Sea; East China Sea; South China Sea. (28, 30)
276. *R. crassus* Zheng: East China Sea. (28, 30)
277. *R. gigas* Zheng: East China Sea; South China Sea. (28, 30)
278. *R. laevigatum* Höglund: Southern Yellow Sea; East China Sea; South China Sea. (28, 30)
279. *R. trochamminiformis* Saidova: East China Sea. (28, 30)
280. *R. turbinatus* (Brady): Central South China Sea. (42)
281. *Thalmannammina parkerae* (Uchio): East China Sea. (28, 30)

HADDONIIDAE

282. *Haddonia torresiensis* Chapman: Xisha (=Paracel Islands); Zhongsha (=Macclesfield Bank); Nansha (=Spratly Islands). (26, 27, 30)

CYCLAMMINIDAE

283. *Alveolophragmium orbiculatum* Stschedrina: East China Sea. (19)
284. *A. wiesneri* (Parr) [356]: East China Sea; South China Sea. (28, 30)
285. *A. pauciloculata* (Brady) [272]: East China Sea; South China Sea. (28, 30, 66)
286. *A. ringens* (Brady) [194]: East China Sea; South China Sea. (28, 30, 48, 66)
287. *A. subglobosum* (G. O. Sars) [205]: East China Sea; South China Sea. (28, 30, 66)
288. *Reticulophragmium reticulatum* (Zheng): East China Sea; South China Sea. (28, 30, 74)
289. *R. sintikuensis* (Nakamura): Taipei (Taiwan). (44)
290. *Cyclammina apenninica* Emiliani: East China Sea continental shelf (Eocene). (3)
291. *C. asanoi* Takayanagi: East China Sea continental shelf (Paleocene). (3)

292. *C. cancellata* Brady: East China Sea. (28, 30)
293. *C. compressa* Cushman: East China Sea; South China Sea; Taiwan (Tertiary). (28, 30, 44)
294. *C. pusilla* Brady: East China Sea; South China Sea. (28, 30, 74)
295. *C. reticulata* Zheng [288]: East China Sea; South China Sea. (28, 30, 74)
296. *C. samanica* Berry: East China Sea continental shelf (Paleocene). (3)
297. *C. tani* Ishizaki: Taiwan (Tertiary). (44, 46)
298. *C. trullisata* (Brady) [232]: East China Sea; South China Sea. (28, 30, 66)

SPIROPLECTAMMINACEA
SPIROPLECTAMMINIDAE

299. *Bolivinopsis bulbosa* (Cushman): East China Sea. (19, 64)
300. *B. elongata* Zheng: East China Sea. (28, 30)
301. *Spiroplectammina adamsi* Lalicker: Kashi area (Xinjiang) (Lower Tertiary). (38)
302. *S. atrata* (Cushman): Yellow Sea; East China Sea; Taiwan Strait. (64, 65)
303. *S. biformis* (Parker & Jones). Northern Yellow Sea. (30, 75)
304. *S. carinata* Subbotina: Kashi Basin (Xinjiang) (Lower Tertiary). (38)
305. *S. compta* (Finlay): East China Sea continental shelf (Paleocene). (3)
306. *S. cylindroides* He & Lin: Leizhou (Guangdong) (Pliocene). (33)
307. *S. esnaensis* LeRoy: Kashi area (Xinjiang) (Tertiary). (3, 38, 46)
308. *S. fistulosa* (Brady) [323]: East China Sea; South China Sea. (19, 28, 30, 66)
309. *S. floridana* Cushman [331]: South China Sea; Zhongsha (=Macclesfield Bank). (27, 30, 67)
310. *S. foliosa* Hao & Zeng: Kashi Basin (Xinjiang) (Lower Tertiary). (38)
311. *S. howei* Stuckey: Kashi Basin (Xinjiang) (Lower Tertiary). (38)
312. *S. mexiaensis* Lalicker: East China Sea continental shelf (Paleocene). (3)
313. *S. monetalis* Bykova: Kashi Basin (Xinjiang) (Lower Tertiary). (38, 46)
314. *S. nuttalli* Lalicker: Kashi Basin (Xinjiang) (Lower Tertiary). (38)
315. *S. phoxa* Hao & Zeng: Kashi Basin (Xinjiang) (Lower Tertiary). (38)
316. *S. plummerae* Cushman: East China Sea; East China Sea cotinental shlef (Paleocene). (3, 19)
317. *S. pseudocarinata* (Cushman) [326]: East China Sea; South China Sea. (28, 30, 67)
318. *S. sicula* Hao & Zeng: Kashi Basin (Xinjiang) (Lower Tertiary). (38)
319. *S. taiwanica* Chang: Taiwan (Miocene). (44)
320. *S. typica* Lacroix: Northern Yellow Sea. (30, 75)
321. *S. wrightii* (Silvestri) [322]: Southern Yellow Sea; East China Sea; Taiwan Strait; South China Sea. (50, 66)
322. *Spiroplectinella wrightii* (Sijvestri): Southern Yellow Sea; East China Sea; South China Sea. (28, 30, 74)
323. *Spirorutilis fistulosa* (Brady): East China Sea; South China Sea. (19, 28, 30, 66)
324. *S. kerimbaensis* (Said): Southern Yellow Sea; East China Sea; South China Sea. (28, 30, 66)
325. *S. marielensis* (Lalicker & Bermudez): Southern Yellow Sea; East China Sea; South China Sea. (28, 30)
326. *S. pseudocarinata* (Cushman): East China Sea; South China Sea. (28, 30, 48, 66, 74)
327. *S. wrightii* (Silvestri) [322]: Southern Yellow Sea; East China Sea; South China Sea. (28, 30, 66, 74)
328. *Vulvulina arenacea* (Bagg): East China Sea. (28, 30)
329. *V. pennatula* (Batsch): Xisha (=Paracel Islands); Zhongsha (=Macclesfield Bank). (48)
330. *V. sinensis* Zheng: East China Sea; South China Sea. (28, 30, 74)
331. *Spirotextularia floridana* (Cushman): South China Sea; Zhongsha (=Macclesfield Bank). (27, 30, 67)
332. *S. ornatissima* (Said): South China Sea. (30)
333. *Morulaeplecta bulbosa* Höglund: Bohai; northern Yellow Sea; Nansha (=Spratly Islands). (30)
334. *M. inflata* Zheng: Zhongsha (=Macclesfield Bank). (27, 30)
335. *Bimonilina sinensis* Zheng: East China Sea. (28, 30)

PSEUDOBOLIVINIDAE

336. *Pseudobolivina antarctica* (Wiesner): Bohai; northern Yellow Sea; Nansha (=Spratly Islands). (30, 75)
337. *P. brevis* Zheng: Xisha (=Paracel Islands); Zhongsha (=Macclesfield Bank); Nansha (=Spratly Islands). (26, 30, 48)
338. *P. nasostoma* Zheng: East China Sea; South China Sea. (28, 30)
339. *P. torquata* (Parker): Bohai; northern Yellow Sea. (30, 75)

NOURIIDAE

340. *Nouria armata* Collins: Southern Yellow Sea; East China Sea; South China Sea. (28, 30)
341. *N. atlantica* (Cushman): East China Sea; Nansha (=Spratly Islands). (28, 30)
342. *N. carinata* Zheng: Nansha (=Spratly Islands). (30)
343. *N. foliacea* Zheng: East China Sea. (28, 30)
344. *N. gracilenta* Zheng: East China Sea; South China Sea. (28, 30)
345. *N. harrisii* Heron-Allen & Earland: East China Sea; South China Sea. (28, 30)
346. *N. polymorphinoides* Heron-Allen & Earland: Southern Yellow Sea; East China Sea; South China Sea. (28, 30)

347. *N. rhombiformis* Zheng: Xisha (=Paracel Islands); Nansha (=Spratly Islands). (28, 30)
348. *N. sinensis* Zheng: Northern Yellow Sea. (30)
349. *N. tenuis* Hada: Southern Yellow Sea. (19, 30)

TROCHAMMINACEA
TROCHAMMINIDAE

350. *Ammoglobigerina globigeriniformis* (Parker & Jones): East China Sea; South China Sea. (19, 28, 30, 66, 74)
351. *A. globulosa* (Cushman): East China Sea; South China Sea. (28, 30, 74)
352. *Paratrochammina charlottensis* (Cushman): Bohai; Yellow Sea; East China Sea; South China Sea. (28, 30, 75)
353. *P. simplissima* (Cushman & McCulloch): East China Sea. (28, 30)
354. *Portatrochammina eltaninae* Echols: East China Sea; Nansha (=Spratly Islands). (28, 30)
355. *P. wiesneri* (Parr): Yellow Sea; East China Sea. (28, 30)
356. *Tritaxis conica* (Parker & Jones): South China Sea. (30)
357. *T. fusca* (Williamson): East China Sea; South China Sea. (28, 30)
358. *Trochammina boltovskoyi* Brönnimann: Bohai; Yellow Sea; East China Sea. (30)
359. *T. carinata* Cushman & McCulloch: Northern Yellow Sea; Xisha (=Paracel Islands); Zhongsha (=Macclesfield Bank). (26, 27, 30)
360. *T. challengeri* Hedley, Hurdle & Burdett: East China Sea. (28, 30)
361. *T. charllottensis* Cushman [352]: Bohai; Yellow Sea; East China Sea; South China Sea. (28, 30, 75)
362. *T. globorotaliformis* Zheng: East China Sea; South China Sea. (28, 30)
363. *T. globigeriniformis* (Parker & Jones) [350]: East China Sea; South China Sea. (19, 28, 30, 66, 74)
364. *T. globulosa* Cushman [351]: East China Sea; South China Sea. (28, 30, 74)
365. *T. hadai* Uchio: Bohai; Yellow Sea. (30, 65, 75)
366. *T. inflata* (Montagu): Bohai; Yellow Sea; East China Sea; Jiangsu (Quaternary). (8, 21, 29, 30, 35, 37, 49, 62)
367. *T. japonica* Ishiwada: Bohai; northern Yellow Sea. (30)
368. *T. macrescens* Brady: Shandong (Quaternary). (29, 30)
369. *T. malovensis* Heron-Allen & Earland: Bohai. (30)
370. *T. minuta* Stschedrina: Yellow Sea. (65)
371. *T. nobensis* Asano: East China Sea. (28, 30)
372. *T. ochracea* (Williamson) [398]: Northern Yellow Sea; South China Sea. (30, 65)
373. *T. ovata* Stschedrina: Southern Yellow Sea. (65)
374. *T. pygmaea* Höglund: South China Sea. (30)
375. *T. quadriloba* Höglund: East China Sea; Xisha (=Paracel Islands); Nansha (=Spratly Islands). (28, 30)
376. *T. rotaliformis* Wright: Northern Yellow Sea. (30)
377. *T. squamata* Jones & Parker: Bohai; East China Sea. (12, 19, 30, 31)
378. *T. trapeziformis* Zheng: Xisha (=Paracel Islands). (26, 30)
379. *T. tricamerata* Earland: South China Sea. (30)
380. *T. triloculina* Wang, He, & Lu: East China Sea continental shelf (Eocene). (3)
381. *T. vesicularis* Goës: East China Sea. (28, 30)
382. *T. wiesneri* Parr: Yellow Sea; East China Sea; South China Sea. (28, 30)
383. *T. xishaensis* Zheng: Xisha (=Paracel Islands). (26, 30)
384. *Trochamminopsis globulosa* Cushman [351]: East China Sea; South China Sea. (28, 30)
385. *T. pusilla* (Höglund): South China Sea. (30)
386. *Polskiammina asiatica* (Polski): Bohai; Yellow Sea; East China Sea; South China Sea. (19, 28, 29, 30, 31, 65, 74, 75)
387. *Rotaliammina carinata* Wang, He, & Lu: East China Sea continental shelf (Eocene). (3)
388. *R. chitinosa* (Collins): Xisha (=Paracel Islands); Zhongsha (=Macclesfield Bank). (26, 27, 30)
389. *Tiphotrocha convexoconcava* Stschedrina: Yellow Sea. (65)
390. *T. kellettae* (Thalmann): South China Sea; Zhongsha (=Macclesfield Bank). (27, 30)
391. *T. minuta* Zheng: Xisha (=Paracel Islands). (26)
392. *Jadammina macrescens* (Brady): Yellow Sea; East China Sea. Salt marsh. (16, 37)
393. *J. planata* He & Li: Haihe (Tianjin). (21)
394. *Arenoparrella asiatica* Polski [386]: Bohai; Yellow Sea; East China Sea; South China Sea. (19, 28, 29, 30, 31, 65, 74, 75)
395. *Trochamminula asiatica* (Polski) [386]: Bohai; Yellow Sea; East China Sea; South China Sea. (19, 28, 29, 30, 31, 65, 74, 75)
396. *T. elongata* Zheng [399]: East China Sea; Xisha (=Paracel Islands); Zhongsha (=Macclesfield Bank). (26, 27, 28, 30)
397. *T. lobatula* Zheng [400]: Xisha (=Paracel Islands); Nansha (=Spratly Islands). (26, 30)
398. *Lepidodeuterammina ochracea* (Willismson): Northern Yellow Sea; South China Sea. (30, 65)
399. *Polystomammina elongata* (Zheng): East China Sea; Xisha (=Paracel Islands); Zhongsha (=Macclesfield Bank). (30, 65)
400. *P. lobatula* (Zheng): Xisha (=Paracel Islands); Nansha (=Spratly Islands). (26, 30)

REMANEICIDAE

401. *Septotrochammina plicata* (Terquem): Xisha (=Paracel Islands). (26, 30)

VERNEUILINACEA
PROLIXOPLECTIDAE

402. *Karrerulina apicularis* (Cushman): Nansha (=Spratly Islands). (30)

403. *K. cylindrica* (Finlay): East China Sea; South China Sea. (28, 30)
404. *Prolixoplecta exilis* (Cushman): East China Sea; Nansha (=Spratly Islands). (28, 30, 62)

VERNEUILINIDAE

405. *Gaudryina aequa* Cushman: East China Sea; South China Sea. (28, 30, 66)
406. *G. arenaria* Galloway & Wissler: East China Sea continental shelf (Quaternary). (3)
407. *G. asanoi* Huang: Taiwan (Pleistocene). (44, 56)
408. *G. chileana* Todd & Kniker: East China Sea. (28, 30)
409. *G. collinsi* Cushman: East China Sea; Taiwan (Oligocene). (28, 30, 44)
410. *G. contorta* Höglund: Xisha (=Paracel Islands). (26, 27, 30)
411. *G. convexa* Cushman: Bohai; Yellow Sea; East China Sea. (28, 30)
412. *G. flintii* Cushman [613]: East China Sea; South China Sea; Nansha (=Spratly Islands). (28, 30)
413. *G. haeringensis* Cushman: East China Sea. (19)
414. *G. hastata* Parr: Zhongsha (=Macclesfield Bank); Nansha (=Spratly Islands). (27, 30, 66)
415. *G. hayasakai* Chang: Taiwan (Lower Oligocene). (44)
416. *G. inflata* Israelsky: East China Sea; South China Sea; East China Sea continental shelf (Eocene). (3, 28, 30, 74)
417. *G. karihaensis* Asano: East China Sea; South China Sea. (44)
418. *G. kokuseiensis* Ishizaki: Taiwan (Miocene). (44)
419. *G. koreaensis* Hornibrook: Xisha (=Paracel Islands); Nansha (=Spratly Islands). (25, 30, 66)
420. *G. mai* Huang: Taiwan (Tertiary). (44)
421. *G. niigataensis* Asano: East China Sea continental shelf (Quaternary). (3)
422. *G. oinomikadoi* Huang: Taiwan (Pleistocene). (44)
423. *G. pacifica* Cushman & McCulloch: East China Sea; South China Sea. (28, 30, 66, 74)
424. *G. parva* Ho & Hu: Guangdong (Pleiocene). (33)
425. *G. pliocenica* Cushman, Stewart & Stewart: East China Sea continental shelf (Quaternary and Pliocene). (3)
426. *G. pseudohayasakai* Chang: Taiwan (Miocene). (44, 46)
427. *G. pyramidata* (Cushman): East China Sea; South China Sea. (28, 30, 66)
428. *G. quadrangularis* Bagg: East China Sea; South China Sea. (28, 30, 66)
429. *G. siphonifera* (Brady) [624]: Xisha (=Paracel Islands); Zhongsha (=Macclesfield Bank); Nansha (=Spratly Islands). (4, 25, 27, 30, 44, 66)
430. *G. subglabrata* Cushman & McCulloch: South China Sea. (30)
431. *G. taiwanica* Huang: Taiwan (Pleistocene). (44, 56)
432. *G. transversaria* (Brady): Xisha (=Paracel Islands); Zhongsha (=Macclesfield Bank). (25, 27, 30, 50, 66)
433. *G. triangularis* Cushman: Northeen Yellow Sea; Nansha (=Spratly Islands). (30, 66)
434. *G. trullisata* Todd: South China Sea. (25, 30, 74)
435. *G. wrightiana* Millett: South China Sea. (42, 66, 67)
436. *Siphogaudryina huanghaiensis* Zheng: Southern Yellow Sea. (30)
437. *Verneuilina advena* Cushmn [442]: Bohai; Yellow Sea; East China Sea; South China Sea. (11, 19, 28, 30, 31, 65)

TRITAXIIDAE

438. *Tritaxia changi* Huang: Taiwan (Pliocene). (44)
439. *T. donghaiensis* Zhang: East China Sea. (19)
440. *T. orientalis* (Cushman): East China Sea. (19)

ATAXOPHRAGMIACEA
GLOBOTEXTULARIIDAE

441. *Globotextularia anceps* (Brady): East China Sea; South China Sea. (28, 30, 66)
442. *Verneuilinulla advena* (Cushman): Bohai; Yellow Sea; East China Sea; South China Sea. (11, 19, 28, 30, 31, 65)
443. *V. polita* (Collins): Southern Yellow Sea; East China Sea. (30)

TEXTULARIELLIDAE

444. *Textulariella cushmani* ten Dam & Sigal: East China Sea continental shelf (Paleocene). (3)
445. *T. simplex* Cushman: East China Sea. (28, 30)

TEXTULARIACEA
EGGERELLIDAE

446. *Dorothia arenata* Cushman: East China Sea; South China Sea. (28, 30)
447. *D. exilis* Cushman [404]: East China Sea; Nansha (=Spratly Islands). (28, 30, 62)
448. *D. nagaoi* Asano & Murata: East China Sea; East China Sea continental shelf (Paleocene). (3)
449. *D. nammalensis* (Haque): East China Sea continental shelf (Paleocene). (3)
450. *D. oxyconoides* Zheng: East China Sea; South China Sea. (28, 30, 66)
451. *D. paupercula* (Cushman): East China Sea; South China Sea. (28, 30, 48)
452. *D. pseudoturris* (Cushman): South China Sea. (30)
453. *D. retusa* (Cushman): East China Sea continental shelf (Paleocene). (30)
454. *D. scabra* (Brady): East China Sea; South China Sea. (28, 30, 66)
455. *D. scrupulosa* Zheng: East China Sea; South China Sea. (28, 30)

456. *D. tortilis* Parr: Xisha (=Paracel Islands); Zhongsha (=Macclesfield Bank). (26, 27, 30)
457. *Eggerella advena* (Cushman) [442]: Bohai; Yellow Sea; East China Sea; South China Sea. (11, 19, 28, 30, 31, 65)
458. *E. bradyi* (Cushman): East China Sea; South China Sea; Okinawa Trough (Middle Pleistocene to Holocene). (28, 30, 39, 45, 66)
459. *E. conica* Zheng: East China Sea; South China Sea. (28, 30, 66)
460. *E. decepta* Finlay: South China Sea. (66)
461. *E. nitens* (Wiesner): East China Sea; Okinawa Trough (Middle Pleistocene to Holocene). (28, 30, 45)
462. *E. polita* Collins [443]: Southern Yellow Sea; East China Sea. (28, 30)
463. *E. subconica* Parr: Nansha (=Spratly Islands). (30)
464. *Karreriella bradyi* (Cushman): East China Sea; South China Sea; East China Sea (Quaternary); South China Sea (Upper Tertiary). (19, 28, 30, 39, 45, 48, 66, 74)
465. *K. chilostoma* (Reuss): East China Sea. (19)
466. *K. cylindrica* Finlay [430]: East China Sea; South China Sea. (28, 30, 66)
467. *K. hantkeniana* Cushman: Taiwan (Oligocene). (44)
468. *K. parkerae* Uchio: East China Sea; South China Sea. (28, 30, 74)
469. *K. pupiformis* Zheng: East China Sea; South China Sea. (28, 30, 74)
470. *K. shangtaoensis* Chang: Taiwan (Miocene). (44)
471. *Martinotiella antillarum* (Cushman): Okinawa Trough (Upper Pleistocene). (45)
472. *M. bradyana* (Cushman): East China Sea; South China Sea. (28, 30, 66)
473. *M. communis* (d'Orbigny): East China Sea; South China Sea; Okinawa Trough (Upper Pleistocene); Taiwan (Tertiary). (28, 30, 44, 45, 66)
474. *M. cylindrica* (Cushman): East China Sea; South China Sea. (28, 30)
475. *M. inculta* Zheng: East China Sea; South China Sea. (28, 30, 66)
476. *M. milletti* (Cushman): East China Sea; South China Sea. (28, 30)
477. *M. minuta* Hofker: East China Sea; South China Sea. (28, 30, 74)
478. *M. nodulosa* (Cushman) [484]: Nansha (=Spratly Islands). (66)
479. *M. occidentalis* (Cushman): Southern Yellow Sea; East China Sea; South China Sea. (28, 30, 66)
480. *M. okinawaensis* Zhang: East China Sea. (19)
481. *M. primaeva* (Cushman): South China Sea; Xisha (=Paracel Islands); Zhongsha (=Macclesfield Bank). (19, 30, 48)
482. *M. victoriensis* (Cushman): Taiwan (Neogene). (44)
483. *M. wrightii* (Cushman): East China Sea; Okinawa Trough (Upper Pleistocene). (45)
484. *Multifidiella nodulosa* (Cushman): Nansha (=Spratly Islands). (66)
485. *Schenckiella victoriensis* Cushman [482]: Taiwan (Neogene). (44)
486. *Rudigaudryina inepta* (Cushman & McCulloch): South China Sea. (30)

TEXTULARIIDAE

487. *Bigenerina ammobaculitoidea* Chang: Taiwan (Miocene). (44)
488. *B. curta* Zheng: East China Sea. (28, 30)
489. *B. irregularis* Phleger & Parker: Taiwan Strait. (55)
490. *B. lytta* Lalicker & Bermudez: Xisha (=Paracel Islands); Zhongsha (=Macclesfield Bank). (48)
491. *B. nodosaria* d'Orbigny: East China Sea; Taiwan Strait; South China Sea. (19, 28, 30, 50, 59, 66, 74)
492. *B. shihtiensis* Chang: Taiwan (Miocene). (44)
493. *B. taiwanica* Nakamura: East China Sea; Taiwan Strait; South China Sea; East China Sea (Paleocene); Taiwan (Tertiary). (3, 30, 44, 59, 66)
494. *Textularia abbreviata* d'Orbigny: Southern Yellow Sea; East China Sea; South China Sea. (28, 30, 74)
495. *T. agglutinans* d'Orbigny: East China Sea; South China Sea. (28, 30, 44, 62, 66)
496. *T. akaminei* Ishizaki: Taiwan (Miocene). (44)
497. *T. alishanensis* Chang: Taiwan (Miocene). (44)
498. *T. articulata* d'Orbigny: Penghu (Taiwan). Beach sand. (51)
499. *T. astutia* Lalicker & McCulloch: Nansha (=Spratly Islands). (66)
500. *T. atrata* Cushman [302]: Shidao (Xisha=Paracel Islands) (Upper Pleistocene). (41, 64, 65)
501. *T. aura* Lalicker & McCulloch: Nansha (=Spratly Islands). (30)
502. *T. australis* Parr: South China Sea. (30)
503. *T. awazea* Finlay: East China Sea continental shelf (Quaternary). (3)
504. *T. bocki* Höglund [574]: East China Sea; South China Sea. (25, 26, 30, 74)
505. *T. bradyi* (Moller): Taizhou Bay (Zhejiang) (Quaternary). (47)
506. *T. calva* Lalicker: South China Sea. (30)
507. *T. candeiana* d'Orbigny: Southern Yellow Sea; East China Sea; South China Sea. (26, 30, 45, 62, 66, 67, 74)
508. *T. concava* (Karrer) [581]: Nansha (=Spratly Islands). (66)
509. *T. conica* d'Orbigny: Yellow Sea; East China Sea; Taiwan Strait; South China Sea; Jiangsu (Quaternary); Guangdong (Pliocene); Xisha (=Paracel Islands) (Upper Pleistocene). (19, 30, 41, 44, 62, 65, 66, 67, 74)
510. *T. contorta* Höglund: South China Sea. (30)
511. *T. corrugata* Heron-Allen & Earland: South China Sea. (30, 66)
512. *T. crenata* Zheng: Xisha (=Paracel Islands); Nansha (=Spratly Islands). (25, 30, 66)
513. *T. crassisepta* Cushman [584]: East China Sea; South China Sea; Taiwan (Tertiary). (28, 30, 44, 66)
514. *T. curtata* Zheng: East China Sea. (28, 30)

515. *T. cuneata* Hada: Shandong (Quaternary). (29)
516. *T. dupla* Todd: South China Sea. (25, 30, 74)
517. *T. excavata* Cushman: Taiwan (Miocene). (44)
518. *T. farafraensis* LeRoy: Kashi Basin (Xinjiang) (Lower Tertiary). (38)
519. *T. foliacea* Heron-Allen & Earland: Bohai; Yellow Sea; East China Sea; Taiwan Strait; South China Sea. (19, 22, 28, 30, 48, 50, 66, 67, 74)
520. *T. gramen* d'Orbigny: Yellow Sea; Taiwan Strait. (44, 63)
521. *T. halkyardi* Lalicker: Kashi Basin (Xinjiang) (Lower Tertiary). (38)
522. *T. hakusikeiensis* Nakamura: Taiwan (Tertiary). (44)
523. *T. hoppoensis* Nakamura: Taiwan (Tertiary); Guangdong (Pliocene). (33, 44)
524. *T. hosonoi* Ishizaki: Taiwan (Tertiary). (44)
525. *T. howei* Puri: East China Sea. (28, 30)
526. *T. intosiana* Nakamura: Taiwan (Tertiary). (44)
527. *T. kansaiensis* Ishizaki: Taiwan (Tertiary). (44, 46)
528. *T. kansireinsis* Nakamura: Taiwan (Tertiary). (44)
529. *T. kapitea* Finlay: Guangdong (Pliocene). (33)
530. *T. kerimbaensis* Said [324]: Southern Yellow Sea; East China Sea; South China Sea. (28, 30, 44, 45, 59, 62, 66)
531. *T. lata* Germeraad: South China Sea. (30)
532. *T. lateralis* Lalicker: Southern Yellow Sea; East China Sea; South China Sea. (28, 30, 48, 66)
533. *T. magallanica* Todd & Kniker: Southern Yellow Sea; East China Sea; South China Sea. (28, 30, 48)
534. *T. megaloculata* Copeland: Beijing area (Lower Quaternary); Leizhou (Guangdong) (Pliocene). (2, 23)
535. *T. magnifica* Lalicker & Bermudez: East China Sea continental shelf (Paleocene) (3).
536. *T. midwayensis* Lalicker: East China Sea continental shelf (Paleocene); Kashi Basin (Xinjiang) (Lower Tertiary). (3, 38)
537. *T. mississippiensis* Cushman: East China Sea continental shelf (Quaternary). (3)
538. *T. neorugosa* Thalmann: Xisha (=Paracel Islands); Zhongsha (=Macclesfield Bank); Nansha (=Spratly Islands). (4, 25, 27, 30, 66)
539. *T. nitens* Earland: Nansha (=Spratly Islands). (30)
540. *T. oceanica* Cushman: East China Sea; South China Sea; East China Sea (Quaternary). (3, 25, 27, 28, 30)
541. *T. orbica* Lalicker & McMulloch: East China Sea; South China Sea. (26, 27, 28, 30, 66)
542. *T. ornatissima* Said: Xisha (=Paracel Islands); Zhongsha (=Macclesfield Bank). (48)
543. *T. paragglutinans* Zheng: East China Sea; South China Sea. (28, 30, 74)
544. *T. parva* Zheng: South China Sea. (30)
545. *T. parvula* Cushman: East China Sea. (28, 30)
546. *T. peritubula* Zheng: East China Sea; South China Sea. (28, 30, 66)
547. *T. pitmani* McLean: Guangdong (Pliocene). (33)
548. *T. porrecta* Brady: Southern Yellow Sea; East China Sea; South China Sea. (28, 30, 74)
549. *T. pseudocarinata* Cushman [326]: East China Sea; South China Sea. (19, 28, 30, 66, 74)
550. *T. pseudogramen* Chapman & Parr: East China Sea; Taiwan Strait; South China Sea. (28, 30, 50, 74)
551. *T. pseudokansaiensis* Chang: Penhu area (Taiwan) (Tertiary); Taiwan (Miocene). (44, 46)
552. *T. pseudosolita* Zheng: East China Sea. (28, 30)
553. *T. pseudotrochus* Cushman: Bohai; Yellow Sea; East China Sea; South China Sea. (28, 30, 48, 66, 74)
554. *T. pseudoturris* Cushman: South China Sea. (30)
555. *T. rokuzyukeiensis* Nakamura: Taiwan (Tertiary). (44)
556. *T. sagittula* Defrance [322]: East China Sea; South China Sea; Okinawa Trough (Upper Pleistocene). (19, 44, 48)
557. *T. scrupula* Lalicker & McCulloch: East China Sea; South China Sea. (27, 28, 30, 48, 66)
558. *T. secasensis* Lalicker & McCulloch: Yellow Sea; East China Sea; Taiwan Strait; Zhongsha (=Macclesfield Bank). (45, 48, 64, 65)
559. *T. semialata* Cushman: East China Sea; Xisha (=Paracel Islands); Zhongsha (=Macclesfield Bank). (28, 30, 48)
560. *T. sineiensis* Nakamura: Taiwan (Tertiary). (44)
561. *T. sintikuensis* Nakamura: Taiwan Strait; Taiwan (Tertiary). (44, 55)
562. *T. stricta* Cushman: Taiwan. (44)
563. *T. subantarctica* Vella: Bohai; Yellow Sea; South China Sea. (30)
564. *T. suttonensis* Lalicker: East China Sea. (28, 30)
565. *T. tainanensis* Nakamura: Taiwan (Miocene). (44)
566. *T. tenuissima* Earland: Nansha (=Spratly Islands). (30)
567. *T. transversaria* Brady [432]: Guangdong (Pliocene); Xisha (=Paracel Islands); Zhongsha (=Macclesfield Bank). (25, 27, 30, 33, 44, 50, 66)
568. *T. truncata* Höglund: East China Sea; South China Sea. (28, 30, 48)
569. *T. truncatiformis* Zheng: East China Sea; South China Sea. (28, 30)
570. *T. tubulosa* Zheng: Zhongsha (=Macclesfield Bank). (27)
571. *T. valentula* L. Zheng & Zhang: East China Sea. (19)
572. *T. vola* Lalicker & McColloch: Yellow Sea. (65)
573. *T. zeaggluta* Finlay: East China Sea; South China Sea. (28, 30)
574. *Textilina bocki* Höglund: East China Sea; South China Sea. (28, 30, 74)
575. *T. crassaformis* Zheng: East China Sea; South China Sea. (28, 30, 66)
576. *T. lythostrota* (Schwager): East China Sea. (28, 30)
577. *T. semialata* (Cushman): East China Sea. (28, 30)
578. *Siphoscutula leroyi* Loeblich & Tappan: East China Sea; South China Sea. (28, 30, 66)
579. *S. pacifica* (LeRoy): East China Sea; Nansha (=Sprtly Islands). (28, 30, 66)
580. *Siphotextularia carinata* Zheng: East China Sea. (28, 30)
581. *S. concava* (Karrer): Nansha (=Spratly Islands). (30)
582. *S. cordis* Hornibrook: Nansha (=Spratly Islands). (30)
583. *S. crassaformis* Zheng: East China Sea. (28, 30)

584. *S. crassisepta* (Cushman): East China Sea; South China Sea; Taiwan (Tertiary). (28, 30, 44, 66)
585. *S. crispata* (Brady) [1907]: Xisha (=Paracel Islands); Nansha (=Spratly Islands). (25, 30, 48, 66)
586. *S. curta* (Cushman): East China Sea. (28, 30)
587. *S. differens* McCulloch: Nansha (=Spratly Islands). (30)
588. *S. flintii* (Cushman): East China Sea; South China Sea; Okinawa Trough (Pleistocene to Holocene). (28, 30, 45)
589. *S. foliosa* Zheng: East China Sea; South China Sea. (28, 30, 66)
590. *S. glabrata* Zheng: East China Sea. (28, 30)
591. *S. heterostoma* (Fornasini): East China Sea; South China Sea. (28, 30, 66)
592. *S. masudai* Asano: East China Sea continental shelf (Quaternary). (3)
593. *S. mestayerae* Vella: Southern Yellow Sea; East China Sea; South China Sea. (28, 30)
594. *S. miniacea* Cai & Tu: Xisha (=Paracel Islands); Zhongsha (=Macclesfield Bank). (48)
595. *S. miocenica* Cushman: East China Sea; Okinawa Trough (Upper Pleistocene to Holocene). (28, 30, 39)
596. *S. obesa* Parr [605]: East China Sea; Okinawa Trough. (28, 30, 39)
597. *S. pacifica* LeRoy [579]: East China Sea; Nansha (=Spratly Islands). (28, 30, 66)
598. *S. philippinensis* (Keijzer): East China Sea. (28, 30)
599. *S. pseudoconcava* Zheng: East China Sea; Nansha (=Spratly Islands). (28, 30, 66)
600. *S. pulchra* Zheng: East China Sea. (28, 30)
601. *S. saulcyana* (d'Orbigny): Taiwan Strait. (50)
602. *S. subplana* (Cushman): East China Sea; South China Sea. (28, 30, 66)
603. *S. subplanoides* Zheng: East China Sea; South China Sea. (28, 30, 66)
604. *S. wairoana* Finlay: East China Sea; South China Sea; Okinawa Trough (Middle Pleistocene). (19, 30, 33, 39)
605. *Textulina obesa* (Parr): East China Sea; Okinawa Trough (Upper Pleistocene). (28, 30, 39)
606. *Cribrobigenerina reophaciformis* (Cushman): East China Sea; South China Sea. (28, 30)
607. *C. robustiformis* Zheng: East China Sea; South China Sea. (28, 30, 66)
608. *C. taiwanica* (Nakamura) [493]: East China Sea; Taiwan Strait; South China Sea; East China Sea continental shelf (Paleocene); Taiwan (Tertiary). (3, 30, 44, 59, 66)
609. *C. textularioidea* (Göes): East China Sea; South China Sea. (28, 30, 64, 66, 74)
610. *Septotextularia rugulosa* (Cushman): Xisha (=Paracel Islands); Zhongsha (=Macclesfield Bank); Nansha (=Spratly Islands). (4, 25, 27, 28, 30)
611. *Textularioides inflata* Cushman: East China Sea. (28, 30)

PSEUDOGAUDRYINIDAE

612. *Clavulinoides szaboi* (Hantken): Taiwan (Tertiary). (44)

613. *Migros flintii* (Cushman): East China Sea; South China Sea. (28, 30)
614. *M. spiritensis* (Stelck & Wall): Kashi Basin (Xinjiang) (Lower Tertiary). (38)
615. *Pseudoclavulina gracilis* Zheng: East China Sea; South China Sea. (28, 30)
616. *P. humilis* (Brady): South China Sea. (30, 66)
617. *P. juncea* Cushman: East China Sea; South China Sea. (28, 30, 74)
618. *P. mexicana* (Cushman): South China Sea. (30)
619. *P. robusta* Zheng: Southern Yellow Sea; East China Sea; South China Sea. (28, 30)
620. *P. scabra* Cushman: Nansha (=Spratly Islands). (30)
621. *P. serventyi* (Chapman & Parr): East China Sea; South China Sea. (28, 30, 48, 66)
622. *Pseudogaudryina atlantica* (Bailey): South China Sea. (30)
623. *Plotnikovina compressa* (Cushman): South China Sea. (30)
624. *Siphoniferoides siphonifera* (Brady): Xisha (=Paracel Islands); Zhongsha (=Macclesfield Bank); Nansha (=Spratly Islands). (4, 25, 27, 30, 44, 66)

VALVULINIDAE

625. *Clavulina crustata* (Cushman): East China Sea. (28, 30)
626. *C. difformis* Brady: Xisha (=Paracel Islands); Nansha (=Spratly Islands). (42, 66)
627. *C. multicamerata* Chapman: Central South China Sea; Nansha (=Spratly Islands). (42, 66)
628. *C. pacifica* Cushman: Xisha (=Paracel Islands); Nansha (=Spratly Islands). (25, 30)
629. *Cylindroclavulina bradyi* (Cushman): East China Sea; South China Sea. (28, 30, 48, 66)
630. *C. elongata* Zheng: East China Sea; Nansha (=Spratly Islands). (28, 30)
631. *C. ovata* Zheng: East China Sea; South China Sea; Nansha (=Spratly Islands). (28, 30)
632. *Goesella rotundata* (Cushman): South China Sea. (30, 67)
633. *Valvulina davidiana* Chapman: Xisha (=Paracel Islands). (25)
634. *Siphobigenerina compressa* Zheng: Xisha (=Paracel Islands); Zhongsha (=Macclesfield Bank); Nansha (=Spratly Islands). (27, 30)
635. *Tritaxilina atlantica* Cushman: East China Sea; South China Sea. (28, 30, 66)
636. *T. caperata* (Brady): East China Sea; South China Sea. (28, 30, 48, 74)
637. *Glaucoammina trilatera* (Cushman): South China Sea. (30)

INVOLUTININA
INVOLUTINIDAE

638. *Involutina anguillae* (Höglund): Central South China Sea. (42)

639. *I. intermedia* (Höglund) [93]: East China Sea; South China Sea. (28, 30)
640. *I. tenuis* (Brady) [96]: East China Sea; South China Sea. (28, 30, 42)

PLANISPIRILLINIDAE

641. *Conicospirillinoides semidecoratus* (Heron-Allen & Earland): Xisha (=Paracel Islands). (25, 30)
642. *Planispirillina denticulata* (Brady): Xisha (=Paracel Islands). (25, 30)
643. *P. paucispira* Zheng: Xisha (=Paracel Islands). (26, 30)
644. *P. tuberculatolimbata* (Chapman): Xisha (=Paracel Islands); Zhongsha (=Macclesfield Bank). (25, 27, 30, 48, 66)

SPIRILLININA
SPIRILLINIDAE

645. *Mychostomina carinata* Zheng: Xisha (=Paracel Islands). (25, 30)
646. *M. peripora* Zheng: Xisha (=Paracel Islands). (26, 30)
647. *M. revertens* (Rhumbler): Xisha (=Paracel Islands); Nansha (=Spratly Islands). (26, 30)
648. *M. truncata* Zheng: Xisha (=Paracel Islands). (26, 30)
649. *Sejunctella laticarinina* Zheng: Xisha (=Paracel Islands). (26, 30)
650. *Spirillina compressa* Zheng: Xisha (=Paracel Islands). (26, 30)
651. *S. grosseperforata* Zheng: Xisha (=Paracel Islands). (26, 30)
652. *S. limbata* Brady: Xisha (=Paracel Islands); Zhongsha (=Macclesfield Bank). (4, 25, 30, 42)
653. *S. mediospinosa* Zheng: Zhongsha (=Macclesfield Bank). (27, 30)
654. *S. minima* Wang, Min & Bian: Hebei (Miocene). (11)
655. *S. pectinimarginata* Chapman, Parr & Collins: Xisha (=Paracel Islands); Nansha (=Spratly Islands). (26, 30)
656. *S. planoconcava* Zheng: Taiwan Strait; Zhongsha (=Macclesfield Bank). (27, 30)
657. *S. porisuturalis* Zheng: Xisha (=Paracel Islands). (26, 30)
658. *S. scalaris* Zheng: Xisha (=Paracel Islands). (26, 30)
659. *S. tuberculata* Chapman: Nansha (=Spratly Islands). (66)
660. *S. vivipara* Ehrenberg: South China Sea; Okinawa Trough (Upper Pleistocene to Holocene); Guangdong (Pliocene). (27, 30, 33, 45)
661. *Turrispirillina altispira* Zheng: Xisha (=Paracel Islands). (26, 30)

PATELLINIDAE

662. *Patellina advena* Cushman: Taiwan Strait; South China Sea. (26, 30, 50)
663. *P. altiformis* Cushman: Xisha (=Paracel Islands). (26, 30)
664. *P. corrugata* Williamson: South China Sea. (26, 30, 66)
665. *P. spinosa* Zheng: Xisha (=Paracel Islands). (26, 30)

CARTERINIDAE

666. *Carterina spiculotesta* (Carter): Zhongsha (=Macclesfield Bank); Nansha (=Spratly Islands). (27, 30)

MILIOLINA
CORNUSPIRACEA
CORNUSPIRIDAE

667. *Cornuspira carinata* (Costa): East China Sea; South China Sea. (28, 30, 66)
668. *C. crassienpta* Brady: East China Sea; Nansha (=Spratly Islands). (28, 30, 66)
669. *C. involvens* (Reuss): East China Sea; South China Sea. (28, 30, 66)
670. *C. planorbis* Schultze: East China Sea; South China Sea. (28, 30, 66)
671. *C. selseyensis* Heron-Allen & Earland: East China Sea; Nansha (=Spratly Islands). (28, 30, 66)
672. *Cyclogyra carinata* (Costa) [667]: East China Sea; South China Sea. (28, 30, 66)
673. *C. crassisepta* (Brady) [668]: East China Sea; Nansha (=Spratly Islands). (28, 30, 66)
674. *C. involvens* (Reuss) [669]: East China Sea; South China Sea. (28, 30, 66)
675. *C. planorbis* (Schultze) [670]: East China Sea; South China Sea. (28, 30, 66)
676. *C. selseyensis* (Heron-Allen & Earland) [671]: East China Sea; Nansha (=Spratly Islands). (28, 30, 60)
677. *Cornuspiroides foliaceus* (Philippi): East China Sea; Taiwan Strait; South China Sea. (28, 30, 50)
678. *C. lacunosus* (Brady): East China Sea; South China Sea. (28, 30)

FISCHERINIDAE

679. *Planispirinella exigua* (Brady): South China Sea. (30)
680. *Fischerinella helix* Heron-Allen & Earland: Nansha (=Spratly Islands). (30)
681. *F. pellucida* (Millett): South China Sea. (30)
682. *F. trochoides* Wang, He & Lu: East China Sea continental shelf (Eocene). (3)
683. *Nodobaculariella rustica* Cushman & Todd: Xisha (=Paracel Islands); Zhongsha (=Macclesfield Bank). (26, 27, 30)
684. *Vertebralina insignis* Brady: Xisha (=Paracel Islands). (26, 30)
685. *V. striata* d'Orbigny: South China Sea. (7, 25, 26, 27, 30, 66)
686. *Wiesnerella auriculata* (Egger): South China Sea; Okinawa Trough (Upper Pleistocene). (26, 27, 30, 45)

NUBECULARIIDAE

687. *Nodophthalmidium carinata* Zheng: Xisha (=Paracel Islands). (26, 30)

688. *N. compressum* (Rhumbler): Xisha (=Paracel Islands). (26, 30)

689. *N. simplex* Cushman & Todd: East China Sea; South China Sea. (28, 30, 48)

690. *N. zhongshaensis* Zheng: Zhongsha (=Macclesfield Bank). (27, 30)

691. *Nubeculina divaricata* (Brady): Nansha (=Spratly Islands). (66)

692. *Nubeculopsis queenslandica* Collins: Xisha (=Paracel Islands). (25, 30)

693. *Nubecularia lucifaga* Defrance: Xisha (=Paracel Islands); Nansha (=Spratly Islands). (25, 30)

OPHTHALMIDIIDAE

694. *Cornuloculina inconstans* (Brady): East China Sea; Nansha (=Spratly Islands). (28, 30, 66)

695. *Edentostomina cultrata* (Brady): Bohai; Yellow Sea; East China Sea; South China Sea; Shandong; Jiangsu (Quaternary); Guangdong (Pliocene). (1, 8, 28, 29, 30, 33, 49, 65, 66)

696. *E. elongata* Zheng: East China Sea; South China Sea. (28, 30)

697. *E. milletti* (Cushman) [967]: Bohai; East China Sea; South China Sea. (28, 30, 48)

698. *E. pseudodepressa* (Mangin): East China Sea; South China Sea. (28, 30, 66)

699. *E. rupertiana* (Brady) [1140]: South China Sea. (30, 66)

700. *E. vulgaris* Ho & Hu: Guangdong (Pliocene). (33)

701. *Ophthalmidium inconstans* (Brady) [694]: East China Sea; Nansha (=Spratley Islands). (28, 30, 66)

702. *O. pusillum* (Earland) [1106]: East China Sea; South China Sea; Okinawa Trough (Upper Pleistocene). (28, 30, 45, 66)

703. *O. tenuimargo* (Cushman): Okinawa Trough (Upper Pleistocene to Holocene). (45)

704. *O. tenuiseptatum* (Brady): East China Sea; South China Sea. (28, 30)

MILIOLACEA
SPIROLOCULINIDAE

705. *Cribrolinoides leizhoensis* Ho & Lu: Guangdong (Upper Pleistocene). (33)

706. *Inaequalina concavo-convexa* Zheng: South China Sea. (30)

707. *I. disparilis* (Terquem): East China Sea; South China Sea. (28, 30, 66)

708. *Nummulopyrgo paraglobulus* (Zheng): East China Sea; Nansha (=Spratly Islands). (28, 30, 66)

709. *Planispirinoides bucculentus* (Brady): East China Sea; South China Sea. (28, 30)

710. *Spiroloculina affixa* Terquem: Guangdong (Pliocene). (33)

711. *S. anderseni* Todd & Brönnimann: Yellow Sea; East China Sea; South China Sea; Guangdong (Pliocene) (3, 33)

712. *S. angulata* Cushman: Xisha (=Paracel Islands); Zhongsha (=Macclesfield Bank). (26, 30, 48)

713. *S. antillarum* d'Orbigny: Nansha (=Spratly Islands). (66)

714. *S. arenaria* Brady [1085]: Yellow Sea; East China Sea; Taiwan Strait; Taiwan (Miocene). (44, 48, 64, 67)

715. *S. bicarinata* d'Orbigny: Mt. Jolmo Lungma (=Mt. Everest) area (Tibet) (Lower Tertiary). (9)

716. *S. biformis* Stschedrina: Southern Yellow Sea. (65)

717. *S. bohaiensis* Zheng [787]: Bohai; southern Yellow Sea; East China Sea; Shandong (Quaternary). (28, 29, 30, 47)

718. *S. cavernosa* Karrer: Okinawa Trough (Holocene). (45)

719. *S. clara* Cushman: Xisha (=Paracel Islands); Zhongsha (=Macclesfield Bank). (25, 27, 30)

720. *S. communis* Cushman & Todd: Bohai; Yellow Sea; East China Sea; Taiwan Strait; South China Sea; eastern Hebei; East China Sea continental shelf (Pliocene); Okinawa Trough (Upper Pleistocene); Taizhou Bay (Zhegjiang) (Quaternary). (3, 19, 28, 30, 34, 44, 45, 47, 48, 50, 59, 65, 66)

721. *S. concava* Petri: Okinawa Trough (Upper Pleistocene). (45)

722. *S. corrugata* Cushman & Todd: Taiwan Strait; Xisha (=Paracel Islands); Zhongsha (=Macclesfield Bank); Nansha (=Spratly Islands). (4, 25, 27, 30, 50, 66)

723. *S. dentata* Cushman & Todd: Yellow Sea. (65)

724. *S. depressa* d'Orbigny: Shandong; shoreline of Bohai (Quarternary). (1, 29)

725. *S. excavata* d'Orbigny: Yellow Sea; East China Sea; South China Sea. (28, 30)

726. *S. excisa* Cushman & Todd: East China Sea; South China Sea. (4, 25, 27, 28, 30)

727. *S. elongata* d'Orbigny: Guangdong (Pliocene). (33)

728. *S. eximia* Cushman: Jiangsu; South China Sea; Taiwan (Quaternary). (8, 30, 37)

729. *S. foveolata* Egger: South China Sea. (25, 27, 30, 48)

730. *S. hadai* Thalmann: Nansha (=Spratly Islands). (66)

731. *S. henbesti* Petri: Shandong (Quaternary). (29, 30)

732. *S. huanghaiensis* Zheng: Taizhou Bay shoreline (Zhejiang) (Upper Tertiary). (47)

733. *S. indica* Cushman & Todd: Shandong (Quaternary). (29, 30)

734. *S. jucunda* Cushman & Todd: East China Sea. (19)

735. *S. laevigata* Cushman & Todd [784]: Bohai; East China Sea; South China Sea; Shandong (Quaternary). (19, 28, 29, 30, 47, 49)

736. *S. limbata* d'Orbigny: Taiwan. (44)

737. *S. lucida* Cushman & Todd: Southern Yellow Sea; Taiwan Strait; Jiangsu (Quaternary). (5, 8, 13, 50)

738. *S. manifesta* Cushman & Todd: East China Sea. (28, 30)

739. *S. norvegica* Cushman & Todd: Shandong; Taizhou Bay (Zhejiang); Jiangsu (Quaternary). (8, 27, 30, 47)

740. *S. planulata* (Lamarck): Shoreline of Bohai (Quaternary). (1)

741. *S. planoconvexa* Zheng: Xisha (=Paracel Islands). (26, 30)

742. *S. pauciloculata* He & Lin: Guangdong (Pliocene). (33)

743. *S. pulchra* Stschedrina: Southern Yellow Sea. (65)

744. *S. regularis* Cushman & Todd: Xisha (=Paracel Islands); Zhongsha (=Macclesfield Bank). (48, 66)
745. *S. robusta* Brady: East China Sea; South China Sea. (28, 30)
746. *S. rotunda* d'Orbigny: Xisha (=Paracel Islands); Zhongsha (=Macclesfield Bank). (48)
747. *S. scita* Cushman & Todd: South China Sea. (4, 26, 27, 30, 48)
748. *S. scrobiculata* Cushman: Southern Yellow Sea; Xisha (=Paracel Islands). (4, 65)
749. *S. soldani* Fornasini: Bohai; southern Yellow Sea; Taizhou Bay (Zhejiang) (Quaternary). (8, 47, 49)
750. *S. stabilis* Zheng: Bohai; Yellow Sea; East China Sea; South China Sea. (28, 30)
751. *S. terquemiana* Fornasini: Shandong; Bohai (Quaternary). (1, 29, 30)
752. *S. valida* Cushman: South China Sea. (67)

HAUERINIDAE

753. *Agglutinella reinemundi* (Haque): East China Sea. (28, 30)
754. *Ammomassilina alevoliniformis* (Millett): Okinawa Trough (Upper Pleistocene to Holocene); central South China Sea. (42, 45)
755. *A. arenarium* (Nakamura): Taiwan (Tertiary). (44)
756. *Dentostomina agglutinans* (d'Orbigny): Taiwan Strait. (59)
757. *Pseudoflintina bulbosa* Zheng: Yellow Sea; East China Sea. (28, 30, 66)
758. *P. triquetra* (Brady): East China Sea; Nansha (=Spratly Islands); Taiwan (Miocene). (28, 30, 44, 66)
759. *Schlumbergerina alveoliniformis* (Brady): Mt. Jolmo Lungma (=Mt. Everest) area (Tibet) (Lower Tertiary). (9)
760. *S. occidentalis* Cushman: Xisha (=Paracel Islands); Nansha (=Spratly Islands). (25, 30)
761. *Siphonaperta agglutinans* (d'Orbigny): East China Sea; South China Sea (Pliocene to Holocene). (3, 19, 33, 66)
762. *S. agglutinata* (Cushman): Dongshan Bay (Fujian). (5)
763. *S. ammophila* (Parr): East China Sea. (28, 30)
764. *S. crassatina* (Brady): East China Sea; Nansha (=Spratly Islands). (28, 30, 66)
765. *S. formosana* (Nakamura): Taiwan Strait; Taiwan (Tertiary). (44, 56, 59)
766. *S. inculta* Zheng: East China Sea. (28, 30)
767. *S. macbeathi* Vella. Guangdong (Pliocene). (33)
768. *S. petrophila* (Bermudez & Seiglie): East China Sea. (28, 30)
769. *S. prominentis* Zheng: East China Sea; South China Sea. (28, 30)
770. *Cycloforina contorta* (d'Orbigny): Bohai; Yellow Sea; East China Sea; Shandong; Taizhou Bay (Zhejiang); Jiangsu (Quaternary). (3, 19, 20, 21, 28, 30, 31, 47, 65)
771. *Hauerina atlantica* Cushman: Xisha (=Paracel Islands); Zhongsha (=Macclesfield Bank). (48)
772. *H. bradyi* Cushman [1082]: South China Sea. (4, 27, 30, 48, 62, 66)
773. *H. diversa* Cushman: Xisha (=Paracel Islands); Nansha (=Spratly Islands). (25, 30, 66)
774. *H. involuta* Cushman: South China Sea. (25, 30, 66)
775. *H. orientalis* Cushman: South China Sea. (26, 30, 48, 66)
776. *H. pacifica* Cushman: Xisha (=Paracel Islands); Zhongsha (=Macclesfield Bank). (48)
777. *H. speciosa* Karrer: Xisha (=Paracel Islands); Zhongsha (=Macclesfield Bank); Nansha (=Spratly Islands). (48, 66)
778. *H. trilocularis* Zheng: Xisha (=Paracel Islands). (25, 30)
779. *Massilina corrugata* Collins: Xisha (=Paracel Islands); Zhongsha (=Macclesfield Bank). (25, 27, 30)
780. *M. hachijensis* (Uchio): Zhongsha (=Macclesfield Bank). (27, 30)
781. *M. humblei* Cushman & Ellisor: Eastern Hebei (Quaternary). (34)
782. *M. inaequalis* Cushman & Ellisor: Eastern Hebei; Taizhou Bay (Zhejiang); Jiangsu (Quaternary). (8, 34, 47, 49)
783. *M. intermedia* Cheng & Zheng: Bohai; Xisha (=Paracel Islands); Zhongsha (=Macclesfield Bank). (4, 21, 25, 30)
784. *M. laevigata* (Cushman & Todd): Bohai; East China Sea; South China Sea; Shandong (Quaternary). (19, 28, 29, 30, 47, 49)
785. *M. magna* Zheng: Xisha (=Paracel Islands); Zhongsha (=Macclesfield Bank). (26, 27, 30)
786. *M. milletti* (Wiesner): East China Sea. (19)
787. *M. penglaiensis* (Jacot): Bohai; southern Yellow Sea; East China Sea; Shandong (Quaternary). (28, 29, 30, 47)
788. *M. pratti* Cushman & Ellisor: Bohai; Yellow Sea; East China Sea; eastern Hebei; Shandong (Quaternary). (8, 20, 21, 28, 30, 34, 49)
789. *M. pyristoma* Zheng: Shandong (Quaternary). (29, 30)
790. *M. quadrans* (Cushman & Ponton): Bohai; East China Sea; Hebei (Quaternary). (21, 28, 30, 34)
791. *M. secans* (d'Orbigny): East China Sea; Shenzhen Bay (Guangdong); Xisha (=Paracel Islands); Jiangsu (Quaternary); Guangdong (Pliocene). (7, 8, 19, 25, 33, 49)
792. *Quinqueloculina agglutinans* d'Orbigny: South China Sea. (62, 66)
793. *Q. agglutinata* Cushman: South China Sea. (67)
794. *Q. akneriana* (d'Orbigny): Bohai; Yellow Sea; East China Sea; South China Sea. (28, 30, 50, 66, 67)
795. *Q. anguina* (Terquem): Nansha (=Spratly Islands). (62, 66)
796. *Q. angulostoma* Zheng: Xisha (=Paracel Islands). (26, 30)
797. *Q. arctica* Cushman: Bohai; Nansha (=Spratly Islands); Shandong (Quaternary); Hainan (Pliocene). (29, 30, 33, 66)
798. *Q. arenata* Said: Xisha (=Paracel Islands); Zhongsha (=Macclesfield Bank); Nansha (=Spratly Islands). (25, 27, 30, 66)
799. *Q. argunica* (Gerke): Shandong; Jiangsu; Taizhou Bay (Zhejiang) (Quaternary). (6, 8, 29, 47, 49)
800. *Q. artusoris* Zheng: East China Sea, South China Sea; shoreline of Boha (Quaternary). (28, 30, 66)
801. *Q. aspera* d'Orbigny: Mt. Jolmo Lungma (=Mt. Everest) area (Tibet) (Lower Tertiary). (9)
802. *Q. asperula* Sequenza: Okinawa Trough. (45)

803. *Q. auberiana* d'Orbigny: Southern Yellow Sea; East China Sea; South China Sea. (28, 30, 45, 67)
804. *Q. bella* Stschedrina: Southern Yellow Sea. (65)
805. *Q. bellatula* Bandy: Bohai; Yellow Sea; East China Sea; South China Sea; Shandong (Quarternary). (21, 28, 29, 30)
806. *Q. berthelotiana* d'Orbigny: Xisha (=Paracel Islands); Nansha (=Spratly Islands). (25, 27, 30, 66)
807. *Q. bicarinata* d'Orbigny: Taiwan (Tertiary). (44, 62)
808. *Q. bicornis* (Walker & Jacob): Yellow Sea. (30)
809. *Q. bicostata* d'Orbigny: East China Sea; South China Sea; South China Sea (Quaternary); Okinawa Trough (Holocene). (3, 19, 26, 28, 30, 45, 66)
810. *Q. bicostoides* Vella: East China Sea. (28, 30)
811. *Q. bidentata* d'Orbigny: Xisha (=Paracel Islans); Nansha (=Spratly Islands). (25, 30)
812. *Q. bifossula* He & Hu: Guangdong (Pliocene). (33)
813. *Q. bosciana* d'Orbigny: East China Sea; Taiwan Strait. (28, 30, 59)
814. *Q. boueana* d'Orbigny: South China Sea. (66, 67)
815. *Q. bradyana* Cushman: Okinawa Trough (Holocene); Xisha (=Paracel Islands); Zhongsha (=Macclesfield Bank); Nansha (=Spratly Islands). (25, 30, 48, 62, 66)
816. *Q. candeiana* d'Orbigny: Southern Yellow Sea; Taiwan Strait; South China Sea. (27, 30, 45, 50, 66)
817. *Q. carinatastriata* (Wiesner): Xisha (=Paracel Islands). (26, 30)
818. *Q. centrostriata* Mangin: East China Sea. (28, 30)
819. *Q. collumnosa* Cushman: Nansha (=Spratly Islands). (26, 30)
820. *Q. compressa* (Wiesner): East China Sea; South China Sea. (28, 30, 66)
821. *Q. complanata* (Gerke & Issaeva): Bohai; East China Sea; Taizhou Bay (Zhejiang); Jiangsu (Quaternary). (8, 19, 21, 47, 49)
822. *Q. compressiostoma* Zheng: East China Sea; South China Sea. (28, 30)
823. *Q. compta* Cushman: Southern Yellow Sea; East China Sea; South China Sea. (28, 30)
824. *Q. contorta* d'Orbigny [770]: Bohai; Yellow Sea; East China Sea; South China Sea; Shandong; Jiangsu (Quaternary). (3, 19, 20, 21, 28, 29, 30, 47, 49, 65)
825. *Q. costata* d'Orbigny: Bohai (Quaternary); Okinawa Trough (Holocene); Nansha (=Spratly Islands). (1, 45, 66)
826. *Q. crassa* Cushman: South China Sea. (25, 30)
827. *Q. crassicarinata* Collins: Xisha (=Paracel Islands). (26, 30)
828. *Q. cultrata* (Brady) [695]: Bohai; East China Sea; South China Sea; Shandong; Jiangsu (Quaternary); Guangdong (Pliocene). (1, 8, 28, 29, 30, 33, 65, 66)
829. *Q. curta* Cushman: Yellow Sea; East China Sea; Nansha (=Spratly Islands). (28, 30, 66)
830. *Q. cuvieriana* d'Orbigny: South China Sea. (25, 27, 30)
831. *Q. decora* Zheng: Xisha (=Paracel Islands). (26, 30)
832. *Q. dimidiata* Terquem: Zhongsha (=Macclesfield Bank). (27, 30)
833. *Q. disparilis* d'Orbigny: Southern Yellow Sea; South China Sea; Guangdong (Pliocene). (33, 48, 65, 67)
834. *Q. distorqueata* Cushman: Penghu (Taiwan). Beach sand. (51)
835. *Q. donghaiensis* Zheng: East China Sea. (19)
836. *Q. donghaiensis* Zheng [900]: East China Sea. (28, 30)
837. *Q. elongata* Natland: Southern Yellow Sea; South China Sea; Xisha (=Paracel Islands); Guangdong (Pliocene). (33, 38, 66)
838. *Q. ferrussacii* Cushman: Xisha (=Paracel Islands); Zhongsha (=Macclesfield Bank). (48)
839. *Q. fichteliana* (d'Orbigny): East China Sea; South China Sea. (28, 30, 66)
840. *Q. flavescens* d'Orbigny: Xisha (=Paracel Islands); Zhongsha (=Macclesfield Bank). (48)
841. *Q. frigida* Parker: East China Sea; South China Sea. (28, 30)
842. *Q. fulgida* Todd: Mt. Jolmo Lungma (=Mt. Everest) area (Tibet) (Lower Tertiary). (9)
843. *Q. funafutiensis* (Chapman): Xisha (=Paracel Islands); Nansha (=Spratly Islands). (25, 30)
844. *Q. fusiformis* Zheng: Xisha (=Paracel Islands). (26, 30)
845. *Q. gigas* Natland: Xisha (=Paracel Islands); Zhongsha (=Macclesfield Bank). (48)
846. *Q. granuliformis* Zheng: East China Sea; South China Sea. (28, 30)
847. *Q. granulocostata* Germeraad: East China Sea; South China Sea. (4, 25, 27, 28, 30, 66)
848. *Q. granulosa* Natland: Okinawa Trough (Upper Pleistocene); Nansha (=Spratly Islands). (45, 66)
849. *Q. grossa* Hu: Mt. Jolmo Lungma (=Mt. Everest) area (Tibet) (Lower Tertiary). (9)
850. *Q. imperialis* Hanna: East China Sea. (28, 30)
851. *Q. implexa* Terquem: Southern Yellow Sea. (65)
852. *Q. impolita* Zheng: Xisha (=Paracel Islands). (26, 30)
853. *Q. inculta* Zheng: Xisha (=Paracel Islands). (26, 30)
854. *Q. juani* Huang: Taiwan Strait. (54, 59)
855. *Q. jugosa* Cushman: Bohai; East China Sea; Taiwan Strait. (1, 28, 30, 47, 54, 65)
856. *Q. kansireiensis* Nakamura: Taiwan (Tertiary). (44)
857. *Q. kerimbatica* (Heron-Allen & Earland): Taiwan. (44)
858. *Q. kuromatunaiensis* Asano: Yongxing Island (Xisha=Paracel Islands) (Cenozoic). (4)
859. *Q. laevigata* d'Orbigny: East China Sea; South China Sea; Okinawa Trough (Upper Pleistocene). (28, 30, 45, 66)
860. *Q. lamarckiana* d'Orbigny: Bohai; Yellow Sea; East China Sea; South China Sea; Jiangsu (Quaternary); East China Sea continental shelf (Miocene); Okinawa Trough (Upper Pleistocene); Guangdong (Pliocene). (1, 3, 8, 19, 33, 34, 44, 47, 49, 50, 59, 64, 66)
861. *Q. lata* Terquem: East China Sea. (28, 30)
862. *Q. linoreticulata* Ho & Hu: Guangdong (Pliocene). (33)
863. *Q. longidentata* Terquem: South China Sea. (26, 27, 30)
864. *Q. longirostra* d'Orbigny: Hong Kong; Mt. Jolmo Lungma (=Mt. Everest) area (Tibet) (Lower Tertiary). (9, 62)

865. *Q. mauricensis* Howe: Hong Kong. (9, 62)
866. *Q. microcostata* Natland: Okinawa Trough (Upper Pleistocene to Holocene). (45)
867. *Q. miles* Vella: East China Sea; South China Sea. (28, 30, 66)
868. *Q. minuta* Beck: East China Sea continental shelf (Eocene). (3)
869. *Q. multicostata* Stschedrina: Southern Yellow Sea. (65)
870. *Q. najaeformis* Stschedrina: Yellow Sea. (65)
871. *Q. neosigmoilinoides* Kennett: East China Sea. (28, 30)
872. *Q. neostriatula* Thalmann: Xisha (=Paracel Islands); Zhongsha (=Macclesfield Bank); Nansha (=Spratly Islands). (4, 25, 27, 30)
873. *Q. notata* Wang: Shoreline of Bohai (Quaternary). (1)
874. *Q. oblonga* (Montagu): Yellow Sea; East China Sea; Nansha (=Spratly Islands). (28, 30, 37, 66)
875. *Q. paravulgaris* Stschedrina: Yellow Sea. (65)
876. *Q. parkeri* (Brady): Taiwan Strait; Xisha (=Paracel Islands); Zhongsha (=Macclesfield Bank); Nansha (=Spratly Islands). (4, 25, 27, 44, 62, 66)
877. *Q. paucilocula* Zheng: Xisha (=Paracel Islands). (26, 30)
878. *Q. pauperata* d'Orbigny: Xisha (=Paracel Islands); Nansha (=Spratly Islands); Mt. Jolmo Lungma (=Mt. Everest) area (Tibet) (Lower Tertiary) (9, 27, 30)
879. *Q. pentagona* Giunta: Okinawa Trough (Upper Holocene). (45)
880. *Q. philippinensis* Cushman: South China Sea; East China Sea (Eocene); South China Sea (Pliocene). (3)
881. *Q. poeyana* d'Orbigny: Hong Kong. (62)
882. *Q. polygona* d'Orbigny: Penghu (Taiwan); Hong Kong; Xisha (=Paracel Islands) (Upper Pleistocene). (39, 51, 62)
883. *Q. praelonga* He & Lin: Guangdong (Pliocene). (33)
884. *Q. procera* Cushman: Zhongsha (=Macclesfield Bank); Xisha (=Paracel Islands). (48)
885. *Q. pseudocandeiana* Stschedrina: Yellow Sea. (65)
886. *Q. pseudoproxima* Zhang: East China Sea. (19)
887. *Q. pseudoreticulata* Parr: East China Sea; South China Sea; Yellow Sea (Pliocene); East China Sea continental shelf (Eocene); Guangdong (Pliocene). (3, 22, 28, 30, 33, 59, 66)
888. *Q. pulchella* d'Orbigny: Hainan (Pliocene). (33)
889. *Q. reticulata* (d'Orbigny): South China Sea. (62, 67)
890. *Q. riveroae* Bermudez & Seiglie: Okinawa Trough (Holocene). (45)
891. *Q. rodolphina* d'Orbigny: Okinawa Trough (Holocene). (45)
892. *Q. rugosa* d'Orbigny: Taizhou Bay (Zhejiang) (Quaternary). (47)
893. *Q. sabulosa* Cushman: Yellow Sea; East China Sea; South China Sea; Jiangsu (Quaternary). (3, 8, 19, 31, 50)
894. *Q. sagamiensis* Asano: East China Sea; South China Sea. (28, 30, 44, 66, 67)
895. *Q. sawanensis* Asano: Taiwan. (44)
896. *Q. seminula* (Linné): Bohai; Yellow Sea; East China Sea; Taiwan; South China Sea; Shandong (Quaternary); South China Sea (Pliocene). (1, 3, 8, 19–22, 31, 41, 47, 49, 59)
897. *Q. seminulangulata* McLean: Bohai; southern Yellow Sea; East China Sea; South China Sea; Shandong (Quaternary). (13, 21, 29–31)
898. *Q. septuosa* Stschedrina: Southern Yellow Sea. (65)
899. *Q. sigmoilinoides* Gianotti: East China Sea; Okinawa Trough (Upper Pleistocene). (28, 30, 45)
900. *Q. sinensis* Zheng, nom. nov.: East China Sea. (28, 30)
901. *Q. stalkeri* Loeblich & Tappan: Taizhou Bay (Zhejiang) (Quaternary). (47)
902. *Q. subarenaria* Cushman: Jiangsu (Quaternary). (8)
903. *Q. subcurta* Zheng: East China Sea; South China Sea. (28, 30)
904. *Q. subdecorata* Cushman: East China Sea. (28, 30)
905. *Q. suborbicularis* d'Orbigny: East China Sea continental shelf (Pliocene); Okinawa Trough (Upper Pleistocene). (3, 45)
906. *Q. subpolygona* Parr: Guangdong (Pliocene). (33)
907. *Q. subquadra* Hada: Huanghua (Hebei) (Quaternary). (34)
908. *Q. subungeriana* Serova: South China Sea; Xisha (=Paracel Islands); Shandong (Quaternary); Taizhou Bay (Zhejiang) (Quarternary); Okinawa Trough (Holocene). (8, 29, 30, 39, 42, 45, 47, 49)
909. *Q. sulcata* d'Orbigny: Okinawa Trough (Upper Pleistocene); Xisha (=Paracel Islands); Zhongsha (=Macclesfield Bank); Nansha (=Spratly Islands). (4, 25, 27, 30, 45, 66)
910. *Q. tikutoensis* Nakamura: East China Sea; East China Sea (Quaternary); Taiwan (Tertiary). (3, 19)
911. *Q. tôtômiensis* Asano: East China Sea; South China Sea. (28, 30)
912. *Q. tricarinata* d'Orbigny: Nansha (=Spratly Islands). (66)
913. *Q. tropicalis* Cushman: East China Sea; South China Sea. (28, 30, 62)
914. *Q. tubilocula* Zheng: Xisha (=Paracel Islands). (26, 30, 39)
915. *Q. ungeriana* d'Orbigny: Bohai; Yellow Sea; East China Sea; South China Sea; Mt. Jolmo Lungma (=Mt. Everest) area (Tibet) (Lower Tertiary). (9, 28, 30)
916. *Q. venusta* Karrer: Southern Yellow Sea; East China Sea; South China Sea; Xisha (=Paracel Islands) (Upper Pleistocene); Jiangsu (Quaternary). (8, 28, 30, 39, 66)
917. *Q. vulgaris* d'Orbigny: Taiwan Strait; South China Sea; Shandong (Quaternary); Okinawa Trough (Upper Pleistocene). (29, 30, 45, 59, 66)
918. *Q. yezoensis* Asano: Zhongsha (=Macclesfield Bank). (27, 30)
919. *Biloculinella globula* (Bornemann): East China Sea; Xisha (=Paracel Islands); Zhongsha (=Macclesfield Bank). (19, 26, 30, 48, 66)
920. *B. inflata* (Schlumberger): East China Sea. (28, 30)
921. *B. irregularis* d'Orbigny: Okinawa Trough (Upper Pleistocene to Holocene). (45)

922. *B. isabelleana* (d'Orbigny): Southern Yellow Sea; East China Sea; South China Sea. (28, 30)
923. *B. labiata* (Schlumberger): Southern Yellow Sea; East China Sea; South China Sea. (28, 30)
924. *B. subsphaerica* (Wiesner): East China Sea. (28, 30)
925. *B. taiwanica* Huang: Taiwan. (44)
926. *B. tenuiaperta* Huang [1014]: Taiwan. (44)
927. *B. toddae* Andersen: East China Sea; Okinawa Trough (Upper Pleistocene to Holocene). (28, 30, 45)
928. *Cribromiliolinella ericsoni* (Loeblich & Tappan): East China Sea. (28, 30)
929. *C. striata* (Loeblich & Tappan): East China Sea. (28, 30)
930. *Sigmamiliolinella australis* (Parr): East China Sea. (28, 30)
931. *S. denticollaris* Zheng: East China Sea. (28, 30)
932. *Ammosigmoilinella eximia* Zheng: East China Sea. (28, 30)
933. *Cruciloculina asanoi* Loeblich & Tappan: Okinawa Trough (Upper Pleistocene to Holocene). (45)
934. *C. japonica* Asano: East China Sea. (28, 30)
935. *C. triangularis* d'Orbigny: East China Sea. (28, 30)
936. *Flintina bradyana* Cushman: East China Sea; South China Sea; East China Sea (Quaternary); Taiwan (Miocene); Guangdong (Pliocene). (3, 28, 30, 33, 66)
937. *F. crassatina* (Brady): Taiwan; Nansha (=Spratly Islands). (44, 59, 66)
938. *F. hainanensis* N. W. Wang & He: Hainan (Upper Pleistocene). (33)
939. *F. triquetra* (Brady) [758]: East China Sea; Nansha (=Spratly Islands); Taiwan (Miocene). (28, 30, 44, 66)
940. *Flintinoides labiosa* (d'Orbigny): East China Sea; South China Sea. (4, 7, 28, 30, 48, 66)
941. *Involvohauerina cribrostoma* (Heron-Allen & Earland): Xisha (=Paracel Islands); Zhongsha (=Macclesfield Bank); Nansha (=Spratly Islands). (25, 27, 30)
942. *Miliolinella australis* (Parr) [930]: East China Sea. (28, 30)
943. *M. californica* Rhumler: East China Sea; Okinawa Trough (Upper Pleistocene to Holocene). (28, 30, 45)
944. *M. chaoyii* Huang: Taiwan Strait. (54, 59)
945. *M. chui* Huang: Taiwan Strait. (54)
946. *M. chukchiensis* Loeblich & Tappan: East China Sea; South China Sea. (19, 28, 30)
947. *M. circularis* (Bornemann): Bohai; Yellow Sea; East China Sea; South China Sea; Guangdong (Pliocene). (19, 28, 30, 33, 66)
948. *M. corrugata* Zheng: East China Sea; South China Sea. (28, 30)
949. *M. hornibrooki* (Vella): East China Sea; South China Sea. (28, 30)
950. *M. iongchuanae* Huang: Taiwan Strait. (54)
951. *M. labiosa* (d'Orbigny) [940]: East China Sea; South China Sea. (4, 7, 28, 30, 48, 66)
952. *M. natchitochensis* (Howe): Mt. Jolmo Lungma (=Mt. Everest) area (Tibet) (Lower Tertiary). (9)
953. *M. oblonga* (Montagu): Yellow Sea; East China Sea; South China Sea. (28, 30, 59)
954. *M. oceanica* (Cushman): Taiwan Strait; South China Sea. (25, 27, 30, 50)
955. *M. pseudoblonga* Zheng: Zhongsha (=Macclesfield Bank); Okinawa Trough (Upper Pleistocene). (27, 30, 45)
956. *M. robusta* Cushman & Todd: East China Sea. (28, 30)
957. *M. striata* (Chapman): East China Sea. (28, 30)
958. *M. subrotunda* (Montagu): East China Sea; South China Sea; Shandong (Quaternary); East China Sea continental shelf (Pliocene). (3, 28, 29, 30)
959. *M. temeii* Huang: Taiwan Strait. (54)
960. *M. vigilax* Vella: East China Sea; South China Sea. (28, 30)
961. *M. webbiana* (d'Orbigny): Taiwan Strait; South China Sea. (30, 50, 66)
962. *Pateoris hauerinoides* (Rhumbler): Southern Yellow Sea. (65)
963. *Nevillina coronata* (Millett): Nansha (=Spratly Islands). (30)
964. *Pseudomassilina australis* (Cushman): Xisha (=Paracel Islands); Zhongsha (=Macclesfield Bank); Nansha (=Spratly Islands). (48, 66)
965. *P. macilenta* (Brady): Xisha (=Paracel Islands). (26, 30)
966. *P. reticulata* (Heron-Allen & Earland): Nansha (=Spratly Islands). (66)
967. *Pseudopyrgo milletti* (Cushman): Bohai; East China Sea; South China Sea. (28, 30, 42, 48)
968. *P. paraglobula* Zheng [708]: East China Sea; Nansha (=Spratly Islands). (28, 30, 66)
969. *P. toddae* (Andersen) [927]: East China Sea; Okinawa Trough (Upper Pleistocene to Holocene). (28, 30, 45)
970. *Pseudosigmoilina minuta* Zheng: Xisha (=Paracel Islands). (26, 30)
971. *Pseudotriloculina cyclostoma* (Reuss): East China Sea; South China Sea. (28, 30)
972. *P. lecalvezae* (Kaaschieter): Guangdong; Xisha (=Paracel Islands); Zhongsha (=Macclesfield Bank); Nansha (=Spratly Islands). (7, 25, 30, 48, 66)
973. *P. lunata* (Zheng): East China Sea; South China Sea. (28, 30)
974. *P. subglobiformis* (Zheng): East China Sea; South China Sea. (28, 30)
975. *P. subsphaeroides* (Zheng): East China Sea. (28, 30)
976. *Sinuloculina cyclostoma* (Reuss) [971]: East China Sea; South China Sea. (28, 30)
977. *S. lunata* Zheng [973]: East China Sea; South China Sea. (28, 30)
978. *S. subglobiformis* Zheng [974]: East China Sea; South China Sea. (28, 30)
979. *S. subsphaeroides* Zheng [975]: East China Sea. (28, 30)
980. *Pyrgo anomala* (Schlumberger): East China Sea; South China Sea; Okinawa Trough (Upper Pleistocene). (28, 30, 45)
981. *P. bougainvillei* (d'Orbigny): East China Sea; South China Sea. (28, 30)
982. *P. bulloides* (d'Orbigny): Mt. Jolmo Lungma (=Mt. Everest) area (Tibet) (Lower Tertiary). (9)

983. *P. calostoma* (Karrer): East China Sea; South China Sea. (28, 30)
984. *P. comata* (Brady): East China Sea; South China Sea. (28, 30)
985. *P. compressioblonga* Zheng: Xisha (=Paracel Islands); Nansha (=Spratly Islands). (28, 30)
986. *P. denticulata* (Brady): Xisha (=Paracel Islands); Zhongsha (=Macclesfield Bank); Nansha (=Spratly Islands). (25, 27, 30, 48, 66)
987. *P. depressa* (d'Orbigny): East China Sea; South China Sea; Okinawa Trough (Upper Pleistocene); Xisha (=Paracel Islands) (Upper Pleistocene). (28, 30, 39, 48, 66)
988. *P. elongata* (d'Orbigny): East China Sea; Nansha (=Spratly Islands); Okinawa Trough (Upper Pleistocene). (19, 45, 66)
989. *P. fornasini* Chapman & Parr: East China Sea; South China Sea. (28, 30, 66)
990. *P. grinzingensis* (Karrer): Xisha (=Paracel Islands) (Middle Pleistocene). (39)
991. *P. inornata* (d'Orbigny): Mt. Jolmo Lungma (=Mt. Everest) area (Tibet) (Lower Tertiary). (9)
992. *P. irregularis* (d'Orbigny): Nansha (=Spratly Islands); Guangdong (Pliocene). (33, 66)
993. *P. johnsoni* Cushman: Central South China Sea. (42)
994. *P. laevis* Defrance: East China Sea; South China Sea; Okinawa Trough (Upper Pleistocene). (28, 30, 45, 68)
995. *P. lucernula* (Schwager): South China Sea; Xisha (=Paracel Islands); Taiwan (Miocene); Guangdong (Pliocene). (33, 39, 44, 64)
996. *P. lunula* (d'Orbigny): East China Sea; South China Sea. (26, 30)
997. *P. megastoma* Zheng: Xisha (=Paracel Islands). (26, 30)
998. *P. murrhina* (Schwager): East China Sea; South China Sea; Okinawa Trough (Upper Pleistocene); Xisha (=Paracel Islands) (Middle Pleistocene). (28, 30, 39, 45, 48, 66)
999. *P. oblonga* (d'Orbigny): Xisha (=Paracel Islands) (Upper Pleistocene). (39)
1000. *P. pacifica* Asano: East China Sea; Guangdong (Pliocene). (28, 30, 33)
1001. *P. rotalaria* Loeblich & Tappan: Okinawa Trough; central South China Sea; Xisha (=Paracel Islands) (Upper Pleistocene). (39, 42, 45)
1002. *P. sagittioris* Zheng: East China Sea. (28, 30)
1003. *P. sarsii* (Schlumberger): East China Sea; South China Sea; Taiwan (Miocene). (19, 28, 30, 44, 66, 67)
1004. *P. serrata* (Bailey): East China Sea; South China Sea; East China Sea (Quaternary); Okinawa Trough (Upper Pleistocene); Xisha (=Paracel Islands) (Middle Pleistocene). (3, 28, 30, 39, 45)
1005. *P. simplex* (d'Orbigny): East China Sea. (28, 30)
1006. *P. spinidorsa* Zheng: Zhongsha (=Macclesfield Bank); Nansha (=Spratly Islands). (27, 30)
1007. *P. striolata* (Brady): South China Sea. (25, 30)
1008. *P. tainanensis* Ishizaki: East China Sea; South China Sea; Taiwan (Pliocene to Pleistocene). (3)
1009. *P. tenuis* (Karrer): Xisha (=Paracel Islands) (Middle Pleistocene). (39)
1010. *P. ventruosa* (Reuss): Xisha (=Paracel Islands) (Upper Pleistocene). (39)
1011. *P. vespertilio* (Schlumberger) [1103]: East China Sea; Guangdong (Pliocene). (3, 28, 30)
1012. *P. williamsoni* (Silvestri): East China Sea; South China Sea; Okinawa Trough (Upper Pleistocene); Xisha (=Paracel Islands) (Upper Pleistocene). (3, 28, 30, 39, 45)
1013. *Pyrgoella sphaera* (d'Orbigny): East China Sea; Taiwan Strait; South China Sea. (28, 30, 50, 66)
1014. *P. tenuiaperta* (Huang): East China Sea; Taiwan Strait; South China Sea. (28, 30, 44, 50, 66)
1015. *Triloculina affinis* d'Orbigny: East China Sea; South China Sea. (28, 30)
1016. *T. alabamensis* Cushman: East China Sea continental shelf (Eocene). (3)
1017. *T. basispinata* Zhang: East China Sea. (19)
1018. *T. bertheliniana* (Brady): Taiwan; Hong Kong; Xisha (=Paracel Islands); Zhongsha (=Macclesfield Bank); Nansha (=Spratly Islands). (4, 25, 27, 30, 44, 62, 66)
1019. *T. bicarinata* d'Orbigny: Xisha (=Paracel Islands); Zhongsha (=Macclesfield Bank); Nansha (=Spratly Islands). (25, 30, 48, 66)
1020. *T. bradyana* (Cushman): Nansha (=Spratly Islands). (30)
1021. *T. circularis* Bornemann [947]: Bohai; Yellow Sea; East China Sea; South China Sea; Guangdong (Pliocene). (19, 28, 30, 33, 66)
1022. *T. compressa* Wang, He, & Lu: East China Sea continental shelf (Eocene). (3)
1023. *T. complanata* Hu: Mt. Jolmo Lungma (=Mt. Everest) area (Tibet) (Lower Tertiary). (9)
1024. *T. consobrina* d'Orbigny: East China Sea. (28, 30)
1025. *T. costifera* Terquem: Xisha (=Paracel Islands); Zhongsha (=Macclesfield Bank); Nansha (=Spratly Islands). (25, 30, 48, 66)
1026. *T. cuneata* Karrer: Nansha (=Spratly Islands). (30)
1027. *T. earlandi* Cushman: Okinawa Trough (Upper Pleistocene); South China Sea. (4, 25, 30, 45, 66)
1028. *T. elongata* d'Orbigny: Okinawa Trough (Holocene); Nansha (=Spratly Islands). (9)
1029. *T. foveata* He & Lin: Guangdong (Pliocene). (33)
1030. *T. gibba* d'Orbigny: Mt. Jolmo Lungma (=Mt. Everest) area (Tibet) (Lower Tertiary). (33)
1031. *T. hartingi* Bosquet: Guangdong (Pliocene). (33)
1032. *T. inflata* d'Orbigny: Bahai; East China Sea; South China Sea; Jiangsu (Quaternary). (3, 8, 49)
1033. *T. insignis* (Brady): Xisha (=Paracel Islands); Zhongsha (=Macclesfield Bank); Nansha (=Spratly Islands). (25, 27, 30, 66)
1034. *T. involuta* Cushman: Xisha (=Paracel Islands); Zhongsha (=Macclesfield Bank). (48)
1035. *T. irregularis* (d'Orbigny): Xisha (=Paracel Islands); Zhongsha (=Macclesfield Bank). (25, 30, 48)

1036. *T. kerimbatica* (Heron-Allen & Earland): Xisha (=Paracel Islands); Zhongsha (=Macclesfield Bank); Nansha (=Spratly Islands). (25, 27, 30, 66)

1037. *T. labiosa* d'Orbigny [940]: East China Sea; South China Sea. (4, 7, 28, 30, 48, 66)

1038. *T. laevigata* d'Orbigny: East China Sea; Xisha (=Paracel Islands); Nansha (=Spratly Islands). (26, 28, 30, 66)

1039. *T. laevis* Cai & Tu: Xisha (=Paracel Islands); Zhongsha (=Macclesfield Bank). (48)

1040. *T. lecalvezae* Kaaschieter [972]: Guangdong; Xisha (=Paracel Islands); Zhongsha (=Macclesfield Bank); Nansha (=Spratly Islands). (7, 25, 30, 48, 66)

1041. *T. linneiana* d'Orbigny: Zhongsha (=Macclesfield Bank); Nansha (=Spratly Islands). (27, 30)

1042. *T. marshallana* Todd: East China Sea; South China Sea. (28, 30)

1043. *T. mindenensis* Howe: Mt. Jolmo Lungma (=Mt. Everest) area (Tibet) (Lower Tertiary). (9)

1044. *T. oblonga* (Montagu) [953]: Shenzhen (Guangdong); Hong Kong; Nansha (=Spratly Islands); shoreline of Bohai (Quaternary). (1, 7, 30, 59, 62, 66)

1045. *T. paradox* Qiu & Lin: Guangdong (Pliocene). (33)

1046. *T. parallela* Zheng: Xisha (=Paracel Islands); Nansha (=Spratly Islands). (26, 30)

1047. *T. paratrigonula* Stschedrina: Southern Yellow Sea. (65)

1048. *T. pectinata* Stschedrina: Southern Yellow Sea. (65)

1049. *T. pentagonalis* Wang: Bohai; southern Yellow Sea; East China Sea; South China Sea. (11, 19, 28, 30)

1050. *T. peregrina* He & Lin: Guangdong (Pliocene). (33)

1051. *T. pinguicula* Zheng: Xisha (=Paracel Islands); Nansha (=Spratly Islands). (25, 30, 66)

1052. *T. pinguis* Bandy: East China Sea continental shelf (Eocene). (3)

1053. *T. reticulostriata* (Cushman): Taiwan (Miocene). (44)

1054. *T. rectilocula* Zheng: Southern Yellow Sea; South China Sea; Shandong; Taizhou Bay (Zhejiang) (Quaternary). (20, 29, 30, 47)

1055. *T. rupertiana* (Brady) [1140]: South China Sea. (67)

1056. *T. sidebottomi* (Martinotti): Southern Yellow Sea. (65)

1057. *T. subcylindrica* Zheng: East China Sea; South China Sea. (28, 30)

1058. *T. subglobosa* Cai & Tu: Xisha (=Paracel Islands); Zhongsha (=Macclesfield Bank). (28, 30)

1059. *T. subgranulata* Cushman: Xisha (=Paracel Islands); Zhongsha (=Macclesfield Bank); Nansha (=Spratly Islands). (25, 27, 30)

1060. *T. subplanciana* Cushman: Xisha (=Paracel Islands); Nansha (=Spratly Islands). (25, 30, 66)

1061. *T. transversestriata* (Brady): Xisha (=Paracel Islands); Zhongsha (=Macclesfield Bank); Nansha (=Spratly Islands). (25, 27, 30)

1062. *T. transvoluta* Todd: Xisha (=Paracel Islands); Zhongsha (=Macclesfield Bank). (48)

1063. *T. tricarinata* d'Orbigny: Yellow Sea; East China Sea; South China Sea; Yellow Sea (Quaternary); Okinawa Trough (Holocene); Taiwan (Tertiary); Guangdong (Pliocene); Xisha (=Paracel Islands) (Upper Pleistocene). (4, 19, 25, 27, 30, 31, 33, 44, 45, 46, 50, 62, 66, 67)

1064. *T. trigonula* (Lamarck): Bohai; Yellow Sea; East China Sea; South China Sea; Shandong; Zhejiang (Upper Tertiary); Taiwan (Quaternary); South China Sea (Miocene). (3, 7, 19, 20, 25, 27, 29, 33, 47, 66, 67)

1065. *T. trotusa* Cushman: Xisha (=Paracel Islands); Zhongsha (=Macclesfield Bank). (48)

1066. *T. vespertilio* Zheng: East China Sea; South China Sea. (28, 30)

1067. *T. vicina* Stschedrina: Southern Yellow Sea. (65)

1068. *Triloculinella asymmetrica* Said: East China Sea. (28, 30)

1069. *T. laevigata* (Bornemann): East China Sea; South China Sea. (26, 28, 30, 66)

1070. *T. oblongus* (Montagu): South China Sea. (67)

1071. *T. patens* (Stschedrina): South China Sea. (65)

1072. *T. tegminis* (Loeblich & Tappan): Yellow Sea; East China Sea; Taiwan Strait; Okinawa Trough (Upper Pleistocene). (45, 65)

1073. *T. translucens* (Stschedrina): Yellow Sea. (65)

1074. *Scutuloris oblongus* (Montagu) [1070]: South China Sea. (67)

1075. *S. patens* Stschedrina [1071]: Southern Yellow Sea. (65)

1076. *S. tegminis* Loeblich & Tappan [1072]: Yellow Sea; East China Sea; Taiwan Strait. (45, 65)

1077. *S. translucens* Ststchedrina [1073]: Yellow Sea. (65)

1078. *Mesosigmoilina minuta* Zheng: Xisha (=Paracel Islands). (26, 30, 73)

1079. *Pseudosigmoilina minuta* Zheng [1078]: Xisha (=Paracel Islands). (26, 30, 73)

1080. *Nummoluculina contraria* (d'Orbigny): East China Sea; South China Sea. (28, 30)

1081. *Polysegmentina circinata* (Brady): Xisha (=Paracel Islands). (25, 30)

1082. *Sigmoihauerina bradyi* (Cushman): South China Sea. (4, 27, 28, 30, 62, 66)

1083. *S. fragillissima* (Brady): South China Sea. (26, 30)

1084. *Sigmoilina amygdaloides* (Brady): South China Sea. (67)

1085. *S. arenaria* (Brady): Yellow Sea; East China Sea; Taiwan Strait; South China Sea; Taiwan (Miocene). (44, 48, 64, 67)

1086. *S. elliptica* Galloway & Wissler: East China Sea; South China Sea; Taiwan (Miocene). (3)

1087. *S. foliacea* Zheng: East China Sea; South China Sea. (28, 30)

1088. *S. formosana* Nakamura: Taiwan (Miocene). (44)

1089. *S. inculta* Wang: Shoreline of Bohai (Quaternary). (1)

1090. *S. minutissima* Zheng: Shandong; southern Yellow Sea; Taizhou Bay (Zhejiang) (Quaternary). (29, 30, 31, 47)

1091. *S. obesa* Heron-Allen & Earland: East China Sea. (28, 30)

1092. *S. schlumbergeri* Silvestri: Eastern Hebei (Quaternary); Taiwan (Miocene). (34, 44)

1093. *S. sigmoidea* (Brady): East China Sea. (28, 30)

1094. *S. subtenuis* Ho, Hu & Wang: Jiangsu (Quaternary); Guangdong (Pliocene). (8, 33)

1095. *S. tenuis* (Czjzek) [1107]: East China Sea; Taiwan Strait; Nansha (=Spratly Islands); Jiangsu (Quaternary); Guangdong (Pliocene). (8, 9, 19, 28, 30, 33, 50, 66)

1096. *S. tenuissima* Stschedrina: Southern Yellow Sea. (65)

1097. *S. syrtica* Martinotti: Eastern Hebei (Quaternary). (34)

1098. *S. victoriensis* Cushman: East China Sea; Nansha (=Spratly Islands). (28, 30, 66)

1099. *Sigmoilinella tortuosa* Zheng: Xisha (=Paracel Islands); Zhongsha (=Macclesfield Bank); Nansha (=Spratly Islands). (25, 27, 30)

1100. *Sigmoilinita asperula* (Karrer): Yellow Sea; East China Sea; South China Sea; Jiangsu (Quaternary). (19, 28, 30, 31, 48, 66)

1101. *S. granulifera* Zheng: East China Sea; South China Sea. (28, 30)

1102. *S. tenuis* (Czjzek) [1107]: East China Sea; Taiwan Strait; Nansha (=Spratly Islands); Jiangsu (Quaternary); Guangdong (Pliocene). (8, 9, 19, 28, 30, 33, 50, 66)

1103. *Sigmopyrgo vespertilio* (Schlumberger): East China Sea; Guangdong (Pliocene). (3, 28, 30)

1104. *Spirosigmoilina crenata* (Karrer): Taiwan Strait. (50)

1105. *S. pulchra* Zheng: South China Sea. (26, 27, 30, 66)

1106. *S. pusilla* (Earland): East China Sea; South China Sea; Okinawa Trough (Upper Pleistocene). (28, 30, 45, 66)

1107. *S. tenuis* (Czjzek) [1107]: East China Sea; Taiwan Strait; Nansha (=Spratly Islands); Jiangsu (Quaternary); Guangdong (Pleistocene). (8, 9, 19, 28, 30, 33, 50, 60)

1108. *Sigmoilopsis asperula* (Karrer) [1100]: Yellow Sea; East China Sea; South China Sea; Jiangsu (Quaternary). (19, 28, 30, 31, 48, 64)

1109. *S. carinata* Zheng: East China Sea; South China Sea. (28, 30, 66)

1110. *S. chapmani* Cushman: East China Sea. (28, 30)

1111. *S. finlayi* Vella (Cushman): East China Sea. (19)

1112. *S. flintii* (Cushman): Jiangsu (Quaternary). (8)

1113. *S. herzensteini* (Schlumberger): East China Sea; South China Sea. (28, 30)

1114. *S. moyi* Atkinson: East China Sea. (28, 30)

1115. *S. obesa* Zheng: East China Sea. (28, 30)

1116. *S. orientalis* Zheng: East China Sea. (28, 30)

1117. *S. schlumbergeri* (Silvestri): Yellow Sea; East China Sea; South China Sea; East China Sea continental shelf (Quaternary); Okinawa Trough (Holocene). (3, 28, 30, 45, 48, 64)

1118. *Articulina alticostata* Cushman: Nansha (=Spratly Islands). (66)

1119. *A. carinata* Cushman: Xisha (=Paracel Islands); Zhongsha (=Macclesfield Bank); Nansha (=Spratly Islands). (25, 30, 38, 66)

1120. *A. curta* Zheng: Xisha (=Paracel Islands). (26, 30)

1121. *A. elongata* Cushman: Xisha (=Paracel Islands); Zhongsha (=Macclesfield Bank). (48)

1122. *A. lineata* Brady: Xisha (=Paracel Islands); Nansha (=Spratly Islands). (26, 30)

1123. *A. mucronata* (d'Orbigny): Xisha (=Paracel Islands); Nansha (=Spratly Islands). (26, 30)

1124. *A. pacifica* Cushman: Xisha (=Paracel Islands); Zhongsha (=Macclesfield Bank); Nansha (=Spratly Islands). (25, 27, 30, 66)

1125. *A. parallela* Zheng: Zhongsha (=Macclesfield Bank). (27, 30)

1126. *A. queenslandica* Collins: Xisha (=Paracel Islands). (26, 30)

1127. *A. ricta* Zheng: Xisha (=Paracel Islands). (26, 30)

1128. *A. schauinslandi* (Rhumbler): Xisha (=Paracel Islands); Zhongsha (=Macclesfield Bank). (25, 30)

1129. *A. sulcata* Reuss: Nansha (=Spratly Islands). (66)

1130. *Parrina bradyi* (Millett): Xisha (=Paracel Islands). (25, 30)

1131. *Tubinella funalis* (Brady): Xisha (=Paracel Islands). (26, 30)

MILIOLIDAE

1132. *Miliola costata* Hu: Mt. Jolmo Lungma (=Mt. Everest) area (Tibet) (Lower Tertiary). (9)

1133. *M. inflata* Zheng: Xisha (=Paracel Islands); Zhongsha (=Macclesfield Bank). (25, 27, 30)

1134. *M. natchitochensis* (Howe): Mt. Jolmo Lungma (=Mt. Everest) area (Tibet) (Lower Tertiary). (9)

1135. *M. prisca* (d'Orbigny): Mt. Jolmo Lungma (=Mt. Everest) area (Tibet) (Lower Tertiary). (9)

1136. *M. pseudocarinata* Le Calvez: Mt. Jolmo Lungma (=Mt. Everest) area (Tibet) (Lower Tertiary). (9)

1137. *M. rostrata* (Terquem): Mt. Jolmo Lungma (=Mt. Everest) area (Tibet) (Lower Tertiary). (9)

1138. *M. saxorum* (Lamarck): Mt. Jolmo Lungma (=Mt. Everest) area (Tibet) (Lower Tertiary). (9)

1139. *M. sublineata* (Brady): Xisha (=Paracel Islands). (25, 30)

1140. *Rupertianella rupertiana* (Brady): South China Sea. (30, 66)

ALVEOLINACEA
ALVEOLINIDAE

1141. *Alveolinella quoyi* (d'Orbigny): Nansha (=Spratly Islands). (66)

1142. *Fasciolites ellipsoidalis* (Schwager): Mt. Jolmo Lungma (=Mt. Everest) area (Tibet) (Lower Tertiary). (9)

1143. *F. boscii* Defrance: Mt. Jolmo Lungma (=Mt. Everest) area (Tibet) (Lower Tertiary). (9)

1144. *F. cylindratus* Hottinger: Mt. Jolmo Lungma (=Mt. Everest) area (Tibet) (Lower Tertiary). (9)

1145. *F. globosus* (Leymerie): Mt. Jolmo Lungma (=Mt. Everest) area (Tibet) (Lower Tertiary). (9)

1146. *F. himalayensis* Sheng & Zhang: Mt. Jolmo Lungma (=Mt. Everest) area (Tibet) (Lower Tertiary). (9)

1147. *F. nuttalli* Davies: Mt. Jolmo Lungma (=Mt. Everest) area (Tibet) (Lower Tertiary). (9)

1148. *F. oblongus* (d'Orbigny): Mt. Jolmo Lungma (=Mt. Everest) area (Tibet) (Lower Tertiary). (9)
1149. *F. oliviformis* Sheng & Zhang: Mt. Jolmo Lungma (=Mt. Everest) area (Tibet) (Lower Tertiary). (9)
1150. *F. rutimereri* (Hottinger): Mt. Jolmo Lungma (=Mt. Everest) area (Tibet) (Lower Tertiary). (9)
1151. *F. subtilis* (Hottinger): Mt. Jolmo Lungma (=Mt. Everest) area (Tibet) (Lower Tertiary). (9)
1152. *F. tibeticus* Sheng & Zhang: Mt. Jolmo Lungma (=Mt. Everest) area (Tibet) (Lower Tertiary). (9)
1153. *Borelis melo* (Fichtel & Moll): Taiwan. (44)
1154. *B. pulchra* (d'Orbigny): Xisha (=Paracel Islands); Zhongsha (=Macclesfield Bank). (25, 27, 30)

SORITACEA
PENEROPLIDAE

1155. *Dendritina pacifica* Zheng: Xisha (=Paracel Islands). (7, 25, 30)
1156. *D. striata* Hofker: Xisha (=Paracel Islands); Zhongsha (=Macclesfield Bank). (4, 7, 25, 27, 30)
1157. *Laevipeneroplis bulloides* (d'Orbigny): Xisha (=Paracel Islands) (Middle Pleistocene). (39)
1158. *L. crassa* (Hofker): Nansha (=Spratly Islands). (66)
1159. *L. malayensis* (Hofker): Xisha (=Paracel Islands); Zhongsha (=Macclesfield Bank). (4, 25, 27, 30, 66)
1160. *Puteolina bulloides* (d'Orbigny) [1157]: Xisha (=Paracel Islands) (Middle Pleistocene). (39)
1161. *P. crassa* Hofker [1158]: Nansha (=Spratly Islands). (66)
1162. *P. malayensis* Hofker [1159]: Xisha (=Paracel Islands); Zhongsha (=Macclesfield Bank); Nansha (=Spratly Islands). (4, 25, 27, 30, 66)
1163. *Monalysidium politum* Chapman: Xisha (=Paracel Islands). (25, 30)
1164. *Peneroplis pertusus* (Forskål): Xisha (=Paracel Islands); Zhongsha (=Macclesfield Bank); Nansha (=Spratly Islands). (4, 25, 27, 30, 66)
1165. *P. planatus* (Fichtel & Moll): Hong Kong; Nansha (=Spratly Islands). (30, 60)
1166. *Spirolina acicularis* (Batsch): Nansha (=Spratly Islands). (30, 66, 67)
1167. *S. arietina* (Batsch): Xisha (=Paracel Islands); Zhongsha (=Macclesfield Bank); Nansha (=Spratly Islands). (25, 27, 30, 48)
1168. *S. mariei* Le Calvez: Mt. Jolmo Lungma (=Mt. Everest) area (Tibet) (Lower Tertiary). (9)

SORITIDAE

1169. *Archaias angulatus* (Fichtel & Moll): Xisha (=Paracel Islands); Zhongsha (=Macclesfield Bank). (48)
1170. *Amphisorus hemprichii* Ehrenberg: Taiwan Strait; Xisha (=Paracel Islands); Zhongsha (=Macclesfield Bank); Nansha (=Spratly Islands). (4, 25, 27, 30, 50)
1171. *Orbitolites cotentinensis* Lehman: Mt. Jolmo Lungma (=Mt. Everest) area (Tibet) (Lower Tertiary). (9)
1172. *O. complanatus* Lamarck: Mt. Jolmo Lungma (=Mt. Everest) area (Tibet) (Lower Tertiary). (9)
1173. *Sorites marginalis* (Lamarck): Taiwan Strait. (50)
1174. *S. orbiculus* (Forskål): Xisha (=Paracel Islands); Zhongsha (=Macclesfield Bank); Nansha (=Spratly Islands). (4, 25, 27, 30, 60)

KERAMOSPHAERIDAE

1175. *Keramosphaera tergestina* (Stache): Mt. Jolmo Lungma (=Mt. Everest) area (Tibet) (Lower Tertiary). (9)

LAGENINA
NODOSARIACEA
NODOSARIIDAE

1176. *Alfredosilvestris levinsoni* Andersen: Okinawa Trough (Upper Pleistocene). (45)
1177. *Botuloides pauciloculus* Zheng: Xisha (=Paracel Islands). (26, 30)
1178. *B. perlucida* Zheng & Zhang: East China Sea. (19)
1179. *Dentalina advena* (Cushman): East China Sea; South China Sea; Okinawa Trough (Upper Pleistocene). (30, 45)
1180. *D. basiplanata* Cushman: East China Sea; Okinawa Trough (Upper Pleistocene); Guangdong (Pliocene). (19, 33, 45)
1181. *D. beyrichana* Neugeboren: Okinawa Trough (Upper Pleistocene). (45)
1182. *D. californica* Cushman & Gray: Southern Yellow Sea; East China Sea, South China Sea, Okinawa Trough (Holocene). (30, 45)
1183. *D. catenulata* Brady: Guangdong (Pliocene). (33)
1184. *D. communis* d'Orbigny: Yellow Sea; East China Sea; South China Sea; Okinawa Trough (Upper Pleistocene); Taiwan (Middle Pleistocene). (19, 30, 44, 45, 48, 65, 66)
1185. *D. consobrina* d'Orbigny: East China Sea; South China Sea; Taiwan (Miocene). (30, 44)
1186. *D. cuvieri* d'Orbigny: Xisha (=Paracel Islands); Zhongsha (=Macclesfield Bank). (48)
1187. *D. decepta* (Bagg): Southern Yellow Sea; East China Sea; Xisha (=Paracel Islands); Zhongsha (=Macclesfield Bank); Guangdong (Pliocene). (3, 19, 33, 48)
1188. *D. emaciata* Reuss: East China Sea; South China Sea. (30, 48)
1189. *D. extensa* Zheng & Zhang: East China Sea. (19)
1190. *D. farcimen* Reuss: East China Sea. (30)
1191. *D. filiformis* (d'Orbigny): Bohai; Yellow Sea; East China Sea; South China Sea; Okinawa Trough (Upper Pleistocene). (19, 30, 45, 66)
1192. *D. frobisherensis* Loeblich & Tappan: Xisha (=Paracel Islands); Zhongsha (=Macclesfield Bank). (48)
1193. *D. gracilis* d'Orbigny: Guangdong (Pliocene). (33)

1194. *D. guttifera* d'Orbigny [1210]: Xisha (=Paracel Islands); Zhongsha (=Macclesfield Bank); Nansha (=Spratly Islands). (48, 66)

1195. *D. insecta* (Schwager): Taiwan (Miocene). (44)

1196. *D. intorta* (Dervieux): East China Sea; Xisha (=Paracel Islands); Zhongsha (=Macclesfield Bank); Okinawa Trough (Holocene). (19, 45, 48)

1197. *D. kreyenhagensis* Mallory: Guangdong (Pliocene). (33)

1198. *D. monroei* Todd: Kashi Basin (Xinjiang) (Lower Tertiary) . (38)

1199. *D. mucronata* Neugeboren: Bohai; Yellow Sea; East China Sea; South China Sea. (30)

1200. *D. mutsui* Hada: Bohai; Yellow Sea; East China Sea; South China Sea. (30)

1201. *D. nasuta* Cushman: South China Sea. (30)

1202. *D. roemeri* Neugeboren: Xisha (=Paracel Islands). (26, 30)

1203. *D. sidebottomi* Cushman: Nansha (=Spratly Islands). (66)

1204. *D. subsoluta* (Cushman) [1228]: Nansha (=Spratly Islands). (66)

1205. *D. tortilis* Franke: Bohai; South China Sea. (30)

1206. *D. vertebralis* (Batsch): East China Sea; South China Sea. (30, 44, 67)

1207. *D. vistulae* Pozaryska: Guangdong (Pliocene). (33)

1208. *Enantiodentalina basitorta* (Cushman): East China Sea; South China Sea. (30)

1209. *Paradentalina muraii* Uchio: Xisha (=Paracel Islands); Zhongsha (=Macclesfield Bank). (48)

1210. *Grigelis guttifera* (d'Orbigny): Xisha (=Paracel Islands); Zhongsha (=Macclesfield Bank); Nansha (=Spratly Islands). (48)

1211. *G. pyrula* (d'Orbigny): Bohai; Yellow Sea; East China Sea; South China Sea; Okinawa Trough (Upper Pleistocene to Holocene); Beijing area (Quaternary). (1, 30, 45, 66)

1212. *G. semirugosa* (d'Orbigny): Southern Yellow Sea; East China Sea; South China Sea; southern Yellow Sea (Quaternary); Okinawa Trough (Holocene). (30, 31, 45)

1213. *Nodosaria affinis* d'Orbigny: East China Sea continental shelf (Quaternary); Kashi Basin (Xinjiang) (Lower Tertiary). (3, 38)

1214. *N. albatrossi* Cushman: Taiwan; Nansha (=Spratly Islands). (44, 66)

1215. *N. calomorpha* Reuss: East China Sea. (30)

1216. *N. cephalota* Wang, He, & Lu: East China Sea continental shelf (Eocene). (3)

1217. *N. flintii* Cushman: East China Sea; South China Sea. (30)

1218. *N. inflexa* Reuss: Okinawa Trough (Holocene). (45)

1219. *N. ittai* Loeblich & Tappan: Northern Yellow Sea; East China Sea. (30)

1220. *N. koina* Schwager: East China Sea. (19)

1221. *N. nitida* Terquem: East China Sea; South China Sea. (30)

1222. *N. propinqua* Costa: East China Sea; South China Sea; South China Sea (Pliocene). (3)

1223. *N. pyrula* d'Orbigny [1211]: Bohai; Yellow Sea; East China Sea; Beijing area (Quaternary). (1, 30, 45, 66)

1224. *N. radicula* (Linné) [1236]: Nansha (=Spratly Islands). (66)

1225. *N. raphana* (Linné) [1239]: East China Sea; South China Sea; Taiwan (Miocene). (30, 44)

1226. *N. semirugosa* d'Orbigny [1212]: Southern Yellow Sea; East China Sea; South China Sea; southern Yellow Sea (Quaternary). (30, 31, 45)

1227. *N. simplex* Silvestri: East China Sea; Xisha (=Paracel Islands); Zhongsha (=Macclesfield Bank); Nansha (=Spratly Islands). (30, 48, 66)

1228. *N. subsoluta* Cushman: East China Sea; South China Sea; Okinawa Trough (Holocene). (30, 45, 48, 66)

1229. *N. tainanensis* Nakamura: Taiwan. (44)

1230. *N. vertebralis* (Batsch) [1206]: East China Sea; Taiwan; South China Sea. (30, 44, 67)

1231. *Pandaglandulina dinapolii* Loeblich & Tappan: Guangdong (Pliocene). (33)

1232. *Pseudonodosaria comatula* (Cushman): East China Sea (Quaternary). (3)

1233. *P. glanduliniformis* (Dervieux): Okinawa Trough (Upper Pleistocene). (45)

1234. *P. japonica* (Asano): Southern Yellow Sea. (65)

1235. *P. laevigata* (d'Orbigny): Taiwan (Miocene). (44)

1236. *P. radicula* (Linné): Taiwan (Miocene). (44)

1237. *P. torrida* (Cushman): Okinawa Trough (Upper Pleistocene to Holocene). (45)

1238. *P. virginiana* (Cushman & Cederstrom): East China Sea continental shelf (Paleocene). (3)

1239. *Pyramidulina raphana* (Linné): East China Sea; South China Sea; Taiwan (Miocene). (30, 44)

1240. *Lingulina costata* d'Orbigny: Taiwan (Miocene). (44)

1241. *L. kansireiensis* Nakamura: Taiwan (Miocene). (44)

1242. *L. paucicostata* Chang: Taiwan (Miocene). (44)

1243. *L. polymorpha* Costa [1390]: Zhongsha (=Macclesfield Bank). (27, 30)

1244. *L. seminuda* Hantken: Xisha (=Paracel Islands); Zhongsha (=Macclesfield Bank). (48)

1245. *Lingulinopsis basicostatum* Zheng: Zhongsha (=Macclesfield Bank). (27, 30)

1246. *L. carlofortensis* Bornemann [1389]: Zhongsha (=Macclesfield Bank). (27, 30)

1247. *L. folium* Zheng: Zhongsha (=Macclesfield Bank). (27, 30)

1248. *Frondicularia inaequalis* Costa: East China Sea. (30)

1249. *F. nitida* Terquem: Nansha (=Spratly Islands). (30)

1250. *Plectofrondicularia advena* (Cushman): East China Sea. (19)

VAGINULINIDAE

1251. *Dimorphina mutsuensis* Stschedrina: Northern Yellow Sea. (65)

1252. *Lenticulina angulata* (Reuss): Southern Yellow Sea. (30)

1253. *L. asterizans* Parr: South China Sea. (30)

1254. *L. calcar* (Linné): Yellow Sea; East China Sea; Taiwan Strait; Okinawa Trough (Holocene). (3, 19, 26, 30, 44, 45, 48, 66, 67)
1255. *L. canariensis* (d'Orbigny): Okinawa Trough (Upper Pleistocene). (45)
1256. *L. changi* Huang: Taiwan (Pleistocene). (44)
1257. *L. chiriguanoi* (Boltovskoy): Okinawa Trough (Pleistocene). (45)
1258. *L. comptoni* (Sowerby): Guangdong (Pliocene). (33)
1259. *L. convergens* (Bornemann): Yellow Sea; East China Sea; South China Sea; Okinawa Trough (Holocene). (30, 45)
1260. *L. costata* (Fichtel & Moll) [1331]: East China Sea; South China Sea; East China Sea continental shelf (Quaternary); Okinawa Trough (Holocene); Taiwan (Miocene); South China Sea (Pliocene). (3, 30, 44, 45, 66)
1261. *L. depressa* (Asano): East China Sea; South China Sea. (30)
1262. *L. dorso-costata* (Cushman): Taiwan (Miocene). (44)
1263. *L. exquisita* He & Lin: Guangdong (Pliocene). (33)
1264. *L. gibba* (d'Orbigny): South China Sea. (42, 44, 66)
1265. *L. hoppoensis* (Nakamura): Taiwan. (44)
1266. *L. incavata* Zheng: Zhongsha (=Macclesfield Bank). (27, 30)
1267. *L. iota* (Cushman): East China Sea; Taiwan; Nansha (=Spratly Islands). (19, 44, 66)
1268. *L. kotiensis* (Asano): East China Sea; South China Sea. (30)
1269. *L. limbosa* (Reuss): East China Sea; Xisha (=Paracel Islands); Zhongsha (=Macclesfield Bank); Nansha (=Spratly Islands). (26, 27, 30)
1270. *L. lobata* (Reuss): Guangdong (Pliocene). (33)
1271. *L. lucida* (Cushman): Taiwan (Miocene). (44)
1272. *L. nicobarensis* (Schwager): East China Sea; South China Sea; Taiwan (Miocene). (67)
1273. *L. orbicularis* (d'Orbigny): East China Sea; South China Sea; Taiwan (Miocene); Xisha (=Paracel Islands) (Pleistocene). (30, 41, 44, 66)
1274. *L. papillosa* (Fichtel & Moll): Nansha (=Spratly Islands). (66)
1275. *L. peregrina* (Schwager) [1321]: East China Sea; South China Sea; Okinawa Trough (Upper Pleistocene to Holocene); Guangdong (Pliocene). (30, 33, 45, 64)
1276. *L. platyrhinos* Zheng: Zhongsha (=Macclesfield Bank). (27, 30)
1277. *L. pseudorotulatus* (Asano): East China Sea; South China Sea. (30)
1278. *L. sintikuensis* Nakamura: Taiwan (Miocene). (44)
1279. *L. stachei* Huang: Taiwan (Miocene). (44)
1280. *L. subangulata* (Reuss): Okinawa Trough (Upper Pleistocene). (45)
1281. *L. subgibba* (Parr): Zhongsha (=Macclesfield Bank). (27, 30)
1282. *L. submamilligera* (Cushman): South China Sea; Xisha (=Paracel Islands) (Cenozoic). (4, 30)
1283. *L. suborbicularis* Parr: East China Sea; East China Sea continental shelf (Quaternary). (27, 30)
1284. *L. surugaensis* (Asano): Taiwan (Miocene). (44)
1285. *L. taluensis* (Chang): Taiwan Strait (Neogene). (44)
1286. *L. tôtômiensis* Makiyama: Taiwan (Miocene). (44)
1287. *L. tumida* (Asano): East China Sea; Taiwan Strait; South China Sea. (19, 44, 66, 67)
1288. *L. xuwenensis* He & Lin: Guangdong (Pliocene). (33)
1289. *Robulus asterizans* Parr [1253]: Xisha (=Paracel Islands); Zhongsha (=Macclesfield Bank). (48)
1290. *R. atlanticus* Thalmann: Xisha (=Paracel Islands); Zhongsha (=Macclesfield Bank). (48)
1291. *R. australis* Parr: Xisha (=Paracel Islands); Zhongsha (=Macclesfield Bank). (48)
1292. *R. calcar* (Linné) [1254]: Southern Yellow Sea; East China Sea; South China Sea; Xisha (=Paracel Islands). (3, 19, 26, 30, 44, 45, 48, 66, 67)
1293. *R. convergens* (Bornemann) [1259]: Yellow Sea; East China Sea; South China Sea; Okinawa Trough (Holocene). (30, 45)
1294. *R. costatus* (Fichtel & Moll) [1331]: East China Sea; South China Sea; East China Sea continental shelf (Quaternary); Okinawa Trough (Holocene); Taiwan (Miocene); South China Sea (Pliocene). (3, 30, 44, 45, 66)
1295. *R. crassus* (d'Orbigny): Xisha (=Paracel Islands); Zhongsha (=Macclesfield Bank). (48)
1296. *R. cultus* Israelsky: East China Sea continental shelf (Paleocene). (3)
1297. *R. echinatus* (d'Orbigny): East China Sea; South China Sea. (30, 66)
1298. *R. expansus* (Cushman): South China Sea. (67)
1299. *R. kimituensis* Asano: East China Sea continental shelf (Quaternary); Xisha (=Paracel Islands) (Cenozoic). (3, 4)
1300. *R. knighti* Toulmin: East China Sea continental shelf (Eocene). (3)
1301. *R. limbatus* (Reuss): Nansha (=Spratly Islands). (66)
1302. *R. mammilligera* (Cushaman): Nansha (=Spratly Islands). (66)
1303. *R. nitidus* (d'Orbigny): Nansha (=Spratly Islands). (66)
1304. *R. nicobarensis* (Schwager) [1272]: East China Sea; South China Sea; Taiwan (Miocene). (67)
1305. *R. orbicularis* (d'Orbigny) [1273]: East China Sea; South China Sea; Taiwan (Miocene); Xisha (=Paracel Islands) (Upper Pleistocene). (30, 41, 44, 66)
1306. *R. papillosa* (Fichtel & Moll) [1274]: Nansha (=Spratly Islands). (66)
1307. *R. pliocenicus* Silvestri: Xisha (=Paracel Islands); Zhongsha (=Macclesfield Bank). (48)
1308. *R. rotulatus* (Lamarck): Nansha (=Spratly Islands). (66)
1309. *R. socia* (Israelsky): East China Sea continental shelf (Paleocene). (3)
1310. *R. submamilligerus* (Cushman) [1282]: South China Sea; Xisha (=Paracel Islands) (Cenozoic). (4, 30)
1311. *R. suborbicularis* (Parr) [1283]: South China Sea; East China Sea continental shelf (Quaternary). (27, 30)

1312. *R. tasmanica* Parr: Xisha (=Paracel Islands); Zhongsha (=Macclesfield Bank). (48)
1313. *R. thalmanni* Hessland: Xisha (=Paracel Islands); Zhongsha (=Macclesfield Bank). (48)
1314. *R. tumeyensis* Israelsky: East China Sea continental shelf (Eocene). (3)
1315. *R. tumidus* Asano [1287]: East China Sea; Taiwan Strait; South China Sea. (19, 44, 66, 67)
1316. *R. vortex* (Fichtel & Moll): Xisha (=Paracel Islands); Zhongsha (=Macclesfield Bank); Nansha (=Spratly Islands). (48, 66)
1317. *Marginulinopsis bradyi* (Goës): Taiwan. (44)
1318. *M. costata* (Batsch): East China Sea; South China Sea. (30, 48)
1319. *M. marshalli* Finlay: East China Sea continental shelf (Eocene). (3)
1320. *M. waiparensis* Finlay: East China Sea continental shelf (Eocene). (3)
1321. *Neolenticulina peregrina* (Schwager): East China Sea; South China Sea; Okinawa Trough (Upper Pleistocene to Holocene); Guangdong (Pliocene). (30, 33, 45, 66)
1322. *Saracenaria acutauricularis* (Fichtel & Moll): East China Sea; South China Sea; East China Sea continental shelf (Quaternary). (3, 30)
1323. *S. angularis* Natland: East China Sea; South China Sea; Okinawa Trough (Upper Pleistocene). (30, 45)
1324. *S. hannoverana* (Franke): East China Sea; South China Sea. (30)
1325. *S. italica* Defrance: East China Sea; South China Sea; Taiwan Strait; Xisha (=Paracel Islands) (Upper Pleistocene). (19, 30, 39, 48, 66)
1326. *S. latifrons* (Brady): East China Sea; Nansha (=Spratly Islands). (30, 66)
1327. *S. limbata* Hussey: East China Sea continental shelf (Paleocene). (3)
1328. *S. midwayensis* Kline: East China Sea. (19)
1329. *S. perforata* Hussey: Guangdong (Pliocene). (33)
1330. *S. schencki* Cushman & Hobson: Taiwan (Miocene). (44)
1331. *Spincterules costatus* (Fichtel & Moll): East China Sea; South China Sea; East China Sea continental shelf (Quaternary); Okinawa Trough (Holocene); Taiwan (Miocene); South China Sea (Pliocene). (3, 30, 44, 45, 66)
1332. *Palmula latifolia* Zheng: Xisha (=Paracel Islands); Zhongsha (=Macclesfield Bank). (27, 30, 48)
1333. *Amphicoryna hirsuta* (d'Orbigny): East China Sea; South China Sea; Okinawa Trough (Upper Pleistocene). (30, 45, 48, 66)
1334. *A. pauciloculata* (Cushman): Yellow Sea; East China Sea; South China Sea. (19, 30, 48, 66, 67)
1335. *A. scalaris* (Batsch): Yellow Sea; East China Sea; South China Sea. East China Sea continental shelf; Jiangsu; Zhejiang (Quaternary). (3, 8, 19, 30, 45, 47, 48, 65, 66)
1336. *A. separans* (Brady): Xisha (=Paracel Islands); Zhongsha (=Macclesfield Bank). (48)
1337. *A. proxima* (Silvestri): East China Sea; Xisha (=Paracel Islands); Zhongsha (=Macclesfield Bank). (19, 48)
1338. *A. sublineata* (Brady): East China Sea; South China Sea. (19, 30, 66)
1339. *A. subscalaris* (Cushman): East China Sea. (30)
1340. *Lagenonodosaria bilocularis* (Rhumbler): Okinawa Trough (Upper Pleistocene). (45)
1341. *L. hirsuta* (d'Orbigny): East China Sea; South China Sea; East China Sea continental shelf (Upper Pleistocene). (30, 45, 48, 66)
1342. *L. hispida* (d'Orbigny): East China Sea; South China Sea. (67)
1343. *L. leizhouensis* He & Hu: Xisha (=Paracel Islands); Zhongsha (=Macclesfield Bank). (8, 48)
1344. *L. pyrula* (d'Orbigny) [1211]: Bohai; Yellow Sea; East China Sea; South China Sea; Okinawa Trough (Upper Pleistocene to Holocene); Beijing area (Quaternary). (1, 30, 45, 66)
1345. *L. pauciloculata* (Cushman) [1334]: Yellow Sea; East China Sea; South China Sea. (19, 30, 48, 66, 67)
1346. *L. scalaris* (Batsch) [1335]: Yellow Sea; East China Sea; South China Sea; Jiangsu; Zhejiang (Quaternary). (3, 8, 19, 30, 45, 47, 48, 65, 66)
1347. *L. semirugosa* (d'Orbigny) [1212]: Southern Yellow Sea; East China Sea, South China Sea; southern Yellow Sea (Quaternary); Okinawa Trough (Holocene). (30, 31, 45)
1348. *L. variabilis* (Terquem & Berthelin): Okinawa Trough (Holocene). (45)
1349. *Astacolus costaus* (Fichtel & Moll) [1331]: East China Sea; South China Sea; East China Sea continental shelf (Quaternary); Okinawa Trough (Holocene); Taiwan (Miocene); South China Sea (Pliocene). (3, 30, 44, 45, 64, 65, 67)
1350. *A. crepidulus* (Fichtel & Moll): East China Sea; South China Sea; Guangdong (Piocene). (30, 33, 66)
1351. *A. gemmatus* (Brady) [1358]: South China Sea. (67)
1352. *A. inflata* (Parr): Okinawa Trough (Holocene). (45)
1353. *A. planulatus* Galloway & Wisseler: Taiwan (Miocene). (44)
1354. *A. pseudoplanulata* Zheng: Zhongsha (=Macclesfield Bank). (27, 30)
1355. *A. schloenbachi* (Reuss): East China Sea; South China Sea; Guangdong (Pliocene). (30, 33)
1356. *A. tricarinellus* (Reuss): South China Sea. (67)
1357. *Hemicristellaria bradyi* (Cushman): East China Sea; South China Sea. (30, 66)
1358. *H. gemmata* (Brady): East China Sea; South China Sea. (30, 48, 66)
1359. *Marginulina costata* Batsch [1318]: East China Sea; South China Sea. (30, 48)
1360. *M. hamuloides* Brotzen: Guangdong (Pliocene). (33)
1361. *M. hanzawai* (Asano): East China Sea. (19)
1362. *M. glabra* d'Orbigny: East China Sea; South China Sea. (30)
1363. *M. moodysensis* Cushman & Todd: East China Sea continental shelf (Eocene). (3)

1364. *M. obesa* Cushman: Taiwan Strait; South China Sea; Okinawa Trough (Holocene). (44, 45, 66)

1365. *M. perobliqua* Reuss: East China Sea; South China Sea. (30)

1366. *M. philippinensis* Cushman: East China Sea; South China Sea. (30, 66)

1367. *M. richteri* (Brotzen): East China Sea continental shelf (Paleocene). (3)

1368. *M. robusta* Wang, He, & Lu: East China Sea continental shelf (Paleocene). (3)

1369. *M. semicostata* Reuss: East China Sea. (19)

1370. *M. sestrona* Loeblich & Tappan: East China Sea continental shelf (Paleocene). (3)

1371. *M. striatula* Cushman: East China Sea; South China Sea. (30, 66, 67)

1372. *M. tenuis* Bornemann: East China Sea; South China Sea. (30, 66)

1373. *M. terquemi* d'Orbigny: East China Sea; South China Sea. (30)

1374. *M. varipapillata* (Israelsky): East China Sea continental shelf (Paleocene). (3)

1375. *Vaginulinopsis marwicki* Finlay: Xisha (=Paracel Islands); Zhongsha (=Macclesfield Bank). (48)

1376. *V. pacifica* (Cushamn & Hanzawa): Xisha (=Paracel Islands); Zhongsha (=Macclesfield Bank). (48)

1377. *V. sinuata* Zheng: Xisha (=Paracel Islands); Zhongsha (=Macclesfield Bank). (27, 30)

1378. *V. sublegumen* Parr: Okinawa Trough (Holocene). (45)

1379. *V. tasmanica* Parr: Xisha (=Paracel Islands); Zhongsha (=Macclesfield Bank); Nansha (=Spratly Islands). (48, 66)

1380. *Planularia tricarinella* (Reuss): East China Sea; South China Sea; Okinawa Trough (Holocene). (19, 30, 45)

1381. *Pseudarcella semisphaerea* Wang, He & Lu: East China Sea continental shelf (Eocene). (3)

1382. *Vaginulina bradyi* Cushman: Taiwan (Neogene). (44)

1383. *V. legumen* (Linné): East China Sea; South China Sea. (30)

1384. *V. margaritifera* (Batsch): Okinawa Trough (Holocene). (45)

1385. *V. patens* Brady: Nansha (=Spratly Islands). (66)

1386. *V. protumida* (Schwager): East China Sea. (30)

1387. *V. spinigera* Brady: East China Sea; South China Sea. (30, 48, 66)

1388. *V. takaoensis* Ishizaki: Taiwan (Neogene). (44)

1389. *Spirolingulina carlofortensis* (Bornemann): Zhongsha (=Macclesfield Bank). (27, 30)

1390. *S. polymorpha* (Costa): Zhongsha (=Macclesfield Bank). (27, 30)

LAGENIDAE

1391. *Hyalinonetrion elongata* (Ehrenberg): Bohai; Yellow Sea; East China Sea; South China Sea; Okinawa Trough (Upper Pleistocene to Holocene); Jiangsu (Quaternary). (8, 19, 30, 31, 45, 48, 66)

1392. *Lagena acuticosta* Reuss: East China Sea; Okinawa Trough (Upper Pleistocene). (19, 45)

1393. *L. alcocki* White: Okinawa Trough (Upper Pleistocene). (45)

1394. *L. amphora* Reuss: Bohai; Yellow Sea; East China Sea; South China Sea. (30)

1395. *L. apiopleura* Loeblich & Tappan: East China Sea. (30)

1396. *L. aspera* Reuss: Nansha (=Spratly Islands); Okinawa Trough (Holocene). (45, 64)

1397. *L. clavata* Williamson: Bohai; Yellow Sea; South China Sea; Jiangsu (Quaternary). (8, 30)

1398. *L. costata* (Williamson): East China Sea; South China Sea. (19, 66)

1399. *L. crenata* Parker & Jones: East China Sea; South China Sea. (30, 66)

1400. *L. desmophora* Rymer & Jones: Nansha (=Spratly Islands). (66)

1401. *L. dentaliniformis* Bagg: East China Sea. (30)

1402. *L. distoma* Parker & Jones: Southern Yellow Sea; East China Sea; Nansha (=Spratly Islands); southern Yellow Sea (Quaternary). (19, 30, 31, 66)

1403. *L. distorta* (Seguenza): Okinawa Trough (Holocene). (45)

1404. *L. doveyensis* Haynes: Southern Yellow Sea; near sea section of Haihe and Ji Canal (Hebei); East China Sea; South China Sea; Shandong (Quaternary). (19, 20, 29, 30, 66)

1405. *L. excentrica* Sidebottom: Okinawa Trough (Upper Pleistocene). (39)

1406. *L. elegantissima* (Bornemann): Nansha (=Spratly Islands). (66)

1407. *L. elongata* (Ehrenberg) [1391]: Bohai; Yellow Sea; East China Sea; South China Sea; Okinawa Trough (Upper Pleistocene to Holocene); Jiangsu (Quaternary). (8, 19, 30, 31, 45, 48, 66)

1408. *L. exsculpta* Brady [1453]: Xisha (=Paracel Islands) (Upper Pleistocene). (39)

1409. *L. flintiana* Cushman: Xisha (=Paracel Islands) (Upper Pleistocene). (39)

1410. *L. feildeniana* Brady: Nansha (=Spratly Islands). (66)

1411. *L. flatulenta* Loeblich & Tappan: East China Sea; South China Sea. (30)

1412. *L. gibbera* Buchner: Nansha (=Spratly Islands). (66)

1413. *L. globosa* (Montagu): East China Sea; South China Sea. (30)

1414. *L. gracilis* Williamson [1451]: Bohai; northern Yellow Sea; Nansha (=Spratly Islands). (30, 66)

1415. *L. gracillima* (Sequenza): Northern Yellow Sea; Nansha (=Spratly Islands); Jiangsu (Quaternary). (8, 30, 66)

1416. *L. gortanii* Selli: Xisha (=Paracel Islands) (Middle Pleistocene). (39)

1417. *L. hispida* Reuss: Southern Yellow Sea; East China Sea; South China Sea; Okinawa Trough (Upper Pleistocene); Jiangsu (Quaternary). (19, 30, 39, 45, 48)

1418. *L. hispidula* Cushman [1452]: East China Sea; Okinawa Trough (Upper Pleistocene to Holocene). (19, 30, 45)

1419. *L. hyugaensis* Oinomikado: East China Sea. (19)
1420. *L. interrupta* Williamson: Southern Yellow Sea (Quaternary). (31)
1421. *L. laevicostata* Cushman & Gray: Okinawa Trough (Upper Pleistocene). (45)
1422. *L. laevis* (Montagu): Bohai; Yellow Sea; East China Sea; Shandong (Quaternary). (29, 30, 31, 48, 66)
1423. *L. longispina* Brady: Xisha (=Paracel Islands); Zhongsha (=Macclesfield Bank). (48)
1424. *L. meridionalis* Wiesner: Xisha (=Paracel Islands); Zhongsha (=Macclesfield Bank). (48)
1425. *L. mississippiensis* Cushman & Todd: East China Sea continental shelf (Paleocene). (3)
1426. *L. nebulosa* Cushman: East China Sea; Okinawa Trough (Holocene); Xisha (=Paracel Islands) (Middle Pleistocene). (39, 45)
1427. *L. oxystoma* Reuss: Xisha (=Paracel Islands) (Middle Pleistocene). (39)
1428. *L. pellita* (Heron-Allen & Earland): Okinawa Trough (Upper Pleistocene to Holocene). (45)
1429. *L. perlucida* (Montagu): Bohai; Yellow Sea; East China Sea; South China Sea; Jiangsu (Quaternary). (8, 19, 30, 45, 65)
1430. *L. pliocenica* Cushman & Gray: Southern Yellow Sea; East China Sea; South China Sea; Jiangsu (Quaternary). (8, 30, 67)
1431. *L. plumigera* Brady: Nansha (=Spratly Islands). (66)
1432. *L. pseudocostata* Olsson: East China Sea continental shelf (Eocene). (3)
1433. *L. quadrilatera* Earland [1457]: East China Sea. (19)
1434. *L. semilineata* Wright [1455]: Southern Yellow Sea; Okinawa Trough (Upper Pleistocene to Holocene). (45, 65)
1435. *L. semistriata* (Williamson): East China Sea; South China Sea. (30, 66)
1436. *L. sesquistriata* Bagg: East China Sea. (30)
1437. *L. setigera* Millett: Shoreline of Bohai (Quaternary); Okinawa Trough (Upper Pleistocene). (45, 49)
1438. *L. spicata* Cushman & McCulloch: East China Sea; southern Yellow Sea (Quaternary); Okinawa Trough (Upper Pleistocene to Holocene). (19, 31, 45)
1439. *L. spinigera* Earland [1456]: Okinawa Trough (Upper Pleistocene). (45)
1440. *L. stavensis* Bandy: Xisha (=Paracel Islands) (Upper Pleistocene). (39)
1441. *L. striata* (d'Orbigny): Bohai; Yellow Sea; East China Sea; South China Sea; East China Sea (Quaternary to Pliocene); Jiangsu (Pliocene). (3, 8, 30, 48, 66)
1442. *L. striatifera* Tappan: East China Sea. (19)
1443. *L. striatopunctata* (Parker & Jones): East China Sea; Xisha (=Paracel Islands). (28, 30)
1444. *L. subamphora* Asano: Southern Yellow Sea. (65)
1445. *L. substriata* Williamson: Bohai; southern Yellow Sea; East China Sea; South China Sea; Jiangsu; Okinawa Trough (Upper Pleistocene to Holocene); Leiqiong (southern Guangdong and Hainan) (Quaternary); Xisha (=Paracel Islands) (Upper Pleistocene). (3, 8, 19, 30, 31, 45, 66)
1446. *L. sulcata* (Walker & Jacob): Bohai; southern Yellow Sea; East China Sea; Taiwan Strait. (30, 44, 46)
1447. *L. sulcatospicata* Cushman & McCulloch: Yellow Sea; Xisha (=Paracel Islands); Zhongsha (=Macclesfield Bank); Nansha (=Spratly Islands). (65, 66)
1448. *L. tenuis* (Bornemann): Bohai. (30)
1449. *L. wiesneri* Parr: Yellow Sea; East China Sea; Xisha (=Paracel Islands) (Middle Pleistocene). (19, 39)
1450. *L. williamsoni* Alcock: Beijing area (Quaternary). (2)
1451. *Procerolagena gracilis* (Williamson): Bohai; northern Yellow Sea; Nansha (=Spratly Islands). (30, 66)
1452. *Pygmaeoseistron hispidula* (Cushman): East China Sea; Okinawa Trough (Upper Pleistocene to Holocene). (19, 30, 45)
1453. *Exculptina exculpta* (Brady): Xisha (=Paracel Islands) (Upper Pleistocene to Holocene). (39)
1454. *E. pliocenica* (Cushman & Gray): Southern Yellow Sea; East China Sea; South China Sea; Jiangsu (Quaternary). (8, 30, 67)
1455. *E. semilineata* (Wright): Southern Yellow Sea; Okinawa Trough (Upper Pleistocene to Holocene). (45, 65)
1456. *E. spinigera* (Earland): Okinawa Trough (Upper Pleistocene). (45)
1457. *Laculatina quadrilatera* (Earland): East China Sea. (19)

POLYMORPHINIDAE

1458. *Falsoguttulina laevigata* Wang, He, & Lu: East China Sea continental shelf (Paleocene). (3)
1459. *Globulina gibba* (d'Orbigny): Guangdong (Pliocene). (33)
1460. *G. landesi* (G. D. Hanna & M. A. Hanna): East China Sea; South China Sea. (26, 28, 30)
1461. *G. minuta* (Roemer): Jiangsu (Quaternary). (8, 47)
1462. *G. trigona* Zheng: Zhongsha (=Macclesfield Bank). (27, 30)
1463. *Guttulina austriaca* (d'Orbigny): Bohai; Yellow Sea; East China Sea; South China Sea. (30, 48)
1464. *G. basalis* Galloway & Hemingway: Guangdong (Miocene). (33)
1465. *G. dawsoni* Cushman & Ozawa: Guangdong (Pliocene). (33)
1466. *G. lactea* (Walker & Jacob): Bohai; East China Sea; South China Sea; Shandong (Quaternary). (29, 30)
1467. *G. hantkeni* Cushman & Ozawa: Jiangsu (Quaternary). (8)
1468. *G. kishinouyi* Cushman & Ozawa: Jiangsu (Quaternary). (8)
1469. *G. orientalis* (Cushman & Ozawa): East China Sea; Jiangsu (Quaternary); Guangdong (Pliocene). (8, 21, 33)
1470. *G. pacifica* (Cushman & Ozawa): East China Sea; Taiwan Strait; South China Sea; shoreline of Bohai (Quaternary). (19, 30, 44, 48, 49, 66)
1471. *G. praelonga* (Egger): Jiangsu (Quaternary). (8)

1472. *G. problema* (d'Orbigny): East China Sea; South China Sea; Xisha (=Paracel Islands) (Upper Pleistocene). (30, 41, 48, 67)
1473. *G. pulchella* (d'Orbigny): South China Sea. (30)
1474. *G. regina* (Brady, Parker, & Jones): East China Sea; Taiwan Strait; East China Sea continental shelf (Miocene). (3, 19)
1475. *G. sadoensis* (Cushman & Ozawa): Guangdong (Pliocene). (33)
1476. *G. silvestri* Cushman & Ozawa: East China Sea; South China Sea. (30)
1477. *Polymorphina elegantissima* Parker & Jones [1493]: East China Sea; South China Sea; Guangdong (Pliocene). (30, 33, 66, 67)
1478. *P. ovata* d'Orbigny: Nansha (=Spratly Islands). (66)
1479. *Polymorphinella vaginulaeformis* Cushman & Hanzawa: Nansha (=Spratly Islands). (30)
1480. *Pseudopolymorphina indica* (Cushman): East China Sea; South China Sea. (19, 26, 30)
1481. *P. netroformis* Hao & Zeng: Kashi Basin (Xijiang) (Lower Tertiary). (38)
1482. *P. ovalis* Cushman & Ozawa: Xisha (=Paracel Islands); Zhongsha (=Macclesfield Bank); Beijing area (Lower Quaternary). (2, 48)
1483. *P. suboblonga* Cushman & Ozawa: Beijing area (Lower Quaternary); Jiangsu (Quaternary). (2, 8)
1484. *P. soldani* (d'Orbigny): Bohai; Yellow Sea; East China Sea. (30)
1485. *Pyrulina albatrossi* Cushman & Ozawa: East China Sea. (30)
1486. *P. angusta* (Brady): Nansha (=Spratly Islands). (66)
1487. *P. cylindroides* (Roemer): East China Sea; South China Sea. (30)
1488. *P. extensa* (Cushman): Nansha (=Spratly Islands). (66)
1489. *P. fusiformis* (Roemer): East China Sea; Xisha (=Paracel Islands); Zhongsha (=Macclesfield Bank). (30, 48)
1490. *P. gutta* d'Orbigny: Nansha (=Spratly Islands). (66)
1491. *Pyrulinoides rasilis* Hao & Zeng: Kashi Basin (Xijiang) (Lower Tertiary). (38)
1492. *Sigmoidella bacomensis* Yabe & Asano: Taiwan (Neogene). (44)
1493. *S. elegantissima* (Parker & Jones): East China Sea; South China Sea; Guangdong (Pliocene). (30, 33, 66, 67)
1494. *S. kagaensis* Cushman & Ozawa: East China Sea; South China Sea. (33)
1495. *S. pacifica* Cushman & Ozawa: Guangdong (Pliocene). (33)
1496. *S. subtaiwanica* Nakamura: East China Sea continental shelf (Pliocene); Taiwan (Tertiary); South China Sea. (33)
1497. *S. taiwanensis* Nakamura: Taiwan (Tertiary). (44)
1498. *Sigmomorphina basistriata* Zheng: Xisha (=Paracel Islands). (25, 26, 30)
1499. *S. gallowayi* Cushman & Ozawa: Shoreline of Bohai (Quaternary). (49)
1500. *S. ozawai* (Hada): Southern Yellow Sea; Okinawa Trough (Holocene); Guangdong (Pliocene). (30, 33, 45)
1501. *S. semitecta* (Reuss): Bohai; East China Sea; South China Sea. (30)
1502. *S. subcircularis* Zheng: Xisha (=Paracel Islands). (26, 30)
1503. *S. trilocularis* Bagg: Beijing area (Lower Quaternary). (2)
1504. *S. undulosa* (Terquem): Shoreline of Bohai (Quaternary). (49)
1505. *S. williamsoni* (Terquem): South China Sea. (30)
1506. *S. yokoyamai* Cushman & Ozawa: Bohai; southern Yellow Sea; East China Sea; South China Sea. (30)
1507. *Ramulina globulifera* Brady: East China Sea; South China Sea. (30, 66)

ELLIPSOLAGENIDAE

1508. *Cushmanina desmophora* Rymer & Jones: Nansha (=Spratly Islands). (66)
1509. *Galwayella trigonomarginata* Parker & Jones: Northern Yellow Sea. (30)
1510. *Favulina hexagona* (Williamson): East China Sea; South China Sea; southern Yellow Sea (Quaternary); East China Sea continental shelf (Tertiary). Guangdong (Pliocene). (3, 19, 30, 33, 45, 48, 65)
1511. *F. melo* (d'Orbigny): Southern Yellow Sea; East China Sea; South China Sea; East China Sea (Quaternary); South China Sea (Pliocene). (3, 19, 33, 48, 66)
1512. *F. squamosa* (Montagu): Southern Yellow Sea; East China Sea; Taiwan Strait; South China Sea. (30, 50)
1513. *Lagnea radiata* (Sequenza): Okinawa Trough (Holocene). (45)
1514. *Oolina ampullodistoma* (Jones): Xisha (=Paracel Islands). (26, 30)
1515. *O. borealis* (Williamson): Beijing area (Lower Quaternary). (2)
1516. *O. brevisolenia* Zheng: Xisha (=Paracel Islands). (26, 30)
1517. *O. caudigera* (Wiesner): Xisha (=Paracel Islands). (26, 30)
1518. *O. costata* (Williamson): Guangdong (Pliocene). (33)
1519. *O. favopunctata* (Brady): East China Sea. (30)
1520. *O. globosa* (Montagu): East China Sea; Taiwan Strait; Nansha (=Spratly Islands). (30, 44, 66)
1521. *O. hertwigiana* (Brady): East China Sea. (30)
1522. *O. hexagona* (Williamson) [1510]: East China Sea; South China Sea; southern Yellow Sea (Quaternary); East China Sea continental shelf (Tertiary); Guangdong (Pliocene). (3, 19, 30, 33, 45, 48, 67)
1523. *O. laevigata* d'Orbigny: East China Sea; Guangdong (Pliocene). (19, 33)
1524. *O. lineata* (Williamson): East China Sea; South China Sea. (30)
1525. *O. longispina* (Brady): Nansha (=Spratly Islands). (66)
1526. *O. melo* d'Orbigny [1511]: Southern Yellow Sea; East China Sea; South China Sea; East China Sea continental shelf (Quaternary); South China Sea (Pliocene). (3, 19, 33, 48, 66)

1527. *O. shunyiense* Wang & He: Beijing area (Lower Quaternary). (2)
1528. *O. spiralis* (Brady): Xisha (=Paracel Islands). (26, 30)
1529. *O. squamosa* (Montagu) [1512]: Yellow Sea; East China Sea; Taiwan Strait; South China Sea. (30, 50)
1530. *O. taiyanggonense* Wang & Hu: Beijing area (Lower Quaternary). (2)
1531. *O. variata* (Brady): East China Sea. (30)
1532. *Fissurina annectens* (Burrows & Holland): Bohai. (30)
1533. *F. aperta* Sequenza: Okinawa Trough (Holocene); Guangdong (Pliocene). (33, 45)
1534. *F. aradasii* Sequenza: Bohai; Shandong (Quaternary); Xisha (=Paracel Islands) (Upper Pleistocene). (21, 29, 30, 39)
1535. *F. atlantica* Redmond: East China Sea continental shelf (Quaternary). (3)
1536. *F. barri* (Cushman & Stainforth): East China Sea continental shelf (Quaternary). (3)
1537. *F. biancae* Sequenza: Okinawa Trough (Upper Pleistocene). (45)
1538. *F. bicarinata* Terquem: East China Sea; South China Sea. (30)
1539. *F. bicaudata* Sequenza: Okinawa Trough (Upper Pleistocene). (45)
1540. *F. carinata* Reuss: Nansha (=Spratly Islands). (66)
1541. *F. clathrata* (Brady): East China Sea. (30)
1542. *F. circularis* Todd: Xisha (=Paracel Islands); Zhongsha (=Macclesfield Bank); Okinawa Trough (Upper Pleistocene to Holocene). (45, 48)
1543. *F. compressioblonga* Zheng: Xisha (=Paracel Islands); Okinawa Trough (Upper Pleistocene). (26, 30, 45)
1544. *F. contusa* Parr: Xisha (=Paracel Islands); Zhongsha (=Macclesfield Bank). (26, 30, 48)
1545. *F. crebra* (Matthes): Shoreline of Bohai (Quaternary); Okinawa Trough (Holocene). (45)
1546. *F. crenulata* Coryell & Rivera: Nansha (=Spratly Islands); Xisha (=Paracel Islands) (Pleistocene). (39, 66)
1547. *F. cucullata* Silvestri: Xisha (=Paracel Islands); Zhongsha (=Macclesfield Bank). (48)
1548. *F. cucurbitasema* Loeblich & Tappan: Bohai; shoreline of Bohai (Quaternary); Okinawa Trough (Upper Pleistocene). (30, 45, 49)
1549. *F. dendiculifera* Buchner: Okinawa Trough (Holocene). (45)
1550. *F. diaphana* (Buchner): Xisha (=Paracel Islands). (26, 30)
1551. *F. disjuncta* Zheng: Xisha (=Paracel Islands). (26, 30)
1552. *F. disjungens* (Buchner) [1594]: Xisha (=Paracel Islands). (26, 30)
1553. *F. earlandi* Parr: Okinawa Trough (Upper Pleistocene to Holocene). (45)
1554. *F. eburnea* (Buchner): Okinawa Trough (Holocene). (45)
1555. *F. elliptica* (Cushman): Okinawa Trough (Holocene). (45)
1556. *F. fasciata* (Egger): Xisha (=Paracel Islands); Okinawa Trough (Upper Pleistocene to Holocene). (26, 30, 45)
1557. *F. fimbriata* (Brady): East China Sea. (30)
1558. *F. formosa* (Schwager): Nansha (=Spratly Islands). (66)
1559. *F. furcicolla* Zheng: Xisha (=Paracel Islands). (26, 30)
1560. *F. granulocostulata* Zheng: Xisha (=Paracel Islands). (26, 30)
1561. *F. jasolini* (Buchner): Xisha (=Paracel Islands). (26, 30)
1562. *F. kerimbatica* (Heron-Allen & Earland): Xisha (=Paracel Islands). (26, 30)
1563. *F. lacunata* (Burrows & Holland): East China Sea; East China Sea continental shelf (Quaternary); Okinawa Trough (Holocene). (3, 19, 45)
1564. *F. laevigata* Reuss: Bohai; southern Yellow Sea; East China Sea; South China Sea; Jiangsu (Quaternary). (19, 20, 29, 30, 49)
1565. *F. laevis* Sequenza: Shoreline of Bohai (Quaternary); East China Sea continental shelf (Quaternary). (3)
1566. *F. lucida* (Williamson): Bohai; Yellow Sea; East China Sea; South China Sea; Shandong (Quaternary). (19, 20, 29, 30, 49)
1567. *F. marginata* Sequenza: Bohai; southern Yellow Sea; Taiwan Strait; South China Sea; Okinawa Trough (Upper Pleistocene to Holocene). (30, 44, 45, 68)
1568. *F. milletti* Todd: Xisha (=Paracel Islands). (26, 30)
1569. *F. minipora* Zheng: Xisha (=Paracel Islands). (26, 30)
1570. *F. occlusa* (Heron-Allen & Earland): Okinawa Trough (Holocene). (45)
1571. *F. orbignyana* (Sequenza): Yellow Sea; East China Sea; Taiwan Strait; South China Sea; shoreline of Bohai (Quaternary); Okinawa Trough (Holocene); Guangdong (Upper Pleistocene). (45)
1572. *F. propinqua* Sequenza: Okinawa Trough (Holocene). (45)
1573. *F. pulchella* (Brady): Beijing area (Lower Quaternary). (2)
1574. *F. pseudoglobosa* (Buchner): Xisha (=Paracel Islands); Nansha (=Spratly Islands). (26, 30)
1575. *F. pseudolucida* Zheng: Xisha (=Paracel Islands). (26, 30)
1576. *F. pseudoorbignyana* (Buchner): Xisha (=Paracel Islands). (45)
1577. *F. quadrata* (Williamson): Okinawa Trough (Holocene). (45)
1578. *F. quadricostulata* (Reuss): South China Sea. (30)
1579. *F. radiata* Sequenza [1513]: Xisha (=Paracel Islands); Okinawa Trough (Holocene). (45)
1580. *F. radiatomarginata* (Parker & Jones): Xisha (=Paracel Islands); Zhongsha (=Macclesfield Bank); Okinawa Trough (Holocene). (26, 30, 45, 48)
1581. *F. radiatostoma* Zheng: Xisha (=Paracel Islands). (26, 30)
1582. *F. robusta* Zheng: Xisha (=Paracel Islands). (26, 30)
1583. *F. sacculus* Fornasini: Nansha (=Spratly Islands). (66)
1584. *F. semimarginata* (Reuss): Yellow Sea; Xisha (=Paracel Islands); Zhongsha (=Macclesfield Bank); Shandong; southern Yellow Sea (Quaternary). (31, 48, 65)
1585. *F. sexcostulata* Silvestri: Okinawa Trough (Holocene). (45)
1586. *F. simplicita* (Matthes): Shoreline of Bohai (Quaternary). (49)

1587. *F. sigmoidella* (Cushman): Okinawa Trough (Holocene). (45)

1588. *F. subglobosa* (Buchner): Nansha (=Spratly Islands). (30)

1589. *F. tricaudata* Silvestri: Nansha (=Spratly Islands); Okinawa Trough (Holocene). (30, 45)

1590. *F. tropicalis* Zheng: Xisha (=Paracel Islands); Okinawa Trough (Holocene). (26, 30, 45)

1591. *F. walleriana* (Wright): Xisha (=Paracel Islands). (26, 30)

1592. *F. wrightiana* (Brady): Xisha (=Paracel Islands); Guangdong (Pliocene). (26, 30, 33)

1593. *F. unicospina* (Cushman & Renz): East China Sea. (19)

1594. *Lagenosolenia disjungens* (Buchner): Xisha (=Paracel Islands). (26, 30)

1595. *Parafissurina agassizi* (Todd & Brönnimann): Xisha (=Paracel Islands) (Pleistocene). (39)

1596. *P. cuniculifera* (Buchner): Okinawa Trough (Upper Pleistocene). (45)

1597. *P. himatiostoma* Loeblich & Tappan: Okinawa Trough (Upper Pleistocene to Holocene); Xisha (=Paracel Islands) (Pleistocene). (39, 45)

1598. *P. minuta* Zheng: Xisha (=Paracel Islands). (26, 30)

1599. *P. oblonga* Zheng: Xisha (=Paracel Islands). (26, 30)

1600. *P. quadrispinata* Zheng: Xisha (=Paracel Islands). (26, 30)

1601. *P. reflecta* Zheng: Xisha (=Paracel Islands). (26, 30)

1602. *P. subcarinata* Parr: Okinawa Trough (Upper Pleistocene to Holocene). (45)

1603. *P. subcircularis* Parr: Okinawa Trough (Upper Pleistocene to Holocene). (45)

1604. *P. sublata* Parr: Okinawa Trough (Upper Pleistocene to Holocene). (45)

1605. *P. subovata* Parr: Okinawa Trough (Holocene). (45)

GLANDULINIDAE

1606. *Esosyrinx curta* (Cushman & Ozawa): Yellow Sea; Jiangsu (Quaternary). (8, 65)

1607. *E. guttuliniformis* Stschedrina: Northern Yellow Sea. (65)

1608. *Glandulina aequalis* Reuss: East China Sea; South China Sea. (30)

1609. *G. apiculata* Costa: East China Sea continental shelf (Eocene). (3)

1610. *G. dimorpha* (Bornemann): Southern Yellow Sea; East China Sea; South China Sea. (30)

1611. *G. echinata* Millet: Taiwan Strait; South China Sea. (50, 66)

1612. *G. elliptica* Reuss: South China Sea. (67)

1613. *G. globosa* Ho & Hu: Xisha (=Paracel Islands); Zhongsha (=Macclesfield Bank); Nansha (=Spratly Islands). (48, 66)

1614. *G. granulosa* Zheng: Xisha (=Paracel Islands). (26, 30)

1615. *G. hainanensis* N. W. Wang & Hu: Guangdong (Pliocene). (33)

1616. *G. laevigata* (d'Orbigny): Yellow Sea; East China Sea; Okinawa Trough (Holocene); South China Sea; Guangdong (Pliocene). (19, 30, 33, 45, 48, 66)

1617. *G. rotundata* Reuss: East China Sea; South China Sea. (30, 66)

1618. *Globulotuba entosoleniformis* Collins: Southern Yellow Sea. (65)

1619. *Laryngosigma caudata* Zheng: Xisha (=Paracel Islands). (26, 30)

1620. *L. hyalascidia* Loeblich & Tappan: Southern Yellow Sea. (65)

1621. *L. lauta* Stschedrina: Northern Yellow Sea. (65)

1622. *L. ovata* Stschedrina: Southern Yellow Sea. (65)

1623. *L. williamsoni* (Terquem): Xisha (=Paracel Islands); Nansha (=Spratly Islands). (25, 26, 30)

1624. *Entolingulina inarimensis* (Buchner): Xisha (=Paracel Islands). (26, 30)

1625. *Obliquilingulina oblonga* Zheng: Xisha (=Paracel Islands). (26, 30)

ROBERTININA
CERATOBULIMINACEA
CERATOBULIMINIDAE

1626. *Ceratobulimina pacifica* Cushman & Harris: East China Sea; South China Sea; Okinawa Trough (Upper Pleistocene). (19, 45, 48, 66)

1627. *Lamarckina fuchsis* Rvehok: Xisha (=Paracel Islands); Zhongsha (=Macclesfield Bank). (48)

1628. *L. scabra* (Brady): East China Sea; South China Sea. (19, 30, 48, 66)

1629. *L. ventricosa* (Brady): Nansha (=Spratly Islands). (66)

EPISTOMINIDAE

1630. *Höglundina bradyi* (Galloway & Wissler): Okinawa Trough (Upper Pleistocene). (45)

1631. *H. elegans* (d'Orbigny): East China Sea; South China Sea; Okinawa Trough (Upper Pleistocene to Holocene); East China Sea continental shelf (Quaternary); Xisha (=Paracel Islands) (Pleistocene to Holocene). (19, 30, 39, 45, 65, 67)

ROBERTINACEA
ROBERTINIDAE

1632. *Alliatina nitida* (Millett): South China Sea; Okinawa Trough (Upper Pleistocene to Holocene). (26, 30, 45, 66)

1633. *A. oinomikadoi* (Huang): Taiwan (Neogene). (44)

1634. *Pseudononionella variabilis* Zheng: Bohai; Yellow Sea; East China Sea; Hebei; Shandong; Taiwan; East China Sea (Quaternary). (3, 19, 20, 21, 29, 30, 31, 34, 37, 47)

1635. *Cushmanella primitiva* Cushman & McCulloch: South China Sea. (30)

1636. *Geminospira bradyi* Bermudez: Xisha (=Paracel Islands); Zhongsha (=Macclesfield Bank); Nansha (=Spratly Islands). (27, 30, 66)

1637. *G. mayori* (Cushman): Xisha (=Paracel Islands); Zhongsha (=Macclesfield Bank). (48)

1638. *G. simaensis* Makiyama & Nakagawa: East China Sea. (19)
1639. *Pseudobulimina convoluta* (Williamson): East China Sea; South China Sea. (30)
1640. *Robertina bradyi* Cushman & Parker [1643]: Xisha (=Paracel Islands); Zhongsha (=Macclesfield Bank); Nansha (=Spratly Islands); Okinawa Trough (Upper Pleistocene). (45, 48, 66)
1641. *R. tasmanica* Parr: Xisha (=Paracel Islands); Zhongsha (=Macclesfield Bank). (48)
1642. *R. translucens* Cushman & Parker: Okinawa Trough (Holocene). (45)
1643. *Robertinoides bradyi* Cushmand & Parker: Xisha (=Paracel Islands); Zhongsha (=Macclesfield Bank); Nansha (=Spratly Islands). (48, 66)
1644. *R. declivis* (Reuss): East China Sea; South China Sea. (30, 66)
1645. *R. normanni* (Goës): Nansha (=Spratly Islands). (66)

GLOBIGERININA
HETEROHELICACEA
GUEMBELITRIIDAE

1646. *Gallitellia vivans* (Cushman): Yellow Sea; East China Sea; Taiwan Strait; South China Sea. (19, 30, 64, 65, 67)
1647. *Guembelitria vivans* Cushman [1646]: Yellow Sea; East China Sea; Taiwan Strait; South China Sea. (19, 30, 64, 65, 67)

HETEROHELICIDAE

1648. *Bifarina mckinnoni* Millet: Nansha (=Spratly Islands). (30)

CHILOGUEMBELINIDAE

1649. *Laterostomella spinosa* Zheng: Zhongsha (=Macclesfield Bank). (27, 30)

GLOBOROTALIACEA
GLOBOROTALIIDAE

1650. *Globorotalia acostaensis* Blow: Beijing area (Lower Quaternary); Taiwan (Pliocene); South China Sea (Neogene). (2, 52, 72)
1651. *G. abundocamerata* Bolli: Kashi Basin (Xinjiang) (Tertiary); Tarim Basin (Kasin area, Xinjiang) (Paleocene). (38, 40)
1652. *G. anfracta* Parker: South China Sea. (70)
1653. *G. angulata* (White) [1730]: Kashi Basin (Xinjiang) (Tertiary). (38)
1654. *G. archeomenardii* Bolli: Taiwan (Miocene). (52)
1655. *G. bermudezi* Rogl & Bolli: Okinawa Trough (Upper Pleistocene to Holocene); Xisha (=Paracel Islands) (Upper Pleistocene to Holocene). (39, 45)
1656. *G. clarkei* (Rogl & Bolli) [1820]: East China Sea; Okinawa Trough (Upper Pleistocene to Holocene); South China Sea (Upper Tertiary). (19, 45, 72)
1657. *G. compressa* (Plummer): Kashi Basin (Xinjiang) (Lower Tertiary). (38)
1658. *G. conoidea* Walter: Taiwan (Upper Tertiary). (38)
1659. *G. conicotruncata* Subbotina: Kashi Basin (Xinjiang) (Lower Tertiary). (38)
1660. *G. continuosa* Blow: Beijing area (Lower Quaternary). (2)
1661. *G. crasaformis* (Galloway & Wissler): Taiwan Strait; South China Sea; Okinawa Trough (Upper Pleistocene to Holocene); Xisha (=Paracel Islands) (Middle Pleistocene to Holocene). (24, 30, 38, 41, 43, 45, 48, 50, 72)
1662. *G. crassula* Cushman & Steward: Okinawa Trough (Upper Pleistocene to Holocene). (45)
1663. *G. crotonensis* Conato & Follador: East China Sea. (12)
1664. *G. cultrata* (d'Orbigny): South China Sea. (67)
1665. *G. eastropacia* Boltovskoy: Okinawa Trough (Upper Pleistocene to Holocene). (45)
1666. *G. ehrenbergi* Bolli: East China Sea continental shelf (Paleocene). (3)
1667. *G. exilis* Blow: South China Sea (Neogene). (72)
1668. *G. fimbriata* (Brady): Okinawa Trough (Holocene). (45)
1669. *G. flexuosa* (Koch): Okinawa Trough (Upper Pleistocene); Xisha (=Paracel Islands) (Middle Pleistocene to Holocene); South China Sea (Neogene). (39, 45, 72)
1670. *G. hirsuta* (d'Orbigny): East China Sea; South China Sea. (23, 24, 30, 43, 45)
1671. *G. humerosa* (Takayanagi & Saito): Beijing area (Lower Quaternary); South China Sea (Neogene). (2, 72)
1672. *G. inflata* (d'Orbigny): East China Sea; South China Sea; shoreline of Taizhou Bay (Zhejiang) (Upper Quaternary); Beijing area (Lower Quaternary). (2, 23, 24, 30, 43, 45, 47, 72)
1673. *G. lata* Bronnimann & Resig: Okinawa Trough (Upper Pleistocene). (45)
1674. *G. margaritae* Bolli & Bermudez: Xisha (=Paracel Islands) (Upper Miocene to present); South China Sea (Neogene). (36, 72)
1675. *G. mayeri* Bolli: Taiwan (Miocene). (61)
1676. *G. menardii* (d'Orbigny): East China Sea; South China Sea; East China Sea (Miocene); Okinawa Trough (Upper Pleistocene to Holocene); Taiwan (Tertiary); Guangdong (Miocene to Pliocene); Leiqiong area (southern Guangdong and Hainan) (Tertiary); Xisha (=Paracel Islands) (Upper Pleistocene). (2, 3, 23, 24, 30, 33, 36, 39, 41, 43, 45, 46, 67, 72)
1677. *G. miocenica* Palmer: South China Sea (Neogene). (72)
1678. *G. multicamerata* Cushman & Jarvis: South China Sea (Neogene). (72)
1679. *G. obesa* Bolli: Okinawa Trough (Upper Pleistocene to Holocene); Taiwan (Miocene). (45, 52)
1680. *G. oscitans* (Todd): Okinawa Trough (Upper Pleistocene to Holocene). (45)
1681. *G. padana* Dondi & Papetti: Taiwan (Neogene). (58)

1682. *G. paekerae* Brönnimann & Resig: Okinawa Trough (Upper Pleistocene to Holocene). (45)
1683. *G. peripheroacuta* Blow & Banner: Taiwan (Miocene). (61)
1684. *G. perpheroronda* Blow & Banner: Taiwan (Miocene); South China Sea (Neogene). (46, 61, 72)
1685. *G. plesiotumida* Blow & Banner: Xisha (=Paracel Islands) (Cenozoic). (4)
1686. *G. praemenardii* Cushman & Stainforth: Taiwan (Miocene); South China Sea (Neogene). (61, 72)
1687. *G. pseudobulloides* Plummer: Kashi Basin (Xinjiang) (Lower Tertiary). (38)
1688. *G. pseudomenardii* Bolli: East China Sea continental shelf (Paleocene). (3)
1689. *G. pumilio* Parker: East China Sea; South China Sea. (12, 43, 45)
1690. *G. puncticulata* (d'Orbigny): Okinawa Trough (Upper Pleistocene to Holocene). (45)
1691. *G. redunca* Cheng & Cheng: East China Sea; Okinawa Trough (Upper Pleistocene to Holocene). (23, 24, 45)
1692. *G. robusta* Bolli: Taiwan (Neogene). (58)
1693. *G. scitula* (Brady): East China Sea; South China Sea; Taiwan Strait; Okinawa Trough (Upper Pleistocene to Holocene); Xisha (=Paracel Islands) (Middle Pleistocene to Upper Pleistocene). (24, 39, 43, 45, 50)
1694. *G. siakensis* Leroy: South China Sea (Neogene). (46, 72)
1695. *G. tosaensis* (Takayanagi & Saito): Okinawa Trough (Upper Pleistocene); South China Sea (Neogene). (45, 72)
1696. *G. trigonula* (d'Orbigny): South China Sea. (67)
1697. *G. truncatulinoides* (d'Orbigny): East China Sea; Taiwan Strait; South China Sea; Okinawa Trough (Upper Pleistocene to Holocene); East China Sea (Pleistocene); Xisha (=Paracel Islands) (Pleistocene). (3, 23, 24, 30, 36, 39, 41, 43, 45)
1698. *G. tumida* (Brady): East China Sea; South China Sea; Okinawa Trough (Upper Pleistocene to Holocene); Jiangsu (Quaternary); Xisha (=Paracel Islands) (Pleistocene); Guangdong (Pliocene). (4, 8, 23, 24, 33, 39, 41, 43, 45)
1699. *G. ungulata* Bermudez: Okinawa Trough (Upper Pleistocene to Holocene); Xisha (=Paracel Islands) (Middle Pleistocene to Upper Pleistocene). (39, 45)
1700. *G. velascoensis* (Cushman): Kashi Basin (Xinjiang) (Tertiary). (38)
1701. *Neogloboquadrina dutertrei* (d'Orbigny): East China Sea; South China Sea; Okinawa Trough (Pliocene); Xisha (=Paracel Islands) (Middle Pleistocene to Holocene). (12, 39, 45, 70)
1702. *N. eggeri* (Rhumbler): East China Sea; South China Sea; Xisha (=Paracel Islands) (Middle Pleistocene). (23, 24, 30, 39)
1703. *N. pachyderma* (Ehrenberg): East China Sea; South China Sea; East China Sea continental shelf (Pliocene); Okinawa Trough (Holocene); Xisha (=Paracel Islands) (Middle Pleistocene to Holocene). (3, 12, 39, 45, 72)
1704. *Turborotalia anticompressa* Wang, He, & Lin: East China Sea continental shelf (Paleocene). (3)
1705. *T. chapmani* Parr: East China Sea continental shelf (Paleocene). (3)
1706. *T. crassaformis* (Galloway & Wissler) [1661]: Taiwan Strait; South China Sea; Xisha (=Paracel Islands) (Upper Pleistocene). (24, 30, 38, 41, 43, 45, 48, 50, 72)
1707. *T. dutertrei* (d'Orbigny) [1672]: East China Sea; South China Sea; Okinawa Trough (Upper Pleistocene to Holocene); Xisha (=Paracel Islands) (Middle Pleistocene to Holocene). (12, 39, 45, 70)
1708. *T. inflata* (d'Orbigny) [1672]: East China Sea; South China Sea; Beijing area (Lower Quaternary); Taizhou Bay (Zhejiang) (Upper Quaternary); East China Sea continental shelf (Pliocene to present). (2, 3, 12, 23, 24, 30, 43, 45, 47, 72)
1709. *T. pseudogriffinae* Wang, He, & Lu: East China Sea continental shelf (Paleocene). (3)
1710. *T. pseudoimitata* Blow: East China Sea (Paleocene). (3)
1711. *T. mayeri* Cushman & Ellisor: Guangdong (Pliocene). (33)
1712. *T. peripheroronda* Blow & Banner [1684]: Taiwan (Miocene); South China Sea (Neogene). (46, 61, 72)
1713. *T. pumilio* (Parker) [1689]: East China Sea; South China Sea; Okinawa Trough (Upper Pleistocene). (12, 43, 45)
1714. *T. siakensis* (LeRoy) [1694]: South China Sea (Neogene). (46, 72)
1715. *T. tripartita* Wang, He, & Lu: East China Sea (Paleocene). (3)

TRUNCOROTALOIDIDAE

1716. *Acarinina appressocamerata* Blow: East China Sea continental shelf (Eocene). (3)
1717. *A. broedermanni* Cushman & Bermudez: East China Sea continental shelf (Eocene). (3)
1718. *A. bullbrooki* Bolli: East China Sea continental shelf (Eocene). (3)
1719. *A. camerata* (Khalilov): East China Sea continental shelf (Eocene). (3)
1720. *A. compacta* Wang, He, & Lu: East China Sea continental shelf (Paleocene). (3)
1721. *A. decepta* (Martin): East China Sea continental shelf (Lower Eocene to Middle Eocene). (3)
1722. *A. hansbolli* Banner & Blow: East China Sea continental shelf (Lower and Middle Paleocene). (3)
1723. *A. interposita* (Subbotina): East China Sea continental shelf (Eocene). (3)
1724. *A. lingfengensis* Wang, He & Lu: East China Sea continental shelf (Paleocene to Eocene). (3)
1725. *A. mattewsae* Blow: East China Sea continental shelf (Eocene). (3)
1726. *A. praeangulata* Blow: East China Sea continental shelf (Paleocene). (3)
1727. *A. primitiva* (Finlay): East China Sea continental shelf (Eocene). (3)

1728. *A. pseudotopilensis* (Subbotina): East China Sea continental shelf (Eocene); Kashi Basin (Xinjiang) (Lower Tertiary). (3, 38)

1729. *Globigerapsis index* (Finlay): East China Sea continental shelf (Paleocene). (3)

1730. *Morozovella angulata* (White): East China Sea continental shelf (Paleocene). (3)

1731. *M. velascoensis* (Cushman) [1755]: Kashi Basin (Xinjiang) (Tertiary). (46)

1732. *Muricoglobigerina soldadoensis* (Brönnimann): East China Sea continental shelf (Upper Paleocene to Lower Eocene). (3)

PULLENIATINIDAE

1733. *Pulleniatina finalis* Banner & Blow: Okinawa Trough (Upper Pleistocene to Holocene); Xisha (=Paracel Islands) (Pleistocene). (39, 45)

1734. *P. obliquiloculata* (Parker & Jones): East China Sea; South China Sea; East China Sea continental shelf (Quaternary); Okinawa Trough (Upper Pleistocene to Holocene); Xisha (=Paracel Islands) (Middle Pleistocene to Holocene). (3, 12, 23, 24, 30, 41, 43, 45, 46, 50, 66, 67, 70)

1735. *P. praecursor* Banner & Blow: Xisha (=Paracel Islands) (Middle Pleistocene to Holocene). (39)

1736. *P. primalis* Banner & Blow: Xisha (=Paracel Islands) (Middle Pleistocene to Holocene). (4, 39)

CANDEINIDAE

1737. *Globigerinita bradyi* (Wiesner): Taiwan Strait. (50)

1738. *G. glutinata* Egger: East China Sea; Taiwan Strait; South China Sea. (12, 39, 43, 45, 66, 68)

1739. *G. incrusta* Akers: Okinawa Trough (Upper Pleistocene). (45)

1740. *G. iota* Parker: Okinawa Trough (Upper Pleistocene to Holocene). (12, 45)

1741. *Tinophodella ambitacrena* Loeblich & Tappan: East China Sea; South China Sea. (23, 24, 30)

1742. *Candeina nitida* d'Orbigny: East China Sea; South China Sea; Okinawa Trough (Upper Pleistocene). (23, 24, 43, 45, 66)

CATAPSYDRACIDAE

1743. *Globoquadrina altispira* (Cushman & Jarvis): Taiwan area (Tertiary); Xisha (=Paracel Islands) (Upper Miocene to to Middle Pliocene). (36, 46)

1744. *G. conglomerata* (Schwager): South China Sea. (43)

1745. *G. dehiscens* (Chapman, Parr, & Collins): Taiwan (Upper Neogene). (46, 58)

1746. *G. dutertrei* (d'Orbigny) [1701]: South China Sea. (43)

1747. *G. hexagona* (Natland): South China Sea. (43)

1748. *G. venezuelana* (Hedberg): Taiwan (Miocene). (52)

1749. *Globorotaloides hexagona* (Natland): Okinawa Trough (Upper Pleistocene). (45)

1750. *Subbotina bakeri* (Cole): East China Sea continental shelf (Eocene). (45)

1751. *S. inaequispira* (Subbotina): East China Sea continental shelf (Eocene). (3)

1752. *S. linaperta* (Finlay): East China Sea continental shelf (Paleocene). (3)

1753. *S. triangularis* (White): East China Sea continental shelf (Paleocene). (3)

1754. *S. triloculinoides* (Plummer): East China Sea continental shelf (Paleocene); Jiangsu (Quaternary); Guangdong (Pliocene). (3)

1755. *S. velascoensis* (Cushman): East China Sea continental shelf (Paleocene); Kashi Basin (Xinjiang) (Lower Tertiary). (3)

HANTKENINACEA
GLOBANOMALINIDAE

1756. *Globanomalina wilcoxensis* (Cushman & Ponton): East China Sea continental shelf (Eocene). (3)

1757. *Pseudohastigerina wilcoxensis* (Cushman & Ponton) [1756]: East China Sea continental shelf (Eocene). (3)

1758. *Hastigerinella digitata* (Brady) [1760]: South China Sea. (24)

1759. *H. digitata* (Rhumbler) [1761]: South China Sea. (23)

GLOBIGERINACEA
GLOBIGERINIDAE

1760. *Beella digitata* (Brady): Xisha (=Paracel Islands) (Middle Pleistocene to Holocene). (23, 24, 30, 39, 66, 70)

1761. *Bolliella adamsi* (Banner & Blow): East China Sea; Taiwan Strait; Okinawa Trough (Upper Pleistocene); South China Sea. (12, 23, 24, 43, 45, 50, 66)

1762. *Globigerina ampliapertura* Bolli: Guangdong (Pliocene). (33)

1763. *G. angustiumbilicata* Bolli: Beijing area (Lower Quaternary); Kashi Basin (Xinjiang) (Tertiary); Taiwan (Miocene). (2, 38, 52)

1764. *G. apertura* Cushman: Taiwan (Miocene). (52)

1765. *G. atlantica* Berggren: Beijing area (Lower Quaternary). (2)

1766. *G. borealis* Brady: Taiwan Strait. (59)

1767. *G. bradyi* Wiesner: East China Sea; Okinawa Trough (Upper Pleistocene to Holocene). (23, 45, 52)

1768. *G. bulloides* d'Orbigny: Southern Yellow Sea; East China Sea; Taiwan Strait; South China Sea; Beijing area; East China Sea continental shelf; Jiangsu (Quaternary); Taiwan (Miocene); Xisha (=Paracel Islands) (Middle Pleistocene to Holocene). (2, 3, 8, 12, 23, 24, 39, 43, 45, 49, 52, 59, 66)

1769. *G. calida* Parker [1790]: Taiwan Strait; South China Sea; Okinawa Trough (Upper Pleistocene). (24, 39, 43, 45, 50, 66)

1770. *G. ciperoensis* Bolli: Leiqiong area (southern Guangdong and Hainan) (Tertiary). (46)

1771. *G. decoraperta* Takayanagi & Saito: Okinawa Trough (Upper Pleistocene). (45)

1772. *G. digitata* Brady [1760]: South China Sea. (70)

1773. *G. diplostoma* Reuss: Taiwan (Miocene). (52)

1774. *G. druryi* Akers: South China Sea; Guangdong area (Pliocene). (33, 67)

1775. *G. eggeri* Rhumbler [1702]: East China Sea; South China Sea; Taiwan (Pliocene to Pleistocene); Guangdong area (Pliocene). (23, 24, 52)

1776. *G. falconensis* Blow: East China Sea; Taiwan Strait; South China Sea; Beijing area; East China Sea continental shelf (Quaternary and Pliocene); Okinawa Trough (Upper Pleistocene); Taiwan (Miocene to Pleistocene). (2, 3, 12, 43, 45, 50, 52)

1777. *G. fringa* Subbotina: Kashi Basin (Xinjiang) (Lowe Tertiary). (38)

1778. *G. hexagona* Natland [1749]: Okinawa Trough (Upper Pleistocene); Taiwan (Miocene); Leiqiong area (southern Guangdong and Hainan) (Tertiary). (45)

1779. *G. nepenthes* Todd: Yellow Sea; South China Sea; Beijing area (Lower Quaternary). (46, 52)

1780. *G. pachyderma* (Ehrenberg) [1703]: East China Sea; South China Sea; Okinawa Trough (Upper Pleistocene); Xisha (=Paracel Islands) (Middle Pleistocene to Holocene). (2, 3, 12, 39, 43, 45, 65, 70)

1781. *G. quinqueloba* Natland: East China Sea; South China Sea; Okinawa Trough (Upper Pleistocene); Xisha (=Paracel Islands) (Middle Pleistocene to Holocene). (12, 39, 43, 45, 67, 70)

1782. *G. rubescens* Hofker [1816]: East China Sea; South China Sea; Okinawa Trough (Upper Pleistocene to Holocene); Guangdong area (Pliocene). (23, 24, 33, 42, 45, 50, 66)

1783. *G. subcretacea* Lomnicki: South China Sea; Taiwan (Miocene to Pleistocene). (52, 67)

1784. *G. triloculinoides* Plummer [1754]: Beijing area (Lower Quaternary); Kashi Basin (Xinjiang) (Lower Tertiary); Guangdong area (Pliocene). (2, 3, 38)

1785. *G. varianta* Subbotina: Kashi Basin (Xinjiang) (Lower Tertiary); Jiangsu (Quaternary). (8, 38)

1786. *G. woodi* Jenkins: Taiwan (Miocene). (52)

1787. *G. velascoensis* (Cushman) [1755]: East China Sea continental shelf (Paleocene); Kashi Basin (Xinjiang) (Lower Tertiary). (3, 38)

1788. *Globigerinella adamsi* (Banner & Blow) [1761]: East China Sea; South China Sea; Okinawa Trough (Upper Pleistocene); Guangdong area (Pliocene); Xisha (=Paracel Islands) (Middle Pleistocene to Holocene). (12, 23, 24, 43, 45, 50, 66)

1789. *G. aequilateralis* (Brady): East China Sea; South China Sea; Okinawa Trough (Upper Pleistocene); Guangdong area (Pliocene); Xisha (=Paracel Islands) (Middle Pleistocene to Holocene). (3, 12, 23, 24, 33, 39, 43, 45, 66, 67)

1790. *G. calida* (Parker): Taiwan Strait; South China Sea; Okinawa Trough (Upper Pleistocene); Xisha (=Paracel Islands) (Middle Pleistocene to Holocene). (24, 39, 43, 45, 49, 64)

1791. *G. siphonifera* (d'Orbigny): East China Sea; Taiwan Strait; South China Sea. (50)

1792. *Alloglobigerinoides conglobatus* (Brady): Taiwan (Miocene). (60)

1793. *Globigerinoides bolli* Blow: Taiwan (Miocene). (52)

1794. *G. conglobatula* Cheng & Cheng: East China Sea; South China Sea; Okinawa Trough (Upper Pleistocene). (23, 24, 45)

1795. *G. conglobatus* (Brady): East China Sea; Taiwan Strait; South China Sea; Xisha (=Paracel Islands); East China Sea continental shelf (Quaternary and Miocene); Okinawa Trough (Upper Pleistocene). (12, 23, 24, 33, 36, 41, 43, 45, 46, 49, 66, 67)

1796. *G. cyclostomus* (Galloway & Wissler): Okinawa Trough (Upper Pleistocene); Xisha (=Paracel Islands) (Middle Pleistocene to Holocene). (39, 45)

1797. *G. elongatus* (d'Orbigny): East China Sea; South China Sea; Okinawa Trough (Upper Pleistocene); Taiwan (Miocene to Pleistocene); Xisha (=Paracel Islands) (Middle Pleistocene to Holocene). (30, 39, 45, 51)

1798. *G. extremus* Bolli & Bermudez: Leiqiong area (southern Guangdong and Hainan) (Tertiary). (46)

1799. *G. fistulosus* (Schubert) [1815]: Taiwan (Upper Tertiary). (58)

1800. *G. helicina* (d'Orbigny): South China Sea; Guangdong area (Pliocene); Xisha (=Paracel Islands) (Middle Pleistocene to Holocene). (33, 39, 42)

1801. *G. immaturus* LeRoy: Taiwan (Tertiary); Guangdong area (Pliocene); Xisha (=Paracel Islands) (Middle Pleistocene to Holocene); Xisha (=Paracel Islands) (Cenozoic). (4, 33, 39, 46)

1802. *G. obliquus* Bolli: Taiwan (Miocene to Pleistocene); Guangdong area (Pliocene); Xisha (=Paracel Islands) (Upper Miocene to Pleistocene). (4, 33, 39, 46)

1803. *G. parkerae* Bermudez: Okinawa Trough (Upper Pleistocene to Holocene). (45)

1804. *G. pyramidalis* (Van den Broeck): Okinawa Trough (Upper Pleistocene); Xisha (=Paracel Islands) (Middle Pleistocene to Holocene). (39, 45)

1805. *G. ruber* (d'Orbigny): Southern Yellow Sea; East China Sea; South China Sea; Zhejiang (Quaternary); Okinawa Trough (Upper Pleistocene); Taiwan (Miocene to Pleistocene); Xisha (=Paracel Islands) (Middle Pleistocene to Holocene). (3, 4, 8, 12, 24, 31, 33, 39, 41, 45, 47, 49, 50, 66, 67)

1806. *G. sacculifera* (Brady): East China Sea; Taiwan Strait; South China Sea; South China Sea (Pleistocene to Holocene). (3, 4, 8, 23, 24, 33, 39, 41, 42, 45, 49, 66, 67)

1807. *G. quadrilobatus* (d'Orbigny): Xisha (=Paracel Islands) (Middle Pleistocene to Holocene). (33)

1808. *G. sicanus* de Stefani: Taiwan (Tertiary). (46)

1809. *G. sinensis* Cheng & Cheng: East China Sea; South China Sea. (23, 24, 42)
1810. *G. tenellus* Parker: Taiwan Strait; South China Sea; Okinawa Trough (Upper Pleistocene). (24, 42, 45, 50)
1811. *G. trilobus* (Reuss): East China Sea; South China Sea; Okinawa Trough (Upper Pleistocene); Guangdong area (Pliocene); Xisha (=Paracel Islands) (Middle Pleistocene to Holocene). (4, 23, 24, 33, 39, 41, 42)
1812. *G. trilocularis* (d'Orbigny): South China Sea. (24)
1813. *G. triloculinoides* Plummer [1754]: East China Sea continental shelf (Paleocene); Jiangsu (Quaternary); Guangdong area (Pliocene). (3, 8, 33, 38)
1814. *G. varianta* Subbotina [1785]: Kashi Basin (Xinjiang) (Lower Tertiary); Jiangsu (Quaternary). (8, 38)
1815. *Globigerinoidesella fistulosa* (Schubert): Taiwan (Upper Tertiary). (58)
1816. *Globoturborotalita rubescens* (Hofker): East China Sea; South China Sea; Okinawa Trough (Upper Pleistocene); Guangdong area (Pliocene). (4, 23, 24, 39, 45, 66, 67)
1817. *Sphaeroidinella dehiscens* (Parker & Jones): East China Sea; South China Sea; Okinawa Trough (Upper Pleistocene); Xisha (=Paracel Islands) (Middle Pleistocene to Holocene). (4, 23, 24, 39, 45, 66, 67)
1818. *Sphaeroidinellopsis seminulina* (Schwager): Taiwan (Tertiary); Xisha (=Paracel Islands) (Middle Pleistocene to Holocene). (36, 46, 52)
1819. *S. subdehiscens* Blow: Taiwan (Tertiary); Xisha (=Paracel Islands) (Cenozoic). (4, 46, 52)
1820. *Turborotalita clarkei* (Rogl & Bolli): East China Sea; Okinawa Trough (Upper Pleistocene); South China Sea (Neogene). (19, 45, 72)
1821. *T. humilis* (Brady): East China Sea; South China Sea; Okinawa Trough (Upper Pleistocene). (12, 13, 45)
1822. *Candorbulina suturalis* (Brönnimann): East China Sea continental shelf (Quaternary, Miocene to present); Leiqiong area (southern Guangdong and Hainan) (Tertiary). (3, 4, 46)
1823. *Orbulina bilobata* (d'Orbigny): South China Sea; Okinawa Trough (Upper Pleistocene); Guangdong area (Pliocene); Xisha (=Paracel Islands) (Middle Pleistocene to Holocene). (3, 23, 39, 42, 43, 45)
1824. *O. circularosuturalis* N. W. Wang & He: Guangdong area (Pliocene). (33)
1825. *O. porosa* Terquem: South China Sea. (42)
1826. *O. suturalis* Brönniman [1822]: East China Sea continental shelf (Quaternary, Miocene to present); Taiwan (Tertiary); Leiqiong area (southern Guangdong and Hainan) (Tertiary); Xisha (=Paracel Islands) (Cenozoic). (3, 4, 46)
1827. *O. universa* d'Orbigny: East China Sea; Taiwan Strait; South China Sea; Beijing area (Lower Quaternary); East China Sea continental shelf (Miocene to present); Okinawa Trough (Upper Pliocene); Taiwan (Tertiary); Xisha (=Paracel Islands) (Cenozoic). (2, 3, 4, 23, 24, 33, 43, 45, 66, 67)

HASTIGERINIDAE

1828. *Hastigerina aequilateralis* (Brady) [1789]: East China Sea; South China Sea. (67)
1829. *H. pelagica* d'Orbigny: East China Sea; Taiwan Strait; South China Sea. (23, 24, 43, 50)
1830. *H. siphonifera* (d'Orbigny) [1789]: East China Sea; South China Sea; Okinawa Trough (Upper Pleistocene); Guangdong area (Pliocene); Xisha (=Paracel Islands) (Middle Pleistocene to Holocene). (3, 12, 23, 24, 33, 39, 43, 45, 66)

ROTALIINA
BOLIVINACEA
BOLIVINIDAE

1831. *Bolivina abbreviata* Heron-Allen & Earland [1861]: South China Sea. (4, 25, 27, 30, 66, 67)
1832. *B. acerosa* (Cushman): Taiwan Strait; Shandong (Quaternary). (29, 30, 50)
1833. *B. acutula* Bandy [1863]: Okinawa Trough (Upper Pleistocene). (45, 66)
1834. *B. aenariensis* (Costa): Nansha (=Spratly Islands). (66)
1835. *B. alata* (Sequenza) [1864]: Nansha (=Spratly Islands). (66)
1836. *B. albatrossi* Cushman: Bohai; Yellow Sea; East China Sea; Shandong (Quaternary). (19, 21, 29)
1837. *B. bassensis* Parr: East China Sea; Taiwan; Xisha (=Paracel Islands) (Middle Pleistocene to Holocene). (39, 64)
1838. *B. capitata* Cushman [1867]: East China Sea; Okinawa Trough (Upper Pleistocene). (30, 45)
1839. *B. cochei* Cushman & Adams: East China Sea; South China Sea; Jiangsu (Quaternary). (8, 12, 22, 48)
1840. *B. compacta* Sidebottom: Bohai; Xisha (=Paracel Islands); Xisha (=Paracel Islands) (Cenozoic); Zhongsha (=Macclesfield Bank); Nansha (=Spratly Islands). (4, 30, 48)
1841. *B. dilatata* (Reuss): Shoreline of Bohai (Quaternary). (49)
1842. *B. goësi* Cushman: South China Sea. (27, 66)
1843. *B. incertum* Hofker: Okinawa Trough (Upper Pleistocene). (45)
1844. *B. obscura* Ho He & Wang: Jiangsu (Quaternary). (8)
1845. *B. ovata* Egger: East China Sea. (64)
1846. *B. pseudoplicata* Heron-Allen & Earland: Yellow Sea; East China Sea. (19, 62)
1847. *B. pseudopunctata* Höglund [1877]: East China Sea; Nansha (=Spratly Islands); shoreline of Bohai (Quaternary); Jiangsu (Quaternary). (8, 30, 51, 66)
1848. *B. pusilla* Schwager: Shoreline of Bohai (Quaternary). (51)
1849. *B. robusta* Brady: Bohai; Yellow Sea; East China Sea; South China Sea; southern Yellow Sea (Quaternary); East China Sea continental shelf (Pliocene); Okinawa Trough (Upper Pleistocene). (3, 19, 31, 33, 44, 45, 49, 50, 59, 66)
1850. *B. semicostata* Cushman: Nansha (=Spratly Islands). (66)
1851. *B. spathulata* (Williamson): East China Sea; Taiwan Strait; South China Sea; southern Yellow Sea (Quaternary); East China Sea continental shelf (Quaternary). (3, 30, 64, 67)

1852. *B. spinata* (Cushman): Okinawa Trough (Holocene); Xisha (=Paracel Islands) (Upper Pleistocene). (39, 45)
1853. *B. spinea* Cushman: Nansha (=Spratly Islands). (66)
1854. *B. spissa* Cushman: East China Sea; South China Sea. (30, 66)
1855. *B. striatula* Cushman [1866]: Bohai; Yellow Sea; East China Sea; South China Sea; Shandong (Quaternary); Guangdong (Pliocene). (19, 21, 29, 33, 37, 42, 62)
1856. *B. subangularis* Brady [1909]: Taiwan Strait; South China Sea. (50)
1857. *B. subspinescens* Cushman [1892]: East China Sea; Okinawa Trough (Upper Pleistocene). (30, 45)
1858. *B. subtenuis* (Cushman) [1891]: Xisha (=Paracel Islands); Zhongsha (=Macclesfield Bank); Nansha (=Spratly Islands). (27, 30, 48)
1859. *B. vadescens* Cushman: Guangdong (Pliocene). (33)
1860. *B. variabilis* (Williamson): Xisha (=Paracel Islands); Nansha (=Spratly Islands). (27, 30)
1861. *Brizalina abbreviata* (Heron-Allen & Earland): South China Sea. (4, 25, 27, 30, 64, 65)
1862. *B. acerosa* (Cushman) [1832]: Taiwan Strait. (49)
1863. *B. acutula* (Brady): Okinawa Trough (Upper Pleistocene to Holocene). (45)
1864. *B. alata* (Sequenza): East China Sea; South China Sea. (19, 30)
1865. *B. anglica* (Cushman): Kashi Basin (Xinjiang) (Lower Tertiary). (38)
1866. *B. britannica* Macfadyan: Xisha (=Paracel Islands). (26, 30)
1867. *B. capitata* Cushman: East China Sea; South China Sea; Guangdong (Pliocene). (3, 30)
1868. *B. doniezi* (Cushman & Wickenden): Xisha (=Paracel Islands); Okinawa Trough (Upper Pleistocene to Holocene). (26, 30, 45)
1869. *B. durrandii* (Millett): South China Sea. (26, 30)
1870. *B. earlandi* (Parr): Okinawa Trough (Upper Pleistocene to Holocene). (45)
1871. *B. hantkeniana* (Brady) [1895]: Taiwan Strait; Guangdong (Pliocene). (33, 44, 59)
1872. *B. interjuncta* (Cushman): Guangdong (Pliocene). (33)
1873. *B. kiiensis* (Asano): Okinawa Trough (Upper Pleistocene to Holocene). (45)
1874. *B. limbata* (Brady) [1900]: Hong Kong. (62)
1875. *B. lowmani* (Phleger & Parker): Hong Kong. (62)
1876. *B. paula* (Cushman & Cahill): Xisha (=Paracel Islands). (26, 30)
1877. *B. pseudopunctata* (Höglund): East China Sea; South China Sea; Jiangsu (Quaternary). (8, 30, 64)
1878. *B. pseudoseminuda* Zheng: South China Sea; Xisha (=Paracel Islands). (26, 30, 66)
1879. *B. punctata* d'Orbigny: South China Sea. (66)
1880. *B. punctatostriata* (Kraeuzberg): East China Sea. (19)
1881. *B. robusta* (Brady) [1849]: Bohai; Yellow Sea; East China Sea; South China Sea; southern Yellow Sea (Quaternary); East China Sea continental shelf (Pliocene); Okinawa Trough (Upper Pleistocene to Holocene). (3, 19, 31, 33, 44, 45, 50, 66)
1882. *B. rhomboidalis* (Millet): East China Sea; South China Sea. (26, 30)
1883. *B. seminuda* (Cushman): Yellow Sea; East China Sea; South China Sea. (19)
1884. *B. semicostata* (Cushman): Taiwan Strait. (44)
1885. *B. spinescens* (Cushman): Xisha (=Paracel Islands). (26, 30)
1886. *B. striatula* (Cushman): Bohai; Yellow Sea; East China Sea; South China Sea; Shandong (Quaternary); Guangdong (Pliocene). (19, 21, 29, 33, 37, 62)
1887. *B. spathulata* (Williamson) [1851]: Yellow Sea; East China Sea. (19)
1888. *B. subcapitata* Zheng: South China Sea. (27, 66)
1889. *B. subreticulata* Parr: East China Sea; South China Sea. (19, 30, 66)
1890. *B. substriatula* Zheng: Xisha (=Paracel Islands); Nansha (=Spratly Islands). (26, 30, 66)
1891. *B. subtenuis* (Cushman): Xisha (=Paracel Islands); Zhongsha (=Macclesfield Bank); Nansha (=Spratly Islands). (27, 30)
1892. *B. subspinescens* (Cushman): Yellow Sea; East China Sea; South China Sea. (19, 30)
1893. *B. tainanensis* (Nakamura): Taiwan. (44)
1894. *B. zanzibarica* (Cushman): Xisha (=Paracel Islands); Zhongsha (=Macclesfield Bank); Nansha (=Spratly Islands); East China Sea (Quaternary). (3, 25, 30, 37)
1895. *Lugdunum hantkenianum* (Brady): Taiwan; Guangdong (Pliocene). (33, 44)

LOXOSTOMATACEA
LOXOSTOMATIDAE

1896. *Loxostomum amygdalaeformis* (Brady): East China Sea; South China Sea. (19, 30, 66, 67)
1897. *L. convallarium* (Millett) [1972]: East China Sea; South China Sea. (30, 67)
1898. *L. formosana* (Nakamura): Taiwan. (44)
1899. *L. karreianum* (Brady) [1971]: Taiwan; Okinawa Trough (Upper Pleistocene to Holocene). (44, 45)
1900. *L. limbatum* (Brady): South China Sea. (66)
1901. *L. mayori* (Cushman): East China Sea; South China Sea. (67)
1902. *L. spinosum* Wang, He, & Lu: East China Sea continental shelf (Eocene). (3)

BOLIVINELLIDAE

1903. *Bolivinella basicostata* Zheng: Zhongsha (=Macclesfield Bank). (27, 30)

1904. *B. elegans* Parr: Xisha (=Paracel Islands). (26, 30)
1905. *B. folia* (Parker & Jones): Xisha (=Paracel Islands); Zhongsha (=Macclesfield Bank); Nansha (=Spratly Islands). (48)
1906. *B. leizhouensis* He & Lin: Guangdong (Pliocene). (33)

TORTOPLECTELLIDAE

1907. *Tortoplectella crispata* (Brady): Xisha (=Paracel Islands); Nansha (=Spratly Islands). (25, 30, 48, 66)

BOLIVINITACEA
BOLIVINITIDAE

1908. *Bolivinita quadrilatera* (Schwager): Okinawa Trough (Upper Pleistocene to Holocene). (44, 45, 66)
1909. *B. subangularis* (Brady): Taiwan Strait; South China Sea. (44, 66)
1910. *B. suturornata* Zhang: Okinawa Trough. (19)

CASSIDULINACEA
CASSIDULINIDAE

1911. *Cassidulina alternans* Yabe & Hanzawa [1937]: Nansha (=Spratly Islands); Okinawa Trough (Upper Pleistocene to Holocene). (45, 66)
1912. *C. bradshawi* Uchio: Okinawa Trough (Holocene). (45)
1913. *C. carinata* Silvestri: Yellow Sea; East China Sea; Taiwan Strait; South China Sea; Okinawa Trough (Upper Pleistocene to Holocene); Xisha (=Paracel Islands) (Middle Pleistocene to Upper Pleistocene). (19, 39, 44, 45, 66)
1914. *C. crassa* d'Orbigny: Nansha (=Spratly Islands). (66)
1915. *C. cuneata* Finlay: Xisha (Upper Pleistocene to Holocene). (39)
1916. *C. delicata* Cushman [1949]: Xisha (=Paracel Islands); Nansha (=Spratly Islands); Okinawa Trough (Upper Pleistocene to Holocene). (26, 45, 64)
1917. *C. jonesiana* Brady: Xisha (=Paracel Islands) (Upper Pleistocene to Holocene). (39)
1918. *C. laevigata* d'Orbigny: East China Sea; Beijing area (Quaternary); Okinawa Trough (Upper Pleistocene to Holocene). (33, 45)
1919. *C. laticamerata* Voloshinova: Xisha (=Paracel Islands) (Middle Pleistocene). (39)
1920. *C. minuta* Cushman [1947]: Okinawa Trough (Middle Pleistocene to Holocene). (45)
1921. *C. moluccensis* Germeraad: Xisha (=Paracel Islands) (Middle Pleistocene). (39)
1922. *C. neocarinata* Thalmann: Southern Yellow Sea; East China Sea; South China Sea; Okinawa Trough (Upper Pleistocene to Holocene). (30, 45, 66, 67)
1923. *C. oblonga* Reuss [1934]: East China Sea; Taiwan Strait; South China Sea. (42, 44, 66)
1924. *C. pacifica* Cushman [1950]: East China Sea; Taiwan Strait; South China Sea. (42, 44, 66)
1925. *C. pulchella* d'Orbigny: East China Sea; South China Sea. (30)
1926. *C. reniforme* Norvang: Okinawa Trough (Upper Pleistocene to Holocene). (45)
1927. *C. striatostoma* Zheng: Zhongsha (=Macclesfield Bank); Nansha (=Spratly Islands). (27, 30, 66)
1928. *C. subcarinata* Uchio: Okinawa Trough (Upper Pleistocene to Holocene). (45)
1929. *C. subglobosa* Brady: South China Sea; Xisha (=Paracel Islands) (Middle Pleistocene). (66, 67)
1930. *C. teretis* Tappan: Nansha (=Spratly Islands); Xisha (=Paracel Islands) (Middle Pleistocene). (39)
1931. *Cassidulinoides braziliensis* (Cushman): Okinawa Trough (Upper Pleistocene to Holocene). (45)
1932. *C. chapmani* Parr: Xisha (=Paracel Islands). (26, 30)
1933. *C. orientale* (Cushman): Okinawa Trough (Holocene). (45)
1934. *Evolvocassidulina belfordi* Nomura: Okinawa Trough (Upper Pleistocene to Holocene); Xisha (=Paracel Islands) (Middle Pleistocene). (39, 45)
1935. *Favocassidulina favus* (Brady): Xisha (=Paracel Islands); Zhongsha (=Macclesfield Bank); Xisha (=Paracel Islands) (Upper Pleistocene). (39, 66)
1936. *Globocassidulina algida* (Cushman): Zhongsha (=Macclesfield Bank). (27, 30)
1937. *G. alternans* (Yabe & Hanzawa): Okinawa Trough (Upper Pleistocene to Holocene). (45)
1938. *G. complanata* (Ujiie & Kusukawa): Xisha (=Paracel Islands). (26, 30)
1939. *G. gemma* (Todd): Okinawa Trough (Upper Pleistocene to Holocene). (45)
1940. *G. producta* (Chapman & Parr): Okinawa Trough (Upper Pleistocene to Holocene). (45)
1941. *G. subglobosa* (Brady): East China Sea; South China Sea; East China Sea continental shelf (Quaternary); Okinawa Trough (Upper Pleistocene to Holocene); Xisha (=Paracel Islands) (Upper Pleistocene). (3, 19, 41, 45, 66)
1942. *Islandiella californica* (Cushman & Hughes): Nansha (=Spratly Islands). (19, 29, 30, 66)
1943. *I. islandica* (Norvang): East China Sea; Nansha (=Spratly Islands); Shandong (Quaternary). (45)
1944. *I. kattoi* (Takayanagi): Okinawa Trough (Uppler Pleistocene). (45)
1945. *I. seranensis* (Germeraad): Okinawa Trough (Upper Pleistocene). (45)
1946. *Lernella inflata* LeRoy: South China Sea. (30)
1947. *Paracassidulina minuta* (Cushman): Okinawa Trough (Upper Pleistocene to Holocene). (45)
1948. *P. nipponensis* (Eade): Okinawa Trough (Holocene). (45)
1949. *Takayanagia delicata* (Cushman): Xisha (=Paracel Islands); Nansha (=Spratly Islands); Okinawa Trough (Upper Pleistocene to Holocene). (26, 45, 64)
1950. *Burseolina pacifica* (Cushman): East China Sea; Taiwan Strait; South China Sea. (42, 44, 64)

1951. *Ehrenbergina pacifica* Cushman: Taiwan Strait; South China Sea; Okinawa Trough (Upper Pleistocene to Holocene). (42, 44, 45, 48, 66)
1952. *E. trigona* Goës: Nansha (=Spratly Islands). (66)
1953. *E. undulata* Parker: Okinawa Trough (Upper Pleistocene). (46)

EOUVIGERINACEA
STAINFORTHIIDAE

1954. *Hopkinsina pacifica* Cushman: Bohai; Yellow Sea; East China Sea; southern Yellow Sea (Quaternary). (19, 21, 30, 31, 35, 65)
1955. *Stainforthia complanata* (Egger): East China Sea. (19)
1956. *S. concava* Hoglund: East China Sea.
1957. *S. spinosa* (Heron-Allen & Earland): East China Sea. (19)
1958. *Virgulopsis orientalis* Ho, Hu, & Wang: East China Sea; South China Sea; East China Sea continental shelf (Quaternary); Guangdong (Pliocene). (3, 19, 33, 48)

BULIMINACEA
SIPHOGENERINOIDIDAE

1959. *Euloxostomum bradyi* (Asano): South China Sea. (30)
1960. *Loxostomina mayori* (Cushman): South China Sea. (67)
1961. *Rectobolivina aculeata* Zheng: Zhongsha (=Macclesfield Bank). (27, 30)
1962. *R. altilocula* Zheng: Xisha (=Paracel Islands); Zhongsha (=Macclesfield Bank); Nansha (=Spratly Islands). (28, 30, 66)
1963. *R. bifrons* (Brady): East China Sea; Taiwan Strait; South China Sea; Okinawa Trough (Pliocene); Xisha (=Paracel Islands) (Upper Pleistocene). (41, 44, 45, 48, 66)
1964. *R. digitata* Parr: Xisha (=Paracel Islands); Zhongsha (=Macclesfield Bank); Okinawa Trough (Holocene); Guangdong (Pliocene). (33, 45, 48)
1965. *R. raphana* (Parker & Jones): Yellow Sea; East China Sea; South China Sea; East China Sea continental shelf (Quaternary); South China Sea (Pliocene); Xisha (Upper Pleistocene). (33, 45, 48)
1966. *R. laevis* Zheng: Xisha (=Paracel Islands); Zhongsha (=Macclesfield Bank). (26, 30, 48)
1967. *R. subbifrons* Zheng: Xisha (=Paracel Islands). (26, 30)
1968. *R. subraphanus* Zheng: Xisha (=Paracel Islands). (26, 30)
1969. *R. virgula* (Brady): Xisha (=Paracel Islands); Nansha (=Spratly Islands). (26, 30)
1970. *R. xuwenensis* Ho & Lu: Guangdong (Pliocene). (33)
1971. *Saidovina karreriana* (Brady): Taiwan Strait; Okinawa Trough (Upper Pleistocene to Holocene). (44, 45)
1972. *Bitubulogenerina convallaria* (Millett): East China Sea; South China Sea. (30, 67)
1973. *Siphogenerina bifrons* (Brady) [1963]: East China Sea; South China Sea; Okinawa Trough (Holocene); Xisha (=Paracel Islands) (Middle Pleistocene). (41, 44, 45, 48, 66)

1974. *S. pacifica* Cushman: Xisha (=Paracel Islands); Zhongsha (=Macclesfield Bank); Okinawa Trough (Holocene). (45, 48)
1975. *S. raphana* (Parker & Jones) [1965]: East China Sea; South China Sea; East China Sea continental shelf (Quaternary); South China Sea (Pliocene); Xisha (=Paracel Islands) (Upper Pleistocene). (3, 19, 25, 44, 48, 65, 66)
1976. *S. striata* (Schwager): Taiwan Strait; South China Sea. (44, 66)
1977. *S. virgula* (Brady) [2143]: Nansha (=Spratly Islands); Okinawa Trough (Holocene). (45, 66)

BULIMINIDAE

1978. *Bulimina aculeata* d'Orbigny: Bohai; Yellow Sea; East China Sea; Taiwan Strait; Okinawa Trough (Upper Pleistocene); East China Sea continental shelf (Quaternary); Xisha (=Paracel Islands) (Middle Pleistocene to Holocene). (3, 19, 30, 39, 44, 45, 66)
1979. *B. barbata* Cushman: Nansha (=Spratly Islands). (30)
1980. *B. clava* Cushman & Parker: Xisha (=Paracel Islands) (Upper Pleistocene). (39)
1981. *B. costata* d'Orbigny: South China Sea. (30)
1982. *B. delreyensis* Cushman & Galliher: Xisha (=Paracel Islands) (Holocene). (39)
1983. *B. ecuadorana* Cushman & Stevenson: Xisha (=Paracel Islands) (Holocene). (39)
1984. *B. elegans* d'Orbigny: East China Sea. (30)
1985. *B. exilis* Brady: Bohai; Yellow Sea; East China Sea; South China Sea; Okinawa Trough (Upper Pleistocene to Holocene). (30, 45, 66)
1986. *B. fissura* Weiss: East China Sea continental shelf (Quaternary). (3)
1987. *B. fossa* Cushman & Parker: Xisha (=Paracel Islands) (Middle Pleistocene). (39)
1988. *B. gibba* Fornasini: Okinawa Trough (Upper Pleistocene). (45)
1989. *B. inflata* Sequenza: South China Sea. (30)
1990. *B. mapiria* Finlay: Okinawa Trough (Upper Pleistocene). (45)
1991. *B. marginata* d'Orbigny: Yellow Sea; East China Sea; Taiwan Strait; South China Sea; Beijing area (Lower Quaternary); Xisha (=Paracel Islands) (Pleistocene). (2, 3, 8, 19, 30, 33, 39, 49, 65, 66, 67)
1992. *B. marginospinata* Cushman & Parker: Guangdong (Pliocene). (33)
1993. *B. mauricensis* Howe: East China Sea continental shelf (Eocene). (3)
1994. *B. mexicana* Cushman: Okinawa Trough (Middle Pleistocene). (45)
1995. *B. midwayensis* Cushman & Parker: East China Sea continental shelf (Paleocene). (3)
1996. *B. nipponica* Asano: Taiwan. (44)
1997. *B. notoensis* (Asano): Xisha (=Paracel Islands) (Middle Pleistocene). (39)

1998. *B. notovata* Chapman [2021]: Nansha (=Spratly Islands); Okinawa Trough (Upper Pleistocene to Holocene). (45, 66)
1999. *B. ovata* d'Orbigny [2022]: East China Sea continental shelf (Paleocene and Eocene); Okinawa Trough (Holocene). (3, 45)
2000. *B. pagoda* Cushman: Okinawa Trough (Upper Pleistocene to Holocene); Xisha (=Paracel Islands) (Middle Pleistocene). (39, 45)
2001. *B. pulchella* d'Orbigny: Okinawa Trough (Holocene). (45)
2002. *B. pupoides* d'Orbigny [2024]: East China Sea; Nansha (=Spratly Islands). (66)
2003. *B. pyrula* d'Orbigny [2019]: South China Sea. (66)
2004. *B. rostrata* Brady: East China Sea; South China Sea; Okinawa Trough (Upper Pleistocene to Holocene). (19, 30, 45, 64, 66)
2005. *B. spicata* Phleger & Parker: Nansha (=Spratly Islands); Xisha (=Paracel Islands) (Upper Pleistocene). (39, 66)
2006. *B. squammigera* d'Orbigny: Okinawa Trough (Holocene). (45)
2007. *B. submarginata* Parr: Okinawa Trough (Upper Pleistocene to Holocene). (45)
2008. *B. subornata* Brady: Okinawa Trough (Upper Pleistocene to Holocene). (45)
2009. *B. subulata* Cushman & Parker: Xisha (=Paracel Islands) (Upper Pleistocene to Holocene). (39)
2010. *B. winniana* Howe: East China Sea continental shelf (Eocene). (3)
2011. *Globobulimina affinis* (d'Orbigny): Xisha (=Paracel Islands) (Upper Pleistocene). (39)
2012. *G. arctica* (Höglund): Xisha (=Paracel Islands) (Upper Pleistocene). (30)
2013. *G. auriculata* (Bailey): East China Sea; South China Sea; Xisha (=Paracel Islands) (Upper Pleistocene). (30, 39)
2014. *G. glabra* Cushman & Parker: Xisha (=Paracel Islands) (Pleistocene). (39)
2015. *G. hanzawaia* Asano: East China Sea. (30)
2016. *G. notovata* (Chapman) [2021]: Okinawa Trough (Upper Pleistocene to Holocene). (19, 45)
2017. *G. pacifica* Cushman: East China Sea; South China Sea; Xisha (=Paracel Islands) (Upper Pleistocene). (39, 66)
2018. *G. perversa* (Cushman): Okinawa Trough (Upper Pleistocene); Taiwan (Miocene). (44, 45)
2019. *G. pyrula* (d'Orbigny): South China Sea. (66)
2020. *G. torta* Cushman: Xisha (=Paracel Islands) (Middle Pleistocene to Holocene). (39)
2021. *Praeglobobulimina notovata* (Chapman): Nansha (=Spratly Islands); Okinawa Trough (Upper Pleistocene to Holocene). (45, 66)
2022. *P. ovata* (d'Orbigny): Okinawa Trough (Holocene); East China Sea continental shelf (Paleocene and Eocene). (3, 45)
2023. *P. spinescens* (Brady): East China Sea. (19)
2024. *Protoglobobulimina pupoides* (d'Orbigny): Nansha (=Spratly Islands). (66)

BULIMINELLIDAE

2025. *Buliminella elegantissima* (d'Orbigny): Bohai; Yellow Sea; East China Sea; South China Sea; Okinawa Trough (Upper Pleistocene); South China Sea (Quaternary). (3, 19, 21, 30, 31, 45, 65)
2026. *B. madagascariensis* (d'Orbigny): Xisha (=Paracel Islands); Zhongsha (=Macclesfield Bank); Guangdong (Pliocene). (33, 48)
2027. *B. milletti* Cushman: Nansha (=Spratly Islands). (66)
2028. *B. seminuda* (Terquem): Guangdong (Pliocene). (33)

UVIGERINIDAE

2029. *Euuvigerina aculeata* (d'Orbigny): East China Sea. (19)
2030. *E. peregrina* (Cushman): Taiwan. (44)
2031. *Neouvigerina ampullacea* (Brady): East China Sea; Taiwan; South China Sea; Okinawa Trough (Upper Pleistocene to Holocene). (30, 44, 45, 66)
2032. *Siphouvigerina ampullacea* (Brady) [2031]: East China Sea; Taiwan Strait; South China Sea (Upper Pleistocene to Holocene). (30, 44, 45, 66)
2033. *S. interrupta* (Brady): South China Sea; East China Sea; Taiwan; South China Sea (Pliocene). (3, 30, 33, 44, 48)
2034. *S. porrecta* (Brady): East China Sea; Xisha (=Paracel Islands); Zhongsha (=Macclesfield Bank); Nansha (=Spratly Islands); Okinawa Trough (Upper Pleistocene to Holocene). (19, 27, 30, 45)
2035. *S. proboscidea* (Schwager): East China Sea; Taiwan Strait; South China Sea; East China Sea continental shelf (Quaternary); South China Sea (Upper Pleistocene); Xisha (=Paracel Islands) (Middle Pleistocene to Holocene). (3, 30, 33, 39, 44)
2036. *S. pseudoampullacea* (Asano): East China Sea; Zhongsha (=Macclesfield Bank); Okinawa Trough (Holocene). (19, 27, 30, 45)
2037. *S. succincta* Qiu & Lin: Guangdong (Pliocene). (33)
2038. *Uvigerina aculeata* d'Orbigny [2029]: Taiwan Strait; South China Sea; Okinawa Trough (Upper Pleistocene to Holocene). (44, 45, 66)
2039. *U. adiposa* d'Orbigny: Xisha (=Paracel Islands) (Upper Pleistocene to Holocene). (39)
2040. *U. ampullacea* Brady [2031]: East China Sea; Taiwan Strait; South China Sea. (44, 66)
2041. *U. asperula* Czjzek: Taiwan Strait; South China Sea. (44, 48, 66)
2042. *U. attennata* (Coryell & Mossman): Xisha (=Paracel Islands) (Upper Pleistocene). (39)
2043. *U. attenuata* Cushman & Renz: Okinawa Trough (Upper Pleistocene to Holocene). (45)
2044. *U. auberiana* d'Orbigny: Taiwan Strait; South China Sea. (64, 65, 67)
2045. *U. bassensis* Parr: South China Sea; Xisha (Paracel Islands) (Middle Pleistocene to Holocene). (39, 42)

2046. *U. bifurcata* d'Orbigny: Taiwan Strait. (44)
2047. *U. bradyana* Fornasini: Xisha (=Paracel Islands); Zhongsha (=Macclesfield Bank) Xisha (=Paracel Islands) (Upper Pleistocene). (45, 48)
2048. *U. brunnensis* Karrer: Nansha (=Spratly Islands). (66)
2049. *U. bullata* Ho & Hu: South China Sea. (42)
2050. *U. canariensis* d'Orbigny: East China Sea; South China Sea; Jiangsu (Quaternary). (8, 19, 66)
2051. *U. chirana* Cushman & Stone: Xisha (=Paracel Islands); Zhongsha (=Macclesfield Bank). (48)
2052. *U. crassa* Egger: Xisha (=Paracel Islands) (Middle Pleistocene to Upper Pleistocene). (39)
2053. *U. cushmani* Todd: Nansha (=Spratly Islands). (66)
2054. *U. dirupta* Todd: East China Sea; South China Sea; Okinawa Trough (Upper Pleistocene to Holocene); Xisha (=Paracel Islands) (Middle Pleistocene). (12, 19, 39, 45)
2055. *U. finisterrensis* Colom: South China Sea; Okinawa Trough (Upper Pleistocene to Holocene). (42, 45)
2056. *U. globulosa* Egger: Okinawa Trough (Upper Pleistocene to Holocene). (45)
2057. *U. graciliformis* Papp & Turnovsky: Leiqiong area (southern Guangdong and Hainan) (Tertiary). (46)
2058. *U. hispida* Schwager: South China Sea; Xisha (=Paracel Islands) (Middle Pleistocene to Upper Pleistocene). (39)
2059. *U. hispido-costata* Cushman & Todd: Xisha (=Paracel Islands) (Middle Pleistocene to Upper Pleistocene). (39)
2060. *U. interrupta* Brady [2033]: East China Sea; South China Sea. (30)
2061. *U. kernensis* Barbat & Von Estorff: Nansha (=Spratly Islands). (66)
2062. *U. mediterranea* Hofker: Taiwan Strait (Neogene). (44)
2063. *U. miozea* Finlay: Xisha (=Paracel Islands); Zhongsha (=Macclesfield Bank). (48)
2064. *U. nitidula* Schwager: Taiwan. (44)
2065. *U. peregrina* Cushman [2030]: South China Sea; Okinawa Trough (Upper Pleistocene to Holocene); Guangdong (Pliocene). (33, 45, 64)
2066. *U. pigmea* d'Orbigny: Taiwan Strait. (44)
2067. *U. porrecta* (Brady) [2034]: Nansha (=Spratly Islands). (66)
2068. *U. proboscidea* Schwager [2035]: East China Sea; South China Sea; Xisha (=Paracel Islands) (Middle Pleistocene to Upper Pleistocene). (39, 44, 66)
2069. *U. rugosa* d'Orbigny: Xisha (=Paracel Islands) (Upper Pleistocene). (39)
2070. *U. schencki* Asano: East China Sea; Taiwan Strait; South China Sea. (19, 44)
2071. *U. schwageri* Brady: East China Sea; South China Sea; East China Sea continental shelf (Quaternary); South China Sea (Pliocene). (3, 19, 30, 44, 66, 67)
2072. *U. segundoensis* Cushman & Galliher: Taiwan Strait. (44)
2073. *U. senticosa* Cushman: East China Sea; South China Sea. (30, 66)
2074. *U. shukrii* Said: Xisha (=Paracel Islands) (Upper Pleistocene). (39)
2075. *U. tenuistriata* Reuss: Taiwan Strait; South China Sea. (44)
2076. *U. torquata* Wezzel: Guangdong (Pliocene). (33)
2077. *U. vadescens* Cushman: Okinawa Trough (Upper Pleistocene to Holocene). (45)
2078. *Angulogerina angulosa* (Williamson): East China Sea; Beijing (Lower Quaternary); southern Yellow Sea (Quaternary); Okinawa Trough (Upper Pleistocene to Holocene). (2, 19, 31, 45, 64)
2079. *A. carinata* Cushman: Nansha (=Spratly Islands). (66)
2080. *A. fluens* Todd: Southern Yellow Sea; Beijing (Lower Quaternary). (2, 30)
2081. *Trifarina angulosa* (Williamson) [2078]: East China Sea; Beijing (Lower Quaternary); southern Yellow Sea (Quaternary); Okinawa Trough (Upper Pleistocene to Holocene). (2, 19, 31, 45, 64)
2082. *T. bradyi* Cushman: East China Sea; Taiwan Strait; South China Sea; East China Sea continental shelf (Quaternary); Okinawa Trough (Upper Pleistocene to Holocene); South China Sea (Pliocene). (3, 19, 33, 45, 48, 50)
2083. *T. costornata* (Hornibrook): East China Sea continental shelf (Quaternary). (3)
2084. *T. fluens* (Todd) [2088]: Beijing (Lower Quaternary). (2, 30)
2085. *T. esnaensis* Leroy: East China Sea continental shelf (Paleocene). (3)
2086. *T. granulosa* Wang & He: Beijing (Lower Quaternary). (2)
2087. *T. guangdongensis* Ho & He: Xisha (=Paracel Islands); Zhongsha (=Macclesfield Bank). (48)
2088. *T. lepida* (Brady): Xisha (=Paracel Islands). (26, 30)
2089. *T. occidentalis* (Cushman): Southern Yellow Sea. (13)
2090. *T. reussi* Cushman: Nansha (=Spratly Islands). (66)

REUSSELLIDAE

2091. *Acostina piramidale* Acosta: East China Sea. (30)
2092. *Chrysalidinella dimorpha* (Brady): Hong Kong. (62)
2093. *C. earlandi* Cushman: Xisha (=Paracel Islands); Zhongsha (=Macclesfield Bank). (26, 27, 30, 48)
2094. *C. fijiensis* Cushman: Xisha (=Paracel Islands); Zhongsha (=Macclesfield Bank); Nansha (=Spratly Islands). (27, 30, 48, 66)
2095. *C. pacifica* Uchio: Xisha (=Paracel Islands); Zhongsha (=Macclesfield Bank). (48)
2096. *Fijiella simplex* (Cushman): Xisha (=Paracel Islands); Zhongsha (=Macclesfield Bank); Nansha (=Spratly Islands). (25, 30, 48)
2097. *Reussella aculeata* Cushman: Taiwan Strait; Xisha (=Paracel Islands) (Neogene). (4, 44)
2098. *R. atlantica* Cushman: Okinawa Trough (Holocene). (45)

2099. *R. costulata* Zheng: Xisha (=Paracel Islands); Zhongsha (=Macclesfield Bank). (25, 27, 30)

2100. *R. haizumensis* Asano: South China Sea. (67)

2101. *R. pulchra* Cushman: Xisha (=Paracel Islands); Zhongsha (=Macclesfield Bank). (27, 30, 48)

2102. *R. simplex* Cushman [2096]: Xisha (=Paracel Islands); Zhongsha (=Macclesfield Bank); Xisha (=Paracel Islands) (Upper Pleistocene). (41, 48)

2103. *R. spinosa* Zheng: Zhongsha (=Macclesfield Bank). (27, 30)

2104. *R. spinulosa* (Reuss): South China Sea; East China Sea (Quaternary). (3, 48)

2105. *Valvobifarina mackinnoi* (Millett): Xisha (=Paracel Islands); Zhongsha (=Macclesfield Bank). (48)

TRIMOSINIDAE

2106. *Mimosina orientalis* (Cushman): Xisha (=Paracel Islands); Nansha (=Spratly Islands). (26, 30, 66)

2107. *M. sidebottomi* Dollfuss: Xisha (=Paracel Islands). (26, 30)

2108. *Trimosina milletti* Cushman: Nansha (=Spratly Islands). (66)

2109. *T. orientalis* Cushman [2106]: Xisha (=Paracel Islands); Nansha (=Spratly Islands). (26, 30, 66)

PAVONINIDAE

2110. *Bifarinella ryukyuensis* (Cushman & Hanzawa): Dongsha (=Pratas Islands); Xisha (=Paracel Islands); Nansha (=Spratly Islands). (26, 30, 48)

2111. *Pavonina flabelliformis* d'Orbigny: Xisha (=Paracel Islands); Zhongsha (=Macclesfield Bank). (25, 30, 48)

2112. *P. ryukyuensis* Cushman & Hanzawa [2110]: Xisha (=Paracel Islands); Zhongsha (=Macclesfield Bank); Nansha (=Spratly Islands). (26, 30, 48)

2113. *P. tasmanensis* (Carter): East China Sea. (12)

MILLETTIIDAE

2114. *Millettia tessellata* (Brady): Zhongsha (=Macclesfield Bank). (27, 30)

FURSENKOINACEA
FURSENKOINIDAE

2115. *Coryphostoma amygdalaeformis* (Brady) [1896]: Xisha (=Paracel Islands) (Pleistocene). (4)

2116. *C. limbata* (Brady) [1900]: Xisha (=Paracel Islands); Zhongsha (=Macclesfield Bank); Guangdong (Pliocene). (33, 48)

2117. *C. lobata* (Brady): Guangdong (Pliocene). (33)

2118. *Fursenkoina bradyi* (Cushman): Nansha (=Spratly Islands). (66)

2119. *F. carinata* (Heron-Allen & Earland): East China Sea; South China Sea. (30)

2120. *F. compressa* (Bailey): East China Sea. (30)

2121. *F. cornuta* (Cushman): East China Sea. (30)

2122. *F. exilis* (Cushman & Ellisor): Kashi Basin (Xinjiang) (Tertiary). (38)

2123. *F. pauciloculata* (Brady): Yellow Sea; East China Sea; Taiwan Strait; South China Sea; Jiangsu (Quaternary); Guangdong (Pliocene). (3, 8, 19, 30, 33, 44, 50, 65)

2124. *F. rotundata* Parr: Nansha (=Spratly Islands). (30)

2125. *F. squammosa* (d'Orbigny): South China Sea. (30)

2126. *F. schreibersiana* (Czjzek): East China Sea; South China Sea; East China Sea (Quaternary); South China Sea (Pliocene). (3, 19, 30, 33, 66, 67)

2127. *F. taiwanica* (Nakamura): Taiwan. (44)

2128. *F. texturata* (Brady): South China Sea. (44)

2129. *F. tikutoensis* (Nakamura): Taiwan. (44)

2130. *Virgulina bradyi* Cushman [2118]: Nansha (=Spratly Islands). (66)

2131. *V. pauciloculata* Brady [2123]: East China Sea; Taiwan Strait; South China Sea. (30, 50, 66)

2132. *V. rotundata* Parr [2124]: Nansha (=Spratly Islands). (66)

2133. *V. schreibersiana* Czjzek [2126]: East China Sea; South China Sea; East China Sea continental shelf (Quaternary); South China Sea (Pliocene). (3, 19, 30, 33, 66, 67)

2134. *V. texturata* (Brady) [2128]: South China Sea. (42)

2135. *Sigmavirgulina basistriata* Zheng: Xisha (=Paracel Islands). (26, 30)

2136. *S. lanceolata* Zheng: Xisha (=Paracel Islands). (26, 30)

2137. *S. tortuosa* (Brady): Xisha (=Paracel Islands); Zhongsha (=Macclesfield Bank); Nansha (=Spratly Islands). (48, 64)

VIRGULINELLIDAE

2138. *Virgulinella fragilis* Grindell & Collen: Southern Yellow Sea. (30)

DELOSINACEA
DELOSINIDAE

2139. *Delosina complanata* Earland: Xisha (=Paracel Islands); Nansha (=Spratly Islands). (26, 30)

STILOSTOMELLACEA
STILOSTOMELLIDAE

2140. *Nodogenerina antillea* (Cushman): Okinawa Trough (Holocene). (45)

2141. *N. aperturata* Boomgaart: Guangdong (Pliocene). (33)

2142. *N. lepida* (Schwager): Guangdong (Pliocene). (33)

2143. *N. virgula* (Brady): Nansha (=Spratly Islands); Okinawa Trough (Holocene). (45, 66)

2144. *Siphonodosaria insoluta* (Schwager): Taiwan (Miocene). (44)

DISCORBACEA
PLACENTULINIDAE

2145. *Ashbrookia compressa* (Zheng): Zhongsha (=Macclesfield Bank). (27, 30)

2146. *Pseudopatellina compressa* Zheng [2145]: Zhongsha (=Macclesfield Bank). (27, 30)

2147. *Patellinella ambulacrata* (Moebius): Zhongsha (=Macclesfield Bank). (27, 30)

2148. *P. hanzawai* Asano: Xisha (=Paracel Islands); Zhongsha (=Macclesfield Bank). (48)

2149. *P. inconspicua* (Brady): Nansha (=Spratly Islands); Okinawa Trough (Holocene). (45)

2150. *P. jugosa* (Brady): Xisha (=Paracel Islands); Zhongsha (=Macclesfield Bank); Guangdong (Pliocene). (33, 48)

2151. *P. spinosa* Zheng: Xisha (=Paracel Islands); Zhongsha (=Macclesfield Bank); Okinawa Trough (Upper Pleistocene to Holocene). (26, 27, 30)

BAGGINIDAE

2152. *Baggina compressa* Hao & Zeng: Kashi Basin (Xinjiang) (Tertiary). (38)

2153. *B. inaperta* Hao & Zeng: Kashi Basin (Xinjiang) (Tertiary). (38)

2154. *B. indica* (Cushman): Southern Yellow Sea; East China Sea; South China Sea; East China Sea continental shelf (Quaternary); Guangdong (Pliocene). (3, 27, 33, 48, 65, 66)

2155. *B. longovata* Hao & Zeng: Kashi Basin (Xinjiang) (Tertiary). (38)

2156. *B. philippinensis* (Cushman): Taiwan. (44)

2157. *B. trapezoida* Hao & Zeng: Kashi Basin (Xinjiang) (Tertiary). (38)

2158. *B. turgidus* Hao & Zeng: Kashi Basin (Xinjiang) (Tertiary). (38)

2159. *Cancris auriculus* (Fichtel & Moll): Southern Yellow Sea; East China Sea; South China Sea; Jiangsu; East China Sea continental shelf (Quaternary); Okinawa Trough (Upper Pleistocene to Holocene); Taiwan (Miocene); Guangdong (Pliocene). (3, 8, 19, 26, 30, 31, 33, 44, 45, 47, 50, 66)

2160. *C. indicus* (Cushman): Southern Yellow Sea; East China Sea; South China Sea; Guangdong (Pliocene). (33, 48, 65)

2161. *C. intermedius* Cushman & Todd: East China Sea; South China Sea. (19, 48, 66)

2162. *C. maoricus* Finlay: Okinawa Trough (Upper Pleistocene). (45)

2163. *C. oblongus* (Williamson): Southern Yellow Sea; East China Sea; South China Sea. (30, 66, 67)

2164. *C. obesus* Qiu & Lin: Guangdong (Pliocene). (33)

2165. *C. ovalis* Hao & Zeng: Kashi Basin (Xinjiang) (Tertiary). (38)

2166. *C. peroblongus* (Cushman): East China Sea; South China Sea. (30)

2167. *C. sarga* (d'Orbigny): South China Sea; southern Yellow Sea (Quaternary). (31, 67)

2168. *C. segmentalis* Hao & Zeng: Kashi Basin (Xinjiang) (Tertiary). (38)

2169. *C. torquertus* Cushman & Todd: Southern Yellow Sea; East China Sea; South China Sea. (30)

2170. *Cribrobaggina reniformis* (Heron-Allen & Earland): Xisha (=Paracel Islands); Zhongsha (=Macclesfield Bank); Nansha (=Spratly Islands). (25, 30, 48)

2171. *Latecella reniformis* (Heron-Allen & Earland) [2170]: Xisha (=Paracel Islands); Zhongsha (=Macclesfield Bank); Nansha (=Spratly Islands). (25, 30, 48)

2172. *Rugidia corticata* (Heron-Allen & Earland): East China Sea; Xisha (=Paracel Islands); Zhongsha (=Macclesfield Bank). (25, 27, 30)

2173. *Valvulineria laevigata* Phleger & Parker: Jiangsu; shoreline of Taizhou Bay (Zhejiang) (Quaternary). (8, 47)

2174. *V. mexicana* Parker: Okinawa Trough (Holocene). (45)

2175. *V. minuta* Parker: Okinawa Trough (Upper Pleistocene to Holocene). (45)

2176. *V. polita* Parr: Nansha (=Spratly Islands). (66)

2177. *V. rugosa* Earland: Nansha (=Spratly Islands). (66)

2178. *V. sadonica* Asano: Shoreline of Bohai (Tertiary). (49)

EPONIDIDAE

2179. *Eponides berthelotianus* (d'Orbigny): South China Sea; East China Sea continental shelf (Quaternary); South China Sea (Pliocene to Holocene). (3, 33, 66)

2180. *E. blancoensis* Brady: Shoreline of Bohai (Tertiary). (49)

2181. *E. cribroconcameratus* (Asano & Uchio) [2192]: Southern Yellow Sea; East China Sea. (30, 65)

2182. *E. pacifica* (LeRoy): East China Sea continental shelf (Quaternary). (3)

2183. *E. parvus* Stschedrina: Southern Yellow Sea. (65)

2184. *E. procerus* (Brady) [2232]: East China Sea; South China Sea; Guangdong (Pliocene). (30, 66)

2185. *E. pusillus* Parr: Yellow Sea. (65)

2186. *E. repandus* (Fichtel & Moll): East China Sea; Taiwan; Xisha (=Paracel Islands); Zhongsha (=Macclesfield Bank); Jiangsu (Quaternary). (8, 30, 44, 48)

2187. *E. subornatus* (Cushman) [2233]: East China Sea; Taiwan; South China Sea; Taiwan (Miocene). (19, 44)

2188. *E. sulcata* Hu: Mt. Jolmo Lungma (=Mt. Everest) area (Tibet) (Tertiary). (9)

2189. *E. umbonatus* (Reuss): Taiwan. (44)

2190. *E. varvus* Stschedrina: Southern Yellow Sea. (65)

2191. *E. xishaensis* Cai & Tu: Xisha (=Paracel Islands); Zhongsha (=Macclesfield Bank). (48)

2192. *Cribroeponides cribroconcameratus* (Asano & Uchio): Southern Yellow Sea; East China Sea. (30, 65)
2193. *Poroeponides calida* Cai & Tu: Xisha (=Paracel Islands); Zhongsha (=Macclesfield Bank). (48)
2194. *P. cribrorepandus* Asano & Uchio: Yellow Sea; East China Sea; Taiwan Strait; South China Sea; Jiangsu; shoreline of Taizhou Bay (Zhejiang) (Quaternary); South China Sea (Pliocene). (3, 4, 8, 19, 22, 30, 47, 59, 65, 66, 67)
2195. *P. incrassatus* Ho, Hu & Wang: East China Sea continental shelf; Jiangsu (Quaternary). (3, 8)
2196. *P. lateralis* (Terquem): Northern Yellow Sea; East China Sea; South China Sea; shoreline of Taizhou Bay (Zhejiang) (Quaternary); Guangdong (Pliocene). (30, 33, 47, 62, 66)
2197. *P. repandus* (Fichtel el Moll) [2866]: East China Sea; Taiwan; Xisha (=Paracel Islands); Zhongsha (=Macclesfield Bank); Jiangsu (Quaternary). (8, 30, 44, 48)
2198. *P. speciosus* Stschedrina: Southern Yellow Sea. (65)
2199. *Vernonina gravida* Wang, He, & Lu: East China Sea continental shelf (Paleocene). (3)
2200. *Sestronophora arnoldi* Loeblich & Tappan: South China Sea. (30)

HELENINIDAE

2201. *Helenina anderseni* (Warren): East China Sea; Zhejiang; Taiwan. (19, 37)
2202. *Mississippina concentrica* Parker & Jones: Xisha (=Paracel Islands); Zhongsha (=Macclesfield Bank); Nansha (=Spratly Islands). (27, 30, 48, 66)

MISSISSIPPINIDAE

2203. *Ungulatella conoides* Cushman: Xisha (=Paracel Islands). (26, 30)
2204. *Ungullatelloides imperialis* Seiglie: Xisha (=Paracel Islands). (26, 30)
2205. *U. pagoda* Zheng: Xisha (=Paracel Islands). (26, 30)

PEGIDIIDAE

2206. *Pegidia dubia* (d'Orbigny): Zhongsha (=Macclesfield Bank); Nansha (=Spratly Islands). (27, 30)
2207. *Sphaeridia papillata* Heron-Allen & Earland: Zhongsha (=Macclesfield Bank); Nansha (=Spratly Islands). (27, 30)

DISCORBIDAE

2208. *Discorbis asterocides* Hao & Zeng: Taiwan (Tertiary); Kashi Basin (Xinjiang) (Lower Tertiary). (38, 46)
2209. *D. australis* Parr [2241]: Taiwan. (44)
2210. *D. biaperturata* Pokorny: Xisha (=Paracel Islands); Zhongsha (=Macclesfield Bank). (48)
2211. *D. bradyi* Cushman [2242]: Nansha (=Spratly Islands). (66)
2212. *D. bullatus* Hao & Zeng: Kashi Basin (Xinjiang) (Lower Tertiary). (38)
2213. *D. candeianus* (d'Orbigny): Nansha (=Spratly Islands). (66)
2214. *D. chinensis* Huang: Taiwan Strait. (59)
2215. *D. distinctus* Hao & Zeng: Kashi Basin (Xinjiang) (Lower Tertiary). (38)
2216. *D. granulosus* (Heron-Allen & Earland): Hong Kong. (62)
2217. *D. latestoma* Zheng: Xisha (=Paracel Islands). (26, 30)
2218. *D. mira* Cushman: Xisha (=Paracel Islands). (26, 30)
2219. *D. minuta* Schubert: Xisha (=Paracel Islands). (26, 30)
2220. *D. murrayi* (Heron-Allen & Earland) [2279]: Hong Kong. (62)
2221. *D. obvelatus* Hao & Zeng: Kashi Basin (Xinjiang) (Lower Tertiary). (38)
2222. *D. orbicularis* (Terquem) [2235]: East China Sea; Nansha (=Spratly Islands). (30, 66)
2223. *D. placoides* Hao & Zeng: Kashi Basin (Xinjiang) (Lower Tertiary). (38)
2224. *D. rugosus* (d'Orbigny): Xisha (=Paracel Islands); Zhongsha (=Macclesfield Bank); Nansha (=Spratly Islands). (25, 27, 30, 48, 66)
2225. *D. subquatus* Hao & Zeng: Kashi Basin (Xinjiang) (Lower Tertiary). (38)
2226. *D. subvesicularis* Collins: Xisha (=Paracel Islands). (25, 30)
2227. *D. taiwanensis* Huang: Taiwan Strait. (44, 59)
2228. *D. tuberculatus* Hao & Zeng: Kashi Basin (Xinjiang) (Lower Tertiary). (38)
2229. *D. vilardeboana* (d'Orbigny) [2256]: Taiwan Strait. (50)
2230. *D. williamsoni* Chapman & Parr: South China Sea. (30)
2231. *Neoeponides berthelinianus* (d'Orbigny): East China Sea. (12)
2232. *N. procera* (Brady): East China Sea; South China Sea; northern South China Sea (Quaternary). (19, 30, 48, 66)
2233. *N. subornatus* (Cushman): East China Sea; South China Sea; Taiwan (Miocene). (19, 44)

ROSALINIDAE

2234. *Gavelinopsis praeggeri* (Heron-Allen & Earland): East China Sea; South China Sea; East China Sea continental shelf (Quaternary); Guangdong (Pliocene). (3, 19, 27, 30, 33, 66)
2235. *Neoconorbina orbicularis* (Terquem): Okinawa Trough (Holocene). (45)
2236. *N. pacifica* Hofker: Nansha (=Spratly Islands). (30)
2237. *N. terquemi* (Rzehak): Southern Yellow Sea; Guangdong (Pliocene). (33, 65)
2238. *Planodiscorbis grossepunctatus* (Parr): Xisha (=Paracel Islands). (26, 30)
2239. *P. lingi* Huang: Taiwan Strait. (44)
2240. *P. rarescens* (Brady): East China Sea; South China Sea. (19, 30, 66)

2241. *Rosalina australis* (Parr): East China Sea. (19)
2242. *R. bradyi* (Cushman): Bohai; Yellow Sea; East China Sea; South China Sea; Jiangsu (Quaternary); South China Sea (Pleistocene). (3, 8, 19, 44, 48, 62, 65, 66)
2243. *R. concinna* (Brady) [2258]: East China Sea; Xisha (=Paracel Islands); Zhongsha (=Macclesfield Bank); Nansha (=Spratly Islands). (25, 30, 48, 67)
2244. *R. crustata* (Cushman): Zhongsha (=Macclesfield Bank); Nansha (=Spratly Islands). (27, 30)
2245. *R. floridana* (Cushman): Hong Kong; shoreline of Bohai (Quaternary). (34, 44, 66)
2246. *R. globularis* d'Orbigny: Taiwan Strait; eastern Hebei (Quaternary). (34, 44, 66)
2247. *R. neapolitana* (Cushman): Xisha (=Paracel Islands); Nansha (=Spratly Islands). (25, 30)
2248. *R. obtusa* d'Orbigny: Okinawa Trough (Holocene). (45)
2249. *R. opima* (Cushman): Zhongsha (=Macclesfield Bank). (27, 30)
2250. *R. orientalis* (Cushman): Xisha (=Paracel Islands); Nansha (=Spratly Islands). (25, 30)
2251. *R. pacifica* (Hofker) [2236]: Xisha (=Paracel Islands); Zhongsha (=Macclesfield Bank); Nansha (=Spratly Islands); Xisha (=Paracel Islands) (Cenozoic). (4, 25, 27, 30, 48, 66)
2252. *R. petasiformis* Zheng: Xisha (=Paracel Islands); Zhongsha (=Macclesfield Bank); Nansha (=Spratly Islands). (25, 30, 48)
2253. *R. subcomplanata* (Parr): Xisha (=Paracel Islands). (26, 30)
2254. *R. terquemi* (Rzehak) [2237]: Zhongsha (=Macclesfield Bank); Nansha (=Spratly Islands). (27, 30, 66)
2255. *R. tuberocapitata* (Chapman): Zhongsha (=Macclesfield Bank). (27, 30)
2256. *R. vilardeboana* d'Orbigny: Southern Yellow Sea; East China Sea; South China Sea; South China Sea (Quaternary). (3, 19, 26, 30, 48, 66, 67)
2257. *R. williamsoni* (Chapman & Parr): Zhongsha (=Macclesfield Bank). (27, 30)
2258. *Tretomphaloides concinnus* (Brady): East China Sea; Xisha (=Paracel Islands); Zhongsha (=Macclesfield Bank); Nansha (=Spratly Islands). (25, 30, 48, 67)
2259. *Tretomphalus bulloides* d'Orbigny: Xisha (=Paracel Islands) (Cenozoic). (4)
2260. *T. concinnus* (Brady) [2258]: East China Sea; Xisha (=Paracel Islands); Nansha (=Spratly Islands). (25, 30, 48, 67)
2261. *T. grandis* Cushman: Xisha (=Paracel Islands); Zhongsha (=Macclesfield Bank); Nansha (=Spratly Islands); Xisha (=Paracel Islands) (Cenozoic). (4, 25, 30, 48)
2262. *T. milletti* (Heron-Allen & Earland) [2416]: Xisha (=Paracel Islands); Nansha (=Spratly Islands). (25, 30)
2263. *T. planus* Cushman: Xisha (=Paracel Islands); Zhongsha (=Macclesfield Bank); Nansha (=Spratly Islands). (25, 30, 48)

SPHAEROIDINIDAE

2264. *Sphaeroidina bulloides* d'Orbigny: East China Sea; Taiwan Strait; South China Sea; East China Sea continental shelf (Quaternary); Taiwan (Miocene); Okinawa Trough (Upper Pleistocene to Holocene). (3, 19, 30, 33, 44, 45, 48, 59, 66, 67)
2265. *S. chilostomata* Galloway & Morrey: Guangdong (Pliocene). (33)
2266. *S. variabilis* Reuss: Guangdong (Pliocene). (33)

GLABRATELLACEA
GLABRATELLIDAE

2267. *Angulodiscorbis corrugata* (Millett): Xisha (=Paracel Islands); Nansha (=Spratly Islands). (25, 30)
2268. *Conorbella pulvinata* (Brady): South China Sea. (66)
2269. *Glabratella crassa* Dorreen: Xisha (=Paracel Islands). (26, 30)
2270. *G. elegans* Zheng: Xisha (=Paracel Islands). (26, 30)
2271. *G. limbata* Zheng: Xisha (=Paracel Islands); Zhongsha (=Macclesfield Bank). (26, 27, 30)
2272. *G. makinoi* Uchio: Xisha (=Paracel Islands). (26, 30)
2273. *G. opercularis* (d'Orbigny): Southern Yellow Sea. (65)
2274. *G. patelliformis* (Brady): Yellow Sea; Taiwan Strait; Xisha (=Paracel Islands); Zhongsha (=Macclesfield Bank); Nansha (=Spratly Islands). (25, 27, 30, 48, 50)
2275. *G. pileolus* (d'Orbigny): Southern Yellow Sea. (65)
2276. *G. pulchella* Zheng: Xisha (=Paracel Islands). (26, 30)
2277. *G. scabra* Zheng: Xisha (=Paracel Islands); Nansha (=Spratly Islands). (26, 30)
2278. *G. tabernacularis* (Brady): Xisha (=Paracel Islands); Nansha (=Spratly Islands). (26, 30)
2279. *Murrayinella murrayi* (Heron-Allen & Earland): Southern Yellow Sea; East China Sea; Okinawa Trough; South China Sea. (19, 22, 45, 65)
2280. *Schackoinella globosa* (Millett) [2279]: Southern Yellow Sea; East China Sea; Okinawa Trough; South China Sea. (19, 22, 45, 65)
2281. *S. lepida* He & Hu: Shoreline of Bohai (Quaternary). (49)

HERONALLENIIDAE

2282. *Heronallenia laevis* Parr: Xisha (=Paracel Islands). (26, 30)
2283. *H. parva* Parr: Xisha (=Paracel Islands). (26, 30)

BULIMINOIDIDAE

2284. *Buliminoides milletti* (Cushman): Xisha (=Paracel Islands); Zhongsha (=Macclesfield Bank); Nansha (=Spratly Islands). (27, 30, 48, 66)
2285. *B. parallela* (Cushman & Parker): Xisha (=Paracel Islands). (26, 30)

2286. *B. williamsonianus* (Brady): Xisha (=Paracel Islands); Zhongsha (=Macclesfield Bank). (25, 27, 30, 48)

SIPHONINACEA
SIPHONINIDAE

2287. *Pulsiphonina elegans* Brotzen: East China Sea; South China Sea; Xisha (=Paracel Islands) (Upper Pleistocene). (19, 41, 66)

2288. *Siphonina australis* Cushman: Okinawa Trough (Upper Pleistocene); Guangdong (Pliocene). (33, 45)

2289. *S. bradyana* Cushman: East China Sea; South China Sea. (19)

2290. *S. pulchra* Cushman: East China Sea; Xisha (=Paracel Islands); Zhongsha (=Macclesfield Bank). (25, 27, 30, 48)

2291. *S. reticulata* (Czjzek): South China Sea. (67)

2292. *S. tubulosa* Cushman: East China Sea; Taiwan Strait; South China Sea; East China Sea continental shelf (Quaternary). (3, 44, 66)

2293. *Siphoninoides echinatus* (Brady): Xisha (=Paracel Islands); Zhongsha (=Macclesfield Bank); Nansha (=Spratly Islands); Xisha (=Paracel Islands) (Cenozoic). (4, 25, 27, 30, 44, 48, 50, 62, 66)

2294. *S. glabra* (Heron-Allen & Earland): Taiwan Strait. (44)

DISCORBINELLACEA
PARRELLOIDIDAE

2295. *Cibicidoides amygdaliformis* Hao & Zeng: Kashi Basin (Xinjiang) (Lower Tertiary). (38)

2296. *C. anomalos* Hao & Zeng: Kashi Basin (Xinjiang) (Lower Tertiary). (38)

2297. *C. bellus* Qiu & Lin: Guangdong (Pliocene). (33)

2298. *C. bradyi* (Trauth): Xisha (=Paracel Islands); Zhongsha (=Macclesfield Bank). (48)

2299. *C. compressa* (Cushman & Renz): Xisha (=Paracel Islands) (Upper Pleistocene). (39)

2300. *C. haitiansis* (Coryell & Rivera): Xisha (=Paracel Islands) (Middle Pleistocene). (39)

2301. *C. hyalinus* (Hofker) [2312]: Xisha (Paracel Islands); Zhongsha (=Macclesfield Bank); Nansha (=Spratly Islands); Okinawa Trough (Upper Pleistocene to Holocene); Xisha (=Paracel Islands) (Middle Pleistocene to Upper Pleistocene). (39, 45, 48, 66)

2302. *C. midwayensis* (Plummer): East China Sea continental shelf (Eocene). (3)

2303. *C. mundulus* (Brady, Parker, & Jones): Xisha (=Paracel Islands); Zhongsha (=Macclesfield Bank). (48)

2304. *C. ovaliformis* Hao & Zeng: Kashi Basin (Xinjiang) (Lower Tertiary). (38)

2305. *C. phaseoliformis* Hao & Zeng: Kashi Basin (Xinjiang) (Lower Tertiary). (38)

2306. *C. pseudoungerianus* (Cushman) [2623]: Kashi Basin (Xinjiang) (Lower Tertiary). (38)

2307. *C. robertsonianus* (Brady): Xisha (=Paracel Islands) (Upper Pleistocene). (39)

2308. *C. subplanospirolus* Hao & Zeng: Kashi Basin (Xinjiang) (Lower Tertiary). (38)

2309. *C. ungerianus* (d'Orbigny): Kashi Basin (Xinjiang) (Lower Tertiary). (38)

2310. *C. succedens* (Brotzen): Xinjiang (Early Tertiary); Guangdong (Paleocene). (3, 38)

2311. *C. temperatus* (Vella): East China Sea continental shelf (Quaternary). (3)

2312. *Parrelloides hyalinus* (Hofker): Xisha (=Paracel Islands); Zhongsha (=Macclesfield Bank); Nansha (=Spratyl Islands); Okinawa Trough (Upper Pleistocene to Holocene); Xisha (=Paracel Islands) (Middle Pleistocene to Upper Pleistocene). (39, 45, 48, 66)

PSEUDOPARRELLIDAE

2313. *Epistominella amakusaensis* Asano & Murata: East China Sea continental shelf (Eocene). (3)

2314. *E. exigua* (Brady): Yellow Sea; East China Sea; Taiwan Strait; South China Sea. (59, 64, 65, 66)

2315. *E. naraensis* (Kuwano): Southern Yellow Sea; South China Sea; East China Sea continental shelf (Quaternary); Okinawa Trough (Upper Pleistocene to Holocene). (3, 19, 31, 33, 45)

2316. *E. nipponica* Kuwano: East China Sea. (19)

2317. *E. pacifica* (Cushman): East China Sea. (10)

2318. *E. pulchra* (Cushman): East China Sea; Zhongsha (=Macclesfield Bank); Nansha (=Spratly Islands). (27, 30, 64)

2319. *E. takayanagii* Iwasa: Okinawa Trough (Upper Pleistocene to Holocene). (45)

2320. *E. tubulifera* (Heron-Allen & Earland) [2430]: East China Sea; Taiwan Strait; South China Sea. (25, 27, 30, 44, 50, 66)

2321. *Stetsonia altilis* Hao & Zeng: Kashi Basin (Xinjiang) (Lower Tertiary). (38)

2322. *S. minuta* Parker: Okinawa Trough (Holocene). (45)

PLANULINOIDIDAE

2323. *Planulinoides planoconcavus* (Chapman, Parr, & Collins): Xisha (=Paracel Islands). (26, 30)

DISCORBINELLIDAE

2324. *Discorbinella advena* (Cushman): Hong Kong. (62)

2325. *D. bertheloti* (d'Orbigny): East China Sea; South China Sea. (19, 30, 66)

2326. *D. floridensis* Cushman: Nansha (=Spratly Islands). (66)

2327. *D. kerimbatica* Zheng: Xisha (=Paracel Islands). (26, 30)

2328. *D. montereyensis* Cushman & Martin: East China Sea continental shelf (Quaternary); Guangdong (Pliocene). (3, 33)

2329. *Laticarinina halophora* (Stache): Xisha (=Paracel Islands); Zhongsha (=Macclesfield Bank). (48)

2330. *L. pauperata* (Parker & Jones): East China Sea; South China Sea; Okinawa Trough (Upper Pleistocene to Holocene); Xisha (=Paracel Islands) (Upper Pleistocene to Holocene). (30, 39, 45, 66)

PLANORBULINACEA
PLANULINIDAE

2331. *Hyalinea balthica* (Schroeter): Southern Yellow Sea; East China Sea; South China Sea; Jiangsu; East China Sea continental shelf (Quaternary); Okinawa Trough (Upper Pleistocene to Holocene); Guangdong (Pliocene). (3, 8, 19, 30, 33, 36, 44, 45, 59, 66)

2332. *Planulina alticamera* Qiu & Lin: Guangdong (Pliocene). (3)

2333. *P. ariminensis* d'Orbigny: East China Sea; Taiwan Strait; South China Sea. (30, 44, 66)

2334. *P. bradyi* Tolmachoff: Xisha (=Paracel Islands); Zhongsha (=Macclesfield Bank). (48)

2335. *P. costata* (Hantken): Okinawa Trough (Upper Pleistocene). (39)

2336. *P. dohertyi* (Gallowet & Morrey): Xisha (=Paracel Islands) (Middle Pleistocene to Upper Pleistocene). (39)

2337. *P. elegans* Tolmachoff: Yellow Sea; Taiwan Strait; South China Sea. (64)

2338. *P. foveolata* (Brady): Nansha (=Spratly Islands). (66)

2339. *P. subdepressus* Asano: East China Sea continental shelf (Quaternary). (3)

2340. *P. subtenuissima* (Nuttall): Xisha (=Paracel Islands) (Middle Pleistocene). (39)

2341. *P. wuellerstorfi* (Schwager) [2393]: East China Sea; Taiwan Strait; South China Sea; Okinawa Trough (Upper Pleistocene to Holocene); Xisha (=Paracel Islands) (Middle Pleistocene to Upper Pleistocene). (30, 39, 44, 45, 48, 66)

CIBICIDIDAE

2342. *Cibicides artemi* Bykova: Kashi Basin (Xinjiang) (Lower Tertiary). (38)

2343. *C. beatus* Martin: East China Sea continental shelf (Eocene). (3)

2344. *C. boueanus* (d'Orbigny): Guangdong (Pliocene). (33)

2345. *C. borislavensis* Aisenstat: Kashi Basin (Xinjiang) (Lower Tertiary); Tarim Basin (Xinjiang) (Tertiary). (38, 40)

2346. *C. bradyi* (Trauth) [2298]: South China Sea; Xisha (=Paracel Islands) (Middle Pleistocene to Holocene). (39, 66)

2347. *C. celebrus* Bandy: Kashi Basin (Xinjiang) (Lower Tertiary). (38)

2348. *C. compressus* (Cushman & Renz) [2299]: Xisha (=Paracel Islands) (Middle Pleistocene to Upper Pleistocene). (39)

2349. *C. cookei* Cushman & Garrett: East China Sea continental shelf (Eocene). (3)

2350. *C. cunobelini* Haynes: East China Sea continental shelf (Eocene). (3)

2351. *C. cushmani* Ujiie & Kusukawa: Yellow Sea; East China Sea; South China Sea. (30, 66)

2352. *C. deusseni* (Weinzieri & Applin): Kashi Basin (Xinjiang) (Lower Tertiary). (38)

2353. *C. decoratus* LeRoy: East China Sea continental shelf (Paleocene). (3)

2354. *C. dorsitubera* Hao & Zeng: Kashi Basin (Xinjiang) (Lower Tertiary). (38)

2355. *C. djaffaensis* Sigal: Guangdong (Pliocene). (33)

2356. *C. deprimus* Phleger & Parker: Yellow Sea; East China Sea; South China Sea; Okinawa Trough (Upper Pleistocene). (45, 48, 64, 65)

2357. *C. entendus* Hao & Zeng: Kashi Basin (Xinjiang) (Lower Tertiary). (38)

2358. *C. farafraensis* Leroy: East China Sea continental shelf (Paleocene). (3)

2359. *C. fletcheri* Galloway & Wissler: Okinawa Trough (Upper Pleistocene). (45)

2360. *C. howelli* Toulmin: Mt. Jolmo Lungma (=Mt. Everest) (Tibet) (Lower Tertiary). (9)

2361. *C. laxus* He & Lin: Guangdong (Pliocene). (33)

2362. *C. lectus* Vassilenko: East China Sea continental shelf (Eocene). (3)

2363. *C. lobatulus* (Walker & Jocob) [2394]: Bohai; Yellow Sea; East China Sea; South China Sea; Taiwan Strait; East China Sea continental shelf (Quaternary and Pliocene); Kashi Basin (Xinjiang) (Lower Tertiary). (1, 3, 19, 38, 44, 48, 59, 66)

2364. *C. lobatus* (d'Orbigny): Kashi Basin (Xinjiang) (Lower Tertiary). (38)

2365. *C. margaritiferus* (Brady): East China Sea; Taiwan Strait; South China Sea. (3, 19, 30, 44, 59, 66, 67)

2366. *C. mayori* (Cushman): Xisha (=Paracel Islands); Zhongsha (=Macclesfield Bank); Nansha (=Spratly Islands). (25, 27, 30, 66)

2367. *C. mendesi* Petri: Tarim Basin (Xinjiang) (Tertiary). (40)

2368. *C. mollis* Phleger & Parker: Southern Yellow Sea; East China Sea. (19, 65)

2369. *C. multicameratus* Stschedrina: Southern Yellow Sea. (65)

2370. *C. nagaoi* Asano & Murata: Kashi Basin (Xinjiang) (Lower Tertiary). (38)

2371. *C. praecursorius* (Schwager): Mt. Jolmo Lungma (=Mt. Everest) (Tibet) (Lower Tertiary). (9)

2372. *C. praecinctus* (Karrer) [2622]: East China Sea; South China Sea; Xisha (=Paracel Islands). (4, 25, 30, 48, 66, 67)

2373. *C. pseudoungerianus* (Cushman) [2623]: Yellow Sea; East China Sea; Taiwan Strait; South China Sea; Jiangsu; East China Sea continental shelf (Quaternary and Miocene). (3, 8, 19, 44, 50, 59, 64, 66)

2374. *C. punjabensis* Haque: East China Sea continental shelf (Eocene). (3)
2375. *C. reinholdi* Dam: Tarim Basin (Xinjiang) (Tertiary); Kashi Basin (Xinjiang) (Lower Tertiary). (40)
2376. *C. refulgens* de Montfort: Guangdong (Pliocene); Xisha (=Paracel Islands) (Upper Pleistocene). (41, 44, 48, 66, 67)
2377. *C. robertsonianus* (Brady) [2307]: South China Sea; Xisha (=Paracel Islands) (Middle Pleistocene to Upper Pleistocene). (42)
2378. *C. robustus* LeCalvez: Mt. Jolmo Lungma (=Mt. Everest) (Tibet) (Lower Tertiary). (9)
2379. *C. rugosus* Phleger & Parr: South China Sea. (42)
2380. *C. succedens* Brotzen [2310]: Tarim Basin (Xinjiang) (Tertiary). (40)
2381. *C. sintikuensis* Nakamura: Taiwan Strait; East China Sea continental shelf (Quaternary); Taiwan (Upper Tertiary). (3, 59)
2382. *C. subhaidingeri* Parr: East China Sea; Taiwan Strait; South China Sea; northern South China Sea (Quaternary); Xisha (=Paracel Islands) (Upper Pleistocene). (3, 30, 39, 48, 50, 66)
2383. *C. tallahattensis* Bandy: Kashi Basin (Xinjiang) (Lower Tertiary); East China Sea continental shelf (Eocene). (3, 38)
2384. *C. tani* Iwasa & Kikuchi: Guangdong (Pliocene). (3)
2385. *C. tenuimargo* (Brady): East China Sea; South China Sea. (25, 30)
2386. *C. umbonatus* Phleger & Parker: Nansha (=Spratly Islands). (66)
2387. *C. ungerianus* (d'Orbigny) [2309]: East China Sea; Nansha (=Spratly Islands). (30, 66)
2388. *C. wuellerstorfi* (Schwager) [2393]: East China Sea; South China Sea; Okinawa Trough (Upper Pleistocene to Holocene). (30, 39, 45)
2389. *Cibicidina expansus* Hao & Zeng: Kashi Basin (Xinjiang) (Lower Tertiary). (38)
2390. *C. minuta* Zheng: Zhongsha (=Macclesfield Bank). (27, 30)
2391. *C. patinaris* Hao & Zeng: Kashi Basin (Xinjiang) (Lower Tertiary). (38)
2392. *C. platumbilica* Hao & Zeng: Kashi Basin (Xinjiang) (Lower Tertiary). (38)
2393. *Fontbotia wuellerstorfi* (Schwager): East China Sea; South China Sea; Okinawa Trough (Upper Pleistocene to Holocene); Xisha (=Paracel Islands) (Middle Pleistocene to Upper Pleistocene). (30, 39, 45)
2394. *Lobatula lobatula* (Walker & Jacob): Bohai; Yellow Sea; East China Sea; South China Sea; East China Sea continental shelf (Quaternary and Eocene); Kashi Basin (Xinjiang) (Lower Tertiary). (1, 3, 19, 30, 38, 44, 48, 66)
2395. *Dyocibicides biserialis* Cushman & Valentine: East China Sea; South China Sea. (30)
2396. *D. epicharis* Zhang: East China Sea. (19)
2397. *D. perforata* Cushman & Valentine: East China Sea; South China Sea. (30)
2398. *Pyropiloides elongatus* Zheng: Xisha (=Paracel Islands). (26, 30)
2399. *Planorbulinoides retinaculata* (Parker & Jones): Nansha (=Spratly Islands). (30)

PLANORBULINIDAE

2400. *Caribbeanella cuspidata* Ho & Hu: Xisha (=Paracel Islands); Zhongsha (=Macclesfield Bank); Nansha (=Spratly Islands). (48, 66)
2401. *C. depressa* Zhang: East China Sea. (19)
2402. *C. incerta* Ho & Hu: Xisha (=Paracel Islands); Zhongsha (=Macclesfield Bank); Nansha (=Spratly Islands). (48, 66)
2403. *C. irregularis* Cai & Tu: Xisha (=Paracel Islands); Zhongsha (=Macclesfield Bank). (48)
2404. *C. katasensis* (Ujiie): Nansha (=Spratly Islands). (66)
2405. *C. ogiensis* (Matsugana): Xisha (=Paracel Islands); Zhongsha (=Macclesfield Bank); Nansha (=Spratly Islands); Guangdong (Pliocene). (33, 48, 66)
2406. *Planorbulina acervalis* Brady: Hong Kong; Xisha (=Paracel Islands); Nansha (=Spratly Islands). (25, 30, 66)
2407. *P. mediterranensis* d'Orbigny: Taiwan Strait. (44)
2408. *P. rubra* d'Orbigny: Xisha (=Paracel Islands); Nansha (=Spratly Islands). (25, 30, 66)
2409. *P. variabilis* (d'Orbigny): Xisha (=Paracel Islands). (26, 30)
2410. *Cibicidella variabilis* (d'Orbigny) [2409]: Xisha (=Paracel Islands). (26, 30)
2411. *Planorbulinella larvata* (Parker & Jones): East China Sea; South China Sea; Guangdong (Pliocene); Xisha (=Paracel Islands) (Tertiary and Upper Pleistocene). (4, 19, 25, 27, 30, 33, 41, 46, 48, 66)

CYMBALOPORIDAE

2412. *Cymbaloporella tabellaeformis* (Brady): Xisha (=Paracel Islands); Zhongsha (=Macclesfield Bank). (25, 27, 30, 66)
2413. *Cymbaloporetta bradyi* (Cushman): Xisha (=Paracel Islands); Zhongsha (=Macclesfield Bank); Nansha (=Spratly Islands); Shenzhen Bay (Guangdong). (7, 25, 30, 48, 66)
2414. *C. solida* Stschedrina: Southern Yellow Sea. (65)
2415. *C. squammosa* (d'Orbigny): Xisha (=Paracel Islands); Zhongsha (=Macclesfield Bank); Nansha (=Spratly Islands); Xisha (=Paracel Islands) (Cenozoic and Tertiary). (4, 25, 27, 30, 46, 48, 66)
2416. *Millettiana milletti* (Heron-Allen & Earland): Xisha (=Paracel Islands); Nansha (=Spratly Islands). (25, 30)
2417. *Pyropilus rotundatus* Cushman: Xisha (=Paracel Islands); Zhongsha (=Macclesfield Bank); Nansha (=Spratly Islands); Xisha (=Paracel Islands) (Cenozoic). (4, 25, 27, 30, 48)

VICTORIELLIDAE

2418. *Carpenteria monticularis* Carter: East China Sea; South China Sea. (30)
2419. *C. proteiformis* Göes: East China Sea. (30)
2420. *C. utricularis* (Carter): Xisha (=Paracel Islands). (26, 30)
2421. *Rupertina sinensis* Zhang: East China Sea. (19)

ACERVULINACEA
ACERVULINIDAE

2422. *Acervulina inhaerens* Schultze: Xisha (=Paracel Islands); Zhongsha (=Macclesfield Bank); Nansha (=Spratly Islands). (25, 27, 30)
2423. *Gypsina globula* (Reuss) [2426]: Xisha (=Paracel Islands); Zhongsha (=Macclesfield Bank); Nansha (=Spratly Islands); Xisha (=Parael Islands) (Cenozoic). (4, 25, 27, 30, 48, 66)
2424. *G. vesicularis* (Parker & Jones): Xisha (=Paracel Islands) (Cenozoic). (4)
2425. *Planogypsina squamiformis* (Chapman): Xisha (=Paracel Islands); Zhongsha (=Macclesfield Bank). (27, 30)
2426. *Sphaerogypsina globulus* (Reuss): Xisha (=Paracel Islands); Zhongsha (=Macclesfield Bank); Nansha (=Spratly Islands); Xisha (=Paracel Islands) (Cenozoic). (4, 25, 27, 30, 48, 66)

HOMOTREMATIDAE

2427. *Homotrema rubrum* (Lamarck): South China Sea; Xisha (=Paracel Islands). (25, 30)
2428. *Miniacina miniacea* (Pallas): Xisha (=Paracel Islands); Zhongsha (=Macclesfield Bank); Nansha (=Spratly Islands). (25, 27, 30)

ASTERIGERINACEA
EPISTOMARIIDAE

2429. *Asanonella shojii* Huang: Taiwan Strait. (44)
2430. *A. tubulifera* (Heron-Allen & Earland): East China Sea; Taiwan Strait; South China Sea. (25, 27, 30, 44, 50, 66)
2431. *Epistomaria annectens* (Parker & Jones): Yellow Sea. (65)
2432. *Pseudoeponides angulatus* Stschedrina: Yellow Sea. (65)
2433. *P. anderseni* Warren: Yellow Sea; East China Sea; South China Sea; near sea section of Haihe and Ji Canal (Hebei); Shandong (Quaternary). (20, 29, 30, 37, 66)
2434. *P. compressum* Zheng: Near sea section of Haihe and Ji Canal (Hebei); Shandong (Quaternary). (20, 29, 30)
2435. *P. heterogeneus* Stschedrina: Yellow Sea. (65)
2436. *P. japonicus* Uchio: Southern Yellow Sea; East China Sea; South China Sea; Jiangsu; East China Sea continental shelf (Quaternary); Okinawa Trough (Upper Pleistocene). (3, 8, 30, 31, 44, 45)
2437. *P. nakazatoensis* (Kuwano): Yellow Sea; East China Sea; Taiwan Strait. (19, 30, 37, 59, 67)
2438. *Eponidella sinensis* (Zheng): Near sea section of Haihe and Ji Canal (Hebei); Shandong (Quaternary); Taizhou Bay (Zhejiang) (Quaternary). (20, 29, 30, 47)
2439. *Pseudogyroidina sinensis* Zheng [2438]: Near sea section of Haihe and Ji Canal (Hebei); Shandong (Quaternary); Taizhou Bay (Zhejiang) (Quaternary). (20, 29, 30, 47)

ALFREDINIDAE

2440. *Epistomaroides polystomelloides* (Parker & Jones): Xisha (=Paracel Islands); Zhongsha (=Macclesfield Bank); Nansha (=Spratly Islands). (25, 27, 39)

ASTERIGERINATIDAE

2441. *Asterigerinata bashbulakensis* Hao & Zeng: Kashi Basin (Xinjiang) (Lower Tertiary). (38)

AMPHISTEGINIDAE

2442. *Amphistegina exquiseta* Ho & Hu: Xisha (=Paracel Islands); Zhongsha (=Macclesfield Bank). (48)
2443. *A. gibbosa* d'Orbigny: Taiwan Strait; Xisha (=Paracel Islands); Zhongsha (=Macclesfield Bank). (44, 48)
2444. *A. lessoni* (d'Orbigny): Taiwan Strait; Xisha (=Paracel Islands); Zhongsha (=Macclesfield Bank); Okinawa Trough (Holocene); Guangdong (Pliocene). (33, 45, 48, 62, 67)
2445. *A. madagascariensis* d'Orbigny: Taiwan Strait; Xisha (=Paracel Islands); Zhongsha (=Macclesfield Bank); Nansha (=Spratly Islands); Xisha (=Paracel Islands) (Upper Pleistocene, Cenozoic, and Tertiary). (4, 25, 27, 30, 41, 44, 46, 48, 59)
2446. *A. papillosa* Said: Nansha (=Spratly Islands). (66)
2447. *A. quoyi* d'Orbigny: Xisha (=Paracel Islands); Zhongsha (=Macclesfield Bank). (48)
2448. *A. radiata* (Fichtel & Moll): Taiwan Strait; Xisha (=Paracel Islands); Zhongsha (=Macclesfield Bank); Nansha (=Spratly Islands); Xisha (=Paracel Islands) (Upper Pleistocene and Cenozoic); Guangdong (Pliocene). (4, 25, 27, 30, 41, 44, 46, 48, 50)
2449. *A. venosa* (Fichtel & Moll): Xisha (=Paracel Islands); Zhongsha (=Macclesfield Bank); Nansha (=Spratly Islands); Xisha (=Paracel Islands) (Upper Pleistocene). (4, 25, 27, 30, 41, 48, 66)

LEPIDOCYCLINIDAE

2450. *Nephrolepidina rutteni* (Fichtel & Moll): Xisha (=Paracel Islands) (Upper Pleistocene). (36)
2451. *Lepidocyclina formosa* Schlumberger: Taiwan (Tertiary). (46)

NONIONICEA
NONIONIDAE

2452. *Evolutononion shansiensis* N. W. Wang: East China Sea; eastern Hebei (Quaternary). (11, 34)

2453. *Florilus atlanticus* (Cushman) [2507]: Okinawa Trough (Upper Pleistocene to Holocene). (45)

2454. *F. belridgensis* (Barbat & Johnson) [2473]: Guangdong (Pliocene). (33)

2455. *F. boueanus* (d'Orbigny) [2475]: Southern Yellow Sea; East China Sea (Tertiary); Taiwan (Miocene). (3, 44)

2456. *F. carinatus* Wang, He, & Lu: East China Sea continental shelf (Eocene). (3)

2457. *F. decorus* (Cushman & McCulloch) [2540]: Bohai; Yellow Sea; East China Sea; South China Sea; Yellow Sea (Quaternary). (3, 8, 19, 21, 30, 33, 47, 48, 66)

2458. *F. extensus* (Cushman) [2480]: Guangdong (Pliocene). (33)

2459. *F. floriensis* (Cole): Kashi Basin (Xinjiang) (Lower Tertiary); East China Sea (Quaternary). (3, 38)

2460. *F. fuscus* Wang, Min, & Bian: East China Sea. (11)

2461. *F. hantkeni* Cushman & Applin: Xisha (=Paracel Islands); Zhongsha (=Macclesfield Bank). (48)

2462. *F. japonicus* (Asano) [2525]: East China Sea; Taiwan Strait; South China Sea; Guangdong (Pliocene); South China Sea (Quaternary). (3, 19, 22, 33, 44, 48, 59)

2463. *F. labradoricum* (Dawson) [2526]: Okinawa Trough (Upper Pleistocene to Holocene); Kashi Basin (Xinjiang) (Lower Tertiary). (39, 45)

2464. *F. limbatostriatus* (Cushman) [2543]: Yellow Sea; East China Sea; shoreline of Taizhou Bay (Zhejiang); Jiangsu (Quaternary). (8, 19, 47)

2465. *F. manpukujiense* (Ostsuka): Taiwan Strait. (44)

2466. *F. scaphum* (Fichtel & Moll) [2494]: Southern Yellow Sea; East China Sea; Xisha (=Paracel Islands); Zhongsha (=Macclesfield Bank); Guangdong; Hainan (Pliocene). (18, 19, 33, 44, 48)

2467. *F. sloanii* (d'Orbigny): Kashi Basin (Xinjiang) (Lower Tertiary). (38)

2468. *F. subgrateloupi* (Galloway & Heminway): Kashi Basin (Xinjiang) (Lower Tertiary). (38)

2469. *Nonion advenum* (Cushman): Kashi Basin (Xinjiang) (Lower Tertiary). (38)

2470. *N. affine* (Reuss) [2561]: Okinawa Trough (Upper Pleistocene to Holocene); Xisha (=Paracel Islands) (Middle Pleistocene to Holocene); Mt. Jolmo Lungma (=Mt. Everest) (Tibet) (Lower Tertiary). (9, 39, 45)

2471. *N. akitaense* Asano: Bohai; Yellow Sea; East China Sea; South China Sea; Shandong; Jiangsu; East China Sea continental shelf (Quaternary); Taizhou Bay (Zhejiang) (Upper Pleistocene); Okinawa Trough (Upper Pleistocene); Guangdong (Pliocene). (3, 8, 19, 21, 29, 30, 33, 45, 47)

2472. *N. anomalinoidea* Gerke: Shoreline of Bohai; Jiangsu (Quaternary). (8, 49)

2473. *N. belridgense* Barbat & Johnson: Jiangsu (Quaternary). (8)

2474. *N. bogdanowiczi* Voloshinova: Tarim Basin (Xinjiang) (Tertiary). (40)

2475. *N. boueanum* (d'Orbigny): Southern Yellow Sea; East China Sea; South China Sea; Jiangsu (Quaternary). (8, 30, 66)

2476. *N. browni* Cole: East China Sea continental shelf (Oligocene). (3)

2477. *N. decorum* (Cushman & McCulloch) [2540]: Bohai; Yellow Sea; East China Sea; Xisha (=Paracel Islands); Zhongsha (=Macclesfield Bank); Nanhsa (=Spratly Islands). (19, 21, 30, 48, 66)

2478. *N. delicatum* Wang: Shoreline of Bohai (Quaternary). (1)

2479. *N. depressulum* (Walker & Jocob): Eastern Hebei; shoreline of Bohai (Quaternary). (1, 34, 49)

2480. *N. extensum* (Cushman): Jiangsu (Quaternary). (8)

2481. *N. glabrum* Ho, Hu, & Wang [2553]: Jiangsu (Quaternary); Taizhou Bay (Zhejiang) (Upper Quaternary). (8, 47)

2482. *N. goudkoffi* Kleinpell: Xisha (=Paracel Islands) (Middle Pleistocene to Holocene). (39)

2483. *N. graniferum* (Terquem): East China Sea continental shelf (Eocene). (33)

2484. *N. granulosum* Dam & Reinhold: Guangdong (Pliocene). (33)

2485. *N. grateloupi* (d'Orbigny) [2529]: South China Sea; Mt. Jolmo Lungma (=Mt. Everest) (Tibet) (Lower Tertiary). (9, 67)

2486. *N. ibericum* Cushman: Okinawa Trough (Upper Pleistocene). (45)

2487. *N. inexcavatum* (Cushman & Applin): Kashi Basin (Xinjiang) (Lower Tertiary). (45)

2488. *N. japonicum* Asano [2525]: Southern Yellow Sea; East China Sea; Taiwan Strait; South China Sea; Okinawa Trough (Holocene). (27, 30, 44, 50, 66, 67)

2489. *N. laevis* (d'Orbigny): Kashi Basin (Xinjiang) (Lower Tertiary). (45)

2490. *N. ornatissimum* Cushman: Kashi Basin (Xinjiang) (Lower Tertiary). (38)

2491. *N. roemeri* Cushman: East China Sea continental shelf (Oligocene). (3)

2492. *N. rolshauseni* Bandy: Kashi Basin (Xinjiang) (Lower Tertiary). (38)

2493. *N. rotulum* Chalilov [2568]: East China Sea continental shelf (Eocene); Kashi Basin (Xinjiang) (Lower Tertiary). (3, 38)

2494. *N. scaphum* (Fichtel & Moll): Bohai; East China Sea; South China Sea. (30, 66)

2495. *N. schwageri* Cushman: Jiangsu (Quaternary). (8)

2496. *N. shansiensis* (Wang) [2452]: Eastern Hebei (Quaternary). (8)

2497. *N. sorachiensis* Asano: East China Sea continental shelf (Eocene). (3)

2498. *N. subdilatatum* Zheng: Xisha (=Paracel Islands). (26, 30)

2499. *N. sublaeve* ten Dam [2537]: Bohai; Jiangsu (Quaternary). (8, 21)

2500. *N. tallahattensis* Bandy: East China Sea continental shelf (Quaternary). (3)

2501. *N. tuberculatum* (d'Orbigny) [2539]: Jiangsu (Quaternary). (8)

2502. *N. turgida* (Williamson): Bohai; East China Sea; South China Sea. (30)

2503. *N. usbekistanensis* Bykova: Jiangsu (Quaternary). (8)

2504. *Nonionella africana* LeRoy: Kashi Basin (Xinjiang) (Lower Tertiary). (38)

2505. *N. alabamensis* Cushman & Ponton: East China Sea continental shelf (Eocene). (3)

2506. *N. ansata* Cushman: Guangdong (Pliocene). (3)

2507. *A. atlantica* Cushman: East China Sea continental shelf; Jiangsu (Quaternary). (3, 8)

2508. *N. auricula* Heron-Allen & Earland: Shoreline of Bohai; Jiangsu (Quaternary). (3, 8)

2509. *N. basiloba* Cushman & McCulloch: East China Sea. (30)

2510. *N. coarseoerforata* Qiu & Lin: Guangdong (Pliocene). (33)

2511. *N. decora* Cushman & McCulloch [2540]: Taizhou Bay (Zhejiang) (Quaternary); East China Sea continental shelf; Jiangsu (Quaternary); Guangdong (Pliocene). (3, 8, 19, 21, 30, 47, 48, 66)

2512. *N. jacksonensis* Cushman: Yellow Sea; East China Sea; Jiangsu (Quaternary). (8, 19)

2513. *N. limbatostriata* Cushman [2543]: Yellow Sea; East China Sea; Nansha (=Spratly Islands); Jiangsu; shoreline of Taizhou Bay (Zhejiang) (Quaternary). (8, 19, 26, 30, 47, 66)

2514. *N. magnalingua* Finlay: Jiangsu (Quaternary). (8)

2515. *N. modesta* Galloway & Heminway: Kashi Basin (Xinjiang) (Tertiary). (38)

2516. *N. opima* Cushman: Southern Yellow Sea; East China Sea; Jiangsu (Quaternary). (8, 19, 65)

2517. *N. penghuensis* Huang: Taiwan. (44)

2518. *N. pulchella* Hada: Bohai; Yellow Sea; East China Sea; Okinawa Trough (Upper Pleistocene). (45)

2519. *N. soldadoensis* Cushman & Renz: East China Sea continental shelf (Eocene). (3)

2520. *N. reussana* Cushman: Kashi Basin (Xinjiang) (Tertiary). (38)

2521. *N. stella* Cushman & Moyer: Bohai; Yellow Sea; East China Sea; Xisha (=Paracel Islands); southern Yellow Sea (Quaternary). (19, 30, 31)

2522. *N. translucens* Cushman: Nansha (=Spratly Islands). (66)

2523. *N. tredeca* (Asano): Yellow Sea; East China Sea; Taiwan Strait. (64, 65)

2524. *Nonionellina frankei* (Cushman): Kashi Basin (Xinjiang) (Tertiary). (38)

2525. *N. japonica* (Asano): Southern Yellow Sea; Taiwan Strait; East China Sea; South China Sea; Okinawa Trough (Upper Pleistocene). (27, 30, 44, 50, 66, 67)

2526. *N. labradorica* (Dawson): Okinawa Trough (Upper Pleistocene to Holocene); Kashi Basin (Xinjiang) (Tertiary). (38, 45)

2527. *N. pizarrense* (Berry): Kashi Basin (Xinjiang) (Tertiary). (38)

2528. *N. reniformis* Hao & Zeng: Kashi Basin (Xinjiang) (Tertiary). (38)

2529. *Nonionoides grateloupi* d'Orbigny: South China Sea; Mt. Jolmo Lungma (=Mt. Everest) (Tibet) (Tertiary). (9, 67)

2530. *Protelphidium anglicum* Murray: Yellow Sea; East China Sea. (37)

2531. *P. compressum* Zheng: Bohai; East China Sea; near sea section of Haihe and Ji Canal (Hebei). (20, 21, 29, 30)

2532. *P. fulvofusculus* Stschedrina: Southern Yellow Sea. (65)

2533. *P. glabrum* (Ho, Hu, & Wang): East China Sea; southern Yellow Sea; Jiangsu (Quaternary). (8, 19, 31)

2534. *P. granosum* (d'Orbigny): Near sea section of Haihe and Ji Canal (Hebei); Bohai; Yellow Sea; East China Sea; South China Sea; Jiangsu; shoreline of Taizhou Bay (Zhejiang) shoreline (Quaternary). (8, 20, 21, 47, 49)

2535. *P. luridus* Stschedrina: Yellow Sea. (65)

2536. *P. pauperatum* (Balkwill & Wright): Shoreline of Bohai (Quaternary). (49)

2537. *P. sublaeve* (ten Dam): Bohai; Jiangsu (Quaternary). (8, 21, 49)

2538. *P. tersum* Ho, Hu, & Wang: Eastern Hebei; Jiangsu (Quaternary). (8, 34)

2539. *P. tuberculatum* (d'Orbigny): Bohai; Yellow Sea; East China Sea; East China Sea (Quaternary). (3, 13, 14, 19, 31, 35, 38)

2540. *Pseudononion decorum* (Cushman & McCulloch): Bohai; Yellow Sea; East China Sea; South China Sea; Yellow Sea (Quaternary). (3, 8, 19, 21, 30, 33, 47, 48, 66)

2541. *P. granuloumbilicatum* Zheng: Zhongsha (=Macclesfield Bank). (27, 30)

2542. *P. japonicus* Asano: East China Sea; Nansha (=Spratly Islands). (3, 19, 22, 30, 33, 44, 48, 59, 66)

2543. *P. limbatostriatum* (Cushman): Yellow Sea; East China Sea; Xisha (=Paracel Islands); Nansha (=Spratly Islands); shoreline of Taizhou Bay (Zhejiang); Jiangsu (Quaternary). (8, 19, 26, 30, 47, 66)

2544. *P. minutum* Zheng: Bohai; Yellow Sea; East China Sea; near sea section of Haihe and Ji Canal (Hebei); Shandong (Quaternary). (20, 21, 29, 31)

2545. *P. oinomikadoi* Matsunaga: Bohai. (21)

2546. *Astrononion australe* Cushman: East China Sea; South China Sea. (30)

2547. *A. fijiense* Cushman & Edwards [2558]: Nansha (=Spratly Islands). (66)

2548. *A. gallowayi* Loeblich & Tappan: Nansha (=Spratly Islands); Okinawa Trough (Upper Pleistocene); Jiangsu (Quaternary). (8, 45, 66)

2549. *A. glabrum* He & Hu: Shoreline of Bohai (Quaternary). (49)

2550. *A. italicum* Cushman & Edwards: Yellow Sea; Taiwan Strait; South China Sea; shoreline of Bohai (Quaternary); Okinawa Trough (Upper Pleistocene to Holocene). (45, 62, 63)
2551. *A. minimum* Zheng: Xisha (=Paracel Islands). (26, 30)
2552. *A. novozealandicum* Cushman & Edwards [2560]: Yellow Sea; East China Sea; South China Sea; Jiangsu (Quaternary). (8, 30)
2553. *A. pressus* Stschedrina: Southern Yellow Sea. (65)
2554. *A. stelligerum* (d'Orbigny): South China Sea; East China Sea (Quaternary). (3, 66, 67)
2555. *A. tasmanensis* Carter: Yellow Sea; East China Sea; East China Sea continental shelf (Quaternary). (3, 66, 67)
2556. *A. tumidum* Cushman & Edwards [2559]: Nansha (=Spratly Islands). (66)
2557. *A. umbilicatulum* Uchio: Taiwan Strait. (50)
2558. *Fijinonion fijiensis* (Cushman & Edwards): Nansha (=Spratly Islands). (66)
2559. *Laminononion tumidum* (Cushman & Edwards): Nansha (=Spratly Islands). (66)
2560. *Pacinonion novozealandicum* (Cushman & Edwards): Yellow Sea; East China Sea; South China Sea; Jiangsu (Quaternary). (8, 30)
2561. *Melonis affinis* Reuss: Okinawa Trough (Upper Pleistocene to Holocene); Xisha (=Paracel Islands) (Middle Pleistocene to Holocene); Mt. Jolmo Lungma (=Mt. Everest) (Tibet) (Tertiary). (9, 38, 39, 45)
2562. *M. barleeanum* (Williamson): East China Sea; Taiwan Strait; South China Sea; Xisha (=Paracel Islands) (Middle Pleistocene to Holocene). (19, 39, 44, 48, 59, 64, 66)
2563. *M. cyrtomatus* Hao & Zeng: Kashi Basin (Xinjiang) (Tertiary). (38)
2564. *M. formosa* (Seguenza): Kashi Basin (Xinjiang) (Tertiary). (38)
2565. *M. minutus* Zheng: Okinawa Trough (Upper Pleistocene to Holocene); Xisha (=Paracel Islands) (Middle Pleistocene to Holocene). (25, 30, 39, 45)
2566. *M. pompilioides* (Fichtel & Moll): Okinawa Trough (Upper Pleistocene); Kashi Basin (Xinjiang) (Tertiary); Xisha (=Paracel Islands) (Middle Pleistocene to Holocene). (38, 39, 45, 48, 64)
2567. *M. probliquus* Hao & Zeng: Kashi Basin (Xinjiang) (Tertiary). (38)
2568. *M. rotulum* (Chalilov): Kashi Basin (Xinjiang) (Tertiary). (38)
2569. *M. suni* Hao & Zeng: Kashi Basin (Xinjiang) (Tertiary). (38)
2570. *M. umbilicatula* (Montagu): East China Sea; Taiwan Strait; South China Sea. (30, 44)
2571. *M. zaandami* (Vorthuysen): South China Sea. (42)
2572. *M. zeobesus* Vella: Guangdong (Pliocene). (33)
2573. *M. zorillus* Qiu & Lin: Guangdong (Pliocene). (33)
2574. *Pullenia apertula* Cushman: Taiwan. (44)
2575. *P. bulloides* (d'Orbigny): East China Sea; Taiwan Strait; South China Sea; East China Sea continental shelf (Quaternary); Okinawa Trough (Upper Pleistocene to Holocene); Xisha (=Paracel Islands) (Pleistocene). (3, 19, 33, 41, 44, 45, 48, 66)
2576. *P. eocenica* Cushman & Siegfus: East China Sea continental shelf (Paleocene). (3)
2577. *P. jarvisi* Cushman: Kashi Basin (Xinjiang) (Tertiary). (38)
2578. *P. malkinae* Coryell & Mossman: Okinawa Trough (Upper Pleistocene). (45)
2579. *P. marssoni* Cushman & Todd: East China Sea continental shelf (Paleocene). (3)
2580. *P. quaternaria* (Reuss): Kashi Basin (Xinjiang) (Tertiary). (38)
2581. *P. quinqueloba* Reuss: Southern Yellow Sea; East China Sea; South China Sea; Jiangsu; East China Sea continental shelf (Quaternary); Okinawa Trough (Upper Pleistocene to Holocene); Kashi Basin (Xinjiang) (Tertiary). (3, 8, 19, 38, 45, 66)
2582. *P. subcarinata* (d'Orbigny): South China Sea; Okinawa Trough (Upper Pleistocene). (42, 45)
2583. *P. subsphaerica* Parr: Nansha (=Spratly Islands). (66)

ALMAENIDAE

2584. *Anomalinella rostrata* (Brady): Xisha (=Paracel Islands); Zhongsha (=Macclesfield Bank); Nansha (=Spratly Islands); Xisha (=Paracel Islands) (Neogene). (4, 25, 30, 48, 66)
2585. *A. tennesseensis* Berry: East China Sea continental shelf (Paleocene). (3)
2586. *A. ungeriana* (d'Orbigny): Taiwan. (44)

CHILOSTOMELLACEA
CHILOSTOMELLIDAE

2587. *Allomorphina paleocenica* Cushman: East China Sea continental shelf (Paleocene). (3)
2588. *Chilostomella cushmani* Chapman: East China Sea; South China Sea; Okinawa Trough (Holocene). (30, 45)
2589. *C. oolina* Schwager: Nansha (=Spratly Islands); Taiwan (Miocene); Xisha (=Paracel Islands) (Middle Pleistocene to Holocene). (39, 44, 66)
2590. *C. ovoidea*: East China Sea; South China Sea. (19, 66)
2591. *Chilostomelloides macrostoma* ten Dam & Sigal: East China Sea continental shelf (Paleocene). (3)
2592. *C. plummerae* ten Dam & Sigal: East China Sea continental shelf (Paleocene). (3)

QUADRIMORPHINIDAE

2593. *Quadrimorphina pacifica* Cushman & Todd: Okinawa Trough (Upper Pleistocene). (45)

OSANGULARIIDAE

2594. *Osangularia bengalensis* (Schwager): Xisha (=Paracel Islands); Zhongsha (=Macclesfield Bank); Nansha (=Spratly Islands); Okinawa Trough (Upper Pleistocene to Holocene). (45, 48, 66)

2595. *O. brunswickernsis* Todd & Kniker: East China Sea continental shelf (Paleocene). (3)

2596. *O. culter* (Parker & Jones): East China Sea; South China Sea. (30, 66)

ANOMALINIDAE

2597. *Anomalina ammonoides* (Reuss): East China Sea; Nansha (=Spratly Islands); shoreline of Bohai (Quaternary). (1, 66)

2598. *A. bradyi* Said: East China Sea; South China Sea. (30, 42)

2599. *A. colligera* Chapman & Parr: East China Sea; Taiwan Strait; Xisha (=Paracel Islands); Zhongsha (=Macclesfield Bank); Nansha (=Spratly Islands); Xisha (=Paracel Islands) (Cenozoic). (4, 19, 44, 48, 66)

2600. *A. glabrata* (Cushman): Southern Yellow Sea; East China Sea; Nansha (=Spratly Islands); Taiwan (Miocene). (30, 44, 66)

2601. *A. globulosa* Chapman & Parr: Xisha (=Paracel Islands); Zhongsha (=Macclesfield Bank); Nansha (=Spratly Islands). (48, 62, 66)

2602. *A. insecta* Schwager: Kashi Basin (Xinjiang) (Tertiary). (38)

2603. *A. pinghuensis* Wang, He, & Lu: East China Sea continental shelf (Eocene). (3)

2604. *A. praespissiformis* Cushman & Bermudez: Kashi Basin (Xinjiang) (Tertiary). (38)

2605. *A. punctulata* Cushman: Hong Kong. (62)

2606. *A. robertsonianus* (Brady) [2307]: Shoreline of Bohai (Quaternary). (49)

2607. *A. semiteres* Finlay: East China Sea continental shelf (Eocene). (3)

2608. *A. semipunctata* (Bailey) [2633]: South China Sea. (48, 66)

2609. *A. tasmanica* Parr: Xisha (=Paracel Islands); Zhongsha (=Macclesfield Bank). (48)

ORIDORSALIDAE

2610. *Oridorsalis stellata* (Silvestri): Xisha (=Paracel Islands) (Middle Pleistocene to Holocene). (39)

2611. *O. tenera* (Brady): East China Sea; South China Sea; Okinawa Trough (Upper Pleistocene to Holocene). (3, 19, 45, 48, 65)

2612. *O. umbonatus* (Reuss): Taiwan Strait; Xisha (=Paracel Islands); Zhongsha (=Macclesfield Bank); Okinawa Trough (Upper Pleistocene to Holocene); Xisha (=Paracel Islands) (Middle Pleistocene to Holocene). (39, 45, 48, 66)

2613. *O. westi* Andersen: Okinawa Trough (Upper Pleistocene to Holocene). (45)

2614. *Schwantzia elegantissima* McCulloch: East China Sea. (19)

HETEROLEPIDAE

2615. *Anomalinoides centrabullus* Hao & Zeng: Kashi Basin (Xinjiang) (Tertiary). (38)

2616. *A. petaliformis* Hao & Zeng: Kashi Basin (Xinjiang) (Tertiary). (38)

2617. *A. vialovi* Bykova: Kashi Basin (Xinjiang) (Tertiary). (38)

2618. *A. welleri* (Plummer): East China Sea. (19)

2619. *Heterolepa dutemplei* (d'Orbigny): East China Sea; East China Sea continental shelf (Quaternary and Eocene); Guangdong (Pliocene). (3, 19, 33)

2620. *H. kezloyensis* Hao & Zeng: Kashi Basin (Xinjiang) (Tertiary). (38)

2621. *H. loxisutura* Hao & Zeng: Kashi Basin (Xinjiang) (Tertiary). (38)

2622. *H. praecincta* (Karrer): East China Sea; Taiwan Strait; Xisha (=Paracel Islands); Zhongsha (=Macclesfield Bank); Okinawa Trough (Upper Pleistocene to Holocene). (19, 44, 45, 48, 59)

2623. *H. pseudoungeriana* (Cushman): Xisha (=Paracel Islands) (Pleistocene). (41)

2624. *H. sigmasutura* Hao & Zeng: Kashi Basin (Xinjiang) (Tertiary). (38)

2625. *H. subpraecincta* (Asano): East China Sea. (19)

2626. *H. tenera* (Brady) [2611]: Xisha (=Paracel Islands); Zhongsha (=Macclesfield Bank); East China Sea continental shelf (Quaternary). (3, 19, 45, 48, 66)

2627. *H. virosus* Hao & Zeng: Xisha (=Paracel Islands); Zhongsha (=Macclesfield Bank). (48)

2628. *H. xystrota* Hao & Zeng: Kashi Basin (Xinjiang) (Tertiary). (38)

GAVELINELLIDAE

2629. *Gyroidinoides nipponicus* (Ishizaki): Southern Yellow Sea; East China Sea; Taiwan Strait. (19, 30, 44)

2630. *G. subzelandica* Hornibrook: Okinawa Trough (Upper Pleistocene); Xisha (=Paracel Islands) (Middle Pleistocene to Holocene). (39, 45)

2631. *Rotaliatina buliminoides* (Reuss): South China Sea. (42)

2632. *Discanomalina coronata* (Parker & Jones) [2632]: Okinawa Trough (Upper Pleistocene). (45)

2633. *D. semipunctata* (Bailey): Okinawa Trough (Upper Pleistocene). (45)

2634. *Paromalina coronata* (Parker & Jones) [2632]: Okinawa Trough (Pleistocene). (45)

2635. *P. semipunctata* (Bailey) [2633]: Okinawa Trough (Upper Pleistocene). (45)

2636. *Gyroidina altiformis* R. E. & K. C. Stewart: Okinawa Trough (Upper Pleistocene); Xisha (=Paracel Islands) (Middle Pleistocene to Holocene). (39, 45)

2637. *G. broeckhiana* (Karrer): East China Sea; Nansha (=Spratly Islands). (30, 66)
2638. *G. depressa* (Alth): Eastern Hebei; Jiangsu; shoreline of Taizhou Bay (Zhejiang) (Quaternary). (8, 34, 47)
2639. *G. gemma* Brady: Xisha (=Paracel Islands) (Holocene). (39)
2640. *G. lottensis* Garret: East China Sea continental shelf (Eocene). (3)
2641. *G. multilocula* Coryell & Mossma: Xisha (=Paracel Islands) (Holocene). (39)
2642. *G. neosoldani* Brotzen: Taiwan Strait; South China Sea; Okinawa Trough (Upper Pleistocene to Holocene); Xisha (=Paracel Islands) (Middle Pleistocene to Holocene). (39, 42, 44, 59, 66)
2643. *G. nipponica* Ishizaki [2629]: Southern Yellow Sea; East China Sea; Taiwan Strait. (19, 30, 44)
2644. *G. orbicularis* d'Orbigny: Taiwan Strait; Nansha (=Spratly Islands); Okinawa Trough (Holocene); Guangdong (Pliocene). (33, 44, 45, 66)
2645. *G. quinqueloba* Uchio: Okinawa Trough (Pleistocene). (45)
2646. *G. soldanii* d'Orbigny [2648]: East China Sea; Taiwan Strait; Guangdong (Pliocene). (33, 50)
2647. *G. tainanensis* Nakamura: Taiwan. (44)
2648. *Hansenisca soldani* (d'Orbigny): East China Sea; Taiwan Strait; Guangdong (Pliocene). (33, 50)
2649. *Hanzawaia boueana* (d'Orbigny): Zhongsha (=Macclesfield Bank); Nansha (=Spratly Islands). (27, 30, 66)
2650. *H. convexa* Ho, Hu, & Wang: Dongshan Bay (Fujian); Taiwan Strait; Jiangsu (Quaternary); Guangdong (Pliocene). (5, 8, 33, 50)
2651. *H. hoppoensis* (Nakamura): Taiwan Strait. (44)
2652. *H. mantaensis* Galloway & Morrey: Southern Yellow Sea; East China Sea; Taiwan Strait; South China Sea; Okinawa Trough (Upper Pleistocene to Holocene); Jiangsu (Quaternary); Guangdong (Pliocene). (8, 12, 13, 33, 48, 50, 66)
2653. *H. nipponica* Asano: Yellow Sea; East China Sea; South China Sea; southern Yellow Sea (Quaternary); East China Sea continental shelf (Pliocene). (3, 19, 22, 31, 64, 65, 67)
2654. *H. nitidula* (Bandy): South China Sea; Guangdong (Pliocene). (33, 42)
2655. *H. rhodiensis* (Parker): East China Sea; South China Sea. (30, 66)
2656. *H. sumitomoi* Asano & Murata: East China Sea; Taiwan Strait. (64)

KARRERIIDAE

2657. *Karreria fallax* Rzehak: Mt. Jolmo Lungma (=Mt. Everest) (Tibet) (Tertiary). (9)

TRICHOHYALIDAE

2658. *Buccella decora* Stschedrina: Southern Yellow Sea. (65)
2659. *B. frigida* (Cushman): Bohai; Yellow Sea; East China Sea; shoreline of Bohai (Quaternary); Okinawa Trough (Pleistocene); East China Sea continental shelf (Pliocene); South China Sea (Quaternary). (8, 19, 21, 29, 30, 31, 35, 45, 49, 67)
2660. *B. inculta* Ho, Hu, & Wang: Jiangsu (Quaternary). (8)
2661. *B. inusitata* Andersen: Northern and southern Yellow Sea; East China Sea continental shelf; Jiangsu; Beijing area (Quaternary). (2, 3, 8, 31, 65)
2662. *B. modica* He & Hu: Shoreline of Bohai (Quaternary). (49)
2663. *B. radiata* Stschedrina: Yellow Sea. (65)
2664. *B. tenerrima* (Brady): Bohai; Yellow Sea; East China Sea; South China Sea. (30)
2665. *B. tunicata* Ho, Hu, & Wang: East China Sea; Shandong; Jiangsu (Quaternary). (8, 19, 29, 30)

ORBITOIDACEA
LEPIDORBITOIDIDAE

2666. *Actinosiphon tibetica* (Douville): Mt. Jolmo Lungma (=Mt. Everest) area (Tibet) (Tertiary). (9)
2667. *Daviesina langhami* Smout: Mt. Jolmo Lungma (=Mt. Everest) area (Tibet) (Tertiary). (9)

ROTALIACEA
ROTALIIDAE

2668. *Thalmannita coronata* Hu: Mt. Jolmo Lungma (=Mt. Everest) area (Tibet) (Tertiary). (9)
2669. *Pararotalia armata* (d'Orbigny): Jiangsu; shoreline of Taizhou Bay (Zhejiang); Tarim Basin (Xinjiang) (Tertiary); Lianyungang (Jiangsu) (Quaternary); Okinawa Trough (Holocene). (8, 9, 32, 40, 45, 47)
2670. *P. audouni* (d'Orbigny): Bohai; Yellow Sea; East China Sea; South China Sea. (30)
2671. *P. bellatula* Stschedrina: Southern Yellow Sea. (65)
2672. *P. calcar* (d'Orbigny): Xisha (=Paracel Islands); Nansha (=Spratly Islands); Xisha (=Paracel Islands) (Cenozoic). (4, 25, 30, 66)
2673. *P. fungiformis* Ho, Hu, & Wang: Southern Yellow Sea; shoreline of Taizhou Bay (Zhejiang); Jiangsu (Quaternary). (8, 47, 65)
2674. *P. inermis* (Terquem): Bohai; Jiangsu; East China Sea continental shelf (Pliocene); South China Sea (Quaternary). (3, 8, 21, 47, 49)
2675. *P. minuta* (Takayanagi): Taiwan Strait. (54)
2676. *P. murrayi* (Heron-Allen & Earland) [2279]: Southern Yellow Sea; East China Sea; South China Sea; Okinawa Trough (Pleistocene). (19, 22, 45, 65)
2677. *P. nipponica* (Asano): East China Sea. (19)
2678. *P. orientalis* (Cushman & Bermudez): East China Sea continental shelf; Jiangsu; shoreline of Taizhou Bay (Zhejiang); South China Sea (Quaternary). (3, 8, 47)

2679. *P. ozawai* (Asano): Southern Yellow Sea; East China Sea; Taiwan Strait; Dongshan Bay (Fujian); Taiwan (Pleistocene). (5, 52, 65)

2680. *P. rosea* (d'Orbigny): Okinawa Trough (Upper Pleistocene to Holocene). (45)

2681. *P. taiwanica* (Nakamura): Taiwan Strait; Taiwan (Miocene to Pliocene). (53)

2682. *P. tuberculata* Ko-Papanicolaou: Hong Kong. (62)

2683. *P. xincunensis* Sun: Lingshui (Hainan). (7)

2684. *P. umbonata* (LeRoy): Kashi Basin (Xinjiang) (Tertiary). (38)

2685. *P. venusta* (Brady): Hong Kong. (62)

2686. *Lockhartia conditi* (Nuttall): Mt. Jolmo Lungma (=Mt. Everest) area (Tibet) (Tertiary). (9)

2687. *L. haimei* (Davies): Mt. Jolmo Lungma (=Mt. Everest) area (Tibet) (Tertiary). (9)

2688. *L. hunti* Ovey: Mt. Jolmo Lungma (=Mt. Everest) area (Tibet) (Tertiary). (9)

2689. *L. megapapulata* Hu: Mt. Jolmo Lungma (=Mt. Everest) area (Tibet) (Tertiary). (9)

2690. *L. tipperi* (Davies): Mt. Jolmo Lungma (=Mt. Everest) area (Tibet) (Tertiary). (9)

2691. *Rotalia calcarinoides*: Xisha (=Paracel Islands); Zhongsha (=Macclesfield Bank); Nansha (=Spratly Islands); Xisha (=Paracel Islands) (Cenozoic). (4, 25, 30, 48, 66)

2692. *R. decipiens* Hu: Mt. Jolmo Lungma (=Mt. Everest) area (Tibet) (Tertiary). (9)

2693. *R. hensoni* Smout: Mt. Jolmo Lungma (=Mt. Everest) area (Tibet) (Tertiary). (9)

2694. *R. microannectens* Wang: Shoreline of Bohai (Quaternary). (1)

2695. *R. orientalis* Cushman & Bermudez [2678]: Mt. Jolmo Lungma (=Mt. Everest) area (Tibet) (Tertiary). (9)

2696. *R. ovata* Hu: Mt. Jolmo Lungma (=Mt. Everest) area (Tibet) (Tertiary). (9)

2697. *R. saxorum* d'Orbigny: Mt. Jolmo Lungma (=Mt. Everest) area (Tibet) (Tertiary). (9)

2698. *R. suessonensis* d'Orbigny: Mt. Jolmo Lungma (=Mt. Everest) area (Tibet) (Tertiary). (9)

2699. *R. trochidiformis* Lamarck: Mt. Jolmo Lungma (=Mt. Everest) area (Tibet) (Tertiary). (9)

2700. *R. venusta* Hu: Mt. Jolmo Lungma (=Mt. Everest) area (Tibet) (Tertiary). (9)

2701. *Smoutina corpuscula* Hu: Mt. Jolmo Lungma (=Mt. Everest) area (Tibet) (Tertiary). (9)

2702. *Ammonia altispira* Qiu & Lin: Guangdong (Pliocene); Leiqiong area (southern Guangdong and Hainan) (Tertiary). (33, 46)

2703. *A. annectens* (Parker & Jones) [2767]: Bohai; Yellow Sea; East China Sea; Taiwan Strait; South China Sea; near sea section of Haihe and Ji Canal (Hebei); Shandong (Quaternary); Okinawa Trough (Pleistocene); East China Sea continental shelf (Miocene); Taiwan (Miocene). (3, 8, 20, 29, 30, 31, 45, 47, 50, 66)

2704. *A. aomoriensis* (Asano): Shoreline of Bohai (Quaternary). (49)

2705. *A. astera* Stschedrina: Southern Yellow Sea. (65)

2706. *A. batava* (Hofker): Shoreline of Bohai (Quaternary). (49)

2707. *A. beccarii* (Linné): Bohai; Yellow Sea; East China Sea; South China Sea; shoreline of eastern Hebei; Jiangsu (Quaternary); Okinawa Trough (Upper Pleistocene to Holocene); East China Sea continental shelf (Miocene); Tarim Basin (Xinjiang) (Tertiary); Taiwan (Miocene); Guangdong (Pliocene). (3, 4, 8, 25, 30, 37, 44, 45, 46, 48, 49, 66)

2708. *A. compressiuscula* (Brady) [2759]: Yellow Sea; East China Sea; Hong Kong; southern Yellow Sea (Quaternary). (19, 22, 31)

2709. *A. confertitesta* Zheng: Bohai; Yellow Sea; East China Sea; South China Sea; eastern Hebei; Shandong; Dongshan Bay (Fujian); shoreline of Taizhou Bay (Zhejiang) (Quaternary). (3, 5, 7, 29, 30, 34, 47)

2710. *A. convexidorsa* Zheng: Yellow Sea; Shandong; shoreline of Taizhou Bay (Zhejiang); eastern Hebei (Quaternary). (12, 13, 19, 20, 29, 30, 31)

2711. *A. crebera* Stschedrina: Southern Yellow Sea. (65)

2712. *A. dilucida* Stschedrina: Southern Yellow Sea. (65)

2713. *A. equatoriana* (LeRoy): Taiwan (Miocene to Pliocene). (53)

2714. *A. faceta* Ho, Hu, & Wang: Jiangsu (Quaternary). (8)

2715. *A. flevensis* (Hofker): Bohai; Shandong (Quaternary). (21, 29, 30)

2716. *A. floscula* (Todd & Post): Southern Yellow Sea. (65)

2717. *A. formosa* Stschedrina: Yellow Sea. (65)

2718. *A. glans* Stschedrina: Northern Yellow Sea. (65)

2719. *A. globosa* (Millett): Bohai; Yellow Sea; East China Sea; near sea section of Haihe and Ji Canal (Hebei); South China Sea (Quaternary). (3, 20, 21, 33, 35)

2720. *A. granuloumbilica* Zheng: Bohai; near sea section of Haihe and Ji Canal (Hebei); Shandong (Quaternary). (20, 21, 29, 30)

2721. *A. hatatatensis* (Takayanagi): Tarim Basin (Xinjiang) (Tertiary). (40)

2722. *A. honyaensis* (Asano): Tarim Basin (Xinjiang) (Tertiary). (40)

2723. *A. hozanensis* (Nakamura): Taiwan (Pliocene to Pleistocene). (44, 53)

2724. *A. indica* (LeRoy): Bohai; Yellow Sea; Taiwan (Tertiary). (44, 46)

2725. *A. inflata* (Sequenza): East China Sea continental shelf (Quaternary); Taiwan (Pliocene and Pleistocene); South China Sea (Pliocene). (3, 33, 54)

2726. *A. japonica* (Hada): Bohai; southern Yellow Sea; East China Sea; Tarim Basin (Xinjiang) (Tertiary); shoreline of Bohai (Quaternary); Taiwan (Miocene); Guangdong (Pliocene). (33, 40, 49, 53)

2727. *A. ketienziensis* (Ishizaki): Yellow Sea; East China Sea; Taiwan Strait; Xisha (=Paracel Islands); Zhongsha (=Macclesfield Bank); southern Yellow Sea (Quaternary); Okinawa Trough (Upper Pleistocene to Holocene);

East China Sea continental shelf (Pliocene); Taiwan (Miocene and Pleistocene). (3, 31, 45, 58, 59, 64, 65)

2728. *A. koeboeensis* (LeRoy): Lianyungang (Jiangsu) (Quaternary). (32)

2729. *A. limbatobeccarii* (McLean): Shandong (Quaternary). (21, 29, 30)

2730. *A. limnetes* (Todd & Brönnimann): Southern Yellow Sea; near sea section of Haihe and Ji Canal (Hebei); Shandong (Quaternary); Tarim Basin (Xinjiang) (Tertiary). (20, 29, 30, 40, 47)

2731. *A. maruhasii* (Kuwano): Yellow Sea; East China Sea; southern Yellow Sea (Quaternary). (12, 31)

2732. *A. multicella* Zheng: Bohai; East China Sea; South China Sea; Shandong; shoreline of Taizhou Bay (Zhejiang) (Quaternary). (21, 29, 30, 47)

2733. *A. nakamurai* (Ishizaki): Taiwan. (44)

2734. *A. nantongensis* Hom Hu, & Wang: Shoreline of Bohai; Jiangsu (Quaternary). (8, 49)

2735. *A. nipponica* (Asano): Taiwan (Upper Miocene). (53)

2736. *A. parkinsoniana* (d'Orbigny): Bohai; East China Sea; South China Sea. (21, 30)

2737. *A. pauciloculata* (Phleger & Parker): Bohai; Yellow Sea; East China Sea; South China Sea; near sea section of Haihe and Ji Canal (Hebei); East China Sea continental shelf (Quaternary and Eocene). (3, 19, 20, 22)

2738. *A. paucipora* Zheng: Xisha (=Paracel Islands). (26, 30)

2739. *A. rolshauseni* (Cushman & Bermudez): Eastern Hebei (Quaternary). (34)

2740. *A. sikokuensis* (Ishizaki): Taiwan (Miocene to Pleistocene). (44, 53)

2741. *A. sobrina* (Shupack): Shoreline of Bohai (Quaternary). (49)

2742. *A. takanabensis* (Ishizaki): Yellow Sea; East China Sea; South China Sea; Shandong; shoreline of Taizhou Bay (Zhejiang) (Quaternary); East China Sea continental shelf (Miocene); Guangdong (Pliocene). (3, 29, 30, 33, 47, 49, 53, 66)

2743. *A. tepida* (Cushman): Bohai; Yellow Sea; East China Sea; Taiwan Strait; South China Sea; Shandong; Taiwan; eastern Hebei (Quaternary); East China Sea continental shelf (Miocene); Tarim Basin (Xinjiang) (Tertiary). (3, 4, 9, 29, 30, 34, 40, 47, 48, 66)

2744. *A. tikutoensis* [2765]: Taiwan (Tertiary). (46)

2745. *A. togopiensis* Whittaker & Hodgkinson: Shenzhen Bay (Guangdong); Hainan. (7)

2746. *A. yabei* (Ishizaki) [2766]: Taiwan area (Tertiary). (46)

2747. *Asterorotalia binhaiensis* Hu, He, & Wang: Jiangsu; shoreline of Taizhou Bay (Zhejiang) (Quaternary); Guangdong (Pliocene). (8, 33, 47)

2748. *A. compressiuscula* (Brady): Southern Yellow Sea. (65)

2749. *A. diplocava* Ho, Hu, & Wang: Jiangsu; shoreline of Taizhou Bay (Zhejiang) (Quaternary). (8, 47)

2750. *A. hexaspinosa* Wang, He, & Lin: Hainan (Pliocene). (33)

2751. *A. inflata* (Millett): Southern Yellow Sea; Taiwan Strait. (44, 65)

2752. *A. inspinosa* Huang: East China Sea continental shelf (Quaternary and Pliocene); Taiwan (Miocene and Pliocene). (3)

2753. *A. multispinosa* (Nakamura): Taiwan (Miocene to Pliocene). (33, 44, 53)

2754. *A. pulchella* (d'Orbigny): East China Sea; South China Sea; Shandong (Quaternary); Hainan (Pliocene). (29, 30, 33)

2755. *A. substrispinosa* (Ishizaki): Shoreline of Bohai; shoreline of Taizhou Bay (Zhejiang); East China Sea continental shelf; South China Sea; Jiangsu (Quaternary); Hainan (Pliocene). (3, 8, 31, 33, 49)

2756. *A. tetraspinosa* Wang, He, & Lin: Hainan (Pliocene). (33)

2757. *A. trispinosa* (Thalmann): Taiwan (Pleistocene). (44, 53)

2758. *A. venusta* Ho, Hu, & Wang: East China Sea continental shelf; Jiangsu; shoreline of Taizhou Bay (Zhejiang) (Quaternary); Hainan (Pliocene). (3, 8, 33, 47)

2759. *Pseudorotalia compressuiscula* (Brady): Nansha (=Spratly Islands). (66)

2760. *P. gaimardii* (d'Orbigny) [2769]: Bohai; Yellow Sea; East China Sea; Taiwan Strait; South China Sea; Shandong (Quaternary); Okinawa Trough (Pleistocene); East China Sea continental shelf (Quaternary); Taiwan (Miocene and Pleistocene); Hainan (Pliocene). (3, 29, 30, 33, 45, 50, 66)

2761. *P. indopacifica* (Thalmann): Bohai; southern Yellow Sea; East China Sea; Taiwan Strait; Taiwan; South China Sea; Hainan (Pliocene); South China Sea (Miocene). (3, 19, 30, 33, 44, 50, 52, 59)

2762. *P. leiqiongensis* Wang & He: Hainan (Pliocene). (3, 33)

2763. *P. papillosa* (Brady): Taiwan. (44)

2764. *P. schroeteriana* (Carpenter, Parker & Jones): Southern Yellow Sea; East China Sea; Taiwan Strait; South China Sea; eastern Hebei; Shandong; Taiwan; southern Yellow Sea (Miocene); Jiangsu (Quaternary). (3, 4, 8, 22, 29, 30, 33, 44, 47, 65, 66)

2765. *P. tikutoensis* (Nakamura): Taiwan; Guangdong (Pliocene). (33, 44)

2766. *P. yabei* (Ishizaki): Taiwan; Taiwan Strait; East China Sea continental shelf (Quaternary); South China Sea (Pliocene). (33, 44, 53)

2767. *Rotalidium annectens* (Parker & Jones): Bohai; Yellow Sea; East China Sea; Taiwan Strait; South China Sea; near sea section of Haihe and Ji Canal (Hebei); Shandong; Jiangsu; shoreline of Taizhou Bay (Zhejiang) (Quaternary); East China Sea continental shelf (Miocene); Okinawa Trough (Pleistocene); Taiwan (Miocene). (3, 8, 20, 29, 30, 31, 45, 47, 50)

2768. *Cavarotalia annectens* (Parker & Jones) [2767]: Bohai; Yellow Sea; East China Sea; Taiwan Strait; South China Sea; near sea section of Haihe and Ji Canal (Hebei); Shandong; Jiangsu; shoreline of Taizhou Bay (Zhejiang)

(Quaternary); East China Sea continental shelf (Miocene); Okinawa Trough (Pleistocene); Taiwan (Miocene). (3, 4, 8, 20, 29, 31, 45, 47, 50)

2769. *Rotalinoides gaimardii* (d'Orbigny): Shandong; Bohai; Yellow Sea; East China Sea; Okinawa Trough; Taiwan Strait; South China Sea; East China Sea continental shelf (Quaternary); Taiwan (Pleistocene and Miocene). (3, 29, 30, 33, 45, 50, 66)

2770. *Streblus compressuiscula* (Brady) [2759]: South China Sea. (67)

2771. *S. tepidus* (Cushman) [2743]: South China Sea. (67)

2772. *S. trispinosus* (Thalmann) [2757]: South China Sea. (67)

CALCARINIDAE

2773. *Baculogypsina sphaerulata* (Parker & Jones): Taiwan. (44)

2774. *Baculogypsinoides spinosus* Yabe & Hanzawa: Taiwan; Zhongsha (=Macclesfield Bank); Nansha (=Spratly Islands). (27, 30, 44, 66)

2775. *Calcarina calcar* d'Orbigny: Taiwan. (44)

2776. *C. hainanensis* Li: South China Sea. (63)

2777. *C. hispida* Brady: Taiwan Strait; Xisha (=Paracel Islands); Zhongsha (=Macclesfield Bank); Nansha (=Spratly Islands); Xisha (=Paracel Islands) (Cenozoic). (4, 25, 27, 30, 44, 48, 66)

2778. *C. spengleri* (Gmelin): Taiwan Strait; South China Sea; Xisha (=Paracel Islands) (Cenozoic). (4, 7, 44, 46)

2779. *C. viennoti* (Greig): Xisha (=Paracel Islands) (Upper Pleistocene). (41)

ELPHIDIIDAE

2780. *Cribroelphidium articulatum* (d'Orbigny) [2808]: Shoreline of Bohai (Quaternary). (49)

2781. *C. neominutum* (Wang): Shoreline of Bohai (Quaternary). (1)

2782. *C. pacificum* Ujiie: Xisha (=Paracel Islands); Zhongsha (=Macclesfield Bank). (48)

2783. *Elphidiononion incertum* (Williamson) [2791]: Shoreline of Bohai (Quaternary). (1)

2784. *E. neominutum* Wang [2781]: Shoreline of Bohai (Quaternary). (1)

2785. *Cribrononion albiumbilicatulus* (Weiss): Yellow Sea. (65)

2786. *C. asiaticum* (Polski): Eastern Hebei and near sea section of Haihe and Ji Canal (Hebei); Shandong; Jiangsu; southern Yellow Sea; East China Sea; South China Sea (Quaternary). (3, 20, 29, 30, 31, 34)

2787. *C. frigidum* (Cushman): East China Sea; Jiangsu (Quaternary). (8, 19)

2788. *C. fronto* Stschedrina: Southern Yellow Sea. (65)

2789. *C. gnythosuturatum* Ho, Hu, & Wang: Yellow Sea; East China Sea; eastern Hebei; Jiangsu (Quaternary). (8, 34, 37, 49)

2790. *C. heterocameratus* Stschedrina: Southern Yellow Sea. (65)

2791. *C. incertum* (Williamson): Near sea section of Haihe and Ji Canal (Hebei); Bohai; Yellow Sea; East China Sea; South China Sea; Shandong; shoreline of Taizhou (Zhejiang) (Late Quaternary); Okinawa Trough (Upper Pleistocene to Holocene); South China Sea (Quaternary). (3, 20, 29, 30, 31, 45, 57)

2792. *C. limpidus* Stschedrina: Southern Yellow Sea. (65)

2793. *C. minutum* (Reuss): Hainan (Pliocene). (33)

2794. *C. pellucens* Stschedrina: Yellow Sea. (65)

2795. *C. poeyanum* (d'Orbigny): Bohai; Jiangsu (Quaternary). (8, 21)

2796. *C. porisuturalis* Zheng: Near sea section of Haihe and Ji Canal (Hebei); Bohai; Yellow Sea; East China Sea; South China Sea; Shandong; southern Yellow Sea (Quaternary); Taiwan (Late Quaternary). (19, 20, 21, 29, 30, 33, 35, 37, 47)

2797. *C. rhomboidale* Ho, Hu, & Wang: Jiangsu (Quaternary). (8)

2798. *C. simplex* (Cushman) [2851]: Southern Yellow Sea. (65)

2799. *C. subincertum* (Asano): Bohai; Yellow Sea; East China Sea (Tertiary and present); Okinawa Trough (Upper Pleistocene). (3, 19, 45, 64, 65)

2800. *C. vitreum* Wang, Zeng, & Min: Bohai; Yellow Sea; East China Sea. (12, 19, 20, 37)

2801. *Elphidiella biseripora* Sun: Shenzhen Bay (Guangdong). (7)

2802. *E. brevicanalis* Zheng: Shandong (Quaternary). (29, 30)

2803. *E. hannai* (Cushman & Grant): Shenzhen Bay (Guangdong); shoreline of Taizhou Bay (Zhejiang) (Late Quaternary). (7, 47)

2804. *E. kiangsuensis* (He, Hu, & Wang): Bohai; East China Sea; South China Sea; near sea section of Haihe and Ji Canal (Hebei); Shandong; Jiangsu (Quaternary). (3, 8, 19, 20, 21, 29, 30, 31, 47)

2805. *E. nutovaensis* (Borovleva): Jiangsu; eastern Hebei (Quaternary). (8, 34)

2806. *E. papilli* Ko-Papanicolaou: Hong Kong. (62)

2807. *Elphidium advenum* (Cushman): Bohai; Yellow Sea; East China Sea; Taiwan Strait; South China Sea; Hebei; Shandong; Jiangsu; Okinawa Trough (Upper Pleistocene to Holocene); Taiwan (Quaternary); Xisha (=Paracel Islands) (Pleistocene). (3, 8, 19, 22, 28, 29, 45, 47, 48, 49, 50, 67)

2808. *E. articulatum* (d'Orbigny): Yellow Sea; East China Sea. (37)

2809. *E. asiaticum* Polski [2786]: Bohai; Yellow Sea; East China Sea; Nansha (=Spratly Islands). (19, 30, 66)

2810. *E. chapmani* (Cushman): Xisha (=Paracel Islands); Zhongsha (=Macclesfield Bank). (48)

2811. *E. clavatum* Cushman: Shoreline of Bohai; Jiangsu; Taiwan (Quaternary). (3, 8, 47, 49)

2812. *E. concinnum* Nicol: Xisha (=Paracel Islands); Zhongsha (=Macclesfield Bank). (48)

2813. *E. crassimargo* Stschedrina: Yellow Sea. (65)

2814. *E. craticulatum* (Fichtel & Moll) [2863]: Taiwan Strait. (44, 67)

2815. *E. crispum* (Linné): Bohai; Taiwan; Xisha (=Paracel Islands); Zhongsha (=Macclesfield Bank); Nansha (=Spratly Islands); Jiangsu (Quaternary); Xisha (=Paracel Islands) (Cenozoic). (1, 4, 8, 25, 27, 29, 44, 48, 62, 66)
2816. *E. decorum* Qiu & Lin: Guangdong (Pliocene). (33)
2817. *E. dellum* Wang: Shoreline of Bohai (Quaternary). (1)
2818. *E. excavatum* (Terquem): Bohai; eastern Hebei (Quaternary). (21, 34, 49)
2819. *E. fissurisuturatum* Wang, He, & Lu: East China Sea (Eocene). (3)
2820. *E. frigidum* Cushman [2787]: Bohai; southern Yellow Sea. (30)
2821. *E. granti* Kleinpell: Guangdong (Pliocene). (33)
2822. *E. guandongensis* (He & Lin): Shenzhen Bay (Guangdong). (33)
2823. *E. gunteri* Cole: East China Sea. (15)
2824. *E. hispidulum* Cushman: Bohai; Yellow Sea; East China Sea; South China Sea; Shandong; shoreline of Taizhou Bay (Zhejiang); Jiangsu (Quaternary). (8, 19, 29, 30, 35, 49, 67)
2825. *E. hokkaidoense* Asano: Taiwan. (44)
2826. *E. hughesi* Cushman & Grant: Yellow Sea; shoreline of Bohai (Quaternary). (1, 35)
2827. *E. ibericum* (Schrodt): Yellow Sea; South China Sea. (37, 49)
2828. *E. incertum* (Williamson) [2791]: Shoreline of Bohai; shoreline of Taizhou Bay (Zhejiang) (Quaternary); Okinawa Trough (Holocene). (45, 47, 49)
2829. *E. jenseni* (Cushman): Southern Yellow Sea; East China Sea; South China Sea; Shandong (Quaternary); East China Sea (Pleistocene). (19, 29, 30, 66)
2830. *E. kusiroense* Asano: Southern Yellow Sea. (65)
2831. *E. latispatium* Poag: Guangdong (Pliocene). (33)
2832. *E. limpidum* Ho, Hu, & Wang: Bohai; Yellow Sea; East China Sea; Shandong; Jiangsu; shoreline of Taizhou Bay (Zhejiang) (Quaternary). (8, 19, 21, 29, 30, 47)
2833. *E. macellum* (Fichtel & Moll): East China Sea; South China Sea; Okinawa Trough (Pleistocene to Holocene). (25, 30, 45, 62, 66)
2834. *E. magellanicum* Heron-Allen & Earland: Bohai; Yellow Sea; East China Sea; Shandong; shoreline of Taizhou Bay (Zhejiang); East China Sea (Quaternary). (3, 19, 20, 29, 30, 31, 47)
2835. *E. margaritaceum* Cushman: Shandong (Quaternary). (29, 30)
2836. *E. milletti* (Heron-Allen & Earland): Nansha (=Spratly Islands). (66)
2837. *E. nakanokawaense* Shirai: Bohai; Yellow Sea; East China Sea; South China Sea; Shandong (Quaternary). (5, 13, 21, 29, 30)
2838. *E. nigarense* Cushman: Xisha (=Paracel Islands); Zhongsha (=Macclesfield Bank). (48)
2839. *E. oceanicum* Cushman: Hong Kong; Xisha (=Paracel Islands); Zhongsha (=Macclesfield Bank); Nansha (=Spratly Islands). (25, 27, 30, 66)
2840. *E. owenianum* (d'Orbigny): South China Sea. (67)
2841. *E. pacificum* Collins: Xisha (=Paracel Islands); Zhongsha (=Macclesfield Bank); Nansha (=Spratly Islands). (25, 27, 30, 66)
2842. *E. panamense* Cushman: South China Sea. (67)
2843. *E. parvum* Zheng: Zhongsha (=Macclesfield Bank); Nansha (=Spratly Islands). (27, 30)
2844. *E. poeyanum* d'Orbigny [2795]: Lingshui (Hainan). (7, 8, 21)
2845. *E. pulvereum* Todd: Hong Kong. (62)
2846. *E. pustulatum* Todd: Hong Kong. (62)
2847. *E. reticulosum* Cushman: Xisha (=Paracel Islands); Zhongsha (=Macclesfield Bank); Nansha (=Spratly Islands). (25, 30, 48, 66)
2848. *E. rischtanicum* Bykova: East China Sea (Eocene). (3)
2849. *E. sagrum* (d'Orbigny): South China Sea. (67)
2850. *E. shandongensis* He & Hu: Shoreline of Bohai (Quaternary). (49)
2851. *E. simplex* Cushman: Near sea section of Haihe and Ji Canal (Hebei); Bohai; Yellow Sea; East China Sea; South China Sea; Shandong (Quaternary). (19, 20, 21, 29, 30, 37).
2852. *E. subadvenum* Wang: Shoreline of Bohai (Quaternary). (1)
2853. *E. subcrispum* Nakamura: Bohai; Shandong; Dongshan Bay (Fujian); shoreline of Taizhou Bay (Zhejiang) (Quaternary). (5, 21, 29, 30, 47)
2854. *E. subincertum* Asano [2799]: Yellow Sea; East China Sea; South China Sea. (3, 19, 45, 64, 65)
2855. *E. taiwanum* Nakamura: Taiwan. (44)
2856. *E. texanum* (Cushman & Applin): Mt. Jolmo Lungma (=Mt. Everest) (Tibet) (Tertiary). (9)
2857. *E. tikutoensis* Nakamura: East China Sea continental shelf (Paleocene). (3)
2858. *E. translucens* Natland: Hong Kong. (62)
2859. *E. tsudai* Chiji & Nakaseko: Yellow Sea; East China Sea; Taiwan Strait. (64, 65)
2860. *E. tungliangensis* (Huang): Penghu (Taiwan). (58)
2861. *E. verriculata* (Brady): Okinawa Trough (Holocene). (45)
2862. *Cellanthus chapmani* Cushman [2801]: Xisha (=Paracel Islands); Zhongsha (=Macclesfield Bank). (48)
2863. *C. craticulatum* (Fichtel & Moll): Taiwan Strait; Xisha (=Paracel Islands); Zhongsha (=Macclesfield Bank); Nansha (=Spratly Islands); Nansha (=Spratly Islands) (Tertiary). (4, 25, 27, 30, 46, 48, 66)
2864. *C. guangdongensis* He & Lin [2822]: Guangdong (Pliocene). (33)
2865. *C. ibericum* (Schrodt) [2872]: Shoreline of Bohai (Quaternary). (37, 49)
2866. *C. taiwanus* (Nakamura) [2855]: Xisha (=Paracel Islands); Zhongsha (=Macclesfield Bank); Guangdong (Pliocene); Xisha (=Paracel Islands) (Cenozoic). (4, 33, 48)
2867. *C. tikutoensis* (Nakamura) [2857]: Taiwan Strait. (44)
2868. *C. tungliangensis* Huang [2860]: Penghu (Taiwan). (58)
2869. *Ozawaia tongaensis* Cushman: Northern Yellow Sea. (30)

2870. *Rectoelphidiella aplata* Ho, Hu, & Wang: Shoreline of Bohai; Shandong; Yellow Sea; East China Sea; Jiangsu; shoreline of Taizhou Bay (Zhejiang); East China Sea continental shelf (Pliocene); South China Sea (Quaternary). (3, 8, 29, 30, 31, 47, 49)

2871. *R. lepida* Ho, He, & Wang: Yellow Sea; East China Sea; shoreline of Taizhou Bay (Zhejiang); Jiangsu (Quaternary). (8, 19, 47)

2872. *Stomoloculina lobata* He & Hu: Shoreline of Bohai (Quaternary). (49)

2873. *S. multangula* Ho, He, & Wang: Bohai; Yellow Sea; East China Sea; near sea section of Haihe and Ji Canal (Hebei); Jiangsu; Taizhou Bay (Zhejiang); eastern Hebei (Quaternary). (8, 19, 20, 29, 30, 34, 47, 49)

2874. *S. symmetrica* Ho & Hu: Xisha (=Paracel Islands); Zhongsha (=Macclesfield Bank). (48)

2875. *Parrellina hispidula* (Cushman) [2824]: Bohai; Yellow Sea; East China Sea; South China Sea; Shandong; shoreline of Taizhou Bay (Zhejiang); Jiangsu (Quaternary). (8, 19, 29, 30, 35, 49, 67)

MIOGYPSINIDAE

2876. *Polystomellina discorbinoides* Yabe & Hanzawa: Shoreline of Bohai (Quaternary). (49)

2877. *Miogypsina polymorpha* Rutten: Xisha (=Paracel Islands) (Cenozoic). (4)

NUMMULITACEA
PELLATISPIRIDAE

2878. *Miscellanea complanata* Sheng & Zhang: Mt. Jolmo Lungma (=Mt. Everest) (Tibet) (Tertiary). (9)

2879. *M. minor* Sheng & Zhang: Mt. Jolmo Lungma (=Mt. Everest) (Tibet) (Tertiary). (9)

2880. *M. miscella* (d'Archiac & Haima): Mt. Jolmo Lungma (=Mt. Everest) (Tibet) (Tertiary). (9)

2881. *M. multicolumnata* Sheng & Zhang: Mt. Jolmo Lungma (=Mt. Everest) (Tibet) (Tertiary). (9)

2882. *M. stampi* (Davies): Mt. Jolmo Lungma (=Mt. Everest) (Tibet) (Tertiary). (9)

NUMMULITIDAE

2883. *Assilina dandotica* Davies: Mt. Jolmo Lungma (=Mt. Everest) (Tibet) (Tertiary). (9)

2884. *A. formosensis* Hanzawa: Taiwan area (Tertiary). (46)

2885. *A. granulosa* (d'Archiac): Mt. Jolmo Lungma (=Mt. Everest) (Tibet) (Tertiary). (9)

2886. *A. levis* Sheng & Zhang: Mt. Jolmo Lungma (=Mt. Everest) (Tibet) (Tertiary). (9)

2887. *A. samashanica* Sheng & Zhang: Mt. Jolmo Lungma (=Mt. Everest) (Tibet) (Tertiary). (9)

2888. *A. subspinosa* Davies: Mt. Jolmo Lungma (=Mt. Everest) (Tibet) (Tertiary). (9)

2889. *Cycloclypeus carpenteri* Brady: Zhongsha (=Macclesfield Bank). (27, 30)

2890. *Heterostegina depressa* d'Orbigny: Xisha (=Paracel Islands); Zhongsha (=Macclesfield Bank); Nansha (Spratly Islands); South China Sea (Tertiary). (4, 25, 27, 30, 44, 46, 48, 66)

2891. *H. longispina* Zheng: Xisha (=Paracel Islands); Zhongsha (=Macclesfield Bank). (27, 30, 48)

2892. *H. suborbicularis* d'Orbigny: Taiwan; Xisha (=Paracel Islands); Zhongsha (=Macclesfield Bank); Nansha (=Spratly Islands); Xisha (=Paracel Islands) (Cenozoic). (4, 25, 27, 30, 44, 48)

2893. *Nummulites ammonoides* (Gronovius): Taiwan Strait; Hong Kong; Xisha (=Paracel Islands); Zhongsha (=Macclesfield Bank). (25, 27, 30, 50, 62)

2894. *N. ataticus* Leymerie: Mt. Jolmo Lungma (=Mt. Everest) (Tibet) (Tertiary). (9)

2895. *N. baguelensis* Verbeek: East China Sea continental shelf (Eocene). (3)

2896. *N. complanata* (Defrance): Nansha (=Spratly Islands). (66)

2897. *N. donghaiensis* Wang, He, & Lu: East China Sea continental shelf (Eocene). (3)

2898. *N. duimiensis* Sheng & Zhang: Mt. Jolmo Lungma (=Mt. Everest) (Tibet) (Tertiary). (9)

2899. *N. georgiensis* Renngarten: East China Sea continental shelf ((Eocene). (3)

2900. *N. guayabalensis* (Barker): East China Sea continental shelf (Eocene). (3)

2901. *N. laevigatus* (Bruguiere): Mt. Jolmo Lungma (=Mt. Everest) (Tibet) (Tertiary). (9)

2902. *N. mamilla* (Fichtel & Moll): Mt. Jolmo Lungma (=Mt. Everest) (Tibet) (Tertiary). (9)

2903. *N. nuttalli* Davies: East China Sea continental shelf (Eocene). (3)

2904. *N. obesus* Sheng & Zhang: Mt. Jolmo Lungma (=Mt. Everest) (Tibet) (Tertiary). (9)

2905. *N. parvulus* Zheng & Zhang: Mt. Jolmo Lungma (=Mt. Everest) (Tibet) (Tertiary). (9)

2906. *N. pengaronensis* Verbeek: Mt. Jolmo Lungma (=Mt. Everest) (Tibet) (Tertiary). (9)

2907. *N. pernotus* Schaub: East China Sea continental shelf (Eocene). (3)

2908. *Operculina ammonoides* (Gronovius) [2893]: Xisha (=Paracel Islands); Zhongsha (=Macclesfield Bank); East China Sea continental shelf (Pliocene); Taiwan (Tertiary). (3, 46, 48)

2909. *O. bartschi* Cushman: South China Sea; Guangdong (Pliocene); Xisha (=Paracel Islands) (Cenozoic). (4, 33, 67)

2910. *O. complanata* (Defrance) [2896]: East China Sea continental shelf (Pliocene); Guangdong (Pliocene); Xisha (=Paracel Islands) (Cenozoic). (3, 4, 33)

2911. *O. canalifera* d'Archiac: Mt. Jolmo Lungma (=Mt. Everest) (Tibet) (Tertiary). (9, 46)

2912. *O. gaimardii* d'Orbigny: South China Sea. (48, 67)

2913. *O. granulosa* Leymerie: Xisha (=Paracel Islands) (Later Pleistocene). (41)
2914. *O. madagascariensis* d'Orbigny: Xisha (=Paracel Islands); Zhongsha (=Macclesfield Bank). (48)
2915. *O. philippinensis* Cushman: East China Sea continental shelf (Pliocene); Guangdong (Pliocene). (3, 33)
2916. *O. subsalsa* Davies: Mt. Jolmo Lungma (=Mt. Everest) (Tibet) (Lower Tertiary). (9, 46)
2917. *O. subgranulosa* d'Orbigny: South China Sea. (67)
2918. *O. tuberculata* Vaughan & Cole: Guangdong (Pliocene). (33)
2919. *O. umbona* He & Lin: Guangdong (Pliocene). (33)
2920. *O. venosa* (Fichtel & Moll): Xisha (=Paracel Islands); Zhongsha (=Macclesfield Bank). (48)

DISCOCYCLINIDAE

2921. *Discocyclina pygmaea* Henrici: Mt. Jolmo Lungma (=Mt. Everest) (Tibet) (Lower Tertiary). (9)
2922. *D. sowerbyi* Nuttall: East China Sea continental shelf (Eocene); Mt. Jolmo Lungma (=Mt. Everest) (Tibet) (Lower Tertiary). (3, 33)

REFERENCES*

(1). Wang, N. and X. He, 1978. Quaternary taxonomic foraminifera of the coastal region of the Bohai Sea and a preliminary study of its paleogeographic problems. Geology and Mineral Industry Research 4:90–136, pls. 1–9.

(2). Wang, N. and X. He, 1985. A study on the early Pleistocene foraminifera from Beijing Plain. Proc. Stratigraphic Paleontology 12:47–66, pls. 1–7.

(3). Wang, N., X. He, and S. Lu, 1989. Foraminifera. In: Cenozoic paleobiota of the continental shelf of the East China Sea (Donghai). Geology Press, pp:6–136, pls. 108–163.

(4). Wang, Z., X. He, and S. Qiu, 1980. Calcium carbonate strata of eastern Yongxing Island Well No. 1 and preliminart study of microfossils. Petroleum Experimental Geology 1:23–38, pls.1–4.

(5). Fang, H., 1990. Environmental implications of foraminifera from surface sediments in Dongshan Bay. Journal of Oceanography in Taiwan Strait 9(2):141–147.

(6). Bian, Y. and P. Wang, 1980. On the Late Quaternary microfauna of the Jiaozhou Bay, Shandong Province and its significance. Papers in Marine Micropaleontology of China. China Ocean Press, Beijing, pp:146–151.

(7). Sun, X. 1991. Foraminifera from sediments of the Shenzhen Bay, Guangdong and the Xinchun Harbour, Hainan Island. Acta Micropaleontologica Sinica 8(3): 325–337, pls.1–2.

(8). He, Y., L. Hu, and K. Wang, 1965. The Quaternary foraminifera of the eastern part of Jiangsu. Mem. Inst. Geol. and Palaeont. Acad. Sinica 4:51–162, pls. 1–16.

(9). He, Y., B. Zhang, L. Hu, and J. Sheng, 1976. Mezozoic foraminifera of Mt. Jomol Lungma region. Report on Mt. Jomol Lungma region Scientific Investigation: Paleontology, pp:1–76, pls. 1–36.

(10). Wang, P. and J. Lin, 1974. Discovery of Paleogene brackish-water foraminifera in a certain basin in central China and its significance. Acta Geologica Sinica 2:175–183, pl. 1.

(11). Wang, P., M. Qiu, and Y. Bian, 1980. On marine-continental transitional faunas in Cenozoic deposits of East China. Papers in Marine Micropaleontology of China, China Ocean Press, Beijing, pp:9–19, pls.1, 2.

(12). Wang, P., J. Zhang, and M. Qiu, 1980. Distribution of foraminifera in surface sediments of the East China Sea. Papers in Marine Micropaleontology of China, China Ocean Press, Beijing, pp:20–38, pls. 3–5.

(13). Wang, P., M. Qiu, and Y. Bian, 1980. Distribution of foraminifera and ostracoda in bottom sediments of the northwestern part of the South Huanghai (Yellow Sea) and its geological significance. Papers in Marine Micropaleontology of China, China Ocean Press, Beijing, pp:61–83, pls.7, 8, 9.

(14). Wang, P., M. Qiu, and J. Gao, 1980. A preliminary study of foraminiferal and ostracod assemblages in the Huanghai Sea. Papers in Marine Micropaleontology of China, China Ocean Press, Beijing, pp:84–100, pl. 11.

(15). Wang, P., M. Qiu, and Y. Bian, 1980. Characteristics of foraminiferal and ostracod thanatocoenoses from some Chinese estuaries and their geological significance. Papers in Marine Micropaleontology of China, China Ocean Press, Beijing, pp:101–111, pls. 12, 13.

(16). Wang, P., J. Zhang, and J. Gao, 1980. A close-up view of lowered sea-level microfauna from the East China and Huanghai Seas in the Late Pleistocene. Papers in Marine Micropaleontology of China, China Ocean Press, Beijing, pp:112–119, pl. 14.

(17). Wang, P. and S. Gu, 1980. Quaternary transgression in the lower reaches of the Xialiaohe River plain. Papers in Marine Micropaleontology of China, China Ocean Press, Beijing, pp:130–139, pl. 16.

(18). Wang, P., L. Xia, and F. Zheng, 1980. A preliminary note on the Late Quaternary microfauna of the Beibu Gulf and its bearing on sea-level changes. Papers in Marine Micropaleontology of China, China Ocean Press, Beijing, pp:140–145, pl. 17.

(19). Wang, P., J. Zhang, H. Zhao, and M. Qiu, 1980. Foraminifera and ostracod in bottom sediments of the East China Sea. China Ocean Press, Beijing, pp:1–438, pls. 1–34.

(20). Li, Y. 1986. Taphonomic foraminiferal communities in the reaches of the Haihe River and Jiyunhe River near the sea and its paleogeographic significance. Acta Micropaleontologica Sinica 3(3):251–259.

(21). Li, Y. and X. He, 1983. Preliminary studies on foraminifera in surface layer of sediments along the north coast of the Bohai Bay and its habitat. Geographical Research 2(1):65–72, pl. 1.

(22). Li, Y. and W. Yan, 1988. Taphonomic foraminifera in surficial sea-floor sediments of Hong Kong. Acta Micropaleontologica Sinica 5(3):221–236, pl. 1.

(23). Cheng, T. and S. Cheng, 1960. The planktonic foraminifera of the Yellow Sea and the East China Sea. Oceanologia & Limnologia Sinica 3(3):125–156, pls. 1–11.

(24). Cheng, T. and S. Cheng, 1964. The planktonic foraminifera of the northern South China Sea. Oceanologia & Limnologia Sinica 6(1):38–77, pls. 1–6.

(25). Cheng, T. and S. Zheng, 1978. The recent foraminifera of the Xisha Islands, Guangdong Province, China. I. Studia Marina Sinica 12:149–266, pls. 1–33.

(26). Zheng, S. 1979. The recent foraminifera of the Xisha Islands, Guangdong Province, China. II. Studia Marina Sinica 13:16–78, pls. 1–10.

(27). Zheng, S. 1980. The recent foraminifera of the Zhongsha Islands. I. Studia Marina Sinica 16:143–182, pls. 1–8.

(28). Zheng, S. 1988. The agglutinated and porcelaneous foraminifera of the East China Sea. China Science Press, Beijing, pp:1–337, pls. 1–54.

(29). Zheng, S., T. Cheng, X. Wang, and Z. Fu, 1978. The Quaternary foraminifera of the Dayuzhang Irrigation Area, Shandong Province, and a preliminary attempt at an interpretation of its depositional environment. Studia Marina Sinica 13:16–78, pls. 1–10.

(30). Zheng, S. and Z. Fu, Foraminifera. Fauna Sinica. (manuscript)

(31). Zheng, G. (Editor-in-Chief) 1989. Stratigraphic correlation of southern Yellow Sea Quaternary stratotype. Science Press, Beijing, pp:1–276, pls. 1–6.

(32). Zheng, L. 1980. Late Quaternary microfauna and marine transgressions of Lianyungang Harbour, Jiangsu Province. Papers in Marine Micropaleontology of China, China Ocean Press, pp:152–163.

(33). Lin, J., N. Wang, X. He, and S. Qiu, 1978. Tertiary foraminifera. In: Paleontological Atlas of Central and Southern Regions of China. Geological Publishing House, Beijing, pp:49–115, pls. 11–29.

(34). Lin, J. 1979. The Quatenary foraminifera in eastern Hebei. Chinese Academy of Geological Sciences Report 1(1):67–82, pls.1, 2, 3.

(35). Lin, H. and X. Zhu, 1986. Study of foraminifera and ostracoda from surface sediments in a small bay of northeastern Jiaozhou Bay, Shandong. Marine Geology and Quaternary Geology 6(3):67–78, pls. 1–3.

(36). Meng, X. 1986. Biostratigraphical boundaries and environmental changes of Xisha Islands since Late Miocene shown by foraminiferal fauna. Acta Micropaleontologica Sinica 6(4):345–356, pl. 1.

(37). Hong, X. 1985. The distribution of recent foraminifera in marshes along the coasts of the Huanghai Sea and the East China Sea and its geological significance. Proc. China Quaternary Symposium, pp:66–75.

(38). Hao, Y. and X. Zheng, 1980. Early tertiary foraminifera from the Kashi Basin of Xinjiang. Acta Paleontologica Sinica 19(2):152–167, pls.1, 2, 3.

(39). Hao, Y. et al. 1989. Quaternary microbiotas and their geological significance from northern Xisha Trench of South China Sea. China University of Geosciences, Wuhan, pp:59–93, pls. 1–11.

(40). Hu, L. 1982. Paleoecology of Tarim Basin Late Tertiary foraminifera and their geological significance. Kexue Tongbao 15:938–941.

(41). Qin, G. and X. Zhu, 1982. Foraminifera assemblage of Late Pleistocene of Shidao, Xisha Islands and its geological significance. Marine Geological Research 2(4):108–112.

(42). Tu, X. and H. Cai, 1980. The distribution of foraminifera and ostracoda in the surface sediments of the central area of the South China Sea. Report on the comprehensive investigation of the South China Sea region. I. pp:99–118, pls.1–4.

(43). Tu, X., F. Zheng, and M. Chen, 1988. Planktonic foraminifera of the central and northern areas of South China Sea. Tropic Oceanology 1:28–38.

(44). Huang, T. 1975. Atlas of fossil benthonic foraminifera. Taiwan Oil and Mineral Prospecting Division and Chinese Petroleum Institute, pp:1–343.

(45). Zeng, X., Y. Hao, D. Sun, and Z. Pan, 1988. Quaternary microbiotas in the Okinawa Trough and their geological significance. Geological Publishing House, Beijing, pp:145–226, pls.1–32.

(46). Qiu, S. and J. Lin, 1980. Tertiary foraminiferal faunas of China and their significance in oil and gas prospecting. Oil and Gas Geology 1(3):208–219.

(47). Cheng, G. 1980. The Late Quaternary foraminifera of the coast of the Taichou Gulf and its paleogeographic significance. J. Shandong College of Oceanology 10(4):91–102.

(48). Cai, H. and X. Hu, 1983. Distribution of foraminifera and ostracoda from the surface of the bottom sediments off the Xisha-Zhongsha Islands, South China Sea. South China Sea Studia Marina Sinica 4:25–58, pls. 1–4.

(49). Ministry of Petro-chemical Industry, Petroleum Exploration and Exploitation Program Institute and Nanjing Geological and Paleontological Institute, Academia Sinica, 1978. The Cenozoic foraminifera from the coastal region of Bohai. China Science Press, pp:1–48, pls. 1–10.

(50). Fujian Institute of Oceanology, 1988. A comprehensive oceanographic survey of the central and northern part of the Taiwan Strait. China Science Press, Beijing, pp:1–423, pls. 1–8.

(51). Huang, T., 1961. Smaller foraminifera from the beach sands at Tanmenkang, Pachao-Tao, Penghu. Proc. Geol. Soc. China 4:80–90, pls. 1–5.

(52). Huang, T., 1963. Planktonic foraminifera from the Peikang PK–3 well in the Peikang Shelf Area, Yunlin, Taiwan, Mr. H. H. Ling's 70th Birthday Jubilee Volume, Petroleum Geology of Taiwan 2:153–181, pls. 1–6.

(53). Huang, T., 1964, "Rotalia" group from the Upper Cenozoic of Taiwan. Micropaleontology 10 (1):49–62, pls. 1–3.

(54). Huang, T., 1970. New Foraminiferida from the Taiwan Strait, Taiwan, China. Proc. Geol. Soc. China 13:108–114, pl. 1.

(55). Huang, T., 1971. Some foraminiferal lineages in Taiwan. Proc. Geol. Soc. China 14:76–85, pl. 1.

(56). Huang, T., 1974. Four new species of *Gaudryina* and *Tritaxia* (Foraminiferida) from southern Taiwan. Proc. Geol. Soc. China 17: 135–138, pl. 1.

(57). Huang, T., 1974. A new *Lenticulina* (Foraminiferida) from south-western Taiwan. Bull. Geol. Survey of Taiwan 24:31–33, pl. 1.

(58). Huang, T., 1975. Late Neogene foraminifera zonation of southwestern Taiwan. Late Neogene Epoch Boundaries, pp:106–114.

(59). Huang, T., 1983. Foraminiferal biofacies of the Taiwan Strait, ROC. Bulletin della Societa Paleontologica Italiana, 22 (102):151–177.

(60). Huang, T., 1986. *Alloglobigerinoides*, a new planktonic foraminiferai genus. Petroleum Geology of Taiwan 3: 93–102, pls. 1–2.

(61). Huang, T., and H. T. Chiu, 1973. Some diagnostic Miocene planktonic foraminifera from west central Taiwan. Petroleum Geology of Taiwan 2:93–102, pls. 1–2.

(62). Ko-panicolaou, K. J. 1981. A study of the distribution of coastal foraminifera from seleceted areas in Hongkong. Univ. Microfilms International, New York Univ., pp:1–195, pls. 1–11.

(63). Li, Q. and P. Wang, 1985. Distribution of larger foraminifera in the northwestern part of the South China Sea, in Wang, P. et al., Marine Micropaleontology of China, China Ocean Press, Beijing Springer-Verlag, Berlin Heidelberg, pp:176–195.

(64). Polski, W., 1959. Foraminiferal biofacies of the North Asiatic coast. Jour. Paleontology 33(4):569–587, pl. 78.

(65). Stschedrina, Z. G. and T. G. Lukina, 1984. Formaminifera of Pacific Shelf in Southeastern Asia. USSR Academy of Sciences, Zoological Institute, pp:1–139, pls. 1–31.

(66). Tu, X. and F. Zheng, 1991. Foraminifera in Surface Sediments of the Nansha Sea Area in Quaternary Biological Groups of the Nansha Islands and the Neighbouring Waters, Edited by the Multidisciplinary Oceanography Expedition Team of Academia Sinica to the Nansha Islands, Zhongshan University Publishing House, pp:129–198, pls. 1–10.

(67). Waller, H. O., 1960. Foraminiferal biofacies of the South China Coast. Jour. Pal. 34(6):1164–1182.

(68). Waller, H. O., and W. Polski, 1959. Planktonic Foraminifera of the Asiatic Shelf. Contr. Cushman Foundation Foram. Res. 10(4):123–126, pl. 10

(69). Wang, P., Q. Min, and Y. Bian, 1985. Foraminifera and ostracoda in bottom sediments of Bohai Gulf and their bearings on Quaternary paleoenvironment. In: Wang, P. et al., Marine Micropaleontology of China, China Ocean Press, Beijing Springer-Verlag Berlin Heidelberg, pp:176–195.

(70). Wang, P., Q. Min, and Y. Bian, 1985. Foraminiferal biofacies in the northern continental shelf of the South China Sea. In: Wang, P. et al., Marine Micropaleontology of China, China Ocean Press, Beijing Springer-Verlag Berlin Heidelberg, pp:151–175.

(71). Wang, P., X. Hong, and Q. Zhao, 1985. Living foraminifera and ostracoda: distribution in the coastal areas of the East China Sea and the Huanghai Sea. In: Wang, P. et al., Marine Micropaleontology of China, China Ocean Press, Beijing Springer-Verlag Berlin Heidelberg, pp:243–255.

(72). Wang, P., L. Xia, and X. Cheng, 1985. Neogene biostratigraphy in the northern shelf of the South China Sea. In: Wang, P. et al., Marine Micropaleontology of China, China Ocean Press, Beijing Springer-Verlag Berlin Heidelberg, pp:291–303.

(73). Zheng, S., 1981. *Mesosigmoilina*, a new name for *Pseudosigmoilina* Zheng, 1979. Jour. of Pal., 55(2):483.

(74). Zheng, S. and Z. Fu, 1990. Faunal trends and assemblages of the northern South China Sea agglutinated foraminifera. Paleoecology, Biostragraphy, Paleoceanography and Taxonomy of Agglutinated Foraminifera. NATO ASI Ser. C: Mathematical and Physical Sciences, vol. 327, pp:541–563, pls. 1–3.

(75). Zheng, S. and Z. Fu, 1991. The Agglutinated foraminifera of the Bohai Sea and Huanghai Sea. Studies in Benthic Foraminifera. BENTHOS'90, Sendai, Tokai University Press, pp:183–197, pls. 1–2.

*: (1)–(50) in Chinese, (65) in Russian

Note: The number in the parenthesis following a species name denotes that the species is synonymous with the numbered species according to Loeblich and Tappan (Foraminiferal Genera and Their Classification, 1987, Van Nostrand Reinhold Company, New York). For example, No. 29. *Rhabdammina cornuta* (Brady) [6] means that *Rhabdammina cornuta* is synonymous with No. 6 *Astrorhizoides cornuta* (Brady). Subtracting these 316 synonymous cases, the actual recorded number of foraminiferal species in China from Cenozoic to present is 2,606 species.

Compiled and edited by Zheng Shouyi and Fu Zhaoxian

FUNGI

YEAST

ENDOMYCETATALES
CRYTOCOCCACEAE

Candida albicans: Bohai. (3)
C. amylolenta: Bohai. (3)
C. azyma: Bohai; Yellow Sea. (2, 3, 4, 5)
C. curvata: Bohai. (2, 4)
C. domerogii: Bohai. (3)
C. edax: Bohai. (2, 4)
C. famata: Bohai; Yellow Sea. (2, 3, 4, 5)
C. guilliermondii: Bohai; Yellow Sea. (2, 3, 4, 5)
C. humicola: Bohai. (3)
C. hylophilia: Bohai. (3)
C. insectorum: Bohai. (2, 4)
C. intermedia: Yellow Sea. (5)
C. krusei: Bohai; Yellow Sea. (2, 3, 4, 5)
C. krusii: Bohai. (3)
C. lambica: Bohai. (3)
C. membranaefaciens: Yellow Sea. (5)
C. multisgemmis: Bohai; Yellow Sea. (2, 3, 4, 5)
C. parapsilosis: Bohai; Yellow Sea. (2, 3, 4, 5)
C. pelliculosa: Bohai. (3)
C. sake: Bohai; Yellow Sea. (2, 3, 4, 5)
C. silvanorum: Yellow Sea. (5)
C. tenuis: Yellow Sea. (5)
C. tepae: Bohai. (3)
C. terebra: Bohai. (2, 4)
C. tropicalis: Bohai; Yellow Sea. (2, 3, 4, 5)
C. valdiviana: Bohai. (2, 4)
C. veronae: Yellow Sea. (5)
C. wiswanathii: Bohai. (5)
C. zeylancides: Bohai. (3)
Cryptococcus albidus: Bohai. (1, 3, 4)
C. ates: Bohai. (4)
C. hungaricus: Bohai. (1, 4)
C. infirmo-miniatus: Bohai. (4)
C. laurentii: Bohai. (1, 3, 4)
C. luteolus: Bohai. (1, 4)
C. magnus: Bohai. (4)
C. uniguttulatus: Bohai. (1, 4)
Rhodotorula aurantiaca: Bohai. (1, 4)
R. glutinis: Bohai. (1, 3, 4)
R. graminis: Bohai. (1, 4)
R. minuta: Bohai. (1, 4)
R. pilimanae: Yellow Sea. (5)
R. rubra: Bohai; Yellow Sea. (1, 3, 4, 5)
Trichosporon cutaneum: Bohai. (4)
T. figueirae: Yellow Sea. (5)
Torulaspora delbrueckii: Bohai. (4)

SACCHAROMYCETACEAE

Aureobasidium pullalans: Yellow Sea. (4, 5)
Debaryomyces hansenii: Bohai; Yellow Sea. (3, 4, 5)
D. polymorphus: Bohai. (4)
Pichia burtonii: Bohai. (4)
P. carsonii: Bohai. (4)
P. etchellsii: Bohai; Yellow Sea. (4, 5)
P. guilliermondii: Bohai; Yellow Sea. (4, 5)
P. heimii: Bohai. (4)
P. kluyveri: Yellow Sea. (5)
P. naganishii: Yellow Sea. (5)
P. ohmeri: Bohai; Yellow Sea. (4, 5)
P. philogaea: Bohai. (4)
P. scolyti: Bohai. (4)
Saccharomyces cerevisiae: Bohai. (4)
S. kluyveri: Bohai. (4)

REFERENCES*

(1). Zhou Yuliang, Chen Houde and Ma Shengwu, 1983. Investigation on marine-occurring yeasts in Bohai (Bo Sea), I. Identification and distribution of Genus *Rhodotorula* and *Cryptococcus* isolated from seawater in Bohai (Bo Sea). Journal of Nankai University, 1:89–98.
(2). Zhou Yuliang et al., 1989. Investigation on marine-occurring yeasts in Bohai (Bo Sea), I. Identification and distribution of Genus *Candida* Berkhout isolated from seawater in Bohai (Bo Sea). Journal of Nankai University, 2:49–54.
(3). Zhou Yuliang et al., 1990. Yeasts from different substrate along Bohai Sea coast. Journal of Nankai University, 4:57–61.
(4). Zhou Yuliang et al., 1991. Yeasts of Bohai seawater. Acta Mycologica Sinica, 10(1): 36–42.
(5). Zhou Yuliang et al. Yeasts of Huanghai seawater. (manuscript)

*: all in Chinese
Compiled by Ni Chunzhi; Edited by Xu Huaishu

OTHER FUNGI

North of Xiamen (Fujian)

EUROTIALES
EUROTIACEAE

Aspergillus (Mick): Bohai; Yellow Sea; East China Sea. (2, 3)
A. flavus Link: Xiamen harbor (Fujian). (8)
A. niger van Tieghem: Xiamen harbor (Fujian). (8)
A. sydowii (Bainer & Sartory) Thom & Church: Xiamen harbor (Fujian). (8)
A. terreus Thom: Xiamen harbor (Fujian). (8)
A. ustus (Baimier) Thom & Church: Xiamen harbor (Fujian). (8)
Penicillium: Bohai; Yellow Sea; East China Sea. (2, 3)
P. chrysogerum Thom: Xiamen harbor (Fujian). (8)

GYMNOASCACEAE

Paecilomyces varioti Bainer: Xiamen harbor (Fujian). (8)

DOTHIDEALES
DOTHIDEACEAE

Ramu lispora Miura: Bohai; Yellow Sea. (2, 4, 5)

XYLARIALES
CEPHALOSPORIUM

Cephalosporium Corda: Bohai; Yellow Sea. (2, 4)

CHAETOMIACEAE

Chaetomium globosum Kunze: Xiamen harbor (Fujian). (2, 4)

MONILIALES
DEMATIACEAE

Alternaria Nees: Bohai; Yellow Sea. (2, 4)

TUBERCULARIACEAE

Fusarium: Bohai. (4)

F. graminearum: Xiamen (Fujian). Pathogen responsible for the death of overwintering *Penaeus penicillatus* broodstock. (7)

F. solani (Mart): Shrimp farm ponds in Zhoushan (Zhejiang). Pathogen for the black gill disease in *Penaeus chinensis*. (6)

F. triciuetum: Xiamen (Fujian). Pathogen responsible for the death of overwintering *Penaeus penicillatus* broodstock. (7)

REFERENCES*

(1). Division of Microbiology, Institute of Oceanology, Acdemia Sinica, 1978. A preliminary study on feeding larvae of *Stichopus japonicus* Selenka with marine yeast. Marine Sciences, 2:37–40.

(2). Wang Wenxing et al., 1980. The concentration of petroleum hydrocarbon and petroleum degradation bacteria in Yellow Sea. Marine Study, 12(4):64–70.

(3). Wang Wenxing et al., 1980. A preliminary analysis of microorganisms from sediment column-like samples in the East China Sea. Marine Study, 12(1): 22–28.

(4). Wang Wenxing et al., 1983. Some biological characteristics of petroleum degrading microorganisms in the Bohai Sea. Marine Environmental Science, 2(4):11–24.

(5). Sun Yi and Chen Dou, 1989. The microbial composition of *Stichopus japonicus* and its physiological property. Oceanologia & Limnologia Sinica, 20(4):300–307.

(6). Chen Bo et al., 1992. Studies on pathogenicity of a species of *Fusarium* in the cultured adult prawn (*Penaeus chinensis*). East China Sea Marine Science, 10(4):7–15.

(7). Hong Xin, Wu Dinghu and Ceng Weihua, 1991. Studies on *Fusarium*-caused disease of shrimps and the pathogenic biology. Xiamen Fisheries Science & Technology, 2:21–25.

(8). Liang Ziyuan et al. 1986. Studies on the erosion of metals by moulds. Acta Oceanlogica Sinica, 8(2):251–154.

*: all in Chinese
Compiled by Ni Chunzhi; Edited by Xu Huaishu

South of Xiamen (Fujian)

ASCOMYCOTINA

Aniptodera chesapeakensis Shearer & Miller: Shenzhen (Guangdong); Hong Kong. Decayed mangrove. (8, 9, 14)

A. haispora Vrijmoed, Hyde & Jones: Hong Kong. Decayed mangrove. (9, 12, 14)

A. lignatilis Hyde: Hong Kong. Decayed mangrove. (9, 12, 14)

A. mangrovei Hyde: Shenzhen (Guangdong). Decayed mangrove. (9, 14)

Antennospora quadricornuta Cribb & Cribb: Hong Kong; South China Sea. Pine wood, teak wood, mangrove, and decayed woods. (1, 2, 3, 4, 9, 11, 14)

Arenariomyces trifurcata Höhnk & E. B. G. Jones: Hong Kong. Decayed mangrove. (8)

Belizeana tuberculata Kohlm. & Vok.-Kohlm: Hong Kong. Decayed mangrove. (9, 12, 14)

B. cf. *tuberculata*: Hong Kong. Decayed mangrove. (9, 12, 14)

Ceriosporopsis halima Linder: Hong Kong. Pine wood. (1, 2, 3, 4)

Chaetomium sp.: Hong Kong. Decayed mangrove leave. (5)

Corollospora maritima Werdermann: Hong Kong. Pine wood and decayed mangrove. (1, 2, 4, 8)

C. pulchella Kohlm., Schmidt & Nair: Hong Kong; South China Sea. Pine wood and decayed wood. (1, 2, 11)

Dactylospora haliotrepha (Kohlm. & Kohlm) Hafellner: Shenzhen (Guangdong); Hong Kong. Decayed mangrove. (9, 12, 14)

Diaporthe salsuginosa Vrijmoed, Hyde & Jones: Hong Kong. Decayed mangrove. (9, 12, 14)

Diatrype sp.: Shenzhen (Guangdong). Decayed mangrove. (9, 14)

Eutypa sp.: Shenzhen (Guangdong); Hong Kong. Decayed mangrove. (9, 12, 14)

Haligena viscidula Kohlm. & Kohlm: Hong Kong. Pine wood. (1, 2)

Halonectria milfordensis E. B. G. Jones: Hong Kong. Pine wood, tung wood, and decayed mangrove. (1, 2, 4, 8)

Halosarpheia abonnis Kohlm: Shenzhen (Guangdong); Hong Kong. Decayed mangrove. (9, 12, 14)

H. fibrosa Kohlm & Kohlm: Hong Kong. Decayed mangrove. (9, 12, 14)

H. indica: Shenzhen (Guangdong). Decayed mangrove. (9, 14)

H. marina (Cribb & Cribb) Kohlm: Shenzhen (Guangdong); Hong Kong. Decayed mangrove. (9, 14)

H. minuta Leong: Hong Kong. Decayed mangrove. (9, 14)

H. ratnagiriensis Patil & Borse: Shenzhen (Guangdong). Decayed mangrove. (9, 14)

H. retorquens Shearer & Crane: Shenzhen (Guangdong); Hong Kong. Decayed mangrove. (9, 12, 14)

H. viscosa [*Isthnoles schmidt*] Shearer & Crane: Shenzhen (Guangdong). Decayed mangrove. (9, 14)

Halosarpheia sp.: Hong Kong. Decayed mangroves. (8, 9, 12, 14)

Halosphaeria hamata (Hohnk) Kohlm: Hong Kong. Pine wood. (1, 2, 3, 4)

H. salina (Meyers) Kohlm: Hong Kong; South China Sea. Pine wood and decayed wood. (1, 4, 11)

Hydronectria tethys Kohlm & Kohlm: Hong Kong. Decayed mangrove. (1, 4, 11)

Hypoxylon oceanicum Schatz: Shenzhen (Guangdong); Hong Kong. Decayed mangrove. (9, 12, 14)

Leptosphaeria australiensis (Cribb & Cribb) G. C. Hughes: Hong Kong. Decayed mangrove. (9, 12, 14)

L. avicenniae Kohlm. & Kohlm: Hong Kong. Decayed mangrove. (9, 12, 14)

L. oraemaris Linder: Decayed mangrove. (8)

Lignincola laevis Hohnk: Shenzhen (Guangdong); Hong Kong. Pine wood and decayed mangrove. (1, 2, 3, 4, 8, 9, 12, 14)

L. longirostris (Cribb & Cribb) Kohlm: Shenzhen (Guangdong); Hong Kong. Decayed mangrove. (9, 12, 14)

Lulworthia grandispora Meyers: Hong Kong. Decayed mangrove. (9, 14)

Lulworthia sp.: Hong Kong. Pine wood, teak wood, and decayed mangrove. (1, 2, 3, 4, 8, 9, 12, 14)

Marinosphaera mangrovei Hyde: Shenzhen (Guangdong); Hong Kong. Decayed mangrove. (9, 12, 14)

M. cf. *mangrovei*: Hong Kong. Decayed mangrove. (9, 12, 14)

Massarina armatispora Hyde, Vrijmoed, Chinnaraj & Jones: Hong Kong. Decayed mangrove. (9, 10, 12, 14)

M. thalassiae Kohlm. & Volk.-Kohlm: Hong Kong. Decayed mangrove. (9, 12, 14)

M. velatospora Hyde & Borse: Hong Kong. Decayed mangrove. (9, 12, 14)

Massarina sp.: Hong Kong. Decayed mangrove. (9, 12, 14)

Melaspilea sp.: Hong Kong. Decayed mangrove. (9, 12, 13, 14)

Nais glitra Crane & Shearer: Hong Kong. Decayed mangrove. (9, 12, 14)

N. inornata Kohlm: Hong Kong. Pine wood and decayed mangrove. (1, 2, 4, 8)

Passeriniella obiones (Crouan & Crouan) Hyde & Mouzouras: Hong Kong. Pine wood. (1, 2)

Passeriniella sp.: Hong Kong. Decayed mangrove. (9, 12, 14)

Petriella setifera (Schm.) Curzi: Hong Kong. Pine wood and teak wood. (1, 2, 3, 4)

Petriellidium ellipsodeum V. Arx & Fassatiova: Hong Kong. Pine wood and teak wood. (1, 2, 3, 4)

Pleospora sp.: Shenzhen (Guangdong); Hong Kong. Decayed mangrove. (8, 9, 12, 14)

Remispora quadriremis (Hohnk) Kohlm: Hong Kong. Pine wood. (1, 2, 3, 4)

Rosellinia sp.: Hong Kong. Decayed mangrove. (9, 12, 14)

Saccardoella marinospora Hyde: Shenzhen (Guangdong); Hong Kong. Decayed mangrove. (8, 9, 12, 14)

Savoryella lignicola Jones & Eaton: Shenzhen (Guangdong); Hong Kong. Decayed mangrove. (8, 9, 12, 14)

S. paucispora (Cribb & Cribb) Koch: Hong Kong. Pine wood and decayed mangrove. (1, 4, 9, 12, 14)

Tpedospora radiata Meyers: Hong Kong. Pine wood. (1, 2)

Trematosphaeria lineolastispora Hyde: Hong Kong. Decayed mangrove. (9, 12, 14)

Verruculina enalia Kohlm. & Kohlm: Shenzhen (Guangdong); Hong Kong. Decayed mangrove. (9, 12, 14)

DEUTEROMYCOTINA

Acremonium sp.: Hong Kong. Sediment of mangrove forest. (6)

Alternaria alternata Kyde: Hong Kong. Decayed mangrove leave. (7)

A. maritima Sutherland: Hong Kong. (1)

Alternaria spp.: Shenzhen (Guangdong). Pine wood and decayed mangrove. (4, 9, 14)

Arthrobotrys sp.: Hong Kong. Teak wood. (1, 4)

Aspergillus spp.: Hong Kong. Pine wood, teak wood, sediment of mangrove forest. (1, 2, 3, 4, 6)

Beltrandia sp.: Shenzhen (Guangdong). Decayed mangrove. (9, 14)

Cirrenalia macrocephala (Kohlm.) Meyers & Moore: Hong Kong. Pine wood and decayed mangrove. (1, 2, 3, 4, 9, 12, 14)

C. pseudomacrocephala Kohlm: Hong Kong. Pine wood. (1, 2)

C. pygmea Kohlm: Shenzhen (Guangdong). Decayed mangrove. (7)

Cirrenalia sp.: Shenzhen (Guangdong). Decayed mangrove. (9, 14)

Clavatospora bulbosa (Anast.) Nakagira & Tubaki: Hong Kong; South China Sea. Decayed mangrove and decayed wood. (8, 9, 11, 14)

Cladosporium cladosporiodes: Hong Kong. Decayed mangrove. (7)

Cladosporium sp.: Hong Kong. Pine wood, teak wood, decayed mangrove leave, and sediment of mangrove forest. (1, 4, 5, 6)

Cytospora spp.: Hong Kong. Pine wood and teak wood. (1, 2, 4)

Dictyosporium pelagicum (Linder) G. C. Hughes: Hong Kong. Pine wood, teak wood, and decayed mangrove. (1, 2, 3, 4, 8, 9, 12, 14)

Dictyosporium sp.: South China Sea. Decayed wood. (11)

Graphium sp.: Hong Kong. Pine wood. (1, 2, 3, 4)

Humicola alopallonella Meyers & Moore: Hong Kong; South China Sea. Pine wood, decayed mangrove, and decayed wood. (1, 2, 3, 4, 9, 11, 12, 14)

H. fuscoatra: Hong Kong. Decayed mangrove leave. (7)

Humicola sp.: Hong Kong. Pine wood and decayed mangrove leave. (1, 2, 5)

Monodictys pelagica (Johnson) E. B. G. Jones: Hong Kong. Pine wood, teak wood, and decayed mangrove. (1, 2, 3, 4, 8)

Monodictys sp.: South China Sea. Decayed wood. (11)

Monodictys sp.: Shenzhen (Guangdong). Decayed mangrove. (9, 14)

Monodictys-like sp.: Shenzhen (Guangdong). Decayed mangrove. (9, 14)

Oidiodendron sp.: Hong Kong. Teak wood. (1, 4)

Papulaspora halima Anastasiou: Hong Kong. Pine wood and decayed mangrove. (1, 2, 3, 4, 8)

Penicillium spp.: Hong Kong. Pine wood, teak wood, sediment of mangrove forest, and decayed mangrove leave. (1, 2, 4, 6, 7)

Periconia prolifica Anastasiou: Shenzhen (Guangdong); Hong Kong; South China Sea. Pine wood, teak wood, decayed mangrove, and decayed wood. (1, 2, 3, 4, 8, 9, 11, 14)

Periconia sp.: Hong Kong. Pine wood. (1, 2, 3, 4)

Periconiella sp.: Hong Kong. Decayed mangrove leave. (7)

Phialophora sp.: Hong Kong. Pine wood and teak wood. (1, 2, 3, 4)

Phialophorophoma litoralis Linder: Hong Kong. Decayed mangrove. (8)

Phoma sp.: Shenzhen (Guangdong); Hong Kong. Sediment of mangrove forest, and decayed mangrove. (6, 8, 9, 14)

Phomopsis sp.: Hong Kong. Decayed mangrove. (9, 12)

Pyrenochaeta sp.: Hong Kong. Pine wood. (1, 3)

Rhabdospora sp.: Shenzhen (Guangdong). Decayed mangrove. (9, 14)

Sporotrichum sp.: Hong Kong. Teak wood. (1, 4)

Stachybotrys-like sp.: South China Sea. Decayed wood. (11)

Stagnospora sp.: Shenzhen (Guangdong). Decayed mangrove. (9, 14)

Trichocladium achrasporum (Meyers & Moore) Dixon in Shearer & Crane: Shenzhen (Guangdong); Hong Kong; South China Sea. Pine wood, teak wood, and decayed mangrove. (1, 2, 3, 4, 8, 9, 11, 14)

T. linderii Shearer: Shenzhen (Guangdong); Hong Kong. Decayed mangrove. (9, 12, 14)

Trichocladium sp.: Shenzhen (Guangdong). Decayed mangrove. (9, 14)

Trichoderma sp.: Hong Kong. Teak wood, sediment of mangrove forest. (1, 4, 6)

Verticillium sp.: Hong Kong. Teak wood. (1, 4)

Zalerion maritimum (Linder) Anastasiou: Hong Kong. Pine wood. (1, 2, 4)

Z. varium Anastasiou: Hong Kong. Pine wood. (1, 2, 3, 4)

Zalerion sp.: Hong Kong. Pine wood. (1, 2, 9, 14)

BASIDIOMYCOTINA

Agerita sp.: Shenzhen (Guangdong). Decayed mangrove. (9, 14)

Halocyphina villosa Kohlm. & Kohlm: Shenzhen (Guangdong); Hong Kong. Decayed mangrove. (9, 12, 14)

REFERENCES

(1). Vrijmoed, L. L. P., I. J. Hodgkiss and L. B. Thrower, 1982. Factors affecting the distribution of lignicolous marine fungi in Hong Kong. Hydrobiologia 87:143–160.

(2). Vrijmoed, L. L. P., I. J. Hodgkiss, and L. B. Thrower, 1982. Seasonal patterns of primary colonization by lignicolous marine fungi in Hong Kong. Hydrobiologia 87:253–262.

(3). Vrijmoed, L. L. P., I. J. Hodgkiss, and L. B. Thrower, 1986. Effects of surface fouling organisms on the occurence of fungi on submerged pine blocks in Hong Kong coastal waters. Hydrobiologia 135: 123–130.

(4). Vrijmoed, L. L. P., I. J. Hodgkiss, and L. B. Thrower, 1986. Occurrence of fungi on submerged pine and teak blocks in Hong Kong coastal waters. Hydrobiologia, 135:109–122.

(5). Vrijmoed, L. L. P. and N. F. Y. Tam, 1989. Higher fungi associated with mangrove leaves on soil surface in subtropical mangrove community. Abstracts of the 5th International Symposium on Microbial Ecology, Kyoto, Japan, p.63.

(6). Vrijmoed, L. L. P. and G. C. Hughes, 1990. Observations of higher fungi in soils of small tropical mangrove community. Abstracts of the 4th International Mycological Congress, Regensburg, Germany, p.165.

(7). Vrijmoed, L. L. P. and N. F. Y. Tam, 1990. Fungi associated with leaves of *Kandelia candel* (L.) Druce in litter bags on the mangrove floor of a small subtropical mangrove community in Hong Kong. Bulletin of Marine Science 47:261–262.

(8). Vrijmoed, L. L. P. 1990. Preliminary observations of lignicolous marine fungi from mangroves in Hong Kong. Proceedings of the Second International Marine Biological Workshop: The Marine Flora and Fauna of Hong Kong and Southern China. Hong Kong, 1986. (Ed. B. Morton), Hong Kong University Press, pp:701–706.

(9). Vrijmoed, L. L. P., E. B. G. Jones and K. D. Hyde, 1992. Observations on subtropical mangrove fungi in the Pearl River Estuary. Research Report, AP-92-15, Department of Applied Science, City Polytechnic of Hong Kong, pp:1–13.

(10). Hyde, K. D., L. L. P. Vrijmoed, S. Chinnaraj, and E. B. G. Jones, 1992. *Massarina armatispora* sp. nov., a new intertidal ascomycete from mangroves. Botania Marina 35:325–328.

(11). Vrijmoed, L. L. P., C. S. W. Kueh, H. Q. Shen, C. H. Cai, and Y. P. Zhou, 1990. A preliminary investigation of marine fungi in the South China Sea. Proceedings of the First International Conference on the Marine Biology of Hong Kong and the South China Sea, Hong Kong University Press.

(12). Vrijmoed, L. L. P., K. D. Hyde and E. B. G. Jones, Observations on mangrove fungi from Macau and Hong Kong, with the description of two Ascomycetes: *Diaharborhe salsuginosa* and *Aniptodera haispora*. (manuscript).

(13). Vrijmoed, L. L. P., K. D. Hyde and E. B. G. Jones, A new *Melaspilea* species from Australia and Hong Kong mangroves. (manuscript).

(14). Jones, E. B. G., and L. L. P. Vrijmoed, Mangrove fungi from the subtropics, with the description of a new *Pleospora* species. (manuscript)

Compiled by Guan Liping (L. L. P. Vrijmoed); Edited by Zhou Zhongcheng

MYCOPHYCOPHYTA [LICHENS]

Verrucariam aura: Hong Kong. Rock surface in supralittoral zone. (1)

REFERENCE

(1). Morton, B. and J. Morton, 1983. The sea shore ecology of Hong Kong. Hong Kong University Press, 359 pp.

Compiled by Huang Zongguo; Edited by B. Morton

PLANTAE

ALGAE

RHODOPHYTA

PORPHYRIDIALES
GONIOTRICHACEAE

Asterocytis ornata (C. Ag.) Hamel: Hong Kong. (42)
Stylonema alsidii (Zanard.) Drew [*Goniotrichum alsidii*]: Fujian; Hong Kong. (60, 120)

BANGIALES
ERYTHROPELTIDACEAE

Erythrocladia irregularis Rosenv. [*E. subintegra*]: Shengsi Islands (Zhejiang); Hong Kong. (67, 120)
Erythrotrichia carnea (Dillw.) J. Ag.: Fujian. (60)
Porphyropsis coccinea (J. Ag. & Aresch.) Rosenv.: Fujian. (24)

BANGIACEAE

Bangia atropurpurea (Roth) C. Ag.: Hong Kong. (124)
B. breviaticulata Tseng: Hong Kong. (123)
B. fusco-purpurea: Zhejiang; Fujian; Hong Kong. (17, 67, 122, 124)
B. gloiopeltidicola Tanaka: Zhejiang seashore. (55, 67)
B. siliaris Carm.: Zhoushan and Shengsi (Zhejiang). (67)
B. yamadai Tanaka: Fujian. (60)
Porphyra crispata Kjellm: Zhejiang; Fujian; Guangdong. (17, 55, 67, 122, 124)
P. dentata Kjellm: Zhejiang; Fujian; Guangdong. (17, 55, 67, 122, 124)
P. dentimarginata C. Y. Chu & S. C. Wang: Zhejiang. (3)
P. fujianensis Wang & Zhang: Fujian. (3)
P. guangdongensis Tseng & T. J. Chang: Fujian; Guangdong. (122)
P. haitanensis Chang & Zheng: Zhoushan, Shengsi and Nanji (Zhejiang); Fujian; Guangdong. (17, 55, 67, 122)
P. ishigecola Miura: Zhejiang. (55)
P. katadai Miura v. *hemiphylla* Tseng & T. J. Chang: Dalian (Liaoning); Qingdao (Shandong). (122)
P. marginata Tseng & T. J. Chang: Dalian (Liaoning); Rongcheng and Qingdao (Shandong); western shore of Yellow Sea. (63, 78, 79, 88, 91, 122)
P. monosporangia Wang & Zhang: Fujian. (1)
P. oligospermatangia Tseng B. F. Zheng: Qingdao (Shandong). (65)
P. pseudocrispata Wang & Zhang: Guangdong. (2)
P. ramosissima Pan & Wang: Guangdong. (104)
P. seriata Kjellm.: Dalian (Liaoning). (91)
P. suborbiculata Kjellm.: Shandong; Jiangsu; Zhejiang; Fujian; Guangdong; Hong Kong. (17, 55, 67, 84, 91, 122)
P. tenera Kjellm.: Liaoning; Shandong; Zhejiang. (79, 84, 87, 91, 122)
P. umbilicalis (L.) J. Ag.: Western shore of Yellow Sea. (88)
P. vietnamensis Tanka & Ho: Hainan. (122)
P. yezoensis Ueda: Liaoning; Shandong; Jiangsu; Zhejiang. (17, 55, 67, 79, 87, 88, 122)

NEMALIALES
ACROCHAETIACEAE

Audouinella daviesii (Dillwyn) Woelkerling [*Acrochaetium crassipes*, *A. daviesii*]: Fujian; Hong Kong. (60, 120)
A. densa (Drew) Garbary [*Acrochaetium arcuatum*]: Hong Kong. (120)
A. microscopica (Naeg. in Kuetz.) Woelkerling [*Acrochaetium microscopicum*]: Dalian (Liaoning). (73)
A. robusta (Boeg.) Garbary [*Acrochaetium robusta*]: Shandong; Hong Kong. (114, 120, 122)
A. subimmersum (S. & G.) Garbary & Rueness [*Acrochaetium subimmersum*]: Dalian (Liaoning). (73)

NEMALIACEAE

Nemalion helminthiodes (Valley) Batt. v. *vermiculare* (Sur.) Tseng: Bohai; Yellow Sea; Hong Kong. (87, 88, 91, 122, 124)

HELMINTHOCLADIACEAE

Dermonema frappieri (Mont. & Mill) Boerg. [*D. gracilis*]: Taiwan; Guangdong; Hong Kong. (122, 124)
D. pulvinata (Grunow) Fan [*Nemalion pulvinatum*]: Zhejiang; Fujian; Taiwan; Guangdong; Hong Kong. (54, 67, 122, 124)
Ganonema farionsa (Lamx.) Fan & Wang [*Liagora farinosa*]: Hainan; Xisha (=Paracel Islands). (124)
G. pinnatiramosa (Yamada) Fan & Wang: Xisha (=Paracel Islands). (107)
Helminthocladia australis Harv.: Hainan. (122)
H. yendoana Narita: Jinzhou (Liaoning); Qingdao (Shandong). (91)
Liagora boergesenii Yam.: Taiwan. (141)
L. ceranoides Lamx.: Hainan. (122)
L. divaricata Tseng: Hainan. (122)
L. filiformis Fan & Li: Xisha (=Paracel Islands). (107, 122)

L. hainanensis Tseng & Li: Hainan. (122)
L. pinnata Harv.: Dongsha (=Pratas Islands); Xisha (=Paracel Islands). (107, 122)
L. rubra Tseng & Li: Hainan. (122)
L. samaensis Tseng: Hainan. (108)
L. setchellii Yamada: Xisha (=Paracel Islands). (107)
L. sinensis Fan, Wang & Pan: Xisha (=Paracel Islands). (107, 122)
Trichogloeopsis hawaiiana Abbott & Doty: Xisha (=Paracel Islands). (107, 122)
Yamadaella caenomyce (Decaisne) Abbott [*Liagora caenomyce*]: Taiwan. (110)

CHAETANGIACEAE

Actinotrichia fragilis (Forssk.) Boerg: Hainan; Xisha (=Paracel Islands). (37, 122, 123)
Galaxaura apiculata Kjellm.: South China Sea. (59)
G. elongata J. Ag.: Hong Kong. (123)
G. fasciculata Kjellm.: Hainan; Nansha (=Spratly Islands). (116, 122)
G. filamentosa Chou in Taylor [*G. rudis*]: Xisha (=Paracel Islands); Nansha (=Spratly Islands). (70, 107, 122)
G. glabriuscula Kjellm.: South China Sea. (59)
G. oblongata (Ell. & Sol.) Lamx.: Fujian; Guangdong; Hong Kong. (54, 122, 123, 124)
G. obtusata (Ell. & Sol.) Lamx.: Hong Kong. (116)
G. pacifica Tanaka: Guangdong. (122)
G. robustsa Kjellm.: Guangdong. (122)
G. stupocaula Kjellm. [*G. arborea*]: Hong Kong; Hainan. (116, 122)
G. subfruticulosa Chou in Taylor [*G. fruticulosa*]: Guangdong; Hong Kong. (122)
G. umbellata (Esp.) Lamx.: South China Sea. (59)
G. ventricosa Kjellm.: Hainan; Xisha (=Paracel Islands). (122)
G. veprecula Kjellm.: Taiwan; Xisha (=Paracel Islands). (122)
Gloiophloea chinensis Tseng: Guangdong; Hainan. (122)
Scinaia boergesenii Tseng: Guangdong; Hainan. (122)
S. japonica Setch: Fujian. (60)
S. latifroms Howe [*S. cottonii*]: Fujian; Hainan. (122)
S. tsinglanensis Tseng: Guangdong; Hainan. (122)

BONNEMAISONIACEAE

Asparagopsis taxiformis (Delile) Collins & Harv. [*A. sanfordiana, Falkenbergia hillebrandii*]: Hong Kong; Hainan; Xisha (=Paracel Islands); Nansha (=Spratly Islands). (37, 70, 122, 123, 124)
Bonnemaisonia hamifera Hariot [*B. nootkana, Asparagopsis hamifera*]: Dalian (Liaoning); Qingdao (Shandong). (103)

GELIDIACEAE

Gelidiella acerosa (Forssk.) Feldm. & Hamel: Hainan; Xisha (=Paracel Islands); Nansha (=Spratly Islands). (70, 122)
Gelidium amansii (Lamx.) Lamx.: China's coast. (24, 109)
G. amansii f. *elegans*: Nansha (=Spratly Islands). (44)
G. amansii f. *radicans*: Common species from Bohai to South China Sea. (17, 55, 67, 79, 84, 88, 124)
G. crinale (Turn.) Lamx.: Widely distributed along China's seashore. (55, 63, 67, 84, 88, 124)
G. divaricatum Martens: Widely distributed along China's seashore. (17, 55, 63, 67, 84, 88, 124)
G. japonicum (Harv.) Okam.: Taiwan. (110)
G. johnstonii Setchell and Gardner: Hong Kong. (123)
G. kintaroi (Okam.) Yam. [*G. clavatum*]: Xiamen (Fujian). (14)
G. latifolium Bormet: Hong Kong. (121, 124)
G. pacificum Okam.: Widely distributed in Zhejiang and Fujian. (17, 55, 67, 122)
G. pusillum (Stackh.) Le Jol.: Beidaihe (Hebei); Qingdao (Shandong); western shore of Yellow Sea; Fujian; Hong Kong. (32, 55, 91, 122, 123)
G. tsengii Fan [*G. grubbae*]: Beidaihe (Hebei); Weihai (Shandong); Hong Kong. (91, 123)
G. vagum Okam.: Yellow Sea. (44, 91, 122)
Pterocladia capillacea (Gmelin) Bornet [*P. temuis*]: China's coast. (17, 55, 63, 67, 84, 88, 122, 124)
P. nana Okam.: Hong Kong. (123)

CRYPTONEMIALES
DUMONTIACEAE

Dumontia simplex Cotton: Yellow Sea shoreline. (88, 122)
Hyalosiphonia caespitosa Okam.: Yellow Sea shoreline. (88, 91, 122)

PEYSSONNELIACEAE

Peyssonnelia rubra (Grev.) J. Ag. v. *orientalis* W.-v. Bosse: Hainan. (122)
Ramicrusta nanhaiensis D. Zhang & J. Zhou: Xisha (=Paracel Islands). (50, 122)
Rhododiscus pulcherrimus Crouan: Fujian. (60)

HILDENBRANDIACEAE

Hildenbrandia prototypus Nardo: Widely distributed along China's seashore. (67, 103, 122)
H. rivularis (Lieb.) J. Ag.: Nanji (Zhejiang). (17)

CORALLINACEAE

Amphiroa aberrans Yendo: Shengsi (Zhejiang). (67)
A. anastomosans W.-v. B.: Hainan. (122)

A. f. *cyathifera* (Lamx.) W.-v. Bosse: Xisha (=Paracel Islands). (48)

A. dilatata Lamx.: East China Sea shoreline; Xisha (=Paracel Islands). (67, 115, 122)

A. ephedraea Dec.: Shengsi and Nanji (Zhejiang); Fujian; Hong Kong. (67, 124)

A. fragilissima (L.) Lamx.: Hainan; Xisha (=Paracel Islands). (47, 48, 122)

A. foliacea Lamx.: South China Sea shoreline. (122)

A. rigida Lamx.: Qingdao (Shandong); Shengsi (Zhejiang); Xiamen (Fujian). (67, 91, 115)

A. valonioides Yendo: Hong Kong. (124)

A. zonata Yendo: Yellow Sea; East China Sea. (55, 88, 122)

Bossiella cretacea (P. & R.) Johansen: Yellow Sea shoreline. (122)

Cheilosporum jungermanniodes Rupr.: Taiwan; Hainan. (122)

Corallina gracilis Lamx.: Fujian. (60)

C. officinalis L.: Bohai; Yellow Sea; East China Sea. (17, 67, 88, 122)

C. pilulifera Post & Rupr.: Yellow Sea; East China Sea; Taiwan Strait; Hong Kong. (17, 67, 88, 122, 124)

C. pilulifera f. *filiformis* Ruprecht: Widely distributed along Fujian's seashore. (115)

C. pilulifera f. *sororia* Ruprecht: Widely distributed in the islands along Fujian's coastline. (115)

C. sessilis Yendo: Nanji (Zhejiang); Fujian; Hong Kong. (122, 124)

Dermatolithon corallinae (Grouan.) Fosl.: Yellow Sea shoreline. (122)

D. tumidulum Fosl.: Shengsi (Zhejiang). (67)

Fosliella farinosa (Lamx.) Howe: From Bohai to South China Sea; Taiwan; Hong Kong; Xisha (=Paracel Islands); Nansha (=Spratly Islands). (48, 122, 124)

Jania adhaerens Lamx. [*J. decussato-dichotoma*]: East China Sea and South China Sea shoreline; Xisha (=Paracel Islands); Nansha (=Spratly Islands). (47, 122, 124)

J. capillacea Harv.: Xisha (=Paracel Islands). (51)

J. crassa Lamx.: Fujian; Hainan. (122)

J. radiata Yendo: Xisha (=Paracel Islands). (47)

J. ungulata Yendo: Hong Kong. (123)

J. ungulata f. *breviov* Yendo: Zhejiang; Fujian; Guandong; Hong Kong. (122, 124)

Leptophytum asperulum (Fosl.) Adey: Xisha (=Paracel Islands). (51)

Lithophyllum kotschyanum (Unger) Fosl.: Hainan. (122)

L. moluccense Fosl.: Hainan. (122)

L. okamurae Fosl.: Zhejiang; Xisha (=Paracel Islands); Nansha (=Spratly Islands). (51, 55, 61)

Lithoporella melobesioides Fosl.: Hainan; Xisha (=Paracel Islands). (48, 122)

L. pacifica (Heydr.) Fosl.: Hainan; Xisha (=Paracel Islands); Nansha (=Spratly Islands). (61, 122)

Lithothamnium aculeiferum Mason in Setch. & Mason: Dalian (Liaoning). (52)

L. glaciale Kjellm: Shandong. (53)

L. intermedium Kjellm.: Liaoning; Shandong. (52, 122)

L. japonicum Fosl. [*L. fretense*]: Dalian (Liaoning); Shandong. (52)

L. pacificum (Fosl.) Fosl.: Liaoning; Shandong. (52)

Marginiosporum aberrans (Yendo) Johansen & Chihara: Zhejiang; Fujian. (122)

M. crassissimum (Yendo) Johansen & Chilara: Zhejiang; Fujian. (122)

Mastophora rosea (C. Ag.) Setch.: Taiwan; Dongsha (=Pratas Islands); Xisha (=Paracel Islands); Nansha (=Spratly Islands). (61)

Mesophyllum erubescens (Fosl.) Lem.: Guangdong. (122)

M. mesomorphum (Fosl.) Adey: Xisha (=Paracel Islands). (48)

M. simulans (Fosl.) Lem.: Hainan. (122)

Neogoniolithon conicum (Dawson) Gordon: Xisha (=Paracel Islands). (49)

N. fosliei (Heyd.) Setch. & Mason: Xisha (=Paracel Islands); Nansha (=Spratly Islands). (49, 61)

N. frutescens (Fosl.) Setch. & Mason: Xisha (=Paracel Islands). (47, 49, 122)

N. frutescens f. *flabelliformis* (Fosl.) Lee: Xisha (=Paracel Islands). (49)

N. megalocystum (Fosl.) Setch. & Mason: Xisha (=Paracel Islands). (49)

N. pacificum (Fosl.) Setch. & Mason: Xisha (=Paracel Islands). (49, 122)

N. trichotomum (Heydr.) Setch. & Mason: Guangdong; Xisha (=Paracel Islands); Nansha (=Spratly Islands). (48, 61, 122)

N. variabile Zhang & Zhou: Hainan; Xisha (=Paracel Islands); Nansha (=Spratly Islands). (49, 61, 122)

Pneophyllum lejolisii (Rosanoff) Chamberlain [*Fosliella lejolisii*]: From Yellow Sea to Hong Kong. (122, 124)

P. zostericolum (Fosl.) Fujita [*Leptophytum zostericolum*, *Heteroderma zostericola*]: Yellow Sea shoreline. (122)

Porolithon bobergesenii (Fosl.) Lemoine in Boerg. [*Hydrolithon boergesenii*]: Xisha (=Paracel Islands). (48)

P. onkodes (Heydr.) Foslie: Hainan; Xisha (=Paracel Islands). (47, 61, 122)

P. onkodes f. *devia* Fosl.: Xisha (=Paracel Islands). (48)

P. onkodes f. *subramosa* Fosl.: Xisha (=Paracel Islands). (48)

Pseudolithophyllum samoense (Fosl.) Adey: Yellow Sea shoreline. (122)

P. yendoi (Fosl.) Adey: Guangdong. (122)

Spongites reinboldii (W.-v. B. & Fosl.) Penrose & Woelkerling [*Hydrolithon reinboldii*]: Hainan; Xisha (=Paracel Islands). (47, 122)

GLOIOSIPHONIACEAE

Gloiosiphonia capillaris (Huds.) Carm.: Lushun and Dalian (Liaoning); Yantai and Qingdao (Shandong); Shengsi (Zhejiang). (67, 91, 122)

ENDKOCLADIACEAE

Gloiopeltis complanata (Harv.) Yamada: Hong Kong. (124)

G. furcata (P. & R.) J. Ag.: Lushun, Dalian, and Hulu Island (Liaoning); Yantai and Qingdao (Shandong); Shengsi and Nanji Island (Zhejiang); Fujian; Hong Kong. (17, 55, 63, 67, 79, 84, 87, 88, 122)

G. furcata v. *coliformis* Okam.: Western shore of Yellow Sea. (91)

G. tenax (Turn.) J. Ag.: From East China Sea shoreline to Guangdong; Hong Kong. (17, 122, 124)

HALYMENIACEAE

Carpopeltis formosana Okam.: Taiwan. (110)

C. rigida (Harv.) Schm.: Taiwan. (110)

Grateloupia divaricata Okam.: Lushun and Dalian (Liaoning). (91)

G. filicina C. Ag.: Widely distributed along China's seashore. (17, 55, 63, 67, 84, 87, 88, 122, 123, 124)

G. filicina v. *lomentaria* Howe: Beidaihe (Hebei); Yantai (Shandong). (91)

G. gelatinosa Grun. & Holms: Shengsi (Zhejiang). (67)

G. imbricata Holm.: Fujian; Hong Kong. (60, 124)

G. latissima Okam.: Fujian. (60)

G. livida (Harv.) Yamada: From Bohai to South China Sea; Hong Kong seashore. (17, 55, 63, 67, 87, 122, 123, 124)

G. okamurae Yamada: Zhejiang. (55)

G. prolongta J. Ag.: Eastern Liaoning Peninsula; Yantai (Shandong); Zhejiang. (55, 91)

Halymenia dilatata Zanard.: Fujian. (60)

H. durvillaei Bory v. *ceylanid*: Taiwan. (79)

H. durvillaei Bory v. *formosa* (Harv. & Kuetz.) Reinbold in Reinecke: Taiwan. (79)

H. sinensis Tseng & C. F. Chang: Bohai; Yellow Sea; East China Sea. (55, 67, 122)

Polyopes polyideoides Okam.: Fujian. (122)

Prionitis cornea (Okam.) Dawson [*Carpopeltis cornea*]: Taiwan. (110)

P. prolifera (Hariot) Kawaguchi & Masuda [*Carpopeltis affinis*]: Bohai; Yellow Sea; Zhejiang. (55, 64, 67, 84, 88, 91)

P. ramosissima (Okam.) Kawaguchi [*Grateloupia ramosissima*]: Beidaihe (Hebei); Shengsi and Nanji Islands (Zhejiang); Fujian. (17, 55, 67, 91, 122)

GIGARTINALES
NEMASTOMACEAE

Schizymenia dubyi (Chauv.) J. Ag.: Fujian; Hong Kong. (60, 121)

Tsengia nakamurae (Yendo) Fan & Fan: Yellow Sea shoreline. (122)

HYPNEACEAE

Hypnea boergesenii Tanaka: Zhejiang; Taiwan; Dongshan (Fujian); Xisha (=Paracel Islands). (17, 54, 55, 67, 87, 122)

H. cervicornis J. Ag.: Zhejiang; Fujian; Hong Kong; Xisha (=Paracel Islands). (37, 55, 87, 122, 123)

H. charoides Lamx. [*H. seticulosa*]: Zhejiang; Fujian; Taiwan; Guangdong; Hong Kong. (55, 87, 122, 124)

H. chordacea Kueta.: Taiwan. (110)

H. japonica Tanaka: Taiwan; Dongshan (Fujian); Hong Kong; Hainan. (54, 66, 87, 122, 123)

H. musciformis (Wulfen) Lamx. v. *esperi* J. Ag.: Nansha (=Spratly Islands). (1a)

H. pannosa J. Ag. [*H. nidulans*]: Taiwan; Hainan; Xisha (=Paracel Islands); Nansha (=Spratly Islands). (37, 70, 122)

PLOCAMIACEAE

Plocamium telfairiae Harv.: Lushun and Dalian (Liaoning); Qingdao (Shandong); Shengsi and Nanji Islands (Zhejiang); Hong Kong. (17, 55, 67, 88, 91, 122)

P. telfairiae f. *uncinatum* Okam.: Fujian. (60)

GRACILARIACEAE

Ceratodictyon spongiosum Zan.: Guangdong; Hong Kong; Xisha (=Paracel Islands). (108, 122, 123)

Gelidiopsis intricata (Ag.) Vickers: Hainan; Xisha (=Paracel Islands); Nansha (=Spratly Islands). (37, 70, 122)

Gracilaria arcuata Zan.: East China Sea; Hainan; Xisha (=Paracel Islands). (14, 50)

G. articulata Chang & Xia: Hainan. (86)

G. asiatica Zhang & Xia [*G. confervoides, G. verrucosa*]: From Liaoning to Guangdong seashore. (39)

G. asiatica v. *zhengii* Zhang & Xia: Fujian. (126)

G. bangmeiana Zhang & Abbott [*Polycarvernosa remulosa*]: Hainan. (112)

G. belinae Zhang & Xia [*G. heteroclata*]: China's seas. (112)

G. blodgettii Harv.: Fujian; Guangdong; Hainan; Hong Kong. (28, 122, 124)

G. chouae Zhang & Xia [*G. bursa-pastoris*]: Zhejiang; Fujian; Hong Kong; Hainan. (17, 28, 55, 67, 122, 123)

G. constricta C. F. Chang & B. M. Xia: Hainan. (28)

G. coronopifolia J. Ag.: Taiwan; Hainan. (28, 122)

G. corticata J. Ag.: Hong Kong. (116)

G. edulis (Gmelin) Silva [*G. lichenoides, Polycavernosa fastigiata*]: Taiwan; Hainan; Xisha (=Paracel Islands). 112)

G. eucheumoides Harv.: Taiwan; Hainan; Xisha (=Paracel Islands). (28, 122)

G. firma C. F. Chang & B. M. Xia: Guangdong. (122)

G. gigas Harv. [*G. foliifera* f. *textorii*]: Haifeng (Guangdong). (28)

G. hainanensis C. F. Chang & B. M. Xia: Hainan. (28, 122)
G. lingula J. Ag.: Xiamen (Fujian). (28)
G. megaspora (Daws) Papenf.: Fujian. (31)
G. mixtra Abbott, Zhang & Xia: Guangdong. (112)
G. punctata (Okam.) Yamada: Dali and Eluanbi (Taiwan). (28)
G. rubra C. F. Chang & B. M. Xia: Guangdong; Hainan. (28, 31)
G. salicornia (Ag.) Dawson [*G. cacalia, G. crassa, G. minor*]: Taiwan; Hainan. (28, 122)
G. sjoestedtii Kylin: Shandong. (122)
G. spinulosa (Okam.) Chang & Xia [*G. purpurascens* f. *spinulosa*]: Taiwan; Hainan. (122)
G. tenuistipitata C. F. Chang & B. M. Xia [*G. tsengiana*]: Guangdong; Guangxi. (122)
G. tenuistipitata v. *liui* Zhang & Xia: Fujian; Guangdong; Guangxi; Hainan. (125)
G. textorii (Suring.) De Teni. [*G. foliifera*]: Bohai; Yellow Sea; Xiamen (Fujian). (30, 87, 91, 122)
G. vieillardii Silva Schm. & Mazza [*G. denticulata*]: Taiwan. (110)
Gracilariopsis chorda (Holmes) Ohmi [*Gracilaria chorda*]: Xiamen (Fujian); Hainan. (112, 115, 122)
G. lemaneiformis (Bory) Daws. Acleto & Foldvit: China's seas. (112)
Polycarvernosa fastigiata Chang & Xia: Hainan; Xisha (=Paracel Islands). (122)
P. ramulosa Chang & Xia: Hainan. (122)

SPHAEROCOCCACEAE

Caulacanthus ustulatus (Mertens) Kuetz [*C. okamurai*]: Widely distributed along China's seashore. (67, 87, 122)
Catenella impudica (Ment.) J. Ag.: Hong Kong. (117)
C. nipae Zan: Hong Kong. (117)
C. subumbellata Tseng: Hong Kong. (117)

SARCODIACEAE

Trematocarpus pygmaeus Yendo [*Eucheuma denticulatum*]: Dongsha (=Pratas Islands); Xisha (=Paracel Islands). (35, 122)

SOLIERIACEAE

Eucheuma arnoldii Webervan Bosse: Xisha (=Paracel Islands). (87)
E. denticulatum (Burman) Collins & Harvey [*E. muricatum*]: Taiwan. (87)
E. gelatinae (Esper) J. Ag.: Taiwan; Hainan. (122)
E. okamurai Yamada: Hainan. (87)
E. serra J. Ag.: Taiwan. (143)
Meristotheca papulosa (Mont.) J. Ag.: Taiwan. (87, 122)
Sarconema filiforme (Sond.) Kylin: Hainan. (122)
S. gracilarioides Chang & Xia: Hainan. (43, 122)
Solieria pacifica (Yam.) Yoshida [*S. robusta*]: Fujian; Hainan. (122)
S. tenuis Zhang & Xia in Xia & Zhang [*S. mollis*]: Widely distributed along China's seashore. (79, 84, 87, 88)

DICRANEMACEAE

Tylotus lichenoides Okam.: Xisha (=Paracel Islands). (122)

PHYLLOPHORACEAE

Ahnfeltia furcellata Okam.: Hainan. (122)
Gymnogongrus flabelliformis Harv.: Widely distributed along China's seashore. (17, 55, 67, 122, 123)

SEBDENIACEAE

Sebdenia agardhii (De-Toni) Codomier: Hong Kong. (121)

GIGARTINACEAE

Chondrus ocellatus Holm.: Zhejiang; Fujian; Taiwan. (55, 122)
Gigartina intermedia Sur.: Widely distributed along China's seashore. (55, 67, 91, 122, 123)
G. pacifica Kjellm.: Fujian. (60)
G. teedii (Roth) Lamx.: Qingdao (Shandong). (91)
G. tenella Harv.: Yellow Sea; East China Sea; Fujian. (55, 67, 91, 122)

POLYIDEACEAE

Rhodopeltis borealis Yam: Taiwan. (143)

RHIZOPHYLLIDACEAE

Portieria hornemannii (Lyngb.) Silva [*Chondrococcus hornemannii*]: Hong Kong; Xisha (=Paracel Islands). (122, 123)

WURDEMARNIACEAE

Wurdemannia miniata (Dudy) Feldm. & Hamel: Dongshan (Fujian). (31, 54)

PALMARIALES
PALMARIACEAE

Rhodophysema georgii Batters [*Rhododermis georgii*]: Dalian (Liaoning). (55)

RHODYMENIALES
RHODYMENIACEAE

Botryocladia connexa Chang & Xia: Xisha (=Paracel Islands). (37)
B. pyriformis (Boerg.) Kylin: Xisha (=Paracel Islands). (35)
B. skottsbergii (Boerg.) Levring: Xisha (=Paracel Islands); Nansha (=Spratly Islands). (37, 70, 122)
Chrysymenia wrightii (Harv.) Yamada: Dalian (Liaoning); Qingdao, Yantai, and Weihai (Shandong); Zhejiang. (55, 67, 88, 91, 122)
Coelarthrum boergesenii W.-v. B. [*C. coactum*]: Xisha (=Paracel Islands). (35, 122)
Coelothrix irregularis (Harv.) Boerg: Xisha (=Paracel Islands). (108)
Cryptarachne okamurai (Yamada & Segawa) Zhang & Xia: Xisha (=Paracel Islands). (37, 122)
Erythrocolon podagricum J. Ag. in Grunow: Eluanbi (Taiwan); Guanghua Reef (Xisha=Paracel Islands). (67, 122, 123)
Rhodymenia intricata (Okam.) Okam.: Yellow Sea; Zhejiang; Fujian; Hong Kong. (67, 122, 123)

CHAMPIACEAE

Champia bififda Okam.: Zhejiang seashore. (55)
C. japonica Okam.: Fujian; Guangdong. (122)
C. parvula (C. Ag.) Harv.: Widely distributed along China's seashore. (55, 122)
Gastroclonium xishaensis Chang & Xia: Xisha (=Paracel Islands). (122)
Lomentaria catenata Harv. in Perry: Zhejiang. (55)
L. hakodatensis Yendo: Qingdao and Yantai (Shandong); Zhoushan, Shengsi, and Nanji Island (Zhejiang). (17, 67, 88, 91, 122)
L. pinnata Segara: Zhejiang. (55)

CERAMIALES
CERAMIACEAE

Aglaothamnion callophyllidicola (Yam.) Boo, Lee, Rueness & Yoshida [*Callithamnion callophylloidicola*]: Zhejiang seashore. (67)
Anotrichium tenue (J. Ag.) Naegeli [*Griffithsia tenuis*]: Dongshan (Fujian); Hainan; Xisha (=Paracel Islands). (62, 122)
Antithamnion cruciatum (Ag.) Naeg.: Beidaihe (Hebei); Zhoushan and Shengsi (Zhejiang). (55, 67, 91)
A. defectum Kylin: Shandong seashore. (15, 122)
A. hubbsii Dawson: Qingdao (Shandong). (15)
A. nipponicum Yamada & Inagaki: Zhoushan and Shengsi (Zhejiang). (67)
A. pylaisei (Mont.) Kjellm: Fujian. (60)
Antithamnionella elegans (Berthold) Price & John [*A. breviramosa*]: Qingdao (Shandong). (16)
A. sarniensis Lyle: Qingdao (Shandong). (16)
A. spirographidis (Schif.) Wollaston: Qingdao (Shandong). (16)
Callithamnion corymbosum (Smith) Lyngb.: Liaoning; Beidaihe (Hebei); Shandong. (122)
Campylaephora crassa (Okam.) Nakamura [*Ceramium crassum*]: Zhejiang; Guangdong; Hong Kong. (55, 122)
C. hypnaeoides J. Ag.: Changhai, Lushun, and Dalian (Liaoning); Yantai and Qingdao (Shandong); Zhejiang. (55, 63, 84, 87, 88, 122)
Centroceras apioulatum Yamada: Xisha (=Paracel Islands). (62)
C. clavulatum (C. Ag.) Montagne: Zhejiang; Fujian; Hong Kong; Hainan; Xisha (=Paracel Islands). (55, 62, 67, 122, 123)
C. japonicum Itono: Xisha (=Paracel Islands). (62)
C. miniatum Yamada: Xisha (=Paracel Islands). (62)
Ceramium aduncum Nakamura: Zhejiang. (55)
C. boydenii Gepp: Bohai; East China Sea. (55, 63, 79, 84, 87, 88, 122)
C. byssoideum Harv.: Hong Kong; Zhongsha (=Mcclesfield Bank). (74, 76)
C. fastigiramosum Boo & Lee [*C. fastigiatum*]: Fujian. (60)
C. japonicum Okam.: Beidaihe (Hebei); Yantai (Shandong); Shengshan (Zhejiang); Fujian. (55, 67, 91, 122)
C. kondoi Yendo [*C. rubrum*]: Bohai; Yellow Sea. (55, 63, 79, 84, 87, 88, 122)
C. masonii Dawson: Fujian. (60)
C. nakamurae Dawson: Zhejiang. (55)
C. paniculatum Okam.: Zhejiang; Fujian. (17, 122)
C. tenerrimum (Mart.) Okam.: Widely distributed along China's seashore. (17, 91, 122, 124)
C. tenuissimum (Roth) J. Ag.: Shandong; Fujian; Taiwan; Hong Kong; Xisha (=Paracel Islands). (122)
Crouania attenuata (C. Ag.) J. Ag.: Fujian. (60)
Gordoniella yonakuniensis (Yamada & Tanaka) Itono [*Spermothammion yonakuniensis*]: Xisha (=Paracel Islands). (55)
Griffithsia coacta Okam.: Xisha (=Paracel Islands). (62)
G. japonica Okam.: Zhejiang. (55)
G. metcalfii Tseng: Hainan; Xisha (=Paracel Islands). (62, 122)
G. rhizophora Grunow ex Web.-v. Bos.: Hong Kong. (116)
G. subcylindrica Okam.: Hainan. (122)
G. tenuis Ag.: Hong Kong. (116)
Haloplegma duperreyi Mont.: Taiwan; Xisha (=Paracel Islands). (37, 122)
H. polyspora Chang & Xia: Xisha (=Paracel Islands). (37, 122)
Herpochondria elegans (Okam.) Itono [*Microcladia elegans*]: Dalian (Liaoning). (84)
Platythamnion yezoense Inagaki: Liaoning; Hebei; Shandong. (122)
Pleonosporium venustissimum (Mont.) De Toni: East China Sea. (110)

Spyridia filamentosa (Wulf.) Harv.: China's seashore. (91, 122, 123)
Tiffaniella suychiroi (Okam.) Kaneko [*Spermothammion suychiroi*]: Zhejiang. (55)
Wrangelia argus (Mont.) Montagne: Xisha (=Paracel Islands). (37, 122)
W. hainanensis Tseng: Fujian; Guangdong. (122)
W. tagoi (Okam.) Kamura & Segawa in Jegawa: Xisha (=Paracel Islands); Nansha (=Spratly Islands). (37, 70, 122)
W. tayloriana Tseng: Fujian. (37, 122)

DELESSERIACEAE

Acrosorium polyneurum Okam.: Zhejiang. (67)
A. reptans Kylin: Pingtan (Fujian). (58, 60)
A. venulosum (Zanard.) Kylin [*A. unicinatum*]: Zhejiang; Fujian. (55, 122)
A. yendoi Yamada: Yellow Sea; Zhejiang; Hong Kong. (17, 55, 67, 88, 122, 124)
Caloglossa chuanshiensis Chen & Zheng: Fujian. (26)
C. leprieurii (Mont.) J. Ag.: Zhejiang; Fujian; Guangdong; Hainan; Hong Kong. (55, 122, 124)
C. ogasawaraensis Okam.: Hong Kong. (124)
Erythroglossum pinnatum Okam.: Zhejiang. (67)
E. repens Okam.: Zhejiang. (17, 67)
Hypoglossum attenuatum Gard: Xisha (=Paracel Islands). (35, 122)
H. geminatum Okam.: Fujian seashore. (14)
Martensia denticulata Harv.: Pingtan and Jinjiang (Fujian). (60)
M. fragilis Harv.: Xisha (=Paracel Islands). (108)
Phycodrys fimbriata (De La Pyl. ex J. Ag.) Kylin: Zhejiang. (55)
P. radicosa (Okam.) Yamada & Inagaki in Yamada: Yellow Sea shoreline; Zhejiang. (17, 55, 91, 103, 122)
Taenioma nanum (Kuetz.) Papernf. [*T. macrourum*]: Hong Kong. (123)
T. perpusillum (J. Ag.) J. Ag.: Hong Kong. (123)
Tsengiella spinulosa Zhang & Xia: Dalian (Liaoning). (120)
Zellera tawallina Martens [*Claudea batanensis*]: Eluanbi and Lu Island (Taiwan); Xisha (=Paracel Islands). (34, 122)

DASYACEAE

Dasya mollis Harv.: Xisha (=Paracel Islands). (37)
D. pedicellata C. Ag.: Xiamen (Fujian). (14, 115)
D. rigidula (Kuetz.) Ardisson: Fujian. (60)
D. scoparia Harv. ex J. Ag.: Fujian; Guangdong. (122)
D. sessilis Yamada: Zhejiang. (55)
D. villosa Harv.: Liaoning; Shandong; Zhejiang. (55, 67, 88, 91, 122)
Dasyopsis pilosa Weber-van Bosse: Xisha (=Paracel Islands). (108)
Heterosiphonia japonica Yamada: Western shore of Yellow Sea; Zhejiang. (17, 55, 67, 88, 122)
H. pulchra (Okam.) Falk: Zhejiang; Fujian. (55, 67, 122)

RHODOMELACEAE

Acanthophora muscoides (L.) Bory: Dongshan (Fujian); Guangdong. (54, 122)
A. spicifera (Vahl.) Boerg. [*A. orientalis*]: Taiwan; Guangdong; Hong Kong; Hainan; Xisha (=Paracel Islands). (35, 87, 122, 123)
Acrocystis nana Zan: Taiwan; Hainan. (122)
Amansia glomerata C. Ag.: Hainan. (122)
Bostrychia binderi Harv.: Hong Kong; Hainan. (118, 122)
B. hongkongensis Tseng: Hong Kong. (118, 122)
B. intricata (Bory) Mont: Hong Kong. (118)
B. kelanensis Grun.: Hong Kong. (118, 139)
B. radicans (Mont.) Mont. in Kuetz.: Hong Kong. (118)
B. simpliciuscula Harv. ex J. Ag.: Hong Kong. (118)
B. tenella (Lamx.) J. Ag.: Hong Kong; Hainan. (122, 123)
Chondria armata (Kuetz.) Okam.: Taiwan; Hainan. (122)
C. crassicaulis Harv.: China's seashore. (17, 55, 67, 88, 115, 122)
C. dasyphylla (Woodward) C. Ag.: Zhejiang; Fujian. (67, 115)
C. expansa Okam.: Xisha (=Paracel Islands). (35)
C. lancifolia Okam.: Hong Kong. (124)
C. repens Boberg. [*C. hapteroclada*]: Hong Kong; Xisha (=Paracel Islands). (35, 124)
C. ryukyuensis Yam.: Pingtan and Jinjiang (Fujian). (60)
C. succulenta (J. Ag.) Fkbg.: Beidaihe (Hebei); Qingdao (Shandong). (91)
C. tenuissima (Good. & Wood.) C. Ag.: Western shore of Yellow Sea; Zhejiang seashore. (55, 67, 88)
C. xishaensis Chang & Xia: Xisha (=Paracel Islands). (35)
Digenea simplex (Wulf.) C. Ag.: Taiwan; Dongsha (=Pratas Islands). (122)
Endosiphonia clavigera (Wolny) Fkbg.: Xisha (=Paracel Islands). (122)
Herposiphonia basilaris Zhang & Chen: Fujian (including Pingtan). (66)
H. caespitosa Tseng: Hong Kong. (123)
H. fissidentoides (Holm.) Okam.: Fujian. (60)
H. fujianensis Zhang & Chen: Fujian. (66)
H. insidiosa (Grev.) Fkbg.: Xiamen (Fujian). (115)
H. parca Setch. [*H. tenella*, *H. terminalis*]: Zhejiang; Xiamen (Fujian); Hong Kong; Xisha (=Paracel Islands). (55, 115, 123)
H. pecten-veneris (H.) Fkbg.: Hong Kong. (123)
H. ramosa Tseng: Hong Kong. (123)
H. secunda (C. Ag.) Anbronn.: Pingtan (Fujian). (60)
H. subdisticha Okam.: Fujian. (60)
Janczewskia ramiformis Chang & Xia: Shandong. (122)
Laurencia articulata Tseng: Hong Kong; Xisha (=Paracel Islands). (38, 124)
L. capituliformis Yamada: Penglai (Shandong). (91)
L. cartilaginea Yamada: Hong Kong; Xisha (=Paracel Islands). (38, 124)
L. changii Xia & Bartsch.: Nansha (=Spratly Islands). (1a)
L. chinensis Tseng: Hong Kong. (122, 124)
L. composita Yamada: Zhejiang. (67)

L. decumbens Kuetz.: Xisha (=Paracel Islands). (38)
L. flexilis setch v. *tropica* (Yam.) Xia & Zhang: Taiwan; Hainan. (122)
L. galtsoffii Howe: Xisha (=Paracel Islands). (38)
L. hongkongensis Tseng, Chang & Xia: Hong Kong. (122, 123, 124)
L. intermedia Yamada: Western shore of Yellow Sea; northern East China Sea. (88, 91, 122)
L. jejuna Tseng: Hong Kong. (119, 122)
L. longicaulis Tseng: Hong Kong. (119, 122)
L. majuscula (Harv.) Lucas: Hong Kong; Xisha (=Paracel Islands); Nansha (=Spratly Islands). (23, 38, 122, 123)
L. mariannensis Yamada: Xisha (=Paracel Islands). (38)
L. microcladia Kuetz.: Pingtan (Fujian). (60)
L. nipponica Yamada: Yantai (Shandong). (91)
L. obtusa (Hudson) Lamx.: Beidaihe (Hebei); Yantai (Shandong); Hong Kong; Xisha (=Paracel Islands); Nansha (=Spratly Islands). (23, 38, 91, 122, 124)
L. obtusa v. *divaricata* Ag.: Beidaihe (Hebei). (91)
L. okamurai Yamada [*L. japonica*]: China's seashore. (91, 110, 122, 124)
L. paniculata (Ag.) J. Ag.: Fujian; Hong Kong. (119, 122)
L. papillosa (Forsk.) Grev.: Xiamen (Fujian). (111, 115)
L. parvipapillata Tseng: Hong Kong; Xisha (=Paracel Islands); Nansha (=Spratly Islands). (23, 38, 123)
L. pinnata Yamada: Shengsi and Nanji Islands (Zhejiang); Xisha (=Paracel Islands). (17, 38)
L. silvai Zhang & Xia [*L. fasciculata*]: Xisha (=Paracel Islands). (38, 122)
L. subsimplex Tseng: Hong Kong. (120)
L. surculigera Tseng: Hong Kong. (120)
L. tenera Tseng: Hong Kong. (120)
L. tristicha Tseng, Chang & Xia: Hong Kong. (122, 124)
L. undulata Yamada: Hong Kong. (120, 122, 124)
L. venusta Yamada: Beidaihe (Hebei). (91)
L. verticillata Zhang & Xia: Xisha (=Paracel Islands). (122)
Levellea jungermannioides (Mert. & Her.) Harv.: Taiwan; Hong Kong; Hainan; Dongsha (=Pratas Islands); Xisha (=Paracel Islands). (34, 122)
Polysiphonia abscissa Hook. & Harv.: Zhejiang; Fujian. (68)
P. coacta Tseng: Hong Kong. (123)
P. crassa Okam.: Zhejiang. (68)
P. decumbens Segi.: Zhejiang; Fujian. (68)
P. elongata (Huds.) Sprengel: Zhejiang; Fujian. (68)
P. ferulacea Suhr. ex J. Ag.: Hong Kong. (123)
P. fragilis Sur.: Zhejiang; Fujian. (68)
P. gracilis Tseng: Hong Kong. (123)
P. harlandii Harv.: Taiwan; Hong Kong; Hainan. (122, 123)
P. japonica Harv. [*P. akkeshiensis*]: Beidaihe (Hebei); Yantai, Weihai, and Qingdao (Shandong); Zhejiang; Fujian. (17, 55, 67, 88, 122)
P. mollis Hook. & Harv.: Shengsi (Zhejiang); Fujian. (55, 68)
P. mollis v. *tongatensis* Harvey & Kuetzing: East China Sea. (28)
P. morrowii Harv.: Yantai and Qingdao (Shandong); western shore of Yellow Sea. (88, 91, 122)
P. patens Harv.: Zhejiang; Fujian. (68)
P. richardsoni Hook.: Zhejiang; Fujian. (68)
P. savatieri Hariot: Zhejiang; Fujian; Hong Kong. (68, 123, 124)
P. senticulosa Harv.: Zhejiang; Fujian. (68)
P. simplex Hollenb.: Zhejiang; Fujian. (68)
P. spiralis Batten: Fujian (including Pingtan). (58, 60)
P. subtilissibma Mont.: East China Sea shoreline; Hong Kong. (68, 124)
P. upolensis (Grun.) Holl.: Zhejiang; Fujian. (68)
P. urceolata Grev.: Weihai and Qingdao (Shandong); western shore of Yellow Sea; East China Sea. (17, 55, 67, 68, 88, 91, 103, 122)
P. yendoi Segi: Zhejiang; Fujian. (68)
Pterosiphonia pennata (Roth.) Fkbg.: Xiamen and Huian (Fujian). (60, 115)
Rhodomela confervoides (Huds.) Silva [*R. subfusca*]: Beidaihe (Hebei); western shore of Yellow Sea. (91, 122)
Sinosiphonia elegans Tseng & B. L. Zheng: Shandong. (122)
Symphyocladia latiuscula (Harv.) Yam. [*S. gracilis*]: Beidaihe (Hebei); western shore of Yellow Sea; Zhejiang. (17, 55, 67, 88, 91, 122)
S. marchantioides (Harv. in Hook.) Fkbg. in Engler & Prantl: Zhejiang; Fujian; Taiwan. (17, 55, 67, 122)
S. pennata Okam.: Zhejiang. (55, 67)
Tolypiocladia calodictyon (Harv.) Kuetz.: Xisha (=Paracel Islands). (108)
T. glomerulata (C. Ag.) Schmitz in Engler & Prantl: Hainan; Dongsha (=Pratas Islands); Xisha (=Paracel Islands). (34, 122)

PHAEOPHYTA

ECTOCARPALES
ECTOCARPACEAE

Acinetospora crinita (Carm. ex Harv. in Hook.) Korn.: Dalian (Liaoning). (72)
Ectocarpus acutus S. & G.: Yellow Sea. (72)
E. arctus Kuetz. [*E. confervoides*]: Seashore from Liaoning to Xiamen (Fujian). (55, 67, 91, 115, 122)
E. commensalis S. & G.: Dalian (Liaoning). (72)
E. corticulatus Saund.: Dalian (Liaoning). (72)
E. rallsiae Vickers: Dalian (Liaoning). (72)
E. rhodochortonoides Boerg.: Hong Kong. (123)
E. siliculosus (Dillw.) Lyngb. [*Giffordia indica*]: Yellow Sea; East China Sea; Xiamen (Fujian); Hong Kong. (55, 91, 115, 122, 124)
E. simplex Crouan: Fujian. (60)
E. tenellus Noda: Yellow Sea. (4)
E. terminalis Kütz: Yellow Sea. (4)
E. tomentosus (Huds.) Lyngb.: Yellow Sea. (4)
E. yezoensis Yam. & Tan.: Dalian (Liaoning). (72)
Feldmannia indica (Sonder) Womersley & Bailey [*Ectocarpus indicus*]: Shengshan (Zhejiang). (55, 67)

F. irregularis (Kuetz.) Hamel [*Giffordia irregularis*]: Yellow Sea. (122)
Giffordia acuto-ramuli (Noda) Luan & Luan: Dalian (Liaoning). (72)
G. fuscata (Zanard.) Kuck.: Dalian (Liaoning). (72)
G. sandriana (Zanardini) Hamel: Dalian (Liaoning). (72)
Hincksia breviarticulata (J. Ag.) Silva [*E. breviarticulatus*]: Zhoushan Islands (Zhejiang); Taiwan; Guangdong; Xisha (=Paracel Islands). (21, 67, 122)
H. granulosus (J. E. Smith) Silva [*Ectocarpus granulosus*]: Fujian. (60)
H. mitchellae (Harv.) Silva [*Giffordia mitchellae*]: Taiwan; Xisha (=Paracel Islands). (21, 122)

SOROCARPACEAE

Botrytella micromora Bory [*Sorocarpus micromorus, S. uvaeformis*]: Qingdao (Shandong). (122)
Polytretus reinboldii (Reinke) Sauv. [*Ectocarpus reinboldii*]: Yellow Sea. (122)
Sorocarpus pacifica Holl: Yellow Sea. (4)
Streblonema anomalum S. & G.: Dalian (Liaoning). (72)
S. bohaiense R. X. Luan: Dalian (Liaoning). (72)
S. codii Borton: Yellow Sea. (4)
S. corymbiferum S. & G.: Dalian (Liaoning). (72)
S. crasscaule R. X. Luan: Dalian (Liaoning). (72)
S. fasciculatum Thuret in Le Jolis: Yellow Sea. (4)

RALFSIALES
RALFSIACEAE

Heteroralfsia saxicola (Okam. & Yamada) Kawai [*Saundersella saxicola*]: Liaoning; Shandong. (122)
Ralfsia fungiformis (Grun.) S. & G.: Fujian. (60)
R. verrucosa (Aresch.) J. Ag.: Qingdao (Shandong); Zhejiang; Hong Kong. (55, 67, 122, 124)

CHORDARIALES
LEATHESIACEAE

Leathesia difformes (L.) Aresch.: Qingdao and Yantai (Shandong); Zhejiang. (55, 67, 122)
L. nana S. & G.: Yellow Sea. (122)
L. saxicola Takamatsu: Yellow Sea. (122)
Petrospongium rugosum (Okam.) S. & G. [*Cylindrocarpus rugosus*]: Guangdong; Hong Kong. (122)

ELACHISTACEAE

Elachista fucicola (Vell.) Aresh.: Yellow Sea shoreline. (122)
Halothrix lumbricalis (Kuetz.) Reinke: Yellow Sea shoreline. (88, 122)

MYRIONEMATACEAE

Myrionema tenue Noda & Honda in Honda & Noda: Dalian (Liaoning). (72)

CHORDARIACEAE

Eudesme virescens (Carm. ex Harv. in Hooker) J. Ag.: Qingdao (Shandong). (88, 103, 122)
Papenfussiella kuromo (Yendo) Inagaki [*Myriocladia kuromo*]: Qingdao (Shandong); Zhejiang. (55, 91, 122)
Sphaerotrichia firma (Gepp) A. Zin. [*Chordaria firma*]: Yellow Sea shoreline. (91, 122)
Tinocladia crassa (Sur.) Kylin: Yellow Sea shoreline. (122)

ACROTHRICACEAE

Acrothrix pacifica Okam & Yamada: Liaoning; Qingdao (Shandong). (91, 122)

SPERMATOCHNACEAE

Nemacystus decipiens (Sur.) Kuck.: Yellow Sea shoreline. (63, 87, 122)

ISHIGEACEAE

Ishige okamurai Yendo: Zhejiang; Guangdong; Hong Kong. (17, 55, 67, 122, 124)
I. sinicola (S. & G.) Chihara [*I. foliacea*]: Zhejiang; Fujian; Guangdong; Hong Kong. (17, 55, 67, 122, 124)

SPOROCHNALES
SPOROCHNACEAE

Sporochnus radiciformis (R. Brown ex Turn.) C. Ag.: Fujian. (122)

DICTYOSIPHONALES
ASPEROCOCCACEAE

Myelophycus simplex (Harv.) Papernf [*M. caespitosus*]: Lushun and Dalian (Liaoning). (32, 122)

DICTYOSIPHONACEAE

Dictyosiphon foeniculaceus (Huds.) Grev.: Common species in Yellow Sea. (122)

PUNCTARIACEAE

Punctaria latifolia Grev.: Lushun and Dalian (Liaoning); Qingdao (Shandong); Zhejiang. (91, 122)

P. plataginea (Roth) Grev.: Western shore of Yellow Sea. (91, 92, 122)

SCYTOSIPHONALES
CHNOOSPORACEAE

Chnoospora implex J. Ag.: Taiwan; Hainan; Xisha (=Paracel Islands). (122)

C. minima (Hering) Papenfuss: Taiwan; Hainan. (122)

SCYTOSIPHONACEAE

Colpomenia bullosa (Saunders) Yamada [*C. bullosa*]: Qingdao (Shandong). (88, 103, 122)

C. sinuosa (Mertens ex Roth) Derb. & Sol.: Widely distributed from Bohai to South China Sea. (17, 55, 67, 88, 122)

Endarachne binghamiae J. Ag.: Zhejiang; Fujian; Taiwan; Guangdong; Hong Kong. (17, 55, 67, 122, 124)

Hydroclathrus clathratus (Bory) Howe: Taiwan; Hong Kong; Hainan; Xisha (=Paracel Islands). (110, 122, 124)

H. tenuis Tseng & Lu: Xisha (=Paracel Islands). (122)

Petalonia fascia (O. F. Muellêr) Kuntze: Dalian (Liaoning); Qingdao (Shandong); Zhejiang; Fujian. (17, 55, 67, 88, 91, 122)

P. zosterifolia (Reinke) Kuetz.: Shandong. (122)

Rosenvingea intricata (J. Ag.) Boerg.: Hong Kong; Hainan. (122)

R. orientalis (J. Ag.) Boerg.: Taiwan; Guangdong; Guangxi; Hainan. (122)

Scytosiphon dotyo Wynne: Zhejiang. (55)

S. lomentarius (Lyngb.) J. Ag.: Widely distributed along China's seashore. (17, 55, 67, 84, 87, 88, 91, 122, 124)

DICTYOTAKES
DICTYOTACEAE

Dictyopteris divaricata (Okam.) [*Haliseris divaricata, Neurocarpus divaricatus*]: Lushun and Dalian (Liaoning); Weihai and Qingdao (Shandong); Zhejiang. (17, 55, 67, 84, 88, 91, 122)

D. latiuscula (Okam.) Okam.: Zhejiang; Fujian; Guangdong. (17, 55, 122)

D. prolifera (Okam. in De Toni & Okam.) Okam.: Zhejiang; Fujian. (122)

D. repens (Okam.) Boerg.: Taiwan; Guangdong; Xisha (=Paracel Islands); Nansha (=Spratly Islands). (22, 23, 122)

D. undulata Holm.: Liaoning; Fujian; Taiwan; Guangdong. (85, 122)

Dictyota cervicornis Kuetz: Hainan; Xisha (=Paracel Islands); Nansha (=Spratly Islands). (22, 23, 122)

D. dichotoma (Huds.) Lamx.: Bohai; Shandong; Zhejiang; Fujian; Hong Kong; Hainan; Xisha (=Paracel Islands). (17, 21, 22, 23, 55, 91, 122)

D. divaricata Lamx.: Lushun and Dalian (Liaoning); Qingdao, Yantai, and Weihai (Shandong); Zhejiang; Fujian; Hong Kong; Xisha (=Paracel Islands). (22, 67, 88, 122, 124)

D. friabilis Lamx. Setch: Hainan; Xisha (=Paracel Islands); Nansha (=Spratly Islands). (17, 21, 22, 23, 122, 124)

D. indica Sond. in Kuetz.: Yellow Sea shoreline; Zhejiang. (67, 88, 122)

D. linearis (C. Ag.) Grev.: Hainan. (122)

D. liqulata Kuetz.: Qingdao (Shandong). (91)

D. patens J. Ag.: Guangdong; Hong Kong; Xisha (=Paracel Islands). (22, 122)

Dilophus okamurae Dawson [*D. maginatus*]: Zhejiang; Fujian; Taiwan. (55, 122)

Lobophora variegata (Lamx.) Womer [*Pocockiella variegata*]: Taiwan; Hong Kong; Xisha (=Paracel Islands); Nansha (=Spratly Islands). (22, 23, 122, 124)

Pachydictyon coriaceum (Holm.) Okam.: Nanji Island (Zhejiang); Fujian; Hong Kong. (17, 55, 122, 124)

Padina arborescens Holm.: Fujian; Guangdong; Hong Kong. (122, 124)

P. australis Hauck: Taiwan; Guangdong; Hong Kong; Hainan; Xisha (=Paracel Islands). (122, 124)

P. boryana Thivy in Taylor [*P. commersonii, Dictyota radicans*]: Xiamen (Fujian); Taiwan; Hong Kong; Guangdong; Xisha (=Paracel Islands). (22, 111, 122, 124)

P. crassa Yamada: Qingdao; Nanji Island (Zhejiang); Hong Kong. (17, 55, 67, 88, 122, 124)

P. jonesii Tsuda: Xisha (=Paracel Islands). (22, 122)

P. minor Yamada: Taiwan; Guangdong; Xisha (=Paracel Islands). (22, 122)

P. tetrastromatica Hauck: Dongshan (Fujian); Hong Kong; Hainan; Dongsha (=Pratas Islands). (122, 124)

P. xishaensis Tseng & Lu: Xisha (=Paracel Islands). (22)

Spatoglossum dichotomum Tseng & Lu: Hong Kong. (122)

S. pacificum Yendo: Zhejiang; Fujian; Guangdong; Hong Kong. (55, 122)

S. solieri (Chauv. ex Mont.) Kuetz.: Fujian. (60)

Stypopodium zonale (Lamx.) Papebfuss: Xisha (=Paracel Islands). (22)

Zonaria coriacea Yamada: Hong Kong. (123)

Z. diesingiana J. Ag.: Fujian; Taiwan; Guangdong; Hong Kong. (122)

Z. nigrescens Sondr.: Xisha (=Paracel Islands). (108)

SPHACELARIALES
SPHACELARIACEAE

Sphacelaria carolinensis Trono: Taiwan; Xisha (=Paracel Islands). (21)

S. fusca (Hudson) C. Ag.: Zhejiang; Fujian; Hong Kong; Xisha (=Paracel Islands). (21, 55, 122)

S. novae-hollandiae Sonder: Fujian; Guangdong; Hong Kong; Xisha (=Paracel Islands). (21, 122, 124)

S. rigidula Kuetz. [*S. furcigera*]: Qingdao (Shandong); Zhejiang; Fujian; Guangdong; Hong Kong; Xisha (=Paracel Islands). (5, 7, 21, 91, 122, 124)

S. subfusca S. & G.: Qingdao (Shandong); western shore of Yellow Sea; Nanji Island (Zhejiang). (17, 67, 88, 122)

S. tribuloides Menegh.: Taiwan; Guangdong; Hong Kong; Xisha (=Paracel Islands). (21, 122, 124)

S. variabilis Sauv.: Fujian. (60)

DESMARESTIALES
DESMARESTIACEAE

Desmarestia viridis (Mueller) Lamour: Lushun and Dalian (Liaoning); Qingdao (Shandong); western shore of Yellow Sea. (88, 91, 122)

LAMINARIALES
CHORDACEAE

Chorda filum (L.) Stackh.: Bohai; Yellow Sea shoreline. (84, 85, 87, 122)

LAMINARIACEAE

Laminaria japonica Aresch. [*L. ochotensis*]: Bohai; Yellow Sea; East China Sea. (17, 55, 63, 67, 79, 84, 87, 88, 122)

ALARIACEAE

Ecklonia kurome Okam.: Zhoushan (Zhejiang); Pingtan and Putian (Fujian). (67, 122)

Undaria pinnatifida (Harv.) Suringar: Bohai and Yellow Sea shoreline; Zhoushan Islands and Nanji Island (Zhejiang). (17, 55, 67, 79, 84, 87, 88, 91, 122)

FUCALES
FUCACEAE

Pelvetia siliquosa Tseng & C. F. Chang [*P. minor*]: Bohai; Yellow Sea. (63, 79, 84, 87, 88, 122)

CYSTOSERIACEAE

Hormophysa articulata Kuetz. [*Hormosiar articulata*]: Hainan; Dongsha (=Pratas Islands); Xisha (=Paracel Islands). (122)

Myagropsis myagroides (Mert. ex Turn.) Fensholt [*Cystophyllum caespitosum*]: Dalian and Lushun (Liaoning). (88, 91, 122)

SARGASSACEAE

Sargassum angustifolium (Turn.) C. Ag.: Guangdong; Hong Kong. (122, 124)

S. assimile Harv.: Guangdong; Hong Kong. (122)

S. cinereum J. Ag.: Guangdong; Hong Kong. (122)

S. confusum Ag. [*S. pallidum*]: Bohai; Yellow Sea. Common species. (63, 79, 80, 82, 84, 87, 88, 122)

S. crassifolium J. Ag.: Taiwan; Guangdong; Xisha (=Paracel Islands). (93, 122)

S. dazhouense sp. nov.: China's coast. (99a)

S. dotyi Trono.: China's coast. (99a)

S. duplicatum J. Ag.: Guangdong; Hainan; Xisha (=Paracel Islands); Nansha (=Spratly Islands). (23, 92, 93, 122)

S. emarginatum Tseng & Lu: Xisha (=Paracel Islands). (92)

S. fulvellum (Turn.) C. Ag. [*S. enerve*]: Guangdong; Hong Kong. (122, 124)

S. fusiforme (Harv.) Setch: Liaoning; Shandong; Fujian; Guangdong; Hong Kong. (17, 55, 67, 79, 84, 88, 91, 122, 124)

S. glaucescens J. Ag.: Hong Kong; Xisha (=Paracel Islands). (93, 122)

S. graminifolium (Journ.) J. Ag.: Zhejiang; Guangdong; Hong Kong. (17, 55, 122)

S. hemiphyllum (Turn.) C. Ag.: East China Sea; Hong Kong; South China Sea. (17, 55, 67, 122, 124)

S. henslowianum C. Ag.: Fujian; Guangdong; Hong Kong. (122)

S. herklotsii Setchell: Guangdong; Hong Kong; Hainan. (122)

S. horneri (Turn.) C. Ag.: Liaoning; Zhejiang; Fujian; Guangdong; Hong Kong. (17, 55, 67, 122, 124)

S. ilicifolium (Turn.) C. Ag.: Taiwan; Hainan; Xisha (=Paracel Islands). (92, 122)

S. kuetzingii Setchell: Guangdong; Hong Kong; Guangxi; Xisha (=Paracel Islands). (92, 122)

S. laxifolium Tseng & Lu: Hong Kong. (95)

S. longifructum Tseng & Lu: Guangdong. (95)

S. maclurei Setch.: Guangdong; Hong Kong; Hainan. (122)

S. megalocystum sp. nov.: China's coast. (99a)

S. microcanthum (Kuetz.) Endlicher: Yantai (Shandong). (91)

S. miyabei Yendo [*S. kjellmaniamum*]: Yellow Sea shoreline; Zhejiang; Fujian; Guangdong; Hong Kong. (17, 55, 63, 67, 84, 87, 88, 122)

S. naozhouense Tseng & Lu: Nanji Island (Zhejiang). Guangdong. (95)

S. nigrifoloides Tseng & Lu: Nanji Island (Zhejiang). (67, 94).

S. parvivesiculosum Tseng & Lu: Xisha (=Paracel Islands). (93, 122)

S. patens C. Ag.: Fujian; Guangdong; Hong Kong. (122, 124)

S. phyllocystum Tseng & Lu: Hainan; Xisha (=Paracel Islands). (93, 122)

S. ploycystum C. Ag.: Taiwan; Guangdong; Guangxi; Hainan; Xisha (=Paracel Islands); Nansha (=Spratly Islands). (23, 92, 122)

S. siliquastrum (Turn.) Ag. [*S. macrocarpum, S. serratifolia, S. tortile*]: Liaoning; Zhejiang; Fujian; Guangdong; Hong Kong. (55, 83, 87, 88, 91, 122)

S. subtilissimum Tseng & Lu: Xisha (=Paracel Islands). (80, 122)

S. swartzii (Turn.) C. Ag.: Guangdong; Hong Kong; Guangxi. (122)

S. tenerrimum J. Ag.: Guangdong; Hong Kong. (122)

S. thunbergii (Mert. ex Roth) O. Kuetz.: China's seashore. Common species. (17, 55, 61, 67, 79, 84, 87, 88, 122)

S. turbinatifolium Tseng & Lu: Xisha (=Paracel Islands). (93, 122)

S. vachellianum Grev.: Zhejiang; Fujian; Guangdong; Hong Kong. (17, 55, 67, 122, 124)

S. xishaense Tseng & Lu: Xisha (=Paracel Islands). (93, 122)

Turbinaria conoides (J. Ag.) Kuetz.: Taiwan; Hainan; Xisha (=Paracel Islands). (122)

T. ornata (Turn.) J. Ag.: Taiwan; Hainan; Xisha (=Paracel Islands); Nansha (=Spratly Islands). (23, 110, 122)

T. parvifolia Tseng & Lu: Xisha (=Paracel Islands). (122)

T. reialata Kuetz.: Hong Kong. (124)

CHLOROPHYTA

TETRASPORALES
COLLINSIELLACEAE

Collinsiella cava (Yendo) Printz.: Yellow Sea. (122)
C. tuberculata Setch. & Gardn.: Yellow Sea; Fujian. (60, 122)

CHLORANGIACEAE

Prasinocladus lubricus Kuck: Fujian. (26)

ULOTRICHALES
ULOTRICHACEAE

Ulothrix flaca (Dillw.) Thur. in Le Jolis: Widely distributed along China's seashore. (17, 62, 63, 79, 84, 122)
U. implexa Kuetz. [*U. pseudoflacca*]: Fujian. (24, 60)

ULVALES
MONOSTROMATACEAE

Monostroma angicava Kjellm: Bohai; Yellow Sea. Common species. (87)

M. arcticum Wittr.: Dalian and Changhai (Liaoning); Yantai, Weihai, and Qingdao (Shandong). (63, 79, 84, 87, 122)

M. crassifolia Tseng & C. F. Chang: Zhejiang. (55)

M. latissimum (Kütz) Wittr.: Hong Kong. (123)

M. nitidum Wittr.: Zhejiang; Fujian; Taiwan; Hong Kong; Hainan. (17, 87, 122, 123)

Protomonostroma undulatum (Wittr.) Vinogradova [*Monostroma undulatum*]: Fujian. (60)

KORMANNIACEAE

Kormannia leptoderma (Kjellm.) Bliding: East China Sea. (110)

ULVACEAE

Blidingia micrococea (Kuetz.) Tseng & Chang [*Enteromorpha micrococea*]: Lushun and Dalian (Liaoning). (36)

B. minima (Naeg. & Kuetz.) Kylin: Yellow Sea; East China Sea. (122)

Enteromorpha clathrata (Roth) Grev. emend Bliding: Common species along China's seashore. (17, 62, 63, 79, 84, 122, 123)

E. compressa (L.) Grev.: Common species along China's seashore. (17, 63, 68, 79, 122, 123)

E. crinita (Roth) J. Ag.: Hong Kong. (124)

E. cruciata Collins: Hainan. (74)

E. erecta (Lyngb.) J. Ag.: Xiamen (Fujian). (12)

E. flexuosa (Wulf.) J. Ag.: Hong Kong; South China Sea. (14, 123)

E. intestinalis (L.) Link: Common species along China's seashore. (17, 63, 70, 79, 84, 88, 122)

E. kylinii Bliding: Yellow Sea. (91)

E. lingulata J. Ag.: South China Sea. (74)

E. linza (L.) J. Ag. [*Ulva linza*]: Common species along China's seashore. (17, 67, 79, 122, 123)

E. plumosa Kuetz: Yellow Sea. (91)

E. prolifera (Muell.) J. Ag.: Widely distributed along China's seashore. (17, 63, 67, 68, 70, 79, 87, 122, 123)

E. salina Kuetz v. *polyclados* Kuetz: Yellow Sea. (91)

E. spiralis Tseng & C. F. Chang: East China Sea. (79)

E. torta (Mert.) Rbd.: Hainan. (74)

E. tubulosa Kuetz: East China Sea; Hong Kong; South China Sea. (17, 91, 122)

Ulva conglobata Kjellm: Common species along China's seashore. (17, 63, 67, 68, 79, 84, 87, 122, 123)

U. fasciata Delile: Fujian; Taiwan; Guangdong; Hong Kong. (122, 123)

U. lactuca L.: Widely distributed along China's seashore. (17, 63, 67, 79, 84, 87, 122, 123)

U. lens Grou.: East China Sea. (67)

U. pertusa Kjellm.: Bohai; Yellow Sea; East China Sea. (17, 63, 79, 84, 122)

U. reticulata Forssk.: Hong Kong. (123)

U. rigida C. Ag.: Yantai (Shandong). (88)

ACROSIPHONIALES
ACROSIPHONIACEAE

Spongomorpha arcta (Dillw.) Kuetz.: Yellow Sea. (122)
Urospora doliifera (Setch. & Gardn.) Doty: Yellow Sea. (122)
U. penicilliformis (Roth.) Aresch.: Yellow Sea. (88, 122)

CLADOPHORALES
CLADOPHORACEAE

Chaetomorpha aerea (Dillw.) Kuetz.: Lushun and Dalian (Liaoning); Qingdao (Shandong); Xiamen (Fujian). (63, 88, 115, 157)
C. aerea f. *versata* Heydr.: Qingdao (Shandong). (88)
C. antennina (Bory.) Kuetz.: Dongshan (Fujian); Hong Kong; South China Sea. (54, 121, 122)
C. brachygona Harv.: Hong Kong. (121)
C. linum (Mull.) Kuetz.: Yellow Sea; East China Sea; Xiamen (Fujian). (115, 122)
C. media (Ag.) Kuetz.: Zhejiang. (17)
C. pachynema (Mont.) Kuetz.: Xiamen (Fujian). (14)
C. spiralis Okam: East China Sea; Hainan. (67, 122)
Cladophora albida (Huds.) Kuetz.: East China Sea. (55, 67)
C. aokii Yam.: Hainan. (122)
C. boodleoides Boerg.: Xisha (=Paracel Islands). (13)
C. catenata (L.) Kuetz. [*C. fuliginosa*]: Taiwan; Hainan; Xisha (=Paracel Islands). (13, 55, 67, 110, 122)
C. cymopoliae Boerg.: Xisha (=Paracel Islands). (13)
C. delicatula Ment: Fujian; Hong Kong. (60, 123)
C. divergens Kjellm.: Hong Kong. (123, 124)
C. echinus (Bias.) Kuetz.: Xisha (=Paracel Islands). (13)
C. fascicularis (Mert. ex C. Ag.) Kuetz.: Widely distributed along China's seashore. (17, 122)
C. flexuosa (Griff.) Harv.: Yantai (Shandong); Hainan; Xisha (=Paracel Islands). (13, 55, 67)
C. glaucescens Harv.: Zhejiang. (55, 67)
C. gracilis Kuetz.: Xiamen (Fujian). (14)
C. montagnei Kuetz. v. *radicans* Yamada: Taiwan. (110)
C. ohkuboana Holm.: Fujian. (122)
C. oligoclada Harv.: Dalian (Liaoning). (73)
C. patentiramea (Mont.) Kuetz.: Taiwan; Xisha (=Paracel Islands). (13, 110, 122)
C. rudolphiana (C. Ag.) Kuetz.: Dalian (Liaoning). (73)
C. rugulosa Mart.: Hainan. (122)
C. saviniana Boerg.: Xisha (=Paracel Islands). (13)
C. scitula (Suhr.) Kuetz.: Taiwan. (110)
C. sibogae Reinb.: Taiwan; Xisha (=Paracel Islands). (13)
C. socialis Kuetz.: Xisha (=Paracel Islands). (13)
C. stimpsonii Harv.: Zhejiang. (67)
C. utriculosa Kuetz.: Zhejiang. (55, 67)
C. wrightiana Harv.: Lushun and Dalian (Liaoning). (91)
Rhizoclonium hookeri Kuetz.: Hong Kong. (123, 124)
R. implexum (Dillw.) Kuetz.: East China Sea; Fujian seashore. (55, 122)
R. riparium (Roth.) Harv.: Liaoning; Yellow Sea; Hong Kong. (50, 53)

DASYCLADALES
DASYCLADACEAE

Acetabularia calyculus Lamx. in Quoy & Gaimard: Guangdong. (122)
A. clavata Yamada: Xisha (=Paracel Islands). (100)
A. parvula Solms-Laubach [*A. möbii*]: Guangdong; Hainan; Xisha (=Paracel Islands). (100, 122)
A. tsengiana Egerod: Xisha (=Paracel Islands). (35)
Bornetella nitida Munier-Chalmas ex Sonder: Xisha (=Paracel Islands); Nansha (=Spratly Islands). (23, 35, 122)
B. oligospora Solms-Laubach: Guangdong. (122)
B. sphaerica (Zanardini) Solms-Laubach: Hainan. (122)
Neomeris annulata Dickie: Guangdong; Xisha (=Paracel Islands). (100, 122)
N. bilimbata Koster: Hainan; Xisha (=Paracel Islands); Nansha (=Spratly Islands). (23, 122)
N. van-bosseae Howe: Guangdong; Hainan; Xisha (=Paracel Islands); Nansha (=Spratly Islands). (11, 50)

SIPHONOCLADIALES
VALONIACEAE

Dictyosphaeria bokotensis Yamada: Dongshan (Fujian); Taiwan; Xisha (=Paracel Islands); Nansha (=Spratly Islands). (23, 42, 54, 110)
D. cavernosa (Forssk.) Boerg.: Fujian; Taiwan; Guangdong; Hong Kong; Xisha (=Paracel Islands). (42, 54, 57, 122, 123)
D. fujianenesis Chen & Chou: Fujian. (60)
D. intermedia Weber-Van Bosse: Hainan. (42)
D. spinifera Tseng & C. F. Chang: Guangdong; Hainan; Xisha (=Paracel Islands). (42, 77, 122)
D. versluysii Webber-Van Bosse: Guangdong; Hainan; Xisha (=Paracel Islands). (23, 42, 122)
Valonia aegagropila C. Ag.: Taiwan; Guangong; Hainan; Xisha (=Paracel Islands); Nansha (=Spratly Islands). (23, 42, 122)
V. fastigiata Harv. ex J. Ag.: Taiwan. (105)
V. utricularis (Roth) C. Ag. [*Conterva utricularis*]: Taiwan; Guangdong; Xisha (=Paracel Islands); Nansha (=Spratly Islands). (23, 42, 122)
Ventricaria ventricosa (J. Ag.) Olsen & West [*Valonia ventricosa*]: Guangdong; Xisha (=Paracel Islands); Nansha (=Spratly Islands). (23, 42, 122)

SIPHONOCLADACEAE

Boergesenia forbesii (Harv.) Feldm.: Taiwan; Xisha (=Paracel Islands). (13, 50)

Clodophoropsis herpestica (Mont.) Howe: Shi Island (Shandong); Xisha (=Paracel Islands). (42, 122)
C. sundanesis Rbd.: Hainan. (122)
C. vaucheriaeformis (Aresch.) Papenf. [*Spongocladia vaucheriaeformis*]: Xisha (=Paracel Islands). (122)
C. zollingeri (Kuetz.) Rbd. [*C. fasciculatus*]: Fujian; Taiwan; Guangdong; Hong Kong. (122, 123)
Siphonocladus tropicus J. Ag.: Hainan. (105)
S. xishaensis C. F. Chang & B. M. Xia: Xisha (=Paracel Islands). (30, 42, 122)

BOODLEACEAE

Boodleopsis aggregata Tseng & Dong: Fujian; Taiwan; Guangdong; Hong Kong. (122)
Boodlea coacta (Dickie) Murr. & De Toni: Fujian; Taiwan. (105)
B. composita (Harvey & Hooker) Brand [*Cladophora composita*]: Fujian; Taiwan; Guangdong; Hong Kong. (122, 123)
B. montagnei (Harv. ex Gray) Egerod: Taiwan. (105)
B. paradoxa Rbd.: Tainan (Taiwan). (110)
B. siamensis Rbd.: Taiwan. (110)
B. struveoides Howe: Yongxing Island (Xisha=Paracel Islands). (42)
B. vanbosseae Reinbold [*Cladophora montagnei*]: Xiamen (Fujian); Yongxing Island (Xisha=Paracel Islands). Low intertidal. (42, 110)
Struvea anastomosans (Harv.) Picc & Grun. ex Picc. [*S. delicatula*]: Taiwan; Hong Kong; Hainan; Xisha (=Paracel Islands). (42, 122, 123)
S. intermedia C. F. Chang & E. Z. Xia: Xisha (=Paracel Islands). (42, 122)

ANADYOMENACEAE

Anadyomene wrightii Gray: Taiwan; Guangdong. (42, 122)
Microdictyon japonicum Setch. [*Rhipidiphyllin reticulatum*]: Taiwan; Guangdong; Hainan; Xisha (=Paracel Islands). (42, 122)
M. nigrescens (Yam.) Setch.: Taiwan. (105)
M. okamurae Setch.: Taiwan; Guangdong; Xisha (=Paracel Islands). (42, 122)
M. pseudohapteron A. & E. S. Gepp: Xisha (=Paracel Islands). (42, 122)
Valoniopsis pachynema (Mart.) Boerg.: Taiwan; Guangdong; Xisha (=Paracel Islands). (23, 42, 77, 122)

CADIACEAE
BRYOPSIDACEAE

Bryopsis arbuscula Ag.: East China Sea. (110)
B. caespitosa Shur.: East China Sea. (55, 67)
B. corticulans Setch.: Yellow Sea; East China Sea. (55, 67, 122)
B. harveyana J. Ag.: Taiwan; Guangdong; Xisha (=Paracel Islands). (101, 122)
B. hypnoides Lamx.: Bohai; Yellow Sea; East China Sea. (55, 67, 88, 91)
B. muscosa Lamx.: Hong Kong. (123)
B. pennata Lamx.: Bohai; East China Sea. (55, 67, 91)
B. plumosa (Huds.) C. Ag.: Widely distributed along China's seashore. (55, 67, 91, 122)
Pseudobryopsis hainanensis Tseng: Hainan. (122)

CAULERPACEAE

Caulerpa brachypus Harv.: Xisha (=Paracel Islands). (100)
C. cupressoides (Vohl) C. Ag.: Hong Kong; Hainan; Xisha (=Paracel Islands). (101, 122)
C. cupressoides (Vahl) Ag. v. *lycopodium* W. -v. B. f. *amicorum* (Harv.) W. -v. B.: Hong Kong. (123)
C. nummularia Harv. ex J. Ag. [*C. peltata* v. *mummularia*]: Hainan; common species in Xisha (=Paracel Islands). (101, 122)
C. okamurae W. -v. Bosse: Dongshan (Fujian). (54, 122)
C. peltata Lamx. [*C. racemosa* v. *peltata*, *C. peltata* Lamx. v. *typica*]: Taiwan; Hong Kong; Hainan; South China Sea; Xisha (=Paracel Islands). (101, 122, 124)
C. prolifera (Forsk.) Lamx. v. *xishaensis* v. nov.: Nansha (=Spratly Islands). (27a)
C. racemosa (Forssk.) W.-v. Bos. v. *clavifera* (Turn.) W. -v. Bos.: Dongshan (Fujian); Taiwan; Xisha (=Paracel Islands). (54, 122)
C. racemosa (Forssk.) J. Ag. v. *occidentalis* (J. Ag.) Boerg.: Hainan; Xisha (=Paracel Islands). (23)
C. serrulata (Forssk.) J. Ag.: Taiwan; Hainan; Xisha (=Paracel Islands); Nansha (=Spratly Islands). (23, 100, 122)
C. sertularioides (Gmel.) Howe f. *longiseta* (Bory) Svedelius: Hong Kong; Hainan; Xisha (=Paracel Islands). (122, 123, 124)
C. taxifolia (Vahl.) C. Ag. [*C. taxifolia* (Vahl.) C. Ag. f. *typica* Sved]: Hong Kong; Hainan; Xisha (=Paracel Islands). (100, 122, 124)
C. verticillata J. Ag.: Hainan. (122)
C. webbiana Mont.: Dongshan (Fujian); Taiwan; Xisha (=Paracel Islands). (54, 100)

DERBESIACEAE

Derbesia marina (Lyngb.) Kjellm: Dongshan (Fujian). (54)
Pedobesia ryukyuensis (Yam. & Tan.) Kobara & Chihara [*Derbesia ryukyuensis*]: Fujian. (64)

VAUCHCRIACEAE

Pseudodichomosiphon constriata Mada: Dongshan (Fujian). (54, 90)

CODIACEAE

Avrainvillea amadelpha (Mont.) A. & E. S. Gepp: South China Sea. (102)

A. erecta (Berkl.) A. & E. S. Gepp: Hainan. (122)

A. lacerata Harv. ex J. Ag.: Hainan; Xisha (=Paracel Islands); Nansha (=Spratly Islands). (122)

A. obscura (C. Ag.) J. Ag. [*A. capituliformis*]: Hainan. (102)

A. ryukyuensis Yam.: Fujian. (60)

A. xishaensis Tseng & Dong: South China Sea. (102)

Chlorodesmis hildebrandtii A. & E. S. Gepp: Hainan. (122)

C. sinensis Tseng & Dong: Xisha (=Paracel Islands). (100, 122)

Codium adhaerens (Cabr.) C. Ag.: Hong Kong. (123)

C. arabicum Kuetz: Taiwan; Hainan; Xisha (=Paracel Islands); Nansha (=Spratly Islands). (23, 101, 122)

C. barlettii Tseng & Dong: Hainan. (122, 123)

C. cylindricum Holm.: Fujian; Guangdong; Hong Kong. (122, 123)

C. divaricatum Holm.: Qingdao (Shandong); East China Sea. (91, 122)

C. fragile (Sur.) Heriot [*C. mucronatum*]: Bohai; Yellow Sea; East China Sea. Widely distributed species. (17, 55, 67, 91, 122)

C. geppii O. C. Schmidt: Hainan; Xisha (=Paracel Islands); Nansha (=Spratly Islands). (23, 101, 122)

C. intricatum Okam.: Hong Kong. (123, 124)

C. ovale Zan.: Jinqing Island (Xisha=Paracel Islands). (100)

C. papillatum Tseng & Gilb v. *hainanense* Tseng: Guangdong; Hainan. (122)

C. repens Crouan frat. in Vicker: Taiwan; Guangdong; Xisha (=Paracel Islands). (101, 122)

C. taitense Setchell [*C. geppei*]: Xisha (=Paracel Islands). (101)

Geppella echinocaulos Cribb: Xisha (=Paracel Islands). (101)

G. japonica Tanaka & Itono: Xisha (=Paracel Islands). (101)

G. prolifera Tseng & Dong: Xisha (=Paracel Islands). (122)

Halimeda cylindraceae Dec.: Nansha (=Spratly Islands). (23)

H. discoidea Decaisne: Taiwan; Xisha (=Paracel Islands); Nansha (=Spratly Islands). (23, 75, 122)

H. gigas W. R. Taylor: Nansha (=Spratly Islands). (23)

H. incrassata (Ellis.) Lamx.: Hainan; Xisha (=Paracel Islands); Nansha (=Spratly Islands). (23, 122)

H. macroloba Dec.: Hainan; Xisha (=Paracel Islands). (75, 122)

H. micronesica Yam.: Xisha (=Paracel Islands); Nansha (=Spratly Islands). (23, 122)

H. opuntia (Linn.) Lamx.: Hainan; Xisha (=Paracel Islands); Nansha (=Spratly Islands). (75, 122)

H. opunita (L.) Lamx. f. *triloba* (Dec.) Barton: Hainan; Xisha (=Paracel Islands). (75, 122)

H. simulans Howe: Nansha (=Spratly Islands). (23)

H. taenicola Taylor: Xisha (=Paracel Islands); Nansha (=Spratly Islands). (23, 75, 122)

H. velasquezii Taylor: Xisha (=Paracel Islands); Nansha (=Spratly Islands). (23, 75, 122)

H. xishaensis Tseng & M. L. Dong: Xisha (=Paracel Islands). (75, 122)

Tydemania expeditionis Web. -v. Bos.: Xisha (=Paracel Islands); Nansha (=Spratly Islands). (23, 100, 122)

Udotea argentea Zanardini v. *spumosa* A. & E. S. Gepp: Nansha (=Spratly Islands). (23)

U. flabellum (Ell. & Sol.) Howe: Hainan; Nansha (=Spratly Islands). (23, 122)

U. fragilifolia Tseng & Dong: Xisha (=Paracel Islands). (99, 122)

U. javensis (Mont.) A. & E. S. Gepp: Hainan; Xisha (=Paracel Islands). (100, 122)

U. reniformis Tseng & Dong: Xisha (=Paracel Islands). (99, 122)

U. tenax Tseng & Dong: Xisha (=Paracel Islands). (99, 122)

U. tenuifolia Tseng & Dong: Xisha (=Paracel Islands). (99, 122)

U. velutina Tseng & Dong: Xisha (=Paracel Islands). (99, 122)

U. xishaensis Tseng & Dong: Xisha (=Paracel Islands). (99, 122)

CHLOROCOCCALES
SCENEDESMACEAE

Scenedesmus brasilieasis Bohl.: Nansha (=Spratly Islands). Planktonic species. (56)

S. javensis Chod.: Nansha (=Spratly Islands). Planktonic species. (56)

S. quadricauda (Turp.) Breb.: Nansha (=Spratly Islands). Planktonic species. (56)

REFERENCES*

(1). Wang Sujuan and Zhang Jingrong, 1980. A new species of *Porphyra* from China—*Porphyra monosporangia* Wang & Zhang. Oceanologia & Limnologia Sinica, 11(2):141–149.

(1a). Wang, Y. Q. and B. M. Xia, 1995. Studies on some marine algae of Nansha Islands, Hainan Province, China, II. Proceedings of Meeting of the Chinese Phycology Society No.3. (abstract)

(2). Wang Sujuan and Zhang Jingyong, 1981. On *Porphyra pseudocruspata*—a new species from China. Journal of Xiamen Fisheries College, 1:27–34.

(3). Zhang Jingrong and Wang Sujuan, 1993. A new species of *Porphyra* from China—*Porphyra fujianensis* sp. nov. Oceanologia & Limnologia Sinica, 24(4):356–359.

(4). Wang Shubo and Luan Rixiao, 1984. A preliminary study on Ectocarpaceae along the coast of Yellow Sea; China. In: Proceedings on the Fourth Workshop of Oceanology and Limnology Society.

(5). Hua Maosen, 1978. Taxonomic studies on marine blue-green algae of Xisha Islands, I. Studia Marina Sinica, 12:59–66.

(6). Hua Maosen, 1981. A new species of the genus *Borzia* from Xisha Islands. Oceanol. Limnol. Sinica, 12(3):265–269.

(7). Hua Maosen, 1983. Studies on some blue-green algae of Xisha Islands, II. Studia Marina Sinica, 20:55–67.

(8). Hua Maosen and Tseng Chengkuei, 1985. Studies on some marine blue-green algae of Xisha Islands, III. Studia Marina Sinica, 24:1–9.

(9). Hua Maosen and Tseng Chengkuei, 1985. Studies on some marine blue-green algae of Xisha Islands, IV. Studia Marina Sinica, 24:11–26.

(10). Hua Maosen and Tseng Chengkuei, 1985. Studies on some blue-green algae of Xisha Islands, V. Studia Marina Sinica, 24:27–50.

(11). Chu Chiayen and Wang Suchuan, 1960. Study on *Porphyra dentimarginata* sp. nov. Acta Botanica Sinica, 9(1):37–41.

(12). Chu Haolan, 1959. Studies on microscopic marine algae from Yellow Sea; I. Blue algae. Journal of Nanjing University, 2:1–22.

(13). Chu Haolan and Liu Xuexian, 1980. Studies on Cladophoraceae from Xisha Islands. Studia Marina Sinica, 17:11–20.

(14). Chu Zhongjia, 1981. The Marine Algae of Xiamen. Journal of Xiamen Fisheries College, 1:35–45.

(15). Liu Jianhua, 1984. Studies on *Antithamnion nageli* of Qingdao. Journal of Shandong College of Oceanology, 14(2):40–47.

(16). Liu Jianhua, 1984. Taxonomic study on *Antithamnion lyle* of Qingdao. Journal of Shandong College of Oceanology, 14(4):68–72.

(17). Sun Jianzhang and Hang Jinxin, 1976. A preliminary survey on benthic algae around Nanji Islands. Acta Phytotaxonomic Sinica, 14(1):51–55.

(18). Li Weixin, 1980. Studies on the reproductive organs of two species of *Liagroa*. Marine Science, 3:25–28.

(19). Li Weixin and Liu Sijian, 1980. Economic seaweeds of Wanshan Islands. Journal of Zhanjiang Fisheries College, 2:11–17.

(20). Li Minren et al., 1984. Laboratory culture of a new species of planktonic blue-green algae—*Trichodesmium maccii* from Qingdao coast. Journal of Shandong College of Oceanology, 14(1):93–100.

(21). Lu Baoren and Tseng Chengkuei, 1980. Studies on some brown algae of Xisha Islands, I. Studia Marina Sinica, 17:21–35.

(22). Lu Baoren and Tseng Chengkuei, 1980. Studies on Dictyotaceae of Xisha Islands. Studia Marina Sinica, 24:69–86.

(23). Lu Baoren et al., 1991. Studies on some brown and green algae of Nasha Islands, I. Transactions of marine organisms nearby Nansha Islands (I). China Ocean Press, 5:1–4.

(24). Chen Zhenfen et al., 1982. Distribution of the intertidal algae in the mouth of Jiulongjiang River. Journal of Oceanography in Taiwan Strait, 1(2):91–97.

(25). Chen Yaqu, 1982. Analysis of the red tide of *Trichodesmium* in East China Sea in the year 1972. Journal of Fisheries 6(2):181–189.

(26). Chen Changsheng and Zhang Jinrong, 1988. An investigation on the resources of major economic seaweeds in the intertidal zone of Fujian Province. Fujian Fisheris, 4:34–42.

(27). Chen Changsheng and Zhang Jinrong, 1991. A study on marine algae in the intertidal zone of Fujian Province, I. The species and its distribution. Journal of Xiamen Fisheris College, 13(2):26–35.

(27a). Wu, X. C., B. L. Lu, and C. K. Tseng, 1995. Studies on *Caulerpa* (green algae) of Nansha Islands, Hainan Province, China. Proceedings of Meeting of the Chinese Phycology Society No. 2. (abstract)

(28). Chang, C. F. and B. M. Xia, 1962. A preliminary phytogeographical study on the Chinese species of *Gracilaria*. Oceanologia & Limnologia, 4(3–4):189–198.

(29). Chang, C. F. and B. M. Xia, 1963. A new genus of Gracilariaceae—*Polycavernosa*. Studia Marina Sinica, 3:119–128.

(30). Chang, C. F. and B. M. Xia, 1964. A comparative study of *Gracilaria foliifera* (Forssk) Boergs. and *Gracilaria textorii* (Suring) De Toni. Acta Bot. Sinica, 12(2):201–209.

(31). Chang, C. F. and B. M. Xia, 1876. Studies on Chinese species of *Gracilaria*. Studia Marina Sinica, 11:91–166.

(32). Chang, C. F. and B. M. Xia, 1978. Studies on some red algae of Xisha Islands, I. Studia Marina Sinica, 12:27–40.

(33). Chang, C. F. and B. M. Xia, 1978. A new species of *Gastroclonium* from Xisha Islands. Oceanologia & Limnologia Sinica, 9(2): 209–214.

(34). Chang, C. F. and B. M. Xia, 1979. Studies on some red algae of Xisha Islands, II. Studia Marina Sinica, 15:21–47.

(35). Chang, C. F. and B. M. Xia, 1980. Studies on some red algae of Xisha Islands, III. Studia Marina Sinica, 17:49–70.

(36). Chang, C. F. and B. M. Xia, 1980. Two new species of *Laurencia* from Xisha Islands. Oceanol. & Limnol. Sinica, 11(3):267–274.

(37). Chang, C. F. and B. M. Xia, 1983. Studies on some marine red algae of Xisha Islands, IV. Studia Marina Sinica, 20: 123–140.

(38). Chang, C. F. and B. M. Xia, 1985. Studies on the genus *Laurencia* of Xisha Islands. Studia Marina Sinica, 24:51–68.

(39). Chang, C. F. and B. M. Xia, 1985. On *Gracilaria asiatica* sp. nov. and *Gracilaria verrucosa* (Huds.) Papenfuss. Oceanologia & Limnologia Sinica, 16(3):175–179.

(40). Chang, C. F. and B. M. Xia, 1985. A new genus of Delesseriaceae—*Tsengiella*. Proceedings of the Secong Chinese Phycological Symposium.

(41). Chang, C. F., E. Z. Xia and B. M. Xia, 1963. A comparative study of *Hypnea musciformis* (Wulf.) Lamouroux and *Hypnea japonica* Tanaka. Oceanologia & Limnologia Sinica, 5(1):35–45.

(42). Chang, C. F., E. Z. Xia and B. M. Xia, 1978. Taxonomic studies on Siphonocladales of Xisha Islands. Studia Marina Sinica, 10:20–60.

(43). Chang, C. F. and E. Z. Xia, 1981. A new species of *Sarconema* from Hainan Island. Oceanologia & Limnologia Sinica, 12(3):259–265.

(44). Chang, C. F. and E. Z. Xia, 1986. On *Gelidium vagum* Okam. and *G. grubbae* Fan. Oceanologia & Limnologia Sinica, 17(6):521–525.

(45). Chang Tejui and Zheng Baofu, 1962. The Chinese *Porphyra* and their geographical distribution. Oceanologia & Limnologia Sinica, 4(3–4):183–188.

(46). Chang Tejui and B.F. Zheng, 1960. *Porphyra haitanensis*, a new species of *Porphyra* from Fujia. Acta Bot. Sinica, 9(1):32–36.

(47). Chang Tejui and Zhou Jinhua, 1978. Studies on Corallinaceae of Xisha Islands, I. Studia Marina Sinica, 12:17–26.

(48). Chang Tejui and Zhou Jinhua, 1980. Studies on Corallinaceae of Xisha Islands, II. Studia Marina Sinica, 12:71–76.

(49). Chang Tejui and Zhou Jinhua, 1980. Studies on Corallinaceae of Xisha Islands, III. Oceanologia & Limnologia Sinica, 11(4): 351–357.

(50). Chang Tejui and Zhou Jinhua, 1981. *Ramicrusta*, a new genus of Peyssonneliaceae. Oceanologia & Limnologia Sinica. 12(6): 538–544.

(51). Chang Tejui and Zhou Jinhua, 1985. Studies on Corallinaceae of Xisha Islands, IV. Studia Marina Sinica, 24:39–50.

(52). Chang Tejui and Zhou Jinhua, 1989. On the taxonomy of some species of Northern China *Lithothamnium*. Studia Marina Sinica, 30:93–102.

(53). Chang Tejui and Zhou Jinhua. The "Hai Fu Shi" from Huangxian of Shandong Province—*Lithothamnium* spp. Proceedings of the First Phycological Symposium.

(54). Zhang Shuijin, 1981. A preliminary study on rocky intertidal algae ecology in Dongshan County and its adjacent islets, Fujian. Acta Ecologica Sinica, 1(4):361–368.

(55). Hang Jinxin and Sun Jianzhang, 1983. The Pictorial Handbook of Marine Algae from Zhejiang Coast. Zhejiang Science Press.

(56). Lin Yongshui and Lin Qiuyan, 1991. Distribution feature of phytoplankton in the waters around Nansha Islands. Transactions of Marine Organisms nearby Nansha Islands (II). China Ocean Press, pp:66–87.

(57). Zhou Nansheng, 1950. The marine algae from Dongshan Island, Fujian. Journal of Xiamen Fisheries College, 1(3):19–28.

(58). Chou Chenying and Chen Zhuohua, 1965. The marine algae of Pingtan Island. Journal of Fujian Teachers College, 1:1–12.

(59). Chou Chenying and Chen Zhuohua. On studies of Chinese species of *Galaxaura*. Proceedings of the First Chinese Phycological Symposium.

(60). Chou Chenying and Chen Zhuohua, 1983. A list of marine algae from Fujian Coast. Journal of Oceanography in Taiwan Strait, 2(1):91–102.

(61). Zhou Jinhua and Zhang Derui, 1991. Studies on corallinaceae of Nansha Islands. Transactions of Marine Organisms nearby Nansha Islands (I). 5:15–19. China Ocean Press.

(62). Zheng Bolin, 1980. Studies on Ceramiaceae of Xisha Islands, I. Studia Marina Sinica, 17:37–48.

(63). Zheng Bolin and Yao Gendi, 1960. The economic marine algae from Yellow Sea and Bohai Sea. Journal of Shandong College of Oceanology, 1:120–210.

(64). Zheng Bailin et al., 1985. A study of the benthic algae of Mati Reef in Qingdao. Journal fo Shandong College of Oceanology, 15(2):46–53.

(65). Zheng Baofu, 1981. A new species of *Porphyra* from China—*Porphyra oligospermatangia* sp. nov. Oceanologia & Limnologia Sinica, 12(5):447–451.

(66). Zheng Yi and Chen Zhuhua, 1992. Two new species of *Herposiphonia* (Rhodophyta) from the coast of Fujian. Oceanologia & Limnologia Sinica, 23(1):95–100.

(67). Xiang Siduan, 1979. The marine algae of Zhejiang. Journal of Hangzhou University, 54–60.

(68). Xiang Siduan, 1979. Studies on the *Polysiphonia* from Zhejiang. Journal of Hangzhou University, 53–54.

(69). Xia, E. Z. 1963. A preliminary phytogeopraphical study on Chinese species of *Eucheuma*. Oceanologia & Limnologia Sinica, 5(1):52–55.

(70). Xia, E. Z. and B. M. Xia, 1991. Studies on some marine red algae of Nansha Islands, I. Transactions of Marine Organisms nearby Nansha Islands (I). 5:20 24. China Ocean Press.

(71). Guo Yujie and Yang Zheyu, 1982. Studies on the ecology of phytoplankton from the continental shelf of East China Sea in the year 1976. Oceanologia & Limnologia Sinica, 19:11–32.

(72). Luan Rixiao and Luan Shujun, 1990. Studies on new recorded species of Chinese Ectocarpaceae II. Proceedings of the Third Phycological Symposium, No. 17.

(73). Luan Shujun, 1990. Studies on *Cladophora* from Dalian. Proceedings of the Third Chinese Phycological Symposium, No. 16.

(74). Dong Meiling, 1963. A preliminary phytogeographical study on Chinese species of *Enteroporpha*. Oceanologia & Limnologia Sinica, 5(1):46–51.

(75). Dong Meiling and Tseng Chengkuei, 1980. Studies on marine green algae from Xisha Islands, I. Studia Marina Sinica, 17:1–10.

(76). Jiang Fukang, 1982. Studies on the marine algae from Huangyan Island. Reports of the Comprehensive Survey on South China Sea, pp:279–292.

(77). Jiang Fukang and Shan Guozhao, 1981. The community structure of the marine algae on Porites in Zhaoshu Island. South China Sea Studia Marina Sinica, 2:93–99.

(78). Tseng Chengkuei and T. J. Chang, 1958. On *Porphyra marginata* sp. nov. and its systematic position. Acta Bot. Sinica, 7(1):15–25.

(79). Tseng Chengkuei and T. J. Chang, 1952. The economic seaweeds of North China. Journal of Shandong University, 2:57–82.

(80). Tseng Chengkuei and T. J. Chang, 1952. Studies on Chinese *Sargassum*, I. *Sargassum confusum* Ag. Acta Bot. Sinica, 3(2): 235–260.

(81). Tseng Chengkuei and T. J. Chang, 1952. On the distribution of *Pelvetia siliquosa* Tseng & C. F. Chang. Acta Bot. Sinica, 2(2):280–292.

(82). Tseng Chengkuei and T. J. Chang, 1954. Studies on Chinese *Sargassum*, II. *Sargassum kjellmanianum* Yendo. Acta Bot. Sinica, 4:353–370.

(83). Tseng Chengkuei and T. J. Chang, 1958. On the geographical distribution of *Pelvetia siliquosa* Tseng & C. F. Chang. Oceanologia & Limnologia Sinica, 1(2):209–215.

(84). Tseng Chengkuei and T. J. Chang, 1959. On the economic marine algae flora of Yellow Sea and East China Sea. Oceanologia & Limnologia Sinica, 2(1):43–52.

(85). Tseng Chengkuei and T. J. Chang, 1959. On the discontinuous distribution of some brown algae along China coast. Oceanologia & Limnologia Sinica, 2(2):86–92.

(86). Tseng Chengkuei and T. J. Chang, 1959. On the regional division of the marine algae of Western North Pacific. Oceanologia & Limnologia Sinica, 2(4):244–267.

(87). Tseng Chengkuei and T. J. Chang et al., 1962. Economic Seaweeds of China. China Science Press.

(88). Tseng Chengkuei and T. J. Chang, 1962. An analytical study of the marine algae flora of the western Yellow Sea Coast, I. Oceanologia & Limnologia Sinica, 4(1–2): 49–59.

(89). Tseng Chengkuei and T. J. Chang, 1962. Studies on Chinese Dictyosphaeria. Acta Bot. Sinica, 10(2):120–134.

(90). Tseng Chengkuei and T. J. Chang, 1963. A preliminary analytical study of the Chinese marine algae flora. Oceanologia & Limnologia Sinica, 5(3):245–253.

(91). Tseng Chengkuei and T. J. Chang, 1964. A critical review of the records of the benthic marine algae as reported from the western Yellow Sea coast. Studia Marina Sinica, 6:1–26.

(92). Tseng Chengkuei and Lu Baoren, 1978. Studies on the Sargassaceae of Xisha Islands, I. Studia Marina Sinica, 12:1–16.

(93). Tseng Chengkuei and Lu Baoren, 1979. Studies on the Sargassaceae of Xisha Islands, II. Studia Marina Sinica, 15:1–20.

(94). Tseng Chengkuei and Lu Baoren, 1985. On a new species of *Sargassum* from East China Sea—*Sargassum nigrifoloides* sp. nov. Oceanologia & Limnologia Sinica, 16(3):169–175.

(95). Tseng Chengkuei and Lu Baoren, 1987. Three new species of *Sargassum* (Fucales, Phaeophyta) form South China Sea. Oceanolgia & Limnologia Sinica, 18(6):515–520.

(96). Tseng Chengkuei and Lu Baoren, 1990. Studies on the Chinese *Sargassum*, III. Proceedings of the Third Chinese Phycological Symposium, No. 228.

(97). Tseng Chengkuei and Lu Baoren, 1990. Studies on the Chinese *Sargassum*, IV. Proceedings of the Third Chinese Phycological Symposium, No. 229.

(98). Tseng Chengkuei and Lu Baoren, 1990. Studies on the Chinese *Sargassum*, V. Proceedings of the Third Chinese Phycological Symposium, No. 230.

(99). Tseng Chengkuei and Dong Meilin, 1975. Some new species of *Udotea* from Xisha Islands. Studia Marina Sinica, 10:1–19.

(99a). Tseng, C. K. and B. L. Lu, 1995. Studies on *Sargassum* in China's seas: some new species and records. Proceedings of Meeting of the Chinese Phycology Society No. 3. (abstract)

(100). Tseng Chengkuei and Dong Meilin, 1978. Studies on some marine green algae from Xisha islands, I. Studia Marina Sinica, 12:41–52.

(101). Tseng Chengkuei and Dong Meilin, 1983. Studies on some marine green algae from Xisha Islands, III. Studia Marina Sinica, 20:109–122.

(102). Tseng Chengkuei and Dong Meilin. Studies on New and Unrecorded Species of *Avrainvillea* from South China Sea. Proceedings of The First Chinese Phycological Symposium. No. 73–74.

(103). Tseng Chengkuei and Zheng Bolin, 1954. Studies on marine algae of Qingdao. Acta Bot. Sinica, 3(1):105–123.

(104). Tseng Chengkuei and Zhang Derui, 1978. Two new species of *Porphyra* from China. Oceanologia & Limnologia Sinica, 9(1):78–80.

(105). Fan Kungchu, 1963. On the phytogeographical distribution of the Siphonocladales of China. Oceanologia & Limnologia Sinica, 5(2):165–171.

(106). Fan Kungchu et al., 1974. Studies on marine algae of Xisha Islands, I. *Ganonema*. Acta Phytotaxonomic Sinica, 12(4):489–495.

(107). Fan Kungchu et al., 1975. Studies on marine algae of Xisha Islands, II. New species and new records of the family Newalionaceae (Rhodophyta). Acta Phytotaxonomic Sinica, 13(2):71–77.

(108). Fan Kungchu et al. Studies on the marine algae of Xisha Islands, III. A list of marine algae from Jinyin Island and its vicinity. In: Collections of Papers on Marine Biological Investigation of Xisha Islands, Zhongsha Islands. China Science Press, pp:55–62.

(109). Pan Guoying and Wang Yungchuan. Tropic Oceanology. China Science Press.

(110). Okayama Kintarou, 1936. Seaweeds of Japan. Uchida Ro Kaku Ho, Tokyo, Japan.

(111). Cotton, A. D. 1915. Some Chinese marine algae. Kew Bull. Misc. Inform., 3:107–113.

(112). Abbott, I. A., Zhang Junfu and Xia Bangmet, 1991. *Gracilaria mixta* sp. nov. and other western Pacific species of the genus *Gracilaria* (Glaciariaceae, Rhodophyta). Pac. Sci., 45(1):12–27.

(113). Tseng, C. K. 1925. Economic seaweeds of Kwangtung Province, South China. Lingn. Sci. Jour., 14(1):93–104.

(114). Tseng, C. K. 1936. On marine algae new to China. Bull. Fan. Mem. Inst. Biol. (Bet), 7(5):169–196.

(115). Tseng, C. K. 1936. Notes on the marine algae from Amoy. Amoy Mar. Biol. Bull., 1(1):1–86.

(116). Tseng, C. K. 1940. Marine algae of Hong Kong, historical survey and list of recorded species. Jour. Hong Kong Fish. Res. Stat., 1(2):194–210.

(117). Tseng, C. K. 1942. Marine algae of Hong Kong, II. The genus *Catenella*. Washington Acad. Sci., 32(5):142–146.

(118). Tseng, C. K. 1943. Marine algae of Hong Kong, III. The genus *Bostrychia*. Papers of the Michigan Academy of Science, Arts and Letters, 28:165–183.

(119). Tseng, C. K. 1943. Marine algae of Hong Kong, IV. The genus *Laurencia*. Papers of the Michigan Academy of Science, Arts and Letters, 28:185–208.

(120). Tseng, C. K. 1944. Marine algae of Hong Kong, VI. The genus *Polysiphonia*. Papers of the Michigan Academy of Science, Arts and Letters, 29:67–82.
(121). Tseng, C. K., C. F. Chang, Enchan Xia and Bangmei Xia, 1980. Studies on some marine red algae from Hong Kong. The Marine Flora and Fauna of Hong Kong and Southern China. Hong Kong University Press.
(122). Tseng, C. K. 1983. Common Seaweeds of China. China Science Press.
(123). Hodgkiss, I. J. and K. Y. Lee, 1981. Checklist and keys to the marine algae of Hong Kong. Memoirs of the Hong Kong Natural History Society, No. 16.
(124). Hodgkiss, I. J. 1983. Hong Kong Seaweed. An Urban Council Publication.
(125). Zhang Junfu and Xia Bangmei, 1988. On two new *Gracilaria* (Gracilariaceae, Rhodophyta) from south China. In: Abbott, I. A. (ed.), Taxonomy of economic seaweeds: with reference to some Pacific and Caribbean species, Vol. 2. California Sea Grant College Program, La Jolla, California, USA, pp:131–136.
(126). Zhang Junfu and Xia Bangmei, 1988. Studies on two new *Gracilaria* from south China and a summary of *Gracilaria* species in China. In: Abbott, I. A. (ed.), Taxonomy of economic seaweeds: with reference to some Pacific and Caribbean species, Vol. 2. California Sea Grant College Program, La Jolla, California, USA, pp:137–139.

*: (1)–(109) in Chinese, (110) in Japanese
Compiled by Xu Zhiming, Jiang Jialun, Zhang Junfu (C. F. Chang), Chen Changsheng, Zhang Shuijin, and Yin Weiping; Edited by Zheng Chengkui (C. K. Tseng), Zhang Junfu (C. F. Chang), and Zhang Shuijin.

TRACHEOPHYTA

PTERIDOPHYTA

LYGODIACEAE

Lygodium microstachyum Desr.: Zhejiang; Fujian; Taiwan; Guangdong; Guangxi; Hainan; and inland area. Coastal rocky shore crevices. (5, 9, 11, 33, 48)

SCHIZAEACEAE

Schizaea dichotoma (L.) Sm.: Hainan. Coastal sandy shore. (40)

LINDSAEACEAE

Stenoloma biflora (Kaulf.) Ching: Fujian; Taiwan; Guangdong. Coastal rocky shore crevices or wall. (11, 36, 48)

GYMNOGRAMMACEAE

Pityrogramma calomelanos (L.) Link.: Guangdong; Hainan. Mangrove forest interior and inland area. (3, 9, 10, 11, 28)

THELYPTERIDACEAE

Cyclosorus interruptus (Willd.) H. Ito: Guangdong; Hainan. Under mangrove canopy and inland area. (3, 9, 11, 28)

BLECHNACEAE

Stenochlaena hainanensis Ching & Chieu: Hainan. Coastal bushes and inland area. (3, 8:3, 9, 11)
S. palustris (Burm. f.) Bedd.: Taiwan; Guangdong; Guangxi; Hainan. Coastal and inland area. (3, 8:3, 9, 14, 45)

DRYOPTERIDACEAE

Cyrtomium falcatum (L. f.) Presl.: Shandong; Zhejiang; Fujian; Taiwan; Guangdong. Coastal rocky shore crevices and inland area. (10, 11, 33, 36, 48)
Dryopteris decipiens (HK.) O. Ktze.: Fujian; Guangdong; Guangxi. Island shoreline and inland area. (11, 14, 48)

ACROSTICHACEAE

Acrostichum aureum L.: Fujian; Taiwan; Guangdong; Guangxi; Hainan. Intertidal mud flat or mangrove forest. (8:3, 9, 10, 24, 30, 32, 35, 45)
A. speciosum Willd.: Hainan. Intertidal mud flat or mangrove forest. (3, 11, 25, 35)

GYMNOSPERMAE

PINACEAE

Pinus elliottii Engelm.: Zhejiang; Fujian; Taiwan; Guangdong; Guangxi. Species for planting in coastal area. (11, 14, 33, 48)
P. massoniana Lamb.: South of mouth of Changjiang (Yangtze River) and inland area. Species for planting in coastal area. (10, 11, 48)
P. thunbergii Parl.: Eastern Liaoning Peninsula; Shandong; Jiangsu; Zhejiang; Fujian; Taiwan. Species for planting in coastal area. (10, 11, 33, 48)

ANGIOSPERMAE

DICOTYLEDONEAE
CASUARINACEAE

Casuarina cunninghamiana Miq.: Along the southeastern China coast. Sandy area, an species used for shoreline protection. (11, 33, 48)
C. equisetifolia L: Along the southeastern China coast. Sandy area, an ideal species used for shoreline protection. (1, 3, 8:20, 11, 33, 48)
C. glauca Sieb.: Along the southeastern China coast. A species used for shoreline protection. (11, 33, 48)

MORACEAE

Taxotrophis ilicifolius Vidal. [*T. macrophylla* Merr.]: Hainan and inland area. Sandy area. (9, 11, 46)

OLACACEAE

Ximenia americana L.: Hainan. Coastal sandy area. (3, 8:24, 9, 11, 12, 46)

POLYGONACEAE

Polygonum sibiricum Lamx.: Coastal and inland area north of Shandong. Coastal salt flat. (10, 11, 40)
Rumex japonicus Houtt.: Shandong; Jiangsu; Zhejiang; Fujian; Taiwan; Guangdong; Guangxi; and inland area. Coastal wetland or rock crevices. (11, 33, 48)

CHENOPODIACEAE

Arthrocnemum indicum (Willd.) Miq.: Hainan. Coastal sand flat. (9, 11, 17, 45)
Atriplex centralasiatica Iljin: Liaoning; Hebei; and inland area. Coastal salt flat. (10, 11, 17, 38)
A. maximowicziana Makino: Fujian; Hainan. Coastal sand flat, muddy sand, or shoreline. (29, 31, 33, 46, 48)
A. patens (Litw.) Iljin: Coastal and inland area north of Shandong. Coastal salt flat. (10, 11)
A. repens Roth.: Guangxi; Hainan. Coastal sand flat or open flat. (3, 9, 10, 14)
A. sibirica L.: Northeastern China; northern coast of China and inland area. Coastal salt flat. (10, 11)
Chenopodium acuminatum Willd. subsp. *virgatum* (Thunb.) Kitan.: Liaoning; Hebei; Jiangsu; Zhejiang; Fujian; Taiwan; Guangdong; Guangxi. Coastal sand flat and open grassy flat. (10, 11, 33, 48)
C. album L.: Along China's coast and inland area. Coastal sand flat or open flat. (10, 11, 13, 33, 34, 48)
C. ambrosioides L.: Jiangsu; Zhejiang; Fujian; Taiwan; Guangdong; Guangxi; and inland area. Coastal bank or open flat. (10,11, 33, 34, 48)
C. glaucum L.: Northeastern China; northern China coast; Jiangsu; Zhejiang; and inland area. Coastal salt flat. (10, 11, 47)
C. serotinum L.: Liaoning; Hebei; Shandong; Fujian; Taiwan; Guangdong; Guangxi; and inland area. Coastal sand flat or open flat. (10, 11, 14, 33, 48)
Corispermum puberulum Iljin: Northern China; northeastern coast of China; Jiangsu. Coastal bank. (11, 16, 18, 20)
Kochia scoparia (L.) Schrad.: Occur almost everywhere in China's coastal and inland area. Coastal bank. (10, 11, 17, 48)
Salicornia europaea L.: Liaoning; Hebei; Shandong; Jiangsu; Guangxi; and inland area. Coastal sand or mud flat. (10, 11, 14, 20, 38)

Salsola collina Pall.: Northeastern China coast; northern China; Jiangsu; and inland area. Coastal sandy bank. (10, 11, 16, 18)
S. komarovii Iljin: Liaoning; Hebei; Shandong; Jiangsu; Zhejiang. Coastal sandy area. (10, 11, 20)
S. pestifer A. Nelson: Liaoning; Hebei; Jiangsu; and inland area. Coastal sandy bank. (10, 11, 18)
Suaeda australis (R. Br.) Moq.: Fujian; Taiwan; Guangdong; Guangxi; Hainan. Mud or sand flat, mangrove forest fringe. (3, 4, 8:25, 9, 10, 24, 29, 31, 34, 48)
S. glauca Bunge: Northern China; northeastern China coast; Shandong; Jiangsu; Zhejiang; and inland area. Coastal salt flat. (10, 11, 17, 20, 22)
S. heteroptera Kitagawa [*S. salsa* (L.) Pall.]: Northestern China coast; Hebei; Jiangsu; Zhejiang; and inland area. Coatal salt flat. (10, 16, 17, 20, 22, 47)
S. laevissima Kitag.: Liaoning; Hebei. Coastal salt flat. (16, 17)
S. liaotungensis Kitag.: Liaoning. Coastal salt flat. (16)

AMARANTHACEAE

Achyranthes aspera L.: Fujian; Taiwan; Guangdong; Guangxi; Hainan; and inland area. Coastal bank, open area. (3, 4, 8:25, 10, 48)
Allmania nodiflora (L.) R. Br.: Hainan. Coastal sandy area. (3, 8:25, 9, 11)
Alternanthera pungens H. B. K.: Southern Fujian coast. Coastal muddy shore or sandy bank. (29, 31, 33, 48)
A. sessilis (L.) DC.: Coastal provinces south of Changjiang (=Yangtze River) and inland area. Coastal sandy wetland. (3, 4, 8:25, 10, 48)
Gomphrena celosioides Mart.: Taiwan; Guangdong; Hainan; and inland area. Coastal sandy area. (3, 8:25, 9, 11)
Trichurus monsoniae (L. f.) C. C. Townsed: Hainan. Coastal sand flat. (3, 8:25, 9, 11)

NYCTAGINACEAE

Boerhavia diffusa L.: Fujian; Taiwan; Guangdong; Guangxi; and inland area. Coast or open area. (3, 4, 9, 10, 36, 48)
B. erecta L.: Xisha (=Paracel Islands). Sandy area. (5)
Pisonia aculeata L.: Taiwan; Guangdong; Hainan. Bushes in coastal flat. (3, 9, 11, 42)
P. grandis R. Br.: Xisha (=Paracel Islands); Nansha (=Spratly Islands). Evergreen coppice. (7, 12)

AIZOACEAE

Gisekia pharnaceoides L.: Hainan. Coastal sandy area. (3, 8, 9, 11)
Mollugo costata Y. T. Chang & C. F. Wei.: Fujian; Taiwan; Guangdong; Guangxi; Hainan. Coastal sandy area. (3, 9, 12, 14)

M. nudicaulia Lamx.: Taiwan; Hainan. Coastal sandy area or open grassy area. (3, 9, 11, 12)

M. oppositifolia L.: Taiwan; Guangdong; Guangxi; Hainan. Coastal sandy area or open flat. (3, 9, 11, 12, 14)

M. pentaphylla L.: Coastal and inland area south of Shandong. Coastal sandy area. (3, 4, 9, 10, 11, 12)

Sesuvium portulacastrum L.: Fujian; Taiwan; Guangdong; Hainan. Sandy beach or mangrove canopy. (3, 9, 10, 31, 33, 45)

Tetragonia tetragonoides (Pall.) Kuntze: Jiangsu; Zhejiang; Fujian; Taiwan; Guangdong. Coastal sandy beach. (12, 29, 31, 33, 48)

Trianthema portulacastrum L.: Taiwan; Hainan; Xisha (=Paracel Islands). Coastal open sand flat. (9, 33, 48)

PORTULACACEAE

Portulaca pilosa L.: Fujian; Guangdong; Hainan; Xisha (=Paracel Islands). Coastal sandy area, mud bank, or rock crevices. (9, 33, 48)

CARYOPHYLLACEAE

Melandrium apricum (Turcz.) Rohrb.: Distributed all over China. Coastal sandy beach and inland area. (9, 11, 16, 31, 33)

Polycarpaea corymbosa (L.) Lamx.: Fujian; Guangdong; Guangxi; Hainan; and inland area. Coastal sandy area. (3, 4, 9, 10, 11, 14, 48)

P. gaudichaudii Gagnep.: Guangdong; Hainan. Coastal sandy area. (9, 12)

Spergularia salina J. & C. Presl.: China's coastal and inland area. Coastal sandy area, salty area, and mud bank. (6, 10, 11, 22, 33, 48)

MENISPERMACEAE

Coculus orbiculatus (L.) DC. [*C. tribolus*]: China's coastal and inland area (except for northwestern China). Bushes on rocky shore. (10, 11, 14, 33, 48)

LAURACEAE

Cassytha filiformis L.: Zhejiang; Fujian; Taiwan; Guangdong; Guangxi; and inland area. Coastal bush or sandy gravel area. (10 11, 14, 33, 36, 48)

Cinnamomum camphora (L.) Presl.: Coastal area south of Changjiang (=Yangtze River) and inland area. Planting species for coastal area. (8:31, 11, 14, 48)

Litsea glutinosa (Lour.) C. B. Rob.: Fujian; Guangdong; Guangxi; Hainan; and inland area. Coastal area. (3, 4, 9, 10, 11, 14)

L. rotundifolia Hemsl. v. *oblongifolia* (Nees) Allen: Zhejiang; Fujian; Taiwan; Guangdong; Guangxi; Hainan; and inland area. Rocky shore crevices. (9, 10, 14, 33, 48)

HERNANDIACEAE

Hernandia nymphaeifolia (Presl.) Kubitzki: Hainan. Coastal flat, forest. (13)

H. sonora L.: Taiwan; Hainan. Beach, coastal forest. (3, 8:31, 10, 11, 30, 36)

CRUCIFERAE

Coronopus didymus (L.) J. E. Smith: Shandong; Jiangsu; Zhejiang; Fujian; Taiwan; Guangdong; and inland area. Coastal bank or sandy gravel area. (10, 33, 36, 48)

NEPENTHACEAE

Nepenthes mirabilis (Lour.) Druce: Guangdong; Hainan. Wetland inside coastal sandy bank. (9, 10, 11, 48)

DROSERACEAE

Drosera burmannii Vahl: Fujian; Taiwan; Guangdong; Guangxi; Hainan; and inland area. Coastal wetland. (9, 10, 11, 14, 33, 48)

D. indica L.: Fujian; Taiwan; Guangdong; Guangxi; Hainan. Coastal wetland, coastal beach. (9, 10, 11, 14, 33, 48)

CRASSULACEAE

Bryophyllum pinnatum (Lamx.) Oken: Fujian; Taiwan; Guangdong; Guangxi; Hainan; and inland area. Coastal rocky crevices or sandy beach. (10, 11, 14, 33, 48)

PITTOSPORACEAE

Pittosporum tobira (Thunb.) Ait.: Coastal provinces south of Changjiang (=Yangtze River). Coastal rocky crevices, coastal forest. (10, 11, 33, 36, 48)

ROSACEAE

Rubus parvifolius L.: Almost all over China's coastal and inland area. Coastal rocky crevices. (10, 11, 14, 33, 48)

LEGUMINOSAE

Abrus precatorius L.: Taiwan; Guangdong; Guangxi; Hainan; and inland area. Coastal bank, thin forest, or bushes. (9, 10, 11, 14, 34)

Acacia auriculiformis A. Cunn. ex Benth.: Southern China. Coastal sandy area, planting species. (11, 14, 33, 48)

A. confusa Merr.: Fujian; Taiwan; Guangdong; Guangxi; Hainan. Planting species for coastal area. (11, 14, 33, 48)

Albizzia corniculata (Lour.) Druce: Fujian; Guangdong; Guangxi; Hainan. Coastal bank or inland area. (9, 10, 11, 14, 34, 48)

Alysicarpus vaginalis (L.) DC.: Fujian; Taiwan; Guangdong; Guangxi; Hainan. Coastal open area and inland area. (5, 9, 10, 14, 33, 34)

Amorpha fruticosa L.: Northern China; northeastern China's coast; Shandong; Jiangsu; and inland area. Shoreline protection forest (sandy area). (10, 11, 20)

Atylosia scarabaeoides (L.) Benth.: Fujian; Taiwan; Guangdong; Guangxi; and inland area. Coastal sandy gravel area or sandy bank. (10, 14, 33, 34, 48)

Caesalpinia crista L.: Fujian; Taiwan; Guangdong; Hong Kong; Guangxi; Hainan. Coastal open area and inland area. (1, 3, 9, 10, 14, 34, 36, 44, 48)

Canavalia cathartica Thou. [*C. microcarpa*]: Fujian; Taiwan; Guangdong; Guangxi; Hainan. Rocky shore or bushes. (9, 14, 34, 48)

C. maritima (Aubl.) Thou.: Fujian; Taiwan; Guangdong; Hong Kong; Guangxi; Hainan. Coastal beach or bank. (3, 9, 14, 33, 34, 44)

Derris trifoliata Lour.: Fujian; Taiwan; Guangdong; Hong Kong; Guangxi; Hainan. Intertidal wetland, mangrove forest. (3, 9, 10, 30, 33, 34, 44, 48)

Desmodium gangeticum (L.) DC.: Fujian; Taiwan; Guangdong; Guangxi; Hainan; and inland area. Coastal bank. (9, 10, 14, 34, 48)

D. podocarpum DC. ssp. *oxyphyllum* (DC.) Ohashi: Jiangsu; Zhejiang; Fujian; Guangdong; Guangxi; and inland area. Coastal sandy bank. (11, 14, 48)

D. rubrum (Lour.) DC.: Guangdong; Guangxi; Hainan. Coastal sandy area. (3, 9, 11, 14)

Indigofera hirsuta L.: Fujian; Taiwan; Guangdong; Hainan. Coastal sandy area. (9, 33, 34, 48)

Lathyrus maritimus (L.) Bigel.: Northestern China; Hebei; Shandong; Jiangsu; Zhejiang. Coastal sandy area. (10, 11, 16, 20, 40)

Leucaena leucocephala (Lamx.) de Wit [*L. glauca* (L.) Benth.]: Fujian; Taiwan; Guangdong; Guangxi; Hainan; and inland area. Islands. (9, 10, 11, 14, 33, 48)

Maackia australis (Dunn) Takeda: Guangdong; Hainan. Island bushes. (3, 11, 34)

Pongamia pinnata (L.) Merr.: Fujian; Taiwan; Guangdong; Guangxi; Hainan. Intertidal or mangrove forest interior. (1, 3, 9, 10, 14, 30, 34, 36, 48)

Rhynchosia volubilis Lour.: Jiangsu; Fujian; Taiwan; Guangdong; Guangxi; Hainan; and inland area. Coastal sandy area. (9, 10, 14, 33, 34, 48)

Robinia pseudoacacia L.: Introduced to all over China. Shoreline protection forest (sandy area). (10, 18, 20, 48)

Sesbania cannabina (Retz.) Pers.: Jiangsu; Zhejiang; Fujian; Taiwan; Guangdong; Hainan. Mud flat and inland area. (9, 10, 11, 47, 48)

OXALIDACEAE

Oxalis corniculata L.: Widely distributed in China. Coastal bank or open area. (9, 10, 11, 14, 33, 48)

GERANIACEAE

Erodium cicutarium (L.) L'Herit. ex Ait.: Hebei; Shandong; Jiangsu; Fujian; and inland area. Coastal sandy or grassy area. (10, 11, 33, 48)

ZYGOPHYLLACEAE

Nitraris sibirica Pall.: Coastal and inland area north of Changjiang (=Yangtze River). Coastal mud flat. (10, 11, 13, 38)

Tribulus cistoides L.: Hainan. Coastal sandy beach. (1, 11, 17)

T. terrestris L.: All over China. Coastal sandy beach or grassy wetland. (1, 3, 9, 10, 11, 48)

RUTACEAE

Atalantis buxifolia (Poir.) Oliv.: Fujian; Guangdong; Guangxi; Hainan. Coastal sandy or low elevation flat area. (1, 3, 9, 11, 12)

Evodia meliaefolia (Hance) Benth.: Fujian; Taiwan; Guangdong; Guangxi; Hainan; and inland area. Coastal flat. (9, 10, 11, 14, 48)

Micromelum falcatum (Lour.) Tanaka: Guangdong; Guangxi; Hainan; and inland area. Coastal flat or dry area. (3, 9, 11, 12, 14)

Murraya alata Drake v. *hainanensis* Swingle: Guangdong; Guangxi; Hainan. Coastal dry sandy area. (9, 11, 14)

M. microphylla (Merr. & Chun) Sw.: Hainan. Coastal low hill or bushes. (3, 9, 11, 12)

Zanthoxylum nitidum (Roxb.) DC.: Fujian; Taiwan; Guangdong; Guangxi; Hainan; and inland area. Coastal rocky crevices. (9, 10, 11, 14, 33, 48)

SIMARUBACEAE

Brucea javanica (L.) Merr.: Southern China. Coastal flat. (9, 10, 11, 14, 33, 48)

MELIACEAE

Turraea pubescens Hellen: Guangdong; Guangxi; Hainan. Coastal bushes. (3, 9, 11, 12, 14)

Xylocarpus granatum Koenig: Hainan. Mangrove forest or shallow muddy shore, mangrove species (3, 9, 25, 30, 35, 45)

EUPHORBIACEAE

Breynia officinalis Hemsl.: Fujian; Taiwan. Coastal low hill or rocky shore. (33, 48)

Bridelia tomentosa Bl.: Fujian; Taiwan; Guangdong; Guangxi; Hainan; and inland area. Coastal low hill. (9, 10, 11, 14, 33, 48)

Croton crassifolius Geisel.: Southern China and inland area. Coastal low hill. (9, 10, 11, 14, 33, 48)

C. laui Merr. & Meto.: Hainan. Coastal sandy beach and inland area. (3, 9, 11)

Dimorphocalyx poilanei Gagnep.: Hainan. Coastal bushes or under thin forest canopy. (3, 9, 11)

Euphorbia atoto Forst. f.: Taiwan; Guangdong; Hainan; Nansha (=Spratly Islands). Coastal sandy area or sandy gravel beach. (3, 7, 9, 11, 36)

E. indica Lamx.: Fujian; Guangdong; Guangxi; Hainan; and inland area. Coastal sandy area. (9, 10, 11, 14, 33, 48)

E. microphylla Heyne.: Fujian; Guangdong. Coastal sandy area. (33, 48)

E. tirucalli L.: Fujian; Taiwan; Guangdong; Guangxi; Hainan. Garden species, occasinally found in coastal area. (3, 9, 11, 33, 36, 48)

Excoecaria agallocha L.: Fujian; Taiwan; Guangdong; Hong Kong; Guangxi; Hainan. Mud flat and mangrove forest. (3, 9, 14, 24–26, 30, 33, 35, 44, 45, 48)

Glochidion obovatum Sieb. & Zucc.: Fujian. Coastal rocky crevices and low elevation hill. (33, 48)

Jatropha curcas L.: Fujian; Guangdong; Guangxi; Hainan; and inland area. All over coastal area. (9, 10, 11, 14, 33, 48)

Phyllanthus leptoclados Benth.: Fujian; Guangdong. Coastal rocky crevices. (33, 48)

Synostemon bacciformis (L.) G. L. Websten [*Agyneia bacciformis* (L.) A. Juss]: Fujian; Taiwan; Guangdong; Hainan. Coastal open area, sandy beach, or mud bank. (9, 10, 11, 29, 31, 33, 48)

ANACARDIACEAE

Pistacia chinensis Bunge: Northern China; northwestern China; South of Changjiang (=Yangtze River). Coastal bank and inland area. (9, 10, 11, 33, 48)

Rhus chinensis Mill. v. *roxburghii* (DC.) Rehd.: Taiwan; Guangdong; Guangxi; Hainan; and inland area. Coastal low hill and flat. (10, 11, 14)

CELASTRACEAE

Maytenus diversifolia (Gray) Hou [*Gymnosporia diversifolia* (Gray) Maxim]: Fujian; Taiwan; Guangdong; Guangxi; Hainan. Coastal rocky crevices, coastal bank, and inland area. (9, 10, 11, 14, 33, 48)

SALVADORACEAE

Azima sarmentosa (Bl.) Benth. & Hook. f.: Hainan. Coastal sandy bank. (10, 11, 46)

SAPINDACEAE

Dodonaea viscosa (L.) Jacq.: Fujian; Taiwan; Guangdong; Guangxi; Hainan; and inland area. Coastal low hill or thin forest. (1, 3, 8:47, 9, 11, 12, 14, 33, 48)

Erioglossum rubiginosa (Roxb.) Bl.: Guangdong; Guangxi; Hainan. Coastal thin forest or sandy bushes. (1, 3, 9, 11, 12, 46)

Mischocarpus sundaicus Bl.: Guangxi; Hainan. Coastal forest. (9, 11, 14, 46)

RHAMNACEAE

Colubrina asiatica (L.) Brongn.: Taiwan; Guangdong; Guangxi; Hainan. Coastal thin forest or low hill. (3, 9, 11, 14)

Paliurus ramosissimus (Lour.) Poir.: Jiangsu; Zhejiang; Fujian; Taiwan; Guangdong; Guangxi; and inland area. Coastal open area or rocky crevices. (10, 11, 14, 33, 48)

Sageretia thea (Osbeck) Johnst. [*S. theezans* (L.) Brongn.]: Jiangsu; Zhejiang; Fujian; Taiwan; Guangdong; Guangxi; and inland area. Coastal low hill or rocky crevices. (10, 11, 14, 33, 49)

TILIACEAE

Grewia piscatorum Hance: Fujian; Taiwan; Hainan. Coastal bushes and inland area. (3, 8:49, 9, 11, 48)

Triumfetta procumbens Forst.: Dongsha (=Pratas Islands); Xisha (=Paracel Islands). (8:49)

MALVACEAE

Hibiscus surattensis L.: Guangdong; Hainan; and inland area. Coastal sandy area, bush, or thin forest canopy. (9)

H. tiliaceus L.: Fujian; Taiwan; Guangdong; Guangxi; Hainan. High intertidal or coastal flat. (3, 9, 10, 12, 14, 25, 26, 30, 35, 48)

Malvastrum coromandelianum (L.) Gurke: Fujian; Taiwan; Guangdong; Guangxi; Hainan; and inland area. Coastal bank or open area. (9, 10, 11, 14, 33, 48)

Sida alnifolia L. v. *microphylla* (Car.) S. Y. Hu: Fujian; Guangdong; Guangxi; Hainan; and inland area. Coastal sandy or open area. (9, 11, 14, 33, 48)

S. chinensis Retz.: Taiwan; Guangdong; Guangxi; Hainan; and inland area. Coastal flat. (3, 9, 11, 12, 14)

S. cordifolia Wall.: Fujian; Taiwan; Guangdong; Guangxi; Hainan; and inland area. Coastal bank or sandy area. (3, 9, 11, 14, 33, 48)

S. corylifolia Wall. ex Master: Guangdong; Guangxi; Hainan; and inland area. Coastal grassy area. (9, 11)

S. rhombifolia L.: Fujian; Taiwan; Guangdong; Guangxi; Hainan; and inland area. Island open area. (1, 3, 9, 11, 12, 33, 48)

S. szechuensis Matsuda: Guangxi and inland area. Coastal sandy area. (10, 11, 14)

Thespesia populnea (L.) Soland. ex Corr.: Taiwan; Guangdong; Guangxi; Hainan. Mangrove forest interior or mud flat interior, mangrove species. (3, 9, 11, 12, 14, 25, 26, 30, 35)

Urena lobata L.: All provinces south of Changjiang (=Yangtze River). Coastal open area and inland area. (1, 3, 9, 10, 11, 14, 33, 48)

STERCULIACEAE

Helicteres angustifolia L.: Fujian; Taiwan; Guangdong; Guangxi; Hainan; and inland area. Coastal low hill, open area. (9, 10, 11, 14, 33, 48)

Heritiera littoralis (Dryand.) Ait.: Taiwan; Guangdong; Hong Kong; Guangxi; Hainan. Mangrove forest, mangrove species. (3, 9–12, 14, 25, 30, 35, 36, 44, 45)

Waltheria indica L. [*W. americana* L.]: Fujian; Taiwan; Guangdong; Guangxi; Hainan; and inland area. Coastal beach, sandy and sandy gravel area. (1, 3, 9, 10, 12, 33, 48)

DILLENIACEAE

Tetracera asiatica (Lour.) Hoogl.: Guangdong; Guangxi; Hainan. Coastal bank and inland area. (9, 10, 11, 14)

THEACEAE

Eurya chinensis R. Br.: Fujian; Taiwan; Guangdong; Guangxi; and inland area. Coastal low elevation hill. (10, 11, 14, 48)

E. emarginata (Thunb.) Makino: Zhejiang; Fujian; Taiwan. Coastal bush in low elevation hill area. (10, 11, 33)

Ternstroemia microphylla Merr. [*T. pseudoverticillata* Merr. & Chun]: Fujian; Guangdong; Guangxi; Hainan. Coastal rocky crevices and inland area. (9, 10, 11, 33, 48)

GUTTIFERAE

Calophyllum inophyllum L.: Taiwan; Guangdong; Guangxi; Hainan. Coastal sandy area and forest, inland area. (3, 8:50, 9, 10, 11, 46)

DIPTEROCARPACEAE

Vatica astrotricha Hance: Hainan. Coastal sandy shore and forest, inland area. (9, 10, 11, 42, 46)

ELATINACEAE

Bergia serrata Blanco: Southern China. Coastal sandy beach and inland area. (9, 33, 48)

TAMARICACEAE

Tamarix chinensis Lour.: Northeastern China; northern China; central and lower basin of Changjiang (=Yangtze River); Guangdong; Guangxi. Coastal salty area and inland area. (10, 11, 15, 17)

T. juniperina Bunge: Northeastern China; northern China; eastern China. Coastal salty area and inland area. (10, 15, 40)

FLACOURTIACEAE

Flacourtia indica (Burm. f.) Merr.: Guangdong; Guangxi; Hainan. Coastal area. (1, 3, 9, 10, 11)

Scolopia chinensis (Lour.) Clos: Southern China. Coastal bush or forest edge. (9, 10, 11, 14, 33, 48)

S. hainanensis Sleum.: Hainan. Coastal open sandy area. (3, 9, 11, 46)

S. henryi Sleum.: Hainan. Coastal sandy area. (3, 9)

CACTACEAE

Opuntia dillenii (Ker-Gawl.) Haw.: Southern China. Coastal dry rocks or bush. (1, 3, 9, 10, 14, 33, 48)

THYMELAEACEAE

Wickstroemia indica (L.) C. A. Mey.: All provinces south of Changjiang (=Yangtze River). Coastal low hill, bank, open area, and inland area. (9, 10, 14, 33, 46, 48)

ELAEAGNACEAE

Elaeagnus oldhami Maxim.: Fujian; Taiwan; Guangdong; Hainan. Coastal bush and inland area. (8:52, 10, 11, 33, 48)

LYTHRACEAE

Pemphis acidula J. R. & Forst.: Taiwan; Hainan; Xisha (=Paracel Islands). Coastal mud flat, mangrove species. (8:52, 19, 25, 35, 36)

SONNERATIACEAE

Sonneratia alba Sm.: Hainan. Coastal salty area or mangrove forest, mangrove species. (8:52, 9, 10, 12, 25, 30, 35, 45)

S. caseolaria (L.) Engl.: Hainan. Coastal salty area or mangrove forest, mangrove species. (8:52, 9, 10, 12, 25, 30, 35, 45)

S. hainanensis Ko. E. Y. Chen & W. Y. Chen.: Hainan. Coastal salty area or mangrove forest, mangrove species. (45)

S. ovata Backer.: Hainan. Coastal salty area or mangrove forest. (3, 11, 30, 45)

LECYTHIDACEAE

Barringtonia asiatica (L.) Kurz.: Taiwan. Coastal forest, mangrove species. (8:52, 36)

B. racemosa (L.) Spreng.: Taiwan; Hainan. High intertidal mud flat or shore, mangrove speceies. (3, 8:52, 9, 25, 30, 35, 45)

RHIZOPHORACEAE

Bruguiera cylindrica (L.) Bl.: Hainan. Mangrove forest, mangrove species. (8:52, 9, 12, 25, 30, 35, 41)

B. gymnorhiza (L.) Savingny: Fujian; Taiwan; Guangdong; Hong Kong; Guangxi; Hainan. Mangrove forest, mangrove species. (8:52, 9, 12, 14, 24, 25, 30, 35, 41, 44, 45, 48)

B. sexangula (Lour.) Poir.: Hainan. Shallow salty area or mangrove forest, mangrove species. (8:52, 9, 12, 25, 30, 35, 41, 45)

B. sexangula (Lour.) Poir v. *rhynchopetala* Ko: Hainan. Shallow salty water or mangrove forest, mangrove species. (3, 8:52, 12, 25, 35)

Ceriops tagal (Perr.) C. B. Rob.: Taiwan; Guangdong; Guangxi; Hainan. Mud flat or mangrove forest, mangrove species. (8:52, 9, 12, 25, 26, 35, 41)

Kandelia candel (L.) Druce: Fujian; Taiwan; Guangdong; Hong Kong; Guangxi; Hainan. Mud flat or mangrove forest, mangrove species. (3, 8:52, 9, 12, 14, 25, 26, 35, 41, 44, 48)

Rhizophora apiculata Bl.: Hainan. Mud flat or mangrove forest, mangrove species. (3, 8:52, 12, 25, 35, 41)

R. mucronata Poir.: Taiwan. Mud flat or mangrove forest, mangrove species. (8:52, 25, 26, 35, 41)

R. stylosa Griff.: Taiwan; Guangdong; Guangxi; Hainan. Mud flat or mangrove forest, mangrove species. (8:52, 9, 12, 25, 35, 41)

COMBRETACEAE

Lumnitzera littorea (Jack.) Voight: Hainan. Mud flat, mangrove forest, or high intertidal; mangrove species. (8:53, 9, 25, 30, 35, 45)

L. racemosa Willd.: Taiwan; Guangdong; Hong Kong; Guangxi; Hainan. Mud flat or mangrove forest interior, mangrove species. (8:53, 9, 24, 25, 26, 30, 35, 44, 45)

Terminalia catappa L.: Fujian; Taiwan; Guangdong; Guangxi; Hainan. Coastal sandy shore, sand flat, and inland area. (8:53, 9, 14, 33, 36)

MYRTACEAE

Eucalyptus citriodora Hook. f.: Southern China. Species for coastal planting. (9, 10, 14, 33, 48)

E. exserta F. V. Muell.: Southern China. Species for coastal planting. (9, 10, 14, 33, 48)

E. robusta Sm.: Southern China. Species for coastal planting. (9, 10, 14, 33, 48)

ONAGRACEAE

Oenothera littoralis Schlect.: Fujian; Guangdong; Hainan. Sandy beach. (3, 29, 31, 33, 37, 48)

UMBELLIFERAE

Glehnia littoralis F. Schmidt ex Miq.: Liaoning; Hebei; Shandong; Jiangsu; Zhejiang; Fujian; Taiwan; Guangdong; Hainan. Sandy beach. (10, 11, 31, 33, 40, 46, 48)

Peucedanum japonicum Thunb.: Shandong; Zhejiang; Fujian; Taiwan. Sandy shore. (33, 36, 48)

MYRSINACEAE

Aegiceras corniculatum (L.) Blanco: Fujian; Taiwan; Guangdong; Hong Kong; Guangxi; Hainan. Mud flat or mangrove forest, mangrove species. (8:58, 9, 10, 12, 24, 25, 35, 44, 48)

Embelia laeta (L.) Mez: Fujian; Taiwan; Guangdong; Guangxi; Hainan; and inland area. Coastal rocky crevices or bush. (9, 10, 14, 33, 48)

Rapanea linearis (Lour.) S. Moore: Guangdong; Guangxi; Hainan. Coastal sandy bank and inland area. (9, 10, 11, 14, 46)

PRIMULACEAE

Anagallis coerulea Schreb.: Fujian; Taiwan; Guangdong; Guangxi. Coastal sandy area, rocky crevices, road side, or open area. (11, 14, 33, 36, 48)

Lysimachia mauritiana Lamx.: Liaoning; Shandong; Jiangsu; Zhejiang; Fujian; Taiwan. Coastal rocky crevices and sandy beach. (10, 11, 33, 48)

PLUMBAGINACEAE

Limonium bicolor (Bunge) O. Kuntze: Coastal and inland area of Hebei. Coastal salty area or sandy hill. (10, 11, 40)

L. sinense (Girard) Kuntze: Liaoning; Hebei; Shandong; Jiangsu; Fujian; Guangdong; Guangxi; Hainan; and inland area. Coastal sandy area, sandy beach, or salty area. (9, 10, 14, 20, 29, 31, 33, 37, 48)

Plumbago zeylanica L.: Fujian; Taiwan; Guangdong; Guangxi; Hainan; and inland area. Coastal bush or sandy area. (8:60, 9, 10, 33, 48)

APOCYNACEAE

Cerbera manghas L.: Taiwan; Guangdong; Guangxi; Hainan. Costal sand flat, mud flat, or mangrove forest; mangrove species. (3, 8:63, 9, 10, 12, 14, 25, 26, 35)

ASCLEPIADACEAE

Calotropis gigantea (L.) Dryand.: Guangdong; Guangxi; Hainan; and inland area. Coastal open area. (8:63, 9, 10, 14)

Cynanchum insulanum (Hance) Hemsl.: Guangdong; Guangxi; Hainan. Coastal sandy area or open area. (8:63, 9, 12, 14, 34)

C. insulanum (Hance) Hemsl. v. *lineare* (Tsiang & Zhang) Tsiang & Zhang: Guangdong; Hainan. Coastal open flat. (8:63, 9, 11, 12)

Sarcostemma acidum (Roxb.) Voigt.: Guangdong; Hainan. Forest. (9, 10, 11)

CONVOLVULACEAE

Calystegia soldanella (L.) R. Br.: Liaoning; Hebei; Shandong; Jiangsu; Zhejiang; Fujian; Taiwan; and inland area. Coastal sandy beach or rocky bank crevices. (5, 10, 11, 33, 36, 40)

Evolvulus alsinoides L. v. *decumbens* (R. Br.) Coststr.: Southern China. Coastal sandy area. (8:64, 9, 11, 33, 48)

Ipomoea maxima (L. f.) Don ex Sweet: Taiwan; Hainan. Coastal bush or open area. (8:64, 9, 11)

I. pes-caprae (L.) Sweet: Zhejiang; Fujian; Taiwan; Guangdong; Guangxi; Hainan. Coastal sandy beach and bank. (1, 8:64, 9, 10, 14, 29, 31, 33, 36, 37, 48)

I. stolonifera (Cyrillo) J. F. Gmel.: Fujian; Taiwan; Guangdong; Hainan. Coastal sandy area. (9, 11, 33, 36, 48)

Merremia hirta (L.) Merr. v. *retusa* Coststr.: Guangdong. Coastal sandy area. (3, 11)

Stictocardis tiliaefolia (Desr.) Hallier f.: Taiwan; Guangdong; Hainan. Coastal bush. (8:64, 9, 11)

BORAGINACEAE

Coldenia procumbens L.: Taiwan; Hainan. Intertidal sandy beach. (8:64, 9, 11)

Cordia subcordata Lamx.: Hainan; Nansha (=Spratly Islands). Coastal evergreen coppice. (7, 9)

Heliotropium marifolium Retz.: Hainan. Coastal grassland. (8:64, 9)

H. strigosum Willd.: Fujian; Guangdong; Hainan. Coastal sandy area. (5, 8:64, 9, 11, 33, 48)

Messerschmidia argentea (L.) John. [*Tournefortia argentea* L.]: Taiwan; Hainan; Xisha (=Paracel Islands). Coastal sandy beach. (8:64, 9)

M. sibirica L. ssp. *angustiior* (DC.) Kitag.: Coastal and inland area north of Changjiang (=Yangtze River). Coastal sandy area and salty flat. (10, 11, 37)

VERBENACEAE

Avicennia marina (Forsk.) Vierh.: Fujian; Taiwan; Guangdong; Hong Kong; Guangxi; Hainan. Mud flat or mangrove forest, mangrove species. (8:65, 9, 10, 14, 24, 25, 26, 29, 31, 35, 44, 48)

Clerodendron inerme (L.) Gaertn.: Fujian; Taiwan; Guangdong; Hong Kong; Guangxi; Hainan. Intertidal sandy beach and bank. (8:65, 9, 10, 14, 29, 30, 31, 44, 45, 48)

Phyla nodiflora (L.) Greene: Jiangsu; Fujian; Taiwan; Guangdong; Hainan; and inland area. Coastal sandy area. (3, 6, 9, 10, 33, 34, 48)

Premna obtusifolia R. Br.: Taiwan; Guangdong; Guangxi; Hainan. Coastal bank, mangrove forest interior, and inland area. (3, 9, 11, 14)

Vitex trifolia L. v. *simplicifolia* Cham.: Coastal provinces and inland area. Coastal sandy beach. (3, 8:65, 9, 10, 14, 29, 31, 37, 47, 48)

LABIATAE

Leucas chinensis (Retz.) R. Br.: Taiwan; Hainan. Coastal flat area. (3, 8:65, 9, 11)

L. zeylanica (L.) R. Br.: Guangdong; Guangxi; Hainan. Coastal flat. (3, 8:65, 9, 10, 11, 14)

Scutellaria strigillosa Hemsl.: Liaoning; Hebei; Shandong; Jiangsu. Coastal sandy beach. (10, 11, 16, 20)

SOLANACEAE

Datura metel L.: Coastal provinces and inland area. Coastal bank or road side. (9, 10, 11, 14, 33)

Lycium chinense Mill.: Coastal provinces and inland area. Coastal bush or bank. (5, 9, 10, 11, 33)

SCROPHULARIACEAE

Bacopa monnieri (L.) Wettst.: Fujian; Taiwan; Guangdong; Guangxi; Hainan; and inland area. Coastal wetland. (5, 8:67, 9, 10, 11, 14, 33)

Veronica undulata Wall.: All over China except for Tibet, Qinghai, and Ningxia. Sandy beach, swamp. (3, 8:67, 10, 33)

BIGNONIACEAE

Dolichandrone spathacae (L. f.) K. Schum.: Hainan. Coastal flat and river mouth area, mangrove species. (11, 19, 39)

ACANTHACEAE

Acanthus ebracteatus Vahl.: Hainan. Mud flat and mangrove forest, mangrove species. (3, 9, 25, 30, 34, 35, 45)

A. ilicifolius L.: Fujian; Guangdong; Hong Kong; Guangxi; Hainan. Mud flat and mangrove forest, mangrove species. (9, 10, 14, 24, 25, 26)

A. xiamenensis R. T. Zhang: Fujian. Mud flat and mangrove forest, mangrove species. (27, 31, 33)

MYOPORACEAE

Myoporum bontiodes (S. & Z.) A. Gray: Zhejiang; Fujian; Taiwan; Guangdong; Guangxi; Hainan. Coastal flat and mangrove forest interior. (3, 10, 14, 33, 34, 47)

RUBIACEAE

Guettarda speciosa L.: Taiwan; Guangdong; Hainan; Xisha (=Paracel Islands). Tropical coastal plant, often form evergreen coppice in Xisha (=Paracel Islands). (7, 9, 11)

Hedyotis pinifolia Wall.: Guangdong; Guangxi; Hainan. Coastal sandy area and inland area. (1, 3, 9, 10, 11)

Morinda citrifolia L.: Taiwan; Hainan; Xisha (=Paracel Islands). Coastal plant. (3, 9, 10, 11, 36)

M. parvifolia Benth.: Southern and southeastern China. Coastal low hill and rocky crevices. (9, 10, 11, 14, 33)

Randia accedens Hance: Guangdong; Guangxi. Coastal sandy beach and inland area. (9, 10, 33)

Scyphiphora hydrophyllacea Gaertn. f.: Hainan. Mud flat and mangrove forest interior, mangrove species. (9, 10, 11, 25, 30, 35, 45)

GOODENIACEAE

Calogyne pilosa R. Br.: Fujian; Guangdong. Coastal hill, riverside grassy area. (3, 8:73, 10, 11)

Scaevola hainanensis Hance: Fujian; Taiwan; Guangdong; Guangxi; Hainan. Coastal rocky crevices, mud flat, or mangrove forest canopy. (8:73, 9, 10, 11, 14, 30, 35, 45)

S. sericea Vahl: Fujian; Taiwan; Guangdong; Guangxi; Hainan; Xisha (=Paracel Islands). Coastal rocky crevices, gravel sandy area, or shoreline. (7, 8:73, 9, 10, 12, 14, 29, 30, 31)

COMPOSITAE

Artemisia capillaris Thunb.: Coastal and inland area all over China. Coastal sandy area, bank, or open area. (10, 11, 14, 20, 33)

A. eriopoda Bunge: Northeastern, northern, and inland China. Coastal gravel and rocky area. (10, 11, 16)

A. scoparia Waldst. & Kit.: Widely distributed in China. Coastal salty area. (10, 11, 17, 40)

A. stelleriana Besse: Northeastern China coast. Coastal sandy area. (10)

A. verbenacea (Kom.) Kitag. [*A. valgaris* L. v. *verbenacea* Kom.]: Northeastern China coast. (44)

Aster subulatus Michx.: Jiangsu; Zhejiang; Fujian; and inland area. Coastal bank and open area. (5, 21, 47)

Blainvillea acmella (L.) Philipson [*B. latifolia* (L. f.) DC.]: Guangdong; Guangxi; Hainan; and inland area. (9, 10, 11, 14)

Conyza canadensis (L.) Cronq.: Almost all over China. Coastal open area and bank. (10, 11, 14, 22, 33)

Glossogyne tenuifolia Cass.: Fujian; Taiwan; Guangdong; Guangxi; Hainan. Coastal sandy area. (5, 9, 10, 14, 33, 34)

Gynura divaricata (L.) DC.: Southern and inland China. Shoreline. (9, 11, 14)

Hemistepta lyrata (Bunge) Bunge: All over China. Shoreline. (9, 10, 11, 14)

Heteropappus hispidus (Thunb.) Less.: Almost all over China. Coastal open area, bank, or grassland (10, 11, 33)

Ixeris debilis A. Gray: Northeastern, eastern, and southern central China. Coastal wetland. (10, 11, 14, 33)

I. japonica Nakai subsp. *salsuginosa* Kitag: Northeastern China coast. Coastal wetland. (16)

I. repens A. Gray: Northeastern, northern, eastern, and southern China. Coastal sandy beach. (3, 10, 14, 31, 33, 37, 40)

Lactuca indica L.: Almost all over China (except for northwestern China). Coastal sandy area. (10, 11, 33)

Launaea sarmentosa (Willd.) Merr. & Chun: Southern China. Coastal sandy beach. (3, 9, 10, 11, 33, 34)

Picris japonica Thunb. v. *koreana* Kitag: Northeastern China. Shoreline. (16)

Pluchea indica (L.) Less: Taiwan; Guangdong; Guangxi; Hainan. Coastal sandy area, intertidal or shoreline. (8:75, 9, 10, 14, 45)

Scorzonera mongolica Maxim. v. *putjatae* C. Winkler: Northeastern China; Hebei; Shandong; Jiangsu; and inland area. Coastal sandy bank. (10, 16, 18)

Senecio pseudo-arnica Less.: Northeastern China. Shoreline. (16)

Synedrella nodiflora Gartn.: Southeastern and inland China. (8:75, 9, 10, 11)

Tridax procumbens L.: Fujian; Taiwan; Guangdong; Guangxi; Hainan. Coastal sandy area. (9, 10, 11, 14, 33)

Tripolium vulgare Ness.: Northeastern, northern, and inland China. Coastal salty area. (10, 11, 17, 22)

Wedelia biflora (L.) DC.: Taiwan; Guangdong; Hainan. Coastal sandy area and inland area. (9, 10, 11)

W. prostrata (Hook. & Arn.) Hemsl.: Jiangsu; Zhejiang; Fujian; Taiwan; Guangdong; Guangxi; Hainan. Coastal sandy beach. (5, 8:75, 14, 29, 31, 37)

Xanthium strumarium L.: Northeastern China; Hebei; Jiangsu. Coastal sandy hill, salty area, or open grassland. (16, 20, 40)

MONOCOTYLEDONEAE
TYPHACEAE

Typha angustifolia L.: Liaoning; Hebei; Shandong; Jiangsu; Zhejiang; Fujian; and inland area. Coastal wetland. (10, 11, 16, 20, 33, 40)

PANDANACEAE

Pandanus forceps Martelli: Guangdong; Hainan. (9, 10)

P. tectorius Sol.: Fujian; Taiwan; Guangdong; Guangxi; Hainan. Coastal sandy area. (3, 9, 10, 14, 29, 31, 33)

POTAMOGETONACEAE

Cymodocea rotundata Aschers. & Schweinf.: Guangdong; Hainan. In the water under mangrove canopy. (9, 21, 42)

Halodule pinifolia (Miki) Hart.: Guangdong; Guangxi; Hainan. Low intertidal to subtidal. (21)

H. uninervis (Forsk.) Aschers.: Taiwan; Guangdong; Guangxi; Hainan. Shallow sea. (9, 21, 42)

Phyllospadix iwatensis Makino: Liaoning; Hebei; Shandong. Rocky shore. (21)

P. japonica Makino: Liaoning; Hebei; Shandong. Rocky shore. (10, 21)

Ruppia rostellata Koch [*R. maritima* L.]: Liaoning; Hebei; Shandong; Jiangsu; Zhejiang; Taiwan; Guangdong; Hainan; and inland area. Coastal flat and salt marsh. (1, 2, 3, 9, 10, 11, 16, 17, 23, 44, 45)

Syringodium isoetifolium (Aschers) Dand: Guangdong; Guangxi. Shallow sea. (21)

Zannichellia palustris L.: Most of China. Coastal and inland salt water and freshwater swamp. (6, 10, 11, 16, 42)

Zostera caespitosa Miki: Liaoning; Hebei; Shandong. Shallow sea. (44, 46)

Z. marina L.: Liaoning; Hebei; Shandong. Shallow sea and beach. (10, 11, 21)

Z. nana Roth [*Z. japonica* Aschers & Greebn]: Hebei; Shandong; Hong Kong. Shallow sea. (10, 44)

Z. pacifica S. Watson: Northeastern China. Shallow sea. (16)

JUNCAGINACEAE

Triglochin maritimum L.: Northern China; northeastern China; Shandong; and inland area. Wet sandy area, salt flat. (44, 46)

HYDROCHARITACEAE

Enhalus acoroides (L. f.) Steud. [*E. koenigii* Rich.]: Eastern China coast; Hainan. Mid-intertidal to subtidal. (10, 21, 42)

Halophila beccarii Aschers.: Guangdong; Hong Kong; Hainan; Xisha (=Paracel Islands). Mangrove forest fringe or low intertidal. (21, 44)

H. ovata (R. Br.) Hook. f.: Taiwan; Guangdong; Hong Kong; Hainan; Xisha (=Paracel Islands). Mid-intertidal to subtidal. (21, 42, 44)

Thalassia hemprichii (Ehrenb.) Aschers: Guangdong; Hainan. Mid-intertidal to subtidal. (21, 42)

GRAMINEAE

Aeluropus littoralis (Gouan) Parl.: Southern part of northeastern China. Coastal sandy area. (10, 16)

A. littoralis (Gouan) Parl. v. *sinensis* Debeaux: Coastal and inland area north of Changjiang (=Yangtze River). Salt flat. (10, 11, 17, 20, 22, 38, 40, 47)

Aristida chinensis Munro: From Fujian to Hainan. Coastal sandy shore or low hill grassland. (10, 11, 14, 33, 46)

Brachiaria subquadripara (Trin.) Hitchc.: Fujian; Taiwan; Guangdong; Guangxi; Hainan. Coastal sandy area. (9, 10)

Brachypodium monshuricum Kitag.: Southern part of northeastern China. Islands. (16)

Calamagrostis epigejos (L.) Roth.: Almost all over China. Coastal flat or wetland. (10, 11, 20)

Cenchrus calyculata Cavan.: Fujian; Taiwan; Guangdong; Hainan. Coastal sandy or sandy gravel area. (3, 9, 10, 11, 33)

Chloris barbata Swartz: Taiwan; Guangdong. Coastal sandy area. (10, 36)

C. formosana (Honda) Keng: Fujian; Taiwan; Guangdong; Guangxi; Hainan. Coastal sandy area or bank. (9, 11, 14, 33, 34)

C. virgata Sw.: Almost all over China. Coastal sandy area or bank. (10, 11, 16, 33)

Chrysopogon orientalis (Desv.) A. Camus: Fujian; Guangdong; Guangxi; Hainan. Coastal sandy area. (9, 10, 11, 14)

Coelachne simpliciuscula (Wight & Arn.) Murro ex Benth.: Southern China. (9, 10)

Cymbopogon tortilis (Presl.) A. Camus: Fujian; Taiwan; Guangdong; Guangxi; Hainan. Coastal low hill. (9, 10, 11, 14, 33)

Cynodon dactylon (L.) Pers.: All provinces south of Yellow River basin. Coastal sandy area, mud flat, bank, and inland. (9, 10, 11, 14, 22)

Dactyloctenium aegyptiacum (L.) Willd.: Zhejiang, Taiwan, southern China. Coastal bank, sandy shore, and inland. (9, 10, 11, 14, 27, 33, 37)

Desmostachya bipinnata (L.) Stapf.: Hainan. Coastal sandy area. (3, 9, 10, 11)

Digitaria heterantha (Nees & Meyen) Merr.: Fujian; Taiwan; Guangdong; Hainan. Coastal sandy beach. (9, 11)

D. heterantha (Nees & Meyen) Merr. subsp. *aequabilis* Henr. v. *holosericea* Henr.: Hainan. Coastal sandy area. (9)

Echinochloa crusgallii (L.) Beauv. v. *formosensis* Ohwi.: Taiwan; Hainan. (9)

Eragrostis cylindrica (Roxb.) Nees: Southern China. Coastal sandy area and bank. (9, 10, 11, 14, 33, 37)

E. nevinii Hance: Fujian; Guangdong; Guangxi; Hainan. Coastal sandy area. (9, 10, 11, 14)

Eustachys tener (Presl.) A. Camus: Fujian; Taiwan; Guangdong; Hainan. Coastal sandy area and inland. (10, 42, 46)

Hymenachne assamica (Hook. f.) Hitche.: Taiwan; Guangdong; Guangxi; Hainan. Coastal sandy area. (9, 10, 11, 14)

Imperata cylindrica (L.) Beauv. v. *major* (Ness) C. E. Hubb: Almost all over China. Coastal sandy beach, sandy bank, high intertidal mud flat, and inland. (3, 9, 10, 14, 20, 22)

Ischaemum barbatum Retz. [*I. aristatum* L. v. *meyenianum* (Nees) A. Camus] : Southeastern China. Coastal bush or low hill grassland. (9, 11, 14, 46)

I. indicum (Houtt.) Merr.: Southeastern and southern China; Hong Kong. Coastal low elevation hill. (9, 11, 14, 33, 48)

Leymus mollis (Trin.) Hochst. v. *coreensis* (Hack.) Keng f.: Northern China coast. Beach. (10, 11, 16)

Microchloa indica (L. f.) Beauv.: Guangdong; Hainan; and inland. Coastal sandy area. (3, 9, 10)

Panicum repens L.: Southeastern China and Hong Kong. Coastal sandy area, wetland, and inland. (2, 3, 9, 10, 14, 33, 44)

Paspalum distichum L.: Jiangsu; Fujian; Taiwan; Guangdong; Guangxi; Hainan; and inland. Coastal sandy area or bank. (2, 3, 9, 10, 14, 33, 45)

Perotis hordeiformis Nees ex Hook. & Arn.: Taiwan; Guangdong; Guangxi; Hainan. Coastal sandy area or sandy mud. (9, 11, 14)

P. indica (L.) O. Kuntze: Hebei; Fujian; Gunagdong; Guangxi; Hainan; and inland. Coastal sandy area. (9, 10, 11, 14, 46)

Phacelurus latifolius (Steud.) Ohwi: From Hebei to Zhejiang. Costal salty area. (6, 10, 11, 20)

P. latifolius (Steud.) Ohwi v. *angustifolia* (Debeaux) Kitag.: Southern part of northeastern China; from Hebei to Zhejiang. Along the ditches in the coastal area. (10, 11, 16)

Phragmites communis Trin.: Almost all over China. Coastal wetland, mud flat, or riverside. (3, 10, 20, 22, 33, 40, 42, 47)

Puccinellia chinampoensis Ohwi: Northeastern and northern China. Coastal mud flat and inland. (10, 11, 16, 17)

P. distans (L.) Parl.: Northeastern China. Coastal salty area. (10, 11, 40)

Saccharum arundinaceum Retz.: Jiangsu; Zhejiang; Fujian; Guangdong; Guangxi; Hainan; and inland. Sandy beach. (9, 10, 11, 14, 33, 37)

S. spontaneum L.: Southern China. Coastal sandy shore and inland. (9, 10, 11, 46)

Spartina alterniflora Loisel.: Jiangsu; Fujian. Mud flat. (8, 11, 33)

S. anglica C. E. Hubb.: Hebei; Jiangsu; Zhejiang; Fujian; Guangxi. Mud flat. (10, 11, 14, 20, 38)

Spinifex littoreus (Burm. f.) Merr.: Fujian; Taiwan; Guangdong; Guangxi; Hainan. Coastal sandy beach. (2, 3, 9, 10, 14, 31, 33, 36, 37)

Sporobolus virginicus (L.) Kunth: Zhejiang; Fujian; Taiwan; Guangdong; Guangxi; Hainan. Coastal salty area. (3, 5, 9, 10, 14, 24, 29, 31, 37, 45)

Thuaria involuta (G. Forst.) R. Br. ex Roem. & Schnlt.: Taiwan; Guangdong; Hainan; Xisha (=Paracel Islands); Nansha (=Spratly Islands). Coastal sandy area or sandy shore. (7, 9, 10, 46)

Zoysia japonica Steud.: Northeastern to eastern China. High elevation mud flat, sand flat. (10, 11, 16, 22, 40, 47)

Z. macrostachys Tranch. & Sav.: Shandong; Jiangsu; Zhejiang; and inland. Coastal sandy area. (8:10, 11, 20)

Z. matrella (L.) Merr.: Taiwan; Guangdong; Guangxi; Hainan. Coastal sandy area or beach. (3, 9, 10, 14, 24, 30, 45)

Z. sinica Hance: Southern, northern, and northeastern China. Coastal sandy beach or rocky crevices. (5, 10, 11, 16, 31, 33, 44)

Z. tenuifolia Willd. & Trin: Southern China. Coastal sandy area. (10, 11)

CYPERACEAE

Bulbostylis barbata (Rottb.) C. B. Clarke: Commonly found all over China. Beach grassland. (2, 3, 9, 10, 11)

B. densa (Wall.) Hand-Mazz.: Hebei; southern and eastern China. Coast and inland. (2, 3, 10, 11)

B. puberula (Poir.) Clarke: Guangdong; Hainan. Sandy beach or grassland. (9, 11, 34)

Carex commixta Steud.: Hainan. Coastal sandy area. (9, 11)

C. kobomugi Ohwi.: Southern part of northeastern China; Hebei; Shandong; Jiangsu; Zhejiang; Taiwan. Coastal sandy beach. (10, 11, 16, 20, 40, 47)

C. pumila Thunb.: Southern part of northeastern China; Hebei; Shandong; Jiangsu; Zhejiang; Taiwan. Coastal sandy beach or sandy area. (10, 11, 16, 20, 33, 40)

C. scabrifolia Steud.: Liaoning; Hebei; Shandong; Jiangsu; Zhejiang; Taiwan. Coastal sandy area or beach. (10, 11, 20, 22)

Cladium chinense Nees: Taiwan; Guangdong; Guangxi; Hainan; and inland. Coastal salty area or swamp. (9, 10, 11)

Courtoisia cyperoides (Roxb.) Nees: Guangxi and inland. Coastal dry sandy area. (10, 14)

Cyperus malaccensis Lamx.: Taiwan; Guangdong; Guangxi; Hainan. Coastal mud flat. (3, 9, 11, 14, 34)

C. malaccensis Lamx. v. *brevifolius* Bocklr.: Fujian; Guangdong; Guangxi. Coastal mud flat and inland riverside. (3, 10, 14, 29, 31, 34, 45)

C. rotundus L.: Almost all over China. Coastal dry sandy beach and inland. (3, 9, 10, 14, 16, 33, 48)

C. stoloniferus Retz.: Fujian; Taiwan; Guangdong; Guangxi; Hainan. Coastal wet sandy area and sandy beach. (9, 10, 11, 14, 33, 34)

Eleocharis acutangula (Roxb.) Schut.: Fujian; Taiwan; Guangdong; Hainan. Coastal swamp and inland rice field. (9, 45)

E. geniculata (L.) Roem & Schult.: Fujian; Taiwan; Guangdong; Hainan. Coastal sandy area. (9)

Fimbristylis aesfivalis (Retz.) Vahl.: Zhejiang; Fujian; Taiwan; Guangdong; Guangxi; Hainan; and inland. Coastal sandy area or salt marsh. (2, 3, 9, 10, 14, 33)

F. dichotoma (L.) Vahl: Almost all over China. Coastal dry sandy beach. (9, 10, 11, 14, 34)

F. ferruginea (L.) Vahl: Fujian; Taiwan; Guangdong; Guangxi; Hainan. Coastal sandy area or salt marsh. (3, 5, 9, 11, 14, 33, 34)

F. longispica Steud.: Northeastern China coast; Jiangsu; Zhejiang; Fujian; and Guangxi. Coastal sandy area or salt marsh. (10, 11, 14, 16, 33)

F. longistipitata Tang & Wang: Hainan. Coastal sandy area. (3, 9, 11)

F. ovata (Burm. f.) Kern.: Fujian; Taiwan; Guangdong; Guangxi; Hainan; and inland. Coastal sandy area. (9, 10, 11, 14, 33)

F. pierotii Miq.: Shandong; Jiangsu; Zhejiang; Fujian; Guangxi; and inland. Coastal sandy area. (10, 11, 14)

F. polytrochoides (Retz.) Vahl: Fujian; Taiwan; Hainan. Coastal wet and salty area. (9, 11)

F. rigidula Nees: Jiangsu; Zhejiang; Guangdong; Guangxi; and inland. Coastal sandy area. (10, 11, 14)

F. sericea (Poir.) R. Br.: Zhejiang; Fujian; Taiwan; Guangdong; Guangxi; Hainan. Coastal sandy beach. (5, 9, 10, 11, 14, 31, 33, 34)

F. schoenoides (Retz.) Vahl: Zhejiang; Fujian; Taiwan; Guangdong; Guangxi; Hainan; and inland. Coastal salt marsh. (5, 9, 10, 11, 33)

F. spathacea Roth: Fujian; Taiwan; Guangdong; Hong Kong; Guangxi; Hainan. Coastal sandy beach. (2, 3, 9, 11, 29, 33, 44)

F. subbispicata Nees & Meyen: Northeastern China; Hebei; eastern China; Guangdong; Guangxi; and inland. Coastal salt marsh. (10, 11, 14, 16, 33, 34)

F. tristachya R. Br.: Guangdong; Hainan. Coastal salt marsh. (9)

Juncellus serotinus (Rottb.) C. B. Clarke: Northeastern, northern, central, and eastern China; Guangdong; and inland. Coastal mud flat. (10, 11, 22)

Kullinga brevifolia (Rottb.): Zhejiang; Fujian; Guangdong; Guangxi; Hainan; and inland. Coastal sandy beach. (1, 2, 17)

Mariscus dubius (Rottb.) Hutch.: Hainan. Coastal sandy area. (9)

M. hainanensis Chun & How: Hainan. Coastal sandy area. (9)

M. monospermus S. M. Huang: Hainan. Coastal grassland. (9)

M. radians (Nees & Meyen) Tang & Wang: Shandong; Zhejiang; Guangdong; Hainan. Coastal grassland. (3, 9, 11)

M. radians (Nees & Meyen) Tang & Wang v. *floribundus* (Camus) S. M. Huang: Guangdong; Hainan. Coastal open area. (9)

Pycreus polystachyos (Rottb.) P. Beauv.: Fujian; Taiwan; Guangdong; Guangxi; Hainan. Coastal mud flat. (9, 10, 14, 29, 33, 45)

Remirea maritima Aubl.: Taiwan; Guangdong; Hainan. Coastal open area and sandy flat. (3, 9, 10, 11)

Scirpus mariqueter Tang & Wang: Jiangsu; Zhejiang. Coastal mud flat. (8:11, 11, 22, 47)

S. planiculmis Fr. Schmidt: Northeastern China; Hebei; Shandong; Jiangsu; Zhejiang; and inland. River mouth area. (10, 11, 17, 20, 38)

S. tabernaemontani Gmel: Northern China; northeastern China; Jiangsu; and inland. Coastal salty area. (10, 11, 38, 40)

S. triqueter L.: Almost all over China (except for Guangdong). Coastal mud flat and inland. (10, 11, 22, 40)

PALMAE

Caryota ochlandra Hance.: Southeastern China and inland. Coastal sandy shore. (9, 10, 11, 46)

Cocos nucifera L.: Taiwan; Guangdong; Guangxi; Hainan. Coastal flat. (1, 3, 9–12)

Nypa fruticans Wurmb: Hainan. Mud flat or mangrove forest, mangrove species. (3, 9, 10, 12, 25, 35)

Rhapis excelsa (Thunb.) Henry ex Rehd.: Southeastern China and inland. Coastal sandy shore. (9, 10, 11, 46)

FLAGELLARIACEAE

Flagellaria indica L.: Taiwan; Guangdong; Hainan. Coastal sandy shore. (9, 10, 11)

RESTIONACEAE

Leptocarpus disjunctus Mast.: Hainan. Coastal sandy area. (9, 10, 11, 42, 46)

CENTROLEPIDACEAE

Centrolepis banksii (R. Br.) Roem. & Schult [*C. hainanensis* Merr. Metc.]: Hainan. Coastal sandy shore. (9, 10, 46)

XYRIDACEAE

Xyris indica L.: Southern China. Coastal low sandy area. (9, 10, 11, 14, 33)
X. pauciflora Willd.: Fujian; Taiwan; Guangdong; Guangxi; Hainan; and inland. Coastal sandy area. (9, 10, 11, 33, 34)

ERIOCAULACEAE

Eriocaulon buergerianum Koern.: Jiangsu; Zhejiang; Fujian; Taiwan; Guangdong; Guangxi; and inland. Coastal sandy area. (10, 11, 14)

COMMELINACEAE

Murdannia kainantensis (Masam.) D. Y. Hong [*Aneilema kainantense* Masam.]: Fujian; Guangdong; Hainan. Coastal islands. (9, 11, 43)

PHILYDRACEAE

Philydrum lanuginosum Banks.: Fujian; Taiwan; Guangdong; Guangxi; Hainan. Coastal sandy shore. (9, 10, 11, 33, 34)

JUNCACEAE

Juncus effusus L.: Coastal and inland area of China. Coastsal salt marsh. (5, 10, 11, 29)
J. gracillimus (Buch.) V. Krecz. & Gontsch.: Jiangsu and the provinces north of Changjiang (=Yangtze River). Coastal salt marsh. (10, 11, 40)
J. haenkei E. Mey.: Northeastern China. Shoreline, shallow sea. (16)

LILIACEAE

Allium anisopodium Ledeb.: All provinces north of Changjian (=Yangtze River) basin. Coastal sandy hill. (2, 17, 36)
Asparagus brachyphyllus Turcz.: Liaoning; Hebei; and inland. Coastal salty area. (10, 11, 16)
A. dauricus Fisch. ex Link: Northeastern China; Hebei; Shandong; Jiangsu; and inland. Coastal sandy beach. (10, 11, 20)
A. officinalis L.: All over China. Coastal sandy area. (10, 11, 16, 33)
A. polyphyllus Stev.: Northeastern China. Coastal salty area. (16)
Dianella ensifolia (L.) DC.: Jiangsu; Zhejiang; Fujian; Taiwan; Guangdong; Guangxi; Hainan; and inland. Coastal low hill bush. (9, 10, 11, 33, 34)

AMARYLLIDACEAE

Agave americana L.: Southern China. Coastal area and coastal rocky crevices. (9, 11, 33, 37)
A. angustifolia Haw: Southern China. Coastal area. (8:16, 10, 11, 33)
Crinum asiaticum L. v. *sinicum* (Roxb. ex Herb.) Baker: Fujian; Taiwan; Guangdong; Guangxi; Hainan. Coastal area. (3, 8:16, 9, 10, 11, 14, 33)

IRIDACEAE

Belamcanda chinensis (L.) DC.: All over China. Coastal area. (5, 9, 10, 11, 46)
Iris ensata Thunb.: Northern, northeastern and eastern China. Coastal salty area. (10, 11, 40)

ZINGIBERACEAE

Costus speciosus (Koenig) Smith: Taiwan; Guangdong; Guangxi; Hainan; and inland. Coastal sandy shore. (10, 11, 14, 33, 34)

REFERENCES*

(1). Guangxi Institute of Botany, 1971. Index florae Guangxi, Tomus 2, Dicotyledoneae.
(2). Guangxi Institute of Botany, 1973. Index Florae Guangxi, Tomus 3, Monocotyledoneae.
(3). Vegetation speciality of comprehensive survey in coastal and tidal flat resources from Guangdong Province, 1987. List of Coastal Plant from Guangdong Province, China. China Ocean Press.
(4). Guangxi Institute of Botany, Academia Guangxi, 1991. Flora of Guangxi, Vol. 1. Guangxi Science and Technology Press.
(5). Wei Xinming, Chen Zhehai, and Lin Zhongxin, 1982. Coastal vegetation on islands along the central Fujian coast (Shengfu Bay to Minjiang Estuary). Journal of Oceanography in Taiwan Strait, 1(2):111–125.
(6). Wang Ningzhu et al., 1983. The atlas of vascular plant aquaticae in China. The People's Publishing House, Hubei, China.
(7). Wang Xianbo et al., 1989. The theory and performance of the natural reservations. The China Environmental Science Press.
(8). Delectis Florae Reipublicae Popularis Sinicae Ag endaae Academiae Sinicae Edita, 1959–1989. Flora Reipublicae Popularis Sinicae, Vol. 3, 10, 11, 16, 20, 24, 25, 31, 47, 49, 50, 52, 53, 58, 60, 63, 64, 65, 67, 73, 75. China Science Press.

(9). Instituti Botanici Aqustro-sinensis Academiae Sinicae, 1964–1977. Flora Hainanica, Tome 1–4. China Science Press.
(10). Instituti Botanici Academiae Sinicae, 1972–1976. Iconographia Cormophytorum Sinicorum, Tomus 1–5. China Science Press.
(11). Editoral Group of Coastal Vegetation in China, 1987. List of coastal plant in China. Vegetation speciality of comprehensive survey in coastal and tidal flat resources from China.
(12). South China Institute of Botany, Academia Sinica, 1987–1991. Flora of Guangdong, Vol. 1–2. Guangdong Science and Technology Press, China.
(13). Qiu Huaxing and Fu Guoai, 1987. *Hernandia* Linn.—A Genus new to Guangdong. Tropical Forestry Science and Technology 4:71–72.
(14). Li Zhiji et al., 1986. Reports of comprehensive survey in coastal and tidal flat resources from Guangxi Province, Vol.7, Plant, 119–174. Investigation Set of Coastal Vegetation from Guangxi.
(15). Li Yansheng et al., Silva Liaoningica. (manuscript)
(16). Liu Shene et al., 1959. Claves Plantarum Chinae Boreali-olientalis. Editio Academiae Scientiarum Sinicae.
(17). Liu Jiaguang and Han Qing, 1985. Studies on the vegetation in the saline-alkali soil and its indicative functions in the Dagang District of Tianjin. Acta Phytoecologica & Geobotanica Sinica, 9(2):158–164.
(18). Liu Fang-Xun et al., 1986. A study of the psammophilous vegetation along the coast of Jiangsu Province. Acta Phytoecologica & Geobotanica Sinica, 9(3):221–232.
(19). Liu Dongfang and Tang Shaoqing, 1989. Pollen morphology of the mangrove plants in China. Guihaia, 9(3):221–232.
(20). Team of Comprehensive Survey in Coastal and Tidal Flat resources from Jiangsu Province, 1985. Reports on comprehensive survey in coastal and tidal flat resources from Jiangsu Province, Chapter 11, Vegetation and plant resources, 336–349.
(21). Yang Zongdai, 1979. The geographical distribution of sea-grasses in China. Transactions of Oceanology and Limnology, 2: 41–46.
(22). Yang Qilun et al., 1988. Reports of comprehensive survey in coastal and tidal flat resources from Fujian Province, Chapter 10, Vegetation. Shanghai Science and Technology Press, pp:151–154.
(23). Wu Zhangzhong et al., 1990. Reports of comprehensive survey in coastal and tidal flat resources from Fujian Province, Chapter 10, Vegetation. China Ocean Press, pp:224–238.
(24). Zhang Hongda, Zhang Chaochang and Wang Bosun, 1957. Mangrove communities of Leizhou Peninsula. Acta Scientianum Naturalium Universitatis Sunyatseni, 1:122–145.
(25). Zhang Raoting and Lin Peng, 1984. Studies on the flora of mangrove plants from the coast of China. Journal of Xiamen University (Natural Science), 23(2):232–239.
(26). Zhang Raoting, 1984. The species composition and geographical distribution of the mangrove plants in Taiwan Strait. Journal of Oceanography in Taiwan Strait, 3(1):112–117.
(27). Zhang Raoting, 1985. A new species of *Acanthus* Linn. from Fujian (Family Acanthaceae). Wuyi Science Journal, 5:237–239.
(28). Chang Hungta et al., 1985. Mangroves of Hong Kong. Eoclogical Science, 2:1–8.
(29). Zhang Raoting and Zheng Peizhong, 1986. On the distribution of higher plants along Jiulong River Estuary. Journal of Xiamen University (Natural Science), 25(3):354–361.
(30). Chen Shupei, Liang Zhixian and Deng Yi, 1986. Mangrove and its reservation in Hainan Island. Ecologic Science, 1:12–19.
(31). Zhang Raoting and Gu Li, 1991. Studies on coastal higher plants in Xiamen and its adjacent areas. Journal of Oceanogrphy in Taiwan Strait, 10(4):386–391.
(32). Zhang Raoting and Huang Dunqing, 1992. Notes on mangrove plant *Acrostichum aureum* L. in Fujian. Journal of Oceanography in Taiwan Strait, 11(4):372–374.
(33). Zhang Raoting, 1992. Index of coastal higher plants in Fujian. (manuscript)
(34). Chen Pangyu and Li Zexian, 1986. An analysis of the flora in the coast region and islands around the mouth of Pearl River. Acta Botanica Austro Sinica, 2:51–70.
(35). Lin Peng, 1984. Mangrove Vegetation. China Ocean Press.
(36). Zhang Yuanchun. 1984. Beachy plants of Taiwan (Nature of Taiwan, Series, 10). Make Holiday Press.
(37). Lin Peng, Qiu Xizhao and Zhang Raoting, 1984. Studies on the vegetation of three islands of Pingtan, Nanri and Meizhou in the middle coast of Fujian Province. Acta Phytoecologica & Geobotanica Sinica, 8(1):303–304.
(38). Biology Deparment of Nankai University et al. 1987. Reports of comprehensive survey in coastal and tidal flat resources from Tianjin, Chapter 3, Vegetation, Resources silvae. China Ocean Press, pp:118–127.
(39). Hu Chiming, Li Zexian and Xing Fuwu, 1987. *Dolichandrone spathacea*—A mangrove plant recently discovered in China. Guihaia, 7(4):303–304.
(40). Hou, H. Y. et al., 1953. A preliminary study on the plant associations of Peitaiho, Hopei Province. Acta Botanica Sinica, 2(4):431–465.
(41). How Foon-chew and Ho Chun-Nien, 1953. Rhixophoraceae in Chinese flora. Acta Phytotaxonomica Sinica, 2(2):133–157.
(42). How Foon-chew (editor; revised by Wu Te-lin, Ko Wan-cheung and Chen Te-Choa et al.), 1982. A dictionary of the families and genera of Chinese seed plants (second edition). China Science Press.
(43). Hong De-yuang, 1974. Revisio Commelinacearum Sinicarum. Acta Phytotaxonomica Sinica, 12(4):474.
(44). Morton, B. and Morton, J., 1983. The sea shore ecology of Hong Kong. Hong Kong University Press.
(45). Ko Wan-cheung, 1985. Mangrove of Guangdong. Tropical Geography, 5(1):1–8.
(46). Huang Peiyou, 1983. The sandy sea shore vegetation of Hainan. Ecologic Science, 2:1–14.
(47). Cai Wanghou et al., 1988. Reports of comprehensive survey in coastal and tidal flat resources from Zhejiang Province, Chapter 10, Vegetation. China Ocean Press, pp:175–184.
(48). Editorial Group of Flora Fujianica, 1982–1989. Flora Fujianica, Tomus 1–4. Fujian Science and Technology Press.

*: all in Chinese except for (44)

Compiled by Zhang Yaoting and Zhang Chaochang; Edited by Xue Deqing

ANIMALIA

PORIFERA [SPONGIA]

DEMOSPONGIAE
AXINELLIDA
AXINELLIDAE

Acanthella hispida Pulitzer-Finali: Hong Kong. (5)
Axinella echidnaea Ridley: Xisha (=Paracel Islands). (4)
Phakellia fusca Thiele: South China Sea. (4)
P. pygmaca Thiele: East China Sea. (4)
Phycopsis terpnis de Laubenfels: South China Sea. (4)
Spongosorites porites de Laubenfels: Xisha (=Paracel Islands). (4)

ASTROPHORIDA
THENEIDAE

Thenea grayi Sollas: East China Sea. (4)

GEODIIDAE

Caminus tunghainensis Li: East China Sea. (4)
Geodia japonica (Sollas): East China Sea. (4)
Sidonops acanthostyles Li: Xisha (=Paracel Islands). (4)

STELLETTIDAE

Pilochrota haeckeli Sollas: East China Sea. (4)
Stelletta tenuis Lindgren: South China Sea. (4)
S. validissina Thiele: East China Sea. (4)

DICTYOCERATIDA
SPONGIIDAE

Ircinia pinna Hentschel: Hong Kong. (6)
I. variabilis (Schmidt): South China Sea. (4)
Phyllospongia foliascens (Pallas): Xisha (=Paracel Islands). (4)
Spongia ceylonensis Dendy: Hong Kong. (6)
S. oblique Duchassain & Micheloti: East China Sea. (4)
S. officinalis L.: South China Sea. (4)

DYSIDEIDAE

Dysidea fragilis (Montagu): Xisha (=Paracel Islands). (4)
Spongionella carteri (Burton): East China Sea. (4)

VERONGIIDAE

Psammaplysilla purpurea (Carter): Hong Kong. (6)

HADROMERIDA
SPIRASTRELLIDAE

Spirastrella aurivilli Lindgren: South China Sea. (4)
S. cuntarix (Schmidt): Hong Kong. (6)
S. semilunaris Lindgren: Xisha (=Paracel Islands). (4)
S. vagabunda Ridley: South China Sea. (4)

SUBERITIDAE

Suberites carnosa (Johnston): Yellow Sea; Xiamen (Fujian). (4)
S. domuncula (Olivi): Yellow Sea. (4)
S. ficus (Nardo): Yellow Sea. (4)

CLIONIDAE

Cliona celata Grant: China's coast. (4)
C. vastifica Hancock: Qingdao (Shandong). (4)

TETHYIDAE

Tethya aurantium (Pallas): Guangxi; Hainan; Xisha (=Paracel Islands). (4)
T. japonica Solla: Hong Kong. (5)
T. robusta Bowerbank: Hong Kong. (6)

HALICHONDRIDA
HALICHONDRIDAE

Halichondria panicea (Pallas): China's coast. (4)
Xestospongia exiqua (Kirkpatrick): Xisha (=Paracel Islands). (4)

HYMENIACIDONIDAE

Hymeniacidon sanguinea (Grant): Xisha (=Paracel Islands). (4)
Leucophloeus massalis Carter: Xisha (=Paracel Islands). (4)

HOMOSCLEROPHORIDA
PLAKINIDAE

Plakortis simplex Schulze: Xisha (=Paracel Islands). (4)

HAPLOSCLERIDA
RENIERIDAE

Gellius microxes Li: Hainan. (2)
G. toxius Topsent: Beibuwan (=Gulf of Tongking). (4)
G. varius (Bowerbank): Hainan. (4)
Pachychalina brevispiculifera Dendy: Fujian. (4)
P. elongata Ridley & Dendy: Xisha (=Paracel Islands). (4)
P. punctata Ridley & Dendy: Nansha (=Spratly Islands). (4)

P. renieroides (Lendenfeld): Xiamen (Fujian); Leizhou (Guangdong). (4)
P. similis Ridley & Dendy: Xisha (=Paracel Islands). (4)
P. variabilis Dendy: Fujian; Hong Kong. (4)
Petrosia testudinaria (Lamarck): Guangxi; Xisha (=Paracel Islands). (4)
Reniera implexa v. *baeri* Wilson: Hong Kong. (5)
Sigmadocia symbiotica Bergquist: Hong Kong; Guangxi; Hainan. (6)

HALICLONIDAE

Callyspongia diffusa (Ridley): Hainan. (4)
C. globosa Pulitzer-Finali: Hong Kong. (5)
C. orieminens Pulitzer-Finali: Hong Kong. (5)
C. ramosa (Gray): Xisha (=Paracel Islands). (4)
Cribochalina chinensis Pulitzer-Finali: Hong Kong. (5)
Gelliodes incrustan Dendy: Xisha (=Paracel Islands). (4)
G. pumila (Von Lendenfeld): Hong Kong. (6)
G. spinosella Thiele: South China Sea. (4)
Haliclona crassiloba (Lamarck): South China Sea. (4)
H. palmata (Lamarck): Nansha (=Spratly Islands). (4)
H. permolis (Bowerbank): Leizhou (Guangdong). (4)
H. subarmigera (Ridley): Guangxi. (4)
Siphonochalina flexa Pulitzer-Finali: Hong Kong. (5)
S. intermedia Ridley & Dendy: Xisha (=Paracel Islands). (4)

POECILOSCLERIDA
CLATHRIIDAE

Clathria lipochela Burton: Xisha (=Paracel Islands). (4)
Microciona prolifera (Ellis & Solander): East China Sea. (4)

AGELASIDAE

Agelas robusta Pulitzer-Finali: Hong Kong. (5)

BIEMNIDAE

Biemna fistulosa (Topsent): Hong Kong. (5)
Tylodesma tylostrongyla Li: Hong Kong. (2)

MYXILLIDAE

Lissodendoryx isodictyalis (Carter): Hainan. (4)
L. tylostyla Li: Hainan. (4)
Tedania anhelans (Lieberkuhn): Hainan. (4)
T. strongyla Li: Hainan. (2)

MYCALIDAE

Mycale macilenta (Bowerbank): Hainan. (4)
M. philippensis (Dendy): Xisha (=Paracel Islands). (2)
M. phylophila Hentschel: Hong Kong. (4)
M. vermistyla Li: Hainan. (2)

ESPERIOPSIDAE

Neofibularia chinensis Pulitzer-Finali: Hong Kong. (5)

SPIROPHORIDA
TETILLIDAE

Cinachyra porosa (Lendenfeld): Hainan. (4)
C. pseudobarbata Li: Nansha (=Spratly Islands). (4)
Craniella spatuliformis Li: East China Sea. (4)
Tetilla lituiformis Li.: East China Sea. (4)
T. ternatensis Kieschnick: Xisha (=Paracel Islands). (4)

CALCAREA

Grantia nipponica Hozawa: Bohai; Shandong (including Qingdao). (5)
Leuconia aspera (Schmidt): Guangxi. (4)
L. solida (Schmidt): Hong Kong. (6)
Leucosolenia botryodes (Ellis & Solander): Yellow Sea. (4)
L. tenui (Schffner): Qingdao (Shandong). (4)
L. variabilis Haeckel: Hong Kong. (4)
Scypha coronatum (Ellis & Solander): Qingdao (Shandong). (4)

HEXACTINELLIDA
AMPHIDISCOPHORIDA
HYALONEMATIDAE

Hyalonema apertum Schulze: East China Sea. (4)
H. proximum Schulze: East China Sea. (4)
H. toxeres Thomson: East China Sea. (4)

PHERONEMATIDAE

Pheronema annae Leidy: East China Sea. (4)
P. ijimai Okada: East China Sea. (1)
P. raphanus Schulze: East China Sea. (4)
P. weberi Ijima: East China Sea. (4)

MONORHAPHIIDAE

Monorhaphis intermedia Li: East China Sea. (3)

SEMPERELLIDAE

Semperella cucumis Schulze: East China Sea. (4)

HEXASTEROPHORIDA
EURETIDAE

Farrea occa (Bowerbank): East China Sea. (4)

APHROCALLISTIDAE

Aphrocallistes beatrix Gray: East China Sea. (4)
A. vastus Schulze: East China Sea. (4)

EUPLECTELLIDAE

Euplectella imperialist Ijima: East China Sea. (4)
E. marshalli Ijima: East China Sea. (4)
E. oweni Herklot & Marshall: East China Sea. (4)

REFERENCES*

(1). Li Jinhe, 1984. Studies on the Hexactinellida of the continental shelf of the East China Sea. Studia Marina Sinica, 23:105–118.
(2). Li Jinhe, 1986. Sponges of marine fouling organisms in China waters, I. Studia Marina Sinica, 26:73–116.
(3). Li Jinhe, 1987. *Monorhaphis intermedia* - a new species of Hexactinellida. Oceanologia & Limnologia Sinica, 18(2):130–137.
(4). Lin Jinhe, 1992. Supplement records of Hexactinellida in China. (manuscript)
(5). Pulitzer-Finali, G. 1980. Some shallow-water sponges from Hong Kong. Proceedings of the First International Marine Biological Workshop: The Marine Flora and Fauna of Hong Kong and Southern China. Hong Kong University Press, pp:97–100.
(6). Van Soest, R. W. M. 1980. A small collection of sponges from Hong Kong. Proceedings of the First International Marine Biological Workshop: The Marine Flora and Fauna of Hong Kong and Southern China. Hong Kong University Press, pp:85–96.

*: (1)-(4) in Chinese
Compiled by Li Jinhe

COELENTERATA [CNIDARIA]

HYDROZOA

HYDROIDOMEDUSAE [HYDROPOLYPSE, HYDROIDSI, HYDROMEDUSAE]
ANTHOMEDUSAE [GYMNOBLA, ATHECATA, ATHECATAE-ANTHOMEDUSAE]
AUSTRALOMEDUSIDAE

Platystoma bitentaculata Xu, Huang & Chen: West of Taiwan Bank. (19, 20)
P. nanhainensis Zhang: Taiwan Strait; offshore of eastern Guangdong; Nansha (=Spratly Islands). (65, 73, 78)

BOUGAINVILLIDAE

Bimeria francisana [*Garveia francisana*]: Bohai. (71)
B. vestita Wright: East China Sea. (83)
Bougainvillia bitentaculata Uchida: From Xiamen (Fujian) to South China Sea. (8, 73, 78)
B. britannica (Forbes) [*B. flavida*]: Bohai; Yellow Sea; East China Sea; Taiwan Strait; northern South China Sea. (27, 91)
B. fulva Agassiz & Mayer: Taiwan Strait; central South China Sea. (5, 73)
B. niobe Mayer: Taiwan Strait; offshore of eastern Guangdong; Beibuwan (=Gulf of Tongking). (5, 8, 73, 74)
B. paraplatygaster Xu, Huang & Chen: Taiwan shallow shore. (20)
B. platygaster (Haeckel): Taiwan Strait; near shore of Guangdong and Hainan; central South China Sea. (5, 14)
B. principis (Steenstrup): Yellow Sea. (27)
B. ramosa (Van Beneden) [*B. autumnalis*]: China's seashore. (9, 11, 49)
B. superciliaris (L. Agassiz): Yellow Sea. (27)
Koellikerina constricta (Menon): Taiwan Strait; Guangdong near shore. (11)
K. diforficulata Xu & Zhang: Southern Taiwan Strait; Guangdong near shore. (11)
K. fasciculata (Péron & Lesueur): Taiwan Strait; near shore of Guangdong and Hainan. (9, 11)
K. heteronemalis Xu, Huang & Chen: South of Taiwan Bank. (20)
K. multicirrata (Kramp): Guangdong near shore. (11)
K. octonemalis (Maas): Guangdong near shore; South China Sea. (12)
K. taiwanensis Xu, Huang & Chen: Taiwan shallow shore. (19, 20)
Lizzia gracilis (Mayer): South China Sea. (78)
Nemopsis bachei L. Agassiz: Bohai; Yellow Sea; East China Sea; mouth of Zhujiang (=Pearl River) (Guangdong). (7)

CYTAEIDIDAE

Cytaeis tetrastyla Eschscholtz: Mouth of Changjiang (=Yangtze River); Taiwan Strait; South China Sea. (5, 10, 11, 74)

CLAVIDAE

Turritopsis lata Von Lendenfeld: Yellow Sea; East China Sea; Taiwan Strait; Beibuwan (=Gulf of Tongking); Donghang (Hainan). (9, 11, 49)
T. nutricula McCrady: Bohai; Yellow Sea; East China Sea; Taiwan Strait; eastern Guangdong near shore. (11, 27, 58, 91)

EUDENDRIIDAE

Eudendrium californicum Torrey: South China Sea. (83)
E. capillare Alder: Bohai; Yellow Sea; Putuoshan (Zhejiang). (47, 71, 80)

E. pusillum amoyicum Hargill: East China Sea. (83, 93)
E. racemosum (Pallas): South China Sea. (71)
E. rameum (Pallas): Bohai. (71)
E. ramosum Linnaeus: Xiamen harbor (Fujian). (15)

HYDRACTINIIDAE

Podocoryne apicata Kramp: Xiamen harbor (Fujian); Taiwan shallow shore; South China Sea. (17, 35)
P. carnea M. Sars: Bohai; Beibuwan (=Gulf of Tongking); South China Sea. (74, 78)
P. gracilis Mayer: Nansha (=Spratly Islands). (79)
P. minima (Trinci): Bohai; Yellow Sea; Taiwan Strait; South China Sea. (5, 27, 73, 78)
P. minuta (Mayer): Daya Bay (Guangdong). (35)

STYLASTERIDAE

Allopora scabiosa (Broch): Dongshan (Fujian). (24)
Stylaster elegans Verrill: Xisha (=Paracel Islands). (23)

CALYCOPSIDAE

Bythocellata cruciformis Nair: South China Sea. (78)
Bythotiara depressa Naumor: Central South China Sea. (56)
B. murrayi Gunfher: Nansha (=Spratly Islands). (78)
Calycopsis bigelowi Vanhöffen: Nansha (=Spratly Islands). (78)
C. papillata Bigelow: South China Sea. (78)
Heterotiara anonyma Maas: South China Sea. (78)
H. minor Vanhöffen: Taiwan Strait; South China Sea. (5, 11)
Kanaka pelagica Uchida: Central South China Sea. (56)
Pseudotiara tropica (Bigelow): Central South China Sea. (56)

NIOBIIDAE

Niobia dendrotentacula Mayer: Taiwan Strait; offshore of eastern Guangdong. (5, 11)

PANDEIDAE

Amphinema australis ((Mayer): East China Sea; South China Sea. (78)
A. dinema (Péron & Lesueur): Yellow Sea; Taiwan Strait; near shore of Guangdong and Hainan; central South China Sea. (5, 9, 47, 78)
A. physophorum (Uchida) [*Stomotoca physophorum*]: Southern Zhejiang near shore. (58, 91)
A. rugosum (Mayer) [*A. rugosum* v. *shantungensis*, *A. rugosa* v. *tsingtauensis*]: Yellow Sea; Taiwan Strait; Guangdong near shore; Nansha (=Spratly Islands). (11, 27, 58, 78)
A. turrida (Mayer): Northern Jiangsu near shore; northern and central Taiwan Strait. (5, 7, 58)
Annatiara affinis (Hartlaub): South China Sea. (78)
Catablema vesicarium (A. Agassiz) [*Turris vesicaria*]: Qingdao (Shandong) near shore. (47)
Cirrhitiara simplex Xu, Huang & Chen: South of Taiwan Bank. (20)
Halitholus pauper Hartlaub: Nansha (=Spratly Islands). (78)
Leuckartiara gardineri Browne: Nansha (=Spratly Islands). (78)
L. hoepplii Hsu [*L. octona* v. *minor*]: Bohai; Yellow Sea; East China Sea; Taiwan Strait; Guangdong near shore. (5, 11, 84, 91)
L. octona (Fleming): Shengshan (Zhejiang); Fujian near shore; northern and central Taiwan Strait; Daya Bay (Guangdong). (1, 10)
L. orientalis Xu, Huang & Chen: West of Taiwan Bank. (20)
L. zacae Bigelon: South China Sea. (78)
Merga bulbosa Bouillon: Central South China Sea. (56)
M. macrobulbosa Xu, Huang & Chen: West of Taiwan Bank. (20)
M. tergestina (Neppi & Stiasng): Xiamen harbor (Fujian); central and northen Taiwan Strait; offshore of eastern Guangdong; central South China Sea. (11, 56)
Neoturris papua (Lesson): Central South China Sea. (56)
Octotiara russelli Kramp: South China Sea. (59)
Pandea conica (Quoy & Gaimard): Xiamen harbor (Fujian); western Guangdong near shore; Nansha (=Spratly Islands). (1, 11, 78)
Pandeopsis ikarii Uchida: Taiwan Strait; western Guangdong near shore; northern Beibuwan (=Gulf of Tongking); central South China Sea. (5, 11, 56)

PROTIARIDAE

Halitiara formosa Fewkes: South China Sea. (78)
Halitiarella ocellata Bouillon: Southwestern Taiwan shallow shore. (20)
H. minutum Xu, Huang & Chen: Taiwan shallow shore. (20)
Latitiara orientalis Xu & Huang: Taiwan shallow shore. (20)
Paratiara digitalis Kramp & Damas: Taiwan Strait; offshore of eastern Guangdong; South China Sea. (5, 11)

RATHKEIDAE

Pseudorathkea macrogastrica Xu & Huang: Luoyuan Bay (Fujian). (18)
Rathkea octopunctata (M. Sars): Bohai; Yellow Sea; southern Zhejiang; Fujian near shore. (10, 27, 58)

CLADONEMATIDAE

Cladonema radiatum Dujardin [*C. mayersi*]: Northern Yellow Sea. (27)

CORYNIDAE

Coryne crassa Fraser: Qingdao (Shandong). (46)
C. pusilla Gaertner: Putoushan (Zhejiang); South China Sea. (71, 80)
Dicodonium jeffersoni (Mayer): Southwestern Taiwan; offshore of eastern Guangdong. (11)
Dipurena ophiogaster Haeckel: Eastern Hainan. (9)
D. strangulata McCrady: Xiamen harbor (Fujian); Daya Bay (Guangdong). (10, 35)
Sarsia nipponica Uchida: Bohai; Yellow Sea. (27)

HYDROCORYNIDAE

Hydrocoryne miurensis Stechow [*Sarsia resplendens*]: Putou (Zhejiang); central and southern Fujian near shore. (8, 80)

MOERISIIDAE

Moerisia inkermanica Palschikowa-Ostroumova [*Ostroumovia inkermanica*]: Mouth of Jiulong River (Fujian) and of Zhujiang (=Pearl River) (Guangdong); Nansha (=Spratly Islands). (9, 78)
Odessia microtentaculata Xu, Huang & Chen: West of Taiwan Bank. (20)
Tiaricodon coeruleus Browne [*Moerisia lyonsi, Urashimea globosa*]: Weihai (Shandong); mouth of Jiulong River (Fujian); northern Beibuwan (=Gulf of Tongking). (1, 11)

CORYMORPHIDAE

Cnidocodon xiamenensis (Zhang & Wu) [*Ramus xiamenensis*]: Xiamen harbor (Fujian); northern Beibuwan (=Gulf of Tongking); South China Sea. (63, 74)
Euphysilla pyramidata Kramp: Off of mouth of Zhujiang (=Pearl River) (Guangdong); South China Sea. (12, 78)
Euphysomma brevia (Uchida): Mouth of Jiulong River (Fujian). (17)
Euphysora annulata Kramp: Central South China Sea. (48)
E. bigelowi Maas [*Euphysa bigelowi*]: Yellow Sea; East China Sea; Taiwan Strait; Hong Kong; South China Sea. (5, 11, 27, 58, 78)
E. furcata Kramp: Central South China Sea. (56)
Gotoea typica Uchida: Northern South China Sea. (12)
Steenstrupia nutans (M. Sars): Pingtan (Fujian) near shore. (7)
Vannuccia forbesii (Mayer) [*Hyxbocodon forbesii*]: Taiwan Strait; Guangdong near shore; Sanya (Hainan); northern Beibuwan (=Gulf of Tongking); Nansha (=Spratly Islands). (5, 7, 9, 35, 78)

EUPHYSIDAE

Euphysa aurata Forbes: Bohai; Yellow Sea; East China Sea; Taiwan Strait; South China Sea. (5, 27, 73)

HALOCORDYLIDAE

Halocordyle cavolinii Ehrenberg [*Pennaria cavolinii*]: Putou and Qingbin (Zhejiang); Fuding and Xiamen (Fujian). Attached to rocks or floats. (15, 80)
H. disticha Goldfuss: East China Sea; Daya Bay (Guangdong). (31, 71, 83)
H. grandis Kramp [*Pennaria grandis*]: South China Sea. (78)
H. tiarella Agres [*Pennaria tiarella*]: Northern and central Taiwan Strait; Jinjiang and Xiamen (Fujian) near shore; eastern Guangdong near shore; South China Sea. (5, 10, 11)

TUBULARIIDAE

Ectopleura dumontieri (Van Beneden): From Bohai to South China Sea. (5, 11, 47)
E. guangdongensis Xu, Huang & Chen: West of Taiwan Bank. (20)
E. latitaeniata Xu & Zhang: Southwestern Taiwan; offshore of eastern Guangdong; off mouth of Zhujiang (=Pearl River) (Guangdong); central South China Sea. (11, 56)
E. minerva Mayer: Bohai; Yellow Sea; Taiwan Strait; Guangdong nearshore; central South China Sea; Nansha (=Spratly Islands). (5, 8, 11, 78)
E. sacculifera Kramp: Southwestern Taiwan shallow shore. (20)
E. xiamenensis Zhang & Ling: Xiamen harbor (Fujian). (66)
Hybocodon amoyensis Hargitt: East China Sea. (83)
H. octopleurus Kao: Yantai (Shandong). (47)
H. prolifer L. Agassiz: Zhangpu (Fujian) nearshore. (8)
Tubularia larynx Ellis & Solander: Putou (Zhejiang). Tide pools and boats. (80)
T. marina Torrey: Bohai; Yellow Sea. (46, 71)
T. mesembryanthemum Allman: Bohai; Yellow Sea; East China Sea; South China Sea. (15, 46, 71, 80)

VELELLIDAE

Porpita porpita (Linné) [*P. pacifica, P. umbrella*]: Southern Zhejiang nearshore; Fujian nearshore; Taiwan Strait; Guangdong nearshore; Hong Kong; South China Sea. (3, 49, 80)
Velella velella (Linné) [*V. lata*]: Offshore of Zhoushan (Zhejiang) and Pingtan (Fujian) offshore; Taiwan shallow shore; Sanya (Hainan); South China Sea. (54, 73)

MILLEPORIDAE

Millepora dichotoma Forskål: Xisha (=Paracel Islands). (22)
M. etaesa Forskål: Xisha (=Paracel Islands). (22)
M. intricata Milne-Edwards: Taiwan; Sanya (Hainan). (6)
M. latifolia Boschma: Sanya (Hainan); Xisha (=Paracel Islands). (6, 22)
M. murrayi Quelch: Taiwan; Sanya (Hainan); South China Sea. (6)

M. platyphylla Hemprich & Ehrenberg: Taiwan; Sanya (Hainan); Xisha (=Paracel Islands); Nansha (=Spratly Islands). (6, 22)

M. tenera Boschma: Taiwan; Xisha (=Paracel Islands); Zhongsha (=Macclesfield Bank). (22)

M. xishanensis Zou: Xisha (=Paracel Islands). (22)

TEISSIERIDAE

Teissiera australe Bouillon: Southwestern Taiwan shallow shore. (20)

T. macrocystae Xu, Huang & Chen: Southwestern Taiwan shallow shore. (20)

T. medusifera Bouillon: West of Taiwan Bank. (20)

T. polypofera Xu, Huang & Chen: Southwestern Taiwan shallow shore. (20)

ZANCLEIDAE

Zanclea costata Gegenbaur: Yellow Sea; Zhoushan (Zhejiang); Taiwan Strait; South China Sea. (27, 47)

Z. orientalis Browne: Nansha (=Spratly Islands). (78)

LEPTOMEDUSAE
[CALYPTOBLASTEA, THECATA, THECATAE-LEPTOMEDUSAE]
AEQUOREIDAE

Aequorea aequorea (Forskål): Eastern and southwestern Taiwan; northern South China Sea. (11)

A. australis Uchida: Yellow Sea; Taiwan Strait; Hong Kong; South China Sea. (5, 27, 58)

A. coerulescens (Brandt): Northern Yellow Sea. (27)

A. conica Browne: Yellow Sea; East China Sea; Taiwan Strait; South China Sea. (1, 5, 9, 58, 74)

A. floridana (Agassiz): Nansha (=Spratly Islands). (79)

A. globosa Eschscholtz: Taiwan Strait; northern South China Sea. (11, 35)

A. macrodactyla (Brandt): Taiwan Strait; northern South China Sea. (11, 35)

A. parva Browne: Southern Yellow Sea; southern Zhejiang nearshore; Taiwan Strait; northern South China Sea. (1, 5, 11, 35, 58)

A. pensilis (Eschscholtz): Fujian; southwestern Taiwan; eastern and western Guangdong; Haikou (Hainan). (1, 9, 16)

Gangliostoma guangdongensis Xu: Northern South China Sea. (14)

Zygocanna vagans Bigelow: Taiwan; South China Sea. (86)

BLACKFORDIIDAE

Blackfordia manhattensis Mayer: Southern Yellow Sea; southern Zhejiang nearshore; Xiamen harbor (Fujian); Daya Bay (Guangdong); northern Beibuwan (=Gulf of Tongking). (49, 58, 74)

B. polytentaculata Hsu & Chin: Mouth of Jiulong River (Fujian); Zhangpu (Fujian) nearshore; mouth of Zhujiang (=Pearl River); northern Beibuwan (=Gulf of Tongking). (7, 11)

B. virginica Mayer: Mouth of Min River and Jiulong River (Fujian) and Zhujiang (=Pearl River) (Guangdong). (7, 11)

CAMPANULINIDAE

Calycella syringa (Linneaus): Yellow Sea. (95)

MALAGAZZIIDAE

Malagazzia carolinae (Mayer) [*Phialucium carolinae, P. virens*]: From Bohai to northern South China Sea. (5, 9, 27, 58, 74)

M. condensum (Kramp) [*Phialucium condensum*]: Bohai; southern Fujian nearshore; southwestern Taiwan; eastern Guangdong nearshore; central South China Sea; Nansha (=Spratly Islands). (10, 11, 73, 78)

M. curviductum (Xu & Zhang) [*Phialucium curviductum*]: Taiwan Strait; eastern and western Guangdong nearshore. (11)

M. cyphogonia (He & Xu) [*Phialucium cyphogonia*]: Yantai (Shandong). (29)

M. taeniogonia (Chow & Huang) [*Phialucium taeniogonia*]: Bohai; Yellow Sea; Fujian nearshore; northern Beibuwan (=Gulf of Tongking). (8, 27)

Octocannoides ocellata Menon: Taiwan Strait; eastern Guangdong nearshore; northern Beibuwan (=Gulf of Tongking); Sanya (Hainan). (8, 9, 11, 74)

Octophialucium funerarium (Quoy & Gaimard): Southeastern Taiwan; offshore of eastern Guangdong. (11, 16)

O. indicum Kramp [*Octocanna polynema*]: Yellow Sea; East China Sea; Taiwan Strait; northern and southern South China Sea. (1, 11, 56)

O. medium Kramp: Taiwan Strait; South China Sea. (9, 11, 78)

O. solidum (Menon): Northern Jiangsu; Zhoushan (Zhejiang); Xiamen harbor (Fujian). (58)

PHIALELLIDAE

Phialella macrogona Xu, Huang & Wang: Mouth of Jiulong River (Fujian); Taiwan shallow shore. (17, 73)

SUGIURIDAE

Sugiura chengshanense (Ling) [*Gastroblasta raffaelei v. chengshanensis, G. chengshanensis, Phialidium chengshanensis*]: From Bohai to northern South China Sea. (11, 27, 58, 74, 91)

DIPLEUROSOMATIDAE

Dichotomia cannoides Brooks: South of Taiwan Bank; offshore of eastern Guangdong; Nansha (=Spratly Islands). (11, 16, 78)

MELICERTIDAE

Melicertoide octolabiatis Xu, Huang & Chen: West of Taiwan Bank. (20)
Melicertum octocostatum (M. Sars): Southwestern Taiwan shallow shore. (20)

EIRENIDAE

Eirene brevigona Kramp: Southern Zhejiang nearshore; Taiwan Strait; South China Sea. (5, 9, 11, 57, 73, 78)
E. ceylonensis Browne [*Phortis ceylonensis*]: Bohai; Yellow Sea; East China Sea; Taiwan Strait; northern South China Sea. (1, 5, 11, 27, 58)
E. chiaochowensis (Kao & Li) [*Phortis lactea v. chiaochowensis*]: Yantai (Shandong); southern Zhejiang nearshore; Taiwan Strait; Hong Kong; northern and central South China Sea. (1, 5, 9, 11, 12, 39, 47, 57)
E. hexanemalis (Goette) [*Irenopsis hexanemalis*]: Yantai (Shandong); southern Zhejiang nearshore; Taiwan Strait; Hong Kong; northern and central South China Sea. (1, 5, 9, 11, 12, 39, 47)
E. kambara Agassiz & Mayer [*Phortis kambara*]: Jiaozhou Bay (Shandong); Taiwan Strait; Daya Bay (Guangdong); northern Beibuwan (=Gulf of Tongking); eastern Hainan. (5, 8, 35, 47)
E. menoni Kramp [*Phortis lactea*]: Mouth of Liao River (Bohai); Yellow Sea; East China Sea nearshore; Taiwan Strait; northern South China Sea. (8, 9, 11, 35, 58)
E. palkensis Browne: Xiamen harbor (Fujian); Taiwan shallow shore; Hong Kong. (7, 73)
E. pyramidalis (L. Agassiz) [*Phortis pyramidalis*]: Bohai; Yellow Sea; Zhoushan (Zhejiang); mouth of Jiulong River (Fujian); Shilinzhou (Hainan). (9, 47, 49)
E. tenuis (Browne): Mouth of Jiulong River (Fujian); northern Beibuwan (=Gulf of Tongking). (13, 74)
E. varidula (Péron & Lesueus): Xiamen harbor (Fujian); central South China Sea. (56)
Eutima curva Browne: Xiamen harbor (Fujian); northern Beibuwan (=Gulf of Tongking); central South China Sea. (13, 56, 57, 74)

E. gegenbauri (Haeckel) [*Octorchis gegenbauri*]: Jiaozhou Bay (Shandong); northern Jiangsu nearshore; South China Sea. (47, 58)
E. gentiana (Haeckel): Xiamen harbor (Fujian). (59)
E. gracilis (Forbes & Goodsir): northern Beibuwan (=Gulf of Tongking); Sanya (Hainan). (9, 74)
E. japonica Uchida: Xiamen harbor (Fujian); Taiwan shallow shore; Daya Bay (Guangdong); northern Beibuwan (=Gulf of Tongking). (13, 35, 73, 74)
E. levuka (Agassiz & Mayer) [*E. campanulata*]: Taiwan Strait; Hong Kong; northern South China Sea. (1, 5, 9, 10, 12)
E. mira McCrady: Xiamen harbor (Fujian). (59)
E. modesta (Harflaub): Xiamen harbor (Fujian). (59)
E. neucaledonia Uchida: Xiamen harbor (Fujian); central South China Sea. (56, 90)
E. orientalis (Browne): Northern Jiangsu; Taiwan Strait; offshore of eastern Guangdong; Daya Bay (Guangdong); northern Beibuwan (=Gulf of Tongking). (5, 8, 11, 35, 74)
E. variabilis McCrady: Central Fujian nearshore; Taiwan shallow shore; Daya Bay (Guangdong); central South China Sea. (10, 35, 56, 73)
Eutonina scientillans (Bigelow): Xiamen harbor (Fujian). (90)
Helgicirrha brevistyla Xu & Huang: Mouth of Jiulong River (Fujian); Daya Bay (Guangdong); northern Beibuwan (=Gulf of Tongking). (13, 35, 73)
H. cornelii Bouillon: Dongshan (Fujian); northern Beibuwan (=Gulf of Tongking). (8, 74)
H. malayensis (Stiasny): Yellow Sea; southern Zhejiang nearshore; Taiwan Strait; northern and central South China Sea. (5, 10, 27, 35, 58)
Tima formosa L. Agassiz: Bohai; Yellow Sea. (27)

LAODICEIDAE

Laodicea indica Browne: Taiwan Strait; South China Sea. (5, 11, 35, 56, 78)
L. undulata (Forbes & Goodsir): Taiwan Strait; South China Sea. (5, 8, 11, 56, 78)
Staurodiscus gotoi (Uchida): Northern and central Taiwan Strait. (5, 10)
S. vietnanensis Kramp: Nansha (=Spratly Islands). (78)
Toxorchis arcuatus Haeckel: Nansha (=Spratly Islands). (78)
T. polynema Kramp: From Taiwan Strait to South China Sea. (5, 59)

CIRRHOLOVENIIDAE

Cirrholovenia polynema Kramp: Taiwan Strait; South China Sea. (11, 56)
C. tetranema Kramp: Southern Yellow Sea; Taiwan shallow shore; South China Sea. (11, 56, 78)
Paralovenia bitentaculata Bouillon: Central South China Sea. (56)

EUCHEILOTIDAE

Eucheilota diademada Kramp: Nansha (=Spratly Islands). (78)
E. duodecimalis A. Agassiz: Zhoushan (Zhejiang); South China Sea. (56, 58, 78)
E. macrogona Zhang & Ling: Xiamen harbor (Fujian); Taiwan shallow shore; South China Sea. (66, 73, 74, 78)
E. menoni Kramp: Southern Yellow Sea; Taiwan Strait; offshore of eastern Guangdong; Nansha (=Spratly Islands). (5, 11, 58, 78)
E. multicirrs Xu & Huang: Luoyuan Bay (Fujian). (18)
E. paradoxia Mayer: Daya Bay (Guangdong); Nansha (=Spratly Islands). (35, 78)
E. taiwanensis Xu & Huang: Taiwan shallow shore. (20)
E. tropica Kramp: Xiamen harbor (Fujian); northern Beibuwan (=Gulf of Tongking); Nansha (=Spratly Islands). (78, 90)
E. ventricularis MaCrady: From Yellow Sea to South China Sea. (5, 8, 47, 78)

LOVENELLIDAE

Lovenella assimilis (Browne) [*Eucheilota menoni*]: From Yellow Sea to South China Sea. (11, 27, 56, 58, 78)
L. haichangensis Xu & Huang: Ronghai and mouth of Jiulong River (Fujian); northern Beibuwan (=Gulf of Tongking). (13, 74)
Tiaropsis multicirrata (M. Sars): Yantai (Shandong). (27)

LAFOEIDAE

Filellum serratum (Clarke) [*F. serpens*]: Bohai; Yellow Sea; East China Sea. (15, 46, 82)
Hebella laterocaudata Billard: Nansha (=Spratly Islands). (52)
H. neglecta Stechow: Yellow Sea. (76)
H. parasitica (Ciamician): Huian (Fujian). (15)
Lafoea dumosa (Fleming): China's seashore. (53, 95)
Zygophylax rufa (Bale): South China Sea. (94)
Z. tizardensis Kirkpatrik: South China Sea. (52)

HALECIIDAE

Halecium cymiforme Allman: East China Sea. (85)
H. flexile Allman: South China Sea. (93)
H. humile Pictet: South China Sea. (88)
H. sessile Norman: East China Sea. (83)
H. tenellum Hincks: East China Sea. (83)

AGLAOPHENIIDAE

Aglaophenis amoyensis Hargitt: East China Sea. (83)
A. cupressina Lamouroux: South China Sea. (52, 85, 94)
A. pluma (Linnaeus): Nansha (=Spratly Islands). (52)
A. suensonii Jäderholm: Xiamen harbor and Shacheng Bay (Fujian). (15)
A. whiteleggei Bale: Yellow Sea; East China Sea; South China Sea. (15, 48, 75, 80, 83, 93)
Gymnangium expansa (Jäderholm): East China Sea. (85)
G. gracilicaulis (Jäderholm): East China Sea. (85)
G. hians (Busk): East China Sea; South China Sea. (48)
G. speciosa (Allman): East China Sea. (83)
G. vagae (Jäderholm): East China Sea; South China Sea. (52, 85)
Lytocarpus niger (Nutting): South China Sea continental shelf. (77)
L. nutlingi Hargitt: East China Sea; South China Sea. (15, 31)
L. philippinus (Kirchenpauer): Taiwan Strait; South China Sea. (15, 48)
L. phoeniceus (Busk): East China Sea. (85, 94)
Monoserius pennarius (Linnaeus): East China Sea; South China Sea. (53, 85, 94)
Thecocarpus brevirostris (Busk): Nansha (=Spratly Islands). (52)
T. myriophyllum orientalis Billar: East China Sea; South China Sea. (53, 88)

HALOPTERIIDAE

Antennella recta Nutting: South China Sea. (94)
A. secundaria (Gmelin): South China Sea. (88)
Halopteris diaphana (Heller): South China Sea. (52)
Monostaechas quadridens (McCrady): East China Sea. (85)

KIRCHENPAUERIIDAE

Pycnotheca mirabilis (Allman): Yellow Sea. (46, 48)

PLUMULARIDAE

Plumularia badia Kirchenpauer: East China Sea. (71)
P. lagenifera Allman: East China Sea. (83)
P. setacea (Linnaeus): From Bohai to South China Sea. (15, 46, 48, 52, 80)
P. setaceoides Bale: From Yellow Sea to South China Sea. (15, 46, 48, 71, 83, 93)

SERTULARIIDAE

Diphasia dubia Hargitt: East China Sea. (83)
D. heurteli Billard: South China Sea. (94)
D. minuta Billard: South China Sea. (53)
D. scalariformis Kirkpatrik: East China Sea. (85)
D. thornelyi Ritchie: East China Sea; South China Sea. (51)
Dynamena crisioides Lamouroux [*Sertularia tubuliformis*]: East China Sea; South China Sea. (15, 52, 93)
D. nanshaensis Tang: Nansha (=Spratly Islands). (52)

D. quadridentata (Ellis & Solander): East China Sea. (83)
Idiellana pristis (Lamouroux): East China Sea; South China Sea. (53, 94, 95)
Selaginopsis trilateralis Fraser: Offshore of Shi Island (Shandong). (46)
Sertularella gayi (Lamouroux): South China Sea. (88)
S. inabai Stechow: Yellow Sea. (76)
S. indivisa bidentata Ling: East China Sea. (93)
S. mirabilis (Jäderholm): East China Sea; South China Sea. (85, 94)
S. miurrensis Stechow: Yellow Sea; East China Sea. (15, 46, 80, 93)
S. philippinensis Hargitt: South China Sea. (94)
S. sinensis Jäderholm: Yellow Sea; East China Sea. (46, 85)
S. tenella (Alder): Huian (Fujian). (15)
Sertularia cornicina (McCrady): Ningde and Lianjiang (Fujian). (15)
S. distans (Lamouroux): South China Sea. (85)
S. fabricii Levinsen: Xiamen harbor (Fujian). (15)
S. furcata Trask: Yantai and Qingdao (Shandong); Putou (Zhejiang); Daya Bay (Guangdong). (31, 46, 80)
S. similis Clark: Yellow Sea. (46, 76)
S. westindica (Stechow): Nansha (=Spratly Islands). (52)
Symplectoscyphus hozawai Stechow: East China Sea. (93)
S. huanghaiensis Tang & Huang: Yellow Sea. (83)
Thuiaria lonchitis (Ellis & Solander): East China Sea. (85)
T. marktanneri Stechow: Yellow Sea. (95)
T. similis (Clark): Offshore of Shi Island (Shandong). (46)
Thyroscyphus torresi (Busk) [*Cnidoscyphyus torresi*]: South China Sea. (53)

SYNTHECIIDAE

Synthecium campylocarpum Allman: Qingdao (Shandong); Putou (Zhejiang); Daya Bay (Guangdong). (31, 46, 80)
S. patulum (Busk): Yellow Sea; East China Sea. (46, 76, 80)
S. tubithecum Allman: South China Sea. (94)

CAMPANULARIIDAE

Campanularia calceolifera Hincks: Jiaozhou Bay (Shandong); Xiamen harbor (Fujian). (15, 46)
C. denticulata Clarke: East China Sea. (71)
C. hincksii Alder: South China Sea. (88)
C. verticillata (Linné): Offshore of Shi Island (Shandong); East China Sea. (46, 71)
C. volubjlis (Linnaeus): South China Sea. (57, 83)
Clytia ambiguum (Agassiz & Mayer) [*Phialidium ambigumm*]: South China Sea; northern Beibuwan (=Gulf of Tongking). (74)
C. cylindrica Agassiz: Yellow Sea; East China Sea; South China Sea. (46, 71)
C. discoida (Mayer) [*Phialidium discoidum*]: Bohai; Taiwan Strait; northern South China Sea. (5, 11, 27)
C. folleata (McCrady) [*Phialidium folleatum*]: Yellow Sea; Taiwan Strait; South China Sea. (5, 11, 47, 58, 74, 78)
C. hemisphaerica (Linné) [*Phialidium hemisphaericium, C. edwardsi, C. minuta*]: From Bohai to South China Sea. (10, 46, 47, 56, 58, 71, 80)
C. hexacanalis Xu, Huang & Chen: West of Taiwan Bank. (20)
C. languida (Agassiz): Nansha (=Spratly Islands). (70)
C. malayense (Kramp) [*Phialidium malayense*]: South China Sea. (12, 56, 78)
C. mccradyi (Brooks) [*Phialidium mccradyi*]: Mouth of Jiulong River (Fujian); Nansha (=Spratly Islands). (17, 79)
C. ovalis (Mayer) [*Phialidium ovale*]: Mouth of Jiulong River (Fujian); Nansha (=Spratly Islands). (13, 79)
C. stechowi Hargiff: East China Sea. (83)
C. uchidai Kramp [*Phialidium uchidia*]: Nansha (=Spratly Islands). (78)
Gonothyraea clarki (Maktanner-Turneretscher): Bohai; Yellow Sea; South China Sea. (46, 71)
G. inorata Nutting: East China Sea. (83)
Laomedea calceolifera (Hincks): Yellow Sea. (46)
L. flexuosa (Hincks): East China Sea. (93)
Obelia bidentala Clark [*O. bicuspidata*]: Yantai (Shandong); Daya Bay (Guangdong). (31, 46)
O. dichotoma (Linnaeus) [*O. borealis, O. gelatinosa, O. gracilis, O. longissima, Campanularia gelatinosa*]: From Yellow Sea to South China Sea. (15, 46, 47, 71)
O. geniculata (Linné): From Bohai to northern South China Sea. (15, 34, 46, 71)
Orthopyxis compressa (Clark) [*Agastra rubra*]: Mouth of Jiulong River (Fujian). (17)
O. integra (Macgillivray) [*Campanularia integra*]: Jiaozhou Bay (Shandong). (46, 82)
O. platycarpa Bale: Putou (Zhejiang). (80)
Rhizocaulus chinensis (Marttanner-Turneretscher): Yellow Sea; East China Sea. (76, 95)
R. verticillatus (Linnaeus): Bohai; Yellow Sea. (46)

PHIALUCIIDAE

Phialucium mbenga (Agassiz & Mayer): Xiamen harbor (Fujian); South China Sea. (10, 56, 78)

LIMNOMEDUSAE
[LIMNOPOLYPAE-LIMNOMEDUSAE]
OLINDIASIDAE

Gonionemus vertens A. Agassiz [*G. murbachii* v. *chekiangensis, G. m.* v. *oshoro*]: Bohai; Yellow Sea; East China Sea. (27, 47, 49, 91)
Maeotia inexpectata Ostroumaff: Mouth of Jiulong River (Fujian). (17)
Scolionema survaense (Agassiz & Mayer): Taiwan Strait; offshore of eastern Guangdong; Nansha (=Spratly Islands). (5, 8, 10, 11, 78)

PROBOSCIDACTYLIDAE

Proboscidactyla flavicirrata Brandt: Bohai; Yellow Sea. (27, 58)
P. mutabilis (Browne) [*Willsia mutabilis*]: Yantai (Shandong) nearshore. (47)
P. ornata (MaCrady) [*P. ornata* v. *gemmifera*]: From Yellow Sea to South China Sea. (1, 9, 27, 47, 56, 57, 58)
P. stellata (Forbés): Yellow Sea; Nansha (=Spratly Islands). (27, 58, 78)

TRACHYMEDUSAE
GERYONIIDAE

Geryonia proboscidalis (Forskål): Taiwan Strait; South China Sea. (5, 9, 11, 56, 57, 78)
Liriope tetraphylla (Chamisso & Eysenhardt): From Yellow Sea to South China Sea. (5, 10, 12, 27, 35, 47, 56, 58, 73, 75, 78)

HALICREATIDAE

Halitrephes maasi Bigelow: Nansha (=Spratly Islands). (78)
Halicreas conica Vanhöffen: Nansha (=Spratly Islands). (79)
H. minimum Fewkes: South China Sea. (59)
H. racovitzae (Maas): Central South China Sea. (56)
Varitentaculata yantaiensis He: Bohai; Yellow Sea. (30)

PETASIDAE

Petasiella asymmetrica Uchida: Taiwan Strait; northern and central South China Sea. (5, 12, 56)

RHOPALONEMATIDAE

Aglantha elata (Haeckel): Nansha (=Spratly Islands). (78)
Aglaura hemistoma Péron & Lésueur: East China Sea; Taiwan Strait; South China Sea. (2, 4, 5, 7, 9, 12, 16, 56, 58, 70, 74, 78)
Amphogona apicata Kramp: Taiwan Strait; South China Sea. (12, 56, 73, 78)
A. apsteini (Vanhöffen): Taiwan Strait; northern and southern South China Sea. (9, 11, 73, 78)
A. pusilla Hartlaub: Taiwan shallow shore; northern and central South China Sea. (12, 35, 56, 73, 74)
Colobonema igneum (Vanhöffen): Nansha (=Spratly Islands). (78)
C. sericeum Vanhöffen: Taiwan Strait; South China Sea. (59)
Crossota brunnea Vanhöffen: Taiwan Strait; South China Sea. (59)
Rhopalonema funerarium Vanhöffen: Off mouth of Zhujiang (=Pearl River) (Guangdong); Nansha (=Spratly Islands). (12, 78)
R. velatum Gegenbaur: East China Sea; Taiwan Strait; South China Sea. (5, 11, 56, 73, 78)
Smithea eurygaster Gegenbaur: Nansha (=Spratly Islands). (78)

NARCOMEDUSAE
AEGINIDAE

Aegina citrea Eschscholtz: Taiwan Strait; eastern Taiwan; South China Sea. (11, 16, 56, 73, 78)
Aeginura grimaldii Maas: From southern Yellow Sea to South China Sea. (2, 4, 5, 7, 9, 11, 12, 16, 56, 58, 73, 74, 78)
Otoporpa polystriata Xu & Zhang: Taiwan Strait; northern and southern South China Sea. (11, 16, 78)
Solmundella bitentaculata (Quoy & Gaimard): From southern Yellow Sea to South China Sea. (1, 5, 8, 9, 11, 12, 16, 50, 56, 58, 73, 74, 78)

SOLMARISIDAE

Pegantha martagon Haeckel: South China Sea. (56, 78)
P. triloca Haeckel: Xiamen (Fujian); South China Sea. (59, 78)
Solmaris flavecens Kolliker: Central South China Sea. (56)
S. leuocostyla (Will): Taiwan Strait; South China Sea. (5, 8, 12, 56, 78)

CUNINIDAE

Cunina frugifera Kramp: Nansha (=Spratly Islands). (78)
C. octonaria MaCrady: East China Sea; Taiwan Strait; Nansha (=Spratly Islands). (59, 78)
C. peregrina Bigelow: Nansha (=Spratly Islands). (78)
Solmissus marshalli Agassiz & Mayer: South China Sea. (56, 59, 78)

SIPHONOPHORAE
CYSTONECTAE
PHYSALIIDAE

Physalia physalis (Linné): East China Sea; South China Sea. (54, 80)

RHIZOPHYSIDAE

Rhizophysa filiformis (Forskål): Taiwan Strait; Dongsha (=Pratas Islands). (45, 54, 61)

PHYSONECTAE
AGALMIDAE

Agalma elegans (Sars): Southern Yellow Sea; East China Sea; Taiwan Strait; northern and central South China Sea. (5, 8, 11, 34, 41, 42, 54, 56)

A. okeni Eschoschohz: East China Sea; Taiwan Strait; South China Sea. (9, 11, 41, 42, 54, 56)

Halistemma rubrum (Voge): Central and southern South China Sea. (41, 61)

Nanomia bijuga (Chiaje): East China Sea; Taiwan Strait; central South China Sea. (34, 39, 41, 42, 56)

PYROSTEPHIDAE

Bargmannia elongata Totton: Central and southern South China Sea. (41, 56, 57, 67)

PHYSOPHORIDAE

Physophora hydrostatica Forskål: From southern Yellow Sea to South China Sea. (2, 4, 5, 11, 39, 41, 42, 56, 58, 67)

ATHORYBIIDAE

Athorybia rosacea (Forskål): Central South China Sea. (38, 56)

Melophyes melo (Quoy & Gaimard): Central South China Sea. (38, 55, 56)

NECTALIDAE

Nectalia loligo Haeckel: Central and southern South China Sea. (54, 56, 61)

FORSKALIDAE

Forskalia cuneata Chun: Central South China Sea. (56)

F. edwardsi Kölliker: East China Sea; Taiwan Strait; central and southern South China Sea. (35, 41, 42, 54, 61)

CALYCOPHORAE
PRAYIDAE

Amphicaryon acaule Chun: East China Sea; Taiwan; South China Sea. (5, 11, 39, 41, 54, 55, 56)

A. ernesti Totton: East China Sea; Taiwan shallow shore; central and southern South China Sea. (56, 67, 73)

A. peltifera (Haeckel): Taiwan Strait; South China Sea. (11, 54, 56, 73)

Praya dubia (Quoy & Gaimard): Central South China Sea. (36, 56)

P. reticulata (Bigelow): East China Sea; central and southern South China Sea. (39, 41, 54, 56)

Rosacea plicata Quoy & Gaimard: East China Sea; Taiwan Strait; South China Sea. (5, 11, 39, 41, 54, 55)

HIPPOPODIIDAE

Hippopodius hippopus Forskål: East China Sea; Taiwan Strait; South China Sea. (5, 11, 39, 41, 42, 54, 56)

Vogtia glabra Bigelow: East China Sea; Taiwan Strait; offshore of eastern Guangdong; central South China Sea. (5, 11, 34, 41, 42, 54, 56)

V. kuruae Alvarino: Central and southern South China Sea. (54, 61)

V. microsticella Zhang & Lin: Southern East China Sea; central South China Sea. (56, 69, 97)

V. pentacantha Kölliker: Taiwan shallow shore; central and southern South China Sea. (41, 54, 73)

V. serrata (Moser): East China Sea; Taiwan Strait; central South China Sea. (34, 39, 56)

V. spinosa Kefferstein & Ehlers: Taiwan shallow shore; offshore of eastern Guangdong; central South China Sea. (11, 41, 56, 73)

DIPHYIDAE

Chelophyes appendiculata (Eschsholtz): From East China Sea to central South China Sea. (5, 11, 54, 55, 56)

C. contorta (Lens & Van Riemsdijk): From East China Sea to central South China Sea. (5, 9, 41, 54)

Dimophyes arctica (Chun): Central and southern South China Sea. (54, 67)

Diphyes bojani (Eschscholtz) [*Diphyopsis bojani*]: From East China Sea to central South China Sea. (5, 7, 39, 41, 54, 56)

D. chamissonis Huxley [*Diphyopsis chamissonis*]: From southern Yellow Sea to central South China Sea. (1, 5, 9, 11, 34, 39, 41, 54, 56, 73)

D. dispar Chanisso & Eysenhardt: From East China Sea to central South China Sea. (5, 8, 34, 39, 41, 56)

Eudoxia macra Totton: East China Sea; Taiwan Strait; central South China Sea. (34, 39, 56, 67, 73)

Eudoxoides mitra (Huxley): East China Sea; Taiwan Strait; South China Sea. (2, 4, 11, 34, 39, 41, 42, 54, 56)

E. spiralis (Bigelow): From East China Sea to central South China Sea. (5, 9, 34, 41, 56)

Lensia baryi Totton: Central South China Sea. (56)

L. campanella (Moser): From East China Sea to central South China Sea. (2, 4, 11, 34, 39, 41, 54, 56, 73)

L. canopusi Stepanjants: Taiwan shallow shore; central South China Sea. (37, 56, 73)

L. challengeri Totton: East China Sea; Taiwan shallow shore; central South China Sea. (41, 54, 56, 73)

L. conoides (Keferstein & Ehlers): From East China Sea to central South China Sea. (11, 34, 39, 56, 73)

L. cordata Totton: Central South China Sea. (32)

L. cossack Totton: From East China Sea to central South China Sea. (11, 34, 41, 42, 56, 73)

L. fowleri (Bigelow): East China Sea; Taiwan shallow shore; northern and central South China Sea. (56, 67, 73)

L. grimaldi (Leloup): Central South China Sea. (56)

L. havock Totton: Central South China Sea. (32)

L. hotspur Totton: From East China Sea to central South China Sea. (5, 11, 54, 55, 73)

L. leloupi Totton: From Taiwan Strait to central South China Sea. (5, 9, 54, 55, 56, 73)

L. lelouveteau Totton: Central South China Sea. (37, 56)

L. meteori (Leloup): East China Sea; Taiwan shallow shore; central and southern South China Sea. (67, 73)

L. multicristata (Moser): Taiwan Strait; central South China Sea. (41, 54, 55)

L. multicristatoides Zhang & Ling: Central South China Sea. (68)

L. reticulata Totton: Central South China Sea. (37, 56)

L. subtilis (Chun): From East China Sea to central South China Sea. (5, 9, 34, 41, 54, 56, 73)

L. subtiloides (Lens & Van Riemsdijk) [*Diphyes subtiloides*, *D. truncata*]: From East China Sea to South China Sea. (7, 34, 41, 54, 56)

L. tottoni A. Daniel & R. Daniel: Central South China Sea. (56)

Muggiaea atlantica Cunningham: From Bohai to central South China Sea. (1, 27, 34, 54)

M. delsmani Totton: Central South China Sea. (56)

Sulculeolaria angusta Totton: Western Taiwan Strait; central and southern South China Sea. (34, 56, 67)

S. bigelowi (Sears): Northern and central South China Sea. (41, 54, 56)

S. biloba (Sars): East China Sea; Taiwan Strait; northern and central South China Sea. (9, 11, 34, 41, 54, 56, 73)

S. brintoni Alvarino: Central and southern South China Sea. (54, 56, 61)

S. chuni (Lens & Van Riemsdijk): East China Sea; Taiwan Strait; northern and central South China Sea. (5, 11, 34, 41, 54, 58)

S. monoica (Chun): East China Sea; Taiwan Strait; northern and central South China Sea. (5, 11, 41, 54, 73)

S. quadrivalvis Blainville: From East China Sea to central South China Sea. (2, 4, 5, 8, 11, 34, 39, 41, 54, 56)

S. tropica Zhang: From Taiwan Strait to central South China Sea. (34, 56, 60)

S. turgida (Gegenbaur): From East China Sea to central South China Sea. (5, 8, 11, 39, 54)

S. xishaensis Hong & Zhang: Taiwan Strait; central South China Sea. (5, 41, 73)

CLAUSOPHYIDAE

Chuniphyes moserae Totton: Central and southern South China Sea. (56)

C. multidentata Lens & Van Riemsdijk: East China Sea; central South China Sea. (54, 62)

Clausophyes galeata Lens & Van Riemadijk: East China Sea; central South China Sea. (62)

C. ovata (Keferstein & Ehlers): South China Sea. (54)

Crystallophyes amygdalina Moser: Central South China Sea. (36, 56)

Heteropyramis maculata Moser: Central South China Sea. (36, 56)

SPHAERONECTIDAE

Sphaeronectes gracilis (Claus) [*S. truncata*]: From East China Sea to central South China Sea. (5, 8, 34, 39, 42, 54, 56)

ABYLIDAE

Abyla bicarinata Moser: East China Sea; central South China Sea. (67)

A. brownia Sears: Central and southern South China Sea. (67)

A. carina Haeckel: Central and southern South China Sea. (61)

A. haeckeli Lens & Van Riemsdijk: East China Sea; South China Sea. (67)

A. ingeborgae Sears: Central South China Sea. (37, 56)

A. peruana Sears: Central South China Sea. (37, 56)

A. schmidti Sears: Taiwan Strait; South China Sea. (11, 54, 56, 73)

A. trigona Quoy & Gaimard: East China Sea; Taiwan Strait; South China Sea. (33, 73)

Abylopsis eschscholtzi Huxley: From East China Sea to central South China Sea. (5, 11, 34, 41, 54)

A. tetragona Otto: East China Sea; Taiwan Strait; South China Sea. (9, 34, 41, 49, 54, 56)

Bassia bassensis (Quoy & Gaimard): From East China Sea to central South China Sea. (5, 8, 9, 11, 34, 39, 41, 49, 54, 56)

Ceratocymba dentata (Bigelow): Central and southern South China Sea. (67)

C. intermedia Sears: Central and southern South China Sea. (61)

C. leuckarti (Huxley): East China Sea; Taiwan Strait; South China Sea. (5, 11, 39, 41, 42, 56)

C. sagitta (Quoy & Gaimard): East China Sea; central and southern South China Sea. (50, 67)

Enneagonum hyalinum (Quoy & Gaimard) [*Cuboides crystallus*, *C. vitreus*, *Cymba crystallus*]: From East China Sea to central South China Sea. (5, 8, 9, 11, 34, 49, 54, 56)

E. searsae Alvarino: From Taiwan Strait to central South China Sea. (11, 41, 54, 73)

SCYPHOMEDUSAE
PTEROMEDUSAE

Tetraplatia volitans Bush: Central South China Sea. (33)

STAUROMEDUSAE
ELEUTHEROCARPIDAE

Haliclystus auricula (Rathke): Yantai (Shandong). (92)

H. steinegeri Kishinouye: Dalian (Liaoning); Yantai (Shandong). Attached to algae. (25, 92)

Kishinouyea nagatensis (Oka): Northern China coast. (92)
Sasakiella cruciformis Okubo: Dalian (Liaoning); Yantai (Shandong). Attached to algae. (25, 92)
S. tsingtaoensis Ling: Qingdao (Shandong). (92)
Stenoscyphus inabai (Kishinouye): Qingdao (Shandong). (92)

CUBOMEDUSAE
CARYBDEIDAE

Carybdea rastoni Haacke: East China Sea; Fujian near shore. (7)
C. sivickisi Stiasny: Southern Fujian near shore; northern South China Sea. (11)
Tamoya gargantua Haeckel [*T. alata*]: Zhoushan (Zhejiang); Fujian near shore; Nanao (Guangdong). (44)

CORONATAE
ATOLLIDAE

Atolla wyvillei Haeckel: Offshore of East China Sea. (45)

ATORELLIDAE

Atorella arcturi Bigelow: Central South China Sea. (56)
A. subglobosa Vanhöffen: Central South China Sea. (56)
A. vanhoeffeni Bigelow: Central South China Sea. (56)

LINUCHIDAE

Linuche draco (Haeckel): Xisha (=Paracel Islands). (44)

NAUSITHOIDAE

Nausithöe punctata Kölliker: From East China Sea to central South China Sea. (11, 44, 56)

PERIPHYLLIDAE

Periphylla periphylla (Péron & Lésueur): East China Sea offshore. (44)

SEMAEOSTOMEAE
PELAGIIDAE

Chrysaora helovola Brandt: Fujian near shore; Daya Bay (Guangdong); Hong Kong; Xisha (=Paracel Islands). (7, 35, 44)
Pelagia noctiluca (Forskål) [*P. panopyra, P. perla*]: From East China Sea to central South China Sea. (11, 35, 44, 55, 56, 79)
Sanderia malayensis Goette: Xiamen harbor (Fujian); offshore of eastern Guangdong; Xisha (=Paracel Islands). (11, 44)

CYANEIDAE

Cyanea capillata (Linné): Southern Yellow Sea; northern East China Sea. (45)
C. ferruginea Eschscholtz: Southern Yellow Sea; northern East China Sea. (45)
C. nozakii Kishinouye: Bohai; Yellow Sea; Fujian; Guangdong near shore; Hong Kong. (7, 26, 35)
C. purpurea Kishinouye: Xiamen (Fujian). (45)

ULMARIDAE

Aurelia aurita (Linné): Yantai (Shandong); Zhejiang; Pingtan (Fujian); Hong Kong. (7, 70)

RHIZOSTOMEAE
CEPHEIDAE

Cephea conifera Haeckel: Southern Zhejiang. (45)
Netrostoma coerulescens Maas: Southern Zhejiang. (44)
N. setouchianum (Kishinouye): Southern Zhejiang. (44)

MASTIGIIDAE

Mastigias ocellatus (Modeer): Hong Kong; Hainan. (44)
M. papua (Lesson): Xiamen harbor (Fujian); Hong Kong; Guangdong near shore; Hainan. (44, 86)

THYSANOSTOMATIDAE

Thysanostoma flagellatum (Haeckal): Central South China Sea. (56)

CATOSTYLIDAE

Acromitus flagellatus (Maas): Southern Fujian near shore; Taiwan. (44)
A. tankahkeei Light: Southern Fujian near shore; Shantou (Guangdong). (7, 89)

LOBONEMATIDAE

Lobonema smithi Mayer: From Xiamen harbor (Fujian) to northern South China Sea. (40)
Lobonemoides gracilis Light: From Xiamen harbor (Fujian) to northern South China Sea. (40)

RHIZOSTOMATIDAE

Rhopilema esculentum Kishinouye: From Bohai to northern South China Sea. (7, 21, 40)
R. hispidum Vanhöeffen: From Xiamen harbor (Fujian) to northern South China Sea. (7, 40, 43)
R. rhopalophorum Haeckel: Xiamen harbor (Fujian); Shaoan (Fujian) near shore. (44)

STOMOLOPHIDAE

Stomolophus meleagris L. Agassiz: Yellow Sea; East China Sea; Hong Kong. (40)

REFERENCES*

(1). Qiu Shuyuan, 1954. Notes on zooplankton, I. Hydromedusae. Acta Zool. Sinica, 6(1):41–46.
(2). Qiu Shuyuan, 1954. On the medusae of southeast coast of China. Acta Zool. Sinica, 6(1):49–51.
(3). Qiu Shuyaun, 1957. On the *Porpita* of China coast. Scientific Advance of Xiamen University, 1:85–93.
(4). Liu Yuai and Ye Huachen, 1979. Studies on the planktonic medusae from continental shelf of northern part of South China Sea. In: Collections of survey reports of fish resources of bottom nets of foreign sea from the continental shelf of northern part of South China Sea, pp:565–586.
(5). Zhu Changshou et al., 1988. The species composition and distribution of the plankton. In: Comprehensive oceanographic survey of the central and northern part of Taiwan Strait. China Science Press, pp:259–305.
(6). Tsi Chungyen and Zou Renlin, 1965. Notes on the species of *Millepora* of Hainan. Acta Zoologica Sinica, 17(2):185–188.
(7). Xu Zhenzu and Jin Dexiang, 1962. Studies on the medusae from Fujian coast, I.. Journal of Xiamen University (Natural Science), 2(3):206–224.
(8). Xu Zhenzu and Zhang Jinbiao, 1964. Studies on the medusae from Fukien coast II. On the taxonomy of Hydromedusae, Siphonophores and Ctenophores off South Fukien. Journal of Xiamen University (Natural Science), 11(3):120–149.
(9). Xu Zhenzu, 1965. Studies on zooplankton of Hainan Island and adjacent waters, I. Hydromedusae. Journal of Xiamen University (Natural Science), 12(1):90–110.
(10). Xu Zhenzu and Zhang Jinbiao, 1974. Studies on the medusae from Fukien coast, III. Studies on classification of the medusae from the central and northern part of the Fukien Coast. Oceanol. Technol. Sinica, 2: 17–32.
(11). Xu Zhenzu and Zhang Jinbiao, 1978. On Hydromedusae, Siphonophorae and Scyuphomedusae from coast of the East Guangdong Province and south Fukien Province. Journal of Xiamen University (Natural Science), 17(4):19–63.
(12). Xu Zhenzu and Zhang Jinbiao, 1981. On Hydromedusae from coastal waters of northern part of South China Sea. Journal of Xiamen University (Natural Science), 20(3):373–382.
(13). Xu Zhenzu and Huang Jiachi, 1983. On Hydromedusae, Siphonophorae, Scyphomedusae and Ctenophorae from the Jiulong River estuary of Fujian, China. Journal of Oceanography in Taiwan Strait, 2(2):99–110.
(14). Xu Zhenzu, 1982. On a new genus and species of Leptomedusae from northern part of South China Sea. Acta Zool. Sinica, 8(1):4–6.
(15). Xu Zhenzu. Studies on the Hydrozoa of the Fukien coast. (manuscript)
(16). Xu Zhenzu, 1983. Ecological studies on Hydromedusae in the southwestern Taiwan Strait of China. Acta Oceanol. Sinica, 2(1):129–139.
(17). Xu Zhenzu, Huang Jiachi and Wang Wenqiao, 1985. On new species and records of Hydromedusae from the Jiulong River Estuary of Fujian, China. Journal of Xiamen University (Natural Science), 24(1):102–110.
(18). Xu Zhenzu and Huang Jichi, 1990. A new genus and two new species of Hydropolypae - Hydromedusae from Luoyuan Bay, Fujian Province. Acta Zootaxon. Sinica, 15(3):262–266.
(19). Xu Zhenzu and Huang Jiachi, 1990. A new genus and two species of Hydromedusae from China. Acta Zootaxon. Sinica, 15(4):401–405.
(20). Xu Zhenzu, Huang Jiachi and Chen Xu, 1991. On new species and record of Hydromedusae in the upwelling region off the Minnan - Taiwan bank fishing ground. In: Minnan - Taiwan bank fishing ground upwelling ecosystem study. China Science Press, pp:469–486.
(21). Wu Baoling, 1955. Haitese (Rhopinema). Bull. Biol., 4:34–40.
(22). Zou Renlin, 1978. Studies on the corals of the Xisha Islands, Guangdong Province, II. The genus *Millepora*, with the description of a new species. In: A survey of the marine biology of Xisha and Zhongsha Islands, China. China Science Press, pp: 81–90.
(23). Zou Renlin, 1978. Studies on the corals of the Xisha Islands, III. An illustrated catalogue of Scleractinean, Hydrocorallinian, Heliporina and Tubiporina. In: A survey of the marine biology of Xisha and Zhongsha Islands, China. China Science Press, pp:91–111.
(24). Zou Renlin, 1983. On the corals and its natural compound. J. Mar. Drugs, 3:125–129.
(25). Zao Ruyi, 1990. Coelenterata. In: The guide of invertebrate collection in the China Coast. China Ocean Press, pp:39–40.
(26). Zhow Taihsuan and Huang Mingchian, 1956. A fisheries enemy of Yellow Sea and Bohai Sea-*Cyanea nozakii*. Biol. Bull., 6:9.
(27). Chow Taihsuan and Huang Mingchian, 1958. A study on hydromedusae of Chefoo. Acta Zool. Sinica, 10(2):173–191.
(28). He Zhenwu, 1964. Hydromedusae of China. Chinese Jour. Zool., 6(2):53–57.
(29). He Zhenwu and Xu Renhe, 1982. Studies on the hydromedusae and a new species of Leptomedusae from Yantai coast. J. Xinxiang Normal College, 4:33–44.
(30). He Zhenwu, 1980. A new genus and species of Tranymedusae from Yantai. Acta Zootaxonomica Sinica, 5(4):327–329.
(31). Lin Sheng, 1990. A preliminary study on Hydroids ecology in Daya Bay. In: Collections of papers on marine ecology in Daya Bay, II. China Ocean Press, pp: 315–319.
(32). Lin Mao and Zhang Jinbiao, 1987. Description of two species of deepwater *Lensia* (Siphonophora) in the central South China Sea. Mar. Sci. Bull., 6(2):105–106.
(33). Lin Mao and Zhang Jinbiao, 1988. A strange medusa - *Tetraplatia volitans* for the first discovery in China. Chinese Jour. Zool., 23(2):1–2.
(34). Lin Mao and Zhang Jinbiao, 1989. Ecological studies on the Siphonophores in the western Taiwan Strait. Mar. Sci. Bull., 8(3):65–71.
(35). Lin Mao, 1989. Taxonomy and fauna on the medusae from the waters of Daya Bay. In: Collections of papers on marine ecology in Daya Bay, I. China Ocean Press, pp:59–65.
(36). Lin Mao, 1990. Two rare species of deep-water Siphonophora (Clausophyidae) for the first discovery in China. Mar. Sci. Bull., 9(1):93–95.
(37). Lin Mao and Zhang Jinbiao, 1991. New records of Siphonophores from China Sea. Acta Zootaxonomica Sinica, 16(4):496.
(38). Lin Mao and Zhang Jinbiao, 1992. A preliminary study on Siphonophora ecology in central part of South China Sea. Acta Oceanologica Sinica, 14(2):99–105.
(39). Hong Hue-shin, 1964. Studies on Medusae of East China Sea, I. Siphonophora. Contribtions of Shanghai Fisheries College, 1:111–120.
(40). Hong Hue-shin, Zhang Shimei and Wang Jinchi, 1978. Haitese (Rhopilema). China Scence Press.
(41). Hong Hue-shin and Zhang Shimei, 1981. Systematic studies on the Siphonophores of Xisha Islands. Journal of Xiamen Fisheries College, 1:1–25.
(42). Hong Hue-shin and Zhang Shimei, 1981. A preliminary study on the fauna of Siphonophora in China Seas. Journal of Xiamen Fisheries College, 1:46–55.
(43). Hong Hue-shin and Zhang Shimei, 1982. Edible Medusae along China coast. Journal of Xiamen Fisheries College, 1:5–18.
(44). Hong Hue-shin, Lin Limin and Zhang Shimei, 1985. Systematic studies on the Scyphomedusae of China Seas. Journal of Xiamen Fisheries College, 2:7–18.
(45). Hong Hue-shin and Zhang Shimei, 1989. Some new records of Scyphomedusae from China Seas. Journal of Xiamen Fisheris College, 11(2): 11–15.

(46). Kao Chih-sheng, 1956. On the Hydrozoa from Shantung coast. Journal of Shantung University, 2(4):70–103.

(47). Kao Chih-sheng, Li Fung-lu, Chang Un-mei and Li Hien-lun, 1958. On the Hydromedusae from Shantung coast. Journal of Shantung University, 1:75–118.

(48). Kao Chih-sheng and Zhang Zhinan, 1965. Studies on the Aglao-pheniidae from China coast. In: Transactions of 30th anniversary of the Chinese Society of Zoology. China Science Press, p:29.

(49). Kao Chih-sheng and Zhang Zhinan, 1962. On the Hydromedusae from Zhoushang Coast. Journal of Shandong College of Oceanology, 1:65–91.

(50). Gao Shangwu, 1982. The medusae of the East China Sea. Studia Marina Sinica, 19:33–42.

(51). Tang Zhican, and Huang Meijun, 1987. The morphological study of *Symplectoscyphus huanghaiensis* Tang & Huang (Hydroida) in Yellow Sea. Oceanologia & Limnoligia Sinica, 18(3):291–294.

(52). Tang Zhican, 1991. On the Hydrozoa from the Nansha Islands. In: Contributions on marine biological research of the Nansha Islands and adjacent waters (I). China Ocean Press, pp:24–37.

(53). Tang Zhican, 1991. *Monoserius pennarius* assemblage and its ecological and geographical studies on the continental shelf of waters around the Nansha Islands. In: Contributions on marine ecological research of Nansha Islands and adjacent waters (II). China Ocean Press, pp:255–261.

(54). Chen Qingchao, 1983. Siphonophores in the central and northern parts of the South China Sea. In: Contributions on marine biological research of the South China Sea (I). China Ocean Press, pp:7–17.

(55). Lian Guangshan et al., 1981. Plankton (Chapter 9). In: R/V "Shijian" observational report of the western central pacific. China Ocean Press, pp:74–88.

(56). State Oceanic Administration, 1988. List of Zooplankton: Hydromedusae, Siphonophora and Scyphomedusae. In: Reports of the comprehensive investigation on environmental resources of central South China Sea. China Ocean Press, pp:230–234.

(57). Zhang Jinbiao and Xu Zhenzu, 1975. Studies on medusae from Fujian Coast, IV. Distribution of the planktonic medusae in southern part of Fujian. Oceanologia & Technologia Sinica, 5:1–14.

(58). Zhang Jinbiao, 1977. Studies on the Hydromedusae, Siphonophora and Ctenophora from the Coast of Jiangsu and Zhejiang provinces. Oceanologia & Technologia Sinica, 7:95–107.

(59). Zhang Jinbiao, 1979. A preliminary analyses on the Hydromedusae fauna of the China Seas. Acta Oceanologica Sinica, 1(1):127–137.

(60). Zhang Jinbiao, 1980. Discovery of a new species of Siphonophora in Pacific Ocean. Acta Oceanologica Sinica, 2(1):152–155.

(61). Zhang Jinbiao and Xu Zhenzu, 1980. On the geographical distribution of Siphonophora in the China Seas. Journal of Xiamen University (Natural Science), 19(3):100–109.

(62). Zhang Jinbiao and Zhang Xilie, 1980. Description of two deep-water Siphonophora of the northern East China Sea. Journal of Xiamen University (Natural Science), 19(3):121–125.

(63). Zhang Jinbiao and Wu Yuqing, 1981. On a new genus and species of the Hydromedusae from Xiamen Harbour, Fujian Province. Acta Oceanologica Sinica, 3(1):184–187.

(64). Zhang Jinbiao et al., 1981. Plankton (Chapter 7). In: R/V "Xiangyang Hong 09" observational data of the western central Pacific. China Ocean Press, pp:118–136.

(65). Zhang Jinbiao, 1982. A new family, genus and species of Anthomedusae from the northern China seas. Acta Oceanologica Sinica, 4(2):209–214.

(66). Zhang Jinbiao and Lin Mao, 1984. Two new species of Hydromedusae from the Xiamen Harbour and adjacent waters. Acta Zootaxonomica Sinica, 9(4):343–347.

(67). Zhang Jinbiao, 1984. The Calycophorae (Siphonophora) from tropical waters of western Pacific Ocean. In: Proceedings of the plankton from the tropical waters of the western Pacific Ocean. China Ocean Press, pp:52–85.

(68). Zhang Jinbiao and Lin Mao, 1987. A new species of Siphonophora genus *Lensia* from the deep waters of the central part of South China Sea. Acta Oceanologica Sinica, pp:603–606.

(69). Zhang Jinbiao and Lin Mao, 1990. A new species of Siphonophora from East China Sea and South China Sea. Acta Oceanologica Sinica, 12(3):352–354.

(70). Morton, B. and J. Morton, 1983. The seashore ecology of Hong Kong. Hong Kong University Press, pp:54–55, 98–99.

(71). Huang Zongguo and Cai Ruxing, 1984. The list of the species of fouling organisms along the coast of China. In: Marine biofouling and its prevention. China Ocean Press, pp:144–145.

(72). Huang Zongguo, Li Shaoqing and Chen Yaping, 1989. On the species composition and distribution of zooplankton at Luoyuan Bay of Fujian. Journal of Xiamen University (Natural Science), 28 (sup.):87–95.

(73). Huang Jiaqi, Chen Xu and Xu Zhenzu, 1991. Ecological studies on the medusae in Minnan-Taiwan bank fishing ground upwelling region. In: Minnan-Taiwan bank fishing groun upwelling ecosystem study. China Science Press, pp:456–468.

(74). Huang Liping, 1987. On the planktonic medusae from the northern coast of Beibu Bay. Guangxi Ocean, 1:1–11.

(75). Huang Meijun, 1988. New records of *Aglaophenia whiteleggei* Bale from Qingdao Coast. Journal of Shandong College of Oceanology, 18(2):44–47.

(76). Huang Meijun and Liu Heng, 1987. Studies on the Hydrozoa in the waters around Jiaozhou Bay. Journal of Shandong College of Oceanology, 17(3):61–69.

(77). Meng Zhimin, 1991. On benthics from the continental shelf of Nansha Islands. In: Reports of the comprehensive investigation of the demersal fishery resources from the continental shelf of Nansha Islands. China Ocean Press, pp:56–67.

(78). Li Aishao and Chen Qingchao, 1991. The Hydromedusae in the waters around Nansha Islands-the species composition, faunastical characteristics and zoogeography of Hydromedusae around Nansha Islands waters. In: Contributions on faunastical and zoogeographical research of Nansha Islands waters. China Ocean Press, pp:1–63.

(79). Li Aishao and Chen Qingchao, 1991. The medusae of Nansha Islands waters, I. The composition and distribution of the Hydromedusae and Scyphomedusae. In: Contributions on marine biological research of Nansha Islands and adjacent waters (II). China Ocean Press, pp:89–102.

(80). Wei Chongde, 1959. A preliminary survey of the Hydrozoa and Hydromedusae from Zhoushan coast. Journal of Hangzhou University, 2:187–212.

(81). Bouilin, J., et al., 1985. Essai de classification des Hydropolypes-Hydromeduses (Hydrozoa-Cnidaria). Indo-Malayan Zool., 2:27–243.

(81a). Bouilin, J., et al., 1992. No Sinphonophoran Hydrozoa: what are we talking about? Sci. Mar., 56: 270- 284.

(82). Calder, D. R., 1991. Shallow-water Hydroid of Bermuda, The Thecatac, Exclusive of Plumulariodea. Life Science Contributions, 154:1–140.

(83). Hargitt, C.W., 1927. Hydroids of South China. Bull. Mus. Comp. Harv. Coll., 67(16):491–520.

(84). Hsu, H. F., 1928. On a new species of Hydromedusae. Contri. Biol. Lab. Sci. Soc. China Nanking, 4:1–7.

(85). Jäderholm, E., 1920. On some exotic Hydroids in Swedish Zoological State Museum. Ark, Zool., 13(3):1–11.

(86). Kramp, P. L., 1961. Synopsis of the medusae fo the world. Jour. Mar. Biol. Ass. U.K., 40: 1–459.

(87). Kirkpatrick, R., 1890. Report upon the Hydrozoa and Polyzoa Collected by P. W. Basselt-Smith, Esq., Surgeon R. N., Moore., during the survey of the Tizard and Macclesfield Bank, in the China Sea, by M. S."Ranbler" Commander W. R. Moore. Ann. Mag. Nat. Hist., 6(5):11–24.

(88). Leloup, E., 1937. Hydropolypes and Scyphopolypes recuueillis par Dawydoff sur les coles de I' Idochine francaise. Mem. Mus. Hist. Nat. Belg., 12(2):3373.

(89). Light, S.F., 1924. A new species of Scyphomedusae jellyfish in Chinese Water, China. J. Sci. & Art., 2:449–450.

(90). Lin, M. and Zhang, I. B., 1990. Ecological studies on the Hydromedusae, Siphonophores and Ctenophores in the Xiamen Harbour and adjacent waters. Acta Oceanologica Sinica, 9(3):429–438.

(91). Ling, S. W., 1937. Studies on Chinese Hydrozoa, I. On some Hydromedusae from Chekiang coast. Peking Nat. Hist. Bull., 11(4):351–365.
(92). Ling, S. W., 1939. Studies on Chinese Stauromedusae, II. Further studies on some Stauromedusae from China. Lingnan Sci. J., 2(4):281–291; 495–504.
(93). Ling, S. W., 1938. Studies on Chinese Hydrozoa, II. Report on some common hydroides from the East Saddle Island. Lingnan Sci. J., 17(2):175–184; 17(3):357- 366.
(94). Nuting, C. C., 1927. Report on the Hydroida collected by the United State Fisheries Steamer "Albatross" in Philippine Region, 1907–1910. Bull. US Nat. Mus., 6(3):195–242.
(95). Stechow, E., 1913. Die Hydroiden fauna der japanicschen Region. J. Coll. Sci. Tokyo Imp. Univ., 44(8):1–23.
(96). Totton, A.K., 1965. A synopsis of the Siphonophora. Trust. Brit. Mus. Hist., London, p:230.
(97). Zhang, J. B. and Lin, M., 1991. On a new Siphonophora from the East China Sea and South China Sea. Acta Oceanologica Sinica, 10(4):609–611.

*: (1)-(80) in Chinese
Hydropolypes compiled by Xu Zhenzhu and Lin Mao; Hydromedusae compiled by Tang Zhichan and Huang Meijun; Milleporidae compiled by Zhou Renlin; Hydropolypes and Siphomedusae edited by Xu Zhenzhu; Hydromedusae edited by Zhang Jinbiao and Hong Huixin.

ANTHOZOA
ZOANTHIDEA
EPIZOANTHIDAE

Epizoanthus fatuus Schultze: East China Sea. (10, 18, 26)
E. glacialis Verrill: East China Sea (10, 26)
E. illorieatus Tischbierek: East China Sea. (10, 26)
E. paguriphilus Verrill: East China Sea. (10, 27)
E. planus Carlgren: East China Sea. (10, 18)
E. ramosus Carlgren: East China Sea. (10, 27)
E. similis Carlgren: East China Sea. (10, 18, 27)
E. sinensis Pei: East China Sea. (10, 26)
E. tube Carlgren: East China Sea. (10, 26, 27)
Palythoa anthoplax Pax & Müller: Xisha (=Paracel Islands). (12, 19, 27)
P. australiae Carlgren: Xisha (=Paracel Islands). (12, 19, 27)
P. capensis Haddon & Shackleton: Xisha (=Paracel Islands). (12, 19, 27)
P. haddoni Carlgren: Xisha (=Paracel Islands). (12, 19, 27)
P. hainanensis Pei: Xisha (=Paracel Islands). (12, 19)
P. liscia Haddon & Duerden: Xisha (=Paracel Islands). (12, 19, 27)
P. natalensis Carlgren: Xisha (=Paracel Islands). (12, 19, 27)
P. nelliae Carlgren: Xisha (=Paracel Islands). (12, 27)
P. sinensis Pei: Xisha (=Paracel Islands). (12, 19, 27)
P. singaporensis Pax & Müller: Xisha (=Paracel Islands). (12, 27)
P. stephensoni Carlgren: Xisha (=Paracel Islands). (12, 19, 26, 27)
P. titanophila Pax & Müller: Xisha (=Paracel Islands). (12, 27)
P. xishaensis Pei: Xisha (=Paracel Islands). (12, 19, 27)
P. yongei Carlgren: Xisha (=Paracel Islands). (12, 19)

ZOANTHIDAE

Zoanthus barnardi Carlgren: Xisha (=Paracel Islands). (11, 19, 26)
Z. cavernarum Pax & Müller: Xisha (=Paracel Islands). (11, 19, 26)
Z. cyanoides Pax & Müller: Xisha (=Paracel Islands). (11, 19, 26)
Z. erythrochlores Pax & Müller: Xisha (=Paracel Islands). (11, 19, 26)
Z. sinensis Pei: Xisha (=Paracel Islands). (11, 19)
Z. vietnamensis Pax & Müller: Xisha (=Paracel Islands). (11, 19)
Z. xishaensis Pei: Xisha (=Paracel Islands). (11, 19, 26)

ACTINIARIA
EDWARDSIDAE

Edwardsia japonica Carlgren: Bohai; Yellow Sea; Hong Kong. (17, 25, 28)
E. sipunculoides Stimpson: Bohai; Yellow Sea. (17, 28)

HORMATHIIDAE

Calliactis argentacoloratus Pei: Xisha (=Paracel Islands). (6, 13, 24)
C. conchicola Parry: Xisha (=Paracel Islands). (6, 13, 24)
C. miriam Haddon & Shackleton: Xisha (=Paracel Islands). (6, 13, 24)
C. polypus (Forskål): Xisha (=Paracel Islands). (6, 13, 24)
C. reticulata Stephenson: Xisha (=Paracel Islands). (6, 13, 24, 28)
C. rosea (Hand): Xisha (=Paracel Islands). (6, 13, 23, 24)
C. xishaensis Pei: Xisha (=Paracel Islands). (6, 13, 24)
Paracalliactis sinica Pei: East China Sea. (6)

METRIDIIDAE

Metridium huanghaiensis Pei: Bohai; Yellow Sea. (14, 28)
M. senile fimbriatum Verrill: Bohai; Yellow Sea. (2, 3, 14, 28)
M. sinensis Pei: Bohai; Yellow Sea. (14, 28)

ACTINIIDAE

Actinia equina (Linné): China's coast. (1, 6, 7, 25, 28)
Anthopleura asiatica Uchida & Murmatsu: China's coast. (15, 28)
A. ballii Stephenson: China's coast. (15, 28)
A. chinensis England: Hong Kong. (22)
A. dixoninana: Hong Kong. (25)
A. incerta England: Hong Kong. (22)
A. japonica (Verrill): China's coast. Intertidal. (2, 3, 15, 22, 25, 28)
A. midori Uchida & Murmastu: China's coast. (15, 28)

A. mortoni England: Hong Kong. (22)

A. nigrescens (Verrill) [*A. pacifica*]: China's coast. Intertidal to shallow sea. (2, 3, 5, 6, 15, 22, 25, 28)

A. qingdaoensis Pei: Yellow Sea (15, 18, 28)

A. stella (Verrill): Fujian's coast. Rocky crevices in mid- and low-intertidal. (6)

A. xanthogrammia (Brandt): China's coast. (2, 3, 6, 15, 28)

Cancrisocia expansa Stimpson: Hong Kong; South China Sea. (9, 29, 30)

Gymnophellia hutchingsae England: Hong Kong. (22)

Gyractis japonica (Verrill): China's coast. Intertidal. (2, 3, 15, 22, 25, 28)

G. stimpsoni (Verrill): Hong Kong. (22)

Haliplanella luciae Hand: China's coast. Intertidal. (1, 2, 3, 5–8, 25, 28, 31)

Spheractis cheungae England: Hong Kong. (22)

Synantheopsis primus England: Hong Kong. (22)

BOLOCEROIDIDAE

Boloceroides mcmurrichi (Kweitniewski): Hong Kong. (22, 25)

AIPTASIDAE

Paraiptasia radiata (Stimpson): Hong Kong. Attached to Nassa snail (*Nassarius*). (22)

SAGARTIIDAE

Actinothoe qingdaoensis Pei: Qingdao (Shandong). (16, 28)

Sagartia carcinophilus Verrill: China's seas. (4, 28)

S. rosea Gosse: Bohai. (4, 28)

CERIANTHARIA
CERIANTHIDAE

Cerianthus filiformis Carlgren: Bohai; Yellow Sea; South China Sea. (25, 28)

C. orientalis Verrill: Hong Kong; South China Sea. (28)

REFERENCES*

(1). Xu Zhenzhu, 1992. Investigation of sea anemones along Fujian's coast. (manuscript)

(2). Song Pengdong et al. 1988. Guidebook on invertebrates along the Dalian coast. Higher Education Press, pp:43–47.

(3). Zhao Ruyi et al., 1990. Guidebook on collection of invertebrate from China coast (in North China). China Ocean Press, pp:42–44.

(4). Lu Zhangkeng et al. 1991. Preliminary study on the pharmaceutical characteristics of red sea anemone (*Sagartia rosea*). Procedings of Marine Drug Development Conference, pp:1–14.

(5). Huang Zongguo and Cairuxing 1984. Marine fouling organism and its prevention (I). China Ocean Press, 352pp.

(6). Pei Zunan, 1982. A new species of the genus *Paracalliactis* (Hormathiidae: Actiniatia) from the East China Sea. Studia Marina Sinica, 19: 65–71.

(7). Pei Zunan, 1982. Introduction of the Coelenterata drugs. Journal of Marine Drugs, 2:53–55.

(8). Pei Zunan, 1984. On the appearance and habit of *Haliplanells luciae*. Chinese Journal of Zoology, 5:1–5.

(9). Pei Zunan, 1990. A sea anemone secreting chitinous membrane *Cancrisocia expansa* (Coelentera, Anthozoa, Actiniidae). Chinese Journal of Zoology, 25(5):4–6.

(10). Pei Zunan. On the Epizoanthus from the East China Sea. (manuscript)

(11). Pei Zunan. Zoantharia (Coelenterata: Zoanthidea) of Xisha Islands: *Zoanthus* (Epizoanthidae). (manuscript)

(12). Pei Zunan. Zoantharia (Coelenterta: Zoanthidea) of Xisha Islands: *Palythoa* (Epizoanthidae). (manuscript)

(13). Pei Zunan, 1996. Three new species of the genus *Calliactis*. Studia Marina Sinica, 37:177–187.

(14). Pei Zunan. Study on the genus *Matridium* in the Yellow Sea and Bohai Sea. Studia Marina Sinica, (in press)

(15). Pei Zunan, 1995. Description of *Anthopleura qingdaoensis* sp. nov. Studia Marina Sinica, 36:227–234.

(16). Pei Zunan, 1993. A new species of the genus *Actinthoe* (Sagartiidae: Actinaria). Studia Marina Sinica, 34:170–174.

(17). Pei Zunan, Study of sea anemone *Edwardsia* in China. (manuscript)

(18). Pei Zunan, Guidebook on invertebrates of Beibuwan (=Gulf of Tongking). Higher Education Press. (in press)

(19). Carlgren, O., 1938. South African Actiniaria and Zoantharia. Kundl. Svenska Vetenskapaakad. Handl. 3 ser., 17(3):1–148.

(20). Carlgren, O., 1954. Actiniaria and Zoantharia from south and west Australia with comments upon some Actiniaria from New Zealand. Ark. F. Zool. Ser. 2, 6(34):571–595

(21). Dunn, D. F., 1980. *Stylobaters*: a shell-forming sea Anemone (Coelenterata, Anthozoa, Actiniidae). Pacific Science, 34(4):379–388.

(22). England, K., 1992. Hong Kong anemones. Proceedings of Fourth International Marine Biological Workshop: The Marine Flora and Fauna of Hong Kong and Southern China. Hong Kong University Press, pp:49–95

(23). Hand, C., 1955. The sea anemones of central California. Wasmann J. Biol., 13(1):37–99, (2):192–203.

(24). Hazlett, B. A., 1970. Interspecific shell fighting in three sympatric species of hermit crabs in Hawaii, Pacfic Science, 24(4):472–482.

(25). Morton, B. and J. Morton, 1983. The sea shore ecology of Hong Kong. Hong Kong University Press.

(26). Pax, F. and Moller, I., 1957. Zoantharien aus Vietnam. Mem. Mus. Hist. Nat. Paris N. S., 16A(1):1–40.

(27). Roule, L., 1900. Sur les genres *Palythoa* and *Epizoanthus* C.R. Acad. Sci., 131:279- 281.

(28). Stephenson, T. A., 1935. The British sea anemones Vol. 2, 426 pp. 19 pls. (Royal Society) London.

(29). Stimpson, W., 1855. Description of some of the new marine invertebrates from the Chinese and Japanese seas. Proceedings of the Academy of Natural Scicences, Philadelphia, 7:375–384.

(30). Uchida, T., 1960. *Carinactis ichikawai*, n. gen., n. sp. an Actiniarian commensal with the crab *Dorippe granulata*. Jap. J. Zool., 12(4): 595–601.

(31). Williams, R. B., 1975. Catch-tentacles in anemones: occurrence in *Haliplanella luciae* (Verril) and a review of current knowledge. Jour. Nat. Hist., 9(3):241–248.

*: (1)-(18) in Chinese
Compiled by Pei Zhunan and Huang Zongguo

SCLERACTINIA

REEF BUILDERS

THAMNASTERIIDAE

Psammocora contigua (Esper): Taiwan; Daya Bay (Guangdong); Weizhou Island (Guangxi); Hainan; Xisha (=Paracel Islands); Nansha (=Spratly Islands). (17, 19, 23, 24, 26, 27, 37)
P. contigua pulchra Nemenzo: Hainan; Xisha (=Paracel Islands). (17, 19, 37)
P. divaricata Gardiner: Hainan; Xisha (=Paracel Islands). (17, 19)
P. nierstrazi van der Horst: Xisha (=Paracel Islands). (19)
P. profundacella Gardiner: Daya Bay (Guangdong). (26)

POCILLOPORIDAE

Pocillopora brevicornis Lamarck: Hainan; Xisha (=Paracel Islands); Nansha (=Spratly Islands). (17, 19, 27, 37)
P. damicornis (Linnaeus): Taiwan; Hainan; Xisha (=Paracel Islands); Nansha (=Spratly Islands). (17, 19, 24, 27, 37)
P. danae Verill: Hainan; Xisha (=Paracel Islands); Nansha (=Spratly Islands. (17, 19, 27, 37)
P. ligulata Dana: Hainan; Xisha (=Paracel Islands); Nansha (=Spratly Islands). (17, 19, 27, 37)
P. meandrina nobilis Verrill: Xisha (=Paracel Islands). (19)
P. verrucosa (Ellis & Solander): Taiwan; Hainan; Xisha (=Paracel Islands); Nansha (=Spratly Islands). (17, 19, 24, 27)
Seriatopora angulata Klunzinger: Xisha (=Paracel Islands); Nansha (=Spratly Islands). (19, 27)
S. caliendrum Ehrenberg: Xisha (=Paracel Islands). (21)
S. hystrix Dana: Taiwan; Xisha (=Paracel Islands); Nansha (=Spratly Islands). (19, 24, 27)
S. stellata Quelch: Xisha (=Paracel Islands); Nansha (=Spratly Islands). (21)
Stylophora danae Mile-Edwards & Haime: Xisha (=Paracel Islands). (19)
S. pistillata (Esper): Xisha (=Paracel Islands); Nansha (=Spratly Islands). (19, 24, 27)

ACROPORIDAE

Acropora aduncata Zou: Xisha (=Paracel Islands). (22)
A. affinis (Brook): Hainan; Xisha (=Paracel Islands); Nansha (=Spratly Islands). (19, 27)
A. armata (Brook): Hainan. (17)
A. brueggemanni (Brook): Weizhou Island (Guangxi); Xisha (=Paracel Islands). (19, 23)
A. brueggemanni uncinata (Brook): Hainan; Xisha (=Paracel Islands); Nansha (=Spratly Islands). (17, 19, 27, 37)
A. caroliniana Nemenzo: Nansha (=Spratly Islands). (28)
A. conferta (Quelch): Hainan. (17)
A. corymbosa (Lamarck): Hainan; Xisha (=Paracel Islands); Nansha (=Spratly Islands). (17, 19, 24, 27, 37)
A. cytherea (Dana): Daya Bay (Guangdong); Weizhou Island (Guangxi). (23, 26)
A. decipiens (Brook): Hainan. (17)
A. delicatula (Brook): Hainan. (17)
A. dissimilis Verrill: Hainan. (17)
A. echinata (Dana): Xisha (=Paracel Islands). (19)
A. formosa (Dana): Hainan; Xisha (=Paracel Islands); Nansha (=Spratly Islands). (17, 9, 24, 27)
A. granulosa (Edward & Haime): Xisha (=Paracel Islands); Nansha (=Spratly Islands). (22, 24)
A. haimei (Milne-Edwards & Haime): Hainan. (17)
A. horrida (Dana): Xisha (=Paracel Islands); Nansha (=Spratly Islands). (19, 27)
A. humilis (Dana): Daya Bay (Guangdong); Weizhou Island (Guangxi); Hainan; Xisha (=Paracel Islands); Nansha (=Spratly Islands). (17, 19, 23, 24, 26, 27)
A. hystrix (Dana): Xisha (=Paracel Islands); Nansha (=Spratly Islands). (19, 27)
A. irregularis (Brook): Dongsha (=Pratas Islands); Xisha (=Paracel Islands). (19)
A. lutkeni Crossland: Hainan. (17)
A. nasuta (Dana): Hainan; Xisha (=Paracel Islands). (17, 19)
A. nasuta crassilabia (Brook): Hainan; Xisha (=Paracel Islands); Nansha (=Spratly Islands). (17, 19, 27)
A. pacifica (Brook): Hainan; Nansha (=Spratly Islands). (17, 27)
A. palifera (Lamarck): Xisha (=Paracel Islands); Nansha (=Spratly Islands). (19, 27)
A. prostrata (Dana): Weizhou Island (Guangxi); Hainan; Xisha (=Paracel Islands); Nansha (=Spratly Islands). (17, 19, 23, 24, 27, 37)
A. pruinosa (Brook): Daya Bay (Guangdong), Weizhou Island (Guangxi). (23, 26)
A. pulchra (Brook): Hainan; Xisha (=Paracel Islands); Nansha (=Spratly Islands). (17, 19, 24, 27, 37)
A. pulchra stricta (Brook): Hainan; Xisha (=Paracel Islands); Nansha (=Spratly Islands). (17, 19, 27)
A. rosaria (Dana): Xisha (=Paracel Islands); Nansha (=Spratly Islands). (19, 27)
A. rotumana (Gardiner): Xisha (=Paracel Islands); Nansha (=Spratly Islands). (19, 24, 27)
A. surculosa (Dana): Hainan. (17)
A. tenella (Brook): Hainan; Xisha (=Paracel Islands). (17, 27)
A. tizardi (Brook): Xisha (=Paracel Islands); Nansha (=Spratly Islands). (17, 27)
A. valida (Dana): Hainan; Xisha (=Paracel Islands). (17, 19)
A. variabilis (Klunzinger): Xisha (=Paracel Islands); Nansha (=Spratly Islands). (19, 27)
Anacropora tapera Zou, Song & Ma: Weizhou Island (Guangxi). (18)
Astreopora myriophthalma (Lamarck): Hainan; Xisha (=Paracel Islands); Nansha (=Spratly Islands). (17, 19, 24, 27)

Montipora aenigmatica Bernard: Hainan; Nansha (=Spratly Islands). (17, 27)

M. cristagalli (Ehrenberg): Hainan. (17)

M. danae Milne-Edwards & Haime: Xisha (=Paracel Islands), Zhongsha (=Macclesfield Bank). (19)

M. foliosa (Pallas): Taiwan; Hainan; Dongsha (=Pratas Islands); Xisha (=Paracel Islands); Nansha (=Spratly Islands). (17, 19, 24, 28)

M. foveolata (Dana): Xisha (=Paracel Islands). (19)

M. fragilis Qouelch: Hainan. (17)

M. fruticosa Bernard: Hainan; Xisha (=Paracel Islands). (17, 19)

M. gaimardi Bernard: Hainan. (17)

M. hispida (Dana): Taiwan; Hainan; Xisha (=Paracel Islands); Nansha (=Spratly Islands). (17, 19, 27, 28).

M. ramosa Bernard: Taiwan; Hainan; Xisha (=Paracel Islands); Nansha (=Spratly Islands). (17, 19, 24, 28)

M. sinensis Bernard: Daya Bay (Guangdong); Weizhou Island (Guangxi); Hainan; Xisha (=Paracel Islands); Nansha (=Spratly Islands). (17, 19, 24, 28)

M. solanderi Bernard: Hainan. (17)

M. striata Bernard: Hainan. (17)

M. trabeculata Bernard: Hainan; Xisha (=Paracel Islands); Nansha (=Spratly Islands). (17, 19, 27, 37)

M. truncata Zou, Song & Ma: Hainan. (17)

M. turgescens Bernard: Hainan; Xisha (=Paracel Islands). (17, 19)

AGARICIIDAE

Leptoseris gardineri van der Horst: Hainan. (17)

L. papyracea (Dana): Zengmuansha (southern tip of Nansha (=Spratly Islands)). (24)

Pachyseris involuta Studer: Xisha (=Paracel Islands). (21)

P. rugosa (Lamarck): Hainan; Dongsha (=Pratas Islands); Xisha (=Paracel Islands); Nansha (=Spratly Islands), (17, 19, 24, 28)

P. speciosa (Dana): Taiwan; Hainan; Dongsha (=Pratas Islands); Xisha (=Paracel Islands). (17, 19, 28)

Pavona cactus (Forskål): Xisha (=Paracel Islands). (19)

P. decussata Dana: Taiwan; Daya Bay (Guangdong); Weizhou Island (Guangxi); Hainan. (17, 23, 26, 28)

P. frandifera Lamarck: Taiwan; Hainan. (17, 28)

P. lata Dana: Hainan. (17)

P. (Polyastra) minikoiensis (Gardiner): Hainan; Nansha (=Spratly Islands). (17, 27)

P. minuta Wells: Xisha (=Paracel Islands); Nansha (=Spratly Islands). (21, 27)

P. praetorta Dana: Taiwan; Hainan; Xisha (=Paracel Islands). (17, 19, 28)

P. seriata Brüggemans: Xisha (=Paracel Islands). (19)

P. varians Verrill: Hainan; Dongsha (=Pratas Islands); Xisha (=Paracel Islands); Nansha (=Spratly Islands). (17, 19, 27, 28, 37)

FUNGIIDAE

Cycloseris cyclolites (Lamarck): Nansha (=Spratly Islands). (27)

Fungia danai Milne-Edwards & Haime: Dongsha (=Pratas Islands); Xisha (=Paracel slands). (19, 28)

F. echinata (Pallas): Hainan; Xisha (=Paracel Islands); Nansha (=Spratly Islands). (17, 19, 27)

F. fieldi Gardiner: Xisha (=Paracel Islands). (19)

F. fungites (Linnaeus): Hainan; Dongsha (=Pratas Islands); Xisha (=Paracel Islands); Nansha (=Spratly Islands). (17, 19, 27, 28, 37)

F. paumotensis Stutchbury: Hainan. (17)

F. repanda Dana: Dongsha (=Pratas Islands); Xisha (=Paracel Islands). (19, 28)

F. scutaria Lamarck: Xisha (=Paracel Islands); Nansha (=Spratly Islands). (19, 27)

Halomitra philippinensis (Studer): Weizhou Island (Guangxi); Xisha (=Paracel Islands). (19, 23)

Heliofungia actiniformis (Quoy & Gaimard): Xisha (=Paracel Islands). (19)

H. actiniformis palawensis Döderlein: Xisha (=Paracel Islands). (19)

Herpolitha limax (Esper): Dongsha (=Pratas Islands); Xisha (=Paracel Islands). (19, 28)

H. weberi (van der Horst): Xisha (=Paracel Islands); Nansha (=Spratly Islands). (19, 27)

Podabacia crustacen (Pallas): Hainan; Nansha (=Spratly Islands). (17, 27)

P. elegans lobata (van der Horst): Hainan. (17)

Polyphyllia talpina Lamarck): Hainan; Xisha (=Paracel Islands); Nansha (=Spratly Islands). (17, 19, 27)

Sandalolitha robusta (Quelch): Hainan; Dongsha (=Pratas Islands); Xisha (=Paracel Islands). (17, 19, 28, 37)

PORITIIDAE

Alveopora excelsa Verrill: Xisha (=Paracel Islands); Zengmuansha (southern tip of Nansha (=Spratly Islands)). (21, 24)

A. japonica Eguchi: Zengmuansha (southern tip of Nansha (=Spratly Islands)). (24)

A. polyformis Zou: Xisha (=Paracel Islands). (20)

Goniopora columna Dana: Daya Bay (Guangdong); Weizhou Island (Guangxi). (23, 26)

G. duofasciata Thiel: Daya Bay (Guangdong); Hainan. (17, 26)

G. gracilis (Bassett-Smith): Xisha (=Paracel Islands); Nansha (=Spratly Islands). (19, 27)

G. minor Crossland: Xisha (=Paracel Islands). (19)

G. wotouensis Zou, Song & Ma: Autou (Guangdong). (18)

Porites andrewsi Vaughan: Hainan; Xisha (=Paracel Islands); Nansha (=Spratly Islands). (17, 19, 27, 37)

P. compressa Dana: Daya Bay (Guangdong). (26)

P. iwqyamaensis Eguchi: Hainan; Xisha (=Paracel Islands); Nansha (=Spratly Islands). (17, 19, 27)

P. lichen Dana: Xisha (=Paracel Islands); Nansha (=Spratly Islands). (19, 27)

P. lutea Milne-Edwards & Haime: Daya Bay (Guangdong); Weizhou Island (Guangxi); Hainan; Xisha (=Paracel Islands); Nansha (=Spratly Islands). (17, 19, 23, 24, 26)

P. mattaii Wells: Hainan; Xisha (=Paracel Islands); Nansha (=Spratly Islands). (17, 19, 27)

P. nigrescens Dana: Hainan; Nansha (=Spratly Islands). (17, 27)

P. pukoensis Vaughan: Weizhou Island (Guangxi); Hainan; Xisha (=Paracel Islands); Nansha (=Spratly Islands), (17, 19, 23, 27)

DENDROPHYLLIDAE

Tugastraea aurea (Quoy & Gaimard): Xiamen harbor (Fujian); Taiwan; Hainan; Xisha (=Paracel Islands). (25, 28)

T. coccinea (Ehrenberg): Guangdong seashore. (25)

T. diaphana (Dana): Dongshan (Fujian); Hainan; Xisha (=Paracel Islands). (25)

T. tenuilamellosa (Milne-Edwards & Haime): Hainan; Xisha (=Paracel Islands). (25)

T. titijimaensis Yabe & Eguchi: Hainan; Xisha (=Paracel Islands). (25)

Turbinaria agaricia Bernard: Daya Bay (Guangdong); Hainan. (17, 26)

T. contorta Bernard: Taiwan; Guangdong; Weizhou Island (Guangxi); Dongsha (=Pratas Islands); Nansha (=Spratly Islands). (23, 26, 27, 28).

T. crater (Pallas): Daya Bay (Guangdong); Hainan; Xisha (=Paracel Islands); Nansha (=Spratly Islands). (17, 19, 24, 26)

T. elegans Bernard: Penghu Islands (Taiwan); Weizhou Island (Guangxi); Dongsha (=Pratas Islands). (23, 28)

T. foliosa Bernard: Taiwan; Naozhou Island (Guangdong). (23, 28)

T. globularis Bernard: Weizhou Island (Guangxi). (23)

T. irregularis Bernard: Weizhou Island (Guangxi); Xisha (=Paracel Islands). (20, 23)

T. mantonae Crossland: Hainan. (17)

T. mesenterina (Lamarck): Daya Bay (Guangdong); Hong Kong; Weizhou Island (Guangxi). (23, 26, 28)

T. peltata (Esper): Taiwan; Guangdong; Hong Kong; Weizhou Island (Guangxi); Hainan; Dongsha (=Pratas Islands); Nansha (=Spratly Islands). (17, 23, 24, 26, 28)

T. tizimeensis Yabe & Sugiyama: Naozhou Island (Guangdong). (26)

T. undata Bernard: Naozhou Island and Daya Bay (Guangdong); Weizhou Island (Guangxi); Hainan. (17, 23, 26)

SIDERASTREIDAE

Coeloseris mayeri Vaughan: Xisha (=Paracel Islands); Nansha (=Spratly Islands). (19, 27)

Coscinaraea exesa (Dana): Xisha (=Paracel Islands); Zhongsha (=Macclesfield Bank); Nansha (=Spratly Islands). (19, 27)

Pseudosiderastrea tayamai Yabe & Sugiyama: Guangdong; Guangxi. (21)

OCULINIDAE

Acrhelia horrescens Dana: Xisha (=Paracel Islands); Nansha (=Spratly Islands). (21, 24, 27)

Galaxea aspera Quelch: Hainan; Xisha (=Paracel Islands); Nansha (=Spratly Islands). (17, 19, 24, 27)

G. fascicularis (Linnaeus): Taiwan; Weizhou Island (Guangxi); Dongsha (=Pratas Islands); Xisha (=Paracel Islands), Nansha (=Spratly Islands). (17, 19, 23, 27, 28)

G. lamarcki Milne-Edwards & Haime: Hainan. (17)

FAVIIDAE

Caulastrea furcata Dana: Hainan; Dongsha (=Pratas Islands); Nansha (=Spratly Islands). (17, 27, 28)

Cyphastrea serailia (Forskål): Taiwan; Daya Bay (Guangdong); Hainan; Dongsha (=Pratas Islands); Xisha (=Paracel Islands); Nansha (=Spratly Islands). (17, 19, 26, 27, 28)

C. zhongjianensis Zou: Xisha (=Paracel Islands). (20)

Diploastrea heliopora (Lamarck): Hainan. (17)

Echinopora gemmacea (Lamarck): Nansha (=Spratly Islands). (24, 27)

E. horrida Dana: Nansha (=Spratly Islands). (27)

E. lamellosa (Esper): Hainan; Dongsha (=Pratas Islands); Xisha (=Paracel Islands); Nansha (=Spratly Islands). (17, 19, 24, 27, 28)

Favia matthaii Vaughan: Hainan; Xisha (=Paracel Islands); Nansha (=Spratly Islands). (17, 19, 24)

F. palauensis Yabe & Sugiyama: Hainan; Xisha (=Paracel Islands). (17, 19)

F. rotumana (Gardiner): Daya Bay (Guangdong); Weizhou Island (Guangxi); Hainan; Dongsha (=Pratas Islands); Xisha (=Paracel Islands); Nansha (=Spratly Islands). (17, 19, 23, 26, 27, 28)

F. speciosa (Dana): Xiamen harbor (Fujian); Taiwan; Daya Bay (Guangdong); Weizhou Island (Guangxi); Hainan; Dongsha (=Pratas Islands); Xisha (=Paracel Islands); Nansha (=Spratly Islands). (17, 19, 23, 27, 28)

F. stelligera (Dana): Xisha (=Paracel Islands); Zhongsha (=Macclesfield Bank); Nansha (=Spratly Islands). (19, 27, 28)

Favites abdita (Ellis & Solander): Taiwan; Daya Bay (Guangdong); Weizhou Island (Guangxi); Hainan; Dongsha (=Pratas Islands); Nansha (=Spratly Islands). (17, 23, 26, 27, 28)

F. halicora (Ehrenberg): Dongsha (=Pratas Islands); Xisha (=Paracel Islands); Nansha (=Spratly Islands). (19, 27, 28)

F. pentagona (Esper.): Daya Bay (Guangdong); Hainan; Dongsha (=Pratas Islands). (17, 26, 28)

F. virens (Dana): Dongsha (=Pratas Islands); Xisha (=Paracel Islands); Nansha (=Spratly Islands). (19, 24, 27, 28)

Goniastrea aspera Verrill: Xiamen harbor (Fujian); Daya Bay (Guangdong); Weizhou Island (Guangxi); Hainan; Dongsha (=Pratas Islands); Xisha (=Paracel Islands); Nansha (=Spratly Islands). (17, 19, 24, 26–28)

G. pectinata (Ehrenberg): Taiwan; Hainan; Dongsha (=Pratas Islands); Xisha (=Paracel Islands); Nansha (=Spratly Islands). (17, 19, 24, 28)

G. retiformis (Lamarck): Weizhou Island (Guangxi); Hainan; Dongsha (=Pratas Islands); Xisha (=Paracel Islands); Nansha (=Spratly Islands). (17, 19, 24, 28)

G. yamanarii (Yabe & Sugiyama): Daya Bay (Guangdong); Hainan. (17, 26)

G. yamanarii profunda (Umbgrove): Hainan; Xisha (=Paracel Islands). (17, 19)

Hydnophora contignatio (Forskål): Hainan. (17)

H. exesa (Pallas): Daya Bay (Guangdong); Weizhou Island (Guangxi); Hainan; Dongsha (=Pratas Islands); Xisha (=Paracel Islands); Nansha (=Spratly Islands). (17, 23, 26, 27, 28)

H. microconos (Lamarck): Hainan; Dongsha (=Pratas Islands); Xisha (=Paracel Islands); Nansha (=Spratly Islands). (17, 19, 24, 27, 28)

H. rigida (Dana): Xisha (=Paracel Islands). (19)

Leptastrea bottae (Milne-Edwards & Haime): Xisha (=Paracel Islands). (19)

L. immersa Klunzinger: Xisha (=Paracel Islands); Nansha (=Spratyl Islands). (19, 27)

L. purpurea (Dana): Taiwan; Hainan; Dongsha (=Pratas Islands); Xisha (=Paracel Islands); Nansha (=Spratly Islands). (17, 19, 24, 26, 28)

L. transversa Klunzinger: Xisha (=Paracel Islands); Nansha (=Spratyl Islands). (19)

Platygyra astreiformis (Milne-Edwards & Haime): Dongsha (=Pratas Islands); Xisha (=Paracel Islands); Nansha (=Spratly Islands). (19, 27)

P. crosslandi (Matthai): Hainan. (17)

P. gracilis (Dana): Hainan; Xisha (=Paracel Islands); Nansha (=Spratly Islands). (17, 19, 27)

P. phrygia (Ellis & Solander): Taiwan; Hainan; Dongsha (=Pratas Islands); Xisha (=Paracel Islands); Nansha (=Spratly Islands). (17, 19, 27, 28)

P. rustica (Dana): Taiwan; Daya Bay (Guangdong); Weizhou Island (Guangxi); Dongsha (=Pratas Islands); Xisha (=Paracel Islands); Nansha (=Spratly Islands). (17, 19, 23, 26, 27, 28)

P. ryukyuensis Yabe & Sugiyama: Hainan; Xisha (=Paracel Islands). (17, 19)

Plesidastrea vacua (Crossland): Xisha (=Paracel Islands); Nansha (=Spratly Islands). (19, 27)

P. versipora (Lamarck): Taiwan; Daya Bay (Guangdong); Hainan; Dongsha (=Pratas Islands); Xisha (=Paracel Islands); Nansha (=Spratly Islands). (19, 24, 26, 27, 28)

Scapophyllia cylindrica Milne-Edwards & Haime: Hainan; Nansha (=Spratyl Islands). (17, 27)

MUSSIDAE

Acanthastrea echinata (Dana): Weizhou Island (Guangxi); Hainan; Xisha (=Paracel Islands); Nansha (=Spratly Islands). (17, 19, 23, 27)

Lobophyllia corymbosa (Forskål): Hainan; Dongsha (=Pratas Islands). (17, 28)

L. costata (Dana): Hainan; Xisha (=Paracel Islands); Nansha (=Spratyl Islands). (17, 19, 27)

Symphyllia agaricia Milne-Edwards & Haime: Hainan; Xisha (=Paracel Islands). (17, 19)

S. gigantea (Yabe & Sugiyama): Hainan. (17)

S. nobilis (Dana): Taiwan; Hainan; Dongsha (=Pratas Islands); Nansha (=Spratly Islands). (17, 27, 28)

S. radians Milne-Edwards & Haime: Hainan; Nansha (=Spratly Islands). (17, 27)

MERULINIDAE

Merulina ampliata (Ellis & Solander): Hainan; Dongsha (=Pratas Islands); Xisha (=Paracel Islands); Nansha (=Spratly Islands). (17, 19, 27, 28)

M. laxa Dana: Hainan; Xisha (=Paracel Islands); Nansha (=Spratly Islands). (17, 19, 27)

M. scabricula Dana: Hainan. (17)

PECTINIIDAE

Echinophyllia aspera Ellis & Solander: Daya Bay (Guangdong); Weizhou Island (Guangxi); Hainan. (17, 23, 26)

Mycedium elephantotus (Pallas): Nansha (=Spratly Islands). (27)

Pectinia lactuca (Pallas): Taiwan; Hainan; Dongsha (=Pratas Islands); Nansha (=Spratly Islands). (17, 27, 28)

CARYOPHYLLIIDAE

Cataliphyllia fimbriata (Spengler): Taiwan; Hainan; Nansha (=Spratly Islands). (17, 27, 28)

Euphyllia turgida Dana: Xisha (=Paracel Islands). (20)

NON-REEF BUILDERS

FUNGIIDAE

Diaseris fragilis Alcock: Southern South China Sea. (29)

Fungiacyathus stephanus (Alcock): East China Sea; southern South China Sea. (29, 30)

F. symmetricus (Pourtales): Northeastern South China Sea. (29, 30)

MICRABACIIDAE

Stephanophyllia formosissima Moseley: Southern South China Sea. (29)

S. fungulus Alcock: East China Sea; northeastern South China Sea. (29, 30)

S. japonica Yabe & Eguchi: Northeastern and southern South China Sea. (29)

ANTHEMIPHYLLIIDAE

Anthemiphyllia dentata (Alcock): East China Sea; northeastern South China Sea. (29)

CARYOPHYLLIIDAE

Acanthocyathus grayi Milne-Edwards & Haime: Southern South China Sea. (29)

A. spiniger Kent: East China Sea; northeastern and southern South China Sea. (29)

Anomocora fecunda (Pourtales): East China Sea. (31)

Balanophyllia rediviva Moseley: South China Sea. (30)

Caryophyllia compressa Yabe & Eguchi: East China Sea. (31)

C. japonica Marenzeller: East China Sea; northeastern and southern South China Sea. (29)

C. pacifica (Yabe & Eguchi): East China Sea. (31)

C. paucipaliata Yabe & Eguchi: Southern South China Sea. (29)

C. scobinosa Alcock: East China Sea. (31)

Ceratotrochs brunneus (Moseley): Central South China Sea. (29)

C. funicolumna Alcock: Northeastern South China Sea. (29)

Deltocyathus lens Alcock: Northeastern and southern South China Sea. (29)

D. murrayi Gardiner & Vaugh: Central South China Sea. (29)

Desmophyllum alabastrum Alcock: South China Sea. (36)

Heterocyathus aequicostatus Milne-Edwards & Haime: Taiwan; northeastern, central, and southern South China Sea. (20)

H. alternatus Verrill: East China Sea; northeastern, central and southern South China Sea. (29)

H. japonicus (Verrill): East China Sea; northeastern and southern South China Sea. (29)

Notocaythus conicus (Alcock): East China Sea; northeastern and southern South China Sea. (29)

Paracyathus pruinosus Alcock: Southern South China Sea. (29)

Stephanocyathus nobilis (Moseley): Central South China Sea. (29)

S. spiniger (Marenzeller): Northeastern and southern South China Sea. (29)

Trochocyathus caryophylloides Alcock: East China Sea; northeastern and southern South China Sea. (29)

T. pileus Alcock: East China Sea; northeastern and southern South China Sea. (29)

T. pseudowingedus Zou & Chen: East China Sea. (31)

Tropidocyathus lessonii Milne-Edwards & Haime: East China Sea; northeastern and southern South China Sea. (29)

RHIZANGIIDAE

Culicia rubeola (Quoy & Gaimard): South China Sea. (31)

Oulangia stokesiana Milne-Edwards & Haime: Northeastern and southern South China Sea. (29)

FLABELLIIDAE

Flabellum deludens Marenzeller: East China Sea. (31)

F. distinctum Milne-Edwards & Haime: East China Sea; northeastern and southern South China Sea. (29)

F. pavoninum Lesson: Northeastern and southern South China Sea. (29)

F. rubrum (Quoy & Gaimard): East China Sea; northeastern and southern South China Sea. (29)

DENDROPHYLLIIDAE

Balanophyllia compressa Yabe & Eguchi: East China Sea. (31)

B. conica (Horst): East China Sea; southern South China Sea. (29)

B. cornu Moseley: East China Sea; northeastern South China Sea. (29)

B. cumingii Milne-Edwards & Haime: Xiamen (Fujian). (38)

B. imperialis Kent: Southern South China Sea. (29)

B. rediviva Moseley: South China Sea. (30)

Endopachys grayi Milne-Edwards & Haime: East China Sea; northeastern and southern South China Sea. (29)

Heteropsammia cochlea (Spengler): East China Sea; southern South China Sea. (29)

H. michelini Milne-Edwards & Haime: East China Sea; southern South China Sea. (29)

ANTIPATHARIA
BATHYPATHIDAE

Bathypathes affinis (Brook): Zhongsha (=Macclesfield Bank). (13, 14)

B. patulis Brook: Zhongsha (=Macclesfield Bank). (13, 14)

B. tenuis Brook: Nansha (=Spratly Islands). (12, 13)

ANTIPATHIDAE

Antipathes ceylonensis Thomson & Simpson: Hainan. (10, 13)

A. chotis Cooper: Nansha (=Spratly Islands). (12, 13)

A. crispis (Brook): Nansha (=Spratly Islands). (12, 13)

A. dichotomis Pallas: Mouth of Zhujiang (=Pearl River) (Guangdong); Hainan; Zhongsha (=Macclesfield Bank); Nansha (=Spratly Islands). (10, 12, 13)

A. flabellum (Pallas): Mouth of Zhujiang (=Pearl River, Guangdong). (10, 13)

A. grandis Verrill: Mouth of Zhujiang (=Pearl River, Guangdong). (10, 13, 16)

A. herdmani Cooper: Zhongsha (=Macclesfield Bank). (13, 14)

A. japonica Brook: Taiwan; Hainan; Zhongsha (=Macclesfield Bank). (10, 12, 13)

A. lentis Pourtalés: Nansha (=Spratly Islands). (12, 13)

A. myriephyllis Pallas: Zhongsha (=Macclesfield Bank); Nansha (=Spratly Islands). (12, 13)

A. planis Cooper: Zhongsha (=Macclesfield Bank). (13, 14)

A. viminalis Roule: Mouth of Zhujiang (=Pearl River, Guangdong). (10, 13, 15)

A. virgatis Esper: Hainan. (10, 13)

Aphanipathes sarothamnoides Brook: Zhongsha (=Macclesfield Bank). (13, 14)

A. somervillei Cooper: Zhongsha (=Macclesfield Bank). (13, 14)

Cirripathes anguinis Dana: Mouth of Zhujiang (=Pearl River, Guangdong); Hainan. (15, 16)

C. hainanensis Zou & Zhou: Mouth of Zhujiang (=Pearl River, Guangdong); Hainan. (13, 15, 16)

C. musculosis (Pesch): Mouth of Zhujiang (=Pearl River, Guangdong). (14, 15)

C. rumphii (Pestch): Mouth of Zhujiang (=Pearl River, Guangdong); Hainan; Xisha (=Paracel Islands). (13, 15, 16)

C. sinensis Zou & Zhou: Mouth of Zhujiang (=Pearl River, Guangdong); Hainan. (13, 15, 16)

C. spiralis (Linnaeus): Xisha (=Paracel Islands); Zhongsha (=Macclesfield Bank). (13, 14)

Parantipathes cylindrices (Brook): Zhongsha (=Macclesfield Bank). (13, 14)

Stichopathes abyssicolis Roule: Nansha (=Spratly Islands). (11, 12, 13)

S. bournei Cooper: Nansha (=Spratly Islands). (11, 12, 13)

S. ceylonensis Thomson & Simpson: Nansha (=Spratly Islands). (11, 12, 13)

S. contortis Thomson & Simpson: Nansha (=Spratly Islands). (11, 12, 13)

S. desbonni (D. & M.): Zhongsha (=Macclesfield Bank). (11, 13, 14)

S. filiformis (Gray): Zhongsha (=Macclesfield Bank). (11, 13, 14)

S. flagellum Roule: Xisha (=Paracel Islands); Nansha (=Spratly Islands). (11–14)

S. gracilis (Gray): Nansha (=Spratly Islands). (11, 12, 13)

S. maldivensis Cooper: Nansha (=Spratly Islands). (11, 12, 13)

S. papillosis Thomson & Simpson: Nansha (=Spratly Islands). (12, 13)

S. regularis Cooper: Nansha (=Spratly Islands). (11, 12, 13)

S. sacculis (Pesch): Nansha (=Spratly Islands). (11, 12, 13)

S. semiglabris (Pesch): Nansha (=Spratly Islands). (11, 12, 13)

S. variabilis (Pesch): Nansha (=Spratly Islands). (11, 12, 13)

HELIOPORACEA
HELIOPORIDAE

Heliopora coerulea (Pallas): Taiwan; Hainan; Dongsha (=Pratas Islands); Xisha (=Paracel Islands); Nansha (=Spratly Islands). (17, 27, 28)

STOLONIFERA
TUBIPORIDAE

Tubipora musica Linnaeus: Taiwan; Hainan; Dongsha (=Pratas Islands); Xisha (=Paracel Islands); Nansha (=Spratly Islands). (17, 27, 28)

ALCYONACEA
ANTHOTHELIDAE

Semperina brunnea Nutting: Hong Kong. (32)

SUBERGORGIIDAE

Subergorgia kollikeri Wright & Studer: Hong Kong; Hainan. (32, 33)

S. ornata Thomson & Simpson: Hainan. (33)

S. reticulata (Ellis & Solander): Guangdong; Hong Kong; Lingaojiao (Hainan). (32, 33)

S. suberosa (Pallas): Guangdong; Hainan. (33)

MELITODIDAE

Acabaria formosa Nutting: Hainan. (33)

A. philippinensis (Wright & Studer): Jinmen (Fujian). (38)

Melitodes modesta Nutting: Shaoan (Fujian). (38)

M. ocracea (Linnaeus): Hainan. (33)

M. squarnata Nutting: Hainan. (33)

Mopsella rubeola (Wright & Studer): Hainan. (33)

ACANTHOGORGIIDAE

Acalycigorgia inermis (Hedlung): Taiwan; Hong Kong. (32)

Acanthogorgia armata Verrill: Xiamen (Fujian). (38)

A. aspera Pourtalis: Xiamen (Fujian). (38)

A. muriata Verrill: Dongshan (Fujian). (38)

A. vegae Aurivillius: Guangdong; Hong Kong. (32, 33)

Anthogorgia bocki Aurivillius: Hong Kong. (32)

PARAMURICEIDAE

Echinogorgia aurantiaca (Valenciennes): Jinmen (Fujian); Guangdong. (33, 38)

E. coccinea (Stimpson): Guangdong; Hong Kong. (32, 33)

E. complexa Nutting: Xiamen harbor (Fujian). (38)

E. flora Nutting: Dongshan (Fujian); Guangdong; Hong Kong. (32, 33, 38)

E. lami Stiasny: Guangdong; Hong Kong. (32, 33)
E. mertoni Kükenthal: Guangdong. (33)
E. pseudosassapo Kölliker: Taiwan Strait; Guangdong; Hong Kong. (32, 33, 38)
E. sassapo reticulata (Esper): Guangdong; Hong Kong; Hainan. (32, 33)
Echinomuricea indomalaccensis Ridley: Hong Kong. (32)
Menella praelonga (Ridley): Guangdong; Hong Kong. (32, 33)
M. rubescens Nutting: Hong Kong. (32)
M. spinifera Kükenthal: Guangdong. (33)
M. verrucosa Brundin: Guangdong. (33)
Muricella abnormalis Nutting: Guangdong; Hong Kong. (32, 33)
M. flexuosa (Verrill): Guangdong; Hong Kong. (32, 33)
M. rubra Thomson: Xiamen (Fujian). (38)
M. sibogae (Nutting): Guangdong; Hong Kong. (32, 33)
M. sinensis (Verrill): Guangdong; Hong Kong. (32, 33)
Villogorgia compressa Hiles: Guangdong; Hong Kong. (32, 33)
V. intricata (Gray): Guangdong. (33)

PLEXAURIDAE

Euplexaura albida Kükenthal: Zhoushan (Zhejiang); Fuding and Lianjiang (Fujian). (38)
E. attenuata (Nutting): Xiapu, Huian, and Xiamen (Fujian). (38)
E. curvata Kükenthal: Guangdong; Hong Kong. (32, 33)
E. erecta Kükenthal: Guangdong; Hong Kong. (32, 33)
E. robusta Kükenthal: Hong Kong. (32)
Hicksonella guishanensis Zou: Qingdao (Shandong); Wenzhou (Zhejiang); Xiamen (Fujian); mouth of Zhujiang (=Pearl River, Guangdong). (33)
H. princeps Nutting: Guangdong; Hong Kong. (32, 33)

ELLISELLIDAE

Ctenocella pectinata (Pallas): Guangdong; Hainan. (33)
Ellisella gracilis (Wright & Studer): Guangdong. (35)
E. laevis (Verrill): Guangdong; Hong Kong. (32, 33)
Isis hippuris Linnaeus: Taiwan; Zhongsha (=Macclesfield Bank). (34)
I. minorbrachyblasta Zou & Huang: Nansha (=Spratly Islands). (34)
I. reticulata Nutting: Xisha (=Paracel Islands). (34)
Junceella fragilis (Ridley): Guangdong; Hainan. (33)
J. gemmacea (Valenciennes): Guangdong. (33)
J. juncea (Pallas): Hainan. (33)
J. racemosa Wright & Studer: Guangdong. (33)
J. squamata Toeplitz: Taiwan; Guangdong; Hainan. (33)
Scirpearia erythraea Kükenthal: Hong Kong; Hainan. (32, 33)
S. gracilis (Wright & Studer): Guangdong. (33)

ALCYONACEA
ALCYONIIDAE

Alcyonium gracillimum Kükenthal: Southern Nansha (=Spratly Islands). (9)
Cladiella humesi Verseveldt: Dapeng Bay (Hong Kong); Banyue Reef (Nansha=Spratly Islands). (6, 8)
C. krempfi Hickson: Yalong Bay (Hainan). (2)
C. madagascarensis Tixier-Durivault: Dapeng Bay (Hong Kong). (6)
C. subtilis Tixier-Durivault: Dapeng Bay (Hong Kong); Xianbin Reef (Nansha=Spratly Islands). (6, 8)
Lobophytum angulatum Tixier-Durivault: Zhongjian Island (Xisha=Paracel Islands). (4)
L. anomalum Li: Panshi Reef (Xisha=Paracel Islands). (4)
L. caledonense Tixier-Durivault: Haikou Reef (Nansha=Spratly Islands). (8)
L. caputospiculatum Li: Zhaoshu Island (Xisha=Paracel Islands). (4)
L. chevalieri Tixier-Durivault: Zhaoshu Island (Xisha=Paracel Islands). (4)
L. compactum Tixier-Durivault: Zhaoshu Island (Xisha=Paracel Islands); Renai Reef (Nansha=Spratly Islands). (4, 8)
L. crassodigitum Li: Zhongjian Island (Xisha=Paracel Islands). (4)
L. crassospiculatum Moser: Zhaoshu Island and Panshi Reef (Xisha=Paracel Islands). (4)
L. cristaglli Marenzeller: Xiane Reef (Nansha=Spratly Islands). (8)
L. cristatum Tixier-Durivault: Huangyan Island (Zhongsha=Macclesfield Bank); Panshi Reef and Yongxin Island (Xisha=Paracel Islands). (2, 4)
L. delectum Tixier-Durivault: Panshi Reef (Xisha=Paracel Islands). (4)
L. denticulatum Tixier-Durivault: Dapeng Bay (Hong Kong). (6)
L. depressum Tixier-Durivault: Dapeng Bay (Hong Kong). (6)
L. gazellae Moser: Zhaoshu Island (Xisha=Paracel Islands). (4)
L. hirsutum Tixier-Durivault: Zhaoshu Island (Xisha=Paracel Islands); Meiji Reef (Nansha=Spratly Islands). (4, 8)
L. irregulare Tixier-Durivault: Zhaoshu Island (Xisha=Paracel Islands). (4)
L. longispiculatum Li: Huaguang Reef (Xisha=Paracel Islands). (4)
L. microspiculatum Tixier-Durivault: Xiane Reef (Nansha=Spratly Islands). (8)
L. oblongum Tixier-Durivault: Shanhu Island (Xisha=Paracel Islands). (4)
L. oligoverrucum Li: Panshi Reef (Xisha=Paracel Islands). (4)
L. pauciflorum (Ehrenberg): Dong Island, Shanhu Island, and Jinqing Island (Xisha=Paracel Islands). (4)
L. pulchellum Tixier-Durivault: Panshi Reef (Xisha=Paracel Islands). (4)
L. pygmapedium Li: Dong Island (Xisha=Paracel Islands). (4)

L. ransoni Tixier-Durivault: Panshi Reef (Xisha=Paracel Islands). (4)

L. salvati Tixier-Durivault: Zhaoshu Island (Xisha=Paracel Islands). (4)

L. schoedei Moser: Zhongjian Island (Xisha=Paracel Islands). (4)

L. spicodigitum Li: Zhaoshu Island (Xisha=Paracel Islands). (4)

L. spissum Tixier-Durivault: Zhaoshu Island, Chenhang Island, and Zhongjian Island (Xisha=Paracel Islands). (4)

L. venustum Tixier-Durivault: Dapeng Bay (Hong Kong). (6)

L. verrucosum Li: Huaguang Reef (Xisha=Paracel Islands). (4)

Sarcophyton acutangulum (Marenzeller): Yagong Island (Xisha=Paracel Islands). (4)

S. boletiforme Tixier-Durivault: Southern Nansha (=Spratly Islands). (8)

S. crassocaule Moser: Huangyan Island (Zhongsha=Macclesfield Bank); Xianbin Reef, Renai Reef, Meiji Reef, and Jianzhang Reef (Nansha=Spratly Islands). (1, 8)

S. elegans Moser: Yagong Island (Xisha=Paracel Islands). (4)

S. furcatum Li: Zhongjian Island (Xisha=Paracel Islands). (4)

S. glaucum (Quoy & Gaimard): Yalong Bay (Hainan). (2)

S. infundibuforme Tixier-Durivault: Yongxin Island (Xisha=Paracel Islands). (4)

S. latum (Dana): Huangyan Island (Zhongsha=Macclesfield Bank). (1)

S. molle Tixier-Durivault: Xianbin Reef and Renai Reef (Nansha=Spratly Islands). (8)

S. trocheliophrum Marenzeller: Taiwan; Huangyan Island (Zhongsha=Macclesfield Bank); Meiji Reef (Nansha=Spratly Islands). (1, 9)

Sinularia capillosa Tixier-Durivault: Yalong Bay (Hainan). (2)

S. compressa Tixier-Durivault: Huangyan Island (Zhongsha=Macclesfield Bank). (1)

S. corpulenta Li: Yalong Bay (Hainan). (2)

S. fibrillosa Li: Yalong Bay (Hainan). (2)

S. granosa Tixier-Durivault: Yalong Bay (Hainan). (2)

S. inexplicita Tixier-Durivault: Huangyan Island (Zhongsha=Macclesfield Bank); Renai Reef (Nansha=Spratly Islands). (1, 8)

S. microclavata Tixier-Durivault: Hainan. (2)

S. monstrosa Li: Yalong Bay (Hainan). (2)

S. nanolobata Verseveldt: Xianbin Reef (Nansha=Spratly Islands). (8)

S. papillosa Li: Yalong Bay (Hainan). (2)

S. partia Tixier-Durivault: Yalong Bay (Hainan).)2)

S. polydactyla (Ehrenberg): Yalong Bay (Hainan). (2)

S. prattae Verseveldt: Zengmuanshan (southern tip of Nansha=Spratly Islands). (7)

S. quarciformis (Pratt): Southern Nansha (=Spratly Islands). (8)

S. ramulosa Tixier-Durivault: Yalong Bay (Hainan). (2)

S. renei Tixie-Durivault: Yalong Bay (Hainan). (2)

S. tenella Li: Yalong Bay (Hainan). (2)

NEPHTHEIDAE

Capnella imbricata Quoy & Gaimard: Banyue Reef (Nansha=Spratly Islands). (8)

C. philippinensis Light: Meiji Reef (Nansha=Spratly Islands). (9)

Lemnalia bournei Roxas: Banyue Reef (Nansha=Spratly Islands). (8)

Morchellana dollfusi Tixier-Durivault: Zengmuansha (southern tip of Nansha=Spratly Islands). (7)

M. elongata (Henderson): Zengmuansha (southern tip of Nansha=Spratly Islands). (7)

M. pulchella (Utinomi): Zengmuansha (southern tip of Nansha=Spratly Islands). (7)

M. rubra (May) [*Dendronephya rubra*]: Taiwan Strait. (38)

Nephthea brassica Kükenthal: Niuchelun Island and Xinyi Reef (Nansha=Spratly Islands). (8)

N. capnelliformis Thomson & Dean: Xianbin Reef (Nansha=Spratly Islands). (9)

N. erecta Kükenthal: Zengmuansha (southern tip of Nansha=Spratly Islands). (7)

N. simulata Verseveldt: Southern Nansha (=Spratly Islands). (8)

Paralemnalia eburnea Kükenthal: Renai Reef (Nansha=Spratly Islands). (9)

P. thyrsoides (Ehrenberg): Taiwan; Xinyi Reef (Nansha=Spratly Islands). (9)

Roxasia cervicornis Wright & Studer: Yalong Bay (Hainan); Nansha (=Spratly Islands). (2, 9)

Scleronephthya corymbosa Verseveldt & Cohen: Dapeng Bay (Hong Kong). (6)

S. pustulosa Wright & Studer: Dapeng Bay (Hong Kong). (6)

Spongodes gigantea Verrill [*Dendronephya gigantea*]: Dongshan (Fujian); Hong Kong. (38)

S. guggenheimi (Roxas) [*Dendronephya guggenheimi*]: Huian (Fujian). (38)

S. hadzii Tixier-Durivault & Prevorsek: Yalong Bay (Hainan). (2)

S. spinifera Holm: Dapeng Bay (Hong Kong). (6)

S. studeri Ridley: Dapeng Bay (Hong Kong). (6)

Stereonephthya inordinata Tixier-Durivault: Xinyi Reef (Nansha=Spratly Islands). (8)

S. pumilia Li: Northeastern South China Sea. (5)

S. rubiflora Utinomi: Southern Nansha (=Spratly Islands). (8)

Umbellulifera formosa Li: Beibuwan (=Gulf of Tongking); South China Sea. (3)

U. oreni Verseveldt: Southern Nansha (=Spratly Islands). (8)

U. striata (Thomson & Henderson): Southern Nansha (=Spratly Islands). (8)

NIDALIIDAE

Siphonogorgia cylindrata Kükenthal: Southern Nansha (=Spratly Islands). (8)

S. gracilis (Herrison): Southern Nansha (=Spratly Islands). (8)

S. variabilis (Hickson): Southern Nansha (=Spratly Islands). (8)

VIGUIERIOTIDAE

Studeriotes debilis Thomson & Dean: Southern Nansha (=Spratly Islands). (8)

S. spinosa Thomson & Dean: Southern Nansha (=Spratly Islands). (7, 8)

XENIIDAE

Xenia elongata Dana: Xianbin Reef (Nansha=Spratly Islands). (9)

X. spicata Li: Huangyan Island (Zhongsha=Macclesfield Bank). (3)

REFERENCES*

(1). Li Chupu, 1982a. Studies on Alcyonacea of Huangyan Island. Reports of the comprehensive investigation and research of the South China Sea (I). China Science Press, pp:293–300.

(2). Li Chupu, 1982b. Studies on the Alcyonacea of the South China Sea, I. Alcyonacea from Yanglong Bay. Tropic Oceanology, 1(2):156–169.

(3). Li Chupu, 1982c. Two new species of Alcyonacea from South China Sea (Coelenterata: Octocorallia). Acta Zootaxonomica Sinica, 7(3):229–232.

(4). Li Chupu, 1984. Studies on the Alcyonacea of the South China Sea, II. Lobophytum and Sarcophyton of Xisha Islands. South China Sea Studia Marina Sinica, 6:103- 119.

(5). Li Chupu, 1986. A new species of Nephthyidae. Tropic Oceanology, 5(3):63–65.

(6). Li Chupu, 1986. The Alcyonacea in Hong Kong waters. Tropic Oceanology, 5(4):19- 25.

(7). Li Chupu, 1987. Alcyonacea. In: Research reports on the comprehensive oceanographic investigation of Chinese Southern Territory: Zengmu Shoal. China Science Press, pp:200–203.

(8). Li Chupu, 1989. Alcyonacea. In: Research reports on the comprehensive oceanographic investigation of Nansha Islands and adjacent waters (I). China Science Press, Vol. 2, pp: 746–755.

(9). Li Chupu, 1991. The Alcyonacea of Nansha Islands, II. In: Collections of research papers on marine organisms in the waters around and near Nansha Islets (I). China Ocean Press, pp:48–55.

(10). Zhou Jinming and Zou Renlin, 1984. Studies on the Antipatharians of China, II. The genus *Antipathes*. Tropic Oceanology, 3(2):56–61.

(11). Zhou Jinming and Zou Renlin, 1987. Studies on the Antipatharians of China, III. The genus *Stichopathens*. Tropic Oceanology, 6(3):63–70.

(12). Zhou Jinming and Zou Renlin, 1991. Antipatharians of Nansha Islands waters. In: Collection of research papers on marine organisms in the waters around and near Nansha Islets (I). China Ocean Press, pp:37–47.

(13). Zhou Jinming and Zou Renlin, 1991. A preliminary study on Antipatharians of Zhongsha Islands waters. In: The Study Collections of Fauna and Geography of Marine Animals in Nansha Islands. China Ocean Press, pp:295–301.

(14). Zhou Jinming and Zou Renlin, 1992. Studies on the Antipatharians from Zhongsha Islands waters. Tropic Oceanology, 11(3):45–52.

(15). Zou Renlin and Zhou Jinming, 1982. Studies on the Antipatharians of China, I. The genus *Cirrhipathes* with description of a new species. Tropic Oceanology, 11(1): 82–91.

(16). Zou Renlin and Zhou Jinming, 1984. Antipatharians from Hong Kong waters with a description of a new species. Asian Marine Biology 1:101–105.

(17). Zou Renlin, Song Shanwen, and Ma Jianghu, 1975. Shallow-water Scleractinica corals of Hainan Island. China Science Press, pp:1–65.

(18). Zou Renlin, 1975. Two new species of shallow-water stony corals from the coast of Guangdong and Guangxi. Acta Zoologica Sinica, 21(3):241–243.

(19). Zou Renlin, 1978. Studies on corals from Xisha Islands, III. Lists of Scleractinia, Hydrocorallina, Heliopora and Tubipora corals. China Science Press, pp:91–124.

(20). Zou Renlin, 1980. Studies on corals from Xisha Islands, IV. Scleractinia corals with description of two new species. South China Sea Studia Marina Sinica, 1:113- 118.

(21). Zou Renlin and Chen Youzhang, 1983. Preliminary studies on the geographical distribution of shallow-water Scleractinia corals from China. South China Sea Studia Marina Sinica, 4:89–95.

(22). Zou Renlin, 1984. Studies on corals from Xisha Islands, V. The deep-water Acropora with description of a new species. Tropic Oceanology, 3(2):52–55.

(23). Zou, Renlin, et al. 1988. An ecological study of reef corals around Weizhou Island. Proc. Mar. Biol. South China Sea, pp:201–211.

(24). South China Sea Institute of Oceanology, Academia Sinica, 1983. Report on comprehensive survey and research about Zengmu Ansha-the southern boundary of China. China Science Press, pp:203–205.

(25). Zhang Yuanlin, 1984. Study on numerical taxonomy of Tubastraea. Tropic Oceanology, 3(1):56–62.

(26). Zhang Yuanlin and Zou Renlin, 1987. Community structure of shallow-water stony corals in Daya Bay. Tropic Oceanology, 6(1):12–18.

(27). Zhang Yuanlin et al., 1989. Scleractinia Corals. Research reports on the comprehensive oceanographic investigation of Nansha Islands and adjacent waters (I). China Science Press, Vol.1:100–103.

(28). Ma, T. Y. H., 1959. Effect of water temperature on growth rate of reef corals. Oceanographic Sinica Spec., 1:1–116, 321 pls. 12 figs.

(29). Zou Renlin and Chen Youzhang, 1988. Studies on ahermatypic corals of the South China Sea, II. Species and genus records, and distribution characteristics. Tropic Oceanology, 7(1):74–83.

(30). Zou Renlin et al., 1983. Ecological analyses of ahermatypic corals from the northern shelf of South China Sea. Tropic Oceanology, 2(3):1–6.

(31). Zou Renlin et al., 1982. Preliminary study on deep-water Scleractinia corals from the East China Sea, I-II. Marine Science Bulletin, 4:51–67; 5:36–42.

(32). Zou Renlin and P. J. B. Scott, 1982. The Gorgonacea of Hong Kong. Procist Internat. Mar. Biol. Workshop 1:135–158.

(33). Zou Renlin and Chen Youzhang, 1984. Study on the shallow-water Gorgonacea from coast of Guangdong. South China Sea Studia Marina Sinica, 5:67–75.

(34). Zou Renlin, Huang Baochao, and Wang Xiangzhen, 1990. Studies on Gorgonacea of China, I. *Isis* with description of a new species. Acta Oceanologica Sinica, 12(1):83–90.

(35). Huang Baochao, 1989. Cluster analysis and numerical distribution of Gorgonian in the shallow water of Nanao Island. South China Sea Studia Marina Sinica, 9:137- 149.

(36). Zou Renlin and Zhang Yuanlin, 1989. Ecological characteristics of deep-water stony corals from northern waters of South China Sea. In: Research reports on the comprehensive oceanographic investigation of Nansha Islands and adjacent waters (I). China Science Press,Vol. 2:742–746.

(37). Zou Renlin, 1983. Corals and their natural products. Journal of Marine Drugs, 2(3): 125–129.

(38). Xu Zhenzu, 1992. Studies on shallow-water stony corals, Gorgonacea and Alcyonacea along the coast of Fujian. (manuscript)

*: All in Chinese except for (16), (23), (28), and (32)
Compiled by Zou Relin, Li Chupu, Zhou Jiming, and Xu Zhenzu; Edited by Zou Renlin

PENNATULACEA
VERETILLIDAE

Cavernularia habereri Moroff: Fujian. Low intertidal muddy sand flat. (6)
C. obesa Milne-Edwards & Haime: Dalian (Liaoning); Qingdao (Shandong); Hong Kong. Intertidal muddy sand flat. (1, 7)
Lituaria amoyensis Koo: Xiamen (Fujian). Intertidal muddy sand flat. (5)

STACHYPTILIDAE

Stachyptilum doflemi Bass: Hong Kong. Intertidal muddy sand flat. (7)

VIRGULARIIDAE

Virgularia gustaviana (Herclots): Fujian coast. (6)
V. reinwardtii Herclots: Xiamen harbor (Fujian). Intertidal muddy sand flat. (6)

PENNATULIDAE

Pennatula fimbriata Hercolts: Northwestern continental shelf off Nansha (=Spratly Islands). (4)
P. murrayi Kolliker: Continental shelf off northwestern Nansha (=Spratly Islands). (4)
P. phosphorea Linné: Continental shelf off northwestern Nansha (=Spratly Islands). (4)

PTEROEIDIDAE

Pteroeides chinense Herclots: Fujian. Muddy sand flat below mid-intertidal. (6)
P. sparmannii Kolliker: Hong Kong. Intertidal muddy sand flat. (7)

TELESTACEA
TELESTAIDAE

Telesto cf. *rubra*: Fujian; Guangdong. Low intertidal to subtidal; on rocks, docks, and buoys. (1, 2, 3)

REFERENCES*

(1). Zhao Ruyi et al., 1990. Guidebook on collection of invertebrate from China coast (in North China). China Ocean Press, 393pp.
(2). Huang Zongguo, Cairuxing et al., 1960. Studies on the littoral ecology of Amoy and its vicinity. Journal of Xiamen University (Natural Science), 7(3):74–95.
(3). Huang Zongguo and Cairuxing 1984. Marine fouling organism and its prevention (I). China Ocean Press, 352pp.
(4). Meng Zhimin, 1991. Benthos in the shelf waters of Nansha Islands. In: Bottom trawling fishing resources investigation in southwestern continental shelf sea waters of Nansha Islands. China Ocean Press, pp:56–72.
(5). Koo, S. Y. 1935. On a new Penatulid (Lituaria) from Amoy. Nat. Sci. Bull. Amoy University. 1(2):157–164.
(6). Koo, S. Y. 1940. Some sea pens (Pennatulacea) from Amoy Island. China Jour. 32(3):113–118.
(7). Morton, B. and J. Morton, 1983. The sea shore ecology of Hong Kong. Hong Kong University, 350pp.

*: (1)-(4) in Chinese
Compiled by Xu Zhenzhu and Huang Zongguo; Edited by Li Chupu.

CTENOPHORA

TENTACULATA
CYDIPPIDA
PLEUROBRACHIDAE

Euchlora rubra (Kölliker): Xiamen (Fujian). (4)
Hormiphora palnata Chun: East China Sea. (14)
Hormiphora sp.: Taiwan Strait. (11)
Pleurobrachia globosa Moser: Jiangsu; Zhejiang; Fujian; Guangdong; Guangxi; central South China Sea. (2, 5–14, 16)

LOBATE
OCYROPSIDAE

Ocyropsis crystallina (Rang): Jiangsu; Zhejiang; Fujian; Guangdong; Guangxi. (1, 2, 5- 9)

BOLINOPSIDAE

Bolinopsis vitrea (L. Agassiz): Zhejiang; Fujian; Guangdong. (3, 5, 7, 9, 12)

CESTIDA
CESTIDAE

Cestum sp.: East China Sea. (14)

NUDA
BEROIDA
BEROIDAE

Beroe cucumis Fabricius: Jiangsu; Zhejiang; Fujian; Guangdong; Guangxi; central South China Sea. (2, 5–9, 11–16)
B. ovata Chamisso & Egenhardt: Fujian. (5, 7)

REFERENCES*

(1). Chiu, S. Y., 1954a. On the occurrence of a tropical Ctenophores in Amoy Harbour. Acta Zoologica Sinica, 6(1):37–39.

(2). Chiu, S. Y., 1957. Preliminary notes on the Ctenophores of the South China Sea. Acta Zoologica Sinica, 9(1):85–100.

(3). Chiu, S. Y. 1962. Notes on Ctenophore *Bolinopsis vitrea* (L. Agassiz) in Amoy Harbour. Journal of Xiamen University (Natural Science), 9(3): 255–258.

(4). Qiu Suyuan, 1980. On the Nematocyst-bearing Ctenophore *Euchlora rubra* Kölliker from Xiamen Harbour. Oceanologia & Limnologia Sinica, 11(3):255–258.

(5). Hsu Chen-tsu and Chang Chin-piao, 1964. Studies on the Medusae from the Fukien coast, II. On the taxonomy of the Hydromedusae, Siphonophores and Ctenophores off south Fukien. Journal of Xiamen University (Natural Science), 11(3):120–129.

(6). Xu Zhenzu and Zhang Jinbiao, 1974. Studies on Medusae from Fujian coast, III. On the taxonomy of Medusae from the central and northern coastal waters. Journal of Marine Science & Technology, 2:17–32.

(7). Hsu Chen-tsu and Chin, T. G., 1962. Studies on the Medusae from the Fukien coast. Journal fo Xiamen University (Natural Science), 9(3):206–224.

(8). Xu Zhenzu and Huang Jiaqi, 1983. On the Hydromedusae, Siphonophora, Scyphomedusae and Ctenophora from the Jiulong River Estuary of Fujian, China. Taiwan Strait, 2(2):99–110.

(9). Zhang Jinbiao, 1977. Investigation and research on Hydromedusae and Ctenophora from the Zhejiang coast. Marine Science & Technology, 7:95–107.

(10). Zhang Jinbiao, 1984. Distribution and abundance of *Pleurobrachia globosa* from the coast of southeastern China. Acta Oceanologica Sinica, 5 (suppl.):841–846.

(11). Li Mao and Zhang Jinbiao, 1989. Ecological studies on Hydromedusae and Ctenophores from the western waters of Taiwan Strait. Acta Oceanologica Sinica, 11(5):621–628.

(12). Lin Mao, 1989. Taxonomy and fauna on the medusaes from the waters of the Daya Bay. In: Collections of Papers on Marine Ecology in the Daya Bay (I). China Ocean Press, pp:59–65.

(13). State Oceanic Administration, 1988. The achievement in comprehensive investigation on environmental resources in the central part of the South China Sea. China Ocean Press, pp:162–252.

(14). Gao Shangwu, 1982. Studies on Medusae of the East China Sea. Studia Marina Sinica, 19:33–42.

(15). Huang Liping, 1987. On planktonic Medusaes from the northern coast of Beibu Bay. Guangxi Oceanology 1:1–9

(16). Lin, Mao, and Zhang Jinbiao, 1990. Ecological studies on the Hydromedusae, Siphonophoras and Ctenophoras in the Xiamen Harbour and adjacent waters, Acta Oceanologica Sinica. 9, 3:429–438.

*: (1)-(15) in Chinese
Compiled by Lin Mao; Edited by Qiu Shuyuan (S. Y. Chiu)

PLATYHELMINTHES

TURBELLARIA
TRICLADIDA
PROCERODIDAE

Procerodes graciliceps (Stimpson): China's coast. Intertidal, muddy sand substrate. (11, 12)

POLYCLADIDA
STYLOCHIDAE

Stylochus corniculatus Stimpson: China's coast. Water depth about 10 m, muddy sand substrate, inside bivalve shells. (10, 11)

S. orientalis Bock: Taiwan Strait; Taiwan. (6, 8, 12)

S. pusilla Bock: China's coast. (6, 12)

Stylochus sp.: Fujian; Guangdong. Inside oyster and barnacle shells. (5)

LEPTOPLANIDAE

Leptoplana delicatula Stimpson: Hong Kong. Intertidal, sandy substrate. (11)

L. fusca Stimpson: Hong Kong. Intertidal rocky substrate, under rocks. (11)

L. trullaeformis Stimpson: China's coast. Under rocks, water depth about 45 m. (10, 11)

Notoplana humilis (Stimpson): Dalian (Liaoning); Qingdao (Shandong); Bohai; Yellow Sea. Under rocks. (1, 2, 4)

Plagiotata promiscus Plehn: China's coast. (9, 12)

Stylochoplana suoensis Kato: Taiwan. (7)

S. taiwanica Kato: Taiwan. (7)

S. utunomii Kato: Taiwan. (7)

PLANOCERIDAE

Elasmodes acutus (Stimpson): China's coast. Water depth about 10 m, muddy substrate. (10, 11)

Paraplanocera oligoglena (Schmarda): Suao and Nanpuao (Taiwan). (7)

P. reticulata (Stimpson): Dalian (Liaoning); Yantai (Shandong); Bohai; Yellow Sea. Intertidal. (1, 3, 4)

DIPLOSOLENIIDAE

Pseudostylochus obscuras (Stimpson): Dalian (Liaoning). Muddy sand flat. (1, 4)

PSEUDOCERIDAE

Pseudoceros exoplatus Kato: Yellow Sea; East China Sea; South China Sea. Inside oyster and barnacle shells. (5)

PROSTHIOSTOMIDAE

Enchiridium japonicum Kato: Taiwan. (7)

Prosthiostomum affine Stimpson: Hong Kong. Intertidal sandy substrate. (11)

P. formosum Kato: Taiwan. (7)

P. obscurum (Stimpson): China's coast. Subtidal sandy substrate. (10, 11, 12)

P. tenebrosum Stimpson: Hong Kong. Under rocks in intertidal sandy shore. (11)

TYPHLOLEPTIDAE

Cryptocoelum opacum Stimpson: Hong Kong. Water depth about 10 m, parasitic. (7)

REFERENCES*

(1). Song Pengdong, Li, Yingxi, Wang Guiyun, and Li Taiwu, 1985. Guidebook for Practice in Invertebrate from Dalian Coast. China Higher Education Press, 338pp.
(2). Chen Xintao, Tang Zhonzhang, and Jiang Jingbou, 1963. A Collection of Illustrative Plates of Animals in China, Platyhelminthes (Nemertea suppl.). China Science Press.
(3). Jiang Zaijie and Liu Lingyun, 1986. Guidebook for Practice in Invertebrate from Yantai Coast. Beijing Normal University Press, 465pp.
(4). Zhao Ruyi et al., 1990. Guidebook on collection of invertebrate from China coast (in North China). China Ocean Press, 393pp.
(5). Huang Zongguo and Cai Ruxing, 1984. Marine fouling organism and its prevention (I). China Ocean Press, 352pp.
(6). Bock, S. 1913. Studien über Polycladen. Zool. Bidr. Upps., 2:31–344.
(7). Kato, K. 1943. Palyclads from Formosa. Bulletin of the Biogeographical Society of Japan, 13:69–77.
(8). Lue, K. Y. and M. Kawakatsu, 1986. History of the study of Turbellaria in China. Part 1: Ages of Materia Medica and early expeditions by westerners. Hydrobiologia, 132:317–322.
(9). Plehn, M. 1896. Neue Polycladen, gesammelt von Hernn Kapitän Chierchia bei der Erdumschiffung de korvette Vettor Pisani, von Herrn Prof. Dr. Kükenthal in nordlichen Eismeer und von Herrn Prof. Dr. Semon in Java. Z. Naturwiss, 30:137–176.
(10). Stimpson, W. 1855. Descriptions of some of the new marine invertebrate from the Chinese and Japanese Seas. Proc. Acad. Nat. Sci. Philad. 7:375–384.
(11). Stimpson, W. 1857. Prodromus descriptionis animalium evertebratorum quae in expeditione ad Oceanum, Pacificum septentrionalem a Republica Federata missa, Johanne Rodgers Duce, observavit et descripsit. Pars. 1. Turbellaria Dendrocoela. Proc. Acad. Nat. Sci. Philad. 9:19–31.
(12). Tu, T. J. 1940. Geschichtlicher überblick über das Studium der Turbellarien in Ostasien und Stand unserer kenntnisse von Diesen. Zool. Jb. Syst. Okol. Geogr. Tiere, 73:201–260.

*: (1)-(5) in Chinese
Compiled by Pan Haihong; Addition and edited by Sun Shichun.

TREMATODA

MONOGENA
ANCYROCEPHALIDAE

Haliotrema thysanophridis (Yamaguti): Qingdao (Shandong). Host: flathead (*Cociella crocodilus*). (26)
Hamatopeduncularia elegans Bychowsky & Nagibina: Yantai (Shandong). Host: sea catfish (*Arius thalassinus*). (26)
H. simplex Bychowsky & Nagibina: Yellow Sea. Host: sea catfish (*Arius thalassinus*). (26)
Ligophorus leporinus (Zhang & Ji): Huian and Xiamen (Fujian). Host: striped mullet (*Mugil cephalus*). (38)
L. vanbenedeni (Parona & Perugia): Huian and Xiamen (Fujian). Host: striped mullet (*Mugil cephalus*). (26, 38)
Murraytrema pricei Bychowsky & Nagibina: Qingdao (Shandong). Host: drum (*Nibea albiflora*). (26)
Triacanthinella principale Bychowsky & Nagibina: Yantai (Shandong). Host: triplespines (*Triacanthus brevirostris*). (26)

DIONCHIDAE

Dionchus sp.: Yantai (Shandong). Host: Remora (*Remora remora*). (26)

HEXABOTHRIIDAE

Erpocotyle sp.: Qingdao (Shandong). Host: houndshark (*Triakis scyllium*). (26)

MICROCOTYLIDAE

Metamicrocotyla gracilis Li: Penglai (Shandong). Host: striped mullet (*Mugil cephalus*). (26, 35)

DICLIDOPHORIDAE

Heterobothrium praeorchis Bychowsky & Nagibina: Qingdao (Shandong). Host: puffer (*Fugu alboplumbeus*). (26)

ANCHOROPHORIDAE

Anchorophorus sinensis Bychowsky & Nagibina: Qingdao and Yantai (Shandong). Host: tonguefish (*Cynoglossus semilaevis*). (26, 76)

GOTOCOTYLIDAE

Gotocotyle sawara Ishii: Qingdao (Shandong). Host: Spanish mackerel (*Scombermorus niphonius*). (26)

AXINIDAE

Zeuxapta japonica Yamaguti: Qingdao (Shandong). Host: amberjack (*Seriola aureovittata*). (26)

PTERINOTREMATIDAE

Pterinotrema mirabilis Bychowsky & Nagibina: Sanya (Hainan). Host: bonefish (*Albula vulpes*). (77)

DIGENEA
BUCEPHALIDAE

Alcicornis hainanensis Gu & Shen: Baimajing (Hainan). Host: jack (*Zonichthys nigrofasciata*). (21, 55)

Bucephalopsis ablennus Gu & Shen: Sanya (Hainan). Host: needlefish (*Ablennes anastomella*). (21, 55)

B. arcuatus (Linton): Pingtan (Fujian). Host: Spanish mackerel (*Scomberomorus guttatus*). (45)

B. bennetti (Hopkins & Sparks): Haikou (Hainan). Host: needlefish (*Tylosurus leiurus*). (21, 55)

B. collichthydis Qiu & Li: Huanghua (Hebei). Host: drum (*Collichthys lucidus*). (26)

B. obpyriformis Gu & Shen: Sanya (Hainan). Host: needlefish (*Tylosurus melanotus*). (21, 55)

B. rhynchobati Wang: Fuqing (Fujian). Host: guitarfish (*Rynchobatus djiddensis*). (45)

B. scombropsis Yamaguti: Pingtan (Fujian). Host: cutlassfish (*Trichiurus haumela*). (39)

Bucephalus harpodontis Wang: Pingtan (Fujian). Host: lizardfish (*Harpodon nehereus*). (39)

B. margaritae Ozaki & Ishibashi: Dongshan (Fujian). Host: flathead (*Platycephalus indicus*). (79)

B. polymorphus Baer: Shenjiamen (Zhejiang). Host: flathead (*Platycephalus indicus*). (48)

B. retractilis Yamaguti: Jinjiang and Pingtan (Fujian). Host: jack (*Atropus atropus*) and herring (*Herklotsichthys*). (45)

B. trifurcatus Wang: Pingtan (Fujian). Host: barracuda (*Sphyraena pinguis*). (39)

B. varicus Manter: Sanya (Hainan). Host: jack (*Zonichthys nigrofasciata*). (55, 77)

Cercaria pernaviridis Tang: Hong Kong. Host: green mussel (*Perna viridis*). (84)

Dollfustrema sinica Gu & Shen: Haikou (Hainan); Dongsha (=Pratas Islands). Host: flathead (*Platycephalus indicus*). (21, 55)

Folliculovarium gymnothoracis Gu & Shen: Xisha (=Paracel Islands). Host: moray eel (*Gymnothorax polyuranodon*). (61)

F. xishaense Gu & Shen: Xisha (=Paracel Islands). Host: grouper (*Epinephelus fasciatus*). (61)

Heterobucephalopsis gymnothoracis Gu & Shen: Xisha (=Paracel Islands). Host: moray eel (*Gymnothorax polyuranodon*). (61)

Neodollfustrema xishaense Gu & Shen: Xisha (=Paracel Islands). Host: moray eel (*Gymnothorax melanospilus*). (61)

Neoprosorhynchus xishaensis Gu & Shen: Xisha (=Paracel Islands). Host: jack (*Selar crumenophthalmus*). (55)

Prosorhynchus clavatum Wang: Pingtan (Fujian). Host: barracuda (*Sphyraena pinguis*). (39)

P. crucibulum (Rud): Qingdao (Shandong). Host: conger (*Conger myriaster*). (26, 70)

P. eleutheronemae Wang: Pingtan (Fujian). Host: threadfin (*Eleutheronema tetradactylus*). (44)

P. epinepheli Yamaguti: Xisha (=Paracel Islands). Host: grouper (*Epinephelus*). (61)

P. facilis (Ozaki): Sanya (Hainan). Host: amberjack (*Seriola dumerili*). (21, 55)

P. fujianensis Wang: Putian (Fujian). Host: eel (*Anguilla marmorata*). (44)

P. ozakii Manter: Dalian (Liaoning). Host: Spanish mackerel (*Scrombermorus iphonius*). (26)

P. sphraenae Gu & Shen: Sanya (Hainan). Host: barracudas (*Sphyraena jello* and *S. pinguis*). (21, 55)

P. synanceiae Wang: Pingtan (Fujian). Host: stonefish (*Synanceia verrucosa*). (44)

P. tsengi Tsin: Beidaihe (Hebei); Qingdao (Shandong); Dongshan and Xiamen (Fujian). Host: flathead (*Platycephalus indicus*). (26, 71)

Pseudobucephalopsis belonnis Gu & Shen: Xisha (=Paracel Islands). Host: needlefish (*Belone platyura*). (61)

Pseudoprosorhynchus fusiformis Wang: Pingtan (Fujian). Host: scorpionfish (*Inimicus*). (44)

Rhipidocotyle adbaculum Mater: Sanya (Hainan). Host: Spanish mackerel (*Scombermorus commersoni*). (21, 54)

R. anguillae Wang: Putian (Fujian). Host: eel (*Anguilla marmorata*). (45)

R. baculum (Linton): Sanya (Hainan). Host: Spanish mackerel (*Scombermorus ommersoni*). (21, 54)

R. clavivesiculum Gu & Shen: Haikou (Hainan). Host: grouper (*Plectropomus leopardus*). (21, 55)

R. croceae Gu & Shen: Dongsha (=Pratas Islands). Host: drum (*Pseudosciaena crocea*). (55)

R. pentagonum (Ozaki): Pingtan (Fujian). Host: Spanish mackerel (*Scomberomorus guttatus*). (39)

R. xishaensis Gu & Shen: Xisha (=Paracel Islands). Host: jack (*Carangoides ignobilis*). (61)

Telorhynchus astrocongeri Shen: Ningbo (Zhejiang). Host: conger eel (*Conger myriaster*). (13)

T. cociellae Gu & Shen: Baimajing (Hainan). Host: flathead (*Cociella crocodilus*). (21, 55)

T. hippocampi Shen: Jieshi (Guangdong). Host: seahorse (*Hippocampus trimaculatus*). (1)

HAPLOPORIDAE

Hapalotrema flecterotestis Zhukov: Hangu (Tianjin). Host: mullet (*Liza haematocheila*). (34)

HAPLOSPANCHNIDAE

Haplosplanchnus cuneatus Tang & Lin: Mouth of Min River (Fujian). Host: striped mullet (*Mugil cephalus*). (66)

H. elongatus Tang & Lin: Mouth of Min River (Fujian). Host: striped mullet (*Mugil cephalus*). (66)

H. purii Srivastava: Fuqing (Fujian). Host: striped mullet (*Mugil cephalus*). (40)

Prohaplosplanchnus diorchis Tang & Lin: Mouth of Min River (Fujian). Host: striped mullet (*Mugil cephalus*). (66)

Provirellotrema crenimugilas Pan: Guangzhou (Guangdong). Host: mullet (*Crenimugil crenilabis*). (74)

Schikhobalotrema acuta (Linton): Sanya (Hainan). Host: needlefish (*Tylosurus giganteus*). (21)

GYLIAUCHENIDAE

Apharyngogyliauchen opisthovarius Gu & Shen: Xisha (=Paracel Islands). Host: wrasses (Labridae). (61)
A. scorustis Gu & Shen: Xisha (=Paracel Islands). Host: parrotfish (*Scarus sordidus*). (61)
Gyliauchen oligoglandulosus Gu & Shen: Sanya (Hainan). Host: rabbitfish (*Siganus guttatus*). (21, 58)

PARAMPHISTOMIDAE

Allassostomoides chelydrae (Maccallum): Pingtan (Fujian). Host: sea turtles. (43)

OPISTHOLEBETIDAE

Heterolebes immaculosus Ku & Shen: Weihai (Shandong). Host: puffers *(Fugu niphobles* and *F. vermicularis*). (26, 53)
H. sinensis Gu & Shen: Qingdao (Shandong). Host: puffer (*Fugu alboplumbeus*). (26, 58)
H. spari Shen: Longkou (Shandong). Host: porgy (*Sparus macrocephalus*). (26)
Macalifer aprionis Shen: Ningbo (Zhejiang). Host: searobin (*Chelidonichthys kumu*). (11)
M. dayawanensis Shen & Tong: Daya Bay (Guangdong). Host: puffer (*Lagocephalus inermis*). (24)
M. pacificus Yamaguti: Pingtan (Fujian). Host: dolphins. (41)
M. zhoushanensis Shen: Zhoushan (Zhejiang). Host: puffer (*Fugu oblongus*). (11)
Opistholebes microovus Ku & Shen: Weihai (Shandong). Host: puffer (*Fugu niphobles*). (26, 53)

ANGIODICTYIDAE

Hexangium sigani Goto & Ozaki: Haikou (Hainan). Host: rabbitfish (*Siganus guttatus*). (21, 53)

RHYTIDODIDAE

Microscaphidium japonicum Oguro: Pingtan (Fujian). Host: sea turtles. (43)
Polyangium linguatula (Looss) Looss: Pingtan (Fujian). Host: sea turtles. (43)
Rhytidodoides cheloniae Wang: Pingtan (Fujian). Host: sea turtles. (43)

ACCACOELIIDAE

Accacladium arii Wang: Putian (Fujian). Host: sea catfish (*Arius thalassinus*). (40)
Rhynchopharynx formionis Shen: Sanya and Baimajing (Hainan). Host: jack (*Formio niger*). (11, 21)

Tetrochetus coryphaenae Yamaguti: Sanya (Hainan), Ningbo (Zhejiang). Host: dolphinfish (*Coryphaena hippurus*). (11, 21)
T. hainanensis Shen: Haikou (Hainan). Host: filefish (*Alutera*). (11, 12)
T. navodonis Shen & Tong: Baimajing (Hainan). Host: puffer (*Navodon septentrionalis*). (21, 22)
T. zhoushanensis Shen: Shenjiamen (Zhejiang). Host: puffers (*Fugu*). (11)

BIVESCULIDAE

Bivesicula auxisae Gu & Shen: Xisha (=Paracel Islands). Host: tuna (*Auxis thazard*). (61)
B. fistulariae Shen: Ningbo (Zhejiang). Host: cornetfish (*Fistularia petimba*). (3)
B. lutiani Gu & Shen: Xisha (=Paracel Islands). Host: snapper (*Lutjanus kasmira*). (61)
B. megalopis Shen: Sanya (Hainan). Host: tarpon (*Megalops cyprinoides*). (3)
B. ostichthydis Shen: Ningbo (Zhejiang). Host: soldierfish (*Myripristis murdjan*). (3)
B. xishaensis Gu & Shen: Xisha (=Paracel Islands). Host: grouper (*Epinephelus fasciatus*). (61)

AZYGIIDAE

Azygia acuminata Goldberger: Qingdao (Shandong). Host: eel (*Anguilla*). (26, 79)
A. micropteri (Maccallum): Fuding (Fujian). Host: temperate bass (*Lateolabrax japonicus*). (41)

FELLODISTOMIDAE

Bacciger lizae Shen: Beitang (Tianjin). Host: mullet (*Liza haematocheila*). (26)
B. mugilis Shen: Shenjiamen (Zhejiang). Host: striped mullet (*Mugil cephalus*). (13)
Benthotrema pyriformis Wang: Pingtan (Fujian). Host: porgy (*Sparus latus*). (45)
Discogasteroides hainanensis Shen: Sanya (Hainan). Host: filefishes (*Alutera* and *Paraluteres*). (21)
Faustula qikouensis Qiu & Li: Qikou (Hebei). Host: goby (*Synechogobius ommaturus*). (26)
Neobenthotrema pyriformis Wang: Pingtan (Fujian). Host: triplespines (*Triacanthus brevirostris*). (47)
Proctoeces longisaccatus Wang: Pingtan (Fujian). Host: medusafish (*Psenopsis anomala*), triplespines (*Triacanthus brevirostris*) and filefish (*Paramonacanthus nipponensis*). (47)
P. maculatus (Looss): Qingdao (Shandong). Host: sole (*Zebrias zebra*) and tonguefishes (*Cynoglossus semilaevis* and *C. joyneri*). (26)

P. orientalis Cao: Xiamen (Fujian); Hong Kong. Host: porgy (*Sparus latus*) and pearl oyster (*Pteria penguin*). (72, 82)

P. parapistipomae Wang: Pingtan (Fujian). Host: grunt (*Parapristipoma trilineatus*). (45)

Prudhnoeus oligolecithosum Wang: Pingtan (Fujian). Host: grunt (*Plectorhynchus cinctus*). (43)

Pseudoantorchis thalassomae Wang: Pingtan (Fujian). Host: wrasse (*Thalassoma hardwicki*). (41)

Pseudofellodistomum plagiorchis Wang: Pingtan (Fujian). Host: threadfin (*Eleutheronema tetradactylus*). (45)

Pseudosteringophorus holognathi Yamaguti: Pingtan (Fujian). Host: grunt (*Plectorhynchus cinctus*). (41)

Steringophorus congeri Shen: Ningbo (Zhejiang). Host: conger eel (*Conger*). (13)

Steringotrema sinensis Cao: Xiamen (Fujian). Host: porgy (*Sparus latus*) and searobin (*Chelidonichthys*). (45)

Tergestia atropi Shen: Sanya (Hainan). Host: jack (*Atropus atropus*). (21)

T. atulis Shen: Sanya (Hainan). Host: jack (*Carangoides djeddaba*). (21)

T. hainanensis Shen: Sanya (Hainan). Host: porgy (*Paragyrops edita*) and jack (*Selar boops*). (21)

T. laticollis (Rud.): Sanya (Hainan). Host: jack (*Selaroides leptolepis*). (21)

T. triacanthi Wang: Pingtan (Fujian). Host: triplespines (*Triacanthus brevirostris*). (45)

MONASCIDAE

Monascus filiformis (Rud.): Baimajing (Hainan). Host: driftfish (*Nomeus gronovii*). (21)

M. orientalis (Srivastava): Longhai (Fujian). Host: butterfish (*Pampus argenteus*). (41)

DIPLANGIDAE

Diplangus hainanensis Shen: Sanya (Hainan). Host: flyingfish (*Cypselurus bahiensis*). (21)

GORGODERIDAE

Phyllodistomum folium Braun: Yingkou (Liaoning). Host: temperate bass (*Lateolabrax japonicus*). (26)

P. pearsei Holi: Sanya (Hainan). Host: needlefishes (*Ablennes anastomella* and *Tylosurus leiurus*). (53)

MONORCHIIDAE

Allolasiotocus pseudosciaenae Wang: Putian (Fujian). Host: drums (*Miichthys miiuy* and *Pseudosciaena polyactis*). (41)

Genolopa sparui Shen: Sanya (Hainan), Tanggu (Tianjin). Host: porgy (*Sparus berda*) and goby (*Odontamblyopus rubicundus*). (21, 26)

Hurleytrema hainanensis Shen: Sanya (Hainan). Host: needlefish (*Tylosurus melanotus*). (21)

Lasiotocus crytostoma (Oshmarin): Sanya (Hainan). Host: grunt (*Pomadasys hasta*). (21)

Longimonorchis pampi Wang: Longhai (Fujian). Host: butterfish (*Pampus argenteus*) and temperate bass (*Lateolabrax japonicus*). (40)

Monorcheides xishaensis Shen: Xisha (=Paracel Islands). Host: snapper (*Lutjanus argentimaculatus*). (7)

Monorchicestrahelmins branchiostegi Shen: Ningbo (Zhejiang). Host: tilefish (*Branchiostegus argentatus*). (13)

Monorchis fusiformis Wang: Pingtan (Fujian). Host: snake eel (*Pisodonophis*). (41)

Opisthomonorcheides decaptei Parukhin: Sanya (Hainan). Host: jack (*Carangoides mate*). (21)

O. indicus Karyokarate: Daya Bay (Guangdong); Sanya (Hainan). Host: jack (*Formio niger*). (21, 24)

Proctotrematoides anguillae Qiu & Tong: Qinhuangdao (Hebei). Host: eel (*Anguilla*). (26)

P. gymnothoracis Shen: Sanya (Hainan). Host: moray eel (*Gymnothorax undulatus*). (21)

Pseudomonorcheides ditrematis Wang: Pingtan (Fujian). Host: surfperch (*Ditrema temmincki*). (41)

ZOOGONIDAE

Cypseluritrematoides microvaei Shen: Ningbo (Zhejiang). Host: searobin (*Chelidonichthys kumu*). (13)

C. minor Gu & Shen: Haikou (Hainan). Host: sailfin flyingfish (*Parexocoetus brachypterus*). (21, 58)

C. triangularis Yamaguti: Beidaihe (Hebei). Host: flyingfish (*Prognichthys agoo*). (26, 87)

Lecithostaphylus ahaaha Yamaguti: Sanya (Hainan); Xisha (=Paracel Islands). Host: needlefish (*Belone platyura*). (21, 51)

L. fugus Zhang, Qiu & Li: Tanggu (Tianjin). Host: puffer (*Fugu niphobles*). (26, 37)

ACANTHOCOLPIDAE

Acanthocolpus acanthocepolae Shen: Sanya (Hainan). Host: bandfish (*Cepola schlegeli*). (21)

A. liodorus Luhe: Sanya (Hainan). Host: wolf herring (*Chirocentrus dorab*). (21)

A. sanyaensis Shen: Sanya (Hainan). Host: jack (*Caranx malabaricus*). (21)

Deropristis paurosoma Wang: Pingtan (Fujian). Host: threatfin (*Eleutheronema tetradactylus*). (45)

Pseudocaenodera alectis Shen: Sanya (Hainan). Host: pompano (*Alectis indica*). (21)

P. nibeae Shen: Tanggu (Tianjing); Longhai (Fujian). Host: drum (*Nibea albiflora*). (16, 26)

Skrjabinopsolus sanyaensis Shen: Sanya (Hainan). Host: jack (*Formio niger*). (21)

Stephanostomoides dorabi Mamaev & Oshmarin: Sanya (Hainan). Host: wolf herring (*Chirocentrus dorab*). (21)

Stephanostomum argyrosomi Shen: Qingdao (Shandong). Host: drum (*Argyrosomus argentatus*). (26)

S. baccatum Nicoll: Putian (Fujian). Host: sea catfish (*Arius thalassinus*) and gizzard shad (*Nematalosa come*). (41).

S. bicoronatum (Stossich): Hangu (Tianjin); Qingdao (Shandong). Host: temperate bass (*Lateolabrax japonicus*) and tonguefish (*Cynoglossus semilaevis*). (26, 56)

S. dentatum (Linton): Pingtan (Fujian). Host: houndshark (*Mustelus griseus*). (41)

S. ditrematis (Yamaguti): Pingtan (Fujian). Host: houndshark (*Mustelus griseus*). (42)

S. fistulariae (Yamaguti): Yingkou (Liaoning). Host: drum (*Nibea albiflora*). (26)

S. lebourae Cabellero: Yingkou (Liaoning); Tanggu (Tianjin). Host: tonguefish (*Cynoglossus semilaevis*). (26)

S. rachycentronis Shen: Sanya (Hainan). Host: cobia (*Rachycentron canadum*). (21)

S. seriolae Yamaguti: Sanya (Hainan). Host: amberjack (*Seriola dumerili*). (21)

S. sphyraenae Wang: Pingtan (Fujian). Host: barracuda (*Sphyraena pinguis*). (41)

Tormopsolus fuzhouensis Wang: Guantou (Fujian). Host: drum (*Nibea albiflora*). (40)

HOMALOMETRIDAE

Crassicutis karwarensis Hafeezullan: Fuding (Fujian). Host: temperate bass (*Lateolabrax japonicus*). (41)

OMPHALOMETRIDAE

Dichiloseccus cheloniae Wang: Pingtan (Fujian). Host: sea turtles. (43)

LEPOCREADIIDAE

Aephnidiogenes hainanensis Shen & Tong: Baimajing (Hainan). Host: grunt (*Plectorhynchus pictus*). (25, 26)

Allolepidapedon pristipomoidis Shen: Xisha (=Paracel Islands). Host: snapper (*Pristipomoides microlepis*). (7)

Bianium dayawanense Shen & Tong: Daya Bay (Guangdong). Host: puffer (*Lagocephalus*). (24)

B. hemistoma (Ozaki): Qinhuangdao (Hebei); Qingdao (Shandong); Pingtan and Xiamen (Fujian). Host: puffers (*Fugu alboplumbeus*, *F. oblongus*, and *F. xanthopterus*). (26, 41, 71)

B. lianyungangense Shen: Lianyungang (Jiangsu). Host: tonguefish (*Cynoglossus semilaevis*). (20, 26)

B. plicitum (Linton): Putian and Pingtan (Fujian). Host: grouper (*Epinephelus akaara*) and sea catfish (*Arius thalassinus*). (41)

Callogotrema fistulariae Oshmarin: South China Sea. Host: red cornetfish (*Fistularia petimba*). (45)

Cephalolepidapedon seba Yamaguti: Beidaihe (Hebei); Ningbo (Zhejiang). Host: Pacific mackerel (*Pneumatophorus japonicus*). (26, 86)

Cercaria elegans Tang: Hong Kong. Host: Philippine clam (*Ruditapes philippinarum*). (84)

Diploproctodaeum waki Shen: Sanya (Hainan). Host: drum (*Wak sina*). (21)

Hypocreadium drepanei Shen: Sanya (Hainan). Host: fish (*Drepane longimana* and *D. punctata*). (21)

Hypoporus phylloides Wang: Pingtan (Fujian). Host: medusafish (*Psenopsis anomala*). (45)

Intusatrium crassum Wang: Pingtan (Fujian). Host: tonguefish (*Cynoglossus*). (45)

Koseiria xishaense Gu & Shen: Xisha (=Paracel Islands). Host: sea chub (*Kyphosus cinerascens*). (61)

Labrifer gymnocrini Shen: Xisha (=Paracel Islands). Host: emperor bream (*Gymnocrinius griseus*). (7)

Lepidapedon aphareι Shen: Ningbo (Zhejiang). Host: snapper (*Aphareus furcatus*). (11)

L. golphick Oschmarine: South China Sea. Host: snapper (*Pristipomoides typus*). (45)

L. longivesculum Hafeeullah: Pingtan (Fujian). Host: butterfish (*Pampus argenteus*). (41)

L. sphyraenae Shen: Haikou (Hainan). Host: barracuda (*Sphyraena pinguis*). (21)

Lepocreadioides discum Wang: Pingtan (Fujian). Host: tonguefish (*Cynoglossus robustus*). (45)

L. hunghuaensis Qiu, Zhang & Li: Qingdao (Shandong); Huanghua (Hebei). Host: tonguefish (*Cynoglossus semilaevis*). (15, 26, 27)

L. indicum Srivastava: Qinhuangdao (Hebei); Fuzhou (Fujian). Host: tonguefish (*Cynoglossus robustus*). (26, 40)

L. zebrini Yamaguti: Yantai (Shandong); mouth of Changjiang (=Yangtze River). Host: sole (*Zebrias zebra*). (17, 26)

Lepocreadium dongxiangensis Wang: Pingtan (Fujian). Host: jack (*Caranx*). (45)

L. drepanei Shen: Sanya (Hainan). Host: fish (*Drepane longimana*). (21)

L. navodoni Shen: Ningbo (Zhejiang); Yantai (Shandong). Host: filefish (*Navodon septentrionalis*). (11, 26)

L. trachinoti Wang: Pingtan (Fujian). Host: pompano (*Trachinotus ovatus*). (45)

Lobatovitelliovarium fusiforme Yamaguti: Xisha (=Paracel Islands). Host: needlefish (*Ablennes*). (45)

Neallolepidapedon hawaiiense Yamaguti: Ningbo (Zhejiang); Sanya (Hainan). Host: red cornetfish (*Fistularia petimba*). (11, 87)

Neomultitestis bengalensis (Madhav): Sanya (Hainan). Host: spadefish (*Platax orbicularis*). (21)

Neonotoporus kareii Qiu & Li: Longkou (Shandong). Host: righteye flounder (*Kareius bicoloratus*). (26)

Notopours astrocongeris Qiu & Li: Dalian (Liaoning). Host: conger eel (*Conger myriaster*). (26)

Opochona glossoides Wang: Pingtan (Fujian). Host: jack (*Megalaspis cordyla*). (45)

Phyllotrema bicaudatum Yamaguti: Pingtan (Fujian). Host: snake eel (*Pisodonophis*). (41)

P. microrchis Jin, Zhang & Ji: Fuding and Huian (Fujian). Host: eel (*Anguilla marmorata*). (41, 65)

P. quadricaudatum Gu & Shen: Pingtan (Fujian); Haikou (Hainan). Host: pike conger (*Muraenesox cinereus*) and snake eel (*Pisodonophis*). (21, 51)

Preptetos cylindricus Wang: Pingtan (Fujian). Host: filefish (*Monacanthus chinensis* and *Navodon septentrionalis*). (45)

Pseudocreadim hainanensis Shen: Sanya (Hainan). Host: triggerfish (*Balistes capistretus*). (21)

P. monacanthi Yamaguti: Yantai (Shandong). Host: filefish (*Navodon septentrionalis*). (26)

Sphincetrostoma japonicus Yamaguti: Ningbo (Zhejiang). Host: tilefish (*Branchiostegus argentatus*). (11)

Trigonotrema alatum Goto & Ozaki: Baimajing (Hainan). Host: tilefishes (*Branchiostegus argentatus* and *B. auratus*). (21)

OPECOELIDAE

Anisoporus cobraeformis Ozaki: Baimajing (Hainan). Host: armored searobin (*Peristedion orientale*). (21)

Cainocreadium epinepheli Yamaguti: Xisha (=Paracel Islands). Host: emperor bream (*Lethrinus haematopterus*). (7)

C. gullella Linton: Xisha (=Paracel Islands). Host: emperor bream (*Gymnocranius griseus*). (7)

Coitocaecum sigani Shen: Sanya (Hainan). Host: rabbitfish (*Siganus oramin*). (21)

Dactylostomum epinepheli Wang: Pingtan (Fujian). Host: grouper (*Epinephelus akaara*). (41)

Decemtestis asymmetricum Wang: Pingtan (Fujian). Host: porgy (*Paragyrops edita*). (47)

D. cynoglossi Li, Zhang, Qiu, Chen & Liang: Tanggu (Tianjin). Host: tonguefish (*Cynoglossus semilaevis*). (26, 30)

Hamacreadium hainanensis Shen: Sanya (Hainan). Host: snapper (*Lutjanus lineolatus*) and drum (*Wak sina*). (21)

H. lethrini Yamaguti: Xisha (=Paracel Islands). Host: emperor breams (*Gymnocranius griseus* and *Lethrinus miniatus*). (7)

H. matabile Linton: Sanya (Hainan). Host: porgy (*Sparus macrocephalus*). (21)

H. synechogobii Qiu & Li: Longkou (Shandong). Host: goby (*Synechogobius ommaturus*). (26)

H. xishaene Shen: Xisha (=Paracel Islands). Host: snapper (*Lutjanus kasmira*). (7)

Helicometra epineli Yamaguti: Pingtan (Fujian). Host: grouper (*Epinephelus akaara*). (41)

H. scorpaenae Wang: Pingtan (Fujian). Host: scorpionfish (*Minous*). (41)

H. selaroidis Shen: Sanya (Hainan). Host: jack (*Selaroides leptolepis*). (10, 21)

Manteriella chanis Shen: Sanya (Hainan). Host: milkfish (*Chanos chanos*). (10, 21)

M. hainanensis Shen: Sanya (Hainan). Host: tilefish (*Branchiostegus argentatus*). (21)

Neohelicometra dalianensis Li, Qiu & Zhang: Dalian (Liaoning). Host: greenling (*Hexagrammos otakii*). (26, 32)

Nicolla ditrematis (Wang): Fuding (Fujian). Host: surfperch (*Ditrema termmincki*). (41)

N. epinepheli (Wang): Pingtan (Fujian). Host: grouper (*Epinephelus akaara*). (41)

Opecoelina pacifica Manter: Sanya (Hainan). Host: cardinalfish (*Apogonichthys carinatus*). (21, 78)

Opecoeloides fugus Li, Qiu & Zhang: Tanggu (Tianjin). Host: puffer (*Fugu vermicularis*). (26, 31)

Opecoelus arii Wang: Fujian. Host: sea catfish (*Arius thalassinus*) and temperate bass (*Lateolabrax japonicus*). (40)

O. bohaiensis Li, Qiu & Zhang: Yingkou (Liaoning). Host: flathead (*Platycephalus indicus*). (26)

O. himezi Yamaguti: Yingkou (Liaoning). Host: flathead (*Platycephalus indicus*). (26, 32)

O. inimici Yamaguti: Dalian (Liaoning). Host: conger eel (*Conger myriaster*). (26, 32)

O. lateolabracis Qiu & Liang: Tanggu (Tianjin). Host: temperate bass (*Lateolabrax japonicus*). (26)

O. lobatus Ozaki: Xisha (=Paracel Islands). Host: mojarra (*Gerres lucidus*). (61)

O. nipponicus Yamaguti: Dalian (Liaoning). Host: scorpionfish (*Sebastodes fuscesens*). (26, 31)

O. pteroisi Shen: Sanya (Hainan). Host: scorpion fish (*Pterois lunulata*). (21)

O. sebastisci Yamaguti: Dalian (Liaoning). Host: scorpionfish (*Sebastodes fuscesens*). (26, 31)

O. spaericus Ozaki: Yingkou and Dalian (Liaoning); Qingdao (Shandong). Host: rockfish (*Sebastodes fuscesens*), conger eel (*Conger myriaster*), and greenling (*Hexagrammos*). (15, 26, 71)

O. zhifuensis Qiu & Li: Yantai (Shandong). Host: temperate bass (*Lateolabrax japonicus*). (26)

Opegaster ditrematis Yamaguti: Longkou (Shandong). Host: halibuts (Paralichthyidae). (26, 31)

O. hippocampi Shen: Jieshi (Hebei). Host: seahorse (*Hippocampus trimaculatus*). (1)

O. synodi Manter: Sanya (Hainan). Host: porgy (*Rhabdosargus sarba*). (21)

O. tamori Yamaguti: Jieshi (Hebei). Host: seahorse (*Hippocampus trimaculatus*). (1)

Ozakia callyodontis Yamaguti: Qingdao (Shandong). Host: porgy (*Sparus macrocephalus*). (15, 21)

O. gerris Shen: Sanya (Hainan). Host: mojarra (*Gerres filamentosus*). (21)

O. lateolabracis Yamaguti: Yingkou (Liaoning). Host: temperate bass (*Lateolabrax japonicus*). (61)

O. sillaginis Shen: Sanya (Hainan). Host: smelt-whiting (*Sillago sihama*). (10, 21)

Paramanteriella cantherini Li, Qiu & Zhang: Dalian (Liaoning). Host: filefish (*Navodon septentrionalis*). (26, 31)

Plagioporus acanthogobii Yamaguti: Qikou (Hebei). Host: goby (*Chaeturichthys hexanema*). (26)

P. apogonichthydis Yamaguti: Huanghua (Hebei). Host: drum (*Collichthys lucidus*). (26)

P. dorosomatis Yamaguti: Huanghua (Hebei). Host: temperate bass (*Lateolabrax japonicus*). (26)

P. epinepheli Shen: Xisha (=Paracel Islands). Host: grouper (*Epinephelus*). (7)

P. hunghaensis Qiu & Li: Huanghua (Hebei). Host: drum (*Collichthys lucidus*). (26)

P. issaitschikowi Price: Huanghua (Hebei); Longkou (Shandong). Host: halibut (Paralichthyidae) and drums (*Collichthys lucidus* and *Nibea albiflora*). (26, 32)

P. parathalassomatis Wang: Pingtan (Fujian). Host: wrasse (*Thalassoma hardwicki*). (41)

P. rhabdosargi Wang: Pingtan (Fujian). Host: porgy (*Rhabdosargus sarba*). (41)

P. sillagonis Yamaguti: Longkou (Shandong). Host: gunnel (*Enedrias nebulosus*). (26)

Podocotyle lethrini Yamaguti: Sanya (Hainan). Host: emperor bream (*Lethrinus ornatus*). (21)

P. lizae Qiu & Liang: Tanggu (Tianjin). Host: mullet (*Liza haematocheila*). (26)

P. lutiani Shen: Sanya (Hainan). Host: snapper (*Lutjanus erythopterus*). (21)

P. nibeae Qiu & Li: Yingkou (Liaoning). Host: drum (*Nibea albiflora*). (26)

P. tetrastyla Tang: Hong Kong. Host: turban snail (*Lunella coronata*). (82)

Podocotyloides plageorchis Shen: Ningbo (Zhejiang). Host: pike conger (*Muraenesox cinereus*). (14)

Pseudopecoelina chirocentrosus Shen: Sanya and Haikou (Hainan). Host: wolf herring (*Chirocentrus dorab*). (10, 21)

P. platycephali Shen: Sanya and Haikou (Hainan). Host: flathead (*Platycephalus indicus*). (10, 21)

P. xishaense Gu & Shen: Xisha (=Paracel Islands). Host: grouper (*Epinephelus*). (61)

Pseudopecoeloides astrocongeris Shen: Ningbo (Zhejiang). Host: conger eel (*Conger myriaster*). (14)

P. carcngis (Yamaguti): Huanghua (Hebei). Host: drum (*Collichthys lucidus*). (26)

P. dayawanensis Shen & Tong: Daya Bay (Guangdong). Host: jack (*Caranx malabaricus*). (24)

P. mugilis Shen: Sanya (Hainan). Host: mullet (*Liza macrolepis*). (21)

Pseudopercoelus alectis Shen: Sanya and Baimajing (Hainan). Host: pompano (*Alectis ciliaris*). (21)

P. elongatus (Yamaguti): Sanya (Hainan). Host: grouper (*Plectropomus leopardus*). (21)

P. epinepheli Wang: Pingtan (Fujian). Host: grouper (*Epinephelus akaara*). (41)

Vesicocoelium solenphagum Tang, Hsu Huang & Lu: Longhai (Fujian). Host: gobies (*Glossogobius giuris*, *Synechogobius hasta*, and *Triaenopogon barbatus*) and sleeper (*Brionobutis koilomatodon*). (69)

PLEORCHIIDAE

Pleorchis hainanensis Shen: Baimajing (Hainan). Host: drum (*Argyrosomus aneus*). (5, 21)

P. nibeae Shen: Huanghua and Qinhuangdao (Hebei). Host: drum (*Nibea albiflora*). (5, 26)

P. sciaenae Yamaguti: East China Sea; Sanya (Hainan). Host: drums (*Nibea albiflora* and *Wak*). (5, 26, 85)

P. uku Yamaguti: Xisha (=Paracel Islands). Host: searobin (*Chelidonichthys kumu*), snappers (*Caesio xanthonotus* and *Pristipomoides microlepis*). (5, 87)

SPIRORCHIIDAE

Haemoxenicon elongatus Wang: Pingtan (Fujian). Host: sea turtles. (43)

PLAGIORCHIIDAE

Cercaria armata Tang: Hong Kong. Host: horn snail (Cerathidea) and cerith snail (*Cerithium*). (82)

C. minus Tang: Hong Kong. Host: snail (*Nodilittorina trochoides*). (82)

C. spelotremoides Tang: Hong Kong. Host: cerith snail (*Clypemorus*) and horn snails (*Batillaria*, *Cerithidea cingulata* and *C. rhizophorarum*). (82)

Enodiotrema hainanensis Chen, Liang, Qiu & Li: Yueqing (Zhejiang). Host: sea turtles. (36)

MACRODEROIDIDAE

Pseudostiotrema rhynchobati Wang: Pingtan (Fujian). Host: guitarfish (*Rhynchobatus*). (40)

ATRACTOTREMATIDAE

Atractotrema fusum Goto & Ozaki: Sanya (Hainan). Host: rabbitfish (*Siganus oramin*). (21)

ACANTHOSTOMIDAE

Paraisocoelium platycephali Shen: Qingdao (Shandong). Host: flathead (*Platycephalus indicus*). (15, 21)

Pseudoisocoelium bigonopori Pen: Guangzhou (Guangdong). Host: temperate bass (*Lateolabrax japonicus*). (75).

Terminoisocoelium laterolecithale Gu & Shen: Shenjiamen (Zhejiang), Tanggu (Tianjin). Host: temperate bass (*Lateolabrax japonicus*). (26, 55).

CRYPTOGONIMIDAE

Biovarium cryptocotyle Yamaguti: Qingdao (Shandong); Pingtan (Fujian). Host: temperate bass (*Lateolabrax japonicus*). (26, 56)

B. pomadasydis Shen & Tang: Longhai (Fujian). Host: grunt (*Pomadasys hasta*). (23)

B. schistolecithale Gu & Shen: Yingkou (Liaoning); Qingdao and Wendeng (Shandong). Host: temperate bass (*Lateolabrax japonicus*). (26, 55)

B. tsingtaoensis Gu & Shen: Qingdao (Shandong). Host: temperate bass (*Lateolabrax japonicus*). (26, 56)

Diplopharyngotrema bipapillosa (Gu & Shen): Shenjiamen (Zhejiang). Host: temperate bass (*Lateolabrax japonicus*). (55)

D. latelabraois Yamaguti: Huanghua (Hebei); Longhai and Fuding (Fujian). Host: temperate bass (*Lateolabrax japonicus*). (26, 41)

Paracryptogonimus apharei Yamaguti: Sanya (Hainan). Host: snapper (*Pristipomoides microlepis*). (21, 87)

P. elongatus Gu & Shen: Sanya (Hainan). Host: threatfin bream (*Nemipterus virgatus*). (21, 58)

P. lutiani Wang: Pingtan (Fujian). Host: snapper (*Lutjanus*).

P. ovatus Yamaguti: Sanya and Baimajing (Hainan). Host: snapper (*Lutjanus vaigiensis*) and wolf herring (*Chirocentrus dorab*). (21)

P. ula-ula Yamaguti: Xisha (=Paracel Islands). Host: snapper (*Pristipomoides microlepis*). (60, 87)

Pseudosiphoderoides lutiani (Yamaguti): Sanya (Hainan). Host: snapper (*Lutjanus vaigiensis*). (21)

P. opakapaka Yamaguti: Xisha (=Paracel Islands). Host: snapper (*Pristipomoides microlepis*). (60, 87)

P. rotivarijera Shen: Ningbo (Zhejiang). Host: searobin (*Chelidonichthys kumu*). (11)

P. xishaensis Gu & Shen: Xisha (=Paracel Islands). Host: snapper (*Caesio xanthonotus*). (61)

Siphodera vinaledwaldsi (Linton): Sanya (Hainan). Host: barracuda (*Sphyraena pinguis*). (21)

Siphoderina asiatica Gu & Shen: Sanya (Hainan). Host: snappers (*Lutjanus fulviflamma* and *L. erythopterus*). (21, 58)

Siphoderooides lutiani Shen: Sanya (Hainan). Host: snapper (*Lutjanus erythopterus*). (21)

DIDYMOZOIDAE

Allodidymocodium unitubulare Yamaguti: South China Sea. Host: rabbitfish (*Siganus fuscescens*). (85)

Allonematobothrium epinepheli Yamaguti: Xisha (=Paracel Islands). Host: grouper (*Epinephelus tauvina*). (19, 61)

A. xishaense Gu & Shen: Xisha (=Paracel Islands). Host: grouper (*Epinephelus tauvina*). (19, 61)

Coeliodidymocystis abdominalis Yamaguti: Xisha (=Paracel Islands). Host: tuna (*Auxis thazard*). (19, 87)

Colocyntotrema auxis Yamaguti: Dongshan (Fujian). Host: tuna (*Auxis thazard*). (6)

Didymocystis bifurcata Yamaguti: Ningbo (Zhejiang). Host: barracuda (*Sphyraena pinguis*). (6, 87)

D. crassa Ishil: Ningbo (Zhejiang). Host: tuna (*Auxis tapeinosoma*). (6)

D. philobranchia Yamaguti: South China Sea. Host: tuna (*Auxis thazard*). (85)

D. radiatus Ku & Shen: Sanya (Hainan). Host: barracudas (*Sphyraena jello* and *S. pinguis*). (21, 52)

D. soleiformis Ishil: Ningbo (Zhejiang). Host: tuna (*Auxis tapeinosoma*). (6)

D. wedli Ariola: Ningbo (Zhejiang). Host: tuna (*Auxis tapeinosoma*). (6)

Didymocystoides buccalis Yamaguti: South China Sea. Host: tuna (*Auxis thazard*). (85)

D. dolichorchis Gu & Shen: Sanya (Hainan). Host: bigeye snapper (*Priacanthus tayenus*). (21, 60)

D. xishaensis Shen: Xisha (=Paracel Islands). Host: tuna (*Auxis thazard*). (9)

Didymozoon biramus Ku & Shen: Lianyungang (Jiangsu); Ningbo (Zhejiang); Sanya (Hainan). Host: barracudas (*Sphyraena jello* and *S. pinguis*). (6, 9, 21, 26, 52)

D. gigas Ku & Shen: Sanya (Hainan). Host: spadefish (*Platax orbicularis*). (12, 52)

D. longicollopsis Gu & Shen: Sanya (Hainan). Host: bigeye snapper (*Priacanthus tayenus*). (19, 21, 60)

D. platycephali Ku & Shen: Yantai (Shandong). Host: flathead (*Platycephalus indicus*). (19, 26, 52)

D. pneumatophori Gu & Shen: Lianyungang (Jiangsu); Ningbo (Zhejiang). Host: Pacific mackerel (*Pneumatophorus japonicus*). (19, 26, 60)

D. priacanthus Ku & Shen: Ningbo (Zhejiang); Sanya (Hainan). Host: bigeye snappers (*Priacanthus macracanthus* and *P. tayenus*). (19, 21, 52).

D. scomberomori Ku & Shen: Sanya (Hainan). Host: Spanish mackerel (*Scombermorus commersoni*). (19, 21, 52)

D. sparial Yamaguti: Yantai (Shandong). Host: flathead (*Platycephalus indicus*). (19, 21, 52)

Gonapodasmius branchialis Yamaguti: Sanya (Hainan). Host: grouper (*Cephalopholis pachycentron*). (21, 52)

G. hainaensis Gu & Shen: Baimajing (Hainan). Host: grouper (*Tristotropis ermopterus*). 21, 60)

G. pacificus Yamaguti: Xisha (=Paracel Islands). Host: groupers (*Epinephelus awoara* and *E. tauvina*). (61)

Kollikeria orientalis Yamaguti: Sanya (Hainan). Host: tuna (*Auxis tapeinosoma*). (21, 52)

K. sphyraenae Shen: Ningbo (Zhejiang). Host: barracuda (*Sphyraena pinguis*). (16, 19)

Lepidodidymozoon sinicum Shen: Qingdao (Shandong); Ningbo (Zhejiang); Lingao (Hainan). Host: barracuda (*Sphyraena pinguis*). (6, 26)

Maccallozum platycephali Ku & Shen: Zhoushan (Zhejiang). Host: flathead (*Platycephalus indicus*). (19, 21, 52)

Metanematobothrium serilae Ku & Shen: Sanya (Hainan). Host: amberjack (*Seriola dumerili*). (19, 21, 52)

Nematobothrium filiforme Yamaguti: Ningbo (Zhejiang). Host: Pacific mackerel (*Pneumatophorus japonicus*). (19, 60)

N. schistogonimum Ku & Shen: Sanya (Hainan). Host: jack (*Megalaspis cordyla*). (21, 52)

Neometadidymozoon zhejiangensis Shen: Ningbo (Zhejiang). Host: bigeye snapper (*Priacanthus macracanthus*). (6)

Oesophogocystis dissimilis Yamaguti: South China Sea. Host: bluefin tuna (*Thunnus thynnus*). (85)

Osteodidymocodium kawakawa Yamaguti: South China Sea. Host: rabbitfish (*Siganus fuscescens*). (85)

Paragonapodasmius huanghaiensis Shen: Qingdao, Yantai, and Wendeng (Shandong). Host: temperate bass (*Lateolabrax japonicus*). (26)

Phacelotrema claviforme Yamaguti: Dongshan (Fujian). Host: tuna (*Auxis thazard*). (19)

Pseudocoeliodidymocystis xishaensis Shen: Xisha (=Paracel Islands). Host: tuna (*Auxis thazard*). (19)

Syncorpozoum hainanensis Ku & Shen: Sanya (Hainan), Ningbo (Zhejiang). Host: bigeye snapper (*Priacanthus macracanthus*). (21, 52)

Univitellodidymocystis miliaris (Yamaguti): Xisha (=Paracel Islands). Host: tuna (*Euthynnus yaito*). (19)

HEMIURIDAE

Aphanurus mugilus Tang: Penglai (Shandong); Fujian. Host: striped mullet (*Mugil cephalus*). (12, 26, 35, 68)

A. multiprostatus Pan: Guangzhou (Guangdong). Host: anchovy (*Coilia grayi*). (74)

A. stossichi (Monticelli): Longhai and Fuqing (Fujian). Host: gizzard shad (*Clupanodon punctatus*). (12, 41, 70)

Aponurus clupanodontis Qiu & Liang: Beidaihe (Hebei). Host: gizzard shad (*Clupanodon punctatus*). (26)

A. collichthydis Qiu & Liang: Tanggu (Tianjin). Host: drum (*Collichthys lucidus*). (26)

A. eleutheronematis Shen: Ningbo (Zhejiang). Host: threadfin (*Eleutheronema tetradactylus*). (12)

A. halieutae Shen: Ningbo (Zhejiang). Host: batfish (*Halieutaea stellata*). (12)

A. laguncula Looss: Longkou (Shandong), Tanggu (Tianjin). Host: herring (Clupeidae) and tonguefish (*Cynoglossus semilaevis*). (26)

A. lizae Shen: Beitang (Zhejiang). Host: mullet (*Liza haematocheila*). (26)

A. rhinoplagusiae Yamaguti: Huanghua (Hebei). Host: temperate bass (*Lateolabrax japonicus*). (26)

A. uraspis Shen: Baimajing (Hainan). Host: jack (*Caranx helvolus*). (21)

A. vitellograndis Layman: Longkou (Shandong); Fuzhou (Fujian). Host: butterfish (*Pampus argenteus*) and halibut (Paralichthyidae). (26, 42)

Chenia cheni Wang: Pingtan (Fujian). Host: tonguefish (*Cynoglossus*). (74)

Derogenes bohaiensis Qiu & Liang: Tanggu (Tianjin); Longkou (Shandong). Host: rockfish (*Sebastodes fuscesens*) and drum (*Argyrosomus argentatus*). (26)

D. chelidonichthydis Shen: Qingdao (Shandong). Host: searobin (*Chelidonichthys kumu*). (26, 15)

D. epinepheli Wang: Pingtan (Fujian). Host: grouper (*Epinephelus akaara*). (42)

D. gadi Shen: Qinhuangdao (Hebei); Yantai (Shandong). Host: cods (Gadidae) and temperate bass (*Lateolabrax japonicus*). (20, 26)

D. macrostoma Yamaguti: Huanghua (Hebei). Host: drum (*Nibea albiflora*). (26)

D. magnus Wang: Pingtan (Fujian). Host: striped mullet (*Mugil cephalus*). (47)

D. minoi Shen: Mouth of Changjiang (=Yangtze River). Host: scorpionfish (*Minous monodactylus*). (17)

D. varicus (Müller): Qingdao and Yantai (Shandong); Lianyungang (Jiangsu); Zhoushan (Zhejiang); Xiamen (Fujian). Host: goosefish (*Lophius litulon*), grunt (*Hapalogenys mucronatus*), flathead (*Platycephalus indicus*), halibuts (Paralichthyidae), and anchovy (*Setipinna taty*). (12, 15, 20, 26)

Dinosoma setipinae Wang: Fuding (Fujian). Host: anchovy (*Setipinna taty*). (42)

Dinurus longissimus Looss: Sanya (Hainan). Host: dolphinfish (*Coryphaena hippurus*). (7, 21, 57)

D. megnacetabalum Gu & Shen: Sanya (Hainan). Host: jack (*Carangoides mate*). (21, 50)

D. scombri Yamaguti: Ningbo (Zhejiang). Host: Pacific mackerel (*Pneumatophorus japonicus*). (12)

Duosphincter zancli Manter & Pritchard: Haikou (Hainan). Host: ladyfish (*Elops saurus*). (21)

Ectenurus carangis Gu & Shen: Sanya (Hainan). Host: jack (*Carangoides kalla*). (21, 57)

E. lepidus Looss: Beidaihe (Hebei); Zhoushan (Zhejiang). Host: jack (*Atropus atropus*) and drum (*Pseudosciaena crocea*). (12, 26)

E. megalaspis Gu & Shen: Haikou (Hainan). Host: jack (*Megalaspis cordyla*). (21, 57)

E. nibeae Shen: Qingdao (Shandong). Host: drum (*Nibea albiflora*). (15, 26)

E. pseudosciaenae Gu & Shen: Zhoushan (Zhejiang). Host: drum (*Pseudosciaena crocea*). (12, 57)

E. theraponae Oshmarin: Sanya (Hainan). Host: grunter (*Therapon theraps*). (21)

E. tianjinensis Li, Zhang & Qiu: Tanggu (Tianjin). Host: cutlassfish (*Eupleurogrammus muticus*). (26, 29)

E. trachuri Yamaguti: Weihai (Shandong). Host: jack (*Trachurus japonicus*). (26, 57)

E. trichiuri Shen: Mouth of Changjiang (=Yangtze River). Host: cutlassfish (*Trichiurus haumela*). (17)

E. zonichthyi Gu & Shen: Sanya (Hainan). Host: jack (*Zonichthys nigrofasciata*). (21, 57)

Elytrophallus coiliae Wang: Pingtan (Fujian). Host: anchovy (*Coilia ectenes*). (12, 42)

Erilepturus bohaiensis Li, Zhang & Qiu: Tanggu (Tianjin). Host: drum (*Collichthys niveatus*). (26, 29)

E. collichthydis Wang: Guantou (Fujian). Host: drum (*Collichthys lucidus*). (12, 42)

E. formosae Reid, Coil & Kuntz: Taiwan. Host: pompano (*Alectis indica*). (80)

E. pisodonophis Shen: Sanya (Hainan). Host: snake eel (*Pisodonophis*). (21)

E. trichiuri Gu & Shen: Baimajing (Hainan). Host: cutlassfish (*Trichiurus haumela*). (21, 57)

Genarchopsis clupeae Qiu & Li: Longkou (Shandong). Host: herring (Clupeidae). (26)

Genolina laticauda Manter: Huanghua (Hebei). Host: rockling (*Enchelyopus*). (26)

Hypohepaticola setipinnae Li, Zhang & Qiu: Tanggu (Tianjin). Host: anchovy (*Setipinna taty*). (26, 28)

Hysterolecitha arii Wang: Putian (Fujian). Host: sea catfish (*Arius thalassinus*). (12, 42)

H. chirocentri Ku & Shen: Sanya (Hainan). Host: wolf herring (*Chirocentrus dorab*). (21, 50)

H. progonimus Ku & Shen: Sanya (Hainan). Host: bonefish (*Albula vulpes*). (21, 50)

H. rosea Linton: Pingtan (Fujian). Host: drum (*Nibea albiflora*). (40)

Hysterolecithoides epinepheli Yamaguti: Sanya (Hainan). Host: rabbitfish (*Siganus oramin*). (21)

H. multiglandularis Tang, Shi & Guan: Longhai (Fujian). Host: ponyfish (*Leiognathus brevirostris*). (70)

Johinophyllum formiae Wang: Pingtan (Fujian). Host: jack (*Formio niger*). (42)

J. qingdaoensis Shen: Qingdao (Shandong). Host: drums (*Argyrosomus argentatus* and *Nibea albiflora*). (15, 26)

Lecithaster atropi Shen: Ningbo (Zhejiang). Host: jack (*Atropus atropus*). (12)

L. confusus Odhner: Guangzhou (Guangdong). Host: anchovy (*Coilia grayi*). (73)

L. fusiformis Wang: Pingtan (Fujian). Host: rabbitfish (*Siganus fuscescens*). (47)

L. sayori Yamaguti: Qinhuangdao (Hebei). Host: needlefish (*Ablennes anastomella*). (26)

L. setipinnae Qiu & Liang: Tanggu (Tianjin). Host: anchovy (*Setipinna taty*). (26)

L. tylosuri Li, Qiu, Chen & Liang: Qinhuangdao (Hebei). Host: needlefish (*Ablennes anastomella*). (26, 33)

Lecithochirium albulae Yamaguti: Xisha (=Paracel Islands). Host: moray eel (*Gymnothorax*). (7)

L. apharei Yamaguti: Ningbo (Zhejiang). Host: threadfin (*Eleutheronema tetradactylus*). (12)

L. branchialis (Stumkard & Nigrelli): Pingtan (Fujian). Host: cutlassfish (*Trichiurus haumela*). (12, 42)

L. caudiparum (Rud.): Tanggu (Tianjin). Host: anchovy (*Coilia ectenes*). (26)

L. dongshanensis Shen: Beidaihe (Hebei); Dongshan (Fujian). Host: drum (*Pseudosciaena crocea*) and righteye flounder (*Kareius bicoloratus*). (9, 12, 26)

L. fistulariae Yamaguti: Ningbo (Zhejiang). Host: red cornetfish (*Fistularia petimba*). (12)

L. fusiforme Luhe: Baimajing (Hainan). Host: cutlassfish (*Trichiurus haumela*). (21)

L. holocentri Yamaguti: Ningbo (Zhejiang); Baimajing (Hainan). Host: porgy (*Paragyrops edita*) and cutlassfish (*Trichiurus haumela*). (12, 21)

L. kawakawa Yamaguti: Xisha (=Paracel Islands). Host: tuna (*Euthynnus yaito*). (7)

L. kawalea Yamaguti: Ningbo (Zhejiang). Host: batfish (*Halieutaea stellata*). (12)

L. keokeo Yamaguti: Sanya (Hainan). Host: emperor bream (*Lethrinus nebulosus*). (21)

L. magnaporum Manter: Dongshan and Pingtan (Fujian). Host: remora (*Echeneis naucrates*) and amberjack (*Seriola aureovittata*). (12, 42)

L. michthydis Wang: Putian (Fujian). Host: drum (*Miichthus miiuy*). (12, 42)

L. microstomum Chandler: Sanya (Hainan). Host: cobia (*Rachycentron canadum*). (21)

L. monticolli (Linton): Pingtan (Fujian). Host: snake eel (*Cirrhimuraena chinensis*). (12, 42)

L. muraenesocis Wang: Pingtan (Fujian). Host: pike conger (*Muraenesox cinereus*) and Spanish mackerel (*Scomberomorus guttatus*). (12, 42)

L. pacificum Tang: Fujian. Host: pike conger (*Muraenesox cinereus*). (12, 68)

L. paracanthi Yamaguti: Dalian (Liaoning); Qidong (Jiangsu); Zhoushan and Ningbo (Zhejiang); Baimajing (Hainan); Xisha (=Paracel Islands). Host: jacks (*Carangoides kalla* and *Caranx helvolus*), soldierfish (*Mystipristis seychellensis*), Pacific mackerel (*Pneumatophorus japonicus*), Spanish mackerel (*Scomberomorus niphonius*), and drum (*Pseudosciaena crocea*). (7, 12, 21, 26)

L. poecilopsettae Shen: Baimajing (Hainan). Host: righteye flounder (*Poecilopsetta*). (21)

L. pseudosciaenae Shen: Ningbo (Zhejiang). Host: drum (*Pseudosciaena polyactis*). (9)

L. rufoviride (Rud.): Ningbo (Zhejiang). Host: conger eel (*Conger*). (12)

L. savalae Shen: Xiamen (Fujian). Host: cutlassfish (*Lepturacanthus savala*). (12)

L. scomberomori Wang: Pingtan (Fujian). Host: Spanish mackerel (*Scombermorus niphonius*). (12, 42)

L. texanum (Chandler): Xisha (=Paracel Islands). Host: tuna (*Euthynnus yaito*). (7)

L. trichiuri Gu & Shen: Yantai and Qingdao (Shandong); Ningbo (Zhejiang); Dongshan (Fujian); Daya Bay (Guangdong); Sanya (Hainan). Host: cutlassfish (*Trichiurus haumela*). (12, 24, 26, 59)

Lecithocladium bulbolabrum Reid, Coil & Kuntz: Dongshan (Fujian); Taiwan. Host: mackerel (*Rastrelliger kanagurta*). (80)

L. dongshanensis Shen: Dongshan (Fujian). Host: drum (*Pseudosciaena crocea*). (9, 12)

L. excisiforme Chen: Longhai (Fujian); Daya Bay (Guangdong); Sanya (Hainan). Host: butterfish (*Pampus chinensis*), jack (*Decapterus maruadsi* and *Formio niger*), and medusafish (*Psenopsis anomala*). (21, 24, 42, 48)

L. excisum (Rud.): Qingdao (Shandong); Pingtan (Fujian); Sanya (Hainan). Host: threadfin bream (*Scolopsis taeniopterus*); jack (*Carangoides macrurus*), Pacific mackerel (*Pneumatophorus japonicus*), Spanish mackerel (*Scombermorus guttatus*). (15, 21, 26, 42)

L. glandulum Chauhan: Pingtan (Fujian); Haikou (Hainan). Host: jack (*Megalaspis cordyla*), righteye flounder (*Pleuronichthys cornutus*), and tonguefish (*Cynoglossus*). (21, 48)

L. harpedontis Srivastava: Qingdao (Shandong); Lianyungang (Jiangsu); Zhoushan (Zhejiang); Sanya (Hainan). Host: Pacific mackerel (*Pneumatophorus japonicus*), and butterfish (*Pampus argenteus*). (12, 15, 26)

L. ilishae Mamaev: Pingtan (Fujian). Host: herring (*Illisha elongata*). (12, 42)

L. leiognathi Wang: Pingtan (Fujian). Host: mojarra (*Leiognathus ruconius*), jacks (*Atropus atropus* and *Caranx*), and drum (*Collichthys lucidus*). (48)

L. lutiani Gu & Shen: Sanya (Hainan). Host: snapper (*Lutjanus erythopterus*). (21, 57)

L. megalaspis Yamaguti: Sanya (Hainan). Host: jack (*Carangoides kalla*). (21)

L. miichthydis Wang: Tanggu (Tianjin); Putian (Fujian). Host: drums (*Argyrosomus argentatus* and *Miichthys miiuy*). (12, 42)

L. pagrosomi Yamaguti: Sanya (Hainan). Host: cutlassfish (*Trichiurus haumela*). (21)

L. prariovum Yamaguti: Qingdao (Shandong); Dongshan and Xiamen (Fujian); Sanya and Baimajing (Hainan). Host: mackerels (*Pneumatophorus japonicus* and *Rastrelliger kanagurta*), and cutlassfish (*Trichiurus haumela*). (12, 15, 21)

L. putianensis Wang: Putian (Fujian). Host: butterfish (*Pampus argenteus*). (12, 42)

L. xishaense Shen: Xisha (=Paracel Islands). Host: surgeonfish (Acantharidae). (7)

Lecithomonoium harpodi Shen: Zhoushan (Zhejiang). Host: Bombay duck (*Harpadon nehereus*). (12)

Lecithophyllum intermedium (Manter): Qinhuangdao (Hebei). Host: rockling (*Enchelyopus*). (26)

Monolecithotrema lizae Shen: Wendeng (Shandong). Host: mullet (*Liza haematocheila*). (20, 26)

Musculovesicula trichiuri Gu & Shen: Zhoushan (Zhejiang). Host: cutlassfish (*Trichiurus haumela*). (12, 59)

M. zonichthydis Shen: Baimajing (Hainan). Host: jack (*Zonichthys nigrofasciata*). (21)

Neoaphanurus magniprotesticus Tang, Shi & Pan: Longhai (Fujian). Host: anchovy (*Coilia mystus*). (12, 70)

Neodichadena mugilis Tang, Shi & Cao: Longhai (Fujian). Host: mullet (*Osteomugil stronylocephalus*). (12, 70)

Oligolecithoides trilobatus Shen: Sanya (Hainan). Host: jack (*Caranx chrysophrys*). (2, 21)

Opisthadena fujianensis Tang, Shi & Pan: Longhai (Fujian). Host: mullet (*Osteomugil stronylocephalus*). (12, 70)

O. marina Tang, Shi & Cao: Longhai (Fujian). Host: Asiatic glassfish (*Ambasis gymnocephalus*). (12, 70)

O. setipinnae Qiu & Liang: Tanggu (Tianjin). Host: anchovy (*Setipinna taty*). (26)

Parahemiurus ambassicola Tang Shi & Cao: Longhai (Fujian). Host: Asiatic glassfish (*Ambassis gymnocephalus*). (70)

P. clupeae Yamaguti: Sanya (Hainan). Host: round herring (*Dussumieria hasselti*). (21)

P. coiliae Wang: Fuzhou (Fujian). Host: anchovy (*Coilia grayi*). (42)

P. collichthydis Li, Zhang & Qiu: Tanggu (Tianjin). Host: drum (*Collichthys niveatus*). (26, 28)

P. harengulae Yamaguti: Qingdao (Shandong); Daya Bay (Guangdong). Host: herring (*Herklotsichthys*) and anchovy (*Thryssa setirostris*). (26)

P. hemirhamphi Wang: Pingtan (Fujian). Host: halfbeak (*Hemiramphus georgii*). (42)

P. merus (Linton): Dongshan and Pingtan (Fujian). Host: herrings (*Etrumeus micropus* and *Ilisha elongata*). (12, 42)

P. pseudoscianae Shen: Zhoushan (Zhejiang). Host: drum (*Pseudosciaena crocea*). (9, 12)

P. qingdaoensis Shen: Qingdao (Shandong). Host: flathead (*Platycephalus indicus*) and drum (*Nibea albiflora*). (15, 26)

P. sardiniae Yamaguti: Daya Bay (Guangdong). Host: Bombay duck (*Harpadon nehereus*). (24)

P. seriolae Yamaguti: Tanggu (Tianjin). Host: flathead (*Platycephalus indicus*). (26)

Paramacradenina xiamensis Tang, Shi & Guan: Longhai (Fujian). Host: mullet (*Osteomugil stronylocephalus*). (12, 70)

Plerurus asterocongeri Shen: Ningbo (Zhejiang). Host: conger eel (*Conger myriaster*). (12)

P. atulis Shen: Baimajing (Hainan). Host: jack (*Caranogoides mate*). (21)

P. carangis Yamaguti: Yingkou (Liaoning); Zhoushan and Ningbo (Zhejiang). Host: flathead (*Platycephalus indicus*) and cutlassfish (*Trichiurus haumela*). (12, 26)

P. cynoglossi Wang: Longhai (Fujian). Host: tonguefish (*Cynoglossus robustus*). (12, 42)

P. hainanensis Gu & Shen: Sanya (Hainan). Host: cutlassfish (*Trichiurus haumela*). (21, 59)

P. longicaudatus Yamaguti: Dongshan and Xiamen (Fujian). Host: cutlassfish (*Trichiurus haumela*). (12)

P. pomadasydis Shen & Tong: Baimajing (Hainan). Host: grunt (*Pomadasys hasta*). (21, 33)

P. pseudosciaenae Shen: Zhoushan (Zhejiang). Host: drum (*Pseudosciaena crocea*). (9)

P. scomberomori Shen: Sanya (Hainan). Host: Spanish mackerel (*Scombermorus niphonius*). (21)

P. spatulocirrus Gu & Shen: Baimajing (Hainan). Host: cutlassfish (*Trichiurus haumela*). (21, 59)

P. trichiuri Gu & Shen: Dongshan and Xiamen (Fujian); Daya Bay (Guangdong); Haikou (Hainan). Host: cutlassfish (*Trichiurus haumela*). (21, 59)

Prosorchiopsis plataxonis Gu & Shen: Sanya (Hainan). Host: spadefish (*Platax teira*). (21, 58)

Prosorchis provitelletus Ku & Shen: Zhoushan (Zhejiang); Fuzhou (Fujian). Host: butterfish (*Pampus argenteus*). (12, 42, 51)

P. rastrelligi Ku & Shen: Sanya (Hainan). Host: mackerel (*Rastrelliger kanagurta*). (21, 51)

P. xishaensis Shen: Xisha (=Paracel Islands). Host: surgeonfish (Acanthuridae). (7)

Pseudobunocotyla clupanodonis Qiu & Tong: Beidaihe (Hebei). Host: lizard shad (*Clupanodon punctatus*). (26)

Separogermiductus sinicus Tang: Fujian. Host: temperate bass (*Lateolabrax japonicus*). (68)

Sterrhurus chirocentri Shen: Sanya (Hainan). Host: wolf herring (*Chirocentrus dorab*). (21)

S. gymnothoracis Yamaguti: Qingdao (Shandong); Sanya (Hainan). Host: conger eel (*Conger myriaster*) and barracuda (*Sphyraena jello*). (21, 26)

S. inimici Yamaguti: Qingdao (Shandong). Host: flathead (*Platycephalus indicus*). (15, 26)

S. longicaudatus Shen: Zhoushan (Zhejiang). Host: goosefish (*Lophius litulon*). (12)

S. microvatus Yamaguti: Baimajing (Hainan). Host: batfish (*Halieutaea stellata*). (21)

S. pisodonophis Shen: Sanya (Hainan). Host: snake eel (*Pisodonophis*). (21)

Stomachicola hainanensis Shen: Sanya (Hainan). Host: wolf herring (*Chirocentrus dorab*). (21)

S. muraenesocis Yamaguti: Qingdao, Yantai, and Longkou (Shandong); Lianyungang (Jiangsu); Zhoushan and Ningbo (Zhejiang); Dongshan (Fujian); Taiwan; Daya Bay (Guangdong); Sanya and Baimajing (Hainan). Host: pike conger (*Muraenesox cinereus*). (12, 15, 21, 26, 41, 68, 80)

Tubulovesicula anguillae Yamaguti: Baimajing (Hainan). Host: righteye flounder (*Poecilopsetta*) and cutlassfish (*Trichiurus haumela*). (21)

T. lindbergi (Layman): Yingkou (Liaoning); Sanya (Hainan). Host: barracuda (*Sphyraena pinguis*) and lizardfish (*Saurida tumbil*). (21, 26, 57)

T. longicorporis Shen: Sanya (Hainan). Host: Spanish mackerel (*Scombermorus koreana*). (21)

T. maraenesocis Yamaguti: Baimajing (Hainan). Host: pike conger (*Muraenesox cinereus*). (21, 57)

T. sauridia Gu & Shen: Weihai (Shandong); Lianyungang (Jiangsu). Host: lizardfish (*Saurida elongata*). (26, 57)

T. serpentis Wang: Pingtan (Fujian). Host: sea snake (*Hydrophis cyanocinctus*). (46)

T. spari Yamaguti: Sanya and Baimajing (Hainan). Host: lizardfish (*Trachinocephalus myops*) and moray eel (*Gymnothorax*). (21)

Uterovesiculurus fujianensis Wang: Pingtan (Fujian). Host: sea bass (*Niphon spinosus*) and temperate bass (*Lateolabrax japonicus*). (42)

U. hamati (Yamaguti): Sanya (Hainan). Host: lizardfishes (*Saurida elongata* and *S. filamentosa*). (21, 57)

U. lutianius Gu & Shen: Sanya (Hainan). Host: snapper (*Lutjanus erythopterus*). (21, 57)

U. sinensis Gu & Shen: Sanya (Hainan). Host: jacks (*Chorinemus lysan* and *Zonichthys nigrofasciata*). (21, 57)

U. spindis Shen: Sanya (Hainan). Host: snapper (*Lutjanus vaigiensis*). (21)

HIRUDINELLIDAE

Hirudinella beebei Chandler: Xisha (=Paracel Islands). Host: wahoo (*Acanthocybium solandi*). (62)

H. xishaensis Gu & Shen: Xisha (=Paracel Islands). Host: tuna (*Euthynnus yaito*). (62)

PROSOGONOTREMATIDAE

Prosogonotrema caesionis Gu & Shen: Sanya (Hainan). Host: snapper (*Caesio erythrogaster*). (21, 58)

P. plataxum Gu & Shen: Sanya (Hainan). Host: spadefish (*Platax orbicularis*). (21, 58)

MONODHELMINTHIDAE

Mehratrema arii Gu & Shen: Sanya and Haikou (Hainan). Host: sea catfish (*Arius thalassinus*). (21, 58)

M. fujianensis Wang: Putian (Fujian). Host: drum (*Miichthus miiuy*). (41).

Monodhelmis philippinensis Velasquey: Putian (Fujian). Host: sea catfish (*Arius thalassinus*). (41)

ISOPARORCHIIDAE

Flongoparorchis arii Shen & Tong: Baimajing (Hainan). Host: sea catfish (*Arius thalassinus*). (21, 22)

AEROBIOTREMATIDAE

Aerobiotrema acinovaria Tang: Fujian. Host: pike conger (*Muraenesox cinereus*). (68)

ALBULATREMATIDAE

Albulatrema ovale Yamaguti: Sanya (Hainan). Host: bonefish (*Albula vulpes*). (21)

HARMOTREMATIDAE

Harmotrema linguiforme Wang: Pingtan (Fujian). Host: sea snake (*Hydrophis cyanocinctus*). (46)

PHILOPHTHALMIDAE

Cercaria cloacicola Tang: Hong Kong. Host: horn snails (*Terebralia sulcata*). (81)
C. gedoelsti Tang: Hong Kong. Host: horn snail (*Cerithidea rhizophorarum*). (81).
C. hongkongensis Tang: Hong Kong. Host: horn snail (*Cerithidea cingulata*). (81)
C. longicauda Tang: Hong Kong. Host: horn snails (*Cerithidea rhizophorarum* and *Terebralia sulcata*). (81).
C. mortoni Tang: Hong Kong. Host: snail (*Nodilittorina trochoides*). (81)
Neopygorchis exvitellina Tang & Tang: Fuzhou (Fujian). Final host: curlew (*Numenius arquata orientalis*). (67)
Parorchis acanthus (Nicoll, 1906) Nicoll: Fuzhou and Pingtan (Fujian). Final host: dunlin (*Calidris alpina*), greater sand plover (*Charadrius leschenaultii*), and redshank (*Tringa t. totanus*). (67)
P. gedoelsti Skrjabin: Pingtan (Fujian). Final host: dunlin (*Calidris alpina*), European dotterel (*Charadrius morinellus*), greater sand plover (*Charadrius leschenaultii*), and curlew (*Numenius arquata orientalis*). (67)
Pittacium egrettum Tang & Tang: Pingtan (Fujian). Final host: egret (*Egretta*). (67)
P. pittacium (Buaum, 1901) Szidat: Pingtan (Fujian). Final host: European dotterel (*Charadrius morinellus*), golden plover (*Pluvialis dominica*), curlew (*Numenius arquata orientalis*), and snowy plover (*Charadrius alexandrinus dealbatus*). (67)

HETEROPHYIDAE

Cercaria magnicaudata Tang: Hong Kong. Host: cerith snail (*Cerithium*) and horn snails (*Batillaria, Cerithidea cingulata*, and *Terebralia sulcata*). (82)
Cercarioides (*C.*) *hoepplii* Tang & Tang: Fuqing (Fujian). Host: turban snail (*Lunella coronata*). (83)

Heterophyes heterophyes (Siebold, 1853) Stiles & Hassal: Hong Kong. Host: cerith snail (*Cerithium*). (82)

CYATHOCOTYLIDAE

Cercaria mesostephanus Tang: Hong Kong. Host: cerith snail (*Cerithium*). (82)

ECHINOSTOMATIDAE

Cercaria himasthloides Tang: Hong Kong. Host: horn snail (*Cerithidea rhizophorarum*). (82)

REFERENCES*

(1). Shen Jiwei, 1982. Two new digenetic trematodes from the sea horse, *Hippocampus trimaculatus*. Oceanologia & Limnologia Sinica, 13(3):285–288.
(2). Shen Jiwei, 1982. On *Oligolecithoides trilobatus* (Hemiuridae trematode) gen. et sp. nov. from China. Oceanologia & Limnologia Sinica, 13(5): 473–476.
(3). Shen Jiwei, 1982. Three new species of genus *Bivesicula* Yamaguchi, 1934 (Trematodes: Bivesiculidae) from marine fishes of China. Oceanologia & Limnologia Sinica, 13(6):570–576.
(4). Shen Jiwei, 1983. A new species of *Paragonapodasmius* Yamaguchi, 1938 - *Paragonapodasmius huanghaiiensis*. Oceanologia & Limnologia Sinica, 14(1): 71–73.
(5). Shen Jiwei, 1983. Two new species to the family Pleorchiidae Poche, 1926 (trematode) from some marine fishes in China. Oceanologia & Limnologia Sinica, 14(4):369–401.
(6). Shen Jiwei, 1984. Digenetic trematodes of Didymozoidae Poche, 1907 from marine fishes in the East China Sea. Studia Marina Sinica, 23:121–129.
(7). Shen Jiwei, 1985. Digenetic trematodes of fishes from the Xisha Islands II. Studia Marina Sinica, 24:167–180.
(8). Shen Jiwei, 1985. Some digenetic trematodes from the *Psuedosciaena crocea* (Richardson) from the East China Sea, Trematodes: Hemiuridae, Bucephalidae. Acta Zootaxonomica Sinica, 10(2):129–136.
(9). Shen Jiwei, 1986. Opecoelidae tremapedes of marine fishes from Hainan Island. Oceanologia & Limnologia Sinica, 17(5):879–885.
(10). Shen Jiwei, 1986. Some digenetic trematodes of Accaoeliiidae Ohhner, 1911 from China. Oceanologia & Limnologia Sinica, Suppl.:191–196.
(11). Shen Jiwei, 1986. Digenetic trematodes of fishes from the East China Sea I. Species of the families Opistolebetidae, Lepocreadiidae and Cryptogonimidae. Studia Marina Sinica, 27:209–219.
(12). Shen Jiwei, 1987. Digenetic trematodes of fishes from the East China Sea II. Description of a new genus and seven species of Hemiuridae. Studia Marina Sinica, 125–139.
(13). Shen Jiwei, 1987. Digenetic trematodes of fishes from the East China Sea III. Five new species of the families Bucephalidae, Fellodistomidae, Zoogonidae and Monorchiidae. Studia Marina Sinica, 28:141–149.
(14). Shen Jiwei, 1989. Digenetic trematodes of fishes from the East China Sea IV. Two new species of parasitic trematodes of eels. Studia Marina Sinica, 30:149–152.
(15). Shen Jiwei, 1989. Studies on the digenetic trematodes of fishes from Jiaozhou Bay. Studia Marina Sinica, 30:153–162.
(16). Shen Jiwei, 1990. Digenetic trematodes of fishes from the East China Sea V. Two new parasitic trematodes from scieaenid fishes. Studia Marina Sinica, 31:109- 114.

(17). Shen Jiwei, 1990. Digenetic trematodes of fishes from Changjiang River Estuary. Studia Marina Sinica, 31:115–120.
(18). Shen Jiwei, 1990. Description of three new species of Hemiuridae. Marine Science Bulletin, 9(5):53–57.
(19). Shen Jiwei, 1990. Didymozoidae trematodes from marine fishes offshore of China. Marine Science Bulletin, 9(3):46–54.
(20). Shen Jiwei, 1990. Description of four new species (Lepocreadiidae and Hemiuridae) and a list of digenetic trematodes of fishes from Yellow Sea. Marine Science Bulletin, 9(4):54–63.
(21). Shen Jiwei, 1990. Digenetic trematodes of marine fishes form Hainan Island. China Science Press, 228pp.
(22). Shen Jiwei and Tong Yungyung, 1984. Two new digenetic trematodes of marine fishes from Hainan Island. Acta Zoologica Sinica, 30(3):243–246.
(23). Shen Jiwei and Tong Yungyung, 1985. Two new species of digenetic trematodes from *Pomadasys hasta* (Bloch). Oceanologia & Limnologia Sinica, 16(6):514- 517.
(24). Shen Jiwei and Tong Yungyung, 1990. Studies on the digenetic trematodes of fishes from the Daya Bay. Acta Zootaxonomica Sinica, 15(4):385–392.
(25). Shen Jiwei and Tong Yungyung, 1990. A new species of *Aephnidionenes* Nicoll. (trematodes: Lepocreadiidae). Acta Zootaxonomica Sinica, 15(3):267–270.
(26). Shen Jiwei and Qiu Zhaozhi, 1995. Studies on the trematodes of fishes from Yellow Sea and Bohai Sea. China Science Press, 207pp.
(27). Qiu Zhaozhi, Zhang Lunshen, and Li, Qinkui, 1987. Digenetic tramatodes of fishes from Bohai Sea, China IV. A new species of Lepocreadiidae. Sichuan Journal of Zoology, 6(1):7–9.
(28). Li Qinkui, Zhang Lunshen, and Qiu Zhaozhi, 1986. Digenetic trematodes of fishes from Bohai Sea, China I. Two new species of Hemiuridae. Acta Zootaxonomica Sinica, 11(2):126–130.
(29). Li Qingkui, Zhang Lunshen, and Li Qinkui, 1986. Digenetic tramatodes of fishes from Bohai Sea, China II. Trematoda: Hemiuridae. Journal of Nankai University, 2:133–136.
(30). Li Qingkui, Zhang Lunshen, Qiu Zhaozhi, Chen Xixin, and Liang Zhong, 1987. Studies on the genus *Decemtestia* Yamaguchi, 1934. Contributions from Tianjin Natural History Museum, 4:5–8.
(31). Li Qingkui, Qiu Zhaozhi, and Zhang Lunshen, 1988. Digenetic trematodes of fishes from Bohai Sea, China V. Trematoda: Opecoelidae. Acta Zootaxonomica Sinica, 13(4):329–336.
(32). Li Qingkui, Qiu Zhaozhi, and Zhang Lunshen, 1989. Digenetic trematodes of fishes from Bohai Sea, China VI. Trematoda: Opecoelidae. Acta Zootaxonomica Sinica, 14(1):12–16.
(33). Li Qingkui, Zhang Lunshen, Qiu Zhaozhi, Chen Xixin, and Liang Zhong, 1989. Digenetic trematodes of fishes from Bohai Sea, China VII. A new species of Lecithastriidae. Marine Science Bulletin, 8(1): 119–120.
(34). Li Minmin, 1984. Parasites of the mullets *Mugil cephalus* (Linnaeus) and *Liza haematochela* (Temminch & Schlegel) in the area of Bohai Gulf I: Haigu area. Acta Zoologica Sinica, 30(2):153–158.
(35). Li Minmin, 1984. Parasites of the mullets *Mugil cephalus* (Linnaeus) and *Liza haematochela* (Temminch & Schlegel) in the area of Bohai Gulf II: Penglai area. Acta Zoologica Sinica, 30(3):231–242.
(36). Chen Xixin, Liang Zhong, Qiu Zhaozhi, and Li Qinkui, 1990. Studies on the subfamily Enodiotrematinae Baer 1924 (Plagiorchiidae, trematoda). Contributions from Tianjin Natural History Museum, 7:10–12.
(37). Zhang Runsheng, Qiu Zhaozhi, and Li Qinkui, 1986. Digenetic trematodes of fishes from Bohai Sea in China III. A new species of Zoogonidae. Acta Zootaxonomica Sinica, 11(4):348–350.
(38). Zhang Jianying and Ji Guoliang, 1981. Monogenetic trematodes of Chinese marine fishes of *Mugil cephalus* with description of a new species. Oceanologia & Limnologia Sinica, 12(4):249–355.
(39). Wang Puqin, 1980. Gastrostomatous trematodes from some fishes in Fujian Province. Acta Zootaxonomica Sinica, 5(4):330–336.
(40). Wang Puqin, 1982. Some digenetic trematodes of marine fishes form Fujian Province. Wuyi Science , 1(2):65–74.
(41). Wang Puqin, 1982. Hemiurid trematodes of marine fishes from Fujian Province. Oceanologia & Limnologia Sinica, 13(2):179–194.
(42). Wang Puqin, 1982. Hemiurid trematodes of marine fishes from Fujian Province. Journal of Fujian Normal University (Natural Science), 1982(2): 73–83.
(43). Wang Puqin, 1982. Some parasitic digenetic trematodes of vertebrate. Wuyi Science, 3:40–49.
(44). Wang Puqin, 1985. Notes on some species of Gasterostome Trematodes of fishes mainly from Fujian Province, China. Journal of Fujian Normal University (Natural Science), 4:73–83.
(45). Wang Puqin, 1987a. Digenetic trematodes of marine fishes in Pingtan County, Fujian Province, South China. Wuyi Science Journal, 7:151–163.
(46). Wang Puqin, 1987b. Four new species of digenetic trematodes in Amphibia and Reptilia. Acta Amphibia and Reptilia, 6(3):71–77.
(47). Wang, Puqin, 1991. One new genus and six new species of digenetic trematodes from Pingtan, Fujian Province. Wuyi Science Journal, 13:40–49.
(48). Wang Yangin, 1987. Digenetic trematodes of marine fishes from Fujian Province. Wuyi Science Journal, 7:165–179.
(49). Ku Changtung and Shen Jiwei, 1964. Two new record of trematodes in marine fishes. Journal of Nankai University (Natural Science), 5(1):69–73.
(50). Ku Changtong and Shen Jiwei, 1964. Studies on two species of hemiurid trematodes belonging to the genus *Hysterolecitha* Linton, 1910. Journal of Nankai University (Natural Science), 5(1):36–43.
(51). Ku Changtung and Shen Jiwei, 1964. Studies on new species of the genus *Prosorchis* Yamaguchi, 1934. Journal of Nankai University (Natural Science), 45–50.
(52). Ku Changtung and Shen Jiwei, 1965. Taxonomic study on the family Didymozoidae Poche, 1907. Journal of Nankai University (Natural Science), 6(1):21–48.
(53). Ku Changtung and Shen Jiwei, 1965. Studies on some digenetic trematodes from marine fishes in China. Acta Parasitologica Sinica, 2(4):355–365.
(54). Ku Changtung and Shen Jiwei, 1975. Studies on the genus *Phipidocotyle* Diesing (Bucephalidae trematodea) from some marine fishes of China. Acta Zoologica Sinica, 21(2):205–221.
(55). Gu Changdong and Shen Jiwei, 1976. Report on some Gastrostomatous trematodes (family Bucephalidae Poche, 1907) from marine fishes in Dong Hai and Nan Hai China. Acta Zoologica Sinica, 22(4):371–384.
(56). Gu Changdong and Shen Jiwei, 1978. Some digenetic trematodes from the sea perch *Lateolabrax japonicus*. Acat Zoologica Sinica, 24(2):170–178.
(57). Gu Changdong and Shen Jiwei, 1978. Some dinurid trematodes (subfamily Dinuridnae Looss, 1907) from marine fishes of economic importance of China. Acta Zoologica Sinica, 24(4):373–387.
(58). Gu Changdong and Shen Jiwei, 1979. Ten new species of digenetic trematodes of marine fishes. Acta Zootaxonomica Sinica, 4(4):342–355.
(59). Gu Changdong and Shen Jiwei, 1981. Digenetic trematodes of ribbonfish, *Trichurus haumela* (Forskål) and their distribution in the fishing groud of China Sea. Acta Zoologica Sinica, 27(1):53–63.
(60). Gu Changdong and Shen Jiwei, 1983. Four new species of didymozoid trematodes from China. Acta Zootaxonomica Sinica, 8(1):17–23.
(61). Gu Changdong and Shen Jiwei, 1983. Digenetic trematodes of fishes from the Xisha Islands, Guangdong Province, China. I. Studia Marina Sinica, 20: 157–184.
(62). Gu Changdong and Shen Jiwei, 1985. Digenetic trematodes of Hirudinellidae Dollfus, 1932 from the Xisha Islands. Studia Marina Sinica, 24:161–166.
(63). Lin Jinlan, 1981. Notes on some vermes and three new species in marine fishes from Fujian Province. Journal of Fujian Normal University (Natural Science), 1:113–118.
(64). Jin Daxion, 1974. One new genus and new species in digenetic trematodes, China. Acta Zoologica Sinica, 20(1):35–40.
(65). Jin Daxiong et al., 1979. A new species of *Phyllotrema* Yamaguchi, 1934 from marine fishes. Oceanogia & Limnologica Sinica, 10(1): 282–284.

(66). Tang Zhongzhang and Lin Xiuming, 1978. Three new species and one new genus of trematodes belonging to the family Haplosplanchnidae Poche, 1925. Acta Zoologica Sinica, 24(4): 203–211.
(67). Tang Zhongzhang and Tang Chongti, 1979. Investigation on trematodes belonging to the family Philophthalmidae Travassos, 1918. Journal of Xiamen University (Natural Science), 1979 (1):99–106.
(68). Tang Zhongzhang, 1981. Description of four species of trematodes hemiurid. Acta Zoologica Sinica, 27(3):254–264.
(69). Tang Chongti et al., 1975. Studies on the parasitic disease of Chinese Razor clam (*Sinonovacula constricta* Lamarck) in the Northern Estauary of Jiulong River, Fujian. Journal of Xiamen University (Natural Science), 1975(2):161–177.
(70). Tang Chongti et al., 1983. Studies on the trematodes of marine fishes from Fujian I. Hemiurid trematodes. Acta Zootaxonomica Sinica, 8(1):33–42.
(71). Tsin Swumei, 1933. Trematode parasites in the fishes. J. Sci. Nat. Univ. Shantung, Tsingtao, China, 1(2):379–392.
(72). Cao Hua, 1989. The life history of *Proctoeces orientalis* sp. nov. from marine bivalves. Acta Zoologica Sinica, 35(1):58–65.
(73). Pan Jinpei, 1984. A study of the family Cryptogonimidae (trematoda Digene) of China, with a description of one new genus and four new species. In: Parasitic Organisms of Fresh-water Fish of China. China Agricultural Publishing House, pp:115–124.
(74). Pan Jinpei, 1984. Two new genera and three new species of hemiurid and Maseniidae from Kwangtung fishes. China Agricultural Publishing House, pp:125–132.
(75). Pan Jinpei, 1984. Description of a new subfamily Pseudoisocoeliinae with a discussion on its classification. China Agricultural Publishing House, pp:133–138.
(76). Bychowsky and Nagibina, 1958. A new species in the family of the Monogenea- *Anchorophorus sinensis* n.g. n. sp. Acta Zoologica Sinica, 10(1):1–18.
(77). Bychowsky and Nagibina, 1959. New deputy of Monogenea from the South China Sea. Acta Zoologica Sinica, 11(2): 21–234.
(78). Mamter, H.W., 1940. Digenetic trematodes of fishes from the Galapagos Islands and neighbouring Pacific. Rep. Allan. Hancock Pacif. Exp., 2(14):35–479.
(79). Dzaki and Ishibashi, 1934. Notes on the cercaria of the pearl oyster. Proc. Jap. Acad. Tokyo, 10:439–442.
(80). Reid, Coil and Kuntz, 1966. Hemiuridae trematodes of Formosan marine fishes I. Subfamilies Dinurnae and Stomachicoline. J. Parasit., 52(1):39–45.
(81). Tang Chongti, 1990. Philophthalmid larval trematodes from Hong Kong and the coast of South China. In : Proceedings of the Second International Marine Biological Workshop: The Marine Flora and Fauna of Hong Kong and Southern China. Hong Kong University Press, pp:213–232.
(82). Tang Chongti, 1990. Further studies on some cercariae of molluscs collected from the shores of Hong Kong. In: Proceedings of the Second International Marine Biological Workshop: The Marine Flora and Fauna of Hong Kong and Southern China. Hong Kong University Press, pp:233–257.
(83). Tang Chongti, 1992. A new species of Cercarioidae (Trematoda: Heterophyidae) from Fujian with a discussion on its distribution. In: Proceedings of the Fourth International Marine Biological Workshop: The Marine Flora and Fauna of Hong Kong and Southern China. Hong Kong University Press, pp:29–35.
(84). Tang Chongti, 1992. Some larval trematodes form marine bivalves of Hong Kong and freshwater bivalves of China coast. In: Proceedings of the Fourth International Marine Biological Workshop. Hong Kong University Press, pp:17- 28.
(85). Yamaguchi, S. , 1938. Studies on the Helminth Fauna of Japan. Part 21. Trematodes of fishes IV. (Published by author), 55–57.
(86). Yamaguchi, S., 1952. Parasitic worms mainly from celebes I. New digenetic trematodes of fishes. Acta Med. Okayama, 8(2):146–198.
(87). Yamaguchi, S., 1970. Digenetic trematodes of Hawaiian fishes. Keigaku Publishing Co. Tokyo, 9–246.

*: (1)-(76) in Chinese

Compiled by Shen Jiwei; Addition and edited by Tang Zhongzhang, Tang Chongti, and Wang Puqin.

CESTOIDEA

PSEUDOPHYLLIDEA
BOTHRIOCEPHALIDAE

Bothriocephalus japonicus Yamaguti: Fujian. Host: Japanese eel (*Anguilla japonica*). (2)

B. scorpi (Müller): Qingdao (Shandong); Fujian. Host: temperate bass (*Lateolabrax japonicus*). (2, 3)

Oncodiscus sauridae Yamaguti: East China Sea. Host: lizardfish (*Saurida*). (2)

Taphrobothrium japonense Luhe: Fujian. Host: eel (*Anguilla*). (2)

TETRAPHYLLIDEA
PHYLLOBOTHRIIDAE

Anthobothrium bifidum Yamaguti: Qingdao (Shandong). Host: stingray (*Dasyatis zugei*). (2)

A. parvum Yamaguti: Pingtan (Fujian). Host: houndshark (*Mustelus manazo*). (2)

Dinobothrium septaria Beneden: Qingdao (Shandong). Host: basking shark (*Cetorhinus maximus*). (2)

Echeneibothrium hui Tseng: Qingdao (Shandong). Host: stingray (*Dasyatis akajei*). (2, 3)

E. variabile Beneden: Qingdao (Shandong). Host: skate (*Raja porosa*). (2)

Phyllobothrium laciniatum (Linton): East China Sea. Host: dogfish shark (*Squalus*). (2)

P. lactuca Beneden: Qingdao (Shandong). Host: dogfish shark (*Squalus*). (4)

P. loculatum Yamaguti: East China Sea. Host: bullhead shark (*Heterodontus zebra*). (4)

P. ptychocephalum Wang: Pingtan (Fujian). Host: stingray (*Dasyatis kuhli*). (2)

P. tumidum Linton: Qingdao (Shandong). Host: houndsharks (*Mustelus* and *Triakis*) and stingray (*Dasyatis akajei*). (2)

Pithophorus musculosus Subhaparadha: Pingtan (Fujian). Host: houndshark (*Mustelus manazo*). (2)

ONCOBOTHRIIDAE

Acanthobothrium benedeni Loennberg: Qingdao (Shandong). Host: guitarfish (*Rhinobatos*). (2)

A. coronatum (Rudolphi): Qingdao (Shandong); Pingtan (Fujian). Host: stingray (*Dasyatis akajei*). (4)

A. grandiceps Yamaguti: East China Sea. Host: stingray (*Dasyatis akajei*). (4)

A. ijimai Yoshida: East China Sea. Host: stingray (*Dasyatis akajei*). (2)

A. micracantha Yamaguti: East China Sea. Host: stingrays (*Dasyatis akajei* and *D. zugei*). (4)

A. pingtanensis Wang: Pingtan (Fujian). Host: stingray (*Dasyatis kuhli*). (2)

A. tsingtaoense Tseng: Qingdao (Shandong). Host: stingray (*Dasyatis akajei*). (3)

LECANICEPHALIDAE

Cephalobothrium longisegmentum Wang: Lianjiang (Fujian). Host: stingray (*Dasyatis kuhli*). (2)

TRYPANORHYNCHA
OTOBOTHRIIDAE

Otobothrium linstowi Southwell: Qingdao (Shandong). Host: cow shark (*Notorynchus platycephalus*). (3)

Poecilancistrum sp. blastocyst: Penglai (Shandong). Host: striped mullet (*Mugil cephalus*). (1)

REFERENCES*

(1). Li Minmin, 1984. Parasites of mullets *Mugil cephalus* (Linnaeus) and *Lita haematocheila* (Temminck & Schlegel) in the areas of Bohai Gulf, II. Penglai Area. Acta Zoologica Sinica, 30(3): 231–242

(2). Wang Puqin, 1984. Lists of Cestoidea from the fishes of Fujian, China. Wuyi Science, 4:71–83.

(3). Tseng Shen, 1933. Study on some cestodes from fishes. J. Sci. Nat. Univ. Shantung, China, 2:1–21

(4). Yamaguti, 1952. Studies on the helminth fauna of Japan. Part 49. Cestodes of fishes II. Acta Med. Okayama, 8(1):1–78.

*: (1) and (2) in Chinsese

Compiled by Huang Zongguo and Shen Jiwei; Edited by Wang Puqin

NEMERTINEA

ANOPLA
ARCHINEMERTEA
CEPHALATHRICIDAE

Cephalathrix linearis Rathke: Qingdao, Yantai, and Rushan (Shandong). Intertidal. (1)

Cephalotrichella alba Gibson & Sundberg: Dapeng Cove (Hong Kong). Shallow sea. (9)

Procephalathrix arenarius Gibson: Dapeng Cove (Hong Kong). Mid-intertidal to subtidal 12 m. (8)

P. orientalis Gibson: Shenzhen (Guangdong). Mangrove forest muddy substrate. (8)

P. spiralis Coe: Qingdao and Yantai (Shandong). (1)

PALAEONEMERTEA
TUBULANIDAE

Carinesta tubulanoides Gibson: Hong Kong. Low intertidal. (8)

Tubulanus punctatus (Takakura): Qingdao (Shandong). (1)

HETERONEMERTEA
CEREBRATULIDAE

Cerebratulina communis Takakura: Qingdao (Shandong). Intertidal, within mud and sand. (7)

C. darvelli Gibson: Dapeng Cove (Hong Kong). Water depth 15 m. (8)

C. natans Punnet: Shenzhen Bay (Guangdong). Mangrove forest muddy substrate and oyster farm. (8)

C. nigra (Stimpson): Hong Kong. (12)

C. rubella (Stimpson): Hong Kong. (12)

C. sinensis (Stimpson): Hong Kong. (12)

Iwatanemertes piperatus (Stimpson): Hong Kong. In intertidal sedentary oranisms. (12)

Perissonemertes pyrrhocephalus Gibson: Hong Kong. Water depth 5–6m. (8)

Quasilineus sinicus Gibson: Hong Kong. Mid-intertidal. (8)

LINEIDAE

Eousia verticivarius Gibson: Dapeng Cove (Hong Kong). Low-intertidal to 10 m water depth. (8)

Lineopselloides albilineus Gibson: Dapeng Cove (Hong Kong). Water depth 5–10 m. (8)

Lineus alborostratus Takakura: Qingdao and Yantai (Shandong). Mid- and low- intertidal. (1)

L. bilineatus (Renier): Dalian (Liaoning). (5)

L. binigrilinearis Gibson: Hong Kong. Low-intertidal. (8)

L. cingulatus (Stimpson): Hong Kong. (12)

L. fuscovirids Takakura: Dalian (Liaoning); Qingdao (Shandong). (5, 7)

L. geniculatus (Delle Chiaje): Qingdao (Shandong). Low-intertidal. (1)

L. torquatus Coe: Northern China. Under rocks in intertidal area. (6)

L. vegetus Coe: Qingdao (Shandong). (1)

Micrura verrilli Coe: Qingdao (Shandong). Under rocks or on seaweed roots in intertidal area. (1)

Paramicrura borborophila Gibson & Sundberg: Hong Kong. Mangrove forest muddy substrate. (9)

Utolineus uberis Gibson: Dalian (Liaoning); Hong Kong. Mid-intertidal mangrove forest. (5, 8)

POLYBRACHIORHYNCHIDAE

Dendrorhynchus sinensis Yin & Zeng: Zhanjiang (Guangdong); Hong Kong. (2)

D. zhanjiangensis Yin & Zeng: Zhanjiang (Guangdong). (2)
Polydendrorhynchus papillaris Yin & Zeng: Zhanjiang (Guangdong). (4)

BASEODISCIDAE

Baseodiscus curtus Hubrecht: Dalian (Liaoning). Seaweed or under rocks in intertidal area. (6)

ENOPLA
HOPLONEMERTEA
MONOSTILIFERA
EMPLECTONEMERIIDAE

Cephalonema brunniceps Stimpson: Hong Kong. (13)
Emplectonema gracile Johnston: Northern China. Intertidal. (1)
E. mitsuii Yanaoka: Dalian (Liaoning). (5)
Nemertopsis gracilis Coe: Hong Kong. (11)
N. quadripunctatus Quoy & Gaimard: Hong Kong. (8)
N. tetraclitophila Gibson: Hong Kong. (8)
Paranemertes katoi Yamaoka: Dalian (Liaoning). (5)
P. peregrina Coe: Dalian (Liaoning); Qingdao and Yantai (Shandong). Intertidal. (1, 5)

PROSORHOCHMIDAE

Eonemertes macrophthalma Gibson: Hong Kong. Under sand or rocks in low intertidal area. (8)
Pantinonemertes daguilarensis Gibson & Sundberg: Hong Kong. Low-intertidal. (9)
P. mortoni Gibson: Hong Kong. Inside volcano barnacle (*Tetraclita*). (8)
P. spectaculum (Yamaoka): Hong Kong. High intertidal rocky crevices. (8)

AMPHIPORIDAE

Amphiporus punctatulus Coe: Dalian (Liaoning); Qingdao and Yantai (Shandong). Under rocks and on seaweed in intertidal area. (8)
A. sinuosum (Stimpson): Hong Kong. (13)
Zygonemertes grandulosa Yamaoka: Dalian (Liaoning). (5)

TETRASTEMMIDAE

Tetrastemma nigrifrons Coe: Yantai (Shandong). (11)
T. verinigrum Iwata: Hong Kong. Subtidal 1.5 m water depth. (8)

PLECTONEMERTIDAE

Plectonemertes sinensis Gibson: Dapeng Cove (Hong Kong). Water depth 21 m. (8)

CARCINONEMERTIDAE

Carcinonemertes mitsukurii Takakura: Hong Kong. (10)

REFERENCES*

(1). Yin Zuofen, Si Jihua and Li Nuo, 1986. Preliminary survey of Nemertiean from Shandong coast. Marine Science Bulletin, 6(1):67–71.
(2). Yin Zuofen and Zeng Fen, 1984. The study of a new species of genus *Dendrorhynchus* - *Dendrohynchus zhanjianensis*, of Lineid heterone-mertean, possessing multibranched probosics. Marine Science Bulletin, 3(6):51–68
(3). Yin Zuofen and Zeng Fen, 1985. On *Dendrorhynchus sinensis* Gen. & sp. Nov. of Lineid heteronemertean. Oceanologia & Limnologia Sinica, 16(4):323–335.
(4). Yin Zuofen and Zeng Fen, 1986. The study of a new genus and species - *Polydendrorhynchus papillaris* of Lineid heteronemertean possessing multibranched probosics. Journal of Shandong College of Oceanology,16(4):323- 335.
(5). Song Pengdong et al., 1985. Guidebook for Practice in Invertebrate from Dalian Coast. China Higher Education Press.
(6). Zhao Ruyi et al., 1990. Guidebook on collection of invertebrate from China coast (in North China). China Ocean Press, 393pp.
(7). Chen Xintao et al., 1963. A Collection of Illustrative Plates of Animals in China, Platyhelminthes (Nemertiea suppl.). China Science Press.
(8). Gibson, R. 1986. The macrobenthic nemeatean fauna of Hong Kong. Proceedings of the Second International Marine Biological Workshop: The Marine Flora and Fauna of Hong Kong and Southern China (Ed. B. Morton), Hong Kong University Press.
(9). Gibson, R. and P. Sundberg, 1992. Three new nemerteans from Hong Kong. Proceedings of the Fourth International Marine Biological Workshop: The Marine Flora and Fauna of Hong Kong and Southern China. Hong Kong University Press, pp:97–129.
(10). Humes, A. G. 1942. The morphology, taxonomy, and bionomics of nemertean genus *Carcinonemertes*. Illinois Biological Monographs 18:1–105.
(11). Morton, B. and J. Morton, 1983. The Sea Shore Ecology of Hong Kong. Hong Kong University Press.
(12). Stimpson, W. 1855. Descriptions of some of the new marine invertebrata from the Chinese and Japanese seas. Proceedings of the Academy of Natural Sciences of Philadelphia, 7:375–384
(13). Stimpson, W. 1857. Prodromus descriptionis animalium evertebratorum quae in Expeditione and Oceanum Pacificum Septemtrionalem a Republica Federata missa. Pars II. Turbellarieorum Nemertineorum. Proceedings of the Academy of Natural Sciences of Philadelphia, 1857, pp:159–165.

*: (1)-(7) in Chinsese
Compiled by Sun Shichun and Pan Haihong; Edited by Shi Jihua and Huang Zongguo

KINORHYNCHA

CYCLORHAGIDA
ECHINODERIDAE

Centroderes sp.: East China Sea continental shelf. (2)
Condyloderes sp.: East China Sea continental shelf. (2)
Echinoderes tchefouensis Lou: Dalian (Liaoning); Yantai and Qingdao (Shandong). (1, 3)
Echinoderes sp. 1: East China Sea continental shelf and sea bottom at the mouth of Changjiang (=Yangtze River). (2)

Echinoderes sp. 3: East China Sea continental shelf. (2)
Echinoderes sp. 4: East China Sea continental shelf. (2)
Echinoderes sp. 5: East China Sea continental shelf. (2)
Kinorhyncha sp.: The mouth of Changjiang (=Yangtze River). (2)
Pycnophyes sp.: Sea bottom at the mouth of Changjiang (=Yangtze River). (2)
Semnoderes sp.: East China Sea continental shelf. (2)

REFERENCES*

(1). Zhang Zhinan, 1992. Kinorhynchia in Yellow Sea and Bohai Sea. (manuscript)
(2). Fleeger, J. W., Tang Zhican and R. P. Higgins. 1986. Preliminary study of the ecology of meiobenthic copepoda and Kinorhyncha in the Changjiang Estuary and adjacent waters. Studia Marina Sinica, 27:19–208
(3). Lou, T. H. 1934. Sur la presence dun nouveau Kinorhynque a Tchefou: *Echinoderes tchefouensis* sp. nov. contributims du Laboratoire de zoologie, Academie Nationale de Peiping, 1(4):1–9, pl. I.

*: (1) and (2) in Chinese
Compiled by Zhang Zhinan and Huang Zongguo; Edited by Yang Dejian

NEMATODA

(PARASITIC)

ASCARIDIDEA
GOEZIIDAE

Goezia gobia Wang: Fuzhou (Fujian). Host: temperate bass (*Lateolabrax japonicus*) and goby (*Synechogobius ommaturus*). (7)
G. parvus Wang & Wu: Fujian. Host: temperate bass (*Lateolabrax japonicus*). (17)

ANISAKIDAE

Anisakis alexandri Hsu & Hoeppli: Fujian. Host: dolphins. (21)
A. simplex Rudolphi: Hebei. Host: minke whale (*Balaenoptera acutorostrata*). (17)
Contracaecum amoyensis Hsü: Dongshan, Longhai, and Fuzhou (Fujian). Host: pike conger (*Muraenesox yamaguichiensis*) and drum (*Miichthys miiuy*). (6, 10, 11, 12)
C. clavatum (Rudolphi) Skrjabin: Huian, Longhai, and Dongshan (Fujian). Host: grunt (*Pomadasys kaakan*). (4)
C. marinum (Linnaeus): Jiangsu. Host: goosefish (*Lophius litulon*). (17)
C. paralichthydis Yamaguti: Qingdao (Shandong). Host: righteye flounder (*Pseudopleuronectes herzensteini*) and tonguefish (*Cynoglossus*). (20, 21)
C. zenopsis Yamaguti: Baimajing (Hainan). Host: armored searobin (*Peristedion orientale*). (1)
Controcaecum sp.: Xiamen (Fujian). Host: bigeye snapper (*Priacanthus macracanthus*), drum (*Johnius*), and porgy (*Pagrosomus major*). (21, 22)
Phocascaris longispiculum: Wang & Wu: Xiamen (Fujian). Host: cobia (*Rachycentron canadum*). (17).
Raphidascaris acentrogobii Wang & Wu: Ningde (Fujian). Host: goby (*Acentrogobius viridipunctatus*). (17)
R. lophii Wang & Wu: Pingtan and Xiamen (Fujian). Host: goosefish (Lophidae). (17)
Raphidascaroides halicutaeae Yin: Sanya (Hainan). Host: batfish (*Halieutaea stellata*). (2)
R. myliobatum Yin & Zhang: Dongshan and Longhai (Fujian). Host: eagle ray (*Myliobatis tobijei*). (4)

QUIMPERIDAE

Neoomeia sphyrna Yin & Zhang: Jinjiang (Fujian); Hainan. Host: hammerhead shark (*Sphyrna*). (5)

HETEROCHEILIDAE

Cloeoascaris trichiuri Yin & Zhang: Zhoushan (Zhejiang). Host: cutlassfish (*Trichiurus haumela*). (3)
Filocapsularis alexandri (Hsu & Hoeppli): Fujian. Host: dolphins. (16)
Ichthyascaris lophii Wu: Shanghai. Host: goosefish (*Lophius litulon*). (21, 25)

SPIRURIDEA
CAMALLANIDAE

Camallanus marinus Schmidt & Kuntz: Sanya (Hainan). Host: Spanish mackerel (*Scombermorus commersoni*). (2)
C. mustelus Wang: Pingtan (Fujian). Host: houndshark (*Mustelus*). (12, 16)
Spirocamallanus pereirai (Annereaux) Olsen: Sanya (Hainan). Host: smelt-whitings (*Sillago maculata* and *S. sihama*). (2)

CUCULLANIDAE

Cucullanus amadai Yamaguti: Baimajing (Hainan). Host: snapper (*Lutjanus erythopterus*). (1)
C. anguillae Wang & Ling: Fujian. Host: Japanese eel (*Anguilla japonica*). (16)
C. cirratus Müller: Sanya (Hainan). Host: snake eel (*Pisodonophis*). (2)
C. spirocaudus Li: Penglai (Shandong). Host: mullet (*Liza haematocheila*). (6)
Dichelyne longispiculus Wang & Ling: Zhoushan (Zhejiang); Fujian. Host: gizzard shad (*Clupanodon punctatus*). (16)
Indocucullanus muraenecis Yin & Zhang: Zhoushan (Zhejiang); Fujian. Host: pike conger (*Muraenesox*). (4, 10)

PHYSALOPTERIDAE

Heliconema brevispiculum Baylis: Fujian. Host: Japanese eel (*Anguilla japonica*). (11, 16)

H. heliconema Travassos: Fujian. Host: Japanese eel (*Anguilla japonica*). (11, 16)

H. longissima (Ortlepp): Fujian. Host: Japanese eel (*Anguilla japonica*). (11, 16)

Paraleptus chiloscyllii Yin & Zhang: Dongshan and Longhai (Fujian). Host: whitespotted bambooshark (*Chiloscyllium plagiosum*). (4)

P. syclii Wu: Xiamen (Fujian). Host: whitespotted bamboo shark (*Chiloscyllium plagiosum*). (21, 24)

RHABDOCHONIDAE

Rhabdochona minjiangensis Wang: Fujian. Host: Japanese eel (*Anguilla japonica*). (9)

R. synechogobii Wang: Fujian. Host: goby (*Synechogobius ommaturus*). (7)

SPIRURIDAE

Ascarophis epinepheli Wang: Pingtan (Fujian). Host: grouper (*Epinephelus awoara*). (15)

GNATHOSTOMATIDAE

Echinocephalus pleroplatae Wang: Fujian. Host: Japanese butterfly ray (*Gymnura japonica*). (11, 16)

PHILOMETRIDEA
ANGUILLICOIDAE

Anguillicola crassa Kuwahara, Niimi & Itagal: Fujian. Host: Japanese eel (*Anguilla japonica*). (13, 16).

A. globiceps Yamaguti: Fujian. Host: Japanese eel (*Anguilla japonica*). (13, 16).

Capillaria fudingensis Wang: Fuding (Fujian). Host: butterfish (*Pampus argenteus*). (14)

OXYURIDEA
KATHANIDAE

Kathlania leptura (Rudolphi): Fujian. Host: sea turtles. (16, 18)

REFERENCES*

(1). Yin Wenzhen and Zhang Naixin, 1982. New record of parasitic nematodes from marine fishes in China. Acta Zootaxonomica Sinica, 7(4):371.
(2). Yin Wenzhen, 1983. On some mematodes from marine fishes in Hainan Island, China. Acta Zootaxonomica Sinica, 8(3):225–228.
(3). Yin Wenzhen and Zhang Naixin, 1983a. On two new species of parasitic nematodes from marine fishes from Zhoushan Islands, China. Acta Zootaxonomica Sinica, 8(1):7–10.
(4). Yin Wenzhen and Zhang Naixin, 1983b. Notes on some nematodes of marine fishes from Fujian, China. Acta Zootaxonomica Sinica, 8(1):11–16.
(5). Yin Wenzhen and Zhang Naixin, 1984. A new genus and species of Quimperidae (Nematoda). Acta Zootaxonomica Sinica, 9(4):347–350.
(6). Li Minmin, 1984. Parasites of the mullets *Mugil cephalus* (Linnaeus) and *Liza hamatocheila* (Temminck & Schlegel) in the areas of Bohai Gulf II: Penglai area. Acta Zoologica Sinica, 30(3):231–242.
(7). Wang Fuqin, 1965. Notes on some Nematodes of the suborder Ascaridata from Fukien, China. Acta Parasitologica Sinica, 2(4):366–379.
(8). Wang Puqian, 1975. Some nematodes, Camallanidae from Fujian Province, China. Acta Zoologica Sinica, 21(4):350–358.
(9). Wang Puqian, 1976. Notes on new species of nematodes (Spiruridea) from Fujian Province. Acta Zoologica Sinica, 23(1):88–96.
(10). Wang Puqin, 1978. Some nematodes in vertebrates from South China. Journal of Fujian Normal University, 2:75–95.
(11). Wang, Puqin, 1979a. A preliminary survey on fish parasites from Fujian Province. Journal of Fujian Normal University (Natural Science), 2:53–63.
(12). Wang, Puqin, 1979b. The parasitic nematodes from southeastern China. Journal of Fujian Normal University, 2:78–92.
(13). Wang Puqin, 1980. Studies on life history of *Anguillicola globiceps* (Nematoda). Acta Zoologica Sinica, 26(3):243
(14). Wang Puqin, 1982. Notes on Anisakidae (Nematoda) from Fujian Province. Acta Zootaxonomica Sinica, 7(2):117–126.
(15). Wang Puqin, 1984a. Some fish parasites from Fujian Province. Acta Zootaxonomica Sinica, 9(3):228–237.
(16). Wang Puqin, 1984b. Descriptions of three new species and a list of parasitic nematodes from vertebrates in Fujian Province. Wuyi Science Journal, 4:113- 132.
(17). Wang Yangin, 1991. Some parasitic nematodes and four new species in the sea turtles from Fujian Province. Journal of Fujian Normal University (Natural Science), 7(1):92–97.
(18). Lin Jinlan, 1981. Notes on some parasitic vermes and three new species in the sea turtles from Fujian Province. Journal of Fujian Normal University (Natural Science), 11:113–118.
(19). Zhang Haixin et al., 1983. The discovery of *Contracaecum marinum* (Linnaeus) in China. Acta Zootaxonomica Sinica, 8(2):119.
(20). Xu Huinan, 1957. The discovery of *Contracaecum paralichthydis* in China . Chinese Journal of Zoology, 1(2):118–119.
(21). Xu Huinan, 1975. The parasites Nematology of vertebrates. China Science Press.
(22). Hsu, H. F., 1933–1934. On some parasitic nematodes collected in Amoy. Pek. Nat. Hist. Bull., 8(2):155–168.
(23). Hsu, H. F., 1934. *Contracaecum amoyense*, correction of the specific name. Addenda to paper on some species of parasitic nematodes from fishes in China. Pek. Nat. Hist. Bull., 8(3):295.
(24). Wu, H. W., 1927. A new nematodes from the stomach of a scylloid shark. Contri. Hab. Sci. Soc. China, 3(2):1–3.
(25). Wu, H. W., 1949. A note on two parasitic nematodes of fishes. Sinensis, 20 (1–6):51–57.

*: (1)-(21) in Chinese
Compiled by Shen Jiwei; Edited by Wang Puqin

(NON-PARASITIC)

PHASMIDEA
RHABDITIDA
RHABDITIDAE

Rhabditis marina Bastian: Qingdao (Shandong); Yellow Sea. (5)

APHASMIDEA
ARAEOLAIMIDA
AXONOLAIMIDAE

Campylaimus gerlachi Timm: Bohai. (5)
Odontophora axonolaimoides Timm: Qingdao (Shandong); Yellow Sea. (5)
O. littoralis Zhang: Dalian Bight (Liaoning). (5)
O. paraxonolaimoides Zhang: Bohai. (5)
Paradontophora bohaiensis Zhang: Bohai. (5)
P. cobbi (Timm): Bohai. (5)
P. marina Zhang: Bohai. (3)

LEPTOLAIMIDAE

Leptolaimus venustus Lorenzen: Qingdao (Shandong). (5)

DESMOSCOLECIDA
DESMOSCOLECIDAE

Desmoscolex americanus Chitwood: Bohai. (5)

MONHYSTERIDA
SIPHONOLAIMIDAE

Siphonolaimus longicaudata Zhang: Bohai. (5)

LINHOMOEIDAE

Desmolaimus zealandicus De Man: Qingdao (Shandong); Yellow Sea. (5)
Eleutherolaimus stenosoma (De Man): Bohai. (5)
Paralinhomoeus attenuatus De Man: Jiaozhou Bay (Shandong); Yellow Sea. (5)
Terschellingia longicaudata De Man: Bohai. (5)
T. longissimicaudata Timm: Bohai. (5)

MONHYSTERIDAE

Paramonhystera pellucida (Cobb): Qingdao (Shandong); Yellow Sea. (5)
Steineria pulchra Mawson: Bohai. (5)
Theristus littoralis Filipjev: Dalian Bight (Liaoning). (5)
T. macroflevensis Gerlach: Daya Bay (Guangdong). (5)
T. metaflevensis Gerlach: Qingdao (Shandong); Yellow Sea. (5)

XYALIDAE

Daptonema alternum (Wiesert): Dalian Bight (Liaoning). (5)
D. fissidens Cobb: Bohai. (5)
D. fistulatum (Wieser & Hopper): Qingdao (Shandong); Yellow Sea. (5)
D. maeoticum (Filipjev): Wuleidao Bay (Shandong); Yellow Sea. (5)
Gnomoxyala breviseta Zhang: Bohai. (5)
Gonionchus metavillosus Zhang: Dalian Bight (Liaoning). (5)

SPHAEROLAIMIDAE

Sphaerolaimus dispar Filipjev: Bohai. (5)
S. pacifica Allgen: Bohai. (5)

DESMODORIDA
DESMODORIDAE

Bolbolaimus denticulatus Cobb: Dalian Bight (Liaoning). (5)

CERAMONEMATIDAE

Metadasynemella cassidiniensis Vitiello & Haspeslagh: Bohai. (5)

MONOPOSTHIIDAE

Monoposthia costata (Bastian): Outside Jiaozhou Bay (Shandong); Yellow Sea. (5)

CHROMADORIDA
CYATHOLAIMIDAE

Acanthonchus (Seuratiella) tridentatus Kito: Dalian Bight (Liaoning); Jiaozhou Bay (Shandong). (5)
Marilynia macrodentatus (Wieser): Bohai. (5)
Paracanthonchus macrodon (Ditlevsen): Qingdao (Shandong); Yellow Sea. (5)

CHROMADORIDAE

Actinonema pachydermatum Cobb: Bohai. (5)
Chromadora quadrilinea Chitwlld & Chitwood: Dalian Bight (Liaoning). (5)
Chromadorina miro (Cobb): Qingdao (Shandong); Yellow Sea. (5)
Graphonema amokurae (Ditlevsen): Jiaozhou Bay (Shandong); Yellow Sea. (5)
Prochromadora orleji (De Man): Jiaozhou Bay (Shandong); Yellow Sea. (5)
Spilophorella paradaxa (De Man): Bohai. (5)
Steineridora borealis Ktio: Dalian Bight (Liaoning); Jiaozhou Bay (Shandong). (5)

COMESOMATIDAE

Cervonema delta Zhang & Hope: Bohai. (5)
Dorylaimopsis rabalaisi Zhang: Bohai. (5)

D. turneri Zhang: Bohai. (4, 5)
Hopperia hexadentata Zhang & Hope: Bohai. (5)
Paracomesoma heterosetosum Zhang: Bohai; Dalian (Liaoning). Sandy beach. (3, 5)
Paradoralaimopsis bohaiensis Zhang: Bohai. (5)
Sabatieria alata Warwick: Wuleidao Bay (Shandong); Yellow Sea. (5)
S. ancudiana Wieser: Bohai. (5)
S. parabyssalis Wieser: Bohai. (5)
S. pulchra (Schnerider): Qingdao (Shandong), Yellow Sea. (5)
S. punctata (Kreis): Qingdao (Shandong), Yellow Sea. (5)
S. trivialis Tchesunov: Wuleidao Bay (Shandong), Yellow Sea. (5)
Setosabatieria fibulata (Wieser): Bohai. (5)
S. hilarula (De Man): Bohai. (5)

CHOANOLAIMIDAE

Halichoanolaimus duodecimpapillatus: Qingdao (Shandong). (5)
H. wulaidaoensis Zhang: Wuleidao Bay (Shandong); Yellow Sea. (5)

ENOPLIDA
IRONIDAE

Thalassironus bohaiensis Zhang: Bohai. (2)

TRIPYLOIDIDAE

Bathylaimus stenolaimus Schuurmans Stekhoven & de Coninck: Dalian (Liaoning). Sandy intertidal. (5)

OXYSTOMINIDAE

Halalaimus alatus Timm: Bohai. (5)
H. gracilis De Man: Bohai. (5)
H. inflata Zhang: Bohai. (5)
Oxystomina delta Zhang: Bohai. (5)
O. miranda Wieser: Bohai. (5)

LEPTOSOMATIDAE

Tharacostoma setosum (Linstow): Jiaozhou Bay (Shandong); Yellow Sea. (5)

ANTICOMIDAE

Paranticoma longicaudata Chitwood: Bohai. (5)

PHANODERMATIDAE

Phanoderma cocksi Bastian: Jiaozhou Bay (Shandong); Yellow Sea. (5)

P. daiyawanensis Zhang: Daya Bay (Guangdong); South China Sea. (5)
P. macrophallum Steiner: Jiaozhou Bay (Shandong); Yellow Sea. (5)

ENOPLIDAE

Enoplus communis Bastian: Dalian Bight (Liaoning); Jiaozhou Bay (Shandong). (5)
Enoploides delamarei Guy Boucher: Daya Bay (Guangdong); South China Sea. (5)
Mesacanthion littoralis Zhang: Daya Bay (Guangdong); South China Sea. (5)

ANOPLOSTOMATIDAE

Anoplostoma copano Chitwood: Wuleidao Bay (Shandong); Yellow Sea. (5)
A. exceptum Schulz: Bohai. (5)

ONCHOLAIMIDAE

Meoncholaimus moles Zhang & Platt: Qingdao (Shandong); Yellow Sea. (1)
Oncholaimus oxyuris Ditlevsen: Wuleidao Bay (Shandong); Yellow Sea. (5)
O. qingdaoensis Zhang & Platt: Yellow Sea. (1)
O. sinensis Zhang & Platt: Qingdao (Shandong); Yellow Sea. (1)

ENCHELIDIIDAE

Belbolla bohaiensis Zhang: Bohai. (5)
Eurystomina ophthalmophora (Steiner): Dalian Bight (Liaoning); Jiaozhou Bay (Shandong). (5)

REFERENCES*

(1). Zhang, Z. N. and H. M. Platt, 1983. New species of marine nematodes from Qingdao, China. Bull. Br. Mus. nat. Hist. (Zool) 45(5):253–261.
(2). Zhang, Z. N. 1990. A new species of the genus *Thalassironus* De Man 1889 (Nematoda, Adenophora, Ironidae). J. Ocean. Univ. Qingdao 20(3):103–108.
(3). Zhang, Z. N. 1991. Two new species of marine nematodes from the Bohai Sea, China. J. Ocean. Univ. Qingdao 21(2):49–60.
(4). Zhang, Z. N. 1992. Two new species of the genus *Dorylaimopsis* Ditlevsen, 1918 (Nematoda, Adenophora, Comesomatidae) from the Bohai Sea. Chin. J. Ocean. Limnol. 10 (1):31–39.
(5). Zhang Zhinan, 1992. Marine Nematode of China (I). (manuscript)

*: (1)-(4) in Chinese
Compiled by Zhang Zhinan

ACANTHOCEPHALA

EACANTHOCEPHALA
NEOECHINORHYNCHIDA
NEOECHINORHYNCHIDAE

Eocollis harengulae Wang: Huian (Fujian). Host: herring (*Herklotsichthys*). (1)

Neoechinorhynchus agilis Rudolphi: Bohai. Host: mullet (*Osteomugil*). (1, 2)

N. coiliae Yamaguti: Pingtan (Fujian). Host: small intestine of anchovies (*Coilia grayi* and *C. mystus*) and herring (*Ilisha elongata*). (1)

N. johnii Yamaguti: East China Sea; Pingtan (Fujian). Host: drums (*Johnius* and *Nibea semifasciata*). (1)

N. topseyi Podder: Pingtan (Fujian). Host: threadfin (Polynemidae), round herring (*Dussumieria hasselti*), etc. (1)

N. tylosuri Yamaguti: Pingtan (Fujian). Host: needlefish (*Tylosurus strongylurus*), striped mullet (*Mugil cephalus*), etc. (1)

GYRACANTHOCEPHALA
QUADRIGYRIDAE

Acanthosentis lizae Wang: Pingtan (Fujian). Host: mullet (*Liza carinatus*). (1)

A. periophthalmi Wang: Fuding (Fujian). Host: mudskippers (Periophthalmidae). (1)

A. scomberomori Wang: Pingtan (Fujian). Host: Spanish mackerel (*Scombermorus niphonius*). (1)

A. similis Wang: Fuzhou (Fujian). Host: anchovy (*Coilia*). (1)

Quadrigyrus polyspinosus Li: Bohai. Host: mullets (*Liza carinatus*, *Mugil cephalus*, and *Osteomugil*). (2)

PALAEACANTHOCEPHALA
ECHINORHYNCHIDEA
ECHINORHYNCHIDAE

Acanthocephalus luzus Li: Penglai (Shandong). Host: mullet (*Liza haematocheila*). (3)

Echinorhynchus gadi Zoega in Müller: Seas and river mouth in northern China. Host: cods (Gadidae), salmon (*Oncorhynchus keta*), and many other fish. (1)

E. salmonis Müller [*E. manaenae*]: Northern China; Pingtan (Fujian). Host: salmon and many other fish. (1)

CAVISOMATIDAE

Filisoma atropi Wang & Wang: Pingtan (Fujian). Host: jack (*Atropus atropus*). (1)

F. indicum Van Cleave: Pingtan (Fujian). Host: lizardfish (*Harpodon nehereus*) and threadfin bream (*Nemipterus virgatus*). (1)

F. oplegnathi Wang & Wang: Pingtan (Fujian). Host: knifejaw (*Oplegnathus fasciatus*). (1)

Neorhadinorhynchus nudum (Harada): Taiwan. Host: jack (*Trachurus japonicus*). (1)

POLYMORPHIDA
POLYMORPHIDAE

Arhythmorhynchus frassoni (Molin): Pingtan (Fujian). Host: sea gulls and many other sea birds. (1)

A. teres Van Cleave: Fuzhou (Fujian). Host: sea gulls and many other sea birds. (1)

Bolbosoma scomberomuri Wang: Pingtan (Fujian). Host: cutlassfish (*Trichiurus haumela*) and many other fish. (1)

B. vasculosum (Rudolphi): Taiwan. Host: bluefin tuna (*Thunnus thynnus*). (1)

Polymorphus minutus (Goeze): Fujian; Taiwan. Host: gull and many other waterfowl; larvae in silversides (Atherinidae), gobies (Gobiidae) and many other fish. (1)

RHADINORHYNCHIDAE

Gorgorhynchoides epinepheli Wang: Pingtan (Fujian). Host: grouper (*Epinephelus awoara*). (1)

G. satoi (Morisita): Qingdao (Shandong); Fuzhou (Fujian); Taiwan. Host: goby (*Glossogobius giuris*) and other freshwater fish. (1)

Micracanthorhynchus dakusuiensis Harada: Taiwan. Host: goby (*Tridentiger trigonocephalus*) and other freshwater fish. (1)

Serrasentis longus Tripathi: Pingtan (Fujian). Host: drum (*Nibea albiflora*) and many other marine fish. (1)

S. sagittifer (Linton): Xiamen and Huian (Fujian). Host: threadfin bream (*Nemipterus virgatus*) and many other marine fish. (1)

ILLIOSENTIDAE

Brentisentis uncinus Leotta: Taiwan. Host: goby (*Glossogobius giuris*). (1)

Indorhynchus arii (Wang): Putian (Fujian). Host: sea catfish (*Arius thalassimus*). (1)

ARHYTHMACANTHIDAE

Arhythmacanthus cynoglosi (Wang): Pingtan (Fujian). Host: tonguefish (*Cynoglossus robustus*). (1)

A. septacanthus Sita in Golvan: Huian (Fujian). Host: sea catfish (*Arius thalassinus*). (1)

REFERENCES*

(1). Wang Puqin, 1991. Fauna of Acanthocephala in Fujian. Fujian Science & Technolgy Press, pp:1–170.

(2). Li Minmin, 1984. Parasites of the mullets *Mugil cephalus* (Linnaeus) and *Liza haematocheila* (Temminck & Echlegel) in the areas of Bohai Gulf, I. Hangu Area. Acta Zoologica Sinica, 30(3):153–158.

(3). Lin Minmin, 1984. Parasites of the mullets *Mugil cephalus* (Linnaeus) and *Liza haematocheila* (Temminck & Echlegel) in the areas of Bohai Gulf, II. Penglai Area. Acta Zoologica Sinica, 30(3):231–242.

*: all in Chinese

Compiled by Huang Zongguo and Shen Jiwei; Edited by Wang Puqin

ROTIFERA

MONOGONONTA
PLOIMA
BRACHIONIDAE

Brachionus angularis Gosse: Mouth of Changjiang (=Yangtze River). (3)

B. calyciflorus Pallas: Mouth of Changjiang (=Yangtze River). (3)

B. leydigi Cohn: Mouth of Changjiang (=Yangtze River). (3)

B. ureeus (Linnaeus): River mouths along China's coast. (1)

Keratella cochlearis (Gosse): Mouth of Changjiang (=Yangtze River). (4)

K. quadrata (Müller): Mouth of Changjiang (=Yangtze River). (3)

K. valga (Ehrenberg): Mouth of Changjiang (=Yangtze River). (4)

Notholca sp.: River mouths along China's coast. (2)

Schizocerca diversicornis Daday: Mouth of Changjiang (=Yangtze River). (4)

SYNCHAETIDAE

Polyarthra trigla Ehrenberg: Mouth of Changjiang (=Yangtze River). (3)

Synchaeta tremula Müller: River mouths along China's coast. (1)

ASPLANCHNIDAE

Asplanchna brightwelli Goose: Mouth of Changjiang (=Yangtze River). (4)

A. priodonta Gosse: Mouth of Changjiang (=Yangtze River). (4)

A. sieboldi (Leydig): Mouth of Changjiang (=Yangtze River). (4)

FLOSCULARIACEA
CONOCHILIDAE

Conochilus hippocrepis (Schrank): Mouth of Changjiang (=Yangtze River). (4)

TESTUDINELLIDAE

Filinia maior (Colditz): Mouth of Changjiang (=Yangtze River). (3)

Testudinella mucronata (Gosse): Mouth of Changjiang (=Yangtze River). (3)

REFERENCES*

(1). Zheng Zhong et al., 1984. Marine Planktonology. China Ocean Press, 865pp.

(2). Group of National Reports on Marine Organisms, 1990. Reports on Marine Organisms. In: Comprehensive investigation on the resources of coastal zone and intertidal zone of China.

(3). Hong Junchao and Pan Haihong et al., 1991. Reports on investigation of the waters around Chongming Island. Zhejiang Science & Technology Press.

(4). Yang Qilun et al., 1988. Reports on comprehensive survey of the resources on the coastal zone and intertidal zone of Shanghai. Shanghai Science & Technology Press, 390pp.

*: all in Chinese

Compiled by Pan Haihong; Edited by Yang Hequan

PRIAPULIDA

PRIAPULIDAE

Priapulopsis cf. *australis* (de Guerne): East China Sea. (2)

Priapulus caudatus Lamarck: Bohai; Yellow Sea. (1, 2, 3)

REFERENCES*

(1). Yang Dejian and Sun Ruiping, 1984. A new record of Priapula to China Sea - *Priapulus caudatus* Lamarck. Marine Science Bulletin, 3(5):95

(2). Yang Dejian and Sun Ruiping, 1986. A brief account about Priapulida. Bulletin of Biology, 8:5–6

(3). Zhang Zhinan, 1992. Priapulida of Yellow Sea and Bohai Sea. (manuscript)

*: all in Chinese

Compiled by Zhang Zhinan; Edited by Yang Dejian

ANNELIDA

POLYCHAETA

ARCHIANNELIDA
PROTODRILIDAE

Protodrilus huanghaiensis Wu, Sun & Chen: Qingdao (Shandong). Intertidal. (44)

P. rubropharzngeus Jagersten: Qingdao (Shandong). In subtidal fine sand. (44)

SACCOCIRRIDAE

Saccocirrus gabriellae Marcus [*S. major*]: Dalian (Liaoning); Yantai and Qingdao (Shandong). Intertidal. (26, 27, 44, 59)

NERILLIDAE

Nerilla sinica Wu & Chen: Yellow Sea. (59)

DINOPHILIDAE

Dinophilus gyrociliatus O. Schmidt: Yellow Sea. (59)

PHYLLODOCIDA
PHYLLODOCIDAE

Clavadoce nigrimaculata (Moore) [*C. splendida, Eulalia nigrimaculata*]: Daya Bay (Guangdong). (54)

Eteone delta Wu & Chen: Shanghai; Fujian. Endemic brackish and freshwater species in China. (28, 58b, 74, 110)

E. longa (Fabricius): Jiaozhou Bay (Shandong). (7, 20, 74, 110)

E. (Mysta) maculata (Treadwell): Qingdao (Shandong); Hainan. Sandy intertidal. (110)

E. (M.) ornata (Grube): Beibuwan (=Gulf of Tongking); South China Sea. (58b, 74, 90)

E. (M.) tchangsii Uschakov & Wu: From Yellow Sea to South China Sea. Intertidal and subtidal. (4, 7, 12, 13, 20, 58b, 74, 110)

Eulalia bilineata (Johnston): Yellow Sea. (74, 110)

E. viridis (Linné): Yellow Sea; East China Sea; South China Sea. (20, 54, 74, 75, 110)

Eumida albopicta (Marenzeller) [*Eulalia albopicta*]: Yellow Sea. (4, 7, 20, 74, 110)

E. sanguinea (Oersted) [*Eulalia sanguinea*]: Widely distributed species, commonly found in northern China; Qingdao and Yantai (Shandong), intertidal and shallow sea soft bottom. (4, 74, 110)

E. tubiformis Moore [*Eulalia tubiformys*]: Yellow Sea. Intertidal to subtidal 20 m water depth. (4, 7, 20, 74)

Hesionura shandongensis Zhao & Wu: Yellow Sea. (111)

Nereiphysa castanea (Marenea) [*Phyllodoce castanea, Genetyllis castanea*]: From Yellow Sea to South China Sea. (4, 12)

Notophyllum foliosum (Sars): Qingdao (Shandong). Intertidal. (4, 20, 74, 110)

N. imbricatum Moore: Bohai; Yellow Sea. Water depth 13–38 m. (4, 74, 110)

N. splendens (Schmarda): Yellow Sea; Fujian; South China Sea; Xisha (=Paracel Islands). (4, 7, 9, 13, 43, 74)

Phyllodoce (Anaitides) chinensis (Uschakov & Wu): Northern China: intertidal, shallow sea soft bottom; South China Sea shoreline. (4, 20)

P. fristedti Bergstrom: Xisha (=Paracel Islands). (43)

P. gracilis Kinberg: Yellow Sea. (10, 74)

P. (A.) groenlandica Oersted: Jiaozhou Bay (Shandong). (20)

P. laminosa laminosa Savigny [*P. laminosa*]: East China Sea; Taiwan Strait; South China Sea. (58b)

P. (A.) madeirensis (Langerhans) [*P. madeirensis*]: Daya Bay (Guangdong); Hong Kong; Hainan. (54, 74, 103)

P. malmgreni Gravier: East China Sea; South China Sea. (74)

P. (A.) papillosa (Uschakov & Wu) [*P. papillosa*]: China's seashore. Intertidal and shallow sea soft bottom. (4, 12, 74)

Pterocirrus macroceros (Oersted) [*Eulalia (P.) macroceros*]: South China Sea. (4, 74)

P. notoensis (Imajima) [*Eulalia (P.) notoensis*]: Xisha (=Paracel Islands). (74)

LOPADORHYNCHIDAE

Lopadorhynchus brevis Grube: Daya Bay (Guangdong); Xisha (=Paracel Islands). (43, 54, 66, 74)

L. krohni (Claparede): Xisha (=Paracel Islands). (14, 15, 74)

L. nationaeis Reibish: Southern Fujian; Taiwan. Planktonic. (63)

L. tuberculus Shen: Xisha (=Paracel Islands). Planktonic. (43, 66)

L. unicinatus Fauvel: Xisha (=Paracel Islands). Planktonic. (43, 66, 74, 110)

Maupasia caeca Viguier: Xisha (=Paracel Islands). Planktonic. (14, 43, 74)

Pedinosoma curtum Reibisch: Xisha (=Paracel Islands). Planktonic. (43, 74)

Pelagobis longicirrata Greeff: Daya Bay (Guangdong); Xisha (=Paracel Islands). Planktonic. (38, 54, 74)

PONTODORIDAE

Pontodora pelagica Greeff: Xisha (=Paracel Islands). Planktonic. (66)

IOSPILIDAE

Phalacrophorus pictus Greeff: Xisha (=Paracel Islands). Planktonic. (43, 66, 74)

P. uniformis Reibisch: Xisha (=Paracel Islands). Planktonic. (43, 66, 74)

ALCIOPIDAE

Alciopina parasitica Cladparede & Panaceri: Southern Fujian; Taiwan; Zhongsha (=Macclesfield Bank); Xisha (=Paracel Islands). Planktonic. (43, 63, 64, 66, 74)

Naiades cantrainii Delle Chiaji: Taiwan; Xisha (=Paracel Islands); Zhongsha (=Macclesfield Bank). Planktonic. (43, 63, 64, 66, 74)

Plotohelmis alata Chamberlin: Southern Fujian, Taiwan. Planktonic. (52)

P. capitata (Greeff): Xisha (=Paracel Islands). Planktonic. (66, 74)

Rhynchonerella gracilis Costa: Southern Fujian; Taiwan; Xisha (=Paracel Islands); Zhongsha (=Macclesfield Bank). Planktonic. (43, 63, 64)

R. petersii (Langerhans): Xisha (=Paracel Islands). Planktonic. (14, 43)

R. xishaensis Shen: Xisha (=Paracel Islands). Planktonic. (66, 74)

Vanadis crystallina Greeff: Xisha (=Paracel Islands). Planktonic. (43, 64, 74)

V. fuscapunctata Treadwill: Xisha (=Paracel Islands). (66, 74)

V. minuta Trezewill: Southern Fujian; Taiwan; Xisha (=Paracel Islands). Planktonic. (43, 63, 64, 66, 74)

TOMOPTERIDAE

Tomopteris apsteini Rosa: Xisha (=Paracel Islands). Planktonic. (14, 43, 74)

T. cavallii Rosa: Xisha (=Paracel Islands). Planktonic. (14, 43, 74)

T. (Johnstonella) duccii Rosa [*T. duccii*]: Xisha (=Paracel Islands). Planktonic. (14, 43, 74)

T. (J.) dunckeri Rosa [*T. dunckeri*]: Xisha (=Paracel Islands). Planktonic. (43)

T. elegans Chun: Yellow Sea; southern Fujian; Taiwan; Daya Bay (Guangdong); Xisha (=Paracel Islands). Planktonic. (63, 66, 74)

T. ligulata Rosa: Xisha (=Paracel Islands). Planktonic. (66, 74)

T. mariana Greeff: Xisha (=Paracel Islands). Planktonoic. (66, 74)

T. nationalis Apatein: Taiwan. Planktonic. (63)

T. (J.) pacifica Izuka [*T. pacifica*]: Southern Fujian; Taiwan; Daya Bay (Guangdong). Planktonic. (54, 63)

T. planktonis Apatein: South China Sea. Planktonic. (66, 74)

T. rolasi Greeff: Southern Taiwan Strait; Daya Bay (Guangdong); Xisha (=Paracel Islands); Zhongsha (=Macclesfield Bank). Planktonic. (14, 43, 54, 63, 64, 66, 74)

T. septentrionalis Quatrefages: Xisha (=Paracel Islands). Planktonic. (14, 43, 74)

TYPHLOSCOLECIDAE

Sagitella kowalevskii Wagner: Southern Fujian; Taiwan; Xisha (=Paracel Islands); Zhongsha (=Macclesfield Bank). Planktonic. (14, 43, 63, 64, 66, 74)

Travisiopsis dubia Stp Bowitz: Taiwan shallow shore. Planktonic. (43, 63)

T. levinseni Southern: Zhongsha (=Macclesfield Bank). Planktonic. (14, 43, 64, 74)

T. lobifera Levisen: Zhongsha (=Macclesfield Bank). Planktonic. (14, 43, 64, 66, 71, 74)

Typhloscolex meulleri Busch: Taiwan shallow shore. Planktonic. (43, 63)

LACYDONIIDAE

Paralacydonia paradoxa Fauvel: Yellow Sea (water depth 7–25 m); East China Sea; South China Sea. (3, 14, 53, 54, 63, 64, 66, 73)

APHRODITIDAE

Aphrodita aculeata Linné: Fujian seashore. (58b)

A. australis Baird: From Bohai and Yellow Sea to South China Sea. Subtida 10–100 m water depth, soft bottom. (4, 20, 74, 87)

A. japonica Marenzeller: East China Sea continental shelf; Fujian seashore. (21, 87)

A. talpa Quatrefages: Jiaozhou Bay (Shandong), Fujian; South China Sea. Subtidal 10–850 m water depth, soft bottom. (20, 74, 83, 87)

Aphroditella sibogae Horst: South China Sea. (43)

Hermonia acantholpis (Grube): Taiwan Strait; Xisha (=Paracel Islands). (43, 58b)

H. hystrix (Savigny): Hainan; South China Sea. (74)

Laetmonice brevepinnata Horst: Beibuwan (=Gulf of Tongking). (74)

L. filicornia Kinberg: Taiwan Strait; South China Sea. (58b)

L. japonica McIntosh: Yellow Sea; East China Sea. Subtidal. (21, 72, 74, 87)

Pontogenia nuda Horst: Hainan; South China Sea. (74)

POLYNOIDAE

Eunoe cf. *barbata* Moore [*E. oerstedi*, *E.* cf. *oerstedi*]: Bohai subtidal; Yellow Sea rocky intertidal and muddy sand subtidal. (7, 20, 74)

Gastrolepidia clavigera Schmarda: South China Sea. Ectoparasite of sea cucumbers. (43, 73, 74)

Gattyana deludens Fauvel: Fujian. Subtidal. (4, 7, 13, 20, 58b)

G. pohaiensis Uschakov & Wu: Bohai; Qingdao (Shandong). Subtidal. (12, 20, 58b, 74, 84)

Halosydna brevisetosa Kinberg [*H. nebulosa*]: China's seashore. Widely distributed on rocky shores (intertidal and subtidal). (4, 20, 54, 58b, 74)

Halosydnopsis pilosa (Horst): Bohai; Yellow Sea; Taiwan Strait; Hainan. Intertidal and subtidal. (4, 20, 23, 58b, 74)

Harmothoe asiatica Uschakov & Wu: Yellow Sea. Intertidal. (4, 20)

H. dictyophora (Grube): South China Sea. Intertidal. (54, 58b, 74)

H. imbricata (Linnaeus): South China Sea. Intertidal. (4, 20, 23, 58b, 74, 87)

H. indica Kinberg: Beibuwan (=Gulf of Tongking). (74)

H. minuta (Potts): Beibuwan (=Gulf of Tongking); Hainan; Xisha (=Paracel Islands). (43, 58b, 87)

H. nigricansa Horst: Xisha (=Paracel Islands). (43, 54, 58b)

Hermadionella truncata (Moore) [*Hermadion truncata*]: Yellow Sea. Subtidal soft bottom. (7)

Hesperonoe hwanhaiensis Uschakov & Wu: Jiaozhou Bay (Shandong). (12, 20)

H. sinagawaensis (Izuka): Taiwan Strait. (58b)

Hololapidella commensalis Willey: Beibuwan (=Gulf of Tongking); Xisha (=Paracel Islands). (43, 73, 74)

Iphione muricata (Savigny) [*I. hirotai*]: Hong Kong; Hainan; Beibuwan (=Gulf of Tongking); Xisha (=Paracel Islands). Dominate species in coral reefs. (37, 43, 74, 92)

Lepidasthenia interrupta (Marenzeller): Xiamen (Fujian). (23, 58b)

L. izukai Imajima & Hartman: Fujian; Taiwan Strait. (58b, 74)

L. longicirrata Berkeley [*L. longissima*]: Xiamen and Pingtan (Fujian). (9, 23, 105)

L. maculata Potts: Beibuwan (=Gulf of Tongking); Hainan intertidal. (74)

L. microlepis Potts: Beibuwan (=Gulf of Tongking); Hainan; Xisha (=Paracel Islands). (43, 74)

L. ocellata (McIntosh) [*L. elegans*]: Dalian (Liaoning); Qingdao, Yantai, and Jiaozhou Bay (Shandong); central and southern Fujian. Muddy sand flat along the seashore. (4, 20, 12, 58b, 74)

L. ohshimai Okuda: Beibuwan (=Gulf of Tongking). (74)

Lepidonotus (*L.*) *carinulatus* Grube [*L. carinulatus*]: Beibuwan (=Gulf of Tongking). (74)

L. (*L.*) *cristatus* Grube [*L. cristatus*]: Xisha (=Paracel Islands). (43, 74)

L. (*L.*) *dentatus* Okuda & Yanada [*L. chinensis, L. dentatus*]: Qingdao (Shandong); Xiamen (Fujian); Hainan. Intertidal. (4, 12, 20, 74)

L. (*L.*) *helotypus* (Grube) [*L. helotypus*]: Dalian (Liaoning); Qingdao and Yantai (Shandong); Dongshan (Fujian). Rocky intertidal. (4, 7, 13, 23, 54, 58b, 74)

L. (*L.*) *jacksoni* Kinberg [*L. jacksoni*]: Hainan. Intertidal. (74)

L. (*Thormora*) *jukesii* (Baird) [*Thormora jukesii*]: Hainan; South China Sea; Xisha (=Paracel Islands). Intertidal. (54, 63, 74)

L. (*L.*) *melanogrammus* Haswell: Taiwan Strait. (58b).

L. (*L.*) *sagamiana* (Izuka) [*Polynoe sagamiana*]: Yellow Sea; Fujian. Intertidal. (4, 13, 20, 74)

L. (*L.*) *squamatus* (Linnaeus) [*L. squamatus*]: Xiamen (Fujian). (23, 58b, 74)

L. (*L.*) *tenuisetosus* (Gravier) [*L. tenuisetosus*]: Fujian; Daya Bay (Guangdong); Hong Kong; Hainan. Intertidal. (54, 58b, 74)

Nonparahalosydna pleiolepis (Marenzeller) [*Polynoe pleiolepis, Parahalosydna pleiolepis*]: Northern Yellow Sea (water depth 14–25 m); Jiaozhou Bay (Shandong), intertidal and subtidal; Xiamen (Fujian); Daya Bay (Guangdong); Hainan. (4, 13, 20, 54, 58b, 74)

Parahalosydnopsis hartmanae Pettbone: Hong Kong. (108)

Paralepidonotus ampulliferus (Grube) [*Malmgrenia ampulliferoides*]: Bohai; Qingdao (Shandong). Subtidal. (4, 12, 20, 58b, 74)

ACOETIDAE

Acoetes flagelliformis (Wesenberg-Lund) [*Polyodontes flagelliformis*]: Beibuwan (=Gulf of Tongking). (74)

A. jogasimae (Izuka) [*Panthalis jogasimae*]: Taiwan Strait; Beibuwan (=Gulf of Tongking); Xisha (=Paracel Islands). (43, 58b)

A. melanonota (Grube) [*Panthalis melanonotus, Polyodontes melanonotus, P. gracilis*]: Yellow Sea intertidal; East China Sea and South China Sea intertidal and subtidal. (4, 5, 7, 13, 20, 74, 110)

Euarche maculosa (Treadwell) [*Eupanthalis maculata, Macellicephala maculosa, Panthalis nigromaculata*]: Beibuwan (=Gulf of Tongking). (74)

Eupanthalis edriophthalma (Treadwell) [*Panthalis edriophthalma*]: Beibuwan (=Gulf of Tongking). (74)

E. kinbergi McIntosh: Beibuwan (=Gulf of Tongking). (74)

Eupolyodontes mitsukurii (Izuka) [*Panthalis mitsukuri*]: Beibuwan (=Gulf of Tongking). (74)

Neopanthalis lepidus Shen & Wu: South China Sea. (74)

N. muricatus Shen & Wu: South China Sea. (74)

Panthalis oerstedi Kinbergii: Beibuwan (=Gulf of Tongking). (74)

Parapanthalis maculata Sun: South China Sea. (74)

P. muricatus Sun: South China Sea. (74)

Polyodontes atromarginatus Horst: Beibuwan (=Gulf of Tongking). (74)

P. maxillosus Ranzani: Jiaozhou Bay (Shandong); Xiamen (Fujian); Beibuwan (=Gulf of Tongking); Xisha (=Paracel Islands). (20, 43, 58b, 110)

Zachsiella nigromaculata (Grube) [*Eupanthalis nigromaculata, E. oculata, Panthalis nigromaculata*]: Beibuwan (=Gulf of Tongking). (74)

SIGALIONIDAE

Ehlersileanira hwanhaiensis (Uschakov & Wu) [*E. izuensis hwanghaiensis, Leanira izuensis hwanghaiensis*]: Yellow Sea; East China Sea. (10, 74)

E. incisa (Grube) [*Leanira izuensis*]: Yellow Sea; East China Sea; Taiwan Strait. Soft bottom. (10, 21, 58b, 74, 87)

E. tentaculata (Augener) [*Sthenolepis tentaculata*]: Beibuwan (=Gulf of Tongking); Xisha (=Paracel Islands). (43)

Euthalenessa digitata (McIntosh): Daya Bay (Guangdong). (54)

E. oculata (Peters) [*Thalenessa oculata*]: Daya Bay (Guangdong). (54)

Fimbriosthenelais minor (Pruvot & Racoritza): Taiwan Strait; South China Sea. (58b)

Leanira japonica McIntosh: Jiaozhou Bay (Shandong); Fujian. Subtidal. (4, 7, 58b)

Pholoe chinensis Wu, Ding & Zhao: Yellow Sea. (89)
P. minuta (Fabricius): Jiaozhou Bay (Shandong); Daya Bay (Guangdong). (20, 54)
Psammolyce fijiensis McIntosh: South China Sea. (56)
P. flava Kinberg: Beibuwan (=Gulf of Tongking). (54)
P. malayana Horst: Western Taiwan Strait. (58b, 83)
Sigalion asiatica (Ushakov & Wu) [*Thalemessa spinosa asiatica*]: Qinhuangdao (Hebei); Yantai, Qingdao, and Jiaozhou Bay (Shandong). Intertidal and subtidal. (4, 20, 74)
S. mathildae Aud. & Milne Edwards: Taiwan Strait. (58b)
S. spinosa (Hartman) [*Thalemessa spinosa*]: Yellow Sea; East China Sea. Intertidal. (74)
Sthenelais fusca Johnson [*S. boa*]: Yellow Sea; East China Sea; South China Sea. (4, 12, 20, 58b, 74, 83)
Sthenolepis japonica (McIntosh) [*Leanira japonica*]: Yellow Sea; East China Sea; Beibuwan (=Gulf of Tongking). (10, 20, 54, 74)
Willetsthenelais diplecirrus Pettibone: Taiwan Strait; South China Sea. (58b)
W. heterochela (Horst): South China Sea. (56)

EULEPETHIDAE

Eulepethus hamifera (Grude): Daya Bay (Guangdong). (54)
Mexiculepis sineca Wu & Sun: South China Sea. (56)
Pareulepis malayana (Horst): South China Sea. (36)

CHRYSOPETALIDAE

Bhawania brevis Gallardo: Daya Bay (Guangdong); Beibuwan (=Gulf of Tongking). Subtidal. (54, 90)
B. cryptocephata Gravier: South China Sea. (74)
B. goodei Imajima & Hartman: Daya Bay (Guangdong); Beibuwan (=Gulf of Tongking). Under rocks in subtidal area. (54, 74)
Chrysopetalum ehlersi Gravier: South China Sea. Water depth 38 m, muddy sand bottom. (74)
C. occidentale Johnson: Jiaozhou Bay and Yantai (Shandong), on seaweed in rocky intertidal; western Taiwan Strait. (6, 20, 83)
Paleanotus chrysolepis Schmarda: Yellow Sea soft bottom; Guangxi rocky intertidal. (6, 74)
P. debilis (Grube): Xisha (=Paracel Islands). (43, 74)

PISIONIDAE

Pisione africana Day: Jiaozhou Bay (Shandong). (20)
P. levisetosa Zhao & Wu: Yellow Sea. (78)

HESIONIDAE

Gyptis capensis (Day): Jiaozhou Bay (Shandong). (20)
G. labatus (Hessle) [*Oxydromus labatus*]: Guangxi. Intertidal. (74)
G. pacificus (Hessle): Taiwan Strait. (74, 83)
Hesione genetta Grube: Xisha (=Paracel Islands). (43, 74)
H. intertexta Grube: Guangxi; Xisha (=Paracel Islands). (43, 74)
H. pantherina Risso: Xisha (=Paracel Islands). (37, 74)
H. splendida Savigny: Fujian; Xisha (=Paracel Islands). (43, 74, 83)
Hesionides arenaria Friedrich: Yantai (Shandong). Intertidal. (8)
Heteropodarke heteromorpha africana Hartman & Schröder: Yellow Sea. (8)
Leocrates chinensis Kinberg: Fujian; Guangxi; Xisha (=Paracel Islands). (74, 83)
L. claparedi Coasta: Guangxi; Beibuwan (=Gulf of Tongking); Xisha (=Paracel Islands). (37, 74)
Microphthalmus biantenna Zhao & Wu: Yellow Sea. (97)
M. hartmanae pacificus Yamanishi: Dalian (Liaoning); Shandong. Intertidal. (8)
Micropodarke dubia (Hessle) [*Kefersteima dubia, Micropodarke amemiyai*]: Yellow Sea, intertidal; Jiaozhou Bay (Shandong). (8, 19, 71)
Ophiodromus angustifrons (Rube) [*Podarke angustifrons*]: Taiwan Strait; Daya Bay (Guangdong). (54, 58b, 74, 83)
O. berriofordi Day: Fujian; Hainan. Sandy intertidal. (56, 83)
O. cf. *obscura* (Verrill) [*Podarke obscura*]: Beibuwan (=Gulf of Tongking). (74)
O. pugettensis Imajima & Hartman: Hong Kong. (108)

PILARGIIDAE

Ancistrosyllis groenlandica McIntosh: Jiaozhou Bay (Shandong). (20)
A. pilargiformis Uschakov & Wu: Qingdao (Shandong); Daya Bay (Guangdong). (8, 54)
Cabira incerta Webster: Jiaozhou Bay (Shandong). (20)
Pilargis berkeleyi Monro: Jiaozhou Bay (Shandong). (20)
P. verrucosa pacifica Uschakov: Beibuwan (=Gulf of Tongking). Water depth 14 m, sandy mud bottom. (74)
Sigambra bassi (Hartman): Yellow Sea. Water depth 22 m, muddy sand bottom. (74)
S. hanaokai Kitamoni: Daya Bay (Guangdong); South China Sea. (54, 74)
S. tentaculata (Treadwell): Jiaozhou Bay (Shandong). (18, 20)
Synelmis albini (Langerhans) [*Ancistrosyllis bassi*]: East China Sea; South China Sea. (37, 74, 83, 90)
S. annamita Gallardo: Fujian; Beibuwan (=Gulf of Tongking); Xisha (=Paracel Islands). (27, 37)
S. sinica Sun & Chen: Jiaozhou Bay (Shandong); East China Sea. (19, 20)

SYLLIDAE

Autolytus cf. *magnus* Berkely: Yellow Sea. (74)

A. purpreimaculatus Okuda: Taiwan Strait; Daya Bay (Guangdong). (54, 58b).

A. rubustissetys Wu & Shen: Xisha (=Paracel Islands); Zhongsha (=Macclesfield Bank). Planktonic. (43, 64, 74)

A. setoensis Imajima: Beibuwan (=Gulf of Tongking). Intertidal. (74)

A. spinoculatus Imajima: Yellow Sea. (101)

Brania clavata (Claparede): Bohai; Yellow Sea. (16, 20, 76)

Campesyllis longicirrus Wu & Ding: Yellow Sea. (76)

Ehlersia anops Ehlers [*Syllis* (*Langerhansia*) *anops*]: Daya Bay (Guangdong). (54)

E. cornuta Rathke [*Langerhansia cornuta, Syllis cornuta*]: Yellow Sea intertidal; Hong Kong subtidal (muddy sand or sandy mud). (8, 20, 74, 108)

E. cf. *rosea* Imajima [*Langerhansia rosea*]: South China Sea. Subtidal sandy mud. (74)

Eusyllis blomstrandi Malmgren: Yellow Sea. (74)

E. inflata (Marenzeller): Yellow Sea. Intertidal. (74)

E. cf. *irregulata* (Augener): South China Sea. Subtidal 95 m water depth. (74)

Exogone dispar (Webster): Yellow Sea. (76)

E. fangopapillata Zhao & Wu: Shandong seashore. Water depth 15–40 m. (12)

E. gemmifera Pagenstecher: Bohai; Yellow Sea. Seaweed holdfast or muddy sand. (8, 74, 76)

E. naidina Oersted: Yellow Sea. (76)

E. verugera Claparede: Yellow Sea; South China Sea. Seaweed holdfast or muddy sand in intertidal area. (74, 76)

Exogonoides antennata Day: Daya Bay (Guangdong). Rocky intertidal. (54)

Haplosyllis spongicala (Grube): Daya Bay (Guangdong), rocky intertidal; Beibuwan (=Gulf of Tongking), water depth 35 m. (54, 74)

H. spongicala tentaculata (Mariom): Guangdong; Guangxi. Intertidal and subtidal. (74)

Myrianida cf. *pachycerus* (Augener) [*Autolytus pachycerus*]: Xiamen (Fujian); Guangxi. Intertidal. (23, 74)

Odontosyllis enopla Verill: Yellow Sea. (76)

O. gibba Claparede: Xisha (=Paracel Islands); Zhongsha (=Macclesfield Bank). Planktonic. (43, 64, 74)

O. maculata Uschakov: Yellow Sea. Rocky intertidal. (8, 20, 74)

O. rubofasciata Grube: Xisha (=Paracel Islands). (43)

Opisthosyllis brunnea Langerhans: Daya Bay (Guangdong). Rocky intertidal. (54)

O. laevis Day: Daya Bay (Guangdong). Rocky intertidal. (54)

Pettia amphophthalma huanghaiensis Wu, Ding & Zhao: Yellow Sea. (88)

Pionosyllis compacta Malmgren: Yellow Sea. Intertidal and subtidal. (74)

P. magnifica Moore: Yellow Sea. (8)

Sphaerosyllis chinensis Zhao & Wu: Shandong seashore. Water depth 15–40 m. (12)

S. erinaceus Claparede: Yellow Sea. Oyster bed, seaweed holdfast, or muddy sand in intertidal area. (20, 64, 74)

S. glandulata Perkins: Yellow Sea. (76)

S. hirsuta Ehlers: Yellow Sea. On seaweed (*Sargassum*) in low rocky intertidal. (74)

S. hystrix Claparede: Xisha (=Paracel Islands). Coraf reef, seaweed holdfast, or muddy sand. (20, 64, 74)

S. longicauda Webster & Bendict: Yellow Sea. (76)

S. periferopsis Perkins: Yellow Sea. (76)

S. pirifera Claparede: Bohai; Yellow Sea. (20, 64)

Syllis amica Quatrefages: Shandong; Daya Bay (Guangdong); Xisha (=Paracel Islands). Rocky intertidal. (20, 54, 74)

S. gracilis Grube: Daya Bay (Guangdong); Xisha (=Paracel Islands). Intertidal. (54, 74)

S. longissima Gravier: Daya Bay (Guangdong). Intertidal. (54)

S. spongiphila Verrill [*S. sclerolaema*]: East China Sea. Sandy mud, water depth 51m. (74)

Trypanosyllis (*Trypanedenta*) *taeniaformis* (Haewell) [*T. taeniaformis*]: South China Sea. Intertidal and subtidal. (54, 74)

T. (*T.*) *zebra* (Grube) [*T. zebra*]: South China Sea. Common species in coral reefs, from intertidal to water depth 30 m. (37, 43, 74)

Typosyllis aciculata orientalis Imajima & Hartman: Daya Bay (Guangdong); Hong Kong; Xisha (=Paracel Islands). (43, 54, 103)

T. adamantens kurilensis Chlebovitsch [*Syllis decorus*]: Yellow Sea. Oyster bed and seaweed holdfast in rocky intertidal. (8, 20, 74)

T. alterata (Moore): Xisha (=Paracel Islands). (43)

T. armillaris (Müller) [*Syllis armillaris*]: Bohai; Yellow Sea; Daya Bay (Guangdog). Intertidal and subtidal. (8, 20, 54)

T. benguellana Day: Daya Bay (Guangdong). Rocky intertidal. (54)

T. cirropunctata Michel: Daya Bay (Guangdong). Rocky intertidal. (54)

T. fasciata Malmgren [*Syllis fasciata*]: Yellow Sea; Daya Bay (Guangdong). Rocky intertidal. (20, 54, 74)

T. hyalina Grube: Yellow Sea. (8)

T. inflata Marenaeller [*Syllis inflata*]: Jiaozhou Bay (Shandong). (74)

T. lutea Hartman & Schröder: Yellow Sea. (8)

T. maculata Imajima: South China Sea. Intertidal and subtidal, muddy sand. (74)

T. monilata Imajima: Xisha (=Paracel Islands). (43)

T. nigropharynea Day: Daya Bay (Guangdong). Rocky intertidal. (54)

T. prolifera (Krohn): Xisha (=Paracel Islands). (43)

T. variegata (Grube): Yellow Sea; East China Sea; Hong Kong. Rocky intertidal. (20, 54, 74, 103)

T. vitata Grube: Daya Bay (Guangdong). Rocky intertidal. (54)

NEREIDAE

Ceratonereis anchylochaeta (Horst) [*Nereis anchylochaeta, C. tongicanda*]: South China Sea. (46)

C. burmensis Monro [*Nereis (C.) burmensis*]: East China Sea; South China Sea. (43, 46, 54, 58b)

C. costae (Grube) [*Nereis (C.) costae*]: East China Sea; Hong Kong; South China Sea. (43, 46, 58b, 108)

C. erythraeensis Fauvel [*Nereis (C.) erythraeensis*]: China's seashore. (12, 20, 46, 53, 54, 55, 73)

C. hircinicola (Eisig): South China Sea. (46)

C. japonica Imajima: Xisha (=Paracel Islands). (43, 46)

C. marmorata Horst: South China Sea. (46, 54)

C. mirabilis Kinberg [*Nereis mirabilis, N. (C.) tentaculata*]: Hong Kong; Xisha (=Paracel Islands). (43, 46)

C. pachychaeta Fauvell [*Nereis pachychaeta*]: Xisha (=Paracel Islands). (43, 46)

C. tripartita Horst: Xisha (=Paracel Islands). (43)

Cheillonereis cyclurus (Harrington): Bohai; Yellow Sea; East China Sea. (46, 58b)

Dendronereis pinnaticirris Grube: Hainan. Brackish and freshwater species. (46)

Laevispinereis fujianensis He & Wu: Western Taiwan Strait. (58a, 58b)

Leonnates decipiens Fauvel: Guangdong; Guangxi. (46, 54)

L. jousseaumei Gravier: Daya Bay (Guangdong); Beibuwan (=Gulf of Tongking); Hainan. (46)

L. niponicus Imajima: Hong Kong. (108)

L. persica Wesenberg-Lund: From Bohai and Yellow Sea to South China Sea. (7, 12, 20, 46, 54, 58b, 83)

Lycastopsis augenari Okuda: Yellow Sea. Around high tide mark, on rocks or gravel. (7, 20, 46)

Namalycastis aibiuma (Müller): Shanghai; Fujian; Sanya (Hainan). River mouth area. (46, 54, 55, 74)

N. longicirris (Takahashi): Taipei (Taiwan). Freshwater stream. (46, 74)

Neanthes donghaiensis Wu: East China Sea. (69)

N. flava Wu & Sun: Yellow Sea; East China Sea. (46, 58b, 83)

N. glandicincta (Southern) [*Nereis glandicincta*]: Lianjiang (Fujian); Daya Bay (Guangdong); Haikou and Sanya (Hainan). (46, 54)

N. japonica (Izuka): China's seashore. (20, 46, 54, 58b, 83)

N. maculata Wu: Daya Bay (Guangdong); Dapeng Cove (Hong Kong); Hainan. (54)

N. nanhaiensis Wu: South China Sea. (46)

N. succinea (Frey & Leuckart): Tianjin; Daya Bay (Guangdong). (46, 54)

N. unifasciata Willey [*Nereis unifasciata*]: Sanya and Lingshui (Hainan); Zhongsha (=Macclesfield Bank). (46)

Nectoneanthes alatopalpis (Wesenberg-Lund) [*Nereis alatopalpis*]: East China Sea. (46, 58b, 83)

N. donghaiensis He: East China Sea. (70)

N. fujianensis Zheng & Wu: East China Sea. (75)

N. ijimai (Izuka) [*Nereis ijimai, Neanthes ijimai*]: East China Sea; South China Sea. (46, 54, 58b, 83)

N. multignatha Wu: Yellow Sea; East China Sea; South China Sea. (24, 46, 54, 58b, 83)

N. oxypoda (Marenzeller) [*Neanthes oxypoda, Nereis oxypoda, Nereis (Neanthes) oxypoda*]: China's seashore. (12, 20, 46, 54, 58b, 83)

Nereis coutieri Gravier: Daya Bay (Guangdong); Dapeng Cove (Hong Kong); Beihai and Weizhou Island (Guangxi); Hainan. (46, 54)

N. dayana Sun & Shen: Xisha (=Paracel Islands); Zhongsha (=Mcclesfield Bank). (43, 46)

N. denhamensis Augener: Xisha (=Paracel Islands); Zhongsha (=Macclesfield Bank). (43, 46, 64)

N. donghaiensis He & Wu: East China Sea. (69)

N. falcaria (Willey) [*Ceratonereis falcaria, N. kauderi*]: Sanya (Guangdong); Xisha (=Paracel Islands). (43, 47)

N. falcaria multignatha Wu & Sun: Bohai; Yellow Sea; East China Sea. (46, 54)

N. grubei (Kinberg): Bohai; Yellow Sea. Intertidal. (46)

N. guandongensis Wu: Daya Bay (Guangdong); Dapeng Cove (Hong Kong); Hainan. (46, 54)

N. hainanica Chlebovitsch: Haikou and Sanya (Hainan). (47)

N. heterocirrata Greadwell: Yantai and Qingdao (Shandong); Shengshan (Zhejiang); Taiwan. (20, 46)

N. heteromorpha Horst: Xisha (=Paracel Islands); Zhongsha (=Macclesfield Bank). (43, 46)

N. huanghaiensis Wu: Yellow Sea. (46)

N. jacksoni Kinberg: Day Bay (Guangdong); Dapeng Cove (Hong Kong). (46, 54)

N. longior Chlebovitsch & Wu: Bohai; Yellow Sea; East China Sea. (20, 46)

N. multignatha Imajima & Hartman: Yantai, Qingdao, and Weihai (Shandong); Dachen Island (Zhejiang), Intertidal; Bohai; Yellow Sea; East China Sea; South China Sea. (20, 46, 54)

N. neoneanthes Hartman: Qingdao and Yantai (Shandong); Taishan Islands (Zhejiang); Daya Bay (Guangdong). (20, 24, 46, 54)

N. nichollsi Kott: Daya Bay (Guangdong); Beibuwan (=Gulf of Tongking); Xisha (=Paracel Islands). (43, 46, 54)

N. pelagica Linnaeus: Bohai; Yellow Sea; East China Sea. (7, 20, 46, 58b)

N. persica Fauvel [*N. zonata persica*]: East China Sea; South China Sea. Intertidal and subtidal. (46, 54, 58b).

N. sinensis Wu: Yellow Sea; East China Sea. (46)

N. surugaense nanhaiensis Sun & Shen: Zhongsha (=Macclesfield Bank). (43, 46)

N. trifasciata Grube: Yellow Sea; South China Sea. (43, 46)

N. vexillosa Grube [*N. ezoensis*]: Dalian (Liaoning); Yantai and Qingdao (Shandong). (46)

N. zhongshaensis Shen & Sun: Xisha (=Paracel Islands); Zhongsha (=Macclesfield Bank). (43, 46)

N. zonata Malmgren [*N. zonata tigrina*]: Yellow Sea; Jiaozhou Bay (Shandong); Xisha (=Paracel Islands). (43, 46)

Nicon japonica Imajima: South China Sea. Water depth 30–50 m. (46)

N. maculata Kinberg [*N. benhami*]: East China Sea; South China Sea near shore. (43, 46)

N. moniloceras (Hartman) [*Leptonereis glauca moniloceras*]: Jiaozhou Bay (Shandong); East China Sea continental shelf and near shore; South China Sea near shore. (7, 20, 46, 58b, 85)

N. sinica Wu & Sun: Yellow Sea; South China Sea. (43, 46)

Paraleonnates uschkovi Chelbovitsch & Wu: Jiaozhou Bay (Shandong); Fujian; Hainan. (20, 46)

Perinereis aibuhitensis Grube [*Nereis aibuhitensis, N. (Perinereis) aibuhitensis, N. (Neanthes) linea, N. (Neanthes) orientalis*]: China's seashore. (20, 46, 54, 58b)

P. camiguinoides Augener [*Nereis (P.) camiguinoides*]: Yellow Sea; East China Sea; South China Sea. (46, 54, 58b)

P. cavifrons Ehlers [*Nereis (P.) cavifrons*]: Aotou (Guangdong); Beihai (Guangxi). (46, 54)

P. cultrifera Grube: East China Sea; South China Sea. (46, 58b)

P. c. floridana Ehlers: Yellow Sea; East China Sea; South China Sea. (3, 7, 46, 54, 58b)

P. c. helleri Grube: Yellow Sea; East China Sea; South China Sea. (46, 54, 58b)

P. c. obfuscata Grube [*Nereis (P.) obfascata, P. obfuscata*]: Hainan. Coral reefs. (47)

P. c. perspicillata Grube: Hainan. Coral reefs. (46)

P. c. striolata Grube: Hainan. Coral reefs. (46)

P. c. typica Grube: Yellow Sea; East China Sea; South China Sea. (20, 54, 58b)

P. nuntia (Savigny): East Chinas Sea; South China Sea. (46, 58b)

P. n. brevicirris (Grube) [*P. brevicirris*]: Yellow Sea; East China Sea; South China Sea. (20, 46, 58b)

P. n. majungaensis Fauvel: South China Sea. (46)

P. n. typica (Savigny): East China Sea; South China Sea. (46, 58b)

P. n. vallata (Grube) [*Perinereis vallata*]: Yellow Sea; East China Sea; South China Sea. (7, 20, 46, 54, 58b)

P. rhombodonta Wu & Sun: Daya Bay (Guangdong); Hainan. (46, 54)

P. suluana Horst [*Nereis (P.) suluana*]: Hainan; Xisha (=Paracel Islands). (43, 46)

P. vancaurica (Ehlers): East China Sea; South China Sea. (46, 56, 58b)

P. weizhouensis Wu & Sun: Daya Bay (Guangdong); Weizhou Island (Guangxi). (46, 54)

Platynereis abnormis (Horst): Hainan; Dongsha (=Pratas Islands); Xisha (=Paracel Islands); Nansha (=Spratly Islands). (43, 46)

P. bicanaliculata (Baird) [*Nereis agassizi, N. dumerilli, N. kobiensis, Platynereis agassizi*]: China's seashore. (20, 43, 46, 54, 58b)

P. dumerilii Audouin & Milne: Taiwan; Aotou (Guangdong); Guangxi; Haikou, Lingshui, and Sanya (Hainan); Xisha (=Paracel Islands). (43, 58b, 64)

P. pulchella Gravier: Beibuwan (=Gulf of Tongking); Hainan. (46)

P. sinica Sun & Shen: Central South China Sea. Endemic species. (43, 46)

Pseudonereis anomala Kinberg: Hainan. Coral reefs. (86)

P. gallapagensis Kinberg: Sanya (Hainan). (46)

P. variegata (Grube): China's seashore. On seaweed, oyster shell or barnacle shell. (20, 46, 54)

Rullierinereis elytrocirra Wu & Sun: Yellow Sea near shore. Endemic species. (46)

R. misakiensis Imagina & Hayashi: Xisha (=Paracel Islands). (43, 46)

Sinonereis heteropoda Wu & Sun: Yellow Sea; East China Sea; South China Sea. (46)

Tambalagamia fauveli Pillai [*Ceratocephala sibogae*]: From Yellow Sea to South China Sea. Subtidal. (7, 46, 54, 58b, 74)

Tylonereis bogoyawleskyi Fauvel: Fujian; Guangdong; Guangxi; Hainan. (46, 54, 58b)

Tylorrhynchus heterochaetus (Quatrefages): Nanjing (Jiangsu); Shanghai; Fujian; Guangdong. (46, 58b)

GLYCERIDAE

Glycera alba (Müller): Yellow Sea; East China Sea; South China Sea. Intertidal and subtidal. (6, 7, 51, 58b, 72, 74, 83, 90, 108)

G. armigera (Moore): East China Sea. (87)

G. capitata Oersted: Bohai; Yellow Sea; East China Sea; South China Sea. (20, 58b, 74)

G. chirori Izuka: China's seashore. Intertidal and subtidal. (6, 12, 21, 51, 54, 58b, 72, 90)

G. clecipiens Marenzeller: Bohai; Yellow Sea; East China Sea continental shelf; Beibuwan (=Gulf of Tongking). (20)

G. convoluta Keferstein: China's seashore. (10, 51, 54, 58b, 74, 83)

G. gigantea Quatrefages: East China Sea; western Taiwan Strait. (58b)

G. lancadivae Schmarda: East China Sea; Beibuwan (=Gulf of Tongking); Xisha (=Paracel Islands). (43, 58b)

G. onomichiensis Izuka: Yellow Sea intertidal and subtidal; East China Sea; Hong Kong; Beibuwan (=Gulf of Tongking) intertidal. (12, 20, 21, 23, 36, 43, 58b, 83)

G. papillosa nigticans Wu: Daya Bay (Guangdong); Beibuwan (=Gulf of Tongking); Xisha (=Paracel Islands). (37, 58b)

G. prashadi Fauvel: East China Sea; Beibuwan (=Gulf of Tongking); Xisha (=Paracel Islands). (37, 43, 58b, 83)

G. robusta Ehlers: Yellow Sea; East China Sea. (20, 58b)

G. rouxu Aud. & Milne Edwards: China's seashore. (12, 20, 36, 51, 54, 58b, 71, 83)

G. sagittariae McIntosh: Xiamen (Fujian), intertidal; Beibuwan (=Gulf of Tongking). (9)

G. siphonostoma (Delle Chiaje): East China Sea continental shelf; western Taiwan Strait; Beibuwan (=Gulf of Tongking). (58b)

G. subaenea Grube: From Qinhuangdao (Hebei) to Daya Bay (Guangdong). Intertidal and subtidal. (12, 20, 54, 58b, 83)

G. tenuis Hartman: Yellow Sea. Intertidal. (20, 74)

G. tesselata Grube: East China Sea; Fujian; Daya Bay (Guangdong); Beibuwan (=Gulf of Tongking); Xisha (=Paracel Islands). (37, 43, 54, 55, 58b, 83)

G. unicornis Savigny: East China Sea; Fujian. (58b)

Hemipodus yenourensis Izuka: Dalian (Liaoning); Jiaozhou Bay and Yantai (Shandong). Intertidal. (20)

GONIADIDAE

Glycinde gurjanovae Uschakov & Wu: Bohai; Yellow Sea; East China Sea; South China Sea. (12, 20, 58b, 74)

G. kameruniana Angener: Western Taiwan Strait; Daya Bay (Guangdong). (54, 58b)

G. oligodon Southern: East China Sea; South China Sea. (54, 58b)

G. pacifica Monro: Western Taiwan Strait. (58b)

Goniada annulata Moore: East China Sea; South China Sea. (21, 43, 54, 58b)

G. emerita Audouin & Milne Edwards: East China Sea; South China Sea. (43, 58b, 83)

G. japonica Izuka: Bohai; Yellow Sea; East China Sea, intertidal and subtidal; Beibuwan (=Gulf of Tongking). (12, 20, 74, 87)

G. littorea Hartman: Daya Bay (Guangdong). (54)

G. maculata Oersted: Bohai; Yellow Sea; East China Sea; South China Sea. (10, 12, 20, 43, 58b, 74, 83)

G. multidentata Arwidsson: Beibuwan (=Gulf of Tongking). (90)

G. paucidens Grube: Hong Kong. (103)

G. uncinigera Ehlers: Daya Bay (Guangdong). (54)

G. vorax (Kinberg): East China Sea. (58b)

Goniadopsis incerta Fauvel: Daya Bay (Guangdong). (54)

Ophioglycera eximia (Ehler): Taiwan Strait. (58b)

NEPHTYIDAE

Aglaophamus dibranchis Grube: East China Sea; South China Sea. (51, 54, 58b, 73)

A. dicirris Hartman: East China Sea. (58b, 87)

A. jeffreysii (McIntosh): East China Sea; South China Sea. (21, 36, 43, 58b)

A. lyrochaeto (Fauvel): East China Sea; South China Sea. (54, 74, 83, 108)

A. macroura (Schmarda): East China Sea continental shelf. (21)

A. munamaorii (Gibbs): South China Sea. (36, 43)

A. orientalis Fauchald: Western Taiwan Strait; South China Sea. (58b, 90)

A. sinensis Fauvel [*Nephtys sinensis, N. (A.) sinensis*]: From Bohai and Yellow Sea to South China Sea. (6, 10, 20, 53, 58b, 71, 90)

A. tepens Fauchald: East China Sea; South China Sea. (54, 58b, 90)

A. toloensis Ohwada: Dapeng Cove (Hong Kong). Water depth 6–23 m. (106)

A. vietnamensis Fauchald: East China Sea; South China Sea. (54, 58b, 90)

Inermonephtys gallardi Fauchald: Taiwan Strait. (58b)

I. cf. *inermis* (Ehlers) [*Nephtys (Aglaophamus) inermis*]: Yellow Sea; East China Sea; South China Sea. (20, 36, 43, 51, 58b, 83, 108)

Micronephtys sphaerocirrata (Wesenberg-Lund): Yellow Sea; East China Sea; South China Sea. (54, 58b)

Nephtys caeca (Fabricius): Bohai, Yellow Sea. (10, 74)

N. californiensis Hartman: Bohai; Yellow Sea; East China Sea; South China Sea. (12, 20, 51, 54, 58b, 74)

N. ciliata (Müller): Bohai; Yellow Sea; East China Sea; South China Sea. (12, 20, 51, 54, 58b, 74)

N. longosetosa Oersted: Yellow Sea; East China Sea. (58b)

N. oligobranchia Southern: Bohai; Yellow Sea; East China Sea; South China Sea. (20, 54, 58b, 74, 83)

N. paradoxa Malmgren: Yellow Sea; Guangxi. (6)

N. polybranchia Southern: Bohai; Yellow Sea; East China Sea; South China Sea. (12, 20, 51, 54, 58b, 74)

N. tulearensis Fauvel: East China Sea. (58b)

ORBINIIDA
ORBINIIDAE

Haploscoloplos elongatus (Johnson): Bohai; Yellow Sea; East China Sea; South China Sea. (20, 22, 54, 58b, 74, 83)

H. kerguelensis (McIntosh): Taiwan Strait; Daya Bay (Guangdong). (54, 58b)

Microbinia linea Hartman: Taiwan Strait. (58b)

Naineris dendritica (Kinberg): Yellow Sea; East China Sea. (22)

N. hainanensis Wu: Hainan. Sandy intertidal. (58)

N. laevigata (Grube): Yellow Sea; East China Sea; South China Sea. (20, 54, 74)

Orbinia curieri (Audouin & Milne-Edwards): Daya Bay (Guangdong). (54)

O. dicrochaeta Wu: Bohai; Yellow Sea; East China Sea. (20, 22)

O. exarmata (Fauvel): Yellow Sea; East China Sea; South China Sea. (74)

O. vietnamensis Gallardo: South China Sea. Subtidal sandy bottom. (74, 90)

Phylo felix Kinberg [*P. felix asiatieus*]: Bohai; Yellow Sea; East China Sea; South China Sea. (20, 22, 54, 58b, 74, 83)

P. kupfferi Ehlers: Hong Kong; Beibuwan (=Gulf of Tongking). (87)

P. nudus (Moore): East China Sea; South China Sea. (74)

P. ornatus (Verrill): Bohai; Yellow Sea; South China Sea. Intertidal and subtidal. (74)

Scoloplos (*S.*) *acmeceps* Chamberlin [*S.* (*S.*) *uschakovi*]: Jiaozhou Bay (Shandong). (20, 22)

S. (*S.*) *armiger* (Müller): East China Sea; South China Sea. (43, 54, 58b)

S. (*S.*) *chrysochaeta* Wu: Yellow Sea; East China Sea. (22, 58b)

S. (*Leodamas*) *dubia* Tebble: East China Sea; South China Sea. (58b, 90)

S. (*L.*) *gracilis* Pillai: East China Sea; South China Sea. (54, 58b, 83)

S. (*L.*) *johnstonei* Day: East China Sea. (58b)

S. (*S.*) *marsupialis* Southern: Yellow Sea; East China Sea; South China Sea. Intertidal and subtidal. (20)

S. (*L.*) *rubra* (Webster) [*S.* (*L.*) *rubra pacifica*]: Yellow Sea; East China Sea; South China Sea. (20, 22, 58b, 74)

S. (*L.*) *rubra orientalis* Gallardo: East China Sea; South China Sea. (58b, 90)

S. (*S.*) *spiniferus* Gallardo: South China Sea. (90)

S. (*S.*) *tumidus* Mackie: Hong Kong. (101)

S. (*L.*) *uniramus* Say: Daya Bay (Guangdong). (54)

PARAONIDAE

Aedicira pacifica Hartman: Yellow Sea; East China Sea; South China Sea. (20, 54, 58b, 83)

Aricidea capensis Day: East China Sea. (58b)

A. curviseta Day: East China Sea. (58b)

A. fragilis Webster [*A. fragilis caeca*]: Bohai; Yellow Sea; East China Sea; South China Sea. (17, 20, 58b, 74)

A. (*Allia*) *nolani* Webster & Benedict: Yellow Sea. (101)

A. (*Acesta*) *simplex* Day: East China Sea. (83)

Cirrophorus branchiatus Ehlers: Yellow Sea; East China Sea; South China Sea. (54, 101)

C. furcatus (Hartman): Yellow Sea; East China Sea. (20)

C. neapolitanus pacificus Zhao & Wu: Yellow Sea. (101)

Paraonis gracilis (Tauber): East China Sea; South China Sea. (54, 58b)

Taubera gracilis (Tauber): Yellow Sea (54, 101)

COSSURIDA
COSSURIDAE

Cossurella dimorpha Hartman [*Cossura coasta, Heterocossura aciculata*]: Bohai; Yellow Sea; East China Sea; South China Sea. (20, 37, 54, 58b, 74)

Cossura longioirrata Webster & Benedict: Daya Bay (Guangdong). (59)

SPIONIDA
SPIONIDAE

Aonides oxycephala (Sars): Bohai; Yellow Sea; East China Sea; South China Sea. (20, 21, 53, 54, 55, 74)

Boccardia polychranchia (Haswell): Daya Bay (Guangdong). (54, 74)

B. proboscidea Hartman: Qingdao (Shandong). Intertidal. (74)

Boccardiella hamata (Webster): Qingdao (Shandong). Intertidal. (74)

Laonice cirrata (Sars): Bohai; Yellow Sea; East China Sea; South China Sea. (20, 36, 51, 54, 58b, 73)

Malacoceros indicus (Fauvel): East China Sea; South China Sea. (36, 43, 53, 54, 55, 58b)

Minuspio cirrifera (Wiren): East China Sea; Hong Kong; South China Sea. (18, 54, 58b, 74, 108)

M. polybranchiata: Day Bay (Guangdong). (54)

Paraprionospio pinnata (Ehlers) [*Prionospio pinnata, P.* (*Paraprionospio*) *pinnata*]: Yellow Sea; East China Sea; South China Sea. (20, 28, 51, 53, 54, 58b, 73, 74, 108)

Polydora armata Langerhans: South China Sea. (74)

P. ciliata (Johnston): Yellow Sea; East China Sea. (17)

P. flava Claparede: East China Sea; South China Sea. (74)

P. giardi Mesnil: Yellow Sea. (77)

P. kempi Southern: Yellow Sea; East China Sea. (28, 58b)

P. cf. *pilikia* (Ward): South China Sea. Intertidal. (74)

P. socialis (Schmarda): Hainan. Intertidal and coral reefs. (74, 96)

P. cf. *tentaculata* Blake & Kudenov: South China Sea. (108)

Polydorella novaegeorgiae Gibbs: Daya Bay (Guangdong). (54)

Prionospio cirrifera Wiren: Jiaozhou Bay (Shandong); Daya Bay (Guangdong). (20, 54)

P. ehlersi Fauvel: Yellow Sea; East China Sea; South China Sea. (20, 36, 43, 54, 58b, 90, 99)

P. japonica Okuda: Yellow Sea; East China Sea; South China Sea. (20, 54, 58b)

P. krusadensis Fauvel: East China Sea. (58b)

P. malayensis Caulley: East China Sea; South China Sea. (43, 58b, 83)

P. malmgreni Claparede: Yellow Sea; East China Sea; South China Sea. (20, 43, 54, 58b, 83, 108)

P. (*Apoprionospio*) *pygmaea* (Hartman) [*P. pygmaeus, Apoprionospio pygmaea*]: Yellow Sea. (74)

P. (*P.*) *queenslandica* Blake & Kudenov: Yellow Sea; East China Sea. (74)

P. saccifera Mackie & Hartley: Hong Kong. (99)

P. xishaensis Wu & Chen: Daya Bay (Guangdong); Xisha (=Paracel Islands). (43, 54, 56)

Pseudopolydora antennata (Claparede) [*Polydora antennata*]: East China Sea; South China Sea. (43, 58b)

P. kempi (Southern): Bohai; Yellow Sea; South China Sea. (54, 74)

P. kempi japonica Imajima & Hartman: Yellow Sea; East China Sea. (20)

P. paucibranchiata (Okuda): East China Sea; South China Sea. (20, 74)

Pygospio elegans Claparede: Xisha (=Paracel Islands). (43)

Rhynchospio glutaea (Ehlers): Yellow Sea. (5)

Scolelepis (Nerinides) globosa Wu & Chen: East China Sea. (30)

S. lefebvrei Gravier: Daya Bay (Guangdong). (54)

S. squamata (Müller) [*Nerine cirratulus*]: Bohai; Yellow Sea. (20, 74)

Spio filicornis (Müller): Daya Bay (Guangdong). (54)

S. martinensis Mesnil: Bohai; Yellow Sea. Intertidal. (74)

Spiophanes bombyx (Claparede): Jiaozhou Bay (Shandong). (20)

S. soederstromi Hartman: Taiwan Strait. (83)

MAGELONIDAE

Magelona cineta Ehlers: Yellow Sea; East China Sea; South China Sea. (54, 58b, 74)

M. crenulifrons Gallardo: East China Sea; South China Sea. (54, 58b, 90, 108)

M. japonica Okuda: Bohai; Yellow Sea; East China Sea; South China Sea. (20, 58b, 74)

M. pacifica Monro: Yellow Sea; East China Sea. (74)

M. papilliorni Müller: Yellow Sea; Taiwan Strait. (58b)

TROCHOCHAETIDAE

Trochochaeta diverapoda (Hoagland) [*Aonides diverapoda*]: Hong Kong. (100)

T. multisetosum (Oersted): East China Sea. (72)

T. cf. *orissae* (Fauvel): Hong Kong. (108)

POECILOCHAETIDAE

Poecilochaetus hystricosus Mackie: Hong Kong. (100)

P. johnsoni Hartman: Bohai; Yellow Sea; East China Sea. (20, 58b, 74)

P. paratropicus Gallardo: East China Sea; South China Sea. (54, 58b, 73)

P. serpens Allen: Yellow Sea; East China Sea; South China Sea. (36, 43, 58b, 74)

P. spinulosus Mackie: Hong Kong. (100)

P. tricirratus Mackie: Hong Kong. (100)

P. tropicus Okuda: South China Sea. (43, 54, 74)

HETEROSPIONIDAE

Heterospio catalinensis (Hartman): Taiwan Strait. (58b)

H. sinica Wu & Chen: Yellow Sea; East China Sea; South China Sea. (20, 31, 58b, 74)

CHAETOPTERIDAE

Chaetopterus variopedatus (Renier): Yellow Sea. Subtidal. (74)

Mesochaetopterus japonicus Fujiwara: Bohai; Yellow Sea; East China Sea. (20, 74)

Phyllochaetopterus claparedii McIntosh: Yellow Sea; East China Sea. (20, 58b, 83)

Spiochaetopterus costarum (Claparede): Hong Kong; South China Sea. (108)

CIRRATULIDAE

Caulleriella cf. *typhlops* (Willey): Xisha (=Paracel Islands). (43)

Chaetozone maotienae Gallardo: East China Sea; South China Sea. (58b, 90)

C. setosa Malmgren: Bohai; Yellow Sea; East China Sea; South China Sea. (54, 58b, 74)

C. spinosa Hartman: East China Sea; Taiwan Strait. (58b)

Cirratulus annamensis Gallardo: East China Sea; South China Sea. (58b, 90)

C. chrysoderma Claparede: Yellow Sea; East China Sea. (20, 51, 58b, 74, 83)

C. cirratus (Müller): Yellow Sea; East China Sea. (20, 58b)

C. filiformis Keferstein: Yellow Sea; East China Sea; South China Sea. (51, 58b, 74, 83, 90)

Cirriformia capensis (Sehmarda): Daya Bay (Guangdong). (54)

C. chefooensis (Grube): East China Sea. (23)

C. dasylophia (Marenzeller): East China Sea. (23)

C. filigera (Delle Chiaje): Bohai; Yellow Sea; East China Sea; South China Sea. (20, 43, 58b, 74)

C. puctata (Grube): South China Sea. (36, 43, 74)

C. semicincta (Ehlers): Yellow Sea. (20)

C. tentaculata (Montaau): Yellow Sea; East China Sea; South China Sea. (20, 58b, 74)

Dodecaceria concharum Oersted: Yellow Sea. (74)

D. fewkesi Berkeley & Bakeley: Yellow Sea; South China Sea. (74)

Tharyx acutus Webster & Benedid: Daya Bay (Guangdong). (54)

T. filibranchia Day: Hong Kong; South China Sea. (108)

T. marioni (Saint-Joseph): Yellow Sea; East China Sea; South China Sea. (20, 51, 54, 58b, 83)

T. multifilis Moore: Bohai; Yellow Sea; East China Sea. (20, 58b, 74, 83)

T. tesselata Hartman: Yellow Sea; East China Sea. (18, 20)

ACROCIRRIDAE

Acrocirrus validus Marenzeller: Yellow Sea. (74)

CAPITELLIDA
CAPITELLIDAE

Capitella capitata (Fabriceus): Bohai; Yellow Sea; East China Sea. (20, 25, 58b, 74)

Dasybranchus caducus (Grube): Daya Bay (Guangdong); Hainan; Xisha (=Paracel Islands). (37, 54)

D. lumbricoides Grube: East China Sea. (51)

Heteromastus filiformis (Claparede): Bohai; Yellow Sea; East China Sea; South China Sea. (20, 55, 58b, 74, 83)

H. simillis Southern: East China Sea. (58b)

Leiochrides australis Auqener: East China Sea; South China Sea. (54, 58b)

Mastobranchus indicus Southern: East China Sea. (28, 58b)

Neoheteromastus lineus Haitmer: Taiwan Strait. (58b)

Neomediomastus glabrus (Hartman): South China Sea. (58b)

Notomastus cf. *aberans* Day: East China Sea; South China Sea. (54, 58b, 74, 83)

N. fauvel Day: East China Sea; South China Sea. (54, 58b, 74, 83)

N. giganteus Moore: Sanya (Hainan). Intertidal. (56)

N. latericeus Sars: Yellow Sea; East China Sea; South China Sea. (20, 36, 43, 58b, 83)

N. polyodon Gallardo: Taiwan Strait. (58b)

Paraleiocapitella mossambica Thomassin: East China Sea; South China Sea. (54, 58b)

Parheteromastus tenuis Monro: Western Taiwan Strait; Daya Bay (Guangdong). (54, 58b)

Rashgua rubrocincta Wesenberg-Lund: East China Sea; South China Sea. (54, 58b)

ARENICOLIDAE

Abarenicora pacifica Healy & Wells: Daya Bay (Guangdong). (54)

Arenicola brasiliensis Monato [*A. cristata*]: Bohai; Yellow Sea. Intertidal. (20)

MALDANIDAE

Asychis disparidentata (Moore): Bohai; Yellow Sea; East China Sea; South China Sea. (54, 58b, 74, 87)

A. cf. *gangeticus* Fauvel: East China Sea; South China Sea. (23, 58b, 74, 83)

A. gotoi (Izuka): Bohai; Yellow Sea; East China Sea; South China Sea. (20, 36, 43, 58b, 73, 74, 83)

A. pigmentata Imajima: Taiwan Strait. (58b)

Axiothella australis Augener: East China Sea. (83)

A. rubrocincta (Johnson): East China Sea; South China Sea. (54, 58b)

Chymenura (*Cephalata*) *aciculata* Imajima: East China Sea. (58b)

C. (*Cephalata*) *longicauda* Imajima: East China Sea; South China Sea. (54, 58b)

Clymenella cincta (Saint-Joseph): Yellow Sea; East China Sea; South China Sea. (20, 36, 43, 83)

C. complanata Hartman: Taiwan Strait. (58b)

Euclymene annandalei Southern: Bohai; Yellow Sea; East China Sea; South China Sea. (23, 58b, 74)

E. insecta (Ehlers): East China Sea. (23)

E. lombricoides (Quatrefages): Bohai; Yellow Sea; East China Sea. (58b, 74)

E. oerstedii (Claparede): Yellow Sea; East China Sea; South China Sea. (20, 54, 58b)

E. unicinata Imajima: East China Sea. (58b)

Isocirrus planiceps (Sars): East China Sea. (58b)

I. cf. *watsoni* (Gravier) [*Clymene* (*Euclymene*) *watsoni*]: Yellow Sea. Muddy sand intertidal. (74)

Maldane cristata Treadwell: East China Sea; South China Sea. (54, 56)

M. globifex Grube: East China Sea. (87)

M. sarsi Malmgren: Yellow Sea; East China Sea; South China Sea. Muddy sand or sandy mud substrate. (20, 36, 43, 58b, 74, 83)

Microclymene caudata Imajima & Shiraki: South China Sea. (54)

Nicomache inormata Moore: East China Sea. (43)

N. lumbricalis (Fabricius): East China Sea. (58b, 83)

N. personata Johnson: Yellow Sea. Intertidal. (74)

Notoproctus pacificus (Moore): Taiwan Strait. (58b)

Petaloproctus cf. *terricolus* Quatrefages: South China Sea. Muddy sand or sandy substrate. (74)

Praxillella cf. *affinis* (Sars) [*Clymene* (*Praxillella*) *affinis*]: South China Sea. (74)

P. gracilis (Sars): East China Sea; South China Sea. (54, 58b, 83)

P. pacifica Berkeley: Southern Yellow Sea; East China Sea; South China Sea. (54, 58b, 74)

P. praetermissa (Malmgren) [*Praxilla praetermissa*]: Bohai; Yellow Sea; East China Sea; South China Sea. (20, 54, 58b, 74, 83)

OPHELIIDA
OPHELIIDAE

Antiobactrum brasiliensis (Hansen): Taiwan Strait. (58b)

Armandia intermedia Fauvel: Yellow Sea; East China Sea; South China Sea. (54, 58b, 74)

A. lanceolata Willey: Yellow Sea. (20)

A. leptocirrs Grube: Yellow Sea; East China Sea; South China Sea. (23, 54, 58b, 74)

A. longicaudata (Caullery): Daya Bay (Guangdong). (54)

Euzonus arcticus Grube: Yellow Sea. Intertidal. (74)

E. dillonesis (Hartman): Yellow Sea. (20, 74)

E. ezoensis (Okuda): Yellow Sea. Fine sand intertidal. (74)

Ophelia acuminata Oersted [*Ammotrypane aulogaster*]: Yellow Sea; East China Sea; South China Sea. Subtidal. (20, 74, 83)

O. cf. *limacina* (Rathke): Yellow Sea. (74)
Ophelina acuminata Oersted: Taiwan Strait. (58b)
O. anlogaster (Rathke): East China Sea; South China Sea. (54, 58b)
O. grandis Pillai: Taiwan Strait. (58b)
Polyophthalmus pictus Dujardin: Yellow Sea; South China Sea. Intertidal. (20, 37, 54, 74)
Travisia japonica Fujiwara: Yellow Sea; East China Sea. (20, 58b, 74, 83)
T. pupa Moore: Yellow Sea. (74)

SCALIBREGMIDAE

Hyboscolex pacificus (Moore): Bohai; Yellow Sea. (20, 74)
Scalibregma inflatum Rathke: Yellow Sea; East China Sea; South China Sea. Subtidal. (20, 54, 58b, 74, 83)
Scalisetosus fragilis (Claparede): Daya Bay (Guangdong). (54)
S. longicirrus (Schmarda): Daya Bay (Guangdong); Xisha (=Paracel Islands). (37, 43, 54)

AMPHINOMIDA
AMPHINOMIDAE

Amphinome pulchra Horst: South China Sea. (43, 54, 74)
A. rostrata (Pallas): Xisha (=Paracel Islands). (43, 74)
Chloeia flava (Pallas): Yellow Sea; East China Sea; South China Sea. (20, 23, 58b, 108)
C. fusca McIntosh: East China Sea; South China Sea. (58b, 74)
C. inermis Quatrefages: Hainan. Intertidal. (56)
C. parva Baird: East China Sea; South China Sea. (31, 54, 58b)
C. rosea Potts: East China Sea. (58b)
C. violacea Horst: East China Sea; South China Sea. (43, 54, 58b, 74, 83)
Eurythoe chilensis Kinberg: East China Sea; South China Sea. (54, 58b)
E. complanata (Pallas): East China Sea; South China Sea. (37, 54, 58b)
E. parvecarunculata Horst: Bohai; Yellow Sea; East China Sea; South China Sea. (20, 43, 46, 51, 54, 58b)
Linopherus ambigua (Monro) [*Pseudeurythoe ambigua*]: Bohai; Yellow Sea; East China Sea. (12, 54, 58b, 74)
L. hirsura (Wesenderg-Lund) [*P. hirsura*]: East China Sea; South China Sea. (51, 54)
L. oligobranchia (Wu, Shen & Chen) [*P. oligobranchia*]: South China Sea. (37)
L. microcephala Fauvel: South China Sea. (54)
L. pancibranchiata (Fauvel) [*P. pancibranchiata*]: Yellow Sea; South China Sea. (20, 54, 74)
L. spiralis (Wesenberg-Lund) [*P. spiralis*]: South China Sea. (54)
Notopygos gigas Horst: South China Sea. (37)
N. sibogae Horst: South China Sea. (37)
N. supragigas Uschakov & Wu: Xisha (=Paracel Islands). (11, 43, 74)

Paramphinome indica Fauvel: South China Sea. (74)
Pareurythoe borealis (Sars): South China Sea. (74)
Pherecardia striata (Kinberg): South China Sea. (37, 74)

EUPHROSINIDAE

Euphrosine foliosa Audouin & Milne Edwards: Xisha (=Paracel Islands). (43)
E. myrtosa Savigny: South China Sea. (74)

EUNICIDA
ONUPHIDAE

Diopatra amboinensis Audouin & Milne Edwards: East China Sea. (58b, 83)
D. chilienis Quatrefages [*D. bilogata, D. neapolitana*]: Yellow Sea; East China Sea; South China Sea. (8, 12, 51, 52, 54, 58b, 74)
D. cuprea (Bosc): East China Sea; South China Sea. (54, 58b)
D. dentata Kinberg: East China Sea. (23)
D. neotridens Hartman: East China Sea. (74)
D. oblique Hartman: Xiamen (Fujian). (58b, 74)
D. cf. *ornata* Moore: East China Sea; South China Sea. (74, 87)
D. sugokai Izuka: Bohai; Yellow Sea. (20)
D. variabilis Southern: East China Sea; South China Sea. (36, 54, 58b, 90, 108)
Epidiopatra hupferiana Augener: East China Sea. (74)
Hyalinoecia tubicola (Müller): South China Sea. (36, 74)
Nothria atlantisa Hartman: Taiwan Strait. (87)
N. conchylega (Sars): East China Sea. (87)
N. holobranchiata (Marenzeller) [*Onuphis holobranchiata*]: Daya Bay (Guangdong). (54)
N. pallida Moore: East China Sea. (87)
Onuphis chinensis Uschakov & Wu: Yellow Sea. (74)
O. cirrobranchiata Moore: East China Sea. (58b, 83)
O. eremita Audouin & Milne Edwards: Yellow Sea; East China Sea; South China Sea. (20, 36, 43, 54, 58b, 71, 73, 74)
O. fujianensis Uschakov & Wu: East China Sea. (9, 58b)
O. geophiliformis (Moore) [*Nothria geophiliformis, O. (N.) geophiliformis*]: Yellow Sea. Subtidal. (10, 74)
O. irenuta Audouin & Milne Edwards: East China Sea. (87)
O. pseudodibranchiata Gallardo: South China Sea. (36, 43, 54)
O. willemoesis Moore: East China Sea. (36, 43, 58b, 87)
Rhamphobrachium chuni Ehlers: Taiwan Strait. (58b)
R. diversosetosum Monro: Taiwan Strait. (58b)

EUNICIDAE

Eunice afra Peters: Daya Bay (Guangdong); Xisha (=Paracel Islands). (43, 54, 56, 74)
E. antennata (Savigny): East China Sea; South China Sea. (43, 54, 58b)

E. aphroditois (Palla): South China Sea. (51, 54, 56, 74)
E. australis Quatrefages: East China Sea; South China Sea. (43, 54, 56, 58b)
E. coccinea Gravier: South China Sea. (43)
E. gracilis Crossland: East China Sea. (23)
E. indica Kinberg: East China Sea; South China Sea. (9, 21, 36, 43, 51, 52, 54, 58b, 74, 83)
E. kobiensis McIntosh: East China Sea. (74)
E. longicirrata Webster: East China Sea. (51)
E. marenzelleri Gravier: South China Sea. (43)
E. medicina Moore: East China Sea. (58b)
E. multipectinata Moore: East China Sea. (80)
E. northioidea Moore: East China Sea. (58b)
E. pennata (Müller): Yellow Sea. (23)
E. vittata (Delle Chiaje): East China Sea. (58b)
Lysidice collaris Grube [*L. ninetta collaris*]: East China Sea. (23)
L. ninetta Audouin & Milne Edwards: East China Sea; South China Sea. (37, 56, 74)
Marphysa belli Audouin & Milne Edwards: East China Sea; Hong Kong; Beibuwan (=Gulf of Tongking). (79, 94)
M. bifurcata Kote: South China Sea. (54)
M. depressa (Schmarda): South China Sea. (54, 74)
M. macintoshi Crossland: Xiamen (Fujian); Daya Bay (Guangdong); Xisha (=Paracel Islands). (36)
M. mossanbica (Peters): South China Sea. (56)
M. sanguinea (Montagu) [*M. iwamusi*]: Bohai; Yellow Sea; East China Sea; South China Sea. (8, 12, 17, 20, 51, 54, 58a, 58b, 74, 108)
M. sinensis Monro: East China Sea. (23, 54, 58b, 83)
M. stragulun (Grube): East China Sea; South China Sea. (36, 43, 51, 58b, 83)
M. tamurai Okuda: South China Sea. (54)
Nematonereis unicornis (Grube): East China Sea; South China Sea. (43, 58b)
Palola siciliensis Grube: South China Sea. (43)

EUNIPHYSIDAE

Euniphysa aculeata Wesenberg-Lund: East China Sea; South China Sea. (53, 54, 58b, 71, 74, 83)
E. oculata Wu, Sun & Chen: South China Sea. (45, 74)
E. unicusa Sen & Wu: Beibuwan (=Gulf of Tongking). (65)
Heterophysa tridontesa Sen & Wu: South China Sea. (65)
Paraeuniphysa falciseta Sen & Wu: Southern Nansha (=Spratly Islands). (65)
P. spinea (Miura) [*Euniphysa spinea, Eunice spinea*]: Southern Nansha (=Spratly Islands). (65)
P. taiwanensis Wu & He: Taiwan Strait. (57)

LUMBRINERIDAE

Lumbrineris acutiformis Gallardo: Taiwan Strait. (58b)
L. amboinensis (Grube): East China Sea; South China Sea. (58b, 90)
L. bifurcata (McIntosh): East China Sea; South China Sea. (43, 54, 58b)
L. caudaensis Gallardo: East China Sea; South China Sea. (58b, 90)
L. cruzensis Hartman: Yellow Sea; East China Sea; South China Sea. (8, 20, 36, 43, 58b, 73, 83)
L. heteropoda (Marenzeller): Bohai; Yellow Sea; East China Sea; South China Sea. Intertidal and subtidal. (8, 12, 20, 51, 52, 71, 72, 74)
L. inflata (Moore): Yellow Sea; East China Sea; South China Sea. (20, 54, 58b, 83)
L. japonica (Marenzeller): Yellow Sea; East China Sea. (20, 58b, 83)
L. latreilli Audouin & Milne Edwards: Bohai; Yellow Sea; East China Sea; South China Sea. Intertidal. (20, 36, 51, 54, 58b, 73, 74)
L. longiforlia Imajima & Hartman [*Lumbriconereis debilis*]: Yellow Sea; East China Sea. (8, 58b, 74)
L. meteorana Augener: Daya Bay (Guangdong). (54)
L. mucronata (Ehlers): Daya Bay (Guangdong). (54)
L. nagae Gallardo: East China Sea; South China Sea. (54, 58b, 74, 90)
L. ocellata Grube: East China Sea. (23)
L. pterignatha Gallardo: Taiwan Strait. (58b)
L. shiinoi Gallardo: East China Sea; South China Sea. (51, 52, 54, 58b, 108)
L. simplex Southern: Daya Bay (Guangdong). (54)
L. sphaerocephala (Schmarda): South China Sea. (37)
L. tetraura (Schmarda) [*Lumbriconereis impatiens*]: Yellow Sea; East China Sea. Intertidal. (12, 20, 52, 54, 58b, 74, 83)
Ninoe bruuni Gallardo: South China Sea. (90)
N. gemma Moore: East China Sea. (87)
N. palmata Moore: Yellow Sea, East China Sea. (10, 54, 74)

ARABELLIDAE

Arabella iricolor (Montagu): Yellow Sea; South China Sea. Intertidal and subtidal. (20, 54, 74)
A. mutans (Chamberlin): East China Sea; South China Sea. (56, 83)
A. novecrinita Crossland: Taiwan Strait; Daya Bay (Guangdong). (54, 83)
Drilonereis filum (Claparede): Bohai; Yellow Sea; East China Sea; South China Sea. (20, 51, 54, 58b)
D. logani Crossland: East China Sea; South China Sea. (36, 43, 54, 58b)
D. robustus (Moore): Yellow Sea; Taiwan Strait. (83)
Labrorostratus parasiticus Saint Josep: Yellow Sea. (20, 74)
Notocirrus japonicus (Okuda): East China Sea; South China Sea. (54, 58b)

DORVILLEIDAE

Dorvillea cf. *pseudorubrovittata* Berkeley [*D. moniloceras*]: Yellow Sea. (8, 20, 74)

Ophryotrocha puerilins Claparede de Metschnikov: Yellow Sea. (20, 74)

Schistomeringos incerta (Schmarda) [*Dorvillea japonica*]: East China Sea; South China Sea. (54, 58b, 74, 90, 108)

S. japonica (Annenkova) [*Dorvillea japonica*]: Yellow Sea; East China Sea. (8, 20, 58b)

S. rudolphi (Chiaja) [*Dorvillea rudolphi*]: Daya Bay (Guangdong); Guangxi. (54, 74)

HARTMANIELLIDAE

Hartmaniella fujianensis He & Wu [*H.* cf. *crecta*]: Taiwan Strait. (67, 83)

HISTRIOBDELLIDAE

Myriochele picta Southern: Taiwan Strait. (58b)

STERNASPIDA
STERNASPIDAE

Sternaspis scutata (Renier): China's seas. (18, 23, 36, 43, 51, 52, 53, 54, 58b, 83)

OWENIIDA
OWENIIDAE

Owenia fusiformis Delle Chiaje: Yellow Sea; East China Sea; South China Sea. (20, 43, 51, 52, 53, 54, 58b)

FLABELLIGERIDA
FLABELLIGERIDAE

Brada ferruginea Gallardo: Taiwan Strait. (58b, 83)

B. mammillata Grube: Yellow Sea. (20)

B. talchsapensis Fauvel: Taiwan Strait. (58b, 83)

B. villosa (Rathke): Yellow Sea; East China Sea; South China Sea. (20, 54, 58b, 83)

Diplocirrus erythroporus Gallardo: South China Sea. (90)

Pherusa cf. *bengalensis* (Fauvel) [*Stylarioides bengalensis*]: Yellow Sea, East China Sea. (58b, 74)

P. eruca (Claparede) [*Stylarioide eruca*]: East China Sea. (58b)

P. granulosus (Caullery) [*Stylarioides granulosus*]: Taiwan Strait. (78)

P. parmata (Grube) [*Stylarioides parmata*]: East China Sea; South China Sea. (58b, 74, 83)

P. plumosa (Müller) [*Stylarioides plumosa*]: Yellow Sea; East China Sea. (20, 58b, 83)

Piromis congoensis (Grube) [*Pycnoderma conogensis*]: Daya Bay (Guangdong). (54)

TEREBELLIDA
SABELLARIDAE

Idanthyrsus pennatus (Peters) [*Pallasia pennata*]: Xisha (=Paracel Islands). Coraf reefs. (43, 74)

Lygdamis giardi (McIntosh): Yellow Sea; East China Sea. (20, 39, 58b)

L. indicus Kinberg: East China Sea; South China Sea. (73)

L. nesiotes (Chamberlin) [*Tetreres nesiotes*]: South China Sea. (37)

L. porrectus Ehlers: South China Sea. (43)

Sabellaria alcocki Gravier: South China Sea. Water depth 30 m, muddy sand substrate. (74)

S. cementarium Moore: South China Sea. Water depth 10 m, sandy mud substrate. (58b, 74)

S. ishikawai Okuda: Yellow Sea. (20, 39, 74)

PECTINARIDAE

Amphictene capensis (Pallas) [*Pectinaria (A.) capensis*]: East China Sea; South China Sea. (43, 58b)

A. japonica Nilsson [*Pectinaria japonica, P. (A.) japonica*]: Yellow Sea, intertidal; South China Sea, subtidal. (20, 58b, 74)

Lagis bocki (Hessle) [*Pectinaria (L.) bockis*]: Yellow Sea. (20, 74)

L. neapolitana Claparede [*Pectinaria (L.) neapolitana*]: Bohai; Yellow Sea. (74)

Pectinaria aegyptia Marenzeller: East China Sea. (58b)

P. conchilega Grube: South China Sea. (43)

P. dimai Zachs: Yellow Sea. (20)

P. papillosa Caullery: East China Sea. (58b)

AMPHARETIDAE

Ampharete acutifrons (Grube): Qingdao and Yantai (Shandong), intertidal; Daya Bay (Guangdong), subtidal. (20, 54)

A. arctica (Malmgren): Yellow Sea; East China Sea. (20, 87)

A. reducta Chamgelin: Yellow Sea; East China Sea. (20, 83)

Amphicteis glabra Moore: East China Sea. (58b)

A. gunneri (Sars): Yellow Sea; East China Sea; South China Sea. (20, 21, 47, 58b, 64, 74, 87)

A. mederi Annenkova: Jiaozhou Bay (Shandong). (20)

A. scophrobranchiata Moore: South China Sea. (74)

Amphisamytha japonica Hessle: East China Sea. (58b, 87)

Auchenoplex crinita Ehlers: East China Sea. (83)
Isolda pulchella Müller: East China Sea; South China Sea. (45, 58b)
I. whydahaensis Augener: Jiaozhou Bay (Shandong). (20)
Melinna aberrans: South China Sea. (45)
M. cristata (Sars): Yellow Sea. (74)
M. elisabethae McIntosh: Yellow Sea. (20)
Paramphicteis angustifolia (Grube): South China Sea. (74)
Sabellides octocirrata (Sars): Taiwan Strait. (58b)
Samytha californiensis: South China Sea. (45)
S. gurjanovae Uschakov: South China Sea. (74)
S. oculata Grube: East China Sea. (58b, 83)
S. sexcirrata Sars: South China Sea. (45)
S. sinica Wu & Sun: East China Sea. (58b)
Schistocomus hiltoni Chamberlin: Yellow Sea. (74)
S. sovjeticus Annekova: Jiaozhou Bay (Shandong). (74)

TRICHOBRACHIDAE

Filibranchus roseus Malmgren: Taiwan Strait; Daya Bay (Guangdong). (54)
Terebellides stroemii Sars: Bohai; Yellow Sea; East China Sea; South China Sea. Subtidal. (20, 21, 43, 52, 53, 54, 58b, 71, 74, 108)
Trichobranchus bibranchiatus Moore: Yellow Sea; East China Sea. (20, 83)

TEREBELLIDAE

Amaeana accraensis: South China Sea. (54)
A. antipoda Augener: Yellow Sea. (20)
A. occidentalis (Hartman): Yellow Sea; East China Sea. (74)
A. trilobata (Sars): Yellow Sea; East China Sea; South China Sea. (20, 54, 58b, 113)
Amphitrite cirrate (Müller): Taiwan Strait. (58b)
A. oculata Hessle: Hong Kong. (113)
Artacama proboscidea Malmgren: Yellow Sea. (74)
Eupolymnia marenzeller (Caullery): South China Sea. (37)
E. nebulosa (Montagu): East China Sea; South China Sea. (37, 58b)
E. umbonis Hutchings: Hong Kong. (113)
Lanassa capensis Day: Taiwan Strait. (52)
Lanice auricula Hutchings: Hong Kong. (113)
L. conchilega (Palla): Yellow Sea; East China Sea. (74)
L. socialis (Müller): Taiwan Strait. (58b)
Longicarpus nodus Hutchings: Hong Kong. (113)
Loimia arborea Moore: East China Sea. (58b)
L. bandera Hutchings: Hong Kong. (113)
L. ingens (Grube) [*Terebella ingens*]: Hong Kong. (113)
L. medusa (Savigny): Yellow Sea; East China Sea; South China Sea. (20, 23, 43, 54, 58b, 83, 108)
L. montagui (Grube): East China Sea. (58b)
Lysilla pacifica Hessle: East China Sea; South China Sea. (54, 58b, 83, 113)
L. sinensis Wu & Sen: East China Sea. (58b)
L. ubianensis Caullery: Taiwan Strait. (58b)
Neoamphitrite robusta (Johnson): Bohai; Yellow Sea. (74)
Pista brevibranchia Caullery: Yellow Sea; East China Sea. (58b, 76)
P. cristata (Müller): Yellow Sea; East China Sea. (20, 54, 58b, 83, 87)
P. elongata Moore: Yellow Sea. (20)
P. fasciata (Grube): Yellow Sea; East China Sea. (20, 74)
P. foliigera Caullery: East China Sea; South China Sea. (54, 58b)
P. macrolobata Hessle: Taiwan Strait. (58b)
P. pachybranchiata Fauvel: East China Sea. (58b)
P. pacifica Berkeley: Yellow Sea. (74)
P. robustiseta Caullery: Yellow Sea. (20)
P. typha Grube: East China Sea; South China Sea. (54, 58b, 90, 113)
P. violacea Hartman-Schröder: Hong Kong. (113)
P. zachsi Annekova: Yellow Sea. (20)
Polycirrus dodeka Hutchings: Hong Kong. (113)
P. multus Hutchings: Hong Kong. (113)
P. plumosus (Wolleback): South China Sea. (54)
P. quadratas Hutchings: Hong Kong. (113)
Rhinothelepus occabus Hutchings: Hong Kong. (113)
Scionella japonica Moore: East China Sea. (58b)
Streblosoma duplicata Hutchings: Taiwan Strait; Hong Kong. (58b, 113)
Terebella copia Hutchings: Hong Kong. (113)
T. ehrenbergi Grube: Yellow Sea; South China Seas. (20, 43, 54, 108, 113)
Thelepus opimus Hutchings: Hong Kong. (113)
T. plagiostoma (Schmardo): East China Sea. (74)
T. pulvinus Hutchings: Hong Kong. (113)

SABELLIDA
SABELLIDAE

Bispira vancouveri (Kinberg): Yellow Sea; Jiaozhou Bay (Shandong). (20)
Branchiomma cingulata (Grube) [*Dasychone cingulata*]: Daya Bay (Guangdong), rocky shore; Hainan, intertidal and coral reefs; South China Sea, water depth 20–40 m. (54, 56, 74, 83)
B. nigromaculata (Baird): Daya Bay (Guangdong). Rocky intertidal. (54)
B. orientalis (McIntosh): East China Sea. (23, 109)
B. pacificum (Johansson): Daya Bay (Guangdong). Rocky intertidal. (54)
B. serratibranchis (Grube): Zhejiang. Subtidal. (74)
Chone filicaudata Southern: East China Sea. Shallow water, muddy sand substrate with shells. (109)

C. infundibuliformis Kröyer: Yellow Sea; East China Sea. Water depth 20–50 m, muddy sand substrate or muddy substrate with shells. (74)

C. teres Bush: Yellow Sea. Water depth 7–27 m, muddy sand substrate with shells. (20)

Demonax microphthalmus (Verrill): Daya Bay (Guangdong). (1, 54)

Eudistylis vancouveri (Kinberg) [*Potamilla chiaochouensis*]: Yellow Sea. Muddy sand intertidal. (54)

Hypsicomus phaeotaenia (Schmarda): Xisha (=Paracel Islands). (37, 43, 74)

Megalomma vesiculosum (Montagu) [*Branchiomma vesiculosm*]: Guangxi; South China Sea. Muddy sand intertidal. (54, 74)

Myxicola infundibulum (Renier): Yellow Sea. Muddy sand intertidal. (20, 74)

Potamilla cf. *acuminata* Moore & Bush [*P. acuminata*]: Bohai. Muddy sand intertidal. (74)

P. cf. *myriops* Marenzeller: Yellow Sea. Muddy sand intertidal. (20, 74)

P. polyophtham Grube: Xiamen (Fujian). (23, 105)

P. reniformis (Müller): Yellow Sea; East China Sea; South China Sea. (54, 58b, 71, 74)

P. torelli Malmgren: China's coast. (23, 54)

Pseudobranchiomma emersoni Jones: South China Sea. (54)

Sabella penicillus Linnaeus [*S. pavonina*]: Yellow Sea; East China Sea; Xisha (=Paracel Islands). (20, 37, 43, 74)

Sabellastarte indica (Savigny): Beibuwan (=Gulf of Tongking), Xisha (=Paracel Islands). Water depth 19–41 m. (9, 23, 37, 105)

S. sanctijosephi (Gravier): South China Sea; Xisha (=Paracel Islands). (36, 43)

S. zebuensis (McIntosh): East China Sea; South China Sea. (9, 23, 54, 105)

SERPULIDAE

Apomatus enosimae Marenzeller: South China Sea. Subtidal. (74)

Ditrupa arietina (Müller): Yellow Sea. Subtidal. (74)

Ficopomatus macrodon Southern: East China Sea continental shelf. (58b)

Hydroides albiceps (Grube): East China Sea; South China Sea. (1, 54, 74, 84, 95)

H. centrospina Wu & Chen: Yulin harbor (Hainan). (34)

H. dirampha Mörch [*H. lunulifera*]: Bohai; Yellow Sea; East China Sea; South China Sea. (1, 43, 54, 74, 97, 103)

H. elegans (Haswell) [*H. norvegica*]: China's seashore, more abundant in the south. (1, 2, 54, 74, 93, 95)

H. exaltatus (Marenzeller): Daya Bay (Guangdong); Hong Kong. (1, 43, 54, 74, 103)

H. ezoensis Okuda: From Bohai and Yellow Sea to Daya Bay (Guangdong). (1, 2, 20, 54, 74, 93)

H. fusca Imajima: South China Sea. (43)

H. fusicola Mörch [*H. uncinata*]: Bohai; Yellow Sea; Hong Kong. (20, 74, 93, 107)

H. helmatus (Irosa): South China Sea. (81)

H. inornata Pillai: Shandong; Daya Bay (Guangdong); Hong Kong; Xisha (=Paracel Islands). (1, 54, 103, 107)

H. longispinosa Imajima: Xielang Bay (Hainan); Yongxing Island (Xisha=Paracel Islands). (93)

H. longistylaris Chen & Wu: Zhangpu (Fujian); Weizhou Island (Guangxi). (61)

H. minax (Grube): Weizhou Island (Guangxi); Xielang Bay (Hainan). (93)

H. multispinosa Marenzeller: South China Sea. (81, 93)

H. nanhaiensis Wu & Chen: South China Sea. (34)

H. prisca Pillai: Yantai (Shandong); Daya Bay (Guangdong). (1, 54, 95)

H. protulicola Benedict: East China Sea. (73)

H. rhombobulus Chen & Wu: East China Sea; Taiwan Strait; South China Sea. (1, 54, 61, 103)

H. tambalalmensis Pillai: Dongshan (Fujian); Daya Bay (Guangdong). (1, 54, 95)

H. trilobulus Wu & Chen: Yongxing Island (Xisha=Paracel Islands). (60)

H. tuberculata Imajima: Weizhou Island (Guangxi). (95)

H. xishaensis Shen & Wu: Yongxing Island (Xisha=Paracel Islands). (60)

Metavermilia acanthophora (Augener) [*Vermiliopus acanthophora*]: South China Sea. (54, 95, 103)

M. annobonensis Zibrowius: South China Sea. (43)

M. inflata Imajima: Hong Kong. (96, 103)

M. cf. *taenia* Zibrowius: South China Sea. (43)

Pomatoleios kraussii (Baird) [*P. crosslandi*]: East China Sea; South China Sea. Mid- and low-intertidal. (1, 20, 54, 81)

Pomatostegus stellatus (Abildgaard): Hong Kong. (18, 96)

Protula tubularia (Montagu): Bohai; Yellow Sea; East China Sea; South China Sea. (37, 74, 107)

Salmacina dysteri (Huxley): Bohai; Yellow Sea; East China Sea; South China Sea. (74, 81)

Serpula (Paraserpula) sinica Wu & Chen: Xisha (=Paracel Islands). (45)

S. vermicularis Linnaeus: China's seashore. Intertidal and subtidal. (1, 2, 54, 58b, 74, 107)

S. watsoni Willey: Hong Kong. (96, 103)

Spirobranchus giganteus (Pallas): South China Sea. Intertidal to 50 m water depth. (37, 58b, 74)

S. jousseaumei (Gravier): East China Sea; South China Sea. Water depth 40–60 m. (74)

S. latiscapus (Marenzeller): South China Sea. Water depth 39–205 m. (74)

S. maldivensis Pixell [*Pomatoceros triqueter*]: Bohai; Hainan; South China Sea. (1, 2, 54, 74)

S. polytrema (Philippi): Dongshan (Fujian); Daya Bay (Guangdong); Hong Kong; Weizhou Island (Guangxi). (1, 54, 96, 107)

S. semperi Mörch: Guangxi, intertidal; South China Sea, water depth 40–50 m. (1, 54, 74, 103)

S. sinensis Wu & Chen: South China Sea. (35)
S. tetraceros Schmarda: South China Sea. (54)
S. tricornis (Mörch) [*S. tricornigerus*]: Bohai; Yellow Sea; East China Sea; South China Sea. (1, 2, 54, 74, 103)
Vermiliopsis infundibulum glandigera-group Imajima [*V. ctenophora, V. glandigerus, V. infundibulum,*]: East China Sea; Daya Bay (Guangdong); Hong Kong. (1, 2, 54, 58b, 95, 103)
V. pygidialis (Willey): East China Sea; South China Sea. (43, 58b)

SPIRORBIDAE

Bushiella argutus (Bush): Yellow Sea. Intertidal. (74)
Circeis spirillum (Linnaeus): China's seashore. Subtidal, water depth 46–51 m. (74)
Dexiospira alveolatus Zachs [*Spirorbis (Dexiospira) nipponicus*]: Bohai; Yellow Sea. Mid-intertidal to upper subtidal. (74)
D. foraminosus Busch [*Neodexiospira foraminosus, Spirorbis foraminosus*]: Bohai; Yellow Sea; East China Sea; South China Sea. (1, 2, 54, 74, 81)

CTENODRILIDA
CTENODRILIDAE

Ctenodrilus serratus Schmidt: Yellow Sea. (49)

MYZOSTOMIDA
MYZOSTOMIDAE

Myzostoma antennulata Graff: Southern China. (12)
M. attenuatum Grygier: Hong Kong. (91)
M. bocki (Jägersten): Hong Kong. Host: crinoid (*Tropiometra afra*) etc. (91)
M. dodecaphalcis Grygier: Hong Kong. Host: crinoids (*Amphimetra*). (91)
M. lobatum von Graff: Dapeng Cove (Hong Kong). Host: crinoid (*Tropiometra afra*) etc. (91)
M. nasonovi (Fedotov): Hong Kong. Host: crinoid (*Tropiometra afra*) etc. (91)
M. cf. *pallidum* von Graff: Hong Kong. (91)

REFERENCES*

(1). Wang Jianjun, 1989. Ecology on fouling Polychaeta in Daya Bay. Collections of papers on marine ecology in the Daya Bay (I). China Ocean Press, pp:74–81.
(2). Wang Jianjun, 1991. Polychaeta foulers along coast of the Yellow Sea and Bohai. Ocean Bulletin 10(5):52–58.
(3). Uschakov, P. 1958. Polychaete *Paralacydonia paradoxa* Fauvel (Phyllodocidae) in Yellow Sea, China. Acta Zoologica Sinica, 10(4):416–420.
(4). Uschakov, P. and B. L. Wu, 1959. Polychaete of Yellow Sea, Phyllodocidae and Aphroditidae. The Collection of Ocean Institute, Academia Sinica, 1(4):1–40.
(5). Uschakov, P. and B. L. Wu, 1960. Fauna of Polychaete, China. Oceanologia & Limnologia Sinica, 3(2):86–93.
(6). Uschakov, P. and B. L. Wu, 1962a. Polychaete of Yellow Sea. II: Chrysopetalidae, Glyceridae, and Nephtyidae. Studia Marina Sinica, 1:1–32.
(7). Uschakov, P. and B. L. Wu, 1962b. Polychaete of Yellow Sea. III: Nereidae. Studia Marina Sinica, 1:33–56.
(8). Uschakov, P. and B. L. Wu, 1962c. Polychaete of Yellow Sea. III: Syllidae, Hesionidae, Pilargiidae, Amphinomidae, and Eunicidae. Studia Marina Sinica, 1:61–88.
(9). Uschakov, P. and B. L. Wu, 1962e. Notes on polychaete in Fujian and Zhejiang. Studia Marina Sinica, 1:89–109.
(10). Uschakov, P. and B. L. Wu, 1962f. Polychaete of Yellow Sea. VI: Polychaeta, Errantia. Studia Marina Sinica, 2:110–138.
(11). Uschakov, P. and B. L. Wu, 1962g. A new species of polychaete (*Notopygos supragigas*) from Xisha, South China Sea. Acta Zoologica Sinica, 14(2):261–265.
(12). Uschakov, P. and B. L. Wu, 1963a. Studies on ecology and zoogeography of polychaete in Yellow Sea, China. Studia Marina Sinica, 3:1–50.
(13). Uschakov, P. and B. L. Wu, 1963b. Study on polychaete zoogeography of Yellow Sea, China. Oceanologia & Limnologia Sinica, 5(2):154–162.
(14). Sun Ruiping and Wu Baoling, 1979. Preliminary report on the pelagic Polychaeta from Xisha Islands, Guangdong Province, China. Studia Marina Sinica, 15:57–70.
(15). Sun Ruiping, Wu Baoling and Yang Dejian, 1980. A study of the morphology and larval development of the *Neanthes japonica* (Isuka) from the China Sea. Journal of Shandong College of Oceanology, 10(3):100–110.
(16). Sun Ruiping and Wu Baoling, 1981. Preliminary studies on species of exogoninae (Polychaeta: Syllidae) of the China Coasts. Acta Oceanologica Sinica, 3(4):611- 628.
(17). Sun Ruiping and Yang Dejian, 1987. Studies of Orbiniidae (Polychaeta) from the Yellow Sea and the East China Sea. Studia Marina Sinica, 28:151–168.
(18). Sun Daoyuan and Dong Yongling, 1986. Ecological features of the Polychaetes in the Changjiang Estuary and adjacent waters. Studia Marina Sinica, 27:158–183.
(19). Sun Daoyuan and Chen Bida, 1990. A new species of *Synelmis* from offshore waters of China. Studia Marina Sinica, 31:129–132.
(20). Sun Daoyuan, 1990. New records and list of Polychaeta from Jiaozhou Bay. Studia Marina Sinica, 31:133–146.
(21). Jiang Jinxiang and Huang Liqiang, 1985. Species composition and distribution of benthic animals on the continental shelf of the East China Sea and its adjacent waters. Taiwan Strait, 4(1):89–98.
(22). Wu, B. L. 1962a. Notes on new species (Orbiniidae and Paraonidae, Polychaeta) from Yellow Sea, China. Acta Zoologica Sinica, 14(3):421–428.
(23). Wu, B. L. 1962b. Polychaete fauna feature in Fujian, China. Oceanologia & Limnologia Sinica, 4(1–2):87–93.
(24). Wu, B. L. 1963. Economic fauna of China (Annelida, Polychaeta). Chinese Science Press, 47pp.
(25). Wu, B. L. 1964. Sub-species and ecological feature of polychaete Capitellidae. Oceanologia & Limnologia Sinica, 6(3):261–271.
(26). Wu, B. L. and Yang, D. Z. 1962. *Saccocirrus* (Polychaeta) and its distribution in China's seas. Oceanologia & Limnologia Sinica, 4(3–4):169–179.
(27). Wu, B. L. and Yang, D. Z. 1963. Notes on *Saccocirrus* (Polychaeta) in China's seas. Shandong College of Oceanography Acta, 1:75–90.
(28). Wu Baoling and Chen Mu, 1963. Some fresh-water and mixohaline water Polychaeta from China. Oceanologia & Limnologia Sinica, 8(2):163–167.
(29). Wu Baoling and Chen Mu, 1963. A new species of Polychaeta worm of the Family Spionidae from Xisha Islands, with a review of the genus *Prionospio* Malmgren, 1867. Acta Zoologica Sinica, 16(1):54–58.
(30). Wu Baoling and Chen Mu, 1964. A new species of Polychaeta worm of the family Spionidae from Chushan Archipelago, East China Sea. Acta Zootaxonomica Sinica, 16(1):54–58.
(31). Wu Baoling and Chen Mu, 1966. A new and interesting species of the genus *Heterispio* (Polychaeta, Heterospionidae). Oceanologia & Limnologia Sinica, 8(2):164–168.

(32). Wu Baoling and Chen Mu, 1977. *Heterocossura*, A new genus of the Cossuridae (Polychaeta: Sedentaria). Acta Zoologica Sinica, 23(1):97–101.
(33). Wu Baoling and Chen Mu, 1980. Morphology, ecology, reproduction and larval development of *Pseudopolychaeta paucibranchiata* (Okuda). Acta Zoologica Sinica, 26(4):356–364.
(34). Wu Baoling and Chen Mu, 1981. Two new species of Hydroides (Polychaeta: Serpulidae) from South China Sea. Oceanologia & Limnologia Sinica, 12(4):354-357.
(35). Wu Baoling and Chen Mu, 1981. Two new species of the Family Serpulidae from the South China Sea. Acta Zootaxonomica Sinica, 6(3):245–249.
(36). Wu Baoling and Chen Mu, 1985. Studies on Polychaeta from the Xisha Islands and its adjacent waters (I-III). Journal of Oceanography of Huanghai and Bohai Seas, 3(2):52–61, 3(3):74–87.
(37). Wu Baoling, Shen Shoupeng and Chen Mu, 1975. Preliminary report of Polychaeta annelids from Xisha Islands, Guangdong Province, China. Studia Marina Sinica, 10:65–104.
(38). Wu Baoling and Sun Ruiping, 1978. Planktonic polychaete zoogeography and development in the islets of South China Sea. Oceanologia & Limnologia Sinica, 9(2):215–223.
(39). Wu Baoling and Sun Ruiping, 1979. On the occurrence of two Sabellariio worms in the Yellow Sea, with notes on their larval development. Acta Zootaxonomica Sinica, 25(2):130–142.
(40). Wu Baoling and Sun Ruiping, 1979. Studies of *Arenicola brasiliensis* Monato in the Bohai Sea and the Yellow Sea. Oceanologia & Limnologia Sinica, 10(3):257–272.
(41). Wu Baoling and Sun Ruiping, 1979. Revision of the genera *Nicon* and *Rullierinereis*, with description of a new genus *Sinonereis* (Polychaeta: Nereidae). Oceanic Selections, 2(2):95–112.
(42). Wu Baoling and Sun Ruiping, 1981. The life history of the Polychaetous annelid *Playnereis bicanaliculata* (Baird). Oceanologia & Limnologia Sinica, 12(3):270-278.
(43). Wu Baoling, Sun Ruiping and Chen Mu, 1980. Zoogeographical studies on Polychaeta from the Xisha Islands and its adjacent waters. Acta Oceanologica Sinica, 2(1):111–130.
(44). Wu Baoling, Sun Ruiping and Chen Mu, 1980. On Archiannelids of the Yellow Sea, I. Protodrilidae. Acta Oceanolgica Sinica 2(2);132–148.
(45). Wu Baoling, Sun Ruiping and Chen Mu, 1981. Two new species of Polychaete (Polychaeta). Acta Zootaxonomica Sinica, 6(1):22–26.
(46). Wu Baoling, Sun Ruiping and Yang Dejian, 1981. Study on Nereidae (Polychaeta) in China's seas. China Ocean Press, 228pp.
(47). Wu Baoling, Sun Ruiping and Yang Dejian, 1982. On the occurrence of endoparasitic Polychaeta annelids in Chinese waters. Oceanologia & Limnologia Sinica, 13(2):202–204.
(48). Wu Baoling, Chen Mu and Sun Ruiping, 1980. On Archiannelids of the Yellow Sea, II. Dinophilidae and Nerillidae. Acta Oceanologica Sinica, 2(3):90–97.
(49). Wu Baoling, Qiu Jianwen and Qian Peiyuan, 1991. New record of *Ctenodrilus serratus* (Polychaeta) in China's seas. Journal of Ocean University of Qingdao, 21(3):101–103.
(50). Wu Baolin and Zhao Jing, 1992. Preliminary studies on species of Hesionidae (Polychaeata) from Yellow Sea. Journal of Oceanography of Huanghai and Bohai Seas, 10(2):36–41.
(51). Wu Qiquan et al., 1984. The pollution survey in East China Sea (biology). China Ocean Press, pp:101–145.
(52). Wu Qiquan et al., 1985. Study on benthic ecology in Taiwan Strait. Acta Oceanologica Sinica, 7(3):378–387.
(53). Wu Qiquan et al., 1988. Comprehensive survey report on marine environment of the Xiamen harbour (benthos). Journal of Oceanography in Taiwan Strait, 7(1):15–35.
(54). Wu Qiquan et al., 1990. Distribution of Daya Bay's Polychaeta. Collections of Papers on Marine Ecology in Daya Bay (II). China Ocean Press, pp:320–332.
(55). Wu Qiquan et al., 1990. Community in sandy zone in Daya Bay. Collections of Papers on Marine Ecology in the Daya Bay (II). China Ocean Press, pp:290–297.
(56). Wu Qiquan and Wu Baoling, 1987. Preliminary study of Polychaeta Ecology in Luhuitou tidal zone around the Hainan Island. Journal of Oceanography in Taiwan Strait, 6(1):78–81.
(57). Wu Qiquan and He Minghai, 1988. A new genus and new species of Eunicidae from Taiwan Strait. Acta Zootaxonomica Sinica, 13(2):123–126.
(58a). Wu Qiquan, 1984. A new species of Orbiniidae (Polychaeta) from Hainan Island. Journal of Oceanography in Taiwan Strait, 3(2):203–207.
(58b). Wu Qiquan et al., 1984. Biodiversity of polychaetes in Taiwan Strait. (manuscript)
(59). Chen Mu, 1959. Zoonic illustration of China (Polychaete). China Science Press.
(60). Chen Mu and Wu Baoling, 1978. Two new species of genus *Hydroides* (Polychaeta, Serpulidae) from the Xisha Islands, Guangdong Province, China. Studia Marina Sinica, 12:141–147.
(61). Chen Mu and Wu Baoling, 1980. Two new species of the genus *Hydroides* (Polychaeta, Serpulidae). Oceanologia & Limnologia Sinica, 11(3):247–249.
(62). Chen Mu and Wu Baoling, 1980. Ecological distribution of three species of Nephtyidae in North China. Acta Zoologica Sinica, 26(3):250–254.
(63). Chen Mu and Wu Baoling, 1983. Pelagic Polychaetes from the Taiwan Bank. Marine Science Bulletin, 2(1):42–50.
(64). Shen Shoupeng and Wu Baoling, 1978. Preliminary report on the pelagic Polychaetes from Zhongsha Islands, Guangdong Province, China. Oceanologia & Limnologia Sinica, 9(1):99–107.
(65). Shen Shoupeng and Wu Baoling, 1990. A new family of Polychaeta (Euniphysidae). Acta Oceanologica Sinica, 12(6):765–772.
(66). Shen Shoupeng and Wu Baoling, 1980. Survey of planktonic polychaete in Xisha islets. Report of marine organisms survey in Xisha islets, South China Sea, pp:201–205.
(67). He Minghai and Wu Qiquan, 1986. A new species of Hartmanidllidae (Polychaeta) from Taiwan Strait. Journal of Oceanography in Taiwan Strait, 5(1):65–69.
(68). He Minghai and Wu Qiquan, 1988. A new genus and a new species of Nereidae (Polychaeta). Acta Oceanologica Sinica, 10(4):490–491.
(69). He Minghai and Wu Qiquan, 1988. Description of a new species of Nereidae (Polychaeta). Acta Oceanologica Sinica, 10(4):490–491.
(69). He Minghai and Wu Qiquan, 1988. Description of a new species of *Nereis* from the East China Sea. Acta Zootaxonomica Sinica, 13(4):337–339.
(70). He Minghai, 1987. A new species of Nereidae from the East China Sea. Acta Zootaxonomica Sinica, 12(4):346–349.
(71). He Minghai, 1990. Distribution of Polychaeta in subtidal zone, Dongshan Bay, Fujian Province. Journal of Oceanography in Taiwan Strait, 9(3):206–211.
(72). He Minghai, 1990. Preliminary study on intertidal ecology of Polychaeta in Dongshan Bay. Marine Science Bulletin, 9(2):48–52.
(73). He Minghai, Wu Qiquan and Zheng Fengwu, 1987. Species composition and number distribution of Polychaeta in Xiamen Harbour. Journal of Oceanography in Taiwan Strait, 6(3):251–259.
(74). Yang Dejian and Sun Ruiping, 1988. Polychaetous annelids commonly seen from the Chinese waters. China Agricultural Press, 352pp.
(75). Zheng Fengwu and Wu Qiquan, 1987. A new species of Mereidae (Polychaeta) from Taiwan Strait. Journal of Oceanography in Taiwan Strait, 9(2):103–106.
(76). Meng Fan and Ding Zhihu. Preliminary study on small-size species of Syllidae (Polychaeta). (manuscript)
(77). Zhao Jing and Wu Baoling, 1991. A preliminary study on the family Paraonidae (Polychaeta) from Huanghai Sea. Journal of Oceanography of Huanghai and Bohai Seas, 9(2):26–35.
(78). Zhao Jing and Wu Baoling, 1991. A new interstitial species of the genus *Pisione* (Polychaeta: Pisionidae) from Yellow Sea, China. Oceanologia & Limnologia Sinica, 22(4):304–308.
(79). Zhao Jing and Wu Baoling, 1992a. A new species of *Pisione* (Pisionidae, Polychaeta) from Yellow Sea, China. Acta Zootaxonomica Sinica, 17(2):181–183.
(80). Zhao Jing and Wu Baoling, 1992b. A new species of *Chone* (Sabellidae, Polychaeta) from Yellow Sea, China. Acta Zootaxonomica Sinica, 17(4):121–123.

(81). Huang Zongguo and Cai Ruxing, 1984. Marine fouling organisms and its prevention (I). China Ocean Press, 423pp.

(82). Dong Yongting and Wang Yonghong, 1985. A new record of three species of *Lumbrineris* (Polychaeta) along the coastal waters of Zhejiang Province. Marine Science Bulletin, 4(2):38–42

(83). Fujian Institute of Oceanography, 1988. A comprehensive oceanographic survey of the central and northern part of the Taiwan Strait. China Science Press, 423pp.

(84). Chlebovitsch and Wu, B. L. 1962a. Studies on annelid (Polychaeta). III: Nereidae. Studia Marine Sinica, 1:33–53.

(85). Chlebovitsch and Wu, B. L. 1962b. Studies on annelid (Polychaeta). V. Nereidae (addenda) from Yellow Sea, China. Acta Zoologica Sinica, 14(2):267–378.

(86). Chlebovitsch and Wu, B. L. 1963. Sudies on Nereidae (Polychaeta) from Hainan Island intertide zone. Studia Marine Sinica, 4:48–60.

(87). Yamashita, Hideo, 1977. Research on benthos of East China Sea and Yellow Sea IV: distribution of polychaetes. Western Sea Region Fishery Research Institute, 49:29–67.

(88). Wu, B. L., Ding, and J. Zhao, 1992. A new subspecies *Petitia* (Polychaeta Syllidae) from the Huanghai Sea. (manuscript)

(89). Wu, B. L., J. Zhao, and Ding, 1992. A new meiofauna Polychaeta *Pholoe* (Polychaeta: Sigalionidae) from the Huanghai Sea. Acta Oceanologia Sinica, in press.

(90). Gallardo, V. A., 1967. Polychaeta from the Nha Tram, South Viet Nam. Nagta Rep. 4(3):35–279

(91). Grygier, M. J. 1992. Hong Kong Myzostomida and their Indo-Pacific distribution. In: Proceeding of the Fourth International Marine Biological Workshop: The Marine Flora and Fauna of Hong Kong and Southern China, 1989. Hong Kong University Press, pp:131–147.

(92). Hanley, R. 1992. Hong Kong scaleworms (Polychaeta: Polynoidae). In: Proceeding of the Fourth International Marine Biological Workshop: The Marine Flora and Fauna of Hong Kong and Southern China, 1989. Hong Kong University Press, pp:361–369.

(93). Imajima, M. 1967. Serpulinae from Japan. Bull. Nat. Sci. Mus. Ser. A (Zool.), 2(4):229–248.

(94). Imajima, M. 1975. Lumbrinereidae of polychaetous annelids from Japan, with descriptions of six new species. Bull. Nat. Sci. Mus. Ser. A (Zool.), 1(1):5–37.

(95). Imajima, M. 1976. Serpulid Polychaetes from Tanega-shima, southwest Japan. Mem. Nat. Sci. Mus., Tokyo, 9:123–143.

(96). Imajima, M. 1977. Serpulidae collected around Chichi-jima. Mem. Nat. Sci. Mus., 10:89–111.

(97). Imajima, M. 1978. Serpulidae collected around Nii-jima and O-shima, Izuland. Mem. Nat. Sci. Mus., Tokyo, 11:49–72.

(98). Imajima, M. and Taked Y., 1987. Nephytidae (Polychaeta) from Japan II. The Genera *Dentinephtys* and *Nepthys*. Bull. Nat. Sci. Mus. A (Zool.), 13(2):41–77.

(99). Mackie, A. S. Y., 1986. *Prionospio saccifera* sp. Nov. (PolychaetaL Spionidae) from Hong Kong and the Red Sea, with a description of *Prionospio ehlersi* Fauvel. Proceedings of the Second International Marine Biological Workshop: The Marine Flora and Fauna of Hong Kong and Southern China Sea. Hong Kong University Press, pp:363–375.

(100). Mackie, A. S. Y., 1986. The Poecilochaetidae and Trochochaetidae (Polychaetidae) of Hong Kong. Proceedings of the Second International Marine Biological Workshop: The Marine Flora and Fauna of Hong Kong and Southern China Sea. Hong Kong University Press, pp:337–362.

(101). Mackie, A. S. Y., 1991. A new species of *Scolplos* (Polychaeta: Orbiniidae) from Hong Kong and a comparison with the closely related *Scolopos marsupialis* Southern, 1921 from India. Asian Marine Biology, 8:35–44.

(102). Mackie, A. S. Y., 1992. Hong Kong Orbiginiidae (Annelida: Polychaeta). Proceedings of International Marine Biological Workshop: The Marine Flora and Fauna of Hong Kong and Southern China, 1989. Hong Kong University Press.

(103). Mak, P. M. S., 1980. The coral associated Polychaetes of Hong Kong, with species reference to the serpulids. Proceedings of the First International Marine Flora and Fauna of Hong Kong and Southern China. Hong Kong University Press, pp:595–617.

(104). Pettibone, Marian H. 1989. Revision of the Aphroditoid Polychaetes of the family Acoetidae kinberg (=Polydontidae Augenger) and the Re-establishment of *Acoetes* Audouin and Milne-Edwards, 1832, and *Euarche* Ehlers, 1887. Smithsonian Institution Press, Smithsonian Contribution to Zoology, 464:1–138.

(105). Monro, C. C. A., 1934. On a collection of Polychaeta from the coast of China. Ann. Mag. Nat. Hist., 13(10):353–380.

(106). Ohwada, T., 1992. A new species of *Aglaophanus* (Polychaeta: Nephtyidae) from Hong Kong. Proceedings of International Marine Biological Workshop, The Marine Flora and Fauna of Hong Kong and Southern China, 1989. Hong Kong University Press, pp:149–155.

(107). Pillai, T. G. 1971. Studies on a collection of marine and brackish water polychaete annelids of the family Serpulidae from Ceylon. Ceylon J. Sci. (Biol. Sci.), 9(2):88–130.

(108). Shin, P. K. S. 1980. Some polychaetous annelids from Hong Kong waters. Proceedings of the First International Marine Biological Workshop: The Marine Flora and Fauna of Hong Kong and Southern China Sea. Hong Kong University Press, pp:161–172.

(109). Treadwell, A. L. 1936. Polychaetous annelids from Amoy, China. Proc. U.S. Nat. Mus., 83(2984):261–279.

(110). Uschakov, P. 1972. Polychaetes of the Suborder Phyllodociformia of the Polar Basin and the northwestern part of the Pacific. Akad. Nauk SSSR. Zool. Inst. Fauna of the SSSR. 102, 271pp.

(111). Zhao, J. and Wu., B. L. 1991. A new species of interstitial Polychaete *Hesionura shandongensis* sp. N. from Yantai, the Huanghai Sea. Acta Oceanologica Sinica, 10(3):447–450.

(112). Zhao, J. and Wu, B. L. 1992. Two new species of Exogone and Sphaerosyllis from the Huang Sea. Acta Oceanologica Sinica, 11(1):131–137.

(113). Zhang, X., F. Y. Zhang, and B. L. Wu, 1963. Economic fauna of China, Annelida, Echinodermata, Hemichordata and Urochordata. China Science Press, 3:29–35.

*: (1)-(86) and (113) in Chinese, (87) in Japanese
Compiled by Wu Qiquan, Wang Jiangjun, and He Minghai; Edited by Wu Baoling, Sun Daoyuan, and Qiu Jianwen.

OLIGOCHAETA

TUBIFICIDAE

Ainudrilus gibsoni Erséus: Hong Kong. Intertidal. (1)

A. lutulentus (Erséus): Jiaozhou Bay (Shandong); Hong Kong. Intertidal. (1, 2)

A. taitamensis Erséus: Hong Kong. Intertidal brackish and freshwater area. (1)

Aktedrilus cavus Erséus: Hong Kong. Intertidal. (1)

A. cuneus Erséus: Hong Kong. Under rocks in low intertidal, sandy mid intertidal. (3)

A. locyi Erséus: Hong Kong. Under rocks in low intertidal. (1)

A. longitubularis Finogenova & Shurova: Hong Kong. Intertidal. (1)

A. mortoni Erséus: Hong Kong. Intertidal. (1)

A. parviprostatus Erséus: Hong Kong. Intertidal. (1)

A. parvithecatus (Erséus): Hong Kong. Intertidal. (1)

A. sinensis Erséus: Hong Kong. Sandy intertidal. (1)

Bathydrilus edwardsi Erséus: Jiaozhou Bay (Shandong); Hong Kong. Intertidal to water depth 13 m. (1, 2)

Doliodrilus tener Erséus: Jiaozhou Bay (Shandong); Hong Kong. Intertidal and freshwater. (1, 2)

Duridrilus piger Erséus: Hong Kong. Low intertidal to water depth 21 m. (1)

D. tardus Erséus: Hong Kong. Water depth 7–39 m. (1)

Heronidrilus bihamis Erséus & Jamieson: Jiaozhou Bay (Shandong); Hong Kong. Intertidal to 14 m water depth. (1, 2)

H. fastigatus Erséus & Jamieson: Hong Kong. Water depth 1 m, sandy substrate. (3)

H. hutchingsae Erséus: Hong Kong. Shallow sea, water depth 3–4 m. (1)

H. keenani Erséus: Hong Kong. Intertidal to 18 m water depth. (1)

H. virilis Erséus: Hong Kong. Water depth 1 m, sandy substrate. (3)

Jamiesoniella athecata Erséus: Hong Kong. Sandy mud. (3)

J. enigmatica Erséus: Hong Kong. Intertidal. (1)

Limnodriloides agnes Hrabê: Jiaozhou Bay (Shandong); Hong Kong. Intertidal to water depth 6 m. (1, 2)

L. biforis Erséus: Brackish water and freshwater, 0.3–0.9 m. (1)

L. fraternus Erséus: Hong Kong. Muddy intertidal. (1)

L. fuscus Erséus: Hong Kong. Water depth 4–12 m. (1)

L. macinnesi Erséus: Hong Kong. Water depth 3–7 m. (1)

L. parahastatus Erséus: Hong Kong. Low intertidal to subtidal. (1)

L. rubicundus Erséus: Hong Kong. Water depth 2 m, fine sand substrate. (3)

L. tenuiculus Erséus: Hong Kong. Water depth 3–14 m. (1)

L. tenuiductus Erséus: Hong Kong. Water depth 0.5–3 m. (1)

L. toloensis Erséus: Hong Kong. Low intertidal to 15 m. (1)

L. uniampullatus Erséus: Hong Kong. Intertidal to water depth 6 m. (1)

L. victoriensis Brinkhurst & Baker: Jiaozhou Bay (Shandong). Intertidal and subtidal, freshwater area. (2)

L. virginiae Erséus: Hong Kong. Water depth 0.5–4 m. (1)

Marcusaedrilus irregularis Erséus: Hong Kong. Intertidal to subtidal water depth 2 m. (1)

M. tuber Erséus: Hong Kong. Low intertidal to water depth 9 m. (1)

M. vesiculatus Erséus: Hong Kong. Low intertidal to water depth 9 m. (1)

Monopylephorus parvus Ditlevsen: Hong Kong. Intertidal (river mouth). (1)

M. rubroniveus Levinsen: Jiaozhou Bay (Shandong). (2)

Paupidrilus breviductus Erséus: Hong Kong. Intertidal. (1)

Pectinodrilus disparatus Erséus: Hong Kong. Sandy mid intertidal. (3)

P. hoihaensis Erséus: Hong Kong. Low intertidal. (3)

P. molestus (Erséus): Hong Kong. Sandy low intertidal. (3)

Phallodrilus darvelli Erséus: Hong Kong. Water depth 5–10 m. (1)

P. vanus Erséus: Hong Kong. Low intertidal to water depth 12 m. (1)

Rhizodrilus russus Erséus: Hong Kong. Muddy intertidal. (1)

Smithsonidrilus minusculus Erséus: Hong Kong. Water depth 6–9 m, sandy substrate. (3)

Tectidrilus achaetus Erséus & Qi: Hong Kong. Brackish water and freshwater area. (1)

T. pictoni (Erséus): Jiaozhou Bay (Shandong); Hong Kong. Intertidal to water depth 15 m. (1, 2)

Thalassodrilides briani Erséus: Dapeng Cove (Hong Kong). Water depth 3 m. (1)

T. gurwitschi (Hrabê): Hong Kong. Brackish water and freshwater. (1)

Tubificoides imajimai Brinkhurst: Hong Kong. Low intertidal to water depth 12.5 m. (1)

Uniporodrilus furcatus Erséus: Hong Kong. Sandy mid intertidal. (3)

NAIDIDAE

Paranais frici Hrabê: Hong Kong. Marine and freshwater. (1)

P. litoralis (Müller): Hong Kong. Intertidal and subtidal, brackish water and freshwater. (1)

P. plenus Erséus: Hong Kong. Sandy intertidal. (1)

ENCHYTRAEIDAE

Enchytraeus kincaidi Eisen: Jiaozhou Bay (Shandong). Sandy high intertidal. (2)

Grania hongkongensis Erséus: Hong Kong. Intertidal to water depth 15 m. (1)

G. inermis Erséus: Dapeng Cove (Hong Kong). Water depth 7–14 m. (1)

G. stilifera Erséus: Dapeng Cove (Hong Kong). Water depth 5–8 m. (1)

Lumbricillus sp.: Jiaozhou Bay (Shandong). (2)

Marionina coatesae Erséus: Jiaozhou Bay (Shandong); Hong Kong. Intertidal. (1, 2)

M. levitheca Erséus: Jiaozhou Bay (Shandong); Hong Kong. Intertidal. (1, 2)

M. nevisensis Righi & Kanner: Jiaozhou Bay (Shandong); Hong Kong. Intertidal to subtidal water depth 3–4 m. (1, 2)

M. vancouverensis Coates: Hong Kong. Intertidal (river mouth area). (1)

Stephensoniella marina (Modre): Hong Kong. River mouth and intertidal. (1)

S. sterreri (Lasserre & Erséus): Jiaozhou Bay (Shandong); Hong Kong. Intertidal. (1, 2)

MEGASCOLECIDAE

Pontodrilus litoralis (Grube): Jiaozhou Bay (Shandong); Hong Kong; Hainan. River mouth area. (1, 2)

REFERENCES

(1). Erséus, C. 1990. Marine Oligochaeta of Hong Kong. In: Proeedings of the Second International Marine Biological Workshop: The Marine Flora and Fauna of Hong Kong and Southern China, pp:259–335.

(2). Erséus, C. Sun Daoyuan, Liang Yanling, and Sun Bin, 1990. Marine Oligochaeta of Jiaozhou Bay, Yellow Sea Coast of China. Hydrobiologia 202:107–124.

(3). Erséus, C. 1992. Hong Kong Marine Oligochaeta: A Supplement. In: Proceedings of the Fourth International Marine Biological Workshop: The Marine Flora and Fauna of Hong Kong and Southern China. Hong Kong University Press, pp:157- 180.

Compiled by Huang Zongguo and Pan Haihong; Edited by Sun Daoyuan

HIRUDINEA

RHYNCHOBDELLIDA
PISCICOLIDAE

Limnotrachelobdella fujianensis Yang: Xiamen (Fujian); Zhuhai (Guangdong). (1)

L. okae (Moore): Qinghuangdao (Hebei); Qingdao (Shandong). (1)

L. sinensis (Blanchard): Jiaozhou Bay (Shandong). (1)

Piscicola magna Yang: Yellow Sea. (1)

Pontobdella loricata Harding [*P. moorei, Pontobdellina macrothela*]: Yantai (Shandong). (1)

Pontobdellina macrothela Schmarda: China's near shore. (1)

REFERENCE

(1). Yang Tong, 1996. Fauna Sinica: Annelida, Hirudinea. China Science Press. (in Chinese)

Compiled by Huang Zongguo

SIPUNCULA

PHASCOLOSOMATIDEA
ASPIDOSIPHONIFORMES
ASPIDOSIPHONIDAE

Aspidosiphon (*A.*) *elegans* (Chaamisso & Eysenhardt): Taiwan; Hainan. Coral reef intertidal. (2, 12, 20)

A. (*Paraspidosiphon*) *grandis* Sato: Taiwan. Coral reef intertidal. (12, 20)

A. (*A.*) *muelleri* Diesing [*A. corallicola*]: Beibuwan (=Gulf of Tongking). Water depth 30 m. (12, 20)

A. (*P.*) *steenstrupii* Diesing [*A. formosanus, A. makoensis*]: Taiwan; Hainan; Xisha (=Paracel Islands). Coral reefs. (3, 12, 20)

A. (*A.*) *torus* Selenka & De Man: South China Sea. (23)

Aspidosiphon (*P.*) sp. (cf. *parvulus* Gerould): South China Sea. Water depth 0–40m. Muddy and shelly substrate. (23)

Cloeosiphon aspergillus (Quatrefages): Xisha (=Paracel Islands). Coral reef intertidal. (3, 12, 20)

Lithacrosiphon maldivensis Shipley: Xisha (=Paracel Islands). Coral reef intertidal. (7, 13)

PHASCOLOSOMALIFORMES
PHASCOLOSOMATIDAE

Antillesoma antillarum (Grube & Oersted) [*Phascolosoma onomichianum, P. similis*]: Qingdao and Rizhao (Shandong); Lianyungang (Jiangsu); Wenzhou (Zhejiang); Taiwan; Shantou and Zhanjiang (Guangdong); Beihai and Weizhou Island (Guangxi); Baimajing (Hainan); Xisha (=Paracel Islands). Intertidal; muddy sand, under gravel, or rocky crevices. (1, 9, 12, 13, 17)

Apionsoma trichocephala Sluiter: South China Sea. Water depth 10–100 m. (23)

Phascolosoma albolineatum Baird: Taiwan; Xisha (=Paracel Islands). Coral reef intertidal. (3, 9, 12, 17, 20)

P. esculenta (Chen & Yeh): Jingxiang, Yuhuan, and Bachao (Zhejiang); Pingtan, Fuqing, and Xiamen (Fujian); Zhanjiang (Guangdong); Beihai, Qinzhou, Fangchenggang, and Weizhou Island (Guangxi); Haikou and Baimajing (Hainan). Mid- and high-intertidal and mangrove fringe. (1, 2, 9, 12, 13, 19)

P. formosense (Sato): Taiwan. Intertidal. (9, 12, 20)

P. japonicum Grube: Taiwan. Coral reef intertidal or rocks. (9, 12, 17, 20)

P. nigrescens Keferstein: Taiwan; Dapeng Cove (Guangdong); Hainan; Xisha (=Paracel Islands). Coral reef intertidal and oysters. (3, 9, 12, 17)

P. pacificum Keferstein: Taiwan; Hainan; Xisha (=Paracel Islands). Intertidal coral crevices. (2, 3, 9, 12, 17)

P. parvum Chen & Yeh: Hainan. Intertidal, on seaweed or oyster shells. (2, 9, 12, 13)

P. (*Edmondsius*) *pectinatum* Keferstein: East China Sea. Rocky reefs or muddy substrate. (16)

P. perlucens Baird [*P. dentigerum*]: Dapeng Cove (Guangdong); Hainan; Xisha (=Paracel Islands). Intertidal muddy sand or coral reefs. (6, 7, 9, 12)

P. rottnesti Edmonds: Fujian; Hainan. Intertidal rocky crevices. (2, 9, 12, 13)

P. scolops (Selenka, de Man & Bulow) [*P. dunwichi, P. rueppelii*]: Taiwan; Shantou, Dapeng Cove, and Daya Bay (Guangdong); Weizhou Island (Guangxi); Hainan. Coral reef intertidal, rocks, oysters, or symbiotically with sponges. (2, 6, 9, 12, 13, 17, 20)

P. sinense Chen & Yeh: Southern Hainan; Xisha (=Paracel Islands). Intertidal coral or rocky crevices. (2, 3, 9, 12)

P. varians Keferstein [*P. hainanicum, P. uncatum*]: Taiwan; Hainan; Xisha (=Paracel Islands). Coral reef intertidal. (2, 9, 12, 20)

SIPUNCULIDEA
GOLFINGIIFORMES
GOLFINGIIDAE

Golfingia elongata (Keferstein): Xiamen (Fujian). Intertidal rocks. (12, 15, 17, 18)

G. margaritacea margaritacea (Sars) [*Phascolosoma japonicum*]: Fujian. Intertidal muddy or gravel substrate. (12, 14, 15, 17, 18)

G. vulgaris vulgaris (de Blainville) [*Phascolosoma vulgare*]: Fujian; Taiwan. Intertidal, muddy sand and gravel substrate. (12, 13, 15, 18, 20)

Thysanocardia nigra (Ikea) [*Phascolosoma pyriformis*]: Changshan Islands and Lanshan (Shandong); Taiwan; Guangdong. Intertidal to water depth 30 m. (12, 17, 20)

PHASCOLIONIDAE

Onchnesoma intermedium Murina: East China Sea. Subtidal muddy sand substrate. (21, 22)

Phascolion moskalevi Murina: Guangdong. Intertidal, in gastropod shells. (12, 19)

THEMISTIDAE

Themiste cymodoceae (Edmonds): Weizhou Island (Guangxi); Baimajing (Hainan). Intertidal, oyster shells. (4, 12).

T. dehamata (Kesteven): Xiamen (Fujian). Intertidal muddy sand substrate (vertical burrow, 60–100 cm deep). (8, 12, 13)

T. lageniformis Baird [*Dendrostoma signifer*]: Xiamen (Fujian); Taiwan; Shantou and Dapeng Cove (Guangdong); Hainan. Coral reef intertidal, rocky crevices, oysters, or symbiotically with sponges. (4, 12, 13, 17, 20)

T. minor (Ikeda): Fujian. Often under gravel in intertidal area. (12, 14, 17)

T. spinulum (Chen & Yeh) [*Dendrostomum spinulum*]: Hainan. Intertidal coral reef or rocky crevices. (2, 12)

SIPUNCULIFORMES
SIPUNCULIDAE

Siphonosoma australe (Keferstein) [*S. pescadolense*]: Taiwan; Zhanjiang, Haifeng, and Dapeng Cove (Guangdong); Hainan. Intertidal, muddy or muddy sand substrate, burrow depth more than 50 cm. (2, 6, 12, 15, 20)

S. cumanense (Keferstein) [*S. edule, S. formosa*]: Taiwan; Hainan; Xisha (=Paracel Islands). Intertidal, under rocks or coral sand, burrow depth about 10 cm. (3, 12, 15, 17, 20)

S. funatuti (Shipley): Xisha (=Paracel Islands). Intertidal, in coral sand. (11, 12, 15, 17)

S. rotumanum (Shipley) [*S. nanhaiensis*]: Hainan; Xisha (=Paracel Islands). Coral reef intertidal, sandy and muddy sand substrate, burrow depth 10–15 cm. (5, 12, 15)

Sipunculus angasoides Chen & Yeh: Hainan. Intertidal muddy sand substrate. (2, 10, 12, 13)

S. indicus Peters: Taiwan; Lingshui (Hainan); Xisha (=Paracel Islands). Intertidal, coral sand and muddy sand substrate. (10, 12, 16, 19, 20)

S. norvegicus Danielssen: Xiamen (Fujian); Baimajing (Hainan). Low intertidal and subtidal. (10, 12, 16, 17)

S. nudus Linnaeus: Yantai and Qingdao (Shandong); Pingtan and Xiamen (Fujian); Taiwan; Dapeng Cove and Zhanjiang (Guangdong); Weizhou Island (Guangxi); Haikou, Baimajing, and Qinglan Bay (Hainan). Intertidal and shallow water muddy sand or sandy substrate, burrow depth 20–40 cm. (1, 2, 6, 10, 12, 13, 16, 17)

S. robustus Keferstein [*S. angasii*]: Xisha (=Paracel Islands). Intertidal, mostly among coral rocks. (3, 10, 12, 16, 19, 20)

REFERENCES*

(1). Chen, Y. and Z. C. Yeh, 1958. Notes on some Gephyrea of China with description of four new species. Acta Zoologica Sinica, 10(3):265–277.

(2). Chen, Y., 1963. A preliminary report on the Gephyrea fauna of Hainan Province. Studia Marina Sinica, 4:1–13.

(3). Li, F. L., 1982. On the peanut worms (Sipunculida) of Xisha Islands, Guangdong Province, China. Journal of Shandong College of Oceanology, 12(2): 57–71.

(4). Li, F. L., 1983. Two new records of genus *Themiste* (Sipuncula) found in Hainan Island, Guangdong Province, China. Journal of Shandong College of Oceanology, 13(2):61–66.

(5). Li, F. L., 1985a. On the peanut worms (Sipunculida) of Xisha Islands, Guangdong Province, China. Journal of Shandong College of Oceanology, 15(3):55–63.

(6). Li, F. L., 1985b. Preliminary studies on the Sipuncula of the Dapeng Cove, Guangdong Province, China. Journal of Shandong College of Oceanology, 15(3):59–66.

(7). Li., F. L., 1987. On the peanut worms (Sipuncula) of Xisha Islands, Guangdong Province, China, III: the genus *Lithacrosiphon*. Journal of Shandong College of Oceanology, 17(2):77–80.

(8). Li, F. L., 1988. Some problems concerning the study of the *Themiste dehamata* (Kesteven), 1903. Journal of Shandong College of Oceanology,18(2):40–43.

(9). Li, F. L., 1989. Studies on the genus *Phascolosoma* (Sipuncula) off the China coasts. Journal of Ocean University of Qingdao, 19(3):78–90.

(10). Li, F. L. et al. , 1990. Studies on the Genus *Sipunculus* (Sipuncula) off the China Coasts. Journal of Ocean University of Qingdao, 20(1)93–99.

(11). Li, F.L., H., Zhou, and W. Wang, 1992. Studies on the Genus *Siphonosoma* (Sipuncula) off the China coasts. Journal of Ocean University of Qingdao, 22(1): 97–102.

(12). Li, F. L., H. Zhou, and W. Wang, 1992. A checklist of Sipuncula from the China coasts. Journal of Ocean University of Qingdao, 22(2):72–88.

(13). Lin L. M. et al., 1985. A preliminary study on the species of the Sipunculida and their distribution in the intertidal zone of Fujian Province. Journal of Xiamen Fisheries College, 2:19–33.

(14). Jin, D. X., 1947. Studies of Fukien Sipunculoidea. Biological Bulletin of Fukien Christian University, 3:97–014.
(15). Cutler, E. B. and N. J. Cutler, 1982. A revision of the genus *Siphonosoma* (Sipuncula). Proc. Biol. Soc. Wash., 95(4):748–762.
(16). Cutler, E. B. and N.J. Cutler, 1985. A revision of the genera *Sipunculus* and *Xenosiphon* (Sipuncula). Zool. J. Linn. Soc., 85:219–246.
(17). Cutler, E. B., N. J. Cutler, and T. Nishikawa, 1984. The Sipuncula of Japan: their systematics and distribution. Publ. Seto Mar. Bol. Lab., 29(4/6):249–332.
(18). Cutelr, E. B. and N. J. Cutler, 1987. Revision of the genus *Golfingia* (Sipuncula: Golfinglidae). Proc. Boc. Wash., 100(4):735–761.
(19). Murina V. V., 1964. Report on the sipunculid worms from the coast of South China Sea. Trudy Inst. Ocean., 69:245–270.
(20). Sato, H. 1939. Studies on the Echiuroidea, Sipunculoidea, and Priapuloidea of Japan. Sci. Rep. Tohoko Univ. Ser. 4, 14:339–459.
(21). Murina, V. V., 1976. New and rare species of Sipunculids from the East China Sea. Vestnik Zool., 2:62–67.
(22). Murina, V. V., 1977. Sipunculans of the Arctic and boreal parts of Eurasian Sea. Zool. Inst. Acad. Sci. USSR, 1–282.
(23). Murina V. V., 1989. The fauna of sipunculids for the coastal waters of southern Vietnam. In: Biology of the coastal waters of Vietnam: Benthic invertebrates of southern Vietnam. Vladivostok: Far East Branch. Academy of Sciences of the USSR, 73–83.

*: (1)-(14) in Chinese
Compiled and edited by Li Fenglu and Zhou Hong

ECHIURA

ECHIURIDA
ECHIUROINEA
ECHIURIDAE

Anelassorhynchus inanensis (Ikeda): Xisha (=Paracel Islands). Low intertidal, under large coral rocks, burrow depth 10 cm. (6, 12)

A. subinus (Lanchester): Daya Bay (Guangdong). Intertidal. (12)

Arhynchite rugosum Chen & Yeh: Qingdao, Jiaozhou Bay, and Chengshanjiao (Shandong). Intertidal to water depth 20 m, sandy substrate. (1)

Ikedosoma qingdaoense Li, Wang, & Zhou: Qingdao, Jiaozhou Bay, and Xuejia Island (Shandong). Intertidal muddy sand, burrow (L shape) depth 50–60 cm. (4, 5)

Listriolobus brevirostris Chen & Yeh: Lianyungang (Jiangsu); Qingdao and Jiaozhou Bay (Shandong); Baimajing (Hainan). Intertidal rocky crevices and shallow water muddy sand substrate. (1)

Ochetostoma erythrogrammon Leukart & Rüppell: Taiwan; Hong Kong; Beihai (Guangxi); Hainan; Xisha (=Paracel Islands). Coral rocks or muddy sand. (2, 6, 8)

O. formolosum (Lampert): Shanghai. Subtidal muddy substrate, water depth 20 m. (6, 12)

Paraarhynchite hexorenale Chen & Yeh: Sanya (Hainan). Intertidal. (2)

Thalassema fuscum Ikeda: Guangdong. Shallow sea, especially low intertidal with firm muddy sand substrate; burrow short and vertical. (8, 11)

T. mortenseni Fischer: Guangdong; Hong Kong. (5, 12)

XENOPNEUSTA
URECHIDAE

Urechis unicinctus (Von Drasche): Dalian (Liaoning); Yantai and Jiaozhou Bay (Shandong); Fujian. Mostly subtidal muddy sand substrate, burrow U shape. (1, 5, 12)

HETEROMYOTA
IKEDAIDAE

Ikeda taenioides (Ikeda): Jiaozhou Bay (Shandong). Low intertidal to subtidal, muddy substrate; vertical burrow. (4, 5)

REFERENCES*

(1). Chen Yi and Yeh Zhengchang, 1958. Note on some Gephyrea of China with description of four new species. Acta Zoologica Sinica, 10(3): 265–278.
(2). Chen Yi and Yeh Zhengchang, 1963. A preliminary report on the Gephyrean fauna of Hainan. Studia Marina Sinica, 4:1–13.
(3). Edmonds, S. J. 1982. Echiura. In Parker, S. P. (ed), Synopsis and classification of living organisms. 2:65–66. McGraw-Hill, New York.
(4). Fisher, W. K., 1964. Echiuroid worms of the North Pacific Ocean. Proc. U. S. Nat. Mus., 96:215–292.
(5). Ikeda, I., 1904. The Gephyrea of Japan. Jour. Coll. Sci. Imp. Univ. Tokyo, Japan, 20(4):1–87.
(6). Ikeda, I., 1907. On three new and remarkable species of Echiuroids. Jour. Coll. Sci. Imp. Univ. Tokyo, Japan, 21(8):1–46.
(7). Ikeda, I., 1924. Further notes on the Gephyrea of Japan, with description of some new species from the Marshall, Caroline, and Palau Island. Jap. Jour. Zool., 1:23–44.
(8). Sato, H., 1934. Sipunculoidea and Echiuroidea obtained in Onomichi Bay. Zool. Mag. Tokyo, 46:245–252.
(9). Sato, H. 1937. Echiuroidea, Sipunculoidea, and Priapuloidea obtained in northeast Honshu, Japan, Res. Bull. Saito Ho-on Kai Mus, 12:137–176.
(10). Sato, H. 1939. Studies on the Echiuroidea, Sipunculoidea, and Pariapuloidea of Japan. Sci. Rep. Tohoku Univ., 14(4):339–459.
(11). Morton, B. and Morton, J., 1983. The sea shore ecology of Hong Kong. Hong Kong University Press, 350pp.
(12). Stephen, A. C. and S. J. Edmonds, 1972. The phyla Sipuncula and Echiura. London (British Museum - natural history), 1–528. Publication No. 717.

*: (1) in Chinese and (8) in Japanese
Compiled and edited by Li Fenglu and Wang Wei

MOLLUSCA

POLYPLACOPHORA [AMPHINEURA]

LEPIDOCHITONIDAE

Hanleya sinica Xu: East China Sea. Water depth 1,680–1,950 m. (8)

Lepidochiton rugatus Carpenter [*Lepidopleurus assimilis*]: Haiyang Island (Liaoning); Yellow Sea. Low intertidal to 40 m water depth. (2)

ISCHNOCHITONIDAE

Ischnochiton alatus (Sowerby): Taiwan. (6)
I. boninensis Bergenhayn: Dapeng Bay (Hong Kong) and Hong Kong. Mid intertidal to 5 m water depth. (5)
I. comptus (Gould): Along China's coast. Intertidal. (1, 3)
I. hakodadensis Pilsbry: Bohai; Yellow Sea. Intertidal. (2, 3)
Lepidozona christiaensis Van Belle: Hong Kong. Intertidal. (5)
L. coreanica (Reeve): Along China's coast. Low intertidal to water depth 5 m. (1, 2, 3, 5)
L. nipponica (Berry): Yellow Sea. Intertidal to water depth 57 m. (2)

CALLISTOPLACIDAE

Callistochiton savigni Pilsbry: Lanyu (Taiwan). (1)
Nuttallina alternata (Sowerby): Lanyu (Taiwan). (1)

MOPALIIDAE

Mopalia retifera Thiele: Along Fujian's coast (north of Dongshan). Intertidal. (2, 3)
M. schrencki Thiele: Xiaochangshan Island (Liaoning); Changsangba Island, Li Island, and Rongcheng (Shandong). (2, 3)
Placiphorella japonica (Dall): Along the coast of southeastern China. Intertidal. (3)

SCHIZOCHITONIDAE

Sinolorica scissurata Xu: East China Sea. Water depth 220 m. (8)

CHITONIDAE

Acanthopleura spinosa Bruguiere: Taiwan. (1, 6)
Acanthozostera gemmata Blaunville: Lanyu (Taiwan). (1, 7)
Chiton (Rhissoplax) komaianus Is & Iw Taki [*Rhissoplax komaiana*]: Taiwan; Dapeng Bay (Hong Kong) and Hong Kong. Intertidal. (1, 5)
Liolophura gaimardi (Blainville): Taiwan. (6)

L. japonica (Lischke): Along the coast of southeastern China; Taiwan; Dapeng Bay (Hong Kong) and Hong Kong. Intertidal. (1, 3, 4, 5)
L. loochooana (Broderip): Taiwan; Hainan. (1)
Lucilina amanda Thiele: Hong Kong; Xisha (=Paracel Islands). On corals. (3, 5)
Onithochiton hirasei Pilsbry: Along the coast of southeastern China. Intertidal. (3)
Onithochiton sp.: Dapeng Bay (Hong Kong). Intertidal. (5)
Squamopleura curtisiana Smith: Lanyu (Taiwan). (7)
Tonicia sp.: Hong Kong. Rocky shore. (5)

ACANTHOCHITONIDAE

Acanthochiton armatus (Pease): Taiwan. (6)
A. bednalli Pilsbry: Xisha (=Paracel Islands). (3)
A. dissimilis Is & Iw Taki: Xiaochangshan Island (Liaoning), Rongcheng and Moye Island (Shandong). Intertidal. (2)
A. rubrolineatus (Lischke): Along China's coast. Intertidal. (2, 3)
A. scutiger (Reeve): Shicheng Island (Liaoning); Rongcheng, Shi Island, Wuleidao Bay, Qingdao, and Shijiushou (Shandong). Intertidal. (2)
A. zeylandicus Quoy & Gaimard: Taiwan. (1)
Craspedochiton laqueatus (Sowerby): Hong Kong. Intertidal to water depth 4 m. (5)
Cryptoplax eleioti Pilsbry: Xisha (=Paracel Islands). (3)
C. larvaeformis (Burrow): Xisha (=Paracel Islands). Coral reefs. (3)
C. oculata (Quoy & Gaimard): Southern Hainan; Xisha (=Paracel Islands). Coral reefs. (3)
C. striatus (Lamarck): Taiwan. (6)
Notoplax doederleini Thiele: Dapeng Bay (Hong Kong). Intertidal. (5)
Notoplax sp.: Dapeng Bay (Hong Kong). Intertidal. (5)

REFERENCES*

(1). Mollusc species-determination group of Taiwan, 1982. Shells of Taiwan (II), pp:4–5, Molluscan Friend.
(2). Qi Zhongyan et al., 1987. Molluscs of the Yellow Sea and Bohai Sea. China Agriculture Press, pp:4–13.
(3). Huang Xiuming and Xu Fengshan, 1964. Polyplacophora. In: A Collection of Illustrative Plates of Animals in China, Molluscs (I). China Ocean Press, pp:1–10.
(4). Van Belle, R. A. 1980. On a Small Collection of Chitons from Hong Kong (Mollusca: Polyplacophora). Proceedings of the First International Marine Biological Workshop: The Marine Flora and Fauna of Hong Kong and Southern China, Hong Kong University Press, pp:33–35.
(5). Van Belle, R. A., 1980. Supplementary notes on Hong Kong Chitons (Mollusca: Polyplacophora). Proceedings of the First International Marine Biology Workshop: The Marine Flora and Fauna of Hong Kong and Southern China, Hong Kong University Press, pp:469–483.
(6). Wu Shi-Kuei, 1969. Some Chitons from Taiwan (Formosa). Malac. Rev., 2:103–111.

(7). Wu Shi-Kuei, 1975. The Chitons of Lanhsu, Taiwan. Bull. China Malac. Soc., 2:69–75.
(8). Xu Fenshan, 1990. New genus and species of Polyplacophora (Mollusca) from the East China Sea. China Joul. Oceanol. Limnol., 8(4):375–377.

*: (1)-(3) are in Chinese
Compiled by Huang Xiuming; Addition by Huang Zongguo

BIVALVIA [LAMELLIBRANCHIA]

SOLEMYOIDA
SOLEMYIDAE

Acharax japonica (Dunker): South China Sea. Water depth 48 m. (76a)

A. johnsoni (Dall): Nansha (=Spratly Islands). Water depth 2,830 m. (76a)

NUCULOIDA
NUCULIDAE

Acila divaricata (Hinds): East China Sea; South China Sea. Water depth 70–160 m. (68)

A. mirabilis (Adams & Reeve): Central Yellow Sea. Water depth 40–85 m. (68)

A. schencki Kuroda: East China Sea. Water depth 200 m. (68)

Nucula (*Leionucula*) *convexa* (Sowerby): South China Sea. Water depth within 40 m. (68)

N. (*L.*) *cumingii* Hinds: East China Sea; South China Sea. Water depth 120–200 m. (68)

N. (*Sinonucula*) *cyrenoides* Kuroda: East China Sea; South China Sea. Water depth 60- 150 m. (68)

N. (*Nucula*) *donaciformis* Smith: Nansha (=Spratly Islands). Water depth 671 m. (76a)

N. (*L.*) *exodonta* Prashad: Nansha (=Spratly Islands). Water depth 1,604 m. (76a)

N. (*Lamellihucula*) *izushotoensis* (Okutani): Okinawa Trough. Water depth 2,000–2,150 m. (68)

N. (*N.*) *nimbosa* Prashad: Nansha (=Spratly Islands). Water depth 1,655 m. (76a)

N. (*L.*) *nipponica* Smith: Southern Yellow Sea. Water depth 44–80 m. (68)

N. (*L.*) *pachydonta* Prashad: South China Sea. Water depth 1,100 m. (68)

N. (*L.*) *parvula* (Gould): Hong Kong. (68)

N. (*N.*) *paulula* A. Adams: Southern Yellow Sea; northern East China Sea. Water depth 20+ meters. (68)

N. (*L.*) *tenuis* (Montagu): Bohai; Yellow Sea. Water depth 10–80 m. (68)

N. (*L.*) *tokyoensis* Yokoyama: Southern Yellow Sea. Water depth within 100 m. (68)

MALLETIIDAE

Malletia conspicua Smith: East China Sea. Water depth 550 m. (73)

M. sumatrensis Thiele: South China Sea. Water depth 1,100 m. (73)

Neilo bisculpta Xu: Nansha (=Spratly Islands). (74)

N. humilior (Prashad): South China Sea. Water depth 1,100 m. (73)

Neilonella aequatorialis Thiele: South China Sea. Water depth 270 m. (73)

N. dubia Prasha: East China Sea. Water depth 520 m. (73)

N. guineensis (Thiele): Central South China Sea. (34)

TINDARIIDAE

Tindaria jinxingae Xu: East China Sea. Water depth 550 m. (73)

NUCULANIDAE

Nuculana (*Thestyleda*) *forticostata* Xu: Nansha (=Spratly Islands). Water depth about 200 m. (74)

N. (*Nuculana*) *sadoensis* (Yokoyama): Northern Yellow Sea. Water depth 30–80 m. (67)

N. (*Sinoleda*) *sinensis* Xu: East China Sea; South China Sea. Water depth about 60 m. (67)

N. (*T.*) *tamseimaruae* Tsuchida & Okutani: East China Sea. Water depth 520 m. (76a)

N. (*T.*) *yokoyamai* Kusoda: Central Yellow Sea. Water depth 58 m. (67)

Portlandia japonica (Adams & Reeve): Yellow Sea. Water depth 25–80 m. (67)

Propeleda soyomaruae (Okutani): South China Sea. Water depth 168 m. (67)

Saccella confusa (Hanley): South China Sea. Water depth 16–42 m. (67)

S. cuspidata (Gould): South China Sea. Shallow water. (67)

S. gordonis (Yokoyama): Southern Yellow Sea; northern East China Sea. Water depth about 70 m. (67)

S. sematensis (Suznki & Isizuka): East China Sea. Water depth 100–400 m. (67)

Sarepta speciosa A. Adams: East China Sea; South China Sea. Water depth 66–150 m. (67)

Spinula tasmanica Knudsen: South China Sea. Water depth 1,100 m. (67)

Yoldia lepidula H. & A. Adams: South China Sea. Water depth 10–65 m. (67)

Y. notabilis Yokoyama: Yellow Sea. Water depth 30–80 m. (67)

Y. serotina (Hinds): Beibuwan (=Gulf of Tongking). Water depth 12–16 m. (67)

Y. similis Kuroda & Habe: Bohai; Yellow Sea; East China Sea. Water depth within 30 m. (67)

ARCOIDA
ARCIDAE

Acar plicata (Dillwyn): Pingtan (Fujian); Daya Bay and Baoan (Guangdong); Dongxing (Guangxi); Hainan; Xisha (=Paracel Islands). Intertidal to subtidal 300 m. (42)

Anadara antiquata (Linnaeus): Taiwan; Haimen (Guangdong); Hainan; Xisha (=Paracel Islands). Subtidal. (40)

A. clathrata (Reeve): Beibuwan (=Gulf of Tongking); South China Sea. Subtidal 22- 51.5 m. (40)

A. crebricostata (Reeve): Daya Bay, Baoan, and Wushi (Guangdong); Weizhou Island (Guangxi); Hainan. Subtidal. (40)

A. ferruginea (Reeve): East China Sea; Taiwan; Jiezhou Island (Guangdong); South China Sea. Subtidal 10–118 m. (40)

A. uropigmelana (Bory): Lingshui (Hainan). Subtidal. (40)

Arca avellana Lamarck: Shandong; Zhejiang; Fujian; Guangdong; Guangxi; Beibuwan (Gulf of Tongking); Hainan; Xisha (=Paracel Islands). Intertidal to subtidal 80 m. (42)

A. boucardi Jousseaume: Liaoning; Shandong; Zhejiang; Fujian; Guangdong; Hainan. Intertidal to subtidal 68 m. (42)

A. navicularis Bruguiere: Nanji (Zhejiang); Guangdong; Weizhou Island, Hepu, and Beihai (Guangxi); Beibuwan (=Gulf of Tongking); Sanya (Hainan). Intertidal to subtidal 55 m. (42)

A. ventricosa Lamarck: Guangdong; Lingshui and Sanya (Hainan); Zhongjie Island and Chenhang Islands (Xisha=Paracel Islands). Intertidal to subtidal. (42)

Arcopsis interplicata (Grabau & King) [*Striarca interplicata*]: Bohai; Yellow Sea; Zhejiang; Fujian; Taiwan Strait. Subtidal to 100 m. (43)

A. minabensis Habe: Beibuwan (=Gulf of Tongking). Subtidal. (43)

A. sculptilis (Reeve): Lingshui and Sanya (Hainan). Intertidal. (43)

A. sinensis (Thiele & Jaeckel): Hong Kong; Beibuwan (=Gulf of Tongking, 22–51 m); South China Sea (12–37 m). Subtidal. (40, 100)

A. symmetrica (Reeve): Hebei; Shandong; Zhejiang; Fujian; Taiwan; Guangdong; Guangxi; Hainan. Intertidal to subtidal. (43)

Barbatia bistrigata (Dunker): Liaoning; Hebei; Shandong; Jiangsu; Zhejiang; Fujian; Guangdong; Guangxi. Intertidal to subtidal 36 m. (42)

B. cometa (Reeve): Guangdong; Guangxi; Lingshui and Sanya (Hainan). Intertidal to subtidal. (42)

B. decussata (Sowerby): Dongshan (Fujian); Guangdong; Hong Kong; Guangxi; Beibuwan (=Gulf of Tongking); Hainan; Xisha (=Paracel Islands). Intertidal to subtidal. (42)

B. fusca (Bruguiere): Dongshan (Fujian); Nanao (Guangdong); Weizhou Island (Guangxi); Hainan; Xisha (=Paracel Islands). Intertidal to subtidal. (42)

B. matsumotoi (Habe): Beibuwan (Gulf of Tongking); South China Sea. Subtidal 54–83 m. (42)

B. parva (Sowerby): Weizhou Island (Guangxi); Xisha (=Paracel Islands). Intertidal to subtidal. (42)

B. signata (Dunker): Sansha Bay, Pingtan, and Xiamen (Fujian); Jiezhou Island (Guangdong); Beibuwan (=Gulf of Tongking); Hainan; South China Sea. Intertidal to subtidal 112 m. (42)

B. stearnsi (Pilsbry): Nanji (Zhejiang); Taiwan; Guangdong; Hainan; Xisha (=Paracel Islands). Intertidal to subtidal 83 m. (42)

B. tenella (Reeve): Guangdong; Guangxi; Hainan; Xisha (=Paracel Islands). (42)

B. uwaensis (Yokoyama): East China Sea continental shelf; Taiwan Strait; Beibuwan (=Gulf of Tongking); South China Sea. Subtidal 20–50 m. (42)

B. virescens (Reeve) [*Arca (B.) virescens* Reeve]: Zhejiang; Fujian; Guangdong; Hong Kong; Guangxi; Hainan. Intertidal to subtidal. (42)

B. yamamotoi Sakurai & Habe: Zhongjie Island (Xisha=Paracel Islands). Subtidal 50- 100 m. (42)

Bathyarca kyurokusimana Nomura & Hatai: East China Sea. Subtidal 20–500 m. (42)

Bentharaca asperula (Dall): East China Sea; South China Sea. Subtidal 400–5,000 m. (42)

B. rubrotincta Kuroda & Habe: East China Sea; South China Sea. Subtidal 100–260 m. (42)

B. xenophoricola (Kuroda): East China Sea; South China Sea. Subtidal 50–2,150 m. (42)

Didimacar tenebrica (Reeve) [*Arca tenebrica*]: Hebei; Shandong; Zhejiang; Fujian; Taiwan; Guangdong; Guangxi; Hainan. Intertidal to subtidal about 20 m. (43)

Estellarca olivacea (Reeve) [*Arca (Barbatia) olivacea*]: Shandong; Jiangsu; Zhejiang; Fujian. Intertidal to subtidal about 20 m. (43)

Mabellarca consociata (Smith): Fujian; Taiwan; Guangdong; Guangxi; Beibuwan (=Gulf of Tongking, water depth 12–86 m); South China Sea (5.5–68 m). Subtidal. (40)

M. dautzenbergi (Lamy): Beibuwan (=Gulf of Tongking, water depth 26–49.6 m); South China Sea (water depth 20–113 m). (40)

Noetiella vivianae Oliver. Hong Kong. (100)

Potiarca pilula (Reeve) [*Arca pilula*]: Taiwan; Guangdong; Hainan; South China Sea. (water depth 11–14 m). (40)

Samacar pacifica (Nomura & Zinbo): East China Sea. Subtidal 20–300 m. (42)

Scapharca anomala (Reeve): Changle (Fujian); Guangdong; Weizhou Island (Guangxi). Subtidal. (40)

S. broughtonii (Schrenck) [*Arca inflata*]: Liaoning; Hebei; Shandong; Zhejiang. Intertidal to subtidal 67 m. (40)

S. cornea (Reeve): Taiwan; Baoan and Haikan (Guangdong); Qisha (Guangxi); Hainan. Subtidal. (40)

S. globosa (Reeve) [*Arca binakayanensis*]: Fujian; Guangdong; Hainan; South China Sea (water depth 17–25 m). Subtidal to about 20 m. (40)

S. gubernaculum (Reeve) [*Arca (Anadara) jousseaumei* Lamy]: Taiwan; Daya Bay, Zhapo, and Wushi (Guangdong); Beihai and Qisha (Guangxi); Hainan. Subtidal. (40)

S. inaequivalvis (Bruguiere): Fujian. Subtidal to about 20 m. (40)

S. indica (Gmelin): Yangjiang and Dianbai (Guangdong); Weizhou Island (Guangxi). (86)

S. japonica (Reeve): Pingtan (Fujian); Shangchuan (Guangdong); Hainan; South China Sea. Subtidal, water depth 13 m. (40)

S. labiosa (Sowerby): Fujian; Guangdong; Hainan. Subtidal. (40)

S. satowi Dunker: Sansha, Pingtan, and Lianjiang (Fujian); Haimen, Shangchuan, and Dianbai (Guangdong); Hainan; East China Sea (water depth 17–27 m); South China Sea (water depth 21–23 m). Subtidal. (40)

S. subcrenata (Lischke) [*Arca subcrenata*]: Liaoning; Hebei; Shandong; Jiangsu; Zhejiang; Fujian; Taiwan; Guangdong; Guangxi. Low intertidal to subtidal 56 m. (40)

S. troscheli Dunker: Beihai (Guangxi); Wanning (Hainan). Subtidal. (40)

S. vellicata (Reeve): Taiwan; South China Sea (water depth 23 m). Rare species. (40)

Striarca thielei Schenck & Reinhart: Taiwan; Beibuwan (=Gulf of Tongking, water depth 42–78 m); South China Sea (41–69 m). Subtidal. (43)

Tegillarca granosa (Linnaeus) [*Arca granosa* Linné]: Liaoning; Hebei; Shandong; Zhejiang; Fujian; Taiwan; Guangdong; Guangxi. Intertidal. (40)

T. nodifera (Martens): Zhejiang; Fujian; Taiwan; Guangdong; Guangxi. Intertidal to subtidal. (40)

Trisidos kiyonoi (Kuroda): Pingtan, Xiamen, and Dongshan (Fujian); Guangdong; South China Sea. Intertidal to subtidal 39 m. (42)

T. semitorta (Lamarck) [*Arca (Parallelpipedum) semitorta* Lamarck]: Daya Bay and Baoan (Guangdong); Hong Kong; Beibuwan (=Gulf of Tongking); Hainan; South China Sea. Subtidal 24–30 m. (42)

T. tortuosa (Linnaeus) [*Arca tortuosa*]: Shaoan Bay (Fujian); Guangdong; Sanya, Yacheng, and Yinggehai (Hainan). Subtidal. (42)

CUCULLAEIDAE

Cucullaea labiosa granulosa Jonas: Southern Taiwan Strait; Zhanjiang (Guangdong); Nansha (=Spratly Islands). (49, 83, 92)

GLYCYMERIDIDAE

Glycymeris albolineata (Lischke): Taiwan. Subtidal about 5–20 m. (44)

G. pilsbryri (Yokoyama): Taiwan Strait. Subtidal. (37, 92)

G. reevei (Mayer): Taiwan Strait; Bei Island and Quanfu Island (Xisha=Paracel Islands). Subtidal. (44)

G. rotunda (Dunker): East China Sea continental shelf. Water depth 78–147 m. (44)

G. vestita (Dunker): Pingtan, Huian, and Dongshan (Fujian); Nanao, Lufeng, Shangchuan, and Wushi (Guangdong); Beibuwan (=Gulf of Tongking); East China Sea continental shelf; South China Sea. Subtidal to 70 m. (44)

G. yessoensis (Sowerby): Yellow Sea. Water depth 58 m. (44)

Tucetilla amboinensis (Gmelin): Taiwan; Jinying Island, Shanhu Island, Chenhang Island, Bei Island, Quanfu Island, and Xishazhou (Xisha=Paracel Islands). Intertidal to subtidal about 20 m. (44)

T. tenuicostata (Reeve): South China Sea. Water depth 12–38 m. (44)

LIMOPSIDAE

Crenulilimopsis oblonga (Adams): East China Sea (water depth 100–147 m); South China Sea (water depth 290 m). (45)

Limopsis tajimae Sowerby: East China Sea. Subtidal 100–800 m. (45)

Oblimopa japonica (A. Adams): East China Sea (water depth 95–130m); Beibuwan (=Gulf of Tongking, water depth 26–72 m); South China Sea (water depth 31–290 m); Taiwan. (45)

MYTILOIDA
MYTILIDAE

Amygdalum peasei (Newcomb): Daya Bay (Guangdong) to offshore off southern Hainan. Subtidal 26–270 m. (11, 16a)

A. soyoae Habe: South China Sea (19° N, 112°30' E). Water depth 270 m. (11, 16a)

A. watsoni (Smith) [*Modiolus watsoni*]: South of Changjiang (=Yangtze River); nearshore off Guangdong; Beibuwan (=Gulf of Tongking); Hainan. Subtidal 11- 177 m. (11, 16a)

Arcuatula elegans (Gray): Daya Bay (Guangdong); Hong Kong. Subtidal. (16a, 97)

Arvella sinica (Wang & Tsi) [*Crenella sinica* Wang & Tsi]: Yellow Sea; East China Sea. Subtidal 50–90 m. (16a, 21)

Botula silicula (Lamarck): Guangdong; Weizhou Island (Guangxi); Hainan; Xisha (=Paracel Islands). Intertidal. (11, 16a)

Brachidontes emarginatus (Reeve): Xiamen (Fujian); Daya Bay (Guangdong); Hong Kong. Intertidal. (16a, 33, 97)

B. striatulus (Hanley) [*Gregariella striatus*]: Taiwan, Hong Kong. (11, 16a, 92, 97).

B. variabilis (Krauss): Xiamen and Dongshan (Fujian); Daya Bay (Guangdong); Hong Kong; Beibuwan (=Gulf of Tongking). Intertidal. (11, 16a, 92)

Gregariella coralliophaga (Gmelin): Shandong; Fujian; Guangxi; Hainan; Xisha (=Paracel Islands). Intertidal to subtidal 30 m. (11, 16a)

G. splendida (Dunker): Pingtan and Xiamen (Fujian). Intertidal. (11, 16a)

Hormomya mutabilis (Gould) [*Brachidontes curvatus*]: Xiamen (Fujian); Aotou and Daya Bay (Guangdong); Hong Kong; Qisha (Guangxi); Beibuwan (=Gulf of Tongking); Hainan. Intertidal. (11, 16a, 92)

H. sinensis Wang: Xisha (=Paracel Islands). (8, 16a)

Idasola japonica Habe: East China Sea (26°20' N, 125°E). Water depth 550 m. (11, 16a)

Limnoperna fortunei (Dunker): Rivers south of Changjiang (=Yangtze River); also river mouth brackish and freshwater areas in Fujian and Guangdong. (11, 16a) Tongking). Intertidal to subtidal 64 m. (11, 16a)

Lithophaga (*Lithophaga*) *antillarum* d'Orbigny: Hong Kong; Sanya (Hainan). Subtidal. (16a, 97)

L. (*Stumpiella*) *calyculatus* (Carpenter): Daya Bay (Guangdong); Hainan; Xisha (=Paracel Islands). Coral reef intertidal. (11, 16a)

L. (*Diberus*) *canaliferus* (Hanley): Hainan. Intertidal. (16a, 92)

L. (*Leiosolenus*) *curta* Lischke: Chengshan and Yuhuan (Zhejiang); Fujian; Guangdong. Coral reef intertidal. (11, 16a)

L. (*D.*) *divaricalx* Iredale: Taiwan. (16a)

L. (*Leiosolenus*) *hanleyma* (Reeve): Guangdong; Hong Kong; Guangxi; Hainan.Subtidal. (16a, 97)

L. (*Labis*) *lepteces* Wang: Pingtan and Dongshan (Fujian). (16a)

L. (*Leiosolenus*) *lessepsiana* (Vaillant) [*L. simplex*]: Hong Kong. Subtidal. (16a, 97)

L. (*D.*) *lima* Lamy [*L.* (*Leiosolenus*) *lima* Lamy]: Dongshan (Fujian); Nanao and Daya Bay (Guangdong); Hong Kong; Weizhou Island (Guangxi); Hainan. Coral reef intertidal. (11, 16a)

L. (*S.*) *lithura* Pilsbry: Hainan; Xisha (=Paracel Islands). Coral reef intertidal. (11, 16a)

L. (*D.*) *malaccana* Reeve: Dongshan (Fujian); Shantou and Shanwei (Guangdong); Weizhou Island (Guangxi); Hainan, Xisha (=Paracel Islands). Coral reef intertidal. (11, 16a)

L. (*Labis*) *mucronata* (Philippi): Along China's coast. Intertidal. (16a, 97)

L. (*Myapalmula*) *nasuta* Philippi: Taiwan. (16a)

L. (*Leiosolenus*) *obesa* (Philippi) [*L. obesa* (Philippi)]: Weizhou Island (Guangxi); Hainan. Coral reefs. (11)

L. (*Lithophaga*) *teres* (Philippi) [*L. teres*]: Aotou and Daya Bay (Guangdong); Hong Kong; Weizhou Island (Guangxi); Hainan. Coral reef intertidal. (11, 16a)

L. (*Lithophaga*) *zitteliana* Dunker [*L. zitteliana*]: Nanao and Daya Bay (Guangdong); Weizhou Island (Guangxi); Hainan. Coral reef intertidal. (11, 16a)

Modiolus (*Modiolus*) *auriculatus* (Krauss): Nanji (Zhejiang); Pingtan (Fujian); Hainan; Xisha (=Paracel Islands). Intertidal. (11, 16a)

M. (*Modiolus*) *comptus* Sowerby [*M. barbatus*]: Along China's coast. Intertidal to subtidal 62 m. (11, 16a)

M. (*Modiolusia*) *elongatus* (Swainson) [*M. subrugosa*, *Volsella subrugasa*]: Along China's coast. Subtidal 10–90 m. (11, 16a, 92)

M. (*Fulgida*) *flavidus* (Dunker): Xiamen (Fujian); Nanao (Guangdong); Beibuwan (=Gulf of Tongking); Hainan. Subtidal within 75 m. (11, 16a, 92)

M. (*Modiolusia*) *hanleyi* (Dunker): Hainan. (16a)

M. (*Modiolus*) *metcalfei* Hanley: Along China's coast. Intertidal to subtidal. (11, 16a)

M. (*Modiolus*) *modiolus* (Linnaeus): Liaoning; Shandong; Wenzhou (Zhejiang). Subtidal to 80 m. (11, 16a)

M. nipponicus (Oyama): Daya Bay (Guangdong); Hong Kong. Subtidal. (16a, 38, 97)

M. (*Modiolus*) *philippinarum* (Hanley): Xiamen (Fujian); Aotou, Daya Bay, and Shenzhen (Guangdong); Hainan. Intertidal. (11, 16a)

M. (*Lioberus*) *plicatus* (Lamarck): Offshore off Zhujiang (=Pearl River, Guangdong), western shore of Leizhou Peninsula (Guangdong); Hainan. Subtidal within 30 m. (11, 16a, 92)

M. (*Modiolusia*) *sirahensis* (Jousseaume): Beibuwan (=Gulf of Tongking). (16a)

M. (*Modiolusia*) *vaginus* (Lamack) [*M. vagina*]: Xinying (Hainan). Intertidal. (11, 16a, 92)

Musculista japonica (Dunker) [*Brachydontes japonica*]: Xiamen (Fujian); Daya Bay (Guangdong); Weizhou Island (Guangxi); Beibuwan (=Gulf of Tongking); Hainan. Subtidal to 65.7 m. (11, 16a, 92)

M. perfragilis (Dunker): Taiwan Strait. Subtidal. (16a, 92)

M. senhausia (Benson) [*Brachydontes aquarius*, *B. senhousei*, *Musculus senhousei*]: Along China's coast. Intertidal to subtidal. (11, 16a, 92, 97).

Musculus cumingiana (Dunker): Pingtan and Xiamen (Fujian). Intertidal. (11, 16a)

M. (*Modiolarca*) *cupreus* (Gould) [*M. marmoratus*]: Liaoning; Shandong; Daya Bay (Guangdong); Hong Kong. Subtidal within 100 m. (16a, 92)

M. mirandus (Smith): Xiamen (Fujian); Daya Bay (Guangdong); Beibuwan (=Gulf of Tongking). Intertidal to subtidal 64 m. (11)

M. (*Modiolarca*) *nanus* (Dunker): Daya Bay (Guangdong); Hong Kong; Hainan Subtidal. (11, 16a, 92)

M. (*Musculus*) *nigrus* (Gray): Yellow Sea. Subtidal 44–95 m. (11, 16a, 92)

Mytilus coruscus Gould: Liaoning; Shandong; Zhejiang; Pingtan and Xiamen (Fujian). Intertidal to subtidal 20 m. (11, 16a)

M. galloprovincialis Lamarck [*M. edulis*]: Bohai; Yellow Sea; Zhejiang; Fujian; Guangdong; Hong Kong. Low intertidal to shallow sea, aquaculture species. (11, 16a, 92)

Perna viridis (Linnaeus) [*Mytilus smaragdinus*]: Fujian; Guangdong; Hainan. Low intertidal to subtidal 17 m, aquaculture species. (11, 16a)

Septifer bilocularis (Linnaeus): Aotou and Daya Bay (Guangdong); Beibuwan (=Gulf of Tongking); Hainan; Xisha (=Paracel Islands). Intertidal. (11, 16a, 92)

S. excisus (Wiegmann): Xiamen and Dongshan (Fujian); Nanao, Haimen, and Shantou (Guangdong); Qisha (Guangxi); Hainan; Xisha (=Paracel Islands). Intertidal. (11, 16a)

S. keenae Nomura: Nanji (Zhejiang); Daya Bay (Guangdong); Hong Kong. Intertidal. (16a, 97)

S. pulcher Wang: Xisha (=Paracel Islands). (8, 16a)

S. virgatus (Wiegmann): Zhejiang; Fujian; Guangdong. Intertidal. (11, 16a)

S. xishaensis Wang: Xisha (=Paracel Islands). (8, 16a)

Solamen spectabilis (A. Adams): Yellow Sea; East China Sea. Subtidal 44–83 m. (11, 16a)

Stavelia subdistorta (Récluz): Qisha (Guangxi). (11, 16a, 92)

Trichomusculus barbatus (Reeve): Xinying (Hainan). (11, 16a)

T. semigranata (Reeves): Lingshui (Hainan). (16a)

Trichomya hirsuta (Lamarck) [*Brachydontes* (*Hormomya*) *hirsutus* (Lamarck)]: Nanji (Zhejiang); Fujian; Guangdong; Qisha and Weizhou Island (Guangxi). Intertidal. (11, 16a)

Xenostrobus atratus (Lischke) [*Vignadula atrata*]: Along China's coast. Intertidal, river mouth, and inner bay. (11, 16a, 97)

PINNIDAE

Atrina (*Servatrina*) *pectinata* (Linnaeus) [*A. pectinata japonica, A. pectinata lischkeana, Pinna inflata, P. pectinata*]: Liaoning; Shandong; Dongtou (Zhejiang); Fujian; Guangdong; Hong Kong; Guangxi; Sanya and Yulin (Hainan). Intertidal to subtidal 75 m. (1, 79)

A. (*Servatrina*) *penna* Reeve: East China Sea; South China Sea. Subtidal about 70 m. (1)

A. (*Atrina*) *vexillum* (Born) [*A. strangei, Pinna strangei, P. vexillum*]: Dongshan (Fujian); Aotou, Daya Bay, and Wushi (Guangdong); Hong Kong; Qisha (Guangxi); Xinying and Sanya (Hainan); Xisha (=Paracel Islands). Intertidal to subtidal 50 m. (1, 79)

Pinna bicolor Gmelin [*Atrina serrata, P. atropurpurea, P. bullata*]: Xiamen and Dongshan (Fujian); Taiwan; Nanao, Aotou, and Shenzhen (Guangdong); Hong Kong; Sanya and Yinggehai (Hainan). Intertidal to subtidal 100 m. (1, 79)

P. incurvata Schröder [*P. attenuata*]: East China Sea; South China Sea. Intertidal to subtidal 75 m. (1)

P. muricata Linnaeus: Taiwan; Lingshui and Sanya (Hainan); Dengqing Island and Yongxing Island (Xisha=Paracel Islands). Intertidal. (1, 79)

Pinna (*Streptopinna*) *saccata* Linnaeus: Taiwan; Lingshui (Hainan). Subtidal. (1, 79)

PTERIOIDA
PTERIIDAE

Electroma ovata (Quoy & Gaimard) [*Pteria alacorvi*]: Hainan; Xisha (=Paracel Islands). (2)

Pinctada anomioides (Reeve): Xinying, Lingshui, and Sanya (Hainan). (2)

P. chemnitzi (Philippi) [*Pteria tegulata*]: Yushan and Nanji (Zhejiang); Dongshan (Fujian); Zhapo and Daya Bay (Guangdong); Dongxing (Guangxi); Hainan. Subtidal 10–60 m. (2)

P. maculata (Gould): Hainan; Xisha (=Paracel Islands); Nansha (=Spratly Islands).

P. margaritifera (Linnaeus) [*Pteria margaritifera*]: Naozhou Island (Guangdong); Weizhou Island (Guangxi); Hainan; Xisha (=Paracel Islands). (2)

P. martensi (Dunker) [*Pteria martensii*]: Xiamen and Dongshan (Fujian); Nanao, Haimen, Daya Bay, and Shenzhen (Guangdong); Hepu and Beihai (Guangxi); Hainan. Intertidal to subtidal. (2)

P. maxima (Jameson): Guangdong; Lingshui and Yulin (Hainan). (2)

P. nigra (Gould): Guangdong; Lingshui and Yulin (Hainan). (2)

P. radiata (Leach): Daya Bay (Guangdong); Lingshui (Hainan). (2)

Pterelectroma zebra (Reeve): Zhoushan and Wenzhou (Zhejiang) to Beibuwan (=Gulf of Tongking); Hainan. Water depth 14–85 m. (2)

Pteria (*Austropteria*) *antelata* (Iredale): Xisha (=Paracel Islands). Intertidal, mostly attached to gorgonians. (2)

P. (*A.*) *brevialata* (Dunker) [*P. brevialata*]: South of Nanji (Zhejiang). Intertidal to subtidal 59 m. (2)

P. (*A.*) *chinensis* (Leach): Offshore off Wenzhou (Zhejiang), water depth 95–100 m; Offshore from Jiazi (Guangdong) to Yulin (Hainan), water depth 68–108 m; Xisha (=Paracel Islands). (2)

P. (*A.*) *coturnix* (Dunker) [*P. coturnix*]: East China Sea; Shanwei (Guangdong); Offshore from Daya Bay (Guangdong) to Yulin (Hainan), water depth 19.5–180 m; Beibuwan (=Gulf of Tongking), water depth 14–92 m; Nansha (=Spratly Islands). (2)

P. (*A.*) *cypsellus* (Dunker): Hainan; Xisha (=Paracel Islands). (2)

P. (*A.*) *dendronephthya* Habe: Northern Hainan. Water depth 55 m. (2)

P. (*A.*) *loveni* (Dunker) [*P. lata*]: Guangdong; Beihai (Guangxi); Hainan. (2)

P. (*Magnavicula*) *penguin* (Röding): Daya Bay (Guangdong); Weizhou Island (Guangxi); Sanya (Hainan). (2)

ISOGNOMONIDAE

Crenatula modiolaris Lamarck: Hainan. (3)

C. nigrina Lamarck: Shenzhen (Guangdong); Hainan. Inside sponges. (3)

Isognomon ephippium (Linnaeus): Taiwan; Shenzhen (Guangdong); Hainan. Intertidal. (3)

I. isognomun (Linnaeus) [*Pedalion isognomum, P. padibulum*]: Hainan; Xisha (=Paracel Islands). (3)

I. legumen (Gmelin) [*Pedalion legumen*]: Pingtan, Xiamen, and Dongshan (Fujian); Nanao and Naozhou Island (Guangdong); Weizhou Island (Guangxi); Hainan; Xisha (=Paracel Islands). Intertidal. (3)

I. nucleus (Lamarck) [*I. acutirostris, Pedalion quadrangularia*]: Xiamen (Fujian) to Xisha (=Paracel Islands). Intertidal. (16)

I. pernum (Linnaeus): Zhapo (Guangdong); Hainan; Xisha (=Paracel Islands). Intertidal to subtidal. (3)

VULSELLIDAE

Vulsella minor Röding: Beibuwan (=Gulf of Tongking). Water depth 21–110 m, inside sponges. (3)

V. vulsella (Linnaeus): Taiwan; Naozhou Island (Guangdong); Hainan. Inside sponges. (3)

MALLEIDAE

Malleus albus Lamarck [*M. malleus*]: Nanao and Daya Bay (Guangdong); Hong Kong; Beibuwan (=Gulf of Tongking); Hainan; Nansha (=Spratly Islands). Subtidal to 78 m. (3, 16)

M. daemoniacus Reeve [*M. regula*]: Hainan. Intertidal. (3)

M. regula (Forskål): Nansha (=Spratly Islands). Intertidal to subtidal. (3, 16)

Parvimalleus irregularis (Jousseaume): Yushan and Nanji (Zhejiang). (17)

PECTINIDAE

Amusium japonicum balloti (Bernardi): Baoan (Guangdong); Beihai and Weizhou Island (Guangxi); Beibuwan (=Gulf of Tongking); Nansha (=Spratly Islands). Subtidal 10–40 m. (10a, 16).

A. japonicum formosum Habe [*A. japonica*]: Nanao, Haimen, and Shanwei (Guangdong).Subtidal 50 m. (10a).

A. plueronectes nanshaensis Wang: Nansha (=Spratly Islands). Subtidal 62 m. (16)

A. pleuronectes pleuronectes (Linnaeus) [*A. pleuronectes*]: Guangdong; Hong Kong; Guangxi; Beibuwan (=Gulf of Tongking); Hainan. Subtidal 5–85 m. (10a)

Annachlamys macassarensis (Chenu): Beibuwan (=Gulf of Tongking); southern and western Hainan; Nansha (=Spratly Islands). Subtidal within 50 m. (16)

Bathyamussium jeffreysi (Smith): East China Sea continental shelf. Water depth 520 m. (72)

Bractechlamys elegans Wang: Yulin (Hainan); Xisha (=Paracel Islands); Nansha (=Spratly Islands). Coral reefs. (9)

B. quadrilirata (Lischke): Taiwan; southwestern Hainan. Water depth 45–90 m. (9)

B. schemeltzii (Kobelt): Sanya (Hainan); Xisha (=Paracel Islands). Intertidal to subtidal 20 m. (9)

Chlamys (Mimachlamys) albolineata (Sowerby): Hainan. Intertidal to subtidal 90 m. (6)

C. (M.) asperulata Adams & Reeve: Taiwan; Nanao Island (Guangdong) to Beibuwan (=Gulf of Tongking). Water depth 14.5–93 m, muddy sand bottom. (6)

C. (Azumapecten) farreri (Jones & Preston): Liaoning; Hebei; Shandong; Zhejiang; Fujian. Intertidal to subtidal 60 m. (6)

C. (M.) gloriosa (Reeve): Guanghai and Zhapo (Guangdong); Beihai and Qisha (Guangxi); from mouth of Zhujiang (=Pearl River, Guangdong) to Hainan. Water depth 73–110 m. (6)

C. (Coralichlamys) irregularis (Sowerby): Fujian; Taiwan; Guangdong; Weizhou Island (Guangxi); Hainan; Nansha (=Spratly Islands). Intertidal to subtidal about 20 m. (6, 16)

C. (C.) jousseaumei (Bavay): Hainan. Water depth 165–290 m, muddy sand bottom. (6)

C. (C.) larvatus (Reeve): Lingshui and Sanya (Hainan). (6)

C. (C.) lemniscata (Reeve): Wenzhou (Zhejiang); Taiwan Strait; Hainan. Water depth 100 m, fine sand bottom. (6)

C. (C.) madreporarum (Sowerby): Aotou (Guangdong); Lingshui and Sanya (Hainan). Around intertidal mean low water line. (6)

C. (M.) nobilis (Reeve): Xiamen (Fujian); Taiwan Strait; Haimen, Baoan, Zhapo, and Daya Bay (Guangdong); Qisha (Guangxi); Beibuwan (=Gulf of Tongking); Lingshui (Hainan); Nansha (=Spratly Islands). Subtidal to 147 m. (6)

C. (A.) squamata (Gmelin): Nanji (Zhejiang); Taiwan Strait; Baoan (Guangdong); Qiongzhou Strait (Guangdong); Beibuwan (=Gulf of Tongking); Qinglan (Hainan). Subtidal within 100 m. (6)

C. (M.) valdecostatus (Melvill): Daya Bay (Guangdong); Hong Kong; Qiongzhou Strait (Guangdong); mouth of Zhujiang (=Pearl River, Guangdong) to Beibuwan (=Gulf of Tongking). Water depth 32–42.5 m. (6)

Cryptopecten complanus Wang: Yellow Sea (31°5' N, 128°E). Water depth 1 m, fine sand bottom. (7)

C. nux (Reeve): Mouth of Zhujiang (=Pearl River, Guangdong), water depth 30–100 m; Zhongsha (=Macclesfield Bank), water depth 39–41 m. Coarse sand bottom. (7)

C. tissotii (Bernardi): South China Sea (20°N, 113°E). Water depth 104 m, coarse sand bottom. (7)

C. vesiculosus (Dunker): Jizhou Island (=Cheju Do, Korea) to north of Taiwan Strait. Water depth 82–150 m, muddy sand or fine sand bottom. (7)

Ctenamussium sinensis Wang: East China Sea continental shelf (26°30' N, 124°E). Water depth 104–126 m. (4)

Decatopecten striatus (Schumacher): Taiwan Strait. Intertidal to subtidal 50 m. (37)

Palliolum macrocheiricola Habe: East China Sea continental shelf. Water depth 520 m. (72)

Pecten (*Notovola*) *albicans* (Schröder): Yellow Sea; East China Sea. Water depth 39–116 m, muddy sand and soft mud bottom. (13)

P. (*Oppenheimopecten*) *excavatus* Anton: Nanao and Daya Bay (Guangdong); Hong Kong. Subtidal. (13)

P. (*Minnivola*) *pyxidatus* (Born): Pingtan and Dongshan (Fujian); Taiwan Strait; Guangdong; Guangxi; Hainan; Beibuwan (=Gulf of Tongking). Subtidal 12–59 m. (13)

Propeamussium (*P.*) *caducum* (Smith): South China Sea (19°N, 112.5°30'E). Water depth 300–472 m, coarse sand bottom. (10)

P. (*Parvamussium*) *gracilis* Wang: South China Sea (19°30' N, 113°E). Water depth 210–242 m, sandy bottom. (10)

P. (*Propeamussium*) *sibogai* (Dautzenberg & Bavay): East of Hainan (19°N, 112.5°E). Water depth 472 m, coarse sand bottom. (10)

P. (*Propeamussium*) *stellus* Wang: South China Sea (19°N, 112.5°E). Water depth 290 m, muddy sand. (10)

P. watsoni (Smith): East China Sea continental shelf. Water depth 850 m. (36, 72).

Semipallium (*S.*) *fulvicostum* (Adams & Reeve): Weizhou Island (Guangxi); Lingshui and Sanya (Hainan). (12)

S. (*Excellichlamys*) *spectabilis* (Reeve): Nanji (Zhejiang) to Nansha (Spratly Islands). (12)

S. (*S.*) *tigris* (Lamarck): Sanya (Hainan); Xisha (=Paracel Islands). (12)

S. (*E.*) *xishaensis* Wang: Jinqing Island (Xisha=Paracel Islands). (12)

Serratovola tricarnatus (Anton): South China Sea; Nansha (=Spratly Islands). Subtidal 50–120 m. (13)

Volachlamys hirasei (Bavay) [*Chlamys solaris*, *C. teilhardi*]: Liaoning; Hebei; Shandong; Zhoushan and Xiangshan Bay (Zhejiang). Intertidal to subtidal 50 m. (14)

V. singaporinus (Sowerby) [*Chlamys pica*]: Dongtou (Zhejiang); Fujian; Guangdong; Guangxi; Beibuwan (=Gulf of Tongking); Hainan. Subtidal to 87 m, muddy sand bottom. (14)

PLICATULIDAE

Plicatula australis Lamarck: Zhanjiang (Guangdong); Lingshui (Hainan). Intertidal to subtidal. (5)

P. philippinarum (Hanley): Shanwei (Guangdong); Guangxi. Coral reef intertidal. (5).

P. plicata (Linnaeus): Aotou and Shenzhen (Guangdong); Guangxi; offshore off northwestern Hainan (water depth 38–71 m). Intertidal to subtidal. (5)

P. simplex Gould: Chaoyang and Nanao (Guangdong); Beibuwan (=Gulf of Tongking); Lingshui (Hainan). Subtidal 18–103 m. (5)

Spiniplicatula muricata (Sowerby): South of Changjiang (=Yangtze River) mouth; southeastern and southern Hainan. Subtidal 30–195 m. (5)

LIMIDAE

Acesta marissinica Yamashita & Habe: East China Sea; South China Sea (183°N, 111- 117°E, water depth 150–200 m). (15)

Ctenoides ales (Finlcy): Nansha (=Spratly Islands) (114°35' E, 5°43' N). Water depth 111 m, sandy bottom. (16)

C. lischkei (Lamy): Nansha (=Spratly Islands) (114°35' E, 5°43' N). Intertidal to subtidal about 100 m, sandy bottom. (16)

Isolima limopsis (Nomura & Zinbo): East China Sea. Water depth 118–228 m, muddy sand bottom. (15)

Lima vulgaris (Link) [*L. sowerby*]: Nanao and Naozhou Island (Guangdong); Weizhou Island (Guangxi), Lingshui and Sanya (Hainan); Nansha (=Spratly Islands). Intertidal to subtidal 46 m. (15)

Limaria (*Limatulella*) *amakusaensis* (Habe): Offshore off Hainan. Water depth 96.5 m, muddy sand bottom, rare species. (15)

L. (*Limaria*) *basilanica* (Adams & Reeve): Lingshui, Sanya and Haikou (Hainan); Yongxing Island (Xisha=Paracel Islands). Intertidal to subtidal 20 m. (15)

L. (*Platilimaria*) *dentata* (Sowerby): Lingshui, Sanya and Xinying (Hainan). Intertidal to subtidal 20 m. (15)

L. (*P.*) *fragilis* (Gmelin): Nanao and Aotou (Guangdong); Lingshui, Sanya and Yulin (Hainan); Yongxing Island and Jinyin Island (Xisha=Paracel Islands). Intertidal to subtidal 20 m. (15)

L. (*Limaria*) *hakodatensis* (Tokunaga): Yellow Sea; East China Sea. Intertidal to subtidal 50 m. (15)

L. (*P.*) *perfragilis* Habe & Kosuge: South China Sea. Water depth 65–94 m, sandy mud bottom. (15)

Limatula (*Stubilima*) *bullata* (Born): Shu Island (Xisha=Paracel Islands). Subtidal only, rare species. (15)

L. (*L.*) *choshiensis* Kuroda & Habe: East China Sea. Water depth 100–300 m, muddy sand bottom, rare species. (15)

L. (*L.*) *nippona* Habe: East China Sea. Water depth 83 m, muddy sand bottom. (15)

L. (*S.*) *textilis* Wang: East China Sea. Water depth 260 m, sandy bottom. (15)

SPONDYLIDAE

Spondylus albibarbatus Reeve: Taiwan. (103)

S. anacanthus Mawe: Nansha (=Spratly Islands). Intertidal mean low water to subtidal about 100 m. (16)

S. barbatus Reeve: Fujian; Guangdong. Intertidal to subtidal. (92)

S. butleri Reeve: Gaoxiong and Hualien (Taiwan). (103)

S. candidus Lamarck: Hengchun (Taiwan). (103)

S. castus Reeve: Penghu Islands (Taiwan). (103)

S. cruentus Lischke: Taiwan. (103)
S. ducalis Röding: Nansha (=Spratly Islands). Subtidal. (16)
S. flabellum Reeve: Taiwan. (103)
S. fragum Reeve: Fujian; Guangdong; Guangxi; Beibuwan (=Gulf of Tongking), Xinying and Lingshui (Hainan). Intertidal intertidal mean low water to subtidal. (60)
S. imperialis Chen: Guangdong; Lingshui and Sanya (Hainan); Nansha (=Spratly Islands). Subtidal 5–107 m. (16, 60)
S. microlepos Lamarck: Nansha (=Spratly Islands). Subtidal. (16)
S. nicobaricus Schreibers: Yushan, Dachen, and Nanji (Zhejiang); Longhai and Dongshan (Fujian); Guangdong; Weizhou Island (Guangxi); Beibuwan (=Gulf of Tongking); Xinying and Lingshui (Hainan); Xisha (=Paracel Islands). Intertidal to subtidal. (60)
S. regius (Linnaeus): Taiwan. (103)
S. serraticosta Prashad: Taiwan. (103)
S. sinensis Schreibers: Taiwan. (103)
S. spinosus Schreibers: Penghu Islands and Jilong (Taiwan). (103)
S. squamosus Schreibers: Taiwan. (103)
S. vesicolor Schreibers: Taiwan. (103)
S. wrightianus Crosse: Nansha (=Spratly Islands). Subtidal about 100 m. (16)

DIMYIDAE

Dimya japonica Habe: East China Sea (124°40' E, 26°10' N). Water depth 220 m, soft mud. (72)

ANOMIIDAE

Anomia chinensis Philippi [*A. cytaeum*]: Liaoning; Hebei; Shandong; Jiangsu; Zhejiang; Fujian; Guangdong; Hong Kong; Guangxi; Hainan. Intertidal. (21, 92).
Enigmonia aenigmatica (Holton) [*Anomia aenigmatica* (Chemnits)]: Fujian; Guangdong, Hong Kong; Xinying and Sanya (Hainan). Intertida, attached to mangrove branch. (19, 20, 92).
Monia umbonata (Gould): Bohai; Yellow Sea; East China Sea. Subtidal to 80 m. (21, 92)

PLACUNIDAE

Placuna (Ephippium) ephippium (Philipsson) [*P. sella*]: Guangdong; Qisha (Guangxi); Beibuwan (=Gulf of Tongking); Xinying and Lingshui (Hainan); Nansha (=Spratly Islands). Intertidal to subtidal. (60, 92)
P. (Placuna) placenta (Linnaeus) [*P. placenta*]: Zhejiang; Fujian; Guangdong; Guangxi; Beibuwan (=Gulf of Tongking); Lingshui and Sanya (Hainan). Intertidal. (60, 92)

GLYPHAEIDAE

Hyotissa hyotis (Linnaeus) [*Ostrea hyotis*]: Guangdong; Beibuwan (=Gulf of Tongking); Lingshui (Hainan). Intertidal. (48, 60, 92)
Parahyotissa imbricata (Lamarck) [*Ostrea imbricata*]: East China Sea; Guangdong; Xinying and Lingshui (Hainan). Intertidal to subtidal. (48, 60, 92, 102)
P. sinensis (Gmelin) [*Ostrea sinensis*]: Yushan and Nanji (Zhejiang); Guangdong; Hong Kong; Xinying, Lingshui, and Yulin (Hainan). Intertidal to subtidal. (48, 60)

OSTREIDAE

Alectryonella plicatula (Gmelin): Hong Kong; Hainan; Nansha (=Spratly Islands). (48)
Crassostrea ariakensis (Fujita): Taiwan. (103)
C. gigas (Thunberg) [*Ostrea gigas* Thunberg]: Along China's coast. Intertidal to subtidal several m. (21, 48, 60, 92)
C. lineata (Röding): Taiwan. (103)
C. rivularis (Gould) [*Ostrea rivularis* Gould]: Low salinity water along China's coast. Intertidal to subtidal. (21, 48, 60)
C. vitrefacta (Sowerby): Taiwan. (103)
Dendostrea crenulifera (Sowerby) [*Ostrea crenulifera*]: Guangdong; Hong Kong; Lingshui and Sanya (Hainan). Intertidal. (48, 60, 92)
D. folium (Linnaeus) [*Ostrea folium*]: Guangdong; Hong Kong; Sanya and Xinying (Hainan). Intertidal mean low water to subtidal. (48, 60, 92)
Lopha cristagalli (Linnaeus) [*Ostrea cristagalli*]: Yulin (Hainan). Intertidal mean low water to subtidal. (48, 60, 92)
Ostrea denselamellosa Lischke: Along China's coast. Subtidal to water depth about 30 m. (60)
Planostrea pestigris (Hanley) [*Ostrea paulucchiae*]: Guangdong; Guangxi; Beibuwan (=Gulf of Tongking); Lingshui and Sanya (Hainan). Intertidal to subtidal. (48, 60)
Saccostrea cucullata (Born) [*Ostrea cucullata, O. plicatula*]: Along southeastern coast of China. Intertidal. (21, 48, 60)
S. echinata (Quoy & Gaimard) [*Ostrea echinata*]: Zhejiang; Fujian; Guangdong; Hong Kong; Guangxi; Beibuwan (=Gulf of Tongking); Hainan. Mid intertidal. (21, 48, 60, 92)
Talonostrea talonata Li & Qi [*Ostrea pestigris*]: Beidaihe (Hebei); Qingdao (Shandong), Xiangshan Bay (Zhejiang); Fujian; Guangdong; Guangxi; Beibuwan (=Gulf of Tongking). Intertidal. (21, 48, 60)

VENEROIDA
LUCINIDAE

Anodontia edentula (Linnaeus) [*Lucina edentula*]: Xiangshan Bay and Nanji (Zhejiang); Fujian; Guangdong; Lingshui and Sanya (Hainan). Intertidal. (17, 60, 88)

A. stearnsiana Oyama [*Lucina philippiana*]: Yantai, Rongcheng, and Qingdao (Shandong); Fujian; Guangdong; Beibuwan (=Gulf of Tongking); Yinggehai (Hainan). Intertidal to subtidal. (21, 62, 92)

Bellucina civica (Yokoyama): Daya Bay (Guangdong). Subtidal. (38, 92, 96)

Codakia punctata (Linnaeus): Yongxing Island and Luo Island (Xisha=Paracel Islands). Intertidal to 20 m. (60, 92)

C. tigerina (Linnaeus): Southern Hainan; Xisha (=Paracel Islands). Intertidal to 20 m. (60, 92, 96)

Eamesiella corrugata (Deshayes): South of Taiwan. Intertidal to water depth 10 m. (92)

Epicodakia delicatula (Pilsbry) [*Ctena delicatula*]: Nanji (Zhejiang); Xiamen (Fujian); Daya Bay (Guangdong). Subtidal. (18, 38, 91, 92)

Lucinoma annulata (Reeve): East China Sea; Taiwan Strait. Subtidal. (83, 92, 102)

L. yoshidai Habe: East China Sea (128°10' E, 31°30' N). Water depth 520 m, soft mud. (72, 92, 102)

Pillucina pisidia (Dunker): Taiwan; Daya Bay (Guangdong). Intertidal to 30 m. (38, 92, 96)

P. (Sydlorina) yamakawai (Yokoyama): Daya Bay (Guangdong). Subtidal. (38, 92, 96)

THYASIRIDAE

Thyasira (Thyasira) tokunagaii Kurida & Habe: Bohai; Yellow Sea. Subtidal 11–84 m. (21, 92)

FIMBRIIDAE

Fimbria fimbriata (Linnaeus): Taiwan. (103)
F. soverbii (Reeve): Taiwan. (103)

UNGULINIDAE

Cycladicama cumingi (Hanley): Qinghuangdao and Beidaihe (Hebei); Yantai and Qingdao (Shandong); Liangyungang (Jiangsu); Nanji (Zhejiang); Fujian. Intertidal to subtidal 50 m. (21, 92)

C. lunaris Yokoyama: Taiwan Strait. Subtidal. (83, 92)

C. oblonga (Hanley) [*Joannisiella oblonga*]: Fujian; Taiwan; Beibuwan (=Gulf of Tongking). Subtidal 10–30 m. (92, 94)

C. tsuchii Yammamoto & Habe [*Joannisiella tsuchii*]: Along China's coast. (17, 91, 92)

Felaniella usta (Gould): Northern Yellow Sea. Subtidal, 8–75 m. (21, 92)

LASAEIDAE

Borniopsis ariakensis (Habe): Fujian; Hong Kong. Subtidal, about 20 m. (37, 98)

B. nodosa Morton & Scott: Hong Kong. Intertidal. (98)

B. ochetostomae Morton & Scott: Hong Kong. Intertidal. (98)

B. tsurumaru Habe: Jiaozhou Bay (Shandong); mouths of Yellow River and Changjiang (=Yangtze River); Hong Kong. Subtidal, about 10 m. (21, 92, 98)

Kellia japonica Pilsbry: Nanji (Zhejiang). Intertidal to subtidal. (18, 92)

K. porculus Pilsbry: Jiaozhou Bay (Shandong); Nanji (Zhejiang); Daya Bay (Guangdong); Hong Kong. Intertidal to subtidal 20 m. (21, 92, 96)

Lasaea nipponica Keen [*Lasaea undulata*]: Jiaozhou Bay (Shandong); Nanji (Zhejiang); Xiamen (Fujian); Daya Bay (Guangdong). Intertidal. (21, 92)

L. rubra (Montagu): Hong Kong. Intertidal. (98)

Pseudopythina macrophthalmensis Morton & Scott: Hong Kong. Intertidal.

P. maipoensis Morton & Scott: Hong Kong. Subtidal. (98)

Squillaconcha subsinuata (Lischke): Hong Kong. Intertidal to subtidal. (92, 98)

CHAMIDAE

Chama brassica Reeve: Xiamen (Fujian); Guangdong; Guangxi; Beibuwan (=Gulf of Tongking); Hainan. Intertidal to subtidal 53 m. (29)

C. divaricata Reeve: Taiwan. (103)

C. dunkeri Lischke: Fujian; Guangdong; Guangxi; Hainan. Intertidal to subtidal. (29)

C. fraga Reeve: Pingtan and Xiamen (Fujian); Guangdong; Guangxi, Hainan. Intertidal to subtidal. (29)

C. iostoma Conrad: Sanya (Hainan). Coral reefs. (29)

C. jukesi Reeve: Putian (Fujian); Autou and Zapou (Guangdong); Sanya (Hainan); Yongxing Island (Xisha=Paracel Islands). Intertidal to subtidal. (29)

C. lazarus Linnaeus: Guangdong; Sanya, Yulin, and Lingshui (Hainan). Intertidal to subtidal. (29)

C. lobata Broderip: Offshore off Nanao and Haimen (Guangdong); Offshore off Macau; southeastern Hainan. Subtidal 35–101 m. (29)

C. reflexa Reeve: East China Sea continental shelf, Dachen Island and Nanji (Zhejiang); Pingtan, Xiamen, and Dongshan (Fujian); Zapou, Wushi, and Autou (Guangdong); Weizhou Island (Guangxi); Sanya and Lingshui (Hainan). Intertidal to subtidal 108 m. (29)

C. semipurpurata Lischke: Pingtan (Fujian); Guangdong; Guangxi; Hainan; Xisha (=Paracel Islands). Intertidal to subtidal. (29)

Pseudochama retroversa (Lischke): Nanji (Zhejiang); Pingtan and Xiamen (Fujian); Taiwan Strait; Zhanjiang (Guangdong). Intertidal to subtidal. (29)

MONTACUTIDAE

Barrimysia siphonosomae Morton & Scott: Hong Kong. Intertidal. (98)
Curvemysella paula (A. Adams): Nanji (Zhejiang); Hong Kong. Intertidal. (18, 92, 98)
Fronsella fujitaniana (Yokoyama): Jiaozhou Bay (Shandong). Subtidal 13–35 m. (69, 92, 96)
Isoconcha sibogai Prashad: East China Sea (129°30' E, 30°20' N). Water depth 760 m, soft mud. (72)
Montacutona compacta (Gould): Hong Kong. Intertidal to subtidal. (92, 98)
M. mutsuwanensis (Yamamoto & Habe): Hong Kong. Subtidal. (92, 98)
Mysella triangularis (A. Adams): Hong Kong. Intertidal. (92, 98)
Nipponomysella oblongata (Yokoyama): Jiaozhou Bay (Shandong). Subtidal 20 m. (21, 69, 92)

GALEOMMATIDAE

Devonia semperi (Ohshima) [*Entovalva semperi*]: Bohai Bay; Jiaozhou Bay (Shandong); Hong Kong. Intertidal to subtidal 5–10 m. (21, 98)
Ephippodonta oedipus Morton: Hong Kong. Subtidal. (98)
Galeomma polita Deshayes: Hong Kong. Intertidal. (98)
G. takii Kuroda: Hong Kong. Intertidal. (98)
Pseudogaleomma japonica (A. Adams): Guangdong; Hong Kong; Hainan. Intertidal. (19, 92)
Scintilla cf. *cuvieri* Deshayes: Hong Kong. Coral reef subtidal. (98)
S. nitidella Habe: Hong Kong. Coral reef intertidal. (98)
S. cf. *opalinus* Kuroda & Habe: Hong Kong. Intertidal. (92, 98)
Scintillona brissae Morton & Scott: Hong Kong. Subtidal. (98)

CARDITIDAE

Beguina semiorbiculata (Linnaeus): Weizhou Island (Guangxi); Beibuwan (=Gulf of Tongking); Xinying, Sanya, and Lingshui (Hainan). Subtidal. (76a)
Cardita crassicosta (Lamarck): Taiwan. (103)
C. leana Dunker: Yushan and Nanji (Zhejiang); Xiamen (Fujian); Taiwan; Hong Kong. Intertidal. (17, 92)
C. nodulosa Lamarck: East China Sea (124°40' E, 26°10' N). Soft mud. (72)
C. variegata Bruguiére: Yushan, Dachen, and Nanji (Zhejiang); Fujian; Guangdong; Hong Kong; Weizhou Island (Guangxi); Beibuwan (=Gulf of Tongking); Sanya and Lingshui (Hainan). Intertidal to subtidal. (60, 92)
Carditella hanzawai (Nomura): Mouth of Changjiang (=Yangtze River); Taiwan. (103)
Glans hirasei Dall: East China Sea continental shelf. (32)
Megacardita ferruginosa (Adams & Reeve) [*Venericardita ferruginosa*]: Eastern Liaoning Bay; Dalian (Liaoning). Subtidal to 23 m. (21)

CRASSATELLIDAE

Crassatella speciosa A. Adams: Zhejiang. Subtidal within 30 m. (17)
Eucrassatella (*Nipponocrassatella*) *japonica* (Dunker): East China Sea continental shelf; Fujian. Subtidal. (32, 92, 96)
E. (*N.*) *nana* (Adams & Reeve): East China Sea continental shelf; Taiwan Strait; South China Sea. Subtidal. (32, 83, 92, 96)
Indocrassatella oblongata (Yokoyama): Taiwan Strait. Subtidal. (92, 96)

CARDIIDAE

Afrocardium infantile (Nomura & Zinbo): Southeastern Hainan. Water depth 75–100 m. (65)
Clinocardium californiense (Deshayes): Yellow Sea. Water depth 23–77 m. (65)
Corculum cardissa (Linnaeus): Xisha (=Paracel Islands). (65)
C. mundum (Reeve): Taiwan. (76a)
Discors aurantiacum (Adams & Reeve): Nansha (=Spratly Islands). Water depth 36 m. (77)
D. biradiatum (Bruguiere): Nansha (=Spratly Islands). Water depth 46 m. (77)
D. multipunctatum ctatum (Sowerby): South China Sea. Water depth 54–100 m. (65)
Fragum bannoi (Otuka): Taiwan. (65)
F. carinatum (Lynge): Xiamen (Fujian); Taiwan; Guangdong; Haikou, Sanya, Xinying, and Lingshui (Hainan). (65).
F. fragum (Linnaeus): Xisha (=Paracel Islands). (65)
F. peronatum Iredale: Beibuwan (=Gulf of Tongking). Water depth 12–29 m. (65)
F. unedo (Linnaeus): Taiwan; Lingshui (Hainan). (65)
Fulvia australis (Sowerby): Hainan. (65)
F. bullata (Linnaeus): Shanwei and Pinghai (Guangdong); Qisha (Guangxi); Xinying, Sanya, and Lingshui (Hainan). (65)
F. hungerfordi (Sowerby): South China Sea. Water depth 12–36 m. (65)
F. mutica (Reeve): Bohai; Yellow Sea. Water depth 0–57 m. (65)
F. oxygona (Sowerby): China's seas. (76a)
Laevicardium attenuatum (Sowerby): Eastern Hainan. Water depth 32 m. (65)
L. lobulatum (Deshayes): Nansha (=Spratly Islands). Water depth 46 m. (77)
Lunulicardia retusa (Linnaues): Taiwan; Weizhou Island (Guangxi); eastern Hainan. Water depth 29–35 m. (65)
Lyrocardium lyratum (Sowerby): Taiwan. (65)

Maoricardium mansitlii (Otuka): Dongshan (Fujian); Taiwan; Nanao (Guangdong); Hainan. Water depth 33–90 m. (65)

M. setosum (Redfield): Xiamen (Fujian); Taiwan; Haimen (Guangdong). (65)

Microcardium exasperatum (Sowerby): East China Sea; South China Sea. Water depth 42–128 m. (65, 77)

M. nomurai (Kuroda & Habe): Southeastern Hainan. Water depth 150–200 m. (65)

M. torresi (Smith): East China Sea; South China Sea. Water depth 40–150 m. (65, 77)

Nemocardium bechei (Reeve): Taiwan. (65)

N. samarangae (Makiyama): Southern Yellow Sea. Water depth 64–80 m. (65)

Trachycardium angulatum (Lamarck): Taiwan; Xisha (=Paracel Islands). (65)

T. arenicolum (Reeve): Pingtan and Changle (Fujian); Taiwan. (65)

T. enode (Sowerby): Taiwan; Sanya (Hainan). (65)

T. flavum (Linnaeus): Taiwan; Weizhou Island (Guangxi); Sanya, Lingshui, and Qinglan (Hainan). (65)

T. foveilatum (Sowerby): Taiwan. (65)

T. impolitum (Sowerby): South China Sea. Shallow water, fine sand bottom. (65)

T. sewelli (Prashad): Nansha (=Spratly Islands). Water depth 53 m. (77)

T. transcendens (Helvill & Stanclen): Nansha (=Spratly Islands). (76a)

Trigoniocardia fornicata (Sowerby): Nansha (=Spratly Islands). Water depth 36 m. (77)

Vepricardium asiaticum (Bruguiere): From Taiwan to Beibuwan (=Gulf of Tongking). Water depth 16–87 m. (65)

V. coronatum (Spengler): South China Sea. Water depth 11–46 m. (65)

V. multispinosum (Sowerby): Taiwan; Beibuwan (=Gulf of Tongking). Water depth 24–97 m. (65)

V. sinense (Sowerby): South China Sea. Water depth 12–36 m. (65)

TRIDACNIDAE

Hippopus hippopus (Linnaeus): Taiwan; Xisha (=Paracel Islands). Reef flat. (23)

Tridacna (*Chamestrachea*) *crocea* Lamarck: Taiwan; Hainan; Xisha (=Paracel Islands). Coral reefs. (23)

T. (*Persikima*) *derasa* (Röding): Xisha (=Paracel Islands). Coral atoll. (23)

T. (*Tridacna*) *gigas* (Linnaeus) [*T.* (*Dinodaena*) *cookiana*]: Xisha (=Spratly Islands). Coral reef sandy bottom. (23)

T. T. (*C.*) *maxima* (Röding) [*T. elongata*]: Hainan; Xisha (=Paracel Islands). Coral reefs. (23)

T. (*C.*) *squamosa* Lamarck: Taiwan; Hainan; Xisha (=Paracel Islands). Coral reefs. (23)

MACTRIDAE

Coelomactra antiquata (Spengler) [*Mactra antiquata, M. spectabilis*]: Liaoning; Hebei; Shandong; Jiangsu; Zhejiang; Fujian; Haimen, Lufeng, and Dianbai (Guangdong). Intertidal to subtidal about 20 m. (30)

Heterocardia elliptica Zhuang: Nanao (Guangdong). (28)

Lutraria arcuata Reeve: Along the coast of western Guangxi; Beibuwan (=Gulf of Tongking). Subtidal to 30 m. (30)

L. impar Reeve: Sansha and Pingtan (Fujian); Haimen and Shanchuang Island (Guangdong); Beibuwan (=Gulf of Tongking). Subtidal to 40 m. (30)

L. (*Psommophila*) *maxima* Jonas: Pingtan (Fujian); Taiwan; Guangdong; islands along the coast of Guangxi. Intertidal to subtidal 10 m. (30)

L. philippinarum Reeve: Pingtan and Dongshan (Fujian); Xinying, Lingshui, and Sanya (Hainan). Intertidal to subtidal. (30)

L. (*P.*) *sieboldii* Reeve: Zhoushan (Zhejiang); Beibuwan (=Gulf of Tongking); Qinglan Bay (Hainan). Subtidal 42–78 m. (30)

Mactra aphrodina Reeve: Bohai; Yellow Sea; East China Sea; South China Sea. Subtidal 12–65 m. (30)

M. (*M.*) *chinensis* Philippi [*M. sulcataria*]: Gaizhou (Liaoning) to Dongshan (Fujian). Intertidal. (30)

M. (*Telemactra*) *crossei* (Dunker): Fujian. Subtidal. (37, 92)

M. (*T.*) *cuneata* Gmelin: Dongshan (Fujian); Nanao and Baoan (Guangdong); Qinglan, Xinying, and Lingshui (Hainan). Subtidal to 20 m. (30)

M. inaequalis Reeve: Nanji (Zhejiang); Nanao (Guangdong); Xinying (Hainan). Subtidal 12–65 m. (30)

M. (*T.*) *luzonica* Reeve: Haimen and Jieshi (Guangdong); Qinglan Bay (Hainan). Subtidal to 124 m. (30)

M. (*M.*) *maculata* Gmelin: Yulin (Hainan). Subtidal 10–30 m. (30)

M. (*M.*) *mera* Reeve [*M. grandis*]: Nanji (Zhejiang); Huian, Jingjiang, Zhangpu, and Dongshan (Fujian); Autou and Wushi (Guangdong); Weizhou Island (Guangxi); Beibuwan (=Gulf of Tongking); Sanya and Lingshui (Hainan). Intertidal to subtidal. (30)

M. (*M.*) *ornata* Gray: Fujian; Taiwan Strait; Macau; Guangxi. Subtidal 10–60 m. (30)

M. (*M.*) *veneriformis* Reeve [*M. quadriangularis*]: Along China's coast. Intertidal to subtidal. (30)

Mactrimula dolabrata (Reeve): Southern Yellow Sea; East China Sea; Beibuwan (=Gulf of Tongking); South China Sea. Subtidal 20–100 m. (30)

Meropesta capillacea (Reeve) [*Standella capillacea*]: Xiamen and Zhangpu (Fujian); Taiwan; Wushi (Guangdong); Sanya, and Wanning (Hainan). Intertidal. (30)

M. nicobarica (Gmelin): Taiwan; Wushi (Guangdong); Qisha (Guangxi); Beibuwan (=Gulf of Tongking); Xinying and Lingshui (Hainan). Intertidal to subtidal 20 m. (30)

M. pellucida (Gmelin) [*Standella pellucida*]: Pingtan and Xiamen (Fujian); Shantou, Dongping, Wushi, and Naozhou (Guangdong); Sanya (Hainan). Intertidal. (30).

M. sinojaponica Zhuang: Xingcheng and Changxing Island (Liaoning); Nanbao (Hebei); Dingzhi Bay, Laizhou, and Qingdao (Shandong); Fuqing and Longhai (Fujian); Nanao (Guangdong). Intertidal. (28)

Micronactra angulifera (Reeve): Pingtan and Dongshan (Fujian); Daya Bay; Wangshan Islands, Gulf of Guangzhou (Guangdong), Beibuwan (=Gulf of Tongking). Subtidal 12–83 m. (30)

Oxyperas (*O.*) *bernardi* (Pilsbry): Nanji (Zhejiang); Dongshan (Fujian); Taiwan. Subtidal 10–100 m. (30)

Raeta (*Raetina*) *pellicula* (Reeve): Xiaochangshan (Liaoning); Beidaihe (Hebei); Qingdao (Shandong); Lianyungang and Rudong (Jiangsu). Intertidal. (30)

Raetellops pulchella (Adams & Reeve): Bohai; Yellow Sea; East China Sea. Subtidal 10–60 m. (30)

MESODESMATIDAE

Anapella retroconvexa Zhuang: Naozhou (Guangdong); Sanya (Hainan). (24)

Atactodea striata (Gmelin) [*Mesodesma striata*]: Xiamen and Dongshan (Fujian), Nanao, Lufeng, Baoan, and Daya Bay (Guangdong); Hainan, Xisha (=Paracel Islands). Intertidal. (24)

Coecella chinensis Deshayes: Liaoning; Shandong. (24)

C. turgida Deshayes: Xiamen (Fujian); Taiwan; Pinghai and Shanchuang Island (Guangdong); Sanya and Lingshui (Hainan). (24)

Davila plana (Hanley): Sanya (Hainan). (24)

CARDILIIDAE

Cardilia semisulcata (Lamarck): Taiwan Strait. Subtidal. (37)

DONACIDAE

Chion dysoni Deshayes [*Donax incarnatus*]: Taiwan; Daya Bay (Guangdong); Beibuwan (=Gulf of Tongking); Wanning and Sanya (Hainan). Intertidal. (82, 88)

C. semigranosus (Dunker) [*Donax semigranosus tropicus*]: Nanji (Zhejiang); Changle, Pingtan, and Dongshan (Fujian); Taiwan; Nanao, Daya Bay, and Zhanjiang (Guangdong); Hong Kong; Hainan. Intertidal. (82, 92)

C. ticaonicus (Hanley): South of Taiwan. Intertidal. (92)

Latona cuneata (Linnaeus) [*Donax cuneatus*]: Nanji (Zhejiang); Changle, Pingtan, and Longhai (Fujian); Shanwei, Nanao, Daya Bay, and Zhanjiang (Guangdong); Hong Kong; Sanya and Yinggehai (Hainan). Intertidal. (82, 92)

L. faba (Gmelin) [*Donax faba*]: Yunxiao (Fujian); Nanao and Daya Bay (Guangdong); Hong Kong; Weizhou Island (Guangxi); Sanya and Lingshui (Hainan). Intertidal. (82, 92)

Tentidonax kiusiuensis (Pilsbry) [*Donax kiusiuensis*]: Beidaihe (Hebei); Yantai and Qingdao (Shandong); Gaoxiong (Taiwan). Intertidal to subtidal 10 m. (82, 92)

T. nitidus (Deshayes) [*Donax nitidus*]: Haikou (Hainan). Subtidal 6–30 m. (82, 92)

TELLINIDAE

Aeretica tomlini (Smith) [*Strigilla tomlini*]: Taiwan; Sanya and Haikou (Hainan). Intertidal. (82)

Angulus compressissimus (Reeve): Qinghuangdao and Beidaihe (Hebei); Laizhou Bay and Jiaozhou Bay (Shandong); Lianyungang (Jiangsu). Intertidal. (82)

A. emarginatus (Sowerby): Fujian; Taiwan Strait; Guangdong; Guangxi; Beibuwan (=Gulf of Tongking); Wanning (Hainan). Subtidal to 60 m. (82)

A. lanceolatus (Gmelin): Fujian; Taiwan Strait; Guangdong. Subtidal. (82)

A. psammotellus (Lamarck): Haikou and Sanya (Hainan). Debris on shore. (82)

A. vestalioides (Yokoyama): Bohai; Yellow Sea; Zhejiang; Fujian; Haikou (Hainan). Intertidal to subtidal 50 m. (82)

A. vestalis (Hanley): Fujian; Guangdong; Hong Kong; Guangxi; Haikou (Hainan). Intertidal to subtidal. (82)

Apolymetis meyeri (Philippi): Western Guangdong; Hainan. Intertidal. (82)

Arcopaginula inflata (Gmelin): Fujian; Taiwan Strait; Sanya (Hainan). Subtidal, water depth 8–60 m. (82)

Arcopella casta (Hanley): Haikou (Hainan). Subtidal 2–40 m. (82)

A. isseli (H. Adams): Taiwan Strait. Water depth 5–100 m. (92)

Cadella delta (Yokoyama): East China Sea continental shelf. (82)

C. delta hainanensis Scarlato: Xinying (Hainan). Intertidal. (82)

C. narutoensis Habe: Qingdao and Jiaozhou Bay (Shandong); Nanji (Zhejiang). Subtidal within 20 m. (82)

C. semitorta (Reeve): Nanji (Zhejiang); Fujian; Taiwan Strait; Beibuwan (=Gulf of Tongking); Xinying and Haikou (Hainan). Subtidal 2–60 m. (82)

C. smithi (Lynge): Xinying (Hainan). Subtidal to 30 m. (82)

Cyclotellina remies (Linnaeus) [*Arcopagia remies*]: Guangdong; Lingshui, Sanya, Haikou, and Xinying (Hainan). Intertidal to subtidal 20 m. (82)

Exotica clathrata (Deshayes): Guangxi; Sanya and Lingshui (Hainan); Beibuwan (=Gulf of Tongking). Subtidal. (82)

E. obliquistriata (Reeve): Lingshui (Hainan). Subtidal. (82)

Fabulina tsichungyeni Scarlato: Lingshui (Hainan). Intertidal. (82)

Heteromacoma irus (Hanley) [*Gastrana contabulata*]: Haiyang Island and Dalian (Liaoning); Shandong. Intertidal to subtidal. (82)

Leporimetis lacunosus (Chemnitz) [*Apolymetis lacunosa*]: Nanji (Zhejiang); Guangdong; Sanya and Haikou (Hainan). Intertidal to subtidal. (82)

L. spectabilis (Hanley): Taiwan; Haikou and Sanya (Hainan). Intertidal. (82)

Macalia bruguieri (Hanley): Taiwan; western Guangxi; Beibuwan (=Gulf of Tongking); Sanya, Lingshui, and Xinying (Hainan). Intertidal to subtidal 20 m. (82)

Macoma (Psammacoma) awajiensis (Sowerby): Taiwan Strait. Subtidal. (82)

M. (P.) candida (Lamarck) [*M. galathaea, M. candida*]: Fujian; Taiwan Strait; Guangdong; Hong Kong; Guangxi; Beibuwan (=Gulf of Tongking); Sanya (Hainan). Intertidal to subtidal 50 m. (82)

M. (Macoma) incongrua (Martens): Dalian (Liaoning); Qinghuangdao (Hebei); Yantai, Weihai, Chengshanjiao and Qingdao (Shandong). Intertidal mean low water to subtidal about 10 m. (82)

M. (P.) lucerna (Hanley): Putuo (Zhejiang); Quanzhou (Fujian); Guanghai (Guangdong); Haikou and Wanning (Hainan). Intertidal. (82)

M. murrayi (Grabau & King): Chengshanjiao and Qingdao (Shandong); Fujian; Guangxi. Intertidal mean low water to subtidal. (82)

M. (P.) nobilis (Hanley): Guangdong; Haikou, Wanning, and Sanya (Hainan). Intertidal to subtidal. (82)

M. (M.) praetexta (Martens): Qinghuangdao and Beidaihe (Hebei); Penlai Bay and Yantai (Shandong); Fujian; Taiwan. Intertidal mean low water to subtidal within 50 m. (82)

M. (P.) praerupta Salisbury [*M. truncata*]: Nanji (Zhejiang); Fujian; Taiwan; Guangdong; Beibuwan (=Gulf of Tongking); Sanya and Haikou (Hainan). Intertidal to subtidal 49 m. (82)

M. (Rexithaerus) sectior Oyama: Taiwan. Subtidal, water depth 20–30 m. (92)

M. (M.) tokyoensis Makiyama: Dalian (Liaoning); Qinghuangdao (Hebei); Chengshanjiao and Rongcheng (Shandong). Intertidal mean low water to subtidal within 50 m. (82)

Merisca capsoides (Lamarck): Hong Kong; Sanya and Haikou (Hainan). (82)

M. diaphana (Deshayes) [*Tellina diaphana*]: Hangzhou Bay, Xiangshan Bay, and Sanmen Bay (Zhejiang); Fujian; Guangdong; Guangxi; Beibuwan (=Gulf of Tongking); Wanning, Qinglan, Sanya, and Haikou (Hainan). Intertidal. (82)

M. perplexa (Hanley): Guangdong; Xinying, Haikou, and Wanning (Hainan). Intertidal. (82)

Moerella culter (Hanley): Xiangshan Bay (Zhejiang); Fujian; Taiwan; Guangdong; Guangxi; Beibuwan. (=Gulf of Tongking); Haikou, Wanning, and Sanya (Hainan). Subtidal. (82)

M. iridescens (Benson): Hulu Island (Liaoning); Qingdao (Shandong); Dafeng (Jiangsu); Zhejiang; Fujian; Taiwan; Guangdong; Beibuwan (=Gulf of Tongking). Intertidal to subtidal about 20 m. (82)

M. jedoensis (Lischke) [*Tellina jedoensis*]: Bohai; Yellow Sea; Nanji (Zhejiang); Fujian; Guangdong. Intertidal to subtidal 20 m. (82)

M. philippinarum (Hanley): Fujian; Guangdong; Haikou, Wanning, Lingshui, Sanya, and Xinying (Hainan). Intertidal. (82)

M. rutila (Dunker): Xiaochangshan Island, Dalian, and Xincheng (Liaoning); Qinghuangdao and Beidaihe (Hebei); mouth of Yellow River, Yantai, and Qingdao (Shandong); Rudong (Jiangsu); Zhejiang; Fujian; Taiwan; Guangdong. Intertidal. (82)

Nitidotellina iridella (Martens): Changshan Islands (Liaoning); Beidaihe (Hebei); Chengshanjiao and Jiaozhou Bay (Shandong); Xiangshan Bay (Zhejiang); Xiamen (Fujian); Taiwan; Guangdong; Beibuwan (=Gulf of Tongking). Intertidal to subtidal 40 m. (82)

N. minuta (Lischke): Bohai; Yellow Sea; Nanji (Zhejiang); Fujian; Guangdong; Lingshui and Xinying (Hainan); Beibuwan (=Gulf of Tongking). Intertidal to subtidal 30 m. (82)

N. nitidula (Dunker): Fujian; Taiwan; Wanning and Sanya (Hainan). Intertidal to subtidal. (82)

Pharaonella perna (Spengler) [*Tellina perna*]: Western Guangdong; Guangxi; Xinying, Sanya, and Lingshui (Hainan). Intertidal to subtidal 20 m. (82)

P. rostrata (Linnaeus) [*Tellina vulsella*]: Guangxi; Beibuwan (=Gulf of Tongking); Xinying, Sanya, and Lingshui (Hainan). (82)

Phylloda foliacea (Linnaeus): Sanya (Hainan). Subtidal 10–20 m. (82)

Pinguitellina pinguis (Hanley): Lingshui (Hainan). Intertidal to subtidal 30 m. (82)

Pseudarcopagia minuta Scarlato: Haikou (Hainan). Intertidal. (82)

Pulvinus micans (Hanley): Fujian; Taiwan Strait; Guangdong; Haikou and Xinying (Hainan). Intertidal to subtidal 40 m. (82)

Quidnipagus palatam (Martyn) [*Tellina ruglsa*]: Taiwan; Guangdong; Lingshui and Sanya (Hainan). Intertidal to subtidal 20 m. (82)

Scutarcopagia scobinata (Linnaeus) [*Arcopagia scobinata*]: Taiwan; Haifeng (Guangdong); Sanya and Lingshui (Hainan); Dengqing Island and Yongxing Island (Xisha=Paracel Islands). Intertidal to subtidal 20 m. (82)

Tellina rastella Hanley: Xisha (=Paracel Islands). Intertidal to subtidal 20 m. (82)

T. spengleri Gmelin: Taiwan Strait; western coast of Guangxi; Haikou and Lingshui (Hainan). Intertidal. (82)

T. virgata Linnaeus: Taiwan; Guangdong; Wenchang, Lingshui, Sanya, and Xinying (Hainan). Intertidal. (82)

Tellinides chinensis (Hanley): Fujian; Guangdong. Intertidal to subtidal. (82)

T. ovalis (Sowerby): Taiwan; Beibuwan (=Gulf of Tongking); Lingshui and Xinying (Hainan). Subtidal 20–170 m. (82)

T. timorensis Lamarck: Xiangshan Bay (Zhejiang); Fujian; Guangdong; Beibuwan (=Gulf of Tongking); Haikou, Sanya, Wanning, and Haikou (Hainan). Intertidal. (82)

Tellinimactra maluccensis (Martens): Hainan. Intertidal. (82)

SEMELIDAE

Abra soyoae Habe: East China Sea (128°10' E. 31°30' N): Water depth 520 m. (72)

Abrina hainanensis Scarlato: Sanya and Haikou (Hainan). Mangrove forest. (82)

A. kinoshitai Kuroda & Habe: Yellow Sea. Subtidal 50–60 m. (21)

A. lunella (Gould): Bohai; Yellow Sea; Jiaozhou Bay (Shandong); Taiwan Strait. Subtidal 5–100 m. (21, 82)

A. magna Scarlato: Fujian; Guangdong; Beibuwan (=Gulf of Tongking); Haikou and Sanya (Hainan). Intertidal. (82)

Leptomya minuta Habe: Bohai; Yellow Sea. Subtidal within 50 m. (21)

L. nitida (Adams & Reeve): South of Taiwan. (92)

Semele cordiformis (Hotlen) [*Semele sinensis*]: Nanji (Zhejiang); Fujian; Guangdong; Qisha (Guangxi); Sanya and Haikou (Hainan). Intertidal. (82)

S. crenulata (Sowerby): Guangdong; Hong Kong; Lingshui, Sanya, and Xinying (Hainan). Intertidal. (82)

S. scabra (Hanley): Xinying (Hainan). Intertidal to subtidal. (82)

S. zebuensis (Hanley): Taiwan Strait. Subtidal. (92)

Theora fragilis (A. Adams) [*T. lubrica*]: Bohai; Yellow Sea; East China Sea; Zhoushan (Zhejiang). Subtidal 9–50 m. (82)

T. lata (Hinds): Xiangshan Bay (Zhejiang); Fujian; Haikou (Hainan). Intertidal to subtidal. (82)

PSAMMOBIIDAE

Asaphis dichotoma (Anton): Taiwan; Guangdong; Hong Kong; Beibuwan (=Gulf of Tongking); Lingshui, Sanya, and Xinying (Hainan). Intertidal. (82)

Gari anomala (Deshayes): Taiwan Strait. Subtidal. (82)

G. hosoyai Habe [*G. reevei*]: South of Jiaozhou Bay (Shandong); Xiangshan Bay (Zhejiang); Taiwan Strait; Guangdong; Guangxi; Beibuwan (=Gulf of Tongking). Intertidal to subtidal 60 m. (82)

G. maculosa (Lamarck) [*Psammobia maculosa*]: Guangdong; Xinying (Hainan). Intertidal to subtidal. (82)

G. schepmani Prashad [*G. schepmani*]: Western Guangxi; Beibuwan (=Gulf of Tongking); Xinying (Hainan). Subtidal to 77 m. (82)

G. truncata (Linnaeus): Taiwan Strait; Guangdong; Sanya (Hainan). Intertidal to subtidal. (82)

Gobraeus kazusensis (Yokoyama) [*Psammobia kazusensis*]: Haiyang Island, Dalian, and Lushun (Liaoning); Huangcheng Island, Rongcheng, and Qingdao (Shandong). Intertidal. (21)

Grammatomya squamosa (Lamarck) [*Gari squamosa*]: Lingshui (Hainan). Subtidal 10- 20 m. (82)

Hiatula acuta Cai & Zhuang: Longhai and Zhangpu (Fujian); Wuchuang and Zhanjiang (Guangdong). Subtidal. (82)

H. ambigua (Reeve) [*Sanguinolaria ambigua*]: Taiwan; Guangdong; Haikou, Xinying, and Lingshui (Hainan). Intertidal. (82)

H. atrata Reeve [*Sanguinolaria atrata*]: Leizhou Peninsula (Guangdong); Beibuwan (=Gulf of Tongking); Haikou, Sanya, and Lingshui (Hainan); Intertidal. (82)

H. castanea Scarlato [*Sanguinolaria castanea*]: Taiwan Strait; Guangdong; Xinying and Haikou (Hainan). Intertidal. (82)

H. chinensis (Mörch) [*Sanguinolaria chinensis*, *S. planulata*]: Southern and eastern coasts of Shandong; Nanji (Zhejiang); Fujian; Taiwan; Taiwan Strait; Guangdong; Guangxi; Beibuwan (=Gulf of Tongking); Lingshui, Sanya, and Xinying (Hainan). Intertidal. (82)

H. diphos (Linnaeus) [*Sanguinolaria diphos*]: Xiaochangshan Island (Liaoning); Li Island, Rongcheng, and Qingdao (Shandong); Xiangshan Bay (Zhejiang); Fujian; Taiwan; Taiwan Strait; Guangdong; Hong Kong; Guangxi; Beibuwan (=Gulf of Tongking); Wanning, Sanya, and Xinying (Hainan). Intertidal. (82)

H. inflata (Bertin) [*Sanguinolaria inflata*]: Guangdong; Lingshui and Sanya (Hainan). Intertidal. (82)

H. minor (Deshayes) [*Sanguinolaria minor*]: Lingshui, Sanya and Haikou (Hainan). Intertidal. (82)

H. rostrata (Lightfoot): South of Taiwan. Intertidal. (92)

H. togata (Deshayes): Western Guangdong; Haikou, Sanya, Wanning, and Xinying (Hainan). Intertidal. (82)

H. violacea (Lamarck) [*Sanguinolaria violacea*]: Taiwan; Taiwan Strait; Fujian; Guangdong; Beibuwan (=Gulf of Tongking); Haikou, Lingshui, and Sanya (Hainan). Intertidal to subtidal. (82)

H. virescens (Deshayes) [*Sanguinolaria virescens*]: Guangdong; Beibuwan (=Gulf of Tongking); Haikou and Lingshui (Hainan). River mouth and lower river basin. (82)

Psammobia radiata (Dunker) [*Gari radiata*]: Taiwan Strait; Guangdong; Guangxi; Beibuwan (=Gulf of Tongking); Xinying (Hainan). Intertidal to subtidal about 10 m. (82)

Nuttallia olivacea (Jay) [*Sanguinolaria olivacea*]: Liaoning; Qinghuangdao and Beidaihe (Hebei); Shandong; Lianyungang, Rudong, and Qidong (Jiangsu); Fujian; Guangdong. Intertidal. (82)

Sanguinolaria (*Hainania*) *tchangsii* Scarlato: Southern tip of Leizhou Peninsula (Guangdong); Xinying (Hainan). Intertidal. (82)

SOLECURTIDAE

Azorinus abbreviatus (Gould): Western Guangdong. Subtidal. (88, 92)

A. coarctatus (Gmelin): Fujian; Guangdong; Hong Kong; Beibuwan (=Gulf of Tongking); Sanya (Hainan). Subtidal to 55 m. (82)

A. scheepmakeri (Dunker) [*Zozia scheepmakeri*]: Taiwan; Wanning (Hainan). Intertidal. (82)

Solecurtus divaricatus (Lischke) [*Solenocurtus divaricatus*]: Jiaozhou Bay (Shandong); Xiangshan Bay and Nanji (Zhejiang); Taiwan; Guangdong; Beibuwan (=Gulf of Tongking); Xinying (Hainan). Intertidal. (60, 92)

S. exaratus (Philippi) [*Solenocurtus exaratus*]: Guangdong; Sanya and Lingshui (Hainan). Subtidal. (60, 92)

S. wilsoni Tryon: Daya Bay (Guangdong). Subtidal. (91, 92)

SOLENIDAE

Solen arcuatus Zhang & Huang: Qingdao (Shandong); Lianyungang (Jiangsu); Kanmen (Zhejiang); Fujian. Intertidal. (63)

S. canaliculatus Zhang & Huang: East China Sea; South China Sea. Subtidal 31–110 m. (63)

S. dunkerianus Clessin: Bohai; Yellow Sea; East China Sea; South China Sea. Subtidal to 90 m. (63)

S. gordonis Yokoyama: Northern Fujian; Baoan (Guangdong). Subtidal. (63)

S. gracilis Philippi: Liaoning; Shandong; Lianyungang (Jiangsu). Subtidal. (21, 63)

S. grandis Dunker: China's coast. Intertidal. (63)

S. linearis Spengler: Zhejiang; Fujian; Guangdong; Guangxi; Beibuwan (=Gulf of Tongking); Xinying (Hainan). Intertidal to subtidal 60 m. (63)

S. roseomaculatus Pilsbry: Nanji (Zhejiang); Xiamen (Fujian); South China Sea. Subtidal within 100 m. (63)

S. sloanii Hanley: Fujian; Guangdong; Guangxi; Beibuwan (=Gulf of Tongking); Hainan. Intertidal to subtidal. (63)

S. strictus Gould [*S. gouldii*]: China's coast. Intertidal. (21, 63)

CULTELLIDAE

Cultellus attenuatus Dunker: Liaoning; Hebei; Shandong; Zhejiang; Fujian; Guangdong; Guangxi; Beibuwan (=Gulf of Tongking); Hainan. Intertidal to subtidal 98 m. (63)

C. cultellus (Linnaeus): Zhujiang (=Pearl River, Guangdong); Beibuwan (=Gulf of Tongking). Subtidal. (63)

C. scalprum (Gould): Daijuyang, Xiangshan Bay, and Yueqing Bay (Zhejiang); Fujian; Guangdong; Guangxi; Beibuwan (=Gulf of Tongking). Intertidal to subtidal 30 m. (63).

Siliqua fasciata (Spengler): Fujian; Guangdong. Subtidal. (63)

S. grayana (Dunker): South China Sea. Subtidal. Water depth 7–70 m. (63)

S. minima (Gmelin): Liaoning; Hebei; Shandong; Jiangsu; Zhejiang; Fujian; Guangdong; Guangxi; Beibuwan (=Gulf of Tongking). Intertidal to subtidal 30 m. (63)

S. pulchella (Dunker): Hebei; Shandong; Jiangsu. Intertidal to subtidal 31 m. (63)

S. radiata (Linaneus): Fujian; Guangdong; Beihai (Guangxi); Beibuwan (=Gulf of Tongking); Eastern Hainan. Subtidal. (63)

PHARELLIDAE

Pharella acutidens (Broderip & Sowerby): Yueqing Bay (Zhejiang); Fujiang; Guangdong; Beibuwan (=Gulf of Tongking); Wenchang and Lingshui (Hainan). Intertidal. (63)

Sinonovacula consctricta (Lamarck): Liaoning; Hebei; Shandong; Zhejiang; Fujian; Guangdong; Beibuwan (=Gulf of Tongking). Subtidal. (63)

DREISSENIDAE

Mytilopsis sallei (Récluz): Dongshan and Xiamen (Fujian); Hong Kong. Subtidal. (81, 95)

Peregrinamor ohshimai Shoji: Qingdao (Shandong); Lianyungang (Jiangsu). Intertidal. (21)

KELLIELLIDAE

Alvenius ojianus (Yokoyama): Jiaozhou Bay and mouth of Yellow River (Shandong). Subtidal 6–27 m. (21)

TRAPEZIIDAE

Coralliophaga coralliophaga (Gmelin): Guangdong; Hainan. (19, 92)

Trapezium (Neotrapezium) liratum (Reeve): Changxing Island and Dalian (Liaoning); Yantai and Qingdao (Shandong); Zhejiang; Fujian; Guangdong; Guangxi. Intertidal. (21)

T. (N.) sublaevigatum Lamarck: Guangdong. Intertidal. (92)

GLOSSIDAE

Meiocardia lamarckii (Reeve) [*Isocardia lamarckii*]: Zhapo (Guangdong); Lingshui (Hainan). Subtidal. (60, 92)

M. tetragona (A. Adams & Reeve): East China Sea continental shelf. Subtidal 93 m. (32, 92)

M. vulgaris (Reeve) [*Isocardia vulgaris*]: Zhuhai (Guangdong); Lingshui and Sanya (Hainan). Subtidal. (60, 92)

VESICOMYIDAE

Calyptogena nanshaensis Xu & Shen: Nansha (=Spratly Islands) (113°37' E, 6°04' N). Water depth 2,626 m. (78)

CORBICULIDAE

Corbicula fluminea (Müller): Common brackish water species in China. Aquaculture species in the mouth of Min River (Fujian). (92)

Cyrenobatissa subsulcata (Clessin): Southeastern coast of China; Taiwan. Brackish water, low intertidal. (76a)

Geloina coaxans (Gmelin): Longhai (Fujian); western Taiwan; Nanao and Daya Bay (Guangdong); Fangcheng (Guangxi); Qinglan, Lingshui, and Sanya (Hainan). Intertidal. (38, 92)

VENERIDAE

Anomalodiscus squamosus (Linnaeus): Taiwan; Fujian; Guangdong; Hong Kong; Guangxi; Hainan. Intertidal to subtidal. (22)

Antigona lamellaris Schumacher: Taiwan; Guangdong; Macau; Guangxi; Nansha (=Spratly Islands). Subtidal 30–50 m. (22)

Callista chinensis (Holten): Nanji (Zhejiang); Pingtan and Dongshan (Fujian); Jieshi (Guangdong); Guangxi; Beibuwan (=Gulf of Tongking). Intertidal to subtidal 45 m. (22)

C. erycina (Linnaeus): Fujian; Guangdong; Lingshui and Sanya (Hainan). Intertidal to subtidal. (22)

Callithaca staminea (Conrad) [*Protothaca staminea euglypta*]: Liaoning; Shandong; Zhoushan (Zhejiang). (22)

Circe corrugata (Gray): Weizhou Island (Guangxi). (22)

C. (*Laevicirce*) *hongkongensis* Jiang & Xu: Hong Kong. Subtidal. (35)

C. scripta (Linnaeus): Guangdong; Hong Kong; Guangxi; Xinying, Sanya, and Lingshui (Hainan). Intertidal to subtidal. (22)

C. stutzeri (Donovan): Nanji (Zhejiang); Fujian; Nanao (Guangdong); Guangxi; Beibuwan (=Gulf of Tongking). Subtidal. (22)

C. (*Redicirce*) *sulcata* Gray [*Gouldia sulcata*]: Sanya (Hainan). (22)

C. tumefacia Sowerby: Autou (Guangdong); Hong Kong; Guangxi; Hainan. Subtidal. (22)

Clausinella calophylla (Philippi): Nanji (Zhejiang); Fujian; Guangdong; Hong Kong; Guangxi; Beibuwan (=Gulf of Tongking); Hainan. Intertidal to subtidal. (22)

C. isabellina (Philippi): Fujian; Guangdong; Hong Kong; Guangxi; Hainan. Intertidal to subtidal 24–78 m. (22)

C. tiara (Dillwyn): Fujian; Zhejiang; Guangdong; Guangxi; Hainan. Intertidal to subtidal. (22)

Clementia papyracea (Gray): Guangdong; Beibuwan (=Gulf of Tongking); southern Hainan. Subtidal 28–87 m. (22)

C. vatheleti Mabille: Liaoning; Hebei; Shandong; Zhejiang; Taiwan Strait. Intertidal to subtidal. (22)

Cryptonema producta (Kuroda & Habe): Dongshan (Fujian); Guangdong; Beihai (Guangxi); Hainan. Intertidal to subtidal. (22)

Cyclina sinensis (Gmelin): Liaoning; Heibe; Shandong; Jiangsu; Zhejiang; Fujian; Guangdong; Guangxi; Hainan. Common species. (22)

Cyclosunetta concinna (Dunker) [*Sunetta concinna*]: Nanji (Zhejiang); Fujian; Guangdong; Hainan. Intertidal to subtidal 52 m. (22)

C. menstrualis (Menke): Rizhao (Shandong); Nanji (Zhejiang). (22)

Dorisca nana (Melvill) [*Gouldia nana*]: South China Sea. Subtidal 94–105.5 m. (22)

Dosinia (*Pardosinia*) *amphidesmoides* (Reeve): Xiamen (Fujian); Guangdong; Xinying (Hainan). Intertidal to subtidal. (22)

D. (*Phacosoma*) *aspera* (Reeve): Nanji (Zhejiang); Fujian; Haimen (Guangdong); Hainan. (22)

D. (*P.*) *biscocta* (Reeve): Liaoning; Shandong; Zhejiang; Taiwan Strait; Guangdong; Guangxi; Hainan. Intertidal. (22)

D. (*Dosinella*) *corrugata* (Reeve) [*D. laminata, D. penicillata*]: Liaoning; Hebei; Shandong; Zhejiang; Fujian; Guangdong; Guangxi; Hainan. Intertidal to subtidal. (21, 22)

D. (*P.*) *cumingii* (Reeve): Nanji (Zhejiang); Dongshan (Fujian); Nanao (Guangdong); Nansha (=Spratly Islands). Intertidal to subtidal. (22)

D. (*Sinodia*) *derupta* Römer [*D. gibba*]: Bohai; Yellow Sea; Zhejiang; Fujian; Guangdong; Hainan. Intertidal to subtidal 60 m. (22)

D. (*Asa*) *exasperata* (Philippi) [*D. angulosa*]: Beibuwan (=Gulf of Tongking); Nansha (=Spratly Islands). Subtidal 30 m. (22)

D. fibula (Reeve): Zhejiang; Dongshan (Fujian); Nanao and Naozhou (Guangdong). (22)

D. (*Lamellidosinia*) *gruneri* (Philippi): Xiongzhou (Guangdong); Sanya (Hainan). (76a)

D. (*L.*) *hanleyana* H. and Adams: Shanwei, Haimen, Guanghai, and Shangchuan Island (Guangdong). (76a)

D. (*Bonartemis*) *histrio* (Gmelin): Fujian; Guangdong; Hainan. Intertidal to subtidal 40+ m. (22)

D. (*P.*) *japonica* (Reeve): Liaoning; Shandong; Jiangsu; Zhejiang; Fujian; Guangdong; Hong Kong; Guangxi; Hainan. Intertidal to 73 m, common species. (22)

D. (*D.*) *orbiculata* Dunker: Guangdong; Beibuwan (=Gulf of Tongking). Subtidal 26–93 m. (22)

D. traillei Adams: Zhejiang; Hainan. (22)

D. (*P.*) *troscheli* Lischke: Nanji (Zhejiang); Guangdong; Guangxi. Intertidal. (22)

D. (P.) truncata Zhuang: Haimen (Guangdong); Xinying and Sanya (Hainan). (76a)

Gafrarium dispar (Dillwyn): Beibuwan (=Gulf of Tongking); Hainan; Jinyin Island (Xisha=Spratly Islands). Intertidal to subtidal. (22)

G. divaricatum (Gmelin): Shujiang and Nanji (Zhejiang); Fujian; Guangdong; Guangxi; Hainan. Intertidal. (22)

G. pectinatum (Linnaeus): Taiwan; Daya Bay (Guangdong); Sanya and Lingshui (Hainan). Intertidal to subtidal. (22)

Gomphina aequilatera (Sowerby) [*G. melanaegis, G. veneriformis*]: Common species along China's coast. (22)

Irus irus (Linnaeus) [*I. mitis*]: Qingdao (Shandong); Nanji (Zhejiang); Xiamen and Dongshan (Fujian); Hong Kong; Sanya (Hainan). Intertidal. (22)

Lioconcha castrensis (Linnaeus): Xisha (=Paracel Islands). Subtidal 10 m. (22)

L. fastigiata (Sowerby): Hainan. (22)

L. lorenziana (Dillwyn): Sanya (Hainan). (22)

L. ornata (Dillwyn): Hainan; Xisha (=Paracel Islands); Nansha (=Spratly Islands). (22)

Marcia hiantina (Lamarck) [*Katelysia eugibba, K. rimularis*]: Xiamen (Fujian); Guangdong; Hong Kong; Beihai (Guangxi); Hainan. Intertidal. (22)

M. japonica (Gmelin): Guangdong; Hainan. Intertidal. (22)

M. marmorata (Lamarck): Guangdong; Beihai (Guangxi); Hainan. Intertidal. (22)

Meretrix castannea (Lamarck): Zhanjiang (Guangdong). (88)

M. lamarckii Deshayes: Nanji (Zhejiang); Guangdong; Hainan. Intertidal to subtidal. (22)

M. lusoria (Rumphius): Fujian; Taiwan; Guangdong; Beihai (Guangxi). Intertidal to subtidal. (22)

M. meretrix (Linnaeus): Bohai; Yellow Sea; East China Sea; South China Sea. Intertidal to subtidal. (22)

Paphia (Paphia) amabilis (Philippi): Fujian; Taiwan; Guangdong; Guangxi; Hainan. Intertidal to subtidal. (22)

P. (Paphia) euglypta (Philippi): Southern Zhejiang; East China Sea continental shelf; Pingtan (Fujian); Taiwan Strait; Guangdong; Guangxi. Subtidal. (22)

P. (Paphia) exarata (Philippi): Zhejiang; Fujian; Taiwan Strait; Guangdong; Beibuwan (=Gulf of Tongking); Hainan. Intertidal to subtidal. (22)

P. (Protapes) gallus (Gmelin) [*P. sinuosa*]: Fujian; Taiwan Strait; Guangdong; Guangxi; Hainan. Intertidal to subtidal 48 m. (22)

P. (Paphia) lirata (Philippi): Pingtan (Fujian); Shanwei (Guangdong); Hainan. Subtidal. (22)

P. (Paphia) paoilionacea Röding [*P. alapapilionis*]: Southern Yellow Sea; Lianyungang (Jiangsu). Subtidal 10–72 m. (22)

P. (Paphia) textile (Gmelin): Guangdong; Hong Kong; Guangxi; Hainan. Subtidal. (22)

P. (Paratapes) undulata (Born): Fujian; Guangdong; Hong Kong; Guangxi; Beibuwan (=Gulf of Tongking); Hainan. Intertidal to subtidal 44 m. (22)

Periglypta chemnitzi (Hanley): Pingtan, Xiamen, and Dongshan (Fujian); Hong Kong; South China Sea. Subtidal. (22)

P. clathrata (Deshayes): Sanya (Hainan). (22)

P. compressa Zhuang: Xinying and Sanya (Hainan). (22)

P. puerpera (Linnaeus) [*P. crispata, P. lacerata*]: Guangdong; Hainan; Xisha (=Spratly Islands). Intertidal to subtidal. (22)

P. reticulata (Linnaeus): Taiwan; Sanya (Hainan); Yongxing Island (Xisha=Paracel Islands). (22)

Pitar (Pitarinum) affinis (Gmelin): Hong Kong; Lingshui (Hainan). Intertidal. (22)

P. (Costellipitar) chordatum (Romer): Taiwan Strait. Subtidal. (22)

P. (P.) crocea (Deshayes): Xinying (Hainan). Lower subtidal. Sandy mud bottom. (22)

P. (P.) hebraea (Lamarck): Beibuwan (=Gulf of Tongkin); southern tip of Hainan. Subtidal 30–45 m. (22)

P. (P.) limatula (Sowerby): Sanya (Hainan). (22)

P. (P.) nipponicum Kuroda & Habe: Taiwan Strait. Subtidal. (22)

P. (P.) noguchii Habe: East China Sea continental shelf. Subtidal 93 m. (22)

P. (P.) obliquata (Hanley): Xisha (=Paracel Islands). (22)

P. (P.) pellucida (Lamarck): Guangdong; Lingshui (Hainan). Intertidal to subtidal. (22)

P. (P.) striatum (Gray): Taiwan; Hong Kong; Guangdong; Sanya and Lingshui (Hainan); Nansha (=Spratly Islands). Intertidal to subtidal. (22)

P. (P.) sulfurea Pilsbry: Taiwan Strait; Guangdong; Xinying (Hainan). Intertidal to subtidal. (22)

Protothaca jadoensis Lischke: Liaoning; Shandong; Jiangsu; Chengshi and Nanji (Zhejiang). Intertidal. (22)

Ruditapes philippinarum (Adams & Reeve) [*Venerupis philippinarum*]: Liaoning; Hebei; Shandong; Jiangsu; Zhejiang; Fujian; Taiwan; Guangdong; Hong Kong. Intertidal. (22)

R. variegata (Sowerby) [*Venerupis variegata*]: Fujian (south of Pingtan); Guangdong; Hong Kong; Guangxi; Hainan. Intertidal to subtidal. (22)

Saxidomus purpurata (Sowerby): Dalian and Changshanba Island (Liaoning); Yantai (Shandong). Intertidal to subtidal. (22)

Sunettina solanderii (Gray): Zhejiang; Guangdong; Qiongzhou Strait (between Guangdong and Hainan). Subtidal 55 m, coarse sand and gravel bottom. (22)

Tapes belcheri Sowerby [*T. quadriradiata*]: Guangdong; Lingshui (Hainan). Intertidal to subtidal. (22)

T. dorsalus (Lamarck) [*T. turgida*]: Southern Fujian; Guangdong; Hong Kong; Hainan. Intertidal to subtidal. (22)

T. literata (Linnaeus): Guangdong; Sanya, Lingshui, and Qinglan (Hainan). Intertidal to subtidal. (22)

T. platyptycha Pilsbry [*T. araneosa*]: Lingshui (Hainan). (22)

Timoclea imbricata (Sowerby): Fujian; Guangdong; Guangxi. Intertidal to subtidal. (22)

T. lionota (Smith): Taiwan Strait; South China Sea. Shallow sea. (22)

T. marica (Linnaeus): Lingshui (Hainan); Xisha (=Paracel Islands). (22)

T. scabra (Hanley): Nanji (Zhejiang); Taiwan; Guangdong. Subtidal. (22)

T. subnodulosa (Hanley): Beibuwan (=Gulf of Tongking); Sanya (Hainan). Subtidal 20+ m. (22)

Venus (*Ventricola*) *foveolata* Sowerby [*V. albina*]: East China Sea continental shelf; Zhejiang; Guangdong; Sanya (Hainan). Subtidal. (22)

V. (*V.*) *toreuma* Gould: Taiwan; Guangdong; Beibuwan (=Gulf of Tongking); Sanya (Hainan). Intertidal to subtidal. (22)

PETRICOLIDAE

Claudiconcha japonica (Dunker): Fujian; Guangdong; Hong Kong; Sanya (Hainan). Intertidal. (20)

GLAUCONOMIDAE

Glauconome chinensis Gray [*Glaucomya chinensis*]: Xiangshan Bay and Yueqing Bay (Zhejiang); Fujian; Guangdong; Xinying (Hainan). Intertidal. (60, 92)

G. primeana Crosse & Debeaux: Donggang (Liaoning) to Dafeng (Jiangsu). Intertidal. (21)

MYOIDA
MYIDAE

Cryptomya busoensis Yokoyama: Yellow Sea. Water depth 45–56 m. (70)

Mya arenaria Linnaeus [*M. japonica*]: Bohai; Yellow Sea (north of Lianyungang, Jiangsu). Intertidal and shallow sea within 10 m. (70)

Paramya recluzii (A. Adams): Bohai; Jiaozhou Bay (Shandong). (70)

Tugonia huanghaiensis Xu: Qingdao and Jiaonan (Shandong); Rudong (Jiangsu). (70)

T. sinensis Xu: Huian (Fujian). (70)

Venatomya truncata (Gould): Yellow Sea; East China Sea. Shallow water. (70)

CORBULIDAE

Anisocorbula crassa (Hinds): East China Sea; South China Sea. Subtidal 76–98 m. (28)

A. cuneata (Hinds): Fujian; Taiwan Strait; East China Sea; South China Sea; Beibuwan (=Gulf of Tongking). Subtidal 25–174 m. (28)

A. lineata (Lynge): Beibuwan (=Gulf of Tongking). Subtidal 49 m. (28)

A. modesta (Reeve): Xiamen (Fujian); South China Sea. Intertidal to subtidal. (28)

A. pallida (Hinds): Xiamen and Dongshan (Fujian); Guangxi. Intertidal to subtidal. (28)

A. scaphoides (Hinds): Fujian; Taiwan Strait; western Guangdong; Beibuwan (=Gulf of Tongking). Subtidal 12–63 m. (28)

A. venusta (Gould): Bohai; Yellow Sea. Intertidal to 200 m. (28)

Corbula fortisulcata Smith: Dongshan (Fujian); Taiwan Strait; Shantou and Leizhou (Guangdong); Beibuwan (=Gulf of Tongking); Hainan. Intertidal to subtidal. (28)

Minicorbula minutissima (Habe): Taiwan Strait. Subtidal. (37, 83)

Potamocorbula amurensis (Schrenck): Dalian (Liaoning); Tanggu (Tianjing); Rizhao (Shandong); Chongming Island (Shanghai); Zhuhai (Guangdong). Brackish water in river mouth, muddy bottom, water depth about 10 m. (28)

P. laevis (Hinds): Widely distributed along China's coast. Intertidal to subitdal. (28)

P. rubromuscula Zhuang & Cai: Shantou and Chaoyang (Guangdong). Intertidal to subtidal. (28)

P. ustulata (Reeve): Qingdao and Jiaozhou Bay (Shandong); Chongming Island (Shanghai); mouth of Jiulong River (Fujian). Intertidal to subtidal. (28)

Solidicorbula erythrodon (Lamarck) [*Aloidis erythrodon*]: Fujian; Guangdong; Guangxi; Hong Kong; Hainan. Intertidal to subtidal 44 m. (28)

S. tunicata (Hinds) [*Corbula tunicata*]: Shantou, Guanghai, and Zhapo (Guangdong); Beibuwan (=Gulf of Tongking); Sanya (Hainan). Intertidal to subtidal 20–60 m. (28)

HIATELLIDAE

Hiatella orientalis (Yokoyama) [*H. arctica*]: Dalian (Liaoning); Yantai, Weihai, Rongcheng, and Qingdao (Shandong); Daya Bay (Guangdong); Hong Kong. Around intertidal mean low water. (21)

Panopea japonica A. Adams: Bohai; Yellow Sea. Subtidal. (21)

GASTROCHAENIDAE

Eufistulana grandis (Deshayes) [*Gastrochaena grandis*]: Xiangshan Bay and Yueqing Bay (Zhejiang); Xiamen and Dongshan (Fujian); Guangdong; Beibuwan (=Gulf of Tongking). Intertidal to subtidal. (17, 88, 92, 94)

Gastrochaena (*G.*) *cuneiformis* (Spengler): Zhongjieshan (Zhejiang); Longhai (Fujian); Guangdong; Xinying, Sanya, and Lingshui (Hainan); Xisha (=Paracel Islands). Intertidal to subtidal. (60, 92)

G. (*Cucurbitula*) *cymbium* Spengler [*G. ovata*]: Jiaozhou Bay (Shandong); Wushi (Guangdong); Weizhou Island (Guangxi); Beibuwan (=Gulf of Tongking); Sanya (Hainan). Subtidal. (21, 60, 92)

PHOLADIDAE

Aspidopholas obtecta (Sowerby) [*Pholadidea cheveyi*]: Xinying (Hainan). Intertidal. (59, 92)

Barnea (*Barnea*) *candida* (Linnaeus): Pingtan and Dongshan (Fujian); Shantou, Wushi, and Zhanjiang (Guangdong); Beibuwan (=Gulf of Tongking); Sanya (Hainan). Subtidal. (59)

B. (*Anchomasa*) *davidi* (Deshayes) [*B. davidi*]: Liaoning; Hebei; Shandong; Shengsi and Putou (Zhejiang). Around intertidal mean low water. (59)

B. (*Umitakea*) *dilatata* (Souleyet) [*B. dilatata*]: Changli and Beidaihe (Hebei); Dingzhi Bay, Qingdao, and Shijiushou (Shandong); Zhoushan and Xiangshan Bay (Zhejiang); Changle (Fujian); Beibuwan (=Gulf of Tongking); Chengmai (Hainan). Low intertidal. (59)

B. (*A.*) *elongata* Tchang, Tsi & Li: Guangdong, Qukou and Snaya (Hainan). Low intertidal. (59)

B. (*A.*) *manilensis* (Philippi) [*B. fragilis*]: Liaoning; Shandong; Zhejiang; Fujian; Guangdong; Guangxi; Hainan. Low intertidal. (59, 76a)

Jouannetia cumingi (Sowerby): Weizhou Island (Guangxi); Beibuwan (=Gulf of Tongking); Qinglan, Lingshui, Sanya, and Xinying (Hainan). Coral reef intertidal. (59)

J. globulosa (Quoy & Gaimard): South China Sea. Water depth 81 m. (59)

Martesia ova (Wood): Yuhuan (Zhejiang); Xiamen (Fujian); Autou (Guangdong). Low intertidal. (59)

M. pygmaea Thang, Tsi & Li: Huangcheng and Qisha (Guangxi). Rocky intertidal. (59)

M. striata (Linnaeus): Xiamen (Fujian); Shantou, Lufeng, and Zhuhai (Guangdong); Weizhou Island and Huangcheng (Guangxi); Wanning and Sanya (Hainan). Subtidal, burrowing inside woods. (59)

M. tubigera (Valencennes): Weizhou Island (Guangxi); Sanya (Hainan). Coral reef intertidal. (59)

M. yoshimurai (Kuroda & Termachi) [*Aspidopholas yoshimurai*]: Xiongyue (Liaoning); Tanggu and Xingang (Tianjin); Qingdao (Shandong); Xiangshan and Yuhuan (Zhejiang); Autou and Baoan (Guangdong); Beibuwan (=Gulf of Tongking). Intertidal. (59)

Parapholas quadrizonata Spengler: Zhelang (Guangdong); Weizhou Island (Guangxi); Beibuwan (=Gulf of Tongking); Sanya and Lingshui (Hainan). Coral reef intertidal. (59)

Penitella kamakurensis (Yokoyama) [*Pholadidea acutithyra, P. dolichothyra*]: Tanggu and Xingang (Tianjin); Shengsi (Zhejiang). Low intertidal. (59)

Pholas orientalis Gmelin: Zhangpu (Fujian); Haikou, Yinggehai, and Sanya (Hainan). Subtidal. (59)

Zirfaea crispata (Linnaeus): Yantai and Qingdao (Shandong); Shengshi, Xiangshan, and Yuhuan (Zhejiang); Pingtan, Xiamen, and Dongshan (Fujian); Nanao (Guangdong). Low intertidal. (59)

Z. minor Tchang, Tsi & Li: Yuhuan (Zhejiang); Xiamen (Fujian); Qisha (Guangxi). Intertidal. (59)

TEREDINIDAE

Bankia (*Bankiella*) *campanullata* Moll & Roch: Weizhou Island (Guangxi); Yinggehai (Hainan). (55)

B. (*B.*) *oryzaformis* Sivickis: Sheyang (Jiangsu); Shipu (Zhejiang); Huian and Xiamen (Fujian); Shantou and Zhuhai (Guangdong); Qisha and Weizhou Island (Guangxi); Haikou, Sanya, and Yinggehai (Hainan); Xisha (=Spratly Island). (55)

B. (*B.*) *saulii* (Wright): Haimen, Zhuhai, and Baoan (Guangdong). (55)

B. (*Neobankia*) *tenuis* Sivickis: Xiamen (Fujian); Shantou and Jieshi (Guangdong); Qisha (Guangxi); Haikou and Qinglan (Hainan). (55)

Lyrodus singaporeana (Roch): Hong Kong. Mangrove forest. (99)

Teredo (*Bitubuloteredo*) *bitubula* Li: Yulin Harbor (Hainan). (55)

T. (*Teredo*) *clava* Gmelin: Qinglan Harbor (Hainan). (55)

T. (*Teredora*) *diederichseni* Roch: Qinglan Bay (Hainan); Yongxing Island (Xisha=Spratley Islands). (55)

T. (*Lyrodus*) *furcifera* Martens: Yinggehai (Hainan). (55)

T. (*Pseudodicyathifer*) *manni* (Wright): Hengshan and Dongping (Guangdong); Qisha (Guangxi); Beibuwan (=Gulf of Tongking); Qinglan Bay and Sanya (Hainan). (55)

T. (*Teredo*) *massa* Jousseaume: Yinggehai (Hainan). (55)

T. (*Ungoteredo*) *matacotana* Bartsch: Yulin Harbor and Haikou (Hainan). (55)

T. (*Psiloteredo*) *megotara* Hanley. Shanwei (Guangdong). (55)

T. (*Teredo*) *navalis* Linnaeus: China's coast. (53)

T. (*Teredo*) *parksi* Bartsch: Shanwei (Guangdong); Qisha (Guangxi); Yinggehai (Hainan). (55)

T. (*Teredothyra*) *remiformis* Li: Xisha (=Spratley Islands). Coconut trees. (55)

T. (*L.*) *samosensis* Miller: Dalian (Liaoning); Qingdao (Shandong). (55)

T. (*L.*) *schizoderma* Li: Sanya (Hainan). (55)

PHOLADOMYOIDA
PHOLADOMYIDAE

Pholadomya (*Nipponopanacca*) *sinica* Xu: Okinawa Trough. (75)

LYONSIIDAE

Agriodesma navicula (Adams & Reeve) [*A. naviculoides*]: Jinzhou (Liaoning); Changshan Islands, Yantai, Qingdao (Shandong). Water depth 36 m. (105)

Lyonsia kawamurai Habe: Qingdao (Shandong). Water depth 20 m. (105)

L. ventricosa Gould [*L. praetenuis, L. rostrata*]: Yellow Sea. Water depth 5–30 m. (105)

Sinolyonsia sinica Xu: Bohai. Water depth 29 m. (105)

PANDORIDAE

Pandora (Frenamya) elongata Carpenter: South China Sea. Water depth 43 m. (76)

P. (Pandorella) otukai Habe: Southern Yellow Sea. (76)

P. (P.) pseudobilirata Nomura & Hatai: Jiaozhou Bay (Shandong). Water depth 24 m. (76)

P. (F.) sinica Xu: Northern East China Sea. Water depth 11–43 m. (76)

P. (P.) wardiana A. Adams: Southern Yellow Sea. (76)

MYOCHAMIDAE

Myodora fluctuosa Gould [*M. japonica, M. proxima*]: Southern Yellow Sea. (21)

PERIPLOMATIDAE

Periploma otohimeae Ozaki: Yellow Sea; East China Sea. (21)

LATERNULIDAE

Laternula (Laternula) anatina (Linnaeus) [*L. valenciennesii*]: China's coast. Intertidal to subtidal. (26)

L. (L.) boschasina (Reeve): China's coast. Intertidal. (26)

L. (Exolaternula) marilina (Reeve) [*L. pechliensis*]: Liaoning; Hebei; Shandong; Jiangsu; Zhejiang; Fujian; Guangdong; Guangxi. Intertidal to subtidal 20 m. (26)

L. (E.) nanhaiensis Zhuang & Cai: Zhanjiang (Guangdong); Hepu (Guangxi). (26)

L. (E.) truncata (Lamarck): Fujian; Taiwan; Guangdong; Guangxi; Hainan. Intertidal. (26)

THRACIIDAE

Asthenothaerus huanghaiensis Xu: Central Yellow Sea. Water depth 40–80 m. (104)

Cyathodonta granulosa (Adams & Reeve): East China Sea. (104)

Thracia (Eximiothracia) concinna Gould: Southern Yellow Sea. Water depth 30–40 m. (104)

T. (Thracia) hainanensis Xu: Xinying (Hainan). (104)

T. (Crassithracia) ovata Xu: Northern Yellow Sea. Water depth 46 m. (104)

Trigonothracia jinxingae Xu: Eastern Liaoning Bay (Liaoning) to Guangdong. Coastal shallow water. (104)

T. pusilla (Gould): Yellow Sea. Water depth 9–80 m. (104)

CLAVAGELLIDAE

Penicillus penis (Linnaeus): South China Sea. Water depth 93 m. (19)

VERTICORDIIDAE

Euciroa rostrata Thiele & Jaeckel [*E. teramachii*]: East China Sea. Water depth 365–395 m. (72)

Verticordia deshayesiana Fishcher: East China Sea. Water depth 110 m. (76a)

POROMYIDAE

Cetoconcha japonica Habe: East China Sea. (76a)

Poromya castanea Habe: Southern Yellow Sea. (21)

CUSPIDARIIDAE

Cardiomya alcocki (Smith): Okinawa Trough. Water depth 520 m. (71)

C. gouldiana (Hinds): Taiwan Strait. Water depth 35–60 m. (37, 83).

C. reticulata (Kuroda) [*Cardiomya nipponica, Cuspidaria abyssicola nipponica*]: East China Sea. Water depth 510 m. (71)

C. singaporensis (Hinds): South China Sea. Water depth 26–65 m. (71)

C. sinica Xu: East China Sea. Water depth about 100 m. (71)

C. tosaensis (Kuroda) [*C. sagamiana*]: Yellow Sea. Water depth 35–80 m. (71)

Cuspidaria approximata Smith: Okinawa Trough. Water depth 520 m. (71)

C. caduca Smith: East China Sea; South China Sea. Water depth 50–200 m. (71)

C. corrugata Prashad: Beibuwan (=Gulf of Tongking); South China Sea. Water depth 50–120 m. (71)

C. japonica Kuroda: East China Sea; South China Sea. Water depth within 80 m. (71)

C. kawamurai Kuroda: Northeastern East China Sea. Water depth 520 m. (71)

C. kyushuensis Okutani: Okinawa Trough. Water depth 520 m. (71)

C. macrorhynchus Smith: East China Sea; South China Sea. Water depth 300–395 m. (71)

C. nobilis (A. Adams): East China Sea. Water depth 79–150 m. (71)

C. okezoko Okutani: Northern East China Sea. Water depth 365–395 m. (71)

C. prolatissima Poutiers: South China Sea. Water depth 115 m. (71)

C. steindachneri Sturany: East China Sea; South China Sea. Water depth 78–115 m. 71)

Pseudoneaera semipellucida (Kuroda) [*Cuspidaria iridella*]: Central Yellow Sea. Water depth about 70 m. (71)

REFERENCES*

(1). Wang Zhenrui, 1964. Preliminary studies on Chinese Pinnidae. Studia Marina Sinica, 5:30–41.

(2). Wang Zhenrui, 1978. A study of the family Pteriidae (Mollusca) off China. Studia Marina Sinica, 14:101–115.

(3). Wang Zhenrui, 1980. Studies on Chinese species of Isognomonidae (Mollusca). Studia Marina Sinica, 16:13–141.

(4). Wang Zhenrui, 1980. Study on Chinese species of the family Pectinidae (Mollusca, Bivalvia) I. A new species of the subfamily Propeamussiinae. Oceanologia & Limnologia Sinica, 11(3):259–262.

(5). Wang Zhenrui, 1982. Study on species of Plicatulidae off China coast. Marine Science, 1:23–27.

(6). Wang Zhenrui, 1983. Studies on Chinese species of the family Pectinidae III. Chlamydinae (1. *Chalamys*). Transactions of the Chinese Society of Malacology No.1, 1:47–54.

(7). Wang Zhenrui, 1983. Studies on Chinese species of the family Pectinidae III. Chlamydinae (a new species and three new records of the genus *Cryptopecten*). Oceanologia & Limnolgia Sinica, 14(4):402–406.

(8). Wang Zhenrui, 1983. Studies on the Mytilidae of the Xisha Islands, Guangdong Province, China. Studia Marina Sinica, 20:213–221.

(9). Wang Zhenrui, 1983. Studies on Chinese species of the family Pectinidae III. Chlamydinae (genus *Bractechlamys*). Oceanologia & Limnologia Sinica, 14(6):531–535.

(10). Wang Zhenrui, 1984. Studies on Chinese species of the family Pectinidae VI. Subfamily Propeamussiinae. Oceanologia & Limnogia Sinica, 15(6):598–604.

(10a). Wang Zhenrui, 1984. Studies on species of Pectinidae off the Chinese coasts II. Subfamily Amusiinae. Studia Marina Sinica, 22:245–253.

(11). Wang Zhenrui and Qi Zhongyan, 1984. Study on Chinese species of the family Mytilidae (Mollusca, Bivalvia). Studia Marina Sinica, 22:199–242.

(12). Wang Zhenrui, 1985. Studies on Chinese species of the family Pectinidae VII. Chlamydinae (genus *Semipallium*). Oceanologia & Limnologia Sinica, 16(6):502–506.

(13). Wang Zhenrui, 1989. Studies on the Chinese species of the family Limidae from waters off China. Studia Marina Sinica, 31:163–176.

(14). Wang Zhenrui, 1990. A study on species of the family Pectinidae I. Genera *Volachlamys* and *Amachlamys* off the Chinese coasts. Transactions of the Chinese Society of Malacology No. 3, 3:13–16.

(15). Wang Zhenrui, 1990. Study on the species of family Limidae from waters off China. Studia Marina Sinica, 31:163–176.

(16). Wang Zhenrui and Chen Ruiqiu, 1991. The species of the Pterioda from the Nansha Islands waters. Papers on Marine Biological Research of the Nansha Islands and its Adjacent Waters (I), pp:150–160.

(16a). Wang Zhenrui, 1997. Fauna Sinica, Mollusca, Mytiloida, China Science Press, 268pp.

(17). You Zhongjie, Li Jiewei and Hong Junchao, 1985. A record of Bivalvia of Zhejiang coast. Journal of Zhejiang College of Fisheries, 4(2):133–144.

(18). You Zhongjie and Wang Yinong, 1989. Additions to the marine Bivalvia of Nanji Islands of China. Journal of Zhejiang College of Fisheries, 8(1):17–28.

(19). Tsi, C. Y. and S. T. Ma, 1980. A preliminary checklist of the marine Gastropoda and Bivalvia (Mollusca) of Hong Kong and southern China. Proceedings of the First International Marine Biological Workshop: The Marine Flora and Fauna of Hong Kong and Southern China. Hong Kong University Press, pp:431–458.

(20). Qi Zhongyan, Ma Xiutong, Xie Yukan, and Lin Biping, 1984. A preliminary checklist of marine Molluscs from Hainan Island. Tropic Oceanology Study, 1:1–22.

(21). Qi Zhongyan, Ma Xiutong, and Wang Zhenrui 1989. Molluscs of the Yellow Sea and Bohai Sea. China Agriculture Press, pp:1–309.

(22). Zhuang Qiqian, 1964. Studies on Chinese species of Veneridae (class Lamellibranchia). Studia Marina Sinica, 5:43–106.

(23). Zhuang Qiqian, 1978. The Tridacnids of the Xisha Islands, Guangdong Province, China. Studia Marina Sinica, 12:133–139.

(24). Zhuang Qiqian, 1978. Studies on the Mesodesmatidae (Lamellibanchia) off the Chinese coasts. Studia Marina Sinica, 14:69–73.

(25). Zhuang Qiqian, Lin Huiqiong and Liang Xianyuan, 1981. On the species of the genus *Ruditapes* (Mollusca, Lamellibranchia, Veneridae) off China coast. Studia Marina Sinica, 18:207–215.

(26). Zhuang Qiqian, 1982. Studies on the Laternulidae off the Chinese coast. Oceanologia & Limnologia Sinica, 13(6):553–561.

(27). Zhuang Qiqian, 1983. Two new species of Mactnidae (Mollusca, Bivalvia) off the Chinese coast. Oceanologia & Limnologia Sinica, 14(1):88–91.

(28). Zhuang Qiqian and Cai Yingya, 1983. Studies on the Corbulidae (Bivalvia) off Chinese coasts. Transactions of the Chinese Society of Malacology No. 1:57–68.

(29). Zhuang Qiqian, 1984. Studies on the Chamidae (Bivalvia) off the China coast. Studia Marina Sinica, 22:191–198.

(30). Zhuang Qiqian, 1992. Studies on Mactnidae off China coast. (manuscript)

(31). Jiang Jinxiang and Chen Chanzhong et al., 1984. Preliminary study on benthic ecology in the nearshore waters of western Taiwan Strait. Oceanologica Sinica, 6(3):389–398.

(32). Jiang Jinxiang, Huang Liqiang and Meng Fan, 1985. Species composition and distribution of benthic animals on the continental shelf of the East China Sea and its adjacent waters. Journal of Oceanography in Taiwan Strait, 4(1):89–98.

(33). Jiang Jinxiang and Cai Erxi, 1990. Species composition and quantitative distribution on benthic animals in Daya Bay. Collections of Papers on Marine Ecology in Daya Bay, pp:237–247.

(34). Jiang Jinxiang and Cai Erxi, 1988. The achievement in comprehensive investigations on environmental resources in the central part of the South China Sea. China Ocean Press, pp:215–252.

(35). Jiang Jinxiang, Xu Fengshan, 1992. A new species of Venefidae (Bivalvia) from Hong Kong waters. Journal of Oceanography in Taiwan Strait, 11(4): 291–293

(36). Jiang Jinxiang, 1992. Study on Bivalvia in the continental shelf of the East China Sea and its adjacent waters. (manuscript)

(37). Jiang Jinxiang, Lin Rongguan, 1992. Study on Bivalvia in Taiwan Strait and its adjacent area. (manuscript)

(38). Jiang Jinxiang and Lin Rongguan, 1992. Distribution and fauna of Mollusca (Bivalvia) in Daya Bay. (manuscript)

(39). Lee J. M., 1965. A new species and a new record of *Teredo navalis* from the coast of China. Studia Marina Sinica, 8:1–7.

(40). Li Fenglan, 1983. Studies on Chinese species of the family Arcidae II. Anadarinae. Transactions of the Chinese Society of Malacology, 1:31–46.

(41). Li Fenglan, 1983. Studies on the Arcidae of the Xisha Islands, Guangdong Province, China. Studia Marina Sinica, 20:223–230.

(42). Li Fenglan, 1984. A study of the Arcinae from China coast I. Arcinae. Studia Marina Sinica, 23:145–161.

(43). Li Fenglan, 1985. Studies on the Chinese species of the family Arcidae III. Striarcinca. Studia Marina Sinica, 25:153–160.

(44). Li Fenglan, 1986. Studies on species of Glycymeridae off the China coasts. Transactions of the Chinese Society of Malacology, 2:23–29.

(45). Li Fenglan, 1990. Studies on species of the family Limopsidae (Bivalvia) off the China coasts. Transactions of the Chinese Society of Malacology, 3:19–23.

(46). Li Fuxue et al., 1959. Studies on *Pinna* off Fujian Province. Journal of Xiamen University, 2:57–70.

(47). Li Rongguan and Jiang Jinxiang, 1990. Variation of Mollusca fauna in the soft-bottom intertidal zone along northern coast of Hangzhou Bay. Transactions of the Chinese Society of Malacology, 3:36–45.

(48). Li Xiaoxu, 1991. Studies on comparative anatomy, systematic taxonomy and evolution of oysters (abstract). In: Contributions on the second symposium of the Chinese Society of Malacology in Qingdao.

(49). Li Chupu, Li Yingping and Chen Ruiqin, 1987. Species lists of Molluscs from Zhengmu Shoal - The reports of comprehensive survey on the southern territory of China, China Science Press, pp:205–212.

(50). Chen Saiying et al., 1980. Studies on shell fauna in Nanji Islands. Acta Zoologica Sinica, 8(1):64–94.

(51). Chen Ruiqiu and Li Yingping, 1989. Species lists of shells from Nansha Islands and its adjacent waters. The Report of Comprehensive Survey nearby the Waters of Nansha Islands. China Science Press, pp:755–758.

(52). Tchang Si, Tsi Chung-yen and Li Jiemin, 1955. Molluscs in northern seas of China. China Science Press, 98pp.

(53). Tchang Si, Tsi Chung-yen and Li Jiemin, 1955. Teredinidae and its variation off northern China. Acta Zoologica Sinica, 7(1):1–16.

(54). Tchang Si and Lou Tze-Kong, 1956. A study on Chinese oysters. Acta Zoologica Sinica, 8(1):64–94.

(55). Tchang Si, Tsi Chung-yen and Li Jiemin, 1958. A study on Teredinidae off southern China. Acta Zoologica Sinica, 10(3):242–257.

(56). Tchang Si and Lou Tze-Kong, 1959. Oyster. China Science Press, 156pp.

(57). Tchang Si, 1959. Economic Mollusc fauna in Yellow Sea and East China Sea. Oceanologia & Limnologia Sinica, 2(1): 27–34.

(58). Tchang Si and Tsi Chung-yen, 1959. Economic Mollusc fauna in the South China Sea. Oceanologia & Limnologia Sinica, 2(4): 268–277

(59). Tchang Si, Tsi Chung-yen and Li Jiemin, 1960. Pholadidae and the new species in China. Acta Zoologica Sinica, 12(1): 63–87.

(60). Tchang Si et al., 1960. Mollusc (Bivalve) in the South China Sea. China Science Press, 274pp.

(61). Tchang Si et al., 1962. Records of Economic Animals in China - Marine Mollusc. China Science Press, 246pp.

(62). Tchang Si et al., 1964. A preliminary study of the Demarcation of Molluscan Faunal Regions of China and its Adjacent Waters. Oceanologia & Limnologia Sinica, 5(2):124–138.

(63). Tchang Si and Hwang Hsia-Ming, 1964. On the Chinese species of Solenidae. Acta Zoologica Sinica, 16(2):193–206.

(64). Zhao Ruyi, Cheng Jimin and Zhao Dadong, 1982. Records on Molluscs in Dalian Sea Waters. China Ocean Press, 167pp.

(65). Xu Fengshan, 1964. Studies on the Cardiid Molluscans of the China Sea. Studia Marina Sinica, 6:82–98.

(66). Xu Fengshan, 1980. Two new species of Bivalvia (Mollusca) from the East China Sea. Oceanologia & Limnologia Sinica, 11(4):337–340.

(67). Xu Fengshan, 1984. Preliminary study on the Protobranchia (Mollusca) from the shallow waters of China I. Nucalanidae. Studia Marina Sinica, 22:167–177.

(68). Xu Fengshan, 1984. Preliminary study on the Protobranchia (Mollusca) from the shallow waters of China II. Nucalanidae. Studia Marina Sinica, 22:179–188.

(69). Xu Fengshan, 1986. The new records and list of bivalves from the Jiaozhou Bay. Transactions of the Chinese Society of Malacology, 2:31–41.

(70). Xu Fengshan, 1984. New species and new records of Myidae from the China coast. Oceanologia & Limnologia Sinica, 15(4):437–441.

(71). Xu Fengshan, 1990. Study on the Septibranchia from Chinese waters, I. Cuspidariidae. Studia Marina Sinica, 31:177–184.

(72). Xu Fengshan, 1990. The Bivalvia in the deep-water area of the East China Sea. Studia Marina Sinica, 31:185–193.

(73). Xu Fengshan, 1990. Preliminary study on the Protobranchia (Mollusca) from the China Seas, III. Malletiidae and Tinariidae. Oceanologia & Limnologia Sinica, 21(6):559–562.

(74). Xu Fengshan, 1991. Two new species of Protobranchia from Nansha Islands waters. Papers on Marine Biological Research of the Nansha Islands and its adjacent waters (I). China Ocean Press, pp:82–85.

(75). Xu Fengshan, 1992. *Pholaclomya* (*Nipponopanacca*) *sinica*, A new species of Pholaclomyidae from the East China Sea. Oceanologia & Limnologia Sinica, 23(2):211–213.

(76). Xu Fengshan, 1992. A new species, *Pandorid sinica* (Bivalvia, Mollusca) from the China Seas. Oceanologia & Limnologia Sinica, 23(3):285–287.

(76a). Xu Fengshan, 1992. New records on Chinese bivalve. (manuscript)

(77). Xu Fengshan and Chen Ruiqiu, 1991. Family Cardiidae from Nansha Islands waters. Papers on Marine Biological Research of the Nansha Islands and its adjacent waters (I). China Ocean Press, pp:161–163.

(78). Xu Fengshan and Shen Shoupeng, 1991. A new species of Vesicomyidae from Nansha Islands waters. Papers on marine research of the Nansha Islands and its adjacent waters (I). China Ocean Press, pp:164–166.

(79). Liang Xianyuan, Lin Huiqiong and Wu Pingru, 1986. Comparison on morphology of Pinnidae (Mollusca, Lamellibranchia) from the Fujian coast. Tropic Oceanology, 5(1):13–19.

(80). Yang Ruiqiong and Zhong Youping, 1987. Bivalves from the Fujian coast. Journal of Xiamen College of Fisheries, 2:32–50.

(81). Huang Zongguo, 1992. Lots of *Mytiolpsis sollei* found in Cijiao of Dongshan Island, Fujian Province, China. (manuscript)

(82). Skarato, 1965. Tellinidae (Bivalvia: Mollusca) from China Seas. Studia Marina Sinica, 8:27–114.

(83). Fujian Institute of Oceanography, 1988. A comprehensive oceanographic survey of the central and northern part of the Taiwan Strait. China Science Press, pp:410–411.

(84). Xiong Daren and Cai Yingya, 1981. Bivalves in the waters near Zhanjiang. Journal of Zhanjiang College of Fisheries, 1:25–34.

(85). Cai Yingya, 1976. Blood clams off Fujian coast. Acta Zoologica Sinica, 2:29–30.

(86). Cai Yingya, 1980. New records of Arcidae off the coast of Guangdong and Guangxi Provinces. Journal of Zhanjiang College of Fisheries, 1:51–54.

(87). Cai Yingya and Zhuang Qiqian, 1985. A new species of Psammobiidae (Mollusca, Bivalvia). Tropic Oceanology, 4(3):64–66.

(88). Cai Yingya and Liu Guimao, 1989. Shells off western Guangxi Province. Journal of Zhanjiang College of Fisheries, 9(1–2):57–78.

(89). Cai Yingya, Lin Yongmu and Ou Ruimu, 1990. A study on shell fauna in Nanao Island, Guangdong Province. Journal of Zhanjiang College of Fisheries, 10(1):1–13.

(90). Tetsimei Yoshiyoyi, 1974. Shells of Japan in color (supplement.). Hoikusha Publishing Co. Ltd., Japan.

(91). Tadashige Habe, 1974. Shells of the world in color, Vol. 1. The tropical Pacific. Hoikusha Publishing Co. Ltd. Japan.

(92). Tadashige Habe, 1977. Taxonomy of Bivalvia and Scaphopoda of Japan (Mollusca). Hodonkan Publishing Co. Ltd., Japan.

(93). Ansell, A. D., 1985. Species of *Donax* from Hong Kong: morphology, distribution, behaviour, and metabolism. The Macocofauna of Hong Kong and southern China. II. Vol. 1. Hong Kong University Press, pp:19–47.

(94). Cai Yingya and Zhang Ziqian, 1988. Molluscs in Beibu Gulf. Proc. on Marine Biology of the South China Sea. China Ocean Press, pp:121–142.

(95). Huang, Z. G. and B. Morton, 1983. *Mytilopsis sallei* establishes in Victorial Harbour. Malacological Review, 16:99–100.

(96). Kuroda, T. and T. Habe, 1981. A catalogue of Mollusca of Wakayama Prefecture, the Province of Kii, I. Bivalvia Scaphopoda and Cephalopoda, Kyoto, Japan, 301pp.

(97). Lee, S. Y. and B. Morton, 1985. The Malacofauna of Hong Kong and Southern China II. Volume 1, pp:49–76.

(98). Morton, B. and P. H. Scott, 1989. The Hong Kong Galeommatacea (Mollusca: Bivalvia) and their hosts, with descriptions of new species, Asian Marine Biology 6:129–160.

(99). Morton, B., 1991. Notes on the first mangrove shipworm, *Lyrodus singaporeana* recorded from Hong Kong, Research notes, pp:295–296.

(100). Oliver, P. G., 1990. A new species of Noetiella (Bivalvia: Arcacea) of Hong Kong, The Marine Flora and Fauna of Hong Kong and Southern China. II. Vol. 1. Hong Kong University Press, pp:19–47.

(101). Turner, R. D., 1996. A Survey and Illustrated Catalogue of the Teredinidae (Mollusca: Bivalvia). The Museum of Comparative Zoology, Harvard University, Cambridge, Massessusetts, 265pp.

(102). Vaught, K. C. (Editted, Abbott, R. T. and Boss, K. J.), 1989. A Classification of the living Mollusca. American Molacologists, Inc., Melbourne, Florida, USA, Class Bivalvia, pp:113–142.

(103). Wu, Wen-Lun 1980. The list of Taiwan Bivalve fauna. Taiwan Mus., 13(1–2):55–208.

(104). Xu Fengshan, 1989. Family Thraciidae (Mollusca: Bivalvia) from Chinese waters. Chin. J. Oceanol. Limnol. 7(1): 33–36.

(105). Xu Fengshan, 1992. On the family Lyonsiidae (Mollusca: Bivalvia), with description of a new genus and species from the China's seas. Chin. J. Oceanol. Limnol. 10(2): 153–155.

*: (1)-(18) and (20)-(89) in Chinese, (90)-(92) in Japanese
Compiled by Jiang Jinxiang, Xu Fengshan, and Li Rongguan; Editted by Qi Zhongyan (C. Y. Tsi), Zhuang Qiqian, Cai Yaoguo, Wang Zhenrui, and Li Fenglan

SCAPHOPODA

DENTALIIDAE

Anatlis weinkauffi (Dunker): East China Sea; South China Sea. (2, 6)

Anulidentalium bambusum Chistikov: East China Sea; Taiwan Strait; South China Sea. Subtidal to continental shelf. (5)

Calliodentalium crocinum (Dall): East China Sea continental shelf. (2)

Dentalium obtusum Qi & Ma: South of Zhejiang. Intertidal. (6)

D. octangulatum Donovan: South of Fujian. Low intertidal to 100 m. (1, 3, 6)

D. sinuosum Boissevain: South China Sea. Water depth 30–156 m. (6)

Episiphon candelatum (Kira): East China Sea continental shelf. (2)

E. kiaochowwanensis (Tchang & Tsi): Yellow Sea, intertidal to 60 m; East China Sea and South China Sea, water depth 10–80 m. (1–4)

Fissidentalium formosum (A. Adams & Reeve): South China Sea. Water depth 77–145 m. (6)

F. tenuicostatum Qi & Ma: South China Sea. Water depth 61–117 m. (6)

F. vernedei (Sowerby): East China Sea; South China Sea. (2, 6)

F. yokoyamai (Makiyamai): East China Sea continental shelf; South China Sea. Water depth 50–130 m. (1, 2, 6)

Fustiaria nipponica (Yokoyama): East China Sea; South China Sea. (2)

Gadilina insoluta (Smith): East China Sea continental shelf. (2)

Graptacme aciculum (Gould): East China Sea; South China Sea. (2, 6)

G. buccinulum (Gould): East China Sea; South China Sea. Intertidal and subtidal. (6)

Laevidentalium eburneum (Linnaeus): East China Sea to Nansha (=Spratly Islands). Subtidal. (1, 2, 5)

Omniglypta cerina (Pilsbry): East China Sea continental shelf. (2)

Striodentalium chinensis Qi & Ma: East China Sea; South China Sea. (2, 5, 6)

S. plycostatum Qi & Ma: East China Sea continental shelf. (2, 6)

S. rhabdotum (Pilsbry): East China Sea; South China Sea. (2, 6)

S. sedecimcostatum (Boissevain): East China Sea; Taiwan Strait; South China Sea. Water depth to 550 m. (6)

SIPHONODENTALIIDAE

Cadulus clavatus Gould: South China Sea. Water depth 5–36 m. (1)

Entaliunopsis habutae (Kiroda & Kikuchi): East China Sea continental shelf. (2)

E. intercostatum (Boissevain): East China Sea continental shelf. (2)

Siphonodentalium japonica Habe: Bohai; Yellow Sea. Subtidal, water depth 20–38 m. (4)

REFERENCES*

(1). Ma Xiutong, 1982. Marine shellfishes of China and their collection. China Ocean Press, 166pp.

(2). Qi Zhongyan and Ma Xiutong, 1979. Studies on Scaphopoda from the continental shelf of the East China Sea (abstract). In: Transactions of the seminar on invertebrate fauna by Chinese Society of Zoology.

(3). Qi Zhongyan and Ma Xiutong, 1984. Lists of molluscs from Hainan Island coast. Tropic Ocean Study, pp:1–22.

(4). Qi Zhongyan, Ma Xiutong and Wang Zhenrui, 1989. Molluscs in the Yellow Sea and Bohai Sea. China Agriculture Press, pp:144–146.

(5). Qi Zhongyan and Ma Xiutong, 1991. On some species of Scaphopoda in the waters of Nansha Islands. In: Contributions on Marine Organisms around Nansha Islands and the adjacent Waters (I). China Ocean Press, pp:89–92.

(6). Qi Zhongyan and Ma Xiutong, 1989. A study of the Family Dentaliidae (Mollusca) found in China. Chin. J. Oceanol. Limnol. 7(2): 112–122.

*: (1)-(5) in Chinese
Compiled by Huang Zongguo; Editted by Ma Xiutong and Qi Zhongyan

GASTROPODA

PROSOBRANCHIA
ARCHAEOGASTROPODA
PLEUROTOMARIIDAE

Entemnotrochus rumphii Schepman: Jilong (Taiwan). Subtidal. (53)

Mikadotrochus hirasei Pilsbry: Jilong and Suao (Taiwan). Subtidal. (53)

M. salmiana Rolle: Jilong and Suao (Taiwan). Subtidal. (53)

Perotrochus teramachii Kuroda: Jilong and Suao (Taiwan). Subtidal. (53)

HALIOTIDAE

Haliotis asinina Linnaeus: Taiwan; Hainan; Dongsha (=Pratas Islands); Xisha (=Paracel Islands). Below intertidal mean low water. (26, 39, 53)

H. clathrata Reeve: Hainan; Xisha (=Paracel Islands). Subtidal. (26)

H. discus hannai Ino [*H. gigantea discus*]: Liaoning; Shandong; Lianyungang (Jiangsu). Subtidal, water depth 2–10 m. (26, 39)
H. diversicolor Reeve [*H. diversicolor aquatilis*]: Southern Zhejiang; Fujian; Taiwan; Guangdong; Guangxi. Low intertidal to water depth about 10 m. (26, 39, 53)
H. ovina Gmelin: Taiwan; Hainan; Xisha (=Paracel Islands). (26)
H. planata Sowerby: Taiwan; Guangdong. (26, 39)
H. varia Linnaeus [*H. semistriata*]: Taiwan; Guangdong; Guangxi. Subtidal. (26, 39)
H. venusta Adams & Reeve: Nansha (=Spratly Islands). (31)

FISSURELLIDAE

Cranopsis pelex A. Adams: East China Sea. Subtidal, water depth 50–350 m. (73)
Diodora cruciata (Gould): Coastal seas of China. (28)
D. galeata Helbling: Taiwan; Xisha (=Paracel Islands). (28, 29)
D. mus (Reeve): Fujian; Nanao (Guangdong); Hainan; Nansha (=Spratly Islands). Water depth about 50 m. (28, 32, 39)
D. proxima (Sowerby): Xisha (=Paracel Islands). (28, 29)
D. quadriradiata (Reeve): Coastal seas of China. (28)
D. reevei (Schepman): South China Sea. Subtidal, water depth 10's of m. (28, 39)
D. suprapunicea Otuka: Coastal seas of China. (28)
D. ticaonica (Reeve): Coastal seas of China. (28)
Emarginula bellula A. Adams: Coastal seas of China. (28)
E. biangulata Sowerby: Coastal seas of China. (28)
E. bicancellata Montrouzier: Xisha (=Paracel Islands). (28, 29)
E. capuloidea Nevill: Xisha (=Paracel Islands). (28, 29)
E. clypeus A. Adams: Coastal seas of China. (28)
E. concinna A. Adams: East China Sea. Subtidal, water depth 105 m. (27, 28)
E. eximia A. Adams: Xisha (=Paracel Islands). (28, 29)
E. fragilis Yokoyama: Western East China Sea. Subtidal, water depth 50–150 m. (27, 28)
E. genestrella Deshayes: Xisha (=Paracel Islands). (40)
E. hosoyai Habe: East China Sea. Subtidal, water depth 50–200 m. (27, 28)
E. imaizumi Dall: Coastal seas of China. (28)
E. obovata A. Adams: Coastal seas of China. (28)
E. planulata A. Adams: Subtidal, water depth about 12 m. (28, 29)
E. sinica Lu: Zhaoshu Island (Xisha=Paracel Islands). (30)
E. variegata A. Adams: Taiwan. Intertidal. (60)
E. xishaensis Lu: Xisha (=Paracel Islands). (30)
Hemitoma panhi (Quoy & Gaimard): South China Sea; Xisha (=Paracel Islands). (29, 39)
Macroschisma sinensis A. Adams: Taiwan. Subtidal. (54)
Pectinoodonta orientalis Schepman. South China Sea. Subtidal. (30a)
Puncturella dorcas Kira & Habe: East China Sea. Subtidal, water depth 500 m. (27, 28)

P. fastigiata A. Adams: East China Sea. Subtidal, water depth 100–200 m. (27, 28)
P. nobilis (A. Adams): Coastal seas of China. (28)
P. pileolus (A. Adams): Western East China Sea. Subtidal, water depth 100–250 m. (27, 28)
P. sinensis Sowerby: Liaoning; Shandong. Subtidal, water depth 50–150 m. (23, 28)
Rimula exquisita A. Adams: Xisha (=Paracel Islands). Subtidal, water depth 10–15 m. (29, 39)
Scutus emarginatus (Philippi): Fujian; Guangdong; Hainan; Xisha (=Paracel Islands). Intertidal. (28, 29, 53)
S. sinensis (Blainville): Zhejiang; Fujian; Taiwan; Guangdong; Hainan; Xisha (=Paracel Islands). Intertidal. (28, 29, 53)
S. unguis (Linnaeus): Hong Kong. (25a)
Tugalia gigas (V. Martens): Bohai; Yellow Sea. Subtidal to water depth 50 m. (23, 28)

PATELLIDAE

Cellana grata (Gould): Taiwan; Hong Kong. Intertidal to shallow sea. (66, 75)
C. nigrolineata (Reeve): Liuqiu Reef (Taiwan). Intertidal. (54)
C. radiata (Born): Taiwan. Intertidal. (54)
C. testudinaria (Linné): Guangdong; Hainan. Mid to low intertidal. (39)
C. toreuma (Reeve): China's coast. Intertidal. (39, 53)
Patella flexuosa Quoy & Gaimard: Taiwan; Hong Kong. Intertidal. (53, 76)
P. stellaeformis Reeve: Pingtan (Fujian); Taiwan; Guangdong; Hainan; Xisha (=Paracel Islands). Intertidal. (39, 40)

ACMAEIDAE

Acmaea concinna (Lischke): Taiwan. Intertidal. (60)
A. mitra: Daya Bay (Guangdong). Intertidal. (34)
A. nanshaensis Lü: Nansha (=Spratly Islands). Subtidal. (31, 32)
A. pallida (Gould) [*A. (Niveotectura) pallida, Patelloida dorsuosa*]: Bohai; Yellow Sea. Intertidal. (22, 39)
A. strongiana: Daya Bay (Guangdong). Intertidal. (34)
Chiazacmea pygmaea lampanicola (Habe): Bohai; Yellow Sea; Hong Kong. Intertidal. (73)
Collisella cellanica Christiaens: Hong Kong. (75)
C. dorsuosa (Gould): Hong Kong. (25a)
C. formosa Christiaens: Taiwan. (75)
C. heroldi (Dunker): Taiwan; Hong Kong. Intertidal. (66, 75)
C. kolarovai (Grabau & King) [*Patelloida kolarovai*]: Bohai; Yellow Sea. Intertidal. (23)
C. (C.) luchuana (Pilsbry) *ericae* Christiaens: Hong Kong. (75)
C. mortoni Christiaens: Hong Kong. (75)
Notoacmea concinna: Hong Kong. Intertidal. (78c)

N. schrenckii (Lischke) [*Acmaea schrenckii, Collisella (Notoacmea) schrenckii, Patelloida schrenckii*]: China's coast. Intertidal. (23, 39)

Patelloida conoidalis (Pease): Hong Kong. (5)

P. grata (Gould): Yushan (Zhejiang); Taiwan. Intertidal. (13)

P. lampanicola (Habe) [*Chiazacmea pygmaea*]: Bohai; Yellow Sea; Hong Kong. Intertidal. (73)

P. pygmaea (Dunker) [*Acmaea pygmaea, Chiazacmea pygmaea*]: Bohai; Yellow Sea; Taiwan. Intertidal. (23, 39)

P. saccharina lanx (Reeve) [*Acmaea saccharina*]: Taiwan; South China Sea. Intertidal. (39, 53, 60)

P. striata (Quoy & Gaimard): Taiwan. Intertidal. (53)

P. toloeasis Christiaens: Hong Kong. (5)

TROCHIDAE

Angaria delphinus (Linnaeus): Taiwan. Intertidal to subtidal. (53, 54)

A. laciniata (Lamarck): Guangdong; Hainan. Intertidal mean low water line. (39)

Bathybembix argenteonitens (Lischke): Taiwan. Subtidal. (54)

B. crumpii (Pilsbry): Taiwan. Subtidal. (54)

Calliostoma formosense Smith: Taiwan. Subtidal. (54)

C. haliarchus (Melvill): Taiwan Strait. Subtidal. (54)

C. koma (Schikama & Habe): Bohai; Yellow Sea. Subtidal. (23)

C. unicum (Dunker): North of Fujian; Taiwan. Subtidal. (39, 66)

Camitia rotellinus (Gould) [*C. rotellina*]: Xisha (=Paracel Islands). (62)

Cantharidus giliberti (P. Fischer) [*C.* cf. *nitens*]: Xisha (=Paracel Islands). (62)

C. infuscatus (Gould): Zhejiang; Fujian; Taiwan; Guangxi. Intertidal. (2, 13)

Chlorostoma argyrostoma (Gmelin) [*Tegula argyrostoma*]: South of Fujian; Taiwan. Intertidal mean low water line. (39, 53, 54, 60)

C. nigerrima (Gmelin) [*Omphalius nigerrimus, Tegula nigerrima*]: Fujian; Taiwan; Guangdong. Intertidal. (39, 54, 60)

C. rustica (Gmelin) [*C. rusticum, Tegula rustica*]: Yellow Sea to South China Sea; Taiwan. Intertidal. (23, 39, 54)

C. xanthostigmata: Hong Kong. Intertidal. (78c)

Chrysostoma paradoxum (Born): Taiwan; Hainan. Intertidal. (39, 54)

Clanculus denticulatus (Gray): Hainan; Xisha (=Paracel Islands). (39)

C. margaritarius (Philippi): Taiwan; Guangdong. Intertidal. (39, 60)

C. stigmatarius A. Adams [*C.* cf. *stigmatarius*]: Xisha (=Paracel Islands). (40, 62)

Euchelus fussulatus (Sowerby): Yongxing Island (Xisha=Paracel Islands). (62)

E. scaber (Linné): Guangdong; Hong Kong; Hainan. Subtidal. (39)

Gibbula affinis Garrett: Xisha (=Paracel Islands); Nansha (=Spratly Islands). (63)

Lischkeia alwinae (Lischke): Taiwan. Subtidal. (53, 54)

Minolia chinensis Sowerby: East China Sea; South China Sea. Subtidal. (39)

Monilea calliferus (Lamarck) [*Trochus calliferus*]: Guangdong; Hainan. Intertidal low mean water line. (17, 39)

Monodonta labio (Linné) [*M. australis*]: Yellow Sea to South China Sea; Taiwan. Intertidal. (39, 52, 60, 78a)

M. labio confusa Tapparone-Canefri: Taiwan. Intertidal. (60)

M. neritoides (Philippi): Zhejiang to Guangdong. Intertidal. (39)

M. perplexa (Pilsbry): Taiwan. Intertidal. (54)

Tegula xanthostigma (A. Adams): Penghu (Taiwan). Intertidal. (54)

Thalotia elongatus (Woods) [*Tosatrochus attenuata*]: Taiwan; Xisha (=Paracel Islands). Subtidal. (39, 62)

Tristichotrochus formosense (Smith): Taiwan. (43)

Trochus chloromphalus A. Adams: Taiwan. Intertidal. (54)

T. conus Gmelin [*T. (T.) conus*]: Xisha (=Paracel Islands). Intertidal mean low water line. (39, 62)

T. creniferus Kiener: Taiwan; Xisha (=Paracel Islands); Nansha (=Spratly Islands). (62, 63)

T. histrio (Reeve) [*T. calcaratus*]: Taiwan; Weizhou Island (Guangxi); Xisha (=Paracel Islands); Nansha (=Spratly Islands). (54, 62, 63)

T. maculatus Linnaeus [*T. (T.) maculatus*]: Taiwan; Guangdong; Hainan; Nansha (=Spratly Islands). Intertidal mean low water line. (39, 53, 60, 62, 63)

T. niloticus Linnaeus [*T. (Tectus) niloticus maximus, Tectus maximus*]: Taiwan; Day Bay (Guangdong); Hainan; Xisha (=Paracel Islands); Nansha (=Spratly Islands). Intertidal mean low water line to subtidal 10 m. (33, 39, 53, 62)

T. noduliferus Lamarck: Xisha (=Paracel Islands). (62)

T. pyramis Born [*T. (Tectus) pyramis, Tectus pyramis*]: South China Sea. Intertidal, shallow subtidal. (39, 60, 62)

T. sacellus rota Dunker: Taiwan. Intertidal to water depth 5 m. (53, 60)

T. stellaris Gmelin [*T. incrassatus*]: Taiwan. Low intertidal to subtidal. (54)

T. triserialis Lamarck [*Tectus triserialis*]: Xisha (=Paracel Islands). (62)

Turcica chinensis Sowerby: South China Sea. Subtidal. (39)

T. coreensis Pease: Northern China; Liaoning. Intertidal, subtidal. (23, 39)

Umbonium costatum (Kiener): Taiwan; Fujian; Guangdong. Subtidal. (39)

U. monoliferum: Hong Kong. Intertidal. (78c)

U. suturale (Lamarck): Taiwan. (60)

U. thomasi (Crosse): Bohai; Yellow Sea. Intertidal. (39)

U. vestiarium (Linné): Guangdong; Hainan. Intertidal. (39)

STOMATIIDAE

Stomatella lutea (Linnaeus): Xisha (=Paracel Islands); Nansha (=Spratly Islands). (62, 63)

S. lyrata Pilsbry: Southern Fujian. Intertidal. (39)

Stomatia angulata A. Adams [*S. decolorata*]: Hainan; Xisha (=Spratley Islands). (62)

S. phymotis Helbling: Xisha (=Paracel Islands). (62)

S. (Gena) planulata Lamarck: Taiwan. Intertidal to subtidal. (54)

S. speciosa (A. Adams): Xisha (=Spratley Islands). (62)

CYCLOSTREMATIDAE

Circlotoma venusta (Hedley): Xisha (=Paracel Islands). (62)

Gena varia A. Adams [*Stomatella (Gena) varia*]: Taiwan; Xisha (=Paracel Islands). Intertidal. (39, 40, 54)

Liotia peronii (Kiener): Xisha (=Paracel Islands). (62)

TURBINIDAE

Astraea haematraga (Menke) [*Astralium haematragtragum*]: Taiwan; South China Sea. Intertidal mean low water line to subtidal. (39, 54)

A. pectrosum (Martyn) [*Astralium petrosum*]: Xisha (=Paracel Islands). (62)

Astralium petrosum (Martyn): Xisha (=Paracel Islands). (62)

Collonia donghaiensis Dong: East China Sea. Subtidal. (61)

C. pillula (Dunker): Xisha (=Paracel Islands). (62)

Galeoastrea (Harisazaea) modesta (Reeve): East China Sea continental shelf. (14)

Guildfordia triumphans (Philippi): Taiwan. Subtidal. (14, 53)

G. yoca (Jousseaume): Taiwan; Guangdong. Subtidal. (39, 53)

Homalopoma amussitatum (Gould): Dalian (Liaoning). Intertidal mean low water line to water depth 50 m. (49)

Lunella coronata coreensis (Récluz) [*L. coronata*]: Yellow Sea; East China Sea; Taiwan. Intertidal. (39, 54)

L. coronata granulata (Gmelin) [*Turbo (Lunella) coronatus, T. coronatus granulatus*]: East China Sea; South China Sea. Intertidal. (39, 64)

Pseudastralium henicus gloriosum (Kira): East China Sea continental shelf. Subtidal. (14)

Turbo argyrostomus Linnaeus: Taiwan; Xisha (=Paracel Islands); Nansha (=Spratly Islands). Intertidal. (62, 63)

T. brunneum Röding [*T. articulatus*]: Southern Fujian; Taiwan; Weizhou Island (Guangxi); Hainan; Xisha (=Paracel Islands); Nansha (=Spratly Islands). (62)

T. chrysostomus Linnaeus: Taiwan; Hainan; Xisha (=Paracel Islands); Nansha (=Spratly Islands). (62, 63)

T. (Lunella) cinereus (Born): Taiwan. (53)

T. cornutus Solander: Zhejiang; Fujian; Taiwan; Xisha (=Paracel Islands); Nansha (=Spratly Islands). (62)

T. marmoratus Linnaeus: Taiwan; Hainan; Xisha (=Paracel Islands); Nansha (=Spratly Islands). Subtidal. (62)

T. petholatus Linnaeus: Taiwan; Xisha (=Paracel Islands); Nansha (=Spratly Islands). (53, 62, 63)

T. reevei Philippi: Penghu (Taiwan). Subtidal. (54)

T. sarmaticus Linnaeus: Xisha (=Paracel Islands). (62)

T. setosus Gmelin: Taiwan. (53)

T. sparverius Gmelin: Taiwan. Subtidal. (54)

T. stenogyrus Fischer: Taiwan; Xisha (=Paracel Islands). Subtidal. (61)

NERITIDAE

Clithon (Pictoneritina) chlorostoma (Sowerby): Taiwan. Intertidal. (69)

C. (C.) corona (Linnaeus): Taiwan. Intertidal. (69)

C. corona angulosa (Récluz): Taiwan. (58c)

C. faba (Sowerby): Taiwan; Hong Kong. (69, 78)

C. oualaniensis (Lesson): Guangdong; Hainan. Intertidal. (39, 78)

C. retropictus (*V. Martens*): Taiwan; Guangdong. (39, 69, 78)

C. sowerbianus (Récluz): Guangdong; Hainan. Intertidal. (39)

Nerita achatina (Reeve): Fujian; Beibuwan (=Gulf of Tongking). Intertidal. (50, 74)

N. (Theliostyla) albicilla Linné [*N. albicilla*]: Xiyang Island (Fujian); Taiwan; South China Sea. Intertidal. (39, 53, 78)

N. chamaeleon Linnaeus: Taiwan; Guangdong to Nansha (=Spratly Islands). Intertidal. (18, 60, 78)

N. (Ritena) costata Gmelin [*N. costata*]: Taiwan; Guangdong; Hainan; Xisha (=Paracel Islands). Intertidal. (35, 53, 78)

N. exuvia (Linné): Taiwan. (53)

N. incerta (Philippi): Taiwan. (53)

N. insculpta: Taiwan. (53)

N. japonica (Dunker): Zhejiang. Intertidal. (13)

N. lineata Gmelin: Hong Kong. (53, 78)

N. ocellata (Récluz): Taiwan. Intertidal. (53, 60)

N. planospira (Anton): Taiwan. (53)

N. (R.) plicata Linné [*N. plicata*]: Taiwan; Guangdong; Hainan; Xisha (=Spratley Islands). Intertidal. (39, 53, 78)

N. (Amphinerita) polita Linné [*N. polita*]: Taiwan; Hainan; Xisha (=Paracel Islands). Intertidal. (39, 53, 78)

N. reticulata (Karsten): Taiwan. (53)

N. signata Macleay: Taiwan. Intertidal. (60)

N. squamulata (Récluz): Taiwan. (53)

N. (R.) striata Burrow [*N. striata*]: Taiwan; Guangdong; Hainan; Xisha (=Spratley Islands). Intertidal. (39, 53)

N. undata Linnaeus: Taiwan; Hong Kong; Nansha (=Spratly Islands). (53, 78)

N. (R.) yoldii Récluz: East China Sea; South China Sea. Intertidal. (39, 78)

Neritina (Neripteron) auriculata Lamarck: Taiwan. (69)

N. (Vittina) parallela (Röding): Taiwan. (69)

N. (V.) plumbea Sowerby: Taiwan. (69)

N. pulligera (Linné): Taiwan. (58c)

N. (V.) turrita (Gmelin): Taiwan. (69)

N. (Vittoida) variegata Lesson: Taiwan. (69)
N. (Dostia) violacea (Gmelin): South of Zhejiang; Taiwan; Hong Kong. (39, 69, 78)
Septaria janelli (Récluz): Taiwan. (58c)
S. porcellana (Linné): Taiwan. (53)

PHENACOLEPATIDAE

Phenacolepas crenulata (Broderip): Taiwan; Xisha (=Paracel Islands). (27a, 40)
P. senta Hedley: Xisha (=Paracel Islands). (27a, 40)

NERITOPSIDAE

Neritopsis radula (Linné): Taiwan; Xisha (=Paracel Islands). (39, 53)

MESOGASTROPODA

LACUNIDAE

Stenotis loui (Yen): Yantai and Qingdao (Shandong). (23)
S. oxytropis (Pilsbry): Beidaihe (Hebei); Qingdao (Shandong). Subtidal. (23, 39)
Temanella turrita (A. Adams): Qingdao (Shandong). Intertidal. (23)

LITTORINIDAE

Echininus cumingii spinulosus (Philippi): Taiwan. Intertidal. (54, 55)
Littorina (L.) brevicula (Philippi) [*Littorina balteata, L. heterospiralis, L. mandshurica, L. sourverbiana, Litorina brevicula* v. *costulata, Littorivaga brevicula*]: Bohai; Yellow Sea; East China Sea; Xiamen (Fujian); Daya Bay (Guangdong); Hong Kong. Intertidal. (12, 33, 39, 79a)
L. (Littoraria) coccinea (Gmelin) [*L. coccinea, Littorinopsis obesa*]: Taiwan; Hong Kong; Xisha (=Paracel Islands). Intertidal. (7, 12, 53)
Littoraria (Littorinopsis) ardouiniana (Heude) [*Littorina scabra*]: Hong Kong. Intertidal. (79a)
L. (Palustorina) articulata (Philippi) [*L. scabra, Litorina sinėnsis, Litorina strigata, Littorina intermedia, L. (Littorinopsis) scabra, Melaraphe blanfordi, M. scabra intermedia*]: Bohai and Yellow Sea to South China Sea; Taiwan; Hong Kong. Intertidal. (7, 12, 23, 53, 79a)
L. (P.) melanostoma Gray [*Littorina melanostoma* v. *articulata*]: South of Fujian. Intertidal. (17, 79a)
L. (Littorinopsis) pallescens (Philippi) [*Littorina scabra, L. sieboldii, Litorina scabra* v. *suturalis, Littorina arboricola*]: Hong Kong. Intertidal. (79a)
L. (Littoraria) undulata (Gray) [*Litorina columna, Littorina acuminata, L. undulata, L. undulata* v. *contracta, L. undulata* v. *sulcatula*]: Taiwan; Hong Kong; Hainan; Xisha (=Paracel Islands); Nansha (=Spratly Islands). Intertidal. (7, 11, 53, 60, 79a)
Mainwaringia rhizophila Reid: Hong Kong. Intertidal. (79a)
Nodilittorina (N.) radiata (Eydoux & Souleyet) [*Litorina exigua, Littorina granularis, Nodilittorina (Granulittorina) exigua, N. granularis, N. miliaris, Tectarius granularis*]: Bohai and Yellow Sea to South China Sea; Hong Kong. Intertidal. (7, 12, 53, 64, 79a)
N. (N.) trochoides (Gray) [*Litorina cecillei, L. malaccana, L. rubra, L. vllis, Littorina monilifera, L. trochoides, N. pyramidalis, Tectarius villis, Trochus nodulosus, Turbo trochiformis*]: South of Zhejiang to South China Sea; Taiwan; Hong Kong. Intertidal. (7, 12, 53, 79a)
N. (N.) vidua (Gould) [*Granulilittorina philippiana, Litorina chaoi, L. ventricosa, Littorina millerana, L. vidula, N. hawaiiensis, N. millegrana, N. picta, N. vidua*]: Taiwan; Hong Kong; Xisha (=Paracel Islands). Intertidal. (7, 60, 79a)
Peasiella infracostata (Issel) [*Risella (Peasiella) tantillus* v. *subinfracostata, Cyclostrema fuscopiperata, P. roepstorffiana*]: Hong Kong. Intertidal. (79a)
P. lutulenta Reid: Hong Kong. Intertidal. (79a)
P. roepstorffiana (Nevill) [*Risella balteata, R. (Peasiella) templiana, R. (P.) templiana* v. *nigrofasciata, R. (P.) templiana* v. *subimbricata*]: Hong Kong. Intertidal. (79a)
Tectarius coronatus (Valenciennes): Taiwan. Intertidal. (53)
T. pogodus (Linné): Taiwan. Intertidal. (53)

STENOTHYRIDAE

Stenothyra divalis (Gould): Hong Kong. (73a)
S. formosana Pilsbry and Hirase: Taiwan. (58c)
S. glabar A. Adams: Bohai and Yellow Sea to Fujian. Intertidal. (23, 39)

RISSOIDAE

Alvania fusca Gould: Rongcheng (Shandong). (23)
A. ligata Gould: Guangdong; Hainan. Intertidal mean low water line. (39)
Costalynia cf. *costulata* (Dunker): Fujian. Intertidal. (47a)
Iravadia (Fairbankia) bombayana Stoliczka: Hong Kong. Intertidal. (77)
I. (I.) ornata Blanford: Hong Kong. Intertidal. (77)
I. (I.) quadrasi Boettger: Hong Kong. Intertidal. (77)
Onoba elegantula A. Adams: Bohai; Yellow Sea. Intertidal to water depth 20 m. (23)
Pellamora trochlearis (Gould): China's coast. Intertidal. (71)
Rissoina bureri Grabau & King: Bohai; Yellow Sea. Intertidal. (23)
R. dunedini Grabau & King: Dalian (Liaoning). Intertidal. (49)

Rissolina plicatula (Gould): Bohai and Yellow Sea to Guangdong. Intertidal. (23, 39)

Zebina tridentata (Michaud): Taiwan; Xisha (=Paracel Islands); Nansha (=Spratly Islands). Subtidal. (23, 39)

ASSIMINEIDAE

Assiminea brevicula Pfeiffer: Fujian; Guangdong; Hong Kong. Intertidal. (77)

A. latericea H. & A. Adams: Bohai and Yellow Sea to Zhejiang. (23, 39)

A. lutea A. Adams: Bohai; Yellow Sea; Hong Kong. Intertidal. (23, 78c)

A. scalaris: Xiamen (Fujian). Intertidal. (50)

A. subeffusa: Hong Kong. Intertidal. (78c)

A. violacea: Xiamen (Fujian); Hong Kong. Intertidal. (50)

TORNIDAE

Pygmaeorota cingulifera (A. Adams): Shanmen Bay (Zhejiang). Intertidal. (13)

TURRITELLIDAE

Turritella bacillum Kiener: South of Zhejiang. Subtidal. (39)

T. fascialis Manke [*Kurosioia fascialis*]: East China Sea; South China Sea. Subtidal. (39)

T. fortilirata Sowerby [*Neohaustator fortilirata*]: Yellow Sea. Subtidal. (23, 39)

T. terebra (Linné): South of Fujian; Taiwan. Subtidal. (39, 54)

ARCHITECTONICIDAE

Acutitectonica acutissima (Sowerby): Taiwan. Subtidal. (54)

Architectonica maxima (Philippi): Taiwan; Guangdong; Hainan. Subtidal. (39, 53)

A. perdix (Hinds): South of Fujian. Subtidal. (39)

A. perspectiva (Linné): Taiwan; Guangdong; Hainan. Subtidal. (39, 53)

A. trochlearis (Hinds): Taiwan; Guangdong; Hainan. Subtidal. (39, 53)

Heliacus stramineus (Gmelin): Taiwan Strait. Subtidal. (54)

H. variegatus (Gmelin): Taiwan; Hainan; Xisha (=Paracel Islands). Subtidal. (39, 40)

Philippia radiata (Röding): Taiwan; Xisha (=Paracel Islands). Subtidal. (39, 40)

SILIQUARIIDAE

Siliquaria anguina (Linnaeus): Taiwan Strait. (54)

S. cumingii (Mörch): Taiwan; South China Sea. Subtidal. (11, 60)

VERMETIDAE

Serpulorbis imbricata (Dunker) [*Vermetus imbricatus*]: Zhejiang and Fujian to South China Sea. Intertidal. (39, 64)

Siphonium maximum (Sowerby): Taiwan; Hainan; Xisha (=Paracel Islands). Intertidal mean low water line. (11, 39, 40)

Vermetus renisectus (Carpenter): South of Fujian. Intertidal. (39)

PLANAXIDAE

Planaxis sulcatus (Born): Fujian; Guangdong; Xisha (=Paracel Islands). Intertidal. (18, 39)

MODULIDAE

Modulus tectum (Gmelin) [*Aplodon tectum*]: Taiwan; Xisha (=Paracel Islands). Intertidal to subtidal. (11, 39, 54)

POTAMIDIDAE

Batillaria bronii (Sowerby): Fujian to Hainan; Taiwan. (18, 77)

B. cumingi (Crosse): Bohai and Yellow Sea to South China Sea. Intertidal. (23, 39, 77)

B. multiformis (Lischke): Bohai and Yellow Sea to South China Sea. Intertidal. (23, 77)

B. sordida (Gmelin): Hong Kong; South China Sea. Intertidal. (77)

B. zonalis (Bruguiere): Bohai and Yellow Sea to South China Sea; Taiwan. Intertidal. (23, 39, 54, 64, 77)

Cerithidea cingulata (Gmelin) [*Tympanotomus cingulatus*]: Bohai and Yellow Sea to South China Sea; Taiwan. Intertidal. (23, 39, 54, 77)

C. djadjariensis (K. Martin): Xiamen (Fujian); Taiwan; Hong Kong; South China Sea. Intertidal. (50, 54, 77)

C. largillierti (Philippi): Bohai and Yellow Sea to South China Sea; Taiwan. Intertidal. (23, 71)

C. microptera (Kiener): South of Fujian. Intertidal. (39)

C. ornata (A. Adams): Hong Kong; South China Sea. Intertidal. (39, 77)

C. rhizophorarum A. Adams [*C. obtusa*]: Bohai and Yellow Sea to South China Sea; Taiwan. Intertidal. (23, 39, 54, 77)

C. sinensis (Philippi) [*Cerithium sinensis*]: Bohai; Yellow Sea. Intertidal. (23, 39)

Telescopium telescopium (Linnaeus): Hainan. Intertidal. (46a)

Terebralia sulcata (Born): Hong Kong; Guangdong; Hainan. Intertidal. (39, 77)

CERITHIIDAE

Bittium craticulatum Gould: South China Sea; Nansha (=Spratly Islands). Subtidal. (11)

Cerithium alutaceum Reeve: Guangdong; Hong Kong; Xisha (=Paracel Islands). (17)

C. (*Vertagus*) *articulatum* (Adams & Reeve): Taiwan; Xisha (=Paracel Islands). (8)

C. (*Conocerithium*) *bayayi* Vignal: Nansha (=Spratly Islands). Subtidal. (11)

C. (*V.*) *cedonulli* Sowerby [*C. cedonulli*]: Hainan; Xisha (=Paracel Islands). Intertidal. (39, 40)

C. citrinum Sowerby: Taiwan; Hainan; South China Sea; Xisha (=Paracel Islands); Nansha (=Spratly Islands). Intertidal mean low water line. (11, 40)

C. columna Sowerby: Taiwan; Hainan; Xisha (=Paracel Islands). Intertidal to subtidal. (39, 40)

C. echinatum Lamarck: Taiwan; Xisha (=Paracel Islands). (40)

C. (*V.*) *fasciatum* Bruguiere [*C. fasciatum*]: Taiwan; Xisha (=Paracel Islands). Subtidal. (39, 40)

C. kobelti (Dunker): Yushan and Nanji (Zhejiang); Taiwan; Guangdong. Intertidal. (13, 60)

C. (*Semivertagus*) *maillardi* (Crosse): Nansha (=Spratly Islands). Intertidal. (11)

C. (*S.*) *nesioticum* Pilsbry & Vanatta [*Semivertagus nesioticus*]: Taiwan; Nansha (=Spratly Islands). Intertidal to subtidal. (11, 60)

C. nodulosum (Bruguiere): Hainan; Xisha (=Paracel Islands); Nansha (=Spratly Islands). Intertidal and subtidal. (11, 40)

C. pfefferi: Xiamen (Fujian). Intertidal. (25)

C. rostratum Sowerby: Xisha (=Paracel Islands). (40)

C. rubus (Martyn): Nanao (Guangdong) to Guangxi. (17)

C. (*V.*) *vertagus* (Linnaeus) [*C. vertagus* (Linné)]: Taiwan; Hainan. Intertidal. (39, 40)

Clypemorus bifasciatus Sowerby: Taiwan; Guangdong to Hainan. Intertidal. (17, 39, 54, 60)

C. humilis (Dunker): Taiwan; Guangdong. Intertidal. (54, 67)

C. monoliferum: Hong Kong. Intertidal. (78c)

C. morus Lamarck: Fujian to Guangxi. (28)

C. petrosus (Wood): Taiwan. Intertidal. (54)

C. trailli (Sowerby): Hainan. Intertidal. (39)

C. zonatus (Woods): Taiwan. Intertidal. (60)

Pseudorertagus aluco (Linnaeus): Penghu (Taiwan). Intertidal to subtidal. (54)

Rhinoclavis (*R.*) *aspera* (Linnaeus) [*Cerithium asperum, C.* (*Vertagus*) *asperum*]: Xisha (=Spratley slands). Intertidal, subtidal to 28 m water depth. (11, 40)

R. fasciata (Bruguiere): Taiwan Strait. Subtidal. (54)

R. (*Proclava*) *kochi* (Philippi): East China Sea; South China Sea. (20)

R. sinensis (Gmelin) [*Cerithium sinense, C.* (*Vertagus*) *sinense, Ochetoclava sinensis*]: South of Fujian; Taiwan. Intertidal. (11, 39, 40, 60)

R. (*P.*) *sordidula* (Gould): South China Sea. (20)

R. vertagus (Linnaeus): Taiwan Strait. Subtidal. (54)

LITIOPIDAE

Diffalaba picta (A. Adams): Bohai; Yellow Sea. Intertidal to subtidal 20 m. (23)

CERITHIOPSIDAE

Cerithiella subreticulata (Dunker): Beidaihe (Hebei). (23)

Notoseila laqueta (Gould): Qingdao (Shandong). Subtidal. (23)

EPITONIIDAE

Acrilla acuminata (Sowerby) [*Scala acuminata*]: Bohai and Yellow Sea to South China Sea. Subtidal. (17, 23, 64)

Amaea magnifica (Sowerby): Taiwan Strait. Subtidal. (53, 54)

A. secunda Kuroda & Ito: Yellow Sea; East China Sea. (23)

A. thielei (de Boury): Bohai; Yellow Sea; East China Sea. Subtidal. (23)

Asperiscala eximia (A. Adams & Reeve): Shandong. Subtidal. (23)

Cirratiscala irregularis (Sowerby): Bohai Bay. Subtidal. (23)

Cirsotrema perplexum (Pease): South of Fujian to Xisha (=Paracel Islands); Taiwan. Subtidal. (17, 40)

C. varicosum (Lamarck): Taiwan; Xisha (=Paracel Islands). Subtidal. (17, 40)

Depressiscala aurita (Sowerby) [*E. auritum*]: Bohai; Yellow Sea; Guangdong. Subtidal. (23)

Epitonium eboreum Lan: Taiwan. (52)

E. scalare (Linnaeus) [*E. pretiosa minor*]: Taiwan; Guangdong; Hainan. Subtidal. (17, 53)

E. scalare minor Graban & King [*E. neglecta*]: Bohai; Yellow Sea; East China Sea; Taiwan. Subtidal. (23)

Glabriscala stigmatica (Pilsbry): Bohai; Yellow Sea. Subtidal. (23)

Gradatiscala gradata pygmaea (Grabau & King): Bohai; Yellow Sea; East China Sea. Intertidal. (23)

Papyriscala latifasciata (Sowerby) [*Epitonium latifasciata, E. lineolatum*]: South of Shandong. Subtidal. (23)

P. yokoyamai (Suzuki & Ichikawa): Yangma Island (Shandong). Subtidal. (23)

Spiniscala japonica (Dunker) [*Epitonium japonicus*]: Bohai. Intertidal to 60 m water depth. (23)

JANTHINIDAE

Janthina janthina (Linnaeus): Taiwan; South China Sea. Planktonic. (17, 53)

J. prolongata Blainville: East China Sea; South China Sea. Planktonic. (17)

MELANELLIDAE=EULIMIDAE

Balcis grandis (A. Adams): Taiwan Strait. Subtidal. (54)

B. luchuma (Pilsbry): Taiwan. Below intertidal mean low water line. (60)

B. thaanumi (Pilsbry): Xisha (=Paracel Islands). (40)

Eulima bifascialis (A. Adams) [*E. bilineata, Melanella bivittata*]: Bohai and Yellow Sea to South China Sea. Intertidal to subtidal 40 m. (17, 23)

E. maria (A. Adams): Bohai; Yellow Sea; East China Sea. Intertidal to subtidal 20 m. (23)

Melanella martinii A. Adams: Xiamen (Fujian); Guangdong. (67)

VANIKORIDAE

Vanikoro cidaris Récluz: Xisha (=Paracel Islands). Subtidal. (17, 40)

V. delicata (Pease): Xisha (=Paracel Islands). (40)

V. gueriniana Récluz: Taiwan; Xisha (=Paracel Islands). (40)

AMALTHEIDAE [HIPPONICIDAE]

Amalthea conica Schumacher [*A. acuta*]: Hainan; Xisha (=Paracel Islands); Nansha (=Spratly Islands). Intertidal and subtidal. (11, 17, 40)

Hipponix foliacea (Quoy & Gaimard) [*Antisabia foliacea*]: Taiwan. Intertidal. (17, 54, 60)

Pilosabia pilosa (Deshayes) [*P. trigona*]: Taiwan; South China Sea. Intertidal. (17, 54, 60)

Sabia conica (Schumacher): Taiwan. (54, 60)

TRICHOTROPIDAE

Amathina tricarinata (Linnaeus): South of Zhejiang. (9, 17, 64)

Separatista helicoides (Gmelin): Hainan. (9)

Trichotropis (*Trichotripis*) *bicarinata* Sowerby [*T. bicarinata*]: Northern Yellow Sea. Subtidal. (9, 17)

T. (*Iphinoe*) *unicarinata* Broderip & Sowerby: Yellow Sea. Subtidal. (9)

CAPULIDAE

Capulus dilatatus A. Adams: Taiwan; Guangdong; Hainan. Subtidal. (17)

C. sagittifer Gould: Xisha (=Paracel Islands). (40)

CALYPTRAEIDAE

Calyptraea morbida (Reeve): East China Sea; Taiwan; Hainan. Intertidal mean low water line. (17, 18)

C. sakaguchi: Hong Kong. Intertidal. (78c)

Cheilea diaphana (Reeve): Beibuwan (=Gulf of Tongking); Xisha (=Paracel Islands). Subtidal. (17, 40, 74)

C. equestris (Linnaeus): Xisha (=Paracel Islands). Intertidal. (17, 40)

Crepidula gravispinosa (Kuroda & Habe): Southern Zhejiang; Pingtan (Fujian). Mean low water line. (17, 45)

C. onyx: Hong Kong. Intertidal. (52)

Siphopatella walshi (Reeve) [*Crepidula walshi*]: Bohai and Yellow Sea to South China Sea. Intertidal to 10's of m water depth. (17, 22)

XENOPHORIDAE

Xenophora (*Omustus*) *exuta* (Reeve) [*X. exuta*, *Onustus exuta*]: East China Sea; Taiwan Starit; South China Sea. Subtidal. (21, 55)

X. (*O.*) *indica* (Gmelin): Taiwan; Guangdong; Hainan. Subtidal. (21)

X. (*X.*) *minuta* Tsi & Ma: East China Sea; South China Sea. Subtidal. (21)

X. pallida (Reeve): Taiwan; Nansha (=Spratly Islands). Subtidal. (43, 53, 55)

X. (*Stellaria*) *sinensis* (Philippi) [*X. calculifera*]: Taiwan; South China Sea. Subtidal. (17, 21)

X. (*X.*) *solarioides* (Reeve): South of Zhejiang; Taiwan. Subtidal. (21)

X. (*S.*) *solaris* (Linnaeus) [*X. solaris*, *Stellaria solaris*]: Taiwan; South China Sea. Subtidal. (21, 53)

STROMBIDAE

Lambis chiragra (Linnaeus) [*L.* (*Harpago*) *chiragra*]: Taiwan; Hainan; Xisha (=Spratley Islands). Below intertidal mean low water line. (2)

L. (*L.*) *crocata* (Link) [*L. crocata*]: Taiwan; Xisha (=Paracel Islands). Below intertidal mean low water line. (2)

L. lambis Linnaeus [*L.* (*L.*) *lambis*, *Pterocera lambis*]: Taiwan; Hainan; Xisha (=Spratley Islands). Below intertidal mean low water line. (2)

L. millepeda (Linné): Taiwan. (53)

L. scorpius (Linnaeus): Taiwan; Xisha (=Paracel Islands). (2, 6)

L. truncata sebae (Kiener) [*L.* (*L.*) *truncata sebae*]: Taiwan; Xisha (=Paracel Islands). Subtidal. (2, 53)

Rimella canceliata Linné: Zengmuansha (southern tip of Nansha=Spratly Islands). Subtidal, water depth 50 m. (38)

Strombus aratrum (Röding) [*S.* (*Euprotomus*) *aratrum*]: Taiwan; South China Sea. Subtidal. (2, 11)

S. aurisdiane Linnaeus: Taiwan. Subtidal. (2, 54)

S. (*Euprotomus*) *bulla* (Röding) [*S. bulla*]: Taiwan; Xisha (=Paracel Islands). Subtidal. (2, 53)

S. canarium Linnaeus [*S.* (*Laevistrombus*) *canarium*, *S. canarium turturellus*]: Taiwan; South China Sea. Subtidal. (2, 53)

S. (*Canarium*) *dentatus* Linnaeus: Taiwan; Xisha (=Paracel Islands). Subtidal. (2)

S. (*Dolomena*) *dilatatus* Swainsoni & Reeve: Guangdong; Hainan. Subtidal. (2)

S. epidromis Linnaeus: Taiwan. (2)

S. (*Canarium*) *erythrinus* Dillwyn: Taiwan; Xisha (=Paracel Islands). Subtidal. (2)

S. (*Canarium*) *fragilis* (Röding): South China Sea. Subtidal. (6)

S. gibberulus gibbosus (Röding): Taiwan; Hainan; Xisha (=Paracel Islands). Intertidal mean low water line. (2, 53)

S. (*Doxander*) *japonicus* Reeve: Zhejiang; Guangdong; Hainan. Subtidal. (2)

S. (*Canarium*) *labiatus* (Röding): Guangdong; Hainan. Below intertidal mean low water line. (2, 54)

S. labiosus Wood: Nansha (=Spratly Islands). Subtidal. (38)

S. (*Tricornis*) *latissimus* Linnaeus: Taiwan; Xisha (=Paracel Islands). Subtidal. (6, 53)

S. lentiginosus Linnaeus [*S.* (*Lentigo*) *lentiginosus*]: Taiwan; Hainan; Dongsha (=Pratas Islands); Xisha (=Spratley Islands). Subtidal. (2)

S. (*Conomurex*) *luhuanus* Linnaeus [*S. luhuanus*]: Taiwan; Guangdong; Guangxi; Hainan; Xisha (=Paracel Islands). Intertidal to water depth 36 m. (2, 11, 53)

S. (*D.*) *marginatus robustus* Sowerby: South of Fujian; Taiwan. Subtidal. (2, 11)

S. (*Canarium*) *microurceus* (Kira): Guangdong; Xisha (=Paracel Islands). Subtidal. (2)

S. (*D.*) *minimus* Linnaeus: Taiwan. (2)

S. (*Canarium*) *mutabilis* Swainson [*S. mutabilis*]: Taiwan; Guangdong; Hainan; Xisha (=Spratley Islands). Intertidal. (2, 53)

S. (*Lentigo*) *pipus* (Röding) [*S. pipus*]: Taiwan. Subtidal. (2, 53, 54)

S. (*D.*) *plicatus pulchellus* Reeve: Taiwan; Hainan; South China Sea. Subtidal. (2, 11)

S. scalariformis Duclos: South China Sea. (2)

S. sinuatus (Humphrey): Taiwan. (53)

S. terebellatus Sowerby: Hainan. (2)

S. thersites (Swainson): Taiwan. (53)

S. urceus Linnaeus [*S.* (*Canarium*) *urceus*, *S. u. ustulatus*]: Taiwan; South China Sea. Below intertidal mean low water line. (2, 53)

S. vittatus Linnaeus [*S.* (*Doxander*) *vittatus*]: Taiwan Strait; Guangdong; Guangxi; Hainan. Subtidal. (2, 54)

Terebellum terebellum (Linnaeus): Taiwan; South China Sea; Nansha (=Spratly Islands). Subtidal. (2, 51, 53)

Tibia (*Varicospira*) *cancellata* (Lamarck): China's coast. Subtidal. (2)

T. fusus Linnaeus [*T.* (*T.*) *fusus*]: Taiwan; South China Sea. Subtidal. (2, 51, 53)

T. martinii (Marrat): Taiwan; Hainan. Subtidal. (2, 53)

T. melanochilus (A. Adams): China's coast. Subtidal. (2)

T. powisi (Petit): Taiwan; Guangdong; Hainan. Subtidal. (2)

ATLANTIDAE

Atlanta depressa Souleyet: South China Sea. Planktonic. (41)

A. fusca Souleyet: South China Sea. Planktonic. (17, 41)

A. helcinoides Souleyet: East China Sea; South China Sea. Planktonic. (41)

A. inclinata Souleyet: South China Sea. Planktonic. (17, 41)

A. inflata Souleyet: East China Sea; Taiwan Strait; South China Sea. Planktonic. (17, 41)

A. lesueuri Souleyet: East China Sea; Taiwan Strait; South China Sea. Planktonic. (17, 41)

A. peroni Lesueur: East China Sea; Taiwan Strait; South China Sea. Planktonic. (17, 41)

A. rosea Souleyet: East China Sea; Taiwan Strait; South China Sea. Planktonic. (17, 41)

A. turriculata d'Orbigny: East China Sea; Taiwan Strait; South China Sea. Planktonic. (17, 41)

Oxygyrus keraudreni (Lesueur): South China Sea. Planktonic. (41)

Protatlanta souleyeti (Smith): Southeastern East China Sea; South China Sea. Planktonic. (17, 41)

CARINARIIDAE

Cardiapoda richardi Vayssiere: South China Sea. Planktonic. (41)

Carinaria cristata (Linné): South China Sea. Planktonic. (41)

C. galea Benson: South China Sea. Planktonic. (17, 41)

Pterosoma planum Lesson: Taiwan; South China Sea. Planktonic. (17, 41)

PTEROTRACHEIDAE

Firoloida desmaresti Lesueur: East China Sea; South China Sea. Planktonic. (17, 41)

Pterotrachea coronata Forshal: South China Sea. Planktonic. (17, 41)

P. hippocampus Philippi: South China Sea. Planktonic. (17, 41)

P. minuta Bonnevie: South China Sea. Planktonic. (17, 41)

P. scutata Gegenbaur: Easteern Taiwan; South China Sea. Planktonic. (17, 41)

NATICIDAE

Eunaticina papilla (Gmelin) [*Sigaretus papillus*]: Bohai and Yellow Sea to South China Sea. Intertidal mean low water line to 20 m water depth. (17, 23)

Lunatica gilva (Philippi) [*Natica fortunei, Polynices fortuneri*]: Bohai and Yellow Sea to South China Sea. Intertidal. (23)

L. yokoyamai (Kuroda) [*L. pallida*]: Bohai; Yellow Sea; Dalian (Liaoning); Weihai (Shandong). Intertidal. (23, 25)

Natica alapapilionis (Röding): Taiwan; South China Sea. Subtidal. (17, 55)

N. albifasciata Liu: South China Sea. Subtidal. (25)

N. ampla (Philippi) [*N. didyma* v. *ampla, Polinices didyma*]: Bohai; Yellow Sea; East China Sea; Taiwan. Subtidal. (23, 53, 55)

N. arachnoidea (Gmelin): Hainan. Intertidal mean low water line. (17)

N. areolata Récluz: South China Sea. (11a)

N. asellus Reeve: Guangdong; Hainan. Subtidal. (25)

N. bibalteata Sowerby: South China Sea. Subtidal. (11, 25)

N. bougei Sowerby: South China Sea. (11a)
N. clausiformis (Oyama): Bohai and Yellow Sea to South China Sea. (11a)
N. concavoperculata Liu: South China Sea. Subtidal. (25)
N. concinna Dunker: East China Sea. Subtidal. (25)
N. crassoperculata Liu: South China Sea. Subtidal. (25)
N. excellenna (Azuma): South China Sea. Subtidal. (25)
N. gualteriana Récluz [*N. tessellata*]: Taiwan; Beibuwan (=Gulf of Tongking); Hainan. Intertidal to subtidal. (18, 55, 74)
N. hainanensis Liu: South China Sea. Intertidal. (25)
N. hirasei Pilsbry: Bohai; Yellow Sea. (11a)
N. janthostomoides Kuroda & Habe [*N. janthostoma*]: Bohai; Yellow Sea. Subtidal. (17, 23)
N. lineata (Röding) [*Notocochlis lineata*]: South of Fujian; Taiwan. Below intertidal mean low water line. (17, 53, 55)
N. lurida (Philippi): Dongshan (Fujian) to South China Sea. (18)
N. melanoperculata Liu: South China Sea. Intertidal. (25)
N. onca (Röding): Hainan. Below intertidal mean low water line. (17)
N. pygmaea sp. nov.: South China Sea. (11a)
N. qizhongyani sp. nov.: South China Sea. (11a)
N. sagittata Menke: East China Sea; South China Sea. (11a)
N. scopaespira Liu: South China Sea. Subtidal. (25)
N. simplex schepma: South China Sea. (11a)
N. sinensis sp. nov.: South China Sea. (11a)
N. spadiceoides Liu: South China Sea. Subtidal. (25)
N. superaotanta Schepman: Hainan. Subtidal. (25)
N. tabularis Kuroda: East China Sea; South China Sea. Subtidal. (25)
N. tenuipicta Kuroda: South China Sea. Subtidal. (11, 25)
N. tessellata Philippi: Beibuiwan (=Gulf of Tong King); Hainan. (18, 74)
N. tigrina (Röding) [*N. maculosa*]: Bohai and Yellow Sea to South China Sea. Intertidal to water depth 10 m. (17, 23, 55)
N. tossansis Kuroda: Guangdong; Hainan. Subtidal. (25)
N. trilli Reeve: Hainan. (25)
N. unibalteata Liu: South China Sea. Subtidal. (25)
N. unicolor sp. nov.: Bohai; Yellow Sea. (11a)
N. vitellus (Linnaeus) [*N. spadicea*]: South of Fujian; Taiwan. Subtidal. (11, 17, 53, 55)
Neverita didyma (Röding) [*Natica didyma bicolor*]: Bohai and Yellow Sea to South China Sea. Intertidal or subtidal. (23)
Polinices albumen (Linnaeus): Hainan. Below intertidal mean low water line. (17)
P. effusus (Swainson): Guangdong; Guangxi; Hainan. (18, 25)
P. flemingianus (Récluz): Guangdong to Xisha (=Paracel Islands). (18, 80)
P. macrostoma (Philippi): South of Fujian. Subtidal. (17)
P. mammata (Röding) [*Mammilla mammata, Polynices mammata, P. (Mammilla) mammatus*]: Taiwan; South China Sea. Subtidal. (11, 20, 55)
P. maurus (Bruguere): Weizhou Island (Guangxi). (25)
P. melanostomoides (Quoy & Gaimard) [*P. melanostomus*]: Taiwan Strait; Hainan. Subtidal. (25, 55)
P. opacus (Récluz): Fujian to Xisha (=Paracel Islands). (18, 80)
P. peseiphanti: Hong Kong. Intertidal. (78c)
P. pyriformis (Récluz): Taiwan; South China Sea. Below intertidal mean low water line. (17)
P. sagamensis (Pilsbry): Fujian to Hainan. (25, 80)
P. (Mammilla) simiae (Deshayes): Taiwan Strait. Subtidal. (55)
P. tumidus (Swainson): Taiwan Strait. Subtidal. (55)
P. vestitus Kuroda: Guangdong; Hainan; South China Sea. Intertidal to water depth 98 m. (25)
Sinum incisum (Reeve) [*Sigaretus undulatus*]: Nanji (Zhejiang); South China Sea. Subtidal. (17, 18, 25)
S. japonicus (Lischke): Fujian; Hong Kong. Intertidal. (35b, 78c)
S. javanicum (Griffith & Pidgeon): Taiwan; South China Sea. Subtidal. (17, 55)
S. cf. *neritoideum* (Linnaeus): Fujian. Intertidal. (35b)
S. oblongum (Reeve): South China Sea. Subtidal. (25)
S. papilla (Gmelin): Bohai and Yellow Sea to South China Sea. Intertidal. (48)
S. planulatum (Récluz): Taiwan; Hainan. Subtidal. (17)

LAMELLARIIDAE

Lamellaria kiiensis Habe [*L. latens*]: Southern Yellow Sea. Intertidal. (23)
Velutina (Velutella) pusio A. Adams: Central Yellow Sea. Intertidal to water depth 100 m. (23)

ERATOIDAE

Alaerato angulifera (Sowerby): Hainan. Subtidal. (11b)
Hespererato nanhaiensis sp. nov.: South China Sea. Subtidal. (11b)
Proterato callosa (A. Adams & Reeve): Qingdao (Shandong) to Guangxi. Intertidal to water depth 34 m. (23)
P. (Sulcerato) limata sp. nov.: Xisha (=Paracel Islands). (11b)
P. (Eratoena) nana (Sowerby): Hainan. Intertidal to subtidal. (11b)
P. (S.) pura (Kuroda and Habe): China's coast. Intertidal. (11b)
P. sulcifera (Sowerby): Hainan; Xisha (=Paracel Islands). Subtidal. (18, 40)

TRIVIIDAE

Trivirostra exigua (Gray): Taiwan; Xisha (=Paracel Islands). (40)
T. oryza (Lamarck): Taiwan; Hainan; Xisha (=Paracel Islands). Below intertidal mean low water line. (40)

OVULIDAE

Aclyvolva granularis Ma: Hainan. Subtidal. (8)
Calpurnus (Procalpurnus) lacteus semistriatus (Pease): Taiwan; Hainan; Xisha (=Paracel Islands). Subtidal. (10)
C. (C.) verrucosus (Linnaeus): Taiwan; Hainan. (5, 10, 53)

Carpiscula galearis Cate: East China Sea; South China Sea. Subtidal. (10, 79)

Crenavolva (*Cuspivolva*) *chinensis* Qi & Ma: Xiamen (Fujian); Hong Kong. Subtidal. (10)

C. (*C.*) *striatula traillii* (A. Adams): Nanao and Shanwei (Guangdong). Subtidal. (10)

Cymbovula deflexa (Sowerby): Taiwan. (53)

Delonovolva formosa (Adams & Reeve): Xiamen (Fujian). (10)

Globovula margarita (Sowerby): Taiwan; Xisha (=Paracel Islands). Intertidal to water depth 30 m. (10)

G. sphaera Cate: Taiwan. (10)

Margovula pyriformis (Sowerby): Dongshan (Fujian) to Hainan. Subtidal. (10)

M. schilderorum Cate: Fujian. (10)

M. tinctilis Cate: Guangxi. (10)

Ovula ovum (Linnaeus): Taiwan; Hainan; Xisha (=Paracel Islands). Subtidal. (10, 53, 64)

Phenacovolva (*P.*) *birostris* (Linnaeus) [*Volva* (*P.*) *philippinarum*]: China's coast. (8)

P. (*Turbovula*) *brevirostris* (Schumacher): South of Fujian; Taiwan Strait. Subtidal. (8, 55)

P. (*T.*) *dancei* Cate [*Volva birostris* nov]: South of Fujian; Hong Kong. Subtidal. (8, 17)

P. (*Takasagovolva*) *gigantea* Azuma: Southwestern Taiwan. (8)

P. (*Calcaria*) *longirostrata* (Sowerby): Southeastern Hainan. Subtidal. (8)

P. (*P.*) *recuva* (A. Adams & Reeve): China's seas. (8)

P. (*P.*) *rosea* (A. Adams): China's coast. Subtidal. (8, 55)

P. (*P.*) *suberflxa* (A. Adams & Reeve): East China Sea. (8)

Primovula (*Adamantia*) *concinna* (A. Adams & Reeve): Xiamen (Fujian); Taiwan; Xisha (=Paracel Islands). (10)

P. (*P.*) *dautzenbergi* Schilder: South of Fujian. (10)

P. piriei Petuch: Taiwan. (10)

P. (*P.*) *singularis* Cate: China's coast. (10)

Prionovolva bulla (Adams & Reeve): East China Sea. (10)

P. pudica (A. Adams): South China Sea. Subtidal. (10)

P. pudica wilsoniana Cate: Northwestern Taiwan. (10)

Prosimnia semperi (Weinkauff) [*Primovula* (*Prosimnia*) *caarctata*]: Xisha (=Paracel Islands). Subtidal to water depth of 10's of m. (8, 40)

Pseudocypraea adamsonii Sowerby [*P. adamonii*]: Taiwan; Xisha (=Paracel Islands). (3, 10)

Pseudosimnia (*Diminovula*) *alabaster* (Reeve): Fujian; Taiwan; Guangdong. Intertidal to water depth 60 m. (10)

P. (*Inflatovula*) *marginata* (Sowerby): Xiamen (Fujian); Taiwan. Subtidal. (10, 55)

P. (*D.*) *punctata* (Duclos): China's coast. (10)

P. (*I.*) *sinensis* (Sowerby): Taiwan Strait; Hong Kong. Subtidal. (10, 55)

P. (*D.*) *whitworthi* Cate: South of Dongshan (Fujian); Hong Kong. Subtidal. (10)

Sandalia rhodia (A. Adams) [*Primovula rhodia*]: Yellow Sea to Haimen (Guangdong). Intertidal to water depth 230 m. (8)

Simnialena formicaria (Sowerby): East China Sea. (8)

Testudovolva adaminea Cate: China's seas. (10)

T. arientis Cate: China's seas. (10)

T. nipponensis (Pilsbry): South China Sea. Subtidal. (10)

Volva volva (Linnaeus): Taiwan; South China Sea. Subtidal. (8, 53)

V. volva habei (Oyama): Zhejiang to Guangdong; Taiwan Strait. Subtidal. (8, 55)

Xandarovula xanthochila (Kuroda): East China Sea. Subtidal. (10)

CYPRAEIDAE

Blasicrura (*Derstolida*) *coffea* (Sowerby): Taiwan; Hainan; Xisha (=Paracel Islands). Below low intertidal. (3, 18)

B. (*D.*) *hirundo neglecta* (Sowerby): Hainan; South China Sea; Xisha (=Paracel Islands). Low intertidal. (1, 18)

B. (*Blasicrura*) *interrupta* (Gray): South China Sea. (1)

B. (*B.*) *quadrimaculata* (Gray): Hainan; South China Sea. Subtidal. (1, 3, 18)

B. (*D.*) *stolida* Linnaeus [*Cypraea stolida*]: Penghu (Taiwan); Hainan; Xisha (=Paracel Islands). Low intertidal. (3, 18, 55)

B. (*D.*) *ursellus* (Gmelin): Xisha (=Paracel Islands). Subtidal. (3)

Chelycypraea testudinaria (Linnaeus) [*Cypraea testudinaria*]: Taiwan; Xisha (=Paracel Islands). Below low intertidal. (1, 55)

Cribraria chinensis (Gmelin) [*C.* (*Ovatipsa*) *chinensis*]: Taiwan; Xisha (=Paracel Islands). Low intertidal. (1, 40, 53)

C. (*C.*) *cribraria* (Linnaeus) [*Cypraea cribraria*]: Taiwan; Hainan; Xisha (=Paracel Islands). Intertidal. (3, 53)

C. (*Talostolida*) *teres* (Gmelin) [*C. tere*]: Taiwan; Hainan; Xisha (=Paracel Islands). (1, 55)

Cypraea artuffeli Jousseaume: Taiwan. (58a)

C. aurantium Gmelin: Taiwan. (58a)

C. boivinii Kiener: Taiwan. Subtidal. (55)

C. (*Lyncina*) *carneola* (Linnaeus): Taiwan; South China Sea; Xisha (=Paracel Islands). Low intertidal. (1, 18)

C. childreni (Gray): Taiwan. Subtidal. (55)

C. chinensis Gmelin: Taiwan. Subtidal. (55)

C. contaminata Sowerby: Taiwan. (58a)

C. depressa Gray: Penghu (Taiwan). Subtidal. (55)

C. eglantina Duclos: Penghu (Taiwan). Subtidal. (55)

C. guttata Gmelin: Taiwan. Subtidal. (53, 55)

C. hirasei Roberts: Taiwan. (53)

C. longfordi (Kuroda): Taiwan. (53)

C. lulea Gmelin: Taiwan. (55)

C. (*L.*) *lynx* (Linnaeus): Taiwan; Hainan; Xisha (=Paracel Islands). Low intertidal. (1, 53)

C. microdon Gray: Taiwan. (58a)

C. pallidula Gaskoin: Taiwan. Subtidal. (55)

C. talpa Linnaeus: Taiwan. Intertidal to subtidal. (55)

C. teramachii (Kuroda): Taiwan. (53)

C. (*C.*) *tigris* Linnaeus [*C. tigris*]: Taiwan; Hainan; Xisha (=Paracel Islands). Low intertidal. (1, 53)

C. vitellus (Linnaeus) [*C. vitellus*]: Taiwan; South China Sea. Low intertidal. (1, 53)

Erosaria (*Ravitrona*) *caputserpentis* (Linnaeus) [*Cyparea caputserpentis*]: Taiwan; South China Sea. Intertidal. (11, 53, 55)

E. erosa (Linnaeus) [*Cypraea erosa*]: Taiwan; south of Guangdong; Xisha (=Paracel Islands). Below low intertidal. (18, 40, 55)

E. (*R.*) *helvola* (Linnaeus) [*Cypraea helvova*]: Taiwan; Hainan; Xisha (=Paracel Islands). Low intertidal. (11, 40, 53)

E. inocellata: Hong Kong. Intertidal. (78a)

E. (*R.*) *labrolineata* (Gaskoin) [*E. labrolineata helenae, Cypraea labrolineata*]: Taiwan; Hainan; Nansha (=Spratly Islands). Intertidal, subtidal. (11, 40, 55)

E. (*E.*) *miliaris* (Gmelin) [*E. miliaris, Cypraea miliaris*]: South of Zhejiang; Taiwan Strait. Low intertidal to more than 80 m water depth. (11, 40, 55)

E. (*E.*) *poraria* (Linnaeus) [*Cypraea poraria*]: Taiwan; Hainan; Nansha (=Spratly Islands). Intertidal to subtidal. (1, 11, 40, 55)

Erronea (*E.*) *caurica* (Linnaeus) [*Cypraea caurica*]: Taiwan; Guangxi; Hainan; Xisha (=Paracel Islands). (1, 40, 55)

E. (*E.*) *cylindrica* (Born) [*Cypraea cylindrica*]: Penghu (Taiwan); South China Sea. Subtidal. (1, 55)

E. (*E.*) *errones* (Linnaeus) [*Cypraea errones*]: Taiwan; Guangxi; Hainan; Xisha (=Paracel Islands). (1, 40, 55)

E. (*Adusta*) *hungerfordi* (Sowerby): East China Sea; Taiwan. Subtidal. (4, 55)

E. onyx (Linnaeus) [*Cypraea onyx*]: South of Pingtan (Fujian); Taiwan. Below low intertidal. (1, 55)

E. (*A.*) *pulchella* (Swainson) [*Cypraea pulchella*]: Taiwan Strait; South China Sea. Subtidal. (1, 11, 53, 55)

E. (*A.*) *pyriformis* (Gray): South China Sea. Subtidal. (1, 11)

E. (*A.*) *walkeri* (Sowerby) [*Cypraea walkeri*]: Taiwan; South China Sea. Subtidal. (1, 11, 55)

Luria isabella (Linnaeus) [*L.* (*Basilitrona*) *isabella, Cypraea isabella*]: Taiwan; Hainan. Intertidal. (1, 55)

Mauritia (*Arabica*) *arabica* (Linnaeus) [*M. arabica, Cypraea arabica*]: South of Dongshan (Fujian). Low intertidal. (1, 11, 53, 64)

M. (*Leporicypraea*) *mappa* (Linné) [*Cypraea mappa*]: Taiwan; Xisha (=Paracel Islands). Below low intertidal. (1, 53, 55)

M. mauritiana (Linnaeus) [*M.* (*M.*) *mauritiana, Cypraea mauritiana*]: Taiwan; Hainan; Xisha (=Paracel Islands). Below low intertidal. (1, 3, 53, 64)

M. (*A.*) *scurra* (Gmelin) [*Cypraea scurra*]: Penghu (Taiwan); South China Sea; Xisha (=Paracel Islands). Intertidal. (3, 55)

Monetaria annulus (Linnaeus) [*Cypraea annulus*]: Taiwan; Hainan; Xisha (=Paracel Islands). Intertidal. (1, 40, 53)

M. (*M.*) *moneta* (Linnaeus) [*Cypraea moneta*]: Taiwan; Hainan; Xisha (=Paracel Islands). Intertidal. (1, 40, 53)

Palmadusta (*P.*) *asellus* (Linnaeus) [*Cypraea asellus*]: Taiwan; Hainan; Xisha (=Paracel Islands). (3, 40, 53)

P. (*P.*) *clandestina* (Linnaeus): Taiwan; Hainan; Xisha (=Paracel Islands). (3, 40)

P. (*Melicerona*) *felina* (Gmelin) [*Cypraea felina*]: Taiwan; Hainan; Xisha (=Paracel Islands). Intertidal to subtidal. (55)

P. (*M.*) *fimbriata maromorata* (Schroter) [*Cypraea fimbriata*]: Taiwan; Xisha (=Paracel Islands). Subtidal. (55)

P. gracilis japonica Schilder [*Cypraea gracilis*]: South of Zhejiang; Taiwan. Intertidal to subtidal. (1, 55)

P. (*P.*) *punctata* (Linnaeus) [*Cypraea punctata*]: Taiwan. Subtidal. (40, 55)

P. (*P.*) *ziczac* (Linnaeus) [*Cypraes ziczac*]: Taiwan; Xisha (=Paracel Islands). Subtidal. (55)

Paulonaria beckii (Gaskoin): Taiwan; Xisha (=Paracel Islands). Below low intertidal. (3)

Pustularia bistrinotata Schilder & Schilder: Taiwan; Hainan; Xisha (=Paracel Islands). Subtidal. (40, 55)

P. cicercula (Linnaeus): Taiwan; Hainan; Xisha (=Paracel Islands). Subtidal. (40)

P. globulus (Linnaeus) [*Cypraea globulus*]: Taiwan; Hainan; Xisha (=Paracel Islands). Subtidal. (40, 55)

Staphylaea (*S.*) *limacina* (Lamarck): Taiwan; Hainan; Xisha (=Paracel Islands). Low intertidal. (3, 55)

S. nucleus (Linnaeus): Taiwan; Hainan; Xisha (=Paracel Islands). Low intertidal. (40)

S. staphylaea (Linnaeus) [*S. staphylaea staphylaea, Cypraea staphylaea*]: Taiwan; Hainan; Xisha (=Paracel Islands); southern South China Sea. Above low intertidal. (18, 40, 55)

Talparia argus (Linnaeus) [*Cypraea argus*]: Taiwan; Xisha (=Paracel Islands); Nansha (=Spratly Islands). Below low intertidal. (1, 11, 55)

T. (*T.*) *talpa* (Linnaeus): Taiwan; Hainan. (1)

CASSIDIDAE

Casmaria erinaceus (Linnaeus): Taiwan; Xisha (=Paracel Islands. Subtidal. (15, 55)

C. ponderosa nipponensis Abbott: East China Sea. Subtidal. (15)

C. ponderosa ponderosa (Gmelin): Taiwan; Hainan; Xisha (=Paracel Islands). Subtidal. (15, 18, 55)

Cassis cornuta (Linnaeus): Taiwan; Xisha (=Paracel Islands). Below low intertidal. (15, 55)

Cypraecassis rufa (Linnaeus) [*Cassis rufa*]: Taiwan; Dongsha (=Pratas Islands). (15, 53, 58)

Galeodea echinophorella Hirase: Taiwan. Subtidal. (58)

Morum cancellatum (Sowerby) [*M.* (*Onimusiro*) *cancellatum*]: East China Sea; South China Sea; Taiwan. Subtidal. (15, 53)

M. grande (A. Adams): Taiwan. Subtidal. (55)

Phalium areala (Linnaeus): Taiwa; Hainan. Intertidal to more than 10 m water depth. (15, 55)

P. (*P.*) *bandatum* (Perry) [*P. bandatum*]: Taiwan Strait; Guangdong; Beibuwan (=Gulf of Tongking); Xisha (=Paracel Islands). Intertidal. (15, 74)

P. bisulcatum (Schuber & Wangner) [*P. (Semicassis) bisulcatum, P. persimilis, Semicassis japonica, S. pila*]: East China Sea; South China Sea; Taiwan. Subtidal. (15, 55, 58)

P. (P.) decussatum (Linnaeus) [*P. decussatum*]: Taiwan; Guangdong; Hainan. Subtidal. (15, 18, 58)

P. glabratum bulla (Habe) [*P. bulla, P. (Semicassis) glabratum bulla*]: Taiwan Strait. Subtidal. (55, 58)

P. glaucum (Linnaeus) [*P. (P.) glaucum*]: Taiwan; Guangdong; Beibuwan (=Gulf of Tongking); Hainan. Intertidal to water depth 20 m. (15, 53, 55, 74)

P. (Xenophalium) inornatum (Pilsbry): Taiwan Strait; Guangdong; Dongsha (=Pratas Islands). Subtidal. (15, 58)

P. (Echinophoria) kurodai Abbott: Taiwan. Subtidal. (58)

P. (P.) strigatum breviculum Tsi & Ma: Bohai; Yellow Sea. Below low intertidal. (15, 23)

P. strigatum strigatum (Gmelin) [*P. flammiferum*]: East China Sea; Taiwan Strait; South China Sea. Below low intertidal. (15, 55)

P. (E.) wyvillei (Watson) [*P. (E.) coronadoi wyvillei*]: Taiwan. Subtidal. (53, 55, 58)

OOCORYTHIDAE

Galeoocory leucodoma (Dall): East China Sea. Subtidal. (51)
Oocorys donghaiensis Xu: East China Sea. Subtidal. (51)
O. elongata Schepman: Nansha (=Spratly Islands). Subtidal. (11)
O. lineata (Schepman): Taiwan. Subtidal. (51)

CYMATIIDAE

Apollon bitubercularia (Lamarck): South China Sea. Subtidal. (27)

A. olivator rubustus (Fulton): South of Zhejiang. Intertidal. (17)

A. perca (Perry) [*A. (Biplex) perca*]: Taiwan; South China Sea. Subtidal. (53, 55)

Biplex pulchra (Gray) [*B. aculeata, B. microstoma*]: Taiwan Strait. Subtidal. (57)

Chaeomia sauliae (Reeve): Taiwan. Subtidal. (55)

Charonia (Septa) hepatica (Röding): Taiwan. Subtidal. (53, 55)

C. tritonis (Linnaeus): Taiwan; Xisha (=Paracel Islands). Below low intertidal. (53)

Cymatium aquatile (Reeve): Taiwan; south of Guangdong. Below low intertidal. (55, 66)

C. caudatum (Gmelin) [*C. (R.) caudatum*]: Taiwan; South China Sea. Subtidal. (67)

C. cingulatum (Lamarck) [*Linatella cingulata, L. poulseni*]: Taiwan; South China Sea. Subtidal. (55)

C. clandestinum (Lamarck) [*Gelagna clandestina, G. succincta*]: Taiwan; Hainan. Subtidal. (55)

C. (Turritriton) comptum (A. Adams): Taiwan Strait. Subtidal. (57)

C. (Ranularia) dunkeri (Lischke): Taiwan. Subtidal. (57)

C. echo: Hong Kong. Intertidal. (78c)

C. (R.) encausticum (Reeve): Taiwan. Subtidal. (57)

C. (R.) exile (Reeve): Penghu (Taiwan). Subtidal. (57)

C. (Septa) flaveolum (Röding): Taiwan. Subtidal. (57)

C. gemmatum (Reeve): Taiwan; Guangdong; Guangxi; Hainan; Xisha (=Paracel Islands). Intertidal. (17)

C. grandimaculatum (Reeve): Taiwan. Subtidal. (55)

C. gutturnium (Röding): Penghu (Taiwan). Subtidal. (55)

C. (T.) kiiensis (Sowerby): Taiwan. Subtidal. (57)

C. (T.) labiosum (Wood): Taiwan. Subtidal. (57)

C. lotarium (Linnaeus): Hainan. (66)

C. lotorium (Linnaeus): Taiwan. Subtidal. (55)

C. moniliferum (A. Adams & Reeve): South China Sea. Subtidal. (17)

C. (Monoplex) mundum (Gould): Taiwan. Subtidal. (57)

C. muricinum (Röding): Taiwan. Subtidal. (55)

C. nicobaricum (Reeve) [*C. (Monoplex) nicobaricum*]: Taiwan; Xisha (=Paracel Islands); Nansha (=Spratly Islands). Intertidal to subtidal. (11, 40, 55, 57)

C. parthenopus (V. Salis) [*C. (Monoplex) parthenopeum echo*]: Taiwan; Hainan. Subtidal. (57, 66)

C. pferifferianus (Reeve) [*C. (Reticutriton) pferifferia*]: Taiwan; Xisha (=Paracel Islands); Nansha (=Spratly Islands). Subtidal. (11)

C. pileare (Linnaeus) [*C. (Monoplex) pileare*]: Taiwan; Guangdong; Guangxi; Hainan. Below low intertidal. (55, 57, 66)

C. pyrum (Linnaeus) [*C. (Ranularia) pyrum*]: Taiwan; South China Sea. Below low intertidal. (55)

C. rubercula (Linnaeus): Taiwan; Hainan; Xisha (=Paracel Islands). Intertidal to subtidal. (40, 55)

C. (R.) sarcostomum (Reeve): Taiwan. Subtidal. (57)

C. sinense (Reeve): South China Sea. Subtidal. (17)

C. (R.) sinensis (Reeve): Penghu (Taiwan). Subtidal. (57)

C. (T.) tenuiliratum (Lischke): Taiwan. Subtidal. (57)

C. (R.) testudinarium (Adams & Reeve): Taiwan. Subtidal. (57)

C. tuberosum (Lamarck): Hainan; Xisha (=Paracel Islands). (40, 66)

C. (T.) vespaceum (Lamarck): Taiwan. Subtidal. (57)

Cyrineum bituberculare (Lamarck): South China Sea. Subtidal. (11)

Distorsio anus (Linnaeus): Taiwan; Guangdong; Guangxi; Hainan; Xisha (=Paracel Islands). Intertidal and subtidal. (18, 40, 55)

D. perdistorta Fulton: Taiwan Strait. Subtidal. (57)

D. reticulata (Röding): Taiwan; South China Sea. Subtidal. (55)

Fusitriton galea Kuroda & Habe: Taiwan. Subtidal. (57)

Gyrineum gyrinum (Linnaeus): Taiwan. Subtidal. (57)

G. pusillum (Broderip): Taiwan. (57)

G. roseum (Reeve): Taiwan. Subtidal. (57)

Sassia semitorta Kuroda & Habe: Taiwan; Taiwan Strait. Subtidal. (57)

BURSIDAE

Bufonaria albivaricosa (Reeve): Taiwan. Subtidal. (56)
B. crumena (Lamarck): Taiwan. Subtidal. (56)

B. echinata (Link) [*B. spinosa*]: Taiwan. Subtidal. (56)

B. margaritula (Deshayes): Taiwan. Subtidal. (56)

B. nobilis (Reeve): Taiwan. Subtidal. (56)

Bursa bufo (Röding) [*B. lissostoma, Tufufa bufo*]: Taiwan. Subtidal. (55, 56)

B. (*B.*) *bufonia* (Gmelin) [*B. bufonia, B. dunkeri*]: Taiwan; Xisha (=Paracel Islands). Subtidal. (16, 55, 56)

B. (*Lampasopsis*) *cruentata* (Sowerby) [*B. cruentata*]: Taiwan; Hainan; Xisha (=Paracel Islands). Subtidal. (16, 55)

B. (*Gyrineum*) *elegans* (Sowerby): South of Fujian. Coastal sea. (16)

B. (*Colubrellina*) *granularis* (Röding) [*B.* (*Dulcerana*) *currugata, B. granularis*]: Taiwan; Guangdong; Hainan; Xisha (=Paracel Islands); Nansha (=Spratly Islands). Intertidal to water depth 18 m. (11, 16, 40, 55)

B. lamarckii (Deshayes): Taiwan. Subtidal. (56)

B. rana (Linnaeus) [*B.* (*Gyrineum*) *rana, Bufonaria rana*]: South of Zhejiang; Taiwan Strait. Subtidal. (11, 16, 55, 56)

B. (*L.*) *rhodostoma* (Sowerby) [*B. rhodostoma*]: Taiwan; Hainan; Xisha (=Paracel Islands). Intertidal to water depth 18 m. (16, 18, 56)

B. rosa (Perry) [*B. mammata, B. siphonata*]: Taiwan. Subtidal. (55)

B. (*B.*) *tuberosissima* (Reeve) [*B. leo, B. tuberosissima*]: Taiwan; Hainan; Xisha (=Paracel Islands). Below low intertidal. (16, 56)

Tutufa bubo (Linnaeus) [*Bursa bubo, B. lampas*]: Taiwan; Hainan; Xisha (=Paracel Islands). Low intertidal. (16, 40, 56)

T. oyamai Habe [*Bursa oyamai*]: East China Sea; South China Sea; Taiwan. Subtidal. (16, 55, 56)

T. ranelloides (Reeve): Taiwan. Subtidal. (56)

T. rubeta (Linnaeus) [*Bursa rubeta*]: Taiwan; Hainan; Xisha (=Paracel Islands). Low intertidal. (16, 55, 56)

TONNIDAE

Eudolium pyriforme (Sowerby): Taiwan Strait. Subtidal. (55)

Galeoocorys leucodoma (Dall): Taiwan Strait. Subtidal. (55)

Malea pomum (Linnaeus): Taiwan; Xisha (=Paracel Islands). Intertidal to water depth 20 m. (19, 53)

Tonna allium (Dillwyn): Taiwan Strait; Guangdong; Hainan. Subtidal. (19, 55)

T. canaliculata (Linnaeus) [*T. cepa*]: Taiwan; Hainan; Xisha (=Paracel Islands). Intertidal to water depth 34 m. (55)

T. chinensis (Dillwyn): East China Sea; South China Sea; Taiwan. Subtidal. (19, 53)

T. lischkeana (Kuster) [*T. marginata*]: Taiwan; Guangdong; Hainan. Subtidal. (19, 53, 55)

T. luteostoma (Kuster): East China Sea. Subtidal. (19)

T. magnifica (Sowerby): East China Sea; South China Sea. Subtidal. (19)

T. olearium (Linnaeus) [*Dolium zonatum*]: South of Zhejiang; Taiwan Strait. Subtidal. (19, 55)

T. perdix (Linnaeus): Taiwan; Xisha (=Paracel Islands). Intertidal to water depth 50 m. (19, 35, 53)

T. sulcosa (Born) [*Dolium salosum*]: South of Min River (Fujian); Taiwan. Subtidal. (19, 55)

FICIDAE

Ficus ficus (Linnaeus): East China Sea; Taiwan Strait; South China Sea; Taiwan. Subtidal. (55, 65)

F. filosus (Sowerby): Taiwan; South China Sea. Subtidal. (55, 65)

F. gracilis (Sowerby): East China Sea; Taiwan Strait; South China Sea. Subtidal. (55)

F. subintermedius (d'Orbigny): East China Sea; Taiwan Strait; South China Sea; Taiwan. Subtidal. (55, 65)

STENOGLOSSA
MURICIDAE

Boreotrophon candelabrum (Reeve): Yellow Sea. Subtidal. (22)

Ceratostoma fournieri (Crosse) [*Tritonalia emarginatus*]: North of Shandong. Low intertidal to 10+ m water depth. (17, 22).

C. inornata (Récluz) [*Ocenebra japonica*]: Bohai; Yellow Sea. Low intertidal. (17, 22)

C. rorifluum (Adams & Reeve): North of Shandong. Low intertidal to several m water depth. (17, 22)

Chicoreus aculeatus (Lamarck): Taiwan; South China Sea. Subtidal. (42, 53)

C. (*Siratus*) *alabaster* (Reeve): Taiwan. Subtidal. (55)

C. asianus Kuroda: East China Sea; Taiwan Strait; South China Sea; Taiwan. Below low intertidal. (42, 55)

C. axicornis (Lamarck) [*C. kawamurai*]: Taiwan; Hainan. Subtidal. (42, 55)

C. banksi (Sowerby): China's seas. (42)

C. brevifrons (Lamarck): China's seas. (42)

C. brunneus (Link) [*C. adustus*]: Taiwan; South China Sea. Below low intertidal. (42, 53, 55)

C. cnissodus (Euthyme): Taiwan Strait. Subtidal. (55)

C. (*Phyllonotus*) *elliscrossi* (Fair): Taiwan. (58a)

C. laciniatus (Sowerby): China's seas. (42)

C. microphyllus (Lamarck): Hainan. Below low intertidal. (42)

C. orientalis Zhang [*Murex* (*Chicorecus*) *sinensis*]: South China Sea. (42)

C. (*S.*) *pliciferoides* Kuroda: Taiwan. Subtidal. (55)

C. problematicus (Lan) [*C.* (*P.*) *problematicus*]: Taiwan. (58a)

C. ramosus (Linnaeus): Penghu (Taiwan); South China Sea. Below low intertidal. (42, 55)

C. rosarius (Perry) [*C. palmarosae*]: Taiwan. Subtidal. (55)

C. saulii (Sowerby): Taiwan. Subtidal. (55)

C. superbus Sowerby [*C.* (*Phyllonotus*) *superbus*]: Taiwan. Subtidal. (55, 58a)

C. torrefactus (Sowerby): Taiwan; South China Sea. (42, 53)

Drupa ambusta (Dall) [*Morula ambusta*]: Taiwan. Intertidal. (58a)

D. anaxares (Kiener) [*Morula anaxares*]: Taiwan; Hong Kong; Hainan. Below low intertidal. (43, 58a)

D. aspera (Lamarck) [*Morula aspera*]: Taiwan; Hainan; Xisha (=Paracel Islands). Below low intertidal. (43, 58a)

D. biconica (Blainville) [*Morula biconica*]: Taiwan; Hainan. Below low intertidal. (43, 58a)

D. borealis (Pilsbry) [*Morula borealis*]: Xisha (=Paracel Islands). Below low intertidal. (43, 58a)

D. cancellata (Quoy & Gaimard): Hainan. Low intertidal. (55)

D. cariosa (Wood) [*Morula (Cronia) cariosa*]: Taiwan. Intertidal. (58a)

D. cavernosa (Reeve) [*Morula cavernosa*]: Taiwan. Intertidal. (58a)

D. clathrata (Lamarck): Taiwan; Hainan; Nansha (=Spratly Islands). Below low intertidal. (24, 43, 55)

D. cornus Röding [*D. elata, Drupella cornus*]: Taiwan; Hainan; Xisha (=Paracel Islands); Nansha (=Spratly Islands). Below low intertidal. (24, 43, 55)

D. fiscella (Gmelin) [*M. fiscella*]: Fujian; Taiwan; Hainan; Xisha (=Paracel Islands). Below low intertidal. (43, 58)

D. fragum (Blainville) [*Drupella fragum*]: Taiwan. Intertidal. (58a)

D. (Cronia) fusca (Kuster) [*M. (Cronia) fusca*]: Taiwan. Intertidal. (58a)

D. granulata (Duclos) [*Morula granulata, Muricodrupa granulata*]: Taiwan; Hainan; Xisha (=Paracel Islands); Nansha (=Spratly Islands). Below low intertidal. (43, 53, 55)

D. grossularia Röding: Taiwan; Hainan; Xisha (=Paracel Islands); Nansha (=Spratly Islands). Below low intertidal. (24, 43, 53, 55)

D. margariticola (Broderip) [*Cronia margriticola, Morula (Cronia) margariticola*]: Taiwan; Guangdong; Hainan; Xisha (=Paracel Islands). Intertidal. (43, 55)

D. marginatra (Blainville) [*Morula marginatra*]: Fujian; Taiwan; Hong Kong; Xisha (=Paracel Islands). Below low intertidal. (43, 58a)

D. morum Röding: Taiwan; Hainan; Xisha (=Paracel Islands); Nansha (=Spratly Islands). Below low intertidal. (24, 43, 53)

D. musiva (Kiener) [*Morula musiva*]: South of Dongshan (Fujian). Intertidal. (43, 58a)

D. ochrostoma (Blainville): Xisha (=Paracel Islands). Below low intertidal. (43)

D. paucimaculata (Sowerby) [*Morula paucimaculata*]: Taiwan. (58a)

D. ricina (Linnaeus) [*D. ricinus*]: Taiwan; Hainan; Xisha (=Paracel Islands). Below low intertidal. (43, 53)

D. ricina form *albolabris* (Blainville): Taiwan. Intertidal. (55)

D. rubusidaeus Röding: Taiwan; Hainan; Xisha (=Paracel Islands); Nansha (=Spratly Islands). Below low intertidal. (24, 43, 53)

D. rugosa (Born) [*D. concatenata, Drupella concatenata, Drupella rugosa*]: Taiwan; Hainan; Xisha (=Paracel Islands); Nansha (=Spratly Islands). Below low intertidal. (24, 43, 55, 58a)

D. spathulifera (Blainville): Taiwan; Xisha (=Paracel Islands); Nansha (=Spratly Islands). Below low intertidal. (24, 43)

D. spinosa (H. & A. Adams) [*Morula spinosa*]: Nansha (=Spratly Islands). Intertidal. (24, 58a)

D. uva Röding) [*Morula uva*]: Taiwan; Hainan; Xisha (=Paracel Islands); Nansha (=Spratly Islands). Below low intertidal. (24, 43, 55)

Eragalatax contractus (Reeve): East China Sea; Taiwan Strait; South China Sea; Taiwan. Intertidal. (18, 55)

Homalocantha anatomica (Perry): Taiwan; Hainan; Xisha (=Paracel Islands). Low intertidal. (53)

H. zamboi (Burch & Burch): Taiwan. (17, 53)

Maculotriton serrialis (Deshayes): Taiwan. (58a)

Marchia bipinnata (Reeve): Taiwan. (53)

M. elongata (Lightfoot): Taiwan. (53)

M. martinetana (Röding) [*M. fenestrata*]: Taiwan. Intertidal. (53, 55)

M. siro (Kuroda) [*Thais siro*]: Taiwan. Subtidal. (53, 55)

Murex aduncospinosus Beck: South of Zhejiang. Subtidal. (42)

M. haustellum Linnaeus [*M. (Haustellum) haustellum*]: Taiwan. Subtidal. (55, 58a)

M. hirasei Hinase: Taiwan. Intertidal. (58a)

M. kiiensis Kira [*M. (Haustellum) kiiensis*]: Taiwan. Subtidal. (55, 58a)

M. nigrispinosus Reeve: Taiwan Strait. Subtidal. (55)

M. pecten (Lightfoot) [*M. triremis*]: East China Sea; Taiwan; South China Sea. Subtidal. (42, 52, 53, 55)

M. rectirostris Sowerby: East China Sea; South China Sea; Taiwan. Subtidal. (42, 55)

M. senkakuensis Shikama: Taiwan. Intertidal. (58a)

M. sobrinus A. Adams: Taiwan. Subtidal. (55)

M. torrefactsu: Hong Kong. Intertidal. (78c)

M. trapa Röding [*M. martinianus*]: South of Zhejiang; Taiwan; Hainan. Below low intertidal. (34, 42, 53, 55)

M. tribulus Linnaeus [*M. ternispina*]: Western Guangdong; Xisha (=Paracel Islands). Subtidal. (24, 42, 46)

M. troscheli Lischke: Taiwan Strait. Subtidal. (55)

Murexsul umbilicatus (Teniso-Woods): Daya Bay (Guangdong). Subtidal. (46a)

Nassa francolinus (Bruguiere): Taiwan; Hainan; Xisha (=Paracel Islands). Low intertidal. (17)

N. serta (Bruguiere): Taiwan. Intertidal. (55)

Pteropurpura aduncas (Sowerby) [*Ocinebrellus fatcatus aduncus, Tritonalia fatcatus*]: Northern Yellow Sea; Taiwan. Low intertidal to 200 m water depth. (22, 55)

P. aduncas form *expansa* Sowerby: Taiwan. (58a)

P. plorator (Adams & Reeve): Taiwan. Subtidal. (55)

Pterynotus alatus (Röding) [*P. pinnatus*]: Dongshan (Fujian); Taiwan; Hainan; South hina Sea. Subtidal. (17, 18, 42, 55)

P. bipinnatus (Reeve): Taiwan. Subtidal. (55)

P. elongatus (Lightfoot): Taiwan. Subtidal. (55)

P. loebbeckei (Kobelt): Taiwan. Subtidal. (58a)

P. (Naguetia) trigonula (Lamarck): Taiwan. Subtidal. (58a)

P. tripterus (Born): Taiwan. Subtidal. (55)

P. vespertilio (Kina): Taiwan. Subtidal. (58a)

Purpura brunneolabrum (Dall): Hainan; Xisha (=Paracel Islands). (40)

P. persica (Linnaeus): Taiwan. Intertidal. (55)

P. rudolphi Lamarck [*P. panama*]: Taiwan; Hainan. Below low intertidal. (17, 53, 55)

Rapana bezoar (Linnaeus): East China Sea; Taiwan Strait; South China Sea; Taiwan. Subtidal. (44, 55)

R. rapiformis (Born): East China Sea; Taiwan Strait; South China Sea; Taiwan. Below low intertidal. (44, 55)

R. venosa (Valenciennes) [*R. thomasiana, R. venosa peichiliensis*]: Bohai; Yellow Sea; East China Sea; Taiwan Strait. Intertidal to subtidal. (44, 55)

Siratus alabaster (Reeve): Taiwan. (53)

Thais aculeata Deshayes [*T. (Stramonita) aculeata, T. distinguenda*]: Taiwan. Intertidal. (58a, 64)

T. alouina (Röding) [*T. manchinella*]: Taiwan; Hainan. Below low intertidal. (17, 18, 55)

T. armigera (Link) [*T. (Stramonita) armigera, Purpura armigera*]: Taiwan; Hainan; Xisha (=Paracel Islands). Below low intertidal. (17, 40, 55, 58a)

T. bronni Dunker: East China Sea; Taiwan. Intertidal. (17, 55)

T. bufo Lamarck: Taiwan; Hainan. Intertidal. (17, 55)

T. carinifera: Hong Kong. Intertidal. (78c)

T. clavigera Kuster: From Bohai and Yellow Sea to South China Sea; Taiwan. Intertidal. (53, 55)

T. echinata Blainville [*T. (Mancinella) echinata*]: Taiwan; Hainan. Intertidal. (17, 55, 58a)

T. echinulata (Lamarck) [*Mancinella echinulata, T. (M.) echinulata*]: Taiwan; Guangdong; Hainan. Intertidal. (18, 53, 55, 58a)

T. gradata Jonas [*T. trigona*]: South of Zhejiang. Intertidal. (64)

T. hippocastanum (Linnaeus) [*T. (Mancinella) hippocastanum, Purpura hippocastanum*]: Taiwan; Hainan; Xisha (=Paracel Islands). Intertidal. (17, 24, 55, 58a)

T. luteostoma (Holten) [*Purpura bronii* v. *suppressa*]: From Bohai and Yellow Sea to South China Sea. Intertidal. (17)

T. mustabilis (Link) [*Cymia mustabilis*]: Taiwan; Guangdong; Hainan. Intertidal. (17, 55)

T. pseudodiadema (Yokoyama): Taiwan. (58a)

T. squamosa (Pease): Taiwan. (58a)

T. tissoti: Hong Kong. Intertidal. (78c)

T. tuberosa (Röding) [*Mancinella tuberosa, T. (M.) tuberosa*]: Taiwan; Hainan; Xisha (=Paracel Islands). Below low intertidal. (18, 53, 55)

Vexilla vexillum (Gmelin): Taiwan. Intertidal. (55)

MAGILIDAE=CORALLIOPHILIDAE

Coralliobia deformis (Lamarck): Taiwan; Hainan; Xisha (=Paracel Islands). Low intertidal. (17)

C. fimbriata (Hinds): Fujian. Subtidal. (47a)

C. monodonta (Blainville): Taiwan; Xisha (=Paracel Islands). (40)

C. violacea (Kiener) [*Coraliophia neritoidea, C. violacea*]: Taiwan; Hainan; Xisha (=Paracel Islands). Low intertidal. (17, 55)

C. xoaruleia (Lamarck): Hainan. (18)

Latiaxis gyratus (Hinds): Taiwan. Subtidal. (55)

L. lischkeanus (Dunker): Taiwan; South China Sea. Subtidal. (17, 55)

L. mawae (Griffith & Pidgeon): Taiwan; Hainan. Subtidal. (17, 55)

L. (Tolema) pagodus (A. Adams) [*L. gemmatus, L. pagodus*]: Taiwan; Hainan. Subtidal. (24, 55)

Leptoconchus striatus Rüppell: Nansha (=Spratly Islands). (24)

Magilus antiquatus Montfort: Taiwan; Hainan; Xisha (=Paracel Islands). Below low intertidal. (17, 55)

Rapa bulbiformis Sowerby [*Coralliophila bulbiformis*]: Taiwan; Hainan; Xisha (=Paracel Islands). Subtidal. (40, 60)

R. rapa (Linnaeus): Taiwan. Subtidal. (55)

Rhizochilus madreporarum (Sowerby): Xisha (=Paracel Islands); Nansha (=Spratly Islands). Low intertidal. (17, 24)

PYRENIDAE

Anachis cf. *liocyma* (Pilsbry): Fujian. Subtidal. (47a)

Columbella turturina Lamarck: Taiwan; Xisha (=Paracel Islands); Nansha (=Spratly Islands). Intertidal. (24)

C. varians Sowerby: Taiwan; Hainan; Xisha (=Paracel Islands). Intertidal. (17)

C. versicolor Sowerby: Taiwan; Guangdong; Hainan; Xisha (=Paracel Islands). Intertidal. (17)

Euplica scripta (Lamarck): Taiwan. Intertidal. (60)

E. turarina (Lamarck): Taiwan. Intertidal. (60)

Mitrella bella (Reeve) [*Pyrene martensi*]: From Bohai and Yellow Sea to South China Sea. Intertidal to subtidal. (23)

M. burchardi (Dunker) [*Pyrene bicincta*]: Bohai; Yellow Sea; East China Sea; Taiwan. Intertidal to water depth 20 m. (23, 70)

Pyrene marquesa (Gaskoin): Taiwan; Xisha (=Paracel Islands). (40)

P. punctata (Bruguiere): Taiwan; Hainan; Xisha (=Paracel Islands). Intertidal. (17, 60)

P. testudinaria tylerai (Griffth & Pidgeon): Taiwan; Guangdong; Hainan; Xisha (=Paracel Islands). Intertidal. (17, 18)

Zafra pumila (Dunker): Qingdao (Shandong); Xiamen (Fujian); Daya Bay (Guangdong). Intertidal. (23)

BUCCINIDAE

Babylonia areolata (Link): East China Sea; Taiwan Strait; South China Sea; Taiwan. Subtidal. (17, 53, 55)

B. formosae (Sowerby): Taiwan Strait. Subtidal. (53, 55)
B. japonica (Reeve): Taiwan. Subtidal. (55)
B. lutosa (Lamarck): East China Sea; Taiwan Strait; South China Sea. Subtidal. (55)
B. perforata (Sowerby): Taiwan Strait. Subtidal. (53, 55)
B. rirana Habe: Taiwan. Subtidal. (55)
Buccinium pemphigum (Dall): Northern Yellow Sea. Intertidal to water depth 50 m. (23)
B. undatumplectrum Stimpson: Northern Yellow Sea. Subtidal. (23)
B. yokomaruae Yamashita & Hale: Yellow Sea; Jiangsu. Subtidal. (23)
Cantharus cecillei (Philippi): From Bohai and Yellow Sea to South China Sea. Intertidal to water depth 36 m. (17)
C. giliberti (P. Fischer) [*C.* cf. *nitens*]: Xisha (=Paracel Islands). Intertidal. (62)
C. melanostomus Sowerby: Taiwan. (55)
C. undosus (Linnaeus): Taiwan; Hainan; Xisha (=Paracel Islands). Intertidal. (17)
Engina carolinae (Lamarck): Taiwan. Intertidal. (60)
E. lineata (Reeve): Taiwan. Intertidal. (60)
E. mendicaria (Linnaeus) [*Pusiostoma mendicaria*]: Taiwan; Hainan; Xisha (=Paracel Islands). Intertidal. (53)
E. pulchra (Reeve) [*E. elegans*]: Taiwan; Dongsha (=Pratas Islands); Xisha (=Paracel Islands). Intertidal to subtidal. (53, 55)
E. trifasciata (Reeve): Hainan. Low intertidal. (46a)
E. zonalis (Lamarck): Taiwan. Subtidal. (60)
Hindsia sinensis (Sowerby): East China Sea; Taiwan Strait; South China Sea. Subtidal. (55)
H. suturalis (A. Adams): East China Sea; South China Sea. Subtidal. (17)
Japelion latus (Dall): Yellow Sea. Subtidal. (23)
Nassaria gracilis Sowerby: South China Sea. Subtidal. (38)
Neptunea arthritica cumingii Crosse [*N. cumingii*]: Bohai; Yellow Sea. Subtidal. (23)
Phos roseatum (Hinds): Nansha (=Spratly Islands). Subtidal. (24)
P. senticosus (Linnaeus): East China Sea; Taiwan Strait; South China Sea. Low intertidal to 10+ m water depth. (55)
Pisania (*Japeuthria*) *cingulata* (Reeve): Penghu (Taiwan). Intertidal. (55)
P. ignea (Gmelin): Taiwan; Hainan; Xisha (=Paracel Islands). Intertidal to subtidal. (55)
P. truncata (Hinds): Taiwan; Xisha (=Paracel Islands). Intertidal. (40)
Pollia fumosus (Dillwyn) [*Cantharus* (*Pollia*) *fumosus*]: Taiwan. Subtidal. (55)
P. undosa (Linné) [*Cantharus* (*Pollia*) *undosus*]: Taiwan. Intertidal to subtidal. (55)
Siphonalia fusoides (Reeve): East China Sea. Subtidal. (17)
S. spadicea (Reeve): Yellow Sea; East China Sea. Subtidal. (23)
S. subdilatata Yen: Yellow Sea. Subtidal. (23)

Volutharpa ampullacea perryi (Tay): Bohai; Yellow Sea. (17, 23)

COLUBRARIIDAE

Antemetula elongata (Dall): Nansha (=Spratly Islands). Subtidal. (46)
Colubraria castanea Kuroda & Habe: Nansha (=Spratly Islands). Subtidal. (46)
C. compta (Sowerby): South China Sea. Subtidal. (24)

GALEODIDAE

Hemifusus colosseus (Lamarck): Taiwan Strait. Subtidal. (55)
H. ternatanus (Gmelin): East China Sea; Taiwan Strait; South China Sea. Subtidal. (17)
H. tuba (Gmelin): East China Sea; Taiwan Strait; South China Sea. Subtidal. (17, 55, 64)
Pugilina cochlidium (Linnaeus): Hainan. Subtidal. (46a)

NASSARIIDAE

Nassarius acuminatus: Hong Kong. Intertidal. (78c)
N. albescens (Dunker): Taiwan. Subtidal. (60)
N. caelatus (A. Adams): Daya Bay (Guangdong). Subtidal. (46a)
N. (*Ninoth*) *conoidalis* (Deshayes) [*N. clathratus*, *N. conoidalis*]: Fujian; Taiwan; Guangdong; Nansha (=Spratly Islands). Subtidal. (24, 60, 70)
N. conoratus (Bruguiere) [*N. brum*]: Taiwan, Hainan. Intertidal. (17)
N. crematus: Hong Kong. Intertidal. (78c)
N. (*Niotha*) *cumingii* (A. Adams): South China Sea. (20)
N. (*Tritonella*) *dominulus* (Tapparone-Canefri): Daya Bay (Guangdong). Subtidal. (46a)
N. (*Zeuxis*) *dorsatus* (Röding) [*N. rutilans*]: South China Sea. Subtidal. (24)
N. (*Reticunassa*) *festivus* (Powys) [*N. dealbatus*]: From Bohai and Yellow Sea to South China Sea. Intertidal. (22)
N. fidus (Reeve): Taiwan. Low intertidal. (60)
N. glans (Linnaeus): Guangdong; Guangxi. Below low intertidal. (17)
N. graniferus Kiener: Hainan, Xisha (=Paracel Islands). Intertidal. (17)
N. (*R.*) *gregarius* (Grabau & King): Bohai; Yellow Sea; East China Sea. Intertidal to water depth 28 m. (22)
N. (*Z.*) *hepaticus* (Pulteney): East China Sea; South China Sea. Subtidal. (17, 18, 20)
N. (*R.*) *hiradoensis* (Pilsbry): Bohai; Yellow Sea. Intertidal. (22)
N. horrida (Dunker): Hainan; South China Sea. (18, 20)
N. livescens: Hong Kong. Intertidal. (78b)
N. mangelioider (Reeve): Hainan; South China Sea. (18, 20)
N. margaratiferus (Dunker): From Bohai and Yellow Sea to South China Sea. Intertidal. (48)

N. nodiferus: Hong Kong. Intertidal. (78c)

N. (Z.) olivaceus (Bruguiere): East China Sea; South China Sea. (20)

N. papillosus (Linnaeus): Beibuwan (=Gulf of Tongking); Hainan; Xisha (=Paracel Islands). Low intertidal. (17, 24)

N. pullus: Hong Kong. Intertidal. (78c)

N. reeveanus (Dunker): Taiwan. Intertidal. (60)

N. semiplicatus (A. Adams): East China Sea; South China Sea. Intertidal to subtidal. (13)

N. (Z.) siquijorensis (A. Adams): Yellow Sea. Subtidal. (17, 24)

N. (Alectrion) spiratus (A. Adams): Nansha (=Spratly Islands). Subtidal. (24)

N. (R.) spurcus (Gould): Dalian (Liaoning). Intertidal to water depth 20 m. (22)

N. (Z.) succinctus (A. Adams): From Bohai and Yellow Sea to South China Sea. Intertidal to water depth 50 m. (22)

N. sufflatus (Gould): Fujian. Subtidal. (47a)

N. terefiusculas: Hong Kong. Intertidal. (78c)

N. thersites (Bruguiere): East China Sea; South China Sea. Intertidal. (17)

N. (Varicinassa) variciferus (A. Adams) [*Nassa varicifera*, *Nassarius variciferus*]: From Bohai and Yellow Sea to South China Sea. Intertidal to water depth 40 m. (22)

N. velatus (Gould): Hainan. Intertidal. (46a)

N. (Z.) vitiensis (Hombron & Jaquinot): Nansha (=Spratly Islands). Subtidal. (24)

Zeuxis engylptus (Sowerby): East China Sea; Taiwan Strait; South China Sea. Intertidal. (48)

Z. scalaris: South China Sea. Subtidal. (37)

FASCIOLARIIDAE

Fasciolaria filamentosa (Röding) [*Latirus filamentosa*]: Taiwan; Xisha (=Paracel Islands); Nansha (=Spratly Islands). Below low intertidal. (17, 24, 55)

F. trapezium (Linnaeus) [*Pleuroploca trapezium*]: Taiwan; Xisha (=Paracel Islands). Low intertidal to water depth 20 m. (17, 55)

Fusinus (F.) colus (Linnaeus): Nansha (=Spratly Islands). Subtidal. (46)

F. cf. *forceps* (Perry): Taiwan Strait; South China Sea. Subtidal. (17, 24, 55)

F. fragosusw (Reeve): South China Sea. Subtidal. (24)

F. longicaudus (Lamarck): South China Sea. Subtidal. (24)

F. nicobaricus (Röding): Penghu (Taiwan). Subtidal. (55)

F. perplexus (A. Adams): Taiwan Strait. Subtidal. (24)

F. (Simplicifusus) simplex (Smith): Nansha (=Spratly Islands). Subtidal. (46)

Fusolatirus pilsbryi (Kuroda & Habe) [*Peristernia pilsbryi*]: Nansha (=Spratly Islands). Low intertidal to water depth 150 m. (38)

Granulifusus nipponicus (Smith): Taiwan. Subtidal. (55)

Latirus craticulatus (Linnaeus): Taiwan; Hainan; Xisha (=Paracel Islands). Low intertidal. (17)

L. polygonus (Gmelin): Taiwan. Intertidal. (55)

L. (Latirolagena) smaragdula (Linnaeus) [*L. smaragdula*, *Leucozonia smaragdula*]: Taiwan; Xisha (=Paracel Islands); Nansha (=Spratly Islands). Low intertidal. (24, 55)

L. turritus (Gmelin): Taiwan. Intertidal to subtidal. (55)

Peristernia nassatula (Lamarck): Taiwan; Hainan; Xisha (=Paracel Islands); Nansha (=Spratly Islands). Intertidal. (24)

OLIVIDAE

Ancilla rubiginosa (Swainson): South of Changjiang (=Yangtze River). Subtidal. (59)

Barysipira mammilla (Sowerby): Taiwan Strait. Subtidal. (53, 55)

Oliva annulata (Gmelin): Taiwan. (53, 55)

O. caerulea (Röding): Taiwan. Subtidal. (60)

O. emicator (Meuschen): Taiwan; Xisha (=Paracel Islands). Below low intertidal. (40, 59)

O. episcopalis Lamarck: Taiwan; Xisha (=Paracel Islands). Below low intertidal. (40, 59)

O. hirasei (Kira): Taiwan Strait. Subtidal. (53, 55)

O. ispidula (Linné): Taiwan; Guangdong; Hainan; Xisha (=Paracel Islands). Below low intertidal. (59)

O. miniacea (Röding) [*O. erythrostoma*]: Taiwan; Guangdong; Guangxi; Hainan. Low intertidal to shallow sea. (55, 59)

O. multiplicata Reeve: Taiwan Strait. Subtidal. (55)

O. mustelina Lamarck [*O. mustelina mustelina*]: From Bohai and Yellow Sea to northern South China Sea. Around low intertidal. (17, 55)

O. mustelina concavospira Sowerby: Taiwan; Hainan; Xisha (=Paracel Islands). (55)

O. ornata Marrat [*O. lignaria*]: Taiwan; Hainan. Intertidal to water depth 80 m. (55, 59)

O. reticulata (Röding): Taiwan Strait. Subtidal. (55)

O. sericea (Röding): Taiwan Strait. Subtidal. (55)

O. textilina Lamarck: Taiwan; Xisha (=Paracel Islands). Below low intertidal. (40, 59)

O. todosina Duclos: South China Sea. Subtidal. (59)

Olivella lepta (Duclos) [*O. fulgurata*]: South of Shandong. Low intertidal. (18, 59)

O. plana (Marrat): Guangdong; Hainan. Subtidal. (59)

MITRIDAE

Cancilla filaris (Linnaeus): Taiwan Strait. Subtidal. (55)

Mitra (Cancilla) abyssicola Schepman: Nansha (=Spratly Islands). Subtidal. (24)

M. ambigua Swainson: Taiwan; Hainan; Xisha (=Paracel Islands). Subtidal. (17)

M. aurantia (Gmelin): Taiwan; Guangdong; Hainan. Subtidal. (17)

M. cardinalis (Gmelin): Taiwan Strait. Subtidal. (53, 55)

M. chinensis Gray: From southern Yellow Sea to South China Sea. Intertidal. (17)

M. (Cancilla) circula Kiener: East China Sea; South China Sea. (20)
M. clathrus (Gmelin): Taiwan. Subtidal. (60)
M. (Chrysame) coffea Schubbert & Waginer [*M. coffea*]: Taiwan; Xisha (=Paracel Islands). Subtidal. (40, 55)
M. (Strigatella) decurtata Reeve: Taiwan. Intertidal. (55)
M. (Dibaphus) edentula (Swainson): Taiwan. Intertidal. (55)
M. (C.) ferruginea Lamarck: Taiwan; Hainan; Xisha (=Paracel Islands). Intertidal. (40, 53, 55)
M. (Cancilla) flammea Quoy & Gaimard: East China Sea; South China Sea. (20)
M. (C.) imperialis Röding: Taiwan, Xisha (=Paracel Islands). Subtidal. (40, 55)
M. (Cancilla) isabella (Swainson) [*Cancilla isabella*]: Taiwan Strait; Guangdong; Nansha (=Spratly Islands). Subtidal. (17, 24, 55)
M. (Strigatella) litterata Lamarck: Taiwan; Hainan; Xisha (=Paracel Islands). Intertidal. (40, 53, 55)
M. (Strigatella) lutea Quoy & Gaimard: Taiwan; Xisha (=Paracel Islands). (40)
M. mitra (Linnaeus) [*M. (M.) mitra*]: Taiwan; Xisha (=Paracel Islands). Subtidal. (17, 40, 53, 55)
M. morchii A. Adams [*M. taiwanica*]: Taiwan Strait. Subtidal. (55)
M. papalis (Linnaeus): Taiwan Strait; Xisha (=Paracel Islands); Nansha (=Spratly Islands). Below low intertidal. (17, 24, 53, 55)
M. (Scabricola) papillio (Link): Taiwan; Xisha (=Paracel Islands). Subtidal. (40, 60)
M. (Strigatella) paupercula (Linnaeus): Taiwan; Xisha (=Paracel Islands). Intertidal. (40, 53, 55)
M. pellisserpentis Reeve: Taiwan. Subtidal. (60)
M. (Cancilla) praestanstissima Röding: Nansha (=Spratly Islands). Subtidal. (24)
M. proscissa Reeve: Taiwan; Guangdong; Hainan; Xisha (=Paracel Islands). Subtidal. (40)
M. puncticulata Lamarck: Penghu (Taiwan). Intertidal. (55)
M. (Strigatella) retusa Lamarck: Taiwan. Intertidal. (55)
M. scutulata (Gmelin): Taiwan; Guangdong; Hainan; Xisha (=Paracel Islands). Intertidal. (17, 55)
M. stictica (Link): Taiwan. Subtidal. (53, 55)
M. (Strigatella) virgata Reeve: Taiwan; Xisha (=Paracel Islands). (40)
M. (Strigatella) zebra Lamarck: Taiwan. Intertidal. (55)
Mitropifex obeliscus (Reeve): Nansha (=Spratly Islands). Subtidal. (24)
Pterygia crenulata (Gmelin): Taiwan; Guangdong; Hainan. Below low intertidal. (17)
P. dactylus (Linnaeus): Taiwan Strait. Subtidal. (55)
P. fenestrate (Lamarck): Taiwan Strait. Subtidal. (55)
P. nucea (Meuschen): Taiwan Strait; Hainan; Xisha (=Paracel Islands). Subtidal. (40, 55)
P. sinensis (Reeve): Taiwan Strait; Guangdong. Subtidal. (17, 55, 80)
Pusia consanguineum (Reeve): Taiwan. Subtidal. (60)
Vexillum crebriliratum Reeve: Taiwan. Intertidal. (60)
V. formosae (Sowerby): Taiwan Strait. Subtidal. (53, 55, 66)
V. cf. *lubens* (Reeve): Taiwan. Intertidal. (60)
V. ornatum coccineum (Reeve) [*V. ornatum*, *V. coccineum*]: Taiwan; Hainan. Subtidal. (17, 53, 55)
V. patricarchalis (Gmelin): Taiwan. Intertidal to subtidal. (55)
V. plicarium (Linnaeus): Hainan. Subtidal. (55)
V. rugosum (Gmelin): Hainan. Intertidal. (46a)
V. vulpeculum Linnaeus: Hainan. Subtidal. (17)

VASIDAE

Afer cumingii (Reeve): Taiwan Strait. Subtidal. (55)
Vasum ceramicum (Linnaeus): Taiwan; Xisha (=Paracel Islands). Below low intertidal. (17, 55)
V. turbinellum (Linnaeus): Taiwan; Hainan; Xisha (=Paracel Islands). Intertidal to several m water depth. (17, 55)

HARPIDAE

Harpa amouretta Röding: Taiwan; Xisha (=Paracel Islands); Nansha (=Spratly Islands). Below low intertidal. (17, 24, 55)
H. articularis Lamarck: Taiwan Strait; South China Sea. Subtidal. (24, 55)
H. conoidalis Lamarck: Taiwan; Guangdong. Below low intertidal. (17)
H. harpa (Linnaeus): Taiwan Strait. Subtidal. (55)
H. major Röding: Taiwan Strait. Subtidal. (55)
H. nobilis Röding: Xisha (=Paracel Islands). Below low intertidal. (17)

VOLUTIDAE

Cymbiola nobilis (Lightfoot): Taiwan. Subtidal. (53, 55)
Cymbium melo (Solander) [*Melo melo*]: Fujian; Taiwan; Guangdong. Subtidal. (17, 53, 55)
Fulgoraria formosana Azuma: Taiwan Strait. Subtidal. (55)
F. glabra Habe & Kosuge: Taiwan Strait. Subtidal. (55)
F. hamillei (Crosse): Taiwan. Subtidal. (55)
F. rupestris (Gmelin): East China Sea; South China Sea; Taiwan. Subtidal. (17, 53, 55)
Lyria kurodai (Kawamura): South China Sea. Subtidal. (53, 55)
L. taiwanica Lan: Taiwan. Subtidal. (53, 55)
Teramachia dalli (Bartsch): Taiwan. (53)
T. tibiaeformis (Kuroda): Taiwan. (53)

CANCELLARIIDAE

Cancellaria mangeloides Reeve: Bohai; Yellow Sea. Below low intertidal. (60)
Fusiaphera macrospira (Adams & Reeve): South China Sea. Subtidal. (17)
Merica asprella (Lamarck): South China Sea. Subtidal. (17)

M. cf. *oblonga* Sowerby: Fujian. Subtidal. (47a)

M. sinensis (Reeve): South China Sea. Subtidal. (17)

Sydaphera spengleriana (Deshayes): From Bohai and Yellow Sea to South China Sea. Subtidal. (17)

Trigonaphera bocageana (Crosse & Debeaux): From Bohai and Yellow Sea to South China Sea. Subtidal. (17)

T. obliquata (Lamarck): South China Sea. Subtidal. (1)

MARGINIDAE

Marginella dactylus Lamarck: South China Sea. Subtidal. (24)

M. philippinarum Redfield: Xisha (=Paracel Islands). (40)

M. tricincta Hinds: East China Sea; South China Sea. Subtidal. (17, 24)

CONIDAE

Conus achatinus Hwass [*C. monachus*]: Taiwan; Guangdong; Guangxi; Hainan; Xisha (=Paracel Islands). Below low intertidal. (17, 55)

C. aculeiformis Reeve: Nansha (=Spratly Islands). Subtidal. (23, 36, 38)

C. arenatus Hwass: Taiwan; Xisha (=Paracel Islands). Intertidal. (47, 55)

C. armadillo Shikama: Taiwan. (58b)

C. auricomus Huass: Taiwan. (58b)

C. australis Holten [*C. aulicus*]: Taiwan Strait; Nansha (=Spratly Islands). Subtidal. (17, 36, 53, 55)

C. betulinus Linnaeus: Taiwan; Hainan; Xisha (=Paracel Islands). Below low intertidal. (17, 55)

C. bullatus Linnaeus: Taiwan; Xisha (=Paracel Islands). Subtidal. (47, 55)

C. cancellatus Hwass, in Bruguiere [*C. pagodus*]: Taiwan; South China Sea. Subtidal. (23, 36, 55)

C. capitaneus Linnaeus: Taiwan; Hainan; Xisha (=Paracel Islands). Below low intertidal. (40, 55)

C. caracteristicus Fischer: Penghu (Taiwan). Subtidal. (55)

C. catus Hwass: Taiwan; Hainan; Xisha (=Paracel Islands). Below low intertidal. (17, 40, 55)

C. chaldaeus Linnaeus: Taiwan; Hainan; Xisha (=Paracel Islands). Below low intertidal. (40, 55)

C. comatosa Pilsbry: Taiwan; South China Sea. Subtidal. (36)

C. concolor Sowerby: Guangdong. Below low intertidal. (17)

C. coronatus Gmelin: Taiwan; Guangdong; Hainan; Xisha (=Paracel Islands). Intertidal. (55)

C. distant Hwass, in Bruguiere: Taiwan; South China Sea. Below low intertidal. (36, 55)

C. ebraeus Linnaeus [*Virroconus ebraeus*]: Taiwan, Hainan, Xisha (=Paracel Islands). Below low intertidal. (40, 53, 55, 78a)

C. eburneus Hwass: Taiwan; Hainan; Xisha (=Paracel Islands). Intertidal to subtidal. (40, 55)

C. episcopus Hwass: Xisha (=Paracel Islands). Below low intertidal. (40)

C. eximius Reeve: Nansha (=Spratly Islands). Subtidal. (36, 38)

C. figulinus Linnaeus [*C. minimus*]: Taiwan Strait. Subtidal. (55)

C. flavidus Lamarck: Taiwan; Hainan; Xisha (=Paracel Islands). Below low intertidal. (40, 55)

C. fulgetrum Sowerby: Taiwan; Xisha (=Paracel Islands). Intertidal. (47, 55)

C. generalis Linnaeus: Taiwan; Hainan; Xisha (=Paracel Islands). Below low intertidal. (40, 55)

C. geographus Linnaeus: Taiwan; Hainan; Xisha (=Paracel Islands). Below low intertidal. (40, 53, 55)

C. glans Hwass: Taiwan; Guangxi; Hainan; Xisha (=Paracel Islands). Below low intertidal. (40)

C. grangeri Sowerby: Taiwan; Nansha (=Spratly Islands). Subtidal. (36)

C. hirasei (Kira): Taiwan. Subtidal. (55)

C. ichinoseana (Kuroda): Taiwan Strait. Subtidal. (55)

C. imperialis Linnaeus: Taiwan; Xisha (=Paracel Islands). Below low intertidal. (40, 55)

C. insculptus Kiener: South China Sea. Subtidal. (23, 36)

C. kinoshitai (Kuroda): Taiwan Strait. Subtidal. (55)

C. legatus Lamarck: Penghu (Taiwan); Hainan; Xisha (=Paracel Islands). Subtidal. (47, 55)

C. leopardus Röding: Penghu (Taiwan); Xisha (=Paracel Islands). Subtidal. (47, 55)

C. litoglyphus Hwass: Penghu Islands (Taiwan). Subtidal. (55)

C. litteratus Linnaeus: Taiwan; Hainan; Xisha (=Paracel Islands). Below low intertidal. (40, 53, 55)

C. lividus Hwass: Taiwan; Xisha (=Paracel Islands). Below low intertidal. (40, 55)

C. lynceus Sowerby: Taiwan Strait; Nansha (=Spratly Islands). Subtidal. (36, 55)

C. magnificus Reeve: Taiwan; Xisha (=Paracel Islands). (47)

C. magus Linnaeus: Taiwan; Hainan; Xisha (=Paracel Islands). (47)

C. marmoreus cf. *vidua* Reeve [*C. marmoreus*]: Taiwan. Below low intertidal. (40, 55, 58b)

C. miles Linnaeus: Taiwan; Hainan; Xisha (=Paracel Islands). Below low intertidal. (40, 55)

C. miliaris Hwass: Taiwan; Hainan; Xisha (=Paracel Islands). (47)

C. mitratus Hwass: Taiwan. Subtidal. (55)

C. musicus Hwass: Taiwan; Hainan; Xisha (=Paracel Islands). Intertidal. (40, 55)

C. mustelinus Hwass: Taiwan; Hainan; Xisha (=Paracel Islands). Intertidal to subtidal. (47, 55)

C. nussatella Linnaeus: Taiwan; Hainan; Xisha (=Paracel Islands). Intertidal to subtidal. (47, 55)

C. obscurus Sowerby: Taiwan. Intertidal to subtidal. (55)

C. orbignyi Audouin: Taiwan; South China Sea. Subtidal. (17, 55)

C. pennaceus Born [*C. omaria*]: Taiwan Strait. Subtidal. (55)

C. pertusus Hwass: Taiwan; Xisha (=Paracel Islands). (47)

C. planorbis Born: Taiwan. Subtidal. (55)
C. praecellens A. Adams [*C. sowerbii*]: Taiwan; Nansha (=Spratly Islands). Subtidal. (23, 36, 55)
C. pseudorbignyi (Rockel & Lan): Taiwan. Subtidal. (55)
C. pulicarius Hwass: Taiwan; Hainan; Xisha (=Paracel Islands). Intertidal to several m water depth. (40, 55)
C. quercinus Lightfoot: Taiwan Strait. Subtidal. (55)
C. radiatus Gmelin: Penghu (Taiwan). Subtidal. (55)
C. rattus Hwass: Taiwan; Hainan; Xisha (=Paracel Islands). Below low intertidal. (40, 55)
C. recluzianus Bernardi: Taiwan; South China Sea. Subtidal. (36)
C. scabriusculus Dillwyn: Taiwan; Xisha (=Paracel Islands). Below low intertidal. (40, 55)
C. schepmani Fulton: Nansha (=Spratly Islands). Subtidal. (23, 36)
C. sieboldii Reeve: Taiwan. Subtidal. (55)
C. sponsalis Hwass: Taiwan; Hainan; Xisha (=Paracel Islands). Low intertidal. (40, 55)
C. stercusmuscarum Linnaeus: Penghu (Taiwan). Subtidal. (55)
C. striatus Linnaeus: Taiwan; Hainan; Xisha (=Paracel Islands). Below intertidal. (40, 55)
C. stupa (Kuroda): Taiwan. (53)
C. stupella (Kuroda): Taiwan. Subtidal. (53, 55)
C. sugimotonis (Kuroda): Taiwan. Subtidal. (55)
C. sulcatus Hwass: Taiwan Strait; South China Sea. Subtidal. (17, 55)
C. teramachii (Kuroda): Taiwan. Subtidal. (53, 55)
C. terebra Born: Taiwan. Subtidal. (55)
C. tessulatus Born: Taiwan; Hainan. Intertidal to subtidal. (40, 55)
C. textile Linnaeus: Taiwan; Guangdong; Guangxi; Hainan; Xisha (=Paracel Islands). Below low intertidal. (55)
C. tribblei Walls: Taiwan Strait. Subtidal. (55)
C. tulipa Linnaeus: Taiwan; Hainan; Xisha (=Paracel Islands). Intertidal to subtidal. (47, 55)
C. urashimanus Kuroda: Taiwan Strait. Subtidal. (53, 55)
C. varius Linnaeus: Taiwan. Subtidal. (55)
C. vexillum Gmelin: Taiwan; Hainan; Dongsha (=Pratas Islands); Xisha (=Paracel Islands). Low intertidal. (40, 55)
C. virgo Linnaeus: Taiwan; Xisha (=Paracel Islands). Below low intertidal. (40, 55)
C. vitulinus Hwass: Taiwan; Hainan; Xisha (=Paracel Islands). Below low intertidal. (40, 55)
C. voluminalis Reeve: Beibuwan (=Gulf of Tongking); Nansha (=Spratly Islands). Subtidal. (36, 38)

TURRIDAE

Brachystomia kurodai Habe & Kosuge: Taiwan. Subtidal. (72)
B. tawamurai Habe & Kosuge: Taiwan; South China Sea. Subtidal. (37, 72)
B. vexillum Habe & Kosuga [*Brachytoma vexillum*]: East China Sea; Taiwan; Daya Bay (Guangdong). Subtidal. (46a, 72)
Clathrodrillia jeffreysii (Smith): Taiwan. Subtidal. (60)
Clathurella (*Etremopa*) *subauriformis* (Smith): From Bohai and Yellow Sea to mouth of Changjiang (=Yangtze River). Subtidal. (23)
Crassispira pseudoprinciplis (Yokoyama) [*Clavatula pseudoprinciplis*]: From Bohai and Yellow Sea to South China Sea. Subtidal. (23)
Gemmula congener cosmoi (Sykes): South China Sea. Subtidal. (38)
G. cosmoi (Sykes): South China Sea. Subtidal. (24)
G. deshayesii (Doumel): From Bohai and Yellow Sea to South China Sea. Subtidal. (23)
G. diomedea Powell: South China Sea. Subtidal. (24)
G. kieneri (Doumet): South China Sea. Subtidal. (17, 24, 46)
G. speciosa (Reeve): South China Sea. Subtidal. (17, 24)
G. (*Unedogemmula*) *unedo* (Kiener): Nansha (=Spratly Islands). Subtidal. (46)
Inquistor flavidula (Lamarck) [*Brachytoma flavidutus*]: From Bohai and Yellow Sea to South China Sea. Subtidal. (23)
I. vulpionis Kuroda & Oyama: South China Sea. Subtidal. (24)
Lophiotoma (*L.*) *acufa* (Perry): Nansha (=Spratly Islands). Subtidal. (24)
L. leucotropis (Adams & Reeve) [*Turris leucotropis*]: East China Sea; South China Sea. Subtidal. (17, 35)
Mangelia costulata Dunker: Off mouth of Changjiang (=Yangtze River). Subtidal. (23)
Marshallena philippinarum (Watson): South China Sea. Subtidal. (24)
Nihonia australis (Roissy): South China Sea. Subtidal. (13, 17)
Pinquigemmula luzonica Powell: South China Sea. Subtidal. (24)
Pseudoetrema fortilirata (Smith): From Bohai and Yellow Sea to mouth of Changjiang (=Yangtze River). Subtidal. (23)
Thatcheria mirabilis (Angas): Taiwan; South China Sea. Subtidal. (17, 53, 55)
Turricula javana (Linnaeus) [*Surcula javana*]: Taiwan; East China Sea; South China Sea. Subtidal. (17, 72)
T. nelliae spurius (Hedley): Taiwan; East China Sea; South China Sea. Subtidal. (17, 72)
Turridrupa cincta (Lamarck): Taiwan; Xisha (=Paracel Islands). Below low intertidal. (40, 60)
T. lamberti (Montrouzier): Xisha (=Paracel Islands). (40)
Turris crispa (Lamarck): Beibuwan (=Gulf of Tongking); South China Sea. Subtidal. (17, 74)
Vexitomina chinensis Ma: From Yellow Sea to South China Sea. Subtidal. (9)

TEREBRIDAE

Diplomeriza duplicata (Linnaeus) [*Terebra duplicata*]: Taiwan; Guangdong; Hainan. Intertidal to subtidal. (17, 46a)

Duplicaria evoluta (Deshayes): Daya Bay (Guangdong). Subtidal. (46a)

Hastula diversa (Smith): Fujian. Subtidal. (47a)

H. strigilata (Linnaeus): Taiwan Strait. Subtidal. (55)

H. verreauxi (Deshayes): Taiwan; Hainan. Below low intertidal. (17)

Strioterebrum serotinum (Adams & Reeve): Fujian. Subtidal. (47a)

Terebra affinis Gray: Taiwan; Hainan; Xisha (=Paracel Islands). Below low intertidal. (18, 40)

T. amanda Hinds: Taiwan Strait. Subtidal. (55)

T. areolata (Link): Taiwan Strait. Subtidal. (55)

T. bellanodosa Gralau & King: Bohai; Yellow Sea. Intertidal and subtidal. (22)

T. cerithina Lamarck: Xisha (=Paracel Islands). (40)

T. chlorata Lamarck: Xisha (=Paracel Islands). Intertidal to water depth 15 m. (40)

T. crenulata (Linnaeus): Taiwan Strait; Hainan; Xisha (=Paracel Islands). Low intertidal to several m water depth. (18, 40, 55)

T. cumingii Deshayes: Taiwan Strait; South China Sea. Subtidal. (17, 55)

T. dimidiata (Linnaeus): Taiwan; Hainan; Xisha (=Paracel Islands). Low intertidal to several m water depth. (18, 40, 55)

T. (Noditerebra) dussumieri Kiener [*Duplicaria dussumieri*]: From Bohai and Yellow Sea to South China Sea. Below low intertidal. (22, 55)

T. felina (Dillwyn): Taiwan; Xisha (=Paracel Islands). Below low intertidal. (40)

T. fenestrata Hinds: Nansha (=Spratly Islands). Subtidal. (46)

T. funiculata Hinds: Taiwan; Xisha (=Paracel Islands). (40)

T. guttata (Röding): Hainan; Xisha (=Paracel Islands). Low intertidal to several m water depth. (40)

T. (Diplomeriza) koreana (Yoo): Bohai; Yellow Sea; East China Sea. Intertidal to water depth 40 m. (22)

T. lanceata (Linnaeus): Xisha (=Paracel Islands). Below low intertidal. (40)

T. maculata (Linnaeus): Taiwan; Xisha (=Paracel Islands). Low intertidal. (40, 55)

T. penicillata Hinds: Hainan; Xisha (=Paracel Islands). (40)

T. (Triplostephanus) pereoa Nomura: Bohai; Yellow Sea; East China Sea; Taiwan. Intertidal to water depth 50 m. (22)

T. prestiosa Reeve: South China Sea. Subtidal. (17)

T. subulata (Linnaeus): Taiwan; Hainan; Xisha (=Paracel Islands). Low intertidal to 10's of m water depth. (18, 40, 55)

T. torguata Adams & Reeve: East China Sea. Subtidal. (13)

T. triseriata Gray: South of Zhejiang; Taiwan Strait. Subtidal. (17, 55)

TRIPHORIDAE

Triphora alveolatus A. Adams & Reeve: Qingdao (Shandong). Intertidal. (23)

T. (Cautor) hungerfordi (Sowerby): From Bohai and Yellow Sea to South China Sea. Intertidal. (23)

REFERENCES*

(1). Ma Siutung, 1962. Cowrie from the China Coasts. Acta Zoologica Sinixa, Vol. 14 supp.:1–30.

(2). Ma Siutung, 1976. Notes on Chinese species of the Family Strombidae (Prosobranchia, Gastropod). Studia Marina Sinica, 11:355–371.

(3). Ma Siutung, 1979. Some new records of Cypraeacea (Prosobranchia) of Xisha Islands, Guangdong Province, China. Studia Marina Sinica, 15:93–98.

(4). Ma Siutung, 1982. Two new rare species of the superfamily Cypraeacea collected from continental shelf of the East China Sea. Studia Marina Sinica, 19:85–87.

(5). Ma Siutung, 1983. The discovery of *Capurnus verrucosus* (Linnaeus) in Hainan Island, Guangdong Province, China. Tropic Oceanology, 2(1):78–79.

(6). Ma Siutung and Lin Biping, 1983. Supplementary studies of the species of Strombidae off the China's coast. Tropic Oceanology, 2(2):163–166.

(7). Ma Siutung, 1985. Studies on species of Littorunidae of the Xisha Islands. Guangdong Province, China. Studia Marina Sinica, 24:189–194.

(8). Ma Siutung, 1986. Studies on species of Ovulidae off the China Coasts, I. Subfamily Voline and one new species. Tran. Chinese Soc. Malac., 2:10–18.

(9). Ma Siutung, 1989. A new species of Turridae (Mollusca: Prosbranchia) from the Yellow Sea. Studia Marina Sinica, 30:163–165.

(10). Ma Siutung, 1990. Study on species of Ovulidae off the China Coasts, II. Subfamily Sulcocypraeinae, III. Subfamily Ovulinae. Tran. Chinese Soc. Malac., 3:1–12.

(11). Ma Siutung, 1991. Study on Mesogastropoda of Prosobranchia from the waters of Nansha Islands. Transactions of marine organisms nearby Nansha Islands (I). China Ocean Press, pp:93–109.

(11a). Ma, X. T. and S. P. Zhang. 1993. Study on species of Naticidae off the China coast I. Subfamily Naticinae with four new species. Transactions of the Chinese Society of Malacology, No. 4:1–21.

(11b). Ma, X. T. 1994. Studies on Eratoidae (Gastropoda, Mollusca) of China seas (new species and record). Studia Marina Sinica, 35:239–247.

(12). Yi Jiansheng and Li Fuxue, 1988. Distribution and abundance variation of Littorinidae from Jiulongjiang River in Fujian Province. Acta Oceanologica Sinica, 10(4):492–500.

(13). You Zhongjie et al., 1985. Distribution and fauna of Molluscs (Prosobranchia) from Zhejiang coasts. J. of Zhejiang Fish. Coll., 4(1):25–34.

(14). Jiang Jingxiang et al., 1985. The characteristics of composition and distribution of species on macrobenthos nearby the waters of the continental shelf in the East China Sea. Journal of Oceanography in Taiwan Strait, 2(1): 89–98.

(15). Tsi Chunyen and Ma Suitung, 1980. Etude sur les especies des Cassidae de la China. Studia Marina Sinica, 16:83–96.

(16). Tsi Chunyen and Ma Suitung, 1983. Studies on Chinese species of Bursidae (Mollusca: Gastropoda). Tran. Chinese Soc. Malac., 1:12–22.

(17). Tsi Chunyen, Ma Suitung, Lou Tzekong and Zhang Fusui, 1983. Zoological illustration from China: Molluscs (II). China Science Press.

(18). Tsi Chunyen, Ma Siutung, Xie Yukan and Lin Biping, 1984. A preliminary checklist of marine Molluscs from Hainan Island. Tropical Marine Research (III). China Ocean Press, pp:1–22.

(19). Tsi Chunyen and Ma Siutung, 1984. Studies on the Family Tonnidae (Prosobranchia). Studia Marina Sinica, 23:123–135.

(20). Tsi Chunyen, Ma Siutung, Lin Guangyu and Li Biping, 1984. A preliminary survey on the benthic molluscs from Sanya Harbour, Hainan Island. South China Sea Studia Marina Sinica, 5:77–96.

(21). Tsi Chunyen and Ma Siutung, 1986. Studies on Chinese species of Xenophoridae (Mollusca: Gastropoda). Tran. Chinese Soc. Malac., 2:1–9.

(22). Tsi Chunyen, Ma Siutung and Li Fenglan, 1988. Studies of Family Conidae from the coasts of China (I). Tropical Marine Research (III). China Ocean Press, pp:61 –94.

(23). Tsi Chunyen et al., 1989. Molluscs from Yellow Sea and Bohai Sea. China Agriculture Press.

(24). Tsi Chunyen et al., 1991. Neogastropoda and Heterogastropoda (Mollusa: Gastropoda) from Nansha Islands. Transactions of marine organisms nearby Nansha Islands (I). China Ocean Press, pp:110–129.

(25). Liu Xixing, 1977. Records on new species of Family Naticidae (Mollusca: Gastropoda) from China. Acta Zool. Sinica, 23(3): 303–312.

(25a). Liu Jiahua, 1989. The Limpets of Hong Kong. (manuscript)

(26). Lu Duanhua, 1978. A study on the Haliotidae from the coast of China. Studia Marina Sinica, 14:89–100.

(27). Lu Duanhua, 1982. Studies on Fissurellidae from the continental shelf of the East China Sea. Studia Marina Sinica, 19:89–91.

(27a). Lu Duanhua, 1994. Study on Neritacea of Xisha Islands, China. Proceedings of Malacology No. 4:12

(28). Lu Duanhua, 1983. Studies on Fissurellidae off China Coast. Tran. Chinese Soc. Malac., 1:208–209.

(29). Lu Duanhua, 1983. Studies on Fissurellidae from Xisha Islands. Studia Marina Sinica, 20:205–212.

(30). Lu Duanhua, 1986. The new species of Fissurellidae (Mollusca: Archaeogastropoda). Tran. Chinese Soc. Malac., 2:19–22.

(30a). Lu Duanhua, 1990. A new record on deep water *Pectinoondonta orientalis* Schepman from China seas. Transaction of the Chinese Society of Malacology 3:24–25.

(31). Lu Duanhua, 1991. A preliminary report of Archiaeogastropoda (Mollusca) from Nansha Islands. Chinese Society of Zoology, Transactions of the Chinese Society of Malacology at the third Representative and the fifth Discussion Conference on Oceanology and Limnology.

(32). Lu Duanhua, 1991. The shells from Nansha Islands. Transactions of marine organisms nearby Nansha Islands (I). China Ocean Press, pp:86–87.

(33). Li Rongguan, Wu Qiquan and Jiang Jinxiang et al., 1989. Ecology of benthos of rocky intertidal zone in Daya Bay. In: Collection of Papers on the Marine Ecology of Daya Bay (I). China Ocean Press, pp:82–92.

(34). Li Rongguan and Jiang Jinxiang, 1990. Distribution and fauna of Prosobranchia (Mollusca) in Daya Bay. In: Collection of Papers on the Marine Ecology of Daya Bay (II). China Ocean Press, pp:355–363.

(35). Li Rongguan and Jiang Jinxiang, 1993. Ecological study of Prosobranchia (Mollusca) in subtidal zone, Dongshan Bay, Fujian. Journal of Oceanography in Taiwan Strait, 12(2):171–179.

(35a). Li Rongguan, 1992. Survey on littoral benthos of islands in Xiamen harbour, China. State Oceanic Administration, China, pp:119–125.

(35b). Li Rongguan, 1993. Survey on littoral benthos of islands in East Fujian, China. State Oceanic Administration, China, pp:148–178.

(36). Li Fenglan and Chen Ruiqiu, 1991. Studies of Family Conidae from Nansha Islands. Transactions of marine organisms nearby the waters of Nansha Islands (I). China Ocean Press, pp:137–145.

(37). Li Chubu et al., 1986. The characteristics of the distribution and fauna of Mollusca from northeast South China Sea. Tropical Oceanology, 5(1):20–27.

(38). Li Chubu, 1987. Species lists of Molluscs from Zhengmu Shoal-The reports of comprehensive survey on the southern territory of China. China Science Press, pp:205–212.

(39). Zhang Si et al., 1964. Zoological Illustration from China: Molluscs (I). China Science Press.

(40). Zhang Si, Tsi Chunyen, Ma Siutung and Lou Tzekong, 1975. A checklist of Prosobranchia Gastropoda form Xisha Islands. Studia Marina Sinica, 10:105–128.

(41). Zhang Fusui, 1964. The pelagic molluscs off the China coast. Studia Marina Sinica, 5:189–226.

(42). Zhang Fusui, 1965. Studies on the species of Muricidae off the China coast, I. *Murex*, *Pterynotus* and *Chicoreus*. Studia Marina Sinica, 8:11–24.

(43). Zhang Fusui, 1976. Studies on the species of Muricidae off the China coast, II. Genus *Drupa*. Studia Marina Sinica, 11:333–350.

(44). Zhang Fusui, 1980. Studies on the species of Muricidae off the China coast, III. *Rapana*. Studia Marina Sinica, 16:113–123.

(45). Chen Saiying et al., 1980. Studies on Molluscan fauna of Nanji Islands, the East China Sea. Acta Zoologica Sinica, 26(2):171–177.

(46). Chen Ruiqiu et al., 1989. Species lists of shells from the coast of Nansha Islands. The Report of Comprehensive Survey nearby the Waters of Nansha Islands. China Science Press, pp:755–758.

(46a). Xu, Z. C. and Z. G. Chen, et al. 1993. The Shellfish Primary Colours Illustrated Handbook of Hainan Island. China Science Press, 125pp.

(47). Zhou Jinming, 1983. Notes on *Conus* (Molluscua: Gastropoda) from Xisha Islands. South China Sea Studia Marina Sinica, 4:97–108.

(47a). Zhou Shiqiang and Li Rongguan, 1994. Survey on littoral benthos of islands of Fujian, China. China Ocean Press, pp:182–188.

(48). Hong Junchao et al., 1984. A preliminary analysis on the Mollusca fauna of Yushan Islands' tide zone, the East China Sea. Jour. Zhejiang Fish. Coll., 3(1):21–27.

(49). Zhao Ruyi et al., 1982. Marine Molluscs from Dalian. China Ocean Press.

(50). Gao Shihe and Li Fuxue, 1985. Community ecology of ground-dwelling macrobenthos in mangrove swamps around the Mouth of Jiulong River. Journal of Oceanography in Taiwan Strait, 4(2):179–190.

(51). Xu Fengshan, 1989. One new species of Oocorysidae from the East China Sea. Ocean. Limn. Sinica, 20(6):514–526.

(52). Huang Zongguo et al., 1984. The ecological and biological characteristics of the distribution of *Crepidulaonyx* (Mollusca: Gastropoda) in Hong Kong. Acta Oceanologica Sinica, 5:827–839.

(53). Lai Kinyang, 1978. Shells in Taiwan. Nat. Sci. Cult. Press.

(54). Lai Kinyang, 1986. Marine Gastropods of Taiwan (I). Published by Taiwan Provincial Museum.

(55). Lai Kinyang, 1987. Marine Gastropods of Taiwan (II). Published by Taiwan Provincial Museum.

(56). Lai Kinyang, 1987. Bursidae (Gastropoda) in Taiwan. Acta Malacology, 13:15–17.

(57). Lai Kinyang,1989. The Family Cymatiidae of Taiwan. Acta Malacology, 14:107–128.

(58). Lai Kinyang, 1990. The Family Cassidae of Taiwan. Acta Malacology, 15:23–34

(58a). Lai Kinyang, 1987. Studies on Muricidae (Gastropoda) of Taiwan. The Pei-Yu, No. 11:12–19.

(58b). The Malocological Society of China, 1986a. The List of Marginidae (Gastropoda). The Pei-Yu, No. 10:9–15.

(58c). The Malocological Society of China, 1986b. The List of Mollusca (freshwater) in Taiwan. The Pei-Yu, No. 10:1–8.

(59). Lou Tzekong, 1965. A study on the Olividae of the China coast. Studia Marina Sinica, 7:1–12.

(60). Tang Tianxi et al., 1986. A survey of marine shells (Gastropods and Bivalve) along the northeast coast of Taiwan. Acta Malacology, 12: 27–47.

(61). Dong Zhengzhi, 1982. A new species of the genus *Collonia* (Turbinidae) from the continental shelf of the East China Sea. Studia Marina Sinica, 19:81–83.

(62). Dong Zhengzhi, 1983. Taxonomic study of the Trochacea of Xisha Islands. Studia Marina Sinica, 20:185–192.

(63). Dong Zhengzhi, 1991. Family Trochacea from the waters of Nansha Islands. Transactions of marine organisms nearby the waters of Nansha Islands (I). China Ocean Press, pp:146–149.

(64). Cai Yingya et al., 1979. Introduction of Malacology. Shanghai Science and Technology Press.

(65). Cai Yingya et al., 1981. On some species of Ficidae along the coast of the East and South China Seas. Journal of Zhanjiang Fisheries College, 1:11–16.

(66). Cai Yingya et al., 1982. A preliminary study on the Cymatiidae (Prosobranchia) from Guangdong and Guangxi coasts. Journal of Zhanjiang Fisheries College, 1:1–7.

(67). Cai Yingya et al., 1990. Study on the Molluscan fauna of Nanao Island, the South China Sea. Journal of Zhanjiang Fisheries College, 10(1): 1–13.

(68). Division of Resources, Department of Biology, 1975. Marine shells nearby Xisha Islands. Published by South China Sea Institute of Oceanology, Sinica.

(69). Omatsi, Shigemi 1986. Freshwater and seawater Amaotsutegai clam, *Venus*. Jap. Jour. Malac., 45(3):169–176.

(70). Habe, Tadashige and Kiyoshi Ito, 1982. Shells of the world in color, Vol. 1. The northern Pacific. Hoikusha Publishing Co. Ltd., Japan.

(71). Habe, Tadashige and Kiyoshi Ito, 1984. Shells of Japan in color, Vol. 2. Hoikusha Publishing Co. Ltd., Japan.

(71a). Kira, T. 1983. Colored Illustrations of the Shells of Japan (Enlarged and Revised Edition). Haikusha Publishing Co., 240pp.

(72). Habe, Tadashige and Kiyoshi Ito, 1984. Shells of the world in color, Vol. 1. The tropical Pacific. Hoikusha Publishing Co. Ltd., Japan.

(73). Tokubei Kuroda et al., 1971. The sea shells of Sagami Bay. Published by Maruzen Co. Ltd., Tokyo.

(73a). Brandt, A. J., 1977. An annnotated checklist of the non-marine Molluscs of Hong Kong. In: Proceedings of First International Workshop on the Malacofuna of Hong Kong and Southern China.

(74). Cai Yingya and Zhang Ziqian, 1988. Molluscs in Beibu Gulf. In: Proceedings on Marine Biology of South China. China Ocean Press, pp:121–141.

(75). Christiaens, J., 1977. The Limpets of Hong Kong with Description of seven new species and subspecies. In: Proceedings of the First International Workshop on the Malacofauna of Hong Kong and Southern China, pp:61–84.

(76). Christiaens, J., 1980. Supplementary notes on Hong Kong Limpets. In: Proceedings of the First International Workshop on the Malacofauna of Hong Kong and Southern China, pp:459–468.

(77). Lily, K. and Y. Tong, 1986. The microgastropods of Hong Kong mangrove. In: Proceedings of the First International Workshop on the Malacofauna of Hong Kong and Southern China, pp:437–448.

(78). Hill, D. S., 1977. The Neritidae (Mollusca: Prosobranchia) of Hong Kong. In: Proceedings of the First International Workshop on the Malacofauna and Southern China, pp:85–99.

(78a). Morton, B. S. 1979. The future of the Hong Kong sea shore. Oxford University Press, Hong Kong.

(78b). Morton, B. 1993. Perturbated soft intertidal and subtidal marine communities in Hong Kong: the significance of scavenging gastropods. In: Proceeding of the second international conference on the marine biology of the South China Sea. Guanschin, China, pp:1–15.

(78c). Morton, B. and J. Morton, 1983. The sea shore ecology of Hong Kong. Hong Kong University Press, 350pp.

(79). Qi, Z. Y. and X. T. Ma, 1983. A new species of Ovulidae from Hong Kong. Proceedings of the First International Workshop on the Malacofauna of Hong Kong and Southern China, pp:125–126.

(79a). Reid, D. G., 1992. The Gastropod family Littorinidae in Hong Kong. In: Proceedings of the Fourth International Marine Biological Workshop: The Marine Flora and Fauna of Hong Kong and Southern China, Hong Kong University Press, pp:187–210.

(80). Tsi, C. Y. and S. T. Ma 1980. A preliminary checklist of the marine Gastropoda and Bivalve (Mollusca) of Hong Kong and Southern China. In: Proceedings of the First International Marine Biological Workshop: The Marine Flora and Fauna of Hong Kong and Southern China, Hong Kong University Press, pp:431–458.

(81). Wells, F.E., 1983. The potamididae (Mollusca: Gastropoda) of Hong Kong with an examination of habitat segregation in a small mangrove system. In: Proc. Sec. Inter. Work. Malac., Hong Kong, pp: 139–154.

*: (1)-(68) in Chinese, (69)-(73) in Japanese
Compiled by Li Rongguan, Jiang Jinxiang, and Lai Jinyang (Lai Kinyang); Edited by Ma Xiutong (S. T. Ma) and Qi Zhongyan (C. Y. Tsi)

OPISTHOBRANCHIA
ENTOMOTAENIATA
PYRAMIDELLIDAE

Actaeopyramis amaena A. Adams: East China Sea. (15)

A. eximia (Lischke): Bohai; East China Sea; Taiwan Strait. (12, 20, 29, 42)

A. lauta (A. Adams): Taiwan Strait. (20)

Amaura japonica (A. Adams): East China Sea. (15)

A. sagamiensis Kuroda & Habe: East China Sea. (15)

Chemnclzia abseida (Dall & Bartsch): Taiwan Strait. (20)

C. acosmia (Dall & Bartsch): Yellow Sea. (25, 42)

C. multigyra (Dunker): Yellow Sea. (42)

Cingulina cingulata (Dunker): Bohai; East China Sea. (15, 25, 42)

Colsyrnola buxeunnea (Gould) [*Obeliscus buxeus*]: Near shore in China's seas. (32)

C. ornata (Gould) [*Turbonilla ornata*]: Taiwan; Hainan; South China Sea. (24, 36, 42)

Iphiana lischke (Dall & Bartsch): East China Sea. (20)

I. tenuisculpta (Lischke): East China Sea. (20)

Lancella bella (Dall & Bartsch): East China Sea; South China Sea. (42)

Leucotina dianae A. Adams: East China Sea. (42)

Mormula mumia (A. Adams) [*Turbonilla mumia*]: Yellow Sea. (25, 42)

M. philippiana (Dunker): Yellow Sea; Taiwan Strait. (20)

M. terebra (A. Adams): Bohai; Yellow Sea; East China Sea. (4, 25, 42)

Odostomia felix Dall & Bartsch: East China Sea. (15)

O. hilgenoerfi Clessin: Taiwan Strait. (20)

O. lecta Dall & Bartsch: East China Sea. (15)

O. lirata Gould: Near shore in China's seas. (32)

O. mauritiana (Dall & Bartsch): Bohai; East China Sea; Taiwan Strait. (15, 20)

O. omaensis Nomura: Bohai; Taiwan Strait. (15, 20, 25, 42)

O. physoides Gould: Near shore in China's seas. (32)

O. planata Gould: Hong Kong. (32)

O. subangulata A. Adams: Bohai; Taiwan Strait; Nansha (=Spratly Islands). (15, 20, 42)

O. subplanata Gould: Hong Kong. (32)

O. tenera A. Adams: Taiwan Strait. (15, 20, 42)

Oscilla tricordata (Nomura): Taiwan Strait. (20)

Otopleura auris cati (Holten): Taiwan; Hainan; Xisha (=Paracel Islands). (4, 36, 42)

Paramormula aspera Kuroda & Habe: Yellow Sea; Taiwan Strait. (15, 20, 42)

Pyramidella nodocincta A. Adams: South China Sea. (4, 42)

P. ventricosa Guerin: Hainan; South China Sea. (4, 36)

Syrnola brunnea (A. Adams): Taiwan Strait. (20)

S. cinctella A. Adams: East China Sea. (15, 42)

S. cinnamomea (A. Adams): East China Sea. (15)

Tiberia ebarana (Yokoyama): East China Sea; Taiwan Strait. (15, 20)

T. pulchella (A. Adams): South China Sea. (24, 36, 42)
T. pusille (A. Adams): Taiwan Strait. (20)
T. terebolla Müller: South China Sea. (42)
Turbonilla caelata Gould: Hong Kong. (32)

CEPHALASPIDEA [BULLOMORPHA]
ACTEONIDAE [PUPIDAE]

Acteon secale Gould: Near shore in China's seas. (32)
A. siebaldii (Reeve): Dapeng Cove (Hong Kong); South China Sea. (14, 34, 42)
A. teramachii Habe: East China Sea; South China Sea. (14, 17)
Bullina lineata (Gray): Hainan; South China Sea. (12, 14, 23, 24, 36, 42)
B. nobilis Habe: East China Sea. (14, 23, 42)
B. truncatula Lin: Hainan. (14, 36, 42)
Obrussena moeshimaensis Habe: South China Sea. (11, 34, 42)
Pupa affinis (A. Adams): Hainan; South China Sea. (12, 36, 42)
P. alveola (Souverbie): Xisha (=Paracel Islands). (5, 14, 42)
P. coccinata (Reeve): Hainan; South China Sea. (14, 36, 42)
P. sinica Lin: East China Sea; South China Sea. (14, 42)
P. solidula (Linnaeus): Taiwan; Hong Kong; Hainan; Xisha (=Paracel Islands). (3, 14, 34, 36, 42)
P. strigosa (Gould): Taiwan; Hainan; South China Sea; Xisha (=Paracel Islands). (12, 14, 34, 36, 42)
P. sulcata (Gmelin): Taiwan; Hainan. (3, 14, 34, 36, 42)
Punctacteon fabreanus (Crosse): Hainan. (14, 36, 42)
P. flammeus (Gmelin): Hainan. (14, 36, 42)
P. kajiyamai Habe: Hainan; South China Sea. (14, 36, 42)
P. kirai (Habe): Yellow Sea; East China Sea; South China Sea. (14, 36, 42)
P. suturalis (A. Adams): East China Sea; South China Sea. (14, 36)
P. virgatus (Reeve): Yellow Sea; East China Sea; South China Sea. (14, 42)
P. yamamurae Habe: Yellow Sea; East China Sea; South China Sea; Hainan. (14, 20, 23, 25, 34, 36, 42)

HYDATINIDAE [APLUSTRIDAE]

Aplustrum amplustre (Linnaeus): Taiwan; Xisha (=Paracel Islands). (12, 23, 42)
Hydatina (Hydatoria) albocincta (Van der Hoeven): East China Sea; Taiwan. (1, 23, 36, 42)
H. physis (Linnaeus): Taiwan; Hainan; South China Sea. (1, 3, 23, 36, 42)
H. (Hydatoria) zonata (Solander): Taiwan; Hainan. (30)

RINGICULIDAE

Pseudoringicula sinensis Lin: Bohai. (7, 23, 25)
Ringicula (Ringiculina) denticulata Gould: Dapeng Cove (Hong Kong); South China Sea. (34)
R. (Ringiculina) doliaris Gould [*R. arctata*]: Bohai; Yellow Sea; East China Sea; Taiwan Strait; South China Sea; Hainan,. (1, 10, 12, 20, 23, 25, 27, 32, 34, 42)
R. (Ringiculina) kurodai Takeyama: Hainan. (10, 12, 34, 36, 42)
R. (Ringicula) niinoi Nomura: East China Sea. (10, 23, 42)
R. (Ringiculinopsis) orientalis Lin: East China Sea. (10, 34)
R. (Ringiculina) pilula Habe: East China Sea. (10, 42)
R. (Ringiculina) propinguan Hinds: South China Sea. (10)
R. (Ringiculina) shenzhenensis Lin: South China Sea. (34, 42)
R. (Ringiculina) teramachii Habe: East China Sea; South China Sea. (10, 42)
R. (Ringiculina) yokoyamai Takeyama: Yellow Sea; East China Sea; South China Sea; Hong Kong. (36, 42)

BULLIDAE [BULLARIIDAE]

Bulla adamsii Menke: Hainan; Xisha (=Paracel Islands). (3, 5, 36, 42)
B. ampulla Linnaeus: Taiwan; Hong Kong; Hainan; South China Sea. (3, 4, 22, 34, 36, 42)
B. ampulla bifasciata Menke: Hainan. (12, 36, 42)
B. ampulla trifasciata Sowerby: Hainan. (12, 36, 42)
B. difficilis Habe: Xisha (=Paracel Islands). (5, 42)
B. orientalis Habe: Taiwan; Hainan; South China Sea; Xisha (=Paracel Islands). (3, 5, 23, 36, 42)
B. vernicosa Gould: East China Sea; South China Sea; Xisha (=Paracel Islands); Taiwan; Hong Kong; Hainan,. (3, 5, 23, 36, 38, 42)

ATYIDAE

Aliculastrum cylindrica (Helbling): South China Sea; Xisha (=Paracel Islands). (3, 5, 23, 36, 42)
Atys naucum (Linnaeus): South China Sea; Xisha (=Paracel Islands); Nansha (=Spratly Islands). (5, 17, 23, 36)
Bullacta exarata (Philippi): Bohai; Yellow Sea; East China Sea; South China Sea; Hainan. (1, 12, 20, 23, 25, 28, 36, 42)
Cryptophthalmus smargdinus (Rüppell & Luckart) [*Lathophthalmus smaragdinus*]: Xisha (=Paracel Islands). (5, 23, 42)
Cylichnatys angustus (Gould): Bohai; Yellow Sea; East China Sea; South China Sea. (23, 24, 25, 36, 42)
Dinatys monodonta (A. Adams): South China Sea. (42)
Haloa binotata (Pilsbry) [*Haminea binotata*]: Taiwan. (30)
H. flavescens (A. Adams): South China Sea. (42)
H. margaitoides Kuroda & Habe: South China Sea. (34)
H. ovalis (Pease): Xisha (=Paracel Islands). (5)
H. rotundata (A. Adams) [*Haminea japonica*]: Yellow Sea. (25)
H. (Vitreohaminoea) vitrea (A. Adams): Hainan; South China Sea. (12, 36, 42)
H. (Sericohaminoea) yamagutii Habe: Taiwan; Hainan; South China Sea. (12, 23, 34, 36, 42)

Lamprohaminoea cymbalum (Quoy & Gaimard): East China Sea; South China Sea. (23, 34, 42)

Liloa curta (A. Adams): Xisha (=Paracel Islands). (12, 42)

L. porcellana (Gould): East China Sea; Taiwan; Hainan; Xisha (=Paracel Islands). (12, 23, 34, 36, 42)

Limulatys angustatus (Gould): East China Sea. (20a)

L. constrictus Habe: Taiwan; Hainan; Xisha (=Paracel Islands). (23, 24, 36, 42)

L. muscarius (Gould): Near shore in China's seas. (30)

L. okamotoi Habe: East China Sea; Nansha (=Spratly Islands). (17, 23, 42)

L. ooformis Habe: South China Sea. (36, 42)

Nipponatys volvulinus (A. Adams): South China Sea. (23, 34, 42)

CYLINDROBULLIDAE

Cylindrobulla xishaensis Lin: Xisha (=Paracel Islands). (6, 23, 42)

AKERATIDAE [ACERIDAE]

Akera constricta Kuroda: Hainan. (3, 12, 23, 36, 42)

RETUSIDAE

Pyrunculus cylindricus Lin: Nansha (=Spratly Islands). (17)

P. lagenula A. Adams [*Soa lagenula*]: Near shore in China's seas. (30)

P. longiformis Lin: East China Sea; South China Sea. (34, 42)

P. phialus (A. Adams): Yellow Sea; East China Sea; South China Sea. (12, 25, 34, 42)

P. pyriformis obesus (Habe): Dapeng Cove (Hong Kong). (34, 42)

P. teramachii Habe: East China Sea; South China Sea. (42)

P. tokyoensis Habe: Yellow Sea; Taiwan Strait; South China Sea. (20, 23, 25, 34, 42)

Retusa (*Coelophysis*) *boenensis* (A. Adams): East China Sea; Hong Kong. (3, 22, 23, 34, 42)

R. (C.) cocillii (Philippi): East China Sea; Hainan; South China Sea. (34, 36, 42)

R. (C.) concentrica (A. Adams): East China Sea; Taiwan Strait; Nansha (=Spratly Islands). (17, 20, 42)

R. (C.) elegantissima (Habe): East China Sea; Taiwan Strait; Dapeng Cove (Hong Kong). (34, 42)

R. (C.) eumicra Crosse: South China Sea. (34)

R. (C.) minima (Yamakawa): Bohai; Yellow Sea; East China Sea; South China Sea. (12, 17, 23, 25, 29, 36)

R. (C.) ovulina Lin: Nansha (=Spratly Islands). (16)

Rhizorus eburnea (A. Adams) [*Volvulella eburnea*]: East China Sea. (23

R. kinokuniama (Habe): Nansha (=Spratly Islands). (16)

R. ovulina (A. Adams): Taiwan Strait; Dapeng Cove (Hong Kong). (20, 23, 34)

R. radiola (A. Adams): Bohai; Yellow Sea; East China Sea. (23, 25, 42)

R. tokunagai (Makiyama): Bohai; Yellow Sea; Taiwan Strait; Dapeng Cove (Hong Kong); South China Sea. (20, 25, 34, 42)

Volvula rostrata A. Adams [*Rhizorus rostrata*]: Bohai. (29)

Volvulopsis chinensis Lin & Wu: Nansha (=Spratly Islands). (17)

TRICLIDAE [SCAPHANDERIDAE]

Abderospira punctulata (A. Adams): Taiwan Strait; Dapeng Cove (Hong Kong); Nansha (=Spratly Islands). (16, 20, 23, 34, 42)

A. umbilicata Habe: East China Sea. (42)

Adamnestia japonica (A. Adams): East China Sea. (42)

A. protracta (Gould): East China Sea; South China Sea. (22, 23, 34)

A. tosaensis Habe: East China Sea; South China Sea. (34, 42)

Cylichna biplicata A. Adams: Near shore in China's seas. (32)

C. melampoides Gould: Near shore in China's seas. (32, 42)

C. operosa Gould: Near shore in China's seas. (20a, 32, 42)

C. pyramidata A. Adams: Bohai. (29)

C. villica Gould: Near shore in China's seas. (32)

Didontoglossa decoratoides Habe: East China Sea. (42)

D. koyasensis (Yokoyama): East China Sea. (42)

Eocylichna braunsi (Yokoyama) [*E. cylindrella*]: Bohai; Yellow Sea; East China Sea; Taiwan. (20, 34, 42)

E. involuta (A. Adams) [*Cylichna involuta*]: Bohai; Yellow Sea; South China Sea; Taiwan. (29, 30)

E. musashiensis (Tokunaga): South China Sea. (34, 42)

E. sigmolabris Habe: East China Sea. (42)

E. soyoae Habe: Dapeng Cove (Hong Kong). (34)

Nipponoscaphander cumingii (A. Adams): Dapeng Cove (Hong Kong); Nansha (=Spratly Islands). (17, 34, 42)

N. japonica (A. Adams): East China Sea; Nansha (=Spratly Islands). (17, 23, 42)

N. teromachii (Habe): Dapeng Cove (Hong Kong). (34)

VOLVATELLIDAE

Volvatella ayakii Hamatai: Nansha (=Spratly Islands). (17)

ACTEOCINIDAE

Acteocina (*Tornatina*) *avenaria* (Watson) [*Tornatina avenaria*]: Bohai. (28, 29)

A. (Tornatina) biplex (A. Adams): South China Sea. (42)

A. coarctata A. Adams [*Tornatina coarctata*]: Taiwan. (30)
A. (*Tornatina*) *decorata* (Pilsbry): Xisha (=Paracel Islands). (5, 23, 42)
A. (*Tornatina*) *exilis* (Dunker): Taiwan Strait; South China Sea. (20, 34, 42)
A. (*Tornatina*) *gordonis* (Yokoyama): East China Sea; South China Sea. (32, 34, 42)
A. (*Tornatina*) *gracilis* (A. Adams): Bohai. (29, 30)
A. (*Tornatina*) *nanshanensis* Lin: Nansha (=Spratly Islands). (17)
A. (*Tornatina*) *orientalis* Lin: East China Sea. (42)
A. (*Truncacteocina*) *oyamai* Kuroda & Habe: Hong Kong; South China Sea. (23, 42)
A. (*Tornatina*) *simplex* (A. Adams): South China Sea. (32, 34, 42)
Decorifera globosa (Yamakawa): Near shore in China's seas. (30)
D. insignis (Pilsbry): Bohai; Yellow Sea; East China Sea. (23, 42)
D. matusimana (Nomura): Bohai; Yellow Sea. (25, 42)

PHILINIDAE

Globophiline huanghenensis Lin & You: Yellow Sea. (18)
Hermania infantilis Habe: South China Sea. (18, 24, 36, 42)
Philine japonica Lischke [*P. argentata*]: Bohai; Yellow Sea; East China Sea; Taiwan Strait; South China Sea. (17, 18, 23, 25, 42)
P. kinglipini Tchang: Bohai; Yellow Sea; East China Sea. (1, 18, 23, 25, 28, 42, 43)
P. kurodii Habe: East China Sea; South China Sea; Taiwan. (18, 23)
P. orientalis A. Adams: Hong Kong; South China Sea. (3, 18, 23, 36)
P. otukai Habe: Yellow Sea; East China Sea; South China Sea; Taiwan. (18, 23, 25, 42)
P. scalpta A. Adams: Bohai; Yellow Sea; South China Sea. (16, 18, 25)
P. vitrea Gould: Taiwan Strait; Hong Kong; Hainan; Nansha (=Spratly Islands). (17, 18, 20, 24, 34, 36, 42)
Yokoyamaia argentata (Gould): Bohai; Yellow Sea; East China Sea. (18)
Y. (*Cheshiphine*) *orientalis* Lin: Hong Kong. (18, 35, 36)

AGLAJIDAE [DORIDIIDAE]

Chelidonura fulvipunctata Baba: Xisha (=Paracel Islands). (5, 23, 42)
C. hirundinina (Quoy & Gaimard): Xisha (=Paracel Islands). (5, 23, 42)
C. inornata Baba: Xisha (=Paracel Islands). (5, 23, 42)
C. tsurugensis Baba: Xisha (=Paracel Islands). (5, 42)

Philinopsis cyaneum V. Martens: Yellow Sea; South China Sea. (20, 25, 42)
P. gigliolii (Tapparone-Canefri) [*Doridium gigliolii*]: Bohai; East China Sea; South China Sea; Hainan. (3, 20, 23, 25, 36, 42)
P. lineolata (Helbling & A. Adams) [*Doridium lineolata*]: Hong Kong; Hainan; South China Sea; Xisha (=Paracel Islands). (3, 23, 38)
P. minor (Tchang) [*Doridium depictum* v. *minor*]: Yellow Sea. (1, 25, 28, 42)

RUNCINIDAE

Metaruncina setoensis (Bata): Qingdao (Shandong). (20a)

GASTEROPTRIDAE

Gasteropteron meckeli Kosse: Near shore in China's seas. Planktonic. (1, 20a)

ANASPIDEA [APLYSINOMORPHA]
APLYSIIDAE

Aplysia (*Varria*) *cornigera* Sowerby: Hainan. (2, 3, 30, 36, 42)
A. (*V.*) *dactylomela* Rang: Hong Kong; South China Sea. (2, 3, 5, 23, 36, 42, 44)
A. (*A.*) *juliana* Quoy & Gaimard: South China Sea. (2, 20, 23, 42)
A. (*V.*) *kurodai* (Baba): Taiwan; Hong Kong; Hainan. (3, 20, 23, 36, 42, 44)
A. (*V.*) *oculifera* Adams & Reeve: East China Sea. (2, 20, 23, 42)
A. (*Pruvolaplysia*) *porvula* Mörch: East China Sea. (2, 20, 23, 42)
A. (*V.*) *pulmonica* Gould: East China Sea; Hainan. (2, 20, 23, 36, 42)
A. (*V.*) *sagamiana* Baba: Hainan. (2, 3, 36, 42)
A. (*V.*) *sinensis* Sowerby: Near shore in China's seas. (1)
Dolabella ecaudata (Rang): Hainan. (12, 42)
D. scapula (Martyn): Hong Kong; Hainan; Xisha (=Paracel Islands). (2, 3, 5, 23, 36, 42)
Dolabrifera dolabrifera (Rang): Taiwan; Hong Kong; Hainan; Xisha (=Paracel Islands). (2, 3, 23, 36, 42)
D. fusca Pease: Hainan. (2, 3, 36, 42)
Notarchus (*N.*) *indicus armatus* Bergh: South China Sea. (2, 42)
N. (*Bursatella*) *leachii cirrosus* Stimpson: East China Sea; Hong Kong; Hainan. (2, 3, 22, 36, 42, 44)
N. (*Stylocheilus*) *longicaudus* (Quoy & Gaimard): Hainan. (2, 3, 23, 36, 42, 44)
N. (*S.*) *risbeci* (Engel): Hainan. (2, 3, 23, 36, 42)
Petalifera punctulata (Tapparone-Canefri): Bohai; Yellow Sea. (2, 23, 25, 42)

P. qingdaonensis Lin: Yellow Sea. (13, 42)
P. ramosa Baba: South China Sea. (13, 25, 42)
Syphonota geographica scripta (Bergh): Hainan. (2, 3, 23, 36, 42)

THECOSOMATA
LIMACINIDAE [SPIRALELLIDAE]

Agadina syimpsoni A. Adams: Yellow Sea; East China Sea; Taiwan Strait; South China Sea. (21, 27)
Limacina bulimoides (d'Orbigny): East China Sea; Taiwan Strait; South China Sea. (21, 23, 27)
L. inflata (d'Orbigny): East China Sea; Taiwan Strait; South China Sea. (21, 23, 27)
L. trochiformis (d'Orbigny): Yellow Sea; East China Sea; Taiwan Strait; South China Sea. (21, 23, 25, 27)

CAVOLINIIDAE

Cavolinia gibbosa (Rang): South China Sea. (21, 23)
C. globulosa (Rang): East China Sea; Taiwan Strait; South China Sea. (21, 23, 27)
C. inflexa v. *labiata* (d'Orbigny): East China Sea; South China Sea. (21, 25)
C. longirostris (Lesueur): East China Sea; Taiwan Strait; South China Sea. (1, 21, 23, 27)
C. longirostris v. *angulata* (Souleyet): East China Sea; South China Sea. (21)
C. tridentata (Forskål): South China Sea. (21, 23)
C. uncinata (Rang): East China Sea; Taiwan Strait; South China Sea. (21, 23, 27)
Creseis acicula Rang: Yellow Sea; East China Sea; Taiwan Strait; South China Sea. (21, 23, 25, 27)
C. chierchiae Boas: East China Sea; Taiwan Strait; South China Sea. (21, 23, 27)
C. clava Rang: East China Sea; Taiwan Strait; South China Sea. (21, 23, 27)
C. virgula Rang: East China Sea; Taiwan Strait; South China Sea. (21, 23, 27)
C. virgula v. *comica* Eschschotlz: Yellow Sea; East China Sea; Taiwan Strait; South China Sea. (21, 23, 25, 27)
Cuvlerina columella (Rang): South China Sea. (21, 23)
Diacria quadridentata (Lesueur): East China Sea; Taiwan Strait; South China Sea. (21, 23, 27)
D. quadridenlata v. *costata* (Pfeffer): East China Sea; South China Sea. (21, 23)
D. trispinosa (Lesueur): Yellow Sea; Taiwan Strait; Taiwan; South China Sea. (21, 23, 27)
Euclia balantium (Rang): South China Sea. (21, 23)
E. pyramidata v. *lanceolata* (Lesueur): South China Sea. (21, 23)
E. pyramidata v. *microcaudata* Zhang: South China Sea. (21, 23)

Hyalocyliz striata (Rang): Yellow Sea; East China Sea; Taiwan Strait; South China Sea. (21, 23, 25, 27)
Styliola subula (Quoy & Gaimard): South China Sea. (21, 23)

PERACLIDAE

Peraclis reticulata (d'Orbigny): Taiwan Strait; South China Sea. (21, 23, 27)

CYMBULIIDAE

Corolla ovata (Quoy & Gaimard): Yellow Sea; East China Sea; Taiwan Strait; South China Sea. (21, 23, 25, 27)
Cymbulia peroni Blainville: South China Sea. (21, 23)
C. tricavernosa Zhang: South China Sea. (21, 23)

DESMOPTERIDAE

Desmopterus papilio Chun: East China Sea; Taiwan Strait; South China Sea. (21, 23, 27)

GYMNOSOMATA [PTEROTA]
PNEUMODERMALIDAE

Abranchaea chinensis Zhang: Yellow Sea; East China Sea; Taiwan Strait; South China Sea. (21, 23, 27)
Pneumoderma atlanticum (Oken): South China Sea. (21, 23)
P. heterocotylum Tesh: South China Sea. (21, 23)
P. mediterraneum (Beneden): South China Sea. (21, 23)
Pneumodermopsis ciliata (Gegenbaur): South China Sea. (21, 23)
P. macrocotyla Zhang: South China Sea. (21, 23)
P. paucidens (Boas): East China Sea; South China Sea. (21, 23)
P. polycatyla (Boas): Yellow Sea; East China Sea; South China Sea. (21, 23, 25)

CLIONIDAE

Notobranchaea macdonaldi Pelseneer: East China Sea; Taiwan Strait; South China Sea. (21, 23, 27)
Paraclione longicaudata (Souleyet): Yellow Sea; East China Sea; Taiwan Strait; South China Sea. (21, 23, 25, 27)
Thliptodon diaphanus (Meisenbeimer): East China Sea; South China Sea. (21, 23)

SACOGLOSSA [ASCOGLOSSA, MONOSTICHOGLOSSA]
STILIGERIDAE

Alderia modesta Lovén: Yellow Sea. (25, 41)
Hermaea dendritica (Alder & Hancock): Bohai; Yellow Sea. (1, 23, 25, 42)

OXYNOEIDAE

Lobiger sagamiensis Baba: Hong Kong. (3)

ELYSIIDAE [PLAKOBRANCHIDAE]

Costasiella pallida Jensen: Hong Kong. (42)
Elysia (*E.*) *atroviridis* Baba: East China Sea; Hong Kong. (20, 31)
E. chilkensis Eliot: Hong Kong. (31)
E. (*E.*) *grandifolia* Kelaart: South China Sea. (12, 36, 42)
E. japonica Eliot: Hong Kong. (31)
E. nigrocapitata Baba: Hong Kong. (31)
E. (*E.*) *trisinuata* Baba: Hong Kong; South China Sea. (3, 30, 36, 42)
E. verrucosa Jensen: Hong Kong. (31)
E. (*E.*) *viridis* (Montagu): Yellow Sea. (1, 25, 42, 43)
Ercolania coerulea Trinchese: Hong Kong. (31)
E. emarginata Jensen: Hong Kong. (31)
E. gopalai (Rao): Hong Kong. (31)
E. tentaculata (Eliot): Hong Kong. (31)
Placobranchus ocellatus Van Hasselt: Hainan; Xisha (=Paracel Islands). (3, 5, 36, 42)

GASCOIGNELLIDAE

Gascoignella aprica Jensen: Hong Kong. (31)

NOTASPIDEA [PLEUROBRANCHOMORPHA]
UMBRACULIDAE

Umbraculum pulchrum Lin: Hong Kong; Hainan. (8, 36, 42)
U. umbraculum (Lightfoot) [*U. sinicum*]: East China Sea; Hainan. (3, 23, 36, 42)

PLEUROBRANCHIDAE

Berthellina delicata (Prease): Xisha (=Paracel Islands). (4, 5, 23, 42)
Euselenops (*Euselenops*) *luniceps* (Cuvier): Hong Kong; Hainan. Subtidal. (3, 4, 20, 23, 24, 36, 42, 44)
Gigantonotum album Lin & Tchang: Hainan. (4, 23, 36, 42)
Oscanius hilli Hedley: Xisha (=Paracel Islands). (1, 5, 23, 36, 42)
O. hirasei (Baba): Xisha (=Paracel Islands). (4, 5, 23, 42)
O. maricus (Forskål): Xisha (=Paracel Islands). (12, 42)
O. xishaensis Lin & Tchang: Xisha (=Paracel Islands). (4, 5, 23, 42)
Pleurobranchaea brock Bergh: East China Sea; Taiwan Strait; Hong Kong; Hainan. (1, 4, 12, 20, 23, 42, 44)
P. novaezealandiae Cheeseman: Bohai; Yellow Sea; East China Sea; Hong Kong; Xisha (=Paracel Islands). (4, 12, 23, 25, 28, 42, 44)

NUDIBRANCHIA
DORIDIDAE

Doris verrucosa Linnaeus: Hong Kong. (38)

HEXABRANCHIDAE

Hexabranchus marginatus (Quoy & Gaimard): Taiwan; Hainan; Xisha (=Paracel Islands). (1, 3, 5, 23, 25, 26, 36, 42)
H. sanguineus Rüppell & Leuckart: Taiwan. (26)

POLYCERIDAE [EUPHURIDAE]

Polycera fuitai Baba [*Palio fuitai*]: Yellow Sea; East China Sea; Hong Kong. (19, 20, 23, 25, 38, 41)
P. japonica Baba: Bohai. (23, 25, 42)

TRIOPHIDAE

Caloplocamus acutus Baba: Yellow Sea; East China Sea. (25, 42)
C. ramosus (Cantraine): Bohai; Yellow Sea; East China Sea; Taiwan Strait; South China Sea; Hong Kong. (1, 19, 20, 23, 25, 28, 42)
Kalinga ornata Alder & Hancock: East China Sea; South China Sea; Taiwan; Hong Kong. (3, 20, 23, 25, 36, 41)
Plocamopherus ceylonicus (Kelaart): Hong Kong. (41)
P. tilesii Bergh: East China Sea; Taiwan Strait; South China Sea; Hong Kong. (23, 25, 38, 41, 42)
Thecacera pennigera (Montagu): Bohai; Yellow Sea. (25, 42)

GYMNODORIDIIDAE

Gymnodoris albe (Bergh): East China Sea; Hong Kong; Hainan. (3, 19, 38, 41)
G. bicolor (Alder & Hancock): East China Sea; Taiwan; Hainan. (19, 36)
G. inornata (Bergh): Hong Kong; Hainan. (3, 20, 36, 38, 41, 42)
G. japonica Baba [*G. citrina*]: Yellow Sea; Taiwan; Hong Kong. (25, 38, 41)
G. okinawea Baba: South China Sea. (12, 23, 36, 42)
G. striata (Eliot): Hong Kong; Hainan. (3, 33, 38, 41)
Lamellana gymnota Lin: Hong Kong. (37)
Nembrotha cristata Bergh: Xisha (=Paracel Islands). (5, 23, 42)
N. kubaryana Bergh: Xisha (=Paracel Islands). (5, 23, 42)
N. lineolata Bergh: Xisha (=Paracel Islands). (12, 42)
Tambja affinis Eliot: Taiwan. (26)
T. morsa Bergh: Taiwan. (26)

ONCHIDORIDIDAE [ACANTHODORIDIDAE]

Acanthodoris pibosa (Abildganrd): Yellow Sea. (25)

ALDISIDAE

Aldisa snguinea Robillliard & Baba: Yellow Sea. (25)

GONIODORIDIDAE [OKENIIDAE]

Goniodoris castanea (Alder & Hancock): Bohai; Yellow Sea. (23, 25, 42)
G. glabra (Baba): Hong Kong. (38)
Hopkinsiella hiroi Baba: Yellow Sea; Hong Kong. (23, 25, 42)
Okenia (*O.*) *japonica* Baba [*O. distincta*]: East China Sea; Hong Kong. (19, 20, 38, 41, 42)
O. (*O.*) *opuntia* Baba: Yellow Sea; East China Sea. (19, 20, 23, 25, 42)
O. (*O.*) *plana* Baba [*Hopkinsia plana*]: Yellow Sea; Hong Kong. (25, 38, 42)

NOTODORIDIDAE [AEGIRETIDAE]

Aegires villosus Farran: South China Sea. (3, 23, 36, 42)

OKADAIIDAE [VAYSSIEREIDAE]

Okadaia elegans Baba [*Vayssierea elegans*]: Yellow Sea; Hong Kong; Taiwan. (19, 20, 23)

ROSTANGIDAE

Rostanga arbutus (Angas) [*R. orientalis*]: Bohai; Yellow Sea; East China Sea; Hong Kong. (19, 20, 23, 25, 28, 38, 42, 44)

DORIDOPSIDIDAE

Doridopsis aurantiaca (Eliot): East China Sea; South China Sea. (3, 23, 36, 42)
D. granulosa Pease: Hong Kong. (38)
D. viridis Pease: Taiwan; Hainan. (3, 23, 36, 42)

GLOSSODORIDIDAE [CHROMODORIDIDAE]

Cadinella ornatissima Kisbec: Taiwan; Hong Kong. (26, 40)
Casella atromarginata (Cuvier) [*Chromodoris atromarginata*]: Taiwan; Hong Kong; Hainan; Xisha (=Paracel Islands). (3, 5, 23, 26, 36, 38, 41, 42, 44)
C. cincia Bergh [*Glossodoris cincta*]: Hong Kong. (41)
Ceratosoma cornigerum Adams & Reeve [*C. trilobatum*]: South China Sea. (23, 42)
Chromodoris elizabethina Bergh: Taiwan. (26)
C. fidelis (Kelaart): Taiwan; Hong Kong. (26, 38, 41)
C. geometrica Adams & Reeve: Taiwan. (26)
C. odhneri (Risbec): Taiwan. (26)
C. sagamiensis (Baba): Hong Kong; Hainan. (12, 36)
C. striatella (Bergh): Hong Kong. (41)
C. youngbleuthi (Kay & Uoung) [*Casella rufomarginata*]: Taiwan; Hong Kong. (26, 41)
Glossodoris aspersa (Gould) [*Chromodoris inornata*]: East China Sea. (19, 42)
G. aureopurpurea Collingwood [*Chromodoris collingwood*]: East China Sea; Hainan. (3, 19, 36, 42)
G. crossei (Angas): Hainan. (3, 23, 36, 42)
G. festiva (Angas) [*Chromodoris festiva*]: Taiwan; Hong Kong. (38, 40)
G. hilaris (Bergh): Hainan. (3, 36)
G. lineolata (Van Hasselt) [*Chromodoris lineolata*]: Hong Kong; Hainan. (3, 23, 36, 38, 41, 42)
G. marginata (Pease) [*Chromodoris marginata*]: Hong Kong; Hainan. (3, 36, 38, 41, 42, 44)
G. multituberculata Baba [*Mexichromis multituberculata*]: Hong Kong; Hainan. (3, 23, 36, 41, 42)
G. obsoluta (Rüppell & Leuckart) [*Chromodoris obsoluta*]: Taiwan. (26)
G. orientalis Rudman [*Chromodoris pallescens*]: East China Sea; Hong Kong; Hainan. (19, 20, 34, 36, 38, 40, 42)
G. pallesens (Bergh) [*Chromodoris placida, Hypselodoris placida*]: East China Sea; Hong Kong. (20, 38, 41)
G. sibogae (Baba) [*Chromodoris sibogae*]: Taiwan; Hainan; Xisha (=Paracel Islands). (5, 26, 42)
G. sinensis Rudman [*Chromodoris rubrocornuta*]: Hong Kong; Hainan. (36, 41)
G. tenuis (Collingwood): East China Sea; Hainan; Xisha (=Paracel Islands). (5, 36, 41, 42)
G. tincloria (Rüppell & Leuckart) [*Chromodoris alderi*]: East China Sea; Taiwan. (19, 42)
G. tumlifera (Collingwood) [*Chromodoris shiranae*]: East China Sea; South China Sea; Hong Kong. (19, 38, 39, 41, 42)
G. xishaensis Lin: Xisha (=Paracel Islands). (5, 23, 42)
Hypselodoris festiva A. Adams [*Glossodoris festive*]: East China Sea; Taiwan; Hong Kong. (20, 26, 38, 41, 42)
H. kanga Rudman: Hong Kong. (38, 41)
H. maritima Baba: Taiwan; Hong Kong. (26, 38)
Noumea nivilis Baba [*Chromodoris alba, Glossodoris alba*]: East China Sea; Taiwan; Hong Kong. (19, 20, 26)
Verconia verconis (Basedow & Hedley): East China Sea. (19, 42)

DISCODORIDIDAE [PELTODORIDIIDAE]

Discodoris concinna (Alder & Hancock): Hong Kong; South China Sea. (12, 23, 38)
D. fragilis (Alder & Hancock): Taiwan; Hong Kong. (26, 38)
D. mauritiana (Bergh): Hong Kong. (38)
Peltodoris cf. *aurea* Eliot: Hong Kong. (38)
Thordisa maculigera Bergh: East China Sea. (20)

KENTRODORIDIDAE

Jorunna tomentosa (Cuvier): East China Sea. (19)
Kentrodoris maculosa Cuvier [*Jorunna funebris*]: Taiwan; Hainan; Xisha (=Paracel Islands). (3, 5, 23, 26, 36, 42)
K. rubescens Bergh: Taiwan. (26)

ACTINOCYCLLIDAE

Actinocyclus japonicus (Eliot): Hainan. (3, 23, 36, 42)

HOMOIODORIDIDAE

Homoiodoris japonica Bergh: Bohai; Yellow Sea; East China Sea; Taiwan Strait; Hong Kong; Hainan. (3, 19, 20, 23, 25, 28, 36, 38, 42, 44)

CADLINIDAE

Cadlina japonica Baba: Bohai. (23, 25, 42)
C. sagamiensis Baba: Xisha (=Paracel Islands). (5, 23, 42)

ARCHIDORIDIDAE

Geitodoris granulata Lin & You: Nansha (=Spratly Islands). (17)
Trippa intecta (Kellaart): Taiwan; Hong Kong. (3, 5, 23, 36, 38, 42)

ARGIDIDAE [PLATYDORIDIIDAE]

Argus cruentus (Quoy & Gaimard): Hainan; Xisha (=Paracel Islands). (12, 23, 36, 42)
A. esakii Baba: Hong Kong; Hainan. (12, 23, 36, 42)
A. laminea (Risbec): Hainan; Xisha (=Paracel Islands). (3, 5, 23, 36, 42)
A. scabra (Cuvier): Nansha (=Spratly Islands). (17)
A. speciosus (Abraham): Hainan; Xisha (=Paracel Islands); Nansha (=Spratly Islands). (3, 5, 23, 36, 42)
A. tabulatus (Abraham): Hainan; Xisha (=Paracel Islands). (3, 5, 23, 36, 42)

ASTERONOTIDAE

Asteronotus cespitosus (Van Hasselt): East China Sea; Hong Kong; Hainan; Xisha (=Paracel Islands). (3, 5, 23, 36, 42, 44)
Halgerda japonica Eliot: East China Sea; Xisha (=Paracel Islands). (16, 19, 42)
H. orientalis Lin: Nansha (=Spratly Islands). (16)
H. rubicunda Baba: Hong Kong. (38)
H. xishaensis Lin: Xisha (=Paracel Islands). (5, 23, 42)

DENDRODORIDIDAE

Dendrodoris (D.) areolata (Alder & Hancock) [*Doris areolata*]: Taiwan; Hainan; Xisha (=Paracel Islands). (13, 19, 23, 36, 42)
D. (D.) denisoni (Angas) [*D. gemmacea*]: East China Sea; Taiwan; Hong Kong; Hainan; Xisha (=Paracel Islands). (3, 12, 19, 23, 26, 38, 42)
D. elongata Baba: Hainan. (12, 19, 36, 42)
D. (D.) guttata (Odhner): Hong Kong. (38)
D. (D.) miniata (Alder & Hancock): East China Sea; Taiwan Strait; Hong Kong. (20, 38)
D. (D.) nigra (Stimpson): East China Sea; Taiwan; Hong Kong; Hainan; Xisha (=Paracel Islands). (1, 3, 5, 19, 20, 23, 36, 38, 42)
D. (D.) rubra (Kelaart): East China Sea; Hong Kong. (12, 19, 38, 42)
D. (D.) tuberculosa (Quoy & Hancock): Hainan. (3, 23, 36, 42)

DENDRONODIDAE

Dendronotus arborescens (Müller): Bohai; Yellow Sea. (1, 23, 25, 42, 43)

PHYLLIDIDAE

Fryeria ruppelli Berg: Taiwan. (26)
Phyllidia elegans Bergh: Taiwan; Hainan; Xisha (=Paracel Islands); Nansha (=Spratly Islands). (11, 16, 26, 36, 42)
P. gemmata (Pruvot-Fol): Xisha (=Paracel Islands). (11, 42)
P. honloni Risbec: Hainan; Xisha (=Paracel Islands); Nansha (=Spratly Islands). (2, 11, 36, 42)
P. nobilis (Bergh): Taiwan; Hong Kong; Xisha (=Paracel Islands). (5, 23, 26, 42, 44)
P. ocellata Cuvier: Taiwan; Xisha (=Paracel Islands). (5, 11, 23, 36, 38, 42)
P. pustulosa Cuvier: Taiwan; Hainan; Xisha (=Paracel Islands); Nansha (=Spratly Islands). (3, 5, 11, 17, 23, 26)
P. rotunda (Eliot): Xisha (=Paracel Islands). (11, 42)
P. sereni Risbec: Hainan; Xisha (=Paracel Islands). (11, 36, 42)
P. varicosa Lamarck: Taiwan; Hong Kong; Hainan; Xisha (=Paracel Islands). (3, 5, 11, 17, 23, 26, 36, 38, 42)
P. xishaensis Lin: Xisha (=Paracel Islands). (11, 42)

ARMINIDAE [PLEUROPHYLLIDIADAE]

Armina (A.) appendiculata Baba: East China Sea; South China Sea. (9, 19, 42)
A. (Linguella) babai (Tchang) [*Linguella babai*]: Bohai; Yellow Sea; East China Sea; Hainan. (1, 3, 9, 19, 20, 25, 42, 44)
A. (A.) bilanella Lin: Yellow Sea; East China Sea; Beibuwan (=Gulf of Tongking). (9, 19, 25, 42)

A. (A.) comta (Bergh): Bohai; Yellow Sea; Hong Kong. (9, 23, 25, 42, 44)

A. (A.) japonica (Eliot): East China Sea; Hong Kong. (9 20, 26, 42)

A. (A.) longicauda Lin: East China Sea; South China Sea. (9, 19, 42)

A. (A.) magna Baba: South China Sea. (9, 23, 42)

A. (A.) major Baba: Yellow Sea; East China Sea; South China Sea. (9, 23, 42)

A. (A.) papillata Baba: Yellow Sea; East China Sea; South China Sea; Hong Kong. (9, 13, 19, 42, 44)

A. (L.) punctilopsis Lin: Hong Kong. (37)

A. (L.) punctilucens (Bergh) [*Linguella punctilucens*]: Bohai; Yellow Sea; East China Sea; Hainan. (3, 9, 23, 25, 36, 42, 44)

A. (L.) punctulata Lin: Hong Kong. (35)

A. (A.) semperi (Bergh): South China Sea. (3, 9, 36, 42)

A. (A.) sinensis Lin: Bohai; Yellow Sea; East China Sea. (9, 19, 25, 42)

A. (L.) variolosa (Bergh): East China Sea; South China Sea. (3, 9, 19, 25, 36, 42)

Dermatobranchus (D.) marginlatus Lin: South China Sea. (9, 23, 42)

D. (Pleuroleura) multistriatus Lin: South China Sea. (9, 42)

D. (P.) ornatus Bergh: East China Sea; South China Sea. (9, 42)

D. (P.) tongshanensis Lin: East China Sea. (9, 42)

Pleurophyllidiopsis amoyensis Tchang: East China Sea. (9, 20, 21, 23, 42)

P. orientalis Lin: Bohai; Yellow Sea. (9, 20, 21, 23, 42)

P. tsingtaoensis Lin: Yellow Sea. (9, 25, 42)

DUVAUCELIIDAE [TRITONIIDAE]

Duvaucelia (D.) exsulans (Bergh): Near shore in China's seas. (1)

Marionia olivacea Baba: Yellow Sea. (23, 25, 42)

Paratritonia lutea Baba: East China Sea. (20)

SCYLLAEIDAE

Notobryon wardi Odhner: Bohai; Yellow Sea; East China Sea; Xisha (=Paracel Islands). (12, 16, 19, 23, 25, 42)

Scyllaea pelagica Linnaeus: South China Sea. (1, 42)

BORNELLIDAE

Bornella digitata Adams & Reeve: Hong Kong; Hainan; Nansha (=Spratly Islands). (1, 3, 16, 23, 36, 42)

B. japonica Baba: Hainan. (3, 36, 42)

FIMBRIIDAE [TETHYIDAE]

Melibe japonica Eliot [*Fimbria fimbria*]: Hainan; South China Sea. (1, 12, 23, 36, 42)

EUBRANCHIDAE [EMBLETONIIDAE]

Embletonia gracile paucipapillata Baba: Yellow Sea. (25, 42)

Eubranchus misakensis Baba: Yellow Sea. (23, 25, 42)

TERGIPEDIDAE [CUTHONIDAE]

Catriona ornata (Baba): Yellow Sea. (42)

C. cf. *purpureoanulata* (Baba): Hong Kong. (38)

Cratena nigricolor (Baba): Hong Kong. (44)

C. serrata (Baba) [*Hervia serrata*]: East China Sea. (20)

Eolis gracilis Alder & Hancock: Near shore in China's seas. (1)

Sakuraeolis enosimensis (Baba) [*Hervia ceylonica*]: Yellow Sea; East China Sea; Hong Kong. (19, 23, 25, 38, 42)

Shinanoeolis emurai (Baba): Yellow Sea. (25, 42)

AEOLIDIIDAE

Aeolidia cf. *foulisi* (Angan): Hong Kong. (38)

A. papillosa (Linnaeus): Yellow Sea. (25, 42)

Aeolidiella albopunctata Lin: Hong Kong. (37)

A. takanosimensis (Baba): Yellow Sea; Hong Kong. (25, 38, 42)

Baeolidia fusiformis Baba: Hong Kong. (44)

Cerberilla albopunctata Baba: Hainan; South China Sea. (12, 23, 36, 42)

C. asamusiensis Baba: Hainan. (25, 36, 42)

PETERAEOLIDIIDAE

Peteraeolidia ianthina (Angas): Taiwan; Hainan; Xisha (=Paracel Islands). (26, 42)

P. spaperi Bergh: South China Sea. (5, 42)

FACELENIDAE

Facelina zhejiangnensis Lin & You: East China Sea. (19, 42)

Facelinella porilophags Rudman: Hong Kong. (38)

F. quadrilineata (Baba): Yellow Sea. (42)

Herviella yatsui (Baba): Yellow Sea. (42)

PHIDIANIDAE

Phidiana indica Bergh: Taiwan. (26)

FLABELLINIDAE

Flabellina ornata Angas: Hong Kong. (38)

F. rubrolineata (O'Donoghue): Taiwan. (26)

SPURILLIDAE

Berghia japonica (Baba): Hong Kong. (38)

PHESTILLIDAE

Phestilla melanobranchia Bergh: Hong Kong. (38)

GLAUCIDAE

Glaucus (*Glaucus*) *marinus* (Dupont): East China Sea. (19, 42)

REFERENCES*

(1). Tchang Si and Qi Zhanyan, 1961. Outline of Malacology. China Science Press, 387pp.
(2). Tchang Si and Lin Guangyu, 1964. A study on Aplysiidae from China Coast. Studia Marina Sinica, 5:4–24.
(3). Lin Guangyu and Tchang Si, 1965. Opisthobranchia from the inter-tidal zone of Hainan Island, China. Oceanologia & Limnologia Sinica, 7(1):1–20.
(4). Lin Guangyu and Tchang Si, 1965. Etuda sur les Mollusques Pleurobranchidae de la cote de Chine. Oceanologia & Limnologia Sinica, 7(3):265–276.
(5). Lin Guangyu, 1975. Opisthobranchia from the Inter-tidal zone of Xisha Islands, Guangdong Province, China. Studia Marina Sinica, 10:141–154.
(6). Lin Guangyu, 1978. A new species of *Cyclinobulla* (Opisthebranchia) of the Xisha Islands, Guangdong Province, China. Oceanologia & Limnologia Sinica, 9(1):95–98.
(7). Lin Guangyu, 1980. A new genus and species of Ringiculidae. Oceanologia & Limnologia Sinica, 11(3):263–266.
(8). Lin Guagyu, 1981. A new species of the genus *Umbraculum* (Opisthobranchia) from China. Oceanologia & Limnologia Sinica, 12(3): 286–290.
(9). Lin Guangyu, 1981. A study of the family Arminidae (Opisthobranchia) of China Coast. Studia Marina Sinica. 18:181–205.
(10). Lin Guangyu, 1983. Studies on the genus *Ringicula* (Opisthobranchia) from China Coast. Transactions of the Chinese Society of Malacology, 1:23–30.
(11). Lin Guangyu, 1983. A study on the genus *Phyllidia* (Opisthobranchia) in China. Tropic Oceanology, 2(2):148–153.
(12). Lin Guangyu, 1986. Additions to the Opisthobranchia fauna of Hainan and Xisha Islands of China. Studia Marina Sinica. 26:117–128.
(13). Lin Guangyu, 1990. A new species and a new record of the genus *Petalifera* (Opisthobranchia) of China. Acta Zootaxonomica Sinica. 15(1): 21–24.
(14). Lin Guangyu, 1989. A study on Acteonidae (Opisthobranchia) from China coast. Studia Marina Sinica, 30:167–176.
(15). Lin Guangyu, 1992. A study on Pyramidellidae (Opisthobranchia) from the Yangtze River estuary. (manuscript)
(16). Lin Guangyu, 1991. A study of the Opisthobranchia from the waters of Nansha Islands. Contributions on the study of Marine Biology in the Nansha Islands and Neighbouring Waters, China, I:130–136. China Ocean Press.
(17). Lin Guangyu and Wu Pingru, 1994. Study on the Opisthobranchia from Nansha Islands and its adjacent waters (II). Studies on Marine Fauna and Floral and Biogeography of the Nansha Islands and Neighbouring Waters, China. I:42–54.
(18). Lin Guangyu and You Zhongjie, 1990. Studies on Nudibranchia from the Zhejiang coast. Studia Marina Sinica, 31:147–162.
(19). Lin Guangyu and You Zhongjie, 1995. A study on Nudibranchia (Opisthobranchia) from China coast. Transactions of the Chinese Society of Malacology, 5–6:1–7.
(20). Lin Guangyu and Wu Pingru. A study on Opisthobranchia off the Fujian coast, China. (manuscript)
(20a). Lin Guangyu, 1997. FAUNA SINICA, Gastropoda, Opisthobranchia, Cephalaspidea. China Science Press, 216pp.
(21). Zhang Fusui, 1965. The pelagic molluscs off China coast. I. A systematic study of Pteropoda and Janthinidae (Ptenoglosa, Prosobranchia). Studia Marina Sinica, 5:125–189.
(22). Hong Junchao and You Zhongjie, 1983. Record on Nudibranchia from inter-tidal zone of Yusan Island of Zhejiang, China. Journal of Zhejiang College of Fisheries, 2(1):15–19.
(23). Qi Zhongyan et al., 1986. Illustrated Encyclopedia of the Fauna of China, Mollusca (III), 1–98. China Science Press.
(24). Qi Zhongyan and Lin Guangyu et al., 1984. A preliminary survey on benthic Mollusks from Sanya Harbour, Hainan Island. South China Sea Studia Marina Sinica, 5:77–98.
(25). Qi Zhongyan and Lin Guangyu et al., 1989. Mollusca of Yellow Sea and Bohai Sea, China Agriculture Press, 309pp.
(26). Tan Tianshi and Bai Zhengyu et al., 1987. Record on the distribution of Nudibranchia from Taiwan coast, China. Bulletin of Malacology, R. O. C., 13:71–90.
(27). Dai Yanyu, 1989. Distribution of Pelagic Mollusca from west Taiwan Strait, Journal of Oceanography in Taiwan Strait, 8(1):54–59.
(28). Zhao Ruyi and Chen Jimin et al., 1981. Record on Opisthobranchia from inter-tidal zone of Dalian. Journal of Northeastern Normal University (Natural Science), 1:43–50.
(29). Grabe, A. W. and S. C. King, 1928. Shells of Pertaiho. Peking Soc. Nat. Hist., No. 2: 239–243.
(30). Habe, T., 1955. A list of the Cephalaspid Opisthobranchia of Japan. Bull. Biogeogr. Soc. Japan, 16–19:54–79.
(31). Jensen, K. R., 1986. Annotated checklist of Hong Kong Ascoglossa (Mollusca: Opisthobranchia) with descriptions of four new species. Proceedings of the Second International Workshop on the Malacofauna of Hong Kong and Southern China, Hong Kong, 1983. Hong Kong University Press, pp:77–107.
(32). Johnson, B.T., 1964. The Recent Mollusca of Augustus Addison Gould. United States National Museum Bull., 239.
(33). King, S. C. and C. P. Ping, 1931. The Molluscan shells of Hong Kong Naturalist, 2(1): 9–29.
(34). Lin Guangyu and Qi Zhongyan, 1986. A preliminary survey of the Cephalaspids (Opisthobranchia) of Hong Kong and Adjacent Waters. Proceedings of the Second International Workshop on the Malacofauna of Hong Kong and Southern China, Hong Kong 1983. Hong Kong University Press, pp:110–124.
(35). Lin Guangyu and Qi Zhongyan, 1990. Opisthobranchia fauna of Hainan Island, China. Bulletin of Marine Science, 47(1):134–38.
(36). Lin Guangyu and Qi Zhongyan, 1990. Two new species of Opisthobranchia from Hong Kong. In: Proceedings of the Second International Workshop: The Marine Flora and Fauna of Hong Kong and Southern China, 1986. Hong Kong University Press, pp:433–36.
(37). Lin Guangyu and Qi Zhongyan, 1992. Three new species of Opisthobranchia from Hong Kong. Proceedings of the Fourth International Marine Biological Workshop: The Marine Flora and Fauna of Hong Kong and Southern China, 1989. Hong Kong University Press, pp:1–6.
(38). Orr, J. 1981. Hong Kong Nudibranchs. Urban Council of Hong Kong, pp:1–82.
(39). Rudman, W. B., 1978. The Chromodoridiidae (Opisthobranchia: Mullusca) of the Indo-west Pacific. Jour. Linn. Zool., 78:105–73.
(40). Rudman, W.B., 1983. The Chromodoridiidae (Opisthobranchia: Mullusca) of The Indo-west Pacific. Jour. Linn. Zool., 90:305–407.
(41). Rudman, W.B. and B.W. Darvell, 1990. Opisthobranch Molluscs of Hong Kong, Part 1. Asia Marine Biology, 7:31–79.
(42). Qi Zhangyan, Lin Guangyu et al., 1992. Molluscs of China, Vol. 1. Seashells. (in press).
(43). Tchand-Si, 1934–36. Contribuion al etude des Opisthobranche de la Cote de Tsingtao. Sur un nouvear Nudibranche de la Cote D'Amoy. Institute od Zoology National Academy of Peiping, 2(2): 1–49.
(44). Tsi, C.Y., and S. T. Ma, 1983. Proceedings of the First International Marine Biological Workshop: The Marine Flora and a Preliminary Checklist of the Marine Gastropod and Bivalve (Mollusca) of Hong

Kong and Southern China. Fauna of Hong Kong and Southern China. 1980. Hong Kong University Press, pp:431–458.

*: (1)-(28) in Chinese

Compiled by Lin Guangyu; Edited by Qi Zhanyan (C. Y. Tsi)

PULMONATA
BASOMMATOPHORA
ELLOBIIDAE

Auriculastra duplicata (Pfeiffer): Hong Kong. Intertidal. (3)
A. subula (Quoy & Gaimard): Hong Kong. Intertidal. (3)
Cassidula plectorematoides (Möllendorff): Hong Kong. Intertidal. (3)
C. schmackeriana (Möllendorff): Hong Kong. Intertidal. (3)
Ellobium aurismidae (Linnaeus): Hainan. Mangrove forest muddy substrate. (1)
E. chinensis (Pfeiffer): Zhejiang; Guangdong; Taiwan; Hong Kong; Hainan. High intertidal area with freshwater input. (1, 3)
E. poita (Metcalfe): Shenzhen (Guangdong). Mangrove forest mud flat. (3)
Leamodonda exatum (H. & A. Adams): Hong Kong. Intertidal. (3)
L. punctatostriata (H. & A. Adams): Hong Kong. Intertidal. (3)
L. punctigera (H. & A. Adams): Hong Kong. Intertidal. (3)
Melampus triticeus (Küster): Hong Kong. High intertidal. (3)
Pythia cecillei (Philippi): Xiamen (Fujian); Guangdong; Hong Kong; Guangxi. High intertidal. (1, 3)
P. fimbriosa (Möllendorff): Hong Kong. Intertidal. (3)

SIPHONARIIDAE

Siphonaria atra Quoy & Gaimard: South of Fujian. High rocky intertidal. (1, 3)
S. japonica (Donovan): China's coast. High rocky intertidal or dock water line. (1, 3)
S. laciniosa (Linnaeus): Taiwan. Rocky intertidal. (2)
S. sirius Pilsbry: Hong Kong. Low rocky intertidal. (3)

STYLOMMATOPHORA
ONCHIDIIDAE

Onchidium verruculatum Cuvier: East China Sea; South China Sea. High and mid muddy sand intertidal. (1, 3)

REFERENCES*

(1). Qi Zhongyan, Ma Xiutong et al.,1985. Pulmonata. In: A Collection of Illustrative Plates of Animals in China, Molluscs, Vol. 4. China Ocean Press, pp:1–115
(2). Tang Tianxi et al., 1986. A survey of marine shells (Gastropods and Bivalve) along the northeast coast of Taiwan. Acta Malacology, 12:27–47.
(3). Morton, B. and J. Morton. 1983. The sea shore ecology of Hong Kong. Hong Kong University Press, 350pp.

*: (1) and (2) in Chinese

Compiled by Huang Zongguo and Li Rongguan; Edited by Lin Guangyu

CEPHALOPODA

NAUTILOIDEA
NAUTILIDAE

Nautilus pompilius Linnaeus: Eastern Taiwan; Hainan; Xisha (=Paracel Islands) (empty shells). (5)

TEUTHOIDEA
HISTIOTEUTHIDAE

Histioteuthis celetaria pacifica (Voss): Offshore of southern Hainan. Water depth 1,100 m. (5)
H. dofleini (Pfeiffer): Offshore of East China Sea; South China Sea continental slope. (4, 5)
H. meleagroteuthis (Chun): Offshore of East China Sea. (4)

ENOPLOTEUTHIDAE

Abralia andamanica Goodrich: Offshore of East China Sea; northern South China Sea. (4, 5)
A. multihamata Sasaki: East China Sea, water depth 15–80 m; South China Sea. (1, 4, 5)
Enoploteuthis chunii Ishikawa: Offshore of East China Sea; northern South China Sea. Deep water. (4, 5)
Pterygioteuthis giardi Fischer: East China Sea; South China Sea. (4, 5)

ONYCHOTEUTHIDAE

Mototeuthis lonnbergii Ishikawa & Wakiya: Offshore of East China Sea; South China Sea continental shelf and continental slope. (4, 5)
Onychoteuthis banksii (Leach): Offshore of East China Sea; Taiwan Strait. (1, 4)
Onykia carribbaea Lesueur: Offshore of East China Sea; northern South China Sea continental shelf. (1, 4, 5)

OCTOPOTEUTHIDAE

Octopoteuthis sicula Rüppell: Offshore of South China Sea. (5)

CRANCHIIDAE

Cranchia scabra Leach: Northern South China Sea. (5)
Leachia pacifica (Issel): Offshore of East China Sea. (5)
Liocranchia globula (Berry): Offshore of East China Sea. (4)

L. reinhardti (Steenstrup): Northern South China Sea continental slope. (5)

Sandalops melancholicus Chun: Southern Xisha (=Paracel Islands). (5)

Taonius pavo (Lesueur): East China Sea continental slope, water depth 1,180 m; South China Sea continental slope. (4, 5)

MASTIGOTEUTHIDAE

Mastigoteuthis cordiformis Chun: South China Sea continental slope. (5)

CHIROTEUTHIDAE

Chiroteuthis imperator Chun: Offshore of East China Sea; South China Sea continental slope. (4, 5)

BRACHIOTEUTHIDAE

Brachioteuthis riisei (Steenstrup): East China Sea; South China Sea. (5)

CTENOPTERYGIDAE

Ctenopteryx siculus (Vérany): Xisha (=Paracel Islands); Zhongsha (=Macclesfield Bank). (5)

THYSANOTEUTHIDAE

Thysanoteuthis rhombus Troschel: East China Sea; South China Sea. (1, 4, 5)

OMMASTREPHIDAE

Hyaloteuthis pelagica (Bosc): Offshore of East China Sea. (4)

Nototodarus hawaiiensis (Berry) [*Ommastrephes hawaiiensis*]: South China Sea. (5)

N. sloani philippinensis Voss: Offshore of East China Sea; South China Sea. (1, 4, 5)

Ommastrephes bartrami (Lesueur): Xisha (=Paracel Islands). (5)

Ornithoteuthis volatilis (Sasaki): East China Sea; offshore of South China Sea. (4, 5)

Symplectoteuthis luminosa (Sasaki): Offshore of East China Sea. (4)

S. oualaniensis (Lesson): East China Sea and South China Sea continental slope. (1, 2, 4, 5)

Todarodes pacificus Steenstrup [*Ommastrephes sloani pacificus*]: From Yellow Sea (fishery species) to offshore of Hong Kong. (1, 4, 5)

LOLIGINIDAE

Loligo beka Sasaki: Bohai; Yellow Sea; East China Sea; South China Sea. (1, 4, 5)

L. bleekeri Keferstein [*Doryteuthis bleekeri*]: Yellow Sea; East China Sea; South China Sea. (1, 4, 5)

L. chinensis Gray [*L. formosana*]: Southern Fujian; Penghu Islands (Taiwan); Taiwan Strait; Nanpeng Island (Guangdong); northern Beibuwan (=Gulf of Tongking); Hainan; South China Sea. Fishery species. (1, 4, 5)

L. duvaucelii Orbigny: Taiwan Strait; South China Sea. (1, 4, 5)

L. edulis Hoyle: Yellow Sea; East China Sea; South China Sea. (1, 4, 5)

L. gotoi Sasaki: East China Sea; South China Sea. (1, 4, 5)

L. japonica Hoyle: Bohai; Yellow Sea; northern East China Sea. (5)

L. kobiensis Hoyle: East China Sea; South China Sea. (1, 4, 5)

L. oshimai Sasaki: South China Sea. (5)

L. sibogae Adam: Southern Taiwan Strait; South China Sea. (1, 4, 5)

L. singhalensis Ortmann: Southern East China Sea. (4)

L. tagoi Sasaki: East China Sea; South China Sea. (5)

L. uyii Wakiya & Ishikawa: East China Sea; South China Sea. (1, 4, 5)

Sepioteuthis lessoniana Lesson: Yellow Sea; East China Sea; South China Sea. (1, 4, 5)

SEPIOIDEA
SEPIIDAE

Metasepia tullbergi (Appellöf): East China Sea; South China Sea. (1, 4, 5)

Sepia aculeata Orbigny: Taiwan Strait; South China Sea. (4, 5, 6)

S. andreana Steenstrup: Yellow Sea. (5)

S. brevimana Steenstrup: Zengmuansha (southern tip of Nansha=Spratly Islands). (2)

S. elliptica Hoyle: Taiwan Strait; South China Sea. (1, 4, 5)

S. esculenta Hoyle: Bohai; Yellow Sea (fishery species); East China Sea; South China Sea. (1, 4, 5)

S. kobiensis Hoyle: East China Sea; South China Sea; Xisha (=Paracel Islands). (1, 4, 5, 6)

S. latimanus Quoy & Gaimard [*S. hercules*]: Taiwan Strait; South China Sea. Fishery species. (1, 4, 5, 6)

S. longipes Sasaki: Offshore of East China Sea. (4)

S. lycidas Gray [*S. subaculeata*]: East China Sea; Taiwan Strait; South China Sea. Water depth 10–100 m; fishery species. (1, 4, 5, 6)

S. nanshiensis Li & Chen: Nansha (=Spratly Islands). Water depth 40–100 m. (3)

S. omani Adam & Rees: Taiwan Strait; East China Sea; South China Sea. (1, 4, 5)

S. pharaonis Ehrenberg [*S. tigris*]: Taiwan Strait; South China Sea. Fishery species. (1, 4, 5, 6)

S. recurvirostra Steenstrup: Taiwan Strait; South Chinas Sea; Zengmuansha (southern tip of Nansha=Spratly Islands). (1, 2, 4)

S. robsoni Sasaki: East China Sea; South China Sea; Xisha (=Paracel Islands). (1, 4, 6)

S. tenuipes Sasaki: Offshore of East China Sea. (4)

S. torosa Ortmann: South China Sea. (5)

Sepiella maindroni de Rochebrune [*S. japonica*]: From Bohai and Yellow Sea to South China Sea. Fishery species in Zhejiang and Fujian. (1, 4, 5)

SEPIOLIDAE

Euprymna berryi Sasaki: East China Sea; South China Sea. (1, 4, 5)

E. morsei (Verrill): Yellow Sea. (5)

Inioteuthis japonica Verrill: Taiwan Strait; South China Sea. (1, 4, 5)

I. maculosa Goodrich: Zengmuansha (southern tip of Nansha=Spratly Islands). (2)

Rossia bipapillata Sasaki: Offshore of East China Sea. (4)

Sepiadarium kochii Steenstrup: Taiwan Strait; South China Sea. (1, 4, 5)

Sepiola birostrat Sasaki: Bohai; Yellow Sea; East China Sea; South China Sea. (1, 4, 5)

IDIOSEPIIDAE

Idiosepius paradoxa (Ortmann): Yellow Sea; Taiwan Strait; inner bays along shoreline of South China Sea. (1, 4, 5)

OCTOPODA
STAUROTEUTHIDAE

Grimpoteuthis umbellata (Fischer): South China Sea continental slope. (5)

Stauroteuthis albatrossi Sasaki: Offshore of East China Sea. (4)

OPISTHOTEUTHIDAE

Opisthoteuthis depressa Ijima & Ikeda: Continental slope of East China Sea and South China Sea. (4, 5)

BOLITAENIDAE

Japetella diaphana Hoyle: Offshore of South China Sea. (5)

VAMPYROTEUTHIDAE

Vampyroteuthis infernalis Chun: East China Sea continental slope. (4)

AMPHITRETIDAE

Amphitretus pelagicus Hoyle: East China Sea continental slope. (4)

ALLOPOSIDAE

Allopus mollis Verrill: Offshore of East China Sea. (4, 5)

TREMOCTOPODIDAE

Tremoctopus violaceus gracilis (Eydoux & Souleyet): Taiwan Strait and South China Sea. Sea surface. (1, 4, 5)

ARGONAUTIDAE

Argonauta argo Linnaeus: East China Sea; offshore of South China Sea. (4, 5)

A. hians Solander: East China Sea; offshore of South China Sea. (4, 5)

OCYTHDIDAE

Ocythoe tuberculata Rafinesque: Offshore of South China Sea. (5)

OCTOPODIDAE

Callistoctopus arakawai Taki: Dong Island (Xisha=Paracel Islands). Coral reefs. (5)

Cistopus indicus (Orbigny): South China Sea. Shoreline to water depth 50 m. (5)

Octopus aegina Gray: Taiwan Strait. Continental shelf with water depth of within 80 m. (5)

O. berenice Gray: East China Sea; South China Sea. (1, 4, 5)

O. bimaculatus Verrill: East China Sea; South China Sea. (1, 4, 5)

O. dollfusi Robson: Southern Fujian; Guangdong; Hainan. (1, 4, 5)

O. fusiformis Brock: Southern Fujian; South China Sea. (1, 4, 5, 6)

O. guangdongensis Dong: Weizhou Island (Guangxi); Hainan. (5)

O. luteus (Sasaki): East China Sea; South China Sea. (1, 4)

O. maculosa Hoyle [*O. faciatus*]: South China Sea. Water depth 20–74 m. (5)

O. nanhaiensis Dong: Dongshan (Fujian) to Hainan; Xisha (=Paracel Islands). (5, 6)

O. ocellatus Gray [*O. aerolatus*]: Bohai; Yellow Sea; East China Sea; South China Sea. Fishery species in Bohai and Yellow Sea. (1, 4, 5)

O. oshimai (Sasaki): From south of Nanji (Zhejiang) to Xisha (=Paracel Islands). Continental shelf with water depth within 100 m. (1, 4, 5, 6)

O. ovulum (Sasaki): East China Sea; South China Sea; Xisha (=Paracel Islands). Water depth 10's of m. (1, 4, 5, 6)

O. pallida Hoyle: South China Sea. (5)

O. rugosus (Bosc): East China Sea; South China Sea. (1, 4)

O. striolatus Dong: From Haimen (Guangdong) to Xisha (=Paracel Islands). (5)

O. variabilis (Sasaki): From Bohai to South China Sea. Fishery species in northern China. (1, 4, 5)

O. vulgaris Cuvier: South of Zhoushan Islands (Zhejiang) to Zengmuansha (southern tip of Nansha=Spratly Islands). Water depth of several m to over 100 m; fishery species in southern China. (1, 2, 4, 5, 6)

REFERENCES*

(1). Li Fuxue, 1983. Studies on the Cephalopod fauna of the Taiwan Strait. Taiwan Strait, 2(1):103–109.
(2). Li Fuxue and Chen Qingchao, 1987. On small-size Cephalopods and the junveniles. In: Research reports on the comprehensive oceanographic investigation of Chinese Southern Territory: Zengmu Shoal. China Science Press, pp:163–167.
(3). Li Fuxue and Chen Qingchao, 1989. A new species of *Sepia* (Cephalopoda: Sepiidae) in the waters of the Nansha Islands of China. Tropical Oceanology, 8(2):6–12.
(4). Zheng Yushui, 1987. Studies on Cephalopoda of the South China Sea. Fujian Fisheries, 3:13–21.
(5). Dong Zhengzhi, 1988. Cephalopoda. In: Fauna of China, Mollusca. China Science Press, pp:1–201.
(6). Dong Zhengzhi, 1991. Cephalopods in the waters of Nansha Islands. In: Contributions on Marine Organisms around Nansha Islands and the adjacent Waters (I). China Ocean Press, pp:167–168.

*: all in Chinese
Compiled by Huang Zongguo; Initial editing by Li Fuxue; Final editing by Dong Zhengzhi

ARTHROPODA

MEROSTOMATA

XIPHOSURA
TACHYPLEIDAE

Carcinoscopius rotundicauda (Latreille): Leizhou Peninsula (Guangdong); Beibuwan (=Gulf of Tongking), water depth within 20 m; Hainan. (3)

Tachypleus gigas (Müller): Hong Kong; Guangxi shoreline. (3)

T. tridentatus (Leach): Zhejiang (south of Zhoushan and Ningbo); Fujian; Guangdong; Hong Kong; Guangxi; Hainan. Moves to sandy intertidal to spwan in April and May, migrates to deep water in September and October. (1–5)

Tachypleus sp.: Guangxi shoreline.

REFERENCES*

(1). Wang Yihao, 1984. The northern distribution of *Tachypleus tridentatus* Leach in China Seas. Marine Sciences, 4:38.
(2). Chou, N. S. and C. Cheng, 1950. A prliminary study of horse-shoe crabs of Xiamen. Journal of Xiamen Fisheries, 1(4): 29–40.
(3). Liang Guangyuan and Zhou Liju, 1984. A preliminary investigation on the resources of horseshoe crabs in the Chinese Beibu Bay. Journal of Marine Drugs, 6(4):32- 33.
(4). Sekiguchi, K. and K. Nakamura, 1980. Sympatric distribution pattern of three species of Asian horseshoe crabs. Jap. Soc. Syst. Zool. 18:1–4.
(5). Sekiguchi, K. and H. Sugita, 1980. Systematics and hybridization in the four living species of horseshoe crabs. Evolution, 34(4):712–718.

*: (1)-(3) in Chinese
Compiled by Huang Zongguo; Edited by Lian Guanyao

PYCNOGONIDA

PANTOPODA

Achelia superba (Loman): Liaoning; Shandong; Hong Kong. On seaweed in intertidal area. (2, 4)

Anoplodactylus glandulifer Stock: Hong Kong. (4)

Lecythorhynchus hilgendorfi (Böhm): Along China's shoreline. Low intertidal to shallow sea. (1, 3)

Tanystylum sinoabductus Bamber: Hong Kong. Mid intertidal. (4)

REFERENCES*

(1). Lu, D. H., 1936. Studies on Pycnogonida of Kiaochou Bay. Contributions from the Institute of Zoology, National Academy of Peiping, 14:30.
(2). Zhao Ruyi et al., 1990. Guidebook on collection of invertebrate from China coast (in North China). China Ocean Press, 393pp.
(3). Huang Zongguo and Cai Ruxing, 1984. Marine Fouling Organism and its Prevention (I). China Ocean Press, 354pp.
(4). Bamber, R. N., 1992. Some pycnogonids from the south China. Asian Marine Biology 9:193–199.

*: (1)-(3) in Chinese
Compiled by I. Bartsch and Huang Zongguo; Edited by Hu Chengye

ARACHNOIDEA

ACARI
HALACARIDAE
HALACARINAE

Agauopsis ammodytes Bartsch: Hong Kong. (8)

A. arenaria Bartsch: Hong Kong. (8)

A. humilis Bartsch: Hong Kong. (6)
A. sordida Bartsch: Hong Kong. (6)
Arhodeoporus minuscuius Bartsch: Hong Kong. Sandy beach and on seaweed (*Sargassum*). (5)

ACAROCHELOPODINAE

Acarochelopodia lapidaria Bartsch: Dapeng Cove (Hong Kong). Coarse sand (1–5 cm). (4)

ACTACARINAE

Actacarus sinensis Bartsch: Hong Kong. Sandy beach. (4)

COPIDOGNATHINAE

Acarothrix palustris Bartsch: Dapeng Cove (Hong Kong). Mangrove forest substrate. (1)
Copidognathus cephalocanthus Bartsch: Hong Kong. Water depth 7–10 m. (5)
C. cerberoideus Bartsch: Dapeng Cove (Hong Kong). Low intertidal coarse sand substrate. (4)
C. consobrinus Bartsch: Dapeng Cove (Hong Kong). Low intertidal coarse sand substrate. (4)
C. gracilunguis Bartsch: Hong Kong. Water depth 8 m. (5)
C. inconspicuus Bartsch: Hong Kong. Deposited substrate. (3)
C. longiunguis Bartsch: Hong Kong. Water depth 5 m. (2, 5)
C. neptuneus Bartsch: Hong Kong. Water depth 7–10 m. (5)
C. occultans Bartsch: Shenzhen Bay (Guangdong). Subtidal deposited substrate. (4)
C. paluster Bartsch: Hong Kong. Mangrove forest deposited substrate. (3)
C. polyporus Bartsch: Hong Kong. Deposited substrate, seaweed, barnacles, and fan worm (feather duster worm, *Hydroides*). (3)

SIMOGNATHINAE

Simognathus fovealatus Bartsch: Shenzhen Bay (Guangdong). Low intertidal deposited substrate. (4)

LOHMANNELLINAE

Scaptognathides hawaiiensis Bartsch: Hong Kong. Mid- to low-intertidal. (4)
Scaptognathus triunguis Bartsch: Hong Kong. High intertidal coarse sand. (4)

RHOMBOGNATHINAE

Isobactrus luxtoni Bartsch: Hong Kong. (7)
I. obesus Bartsch: Hong Kong. (7)

Rhombognathus arenarius Bartsch: Hong Kong. (7)
R. dictyotus Bartsch: Hong Kong. (7)
R. hirtellus Bartsch: Hong Kong. (7)
R. neptunellus Bartsch: Hong Kong. (7)
R. setellus Bartsch: Hong Kong. (7)
R. sinensis Bartsch: Hong Kong. Low intertidal sandy substrate. (2)
R. sinensoides Bartsch: Hong Kong. (7)
R. verrucosus Bartsch: Hong Kong. (7)

EUPODIDAE

Halotydeus mollis Luxton: Hong Kong. High intertidal rocky crevices. (9)

NEOPHYLLOBIIDAE

Neophyllobius sp.: Dapeng Cove (Hong Kong). On rocks and seaweed, water depth 6 m. (9)

FORTUYNIIDAE

Alismobates reticulatus Luxton: Hong Kong. On seaweed in high intertidal. (10)
A. rotundus Luxton: Hong Kong. Mangrove forest. (10)
Circellobates venustus Luxton: Hong Kong. High intertidal, deposited substrate, oysters, and barnacles. (10)
Fortuynia sinensis Luxton: Dapeng Cove (Hong Kong). On goose barnacle (*Capitulum mitella*), low rocky intertidal barnacles and red algae. (10)
Fortuynia sp.: Dapeng Cove (Hong Kong). High intertidal rocky crevices. (9)

SELENORIBATIDAE

Arotrobates granulatus Luxton: Hong Kong. Rocky crevices. (10)
A. lanceolatus Luxton: Hong Kong. Mangrove forest (on green algae). (10)
Psednobates uncunguis Luxton: Hong Kong. Green algae. (10)

HYADESIIDAE

Amhyadesia bartschae Luxton: Hong Kong. Mid- and high-intertidal seaweed, oysters, and barnacles. (11)
A. heteromorpya Luxton: Hong Kong. Mid- and high-intertidal rocks and barnacles. (11)

RHODACARIDAE

Rhodacaropsis cheungae Luxton: Hong Kong. Sandy beach. (12)

UROPODIDAE

Uroobovella magna Hiramatsu & Hirschmann: Dapeng Cove (Hong Kong). High intertidal crevices. (9)

REFERENCES

(1). Bartsch, I. 1990a. *Acarothrix palustris* gen et. spec. nov. (Halacaroidea, Acari), ein Bewohner der Salzwiesen Södchinas. Zool. Anz. 224:204–210.
(2). Bartsch, I. 1990b. Halacaridae (Acari) of Hong Kong. In: Proceedings of the Second International Marine Biological Workshop: The Marine Flora and Fauna of Hong Kong and Southern China. Hong Kong University Press, pp:661–665.
(3). Bartsch, I. 1991b. Halacariden (Acari) von Hong Kong. Beschreibung von drei Arten der Gattung Copidognathus. Mitt. hamb. zool. Mus. Inst. 88:175–184.
(4). Bartsch, I. 1991b. Arenicolous Halacaridae (Acari) from Hong Kong. Asian Mar. Biol.:57–76.
(5). Bartsch, I. 1992a. Halacariden (Acari) von Hong Kong. Beschreibung von drei Copidognathus - Arten aus dem Sublitoral. Entomol. Mitt. zool. Mus. Hamburg 10(145):187–198.
(6). Bartsch, I. 1992b. Two new species of littoral *Agauopsis* (Acari: Halacaridae) from Hong Kong. In: Proceedings of the Fourth International Marine Biological Workshop: The Marine Flora and Fauna of Hong Kong and Southern China. Hong Kong University Press, pp:243–250.
(7). Bartsch, I. 1992c. Hong Kong rhombognathine mites (Acari: Halacaridae). In: Proceedings of the Fourth International Marine Biological Workshop: The Marine Flora and Fauna of Hong Kong and Southern China. Hong Kong University Press, pp:251–276.
(8). Bartsch, I. 1992d. Two new species of arenicolous *Agauopsis* (Acari: Halacaridae) from Hoi Ha Wan. In: Proceedings of the Fourth International Marine Biological Workshop: The Marine Flora and Fauna of Hong Kong and Southern China. Hong Kong University Press, pp.891–898.
(9). Luxton, M. 1986. Mites (Arachnida: Acari) from the coast of Hong Kong. In: Proceedings of the Fourth International Marine Biological Workshop: The Marine Flora and Fauna of Hong Kong and Southern China Hong Kong University Press, pp:667–671.
(10). Luxton, M. 1992a. Oribatid mites from the marine littoral of Hong Kong (Acari: Cryptostigmata). In: The marine flora and fauna of Hong Kong and southern China III. Proceedings of the Fourth International Marine Biological Workshop: The Marine Flora and Fauna of Hong Kong and Southern China. Hong Kong University Press, pp:221–228.
(11). Luxton, M. 1992b. Hong Kong hyadesiid mites (Acari: Astigmata). In: Proceedings of the Fourth International Marine Biological Workshop. Hong Kong University Press, pp:229–236.
(12). Luxton, M. 1992c. A new species of *Rhodacaropsis* (Acari: Mesostigmata) from interstitial coastal sand in Hong Kong. In: Proceedings of the Fourth International Marine Biological Workshop: The Marine Flora and Fauna of Hong Kong and Southern China. Hong Kong University Press, pp:237–242.

Compiled by I. Bartsch and Huang Zongguo; Edited by Hu Chengye

INSECTA

HEMIPTERA
GNRRIDAE

Halobates germanus White: Offshore of East China Sea and South China Sea. (1, 2)

H. micans Eschscholtz: East China Sea (26°28'–31° N); offshore of South China Sea. (1, 2)
H. sericeus Eschscholtz: East China Sea (26°28'–31° N); offshore of South China Sea. (1, 2)

CORIXIDAE

Trichocorixa sp.: Hong Kong. Brackish water and freshwater ditch. (2)

DIPTERA
CANACEIDAE

Asclepios shiranus Coreanus: Hong Kong. Mangrove forest. (2, 3)
Trichocanace sinensis: Hong Kong. Mangrove forest. (2, 3)

CERATOPOGONIDAE

Culicoides sp.: Hong Kong. Mangrove forest. (2, 3)

CULICIDAE

Aedes togoi Theobald: Hong Kong. Brackish water and freshwater ditch. (2)
Culex sitiens: Hong Kong. Brackish water and freshwater ditch. (2)

TENDIPEDIDAE

Chironomus sp.: Hong Kong. Brackish water and freshwater ditch. (2)

ASILIDAE

Philodicus javanicus: Hong Kong. Sandy beach. (2)

EPHYDRIDAE

Ephydra sp.: Hong Kong. Brackish water and freshwater ditch. (2)

ODONATA
LIBELLULIDAE

Pantala flavescens Fabr: Hong Kong. Brackish water and freshwater ditch. (2)

COLLEMBOLA
NEANURIDAE

Oudemansia esakii: Hong Kong. Sandy beach. (2)

STAPHYLINIDAE

Bryothinusa sp.: Hong Kong. Sandy beach. (2)

COLEOPTERA
TENEBRIONIDAE

Gonocephalum pseudopubens: Hong Kong. Sandy beach. (2)

CICINDELIDAE

Cicindela anchoralis: Hong Kong. Sandy beach. (2)

ORTHOPTERA
TETTIGONIIDAE

Callimenellus fumidus: Hong Kong. Rocky shore. (2)

HOMOPTERA
COCCIDAE

Ceroplastes rubens Maskell: Hong Kong. Mangrove leave. (2)

LEPIDOPTERA
PSYCHIDAE

Hyalarcta sp.: Hong Kong. Mangrove forest. (2)

REFERENCES*

(1). Cheng Lanna 1982. Halobates from the coastal waters of China. Oceanologia & Limnologia Sinica, 13(4):346–349
(2). Cheng L. and D. S. Hill, 1980. Marine insects of Hong Kong. In: Proceedinngs of the First Marine Biological Workshop: The marine flora and fauna of Hong Kong and southern China, Hong Kong University Press, pp:172–183.
(3). Morton, B. 1983. The sea shore ecology of Hong Kong. Hong Kong University Press, 350pp.

*: (1) in Chinese

Compiled by Huang Zongguo and Zeng Guoshou; Edited by Hu Chengye

CRUSTACEA

BRANCHIOPODA
ANOSTRACA
ARTEMIIDAE

Artemia salina (Linnaeus) [*A. parthenogenetica*, *Artemia* sp. (8)]: Along China's coast. Salt pond and other high salinity water. (2)

CLADOCERA
LEPTODORIDAE

Leptodora kindti (Focke): Mouth of Jiulong River (Fujian). (1)

SIDIDAE

Diaphanosoma brachyurun (Lievin): Mouth of Jiulong River (Fujian). (1)
D. leuchtenbergianum Fischer: Mouth of Jiulong River (Fujian). (1)
Latonopsis sp.: Mouth of Jiulong River (Fujian). (1)
Penilia avirostris Dana: From Bohai to South China Sea. (1–5, 7)
Sida crystallina (O. F. Müller): Mouth of Jiulong River (Fujian). (1)

DAPHNIIDAE

Ceriodaphnia cornuta Sars: Mouth of Jiulong River (Fujian). (1)
C. laticaudata P. E. Müller: Mouth of Jiulong River (Fujian). (1)
C. quadrangula (O. F. Müller): Mouth of Jiulong River (Fujian). (1)
Daphnia carinata King: Mouth of Jiulong River (Fujian). (1)
D. longispina O. F. Müller: Mouth of Jiulong River (Fujian). (1)
D. obtusa Kurz: Mouth of Jiulong River (Fujian). (1)
D. pulex Leydig: Mouth of Jiulong River (Fujian). (1)
Scapholeberis kingi Sars: Mouth of Jiulong River (Fujian). (1)
Simocephalus vetulus (O. F. Müller): Mouth of Jiulong River (Fujian). (1)

MOINIDAE

Moina macrocopa (Straus): Mouth of Beitang River (Zhejiang) and Jiulong River (Fujian). (1, 6)
M. micrura Kurz: Mouth of Beitang River (Zhejiang). (6)

BOSMINIDAE

Bosmina longirostris (O. F. Müller): Mouth of Jiulong River (Fujian). (1)
Bosmina sp.: Mouth of Jiulong River (Fujian). (1)

MACROTHRICIDAE

Macrothrix brevicornis Shen, Tai & Chiang: Mouth of Jiulong River (Fujian). (1)
M. laticornis (Jurine): Mouth of Jiulong River (Fujian). (1)
M. rosea (Jurine): Mouth of Jiulong River (Fujian). (1)
M. spinosa King: Mouth of Jiulong River (Fujian). (1)

CHYDORIDAE

Alona quadrangularis (O. F. Müller): Mouth of Jiulong River (Fujian). (1)
A. rectangula Sars: Mouth of Jiulong River (Fujian). (1)
Alonella globulosa Daday: Mouth of Jiulong River (Fujian). (1)
Chydorus ventricosus Daday: Mouth of Jiulong River (Fujian). (1)
Leydigia ciliata (Gauthier): Mouth of Jiulong River (Fujian). (1)
Pleuroxus hamulatus Birge: Mouth of Jiulong River (Fujian). (1)
P. laevis Sars: Mouth of Jiulong River (Fujian). (1)

PODONIDAE

Evadne nordmanni Loven: North of Hangzhou Bay (Zhejiang). (2–5)
E. tergestina Claus: From Bohai to South China Sea. (1, 3, 4, 5, 7)
Podon polyphemoides (Leuckart): Northern China coast; Taiwan Strait. (3, 4, 5)
P. schmackeri Poppe: From Yellow Sea to South China Sea. (1, 3, 4, 5, 7)

REFERENCES*

(1). Chen Yaping and Huang Jiaqi, 1992. Distribution of Cladocera in Jiulong River Mouth. Journal of Oceanography in Taiwan Strait, 11(3):233–237.
(2). Chen Shouzhong et al., 1982. Zoological Illustration from China: Crustacea (I). China Science Press, pp:5–47.
(3). Cheng, C. and S. L. Chen, 1966. Studies on the marine cladocera of China, I. Taxonomy. Oceanologia & Limnologia Sinica, 8(2):168–174.
(4). Cheng, C. and Cao Wenqing, 1982. Studies of cladocera in China Seas, I. Distribution. Acta Oceanologica Sinica, 4(6):731–742.
(5). Cheng C. and Cao Wenqing, 1987. Marine Cladocera Biology. Xiamen University Press, 177pp.
(6). Zhong Yicheng et al., 1984. A preliminary study on the ecology of zooplankton in the estuary of Beitang River. Acta Ecologica Sinica, 4(4):393–400.
(7). Cai Bingji, 1990. Abundance of cladocera in Daya Bay. Collections of Papers on Marine Ecology in the Daya Bay (II). China Ocean Press, pp:369–373.
(8). Sorgeloos P. et al., 1984. Production and use of *Artemia* in aquaculture. CMFRI Special Publication, 74pp.

*: (1)-(7) in Chinese
Compiled by Cai Bingji and Tang Senming; Edited by Zheng Zhong

OSTRACODA
CYPRIDINIFORMES
CYPRIDINIDAE

Amphisiphonostra naviformis Poulsen: East China Sea; western Taiwan Strait; Nansha (=Spratly Islands). (9, 10, 18)
Codonocera elongata Poulsen: East China Sea; western Taiwan Strait. (9, 10)
C. goniacantha Müller: Taiwan shallow shore. (27)
C. mortenseni Poulsen: Eastern Taiwan Strait; northern Taiwan. (27)
C. polygonia (Müller): East China Sea; eastern Taiwan Strait; northern Taiwan. (10, 19, 27)
C. pusilla (Müller): Western Taiwan Strait; Nansha (=Spratly Islands). (9, 10, 18)
C. stellifera (Claus): Eastern East China Sea. (10, 19)
Cypridina acuminata (Müller): Southern Yellow Sea; East China Sea; eastern and western Taiwan Strait; northern Taiwan; Hong Kong; northern and central South China Sea; Nansha (=Spratly Islands). (1, 2, 3, 10, 12, 13, 15–18, 24–29)
C. amphiacantha Müller: East China Sea. (10)
C. dentata (Müller): East China Sea; eastern and western Taiwan Strait; Hong Kong; central South China Sea, Nansha (=Spratly Islands). (1, 2, 3, 10, 12, 13, 16, 18, 27, 28, 29)
C. inermis Müller: East China Sea; eastern and western Taiwan Strait; Taiwan shallow shore; Hong Kong. (1, 2, 3, 26, 27, 28)
C. japonica (Müller): Northeastern East China Sea. (19)
C. lepidophora (Müller): Bashi Channel (Taiwan). (9)
C. nami Chavtur: Nansha (=Spratly Islands). (18)
C. nana Poulsen: East China Sea; Hainan; central South China Sea. (9, 10, 12)
C. natans (Brady): East China Sea. (10)
C. punctata (Müller): Northern South China Sea. (19)
C. serrata (Müller): Eastern Taiwan Strait; northern and southeastern Taiwan; southwestern Yongxing Island (Xisha=Paracel Islands). (7, 9, 27)
C. sinuosa (Müller): Eastern Taiwan Strait; Hong Kong; central and southern South China Sea. (12, 19, 26, 28)

Cypridinodes asymmetrica (Müller): Eastern and western Taiwan Strait; Taiwan shallow shore; Nansha (=Spratly Islands). (1, 9, 18, 26, 29)

C. avis Tseng: East China Sea. (27)

C. bairdi (Brady) [*Cypridina bairdii*]: East China Sea near shore; eastern and western Taiwan Strait. (2, 3, 19)

C. codonocera Tseng: Taiwan shallow shore. (27)

C. galatheae Poulsen: Western Taiwan Strait; Taiwan shallow shore; northern, central and southwestern South China Sea; Nansha (=Spratly Islands). (1, 9, 15, 18, 19, 27, 29)

C. minuta Poulsen: East China Sea; western Taiwan Strait; Daya Bay (Guangdong). (6, 9, 10)

Euphilomedes agilis (Thomason): Hong Kong. (26)

E. corrugata (Brady): Dongshan (Fujian). (9)

E. interpuncta (Baird): East China Sea; Hainan; central South China Sea. (9, 10, 12)

E. japonica (Müller) [*Philomedes japonica*]: Western Taiwan Strait; Taiwan shallow shore. (1, 2, 3, 9)

E. longiseta (Juday): Southern Yellow Sea; East China Sea; western Taiwan Strait. (9, 10, 29)

E. nodosa Poulsen: Western East China Sea; eastern Taiwan Strait. (19, 26)

E. sordida (Müller): Western Taiwan Strait. (9)

Gigantocypris agassizi Müller: Central South China Sea. Water depth 1,000–4,000 m. (11, 12)

G. australis Poulsen: Central South China Sea. (11)

G. danae Poulsen: Southern Taiwan; central and southern South China Sea. (19)

G. dracontovali Cannon: Nansha (=Spratly Islands). (18)

Heterodesmus adamsii Brady: East China Sea near shore; western Taiwan Strait. (2, 3)

Macrocypridina castanea rotunda Poulsen: Offshore of eastern Taiwan; South China Sea. (19)

Melavargula japonica Poulsen: Eastern Taiwan Strait; central South China Sea. (12, 19)

Monopia flaveola Claus: Daya Bay (Guangdong); Hainan; central South China Sea; Nansha (=Spratly Islands). (6, 9, 12, 17, 18, 19)

M. tehani Tseng: Taiwan shallow shore; northern and central South China Sea. (15, 27)

Paradoloria dorsoserrata (Müller): Northeastern East China Sea. (19)

Paraphilomedes tricornuta Poulsen: Southern South China Sea. (19)

P. unicornuta Poulsen: East China Sea; Taiwan shallow shore; eastern Taiwan Strait; central South China Sea. (1, 19, 27)

Paravargula formosana Tseng: Taiwan shallow shore. (27)

P. hirsuta (Müller): Western Taiwan Strait; Taiwan shallow shore; Daya Bay (Guangdong); central South China Sea; Nansha (=Spratly Islands). (1, 6, 9, 12, 13, 16, 17, 18, 19, 29)

P. taiwantuia Tseng: Taiwan shallow shore; northern and central South China Sea; Zengmuansha (southern tip of Nansha=Spratly Islands). (15, 16, 27)

Philomedes eugeniae Skogsberg: Hainan; central South China Sea. (9, 12)

P. lilljeborg (Sars): Western Taiwan Strait. (13, 29)

Pterocypridina alata Poulsen: Southwestern South China Sea. (19)

Skogsbergia crenulata Poulsen: Taiwan shallow shore. (1)

S. curvata Poulsen: Taiwan shallow shore. (1)

S. minuta Poulsen: Taiwan shallow shore; southwestern South China Sea. (1, 19, 27)

Vargula higendorfi (Müller): Western Taiwan Strait; Daya Bay (Guangdong); southwestern South China Sea; Nansha (=Spratly Islands). (6, 9, 18, 19, 29)

V. spinulosa Poulsen: Northeastern East China Sea. (19)

Zeugophilomedes polae (Graf): East China Sea; western Taiwan Strait; Dongshan (Fujian). (9, 10, 29)

SARSIELLIDAE

Eusarsilla longinenna Poulsen: Western Taiwan Strait. (9, 29)

E. parvispinosa Poulsen: Western Taiwan Strait. (9)

E. spinulosa Poulsen: East China Sea near shore; western Taiwan Strait. (2, 3, 29)

E. tumida (Scott): East China Sea near shore; western Taiwan Strait. (2, 3)

Scottiella crispata (Scott): Western South China Sea. (24)

ASTEROPIDAE

Asteropina grimaldi (Skogsberg): East China Sea near shore; western Taiwan Strait; Daya Bay (Guangdong). (2, 3, 6, 29)

A. minuta Poulsen: Western South China Sea. (20)

Asteropteron fuscum Müller: Northeastern East China Sea. (20)

Cyclasterope bisetosa Poulsen: Daya Bay (Guangdong). (6)

C. fascigera Brady: Western Taiwan Strait; Daya Bay (Guangdong). (6, 9)

C. hilgendorfi Müller: Western Taiwan Strait. (9)

Cycloleberis americana (Müller): Western Taiwan Strait. (29)

C. biminiensis Kornicker: Western Taiwan Strait. (9)

C. bradyi Poulsen: Western Taiwan Strait. (29)

C. brevis (Müller): Western Taiwan Strait; Daya Bay (Guangdong); central South China Sea. (6, 9, 12, 13, 29)

C. poani Tseng: Taiwan shallow shore. (1, 27)

C. similis (Brady): East China Sea near shore; western Taiwan Strait; Daya Bay (Guangdong). (1, 2, 3, 6, 9, 29)

Diasterope bisetosa Poulsen: Taiwan shallow shore. (27)

Synasterope knudseni Poulsen: Western South China Sea. (20)

HALOCYPRIFORMES
HALOCYPRIDAE

Alacia alata (Müller): Offshore of East China Sea. (10)

A. alata major (Rudjakov): East China Sea. (10)

A. alata minor (McHardy): Northeastern East China Sea; southeastern Taiwan. (7, 8)

A. belgicae (Müller): Central South China Sea. Water depth 500–1,000 m. (11, 12)

A. hettacro (Müller): Central South China Sea. (12)

A. leptothrix (Culler) [*Conchoecia leptothrix*]: Northeastern East China Sea; northern and central South China Sea; Nansha (=Spratly Islands). (8, 12, 15, 17, 18, 22)

A. valdiviae (Müller) [*Conchoecia valdiviae*]: Northeastern East China Sea; northern and central South China Sea; Nansha (=Sratly Islands). (8, 12, 15, 17, 18, 22)

Archiconchoecia cucullata (Brady): South China Sea; Nansha (=Spratly Islands). (15, 18, 21)

A. falcata Deevey: Central South China Sea. (12)

A. striata Müller: Taiwan Strait; Taiwan shallow shore; southeastern Taiwan; Bashi Channel (Taiwan); northern and central South China Sea; Nansha (=Spratly Islands). (1, 7, 15–18, 24, 27)

A. versicula Deevey: Southeastern Taiwan; Bashi Channel (Taiwan). (7)

Bathyconchoecia angeli George: East China Sea; Nansha (=Spratly Islands). (10, 18)

B. crosnieri Poulsen: East China Sea. (10)

B. galerita Deevey: Nansha (=Spratly Islands). (18)

B. lacunosa (Müller): Central South China Sea. Water depth 1,000–4,0000 m. (11, 12)

B. paulula Deevey: Nansha (=Spratly Islands). (18)

Boroecia barealis (Sars) [*Conchoecia barealis*]: Northern, southeastern, and southern Taiwan. (27)

Conchoecetta acuminata Claus [*Conchoecia acuminata*]: East China Sea; Taiwan Strait; eastern and southeastern Taiwan; Bashi Channel (Taiwan); Hong Kong; northern and central South China Sea; Nansha (=Spratly Islands). (1, 7, 8, 10, 12–15, 17, 18, 22, 24–27, 29)

C. giesbrechti (Müller) [*Conchoecia giesbrechti*]: East China Sea; Taiwan; northern and central South China Sea; Nansha (=Spratly Islands). (1, 10, 11, 12, 15, 17, 18)

Conchoecia hyalophyllum Claus: Northeastern East China Sea; southeastern Taiwan; central South China Sea. (7, 8, 12)

C. lophura Müller: Northeastern East China Sea; Taiwan; Bashi Channel (Taiwan); northern and central South China Sea; Nansha (=Spratly Islands). (1, 7, 8, 12, 15, 17, 18, 22, 24, 27)

C. lophura lissoides Martens: Taiwan. (1, 7)

C. macrocheira Müller: East China Sea; Taiwan; Taiwan Strait; northern and southern South China Sea; Nansha (=Spratly Islands). (1, 8, 10, 12, 14, 15, 17, 18, 25, 26, 27)

C. magna Claus: East China Sea; Taiwan; Taiwan Strait; Bashi Channel (Taiwan); Hong Kong; northern and central South China Sea; Nansha (=Spratly Islands). (1, 7, 8, 10, 12–15, 17, 18, 24, 27, 28, 29)

C. magna rhombica Müller: East China Sea; Taiwan shallow shore; central South China Sea. (1, 8, 10, 12, 14)

C. parvidentata Müller: East China Sea, northern and southeastern Taiwan; Bashi Channel (Taiwan); northern and central South China Sea; Nansha (=Spratly Islands). (7, 8, 10, 12, 14, 15, 17, 18, 24, 27)

C. subarcuata Claus: East China Sea; Taiwan; Bashi Channel (Taiwan); northern and central South China Sea; Nansha (=Spratly Islands). (1, 7, 8, 10, 12, 14, 15, 17, 18, 27)

Conchoecilla daphnoides Claus: East China Sea; northern, eastern, and southern Taiwan; Bashi Channel (Taiwan); central South China Sea. (7, 8, 10, 12, 22, 24, 27)

C. daphnoides minor Müller [*Conchoecia daphnoides*]: East China Sea; Taiwan; northern and central South China Sea; Nansha (=Spratly Islands). (1, 8, 10, 15, 17, 18)

Conchoecissa ametra (Müller) [*Conchoecia ametra*]: Eastern Taiwan; northern and central South China Sea; Nansha (=Spratly Islands). (11, 12, 15, 18, 22, 27)

C. imbricata (Brady) [*Conchoecia imbricata*]: East China Sea; Taiwan; Bashi Channel (Taiwan); northern and central South China Sea; Nansha (=Spratly Islands). (1, 8, 10, 14, 15, 17, 18, 22, 27)

C. plinthina (Müller): Central South China Sea. Water depth 500–1,000 m. (11, 12)

C. symmetrica (Müller) [*Conchoecia symmetrica*]: Northern and central South China Sea; Nansha (=Spratly Islands). (12, 15, 18)

Discoconchoecia discophora (Müller) [*Conchoecia discophora*]: Northern, eastern, and southern Taiwan. (7, 24, 27)

D. elegans (Sars) [*Conchoecia elegans*]: East China Sea; Taiwan Strait; Taiwan; Bashi Channel (Taiwan); Hong Kong; northern and central South China Sea; Nansha (=Spratly Islands). (1, 7, 8, 10–15, 17, 18, 24, 27, 28, 29)

D. pseudodiscophora (Rudjakov): East China Sea; central South China Sea. (10, 12)

D. tamensis (Poulsen) [*Conchoecia tamensis, Paraconchoecia tamensis*]: East China Sea; western Taiwan Strait; Taiwan; northern and central South China Sea; Xisha (=Paracel Islands). (1, 7, 8, 10, 12, 13, 15, 17, 18, 27)

Euconchoecia aculeata (Scott): Southern Yellow Sea; East China Sea; Taiwan Strait; Taiwan; Hong Kong; northern, central, and western South China Sea; Nansha (=Spratly Islands). (1, 2, 8, 10, 12–18, 21, 25, 28, 29)

E. bifurata Chen & Lin: East China Sea; western Taiwan Strait; Taiwan shallow shore; central South China Sea. (1, 4, 7, 10, 12, 29)

E. chierchiae Müller: East China Sea; Taiwan; western Taiwan Strait; Bashi Channel (Taiwan); Hong Kong. (1, 7, 8, 10, 14, 25, 27, 28, 29)

E. elongata Müller: Southern Yellow Sea; East China Sea; Taiwan Strait; Taiwan shallow shore; Bashi Channel (Taiwan); Hong Kong; central South China Sea; Nansha (=Spratly Islands). (7, 10, 27, 29)

E. maimai Tseng: Southern Yellow Sea; East China Sea; Taiwan Strait; Taiwan; central South China Sea; Nansha (=Spratly Islands). (1, 2, 7, 8, 10, 12, 13, 14, 16, 17, 18, 24, 25, 27, 28, 29)

E. shenghwai Tseng: Southern Yellow Sea; East China Sea; western Taiwan Strait; southeastern and northwestern Taiwan. (7, 10, 27, 29)

Fellia bicornis (Müller): South China Sea, water depth 200–500 m; Nansha (=Spratly Islands). (11, 12, 15, 18, 21)

F. cornuta (Müller): Northeastern East China Sea; South China Sea; Nansha (=Spratly Islands). (8, 15, 17, 18, 21)

F. cornuta dispar (Müller): East China Sea. (10)

Gaussicia edentata (Müller): Nansha (=Spratly Islands). (18)

G. gaussi (Müller): Nansha (=Spratly Islands). (18)

G. incisa (Müller) [*Conchoecia incisa*]: East China Sea; Taiwan; South China Sea; Nansha (=Spratly Islands). (1, 10, 12, 15, 17, 18, 22)

Halocypria globosa Claus: East China Sea; western Taiwan Strait; Taiwan; Hong Kong; northern and central South China Sea; Nansha (=Spratly Islands). (1, 8, 10, 12, 14, 15,1 8, 21, 24, 27, 28)

Halocypris brevirostris (Dana) [*H. inflata*]: East China Sea; Taiwan Strait; Taiwan; Bashi Channel (Taiwan); Hong Kong; central and northern South China Sea; Nansha (=Spratly Islands). (1, 2, 7, 8, 10, 12–15, 17, 18, 21, 24, 25, 27, 28, 29)

Loricoecia ctenophora (Müller): Northeastern East China Sea; eastern and southeastern Taiwan; central South China Sea; Nansha (=Spratly Islands). (8, 12, 18, 22)

L. loricata (Claus) [*Conchoecia loricata*]: East China Sea; Taiwan; Bashi Channel (Taiwan); northern, central and southeastern South China Sea; Nansha (=Spratly Islands). (1, 7, 8, 10, 11, 12, 14, 15, 17, 18, 22, 27)

Metaconchoecia abyssalis (Rudjakov): Northeastern East China Sea; central South China Sea; Xisha (=Paracel Islands). (8, 12, 17, 18)

M. glandulosa (Müller) [*Conchoecia glandulosa*]: Northern and southern Taiwan; northern and central South China Sea; Nansha (=Spratly Islands). (15, 18, 27)

M. isocheira (Müller): Northern, eastern, and southeastern Taiwan; Bashi Channel (Taiwan); Nansha (=Spratly Islands). (7, 24, 27)

M. kyrtophora (Müller) [*Conchoecia kyrtophora*]: Northern, eastern, and southern Taiwan; Nansha (=Spratly Islands). (18, 27)

M. macromma (Müller) [*Conchoecia macromma*]: Northern and central South China Sea; Nansha (=Spratly Islands). (12, 15, 18)

M. pusilla (Müller) [*Conchoecia pusilla*]: Bashi Channel (Taiwan); northern and central South China Sea; Nansha (=Spratly Islands). (7, 12, 15, 18)

M. rotundata (Müller) [*Conchoecia rotundata*]: East China Sea; Taiwan; Bashi Channel (Taiwan); northern and central South China Sea; Nansha (=Spratly Islands). (1, 7, 8, 10, 12, 15, 18)

M. skogsbergi (Iles) [*Conchoecia skogsbergi*]: East China Sea; Taiwan. (7, 10, 24, 27)

M. teretivalvata (Iles): Bashi Channel (Taiwan). (7)

Microconchoecia acuticosta (Müller): Southern and southeastern Taiwan. (7, 27)

M. curta (Lubbock) [*Conchoecia curta*]: Southern Yellow Sea; Taiwan Strait; Taiwan; Bashi Channel (Taiwan); northern and central South China Sea; Nansha (=Spratly Islands). (1, 7, 8, 10, 12, 13, 15–18, 27, 29)

M. echinulata (Claus): Southeastern Taiwan; Bashi Channel (Taiwan). (7)

M. stigmatica (Müller): Nansha (=Spratly Islands). (18)

Mollicia acanthophora (Müller): Southern South China Sea. (22)

M. kampta (Müller) [*Conchoecia kampta*]: Northern and southeastern Taiwan; northern and central South China Sea. (15, 27)

M. minki Poulsen: Southeastern Taiwan. (7)

M. mollis (Müller): East China Sea; Bashi Channel (Taiwan); Nansha (=Spratly Islands). (7, 10, 18)

Obtusata antarotica (Müller) [*Conchoecia antarotica*]: Taiwan. (27)

Orthoconchoecia atlantica (Lubbock) [*Conchoecia atlantica*]: East China Sea; Taiwan Strait; Taiwan; northern and central South China Sea; Nansha (=Spratly Islands). (1, 7, 8, 10, 12–15, 17, 18, 22, 24, 25, 27, 29)

O. bispinosa (Claus) [*Conchoecia bispinosa*]: East China Sea; eastern Taiwan Strait; Taiwan; Bashi Channel (Taiwan); northern and central South China Sea; Nansha (=Spratly Islands). (1, 7, 8, 10, 11, 12, 14, 15, 17, 18, 22, 24, 27)

O. haddoni (Brady & Norman) [*Conchoecia haddoni*]: Northeastern East China Sea; Taiwan; Bashi Channel (Taiwan); central and south South China Sea. (7, 8, 22, 24, 27)

O. secernenda (Vavra) [*Conchoecia secernenda*]: East China Sea; northern, eastern, and southern Taiwan; northern and central South China Sea. (7, 8, 10, 12, 24, 27)

O. striola (Müller) [*Conchoecia striola*]: East China Sea; southeastern, northeastern, and southwestern Taiwan; northern and central South China Sea. (7, 8, 10, 12, 15, 22)

Paraconchoecia aequiseta (Müller) [*Conchoecia aequiseta*]: Taiwan; eastern Taiwan Strait; Bashi Channel (Taiwan); northern and central South China Sea; Nansha (=Spratly Islands). (1, 7, 12, 15, 18, 24, 27)

P. allotherium (Müller): East China Sea; southeastern Taiwan; Bashi Channel (Taiwan); central South China Sea. Water depth 100–200 m. (7, 8, 10, 11, 12)

P. brachyaskos (Müller) [*Conchoecia brachyaskos*]: Taiwan; northern and central South China Sea; Nansha (=Spratly Islands). (7, 12, 15, 17, 18, 27)

P. caudata (Müller): Central South China Sea. (12)

P. cophopyga (Müller) [*Conchoecia cophopyga*]: Northern and central South China Sea; Nansha (=Spratly Islands). (15, 18)

P. dasyophthalma (Müller) [*Conchoecia dasyophthalma*]: Northern and central South China Sea; Nansha (=Spratly Islands). (12, 15, 18)

P. decipiens (Müller) [*Conchoecia decipiens*]: East China Sea; Taiwan Strait; Taiwan; Bashi Channel (Taiwan); northern and central South China Sea; Nansha (=Spratly Islands). (1, 7, 8, 10, 12–15, 17, 18, 25)

P. dentata (Müller) [*Conchoecia dentata*]: Northeastern East China Sea; Taiwan; northern and central South China Sea; Nansha (=Spratly Islands). (1, 8, 12, 15, 17, 18)

P. diacanthus Chen & Lin: Northeastern East China Sea. (23)

P. dorsotuberculata (Müller) [*Conchoecia dorsotuberculata*]: Northern and central South China Sea; Nansha (=Spratly Islands). (15, 18)

P. echinata (Müller) [*Conchoecia echinata*]: East China Sea; western Taiwan Strait; Taiwan; Bashi Channel (Taiwan); northern and central South China Sea; Nansha (=Spratly Islands). (1, 7, 8, 10, 12, 14, 15, 17, 18)

P. gerdhartmanni Martens: East China Sea. (10)

P. inermis Claus [*Conchoecia inermis*]: East China Sea; Taiwan; northern and central South China Sea; Nansha (=Spratly Islands). (1, 10, 12, 15, 17, 18)

P. macroprocera (Angel): East China Sea; southeastern Taiwan; Bashi Channel (Taiwan); central South China Sea. (7, 8, 10, 12)

P. mamillata (Müller): Central South China Sea. (12)

P. microprocera (Angel): East China Sea; southeastern Taiwan; Bashi Channel (Taiwan); central South China Sea; Nansha (=Spratly Islands). (7, 8, 10, 12, 14, 18)

P. oblonga Claus [*Conchoecia oblonga*]: Southern Yellow Sea; East China Sea; Taiwan Strait; Taiwan; Bashi Channel (Taiwan); Hong Kong; northern and central South China Sea; Xisha (=Paracel Islands). (1, 7, 8, 10–15, 17, 18, 24, 27, 28, 29)

P. procera (Müller) [*Conchoecia procera*]: East China Sea; Taiwan Strait; Taiwan; Bashi Channel (Taiwan); northern and central South China Sea; Nansha (=Spratly Islands). (1, 7, 8, 10–15, 17, 18, 25, 27, 29)

P. reticulata (Müller) [*Conchoecia reticulata*]: Northern and central South China Sea; Nansha (=Spratly Islands). (12, 15, 18)

P. spinifera Claus [*Conchoecia spinifera*]: East China Sea; Taiwan Strait; Taiwan; Bashi Channel (Taiwan); northern and central South China Sea; Nansha (=Spratly Islands). (1, 7, 8, 10, 11, 12, 14, 15, 17, 18, 22, 24–27)

P. vitjazi (Rudjakov): Central South China Sea. Water depth 1,000–4,000 m. (11, 12)

Paramollicia dichotoma (Müller): Taiwan; Nansha (=Spratly Islands). (18, 27)

P. plactolycos (Müller): Central South China Sea. Water depth 500–1,000 m. (11, 12)

P. rhynchena (Müller) [*Conchoecia rhynchena*]: Eastern, northern, and southern Taiwan; Nansha (=Spratly Islands). (18, 22, 27)

P. siboga (Müller): Northeastern East China Sea. (14)

Platyconchoecia prosadene (Müller): Taiwan shallow shore. (1)

Porroecia porrecta Claus [*Conchoecia porrecta*]: East China Sea; Taiwan Strait; Taiwan; Bahsi Strait (Taiwan); northern and central South China Sea; Nansha (=Spratly Islands). (1, 2, 7, 8, 10–14, 17, 18, 24, 25, 27, 28, 29)

P. spinirostris Claus [*Conchoecia spinirostris*]: Southern Yellow Sea; East China Sea; Taiwan Strait; Taiwan; Bashi Channel (Taiwan); northern and central South China Sea; Nansha (=Spratly Islands). (1, 2, 7, 8, 10–15, 17, 18, 24, 27, 28, 29)

Pseudoconchoecia concentrica (Müller) [*Conchoecia concentrica*]: East China Sea; western Taiwan Strait; Taiwan; Bashi Channel (Taiwan); Guangdong; northern and central South China Sea; Nansha (=Spratly Islands). (1, 2, 7, 8, 9, 12–15, 17, 18, 19, 22, 27, 29)

P. serrulata Claus: Northeastern East China Sea; southeastern Taiwan; Bashi Channel (Taiwan). (7, 8)

Spinoecia crassispina Chen & Lin: Offshore of East China Sea; central South China Sea. (5, 8, 10, 12)

S. parthenoda (Müller) [*Conchoecia parthenoda*]: East China Sea; Taiwan Strait; Taiwan; Bashi Channel (Taiwan); central, eastern, and southern South China Sea; Nansha (=Spratly Islands). (1, 7, 8, 10, 11, 12, 14, 17, 18, 22, 24, 25, 27, 29)

S. pseudoparthenoda (Angel) [*Conchoecia pseudoparthenoda*]: East China Sea; Taiwan; central South China Sea; Nansha (=Spratly Islands). (1, 7, 8, 10, 12, 17)

REFERENCES*

(1). Zhu Changshou, Huang Jiaqi and Li Shaoqing, 1980. Studies of the Ecology of Planktonic Ostracod in the Upwelling Area of Minnan - Taiwan Bank. Tropic Oceanology, 10(4):67–73.

(2). Chen Ruixiang, 1982. The distribution of planktonic Ostracoda along the western coast of Taiwan Straits, Acta Oceanologica Sinica, 1(2):289–298.

(3). Chen Ruixiang , 1982. On planktonic Ostracoda in nearshore waters of the East China Sea. Marine Science Bulletin, 1(6): 45–57.

(4). Chen Ruixiang and Lin Jinhong, 1985. A new species of *Euconchoecia* from the East China Sea. Acta Oceanolgica Sinica, 4(1):131–134.

(5). Chen Ruixiang and Lin Jinhong, 1987. *Spinoecia crassispina* (nov. sp.) - A new species of planktonic Ostrapoda. Acta Oceanologica Sinica, 6(1):153–158.

(6). Chen Ruixiang and Lin Jinhong, 1990. Ostrapoda near inlet of Nuclear Power Station in Daya Bay. Collections of Papers on Marine Ecology in the Daya Bay (II). China Ocean Press, pp:364–368.

(7). Chen Ruixiang and Lin Jinhong, 1994. Notes on the distribution of planktonic Ostrapoda between the source of Kuroshio and west of Taiwan Strait. Acta Oceanologica Sinica, 13(3): 445–452.

(8). Chen Ruixiang and Lin Jinhong, 1984. The comparison between abundance and diversity of Ostracoda in the northeast East China Sea and in the source area of the Kuroshio. Papers on Kuroshio of Survey Research (V). China Ocean Press.

(9). Chen Ruixiang and Lin Jinhong. Planktonic Ostrapoda in the nearshore waters of the East China Sea. (manuscript)

(10). Chen Ruixiang and Lin Jinhong, 1994. Ecological characteristics of Ostrapoda in the South Huanghai Sea and East China Sea. Acta Oceanologica Sinica, 13(3):401–412.

(11). Chen Ruixiang and Lin Jinhong, 1989. Notes on vertical distribution of zooplankton in centre of South China Sea, 8(1):158–160.

(12). Chen Ruixiang et al., 1988. Plankton. In: Reports on Comprehensive Investigations of Environmental Resources of the Central South China Sea. China Ocean Press, 162–215, pp:237–238.

(13). Chen Ruixiang and Lin Jinhong, 1991. Ecological characteristics of planktonic Ostrapoda in the western waters of Taiwan Straits. Acta Oceanologica Sinica, 10(2): 289–296.

(14). Lin Jinhong and Chen Ruixiang, 1994. Ecology of the planktonic Ostrapoda in the Kuroshio area of the East China Sea. Ecology,14(2):174–179

(15). Chen Qingchao, Yin Jianqiang and Zhang Guxian, 1983. Studies on pelagic Ostrapods in the central and northern part of the South China Sea. Contributions on Marine Biological Research of the South China Sea. China Ocean Press, pp:82–132.

(16). Chen Qingchao, Zhang Guxian and Yin Jianqiang, 1987. Species, quantity and biology of zooplankton. In: Report on comprehensive survey and research about Zengmu Ansha-the southern boundary of China. China Science Press, pp:132- 146.

(17). Chen Qingchao, Zhang Guxian and Yin Jianqiang, 1989. Zooplankton. In: Research reports on the comprehensive oceanographic investigation of Nansha Islands and adjacent waters (I). China Science Press, Vol. 2:659–707

(18). Yin Jianqiang and Chen Qingchao, 1991. Species, fauna and geological distribution of planktonic Ostracoda. In: The Study Collections of Fauna and Geography of Marine Animals in Nansha Islands. China Ocean Press, pp:64–139.

(19). Poulsen, E. M. 1962. Ostracoda-Myodocopa. Part I. Cypridiniformes - Cypridinidae. Dana Report, 57:1.

(20). Poulsen, E. M. 1965. Ostracoda-Myodocopa. Part II. Cyprinidiniformes - Rutidermatidae, Sarsiellidae and Asteropidae. Dana Report, 65:1–484.

(21). Poulsen, E. M. 1969. Ostracoda-Myodocopa. Part III A. Halocypriformes - Thaumatocypridae and Halopridae. Dana Report, 75:1–99.

(22). Poulsen, E. M. 1973. Ostracoda-Myodocopa. Part III B. Halocypriformes - Halocypridae Conchoecinae. Dana Report, 84:1–224.

(23). Chen, R. X. and J. H. Lin, 1994. A new species of pelagic Ostracoda - *Paraconchoecia diacanthus*. Acta Oceanologica Sinica, 16(1):89–92.

(24). Tseng Wen-Young, 1970a. Occurrence of Ostracods in the neigbouring seas of Taiwan. Proceedings of the second CSK symposium, 285–295.

(25). Tseng Wen-Young 1970b. The zooplankton communities in the surface waters of Taiwan Strait. Proceedings of the second CSK symposium, 261–268.

(26). Tseng Wen-Young 1970c. A preliminary report on Cypridinids (Ostracoda) from Taiwan Strait. Kuroshio University, Hawaii, pp:339–346.

(27). Tseng Wen-Young 1977. Pelagic Ostracoda of Taiwan, Part 1 - Cypridiniformes. Taiwan Fisheries Research Institute, 30:1–240.

(28). Tseng Wen-Young 1980. The pelagic Ostracoda of Hong Kong. In: Proceedings of the First Marine Biological Workshop: The Marine Flora and Fauna of Hong Kong and southern China. Hong Kong University Press, pp:401–430.

(29). The Third Oceanology Institute, 1987. The comprehensive plankton survey in the western Taiwan Strait, 71pp (internal report).

*: (1)-(18), (23), and (29) in Chinese

Compiled by Chen Ruixiang and Lin Jinghong; Edited by Chen Qingchao

COPEPODA
CALANOIDA
CALANIDAE

Calanus sinicus Brodsky [*C. pacificus* Brodsky, Shen Jiarui, 1956; Zheng Zhong et al. 1965]: From Bohai to northern South China Sea. (1, 2, 4, 12, 14, 17, 25–28, 39)

Calanoides carinatus (Kröyer): East China Sea; South China Sea. (1, 2, 12, 14, 20, 21, 25, 28, 30)

Canthocalanus pauper (Giesbrecht): From Bohai to South China Sea. (1, 2, 12, 14, 17, 18–22, 25–28, 32, 39)

Nannocalanus minor (Claus): From Yellow Sea to South China Sea. (1, 2, 12, 14, 17, 18–21, 25–28, 39)

Neocalanus gracilis (Dana): From Yellow Sea to South China Sea. (1, 12, 14, 18–21, 25, 28, 29, 39)

N. robustior (Giesbrecht): East China Sea and South China Sea. (1, 2, 12, 14, 20, 25, 28, 29, 30, 39)

N. tenuicornis (Dana): From Yellow Sea to South China Sea. (1, 2, 12, 14, 19, 20, 25- 28, 39)

Undinula darwinii (Lubbock): From Yellow Sea to South China Sea. (1, 2, 12, 14, 17, 18–21, 25–28, 39)

U. vulgaris (Dana): From Yellow Sea to South China Sea. (1, 2, 12, 14, 17, 18- 21, 25- 28, 32, 39)

EUCALANIDAE

Eucalanus attenuatus (Dana): From Yellow Sea to South China Sea. (1, 2, 12, 14, 19, 20, 21, 25, 26, 28, 29, 39)

E. crassus Giesbrecht: East China Sea; South China Sea. (1, 2, 12, 14, 17–22, 25–28, 39)

E. elongatus (Dana): From Yellow Sea to South China Sea. (1, 2, 12, 14, 18–21, 28, 30, 39)

E. mucronatus Giesbrecht: From Yellow Sea to South China Sea. (1, 2, 12, 14, 20, 21, 25, 28, 30)

E. pileatus Giesbrecht: East China Sea; South China Sea. (1, 2, 14, 25–28, 30)

E. pseudattenuatus Sewell: East China Sea; South China Sea. (1, 2, 12, 14, 17–19, 20, 25, 28, 30)

E. subcrassus Giesbrecht: From Yellow Sea to South China Sea. (1, 2, 12, 14, 17–22, 25–28, 39)

E. subtenuis Giesbrecht: From Yellow Sea to South China Sea. (1, 2, 12, 14, 17, 18, 19, 21, 25, 28, 39)

Mecynocera clausi Thompson: From Yellow Sea to South China Sea. (1, 2, 12, 14, 20, 21, 25, 39)

Rhincalanus cornutus Dana: From Yellow Sea to South China Sea. (1, 2, 12, 14, 17–21, 25, 26, 28)

R. nasutus Giesbrecht: From Yellow Sea to South China Sea. (1, 2, 12, 14, 19, 20, 21, 25, 26, 28)

PARACALANIDAE

Acrocalanus andersoni Bowman: East China Sea; Taiwan Strait. (31)

A. gibber Giesbrecht: From Bohai to South China Sea. (1, 2, 12, 14, 17, 19, 20, 22, 25- 28, 30, 33)

A. gracilis Giesbrecht: From Yellow Sea to South China Sea. (1, 2, 9, 12, 14, 17, 18–22, 25, 26, 27, 30)

A. indicus Tanaka: Taiwan Strait; South China Sea. (12, 19, 20, 26, 27)

A. longicornis Giesbrecht: From Yellow Sea to South China Sea. (1, 2, 12, 14, 20, 21, 25–28, 30, 39)

A. monachus Giesbrecht: East China Sea; Taiwan Strait; South China Sea. (1, 2, 14, 19, 20, 27, 28, 30)

Bestiola amoyensis Li & Huang: East China Sea; northern South China Sea. (11, 26, 27)

B. sinicus (Shen & Lee) [*Acrocalanus sinicus* Shen & Lee, 1966]: Northern South China Sea. (10, 11)
Calocalanus contractus Farran: East China Sea; South China Sea. (1, 2, 14, 32)
C. gracilis Tanaka: Taiwan Strait, South China Sea. (1, 14, 20, 28, 32, 39)
C. monospinus Chen & Zhang: Taiwan Strait, South China Sea. (1, 14, 27, 32)
C. pavo (Dana): From Yellow Sea to South China Sea. (1, 2, 12, 14, 19, 20, 25, 27, 28, 32, 39)
C. pavoninus Farran: East China Sea; South China Sea. (1, 2, 12, 14, 25, 27, 28, 30, 32)
C. plumulosus (Claus): From Yellow Sea to South China Sea. (1, 2, 12, 14, 19, 20, 25, 27, 28, 30, 32)
C. styliremis Giesbrecht: East China Sea; South China Sea. (1, 2, 14, 20, 25, 27, 30, 32)
Paracalanus aculeatus Giesbrecht: From Yellow Sea to South China Sea. (1, 2, 12, 14, 20, 21, 22, 25–28, 30, 32, 33, 39)
P. crassirostris Dahl: From Bohai to South China Sea. (1, 2, 12, 14, 17, 22, 25–28, 30)
P. gracilis Chen & Zhang: East China Sea; South China Sea. (1, 2, 12, 14, 25–28, 30, 32)
P. nanus Sars: Taiwan Strait; South China Sea. (1, 14, 20)
P. nudus Sewell: East China Sea; Taiwan Strait; South China Sea. (1, 14, 25, 27, 32)
P. parvus (Claus): From Bohai to South China Sea. (1, 2, 4, 9, 12, 14, 17, 20, 21, 22, 25- 28, 32, 39)
P. serrulus Shen & Lee: Northern South China Sea shoreline. (9, 14)

PSEUDOCALANIDAE

Clausocalanus arcuicornis (Dana): From Yellow Sea to South China Sea. (1, 2, 12, 14, 20, 22, 25–28, 30, 32, 39)
C. farrani Sewell [*C. pergens*, Chen Qingchao et al., 1965]: East China Sea; Taiwan Strait; South China Sea. (1, 2, 12, 14, 19, 25, 27, 30, 31, 32)
C. furcatus (Brady): East China Sea; Taiwan Strait; South China Sea. (1, 2, 12, 14, 17, 18, 19, 20, 25–28, 30, 32, 39)
C. laticeps: Taiwan Strait. (30, 45)
C. mastigophorus (Claus): South China Sea. (14, 19)
C. minor Sewell: South China Sea. (14, 19)
C. parapergens Frost & Fleminger: Northern South China Sea. (14)
C. paululus Farran: Taiwan Strait; South China Sea. (2, 14, 32)
C. pergens Farran: Nansha (=Spratly Islands). (19)
Ctenocalanus vanus Giesbrecht: East China Sea; Taiwan Strait; South China Sea. (14, 20, 25, 32)
Drepanopsis frigidus Wolfenden: Central and southern South China Sea. (19, 32)
D. orbus Tanaka: Central South China Sea. (32)
Microcalanus pusillus Sars: Xisha (=Paracel Islands). (14, 20)
Mimocalanus cultrifer Farran: Central South China Sea. (2, 32)
Monacilla typicus Sars: Central South China Sea. (32)
Spinocalanus angusticeps Sars: Central South China Sea. (32)
S. horridus Welfenden: Central and southern South China Sea. (19, 32)
S. magnus Wolfenden: Central and southern South China Sea. (19, 32)
S. oligospinosus Park: Central South China Sea. (32)
S. spinosus Farran: Central South China Sea. (32)

AETIDEIDAE

Aetideus armatus (Boeck): East China Sea; Taiwan Strait; South China Sea. (1, 2, 19, 30, 45)
Chiridiella macrodactyla Sars: Central and southern South China Sea. (19, 32)
Chiridius gracilis Farran: Central South China Sea. (20)
C. poppei Giesbrecht: East China Sea; Taiwan Strait; South China Sea. (1, 2, 12, 19, 20, 30, 32)
Chirundina streetsi Giesbrecht: Central South China Sea. (32)
Euaetideus acutus (Farran): East China Sea; Taiwan Strait; South China Sea. (1, 2, 12, 19, 20, 25, 28, 32)
E. giesbrechti (Claus): East China Sea; Taiwan Strait; South China Sea. (1, 2, 12, 19, 20, 21, 25, 30, 32)
Euchirella amoena Giesbrecht: East China Sea; Taiwan Strait; South China Sea. (2, 19, 20, 21, 28, 30, 32)
E. bella Giesbrecht [*E. areata* Takana, 1957; Chen Qingchao et al., 1965, 1989]: East China Sea; Taiwan Strait; South China Sea. (2, 12, 19, 20, 25, 30, 32)
E. bitumida (With): Central South China Sea. (20)
E. curticauda Giesbrecht: East China Sea; Taiwan Strait; South China Sea. (2, 20, 25, 32, 33)
E. galeata Giesbrecht: Central and southern South China Sea. (20, 32)
E. indica Vervoort: East China Sea; central South China Sea. (21, 25, 32, 38)
E. maxima Wolfenden: Central South China Sea. (32)
E. messinensis (Claus): East China Sea, Nansha (=Spratly Islands). (19, 38)
E. orientalis Sewell: Central and southern South China Sea. (19, 32)
E. pulchra (Lubbock): Central and southern South China Sea. (19, 20, 32)
E. rostrata (Claus): Central South China Sea. (20)
E. unispina Park, 1968 [*E. acuta* Takana & Omori, 1969; Lian Guangshan et al. 1978]: East China Sea; central South China Sea. (20, 25, 32)
E. venusta Giesbrecht: East China Sea; Taiwan Strait; South China Sea. (2, 19, 20, 32, 33, 38)
Gaetanus brevicornis Esterly: Central and southern South China Sea. (19, 32)
G. miles Giesbrecht: Taiwan Strait; South China Sea. (2, 19, 20, 32)
G. minor Farran: East China Sea; Taiwan Strait; South China Sea. (2, 12, 19, 20, 21, 25, 30, 32, 39)
G. pileatus Farran: Central and southern South China Sea. (19, 20, 32)

Gaidius pungens Giesbrecht: Central South China Sea. (32)
Pseudochirella scopularis Sars: Central and southern South China Sea. (19, 32)
Undeuchaeta intermedia A. Scott: Central and southern South China Sea. (19, 32)
U. major Giesbrecht: East China Sea, central and southern South China Sea. (19, 20, 32)
U. plumosa Lubbock: East China Sea; Taiwan Strait; South China Sea. (2, 19, 20, 21, 25, 32, 38)
Undinopsis armatus (Brady) [*Aetideus armatus*, Chen Qingchao et al., 1965]: East China Sea, Taiwan Strait; South China Sea. (25, 28, 30, 31, 32)

EUCHAETIDAE

Euchaeta acutus Giesbrecht: East China Sea. (38)
E. concinna Dana: From Yellow Sea to South China Sea. (1, 2, 12, 15, 18–21, 25–28, 32, 38, 39)
E. longicornis Giesbrecht: From Yellow Sea to South China Sea. (1, 2, 12, 15, 19, 20, 21, 25–28, 32, 38, 39)
E. marina (Prestandrea): From Yellow Sea to South China Sea. (1, 2, 12, 15, 17- 21, 25, 28, 30, 38, 39)
E. media Giesbrecht: East China Sea; Taiwan Strait; South China Sea. (1, 2, 15, 19, 20, 25, 31, 32, 38)
E. plana Mori: From Yellow Sea to South China Sea. (1, 2, 12, 15, 17, 21, 22, 25–28, 30, 32, 33, 38, 39)
E. spinosa Giesbrecht: East China Sea; South China Sea. (2, 15, 19, 20, 32)
E. tenuis Esterly: Taiwan Strait; central and southern South China Sea. (2, 19, 20, 32)
E. wolfendeni A. Scott: East China Sea; Taiwan Strait; South China Sea. (1, 2, 12, 15, 19, 20, 21, 25, 26, 28, 30, 32, 38)
Pareuchaeta flava (Giesbrecht): From Yellow Sea to South China Sea. (1, 30, 44)
P. malayensis Sewell: Central South China Sea. (20)
P. russelli (Farran): From Yellow Sea to South China Sea. (1, 2, 12, 15, 20, 25, 28, 30, 3 38, 39)
P. scaphula (Fountaine): Central South China Sea. (20)
P. tuberculata A. Scott: Central South China Sea. (32)
P. weberi A. Scott: East China Sea. (38)

PHAENNIDAE

Onchocalanus cristatus (Wolfenden): Central South China Sea. (20)
Oothrix bidentata Farran: East China Sea; Taiwan Strait. (1, 12)
Phaenna spinifera Claus: East China Sea; Taiwan Strait; South China Sea. (1, 2, 12, 19, 20, 21, 28, 30, 38, 39)
Xanthocalanus agilis Giesbrecht: Taiwan Strait; central South China Sea. (2, 32)
X. dilatus Grice: Central South China Sea. (32)

X. multispinus Chen & Zhang: East China Sea; Taiwan Strait. (2, 12, 25)

SCOLECITHRICIDAE

Lophothrix frontalis Giesbrecht: Central and southern South China Sea. (19, 20, 32)
L. latipes (T. Scott, 1894) [*Xanthocalanus pulchra*, Lian Guangshan et al., 1978]: East China Sea; Taiwan Strait; South China Sea. (2, 25, 32)
Macandrewella joanae A. Scott: Central and southern South China Sea. (19, 32)
M. tuberculata Chen: Nansha (=Spratly Islands). (18, 19)
Racovitzanus levis Tanaka: Southern Taiwan Strait; central and southern South China Sea. (2, 19, 32)
Scaphocalanus brevicornis Sars: Central and southern South China Sea. (19, 32)
S. echinatus Farran: East China Sea; Taiwan Strait; South China Sea. (2, 25, 30, 32)
S. longifurca (Giesbrecht): East China Sea; central South China Sea. (25, 32)
S. magnus (T. Scott): Central and southern South China Sea. (19, 32)
S. major (T. Scott): Central South China Sea. (32)
Scolecithricella abyssalis (Giesbrecht): East China Sea; Taiwan Strait; South China Sea. (2, 19, 20, 25, 30, 32)
S. altera Farran: Central South China Sea. (32)
S. arcuata (Sars): Central South China Sea. (32)
S. auropecten Giesbrecht: Central and southern South China Sea. (19, 32)
S. bradyi (Giesbrecht): East China Sea; Taiwan Strait; South China Sea. (1, 2, 12, 19, 20, 21, 25, 26, 30, 32, 38, 39)
S. ctenopus (Giesbrecht): East China Sea; Taiwan Strait; South China Sea. (1, 2, 12, 25, 30, 39)
S. dentata (Giesbrecht): East China Sea; Taiwan Strait; South China Sea. (1, 2, 25, 31, 32)
S. fowleri Farran: Central South China Sea. (32)
S. gracilis Sars: Central South China Sea. (32)
S. longispinosa Chen & Zhang: From Yellow Sea to South China Sea. (1, 2, 12, 17, 20, 22, 25, 26, 27, 30, 32, 38)
S. minor (Brady): From Yellow Sea to norththern South China Sea. (1, 2, 12, 20, 25, 30)
S. ovata (Farran): Southern Taiwan Strait; central South China Sea. (2, 32)
S. tenuiserrata (Giesbrecht): East China Sea. (2, 25, 27, 32)
S. timida Tanaka: Central and southern South China Sea. (19, 32)
S. tropica Grice: East China Sea; Taiwan Strait; South China Sea. (2, 25, 31, 32)
S. valens (Farran): Central South China Sea. (32)
S. vittata (Giesbrecht): East China Sea; Taiwan Strait; South China Sea. (1, 2, 12, 19, 25, 31, 32, 39)
Scolecithrix danae Lubbock: From Yellow Sea to South China Sea. (1, 2, 12, 17, 18–21, 25, 26, 28, 32, 38, 39)

S. nicobarica Sewell: From Yellow Sea to South China Sea. (1, 2, 12, 20, 21, 25–28, 31, 32, 33, 38)
Scottocalanus australis Farran: Central South China Sea. (32)
S. farrani A. Scott: Central South China Sea. (20)
S. helenae (Lubbock): East China Sea; South China Sea. (2, 20, 23, 32)
S. persecans Giesbrecht: Central and southern South China Sea. (19, 32)
S. rotundatus Tanaka: Central South China Sea. (32)
S. securifrons (T. Scott): East China Sea; Taiwan Strait; South China Sea. (2, 19, 20, 32, 38)
S. sedatus Farran: Southern Taiwan Strait. (2)
S. thomasi A. Scott: Central South China Sea. (20)

STEPHIDAE

Stephos pentacanthos Chen & Zhang: Western Taiwan Strait. (12, 30)

TEMORIDAE

Eurytemora pacifica Sato: Bohai; northern Yellow Sea. (12)
Temora discaudata Giesbrecht: From Yellow Sea to South China Sea. (1, 2, 12, 18–22, 25–28, 30, 32, 33, 39)
T. stylifera (Dana): From Yellow Sea to South China Sea. (1, 2, 12, 17, 20, 21, 22, 25- 28, 32, 38, 39)
T. turbinata (Dana): From Yellow Sea to South China Sea. (1, 2, 9, 12, 17, 18–22, 25–28, 30, 32, 39)
Temorites brevis Sars: Central and southern South China Sea. (19, 32)
Temoropia mayumbaensis T. Scott: East China Sea; Taiwan Strait; South China Sea. (2, 12, 19, 25, 32)

METRIDIIDAE

Metridia brevicauda (Giesbrecht): Southern Taiwan Strait; central and southern South China Sea. (2, 19, 32)
M. macrura Sars: Central South China Sea. (32)
M. princeps Giesbrecht: Central and southern South China Sea. (19, 32)
M. venusta Giesbrecht: Southern Taiwan Strait; central and southern South China Sea. (2, 19, 32)
Pleuromamma abdominalis (Lubbock): East China Sea; Taiwan Strait; South China Sea. (1, 2, 12, 19, 20, 21, 25, 28, 32, 38, 39)
P. boraelis (Dahl): East China Sea; Taiwan Strait; South China Sea. (1, 2, 12, 20, 21, 25, 28, 32, 38)
P. gracilis (Claus): From Yellow Sea to South China Sea. (1, 2, 12, 19, 20, 21, 25, 26, 28, 32, 38, 39)
Note: Steaer (1932) classified the females of this species into *minima*, *maxima*, and *piseki* types. All three types are found in China's seas.

P. robusta (Dahl): From Yellow Sea to South China Sea. (1, 2, 12, 19, 20, 21, 24, 25, 26, 28, 32, 33, 38, 39)
P. xiphias (Giesbrecht): East China Sea; Taiwan Strait; South China Sea. (1, 2, 12, 19, 20, 21, 25, 28, 30, 32, 38, 39)

CENTROPAGIDAE

Centropages abdominalis Sato [*C. mcmurrichi* Willey, 1921; Chen Qingchao et al, 1965]: From Bohai to Taiwan Strait. (1, 4, 12, 22, 25, 26, 28, 39)
C. brevifurcus Shen & Lee: Northern South China Sea. (9)
C. calaninus (Dana): From Yellow Sea to South China Sea. (1, 2, 12, 17, 18–21, 28, 30, 39)
C. dorsispinatus Thompson & Scott: From Yellow Sea to Taiwan Strait. (1, 2, 12, 22, 25, 28, 33, 39)
C. elongatus Giesbrecht: East China Sea; South China Sea. (30, 31, 32)
C. furcatus (Dana): From Yellow Sea to South China Sea. (1, 2, 12, 17, 18–22, 27, 28, 32, 38, 39)
C. gracilis (Dana): From Yellow Sea to South China Sea. (1, 2, 12, 17, 18–21, 25, 28, 30, 39)
C. longicornis Mori: East China Sea; Taiwan Strait. (1, 12, 30, 44)
C. orsinii Giesbrecht: East China Sea; Taiwan Strait; South China Sea. (1, 2, 12, 17, 18, 19, 27, 28, 30, 39)
C. sinensis Chen & Zhang: From Yellow Sea to South China Sea. (12, 17, 19, 22, 25, 26, 30)
C. tenuiremis Thompson & Scott: From Bohai to South China Sea. (1, 2, 4, 9, 12, 17, 22, 25–28, 30, 39)
C. violaceus (Claus): East China Sea. (44)
Sinocalanus laevidactylus Shen & Tai: Western Taiwan Strait; northern South China Sea. (12, 26, 30)
S. sinensis (Poppe): From Bohai to northern South China Sea. (3, 12, 22, 25, 30, 39)
S. solstitialis Brehm: Northern South China Sea shoreline. (9)
S. tenellus (Kikuchi): From Yellow Sea to Taiwan Strait. (12, 21, 25, 26)

PSEUDODIAPTOMIDAE

Pseudodiaptomus incisus Shen & Lee: Northern South China Sea. (9, 27)
P. marinus Sato: From Bohai to South China Sea. (9, 12, 17, 22, 26, 27, 28, 30, 39)
P. penicillus Li & Huang: Western Taiwan Strait; northern South China Sea. (11, 27, 37)
Schmackeria dubia (Kiefer): East China Sea; South China Sea. (12, 26, 30, 43)
S. forbesi Poppe & Richard: Western Taiwan Strait; northern South China Sea. (9, 36)
S. inopinus Burckhardt: From Bohai to South China Sea. (9, 12, 22, 25, 30)
S. poplesia Shen: From Bohai to South China Sea. (3, 12, 22, 25, 26, 30, 38, 39)

DIAPTOMIDAE

Mongolodiaptomus birulai (Rylov) [*M. formosanus* Kiefer, Shen Jiarui et al., 1963]: Mouth of Jiulong River (Fujian) and Jian River (Guangdong). (9)

Neodiaptomus yangtsekiangensis Mashiko: Mouth of Jian River (Guangdong). (9)

Tropodiaptomus oryzanus Kiefer: Mouth of Jian River (Guangdong). (9)

LUCICUTIIDAE

Isochaeta ovalis Giesbrecht: Central South China Sea. (32)
Lucicutia bicornis: East China Sea. (21)
L. clausi (Giesbrecht): East China Sea; Taiwan Strait; South China Sea. (2, 12, 19, 20, 25, 38)
L. curta Farran: Central and southern South China Sea. (19, 32)
L. flavicornis (Claus): From Yellow Sea to South China Sea. (1, 2, 12, 17, 19, 20, 21, 25–28, 38, 39)
L. gemina Farran: Central and southern South China Sea. (19, 32)
L. longiserrata (Giesbrecht): Central South China Sea. (32)
L. magna Wolfenden: Central and southern South China Sea. (19, 32)
L. maxima Steuer: Central South China Sea. (32)
L. ovalis Welfenden: East China Sea; Taiwan Strait; South China Sea. (2, 12, 17–20, 25- 28, 38, 39)
L. tenuicauda Sars: Central South China Sea. (32)
L. wolfendeni Tanaka: Central South China Sea. (32)

HETERORHABDIDAE

Disseta palumboi Giesbrecht: Central South China Sea. (32)
Heterorhabdus abyssalis (Giesbrecht): Central South China Sea. (32)
H. clausi (Giesbrecht): Central South China Sea. (32)
H. compactus Sars: Central South China Sea. (32)
H. longicornis (Giesbrecht): East China Sea; central South China Sea. (32)
H. papilliger (Claus): East China Sea; Taiwan Strait; South China Sea. (1, 2, 12, 19, 20, 21, 25, 28, 30, 32)
H. spinifrons (Claus): East China Sea; Taiwan Strait; South China Sea. (2, 19, 31, 32)
H. vipera (Giesbrecht): Southern Taiwan Strait; central and southern South China Sea. (2, 19, 32)
Mesorhabdus gracilis Sars: Central South China Sea. (32)

AUGAPTILIDAE

Augaptilus glacialis Sars: Central South China Sea. (32)
A. longicaudatus (Claus): East China Sea; central South China Sea. (25, 32)
A. megalurus Giesbrecht [*A. anceps* Farran, Tanaka, 1964; Lin Yuhui et al., 1988]: Central South China Sea. (32)
A. spinifrons Sars: Southern Taiwan Strait. (2)

Centraugaptilus rattrayi (T. Scott): Central South China Sea. (32)
Euaugaptilus affinis Sars: Central South China Sea. (32)
E. facilis (Farran): Central South China Sea. (32)
E. filigerus (Claus): East China Sea. (25)
E. hecticus (Giesbrecht): East China Sea; Taiwan Strait; South China Sea. (2, 19, 20, 25, 31, 32)
E. mixtus Brodsky: Central South China Sea. (32)
E. nodifrons Sars: Southern Taiwan Strait; central South China Sea. (2, 32)
E. palumbii (Giesbrecht): Southern Taiwan Strait; central and southern South China Sea. (2, 19, 20, 32)
Haloptilus acutifrons Giesbrecht: East China Sea; central and southern South China Sea. (1, 19, 20, 30, 44)
H. austini Grice: Southern Taiwan Strait; central South China Sea. (2, 32)
H. fertilis (Giesbrecht): Central and southern South China Sea. (19, 32)
H. longicirrus Brodsky, 1950, Park, 1970. [*H. setuliger* Tanaka, 1964b; Lin Yuhui et al., 1988]: Central South China Sea. (32)
H. longicornis (Claus): East China Sea; Taiwan Strait; South China Sea. (1, 2, 12, 19, 20, 21, 25, 28, 30, 31, 32)
H. mucronatus (Claus): East China Sea; Taiwan Strait; South China Sea. (1, 2, 19, 20, 25, 30, 32)
H. ornatus (Giesbrecht): East China Sea; Taiwan Strait; South China Sea. (1, 2, 19, 20, 21, 25, 30, 32)
H. oxycephalus (Giesbrecht): East China Sea; Taiwan Strait; South China Sea. (1, 2, 19, 20, 25, 30, 32)
H. paralongicirrus Park: East China Sea; central South China Sea. (32)
H. spiniceps Giesbrecht: East China Sea; Taiwan Strait; South China Sea. (1, 2, 19, 20, 21, 25, 32)
H. tenuis Farran: Central South China Sea. (32)
Pseudaugaptilus orientalis Tanaka: Central South China Sea. (32)

ARIETELLIDAE

Arietellus aculeatus (T. Scott): Central and southern South China Sea. (19, 20, 32)
A. plumifer Sars: Central and southern South China Sea. (19, 32)
A. setosus Giesbrecht: Southern Taiwan Strait; central South China Sea. (2, 20)
Metacalanus aurivilli Cleve: East China Sea; Taiwan Strait; South China Sea. (1, 2, 12, 26, 27, 28, 30, 32)
Paramisophria sinica Lian & Qian: Southern Yellow Sea; northern South China Sea. (27a)
Paraugaptilus buchani Wolfenden: Central South China Sea. (32)
Phyllopus aequalis Sars: Central South China Sea. (32)
P. bidentatus Brady: Central South China Sea. (32)
P. helgae Farran: Southern Taiwan Strait; central and southern South China Sea. (2, 19, 32)
P. impar Farran: Central and southern South China Sea. (19, 32)
P. mutatus Tanaka: Central South China Sea. (32)

CANDACIIDAE

Candacia aethiopica (Dana): From Yellow Sea to South China Sea. (1, 2, 12, 19, 20, 21, 25, 28, 32, 38)

C. bipinnata (Giesbrecht): From Yellow Sea to South China Sea. (1, 2, 19–22, 25, 28, 32, 39)

C. bradyi A. Scott: From Yellow Sea to South China Sea. (1, 2, 12, 18–22, 25, 28, 32, 33, 38, 39)

C. catula Giesbrecht: East China Sea; Taiwan Strait; South China Sea. (1, 2, 12, 17–21, 25, 28, 32)

C. columbiae Campbell [*C. pacifica* Mori, 1937; Tan Tienhsi, 1970; Zheng Zhong (C. Cheng) et al., 1982]: Taiwan Strait; eastern Taiwan. (30, 44, 45)

C. curta (Dana): From Yellow Sea to South China Sea. (1, 2, 12, 19, 20, 21, 25, 28, 30, 32, 39)

C. discaudata A. Scott: East China Sea; Taiwan Strait; South China Sea. (1, 2, 12, 18- 22, 25, 28, 32)

C. guggenheimi Grice & Jones: East China Sea; central South China Sea. (25, 32)

C. longimana (Claus): East China Sea; Taiwan Strait; South China Sea. (2, 19, 20, 21, 25, 28, 32, 39)

C. pachydactyla (Dana): From Yellow Sea to South China Sea. (1, 2, 12, 18–21, 25, 28, 32, 39)

C. tenuimana (Giesbrecht): Central and southern South China Sea. (19, 32)

C. varicans (Giesbrecht): East China Sea; central South China Sea. (20, 25, 32)

Paracandacia bispinosa (Claus) [*Candacia bispinosa* Claus]: East China Sea; Taiwan Strait; South China Sea. (1, 2, 25, 28, 30, 31, 32, 38)

P. simplex (Giesbrecht) [*Candacia simplex* (Giesbrecht)]: East China Sea; Taiwan Strait; South China Sea. (1, 2, 12, 20, 25, 32, 38)

P. truncata (Dana) [*Candacia truncata* (Dana)]: East China Sea; Taiwan Strait; South China Sea. (1, 2, 12, 18–21, 25, 28, 30, 32, 39)

PONTELLIDAE

Calanopia elliptica (Dana): East China Sea; Taiwan Strait; South China Sea. (1, 2, 12, 15, 17, 18–22, 25–28, 32, 39)

C. minor A. Scott: East China Sea; Taiwan Strait; South China Sea. (1, 2, 12, 15, 19–22, 25–28, 32)

C. thompsoni A. Scott: From Bohai to South China Sea. (1, 2, 9, 12, 15, 17, 25- 28, 39)

Labidocera acuta (Dana): From Yellow Sea to South China Sea. (1, 2, 12, 15, 18–21, 25–28, 32, 39)

L. acutifrons (Dana): Southern Taiwan Strait; South China Sea. (28, 30)

L. bataviae A. Scott: Southern Taiwan Strait. (2)

L. bipinnata Tanaka: From Bohai to South China Sea. (1, 2, 4, 9, 12, 15, 19, 25–28, 32, 39)

L. detruncata (Dana): Bohai; East China Sea; Taiwan Strait; South China Sea. (1, 2, 12, 15, 19, 20, 21, 25, 28, 32, 39)

L. euchaeta Giesbrecht: From Bohai to South China Sea. (1, 2, 4, 9, 12, 15, 25–28, 33, 39)

L. kroeyeri (Brady): East China Sea; Taiwan Strait; South China Sea. (1, 2, 12, 15, 25- 28, 32, 39)

L. laevidentata (Brady): Taiwan Strait; central and southern South China Sea. (1, 18, 19, 31, 32)

L. minuta (Giesbrecht): East China Sea; Taiwan Strait; South China Sea. (1, 2, 12, 15, 19, 21, 25–28, 32, 39)

L. pavo Giesbrecht: From Bohai to South China Sea. (9, 12, 15, 20, 21, 26, 27, 28, 32)

L. sinilobata Shen & Lee: Bohai; East China Sea; Taiwan Strait; South China Sea. (9, 12, 15, 20, 22, 25, 28, 37)

Pontella chierchiae Giesbrecht: From Bohai to South China Sea. (1, 2, 12, 15, 20, 22, 25–28, 30, 38)

P. danae Giesbrecht: Taiwan Strait. (28, 30, 37)

P. denticauda A. Scott: Central South China Sea. (32)

P. fera Dana: East China Sea; Taiwan Strait; South China Sea. (1, 2, 12, 15, 19, 22, 25, 28, 32, 39)

P. kieferi Pesta: Southern Taiwan Strait; central South China Sea. (28, 30, 32)

P. latifurca Chen & Zhang: Bohai; Taiwan Strait. (12, 26, 37)

P. princeps Dana: Southern Taiwan Strait; South China Sea. (28, 30)

P. securifer Brady: East China Sea; Taiwan Strait; South China Sea. (2, 12, 15, 19, 25, 28, 30, 39)

P. sinica Chen & Zhang: East China Sea; Taiwan Strait; South China Sea. (2, 12, 15, 18, 19, 20, 28, 30, 32)

P. spinicauda Mori: From Bohai to South China Sea. (12, 22, 25, 26, 28, 30)

P. tridactyla Shen & Lee: Southern Taiwan Strait; northern South China Sea. (2, 9, 15)

Pontellina plumata (Dana): East China Sea; Taiwan Strait; South China Sea. (1, 2, 12, 15, 17, 18–22, 25, 28, 32, 39)

Pontellopsis armatus (Giesbrecht): East China Sea; Taiwan Strait; South China Sea. (2, 12, 19, 28, 32)

P. inflatodigitata Chen & Shen: Taiwan Strait; northern South China Sea. (1, 15)

P. krameri (Giesbrecht): East China Sea; Taiwan Strait; South China Sea. (1, 2, 15, 25, 28, 32)

P. macronyx A. Scott: Taiwan Strait; South China Sea. (1, 2, 15, 28, 32)

P. regalis (Dana): From Yellow Sea to South China Sea. (2, 12, 15, 19, 20, 25, 28, 32, 39)

P. strenua (Dana): East China Sea; Taiwan Strait; South China Sea. (2, 12, 19, 28, 32)

P. tenuicauda (Giesbrecht): From Bohai to South China Sea. (1, 2, 4, 12, 15, 19, 25–28, 33, 39)

P. villosa Brady: East China Sea; Taiwan Strait; South China Sea. (1, 2, 12, 15, 20, 28, 32, 39)

P. yamadae Mori: From Bohai to South China Sea. (1, 2, 12, 19, 22, 25–28, 33, 39)

BATHYPONTIIDAE

Bathypontia longicornis Tanaka: Central South China Sea. (32)
B. minor Welfenden: Central South China Sea. (32)
B. similis Tanaka: Central South China Sea. (32)
B. spinifera A. Scott: Central South China Sea. (32)

ACARTIIDAE

Acartia bifilosa (Giesbrecht): From Bohai to Taiwan Strait. (4, 12, 25, 26, 28, 33)
A. clausi Giesbrecht: From Bohai to South China Sea. (4, 12, 22, 25, 27, 28, 30, 38, 39)
A. danae Giesbrecht: From Yellow Sea to South China Sea. (1, 2, 12, 20, 22, 25, 27, 28, 32)
A. erythraea Giesbrecht: East China Sea; Taiwan Strait; South China Sea. (1, 2, 12, 18, 19, 25–28, 32)
A. hamata Mori: East China Sea. (44)
A. longiremis Lilljebory: Taiwan Strait; northern South China Sea. (30, 44, 45)
A. negligens Dana: East China Sea; Taiwan Strait; South China Sea. (1, 2, 12, 18–21, 25–28, 32, 39)*A. pacifica* Steuer: From Bohai to South China Sea. (1, 2, 9, 12, 18, 19, 22, 25–28, 32, 33)
A. pacifica Steuer: From Bohai to South China Sea. (1, 2, 9, 12, 19, 22, 25, 28, 32, 33)
A. southwelli Sewell: Hangzhou Bay (Zhejiang); northern South China Sea. (16, 22)
A. spinicauda Giesbrecht: East China Sea; Taiwan Strait; northern South China Sea. (1, 12, 25–28, 30, 38)
Acartiella sinensis Shen & Lee: Taiwan Strait; northern South China Sea. (9, 12, 26, 27, 28, 30)

TORTANIDAE

Tortanus barbatus (Brady): Northern South China Sea. (26, 27, 28, 31)
T. denticulatus Shen & Lee: Northern South China Sea. (9)
T. derjugini Smironov: From Bohai to northern South China Sea. (1, 12, 16, 22, 25, 26, 28, 30, 39)
T. dextrilobatus Chen & Zhang: East China Sea; Taiwan Strait; northern South China Sea. (2, 12, 16, 22, 26, 36, 39)
T. forcipatus (Giesbrecht): From Bohai to northern South China Sea. (1, 2, 12, 16, 22, 25–28, 39)
T. gracilis (Brady): From Bohai to South China Sea. (1, 2, 12, 16, 18, 19, 25, 28, 30, 32, 39)
T. murrayi A. Scott: Central South China Sea. (16)
T. scaphus Bowman: Central South China Sea. (16)
T. sinicus Chen: Central South China Sea. (16)
T. spinicaudatus Shen & Bai: From Bohai to northern South China Sea. (2, 4, 12, 16, 22, 25, 28, 39)
T. vermiculus Shen: From Yellow Sea to northern South China Sea. (3, 12, 16, 22, 25, 30, 33, 39)

CYCLOPOIDA
OITHONIDAE

Limnoithona tetraspina Zhang & Li: East China Sea; Taiwan Strait. (24, 35, 37)
Oithona attenuata Farran: From Yellow Sea to South China Sea. (1, 13, 27, 30, 33, 38)
O. brevicornis Giesbrecht: From Bohai to South China Sea. (1, 13, 24, 26, 27, 30, 33)
Note: Nishida et al. (1977) divided *O. brevicornis* into two types: *O. brevicornis* forma *typica* and *O. brevicornis* forma *minor*. Both types are found in China's seas (Chen Qingchao et al. 1974; Lian Guangshan et al, 1978). Ferrari (1981) considered each type as a species: *O. spinulosa* Lindberg, 1950 and *O. wellershausi* Ferrari, 1981.
O. decipiens Farran: From Yellow Sea to South China Sea. (2, 13, 22, 24, 30, 33)
O. fallax Farran: East China Sea; Taiwan Strait; South China Sea. (1, 13, 19, 24, 26, 27, 30)
O. longispina Nishida et al.: East China Sea; Taiwan Strait; South China Sea. (31, 32)
O. nana Giesbrecht: From Yellow Sea to South China Sea. (1, 13, 24, 26, 30, 32, 33)
O. plumifera Baird: From Yellow Sea to South China Sea. (1, 13, 19, 24, 26, 27, 30, 32, 33)
O. rigida Giesbrecht: From Yellow Sea to South China Sea. (1, 13, 19, 22, 26, 27, 30, 32, 42)
O. robusta Giesbrecht: East China Sea; Taiwan Strait; South China Sea. (24, 31, 32)
O. setigera Dana: East China Sea; Taiwan Strait; South China Sea. (1, 13, 18, 19, 24, 26, 30, 32)
O. similis Claus: From Bohai to South China Sea. (1, 4, 13, 24, 27, 28, 30, 32, 33)
O. simplex Farran: East China Sea; Taiwan Strait; South China Sea. (1, 13, 26, 27, 31, 32)
O. tenuis Rosendorn: East China Sea; Taiwan Strait; South China Sea. (1, 13, 18, 19, 24, 30, 32)
O. vivida Farran: East China Sea; Taiwan Strait; South China Sea. (2, 13, 19, 30, 38)
Pontoeciella abyssicola T. Scott: Taiwan Strait; South China Sea. (2, 32)
Ratania flava Giesbrecht: Taiwan Strait. (2)

ONCAEIDAE

Lubbockia aculeata Giesbrecht: East China Sea; Taiwan Strait; South China Sea. (2, 20, 26, 32)
L. marukawai Mori: East China Sea; Taiwan Strait; South China Sea. (2, 20, 31)
L. squillimana Claus: East China Sea; Taiwan Strait; South China Sea. (1, 13, 19, 24, 30, 32, 38)
Oncaea clevei Fruchtl: East China Sea; Taiwan Strait; South China Sea. (1, 13, 20, 24, 26, 27, 30, 38)

O. conifera Giesbrecht: East China Sea; Taiwan Strait; South China Sea. (1, 13, 24, 26, 27, 30, 32, 38)

O. dentipes Giesbrecht: East China Sea; Taiwan Strait; South China Sea. (1, 13, 20, 24, 27, 30, 38)

O. media Giesbrecht: From Yellow Sea to South China Sea. (1, 13, 18, 19, 24, 26, 27, 30, 32, 33)

O. mediterranea Claus: East China Sea; South China Sea. (1, 13, 20, 24, 26, 27, 31, 32, 38)

O. minuta Giesbrecht: East China Sea; Taiwan Strait; South China Sea. (13, 24, 30, 32, 38)

O. ornata Giesbrecht: East China Sea; Taiwan Strait; South China Sea. (13)

O. similis Sars: East China Sea; Taiwan Strait; South China Sea. (1, 13, 24, 38)

O. venusta Phillippi: East China Sea; Taiwan Strait; South China Sea. (1, 13, 19, 24, 26, 27, 30, 32, 38)

Pachysoma dentatum Mori: Taiwan Strait; South China Sea. (28, 31, 32, 44)

P. punctatum Claus: East China Sea; Taiwan Strait; South China Sea. (1, 13, 24, 28, 30, 32)

Pseudanthessius obscurus A. Scott: South China Sea. (27)

P. parvus A. Scott: South China Sea. (27)

SAPPHIRINIDAE

Copilia lata Giesbrecht: East China Sea; Taiwan Strait; South China Sea. (2, 20, 24, 28, 32, 38)

C. longistylis Mori: East China Sea; Taiwan Strait; South China Sea. (20, 24, 28)

C. mirabilis Dana: East China Sea; Taiwan Strait; South China Sea. (1, 13, 18, 19, 20, 24, 27, 30, 31, 32, 38, 43)

C. quadrata Dana: East China Sea; Taiwan Strait; South China Sea. (1, 18, 19, 20, 24, 28, 30, 32, 38, 43)

C. recta Giesbrecht: East China Sea. (24)

C. vitrea Haeckel: Taiwan Strait, South China Sea. (1, 2, 28, 32)

Corina granulosa Giesbrecht: East China Sea; Taiwan Strait; South China Sea. (13, 19)

Sappirina angusta Dana: East China Sea; Taiwan Strait; South China Sea. (1, 13, 20, 24, 30, 32, 38)

S. auronitens Claus: East China Sea; Taiwan Strait; South China Sea. (20, 24, 30, 32)

S. bicuspidata Giesbrecht: East China Sea; Taiwan Strait; South China Sea. (20, 24, 30, 32)

S. darwinii Haeckel: East China Sea; Taiwan Strait; South China Sea. (1, 13, 20, 21, 24, 30, 32, 38)

S. gastrica Giesbrecht: East China Sea; Taiwan Strait; South China Sea. (20, 24, 30, 32, 38)

S. gemma Dana: East China Sea; Taiwan Strait; South China Sea. (1, 13, 19, 20, 24, 30, 32, 38, 43)

S. intestinata Giesbrecht: East China Sea; Taiwan Strait; South China Sea. (1, 13, 19, 24, 30, 32, 38)

S. iris Dana: Central and southern South China Sea. (19, 32)

S. lactens Giesbrecht: East China Sea; Taiwan Strait; South China Sea. (28, 30, 32)

S. maculosa Giesbrecht: Taiwan Strait; South China Sea. (31, 32)

S. metallina Dana: East China Sea; Taiwan Strait; South China Sea. (1, 19, 20, 24, 30, 32)

S. nigromaculata Claus: From Yellow Sea to South China Sea. (1, 13, 19, 20, 24, 27, 30, 32, 33, 38, 43)

S. opalina Dana: East China Sea; Taiwan Strait; South China Sea. (1, 13, 18, 19, 20, 24, 26, 27, 30, 32, 38)

S. ovatolanceolata Dana: East China Sea; Taiwan Strait; South China Sea. (1, 13, 18, 19, 20, 24, 30, 32)

S. scarlata Giesbrecht: From Yellow Sea to South China Sea. (1, 13, 18, 19, 20, 24, 27, 30, 32, 33, 38)

S. sinuicauda Brady: East China Sea; Taiwan Strait; South China Sea. (1, 13, 19, 20, 24, 30, 32, 38)

S. stellata Giesbrecht: East China Sea; Taiwan Strait; South China Sea. (1, 13, 19, 20, 21, 24, 30, 32, 38, 43)

CORYCAEIDAE

Corycaeus affinis Mcmurrichi: From Bohai to South China Sea. (1, 13, 24, 26, 27, 30, 32, 38)

C. agilis Dana: East China Sea; Taiwan Strait; South China Sea. (1, 13, 20, 24, 27, 30, 38)

C. andrewsi Farran: East China Sea; Taiwan Strait; South China Sea. (1, 13, 20, 24, 26, 27, 30, 32, 38)

C. asiaticus F. dahl: East China Sea; Taiwan Strait; South China Sea. (1, 13, 19, 24, 26, 27, 30, 32)

C. calaninus (Claus): Nansha (=Spratly Islands) and nearby water. (19)

C. carinata Giesbrecht: East China Sea; Taiwan Strait; South China Sea. (1, 13, 20, 24, 27, 30, 32, 38)

C. catus F. dahl: From Yellow Sea to South China Sea. (1, 13, 20, 24, 26, 27, 30, 32, 38)

C. clausi F. dahl: Taiwan Strait. (30)

C. concinnus (Dana): East China Sea; Taiwan Strait; South China Sea. (1, 13, 20, 24, 27, 30, 32, 38)

C. crassiusculus Dana: East China Sea; Taiwan Strait; South China Sea. (1, 13, 20, 24, 27, 30, 32, 38)

C. dahli Tanaka: From Yellow Sea to South China Sea. (1, 13, 18, 19, 20, 24, 26, 27, 30, 32, 33)

C. erythraeus Cleve: East China Sea; Taiwan Strait; South China Sea. (1, 13, 19, 24, 26, 27, 30, 32)

C. flaccus Giesbrecht: East China Sea; Taiwan Strait; South China Sea. (1, 13, 20, 24, 30, 32, 38)

C. furcifer Claus: East China Sea; Taiwan Strait; South China Sea. (1, 13, 19, 24, 32, 38)

C. gibbulus (Giesbrecht): East China Sea; Taiwan Strait; South China Sea. (1, 13, 20, 24, 27, 30, 32, 38)

C. giesbrechti F. dahl: East China Sea; Taiwan Strait; South China Sea. (2, 13, 19, 24, 27, 38)

C. lautus Dana: East China Sea; Taiwan Strait; South China Sea. (1, 13, 19, 20, 24, 30, 32, 38)

C. limbatus Brady: East China Sea; Taiwan Strait; South China Sea. (1, 24, 32)

C. longicaudis Dana: East China Sea; Taiwan Strait; South China Sea. (13, 20)

C. longistylis Dana: East China Sea; Taiwan Strait; South China Sea. (1, 13, 20, 24, 30, 32, 38)

C. lubbocki Giesbrecht: East China Sea; Taiwan Strait; South China Sea. (1, 13, 19, 24, 26, 27, 38)

C. ovalis Claus: East China Sea. (44)

C. pacificus F. dahl: East China Sea; Taiwan Strait; South China Sea. (1, 13, 24, 26, 27, 30, 32, 38)

C. pumilus M. dahl: East China Sea; Taiwan Strait; South China Sea. (1, 13, 27, 38)

C. robustus Giesbrecht: East China Sea; Taiwan Strait; South China Sea. (1, 13, 19, 20, 24, 30, 32, 38)

C. rostratus (Claus): East China Sea; Taiwan Strait; South China Sea. (1, 13, 24, 32)

C. speciosus Dana: East China Sea; Taiwan Strait; South China Sea. (1, 13, 18, 19, 20, 24, 26, 27, 30, 32, 38)

C. subtilis M. dahl: East China Sea; Taiwan Strait; South China Sea. (1, 13, 27, 30)

C. typicus Kröyer: East China Sea; Taiwan Strait; South China Sea. (1, 24, 31, 32, 38)

C. viretus Dana: East China Sea; Taiwan Strait; South China Sea. (2, 13, 20, 24, 27, 30, 32, 38)

CLAUSIDIIDAE

Hemicyclopus dilatatus Shen & Bai: Northern Yellow Sea; Taiwan Strait. (4, 31)

CYCLOPINIDAE

Cyclopina heterospina Shen & Bai: Northern Yellow Sea. (4)

CYCLOPIDAE

Halicyclops sinensis Kiefer: Taiwan Strait. (36)

BOMOLOCHIDAE

Bomolochus decipteri Yamaguti: South China Sea. Parasitic species. (5)

B. gassae Shen: South China Sea. Parasitic species. (5)

B. managatuwo Yamaguti: South China Sea. Parasitic species. (5)

Orbitacolax uniunguis Shen: South China Sea. Parasitic species. (5)

Pumiliopes opisthopteri Shen: South China Sea. Parasitic species. (5)

TAENIACANTHIDAE

Taeniacanthus pteroisi Shen: South China Sea. Parasitic species. (5)

HARPACTICOIDA
LONGIPIDIIDAE

Longipedia scotti A. Scott: Taiwan Strait. (31)
L. weberi A. Scott: Taiwan Strait. (31)

ECTINOSOMIDAE

Microsetella norvegica (Boeck): From Bohai to South China Sea. (1, 13, 24, 26, 27, 30, 32, 33, 38)
M. rosea (Dana): From Yellow Sea to South China Sea. (1, 13, 24, 26, 27, 30, 32, 33, 38)

TACHIDIIDAE

Danielssenia typica Boeck: Northern Yellow Sea. (4)
Euterpina acutifrons Dana: From Yellow Sea to South China Sea. (1, 13, 24, 26, 27, 30, 33)

MACROSETELLIDAE

Miracia efferata Dana: Taiwan Strait; South China Sea. (28, 30)
Setella gracilis Dana: East China Sea; Taiwan Strait; South China Sea. (1, 13, 20, 24, 26, 27, 30, 33)

CLYTEMNESTRIDAE

Clytemnestra rostrata (Brady): East China Sea; Taiwan Strait; South China Sea. (1, 13, 24, 26, 27, 30)
C. scutellata Dana: From Yellow Sea to South China Sea. (1, 13, 20, 24, 26, 27, 30, 33)

AEGISTHIDAE

Aegisthus aculeatus Giesbrecht: Central South China Sea. (32)
A. mucronatus Giesbrecht: Taiwan Strait; central South China Sea. (1, 20, 32)

PELTIDIIDAE

Altentha interrupta (Goodsir): Yellow Sea; Taiwan Strait; South China Sea. (4, 26, 27)
Eupelte acutispinis Zhang & Li: Xisha (=Paracel Islands). (34)
Peltidium falcatum A. Scott: Xisha (=Paracel Islands). (34)
P. ovale Thompson & Scott: Xisha (=Paracel Islands). (34)

PORCELLIDIIDAE

Porcellidium acuticaudatum Thompson & Scott: Xisha (=Paracel Islands). (34)
P. ovatum Haller: Xisha (=Paracel Islands). (34)

HARPACTICIDAE

Harpacticus uniremis Kröyer: Yellow Sea; East China Sea. (4, 28)

TISBIDAE

Tisbe bermudensis Willey: Xisha (=Paracel Islands). (34)

LOURINIIDAE

Lourinia armata Claus: Xisha (=Paracel Islands). (34)

LAOPHONTIDAE

Esola longicauda Edwards: Xisha (=Paracel Islands). (34)
Onychocamptus mirabilis (Gurney): Xisha (=Paracel Islands). (34)
Paralaophonte congenera (Sars): Xisha (=Paracel Islands). (34)

ANCORABOLIDAE

Laophontodes himatus Thompson: Xisha (=Paracel Islands). (34)

THALESTRIDAE

Dactylopodia tisboides (Claus): Xisha (=Paracel Islands). (34)
Eudactylopus latipes T. Scott: Xisha (=Paracel Islands). (34)
Microthalestris forficula (Claus): Taiwan Strait. (26)
Rhynchothalestris rufocincta (Brady): Taiwan Strait; Xisha (=Paracel Islands). (31, 34)

PARASTENHELIIDAE

Parastenhelia spinosa littoralis (Sars): Xisha (=Paracel Islands). (34)

DIOSACCIDAE

Amphiascella subdebilis Willeg: Xisha (=Paracel Islands). (34)
Amphiascopsis havelocki Thompson & Scott: Xisha (=Paracel Islands). (34)
Metamphiascopsis hirsutus bermudae Willey: Xisha (=Paracel Islands). (34)

AMEIRIDAE

Ameira longipes Boeck: Xisha (=Paracel Islands). (34)

CANTHOCAMPTIDAE

Parameira pendula Shen & Bai: Northern Yellow Sea. (4)
Pseudomeira brevifurca Shen & Bai: Northern Yellow Sea. (4)

TETRAGONICIPITIDAE

Phyllopodopsyllus furciger Sars: Xisha (=Paracel Islands). (34)

MONSTRILLOIDA
MONSTRILLIDAE

Monstrilla grandis Giesbrecht: From Yellow Sea to South China Sea. (1, 4, 24, 26, 27, 30, 38)

MISOPHRIOIDA
MISOPHRIIDAE

Misophria sinensis Boxshall: Hong Kong near shore. (41)

SIPHONOSTOMATOIDA
ASTEROCHERIDAE

Asterocheres aesthetes Ho: Hong Kong near shore. Symbiotic with sponges. (42)
Dermatomyzon nigripes Brady & Robertson: Hong Kong near shore. Symbiotic with sponges. (42)
Inermocheres quadratus Boxshall: Hong Kong near shore. Symbiotic with sponges. (42)
Sinopontius aesthetascus Boxshall: Hong Kong near shore. Symbiotic with sponges. (42)
S. punctatus Boxshall: Hong Kong near shore. Symbiotic with sponges. (42)

DINOPONTIIDAE

Stenopontius parvus Boxshall: Hong Kong near shore. (42)

ARGULOIDA
ARGULIDAE

Argulus scutiformis Thiele: Bohai. Parasitic. (8a)

CALIGOIDA
CALIGIDAE

Caligus aduncus Shen & Li: Jiaozhou Bay (Shandong). Parasitic. (7b)
C. bifurcus Shen: Bohai. Parasitic. (6)
C. biseriodentatus Shen: Northern South China Sea. Parasitic. (6)

C. brevisoris Shen: Northern South China Sea. Parasitic. (6)
C. communis Shen: Bohai; northern Yellow Sea. Parasitic. (6)
C. confusus Pillai: Hainan. Parasitic. (40)
C. coryphaenae Steenstrup & Lütken: Xisha (=Paracel Islands). Parasitic. (40)
C. costatus Shen & Li: Jiaozhou Bay (Shandong). Parasitic. (7b)
C. eleutheronemi Shen: Northern South China Sea. Parasitic. (6)
C. euthynus Kurian: Xisha (=Paracel Islands). Parasitic. (40)
C. hilsae Shen: Northern South China Sea shoreline. Parasitic. (6)
C. hirsutus Bassett-Smith: Northern South China Sea. Parasitic. (16)
C. laticorpus Shen: Bohai. Parasitic. (6)
C. longicaudus Bassett-Smith: Yinggehai (Hainan). Parasitic. (40)
C. multispinosus Shen: Northern South China Sea. Parasitic. (6)
C. nibeae Shen: Northern South China Sea. Parasitic. (6)
C. rotundigenitalis Yu: Bohai; northern Yellow Sea. Parasitic. (6)
C. undulatus Shen & Li: Jiaozhou Bay (Shandong). Parasitic. (7b)
Lepeophtheirus edwardsi Wilson: Bohai. Parasitic. (7a)
L. kareii Yamaguti: Bohai. Parasitic. (7a)
L. lateolabraxi Shen: Bohai. Parasitic. (7a)
L. scutiger Shiino: Bohai. Parasitic. (7a)
Parapetalus denticulatus (Shen) [*Sinocaligus denticulatus* Shen, 1957]: Northern South China Sea shoreline. Parasitic. (6, 7a)
Synestius caliginus Steenstrup & Lütken: Northern South China Sea. Parasitic. (6)

ANTHOSOMATIDAE

Lernanthropus abitocephalus Tripathi: Haikou (Hainan) shoreline. Parasitic. (40)
L. atrox Heller: Sanya (Hainan) shoreline. Parasitic. (40)
L. chirocentrosus Tripathi: Haikou and Yinggehai (Hainan) shoreline. Parasitic. (40)
L. chrysophrys Shishido: Sanya (Hainan) shoreline. Parasitic. (40)
L. corniger Yamaguti: Yinggehai (Hainan). Parasitic. (40)
L. cornutus Kirtisinghe: Qingdao (Shandong). Parasitic. (40)
L. gazzis Song & Chen: Sanya (Hainan). Parasitic. (40)
L. giganteus Kröyer: Haikou (Hainan) shoreline. Parasitic. (40)
L. gisleri Van Beneden: Sanya (Hainan) shoreline. Parasitic. (40)
L. paenulatus Wilson: Qingdao and Yantai (Shandong). Parasitic. (40)
L. sanguineus Song & Chen: Sanya (Hainan) shoreline. Parasitic. (40)
L. shishidoi Shiino: Qingdao (Shandong); Haikou and Sanya (Hainan) shoreline. Parasitic. (40)
L. sillaginis Pillai: Sanya (Hainan) shoreline. Parasitic. (40)
L. trifoliatus Bassett-Smith: Taishan and Guanghai (Guangdong). Parasitic. (40)
Mitrapus heteropodus (Yü), Comb, nov.: Haikou and Sanya (Hainan) shoreline. Parasitic. (40)
Norion globosus Pillai: Baimajing (Hainan) shoreline. Parasitic. (40)
Sanya equulus Song & Chen: Sanya (Hainan) shoreline. Parasitic. (40)

DICHELESTHIIDAE

Pseudocycnus appendiculatus Heller: Sanya (Hainan) shoreline. Parasitic. (40)
P. armatus (Bassett-Smith): Haikou and Sanya (Hainan) shoreline. Parasitic (40)

PANDARIDAE

Achtheinus chinensis (Yü): Qingdao and Yantai (Shandong) shoreline. Parasitic. (8b)
A. impenderus Shen & Wang: Beidaihe (Hebei) shoreline. Parasitic. (8b)

LERNAEOPODOIDA
CHONDRACANTHIDAE

Acanthochondrites annulatus (Olsson): Yantai (Shandong) shoreline. Parasitic. (40)
Protochondracanthus alatus (Heller): Haikou (Hainan) shoreline. Parasitic. (40)

LERNAEOPODIDAE

Brachiella lata Song & Chen: Sanya (Hainan) shoreline. Parasitic. (40)
B. trichiuri Gnanamuthu: Yinggehai (Hainan). Parasitic. (40)
B. yongxingensis Song & Chen: Yongxing Island (XIsha=Paracel Islands). Parasitic. (40)
Clavellodes macrotrachelus (Brian): Qingdao (Shandong) shoreline. Parasitic. (40)
Clavellopsis appendiculata (Heegaard): Baimajing (Hainan) shoreline. Parasitic. (40)
C. sargi (Kurz): Qingdao (Shandong) shoreline. Parasitic. (40)
Epibrachiella magna Song & Chen: Shanwei (Guangdong) and Sanya (Hainan) shoreline. Parasitic. (40)
Parabrachiella sihama Song & Chen: Sanya (Hainan) shoreline. Parasitic. (40)
Pseudocharopinus markewitschi (Gussev): Qingdao and Yantai (Shandong) shoreline. Parasitic. (40)
Thysanote appendiculata (Steenstrup & Lütken): Baimajing (Hainan) shoreline. Parasitic. (40)

T. fimbriata (Heller): Yongle Island (Xisha=Paracel Islands). Parasitic. (40)

REFERENCES*

(1). Zhu Changshou et al., 1988. Appendix: List of Zooplankton: Copepoda. In: A comprehensive rapid survey of the central and northern part of the Taiwan Strait. China Science Press, pp:399–401.

(2). Zhu Changshou, Huang Jiaqi and Li Shaoqing, 1991. Studies on the ecology of copepods in Minnan-Taiwan Bank Fishing Ground. In: Minna-Taiwan Bank Fishing Ground Upwelling Ecosystem Study. China Science Press, pp:440–455.

(3). Chen Chiajui, 1955. On some marine crustaceans from the coastal water of Fenghsien, Kiangsu Province. Acta Zoologica Sinica, 7(2):75–100, pls1–2.

(4). Chen Chiajui and Bai Syeo, 1956. The marine copepoda from the spawning ground of *Pneumatophorus japonicus* (Honttyn) off Chefoo, China. Acta Zoologica Sinica, 8(2):177–234, pls.1–3.

(5). Shen Chiajui, 1957. Parasitic copepods from fishes of China, Pt. I. Cyclopoida (1). Acta Zoologica Sinica, 9(4):297–327, pls. 1–9.

(6). Shen Chiajui, 1957. Parasitic copepods from fishes of China, Pt. II. Caligidia, Caligidae (1). Acta Zoologica Sinica, 9(4):351–377, pls.1–11.

(7a). Shen Chiajui, 1958. Parasitic copepods from fishes of China, Pt. III. Caligoida, Caligidae (2). Acta Zoologica Sinica, 10(2):131–144, pls.1–5.

(7b). Shen Chiajui and Li Houlon,1959. Parasitic copepods from fishes of China, Pt. IV. Caligoida, Caligidae (3). Acta Zoologica Sinica,11(1):12–19, pls.1–3.

(8a). Shen Chiajui, 1958. A marine argulid found in China's seas. Acta Zoologica Sinica, 10(1):31–33, pl. 1.

(8b). Shen Chiajui and Wu Kangnan, 1958. A new parasitic copepods, *Achtheinus impenderus* (Caligoida, Pandaridae), from a shark taken at Peitaiho, Hopei Province. Acta Zoologica Sinica, 10(1): 27–30, pl.1.

(9). Shen Chiajui and Lee Foosiang, 1963. The estuarine copepoda of Chiekong and Zaikong rivers, Kwangtung Province, China. Acta Zoologica Sinica, 15(4):571- 596.

(10a). Shen Chiajui and Lee Foosiang, 1966. Notes on new copepods from South China Sea. Acta Zoologica Sinica, 3(3):213–223.

(10b). Shen Chiajui, Tai Aiyun et al., 1979. Fauna Sinica, Crustacean, Freshwater Copepoda, China Science Press, pp:1–450.

(11). Li Shaoqing and Huang Jiaqi, 1984. On two new species of planktonic copepoda from the Estuary of Jiulong River, Fujian, China. Journal of Xiamen University (Natural Science), 23(3):381–390.

(12). Chen Qingchao and Zhang Shuzhen, 1965. The planktonic copepoda of the Yellow Sea and the East China Sea (I), Calinoida. Studia Marina Sinica, 7:20–131, pls. 1- 53.

(13). Chen Qingchao, Zhang Shuzhen and Zhu Changshou, 1974. The planktonic copepods of the Yellow Sea and the East China Sea (II). Cyclopoida and Harpacticoida. Studia Marina Sinica, 9:27–76, 24 pls.

(14). Chen Qingchao and Zhang Shuzhen, 1974. The pelagic copepods of the South China Sea (1). Studia Marina Sinica, 9:101–116, 8pls.

(15). Chen Qingchao and Shen Chiajui, 1974. The pelagic copepods of the South China Sea (II). Studia Marina Sinica, 9:125–137.

(16). Chen Qingchao, 1983. The pelagic copepods of the South China Sea (III). In : Contribution on Marine Biological Research of the South China Sea (I). China Science Press. pp:133–138.

(17). Chen Qingchao et al., 1986. List of zooplankton. In: Report on the arine organisms in the subtidal zone of Hainan Island - comprehensive survey of the resources in the coastal waters and beach areas off Guangdong Province. South China Sea Institute of Oceanography, Acdemica Sinica, pp:83–86.

(18). Chen Qingchao, 1987. A new species of *Macandrewella* (Copepoda, Calanoida). In: Research reports on the comprehensive oceanographic investigation of Chinese Southern Territory: Zengmu Shoal. China Science Press, pp:218–221.

(19). Chen Qingchao et al., 1989. Zooplankton. In: Research reports on the comprehensive oceanographic investigation of Nansha Islands and adjacent waters (I). China Science Press, Vol.1:99–100, vol. 2:659–707.

(20). Chen Boyun, 1982. The species composition and distribution of the planktonic copepods in the Xisha and Zhongsha Islands, China. Journal of Xiamen University (Natural Science), 21(2):209–217.

(21). Chen Yaqu, 1986. A study on the zooplankton in deep-sea fishing ground off continental shelf margin and slope in the East China Sea (II): horizontal and vertical distribution of copepods. In: Transactions of the Chinese Crustacean Society. China Science Press, pp:86–99.

(22). He Dehua et al, 1985. Appendix II: List of zooplankton (Copepoda). In: Investigation reports of Oceanography, State Oceanic Administration, Hangzhou, pp:13–16.

(23). He Dehua et al., 1987. Study of zooplanktonic ecology in the Zhejiang coastal upwelling system (II): species distribution and diversity of zooplankton. Acta Oceanologica Sinica, 9(5):617–626.

(24). Lian Guangshan and Lin Jinmei, 1978a. Studies on the planktonic cyclopoid copepods of the Southern Yellow Sea and the East China Sea. Marine Science and Technology, 9: 45–67.

(25). Lian Guangshan and Lin Jinmei, 1978b. Studies on the calanoid copepods of the Southern Yellow Sea and the East China Sea, Marine Science and Technology, 11:59–112.

(26). Lian Guangshan and Lin Yuhui, 1986. Appendix: List of zooplankton (copepoda). In: Research reports on the plankton of the comprehensive investigation of the resources in the coastal water and beach areas off Fujian Province. Third Institute of Oceanography, State Oceanic Administration, Xiamen, pp:24–30.

(27). Lian Guangshan and Lin Yuhui, 1988,1990. Appendix: List of zooplankton: Copepoda. In: Reports on the zero marine ecological investigation in the waters around the Nuclear Power Station in Daya Bay, Guangdong, China. Third Institute of Oceanography, State Oceanic Administration, pp:474–484 (1988), pp:9.35–9.40 (1990)

(27a) Lian Guangshan and Qian Honglin, 1994. A new species of the genus *Paramisophria* (Copepoda, Calanoida, Arietellidae) from the China Seas. Acta Oceanologica Sinica, 13(4):551–555.

(28). Zheng Zhong, Li Song and Li Shaoqing et al., 1965, 1982. The planktonic copepods of China Seas. Shanghai Science and Technology Publisher, Vol. 1, 210pp; Vol. 2, 162pp.

(29). Zheng Zhong, Li Song and Li Shaoqing, 1978. On the species composition and geographic distribution of the planktonic copepods of the China Seas. Journal of Xiamen University (Natural Science), 17(2):51–63.

(30). Zheng Zhong, Li Shaoqing, Li Song et al., 1982. On the distribution of planktonic copepods in the Taiwan Strait. Journal of Oceanography in Taiwan Strait, 1(1): 69–79.

(31). Lin Yuhui and Lian Guangshan, 1988. Ecology of planktonic copepods in the Taiwan Strait. Journal of Oceanography in Taiwan Strait, 7(3):248–255.

(32). Lin Yuhui and Lian Guangshan, 1988. List of Zooplankton: Copepoda. In : Reports of the comprehensive investigation on environmental resources of the central China Sea. China Ocean Press, pp:238–246.

(33). Meng Fan et al., 1987. On the composition and distribution of zooplankton species in the coastal waters off Jiangsu Province. Acta Ecologica Sinica, 7(3):256–266.

(34). Zhang Chongzhou and Li Zhiying, 1976a. Harpacticoida (Copepoda, Crustacea) from Xisha Islands of Guangdong Province, China. Acta Zoologica Sinica, 22(1):66–70.

(35). Zhang Chongzhou and Li Zhiying, 1976b. A new species of *Limnoithona* (Oithonidae, Cyclopoida) from China. Acta Zoologica Sinica, 22(4): 403–405.

(36). Huang Jiaqi and Chen Boyun, 1985. Species composition and distribution of planktonic copepods in the Jiulong River estuary, Fujian Province. Journal of Oceanography in Taiwan Strait, 4(1):79–88.
(37). Huang Jiaqi, Li Shaoqing and Chen Yiapin, 1989. On the species composition and distribution of zooplankton in the Luoyuan Bay of Fujian, China. Journal of Xiamen University (Natural Science), 28 (suppl.):85–95.
(38). Huang Shimei, 1986. Appendix: List of zooplankton (Copepoda). In: Ecological study of zooplankton, Journal of Shandong College of Oceanography, 16(2):75- 79.
(39). The Group of Planktonic Study, Institute of Oceanography, Academia Sinica, 1977. List of neritic plankton in China Seas: Copepoda. In: On the study of neritic plankton in the China Seas. The Oceanographic Group, The Chinese Committee of Science and Technology, pp:153–156.
(40). Song Daxiang and Chen Guoxiao, 1976. Some parasitic copepods from marine fishes of China. Acta Zoologica Sinica, 22(4): 406–424.
(41). Boxshall, G. A., 1986. A new species of *Misophria* (Copepoda: Misophrioida) from Hong Kong. In: Proceedings of the Second International Marine Biological Workshop: The Marine Flora and Fauna of Hong Kong and Southern China. Hong Kong University Press, pp:515–522.
(42). Boxshall, G. A., 1986. Siphonostome copepods associated with sponges from Hong Kong. In: Proceedings of the Second International Marine Biological Workshop: The Marine Flora and Fauna of Hong Kong and Southern China. Hong Kong University Press, pp:523–547.
(43). Chen Qingchao, 1980. The marine zooplankton of Hong Kong. In: Proceedings of the First International Marine Biological Workshop: The Marine Flora and Fauna of Hong Kong and Southern China. Hong Kong University Press, pp:789–799.
(44). Mori, T., 1937. The pelagic copepods from the neighbouring waters of Japan. Tokyo, 150pp., 80 pls.
(45). Tan Tienhsi, 1970. On the distribution of copepods in waters surrounding Taiwan. In: The Kuroshio, a symposium of the Japan Current. Edited by J. C. Marr., University of Hawaii Press, pp:323–332.

*: (1)-(40) in Chinese
Compiled by Lian Guangshan and Lin Yuhui; Edited by Zheng Zhong (C. Cheng), Li Shaoqing, and Chen Qingchao

CIRRIPEDIA
THORACICA
SCALPELLIDAE

Abathescalpellum fissum (Hoek): Offshore of East China Sea. Water depth 580 m. (9)

A. koreanum (Hiro): Bohai; Yellow Sea; East China Sea. Water depth 14–184 m. (7)

Amigdoscalpellum vitreum (Hoek) [*Scalpellum condensum*]: Offshore of East China Sea. Water depth 550–760 m. (7, 10)

Annandaleum gruvelii (Annandale): Nansha (=Spratly Islands). Water depth 1,252 m. (12)

Arcoscalpellum ciliatum Ren: Offshore of northern East China Sea. Water depth 300+ m. (7)

A. pertosum Foster: Nansha (=Spratly Islands). Water depth 1,241 m. (12)

Capitulum mitella (Linnaeus) [*Mitella mitella, Pollicipes mitella*]: South of Zhoushan (Zhejiang) to Beibuwan (=Gulf of Tongking) and Hainan. Mid- and high- intertidal. (7, 14, 16, 18)

Catherinum rossi (Lakshmana Rao & Newman): Offshore of East China Sea; Nansha (=Spratly Islands). Water depth 510–2150 m. (7, 12)

Euscalpellum rostratum (Darwin): Northern South China Sea. Water depth 23–32 m, attached to hydrozoans. (7)

E. stratum (Aurivillius): Northern South China Sea. Water depth 195 m, attached to hydrozoans. (7)

Lithotrya nicobarica Reinhardt: Xisha (=Paracel Islands); Nansha (=Spratly Islands). Coral reef intertidal. (7, 12)

L. valentiana (Gray): Xisha (=Paracel Islands). Coral reef intertidal. (7)

Litoscalpellum intermedium (Hoeck): Northern East China Sea. Water depth 365–510 m, attached to hydrozoans. (7)

L. sinense Ren: Offshore of East China Sea. Water depth 550 m. (7)

Neoscalpellum dicheloplax Pilsbry: East China Sea. (16)

Pilsbryiscalpellum condensum (Nilsson-Cantell) [*Scalpellum condensum*]: Offshore of East China Sea. (7, 10)

Scalpellum rubrum Hoeck: East China Sea continental shelf. (13)

S. stearnsii Pilsbry: East China Sea continental shelf; northern South China Sea; Xisha (=Paracel Islands). Water depth 290 m. (9, 17)

Smilium acutum (Hoek): East China Sea, water depth 105–184 m; Nansha (=Spratly Islands), water depth 167–1241 m, attached to hydrozoans. (7, 12)

S. scorpio (Aurivillius): Bohai; Yellow Sea; northern East China Sea. Water depth 31- 103 m, attached to hydrozoans. (7, 16)

S. sinense (Annandale): Offshore of East China Sea; Nansha (=Spratly Islands). Water depth 55–150 m, attached to seaweed, hydrozoans, and corals. (7, 12)

Tarasovium orientale Ren: Northern East China Sea. Water depth 100 m. (7)

Teloscalpellum album Liu & Ren: Offshore of northern East China Sea. Water depth 300+ m. (7)

Trianguloscalpellum balanoides (Hoek): Offshore of East China Sea. Water depth 220 m, attached to crinoids. (7)

T. michelottianum (Sequenza): Offshore of East China Sea. Water depth 550–1,115 m. (7)

T. uniarticulatum (Nilsson-Cantell): Offshore of East China Sea; northern South China Sea. Water depth 150–300 m. (7)

IBLIDAE

Ibla cumingi Darwin: From Xiamen (Fujian) to Baibuwan (=Gulf of Tongking); Hainan. Mid-intertidal. (7, 14, 16)

LEPADIDAE

Alepas pacifica Pilsbry: From Shanghai to Guangdong. Attached to bells of large jellyfish. (7, 16)

Conchoderma auritum (Linnaeus): Dalian (Liaoning); Qingdao (Shandong); Shanghai; Xiamen (Fujian); Hainan. Attached to boats, occasionally on floating objects. (9, 16, 19)

C. hunteri (Owen) [*C. virgatum hunteri*]: From Qingdao (Shandong) to Qiongzhou Strait (between Guangdong and Hainan). Attached to boats and floating objects. (7, 16)

C. virgata (Spengler): Qingdao (Shandong); Guangdong; Xisha (=Paracel Islands). Attached to floating wood, boats, and sea turtles. (7, 16, 19)

Lepas anatifera anatifera Linnaeus: From Qingdao (Shandong) to Hainan; Xisha (=Paracel Islands). Attached to floating objects. (7, 16)

L. anatifera striata de Graef: Northern East China Sea. (13)

L. anserifera Linnaeus: From Rongcheng (Shandong) to Nansha (=Spratly Islands). Attached to floating objects, boats, and buoys. (7, 12, 16, 21)

L. indica Annandale: Northern South China Sea; Nansha (=Spratly Islands). Attached to floating wood. (11)

L. pectinata Spengler: South China Sea. Attached to floating objects. (7, 16, 17)

L. testudinata Aurivillius: Qingdao (Shandong); Hainan. On floating objects. (7)

POECILASMATIDAE

Heteralepas japonica (Aurivillius): East China Sea; northern South China Sea. Water depth 103–300 m, attached to corals and sea pansies. (7, 17)

H. quadrata (Aurivillius): Zhejiang; Taiwan. (16, 19)

H. smilium Ren: East China Sea; Taiwan Strait; northern South China Sea. Attached to corals and hydrozoans. (7)

Koleolepas avis (Hiro): East China Sea continental shelf. Water depth 100 m. (70)

Megalasma annandalei Pilsbry: East China Sea. Water depth 1,115 m, attached to sponges (*Hyalonema*). (7, 17)

M. hamatum Calman: East China Sea. Water depth 850 m, attached to spines of sea urchins. (7)

M. striatum Hoeck: East China Sea; Taiwan Strait; northern South China Sea; Xisha (=Paracel Islands); Nansha (=Spratly Islands). Water depth 125–400 m, attached to sea urchins. (7, 9, 11, 16)

Octolasmis angulata (Aurivillius): Fujian; Guangdong; Hainan. Attached to mouth parts of lobsters and crabs. (7)

O. aymonini geryonophila Pilsbry: East China Sea. (12)

O. bathynomi (Annandale): Northern South China Sea. Water depth 784–864 m, attached to pleopods (swimming legs) of large isopods. (12)

O. bullata (Aurivillius): Guangdong coast. Attached to mouth parts of lobsters. (7)

O. cor (Aurivillius): Wenchang (Hainan). Attached to gills of mud crab (*Scylla serrata*). (7)

O. geryonophila Pilsbry: Northern South China Sea. Water depth 784–864 m, attached to crabs. (9, 16)

O. neptuni (MacDonald): Shallow sea along China's coast. Attached to gill and mouth parts of crabs. (9, 16)

O. nierstraszi (Hoek): Offshore of northern East China Sea to central South China Sea. Water depth 12–300 m, attached to hydrozoans and corals. (7)

O. orthogonia (Darwin): East China Sea; Taiwan Strait; northern South China Sea; Xisha (=Paracel Islands); Nansha (=Spratly Islands). Water depth 1,430 m, attached to hydrozoans and corals.

O. scuticosta Hiro: Taiwan. (16, 19)

O. tridens (Aurivillius): Dapeng Cove (Hong Kong). Attached to gills of crabs. (17)

O. warwickii Gray: From Shengshan (Zhejiang) to Beibuwan (=Gulf of Tongking); Hainan. Intertidal to 100 m water depth, attached to crabs. (7, 16, 21, 22)

Oxynaspis aurivillii Stebbing: Northern South China Sea. Water depth 200 m. (9)

O. bocki Nilsson-Cantell: Northern South China Sea. Water depth 300 m, attached to gorgonians (Alcyonaceae). (7)

O. faroni Totton: Rengai Reef (Nansha=Spratly Islands). (10)

O. pacifica Hiro: Taiwan. (16, 19)

O. reducens Foster: Dapeng Cove (Hong Kong). Water depth 12 m, attached to horn corals. (18)

O. sinensis Ren: East China Sea continental shelf. Water depth 100 m, attached to corals. (7)

Paralepas nodulosa (Broch): East China Sea; Taiwan Strait; northern South China Sea. Water depth 110–300 m, attached to sea urchins. (7)

Poecilasma kaempferi (Darwin): East China Sea; South China Sea. Water depth 90–1,800 m, attached to dorsal carapace of crabs. (7, 17)

P. litum Pilsbry: Offshore of northern East China Sea. Water depth 360–420 m. (7)

P. obliqua Hoek: Northern South China Sea. Water depth 784–864 m, attached to antennae of crabs. (7)

Temnaspis amygdalum (Aurivillius): From Dongshan (Fujian) to Xisha (=Paracel Islands). Attached to mouth parts of lobsters. (7, 16)

T. donghaiense (Du & Zhu): East China Sea. (12)

T. excavatum (Hoek): Offshore of East China Sea; South China Sea. Attached to mouth parts and appendages of crabs and shrimps. (7, 15)

T. kilepoae Zevina: Weizhou Island (Guangxi). Attached to mouth parts of crabs. (7, 23)

T. tridens (Aurivillius): Offshore of East China Sea; northern South China Sea. Water depth 55–148 m, attached to mouth parts and appendages of crabs. (7)

Trilasmis eburnea Hinds: Northern South China Sea. Water depth 180–300 m, attached to sea urchins. (7)

VERRUCIDAE

Verruca albatrossiana Pilsbry: East China Sea continental shelf. Water depth 365–395 m. (5)

V. cristallina Gruel: Northern South China Sea. Water depth 290 m. (5)

V. gibbosa Hoeck: East China Sea offshore. Water depth 2,000–2,150 m. (5)

V. koehleri Gruvel: East China Sea continental shelf; northern South China Sea; Nansha (=Spratly Islands). Water depth 100–600 m. (5, 9)

V. nitida Hoek: East China Sea continental shelf. Water depth 1,350–1,420 m. (5)

V. sculpta Aurivillius: East China Sea continental shelf. Water depth 114–395 m. (5)

CHTHAMALIDAE

Chinochthamalus scutelliformis (Darwin) [*Chamaesipho scutelliformis*]: From Dongtou (Zhejiang) to Beibuwan (=Gulf of Tongking); Hainan. Mid- and high-intertidal. (4, 14, 15, 16, 18, 22)

Chthamalus antennatus Darwin: Lingshui and Yulin (Hainan). Intertidal and subtidal. (4)

C. challengeri Hoek [*C. dalli*]: Bohai; northern Yellow Sea. High intertidal. (4, 16, 22)

C. intertextus Darwin: Taiwan. (15, 18)

C. malayensis Pilsbry: From Dongshan (Fujian) to Beibuwan (=Gulf of Tongking); Taiwan; Hainan. High intertidal. (4, 16, 17, 22)

C. moro Pilsbry: Taiwan; Xisha (=Paracel Islands). Intertidal. (4, 16)

C. sinensis Ren: South of Dongtou (Zhejiang). Mid- and low-intertidal. (4)

Euraphia caudata (Pilsbry): Southern Hainan; Weizhou Island (Guangxi). Mid- and high-intertidal. (4, 16)

E. withersi (Pilsbry) [*Chthamalus withersi*]: From Yueqing (Zhejiang) to Beibuwan (=Gulf of Tongking); Hainan inner bays. High intertidal. (4, 14, 16, 18, 22)

Octomeris brunnea Darwin: Taiwan. (15, 18)

O. sulcata Nilsson-Cantell: Taiwan. (15, 18)

Tetrachthamalus sinensis Ren: Weizhou Island (Guangxi); Southern Hainan. (4)

TETRACLITIDAE

Astroclita longicostata Ren: Xisha (=Paracel Islands). Coral reef intertidal. (2)

Tesseropora alba Ren: Xisha (=Paracel Islands). Intertidal. (2)

Tetraclita coerulescens (Spengler): Weizhou Island (Guangxi); Hainan; Xisha (=Paracel Islands). Mid intertidal. (2, 16)

T. japonica Pilsbry [*T. squamosa japonica*]: From Chengshan (Zhejiang) to Guangdong. Mid intertidal. (2, 14, 18)

T. squamosa squamosa (Bruguiere) [*T. viridis*]: From Chengshan (Zhejiang) to Xisha (=Paracel Islands). Mid intertidal. (2, 14, 15–18)

Tetraclitella chinensis (Nilssen-Cantell): Dongshan (Fujian); Taiwan; Zhelang (Guangdong); Hong Kong. (2, 16, 18)

T. costata (Darwin): South China Sea. (16)

T. darwini (Pilsbry): Taiwan; Naozhou Island (Guangdong). Subtidal. (2, 16)

T. divisa (Nilsson-Cantell): Taiwan; Hainan. Intertidal. (2, 16)

T. multicostata (Nilsson-Cantell): Lingshui (Hainan). Subtidal. (2, 22)

T. pilsbryi (Utinomi): Dapeng Cove (Hong Kong). (12, 18)

PYRGOMATIDAE

Boscia anglica (Sowerby): Northern South China Sea. Water depth 31–80 m, buried in corals. (8)

B. oulastreae (Utinomi): Dapeng Cove (Hong Kong). Buried in corals. (18)

Cantellius albus Ren: Sanya and Lingshui (Hainan). Buried in corals. (8, 18)

C. gregarea (Sowerby): Xisha (=Paracel Islands). Low intertidal, buried in corals. (8)

C. iwayama (Hiro): Sanya (Hainan); Xisha (=Paracel Islands). Low intertidal, buried in corals. (8)

C. pallidus (Broch) [*Creusia spinulosa pallida*]: Dapeng Cove (Hong Kong); Weizhou Island (Guangxi); Sanya and Yulin (Hainan). Buried in stone corals. (10, 17)

C. secundus (Broch): Dapeng Cove (Hong Kong); Sanya (Hainan); Xisha (=Paracel Islands). Buried in corals. (8, 18)

C. septimus (Hiro): Sanya (Haian); Xisha (=Paracel Islands). Buried in corals. (8)

C. sinensis Ren: Xisha (=Paracel Islands). Buried in corals. (8)

C. spinulosa euspinulosa (Broch) [*Creusia spinulosa euspinulosa*]: Sanya (Hainan). (16)

Creusia indicum (Annandale): Dapeng Cove (Hong Kong); southern Hainan. Buried in corals. (8, 16, 18)

Nobia conjugatum (Darwin): Dongshan (Fujian); Dapeng Cove (Hong Kong); southern Hainan. Buried in corals. (8, 18)

N. grandis Sowerby: Dapeng Cove (Hong Kong). Buried in corals. (18)

N. orbicellae (Hiro): Sanya and Dongfang (Hainan). Buried in corals. (8)

N. sinica Ren: Northern South China Sea. Water depth 145 m, buried in corals. (8)

Pyrgoma cancellata Leach: Naozhou Island (Guangdong); Dapeng Cove (Hong Kong); southern Hainan; northern South China Sea. Water depth 83 m, buried in corals. (8, 18)

P. stellula Rosell: Dapeng Cove (Hong Kong). Buried in corals. (18)

Savignium crenatum (Sowerby): Dapeng Cove (Hong Kong); Sanya and Lingshui (Hainan). Buried in corals. (8, 18)

S. dentatum (Darwin): Dapeng Cove (Hong Kong); Xisha (=Paracel Islands). Buried in corals. (8, 18)

S. elongatum (Hiro): Dapeng Cove (Hong Kong). Buried in stone corals. (20)

S. milleporae (Darwin): Southern Hainan; Xisha (=Paracel Islands). Buried in corals. (8)

S. orientale Ren: Daya Bay (Guangdong). Buried in corals. (8)

ARCHAEOBALANIDAE

Acasta conica Hoek [*A. spinosa*]: Guangdong; Guangxi; Hainan. Low intertidal to 18 m water depth, in sponges. (6, 16)

A. coriobasis Broch: Daya Bay (Guangdong); Dapeng Cove (Hong Kong). In sponges. (6)

A. dofleini Krüger: From Huian (Fujian) to Beibuwan (=Gulf of Tongking); Hainan. Low intertidal to 103 m water depth, in sponges. (6, 18)

A. fenestrata Darwin: Sanya (Hainan). Low intertidal, in sponges. (6)

A. flexuosa (Nilsson-Cantell): Beibuwan (=Gulf of Tongking); Hainan. Low intertidal to 44 m water depth, in sponges. (6)

A. fragilis Ren: Xinying (Hainan). Low intertidal, in sponges. (6)

A. glans Lamarck: Nothern South China Sea. Water depth 62 m, in sponges. (6)

A. haimanensis Ren: Northern South China Sea. Water depth 112 m, in sponges. (6)

A. laevigata Gray: Xisha (=Paracel Islands). Intertidal, in sponge (*Phyllospongia foliascens*). (6)

A. membranacea Barnard: Northern South China Sea. Water depth 46 m, in sponges. (6)

A. pectinipes Pilsbry: From Yellow Sea to South China Sea. Low intertidal to 52 m water depth, in sponges. (6, 17)

A. sinica Ren: Sanya (Hainan). Subtidal, in sponges. (6)

A. spinitergum Broch: Dapeng Cove (Hong Kong). In sponges. (18)

A. spongites (Poli): Northern South China Sea. Water depth 54–112 m. (6)

A. sporillius Darwin: Qiongzhou Strait (between Guangdong and Hainan); Lingshui (Hainan). Low intertidal to 57 m water depth, in sponges. (6)

A. sulcata Lamarck: Northern South China Sea. Water depth 22–50.5 m, in sponges. (6, 18)

A. zuiho Hiro: Weizhou Island (Guangxi); northern South China Sea. Low intertidal to 55 m water depth, in sponges. (6, 16)

Armatobalanus allium (Darwin) [*Balanus allium*]: Southern Hainan; Xisha (=Paracel Islands). Attached to corals. (1, 16, 18)

A. cepa (Darwin) [*Balanus cepa*]: Northern South China Sea. Water depth 21 m. (1, 16)

Chirona amaryllis (Darwin) [*Balanus amaryllis*]: China's seas. Low intertidal to water depth 124 m. (1, 12, 14, 16)

C. cristatus (Ren & Liu) [*Balanus cristatus*]: Tanggu (Tianjin); Shi Island (Shandong). (1, 16)

C. kruegeri (Pilsbry) [*Balanus kruegeri*]: Northern South China Sea. Water depth 118 m. (1, 16)

C. taiwanensis Hiro: Taiwan. (19)

C. tenuis (Hoek) [*Balanus tenuis*]: East China Sea; South China Sea. Water depth 17- 167 m. (1, 12)

Conopea calceopus (Ellis) [*Balanus calceolus*]: Northern Yellow Sea, water depth 29 m; from Jiaozhou Bay (Shandong) to Beibuwan (=Gulf of Tongking) and Hainan, attached to gorgonian corals. (1, 16)

C. canaliculatus (Ren & Liu) [*Balanus canaliculatus*]: South China Sea. Attached to gorgonian corals. (1, 16)

C. cuneiformis (Hiro): Sanya (Hainan). (16)

C. cymbiformis (Darwin) [*Balanus cymbiformis*]: Daya Bay (Guangdong); Beibuwan (=Gulf of Tongking). Attached to gorgonian corals. (1, 16)

C. granulatus Hiro: Taiwan. (19)

C. navicula (Darwin) [*Balanus navicula*]: Guangdong; Guangxi; Hainan. Attached to gorgonian corals. (1, 16)

C. sinensis (Ren & Liu) [*Balanus sinensis*]: Dongshan (Fujian); Daya Bay (Guangdong). Attached to horn corals. (1, 16)

Membranobalanus longirostrum (Hoek) [*Balanus longirostrum*]: Lingshui (Hainan). In sponges. (1, 16)

Solidobalanus cidaricola (Ren & Liu) [*Balanus cidaricola*]: Offshore of East China Sea; Nansha (=Spratly Islands). Water depth 110–167 m, attached to sea urchins. (1, 9, 12, 16)

S. ciliatus (Hoek) [*Balanus ciliatus*]: Offshore of East China Sea; Dapeng Cove (Hong Kong); northern South China Sea. Water depth 50–66 m. (1, 16, 18)

S. socialis (Hoek) [*Balanus socialis*]: Northern South China Sea. Water depth 18–106 m. (1, 16)

CORONULIDAE

Chelonibia patula (Ranzani): From Zhejiang to Guangxi; Hainan. Attached to swimming crabs. (9, 20)

C. patula dentata Henry: Fujian, Guangdong. Attach to crab carapace. (9, 16)

C. testudinaria (Linnaeus): Xisha (=Paracel Islands). Attach to sea turtles. (3, 16)

Coronula diadema (Linnaeus): Hainan. (16)

Cylindrolepas sinica Ren: Xisha (=Paracel Islands). Attached to sea turtles. (3)

Platylepas decorada Darwin: Xisha (=Paracel Islands). Attached to sea turtles. (3)

P. hexastylos (Fabricus): Zhangjiang (Guangdong), attached to sea snakes; Xisha (=Paracel Islands), attached to sea turtles. (3, 16)

Stomatolepas elegans (Costa): Xisha (=Paracel Islands). Attached to sea turtles. (9)

S. pulchra Ren: Xisha (=Paracel Islands). Attached to sea turtles. (3)

BALANIDAE

Balanus albicostatus Pilsbry [*B. amphitrite albicostatus*]: China's inner bays. High intertidal. (1, 14, 16, 18)

B. amphitrite amphitrite Darwin [*B. amphitrite communis, B. hawaiiensis*]: China's seas. Mid intertidal to shallow sea, boats. (1, 14, 16, 17)

B. cirratus Darwin [*B. amphitrite cirratus, B. variegatus cirratus*]: China's seas. Low intertidal to shallow sea. (1, 14, 16, 18)

B. crenatus Bruguiére: Dalian (Liaoning); Qingdao (Shandong). Subtidal and attached to boats. (1, 16)

B. eburneus Gould: Qingdao (Shandong); Hong Kong. (1, 19)

B. improvisus Darwin: Lushun and Dalian (Liaoning); Shanhaiguan (Hebei); Tanggu (Tianjin); Yantai (Shandong). Subtidal. (1, 16)

B. littoralis Ren & Liu: Donghai Island (Guangdong); Fangcheng (Guangxi); Hainan. Attached to mid intertidal mangrove roots. (1, 16)

B. poecilotheca Krügeri: From Dongshan (Fujian) to Beibuwan (=Gulf of Tongking); northern South China Sea. Low intertidal to water depth 95 m. (1, 16)

B. pulchellus Ren: South China Sea. Attached to buoys. (11)

B. reticulatus Utinomi [*B. amphitrite communis*]: Qingdao (Shandong), docks; from Fujian to Nansha (=Spratly Islands), mid- and low-intertidal to shallow sea and boats. (1, 12, 14, 16, 18)

B. rostratus rostratus Hoek: Around Haiyang Island (Liaoning). Water depth 50 m. (1, 18)

B. trigonus Darwin: Yantai (Shandong), docks; from Chengshan (Zhejiang) to Guangdong and Hainan, low intertidal to shallow sea. (1, 14, 16, 18)

B. uliginosus Utinomi [*B. amphitrite kruegeri*]: From Liaoning to Xisha (=Paracel Islands) shoreline. Particularly abundant in low salinity water. (1, 14)

B. zhujiangensis Ren: Off mouth of Zhujiang (=Pearl River, Guangdong). (11)

Megabalanus rosa (Pilsbry) [*Balanus rosa*]: From Zhoushan (Zhejiang) to Guangdong; Hainan. Subtidal. (1, 16)

M. tintinnabulum crispatus (Darwin): South China Sea. (16)

M. tintinnabulum occator (Darwin): Taiwan; Hainan. (16, 19)

M. tintinnabulum tintinnabulum (Linnaeus) [*Balanus tintinnabulum tintinnabulum*]: From Chengshan (Zhejiang) to Guangdong; Taiwan; Hainan. Subtidal, attached to buoys. (1, 16)

M. tintinnabulum zebra (Darwin): Taiwan. (16, 19)

M. volcano (Pilsbry) [*Balanus volcano*]: From Chengshan (Zhejiang) to Xisha (=Paracel Islands). Low intertidal. (1, 14, 16, 18)

M. xishaensis (Ren & Liu) [*Balanus xishaensis*]: Xisha (=Paracel Islands). Low intertidal. (1, 14,1 6, 18)

ACROTHORACICA
SACCULINIDAE

Sacculina confragosa Boschma: East China Sea; South China Sea. Live on crab's abdomen. (16)

PELTOGASTERIDAE

Peltogaster paguri Rathke: Zhoushan (Zhejiang). Live on abdomen of hermit crab (*Pagurus geminus*). (16)

Thompsonia sp.: Zhejiang. Live on abdomen of mitten crab (*Eriocheir sinensis*). (16)

RHIZOCEPHALA
RHIZOLEPADIDAE

Rhizolepas gurjanovae Zevina: Beibuwan (=Gulf of Tongking). (23)

REFERENCES*

(1). Ren Xianqiu and Liu Ruiyu, 1978. Studies on Chinese Cirripedia (Crustacea) I. Genus *Balanus*. Studia Marina Sinica, 13:119–196.

(2). Ren Xianqiu and Liu Ruiyu, 1979. Studies on Chinese Cirripedia (Crustacea) II. Family Tetraclitidae. Ocaenologia & Limnologia Sinica, 10(4): 336–353.

(3). Ren Xianqiu, 1980. Turtle barnacles of the Xisha Islands, Guangdong Province, China. Studia Marina Sinica, 17:187–197.

(4). Ren Xianqiu, 1984. Studies on Chinese Cirripedia (Crustacea) III. Family Chthamalidae. Studia Marina Sinica, 22:145–166.

(5). Ren Xianqiu, 1984. Studies on Chinese Cirripedia (Crustacea) IV. Family Verrucidae. Studia Marina Sinica, 23:165–182.

(6). Ren Xianqiu, 1984. Studies on Chinese Cirripedia (Crustacea) V. Genus *Acasta*. Studia Marina Sinica, 23:183–220.

(7). Liu Ruiyu and Ren Xianqiu, 1985. Studies on Chinese Cirripedia (Crustacea) VI. Suborder Lepadomorpha. Studia Marina Sinica, 25:179–281.

(8). Ren Xianqiu, 1986. Studies on Chinese Cirripedia (Crustacea) VII. Family Pyrgomatidae. Studia Marina Sinica, 26:119–158.

(9). Ren Xianqiu, 1987. Studies on Chinese Cirripedia (Crutacea) VIII. Supplementary Report. Studia Marina Sinica, 28:175–188.

(10). Ren Xianqiu, 1987. A short report on the Cirripedia (Crustacea) from the Nansha Islands. Studia Marina Sinica, 28:189–193.

(11). Ren Xianqiu, 1989. Two new species and one new records of Cirripedia Thoracica from South China Sea. Oceanologia & Limnologia Sinica, 20(5):466–473.

(12). Ren Xianqiu, 1991. Cirripedia (Crustacea) of Nansha Islands, Guangdong Province, China. In: Contributions on Marine Organisms around Nansha Islands and the adjacent Waters (I). China Ocean Press, pp:169–180.

(13). Du Nanshan and Zhu Zhenqin, 1986. Studies on Thoracica cirripedia fauna in the northern part of the East China Sea. In: Transactions on the Chinese Crustacean Society (I). China Science Press, pp:194–195.

(14). Huang Zongguo et al., 1986. The Cirripede foulers of Hong Kong waters. In: Transactions on The Chinese Crustacean Society (I). China Science Press, pp:109–117.

(15). Huang Zongguo, 1992. The habitation and distribution characteristics of Daya Bay Cirripeds. In: Transactions on the Chinese Crustacean Society (III). Qingdao Ocean University Press, pp:6–15.

(16). Dong Yumao et al., 1982. Zoological Illustration from China: Crustacean Vol.I (2nd edition). China Science Press, pp:64–114.

(17). Dong Yumao, 1988. The deep sea crustaceans in the East China Sea. Zhejiang Science & Technology Press, 132pp.

(18). Foster, B. A. 1980. Shallow water barnacles from Hong Kong. In: Proceedings of the First International Marine Biological Workshop: The Marine Flora and Fauna of Hong Kong and Southern China. Hong Kong University Press, pp:207–232.

(19). Hiro, F. 1939. Studies on the Cirripedia Fauna of Japan. IV. Cirripeds of Formosa (Taiwan), with Some Geographical and Ecological Remarks on the Littoral Forms. Mem. Coll. Sci. Imp. Univ. Ser. B15(2):245–284.

(20). Rainbow, P. S. 1990. A new barnacle record from Hong Kong. In: Proceedings of the Second International Marine Biological Workshop: The Marine Flora and Fauna of Hong Kong and Southern China. Hong Kong University Press, pp:567- 568.

(21). Wu Shi-Kue, 1967. Two new records of Octolasmid cirripeds from Taiwan. Crustaceana 12(3):274–278.

(22). Zevina, G. B. and N. I. Taraso, 1963. I. Cirripedia Thoracica of mainland coasts of southeastern Asia. Trud. Inst. Okeanol. 70:76–100.

(23). Zevina, G. B. and N. I. Taraso, 1963. New species of Lepadimorpha (Cirripedia Thoracica) from the Bay of Tonkin. Crustacea 15(1):35–40.

*: (1)-(17) in Chinese and (22) in Russian
Compiled by Huang Zongguo; Edited by Ren Xianqiu and Liu Ruiyu

MALACOSTRACA
MYSIDACEA
LOPHOGASTRIDAE

Gnathophausia ingens (Dohrn): East China Sea. Water depth 200 m. (13)
Lophogaster hawaiensis Fage: South China Sea. (14)
L. pacificus Fage: South China Sea. (14)
Paralophogaster glaber Hansen: South China Sea. (14)

EUCOPIIDAE

Eucopia australis Dana: South China Sea. (14)

MYSIDAE

Acanthomysis aspera Ii: Bohai, Yellow Sea. (8)
A. crassispinosa Liu & Wang: Taiwan Strait; South China Sea. (4, 7, 12)
A. fujinagai Ii: Yellow Sea; East China Sea. (10)
A. hwanhaiensis Ii: Bohai; Yellow Sea. (8)
A. koreana Ii: Bohai; Yellow Sea. (8)
A. laticauda Liu & Wang: Taiwan Strait; South China Sea. (4, 7, 9, 12)
A. leptura Liu & Wang: Taiwan Strait; South China Sea. (4, 7, 12)
A. longirostris Ii: Bohai; Yellow Sea; East China Sea. (7, 8, 10)
A. meridionalis Liu & Wang: South China Sea. (6, 7)
A. okayamaensis Ii: Yellow Sea. (8)
A. platycauda (Pillai): South China Sea. (7)
A. quadrispinosa Nouvel: South China Sea. (7, 11)
A. rotundicauda Liu & Wang: South China Sea. (4, 7)
A. serrata Liu & Wang: South China Sea. (4, 7)
A. sheni Wang & Liu: Yellow Sea. (8)
A. sinensis Ii: Yellow Sea; East China Sea. (10)
A. tenella Liu & Wang: South China Sea. (6, 7)
Afromysis dentisinus Pillai: South China Sea. (7)
Anchialina grossa Hansen: South China Sea. (2, 14)
A. parva Ii: South China Sea. (2, 9, 14)
A. typica (Kröyer): Yellow Sea; East China Sea; Taiwan Strait; South China Sea. (2, 9, 10, 12, 14)
A. zimmer Tattersall: South China Sea. (2)
Anisomysis bipartoculata Ii: Yellow Sea; South China Sea. (9, 10, 14)
A. brevicauda Wang: South China Sea. (3)
A. ijimai Nakazawa: South China Sea. (11)
A. minuta Liu & Wang: South China Sea. (6, 7)
A. quadrispinosa Wang: South China Sea. (3)
Archaeomysis kokuboi Ii: South China Sea. (8, 9)
Boreomysis rostrata orientalis Ii: South China Sea. (14)
Doxomysis littoralis Tattersall: South China Sea. (7)
D. longiura Pillai: South China Sea. (7)
D. quadrispinosa Illig: South China Sea. (9, 14)
Erythrops minuta Hansen: Yellow Sea; East China Sea; Taiwan Strait; South China Sea. (7, 8, 10, 12)
Euchaetomera oculata Hansen: South China Sea. (9, 14)
E. tenuis G. O. Sars: South China Sea. (9)
Euchaetomeropsis merolepis (Illig): South China Sea. (14)
Gastrosaccus dunckeri Zimmer: Taiwan Strait. (12)
G. formosensis Ii: Bohai; Yellow Sea; Taiwan Strait; South China Sea. (2, 8)
G. hibii Ii: Taiwan Strait; South China Sea. (2, 12)
G. indicus Hansen: South China Sea. (2, 11, 14)
G. ohshimai Ii: South China Sea. (2, 11)
G. parvus Hansen: South China Sea. (2, 14)
G. pelagicus Ii: Yellow Sea; East China Sea; Taiwan Strait. (8, 10)
Hemisiriella parva Hansen: South China Sea. (9, 14)
H. pulchra Hansen: Yellow Sea; East China Sea; Taiwan Strait; South China Sea. (9, 10, 12)
Holmesiella affinis Ii: Yellow Sea; East China Sea; South China Sea. (7, 10)
Hypererythrops spinifera (Hansen): Yellow Sea; East China Sea; Taiwan Strait; South China Sea. (7, 10, 12, 14)
H. zimmeri Ii: Yellow Sea, East China Sea. (10)
Lycomysis spinicauda Hansen: South China Sea. (7, 11)
Mysidopsis indica Tattersall: Taiwan Strait; South China Sea. (7, 12)
M. kempi Tattersall: South China Sea. (7)
Neomysis awatschensis (Brandt): From Bohai to South China Sea. (7, 8)
N. japonica Nakazawa: From Bohai to South China Sea. (7, 8)
Nipponomysis quadrispinosa (Ii) [*Proneomysis quadrispinosa* (1, 7)]: South China Sea. (1, 7)
N. sinensis (Wang) [*Proneomysis sinensis* (1, 7)]: South China Sea. (1, 7)
Paracanthomysis hispida Ii: South China Sea. (11)
Paraleptomysis sinensis Liu & Wang: South China Sea. (5, 7)
P. xenops (Tattersall): South China Sea. (5, 7)
Parastilomysis paradoxa Ii: Yellow Sea; East China Sea. (10)
Pleurerythrops inscita Ii: Taiwan Strait; South China Sea. (7, 12)

P. monospinosa Liu & Wang: South China Sea. (7)
Prionomysis aspera Ii: South China Sea. (7)
Promysis orientalis Dana: Yellow Sea; East China Sea; Taiwan Strait; South China Sea. (7, 10, 12)
Pseudanchialina inermis (Illig): South China Sea. (2, 14)
P. pusilla G. O. Sars: Yellow Sea; East China Sea; Taiwan Strait; South China Sea. (2, 9, 10, 12, 14)
Pseudomma brevicaudum Shen & Liu: Yellow Sea. (8)
Pseudomysisdetes cochinensis Panampunnaypil: South China Sea. (7)
Rhopalophthalmus longipes Ii: South China Sea. (14)
R. orientalis O. S. Tattersall: Taiwan Strait; South China Sea. (9, 10, 12)
Siriella aequiremis Hansen: South China Sea. (9, 14)
S. dubia Hansen: South China Sea. (14)
S. gracilis Dana: South China Sea. (9, 14)
S. japonica Ii: Yellow Sea. (11)
S. japonica izuensis Ii: South China Sea. (11, 14)
S. media Hansen: South China Sea. (11, 14)
S. okadai Ii: South China Sea. (11)
S. quadrispinosa Hansen: South China Sea. (14)
S. sinensis Ii: Yellow Sea; East China Sea; Taiwan Strait; South China Sea. (9, 10, 12)
S. thompsoni (H. Milne-Edwards): South China Sea. (9, 14)
S. trispina Ii: Yellow Sea; East China Sea; South China Sea. (9, 11)
S. wadai Ii: South China Sea. (11)
Synerythrops intermedia Hansen: South China Sea. (7)
Tenagomysis orientalis Ii: South China Sea. (11)

REFERENCES*

(1). Wang Shaowu, 1981. On a new species and a new record of the genus *Proneomysis* from the South China Sea. Oceanologia & Limnologia Sinica, 12(2): 142–147.
(2). Wang Shaowu and Liu Ruiyu, 1987. Preliminary study of the subfamily Gastrosacclnae (Crustacea: Mysiacea) of the South China Sea. Studia Marina Sinica, 28:205–231.
(3). Wang Shaowu, 1989. On the two new species of the genus *Anisomysis* (Crustacea: Mysidacea) from the South China Sea. Studia Marina Sinica, 30:229–237.
(4). Wang Shaowu, 1980. Five new species of the genus *Acanthomysis* (Crustacea: Mysidacea) from the South China Sea. Oceanologia & Limnologia Sinica, 11(4):320–334.
(5). Liu Ruiyu and Wang Shaowu, 1983. On a new genus of Mysidacea, *Paraleptomysis* Gen. Nov. from the South China Sea. Oceanologia & Limnologia Sinica, 14(3):203–207.
(6). Liu Ruiyu and Wang Shaowu, 1983. On three new species of Mysidacea (Crustacea) from the coastal waters of Guangdong, China. Oceanologia & Limnologia Sinica, 14(6): 522–530.
(7). Liu Ruiyu and Wang Shaowu, 1986. Studies on *Mysina* (Crustacea: Mysidacea) of the Northern South China Sea. Studia Marina Sinica, 26:159–202.
(8). Shen Jiarui, Liu Ruiyu and Wang Shaowu, 1989. Mysidacea in waters off northern China coasts. Studia Marina Sinica, 30:189–227.
(9). State Oceanic Administration, 1988. A report of comprehensive survey of environments and resources in the central part of the South China Sea. China Ocean Press, 248pp.
(10). Cai Binji, 1980. A preliminary study of the Mysidacea (Crustacea) from the South Yellow Sea and the East China Sea. Oceanologia & Technologia Sinica, 16:40- 56.
(11). Cai Binji, 1989. Mysidacea from inlet waters of nuclear power station in Daya Bay. Collections of papers on marine ecology in Daya Bay (I). China Ocean Press, pp:130–140.
(12). Cai Binji, 1989. The distribution of Mysidacea in the western Taiwan Strait. Journal of Oceanography in Taiwan Strait, 8(2):125–131.
(13). Dong Yumao et al., 1988. A comprehensive survey on abysmal fishing ground between outside-shelf and slope of continent in the East Cina Sea. Zhejiang Science & Techonology Press, 8pp.
(14). Wang Shaowu and Liu Ruiyu, 1994. A faunal study of the Mysidea (Crustacea) from Nansha Islands and its adjacent waters. In: Marine Fauna and Flora and Biogeography of the Nansha Islands and Neibouring Waters, I. China Ocean Press, pp:61–111.

*: all in Chinese
Compiled by Cai Binji; Edited by Liu Ruiyu and Wang Shaowu

CUMACEA
BODOTRIIDAE

Bodotria chinensis Lomadina: Jiazhou Bay (Shandong). Intertidal. (1, 3)
B. ovalis Gamo: From Shandong Peninsula to mouth of Changjiang (=Yangtze River). Water depth 5–64 m. (1)
B. scorpioides (Montagu): Jiazhou Bay (Shandong). Water depth 26.5 m. (3)
Cyclaspis linguiloba Liu & Liu: Jiazhou Bay (Shandong). Water depth 22 m. (1)
Eocuma lata Calman: Jiaozhou Bay and mouth of Yellow River (Shandong). Water depth 4–26 m. (1)
Heterocuma sarsi Miers: Jiazhou Bay (Shandong). Water depth 7–14 m. (1, 3)
Iphinoe gurjanovae Lomakina: Tanggu (Tianjin). Intertidal. (3)
I. tenera Lomakina: Bohai; Yellow Sea; East China Sea. Water depth 4–37 m, common species. (1, 3)

DIASTYLIDAE

Diastylis tricincta (Zimmer): From Bohai to mouth of Changjiang (=Yangtze River). Water depth 6–37 m. (1)
Dimorphostylis asiatica Zimmer: From mouth of Yellow River to mouth of Changjiang (=Yangtze River), water depth 9–40 m; Taiwan. (1, 2)

GYNODIASTYLIDAE

Gynodiastylis anguicephala Harada: Mouth of Yellow River. Water depth 28 m. (1)

LAMPROPIDAE

Lamprops hexaspinula Liu & Liu: East China Sea. Water depth 13–41 m. (1)

LEUCONIDAE

Eudorella pacifica Hart: Mouth of Yellow River. Water depth 8–28 m. (1, 3)

Hemileucon bidentatus Liu & Liu: Bohai; Yellow Sea. Intertidal to 15 m water depth. (1)

NANNASTACIDAE

Campylaspis amblyoda Gamo: East China Sea. Water depth 107 m. (1)

C. fusiformis Gamo: From mouth of Yellow River to mouth of Changjiang (=Yangtze River). Water depth 7–52 m. (1)

Cumella arguta Gamo: Mouth of Yellow River. Water depth 8–28 m. (1)

REFERENCES*

(1). Liu Heng and Liu Ruiyu, 1990. Preliminary studies on nearshore Cumacea in north China. Studia Marina Sinica, 31:195–228

(2). Liu Heng and Liu Ruiyu, 1992. Supplementary report on nearshore Cumacea in north China. (manuscript)

(3). Romaknha, H. B. 1960. Studies on Cumacea (Crustacea: Malacostaca) fauna from the coast of the Yellow Sea. Oceanologia & Limnologia Sinica, 3(2):94–114

*: all in Chinese
Compiled by Huang Zongguo and Liu Heng; Edited by Liu Ruiyu

TANAIDACEA
TANAIDAE

Anatanais normani (Richardson): Daya Bay (Guangdong). (2)
Tanais cavolinii H. Milne-Edwards: Fujian; Guangdong. (1)

PARATANAIDAE

Leptochelia dubia (Kröyer): Fujian; Guangdong. (1)

ASPEUDIDAE

Aspeudes nipponicus Shiino: Yellow Sea. (1)

REFERENCES*

(1). Huang Zongguo and Cai Ruxing, 1984. Marine fouling organism and its prevention (I). China Ocean Press, 352pp.

(2). Huang Zongguo et al., 1990. Biofouling at water inlet of nuclear power station of Daya Bay. In: Collections of papers on marine ecology in the Daya Bay. China Ocean Press, pp:478–488.

*: both in Chinese
Compiled by Huang Zongguo

ISOPODA
CIROLANIDAE

Cirolana albida Richardson: Zhejiang coast. (11, 21)

C. cubensis Hay: East China Sea; Putuo, Chengshi, and Nanji (Zhejiang). (9, 18)

C. harfordi japonica Thielemann: Dalian (Liaoning); Hong Kong. (5, 12)

C. japonensis (Richardson): Bohai; Yellow Sea; Zhoushan (Zhejiang). Water depth 121 m. (5, 9)

C. minuta Hansen: Zhoushan (Zhejiang). Low intertidal. (9, 18)

C. sphaeromiformis Hansen: Zhoushan (Zhejiang). (9, 18)

Excirolana chiltonii Richardson: Taiwan; Hong Kong. (12)

BATHYNOMIDAE

Bathynomus döederlein Ortamann: East China Sea; Dongsha (=Pratas Islands). Deep sea. (9)

B. gigantia H. Milne-Edwards: Northern South China Sea. Water depth 400 m. (1)

CYMOTHOIDAE

Aegathoa oculata (Say): Zhoushan and Zhenhai (Zhejiang). (9, 18, 19)

Ceratothoa oxyrrhynchaena Koelbel: Hong Kong. Parasitic. (11)

Codonophilus trigonocephalus (Leach): East China Sea. (10, 15)

Nerocila acuminata Schioedte & Meinert: East China Sea; Zhejiang. Attached to fish. (9, 18, 20, 21)

N. depressa Edwards: Taiwan; Hong Kong. (12)

N. phaeopleura Bleeker: Hong Kong. Parasitic. (11)

N. sundaica Bleeker: Hong Kong. Parasitic. (11)

AEGIDAE

Aega dofleini Thielemann: East China Sea. Water depth 100 m. (9, 18)

Anilocra dimidiata Bleeker: Hong Kong. (11)

Rocinela propodialis Richardson: Zhoushan (Zhejiang). (9, 18)

CORALLANIDAE

Lanocira gardinieri Stebbing: Hong Kong. Parasite of sponges, water depth 13 m. (11)

Tachaea chinensis Thielemann: Zhoushan (Zhejiang); Fujian; Hong Kong. Brackish water and freshwater. (8, 12)

LIMNORIIDAE

Limnoria lignorum (Rathke): Zhoushan (Zhejiang). (7, 9, 10, 18)

L. tripunctata Menzies: China's coast. Intertidal woods. (5)

SPHAEROMIDAE

Amphoroidella sp.: Hong Kong. (12)

Clianella brucei Harrison & Holdich: Hong Kong. (12)

Cymodoce acuta Richardson: Dapeng Cove (Hong Kong). (11)

C. japonica Richardson: China's coast. Seaweed and shells in low intertidal. (5)

C. pubescens Haswell: East China Sea. Water depth 137 m. (9, 13)

Dynamenopsis sp.: Hong Kong. (12)

Dynoides dentisinus Shen: From Bohai; Yellow Sea; East China Sea. (2, 9, 21)

Exosphaeroma oregonensis (Dana): Zhoushan (Zhejiang). Mid- and low-intertidal. (5, 9, 18–21)

Paracerceis sculpta (Holmes): Hong Kong. (12)

Paradella dianae (Menzies): Hong Kong. (12)

Sphaeroma retrolaevis Richardson: Fujian coast. Brackish water and freshwater, wood boring species. (6)

S. walkeri Stebbing: South of Xiamen (Fujian). Low intertidal to shallow sea. (12, 17)

IDOTHEOIDAE

Cleantiella isopus (Grube): Dalian (Liaoning); Qingdao (Shandong). Intertidal, interstitial or seaweed. (5)

Cleantis plancicauda Benedict: Wenzhou (Zhejiang). (9, 18)

Idotea ochotensis Brandt: Zhejiang. Gravel and seaweed. (9, 10, 18, 20)

I. stenops (Benedict): Xiangshan and Putou (Zhejiang). (9)

Synidotea laevidorsalis Miers: Putou (Zhejiang). Shallow sea. (3, 9, 10)

LIGIIDAE

Ligia exotica (Roux): China's coast. High intertidal and above. (4, 9, 10, 16, 20)

L. occidentalis Dana: Yantai (Shandong). (4, 9, 10, 20)

TYLIDAE

Tylos sp.: Hong Kong. Intertidal. (16)

OLIBRINIDAE

Olibrinus sp.: Hong Kong. Intertidal. (16)

PHILOSCIIDAE

Littorophiloscia aldabrana Ferranra & Taiti: Hong Kong. Intertidal. (16)

SCYPHACIDAE

Armadilloniscus littoralis Budde-Lund: Hong Kong. Intertidal. (16)

BOPYRIDAE

Allokepon goetici (Shiino): Hong Kong, low intertidal. Host: crabs. (14)

A. sinensis (Danforth): Hong Kong. Host: swimming crab (*Lissocarcinus orbicularis*). (14)

Anchiarthrus derelictus Markham: Hong Kong, water depth 1 m. Host: snapping shrimp (*Alpheus*). (15a)

Anisarthrus sp.: Hezui (Hong Kong). Host: snapping shrimps (*Alpheus*). (14)

Apocepen pulcher Nierstrasz et al.: Shengsi (Zhejiang), water depth 61 m. Host: purse crab (*Philyra globulosa*). (9, 10)

Apophrixus constricutus Markham: Hong Kong, coral reefs. Host: snapping shrimps (*Alpheus*). (14)

Aporbopyrus enosteoidis (Markham): Hong Kong. (15a)

A. megacephalon Nierstrasz: Hong Kong. Host: shore crab (*Pachygrapsus*). (15a)

Athelges takanoshimensis Ishii: Zhoushan (Zhejiang) and Hong Kong, water depth 48 m. Host: hermit crabs (*Dardanus* and *Diogenes edwardsii*). (9, 11, 14)

Bopyrella pyriforma (Shiino): Dapeng Cove (Hong Kong), water depth 12 m. Host: hermit crab (*Diogenes edwardsii*). (15a)

B. tanytelson Markham: Dapeng Cove (Hong Kong), water depth 5–10 m. Host: snapping shrimp (*Alpheus*). (15)

Bopyrinella albida Shiino: Hong Kong. Host: snapping shrimp (*Alpheus*). (15a)

Bopyrione albida Shiino: Hong Kong. Host: snapping shrimp (*Athanas*). (15)

B. longicapitata Markham: Hong Kong, intertidal. Host: snapping shrimp (*Alpheus*).

B. toloensis Markham: Hong Kong, coral reefs. Host: snapping shrimp (*Alpheus*). (14)

Bopyrione sp.: Dapeng Cove (Hong Kong), water depth 4–6 m. Host: snapping shrimp (*Alpheus*). (15)

Bopyrissa pyriforma (Shiino): Hong Kong. Host: hermit crab (*Clibannarius arethusa*). (14)

Dicropleon morator Markham: Dapeng Cove (Hong Kong), water depth 6–20 m. Host: caridean shrimp (*Periclimenaeus*). (14)

Eophrixus shojii Shiino: Hong Kong, low intertidal. Host: snapping shrimp (*Alpheus*). (14, 15)

Epipenaeon ingens Nobili: Hong Kong. Host: penaeid shrimp (*Penaeus semisulcatus*). (14)

Hypocepon globosus Markham: Hong Kong. Host: pea crab (*Pinnotheres*). (15a)

Hypophryxus filiformis (Chopra): Hong Kong, water depth 6 m. Host: snapping shrimp (*Alpheus*). (15 a)

Leidya sesarmae Pearse: Mouth of Min River (Fujian). Host: shore crab (*Sesarma dehaani*). (14)

L. ucae Pearse: Mouth of Min River (Fujian). Host: fiddler crab (*Uca*). (14)

Litobopyrus longicaudatus Markham: Hong Kong, intertidal. Host: snapping shrimp (*Alpheus*). (14)

Metacepon choprai George: Hong Kong, mud flat. Host: shore crab (*Sesarma*). (15)

M. pleopodata Bourdon & Stock: Dapeng Cove (Hong Kong), water depth 18–21 m. Host: xanthid crab (*Pilummus*). (15a)

Orbione halipori Nierstrasz & Brenderá Brandis: Mouth of Zhujiang (Pearl River, Guangdong). Host: penaeid shrimp (*Metapenaeus ensis*). (14)

O. penei Bonnier: Hong Kong. Host: penaeid shrimps. (14)

Parabopyrella indica (Chopra): Hong Kong. Host: snapping shrimp (*Prionalpheus*). (14)

P. perplexa Markham: Dapeng Cove (Hong Kong), water depth 55 m. Host: shrimps. (15)

Parapagurion calcinicola Shiino: Dapeng Cove (Hong Kong), water depth 11 m. Host: hermit crabs. (15a)

Parapenaeon consolidata v. *richardsonae* Shiino: Zhejiang. Host: penaeid shrimp (*Trachypenaeus anchoralis*). (9, 10)

P. japonica (Thielemann): Mouth of Zhujiang (Pearl River, Guangdong). Host: Japanese penaeid shrimp (*Penaeus japonica*). (14)

Paraphrixus brevicauda (Chopra): Hong Kong, coral reefs. Host: snapping shrimps (*Alpheus*). (14)

Parione pachychelii Shiino: Dapeng Cove (Hong Kong), water depth 10 m. Host: porcelain crab (*Pachycheles*). (15a)

Parioninella astridae Shiino: Hong Kong. Host: ghost crab (*Paracleistostoma*). (14, 15)

Pleurocryptosa enosteoidis Markham: Dapeng Cove (Hong Kong). Host: crabs. (14)

P. megacephalon Nierstrasz & Brender: Dapeng Cove (Hong Kong). Host: crabs. (14)

Probopyria elliptica Markham: Hong Kong, water depth 6 m, coral reefs. Host: snapping shrimps (*Alpheus*). (15a)

Probopyriscus novempalensis Markham: Dapeng Cove (Hong Kong), water depth 14 m. Host: snapping shrimp (*Alpheus*). (14)

Progebiophilus sinicus Markham: Hong Kong, low intertidal. Host: mud shrimp (*Upogebia major*). (14, 15)

Pseudione longicauda Shiino: Hong Kong. Muddy intertidal. Host: hermit crabs. (15a)

Pseudostegias dulcilaeuum Markham: Hong Kong. Subtidal. Host: hermit crabs. (14)

P. setoensis Shiino: Taiwan; Hong Kong. Host: hermit crabs. (14)

Rhopalione sinensis Markham: Shenzhen Bay (Guangdong). Host: pea crab (*Pinnotheres*). (15)

Stegophryxus minutus Markham: Hong Kong. Host: hermit crabs. (15a)

Tylokepon naxiae (Bonnier): Hong Kong. Host: spider crab (*Hyastenus diacanthus*). (14, 15a)

REFERENCES*

(1). Liu Ruiyu and Wang Yongliang, 1984. On *Bathynomus gigantia* (Crustacea: Isopoda) from China. In: Transactions on the Chinese Society of Zoology in the 50th Anniversary.

(2). Shen Jiarui, 1929. New species of Isopoda, China. Jing Sheng Biology 1(4):65–78.

(3). Shen Chia-jui, 1955. On some marine crustaceans from the coastal waters of Fenghsien, Kiangsu. Acta Zoologica Sinica, 7(2): 75–100.

(4). Chen Guoxiao, 1987. The description of two species of *Ligia*. Chinese Journal of Zoology, 22(2): 8–11.

(5). Zhao Ruyi et al., 1990. Guidebook on collection of invertebrate from China coast (in North China). China Ocean Press, pp:260–265.

(6). Huang Zongguo and Cai Ruxing, 1984. Marine fouling organism and its prevention (I). China Ocean Press, pp:1–354.

(7). Tsai Ru-hsing, Huang Tsung-kuo and Jiang Jin-shan, 1962. On the marine prevention (I). China Ocean Press, pp:1–354.

(8). Zhang Shuzhen, 1991. Benthic in the shelf waters of Nansha Islands. In: Bottom trawling fishing resources investigation in southwestern continental shelf sea waters of Nansha Islands. China Ocean Press, pp:40–55.

(9). Wei Chongde, 1991. Isopoda, Crustacea. In: Fauna of Zhejiang. Zhejiang Science & Technology Press, pp:94–147

(10). Shiino, 1967. Isopoda. New Illustrated Encyclopedia of the Fauna of Japan. Hokuryu Kan, pp:539–555.

(11). Bruce, N. L., 1982. On a small collection of marine isopoda from Hong Kong. In: Proceedings of the First International Marine Biological Workshop: The Marine f Flora and Fauna of Hong Kong and Southern China, Hong Kong University Press, pp:315–324.

(12). Bruce, N. L., 1986. New records of Isopod crustaceans from Hong Kong. In: Proceedings of the Second International Marine Biological Workshop: The Marine Flora and Fauna of Hong Kong and Southern China. Hong Kong University Press, pp:549–566.

(13). Haswell, W. A., 1881a, 1881b. On some new Australian Marine Isopoda, Part I. Proc. Linn. Soc. New South Wales 5:470–481.

(14). Markham, J. C., 1982. Bopyrid Isopods parasitic on decapod crustaceans in Hong Kong and Southern China. In: Proceedings of the First International Marine Biological Workshop: The Marine Flora and Fauna of Hong Kong and Southern China. Hong Kong University Press, pp:326–391.

(15). Markham, J. C., 1986. Further notes on the Isopoda Bopyrida of Hong Kong. In: Proceedings of the Second International Marine Biological Workshop: The Marine Flora and Fauna of Hong Kong and Southern China. Hong Kong University Press, pp:555–566.

(15a). Markham, J. C., 1992. Second list of additions to the Isopoda Bopyridae of Hong Kong. In: Proceedings of the Fourth International Marine Biological Workshop: The Marine Flora and Fauna of Hong Kong and Southern China. Hong Kong University Press, pp:277–302.

(16). Ma, H. H. T., 1986. Hong Kong intertidal isopods, with notes on the feeding and reproduction of *Armadilloniscus litoralis* Budde - Lund, 1885. In: Proceedings of the Fourth International Marine Biological Workshop: The Marine Flora and Fauna of Hong Kong and Southern China. Hong Kong University Press, pp:1,023–1,031.

(17). Mak, P. M. S., Z. G. Huang, and B. S. Morton, 1985. *Sphaeroma walkeri* introduced into and established in Hong Kong. Crustaceana, 49(1):75–82.

(18). Richardson, H. R., 1899a. Key to the isopods of the Pacific coast of North America, with descriptions of twenty two new species. Proc. U.S. Nat. Mus. 21:815–869.

(19). Richardson, H. R., 1907. Descriptions of new isopod crustaceans of the Family Sphaeromidae. Proc. U.S. Nat. Mus. 31:1–22.

(20). Van Name, W. G. 1920. Isopod collected by the American Museum Congo Expedition. Bull. Amer. Mus. Nat. Hist. 43:41–108.

(21). Van Name, W. G., 1936. The American land and freshwater isopod Crustacea. Bull. Am. Mus. Nat. Hist. 71:1–535.

*: (1)-(9) in Chinese, (10) in Japanese

Compiled by Huang Zongguo; Edited by Cai Ruxing, Wei Chongde, and Liu Ruiyu

AMPHIPODA
GAMMARIDEA
AMPELISCIDAE

Ampelisca bocki Dahl: Yellow Sea; East China Sea; South China Sea. (2, 10)
A. brevicornis (Costa): Bohai; Yellow Sea; East China Sea; South China Sea; Hong Kong. Water depth 4–10 m. (2, 9, 10)
A. chinensis Imbach: South China Sea. (10)
A. cyclops Walker: China's seas. (1, 2)
A. hongkongensis Hirayama: Dapeng Cove (Hong Kong). Water depth 16 m. (9)
A. honmungensis Imbach: South China Sea. (10)
A. iyoensis Nagata: Yellow Sea; East China Sea. (2)
A. maia Imbach: South China Sea. (10)
A. miharaensis Nagata: Yellow Sea; East China Sea; South China Sea; Dapeng Cove (Hong Kong). Water depth 15 m. (2, 9, 10)
A. misakiensis Dahl: East China Sea; South China Sea. (2, 10)
A. nanshaensis Ren: South China Sea. (1)
A. orops Imbach: South China Sea. (10)
Byblis calisto Imbach: South China Sea. (10)
B. febris Imbach: South China Sea. (10)
B. io Imbach: South China Sea. (10)
B. japonicus Dahl: Bohai; Yellow Sea; Dapeng Cove (Hong Kong). Water depth 9- 15 m. (2, 9)
B. kallarthra Stebbing: Dapeng Cove (Hong Kong); South China Sea. Water depth 15 m. (1, 9)
B. pilosa Imbach: South China Sea. (10)

AMPHILOCHIDAE

Gitanopsis japonica Hirayama: Yellow Sea; East China Sea. (12)
G. longus Hirayama: Daya Bay (Guangdong). (3)

AMPITHOIDAE

Ampithoe lacertosa (Bate): Bohai; Yellow Sea; East China Sea. (2)
A. zachsi Gurjanova: Hong Kong. (11)
Cymadusa vadosa Imbach: South China Sea. (10)
Paradusa mauritiensis Ledoyer: Hainan. (4)
Peramphithoe orientalis (Dana): Hainan. (4)
Sunamphitoe plumosa Stephensen: Bohai; Yellow Sea; East China Sea. (2, 4)

ANISOGAMMARIDAE

Anisogammarus (*Eogammarus*) *turgimanus* Shen: Yellow Sea. (5)
Eogammarus sinensis Ren: Yellow Sea. (2)

ATYLIDAE

Atylus minikoi (Walker): Hainan. (4)

CALLIOPIIDAE

Paracalliope karitane J. L. Barnard: Hainan. (4)

COROPHIIDAE

Aoroides columbiae Walker: Yellow Sea; East China Sea. (2)
Cerapus tubularis Say: China's seas. (7)
Cheiriphotis megacheles (Giles): Hainan. (4)
Corophium acherusicum Costa: Bohai; Yellow Sea; East China Sea; South China Sea; Dapeng Dove (Hong Kong). Water depth 9 m. (2, 3, 7, 11)
C. baconi Shoemaker: Yellow Sea; Hong Kong. Water depth 2 m. (3, 7)
C. crassicorne Bruzelius: Dapeng Cove (Hong Kong). Water depth 9 m. (7)
C. heteroceratum Yu: Bohai. (5)
C. homoceratum Yu: Bohai. (5)
C. hongkongensis Hirayama: Dapeng Cove (Hong Kong). Water depth 9 m. (7)
C. insidiosum Crawford: Yellow Sea; East China Sea; Dapeng Cove (Hong Kong). Abundant in intertidal tide pools. (2, 7)
C. kitamorii Nagata: Dapeng Cove (Hong Kong), water depth 5–15 m; Hainan. (4, 7)
C. lamellatum Hirayama: Shenzhen Bay (Guangdong). Intertidal. (7)
C. major Ren: Yellow Sea. (2)
C. monospinum Shen: Yellow Sea. (5)
C. mortonii Hirayama: Hong Kong. Water depth 6 m. (7)
C. sextonae miospinulosum Hirayama: Hong Kong. Water depth 2–3 m. (7)
C. sinense Zhang: Yellow Sea. (2, 6)
C. triangulapedarum Hirayama: Shenzhen Bay (Guangdong). Salt marsh. (7)
C. tridentium Hirayama: Hong Kong. Water depth 4 m. (7)
C. uenoi Stephensen: Yellow Sea; East China Sea; Hong Kong, water depth 2 m; Dapeng Cove (Hong Kong), on seaweed. (2, 7, 11)
Ericthonius brasiliensis (Dana): Hainan; South China Sea. (3, 11)
E. pugnax Dana: China's seas. (7)
Gammaropsis japonica (Nagata): Yellow Sea. (2)
G. laevipalmata Ren: Yellow Sea. (2)
G. liuruiyui Ren: Bohai; Yellow Sea. (2)
G. nitida (Stimpson): Yellow Sea. (2)
G. sexdentata (Stephensen): Bohai; Yellow Sea; East China Sea. (2)
G. togoensis (Schellenberg): South China Sea. (3)
G. utinomi (Nagata): Bohai; Yellow Sea; East China Sea. (2)

Grandidierella gilesi Chilton: Hainan. (4)
G. japonica Stephensen: Yellow Sea; East China Sea. (2)
G. megnae (Giles): Hainan. (4)
Photis digitata Earnard: Hainan. (4)
P. longicaudata (Bate & Westwood): Yellow Sea; East China Sea. (2)

DEXAMINIDAE

Guernea (Prinassus) longidactyla Hirayama: Dapeng Cove (Hong Kong). Water depth 18 m. (8)
G. (P.) mackiei Hirayama: Dapeng Cove (Hong Kong). Water depth 6–9 m. (10)
G. (Guernea) sombati Hirayama: Dapeng Cove (Hong Kong). Water depth 6 m. (8)
Paradexamine setigera Hirayama: Hong Kong. Sandy intertidal. (7)

HAUSTORIIDAE

Urothoe carda Imbach: South China Sea. (10)
U. cuspis Imbach: South China Sea. (10)
U. gelasina Imbach: South China Sea. (10)
U. orientalis Gurjanova: South China Sea. (10)
U. spinidigitus Walker: South China Sea. (10)

EUSIRIDAE

Pontogeneia litorea Ren: Yellow Sea. (2)

HYALIDAE

Hyale dollfusi Chevreux: Hainan. (4)
H. grandicornis (Kröyer): Bohai; Yellow Sea; East China Sea; South China Sea. (2, 3)
H. honoluluensis Schellenberg: Hainan. (4)
H. schmidti (Heller): Bohai, Yellow Sea. (2)
Parhyale hawaiensis (Dana): Hainan; South China Sea. (3, 4)
P. plumulosa (Stimpson): Yellow Sea; East China Sea. (2)
Parhyalella pietschmanni Schellenberg: Hainan. (4)

ISCHYROCERIDAE

Jassa falcata (Montagu): Yellow Sea, East China Sea; South China Sea. (2, 3)

LEUCOTHOIDAE

Leucothoe alata J. L. Barnard: Daya Bay (Guangdong). (3)
L. alcyone Imbach: South China Sea. (11)
L. furina (Savigny): South China Sea. (11)
Leucothoella bannwarthi Schellenberg: Hainan. (4)

LILIEBORGIIDAE

Idunella curvidactyla Nagata: Bohai; Yellow Sea; East China Sea. (2)
I. janisae Imbach: South China Sea. (11)
I. pauli Imbach: South China Sea. (11)
I. serra Imbach: South China Sea. (11)
Lilieborgia serrata Nagata: Bohai; Yellow Sea; East China Sea. (11)
L. sinica Ren: Yellow Sea. (2)

LYSIANASSIDAE

Lepidepecreum nudum Imbach: South China Sea. (10)
Lysianassa cinghalensis (Stebbing): South China Sea. (10)
Orchomene breviceps Hirayama: Yellow Sea; East China Sea. (2)
Socarnes dissimulantia Imbach: South China Sea. (10)

GAMMARIDAE

Ceradocus hawaiensis J. L. Barnard: Hainan. (4)
Elasmopus ecuadorensis hawaiensis Schellenberg: Hainan; South China Sea. (3)
E. japonicus Stephensen: China's seas. (7)
E. molokai J. L. Barnard: Hainan. (4)
E. pectenicrus (Bate): Daya Bay (Hong Kong); Hong Kong; Hainan. (2, 8)
E. spinidactylus (Chevreux): Hainan; South China Sea. (3, 4)
E. spinimanus Walker: Hainan; Nansha (=Spratly Islands). (1, 4)
Eriopisa chilkensis (Chilton): Hainan. (4)
E. elongata (Bruzelius): South China Sea. (10)
Eriopisella propagatio Imbach: South China Sea. (10)
E. sechellensis (Chevreux): Bohai; Yellow Sea; East China Sea; South China Sea. (2)
Maera othonides Walker: Hainan. (4)
M. pacifica Schellenberg: Hong Kong. (10)
M. serratipalma Nagata: East China Sea. (2)
Mallacoota insignis (Chevreux): Hainan. (4)
M. subcarinata (Haswell): Hainan. (4)
Megaluropus agilis Hoek: South China Sea. (10)
Melita alluaudi Ledoyer: Hainan. (4)
M. appendiculata (Say): Hainan. (4)
M. denticulata Nagata: Yellow Sea; East China Sea. (2)
M. koreana Stephensen: Yellow Sea; East China Sea; Hainan. (2, 3, 4)
M. longidactyla Hirayama: Yellow Sea; East China Sea. (2)
M. orgasmos Barnard: Hainan. (4)
M. rylovae Bulycheva: Yellow Sea; East China Sea. (2)
M. setiflagella Yamato: Hainan. (4)
M. tuberculata Nagata: Yellow Sea; East China Sea. (2)

OEDICEROTIDAE

Caviplaxus jiaozhouwanensis Ren: Yellow Sea. (2)
C. longiflagella Ren: Yellow Sea. (2)
Monoculodes limnophilus Tattersall: Yellow Sea. (5)
Oediceroides ornithorhynchus Pirlot: South China Sea. (10)
Pontocrates altamarimus (Bate & Westwood): Bohai; Yellow Sea; East China Sea. (2)
Sinoediceros homopalmulus Shen: Bohai; Yellow Sea; East China Sea. (5)
Synchelidium miraculum Imbach: South China Sea. (10)

PHOXOCEPHALIDAE

Harpiniopsis vadiculus Hirayama: Bohai; Yellow Sea; East China Sea. (2)

PODOCERIDAE

Podocerus inconspicuus (Stebbing): Hainan. (4)
P. tuberculosus Ren: Yellow Sea. (2)

STENOTHOIDAE

Stenothoe gallensis Walker: Daya Bay (Guangdong); South China Sea. (3)
S. qingdaoensis Ren: Yellow Sea. (2)

TALITRIDAE

Orchestia amonala Chevreux: Hainan. (4)
O. platensis Kröyer: Hainan. (4)
Platorchestia platensis (Kröyer): Hong Kong. Intertidal. (11)
Talorchestia martensii (Weber): Hainan. (4)

REFERENCES*

(1). Ren Xianqiu, 1991. On five species of Amphipoda (Crustacea) from Nansha Islands, Guangdong Province, China. In: Contributions on Marine Organisms around Nansha Islands and the adjacent Waters (I). China Ocean Press, pp:181–188.
(2). Ren Xianqiu, 1992. Studies on the Gammaridea (Crustacea:Amphipoda) from Jiaozhou Bay (Yellow Sea). In: Transactions on the Chinese Crustacean Society (III). Qingdao Ocean University Press, pp:214–314.
(3). Ren Xianqiu, 1994. Studies on Gammaridea (Crustacea:Amphipoda) from Hong Kong, Daya Bay and adjacent waters. Studia Marina Sinica, 35:249–272.
(4). Ren Xianqiu. Studies on Gammarida (Crustacea: Amphipoda) from Hainan Island. (manuscript)
(5). Shen Chia-jui, 1955. On some marine crustaceans from the coastal water of Fenghsien, Kiangsu. Acta Zoologica Sinica, 7(2): 75–100.
(6). Zhang Weiquan, 1974. A new species of the genus *Corophium* (Crustacea, Amphipoda, Gammaridea) from the southern coast of Shantung Penisula, North China. Studia Marina Sinica, 9:139–146.
(7). Hirayama, A. 1986a. Marine Gammaridean Amphipoda (Crustacea) from Hong Kong. I: The Family Corophiidae, Genus *Corophium*. In: Proceedings of the Second International Marine Biological Workshop: The Marine Flora and Fauna of Hong Kong and Southern China. Hong Kong University Press, pp:449–485.
(8). Hirayama, A. 1986b. Marine Gammaridean Amphipoda (Crustacea) from Hong Kong. II: The Family Dexaminidae. In: Proceedings of the Second International Marine Biological Workshop: The Marine Flora and Fauna of Hong Kong and Southern China. Hong Kong University Press, pp:487–501.
(9). Hirayama, A. 1991. Marine Ampeliscidae from Hong Kong. Asian Marine Biology 13(8):77–93.
(10). Imbach, M. C. 1967. Gammaridean Amphipoda from the South China Sea. Naga Report 4(1):39–167.
(11). Moore, P. G. 1986. Preliminary notes on a collection of Amphipoda from Hong Kong. In: Proceedings of the Second International Marine Biological Workshop: The Marine Flora and Fauna of Hong Kong and Southern China. Hong Kong University Press, pp:503–513.
(12). Yu, Shou-Chie, 1938. Description of two new Amphipod Crustacea from Tangku. Bull. Fan. Mem. Inst. Biol. Zool. Ser. 8(2):83–103.

*: (1)-(6) in Chinese
Compiled by Ren Xianqiu and Huang Zongguo; Edited by Liu Ruiyu

HYPERIIDEA
LANCEOLIDAE

Lanceola pacifica Stebbing: Central South China Sea. (11)

SCINIDAE

Acanthoscina acanthodes (Stebbing): Central South China Sea. (11)
Scina borealis (Sars): East China Sea continental shelf; central South China Sea. (9, 11)
S. crassicornis (Fabricius): East China Sea; southern Fujian; Taiwan; South China Sea. (2–5, 8, 11, 13, 14)
S. curvidactyla Chevreuz: Central South China Sea. (11)
S. incerta Chevreux: Northern East China Sea; central South China Sea. (11, 14)
S. latifrons Wagler: Central South China Sea. (11)
S. nana Wagler: Central South China Sea. (11)
S. similis Stebbing: Central South China Sea. (11)
S. submarginata Tattersall: East China Sea. (13)
S. tullbergi (Bovallius): East China Sea; southern Fujian; Taiwan; central South China Sea. (3, 11, 13)

VIBILIIDAE

Vibilia armata Bovallius: East China Sea; southern Fujian; Taiwan; South China Sea. (2–6, 8, 9, 11, 13, 14)
V. australis Stebbing: East China Sea; southern Fujian; Taiwan; central South China Sea. (2, 3, 11)
V. chuni Behning & Woltereck: East China Sea; southern Fujian; Taiwan; northern and central South China Sea continental shelf. (3, 5, 6, 11, 13, 14)
V. cultripes Vosseler: Central South China Sea. (4, 11)

V. gibbosa Bovallius: East China Sea; northern and central Taiwan Strait; South China Sea. (1, 2, 4, 5, 8)
V. longicarpus Behning: Southern Fujian; Taiwan. (3)
V. propinqua Stebbing: Offshore of East China Sea. (6)
V. pyripes Bovallius: East China Sea; northern Xisha (=Paracel Islands); Nansha (=Spratly Islands). (2, 4, 8, 14)
V. stebbingi Behning & Woltereck: East China Sea; northern and central Taiwan Strait; northern and central South China Sea continental shelf. (1, 5, 6, 11, 13, 14)
V. viatrix Bovallius: East China Sea; southern Fujian; Taiwan; northern Xisha (=Paracel Islands). (2, 3, 4, 9)

PARAPHRONIMIDAE

Paraphronima crassipes Claus: Northern East China Sea; South China Sea. (4, 5, 11, 14)
P. gracilis Claus: East China Sea; South China Sea. (2, 4, 5, 8, 11–14)

HYPERIIDAE

Hyperia galba (Montagu): East China Sea continental shelf. (9)
Hyperiella antarctica Bovallius: Northern East China Sea. (14)
Hyperietta luzoni (Stebbing): East China Sea; western Taiwan Strait, northern and central South China Sea continental shelf. (2–5, 10–14)
H. parviceps Bowman: Northern and central Taiwan Strait. (1)
H. stebbingi Bowman: East China Sea; western Taiwan Strait; central South China Sea. (3, 10, 11, 13)
H. stephenseni Bowman: East China Sea; Taiwan Strait; central South China Sea. (1, 3, 10, 11, 13)
H. vosseleri (Stebbing): East China Sea; Taiwan Strait; central South China Sea. (1, 3, 10, 11, 13)
Hyperioides longipes Chevreus: East China Sea; western Taiwan Strait; central South China Sea. (2, 3, 6, 10–14)
H. sibaginis (Stebbing) [*Hyperia sibaginis*]: Southern Yellow Sea; East China Sea, Dongshan Bay (Fujian); Taiwan Strait; northern and central South China Sea continental shelf. (1, 2, 3, 5, 6, 9–13)
Hyperionyx macrodactylus (Stephensen): Southern Fujian; Taiwan; central South China Sea. (3, 11)
Hyperoche martinezi (Müller): East China Sea; western Taiwan Strait; central South China Sea. (10, 11, 13)
H. mediterranea Senna: East China Sea; Taiwan Strait. (1, 6, 10, 13)
H. medusarum (Kröyer): East China Sea; northern and central Taiwan Strait; northern South China Sea continental shelf. (1, 5, 6, 9, 14)
H. picta Bovallius: East China Sea. (13)
Laxohyperia vespuliformis Vinogradov & Volkov: Southeastern Zhongsha (=Macclesfield Bank). (11)
Lestrigonus bengalensis Giles [*Hyperia bengalensis, H. dysschistus*]: Southern Yellow Sea; East China Sea; Dongshan Bay (Fujian); Taiwan Strait; Daya Bay (Guangdong); South China Sea; Xisha (=Paracel Islands); Zhongsha (=Macclesfield Bank); Nansha (=Spratly Islands). (1, 3, 5–8, 10, 11, 13)
L. crucipes (Bovallius): Central South China Sea. (11)
L. latissimus (Bovallius) [*Hyperia latissimus*]: East China Sea; Taiwan Strait; northern South China Sea continental shelf. (1, 2, 3, 5, 6, 10, 14)
L. macrophthalmus (Vosseler) [*Hyperia macrophthalmus*]: East China Sea; Luoyuan Bay and Meizhou Bay (Fujian); Taiwan Strait; Daya Bay (Guangdong); northern and central South China Sea continental shelf. (1, 3–6, 10, 14)
L. schizogeneios (Stebbing) [*Hyperia schizogeneios*]: East China Sea; Louyuan Bay (Fujian); Taiwan Strait; South China Sea; Xisha (=Paracel Islands); Zhongsha (=Macclesfield Bank); Nansha (=Spratly Islands). (1–14)
L. shoemakeri Bowman: Northern and central Taiwan Strait. (1)
Parathemisto gaudichaudi Guérin [*Euthemisto bispinosa, Themisto gracilipes*]: Bohai; Yellow Sea; East China Sea. (2, 6, 9, 13, 14)
Phronimopsis spinifera Claus: East China Sea; Taiwan Strait; South China Sea. (1–5, 8, 9, 11, 13, 14)
Themistella fusca (Dana): Southern Yellow Sea; East China Sea; Taiwan Strait; central South China Sea. (1, 10, 11, 13)

PHRONIMIDAE

Phronima atlantica Guérin-Méneville: East China Sea; western Taiwan Strait; northern and central South China Sea continental shelf. (2, 4, 5, 10–13)
P. bucephala Giles: Western Taiwan Strait; central South China Sea. (3, 10, 11)
P. colletti Bovallius: Yellow Sea; East China Sea; southern Fujian; Taiwan. (2–5, 8, 11, 13)
P. curvipes Vossler: Southern Fujian; Taiwan; central South China Sea. (3, 11)
P. pacifica Streets: East China Sea; Taiwan Strait; central South China Sea. (1–4, 11, 13)
P. sedentaria (Forskål): East China Sea; southern Fujian; Taiwan; northern and central South China Sea continental shelf. (2–5, 9, 11)
P. solitaria Guérin-Méneville [*P. megalodous*]: Offshore of Wenzhou (Zhejiang); Taiwan Strait; central South China Sea. (1, 4, 10, 11, 13)
P. stebbingi Vosseler: East China Sea; northern South China Sea continental shelf; Zhongsha (=Macclesfield Bank); Xisha (=Paracel Islands). (2, 4, 5, 9, 13, 14)
Phronimella elongata (Claus): East China Sea; Taiwan Strait; South China Sea. (1–5, 8- 11, 13, 14)

PHROSINIDAE

Anchylomera blossevillei Milne-Edwards: East China Sea; Taiwan Strait; northern and central South China Sea continental shelf. (1–6, 9, 10, 11, 13)

Phrosina semilunata Risso: East China Sea; Taiwan Strait; South China Sea. (1–6, 8–11, 13, 14)

Primno brevidens Bowman: East China Sea; Taiwan Strait; central South China Sea. (1, 3, 10, 11, 13)

P. latreillei Stebbing: Southern Yellow Sea; East China Sea; western Taiwan Strait; central South China Sea. (3, 10, 11, 13)

P. macropa Guérin-Méneville: East China Sea; Taiwan Strait; South China Sea. (1–6, 8- 11, 13, 14)

LYCAEOPSIDAE

Lycaeopsis themistoides Claus: East China Sea; Taiwan Strait; from central South China Sea to around Zengmuansha (southern tip of Nansha=Spratly Islands). (1–4, 7, 8, 10–14)

L. zamboangae (Stebbing): East China Sea; western Taiwan Strait; South China Sea. (2, 4, 5, 8–14)

PRONOIDAE

Eupronoe armata Claus [*E. intermedia*]: East China Sea; Taiwan Strait; central South China Sea. (1, 3, 11, 13)

E. laticarpa Stephensen: Central South China Sea. (11)

E. maculata Claus: East China Sea; Taiwan Strait; South China Sea. (1–6, 8–13)

E. minuta Claus: East China Sea; Luoyuan Bay (Fujian); Taiwan Strait; South China Sea. (1–6, 8–14)

Paralycaea gracilis Claus: East China Sea; Taiwan Strait; central South China Sea. (1, 2, 3, 10, 11, 13)

Parapronoe campbelli Stebbing: Central South China Sea. (11)

P. crustulum Claus: Northern East China Sea; Xisha (=Paracel Islands). (4, 14)

P. elongata Semenova: Southern Fujian; Taiwan. (3)

P. parva Claus [*Sympronoe parva*]: East China Sea; western Taiwan Strait; central South China Sea. (10–13)

Pronoe capito Guérin-Méneville: Around Huangyan Island (Zhongsha=Maccslesfield Bank). (11)

LYCAEIDAE

Lycaea bajensis Shoemaker: Taiwan Strait; Hainan. (1, 6, 10)

L. bovalli Chevreux: Northern East China Sea. (14)

L. bovallioides Stephensen: Southern Yellow Sea; East China Sea; western Taiwan Strait; central South China Sea. (10, 11, 13)

L. nasuta Claus: East China Sea continental shelf. (9)

L. pachypoda [*Pseudolycaea pachypoda*]: East China Sea; central South China Sea. (2, 11, 13)

L. pulex Marion: East China Sea; Taiwan Strait; from northern South China Sea continental shelf to around Zengmuansha (southern tip of Nansha=Spratly Islands). (1–11, 14)

L. vincenti Stebbing: East China Sea; Taiwan Strait; Daya Bay (Guangdong); northern and central South China Sea continental shelf. (1, 2, 3, 5, 6, 10, 11, 12)

Simorhychotus antennarius (Claus) [*S. lilljeborgi*]: East China Sea; Taiwan Strait, northern and central South China Sea continental shelf. (1, 2, 3, 5, 6, 9, 10, 11, 13, 14)

TRYPHANIDAE

Tryphana malmi Boeck: East China Sea; central South China Sea. (2, 11, 13)

BRACHYSCELIDAE

Brachyscelus crusculum Bate: East China Sea; Taiwan Strait; South China Sea. (1–6, 8- 14)

B. globiceps (Claus) [*B. latipes*]: East China Sea; Luoyuan Bay (Fujian); Taiwan Strait; from northern South China Sea continental shelf to around Zengmuansha (southern tip of Nansha=Spratly Islands). (1–11, 13, 14)

B. rapax (Claus) [*B. rapacoides*]: East China Sea. (6)

OXYCEPHALIDAE

Calamorhynchus pellucidus Streets [*C. rigidus*]: East China Sea; central South China Sea. (11, 12)

Cranocephalus scleroticus (Streets) [*C. goesi*]: Northern and central Taiwan Strait; South China Sea. (1, 4, 5, 8)

Glossocephalus milne-edwardsi Bovallius: East China Sea; Taiwan Strait; South China Sea. (1–6, 8–14)

Leptocotis tenuirostris (Claus) [*L. ambobus*, *L. spinifera*]: East China Sea; southern Fujian; Taiwan; South China Sea. (2–5, 8, 9, 14)

Oxycephalus clausi Bovallius: East China Sea; Taiwan Strait; Daya Bay (Guangdong); northern and central South China Sea continental shelf. (1–6, 9, 10, 11, 13)

O. latirostris Claus [*O. notabilis*]: East China Sea; southern Fujian; Taiwan; northern and central South China Sea continental shelf. (2–5, 9, 11, 13)

O. longipes Spandl: East China Sea; southern Fujian; Taiwan shallow shore; South China Sea. (3, 4, 8, 11, 13)

O. piscator Milne-Edwards: East China Sea; western Taiwan Strait; central South China Sea. (6, 10–13)

Rhabdosoma armatum (Milne-Edwards): East China Sea; central South China Sea. (2, 9, 11)

R. brevicaudatum Stebbing: East China Sea; southern Fujian; Taiwan; South China Sea. (2, 3, 4, 8, 13)

R. minor Fage: Southern Zhongsha (=Macclesfield Bank). (11)

R. whitei Bate: East China Sea; western Taiwan Strait; South China Sea. (3, 4, 7–11, 13)

Streetsia challengeri Stebbing: East China Sea; central South China Sea. (2, 11–14)

S. mindanaonis (Stebbing): Central South China Sea. (11)

S. porcella (Claus) [*S. intermedia, Oxycephalus porcella*]: East China Sea; Taiwan Strait; South China Sea. (1–6, 8, 10–14)

S. steenstrupi (Bovallius): East China Sea; central South China Sea. (11, 13)

Tullbergella cuspidata Bovallius: Southern Yellow Sea; East China Sea; Taiwan Strait; Xiamen harbor (Fujian); South China Sea. (1–11, 13, 14)

PLATYSCELIDAE

Amphithyrus bispinosus Claus: East China Sea; Taiwan Strait; South China Sea. (1, 3–6, 8–11, 13, 14)

A. glaber Spandl: Northern and central Taiwan Strait. (1)

A. muratus Volkov: East China Sea; Taiwan Strait; central South China Sea. (1, 3, 10, 11, 13)

A. sculpturatus Claus [*A. orientalis*]: East China Sea; Taiwan Strait; South China Sea. (1, 3–10, 13, 14)

A. similis Claus: Northern and central Taiwan Strait. (1)

Hemityphis crustulatus Claus: Central South China Sea. (11)

H. tenuimanus Claus: East China Sea; central South China Sea. (11, 12, 13)

Paratyphis maculatus Claus: Northeastern East China Sea; southern Fujian; Taiwan. (3, 14)

P. parvus Claus: East China Sea; Taiwan Strait; central South China Sea. (1, 3, 10, 11, 13)

P. promonitori Stebbing: Huangyan Island (Zhongsha=Maccslesfield Bank). (11)

P. spinosus Spandl: East China Sea. (2, 12, 13)

Platyscelus armatus (Claus): East China Sea; northern South China Sea continental shelf; Zhongsha (=Macclesfield Bank); Xisha (=Paracel Islands). (4, 5, 12)

P. ovoides (Risso): East China Sea; South China Sea. (4, 5, 8, 11)

P. serratulus Stebbing: East China Sea; Taiwan Strait; from northen South China Sea continental shelf to around Zengmuansha (southern tip of Nansha=Spratly Islands). (1–9, 11–14)

Tetrathyrus arafurae Stebbing: East China Sea; Taiwan Strait; central South China Sea. (1, 6, 10, 11, 13)

T. forcipatus Claus [*T. monceuri*]: East China Sea; Luoyuan Bay (Fujian); Taiwan Strait; Daya Bay (Guangdong); from northern South China Sea continental shelf to around Zengmuansha (southern tip of Nansha=Spratly Islands). (1–14)

PARASCELIDAE

Parascelus edwardsi Claus [*P. zebu*]: East China Sea; Taiwan Strait; South China Sea. (1, 2, 4, 5, 8–11, 13)

P. typhoides Claus [*Thyropus typhoides*]: East China Sea; western Taiwan Strait; Hainan. (2, 6, 9, 10)

Schizoscelus ornatus Claus: East China Sea; central South China Sea. (11, 12, 13)

Thyropus sphaeroma (Claus) [*T. diaphanus*]: East China Sea; Taiwan Strait; from northern South China Sea continental shelf to around Zengmuansha (southern tip of Nansha=Spratly Islands). (1, 2, 4–14)

REFERENCES*

(1). Zhu, C. S., J. Z. Wu, Y. S. Lin, M. Su and C. J. Huang, 1988. Species composition, distribution and abundance of marine plankton. In: Fujian Institute of Oceanology (Ed.). A Comprehensive Oceanographic Survey of the Central and Northern Part of Taiwan Strait. China Science Press, pp:259–305.

(2). Li, Q. L., 1987. Distribution and ecology of the Hypriidea Amphipoda in the Kuroshio region of the East China Sea in the summer of 1997. In: Su. J. I.. (ed.). Selected Research Reports of Kuroshio Investigation II. China Ocean Press, pp:107–113.

(3). Zhu, C. S., J. Q. Huang and S. J. Li, 1991. Distribution of planktonic Amphipoda (Hyperiidea) in Minnan-Taiwan bank fishing ground. Minnan-Taiwan Bank Fishing Ground Upwelling Ecosystem Study. China Science Press, pp:496–501.

(4). Chen Q. C., B. Y. Chen and G. X. Zhang, 1978. Pelagic Amphipoda in the vicinity of Xisha Islands and Zhongsha Islands. In: Research Reports of the Marine Biological Investigation of Xisha and Zhongsha Islands. China Science Press, pp:227–260.

(5). Song, S. X., G. Q. Chen, W. F. Wu, L. R. Ren, K. Z. Liang and Q. L. Guan, 1979. Distribution and variation of abundance of plankton Amphioda on the continental shelf of northern South China Sea. Reports of Investigation on the Demersal Fishery Resources off the Continental Shelf of Northern South China Sea. Bureau of Oceanography and Fisheries Institute of South China Sea. National General Bureau of Fisheries Press, Guangzhou, pp:550–557.

(6). Chen R. X., 1983. The planktonic amphipoda in the East and South China Seas. Collected Oceanic Works, 6(1):76–92.

(7). Chen, Q. C., G. X. Zhang and J. J. Yin, 1987. Species, abundance and biology of zooplankton. Research Reports of Multidisciplinary Investigation on Chinese Southern Territory: Zengmu Shoal. China Science Press, pp:32–146.

(8). Chen, Q. C., G. X. Zhang, Q. Z. Gao and J. J. Yin, 1989. Zooplankton. Research Reports on the Multidiscipline Investigation of Nansha Islands and its Adjacent Seas. China Science Press, pp:659–707.

(9). Lin, M. Y., 1982. A preliminary report on the Hyperiidea amphipods from the continental shelf of the East China Sea. Studia Marina Sinica, 19:43–50.

(10). Lin, J. H. and R. X. Chen, 1988. Distribution of planktonc Amphipoda in western Taiwan Strait. Journal of Taiwan Strait, 7(4):324–330.

(11). Lin, J. H. and R. X. Chen, 1994. Distribution of pelagic Amphipoda in central South China Sea. Acta Oceanologica Sinica, 16(4):113–119.

(12). Lin, Y. M., 1989. The vertical distribution of Amphipoda Hyperiidea in the East China Sea. Studia Marina Sinica, 30:277–285.

(13). Lin, J. H. and R. X. Chen, 1995. Distribution characteristics of the planktonic amphipods in the south Yellow Sea and the East China Sea. Acta Oceanologica Sinica, 14(4):553–561.

(14). Huang. S. M., 1986. Ecological study of zooplankton. Journal of Shandong College of Oceanography, 16:55–87.

*: (6) and (13) in English, others in Chinese

Compiled by Lin Jinghong and Chen Ruixiang; Edited by Chen Qingchao

CAPRELLIDEA
CAPRELLIDAE

Caprella acanthogaster Mayer: Dalian (Liaoning); Tuojin Island and Jiaozhou Bay (Shandong). Commonly found in seaweed and scallop aquaculture areas. (2, 3, 4)

C. bispinosa Mayer: Yellow Sea. (4)

C. californica Mayer: Hong Kong. (4)

C. equilibra Say: China's coast. Very common species in low intertidal and aquaculture cages. (3, 4, 5)

C. iniquilibra Mayer: Taiwan. Water depth 46–92 m. (4)

C. kroyeri De Haan: East China Sea. (4)

C. penantis Leach [*C. acutifrons* (3)]: Tuoji Island (Shandong); Chengshan (Zhejiang); Hong Kong. Common species. (3, 4, 5)

C. rhopalochir Mayer: China's seas. (4)

C. scaura Templeton: China's coast. Common species from intertidal to shallow sea. (2–5)

C. simia Mayer: Hong Kong. (5)

C. vidua Mayer: Yellow Sea. (4)

Hemiaegina minuta Mager: Xiamen harbor (Fujian). Water depth 15–46 m. (4)

Paracaprella crassa Mayer: Northern Yellow Sea. Water depth 51 m. (4)

CYAMIDAE

Cyamus erratiaus de Vauzene: Northern Yellow Sea. On the skin of whales. (1)

C. ovalis de Vauzene: Northern Yellow Sea. On the skin of whales. (1)

C. scammoni Dall: Northern Yellow Sea. On the skin of whales. (1)

REFERENCES*

(1). Zhou Ninqi, 1980. The first record of Cymidae in China. Acta Zootaxonomica Sinica, 5(4): 381

(2). Zhao Ruyi et al., 1990. Cappellidea. In: Guidebook on collection of invertebrate from China coast (in North China). China Ocean Press, pp:268–269

(3). Huang Zongguo and Cai Ruxing, 1984. Marine fouling organism and its prevention (I). China Ocean Press, 352pp.

(4). Arimoto, I. 1976. Taxonomic studies of caprellids found in the Japanese and adjacent waters. Special publications from the Seto marine biological laboratory III. 229pp.

(5). Morton, B. and J. Morton, 1983. The sea shore ecology of Hong Kong. Hong Kong University Press, 350pp.

*: (1)-(3) are in Chinese
Compiled by Huang Zongguo and Ren Xianqiu; Edited by Cai Ruxing

INGOLFIELLIDEA
INGOLFIELLIDAE

Paranthura japonica Richardson: Yantai (Shandong); Zhoushan (Zhejiang). Rocky shore. (1, 2)

REFERENCES*

(1). Zhao Ruyi et al., 1990. Guidebook on collection of invertebrate from China coast (in North China). China Ocean Press, pp:1–393, fig. 9–42.

(2). Huang Zongguo and Cai Ruxing, 1984. Marine fouling organism and its prevention (I). China Ocean Press, 352pp.

*: both in Chinese
Compiled by Huang Zongguo; Edited by Cai Ruxing

EUPHAUSIACEA
BENTHEUPHAUSIIDAE

Bentheuphausia amblyops G. O. Sars: South China Sea. (3, 8)

EUPHAUSIIDAE

Euphausia brevis Hansen: Yellow Sea; East China Sea; South China Sea. (1, 8)

E. diomedeae Ortmann: Yellow Sea; East China Sea; Taiwan Strait; South China Sea. (1–8)

E. gibba G. O. Sars: Yellow Sea. (2, 8)

E. hemigibba Hansen: South China Sea. (1, 3, 4, 5, 8)

E. mutica Hansen: Yellow Sea; East China Sea; Taiwan Strait; South China Sea. (1–8)

E. nana Brinton: Yellow Sea; East China Sea; Taiwan Strait; South China Sea. (1–8)

E. pacifica Hansen: Yellow Sea; East China Sea; South China Sea. (1, 3, 6, 8)

E. paragibba Hansen: Yellow Sea; East China Sea. (1, 8)

E. pseudogibba Ortmann: Yellow Sea; East China Sea; South China Sea. (1, 2, 4, 6, 8)

E. recurva Hansen: Yellow Sea; East China Sea; Taiwan Strait; South China Sea. (1–8)

E. sanzoi Torelli: Yellow Sea; East China Sea; Taiwan Strait; South China Sea. (1–8)

E. sibogae Hansen: South China Sea. (2, 4, 8)

E. similis G. O. Sars: Yellow Sea; East China Sea; South China Sea. (1, 4, 6, 8)

E. tenera Hansen: Yellow Sea; East China Sea; Taiwan Strait; South China Sea. (1–8)

Nematobrachian booepis (Calman): Yellow Sea; East China Sea; South China Sea. (1- 3, 8)

N. flexipes (Ortmann): Yellow Sea; East China Sea; South China Sea. (1, 3, 8)

N. sexspinosus Hansen: South China Sea. (8)

Nematoscelis atlantica Hansen: Yellow Sea; East China Sea; Taiwan Strait; South China Sea. (1, 2, 3, 5, 8)

N. gracilis Hansen: Yellow Sea; East China Sea; Taiwan Strait; South China Sea. (1–6, 8)

N. lobata Hansen: Yellow Sea; East China Sea; South China Sea. (1, 8)

N. microps G. O. Sars: Yellow Sea; East China Sea; Taiwan Strait; South China Sea. (1- 5)

N. tenella G. O. Sars: Yellow Sea; East China Sea; South China Sea. (1–4, 8)

Pseudeuphausia latifrons (S. O. Sars): Yellow Sea; East China Sea; Taiwan Strait; South China Sea. (1–8)

P. sinica Wang & Chen: Yellow Sea; East China Sea; Taiwan Strait; South China Sea. (1–8)

Stylocheiron abbreviatum G. O. Sars: Yellow Sea; East China Sea; Taiwan Strait; South China Sea. (1–5, 7, 8)

S. affine Hansen: Yellow Sea; East China Sea; Taiwan Strait; South China Sea. (1–8)

S. carinatum G. O. Sars: Yellow Sea; East China Sea; Taiwan Strait; South China Sea. (1–8)

S. elongatum G. O. Sars: South China Sea. (2–5, 8)

S. indicus Silas & Mathew: South China Sea. (2, 8)

S. longicorne G. O. Sars: Yellow Sea; East China Sea; South China Sea. (1–5, 8)

S. maximum Hansen: South China Sea. (2, 3, 8)

S. microphthalma Hansen: Yellow Sea; East China Sea; Taiwan Strait; South China Sea. (2–8)

S. robustum Brinton: South China Sea. (8)

S. suhmii G. O. Sars: Yellow Sea; East China Sea; Taiwan Strait; South China Sea. (2–8)

Thysanopoda acutifrons Holt & Tattersall: South China Sea. (2, 3, 4, 8)

T. aequalis Hansen: Yellow Sea; East China Sea; South China Sea. (1, 2, 3, 6, 8)

T. astylata Brinton: Yellow Sea; East China Sea; Taiwan Strait; South China Sea. (3–8)

T. cornuta Illig: South China Sea. (2, 4, 8)

T. cristata G. O. Sars: South China Sea. (8)

T. egregia Hansen: South China Sea. (8)

T. monacantha Ortmann: South China Sea. (2, 3, 4, 8)

T. obtusifrons G. O. Sars: Yellow Sea; East China Sea; South China Sea. (1, 3, 8)

T. orientalis Hansen: South China Sea. (2, 3, 8)

T. pectinata Ortmann: South China Sea. (2, 3, 4, 8)

T. tricuspidata Milne-Edwards: Yellow Sea; East China Sea; Taiwan Strait, South China Sea. (1–8)

REFERENCES*

(1). Shandong College of Oceanology, 1986. A report of comprehensive survey in the mouth of Changjiang River and Jizhou Island and its adjacent waters. Journal of Shandong College of Oceanology, 16(2):80–81.

(2). Comprehensive Scientific Investigation Team on Nansha Islands, Academia Sinica, 1989, Study report of comprehensive survey in the waters around Nansha Islands (I). China Science Press, pp:668–669.

(3). Chen Qingchao and Zhang Guxian, 1983. Studies on Euphausiacea in the central and northern parts of the South China Sea. In: Paper collections of studies on marine organisms in the South China Sea (I). China Ocean Press, pp:139–172.

(4). State Oceanic Administration, 1988. A report of comprehensive survey on environments and resources in the middle South China Sea. China Ocean Press, pp:248–249.

(5). Fujian Institute of Oceangraphy, 1988. The species composition and distribution of zooplankton. The comprehensive survey in the middle and northern pars of Taiwan Strait. China Science Press, 402pp.

(6). Cai Binji, 1978. A preliminary study of taxonomy of Euphausiacea (Crustacea) from south Yellow Sea and East China Sea. Oceanologia & Technologia Sinica, 8:39- 56.

(7). Cai Binji, 1989. The distribution of Euphausiacea in western Taiwan Strait. Acta Oceanologica Sinica, 11(6):763–768.

(8). Zhang Guxian and Chen Qingchao, 1991. Euphausiids in South China Sea and its adjacent waters. The Study Collections of Fauna and Geography of Marine Animals in Nansha Islands. China Ocean Press, pp:140–271.

*: all in Chinese

Compiled by Cai Binji and Li Shaoqing; Edited by Chen Qingchao

DECAPODA
DENDROBRANCHIATA
PENAEOIDEA
ARISTEIDAE

Aristaeomorpha foliacea (Risso) [*Penaeus foliaceus* Risso, 1827; *Aristeus rostridentatus* Bate, 1881; *Aristeus japonicus* Yokoya, 1933]: East China Sea; Taiwan; northern South China Sea continental shelf and slope. (6, 20, 32, 41, 44, 77)

Aristeus virilis (Bate) [*Hemipenaeus tomentosus* Bate, 1881; *H. virilis* Bate, 1881]: East China Sea; Taiwan; northern South China Sea continental slope. (6, 20, 32, 41, 44, 77)

Benthesicymus altus Bate: East China Sea continental slope. Water depth 620–1,030 m. (6)

B. investigaloris Alcock & Anderson: Northern South China Sea continental slope. Water depth 248–1,024 m. (32, 41, 44)

Benthogennema intermedia (Bate) [*Gennadas intermedius* Bate, 1888; *G. alicei* Bouvier, 1906]: Deep water around Xisha (=Paracel Islands) and Zhongsha (=Macclesfield Bank). (32, 40)

Benthonectes filipes Smith: Northern South China Sea continental slope. Water depth 519–524 m. (32, 41)

Gennadas incertus (Balss) [*Amalopenaeus incertus* Balss, 1927; *G. incertus* Timizi, 1960]: Deep water around Xisha (=Paracel Islands) and Zhongsha (=Macclesfield Bank). (32, 40)

G. parvus Bate: Deep water around Xisha (=Paracel Islands) and Zhongsha (=Macclesfield Bank). (32, 40)

G. propinquus Rathbun: Deep water around Xisha (=Paracel Islands) and Zhongsha (=Macclesfield Bank). (32, 40)

G. scutatus Bouvier: Deep water around Xisha (=Paracel Islands) and Zhongsha (=Macclesfield Bank). (32, 40)

Hepomedus tener Bate: Northern South China Sea continental slope. Water depth 664- 934 m. (32, 41, 44)

Parahepomadus vaubani Crosnier: Northern South China Sea continental slope. Water depth 610–934 m. (32, 41, 44)

Plesiopenaeus coruscans (Wood-Mason) [*Aristeus coruscans* Wood-Mason, 1891]: Northern South China Sea continental slope. Water depth 810–1,024 m. (32, 41, 44)

P. edwardsianus (Johnson) [*Penaeus edwardsianus* Johnson, 1867; *Aristeus coralinus* Bate, 1888]: East China Sea; northern South China Sea continental slope. Water depth 310–946 m. (6, 32, 41, 44)

Pseudaristeus crassipes (Wood-Mason) [*Aristeus crassipes* Wood-Mason, 1891; *Hemipenaeus crassipes* Dong et al. 1988]: Western East China Sea, water depth 720–1,080 m; northern South China Sea continental slope. (6, 20, 32, 41, 44, 77)

SOLENOCERIDAE

Cryptopenaeus sinensis (Liu & Zhong) [*Crassipenaeus sinensis* Liu & Zhong, 1983]: Northeastern South China Sea continental slope. (32, 73)

Hadropenaeus lucasii (Bate) [*Philonicus lucasii* Bate, 1888; *Hymenopenaeus lucasii* Kubo, 1949]: Northeastern South China Sea continental slope. Water depth 260- 540 m. (32, 41, 43, 44)

H. spinicauda Liu & Zhong: Continental shelf off Guangdong. (32, 73)

Haliporoides sibogae (de Man) [*Haliporus sibogae* de Man, 1907; *Hymenopenaeus sibogae* Burkenroad, 1936; *Parahaliporus sibogae* Kubo, 1949]: Western East China Sea; Taiwan; northern South China Sea continental slope. Water depth 210–790 m. (6, 7, 32, 41, 43, 44, 70, 77)

Haliporus taprobanensis Alcock & Anderson: Northern South China Sea continental slope. Water depth 510–890 m. (32, 41, 43, 44)

Hymenopenaeus aequalis (Bate) [*Haliporus equalis* Bate, 1888; *H. aequalis* Wood- Mason, 1891]: Taiwan; northern South China Sea continental slope. Water depth 310–590 m. (20, 32, 41, 44, 77)

H. halli Bruce: Northern South China Sea continental slope. Water depth 510–590 m. (32, 41, 44)

H. propinquus (de Man) [*Haliporus propinquus* de Man, 1907]: Northern South China Sea continental slope. Water depth 510–640 m. (32, 41)

Mesopenaeus mariae Perez Farfante & Ivanov: Continental shelf off northeastern Taiwan. Water depth 158–220 m. (77)

Solenocera alticarinata Kubo [*S. choprai* Natarai Yu & Chan, 1986]: East China Sea; Taiwan; South China Sea continental shelf. (13, 14, 20, 32, 26, 45, 77)

S. comata Stebbing [*S. brevipes* Kubo, 1949]: Taiwan; northern South China Sea continental shelf. (32, 77)

S. crassicornis (H. Milne-Edwards) [*Penaeus crassicornis* H. Milne-Edwards, 1837; *Solenocera distincta* Yu, 1935; *S. sinensis* Yu 1937; *S. subnuda* Kubo, 1949; *S. indica* Hunju, 1968; *S. kuboi* Hall, 1961]: Southern Yellow Sea; East China Sea, Taiwan; South China Sea shallow continental shelf. (13, 14, 20, 32, 36, 43, 44, 70, 74, 77)

S. faxoni de Man [*S. brevipes* Kubo, 1944]: Taiwan; northern South China Sea continental slope. Water depth 248–259 m. (32, 77)

S. koelbeli de Man [*S. depressa* Kubo, 1949]: Western East China Sea; Taiwan; northern South China Sea continental shelf. (13, 14, 20, 32, 36, 45, 77)

S. melantho de Man [*S. prominentis* Kubo]: East China Sea; Taiwan; northern South China Sea continental shelf. (20, 36, 77)

S. pectinata (Bate) [*Philonicus pectinatus* Bate, 1888]: Western East China Sea; Taiwan; northern South China Sea continental shelf. (13, 32, 36, 45, 77)

S. pectinulata Kubo: Northern South China Sea continental shelf. (32)

S. rathbunae Ramadan [*S. rathbuni* Ramadan, 1938]: Northern South China Sea shallow continental shelf. (32)

PENAEIDAE

Atypopenaeus stenodactylus (Stimpson) [*Penaeus stenodactylus* Stimpson, 1860; *P. compressipes* Henderson, 1893; *A. compressipes* Alcock, 1905; *Parapenaeopsis brevirostris* Kubo, 1936; *A. stenodactylus* Hall, 1961]: Western East China Sea; Taiwan; northern South China Sea. (8, 13, 14, 20, 32, 36, 45, 70, 77)

Funchalia villosa (Bouvier) [*Hemipenaeopsis villosus* Bouvier, 1905; *F. villosa* Burkenroad, 1936]: Zhongsha (=Macclesfield Bank). (32)

Metapenaeopsis acclivis Rathbun [*Parapenaeus acclivis* Rathbun, 1902; *Penaeus (Metapenaeus) acclivis* de Man, 1907]: Western East China Sea; Taiwan; northern South China Sea continental shelf. (8, 20, 77)

M. andamanensis (Wood-Mason) [*Penaeopsis coniger* v. *andamanensis* Wood-Mason, 1891]: Taiwan; northern South China Sea continental shelf and slope. (20, 70, 77)

M. barbata (de Haan) [*Penaeus affinis barbatus* de Haan, 1850; *Parapenaeus akayebi* Rathbun, 1902]: Western East China Sea; Taiwan; northern South China Sea continental shelf. (13, 14, 18, 20, 32, 36, 45, 70, 77)

M. dalei (Rathbun) [*Parapenaeus dalei* Rathbun, 1902]: Bohai; Yellow Sea; East China Sea; Taiwan; northern South China Sea continental shelf. (13, 20, 21, 32, 36, 45, 70, 77, 79)

M. dura Kubo: Taiwan; northern South China Sea continental shelf. (20, 32, 77)

M. hilarula (de Man) [*Penaeopsis hilarula* de Man, 1911]: Northern South China Sea. Near shore. (13, 31)

M. lamellata (de Haan) [*Penaeus lamellata* de Haan, 1850]: Western East China Sea; from Taiwan to eastern Guangdong. (13, 20, 36, 45, 77)

M. lata Kubo: Northern South China Sea continental shelf. Water depth 103–472 m. (32)

M. liui Crosnier: Northern South China Sea continental shelf. (33, 65)

M. mogiensis (Rathbun) [*Parapenaeus mogiensis* Rathbun, 1902; *Metapenaeus mogiensis* Alcock, 1906]: Taiwan; Guangdong; Guangxi; Hainan. (18, 20, 32, 45, 76, 77, 79)

M. palmensis (Haswell) [*Penaeus palmensis* Haswell, 1879; *P. velutinus* Bate, 1888; *Metapenaeopsis barbeensis* Hall, 1962]: Taiwan; northern South China Sea. (18, 32, 77, 79)

M. philippi (Bate) [*Penaeus philippinensis* Bate, 1881]: Western East China Sea continental shelf and slope. (6)

M. sinica Liu & Zhong: Taiwan; northern South China Sea continental shelf. (32)

M. stridulans (Alcock) [*Metapenaeus stridulans* Alcock, 1905]: Northern South China Sea continental shelf. (32, 45, 79)

M. tenella Zhong & Liu: Northern South China Sea continental slope. (32)

M. toloensis Hall: Northern South China Sea. (32, 45, 79)

M. velutina (Dana) [*Penaeus velutinus* Dana, 1852; *Metapenaeus velutinus* Rathbun, 1906; *Metapenaeopsis insona* Racek & Dall, 1965; *M. caliper* Liu & Zhong, 1988]: Northern South China Sea continental shelf. (32)

Metapenaeus affinis (H. Milne-Edwards) [*Penaeus affinis* H. Milne-Edwards, 1837; *Penaeopsis affinis* Kemp, 1915; *Metapenaeus mutatus* Hall, 1962]: Fujian; Taiwan; northern South China Sea continental shelf. (14, 20, 32, 45, 70, 77, 79)

M. ensis (de Haan) [*Penaeus monoceros ensis* de Haan, 1844; *P. mastersis* Haswell 1879; *P. incisipes* Bate, 1888; *Metapenaeus monoceros* Kubo, 1949]: Western East China Sea; Taiwan; northern South China Sea continental shelf. (13, 14, 18, 20, 32, 36, 45, 70, 76, 77, 79)

M. intermedius (Kishinouye) [*Penaeus intermedius* Kishinouye, 1900]: Western East China Sea; Taiwan; northern South China Sea. (14, 18, 20, 32, 45, 70, 79)

M. joyneri (Miers) [*Penaeus joyneri* Miers, 1880]: Yellow Sea; East China Sea; Taiwan; northern South China Sea. (8, 10, 14, 18, 20, 32, 36, 45, 77, 79)

M. moyebi (Kishinouye) [*P. mastersii* Haswell, 1882; *Penaeus moyebi* Kishinouye, 1896; *Metapenaeus burkenroadi* Kubo, 1954]: Western East China Sea; Taiwan; northern South China Sea. Low salinity water. (14, 18, 20, 32, 45, 77)

M. tenuipes Kubo [*M. spinulatus* Kubo, 1949]: Taiwan. Shallow sea. (77)

Miyadiella podophthalmus (Stimpson) [*Penaeus podophthalmus* Stimpson, 1860; *Miyadiella pedunculata* Kubo, 1949]: Western East China Sea; northern South China Sea. (13, 32, 36, 70)

Parapenaeopsis cornuta (Kishinouye) [*Penaeus cornutus* Kishinouye, 1990]: Western East China Sea; Taiwan, northern South China Sea. (8, 10, 13, 14, 18, 20, 30, 32, 45, 77, 79)

P. cultrirostris (Alcock) [*P. scultptilis* v. *cultrirostris* Alcock, 1906]: Northern Yellow Sea; western East China Sea; Taiwan; northern South China Sea. (13, 14, 20, 30, 32, 36, 45, 70, 77)

P. hardwickii (Miers) [*Penaeus hardwickii* Miers, 1878]: Southern Yellow Sea; western East China Sea; Taiwan; northern South China Sea. (13, 14, 20, 30, 32, 36, 45, 70, 77, 79)

P. hungerfordi Alcock: Western East China Sea; northern South China Sea. (14, 32, 45, 70)

P. incisa Wang & Liu: Northern South China Sea. (13, 32)

P. sinica Liu & Wang: Northern South China Sea. (13, 32)

P. tenella (Bate) [*Penaeus tenellus* Bate, 1888]: Yellow Sea; East China Sea; Taiwan; northern South China Sea. (13, 14, 20, 32, 36, 45, 70, 77, 79)

Parapenaeus fissuroides Crosnier [*P. fissurus* de Man, 1911 (non Bate, 1988) Liu, 1963; Dong et al., 1980, 1988; Yu & Chan, 1986]: Western East China Sea; northern South China Sea continental shelf. (6, 13, 20, 32, 36, 41, 77)

P. investigatoris Alcock & Anderson: Taiwan; northern South China Sea continental shelf. Water depth 125–330 m. (20, 32, 41, 43)

P. lanceolatus Kubo: Taiwan; northern South China Sea continental shelf and slope. Water depth 45–350 m. (32, 41, 45, 70, 77)

P. longipes Alcock: Western East China Sea; Taiwan; northern South China Sea continental shelf. (20, 32, 43, 77)

P. sextuberculatus Kubo: Western East China Sea; Taiwan; northern South China Sea continental slope. (32, 41, 43, 77)

Penaeopsis eduardoi Perez-Farfante: East China Sea; Taiwan; northern South China Sea continental slope. Water depth 310–540 m. (32, 41, 43, 44, 77)

P. rectacutus (Bate) [*Penaeus rectacutus* Bate, 1888; *Penaeopsis challengeri* Balss, 1925]: East China Sea; Taiwan; northern South China Sea continental slope. (20, 32, 41, 43, 44, 77)

Penaeus (*Melicertus*) *canaliculatus* (Oilvier) [*Palaemon canaliculatus* Olivier, 1811]: Taiwan. (20, 32, 70, 77)

P. (*Fenneropenaeus*) *chinensis* (Osbeck) [*Cancer chinensis* Osbeck, 1765; *Penaeus orientalis* Kishinouye, 1918]: Bohai; Yellow Sea; northern East China Sea; western Guangdong and mouth of Zhujiang (=Pearl River) (Guangdong). (1, 5, 9, 24, 37, 46, 70, 76, 77)

P. (*F.*) *indicus* H. Milne-Edwards: Taiwan; central and western South China Sea. (20, 32, 77, 79)

P. (*Marsupenaeus*) *japonicus* Bate: Southern Yellow Sea; East China Sea; Taiwan; northern South China Sea. (10, 13, 14, 20, 32, 35, 36, 45, 70, 77, 79)

P. (*Melicertus*) *latisulcatus* Kishinouye: Northern East China Sea; Taiwan; northen South China Sea. (14, 20, 32, 45, 70, 76, 77, 79)

P. (*Melicertus*) *longistylus* Kubo: Northern South China Sea; offshore of Hainan. (32)

P. (*Melicertus*) *marginatus* Randall [*Penaeus teraei* Kubo, 1949]: Taiwan; northern South China Sea. (20, 32, 77)

P. (F.) merguiensis de Man: Taiwan Strait; northern South China Sea. (14, 32, 45, 77, 79)

P. (P.) monodon Fabricius [*P. tahitensis* Heller, 1862; *P. bubulus* Kubo, 1949]: Western East China Sea; Taiwan, northern South China Sea. (14, 20, 32, 36, 45, 70, 76, 77)

P. (F.) penicillatus Alcock: Western East China Sea; Taiwan Strait; northern South China Sea. (14, 20, 32, 35, 45, 70, 76, 77)

P. (P.) semisulcatus de Haan [*P. monoon* Bate, 1888; Kubo, 1949 (non Fabricius) 1798; *P. ashiaka* Kishinouye, 1900; *P. carinatus* Yu, 1935]: Western East China Sea; Taiwan; northern South China Sea. (14, 20, 32, 36, 45, 70, 76, 77, 79)

Trachypenaeus anchoralis (Bate): Taiwan. (23, 70, 77)

T. curvirostris (Stimpson) [*Penaeus curvirostris* Stimpson, 1860]: From Bohai to South China Sea. (13, 14, 20, 21, 32, 36, 45, 70, 72, 77, 79)

T. longipes (Paulson) [*Penaeus longipes* Paulson, 1875; *T. asper* Alcock, 1906 (non *T. asper* Kubo, 1949]: Northern South China Sea. (32, 45, 79)

T. malaiana Balss [*T. asper* Kubo, 1949; *T. fulvus* Dall, 1957]: Northern South China Sea. (32, 45, 79)

T. pescadoreensis Schmitt [*T. granulosus*, Yu & Chan, 1986]: Taiwan; northern South China Sea. (14, 20, 32, 45, 70, 77, 79)

T. sedili Hall: Northern South China Sea. (32, 45)

SICYONIDAE

Sicyonia cristata (de Haan) [*Hippolyte cristata* de Haan, 1844; *Eusicyonia cristata* Kubo, 1949]: Taiwan; northern South China Sea continental shelf and slope. (13, 20, 32, 45, 58, 70, 77)

S. curvirostris Balss [*Eusicyonia curvirostris* Kubo, 1949]: Taiwan; northern South China Sea continental shelf and slope. (32, 59, 77)

S. formosa Chan & Yu: Continental shelf off Taiwan. (59, 77)

S. japonica Balss [*S. lancifer* v. *japonica* Balss, 1914; *Eusicyonia lancifer japonica* Kubo, 1949]: Taiwan; northern South China Sea continental shelf. (32, 59, 77)

S. lancifer (Olivier) [*Palaemon lancifer* Olivier, 1811; *Eusicyonia lancifer* Kubo, 1949]: Northern South China Sea continental shelf. (13, 32, 45, 70, 79)

S. longicauda Rathbun: Continental shelf off Taiwan. (59, 77)

S. ommanneyi Hall: Northern South China Sea continental shelf. (13, 32, 45)

SERGESTIOIDEA
SERGESTIDAE

Acetes chinensis Hansen: Bohai; Yellow Sea; western East China Sea; northern South China Sea. (13, 21, 22, 25, 70, 72, 75)

A. erythraeus Nobili: Northern South China Sea. (25, 70, 75)

A. intermedius Omori: Taiwan. (70, 75)

A. japonicus Kishinouye: Southern Yellow Sea; East China Sea; South China Sea shore. (13, 15, 21, 22, 25, 38, 70, 72, 75)

A. serrulatus (Kröyer) [*Sergestes serrulatus* Kröyer, 1859; *Acetes insularis* Kemp, 1917]: Northern South China Sea; Hainan. (25, 75)

A. vulgaris Hansen: Taiwan; southern China. (70, 75)

Lucifer faxoni Borradaile: Southern Yellow Sea; East China Sea; South China Sea. (2, 13)

L. hanseni Nobili: Southern Yellow Sea; East China Sea; South China Sea. (1, 2, 13)

L. intermedius Hansen: Southern Yellow Sea; East China Sea; South China Sea. (1, 2, 13, 15, 39)

L. orientalis Hansen: Southern Yellow Sea; East China Sea; South China Sea. (2, 13, 15)

L. penicillifer Hansen: Southern Yellow Sea; East China Sea; South China Sea. (1, 2, 13, 15, 39)

L. typus H. Milne-Edwards: Southern Yellow Sea; East China Sea; South China Sea. (1, 2, 13, 15)

Sergestes gardineri Kemp: Central South China Sea. (15)

S. orientalis Hansen: Central South China Sea. (15)

S. talismani Barnard: South China Sea continental slope. Water depth 510–590 m. (41)

PLEOCYAMATA
STENOPODIDEA
STENOPODIDAE

Microprosthema validum Stimpson: Hong Kong; Hainan; northern South China Sea. (55)

Spongicola venusta de Haan: East China Sea; northern South China Sea continental shelf. Symbiotic with glass sponges (Hexactinellida). (6)

Stenopus hispidus (Olivier) [*Palaemon hispidus*]: Hong Kong; Hainan; northern South China Sea. (74, 78)

CARIDEA
PASIPHAEOIDEA
PASIPHAEIDAE

Eupasiphae latirostris (Wood-Mason & Alcock) [*Parapasiphae latirostris* Wood-Mason & Alcock, 1891]: East China Sea; South China Sea. Continental slope, water depth 850–940 m. (6, 41)

Glyphus marsupialis Filhol [*Sympasiphaea imperialis*]: East China Sea continental slope, water depth 300–400 m; South China Sea continental slope. (6, 41)

Leptochela gracilis Stimpson: Bohai to South China Sea. Continental shelf. (13, 14, 18, 34, 36)

L. hainanensis Yu [*L. aculeocauda* Paulson]: Bohai to South China Sea. Continental shelf. (31, 36, 74)

L. japonicus Hayashi & Miyake: Hainan. Shallow sea. (33)

L. pugnax de Man: East China Sea; South China Sea. Continental shelf. (13, 31)

L. robusta Stimpson: East China Sea; South China Sea. Continental shelf. (31)

Parapasiphae sulcatifrons Smith: East China Sea; South China Sea. Water depth 1,000- 1,049 m. (6, 41, 44)

Pasiphaea japonica Omori: East China Sea continental slope. Water depth about 400 m. (6)

P. pacifica Rathbun: East China Sea; South China Sea. Continental slope, water depth 650–990 m. (6, 41, 43)

P. semispinosa Holthuis: South China Sea continental slope. Water depth 700–749 m. (6, 43)

P. sinensis Hayashi & Miyake: East China Sea. Water depth 780–1,075 m. (31)

OPLOPHOROIDEA
OPLOPHORIDAE

Acanthephyra armata H. Milne-Edwards: East China Sea; South China Sea. Continental slope, water depth 360–1,090 m. (6, 41, 44)

A. carinata Bate: East China Sea continental slope. Water depth about 700 m. (6)

A. curtirostris Wood-Mason: East China Sea; South China Sea. Continental slope, water depth about 600 m. (6, 40, 41, 44)

A. eximia Smith [*A. angusta*]: South China Sea continental shelf. Water depth 110–790 m. (6, 41, 44)

A. purpurea A. Milne-Edwards: Northeastern and central South China Sea deep water. (31, 40)

Notostomus brevirostris Bate: East China Sea continental slope. Water depth 400–500 m. (6)

N. longirostris Bate: East China Sea continental slope. Water depth 400–500 m. (6)

N. patentissimus Bate: South China Sea continental slope. (31)

Oplophorus typus A. Milne-Edwards [*O. brevirostris* Bate, 1888]: South China Sea continental slope. Water depth 510–590 m. (6, 40, 41, 44)

Systellaspis debilis (A. Milne-Edwards) [*Acanthephyra debilis* A. Milne-Edwards, 1881]: Northeastern and central South China Sea continental slope. (31, 40)

BRESILIOIDEA
BRESILIIDAE

Lucaya sp.: South China Sea continental slope. Water depth 660–790 m. (41)

NEMATOCARCINOIDEA
EUGONATONOTIDAE

Eugonatonotus crassus (A. Milne-Edwards) [*Gonatonotus crassus* A. Milne-Edwards, 1881]: South China Sea continental slope. Water depth 210–540 m. (41, 44)

NEMATOCARCINIDAE

Nematocarcinus cursor A. Milne-Edwards: South China Sea continental slope. Water depth 510–740 m. (41, 44)

N. sibogae de Man: South China Sea continental slope. Water depth 660–740 m. (41)

N. undulatipes Bate: East China Sea; South China Sea. Continental slope, water depth 310–840 m. (6, 41, 44)

PHYNCHOCINETIDAE

Phynchocinetes uritai Kubo: Hainan shoreline. (45)

PSALIDOPODOIDEA
PSALIDOPODIDAE

Psalidopus huxleyi Wood-Mason & Alcock: East China Sea; South China Sea. Continental slope, water depth 360–690 m. (6, 41, 44)

P. spiniventris Wood-Mason: South China Sea continental slope. Water depth 660–740 m. (41)

STYLLODACTYLOIDEA
STYLODACTYLIDAE

Neostylodactylus amarhynsus (de Man) [*Stylodactylus amarhynsus*]: South China Sea continental shelf. (31)

Parastylodactylus bimaxillaris (Bate) [*Stylodactylus bimaxillaris*]: South China Sea continental shelf. (31)

Stylodactylus multidentatus Kubo: East China Sea; South China Sea. Continental slope, water depth 300–349 m. (31, 41)

PALAEMONOIDEA
GNATHOPHYLLIDAE

Gnathophyllum americanum Guerin-Meneville: Taiwan; Hong Kong; northern South China Sea. (42)

HYMENOCERIDAE

Hymenocera elegans Heller: Xisha (=Paracel Islands). Coral reefs. (33)

ANCHISTIOIDIDAE

Anchistioides compressus Paulson: Northern South China Sea. Shallow water. (52)

A. willeyi (Borradaile) [*Palaemonopsis willeyi* Borradaile, 1899]: Northern South China Sea. Shallow water. (52)

PALAEMONIDAE

Anchistus custos (Forskål) [*Cancer custos* Forskål, 1775]: Guangdong; Hong Kong; Hainan; northern South China Sea. Commensal with bivalves. (33, 52, 53)

A. miersi (de Man) [*Harpilius miersi* de Man, 1888]: Northern South China Sea. (52)

Conchodytes monodactylus Holthuis: Guangdong; Hong Kong; Hainan; South China Sea. Commensal with bivalves. (33, 52, 53)

C. nipponensis (de Haan) [*Hymenocera nipponensis*]: East China Sea; northern South China Sea. Commensal with bivalves. (33, 36)

C. tridacnae Peters: Northern and central South China Sea. (52)

Coralliocaris brevirostris Borradaile: Hainan. Coral reefs. (26)

C. graminea (Dana) [*Oedipus graminea*]: Hong Kong; Hainan; northern South China Sea; Xisha (=Paracel Islands). Commensal with Scleractinia corals. (33, 53)

C. superba (Dana) [*Oedipus superbus*]: Hainan; northern South China Sea; Xisha (=Paracel Islands). Commensal with corals. (33)

C. venusta Kemp: Hainan; northern South China Sea; Dongsha (=Pratas Islands). (33, 52)

Exopalaemon annandalei (Kemp) [*Leander annandalei* Kemp; *L. annandalei stylirostris* Yu, 1930]: Bohai; Yellow Sea; East China Sea. Near river mouth. (14, 21, 29, 35, 36, 72)

E. carinicauda (Holthuis) [*Leander longirostris* v. *carinatus* Ortmann, 1891; *Palaemon (Exopalaemon) carinicauda* Holthuis, 1950]: Bohai; Yellow Sea; East China Sea; Taiwan Strait; northern South China Sea. Low salinity shallow water. (14, 21, 29, 35, 36, 72)

E. modestus (Heller) [*Leander modestus* Heller, 1862; *Palaemon modestus* Sowerby, 1925]: Northern China; eastern China. Freshwater and river mouth. (14, 21, 29, 35, 72)

E. orientalis (Holthuis) [*Leander longirostris japonicus* Ortmann; *Palaemon orientalis* Holthuis 1950]: East China Sea; South China Sea. Shallow water. (29, 72)

Hamodactylus boschmai Holthuis: Hong Kong. Commensal with gorgonians. (53)

Hamopontonia corallicola Bruce: Hong Kong. (52, 53)

Harpiliopsis beaupressi Audouin (Nobili) [*Palaemon beaupresii* Kemp, 1922]: Hainan; Zhongsha (=Macclesfield Bank). (26, 52)

H. depressa (Stimpson) [*Harpilius depressus* (Stimpson), 1860]: Northern and central South China Sea. (52)

Jocaste japonica (Ortmann) [*Coralliocaris superba* v. *japonica* Ortmann, 1890]: Northern South China Sea. Commensal with Scleractinia corals. (33, 52)

J. lucina (Nobili) [*Coralliocaris lucina* Nobili, 1901]: Northern South China Sea. Commensal with Scleractinia corals. (33, 52)

Leander urocaridella Holthuis: Guangdong; Beibuwan (=Gulf of Tongking); Hainan. (13, 28)

Leandrites antonbruunii (Bruce) [*Periclimenes antonbruunii* Bruce, 1967; *Leandrites cyrtorhynchus* Fujino & Miyake, 1969; *Leandrites longipes* Liu, Liang & Yan, 1990]: Beibuwan (Gulf of Tongking); South China Sea. (28)

L. deschampsi (Nobili) [*Leander deschampsi* Nobili, 1903]: Guandong. (28)

Leptocarpus potamiscus (Kemp) [*Leander potamaiscus* Kemp, 1918; *Palaemon potamiscus* Suivarti, 1937]: Guangdong. Freshwater. (28)

Macrobrachium equidens (Dana) [*Palaemon equidens* Dana, 1852; *P. (Eupalaemon) sundaicus* de Man, 1892]: South of Fujian. Brackish water. (28, 42, 72)

M. grandimanus (Randall): Hainan. Brackish water. (28)

M. hainanense (Parisi) [*Palaemon (Parapalaemon) hainanse* Parisi, 1919; *P. sinmilis* Yu 1931]: South of Wenzhou (Zhejiang). Brackish water. (28)

M. latidactylus (Thallwitz) [*Palaemon (Macrobrachium) lampropus* de Man, 1892]: Hainan shoreline. (28)

M. mammillodactylus (Thallwitz) [*Palaemon idae mammillodactylus* Thallwitz, 1892]: Hainan shoreline. (28)

M. nipponense (de Haan) [*Palaemon sinensis* Heller, 1862]: All over China. River mouth. (21, 28, 36, 45, 72)

M. rosenbergii (de Man) [*Palaemon rosenbergii* de Man, 1879; *P. carcinus* de Man, 1888; *P. carcinus rosenbergii* Ortmann, 1891]: River mouth in southern China. Recently introduced for aquaculture. (28)

M. superbum (Heller) [*Palaemon superbus* Heller, 1862]: South of mid- and lower- Changjiang (=Yangtze River). Freshwater and river mouth. (28, 42)

Mesopontonia gorgoniophila Bruce: Dongsha (=Pratas Islands). (52)

Onycocaris oligodenttata Fujino & Miyake: Hong Kong. Commensal with sponges. (53)

O. quadratophthalma Balss [*Pontonia quadraphthalma*]: Dapeng Cove (Hong Kong). (55a)

Palaemon concinnus Dana: Shoreline from Fujian to Hainan. (29)

P. gravieri (Yu) [*Leander gravieri* Yu, 1930]: Bohai; Yellow Sea; East China Sea; Fujian. Shallow water. (14, 17, 21, 29, 36, 72)

P. guangdongensis Liu, Liang & Yan: Guangdong. Shallow sea. (29)

P. macrodactylus Rathbun: China's shallow sea and river mouth. (14, 17, 21, 29, 36, 72)

P. ortmanni Rathbun: Yellow Sea; East China Sea; Taiwan. Shallow sea. (14, 17, 21, 29, 72)

P. pacificus (Stimpson) [*Leander pacifica*]: South of Zhejiang. Shallow sea. (14, 29, 34, 72)

P. paucidens de Haan: Yellow Sea shoreline. Freshwater and brackish water. (29, 72)

P. serrifer (Stimpson) [*Leander serrifer*]: China's shoreline. (14, 18, 45, 54, 58, 72)

P. sewelli (Kemp) [*Leander sewelli*]: Guangdong; Guangxi. Shoreline, low salinity water. (29)

P. tenuidactylus Liu, Liang & Yan: Bohai; Yellow Sea; East China Sea. River mouth brackish water. (29)

Palaemonella pottsi (Borradaile) [*Periclimenes pottsi* Borradaile, 1898]: Guangdong shoreline. (52)

P. rotumana (Borradaile) [*Periclimenes rotumanus* Borradaile, 1898]: Hong Kong; Hainan; South China Sea; Dongsha (=Pratas Islands). (52, 53, 58, 74)

Periclimenaeus arabicus (Calman) [*Periclimenes arabicus*]: Hong Kong. Commensal with sponges. (53)

P. rastrifer Bruce [*P. spongicola* Bruce, 1980]: Hong Kong. Commensal with sponges. (52, 53)

P. spinicauda Bruce: Northern South China Sea continental shelf. (52)

P. stylirostris Bruce: Northern South China Sea continental shelf. (52)

P. tridentatus (Miers) [*Coralliocaris tridentata*]: Hainan. (26)

Periclimenens affinis (Zehnter) [*Palaemonella affinis* Zehnter, 1894]: Dongsha (=Pratas Islands). (52)

P. amymone de Man: Hainan shoreline. (33)

P. brevicarpalis (Schenkel) [*Ancylocaris brevicarpalis* Schenkel, 1902]: Guangdong; Hong Kong; Hainan; Nansha (=Spratly Islands), commensal with sea anemones. (52, 55, 78)

P. commensalis Borradaile: Hong Kong. Commensal with sea feather (comatulids). (52, 53)

P. cristimanus Bruce: Hong Kong. Commensal with sea urchins. (52, 53)

P. demani Kemp: Hong Kong. (52, 53)

P. denticulatus Nobili: Zhongsha (=Macclesfield Bank). (52)

P. digitalis Kemp: Hong Kong. (53)

P. elegans (Paulson) [*Anchistia elegans* Paulson, 1875]: Guangdong; Hong Kong; Hainan. Shoreline. (26, 52, 53)

P. exederons Bruce: Northern South China Sea. Shallow water. (52)

P. gorgonicola Bruce: Northern South China Sea continental shelf. (52)

P. holthuisi Bruce [*P. aesopius* Bruce]: Hong Kong. Commensal with sea anemones. (52, 53)

P. hongkongensis Bruce [*P. setoensis* Fujino & Miyake]: Hong Kong. (52, 53)

P. imperator Bruce: Hainan shoreline. (33)

P. inornatus Kemp: Central South China Sea. (52)

P. laccadivensis (Alcock & Anderson) [*Palaemonella laccadivensis* Alcock & Anderson, 1894]: Northern South China Sea continental shelf. (52)

P. lanipes Kemp: Zhongsha (=Macclesfield Bank). Shallow water. (52)

P. nilandensis Borraadaile: Northern South China Sea continental shelf. (52)

P. ornatus Bruce: Hong Kong. Commensal with sea anemones. (52, 53)

P. paraparvus Bruce: Northern South China Sea continental shelf. (52)

P. pertubans Bruce: Hong Kong. Commensal with corals. (52, 55)

P. sinensis Bruce: Hong Kong. (52, 53)

P. soror Nobili: Hong Kong. Commensal with echinoderms. (52, 55)

P. spiniferus de Man: Central South China Sea. (52)

P. tenuipes Borradaile: Hainan shoreline. (33)

P. toloensis Bruce: Hong Kong. Commensal with gorgonians. (52, 53)

P. tosaensis Kubo: Western East China Sea; Taiwan Strait continental shelf. (13, 31)

Periclimenoides odontodactylus Bruce: Hong Kong. (55)

Pontonia okai Kemp: Northern South China Sea. Shallow water. (52)

Vir orientalis (Balls) [*Palaemonella orientalis* Dana, 1852]: Zhongsha (=Macclesfield Bank). (52)

PROCESSOIDEA
PROCESSOIDAE

Nikoides sibogae de Man: East China Sea; northern South China Sea continental shelf. (13, 33)

Processa aequimana (Paulson) [*Nike aequimana*]: Hong Kong. (55)

P. demani Hayashi: Hong Kong. (55)

P. japonica (de Haan) [*Nike japonica*]: East China Sea; South China Sea. Shallow water. (13, 33)

P. sulcata Hayashi: Hong Kong. (55)

ALPHEOIDEA
ALPHEIDAE

Alpheus acutocarinatus de Man: Continental shelf off Guangdong and Guangxi. (31, 47)

A. alcyone de Man: Hainan. Shallow sea. (27, 47)

A. avarus Fabricius: Northern South China Sea. Shallow sea. (47)

A. bannerorum Bruce: Gaungdong; Hong Kong. Shallow sea. (55)

A. bidens (Olivier) [*Palaemon bidens*]: Guangdong; Hong Kong. Shallow sea. (26, 47)

A. bisincisus de Haan: Fujian; Guangdong, Hainan. Shallow sea. (26, 55, 72)

A. brevicristatus de Haan: Bohai; Yellow Sea; East China Sea; northern South China Sea. Shallow sea. (14, 18, 21, 23, 24, 35, 36, 45, 55, 72)

A. brevipes Stimpson: Northern South China Sea. Shallow sea. (27)

A. bucephalus Coutiere: Hainan; South China Sea; Xisha (=Paracel Islands). Shallow sea. (27, 47)

A. canaliculatus Banner & Banner: Guangdong; Hainan. Shallow sea. (47)

A. chiragricus H. Milne-Edwards: Guangdong; Guangxi; Hainan. Shallow sea. (31, 47)

A. collumianus Stimpson [*A. seurati* Coutiere, 1905; *A. collumianus medius* Banner, 1956]: Hainan; Xisha (=Paracel Islands). Coral reefs. (26, 27, 47)

A. diadema Dana: Xisha (=Paracel Islands). Coral reefs. (27, 47)

A. distinguendus de Man [*Crangon heterocarpus* Yu, 1935b]: Bohai; East China Sea; Taiwan Strait. Shallow water. (13, 14, 18, 21, 24, 35, 36, 45, 55, 72)

A. edamensis de Man: Guangdong; Xisha (=Paracel Islands). Shallow sea. (31, 47)

A. edwardsii (Audouin) [*Athanas edwardsii* Audouin, 1827]: Northern South China Sea. Intertidal and shallow sea. (26, 47, 55)

A. facetus de Man: Guangdong; Hainan. Shallow sea. (31, 47)

A. frontalis H. Milne-Edwards: Guangdong; Hainan. Shallow sea. (26, 47)

A. gracilipes Stimpson: Hainan. Commensal with corals. (26, 27, 47)

A. hippothoe de Man: Hong Kong; Hainan; northern South China Sea; Xisha (=Paracel Islands). (47, 55)

A. hoplocheles Coutiere: China's shoreline. Shallow sea. (14, 21, 36)

A. inopinatus Holthuid & Gottlidb [synonymous with *A. serenei* Tiwari, Banner & Banner, 1982; a valid species, Chace, 1988]: Northern South China Sea continental shelf. (31, 47)

A. japonicus Miers: China's shoreline. Shallow sea. (13, 14, 21, 27, 28, 72)

A. leptocheles Banner & Banner: Guangdong. Shallow sea. (31)

A. leviusculus Dana: Continental shelf off Guangdong. (27, 47)

A. lobidens de Haan: Guangdong; Hainan. Intertidal and shallow sea. (26, 47)

A. lottini Guerin [*A. ventrosus* H. Milne-Edwards, 1937]: Hainan; South China Sea; Xisha (=Paracel Islands). Commensal with corals. (27, 47, 78)

A. macroskeles Alcock & Anderson: Northern South China Sea continental shelf. (26, 47)

A. malabaricus leptopus de Man: Guangdong. Shallow sea. (26, 31, 47)

A. malleodigitus (Bate) [*Betaeus malleodigitus* Bate, 1888]: Hainan; Xisha (=Paracel Islands). Coral reefs. (26, 27, 47, 55)

A. microstylus (Bate) [*Betaeus microstylus* Bate, 1888]: Hainan; northern South China Sea; Xisha (=Paracel Islands). Coral reefs. (26)

A. mites Dana: Hainan shoreline. (26)

A. nonalter Kensley: Northern South China Sea continental shelf. (26, 47)

A. obesomanus Dana: Northern South China Sea continental shelf. (26, 47)

A. pachychirus Stimpson: Xisha (=Paracel Islands). Coral reefs. (26, 47)

A. pacificus Dana: Guangdong; Hainan. Shoreline shallow sea. (26, 47)

A. paralcyone Coutiere: Guangdong; Hainan; Xisha (=Paracel Islands). Shoreline shallow sea. (26, 47, 55)

A. pareuchirus Coutiere: Guangdong. Shoreline shallow sea. (31, 47, 55)

A. parvirostris Dana: Hainan; Xisha (=Paracel Islands). Shoreline shallow sea. (27, 33, 74)

A. parvus de Man: Northern South China Sea. Shallow water. (47)

A. polyxo de Man: Xisha (=Paracel Islands). Coral reefs. (47)

A. pubenscens de Man: Northern South China Sea continental shelf. (47)

A. pustulosus Banner & Banner: Guangdong. Shoreline shallow sea. (46, 47, 71)

A. rapacida de Man: Guangdong; Hong Kong. Shoreline shallow sea. (31, 47)

A. rapax Fabricius: Guangdong; Hong Kong. Shallow sea. (31, 47)

A. sereni Tiwari: Hainan shallow sea. (31, 47)

A. sibogae de Man: Guangdong; Guangxi. Shallow sea. (31, 47)

A. spatulatus Banner & Banner: Northern South China Sea continental shelf. (46)

A. splendidus Coutiere: Northern South China Sea continental shelf. (47)

A. stanley dearmarus de Man: Guangdong; Xisha (=Paracel Islands). Shallow water. (27, 48)

A. sudara Banner & Banner: Northern South China Sea. Coral reefs. (26, 48)

A. tenuicarpus de Man: Northern South China Sea shallow water. (26, 27)

A. tirmiziae Kasmi: Hong Kong; northern South China Sea. (48)

A. xishaensis Liu & Lan: Xisha (=Paracel Islands). Coral reefs. (27)

Athanas areteformis Coutiere: Northern South China Sea shallow water. (47)

A. dimorphus Ortmann: Hong Kong. (47, 55)

A. djiboutensis Coutiere: Guangdong; Hong Kong. Near shore. (47)

A. dorsalis (Stimpson) [*Arete dorsalis* Stimpson, 1860]: Guangdong; Hong Kong; Hainan. Shoreline, commensal with echinoderms. (26, 48, 55)

A. gracilipes Banner & Banner: South China Sea continental slope. (47)

A. hongkongensis Bruce: Hong Kong. (55)

A. japonicus Kubo [*A. lamelifer* Kubo, 1940]: Southern Yellow Sea; East China Sea; northern South China Sea. Shallow water. (33)

A. ornithorhynchus Banner & Banner: Hong Kong. (55)

A. oshimai Kubo: Southern Yellow Sea; East China Sea; northern South China Sea shallow water. (26)

A. sibogae de Man [*A. parvus* de Man, 1910]: Hong Kong. (55)

Automate anacanthopus de Man: Guangdong; Hong Kong. Shallow sea. (26, 47)

A. dolichognatha de Man [*A. gardineri* Coutiere]: Northern South China Sea shallow water. (26)

Betaeus granulimanus Yokoya [*B. yokoyai* Kubo]: Western East China Sea; northern South China Sea. Shallow water. (55)

Prionalpheus mortoni Bruce: Hong Kong. (55)

P. triarticulatus Banner & Banner: Hong Kong. (55)

Salmoneus serratidigitus (Coutiere) [*Jousseaumia serratidigitus*]: Hong Kong. Water depth 6 m. (55)

S. sibogae (de Man) [*Jousseaumia sibogae*]: Hong Kong. (47)

Synalpheus bispinosus de Man: Northern South China Sea continental shelf. (33)

S. carinatus (de Man) [*Alpheus carinatus* de Man, 1888]: Northern South China Sea continental shelf. (47)

S. charon (Heller) [*Alpheus charon* Heller, 1861]: South China Sea continental shelf. (47)

S. coutiere Banner: Northern South China Sea, coral reefs, Hong Kong. (47, 55)

S. demani Borradaile: Hainan; northern South China Sea. (26)

S. gravieri Coutiere: Northern South China Sea. Shallow water. (26)

S. hastilicrassus Coutiere: Northern South China Sea. Shallow water. (33, 47, 55)

S. iocosta de Man: Hong Kong; northern South China Sea. (33, 47, 55)

S. neomeris (de Man) [*Alpheus neomeris* de Man, 1897]: Northern South China Sea continental shelf. (33, 47, 55)

S. neptunus (Dana) [*Alpheus neptunus* Dana, 1852]: Hong Kong; northern South China Sea. Shallow sea. (26, 47)

S. nilandensis Coutiere: Northern South China Sea continental shelf. (47)

S. odontophorus de Man: Northern South China Sea continental shelf. (26, 47)

S. paraneomeris Coutiere: Northern South China Sea continental shelf. (33, 47)

S. pescadorensis Coutiere: Taiwan Strait; northern South China Sea. Shallow water. (33, 47)

S. stimpsoni (de Man) [*Alpheus stimpsoni* de Man, 1888]: Northern South China Sea continental shelf. (33, 47)

S. stormi de Man: Northern South China Sea continental shelf. (33, 47)

S. streptodactylus Coutiere: Hong Kong; northern South China Sea. (47, 55)

S. theano de Man: South China Sea continental shelf. (26)

S. trispinosus de Man: Western East China Sea; South China Sea continental shelf. (47, 55)

S. tumidomanus (Paulson) [*Alpheus tumidomanus* Paulson, 1875]: South China Sea. Coral reefs. (47)

OGYRIDIDAE

Ogyrides orientalis (Stimpson) [*Ogyris orientalis*]: China's shoreline. Shallow water. (14, 21, 36, 47, 54)

O. striacticauda Kemp: Yellow Sea; East China Sea; northern South China Sea. Intertidal and shallow sea. (25, 36)

HIPPOLYTIDAE

Birulia kishinouyei (Yokoya) [*Paraspirontocaris kishinouyei*]: Yellow Sea. (24, 31)

Eualus gracilirostris (Stimpson) [*Hippolyte gracilirostris* Stimpson, 1860]: Yellow Sea. Deep water. (24, 37)

E. leptognathus (Stimpson) [*Hippolyte leptognathus* Stimpson, 1860]: Yellow Sea. Shallow water. (21)

E. sinensis (Yu) [*Spirontocaris sinensis* Yu, 1930]: Yellow Sea. Shoreline shallow water. (21, 72)

E. spathulirostris (Yokoya) [*Spirontocaris spathulirostris*]: Yellow Sea. Deep water. (24, 31, 32)

Exhippolysmata ensirostris (Kemp) [*Hippolysmata ensirostris* Kemp]: Western East China Sea; northern South China Sea. Shallow sea. (14, 18, 24, 33, 35, 36)

Heptacarpus camptschaticus (Stimpson) [*Hippolyte camtschaticus* Stimpson, 1860]: Yellow Sea. Deep water. (24, 31, 72)

H. futilirostris (Bate) [*Nauticaris futilirostris* Bate, 1888]: Yellow Sea; East China Sea. Shoreline shallow water. (14, 21, 36)

H. geniculatus (Stimpson) [*Hippolyte geniculatus* Stimpson, 1860]: Yellow Sea; East China Sea. Shoreline shallow water, often attached to seaweed. (21)

H. pandaloides (Stimpson) [*Hippolyte pandaloides* Stimpson, 1860]: Yellow Sea; East China Sea. Shoreline shallow water, often attached to seaweed. (21)

H. rectirostris (Stimpson) [*Hippolyte rectirostris, Spirontocens rectirostris*]: Yellow Sea. Deep water. (36)

Hippolyte ventricosa H. Milne-Edwards: Guangdong; Hainan; Xisha (=Paracel Islands). Attached to seaweed. (55)

Latreutes anoplonyx Kemp: From Bohai to northern South China Sea. Shoreline shallow water, often attached to mouthparts of jellyfish. (21, 36)

L. laminirostris Ortmann: Yellow Sea; East China Sea. Shoreline shallow water, often attached to seaweed. (21, 36)

L. mucronatus Stimpson: Northern South China Sea. Shoreline shallow water. (15, 36)

L. planirostris (de Haan) [*Cyclorhynchus planirostris, Hippolyte planirostris*]: Bohai; northern East China Sea. Shoreline shallow water. (14, 21, 36)

Lysmata kuekenthali (de Man) [*Hippolyte keukenthali* de Man]: Northern South China Sea shallow water. (33)

L. vittata (Stimpson) [*Hippolysmata vittata*]: China's shoreline. Shallow water, often attached to seaweed. (21, 36)

Merhippolyte calmani Kemp & Sewell: South China Sea continental slope. (31, 46)

Saron mammoratus (Olivier): Hainan shoreline. (26)

Spirontocaris crassirostris Kubo: Yellow Sea. Deep water. (31, 33)

S. pectinifera (Stimpson) [*Hippolyte pectinifera*]: Yellow Sea. Deep water. (31)

Thor amboinensis (de Man) [*Hippolyte emboinensis*]: Hong Kong; northern South China Sea. Shallow sea, symbiotic with sea anemones. (33)

T. paschalis [*Hippolyte paschalis*]: Guangdong; Hong Kong; Guangxi; Hainan. Shoreline shallow water, often attached to seaweed. (33, 55)

Tozeuma armatum Paulson: East China Sea; South China Sea. Shoreline shallow water, planktonic. (36)

T. lanceolatum Stimpson: East China Sea; South China Sea. Shallow water, planktonic. (13, 18, 31, 36, 54)

T. tomentosum (Baker) [*Angasia tomentosa*]: East China Sea; South China Sea. Shallow water, benthic. (31, 33)

PANDALOIDA
PANDALIDAE

Chlorotocella gracilis Balss: Western East China Sea; northern South China Sea. Shoreline shallow water, planktonic. (13, 14, 35, 36, 55)

Chlorotocoides spinicauda (de Man) [*Chlorotocus spinicauda*]: Northern South China Sea continental shelf. (33)

Chlorotocus crassicornis (Costa) [*Pandalus crassicornis*]: Northern South China Sea continental shelf. (33)

C. incertus Bate: Northern South China Sea continental shelf. (42)

Heterocarpoides laevicarina (Bate) [*Dorodotes laevicarina*]: East China Sea; South China Sea continental shelf. (13, 14, 24, 36, 45)

Heterocarpus alphonsi Bate: Northern South China Sea continental slope. Water depth 500–1,040 m. (31, 41, 43)

H. dorsalis Bate: East China Sea; Taiwan; northern South China Sea continental slope. Water depth 350–949 m. (6, 19)

H. gibbosus Bate: East China Sea; Taiwan; northern South China Sea continental slope.

H. laevigatus Bate: Taiwan; South China Sea continental slope. (19, 41, 44)

H. parvispina de Man: Taiwan. Deep sea. (19)

H. sibogae de Man: East China Sea; South China Sea continental slope. Water depth 250–740 m. (6, 19, 41, 43, 44)

H. tricarinatus Alcock & Anderson: Southern Taiwan, water depth 1,139–1,259 m; South China Sea continental slope, water depth 360–940 m. (19, 41)

H. woodmasoni Alcock: South China Sea continental shelf and slope. Water depth 100- 749 m. (19, 31, 40, 41, 42, 43)

Pandalus meridionalis Balss: Bohai; Yellow Sea. Deep water. (24, 31)

Parapandalus zurstrasseni Balss: Central western South China Sea continental slope. (40)

Plesionika alcocki (Anderson) [*Pandalus alcocki*]: Northern South China Sea. (31)

P. bifurca Alcock & Anderson: Northern South China Sea continental slope. Water depth 350–390 m. (41)

P. binoculus (Bate) [*Nothocaris binoculus* Bate, 1888]: Northern South China Sea continental shelf. (31)

P. crosnieri Chan, Lee & Yu: Southern Taiwan. Deep sea. (63)

P. dentirostris Tung, Wang & Li: East China Sea. Deep water. (6)

P. edwardsii (Brandt): Southern Taiwan. Deep water. (63)

P. ensis (A. Milne-Edwards) [*Hippolyte ensis*]: Northern South China Sea continental slope. (41)

P. indica de Man: Northern South China Sea continental slope. Water depth 250–540 m. (41, 44)

P. izumiae Omori: East China Sea continental shelf. (13, 31, 36)

P. martia (A. Milne-Edwards) [*Pandalus martia*]: East China Sea; South China Sea continental slope. (31, 41, 43, 44)

P. ortmanni Doflein: East China Sea. Deep water. (31)

P. semilaevis Bate: South Chian Sea continental slope. (41, 44)

P. sindoi (Rathbun) [*Pandalus sindoi*]: South China Sea continental slope. Water depth 350–390 m. (41)

P. unidens Bate: Northern South China Sea continental slope. (31)

P. yui Chan & Crosnier: South China Sea continental shlef. (64)

THALASOCARIDIDAE

Thalasocaris crinita Dana: Northern South China Sea continental shelf. (31)

CRANGONOIDEA
CRANGONIDAE

Crangon affinis de Haan [*C. hakodatei* Rathbun, 1902]: Yellow Sea, deep water; northwestern East China Sea continental shelf. (21, 24, 26)

C. cassiope de Man [*C. crangon*]: Yellow Sea; East China Sea. Around river mouths. (21, 31, 36)

Metacrangon sinensis Fujino & Miyaka: Yellow Sea. Deep water. (33)

Paracrangon abei Kubo: Yellow Sea. Deep water. (33)

Philocheras lowisi (Kemp) [*Crangon lowisi* Demp, 1916]: Hong Kong. (55)

Pontocaris pennata Bate: Fujian; northern South China Sea continental shelf. (14, 31)

P. rathbunae Doflein [*Aegion rathbun*]: East China Sea continental shelf. (31)

P. sibogae (de Man) [*Aegion sibogae* de Man]: South China Sea continental shelf. (31)

Pontophilus angustirostris Kemp: South China Sea continental shelf. (31)

P. bidentatus (de Haan) [*Crangon bidentatus* de Haan]: Yellow Sea. Deep water. (31)

P. carinicauda Stimpson: Hong Kong. (33)

P. incisus Kemp: Fujian; Taiwan Strait. (31)

P. parvirostris Kemp: Taiwan Strait; northern South China Sea continental shelf. (31)

GLYPHOCRANGONIDAE

Glyphocrangon granulosis Bate: South China Sea continental slope. Water depth 860–940 m. (41, 44)

G. hastacauda Bate: East China Sea; South China Sea continental slope. Water depth 510–840 m. (6, 41, 44)

G. pugnax de Man: South China Sea continental shelf. Water depth 500–549 m. (41, 44)

G. regalis Bate: East China Sea; South China Sea continental shelf. Water depth 560–740 m. (6, 21, 41)

ASTACIDEA
NEPHROPSIDEA
NEPHROPSIDAE

Acanthacaris tenuimana (Bate) [*Phoberus tenuimanus* Bate, 1888; *P. brevirostris* Tung & Li, 1985]: East China Sea; South China Sea continental slope. Water depth 560–890 m. (11, 71)

Metanephrops andamanicus Wood-Mason [*Nephrops andamanicus* Wood-Mason, 1891]: Northern South China Sea continental slope. Water depth 260–590 m. (31, 71)

M. armatus Chan & Yu: Southeastern Taiwan. Water depth 200–450 m. (71)

M. formosanus Chan & Yu: Taiwan. Water depth 150–450 m. (31)

M. japonicus (Tepparone-Canefri) [*Nephrops japonicus*]: East China Sea continental slope. (31, 49, 71)

M. neptunus (Bruce): Northern South China Sea continental slope. Water depth 510–570 m. (41, 71)

M. sagamiensis (Parisi): Northern South China Sea deep water. (59, 71)

M. sinensis (Bruce): Northwestern South China Sea continental slope. Water depth 210–740 m. (52, 59, 71)

M. thompsoni (Bate): Southeastern Yellow Sea; East China Sea and northern South China Sea, continental shelf and slope. (2, 3, 31, 36, 41, 51)

Nephropsis carpenteri Wood-Mason: South China Sea continental slope. Water depth 510–590 m. (41, 44, 71)

N. stewarti Wood-Mason: East China Sea; Taiwan; South China Sea continental slope. Water depth 360–749 m. (6, 31, 41, 44, 63a, 71)

N. sulcata Macpherson: Northern South China Sea continental slope. (41)

THAUMASTOCHELIDAE

Thaumastocheles japonicus Calman: Eastern Taiwan, water depth 300–700 m; South China Sea continental slope. (41, 44, 63a)

THALASSINIDEA
THALASSINOIDEA
AXIIDAE

Axiopsis harbereri (Balss) [*Axius harbereri*]: Western East China Sea continental shelf. (31)

A. serratifrons (H. Milne-Edwards) [*Axius serratifrons*]: Northern South China Sea continental shelf. (31)

Calocois japonicus (Parisi) [*Oryrhynchaxius japonicus* (Parisi)]: East China Sea continental shelf. (31)

CALLIANASSIDAE

Callianassa japonica Ortmann [*C. harmandi* Bouvier]: Bohai; Yellow Sea. Shoreline. (21, 31, 71)

C. joculatrix de Man: Northern South China Sea. (13, 31)

C. modesta de Man: Northern South China Sea. Shallow water. (31)

C. petalura Stimpson: Bohai; Yellow Sea. Shallow water. (21, 71)

Callianidea typa H. Milne-Edwards: Northern South China Sea continental shelf. (31)

Ctenocheles balssi Kishinouye: East China Sea; South China Sea continental shelf. (31)

LAOMEDIIDAE

Laomedia astacina de Haan: Yellow Sea; East China Sea; northern South China Sea. Intertidal. (31)

THALASSINIDAE

Thalassina anomalna (Herbst) [*Cancer (Astacus) anomalus* Herbst, 1804]: Hainan. Burrowing in muddy banks. (33)

UPOGEBIIDAE

Upogebia barbata (Strahl) [*Gebia barbata*]: South China Sea. Shallow water. (31, 75a)

U. carinicauda (Stimpson) [*Gebia carinicauda*]: South China Sea. Shallow water. (31, 75a)

U. darwinii (Miers) [*Gebiopsis darwini*]: South China Sea. Shallow water. (31, 75a)

U. imperfecta Sakai: Yellow Sea. Shallow water. (31, 75a)

U. issaeffi (Balss) [*Gebia issaeffi*]: Yellow Sea. Shallow water. (31, 75a)

U. major (de Haan) [*Gebia major* de Haan]: Bohai; Yellow Sea. Intertidal and shallow sea. (21, 75a)

U. shenjiajuiii Yu: Yellow Sea. Shallow water. (21, 75a)

U. spinifrons (Haswell) [*Gebia spinifrons*]: South China Sea. Shallow water. (31, 75a)

U. wuhsienweni Yu: Yellow Sea; East China Sea. Intertidal and shallow sea. (21, 75a)

PALINURIDEA
ERYONIDEA
ERYONIDAE

Polycheles baccatus Bate: East China Sea. Water depth about 300–915 m. (6, 61, 63a)

P. enthrix (Bate): Taiwan. Water depth 500–1,153 m. (63a)

P. typholps Heller: East China Sea; Taiwan; South China Sea continental slope. Water depth 360–640 m. (41, 44, 61, 63a)

Stereomastis andamanensis Alcock: East China Sea; South China Sea continental slope. Water depth 350–940 m. (6, 41, 44)

S. phospherus (Alcock): South China Sea continental slope. Water depth 560–940 m. (41, 44)

S. sculpta (Smith): South China Sea continental slope. Water depth 900–940 m. (41, 44)

PALINUROIDEA
PALINURIDAE

Justitia japonica (Kubo) [*Napalirus japonicus*]: Eastern Taiwan. (63a)

J. longimanus (H. Milne-Edwards) [*Palinurus longimanus* H. Milne-Edwards, 1837]: Eastern Taiwan. Water depth 50–150 m. (63a, 64)

Linuparus sordidus Bruce: Taiwan; South China Sea continental slope. Water depth 260–640 m. (41, 44, 50, 63a, 71)

L. trigonus (Von Siebold) [*Palinurus trigonus*]: East China Sea; South China Sea continental slope. Water depth 260–390 m. (23, 24, 33, 35, 36, 41, 44, 63a, 71, 72)

Palinustus waguensis Kubo: Eastern Taiwan. Water depth 180 m. (63a)

Panulirus homarus (Linnaeus) [*Cancer homarus* Linnaeus, 1758; *Palinurus homarus* Fabricius, 1798; *P. dasypus* H. Milne-Edwards, 1837; *P. burgeri* de Haan, 1841]: Taiwan; South China Sea. Shoreline shallow sea. (16, 33, 63a, 71)

P. japonicus (Von Siebold) [*Palinurus japonicus*]: East China Sea; Taiwan; eastern South China Sea. Shoreline shallow sea. (4, 16, 33, 63a, 71)

P. longipes (H. Milne-Edwards) [*Palinurus longipes*]: Taiwan; northern South China Sea. Shallow sea. (4, 16, 33, 63a, 71)

P. ornatus (Fabricius) [*Palinurus ornatus*]: East China Sea; Taiwan; South China Sea. Shallow sea. (4, 16, 35, 36, 63a, 71)

P. penicillatus (Olivier) [*Astacus penicillatus*]: Taiwan; South China Sea; Xisha (=Paracel Islands). Shallow sea. (4, 16, 35, 36, 63a, 71)

P. polyphagus (Herbst) [*Cancer polyphagus*]: Taiwan; South China Sea continental shelf; Xisha (=Paracel Islands); Nansha (=Spratly Islands). (16, 33, 63a, 71)

P. stimpsoni Holthuis: East China Sea; South China Sea. Shoreline shallow water. (16, 33, 35, 37, 63a, 71)

P. versicolor (Latreille) [*Palinurus versicolor*]: Hainan; Xisha (=Paracel Islands). Shoreline shallow sea. (16, 33, 37, 63a, 71)

Puerulus angulatus (Bate) [*Panulirus angulatus* Bate, 1888]: Taiwan; South China Sea. Continental shelf and slope. (4, 33, 41, 60, 63a, 71)

SCYLLARIDAE

Ibacus ciliatus (Von Siebold) [*Scyllarus ciliatus* Von Siebold, 1824]: East China Sea; South China Sea continental shelf. (14, 23, 24, 35, 36, 63a, 71, 72)

I. novemdentatus Gibbes: East China Sea; South China Sea. (14, 35, 36, 63a, 71, 72)

Parribacus antarcticus (Lund) [*Scyllarus antarcticus* Lund]: Taiwan; Hainan; Xisha (=Paracel Islands); Nansha (=Spratly Islands). Rocky shore. (63a)

P. japonicus Holthuis: Taiwan. Rocky shore. (71)

Scyllarides haanii (de Haan) [*Scyllarus haanii*]: East China Sea continental shelf. (33)

S. squamosus (H. Milne-Edwards) [*Scyllarus squamosus*]: Taiwan; South China Sea. 72, 86)

Scyllarus batei Holthuis: East China Sea; Taiwan; South China Sea continental shelf. (59, 63a, 71)

S. bertholdi Paulson: Taiwan; South China Sea continental shelf. (59, 63a, 71)

S. brevicornis Holthuis: East China Sea; Taiwan. (63a, 71)

S. cultrifer (Ortmann): South China Sea continental shelf. (37, 63a, 71)

S. formosanus Chan & Yu: Eastern Taiwan. Sandy substrate, water depth 250 m. (63a)

S. kitanoviriosus Harada: Taiwan; northern South China Sea. Shoreline. (59, 63a)

S. longidactylus Harada: Northern South China Sea continental shelf. (59, 63a)

S. martensii Pfeffer: East China Sea; Taiwan; South China Sea. Shallow sea. (31, 36, 63a, 71, 72)

S. rugosus H. Milne-Edwards: Taiwan; South China Sea. Shoreline. (59, 63a)

S. sordidus (Stimpson) [*Arctus sordidus*]: Northern South China Sea continental shelf. (31, 71)

S. tuberculatus Bate: Northern South China Sea continental shelf. (31)

REFERENCES*

(1). Cai Bingji, 1988. Distribution of *Lucifer* in western Taiwan Strait. Journal of Marine Science Bulletin, 7(4):60–65.

(2). Cai, B. J. and Cheng, Z. 1965. Studies on taxonomy of *Lucifer* in southeast coastal waters of China. Journal of Xiamen University (Natural Science), 12(2):111–122

(3). Chen, T. R., 1990. Crayfishes of Taiwan. Journal of Aquatic Ecology, 22:78–80.

(4). Chen, T. R., 1990. Lobsters of Taiwan. Journal of Aquatic Ecology, 24:90–93.

(5). Chen, T. R. and Chu, J. H., 1991. Lobsters of Hong Kong. China Fishery Monthly 466:13–24.

(6). Dong Yumao, 1988. Reports on Comprehensive survey of deep-sea crustaceans on the outer edge of continental shelf and deep-sea fishing ground on continental slope of the East China Sea. Zhejiang Science & Technology Press, pp: 1–132.

(7). Dong Yumao et al., 1984. Species composition and fauna of crustaceans on the outer edge of continental shelf and on continental slope of the East China Sea. In: Reports on comprehensive survey of the outer edge of continental shelf and deep- sea fishing ground on continental slope of the East China Sea.

(8). Dong Yumao and Hu Yuying, 1980. Reports on nektonic shrimps in the coastal waters of Zhengjiang, II. Journal of Zoology, 1930 (2):20–24.

(9). Dong Y. M. et al., 1958. Reports on crawling shrimps in the coastal waters of Zhejiang. Journal of Zoology, 2(3):166–170

(10). Dong Yumao et al., 1986. Reports on nektonic shrimps in the coastal waters of Zhejiang. Acta Zootaxonomica Sinica, 11(5):4–6.

(11). Dong Yumao, Wang Baoyong and Li Zhicheng, 1985. A new species of Nephropsicea form the deep waters of the East China Sea. Acta Zootaxonomica Sinica, 10(4):379–380.

(12). Dong Y. M. et al., 1959. Reports on nektonic shrimps in the coastal waters of Zhejiang. Journal of Zoology, 3(9):380–394.

(13). Fujian Institute of Oceanography, 1988. A comprehensive oceanographic survey of the central and northern part of the Taiwan Strait. China Science Press, 423pp.

(14). Shrimp resources group of Fujian, 1986. Investigation reports on exploring and fishing shrimp resources of Fujian sea area. In: Research reports on exploring and fishing shrimp resources and fishing technology in Fujian sea area, pp:6–54.

(15). State Oceanic Administration, 1988. A comprehensive environmental resource survey of the central part of the South China Sea. China Ocean Press, 249pp.

(16). He, Y. D. and You, S. P., 1979. Studies on lobsters of Taiwan. Science of Taiwan Provincial Museum, 22:97–133.

(17). Huang, J. J. and You, S. P., 1982. Studies on freshwater palaemonideans of Taiwan. Science of Taiwan Provincial Museum, 25:157–180.

(18). Huang Zongguo, 1990. Ecological studies on shrimps in waters near Daya Bay Nuclear Power Station. Collections of Papers on Marine Ecology in the Daya Bay, II. China Ocean Press, pp:413–421.

(19). Lee, D. A., 1990. Studies on deep-sea shrimps of Taiwan. Experimental reports of Fisheries Institute of Taiwan Province, 49:125–138.

(20). Lee, D. A. and You, S. P., 1977. Prawns of Taiwan, 110pp.

(21). Liu, R. Y., 1955. Economic shrimps in northern China. China Science Press, 73pp, 22 pls.

(22). Liu, J. Y. 1956. Notes on two species of Acetes of the Family Sergestidae (Crustacea: Decapoda) from the coasts of North China. Acta Zoologica Sinica, 8(1):29–42.

(23). Liu, J. Y. 1959. Notes on the econommic Macrurous Crustacean fauna of the Yellow Sea and the East China Sea. Oceanologia & Limnologia Sinica, 2(1):35–42.

(24). Liu, J. Y. 1963. Zoogeographical studies on the Macrurous crustacean fauna of the Yellow Sea and the East China Sea. Oceanologia & Limnologia Sinica, 5(3):230- 244.

(25). Liu, R. Y., 1965. A study on Acetes off China. In: Transactions of the Chinese Society of Zoology at the 30th Anniversary, pp:1–131.

(26). Liu, R. Y. Studies on *Alpheus* of Hainan Island. Studia Marina Sinica, in press.

(27). Liu, R. Y. and Lan, J. Y., 1980. A preliminary study on *Alpheus* around Xisha Islands. Studia Marina Sinica, 17:77–116.

(28). Liu Ruiyu, Liang Xiangqiu and Yan Shengliang, 1990. A study of the Palaemoninae (Crustacea: Decapoda) from China, I. *Macrobrachium, Leander* and *Leandrites*. Transactions of the Chinese Crustacean Society, 2:102–134.

(29). Liu Ruiyu, Liang Xiangqiu and Yan Shengling, 1991. A study of the Palaemoninae (Crustacea: Decapoda) from China, II. *Palemon, Expalaemon, Palaemon* and *Leptoearpus*. Studia Marina Sinica, 31:229–267.

(30). Liu Ruiyu and Wang Yongliang, 1987. Studies on Chinese species of the Genus *Parapen* Aeopsis (Decapoda, Crustacea). Oceanologia & Limnologia Sinica, 18(6):523–529.

(31). Liu Ruiyu et al., 1994. Studies on offshore benthics of China. In: Comprehensive investigation report on China Seas (IV). China Ocean Press.

(32). Liu Ruiyu and Zhong Zhenru, 1988. Prawns in the South China Sea. China Agriculture Press, 278pp, 6 pls.

(33). Liu Ruiyu. Distribution of some shrimps in China seas. Studia Marina Sinica, in press.

(34). Song Haitang et al., 1991. Reports on shrimp resources in nearshore outer waters of the East China Sea, 150pp.

(35). Wang Yihao, 1987. Notes on the shrimp and lobster fauna of the Zhoushan Archipelago Waters. Oceenologia & Limnogia Sinica, 18(1):48–53.

(36). Wei Chongde and Chen Yongshou (ed.), 1991. Fauna of Zhejiang (Crustacea), Zhejiang Science & Technology Press, 481pp.

(37). Wu, C. C., et al., 1989. Resources research on gastropods and shrimps of bottom- trawling fishing ground in northern Taiwan. In: Experimental Reports on Taiwan Fisheries, 46pp.

(38). Cheng, C. 1953. Studies on marine planktonic crustaceans in Amoy, I. *Acetes*. Journal of Xiamen University, 3(2): 29–44.

(39). Cheng, C. 1954. Studies on marine planktonic crustaceans in Amoy, II. *Lucifer*. Journal of Xiamen University (Natural Science), 4(3): 1–12.

(40). Zhong Huitao and Lan Jinyan, 1983. Studies on the deep-water shrimps from northeastern and central parts of the South China Sea. Contributions on Marine Biological Research of the South China Sea, 1:173–191.

(41). Zhong Zhenru, 1989. Studies on shrimps in the outer waters of continental slope of South China Sea. In: Collections of papers on fisheries of the South China Sea (I). Guangdong Science & Technology Press, 33pp.

(42). Zhong Zhenru. New records of shrimps in the South China Sea (Crutacea: Decapoda). (manuscript)

(43). Zhong Zhenru, Jiang Jiyang and Min Xinai, 1979. Investigation reports on shrimps on the edge of continental shelf of northern South China Sea. In: Bottom trawling fishing resources investigation in northern continental shelf sea waters on the south China sea (1978.2–1979.1) (I).

(44). Zhong Zhenru, Jiang Jiyang and Min Xinai, 1981. Shrimps. In: Comprehensive Survey Report on the Fisheries Resources in the Waters on Continental Slope in Northern South China Sea (Chapter 8).

(45). Zhong Zhenru, Jiang Jiyang and Min Xinai, 1982. Investigation reports on shrimp resources in the near-shore of northern South China Sea. Research reports by Fisheries Institute of South China Sea, State Fisheries Administration, 37pp.

(46). Banner, D. M. and A. H. Banner, 1968. Three new species of the genus *Alpheus* (Decapoda: Alpheidae) from the International Indian Ocean Expedition, Crustaceana, 15(2):141–148. (For correction of the title, see Crustaceana, 16(2):207

(47). Banner, D. M. and A. H. Banner, 1978. Annotated Checklist of Alpheid and Ogyridid Shrimp from the Philippine Archipelago and the South China Sea. Micronesika 14(2):215–257.

(48). Banner, D. M. and A. H. Banner, 1982. The Alpheid Shrimp of Australia. Part 3. The remaining alpheid principally the genus *Alpheus* and the family Ogyrididae. Rec. Austr. Mus., 34(1, 2):1–362.

(49). Bruce, A. J. 1965a. On a new species of Nephrops (Decapoda, Retantia) from the South China Sea. Crustaceana 9(3):274–284, pls. 13–15.

(50). Bruce, A. J. 1965b. A new species of the genus *Linuparus* White from the South China Sea (Crustacea, Decapoda). Zool. 41:1–13.

(51). Bruce, A. J. 1966. *Nephrops sinensis* sp. nov., a new species of lobster from the South China Sea. Crustaceana, 10(2):155–156, pls.10–12.

(52). Bruce, A. J. 1979. Records of some Pontoniine shrimps from the South China Sea. Cahiers de Indopacifique, 1(2):215–248.

(53). Bruce, A. J. 1980. The pontoniine shrimp fauna of Hong Kong. In: Proceedings of First International Marine Biological Workshop: The Marine Flora and Fauna of Hong Kong and Southern China. Hong Kong University Press, pp:233–283.

(54). Bruce, A. J. 1986a. Redescription of five Hong Kong carideans first described by William Simpson 1860. In: Proceedings of Second International Marine Biological Workshop: The Marine Flora and Fauna of Hong Kong and South China. Hong Kong University Press, pp:569–610.

(55). Bruce, A. J. 1986b. Additions to the marine shrimp fauna of Hong Kong. In: Proceedings of Second International Marine Biological Workshop: The Marine Flora and Fauna of Hong Kong and South China, pp:611–643.

(55a). Bruce, A. J. 1992. Additions to the marine caridean fauna of Hong Kong, with a description of a new species of *Onyqcocaris* (Crustacea: Decapoda: Palaemonidae) from Turalu. In: Proceedings of the Third International Marine Biological Workshop: The Marine Flora and Fauna of Hong Kong and Southern China. Hong Kong University Press, pp:329–343.

(56). Chace, F. A. Jr. 1976. Shrimps of the Pasiphaeid genus *Leptochela* with descriptions of three new species (Crustacea: Decapoda: Caridea). Smithsonian Cont. Zool., 222:1–51.

(57). Chan, Tin-Yam and Hsiangping Yu, 1985. On the rock shrimps of the family Sicyoniidae (Crustacea: Decapoda) from Taiwan, with a description of one new species. Asian Mar. Biol., 2:93–105.

(58). Chan, Tin-Yam and Hsiangping Yu, 1985. Studies on the shrimp of the genus *Palaemon* (Crustacea: Decapoda: Palaemonidae) from Taiwan. J. Taiwan Mus., 38(1):119–128.

(59). Chan, Tin-Yam and Hsiangping Yu, 1986. A report on the *Scyllarus* lobsters (Crustacea: Decapoda: Scyllaridae) from Taiwan. J. Taiwan Mus., 39(2):147- 174.

(60). Chan, Tin-Yam and Hsiangping Yu, 1989a. A deep-sea lobster of the genus *Puerulus* (Crustacea: Decapoda: Palinuiridae). Bull. Inst. Zool., Academia Sinica, 28(1):1–6.

(61). Chan, Tin-Yam and Hsiangping Yu, 1989b. Two blind lobsters of the genus *Polycheles* (Crustacea: Decapoda: Eryonidae). Bull. Inst. Zool. Academica Sinica, 28(3):165–170.

(62). Chan, Tin-Yam and Hsiangping Yu, 1991a. *Eugonatonotus chacei* sp. nov., second species of the genus (Crustacea: Decapoda: Eugonatonotidae). Bull. Mus. nat. Hist. Paris 4(13), 1991, sec. a. nos 1–2:143–152.

(63). Chan, Tin-Yam and Hsiangping Yu, 1991b. Two similar species: *Plesionika edwardsi* (Brandt, 1851) and *Plesionika crosnieri*, new species (Crustacea: Decapoda: Pandalidae). Proc. Biol. Soc. Wash. 104(3):545–555.

(64). Chan, Tin-Yam, Dinan Lee and Hsiangping Yu, 1991. Short note: two additional lobster species (Crustacea, Mercrura, Reptantia) in Taiwan. Bull. Inst. Zool. Academia Sinica, 30(3):249–255.

(65). Crosnier, A. 1987. Les especes indo-ouest pacifiques d'eau in profounde du genre *Metapenaeopsis* (Crustacea, Decapoda, Penaeidae). Bull. Mus. nat. Hist. nat., Paris (4) 9, sec. A (2):409–453.

(66). Crosnier, A. 1988. Sur les *Heterocarpus* (Crustacea, Decapoda, Pandalida) du sud- ouest de l'Ocean Indien Remarques sur l'outres especes ouest Pacifiques du genre et description de quartr taxa nouves. Bull. Mus. nat. Hist. Paris, 4e ser. 10:57–96, pl. 1.

(67). Crosnier, A. 1991. Crustacea Decapoda: Les *Metapanenaeopsis* indo-ouest- pacifiquessans appareil stridulant (Penaeidae). Deuxieme partie. In: Crosnier, A. (ed.), Resaltatsdes Campangnes MUSORSTOM, vol. 9. Mem. Mus. nat. Hist. Nat., (A), 152:155–297.

(68). Fujino, T. and S. Miyake, 1970. Caridean and stenopodidean shrimps from the East China Sea and the Yellow Sea (Crustacea, Decapoda, Natantia). J. Fac. Agr. Kyushu University, 16(3):237–312.

(69). Hayashi, K. and S. Miyake, 1968. Studies on the hippolytid shrimps from Japan V. Hippolytid fauna of the sea around the Amakusa Marine Biological Laboratory. OHMU 1(6):121–163.

(70). Holthuis, L. B. 1980. FAO Species Catalogue. Vol. 1. Shrimps and prawns of the world. An annotated catalogue of species of interest to fisheries. FAO Fish. Synopsis, No. 125, Vol. 1, 271pp.

(71). Holthuis, L. B. 1991. FAO Species Catalogue. Vol. 13. Marine lobsters of the world. An annotated and illustrated catalogue of species of interest to fisheries known to date. FAO Fish. Synopsis, No. 125, Vol. 13, 292pp.

(72). Kim, Hooh-Soo 1977. Illustrated flora and fauna of Korea. Vol. 19. Macrlula, 414pp.

(73). Liu Ruiyui and Zhong Zhenru, 1983. On a new genus and two species of Solenocerid shrimps (Crustacea, Penaeoidea) from South China Sea. Chin. J. Ocean. Limnol. 1(2):171–176.

(74). Morton, B. 1991. Hong Kong's scleractinian coral - gallery communities. Asian Mar. Biol. 8:103–115.

(75). Omori Makoto, 1975. The systematics, biogeography and fishery of epipelagic shrimps of the genus *Acetes* (Crustacea, Decapoda, Sergestidae). Bull. Ocean Res. Inst. University of Tokyo, No.7–1:91, pl. 1.

(75a). Sakai, K. 1992. Revision of Upogebiidae (Decapoda: Thalassinidea) in the Indo West Pacific Region. Res. on Crustacea, Spec. No. 1–1:106.

(76). Teong Won-Young and Wo-Wing Cheng, 1980. The economic shrimps of Hong Kong. In: Proceedings of the First International Marine Biological Workshop: The Marine Flora and Fauna of Hong Kong and Southern China. Hong Kong University Press, pp:285–313.

(77). Yu Hsiang-Ping and Tin-Yam Chan, 1986. The Illustrated Penaeoid Prawns of Taiwan. Southern Materials, 183pp.

(78). Yu, S. C. 1936. Report on the macrurous Crustacea collected during the Hainan Biological Expedition in 1934. Chin. J. Zool. 2:85–99.

(79). Starobogatov, Y. I. 1972. Penaeidae (Family Ponasidae, Crustacea: Decapoda) Tonkinskogo Zaliva. Penaeidae of Tonking Gulf. Explor. Fauna Seas 10(18):359–415.

*: (1)-(43) in Chinese, (72) in Korean, (79) in Russian
Compiled by Liu Ruiyu and Zhong Zhenru

ANOMURA
COENOBITOIDEA
DIOGENIDAE

Aniculus aniculus (Fabricius): East China Sea; South China Sea; Taiwan. Shallow sea. (13, 59)

Calcinus elegans (H. Milne-Edwards): Taiwan; South China Sea. Intertidal. (15)

C. formosus Neumann: Taiwan. (15)

C. gaimardii (H. Milne-Edwards): Taiwan; South China Sea. Intertidal. (15)

C. laevimanus (Randall): Taiwan; South China Sea. Intertidal. (15)

C. latens (Randall): Taiwan; South China Sea. Intertidal. (15)

Clibanarius aequabilis Dana: South China Sea. Intertidal. (15, 70)

C. arethusa de Man: East China Sea; South China Sea. (15, 54)

C. bimaculatus (de Haan): South China Sea. Intertidal. (15, 40, 70)

C. clibanarius (Herbst): South China Sea. Intertidal. (15)

C. corallinus (H. Milne-Edwards): Taiwan; South China Sea. Intertidal. (15)
C. cruentatus (H. Milne-Edwards): South China Sea. Intertidal. (15)
C. eurysternus Hilgendorf: Taiwan. (15)
C. formosus Ives: Taiwan. (15)
C. humilis Dana: South China Sea. Intertidal. (15, 70)
C. inaequalis (de Haan): East China Sea; Taiwan; South China Sea. (15, 54)
C. infraspinatus Hilgendorf: From Yellow Sea to South China Sea; Taiwan. Intertidal. (3, 13, 15, 52, 70)
C. japonicus Rathbun: East China Sea. Intertidal. (15, 40)
C. longitarsus (de Haan): East China Sea; Taiwan; South China Sea. (15, 40, 70)
C. multipuctatum Wang et al.: East China Sea deep water. (8, 12, 13, 15, 23)
C. padavensis de Man: South China Sea. Intertidal. (2, 14, 15)
C. striolatus Dana: Taiwan; South China Sea. Intertidal and coral reefs. (15, 52, 59, 70)
C. virescens (Krauss): East China Sea; Taiwan; South China Sea. Intertidal. (15, 59)
Dardanus arrosor (Herbst): East China Sea; South China Sea; Taiwan. Shallow sea. (13, 15, 37, 57, 59)
D. aspersus (Berthold): East China Sea; South China Sea; Taiwan. Intertidal and shallow sea. (13, 15, 52, 59)
D. crassimanus (H. Milne-Edwards): South China Sea. (15)
D. deformis (H. Milne-Edwards): Taiwan; South China Sea. Intertidal. (15)
D. guttatus (Olivier): Taiwan; South China Sea. Shallow sea. (15)
D. hessii (Miers): Taiwan; South China Sea. Intertidal. (13, 15, 55, 59)
D. imbricatus (H. Milne-Edwards): South China Sea. Intertidal. (15)
D. impressus (de Haan) [*Pagurus impressus*]: East China Sea; South China Sea; Taiwan. Shallow sea. (13, 15, 59)
D. lagopodes (Forskål): Taiwan. Subtidal. (15, 59)
D. megistos (Herbst): Taiwan; South China Sea. Intertidal and shallow sea. (15, 34, 52, 54, 55, 57)
D. pedunculatus (Herbst): Taiwan; South China Sea. (15, 59)
D. scutellatus (H. Milne-Edwards): South China Sea. (15)
D. tinctor (Forskål): South China Sea. (15)
D. vulnerans (Thallwitz): Taiwan. (15, 34)
Diogenes avarus Heller: East China Sea; South China Sea. (4, 13, 19)
D. custas (Fabricius): South China Sea. Intertidal. (15)
D. deflectomanus Wang & Tung: From Bohai and Yellow Sea to South China Sea. Intertidal. (13, 20, 21)
D. edwardsii (de Haan): From Bohai and Yellow Sea to South China Sea; Taiwan. Intertidal. (13, 52)
D. gordineri Alcock: South China Sea. Intertidal. (15)
D. hainanica Wang & Dong: South China Sea shallow water. (20)
D. investigatoris (Alcock): South China Sea. Intertidal. (15, 34, 59)
D. paracristimanus Wang & Dong: Bohai; Yellow Sea; East China Sea. Shallow water. (13, 20)
D. rectimanus Miers: From Bohai and Yellow Sea to South China Sea. Intertidal. (15, 55)
D. tomentosus Wang & Dong: Around Zhoushan Islands (Zhejiang). (13, 20, 21)
Isocheles sp.: East China Sea. (15)
Paguristes acanthomerus Ortmann: East China Sea. Water depth 64–250 m. (9, 15, 59)
P. barbatus (Ortman): Yellow Sea; East China Sea; South China Sea. (13, 59)
P. ciliatus Heller: South China Sea. Intertidal. (15, 46)
P. hians Henderson: South China Sea. Intertidal. (15, 46)
P. kagochimensis Ortmann: East China Sea shallow water. (15, 59)
P. ortmanni Miyake: From Bohai and Yellow Sea to South China Sea. Low intertidal to shallow water. (15, 59)
P. palythophilus Ortmann: East China Sea. Water depth 110–600 m. (8, 15, 59)
P. pusillus Henderson: South China Sea. (15, 46, 59)
P. pusillus zhejiangensis Wang & Dong: Zhejiang. (22)
P. sinensis Dong & Wang: East China Sea. (13, 15)
Trizopagurus strigatus (Herbst) [*Aniculus strigatus*]: East China Sea; South China Sea. Shallow water. (13, 59)
Troglopagurus jubatus Nobili: Taiwan. (15)

COENOBITIDAE

Birgus latro (Linnaeus): Taiwan; South China Sea. Shoreline, terrestrial. (7, 15, 34, 39)
Coenobita brevimanus Dana [*C. clypeatus*]: South China Sea shoreline. Terrestrial. (15, 34, 38, 39, 61)
C. cavipes Taiwan; South China Sea. Shoreline, terrestrial. (15, 34, 38, 39, 61, 68)
C. perlatus H. Milne-Edwards: Taiwan; South China Sea. Shoreline, terrestrial. (15, 34, 38, 39, 61)
C. rugosus H. Milne-Edwards: East China Sea; South China Sea; Taiwan. Shoreline, terrestrial. (14, 15, 34, 38, 39, 61)

PAGUROIDEA
PARAPAGURIDAE

Parapagurus acutus acutus de Saint Laurent: South China Sea. (15, 59)
P. acutus hirsutus de Saint Laurent: South China Sea. (15, 59)
P. arcustus monstrosus Alcock [*P. monstrosus*]: East China Sea shallow water. (13, 15, 19, 39)
P. pilosimanus Smith: East China Sea deep water. (13, 15, 19, 39)
P. sibogae de Saint Laurent: South China Sea. (15, 59)

PAGURIDAE

Catapaguroides sp.: East China Sea. (15)
Catapagurus ensifer Henderson: Taiwan. (15)

C. japonicus Yokoya: East China Sea. (13, 15, 59, 71)
Cestopagurus misakiensis (Terao): Taiwan. (15, 59, 69)
Munidopagurus macrocheles (H. Milne-Edwards): East China Sea. (13, 15)
Nematopagurus gardimeri Alcock: East China Sea; South China Sea. (13, 15, 59)
N. indicus Alcock: East China Sea. (13, 15, 34)
N. muricatus (Henderson): South China Sea. (15, 59)
N. squamichelis Alcock: East China Sea; South China Sea. (13, 15, 34)
N. vallatus (Melin): East China Sea. (15, 59)
Pagurus brachiomastus (Thallwitz): Bohai; Yellow Sea; East China Sea. Shallow water. (13, 15, 59)
P. carpoforaminatus (Alcock): South China Sea. (13, 34)
P. carpoforaminatus nephromma (Alcock): East China Sea. Intertidal. (13, 15, 34)
P. dubius (Ortmann): From Bohai and Yellow Sea to South China Sea; Taiwan. Inner bay intertidal. (13, 15, 59)
P. geminus Mclaughlin: From Bohai and Yellow Sea to South China Sea; Taiwan. Intertidal. (13, 15, 59)
P. gracilipes (Stimpson): Yellow Sea. (15, 59, 68)
P. janitor Alcock: Bohai; Yellow Sea; South China Sea. Intertidal. (15, 34)
P. japonicus (Stimpson) [*Eupagurus japonicus*]: Bohai; Yellow Sea. Intertidal. (15, 59, 68)
P. lanuginosus de Haan: Bohai; Yellow Sea. Intertidal. (15, 59)
P. megalops (Stimpson): From Bohai and Yellow Sea to South China Sea. (15, 59, 68)
P. ochotensis Brandt: Yellow Sea; northern South China Sea. Benthic. (1, 13, 15, 24, 73)
P. pectinatus (Stimpson): Bohai; Yellow Sea; northern East China Sea. Benthic, symbiotic with sponges. (13, 15, 24)
P. pilosipes (Stimpson): East China Sea. (15, 59)
P. rubricatus Henderson: East China Sea shallow water. (13, 15, 46)
P. sagamiensis Miyake: East China Sea. (15, 59)
P. trigonocheirus (Stimpson): East China Sea shallow water. (13, 15, 46)
P. triserratus (Ortmann): Bohai; Yellow Sea; East China Sea. (15, 59)
P. zebra (Henderson): East China Sea; South China Sea. (15, 46, 59)
Porcellanopagurus japonicus Balss: East China Sea. (15, 59)
Pylopagurus serpulophilus Miyake: East China Sea. (15, 59)
Spiropagurus spiriger (de Haan): East China Sea; South China Sea; Taiwan. Shallow sea. (13, 15, 59)

LITHODIDAE

Dedignathus inermis (Stimpson): East China Sea. (11)
Hapalogaster dentata (de Haan): Bohai; Yellow Sea. Intertidal and shallow sea. (24, 59)
Lithodes longispinus Sakai: East China Sea deep water. (13, 16)
L. turritus Ortmann: East China Sea deep water. (13, 16)
Neolithodes nipponensis Sakai: East China Sea deep water. (13, 16)
Paralomis dofleini Balss: East China Sea deep water. (13, 16)
P. heterotuberculata Tung et al.: East China Sea deep water. (13)
P. hystrix (de Haan): East China Sea deep water. (13, 16)

GALATHEOIDEA
GALATHEIDAE

Allogalathea elegans (Adams & White) [*Galathea elegans*]: Southern East China Sea shallow water. (9, 13, 35)
Cervimunida princeps Benedict: East China Sea. Water depth 360–365 m. (13, 17, 35)
Galathea balssi Miyake & Baba: East China Sea; South China Sea. Shallow water. (9, 13, 35)
G. formosus de Man: Taiwan shallow sea. (9)
G. orientalis Stimpson: South China Sea shallow water. (9, 36, 68)
G. pubescens Stimpson: East China Sea. Intertidal. (9, 13, 35, 68)
G. pusilla Henderson: East China Sea shallow water. (9)
G. subsquamata Stimpson: South China Sea shallow water. (9, 13, 68)
Munida babai Tirmizi & Javed: Hong Kong. Water depth 112–150 m. (35)
M. compressa Baba: Taiwan; South China Sea. Water depth 68–198 m. (35)
M. exigua Baba: Hong Kong. Water depth 68–198 m. (35)
M. incerta Henderson: East China Sea; South China Sea. Shallow water. (9, 13, 35)
M. japonica Stimpson: East China Sea; South China Sea. Water depth 30–900 m. (9, 13, 35, 68)
M. perarmata H. Milne-Edwards & Bouvier: South China Sea shallow water. (9)
M. prominula Baba: Taiwan. Water depth 421 m. (13, 17, 35)
M. sinensis Zhong & Wang: South China Sea. Water depth 504–558 m. (25)
Munidopsis longirostris H. Milne-Edwards & Bouvier: East China Sea; South China Sea. Deep water. (13, 17)
Paramunida scabra (Henderson) [*Munida scabra*]: East China Sea. Water depth 70–1,630 m. (9, 13, 35)

CHIROSTYLIDAE

Eumunida smithii Henderson: Taiwan. Water depth 204–421 m. (35)
Ptychogaster defensa Benedict: South China Sea. Water depth 510 m. (25)
Uroptychus albatrossae Baba: East China Sea. Water depth 265–510 m. (13, 17, 35)
U. gracilimanus (Henderson): East China Sea. Water depth 421–1668 m. (35)
U. granulatus japonicus Balss: East China Sea deep water. (13, 17)

U. naso Van Dam: East China Sea. Water depth 68–439 m. (35)
U. scandens Benedict: East China Sea. Water depth 100 m. (13, 18)

PORCELLANIDAE

Aliaporcellana suluensis (Dana): Hong Kong; Guangxi; Nansha (=Spratly Islands). (30, 31, 41, 45a)
A. telestophila (Johnson): Nansha (=Spratly Islands). Shallow water. (31, 41, 47)
Enosteoides ornata (Stimpson) [*Porcellana ornata*]: Fujian; Hong Kong; Guangxi. Low intertidal. (5, 29, 30, 45a)
E. variabilis (Yang & Sun): Weizhou Island (Guangxi). Subtidal. (28)
Lissoporcellana quadrilobata (Miers): Taiwan Strait; Hong Kong; Nansha (=Spratly Islands). Shallow sea. (31, 45, 45a)
L. spinuligera (Dana) [*Porcellana armata*]: Dapeng Cove (Hong Kong); Hong Kong; Guangxi. Subtidal. (30, 41, 45a, 56)
L. streptochiroides (Johnson): From Taiwan Strait to Hong Kong. (45a)
Neopetrolisthes maculatus (H. Milne-Edwards): Xisha (=Paracel Islands); Nansha (=Spratly Islands). Coral reefs. (5, 26, 31, 43)
Pachycheles garciaensis (Ward): Xisha (=Paracel Islands). Coral reefs. (5, 26, 43)
P. johnsoni Haig: Xisha (=Paracel Islands). Coral reefs. (5, 26, 43, 62)
P. pectinicarpus Stimpson: Hong Kong. Intertidal. (5, 43, 68)
P. pisum (H. Milne-Edwards): Guangxi. Subtidal. (5, 30)
P. sculptus (H. Milne-Edwards): Hong Kong; Xisha (=Paracel Islands); Nansha (=Spratly Islands). Coral reefs. (26, 31, 62)
P. spinipes (H. Milne-Edwards): Xisha (=Paracel Islands); Nansha (=Spratly Islands). Coral reefs. (5, 26, 31, 43)
Petrolisthes asiaticus (Leach): Guangxi. Intertidal. (30, 50)
P. boscii (Audouin): East China Sea; South China Sea. Low intertidal. (5, 6, 13, 45a, 56, 59)
P. carinipes (Heller): Nansha (=Spratly Islands). Coral reefs. (31, 60)
P. coccineus (Owen): Fujian; Taiwan; Hong Kong. Intertidal. (5, 29, 45a, 56, 59)
P. fimbriatus Borradaile: Xisha (=Paracel Islands); Nansha (=Spratly Islands). Coral reefs. (5, 26, 31, 56)
P. hastatus Stimpson: East China Sea; South China Sea. Subtidal. (5, 6, 56, 64)
P. haswelli Miers: Guangxi. Intertidal. (30, 42, 44, 55)
P. indicus de Man: Taiwan. (5, 56)
P. japonicus (de Haan): From Zhejiang to Guangdong and Hong Kong. Intertidal. (5, 13, 29, 45a, 56)
P. lamarckii (Leach): Hong Kong; Guangxi; Xisha (=Paracel Islands). Intertidal. (26, 30, 45a, 56)

P. militaris (Heller): East China Sea; Taiwan Strait; Hong Kong; Xisha (=Paracel Islands); Nansha (=Spratly Islands). Coral reefs. (5, 26, 31, 45a, 59)
P. scabriculus (Dana): Nansha (=Spratly Islands). Coral reefs. (31, 41, 44, 63)
P. tomentosus (Dana) [*P. penicillatus*]: Xisha (=Paracel Islands); Nansha (=Spratly Islands). Coral reefs. (5, 26, 31)
P. unilobatus Henderson: Taiwan. Coral reefs. (5, 56)
Pisidia dispar (Stimpson): Hong Kong. Coral reefs. (45a)
P. gordoni Johnson: Fujian; Hong Kong. Subtidal. (29, 45, 45a, 48)
P. serratifrons (Stimpson): Yellow Sea; Zhejiang; Fujian; Hong Kong; Guangxi. Intertidal to shallow sea. (5, 13, 29, 30, 45a, 56)
P. streptocheles (Stimpson) [*Porcellana streptocheles*]: East China Sea. Intertidal. (5, 6, 13, 45)
P. striata Yang & Sun: Xiamen (Fujian). Low intertidal. (29)
Polyonyx biunguiculata Dana: Taiwan Strait; Hong Kong. Coral reef sponges. (45a)
P. obsesulus Miers: Hong Kong; Guangxi; Nansha (=Spratly Islands). Shallow sea. (30, 31, 41, 44, 45a, 47)
P. sinensis Stimpson [*P. asiaticus*]: Bohai; Yellow Sea. Intertidal. (5, 47, 56, 66, 68)
Porcellana curvifrons Yang & Sun: Fujian. Intertidal, symbiotic with hermit crabs. (29)
P. habei Miyake: Nansha (=Spratly Islands). Coral reefs. (31, 45, 58)
P. pulchra Stimpson: From Bohai and Yellow Sea to South China Sea. Shallow water. (5, 13, 29, 45a, 56)
Porcellanella triloba White: East China Sea; Taiwan; Hong Kong; South China Sea. Intertidal, often commersal with sea pansies (*Pteroeides*). (29, 31, 45, 45a, 56)
Raphidopus ciliatus Stimpson: From Bohai and Yellow Sea to South China Sea. Shallow water. (5, 13, 45, 45a, 56)

HIPPIOIDEA
ALBUNEIDAE

Albunea dayriti Serene & Umali: Southern East China Sea; South China Sea. Shallow water. (10, 27, 65)
A. symnista (Linnaeus): East China Sea; South China Sea. Shallow water. (10, 13, 59)
Blepharipoda liberata Shen: Yellow Sea shallow water. (10, 59, 67)
Lophomastrix japonica (Durufle): Yellow Sea shallow water. (10, 59, 67)

HIPPIDAE

Hippa adactyla Fabricius: Taiwan; South China Sea. Intertidal. (10, 59)
H. truncatifrons (Miers): South China Sea. (10, 55)
Mastigopus gracilis Stimpson: East China Sea; South China Sea. (10, 68)

REFERENCES*

(1). Wang Fuzhen, 1982. Pharmaceutic hermit crabs from China. Journal of Marine Drugs, 4:49–55

(2). Wang Fuzhen, 1983. New records of hermit crabs from China. East China Sea Marine Science, 2:50–57.

(3). Wang Fuzhen, 1984. Hermit crabs of China. Biological Education, 2:5–8.

(4). Wang Fuzhen, 1985. Textual research and investigation of pharmaceutic crabs from China. Journal of Marine Drugs, 1:17–19.

(5). Wang Fuzhen, 1986. The Porcellanidae (Crustacea, Anomura) of China. Transactions of Oceanology and Limnology, 4:51–55.

(6). Wang Fuzhen, 1987. New records of Porcellain crabs (Crustacea, Anomura) from China. Sichuan Journal of Zoology, 6(4):30.

(7). Wang Fuzhen, 1988. Unusual records of coconut crabs (Crustacea) from China. Bulletin of Biology, 12:24.

(8). Wang Fuzhen, 1988. Tribe Anomura. In: Report on crustaceans of the deep East China Sea. Zhejiang Science & Technology Press, pp:52–72.

(9). Wang Fuzhen, 1989. The Squat lobsters (Crustacea, Anomura) of China. Transactions of Oceanology and Limnology, 2:63–65.

(10). Wang Fuzhen, 1989. New records of cicada crabs (Crustacea, Hippa) from China. Sichuan Journal of Zoology, 8(1):31.

(11). Wang Fuzhen, 1990. New records of Lithodidae (Crustacea, Anomura) from China. Sichucan Journal of Zoology, 9(1):31.

(12). Wang Fuzhen, 1990. New records of hermit crabs from China. Sichuan Journal of Zoology, 9(2):31.

(13). Wang Fuzhen, 1991. Tribe Anomura. In: Fauna of Zhejiang, Crustacea. Zhejiang Science & Techonolgy Press.

(14). Wang Fuzhen, 1992. Hermit crabs from Hong Kong. Sichuan Journal of Zoology, 11(1):40.

(15). Wang Fuzhen, 1992. Studies on the hermit crabs of China (Crustacea, Anomura). East China Sea Marine Science, 10(1):59–63.

(16). Wang Fuzhen and Li Zhicheng, 1983. New reocords of Lothodidae (Crustacea, Anomura) from China. Marine Fisheries, 3:117–122.

(17). Wan Fuzhen and Li Zhicheng, 1986. New records of Galatheidae (Crustacea, Anomura) from the East China Sea. Marine Science, 10(5):28–31.

(18). Wang Fuzhen and Hu Yumei, 1983. New records of Galatheidae from China. Marine Science Bulletin, 2(4):79–83.

(19). Wang Fuzhen and Hu Yuemei, 1984. Studies on the hermit crabs from East China Sea. Journal of Marine Drugs, 2:40–45.

(20). Wang Fuzhen and Tung Yumao, 1977. Two new species of Hermit crabs (Crustacea, Anomura) from China. Acta Zoologica Sinica, 23(1):109–112.

(21). Wang Fuzhen and Dong Yumao, 1977. Two new species of hermit crabs from China. Acta Zootaxonomica Sinica, 5(1):35–38.

(22). Wang Fuzhen and Tung Yumao, 1982. New subspecies and new records of hermit crabs (Crustacea, Anomura) from China. Acta Zootaxonomica Sinica, 7(4):369- 371.

(23). Wang Fuzhen, Dong Yumao and Li Zhicheng, 1986. A new species of hermit crab from the East China Sea (Crustacea, Anomura). Acta Zootaxonomica Sinica, 11(1):32–34.

(24). Shen Jiarui and Liu Ruiyu, 1976. Shrimps and Crabs of China. China Science Press, pp:132–135.

(25). Zhong Zhenru and Wang Fuzhen, 1989. A new species and three new records of squat lobster (Crustacea, Anomura) from China. Journal of Fisheries of China, 13(1):65–69.

(26). Yang Siliang, 1983. Preliminary report of porcellainidae (Crustacea, Anomura) from the Xisha Isladns, Guangdong Province, China. In: Research report by Beijing Natural History Museum, 45:1–15

(27). Yang Siliang and Sun Xiumin, 1979. A new record of Albuneidae (Crustacea, Anomura) from China. Acta Zootaxonomica Sinica, 4(3):213.

(28). Yang Siliang and Sun Xiumin, 1985. A new species of Porcellanidea (Crustacea, Anomura) from China. Acta Zoologica Sinica, 31(2) 150–153.

(29). Yang Siliang and Sun Xiumin, 1990. Porcellainidae (Crustacea, Anomura) of Fujian Province. Research Report by Beijing Natural History Museum, 45:1–15.

(30). Yang Siliang and Sun Xiumin, 1992. On the Porcellanid crab (Anomura: Porcellanidae) of Guangxi Province, China. In: Transactions of the Chinese Crustacean Society, No.3.

(31). Yang Siliang and Xu Zhenxiong, 1994. Reports on Porcellanidae from Nansha Islands and its adjacent waters. Marine Fauna and Flora and Biogeography of the Nansha Islands and Neibouring Waters, I:122–124. China Ocean Press.

(32). Tung Yumao and Wang Fuzhen, 1965. Preliminary studies of the hermit crab fauna of China. Acta Zootaxonomica Sinica, 17(4):401–405.

(33). Tung Yumao and Wang Fuzhen, 1984. A new species of Lithodidae (Crustacea, Anomura) from the East China Sea. Zoological Research, 5(4):329–333.

(34). Alcock, A. 1905. Catalogue of the Indian Decapod Crustacea in the Collection of Indian Museum. Pt. II. Anomura, Fasc. I. Paguridae:1–197.

(35). Baba, K. 1988. Chirostylid and Galatheid Crustaceans (Decapoda: Anomura) of the "Albatross" Phillippine Expedition, 1907–1910. Researches on Crustacea. Special Number 2, Shimoda Printing.

(36). Balss, H. 1913. Ostasiatische Decapoden I. Die Galatheiden und Paguriden. Munish, Math. Phys. Kl. 2(9):1–85.

(37). Buitendijk, A. M. 1937. Biological results of the Snellius Expedition IV. Paguridea, Temminekia, 2:251–280, figs. 1–19.

(38). Fize, A. and R. Serene, 1955. Les Pagures du Vietnam. Notes Inst. Oceanograph Nhatranf. Viet-nam, 45:i–ix, 1–228, figs. 1–35, pl. 1–6.

(39). Gordon, J. 1956. A bibliography of Pagurid Crabs, Exclusive of Alcock, 1905. Bull. Amer. Mus. Nat. Hist. 108(3):253–352.

(40). Haan, W. de. 1833–1850 Crustacea, In: P. F. von Siebold, Fauna Japonica (4):ix-xvi, xii-xvii, i-xxxi, 1–244, pls.1–55.

(41). Haig, J. 1964. Porcellanid Crabs from the Indo-West Paficic, Part I. Papers from Dr. Th. Mortensen's Pacific Expedition, 1914–1916. 81. Vidensk Meddr Dansk naturh. Foren 126:355–386.

(42). Haig, J. 1965. The Porcellanidae (Crustacea, Anomura) of western Australia with Descriptions of Four New Australian Species. J. Proc. R. Soc. West Aust. 48:97- 118.

(43). Haig, J. 1966. A review of the Indo-Pacific Genus *Pachycheles* (Porcellanidae, Anomura). Proc. Sympos. Crust. India, I:285–294.

(44). Haig, J. 1979. Expedition Rumphius II (1975), Crustacea Parasites, Commensaux, etc. (Th. Monod & R. Serene, ed.), 5. Porcellanidae (Crustacea, Decapoda, Anomura). Bull. Mus. Natn. Hist. nat. Paris (Zool. Biol. Ecol., Snim.) No. 1:119- 136.

(45). Haig, J. 1981. Porcellanid Crabs from the Indo West Pacific Part 2. Steenstouppia 7(12):269–291.

(45a). Haig, J. 1992. Hong Kong's Porcellanid Crabs. In: Proceedings of International Marine Biological Workshop: The Marine Flora and Fauna of Hong Kong and Southern China. Hong Kong University Press, pp:303–327.

(46). Henderson, J. R. 1888. Report on the Scientific Results of the Voyage of H. M. S. Challenger During the Years 1873–1876, 27(69):48–102.

(47). Johnson, D. S. 1958. The Indo-West Pacific Species of Genus *Polyonyx* (Crustacea Decapoda, Porcellanidae). Ann. of Zool. (Agra., India) 2:95–118, figs.: 1–4.

(48). Johnson, D. S. 1970. The Galatheidea of Singapore and Adjacent Waters. Bull. Natn. Mus. st. Singapore 35(1):1–44, figs. 1–6.

(49). Kamalaveni, S. 1950. On Hermit Crabs in the Collection of the Indian Museum. Rec. Ind. Mus. 47(1):77–85.

(50). Kropp, R. R. 1983. Three New Species of Porcellanidae (Crustacea: Anomura) from the Mariana Islands and a Discussion of Borradaile's *Petrolisthes* Lamarckii Complex. Micronisica 19(1–2):91–106.

(51). Kropp, R. R. 1986. A Neotype Designation for *Petrolisthes tomentosus* (Dana), and Description of *Petrolisthes heterochorus*, New Species from the Mariana Islands (Anomura: Porcellanidae). 99:452–463.
(52). Lee, S. C. 1969. Anomuran Crustacean of Taiwan. Part I. Diogenidae 8:39–57.
(53). Lewinson, C. 1979. Researches on the Coast of Somolia. The Shore and the Dune of Sar Uanle, 23. Porcellanidae (Crustacea: Decapoda: Anomura). Suppl. 39–57.
(54). Man, J. G. de. 1888. Report on the Polophthalmous Crustacea of the Mergina Archipelago 22:1–312, pls. 1–19.
(55). Miers, E. J. 1880. On a Collections of Crustacean from the Malaysian Region, Part III. Ser. 5. 5(20):370–384, pls. 14, 15.
(56). Miyake, S. 1943. Studies on the Crab Shaped Anomura of Nippon and Adjacent Waters. J. Dep. Agr. Kyushu. Tmp. Univ. 7(3):49–158, figs. 1–61.
(57). Miyake, S. 1956. Invertebrate Fauna of the Intertidal Zone of the Tokara Island XIII. Anomura. Pub. Seto Mar. 303–337.
(58). Miyake, S. 1961. Three New Species of the Anomura from Japan (Crustacea, Decapoda). Anomura. Pub. Seto Mar. 237–247.
(59). Miyake, S. 1978. The Crustacean Anomura of Sagami Bay, 1–200.
(60). Miyake, S. and Y. Nakasone, 1966. On Two Species of the Genus *Petrolisthes* (Anomura, Porcellanidae) from Japan.
(61). Nakasone, Y. 1988. Land Hermit Crabs from the Ryukyus, Japan, with a Description of a New Species from the Philippines (Crustacea, Decapoda, Coenobitidae). 5:165–178.
(62). Nakasone, Y. and Miyake, S. 1968. On Six Species of *Pachycheles* (Anomura: Porcellanidae) from the West-Pacific. I:61–83, 8 figs., 2 pls.
(63). Nakasone, Y. 1968. Four Unrecorded Porcellanis Crabs (Anomura: Porcellanidae) from Okinawa, the Ryukyo Island. Ohmu I:97–111.
(64). Nakasone, Y. 1971. Porcellanid Crabs (Anomura: Porcellanidae) from New Caledonia and the Fiji Island 18:1–13.
(65). Serone, R. and A. F. Umali, 1965. A Review of Philippine Albuneidae, with Description of Two New Species. 94:87–116, figs. 1–12, pls.1–6.
(66). Shen, C. J. 1936. Notes on the Genus *Polyonyx* (Porcellanidae) with Description of a New Species. pp:275–286.
(67). Shen, C. J. 1949. Notes on the Genera *Blepharipoda* and *Lophomastrix* of the Family Albuneidae (Crustacea, Anumora) with Description of a New Species *B. liberata* from China.
(68). Stimpson, W. 1907. Report on the Crustacea Collected by the North Pacific Exploring Expedition 1853–1856. 49:(1717):1240.
(69). Terao, A. 1913. A Catalogue of Hermit Crabs found in Japan with Description of Four New Species. 291–355.
(70). Vap Chiongco, J. V. 1938. The Littoral Paguridae in the Collection of the University of the Philippine, 183–219.
(71). Yoloya, Y. 1933. On the Distribution of Decapod Crustacean Inhabiting the Continental Shelf Around Japan. J. Coll. Agr. Tokyo 12:1–226.
(72). Yoloya, Y. 1939. Maorura and Anomura of Decapod Crustacea found in the Neibourhood of Onagawa Miyagi-Ken, 278–289.
(73). Makapob, B. B. 1938. Anomura, fauna of CCCP. 10(3):1–309.

*: (1)-(33) in Chinese and (73) in Russian
Compiled by Wang Fuzhen and Yang Siliang; Edited by Wang Yongliang and Liu Ruiyu

BRACHYURA
DROMIIDAE

Conchoecetes artificiosus (Fabricius): Zhejiang; Fujian; Guangdong. (18, 20, 26)

Cryptodromia areolata Ihle: East China Sea. Water depth 50–500 m. (20, 26)
C. canaliculata Stimpson: Hainan. Low intertidal. (20, 26)
C. coronata Stimpson: Xisha (=Paracel Islands). Coral reefs. (20, 26)
C. dubia Dai & Chen: Hainan. (20, 26)
C. hilgendorfi de Man: Xisha (=Paracel Islands). Coral reefs. (20, 26)
C. planaria Dai & Yang: Xisha (=Paracel Islands). Coral reefs. (20, 26)
C. protuberum Dai & Song: Hainan. (20, 26)
Dromia dehaani Rathbun: Fujian; Taiwan; Guangdong; Hainan. Water depth 8–150 m. (20, 26)
D. intermedia Laurie: Taiwan. (30)
Dromidiopsis cranioides (de Man): Hainan. (20, 26)
D. dromid (Linnaeus): South China Sea. Water depth 20–50 m. (20, 26)
Petalomera granulata Stimpson: Guangdong; Guangxi. Water depth 50–150 m. (20, 26)
P. japonica (Henderson): Fujian. Water depth 10+ m. (20, 26)
P. lateralis (Gray): South China Sea. (20, 26)
P. longipedalis Dai & Yang: Hainan. (20, 26)
P. sheni Yang & Dai [*P. granulata* Shen, 1964]: Shandong Peninsula. (20, 26)

DYNOMENIDAE

Dynomene granulobata Dai & Yang: Xisha (=Paracel Islands). Coral reefs. (20, 26)
D. hispida Desmarest: Taiwan; Xisha (=Paracel Islands). Rocky crevices and coral reefs. (4, 20, 26)
D. sinense Chen: Shi Island and Chenhang Islands (Xisha=Paracel Islands). (7)
D. spinosa Rathbun: Xisha (=Paracel Islands). Coral reefs. (4, 20, 26)
D. tenuilobata Dai & Lan: Xisha (=Paracel Islands). Coral reefs. (20, 26)

TYMOLIDAE

Tymolus japonicus Stimpson: East China Sea. Water depth 50–350 m. (20, 26)
T. uncifer (Ortmann): East China Sea. Water depth 50–300 m. (20, 26)

HOMOLIDAE

Homola orientalis Henderson: East China Sea continental slope; South China Sea. (9, 16, 20, 26)
Homologenus sinensis Chen: Southeastern China. Subtidal. (9)
H. validiviae Doflein: East China Sea. Water depth 200 m. (16)
Hypophrus longirostris Chen: Southeastern China. Subtidal. (9)

Paromola macrochira Sakai: Southeastern China. Continental shelf, water depth 200- 300 m. (9, 16)

Paromolopsis boasi Wood-Mason: South China Sea. Water depth about 300 m. (9, 20, 26)

LATREILLIDAE

Latreillia phalangium de Haan: East China Sea. Water depth 30–300 m. (20, 26)

L. valida de Haan: East China Sea; South China Sea; Xisha (=Paracel Islands); Nansha (=Spratly Islands). (15, 16, 20, 26)

Latreillopsis bispinosa Henderson: Nansha (=Spratly Islands). Water depth 185 m. (15)

L. tetraspinosa Dai & Chen: East China Sea; South China Sea; Nansha (=Spratly Islands). Water depth 50–150 m. (15, 19, 20, 26)

RANINIDAE

Cosmonatus genkaiae Takeda & Miyake: East China Sea offshore. (20, 26)

C. grayii Adams & White: East China Sea; Nansha (=Spratly Islands). Water depth 30–165 m. (15, 20, 26)

Lyreidus stenops Wood-Mason [*L. integra* Shen, 1964]: East China Sea; Guangdong; Hainan. (20, 26)

L. tridentatus de Haan: Guangdong; Nansha (=Spratly Islands). Water depth 84–300 m. (15, 20, 26)

Ranilia horikoshii Takeda: East China Sea. (20, 26)

R. orientalis Sakai: Xisha (=Paracel Islands). Coral reefs. (20, 26)

Ranina ranina (Linnaeus): Taiwan; Guangdong; Guangxi; Xisha (=Paracel Islands). Water depth 10–50 m. (4, 20, 26)

Raninoides hendersoni Chopra: Nansha (=Spratly Islands). Water depth 113 m. (15)

R. intermedius Dai & Xu: Nansha (=Spratly Islands). Water depth 99 m. (24)

R. personatus Henderson: East China Sea; Nansha (=Spratly Islands). Water depth 127 m. (15, 20, 26)

R. serratifrons Henderson: South China Sea; Nansha (=Spratly Islands). (15, 20, 26)

DORIPPIDAE

Dorippe (*Neodorippe*) *callida* (Fabricius) [*D. astuta*]: Fujian; Guangdong; Hainan. (20, 26)

D. (*Dorippides*) *facchino* (Herbst): Fujian; Guangdong; Guangxi. Water depth 15–100 m. (13, 20, 27)

D. (*Dorippe*) *frascone* (Herbst) [*D. quadridens*, *D. dorsipes* Shen, 1964]: East China Sea; Taiwan; Beibuwan (=Gulf of Tongking); South China Sea. Water depth 15- 50 m. (13, 15, 20, 26)

D. (*Paradorippe*) *granulata* de Haan: Shandong Peninsula; Fujian; Taiwan; Guangdong; Hainan. (20, 26, 27)

D. (*Nobilum*) *histirio* (Nobili) [*Nobilum histirio* (Nobili)]: Ningbo (Zhejiang); Fuzhou (Fujian). (13)

D. (*Neodorippe*) *japonica* von Siebold [*Heikea japonica*]: China's coast. (20, 26, 27)

D. (*Paradorippe*) *polita* Alcock & Anderson: Shandong; Zhejiang; Fujian; Guangdong; Hainan. (20, 26, 27)

D. (*Dorippe*) *sinica* Chen: East China Sea; Guangdong; Hainan. Water depth 15–50 m. (13, 20, 26)

D. (*Dorippe*) *tenuipes* Chen: East China Sea; Guangdong; Guangxi; Hainan; South China Sea. (13, 20, 26)

Ethusa indica Alcock: East China Sea. Water depth 30–795 m. (13, 20, 26)

E. izuensis Sakai: East China Sea. Water depth 30–147 m. (13, 20, 26)

E. latidactyla (Parisi): South China Sea. Water depth 168–173 m. (13)

E. minuta Sakai: East China Sea. Water depth 50–760 m. (13, 20, 26)

E. quadrata Sakai: East China Sea. Water depth 116–120 m. (13)

E. sexdentata (Stimpson): East China Sea; South China Sea. Water depth 220–300 m. (13, 20, 26)

Ethusina desciscens Alcock: East China Sea. Water depth 1,115 m. (13)

E. investigatoris Alcock: East China Sea; Nansha (=Spratly Islands). Water depth 1,115- 2,378 m. (13, 15)

E. robusta Miers: East China Sea. Water depth 1,350 m. (13)

Philippidorippe philippinensis Chen: Nansha (=Spratly Islands). Water depth 96–195 m. (15)

LEUCOSIIDAE

Arcania elongata Yokoya: East China Sea; Guangdong; Hainan. Water depth 30–100 m. (20, 26)

A. erinaceus (Fabricius) [*A. globata* Shen, 1964]: Taiwan Strait; Hainan. (18, 20, 26)

A. heptacantha (de Haan): Zhejiang; Fujian; Taiwan; Guangdong; Hainan. Water depth 50–150 m. (17, 18, 20, 26)

A. novemspinosa Adams & White: Guangdong. (20, 26)

A. quinquespinosa Alcock & Anderson: Taiwan; Guangdong; Hainan. Water depth 68–108 m. (20, 26)

A. sagamiensis Sakai: Hainan shallow sea. (20, 26)

A. septemspinosa (Fabricius): Guangdong; Hainan. (20, 26)

A. undecimspinosa de Haan: Fujian; Taiwan; Guangdong; Hainan. (20, 26)

Cryptocnemus obolus Ortmann: South China Sea. Water depth 180–200 m. (20, 26)

Drachiella morum (Alcock): East China Sea; South China Sea. Water depth 55–150 m. (20, 26)

Ebalia longimana Ortmann: East China Sea. (20, 26)

E. malefactrix Kemp: Hainan. Intertidal. (20, 26)

E. scabra Dai: Haikou and Sanya (Hainan). (20, 26)

E. scabriuscula Ortmann: East China Sea. (20, 26)

E. tosaeusis Sakai: East China Sea. (20, 26)

E. tuberculosa (H. Milne-Edwards): East China Sea. (1, 20, 26)
Iphiculus spongiosus Adams & White: Fujian; Guangdong; Hainan. Water depth 25–106 m. (18, 20, 26)
Ixa cylindrus (Fabricius): Taiwan; Guangdong; Hainan. Water depth 10–30 m. (20, 26)
I. edwardsii Lucas: Guangxi; Hainan. Shallow sea. (20, 26)
Ixoides cornutus (McGilchrist): Taiwan; Guangdong; Hainan. Water depth 50–200 m. (20, 26)
Leucosia anatum (Herbst) [*L. longifrons* Shen, 1964]: Dongshan (Fujian); Taiwan Strait; Guangdong. Water depth 10–80 m. (18, 20, 26)
L. biminents Dai & Xu: Nansha (=Spratly Islands). Water depth 50 m. (24)
L. compressa Shen & Chen: Hainan. Water depth 33 m. (2)
L. craniolaris (Linnaeus): Fujian; Taiwan; Guangdong; Guangxi. Water depth 20–80 m. (17, 18, 20, 26)
L. formosensis Sakai: Taiwan Strait; Taiwan. Water depth 2–40 m. (18, 20, 26)
L. haematosticta Adams & White: Fujian; Taiwan Strait; Hainan. Water depth 30–100 m. (18, 20, 26)
L. latirostrata Shen & Chen: Fujian; Guangdong. (2)
L. longibrachia Shen & Chen: Guangdong; Beibuwan (=Gulf of Tongking); Hainan; South China Sea. (2)
L. minuta Chen: Nansha (=Spratly Islands). Water depth 56 m. (15)
L. parapulchella Dai & Xu: Nansha (=Spratly Islands). Water depth 50 m. (24)
L. puchella Ball [*L. pseuolo margaretata* Chen]: Guangxi; Hainan. Water depth 30–90 m. (14)
L. pulchra Chen: Guangxi; Hainan. Water depth 12–24 m. (15)
L. rhomboidalis de Haan: Fujian; Taiwan Strait; Guangdong. Water depth 27–100 m. (17, 18, 20, 26)
L. sinica Shen & Chen: Beihai (Guangxi). Water depth 11–12 m. (2)
L. unidentata de Haan: Taiwan Strait; Guangdong; Hainan. Water depth 30–100 m. (18, 20, 26)
L. vittata Stimpson: Fujian; Taiwan; Guangdong. Water depth 5–53 m. (18, 20, 26)
L. whitei Bell: Nansha (=Spratly Islands). Water depth 50 m. (15)
Merocryptus lambriformis H. Milne-Edwards: East China Sea. Water depth 35–270 m. (20, 26)
Myra affinis Bell: Guangxi; South China Sea. Water depth 30–150 m. (20, 26)
M. coalita Hilgendorf: South China Sea. (1)
M. fugax (Fabricius): Fujian; Taiwan; Guangdong; Hainan. Water depth 10–24 m. (20, 26)
M. kessleri (Paulson): Nansha (=Spratly Islands). Water depth 53 m. (24)
Nucia speciosa Dana: Xisha (=Paracel Islands). Coral reefs. (7, 20, 26)
Nursia abbreviata Bell: Xiamen (Fujian); Hainan. (17, 20, 26)
N. japonica Sakai: East China Sea. (20, 26)
N. minor (Miers): Xiamen and Tongan (Fujian); Hainan. Water depth 20–35 m. (17, 20, 26)

N. rhomboidalis (Miers) [*N. sinica* Shen, 1937, 1964]: China's coast. (20, 26)
Nursilia dentata Bell: Guangdong; Hainan; Nansha (=Spratly Islands). Water depth 15–80 m. (8, 20, 24, 26)
N. sinica Chen: East China Sea; South China Sea. Water depth 100–174 m. (8)
N. tonsor Alcock: South China Sea. Water depth 33–202 m. (8)
Oreophorus reticulatus Adams & White: Nansha (=Spratly Islands). Water depth 50 m. (24)
Parilia ovata Chen: South China Sea. Water depth 160–230 m. (11)
Pariphiculus mariannae (Herklots): Guangdong; Hainan. Shallow sea. (20, 26)
Philyra acutidens Chen: East China Sea. (14)
P. biprotubera Dai & Guan: Guangdong. (18)
P. carinata Bell: Eastern Liaoning Bay; Shandong Peninsula; Fujian; Guangdong. Low intertidal. (20, 26)
P. globulosa H. Milne-Edwards: Guangdong; Hainan. Low intertidal. (20, 26)
P. heterograna Ortmann: Shandong Peninsula; East China Sea, Guangdong. Water depth 7–30 m. (20, 26)
P. olivacea Rathbun [*Pseudophilyra olivacea* Shen, 1964]: Zhejiang; Fujian; Guangdong; Hainan. Shallow sea. (20, 26)
P. pisum de Haan: South of Eastern Liaoning Peninsula. Low intertidal to shallow sea. (20, 26)
P. platychira de Haan: Taiwan; Guangdong. (20, 26)
P. tuberculosa Stimpson: Taiwan; Guangdong. Low intertidal. (20, 26)
Praebebalia longidactyla Yokoya: East China Sea. (20, 26)
Randallia eburnea Alcock: East China Sea; Guangdong. Water depth 30–300 m. (18, 20, 26)

CALAPPIDAE

Calappa gallus (Herbst): Taiwan; Hainan; Xisha (=Paracel Islands); Nansha (=Spratly Islands). Coral reefs and shallow sea. Water depth 15–100 m. (1, 4, 15, 20, 26)
C. hepatica (Linnaeus): Taiwan; Hainan; Xisha (=Paracel Islands). Water depth 10–100 m. (15, 20, 26)
C. lophos (Herbst): East China Sea; Taiwan; Hainan; Nansha (=Spratly Islands). Water depth 30–100 m. (15, 16, 20, 26)
C. philargius (Linnaeus): Fujian; Taiwan; Guangdong; Hainan; Nansha (=Spratly Islands). Water depth 30–100 m. (20, 26)
C. pustulosa Alcock: Hainan; Nansha (=Spratly Islands). Water depth 50–150 m. (15, 20, 26)
C. terraercginae Ward: Guangdong; Guangxi; Hainan; Nansha (=Spratly Islands). Water depth 30–122 m. (15, 20, 26)
Cycloes granulosa de Haan: Taiwan Strait; Taiwan; Hainan; Nansha (=Spratly Islands). Water depth 30–100 m. (15, 18, 20, 26)

Matuta banksii Leach: Taiwan Strait; Taiwan; Guangdong; Hainan. Low intertidal to water depth 70 m. (18, 20, 26)

M. granulosa Miers: Hainan. (20, 26)

M. lunaris Farskål: Fujian; Taiwan; Hainan. Water depth 10–15 m. (20, 26)

M. planipes Fabricius: China's coast. Low intertidal to water depth 15 m, sandy substrate. (20, 26)

Mursia armata de Haan: East China Sea; Guangxi; Hainan; Nansha (=Spratly Islands). Water depth 50–170 m. (15, 20, 26)

M. curtispina Miers: East China Sea; Nansha (=Spratly Islands). Water depth 138–147 m. (15, 16)

M. trispinosa Parisi: East China Sea; Nansha (=Spratly Islands). Water depth 170 m. (15, 16)

Orithyia sinica Linnaeus [*O. mammillaris* Shen, 1964]: From Eastern Liaoning Bay to Guangdong. Shallow sea. (20, 26)

Paracyclois mileneedwardsii Miers: Nansha (=Spratly Islands). Water depth 170 m. (15)

HYMENOSOMATIDAE

Elamenopsis introversus (Kcmp) [*Neorhynchoplax introversus* Dai et al.]: Jiangsu; Hai River (from Zhejiang to Tianjin). Lakes, grass or muddy substrate. (20, 26)

E. sinensis (Shen) [*Neorhynchoplax sinensis* Dai et al.]: Southern Shandong Peninsula. Low intertidal. (20, 26)

Halicarcinus messor Stimpson [*Rhynchoplax messor* Dai et al.]: Zhejiang; Fujian. Low intertidal to shallow sea. (17, 18, 20, 26)

H. setirostris (Stimpson) [*Rhynchoplax setirostris* Dai et al.]: Shandong Peninsula; Fujian; Guangdong. (18, 20, 26)

MAJIDAE

Achaeus japonicus de Haan: East China Sea; Guangdong. Water depth 20–100 m. (20, 26)

A. pugnax de Man: East China Sea. (20, 26)

A. robustus Yokoya: East China Sea. (20, 26)

A. superciliaris Ortmann: East China Sea. (20, 26)

A. tuberculatus Miers: Eastern Shandong Bay; East China Sea. Water depth 30–100 m. (20, 26)

A. varians Takeda & Miyaka: East China Sea. (20, 26)

Camposcia retusa Latreille: Guangdong; Hong Kong; Xisha (=Paracel Islands). Water depth 10–30 m. (7, 20, 26)

Chlorinoides aculeatus (H. Milne-Edwards): Guangdong. Water depth 10–50 m. (20, 26)

C. harmandi (Bouvier): East China Sea. (20, 26)

Choniognothus reini (Balss): East China Sea. (20, 26)

Criocarcinus superciliosus Herbst: Xisha (=Paracel Islands). Coral reefs. (7, 20, 26)

Cyclax spinicinctus Heller: South China Sea. Coral reefs. (7, 20, 26)

C. suborbicularis (Stimpson): Hainan; Xisha (=Paracel Islands). Coral reefs. (7, 20, 26)

Cyrtomaia hispida (Borradaile): East China Sea. Water depth about 100 m. (20, 26)

C. owstoni Terazaki: East China Sea. Water depth 65–900 m. (20, 26)

Doclea canalifera Stimpson: Fujian; Taiwan Strait; Guangdong; Guangxi; Hainan. Water depth 40–69 m. (18, 20, 26)

D. gracilipes Stimpson: Fujian; Taiwan Strait; Guangdong. (17, 18, 20, 26)

D. japonica (Herbst) [*D. ovis* Dai et al.]: Fujian; Taiwan Strait; Guangdong. River mouth and shallow sea. (18, 20, 26)

D. sinensis Dai: Fujian; Guangxi; Hainan. River mouth and shoreline. (20, 26)

D. tetraptera Walker: Fujian; Hainan. Shallow sea off river mouth. (20, 26)

Eurynome orientalis Sakal: East China Sea. Water depth 65–140 m. (20, 26)

Huenia brevifrons Ward: Xisha (=Paracel Islands). On seaweed. (7, 20, 26)

H. proteus de Haan: Guangdong. Low intertidal to water depth 50 m. (20, 26)

Hyas coaractatus alutaceus Brandt: Shanghai; Xiamen (Fujian). Water depth 30–200 m. (20, 26)

H. cocarctatus ursinus Rathbun: Shanghai. (20, 26)

Hyastenus diacanthus (de Haan): East China Sea; Xiamen (Fujian); Taiwan Strait; Guangdong. Water depth 27–100 m. (17, 18, 20, 26)

H. pleione (Herbst): Jiazhou Bay (Shandong); Fujian; Guangdong. (20, 26)

Leptomithrax compressipes Miers: China's seas. (20, 26)

L. edwardsi (de Haan): East China Sea; South China Sea. Continental shelf, water depth 50–200 m. (16)

L. sinensis Rathbun: China's seas. (20, 26)

Macrocheira kaempferi (Temminck): East China Sea. Water depth 300+ m. (16)

Maja brevispinosis Dai: Guangdong. (20, 26)

M. gibba Alcock: East China Sea. Water depth 100–200 m. (20, 26)

M. japonica Rathbun: Taiwan Strait; Taiwan. Water depth 50–100 m. (18, 20, 26)

M. miersi Walker: Taiwan Strait. Water depth 60–72 m. (18)

M. sakaii Takeda & Miyake: East China Sea. (20, 26)

M. spinigera de Haan: Taiwan. Water depth 15–50 m. (20, 26)

Menaethius monoceros (Latreille): Taiwan; Hainan; Xisha (=Paracel Islands). Low intertidal to shallow sea. (7, 20, 26)

Micippa philyra (Herbst): Dapeng Cove (Hong Kong); Hainan. Intertidal to water depth 20 m. (20, 26, 28)

M. platipes Rüppell: Hainan; Xisha (=Paracel Islands). Intertidal to water depth 20 m. (7, 20, 26)

M. thalia (Herbst): Taiwan Strait; Guangdong. Water depth 20–100 m. (18, 20, 26)

M. xishanensis Chen: Xisha (=Paracel Islands). (7)

Micippoides angustifrons H. Milne-Edwards: Xisha (=Paracel Islands). (7)

Naxioides mammillata (Ortmann): East China Sea. Water depth 30–200 m. (16)

N. taurus (Pocock): China's seas. Water depth 30–50 m. (20, 26)

Oncinopus angustefrons (Takeda & Miyake): Offshore of East China Sea. (20, 26)

Oregonia gracilis Dana: Bohai; Yellow Sea. Shallow water to water depth 370 m. (20, 26)

Paratymolus sexspinosus Miers: Hainan shoreline. (20)

Perinea tumida Dana: Xisha (=Paracel Islands). (7, 20, 26)

Phalangipus filiformis (Rathbun): Taiwan Strait; Taiwan; Guangdong; Guangxi. Shallow sea. (18, 20, 26)

P. hystrix (Miers): Taiwan Strait; Taiwan; Hainan. Water depth 50–100 m. (18, 20, 26)

P. longipes (Linnaeus): Taiwan Strait; Guangdong. Water depth 10–100 m. (18, 20, 26)

Platymaia alcocki Rathbun: East China Sea. Water depth 150–400 m. (16)

P. bartschi Rathbun: East China Sea. Water depth 250 m. (16)

P. fimbriala Rathbun: East China Sea. (20, 26)

Pleistacantha japonica (Yokoya): East China Sea. (20, 26)

P. oryx Ortmann: East China Sea; Guangdong. Water depth 100–300 m. (20, 26)

P. simplex Rathbun: East China Sea. (20, 26)

Prosphorachaeus suluensis (Rathbun): East China Sea. (20, 26)

Pseudocollodes demanni Balss: East China Sea. (20, 26)

Pugettia incisa (de Haan): Fujian. Water depth 50–100 m. (20, 26)

P. minor Ortmann: Shandong; Fujian. Low intertidal to shallow sea. (20, 26)

P. nipponensis Rathbun: East China Sea. Water depth 50–150 m. (20, 26)

P. quadridens (de Haan): China's coast. Low intertidal to shallow sea. (20, 26)

Schizophrys aspera (H. Milne-Edwards): Guangdong; Hong Kong; South China Sea; Xisha (=Paracel Islands). Shoreline and coral reefs. (7, 20, 26, 28)

Scyra compressipes Stimpson: Yellow Sea. Water depth 10–168 m. (20, 26)

Tiarinia angusta Dana: Xisha (=Paracel Islands). Coral reefs. (7, 20, 26)

T. depressa Stimpson: Taiwan. (30)

Trigonothir camelus (Klunzinger): Huangyan Island (Zhongsha=Maccslesfield Bank). Coral reefs. (20, 26)

Tylocarcinus sinensis Dai et al.: Xisha (=Paracel Islands). Coral reefs. (20, 26)

T. styx (Herbst): Taiwan; Hainan; Xisha (=Paracel Islands). Low intertidal to shallow water coral reefs and rocks. (7, 20, 26)

Xenocarcinus depressus Miers: Hainan; Xisha (=Paracel Islands). Coral reefs. (7, 20, 26)

PARTHENOPIDAE

Aethra scruposa (Linnaeus): Xisha (=Paracel Islands). (20, 26)

Ceratocarcinus trilobatus (Sakai): Fujian. Commensal with sea pansy (*Pteroeides*). (20, 26)

Cryptopodia collifer Flipse: South China Sea. Water depth 88 m. (20, 26)

C. fronicata (Fabricius): Fujian; Taiwan; Taiwan Strait; Guangdong; Hainan; Nansha (=Spratly Islands). Water depth 12–67 m. (15, 18, 20, 26)

C. sinica Chen: Nansha (=Spratly Islands). Water depth 37 m. (15)

Daldorfia acuta (Klunzinger): Xisha (=Paracel Islands). Coral reefs. (15)

D. horrida (Linnaeus) [*Parthenope horrida*]: Taiwan; Hainan; Xisha (=Paracel Islands). Intertidal to water depth 125 m. (4, 20, 26)

Echinoecus pentagonus (H. Milne-Edwards): Xisha (=Paracel Islands) shallow sea. Commensal with sea urchins. (4, 20, 26)

Harrovia bituberculata Dai & Chen: Hainan. (3, 20, 26)

H. elegans de Man: Taiwan Strait. Water depth 63–70 m. (18)

H. frontodentata Chen & Dai: Guangdong. (3, 20, 26)

H. longipes Chen: Nansha (=Spratly Islands). Water depth 94 m. (15)

H. tuberculata Haswell: Nansha (=Spratly Islands). Water depth 65 m. (15)

Heterocrypta investigatoris Alcock: Guangdong; Beibuwan (=Gulf of Tongking); Nansha (=Spratly Islands). Water depth 40–50 m. (15, 20, 26)

H. transitans Ortmann: Taiwan Strait; South China Sea. Water depth 35–150 m. (18, 20, 26)

Parthenope brevibrachiatus Chen & Dai: South China Sea. (3, 20, 26)

P. calappoides Adams & White: Beibuwan (=Gulf of Tongking); South China Sea. Water depth 30–50 m. (20, 26)

P. curvispinus (Miers): Nansha (=Spratly Islands). Water depth 46 m. (15)

P. erosa (Miers): Nansha (=Spratly Islands). Coral reefs. (15)

P. hayamaensis (Sakai): South China Sea. Water depth 65 m. (20, 26)

P. lamellifrons (Adams & White): Taiwan; Hainan. Water depth 50–100 m. (20, 26)

P. longimanus (Linnaeus): East China Sea; Beibuwan (=Gulf of Tongking); South China Sea. Water depth 60–70 m. (15, 20, 26)

P. longispinis (Miers): East China Sea; Guangdong; Hainan; Nansha (=Spratly Islands). Water depth 20–180 m. (15, 20, 26)

P. sinensis Chen: Beibuwan (=Gulf of Tongking). Water depth 26 m. (3, 20, 26)

P. tuberculosus (Stimpson): Fujian; Taiwan Strait; Guangdong; Guangxi; Hainan. Water depth 30 to 65 m. (18, 20, 26)

P. turriger (Adams & White): Nansha (=Spratly Islands). Water depth 46–65 m. (15, 20, 26)

P. validus de Haan: Bohai; Yellow Sea; East China Sea; South China Sea. Shallow water. (20, 26)

P. validus forma *intermedius* (Miers): East China Sea. Water depth 200 m. (16)

Rhabdontus pictus H. Milne-Edwards: Beibuwan (=Gulf of Tongking); South China Sea. Water depth 48–63 m. (20, 26)

Tutankhamen pteromerus (Ortmann): East China Sea; South China Sea. Water depth 50–210 m. (20, 26)

Zebrida adamsi White: South China Sea shallow water. Commensal with sea urchins. (20, 26)

CORYSTIDAE

Gomeza bicornis Gray: East China Sea. Water depth 30–50 m. (20, 26)

Jonas distincta (de Haan): Fujian; Taiwan Strait; Guangdong. Water depth 27–90 m. (18, 20, 26)

Podocatactes hamifer Ortmann: East China Sea. Water depth 50–550 m. (20, 26)

ATELECYCLIDAE

Kraussia integra (de Haan): Fujian. Intertidal. (20, 26)

K. nitida Stimpson: Fujian; Taiwan Strait. Low intertidal to water depth 44 m. (18, 20, 26)

K. obliquefrons Yang: Daya Bay (Guangdong). Intertidal. (20, 26)

K. rugulosa (Krauss): Taiwan; Xisha (=Paracel Islands). Intertidal and coral reefs. (20, 26)

CANCRIDAE

Cancer amphioetus Rathbun [*C. pygmaeus* Shen, 1932, 1964]: Bohai; northern Shandong Peninsula. Low intertidal. (20, 26)

C. gibbosulus (de Haan): East Liaoning Peninsula. Water depth 30–100 m. (20, 26)

PORTUNIDAE

Caphyra laevis (H. Milne-Edwards): Xisha (=Paracel Islands). Commensal with sea pansy (*Pteroeides*). (4, 20, 26)

C. rotundifrons (H. Milne-Edwards): Xisha (=Paracel Islands). (7, 20, 26)

Carupa laeviuscula Heller: Hainan; Xisha (=Paracel Islands). (4)

C. tenuipes Dana: Xisha (=Paracel Islands); Nansha (=Spratly Islands). Coral reefs, intertidal to water depth 80 m. (4, 20, 24, 26)

Catoptrus nitidus H. Milne-Edwards: Hainan; Xisha (=Paracel Islands). Water depth 10 m. (4, 20, 26)

C. truncatifrons (de Haan): Xisha (=Paracel Islands). (20, 26)

Charybdis acuta H. Milne-Edwards: Fujian; Guangdong. Water depth 10–20 m. (20, 26)

C. affinis Dana: Fujian; Taiwan; Guangdong; Guangxi. Shallow sea. (18, 20, 26)

C. anisodon (de Haan): Fujian; Taiwan; South China Sea. (20, 26)

C. annulata (Fabricius): Fujian; Guangdong; Guangxi. Low intertidal. (20, 26)

C. bimaculata (Miers): Shandong Peninsula; Fujian; Taiwan; Nansha (=Spratly Islands). Water depth 20–430 m. (20, 26)

C. callianassa (Herbst): Zhejiang; Fujian; South China Sea. (20, 26)

C. feriatus (Linnaeus) [*C. cruciata* Shen, 1964]: Fujian; Taiwan; Guangdong; Guangxi. (20, 26)

C. hellerii (H. Milne-Edwards): Fujian; Taiwan Strait; Guangdong; Guangxi. Coral reefs, intertidal to water depth 40 m. (18, 20, 26)

C. hongkongensis Shen: Fujian; Taiwan Strait; Guangdong; Guangxi. Water depth 30- 400 m. (18, 20, 26)

C. japonica (H. Milne-Edwards): China's coast. Low intertidal to water depth 10 m. (18, 20, 26)

C. lucifera (Fabricius): Taiwan. (20, 26)

C. miles (de Haan): Fujian; Taiwan; Guangdong; Guangxi; Nansha (=Spratly Islands). Water depth 10–200 m. (18, 20, 26)

C. natator (Herbst): Fujian; Taiwan; Guangdong; Guangxi. Shallow sea. (20, 26)

C. obtusifrons Leene: Xisha (=Paracel Islands). Coral reefs. (20, 26)

C. orientalis Dana: Hainan. Water depth 10–30 m. (20, 26)

C. riversandersoni Alcock: East China Sea. Water depth 30–100 m. (16, 20, 26)

C. truncata (Fabricius): Fujian; Taiwan Strait; Guangdong; Guangxi; Nansha (=Spratly Islands). Water depth 10–100 m. (18, 20, 24, 26)

C. vadorum Alcock: Fujian; Taiwan Strait; Guangdong; Guangxi. Shallow sea. (18, 20, 26)

C. variegata (Fabricius): Fujian; Taiwan Strait; Guangdong. Water depth 27–65 m. (18, 20, 26)

C. variegata brevispinosa Leene: Hainan; South China Sea. Water depth about 30 m. (20, 26)

Libystes edwardsi Alcock: Guangdong; Nansha (=Spratly Islands). Water depth 53 m. (20, 24, 26)

L. villosus Rathbun: Xisha (=Paracel Islands). (4)

Lissocarcinus arkati Kemp: Guangdong; Xisha (=Paracel Islands). Water depth 10–65 m. (20, 24, 26)

L. orbicularis Dana: Xisha (=Paracel Islands); Nansha (=Spratly Islands). Commensal with sea cucumbers. (4, 20, 24, 26)

L. polybioides Adams & White: Taiwan Strait; Xisha (=Paracel Islands). Water depth 50–72 m. (18, 24, 28)

Lupocyclus inaequalis (Walker): Nansha (=Spratly Islands). Water depth 46 m. (24)

L. philippinensis Semper: Guangxi; Hainan; Nansha (=Spratly Islands). Water depth 35–100 m. (20, 24, 26)

L. rotundatus Adams & White: Nansha (=Spratly Islands). Water depth 67 m. (24)

L. tugelae Barbard: Taiwan Strait; Nansha (=Spratly Islands). Water depth 44–73 m. (18, 24)

Macropipus corrugatus (Pennant): Fujian; Taiwan Strait. Water depth 30–120 m. (18, 20, 26)

Ovalipes iridescens (Miers): East China Sea. Water depth about 200 m. (16, 20, 26)

O. punctatus (de Haan): From southern Yellow Sea to Fujian. Water depth 10–65 m. (16, 20, 26)

Parathranites orientalis (Miers): East China Sea. Water depth 80–230 m. (20, 26)

Podophthalmus nacreus Alcock: Nansha (=Spratly Islands). Water depth 46–97 m. (24)

P. vigil (Fabricius): Taiwan; Guangxi; Hainan. Shallow sea. (20, 26)

Portunus argentatus (White) [*Neptunus argentatus*]: Fujian; Taiwan Strait; Guangdong; Hainan; Nansha (=Spratly Islands). Water depth 30–100 m. (18, 20, 24, 26)

P. brockii (de Man): Hainan. Water depth 30 m. (20, 26)

P. dayawanensis Chen: Daya Bay (Guangdong). (12)

P. gracilimanus (Stimpson): Fujian; Taiwan Strait; Guangdong; Hainan. Shallow sea. (18, 20, 26)

P. granulatus (H. Milne-Edwards): Taiwan; Xisha (=Paracel Islands). Coral reefs. (4, 20, 26)

P. haanii (Stimpson) [*P. gladiator* Shen, 1964]: Fujian; Taiwan Strait; Guangdong; Guangxi; Nansha (=Spratly Islands). Water depth 10–100 m. (18, 20, 24, 26)

P. hainanensis Chen: Hainan. (12)

P. hastatoides (Fabricius): Fujian; Guangdong; Guangxi. Low intertidal to water depth 82 m. (20, 26)

P. iranjae Crosnier: Xisha (=Paracel Islands). Intertidal. (7, 20, 26)

P. orbitosinus Rathbun: Nansha (=Spratly Islands). Water depth 67 m. (24)

P. pelagicus (Linnaeus): Zhejiang; Fujian; Taiwan; Guangdong; Guangxi. Water depth 10–30 m. (20, 26)

P. pubescens (Dana): Hainan. Water depth 20–30 m. (20, 26)

P. pulchricristatus Gordon: Guangdong; Hainan; Nansha (=Spratly Islands). Water depth 10–97 m. (20, 24, 26)

P. sanguinolentus (Herbst): Fujian; Taiwan; Guangdong; Guangxi. Water depth 10–30 m. (20, 26)

P. tenuipes (de Haan): Guangdong. Water depth 15–35 m. (20, 26)

P. tridentatus Yang & Dai: Xisha (=Paracel Islands). Coral reefs. (20, 26)

P. trituberculatus (Miers): China's coast. Water depth 10–30 m. (16, 20, 26)

P. tuberculosus (H. Milne-Edwards): Nansha (=Spratly Islands). Water depth 20–85 m. (20, 24, 26)

P. tweediei Shen: Guangdong shallow sea. (20, 26)

Scylla serrata (Forskål): From Zhejiang to Guangxi. River mouth low salinity intertidal and shallow subtidal. (20, 26)

Thalamita admete (Herbst): Taiwan; Hainan; Xisha (=Paracel Islands); Nansha (=Spratly Islands). Intertidal to water depth 32 m. (20, 24, 26)

T. chaptali (Audouin & Savigny): Hainan. Water depth about 10 m, coral reefs. (20, 26)

T. coeruleipes Jacquinot & Lucas: Hainan; Xisha (=Paracel Islands). Coral reefs. (4, 20, 26)

T. corrugata Stephenson & Rees: Xisha (=Paracel Islands); Nansha (=Spratly Islands). Coral reefs. (20, 24, 26)

T. crenata (Latreille): Zhejiang; Fujian; Guangdong; Guangxi. Low intertidal and coral reefs. (20, 26)

T. danae Stimpson: Guangxi; Hainan; Xisha (=Paracel Islands). Coral reefs and shallow sea. (20, 26)

T. demani Nobili: Xisha (=Paracel Islands). Water depth about 10 m, coral reefs. (20,2 6)

T. edwardsi Borradaile: Xisha (=Paracel Islands). (20, 26)

T. imparimana Alcock: Nansha (=Spratly Islands). Water depth 53–74 m. (24)

T. integra Dana: Hainan; Xisha (=Paracel Islands). Water depth 7–9 m. (20, 26)

T. kagosshimaensis Sakai: Nansha (=Spratly Islands). Water depth 53 m. (24)

T. picta Stimpson: Fujian; Taiwan; Guangdong. Coral reefs and shallow sea. (4, 20, 26)

T. platypedis Dai & Chen: Sanya (Hainan). Water depth about 20 m and coral reefs. (20, 26)

T. procorrugata Dai & Chen: Hainan. On corals of water depth about 20 m. (20, 26)

T. prymna (Herbst): Taiwan; Guangxi; Hainan; Xisha (=Paracel Islands). Low intertidal and coral reefs. (4, 20, 26)

T. sima H. Milne-Edwards: Fujian; Taiwan; Guangdong. Low intertidal and shallow sea. (20, 26)

T. spinifera Borrodaile: Nansha (=Spratly Islands). Water depth 67 m. (24)

T. spinimana Dana: Hainan; Xisha (=Paracel Islands). Water depth about 30 m. (7, 20, 26)

T. stephensoni Crosnier: Hainan. Coral reefs. (20, 26)

T. tenuipes Borradaile: Xisha (=Paracel Islands). Coral reefs. (20, 26)

Thalmitoides quadridens H. Milne-Edwards: Nansha (=Spratly Islands). Coral reefs. (24)

T. tridens H. Milne-Edwards: Xisha (=Paracel Islands). (7, 20, 26)

GERYONIDAE

Geryon trispinosus (Herbst): East China Sea continental shelf. Water depth 80–300 m. (16)

XANTHIDAE

Actaea amoyensis (de Man): Fujian; Taiwan Strait. (20, 26)

A. bocki Odhner: Taiwan. Water depth 35–120 m. (20, 26)

A. calculosa (H. Milne-Edwards): China's coast. (20, 26)

A. cavipes (Dana): Xisha (=Paracel Islands). Coral reefs. (20, 26)

A. modesta (de Haan): Hainan. (20, 26)

A. nodulosa White: Zhongsha (=Macclesfield Bank). (20, 26)

A. polyacantha (Heller): Hainan. (20, 26)

A. pulchella (H. Milne-Edwards): Fujian; Taiwan; Guangxi. Shallow sea or coral reefs. (20, 26)

A. pura Stimpson: Southwestern Dapeng Cove (Hong Kong) shallow water. (28)

A. savignyi (H. Milne-Edwards): Zhejiang; Fujian; Guangdong. Intertidal to water depth 300 m. (20, 26)

A. superciliaris Odhner: Hainan. Low intertidal. (20, 26)

Actaeodes areolatus (Dana): Hainan. Coral reefs. (20, 26)

A. hirsutissimus (Rüppell): Hong Kong; Xisha (=Paracel Islands). Coral reefs. (20, 26, 28)

A. tomentosa (H. Milne-Edwards): Fujian; Taiwan; Hainan; Xisha (=Paracel Islands). Shallow sea and coral reefs. (20, 26)

Actumnus dorsipes (Stimpson): Taiwan Strait; Guangdong; Hong Kong. Water depth 10–50 m, rock crevice or coral reefs. (18, 20, 26)

A. forficigerus (Stimpson): Tongan Bay (Fujian); Taiwan Strait. Water depth about 20 m to 70 m. (17, 18)

A. setifer (de Haan): Tongan Bay (Fujian); Guangdong. Shallow sea rocky crevices or coral reefs. (4, 18, 20, 26, 28)

Atergatis floridus (Linnaeus): Taiwan; Hainan; Xisha (=Paracel Islands). Low intertidal and coral reefs. (20, 26)

A. integerrimus (Lamarck): Guangdong; Hainan. Water depth 10–30 m. (20, 26)

A. reticulatus de Haan: Zhejiang; Fujian; Guangdong. Low intertidal to water depth 20 m. (20, 26)

Banareia banareias (Rathbun): Xisha (=Paracel Islands). On coral branches. (20, 26)

B. subglobosa (Stimpson): Guangdong; Hong Kong. Water depth 10–35 m, on soft corals. (20, 26)

Baptozius vinosus (H. Milne-Edwards): Hainan. River mouth of mangrove forests. (20, 26)

Calvactaea tumida Ward: Hainan. Shallow water, on soft corals. (20, 26)

Carpilius converxus (Forskål): Taiwan; Hainan; Xisha (=Paracel Islands). Coral reefs. (20, 26)

C. maculatus (Linnaeus): Taiwan; Hainan; Xisha (=Paracel Islands). Coral reefs. (20, 26)

Chlorodiella barbata (Borradaile): Xisha (=Paracel Islands). Coral reefs. (20, 26)

C. bidentata (Nobili): Hainan; Xisha (=Paracel Islands). Coral reefs. (20, 26)

C. crispipleopa Yang: Taiwan; Xisha (=Paracel Islands). Coral reef seaweed. (20, 26)

C. cytherea (Dana): Hainan. (20, 26)

C. laevissima (Dana): Taiwan; Xisha (=Paracel Islands). Coral reefs. (20, 26)

C. nigra (Forskål): Taiwan; Hong Kong; Hainan; Xisha (=Paracel Islands). Subtidal under rocks or coral reefs. (20, 26, 28)

C. xishaensis Chen & Lan: Yongxing Island and Zhaoshu Island (Xisha=Paracel Islands). Coral reefs. (20, 26)

Cymo andreossyi (Aueowis): Xisha (=Paracel Islands). (20, 26)

C. deplanatus H. Milne-Edwards [*C. melanodactylus* Chen, 1978]: Xisha (=Paracel Islands). Coral reefs. (20, 26)

C. melanodactylus de Haan: Guangdong; Xisha (=Paracel Islands). Coral reefs. (20, 26)

C. quadrilobatus Miers: Xisha (=Paracel Islands). Coral reefs. (20, 26)

Dacryopiliumnus rathbunae Balss: Hainan. Coral reefs. (20, 26)

Daira perlata (Herbst): Hainan; Xisha (=Paracel Islands). On corals. (20, 26)

Demania baccalipes (Alcock): Guangdong. Water depth 15–35 m. (20, 26)

D. rotundata Serene: Taiwan; Guangdong. Water depth 35–150 m. (20, 26)

D. scaberrima (Walker) [*Xantho reynaudii* Shen, 1964]: Zhejiang; Fujian; Taiwan Strait; Guangdong; Guangxi. Water depth 15–65 m. (18, 20, 26)

Domecia glabra Alcock: Hainan; Xisha (=Paracel Islands). Coral reefs. (20, 26)

D. hispida Eydoux & Souleyet: Hainan; Xisha (=Paracel Islands). Coral reefs. (20, 26)

Epixanthus corrosus H. Milne-Edwards: Xisha (=Paracel Islands). Coral reefs. (20, 26)

E. frontalis (H. Milne-Edwards): Taiwan; Guangdong; Guangxi; Xisha (=Paracel Islands). Low intertidal. (20, 26)

Eriphia scabricula Dana: Taiwan; Xisha (=Paracel Islands). Intertidal rocky crevices or coral reefs. (20, 26)

E. sebana (Shaw & Nodder) [*E. laevimana*]: Guangdong; Xisha (=Paracel Islands). Rock crevice or coral reefs. (20, 26)

E. smithi Macleay [*E. laevimanus* v. *smithi*]: Fujian; Hainan. Low intertidal rocky crevices or coral reefs. (20, 26)

Etisus anaglyptus H. Milne-Edwards: Guangxi; Xisha (=Paracel Islands). Subtidal. (20, 26)

E. bifrontalis Edmondson: Xisha (=Paracel Islands). Coral reefs. (20, 26)

E. demani Odhner: Xisha (=Paracel Islands). Coral reefs. (20, 26)

E. dentatus (Herbst): Hainan; Xisha (=Paracel Islands). Shallow sea and coral reefs. (20, 26)

E. electra (Herbst): Taiwan; Hainan. Coral reefs. (20, 26)

E. frontalis Dana: Xisha (=Paracel Islands). Coral reefs. (20, 26)

E. laevimanus Randall: Hainan; Xisha (=Paracel Islands). Subtidal. (20, 26)

E. splendidus Rathbun: Zhongsha (=Macclesfield Bank). Coral reefs. (20, 26)

E. utilis Lucas: Zhongsha (=Macclesfield Bank). (20, 26)

Euxanthus exsculptus (Herbst): Hainan; Xisha (=Paracel Islands). Coral reefs. (20, 26)

E. sculptilis Dana: Xisha (=Paracel Islands). Coral reefs. (20, 26)

Galene bispinosa (Herbst): Fujian, Taiwan, Guangdong, Guangxi. Shallow sea. (20, 26)

Glabropilumnus dispar (Dana): Xisha (=Paracel Islands). Coral reefs. (16, 20, 26)

G. sodalis (Alcock): East China Sea; Taiwan. Water depth 32–210 m. (16, 20, 26)

Globopilumnus globosus (Dana): Taiwan; Guangdong. Shallow sea and coral reefs. (20, 26)

Halimede ochtodes (Herbst): Guangxi; Hainan. Water depth 20–50 m. (20, 26)

H. tyche (Herbst) [*H. fragifer* Shen, 1964]: Fujian; Taiwan Strait; Hainan. (18, 20, 26)

Heteropanope glabra Stimpson: Guangdong; Xiamen (Fujian). Intertidal. (18, 20, 26)

Heteropilumnus ciliatus (Stimpson): Shandong Peninsula; Zhejiang; Fujian; Hong Kong. (7, 20, 26, 28)

H. subinteger (Lanchester): Fujian; Guangdong; Guangxi. Intertidal to shallow sea. (18, 20, 26)

Lachnopodus bidentatus (H. Milne-Edwards): Xisha (=Paracel Islands). Coral reefs. (20, 26)

L. subacutus (Stimpson): Xisha (=Paracel Islands). Coral reefs. (20, 26)

Leptodius danae (Odhner): Xisha (=Paracel Islands). Coral reefs. (20, 26)

L. exaratus (H. Milne-Edwards): Fujian; Taiwan; Guangdong; Xisha (=Paracel Islands). Low intertidal and coral reefs. (20, 26)

L. gracilis (Dana): Guangdong; Xisha (=Paracel Islands). Low intertidal and coral reefs. (20, 26)

L. nigromaculatus Serene: Guangxi; Xisha (=Paracel Islands). Intertidal and coral reefs. (20, 26)

L. sanguineus (H. Milne-Edwards): Taiwan; Hainan; Xisha (=Paracel Islands). Shallow sea under rocks or on corals. (20, 26)

Liagore rubromaculata (de Haan): Fujian; Taiwan Strait; Hainan. Water depth 15–65 m and coral reefs. (18, 20, 26)

Liomera bella (Dana): Xisha (=Paracel Islands). Shallow sea and coral reefs. (20, 26)

L. caelata (Odhner): China's coast. Coral reefs. (20, 26)

L. cinctimana (White): Xisha (=Paracel Islands). (20, 26)

L. laevis H. Milne-Edwards: Southwestern Dapeng Cove (Hong Kong); Xisha (=Paracel Islands). Coral reefs. (20, 26, 28)

L. margaritata (H. Milne-Edwards): Hong Kong; Guangxi. Coral reefs. (20, 26, 28)

L. monticulosa (H. Milne-Edwards): Xisha (=Paracel Islands). Shallow water coral reefs. (20, 26)

L. rugata (H. Milne-Edwards): Hainan. Coral reefs. (20, 26)

L. stimpsoni (H. Milne-Edwards): Xisha (=Paracel Islands). Low intertidal to water depth 30 m. (20, 26)

L. venosa (H. Milne-Edwards): Hong Kong; Guangxi; Hainan. Shallow sea and coral reefs. (20, 26, 28)

Lophozozymus pictor (Fabricius): Hainan. Low intertidal to water depth 30 m or coral reefs. (20, 26)

L. pulchellus H. Milne-Edwards: Xisha (=Paracel Islands). (20, 26)

Lybia caestifer (Alcock): Xisha (=Paracel Islands). Coral reefs. (20, 26)

L. tessellata (Latreille): Xisha (=Paracel Islands). (20, 26)

Lydia annulipes (H. Milne-Edwards): Taiwan; Xisha (=Paracel Islands). Coral reefs. (20, 26)

Macromedaeus crassimanus (H. Milne-Edwards): Xisha (=Paracel Islands). Coral reefs. (20, 26)

M. distinguendus (de Haan): China's coast. Low intertidal. (20, 26)

Maldivia palmyrensis Rathbun: Xisha (=Paracel Islands). Coral reefs. (20, 26)

M. triunguiculata (Borradaile): Xisha (=Paracel Islands). Coral reefs. (20, 26)

Medaeops granulosus (Haswell): Guangdong. Low intertidal. (20, 26)

Nanocassiops granulipes (Sakai): East China Sea. Water depth 30–120 m. (20, 26)

Neoliomera pubescens (H. Milne-Edwards): Xisha (=Paracel Islands). Shallow water coral reefs. (20, 26)

Neoxanthias impressus (Lamarck): Xisha (=Paracel Islands). Coral reefs. (20, 26)

N. lineatus (H. Milne-Edwards): Nansha (=Spratly Islands). (24)

N. michelae Seréne & Vadon: Nansha (=Spratly Islands). (24)

Ozius guttatus H. Milne-Edwards: Hainan. (20, 26)

O. rugulosus Stimpson: Taiwan; Xisha (=Paracel Islands). Coral reefs. (20, 26)

O. tuberculosus H. Milne-Edwards: Hainan. Low intertidal under rocks or coral reefs. (20, 26)

Paractaea garretti (Rathbun): Hainan. (20, 26)

P. orientalis (Odhner) [*Actaea orientalis*]: Shandong Peninsula; Fujian; Guangdong. Shallow sea. (17, 18, 20, 26)

P. philippinensis (Ward): Xisha (=Paracel Islands). Coral reefs. (20, 26)

P. rueppellii (Krauss) [*Actaea rueppellii*]: Guangdong. Water depth 15–30 m. (20, 26)

P. tumulosa (Odhner): Xisha (=Paracel Islands). Coral reefs. (20, 26)

Paramedaeus simplex (H. Milne-Edwards): Hong Kong. Subtidal. (28)

Parapanope euagora de Man: Jiaozhou Bay (Shandong); Zhejiang; Fujian; Taiwan Strait; Guangdong. Shallow sea and coral reefs. (18, 20, 26)

Paraxanthias elegans (Stimpson): Taiwan; Hainan. (20, 26)

P. notatus (Dana): Xisha (=Paracel Islands). Coral reefs. (20, 26)

P. pachydactylus H. Milne-Edwards: Xisha (=Paracel Islands). Coral reefs. (20, 26)

Phymodius drachi Guinot: Xisha (=Paracel Islands). Coral reefs. (20, 26)

P. granulosus (de Man): Xisha (=Paracel Islands). Coral reefs. (20, 26)

P. laysani Rathbun: Xisha (=Paracel Islands). Coral reefs. (20, 26)

P. monticulosus (Dana): Hainan; Xisha (=Paracel Islands). Coral reefs. (20, 26)

P. nitidus Dana: Xisha (=Paracel Islands). Coral reefs. (20, 26)

P. ungulatus (H. Milne-Edwards): Xisha (=Paracel Islands). Coral reefs. (20, 26)

Pilodius areolatus (H. Milne-Edwards): Hainan; Xisha (=Paracel Islands). Coral reefs. (20, 26)

P. granulatus Stimpson: Guangdong; Hong Kong; Guangxi. Coral reefs. (20, 26, 28)

P. melanospinis (Rathbun): Xisha (=Paracel Islands). Coral reefs. (20, 26)

P. nigrocrinitus Stimpson: Weizhou Island (Guangxi). Coral reefs. (20, 26)

P. pubescens Dana: Xisha (=Paracel Islands). Coral reefs. (20, 26)

P. pugil Dana: Xisha (=Paracel Islands). Coral reefs. (20, 26)

P. scabriculus Dana: Xisha (=Paracel Islands). Coral reefs. (20, 26)

Pilumnopeus eucratoides (Stimpson): Fujian; Guangdong. Shallow sea. (20, 26)

P. indica (de Haan): Guangdong; Guangxi. Shallow sea or coral reefs. (15, 20, 26)

P. makiana (Rathbun): Eastern Liaoning Peninsula; Shandong Peninsula; Fujian; Taiwan. Intertidal and shallow sea. (16, 20, 26)

P. trispinosus Sakai: Hong Kong; Guangxi. Shallow sea or dead corals. (20, 21, 26, 28)

Pilumnus andersoni de Man: Hong Kong. (28)

P. cursor H. Milne-Edwards: Guangxi shallow sea. (20, 26)

P. interifrontus Shen: Jiaozhou Bay (Shandong). Shallow water. (20, 26)

P. longicornis Hilgendorf: Hong Kong. Subtidal. (28)

P. minutus de Haan: Shandong Peninsula; Taiwan Strait; Guangdong; Hong Kong. Water depth 10–58 m. (18, 20, 26, 28)

P. orbitospinis Rathbun: Taiwan Strait. Water depth 30–58 m. (18)

P. rotumanus Borradaile: Xisha (=Paracel Islands). Coral reefs. (20, 26)

P. sinensis Gordon: Fujian; Guangdong; Guangxi. Shallow sea. (20, 26)

P. spinulus Shen: Shandong Peninsula. Low intertidal. (20, 26)

P. tuantaoensis Shen: Shandong Peninsula. Shallow water. (20, 26)

P. vespertilio (Fabricius): Fujian; Hainan; Xisha (=Paracel Islands). Coral reefs or low intertidal to water depth 13 m. (18, 20, 26)

Platypodia anaglypta (Heller): Xisha (=Paracel Islands). Coral reefs. (20, 26)

P. granulosa (Rüppell): Taiwan; Xisha (=Paracel Islands). Coral reefs. (20, 26)

P. semigranosa (Heller): Xisha (=Paracel Islands). Coral reefs, commensal with sponges. (20, 26)

Pseudoliomera granosimana (H. Milne-Edwards): Hainan; Xisha (=Paracel Islands). Coral reefs. (20, 26)

P. helleri (H. Milne-Edwards): South China Sea. Water depth 15–35 m. (20, 26)

P. speciosa (Dana): Fujian; Taiwan Strait; Xisha (=Paracel Islands). Shallow sea and coral reefs. (18, 20, 26)

P. variolosa Borradaile: Xisha (=Paracel Islands). Coral reefs. (20, 26)

Pseudozius caystrus (Adams & White): Xisha (=Paracel Islands). Coral reefs. (20, 26)

Sphaerozius nitidus Stimpson: Shandong Peninsula; Zhejiang; Fujian; Guangdong; Hong Kong. Shallow sea rocks. (20, 26, 28)

Tetralia glaberrima (Herbst): Xisha (=Paracel Islands). On coral branches. (20, 26)

T. heterodactyla Heller: Xisha (=Paracel Islands). On coral branches. (20, 26)

T. heterodactyla fusca Seréne: Xisha (=Paracel Islands). On coral branches. (20, 26)

Trapezia aerolata Dana: Xisha (=Paracel Islands). On corals. (20, 26)

T. cymodoce (Herbst): Hainan; Xisha (=Paracel Islands). On corals. (20, 26)

T. digitalis Latreille: Hainan; Xisha (=Paracel Islands). On corals. (20, 26)

T. ferruginea Latreille: Hainan, Xisha (=Paracel Islands). On corals. (20, 26)

T. flavopunctata Eydoux & Souleyet [*T. maculata*]: Xisha (=Paracel Islands). (20, 26)

T. formosa Smith: Xisha (=Paracel Islands). On corals. (20, 26)

T. guttata Rüppell: Hainan, Xisha (=Paracel Islands). On corals. (20, 26)

T. reticulata Stimpson: Hainan; Xisha (=Paracel Islands). On corals (20, 26)

T. rufopunctata (Herbst): Xisha (=Paracel Islands). Coral reefs. (20, 26)

T. speciosa Dana: Xisha (=Paracel Islands). (20, 26)

Xanthias canaliculatus Rathbun: Xisha (=Paracel Islands). Coral reefs. (20, 26)

X. lamarckii (H. Milne-Edwards): Taiwan; Hainan; Xisha (=Paracel Islands). Coral reefs. (20, 26)

X. samoensis Ward: Xisha (=Paracel Islands). Coral reefs. (20, 26)

Zozymodes cavipes (Dana): Xisha (=Paracel Islands). Coral reefs. (20, 26)

Zozymus aeneus (Linnaeus): Taiwan; Hainan; Xisha (=Paracel Islands). Shallow sea coral reefs. (20, 26)

GONEPLACIDAE

Camatopsis rubida Alcock & Anderson: Fujian; Taiwan Strait. Water depth about 20 to 65 m. (18)

Carcinoplax bispinosa Rathbun: South China Sea. Water depth 160–180 m. (10)

C. longimana (de Haan): Zhejiang; Fujian; Taiwan Strait; Guangdong. Water depth 30–100 m. (16, 18, 20, 26)

C. longipes (Wood-Mason): East China Sea. Water depth 550 m. (10)

C. longispinosa Chen: South China Sea. Water depth 856–1,100 m. (10)

C. meridionalis Rathbun: East China Sea; Fujian; Taiwan Strait. Water depth about 20 m to 133 m. (10, 18)

C. purpurea Rathbun: Taiwan Strait; Guangdong; Guangxi. (18, 20, 26)

C. sinica Chen: Guangdong; Guangxi. Water depth 100–150 m. (20, 26)

C. surugensis Rathbun: East China Sea; Taiwan Strait. Water depth 50–120 m. (18, 20, 26)

C. vestita (de Haan): Shandong Peninsula. Water depth 30–100 m. (20, 26)

Ceratoplax obtusignathus Dai & Song: Longmen (Guangxi). (21)

Chasmocarcinops gelasimoides Alcock: Guangdong; Hainan. Water depth 10 m. (20, 26)

Eucrate alcocki Seréne [*E. maculata* Yang, 1979]: Fujian; Taiwan Strait; Guangdong. Water depth about 11–50 m. (18, 20, 26)

E. costata Yang & Sun: Fujian. (20, 26)

E. crenata de Haan: Bohai; Shandong Peninsula; Fujian; Taiwan Strait; Guangdong. Low intertidal to water depth about 100 m. (18, 20, 26)

E. haswelli Campbell: Fujian. Water depth about 30–100 m. (20, 26)

E. solaris Yang & Sun: Guangdong; Hainan. Water depth about 50 m. (20, 26)

Goneplax renoculis Rathbun: East China Sea. Water depth 50–100 m. (20, 26)

Heteroplax dentata Stimpson: Hong Kong. Water depth 20–30 m. (20, 26)

H. nagasakiensis Sakai: Fujian; Taiwan Strait. Water depth about 10 m to 60 m. (18)

H. nitida Miers: Fujian; Taiwan Strait. Water depth about 10 m to 44 m. (18)

Hexapus anfractus Rathbun: Fujian; Taiwan Strait; eastern Guangdong. Intertidal to water depth 58 m. (17, 18)

H. granuliferus Campbell & Stephenson: Fujian; Guangdong. Shallow sea. (18, 20, 26)

Lophoplax sculpta (Stimpson): Hainan. Water depth 10–20 m. (20, 26)

L. teschi Seréne: Hainan. Water depth about 70 m. (20, 26)

Ommatocarcinus macgillivrayi White: Central and southern Fujian; Taiwan Strait; Hainan. Water depth 12–150 m. (18, 20, 26)

O. pulcher Barnard: Guangdong. (20, 26)

Psopheticus insignis Alcock: Nansha (=Spratly Islands). (24)

Scalopidia spinosipes Stimpson: Fujian; Taiwan Strait; Guangdong. Water depth 12–65 m. (18, 20, 26)

Ser fukiensis Rathbun: Fujian. Mud flat to shallow sea. (18, 20, 26)

Thaumatoplax orientalis Rathbun: Guangdong. Shallow sea, commensal with polychaetes. (20, 26)

Typhlocarcinops canaliculata Rathbun: Shandong; Fujian. (20, 26)

T. denticarpes Dai & Yang: Shanwei (Guangdong). (20, 26)

T. ocularia Rathbun: East China Sea. Water depth 75–90 m. (20, 26)

Typhlocarcinus nudus Stimpson: Jiaozhou Bay (Shandong); Fujian; Taiwan Strait; Guangdong. Intertidal to water depth about 70 m. (18, 20, 26)

T. villosus Stimpson: Fujian; Taiwan Strait; Hong Kong. Water depth about 10 m to 60 m. (18, 20, 26)

Xenophthalmodes moebii Richters: Xiamen (Fujian); Taiwan Strait. Water depth 6–69 m. (17, 18)

PINNOTHERIDAE

Anomalifrons lightana Rathbun: Fujian; Guangdong. Intertidal to water depth about 5 m. (17, 18, 20, 26)

Durckheimia caeca Bürger: Hong Kong. Mantle cavity of file clam (*Lima lima*). (31)

Fabia obtusidentata Dai: Guangdong; Hong Kong; Guangxi; Hainan. Mantle cavity of bivalves. (20, 26, 31)

Indopinnixa mortoni Davie: Hong Kong. Low intertidal, inside polychaete mud tubes. (26a)

Mortensenella forceps Rathbun: Hainan. Muddy sand flat. (20, 26)

Neoxenophthalmus obscurus (Henderson): Fujian; Taiwan Strait; Guangdong; Hainan. Intertidal to water depth 500 m. (17, 18, 20, 26)

Pinnixa penultipedalis Stimpson: Jiaozhou Bay (Shandong); Guangdong. Inside polychaete (Nereidae) tubes. (20, 26)

P. tumida Stimpson: Bohai Bay; Shandong Peninsula. Commensal with sea cucumbers (*Parcaudina*). (20, 26)

Pinnotheres boninensis Stimpson: Hainan. Mantle cavity of bivalves. (20, 26)

P. cyclinus Shen: Shandong Peninsula. Mantle cavity of bivalves. (20, 26)

P. dilatatus Shen: Shandong Peninsula. Mantle cavity of bivalves. (20, 26)

P. excussus Dai: Hainan. Mantle cavity of bivalves. (20, 26)

P. gordoni Shen: Dapeng Cove (Hong Kong). Mantle cavity of pen clams (Pinnidae). (31)

P. haiyangensis Shen: Shandong Peninsula; Dapeng Cove (Hong Kong). Mantle cavity of bivalves. (20, 26, 31)

P. latus Bürger Hainan: Mantle cavity of pen clams (Pinnidae). (20, 26)

P. luminatus Dai: Dapeng Cove (Hong Kong); Hainan. Mantle cavity of bivalves. (20, 26)

P. nigrans Rathbun: Hainan. (20, 26)

P. obscurus Stimpson: Hong Kong. Mantle cavity of bivalves. (31)

P. paralatissinus Dai & Song: Beihai (Guangxi). (21)

P. pholadis de Haan [*P. affinis* Shen, 1933, 1964]: Shandong Peninsula; Xiamen (Fujian); Dapeng Cove (Hong Kong). Mantle cavity of bivalves. (17, 20, 26, 31)

P. pilulus Dai: Guangdong; Hainan. Mantle cavity of bivalves. (20, 26)

P. sinensis Shen: Eastern Liaoning Peninsula; Xiamen (Fujian); Dapeng Cove (Hong Kong). Mantle cavity of bivalves. (18, 20, 26, 31)

P. spinidactylus Gordon: Hainan. Mantle cavity of bivalves. (20, 26)

P. tsingtaoensis Shen: Eastern Liaoning Peninsula; Shandong Peninsula. Mantle cavity of bivalves. (20, 26)

Tritodynamia fujianensis Chen: Pingtang (Fujian). (6)

T. hainaensis Dai: Xiamen (Fujian); Hainan. Intertidal to water depth about 10 m. (17, 18, 20, 26)

T. horvathi Nobili: Shandong Peninsula shallow sea. (20, 26)

T. intermedia Shen: Shandong Peninsula; Xiamen (Fujian). Muddy sand flat, water depth 13 m. (18, 20, 26)

T. longipropodum Dai: Fujian; Guangdong. Intertidal to water depth 20 m. (18, 20, 26)

T. rathbunae Shen: Eastern Liaoning Peninsula; Shandong Peninsula. Inside polychaete (Aphroditidae) tubes. (20, 26)

Xanthasia murigera White: Xisha (=Paracel Islands). Mantle cavity of giant clams (Tridacnidae). (4, 20, 26)

Xenophthalmus pinnotheroides White: Bohai; Shandong Peninsula; Fujian; Guangdong; Hainan. Water depth about 10 m. (17, 18, 20, 26)

RETROPLUMIDAE

Retropluma denticulata Rathbun: Guangdong; Guangxi; Hainan; Nansha (=Spratly Islands). Water depth 50–200 m. (15, 20, 26)

R. quadrata Saint Laurent: Nansha (=Spratly Islands). Water depth 206 m. (15)

PALICIDAE

Crossotonotus gardineri Rathbun: Xisha (=Paracel Islands). (7)

C. spinipes (de Haan): Xisha (=Paracel Islands). Shallow sea. (4, 20, 26)

Parapalicus nanshaensis Dai & Xu: Nansha (=Spratly Islands). Water depth 138 m. (24)

P. trituberculatus (Chen) [*Palicus trituberculatus*]: South China Sea. Water depth 66–202 m. (20, 26)

MICTYRIDAE

Mictyris longicarpus Latreille: Fujian; Taiwan; Guangdong; Guangxi. River mouth mud flat. (20, 26)

OCYPODIDAE

Camptandrium elongatum Rathbun: Fujian; Guangdong. Muddy sand flat. (20, 26)

C. sexdentatum Stimpson: Eastern Liaoning Bay; Shandong Peninsula; Fujian; Guangdong; Hainan. Burrowing in muddy intertidal and shallow sea. (17, 18, 20, 26)

Cleistostoma dilatatum de Haan: Bohai Bay; Eastern Liaoning Peninsula; Shandong Peninsula; Fujian; Guangdong. River mouth low intertidal. (20, 26)

Dotilla wichmanni de Man: Dongshan and Xiamen (Fujian); Hong Kong; Hainan. Muddy sand intertidal. (18, 20, 26, 26a)

Ilyoplax dentimerosa Shen: Shandong Peninsula; Dongshan and Xiamen (Fujian). Muddy low intertidal. (18, 20, 26)

I. deschampsi (Rathbun): Bohai Bay; East Liaoning Peninsula; Shandong Peninsula; Jiangsu; mouth of Zhang River (Fujian). River mouth mud flat. (18, 20, 26)

I. formosensis Rathbun: Dongshan and Xiamen (Fujian); Taiwan. River mouth, inner bay mud flat. (18, 20, 26)

I. ningpoensis Shen: Zhejiang; Fujian; Hong Kong. Sandy intertidal. (20, 26, 26a)

I. pingi Shen: Bohai Bay; Eastern Liaoning Peninsula; Shandong Peninsula. River mouth mud flat. (20, 26)

I. serrata Shen: Zhejiang; Fujian. Intertidal, burrower. (20, 26)

I. tansuiensis Sakai: Zhejiang; Fujian; Taiwan; Guangdong. River mouth mud flat. (20, 26)

Leipocten sordidulum Kemp: Fujian; Taiwan. River mouth mud flat. (20, 26)

L. trigranulum Dai & Song: Guangdong; Hainan. Intertidal. (21)

Macrophthalmus abbreviatus: Hong Kong. (26a)

M. banzai: Hong Kong. (26a)

M. (Mopsocarcinus) boscii Ausouin & Savignyi: Taiwan; Guangdong; Guangxi; Hainan; Xisha (=Paracel Islands). Intertidal. (7, 20, 26)

M. (Hemiplax) boteltobagoe (Sakai): Taiwan; Hong Kong. (25, 26a)

M. (Macrophthalmus) brevis (Herbst): South China Sea. Muddy sand flat burrows. (20, 26)

M. (Macrophthalmus) convexus Stimpson: Dongshan and Xiamen (Fujian); Taiwan; Guangdong; Hong Kong; Hainan. River mouth, inner bay mud flat. (18, 20, 26, 26a)

M. (Macrophthalmus) crassipes H. Milne-Edwards: Hainan. (20, 26)

M. (Mareotis) definitus Adams & White: Fujian; Hong Kong; Hainan. Muddy intertidal. (20, 26, 26a)

M. (Macrophthalmus) dentatus Stimpson: Guangdong shallow sea. (20, 26, 26a)

M. (Mareotis) depressus Rüppell: South China Sea. (25)

M. (Macrophthalmus) dilatum (de Haan): China's coast. River mouth mud flat. (20, 26)

M. (Mareotis) erato de Man: Fujian; Guangdong; Hong Kong; Hainan. Muddy intertidal. (20, 26, 26a)

M. (Mareotis) japonicus de Haan: China's coast. River mouth, inner bay intertidal. (20, 26)

M. (Mareotis) japonicus froequens Dai & Song: East China Sea. (25)

M. (Venitus) latreillei (Desmarest): Xiamen (Fujian); Guangdong; Hong Kong. (17, 18, 20, 26, 26a)

M. (Mareotis) pacificus Dana: Guangdong; Hong Kong. (20, 26, 26a)

M. (*Macrophthalmus*) *telescopicus* (Owen): South China Sea. (25)

M. (*Mareotis*) *tomentosus* (Souleyet): Fujian; Hong Kong. Muddy sand intertidal. (20, 26, 26a)

M. (*Macrophthalmus*) *verreauxi* H. Milne-Edwards: Hainan shallow sea. (20, 26)

Ocypode ceratophthalmus (Pallas): Fujian; Taiwan; Guangdong; Guangxi; Hainan; Xisha (=Paracel Islands). Sandy high intertidal. (4, 20, 26)

O. cordimana Desmarest: Fujian; Taiwan; Guangdong; Hong Kong; Hainan; Xisha (=Paracel Islands). Sandy intertidal. (4, 20, 26)

O. mortoni George [*O. macrocera* Dai et al., 1985]: Guangdong; Guangxi; Hainan. Around high intertidal. (26)

O. sinensis Dai et al.: Fujian; Guangdong; Hainan; Xisha (=Paracel Islands). Above high intertidal. (26)

O. stimpsoni Ortmann: Bohai Bay; Shandong Peninsula; Fujian; Taiwan; Guangdong. Burrowing in muddy intertidal. (20, 26)

Paracleistostoma crassipilum Dai: Hainan. Intertidal mud flat. (20, 26)

P. cristatum de Man: Bohai Bay; Shandong Peninsula; Fujian. River mouth mud flat. (20, 26)

P. depressum de Man: Fujian; Guangxi; Hainan. River mouth mud flat. (20, 26)

Scopimera bitympana Shen: Bohai Bay; Shandong Peninsula; Jiangsu; Fujian; Taiwan; Hong Kong; Hainan. Muddy sand low intertidal, burrower. (20, 26, 26a)

S. curtelsoma Shen: Hainan. Fine sand low intertidal. (20, 26)

S. globosa de Haan: Shandong Peninsula; Fujian; Taiwan; Guangdong. Muddy sand intertidal, burrower. (20, 26)

S. longidactyla Shen: Bohai Bay; Shandong Peninsula; Taiwan. Sandy intertidal, burrower. (20, 26)

S. tuberculata Stimpson: Guangdong. Muddy sand low intertidal, burrower. (20, 26)

Shenius anomalus (Shen): Xiamen (Fujian); Guangdong. Low intertidal mud flat. (18, 20, 26)

Tmethypocoelis ceratophora (Koelbel): Fujian; Taiwan; Guangdong. Fine sandy flat or mangrove forest. (20, 26)

Uca (*Celuca*) *annulipes* (H. Milne-Edwards): Hainan. Mangrove forest mud bank. (20, 26)

U. (*Deltuca*) *arcuata* (de Haan): Shandong Peninsula; Zhejiang; Fujian; Taiwan; Guangdong. Inner bay mud flat. (20, 26)

U. (*Thalassuca*) *borealis* Crane: Fujian; Taiwan; Guangdong; Guangxi; Hainan. River mouth mud flat. (20, 26)

U. (*Amphiuca*) *chlorophthalmus crassipes* (Adams & White): Guangdong. River mouth mud flat or mangrove forest. (20, 26)

U. (*Deltuca*) *dussumieri* H. Milne-Edwards: Fujian; Guangdong. Burrowing in river mouth mud flat. (18, 20, 26)

U. (*Thalassuca*) *formosensis* Rathbun: Taiwan. Intertidal. (20, 26)

U. (*Celuca*) *lactea* (de Haan) [*U. lactea* Shen, 1964]: Fujian; Taiwan; Hainan. River mouth low intertidal. (20, 26)

U. (*Deltuca*) *paradussumieri* (Batt) [*U. dussumieri spinate* Crane]: South China Sea. (20, 26)

U. (*Celuca*) *triangularis* (H. Milne-Edwards): Taiwan. River mouth mud flat off mangrove forest. (20, 26)

U. (*Thalassuca*) *vocans* (Linnaeus): Fujian; Taiwan; Hainan. River mouth, burrowing in mud flat. (20, 26)

GRAPSIDAE

Acmaeopleura balssi Shen: Shandong Peninsula. Shallow sea, commensal with polychaetes. (20, 26)

A. toriumii Takeda: Hong Kong. Commensal with mud shrimp (*Upogebia major*) and spoon worm (*Ochetostoma erythrogrammon*). (26a)

Bresedium brevipes (de Man): Guangdong coast. Brackish water and freshwater. (20, 26)

Chasmagnathus convexus de Haan: Zhejiang; Fujian; Taiwan; Hainan. River mouth marsh. (20, 26)

Clistocoeloma merguiensis de Man: Zhejiang; Taiwan. Low intertidal mud flat. (20, 26)

C. sinensis Shen: Zhejiang. Low intertidal. (20, 26)

Cyclograpsus granulatus Dana: Xisha (=Paracel Islands). (7)

C. incisus Shen: Guangdong. Intertidal. (20, 26)

C. integer H. Milne-Edwards [*C. granulatus* Chen, 1980]: Xisha (=Paracel Islands). High intertidal. (20, 26)

C. intermedius Ortmann: Taiwan. High intertidal. (20, 26)

C. longipes Stimpson: Xisha (=Paracel Islands). High intertidal. (20, 26)

C. lucidus Dai: Haikou and Sanya (Hainan). (20, 26)

Eriocheir japonica de Hann: Fujian; Taiwan; Guangdong. Freshwater. (20, 26)

E. japonica hepuensis Dai: Guangxi. Freshwater. (23)

E. leptognathus Rathbun: Coastal river mouth north of Fujian. (20, 26)

E. rectus Stimpson: Taiwan; Guangdong. Freshwater. (20, 26)

E. sinensis H. Milne-Edwards: China's coast. Freshwater. (20, 26)

Gaetice depressus (de Haan): Shandong Peninsula; Jiangsu; Zhejiang; Fujian; Taiwan; Guangdong. Low intertidal under rocks. (20, 26)

Grapsus albolineatus Lamarck [*G. strigosus* Shen, 1964]: Taiwan; Guangdong; Hainan; Xisha (=Paracel Islands). High intertidal. (20, 26)

G. crinipes (Dana): Xisha (=Paracel Islands). Burrowing on banks or forest. (4, 20, 26)

G. grayi (H. Milne-Edwards): Taiwan. Burrowing on banks or forest. (20, 26)

G. longitarsis Dana: Guangdong; Hong Kong; Xisha (=Paracel Islands); Nansha (=Spratly Islands). (7, 20, 26)

G. tenuicrustatus (Herbst): Taiwan; Hong Kong; Hainan; Xisha (=Paracel Islands); Nansha (=Spratly Islands). High intertidal or coral reefs. (4, 15, 20, 26)

Helice (*Helice*) *formosensis* Rathbun: Taiwan. (26)

H. (Helicana) japonica Sakai & Yatsajuka [*H. tridens wuane* Shen, 1930]: Shandong; Zhejiang; Fujian; Taiwan. (26, 32)

H. (Helice) latimera Parisi [*H. pingi* Rathbun]: Zhejiang; Fujian; Taiwan; Guangdong. Intertidal mud flat or shoreline. (18, 20, 26)

H. (Helice) leachii Hell: Taiwan; Hainan. River mouth marsh. (20, 26)

H. (Helice) tientsinensis Rathbun: China's coast. River mouth mud flat. (20, 26)

H. (Helicana) wuana Rathbun [*H. tridens sheni* Sakai]: China's coast. Mud flat. (20, 26, 32)

Hemigrapsus longitarsis (Miers): Shandong Peninsula shallow sea. (20, 26)

H. peniciillatus (de Haan): China's coast. Intertidal and river mouth. (20, 26)

H. sanguineus (de Haan): China's coast. Low intertidal. (20, 26)

H. sinensis Rathbun: Eastern Liaoning Peninsula; Shandong Peninsula; Fujian; Guangdong. Intertidal. (20, 26)

Labuanium rotundatum (Hess): Taiwan. Mud flat. (20, 26)

Metaplax elegans de Man: Xiamen and Dongshan (Fujian); Guangdong; Hong Kong. Mangrove forest. (18, 20, 26, 26a)

M. longipes Stimpson: Zhejiang; Fujian; Guangdong; Hong Kong. Intertidal mud flat. (20, 26, 26a)

M. sheni Gordon: Fujian. Mud flat. (20, 26)

M. takahashii Sakai: Fujian; Taiwan; Guangdong; Hong Kong. River mouth. (20, 26, 26a)

Metasesarma aubryi H. Milne-Edwards: Taiwan. Mangrove forest. (20, 26)

Metopograpsus frontalis Miers: Hainan. Subtidal under rocks. (20, 26)

M. latifrons (White): Hainan. Mangrove forest mud flat. (20, 26)

M. quadridentatus Stimpson: Southern Shandong Peninsula; Zhejiang; Fujian; Guangdong. Rocky low intertidal. (20, 26)

M. thukuhar (Owen): Guangdong; Weizhou Island (Guangxi); Hainan; Xisha (=Paracel Islands). Intertidal under rocks. (7, 20, 26)

Nanosesarma (Beanium) batavicum (Moreira): Hainan. Mangrove forest. (20, 26)

N. (Nanosesarma) minutum (da Man) [*Sesarma gordonae* Shen, 1935, 1964]: Hebei; Zhejiang; Fujian; Guangdong. Intertidal to shallow sea. (20, 26)

N. (Nanosesarma) pontianacensis (de Man): Hainan. Mangrove forest. (20, 26)

Neoepisesarma (Neoepisesarma) mederi (H. Milne-Edwards): Jiangsu; Hainan. Intertidal mud flat. (20, 26)

N. (Neoepisesarma) versicolor (Tweedie): Hainan. Intertidal mud flat. (20, 26)

Neosarmatium meinerti (de Man): Taiwan. Mangrove forest. (20, 26)

N. smithi (H. Milne-Edwards): Hainan. Intertidal mud flat. (20, 26)

Pachygrapsus crassipes Randall: Zhejiang; Fujian; Guangdong. Intertidal. (20, 26)

P. minutus H. Milne-Edwards: Taiwan; Hainan; Xisha (=Paracel Islands); Nansha (=Spratly Islands). Intertidal or coral reefs. (4, 15, 20, 26)

P. planifrons de Man: Hainan; Xisha (=Paracel Islands). Intertidal. (7, 20, 26)

P. plicatus (H. Milne-Edwards): Taiwan; Hainan; Xisha (=Paracel Islands); Nansha (=Spratly Islands). Intertidal rock crevices or coral reefs. (15, 20, 26)

Percnon abbreviatum (Dana): Hainan; Xisha (=Paracel Islands); Nansha (=Spratly Islands). Coral reefs. (4, 15, 20, 26)

P. guinotae Crosnier: Hainan; Xisha (=Paracel Islands). Coral reefs. (4, 20, 26)

P. planissimum (Herbst): Hainan; Xisha (=Paracel Islands). Intertidal and coral reefs. (4, 20, 26)

P. sinense Chen: Hainan. Coral reefs. (5, 20, 26)

Plagusia dentipes de Haan: Taiwan. Intertidal. (20, 26)

P. depressa tuberculata Lamarck: East China Sea. (16, 20, 26)

P. immaculata Lamarck: Xisha (=Paracel Islands); Nansha (=Spratly Islands). Coral reefs. (4, 15, 20, 26)

P. tuberculata Lamarck [*P. depressa* Shen, 1964]: East China Sea; Taiwan; Guangdong; Hong Kong; Hainan; Xisha (=Paracel Islands). Intertidal and coral reefs. (4, 16, 20, 26)

Planes cyaneus Dana: Xisha (=Paracel Islands). On floating objects. (20, 26)

P. marinus Rathbun: Xisha (=Paracel Islands). On floating objects. (7, 20, 26)

Pseudograpsus albus Stimpson: Xisha (=Paracel Islands). High intertidal or coral reefs. (7, 20, 26)

Ptychognathus barbatus (H. Milne-Edwards): Taiwan; Guangdong. Shallow water. (20, 26)

Pyxidognathus deianira de Man: Hainan. Mangrove forest mud flat. (20, 26)

Sesarma (Chiromantes) bidens (de Haan): Fujian; Taiwan; Guangdong; Guangxi. River mouth mud flat. (20, 26)

S. (Holometopus) dehaani H. Milne-Edwards: China's coast. River mud bank. (20, 26)

S. (Parasesarma) exquisitum Dai & Song: Longmen (Guangxi). (21)

S. (Chiromantes) fasciata Lanchester: Fujian; Hainan. Intertidal under rocks. (20, 26)

S. (Holometopus) haematocheir (de Haan): Jiaozhou Bay (Shandong); Jiangsu; Zhejiang; Fujian; Taiwan; Guangdong. Burrowing on coastal river bank. (20, 26)

S. (Sesarmops) impressum (H. Milne-Edwards): Taiwan. River mouth marsh or bushes. (20, 26)

S. (Parasesarma) pictum (de Haan): Shandong Peninsula; Zhejiang; Fujian; Taiwan; Guangdong. Low intertidal. (20, 26)

S. (Parasesarma) plicata (Latreille): Jiaozhou Bay (Shandong); Zhejiang; Fujian; Taiwan; Guangdong. Mud flat under rocks. (20, 26)

S. (Sesarmops) sinensis H. Milne-Edwards: Zhejiang; Fujian; Guangdong. River mouth mud burrows. (20, 26)

S. (Holometopus) tangi Rathbun: Fujian. Intertidal. (20, 26)

S. (Parasesarma) tripectinis Shen: Jiaozhou Bay (Shandong); Zhejiang; Fujian. Intertidal. (20, 26)

Thalassograpsus harpax (Hilgendorf): Taiwan. Intertidal or coral reefs. (20, 26)

Utica borneensis de Man: Taiwan. River mouth mud flat. (20, 26)

Varuna litterata (Fabricius): Zhejiang; Fujian; Taiwan; Guangdong; Hainan; Xisha (=Paracel Islands); Nansha (=Spratly Islands). (4, 15, 20, 26)

V. yui Hwang J. J. & M. Takeda: Taiwan; Guangdong; Hong Kong; Guangxi. (26a, 29)

GECARCINIDAE

Cardisoma carnifex (Herbst): Taiwan; Hainna. Mud burrower. (20, 26)

C. hirtipes Dana: Taiwan. Mud burrower. (20, 26)

Gecarcoidea lalandii H. Milne-Edwards: Taiwan; Xisha (=Paracel Islands). Lives on land, migrates to sea to spawn. (20, 26)

HAPALOCARCINIDAE

Hapalocarcinus marsupialis Stimpson: Xisha (=Paracel Islands). Coral reefs. (4, 20, 26)

Pseudohapalocarcinus ransoni Fize & Serene: Xisha (=Paracel Islands). Coral reefs. (20, 26)

Troglocarcinus viridis (Hiro) [*Cryptochirus hongkongensis* Shen, 1964]: Guangdong; Hong Kong. Coral reefs. (20, 26)

REFERENCES*

(1). Shen Chia-Jui and Dai Aiyun, 1964. A Collection of Illustrative Plates for Animals in China, Crustaceans, II. Crabs. China Science Press, 142pp.

(2). Shen, Chia-Jui and Chen Huilian, 1978. Description of five new species of *Leucosia* (Crustacea, Decapoda: Leucosiisae) off Chinese waters. Studia Marina Sinica, 14:75–86.

(3). Shen Chia-Jui, Dai Aiyun, and Chen Huilian, 1982. New and rare species of Parthenopidae (Crustacea: Brachyura) from China Seas. Acta Zootoxonomica Sinica, 7(2):139–149.

(4). Chen Huilian, 1975. Studies on the crabs of Xisha Islands I. Guangdong Province, China. Studia Marina Sinica, 10:157–179.

(5). Chen Huilian, 1977. A new species of *Percnon* (Crustacea:Brachyura) from the Hainan Island, Guangdong Province, China. Acta Zoologica Sinica, 23(4):377- 379.

(6). Chen Huilian, 1979. A new species of Tritodynamia (Crustacea, Brachyura) from Fujian Province, China. Oceanologia & Limnologia Sinica, 10(4):334–336.

(7). Chen Huilian, 1980. Studies on the crabs of the Xisha Islands, Guangdong Province, China, II. Studia Marina Sinica, 17:117–147.

(8). Chen, Huilian 1982. On the Genus *Nursilia* (Crustacea: Decapoda: Leucosiidae) of Chinese waters. Oceanologia & Limnologia Sinica, 13(3):267–272.

(9). Chen, Huilian 1983. Preliminary studies on the Homolidae (Brachyura, Crustacea) of Chinese waters. In: Transactions of the Chinese Crustacean Society, No. I, pp: 227–229.

(10). Chen, Huilian 1984. Study of the Genus *Carcinoplax* (Crustacea: Decapoda: Goneplacidae) of Chinese Waters. Oceanologia & Limnologia Sinica, 15(2):188- 201.

(11). Chen Huilian, 1984. A new species of *Parilia* (Crustacea: Brachyura) from the South China Sea. Oceanologia & Limnologia Sinica, 15(5):482–486.

(12). Chen Huilian, 1986. On the new species of the Genus *Portunus* (Crustacea: Brachyura) from the coast of Guangdong, China. Oceanologia & Limnologia Sinica, 17(1): 84–90.

(13). Chen Huilian, 1986. Studies on the Dorippidae (Crustacea, Brachyura) of Chinese Waters. Transactions of the Chinese Crustacean Society. China Science Press, pp:118–139.

(14). Chen Huilian, 1987. On two new species of Leucosiidae (Crustacea, Brachyuura) from the Chinese waters. Studia Marina Sinica, 28:197–203.

(15). Chen Huilian and Xu Zhenxiong, 1991. On the crabs of the Nansha Islands. In: Contribution on Marine Biological Research of Nansha Islands and Its Adjacent Waters Sea (III). China Ocean Press, pp:48–106.

(16). Dong Yumao, 1988. Molluscs in Deep Sea of the East China Sea. Zhejiang Science Press.

(17). Cai Erxi, 1990. Distribution of crabs in Xiamen Harbour, Fujian. Journal of Oceanography in Taiwan Strait, 9(2):166–171.

(18). Cai Erxi, 1992. Crabs along southeast China. (manuscript)

(19). Dai Aiyun and Chen Huilian, 1980. A new species of *Latrelillopsis* (Latreillidae, Brachyura) from the South China Sea. Acta Zootaxonomica Sinica, 5(1):39–41.

(20). Dai Aiyun and Yang Siliang, 1986. Crabs of the China Seas. China Ocean Press, 642pp.

(21). Dai Aiyun and Song Yuzhi, 1986. Intertidal crabs from Beibu Gulf of Guangxi. Transactions of the Chinese Crustacean Society. China Science Press, pp:54–62.

(22). Dai Aiyun and Guan Shiquan, 1986. One new species of *Philyra* from Guangdong Province (Decapoda: Leucosiidae). Acta Zootaxonomica Sinica, 11(2):148–150.

(23). Dai Aiyun, 1991. Studies on the sub-family differentiation of Eriocheir (Decapoda, Brachyura). Collections of Papers on Systematic Evolutionary Zoology, 1:61–71.

(24). Dai Aiyun and Xu Zhenxiong, 1991. A preliminary study of crabs from Nansha Islands. In: Contribution on Marine Biological Research of Nansha Islands and its Adjacent Waters (III). China Science Press, pp:1–47.

(25). Dai Ai-Yun and Song Yu-Ji, 1984. *Macrophthalmus* (Decapoda, Brachyura) of the seas of China. Crustaceana 46(1):76–86

(26). Dai Ai-Yun and Yang Si-Liang, 1991. Crabs of the China Seas. China Ocean Press, 680pp.

(26a). Daive, P. J. F. 1992, A new species and new records of intertidal crabs from Hong Kong. In: Proceedings of International Marine Biological Workshop: The Marine Flora and Fauna of Hong Kong and Southern China, Hong Kong University Press, pp:245–359.

(27). Holthius, L. B. and R. B. Manning, 1990. Crabs of the subfamily Dorippinae Macleay, 1938 from the Indo-West Pacific Region (Crustacea: Decapoda: Dorippidae). Researches on Crustacea, special No. 3:1–151.

(28). Horikoshi, M. and M. Takeda, 1980. An assemblages of small Crustaceans in shallow subtidal depths at the entrance to Tolo channel, Hong Kong Proceedings of the First International Marine Biological Workshop: The Marine Flora and Fauna of Hong Kong and Southern China, pp:619–626.

(29). Hwang, J. J. and M. Takeda, 1986. A New Freshwater Crabs of the Family Grapsidae from Taiwan. Proc. Soc. Japn. Syst. Zool. 33:11–18.

(30). Lin Chao-Chi, 1949. A Catalogue of Brachyura of Taiwan. Quart. J. Taiwan Mus., 2(1):10–33.

(31). Pregebzer, C. and B. Morton, 1986. Hong Kong Pinnotheridae: Pinnitherinae (Crustacea: Decapoda). In: Proceedings of the Second International Marine Biological Workshop: The Marine Flora and Fauna of Hong Kong and Southern China. Hong Kong University Press, pp:649–659.

(32). Sakai, T. and K. Yatsajuka, 1980. Notes on some Japanese and Chinese *Helice* with *Helice* (*Helicana*) subgen., including *Helice* (*Helicana*) *japonica* n. sp. (Crustacea: Decapoda). Senckenbergiana biol. 60(516):393–411.

(33). Shen, C. J. 1932. The Crabs of Hong Kong, Pt. 3, Hong Kong Naturalist 3(1):32–45.

*: (1)-(24) in Chinese

Compiled by Cai Erxi and Huang Zongguo; Edited by Dai Aiyun, Chen Huilian, and Yang Shiliang

STOMATOPODA
SQUILLOIDEA
SQUILLIDAE

Alima hieroglyphica (Kemp): South China Sea. (1)
Anchisquilla fasciata (de Haan): East China Sea; South China Sea. (1, 4, 5, 7, 12)
Carcinosquilla carinata (Serene): South China Sea. (1)
C. multicarinata (White): South China Sea. (5, 8, 12)
Clorida clorida (Brooks): South China Sea. (5)
C. decorata (Wood-Mason): East China Sea; South China Sea. (5)
C. japonica Manning: South China Sea. (1)
C. latispina Manning: East China Sea; South China Sea. (5)
C. latreillei (Eydoux & Souleyet): East China Sea; South China Sea. Shallow sea. (4, 5)
C. microphthalma (H. Milne-Edwards): East China Sea (south of Zhejiang); Taiwan Strait; South China Sea. (4–7)
C. rotundicauda (Miers): Qingdao (Shandong); southern Zhejiang; Fujian; Taiwan. (1, 4, 12)
C. verrucosa (Hansen): South China Sea. (1)
Cloridopsis aquilonaris Manning: East China Sea; South China Sea. Shallow sea. (1, 4)
C. scorpio (Latreille): East China Sea; South China Sea. Intertidal and shallow sea. (4, 5, 7, 8, 12)
Dictyosquilla foveolata (Wood-Mason): East China Sea; South China Sea. Shallow water. (1, 4, 5, 8, 12)
Kempina mikado (Kemp & Chopra): East China Sea; South China Sea. (1, 4–7)
Lavisquilla inermis (Manning): East China Sea; South China Sea. (5)
Lenisquilla lata (Brooks): East China Sea; South China Sea. Shallow water. (1, 4, 5)
Lophosquilla costata (de Haan): South of Zhoushan (Zhejiang); Xiamen (Fujian); Taiwan; Guangdong; Hong Kong; Guangxi; Hainan. Shallow sea. (1, 4–8, 12)
L. tiwarii Blumstein: South China Sea. (5)
Oratosquilla anomala (Tweedie): East China Seal; South China Sea. (11)
O. gonypetes (Kemp): East China Sea; South China Sea. (1, 4, 5)
O. imperialis (Manning): South China Sea. (1)
O. interrupta (Kemp): East China Sea; South China Sea. (1, 4–8, 12)
O. kempi (Schmitt): Shandong; Zhejiang; Fujian; Guangdong. Intertidal and shallow sea. (1, 4, 8, 12)
O. nepa (Latreille): Zhejiang; Fujian; Taiwan; Guangdong; Guangxi; Hainan. Shallow sea. (1, 4, 5, 12)
O. oratoria (de Haan) [*Squilla oratoria*]: Widely distributed along China's coast, very abundant in Bohai and Yellow Sea. Water depth within 60 m. (1, 4–8, 10, 12)
O. ornata Manning: East China Sea; South China Sea. (5, 10)
O. perpensa (Kemp): East China Sea; South China Sea. (6–8, 11, 12)
O. quinquedentata (Brooks): South China Sea. (1)
O. woodmasoni (Kemp): East China Sea; South China Sea. (5, 8, 11, 12)
Squilloides leptosquilla (Brooks): East China Sea, water depth more than 100 m; South China Sea. (4, 7)

HARPIOSQUILLIDAE

Harpiosquilla annandalei (Kemp): East China Sea; South China Sea. (4, 5, 7, 9)
H. harpax (de Haan): South China Sea. (5, 7, 8, 9)
H. japonica Manning: South China Sea. (5)
H. raphidea (Fabricius): South China Sea. (1)

GONODACTYLOIDEA
GONODACTYLIDAE

Gonodactylus chiragra (Fabricius): Southern Fujian; Taiwan; Guangdong; Hainan; Xisha (=Paracel Islands); Nansha (=Spratly Islands). Coral reefs. (1, 2, 4, 6, 7, 12)
G. demani Henderson: South China Sea. (1)
G. platysoma Wood-Mason: Xisha (=Paracel Islands). (2)
G. smithii Pocock: Xisha (=Paracel Islands). (2)
G. ternatensis de Man: Xisha (=Paracel Islands). (2)

EURYSQUILLIDAE

Coronidopsis bicuspis Hansen: South China Sea. (5)
C. medius Blumstein: South China Sea. (5)
Eurysquilloides sibogas (Hansen): South China Sea. (5)
Sinosquilla hispida Liu & Wang: South China Sea. Water depth 260 m. (3)
S. sinica Liu & Wang: South China Sea. Water depth 58–89 m. (3)

ODENTODACTYLIDAE

Odentodactylus cultrifer (White): South China Sea. (1)
O. japonicus (de Haan): South China Sea. (12)
O. scyllarus (Linné): South China Sea. (1)

PSEUDOSQUILLIDAE

Parasquilla haani (Holthuis): East China Sea; South China Sea. (4, 7, 12)
Pseudosquilla oculata (Brulle): Xisha (=Paracel Islands); Nansha (=Spratly Islands). (2)
P. ornata Miers: Xisha (=Paracel Islands). (2)

LYSIOSQUILLOIDEA
CORONIDIDAE

Parvisquilla multituberculata (Borradaile): Xisha (=Paracel Islands). (2)

LYSIOSQUILLIDAE

Heterosquilla latifrons (de Haan): East China Sea; South China Sea. (4)
Lysiosquilla maculata (Fabricius): Taiwan; South China Sea. Shallow sea. (5, 6, 7, 12)
L. sulcirostris Kemp: Taiwan; Guangdong. (6, 7)

NANNOSQUILLIDAE

Acanthosquilla acanthocarpus (Miers): Taiwan Strait; South China Sea. (6, 7, 12)
A. multifasciata (Wood-Mason): Taiwan Strait; South China Sea. (5, 6, 7, 12)
A. multispinosa Blumstein: South China Sea. (1)
Nannosquilla varicosa (Komai & Tung): South China Sea. (1)

REFERENCES*

(1). Liu Ruiyu and Wang Yongliang, 1962. Studies on geography of Stomatopoda in China's seas. Proceedings of Animal Ecology and Geography Conference. China Science Press.
(2). Liu, Ruiyu, 1975. On a collection of Stomatopod Crustacea from the Xisha Islands, Guangdong Province, China. Studia Marina Sinica, 10:183–197.
(3). Liu, Ruiyu and Wang Yongliang, 1978. Description of a new genus and two species of Stomatopod Crustacea from the South China Sea. Oceanologia & Limnologia Sinica 9 (1):89–94.
(4). Dong, Y. M., 1983. Stomatopoda of East China Sea. East China Sea Marine Sciences 1(1):82–94.
(5). Guan Xilian and Shen Shoupeng, 1988. An ecological analysis of Stomatopoda in the continental shelf of the northern part of South China Sea. Proc. Mar. Biol. South China Sea, 193–200.
(6). Komai, T. 1927. Stomatopoda of Japan and adjacent localities. Mem. Coll. Sci. Kyoto Imp. Univ., B3:307–354.
(7). Lee, Sin-Che and Shi-Keui Wu, 1966. The Stomatopoda Crustacean of Taiwan. Bull. Inst. Zool. Academia Sinica, 5:41–58, 8 figs.
(8). Liu, J. Y. 1949. On some species of *Squilla* (Crustacea: Stomatopoda) from China coast. Contr. Inst. Zool. Nat. Acad. Peiping, 5(1):27–474.
(9). Manning, R. B. 1969. A review of the genus *Harpiosquilla* (Crustacea: Stomatopoda), with descriptions of three new species. Smithsonian Contr. Zool., 36:1–41, 43 figs.
(10). Manning, R. B. 1971. Key to the species of *Oratosquilla* (Crustacea: Stomatopoda), with descriptions of two new species. Smithsonian Contr. Zool., 71:1–16, 4 figs.
(11). Maning, R. B. 1978. Further observation on *Oratosquilla*, with accounts of two new genera and nine new species (Crustacea: Stomatopoda: Squillidae). Smithsonian Contr. Zool., 272:1–44, 25 figs.
(12). Schmitt, W. L. 1929. Chinese Stomatopods collected by S. D. Light. Lingnan Sci. Jour., 8:127–148, 4 pls.

*: (1)-(4) in Chinese
Compiled by Wang Yongliang; Edited by Liu Ruiyu

BRYOZOA [ECTOPROCTA]

STENOLAEMATA
TUBULIPORATA [CYCLOSTOMATA]
CRISIIDAE

Bicrisia edwardsiana (d'Orbigny): Zhejiang; Fujian. Water depth 0–70 m. (24, 33)
Crisia crisioides Ortmann: Zhejiang; Fujian; Guangdong. Water depth 0–45 m. (24, 34)
C. cureata Maplestone: Guangdong; Hainan. Water depth 0–60 m. (24)
C. denticulata (Linnaeus): Zhejiang. Water depth 0–10 m. (34)
C. ebuneodenticulata Smitt: East China Sea; Hainan. Water depth 0–80 m. (11, 34, 39)
C. elongata Milne-Edwards: Fujian; Guangdong; Hainan; Nansha (=Spratly Islands). Water depth 0–57 m. (7, 10, 24, 35)
C. geniculata Milne-Edwards: Fujian; Guangdong. Water depth 0–60 m. (24, 32)
C. setosa MacGillivray: Zhongsha (=Macclesfield Bank). Water depth 45–100 m. (24)
Crisidia cornuta (Linnaeus): Zhejiang; Fujian; Guangdong; Hainan. Water depth 20–50 m. (24, 35)

TUBULIPORINA
DIASTOPORIDAE

Berenicea ampulliformis Okada: Zhejiang; Fujian; Guangdong. Water depth 10–45 m. (24, 33)
B. lineata (MacGillivray): Zhongsha (=Maccslefield Bank). Water depth 60–150 m. (8, 24)
B. meanarina (S. Wood): East China Sea. Water depth 80–150 m. (24, 33)
B. sarmiensis Norman: Nansha (=Spratly Islands). Water depth 50–80 m. (8, 24)

TUBULIPORIDAE

Idmonea irregularis Meneghini: East China Sea. Water depth 108 m. (36)
I. parasitica Busk: South China Sea. (34)
I. radiata Meneghini: Nansha (=Spratly Islands). Water depth 50–70 m. (3, 24)
Tubulipora atlantica Johnston: East China Sea; Guangdong; Hainan. Water depth 30–70 m. (24)
T. cortorta Busk: Lingshan Island (Shandong). Water depth 5–20 m. (24)
T. flabellaris (Fabricius): Eastern Liaoning Peninsula; Hebei; Shandong Peninsula; Jiangsu; Zhejiang; Fujian; Guangdong; Hainan. Water depth 0–60 m. (24, 30)
T. flexucosa (Pourlales): East China Sea. (38)

T. lobulata (Hinks): Guangdong; Hainan. Water depth 0–80 m. (24)

T. pulcherrima Kirkpatrick: Guangdong; Hainan; Xisha (=Paracel Islands); Nansha (=Spratly Islands). Water depth 0–60 m. (8, 24)

T. pulchra MacGillivray: Shandong Peninsula; Zhejiang; Fujian; Guangdong; Guangxi; Hainan. Water depth 10–90 m. (1, 8, 24, 30)

ONCOUSOECIIDAE

Filisparsa candeana (d'Orbigny): Hainan. Water depth 0–35 m. (22, 24)

F. rustica (d'Orbigny): Guangdong; Hainan. Water depth 0–35 m. (2, 24)

Proboscina coapta Canu & Bassler: Zhejiang; Fujian. Water depth 0–25 m. (24, 32, 35)

ENTALOPHOROECIIDAE

Entalophoroecia deliculata (Busk): Hebei; Changshan Islands (Liaoning); Shandong Peninsula. Water depth 0–50 m. (24)

E. proboscidea (Milne-Edwards): Guangdong. (38)

DIAPEROECIIDAE

Crisulipora occidentalis Robertson: Daya Bay (Guangdong). Water depth 5 m. (10, 11)

Diaperoecia radiata (Kirpatrick): Daya Bay (Guangdong). (38)

D. scalaris Canu & Bassler: Zhejiang; Fujian; Guangdong. Intertidal. (10, 11)

HETEROPORIDAE

Heteropora pelluculata Waters: Hainan. Water depth 0–25 m. (1, 24)

RECTANGULATA
LICHENOPORIDAE

Lichenopora capillata Kirpatrick: Zhongsha (=Macclesfield Bank). Water depth 40–60 m. (24, 28)

L. imperialis (Ortmann): Eastern Liaoning Peninsula; Hebei; Shandong Peninsula; Jiangsu; Zhejiang; Fujian; Guangdong; Hainan; Xisha (=Paracel Islands); Nansha (=Spratly Islands). Water depth 0–100 m. (11, 24)

L. radiata (Audouin): Bohai; Shandong Peninsula; Zhejiang; Fujian; Guangdong; Hainan; Xisha (=Paracel Islands). Water depth 0–45 m. (38)

L. simplex Busk: Zhongsha (=Macclesfield Bank). Water depth 40–60 m. (24, 28)

GYMNOLAEMATA
CTENOSTOMATA
STOLONIFERA
WALKEROIDEA
WALKERIIDAE

Walkeria tuberosa Heller: Guangdong; Hainan. Water depth 0–15 m. (24)

W. uva (Linnaeus): Hainan; Zhongsha (=Macclesfield Bank). Water depth 0–25 m. (8, 24)

VESICULARIIDAE

Amathia distans Busk: Guangdong; Hainan. Water depth 30–80 m. (10, 11, 24)

Bowerbankia caudata Hincks: Guangdong; Hainan; Xisha (=Paracel Islands). Water depth 0–50 m. (9, 10, 24, 30)

B. imbricata (Adams): Hebei; Changshan Islands (Liaoning); Shandong Peninsula; Zhejiang; Guangdong; Hainan; Xisha (=Paracel Islands); Nansha (=Spratly Islands). Water depth 0–70 m. (7, 9, 10, 24, 30)

Mimosella verticellata (Heller): Guangdong; Hainan. Water depth 20–25 m. (24)

Zoobotryon verticellatum (Delle Chiage): Guangdong; Hainan. Water depth 0–25 m. (9, 10)

BUSKIIDAE

Buskia nitens Alder: Guangdong; Hainan. Water depth 0–15 m. (24)

AEVERRIILIIDAE

Aeverrillia setogera (Hincks): Nansha (=Spratly Islands). Water depth 40–50 m. (8)

CARNOSA
ARACHNIDIIDEA
ARACHNIDIDOIDAE

Nollela papuensis (Busk): Guangdong; Hainan; Nansha (=Spratly Islands). Water depth 15–45 m. (24)

ALCYONIDIOIDEA
ALCYONIDIIDAE

Alcyonidium gelatinosum (Linnaeus): Southern Shandong Peninsula. Water depth 30–70 m. (24)

A. nanum Silen: Changshan Islands (Liaoning). Water depth 0–45 m. (24)

A. polyoum (Hassall): Eastern Liaoning Peninsula; Hebei; Shandong Peninsula; Jiangsu; Zhejiang; Fujian; Guangdong; Guangxi; Hainan; Xisha (=Paracel Islands). Water depth 0–85 m. (9, 11, 24)

FLUSTRELLIDRIDAE

Flustrellidra flabellaris (Kirpatrick): Zhongsha (=Macclesfield Bank). Water depth 40- 60 m. (8)

CHEILOSTOMATA
INOVICELLINA
AETEIDAE

Aetea anguina (Linnaeus): Eastern Liaoning Peninsula; Hebei; Shandong Peninsula; Jiangsu; Zhejiang; Fujian; Guangdong; Guangxi; Hainan; Zhongsha (=Macclesfield Bank); Xisha (=Paracel Islands); Nansha (=Spratly Islands). Water depth 0–150 m. (7, 10, 11, 24, 26, 27, 30, 32, 37)

A. sinica Liu: Northern Yellow Sea. Water depth 33 m. (27)

A. truncata (Lansborough): Shandong Peninsula; Jiangsu; Zhejiang; Fujian; Guangdong; Xisha (=Paracel Islands); Nansha (=Spratly Islands). Water depth 0–150 m. (11, 24, 27, 30)

A. xishaensis Liu: Xisha (=Paracel Islands). Water depth 80 m. (10, 11, 23, 27)

SCRUPARIINA
SCRUPARIOIDEA
SCRUPARIIDAE

Scruparia chelata (Linnaeus): Northern Yellow Sea; Nansha (=Spratly Islands). Water depth 0–45 m. (8, 24, 27)

MALACOSTEGINA
MEMBRANIPOROIDEA
MEMBRANIPORIDAE

Conopeum eriphorum (Lamouroux): Southern Hainan. Water depth 0–30 m. (26, 27)

C. hirtissimum Liu & Ristedt: Southern and southwestern Hainan. Water depth 0–27 m. (26, 27)

Membranipora barstschi (Canu & Bassler): Nansha (=Spratly Islands). Water depth 45 m. (27)

M. bicornica Liu: Southern Hainan. Water depth 37 m. (27)

M. conjunctiva Zhang & Liu: Guangdong. Water depth 0–15 m. (27, 30, 41)

M. eriophoroidea Liu: Southern Hainan. Water depth 15–25 m. (24, 26, 27)

M. falsitenuis Liu: East China Sea; southern Hainan. Water depth 40–60 m. (23, 27)

M. grandicella (Canu & Bassler): Eastern Liaoning Peninsula; Hebei; Shandong Peninsula; Jiangsu; Zhejiang; Fujian; Guangdong; Guangxi; Hainan; Xisha (=Paracel Islands). Water depth 0–45 m. (7, 9, 11, 23, 24, 27, 30, 31, 37)

M. hainanica Liu & Ristedt: Guangdong; Hainan. Water depth 0–5 m. (26, 27)

M. irregulata Liu: Bohai; Changshan Islands (Liaoning); Qingdao (Shandong); Zhejiang; Fujian; Guangdong; Hainan; Zhongsha (=Maccleslesfield Bank); Nansha (=Spratly Islands). Water depth 0–100 m. (22, 23, 24, 27)

M. limosa Waters [*M. quadrata*, Wang and Cai, 1977]: Zhejiang; Fujian; Guangdong; Guangxi; Hainan. Water depth 0–80 m. (24, 27, 29, 30)

M. limosoidea Liu: Hainan; Nansha (=Spratly Islands). Water depth 0–46 m. (22, 24, 27)

M. lingdingensis Liu & Li: Mouth of Zhujiang (=Pearl River) and Shantou harbor (Guangdong). Water depth 2–5 m. (7, 11, 24, 27, 30, 31)

M. paragrandicella Liu & Ristedt: Fujian; Guangdong; Hainan. Water depth 0–5 m. (24, 27, 30, 31)

M. parasavartii Liu: East China Sea. Water depth 70 m. (27, 31)

M. pseudoirregulata Liu & Ristedt: Southern and northwestern Hainan. (6, 27, 30)

M. savartii (Audouin): Hebei; Shandong; Jiangsu; Zhejiang; Fujian; Guangdong; Hainan; Xisha (=Paracel Islands); Nansha (=Spratly Islands). Water depth 0–150 m. (1, 7, 9, 11, 24, 27, 30, 31)

M. serrilamelloides Liu & Li: Fujian; Guangdong; Hainan. Water depth 0–5 m. (24, 27, 30, 31)

M. similus Liu [*Conopeum reticulum* and *M. serrilamella*, Liu Xixing and Androsova (1959, 1963)]: Shandong Peninsula; Jiangsu; Zhejiang; Fujian; Guangdong; Hainan; Xisha (=Paracel Islands); Nansha (=Spratly Islands). Water depth 0–75 m. (23, 27, 30)

M. tachypleusae Liu & Ristedt: Northwestern Hainan. Water depth 0–10 m. (26, 27)

M. tenuis Desor: East China Sea. Water depth 70–85 m. (24, 27)

M. tuberculata (Bosc): Shandong Peninsula; Jiangsu; Zhejiang; Fujian; Guangdong; Guangxi; Hainan; Xisha (=Paracel Islands); Nansha (=Spratly Islands). Water depth 0–5 m. (7, 9, 10, 23, 24, 27, 30, 37)

M. tuberculatoidea Liu: Xisha (=Paracel Islands); Nansha (=Spratly Islands). Water depth 0–5 m. (27, 31)

M. virgata (Canu & Bassler): Nansha (=Spratly Islands). Water depth 30–60 m. (22, 24, 27)

Membraniporopsis bifloris (Wang & Tung): Bohai; Eastern Liaoning Peninsula; Shandong Peninsula; Jiangsu; Zhejiang; Fujian; Guangdong. Water depth 0–25 m. (7, 9, 27–32, 35, 40)

M. bispinosa (Liu): Qingdao (Shandong); Jiangsu. Water depth 0–5 m. (2, 21, 27, 28, 30, 31)

ELECTRIDAE

Aspidelectra bihamata Liu & Wass [*Electra deviensis*, Tung and Wang (1960); Wang and Cai (1977)]: Hebei; Shandong; Jiangsu; Zhejiang; Fujian; Guangdong; Hainan. Water depth 0–70 m. (9, 24, 26, 27, 29, 30)

A. orientalis Liu & Wass [*Membraniporella aragoi zhoushanica*]: Bohai; Changshan Islands (Liaoning); Shandong Peninsula; Jiangsu; Zhejiang; Fujian; Guangdong. Water depth 0–70 m. (23, 24, 27, 30, 31)

Electra axialata Liu: Qingdao harbor (Shandong). Intertidal. (27, 30, 31)

E. bellula (Hincks) [*Membranipora hugliensis*, Tung and Wang, 1960, Wang and Cai, 1977; *E. pilosa*, Wang and Cai, 1997]: Eastern Liaoning Peninsula; Hebei; Shandong Peninsula; Jiangsu; Zhejiang; Fujian; Guangdong; Guangxi; Hainan; Xisha (=Paracel Islands). Water depth 0–100 m. (11, 24, 27–31)

E. bengaliensis (Stoliczka) [*E. anomala*]: Fujian; Guangdong; Hainan. Water depth 0–25 m. (7, 9, 11, 23, 24, 27–32)

E. deviensis Robertson: Fujian; Guangdong; Guangxi; Hainan; Xisha (=Paracel Islands); Nansha (=Spratly Islands). Water depth 0–50 m. (24, 27, 30, 31)

E. gracilis Liu: Xiamen (Fujian). Water depth 2–10 m. (27, 30, 31)

E. inarmata Liu & Ristedt: Northern and western Hainan. Water depth 0–10 m. (26, 27, 30)

E. inermis Liu & Ristedt: Southern Hainan. Water depth 2–10 m. (26, 27)

E. monilophora Liu: Qikou (Hebei). Intertidal. (27, 30)

E. pilosa (Linnaeus): Southern Hainan. Water depth 0–15 m. (27, 30)

E. pseudopilosa Liu & Wass: Zhejiang; Fujian; Guangdong. Water depth 0–5 m. (7, 27, 28, 30)

E. pseudospinosa Liu & Ristedt: Northern Hainan. Water depth 0–5 m. (24, 28)

E. spiculata (Borg): Tanggu harbor (Tianjin); Hebei. Intertidal. (1, 27)

E. spinigera Liu & Wass: Fujian. Water depth 0–10 m. (27, 31)

E. tenella (Hincks): Eastern Liaoning Peninsula; Hebei; Shandong Peninsula; Jiangsu; Zhejiang; Fujian; Guangdong; Guangxi; Hainan; Xisha (=Paracel Islands); Zhongsha (=Macclesfield Bank); Nansha (=Spratly Islands). Water depth 0–100 m. (7, 9, 11, 24, 27, 28, 30, 31, 32)

E. teniuspinosa Liu & Ristedt: Southern Fujian; northern Hainan. Water depth 0–10 m. (26, 27, 29, 30, 31)

E. xiamenensis Liu: Xiamen (Fujian); Shanwei (Guangdong). Water depth 0–10 m. (27, 30)

NEOCHEILOSTOMINA
PSEUDOMALACOSTEGOMORPHA
CALLOPOROIDEA
QUADRICELLARIIDAE

Nellia oculata Busk: Fujian; Guangdong; Hainan; Xisha (=Paracel Islands); Nansha (=Spratly Islands). Water depth 0–200 m. (1, 7, 22, 24, 27, 29, 30)

N. tenuis Harmer: Xisha (=Paracel Islands); Nansha (=Spratly Islands). Water depth 0–50 m. (4, 22, 24, 27)

FLUSTRIDAE

Antropora cuculata Liu: East China Sea. Water depth 112 m. (27)

A. donghainian Liu: East China Sea. Water depth 50 m. (27)

A. errectorostata Liu & Ristedt: Eastern Hainan. Water depth 6–12 m. (26, 27)

A. fenglingensis Liu: Zhejiang. Water depth 0–20 m. (15, 27, 28)

A. laguncula (Canu & Bassler): Zhejiang; Fujian; Guangdong; Hainan; Nansha (=Spratly Islands). Water depth 0–80 m. (15, 24, 27)

A. leucophora (Marcus): Hainan; Nansha (=Spratly Islands). Water depth 0–60 m. (22, 26, 27)

A. longirostrata Liu & Ristedt: Hainan. Water depth 1–15 m. (26, 27)

A. minus (Hincks): Guangdong; Hainan; Nansha (=Spratly Islands). Water depth 0–60 m. (26, 27)

A. ogivalina (Canu & Bassler): Guangdong; Hainan; Nansha (=Spratly Islands). Water depth 0–45 m. (15, 24, 26, 27, 28)

A. ovata (Canu & Bassler): Zhejiang; Fujian; Guangdong; Hainan; Nansha (=Spratly Islands). Water depth 0–130 m. (15, 27)

A. paralaguncula Liu & Wass: East China Sea. Water depth 110–130 m. (27, 28)

A. serrirostrata Liu: Nansha (=Spratly Islands). Water depth 90 m. (27)

A. subminus Liu & Ristedt: Eastern Hainan. Water depth 0–12 m. (26, 27)

A. subovata Liu & Ristedt: Southern Hainan. Water depth 5 m. (26, 27)

A. tincta (Hastings): Shandong Peninsula; Hainan; Nansha (=Spratly Islands). Water depth 0–60 m. (26, 27, 30)

A. trapezoides (Canu & Bassler): Hainan; Nansha (=Spratly Islands). Water depth 0–45 m. (24, 26, 27)

Aplousina falsifilum Liu: Northern South China Sea. Water depth 70 m. (27)

A. laevigata Liu & Ristedt: Southern Hainan. Water depth 0–6 m. (26, 27)

A. minacis Liu & Wass: East China Sea. Water depth 99 m. (27, 28)

A. nanshaensis Liu: Nansha (=Spratly Islands). Water depth 113 m. (27)

A. sulcata Liu: Southern Yellow Sea. Water depth 70 m. (27)

Carbasea carbasea (Ellis & Solander): East China Sea. Water depth 82 m. (14, 24, 27)

C. donghaiensis Liu [*C. episcopalis*, Liu (1982)]: East China Sea. Water depth 80 m. (14, 24, 27)

C. meridionalis Liu: Guangdong. Water depth 40–50 m. (13, 23, 27)

C. orientalis Liu: East China Sea. Water depth 80–90 m. (14, 23, 27)

C. sinica Liu: East China Sea. Water depth 70–115 m. (14, 27)

Cranosina coronata (Hincks): Hainan; Xisha (=Paracel Islands); Zhongsha (=Maccslefield Bank); Nansha (=Spratly Islands). Water depth 0–200 m. (26, 27)

C. latimandibulata Liu: Zhongsha (=Maccslefield Bank). Water depth 100–200 m. (27)

C. philippinensis (Canu & Bassler): Hong Kong. Water depth 60–265 m. (3, 27, 28, 30)

Ellisina canui (Sakakura): East China Sea. Water depth 60–130 m. (26, 27, 28)

E. costigera Liu: East China Sea. Water depth 135 m. (16, 27)

E. huanghainiana Liu: Northern Yellow Sea. Water depth 45 m. (27)

Hincksina curvimandibulata Liu: Zhongsha (=Maccslefield Bank): Water depth 50–200 m. (27)

Hippoflustra bifurcillata Liu & Wass: East China Sea. Water depth 70–150 m. (24, 26, 27)

H. granulata Liu & Ristedt: Southern Hainan. Water depth 7–10 m. (27)

H. inarmata Liu & Ristedt: Southern Hainan. Water depth 5 m. (26, 27)

Retiflustra anatina (Liu): Guangdong. Water depth 80 m. (14, 27)

R. reticulum (Hincks): Nansha (=Spratly Islands). Water depth 0–46 m. (14, 22, 24, 27)

R. schonauii Levinsen: Fujian (south of Pingtan); Guangdong; Guangxi; Hainan; Nansha (=Spratly Islands). Water depth 0–105 m. (14, 22, 27)

R. sinica (Liu): Guangdong. Water depth 160 m. (14, 24, 27)

Sinoflustra amoyensis (Robertson): Zhejiang; Fujian; Guangdong; Hainan. Water depth 0–25 m. (11, 23, 24, 27–32)

S. annae (Osburn): Zhejiang; Fujian; Guangdong; Hainan. Water depth 0–10 m. (10, 22, 23, 27–31)

CUPULADRIIDAE

Cupuladria quineensis (Busk): East China Sea; South China Sea. Water depth 30–150 m. (20, 27)

SETOSELLINIDAE

Setosellina constricta Harmer: East China Sea. Water depth 90–130 m. (27)

CALLOPORIDAE

Alderina qingdaoensis: Jiaozhou Bay near Qingdao harbor (Shandong). Water depth 29 m. (27)

Amphiblestrum granulatum Liu: Nansha (=Spratly Islands). Water depth 90–143 m. (27)

Callopora biarmata Liu & Ristedt: Southern Hainan. Water depth 2–10 m. (26, 27)

C. bifursca Liu: Nansha (=Spratly Islands). Water depth 95–135 m. (27)

C. corniculifera (Hincks): Hainan. Water depth 80 m. (27)

C. craticula (Alder): Northern Yellow Sea. Water depth 80 m. (27)

C. granulata Liu & Ristedt: Southern Hainan. Water depth 2–10 m. (26, 27)

C. horridoidea Androsova: Changshan Islands (Liaoning); Shandong Peninsula. Water depth 0–45 m. (24, 27, 28, 30)

C. inarmata Liu: Jiaozhou Bay (Shandong). Water depth 29 m. (27)

C. inaviculata Liu: Southern Shandong Peninsula. Water depth 15–20 m. (27)

C. inermis Liu & Wass: Eastern Liaoning Peninsula; Shandong Peninsula; Jiangsu; Zhejiang; Fujian. Water depth 10–80 m. (23, 26, 27, 28, 30)

C. laevigata Liu: Bohai. Water depth 29 m. (27)

C. lineata (Linnaeus): Changshan Islands (Liaoning); Shandong Peninsula; Zhejiang. Water depth 0–70 m. (24, 27, 28)

C. spatulata Liu & Wass: East China Sea. Water depth 70–150 m. (27, 28)

Cauloraphus cymbaeformis (Hincks): Southern Yellow Sea. Water depth 70–80 m. (24, 27)

C. spiniferus (Johnston): Changshan Islands (Liaoning); Shandong Peninsula; Zhejiang. Water depth 0–45 m. (24, 27, 32, 35)

Chaperia acanthina (Lamouroux): Fujian; Guangdong; Hainan. Water depth 5–70 m. (23, 27)

Chaperiopsis beniensis Silen: East China Sea. Water depth 70–150 m. (27)

C. hainanica Liu & Ristedt: Southern Hainan. Water depth 10 m. (26, 27)

C. pleuroaviculata Liu: Nansha (=Spratly Islands). Water depth 30–46 m. (22, 27)

Corbullella nanshaensis Liu: Nansha (=Spratly Islands). Water depth 72 m. (27)

C. tenuirostre (Hincks): East China Sea. Water depth 80–100 m. (22, 27)

Crassimarginatella crassimarginata (Hincks): Southern Hainan; Nansha (=Spratly Islands). Water depth 5–60 m. (17, 27)

C. japonica (Ortmann): Southern Zhejiang. (11, 27)

C. kumatae (Okada): Southern Yellow Sea; East China Sea. Water depth 80–110 m. (17, 27)

C. laguncula Liu: Nansha (=Spratly Islands). Water depth 60 m. (22, 26, 27)

C. lunata Liu: Pinyan (Zhejiang). Intertidal. (17, 27)

C. nanshaensis Liu: Zhongsha (=Macclesfield Bank). Water depth 60–100 m. (27)

C. sinica Liu: East China Sea. Water depth 120–150 m. (17, 27)

Parellisina curvirostris (Hincks): Nansha (=Spratly Islands). Water depth 30–80 m. (22, 27)

P. intercatatoporata Liu & Wass: Southern Hainan. Water depth 30–140 m. (27, 28)

P. longirostris Liu & Wass: East China Sea, Hainan. Water depth 40–80 m. (27, 28)

P. sileni Osburn: East China Sea. Water depth 70–130 m. (27, 28)

Ramphotonotus biformis Liu: Northern South China Sea. Water depth 80 m. (27)

R. bispinigera Liu: Taiwan Strait. Water depth 50 m. (27)

R. circulatis (Liu): East China Sea. Water depth 110 m. (16, 27)

R. uniformis Liu: Northern South China Sea. Water depth 65 m. (27)

R. variabillis Liu & Wass: East China Sea. Water depth 70–120 m. (23, 27, 28)

Retevirgula kenozooidata Liu & Wass: East China Sea. Water depth 80–120 m. (27, 28)

R. spiniaviculata Liu & Ristedt: Southern Hainan; Nansha (=Spratly Islands). Water depth 10–76 m. (26, 27)

Sinopora sinica Liu & Wass: East China Sea. Water depth 80–130 m. (24, 27, 28)

Tegella disincrustata Liu & Wass: Southern Yellow Sea; East China Sea. Water depth 25–50 m. (23, 27, 28)

T. lamellatoidea Liu & Wass: East China Sea. Water depth 70–120 m. (23, 27, 28)

T. unicornis (Fleming): Shandong Peninsula. Water depth 40–70 m. (27)

HIANTOPORIDAE

Hiantopora intermedia (Kirpatrick): East China Sea; South China Sea; Nansha (=Spratly Islands). Water depth 120–150 m. (27)

CELLULARIOMORPHA
BUGULOIDEA
BUGULIDAE

Bicellariella gracilis Busk: Southern Yellow Sea; Zhejiang. Water depth 20–80 m. (20, 27)

B. sinica Liu: Northern East China Sea. Water depth 30–60 m. (18, 27)

Brettia mollis Harmer: Nansha (=Spratly Islands). Water depth 70 m. (27)

B. vectifera Liu: Nansha (=Spratly Islands). Water depth 70 m. (27)

Bugula aspinosa Liu: East China Sea. Water depth 61 m. (18, 27)

B. californica Robertson: Eastern Liaoning Peninsula; Hebei; Shandong Peninsula. Water depth 0–20 m. (7, 9, 18, 27, 30)

B. dentata (Lamouroux): Changshan Islands (Liaoning); Shandong Peninsula; Jiangsu; Zhejiang; Fujian; Guangdong; Guangxi; Hainan; Xisha (=Paracel Islands); Nansha (=Spratly Islands). Water depth 0–160 m. (7, 18, 28, 30)

B. flabellata (J. V. Thompson): Liaoning; Hainan. Water depth 0–10 m. (18, 27, 29, 30)

B. intermedia Liu: East China Sea. Water depth 50–70 m. (18, 27)

B. nanshaensis Liu: Nansha (=Spratly Islands). Water depth 108 m. (27)

B. neritina (Linnaeus): Eastern Liaoning Peninsula; Shandong Peninsula; Jiangsu; Zhejiang; Fujian; Guangdong; Guangxi; Hainan; Xisha (=Paracel Islands); Nansha (=Spratly Islands). Water depth 0–45 m. (7, 9, 18, 27, 29, 30)

B. orientalis Liu: East China Sea. Water depth 60–80 m. (18, 27)

B. pacifica Robertson: Eastern Liaoning Peninsula. Water depth 0–15 m. (17, 30)

B. pseudoaspinosa Liu: Nansha (=Spratly Islands). Water depth 80 m. (27)

B. robusta MacGillivray: Guangdong; Hainan; Nansha (=Spratly Islands). Water depth 0–160 m. (12, 18, 22, 27, 30)

B. scaphoides Kirpatrick: Xisha (=Paracel Islands); Nansha (=Spratly Islands). Water depth 30–60 m. (18, 27)

B. stolonifera Ryland: Southern Shandong Peninsula; Jiangsu; Zhejiang; Fujian; Guangdong; Hainan. Water depth 0–15 m. (7, 26, 27, 29)

B. subglobosa Harmer: Zhejiang; Fujian; Hainan; Xisha (=Paracel Islands); Nansha (=Spratly Islands). Water depth 5–17 m. (26, 27)

B. vectifera Harmer: Hainan. Water depth 32–83 m. (18, 27)

Bugulella sinica Liu: East China Sea. Water depth 210 m. (18, 27)

Caulibugula binata Liu: East China Sea. Water depth 75 m. (20, 27)

C. caliculata (Levinsen): Hong Kong; Qiongzhou Strait (between Guangdong and Hainan); northern Hainan. Water depth 0–20 m. (20, 27)

C. ciliata (Robertson): Southern Hainan. Water depth 5–17 m. (22, 27)

C. ciliatoidea Liu: Jiaozhou Bay (Shandong). Water depth 5–17 m. (20, 27)

C. dendrograpta (Waters): Guangdong. Water depth 70 m. (20, 27)

C. gracilenta Liu: Southern Hainan. Water depth 45–86 m. (20, 27)

C. hainanica Liu: Southern Hainan. Water depth 102 m. (20, 27)

C. inermis Harmer: Southern Yellow Sea; East China Sea; South China Sea. Water depth 30–90 m. (20, 22, 27, 29)

C. irregularis Liu: Eastern Hainan. Water depth 79 m. (20, 27)

C. longiconica Liu: East China Sea. Water depth 60–90 m. (20, 27)

C. longirostris Liu: Eastern Hainan. Water depth 164 m. (20, 27)

C. mortenseni (Marcus): Guangdong; Hainan. Water depth 70–100 m. (20, 27)

C. singulata Liu & Ristedt: Southern Hainan. Water depth 0–15 m. (26, 27)

C. sinica Liu: Hainan. Water depth 5–97 m. (20, 27)

C. zanzibariensis (Waters): Southern Hainan. Water depth 97 m. (20, 27)

Falsibugulella sinica Liu: Southern Hainan. Water depth 98 m. (18, 27)

Halophila longicauda (Harmer): East China Sea. Water depth 2150 m. (18, 27)

Kinetoskias oblongata Liu: East China Sea. Water depth 70–80 m. (12, 27)

EPISTOMIDAE

Synnotum aegypticaum (Audouin): Changshan Islands (Liaoning); Shandong Peninsula; Jiangsu; Zhejiang; Fujian; Guangdong; Guangxi; Hainan; Xisha (=Paracel Islands); Nansha (=Spratly Islands). Water depth 0–160 m. (11, 19, 27, 29, 30)

S. pembaense (Waters): Nansha (=Spratly Islands). Water depth 30–60 m. (18, 27)

FARCIMINARIIDAE

Didymozoum simplex (Busk): Zhongsha (=Macclesfield Bank): Water depth 50 m. (8)

BEANIIDAE

Beania aspinosa Liu: Eatsern Hainan. Water depth 106 m. (18, 27)

B. cupulariensis Osburn: Southern Hainan. Water depth 5–17 m. (26, 27)

B. discordermiae (Ortmann): Hainan; Xisha (=Paracel Islands). Water depth 106 m. (18, 27)

B. farreae Liu: East China Sea. Water depth 14–20 m. (15, 27)

B. hirtssima (Heller): Southern Hainan. Water depth 54 m. (18, 27)

B. intermedia (Hincks): Guangdong; Hainan. Water depth 0–60 m. (10, 18, 26, 27, 30)

B. klugei Cook: Southern Hainan. Water depth 5–10 m. (26, 27)

B. mirabilis Johnston: Eastern Liaoign Peninsula; Hebei; Shandong Peninsula; Jiangsu; Zhejiang; Fujian; Guangdong; Guangxi; Hainan; Xisha (=Paracel Islands); Zhongsha (=Macclesfield Bank); Nansha (=Spratly Islands). Water depth 0–100 m. (11, 18, 24, 26, 27, 30)

B. petiolata Harmer: Qiongzhou Strait (between Guangdong and Hainan); southern Hainan; Nansha (=Spratly Islands). Water depth 50–85 m. (18, 22, 27, 29, 30)

B. regularis Thornely: Guangdong; Hainan. Water depth 20–80 m. (18, 27, 29, 31, 33)

B. spinigera (MacGillivray): East China Sea. Water depth 80–120 m. (18, 27)

Dendrobeania longispinosa (Robertson): Bohai; southern Shandong Peninsula. Water depth 5–50 m. (18, 27)

D. sinica Liu: Southern Shandong Peninsula. Water depth 5–45 m. (18, 27, 30)

CANDIDAE

Amastigia rudis (Busk): Eastern Liaoning Peninsula; Hebei; Shandong Peninsula; Jiangsu; Zhejiang; Fujian; Guangdong; Guangxi; Hainan; Xisha (=Paracel Islands); Nansha (=Spratly Islands). Water depth 0–150 m. (19, 27, 29, 30)

A. varians Liu: Zhongjian Island (Xisha=Paracel Islands). Water depth 70–80 m. (19, 27)

A. xishaensis Liu: Zhongjian Island (Xisha=Paracel Islands). Water depth 70–80 m. (19, 27)

Caberea boryi (Audouin): Fujian; Guangdong; Hainan; Xisha (=Paracel Islands); Nansha (=Spratly Islands). Water depth 5–80 m. (19, 27, 29, 30)

C. busifera Harmer: East China Sea. Water depth 100 m. (18, 27)

C. climacina Ortmann: Eastern Hainan. Water depth 0–30 m. (19, 27)

C. lata Busk: Changshan Islands (Liaoning); Hebei; Shandong Peninsula; Jiangsu; Zhejiang; Fujian; Guangdong; Guangxi; Hainan; Xisha (=Paracel Islands); Nansha (=Spratly Islands). Water depth 0–150 m. (11, 19, 27, 29, 30)

C. megaceras Yanagi & Okada: East China Sea. Water depth 75 m. (17, 21, 27)

C. symmetrica Liu: Southern Hainan. Water depth 140 m. (18, 27)

C. transversa Harmer: Hainan; Xisha (=Paracel Islands); Nansha (=Spratly Islands). Water depth 0–130 m. (19, 22, 27)

Canda clypea (Haswell): Southern Hainan. Water depth 30–45 m. (19, 27)

C. pecten Thornely: Fujian; Guangdong; Hainan; Xisha (=Paracel Islands); Nansha (=Spratly Islands). Water depth 30–110 m. (19, 27)

C. scutata Harmer: Xisha (=Paracel Islands); Nansha (=Spratly Islands). Water depth 6–80 m. (19, 27)

Scrupocellaria californica Robertson: Yulin Harbor (Hainan). Water depth 0–5 m. (19, 27)

S. circulata Liu & Ristedt: Southern Hainan. Water depth 6–11 m. (26, 27)

S. curvata Harmer: Guangdong; Nansha (=Spratly Islands). Water depth 60–100 m. (19, 22, 27)

S. delilii (Audouin): Guangdong; Hainan; Xisha (=Paracel Islands); Nansha (=Spratly Islands). Water depth 50–60 m. (19, 22, 27)

S. diadema Busk [*S. pinigera*, Wang and Hu (1982), Li (1992); *S. scrupea* (pars), Wang and Hu (1982), Li (1988); *S. diegensis*, Li (1988)]: Guangdong; Hainan; Xisha (=Paracel Islands); Nansha (=Spratly Islands). Water depth 0–150 m. (19, 22, 27, 29, 30)

S. ferox (Busk): Nansha (=Spratly Islands). Water depth 30–60 m. (19, 27)

S. inarmata Liu: Nansha (=Spratly Islands). Water depth 99 m. (27)

S. maderensis Busk: Fujian; Guangdong; Hainan; Xisha (=Paracel Islands); Nansha (=Spratly Islands). Water depth 0–70 m. (9, 11, 22, 27, 30)

S. meijensis Liu: Meijijiao Island (Nansha=Spratly Islands). Water depth 30 m. (27)

S. nanshaensis Liu: Nansha (=Spratly Islands). Water depth 5–10 m. (19, 27)

S. obtecta Haswell: Fujian; Guangdong; Hainan; Xisha (=Paracel Islands); Nansha (=Spratly Islands). Water depth 0–90 m. (19, 22, 27)

S. obtectoidea Liu & Ristedt: Southern Hainan. Water depth 3–17 m. (26, 27)

S. scabra (Van Beneden) [*S. scrupea* (pars), Wang and Hu (1982), Li (1988)]: Bohai; northern Yellow Sea. Water depth 30–70 m. (19, 27)

S. seculifera Busk: Nansha (=Spratly Islands). Water depth 30–50 m. (19, 22, 27)

S. spatulata (d'Orbigny): Zhejiang; Fujian; Guangdong; Hainan; Xisha (=Paracel Islands); Nansha (=Spratly Islands). Water depth 0–170 m. (19, 22, 27)

S. spatulatoidea Liu: Guangdong; Xisha (=Paracel Islands); Nansha (=Spratly Islands). Water depth 0–170 m. (19, 22, 27)

S. unicornis Liu: Hainan; Xisha (=Paracel Islands); Nansha (=Spratly Islands). Water depth 0–40 m. (11, 22, 27, 29)

S. uniseriata Liu: Eastern Hainan. Water depth 102 m. (19, 27)

Tricellaria gracilis Smitt: Bohai; southern Yellow Sea. Water depth 30–80 m. (9, 22, 27)

T. longispinosa (Yanagi & Okada): Southern Yellow Sea; northern East China Sea. Water depth 30–70 m. (19, 27)

T. multispinosa Liu: East China Sea. Water depth 80–120 m. (19, 27)

T. occidentalis (Trask): Eastern Liaoning Peninsula; Hebei; Shandong Peninsula; Jiangsu; Zhejiang; Fujian; Guangdong; Guangxi; Hainan. Water depth 0–25 m. (7, 19, 22, 27, 29, 30)

T. peachi (Busk): Southern Shandong Peninsula. Water depth 35 m. (27)

T. porata Liu: Southern Shandong Peninsula. Water depth 25 m. (27)

CRYPHCYSTOMORPHA
MICROPOROIDEA
MICROPORIDAE

Micropora granulata Liu: East China Sea. Water depth 80 m. (27)

Monoporella nodulifera (Hincks): East China Sea; Hainan; Zhongsha (=Macclesfield Bank); Nansha (=Spratly Islands). Water depth 5–60 m. (8, 27)

ONYCHOCELLIDAE

Caleschara laxa Canu & Bassler: Southern Hainan. Water depth 0–15 m. (27)

C. levinseni Harmer: East China Sea; Guangdong; Hainan; Nansha (=Spratly Islands). Water depth 10–50 m. (27)

Onychocella angulosa (Reuss): Fujian; Guangdong; Hainan; Xisha (=Paracel Islands), Nansha (=Spratly Islands). Water depth 0–90 m. (22, 27)

O. subsymmetrica Canu & Bassler: Hainan. Water depth 0–10 m. (26, 27)

Smittipora abyssicola (Smitt): Guangdong; Hainan; Nansha (=Spratly Islands). Water depth 0–80 m. (27)

S. cordiformis Harmer: Hainan; Xisha (=Paracel Islands); Nansha (=Spratly Islands). Water depth 0–15 m. (27)

S. harmeriana Canu & Bassler: Hainan; Xisha (=Paracel Islands); Nansha (=Spratly Islands). Water depth 0–15 m. (27)

CHLIDONIDAE

Crepis longipes Jullien: Zhejiang; Fujian; Guangdong. Water depth 0–50 m. (27, 32, 34)

Poricellaria ratoniensis Jullien: Xisha (=Paracel Islands); Nansha (=Spratly Islands). Water depth 0–60 m. (22, 27)

STEGINOPORELLIDAE

Labioporella cornuta Harmer: Yellow Sea; East China Sea; South China Sea. Water depth 40–120 m. (27)

Steginoporella magniblaris (Busk): East China Sea; Guangdong; Hainan; Zhongsha (=Maccslesfield Bank); Nansha (=Spratly Islands). Water depth 30–130 m. (11, 24, 27)

THALAMOPORELLIDAE

Thalamoporella hamata Harmer: Nansha (=Spratly Islands). Water depth 30–60 m. (21, 27)

T. indica (Hincks) [*T. gothica*, Huang et al. (1990)]: Guangdong; Guangxi; Hainan. Water depth 0–50 m. (7, 11, 27, 30)

T. linearis Canu & Bassler: Hong Kong. Water depth 169 m. (3, 27)

T. lioticha (Ortmann): East China Sea; Guangdong; Hainan; Xisha (=Paracel Islands); Nansha (=Spratly Islands). Water depth 30–120 m. (27)

T. rozieri (Savigny & Audouin): Zhongsha (=Macclesfield Bank); Nansha (=Spratly Islands). Water depth 50–70 m. (8, 27)

T. stapiliera Levinsen: Guangdong; Hainan; Nansha (=Spratly Islands). Water depth 0–60 m. (2, 7, 27, 30)

T. tubifera Levinsen: Southern Hainan; Nansha (=Spratly Islands). Water depth 0–50 m. (27)

CELLARIOIDEA
CELLARIIDAE

Cellaria beniensis Silen [*C. veleronis*, Wang and Hu (1984)]: Southern East China Sea; Guangdong; Guangxi; Hainan; Xisha (=Paracel Islands); Nansha (=Spratly Islands). Water depth 20–70 m. (27)

C. punctata (Busk) [*C. gracilis*, Liu (1991)]: Bohai; Shandong Peninsula; Jiangsu; Zhejiang; Fujian; Guangdong; Guangxi; Hainan; Zhongsha (=Macclesfield Bank); Xisha (=Paracel Islands); Nansha (=Spratly Islands). Water depth 0–90 m. (22, 27, 30, 32)

C. sinica Liu: Zhongsha (=Macclesfield Bank). Water depth 100–210 m. (27)

ASCIPHORINA
CRIBRIMORPHA
CRIBRILINOIDEA
CRIBRILINIDAE

Cribrilina annulata (Fabricius): Shandong Peninsula; Zhejiang. Water depth 0–70 m. (24, 30)

C. biaviculariata Liu: Southern Hainan. Water depth 5–25 m. (24)

C. flabellifera (Kirkpatrick): Hong Kong; Nansha (=Spratly Islands). Water depth 0–45 m. (3, 22, 24)

C. harmeri Ristedt: Southern Hainan. Water depth 5–25 m. (24)

C. innominata (Couch): Guangdong; Hainan; Xisha (=Paracel Islands); Nansha (=Spratly Islands). Water depth 0–60 m. (24)

C. radiata (Moll): Zhejiang; Fujian; Guangdong; Hainan; Zhongsha (=Macclesfield Bank); Xisha (=Paracel Islands); Nansha (=Spratly Islands). Water depth 0–130 m. (22, 24, 27)

C. setosa Kirkpatrick: Zhongsha (=Macclesfield Bank); Nansha (=Spratly Islands). Water depth 30–200 m. (24)

Figularia figularis (Johnston): Shandong Peninsula; Zhejiang. Water depth 0–70 m. (24, 30)

F. fissa (Hincks): Nansha (=Spratly Islands). Water depth 30–60 m. (22, 24)

F. fissulata Canu & Bassler: Fujian. Water depth 0–20 m. (24)

Lyrula hippocrepis (Hincks): Shandong Peninsula. Water depth 0–45 m. (27, 30)

L. ortmanni (Silen): East China Sea. Water depth 70–130 m. (24)

Membraniporella aragoi (Audouin): East China Sea; Xisha (=Paracel Islands). Water depth 0–120 m. (24, 27)

Reginella furcata (Hincks): Shandong Peninsula. Water depth 0–45 m. (27, 30)

CATENICELLOIDEA
CATENICELLIDAE

Catenicella elegans Busk: Fujian; Guangdong; Guangxi; Hainan; Zhongsha (=Macclesfield Bank); Xisha (=Paracel Islands); Nansha (=Spratly Islands). Water depth 0–80 m. (11, 30)

C. triangulifera (Harmer): Guangdong; Hainan. Water depth 0–45 m. (30)

C. uberrima (Harmer): East China Sea; Guangdong; Hainan. Water depth 30–80 m. (30)

EURYSTOMELLIDAE

Eurystomella bilabiata (Waters): Shandong Peninsula; Zhejiang; Fujian. Water depth 0–80 m. (24)

SAVIGNIELLIDAE

Halysisis ijimai (Okada): Zhejiang; Fujian; Guangdong; Hainan; Xisha (=Paracel Islands); Nansha (=Spratly Islands). Water depth 45–80 m. (5, 24, 26, 28)

Savigniella lafontii (Audouin): Guangdong; Hainan; Xisha (=Paracel Islands); Nansha (=Spratly Islands). Water depth 0–25 m. (24, 28, 30)

Vasigniella otophora (Kirkpatrick): Nansha (=Spratly Islands). Water depth 40–60 m. (24, 34)

HIPPOTHOOMORPHA
HIPPOTHOOIDEA
HIPPOTHOIDAE

Celleporella expansa Dawson: Changshan Islands (Liaoning); Shandong Peninsula. Water depth 0–45 m. (24)

C. hylina (Linnaeus): Eastern Liaoning Peninsula; Hebei; Shandong Peninsula; Jiangsu. Water depth 0–70 m. (9, 24)

Hippothoa distans MacGillivray: Shandong Peninsula; Jiangsu; Zhejiang; Fujian; Guangdong; Hainan. Water depth 0–75 m. (24)

H. divaricata Lamouroux: East China Sea. Water depth 30–70 m. (24)

H. flagellum Lamouroux: East China Sea; Guangdong; Hainan; Zhongsha (=Macclesfield Bank); Xisha (=Paracel Islands); Nansha (=Spratly Islands). Water depth 0–200 m. (8, 24, 28)

CHORIZOPORIDAE

Chorizopora brongniartii (Audouin): Fujian; Guangdong; Hainan; Xisha (=Paracel Islands); Nansha (=Spratly Islands). Water depth 0–45 m. (24)

UMBONULOMORPHA
ARACHNOPUSIOIDEA
EXECHONELLIDAE

Exechonella discoidea Canu & Bassler: Southern Hainan; Nansha (=Spratly Islands). Water depth 5–45 m. (24)

E. magna (MacGillivray): East China Sea; Guangdong; Hainan; Zhongsha (=Macclesfield Bank); Xisha (=Paracel Islands); Nansha (=Spratly Islands). Water depth 0–80 m. (8, 24)

E. tuberculata (MacGillivray): Hainan. Water depth 0–25 m. (24)

LEPRALIELLOIDEA
UMBONULIOIDEA
UMBONULIDAE

Rhamphostomella costata (Lorenz): Shandong Peninsula. Water depth 0–30 m. (24, 30)

R. scabra (Fabricius): Changshan Islands (Liaoning); Shandong Peninsula; Zhejiang. Water depth 0–45 m. (24)

Umbonula alvareziana (d'Orbigny): East China Sea. (33, 39)

U. arctica (M. Sars): Changshan Islands (Liaoning); northern Shandong Peninsula. Water depth 0–45 m. (24)

U. verrucosa (Esper): Fujian; Guangdong. Water depth 0–60 m. (24)

EXOCHELLIDAE

Escharella immersa (Fleming): Fujian; Guangdong. Water depth 0–45 m. (24)

E. longirostris Jullien: Shandong Peninsula; Jiangsu; Zhejiang; Fujian. Water depth 0- 110 m. (24, 28, 30)

E. ventricosa (Hassall): Changshan Islands (Liaoning). Water depth 0–45 m. (24)

Escharoidea sauroglossa Levinsen: Changshan Islands (Liaoning); Shandong Peninsula. Water depth 0–45 m. (24)

Exochella areolata Okada & Mawatari: Shandong Peninsula; Jiangsu; Zhejiang; Fujian. Water depth 0–60 m. (24)

LEPRALIELLIDAE

Celleporaria asperta (Hincks): Southern Shandong Peninsula; Zhejiang; Fujian; Guangdong; Hainan. Water depth 10–110 m. (24, 30)

C. columnaris (Busk): Guangdong; Hainan. Water depth 0–60 m. (30)

C. erectorostris (Canu & Bassler): East China Sea; Guangdong; Hainan; Xisha (=Paracel Islands); Nansha (=Spratly Islands). Water depth 0–80 m. (11, 24, 30, 34)

C. fusca (Busk): Shandong Peninsula; Zhejiang; Fujian. Water depth 0–50 m. (1, 24, 30)

C. projecta (Okada & Mawatari): Guangdong. Water depth 0–45 m. (24)

C. sinensis (Canu & Bassler): Hong Kong. Water depth 160 m. (3)

C. tridenticulata (Busk): East China Sea; Guangdong; Hainan; Xisha (=Paracel Islands); Nansha (=Spratly Islands). Water depth 70–110 m. (24, 30)

C. umbonatoidea (Liu & Li): Fujian; Guangdong. Water depth 0–20 m. (7, 21, 22, 30)

ADEONOIDEA
ADEONIDAE

Adeonellopsis arculifera (Canu & Bassler): Nansha (=Spratly Islands). Water depth 30- 60 m. (24)

A. yarransis (Waters): East China Sea; Guangdong; Hainan. Water depth 30–110 m. (24, 28, 30)

Reptadeonella joloensis (Bassler): Nansha (=Spratly Islands). Water depth 40–105 m. (11, 34)

ADEONELLIDAE

Adeonella japonica Ortmann: Zhejiang; Fujian; Guangdong. Water depth 5–105 m. (11, 34)

A. lichenoides (Lamarck) [*A. platalea* (Busk, 1854), *A. minutipora* (Canu & Bassler, 1929)]: Zhejiang; Fujian; Guangdong. Water depth 0–65 m. (11, 34)

LEPRALIOMORPHA
SCHIZOPORELLOIDEA
HIPPOPODINIDAE

Cosciniopsis lonchaea (Busk): Guangdong; Hainan; Nansha (=Spratly Islands). Water depth 40–75 m. (8, 24)

C. vestila (Hincks): Guandong; Hainan; Nansha (=Spratly Islands). Water depth 30–70 m. (24)

Hippopodina feegeensis (Busk): Zhejiang; Fujian; Guangdong; Guangxi; Hainan; Zhongsha (=Macclesfield Bank); Xisha (=Paracel Islands); Nansha (=Spratly Islands). Water depth 0–200 m. (7, 10, 11, 24, 26, 29, 30)

H. perforata (Okada & Mawatari): Hong Kong. Water depth 0–5 m. (7, 11, 30)

Robertsonidra ingens (Canu & Bassler): Guangdong; Hainan; Nansha (=Spratly Islands). Water depth 5–30 m. (3, 4, 7, 24, 30)

HIPPOPORINIDAE

Arthropoma cecillii (Audouin): Zhejiang; Fujian; Guangdong; Guangxi; Hainan; Zhongsha (=Macclesfield Bank); Xisha (=Paracel Islands); Nansha (=Spratly Islands). Water depth 0–150 m. (24, 28, 36)

A. circinatum (MacGillivray): Zhejiang; Fujian; Guangdong; Hainan. Water depth 0–70 m. (24, 28)

Calyptotheca capitifera (Canu & Bassler): Guangdong; Hainan. Water depth 20–70 m. (24)

C. nivea (Busk): Zhejiang; Fujian; Guangdong; Hainan. Water depth 20–70 m. (24)

C. parviminuta Harmer: Guangdong. Water depth 0–15 m. (24)

C. wasiensis (Waters): Zhejiang; Fujian; Guangdong; Guangxi; Hainan; Zhongsha (=Macclesfield Bank); Xisha (=Paracel Islands); Nansha (=Spratly Islands). Water depth 0–150 m. (24)

Codonellina biformis Zhang & Liu: Southern Shandong Peninsula. Water depth 5–35 m. (30, 41)

C. montferrandii (Audouin): East China Sea; Guangdong; Hainan; Nansha (=Spratly Islands). Water depth 45–120 m. (24, 30)

C. obtusata (Ortmann): Guangdong. Water depth 0–45 m. (24, 34)

Emballotheca acutirostris Harmer: Guangdong. Water depth 0–30 m. (24)

E. incisa (Busk): Bohai; Changshan Islands (Liaoning); Shandong Peninsula; Jiangsu; Zhejiang. Water depth 0–30 m. (2, 9, 24)

E. pacifica Harmer [*Lepralia squadrata*]: Zhongsha (=Macclesfield Bank). Water depth 20–100 m. (6, 8, 24)

E. subsinuata (Hincks): Hong Kong. (4, 24)

Hippoporella spinigera (Philipps) [*Hippoporina spinifera*]: Guangdong; Hainan; Nansha (=Spratly Islands). Water depth 0–80 m. (3, 6, 8, 24)

Hippoporina porcellana (Busk): East China Sea; Guangdong; Hainan; Nansha (=Spratly Islands). Water depth 30–120 m. (24)

H. verrucosa Canu & Bassler: Nansha (=Spratly Islands). Water depth 30–60 m. (24)

GIGANTOPORIDAE

Gigantopora mutabilis (Canu & Bassler): Hainan; Xisha (=Paracel Islands); Nansha (=Spratly Islands). Water depth 0–25 m. (24)

G. pupa (Jullien): Hainan; Xisha (=Paracel Islands); Nansha (=Spratly Islands). Water depth 0–40 m. (24)

CHEILOPORINIDAE

Cheiloporina haddoni (Harmer): East China Sea; Guangdong. Water depth 80–100 m. (9, 11, 24, 28, 30)

Hippaliosina acutirostris Canu & Bassler: Hainan. Water depth 0–25 m. (24)

H. calyciformis (Phillips): Zhongsha (=Macclesfield Bank). Water depth 45–100 m. (8, 24)

H. hippocrepis (Simitt): Eastern Liaoning Peninsula; Hebei; Shandong Peninsula. Water depth 0–45 m. (2, 24, 30)

H. hongkongensis (Liu & Li): Fujian; Guangdong; Hainan. Water depth 0–15 m. (7, 23, 24, 25, 26, 30)

H. ovicellata Harmer: Guangdong; Hainan; Nansha (=Spratly Islands). Water depth 30–50 m. (24)

H. triformis Canu & Bassler: Hainan; Nansha (=Spratly Islands). Water depth 0–60 m. (24)

CRYPTOSULIDAE

Cryptosula pallasiana (Moll): Eastern Liaoning Peninsula; Hebei; Shandong Peninsula; northern Jiangsu. Water depth 0–15 m. (6, 24, 30, 34)

PHYLACTELLIDAE

Phylactella geometrica Kirkpatrick: Zhongsha (=Macclesfield Bank). Water depth 50–60 m. (8)

Phylactellipora collaris (Norman): Shandong Peninsula; Jiangsu; Zhejiang; Fujian; Guangdong; Hainan; Xisha (=Paracel Islands). (9, 11)

WATERSIPORIDAE

Watersipora subtorguata (d'Orbigny) [*Dakaria typica, W. subovoidea*]: Eastern Liaoning Peninsula; Hebei; Shandong Peninsula; Jiangsu; Zhejiang; Fujian; Guangdong; Guangxi; Hainan; Xisha (=Paracel Islands); Nansha (=Spratly Islands). Water depth 0–20 m. (7, 9, 11, 24, 30, 35)

SCHIZOPORELLIDAE

Characodoma latisinuatum Harmer: Southern Hainan. Water depth 0–25 m. (24)

Escharina grandicella (Canu & Bassler): Guangdong. Water depth 30–70 m. (24)

E. pesanselis (Smitt) [*Mastigophora duterteri*]: Zhejiang; Fujian; Guangdong; Guangxi; Hainan; Xisha (=Paracel Islands); Zhongsha (=Macclesfield Bank); Nansha (=Spratly Islands). Water depth 0–150 m. (11, 24, 28, 39)

Schizomavella australis (Haswell): Southern East China Sea; Guangdong; Hainan. Water depth 30–90 m. (24, 28)

S. granulosa Canu & Bassler: Hong Kong. Water depth 80–120 m. (2, 24)

S. pacifica Liu & Wass: East China Sea. Water depth 70–130 m. (24, 28)

S. porata Liu & Wass: East China Sea; Guangdong; Hainan; Nansha (=Spratly Islands). Water depth 30–130 m. (24, 28)

S. umbonatoidea Liu & Wass: East China Sea. Water depth 100 m. (24, 28)

Schizoporella biaperta (Michelin): Qingdao harbor (Shandong); Liangyungang (Jiangsu); Zhejiang. (33, 34)

S. erratoidea Liu [*S. errata*, Huang et al. (1990)]: Guangdong; Hainan. Water depth 0–15 m. (10, 11, 30)

S. tumulosa Hincks: Zhejiang. (33)

S. unicornis (Johnston): Eastern Liaoning Peninsula; Hebei; Shandong Peninsula; Jiangsu; Zhejiang; Fujian; Guangdong; Guangxi; Hainan. Water depth 0–120 m. (3, 9, 11, 24, 34)

Stylopoma duboisii (Audouin): Guangdong; Hainan; Zhongsha (=Macclesfield Bank); Nansha (=Spratly Islands). Water depth 0–200 m. (24)

S. parviporosum Canu & Bassler: Guangdong; Hainan. Water depth 0–40 m. (24)

STOMACHETOSELLIDAE

Cigclisula occlusa Busk: Xisha (=Paracel Islands); Nansha (=Spratly Islands). Water depth 30–60 m. (24)

C. turrids (Smitt): Guangdong; Hainan. Water depth 5–80 m. (24)

Stomachetosella collipora Liu & Wass: East China Sea. Water depth 80–120 m. (24, 28)

SMITTINIDAE

Aimulosa signata (Waters): Zhejiang; Fujian; Guangdong; Hainan; Nansha (=Spratly Islands). Water depth 0–80 m. (24, 28)

Parasmittina acumenata Liu & Wass: East China Sea; Guangdong; Hainan; Nansha (=Spratly Islands). Water depth 20–120 m. (24, 28)

P. acuta (Canu & Bassler): Zhoushan Islands (Zhejiang). (30, 38)

P. biformis Liu & Wass: East China Sea. Water depth 100–120 m. (24, 28)

P. californica (Robertson): East China Sea. (36)

P. deliculata (Busk) [*Smittina aviculata*]: Zhejiang; Fujian; Guangdong; Hainan. Water depth 0–70 m. (24, 35)

P. donghaiensis Liu & Wass: East China Sea. Water depth 100–120 m. (24, 28)

P. glomerata (Thornely): Changshan Islands (Liaoning); Shandong Peninsula; Jiangsu; Zhejiang; Fujian; Guangdong; Guangxi; Hainan; Xisha (=Paracel Islands); Nansha (=Spratly Islands). Water depth 0–150 m. (1, 22, 24, 30)

P. malleolus (Hincks): Zhongsha (=Macclesfield Bank); Nansha (=Spratly Islands). Water depth 30–150 m. (24)

P. papulata (Harmer): East China Sea. (36)

P. parsenvalii (Audouin): Guangdong; Hainan. Water depth 0–60 m. (24)

P. projecta (Okada & Mawatari): Zhejiang. (35)

P. raigii (Audouin): Zhejiang; Fujian; Guangdong; Hainan; Xisha (=Paracel Islands); Nansha (=Spratly Islands). Water depth 0–120 m. (24)

P. triformis Liu & Wass: East China Sea. Water depth 90–130 m. (24, 28)

P. trispinosa (Johnston): Zhejiang; Fujian; Guangdong. Water depth 0–80 m. (11)

P. tropica (Waters): Zhongsha (=Macclesfield Bank). Water depth 80–100 m. (8)

P. varians Liu & Wass: East China Sea. Water depth 90–130 m. (24, 28)

Porella compressa (J. Sowerby): Hong Kong. (37)

P. rotundirostris Liu: Southern Shandong Peninsula. Water depth 10–20 m. (21)

Smittina landsborovii (Audouin): Changshan Islands (Liaoning); Hebei; Shandong Peninsula; Jiangsu; Zhejiang; Fujian; Guangdong; Hainan. Water depth 0–60 m. (24)

Smittoidea levis (Kirkpatrick): East China Sea; Guangdong; Hainan; Nansha (=Spratly Islands). Water depth 30–130 m. (23, 26)

S. pacifica Soule & Soule: East China Sea; Guangdong; Nansha (=Spratly Islands). Water depth 0–80 m. (24, 28)

S. prolifera Obsburn: Eastern Liaoning Peninsula; Hebei; Shandong Peninsula; Jiangsu; Zhejiang; Fujian; Guangdong; Guangxi; Hainan; Xisha (=Paracel Islands); Nansha (=Spratly Islands). Water depth 0–150 m. (1, 9, 24, 28)

S. reticulata (MacGillivray): East China Sea; Guangdong; Hainan; Nansha (=Spratly Islands). Water depth 30–130 m. (23, 26)

S. spinigera Liu [*S. reticulata*, Wang and Cai (1977)]: Southern Shandong Peninsula; Zhejiang. Water depth 0–20 m. (21, 24, 30)

PETRALIELLIDAE

Hippopetraliella magna (d'Orbigny): Fujian; Guangdong. Water depth 0–45 m. (10, 24, 30)

Mucropetraliella philippinensis (Canu & Bassler): Zhejiang; Fujian; Guangdong. Water depth 0–30 m. (1, 24, 26, 30, 37)

M. robusta (Canu & Bassler): Zhejiang; Fujian; Guangdong. Water depth 0–45 m. (24, 35)

M. thenardii (Audouin): Nansha (=Spratly Islands). (8, 24)

M. verrucosa (Canu & Bassler): Qiongzhou Strait (between Guangdong and Hainan). Water depth 0–15 m. (24)

M. watersi Harmer: Guangdong; Hainan. Water depth 0–70 m. (24)

Petraliella bisinuata (Smitt): Southern Hainan. Water depth 5–25 m. (24)

P. chuakensis (Waters): Southern Hainan. Water depth 5–25 m. (24)

Sinupetraliella umbonata (Okada & Mawatari): Zhejiang; Fujian; Guangdong. Water depth 0–20 m. (10, 35)

CREPIDACANTHIDAE

Crepidacantha crinispina (Linnaeus): Hainan. Water depth 0–25 m. (24)

C. posissoni (Audouin): Hainan; Xisha (=Paracel Islands); Zhongsha (=Macclesfield Bank); Nansha (=Spratly Islands). Water depth 0–150 m. (8, 24)

LANCEOPORIDAE

Lanceopora formosa Harmer: East China Sea. Water depth 70–105 m. (24)

MICROPORELLIDAE

Calloporina sculpta Canu & Bassler: Southern Hainan. Water depth 5–25 m. (24)

Fenestrulina infundibulipora Canu & Bassler: Guangdong; Hainan; Xisha (=Paracel Islands); Nansha (=Spratly Islands). Water depth 0–80 m. (24)

F. malusi (Audouin): Shandong Peninsula; Jiangsu; Zhejiang; Fujian; Guangdong; Hainan; Xisha (=Paracel Islands); Zhongsha (=Macclesfield Bank); Nansha (=Spratly Islands). Water depth 0–150 m. (7, 9, 24, 28, 30, 34)

Microporella ciliata (Pallas): Eastern Liaoning Peninsula; Hebei; Shandong Peninsula; Jiangsu; Zhejiang; Fujian; Guangdong; Guangxi; Hainan; Xisha (=Paracel Islands); Zhongsha (=Macclesfield Bank); Nansha (=Spratly Islands). Water depth 0–200 m. (7, 24, 28, 30, 32, 38)

M. coronata (Audouin): East China Sea. (37)

M. coscinophora Reuss: Zhongsha (=Macclesfield Bank). Water depth 60–150 m. (8, 24)

M. lunifera Haswell: Southern Shandong Peninsula; Zhejiang; Fujian; Guangdong; Hainan; Xisha (=Paracel Islands); Nansha (=Spratly Islands). Water depth 20–150 m. (24, 30)

M. orientalis Harmer: Guangdong; Hainan; Xisha (=Paracel Islands); Nansha (=Spratly Islands). Water depth 0–150 m. (24, 30)

M. vibraculifera (Hincks): Southern Shandong Peninsula; Zhejiang; Fujian; Guangdong; Hainan; Xisha (=Paracel Islands); Nansha (=Spratly Islands). Water depth 0–120 m. (24, 28, 30)

Stephanosella bernardii (Audouin): Nansha (=Spratly Islands). Water depth 30–60 m. (24)

S. bolini Osburn: Zhejiang. Water depth 78 m. (36)

CALWELLIDAE

Onchoporella selenoides Ortmann: Changshan Islands (Liaoning); Shandong Peninsula; Jiangsu; Zhejiang. Water depth 20–80 m. (24, 28, 36)

MARGARETTIDAE

Margaretta cereoides (Ellis & Solander): Guangdong; Xisha (=Paracel Islands); Nansha (=Spratly Islands). Water depth 30–100 m. (3, 8, 24)

M. gracilis (Ortmann): Nansha (=Spratly Islands). Water depth 30–60 m. (24)

SPIROPORINIDAE

Spiroporina longicollis (Canu & Bassler): Nansha (=Spratly Islands). Water depth 30–60 m. (8)

S. vertebralis (Stocliczk): Nansha (=Spratly Islands). Water depth 30–50 m. (8)

ACTISECIDAE

Actisecos regularis d'Orbigny: East China Sea. Water depth 70–110 m. (24)

CELLEPOROIDEA
HIPPOPORIDRIDAE

Hippoporidra calcarea (Smitt): East China Sea. Water depth 70–130 m. (24)

CELLEPORIDAE

Celleporina costazii (Audouin): Shandong Peninsula; Jiangsu; Zhejiang; Fujian; Guangdong; Hainan; Xisha (=Paracel Islands); Zhongsha (=Macclesfield Bank); Nansha (=Spratly Islands). Water depth 0–150 m. (1, 9, 30, 37)

C. geminata (Ortmann): Shandong Peninsula; Jiangsu; Zhejiang; Fujian; Guangdong. Water depth 70–112 m. (24, 28, 30)

C. pisiformis Canu & Bassler: Guangdong. Water depth 20–50 m. (3, 4, 6)

C. porosissima Harmer: Zhejiang; Fujian; Guangdong; Xisha (=Paracel Islands). Water depth 0–110 m. (24, 35, 38)

C. radiata (Ortmann): East China Sea; Guangdong; Hainan. Water depth 30–70 m. (3, 24, 38)

C. spathulata (MacGillivray): Guangdong; Xisha (=Paracel Islands). Water depth 20–70 m. (8, 24)

PHIDOLOPORIDAE

Hippellozoum pectinatum (Kirkpatrick): Zhongsha (=Macclesfield Bank); Nansha (=Spratly Islands). Water depth 40–150 m. (9, 24)

Iodictyum axillare (Ortmann): Zhejiang; Fujian; Guangdong; Hainan. Water depth 0–70 m. (24, 37)

I. gibberosum (Buchner): Nansha (=Spratly Islands). Water depth 40–60 m. (6)

I. polycrenulatum (Okada): East China Sea. (36)

I. willeyi Harmer: Zhongsha (=Macclesfield Bank). Water depth 60–100 m. (6, 8)

Phidolopora pacifica (Robertson): Changshan Islands (Liaoning); Shandong Peninsula. Water depth 5–45 m. (1, 24, 28, 30)

Reteporellina denticulata (Busk): Guangdong; Hainan. Water depth 5–70 m. (24, 30)

Rhynchozoon globosum Harmer: Eastern and southern Hainan; Nansha (=Spratly Islands). Water depth 10–60 m. (24)

R. grandicellum Canu & Bassler: Southern Hainan. Water depth 5–25 m. (24)

R. rostratum (Busk): Nansha (=Spratly Islands). Water depth 30–60 m. (24)

Schizoretepora tumescens (Ortmann): Changshan Islands (Liaoning); Shandong Peninsula. Water depth 0–45 m. (38)

Sertella suluensis Harmer: Zhongsha (=Macclesfield Bank); Nansha (=Spratly Islands). Water depth 30–130 m. (24)

Triphyllozoon bimunitum (Ortmann): Zhejiang. Water depth 99 m. (33, 38)

T. hirsutum (Busk): Northern Hainan. (11)

T. inovicellatum Liu & Li: Hong Kong. Water depth 0–15 m. (7, 11, 25)

T. moniliferum (MacGillivray): Zhongsha (=Macclesfield Bank); Nansha (=Spratly Islands). Water depth 30–100 m. (8, 24)

T. sinicum Liu & Li: Hong Kong. Water depth 0–15 m. (7, 25)

CLEIDOCHASMATIDAE

Cleidochasma bassleri (Calvet): Hainan; Nansha (=Spratly Islands). Water depth 0–60 m. (24)

C. biavicularium (Canu & Bassler): Guangdong; Hainan. Water depth 0–60 m. (24)

C. fallax (Canu & Bassler): Guangdong; Hainan. Water depth 0–60 m. (24)

C. porcellanum (Busk): Southern Hainan. Water depth 0–15 m. (24)

C. protrusum (Thornely): Hainan; Nansha (=Spratly Islands). Water depth 0–45 m. (24)

Drepanophora corrugata (Thornely): Southern Hainan. Water depth 0–15 m. (24, 30)

CONESCHARELLIOIDEA
CONESCHARELLINIDAE

Conescharellina angusta d'Orbigny: Yellow Sea; East China Sea; northern South China Sea. Water depth 30–150 m. (24)

C. breviconica Canu & Bassler: Yellow Sea; East China Sea; northern South China Sea. Water depth 20–80 m. (24)

C. catella Canu & Bassler: Southern Shandong Peninsula; East China Sea; Guangdong; Hainan; Nansha (=Spratly Islands). Water depth 30–150 m. (24)

C. concava Canu & Bassler: Guangdong; Hainan; Nansha (=Spratly Islands). Water depth 30–80 m. (2, 24)

C. crassa (Tenison Wood): East China Sea; northern South China Sea. Water depth 30–110 m. (24)

C. elongata d'Orbigny: Yellow Sea; East China Sea; northern South China Sea. Water depth 20–110 m. (24)

C. jucunda Canu & Bassler: East China Sea. Water depth 50–100 m. (24)

C. milleporacea Canu & Bassler: Nansha (=Spratly Islands). Water depth 30–60 m. (24)

C. obliqua Canu & Bassler: East China Sea. Water depth 50–90 m. (24)

C. ovalis Harmer: Nansha (=Spratly Islands). Water depth 30–60 m. (24)

C. papulifera Harmer: Guangdong; Hainan; Nansha (=Spratly Islands). Water depth 30–150 m. (24)

C. rectilinea Harmer: East China Sea. Water depth 50–80 m. (24)

C. symmetrica Canu & Bassler: Guangdong; Hainan; Nansha (=Spratly Islands). Water depth 30–150 m. (24)

Flabellopora acutirostris Canu & Bassler: East China Sea. Water depth 60–80 m. (24)

F. elegans d'Orbigny: Southern Yellow Sea; East China Sea; South China Sea; Nansha (=Spratly Islands). Water depth 30–150 m. (2, 24, 38)

F. irregularis Canu & Bassler: Southern Yellow Sea; East China Sea; South China Sea; Nansha (=Spratly Islands). Water depth 30–150 m. (2, 24, 32)

F. lenticularis Canu & Bassler: Nansha (=Spratly Islands). Water depth 30–60 m. (24)

F. linguna Silen: Southern Yellow Sea; East China Sea. Water depth 30–110 m. (24)

F. pisiformis Canu & Bassler: East China Sea. Water depth 70–90 m. (24)

F. pusilla Silen: Southern Yellow Sea; East China Sea. Water depth 30–80 m. (24)

F. tuberosa Canu & Bassler: East China Sea. Water depth 70–110 m. (24)

Trochodoson linearis Canu & Bassler: Guangdong; Hainan; Zhongsha (=Macclesfield Bank); Nansha (=Spratly Islands). Water depth 40–150 m. (6, 24)

T. oplatus Harmer: East China Sea; Guangdong; Hainan; Nansha (=Spratly Islands). Water depth 30–80 m. (6, 24)

Zeuglopora lanceolata (Maplestone): Nansha (=Spratly Islands). Water depth 60–250 m. (24)

REFERENCES*

(1). Androsova, H. I. 1959. Some data on the Bryozoa of the Yellow Sea. Arch. Inst. Oceanol. Sin. 1(4):56–70, pls. 1–3.

(2). Androsova, H. I. 1963. Bryozoa from the South China Sea. Stud. Mar. Sin. 4:21–45, pls.1–3.

(3). Canu, F. and R. S. Bassler, 1929. Bryozoa of the Phillippine region. Proc. U.S. Nat. Mus. 100(9):1–685, figs. 1–224, pls. 1–94.

(4). Harmer, S. F. 1926. The Polyzoa of the Siboga Expedition. Part 2. Cheilostomata Anasca. Siboga Exped., 28b:1–360, pls. 13–34.

(5). Harmer, S. F. 1934. The Polyzoa of the Siboga Expedition. Part 3. Cheilostomata Ascophora I. Family Reteporidae, Siboga Exped., 28c:vii and 503–640, pls. 35–41.
(6). Harmer, S. F. 1957. The Polyzoa of the Siboga Expedition. Part 4. Cheilostomata II. Siboga Exped., 28d:641–1147, pls. 42–74.
(7). Huang, Z. G., C. Y. Li, and X. X. Liu, 1990. The Bryozoan foulers of Hong Kong and neibouring waters. In: Proceedings of Second International Marine Biological Workshop: The Marine Flora and Fauna of Hong Kong and Southern China. Hong Kong University Press, pp:737–765.
(8). Kirkpatrick, R. 1890. Report of the Hydrozoa and Bryozoa collected by P. W. Bassett - Smity Esq. (Sargon R. N., during the Survey of the Tizard and Maccslesfield Bank in China Sea). Ann. Mag. Nat. Hist. 6(5):11–24.
(9). Li, C. Y. 1988. Bryozoans in the fouling organisms from the Bohai Bay and the Yellow Sea. Acta Ecol. Sin. 8(2):61–68.
(10). Li, C. Y. 1989. Ecology on Bryozoans in Daya Bay. In: Collections of Papers on Marine Ecology in the Daya Bay (I). Third Institute of Oceanography (SOA) (Ed.), China Ocean Press, pp:106–116.
(11). Li, C. Y. 1992. Bryozoans in fouling organisms from the northern waters of the South China Sea. J. Oceanog. Taiwan, 11(1):61–68.
(12). Liu, X. X. 1980. Two new species of Scrupocellaria (Scrupocellariidae, Anasca) of the Xisha Islands, Guangdong Province. Stud. Mar. Sin. 17:179–185, figs. 1–8.
(13). Liu, X. X. 1982a. A new species of the genus *Kinetoskias* Daniellsen collected from the continental shelf of the East China Sea. Stud. Mar. Sin. 19:72–80.
(14). Liu, X. X. 1982b. Notes on Chinese Bryozoa of the Family Flustridae. Act. Zootax. Sin. 7(2):132–138.
(15). Liu, X. X. 1983a. New species and new records of Antropora and Beania (Anasca, Bryozoa) from China Seas. Oceanol. Limnol. Sin. 13(1):97–101, pls. 1–2.
(16). Liu, X. X., 1983b. On three species of genus *Ellisina* (Bryozoa) from the continental shelf of the East China Sea. Oceanol. Limnol. Sin. 14(2):169–172, fig. 1.
(17). Liu, X. X. 1983c. Two new species and one first record of genus *Crassimarginatella* from the East China Sea. Oceanol. Limnol. Sin. 13(3):289–292, fig. 1.
(18). Liu, X. X. 1984a. On the Species of Family Bicellariellidae (Bryozoa) from Chinese Seas. Stud. Mar. Sin. 22:254–324, figs. 1–49.
(19). Liu, X. X. 1984b. On the Species of Family Scrupocellaridae Collected from Chinese Seas. Stud. Mar. Sin. 23:257–308, figs. 1–43.
(20). Liu, X. X. 1985. On the Genus *Caulibugnla* Verrill, 1900, Collected from Chinese Seas. Stud. Mar. Sin. 25:127–152, figs. 1–29.
(21). Liu, X. X. 1990. Three new Cheilostome bryozoans from the coastal waters of Shandong and Zhejiang provinces, China. Stud. Mar. Sin. 31:121–128.
(22). Liu, X. X. 1991. A study on the Anascan and Cribrimorphan Bryozoa from Nansha (=Spratly Islands). In: Proceedings of Marine Biology of Nansha Islands and Neighbouring Waters I. China Ocean Press, pp:56–81, 5 figs.
(23). Liu, X. X. 1992. On the genus *Membranipora* (Anasca: Chelostomata: Bryozoa) of southern Chinese seas. The Raffles Bulletin of Zoology 40(1):103–144.
(24). Liu, X. X. Catalogue of marine bryozoans from the Chinese seas. (manuscript)
(25). Liu, X. X., C. Y. Li, 1987. Six new species of bryozoans from the waters of Hong Kong and Zhujiang Estuary. J. Oceanogr. Taiwan Strait 6(1):53–66, pls. 1–4.
(26). Liu, X. X. and H. Ristedt. Cheilostome Bryozoans from Hainan Island, China. I. Suborders Inovicellina, Malacostegina and Neocheilostomina (Infraorders Paramalacostegomorpha and Cellulariomorpha). Senckenberg Mar. Fauna. (in press)
(27). Liu, X. X. and F. Z. Wang. Fauna Sinica: Phyllum Bryozoa: Classes Phylactolaemata and Gymnolaemata: Order Cheilostomida: Suborders Inovicellina, Scrupariina, Malacostegina and Neocheilostomina. Committee of Fauna Sinica (Ed.), Chinese Academy of Sciences. China Science Press. (in press)
(28). Liu, X. X. and R. E. Wass. Cheilostomes from Chinese seas. Invertebrate Taxonomy (Australia). (in press)
(29). Liu, X. X. and Z. Y. Yang, 1995. Systematic position of *Membranipora amoyensis* Robertson, 1921 (Membraniporidae: Cheilostomata). In Yu Xiaoqu (Ed.), Proc. Sym. Mar. Sci. in Taiwan Strait and its Adjacent Waters. China Ocean Press, pp:346–355, figs. 1–2, pl. 1.
(30). Liu, X. X., Yin. X. M. and J. H. Ma. Biology on marine bryozoan foulers from the Chinese seas. (manuscript)
(31). Liu, X. X., Yin, X. M. and W. Xia. Significance of Early Astogeny of Cheilostome Bryozoans in Their Evolution. I. The Characteristics of Early Astogeny of Suborder Malacostegina (Membraniporidae and Electridae), with Descriptions of New Genus and Six New Species. Stud. Mar. Sin. 40. (in press)
(32). Tung, Y. M. and F. T. Wang. 1960. Studies on Marine Bryozoa of Chejiang I. Acta Zool. Sin. 12(2):191–200, figs. 1–5.
(33). Wang, F. T. 1988. A new species of *Electra* from China (Cheilostomata: Electridae). Acta Zootax. Sin. 13(2):111–114, figs. 1–5.
(34). Wang, F. Z. 1989. New subspecies and new records of Ectoprocta from China (Cheilostomata: Caribrilinidae). Sichuan Jour. Zool. 8(4):1–3, figs. 1–4.
(35). Wang, T. Z., and R. X. Cai. 1977. Report on Ectoprocta (Bryozoa) from the East China Sea. Mar. Sci. Tech. 7:14–47, figs. 1–47
(36). Wang, F. Z. and R. X. Cai, 1980. New records of Ectoprocta (Bryozoa) from the East China Sea. Contributions of Hangzhou University Natural Science No. 1 (1980):88–94, figs. 1–14.
(37). Wang, F. Z. and R. X. Cai. 1991. Fouling bryozoans from the waters of Gouba, Zhoushan Islands, China. Contributions of Hangzhou University Natural Science No. 12.
(38). Wang, F. Z. and Y. M. He. 1982a. New records of Ectoprocta (Bryozoa) from China. Ocean Practises, No. 1 (1982):47–52, figs. 1–12.
(39). Wang, F. Z. and Y. M. He. 1982b. Bryozoans from the Xisha Islands (Paracel), China. Ocean Practises, No. 2 (1982):43–47, figs. 1–7.
(40). Wang, F. T. and Y. M. Tung. 1976. A new species of Ectopracta from China. Acta Zool. Sin. 22(3):302–303, figs. 1–2.
(41). Zhang, S. L. and X. X. Liu. 1995a. A New species of the Genus *Membranipora* from the southern Chinese seas. Acta Zootax. Sin. 20(2):133–136, fig. 1.
(42). Zhang, S. L. and X. X. Liu. 1995b. A New species of Genus *Codonellina* from the Coastal Waters of Shandong Peninsula (Bryozoa: Cheilostomata: Hipporinidae). Acta Zooltax. Sin. 20(3):257–261, fig. 1.

*: (1) and (2) in both Chinese and Russian; (9), (22), (24), (25), (25), (29)-(42) in Chinese

Compiled by Li Chuanyan and Liu Xixing; Edited by Liu Xixing

ENTOPROCTA

LOXOSOMATIDAE

Loxosoma annulatum Harmer: Guangdong; Hainan; Xisha (=Paracel Islands). Attached to cellularine bryozoans (e.g., *Canda sculata* and *Scrupocellaria diadema*). Water depth 25–85 m. (1, 4)

L. circulare Harmer: Xisha (=Paracel Islands). Attached to cellularine bryozoans (e.g. *Canda scutata* and *Scrupocellaria diadema*). Water depth 30–70 m. (1)

Loxosomella crassicauda (Salensky): Nansha (=Spratly Islands). Attached to corals, gravel, and calcareous bryozoans (e.g., adeonellids). Water depth 30–80 m. (1, 4)

L. harmeri (Schultz): Yellow Sea. Attached to worm tubes. (2)

Loxosomella sp.: Hong Kong. Attached to pea crab (*Pinnixa rathbuni*). (5)

PEDICELLINIDAE

Pedicellina echinata M. Sars: Yellow Sea. Attached to the spines of cut side sea surchin (*Temnopleurus hardwickii*). Water depth 70 m. (1)

BARENTSIIDAE

Barentsia discreta Busk: Fujian; Guangdong; Hainan; Nansha (=Spratly Islands). Attached to corals, gravel, mulluscan shells, barnacles, hydroids or algae. Water depth 0–60 m. (1, 4)

B. gracilis (M. Sars): Nansha (=Spratly Islands). Attached to corals. Water depth 40–70 m. (1, 4)

B. laxa Kirkpatrick: Hainan; Xisha (=Paracel Islands). Attached to shells, corals, and gravel. Water depth 0–60 m. (1)

REFERENCES*

(1). Liu, X. X. Catalogue of Entoprocts from the Chinese Seas. (manuscript)
(2). Sun, R. P. A Entoprocts Species from the Yellow Sea. (manuscript)
(3). Huang, Z. G. and C. Y. Li. 1987. Attaching organisms on artificial reefs for fish. In: Proceedings of Artificial Reefs for Fish. China Ocean Press, pp:118–132.
(4). Kirkpatrick, R. 1890. Report of the Hydrozoa and Polyzoa Collected by P. W. Basset-Smith Esg. S. "Ramller". Commander W. W. Moore. Ann. Mag. Nat. Hist. 6(5):11–24.
(5). Morton, B. and J. Morton. 1983. The Sea Shore Ecology of Hong Kong. Hong Kong University Press, 78pp.

*: (1)-(3) in Chinese
Compiled by Liu Xixing and Sun Ruiping; Edited by Yang Dejian

BRACHIOPODA

INARTICULATA
ATREMATA
LINGULIDAE

Lingula adamsi Dall [*L. shangtangensis*]: Shandong Peninsula; Jiangsu; Zhejiang; Fujian; Guangdong; Guangxi; Hainan. Water depth 0–5 m. (2, 3, 4)

L. anatina Lamarck [*L. unguis*]: Shandong Peninsula; Jiangsu; Zhejiang; Fujian; Guangdong; Guangxi; Hainan. Water depth 0–30 m. (1–4)

NEOTREMATA
DISCINIDAE

Discinisca stella (Gould): Hebei; Shandong; Jiangsu; Zhejiang; Fujian; Guangdong; Hainan. Water depth 5–110 m. (3, 4)

Pelagodiscus atlantica (King): East China Sea. Water depth 1,950 m. (3, 4)

TELOTREMATA
CANCELLOTHYRIDAE

Terebratulina hataina Cooper: East China Sea; Nansha (=Spratly Islands). Water depth 0–150 m. (3, 4)

LAQUEIDAE

Frenulina sanguinolenta (Gmelin): Xisha (=Paracel Islands). Water depth 30–50 m. (3, 4)

TEREBRATALIIDAE

Terebratalia coreanica (Adams & Reeve): Bohai Bay; Eastern Liaoning Peninsula; Shandong Peninsula; northern Jiangsu. Water depth 0–76 m. (1–4)

DALLINIDAE

Compages mariae (A. Adams): East China Sea. Water depth 85–150 m. (3, 4)

REFERENCES*

(1). Invertebrate Department, Institute of Oceanology, Academia Sinica (Ed.), 1959. Common invertebrates in the coastal waters off Qingdao. China Science Press, 161pp.
(2). Hatai, K. M. 1937. On Some Recent Brachiopods from Eastern Shangtung, China. Bull. Biogeogr. Soc. Jap. 7(3):214–318.
(3). Liu, X. X. Catalogue of Brachiopods from the Chinese Seas. (manuscript)
(4). Richardson, J. R., I. R. Stewent, and X. X. Liu. 1989. Brachiopods from China Seas. Chin. J. Oceanal. Limnal. 7(3):211–224.

*: (1) in Chinese
Compiled by Liu Xixing; Edited by Yang Dejian

PHORONIDA

PHORONIDEA
PHORONIDAE

Phoronis anchitecta Andreas: Yellow Sea. (1)
P. australis Haswell: South China Sea. (1)
P. ijimai Oka [*P. hippocrepia, P. vancouvereusis*]: Yellow Sea. Intertidal. (2)
P. muelleri Selys-Longchamps: Yellow Sea. (1)

REFERENCES*

(1). Wu Baoling, Chen Mu et al. 1964. Preliminary study of Phoronids in China's seas. Abstract Book of Chinese Zoological Society 30th Anniversary Conference.

(2). Wu Baoling, Chen Mu et al. 1980. On the occurrence of *Phoronis ijimai* Oka in the Huanghai, with notes on its larval development. Studia Marina Sinica, 16:101- 112.

*: both in Chinese
Compiled by Wu Baoling

CHAETOGNATHA

SAGITTOIDEA
SAGITTIDAE

Eukrohnia bathyantarctica David: Deep water of South China Sea. (2, 3)

E. bathypelagica Alvarino: Deep water of South China Sea. (2, 3)

E. fowleri Rritter-Zahong: Deep water of South China Sea. (2, 3)

E. hamata (Mobius): Deep water of South China Sea. (2, 3)

E. sinica Zhang & Cheng: Deep water of South China Sea. (2)

Krohnitta pacifica (Aida): Southern Yellow Sea; East China Sea; Taiwan Strait; Taiwan; South China Sea. (1–5)

K. subtilis (Grassi): East China Sea; Taiwan Strait; Taiwan; South China Sea. (1–5)

Pterosagitta draco (Krohn): East China Sea; Taiwan Strait; Taiwan; South China Sea. (1–6)

Sagitta bedfordii Doncaster: East China Sea; Taiwan Strait; South China Sea. (1–5)

S. bedoti Beraneck: Southern Yellow Sea; East China Sea; Taiwan Strait; Taiwan; South China Sea. (1–6)

S. bipunctata Quoy & Gaimard: East China Sea; Taiwan Strait; Taiwan; South China Sea. (1–6)

S. bruani Alvarino: East China Sea; Taiwan Strait; South China Sea. (1–5)

S. crass forma naikaiensis Tokioka: Southern Yellow Sea; East China Sea; Taiwan Strait; Guangdong. (1–5)

S. crassa Tokioka: Bohai; Yellow Sea; northern East China Sea; South China Sea. (1, 3, 5)

S. decipiens Fowler: East China Sea; Taiwan Strait; Taiwan; South China Sea. (1–6)

S. delicata Tokioka: Taiwan Strait; Guangdong; South China Sea. (2, 3, 4)

S. enflata Grassi: Yellow Sea; East China Sea; Taiwan Strait; Taiwan; South China Sea. (1–6)

S. ferox Doncaster: Southern Yellow Sea; East China Sea; Taiwan Strait; Taiwan; South China Sea. (1–6)

S. hexaptera d'Orbigny: East China Sea; Taiwan Strait; offshore of southeastern and southwestern Taiwan; South China Sea. (1–6)

S. johorensis Pathansali & Tokioka: Taiwan Strait; South China Sea. (2, 3, 4)

S. lyra Krohn: East China Sea; Taiwan Strait; Taiwan; South China Sea. (1–6)

S. macrocephala Fowler: Offshore of South China Sea. (2, 3)

S. minima Grassi: Southern Yellow Sea; East China Sea; Taiwan Strait; Taiwan; South China Sea. (1–5)

S. nagae Alvarino: Bohai; Yellow Sea; East China Sea; Taiwan Strait; Guangdong; Nansha (=Spratly Islands). (1–5)

S. neglecta Aida: Southern Yellow Sea; East China Sea; Taiwan Strait; South China Sea. (1–6)

S. neodecipiens Tokioka: East China Sea; Taiwan Strait; Taiwan; South China Sea. (1–5)

S. oceania Gray: Nansha (=Spratly Islands). (3)

S. pacifica Tokioka: Southern Yellow Sea; East China Sea; Taiwan Strait; Taiwan; South China Sea. (1–6)

S. planctonis Steinhaus: East China Sea; South China Sea. (2, 3, 5)

S. pseudoserratodentata Tokioka: Offshore of East China Sea; Taiwan Strait; southern Taiwan; South China Sea. (1–6)

S. pulchra Doncaster: Yellow Sea; Taiwan Strait; Taiwan; South China Sea. (1–6)

S. regularis Aida: Southern Yellow Sea; East China Sea; Taiwan Strait; Taiwan; South China Sea. (1–6)

S. robusta Doncaster: East China Sea; Taiwan Strait; Taiwan; South China Sea. (1–6)

S. septata Doncaster: East China Sea; Taiwan Strait; South China Sea. (1–5)

S. tenuis Conant: East China Sea; Taiwan Strait; South China Sea. (1–5)

S. tokiokai Alvarino: East China Sea; Taiwan Strait; South China Sea. (1–5)

S. zetesios Fowler: South China Sea. (2, 3)

REFERENCES*

(1). Zhaung Side and Chen Xiaolin, 1978. A preliminary study on the taxonomy of Chaetognaths in the southern Yellow Sea and East China Sea. Marine Science & Technology, 9:1–44.

(2). Zhang Guxian and Chen Qingchao, 1983. Studies on Chaetognaths in the central and northern parts of the South China Sea. In: Contribution on marine biological research of the South China Sea. China Ocean Press, pp:17–63.

(3). Zhang Guxian and Chen Qingchao, 1991. Studies on Chaetognaths from Nansha Islands in the Spring and Summer. In: Contributions on Marine Organisms around Nansha Islands and the adjacent Waters (II). China Ocean Press, pp: 102–122.

(4). Dai Yanyu, 1989. Ecological study on Chaetognaths from the western Taiwan Strait. Acta Oceanologica Sinica, 11(4): 486–492.

(5). Dai Yanyu, Study on the ecology of Chaetognaths in southern Yellow Sea and East China Sea. (manuscript)

(6). Liaw Wen Kuang, 1970. On the chaetognaths collected from the waters surrounding Taiwan during CSK Cruises. The Kuroshio. University of Hawaii Press, pp:313- 321.

*: (1)-(5) in Chinese
Compiled by Dai Yanyu;Edited by Chen Qingchao

ECHINODERMATA

CRINOIDEA

ARTICULATA
BATHYCRINIDAE

Democrinus japonicus (T. Gislen): Eastern Hainan. Water depth 150 m. (15, 16, 52)

ISOCRINIDAE

Metacrinus interruptus (P. H. Carpenter): Nansha (=Spratly Islands). Water depth 168m. (36).
M. multisegmentatus (Chang & Liao): Eastern Hainan. Water depth 104–230 m. (15, 16, 52)
M. rotundus (P. H. Carpenter): Eastern Hainan. Water depth 104–200m. (15, 16, 52)

COMASTERIDAE

Capillaster multiradiata (Linné): South China Sea; Nansha (=Spratly Islands). (16, 24, 52).
C. sentosa (P. H. Carpenter): Eastern Hainan; Nansha (=Spratly Islands). Water depth 0- 135 m. (16, 52)
C. gracilis (Hartlaub): Xisha (=Paracel Islands); Zhongsha (=Macclesfield Bank). (50)
C. multibranchiata (P. H. Carpenter): Southern Hainan. Water depth 5–83 m. (16)
C. multifidus (Miller): Taiwan. Water depth 20m. (52)
Comantheria delicata (A. H. Clark): South China Sea. Water depth 54–89 m. (16, 25, 52)
C. grandicalyx (P. H. Carpenter): Fujian; Guangdong. (16, 52)
C. intermedia (A. H. Clark): Xiamen (Fujian). (50)
C. polycnemis (A. H. Clark): Taiwan; Xisha (=Paracel Islands). (52)
Comanthina schlegeli (P. H. Carpenter): Hainan. Water depth 0–278 m. (16)
Comanthus bennettic (J. Müller): Taiwan; Xisha (=Paracel Islands). (52)
C. japonicus (J. Müller): Fujian; Taiwan Strait. Water depth 0–256 m. (16, 52)
C. parvicirra (J. Müller): Fujian; Taiwan Strait; Guandong; Xisha (=Paracel Islands); Nansha (=Spratly Islands). (16, 25, 35, 52)
C. pinguis (A. H. Clark): Hong Kong. (45)
C. solaster (A. H. Clark): Taiwan Strait; Shantou (Guangdong). (52)
Comaster distinctus (P. H. Carpenter): Southern Hainan. Water depth 15–110 m. (16)
Comatella maculata (P. H. Carpenter): Xisha (=Paracel Islands). (35, 52)
C. stelligera (P. H. Carpenter): Southern Hainan; Zhongsha (=Macclesfield Bank); Xisha (=Paracel Islands); Nansha (=Spratly Islands). Water depth 0–32 m. (16, 35, 25, 50)
Comatula pectinata (Linné): Taiwan Strait; Guangdong; Beibuwan (=Gulf of Tongking); Hainan; Nansha (=Spratly Islands). Water depth 0–73 m. (16, 24, 25).
C. solaris (Lamarck): Nansha (=Spratly Islands). (26)
Comissia peregrina (Bell): Taiwan; Xisha (=Paracel Islands). (52)

ZYGOMETRIDAE

Catoptometra hartlaubi (A. H. Clark): Offshore of Shantou (Guangdong). Water depth 260 m. (52)
C. magnifica (A. H. Clark): Eastern Hainan; Nansha (=Spratly Islands). Water depth 39- 914 m. (16, 25, 52)
C. rubroflava (A. H. Clark): Southern Fujian; Hong Kong. (17, 52)
Zygometra comata (A. H. Clark): Southern Fujian; Hainan; eastern Leizhou Peninsula (Guangdong); Nansha (=Spratly Islands). Water depth 55–115 m. (16, 24, 50)

EUDIOCRINIDAE

Eudiocrinus indivisus (Semper): Southern Hainan; Nansha (=Spratly Islands). (16, 25, 52)
E. venustulus (A. H. Clark): Nansha (=Spratly Islands). (25, 52)

HEMEROMETRIDAE

Amphimetra laevipinna (P. H. Carpenter): Southern Fujian; Taiwan Strait; Guangdong; southern Hainan; Nansha (=Spratly Islands). (16, 24, 25)
A. tessellata (J. Müller): Taiwan Strait. (52)

MARIAMETRIDAE

Dichrometra doederleini (Loriol): Northern and central Taiwan Strait; Beibuwan (=Gulf of Tongking). (18, 52)
D. flagellata (Müller): Hong Kong. (17)
Lamprometra palmata palmata (J. Müller): Southern Guangdong; southern Hainan. (16, 52)
Mariametra subcarinata (A. H. Clark): Taiwan. (5)
M. vicaria: Hainan; Zhongsha (=Macclesfield Bank). Water depth 125 m. (5)

COLOBOMETRIDAE

Cyllometra manca (P. H. Carpenter): Hong Kong; western Guangxi; Beibuwan (=Gulf of Tongking); Hainan. Water depth 98–174 m. (17, 52)
Decametra laevininna (A. H. Clark): Hong Kong. (50)
D. mylitta (A. H. Clark): Beibuwan (=Gulf of Tongking); Hainan. Water depth 25–85 m. (52)

Oligometra chinensis (A. H. Clark): Central and southern Fujian; eastern Guangdong. (52)

O. serripinna (P. H. Carpenter): Hong Kong; Hainan. (17, 52)

TROPIOMETRIDAE

Tropiometra afra (Hartlaub): Fujian; Guangdong. Water depth 0–110 m. (16)

CALOMETRIDAE

Gephyrometra versicolor (A. H. Clark): Eastern Hainan. Water depth 80–90 m, Often form large aggregations. (16)

Neometra alecto (A. H. Clark): Eastern Hainan. Water depth 156–260 m. (52)

ASTEROMETRIDAE

Asterometra mirifica (A. H. Clark): South China Sea. Water depth 148 m. (39,52)

Pterometra pulcherrima (A. H. Clark): Eastern and western Hainan. Water depth 70- 174 m. (39, 52)

P. trichopoda (A. H. Clark): Southeastern Hainan. Water depth 140 m. (14)

Sinometra acuticirra (Liao): East China Sea. Water depth 139–147 m. (39)

THALASSOMETRIDAE

Parametra orion (A. H. Clark): South China Sea; Nansha (=Spratly Islands). Water depth 128–306 m. (16, 52)

ANTEDONIDAE

Antedon parviflora (A. H. Clark): Hainan. Water depth 66 m. (52)

A. serrata (A. H. Clark): Jiaozhou Bay (Shandong); Fujian; Taiwan Strait. (7, 16)

Dorometra aphrodite (A. H. Clark): Guangdong; Guangxi. Water depth 100 m. (14).

HOLOTHUROIDEA

DENDROCHIROTA
CUCUMARIIDAE

Acolochirus inornata (V. Marenzeller): Qingdao (Shandong); Shengsi Islands (Zhejiang); Xiamen and Dongshan (Fujian). (3, 4, 5, 16, 52, 58)

Allothyone longicauda (Oestergren): Dalian (Liaoning). (8, 16)

Colochirus anceps (Selenka): Taiwan Strait; Fujian; Guangdong; Nansha (=Spratly Islands). (3, 16, 17, 25, 52, 58)

C. quandrangularis (Troschel): Taiwan Strait; Fujian; Guangdong; Hainan; Nansha (=Spratly Islands). (3, 16, 17, 25, 52, 58)

Havelockia versicolor (Semper): Beibuwan (=Gulf of Tongking); Hainan. Water depth 47–59 m. (52)

Leptopentacta imbricata (Semper): Fujian; Guangdong; Beibuwan (=Gulf of Tongking); Hainan; Nansha (=Spratly Islands). (16, 17, 24, 25)

Pentacta nipponensis (H. L. Clark): Taiwan Strait. (18, 52)

Pseudocnus echinata (Marenzeller): Fujian; Taiwan Strait; Guangdong. (16, 52)

Pseudocolochirus violaceus (Theel): Hong Kong. (45)

Sclerodactula meltipes (Theel): Yantai and Qingdao (Shandong). Shallow sea. (7, 16)

Stolus albescens (Liao): From eastern Guangdong to Beibuwan (=Gulf of Tongking). Water depth 16–109 m. (52)

S. buccalis (Stimpson): From Fujian to Beibuwan (=Gulf of Tongking). Water depth 0- 54 m. (52)

S. canescens (Semper): Central Guangdong. Water depth 74–89 m. (52)

S. molpadioides (Semper): From eastern Guangdong to Beibuwan (=Gulf of Tongking). Water depth 17–57 m. (52)

Thorsonia adversaria (Semper): East China Sea; Beibuwan (=Gulf of Tongking). Water depth 16–50 m. (52)

Thyone anomaus (Oestegren): Taiwan Strait, southern Fujian. (52, 58)

T. bicornis (Ohshima): Xiamen (Fujian); Daya Bay (Guangdong); Beibuwan (=Gulf of Tongking); Nansha (=Spratly Islands). (16, 25, 52, 55, 58)

T. fusus v. *chinensis* (Yang): Fujian. (56)

T. papuensis (Theel): Hong Kong. (17, 50)

T. pedata (Semper): Beibuwan (=Gulf of Tongking). Water depth 55 m. (52)

T. pohaiensis (Liao): Bohai. Water depth 0–10 m. (46)

T. spinifera (Liao): From eastern Guangdong to Beibuwan (=Gulf of Tongking). Water depth 15–115 m. (52)

PHYLLOPHORIDAE

Actinocucumis typicus (Ludwig): Fujian; Guangdong. (2, 14, 42)

Afrocucumis africanus (Semper): Southern Hainan; Xisha (=Paracel Islands). (14, 15)

Anthochirus loui (Chang): Jiaozhou Bay (Shandong); Yellow Sea. (7, 16)

Cladolabes aciculus (Semper): Yongxing Island (Xisha=Paracel Islands). (27, 52)

C. crassus (H. L. Clark): Hong Kong. (17)

C. schmeltzi (Ludwig): Hainan. (52)

Euthyonidiella tungshanensis (Yang): Xiamen and Dongshan (Fujian); Taiwan Strait. (52, 58)

Mensamaria intercedens (Lampert): Taiwan Strait; Fujian; Guangdong; Hainan. (2, 14, 42)

Ohshimella ehrenbergi (Selenka): Yongxing (Xisha=Paracel Islands). (27, 52)

Phyllophorus hypsipyrgus (V. Marenzeler): Yellow Sea; East China Sea. (16)

P. konkuliensis (Heding & Panning): Beibuwan (=Gulf of Tongking). Water depth 32- 53 m. (52)

P. ordinatus (Chang): Dalian (Liaoning); Qingdao (Shandong). (7, 8, 16)

P. spiculatus (Chang): Xiamen and Dongshan (Fujian). (4, 16, 58)

Phyrella fragilis (Ohshima): Leizhou Peninsula (Guangdong); Sanya and Lingshui (Hainan); Xisha (=Paracel Islands). (16, 27, 50)

P. liuwushanensis (Yang): Xiamen (Fujian). (52, 58)

YPSILOTHURIIDAE

Yoshithuriz bitentaculata (Ludwig): Nansha (=Spratly Islands). (25)

ASPIDOCHIROTA

SYNALLACTIDAE

Bathyplotes moseleyi (Theel): East China Sea (30°30'N, 128°E). Water depth 365 m. (37)

B. natans (M. Sars): East China Sea (26°30'N, 125°E). Water depth 450 m. (37)

Mesothuria marginata (Sluiter): East China Sea (30°30'N, 128°E). Water depth 365 m. (37)

M. media (Ohshima): East China Sea (26°30'N, 125°E). Water depth 550 m. (37)

Pseudostichopus trachus (Sluiter): East China Sea (26°N, 125°E, water depth 550m; 28°30'N, 128°30'E, water depth 1,420 m). (37)

P. unguiculatus (Ohshima): South China Sea (19°N, 133°30'E). Water depth 1,100 m. (37)

Synallactes discoidalis (Misukuri): East China Sea (30°15'N, 128°E). Water depth 400 m. (37)

HOLOTHURIIDAE

Actinopyga echinites (Jaeger): Baoan and Aotou (Guangdong); Sanya (Hainan); Xisha (=Paracel Islands). (16, 27, 37, 52)

A. lecanora (Jaeger): Nansha (=Spratly Islands); Xisha (=Paracel Islands). (16, 26, 27, 37, 52)

A. mauritianan (Quoy & Gaimard): Southern Hainan; Xisha (=Paracel Islands); Nansha (=Spratly Islands). (16, 26, 27, 37, 52)

A. miliaris (Quoy & Gaimard): Xisha (=Paracel Islands). (27, 37, 52)

Bohadschia argus (Jaeger): Xisha (=Paracel Islands); Nansha (=Spratly Islands). (16, 24, 27, 37, 52)

B. graeffei (Semper): Nansha (=Spratly Islands). (24)

B. marmorata (Jaeger) [*B. bivittata, B. koellikeri*]: Dazhou Island (Hainan); Xisha (=Paracel Islands); Nansha (=Spratly Islands). (24, 27, 37, 52)

Holothuria albiventer (Semper): Southern Hainan. (37, 52)

H. arenicola (Semper): Xisha (=Paracel Islands). (16, 27, 37, 52)

H. atra (Jaeger): Hainan; Xisha (=Paracel Islands); Nansha (=Spratly Islands). (16, 27, 52)

H. axiologa (H. L. Clark): Xisha (=Paracel Islands). (37, 52)

H. cinerascens (Brandt): Hainan; Yongxing Island (Xisha=Paracel Islands). (16, 27, 37, 52)

H. dicrepans (Semper): Yongxing Island (Xisha=Paracel Islands). (27, 37, 52)

H. difficilis (Semper): Taiwan; southern Hainan; Xisha (=Paracel Islands); Nansha (=Spratly Islands). (16, 26, 27, 37, 52)

H. edulis (Lesson): Southern Hainan. (16, 27, 52)

H. flavomaculata (Semper): Sanya (Hainan). Intertidal. (37, 52)

H. fuscocinerea (Jaeger): Baoan and Daya Bay (Guangdong); Hainan; Xisha (=Paracel Islands). (16, 27, 52, 57)

H. gracilis (Semper): Southern Hainan. (37, 52)

H. hilla (Lesson): Hainan; Xisha (=Paracel Islands); Nansha (=Spratly Islands). (16, 24, 27, 37, 52)

H. impatiens (Forskaal): Hainan; Xisha (=Paracel Islands); Nansha (=Spratly Islands). (16, 24, 27, 37, 52)

H. inhabilis (Selenka): Xisha (=Paracel Islands). (27, 37, 52)

H. leucospilota (Brandt): Hainan; Xisha (=Paracel Islands). (3, 16, 27, 37, 52)

H. martensi (Semper): Beibuwan (=Gulf of Tongking). (16, 37, 52)

H. moebi (Ludwig): Southern Fujian; Guangdong; Xisha (=Paracel Islands). (3, 16, 37, 52)

H. multipilula (Liao): Xisha (=Paracel Islands). (27, 37, 52)

H. nobilis (Selenka): Southern Hainan; Xisha (=Paracel Islands). (16, 27, 37, 52)

H. ocellata (Jaeger): Mouth of Zhujiang (=Pearl River, Guangdong); Beibuwan (=Gulf of Tongking); southern Hainan; Nansha (=Spratly Islands). (16, 17, 24, 37, 52)

H. olivacea (Ludwig): Yongxing Island (Xisha=Paracel Islands). (27, 37, 52)

H. pardalis (Selenka): Hainan; Xisha (=Paracel Islands). (16, 27, 37, 52)

H. pervicax (Selenka): Southern Hainan; Xisha (=Paracel Islands). (27, 37, 52)

H. rigida (Selenka): Yongxing Island (Xisha=Paracel Islands). (27, 37, 52)

H. scabra (Jaegera): Western Guangdong; Hainan; Xisha (=Paracel Islands). (16, 37, 52)

H. sinica (Liao): Central and southern Guangdong; southern Hainan. (3, 37, 52)

H. spinifera (Theel): Xisha (=Paracel Islands). (16, 37, 52)

H. verrucosa (Selenka): Xisha (=Paracel Islands). (27, 37, 52)

Labidodemas semperianum (Selenka): Yongxing Island (Xisha=Paracel Islands); Nansha (=Spratly Islands). (26, 27, 37, 52)

STICHOPODIDAE

Apostichopus japonicus (Selenka): Bohai; Yellow Sea. Shoreline. (8, 16, 32, 37)
Stichopus chloronotus (Brandt): Southern Hainan; Xisha (=Paracel Islands). (16, 27, 37, 52)
S. flaccus (Liao): Central and western Beibuwan (=Gulf of Tongking). (37)
S. horrens (Selenka): Hainan; Xisha (=Paracel Islands); Nansha (=Spratly Islands). (16, 24, 27, 37, 52)
S. variegatus (Semper): Leizhou Peninsula (Guangdong); Beibuwan (=Gulf of Tongking); Hainan; Xisha (=Paracel Islands); Nansha (=Spratly Islands). (16, 24, 27, 37, 52)
Thelenota ananas (Jaeger): Xisha (=Paracel Islands); Nansha (=Spratly Islands). (16, 24, 27, 37, 52)
T. anax (H. L. Clark): Xisha (=Paracel Islands). (27, 37, 52)

MOLPADONIA
MOLPADIIDAE

Molpadia changi (Pawson & Liao): South China Sea. Water depth 89–200 m. (52, 56)
M. guangdongensis (Pawson & Liao): South China Sea. Water depth 89–200 m. (52, 56)
M. roretzii (V. Marenzeller): Yellow Sea; East China Sea; South China Sea. Water depth 50–60 m. (3, 16, 52)

CAUDINIDAE

Acaudina leucoprocta (H. L. Clark): Southern Zhejiang; Fujian; Guangdong; Hainan; Nansha (=Spratly Islands). (25, 38, 52)
A. molpadioides (Semper): China's seas. (3, 4, 5, 6, 17, 38, 52, 58)
Caudina atacta (Pawson & Liao): Beibuwan (=Gulf of Tongking). Water depth 63–91 m. (52, 56)
C. intermedia (Liao & Pawson): Eastern Guangdong. Water depth 107 m. (54)
C. similis (Augustin): Yellow Sea. Water depth 36–66 m. (58)
C. zhejiangensis (Pawson & Liao): Zhejiang. Water depth 53–89 m. (56)
Paracaudina chilensis (Müller): China's seas. (3, 5, 7, 16, 52)
P. declicata (Pawson & Liao): Xiamen (Fujian); Beibuwan (=Gulf of Tongking). (52, 56)

APODA

Anapta gracilis (Semper): Beihai (Guangxi). (52)
Euapta godefforyi (Semper): Southern Hainan; Yongxing, Zhaoshu and Shenhang Islands (Xisha=Paracel Islands). (16, 27)
Labidoplax dubia (Semper): China' seas. (16)
L. incerta (Ludwig): From Fujian to Beibuwan (=Gulf of Tongking). Water depth 20- 100 m. (52)
Opheodesome australiensis (Heding): Yongxing and Shenhang Islands (Xisha=Paracel Islands). (27, 52)
O. grisea (Semper): Southern Hainan, including Sanya and Lingshui. (16, 52)
Patinapta ooplax (V. Marenzeller): Bohai; Yellow Sea; Fujian. (3, 5, 7, 8, 16, 52, 58)
Polyplectana kefersteinii (Selenka): Xisha (=Paracel Islands). (27, 52)
Protankyra asymmetrica (Ludwig): China' seas except for Bohai. (7, 16, 25, 52, 58)
P. bidentata (Woodward & Barrett): Bohai; Yellow Sea; Zhejiang; Fujian; Taiwan Strait; Guangdong. (3, 5, 7, 8, 16, 17, 52, 58)
P. magnihanula (Heding): Hong Kong. (17, 52)
P. pseudodigitata (Semper): Fujian; Taiwan Strait; Guangdong; Beibuwan (=Gulf of Tongking). (16, 52, 58)
P. suensoni (S. G. Heding): Central and northern Taiwan Strait. (18, 52)
P. verrilli (Theel): Xiamen (Fujian). (52, 58)
Synapta maculata (Chamisso & Eysenhardt): Xisha (=Paracel Islands). (16, 27, 52)
Synaptula reticulata (Semper): Hainan. (52)

CHIRIDOTIDAE

Chiridota rigida (Semper): Xisha (=Paracel Islands). (27, 52)
C. stuhlmanni (Lampert): Xisha (=Paracel Islands). (27, 52)
Polycheira fusca (Quoy & Gaimard): Fujian; Guangdong; Hainan; Xisha (=Paracel Islands). (3, 16, 27, 52)

ASTEROIDEA

PHANEROZONIA
BENTHOPECTINIDAE

Cheiraster niasicus (Ludwig): Nansha (=Spratly Islands). Water depth 206 m. (26)

ASTROPECTINIDAE

Astropecten monacanthus (Sladen): Western Guangdong; Hong Kong; Hainan; Nansha (=Spratly Islands). (16, 17, 24, 25, 52)
A. phragmorus (Fisher): Nansha (=Spratly Islands). (24, 25)
A. polyacanthus (Müller & Troschel): Fujian; Guangdong; Nansha (=Spratly Islands). (16, 25)

A. vappa (Müller & Troschel): Taiwan Strait; Fujian; Hong Kong. (17,50)

A. velitaris (von Martens): Taiwan Strait; Beibuwan (=Gulf of Tongking); Nansha (=Spratly Islands). (16, 24, 25)

Craspidaster hesperus (Müller & Troschel): Fujian; Guangdong; Nansha (=Spratly Islands). (9, 16, 25)

Ctenophoraster donghaiensis (Liao & Sun): East China Sea. Water depth 147 m. (49)

Ctenopleura astropectinides (Fisher): Nansha (=Spratly Islands). (25)

C. ludwigi (de Lorio): East China Sea; Taiwan. (49)

C. sinica (Döderlein): East China Sea; Taiwan Strait; South China Sea. Water depth 77- 134 m. (52)

Dipsacaster pretiosus (Döderlein): East China Sea. Water depth 60–116 m. (52)

Koremaster evaulus (Fisher): Nansha (=Spratly Islands, 9°50'N, 117°53'E). Water depth 1,241 m. (26)

LUIDIIDAE

Luidia avicularia (Fisher): From eastern Guangdong to Hainan. Water depth 80–158 m. (50)

L. hardwicki (Gray): Taiwan Strait; Hong Kong; Beibuwan (=Gulf of Tongking); Nansha (=Spratly Islands). (16, 17, 25, 52)

L. longispina (Sladen): Hong Kong; Beibuwan (=Gulf of Tongking). (17, 52)

L. maculata (Müller & Troschel): Taiwan Strait; Guangdong; Hainan. (16, 17, 52)

L. orientalis (Fisher): Hong Kong; eastern Hainan. Water depth 200 m. (16,52)

L. quinaria (von Martens): China's seas. (5, 7, 9, 16, 52)

L. yesoensis (Goto): Bohai; Yellow Sea. (7, 8, 16)

ARCHASTERIDAE

Archaster typicus (Müller & Troschel): Daya Bay (Guangdong); Hong Kong; western Guangxi; Hainan. (9, 16, 17, 52)

GONIASTERIDAE

Anthenea aspera (Döderlein): Xiamen (Fujian); Hong Kong. (17)

A. chinensis (Gray): Taiwan Strait; Fujian; Guangdong. (9, 16, 17, 52)

A. difficilis (Liao): From Fujian to Hong Kong. (52)

A. flavescens (Gray): Xiamen (Fujian); Hong Kong. (9, 17,52)

A. viguieri (Döderlein): Zhapo (Guangdong). Water depth 10–200 m. (52)

Anthenoides cristatus (Sladen): Nansha (=Spratly Islands). (25)

A. granulosus (Fisher): Nansha (=Spratly Islands). (25)

A. laevigatus (Liao & Clark): Eastern Hainan. (42, 52)

A. lithosorus (Fisher): Hong Kong. (52)

A. tenuis (Liao & Clark): Nansha (=Spratly Islands). (42, 52)

Astrothauma euphylacteum (Fisher): Eastern Hainan. Water depth 200 m. (16)

Calliaster cf. *childreni* (Gray): Nansha (=Spratly Islands). (25)

C. quadrispinus (Liao): South China Sea (20°30'N, 113°30'E). Water depth 88 m. (40)

Calliderma emma (Gray): Western Guangdong; Hainan. Water depth 156–300 m. (52)

Goniodiscaster forficulatus (Perrier): Nansha (=Spratly Islands). (25)

G. granuliferus (Gray): Taiwan Strait; Nansha (=Spratly Islands). (18,52)

G. scaber (Mobius): Nansha (=Spratly Islands). (25)

Ogmaster capella (Müller & Troschel): Northeastern East China Sea; Nansha (=Spratly Islands). Water depth 60–87 m. (24,25,52)

Paragonaster chinensis (Liao): Yellow Sea (34°N, 124°E). Water depth 50–60 m. (50)

P. ctenipes (Sladen): Nansha (=Spratly Islands). (25)

Rosaster attenuatus (Liao): East China Sea. Water depth 250 m. (47)

R. bipunctus (Sladen): Nansha (=Spratly Islands). (25)

R. symbolicus (Sladen): Eastern Hainan; Xisha (=Paracel Islands). Water depth 200 m. (16, 25, 52)

Stellaster equestris (Retzius): East China Sea; Taiwan Strait; South China Sea; Nansha (=Spratly Islands). (16, 24, 25, 52)

OREASTERIDAE

Choriaster granulatus (Lütken): Xisha (=Paracel Islands). (25, 31, 52)

Culcita novaeguineae (Müller & Troschel): Southern Hainan; Xisha (=Paracel Islands). (16, 25, 31, 52)

Pentaceraster alveolatus (Perrier): Qiongzhou Strait (between Guangdong and Hainan). Water depth 40–60 m. (52)

P. chinensis (Gray): Western Guangdong; Hainan; Nansha (=Spratly Islands). (24, 25, 52)

P. magnificus (Goto): Hong Kong. (17, 52)

P. regulus (Müller & Troschel): Nansha (=Spratly Islands). (25)

P. sibogae (Döderlein): Lingshui (Hainan). (52)

Poraster superbus (Mobius): Nansha (=Spratly Islands). (25)

Protoreaster nodosus (Linné): Lingshui (Hainan). (16, 52)

ASTERODISCIDIDAE

Asterodiscides elegans (Gray): Hainan. Water depth 60–80 m. (52)

ASTEROPSEIDAE

Asteropsis carinifera (Lamarck): Hainan; Yongxing Island (Xisha=Paracel Islands). (16, 31, 52)

OPHIDIASTERIDAE

Bunaster ritteri (Döderlein): Xisha (=Paracel Islands); Nansha (=Spratly Islands). (25, 31, 52)

Celerina heffernani (Livingstane): Zhongsha (=Maccleslesfield Bank). (52)

Dactyosaster cylindricus (Lamarck): Yongxing and Zhaoshu Islands (Xisha=Paracel Islands). (31, 52)

Fromia milleporella (Lamarck): Yongxing Island (Xisha=Paracel Islands). (31, 52)

Gomophia egyptiaca egeriae (A.M. Clark): Zhongsha (=Macclesfield Bank). (52)

Hacelia tuberculata (Liao): Diaoyu Island (= Senkaku Gunto, northeast of Taiwan). Water depth 100–200 m. (49)

Leiaster spiciosus (von Marlins): Xisha (=Paracel Islands). (52)

Linchia gracilis (Liao): Diaoyu Island (=Senkaku Gunto, northeast of Taiwan). Water depth 100–200 m. (49)

L. laevigata (Linnaeus): Southern Hainan; Xisha (=Paracel Islands); Nansha (=Spratly Islands). (16, 25, 31, 52)

L. multifora (Lamarck): Xisha (=Paracel Islands); Nansha (=Spratly Islands). (25, 52)

Nardoa frianti (Koehler): Yongxing Island (Xisha=Paracel Islands). (31, 52)

Neoferdina cumingii (Gray): Yongxing Island (Xisha=Paracel Islands). (31, 52)

Ophidiaster chinensis (Perrier): Guangzhou (Guangdong). (50)

O. granifer (Lütken): Yongxing, Shenhang, Beijiao and Jinyen Islands (Xisha=Paracel Islands). (31, 52)

O. hemprichi (Müller & Troschel): Yongxing Island (Xisha=Paracel Islands); Nansha (=Spratly Islands). (25, 31, 52)

O. multispinus (Liao & Clark): Xiamen (Fujian); Hong Kong; Beibuwan (=Gulf of Tongking). (53)

Sinoferdina gigantea (Liao): East China Sea continental shelf (27°30'N, 126°E). Water depth 162 m. (44)

PORCELLANASTERIDAE

Abyssaster planus (Sladen): Nansha (=Spratly Islands). (25)
Eremicaster crassus (Sladen): Nansha (=Spratly Islands). (25)
Thoracaster cylindratus (Sladen): Nansha (=Spratly Islands). (25)

METRODIRIDAE

Metrodira subulata (Gray): Taiwan Strait; Nansha (=Spratly Islands). (18, 25, 52)

SPINULOSA
ASTERINIDAE

Anseropoda rosacea (Lamarck): Eastern Hainan; Nansha (=Spratly Islands). Water depth 100 m. (16, 25, 52)

Asterina batheri (Goto): Yantai (Shandong). (16)

A. cepheus (Müller & Troschel): Xisha (=Paracel Islands) Zhongsha (=Macclesfield Bank). (31, 52)

A. limboonkengi (G. A. Smith): Taiwan Strait; Fujian; Guangdong. (16, 52)

A. orthodon (Fisher): Hong Kong. (17)

A. pectinifera (Müller & Troschel): Northern China. Shallow water. (5, 7, 8, 9, 16)

Disasterina odontacantha (Liao): Xisha (=Paracel Islands). (27, 52)

Nepanthia belcheri (Perrier): Xuwen (Guangdong). (52)

N. briareus (Bell): Zhongsha (=Macclesfield Bank). Water depth 54–83 m. (52)

N. variabilis (H. L. Clark): Nansha (=Spratly Islands). (25)

SPHAERASTERIDAE

Podosphaeraster polyplax (Clark & Wright): Zhongsha (=Macclesfield Bank). (52)

ECHINASTERIDAE

Echinaster luzonicus (Gray): Southern Hainan; Xisha (=Paracel Islands); Nansha (=Spratly Islands). (16, 25, 31, 52)

E. stereosomus (Fisher): Nansha (=Spratly Islands). (25, 52)

Henricia aspera robusta (Djakonov): Offshore off Shi Island (Shandong). Water depth 70 m. (16)

H. leviuscula (Stimpson): Dalian (Liaoning); Yantai, Rongcheng and Qingdao (Shandong). (7, 8, 9, 16)

H. spiculifera (H. L. Clark): Northern Yellow Sea. (16)

ACANTHASTERIDAE

Acanthaster planci (Linnaeus): Southern Hainan; Xisha (=Paracel Islands); Nansha (=Spratly Islands). (16, 25, 31, 52)

MITHRODIIDAE

Mithrodia clavigera (Lamarck): Xisha (=Paracel Islands). (31, 52)

SOLASTERIDAE

Crossaster papposus (Linnaeus): Yellow Sea. (16)
Solaster dawsoni (Verrill): Yellow Sea. (16)

PTERASTERIDAE

Euretaster insignis (Sladen): Southern Hainan; Nansha (=Spratly Islands). (18, 24, 25, 52)

FORCIPULATA
ZOROASTERIDAE

Zoroaster carinatus philippinensis (Fisher): Hainan. (16, 52)

ASTERIIDAE

Aphelasterias changfengyingi (Baranova & Wu): Bohai Bay; northern Yellow Sea. (1, 16)
A. japonica (Bell): Bohai Bay; northern Yellow Sea. (1, 16)
Asterias argonauta (Djakonov): Liaoning; Shandong. (1,16)
A. rollestoni (Bell): Yellow Sea; East China Sea (northern Fujian). (5, 7, 8, 9, 16, 17)
A. versicolor (Sladen): Bohai; northern Yellow Sea; Taiwan Strait; Fujian; Hong Kong. (1, 16, 52)
Coronaster volsellatus (Sladen): Eastern Hainan. Water depth 200 m. (16, 52)
Coscinasterias acutispina (Stimpson): Fujian; Guangdong. (16, 17, 52)
Distolasterias elegans (Djakonov): Offshore off Shi Island (Shandong). (16)
D. nipon (Döderlein): Central Yellow Sea; Hong Kong. (16)

ECHINOIDEA
CIDAROIDEA
CIDARIDAE

Eucidaris metularia (Lamarck): Southern Hainan; Xisha (=Paracel Islands); Nansha (=Spratly Islands). (16, 22, 24, 29)
Goniocidaris biserialis (Döderlein): East China Sea continental shelf. (45)
Prionocidaris baculosa (Lamarck): Nansha (=Spratly Islands). (25, 52)
P. baculosa v. *annulifera* (Lamarck): Eastern Hainan. Water depth 118 m. (16)
P. bispinosa (Lamarck): China's seas. (52)
P. vrticillata (Lamarck): Taiwan. (52)
Stereocidaris indica philippinensis (Mortensen): East China Sea continental shelf. (45)
Stylocidaris annulosa (Mortensen): Eastern Hainan; Nansha (=Spratly Islands). (2, 10, 12, 16)
S. bracteata (A. Agassiz): East China Sea continental shelf. (45)
S. reini (Döderlein): Northeastern South China Sea; Nansha (=Spratly Islands). (19, 25)
S. reini v. *cladothrix* (Mortensen): Southeast of Qinglan Bay (Hainan). (10, 16)
S. ryukyuensis (Shigei): East China Sea continental shelf. (45)

LEPIDOCENTROIDA
ECHINOTHURIDAE

Araeosoma owstoni (Mortensen): East China Sea continental shelf; Nansha (=Spratly Islands). (26, 45)
A. owstoni v. *nudum* (Mortensen): Southeast of Qinglan Bay (Hainan). (10, 16)
Asthenosoma ijmai (Yoshiwara): East China Sea. (45, 52)
A. varium (Grube): East China Sea continental shelf; South China Sea; Nansha (=Spratly Islands). (10, 16, 24, 43, 45)
Calveriosoma gracile (A. Agassiz): Nansha (=Spratly Islands, 6°24'N,115°35'E). Water depth 145–185m. (26)
Phormosoma bursarium (A. Agassiz): East China Sea continental shelf. (45)

STIRODONTA
PHYMOSOMATIDAE

Glyptocidaris crenularis (A. Agassiz): Yellow Sea. (8, 16)

STOMOPNEUSTIDAE

Stomopneustes variolaris (Lamarck): Southern Hainan; Xisha (=Paracel Islands). (16, 29, 52)

ARBACIIDAE

Coelopleurus maculatus (A. Agassiz & H. L. Clark): East China Sea continental shelf. (45)

AULODONTA
DIADEMATIDAE

Astropyga radiata (Leske): Beibuwan (=Gulf of Tongking); South China Sea; Nansha (=Spratly Islands). (10, 16, 24, 25, 52)
Chaetodiadema granulatum (Mortensen): Central and northern Taiwan Strait; Guangdong; Beibuwan (=Gulf of Tongking). (10, 16, 43, 52)
C. japonicum (Mortensen): East China Sea continental shelf. (45)
Diadema savignyi (Michelin): Hong Kong; Lingshui (Hainan). (10, 17, 52)
D. setosum (Leske): Guangdong; Hainan; Xisha (=Paracel Islands); Nansha (=Spratly Islands). (10, 12, 16, 24, 29, 43, 52, 55)
Echinothrix calamaris (Pallas): Hong Kong; southern Hainan; Xisha (=Paracel Islands); Nansha (=Spratly Islands). (10, 12, 16, 25, 29, 43, 52)
E. diadema (Linnaeus): Southern Hainan; Xisha (=Paracel Islands). (10, 16, 29, 43, 52)

PEDINIDAE

Caenopedina mirabilis (Döderlein): East China Sea continental shelf. (45)

CAMARODONTA
TEMNOPLEURIDAE

Desmechinus anomalus (H. L. Clark): China's seas. (52)
D. rufus (Bell): Zhongsha (=Macclesfield Bank). (52)
D. versicolor (Mortensen): Zhongsha (=Macclesfield Bank). (52)
Mespilia globulus (Linnaeus): Xisha (=Paracel Islands). (16, 29, 43, 52)
Microcyphus olivaceus (Döderlein): East China Sea continental shelf. (45)
Paratrema doederleini (Mortensen): Hong Kong; Beibuwan (=Gulf of Tongking). (17, 43, 52)
Prionechinus forbesianus (A. Agassiz): East China Sea continental shelf. (45)
Salmaciella dussumieri (L. Agassiz): Beibuwan (=Gulf of Tongking); Xinying (Hainan); Nansha (=Spratly Islands). (16, 25, 43, 52)
Salmacis bicolor rarispina (L. Agassiz): Beibuwan (=Gulf of Tongking); South China Sea. (10, 12, 16, 25, 29, 43)
S. bicolor typica (Mortensen): Lingshui (Hainan); Nansha (=Spratly Islands) (10, 16, 25, 43)
S. sphaeroides variegata (Mortensen): Hong Kong. (10, 17, 43, 52)
S. virgulata (L. Agassiz): Weizhou Island (Guangxi). Water depth 10–55 m. (43, 52)
Temnopleurus apodus (A. Agassiz & H. L. Clark): East China Sea continental shelf. (45)
T. hardwickii (Gray): Bohai; Yellow Sea; East China Sea continental shelf; Zhoushan Islands (Zhejiang); Taiwan Strait. (2, 5, 7, 8, 12, 16, 52)
T. reevesii (Gray): East China Sea; Taiwan Strait; South China Sea. Water depth 5–500 m. (10, 16, 24, 25, 35, 43, 52)
T. toreumaticus (Leske): China's seas. (2, 5, 7, 10, 12, 16, 17, 43, 52)
Temnotrema muculatum (Mortensen): Zhujiang (=Pearl River, Guangdong); Qiongzhou Strait (between Guangdong and Hainan); Hainan. Water depth 51–112 m. (43, 52)
T. reticulatum (Mortensen): Guangdong; Beibuwan (=Gulf of Tongking); Hainan. Water depth 12–270 m. (43, 52)
T. sculptum (A. Agassiz): Qingdao (Shandong); Taiwan Strait. (16, 52)
T. siamense (Mortensen): Central Beibuwan (=Gulf of Tongking); southern Hainan. Water depth 33–46 m. (43, 52)
T. xishaensis (Liao): Hainan; Yongxing Island (Xisha=Paracel Islands). (29, 52)

TOXOPNEUSTIDAE

Cyrtechinus verruculatus (Luken): Zhaoshu Island (Xisha=Paracel Islands). (29, 52)
Goniopneustes pentagonus (A. Agassiz): Zhongsha (=Macclesfield Bank). (52)
Nudechinus multicolor (Yoshiwara): Guangdong. (16, 43, 52)
Pseudoboletia maculata (Michelin): Zhaoshu Island (Xisha=Paracel Islands). (29, 43, 52)
Toxopneustes pileolus (Lamarck): Southern Hainan; Xisha (=Paracel Islands). Common species in coral reefs. (10, 16, 29, 43, 52)
Tripneustes gratilla (Linnaeus): Taiwan; Guangdong; Hainan; Xisha (=Paracel Islands). (10, 12, 16, 17, 29, 43, 52)

STRONGYLOCENTROTIDAE

Hemicentrotus pulcherrimus (A. Agassiz): Bohai; Yellow Sea; Zhejiang; Fujian. (2, 5, 7, 8, 12, 16, 52)
Strongylocentrotus nudus (A. Agassiz): Eastern Liaoning Peninsula; northern Shandong Peninsula. (12, 16)

PARASALENIIDAE

Parasalenia gratiosa (A. Agassiz) [*P. gratiosa* v. *boninensis*]: Hong Kong; Hainan; Xisha (=Paracel Islands). (10, 12, 16, 17, 29, 43, 52)
P. pohlii (Pfeffer): Xisha (=Paracel Islands); Nansha (=Spratly Islands). (26, 29, 52)

ECHINOMETRIDAE

Anthocidaris crassispina (A. Agassiz): Zhejiang; Fujian; Guangdong. (2, 10, 12, 16, 17, 43, 52)
Echinometra mathaei mathaei (Blainville) [*E. mathaei*]: Taiwan; Hainan; Xisha (=Paracel Islands). Widely distributed tropical species. (10, 16, 29, 43, 52)
E. mathaei oblonga (Blainville): Eastern Hainan; Sanya (Hainan); Xisha (=Paracel Islands). (10, 12, 29, 43)
Echinostrephus aciculatus (A. Agassiz): Taiwan. (52)
E. molaris (Blainville): Taiwan; intertidal zone of Lingshui (Hainan). (43, 52)
Heterocentrotus mammillatus (Linnaeus): Southern Hainan; Xisha (=Paracel Islands); Nansha (=Spratly Islands). (10, 12, 16, 29, 43, 52)

HOLECTYPOIDA
ECHINONEIDAE

Echinoneus abnormalis (de Loriol): Zhongsha (=Macclesfield Bank). (52)
E. cyclostomus (Leske): Hainan; Xisha (=Paracel Islands). (16, 29)

CLYPEASTEROIDA

Clypeaster miniaceus (H. L. Clark): Zhongsha (=Macclesfield Bank). (52)

C. reticulatus (Linnaeus): Beibuwan (=Gulf of Tongking); southern Hainan; Xisha (=Paracel Islands); Nansha (=Spratly Islands). (10, 12, 16, 25, 29, 43, 52)

C. reticulatus sundaicus (Mortensen): Guangdong; Beibuwan (=Gulf of Tongking). Shallow sea. (29, 43)

C. virescens (Döderlain): Eastern Hainan. (16, 43, 52)

ARACHNOIDAE

Arachnoides placenta (Linnaeus): Fujian; Guangdong. (2, 10, 12, 16, 43, 52)

FIBULARIIDAE

Echinocyamus cripus (Mazzetti): Taiwan Strait; Jinyin Island (Xisha=Paracel Islands); Nansha (=Spratly Islands). (25, 29, 52)

Fibularia acuta (Yoshiwara): Qinghuangdao (Hebei); Qingdao and Laizhou (Shandong); Taiwan Strait. (7, 12, 16)

F. angulipora (Mortensen): Beibuwan (=Gulf of Tongking); Sanya (Hainan). (43, 52)

F. ovulum (Lamarck): Jinyin Island (Xisha=Paracel Islands). (29, 52)

F. volva (L. Agassiz): Northeastern South China Sea (22°37' N,118°20' E). Water depth 35 m. (43, 52)

LAGANIDAE

Laganum decagonale (de Blainville): Guangdong; Beibuwan (=Gulf of Tongking); Hainan. (10, 12, 16, 17, 52)

L. depressum (Lesson): Taiwan Strait; Hong Kong; Hainan; Nansha (=Spratly Islands). (10, 16, 24, 52, 55)

L. fudsiyama (Döderlein): Nanpeng and Shangchuan Islands (Guangdong). Water depth 60 m. (10)

Peronella lesueuri (Valenciennes): Fujian; Guangdong. (2, 10, 12, 16, 52)

P. pellucida (Döderlein): East China Sea continental shelf. (45)

NEOLAGANIDAE

Tetradiella sinica (Liao & Lin): Beibuwan (=Gulf of Tongking). Fossil. (48)

SCUTELLIDAE

Astriclypeus manni (Verrill): Taiwan Strait; Fujian; Guangdong. (10, 12, 16, 52)

Echinodiscus auritus (Leske): Shuixi (Guangdong); Beihai (Guangxi). (12, 16, 52)

E. tenuissimus (L. Agassiz): Zhongjian Island (Xisha=Paracel Islands). (29, 52)

Scaphechinus mirabilis (A. Agassiz): East China Sea continental shelf. (45)

DENORASTERIDAE

Sinaechinocyamus mai (Wang): From Fujian to Hainan, including Taiwan Strait. (52)

S. planus (Liao): Northern Yellow Sea (37°50'N,124°E). Water depth 72 m. (30,52)

SPATANGOIDA
SPATANGIDAE

Maretia planulata (Lamarck): Off Zhujiang (=Pearl River, Guangdong); Xisha (=Paracel Islands); Nansha (=Spratly Islands). (26, 43, 52)

Pseudomaretia alta (A. Agassiz): Taiwan Strait; southern Fujian; Haimen (Guangdong). (10, 16, 52)

LOVENIIDAE

Echinocardium cordatum (Pennant): Bohai; Yellow Sea. (2, 5, 7, 8, 12, 16)

Lovenia elongata (Gray): Fujian; Guangdong; Nansha (=Spratly Islands). (10, 12, 16, 25, 52)

L. subcarinata (Gray): Hong Kong; Beibuwan (=Gulf of Tongking). (16, 17, 25, 52)

L. triforis (Koehler): East China Sea; South China Sea. (16, 52)

SCHIZASTERIDAE

Faorina chinensis (Gray): Beibuwan (=Gulf of Tongking); South China Sea. Water depth 40–220 m. (10, 12, 16, 25)

Moira lachesinella (Mortensen): Shenzhen (Guangdong). (52)

Paraster compactus (Koehler): Beibuwan (=Gulf of Tongking). (43, 52)

Prymnaster rostratus (Smith): Xisha (=Paracel Islands). (43, 52)

Schizaster lacunosus (Linnaeus): East China Sea; Taiwan Strait; South China Sea. (16)

PERICOSMIDAE

Pericosmus melanostomus (Mortensen): Hong Kong. (52)

P. porphyrocardius (McNamara): Southeast of Qinglan Bay (Hainan), water depth 80–200 m; Nansha (=Spratly Islands). (10, 26, 52)

PALAEOPNEUSTIDAE

Limnopneustes murrayi (A. Agassiz): Nansha (=Spratly Islands, 6°40'N, 115°19'E). Water depth 740–1,178 m. (26)

Platybrissus roemeri (Grube): Xisha (=Paracel Islands). (43)

BRISSIDAE

Anametalia sternaloides (Bolau): Beibuwan (=Gulf of Tongking). (43, 52)

Brissopsis luzonica (Gray): Central and western Guangdong; Beibuwan (=Gulf of Tongking); Hainan; Nansha (=Spratly Islands). Water depth 41–220 m. (25, 43, 52)

Brissus latecarinatus (Leske): Xisha (=Paracel Islands). (16, 29, 52)

Gymnopataqus mirabile (A. Agassiz & H. L. Clark): East China Sea continental shelf. (45)

Metalia dierana (H. L. Clark): Xisha (=Paracel Islands). (29, 43, 52)

M. spatagus (Linnaeus): Xisha (=Paracel Islands). (16, 19, 52)

Rhinobrissus pyramidalis (A. Agassiz): Southern Hainan. (10, 16, 52)

PALAEOSTOMATIDAE

Palaeostoma mirabile (Gray): Hong Kong. (17)

OPHIUROIDEA

EURYALAE
EURYALIDAE

Asteronyx loveni (Müller & Troschel): Eastern Hainan; Nansha (=Spratly Islands). Water depth 950–1,100 m. (13, 16, 25, 52)

Astroceras pergamenum (Lyman): Eastern Hainan; South China Sea. (13, 16, 52)

Euryale aspera (Lamarck): Beibuwan (=Gulf of Tongking); Qiongzhou Strait (between Guangdong and Hainan); southern Hainan; Nansha (=Spratly Islands). (13, 16, 25, 52)

E. purpura (Mortensen): Hong Kong; southwestern Hainan. (13)

Sthenocephalus indicus (Koehler): Eastern Hainan. Water depth 217–300 m. (13)

Trichaster acanthifer (Döderlein): Fujian; Taiwan Strait; Beibuwan (=Gulf of Tongking). (13, 16, 52)

T. flagellifer (von Martens): Taiwan Strait; Fujian; Guangdong; Nansha (=Spratly Islands). (13, 16, 26, 52)

T. palmiferus (Lamarck): Taiwan Strait; Fujian; Guangdong. (11, 13, 16, 52)

GORGONOCEPHALIDAE

Asteroporpa hadracantha (H. L. Clark): Hainan. Water depth 64–366 m. (13, 16, 52)

Astroboa albatrossi (Döderlein): Hong Kong (21°30'N, 116°32'E). Water depth 250 m. (13)

A. nuda v. *elegans* (Koehler): Southern Hainan. (13, 16, 52)

Astrocladus exiguus (Lamarck): Taiwan Strait; Fujian; Nanao (Guangdong); eastern Hainan. (13, 16, 52)

Astrodendrum sagaminum (Döderlein): Yellow Sea. (13, 16)

Astroglymma sculptum (Döderlein): Hong Kong; southwestern Hainan. (13, 16, 52)

Gorgonocephalus dolichodactylus (Döderlein): Southeastern Hainan. Water depth 150–1,100 m. (13, 16, 52)

OPHIOMYXIDAE

Astrogymnotes catasticta (H. L. Clark): Western Guangdong; Hainan. Water depth 55–128 m. (52)

Ophiodera neglecta (Koehler): South China Sea. Water depth 64–722 m. (16, 52)

Ophiomyxa australis (Lütken): Xisha (=Paracel Islands). (16, 28, 52)

LAEMOPHIURIDA
OPHIACANTHIDAE

Amphilimna granulosa (Liao): Eastern Hainan. Water depth 280 m. (41, 52)

A. polycacantha (Liao): Southeastern Hainan. Water depth 72–200 m. (34, 52)

A. sinica (Liao): Eastern Hainan. Water depth 280 m. (41, 52)

Ophiacantha composita (Koehler): Eastern South China Sea (19°55'N, 118°4'E). Water depth 3,050 m. (23)

O. pentagona (Koehler): Eastern Hainan; Nansha (=Spratly Islands). Water depth 81–1,698 m. (16, 25, 52)

Ophiocamax rugosa (Koehler): Eastern Hainan; Nansha (=Spratly Islands). Water depth 300–472 m. (16, 25, 52)

Ophiomitra plicata (Lyman): Northeastern South China Sea (17°45'N, 113°5'E). Water depth 2,700 m. (23)

Ophiotreta gratiosa (Koehler): Eastern Hainan; Nansha (=Spratly Islands). Water depth 217–472 m. (16, 25, 52)

O. matura (Koehler): Nansha (=Spratly Islands, 9°50'N, 117°53'E). Water depth 1,241 m. (26)

GNATHOPHIURIDA
AMPHIURIDAE

Amphilycus scripta (Koehler): Beihai (Guangxi). (52)

Amphioplus ancistrotus (H. L. Clark): Yellow Sea; northern Fujian; Taiwan Strait. (7, 16)

A. cyrtacanthus (H. L. Clark): Eastern Guangdong. Water depth 60–105 m. (52)

A. depressus (Ljungman): East China Sea; Taiwan Strait. (16, 24, 25, 52)

A. impressus (Ljungman): East China Sea; Taiwan Strait; South China Sea; Nansha (=Spratly Islands). (16, 24, 25, 52)

A. intermedius (Koehler): From eastern Guangdong to Beibuwan (=Gulf of Tongking). Water depth 90–220 m. (52)

A. japonicus (Matsumoto): Liaoning; Shandong. Shallow sea. (5, 7, 52)

A. laevis (Lyman): East China Sea; Taiwan Strait; South China Sea. (16, 25, 52)

A. lucidus (Koehler): Taiwan Strait; Fujian; Guangdong; Nansha (=Spratly Islands). (18, 24, 25, 52)

Amphipholis duplicata (Koehler): Fujian; Daya Bay (Guangdong). (57)

A. kochii (Lütken): Lushun and Dalian (Liaoning); Miao Islands (Shandong). (16)

A. loripes (Koehler): Nansha (=Spratly Islands). (25, 52)

A. microplax (Burfield): From Fujian to Hainan. Water depth 10–40 m. (52)

A. sobrina (Matsumoto): Nansha (=Spratly Islands). (26, 52)

A. squamata (Delle Chiaje): Yongxing Island (Xisha=Paracel Islands). (28, 52)

Amphiura ceramis (Koehler): Nansha (=Spratly Islands). (36)

A. digitula (H. L. Clark): Taiwan Strait; northeastern South China Sea. (22)

A. divaricata (Ljungman): Northern and central Taiwan Strait. (18, 52)

A. ecnomiotata (H. L. Clark): Beibuwan (=Gulf of Tongking). Water depth 50 m. (52)

A. koreae (Duncan): Jizhou Island (=Cheju Do, Korea). (16)

A. tenuis (H. L. Clark): Zhanjiang (Guangdong); Haikou (Hainan). (52)

A. vadicola (Matsumoto): China's seas. (5, 7, 11, 16, 52)

Dougaloplus echinatus (Ljunman): Taiwan Strait; Fujian. (18, 22, 52, 55)

Ophiocentrus anomalus (Liao): Fujian; Taiwan Strait; Beibuwan (=Gulf of Tongking); Nansha (=Spratly Islands). (18, 24, 25, 36, 52)

O. inaequalis (H. L. Clark): Hong Kong. (17, 52)

O. koehleri (Gislen): From Hong Kong to Beibuwan (=Gulf of Tongking). Water depth 30–84 m. (52)

O. putnami (Lyman): Hong Kong. (17, 52)

Ophiodaphne formala (Koehler): Eastern Hainan; Nansha (=Spratly Islands). (16, 25, 52)

Ophionephthy difficilis (Duncan): Taiwan Strait; Zhejiang; Fujian; Daya Bay (Guangdong). (18, 52, 57)

Paracrocnida sinensis (A. H. Clark): Hong Kong. (17, 52)

Paramphicondrius tetradontus (Guille): Northern and central Taiwan Strait. (18, 52)

OPHIACTIDAE

Ophiactis acosmeta (H. L. Clark): Northern and central Taiwan Strait. (18, 52)

O. affinis (Duncan): Yellow Sea; East China Sea; Taiwan Strait; South China Sea. (7, 11, 16, 26, 52)

O. hexacantha (H. L. Clark): Xisha (=Paracel Islands); Nansha (=Spratly Islands). (24, 25, 28, 52)

O. modesta (Brock): Taiwan Strait; southern Fujian; southern Hainan; Xisha (=Paracel Islands); Nansha (=Spratly Islands). (11, 16, 25, 28, 52)

O. picteci (Loriol): Hong Kong. (17, 52)

O. savignyi (Müller & Troschel): Taiwan Strait; Fujian; Guangdong; Hainan; Xisha (=Paracel Islands); Nansha (=Spratly Islands). (11, 16, 25, 28, 52)

Ophiopholis brachyactis (H. L. Clark): East China Sea. (5)

O. mirabilis (Duncan): Northern and central Yellow Sea. (7, 11, 16)

OPHIOTRICHIDAE

Macrophiothrix capillaris (Lyman): Eastern Hainan; Nansha (=Spratly Islands). (16, 25, 52)

M. demessa (Lyman): Xisha (=Paracel Islands); Nansha (=Spratly Islands). (25, 28, 52)

M. galateae (Lütken): Xisha (=Paracel Islands). (18, 24, 25, 52)

M. longipeda (Lamarck): Guangdong; southern Hainan; Xisha (=Paracel Islands); Nansha (=Spratly Islands). (11, 16, 17, 25, 28, 52)

M. lorioli (A. M. Clark): Xisha (=Paracel Islands); Nansha (=Spratly Islands). (24, 25, 28, 52)

M. propinqua (Lyman): Hainan; Xisha (=Paracel Islands); Nansha (=Spratly Islands). (24, 25, 28, 52)

M. robillardi (de Loriol): Xisha (=Paracel Islands); Nansha (=Spratly Islands). (24, 25, 28, 52)

M. unicolor (Liao): Yongxing Island (Xisha=Paracel Islands). Coral reefs. (28, 52)

M. variabilis (Duncan): Taiwan Strait; Hong Kong. (17, 18, 52)

Ophiocnemis marmorata (Lamarck): East China Sea; South China Sea. (16, 52)

Ophiomaza cacaotica (Lyman): Taiwan Strait; Fujian; Guangdong; Hainan; Xisha (=Paracel Islands); Nansha (=Spratly Islands). (16, 25, 52)

Ophiomymna elegans (Ljungman): Guangdong; Beibuwan (=Gulf of Tongking); Hainan. (52)

O. funesta (Koehler): From eastern Guangdong to Hainan. Water depth 80–184 m. (52)

O. pulchella (Koehler): From eastern Guangdong to Hainan. (52)

Ophiopteron elegans (Ludwig): Guangdong; Beibuwan (=Gulf of Tongking); Hainan; Nansha (=Spratly Islands). (16, 24, 25, 52)

Ophiothela danae (Verrill): Taiwan Strait; Fujian; Guangdong; Hainan; Xisha (=Paracel Islands); Nansha (=Spratly Islands). (11, 16, 25, 52)

Ophiothrix ciliaris (Lamarck): Fujian; Taiwan; Guangdong; Hainan. (17, 52)

O. exigua (Lyman): Fujian; Taiwan Strait; Guangdong; Hainan; Nansha (=Spratly Islands). (11, 16, 24, 25, 52)

O. hybrida (H. L. Clark): Taiwan; Hong Kong; Xisha (=Paracel Islands); Nansha (=Spratly Islands). (17, 24, 25, 28)
O. infirma (Koehler): Nansha (=Spratly Islands). (25)
O. marenzelleri (Koehler): China's seas. (7, 8, 11, 16)
O. melanosticta (Grube): China's seas. (52)
O. nereidina (Lamarck): Xisha (=Paracel Islands). (28, 52)
O. plana (Lyman): Fujian; Taiwan; Guangdong; Hainan; Nansha (=Spratly Islands). (16, 25, 28, 52)
O. proteus (Koehler): From Hong Kong to Zhongsha (=Macclesfield Bank). (52)
O. purpurea (von Martens): Zhongsha (=Macclesfield Bank); Nansha (=Spratly Islands). (25, 52)
O. pusilla (Lyman): Zhongsha (=Macclesfield Bank). (52)
O. cf. *scotiosa* (Murakami): Xisha (=Paracel Islands). (28)
O. signata (Koehler): Northeastern South China Sea. (22)
O. striolata (Grube): Fujian; Taiwan; Guangdong; Nansha (=Spratly Islands). (16, 25, 52)
O. trilineata (Lütken): Taiwan; Xisha (=Paracel Islands); Nansha (=Spratly Islands). (25, 28, 52)
O. virdialba (von Martens): Zhongsha (=Macclesfield Bank). (52)

CHILOPHIURIDA
OPHIOCHITONIDAE

Ophiochiton fastigatus (Lyman): Eastern Hainan. (16,52)
O. megalaspis (H. L. Clark): Nansha (=Spratly Islands). (25)
Ophionereis dubia (Müller & Troschel): Fujian; Guangdong; Hong Kong; Hainan; Xisha (=Paracel Islands); Nansha (=Spratly Islands). (16, 17, 25, 52)
O. dubia amoyensis (A. M. Clark): Hong Kong. (17, 52)
O. porrecta (Lyman): Nansha (=Spratly Islands). (24, 28)
O. variegata (Duncan): Southern Hainan. (16, 52)

OPHIOCOMIDAE

Ophiarthrum elegans (Peters): Southern Hainan; Xisha (=Paracel Islands). (16, 28, 52)
O. pictum (Müller & Troschel): Xisha (=Paracel Islands). (16, 28, 52)
Ophiocoma brevipes (Peters): Xisha (=Paracel Islands). (28, 52)
O. dentata (Müller & Troschel): Hainan; Xisha (=Paracel Islands). (16, 28, 52)
O. erinaceus (Müller & Troschel): Hainan; Xisha (=Paracel Islands). (16, 25, 28, 52)
O. pica (Müller & Troschel): Xisha (=Paracel Islands). (16, 28, 52)
O. pusilla (Brock): Xisha (=Paracel Islands). (25, 28, 52)
O. schoeleini (Müller & Troschel): Taiwan; Xisha (=Paracel Islands). (25,28)
O. scolopendrina (Lamarck): Southern Hainan; Xisha (=Paracel Islands). (16, 25, 28, 52)

Ophiocomella sexradia (Duncan): Xisha (=Paracel Islands); Nansha (=Spratly Islands). (18, 26, 52)
Ophiomastix annulosa (Lamarck): Xisha (=Paracel Islands). (16, 18, 52)
O. caryophyllata (Lütken): Zhongsha (=Macclesfield Bank). (52)
O. mixta (Lütken): Xisha (=Paracel Islands). (28)
O. variabilis (Koehler): Zhongsha (=Macclesfield Bank). (52)
Ophiopsila abscissa (Liao): Beibuwan (=Gulf of Tongking, 19°00'N,108°30'E). Water depth 68 m. (25, 33, 52)
O. pantherina (Koehler): Taiwan Strait; Beibuwan (=Gulf of Tongking). Water depth 10–100 m. (52)
O. polyacantha (H. L. Clark): Northern and central Taiwan Strait. (18).

OPHIODERMATIDAE

Bathypectinura heros (Lyman): Eastern Hainan. (16, 52)
Ophiarachna incrassata (Lamarck): Nansha (=Spratly Islands). (28, 52)
O. ohshimai (Murakami): Xisha (=Paracel Islands); Nansha (=Spratly Islands). (28, 52)
Ophiarachnella elegans (Bell): Zhongsha (=Macclesfield Bank). (52)
O. gorgonia (Müller & Troschel): Guangdong; Hong Kong; Hainan; Xisha (=Paracel Islands). (16, 17, 28, 52)
O. infernalis (Müller & Troschel): Hong Kong; southern Hainan; Nansha (=Spratly Islands). (16, 17, 25, 52)
O. paucispina (Koehler): Zhongsha (=Macclesfield Bank). (52)
O. septeinspinosa (Müller & Troschel): Zhongsha (=Macclesfield Bank). (52)
O. stablis (Koehler): Zhongsha (=Macclesfield Bank). (52)
Ophiochasma stellatum (Ljungman): Nansha (=Spratly Islands). (24, 25)
Ophiopeza spinosa (Ljungman): Nansha (=Spratly Islands). (16, 18, 52)
Ophiopsammus yoldii (Lütken): Nansha (=Spratly Islands). (25)

OPHIURIDAE

Amphiophiura paupera (Koehler): Northeastern South China Sea; Nansha (=Spratly Islands, 17°45'N, 113°05'E). Water depth 2,700 m. (23, 25)
A. radiata (Lyman): Northeastern South China Sea (17°45'N, 113°5'E). Water depth 2,700 m. (22, 23)
A. sculptilis (Lyman): Northeastern South China Sea (19°55'N, 118°4'E). Water depth 3,050 m. (22, 23, 26)
A. sordida (Koehler): Eastern Hainan; Nansha (=Spratly Islands). (16, 25, 52)
A. spatulifera (Koehler): Northeastern South China Sea. Water depth 2,300–3,050 m. (22, 23)
Ophiolepis cincta (Müller & Troschel): Hainan; Xisha (=Paracel Islands). (16, 28, 52)

O. superba (H. L. Clark): Southern Hainan; Xisha (=Paracel Islands). (16, 28, 52)

Ophiolipus granulatus (Koehler): Eastern Hainan. Water depth 270–470 m. (52)

Ophiomastus tegulitius (Lyman): Northeastern South China Sea. Water depth 2,300- 3,050 m. (22, 23, 26)

Ophiomusium corticosum (Lyman): Northeastern South China Sea (17°45'N, 113°5'E). Water depth 2,700 m. (23, 26)

O. facundum (Koehler): Eastern Hainan. Water depth 460 m. (52)

O. lymani (Wyville & Thomson): Northeastern Hainan; southeastern Dongsha (=Pratas Islands); Nansha (=Spratly Islands). Water depth 1,189–3,343 m. (22, 23, 25)

O. morio (Koehler): Northeastern Nansha (=Spratly Islands). (22)

O. scalare (Lyman): Eastern Hainan. Water depth 124–300 m. (52)

O. simplex (Lyman): Nansha (=Spratly Islands, 12°0'N, 116°0'E). Water depth 4,350 m. (22, 25)

Ophioplocus imbricatus (Müller & Troschel): Weizhou Island (Guangxi); southern Hainan. (16)

O. japonicus (H. L. Clark): Taiwan Strait; Fujian; Guangdong. (16, 52)

Ophiotylos brevipes (Liao): Xisha (=Paracel Islands). (28, 52)

Ophiotypa simplex (Koehler): Nansha (=Spratly Islands). (25)

Ophiozonella subtilis (Koehler): Eastern Hainan. Water depth 152–920 m. (26, 28, 52)

Ophiura flagellata (Lyman): Nansha (=Spratly Islands). (34, 52)

O. kinbergi (Ljungman): China's seas. (5, 7, 8, 11, 16, 52)

O. lanceolata (H. L. Clark): Northeastern South China Sea. (22, 52)

O. micracantha (H. L. Clark): Eastern Hainan; Nansha (=Spratly Islands). Water depth 180–360 m. (16, 25, 52)

O. pteracantha (Liao): Daya Bay (Guangdong); Beibuwan (=Gulf of Tongking); Nansha (=Spratly Islands). (24, 25, 33, 52)

O. sarsii (Lütken): Northern and central Yellow Sea. Water depth 10–2,000 m. (7, 8, 11, 16)

Sinophiura multispina (Koehler): Nansha (=Spratly Islands). (34, 52)

Stegophiura hainanensis (Liao): Eastern and southern Hainan. Water depth 140–472 m. (52)

S. sladeni (Duncan): Yellow Sea. (16)

S. sterilis (Koehler): Hong Kong. Water depth 380 m. (52)

S. vivipara (Matsumoto): Southern Yellow Sea. (16)

OPHIOLEUCIDAE

Ophioleuce seminudum (Koehler): East China Sea; eastern and southern Hainan; Nansha (=Spratly Islands). (25, 52)

Ophiopallas paradoxa (Koehler): Eastern Hainan. Water depth 204–470 m. (16, 52)

REFERENCES*

(1). Baranova, Z. I. and Wu, B.-L., 1962. On some asteriid sea-stars from the Yellow Sea. Studia Marina Sinica, 1962 no. 1:109–118.

(2). Chang, F.-Y., 1932. Echinoidea of the China coast. Contr. Inst. Zool. Natn. Acad. Peiping, 1(2): 1–21, 11 figs. 3 pls.

(3). Chang, F.-Y, 1934. Report on the holothurians from the China coast. Contr. Inst. Zool. Natn. Acad. Peiping, 2(10):1–52, 20 figs, 3pls.

(4). Chang, F.-Y., 1935. Additions to the holothurians of the China coast. Contr. Inst. Zool. Natn. Acad. Peiping, 2(3):1–18, 10 figs.

(5). Chang, F.-Y., 1935. The distribution of the echinoderms from the Jiaozhou Bay and its vicinity. Bull. Inst. Zool. Natn. Acad. Peiping, 12:1–12.

(6). Chang, F.-Y., 1943. Notes on some echinoderms from Chengshen, Chowshan Archipelago. J. Natn. Normal Uni. Peiping, 1943:1–8

(7). Chang, F.-Y., 1948. Echinoderms from Tsingtao. Contr. Inst. Zool. Natn. Acad. Peiping, 4:33–104, 24 figs, 11 pls.

(8). Chang, F.-Y. and Wu, B.-L., 1954. On the echinoderms of Dalian and its vicinity. Acta. Zool. Sin., 6(2):123–145.

(9). Chang, F.-Y, 1956. The sea-stars of China. Biol. Bull. China, 6:19–21, 1 pl.

(10). Chang, F.-Y., 1957. The echinoids of Kwangtung Province. Bull. Inst. Mar. Biol. Acad. Sin., 1(1):1–80, 52 figs, 25 pls.

(11). Chang, F.-Y. and Liao, Y., 1958. The ophiurans of China. Biol. Bull. China, 11: 17- 22, 1 pl.

(12). Chang, F.-Y., Wu, B.-L. and Liao, Y., 1957. The sea urchins of China. Biol. Bull. China, 7:18–25, 1 pl.

(13). Chang, F.-Y., 1962. Euryalae of the China sea. Acta Zool. Sin. Supple., 14:53–68, 2 figs, 4pls.

(14). Chang, F.-Y. and Liao, Y., 1963. Echinodermata. In: Tschang, S (Ed.). Economic Fauna of China. (Polychaeta, Echinodermata and Protochordata). China Science Press, pp:42–113, 3pls.

(15). Chang, F.-Y., 1963. On the recent stalked crinoids of China. Acta Zool. Sin., 15(2):282–288, 2pls.

(16). Chang, F.-Y., Liao, Y., Wu, B.-L. and Chen, L., 1964. Echinodermata. In: Illustrated Fauna of China. China Science Press, 142 pp.

(17). Clark, A. M., 1982. Echinoderms of Hong Kong. In: Morton, B. and C. K. Tseng, (eds.). Marine Fauna and Flora of Hong Kong. Hong Kong University Press, pp:485–502.

(18). Fujian Institute of Oceanology, 1988. A comprehensive oceanographic survey of the central and northern parts of Taiwan Strait. China Science Press, pp: 415–416.

(19). Li, G., 1982. Echinoderms from the China coast of the northern part of Beibuwa. South China Sea Oceanic Science and Technology, 6:14–18.

(20). Li, G., 1983. A report on the echinoderms from the Xisha Islands, Guangdong province, China. Contr. Mar. Biol. Res. South China Sea, 1:256–273.

(21). Li, G., 1985. The echinoderms from the mouth of the Pearl River. In: Treatise on the survey of seashore taken at the mouth of the Pearl River. China Science Press, pp:66–82.

(22). Li, G., 1986. Distribution of echinoderms in northeastern waters of the South China Sea. Trop. Oceanol., 5(2):51–59.

(23). Li, G., 1987. Deep-sea ophiurans in the South China Sea. South China Sea Stud. Mar. Sin., no. 8:143–153.

(24). Li, G., 1987. Tsengmuansha. 5. Echinodermata. In: Investigating report on the comprehensive survey of the south China Territory. China Science Press, pp:212- 218.

(25). Li, G., 1989. 5. Echinodermata. In: Investigating report on the comprehensive survey of the Nansha Islands and their neighbouring waters. No. 1:766–774. China Science Press.

(26). Li, G., 1991. Additions to the echinoderms of the Nansha Islands and their neighbouring waters. In: Treatise on the marine biological research of the Nansha Islands and their neighbouring waters. China Ocean Press, pp:189–195.

(27). Liao, Y., 1975. The echinoderms of Xisha Isalnds, Guangdong Province, China. 1. Holothurioidea. Stud. Mar. Sinica, 10:199–230, 27 figs.

(28). Liao, Y., 1978a. The echinoderms of Xisha Isalnds, Guangdong Province, China. 2. Ophiuroidea. Stud. Mar. Sinica, 12:69–104, 4 pls.
(29). Liao, Y., 1978b. The echinoderms of Xisha Isalnds, Guangdong Province, China. 3. Echinoidea. Stud. Mar. Sinica, 12:102–127, 5 pls.
(30). Liao, Y., 1979. A new genus of Clypeasteroid sea-urchin form Huang Hai. Oceanol. Limnol. Sinica, 10(1): 67–73, 1 pl.
(31). Liao, Y., 1980a. The echinoderms of Xisha Islands, Guangdong Province, China. 4. Asteroidea. Stud. Mar. Sinica, 17:153–171, 6 pls.
(32). Liao, Y., 1980b. The aspidochirote holothurians of China, with erection of a new genus. In: Jangoux, M. (Ed.) Echinoderms: present and past. Rotterdam: A. A. Balkema, pp:115–120.
(33). Liao, Y., 1982. Two new ophiurans from the Gulf of Beibuwan. Oceanol. Limnol. Sinica, 13(6):562–569, 2 pls.
(34). Liao, Y., 1983a. On *Amphilimna polyacantha* sp. nov. and the systematic position of *A. multispina* Koehler (Ophiuroidea). Chin. J. Oceanol. Limnol., 1(2):177–184, 2 pls.
(35). Liao, Y., 1983b. The echinoderms of Xisha Islands, Guangdong Provinve, China. 5. Crinoidea. Stud. Mar. Sinica, 20:263–277, 1 pl.
(36). Liao, Y., 1983c. *Ophiocentrus anomalus* sp. nov. of the family Amphiurdae (Ophiuroidea) from southern China. Oceanol. Limnol. Sinica, 14(4):407–410, 1 pl.
(37). Liao, Y., 1984a. The aspidochirote holothurians of China. Stud. Mar. Sinica, 23:221–247.
(38). Liao, Y., 1984b. Notes on the genus *Acaudina* Clark, 1907 (Echinodermata: Holothurioidea) of China. Stud. Mar. Sinica, 23:248–255.
(39). Liao, Y., 1984c. On the family Asterometridae (Crinoidea) of China, with a description of *Sinometra acuticirra* gen. et sp. nov. China. J. Oceanol. Limnol., 2(1): 109–116.
(40). Liao, Y., 1989a. *Calliaster quadrispinus*, a new species of the family Goniasteridae (Asteroidea) from southern China. Oceanol. Limnol. Sinica, 20(1):23–27, 1 pl.
(41). Liao, Y., 1989b. Two new species of the genus *Amphilimna* (Echinodermata: Ophiuroidea) from southern China. Chin. J. Oceanol. Limnol., 7(4): 339–344, 2 figs.
(42). Liao, Y. and Clark, A. M., 1989. Two new species of the genus *Anthenoides* (Echinodermata: Asteroidea) from southern China. Chin. J. Oceanol. Limnol., 7(1):37–42, 2 pls.
(43). Liao, Y. and Li, G., 1985. Additional report on the Echinoidea of Guangdong Province, China. South China Sea Stud. Mar. Sinica, 7: 143–161, 3 pls.
(44). Liao, Y., 1982. A new genus of starfish from the continental shelf of the East China Sea. Stud. Mar. Sinica, 19:93–97, 1 pl.
(45). Liao, Y., 1986. The echinoderms of the continental shelf of the East China Sea. I. Echinoidea. Only a summary in the Proceedings of the Conference on the Oceanographic Survey of the East China Sea.
(46). Liao, Y., 1986. *Thyone pohaiensis*, A new sea cucumber from the Bohai Sea, China. Chin. J. Oceanol. Limnol, 4(3):313–316.
(47). Liao, Y., 1984. *Rosaster attenuatus*, a new species of the family Goniasteridae (Asteroidea) from the East China Sea. Oceanol. Limnol. Sinica, 15(5): 478–481, 1 pl.
(48). Liao, Y. and Lin, C., 1981. A new Echinoid with sexual dimorphism from the late Tertiary deposits of Beibuwan, Guangxi. Acta Paleont. Sinica, 20(5): 482–484, 1 pl.
(49). Liao, Y., 1985. Two new species of Ophidiasterd sea-stars from the vicinity of Diaoyudao, East China Sea. Chin. J. Oceanol. Limnol., 3(1): 30–34, pl.1–2.
(50). Liao, Y., 1983. A new species of goniasterid sea-star from the Huanghai, China. Chin. J. Oceanol. Limnol., 1(2): 367–369.
(51). Liao, Y. and Sun, S., 1989. A new species of astropectinid sea-star from the East China Sea. Chin. J. Oceanol. Limnol., 7(3):225–229.
(52). Liao, Y. and Clark, A. M., 1995. The echinoderms of southern China. China Science Press. 614 pp, 338 figs, 23 pls.
(53). Liao, Y., 1996. On a new species of Ophidiaster (Echinodermata: Asteroidea) from southern China. Bull. Nat. Hist. Mus. Lond (Zool.), 62(1):37–39.
(54). Liao, Y. and Pawson, D. L., 1993. *Caudina intermedia*, A new species of sea cucumber from the South China Sea. Proc. Biol. Soc. Washington, 106(2):366–368.
(55). Lu, X., 1991. Distribution of echinoderms in the central and northern parts of the Taiwan Strait. J. Oceano. Taiwan Strait, 10(2):127–132.
(56). Pawson, D. L. and Liao, Y., 1992. Molpadiid sea cucumber of China, with description of five new species (Echinodermata: Holothuroidea). Pro. Biol. Soc. Wash., 105(2):373–388.
(57). Xu, H., 1989. The composition and the quantity distribution of the echinoderms for the Daya Bay. Collections of paper on marine ecology in the Daya Bay. Vol. 1:148–151. China Ocean Press.
(58). Yang, P. F., 1937. Report on the holothurians from the Fukien coast. Bull. Mar. Biol. Amoy, 2(1): 1–46, 20 figs, 4 pls.

*: all in Chinese except for (17), (53), (54) and (56)
Compiled by Xu Huizhou and Liao Yulin; Edited by Liao Yulin

HEMICHORDATA [STOMOCHORDATA, BRANCHIOTREMA]

ENTEROPNEUSTA
BALANOGLOSSIDA
HARRIMANIIDAE

Saccoglossus hulangtauensis (Tchang & Koo): Jiaozhou Bay (Shandong). High intertidal. (1, 4, 5)

PTYCHODERIDAE

Balanoglossus carnosus (Willey): Lingshui and Xinying (Hainan). (5)
B. misakiensis Kuwano: Jiaozhou Bay (Shandong); Hong Kong; Hepu (Guangxi). (1, 2, 4, 5, 6)
Glossobalanus mortenseni Horst: Sanya (Hainan). (5)
G. polybranchioporus (Tchang & Liang): Northern Bohai; Beidaihe (Hebei); Jiaozhou Bay (Shandong); Dafeng (Jiangsu); southern Yellow Sea. (3, 4, 5)
Ptychodera flava (Eschscholtz): Sanya and Lingshui (Hainan); Zhaoshu Island (Xisha=Paracel Islands). (5)

REFERENCES*

(1). Chang, S. and K. Z. Gu, 1935. On two species of Enteropneusta from Kiaochou Bay. Contributions from the Institute of Zoology, National Academy of Peiping, 13:1–12.
(2). Chang, S. and Y. M. Chang, 1963. Economic Fauna Sinica, Hemichordata. China Science Press, pp:119–121.
(3). Tchang Si and Liang Xian-yuan, 1965. Description of a new species of Enteropneusta, *Glossobalanus polybranchioporus* from China seas. Acta Zootaxonomica Sinica, 2(1): 1–10.
(4). Liang Xian-yuan, 1965. Balanoglossida. Bulletin of Biology, 1:22–71.
(5). Liang Xianyuan, 1984. Studies on Enteropneusta of intertidal zones of the China seas. Studia Marina Sinica, 22: 127–138.
(6). Morton, B. and J. Morton, 1983. The Sea Shore Ecology of Hong Kong. Hong Kong University Press, 350 pp.

*: (1)-(5) in Chinese
Compiled by Huang Xiuming

UROCHORDATA

APPENDICULATA [COPELATA]

OIKOPLEURIDAE

Althoffia tumida Lohmann: East China Sea; northern South China Sea. (18, 19)

Megalocercus huxleyi (Ritter): Jizhou Island (=Cheju Do, Korea); mouth of Changjiang (=Yangtze River); central and southern East China Sea; southern Fujian; Taiwan; northern South China Sea. (1, 5, 6, 14, 19)

Oikopleura albicans (Leuckart): Southern Fujian; Taiwan; northern South China Sea. (14, 19)

O. cophocera Gegenbaur: Northern South China Sea. (2, 19)

O. cornutogastra (Aida): Central and southern East China Sea; northern South China Sea. (5, 17, 19)

O. dioica Fol: Jizhou Island (=Cheju Do, Korea); mouth of Changjiang (=Yangtze River); central and southern East China Sea; northern South China Sea; Xisha (=Paracel Islands); Zhongsha (=Macclesfield Bank). (1, 2, 5, 9, 11, 19)

O. fusiformis Fol: Jizhou Island (=Cheju Do, Korea); mouth of Changjiang (=Yangtze River); East China Sea; southern Fujian; northern and central Taiwan Strait; Taiwan; northern South China Sea. (1, 6, 14, 17, 19, 20)

O. graciloides Lohmann & Buckmann: Nansha (=Spratly Islands) and adjacent water. (4)

O. intermedia Lohmann: Jizhou Island (=Cheju Do, Korea); mouth of Changjiang (=Yangtze River); East China Sea; northern South China Sea. (1, 6, 17, 19)

O. longicauda (Vogt): Bohai; Yellow Sea; Jizhou Island (=Cheju Do, Korea); mouth of Changjiang (=Yangtze River); central and southern East China Sea; southern Fujian; Taiwan; northern, central and western Taiwan Strait; northern and central South China Sea; Nansha (=Spratly Islands). (1–6, 11, 14, 17, 19, 20)

O. megastoma Aida: Zhongsha (=Macclesfield Bank); Xisha (=Paracel Islands). (9)

O. parva Lohmann: Northern South China Sea. (19)

O. rufescens Fol: Jizhou Island (=Cheju Do, Korea); mouth of Changjiang (=Yangtze River); central and southern East China Sea; southern Fujian; northern and central Taiwan Strait; Taiwan; northern South China Sea; Nansha (=Spratly Islands). (1- 6, 14, 17, 19, 20)

Pleagopleura verticalis (Lohmann): Northern South China Sea. (19)

Stegosoma magnum Langerhans: East China Sea; southern Fujian; Taiwan; northern South China Sea. (6, 14, 19)

FRITILLARIDAE

Fritillaria bicornis Lohmann: Southern Fujian; Taiwan. (14)

F. borealis (Lohmann): Northern South China Sea. (10)

F. borealis sargassi (Lohmann): Northern South China Sea. (19)

F. formica Fol: East China Sea; southern Fujian; Taiwan; Nansha (=Spratly Islands) and adjacent water. (4, 6, 14, 17, 19)

F. fraudax Lohmann: Northern South China Sea. (19)

F. haplostoma Fol: Northern South China Sea. (19)

F. megachile Fol: Xisha (=Paracel Islands); Zhongsha (=Macclesfield Bank). (9)

F. pellucida (Busch): Jizhou Island (=Cheju Do, Korea); mouth of Changjiang (=Yangtze River); central and southern East China Sea; southern Fujian; Taiwan; northern South China Sea; Nansha (=Spratly Islands) and adjacent water. (1, 4, 5, 6, 11, 14, 17, 19)

F. tenella Lohmann: Northern South China Sea. (19)

F. venusta Lohmann: Northern South China Sea. (19)

Tectillaria fertilis (Lohmann): Northern South China Sea. (19)

THALIACEA

CYCLOMYARIA
DOLIOLIDAE

Dolioletta gegenbauri Uljanin: Jizhou Island (=Cheju Do, Korea); mouth of Changjiang (=Yangtze River); East China Sea; southern Fujian; northern, central and western Taiwan Strait; Taiwan; northern South China Sea; Nansha (=Spratly Islands) and adjacent water. (1–7, 11–14, 17, 18, 20)

Doliolina mulleri krohni Borgert: East China Sea; central South China Sea. (7, 17)

D. obscura Tokioka & Berner: Northern, central, and western Taiwan Strait; northern South China Sea. (6, 12, 18, 20)

D. separata Tokioka & Berner: South China Sea. (6)

Dolioloides rarum Grobben: East China Sea; South China Sea. (17)

Doliolum denticulatum Quoy & Gaimard: Jizhou Island (=Cheju Do, Korea); mouth of Changjiang (=Yangtze River); central and southern East China Sea; southern Fujian; northern, central, and western Taiwan Strait; Taiwan; northern and central South China Sea; Nansha (=Spratly Islands) and adjacent water. (1–7, 11–14, 17, 18, 20)

D. denticulatum ehrenbergi Uljanin: East China Sea; central South China Sea. (7, 17)

D. nationalis Borgert: East China Sea; South China Sea. (17)

PYROSOMIDA
PYROSOMATIDAE

Pyrosoma atlanticum Peron: Southern Fujian; Taiwan; central South China Sea; Nansha (=Spratly Islands) and adjacent water. (4, 7, 14)

HEMIMYARIA [DESMOMYARIA]
SALPIDAE

Brooksia rostrata (Traustedt): Mouth of Changjiang (=Yangtze River); central and southern East China Sea; northern, central, and western Taiwan Strait; northern and central South China Sea. (1, 5, 6, 7, 12, 13, 16, 18, 20)

Cyclosalpa affinis (Chamisso): Western Taiwan Strait; northern and central South China Sea. (7, 12, 15, 16, 18)

C. bakeri Ritter: Southern Yellow Sea; Jizhou Island (=Cheju Do, Korea); mouth of Changjiang (=Yangtze River); East China Sea; western Taiwan Strait; northern South China Sea. (1, 6, 10, 12, 16, 18)

C. floridana (Apstein): Southern Yellow Sea; East China Sea; northern, central, and western Taiwan Strait; northern and central South China Sea; Nansha (=Spratly Islands) and adjacent water. (4, 6, 7, 11, 12, 16, 18, 20)

C. pinnata (Forskål): Jizhou Island (=Cheju Do, Korea); mouth of Changjiang (=Yangtze River); central and southern East China Sea; southern Fujian; northern, central, and western Taiwan Strait; Taiwan shallow sea; northern and central South China Sea; Nansha (=Spratly Islands) and adjacent water. (1, 4–7, 12, 14, 15, 16, 18, 20)

C. pinnata polae (Sigl.) [*C. polae*]: East China Sea; northern and central Taiwan Strait; northern and central South China Sea. (7, 15, 18, 20)

C. pinnata f. *quadriluminis* Berner [*Cyclosalpa pinnata quadriluminis*]: South China Sea. (15)

C. pinnata sewelli Metcalf: Western Taiwan Strait; central South China Sea. (7, 12)

Helicosalpa virgula (Vogt) [*Cyclosalpa virgula*]: Northern South China Sea. (18)

Iasis zonaria (Pallas): Jizhou Island (=Cheju Do, Korea); mouth of Changjiang (=Yangtze River); southern Fujian; Taiwan; northern and central South China Sea; Nansha (=Spratly Islands) and adjacent water. (1, 4, 6, 7, 14, 15, 16, 18)

Ihlea punctata (Forskål): South China Sea. (6, 15)

Metcalfina hexagona (Quoy & Gaimard): South China Sea. (16)

Pegea confoederata (Forskål): Northern and southern Yellow Sea; Jizhou Island (=Cheju Do, Korea); mouth of Changjiang (=Yangtze River); East China Sea; northern and central South China Sea; Nansha (=Spratly Islands) and adjacent water. (1, 4, 6, 7, 13–16, 18, 20)

Ritteriella amboinensis (Apstein): Jizhou Island (=Cheju Do, Korea); mouth of Changjiang (=Yangtze River); East China Sea; northern and central Taiwan Strait; northern and central South China Sea; Xisha (=Paracel Islands); Zhongsha (=Macclesfield Bank); Nansha (=Spratly Islands). (1, 6, 7, 15, 16, 18, 20)

R. picteti (Apstein): Northern and central South China Sea. (6, 7, 15, 16, 18)

Salpa fusiformis Cuvier: Northern Yellow Sea; Jizhou Island (=Cheju Do, Korea); mouth of Changjiang (=Yangtze River); East China Sea; southern Fujian; western Taiwan Strait; Taiwan; South China Sea; Zhongsha (=Macclesfield Bank); Xisha (=Paracel Islands); Nansha (=Spratly Islands) and adjacent water. (1, 4, 7, 10, 11, 12, 14, 16, 18)

S. fusiformis aspera Chamisso [*S. aspera*]: East China Sea; South China Sea. (7, 13, 15, 16)

S. maxima Forskål: South China Sea. (7, 15, 16, 18)

S. maxima tuberculata Metcalf: South China Sea. (6, 15)

Thalia democratica (Forskål) [*T. democratica democratica*]: Bohai; southern Yellow Sea; Jizhou Island (=Cheju Do, Korea); mouth of Changjiang (=Yangtze River); central and southern East China Sea; southern Fujian; northern, central, and western Taiwan Strait; Taiwan; northern and central South China Sea; Xisha (=Paracel Islands); Zhongsha (=Macclesfield Bank); Nansha (=Spratly Islands) and adjacent water. (1–8, 11–18, 20)

T. democratica echinata Tokioka: East China Sea; northern and central Taiwan Strait. (6, 7, 13, 15, 17, 18, 20)

T. democratica orientalis Tokioka: Southern Yellow Sea; central and southern East China Sea; northern and central Taiwan Strait; South China Sea. (5, 6, 7, 18, 20)

Thetys vagina Tilesius: Xisha (=Paracel Islands). (15)

Traustedtia multitentaculata (Quoy & Gaimard): Jizhou Island (=Cheju Do, Korea); mouth of Changjiang (=Yangtze River); East China Sea; northern and central Taiwan Strait; northern and central Taiwan Strait; South China Sea. (1, 6, 7, 15, 16, 18, 20)

Weelia cylindrica (Cuvier) [*Salpa cylindrica*]: Jizhou Island (=Cheju Do, Korea); mouth of Changjiang (=Yangtze River); central and southern East China Sea; southern Fujian; northern, central, and western Taiwan Strait; Taiwan; South China Sea. (1, 5, 6, 7, 12, 14,1 5, 16, 18, 20)

REFERENCES*

(1). Shandong College of Oceanology, 1986. Comprehensive Survey and Research Report on the Water Areas Adjacent to the Changjiang River Estuary and Chejudo Island. Journal of Shandong College of Oceanology, 16(2):55–84.

(2). South China Sea Institute of Oceanography, Acdemica Sinica, 1984. Investigation reports on plankton around Dapengao and its adjacent waters, pp:7–16, 28–29.

(3). South China Sea Institute of Oceanography, Acdemica Sinica, 1987. Research reports on the comprehensive oceanographic investigation of Chinese Southern Territory: Zengmu Shoal. China Science Press, pp:132–163.

(4). Comprehensive Scientific Investigation Team on Nansha Islands, Academia Sinica, 1989, Study report of comprehensive survey in the waters around Nansha Islands (I). China Science Press, pp:659–707.

(5). Liu Hongbin, He Dehua and Wang Chunsheng, 1991. Preliminary study on distribution and communities of zooplankton in the Kuroshio area of the Middle- southern East China Sea. Papers on the Kuroshio survey research, 3:305–313.

(6). The Committee of Science and Technology, the Oceanographic Group, 1977. Comprehensive investigation report on China Seas (Vol. 8), pp:121–124, 158- 159.

(7). State Oceanic Administration, 1988. A comprehensive environmental resource survey of the central part of the South China Sea. China Ocean Press, 249pp.

(8). Editoral board for marine atlas, 1991. Biology. In: Marine atlas of Bohai Sea, Yellow Sea and East China Sea. China Ocean Press.
(9). Chen, Qingchao et al., 1978. Horizontal distribution of zooplankton in Xisha and Zhongsha Isalnds. In: A survey of the marine biology of Xisha and Zhongsha Islands, China. China Science Press, pp:63–71.
(10). Chen, Jiekang 1978. Two species of Salpidae from the northern Yellow Sea. Chinese Journal of Zoology, 2:13–16.
(11). Zheng Zhong et al., 1984. Marine Planktonology. China Ocean Press, pp:527–546.
(12). Lin Mao, 1988. A preliminary analysis of Thaliacea from western waters of Taiwan Strait. Marine Science Bulletin, 7(4): 66–71.
(13). Lin Mao, 1990. Ecological studies of Thaliaeea in Daya Bay. Collections of Papers on Marine Ecology in the Daya Bay (II). China Ocean Press, pp:390–395.
(14). The Fishery Resources Groups of Southern Fujian, 1980. The Fishery Resources in Minnan-Taiwan Bank, China, pp:177–242. (internal report)
(15). Hu Qingbo, 1985. The pelagic Tunicata off the coast of China, I. Salpidae. Journal of Xiamen Fisheries College, 1:1–22.
(16). Hou Shumin, 1984. The Hemimyaria (Tunicata) in tropical waters of the western Pacific Ocean. In: Proceedings of the Plankton from the Tropical Waters of the Western Pacific Ocean. China Ocean Press, pp:217–244.
(17). Group of National Reports on Marine Organisms, 1990. Reports on Marine Organisms. In: Comprehensive investigation on the resources of coastal zone and intertidal zone of China, pp:1–128.
(18). Yang Guofeng et al.,1979. A preliminary study on planktonic Pelagic tunicata on the continental shelf of northern South China Sea. In: Reports of Investigation on the Demersal Fishery Resources off the Continental Shelf of Northern South China Sea (Vol. 2). Bureau of Oceanography and Fisheries Institute of South China Sea. National General Bureau of Fisheries Press, Guangzhou, pp:558–568.
(19). Yang Guofeng et al., 1988. Preliminary Study on Appendicularia of the Northern Part of South Sea. In: Proceedings on Marine Biology of the South China Sea, China Ocean Press, pp:143–152..
(20). Fujian Institute of Oceanography, 1988. A comprehensive Oceanographic Survey of the Central and Northern Part of the Taiwan Strait (VII). China Science Press, pp:297–298.

*: all in Chinese except for (19)
Compiled by Hou Shuming and Lin Mao; Edited by Hu Qinbo

ASCIDIACEA

ENTEROGONA
POLYCLINIDAE

Amaroucium constellatum Verrill: Lushun (Liaoning); Penglai (Shandong). (1)
Aplidium depressum Sluiter: Hong Kong. (8)
A. multiplicatum Sluiter: Hong Kong. (8)
Polyclinum constellatum Savigny: Daya Bay (Guangdong). (2)
P. festum Hartmeyer: Hong Kong. (8)
P. sundaicum (Sluiter): Zhangjiang (Guangdong). (4, 8)

DIDEMNIDAE

Didemnum areolatum Tokioka: Zhifu Bay (Shandong); Daya Bay (Guangdong). (1, 2)
D. aspersum Tokioka: Hong Kong. (8)
D. fuscum (Oka): Dongshan Bay (Fujian); Daya Bay (Guangdong). (2, 4)
D. granulatum Tokioka: Hong Kong. (8)
D. membranaceum Sluiter: Hong Kong. (8)
D. moseleyi (Herdman): Qingdao (Shandong); Hong Kong. (8)
D. tonga (Herdman): Hong Kong. (8)
Leptoclinides madara Tokoika: Hong Kong. (8)
L. reticulatus (Sluiter): Hong Kong. (8)
L. rufus (Sluiter): Hong Kong. (8)
Leptoclinum mitsukurii (Oka): Penlai harbor, Qingdao harbor, Yantai harbor, Duoji Island, and Zhifu Bay (Shandong); Daya Bay (Guangdong). (1, 2)

POLYCITORIDAE

Eudistama laysami (Sluiter): Zhangjiang (Guangdong). (4)

DIAZONIDAE

Rhopalaea crassa (Herdman): Hong Kong. (8)

CIONIDAE

Ciona intestinalis Linnaeus: Lushun (Liaoning); Penglai, Yantai, Zhifu Bay, and Qingdao (Shandong); Xiamen (Fujian); Daya Bay (Guangdong); Hong Kong. (1, 2, 8)

ASCIDIIDAE

Ascidia alpha Tokioka: East China Sea; Xinying (Hainan). (5)
A. armata Hartmeyer: Yellow Sea; East China Sea; Wushi (Guangdong). (5)
A. gemata Sluiter: Xinying (Hainan). (5)
A. lobata Huang: Xinying (Hainan). (5)
A. longistriata Hartmeyer: Penglai, Yantai, and Qingdao, and Zhifu Bay (Shandong); Daya Bay (Guangdong); Hong Kong; Sanya (Hainan). (1, 2, 5, 8)
A. pacifica Tokioka: Qingdao (Shandong); Sanya (Hainan). (5)
A. papillosa Tokioka: Xiamen harbor (Fujian). (4)
A. rhabdophora Sluiter: Yulin Harbor (Hainan). (5)
A. sydneiensis Stimpson: Zhangjiang (Guangdong); Hong Kong; Yulin harbor (Hainan); Yongxin Island (Xisha =Paracel Islands). (4, 5, 8)

CORELLIDAE

Chelyosoma siboja Oka: Zhifu Bay (Shandong). (1)

PLEUROGONA
BOTRYLLIDAE

Botrylloides perspicuum Herdman: Hong Kong. (8)
B. simodensis Saito & Watanabe: Qingdao harbor and Zhifu Bay (Shandong). (1, 6)

B. violaceulus Oka: Penglai, Yantai, Qingdao, Duoji Island, and Zhifu Bay (Shandong); Liangyungang (Jiangsu). (1, 6)

Botryllus primigenus Oka: Hong Kong. (4)

B. schlosseri (Pallas): Qingdao harbor (Shandong); Liangyungang (Jiangsu); Daya Bay (Guangdong); Hong Kong; Yongxin Island (Xisha=Paracel Islands). (1, 2, 6, 8)

B. tsingtaoensis Ger & Zan: Lushun harbor (Liaoning); Qingdao harbor (Shandong); Liangyungang (Jiangsu). (1, 6)

B. tuberatus Ritter & Forsyth: Zhifu Bay and Qingdao harbor (Shandong); Liangyungang (Jiangsu); Xiamen harbor (Fujian); Daya Bay (Guangdong). (1, 2, 4, 6)

STYELIDAE

Cnemidocarpa areolata (Heller): Hong Kong. (8)

C. chinensis Tokioka: Qingdao (Shandong); Xiamen (Fujian). (4, 7)

Cnemidocarpa sp.: Qingdao (Shandong). (7)

Plyandrocarpa latericius (Sluiter): Xielang Bay (Hainan). (3)

P. monotestis Tokioka: Xiamen harbor (Fujian). (4)

P. sagamiensis Tokioka: Zhangjiang (Guangdong). (4)

Polycarpa circumarta (Sluiter): Zhangjiang (Guangdong). (4)

P. irregularis Herdman: Zhangjiang (Guangdong). (4)

Styela canopus Savigny: Penglai Bay; Duoji Island; and Qingdao harbor (Shandong); Liangyungang (Jiangsu); Xiamen harbor and Dongshan Bay (Fujian); Daya Bay (Guangdong); Hong Kong; Yulin harbor (Hainan); Yongxin Island (Xisha=Paracel Islands). (1–4, 7, 8)

S. clava Herdman: Lushun (Liaoning); Tanggu (Tianjin); Penglai, Yantai, Qingdao, Duoji Island, and Zhifu Bay (Shandong); Liangyungang (Jiangsu); Luoyuan Bay (Fujian). (1, 7)

S. plicata (Lesueur): Xiamen and Dongshan Bay (Fujian); Daya Bay and Shenzhen Bay (Guangdong); Hong Kong. (2, 3, 4, 8)

S. qindaoensis Ger & Zang: Qingdao harbor (Shandong). (7)

S. sinensis Ger & Zang: Qingdao (Shandong). (7)

Styela sp.: Qingdao harbor (Shandong). (7)

Symplegma oceania Tokioka: Xiamen harbor (Fujian); Daya Bay and Zhangjiang (Guangdong); Hong Kong; Yongxin Island (Xisha=Paracel Islands). (2, 4, 8)

S. reptans (Oka): Hong Kong. (8)

PYURIDAE

Herdmania momus (Savigny): Xiamen harbor and Dongshan Bay (Fujian); Daya Bay and Zhangjiang (Guangdong); Hong Kong; Yulin harbor (Hainan); Xisha (=Paracel Islands). (2, 3, 4, 8)

Microcosmus australis Herdman: Daya Bay and Zhangjiang (Guangdong); Hong Kong. (2, 3, 4, 8)

M. exasperatus Heller: Xiamen harbor and Dongshan Bay (Fujian); Shantou harbor, Nanao Island, Daya Bay, and Zhangjiang (Guangdong); Hong Kong; Yulin harbor (Hainan). (2, 3, 4)

Pyura curvigona Tokioka: Hong Kong. (8)

P. elongata Tokioka: Hong Kong. (8)

P. lignosa Michaelsen: Dongshan Bay (Fujian); Daya Bay (Guangdong). (4)

P. mirabilis (Drasche): Hong Kong. (8)

P. vittata (Stimpson): Hong Kong; Xielang Bay (Hainan). (3, 8)

Pyura sp.: Daya Bay (Guangdong). (2)

MOLGULIDAE

Hartmeyeria chinensis Tokioka: Xiamen harbor (Fujian); Daya Bay (Guangdong). (4)

Molgula diversa Kett: Hong Kong. (8)

M. manhattensis (Delay): Lushun (Liaoning); Tanggu (Tianjin); Penglai, Yantai, Qingdao, and Zhifu Bay (Shandong); Liangyungang (Jiangsu); Luoyuan Bay, Xiamen harbor, and Dongshan Bay (Fujian); Shantou harbor and Daya Bay (Guangdong). (1–4)

REFERENCES*

(1). Zheng Chengxing, 1988. The Ascidians among the fouling organisms in the coast of the Yellow Sea and Bohai Bay. Acta Zoologia Sinica, 34(2):180–188.

(2). Zheng Chengxing, 1990. Ecology of Ascidians in Daya Bay. In: Collections of Papers on Marine Ecology in the Daya Bay (II). China Ocean Press, pp:397–403.

(3). Zheng Chengxing, 1984. Ascidians in the fouling organisms off China's coasts (the Newly-recorded species). Marine Science Bulletin, 3(4):104–106.

(4). Zheng Chengxing, 1993. Ecology of Ascidians off China. (manuscript)

(5). Huang Xiuming, 1989. Studies on Ascidians of China Seas, I. The Species of Genus *Ascidia*. Studia Marina Sinica, 30:239–250.

(6). Ge Guochang and Zang Yanlan, 1983. Ascidians of Jiaozhou Bay, I. Botryllidae. Journal of Shandong College of Oceanology, 13(3):93–100.

(7). Ge Guochang et al., 1987. Ascidians of Jiaozhou Bay, II. Styelidae. Journal of Shandong College of Oceanology, 17(4):95–102.

(8). Kott, P. and Ivan Goodbody, 1980. The Ascidians of Hong Kong. In: Proceedings of the First International Marine Biological Workshop: The Marine Flora and Fauna of Hong Kong and Southern China. Hong Kong University Press, pp:503–554.

*: all in Chinese except for (8)

Compiled by Zheng Chengxing; Edited by Huang Xiuming

CHORDATA

CEPHALOCHORDATA

AMPHIOXI
AMPHIOXIFORMES
AMPHIOXIDAE

Asymmefron culfellum (Pefers): Southern Taiwan Strait, water depth 30–60 m; around Hainan. (1, 4, 5)

Branchiostoma belcheri (Gray): From Putian (Fujian) to Nanao (Guangdong), high- to mid-intertidal; southwestern Taiwan Strait, subtidal (<60 m water depth); Suixi (Guangdong); Hong Kong; Hepu (Guangxi); Xinying and Xinying (Hainan). (1–7)

B. belcheri tsingtauense Tchang & Koo: Qinhuangdao (Hebei); Yantai, Jiaozhou Bay, and Dagong Island (Shandong). (5)

REFERENCES*

(1). Fang Shaohua and Lu Xiaomei, 1991. Ecological characteristics of two species of *Amphioxus* in Minnan-Taiwan Bank Fishing Ground. In: Minnan-Taiwan Bank Fishing Ground Upwelling Ecosystem Study. China Science Press, pp:603–607.
(2). Wang Weiyang et al., 1989. An investigation on the distribution of lancelet *Branchiostoma belcheri* Gray in the coastal zone from Minjiang Estuary to Nanao. Fujian Fisheries, 1:14–16.
(3). Kin Te-siang, 1953. Amphioxus of Amoy. Acta Zoologica Sinica, 5(1):65–78.
(4). Tchang-Si, 1962. Sur la presence du genre Asymmetron dans la mer de Chine et la distribution geographique de *Branchiostoma belcheri* (Gray). Acta Zoologica Sinica, 14(4): 525–528.
(5). Tchang-Si et al., 1963. Economic Fauna Sinica: Annelida, Echinodermata and Hemichordata. China Science Press, 141pp.
(6). Cai Yingya, Liu Zhigang et al., 1986. The *Amphioxus* in the Beibu Gulf of China. Tropic Oceanology, 5(2):40–50.
(7). Light, S. F. 1923. Amphioxus fisheries near the Unviersity of Amoy, China. Science 58:57–61.

*: all in Chinese except for (7)
Compiled by Huang Zongguo; Edited by Jin Dexiang

VERTEBRATA

CYCLOSTOMATA [MARSIPOBRANCHII]
PETROMYZONIFORMES
PETROMYOZONIDAE

Lampetra japonica (Martens): Yellow Sea; East China Sea. Migratory species (between freshwater and sea). (1–6)

EPTATRETIDAE

Eptatretus burgeri (Girard): Yellow Sea; East China Sea. Parasitic, generally on gills or cheeks of fish. (1–6)
E. okinoseanus (Dean): East China Sea; Taiwan. Water depth 200–600 m. (7, 8)

MYXINIDAE

Paramyxine cheni Shen & Tao: Gaoxiong and Donggang (Taiwan). (8)
P. fernholmi Huang Kuo, Mok & Huang: Taiwan. (8)
P. nelsoni Kuo, Mok & Huang: Taiwan. (8)
P. sheni Huang, Kuo, Mok & Huang: Taiwan. (8)
P. taiwanae Shen & Tao: Nanao and Donggang (Taiwan). (8)
P. yangi Teng: Northeastern and southwestern Taiwan. Muddy substrate. (8)

CHONDRICHTHYES
HEXANCHIFORMES
HEXANCHIDAE

Heptranchias dakini Whitley: East China Sea; South China Sea. Water depth 27–1,000 m. (1–6)
H. perlo Bonnaterre: Yellow Sea; East China Sea; Taiwan; South China Sea. Water depth 423–450 m. (1–6)
Hexanchus griseus (Bonnaterre): Southern East China Sea; Taiwan Strait; South China Sea. Benthic species, water depth 0–2,480 m. (1–6)
H. vitulus Springer & Waller: Taiwan. (8)
Notorhynchus platycephalus (Tenore) [*N. cepedianus*]: Yellow Sea; East China Sea; Taiwan; South China Sea. (8)

CHLAMYDASELACIDAE

Chlamydalachus anguineus Garman: Taiwan. Deep sea. (8)

HETERODONTIFORMES
HETERODONTIDAE

Heterodontus japonicus Dumeril: Yellow Sea; East China Sea; Taiwan. (1–6)
H. zebra (Gray): East China Sea; Taiwan Strait; South China Sea. Near shore bottom fish. (1–6)

ISURIFORMES [LAMNIFORMES]
CARCHARIIDAE

Carcharias arenarias Ogilby: South China Sea. Near bottom or bottom. (1–6)
C. owstoni Garman: Yellow Sea; East China Sea. Near bottom or bottom. (1–6)
Eugomphodus taurus (Kafinesque) [*C. taurus*]: Taiwan. (8)

ISURIDAE [LAMNIDAE]

Carcharodon carcharias (Gunner): Bohai; Yellow Sea; East China Sea; Taiwan; South China Sea. (1–6)
Isurus glaucus (Müller & Henle) [*I. oxyrinchus* (8)]: East China Sea; Taiwan Strait; South China Sea. (1–6)
I. paucus Guitart Manday: Taiwan. (8)

CETORHINIDAE

Cetorhinus maximus (Gunner): Yellow Sea; East China Sea; Taiwan; South China Sea. Offshore large shark. (1–6)

ALOPIIDAE

Alopias pelagicus Nakamura: East China Sea; Taiwan; South China Sea. Oceanic species. (1–6)

A. profundus Nakamura [*A. superciliosus*]: Suao and Gaoxiong (Taiwan). (1–6)

A. vulpinus (Bonnaterre): Yellow Sea; East China Sea; Taiwan; South China Sea. Oceanic species. (1–6)

ORECTOLOBIFORMES
ORECTOLOBIDAE

Chiloscyllium colax (Meuschen) [*C. indicum* and *C. isabellum* (8)]: Taiwan; South China Sea. (1–6)

C. griseum (Müller & Henle): South China Sea. (1–6)

C. plagiosum (Bennett): East China Sea; Taiwan Strait; South China Sea. Shallow sea or inner bay. (1–6)

C. punctatum (Müller & Henle): Taiwan; South China Sea. (1–6)

Ginglymostoma ferrugineum Lesson: Taiwan Strait; South China Sea. Reefs, benthic species. (1–6)

Nebrius formosanum Teng: Endemic in Taiwan. (8)

N. macrurus (Garman): South China Sea. (1–6)

Orectolobus japonicus Regan: East China Sea; Taiwan Strait; South China Sea. Near shore sandy mud and seaweed substrate. (1–6)

O. maculatus (Bonnaterre): East China Sea; Taiwan; South China Sea. (1–6)

Stegostoma fasciatum (Hermann): East China Sea; South China Sea. Benthic species. (1–6)

CIRRHOSCYLLIIDAE

Cirrhoscylliun expolitum Smith & Radcliffe: South China Sea. Near shore deep water. (1–6)

RHINCODONTIDAE

Rhincodon typus Smith: Yellow Sea; East China Sea; Taiwan Strait; South China Sea. Oceanic species, sometimes migrates to near shore. (1–6)

CARCHARHINIFORMES
SCYLIORHINIDAE

Apristurus abbreviatus Deng, Xiong, & Zhan: East China Sea. Water depth 650–720 m. (7)

A. canutus Springer & Heemstra: South China Sea. Water depth 655–716 m. (9)

A. herklotsi (Fowler): South China Sea. Deep water. (9)

A. internatus Deng, Xiong & Zhan: East China Sea. Deep water. (7)

A. japonicus Nakaya: East China Sea. Water depth 670–980 m. (7)

A. longicephalus Nakaya: East China Sea; South China Sea. Deep water. (1–6)

A. macrorhynchus (Tanaka): Southern East China Sea; northeastern and southwestern Taiwan; South China Sea. Deep water. (1–6)

A. microps (Gilchrist): South China Sea. Deep water. (1–6)

A. nasutus Buen: South China Sea. Deep water. (9)

A. pinguis Deng, Xiong & Zhan: East China Sea. Deep water. (7)

A. platyrhynchus (Tanaka): East China Sea; South China Sea. Water depth 510–655 m. (1–6)

A. sinensis Chu & Wu: East China Sea; South China Sea. Water depth 200–1,000 m. (1-6)

A. verweyi (Fowler): South China Sea. Deep water. (1–6)

Atelomycterus marmoratus (Bennett): Southern East China Sea; Taiwan; South China Sea. (1–6)

Cephaloscyllium fasciatum Chan: Southern East China Sea; South China Sea. (1–6)

C. formosanum Teng [*C. isabellum* (8)]: Southwestern Taiwan. Muddy sand bottom. (8)

C. umbratile Jordan & Fowler: Yellow Sea; East China Sea; Taiwan Strait; South China Sea. Near shore bottom fish. (1–6)

Figaro melanobranchus (Chan): East China Sea; South China Sea. Water depth about 1,000 m. (1–6)

Galeus eastmani (Jordan & Snyder): East China Sea; Taiwan; South China Sea. (1–6)

G. nipponensis Nakaya: South China Sea. (1–6)

G. sauteri Jordan & Richardso: Taiwan Strait; South China Sea. (1–6)

Halaelurus burgeri (Müller & Henle): East China Sea; Taiwan Strait; South China Sea. Near shore benthic species. (1–6)

Proscyllium habereri (Hilgendorf): Southern East China Sea; South China Sea. (1–6)

Scyliorhinus torazame (Tanaka) [*Apristurus macrorhynchus* (8)]: Yellow Sea; East China Sea. (1–6)

PSEUDOTRIAKIDAE

Pseudotriakis acrages Jordan & Snyder [*P. microdon* (8)]: Hualian (Taiwan). (1–6)

TRIAKIDAE

Eridacnis radcliffei Smith: East China Sea; South China Sea. Deep water. (1–6)

Galeorhinus hyugaensis (Miyosi) [*Hypogaleus hyugauensis* (8)]: East China Sea. (1–6)

G. japonicus (Müller & Henle) [*Hemitriakis japonicus*]: Yellow Sea; East China Sea; South China Sea. Water depth 130–180 m. (1–6)

Mustelus griseus Pietschmann: Yellow Sea; East China Sea; Taiwan Strait; South China Sea. Near shore. (1–6)

M. kanekonis (Tanaka): Southern East China Sea; South China Sea. Near shore. (1–6)

M. manazo Bleeker: Yellow Sea; East China Sea; Taiwan. (1–6)

Triakis scyllium Müller & Henle: Yellow Sea; East China Sea; Taiwan; South China Sea. Near shore. (1–6)

T. venustum (Tanaka) [*Calliscyllium venustum, Proscyllium habereri*]: East China Sea; South China Sea. Water depth 100–320 m. (1–6)

CARCHARHINIDAE

Aprionodon brevipenna (Müller & Henle): East China Sea; Taiwan Strait; South China Sea. (1–6)

Carcharhinus albimarginatus (Rüppell): Southern East China Sea; Taiwan Strait; South China Sea. (1–6)

C. altimus (Springer): Taiwan. (1–6)

C. atrodorsus Deng, Xiong & Zhan: East China Sea. (1–6)

C. brachyurus (Günther): Taiwan. (1–6)

C. brevipinna (Müller & Henle): Taiwan. (8)

C. dussumieri (Müller & Henle): Taiwan; South China Sea. (1–6)

C. falciformis (Bibron): Taiwan. (8)

C. gangeticus (Müller & Henle) [*Glyphis gangeticus* (8)]: East China Sea; South China Sea. (1–6)

C. latistomus Fang & Wang: Yellow Sea; East China Sea. (1–6)

C. leucas (Valenciennes): Taiwan. (8)

C. limbatus (Valenciennes): Taiwan. (8)

C. longimanus (Pocy): East China Sea; Taiwan Strait; South China Sea. (1–6)

C. macloti (Müller & Henle): Taiwan. (8)

C. melanopterus (Quoy & Gaimard): Southern East China Sea; Taiwan; South China Sea. (1–6)

C. menisorrah (Müller & Henle) [*C. amblyrhy* (8)]: Yellow Sea; East China Sea; South China Sea. (1–6)

C. microphthalmus Chu: South China Sea. (1–6)

C. obscunella Deng, Xiong & Zhan: East China Sea. (1–6)

C. obscurus (Lesueur): Taiwan. (8)

C. pleurotaenia (Bleeker): Taiwan Strait; Hainan; northern South China Sea; Dongsha (=Pratas Islands); Xisha (=Paracel Islands); Nansha (=Spratly Islands). (1–6)

C. plumbeus (Nardo): Taiwan. (8)

C. remotoides Deng, Xiong & Zhan: East China Sea. (1–6)

C. sorrah (Müller & Henle): Southern East China Sea; Taiwan Strait; South China Sea. (1–6)

Galeocerdo cuvier (Lesueur): Yellow Sea; East China Sea; Taiwan; South China Sea. (1–6)

Hypoprion atripinnis Chu: Taiwan Strait; South China Sea. (1–6)

H. macloti (Müller & Henle) [*Carcharhinus macloti* (8)]: East China Sea; Taiwan Strait; South China Sea. Near shore. (1–6)

Loxodon macrorhinus (Müller & Henle): Taiwan. (8)

Negaprion queenslandicus (Whitley): South China Sea. (1–6)

Negogaleus balfouri (Day): Southern East China Sea; South China Sea. (1–6)

N. brachygnathus Chu: South China Sea. (1–6)

N. macrostoma (Bleeker) [*Chaenogaleus macrostoma* (8)]: East China Sea; South China Sea. Near shore. (1–6)

N. microstoma (Bleeker) [*Hemigaleus microstoma* (8)]: Southern East China Sea; South China Sea. Near shore. (1–6)

Paragaleus acutiventralis Chu: Taiwan Strait; South China Sea. (1–6)

P. tengi (Chen): Taiwan. (8)

Physodon mulleri (Müller & Henle): South China Sea. (1–6)

Prionace glauca (Linnaeus): Southern East China Sea; Taiwan; South China Sea. (1–6)

Rhizopriondon acutus (Rüppell): Taiwan. (8)

Scoliodon dumerili (Bleeker): South China Sea. (1–6)

S. laticaudus (Müller & Henle): Taiwan. (8)

S. palasorrah (Cuvier): Southern East China Sea; South China Sea. (1–6)

S. sorrakowah (Cuvier): Yellow Sea; East China Sea; Taiwan Strait; South China Sea. (1–6)

S. walbeehmi Bleeker: East China Sea; Taiwan Strait; South China Sea. (1–6)

Triaenodon obesus (Rüppell): Taiwan Strait; South China Sea. Near shore reefs. (1–6)

SPHYRNIDAE

Sphyrna blochi (Cuvier): South China Sea. (1–6)

S. lewini (Griffith): Yellow Sea; East China Sea; Taiwan; South China Sea. Oceanic species. (1–6)

S. mokarran (Rüppell): Southern East China Sea; South China Sea. (1–6)

S. zygaena (Linnaeus): Yellow Sea; East China Sea; Taiwan. (1–6)

SQUALIFORMES
SQUALIDAE

Centrophorus acus Garman: Taiwan; South China Sea. Deep water. (1–6)

C. granulosus (Müller & Henle): South China Sea. Deep water. (9)

C. lusitanicus Bocage & Capello: Taiwan. (8)

C. moluccensis Bleeker: Taiwan. (8)

C. niaukang Teng: Taiwan. (8)

C. robustus Deng, Xiong & Zhan: East China Sea. Deep water. (9)
C. squamosus (Bonnaterre): East China Sea; South China Sea. Deep water. (7, 9)
C. squamulosus (Günther): South China Sea. Deep water. (9)
C. tasselatus Günther: South China Sea. Deep water. (9)
C. uyato (Rafinesque): Taiwan. (8)
Centroscyllium fabricii (Reinharat): South China Sea. Deep water. (9)
C. kamoharai Abe: East China Sea. Water depth 520–1,040 m. (7)
C. nigrum Garman: South China Sea. Deep water. (9)
Centroscymnus owstoni Garman: East China Sea; South China Sea. Deep water. (9)
Cirrhigaleus barbifer Tanaka: Taiwan. (8)
Dalatias licha (Bonnaterre): Donggang (Taiwan). (1–6)
Deania aciculata (Garman): Taiwan; South China Sea. Water depth 710–950 m. (1–6)
Etmopterus lucifer Jordan & Snyder: East China Sea; Taiwan; South China Sea. Water depth 270–950 m. (1–6)
E. molleri (Whitley): Taiwan. (8)
E. pusillus (Lowe): Taiwan; South China Sea. Deep water. (8, 9)
E. splendidus Yano: Taiwan. (8)
Isistius brasiliensis (Quoy & Gaimard): Taiwan. (8)
Pseudocentrophorus isodon Chu, Meng & Liu: South China Sea. Deep water. (1–6)
Scymnodon niger Chu & Meng: East China Sea; South China Sea. Water depth about 1,000 m. (7)
S. squamulosus (Günther): East China Sea; South China Sea. Water depth 700–1,000 m. (7, 9)
Squaliolus laticaudus Smith & Radcliffe: Taiwan. (8)
Squalus acanthias Linnaeus: Yellow Sea; East China Sea. (1–6)
S. blainvillei (Risso): Taiwan. (8)
S. brevirostris Tanaka [*S. megalops* (8)]: Yellow Sea; East China Sea; South China Sea. (1–6)
S. japonicus Ishikawa: Taiwan. (8)
S. mitsukurii Jordan: Yellow Sea; East China Sea; South China Sea. (1–6)

ECHINORHINIDAE

Echinorhinus cookei Pietschmann: Taiwan. (8)

SQUATINIFORMES
SQUATINIDAE

Squatina formosa Shen & Ting: Taiwan. (8)
S. japonica Bleeker: Yellow Sea; East China Sea; Taiwan. (1–6)
S. nebulosa Regan: Southern East China Sea; Taiwan Strait; South China Sea. Water depth 280–330 m. (1–6)
S. tergocellatoides Chen: Taiwan. (8)

PRISTIOPHORIFORMES
PRISTIOPHORIDAE

Pristiophorus japonicus Günther: Yellow Sea; East China Sea; South China Sea. Deep water. (1–6)

PRISTIFORMES
PRISTIDAE

Pristis cuspidatus Latham: Southern East China Sea; Taiwan Strait; South China Sea. (1–6)
P. microdon Latham: South China Sea. (1–6)

RAJIFORMES
RHINIDAE

Rhina ancylostoma Bloch & Schneider: Southern East China Sea; Taiwan Strait; South China Sea. Near shore bottom fish. (1–6)

RHYNCHOBATIDAE

Rhynchobatus djiddensis (Forskål): East China Sea; Taiwan Strait; South China Sea. (1- 6)

RHINOBATIDAE

Rhinobatos formosensis Norman: Taiwan. (8)
R. hynnicephalus Richardson: East China Sea; Taiwan; South China Sea. Near shore muddy sand substrate. (1–6)
R. schlegeli Müller & Henle: East China Sea; Taiwan Strait; South China Sea. (1–6)
Scobatus granulatus (Cuvier) [*Rhinobatos granulatus* (8)]: Southern East China Sea; Taiwan; South China Sea. (1–6)
S. halavi (Forskål): South China Sea. (1–6)

PLATYRHINIDAE

Platyrhina limboonkenkengi Tang: East China Sea; Taiwan Strait; South China Sea. (1- 6)
P. sinensis (Bloch & Schneider): Yellow Sea; East China Sea; Taiwan; South China Sea. (1–6)

RAJIDAE

Bathyraja isotrachys (Günther): East China Sea. Water depth 450–1,100 m. (7)
Breviraja tobitukai Hiyama: East China Sea; South China Sea. Water depth 410–1,000 m. (1–6)
Raja acutispina Ishiyama: Taiwan; South China Sea. Deep water. (1–6)

R. boesemani Ishihara: Taiwan. (8)

R. chinensis Basilewsky: East China Sea. (1–6)

R. gigas Ishiyama: East China Sea. Water depth 300–1,000 m. (7)

R. hollandi Jordan & Richardson: East China Sea; Taiwan Strait; South China Sea. (1–6)

R. katsukii Tanaka: Yellow Sea. (1–6)

R. kanojei Müller & Henle: Yellow Sea; East China Sea; Taiwan Strait; South China Sea. (1–6)

R. kwangtungensis Chu: Taiwan; South China Sea. (1–6)

R. macrocauda Ishiyama: East China Sea; Taiwan. Water depth about 350 m. (7, 8)

R. macrophthalma Ishiyama: East China Sea; South China Sea. Water depth 130–285 m. (7, 8)

R. meerdervcortii Bleeker: Taiwan. (8)

R. porosa Günther: Yellow Sea; East China Sea. (1–6)

R. pulchra Liu: Yellow Sea; East China Sea. (1–6)

R. tengu Jordan & Fowler: East China Sea; Taiwan Strait. Water depth about 700 m. (7)

ANACANTHOBATIDAE

Anacanthobatis borneensis Chan: Taiwan; South China Sea. Deep water. (1–6)

A. donghaiensis (Deng, Xiong & Zhan): East China Sea. Water depth 200–1,000 m. (7)

Springeria melanosoma Chan: Taiwan; South China Sea. Deep water. (1–6)

S. nanhaiensis Meng & Li: South China Sea. Deep water. (1–6)

MYLIOBATIFORMES
HEXATRYGONIDAE

Hexatrygon bickelli Heemstra & Smith: South China Sea. (1–6)

H. brevirostra Shen: Taiwan. (8)

H. longirostrum (Chu & Meng): East China Sea; South China Sea. Water depth 200–400 m. (7)

H. taiwanensis Shen: Taiwan. (8)

H. yangi Shen & Liu: Taiwan. (8)

UROLOPHIDAE

Urolophus aurantiacus Müller & Henle: East China Sea; Taiwan. (1–6)

U. marmoratus Chu, Hu & Li: South China Sea. (1–6)

Urotrygon daviesis Wallace: East China Sea. Water depth 200–400 m. (7)

U. mundus Gill: Taiwan. (8)

DASYATIDAE

Dasyatis akajei (Müller & Henle): East China Sea; South China Sea. (1–6)

D. atratus Ishiyama: Xisha (=Paracel Islands). (1–6)

D. bennetti (Müller & Henle): Southern East China Sea; Taiwan; South China Sea. (1–6)

D. gerrardi (Gray): Southern East China Sea; Taiwan Strait; South China Sea. (1–6)

D. imbricatus (Bloch & Schneider): Southern East China Sea; South China Sea. (1–6)

D. kuhli (Müller & Henle): Southern East China Sea; Taiwan Strait; South China Sea. (1–6)

D. laevigatus Chu: Bohai; Yellow Sea; East China Sea; Taiwan. (1–6)

D. lata (Garman): Taiwan. (8)

D. microphthalmus Chen: Southern East China Sea; Taiwan; South China Sea. (1–6)

D. navarrae (Steindachner): Bohai; Yellow Sea; East China Sea; Taiwan Strait, Taiwan. (1–6)

D. sinensis (Steindachner): Bohai; Yellow Sea; East China Sea. (1–6)

D. uarnak (Forskål): Southern East China Sea; Taiwan Strait; South China Sea. (1–6)

D. ushiei Jordan & Hubbs: Taiwan. (8)

D. zugei (Müller & Henle): Yellow Sea; East China Sea; Taiwan Strait; South China Sea. (1–6)

Taeniura melanospilos Bleeker: East China Sea; Taiwan Strait; South China Sea. (1–6)

Urogymnus africanus (Bloch & Schneider): Southern East China Sea; South China Sea. (1–6)

GYMNURIDAE

Aetoplatea zonura Bleeker: East China Sea; Taiwan; South China Sea. (1–6)

Gymnura bimaculata (Norman): East China Sea; Taiwan Strait; South China Sea. (1–6)

G. japonica (Temminck & Schlegel): East China Sea; Taiwan Strait; South China Sea. (1–6)

G. poecilura (Shaw): South China Sea. (1–6)

MYLIOBATIDAE

Aetomylaeus maculatus (Gray): East China Sea; Taiwan Strait; South China Sea. (1–6)

A. milvus (Müller & Henle): East China Sea; Taiwan Strait; South China Sea. (1–6)

A. nichofii (Bloch & Schneider): East China Sea; Taiwan Strait; South China Sea. (1–6)

A. vespertilio (Bleeker): East China Sea; Taiwan Strait; South China Sea. (1–6)

Myliobatis tobijei Bleeker: Yellow Sea; East China Sea; Taiwan Strait; South China Sea. (1–6)

AETOBATIDAE

Aetobatus flagellum (Bloch & Schneider): East China Sea; Taiwan Strait; South China Sea. (1–6)
A. guttatus (Shaw): East China Sea; Taiwan Strait; South China Sea. (1–6)
A. narinari (Euphrasen): Penghu area (Taiwan). (8)
A. reticulatus Teng: Taiwan. Endemic (8)

RHINOPTERIDAE

Rhinopterus hainanica Chu: South China Sea. (1–6)
R. javanica Müller & Henle: Jilong (Taiwan). (1–6)

MOBULIDAE

Manta birostris (Walbaum): Yellow Sea, East China Sea; South China Sea. (1–6)
Mobula diabolus (Shaw): East China Sea; South China Sea. (1–6)
M. formosana Teng: Taiwan. Endemic. (1–6)
M. japonica (Müller & Henle): East China Sea; Taiwan Strait; South China Sea. (1- 6)

TORPEDINIFORMES
TORPEDINIDAE

Benthobatis moresbyi Alcock: Taiwan Strait, South China Sea. Deep water. (1–6)
Narcine lingula Richardson: East China Sea; Taiwan Strait; South China Sea. (1–6)
N. maculata (Shaw): Southern East China Sea; Taiwan; South China Sea. (1–6)
N. timlei (Bloch & Schneider): Southern East China Sea; Taiwan; South China Sea. (1- 6)
Torpedo macnilli (Whitley): South China Sea. Deep water. (1–6)
T. nobiliana Bonaparte: Taiwan. (8)
T. tokionis (Tanaka): East China Sea; Taiwan Strait. Water depth 220–1,100 m. (1–6)

NARKIDAE

Crassinarke dormitor Takagi: East China Sea; Taiwan Strait; South China Sea. (1–6)
Narke japonica (Temminck & Schlegel): Bohai; Yellow Sea; East China Sea; Taiwan; South China Sea. (1–6)

CHIMAERIFORMES
CHIMAERIDAE

Chimaera jordani Tanaka: East China Sea. Water depth 700–800 m. (7)
C. phantasma Jordan & Snyder: Yellow Sea; East China Sea; Taiwan; South China Sea. Water depth 181–500 m. (1–6)
Hydrolagus isengi (Fang & Wang): Yellow Sea; East China Sea. (1–6)
Psychichthys mitsukurii (Dean): East China Sea; South China Sea. Water depth 490–979 m. (7)

RHINOCHIMAERIDAE

Harriotta opisthoptera Deng, Xiong & Zhan: East China Sea. Water depth 200–1,000 m. (7)
Rhinochimaera pacifica (Mitsukuri): East China Sea; South China Sea. Water depth 716–1,054 m. (7)

OSTEICHTHYES
ACIPENSERIFORMES
ACIPENSERIDAE

Acipenser sinensis Gray: Northern and central Yellow Sea; East China Sea; South China Sea. Migratory species. (1–6)

POLYODONTIDAE

Psephurus gladius (Martens): Yellow Sea; East China Sea. (1–6)

ELOPIFORMES
ELOPIDAE

Elops saurus Linnaeus [*E. machnata* (8)]: East China Sea; Taiwan Strait; South China Sea. Near shore species. (1–6)

MEGALOPIDAE

Megalops cyprinoides (Broussonet): East China Sea; Taiwan Strait; South China Sea. Near shore species. (1–6)

ALBULIDAE

Albula vulpes (Linnaeus) [*A. glossodonta* (8)]: East China Sea; Taiwan Strait; South China Sea. (1–6)

GONORHYNCHIFORMES
GONORHYNCHIDAE

Gonorhynchus abbreviatus Temminck & Schlegel: East China Sea; Taiwan Strait; South China Sea. Near shore bottom fish. (1–6)

CHANIDAE

Chanos chanos (Forskål): China's coast. River mouth, important brackish water aquaculture species. (1–6)

CLUPEIFORMES
CLUPEIDAE

Anodontostoma chacunda (Hamilton): East China Sea; South China Sea. (1–6)

Clupanodon punctatus (Tenminck & Schlegel) [*Konosirus punctatus* (8)]: East China Sea; Taiwan Strait; South China Sea. River mouth, estuary, and harbor. (1–6)

C. thrissa (Linnaeus): East China Sea; Taiwan Strait; South China Sea. River mouth and estuaries. (1–6)

Clupea harengus pallasi Valenciennes: Bohai; Yellow Sea. (1–6)

Dussumieria acuta Cuvier & Valenciennes: Taiwan Strait; South China Sea. (1–6)

D. hasselti Bleeker: East China Sea; Taiwan Strait; South China Sea. Near shore. (1–6)

Etrumeus micropus (Temminck & Schlegel) [*E. teres* (8)]: Southern Yellow Sea; East China Sea; Taiwan Strait; South China Sea. Often found around rocky reefs in islands. (1–6)

Herklotsichthys punctatus (Rüppell): Taiwan. (8)

H. quadrimaculatus (Rüppell): Taiwan. (8)

Ilisha elongata (Bennett): China's coast. Shallow sea, economic species. (1–6)

I. indica (Swainson): East China Sea; Taiwan Strait; South China Sea. Shallow water. (1–6)

I. melastoma (Bloch & Schneider): Taiwan. (8)

Kowala coval (Cuvier): South China Sea. Near shore. (1–6)

Macrura kelee (Cuvier) [*Marcrua ilisha*]: East China Sea; Taiwan Strait; South China Sea. Near shore, river mouth, islands. (1–6)

M. reevesi (Richardson): China's coast. Migratory species. (1–6)

Nematalosa come (Richardson): Taiwan. (8)

N. japonica Regan: Taiwan. (8)

N. nasus (Bloch): Bohai; Yellow Sea; East China Sea; South China Sea. (1–6)

Opisthopterus tardoore (Cuvier & Valenciennes): East China Sea; Taiwan Strait; South China Sea. Shallow water. (1–6)

Pellona ditchela Valenciennes: Taiwan. (8)

Sardinella aurita Valenciennes: East China Sea; Taiwan Strait; South China Sea. Near shore, important commercial species. (1–6)

S. brachysoma Bleeker: Taiwan Strait; South China Sea. Around esturies and harbors. (1–6)

S. clupeoides (Bleekers): Hainan; Xisha (=Paracel Islands). (1–6)

S. fimbriata (Cuvier & Valenciennes): Taiwan Strait; South China Sea. (1–6)

S. hualensis (Chu & Tsai): Northeastern and eastern Taiwan shoreline; eastern Guangdong coast. (8, 20)

S. jussieu (Lacépéde) [*S. gibbosa*]: Southern East China Sea; Taiwan Strait; South China Sea. Near shore or around estuaries and harbors. (1–6)

S. melanura (Cuvier): Southwestern Taiwan shoreline; Hainan. (8, 20)

S. nymphaea (Richardson): East China Sea; South China Sea. (1–6)

S. perforata (Cantor) [*S. albella*]: Taiwan Strait; South China Sea. River mouth, harbors, and estuaries. (1–6)

S. richardsoni Wongratana (Richardson) [*Harengula nymphaea, S. nymphaea*]: East China Sea; South China Sea. River mouth or around islands. (1–6)

S. sindensis (Day): East China Sea; Taiwan. (1–6)

S. sirm (Walbaum) [*Amblygaster sima* (8)]: East China Sea; Taiwan; South China Sea. (1–6)

S. zunasi (Bleeker) [*Harengula zunasi*]: China's coast. Coastal and estuarine water. (1- 6)

Sardinops sagax melanosticta (Temminck & Schlegel): South China Sea. (1–6)

ENGRAULIDAE

Coilia ectenes Jordan & Seale: Bohai; Yellow Sea; East China Sea; Taiwan Strait. Coastal species. (1–6)

C. grayi Richardson: East China Sea; Taiwan Strait; South China Sea. River mouth and coast. (1–6)

C. mystus (Linnaeus): East China Sea; South China Sea. Estuary and river mouth. (1–6)

Engraulis japonicus Temminck & Schlegel: Bohai; Yellow Sea; East China Sea; Taiwan Strait. Common fish along the coast. (1–6)

Setipinna taty (Valeinciennes) [*S. giberti*]: China's coast. Near shore small fish. (1–6)

Stolephorus chinensis (Günther) [*Encrosicholina chinensis* (8)]: East China Sea; Taiwan Strait; South China Sea. River mouth, estuary, and harbor. (1–6)

S. commersoni (Lacépéde) [*Anchoviella commersoni, Encrosicholina commersoni* (8)]: East China Sea; Taiwan Strait; South China Sea. Estuary, harbor, and river mouth. (1–6)

S. heteroloba (Rüppell) [*Anchoviella heteroloba, Encrosicholina heteroloba* (8)]: East China Sea; South China Sea. Near shore small fish. (1–6)

S. indicus (van Hasselt) [*Anchoviella indica*]: East China Sea; Taiwan Strait; South China Sea. Near shore. (1–6)

S. insularis Hardenberg: Taiwan. (8)

S. shantungensis Li [*Anchoviella shantungensis, Encrosicholina shantungensis* (8)]: Bohai. (1–6)

S. tri (Bleeker) [*Encrosicholina tri* (8)]: East China Sea; Taiwan Strait; South China Sea. (1–6)

S. zollingeri (Bleeker) [*Encrosicholina zollingeri* (8)]: East China Sea; Taiwan Strait; South China Sea. Shallow sea and river mouth. (1–6)

Thrissa dussumieri (Valenciennes) [*Thryssa dussumieri* (8)]: East China Sea; Taiwan Strait; South China Sea. Near shore small fish. (1–6)

T. hamiltonii (Gray) [*Thryssa hamiltonii* (8)]: East China Sea; South China Sea. Estuary, harbor, and river mouth. (1–6)

T. kammalensis (Bleeker): Bohai; Yellow Sea; East China Sea; South China Sea. Estuary, harbor, and river mouth. (1–6)

T. mystax (Bloch & Schneider): China's coast. Near shore small fish. (1–6)

T. setirostris (Broussonet) [*Thryssa setirostris* (8)]: East China Sea; Taiwan Strait; South China Sea. Shallow sea small fish. (1–6)

T. vitirostris (Gilchrist & Thompson): Southern East China Sea; Taiwan Strait; South China Sea. Near shore small fish. (1–6)

CHIROCENTRIDAE

Chirocentrus dorab (Forskål): Southern East China Sea; Taiwan Strait; South China Sea. (1–6)

C. nudus (Swainson): Taiwan; South China Sea. (1–6)

SALMONIFORMES
PLECOGLOSSIDAE

Plecoglossus altivelis Temminck & Schlegel: China's coast. Anadromus fish. (1–6)

SALANGIDAE

Leucosoma chinensis (Osbeck): Fujian; Guangdong. Along the coast. (1–6)

Neosalanx anderssoni (Rendahl): Bohai; Yellow Sea. Shoreline. (1–6)

N. tangkahkeii (Wu): Fujian; Guangdong. Shoreline river mouth. (1–6)

Protosalanx hyalocranius (Abbott): Bohai; Yellow Sea; East China Sea. Shoreline. (1- 6)

Salanx acuticeps Regan: China's coast. River mouth. (1–6)

S. ariakensis Kishinouye: Bohai; Yellow Sea; East China Sea. Shoreline. (1–6)

S. cuvieri Valenciennes: Yellow Sea; East China Sea; South China Sea. Shoreline. (1- 6)

ARGENTINIDAE

Argentina kagoshimae Jordan & Snyder: East China Sea; Taiwan; South China Sea. Water depth 225–385 m. (1–6)

A. semifasciata Kishinouye: East China Sea. Deep water. (1–6)

Glossanodon semifasciata (Kishinouye): Central East China Sea; Taiwan. Water depth 321–1,017 m. (7, 8)

Nansenia ardesiaca Jordan & Thompson: East China Sea. Water depth 300–1,000 m. (7)

OPISTHOPROCTIDAE

Dolichopteryx longipes (Vaillant): South China Sea. Deep water. (1–6)

Opisthoproctus soleatus Vaillant: South China Sea. Deep water. (1–6)

BATHYLAGIDAE

Bathylagus argyrogaster Norman: South China Sea. Deep water. (1–6)

B. ochotensis Schmidt: East China Sea. Water depth within 6,100 m. (7)

GONOSTOMATIDAE

Cyclothone acclinidens Garman: South China Sea. Deep water. (1–6)

C. alba Brauer: East China Sea; South China Sea. Deep water. (1–6)

C. atraria Gilbort: South China Sea. Deep water. (9, 10, 14)

C. obscura Brauer: South China Sea. Deep water. (1–6)

C. pallida Brauer: East China Sea; South China Sea. Deep water. (1–6)

C. pseudopallida Mukacheva: South China Sea. Deep water. (1–6)

Diplophos pacifica Günther: Taiwan. (8)

D. taenia Günther: South China Sea. Deep water. (8, 9, 10, 14)

Gonostoma atlanticus Norman: South China Sea. Deep water. (1–6)

G. elongatus Günther: East China Sea; Taiwan; South China Sea. Water depth 60–3,292 m. (1–6)

G. gracile Günther: East China Sea; Taiwan; South China Sea. Water depth 20–2,300 m. (7, 8, 10, 14)

Ichthyococcus ovatus (Cocco): South China Sea. Deep water. (9, 10, 14)

Maurolicus muelleri (Gmelin): East China Sea. Water depth 150–1,317 m. (7)

Polymetme elongata (Matsubara): East China Sea. Water depth 250–580 m. (7)

P. illustris (McCulloch) [*Yarrella illustris*]: Dongsha (=Pratas Islands); Xisha (=Paracel Islands). Deep water. (1–6)

Triplophos hemingi (McArdle): Taiwan. (8)

Valenciennellus tripunctulatus (Esmark): South China Sea. Deep water. (1–6)

Vinciguerria attenuata (Cocco): South China Sea. Deep water. (9, 10, 14)

V. nimbaria (Jordan & Williams): South China Sea. Deep water. (1–6)

V. poweriae (Cocco): South China Sea. Deep water. (9, 10, 14)

STERNOPTYCHIDAE

Argyropelecus aculeatus Valenciennes: East China Sea; Taiwan; South China Sea. Water depth 200–1,950 m. (1–6)

A. affinis Garman: South China Sea. Deep water. (1–6)

A. hemigymnus Cocco: Deep water. (1–6)

A. sladeni Regan: Deep water. (1–6)

Polyipnus aquavitus Baird: Nansha (=Spratly Islands). (17)

P. nuttingi Gilbert: East China Sea; South China Sea. Water depth 430–700 m. (7)

P. spinosus Günther: East China Sea; South China Sea. Water depth within 500 m. (1–6)

P. sterope Jordan & Starks: Taiwan; South China Sea. Deep water. (1–6)

P. tridentifer MucCulloch: South China Sea. Deep water. (1–6)

P. triphanos Schultz: Taiwan. (8)

P. unispinus Schultz: Taiwan. (8)

Sternoptyx diaphana Hermann: East China Sea; South China Sea. (1–6)

S. obscura Garman: South China Sea. Water depth about 1,000 m. (1–6)

S. pseudobscura Baird: East China Sea; South China Sea. Water depth within 2,000 m. (1–6)

STOMIATIDAE

Stomias affinis Günther: East China Sea; Donggang (Taiwan); South China Sea. Water depth 300–2,414 m. (1–6)

S. nebulosus Alcock: South China Sea. Deep water. (1–6)

MELANOSTOMIATIDAE

Bathophilus longipinnis (Pappenheim): South China Sea. Deep water. (1–6)

Eustomias longibarba Parr: East China Sea; South China Sea. Deep water. (1–6)

Leptostomias robustus Imai: East China Sea; Donggang (Taiwan). Water depth 692–979 m. (1–6)

Melanostomias melanopogon Regan & Trewavas: Donggang (Taiwan). (1–6)

M. valdiviae Brauer: South China Sea. Deep water. (14)

Opostomias mitsuii Imai: East China Sea. Water depth within 1,080 m. (7)

Pachystomias microdon (Günther): South China Sea. Deep water. (1–6)

Photonectes albipennis (Döderlein): East China Sea; Donggang (Taiwan). Water depth 350–1,100 m. (1–6)

Thysanactis dentex Regan & Trewavas: East China Sae; South China Sea. Water depth 100–1,080 m. (7)

IDIACANTHIDAE

Idiacanthus fasciola Peters: East China Sea; Taiwan; South China Sea. Deep water. (1- 6)

ASTRONESTHIDAE

Astronesthes chrysophekadion (Bleeker): Taiwan. (8)

A. indicus Brauer: South China Sea. Deep water. (1–6)

A. lucifer Gilbert: East China Sea; Donggang (Taiwan). Water depth 270–2,230 m. (1- 6)

Borostomias elucens (Brauer): South China Sea. Deep water. (10, 14)

CHAULIODONTIDAE

Chauliodus sloani Schneider: East China Sea; Taiwan. Water depth 780–3,886 m. (7)

MALACOSTEIDAE

Malacosteus niger Ayres: East China Sea; South China Sea. Water depth 780–3,886 m. (7)

PTEROTHRISSIDAE

Pterothrissus gissus Hilgendorf: East China Sea. Water depth 200–1,000 m. (7)

ALEPOCEPHALIDAE

Alepocephalus bicolor Alcock: Taiwan. (8)

A. longiceps Lioyd: East of Hainan. Water depth 1,100 m. (21)

A. longirostris Okamura & Kawanishi: East China Sea. Water depth 979–1,140 m. (7)

A. owstoni Tanaka: East China Sea. Water depth 500–1,037 m. (7)

A. triangularis Okamura & Kawanishi: East China Sea. Water depth 900–1,140 m. (7)

A. umbriceps Jordan & Thompson: East China Sea. Water depth 500–1,980 m. (7)

Bathyroctes rostratus Günther: South China Sea. Water depth 500–1,037 m. (9, 10, 14)

Leptoderma retropinnum Fowler: East China Sea. Water depth 500–1,786 m. (7)

Narcetes lloydi Fowler: East China Sea. Water depth 716–1,350 m. (7)

Rouleina guentheri (Alcock): East China Sea. Water depth 500–1,240 m. (7)
R. watasei (Tanaka): East China Sea. Water depth 500–1,240 m. (7)
Talismania antillarum (Goode & Bean): East China Sea. Water depth 692–1,055 m. (1- 6)
T. brachycephala Sazonov: East China Sea. Water depth 680–1,167 m. (7)
Xenodermichthys nodulosus Günther: Taiwan. (8)

MYCTOPHIFORMES
SYNODIDAE

Saurida elongata (Temminck & Schlegel): China's coast. River mouth. (1–6)
S. filamentosa Ogilby: East China Sea; Taiwan Strait; South China Sea. Near shore bottom fish. (1–6)
S. gracilis (Quoy & Gaimard): East China Sea; Taiwan Strait; South China Sea. Deep water sandy bottom. (1–6)
S. tumbil (Bloch & Schneider): East China Sea; Taiwan Strait; South China Sea. Shallow water bottom fish. (1–6)
S. undosquamis (Richardson): East China Sea; Taiwan Strait; South China Sea. Water depth 50–120 m, sandy bottom. (1–6)
S. wanieso Shindo & Yamada: East China Sea. Water depth 60–200 m. (1–6)
Synodus englemani Schultz: Xisha (=Paracel Islands). Coral sand, water depth 5–10 m. (1–6)
S. fuscus Tanaka: East China Sea. (1–6)
S. hoshinonis Tanaka: Southern East China Sea; Taiwan Strait; South China Sea. Near shore bottom fish. (1–6)
S. indicus (Day): South China Sea. (1–6)
S. jaculum Russell: Taiwan. Muddy sand or rocky reefs. (8)
S. kaianus (Günther): East China Sea; South China Sea. Water depth 210–325 m. (1–6)
S. macrops Tanaka: Southern East China Sea; Taiwan Strait; South China Sea. Near shore muddy sand substrate. (1–6)
S. rubromarmoratus Russell & Cressey: Taiwan. (8)
S. ulae Schultz: Taiwan. (8)
S. variegatus (Lacépéde): Southern East China Sea; Taiwan Strait; South China Sea. Sandy substrate. (1–6)
Trachinocephalus myops (Bloch & Schneider): East China Sea; Taiwan Strait; South China Sea. Near shore muddy sand and rocks. (1–6)

HARPODONTIDAE

Harpodon microchir Günther: Taiwan; East China Sea. (1–6)
H. nehereus (Hamilton): Yellow Sea; East China Sea; Taiwan Strait; South China Sea. River mouth. (1–6)

AULOPODIDAE

Hime japonicus (Günther) [*Aulopus japonicus* (8)]: East China Sea; South China Sea. (1- 6)

CHLOROPHTHALMIDAE

Chlorophthalmus acutifrons Hiyama: East China Sea; Taiwan; South China Sea. Water depth 260–950 m. (1–6)
C. agassizi Bonaparte: East China Sea. (1–6)
C. albatrossis Jordan & Starks: East China Sea; South China Sea. Water depth 250–620 m. (1–6)
C. bicornis Norman: South China Sea. (1–6)
C. borealis Kuronuma & Yamaguchi: Taiwan. (8)
C. nigromarginatus (Kamohara): East China Sea; Taiwan. Water depth 184–440 m. (1- 6)
C. oblongus Kamohara: South China Sea. (1–6)

NEOSCOPELIDAE

Neoscopelus macrolepidotus Johnson: East China Sea. Water depth 300–1,590 m. (7)
N. microchir Matsubara: East China Sea; Taiwan. Water depth 250–700 m. (1–6)
N. porosus Arai: South China Sea. Water depth 426–600 m. (7)
Scopelengys tristis Alcock: East China Sea; South China Sea. Water depth 400–1,829 m. (1–6)

MYCTOPHIDAE

Benthosema fibulatum (Gilbert & Cramer): East China Sea. Water depth 122–567 m. (7)
B. pterotum (Alcock) [*Myctophum pterotum*]: East China Sea; Taiwan Strait. Water depth 15–40 m. (1–6)
B. simile (Tåning): South China Sea. Deep water. (12, 14)
B. suborbitale [*B. suborbitale*] Gilbert: South China Sea. Deep water(1–6)
Bolinichthys longipes (Brauer): South China Sea. Deep water. (1–6)
B. nanshanensis Yang & Huang: South China Sea. Deep water. (23)
B. pyrsobolus (Alcock) [*B. blacki*]: South China Sea. Deep water. (12, 14)
Centrobranchus andreae (Lütken): South China Sea. Deep water. (1–6)
C. choerocephalus Fowler: South China Sea. Deep water. (1–6)
C. nigroocellatus (Günther): South China Sea. Deep water. (11, 12, 14)
Ceratoscopelus townsendi (Eigenmann & Eigenmann): East China Sea; South China Sea. Deep water. (12, 14)
C. warmingi (Lurken): East China Sea; Taiwan; South China Sea. Deep water. (1–6)
Diaphus aliciae Fowler: Taiwan; South China Sea. Deep water. (8, 12, 14)

D. brachycephalus Tåning: South China Sea. Deep water. (12, 14)
D. burtoni Fowler: South China Sea. Deep water. (1–6)
D. chrysorhynchus Gilbert & Cramer: East China Sea. Deep water. (1–6)
D. coeruleus (Klunzinger): South China Sea. Water depth 465–600 m. (22)
D. diadematus Tåning: South China Sea. Deep water. (12, 13)
D. diademophilus Nafpaktitis: South China Sea. Deep water. (1–6)
D. fragilis Tåning: Nansha (=Spratly Islands). (17)
D. fulgens (Brauer): South China Sea. Deep water. (12, 14)
D. garmani Gilbert: East China Sea; South China Sea. Deep water. (1–6)
D. holti Tåning: South China Sea. Deep water. (12, 14)
D. jenseni Tåning: East China Sea; South China Sea. Deep water. (1–6)
D. lucidus (Gilbert & Brauer): Nansha (=Spratly Islands). (17)
D. luetkeni (Brauer): East China Sea; Taiwan; South China Sea. Deep water. (1–6)
D. malayanus Weber: East China Sea, South China Sea. Deep water. (1–6)
D. megalops Nafpaktitis: South China Sea. Deep water. (12)
D. mollis Tåning: Taiwan; South China Sea. Deep water. (1–6)
D. parri Tåning: South China Sea. Deep water. (1–6)
D. perspicillatus (Ogilby): South China Sea. Deep water. (1–6)
D. problematicus Parr: South China Sea. Deep water. (1–6)
D. regani Tåning: South China Sea. Deep water. (12, 14)
D. richardsoni Tåning: South China Sea. Deep water. (1–6)
D. sagamiensis Gilbert: Taiwan. (8)
D. signatus Gilbert: Taiwan; South China Sea. Deep water. (8, 12, 14)
D. similis Wisner: South China Sea. Deep water. (12, 14)
D. splendidus Brauer: Taiwan. (8)
D. suborbitalis Weber: East China Sea; South China Sea. Deep water. (1–6)
D. taaningi Norman: Taiwan. (8)
D. tanakae Gilbert: Taiwan; South China Sea. Deep water. (12, 14)
D. termophilus Tåning: South China Sea. Deep water. (12, 14)
D. umbroculus Fowler: East China Sea; South China Sea. (12, 14)
D. watasei Jordan & Starks: East China Sea. Deep water. (7)
Diogenichthys atlanticus (Tåning): East China Sea; South China Sea. Deep water. (1–6)
D. laternatus (Garman): South China Sea. Deep water. (1–6)
D. panurgus Bolin: East China Sea; South China Sea. Deep water. (1–6)
Gonichthys coccoi (Cocco): East China Sea. Deep water. (1–6)
Hygophum atratum (Garman): South China Sea. Deep water. (1–6)
H. macrochir (Günther): South China Sea. Deep water. (12, 14)
H. proximum Becker: South China Sea. Deep water. (1–6)

Lampadena luminosa Garman: China's seas. Water depth 50–1,485 m. (1–6)
L. speculingera Goode & Bean: South China Sea. Deep water. (11, 12, 14)
Lampanyctus alatus Goode & Bean: South China Sea. Deep water. (1–6)
L. bensoni (Fowler): East China Sea. Water depth 790–1,051 m. (7)
L. hubbsi Wisner: South China Sea. Deep water. (1–6)
L. macropterus (Brauer): South China Sea. Deep water. (1–6)
L. niger Günther: South China Sea. Deep water. (1–6)
L. nobilis Tåning: Taiwan. (8)
L. omostigma Gilbert: South China Sea. Deep water. (1–6)
L. punctatissimus Gilbert: South China Sea. Deep water. (12, 14)
L. tenuiformis (Brauer): Nansha (=Spratly Islands). (17)
Myctophum affine (Lütken): East China Sea; South China Sea. Deep water. (14, 17, 18)
M. asperum Richardson: East China Sea; South China Sea. Deep water. (1–6)
M. aurolaternatum Garman: South China Sea. Deep water. (1–6)
M. brachygnathum (Bleeker): South China Sea. Deep water. (1–6)
M. lychnobium Bolin: South China Sea. Deep water. (1–6)
M. nitidulum Garman: South China Sea. Deep water. (1–6)
M. obtusirostris Tåning: East China Sea; Taiwan; South China Sea. Deep water. (1–6)
M. selenoides Wisner: South China Sea. Deep water. (11, 12)
M. spinosum (Steindachner): South China Sea. Deep water. (1–6)
Notoscopelus resplendens (Richardson): South China Sea. Deep water. (12, 14)
Symbolophorus boops (Richardson): South China Sea. Deep water. (1–6)
S. evermanni (Gilbert): South China Sea. Deep water. (1–6)
S. rufinum (Taninh): South China Sea. Deep water. (11, 12, 14)
Taanigichthys bathyphilus (Tåning): South China Sea. Deep water. (12, 14)
Triphoturus micropterus (Brauer): East China Sea; South China Sea. Deep water. (1–6)
T. nigrescens (Brauer): South China Sea. Deep water. (1–6)

SCOPELARCHIDAE

Benthalbella linguidens (Mead & Böhlke): South China Sea. Deep water. (12, 14)
Scopelarchoides danae Johnson: South China Sea. Deep water. (1–6)
S. nicholsi Parr: South China Sea. Deep water. (12, 14)
Scopelarchus analis (Brauer): South China Sea. Deep water. (1–6)
S. guentheri Alcock: South China Sea. Deep water. (1–6)

PARALEPIDIDAE

Lestidiops mirabilis (Ege): South China Sea. Deep water. (10, 14)
Lestidium atlanticum Borodin: South China Sea. Deep water. (14)
L. japonica (Tanaka) [*Lestrolepis japonica*]: East China Sea; Taiwan. Water depth 240- 732 m. (7, 8)
L. prolixum Harry: East China Sea. Water depth 234–523 m. (7)
Notolepis rissoi (Bonaparte): South China Sea. Deep water. (14)
Paralepis atlantic indica Ege: South China Sea. Deep water. (12, 14)
P. elongata (Brauer): South China Sea. Deep water. (14)

ALEPISAURIDAE

Alepisaurus ferox Lowe: East China Sea; Taiwan; South China Sea. Water depth within 1,400 m. (8)
Stemonosudis gracile (Ege): South China Sea. Deep water. (10)

IPNOPIDAE

Ipnops pristibrachium (Fowler): Nansha (=Spratly Islands). (17, 18)

OMOSUDIDAE

Omosudis lowei Günther: East China Sea; South China Sea. Water depth 732–3,396 m. (1–6)

BATHYPTEROIDAE

Bathypterois antennatus Gilbert: South China Sea. Deep water. (9)
B. atricolor Alcock: East China Sea. Water depth 260–1,000 m. (7)
B. guentheri Alcock: East China Sea; South China Sea. Water depth 542- 1,315 m. (7)

EVERMANNELLIDAE

Coccorella atrata Alcock: Nansha (=Spratly Islands). (17)
Evermannella indica Brauer: South China Sea. Deep water. (12)
Odontostomps braneri Richardson: Nansha (=Spratly Islands). (17)
O. normalops (Parr): South China Sea. Deep water. (12, 14)

CETOMIMIFORMES
MEGALOMYCTERIDAE

Vitiaziella cubiceps Rass: South China Sea. Deep water. (1–6)

RONDELETIIDAE

Rondeletia loricata Abe & Hotta: Taiwan; East China Sea; South China Sea. Water depth 140–3,444 m. (7, 8)

BARBOURISIDAE

Barbourisia rufa Parr: East China Sea. Water depth 410–1,080 m. (7)

ATELEOPIDAE

Ijimaia dofleini Sauter: East China Sea. (1–6)

ANGUILLIFORMES
SYNAPHOBRANCHIDAE

Synaphobranchus affinis Günther: East China Sea; Taiwan. Water depth 400–1,100 m. (7, 8)
S. brevidorsalis Günther: East China Sea. Water depth 530–1,958 m. (7)
S. kaupii Johnson: East China Sea. Water depth 530–1,280 m. (7)
S. pinnatus (Gray): South China Sea. Deep water. (9, 17)

ANGUILLIDAE

Anguilla bicolor pacifica Schmidt: Taiwan. (8)
A. breviceps Chu & Jin sp. nov.: Mouth of Min River (Fujian). Catadromous fish. (1–6)
A. elphinstonei Sykes: East China Sea; South China Sea. Catadromous fish. (1–6)
A. foochowensis Chu & Jin sp. nov.: Mouth of Min River (Fujian). Catadromous fish. (1–6)
A. japonica Tamminck & Schlegel: China's coast. Catadromous fish. (1–6)
A. marmorata Quoy & Gaimard: Yellow Sea; East China Sea; Taiwan; South China Sea. Catadromous fish. (1–6)
A. nigricans Chu & Wu sp. nov.: Southern Fujian. Catadromous fish. (1–6)
A. sinensis McClelland: China's coast. Catadromous fish. (1–6)

CONGRIDAE

Alloconger anagoides (Bleeker) [*A. shiroanago*]: East China Sea; Taiwan. (1–6)
A. major Asano: South China Sea. (1–6)
Anago anago (Temminck & Schlegel): East China Sea; Taiwan Strait; South China Sea. Near shore bottom fish. (1–6)
Bathymyrus simus Smith: East China Sea; Taiwan. (1–6)

Bathyuroconger vicinas Vaillant: East China Sea. Water depth 122–1,495 m. (7)
Coloconger scholesi Chan: South China Sea. (1–6)
Conger cinereus Rüppell: East China Sea; Taiwan Strait; South China Sea. (1–6)
C. japonicus Bleeker: Yellow Sea; East China Sea; Taiwan Strait. Near shore bottom fish. (1–6)
C. myriaster (Brevoort) [*Astroconger myriaster*]: Bohai; Yellow Sea; East China Sea; Taiwan. Near shore bottom fish. (1–6)
Congrina retrotincta (Jordan & Snyder): East China Sea. (1–6)
Congriscus megestomus (Günther): East China Sea. Water depth 61–830 m. (1–6)
Gorgasia japonica Abe, Mik & Asai: Taiwan. (8)
G. taiwanensis Shao: Taiwan. (1–6)
Parabathymyrus macrophthalmus Kamohara: East China Sea; Taiwan. (1–6)
Rhynchoconger brevirostris Chen & Wang: East China Sea; Taiwan. Near shore bottom fish. (1–6)
R. ectenurus (Jordan & Richardson) [*Rhynchocymba ectenura*]: East China Sea; Taiwan; South China Sea. Near shore bottom fish. (1–6)
Rhynchocymba nystromi (Jordan & Snyder): East China Sea; Taiwan Strait; South China Sea. Near shore bottom fish. (1–6)
R. sivicola (Matsubara & Ochiai): East China Sea. (1–6)
Uroconger lepturus (Richardson): East China Sea; Taiwan Strait; South China Sea. Near shore bottom fish. (1–6)

MURAENESOCIDAE

Muraenesox bagio (Hamilton): Taiwan. (8)
M. cinereus (Forskål): China's coast. Water depth 50–80 m, my sand or rocky crevices. (1–6)
M. talabon Cuvier: East China Sea; South China Sea. (1–6)
M. talabonoides (Bleeker): South China Sea. (1–6)
M. yamaguchiensis Katayama & Takai: East China Sea; South China Sea. (1–6)
Oxyconger leptognatus Bleeker: East China Sea; Taiwan Strait; South China Sea. (1–6)

NETTASTOMIDAE

Chlopsis fierasfer Jordan & Snyder [*Saurenchelys fierasfer* (8)]: Taiwan Strait, South China Sea. Near shore bottom fish. (1–6).
C. taiwanensis Chen & Wang: East China Sea; southern Taiwan. (1–6)
Nettastoma parvviceps Günther: East China Sea. Water depth 600–1,140 m. (7)

MURAENIDAE

Anarchias allardice Jordan & Starks: Taiwan. (8)
Echidna delicatula (Kaup): East China Sea; South China Sea. Coral reefs. (1–6)
E. nebulosa (Ahl): East China Sea; Taiwan Strait; South China Sea. Coral reefs. (1–6)
E. polyzona (Richardson): Taiwan; Hainan; Xisha (=Paracel Islands). Coral reefs. (1–6)
E. zebra (Shaw): Taiwan Strait; South China Sea. Coral reefs. (1–6)
Enchelycore bikiniensis (Schultz): Taiwan. (8)
E. lichenosa (Jordan & Snyder): Taiwan. (8)
E. schismatorhynchus (Bleeker): Taiwan. (8)
Gymnomuraena concolor (Rüppell): East China Sea; South China Sea. Coral reefs. (1- 6)
G. marmorata Lacépéde: East China Sea. (1–6)
G. tigrina (Lesson): South China Sea. (1–6)
G. zebra (Shaw): Southern tip of Taiwan. Rocky shore. (8)
Gymnothorax berndti Snyder: Taiwan. (8)
G. boschi Bleeker: South China Sea. (1–6)
G. brunneus Herre: Taiwan. (8)
G. buroensis (Bleeker): Taiwan. (8)
G. chilospilus Bleeker: Southern tip of Taiwan. Rocky shore shallow water. (1–6)
G. chlamydatus Snyder: Taiwan. (8)
G. eurostus (Abbott): Northern Taiwan. Rocky shore. (8)
G. favagineus (Bloch & Schneider): East China Sea; Taiwan Strait; South China Sea. Shallow water coral reefs. (1–6)
G. fimbriatus (Bennett): East China Sea; Taiwan Strait; South China Sea. Shallow water coral reefs. (1–6)
G. flavimarginatus (Rüppell): East China Sea; Taiwan Strait; South China Sea. Shallow water coral reefs. (1–6)
G. hepaticus (Rüppell): Lanyu (Taiwan). (8)
G. javanicus (Bleeker): Taiwan. (8)
G. kidako (Temminck & Schlegel): East China Sea; Taiwan. Rocky shore shallow water. (1–6)
G. leucostingmus Jordan & Richardson: East China Sea; Taiwan Strait; South China Sea. Shallow water coral reefs. (1–6)
G. margaritophorus Bleeker: Taiwan. (8)
G. melanospilus (Bleeker): East China Sea; Taiwan Strait; South China Sea. Rocky shore. (1–6)
G. melatremus Schultz: Taiwan. (8)
G. meleagris (Shaw & Nodder): East China Sea; Taiwan Strait; South China Sea. Shallow water coral reefs. (1–6)
G. monostigmus (Regan): Southern tip of Taiwan. Rocky shore. (8)
G. neglectus Tanaka: Taiwan. (8)
G. nudivomer (Playfair & Gther): Taiwan. (8)
G. pescadoris Jordan & Evermann: East China Sea. (1–6)
G. petelli (Bleeker): East China Sea; Taiwan Strait; South China Sea. Shallow water coral reefs. (1–6)
G. pictus (Ahl): East China Sea; Taiwan Strait; South China Sea. Coral reefs. (1–6).

G. pindae (Smith): Taiwan. (8)
G. polyuranodon (Bleeker): East China Sea; Taiwan. Rocky shore. (1–6)
G. pseudothyrsoideue Bleeker: East China Sea; Taiwan Strait; South China Sea. (1–6)
G. punctatofasciatus Bleeker: East China Sea; Taiwan; South China Sea. Rocky shore. (1–6)
G. reevesi (Richardson): East China Sea; Taiwan Strait; South China Sea. Shallow water coral reefs. (1–6)
G. reticularis Bloch: East China Sea; Taiwan Strait; South China Sea. Rocky shore. (1- 6)
G. richardsoni (Bleeker): East China Sea; Taiwan Strait; South China Sea. Shallow water coral reefs. (1–6)
G. rueppelliae (McClelland): Taiwan. (8)
G. thyrsoideus (Richardson): East China Sea; Taiwan Strait; South China Sea. Shallow water coral reefs. (1–6)
G. undulatus (Lacépéde): East China Sea; Taiwan Strait; South China Sea. Shallow water coral reefs. (1–6)
G. zonipectis Seale: Taiwan. (8)
Muraena pardalis Temminck & Schlegel: East China Sea; Taiwan. (1–6)
Rhinomuraena amboinensis Barbour: East China Sea. (1–6)
R. quaesita Garman: East China Sea, southern tip of Taiwan and Lu Island (Taiwan). (1- 6)
Siderea picta (Ahl): Taiwan. (8)
S. thyrsoidea (Richardson): Taiwan. (8)
Strophidon brummeri Bleeker: East China Sea. Rocky shore. (1–6)
S. sathete (Hamilton): Taiwan. (8)
S. ui Tanaka: East China Sea. (1–6)
Thyrsoidea macrurus (Bleeker): East China Sea; Taiwan Strait; South China Sea. Near shore bottom fish. (1–6)
Uropterygius macrocephalus Lesson: Taiwan. (8)
U. micropterus (Bleeker): Taiwan. (8)
U. tigrinus Lesson: Taiwan. (8)

NEENCHELYIDAE

Neenchelys parvipectoralis Chu Wu & Jin: East China Sea. (1–6)

MORINGUIDAE

Moringua abbreviata (Bleeker): East China Sea; Taiwan. (1–6)
M. macrocephalus (Bleeker): East China Sea; Taiwan; South China Sea. Sandy mud flat and river mouth. (1–6)
M. macrochir (Bleeker): East China Sea; South China Sea. Sandy mud flat. (1–6)

NEMICHTHYIDAE

Avocettina infans (Günther): East China Sea; Taiwan. Water depth 50–5,033 m. (7, 8)

Nemichthys scolopaceus Richardson: East China Sea, Donggang (Taiwan). Water depth 205–4,335 m. (1–6)

SERRIVOMERIDAE

Serrivomer beani Gill & Ryder: East China Sea. Water depth 692–2,196 m. (1–6)

DYSOMMIDAE

Dysomma anguillaris Barnard: East China Sea; Taiwan Strait; South China Sea. Near shore bottom fish. (1–6)
D. dolichosomatum Karrer: Taiwan. (8)
D. goslinei Robins & Robins: Taiwan. (8)
D. melanurum Chen & Wang: East China Sea; Taiwan Strait. (1–6)
Meadia roseni Mok, Lee & Chan: Taiwan. (8)
Simenchelys parasiticus Gill: Taiwan. (8)

ECHELIDAE

Muraenichthys gymnupterus Bleeker: East China Sea; Taiwan Strait; South China Sea. Mud flat. (1–6)
M. hattae Jordan & Snyder: East China Sea; Taiwan Strait. Rocky shore and sand flat. (1–6)
M. macropterus Bleeker: East China Sea; Taiwan Strait; South China Sea. Near shore bottom fish. (1–6)
M. malabonensis Harre: South China Sea. (1–6)
Myrophis cheni Chen & Wang: East China Sea; Taiwan Strait. (1–6)
M. macrophthalmus (Kamohara): South China Sea. (1–6)

OPHICHTHYIDAE

Bascanichthys kirkii (Günther): East China Sea; Taiwan Strait. (1–6)
Brachysomophis cirrhochilus (Bleeker): East China Sea; Taiwan Strait; South China Sea. Shallow sea bottom fish. (1–6)
B. crocodilinus (Bennett): East China Sea; Taiwan Strait; South China Sea. Near shore bottom fish. (1–6)
Caecula longipinnis Kner & Steindachner: South China Sea. (1–6)
Callechelys maculatus Chu, Wu & Jin: East China Sea; Taiwan Strait. (1–6)
Cirrhimuraena chinensis Kaup: East China Sea; Taiwan Strait; South China Sea. Sandy mud low intertidal. (1–6)
Leiuranus semicinctus (Lay & Bennett): East China Sea; Taiwan Strait; South China Sea. Rocky shore. (1–6)
Myrichthys aki Tanaka: East China Sea; Taiwan Strait; South China Sea. Coral reefs. (1–6)
M. colubrinus (Boddaert): East China Sea; Taiwan Strait; South China Sea. Coral reefs. (1–6)

M. maculosus (Cuvier): East China Sea; Taiwan Strait; South China Sea. Coral reefs. (1–6)

Mystriophis porphyreus (Temminck & Schlegel): East China Sea. (1–6)

Ophichthus altipins (Kner): Nansha (=Spratly Islands). (13, 17)

O. apicalis (Bennett): East China Sea; Taiwan Strait; Taiwan; South China Sea. Near shore sandy substrate. (1–6, 8)

O. asakusae Jordan & Snyder: East China Sea. (1–6)

O. brevicaudatus Chu, Wu & Jin: East China Sea. Near shore bottom fish. (1–6)

O. celebicus Bleeker: Taiwan Strait, South China Sea. Near shore bottom fish. (1–6)

O. cephalozona Bleeker: East China Sea; Taiwan Strait. Near shore bottom fish. (1–6)

O. erabo Jordan & Snyder [*Microdonophis erabo*]: Taiwan Strait. Near shore bottom fish. (1–6)

O. evermanni Jordan & Richardson: Taiwan Strait, South China Sea. Near shore bottom fish. (1–6)

O. fasciatus Chu, Wu & Jin [*Microdonophis fasciatus*]: East China Sea; Taiwan Strait. Near shore bottom fish. (1–6)

O. intermedius (Regan) [*Microdonophis intermedius*]: Taiwan Strait. Near shore bottom fish. (1–6)

O. macrochir (Bleeker): East China Sea. (1–6)

O. polyophthalmus (Bleeker) [*Microdonophis polyophthalmus*]: East China Sea. Near shore bottom fish. (1–6)

O. sternopterus Cope: East China Sea. (1–6)

O. tsuchidae Jordan & Snyder: Taiwan. (8)

O. urolophus (Temminck & Schlegel): East China Sea; Taiwan Strait; South China Sea. (1–6)

Ophichthys tsuchidai Jordan & Snyder: East China Sea. Water depth within 261 m. (7)

Ophisurus macrorhynchus Bleeker: East China Sea continental shelf. Water depth 240–500 m. (1–6)

Pisodonophis boro (Hamilton-Buchanan): East China Sea; Taiwan Strait; South China Sea. Near shore bottom fish. (1–6)

P. cancrivorus (Richardson): East China Sea; Taiwan Strait; South China Sea. Near shore bottom fish. (1–6)

P. rubicandus Chen: South China Sea. (1–6)

Xyrias revulsus Jordan & Snyder: East China Sea; South China Sea. (1–6)

NOTACANTHIFORMES
HALOSAURIDAE

Halosauropsis affinis (Günther): East China Sea. Water depth 383–2,617 m. (1–6)

Halosaurus ovenii Johnson: South China Sea continental slope. (9)

H. sinensis Abe: Dongsha (=Pratas Islands). (1–6)

NOTACANTHIDAE

Notacanthus abbotti Fowler: East China Sea. Water depth 250–1,000 m. (7)

SILURIFORMES
PLOTOSIDAE

Plotosus anguillaris (Bloch): East China Sea; Taiwan Strait; South China Sea. Near shore rocky bottom. (1–6)

P. lineatus (Thunberg): Taiwan. (8)

ARIIDAE

Arius leiotetocephalus Bleeker: South China Sea. Near shore bottom fish. (1–6)

A. maculatus (Thunberg): Taiwan. (8)

A. sinensis (Lacépéde): East China Sea; Taiwan Strait, South China Sea. Near shore bottom fish. (1–6)

A. thalassinus (Rüppell): East China Sea; Taiwan Strait; South China Sea. Near shore bottom fish. (1–6)

ATHERINIFORMES
ATHERINIDAE

Allanetta barnesi (Starus): Nansha (=Spratly Islands). (17)

A. bleekeri (Günther): China's coast. Inner bays. (1–6)

A. forskali (Rüppell): South China Sea. (1–6)

A. woodwardi (Jordan & Starks) [*Hypoatherina woodwardi* (8)]: Taiwan shoreline. (1–6)

Atherinomorus lacunosus (Bloch & Schneider): Taiwan. (8)

Atherion elymus Jordan & Stars: Northern and southern tips of Taiwan. Shoreline. (1–6)

Hypoatherina tsurugae (Jordan & Starks): Taiwan. (8)

H. valenciennei (Bleeker): Taiwan. (8)

ISONIDAE

Iso flosmaris Jordan & Starks: Taiwan. (8)

I. rhothophilus (Ogilby): Taiwan. (8)

BELONIFORMES
SCOMBRESOCIDAE

Cololabis saira (Brevoort): Yellow Sea. (1–6)

BELONIDAE

Ablennes anastomella (Cuvier & Valenciennes) [*Strongylura anastomella*, *Tylosurss anastomella* (8)]: China's coast. Near shore or river mouth. (1–6)

A. hians (Cuvier & Valenciennes): East China Sea; Taiwan Strait; South China Sea. Near shore. (1–6)

Belone platyura Bennett: Xisha (=Paracel Islands). (1–6)

Platybelone argalus platyura (Bennett): Taiwan. (8)

Tylosurus crocodilus (Le Sueur): East China Sea; Taiwan; South China Sea. (1–6)

T. giganteus (Temminck & Schlegel): East China Sea; Taiwan Strait; South China Sea. Near shore. (1–6)

T. leiurus (Bleeker) [*Strongylura leiurus* (8)]: East China Sea; South China Sea. Near shore. (1–6)

T. melanotus (Bleeker) [*T. acus* (8)]: East China Sea; Taiwan Strait; South China Sea. Near shore. (1–6)

T. strongylurus (Van Hasselt) [*Strongylura strongylurus* (8)]: East China Sea; South China Sea. Near shore. (1–6)

HEMIRAMPHIDAE

Euleptorhamphus viridis (van Hasselt): East China Sea; Taiwan Strait; South China Sea. (1–6)

Hemiramphus dussumieri (Cuvier & Valenciennes) [*Hyporhamphus dussumieri*]: Taiwan Strait, South China Sea. (1–6)

H. far Forskål: East China Sea; Taiwan Strait; South China Sea. (1–6)

H. georgii Cuvier & Valenciennes [*Rhynchorhamphus georgii*]: East China Sea; Taiwan Strait; South China Sea. Near shore. (1–6)

H. gernaeti (Cuvier & Valenciennes) [*Hyporhamphus gernaeti*]: East China Sea; Taiwan. Near shore. (1–6)

H. intermedius Cantor [*Hyporhamphus intermedius*]: China's coast. Near shore. (1–6)

H. limbatus (Cuvier & Valenciennes) [*Hyporhamphus limbatus*]: East China Sea; Taiwan. Near shore. (1–6)

H. lutkei Valenciennes: Northeastern Taiwan. (8)

H. marginatu (Forskål): East China Sea. Near shore. (1–6)

H. melanurus Cuvier & Valenciennes [*Hyporhamphus melanurus*]: East China Sea; South China Sea. (1–6)

H. paucirastris Collette & Parin [*Hyporhamphus paucirastris*]: East China Sea; Taiwan Strait. River mouth and estuary. (1–6)

H. quoyi Cuvier & Valenciennes [*Hyporhamphus quoyi*]: East China Sea; South China Sea. Near shore. (1–6)

H. sajori Temminck & Schlegel [*Hyporhamphus sajori*]: Bohai; Yellow Sea; East China Sea. Near shore. (1–6)

H. sinensis (Günther) [*Hyporhamphus sinensis*]: East China Sea; South China Sea. (1–6)

Hyporhamphus taiwanensis Collette & Su: Taiwan. (8)

H. yuri Collette & Parin: Taiwan. (8)

Zenarchopterus buffoni (Cuvier & Valenciennes): Taiwan; South China Sea. (1–6)

OXYPORHAMPHIDAE

Oxyporhamphus convexux (Wevier & de Beaufort): Taiwan; South China Sea. (8, 15)

O. micropterus (Cuvier & Valenciennes): East China Sea; Taiwan Strait; South China Sea. Near shore. (1–6)

EXOCOETIDAE

Cheilopogon agoo (Temminck & Schlegel): South China Sea. (14, 15)

Cypselurus angusticeps Nichols & Breder: Taiwan. (8)

C. arcticeps (Günther): Southern Taiwan Strait; South China Sea. (1–6)

C. atrisignis Jenkins: Taiwan; South China Sea; Xisha (=Paracel Islands); Zhongsha (=Macclesfield Bank). (1–6)

C. bahiensis (Renzani): East China Sea; Taiwan Strait; South China Sea. (1–6)

C. brevis Weber & Beaufort: South China Sea. (1–6)

C. cyanopterus (Cuvier & Valenciennes): East China Sea; South China Sea. (1–6)

C. katoptron (Bleeker) [*C. altipennis*]: East China Sea; South China Sea. (1–6)

C. naresi (Günther): Taiwan. (1–6)

C. nigripennes (Cuvier & Valenciennes): East China Sea. (1–6)

C. oligolepis (Bleeker): Taiwan Strait, South China Sea. (1–6)

C. opisthopus (Bleeker): South China Sea. (1–6)

C. oxycephalus (Bleeker): East China Sea; Taiwan Strait; South China Sea. Near shore. (1–6)

C. pinnatibarbatus japonicus (Franz): East China Sea; South China Sea. (1–6)

C. poecilopterus (Bleeker & Valenciennes): Taiwan; Xisha (=Paracel Islands); Zhongsha (=Macclesfield Bank). (1–6)

C. simus (Cuvier & Valenciennes): South China Sea. (1–6)

C. speculiger (Cuvier & Valenciennes) [*Hirundichthys speculiger* (8)]: South China Sea; Xisha (=Paracel Islands). Oceanic species. (1–6)

C. spilonotopteus (Bleeker): Taiwan. (8)

C. spilopterus (Cuvier & Valencienne): Taiwan; South China Sea. (1–6)

C. suttoni (Whitley & Colefax): Taiwan; South China Sea. (8)

C. unicolor (Valenciennes): Taiwan. (8)

Exocoetus monocirrhus (Richardson): East China Sea; Taiwan Strait; South China Sea. (1–6)

E. volitans Linnaeus: East China Sea; Taiwan Strait; South China Sea. Oceanic species. (1–6)

Fodiator acutus pacificus Brunn: South China Sea. (1–6)

Hirundichthys oxycephalus (Bleeker): Taiwan. (8)

H. rondeletii (Valenciennes): Taiwan. (8)

Parexocoetus brachypterus (Richardson): East China Sea; Taiwan Strait; South China Sea. Near shore. (1–6)

P. mento (Cuvier & Valenciennes): East China Sea; South China Sea. (1–6)

Prognichthys agoo (Temminck & Schlegel) [*Cypselurus agoo* (8)]: Yellow Sea; Taiwan; South China Sea. (1–6)

P. albimaculatus (Fowler): South China Sea. (1–6)

P. brevipinnis (Cuvier & Valenciennes) [*Cypselurus brevipinnis* (8)]: South China Sea. (1–6)

P. rondeleti (Cuvier & Valenciennes) [*Hirundichthys rondeleti* (8)]: East China Sea; Taiwan; South China Sea. (1–6)

P. sealei Abe: South China Sea. (1–6)

GADIFORMES
MORIDAE

Lotella phycis (Temminck & Schlegel): South China Sea. Deep water. (1–6)

Physiculus inbarbatus Kamohara: East China Sea; Taiwan. Deep water. (7, 8)

P. japonicus Hilgendorf: East China Sea; Taiwan; South China Sea. Deep water. (1–6)

P. jordani Böhlke & Mead: South China Sea. Deep water. (1–6)

P. maximowiczi (Herzenstein): East China Sea; Taiwan; South China Sea. Deep water. (1–6)

P. nigrescens Smith & Radcliffe: East China Sea; South China Sea. Deep water. (1–6)

P. roseus Alcock: Taiwan. (8)

GADIDAE

Ciliata pacifica (Temminck & Schlegel): Yellow Sea; East China Sea. (1–6)

Gadus macrocephalus Tilesius: Yellow Sea. (1–6)

Theregra chalogramma (Pallas): Eastern Yellow Sea. (1–6)

BREGMACEROTIDAE

Bregmaceros arabicus D'Ancona & Cavinato: East China Sea. (1–6)

B. atlanticus Goode & Bean: South China Sea. (1–6)

B. bathymaster Jordan & Bollman: South China Sea. (13)

B. japonicus Tanaka: East China Sea; Taiwan. Water depth 1,014–1,037 m. (7, 8)

B. lanceolatus Shen: East China Sea; Taiwan Strait. Near shore. (1–6)

B. macclellandi Thompson [*B. atripinnis*]: East China Sea; Taiwan; South China Sea. Generally live in water depth 20–50 m. (8)

B. nectabanus Whitley: South China Sea. (1–6)

B. pescadorus Shen: Taiwan. (8)

B. rarisquamosus Munro: South China Sea. (1–6)

MACROURIDAE [CORYPHAENOIDIDAE]

Bathygadus antrodes (Jordan & Gilbert): East China Sea. Water depth 950–979 m. (7)

B. garretti Gilbert & Hubbs: East China Sea. Water depth 692 m. (7)

Caelorhynchus anatirostris Jordan & Gilbert: East China Sea; Taiwan. Water depth 520–655 m. (7, 8)

C. asteroides Okamura: East China Sea. Water depth 300–400 m. (7)

C. cingulatus Gilbert & Hubbs: Taiwan; South China Sea. (1–6)

C. commutabilis Smith & Radcliffe: South China Sea. (1–6)

C. formosanus Okamura: East China Sea; Taiwan. Near shore bottom fish. (1–6)

C. intermedius Chu & Lo: East China Sea. Water depth 213–285 m. (7)

C. japonicus (Temminck & Schlegel): East China Sea; Taiwan. Water depth 490–716 m. (7, 8)

C. jordani Smith & Pope: East China Sea. Water depth 213–420 m. (7)

C. kamoharai Matsubara: East China Sea; Taiwan; South China Sea. Water depth 213- 420 m. (1–6)

C. kishinouyei Jordan & Snyder: Taiwan; South China Sea. (1–6)

C. longissimus Matsubara: East China Sea. Water depth 250–450 m. (7)

C. multispinulosus Katayama [*C. commutabilis*]: East China Sea; Taiwan; South China Sea. Near shore bottom fish. (1–6, 8)

C. parallelus (Günther): East China Sea; Taiwan. Water depth 790–979 m. (7, 8)

C. smithi Gilbert & Hubbs: East China Sea. 423–780 m. (7)

C. tokiensis (Steindachner & Döderlein): East China Sea; Taiwan. Water depth 490–550 m. (1–6)

Caelorinchs dorsalis Gilbert & Hubbs: Taiwan. (8)

C. gilberti Jordan & Hubbs: Taiwan. (8)

C. hubbsi Matsubara: Taiwan. (8)

C. matsubarai Okamura: Taiwan. (8)

Cetonurus crassiceps (Günther): South China Sea. Deep water. (9)

Coryphaenoides marginatus Steindachner & Döderlein: East China Sea; Taiwan. Water depth 520–919 m. (7, 8)

C. nasutus Günther: East China Sea. Water depth 940–980 m. (7)

Gadomus colletti Jordan & Gilbert: East China Sea. Water depth 950 m. (7)

G. melanoptorus Gilbert: South China Sea. Deep water. (13)

G. multifilis (Günther): South China Sea. (1–6)

Hymenocephalus gracilis Gilberts & Hubbs: South China Sea. (1–6)

H. lethonemus Jordan & Gilbert: East China Sea; South China Sea. Deep water. (1–6)

H. longiceps Smith & Radcliffe: Taiwan; South China Sea. Deep water. (1–6)

H. striatissimus Jordan & Gilbert: East China Sea; Taiwan; South China Sea. Water depth 420–587 m. (1–6)

Hymenogadus kuronumai (Kamohara): East China Sea. Deep water. (7)

Malacocephalus laevis (Lowe): Taiwan; East China Sea. Deep water. (7, 8)

Nezumia proximus (Smith & Radcliffe): East China Sea. Water depth 979 m. (7)

Trachonurus villosus (Günther): East China Sea. Water depth 850–1,000 m. (7)

Ventrifossa divergens Gilbert & Hubbs: South China Sea. Deep water. (1–6)

V. garmani (Jordan & Gilbert): East China Sea; Taiwan; South China Sea. Water depth 420–780 m. (1–6)

V. nigrodorsalis Gilbert & Hubbs: Taiwan; South China Sea. Deep water. (1–6)

V. rhipidodorsalis Okamura: East China Sea. Deep water. (1–6)

MACROUROIDIDAE

Squalogadus modificatus Gilbert & Hubbs: East China Sea. Water depth 980 m. (7)

CARAPIDAE [FIERASFERIDAE]

Carapus homei (Richardson): Taiwan; South China Sea, Nansha (=Spratly Islands). Often live inside respiratory cavity of sea cucumbers, ascidians, and cockles (Cardiidae). (1–6)
C. kagoshimanus (Steindachner & Döderlein): South China Sea; Xisha (=Paracel Islands). Inside sea cucumbers in coral reefs. (1–6)
C. lumbricoides (Bleeker): South China Sea, Nansha (=Spratly Islands). (1–6)
C. parvipinnis (Kaup): Taiwan; South China Sea, Xisha (=Paracel Islands). Live inside sea cucumbers. (1–6)
Echiodon owasianus (Matsubara): Taiwan. (8)
Encheliophis gracilis (Bleeker): Taiwan. (8)
E. sagamianus (Tanaka): Taiwan. (8)
Onuxodon margaritiferae (Rendahl): Taiwan. (8)
O. parvibrachium (Fowler): Taiwan. (8)
Pyramodon ventralis Smith & Radcliffe: Taiwan. (8)

OPHIDIIDAE [BROTULIDAE]

Bassobythites macropterus (Smith & Radcliffe): East China Sea. Water depth 979–1,006 m. (7)
Bassozetus robustus Smith & Radcliffe: South China Sea. (7)
Brotula multibarbata Temminck & Schlegel: East China Sea; Taiwan; South China Sea. Near shore muddy sand substrate. (1–6)
Dicrolene quinquarius (Günther): East China Sea. Water depth 900–1,140 m. (1–6)
D. tristis Smith & Radcliffe: East China Sea; South China Sea. Water depth 560–1,100 m. (1–6)
Homostolus japonicus Matsubara: South China Sea. (1–6)
Hoplobrotula armata (Temminck & Schlegel): Yellow Sea; East China Sea; Taiwan; South China Sea. Near shore bottom fish. (1–6)
Luciobrotula bartschi Smith & Radcliffe: South China Sea. (1–6)
Monomitopus kumae Jordan & Hubbs: East China Sea. Water depth 250–990 m. (1–6)
M. longiceps Smith & Radcliffe: South China Sea. (1–6)
Neobythites fasciatus Smith & Radcliffe: South China Sea. (1–6)
N. nigromaculatus Kamohara: Taiwan; South China Sea. (1–6)
N. sivicola (Jordan & Snyder): East China Sea; Taiwan. (1–6)
N. stigmosus Machida: Taiwan. (8)
Ophidion asiro (Jordan & Fowler): Taiwan; South China Sea. (1–6)
O. muraenolenis Günther: South China Sea. (1–6)
Pycnocraspedum microlepis (Matsubara): South China Sea. (1–6)
Sirembo imberbis (Temminck & Schelegel): East China Sea; Taiwan; South China Sea. Near shore bottom fish. (1–6)
S. marmoratum (Goode & Bean): South China Sea. Deep water. (1–6)

BYTHITIDAE

Brotulina fusca Fowler: Taiwan. (1–6)
Dinematichthys dasyrhynchus Cohen & Hutchins: Taiwan. (8)
D. iluocoeteoides Bleeker: South China Sea; Xisha (=Paracel Islands). Shallow water coral reefs. (1–6)
D. minyomma Sedor & Cohen: Taiwan. (8)
Diplacanthopoma brunnea Smith & Radcliffe: South China Sea. (1–6)
Oligopus robustus (Smith & Radcliffe): South China Sea. (1–6)
Saccogaster tubercularis Chan: South China Sea. (1–6)

APHYONIDAE

Aphyonus bolinu Nielsen: South China Sea. (1–6)
Barathronus diaphanus Brauer: South China Sea. (1–6)

BERYCIFORMES
MELAMPHAIDAE

Melamphaes leprus Ebeling: South China Sea. Deep water. (13, 14)
M. simus Ebeling: South China Sea. Deep water. (14)
Poromitra crassicepsi (Günther): East China Sea. Water depth 979 m. (7)
P. megalops Lütken: South China Sea. Deep water. (14)
P. oscitons Ebeling: South China Sea. Deep water. (1–6)
Scopeloberyx opisthopterus (Parr): South China Sea. Deep water. (13)
S. robustus (Günther): South China Sea. Deep water. (1–6)
Scopelogadus mizolepis (Günther): South China Sea. Deep water. (1–6)

POLYMIXIIDAE

Polymixia berndti Gilbert: Taiwan; South China Sea. Water depth 375–2,000 m. (1–6)
P. japonicus Günther: East China Sea; Taiwan; South China Sea. Deep water. (1–6)
P. longispina Deng, Xiong & Zhan: East China Sea. Deep water. (7)

DIRETMIDAE

Diretmus argenieus Johnson: East China Sea; South China Sea. Water depth 530–1,030 m. (1–6)

HISPIDOBERYCIDAE

Hispidoberyx ambagiosus Kotlyar: South China Sea. Water depth 1,019 m. (25)

BERYCIDAE

Beryx splendens Lowe: East China Sea; South China Sea. Deep water. (1–6)
Centroberyx lineatus (Cuvier & Valenciennes) [*Trachichthodes lineatus*]: South China Sea. Deep water. (1–6)
C. rubricaudus Liu & Shen: Taiwan. (8)

TRACHICHTHYIDAE

Gephyroberyx japonicus (Döderlein): South China Sea. Deep water. (1–6)
Hoplostethus crassispinus Kotlyar: Taiwan. (8)
H. mediterraneus Cuvier & Valenciennes: East China Sea; Taiwan Strait; South China Sea. Deep water. (1–6)
Paratrachichthys prosthemius Jordan & Fowler: Taiwan; South China Sea. Deep water. (1–6)

ANOPLONGASTERIDAE

Anoplongaster cornuta (Valenciennes): East China Sea. Water depth 1,014–1,037 m. (7)

HOLOCENTRIDAE

Ostichthys japonicus (Cuvier & Valenciennes): East China Sea; South China Sea. Water depth 180–340 m. (1–6)
O. kaianus (Günther): Taiwan. (8)
O. sheni Chen, Shao & Mok: Taiwan. (8)

ANOMALOPIDAE

Adioryx caudimaculatus (Cuvier & Valenciennes) [*Holocentrus caudimaculatus, Sargocentron caudimaculatus* (8)]: Taiwan; South China Sea; Xisha (=Paracel Islands); Nansha (=Spratly Islands). Coral reefs, water depth 10–20 m. (1–6)
A. cornutus (Bleeker) [*Holocentrus cornutus, Sargocentron cornutus* (8)]: Taiwan; South China Sea; Xisha (=Paracel Islands). Coral reefs. (1–6)
A. diadema (Lacépéde) [*Sargocentron diadema* (8)]: South China Sea. (1–6)
A. furcatus (Günther) [*Holocentrus furcatus*]: South China Sea; Xisha (=Paracel Islands). Coral reefs. (1–6)
A. lacteoguttatus (Cuvier & Valenciennes) [*Holocentrus lacteoguttatus, Sargocentron lacteoguttatus* (8)]: Taiwan; South China Sea; Xisha (=Paracel Islands). Coral reefs. (1–6)
A. microstomus (Günther) [*Holocentrus microstomus*]: Taiwan; South China Sea. (1–6)
A. spinifer (Forskål) [*Holocentrus spinifer*]: Southern Taiwan; Xisha (=Paracel Islands); Zhongsha (=Macclesfield Bank); Nansha (=Spratly Islands). Coral reefs. (1–6)
A. spinosissimus (Temminck & Schlegel) [*Sargocentron spinosissimus* (8)]: Southwestern Taiwan. Rocky shore shallow water. (1–6)
A. tiere (Cuvier & Valenciennes) [*Holocentrus tiere, Sargocentron tiere* (8)]: Taiwan; South China Sea; Xisha (=Paracel Islands). Coral reefs. (1–6)
A. violaceus (Bleeker) [*Holocentrus violaceus*]: South China Sea; Xisha (=Paracel Islands). Coral reefs. (1–6)
Anomalops katoptron (Bleeker): Taiwan. (8)
Dispinus ruber (Forskål) [*Holocentrus ruber*]: Taiwan; South China Sea. Coral reef near bottom fish. (1–6)
Flammeo argenteus (Cuvier & Valenciennes) [*Holocentrus laevis*]: Xisha (=Paracel Islands). Coral reefs. (1–6)
F. opercularis (Cuvier & Valenciennes) [*Holocentrus opercularis, Neoniphon opercularis* (8)]: Taiwan; South China Sea; Xisha (=Paracel Islands). Coral reefs. (1–6)
F. sammara (Forskål) [*Holocentrus sammara, Neoniphon sammara* (8)]: Taiwan; Xisha (=Paracel Islands). Coral reefs. (1–6)
F. scythrops (Lacépéde & Ebeling): Nansha (=Spratly Islands). (17)
Holocentrus bleeker Weber: Xisha (=Paracel Islands). Coral reefs. (1–6)
Myripristis adustus Bleeker: East China Sea; Taiwan; South China Sea. (1–6)
M. berndti Jordan & Evermann: Taiwan. (8)
M. chryseres Jordan & Evenmann: Taiwan. (8)
M. hexagonus (Lacépéde): South China Sea. (1–6)
M. kuntee Valenciennes: Taiwan. (8)
M. melanosticyus Bleeker: Taiwan, Xisha (=Paracel Islands); Nansha (=Spratly Islands). Coral reefs. (1–6)
M. murdjan (Forskål): Taiwan; South China Sea; Xisha (=Paracel Islands). Coral reefs. (1–6)
M. parvidens Cuvier & Valenciennes: South China Sea. (1–6)
M. pralinius Cuvier & Valenciennes: Taiwan; South China Sea. Coral reefs. (1–6)
M. seychellensis Cuvier: Taiwan. (8)
M. shultzei Seale: South China Sea, Xisha (=Paracel Islands). (1–6)
M. violaceus Bleeker: Penghu Islands (Taiwan); South China Sea; Xisha (=Paracel Islands); Nansha (=Spratly Islands). Coral reefs. (1–6)
M. vittata Cuvier: Taiwan. (8)
Plectrypops lima (Valenciennes): Taiwan. (8)
Sargocentron ittodai (Jordan & Fowler): Taiwan. (8)
S. melanospilos (Bleeker): Taiwan. (8)
S. praslin (Lacépéde): Taiwan. (8)
S. rubrum (Forskål): Taiwan. (8)
S. spiniferum (Forskål): Taiwan. (8)
S. tiere (Cuvier & Valenciennes): Taiwan. (8)

MONOCENTRIDAE

Monocentrus japonicus (Houttuyn): Yellow Sea; East China Sea; Taiwan; South China Sea. Near shore benthic species. (1–6)

ZEIFORMES
GRAMMICOLEPIDAE

Xenolepidichthys dalgleishi Gilchrist: Taiwan. (8)

ZEIDAE

Cyttomimus affinis Weber: South China Sea. (1–6)
Cyttopsis roseus (Lowe): East China Sea; Taiwan. Water depth 140–553 m. (7)
Parazen pacificus Kamohara: East China Sea; Taiwan; South China Sea. Water depth 140–360 m. (1–6)
Zen cypha (Fowler): South China Sea. (1–6)
Zenion hololepis (Goode & Bean): South China Sea. Deep water. (1–6)
Z. japonicus Kamohara: East China Sea; South China Sea. Water depth 200–620 m. (7)
Zenopsis nebulosa (Temminck & Schlegel): East China Sea; Taiwan. Water depth 210- 270 m. (1–6)
Zeus japonicus Cuvier & Valenciennes: East China Sea; Taiwan; South China Sea. Near shore bottom fish. (1–6)

ANTIGONIDAE [CAPROIDAE]

Antigonia capros Lowe: East China Sea; Taiwan; South China Sea. Water depth 50–750 m. (1–6)
A. rubescens (Günther): East China Sea; Taiwan; South China Sea. Deep water. (1–6)
A. rubicunda Ogilby: East China Sea. Water depth 50–345 m. (1–6)

LAMPRIDIFORMES
VELIFERIDAE

Velifer hypselopterus Bleeker: Taiwan Strait; South China Sea. Generally lives in deep water. (1–6)

REGALECIDAE

Regalecus russelli (Show): South China Sea. Rare species. (1–6)

TRACHIPTERIDAE

Desmodema polystictum Ogilby: Taiwan. (8)
Trachypterus iris (Walbaum): Taiwan Strait. Deep water. (1–6)

T. ishikawae Jordan & Snyder: Taiwan. (8)
Zu cristatus (Bonelli): South China Sea. Deep water. (1–6)

LAMPRIDAE

Lampris guttatus (Brünnich): Taiwan. (8)

ATELEOPODIDAE

Ateleopus japonicus Bleeker: Taiwan. (8)
A. purpureus Tanaka: East China Sea; South China Sea. Water depth 135–587 m. (1–6)

GASTEROSTEIFORMES
FISTULARIIDAE

Fistularia commersonii ppell: Taiwan. (8)
F. petimba Lacépéde: Yellow Sea; East China Sea; Taiwan; South China Sea. Near shore bottom fish. (1–6)
F. villosa Klunzinger: Southern East China Sea; Taiwan Strait; South China Sea. Near shore bottom fish. (1–6)

AULOSTOMIDAE

Aulostomus chinensis (Linnaeus): East China Sea; Taiwan Strait; South China Sea. (1–6)

MACRORHAMPHOSIDAE

Macrorhamphosus gracilis (Lowe): East China Sea. Water depth 182–279 m. (7)
M. japonicus (Günther): East China Sea. (1–6)
M. scolopax (Linnaeus): East China Sea; Taiwan. Water depth 123–420 m. (7)

CENTRISCIDAE

Aeoliscus strigatus (Günther): Southern tip of Taiwan. Rocky shore. (1–6)
Centriscus scutus (Linnaeus): East China Sea; Taiwan Strait; South China Sea. Rocky shore. (1–6)

SOLENOSTOMIDAE

Solenostomus armatus Weber: East China Sea; South China Sea. (1–6)
S. cyanopterus Bleeker: China's coast. (1–6)
S. paegnius Jordan & Thompson: Taiwan. (8)

S. paradoxus Bleeker: Southern and northern Taiwan. Rocky shore. (1–6)

SYNGNATHIDAE

Choeroichthys sculptus (Günther): Taiwan. (8)
Corythoichthys crenulatus (Weber): East China Sea. (1–6)
C. fasciatus (Gray): East China Sea; Taiwan Strait; South China Sea. (1–6)
C. flavofasciatus (Rüppell): Taiwan. (8)
Doryrhamphus dactyliophorus (Bleeker): Taiwan. Rocky reefs. (8)
D. excisus Kaup: Taiwan. (8)
D. japonicus Araga & Yoshino: Taiwan. (8)
D. melanopleura (Bleeker): Taiwan; Xisha (=Paracel Islands). Rocky reefs. (1–6)
Dunckerocampus dactyliophorus (Bleeker): Southern tip of Taiwan. Rocky shore. (8)
Halicampus koilomatodon Bleeker [*H. grayi* (8)]: South China Sea. (1–6)
H. macrorhynchus Bambe: Taiwan. (8)
H. mataafae (Jordan & Seale): Taiwan. (8)
Hippichthys heptagonus Bleeker: Taiwan. (8)
Hippocampus coronatus Temminck & Schelegel: Bohai; Yellow Sea. (1–6)
H. histris Kaup [*H. erinaceus* (8)]: East China Sea; Taiwan Strait; South China Sea. Inner bay, seaweed. (1–6)
H. japonicus Kaup: China's coast. Estuary, seaweed. (1–6)
H. kelloggi Jordan & Snyder: Bohai, East China Sea; South China Sea. Often on seaweed. (1–6)
H. kuda Bleeker: Bohai; East China Sea; Taiwan; South China Sea. Seaweed. (1–6)
H. trimaculatus Leach: East China Sea; Taiwan Strait; South China Sea. (1–6)
Micrognatus brevirostris (Rüppell): Southern tip of Taiwan, rocky shore; Nansha (=Spratly Islands). (8, 17)
M. metaatae (Jordan & Snyder) [*Halicampus metaatae* (8)]: Lanyu (Taiwan). Rocky shore. (8)
Microphis boaja (Bleeker) [*Doryrhamphus boaja* (8)]: East China Sea. (1–6)
M. brachyurus (Bleeker): Taiwan. (8)
M. leiaspis (Bleeker): Taiwan. (8)
M. manadensis (Bleeker): Taiwan. (8)
Phoxocampus belcheri (Kaup): Northeastern Taiwan. Rocky shore. (8)
Solegnathus hardwicki (Gray): East China Sea; Taiwan Strait; South China Sea. Deep water. (1–6)
Syngnathoides biaculeatus (Bloch): East China Sea; Taiwan; South China Sea. (1–6)
Syngnathus acus Linnaeus: China's coast. Seaweed. (1–6)
S. argyrostictus Kaup: East China Sea; South China Sea. (1–6)
S. cyanospilus Bleeker [*Hippichthys cyanospilus* (8)]: East China Sea; South China Sea. (1–6)
S. djarong Bleeker: East China Sea; South China Sea. Estuary and river mouth. (1–6)
S. pelagicus Linnaeus: East China Sea. (1–6)
S. spicifer Rüppell [*Hippichthys spicifer* (8)]: South China Sea. (1–6)
Tchthyocampus belcheri Kaup: South China Sea. (1–6)
Trachyrhamphus serratus Temminck & Schlegel: East China Sea; Taiwan Strait; South China Sea. Seaweed. (1–6)
Urocampus nanus Günther: China's coast. (1–6)
Yozia bicoarctatus Bleeker: South China Sea. (1–6)

MUGILIFORMES
SPHYRAENIDAE

Sphyraena acutipinnis Day: Taiwan. (8)
S. barracuda (Walbaum): Taiwan, Guangdong, Xisha (=Paracel Islands). (1–6)
S. flavicauda Rüppell: Northeastern Taiwan. (8)
S. forsteri Cuvier & Valenciennes: Taiwan Strait; Xisha (=Paracel Islands); Nansha (=Spratly Islands). Coral reefs. (1–6)
S. helleri Jenkins: South China Sea; Xisha (=Paracel Islands). (1–6)
S. japonica Cuvier & Valenciennes: East China Sea; Taiwan Strait; South China Sea. Near shore. (1–6)
S. jello Cuvier & Valenciennes: East China Sea; Taiwan Strait; South China Sea. Near shore. (1–6)
S. nigripinnis Temminck & Schlegel: Taiwan. (8)
S. obtusata Cuvier & Valenciennes: Xisha (=Paracel Islands). (1–6)
S. pinguis Günther: China's coast. Near shore. (1–6)
S. putnamiae Jordan & Seale: Taiwan. (8)

MUGILIDAE

Crenimugil crenilabis (Forskål): Taiwan, Xisha (=Paracel Islands). (1–6)
Ellochelon vaigiensis (Quoy & Gaimard) [*Mugil vaigiensis*]: Hainan; Xisha (=Paracel Islands); Zhongsha (=Macclesfield Bank). (1–6)
Liza alata (Steinderchner): Taiwan. (8)
L. carinatus (Cuvier & Valenciennes) [*Mugil carinatus*]: East China Sea; South China Sea. Shallow sea or river mouth. (1–6)
L. dussumieri (Cuvier & Valenciennes) [*Mugil dussumier*]: East China Sea; Taiwan; South China Sea. Shallow sea or river mouth brackish water. (1–6)
L. haematocheila (Temminck & Schlegel) [*L. soiuy*, *Mugil soiuy*]: China's coast. (1–6)

L. macrolepis (Smith) [*Mugil macrolepis*]: Taiwan; South China Sea. (1–6)

L. melinopterus (Cuvier & Valenciennes) [*Mugil melinopterus*]: Taiwan Strait. (1–6)

L. subviridis (Valenciennes): Taiwan. (8)

L. tade (Forskål): East China Sea; South China Sea. (1–6)

Mugil cephalus Linnaeus: China's coast. Shallow sea or river mouth. (1–6)

Osteomugil ophuyseni (Bleeker) [*Mugil affinis, M. kelaertii*]: East China Sea; South China Sea. Shallow brackish water. (1–6)

O. strongylocephalus (Richardson) [*Mugil engeli*]: East China Sea; South China Sea. Shallow sea or river mouth. (1–6)

Plicomugil labiosus (Cuvier & Valenciennes) [*Oedalechilus labiosus*]: Taiwan; Xisha (=Paracel Islands). (1–6)

Valamugil buchanani (Bleeker): South China Sea. (1–6)

V. cunnesius (Valenciennes): Taiwan. (8)

V. formosae (Oshima): Taiwan. (8)

V. seheli (Forskål) [*Mugil seheli*]: East China Sea; South China Sea. (1–6)

POLYNEMIDAE

Eleutheronema tetradactylus (Shaw): China's coast. (1–6)

Polynemus indicus (Shaw): Northeastern and southwestern Taiwan. (8)

P. microstoma (Bleeker): Southwestern Taiwan and Nanyu (Taiwan) coast. (8)

P. plebeius Broussonet: East China Sea; South China Sea. (1–6)

P. sextarius Bloch & Schneider [*Polydactyeus sextarius*]: East China Sea; South China Sea. (1–6)

PERCIFORMES
AMBASSIDAE

Ambassis gymnocephalus Lacépéde: East China Sea; Taiwan; South China Sea. (1–6)

A. kopsi Bleeker: South China Sea. (1–6)

A. miops Günther: South China Sea. (1–6)

A. urotaenia Bleeker: Taiwan; South China Sea. (1–6)

LATIDAE

Lates calcarifer (Bloch): Taiwan; South China Sea. (1–6)

Psammoperca waigiensis (Cuvier & Valenciennes): Taiwan; South China Sea. (1–6)

SERRANIDAE

Aethaloperca rogaa (Forskål): Taiwan; South China Sea. (1–6)

Anthias dispar (Herre): Lu Island (Taiwan). Rocky reefs. (8)

A. fasciatus (Kamohara): Taiwan. (8)

A. luzonensis Katayama & Masuda: Taiwan. (8)

A. pascalus (Jordan & Tankaa): Lu Island (Taiwan). Rocky reefs. (8)

A. pleurotaenia Bleeker: Lu Island and Penghu Islands (Taiwan). Rocky reefs. (8)

A. squamipinnis (Peters): East China Sea; Taiwan; South China Sea. Deep water. (1–6)

A. truncatus Katayama & Masuda: Taiwan. (8)

Anyperodon leucogrammicus (Cuvier & Valenciennes): Taiwan; Hainan; Dongsha (=Pratas Islands); Xisha (=Paracel Islands); Nansha (=Spratly Islands). Coral reefs. (1–6)

Aulacocephalus temmincki Bleeker: Northern and northeastern Taiwan. Rocky shore. (1–6)

Callanthias japonicus Franz: East China Sea; Taiwan. Rocky reefs. (7, 8)

Caprodon schlegeli (Günther): East China Sea; Taiwan; South China Sea. Water depth 70–200 m. (1–6)

Cephalopholis analis (Valenciennes): Taiwan. (8)

C. argus (Bloch & Schneider): Penghu Islands (Taiwan); Hainan; Dongsha (=Pratas Islands); Xisha (=Paracel Islands); Nansha (=Spratly Islands). Coral reef common species. (1–6)

C. aurantius (Cuvier & Valenciennes): South China Sea. (1–6)

C. boenack (Bloch): Taiwan; South China Sea. Rocky reefs. (1–6)

C. formosa (Shaw & Nodder): Taiwan. (8)

C. igarashiensis Katayama: Taiwan. (8)

C. leopardus (Bloch & Schneider): South China Sea. (1–6)

C. miniatus (Forskål): Taiwan; Hainan; Dongsha (=Pratas Islands); Xisha (=Paracel Islands); Nansha (=Spratly Islands). Coral reefs and rocky reefs. (1–6)

C. pachycentron (Cuvier & Valenciennes) [*C. albomarginatus*]: East China Sea; Taiwan Strait; South China Sea. Coral reef bottom fish. (1–6)

C. sexmaculabus (Rüppell): Taiwan; Nansha (=Spratly Islands). (8, 17)

C. sonnerati (Cuvier & Valenciennes): Penghu Islands (Taiwan); Hainan; Dongsha (=Pratas Islands); Xisha (=Paracel Islands); Nansha (=Spratly Islands). Coral reefs and shoreline. (1–6)

C. spiloparaeus (Valenciennes): Taiwan. (8)

C. urodelus (Cuvier & Valenciennes): South China Sea; Xisha (=Paracel Islands); Nansha (=Spratly Islands). Coral reef common species. (1–6)

C. urodeta (Forster): Taiwan. (8)

Chelidoperca hirundinacea Cuvier & Valenciennes: East China Sea; Taiwan; South China Sea. Rocky reefs. (1–6)

C. margaritifera Weber: South China Sea. Water depth 100 m. (1–6)

C. pleurospilus (Günther): East China Sea; Taiwan. Water depth 80–200 m. (1–6)

Chorististium japonicus (Döderlein) [*Liopropoma japonicus* (8)]: Taiwan Strait; South China Sea. Rocky reefs. (1–6)

C. latifasciata (Tanaka) [*Liopropoma latifasciata* (8)]: Taiwan Strait, South China Sea. Rocky reefs. (1–6)

C. lunulatum (Guichenot): Northeastern Taiwan. Rocky reefs. (8)

Cromileptes altivelis (Cuvier & Valenciennes): Taiwan; South China Sea. (1–6)

Diploprion bifasciatum (Kuhl & van Hasselt): East China Sea; Taiwan Strait; South China Sea. Near shore. (1–6)

Doederleinia berycoides (Hilgendorf): Bohai; Yellow Sea; East China Sea; Taiwan. Deep water. (1–6)

Epinephelus akaara (Temminck & Schlegel): East China Sea; Taiwan Strait; South China Sea. Rocky reefs. (1–6)

E. amblycephalus (Bleeker): East China Sea; Taiwan Strait; South China Sea. Rocky reefs. (1–6)

E. areolatus (Forskål): Taiwan; Guangdong; Hainan; Dongsha (=Pratas Islands); Xisha (=Paracel Islands); Nansha (=Spratly Islands). Rocky reefs. (1–6)

E. awoara (Temminck & Schlegel): East China Sea; Taiwan Strait; South China Sea. Rocky reefs around islands. (1–6)

E. bleekeri (Vaillant & Bocourt): Taiwan; South China Sea. Rocky reefs. (1–6)

E. bontoides (Bleeker): Taiwan. (8)

E. brunneus (Bloch): Taiwan; South China Sea. (1–6)

E. caeruleopunctatus (Bloch): Taiwan; South China Sea. (1–6)

E. chlorostigma (Cuvier & Valenciennes): East China Sea; Taiwan Strait; South China Sea. Rocky reefs. (1–6)

E. coioides (Hamilton): Taiwan. (8)

E. cometae Tanaka: Hainan; Dongsha (=Pratas Islands); Xisha (=Paracel Islands); Nansha (=Spratly Islands). Deep water rocky reefs. (1–6)

E. corallicola (Cuvier & Valenciennes): Taiwan Strait; Hainan; Dongsha (=Pratas Islands); Xisha (=Paracel Islands); Nansha (=Spratly Islands). Near shore and coral reefs. (1–6)

E. cyanopodus (Richardson): Taiwan. (8)

E. diacanthus (Cuvier & Valenciennes): South China Sea. (1–6)

E. epistictus (Temminck & Schlegel): East China Sea; Taiwan Strait; South China Sea. Rocky reefs. (1–6)

E. fario (Thunberg): East China Sea; Taiwan Strait; South China Sea. Near shore rocky reefs. (1–6)

E. fasciatomaculatus (Peters): Taiwan; South China Sea. (1–6)

E. fasciatus (Forskål): Taiwan Strait, South China Sea. Rocky reef shallow water. (1–6)

E. fuscoguttatus (Forskål): Taiwan Strait; South China Sea. (1–6)

E. hexagonatus (Bloch & Schneider): Southern Taiwan; Hainan; Dongsha (=Pratas Islands); Xisha (=Paracel Islands); Nansha (=Spratly Islands). Coral reefs. (1–6)

E. hoedii (Bleeker): Hainan; Dongsha (=Pratas Islands); Xisha (=Paracel Islands); Nansha (=Spratly Islands). Coral reef bottom fish. (1–6)

E. latifasciatus (Temminck & Schlegel): East China Sea; Taiwan Strait; South China Sea. Rocky reefs. (1–6)

E. macrospilos (Bleeker): South China Sea. (1–6)

E. maculatus (Bloch): Taiwan; South China Sea. (1–6)

E. malabaricus (Bloch & Schneider): East China Sea; Taiwan Strait; South China Sea. Rocky reefs. (1–6)

E. megachir (Richardson): East China Sea; Taiwan Strait; South China Sea. Rocky reefs. (1–6)

E. melanostigma Schultz: Taiwan. (8)

E. merra (Bloch): East China Sea; Taiwan Strait; South China Sea. Shallow water rocky reefs. (1–6)

E. microdon (Bleeker): South China Sea. (1–6)

E. moara (Temminck & Schlegel): East China Sea; Taiwan Strait; South China Sea. Rocky reefs. (1–6)

E. morrhua (Cuvier & Valenciennes): Taiwan; South China Sea. (1–6)

E. octofasciatus Griffin: Taiwan. (8)

E. poecilonotus (Temminck & Schlegel): Taiwan. (8)

E. quoyanus (Valenciennes): Taiwan. (8)

E. radiatus (Day): Taiwan. (8)

E. retouti Bleeker: Taiwan. (8)

E. rhyncholepis (Bleeker) [*E. rivulatus* (8)]: South China Sea. (1–6)

E. septemfasciatus (Thunberg): Yellow Sea, East China Sea. Water depth 2–320 m. (1–6)

E. sexfasciatus (Cuvier & Valenciennes): South China Sea. (1–6)

E. spilotocep Schultz: Hainan; Dongsha (=Pratas Islands); Xisha (=Paracel Islands); Nansha (=Spratly Islands). Coral reefs. (1–6)

E. summana (Forskål): Penghu Islands (Taiwan); Hainan; Dongsha (=Pratas Islands); Xisha (=Paracel Islands); Nansha (=Spratly Islands). (1–6)

E. tauvina (Forskål): Taiwan; Hainan; Dongsha (=Pratas Islands); Xisha (=Paracel Islands); Nansha (=Spratly Islands). Shallow water inside rocky reefs. (1–6)

E. trimaculatus (Valenciennes): Taiwan. (8)

E. truncatus Katayama: East China Sea; Taiwan Strait; South China Sea. Rocky reefs. (1–6)

E. tukula Morgans: Taiwan; South China Sea. (1–6)

E. undulosus (Quoy & Gaimard): Taiwan; South China Sea. Rocky reefs. (1–6)

Giganthias immaculatus Katayama: Taiwan. (8)

Gracila albomarginatus (Fowler & Bean): Taiwan; South China Sea. (1–6)

Grammistes sexlineatus Thunberg: Taiwan; Hainan; Dongsha (=Pratas Islands); Xisha (=Paracel Islands); Nansha (=Spratly Islands). (1–6)

Holenthias borbonius (Valenciennes): Taiwan. (8)

H. chrysostictus (Günther): Taiwan. (8)

Lateolabrax japonicus (Cuvier & Valenciennes): China's coast. River mouth. (1–6)

Liopropoma aragai Randall & Taylor: Taiwan. (8)

L. susumi (Jordan & Seale): Taiwan. (8)

Malakichthys elegans Matsubara & Yamaguchi: East China Sea; South China Sea. Water depth 150–400 m. (1–6)
M. griseus Döderlein: South China Sea. Deep water. (1–6)
M. wakiyai Jordan & Hubbs: East China Sea. Water depth 166–235 m. (1–6)
Mirolabrichthys tuka Here: South China Sea. (1–6)
Niphon spinosus Cuvier & Valenciennes: East China Sea; Taiwan. (1–6)
Odontanthias rhodopeplus (Günther): Taiwan. (8)
O. unimaculatus (Tanaka): Taiwan. (8)
Plectranthias anthioides (Günther): Taiwan. (8)
P. helenae Randall: Taiwan. (8)
P. kelloggi (Jordan & Evermann): Taiwan. (8)
P. longimanus (Weber): Taiwan. (8)
P. nanus Randall: Taiwan. (8)
P. wheeleri Randall: Taiwan. (8)
P. whiteheadi Randall: Taiwan. (8)
P. yamakawai Yoshino: Taiwan. (8)
Plectropomus leopardus (Lacépéde): East China Sea; Taiwan; South China Sea. Coral reefs and near shore water with rocky substrate. (1–6)
P. melanoleucus: Southern tip of Taiwan and Penghu (Taiwan). Rocky shore. (8)
P. oligacanthus Bleeker: South China Sea. (1–6)
P. truncatus Fowler & Bean: South China Sea. Coral reefs and shallow water with rocky substrate. (1–6)
Pogonoperca ocellatus Günther: Hainan; Dongsha (=Pratas Islands); Xisha (=Paracel Islands); Nansha (=Spratly Islands). (1–6)
Promicrops lanceolatus (Eloch): Hainan; Dongsha (=Pratas Islands); Xisha (=Paracel Islands); Nansha (=Spratly Islands). Coral reefs and shoreline, water depth 80 m. (1–6)
Pseudanthias cichlops (Bleeker) [*Anthias cichlopso*]: Taiwan Strait; South China Sea. (1–6)
P. elongatus (Franz) [*Anthias elongatus* (8)]: South China Sea. (1–6)
Sacura margaritacea (Higendorf): Taiwan; South China Sea. (1–6)
Saloptia powelli Smith: Taiwan. (8)
Sayonara satsumae Jordan & Seale: East China Sea; South China Sea. (1–6)
Selenanthias analis Tanaka: South China Sea. (1–6)
Serranocrrhitus latus Watanabe: Lu Island (Taiwan). Rocky reefs. (8)
Synagrops japonicus (Steindachner & Döderlein) [*Plectranthias japonicus* (8)]: East China Sea; South China Sea. (1–6)
S. philippinensis (Günther): East China Sea; South China Sea. (1–6)
S. serratospinosus Smith & Radcliffe: South China Sea. (1–6)
Tosana niwae Smith & Pope: East China Sea; Taiwan; South China Sea. (1–6)
Trisotropis dermopterus (Temmick & Schlegel) [*Triso dermopterus* (8)]: East China Sea; Taiwan Strait; South China Sea. Rocky reefs, water depth 30–60 m. (1–6)

Variola albimarginata Baissac: Taiwan. (8)
V. louti (Forskål): East China Sea; Taiwan Strait; South China Sea. Coral reefs. (1–6)
Zalanthias azumanus (Jordan & Richardson): East China Sea; South China Sea. Deep water. (1–6)

GLAUCOSOMIDAE

Glaucosoma faurellii Sauvage: Taiwan Strait; South China Sea. Rocky reefs. (1–6)
G. hebraicum Richardson: Taiwan Strait; South China Sea. Rocky reefs. (1–6)

SCOMBROPIDAE

Pomatomus saltatris (Cinnaeus): Taiwan. (8)
Scombrops boops (Houttuyn): East China Sea. Water depth 150–550 m. (1–6)

SYMPHYSANODONTIDAE

Symphysanodon katayamai Anderson: Taiwan. (8)

PSEUDOCHROMIDAE

Dampieria cyclophthalmus (Miller & Troschel) [*Labracinus cyclophthalmus*]: Lu Island, Lanyu, and Penghu Islands (Taiwan). Rocky reefs. (8)
D. lineata Gastelnau: Taiwan. (8)
D. melanotaenia (Bleeker): Hainan; Dongsha (=Pratas Islands); Xisha (=Paracel Islands); Nansha (=Spratly Islands). Rocky reef shallow water. (1–6)
Pseudochromis cynotaenia Bleeker: Lu Island and Lanyu (Taiwan). Rocky reefs. (8)
P. fuscus Müller & Troschel: Taiwan; Hainan; Dongsha (=Pratas Islands); Xisha (=Paracel Islands); Nansha (=Spratly Islands). (8)
P. kikaii Aoyagi: Southern tip of Taiwan. Rocky reefs. (8)
P. luteus Aoyagi: Southern tip of Taiwan. Rocky reefs. (8)
P. melanotoenia (Bleeker): Southern tip of Taiwan. Intertidal. (8)
P. porphyreus Lubbock & Goldmon: Lanyu (Taiwan). Rocky intertidal. (8)
P. sureus Seale: Southeastern Taiwan and Lanyu (Taiwan). Rocky reefs. (1–6)
P. tapeinosoma Bleeker: Lanyu (Taiwan). Intertidal. (8)
P. xanthochir Bleeker: Taiwan. (8)

PLESIOPIDAE

Assesor randalli Allen & Kuitar: Southern tip of Taiwan. Rocky reefs. (8)

Calloplesiops altivelis (Steindachner) [*Plesiops altivelis*]: Lanyu and Lu Island (Taiwan), southern and southeastern Taiwan. Rocky intertidal. (8)

Plesiops coeruleolineatus Rüppell: Northern and southern Taiwan. Rocky intertidal. (8)

P. corallicola Bleeker: Taiwan. (8)

P. melas Bleeker [*P. coeruleolineatus* (8)]: Hainan; Dongsha (=Pratas Islands); Xisha (=Paracel Islands); Nansha (=Spratly Islands). Near shore. (1–6)

P. nakaharae Tanaka: Taiwan. (8)

PSEUDOGRAMMIDAE

Aporops blinearis Schultz: Lanyu (Taiwan). Rocky reefs. (8)

ACANTHOCLINIDAE

Acanthoplesiops psilogaster Hardy: Taiwan. (8)
Beliops batanensis Smith-Vaniz & Johnson: Taiwan. (8)
Belonepterygion fasciolatum (Ogilby): Taiwan. (8)

KUHLIIDAE

Kuhlia marginata (Cuvier & Valenciennes): Taiwan; South China Sea. (1–6)

K. rupestris (Lacépéde): Taiwan; South China Sea. (1–6)

K. taeniura (Cuvier & Valenciennes) [*K. mugil* (8)]: Taiwan; South China Sea. (1- 6)

PRIACANTHIDAE

Priacanthus blochi Bleeker: Taiwan; South China Sea. (1–6)

P. boops (Bloch & Schneider): East China Sea; Taiwan Strait; South China Sea. Water depth 140–325 m. (1–6)

P. cruentatus (Lacépéde): Taiwan; South China Sea. (1–6)

P. hamrur (Forskål) [*Cookeolus boops* (8)]: Hainan; Dongsha (=Pratas Islands); Xisha (=Paracel Islands); Nansha (=Spratly Islands). (1–6)

P. macracanthus Cuvier & Valenciennes: East China Sea; Taiwan Strait; South China Sea. Near shore bottom fish. (1–6)

P. tayenus Richardson: East China Sea; Taiwan Strait; South China Sea. (1–6)

Pristigenys multifascia Yoshino & Iwai: Taiwan. (8)

Pseudopriacanthus multifasciatus (Yoshino & Iwai) [*Priacanthus multifasciatus* (8)]: South China Sea. (1–6)

P. niphonius (Cuvier & Valenciennes) [*Priacanthus niphontus* (8)]: Yellow Sea; East China Sea; South China Sea. (1–6)

ACROPOMIDAE

Acropoma hanedai Matsubara: Taiwan; South China Sea. (1–6)

A. japonicum Günther: East China Sea; Taiwan Strait; South China Sea. (1–6)

PERCICHTHYIDAE

Bathysphyraenops simplex Parr: Taiwan. (8)

APOGONIDAE

Apogon ambionensis Bleeker: East China Sea; Taiwan Strait; South China Sea. (1–6)

A. angustatus (Smith & Radcliffe): Taiwan. (8)

A. apogonides (Bleeker): Taiwan. (8)

A. aroubiensis Hombron & Jacquinot: Northeastern Taiwan. Rocky shore. (8)

A. aureus (Lacépéde): Taiwan. (8)

A. bandanensis Bleeker: Hainan; Dongsha (=Pratas Islands); Xisha (=Paracel Islands); Nansha (=Spratly Islands). Coral reefs. (1–6)

A. carinatus Cuvire: Taiwan. (8)

A. cheni Hayashi: Taiwan. (8)

A. chrysotaenia Bleeker: Taiwan. (8)

A. compressus (Smith & Radcliffe): Taiwan. (8)

A. cookii Macleay: Taiwan. (8)

A. crassiceps Garman: Taiwan. (8)

A. cyanosoma Bleeker: Northern Taiwan. Rocky reefs. (8)

A. doederleini Jordan & Snyder: Taiwan; South China Sea. (1–6)

A. doryssa (Jordan & Seale): Taiwan. (8)

A. endekataenia (Bleeker): South China Sea. (1–6)

A. erythrinus Snyder [*A. coccineus* (8)]: Taiwan Strait; South China Sea. (1–6)

A. exostigma (Jordan & Starks): Taiwan. (8)

A. fleurieu (Lacépéde): East China Sea; Taiwan Strait; South China Sea. (1–6)

A. fraenatus Cuvier & Valenciennes: South China Sea. (1–6)

A. fusca (Quoy & Gaimard): Taiwan; South China Sea. (1–6)

A. guamensis Valenciennes: Taiwan. (8)

A. hyalosoma Bleeker: Taiwan. (8)

A. kallopterus Bleeker: Taiwan. (8)

A. kiensis Jordan & Snyder: East China Sea; Taiwan Strait; South China Sea. (1–6)

A. lateralis Valenciennes: Taiwan. (8)

A. melas Bleeker: Southern tip of Taiwan. Rocky reefs. (8)

A. moluccensis Valenciennes: Taiwan. (8)

A. nigripinnis Cuvier: Taiwan. (8)

A. nigrofasciatus Lachner: Taiwan. (8)

A. nitidus (Smith): Taiwan. (8)

A. notatus (Houttuyn): Southwestern Taiwan. Rocky reefs. (8)

A. novemfasciatus Cuvier & Valenciennes: Taiwan; South China Sea. (1–6)

A. orbicularis Cuvier & Valenciennes [*Sphaeramia orbicularis* (8)]: South China Sea. (1–6)

A. quadrifasciatus Cuvier & Valenciennes [*A. fasciatus* (8)]: East China Sea; Taiwan Strait; South China Sea. (1–6)

A. robustus (Smith & Radcliffe) [*A. cookii* (8)]: Taiwan; South China Sea. (1–6)
A. savayensis Günther: Southern tip of Taiwan. Rocky reefs. (8)
A. semilineatus Temminck & Schlegel: East China Sea; Taiwan Strait; South China Sea. (1–6)
A. semiornatus Peters: Taiwan. (8)
A. taeniatus Cuvier & Valenciennes [*A. pseudotaeniatus* (8)]: East China Sea; Taiwan Strait; South China Sea. (1–6)
A. taeniophorus Regan: Taiwan. (8)
A. thermalis Cuvier & Valenciennes: Eastern and southwestern Taiwan. Deep sea. (8)
A. timorensis Bleeker: Taiwan. (8)
A. trimaculatus Cuvier & Valenciennes: South China Sea. (1–6)
A. unicolor Döderlein: Taiwan. (8)
Apogonichthys albomarginatus Radcliffe: South China Sea. (1–6)
A. arafurae (Günther): South China Sea. (1–6)
A. brachygrammus (Jenkins): South China Sea. (1–6)
A. carinatus (Cuvier & Valenciennes): East China Sea; Taiwan Strait; South China Sea. (1–6)
A. ellioti (Day) [*Apogon ellioti* (8)]: East China Sea; Taiwan Strait; South China Sea. (1–6)
A. lineatus (Temminck & Schlegel) [*Apogon lineatus* (8)]: Bohai; Yellow Sea; East China Sea; Taiwan; South China Sea. (1–6)
A. niger Döderlein [*Apogon niger* (8)]: East China Sea; Taiwan Strait; South China Sea. (1–6)
A. perdix Bleeker: South China Sea. (1–6)
A. striatus (Smith & Radcliffe) [*Apogon striatus* (8)]: East China Sea; Taiwan Strait; South China Sea. (1–6)
A. waikiki Jordan & Evermann: Taiwan. (8)
Archamia biguttata Lachner: Taiwan. (8)
A. buroensis (Bleeker): Taiwan. (8)
A. dispilus Lachner: Taiwan. (8)
A. fucata (Canter): Taiwan. (8)
A. goni Chen & Shao: Taiwan. (8)
A. lineolata (Cuvier & Valenciennes): South China Sea. (1–6)
A. macropterus (Cuvier & Valenciennes): South China Sea. (1–6)
Cheilodipterus macrodon (Lacépéde): South China Sea. (1–6)
C. quinquelineatus Cuvier: Taiwan. (8)
C. subulatus Weber: Taiwan. (8)
Foa abocellata (Goren & Karplus): Taiwan. (8)
Fowleria isostigma (Jordan & Seale): Southern tip of Taiwan. Rocky reefs. (8)
F. marmorata (Alleyne & Macleay): Taiwan. (8)
F. ocellata (Weber) [*Apogonichthys ocellatus*]: Northern and southern Taiwan. Rocky reefs. (8)
F. variegata (Valenciennes): Taiwan. (8)
Gymnapogon annona (Whitley): Taiwan. (8)
G. japonicus Regan: Taiwan. (8)
G. philippinus (Herre): Taiwan. (8)
G. urospilotus Lachner: Taiwan. (8)
Howella sherborni (Norman): East China Sea. Water depth 552–810 m. (7)
Neamia octospina Smith & Radcliffe: Taiwan. (8)
Papillapogon auritus (Cuvier & Valenciennes) [*Fowleria auritus* (8)]: Taiwan; Hainan; Dongsha (=Pratas Islands); Xisha (=Paracel Islands); Nansha (=Spratly Islands). Coral reefs. (1–6)
Paramina quinquelineatus (Cuvier & Valenciennes) [*Cheilodipterus quinquelineatus* (8)]: Hainan; Dongsha (=Pratas Islands); Xisha (=Paracel Islands); Nansha (=Spratly Islands). (1–6)
Pseudamiops gracilicauda (Lachner): Taiwan. (8)
Pseudemia gelatinosa Smith: Taiwan. (8)
Rhabdamia cypselurus Weber: Taiwan. (8)
R. gracilis (Bleeker): Taiwan. (8)
Siphamia fuscolineata Lachner: Taiwan. (8)
S. majimai Mataubara & Iwia: Taiwan. (8)
S. versicolor (Smith & Radcliffe): Taiwan. (8)
Sphaeramia nematoptera (Bleeker): Lanyu (Taiwan). Coral reefs. (8)
S. orbicularis (Cuvier): Taiwan. (8)
Synagrops japonicus (Steindachner & Döderlein): East China Sea; South China Sea. (7)

LACTARIIDAE

Lactarius lactarius (Bloch & Schneider): East China Sea; Taiwan Strait; South China Sea. (1–6)

SILLAGINIDAE

Sillago (Parasillago) asiatica Mckay: Taiwan. (8)
S. (Sillaginopodys) chondropus Bleeker: Taiwan. (8)
S. (Parasillago) ingenuua Mckay: Taiwan. (8)
S. japonica Temminck & Schlegel: East China Sea; Taiwan Strait; South China Sea. (1–6)
S. maculata Quoy & Gaimard: East China Sea; Taiwan Strait; South China Sea. (1- 6)
S. (Parasillago) microps Mckay: Taiwan. (8)
S. (Sillago) parvisquamis Gill: Taiwan. (8)
S. sihama (Forskål): China's coast. (1–6)

BRANCHIOSTEGIDAE

Branchiostegus albus Dooley: Northern and southwestern Taiwan. Muddy sand substrate. (8)
B. argentatus (Cuvier & Valenciennes): East China Sea; Taiwan; South China Sea. (1- 6)
B. auratus (Kishinonye): Taiwan Strait; South China Sea. (1–6)
B. japonicus (Houttuyn): Bohai; Yellow Sea; East China Sea; Taiwan. (1–6)
Malacanthus brevirostris Guichenot: Southwestern Taiwan. Rocky reefs. (8)
M. hoedti Bleeker: Hainan; Dongsha (=Pratas Islands); Xisha (=Paracel Islands); Nansha (=Spratly Islands). (1–6)

M. latovittatus (Lacépéde): Taiwan Strait; Hainan; Dongsha (=Pratas Islands); Xisha (=Paracel Islands); Nansha (=Spratly Islands). (1–6)

CARANGIDAE

Alectis ciliaris (Bloch): Yellow Sea; East China Sea; Taiwan Strait; South China Sea. (1–6)

A. indica (Rüppell): Yellow Sea; East China Sea; Taiwan Strait; South China Sea. (1–6)

Atropus atropus (Bloch & Schneider): China's coast. Near shore. (1–6)

Carangoides (*C.*) *bucculentus* Alleyne & Macleay: Taiwan; South China Sea. (1–6)

C. (*Longirostirum*) *delicatissimus* (Döderlein): South China Sea. (1–6)

C. (*Atule*) *djeddaba* (Forskål) [*Alepes djeddaba* (8)]: Yellow Sea; East China Sea; South China Sea. (1–6)

C. hedlandensis (Whitley): Taiwan. (8)

C. (*C.*) *hippos* (Linnaeus): South China Sea. (1–6)

C. (*C.*) *ignobilis* (Forskål): Yellow Sea; East China Sea; Taiwan Strait; South China Sea. (1–6)

C. (*C.*) *ishikqwai* Wakiya: South China Sea. (1–6)

C. (*A.*) *kalla* Cuvier & Valenciennes [*Alepes para* (8)]: East China Sea; Taiwan Strait; South China Sea. (1–6)

C. lugubris Poey: Taiwan. (8)

C. (*A.*) *macrurus* (Bleeker) [*Caranx* (*Alepes*) *vari* (8)]: South China Sea. (1–6)

C. (*A.*) *malam* (Bleeker) [*Caranx* (*Alepes*) *melanoptera* (8)]: South China Sea. (1–6)

C. (*A.*) *mate* Cuvier & Valenciennes [*Atule kalla* (8)]: Taiwan; South China Sea. (1–6)

C. (*C.*) *melampygus* Cuvier & Valenciennes: Taiwan; Hainan; Dongsha (=Pratas Islands); Xisha (=Paracel Islands); Nansha (=Spratly Islands). (1–6)

C. orthogrammus (Jordan & Gilbert): Taiwan. (8)

C. (*C.*) *oshimai* Wakiya: South China Sea. (1–6)

C. papuensis (Alleyne & Macleay): Taiwan. (8)

C. (*A.*) *pectoralis* Chu & Cheng: South China Sea. (1–6)

C. (*L.*) *platessa* (Cuvier & Valenciennes): South China Sea. (1–6)

C. (*C.*) *sansum* (Forskål): South China Sea. (1–6)

C. (*C.*) *stellatus* Eydoux & Souleyer: Hainan; Dongsha (=Pratas Islands); Xisha (=Paracel Islands); Nansha (=Spratly Islands). (1–6)

C. tille Cuvier: Taiwan. (8)

Caranx (*Citula*) *armatus* (Forskål) [*Carangoides armatus* (8)]: South China Sea. (1–6)

C. (*Citula*) *chrysophrys* (Bloch & Schneider) [*Carangoides chrysophrys* (8)]: Yellow Sea; East China Sea; Taiwan; South China Sea. Generaly below 60 m water depth. (1–6)

C. (*Citula*) *coeruleopinnatus* [*Carangoides coeruleepinnatus* (8)] Rüppell: East China Sea; Taiwan Strait; South China Sea. (1–6)

C. (*Citula*) *dinema* (Bleeker): South China Sea. (1–6)

C. (*Carangoides*) *equula* Temminck & Schlegel: Southern Yellow Sea; East China Sea; Taiwan Strait; South China Sea. (1–6)

C. (*Carangoides*) *ferdau* (Forskål) [*Carangoides ferdau* (8)]: Taiwan; Guangdong; Hainan; Dongsha (=Pratas Islands); Xisha (=Paracel Islands); Nansha (=Spratly Islands). (1–6)

C. (*Uraspis*) *helvolus* (Forster): East China Sea; Taiwan Strait; South China Sea. (1–6)

C. (*Citula*) *malabaricus* (Bloch & Schneider): Yellow Sea; East China Sea; South China Sea; Taiwan. Near shore, often within 60 m water depth. (1–6)

C. (*Citula*) *oblongus* Cuvier & Valenciennes: South China Sea. (1–6)

C. (*Carangoides*) *plagiotaenia* (Bleeker) [*Carangoides plagiotaenia* (8)]: South China Sea. (1–6)

C. (*Citula*) *plumbeus* (Quoy & Gaimard): East China Sea; Taiwan Strait; South China Sea. (1–6)

C. (*Carangoides*) *praeustus* Bennett: South China Sea. (1–6)

C. (*Caranx*) *sexfasciatus* Quoy & Gaimard: Yellow Sea; East China Sea; Taiwan; Hainan; Dongsha (=Pratas Islands); Xisha (=Paracel Islands); Nansha (=Spratly Islands). (1–6)

C. (*Citula*) *talamparoidas* (Bleeker): South China Sea. (1–6)

C. (*U.*) *uraspis* (Günther): Taiwan. (8)

Chorinemus formosanus (Wakiya): Taiwan Strait; South China Sea. (1–6)

C. hainanensis Chu & Cheng: South China Sea. (1–6)

C. lysan (Forskål) [*Scomberoides lysan* (8)]: Taiwan; South China Sea. (1–6)

C. moadetta Cuvier & Valenciennes: Yellow Sea; East China Sea; South China Sea. (1- 6)

C. orientalis Temminck & Schlegel: Taiwan Strait; South China Sea. (1–6)

C. sanctipetri Cuvier & Valenciennes: South China Sea. (1–6)

C. tala (Cuvier & Valenciennes) [*Scomberoides orientalis*]: South China Sea. (1–6)

C. tolooparah (Rüppell): Taiwan Strait; South China Sea. (1–6)

Decapterus fasciatus Bleeker: South China Sea. (1–6)

D. kurroides Bleeker: East China Sea; Taiwan Strait; South China Sea. (1–6)

D. lajang Bleeker: East China Sea; Taiwan Strait; South China Sea. (1–6)

D. macarellus (Cuvier): Taiwan. (8)

D. macrosoma Bleeker: East China Sea; Taiwan Strait; South China Sea. (1–6)

D. maruadsi (Temminck & Schlegel): China's coast. Near shore. (1–6)

D. muroadsi (Temminck & Schlegel): South China Sea. (1–6)

D. russelli (Rüppell): East China Sea; South China Sea. Water depth 125–415 m. (1–6)

D. tabl (Berry): Taiwan. (8)

Elagatis bipinnulatus (Quoy & Gaimard): East China Sea; Taiwan Strait; South China Sea. (1–6)

Gnathanodon speciosus (Forskål): South China Sea. (1–6)

Megalaspis cordyla (Linnaeus): East China Sea; Taiwan Strait; South China Sea. Near shore. (1–6)

Naucrates ductor (Linnaeus): Taiwan; Xisha (=Paracel Islands). (1–6)

Scomberoides commersonianus Lacépéde: Taiwan. (8)

S. tol (Cuvier): Taiwan. (8)

Selar boops Cuvier & Valenciennes: Taiwan Strait; South China Sea. Near shore. (1–6)

S. crumenophthalmus (Bloch): Yellow Sea; East China Sea; Taiwan Strait; South China Sea. (1–6)

Selaroides leptolepis (Cuvier & Valenciennes): Taiwan Strait; South China Sea. Near shore. (1–6)

Seriola aureovittata Temminck & Schlegel: China's coast. (1–6)

S. dumerili (Risso): China's coast. (1–6)

S. quinqueradiata Temminck & Schlegel: China's coast. (1–6)

Trachinotus bailloni (Lacépéde): Taiwan; South China Sea. (1–6)

T. ovatus (Linnaeus) [*T. blochii* (8)]: China's coast. (1–6)

T. russelli Cuvier & Valenciennes: South China Sea. (1–6)

Trachurus japonicus (Temminck & Schelegel): China's coast. (1–6)

Ulua mandibularis (Macleay) [*U. mentail* (8)]: South China Sea. (1–6)

Zonichthys nigrofasciata (Ruppell) [*Seriolina nigrofasciata* (8)]: East China Sea; South China Sea. (1–6)

MENIDAE

Mene maculata (Bloch & Scheider): East China Sea; Taiwan Strait; South China Sea. Near shore. (1–6)

FORMIONIDAE

Formio niger (Bloch) [*F. parastromateus* (8)]: Yellow Sea; East China Sea; South China Sea; Taiwan. Near shore. (1–6)

RACHYCENTRIDAE

Rachycentron canadum (Linnaeus): Bohai; Yellow Sea; East China Sea; South China Sea; Taiwan. Offshore. (1–6)

CORYPHAENIDAE

Coryphaena hippurus Linnaeus: Bohai; Yellow Sea; East China Sea; South China Sea; Taiwan. Migratory species. (1–6)

BRAMIDAE

Brama japonica Hilgendorf: East China Sea; Taiwan. Water depth 570–780 m. (7)

B. rayi (Bloch): South China Sea. Deep water. (9)

Collybus drachme Snyder: South China Sea. Deep water. (9)

Eumegistus illustris Jordan & Jordan: Taiwan. (8)

Pterycombus petersii (Hilgendorf): Taiwan. (8)

Taractes rubescens (Jordan & Evermann): Taiwan. (8)

Taractichthys steindachneri (Döderlein): Taiwan. (8)

SCIAENIDAE

Argyrosomus aneus (Bloch): East China Sea; Taiwan Strait; South China Sea. Near shore bottom fish. (1–6)

A. argentatus (Houttuyn) [*Pennahia argentatus* (8)]: Yellow Sea; East China Sea; South China Sea. Near shore bottom fish. (1–6)

A. macrocephalus (Tang) [*Pennahia macrocephalus* (8)]: East China Sea; Taiwan Strait; South China Sea. Near shore bottom fish. (1–6)

A. pawak Lin [*Pennahia pawak* (8)]: East China Sea; South China Sea. Near shore bottom fish. (1–6)

Atrobucca nibe (Jordan & Thomson) [*Argyrosomus nibe*]: East China Sea; Taiwan Strait; South China Sea. Near shore bottom fish. (1–6)

Bahaba flavolabiata (Lin): East China Sea; Taiwan Strait; South China Sea. Water depth 50–60 m. (1–6)

Chrysochir aureus (Richardson) [*Nibea acuta*]: East China Sea; Taiwan Strait; South China Sea. Near shore bottom fish. (1–6)

Collichthys lucidus (Richardson) [*C. fragilis*]: Yellow Sea; East China Sea; Taiwan,, South China Sea. Water depth about 20 m. (1–6)

C. niveatus Jordan & Starks: Yellow Sea; East China Sea. (1–6)

Johnius amblycephalus (Bleeker): East China Sea; Taiwan Strait; South China Sea. Near shore bottom fish, water depth 20–40 m. (1–6)

J. belengeri (Cuvier & Valenciennes): China's coast. Near shore, water depth within 20 m. (1–6)

J. carutta Bloch: South China Sea. (1–6)

J. dussumieri (Cuvier & Valenciennes): South China Sea. (1–6)

J. fasciatus Chu, Lo & Wu: Taiwan Strait, South China Sea. Near shore bottom fish. (1- 6)

J. macrorhynus (Mohan): Taiwan. (8)

Macrospinosa cuja (Hamilton): South China Sea. (1–6)

Megalonibes fusca Chu, Lo & Wu: Yellow Sea; East China Sea; Taiwan Strait; South China Sea. Near shore large bottom fish. (1–6)

Miichthys miiuy (Basilewsky): China's coast. Near shore, water depth 15–70 m. (1–6)

Nibea albiflora (Richardson): China's coast. Near shore bottom fish. (1–6)

N. chui Trewavas: East China Sea; South China Sea. Near shore bottom fish. (1–6)

N. diacanthus (Lacépéde) [*Protonibea diacanthus* (8)]: Taiwan Strait; South China Sea. Near shore bottom fish. (1–6)

N. japonica (Temminck & Schlegel) [*Argyrosomus japonica* (8)]: Yellow Sea; East China Sea; Taiwan. (1–6)

N. miichthioides Chu, Lo & Wu: Taiwan Strait; South China Sea. Near shore bottom fish. (1–6)

N. semifasciata Chu, Lo & Wu: East China Sea; Taiwan; South China Sea. (1–6)

Otolithes argenteus (Cuvier & Valenciennes): East China Sea; Taiwan Strait; South China Sea. Near shore bottom fish, water depth within 60 m. (1–6)

O. ruber (Bloch & Schneider): Taiwan; South China Sea. (1–6)

Panna microdon (Bleeker): South China Sea. (1–6)

Paranibea semiluctuosa (Cuvier & Valenciennes): South China Sea. (1–6)

Pennahia macrophthalmus Bleeker: Taiwan. (8)

Pseudosciaena crocea (Richardson) [*Laramichthys crocea* (8)]: Yellow Sea; East China Sea; Taiwan; South China Sea. Near shore, water depth within 60 m. (1–6)

P. polyactis Bleeker [*Laramichthys polyactis* (8)]: Bohai; Yellow Sea; East China Sea. Water depth within 105 m. (1–6)

Umbrina russelli Cuvier & Valenciennes [*Sciaena russelli*]: East China Sea; Taiwan Strait; South China Sea. Near shore bottom fish. (1–6)

Wak coitor (Hamilton): South China Sea. (1–6)

W. sina (Cuvier & Valenciennes) [*Johnius sina* (8)]: East China Sea; Taiwan Strait; South China Sea. Near shore. (1–6)

W. soldado (Lacépéde): South China Sea. (1–6)

W. tingi (Tang) [*Johnius tingi* (8)]: East China Sea; Taiwan Strait; South China Sea. Near shore. (1–6)

LEIOGNATHIDAE

Gazza achlamys Jordan & Starks: Southwestern Taiwan. Sandy substrate. (8)

G. minuta (Bloch): East China Sea; South China Sea. Near shore small fish. (1–6)

Leiognathus berbis (Cuvier & Valenciennes): East China Sea; Taiwan Strait; South China Sea. Near shore small fish. (1–6)

L. bindus (Cuvier & Valenciennes): East China Sea; Taiwan Strait; South China Sea. Near shore small fish. (1–6)

L. brevirostris (Cuvier & Valenciennes): East China Sea; Taiwan Strait; South China Sea. Near shore small fish. (1–6)

L. daura (Cuvier): South China Sea. Near shore small fish. (1–6)

L. dussumieri (Cuvier & Valenciennes): South China Sea. Near shore small fish. (1–6)

L. elongatus (Günther): Taiwan; South China Sea. Near shore small fish. (1–6)

L. equulus (Forskål): Taiwan; South China Sea. Near shore small fish. (1–6)

L. fasciatus Lacépéde: South China Sea. Near shore small fish. (1–6)

L. insidiator (Bloch) [*Secutor insidiator* (8)]: East China Sea; South China Sea. Coastal fish. (1–6)

L. leuciscus (Günther): Taiwan; South China Sea. Near shore small fish. (1–6)

L. lineolatus (Cuvier & Valenciennes): Taiwan Strait, South China Sea. Near shore small fish. (1–6)

L. nuchalis (Temminck & Schlegel): Taiwan. (8)

L. rivulatus (Temminck & Valenciennes): East China Sea; Taiwan Strait; South China Sea. Near shore small fish. (1–6)

L. ruconius (Hamilton-Buchanan) [*Secutor ruconius* (8)]: Yellow Sea, East China Sea; South China Sea. Near shore small fish. (1–6)

L. splendens (Cuvier): Taiwan; South China Sea. Near shore small fish. (1–6)

GERRIDAE

Gerreomorpha japonica (Bleeker): East China Sea; Taiwan; South China Sea. Common bottom fish in estuaries. (1–6)

Gerres abbreviatus Bleeker: East China Sea; Taiwan; South China Sea. (1–6)

G. acinaces Bleeker: Xisha (=Paracel Islands). (1–6)

G. argyreus (Bloch & Schneider): Taiwan; South China Sea. (1–6)

G. filamentosus Cuvier: East China Sea; Taiwan Strait; South China Sea. Common bottom fish in estuaries. (1–6)

G. lucidus Cuvier: East China Sea; Taiwan; South China Sea. Common bottom fish in estuaries. (1–6)

G. macrosoma Bleeker: East China Sea; Penghu (Taiwan); South China Sea. (1–6)

G. oblongus Cuvier: Taiwan; Xisha (=Paracel Islands). (1–6)

G. oyena (Forskål) [*G. equula*]: Taiwan; Hainan; Dongsha (=Pratas Islands); Xisha (=Paracel Islands); Nansha (=Spratly Islands). (1–6)

G. poeti Cuvier & Valenciennes: South China Sea. (1–6)

Pentarion longimanus (Cantor): Taiwan; South China Sea. Near shore small fish. (1–6)

LUTJANIDAE

Aphareus furcatus (Lacépéde): Taiwan; Hainan; Dongsha (=Pratas Islands); Xisha (=Paracel Islands); Nansha (=Spratly Islands). (1–6)

A. rutilans Cuvier: Taiwan. (8)

Aprion virescens Cuvier & Valenciennes: Taiwan; Xisha (=Paracel Islands); Zhongsha (=Macclesfield Bank); Nansha (=Spratly Islands). Water depth within 100 m. (1- 6)

Caesio chrysozona Cuvier & Valenciennes [*Pterocoesio chrysozona* (8)]: Taiwan; South China Sea. (1–6)

C. coerulaureus Lacépéde: Taiwan; Hainan; Dongsha (=Pratas Islands); Xisha (=Paracel Islands); Nansha (=Spratly Islands). Around rocky shore and coral reefs. (1–6)

C. diagramma Bleeker [*Pterocoesio diagramma* (8)]: Taiwan; Xisha (=Paracel Islands). Often live in shallow water around coral reefs. (1–6)

C. erythrogaster Cuvier & Valenciennes: Taiwan; South China Sea. (1–6)

C. lunaris Cuvier & Valenciennes: Taiwan; Xisha (=Paracel Islands). Shallow water around coral reefs. (1–6)

C. tile Cuvier & Valenciennes [*Pterocoesio tile* (8)]: Taiwan; Xisha (=Paracel Islands). Shallow water around coral reefs. (1–6)

C. xanthonotus Bleeker: Taiwan; Hainan; Dongsha (=Pratas Islands); Xisha (=Paracel Islands); Nansha (=Spratly Islands). Shallow water around coral reefs. (1–6)

Etelis carbunculus Cuvier & Valenciennes: Taiwan; Hainan; Dongsha (=Pratas Islands); Xisha (=Paracel Islands); Nansha (=Spratly Islands). Often live in rocky reef deeper water. (1–6)

E. coruscans Valenciennes: Taiwan. (8)

E. radiosus Anderson: Taiwan. (8)

Glabrilutjanus nematophorus (Bleeker): South China Sea. (1–6)

Lutjanus altifrontalis Chan: South China Sea. (1–6)

L. argentimaculatus (Forskål): East China Sea; Taiwan Strait; Hainan; Dongsha (=Pratas Islands); Xisha (=Paracel Islands); Nansha (=Spratly Islands). Coral reef shallow water (to 80 m water depth). (1–6)

L. bengalensis (Bloch): Taiwan. (8)

L. bohar (Forskål): Taiwan; Xisha (=Paracel Islands). Often live around coral reefs or rocky reefs. (1–6)

L. boutton (Lacépéde): Penghu (Taiwan). (8)

L. carponotatus (Richardson): Taiwan. (8)

L. chrysotaenia (Bleeker): South China Sea. (1–6)

L. decussatus (Cuvier): Northeastern Taiwan and Penghu (Taiwan). (8)

L. dodecacanthoides (Bleeker): Taiwan. (8)

L. erythopterus Bloch: East China Sea; Taiwan Strait; South China Sea. Water depth 30–90 m. (1–6)

L. fulviflamma (Forskål): East China Sea; Taiwan Strait; South China Sea. Often live in shallow water around coral reefs. (1–6)

L. fulvus (Forster): Taiwan Strait; South China Sea. (1–6)

L. gibbus (Forskål): Taiwan; Xisha (=Paracel Islands); Nansha (=Spratly Islands). Shallow water around coral reefs. (1–6)

L. johni (Bloch): East China Sea; Taiwan Strait; South China Sea. Near shore bottom fish. (1–6)

L. kasmira (Forskål): East China Sea; Taiwan Strait; South China Sea. Around coral reefs. (1–6)

L. lineolatus (Rüppell): East China Sea; Taiwan Strait; South China Sea. Shallow water near bottom fish. (1–6)

L. lutjanus Bloch: Taiwan Strait; South China Sea. Shallow water near bottom fish. (1–6)

L. malabaricus (Bloch & Schneider): Taiwan. (8)

L. monostigma (Cuvier & Valenciennes): Taiwan; Xisha (=Paracel Islands). Often live around coral reefs. (1–6)

L. niger (Forskål) [*Macolor niger* (8)]: Taiwan; Xisha (=Paracel Islands); Nansha (=Spratly Islands). Often live in coral reef shallow water. (1–6)

L. quinquelineatus (Bloch): Taiwan. (8)

L. rivulatus (Cuvier & Valenciennes): Taiwan Strait; South China Sea. (1–6)

L. russelli Bleeker: East China Sea; Taiwan Strait; South China Sea. Often live in shallow water around coral reefs or rocky reefs. (1–6)

L. sanguineus (Cuvier & Valenciennes): East China Sea; South China Sea. (1–6)

L. sebae (Cuvier & Valenciennes): East China Sea; Taiwan Strait; South China Sea. Juveniles in shallow water, adults in water depth about 100 m. (1–6)

L. spilurus (Bennett): East China Sea; Taiwan Strait; South China Sea. Around coral reefs and rocky reefs. (1–6)

L. stellatus Akuzaki: Northern and southern Taiwan. (8)

L. vaigiensis (Quoy & Gaimard): Taiwan; South China Sea. Often live around coral reefs and rocky reefs. (1–6)

L. vitta (Quoy & Gaimard): East China Sea; Taiwan Strait; South China Sea. Shallow water near bottom fish. (1–6)

Macolor niger (Forskål): South China Sea. (1–6)

Paracaesio caeruleus (Katayama): Northeastern Taiwan. Rocky reefs. (8)

P. kusakarii Abe: Taiwan. (8)

P. xanthurus Bleeker: Taiwan Strait; Hainan; Dongsha (=Pratas Islands); Xisha (=Paracel Islands); Nansha (=Spratly Islands). Often lives around coral reefs. (1-6)

Pinjalo pinjalo (Bleeker): Taiwan; South China Sea. (1–6)

Pristipomoides auricilla (Jordan, Evermann & Tanaka): Taiwan. (8)

P. filamentosus (Valenciennes): Northeastern Taiwan. Rocky reefs. (8)

P. flavipinnis Shinohara: Northeastern Taiwan. (8)

P. microlepis (Bleeker): East China Sea; Taiwan Strait; South China Sea. Near shore. (1–6)

P. multidens (Day): Taiwan; Nansha (=Spratly Islands). (17, 19)

P. sieboldii (Bleeker): Taiwan. (8)

P. typus Bleeker: East China Sea; Taiwan Strait; South China Sea. Near bottom, economic species. (1–6)

Symphorus nematophorus (Bleeker): Taiwan. (8)

S. spilurus Günther: South China Sea. Around coral reefs. (1–6)

Tangia carnolabrum Chan: South China Sea. (1–6)

Tropidinius amoenus (Snyder): Taiwan. (8)

T. zonatus (Valenciennes): Taiwan. (8)

EMMELICHTHYIDAE

Dipterygotus leucogrammicus (Bleeker): East China Sea; Taiwan Strait; South China Sea. (1–6)

Emmelichthys seintillans (Jordan & Thompsons): Taiwan. (8)

E. struhsakeri Heemstra & Randall: East China Sea; Taiwan. Water depth 0–360 m. (7, 8)

Erythrocles schlegeli (Richardson): East China Sea; Taiwan; South China Sea. Water depth 150–300 m. (7)

LETHRINIDAE

Lethrinus haematopterus Temminck & Schlegel: Taiwan Strait; South China Sea. (1–6)
L. harak (Forskål): Taiwan. (8)
L. kalloperus Bleeker: South China Sea. (1–6)
L. leutjanus (Lacépéde) [*L. lentjan* (8)]: Taiwan; South China Sea. (1–6)
L. mahsena (Forskål): Taiwan. (8)
L. mahsenoides Valenciennes: Taiwan. (8)
L. miniatus (Bloch & Schneider): Taiwan; Hainan; Xisha (=Paracel Islands); Zhongsha (=Macclesfield Bank). (1–6)
L. nebulosu (Forskål): Taiwan Strait; South China Sea. Water depth 30–60 m, rocky reefs. (1–6)
L. nematacanthus Bleeker: Taiwan; South China Sea. Benthic fish. (1–6)
L. ornatus Cuvier & Valenciennes: Taiwan; South China Sea. (1–6)
L. reticulatus Valenciennes: Southern tip of Taiwan. Rocky reefs. (8)
L. rhodopterus Bleeker: Taiwan Strait; South China Sea. (1–6)
L. rubrioperculatus Sato: Taiwan. (8)
L. semicinctus Valenciennes: Taiwan. (8)
L. variegatus Cuvier & Valenciennes: Taiwan; South China Sea. (1–6)
L. xanthochilus Klunzinger: Taiwan. (8)

SPARIDAE

Acanthopagrus australis (Günther): Taiwan. (8)
Argyrops bleekeri Oshima: East China Sea; Taiwan; South China Sea. Near shore bottom fish. (1–6)
A. spinifer (Forskål): Taiwan. Rocky reefs. (8)
Evynnis cardinalis (Lacépéde): Northern and southwestern Taiwan. (8)
Monotaxis grandoculis (Forskål): Taiwan; Xisha (=Paracel Islands); Nansha (=Spratly Islands). (1–6)
Pagrosomus major (Temminck & Schlegel) [*Pagrus major* (8)]: China's coast. Near shore bottom fish, water depth 30–90 m. (1–6)
Paragyrops edita Tanaka: East China Sea; South China Sea. Water depth 20–70 m. (1-6)
Rhabdosargus sarba (Forskål) [*Sparus sarba* (8)]: Yellow Sea; East China Sea; Taiwan; South China Sea. Near shore benthic fish. (1–6)
Sparus berda Forskål [*Acanthopagrus berda* (8)]: East China Sea; Taiwan; South China Sea. Shallow water benthic fish, rocky reefs. (1–6)
S. latus Houttuyn [*Acanthopagrus latus* (8)]: East China Sea; Taiwan; South China Sea. Shallow water bottom fish. (1–6)
S. macrocephalus (Basilewsky) [*Acanthopagrus schlegeli* (8)]: China's coast. Shallow water with sandy mud or rocky reefs.
Taius tumifrons (Temminck & Schlegel) [*T. fumifrons*]: Yellow Sea; East China Sea; Taiwan; South China Sea. Water depth 0–90 m. (1–6)

LOBOTIDAE

Lobotes surinamensis (Bloch): China's coast. Shallow sea. (1–6)

BANJOSIDAE

Banjos banjos (Richardson): East China Sea; Taiwan; South China Sea. Water depth 175–400 m. (1–6)

NEMIPTERIDAE

Nemipterus aurorus Russell: Taiwan. (8)
N. bathybius (Snyder): East China Sea; Taiwan Strait; South China Sea. Water depth more than 90 m. (1–6)
N. furcosus (Valenciennes): Taiwan. (8)
N. hexodon (Quoy & Gaimard): Taiwan Strait; South China Sea. (1–6)
N. japonicus (Bloch): East China Sea; Taiwan Strait; South China Sea. Generally in water depth 20–100 m. (1–6)
N. oveni (Bleeker): South China Sea. (1–6)
N. peronii (Valenciennes): Taiwan. (8)
N. thosaporni Russell: Taiwan. (8)
N. tolu (Cuvier & Valenciennes): East China Sea; South China Sea. (1–6)
N. virgatus (Houttuyn): East China Sea; Taiwan; South China Sea. Near shore bottom fish, water depth 30 m. (1–6)
N. zysron (Bleeker): Taiwan. (8)

PENTAPODIDAE

Gnathodentex aurolineatus (Lacépéde): Xisha (=Paracel Islands); Nansha (=Spratly Islands). Coral reefs. (1–6)
Gymnocranius griseus (Temminck & Schlegel): East China Sea; Taiwan Strait; South China Sea. (1–6)
G. japonicus Akazaki: Taiwan. (8)
Pentapus macrurus (Bleeker): Taiwan. (8)
P. nagasakiensis (Tanaka): Taiwan. (8)
P. setosus Cuvier & Valenciennes: South China Sea. (1–6)

SCOLOPSIDAE

Parascolopsis tosensis (Kamohara): Taiwan. (8)
Scolopsis affinis Peters: Taiwan. (8)

S. bilineatus (Bloch): Taiwan Strait; South China Sea. (1–6)

S. bimaculatus Rüppell: South China Sea. (1–6)

S. cancellatus (Cuvier & Valenciennes): Xisha (=Paracel Islands); Nansha (=Spratly Islands). Coral reefs. (1–6)

S. ciliatus (Lacépéde): Taiwan Strait; South China Sea. (1–6)

S. eriomma Jordan & Richardson [*Parascolopsis eriomma* (8)]: Taiwan; South China Sea. (1–6)

S. inermis Rüppell (Temminck & Schlegel) [*Parascolopsis inermis* (8)]: Taiwan Strait; South China Sea. (1–6)

S. lineatus Quoy & Gaimard: Taiwan. (8)

S. margaritifer (Cuvier & Valenciennes): Taiwan; South China Sea. (1–6)

S. monogramma (Cuvier): Taiwan. (8)

S. taeniopterus (Cuvier & Valenciennes): South China Sea. (1–6)

S. temporalis Cuvier: South China Sea. (1–6)

S. vosmeri (Bloch): Taiwan Strait; South China Sea. (1–6)

S. xenochrous Günther: Taiwan. (8)

POMADASYIDAE

Hapalogenys kishinouyei Smith & Pope: East China Sea; Taiwan; South China Sea. Often in areas with rocky reefs or muddy substrate. (1–6)

H. mucronatus (Eydoux & Souleyet): China's coast. Shallow sea. (1–6)

H. nitens (Richardson) [*H. nigripinnis*]: China's coast. Near shore. (1–6)

Parapristipoma trilineatus (Thunberg): East China Sea; Taiwan Strait; South China Sea. (1–6)

Plectorhynchus albovittatus Rüppell: Taiwan. (8)

P. celebicus Bleeker: Taiwan. (8)

P. chaetodonoides Lacépéde: Taiwan; Xisha (=Paracel Islands); Nansha (=Spratly Islands). (1–6)

P. cinctus (Temminck & Schlegel): China's coast. Shallow sea bottom fish, often in rocky reef area. (1–6)

P. diagrammus (Linnaeus): Taiwan; Xisha (=Paracel Islands). (1–6)

P. flavomaculatus (Cuvier): Northern Taiwan. Rocky reefs. (8)

P. foetela (Forskål): South China Sea. (1–6)

P. goldmanni (Bleeker): Taiwan; Xisha (=Paracel Islands). (1–6)

P. lineatus (Cuvier & Valenciennes): Taiwan; South China Sea. (1–6)

P. nigrus (Cuvier): Taiwan; South China Sea. (1–6)

P. orientalis (Bloch): Taiwan; Xisha (=Paracel Islands). Rocky reefs. (1–6)

P. pictus (Thunberg): East China Sea; Taiwan Strait; South China Sea. (1–6)

P. picus (Cuvier): Taiwan. (8)

P. punctatissimus (Playfair): Xisha (=Paracel Islands). (1–6)

P. reticulatus (Günther): South China Sea. (1–6)

P. schotaf (Forskål): Taiwan. (8)

P. sinensis Zhu, Wu & Jin: East China Sea; Taiwan Strait; South China Sea. Around rocky reefs. (1–6)

Pomadasys argenteus (Forskål): Taiwan; South China Sea. Near shore. (1–6)

P. furcatus (Bloch & Schneider): South China Sea. (1–6)

P. grunniens (Bloch & Schneider): South China Sea. Near shore. (1–6)

P. hasta (Bloch): East China Sea; Taiwan Strait; South China Sea. Near shore bottom fish. (1–6)

P. kaakan (Cuvier): Taiwan. (8)

P. maculatus (Bloch): East China Sea; Taiwan; South China Sea. Shallow sea. (1–6)

P. quadrilineatus Shen: Taiwan. (8)

P. umimaculatus Tian: South China Sea. (1–6)

THERAPONIDAE

Helotes cancellatus (Cuvier & Valenciennes) [*Terapon cancellatus* (8)]: Taiwan; South China Sea. (1–6)

H. sexlineatus (Quoy & Gaimard): East China Sea; Taiwan Strait; South China Sea. Near shore bottom fish. (1–6)

Pelates quadrilineatus (Bloch): East China Sea; Taiwan Strait; South China Sea. Near shore shallow sea. (1–6)

Therapon jarbua (Forskål) [*Terapon jarbua* (8)]: East China Sea; Taiwan Strait; South China Sea. Near shore bottom fish. (1–6)

T. oxyrhynchus Temminck & Schlegel: East China Sea; Taiwan Strait; South China Sea. Near shore bottom fish. (1–6)

T. thraps (Cuvier & Valenciennes) [*Terapon thraps* (8)]: East China Sea; Taiwan Strait; South China Sea. Shallow sea bottom fish. (1–6)

KYPHOSIDAE

Kyphosus bigibbus Lacépéde: Taiwan. (8)

K. cinerascens (Forskål): Taiwan; South China Sea. (1–6)

K. lembus (Cuvier & Valenciennes): East China Sea; Taiwan Strait; South China Sea. Near shore. (1–6)

GIRELLIDAE

Girella melanichthys (Richardson): Northern and northeastern Taiwan. (8)

G. mezina Jordan & Starks: Northern Taiwan. Rocky reefs. (8)

G. punctata Gray: East China Sea; Taiwan Strait; South China Sea. Near shore rocky reefs. (1–6)

MULLIDAE

Mulloidichthys flavolineatus (Lacépéde): Southern tip of Taiwan and Lanyu (Taiwan). Rocky reef area with sandy substrate. (8)

M. pflugeri (Steindachner): Southern tip of Taiwan. Rocky reefs. (8)

M. samoensis (Günther): Xisha (=Paracel Islands). Coral reefs. (1–6)

M. suriflamma (Forskål): Taiwan; Xisha (=Paracel Islands); Nansha (=Spratly Islands). Coral reefs. (1–6)

M. vanicolensis Cuvier & Valenciennes: Taiwan; Xisha (=Paracel Islands). Coral reefs. (1–6)

Parupeneus barberinoides (Bleeker): Southern tip of Taiwan. Sandy substrate. (8)

P. barberinus (Lacépéde): Taiwan Strait; South China Sea. (1–6)

P. bifasciatus (Lacépéde): Taiwan; Xisha (=Paracel Islands). (1–6)

P. chryserdros (Lacépéde): Xisha (=Paracel Islands); Nansha (=Spratly Islands). (1–6)

P. chrysopleuron (Temminck & Schlegel): East China Sea; Taiwan; South China Sea. Near shore. (1–6)

P. ciliatus (Lacépéde): Taiwan. (8)

P. cyclostomus (Lacépéde): Taiwan. (8)

P. fraterculus (Cuvier & Valenciennes): South China Sea. Near shore. (1–6)

P. indicus (Shaw): Taiwan Strait; South China Sea. (1–6)

P. luteus (Cuvier & Valenciennes): East China Sea; Taiwan Strait; South China Sea. (1-6)

P. megalops (Tanaka): South China Sea. (1–6)

P. multifasciatus (Quoy & Gaimard): Taiwan. (8)

P. pleurostigma (Bennett): Hainan; Dongsha (=Pratas Islands); Xisha (=Paracel Islands); Nansha (=Spratly Islands). (1–6)

P. spilurus (Bleeker): Southwestern Taiwan and Penghu (Taiwan). Rocky reefs. (8)

P. trifasciatus (Lacépéde): Taiwan; South China Sea. Common species in coral reefs. (1–6)

Upeneus bensasi (Temminck & Schlegel): Yellow Sea; East China Sea; Taiwan; South China Sea. Near shore bottom fish, water depth 20–40 m. (1–6)

U. luzonius Jordan & Seale: South China Sea. Near shore fish. (1–6)

U. moluccensis (Bleeker): Taiwan; South China Sea. (1–6)

U. quadrilineatus Chen & Wang: Taiwan; South China Sea. (1–6)

U. subvittatus (Temminck & Schlegel): South China Sea. Near shore bottom fish. (1–6)

U. sulphureus Cuvier & Valenciennes: East China Sea; Taiwan Strait; South China Sea. Near shore small fish. (1–6)

U. tragula Richardson: East China Sea; Taiwan Strait; South China Sea. Near shore. (1–6)

U. vittatus (Forskål): Taiwan; South China Sea. (1–6)

PEMPHERIDAE

Parapriacanthus compressus White: Taiwan. (8)

P. ransonneti Steindachner: Taiwan; South China Sea. (1–6)

P. vanicolensis Cuvier: Taiwan. (8)

Pempheris japonicus Döderlein: South China Sea. (1–6)

P. nyctereutes Jordan & Evermann: Taiwan Strait; South China Sea. (1–6)

P. oualensis Cuvier & Valenciennes [*P. molucca*]: East China Sea; Taiwan Strait; South China Sea. (1–6)

P. xanthopterus Toninaga: South China Sea. (1–6)

EPHIPPIDAE

Ephippus orbis (Bloch): East China Sea; Taiwan Strait; South China Sea. Rocky reefs or coral reefs. (1–6)

Platax batavianus Cuvier & Valenciennes: East China Sea; Taiwan Strait; South China Sea. (1–6)

P. orbicularis (Forskål): East China Sea; Taiwan; South China Sea. (1–6)

P. pinnatus (Linnaeus): Southern Taiwan. Rocky reefs. (8)

P. teira (Forskål): East China Sea; Taiwan Strait; South China Sea. Often under floating objects. (1–6)

DREPANIDAE

Drepane longimana (Bloch & Schneider): East China Sea; Taiwan Strait; South China Sea. Near shore shallow water. (1–6)

D. punctata (Linnaeus): East China Sea; Taiwan Strait; South China Sea. Near shore shallow water. (1–6)

PSETTIDAE [MONODACTYLIDAE]

Monodactylus argenteus (Linnaeus): East China Sea; Taiwan Strait; South China Sea. Rocky shore or coral reefs. (1–6)

SCATOPHAGIDAE

Scatophagus argus (Linnaeus): East China Sea; Taiwan Strait; South China Sea. Near shore rocky reefs and seaweed. (1–6)

SCORPIDAE

Microcanthus strigatus (Cuvier & Valenciennes): Yellow Sea; East China Sea; Taiwan; South China Sea. Rocky shore. (1–6)

CHAETODONTIDAE

Apolemichthys trimaculatus (Lacépéde): Taiwan. (8)

Centrophyge bicolor (Bloch): Taiwan; South China Sea. Rocky reefs. (1–6)

C. bispinosus (Günther) [*Xiphipops bispinosus*]: Taiwan; Xisha (=Paracel Islands). Rocky reefs. (1–6)

C. ferrugatus Randall & Burgess [*Xiphipops ferrugatus*]: Taiwan; South China Sea. Rocky reefs. (1–6)

C. fisheri (Snyder): South China Sea. (1–6)

C. flavicauda (Fraser-Brunner) [*Xiphipops flavicauda*]: Taiwan. Rocky reefs. (1–6)

C. heraldi Woods & Schultz: Southern Taiwan; South China Sea. Rocky reefs. (1–6)

C. interruptus (Tanake): Taiwan. (8)

C. nox (Bleeker) [*Xiphipops nox*]: Lanyu (Taiwan) and southeastern Taiwan. Rocky reefs. (8)

C. tibicen Cuvier & Valenciennes [*Xiphipops tibicen*]: Taiwan; South China Sea. Rocky reefs. (1–6)

C. vroliki (Bleeker): Southern tip of Taiwan; Xisha (=Paracel Islands). (1–6)

Chaetodon adiergastos Seale: Taiwan; South China Sea. Coral reefs. (1–6)

C. argentatus Smith & Radcliffe: Taiwan; South China Sea. Coral reefs. (1–6)

C. auriga Forskål: Taiwan; Hainan; Xisha (=Paracel Islands); Nansha (=Spratly Islands). Coral reefs. (1–6)

C. auripes Jordan & Snyder: Southern and northern Taiwan. Coral reefs. (8)

C. baronessa Cuvier: Southern tip of Taiwan. Coral reefs. (8)

C. bennetti Cuvier & Valenciennes: Xisha (=Paracel Islands). Coral reefs. (1–6)

C. chrysurus Desjardins: Taiwan; Xisha (=Paracel Islands). Coral reefs. (1–6)

C. citrinellus Cuvier & Valenciennes: Taiwan; Xisha (=Paracel Islands). Coral reefs. (1–6)

C. collare Bloch: Taiwan; Hainan; Dongsha (=Pratas Islands); Xisha (=Paracel Islands); Nansha (=Spratly Islands). Coral reefs. (1–6)

C. ephippium Cuvier & Valenciennes: Taiwan; Xisha (=Paracel Islands); Nansha (=Spratly Islands). Coral reefs. (1–6)

C. falcula Bloch: Taiwan; South China Sea. Coral reefs. (1–6)

C. guencheri Ahl: Southern tip of Taiwan. Coral reefs. (1–6)

C. kleini Bloch: Taiwan; Xisha (=Paracel Islands). Coral reefs. (1–6)

C. lineolatus Cuvier & Valenciennes: South of Penghu Islands (Taiwan); South China Sea. Coral reefs. (1–6)

C. lunula (Lacépéde) [*C. fasciatus*]: Penghu Islands (Taiwan); Xisha (=Paracel Islands). Coral reefs. (1–6)

C. melanotus Bloch & Schneider: Penghu Islands (Taiwan); Xisha (=Paracel Islands). Coral reefs. (1–6)

C. meyeri Schneider: Southern tip of Taiwan. Coral reefs. (8)

C. miliaris Quoy & Gaimard: Taiwan; South China Sea. Coral reefs. (1–6)

C. modestus Temminck & Schlegel: China's coast. Coral reefs. (1–6)

C. nippon Döderlein: Northeastern and southeastern Taiwan. Coral reefs. (8)

C. octofasciatus Bloch: Penghu (Taiwan) and southern tip of Taiwan; South China Sea. Coral reefs. (1–6)

C. ornatissimus Cuvier & Valenciennes: Taiwan; Xisha (=Paracel Islands); Nansha (=Spratly Islands). (1–6)

C. plebeius Broussonet: Taiwan; South China Sea. Coral reefs. (1–6)

C. punctatofasciatus Cuvier & Valenciennes: Taiwan; Xisha (=Paracel Islands). Coral reefs. (1–6)

C. quadrimaculatus Gray: Southern Taiwan. Coral reefs. (8)

C. rafflesi Bennett: Taiwan; South China Sea. (1–6)

C. reticulatus Cuvier & Valenciennes: Taiwan; South China Sea. Coral reefs. (1–6)

C. selene Bleeker: Taiwan; South China Sea. Coral reefs. (8)

C. semeion Bleeker: Taiwan; Xisha (=Paracel Islands). Coral reefs. (1–6)

C. speculum Cuvier & Valenciennes: Taiwan; Hainan; Xisha (=Paracel Islands). Coral reefs. (1–6)

C. strigangulus Gmelin: Taiwan; Xisha (=Paracel Islands); Nansha (=Spratly Islands). Coral reefs. (1–6)

C. triangulum Cuvier & Valenciennes: Taiwan; South China Sea. Coral reefs. (1–6)

C. trifascialis Quoy & Gaimard: Taiwan. (8)

C. trifasciatus Mungo Park: Taiwan; Hainan; Dongsha (=Pratas Islands); Xisha (=Paracel Islands); Nansha (=Spratly Islands). Coral reefs. (1–6)

C. ulietensis Cuvier: Southern Taiwan. Coral reefs. (8)

C. unimaculatus Bloch: Taiwan; Xisha (=Paracel Islands). Coral reefs. (1–6)

C. vagabundus Linnaeus: Taiwan; South China Sea. Coral reefs. (1–6)

C. wiebeli Kaup [*C. bellamaris*]: Penghu Islands (Taiwan); Hainan; Dongsha (=Pratas Islands); Xisha (=Paracel Islands); Nansha (=Spratly Islands). Coral reefs. (1–6)

C. xantharus Bleeker: Taiwan. Coral reefs. (8)

Chaetodontoplus chrysocephalus Bleeker: Taiwan. Coral reefs. (8)

C. duboulayi (Günther): Taiwan; South China Sea. (1–6)

C. melanosoma (Bleeker): Taiwan. Coral reefs. (8)

C. personifer (McCulloch): Donggang (Taiwan). (8)

C. septentrionalis (Temminck & Schlegel): Taiwan Strait; South China Sea. (1–6)

Chelmon rostratus (Linnaeus): Taiwan; Hainan; Dongsha (=Pratas Islands); Xisha (=Paracel Islands); Nansha (=Spratly Islands). Coral reefs. (1–6, 8)

Coradion chrysozonus (Cuvier & Valenciennes): Taiwan; South China Sea. Coral reefs. (1–6)

Euxiphipops sexstriatus (Cuvier & Velenciennes): Taiwan; Xisha (=Paracel Islands). Coral reefs. (1–6)

Forcipiger falvissimus Jordan & Mcgreger: Lanyu (Taiwan) and southern tip of Taiwan. Coral reefs. (8)

F. longirostris (Broussonet) [*Chaetodon longirostris*]: Southern Taiwan; Xisha (=Paracel Islands); Nansha (=Spratly Islands). Coral reefs. (1–6)

Genicanthus lamarck (Lacépéde): Taiwan. Rocky reefs. (1–6)

G. macclesfieldiensis Chen: Zhongsha (=Macclesfield Bank). (1–6)

G. melanospilos (Bleeker): Taiwan; Xisha (=Paracel Islands). (1–6)

G. semifasciatus (Kamohara): Taiwan. Rocky reefs. (1–6)

G. watanabei (Yasuda & Tominara): Southeastern Taiwan. Coral reefs. (8)

Hemitaurichthys polylepis (Bleeker): Southern Taiwan. Coral reefs. (8)

H. zoster (Bennett): Taiwan; Hainan; Dongsha (=Pratas Islands); Xisha (=Paracel Islands); Nansha (=Spratly Islands). Coral reefs. (1–6)

Heniochus acuminatus (Linnaeus): Taiwan; Hainan; northern South China Sea; Dongsha (=Pratas Islands); Xisha (=Paracel Islands); Nansha (=Spratly Islands). Coral reefs. (1–6)

H. chrysostomus Cuvier: Southern Taiwan. Coral reefs. (8)

H. monoceros Cuvier & Valenciennes: Taiwan Strait; Hainan; Dongsha (=Pratas Islands); Xisha (=Paracel Islands); Nansha (=Spratly Islands). Coral reefs. (1–6)

H. permutatus Cuvier & Valenciennes: Taiwan; South China Sea. Coral reefs. (1–6)

H. singularius Smith & Radcliffe: Taiwan; Hainan; Dongsha (=Pratas Islands); Xisha (=Paracel Islands); Nansha (=Spratly Islands). Coral reefs. (1–6)

H. varius (Cuvier): Taiwan; Hainan; Dongsha (=Pratas Islands); Xisha (=Paracel Islands); Nansha (=Spratly Islands). Coral reefs. (1–6)

Holacanthus trimaculatus Cuvier & Valenciennes: Taiwan Strait; South China Sea. (1- 6)

H. venustus Yasuda & Tominoga: Northern Taiwan. (1–6)

Parachaetodon ocellatus (Cuvier & Valenciennes): South China Sea. (1–6)

Pomacanthus annularis (Bloch): Taiwan; Xisha (=Paracel Islands). Coral reefs. (1–6)

P. imperator (Bloch): Taiwan; Xisha (=Paracel Islands). Coral reefs. (1–6)

P. semicirculatus (Cuvier & Valenciennes): Taiwan; Xisha (=Paracel Islands). (1–6)

P. sexstriatus (Cuvier): Taiwan. (8)

Pygoplites diacanthus (Boddaert): Taiwan; Xisha (=Paracel Islands); Nansha (=Spratly Islands). Rocky reefs. (1–6)

HISTIOPTERIDAE

Histiopterus typus Temminck & Schlegel: East China Sea; Taiwan; South China Sea. Rocky reefs. (1–6)

Pentaceros japonicus Döderlein: East China Sea; Taiwan. Water depth 210–270 m. (7)

OPLEGNATHIDAE

Oplegnathus fasciatus (Temminck & Schlegel) [*Hoplegnetus fasciatus*]: Yellow Sea; East China Sea; Taiwan Strait. (1–6)

O. punctatus (Temminck & Schlegel) [*Hoplegnetus punctatus*]: Yellow Sea; East China Sea; Taiwan. (1–6)

CEPOLIDAE

Acanthocepola indica (Day): Taiwan; South China Sea. Benthic species. (1–6)

A. krusensterni (Temminck & Schlegel): Taiwan Strait; South China Sea. Bottom fish. (1–6)

A. limbata Cuvier & Valenciennes: East China Sea; Taiwan; South China Sea. Benthic species. (1–6)

Cepola schlegeli (Day): South China Sea. (1–6)

Pseudocepola taeniosoma Kamohara. Taiwan. (8)

EMBIOTOCIDAE

Ditrema temmincki Bleeker: Bohai; Yellow Sea. (1–6)

LABRIDAE

Anampses caeruleopunctatus Rüppell: Taiwan; Hainan; Dongsha (=Pratas Islands); Xisha (=Paracel Islands); Nansha (=Spratly Islands). Rocky reefs. (1–6)

A. diadematus Rüppell: Taiwan; Hainan; Dongsha (=Pratas Islands); Xisha (=Paracel Islands); Nansha (=Spratly Islands). Rocky reefs. (1–6)

A. geographius Cuvier & Valenciennes: Southwestern Taiwan; South China Sea. Rocky reefs. (1–6)

A. melanurus Bleeker: Taiwan; Hainan; Dongsha (=Pratas Islands); Xisha (=Paracel Islands); Nansha (=Spratly Islands). Rocky reefs. (1–6)

A. meleagrides Cuvier & Valenciennes: Southwestern Taiwan; South China Sea. Rocky reefs. (1–6)

A. neoguinaicus Bleeker: Southwestern Taiwan. Rocky reefs. (1–6)

A. twisti Bleeker: Southwestern Taiwan; Hainan; Dongsha (=Pratas Islands); Xisha (=Paracel Islands); Nansha (=Spratly Islands). Rocky reefs. (1–6)

Bodianus anthioides (Bennett): Southeastern Taiwan. Rocky reefs. (8)

B. axillaris (Bennett): Taiwan; Hainan; Dongsha (=Pratas Islands); Xisha (=Paracel Islands); Nansha (=Spratly Islands). Rocky reefs. (1–6)

B. bilunulatus (Lacépéde): Taiwan; Hainan; Dongsha (=Pratas Islands); Xisha (=Paracel Islands); Nansha (=Spratly Islands). (1–6)

B. cylindriatus (Tanaka): Taiwan. (8)

B. diana (Lacépéde): Taiwan; South China Sea. Rocky reefs. (1–6)

B. hirsutus (Lacépéde) [*B. loxozunus*, *B. macrutus*]: Taiwan; Xisha (=Paracel Islands); Nansha (=Spratly Islands). Rocky reefs. (1–6)

B. izuensis Arga & Yoshino: Taiwan. (8)

B. leucostictus Bennett: Taiwan. (8)

B. luteopunctatus (Smith): Taiwan; South China Sea. Rocky reefs. (8)

B. macrourus (Günther): Taiwan. (8)

B. masudai Arga & Yoshino: Taiwan. (8)

B. mesothorax (Bloch & Schneider): Northeastern and southwestern Taiwan. Rocky reefs. (8)
B. oxycephalus (Bleeker) [*B. vulpinus*]: Taiwan; South China Sea. Rocky reefs. (1–6)
B. pacificus (Kamohara): Taiwan. (8)
B. perditio (Quoy & Gaimard): Taiwan; South China Sea. Rocky reefs. (1–6)
B. prognathus (Günther): Southern tip of Taiwan. Rocky reefs. (8)
B. thoracotaeniatus Yamamoto: Taiwan. (8)
Cheilinus bimaculatus Cuvier & Valenciennes: Southwestern Taiwan; South China Sea. Rocky shore. (1–6)
C. celebieus Bleeker: Taiwan; Hainan; Dongsha (=Pratas Islands); Xisha (=Paracel Islands); Nansha (=Spratly Islands). (1–6)
C. chlorurus (Bloch): Taiwan; Hainan; Dongsha (=Pratas Islands); Xisha (=Paracel Islands); Nansha (=Spratly Islands). Rocky reefs. (1–6)
C. diagrammus (Lacépéde): Taiwan; South China Sea. Rocky reef shallow water. (1–6)
C. fasciatus (Bloch): Taiwan; Hainan; Xisha (=Paracel Islands); Nansha (=Spratly Islands). Often around rocky reefs. (1–6)
C. mentalis Rüppell: East China Sea; Taiwan Strait; South China Sea. (1–6)
C. oxycephalus Bleeker: Southwestern Taiwan; Hainan; Dongsha (=Pratas Islands); Xisha (=Paracel Islands); Nansha (=Spratly Islands). Rocky shore. (1–6)
C. oxyrhynchus Bleeker: Taiwan. (8)
C. rhodochrous Günther: Taiwan; Hainan; Dongsha (=Pratas Islands); Xisha (=Paracel Islands); Nansha (=Spratly Islands). Rocky reef shallow water. (1–6)
C. trilobatus Lacépéde: Taiwan; Xisha (=Paracel Islands); Nansha (=Spratly Islands). Rocky reef shallow water. (1–6)
C. undulatus Rüppell: Taiwan; Hainan; Dongsha (=Pratas Islands); Xisha (=Paracel Islands); Nansha (=Spratly Islands). Rocky reefs. (1–6)
C. unifasciatus Streets: Taiwan. (8)
Cheilio inermis (Forskål): Taiwan; Hainan; Dongsha (=Pratas Islands); Xisha (=Paracel Islands); Nansha (=Spratly Islands). Rocky and sandy area. (1–6)
Choerodon anchorago (Bloch): Taiwan; South China Sea. (1–6)
C. azurio (Jordan & Snyder): Taiwan Strait; South China Sea. Rocky reefs. (1–6)
C. gymnogenys (Günther): Taiwan. (8)
C. jordani (Snyder): Taiwan. (8)
C. melanostigma Fowler & Bean: Taiwan; South China Sea. (1–6)
C. nectemblema (Jordan & Evermann): South China Sea. (1–6)
C. pescadorensis Yu: South China Sea. (1–6)
C. quadrifasciatus Yu: Taiwan. (8)
C. robustus (Günther): Taiwan. (8)
C. schoenleini (Cuvier & Valenciennes): Taiwan; Hainan; Dongsha (=Pratas Islands); Xisha (=Paracel Islands); Nansha (=Spratly Islands). Rocky reefs. (1–6)
C. zamboangae (Seale & Bean): Southern tip of Taiwan. Rocky reefs. (8)
Choerodonoides japonicus Kamohara: South China Sea. (1–6)
Cirrhilabrus cyanopleura (Bleeker): Taiwan. (8)
C. exquistitus Smith: Southwestern tip of Taiwan. Rocky shore. (8)
C. melanomarginatus Randall & Shen: Southwestern tip of Taiwan. Rocky shore. (8)
C. rubimarginatus Randall: Taiwan. (8)
C. solorensis Bleeker [*C. diagrammus*]: Taiwan; Hainan; Dongsha (=Pratas Islands); Xisha (=Paracel Islands); Nansha (=Spratly Islands). Rocky reefs. (1–6)
C. temminckii Bleeker: Southwestern tip of Taiwan. Rocky shore. (8)
Coris aygula Lacépéde: Taiwan; Hainan; Dongsha (=Pratas Islands); Xisha (=Paracel Islands); Nansha (=Spratly Islands). Rocky reefs. (1–6)
C. dorsomacula (Fowler): Taiwan. (8)
C. gaimard (Quoy & Gaimard): Taiwan; Xisha (=Paracel Islands); Nansha (=Spratly Islands). Rocky reefs. (1–6)
C. musume (Jordan & Snyder) [*C. picta* (8)]: Taiwan; South China Sea. (1–6)
C. picta (Bloch & Schneider): Taiwan. (8)
C. pictoides Randall & Kuiter: Taiwan. (8)
Cymolutes lecluse (Quoy & Gaimard): South China Sea. (1–6)
C. torquatus (Valenciennes): Taiwan. (8)
Duymeria flagellifera (Cuvier & Valenciennes): Guangdong; Hainan; Dongsha (=Pratas Islands); Xisha (=Paracel Islands); Nansha (=Spratly Islands). (1–6)
Epibulus insidiator (Pallas): Taiwan; Hainan; Dongsha (=Pratas Islands); Xisha (=Paracel Islands); Nansha (=Spratly Islands). Rocky reefs. (1–6)
Gomphosus varius Lacépéde [*G. tricolor*]: Hainan; Dongsha (=Pratas Islands); Xisha (=Paracel Islands); Nansha (=Spratly Islands). (1–6)
Halichoeres ambionensis (Bleeker): South China Sea. Rocky reefs. (1–6)
H. argus (Bloch & Schneider): Taiwan Strait; South China Sea. Rocky reefs. (1–6)
H. bioceatus Schultz: Southwestern Taiwan. Rocky shore. (8)
H. centiquadrus (Lacépéde): East China Sea; Taiwan; South China Sea; Xisha (=Spartly Islands). Rocky reefs. (1–6)
H. chrysus Randall: Taiwan. (8)
H. cyanopleura (Bleeker): South China Sea. Rocky reefs. (1–6)
H. dussumieri (Valenciennes): Taiwan. (8)
H. hartzfeldi (Bleeker): Hainan; Dongsha (=Pratas Islands); Xisha (=Paracel Islands); Nansha (=Spratly Islands). Rocky reefs. (1–6)
H. hortulanus Lacépéde: Taiwan. (8)
H. hyrili (Bleeker): Hainan; Dongsha (=Pratas Islands); Xisha (=Paracel Islands); Nansha (=Spratly Islands). (1–6)
H. leparensis (Bleeker): Taiwan; South China Sea. Rocky reefs. (1–6)

H. margaritaceus (Cuvier & Valenciennes): Taiwan; Xisha (=Paracel Islands); Nansha (=Spratly Islands). Rocky reefs. (1–6)

H. marginatus Rüppell: Taiwan; Hainan; Dongsha (=Pratas Islands); Xisha (=Paracel Islands); Nansha (=Spratly Islands). Rocky reefs. (1–6)

H. melanochir (Fowler & Bean): Taiwan; Hainan; Dongsha (=Pratas Islands); Xisha (=Paracel Islands); Nansha (=Spratly Islands). Rocky reefs. (1–6)

H. melanurus (Bleeker): Taiwan; Hainan; Dongsha (=Pratas Islands); Xisha (=Paracel Islands); Nansha (=Spratly Islands). Rocky reefs. (1–6)

H. miniatus (Cuvier & Valenciennes): Taiwan; Hainan; Dongsha (=Pratas Islands); Xisha (=Paracel Islands); Nansha (=Spratly Islands). Rocky reefs. (1–6)

H. nebulosus (Cuvier & Valenciennes): Taiwan; Hainan; Dongsha (=Pratas Islands); Xisha (=Paracel Islands); Nansha (=Spratly Islands). Rocky reefs. (1–6)

H. nigrescens (Bloch & Schneider): Taiwan; Hainan; Dongsha (=Pratas Islands); Xisha (=Paracel Islands); Nansha (=Spratly Islands). Rocky reefs. (1–6)

H. ornatissimus (Garrett): Taiwan. (8)

H. pelicieri Randall & Smith: Taiwan. (8)

H. poecilopterus (Temminck & Schlegel): Taiwan; Hainan; Dongsha (=Pratas Islands); Xisha (=Paracel Islands); Nansha (=Spratly Islands). Rocky reefs. (1–6)

H. prosopein (Bleeker): Taiwan. (8)

H. purpurascens (Bleeker): Taiwan; South China Sea. Rocky reefs. (1–6)

H. scapularis (Bennett): Southwestern tip of Taiwan; South China Sea. Rocky reefs. (1- 6)

H. tenuispinis (Günther): Taiwan; South China Sea. Shallow sea around rocky reefs. (1–6)

H. timorensis (Bleeker): Taiwan. (8)

H. trimaculatus (Quoy & Gaimard): Southwestern Taiwan; Xisha (=Paracel Islands); Nansha (=Spratly Islands). Rocky reefs. (1–6)

Hemigymnus fasciatus (Bloch): Southern tip of Taiwan; Hainan; Dongsha (=Pratas Islands); Xisha (=Paracel Islands); Nansha (=Spratly Islands). Rocky reefs. (1–6)

H. melapterus (Bloch): Southwestern Taiwan; Hainan; Dongsha (=Pratas Islands); Xisha (=Paracel Islands); Nansha (=Spratly Islands). Rocky reefs. (1–6)

Hemipteronotus aneitensis (Günther) [*Xyrichthys aneitensis* (8)]: Hainan; Dongsha (=Pratas Islands); Xisha (=Paracel Islands); Nansha (=Spratly Islands). (1–6)

H. caerulopunctatus Yu [*Xyrichthys caerulopunctatus* (8)]: Taiwan Strait; South China Sea. Water depth 30–40 m. (1–6)

H. evides Jordan & Richardson [*Xyrichthys evides* (8)]: South China Sea. (1–6)

H. melanopus (Bleeker) [*Xyrichthys melanopus* (8)]: Taiwan; Hainan; Dongsha (=Pratas Islands); Xisha (=Paracel Islands); Nansha (=Spratly Islands). (1–6)

H. pentadactylus (Linnaeus) [*Xyrichthys pentadactylus* (8)]: South China Sea. (1–6)

H. verrens (Jordan & Evermann) [*Xyrichthys verrens*]: Taiwan; South China Sea. (1–6)

Hologymnosus annulatus (Lacépéde): Taiwan. (8)

H. doliatus (Lacépéde): Taiwan. (8)

H. semidiscus (Lacépéde): Taiwan; Hainan; Dongsha (=Pratas Islands); Xisha (=Paracel Islands); Nansha (=Spratly Islands). Rocky reefs. (1–6)

Iniistius dea Temminck & Schlegel [*Xyrichthys des, Xyrichthys dea* (8)]: Taiwan; South China Sea. Rocky reefs or sandy substrate. (1–6)

I. pavo (Cuvier & Valenciennes) [*Xyrichthys pavo* (8)]: Taiwan Strait; Guangdong; Hainan; Dongsha (=Pratas Islands); Xisha (=Paracel Islands); Nansha (=Spratly Islands). (1–6)

Labrichthys cyanotaenia Bleeker: Hainan; Dongsha (=Pratas Islands); Xisha (=Paracel Islands); Nansha (=Spratly Islands). (1–6)

L. unilineatus (Guichenot): Taiwan. (8)

Labroides bicolor Fowler & Bean: Southwestern tip of Taiwan. Rocky reefs. (1–6)

L. dimidiatus (Cuvier & Valenciennes): Taiwan; Guangdong; Guangxi; Hainan; Dongsha (=Pratas Islands); Xisha (=Paracel Islands); Nansha (=Spratly Islands). Rocky reefs. (1–6)

Labropsis manabei Schmidt: Hainan; Dongsha (=Pratas Islands); Xisha (=Paracel Islands); Nansha (=Spratly Islands). (1–6)

Leptojulis lambdastigma Randall & Ferraris: Taiwan. (8)

Lienardella fasciatus (Günther): Southern tip of Taiwan. Rocky reefs. (8)

Macropharyngodon meleagris (Cuvier & Valenciennes): Taiwan; Hainan; Dongsha (=Pratas Islands); Xisha (=Paracel Islands); Nansha (=Spratly Islands). Rocky reefs. (1–6)

M. negrosensis Herre: Southern tip of Taiwan. Rocky reefs. (8)

Neocirrhilabrus oxyurus Cheng & Wang: Hainan; Dongsha (=Pratas Islands); Xisha (=Paracel Islands); Nansha (=Spratly Islands). (1–6)

Novaculichthys macrolepidotus (Bloch): Taiwan. (8)

N. taeniourus (Lacépéde) [*Xyrichthys taeniourus*]: Taiwan; Hainan; Dongsha (=Pratas Islands); Xisha (=Paracel Islands); Nansha (=Spratly Islands). Rocky reefs. (1–6)

Paracheilinus carpenteri Randall & Lubbok: Taiwan. (8)

Pseudocheilinus evanidus Jenkins: Taiwan. (8)

P. hexataenia (Bleeker): Southern tip of Taiwan; Hainan; Dongsha (=Pratas Islands); Xisha (=Paracel Islands); Nansha (=Spratly Islands). Rocky reefs. (1–6)

P. octotaenia Jenkins: Taiwan. (8)

Pseudocoris awayae (Schmidt) [*Coris awayae*]: Taiwan; Hainan; Dongsha (=Pratas Islands); Xisha (=Paracel Islands); Nansha (=Spratly Islands). (1–6)

P. yamashiroi (Schmidt) [*Coris yamashiroi*]: Hainan; Dongsha (=Pratas Islands); Xisha (=Paracel Islands); Nansha (=Spratly Islands). Rocky reefs. (1–6)

Pseudodax moluccanus (Valenciennes): Taiwan. (8)

Pseudojuloides cerasina (Snyder): Taiwan. (8)
Pseudolabrus gracilis (Steindachner) [*Suezichthys gracilis* (8)]: East China Sea; Taiwan; South China Sea. Rocky reefs. (1–6)
P. japonicus (Houttuyn): East China Sea; Taiwan; South China Sea. Rocky reefs. (1–6)
Pteragogus flagellifera Valenciennes: Taiwan. (8)
Stethojulis axillaris (Quoy & Gaimard): Taiwan; Hainan; Dongsha (=Pratas Islands); Xisha (=Paracel Islands); Nansha (=Spratly Islands). Rocky reefs. (1–6)
S. interrupta (Bleeker): Taiwan; Hainan; Dongsha (=Pratas Islands); Xisha (=Paracel Islands); Nansha (=Spratly Islands). Rocky reefs. (1–6)
S. kalosoma (Bleeker): Taiwan; South China Sea. Rocky reefs. (1–6)
S. linearis (Schltz) [*S. bandanensis*]: Hainan; Dongsha (=Pratas Islands); Xisha (=Paracel Islands); Nansha (=Spratly Islands). (1–6)
S. phekadopleura (Bleeker): Taiwan; South China Sea. Rocky reefs. (1–6)
S. renardi (Bleeker): Taiwan; Hainan; Dongsha (=Pratas Islands); Xisha (=Paracel Islands); Nansha (=Spratly Islands). Rocky reefs. (1–6)
S. strigiventer Bennett: Zhejiang; Taiwan; Hainan; Dongsha (=Pratas Islands); Xisha (=Paracel Islands); Nansha (=Spratly Islands). Rocky reefs. (1–6)
S. trilineata (Bloch & Schneider): Taiwan; South China Sea. Rocky reefs. (1–6)
Thalassoma amblycephalus (Bleeker): Southwestern Taiwan; Hainan; Dongsha (=Pratas Islands); Xisha (=Paracel Islands); Nansha (=Spratly Islands). Rocky reefs. (1–6)
T. cupido (Temmick & Schlegel): Taiwan; Hainan; Dongsha (=Pratas Islands); Xisha (=Paracel Islands); Nansha (=Spratly Islands). Rocky reefs. (1–6)
T. fuscum (Lacépéde): Penghu (Taiwan) and southwestern Taiwan; Hainan; Dongsha (=Pratas Islands); Xisha (=Paracel Islands); Nansha (=Spratly Islands). Rocky reefs. (1–6)
T. hardwicki (Bennett): Taiwan; Xisha (=Paracel Islands); Nansha (=Spratly Islands). Rocky reefs. (1–6)
T. jansenii (Bleeker): Taiwan;Hainan; Dongsha (=Pratas Islands); Xisha (=Paracel Islands); Nansha (=Spratly Islands). Rocky reefs. (1–6)
T. lunare (Linnaeus): Taiwan; Hainan; Dongsha (=Pratas Islands); Xisha (=Paracel Islands); Nansha (=Spratly Islands). Rocky reefs. (1–6)
T. lutescens (Lay & Bennett): Taiwan; Hainan; Dongsha (=Pratas Islands); Xisha (=Paracel Islands); Nansha (=Spratly Islands). Rocky reefs. (1–6)
T. purpureum (Forskål): Southwestern Taiwan; Hainan; Dongsha (=Pratas Islands); Xisha (=Paracel Islands); Nansha (=Spratly Islands). Rocky reefs. (1–6)
T. quinquevittatus Lay & Bennett: Taiwan; Hainan; Dongsha (=Pratas Islands); Xisha (=Paracel Islands); Nansha (=Spratly Islands). Rocky reefs. (1–6)
T. trilobatum Lacépéde: Taiwan. (8)
T. umbrostigma (Rüppell): Hainan; Dongsha (=Pratas Islands); Xisha (=Paracel Islands); Nansha (=Spratly Islands). (1–6)
Wetmorella nigropinnata (Seale): Taiwan. (8)
Xiphocheilus quadrimaculatus Günther: Taiwan Strait; South China Sea. (1–6)
X. typus Bleeker: South China Sea. (1–6)
Xyrichthys twistii (Bleeker): Taiwan. (8)
X. woodi (Jenkins): Taiwan. (8)

SCARIDAE

Bolbometopon bicolor (Rüppell) [*Cetoscarus bicolor* (8), *Chlorurus bicolor*]: Taiwan; South China Sea. Coral reefs or rocky reefs. (1–6)
B. muricatus (Cuvier & Valenciennes) [*Chlorurus muricatus*]: Taiwan; South China Sea. Coral reefs. (1–6)
Calotomus carolinus (Valenciennes): Taiwan. (8)
C. japonicus (Cuvier & Valenciennes): Hainan; Dongsha (=Pratas Islands); Xisha (=Paracel Islands); Nansha (=Spratly Islands). Coral reefs. (1–6)
C. spinidens (Quoy & Gaimard): Taiwan; Hainan; Dongsha (=Pratas Islands); Xisha (=Paracel Islands); Nansha (=Spratly Islands). Coral reefs or rocky reefs. (1–6)
Leptoscarus vaigiensis (Quoy & Gaimard): Taiwan Strait; Guangdong; Hainan; Dongsha (=Pratas Islands); Xisha (=Paracel Islands); Nansha (=Spratly Islands). Coral reefs or rocky reefs. (1–6)
Scarus aeruginosus Cuvier & Valenciennes: Taiwan; Guangdong; Hainan; Dongsha (=Pratas Islands); Xisha (=Paracel Islands); Nansha (=Spratly Islands). Coral reefs or rocky reefs. (1–6)
S. atropectoralis Schultz: Taiwan. Coral reefs. (8, 25)
S. bowersi (Snyder): Taiwan. Coral reefs. (25)
S. chlorodon Jenyns: Hainan; Dongsha (=Pratas Islands); Xisha (=Paracel Islands); Nansha (=Spratly Islands). Coral reefs. (25)
S. dimidiatus Bleeker: Taiwan; Hainan; Dongsha (=Pratas Islands); Xisha (=Paracel Islands); Nansha (=Spratly Islands). Coral reefs. (1–6)
S. fasciatus Cuvier & Valenciennes: Hainan; Dongsha (=Pratas Islands); Xisha (=Paracel Islands); Nansha (=Spratly Islands). Coral reef common species. (1–6)
S. forsteri Cuvier & Valenciennes: Taiwan; Hainan; Dongsha (=Pratas Islands); Xisha (=Paracel Islands); Nansha (=Spratly Islands). Coral reefs. (1–6)
S. frenatus Lacépéde: Fujian; Taiwan; Hainan; Dongsha (=Pratas Islands); Xisha (=Paracel Islands); Nansha (=Spratly Islands). Coral reefs or rocky reefs. (1–6)
S. ghobban Forskål [*S. dussumieri*]: Taiwan; Guangdong; Hainan; Dongsha (=Pratas Islands); Xisha (=Paracel Islands); Nansha (=Spratly Islands). Rocky shore. (1–6)
S. gibbus Rüppell: Taiwan; Xisha (=Paracel Islands); Nansha (=Spratly Islands). Coral reefs or rocky reefs. (1–6)

S. globiceps Cuvier & Valenciennes: Taiwan; Hainan; Dongsha (=Pratas Islands); Xisha (=Paracel Islands); Nansha (=Spratly Islands). Rocky reefs. (1–6)

S. janthochir Bleeker: Hainan; Dongsha (=Pratas Islands); Xisha (=Paracel Islands); Nansha (=Spratly Islands). Rocky reefs. (1–6)

S. lepidus Jenyns: Taiwan; Hainan; Dongsha (=Pratas Islands); Xisha (=Paracel Islands); Nansha (=Spratly Islands). Common species in coral reefs. (1–6)

S. longiceps Cuvier & Valenciennes [*Hipposcarus longiceps* (8)]: Southern Fujian; Taiwan; Hainan; Dongsha (=Pratas Islands); Xisha (=Paracel Islands); Nansha (=Spratly Islands). Coral reefs or rocky reefs. (1–6)

S. lunula (Snyder): Taiwan; Xisha (=Paracel Islands). Common species in coral reefs. (1–6)

S. niger Forskål: Taiwan; Guangdong; Hainan; Dongsha (=Pratas Islands); Xisha (=Paracel Islands); Nansha (=Spratly Islands). (1–6)

S. oedema (Snyder): Taiwan. Coral reefs. (25)

S. oviceps Cuvier & Valenciennes: Taiwan; Hainan; Dongsha (=Pratas Islands); Xisha (=Paracel Islands); Nansha (=Spratly Islands). Coral reefs. (1–6)

S. ovifrons (Temminck & Schlegel): Taiwan; South China Sea. Rocky reefs. (1–6)

S. pyrrhurus (Jordan & Seabe): Taiwan. Coral reefs. (25)

S. scaber Cuvier & Valenciennes: Hainan; Dongsha (=Pratas Islands); Xisha (=Paracel Islands); Nansha (=Spratly Islands). Rocky reefs. (1–6)

S. sordidus Forskål [*S. erythrodon*]: Taiwan; Guangdong; Hainan; Dongsha (=Pratas Islands); Xisha (=Paracel Islands); Nansha (=Spratly Islands). Coral reefs or rocky reefs. (1–6)

S. taeniurus Cuvier & Valenciennes: Taiwan; Hainan; Dongsha (=Pratas Islands); Xisha (=Paracel Islands); Nansha (=Spratly Islands). Common species in coral reefs. (1–6)

S. venosus Cuvier & Valenciennes: Taiwan; Hainan; Dongsha (=Pratas Islands); Xisha (=Paracel Islands); Nansha (=Spratly Islands). Common species in coral reefs. (1–6)

Scarops festivus Valenciennes: Taiwan. (8)

S. forsteni (Bleeker): Taiwan. (8)

S. prasiognathos Valenciennes: Taiwan. (8)

S. psittacus Forssk: Taiwan. (8)

S. rivulatus Valenciennes: Taiwan. (8)

S. rubroviolaceus (Bleeker): Fujian; Taiwan; Hainan; Dongsha (=Pratas Islands); Xisha (=Paracel Islands); Nansha (=Spratly Islands). Coral reefs or rocky reefs. (1–6)

S. schlegeli (Bleeker): Taiwan. (8)

POMACENTRIDAE

Abudefduf anabatoides (Bleeker) [*Neopomacentrus anabatoides* (8)]: Hainan; Dongsha (=Pratas Islands); Xisha (=Paracel Islands); Nansha (=Spratly Islands). (1–6)

A. aureus (Cuvier & Valenciennes): Hainan; Dongsha (=Pratas Islands); Xisha (=Paracel Islands); Nansha (=Spratly Islands). (1–6)

A. bankieri (Richardson) [*Neopomacentrus bankieri* (8)]: Hainan; Dongsha (=Pratas Islands); Xisha (=Paracel Islands); Nansha (=Spratly Islands). (1–6)

A. behni (Bleeker): Hainan; Dongsha (=Pratas Islands); Xisha (=Paracel Islands); Nansha (=Spratly Islands). (1–6)

A. bengalensis (Bloch): Taiwan; South China Sea. (1–6)

A. biocellatus (Quoy & Gaimard) [*Chrysiptera biocellatus* (8)]: Taiwan; Hainan; Xisha (=Paracel Islands); Nansha (=Spratly Islands). (1–6)

A. coelestinus (Cuvier & Valenciennes) [*Pomacentrus coelestinus* (8)]: Taiwan; Guangdong; Xisha (=Paracel Islands); Nansha (=Spratly Islands). (1–6)

A. curacao (Bloch): Taiwan; South China Sea. (1–6)

A. cyaneus (Quoy & Gaimard) [*Chrysiptera cyaneus* (8)]: Xisha (=Paracel Islands); Nansha (=Spratly Islands). (1–6)

A. dicki (Lienard) [*Plectroglyphidodon dicki* (8)]: Taiwan; Hainan; Dongsha (=Pratas Islands); Xisha (=Paracel Islands); Nansha (=Spratly Islands). (1–6)

A. glaucus (Cuvier & Valenciennes) [*Chrysiptera glaucus* (8)]: Taiwan; Hainan; Dongsha (=Pratas Islands); Xisha (=Paracel Islands); Nansha (=Spratly Islands). (1–6)

A. lacrymatus (Quoy & Gaimard): Taiwan; Hainan; Dongsha (=Pratas Islands); Xisha (=Paracel Islands); Nansha (=Spratly Islands). (1–6)

A. leucogaster (Bleeker) [*Amblyglyphidodon leucagaster* (8)]: Taiwan; Hainan; Dongsha (=Pratas Islands); Xisha (=Paracel Islands); Nansha (=Spratly Islands). (1–6)

A. lorenzi Hensley & Allen: Taiwan. (8)

A. melas (Cuvier & Valenciennes) [*Neopomacentrus mela* (8)]: Taiwan; Hainan; Dongsha (=Pratas Islands); Xisha (=Paracel Islands); Nansha (=Spratly Islands). (1–6)

A. notatus (Day): Northern and southern Taiwan. Rocky reefs. (8)

A. richardsoni Snyder [*Pomachromis richardsoni* (8)]: Hainan; Dongsha (=Pratas Islands); Xisha (=Paracel Islands); Nansha (=Spratly Islands). (1–6)

A. septemfascistus (Cuvier & Valenciennes): Taiwan; Hainan; Dongsha (=Pratas Islands); Xisha (=Paracel Islands); Nansha (=Spratly Islands). (1–6)

A. sexfasciatus (Lépéde): Taiwan. Rocky reefs. (8)

A. sordidus (Forskål): Taiwan; Gunagdong; Xisha (=Paracel Islands); Nansha (=Spratly Islands). (1–6)

A. thoraataeniatus Fowler & Bean: Taiwan; Hainan; Dongsha (=Pratas Islands); Xisha (=Paracel Islands); Nansha (=Spratly Islands). (1–6)

A. uniocellatus (Quoy & Gaimard) [*Chrysiptera uniocellatus* (8)]: Taiwan; Hainan; Dongsha (=Pratas Islands); Xisha (=Paracel Islands); Nansha (=Spratly Islands). (1–6)

A. vaigiensis (Quoy & Gaiamard): Taiwan; Guangdong; Hainan; Dongsha (=Pratas Islands); Xisha (=Paracel Islands); Nansha (=Spratly Islands). (1–6)

A. xanthozona (Bleeker): Taiwan; Hainan; Dongsha (=Pratas Islands); Xisha (=Paracel Islands); Nansha (=Spratly Islands). (1–6)

A. zonatus (Cuvier & Valenciennes) [*Chrysiptera zonatus* (8)]: Hainan; Dongsha (=Pratas Islands); Xisha (=Paracel Islands); Nansha (=Spratly Islands). (1–6)

Amblyglyphidodon aures (Cuvier): Southern tip of Taiwan. Rocky reefs. (8)

A. curacao (Bloch): Southern tip of Taiwan. Rocky reefs. (8)

Amphiprion akallopispisus Bleeker: Hainan; Dongsha (=Pratas Islands); Xisha (=Paracel Islands); Nansha (=Spratly Islands). (1–6)

A. bicinctus Rüppell: Taiwan; Guangdong; Xisha (=Paracel Islands); Nansha (=Spratly Islands). (1–6)

A. clarkii (Bennett): Taiwan. Rocky reefs, symbiotic with sea anemone. (8)

A. frenatus Brevoort: Taiwan; Hainan; Dongsha (=Pratas Islands); Xisha (=Paracel Islands); Nansha (=Spratly Islands). Rocky reefs. (1–6)

A. ocellaris Cuvier: Lu Island (Taiwan) and southern Taiwan. Rocky reefs, symbiotic with sea anemone. (8)

A. percula (Lacépéde): Hainan; Dongsha (=Pratas Islands); Xisha (=Paracel Islands); Nansha (=Spratly Islands). (1–6)

A. perideraion Bleeker: Lanyu and Lu Island (Taiwan) and southern Taiwan; Hainan; Dongsha (=Pratas Islands); Xisha (=Paracel Islands); Nansha (=Spratly Islands). Rocky reefs. (1–6)

A. polymnus (Linnaeus): Hainan; Dongsha (=Pratas Islands); Xisha (=Paracel Islands); Nansha (=Spratly Islands). (1–6)

Cheiloprion labiatus (Day): Taiwan; South China Sea. (1–6)

Chromis alleni Randall, Ida & Moyer: Taiwan; South China Sea. (1–6)

C. analis (Cuvier): Taiwan; South China Sea. Rocky reefs. (1–6)

C. atripectoralis Welancler: Southern tip of Taiwan. Rocky reefs. (8)

C. atripes Fowler: Northeastern Taiwan. Rocky reefs. (8)

C. caeruleus (Cuvier & Valenciennes): Hainan; Dongsha (=Pratas Islands); Xisha (=Paracel Islands); Nansha (=Spratly Islands). (1–6)

C. chrysura (Bliss) [*C. isharae*]: Taiwan; Hainan; Dongsha (=Pratas Islands); Xisha (=Paracel Islands); Nansha (=Spratly Islands). (1–6)

C. cinerascens (Cuvier): Taiwan. (8)

C. dimidiatus (Klunzinger): Taiwan; Hainan; Dongsha (=Pratas Islands); Xisha (=Paracel Islands); Nansha (=Spratly Islands). (1–6)

C. elerae Fowler & Bean: South China Sea. (1–6)

C. flavomaculata Kamohara & Schultz: Southern tip of Taiwan. Rocky reefs. (8)

C. fumea (Tanaka): Taiwan; South China Sea. (1–6)

C. lepidolepis Bleeker: Taiwan; Hainan; Dongsha (=Pratas Islands); Xisha (=Paracel Islands); Nansha (=Spratly Islands). (1–6)

C. magaritifer Fowler: Southern tip of Taiwan. Rocky reefs. (8)

C. mirationis Tanaka: East China Sea; Taiwan; South China Sea. (1–6)

C. notatus (Temminck & Schlegel): East China Sea; Taiwan; South China Sea. Rocky reefs. (1–6)

C. ovatiformis Fowler: Taiwan; Hainan; Dongsha (=Pratas Islands); Xisha (=Paracel Islands); Nansha (=Spratly Islands). (1–6)

C. retrofasciatus Weber: Taiwan. (8)

C. ternatensis (Bleeker): Taiwan; Hainan; Dongsha (=Pratas Islands); Xisha (=Paracel Islands); Nansha (=Spratly Islands). Rocky shore. (1–6)

C. vanderbilti (Fowler): Southern tip of Taiwan and Lu Island (Taiwan). Rocky reefs. (8)

C. viridis (Cuvier): Taiwan. (8)

C. weberi Fowler & Bean: Southern tip of Taiwan. Rocky reefs. (8)

C. xanthochir (Bleeker): Taiwan; South China Sea. (1–6)

C. xanthura (Bleeker): Southern tip of Taiwan. Rocky reefs. (8)

C. xanthurus (Bleeker): Taiwan; Hainan; Dongsha (=Pratas Islands); Xisha (=Paracel Islands); Nansha (=Spratly Islands). (1–6)

Chrysiptera glauca (Cuvier): Taiwan. Rocky reefs. (8)

C. leucopoma (Lessen): Taiwan. Rocky reefs. (8)

C. rex (Snyder): Southern tip of Taiwan. Rocky reefs. (8)

C. starcki (Allen): Northern and southern tips of Taiwan. Rocky reefs. (8)

Dascyllus aruanus (Linnaeus): Taiwan; Hainan; Dongsha (=Pratas Islands); Xisha (=Paracel Islands); Nansha (=Spratly Islands). Coral reefs. (1–6)

D. marginatus (Rüppell): Taiwan; Hainan; Dongsha (=Pratas Islands); Xisha (=Paracel Islands); Nansha (=Spratly Islands). Rocky reefs. (1–6)

D. melanurus Bleeker: Taiwan. (8)

D. reticulatus (Richardson): Southern tip of Taiwan. Rocky reefs. (8)

D. trimaculatus (Rüppell): Taiwan; Hainan; Dongsha (=Pratas Islands); Xisha (=Paracel Islands); Nansha (=Spratly Islands). Symbiotic with sea anemone. (1–6)

Daya jordani (Rutter): East China Sea; Taiwan Strait; South China Sea. Near shore, water depth 30–50 m. (1–6)

Hemiglyphidodon plagimetopon (Bleeker): Hainan; Dongsha (=Pratas Islands); Xisha (=Paracel Islands); Nansha (=Spratly Islands). (1–6)

Neoglyphidodon nigroris (Cuvier): Eastern and southern Taiwan. (1–6)

Neopomacentrus azysron (Bleeker): Taiwan. (8)

Plectroglyphidodon imparipennis (Villant & Sauvage): Taiwan. Rocky reefs. (8)

P. johnstonianus Fowler & Ball: Southern tip of Taiwan. Rocky reefs. (8)

P. lacrymatus (Quoy & Gaimard): Northeastern and southern Taiwan. Rocky reefs. (8)

P. leucozona (Bleeker): Southern tip of Taiwan. Rocky reefs. (8)

P. phoenixensis (Schultz): Taiwan. (8)

Pomacentrus albifasciatus Schelegel & Müller [*Stegastes albifasciatus* (8)]: Taiwan; Hainan; Dongsha (=Pratas Islands); Xisha (=Paracel Islands); Nansha (=Spratly Islands). (1–6)
P. amboinensis Bleeker: Southern tip of Taiwan. Rocky reefs. (8)
P. chrysurus Cuvier: Taiwan. (8)
P. coelestis Jordan & Starks: Taiwan. (8)
P. dorsalis Gill [*P. bankanensis* (8)]: South China Sea. (1–6)
P. grammorhynchus Fowler: Taiwan. (8)
P. jenkinsi Jordan & Evermann: Taiwan; Hainan; Dongsha (=Pratas Islands); Xisha (=Paracel Islands); Nansha (=Spratly Islands). (1–6)
P. lepidogenys Fowler & Bean: Taiwan. (8)
P. lividus (Bloch & Schneider) [*Stegastes lividus* (8)]: Hainan; Dongsha (=Pratas Islands); Xisha (=Paracel Islands); Nansha (=Spratly Islands). (1–6)
P. melanopterus Bleeker [*Neopomacentrus brachialis* (8)]: Taiwan; Hainan; Dongsha (=Pratas Islands); Xisha (=Paracel Islands); Nansha (=Spratly Islands). (1–6)
P. moluccensis Bleeker: Taiwan; Hainan; Dongsha (=Pratas Islands); Xisha (=Paracel Islands); Nansha (=Spratly Islands). (1–6)
P. nagasakiensis Tanaka: Taiwan. (8)
P. nigricans (Lacépéde) [*Stegastes nigricans* (8)]: Taiwan; Hainan; Dongsha (=Pratas Islands); Xisha (=Paracel Islands); Nansha (=Spratly Islands). (1–6)
P. nigromarginatus Allen: Taiwan. (8)
P. niomatus De Vis: Hainan; Dongsha (=Pratas Islands); Xisha (=Paracel Islands); Nansha (=Spratly Islands). (1–6)
P. notophathalmus Bleeker [*Dischistodus melanotus* (8)]: Taiwan; Hainan; Dongsha (=Pratas Islands); Xisha (=Paracel Islands); Nansha (=Spratly Islands). (1–6)
P. pavo (Bloch): Taiwan; Hainan; Dongsha (=Pratas Islands); Xisha (=Paracel Islands); Nansha (=Spratly Islands). (8)
P. perspicilliatus Cuvier & Valenciennes [*Dischistodus perspicillatus* (8)]: Taiwan; Hainan; Dongsha (=Pratas Islands); Xisha (=Paracel Islands); Nansha (=Spratly Islands). (1–6)
P. philippinus Evermann & Seale: Taiwan; Hainan; Dongsha (=Pratas Islands); Xisha (=Paracel Islands); Nansha (=Spratly Islands). (1–6)
P. prosopotaenia Bleeker [*Dischistodus prosopotaenia* (8)]: Taiwan; Hainan; Dongsha (=Pratas Islands); Xisha (=Paracel Islands); Nansha (=Spratly Islands). (1–6)
P. stigma Montalbon: Taiwan. (8)
P. taeniurus Bleeker [*Dischistodus taeniurus* (8)]: Taiwan; Hainan; Dongsha (=Pratas Islands); Xisha (=Paracel Islands); Nansha (=Spratly Islands). (1–6)
P. tripunctatus Cuvier & Valenciennes: Guangdong; Guangxi; Hainan; Dongsha (=Pratas Islands); Xisha (=Paracel Islands); Nansha (=Spratly Islands). (1–6)
P. vaiuli Jordan & Seale: Taiwan. Rocky reefs. (8)
P. violascens (Bleeker) [*Neopomacentrus violascens* (8)]: Taiwan; Hainan; Dongsha (=Pratas Islands); Xisha (=Paracel Islands); Nansha (=Spratly Islands). (1–6)
Pristotis jerdoni (Day): Southern tip of Taiwan. Rocky reefs. (8)
Stegastes altus (Okada & Ikeda): Taiwan. (8)
S. apicalis (De Vis): Taiwan. (8)
S. aureus (Fowler): Taiwan. (8)
S. fasciolatus (Ogilby): Northeastern Taiwan. Rocky shore. (8)
S. insularis Allen & Emery: Taiwan. (8)
Teixeirichthys formosana (Fowler & Bean): Taiwan. (8)
T. jordani (Rutter): Northeastern and southwestern Taiwan. Muddy sand substrate. (8)

CIRRHITIDAE

Amblycirrhitus bimacula (Jenkins): Southern tip of Taiwan; South China Sea. Rocky reefs. (8)
A. unimacula (Kamahara): Taiwan. (8)
Cirrhitichthys aprinus (Cuvier & Valenciennes): Taiwan Strait; South China Sea. (1–6)
C. aureus (Temminck & Schlegel): Taiwan; Guangdong; Guangxi; Hainan; Dongsha (=Pratas Islands); Xisha (=Paracel Islands); Nansha (=Spratly Islands). (1–6)
C. falco Randall: Lanyu (Taiwan) and southern tip of Taiwan; South China Sea. Rocky shore. (1–6)
C. oxycephalus (Bleeker): Taiwan; South China Sea. (1–6)
Cirrhitus pinnulatus (Bloch & Schneider): Taiwan; South China Sea. Often in rocky reefs. (1–6)
Cyprinocirrhites polyactis (Bleeker): Southern tip of Taiwan; South China Sea. Rocky reefs. (1–6)
Paracirrhites arcatus (Cuvier & Valenciennes): Lanyu (Taiwan) and southern tip of Taiwan; South China Sea; Dongsha (=Pratas Islands); Xisha (=Paracel Islands); Nansha (=Spratly Islands). Rocky shore. (1–6)
P. forsteri (Bloch & Schneider): Lanyu (Taiwan) and southern tip of Taiwan; South China Sea. Rocky shore. (1–6)

APLODACTYLIDAE

Goniistius quadricoris (Günther): Northern and southern Taiwan; South China Sea. Rocky shore. (1–6)
G. zebra (Döderlein): Taiwan. (8)
G. zonatus (Cuvier & Valenciennes): Taiwan Strait; South China Sea. Rocky reefs. (1-6)

BEMBROPIDAE

Bembrops caudimacula Steindachner: East China Sea; Taiwan. Water depth within 150 m. (1-6)
B. curvatura Okada & Suzuki: South China Sea. Deep water. (1–6)
Chrionenma chlorotaenia McKay: Taiwan. (8)
C. chryseres Gilbert: East China Sea; southwestern Taiwan. (1–6)

PARAPERCIDAE [MAGILODIDAE]

Kochichthys flavofasciata (Kamohara): Taiwan. (8)
Parapercis alboguttata (Günther): South China Sea. (1–6)

P. aurantiaca (Döderlein): Taiwan; South China Sea. (1–6)
P. cephalopunctata (Seale): Southern Taiwan. Rocky reefs. (8)
P. clathrata Ogilby: Northeastern and southern Taiwan. Rocky reefs. (8)
P. cylindrica (Bloch): East China Sea; northeastern and southern Taiwan; South China Sea. Rocky reefs. (1–6)
P. decemfasciata (Franz): Northeastern and southwestern Taiwan. Muddy sand substrate. (8)
P. hexophthalma (Cuvier & Valenciennes): Taiwan, Xisha (=Paracel Islands). Coral reef bottom fish. (1–6)
P. kamoharai Schultz: Southwestern and southern Taiwan. Muddy sand substrate. (8)
P. mimaseana (Kamohara): East China Sea; Taiwan. Water depth 310 m. (7, 8)
P. multifasciata (Döderlein): Notheastern and southwestern Taiwan; South China Sea. Muddy sand substrate. (1–6)
P. multiplicata Randall: Taiwan. (8)
P. muronis Tanaka: Southwestern Taiwan; South China Sea. Muddy sand substrate. (1- 6)
P. ommatura (Jordan & Starks): East China Sea; Taiwan Strait; South China Sea. (1–6)
P. polyophthalma (Cuvier): Taiwan. (8)
P. pulchella (Temminck & Schlegel): Taiwan Strait; South China Sea. Rocky reefs. (1- 6)
P. punctata (Cuvier & Valenciennes): South China Sea. Near shore bottom fish. (1–6)
P. quadrispinosus Weber: Xisha (=Paracel Islands). (1–6)
P. sexfasciata (Temminck & Schlegel): East China Sea; northeastern and southwestern Taiwan. Muddy sand substrate. (8)
P. snyderi (Jordan & Starks): Taiwan; South China Sea. (1–6)
P. somaliensis Schultz: Taiwan. (8)
P. striolata (Weber): Taiwan; South China Sea. (1–6)
P. tetracanthus (Lecépéde): Taiwan; South China Sea. (1–6)
P. xanthozona (Bleeker): Penghu (Taiwan) and southern tip of Taiwan; South China Sea. Sandy substrate. (1–6)

CHIASMODONTIDAE

Chiasmodon niger Johnson: Eastern Taiwan. Deep sea. (8)
Pseudoscopelus scriptus Sagamianus & Tanaka: East China Sea. Deep water. (7)

HEMEROCOETIDAE

Acanthaphritis grandisquamis Günther: South China Sea. (1–6)

TRICHONOTIDAE

Trichonotus filamentosus (Steindachner): South China Sea. (1–6)
T. setiger Bloch & Schneider: Taiwan; South China Sea. (1–6)

OWSTONIDAE

Owstonia tosaensis Kamohara: Taiwan; South China Sea. Water depth about 200 m, bottom fish. (1–6)
O. totomiensis Tanaka: Southwestern Taiwan. (8)

OPISTOGNATHIDAE

Gnathypops evermanni Jordan & Snyder [*Opistognathus evermanni* (8)]: Taiwan; South China Sea. (1–6)
Opistognathus castelnaui Bleeker: Taiwan. (8)
O. hongkongiensis Chan: Taiwan. (8)
Stalix immculatus Xu & Zhan: East China Sea. (1–6)
S. sheni Smith-Vaniz: Taiwan. (8)

URANOSCOPIDAE

Gnathagnus elongatus (Temminck & Schlegel): Yellow Sea; East China Sea; South China Sea. Near shore bottom fish. (1–6)
Ichthyoscopus lebeck (Bloch & Schneider): East China Sea; Taiwan Strait; South China Sea. Near shore bottom fish. (1–6)
Uranoscopus bicinctus Temminck & Schlegel: East China Sea; Taiwan Strait; South China Sea. Near shore bottom fish. (1–6)
U. chinensis Guichnot: Taiwan. (8)
U. japonicus Houttuyn: Yellow Sea; East China Sea; South China Sea. Water depth 7- 10 m. (1–6)
U. oligolepis Bleeker: East China Sea; Taiwan Strait; South China Sea. Near shore bottom fish. (1–6)
Zalescopus tosae Jordan & Hubbs: East China Sea; South China Sea. Bottom fish. (1–6)

CHAMPSODONTIDAE

Champsodon atridorsalis Ohiai & Nakamura: South China Sea. (1–6)
C. guentheri Regan: Taiwan. (8)
C. snyderi Franz [*C. capensis*]: East China Sea; Taiwan Strait; South China Sea. Deep water. (1–6)

CREEDIIDAE

Limnichthys fasciatus Waite: Taiwan. (8)

PERCOPHIDAE

Pteropsaron evolans Jordan & Snyder: Taiwan. (8)
Spinapsaron barbatus Okamura & Kishida: Taiwan. Coral reefs. (8)
S. formosensis (Kao & Shen): Taiwan. (8)

TRIPTERYGIDAE

Enneapterygius hemimelas (Kner & Steindachner): Taiwan. (8)
E. minutus (Günther): Taiwan. (8)
E. nanus (Schultz): Taiwan. (8)
Helcogramma habena William & McCormick: Taiwan. (8)
H. obtusirostre (Klunzinger): Taiwan. (8)
H. striata Hadley Hansen: Taiwan. (8)
Tripterygion etheostoma Jordan & Snyder [*Enneapterygius etheostoma* (8)]: Taiwan; South China Sea. Rocky reefs. (1-6)
T. fuligicauda (Fowler): South China Sea. (1-6)

CLINIDAE

Spriangeratus xanthosome (Bleeker): Northern and southern Taiwan. Rocky shore. (8)

BLENNIIDAE

Alticus saliens (Forster): Taiwan; South China Sea. (1-6)
Andamia pacifica Tomiyama: South China Sea. (1-6)
A. reyi (Sauvage): Taiwan. (8)
A. tetradactylus (Bleeker): Taiwan. (8)
Aspidontus filamentosus (Cuvier & Valenciennes): Xisha (=Paracel Islands). (1-6)
A. taeniatus Quoy & Gaimard: Southern tip of Taiwan; Xisha (=Paracel Islands). Rocky reefs. (1-6)
Atrosalarias fuscus (Rüppell): South China Sea. (1-6)
Blennius yatabei Jordan & Snyder [*Parablennius yatabei* (8)]: Yellow Sea; East China Sea; Taiwan. (1-6)
Cirripectes castaneus (Valenciennes): Taiwan. (8)
C. filamentosus (Alleyne & Macleay): Taiwan. (8)
C. fuscoguttatus Strasburg & Schultz: Taiwan. (8)
C. imitater Williams: Taiwan. (8)
C. perustus Smith: Taiwan. (8)
C. polyzona (Bleeker): Taiwan. (8)
C. quagga (Fowler & Ball): Taiwan. (8)
C. variolosus (Cuvier & Valenciennes): Lanyu (Taiwan); South China Sea. Rocky reefs. (1-6)
Dasson breviceps (Valenciennes): Taiwan. (8)
D. springeri Smith-Vaniz: Taiwan. (8)
D. trossulus (Jordan & Snyder): South China Sea. (1-6)
D. variabilis Cantor: Taiwan. (8)
Ecsenius bicolor (Day): Taiwan. (8)
E. frontalis (Cuvier & Valenciennes) [*E. namiye* (8)]: Northern Taiwan and Penghu (Taiwan); South China Sea. (1-6)
E. lineatus Klauswitz: Southern Taiwan. Coral reefs. (8)
E. oculus Springer: Taiwan. (8)
E. yaeyamaensis (Aoyagi): Taiwan. (8)
Entomacrodus caudofasciatus (Regan): Taiwan. (8)
E. decussatus Bleeker: Taiwan. (8)
E. epalzeocheilos (Bleeker): Taiwan. (8)
E. niuafoouensis (Fowler): Taiwan. (8)
E. stellifer lighti (Herre): Taiwan. (8)
E. striatus (Quoy & Gaimard): Taiwan. (8)
E. thalassinus (Jordan & Seale): Taiwan. (8)
Exallias brevis (Kner): Taiwan. (8)
Istiblennius bilitonensis (Bleeker): Shoreline along southern tip of Taiwan. (8)
I. coronatus (Günther): Taiwan. (8)
I. cynostigma (Bleeker): Shoreline along southern tip of Taiwan. (8)
I. edentulus (Bloch & Schneider): Taiwan. (8)
I. interruptus (Bleeker): Shoreline along southern tip of Taiwan. (8)
I. mulleri (Klunzinger): Taiwan. (8)
Meiacanthus grammistes (Valenciennes): Southern tip of Taiwan. Rocky reefs. (8)
Mimoblennius atrocinctus (Regan): Taiwan. (8)
Omobranchus elegans (Steindachner) [*Enchelyurus kraussi* (8)]: Taiwan; South China Sea. (1-6)
O. fasciolaticeps (Richardson): Taiwan. (8)
O. germaini (Sauvoge): Penghu (Taiwan) and southern Taiwan. Rocky reefs. (1-6)
O. japonicus (Bleeker) [*Dasson japonicus*]: South China Sea. Coral reefs. (1-6)
O. kallosoma (Bleeker) [*Petroscirtes kallosoma*]: South China Sea. Coral reefs. (1-6)
O. kraussi (Klunzinger): South China Sea. (1-6)
O. punctatus (Valenciennes): Taiwan. (8)
O. uekii (Katayama): South China Sea. (1-6)
Parenchelyurus hepburni (Snyder): Taiwan. (8)
Petroscirtes mitratus (Rüppell): Taiwan. (8)
Plagiotremus rhinorhynchos (Bleeker): Southern tip of Taiwan. Rocky shore. (8)
P. spilistius Gill [*Lembeichthys furcocaudalis*]: South China Sea. (1-6)
P. tapeinosoma (Bleeker): Southern tip of Taiwan. Rocky shore. (8)
Praealticus margaritarius (Snyder): Taiwan. (8)
P. striatus Bath: Taiwan. (8)
P. tanegasimae (Jordan & Starks): Taiwan. (8)
Rhabdoblennius ellipes (Jordan & Starks): Taiwan. (8)
Salarias dussumieri (Cuvier & Valenciennes) [*Istiblennius dussumieri* (8)]: Taiwan Strait; South China Sea. Coral reefs. (1-6)
S. edentulus (Bloch & Schneider) [*Istiblennius edentulus*]: Xisha (=Paracel Islands). (1-6)
S. fasciatus (Bloch): Taiwan; South China Sea; Xisha (=Paracel Islands). (1-6)
S. guttatus (Cuvier & Valenciennes): South China Sea. Coral reefs. (1-6)
S. lineatus Cuvier & Valenciennes [*Istiblennius lineatus* (8)]: Taiwan; Xisha (=Paracel Islands). Rocky reefs. (1-6)
S. margaritatus (Kendall & Radcliffe): Xisha (=Paracel Islands). Coral reefs. (1-6)

S. periophthalmus (Cuvier & Valenciennes) [*Istiblennius periophthalmus* (8)]: Taiwan; Xisha (=Paracel Islands). Rocky reefs. (1–6)
Scartela emarginatus (Günther): Taiwan. (8)
Stanulus seychellensis Smith: Taiwan. (8)
Xiphasia setifer Swainson: East China Sea; Taiwan Strait; South China Sea. (1–6)

PHOLIDAE

Azuma emmnion Jordan & Snyder: Bohai; Yellow Sea. (1–6)
Dictyosoma burgeri Van der Hoeven: Bohai; Yellow Sea. (1–6)
Enedrias fangi Wang & Wang: Bohai; Yellow Sea. (1–6)
E. nebulosus (Temminck & Schlegel): Bohai; Yellow Sea; East China Sea. (1–6)

STICHAEIDAE

Alectrias benjamini Jordan & Snyder: Bohai; Yellow Sea. (1–6)
Ernogrammus hexagrammus (Temminck & Schlegel): Bohai; Yellow Sea. (1–6)

CONGROGADIDAE

Congrogadus subducens (Richardson): South China Sea. Coral reefs. (1–6)

CEBIDICHTHYIDAE

Zoarchias uchidai Matsubara: Yellow Sea. (1–6)

ZOARCIDAE

Enchelyopus elongatus Kner [*Zoarces elongatus*]: Bohai; Yellow Sea; East China Sea. (1–6)

SCHINDLERIIDAE

Schindleria pictschmanni (Schindler): South China Sea. (8; 15)
S. praematura (Schindler): South China Sea. (1–6)

AMMODYTIDAE

Ammodytes personatus Girard: Bohai; Yellow Sea. (1–6)
Bleekeria anguilliviridis (Fowler): East China Sea; Taiwan Strait; South China Sea. Near shore, sandy substrate. (1–6)
Embolichthys mitsukurii (Jordan & Evermann): Northeastern Taiwan; South China Sea. Shallow sea muddy sand substrate. (1–6)

DRACONETTIDAE

Draconetta margarostigma Cheng & Tain: South China Sea. (1–6)
D. xenica Jordan et Fowler: South China Sea. (1–6)

CALLIONYMIDAE

Bathycallicaymus formocanus (Fowler): Taiwan; Nansha (=Spratly Islands). (17)
Callionymus altidorsalis Wang & Ye: South China Sea. (1–6)
C. altipinnis Fricke: Taiwan. (8)
C. beniteguri Jordon & Snyder: Bohai; Yellow Sea; East China Sea; Taiwan. Near shore bottom fish. (1–6)
C. curvicornis Valenciennes: Taiwan. (8)
C. enneactis Bleeker: Taiwan. (8)
C. flagris Jordan & Fowler: East China Sea; South China Sea. Near shore. (1–6)
C. formosanus Fricke: Taiwan. (8)
C. hainanensis Li: Taiwan; South China Sea. (1–6)
C. hindsi Richardson: Taiwan. (8)
C. kaianus Günther [*Bathycallionymus kainus*]: South China Sea. (1–6)
C. kitaharae Jordan & Seale: Bohai; Yellow Sea. (1–6)
C. lunaius Temminck & Schlegel: Taiwan; South China Sea. (1–6)
C. marisinensis Fowler: South China Sea. (1–6)
C. meridionalis Suwardji: Taiwan. (8)
C. monofilispinnus Li: South China Sea. (1–6)
C. octostigmatus Fricke: Taiwan. (8)
C. olidus Günther: Bohai; Yellow Sea; East China Sea; Taiwan. Often moves into rivers. (1–6)
C. ornatipinnis Regan: Taiwan. (8)
C. planu Ochiai: Taiwan. (8)
C. richardsoni Bleeker: Bohai; Yellow Sea; East China Sea; South China Sea. Near shore. (1–6)
C. schaapi Bleeker [*Repomucenus schaapi*]: Northeastern Taiwan; South China Sea. Bottom fish. (1–6)
C. valenciennei Temminck & Schlegel: Taiwan. (8)
C. virgis Jordan & Fowler: East China Sea; Taiwan Strait. (1–6)
Calliurichthys belcheri recurvispinnis (Li): Taiwan. (8)
C. dorysus Jordan & Fowler: Southern Yellow Sea; East China Sea; South China Sea. Near shore bottom fish. (1–6)
C. filamentous (Cuvier & Valenciennes): Taiwan; South China Sea. (1–6)
C. japonicus (Houttuyn): East China Sea; Taiwan Strait; South China Sea. Sandy substrate. (1–6)
C. martinae Fricke: Taiwan. (8)
C. recurvispinnis Li: Taiwan; South China Sea. (1–6)
C. variegatus (Temminck & Schlegel): Taiwan; South China Sea. (1–6)
Dactylopus dactylopus (Bennett): Penghu (Taiwan); South China Sea. (1–6)
Diplogrammus goramensis (Bleeker): South China Sea. (1–6)

D. xenicus (Jordan & Thompson): Taiwan. (8)
Draculo mirabilis Snyder: Bohai. (1–6)
Spinicapitichthys draconis (McCulloch): Nansha (=Spratly Islands). (17)
Synchiropus altivelis (Temminck & Schlegel) [*Foetorepus altivelis*]: Taiwan; South China Sea. Deep sea. (1–6)
S. delandi Fowler: Taiwan. (8)
S. grinnelli Fowler: Taiwan. (8)
S. lateralis (Richardson): Taiwan. (8)
S. masudai (Nakabo): Taiwan. (8)
S. ocellatus (Pallas): Taiwan; Xisha (=Paracel Islands). Shallow water in coral reefs and rocky reefs. (1–6, 8)
S. ornatus Fowler: South China Sea. (1–6)
S. picturatus (Peters): Taiwan. (8)
S. splendidus (Herre): Taiwan. (8)

SIGANIDAE

Siganus argenteus (Quoy & Gaimard): Taiwan. (8)
S. canaliculatus (Park): Taiwan. (8)
S. chrysospilos (Bleeker): Taiwan; Xisha (=Paracel Islands). Coral reefs or rocky shore. (1–6)
S. corallinus (Cuvier & Valenciennes): Xisha (=Paracel Islands). Rocky reefs and coral reefs. (1–6)
S. fuscescens (Houttuyn): East China Sea; Taiwan Strait; South China Sea. Near shore. (1–6)
S. guttatus (Bleeker): Taiwan; Hainan; Xisha (=Paracel Islands). Rocky reefs or coral reefs. (1–6)
S. javus (Linnaeus): South China Sea. (1–6)
S. oramin (Bloch & Valenciennes): East China Sea; Taiwan; South China Sea. Rocky reefs or coral reefs. (1–6)
S. puellus (Schlegel): Taiwan Strait; Xisha (=Paracel Islands). Rocky reefs and coral reefs. (1–6)
S. rostratus (Cuvier & Valenciennes): Taiwan; Xisha (=Paracel Islands); Nansha (=Spratly Islands). Rocky reefs or coral reefs. (1–6)
S. spinus (Linnaeus): Taiwan; Xisha (=Paracel Islands). Rocky reefs or coral reefs. (1–6)
S. virgatus (Cuvier & Valenciennes): Taiwan; South China Sea. (1–6)
S. vulpinus (Schlegel & Müller): Taiwan; South China Sea; Xisha (=Paracel Islands); Nansha (=Spratly Islands). Coral reefs. (1–6)

ZANCLIDAE

Zanclus canescens (Linnaeus): Taiwan Strait. Coral reefs. (1–6)
Z. cornutus (Linnaeus): Taiwan; South China Sea. Rocky reef shallow water. (1–6)

ACANTHURIDAE

Acanthurus bariene Lesson: Northeastern Taiwan; South China Sea. Rocky reefs. (1–6)
A. bleekeri Günther: Taiwan Strait; South China Sea. (1–6)
A. dussumieri Cuvier & Valenciennes: Taiwan Strait; Xisha (=Paracel Islands). Coral reefs. (1–6)
A. gahhm (Forskål): Taiwan; Xisha (=Paracel Islands). (1–6)
A. glaucopareius Cuvier: Southern tip of Taiwan; Xisha (=Paracel Islands). Coral reefs. (1–6)
A. guttatus Bloch & Schneider: Taiwan. (8)
A. japonicus (Schmidt): Taiwan. (8)
A. leucopareius (Jenkins): Taiwan. (8)
A. lineatus (Linnaeus): Taiwan; Xisha (=Paracel Islands); Nansha (=Spratly Islands). (1–6)
A. lishenus Shen & Lim: Taiwan. (8)
A. maculiceps (Ahl): Southern tip of Taiwan; South China Sea. (1–6)
A. mata (Cuvier): Taiwan; South China Sea. (1–6)
A. matoides Cuvier & Valenciennes: Taiwan Strait; South China Sea. (1–6)
A. melanopterus Shen & Lim: Taiwan. (8)
A. nigrican (Cinnaeus): Taiwan. (8)
A. nigricanda Dunker, Dunker & Mohr: Taiwan. (8)
A. nigrofuscus (Forskål): Taiwan; Xisha (=Paracel Islands); Nansha (=Spratly Islands). (1–6)
A. olivaceus Bloch & Schneider: Taiwan; Xisha (=Paracel Islands); Nansha (=Spratly Islands). (1–6)
A. pyroferus Kittlitz: Southern tip of Taiwan; South China Sea. (1–6)
A. reticulatus Shen & Lim: Taiwan. (8)
A. thompsoni Fowler: Xisha (=Paracel Islands). (1–6)
A. triostegus (Linnaeus): Taiwan and nearby islands; Xisha (=Paracel Islands). (1–6)
A. xanthopterus Cuvier & Valenciennes: Taiwan; South China Sea. (1–6)
Axinurus thynnoides Cuvier & Valenciennes [*Naso thynnoides*]: Taiwan Strait; South China Sea. (1–6)
Callicanthus hexacanthus (Bleeker) [*Naso hexacanthus* (8)]: Taiwan; South China Sea; Xisha (=Paracel Islands). Economic species. (1–6)
C. lituratus (Bloch & Schneider) [*Naso lituratus*]: Taiwan; Xisha (=Paracel Islands); Nansha (=Spratly Islands). (17)
C. lopezi (Herre) [*Naso lopezi* (8)]: Taiwan; Nansha (=Spratly Islands). (8, 17, 19)
Ctenochatus binotatus Randall: Southern Taiwan. (8)
C. striatus (Quoy & Gaimard): Taiwan; Xisha (=Paracel Islands); Nansha (=Spratly Islands). (1–6)
Naso annulatus (Quoy & Gaimard): South China Sea. (1–6)
N. brachycantron (Valenciennes): Taiwan. (8)
N. brevirostris (Cuvier & Valenciennes): Taiwan; Guangdong; Xisha (=Paracel Islands); Nansha (=Spratly Islands). Coral reefs. (1–6)

N. herrei Smith: South China Sea. (1–6)

N. thynnoides (Valenciennes): Taiwan. (8)

N. unicornis (Forskål): Zhoushan Islands (Zhejiang); Taiwan; Xisha (=Paracel Islands); Zhongsha (=Macclesfield Bank). (1–6)

N. vlamingi (Cuvier & Valenciennes): Lanyu and Lu Island (Taiwan); Xisha (=Paracel Islands). Rocky reefs. (1–6)

Paracanthurus hepatus (Linnaeus): Taiwan; South China Sea. (1–6)

Prionurus scalprus Cuvier & Valenciennes: Taiwan Strait; South China Sea. (1–6)

Zebrasoma flavescens (Bennett): Southern tip of Taiwan; Xisha (=Paracel Islands). (1- 6)

Z. scopas (Cuvier): Southern Taiwan shoreline. (1–6)

Z. veliferum (Bloch): Taiwan; Hainan; Xisha (=Paracel Islands); Nansha (=Spratly Islands). (1–6)

TRICHIURIDAE

Benthodeodesmus tonuis (Günther): East China Sea. Water depth 410–655 m. (7)

Euplerogrammus muticus (Gray) [*Lepterus muticus, Trichiurus muticus*]: China's coast. Near shore shallow sea. (1–6)

Evoxymetopon poeyi Günther: Taiwan. (8)

E. taeniatus Poey: Taiwan. (8)

Lepturacanthus savala (Cuvier) [*Trichiurus savala*]: East China Sea; Taiwan Strait; South China Sea. Near shore. (1–6)

Tentoriceps cristatus (Klunzinger) [*Pseudoxymetopon sinensis*]: East China Sea; Taiwan Strait; South China Sea. Generally in water depth 50–70 m. (1–6)

Trichiurus brevis Wang & You: South China Sea. Near shore. (24)

T. haumela (Forskål) [*Lepters haumela*]: China's coast. Generally near shore muddy substrate, water depth 60–100 m. (1–6)

T. lepturus Linnaeus: Taiwan. (8)

T. nanhaiensis Wang & Xu: South China Sea. Near shore. (24)

GEMPYLIDAE

Epinula orientalis Gilcrist & Von Bonder: South China Sea continental slope. (9)

Gempylus serpens Cuvier & Valenciennes [*G. notha* (8)]: Taiwan; Hainan; Dongsha (=Pratas Islands); Xisha (=Paracel Islands); Nansha (=Spratly Islands). Oceanic species. (1–6)

Lepidocybium flavobrunneum (Smith): Taiwan; Hainan; Dongsha (=Pratas Islands); Xisha (=Paracel Islands); Nansha (=Spratly Islands). Deep sea. (1–6)

Nealotus tripes Johnson: East China Sea. Water depth 655–1,014 m. (7)

Neoepinnula orientalis (Gilchrist & Von Bonder): East China Sea; Taiwan. Water depth about 200–300 m. (7, 8)

Nesiarchus nasutus Johnson: East China Sea. Water depth 950 m. (7)

Promethichthys prometheus (Cuvier & Valenciennes): East China; Taiwan; South China Sea. Deep sea. (1–6)

Rexea prometheoides (Bleeker): East China Sea; Taiwan; South China Sea. Deep sea. (1–6)

R. solandri (Cuvier): East China Sea. Water depth 270–279 m. (7)

Ruvettus pretiosus Cocco: Taiwan. (8)

R. tydemani Weber: Hainan; Dongsha (=Pratas Islands); Xisha (=Paracel Islands); Nansha (=Spratly Islands). Deep sea. (1–6)

Thyrsitoides marleyi Fowler: Taiwan Strait; Hainan; Dongsha (=Pratas Islands); Xisha (=Paracel Islands); Nansha (=Spratly Islands). Oceanic species. (1–6)

SCOMBRIDAE

Pneumatophorus japonicus (Houttuyn) [*Scomber japonicus* (8)]: Yellow Sea; East China Sea; Taiwan; South China Sea. Oceanic migratory species. (1–6)

P. tapeinocephalus (Bleeker): Yellow Sea; East China Sea; Taiwan Strait. Near shore. (1–6)

Rastrelliger faughni Matsui: Taiwan. (8)

R. kanagurta (Cuvier): East China Sea; Taiwan Strait; South China Sea. Common oceanic species. (1–6)

CYBIIDAE

Acanthocybium solandi (Cuvier & Valenciennes): Taiwan; Hainan; Dongsha (=Pratas Islands); Xisha (=Paracel Islands); Nansha (=Spratly Islands). Economic species. (1–6)

Grammatorcynus bicarinatus (Quoy & Gaimard): Taiwan; Hainan; Dongsha (=Pratas Islands); Xisha (=Paracel Islands); Nansha (=Spratly Islands). Offshore off coral reefs. (1–6)

Scomber australasicus Cuvier [*Scombermorus australasicus* (8)]: Taiwan. (8)

Scombermorus cavalla (Cuvier): East China Sea; South China Sea. (1–6)

S. commersoni (Lacépéde): East China Sea; Taiwan Strait; South China Sea. Near shore. (1–6)

S. guttatus (Bloch & Schneider): East China Sea; Taiwan Strait; South China Sea. Near shore. (1–6)

S. koreana (Kinshinouye) [*Sewara koreana*]: Bohai; Yellow Sea; East China Sea. Near shore. (1–6)

S. niphonius (Cuvier & Valenciennes) [*Sewara niphonia*]: Yellow Sea; East China Sea; Taiwan Strait. Near shore. (1–6)

S. sinensis (Lacépéde): Yellow Sea; East China Sea; Taiwan Strait. Near shore. (1–6)

HISTIOPHORIDAE

Histiophorus gladius (Broussonet): Taiwan Strait; northern South China Sea; Hainan; Dongsha (=Pratas Islands); Xisha (=Paracel Islands); Nansha (=Spratly Islands). Oceanic species. (1–6)

H. orientalis Temminck & Schlegel: Yellow Sea; East China Sea; Taiwan Strait; South China Sea. Oceanic species. (1–6)

H. platypterus (Shaw & Nodder) [*Istiophorus platypturus* (8)]: Taiwan. (8)

Makaira formosana (Hirasaka & Nakamura): South China Sea. Oceanic species. (1–6)

M. indica (Cuvier): Taiwan. (8)

M. marlina Jordan & Hill: Taiwan; Hainan; Dongsha (=Pratas Islands); Xisha (=Paracel Islands); Nansha (=Spratly Islands). Oceanic species. (1–6)

M. mazara (Jordan & Snyder): Taiwan; South China Sea. Oceanic species. (1–6)

M. mitsukurii (Jordan & Snyder): South China Sea. Oceanic species. (1–6)

Tetrapturus angustirostris Tanaka: Taiwan; South China Sea. Oceanic species. (1–6)

T. audax (Phillipi): Taiwan. (8)

XIPHIIDAE

Xiphias gladius Linnaeus: East China Sea; Taiwan Strait; South China Sea. Oceanic species. (1–6)

THUNNIDAE

Auxis rochei (Risso): Taiwan. (8)

A. tapeinosoma (Bleeker): Yellow Sea; East China Sea; South China Sea. Offshore. (1–6)

A. thazard (Lacépéde): Yellow Sea; East China Sea; South China Sea. Offshore. (1–6)

Euthynnus affinis (Cuvier): Taiwan. (8)

E. yaito Kishinouye: Taiwan Strait; South China Sea. Oceanic species. (1–6)

Gymnosarda unicolor (Rüppell): Taiwan; Hainan; Dongsha (=Pratas Islands); Xisha (=Paracel Islands); Nansha (=Spratly Islands). Often lives in offshore off rocky reefs. (1–6)

Katsuwonus pelamis (Linnaeus) [*Euthynnus pelamis* (8)]: East China Sea; Taiwan Strait; South China Sea. Oceanic species. (1–6)

Sarda orientalis (Temminck & Schlegel): East China Sea; Taiwan Strait; South China Sea. Oceanic species. (1–6)

Thunnus alalunga (Bonnaterre): East China Sea; Taiwan; South China Sea. (1–6)

T. albacora (Lowe): Taiwan Strait; South China Sea. Oceanic migratory species. (1–6)

T. obesus Lowe: Taiwan; Hainan; Dongsha (=Pratas Islands); Xisha (=Paracel Islands); Nansha (=Spratly Islands). Oceanic migratory species. (1–6)

T. thynnus (Linnaeus): East China Sea; Taiwan; South China Sea. (1–6)

T. tonggol (Bleeker): Taiwan Strait; South China Sea. Oceanic migratory species. (1–6)

ARIOMMIDAE

Ariomma evermanni Jordan & Snyder: East China Sea; South China Sea. (1–6)

A. indica (Day) [*Psenes indicus*]: East China Sea; Taiwan Strait; South China Sea. Near shore. (8)

A. lurida Jordan & Snyder: South China Sea. (9)

NOMEIDAE

Cubiceps squamicepoides Deng, Xiong & Zhan: East China Sea. Water depth 655–670 m. (7)

C. squamiceps (Lloyd): East China Sea; Taiwan Strait. Near shore fish. (1–6)

Nomeus gronovii (Gmelin): Northern Taiwan; South China Sea. Rocky reefs, symbiotic with jellyfish. (8)

Psenes arafurensis Günther: East China Sea; South China Sea. (1–6)

P. cyanophrys Cuvier & Valenciennes: East China Sea; Taiwan. Water depth within 1,500 m. (1–6)

P. pellucidus Lütken [*Icticus pellucidus*]: East China Sea; Taiwan; South China Sea. Often lives around jellyfish. (1–6)

STROMATEIDAE

Pampus argenteus (Euphrasen) [*Stromateus argenteus*]: China's coast. Near shore, water depth 30–70 m. (1–6)

P. chinensis (Euphrasen) [*Stromateus sinensis*]: East China Sea; Taiwan Strait; South China Sea. Near shore. (1–6)

P. nozawae (Ishikawa) [*Stromateus cinereus*]: East China Sea; Taiwan Strait; South China Sea. Near shore migratory species, water depth 30–70 m. (1–6)

CENTROLOPHIDAE

Hyperoglyphe japonicus (Döderlein): East China Sea; South China Sea. (1–6)

Psenopsis anomala (Temminck & Schlegel): East China Sea; Taiwan Strait; South China Sea. Near shore, water depth 45–120 m. (1–6)

Schedophilus maculatus Günther: South China Sea. (1–6)

ELEOTRIDAE

Asterropteryx semipunctatus Rüppell: Hainan; Xisha (=Paracel Islands). Coral reefs. (1–6)
Austrolethops wardi Whitley: Taiwan. (8)
Bostrichthys marmoratus (Bleeker): Taiwan. (8)
B. sinensis (Lacépéde): East China Sea; Taiwan; South China Sea. Mud flat and river mouth. (1–6)
Brionobutis koilomatodon (Bleeker): East China Sea; Taiwan Strait; South China Sea. Rocky shore. (1–6)
Butis butis (Hamilton): South China Sea. (1–6)
B. gymnopomus (Bleeker): Taiwan. (8)
B. koilomatodon (Bleeker): Taiwan. (8)
B. melanostigma (Bleeker): Taiwan. (8)
Eleotriodes immaculatus Ni: South China Sea. (1–6)
E. longipinnis (Bennett): Penghu Islands (Taiwan); Xisha (=Paracel Islands). Coral reefs and rocky reefs. (1–6)
E. muralis (Quoy & Gaimard): South China Sea. (1–6)
E. strigatus (Brousonet): Taiwan; Xisha (=Paracel Islands). Coral reefs. (1–6)
Eleotris acanthopomus Bleeker: Taiwan. (8)
E. melanosoma Bleeker: Taiwan. (8)
Eviota abax (Jordan & Snyder): Taiwan; Hainan; Xisha (=Paracel Islands). Coral reefs. (1–6)
E. prasina (Klunzinger): Taiwan. (8)
Hypseleotris bipartita Herre: Taiwan. (8)
Nemateleotris magnificus Fowler: Taiwan. (8)
Ophiocara aporos (Bleeker): Taiwan; South China Sea. (1–6)
O. porocephala (Cuvier & Valenciennes): Southern tip of Taiwan; South China Sea. Rocky reefs. (1–6)
Oxymetopon compressus Chan: South China Sea. (1–6)
Parioglossus taeniatus Regan: Taiwan. (8)
Pogonoculius zebra Fowler: Taiwan. (8)
Ptereleotris evides (Jordan & Hubbs): Taiwan. (8)
P. heteropterus (Bleeker): Taiwan. (8)
P. microlepis (Bleeker): Southern tip of Taiwan; Xisha (=Paracel Islands). Coral reefs and rocky reefs. (1–6)
Valenciennea heldsdingenii (Bleeker) [*Eleotris heldsdingenii* (8)]: Taiwan. (8)
V. sexguttatus (Valenciennes) [*Eleotris sexguttatus* (8)]: Taiwan. (8)
V. strigatus (Broussonet) [*Eleotris strigatus* (8)]: Taiwan. (8)

GOBIIDAE

Aboma isushimae Jordan & Snyder: Yellow Sea. (1–6)
A. lactipes (Hilgendorf): Bohai; Yellow Sea; East China Sea. (1–6)
Acanthogobius flavimanus (Temminck & Schlegel): Bohai; Yellow Sea. (1–6)
A. jacoti Fowler: Yellow Sea. (1–6)
Acentrogobius bonti (Bleeker): Taiwan Strait; South China Sea. (1–6)
A. campbelli (Jordan & Snyder) [*Istigobius campbelli* (8)]: East China Sea; Taiwan Strait; South China Sea. (1–6)
A. caninus (Cuvier & Valenciennes): East China Sea; Taiwan Strait; South China Sea. (1–6)
A. cauerensis (Bleeker): South China Sea. (1–6)
A. chlorostigmatoides (Bleeker): East China Sea; Taiwan Strait; South China Sea. (1–6)
A. cyanomos (Bleeker): South China Sea. (1–6)
A. hoepplii (Wu): East China Sea; Taiwan Strait. (1–6)
A. hoshinonis (Tanaka): South China Sea. (1–6)
A. masoni (Day): Taiwan; South China Sea. (1–6)
A. microps Chu & Wu: East China Sea. (1–6)
A. ornatus (Rüppell) [*Istigobis ornatus* (8)]: Taiwan Strait; South China Sea. Rocky reefs or coral reefs. (1–6)
A. punctang (Bleeker) [*Exyurias punctang* (8)]: South China Sea. (1–6)
A. triangularis (Weber): South China Sea. (1–6)
A. viganensis (Steindachner): Taiwan. (8)
A. viridipunctatus (Cuvier & Valenciennes): South China Sea. (1–6)
Amblyeleotris fasciata (Herre): Taiwan. (8)
A. guttata (Fowler): Taiwan. (8)
Amblygobius albimaculatus (Rüppell): Taiwan; Hainan; Xisha (=Paracel Islands). Rocky reefs or coral reefs. (1–6)
A. bynoensis (Richardson): South China Sea. (1–6)
A. shatinensis Herre: South China Sea. (1–6)
Apocryptes bato (Hamilton): South China Sea. (1–6)
Apocryptichthys sericus Herre: East China Sea; Taiwan Strait; South China Sea. Mud flat. (1–6)
Apocryptodon glyphisodon (Bleeker): Southern East China Sea. (1–6)
A. madurensis (Bleeker): Yellow Sea; East China Sea; Taiwan; Taiwan Strait; South China Sea. Mud flat. (1–6, 8)
A. malcolmi Smith: South China Sea. (1–6)
Awaous melanocephalus (Bleeker): Taiwan; South China Sea. (1–6)
Bathygobius cocosensis (Bleeker): Taiwan. (8)
B. cotticeps (Steindachner): Taiwan. (8)
B. cyclopterus (Valenciennes): Taiwan. (8)
B. fuscus (Rüppell): East China Sea; Taiwan; South China Sea. Near shore shallow water bottom fish. (1–6)
Callogobius liolepis Kouman: South China Sea. (1–6)
C. sclateri (Steindachner): Xisha (=Paracel Islands). Rocky shore and coral reefs. (1–6)
C. snelliusi Kouman: Taiwan. (8)
Chaenogobius annularis Gill: Bohai; Yellow Sea. (1–6)
Chaeturichthys hexanema Bleeker: China's coast. (1–6)
C. stigmatias Richardson: China's coast. (1–6)
Chasmichthys gulosus (Guichenot): Yellow Sea. (1–6)
Chloea castanea (O'saughnessy): Bohai; Yellow Sea. (1–6)
C. mororana Jordan & Snyder: Yellow Sea. (1–6)
C. sarchynnis Jordan & Snyder: Bohai; Yellow Sea. (1–6)
Cryptocentrus albidorsus (Yanagisawa): Taiwan. (8)
C. cryptocentrus (Valenciennes): Taiwan. (8)

C. filifer (Cuvier & Valenciennes): China's coast. (1–6)
C. gymnocephalus (Bleeker): South China Sea. (1–6)
C. nigrocellatus (Yanagisawa): Taiwan. (8)
C. papuanus (Peters): South China Sea. (1–6)
C. pavontnoides (Bleeker): South China Sea. (1–6)
C. russus (Cantor): East China Sea; South China Sea. (1–6)
C. strigilliceps (Jordan & Seale): Taiwan. (8)
C. yatsui Tomiyama: South China Sea. (1–6)
Ctenogobius aurocingulus (Herre): Taiwan. (8)
C. brevirostris (Günther): East China Sea; Taiwan Strait; South China Sea. Shallow sea and river mouth. (1–6)
C. criniger (Cuvier & Valenciennes): Taiwan Strait; South China Sea. (1–6)
C. crocineus Smith: Taiwan. (8)
C. gymnauehen (Bleeker): China's coast. (1–6)
C. notophthalmus Bleeker: South China Sea. (1–6)
C. pflalumi (Bleeker): Bohai; Yellow Sea. (1–6)
C. puncticeps Deng & Xiong: East China Sea. (1–6)
C. tangaroai Lubbock & Polunin: Taiwan. (8)
Eutaeniichthys gilli Jordan & Snyder: Yellow Sea. (1–6)
Glossogobius biocellatus (Cuvier & Valenciennes): East China Sea; Taiwan; South China Sea. (1–6)
G. brunnoides (Nichols): Taiwan. (8)
G. circumspectus (Macleay): Taiwan. (8)
G. giuris (Hamilton): East China Sea; Taiwan Strait; South China Sea. Mud flat, rocky shore, and river mouth. (1–6)
G. obscuripinnis (Peters): Taiwan. (8)
G. olivaceus (Temminck & Schlegel) [*G. fasciato-punctotus*]: East China Sea; Taiwan Strait; South China Sea. Shallow sea and river mouth. (1–6)
Gnatholepis cauerensis (Bleeker): Taiwan. (8)
G. otakii (Jordan & Snyder) [*Hazeus otakii* (8)]: Taiwan. (1–6)
Gobiodon citrinus (Rüppell): Southern tip of Taiwan. Rocky shore. (8)
G. erythrospilus Bleeker: Hainan; Xisha (=Paracel Islands). Coral reefs. (1–6)
G. multilineatus Wu: Xisha (=Paracel Islands). Coral reefs. (1–6)
G. oculolineatus Wu: Xisha (=Paracel Islands). Coral reefs. (1–6)
G. okinawae Wawada; Arai & Abe: Xisha (=Paracel Islands). Coral reefs. (1–6)
G. quinquestrigatus (Cuvier & Valenciennes): Taiwan; Xisha (=Paracel Islands). Coral reefs and rocky reefs. (1–6)
G. ruvalabis (Rüppell): South China Sea. (17)
G. verticalis Alleyne & Macleay: Nansha (=Spratly Islands). (1–6)
Lophiogobius ocellicauda Günther: Yellow Sea; East China Sea. (1–6)
Luciogobius guttatus Gill: Bohai; Yellow Sea; East China Sea. (1–6)
L. saikaiensis Dotu: Taiwan. (8)
Mugilogobius abei (Jordan & Snyder): Yellow Sea; East China Sea; Taiwan Strait; South China Sea. (1–6)
M. tagala (Herre): Taiwan. (1–6)
Oligolepis acutipinnis (Cuvier & Valenciennes): Taiwan. (8)

O. fasciatus Wu & Lin: Yellow Sea; East China Sea; Taiwan Strait. Near shore muddy sand substrate, water depth 10+ m. (1–6)
Oplopomus caninoides (Bleeker): Taiwan; South China Sea. (1–6)
Oxyurichthys macrolepis Chu & Wu: East China Sea. (1–6)
O. microlepis (Bleeker): East China Sea; South China Sea. (1–6)
O. ophthalmonema (Bleeker): East China Sea; Taiwan Strait; South China Sea. (1–6)
O. papuensis (Cuvier & Valenciennes): East China Sea; Taiwan; South China Sea. (1–6)
O. tentacularis (Cuvier & Valenciennes): East China Sea; Taiwan Strait; South China Sea. (1–6)
Parachaeturichthys polynema (Bleeker): East China Sea; Taiwan Strait; South China Sea. (1–6)
Paragobiodon echinocephalus (Rüppell): Hainan; Xisha (=Paracel Islands). Coral reefs. (1–6)
P. melanosomus (Bleeker): Xisha (=Paracel Islands). Coral reefs. (1–6)
P. xanthosomus (Bleeker): Xisha (=Paracel Islands). Coral reefs. (1–6)
P. zacalles Jordan & Snyder: Yellow Sea. (1–6)
Parapocryptes macrolepis (Bleeker): East China Sea; South China Sea. (1–6)
P. serperaster (Richardson): East China Sea; Taiwan Strait; South China Sea. Mud flat or around river mouth. (1–6)
Pseudapocryptes lanceolatus (Bloch & Schneider): South China Sea. (1–6)
Quisquilius eugenius Jordan & Evermann: Northeastern Taiwan; South China Sea. Rocky reefs. (1–6)
Sicyopterus micrurus (Bleeker) [*S. japonicus* (8)]: Taiwan. (8)
Stenogobius genivittatus (Valenciennes): Taiwan. (8)
S. lacrymosus (Peters): Taiwan. (8)
Stigmatogobius hoeveni (Bleeker): South China Sea. (1–6)
S. javanicus (Bleeker): South China Sea. (1–6)
Synechogobius hasta (Temminck & Schlegel): Bohai; Yellow Sea; East China Sea; Taiwan. Sea bottom with muddy substrate or river mouth. (1–6)
S. ommaturus (Richardson): China's coast. (1–6)
Triaenopogon barbatus (Günther): China's coast. Near shore bottom fish. (1–6)
Tridentiger obscurus (Temminck & Schlegel): China's coast. Near shore bottom fish. (1–6)
T. trigonocephalus (Gill): China's coast. Near shore bottom fish. (1–6)
Zonogobius naraharae (Snyder) [*Priolepis naraharae* (8)]: Taiwan. (8)
Z. semidoliatus (Cuvier & Valenciennes): Taiwan; Hainan; Xisha (=Paracel Islands). Rocky reefs and coral reefs. (1–6)

PERIOPHTHALMIDAE

Boleophthalmus maculatus (Oshima): South China Sea. (1–6)

B. pectinirostris (Linnaeus) [*B. chinensis*]: China's coast. Muddy low intertidal or river mouth mud flat. (1–6)
Periophthalmus argentilineatus Cuvier & Valenciennes: South China Sea. (1–6)
P. cantonensis (Osbeck): China's coast. Muddy or muddy sand high intertidal or mud flat around river mouth. (1–6)
Scartelaos gigas Chu & Wu: East China Sea. (1–6)
S. viridis (Hamilton-Buchanan): East China Sea; Taiwan Strait; South China Sea. Mud flat around river mouth. (1–6)

TAENIOIDIDAE

Amblyotrypauchen arctocephalus (Alcock): South China Sea. (1–6)
Brachyamblyopus anotus (Franz): Taiwan. (8)
B. brachysoma (Bleeker): South China Sea. (1–6)
Ctenotrypauchen chinensis Steindachner: East China Sea. (1–6)
C. microcephalus (Bleeker): China's coast. (1–6)
Odontamblyopus rubicundus (Hamilton-Buchanan): China's coast. Shallow sea, water depth 2–20 m. (1–6)
Taenioides aguillaris (Linnaeus): East China Sea; Taiwan; South China Sea. Intertidal. (1–6)
T. cirratus (Blyth): East China Sea; Taiwan Strait; South China Sea. Near shore and river mouth mud flat. (1–6)
Trypauchen taenlia Koumans: South China Sea. (1–6)
T. vagina (Bloch & Schneider): East China Sea; South China Sea. Brackish water with muddy substrate, water depth 20+ m. (1–6)

ECHENEIDAE

Echeneis naucrates Linnaeus: China's coast. Often attached to ships or large fish. (1–6)
Remora albescens (Temminck & Schlegel) [*Remorina albescen* (8)]: East China Sea; Taiwan; South China Sea. Often lives on gills or inside anus of mantas (Mobulidae). (1–6)
R. australis (Bennett): Taiwan. (8)
R. brachyptera (Lowe): Taiwan; Hainan; Dongsha (=Pratas Islands); Xisha (=Paracel Islands); Nansha (=Spratly Islands). Attached to oceanic fish. (1–6)
R. remora (Linnaeus): China's coast. (1–6)
Rhombochirus osteochir (Cuvier): Taiwan; Hainan; Dongsha (=Pratas Islands); Xisha (=Paracel Islands); Nansha (=Spratly Islands). (1–6)

SCORPAENIFORMES
SCORPAENIDAE

Apistus carinatus Bloch & Schneider [*A. alatus*]: East China Sea; Taiwan Strait; South China Sea. Near shore bottom fish. (1–6)
Brachypterois serrulatus (Richardson): East China Sea; Taiwan Strait; South China Sea. (1–6)
Dendrochirus bellus Jordan & Hubbs [*Brachirus bellus*, *Brachypteros bellus*]: East China Sea; Taiwan Strait; South China Sea. (1–6)
D. biocellatus (Fowler): Northern Taiwan. Rocky shore. (8)
D. zebra (Quoy & Gaimard) [*Brachirus zebra*]: Taiwan; Hainan; Xisha (=Paracel Islands). Rocky reef shallow water. (1–6)
Ebosia bleekeri (Steindachner & Döderlein): East China Sea; Taiwan; South China Sea. (1–6)
Helicolenus hilgendorfi (Steinachner & Döderlein): East China Sea; Taiwan. Water depth 174–655 m. (7, 8)
Hoplosebastes armatus (Cuvier & Valenciennes): East China Sea; Taiwan Strait; South China Sea. (1–6)
Hozukius embremarius (Jordan & Starks): East China Sea; South China Sea. Water depth 650–900 m. (7)
Iracundus signifer Jordan & Evermann: Taiwan. Coral reefs, water depth 11–70 m. (8)
Neomerinthe megalepis (Fowler): Taiwan. (8)
N. procorva Chen: Taiwan. (8)
N. procurva Chen: Northeastern Taiwan. Muddy sand. (8)
N. rotunda Chen: Taiwan. (8)
Neosebastes entaxis Jordan & Starks: Northern Taiwan. (1–6)
Parapterois heterurus Bleeker: East China Sea; Taiwan Strait; South China Sea. (1–6)
Parascorpaena mcadamsi (Fowler): Taiwan. (8)
P. mossambica (Peter): Eastern, western and southern Taiwan. Rocky intertidal and subtidal. (8)
P. picta (Cuvier & Valenciennes): Taiwan; Hainan; Xisha (=Paracel Islands). Rocky reefs. (1–6)
Plectrogenium nanum Gilbert: Taiwan. (1–6)
Pontinus tentacularis (Fowler): East China Sea; Taiwan. Water depth 269 m. (7)
Pteroidichthys amboinensis Bleeker: Taiwan; South China Sea. (1–6)
Pterois antennata (Bloch): Taiwan; South China Sea. (1–6)
P. lunulata Temminck & Schlegel [*P. miles*]: East China Sea; Taiwan Strait; South China Sea. (1–6)
P. radiata Cuvier & Valenciennes: Taiwan; Xisha (=Paracel Islands). Rocky reefs. (1- 6)
P. russelli Bennett: East China Sea; Taiwan Strait; South China Sea. Rocky reefs or coral reefs. (1–6)
P. volitans Linnaeus [*P. miles*]: Taiwan Strait; South China Sea. Rocky reefs. (1–6)
Rhinopias frondosa (Günther): Taiwan Strait. (1–6)
Scorpaena hatizyoensis Matsubara: Xisha (=Paracel Islands). Rocky reefs. (1–6)
S. izensis Jordan & Starks: China's coast. Sandy gravel, rocky reefs, or coral reefs. (1–6)
S. neglecta Temminck & Schlegel [*Scorpaena onaria* (8)]: China's coast. Intertidal to deep water. (1–6)
Scorpaenodes guamensis (Quoy & Gaimard) [*S. kelloggi* (8)]: Taiwan; South China Sea. Rocky reefs and coral reefs. (1–6)
S. hirsutus (Smith): Taiwan. (8)
S. littoralis Tanaka: Northeastern Taiwan. Rocky shore. (8)

S. minor (Smith): Taiwan. (8)

S. parvipinnis (Garrett): Taiwan. (8)

S. scabra (Ramsay & Ogilby): Taiwan; Xisha (=Paracel Islands). Rocky reef crevices. (1–6)

S. varipinnis Smith: Northeastern Taiwan. Rocky reefs. (8)

Scorpaenopsis cirrhosa (Thunberg): East China Sea; Taiwan Strait; South China Sea. Rocky reefs. (1–6)

S. diabolus (Cuvier & Valenciennes): Taiwan; South China Sea. Rocky reefs. (1–6)

S. gibbosa (Bloch & Schneider): East China Sea; Taiwan Strait; South China Sea. (1–6)

S. neglecta Heckel: Taiwan. (8)

Sebastapistes albobrunnea (Günther): Southern tip of Taiwan. Rocky intertidal and subtidal. (1–6)

S. bynoensis (Richardson) [*Scorpaena bynoensis*]: Taiwan; South China Sea. (1–6)

S. megalepis (Fowler): South China Sea. Water depth about 62–82 m, bottom fish. (1–6)

S. nuchalis Günther: Taiwan Strait; Xisha (=Paracel Islands). Rocky reefs. (1–6)

S. vachelli (Richardson): South China Sea. (1–6)

Sebastes hubbsi (Matsubara): Yellow Sea; East China Sea. (1–6)

S. intermis Cuvier & Valenciennes: Yellow Sea. (1–6)

S. itinus (Jordan & Starks): Yellow Sea. (1–6)

S. joyneri Günther: Taiwan. (1–6)

S. nigricans (Schmidt): Yellow Sea. (1–6)

S. nivosus (Hilgendorf): Yellow Sea. (1–6)

S. pachycephalus Temminck & Schlegel: Yellow Sea; East China Sea. (1–6)

S. schlegeli (Hilgendorf): Yellow Sea; East China Sea. (1–6)

S. thompsoni (Jordan & Hubbs): Yellow Sea; East China Sea. (1–6)

S. trivittatus Hilgendorf: Yellow Sea. (1–6)

Sebastiscus albofasciatus (Lacépéde): Northern and northeastern Taiwan; South China Sea. Rocky reefs. (1–6)

S. marmoratus (Cuvier & Valenciennes): China's coast. Near shore rocky reefs. (1–6)

S. tertius Barsukov & Chen: Taiwan. (8)

Setarches fidjiensis (Günther): South China Sea. Deep sea near bottom fish. (1–6)

S. longimanus (Alcock & McGrichrist): East China Sea; Taiwan; South China Sea. Water depth 321–1,054 m. (1–6)

Thysanichthys crossotus Jordan & Starks: Taiwan. (8)

SYNANCEIIDAE

Choridactylus multibarbis Richardson: South China Sea. (1–6)

Erosa erosa (Langsdorf): East China Sea; Taiwan Strait; South China Sea. Coral reefs and seaweed. (1–6)

Inimicus cuvieri (Gray): East China Sea; Taiwan Strait; South China Sea. (1–6)

I. didactylus (Pallas): Taiwan Strait; South China Sea. (1–6)

I. japonicus (Cuvier & Valenciennes): China's coast. Shoreline or around islands. (1–6)

I. sinensis (Valenciennes): Penghu Islands (Taiwan). (8)

Minous coccineus Alcock: Taiwan. (8)

M. inermis Alcock: East China Sea; Taiwan Strait; South China Sea. Near shore bottom fish. (1–6)

M. monodactylus (Bloch & Schneider): China's coast. (1–6)

M. pictus Günther: Taiwan. (8)

M. pusillus Temminck & Schlegel: East China Sea; Taiwan Strait; South China Sea. (1- 6)

M. quincarinatus (Fowler): Northern, northeastern, and southwestern Taiwan. (8)

M. trachycephalus (Bleeker): Taiwan Strait. (1–6)

Polycaulus uranoscopa (Bloch & Schneider): East China Sea; Taiwan Strait; South China Sea. Near shore benthic species. (1–6)

Snyderina yamanokami Jordan & Starks [*Polycaulus yamanokami* (8)]: Taiwan. (8)

Synanceia horrida Linnaeus: South China Sea. (1–6)

S. verrucosa Bloch & Schneider: Taiwan Strait; South China Sea. Coral reefs and seaweed. (1–6)

TRIGLIDAE

Chelidonichthys kumu (Lesson & Garnot): China's coast. Water depth 30–40 m. (1–6)

C. spinosus (McClelland): East China Sea; South China Sea. Water depth 140–435 m. (7)

Lepidotrigla abyssalis Jordan & Starks: East China Sea; Taiwan. (7, 8)

L. alata (Houttuyn): East China Sea; Taiwan Strait; South China Sea. (1–6)

L. guentheri Hilgendorf: East China Sea; Taiwan Strait. (1–6)

L. hime Matsubara & Hiyama: East China Sea. (7)

L. japonica (Bleeker): East China Sea; Taiwan Strait; South China Sea. (1–6)

L. kishinouyi Snyder: East China Sea; Taiwan. (1–6)

L. lepidojugulata Li: Taiwan Strait; South China Sea. (1–6)

L. longifaciata Yatou: East China Sea. (7)

L. longimana Li: South China Sea. (1–6)

L. marisinensis Fowler: Northern South China Sea. (1–6)

L. micropterus Günther: Bohai; Yellow Sea; East China Sea; Taiwan Strait. (1–6)

L. oglina Fowler: Northern South China Sea. (1–6)

L. punctipectoralis Fowler: East China Sea; Taiwan; South China Sea. Often lives in water depth 180–245 m. (1–6)

L. spilopterus Günther: South China Sea. (1–6)

Parapterygotrigla macrorhynchus (Kamohara): Taiwan. (8)

P. multicellata Matsubara: East China Sea; northeastern Taiwan. (1–6)

Pterygotrigla hemisticta (Temminck & Schlegel): Taiwan Strait; South China Sea. (1–6)

P. ryukyuensis Matsubara & Hiyama: South China Sea. (1–6)

PERISTEDIIDAE

Gargariscus prionocephalus (Duméril): Taiwan; South China Sea. Deep sea. (1–6)
Peristedion nierstrasi Weber: Taiwan. (1–6)
P. orientale Temminck & Schlegel: East China Sea; Taiwan. (1–6)
Satyrichthys amiscus (Jordan & Starks): East China Sea. Water depth 275–610 m. (1–6)
S. fowleri Beaufort & Briggs: South China Sea. (1–6)
S. piercei Fowler: South China Sea. (1–6)
S. rieffeli (Kaup): Taiwan Strait; South China Sea. (1–6)
S. welchi (Herre): Taiwan. (1–6)

CONGIOPODIDAE [APLOACTIDAE]

Acanthosphex leurynnis (Jordan & Seale): South China Sea. (1–6)
Amblyapistus macracanthus Bleeker [*Ablabys macracanthus* (8)]: East China Sea; Taiwan Strait; South China Sea. (1–6)
A. taenianotus (Cuvier & Valenciennes) [*Ablabys taenianotus* (8)]: Taiwan Strait; South China Sea. (1–6)
Aploactis aspera Richardson: Southwestern Taiwan; South China Sea. (1–6)
Caratanthus maculatus Günther: Nansha (=Spratly Islands). (8, 17)
C. unipinna (Gray): Taiwan. (8)
Centropogon fuscovirens (Cuvier & Valenciennes): South China Sea. (1–6)
C. urostigma (Bleeker): South China Sea. (1–6)
Cottapistus cottoides (Linnaeus): South China Sea. (1–6)
Erisphex potti (Steindachner): China's coast. Water depth more than 100 m. (1–6)
E. simplex Chen: Taiwan. (8)
Hypodytes indicus (Day) [*Paracentropogon indicus*]: East China Sea; Taiwan Strait; South China Sea. (1–6)
H. longispinis (Cuvier & Valenciennes): South China Sea. (1–6)
Neocentropogon japonicus Matsubara: Taiwan Strait. (1–6)
Ocosia spinosa Chen: Taiwan. (8)
O. vespa Jordan & Starks: Taiwan Strait; South China Sea. (1–6)
Paraploactis kagoshimensis (Ishikawa): Taiwan. (8)
Sthenopus mollis Richardson: South China Sea. (1–6)
Taenianotus triacanthus Lacépède [*Amblyapistus triacanthus* (8)]: Taiwan. (8)
Tetraroge leucogaster (Richardson) [*Gymnapistes leucogaster*]: South China Sea. (1–6)
Vespicula sinensis (Bleeker): South China Sea. (1–6)
V. trachinoides (Cuvier & Valenciennes): South China Sea. Rocky reefs. (1–6)

HEXAGRAMMIDAE

Agrammus agrammus (Temminck & Schlegel): Yellow Sea; East China Sea. (1–6)
Hexagrammos lagocephalus (Pallas): Yellow Sea. (1–6)
H. octogrammus (Pallas): Yellow Sea. (1–6)
H. otakii Jordan & Starks: Yellow Sea; East China Sea. (1–6)

PARABEMBRIDAE

Parabembras curtus (Temminck & Schlegel): Yellow Sea; East China Sea. (1–6)

BEMBRIDAE

Bembras japonicus Cuvier & Valenciennes: East China Sea; Taiwan Strait; South China Sea. Water depth 100–200 m. (1–6)
B. laevis (Nakaya): Nansha (=Spratly Islands). (17)

PLATYCEPHALIDAE

Cociella crocodilus (Tilesius): China's coast. (1–6)
Cymbacephalus nematophthalmus Günther: South China Sea. (1–6)
Elates ransonneti (Steindachner): South China Sea. (1–6)
Grammoplites scaber (Linnaeus): East China Sea; Taiwan Strait; South China Sea. (1–6)
Inegocia guttatus (Cuvier & Valenciennes): East China Sea; Taiwan Strait; South China Sea. (1–6)
I. japonicus (Tilesius): China's coast. (1–6)
Kumococius detrusus (Jordan & Seale): South China Sea. (1–6)
Onigocia macrolepis (Bleeker): East China Sea; Taiwan; South China Sea. (1–6)
O. spinosus (Temminck & Schlegel): East China Sea; Taiwan Strait; South China Sea. (1–6)
O. tuberculatus (Cuvier & Valenciennes) [*Sorsogon tuberculatus* (8)]: East China Sea; Taiwan Strait; South China Sea. (1–6)
Platycephalus indicus (Linnaeus): China's coast. Shoreline to water depth 50 m. (1–6)
Ratabulus megacephalus (Tanaka): East China Sea; Taiwan Strait; South China Sea. (1–6)
Rogadius asper (Cuvier & Valenciennes): East China Sea; Taiwan Strait; South China Sea. (1–6)
R. patriciae Knapp: Taiwan. (1–6)
Suggrundus longirostris Shao & Chen: Taiwan. (8)
S. macracanthus (Bleeker): Taiwan. (1–6)
S. meerdvoorti (Bleeker): East China Sea; Taiwan Strait; South China Sea. (1–6)
S. rodericensis Cuvier & Valenciennes [*Kumococius rodericensis* (8)]: Taiwan. (1–6)
Thysanophrys arenicda Schultz: Taiwan. (8)
T. bataviensis (Bleeker): South China Sea. (1–6)
T. celebica (Bleeker): Taiwan. (8)
T. chiltonae Schultz: Taiwan. (8)
T. otaitensis (Parkinson): Taiwan. (8)

HOPLICHTHYIDAE

Hoplichthys fasciatus Matsubara: Northeastern Taiwan. Deep sea with muddy sand substrate. (8)

H. gilberti Jordan & Richardson: East China Sea; Taiwan; South China Sea. (1–6)

H. langsdorfi Cuvier & Valenciennes: East China Sea; Taiwan Strait; South China Sea. (1–6)

H. regani Jordan & Richardson: Taiwan; South China Sea. (1–6)

COTTIDAE

Ceratocottus diceraus (Pallas): Yellow Sea. (1–6)

Cottiusculus gonez Schmidt: Yellow Sea; East China Sea. (1–6)

Gymnocanthus herzensteini Jordan & Starks: Yellow Sea. (1–6)

Hemitripterus villosus (Pallas): Yellow Sea. (1–6)

Pseudoblennius cottoides (Richardson): Yellow Sea. (1–6)

Trachidermus fasciatus Heckel: Yellow Sea; East China Sea. (1–6)

Vellitor centropomus (Richardson): Yellow Sea. (1–6)

PSYCHROLUTIDAE

Psychrolutes inermis (Valliant): East China Sea. Water depth 520–979 m. (7)

EREUNIIDAE

Ereunias grallator Jordan & Snyder: East China Sea. Water depth 410–670 m. (7)

AGONIDAE

Podothecus sturiodes (Güichenot): Yellow Sea. (1–6)

CYCLOPTERIDAE

Lethotremus awae Jordan & Snyder: Yellow Sea; East China Sea. (1–6)

LIPARIDAE

Liparis chefuensis Wu & Wang: Yellow Sea. (1–6)

L. choanus Wu & Wang: Yellow Sea. (1–6)

L. petschiliensis (Rendahl): Yellow Sea. (1–6)

L. tanakae (Gilbert & Burke): Bohai; Yellow Sea; East China Sea. (1–6)

Paraliparis meridionalis Kido: East China Sea. Water depth 600–932 m. (7)

DACTYLOPTERIDAE

Dactyloptena gilberti (Snyder): Taiwan Strait; South China Sea. (1–6)

D. orientalis (Cuvier & Valenciennes): East China Sea; Taiwan Strait; South China Sea. (1–6)

Daicocus peterseni (Nystrom): East China Sea; Taiwan Strait; South China Sea. (1–6)

Ebisinus cheirophthalmus (Bleeker): Taiwan. (1–6)

PLEURONECTIFORME [HETEROSOMATA, PLEURONECTIDA] PSETTODIDAE

Psettodes erumei (Bloch & Schneider): East China Sea; Taiwan Strait; South China Sea. Near shore benthic species. (1–6)

CITHARIDAE

Brachypleura novaezeelandiae Günther: South China Sea. (1–6)

Citharoides macrolepidotus Hubbs [*C. macrolepis*]: Eastern East China Sea; northern South China Sea. Water depth 115–549 m. (1–6)

Lepidoblepharon ophthalmolepis Weber: East China Sea; South China Sea. Deep water. (1–6)

PARALICHTHYIDAE

Paralichthys olivaceus (Temminck & Schlegel): China's coast. Important economic species. (1–6)

Pseudorhombus arsius (Hamilton-Buchanan): East China Sea; Taiwan Strait; South China Sea. (1–6)

P. cinnamomeus (Temminck & Schlegel): China's coast. Near shore bottom fish. (1–6)

P. ctenosquamis (Oshima): South China Sea. (1–6)

P. dupliocellatus Regan: East China Sea; Taiwan Strait; South China Sea. Near shore bottom fish. (1–6)

P. elevatus Ogilby: Taiwan; South China Sea. (1–6)

P. javanicus (Bleeker): Southern East China Sea; South China Sea. (1–6)

P. levisquamis (Oshima): Taiwan; South China Sea. (1–6)

P. malayanus Bleeker: South China Sea. (1–6)

P. neglectus Bleeker: China's coast. (1–6)

P. oligodon (Bleeker): East China Sea; Taiwan Strait; South China Sea. Near shore bottom fish. (1–6)

P. pentophthalmus Günther: Taiwan Strait; South China Sea. (1–6)

P. quinquocellatus Weber & Beaufort: Taiwan Strait; South China Sea. (1–6)

P. triocellatus (Schneider): South China Sea. (1–6)

Tarphops oligolepis (Bleeker): Yellow Sea; East China Sea; Taiwan Strait; northern South China Sea. (1–6)

Tephrinectes sinensis (Lacépéde): East China Sea; Taiwan Strait; South China Sea. Near shore benthic species. (1–6)

BOTHIDAE

Arnoglossus aspilos (Bleeker): South China Sea. (1–6)

A. japonicus (Hubbs): East China Sea; Taiwan; South China Sea. (1–6)

A. polyspilus (Günther): East China Sea; Taiwan; South China Sea. Water depth 90–252 m. (1–6)

A. scapha (Schneider): South China Sea. (1–6)

A. tapeinosoma (Bleeker): Taiwan; South China Sea. (1–6)

A. tenuis Günther: East China Sea; Taiwan; South China Sea. (1–6)

Asterorhombus intermedius (Bleeker) [*Arnoglossus intermedius*]: South China Sea. (1- 6)

Bothus assimilis (Günther): Northeastern Taiwan; South China Sea. (1–6)

B. mancus (Broussonet): Southern tip of Taiwan and Lanyu (Taiwan); Xisha (=Paracel Islands). Rocky reefs. (1–6)

B. myriaster (Temminck & Schlegel): Taiwan Strait; South China Sea. (1–6)

B. pantherinus (Rüppell): Taiwan; Xisha (=Paracel Islands). Coral reefs or rocky reefs. (1–6)

Chascanopsetta lugubris Alcock: East China Sea; Taiwan Strait; South China Sea. Water depth 183–977 m. (1–6)

Crossorhombus azureus (Alcock): East China Sea; Taiwan Strait; South China Sea. (1- 6)

C. kanekonis (Tanaka): Northeastern Taiwan; South China Sea. (1–6)

C. kobensis (Jordan & Starks): East China Sea; Taiwan Strait; northern South China Sea. Water depth 50–275 m. (1–6)

C. valderostratus (Alcock): East China Sea; South China Sea. (1–6)

Engyprosopon filipennis Wu & Tang [*E. macroptera* (8)]: Taiwan; South China Sea. (1- 6)

E. grandisquama (Temminck & Schlegel): East China Sea; Taiwan Strait; South China Sea. (1–6)

E. latifrons (Regan): South China Sea. (1–6)

E. longipelvis Amaoka: South China Sea. (1–6)

E. mogki (Bleeker): South China Sea. (1–6)

E. multisquama Amaoka: Taiwan; South China Sea. (1–6)

Grammatobothus krempfi Chabanaud: Northeastern and southern Taiwan. Rocky reefs. (8)

G. polyophthalmus (Bleeker): South China Sea. (1–6)

Japonolaeops dentatus Amaoka: Taiwan. (8)

Kamoharia megastoma (Kamohara): Taiwan; South China Sea. (1–6)

Laeops kitakarae (Smith & Pope): East China Sea; Taiwan Strait; South China Sea. (1- 6)

L. lanceolatata Franz: East China Sea; South China Sea. (1–6)

L. parviceps Günther: East China Sea; Taiwan; South China Sea. (1–6)

L. tungkongensis Chen & Weng: Taiwan. (8)

Neolaeops microphthalmus (Von Bonder): Taiwan; South China Sea. (1–6)

Parabothus chlorospilus Gilbert: Northeastern Taiwan. (8)

P. coarctatus (Gilbert): East China Sea; South China Sea. Water depth 210–403 m. (1- 6)

Psettina filimanus Li & Wang: East China Sea; South China Sea. (1–6)

P. gigantea Amaoka: Taiwan. (8)

P. hainanensis (Wu & Tang): South China Sea. (1–6)

P. iijimae (Jordan & Starks): Taiwan; South China Sea. (1–6)

P. tosana Amaoka: Northeastern Taiwan. Muddy sand substrate. (8)

PLEURONECTIDAE

Cleisthenes herzensteini (Schmidt): Yellow Sea; East China Sea. (1–6)

Clidoderma asperrima (Temminck & Schlegel): Bohai; Yellow Sea; East China Sea. (1- 6)

Eopsetta grigorijewi (Herzestein): Bohai; Yellow Sea; East China Sea; Taiwan. (1–6)

Hippoglossoides dubius Schmidt: Yellow Sea; northern East China Sea. (1–6)

Kareius bicoloratus (Basilewsky): Bohai; Yellow Sea; East China Sea. (1–6)

Liopsetta obscura (Herzenstein): Yellow Sea. (1–6)

Microstomus achne (Jordan & Starks): Bohai; Yellow Sea; East China Sea. (1–6)

Plagiopsetta fasciatus (Fowler): South China Sea. (1–6)

P. glossa Franz: East China Sea; Taiwan; South China Sea. (1–6)

Pleuronichthys cornutus (Temminck & Schlegel): China's coast. Near shore bottom fish. (1–6)

Poecilopsetta colorata Günther: South China Sea. (1–6)

P. megalepis Fowler: Taiwan. (8)

P. natalensis Norman: Yellow Sea; East China Sea; Taiwan; South China Sea. (1–6)

P. plinthus (Jordan & Starks): Yellow Sea; East China Sea; Taiwan; South China Sea. (1–6)

P. praelonga Alcock: Taiwan; South China Sea. (1–6)

Pseudopleuronectes herzensteini (Jordan & Snyder): Bohai; Yellow Sea; East China Sea. (1–6)

P. yokohamae (Günther): Bohai; Yellow Sea; northern East China Sea. (1–6)

Samaris cristatus Gray: East China Sea; Taiwan Strait; South China Sea. (1–6)

Samariscus filipectoralis Shen: Northeastern Taiwan. (8)

S. huysmani Weber: South China Sea. (1–6)

S. inornatus (Lloyd): South China Sea. (1–6)

S. latus Matsubara & Takamuki: East China Sea; Taiwan; South China Sea. (1–6)

S. longimanus Norman: South China Sea. (1–6)

Tanakius kitaharae (Jordan & Starks): Bohai; Yellow Sea; East China Sea. (1–6)

Verasper moseri Jordan & Gilbert: Bohai; Yellow Sea. (1–6)

V. variegatus (Temminck & Schlegel): Bohai; Yellow Sea; East China Sea. (1–6)

SOLEIDAE

Aesopia cornuta Kaup: Taiwan Strait; South China Sae. Near shore bottom fish. (1–6)

Aseraggodes kaianus (Günther): Northeastern Taiwan; South China Sea. Muddy sand substrate. (1–6)

A. kobensis (Steindachner): Northern and southwestern Taiwan; South China Sea. Muddy sand substrate. (1–6)

Brachirus annularis Fowler [*Synaptura annularis* (8)]: Southwestern Taiwan; South China Sea. Muddy sand substrate. (1–6)

B. orientalis (Schneider) [*Synaptura orientalis*]: East China Sea; Taiwan Strait; South China Sea. (1–6)

B. pan (Hamilton): South China Sea. (1–6)

B. swinhonis (Steindachner): Northern South China Sea. (1–6)

Heteromycteris japonicus (Temminck & Schlegel): Yellow Sea; East China Sea. (1–6)

Liachirus melanospilus (Bleeker) [*Aseraggodes melanospilus* (8)]: Taiwan Strait; South China Sea. (1–6)

Monochirus trichodactylus (Linnaeus): South China Sea. (1–6)

Pardachirus pavoninus (Lacépède): Southern tip of Taiwan; South China Sea. Rocky reefs. (1–6)

P. xenicus Matsubara & Ochiai: Taiwan. (8)

Solea ovata Richardson: East China Sea; Taiwan Strait; South China Sea. (1–6)

Soleichthys heterorhinos (Bleeker): Northern and southern Taiwan. Rocky reefs. (1–6)

Zebrias crossolepis Cheng & Chang: Taiwan Strait; South China Sea. (1–6)

Z. japonicus (Bleeker): East China Sea; South China Sea. (1–6)

Z. quagga (Kaup): Southwestern Taiwan; South China Sea. Muddy sand substrate. (1–6)

Z. zebra (Bloch): China's coast. Near shore bottom fish. (1–6)

CYNOGLOSSIDAE

Cynoglossus abbreviatus (Gray): China's coast. (1–6)

C. arel (Schneider): Taiwan; South China Sea. (1–6)

C. bilineatus (Lacépède): East China Sea; Taiwan Strait; South China Sea. (1–6)

C. brachycephalus Bleeker: South China Sea. (1–6)

C. gracilis Günther: Northern, northeastern, and southwestern Taiwan; South China Sea. (1–6)

C. interruptus Günther: East China Sea; Taiwan Strait; South China Sea. (1–6)

C. itinus (Snyder): Taiwan Strait; South China Sea. (1–6)

C. joyneri Günther: China's coast. Muddy sand substrate. (1–6)

C. kopsi (Bleeker): Northeastern and southwestern Taiwan. Muddy sand substrate. (1–6)

C. lida (Bleeker): Taiwan; South China Sea. (1–6)

C. lighti Norman: China's coast. (1–6)

C. lineolatus Steindachner: South China Sea. (1–6)

C. macrolepidotus (Bleeker): East China Sea; South China Sea. (1–6)

C. melampetalus (Richardson): East China Sea; Taiwan Strait; South China Sea. Near shore bottom fish. (1–6)

C. monopus (Bleeker): South China Sea. (1–6)

C. nigropinnatus Ochisai: Western East China Sea; northern South China Sea. Water depth within 150 m. (1–6)

C. oligolepis (Bleeker): East China Sea; South China Sea. (1–6)

C. puncticeps (Richardson): East China Sea; Taiwan Strait; South China Sea. (1–6)

C. purpureomaculatus Regan: Bohai; Yellow Sea; East China Sea; northern South China Sea. (1–6)

C. robustus Günther: China's coast. (1–6)

C. roulei Wu: East China Sea; Taiwan Strait; South China Sea. (1–6)

C. semilaevis Günther: China's coast. Near shore muddy sand substrate. (1–6)

C. sibogae (Weber): Taiwan Strait; South China Sea. (1–6)

C. sinicus Wu: Western East China Sea; Taiwan Strait; northern South China Sea. (1–6)

C. suyeni Fowler: Southwestern Taiwan. Muddy sand substrate. (1–6)

C. trigrammus Günther: Southern East China Sea; Taiwan Strait; South China Sea. Near shore or river mouth bottom fish. (1–6)

C. xiphoides Günther: Northern South China Sea. (1–6)

Paraplagusia bilineata (Bloch): East China Sea; Taiwan Strait; South China Sea. Muddy sand substrate. (1–6)

P. blochi (Bleeker): Western and southwestern Taiwan; northern South China Sea. (1–6)

P. guttata Macleay: Yellow Sea; East China Sea; Taiwan; South China Sea. (1–6)

P. japonica (Temminck & Schlegel): East China Sea; Taiwan Strait; South China Sea. Near shore bottom fish. (1–6)

Symphurus novemfasciatus Shen: Southern tip of Taiwan. Sandy intertidal. (8)

S. orientalis (Bleeker): From northeastern Yellow Sea to northern South China Sea. (1–6)

S. strictus Gilbert: Taiwan. (8)

TETRAODONTIFORMES
TRIACANTHODIDAE

Halimochirurgus alcocki Weber: Taiwan; South China Sea. Water depth 309–394 m. (1- 6)

Macrorhamphosodes uradoi (Kamohara): East China Sea. Water depth 183–400 m. (7)

Paratriacanthodes retrospinis Fowler: Taiwan's coast. (1–6)

Triacanthodes anomalus (Temminck & Schlegel): Northeastern Taiwan; South China Sea. Bottom fish. (1–6)

Tydemania japonica Kamohara: South China Sea. Muddy sand substrate with water depth about 420 m.

T. navigatoris Weber: Southwestern Taiwan; South China Sea. Muddy sand substrate with water depth of 60–420 m. (1–6)

TRIACANTHIDAE

Pseudotriacanthus strigilifer (Cantor) [*Triacanthus strigilifer*]: East China Sea; Taiwan Strait; South China Sea. Near shore bottom fish. (1–6)

Triacanthus biaculeatus (Bloch): Northeastern and southwestern Taiwan; South China Sea. (1–6)

T. blochi Bleeker [*Tripodichthys blochi* (8)]: Northeastern and southwestern Taiwan; South China Sea. (1–6)

T. brevirostris Temminck & Schlegel: China's coast. Near shore bottom fish. (1–6)

T. nieuhofi Bleeker: South China Sea. (1–6)

BALISTIDAE

Abalistes stellatus (Lacépéde): East China Sea; Taiwan Strait; South China Sea. Common near shore bottom fish. (1–6)

Balistapus undulatus (Mungo Park): Taiwan; Hainan; Xisha (=Paracel Islands); Nansha (=Spratly Islands). Coral reefs or rocky reefs. (1–6)

Balistes capistretus Shaw: Xisha (=Paracel Islands). Coral reefs or rocky reefs. (1–6)

B. vetula Linnaeus: South China Sea. (1–6)

Balistoides conspicillum (Bloch & Schneider) [*Balistes conspicillum*]: Taiwan Strait; Xisha (=Paracel Islands). Coral reefs. (1–6)

B. viridescens (Bloch & Schneider) [*Balistes viridescens*]: Taiwan; Hainan; Xisha (=Paracel Islands). Rocky reefs. (1–6)

Canthidermis maculatus (Bloch) [*C. rutundatus*]: East China Sea; Taiwan Strait; South China Sea. Coral reefs or rocky reefs. (1–6)

Melichthys niger (Bloch): South China Sea. (1–6)

M. vidula (Solander) [*Balistes vidula*]: Southern tip of Taiwan; Hainan; Dongsha (=Pratas Islands); Xisha (=Paracel Islands); Nansha (=Spratly Islands). Rocky reefs. (1–6)

Odonus niger (Rüppell): Taiwan; Xisha (=Paracel Islands); Nansha (=Spratly Islands). Coral reefs or rocky reefs. (1–6)

Pseudobalistes flavimarginatus (Rüppell) [*Balistes flavimaginatus*]: Northeastern and southern Taiwan, Lu Island (Taiwan); Xisha (=Paracel Islands). Rocky reefs. (1- 6)

P. fuscus (Bloch & Schneider) [*Balistes fuscus*]: Southern Taiwan; Xisha (=Paracel Islands). Rocky shore. (1–6)

Rhinecanthus aculeatus (Linnaeus) [*Balistapus aculeatus*]: Taiwan; South China Sea. Coral reefs or rocky reefs. (1–6)

R. echarpe (Lacépéde) [*Balistapus rectangulus*]: Taiwan; Xisha (=Paracel Islands). Coral reefs or rocky reefs. (1–6)

R. verrucosus (Linnaeus): Southern tip of Taiwan; South China Sea. Rocky reefs. (1–6)

Sufflamen bursa (Bloch & Schneider): Southern tip of Taiwan. Rocky reefs. (8)

S. chrysopterus (Bloch & Schneider): Southern tip of Taiwan; Xisha (=Paracel Islands). Rocky reefs. (1–6)

S. fraenatus (Latreille): Southern tip of Taiwan; South China Sea. Rocky reefs. (8)

Xanthichthys auromarginatus (Bennett): Taiwan. (8)

X. lineopunctatus (Hollard): Hainan; Dongsha (=Pratas Islands); Xisha (=Paracel Islands); Nansha (=Spratly Islands). (1–6)

ALUTERIDAE

Alutera monoceros (Osbeck): Yellow Sea; East China Sea; Taiwan Strait; South China Sea. (1–6)

A. scripta (Osbeck): Taiwan; South China Sea. (1–6)

Amanses scopas (Cuvier): Taiwan. (8)

Arotrolepis sulcatus (Hollard) [*Monacanthus sulcatus*]: Yellow Sea; East China Sea; Taiwan Strait; South China Sea. Water depth more than 20 m. (1–6)

Cantherhines dumerili (Hollard): Lanyu (Taiwan) and southern tip of Taiwan; Hainan; Dongsha (=Pratas Islands); Xisha (=Paracel Islands); Nansha (=Spratly Islands). Rocky reefs. (1–6)

C. fronticinctus (Günther): Taiwan. (8)

C. pardalis (Rüppell): Lanyu (Taiwan) and southern tip of Taiwan; Hainan; Dongsha (=Pratas Islands); Xisha (=Paracel Islands); Nansha (=Spratly Islands). Rocky reefs. (1–6)

Chaetodermis spinosissimus (Quoy & Gaimard) [*C. penicilligerus* (8)]: Taiwan; Xisha (=Paracel Islands). (1–6)

Laputa kneri (Steindachner): Taiwan; South China Sea. (1–6)

Monacanthus chinensis (Osbeck): East China Sea; Taiwan Strait; South China Sea. Rocky reefs. (1–6)

Navodon septentrionalis (Günther) [*N. modestus, Thamnaconus septentrionalis* (8)]: China's coast. Bottom fish, water depth 50–120 m. (1–6)

N. tessellatus (Günther) [*Cantherhines nodestus*]: Northeastern Taiwan; South China Sea. Rocky reefs. (1–6)

N. xanthopterus Xu & Zhen sp. no.: East China Sea; South China Sea. Water depth 100- 225 m. (7)

Oxymonacanthus longirostris (Blocj & Schneider): Southern tip of Taiwan; Hainan; Dongsha (=Pratas Islands); Xisha (=Paracel Islands); Nansha (=Spratly Islands). Rocky reefs. (1–6)

Paraluteres prionurus (Bleeker): Southern tip of Taiwan. Rocky shore. (8)

Paramonacanthus nipponensis (Kamohara) [*Monacanthus nipponensis*]: Taiwan Strait; South China Sea. Near shore. (1–6)

P. oblongus (Temminck & Schlegel): Taiwan Strait; South China Sea. (1–6)

Pervagor aspricaudus (Hollard): Taiwan. (8)

P. janthinosoma (Bleeker): Taiwan. (8)

P. melanocephalus (Bleeker): Southern tip of Taiwan; Hainan; Dongsha (=Pratas Islands); Xisha (=Paracel Islands); Nansha (=Spratly Islands). Rocky reefs. (1–6)

P. nitens (Hollard): Hainan; Dongsha (=Pratas Islands); Xisha (=Paracel Islands); Nansha (=Spratly Islands). (1–6)

P. tomentosus (Linnaeus): South China Sea. (1–6)

Pseudalutarius nasicorinis (Schlegel): South China Sea. (15, 17)

Rudarius ercodes Jordan & Fowler: Northeastern Taiwan; South China Sea. Rocky reefs. (1–6)

Stephanolepis cirrhifer (Temminck & Schlegel) [*Monacanthus setifer*]: Yellow Sea; East China Sea; Taiwan; South China Sea. Near shore benthic species. (1–6)

S. japonicus (Tilesius): East China Sea. (1–6)

PSILOCEPHALIDAE

Psilocephalus barbatus (Gray): East China Sea; South China Sea. (1–6)

ARACANIDAE

Aracana rosapinto (Smith): Taiwan Strait; South China Sea. Benthic species. (1–6)

Kentrocapros aculeatus (Houttuyn): Northeastern Taiwan; South China Sea. (1–6)

OSTRACIONTIDAE

Lactoria cornutus (Linnaeus) [*Ostracion cornutus*]: China's coast. Near shore benthic species. (1–6)

L. diaphanus (Bloch & Schneider): Northeastern Taiwan; South China Sea. (1–6)

L. fronasini (Bianconi): Taiwan. (8)

Ostracion meleagris Shaw: Southern tip of Taiwan. Rocky shore. (8)

O. solorensis Bleeker: Taiwan Strait. (1–6)

O. tuberculatus Linnaeus [*O. cubicus* (8)]: Yellow Sea; East China Sea; Taiwan Strait; South China Sea. Near shore benthic species. (1–6)

Rhinesomus concatenatus Bloch & Schneider [*Tetrosomus concatenatus* (8)]: Taiwan Strait; South China Sea. Near shore benthic species. (1–6)

R. gibbosus Linnaeus [*Lactophry gibbosus*, *Tetrosomus gibbosus* (8)]: Taiwan Strait; South China Sea. Near shore benthic species. (1–6)

Rhynchostracion nasus Bloch: South China Sea. (1–6)

R. rhinorhynchus Bleeker: Southern Taiwan Strait. (1–6)

TETRAODONTIDAE

Amblyrhynchotus hypselogeneion (Bleeker): Taiwan Strait; South China Sea. Near shore bottom fish. (1–6)

A. honckeni (Bloch): South China Sea. (1–6)

A. rufopunctatus Li: South China Sea. (1–6)

A. spinosissimus (Regan): South China Sea. (1–6)

Arothron alboreticulatus (Tanaka): Taiwan. (8)

A. firmamentum (Temminck & Schlegel): Taiwan. (8)

A. hispidus (Linnaeus): Taiwan; Xisha (=Paracel Islands). (1–6)

A. immaculatus (Bloch & Schneider): Taiwan; South China Sea. (1–6)

A. manilensis Marion de Proce: Taiwan. (8)

A. mappa (Lesson): Southern East China Sea; Taiwan. (1–6)

A. meleagris (Lacépéde): Xisha (=Paracel Islands). (1–6)

A. nigropunctatus (Bloch & Schneider): Taiwan; Hainan; Xisha (=Paracel Islands). (1- 6)

A. reticularis (Bloch & Schneider): Taiwan Strait. (1–6)

A. stellatus (Bloch & Schneider): Taiwan Strait; South China Sea. (1–6)

Boesemanichthys firmamentum (Temminck & Schlegel): Taiwan Strait; northern South China Sea shoreline. (1–6)

Canthigaster amboinensis (Bleeker): Southern tip of Taiwan. Rocky shore. (8)

C. bennetti (Bleeker): Taiwan. (8)

C. compressus Marion de Proce: Taiwan. (8)

C. coronata (Vaillant & Sauvage): Taiwan. (8)

C. jactator (Jenkins): Taiwan Strait; Xisha (=Paracel Islands). (1–6)

C. janthinopterus (Bleeker): Southern tip of Taiwan. Rocky shore. (8)

C. rivulatus (Temminck & Schlegel): Taiwan Strait; South China Sea. (1–6)

C. solandri (Richardson): Southern tip of Taiwan. Rocky shore. (8)

C. valentini (Bleeker): Southern tip of Taiwan; South China Sea. Rocky shore. (1–6)

Chelonodon patoca (Hamilton-Buchanan): East China Sea; Taiwan Strait; South China Sea. Near shore bottom fish. (1–6)

Fugu alboplumbeus (Richardson) [*Spheroides alboplumbeus*]: China's coast. Near shore bottom fish. (1–6)
F. basilevskianus (Basiltwsky): Bohai; Yellow Sea. (1–6)
F. bimaculatus (Richardson): Southern Yellow Sea; East China Sea; South China Sea. Near shore bottom fish. (1–6)
F. flavidus Li, Wang & Wang & Wang: China's coast. (1–6)
F. niphobles (Jordan & Snyder) [*Takifugu niphobles* (8)]: China's coast. Near shore bottom fish. (1–6)
F. oblongus (Bloch) [*Takifugu oblongus* (8)]: East China Sea; Taiwan; South China Sea. Near shore bottom fish. (1–6)
F. obscurus (Abe) [*Spheroides obscurus*]: China's coast. Near shore bottom fish. (1–6)
F. ocellatus (Linnaeus) [*Spheroides ocellatus*]: China's coast. Near shore bottom fish. (1–6)
F. pardalis (Temminck & Schlegel): Bohai; Yellow Sea. (1–6)
F. porphyreus (Temminck & Schlegel): Bohai; Yellow Sea; East China Sea. (1–6)
F. pseudommus (Chu): Bohai; Yellow Sea; East China Sea. (1–6)
F. reticularis Tian, Cheng & Wang [*Takifuga poecilonotus* (8)]: Bohai; Yellow Sea; East China Sea. (1–6)
F. rubripes (Timminck & Schlegel): Bohai; Yellow Sea; East China Sea. (1–6)
F. stictonotus (Temminck & Schlegel): East China Sea. (1–6)
F. vermicularis (Temminck & Schlegel) [*Spheroides vermicularis*]: China's coast. Near shore bottom fish. (1–6)
F. xanthopterus (Temminck & Schlegle) [*Spheroides xanthopterus, Takifugu xanthopterus* (8)]: China's coast. Near shore bottom fish. (1–6)
Gastrophysus lunaris (Bloch & Schneider) [*Lagocephalus lunaris*]: East China Sea; Taiwan Strait; South China Sea. Near shore bottom fish. (1–6)
G. spadiceus (Richardson) [*Lagocephalus spadiceus*]: East China Sea; Taiwan Strait; South China Sea. Near shore bottom fish. (1–6)
Lagocephalus gloveri Abe & Tabeta: Taiwan. (8)
L. inermis (Temminck & Schlegel): Southern Yellow Sea; East China Sea; Taiwan Strait; South China Sea. Near shore bottom fish. (1–6)
L. oceanicus Jordan & Evermann: South China Sea. (1–6)
L. wheeleri Abe, Tabeta & Kitahama: Taiwan. (8)
Liosaccus cutaneus (Günther): South China Sea. (1–6)
L. pachygaster (Müller & Troschel): East China Sea. Water depth 25–480 m. (7)
Pleuranacanthus sceleratus (Forster) [*Gastrophysus sceleratus, Lagocephalus sceleratus*]: Taiwan Strait; South China Sea. (1–6)
P. suezensis (Gobar) [*Gastrophysus suezensis, Lagocephalus suezensis*]: Taiwan Strait; South China Sea. Near shore bottom fish. (1–6)
Sphoeroides pachygaster (Müller & Trosche) [*Lagocephalus pachygaster* (8)]: Taiwan. (8)

TRIODONTIDAE

Triodon bursarius Reinhardt: Northeastern Taiwan. (1–6)
T. macropterus Lesson: Taiwan. (8)

DIODONTIDAE

Chilomycterus affinis Günther: Taiwan Strait. (1–6)
C. echinatus (Cronow): South China Sea. (1–6)
C. orbicularis (Bloch): Taiwan Strait; South China Sea. (1–6)
Diodon bleekeri Günther: South China Sea. (1–6)
D. eydiuxii Birssout ded Barneville: Taiwan. (8)
D. holacanthus Linnaeus: Yellow Sea; East China Sea; Taiwan Strait; South China Sea. (1–6)
D. hystrix Linnaeus: Northern and northeastern Taiwan; Xisha (=Paracel Islands). Rocky shore. (1–6)
D. liturosus Shaw: Taiwan. (8)
D. novemmaculatus Cuvier: Dongsha (=Pratas Islands); Xisha (=Paracel Islands). (1–6)

MOLIDAE

Masturus lanceolatus (Lienard) [*M. oxyuropterus*]: Taiwan Strait; South China Sea. Oceanic species. (1–6)
Mola mola (Linnaeus): Yellow Sea; East China Sea; Taiwan; South China Sea. Oceanic species. (1–6)

PEGASIFORMES
PEGASIDAE

Pegasus draconis Linnaeus [*Eurypegasus draconis* (8)]: Northeastern Taiwan; South China Sea. (1–6)
P. laternarius Cuvier: East China Sea; Taiwan Strait; South China Sea. Deep water benthic species. (1–6)
P. volitans Cuvier: East China Sea; Taiwan Strait; South China Sea. Near shore benthic pecies. (1–6)

LOPHIIFORMES
LOPHIIDAE

Lophiodes infrabrunneus Smith & Radcliffe: Okinawa Trough. Water depth 494–1,560 . (7)
L. naresi (Günther): East China Sea. Water depth 183–457 m. (7)
Lophiomus setigerus (Vahl): East China Sea; Taiwan Strait; South China Sea. Water depth 40–50 m, muddy sand substrate. (1–6)
Lophius litulon (Jordan): Bohai; Yellow Sea; East China Sea; Taiwan Strait; South China Sea. (1–6)
Slandenis remiger Smith & Radcliffe: East China Sea. Water depth 979–1294 m. (7)

MELANOCETIDAE

Melanocetus johnoni Günther: East China Sea. Water depth 275–4,475 m. (7)

CERATIIDAE

Ceratias holboelli Kröyer: East China Sea; South China Sea. Water depth 697–4,393 m. (7, 14)
Cryptopsaras couesi Gill: South China Sea. Deep water. (1–6)

DICERATIIDAE

Diceratias bispinosus (Günther): South China Sea. Water depth 300–1,400 m. (16)
Phrynichthys thele Uwate: East China Sea; South China Sea. Water depth 680–979 m. (7, 16)

GIGANTACTINIDAE

Gigantactis garganthus Bertelsen, Pietsch & Lavenberg: East China Sea. Water depth 500–1,850 m. (7)
G. vanhoeffeni Brauer: East China Sea. Water depth 300–5,300 m. (7)

ONEIRODIDAE

Oneirodes appendixus Ni & Xu, sp. nov.: East China Sea. Water depth 432–810 m. (7)

LINOPHRYNIDAE

Linophryne polypogon Regan: East China Sea. Water depth 1,000–5,000 m. (7)

HIMANTOLOPHIDAE

Himantolophus groenlandicus Reinhardt: Taiwan. (8)

ANTENNARIDAE

Antennarius biocellatus (Cuvier): Taiwan. (8)
A. commersonii (Shaw): Taiwan. (8)
A. dorehensis Bleeker: Taiwan; Xisha (=Paracel Islands). Rocky reefs and coral reefs. 1–6)
A. hispidus Bloch & Schneider: East China Sea; Taiwan Strait; South China Sea. Near shore bottom fish. (1–6)
A. melas Bleeker: East China Sea; Taiwan Strait. Near shore bottom fish. (1–6)
A. nummifer Cuvier: Taiwan Strait; South China Sea. Rocky reefs. (1–6)
A. pinniceps Commerson [*A. tridens* (8)]: East China Sea; Taiwan Strait; South China Sea. Mud flat, seaweed, rocky reefs, and coral reefs. (1–6)
A. randalli Allen: Taiwan. (8)
A. striatus (Shaw): Taiwan. (8)
Histiophryne cryptacanthus (Cuvier): Taiwan. (8)
Histrio histrio (Linnaeus) [*Pterophryne marmoratus*]: Taiwan; South China Sea. Rocky shore with dense seaweed. (1–6)

CHAUNACIDAE

Chaunax abei Le Danois: Northeastern Taiwan. (8)
C. fimbriatus Hilgendrof: Northern and northeastern Taiwan; South China Sea. Coral reefs. (1–6)

OGCOCEPHALIDAE

Dibranchus japonicus Amaoka & Toyoshimo: Taiwan. (8)
Halicmetus reticulatus Smith & Radcliffe: Taiwan; Dongsha (=Pratas Islands); Xisha (=Paracel Islands). (1–6)
Halieutaea fitzsmonsi (Gilchrist & Thompson): Southwestern Taiwan. (1–6)
H. fumosa Alcock: Taiwan Strait; South China Sea. (1–6)
H. indica Annandala & Jonkins: Xisha (=Paracel Islands). (1–6)
H. nigra Alcock: South China Sea. (1–6)
H. sinica Tchang & Chang: East China Sea; South China Sea. Near shore bottom fish. (1–6)
H. stellata (Vahl): China's coast. Mud flat, shallow sea with sandy mud substrate. (1–6)
Malthopsis annulifera Tanaka: East China Sea; Taiwan. Water depth 70–778 m. (7, 8)
M. jordeni Gilbert: Taiwan. (8)
M. luteus Alcock: Southwestern Taiwan; South China Sea. Muddy sand substrate. (1–6)

GOBIESOCIFORMES
GOBIESOCIDAE

Aspasma minima (Döderlein): Southern tip of Taiwan. Rocky reefs. (8)
Conidens laticephalus (Tanaka): Northern and northeastern Taiwan. Rocky reefs. (8)
Discoetrema crinophila Briggs: Southern tip of Taiwan. Rocky reefs. (8)
Lepadichthys frenatus Waite: Northern and southern Taiwan. Rocky reefs. (8)

Pherallodus indicus (Weber): Southern tip of Taiwan. Rocky reefs. (8)

REFERENCES*

(1). Cheng Qingtai and Zheng Baoshan (ed.), 1987. Systematic Synopsis of Chinese Fishes, Vol. 1. China Science Press, 1,458pp.
(2). Institute of Zoology, Academy Sinica et al., 1962. The Fishes of South China Sea. China Science Press, 1,111pp.
(3). Chu Yuanting (ed.), 1963. The Fishes of East China Sea. China Science Press, 611pp.
(4). Fisheries Institute of South China Sea, State Fisheries Administration, 1979. The Fishes of the Islands in the South China Sea. China Science Press, 613pp.
(5). Zhang Chunlin et al., 1955. Report of the Survey on the Fishes in Yellow Sea and Bohai Sea. China Science Press, pp:8–332.
(6). Chu Yuanting (ed.), 1984 (Vol.1), 1985 (Vol. 2). The Fishes of Fujian. Fujian Science Press, 528pp (Vol. 1), 700pp (Vol. 2).
(7). Fisheries Institute of East China Sea, 1988. The Deep-Sea Fishes of East China Sea. Xuelin Publishing House, 350pp.
(8). Shen Shichieh et al. 1983. Fishes of Taiwan. Department of Zoology, National Taiwan University, 170pp.
(9). Fisheries Institute of South China Sea, State Fisheries Administration, 1981. Species and Catches of The Fishes. Comprehensive Survey Report on the Fisheries Resources in the Waters on Continental Slope in Northern South China Sea. pp:6.1- 6.63.
(10). Yang Jiaju and Huang Zengyue, 1983. The Deep-Sea Fishes in the Waters around Dongsha Islands, I. Salmoniformes. Contributions on the Marine Organisms in South China Sea (I). China Ocean Press, pp:217–233.
(11). Huang Zengyue and Yang Jiaju, 1983. The Deep-Sea Fishes in the Waters around Dongsha Islands, II. Myctophiformes. Contributions on the Marine Organisms in the South China Sea (I). China Ocean Press, pp:234–255.
(12). Chen Zhenran, 1983. Studies on the Lanternfishes (Myctophidae) from the Central Waters of the South China Sea. Contributions on the Marine Organisms in South China Sea (I). China Ocean Press, pp:199–216.
(13). Huang Zengyue and Yang Jiaju, 1984. The Deep-Sea Fishes in the Waters around Dongsha Islands, III. Macruriformes and Beryciformes etc.. Tropic Oceanology, 3(3):44–50.
(14). Yang Jiaju and Huang Zengyue, 1989. The fauna and geographical distribution of deep-sea fishes in South China Sea. South China Sea Studia Marina Sinica, 9: 123–136.
(15). Multidisciplinary Scientific Survey Group on Nansha Islands, Academy Sinica, 1989. Multidisciplinary Survey Report on Nansha Islands and the Adjacent Waters (I), Vol. 1. The Atoll Fishes. China Science Press, pp:106–116.
(16). Ni Yong, Wu Hanlin, and Li Sheng, 1989. Two anglerfishes (Diceratiidae) new to Chinese fauna. Collections of Study on Fisheries in South China Sea (I):87–96.
(17). Huang Zengyue, Yang Jiaju, and Li Qingxin, 1991. Preliminary study on the fish fauna of Nansha Islands. Contributions on the Fauna and Zoogeography in the Waters around Nansha Islands. China Ocean Press, pp:272–294.
(18). Huang Zengyue et al., 1991. Preliminary study on the fishes of Nansha Islands. Contributions on Marine Organisms around Nansha Islands and the adjacent Waters (I). China Ocean Press, pp:190–197.
(19). Li Qingxin, Yang Jiaju, and Huang Zengyue, 1991. Bottom trawl fishes of Nansha Islands (I). Contributions on Marine Organisms around Nansha Islands and the adjacent Waters (I). China Ocean Press, pp:198–210.
(20). Qiu Shuyuan, 1982. A preliminary study on the Sardines of the South China Sea. Journal of Xiamen University (Natural Science), 21(1):55–67.
(21). Tian Mingcheng, 1986. Alepocephalidae new to Chinese deep-sea fishes. Oceanologia & Limnolgia Sinica, 17(1):91–92.
(22). Chen Suihi and Yang Yurong, 1991. Report on the deep-sea fishes from the adjacent waters of Dongsha Islands. Collected Papers of Zoology, 8:145–155.
(23). Yang Yurong, Zeng Bingguang and John R. Paxton, 1988. Additional Specimens of the Deep-Sea Fish *Hispidoberyx ambagiousus* (Hispidoberycidae, Bercigormes) from the South China Sea, with Comments on the Family Relationships. UO 38:1- 8.
(24). Wang Keling et al., 1993. Studies on the species identification of hairtail fishes (*Trichiurus*) in China coastal waters. Acta Oceanologica Sinica, 15(2):77–83.
(25). Shao Kwang-Tsao and Lih-Wen Chen, 1989. Fish of the Family Scarridae from Taiwan. Bull. Inst. Zool., Acad. Sinica, 28(1):15–39.

*: all in Chinese except for (23) and (25)
Compiled by Lin Shuangdan; Edited by Qiu Shuyan, Yang Jiaju, Lin Fushen and Shao Kwang Tsao
Note: Some fish species are classified differently in mainland China and Taiwan. In those cases, mainland China's are used with Taiwan's as synonyms.

AMPHIBIA
SALIENTIA [ANURA]
RANIDAE

Euphlyctis cancrivora (Gravenhorst) [*Rana cancrivora*]: Taiwan; Macau; Beihai, Fangcheng, and Hepu (Guangxi); Haikou and Wenchang (Hainan). (1, 2)

REFERENCES*

(1). Sichuan Biology Research Institute, 1977. Systematic key of Chinese amphibians. China Science Press, 93pp
(2). Fei, L. et al., 1990. Key to Chinese Amphibia. Chongqing Branch, Scientific and Technological Literature Press, 364pp

*: both in Chinese
Compiled by Huang Zongguo; Edited by Fei Liang

REPTILIA
TESTUDIFORMES
CHELONIIDAE

Caretta c. gigas Deraniyagala [*C. c. olivacea*]: Yellow Sea; East China Sea; South China Sea. (1, 6, 8)

Chelonia mydas (Linnaeus): Yellow Sea; East China Sea; South China Sea. (1, 3)

Eretmochelys imbricata (Linnaeus): Yellow Sea; East China Sea; South China Sea. (1, 3)

Lepidochelys olivacea (Eschscholtz): From Jiangsu to Guangxi; Taiwan. (7, 9)

DERMOCHELYIDAE

Dermochelys coriacea (Linnaeus): Bohai; Yellow Sea; East China Sea; South China Sea. (1)

SERPENTIFORMES
ACROCHORDIDAE

Acrochordus granulatus (Schneider): Sanya (Hainan). River mouth. (1)

HYDROPHIIDAE

Acalyptophis peronii (Dumeril & Bibron): Guangdong coast. (1)

Astrotia stokesi (Gray): Taiwan coast. (1)

Emydocephalus ijimae Stejneger: Taiwan coast. (1)

Hydrophis caerulescens (Shaw): Shandong; Guangdong. Coast. (1)

H. cyanocinctus Daudin: Liaoning; Shandong; Jiangsu; Zhejiang; Fujian; Taiwan; Guangdong; Guangxi; Hainan. Coast. (1, 4)

H. fasciatus atriceps (Günther): Fujian; Guangdong; Guangxi; Hainan. Coast. (1, 2)

H. melanocephalus (Gray): Zhejiang; Fujian; Taiwan; Guangdong. Coast. (1)

H. ornatus (Gray): Shandong; Taiwan; Guangdong; Guangxi; Hainan. Coast. (1)

Kerilia jerdonii siamensis Smith: Taiwan coast. (8)

Lapemis hardwickii (Gray): Shandong; Fujian; Taiwan; Guangdong; Guangxi; Hainan. Coast. (1)

Laticauda colubrina (Schneider): Taiwan coast. (1)

L. laticaudata (Linnaues): Fujian; Taiwan. Coast. (1, 2)

L. semifasciata (Reinwardt): Liaoning; Fujian; Taiwan. Coast. (1, 2, 4)

Microcephalophis gracilis (Shaw): Fujian; Guangdong; Guangxi; Hainan. Coast. (1)

Pelamis platurus (Linnaeus): Zhejiang; Fujian; Taiwan; Guangdong; Hainan. Coast. (1)

Praescutata viperina (Schmidt): Liaoning; Fujian; Guangdong; Guangxi; Hainan. Coast. (1, 5)

COLUBRIDAE

Enhydris bennetti (Gray): Hong Kong. Mangrove forest mud flat and inland. (10)

VIPERIDAE

Agkistrodon qianshanensis shedaoensis Zhao [*A. halys*, *A. shedaoensis*]: She Island (=Snake Island, 7 miles west off Lushun, Liaoning). There were 14,000 individuals in the 0.73 km^2 island. (4)

REFERENCES*

(1). Sichuan Biology Research Institute, 1977. Systematic key of Chinese reptiles. China Science Press, 111pp.

(2). Tian Wanshu, 1986. Identification Manual of Chinese Amphibians and Reptiles. China Science Press, 164pp.

(3). Lee Z. H., 1955. Sea turtles in the coast of southeast China. Bulletin of Biology, 12:36–39.

(4). Ji Daming and Wen Sisheng, 1992. On the taxonomy of Pallas pit viper from northeastern China. In: Transactions on Division of Herpetology, the Chinese Society of Zoology (10th anniversary, Dalian).

(5). Ji Daming et al., 1987. Fauna of Liaoning: Amphibia and Reptilia. Liaoning Science Press, 170pp.

(6). Yang Jiaju, 1975. A preliminary observation of the biological characteristics of sea turtles from Xisha Islands. In: Preliminary Reports on Comprehensive Investigation of Xisha Islands, pp:84–87.

(7). Zhou Kaiya, 1983. Studies on *Chelonia mydas*, *Lepidochelys olivacea* and *Dermochelys coriacea* from the coast of Jiangsu. Acta Herpetologica Sinica, 2(3):57–62.

(8). Hu Shuqin et al., 1987. Zoological Illustration from China: Amphibia and Reptilia. China Science Press, 110pp

(9). Huang Meihua et al., 1990. Fauna of Zhejiang: Amphibia and Reptilia. Zhejiang Science & Technology Press, 305pp.

(10). Morton, B. 1983. The Sea Shore Ecology of Hong Kong. Hong Kong University Press, 350pp.

*: all in Chinese except for (10)

Compiled by Lin Jinmei; Edited by Fei Liang

AVES
GAVIIFORMES
GAVIIDAE

Gavia adameii (G. R. Gray): Eastern Liaoning; Fujian. Migratory sea bird. (2, 5)

G. arctica viridigularis Dwight: Eastern Liaoning; Fujian. Sea bird. (2, 5)

G. s. stellata (Pontoppidan): China's coast. Sea bird. (2, 4, 5)

PODICIPEDIFORMES
PODICIPEDIDAE

Tachybaptus grisegena caspicus (Hablizl): China's coast and inland. (1–5)

T. grisegena cristatus (Linnaeus): China's coast. Migratory species. (1–6)

T. grisegena holboellii Reinhardt: Hebei; Fujian; Hong Kong. Migratory species. (2, 5, 6)

T. ruficollis poggei (Reichenow): Eastern Liaoning; Shanghai; Taiwan; Zhanjiang (Guangdong); Hong Kong; Hainan. (1, 2, 4, 5, 6)

PROCELLARIIFORMES
DIOMEDEIDAE

Diomedea albatrus Pallas: China's coast. Sea bird. (2, 5)

D. nigripes Audubon: China's coast. Sea bird. (2, 5)

PROCELLARIIDAE

Bulmeria bulwerii (Jardine & Selby): Fujian; Taiwan; Guangdong. Islands. (2, 5)

Fulmarius glacialis rodgersii Cassin: Northeastern China's coast. Sea bird. (2, 5)
Pterodroma r. rostrata (Peale): Taiwan. Sea bird. (2, 5)
Puffinus griseus (Gmelin): Fujian; Taiwan. Sea bird. (2, 5)
P. leucomelas (Temminck): China's seas. Islands. (2, 4, 5)
P. pacificus cuneatus Salvin: Penghu (Taiwan). Sea bird. (2, 5)

HYDROBATIDAE

Oceanodroma r. monorhis (Swinhoe): Shandong; Fujian; Taiwan; Guangdong. Islands. (2, 5)

PELECANIFORMES
PHAETHONTIDAE

Phaethon aethereus indicus Hume: Xisha (=Paracel Islands). (2, 5)
P. lepturus dorotheae Mathews: Taiwan. (2, 5)
P. rubricauda rothschildi (Mathews): Taiwan. (2, 5)

PELECANIDAE

Pelecanus p. philippensis Gmelin: Hebei; Shandong; Fujian; Guangdong; Guangxi; Hainan. Coast, islands, and inland. (2–5)

SULIDAE

Sula leucogaster plotus (Forster): From Shanghai to Taiwan; Hainan; Xisha (=Paracel Islands). Sea bird. (2, 5)
S. sula rubripes Gould: Reproduce in Xisha (=Paracel Islands). Sea bird. (2, 5)

PHALACROCORACIDAE

Phalacrocorax carbo sinensis (Blumenbach): China's coast and inland. Migratory species. (2–6)
P. filamentosus (Temminck & Schlegel): From Lushun (Liaoning) to Fujian; Taiwan. Coast and inland, migratory species. (2, 5)
P. p. pelagicus Pallas: China's coast during winter. Migratory species. (2, 5)

FREGATIDAE

Fregata andrewsi Mathews: Guangdong. Islands. (2, 5)
F. a. ariel (G. R. Gray): Fujian; Taiwan; Xisha (=Paracel Islands). Islands. (2, 5)
F. m. minor (Gmelin): Guangdong; Fujian; Xisha (=Paracel Islands); occasionally in Bohai and Yellow Sea. Sea bird. (2, 5)

CICONIIFORMES
ARDEIDAE

Ardea cinerea rectirostris Gould: China's coast. (1–6)
A. purpurea manilensis Meyen: China's coast. (1–6)
Ardeola bacchus (Bonaparte): Coasts south of Changjiang (=Yangtze River). Migratory species. (2, 4, 5, 6)
Botaurus s. stellaris (Linnaeus): China's coast. Migratory species. (1, 2, 4, 5)
Bubulcus ibis coromandus (Boddaert): Coasts south of Changjiang (=Yangtze River); occasionally in Weihai (Shandong). (2, 4, 5, 6)
Butorides striatus connectens Stresemann: China's coast. Migratory species. (2, 3, 4, 5, 6)
Egretta alba modestus (J. E. Gray): China's coast. Migratory species. (1, 2, 4, 5, 6)
E. eulophotes (Swinhole): China's coast; Xisha (=Paracel Islands). Migratory species. (2, 5)
E. g. garzetta (Linnaeus): Coasts south of Changjiang (=Yangtze River); occasionally in Weihai (Shandong). Migratory species. (2, 4, 5, 6)
E. i. intermedia (Wagler): South of Changjiang (=Yangtze River). Migratory species. (2, 4, 5, 6)
E. s. saora (Gmelin): Fujian; Penghu and Lanyu (Taiwan); Guangdong; Hainan. Migratory species. (2, 5)
Ixobrychus cinnamomeus (Gmelin): China's coast. Migratory species. (2, 4, 5, 6)
I. eurhythmus (Swinhoe): China's coast. Migratory species. (1, 2, 4, 5, 6)
I. s. sinensis (Gmelin): China's coast. Migratory species. (1, 2, 4, 5, 6)
Nycticorax n. nycticorax (Linnaeus): China's coast. Migratory species. (2, 4, 5, 6)

CICONIIDAE

Ciconia ciconia boyciana Swinhoe: Coasts south of Changjiang (=Yangtze River). Migratory species. (1–6)
C. nigra (Linnaeus): China's coast. Migratory species. (1, 2, 5, 6)

THRESKIORNITHIDAE

Platalea leucorodia Linnaeus: Coasts south of Changjiang (=Yangtze River). Migratory species. (1, 2, 5)
P. minor Temminck & Schlegel: China's coast. Migratory species. (2, 5)
Threskiornis aethiopicus melanocephalus (Latham): China's coast. Migratory species. (1, 2, 5)

ANSERIFORMES
ANATIDAE

Aix galericulata (Linnaeus): China's coast. (1–5)
Anas a. acuta Linnaeus: China's coast. (2, 5)
A. c. crecca Linnaeus: China's coast. (1–6)
A. clypeata Linnaeus: China's coast. (1–6)
A. falcata Georgi: China's coast. Migratory species. (1–6)
A. formosa Georgi: Coasts south of Jiangsu. Migratory species. (2, 5)
A. p. platyrhynchos Linnaeus: Coasts south of Bohai Bay. (1–6)
A. penelope Linnaeus: China's coast. (1–6)
A. poecilorhyncha zonorhyncha Swinhoe: Coasts south of Changjiang (=Yangtze River). Migratory species. (1–6)
A. querquedula Linnaeus: China's coast. (1–6)
A. s. strepera Linnaeus: Shandong (rarelly); coast south of Changjiang (=Yangtze River). (2–6)
Anser a. albifrons (Scopoli): Coast south of Changjiang (=Yangtze River). Migratory species. (1, 2, 4, 5)
A. anser (Linnaeus): Coast south of Changjiang (=Yangtze River). Migratory species. (2–6)
A. cygnoides (Linnaeus): Fujian; Guangdong. Coast, migratory species. (2, 4, 5)
A. erythropus (Linnaeus): Coast south of Changjiang (=Yangtze River). Migratory species. (2, 4, 5)
A. fabalis sibircus (Latham): Coast from Liaoning to southern China. Migratory species. (1, 2, 4, 5)
Aythya baeri (Radde): Coast of Shandong and south of Changjiang (=Yangtze River). Migratory species. (1–6)
A. ferina (Linnaeus): Coast south of Changjiang (=Yangtze River). Migratory species. (2, 5, 6)
A. fuligula (Linnaeus): China's coast. (1–6)
A. marila Linnaeus: Coast of Shandong Peninsula and south of Changjiang (=Yangtze River). (2–6)
Branta bernicla (Linnaeus): Coast from Lushun (Liaoning) to Fuzhou (Fujian). Migratory species. (2, 5)
Bucephala clangula (Linnaeus): Coast south of Shanghai to Guangdong. (2, 4, 5, 6)
Clangula hyemalis (Linnaeus): Coast from Hebei to Fujian. (2, 5)
Cygnus c. cygnus (Linnaeus): Shanghai; occasionally in Fujian. (2, 4, 5)
C. columbianus (Orl): China's coast. (2, 5)
C. olor (Gmelin): North of Changjiang (=Yangtze River); Taiwan. (2, 5)
Dendrocygna javanica (Horsfield): Coast south of Changjiang (=Yangtze River). Migratory species. (2, 5)
Melanitta fusca stejnegeri (Ridgway): Coast from Hebei to Fujian. (2, 5)
M. nigra americana (Swainson): Fujian coast. (2, 5)
Mergus albellus (Linnaeus): China's coast. (1–5)
M. serrator Linnaeus: China's coast. (1–6)
M. squamatus Gould: Shanghai; Fujian; Guangdong. (2, 4, 5)
Nettapus coromandelianus (Gmelin): Coast south of Changjiang (=Yangtze River). (2, 5, 6)
Tadorna ferruginea (Pallas): North of Changjiang (=Yangtze River). (2, 4, 5)
T. tadorna (Linnaeus): North of Changjiang (=Yangtze River); Fujian; Guangdong; Guangxi; Taiwan (occasionally). Migratory species. (1–6)

FALCONIFORMES
ACCIPITRIDAE

Haliaeetus albicilla (Linnaeus): Coast south of Changjiang (=Yangtze River). Migratory species. (5)
H. leucogaster (Gmelin): Southern China coast. (2, 3, 5, 6)

GRUIFORMES
GRUIDAE

Grus grus lilfordi Sharpe: China's coast. Migratory species. (1–6)
G. japonensis (P. L. S. Müller): Eastern Liaoning Bay; middle and lower Changjiang (=Yangtzc River) basin; northern Jiangsu coast; Taiwan. (1, 3, 5)
G. leucogeranus Pallas: Eastern Liaoning Bay; mouth of Changjiang (=Yangtze River). Migratory species. (1, 3, 5)
G. monacha Temminck: Eastern Liaoning Bay; northern Jiangsu coast. Migratory species. (1, 3, 5)
G. vipio Pallas: China's coast. Migratory species. (2–5)

RALLIDAE

Amaurornis phoenicurus chinensis (Boddaert): Coasts south of Shandong to Xisha (=Paracel Islands). Migratory species. (2, 3, 4, 6)
Fulica a. atra Linnaeus: China's coast. Migratory species. (1–6)
Gallicrex c. cinerea (Gmelin): China's coast. (2–6)
Gallinula chloropus indica Blyth: China's coast. Migratory species. (1–6)
Porzana fusca erythrothorax (Temminck & Schlegel): China's coast. Migratory species. (2, 4, 5, 6)
P. p. pusilla (Pallas): China's coast. Migratory species, (1–6)
Rallus aquaticus indicus Blyth: China's coast. Migratory species. (2, 4, 5, 6)
R. striatus gularis Horsfield: Mouth of Changjiang (=Yangtze River); Fujian; Hainan. Migratory species. (3, 4)

CHARADRIIFORMES
JACANIDAE

Hydrophasianus chirugus (Scopoli): Coasts south of Changjiang (=Yangtze River). Migratory species. (2, 5, 6)

ROSTRATULIDAE

Rostratula b. benghalensis (Linnaeus): Coasts south of Changjiang (=Yangtze River). Migratory species. (2, 4, 5, 6)

HAEMATOPDIDAE

Haematopus ostralegus osculans Swinhole: China's coast. Migratory species. (1–6)

CHARADRIIDAE

Charadrius alexandrinus dealbatus (Swinhoe): China's coast. Migratory species. (1–6)

C. asiaticus veredus Gould: China's coast. Migratory species. (1–6)

C. dubius curonicus Gmelin: China's coast. Migratory species. (1–6)

C. hiaticula placidus J. E. & G. R. Groy: East China Sea. Coast, migratory species. (3, 4, 5)

C. leschenaultii Lesson: China's coast. Migratory species. (1–6)

C. m. mongolus Pallas: China's coast. Migratory species. (1–6)

Pluvialis dominica fulva (Gmelin): China's coast; Xisha (=Paracel Islands). Migratory species. (1–6)

P. squatarola (Linnaeus): China's coast. Migratory species. (1–6)

Vanellus cinereus (Blyth): China's coast. Migratory species. (1, 2, 4, 5, 6)

V. vanellus (Linnaeus): China's coast. Migratory species. (1–6)

SCOLOPACIDAE

Arenaria i. interpres (Linnaeus): China's coast. Migratory species. (2–6)

Calidris acuminata (Horsfield): China's coast. Migratory species. (1–6)

C. alpina sakhalina (Viellot): Coasts south of Changjiang (=Yangtze River). Migratory species. (1–6)

C. canutus rogersi (Mathews): China's coast. Migratory species. (2–6)

C. ferruginea (Poatoppidan): Guangdong; Hainan. Migratory species. (2–6)

C. ruficollis (Pallas): China's coast. Migratory species. (2–6)

C. subminuta (Middlendorff): Taiwan; Guangdong. Migratory species. (3, 5, 6)

C. temminckii (Leisler): Fujian; Taiwan; Guangdong. Migratory species. (5)

C. tenuirostris (Horsfield): China's coast. Migratory species. (2–6)

Capella g. gallinago (Linnaeus): China's coast. Migratory species. (1–6)

C. megala (Swinhoe): Taiwan; Guangdong; Hainan. Migratory species. (3–6)

C. solitaria japonica (Bonaparte): North of Fujian; Guangdong. (3, 5)

C. stenura (Bonaparte): Fujian; Taiwan; Guangdong; Hainan. Migratory species. (3, 5, 6)

Crocethia alba (Pallas): South of Fujian; Taiwan. Migratory species. (2–6)

Eurynorhynchus pygmeus (Linnaeus): Shanghai; Fuzhou (Fujian); Guangdong; Hainan. Migratory species. (2–6)

Limicola falcinellus sibirica Dresser: China's coast. Migratory species. (2–6)

Limnodromus semipalmatus (Blyth): Coasts south of Shanghai. Migratory species. (2–6)

Limosa lapponica novaezealandiae G. R. Groy: China's east coast; Taiwan. Migratory species. (1–6)

L. limosa melanuroides Gould: China's east coast; Taiwan. Migratory species. (2-6)

Lymnocryptes minimus (Brunnich): Fujian; Taiwan; Guangdong. Migratory species. (2, 5)

Numenius arquata orientalis Brehm: China's coast. Migratory species. (1–6)

N. borealis minutus Gould: China's coast. Migratory species. (1–5)

N. madagascariensis (Linnaeus): China's east coast; Taiwan. Migratory species. (1–6)

N. phaeopus variegatus (Scopoli): China's coast. Migratory species. (1–6)

Philomachus pugnax (Linnaeus): China's coast. Migratory species. (2, 4, 5, 6)

Scolopax r. rusticola Linnaeus: China's coast. Migratory species. (3, 5)

Tringa erythropus (Pallas): China's coast. Migratory species. (1–6)

T. glareola Linnaeus: China's coast. Migratory species. (1–6)

T. guttifer (Nordmann): Southeastern China. Coast. (1–6)

T. hypoleucos Linnaeus: China's southern coast. Migratory species. (1–6)

T. incana bervipes (Vieillot): China's coast. Migratory species. (2–6)

T. nebularis (Gunnerus): Coasts south of Changjiang (=Yangtze River). Migratory species. (1–6)

T. ochropus Linnaeus: China's southeastern coast; Taiwan; Hainan. Migratory species. (1–6)

T. stagnatilis (Bechstein): China's coast. Migratory species. (1–6)

T. t. totanus (Linnaeus): China's coast. Migratory species. (1–6)

Xenus cinerea (Culdenstadt): China's coast. Migratory species. (2–6)

RECURVIROSTRIDAE

Himantopus h. himantopus (Linnaeus): China's coast. Migratory species. (1–6)

Recurvirostra avosetta (Linnaeus): China's coast. Migratory species. (1–6)

PHALAROPODIDAE

Phalaropus lobatus (Linnaeus): China's coast. Migratory species. (2, 5, 6)

GLAREOLIDAE

Glareola maldivarum Forster: China's coast. Migratory species. (1–6)

LARIFORMES
LARIDAE

Anous stolidus pileatus (Scopoli): Fujian; Taiwan Strait. (2, 5)
Childonias hybrida swinhoei (Mathews): China's coast. Migratory species. (1, 2, 5)
C. leucoptera (Temminck): China's coast. Migratory species. (1–6)
Gelochelidon n. nilotica (Gmelin): Eastern Liaoning Bay. (1, 5)
Gygis alba candida (Gmelin): Macau; Xisha (=Paracel Islands). (2, 5)
Hydroprogne t. tschegrava (Lepechin): Lushun (Liaoning); Qingdao (Shandong); Fujian; Taiwan; Guangdong; Hainan. Migratory species. (2, 5)
Larus a. cachinnans Pallas: Liaoning; Hong Kong. Coast. (1, 6)
L. argentatus Ponloppida: China's coast. Migratory species. (1–6)
L. brunnicephalus Jerdon: Hong Kong. (5, 6)
L. canus kamtschatschensis (Bonaparte): China's coast. Migratory species. (1–5)
L. crassirostris Vieillot: China's coast. Migratory species. (1–6)
L. glaucescens Naumann: Fujian coast. (2, 5)
L. ichthyaetus Pallas: Hong Kong. (5, 6)
L. ridibundus Linnaeus: All over China's coast during winter. (1–6)
L. saundersi (Swinhoe): China's coast. Migratory species. (1–6)
L. schistisagus Stejneger: China's coast. Migratory species. (2, 5)
Stercorarius pomarinus (Temminck): Guangdong coast. (2, 5, 6)
Sterna a. anaethetus Scopoli: Fujian (islands); Taiwan; Hainan. Coast, migratory species. (2, 5)
S. albifrons sinensis Gmelin: China's coast. Migratory species. (1, 2, 5, 6)
S. dougallii bangsi Mathews: Zhejiang; Fujian; Taiwan; Guangdong. Migratory species. (2, 5)
S. fuscata Linnaeus: China's coast. Migratory species. (2, 5)
S. hirundo longipennis Nordmann: China's coast. Migratory species. (1–6)
S. s. sumatrana Raffles: Hebei; Shandong; southern Zhejiang; Fujian; Taiwan; Guangdong; Hainan. Migratory species. (1, 2, 5, 6)

Thalasseus bergii cristatus (Stephens): Fujian; Guangdong; Hainan. (2, 5)
T. zimmermanni (Reichenow): Shandong; Fujian; Guangdong. Coast. (2, 5)

ALCIDAE

Synthliboramphus antiquus (Gmelin): Lushun (Liaoning); Yantai and Qingdao (Shandong); Taiwan Strait. Migratory species. (2, 5)

APODIFORMES
APODIDAE

Apus affinis subfurcatas (Blyth): China's coast. Migratory species. (2, 5, 6)
A. pacificus (Latham): China's coast. (2, 5, 6)
A. p. kanoi (Yamashina): Fujian; Taiwan; Guangdong; Guangxi. Migratory species. (2, 5)
Collocalia fuciphaga (Thunberg): Dazhou Island (Hainan). (5)
Hirundapus c. caudcutus (Latham): China's coast. Migratory species. (2, 5, 6)
H. cochinchinensis (Oustalet): Hainan; Dongsha (=Pratas Islands); Xisha (=Paracel Islands); Nansha (=Spratly Islands). (2)

REFERENCES*

(1). Liaoning Forestry Reconnaissance Designing Institute, 1989. Lists of Wildlife in Shuangtai Hekou National Natural Reserve.
(2). Cheng Tso-hsio, 1976. Records on Distribution of Chinese Avifauna. China Science Press, 1,218pp.
(3). Cheng Tso-hsio, 1982. Taxonomy of Invertebrates. China Agriculture Press, 557pp.
(4). Zhou Manzhang et al., 1991. Demonstration Report on the Establishment of Shanghai Chongming Dongtai Bird Natural Conservation Area, 9pp.
(5). Cheng Tso-hsio, 1987. A synopsis of the Avifauna of China. China Science Press, 1,222 pp.
(6). World Wildlife Fund, 1975. Checklist of birds in Mai Po Nature Reserve, 15pp.

*: (1)-(4) in Chinese

Compiled by Huang Zongguo, Yan Songkai, and Tang Zhaohe; Edited by Zheng Zuoxin (Cheng Tsohsin) and Zhang Fuyun

MAMMALIA
CETACEA
BALAENIDAE

Eubalaena glacialis (Borowski): Yellow Sea. Oceanic species. (1)

ESCHRICHTIIDAE

Eschrichtius robustus (Lilljeborg): Yellow Sea; East China Sea; Taiwan; South China Sea. Shoreline. (1, 2, 5)

BALAENOPTERIDAE

Balaenoptera acutorostrata Lacépéde: Bohai; Yellow Sea; East China Sea; Taiwan; South China Sea. Oceanic species. (1, 2, 5)

B. borealis Lesson: Yellow Sea; East China Sea; Taiwan; South China Sea. Oceanic species. (1, 2, 5)

B. edeni Anderson: Yellow Sea; East China Sea; Taiwan; South China Sea. Oceanic species. (1, 2, 5)

B. musculus (Linnaeus): Yellow Sea; Taiwan; South China Sea. Oceanic species. (1, 2)

B. physalus (Linnaeus): Bohai; Yellow Sea; East China Sea; Taiwan; South China Sea. Oceanic species. (1, 2, 5)

Megaptera novaeangliae (Borowski): Yellow Sea; East China Sea; Taiwan; South China Sea. Oceanic species. (1, 2, 5)

PHYSETERIDAE

Kogia breviceps (Blainville): Taiwan; South China Sea. Oceanic species. (1, 2, 5)

K. simus Owen: Taiwan. (2)

Physeter macrocephalus Linnaeus: Yellow Sea; East China Sea; Taiwan; South China Sea. Oceanic species. (1, 2, 5)

ZIPHIIDAE

Mesoplodon densirostris (Blainville): Taiwan. Oceanic species. (1, 2, 5)

M. ginkgodens Nishiwaki & Kamiya: Yellow Sea; East China Sea; Taiwan; South China Sea. Shoreline. (1, 2, 5)

Ziphius cavirostris G. Cuvier: Taiwan. Oceanic species. (1, 2, 5)

GLOBICEPHALIDAE

Feresa attenuata Gray: Taiwan. (2)

Grampus griseus (G. Cuvier): East China Sea; Taiwan; South China Sea. Oceanic species. (1, 2, 5)

Orcinus orca (Linnaeus): Bohai; Yellow Sea; East China Sea; Taiwan; South China Sea. Oceanic species. (1, 2, 5)

Peponocephala electra (Gray): Taiwan. Oceanic species. (1, 2, 5)

Pseudorca crassidens (Owen): Bohai; Yellow Sea; East China Sea; Taiwan; South China Sea. Oceanic species. (1, 2, 5)

PHOCOENIDAE

Neophocaena phocaenoides (G. Cuvier): Bohai; Yellow Sea; East China Sea; Taiwan; South China Sea; Changjiang (=Yangtze River). Mouth of major rivers. (1, 2, 5)

DELPHINIDAE

Delphinus delphis Linnaeus: Bohai; Yellow Sea; East China Sea; Taiwan; South China Sea. Oceanic species. (1, 2, 5)

D. tropicalis van Bree: Taiwan; South China Sea. Oceanic species. (1, 2, 5)

Lagenodelphis hosei Fraser: Taiwan; South China Sea. Shoreline. (1, 2, 5)

Lagenorhynchus obliquidens Gill: Yellow Sea; East China Sea; South China Sea Shoreline. (1)

Sousa chinensis Osbeck: East China Sea; South China Sea. Shoreline. (1, 2, 5)

S. plumbea Cuvier: East China Sea; South China Sea. Shoreline. (1)

Stenella attenuata (Gray): Taiwan; South China Sea. (2)

S. coeruleoalba (Meyen): Taiwan. Oceanic species. (1, 2, 5)

S. frontalis (G. Cuvier): Taiwan; South China Sea. Shoreline. (1, 2, 5)

S. longirostris (Gray): Taiwan; South China Sea. Oceanic species. (1, 2)

Steno bredanensis (Lesson): East China Sea; Taiwan; South China Sea. Shoreline. (1, 2, 5)

Tursiops aduncus (Ehrenberg): East China Sea; Taiwan; South China Sea. Oceanic species. (1, 2, 5)

T. truncatus (Montagu): Bohai; Yellow Sea; East China Sea; Taiwan; South China Sea. Shoreline. (1, 2, 5)

PINNIPEDIA
PHOCIDAE [OTARIIDAE]

Erignathus barbatus Erxleben: East China Sea. Shoreline. (1, 4)

Phoca hispida (Schreber): Yellow Sea. Shoreline. (1, 4)

P. largha Pallas [*P. vitulina*]: Bohai; Yellow Sea; East China Sea; South China Sea. Shoreline. (1, 4)

OTARIDAE

Callorhinus ursinus (Linnaeus): Yellow Sea. Shoreline. (1, 4)

Eumetopias jubata (Schreber): Yellow Sea. Shoreline. (1, 4)

SIRENIA
DUGONGIDAE

Dugong dugon Müller: Southern Taiwan; Guangdong (south of Yanjiang); Guangxi; Hainan. Shoreline. (1, 2, 3, 5)

REFERENCES*

(1). Wang Peilie, 1993, Fauna of marine mammals in China. Acta Oceanologica Sinica, 12(2):273–278

(2). Wang Peilie, 1991. Teh cetaceans and its resources protection in Taiwan. Fisheries Science, 10(40:24–28.

(3). Wang Peilie and Sun Jianyun, 1986. Distribution of the Dugong off the coast of China. Acta Theriologica Sinica, 6(3):175–181

(4). Wang Peilie and Wang Zhemao, 1986. Pinnipeds in China. Marine Fisheries, 8(6): 250–253.

(5). Yiang, H. C. 1986. Studies on cetaceans of Taiwan. Science of Taiwan Provincial Museum Monthly, 19:131–178.

*: all in Chinese

Compiled by Lin Jinmei; Edited by Wang Pilie, Wang Song, and Dong Jinhai

INDEX

The names of all the classification levels (except for subgenus) in the primary entry are listed alphabetically. The synonyms are listed within brackets. If there are more than one species in a genus, then from the second species on only the species name(s) is (are) listed. In case of subspecies, variations, and forms, the species name is replaced with a 3-em dash (———). For example:

Chaetoceros abnoumis
　　didymus
　　——— v. *anglica*
　　——— f. *protubernas*

A

Abalistes stellatus 459
Abarenicora pacifica 225
Abathescalpellum fissum 318
　　koreanum 318
Abderospira punctulata 286
　　umbilicata 286
[*Ablabys macracanthus*] 455
　　[*taenianotus*] 455
Ablennes anastomella 418
　　hians 418
Aboma isushimae 451
　　lactipes 451
Abra soyoae 252
Abralia andamanica 294
　　multihamata 294
Abranchaea chinensis 288
Abrina hainanensis 252
　　kinoshitai 252
　　lunella 252
　　magna 252
Abrus precatorius 153
Abudefduf anabatoides 442
　　aureus 442
　　bankieri 442
　　behni 442
　　bengalensis 442
　　biocellatus 442
　　coelestinus 442
　　curacao 442
　　cyaneus 442
　　dicki 442
　　glaucus 442
　　lacrymatus 442
　　leucogaster 442
　　lorenzi 442
　　melas 442
　　notatus 442
　　richardsoni 442
　　septemfasciatus 442
　　sexfasciatus 442
　　sordidus 442
　　thoraataeniatus 442
　　uniocellatus 442
　　vaigiensis 442
　　xanthozona 443
　　zonatus 443
Abyla bicarinata 175
　　brownia 175

Abyla (continued)
　　carina 175
　　haeckeli 175
　　ingeborgae 175
　　peruana 175
　　schmidti 175
　　trigona 175
ABYLIDAE 175
Abylopsis eschscholtzi 175
　　tetragona 175
Abyssaster planus 391
Acabaria formosa 186
　　philippinensis 186
Acacia auriculiformis 153
　　confusa 153
Acalycigorgia inermis 186
Acalyptophis peronii 464
Acanthacaris tenuimana 345
ACANTHACEAE 158
Acanthaphritis grandisquamis 445
ACANTHARIA 61
Acanthaster planci 391
ACANTHASTERIDAE 391
Acanthastrea echinata 184
Acanthella hispida 164
[*Acanthephyra angusta*] 339
　　armata 339
　　carinata 339
　　curtirostris 339
　　[*debilis*] 339
　　eximia 339
　　purpurea 339
Acanthobothrium benedeni 206
　　coronatum 206
　　grandiceps 207
　　ijimai 207
　　micracantha 207
　　pingtanensis 207
　　tsingtaoense 207
ACANTHOCEPHALA 213
Acanthocephalus luzus 213
Acanthocepola indica 438
　　krusensterni 438
　　limbata 438
Acanthochiasma ruboscens 61
ACANTHOCHIASMIDAE 61
Acanthochiton armatus 238
　　bednalli 238
　　dissimilis 238
　　rubrolineatus 238

Acanthochiton (continued)
　　scutiger 238
　　zeylandicus 238
ACANTHOCHITONIDAE 238
Acanthochondrites annulatus 316
ACANTHOCLINIDAE 428
Acanthocolla cruciata 61
　　solidissima 61
ACANTHOCOLPIDAE 195
Acanthocolpus acanthocepolae 195
　　liodorus 195
　　sanyaensis 195
Acanthocorys castanoides 67
　　umbelifera 67
　　variabilis 67
Acanthocyathus grayi 185
　　spiniger 185
Acanthocybium solandi 449
Acanthodesmia viniculata 66
[ACANTHODORIDIDAE] 289
Acanthodoris pibosa 289
Acanthogobius flavimanus 451
　　jacoti 451
Acanthogorgia armata 186
　　aspera 186
　　muriata 186
　　vegae 186
ACANTHOGORGIIDAE 186
Acantholithium dicopum 61
Acanthometra pellucida 61
ACANTHOMETRIDAE 61
Acanthomysis aspera 323
　　crassispinosa 323
　　fujinagai 323
　　hwanhaiensis 323
　　koreana 323
　　laticauda 323
　　leptura 323
　　longirostris 323
　　meridionalis 323
　　okayamaensis 323
　　platycauda 323
　　quadrispinosa 323
　　rotundicauda 323
　　serrata 323
　　sheni 323
　　sinensis 323
　　tenella 323
Acanthonchus tridentatus 211
Acanthopagrus australis 434

Acanthopagrus (continued)
  [*berda*] 434
  [*latus*] 434
  [*schlegeli*] 434
Acanthophora muscoides 139
  [*orientalis*] 139
  spicifera 139
ACANTHOPLEGMIDAE 61
Acanthoplesiops psilogaster 428
Acanthopleura spinosa 238
Acanthoscina acanthodes 330
Acanthosentis lizae 213
  periophthalmi 213
  scomberomori 213
  similis 213
Acanthosphaera aff. *barbati* 63
  capillaris 63
  dodecastyla 63
  insignis 63
  nanhaiensis 63
Acanthosphex leurynnis 455
Acanthosquilla acanthocarpus 369
  multifasciata 369
  multispinosa 369
Acanthostaurus nordgaardi 61
  prupurascens 61
ACANTHOSTOMIDAE 198
Acanthozostera gemmata 238
ACANTHURIDAE 448
Acanthurus bariene 448
  bleekeri 448
  dussumieri 448
  gahhm 448
  glaucopareius 448
  guttatus 448
  japonicus 448
  leucopareius 448
  lineatus 448
  lishenus 448
  maculiceps 448
  mata 448
  matoides 448
  melanopterus 448
  nigrican 448
  nigricanda 448
  nigrofuscus 448
  olivaceus 448
  pyroferus 448
  reticulatus 448
  thompsoni 448
  triostegus 448
  xanthopterus 448
Acanthus ebracteatus 158
  ilicifolius 158
  xiamenensis 158
Acar plicata 240
ACARI 297
Acarinina appressocamerata 100
  broedermanni 100
  bullbrooki 100
  camerata 100
  compacta 100
  decepta 100
  hansbolli 100
  interposita 100

Acarinina (continued)
  lingfengensis 100
  matthewsae 100
  praeangulata 100
  primitiva 100
  pseudotopilensis 101
Acarochelopodia lapidaria 298
ACAROCHELOPODINAE 298
Acarothrix palustris 298
Acartia bifilosa 312
  clausi 312
  danae 312
  erythraea 312
  hamata 312
  longiremis 312
  negligens 312
  pacifica 312
  southwelli 312
  spinicauda 312
Acartiella sinensis 312
ACARTIIDAE 312
Acasta conica 321
  coriobasis 321
  dofleini 321
  fenestrata 321
  flexuosa 321
  fragilis 321
  glans 321
  hainanensis 321
  laevigata 321
  membranacea 321
  pectinipes 321
  sinica 321
  spinitergum 321
  [*spinosa*] 321
  spongites 321
  sporillius 321
  sulcata 321
  zuiho 321
Acaudina leucoprocta 389
  molpadioides 389
Accacladium arii 194
ACCACOELIIDAE 194
ACCIPITRIDAE 466
Acentrogobius bonti 451
  campbelli 451
  caninus 451
  cauerensis 451
  chlorostigmatoides 451
  cyanomos 451
  hoepplii 451
  hoshinonis 451
  masoni 451
  microps 451
  ornatus 451
  puntang 451
  triangularis 451
  viganensis 451
  viridipunctatus 451
[ACERIDAE] 286
Acervulina inhaerens 116
ACERVULINACEA 116
ACERVULINIDAE 116
Acesta marissinica 245
Acetabularia calyculus 145

Acetabularia (continued)
  clavata 145
  [*möbii*] 145
  parvula 145
  tsengiana 145
Acetes chinensis 338
  erythraeus 338
  [*insularis*] 338
  intermedius 338
  japonicus 338
  serrulatus 338
  vulgaris 338
ACETOBACTERACEAE 21
Achaeus japonicus 356
  pugnax 356
  robustus 356
  superciliaris 356
  tuberculatus 356
  varians 356
Acharax japonica 239
  johnsoni 239
Achelia superba 297
ACHNANTHACEAE 45
ACHNANTHALES 44
[*Achnanthes angustata*] 45
  bengalensis 45
  brevipes 45
  ———— v. *angustata* 45
  ———— v. *intermedia* 45
  ———— v. *leudugeri* 45
  ———— v. *parvula* 45
  citronella 45
  clevei 45
  coarctata 45
  conspicus v. *brevistrista* 45
  cotteriensis 45
  curvirostrum 45
  danica 45
  delicatula 45
  dispar v. *angulata* 45
  hauckiana 45
  javanica 45
  ———— v. *subconstricta* 45
  ———— v. *tenuistauros* 45
  kryophila 45
  kuwaitensis 45
  longipes 45
  microcephala 45
  minutissima 45
  oblongella 45
  orientalis 45
  ploenensis 45
  radiata 45
Achtheinus chinensis 316
  impenderus 316
Achyranthes aspera 151
Acila divaricata 239
  mirabilis 239
  schencki 239
Acinetobacter 21
Acinetospora crinita 140
Acipenser sinensis 409
ACIPENSERIDAE 409
ACIPENSERIFORMES 409
Aclyvolva granularis 270

Index

Acmaea concinna 262
   mitra 262
   nanshaensis 262
   pallida 262
   [*pallida*] 262
   [*pygmaea*] 263
   [*saccharina*] 263
   [*schrenckii*] 263
   strongiana 262
ACMAEIDAE 262
Acmaeopleura balssi 365
   toriumii 365
Acoetes flagelliformis 217
   jogasimae 217
   melanonota 217
ACOETIDAE 217
Acolochirus inornata 387
Acostina piramidale 108
Acremonium sp. 131
Acrhelia horrescens 183
Acrilla acuminata 267
Acrobotrissa cribosa 67
Acrocalanus andersoni 306
   gibber 306
   gracilis 306
   indicus 306
   longicornis 306
   monachus 306
   [*sinicus*] 307
ACROCHAETIACEAE 133
   [*Acrochaetium arcuatum*] 133
   [*crassipes*] 133
   [*daviesii*] 133
   [*microscopicum*] 133
   [*robusta*] 133
   [*subimmersum*] 133
ACROCHORDIDAE 464
Acrochordus granulatus 464
ACROCIRRIDAE 224
Acrocirrus validus 224
Acrocystis nana 139
Acromitus flagellatus 176
   tankahkeei 176
Acropoma hanedai 428
   japonicum 428
ACROPOMIDAE 428
Acropora aduncata 181
   affinis 181
   armata 181
   bruggemanni 181
   ——— *uncinata* 181
   caroliniana 181
   conferta 181
   corymbosa 181
   cytherea 181
   decipiens 181
   delicatula 181
   dissimilis 181
   echinata 181
   formosa 181
   granulosa 181
   haimei 181
   horrida 181
   humilis 181
   hystrix 181

Acropora (continued)
   irregularis 181
   lutkeni 181
   nasuta 181
   ——— *crassilabia* 181
   pacifica 181
   palifera 181
   prostrata 181
   pruinosa 181
   pulchra 181
   ——— *stricta* 181
   rosaria 181
   rotumana 181
   surculosa 181
   tenella 181
   tizardi 181
   valida 181
   variabilis 181
ACROPORIDAE 181
ACROSIPHONIACEAE 145
ACROSIPHONIALES 145
Acrosorium polyneurum 139
   reptans 139
   [*uncinatum*] 139
   venulosum 139
   yendoi 139
Acrosphaera collina 62
   spinosa 63
ACROSTICHACEAE 150
Acrostichum aureum 150
   speciosum 150
ACROTHORACICA 322
ACROTHRICACEAE 141
Acrothrix pacifica 141
ACTACARINAE 298
Actacarus sinensis 298
Actaea amoyensis 359
   bocki 359
   calculosa 359
   cavipes 360
   modesta 360
   nodulosa 360
   [*orientalis*] 361
   polyacantha 360
   pulchella 360
   pura 360
   [*rueppellii*] 361
   savignyi 360
   superciliaris 360
Actaeodes areolatus 360
   hirsutissimus 360
   tomentosa 360
Actaeopyramis amaena 284
   eximia 284
   lauta 284
Acteocina avenaria 286
   biplex 286
   coarctata 287
   decorata 287
   exilis 287
   gordonis 287
   gracilis 287
   nanshanensis 287
   orientalis 287
   oyamai 287

Acteocina (continued)
   simplex 287
ACTEOCINIDAE 286
Acteon secale 285
   siebaldii 285
   teramachii 285
ACTEONIDAE 285
Actinia equina 179
ACTINIARIA 179
ACTINIIDAE 179
ACTINOBACTERIA 23
Actinocucumis typicus 387
ACTINOCYCLLIDAE 291
Actinocyclus alienus 31
   appendieulatus 31
   australis 31
   curvatulus 31
   ehrenbergii 31
   ——— v. *crassa* 31
   ——— v. *ralfsii* 31
   ——— v. *tenella* 31
   ellipticus 31
   ——— f. *lanceolata* 31
   fasciculatus 31
   ingens 31
   japonicus 291
   normani 31
   [*ralfsii*] 31
   roperi 31
   subtilis 31
   varicus 31
ACTINODISCACEAE 31
Actinomma arcadophorum 63
   bareale 63
   capillaceum 63
   eriosperma 63
   leptodermum 63
   medianum 63
   mediterranensis 63
   multispinula 63
   popofskii 63
   saccoi 64
ACTINOMYCETALES 23
Actinonema pachydermatum 211
ACTINOPLANACEAE
Actinoplanes 24
Actinoptychus annulatus 31
   ——— v. *minor* 31
   australis 31
   hexagonus 31
   marylandicus 31
   notabilis 31
   pericavatus 31
   splendens 31
   stella v. *thunii* 31
   subangulatus 31
   triacriformis 31
   trilingulatus 31
   undulatus 31
   vulgaris 31
Actinopyga echinites 388
   lecanora 388
   mauritiana 388
   miliaris 388
Actinosiphon tibetica 121

Actinothoe qingdaoensis 180
Actinotrichia fragilis 134
ACTISECIDAE 381
Actisecos regularis 381
Actumnus dorsipes 360
   forficigerus 360
   setifer 360
Acutitectonica acutissima 266
Adamnestia japonica 286
   protracta 286
   tosaensis 286
Adeonella japonica 378
   lichenoides 378
   [*minutipora*] 378
   [*platalea*] 378
ADEONELLIDAE 378
Adeonellopsis arculifera 378
   yarransis 378
ADEONIDAE 378
ADEONOIDEA 378
Adercotryma glomerata 74
Adioryx caudimaculatus 422
   cornutus 422
   diadema 422
   furcatus 422
   lacteoguttatus 422
   microstomus 422
   spinifer 422
   spinosissimus 422
   tiere 422
   violaceus 422
Aedes togoi 299
Aedicira pacifica 223
Aega dofleini 325
Aegathoa oculata 325
Aegiceras corniculatum 156
AEGIDAE 325
AEGINIDAE 173
Aegina citrea 173
Aeginura grimaldii 173
[*Aegion rathbuni*] 344
   [*sibogae*] 344
Aegires villosus 290
[AEGIRETIDAE] 290
AEGISTHIDAE 314
Aegisthus aculeatus 314
   mucronatus 314
Aeluropus littoralis 159
   ——— v. *sinensis* 159
Aeolidia cf. *foulisi* 292
   papillosa 292
Aeolidiella albopunctata 292
   takanosimensis 292
AEOLIDIIDAE 292
Aeoliscus strigatus 423
Aephnidiogenes hainanensis 195
Aequorea aequorea 169
   australis 169
   coerulescens 169
   conica 169
   floridana 169
   globosa 169
   macrodactyla 169
   parva 169
   pensilis 169

AEQUOREIDAE 169
Aeretica tomlini 250
Aerobiotrema acinovaria 204
AEROBIOTREMATIDAE 204
Aeromonas 21
Aesopia cornuta 458
Aetea anguina 371
   sinica 371
   truncata 371
   xishaensis 371
AETEIDAE 371
Aethaloperca rogaa 425
Aethra scruposa 357
AETIDEIDAE 307
Aetideus armatus 307
   [*armatus*] 308
AETOBATIDAE 409
Aetobatus flagellum 409
   guttatus 409
   narinari 409
   reticulatus 409
Aetomylaeus maculatus 408
   milvus 408
   nichofii 408
   vespertilio 408
Aetoplatea zonura 408
Aeverrillia setogera 370
AEVERRIILIIDAE 370
Afer cumingii 279
Afrocardium infantile 248
Afrocucumis africanus 387
Afromysis dentisinus 323
Agadina syimpsoni 288
AGLAJIDAE 287
Agalma elegans 173
   okeni 174
AGALMIDAE 173
AGARICIIDAE 181
[*Agastra rubra*] 172
Agauopsis ammodytes 297
   arenaria 297
   humilis 298
   sordida 298
Agave americana 162
   angustifolia 162
Agelas robusta 165
AGELASIDAE 165
Agerita sp. 132
Aggerostramen rusticum 72
Agglutinella reinemundi 83
[*Agkistrodon halys*] 464
   qianshanensis shedaoensis 464
   [*shedaoensis*] 464
Aglantha elata 173
Aglaophamus dibranchis 222
   dicirris 222
   jeffreysii 222
   lyrochaeto 222
   macroura 222
   munamaorii 222
   orientalis 222
   sinensis 222
   tepens 222
   toloensis 222
   vietnamensis 222

Aglaophenis amoyensis 171
   cupressina 171
   pluma 171
   suensoni 171
   whiteleggei 171
AGLAOPHENIIDAE 171
Aglaothamnion callophyllidicola 138
Aglaura hemistoma 173
AGONIDAE 456
Agrammus agrammus 455
Agriodesma navicula 257
   [*naviculoides*] 257
[*Agyneia bacciformis*] 154
Ahnfeltia furcellata 137
Aimulosa signata 380
Ainudrilus gibsoni 233
   lutulentus 233
   taitamensis 233
AIPTASIDAE 180
Aix galericulata 466
AIZOACEAE 151
Akera constricta 286
AKERATIDAE 286
Aktedrilus cavus 233
   cuneus 233
   locyi 233
   longitubularis 233
   mortoni 233
   parviprostatus 233
   parvithecatus 233
   sinensis 233
Alacia alata 302
   ——— *major* 302
   ——— *minor* 302
   belgicae 303
   hettacro 303
   leptothrix 303
   valdiviae 303
Alaerato angulifera 270
ALARIACEAE 143
Albizzia corniculata 153
Albosporus group 24
Albulatrema ovale 204
ALBULATREMATIDAE 204
[*Albula glossodonta*] 409
   vulpes 409
ALBULIDAE 409
Albunea dayriti 351
   symnista 351
ALBUNEIDAE 351
Alcaligenes 21
Alcicornis hainanensis 192
ALCIDAE 468
ALCIOPIDAE 215
Alciopina parasitica 215
ALCYONACEA 186, 187
ALCYONIDIIDAE 370
ALCYONIDIOIDEA 370
Alcyonidium gelatinosum 370
   nanum 370
   polyoum 370
ALCYONIIDAE 187
Alcyonium gracillimum 187
Alderia modesta 288
Alderina qingdaoensis 373

Index

Aldisa snguinea 290
ALDISIDAE 290
Alectis ciliaris 430
   indica 430
Alectrias benjamini 447
Alectryonella plicatula 246
Alepas pacifica 318
[*Alepes djeddaba*] 430
   [*para*] 430
ALEPISAURIDAE 415
Alepisaurus ferox 415
ALEPOCEPHALIDAE 412
Alepocephalus bicolor 412
   longiceps 412
   longirostris 412
   owstoni 412
   triangularis 412
   umbriceps 412
ALFREDINIDAE 116
Alfredosilvestris levinsoni 90
ALGAE 133
Aliaporcellana suluensis 351
   telestophila 351
Aliculastrum cylindrica 285
Alima hieroglyphica 368
Alismobates reticulatus 298
   rotundus 298
Allanetta barnesi 418
   bleekeri 418
   forskali 418
   woodwardi 418
Allassostomoides chelydrae 194
Alliatina nitida 98
   oinomikadoi 98
Allium anisopodium 162
Allmania nodiflora 151
Alloconger anagoides 415
   major 415
   [*shiroanago*] 415
Allodidymocodium unitubulare 199
Allogalathea elegans 350
Alloglobigerinoides conglobatus 102
ALLOGROMIIDAE 69
ALLOGROMIINA 69
Allokepon goetici 326
   sinensis 326
Allolasiotocus pseudosciaenae 195
Allolepidapedon pristipomoidis 195
Allomorphina paleocenica 119
Allonematobothrium epinepheli 199
   xishaensis 199
Allopora scabiosa 167
ALLOPOSIDAE 296
Allopus mollis 296
Allothyone longicauda 387
ALMAENIDAE 119
[*Aloidis erythrodon*] 256
Alona quadrangularis 301
   rectangula 301
Alonella globulosa 301
Alopias pelagicus 405
   profundus 405
   [*superciliosus*] 405
   vulpinus 405
ALOPIIDAE 405

ALPHEIDAE 341
ALPHEOIDEA 341
Alpheus acutocarinatus 341
   alcyone 341
   avarus 341
   bannerorum 341
   bidens 341
   bisincisus 342
   brevicristatus 342
   brevipes 342
   bucephalus 342
   canaliculatus 342
   [*carinatus*] 343
   [*charon*] 343
   chiragricus 342
   collumianus 342
   [——— *medius*] 342
   diadema 342
   distinguendus 342
   edamensis 342
   edwardsii 342
   facetus 342
   frontalis 342
   gracilipes 342
   hippothoe 342
   hoplocheles 342
   inopinatus 342
   japonicus 342
   leptocheles 342
   leviusculus 342
   lobidens 342
   lottini 342
   macroskeles 342
   malabaricus leptopus 342
   malleodigitus 342
   microstylus 342
   mites 342
   [*neomeris*] 343
   [*neptunus*] 343
   nonalter 342
   obesomanus 342
   pachychirus 342
   pacificus 342
   paralcyone 342
   pareuchirus 342
   parvirostris 342
   parvus 342
   polyxo 342
   pubescens 342
   pustulosus 342
   rapacida 342
   rapax 342
   [*serenei*] 342
   [*seurati*] 342
   sereni 342
   sibogae 342
   spatulatus 342
   splendidus 342
   stanleyi dearmarus 342
   [*stimpsoni*] 343
   sudara 342
   tenuicarpus 342
   tirmiziae 342
   [*tumidomanus*] 343
   [*ventrosus*] 342

Alpheus (continued)
   xishaensis 342
Altentha interrupta 314
Alternanthera pungens 151
   sessilis 151
Alternaria 130
   alternata 131
   maritima 131
Alternaria sp. 131
Althoffia tumida 400
Alticus saliens 446
Alutera monoceros 459
   scripta 459
ALUTERIDAE 459
Alvania fusca 265
   ligata 265
Alvenius ojianus 253
ALVEOLINACEA 89
Alveolinella quoyi 89
ALVEOLINIDAE 89
Alveolophragmium orbiculatum 74
   pauciloculata 74
   ringens 74
   subglobosum 74
   wiesneri 74
Alveopora excelsa 182
   japonica 182
   polyformis 182
Alysicarpus vaginalis 153
Amaea magnifica 267
   secunda 267
   thielei 267
Amaeana accraensis 229
   antipoda 229
   occidentalis 229
   trilobata 229
[*Amalopenaeus incertus*] 335
[*Amalthea acuta*] 268
   conica 268
AMALTHEIDAE 268
Amanses scopas 459
Amansia glomerata 139
AMARANTHACEAE 151
Amaroucium constellatum 402
AMARYLLIDACEAE 162
Amastigia rudis 375
   varians 375
   xishaensis 375
Amathia distans 370
Amathina tricarinata 268
Amaura japonica 284
   sagamiensis 284
Amaurornis phoenicurus chinensis 466
AMBASSIDAE 425
Ambassis gymnocephalus 425
   kopsi 425
   miops 425
   urotaenia 425
Amblyapistus macracanthus 455
   taenianotus 455
   [*triacanthus*] 455
Amblycirrhitus bimacula 444
   unimacula 444
Amblyeleotris fasciata 451
   guttata 451

[*Amblygaster sima*]
Amblyglyphidodon aures 443
   curacao 443
   [*leucagaster*] 442
Amblygobius albimaculatus 451
   bynoensis 451
   shatinensis 451
Amblyotrypauchen arctocephalus 453
Amblyrhynchotus hypselogeneion 460
   honckeni 460
   rufopunctatus 460
   spinosissimus 460
Ameira longipes 315
AMEIRIDAE 315
Amhyadesia bartschae 298
   heteromorpya 298
Amigdoscalpellum vitreum 318
Ammobaculites agglutinans 74
   akabiraensis 74
   calcareus 74
   catenulatus 74
   crassaformis 74
   cylindricus 74
   exiguus 74
   filiformis 74
   foliaceus 74
   formosensis 74
   hayasakai 74
   huanghaiensis 74
   josephi 74
   reophaciformis 74
   robusta 74
AMMODISCACEA 71
AMMODISCIDAE 71
Ammodiscoides turbinatus 71
Ammodiscus ambiguus 71
   anthosatus 71
   argenteus 71
   evolutus 71
   gullmarensis 71
   hoeglundi 71
   incertus 71
   intermedius 71
   pacificus 71
   planus 71
   tenuis 71
Ammodytes personatus 447
AMMODYTIDAE 447
Ammoglobigerina globigeriniformis 76
   globulosa 76
Ammolagena clavata 71
Ammomarginulina ensis 74
   tenerissima 74
Ammomassilina alveoliniformis 83
   arenarium 83
Ammonia altispira 122
   annectens 122
   aomoriensis 122
   astera 122
   batava 122
   beccarii 122
   compressiuscula 122
   contertitesta 122
   convexidorsa 122
   crebera 122

Ammonia (continued)
   dilucida 122
   eguatoriana 122
   faceta 122
   flevensis 122
   floscula 122
   formosa 122
   glans 122
   globosa 122
   granuloumbilica 122
   hatatensis 122
   honyaensis 122
   hozanensis 122
   indica 122
   inflata 122
   japonica 122
   ketienziensis 122
   koeboeensis 123
   limbatobeccarii 123
   limnetes 123
   maruhasii 123
   multicella 123
   nakamurai 123
   nantongensis 123
   nipponica 123
   parkinsoniana 123
   pauciloculata 123
   paucipora 123
   rolshauseni 123
   sikokuensis 123
   sobrina 123
   takanabensis 123
   tepida 123
   tikutoensis 123
   togopiensis 123
   yabei 123
Ammoscalaria agrestis 73
   fontinense 73
   pseudospiralis 73
   tenuimargo 73
Ammosigmoilinella eximia 86
Ammosphaeroidina sphaeroidiniformis 74
AMMOSPHAEROIDINIDAE 74
Ammotium minutum 74
   palustre 74
   stenostomum 74
[*Ammotrypane aulogaster*] 225
Amorpha fruticosa 152
Ampelisca bocki 328
   brevicornis 328
   chinensis 328
   cyclops 328
   hongkongensis 328
   honmungensis 328
   iyoensis 328
   maia 328
   miharaensis 328
   misakiensis 328
   nanshaensis 328
   orops 328
AMPELISCIDAE 328
Ampharete acutifrons 228
   arctica 228
   reducta 228
AMPHARETIDAE 228

Amphiascella subdebilis 315
Amphiascopsis havelocki 315
Amphibelone anomala 61
   hydrotomica 61
AMPHIBIA 463
Amphiblestrum granulatum 373
Amphibrachium sponguroides 64
Amphicaryon acaule 174
   ernesti 174
   peltifera 174
Amphicoryna hirsuta 93
   pauciloculata 93
   proxima 93
   scalaris 93
   separans 93
   sublineata 93
   subscalaris
Amphicteis glabra 228
   gunneri 228
   mederi 228
   scophrobranchiata 228
Amphictene capensis 228
   japonica 228
AMPHIDISCOPHORIDA 165
AMPHILEPTIDAE 57
Amphileptus gutta 57
   litonotiformis 57
Amphilimna granulosa 395
   polyacantha 395
   sinica 395
AMPHILITHIDAE 61
Amphilithium clavarium 61
   concretum 61
AMPHILOCHIDAE 328
Amphilonche elongata 61
Amphilycus scripta 395
Amphimetra laevipinna 386
   tessellata 386
Amphinema australis 167
   dinema 167
   physophorum 167
   rugosum 167
   [—— v. *shantungensis*] 167
   [*rugosa* v. *tsingtauensis*] 167
   turrida 167
[AMPHINEURA] 237
Amphinome pulchra 226
   rostrata 226
AMPHINOMIDA 226
AMPHINOMIDAE 226
Amphiophiura paupera 397
   radiata 397
   sculptilis 397
   sordida 397
   spatulifera 397
Amphioplus ancistrotus 395
   cyrtacanthus 395
   depressus 395
   impressus 395
   intermedius 396
   japonicus 396
   laevis 396
   lucidus 396
AMPHIOXI 403
AMPHIOXIDAE 403

Index

AMPHIOXIFORMES 403
Amphipholis duplicata 396
   kochii 396
   loripes 396
   microplaxa 396
   sobrina 396
  squamata 396
Amphipleura micans 35
  rutilans 35
  —— v. *antarctica* 35
AMPHIPODA 328
AMPHIPORIDAE 208
Amphiporus punctatulus 208
  sinuosum 208
Amphiprion akallopispisus 443
  bicinctus 443
  clarkii 443
  frenatus 443
  ocellaris 443
  percula 443
  perideraion 443
  polymnus 443
Amphiprora alata 35
  paludosa v. *hyalina* 35
  venusta 35
Amphirhopalum ypsilon 64
Amphiroa aberrans 134
  anastomosans 134
  f. *cyathifera* 135
  dilatata 135
  ephedraea 135
  fragilissima 135
  foliacea 135
  rigida 135
  valoniodes 135
  zonata 135
Amphisamytha japonica 228
Amphisiphonostra naviformis 301
Amphisolenia asymmetrica 51
  bidentata 51
  extensa 51
  globifera 51
  palaeatheroides 51
  rectangulata 51
  schauinslandi 51
  schroederi 51
  spinulosa 51
  thrinax 51
AMPHISOLENIACEAE 51
Amphisorus hemprichii 90
Amphisphaera palliatum 63
Amphispyris bria 66
  costata 66
  reticulate 66
  sinensis 66
Amphistaurus tetrapterus 61
Amphistegina exquiseta 116
  gibbosa 116
  lessoni 116
  madagascariensis 116
  papillosa 116
  quoyi 116
  radiata 116
  venosa 116
AMPHISTEGINIDAE 116
Amphistylus xishaensis 63
Amphitholonium pylonium 65
  trasversarum 65
Amphitholus cf. *acanthometra* 65
AMPHITRETIDAE 296
Amphithrix janthina 27
Amphithyrus bispinosus 333
  glaber 333
  muratus 333
  [*orientalis*] 333
  scupturatus 333
  similis 333
Amphitretus pelagicus 296
Amphitrite cirrate 229
  oculata 229
Amphiura ceramis 396
  digitula 396
  divaricata 396
  economiotata 396
  koreae 396
  tenuis 396
  vadicola 396
AMPHIURIDAE 395
Amphogona apicata 173
  apsteini 173
  pusilla 173
Amphora acuta 42
  angusta 42
  —— v. *chinensis* 42
  —— v. *diducta* 42
  —— v. *eulensteinii* 42
  arenaria 42
  arenicola 42
  —— v. *major* 42
  —— v. *subaequalis* 42
  bigibba 42
  coffeaeformis 42
  —— v. *acutiuscula* 42
  commutata 42
  costata 42
  —— v. *inflata* 42
  crassa 42
  —— v. *canpechiana* 42
  —— v. *interrupta* 42
  —— v. *punctata* 42
  cymbelloides 42
  cymbifera 42
  decussata 42
  egregia 42
  eunotia 42
  —— v. *gigantea* 42
  exigua 42
  exsecta 42
  gigantea v. *fusca* 42
  graeffii 42
  grevilleana 42
  grovei 42
  helenensis 42
  hyalina 42
  javanica 42
  laevis 42
  libyca 42
  lineolata v. *chinensis* 42
  lunula 42
  lyrata 42
Amphora (continued)
  macilenta 42
  —— v. *ergadensis* 42
  maria 42
  marina 42
  mexicana 42
  micrometra 42
  obtusa 42
  —— v. *oceanica* 42
  ocellata 42
  oculus 42
  ostrearia 42
  —— v. *vitrea* 42
  ovalis 42
  —— v. *pediculus* 42
  pediculus 42
  proteus 42
  —— v. *oculata* 43
  rhombica 43
  v. *sinica* 43
  robusta 43
  schmidtii 43
  spectabilis 43
  staurophora 43
  terroris 43
  turgida 43
Amphorella brandti 60
Amphorellopsis acuta 60
  pentagona 60
Amphoroidella sp. 325
Ampithoe lacertosa 328
  zachsi 328
AMPITHOIDAE 328
[*Amusium japonica*] 244
  japonicum balloti 244
  —— *formosum* 244
  [*pleuronectes*] 244
  —— *nanshaensis* 244
  —— *pleuronectes* 244
Amygdalum peasei 241
  soyoae 241
  watsoni 241
Anabaena variabilis 26
ANACANTHOBATIDAE 408
Anacanthobatis borneensis 408
  donghaiensis 408
ANACARDIACEAE 154
Anachis cf. *liocyma* 276
Anacropora tapera 181
Anacystis elabens 24
Anadara antiquata 240
  clathrata 240
  crebricostata 240
  ferruginea 240
  uropigmelana 240
ANADYOMENACEAE 146
Anadyomene wrightii 146
Anagallis coerulea 156
Anago anago 415
Anametalia sternaloides 395
Anampses caeruleopunctatus 438
  diadematus 438
  geographius 438
  melanurus 438
  meleagrides 438

Anampses (continued)
   neoguinaicus 438
   twisti 438
Anapella retroconvexa 250
Anapta gracilis 389
Anarchias allardice 416
Anas a. acuta 466
   c. crecca 466
   clypeata 466
   falcata 466
   formosa 466
   p. platyrhynchos 466
   penelope 466
   poecilorhyncha zonorhyncha 466
     querquedula 466
     s. strepera 466
ANASPIDEA 287
Anatanais normani 325
ANATIDAE 466
Anatlis weinkauffi 261
Anchialina grossa 323
   parva 323
   typica 323
   zimmer 323
Anchiarthrus derelictus 326
Anchisquilla fasciata 368
[*Anchistia elegans*] 341
Anchistioides compressus 340
   willeyi 340
ANCHISTIOIDIDAE 340
Anchistus custos 340
   miersi 340
ANCHOROPHORIDAE 192
Anchorophorus sinensis 192
[*Anchoviella commersoni*] 410
   [*heteroloba*] 410
   [*indica*] 410
   [*shantungensis*] 410
Anchylomera blossevillei 332
Ancilla rubiginosa 278
[*Ancistrosyllis bassi*] 218
   groenlandica 218
   pilargiformis 218
ANCORABOLIDAE 315
[*Ancylocaris brevicarpalis*] 341
ANCYROCEPHALIDAE 192
Andamia pacifica 446
   reyi 446
   tetradactylus 446
ANDROSPYRIDAE 66
[*Aneilema kainantense*]
Anelassorhynchus inanensis 237
   subinus 237
Angaria delphinus 263
   laciniata 263
[*Angasia tomentosa*] 344
ANGIODICTYIDAE 194
ANGIOSPERMAE 150
ANGUILLICOIDAE 210
Anguillicola crassa 210
   globiceps 210
ANGUILLIDAE 415
ANGUILLIFORMES 415
Anguilla bicolor pacifica 415
   breviceps 415

Anguilla (continued)
   elphinstonei 415
   foochowensis 415
   japonica 415
   marmorata 415
   nigricans 415
   sinensis 415
Angulodiscorbis corrugata 112
Angulogerina angulosa 108
   carinata 108
   fluens 108
Angulus compressissimus 250
   emarginatus 250
   lanceolatus 250
   psammotellus 250
   vestalioides 250
   vestalis 250
Aniculus aniculus 348
   [*strigatus*] 349
Anilocra dimidiata 325
ANIMALIA 164
Aniptodera chesapeakensis 130
   haispora 130
   lignatilis 130
   mangrovei 130
ANISAKIDAE 209
Anisakis alexandri 209
   simplex 209
Anisarthrus sp. 326
Anisocorbula crassa 256
   cuneata 256
   lineata 256
   modesta 256
   pallida 256
   scaphoides 256
   venusta 256
ANISOGAMMARIDAE 328
Anisogammarus turgimanus 328
Anisomysis bipartoculata 323
   brevicauda 323
   ijimai 323
   minuta 323
   quadrispinosa 323
Anisoporus cobraeformis 197
Annachlamys macassarensis 244
Annandaleum gruvelii 318
Annatiara affinis 167
ANNELIDA 214
Anodontia edentula 246
   stearnsiana 247
Anodontostoma chacunda 410
Anomalifrons lightana 363
Anomalina ammonoides 120
   bradyi 120
   colligera 120
   glabrata 120
   globulosa 120
   insecta 120
   pinghuensis 120
   praespissiformis 120
   punctulata 120
   robertsonianus 120
   semipunctata 120
   semiteres 120
   tasmanica 120

Anomalinella rostrata 119
   tennesseensis 119
   ungeriana 119
ANOMALINIDAE 120
Anomalinoides centrabullus 120
   petaliformis 120
   vialovi 120
   welleri 120
Anomalodiscus squamosus 254
ANOMALOPIDAE 422
Anomalops katoptron 422
[*Anomia aenigmatica*] 246
   chinensis 246
   [*cytaeum*] 246
ANOMIIDAE 246
Anomocora fecunda 185
Anomoeoneis costata 35
ANOMURA 348
ANOPLA 207
Anoplodactylus glandulifer 297
Anoplongaster cornuta 422
ANOPLONGASTERIDAE 422
Anoplostoma copano 212
   exceptum 212
ANOPLOSTOMATIDAE 212
Anorthoneis eurystoma 44
   excentrica 44
ANOSTRACA 300
Anotrichium tenue 138
Anous stolidus pileatus 468
Anser a. albifrons 466
   anser 466
   cygnoides 466
   erythropus 466
   fabalis sibircus 466
ANSERIFORMES 466
Anseropoda rosacea 391
Antedon parviflora 387
   serrata 387
ANTEDONIDAE 387
Antemetula elongata 277
ANTENNARIDAE 462
Antennarius biocellatus 462
   commersonii 462
   dorehensis 462
   hispidus 462
   melas 462
   nummifer 462
   pinniceps 462
   randalli 462
   striatus 462
   [*tridens*] 462
Antennella recta 171
   secundaria 171
Antennospora quadricornuta 130
Anthemiphyllia dentata 185
ANTHEMIPHYLLIIDAE 185
Anthenea aspera 390
   chinensis 390
   difficilis 390
   flavescens 390
   viguieri 390
Anthenoides cristatus 390
   granulosus 390
   laevigatus 390

Index

Anthenoides (continued)
  lihosorus 390
  tenuis 390
[*Anthias cichlopso*] 427
  dispar 425
[*elongatus*] 427
  fasciatus 425
  luzonensis 425
  pascalus 425
  pleurotaenia 425
  squamipinnis 425
  truncatus 425
Anthobothrium bifidum 206
  parvum 206
Anthochirus loui 387
Anthocidaris crassispina 393
ANTHOCYRTIDAE 67
Anthocyrtidium ophirense 67
  zanguebaricus 67
Anthogorgia bocki 186
ANTHOMEDUSAE 166
Anthopleura asiatica 179
  ballii 179
  chinensis 179
  dixoninana 179
  incerta 179
  japonica 179
  midori 179
  mortoni 180
  nigrescens 180
  [*pacifica*] 180
  qingdaoensis 180
  stella 180
  xanthogrammia 180
ANTHOSOMATIDAE 316
ANTHOTHELIDAE 186
ANTHOZOA 179
ANTICOMIDAE 212
Antigona lamellaris 254
Antigonia capros 423
  rubescens 423
  rubicunda 423
ANTIGONIDAE 423
Antillesoma antillarum 235
Antiobactrum brasiliensis 225
ANTIPATHARIA 185
Antipathes ceylonensis 185
  chotis 185
  crispis 185
  dichotomis 186
  flabellum 186
  grandis 186
  herdmani 186
  japonica 186
  lentis 186
  myriephyllis 186
  planis 186
  viminalis 186
  virgatis 186
ANTIPATHIDAE 185
[*Antisabia foliacea*] 268
Antithamnion cruciatum 138
  defectum 138
  hubbsii 138
  nipponicum 138

Antithamnion (continued)
  pylaisei 138
[*Antithamnionella breviramosa*] 138
  elegans 138
  sarniensis 138
  spirographidis 138
Antropora cuculata 372
  donghainian 372
  errectorostata 372
  fenglingensis 372
  laguncula 372
  leucophora 372
  longirostrata 372
  minus 372
  ogivalina 372
  ovata 372
  paralaguncula 372
  serrirostrata 372
  subminus 372
  subovata 372
  tincta 372
  trapezoides 372
Anulidentalium bambusum 261
[ANURA] 463
Anyperodon leucogrammicus 425
[*Aonides diverapoda*] 224
  oxycephala 223
Aoroides columbiae 328
Aphanipathes sarothamnoides 186
  somervillei 186
Aphanothee saxicola 24
Aphanurus mugilus 200
  multiprostatus 200
  stossichi 200
Aphareus furcatus 432
  rutilans 432
Apharyngogyliauchen opisthovarius 194
  scorustis 194
APHASMIDEA 211
Aphelasterias changfengyingi 392
  japonica 392
Aphrocallistes beatrix 166
  vastus 166
APHROCALLISTIDAE 166
Aphrodita aculeata 216
  australis 216
  japonica 216
  talpa 216
Aphroditella sibogae 216
APHRODITIDAE 216
APHYONIDAE 421
Aphyonus bolinu 421
Apionsoma trichocephala 235
[*Apistus alatus*] 453
  carinatus 453
Aplidium depressum 402
  multiplicatum 402
[APLOACTIDAE] 455
Aploactis aspera 455
APLODACTYLIDAE 444
[*Aplodon tectum*] 266
Aplousina falsifilum 372
  lavevigata 372
  minacis 372
  nanshaensis 372

Aplousina (continued)
  sulcata 372
[APLUSTRIDAE] 285
Aplustrum amplustre 285
Aplysia cornigera 287
  dactylomela 287
  juliana 287
  kurodai 287
  oculifera 287
  parvula 287
  pulmonica 287
  sagamiana 287
  sinensis 287
APLYSIIDAE 287
[APLYSINOMORPHA] 287
Apocepen pulcher 326
Apocryptes bato 451
Apocryptichthys sericus 451
Apocryptodon glyphisodon 451
  madurensis 451
  malcolmi 451
APOCYNACEAE 157
APODA 389
APODIDAE 468
APODIFORMES 468
Apogon amboinensis 428
  angustatus 428
  apogonides 428
  aroubiensis 428
  aureus 428
  bandanensis 428
  carinatus 428
  cheni 428
  chrysotaenia 428
  [*coccineus*] 428
  compressus 428
  cookii 428
  [*cookii*] 429
  crassiceps 428
  cyanosoma 428
  doederleini 428
  doryssa 428
  [*ellioti*] 429
  endekataenia 428
  erythrinus 428
  exostigma 428
  [*fasciatus*] 428
  fleurieu 428
  fraenatus 428
  fusca 428
  guamensis 428
  hyalosoma 428
  kallopterus 428
  kiensis 428
  lateralis 428
  [*lineatus*] 429
  melas 428
  moluccensis 428
  [*niger*] 429
  nigripinnis 428
  nigrofasciatus 428
  nitidus 428
  notatus 428
  novemfasciatus 428
  orbicularis 428

Apogon (continued)
   [*pseudotaeniatus*] 429
   quadrifasciatus 428
   robustus 429
   savayensis 429
   semilineatus 429
   semiornatus 429
   [*striatus*] 429
   taeniatus 429
   taeniophorus 429
   thermalis 429
   timorensis 429
   trimaculatus 429
   unicolor 429
Apogonichthys albomarginatus 429
   arafurae 429
   brachygrammus 429
   carinatus 429
   ellioti 429
   lineatus 429
   niger 429
   [*ocellatus*] 429
   perdix 429
   striatus 429
   waikiki 429
APOGONIDAE 428
Apolemichthys trimaculatus 436
Apollon bituberculatus 273
   olivator rubustus 273
   perca 273
   [*perca*] 273
[*Apolymetis lacunosa*] 251
   meyeri 250
Apomatus enosimae 230
Aponurus clupanodontis 200
   collichthydis 200
   eleutheronematis 200
   halieutae 200
   laguncula 200
   lizae 200
   rhinoplagusiae 200
   uraspis 200
   vitellograndis 200
Apophrixus constrictus 326
[*Apoprionospio pygmaea*] 223
Aporbopyrus enosteoidis 326
   megacephalon 326
Aporops blinearis 428
Apostichopus japonicus 389
APPENDICULATA 400
Aprion virescens 432
Aprionodon brevipenna 406
Apristurus abbreviatus 405
   canutus 405
   herklotsi 405
   internatus 405
   japonicus 405
   longicephalus 405
   macrorhynchus 405
   [*macrorhynchus*] 405
   microps 405
   nasutus 405
   pinguis 405
   platyrhynchus 405
   sinensis 405
   verweyi 405

Apus affinis subfurcatas 468
   pacificus 468
   p. kanoi 468
Arabella iricolor 227
   mutans 227
   novecrinita 227
ARABELLIDAE 227
Aracana rosapinto 460
ARACANIDAE 460
ARACHNOIDEA 297
ARACHNIDIDOIDAE 370
ARACHNIDIIDEA 370
ARACHNOIDAE 394
Arachnoides placenta 394
Arachnoidiscus ehrenbergii 31
   —— v. *monteryana* 31
   ornarus 31
ARACHNOPUSIOIDEA 378
Arachnosphaera myriacantha 64
ARAEOLAIMIDA 211
Araeosoma owstoni 392
   —— v. *nudum* 392
ARBACIIDAE 392
Arca avellana 240
   [*binakayanensis*] 240
   boucardi 240
   [*granosa*] 241
   [*inflata*] 240
   [*jousseaumei*] 241
   navicularis 240
   [*olivacea*] 240
   [*pilula*] 240
   [*semitorta*] 241
   [*subcrenata*] 241
   [*tenebrica*] 240
   [*tortuosa*] 241
   ventricosa 240
   [*virescens*] 240
Arcania elongata 354
   erinaceus 354
   [*globata*] 354
   heptacantha 354
   novemspinosa 354
   quinquespinosa 354
   sagamiensis 354
   septemspinosa 354
   undecimspinosa 354
ARCHAEOBALANIDAE 321
ARCHAEOGASTROPODA 261
Archaeomysis kokuboi 323
Archaias angulatus 90
Archamia biguttata 429
   buroensis 429
   dispilus 429
   fucata 429
   goni 429
   lineolata 429
   macropterus 429
Archaster typicus 390
ARCHASTERIDAE 390
ARCHIANNELIDA 214
Archiconchoecia cucullata 303
   falcata 303
   striata 303
   versicula 303
ARCHIDORIDIDAE 291

Archimerismus arenaceus 73
ARCHINEMERTEA 207
Archipera dipleura 67
   triclavigera 67
Archipilium eburneforme 67
Architectonica maxima 266
   perdix 266
   perspectiva 266
   trochlearis 266
ARCHITECTONICIDAE 266
ARCIDAE 240
Arcocellulus cornucer 43
   mammifer 43
ARCOIDA 240
[*Arcopagia remies*] 250
   [*scobinata*] 251
Arcopaginula inflata 250
Arcopella casta 250
   isseli 250
Arcopsis interplicata 240
   minabensis 240
   sculptilis 240
   sinensis 240
   symmetrica 240
Arcoscalpellum ciliatum 318
   pertosum 318
[*Arctus sordidus*] 346
Arcuatula elegans 241
Ardea cinerea rectirostris 465
   purpurea manilensis 465
ARDEIDAE 465
Ardeola bacchus 465
Arenaria i. interpres 467
Arenariomyces trifurcata 130
Arenicola brasiliensis 225
   [*cristata*] 225
ARENICOLIDAE 225
Arenoparrella asiatica 76
Arenoturrispirillina catinus 71
[*Arete dorsalis*] 342
Argentina kagoshimae 411
   semifasciata 411
ARGENTINIDAE 411
ARGIDIDAE 291
Argillotuba argillacea 69
Argonauta argo 296
   hians 296
ARGONAUTIDAE 296
ARGULIDAE 315
ARGULOIDA 315
Argulus scutiformis 315
Argus cruentus 291
   esakii 291
   laminea 291
   scabra 291
   speciosus 291
   tabulatus 291
Argyropelecus aculeatus 412
   affinis 412
   hemigymnus 412
   sladeni 412
Argyrops bleekeri 434
   spinifer 434
Argyrosomus aneus 431
   aregntatus 431
   [*japonica*] 431

Index

Argyrosomus (continued)
    macrocephalus 431
    [*nibe*] 431
    pawak 431
Arhodeoporus minuscuius 298
Arhynchite rugosum 237
ARHYTHMACANTHIDAE 213
Arhythmacanthus cynoglosi 213
    septacanthus 213
Arhythmorhynchus frassoni 213
    teres 213
Aricidea capensis 223
    curviseta 223
    fragilis 223
    [———— *caeca*] 223
    nolani 223
    simplex 223
ARIETELLIDAE
Arietellus aculeatus 310
    plumifer 310
    setosus 310
ARIIDAE 418
Ariomma evermanni 450
    indica 450
    lurida 450
ARIOMMIDAE 450
Aristaeomorpha foliacea 335
ARISTEIDAE 335
    [*Aristeus coralinus*] 336
    [*coruscans*] 336
    [*crassipes*] 336
    [*japonicus*] 335
    [*rostridentatus*] 335
    virilis 335
Aristida chinensis 159
Arius leiotetocephalus 418
    maculatus 418
    sinensis 418
    thalassinus 418
Armadilloniscus littoralis 326
Armandia intermedia 225
    lanceolata 225
    leptocirrs 225
    longicaudata 225
Armatobalanus allium 321
    cepa 321
Armina appendiculata 291
    babai 291
    bilanella 291
    comta 292
    japonica 292
    longicauda 292
    magna 292
    major 292
    papillata 292
    punctilopsis 292
    punctilucens 292
    puntulata 292
    semperi 292
    sinensis 292
    variolosa 292
ARMINIDAE 291
Arnoglossus aspilos 457
    [*intermedius*] 457
    japonicus 457
    polyspilus 457

Arnoglossus (continued)
    scapha 457
    tapeinosoma 457
    tenuis 457
Arothron alboreticulatus 460
    firmamentum 460
    hispidus 460
    immaculatus 460
    manilensis 460
    mappa 460
    meleagris 460
    nigropunctatus 460
    reticularis 460
    stellatus 460
Arotrobates granulatus 298
    lanceolatus 298
Arotrolepis sulcatus 459
Artacama proboscidea 229
[*Artemia parthenogenetica*] 300
    salina 300
[*Artemia* sp.] 300
ARTEMIIDAE 300
Artemisia capillaris 158
    eriopoda 158
    scoparia 158
    stelleriana 158
    [*valgaris* v. *verbenacea*] 158
    verbenacea 158
ARTHACANTHA 61
Arthrobacter 22
Arthrobotrys sp. 131
Arthrocnemum indicum 151
ARTHROPODA 297
Arthropoma cecilii 378
    circinatum 379
Arthrospira maxima 25
ARTICULATA 386
Articulina alticostata 89
    carinata 89
    curta 89
    elongata 89
    lineata 89
    mucronata 89
    pacifica 89
    parallela 89
    queenslandica 89
    ricta 89
    schauinslandi 89
    sulcata 89
Artropilium fusiforme 68
    sitularius 68
ARTOSTROBIIDAE 68
Artostrobium auritum 68
Arvella sinica 241
Asanonella shojii 116
    tubulifera 116
Asaphis dichotoma 252
ASCARIDIDEA 209
Ascarophis epinepheli 210
Ascidia alpha 402
    armata 402
    gemata 402
    lobata 402
    longistriata 402
    pacifica 402
    papillosa 402

Ascidia (continued)
    rhabdophora 402
    sydneiensis 402
ASCIDIACEA 402
ASCIDIIDAE 402
Aschemonella scabra 71
ASCHEMONELLIDAE 71
ASCIPHORINA 377
ASCLEPIADACEAE 157
Asclepios shiranus 299
[ASCOGLOSSA] 288
ASCOMYCOTINA 130
ASCOPHORINA 377
Aseraggodes kaianus 458
    kobensis 458
    [*melanospilus*] 458
Ashbrookia compressa 110
ASILIDAE 299
[*Asparagopsis hamifera*] 134
    [*sanfordiana*] 134
    taxiformis 134
Asparagus brachyphyllus 162
    dauricus 162
    officinalis 162
    polyphyllus 162
Aspasma minima 462
Aspergillus 129
    flavus 129
    niger 129
    sydowii 129
    terreus 129
    ustus 129
Aspergillus spp. 131
Asperiscala eximia 267
ASPEROCOCCACEAE 141
Aspeudes nipponicus 325
ASPEUDIDAE 325
Aspidelectra bihamata 371
    orientalis 372
Aspidisca leptaspos 60
    lynceus 60
    magna 60
    turrita 60
ASPIDISCIDAE 60
ASPIDOCHIROTA 388
Aspidontus filamentosus 446
    taeniatus 446
Aspidopholas obtecta 257
    [*yoshimurai*] 257
[*Aspidosiphon corallicola*] 235
    elegans 235
    [*formosanus*] 235
    grandis 235
    [*makoensis*] 235
    muelleri 235
    steenstrupii 235
    torus 235
Aspidosiphon sp. (cf. *parvulus*) 235
ASPIDOSIPHONIDAE 235
ASPIDOSIPHONIFORMES 235
Asplanchna brightwelli 214
    priodonta 214
    sieboldi 214
ASPLANCHNIDAE 214
Assesor randalli 427
Assilina dandotica 126

Assilina (continued)
   formosensis 126
   granulosa 126
   levis 126
   samashanica 126
   subspinosa 126
Assiminea brevicula 266
   latericea 266
   lutea 266
   scalaris 266
   subeffusa 266
   violacea 266
ASSIMINEIDAE 266
ASTACIDEA 345
Astacolus costata 93
   crepidulus 93
   gemmatus 93
   inflata 93
   planulatus 93
   pseudoplanulata 93
   schloenbachi 93
   tricarinellus 93
[*Astacus penicillatus*] 346
Aster subulatus 158
Asterias argonauta 392
   rollestoni 392
   versicolor 392
ASTERIGERINACEA 116
Asterigerinata bashbulakensis 116
ASTERIGERINATIDAE 116
ASTERIIDAE 392
Asterina batheri 391
   cepheus 391
   imboonkengi 391
   orthodon 391
   pectinifera 391
ASTERINIDAE
Asterionella formosa 43
   glacialis 43
   japonica 43
   kariana 43
   notata 43
Asterocheres aesthetes 315
ASTEROCHERIDAE 315
Asterocytis ornata 133
ASTERODISCIDIDAE 390
Asterodiscides elegans 390
ASTEROIDEA 389
Asterolampra marylandica 32
   vanheurckii 32
Asterometra mirifica 387
ASTEROMETRIDAE 387
Asteromphalus arachne 32
   brookei 32
   cleveanus 32
   elegans 32
   flabellatus 32
   heptactis 32
   robustus 32
   roperianus 32
ASTERONOTIDAE 291
Asteronotus cespitosus 291
Asteronyx loveni 395
ASTEROPIDAE 302
Asteropina grimaldi 302
   minuta 302

Asteropora myriophthalma
Asteroporpa hadracantha 395
Asteropsis carinifera 390
Asteropteron fuscum 302
Asterorhombus intermedius 457
ASTEROPSEIDAE 390
Asterorotalia binhaiensis 123
   compressiuscula 123
   diplocava 123
   hexaspinosa 123
   inflata 123
   inspinosa 123
   multispinosa 123
   pulchella 123
   subtrispinosa 123
   tetraspinosa 123
   trispinosa 123
   venusta 123
Asterropteryx semipunctatus 451
Asthenosoma ijimai 392
   varium 392
Asthenothaerus huanghaiensis
Astracantha paradoxa 68
ASTRACANTHIDAE 68
Astraea haematraga 264
   petrosum 264
[*Astralium haematragtragum*] 264
   petrosum 264
   [*petrosum*] 264
Astrammina rara 70
Astreopora myriophthalma 181
Astriclypeus manni 394
Astroboa albatrossi 395
   nuda v. *elegans* 395
Astroceras pergamena 395
Astrocladus exiguus 395
Astroclita longicostata 320
[*Astroconger myriaster*] 416
Astrodendrum sagaminum 395
Astroglymma sculptum 395
Astrogymnotes catasticta 395
ASTROLITHIDAE 61
Astrolithium bulbiferum 61
Astronesthes chrysophekadion 412
   indicus 412
   lucifer 412
ASTRONESTHIDAE 412
Astrononion australe 118
   fijiense 118
   gallowayi 118
   glabrum 118
   italicum 119
   minimum 119
   novozealandicum 119
   pressus 119
   stelligerum 119
   tasmanensis 119
   tumidum 119
   umbilicatulum 119
Astropecten monacanthus 389
   phragmorus 389
   polyacanthus 389
   vappa 390
   velitaris 390
ASTROPECTINIDAE 389
Astrophaera hexagonalis 64

ASTROPHORIDA 164
Astropyga radiata 392
Astrorhiza arenaria 69
   granulosa 69
   limnicola 69
ASTRORHIZACEA 69
ASTRORHIZIDAE 69
Astrorhizoides cornutus 70
ASTROSPHAERIDAE 63
Astrotia stokesi 464
Astrothauma euphylacteum 390
Asychis disparidentata 225
   cp. *gangeticus* 225
   gotoi 225
   pigmentata 225
Asymmefron culfellum 403
Atactodea striata 250
Atalantis buxifolia 153
ATAXOPHRAGMIACEA 77
ATELECYCLIDAE 358
ATELEOPIDAE 415
ATELEOPODIDAE 423
Ateleopus japonicus 423
   purpureus 423
Atelomycterus marmoratus 405
Atergatis floridus 360
   integerrimus 360
   reticulatus 360
Athanas areteformis 342
   dimorphus 342
   djiboutensis 342
   dorsalis 342
   [*edwardsii*] 342
   gracilipes 342
   hongkongensis 342
   japonicus 343
   [*lamelifer*] 343
   ornithorhynchus 343
   oshimai 343
   [*parvus*] 343
   sibogae 343
[ATHECATA] 166
[ATHECATAE-ANTHOMEDUSAE] 166
Athelges takanoshimensis 326
ATHERINIDAE 418
ATHERINIFORMES 418
Atherinomorus lacunosus 418
Atherion elymus 418
Athorybia rosacea 174
ATHORYBIIDAE 174
Atlanta depressa 269
   fusca 269
   helcinoides 269
   inclinata 269
   inflata 269
   lesueuri 269
   peroni 269
   rosea 269
   turriculata 269
ATLANTIDAE 269
Atolla wyvillei 176
ATOLLIDAE 176
Atorella arcturi 176
   subglobosa 176
   vanhoeffeni 176
ATORELLIDAE 176

Index

Atractotrema fusum 198
ATRACTOTREMATIDAE 198
ATREMATA 384
Atrina pectinata 243
    [————— *japonica*] 243
    [————— *lischkeana*] 243
    penna 243
    [*serrata*] 243
    [*strangei*] 243
    vexillum 243
Atriplex centralasiatica 151
   maximowicziana 151
   patens 151
   repens 151
   sibirica 151
Atrobuca nibe 431
Atropus atropus 430
Atrosalarias fuscus 446
Attheya zachariasi 32
[*Atule kalla*] 430
ATYIDAE 285
ATYLIDAE 328
Atylosia scarabaeoides 152
Atylus minikoi 328
[*Atypopenaeus compressipes*] 336
   stenodactylus 336
[*stenodactylus*] 336
Atys naucum 285
Auchenoplex crinita 229
Audouinella daviesii 133
   densa 133
   microscopica 133
   robusta 133
   subimmersum 133
AUGAPTILIDAE 310
[*Augaptilus anceps*] 310
   glacialis 310
   longicaudatus 310
   megalurus 310
   spinifrons 310
Aulacocephalus temmincki 425
Aulacodiscus affinis 31
   argus 31
   margaritaceus 31
   reticulata 31
Aularia ternaria 68
Auliscus caelatus 31
   incertus 31
   punctatus 31
   sculptus 31
AULODONTA 392
AULOPODIDAE 413
[*Aulopus japonicus*] 413
AULOSPHAERIDAE 68
AULOSTOMIDAE 423
Aulostomus chinensis 423
Aurelia aurita 176
Aureobasidium pullalans 129
Aureus group 24
Auricula complexa 43
   insecta 43
   intermedia 43
   minuta 43
AURICULACEAE 43
Auriculastra duplicata 294
   subula 294

AUSTRALOMEDUSIDAE 166
Austrolethops wardi 451
Autolytus cf. *magnus* 219
   [*pachycerus*] 219
   purpreimaculatus 219
   rubustissetys 219
   setoensis 219
   spinoculatus 219
Automate anacanthopus 343
   dolichognatha 343
   [*gardineri*] 343
Auxis rochei 450
   tapeinosoma 450
   thazard 450
AVES 464
Avicennia manna 157
Avocettina infans 417
Avrainvillea amadelpha 147
   [*capituliformis*] 147
   erecta 147
   lacerata 147
   obscura 147
   ryukyuensis 147
   xishaensis 147
Awaous melanocephalus 451
AXIIDAE
Axinella echidnaea 164
AXINELLIDA 164
AXINELLIDAE 164
AXINIDAE 192
Axinurus thynnoides 448
Axiopsis harbereri 345
   serratifrons 345
Axiothella australis 225
   rubrocincta 225
[*Axius harbereri*] 345
   [*serratifrons*] 345
AXONOLAIMIDAE 211
Aythya baeri 466
   ferina 466
   fuligula 466
   marila 466
Azima sarmentosa 154
Azorinus abbreviatus 253
   coarctatus 253
   scheepmakeri 253
Azotobactera halophilum v. *marinus* 21
AZOTOBACTERACEAE 21
Azuma emmnion 447
Azygia acuminata 194
   micropteri 194
AZYGIIDAE 194

B

Babylonia areolata 276
   formosae 277
   japonica 277
   lutosa 277
   perforata 277
   rirana 277
Bacciger lizae 194
   mugilis 194
BACILLACEAE 22
BACILLALES 22
Bacillaria paradoxa 46

BACILLARIOPHYTA 27
Bacillus 22
B. brevis 22
   Fastidiosus 22
   firmus 22
   sphaerieus 22
Bacopa monnieri 158
BACTERIA 21
Bacteriastrum comosum 28
   ————— v. *hispida* 28
   delicatulum 28
   elongatum 28
   ————— v. *diversum* 28
   hyalinum 28
   ————— v. *princeps* 28
   mediterraneum 28
   minus 28
   seticulosa 28
   varians 28
Baculogypsina sphaerulata 124
Baculogypsinoides spinosus 124
Baeolidia fusiformis 292
Baggina compressa 110
   inaperta 110
   indica 110
   longovata 110
   philippinensis 110
   turgidus 110
   trapezoida 110
BAGGINIDAE 110
Bahaba flavolabiata 431
BALAENIDAE 469
Balaenoptera acutorostrata 469
   borealis 469
   edeni 469
   musculus 469
   physalus 469
BALAENOPTERIDAE 468
BALANIDAE 321
BALANOGLOSSIDA 399
Balanoglossus carnosus 399
   misakiensis 399
Balanophyllia compressa 185
   conica 185
   cornu 185
   cumingii 185
   imperialis 185
   redivia 185
Balanus albicostatus 321
   [*allium*] 321
   [*amaryllis*] 321
   amphitrite amphitrite 322
   [————— *albicostatus*] 321
   [————— *cirratus*] 322
   [————— *communis*] 322
   [————— *kruegeri*] 322
   [*calceolus*] 321
   [*canaliculatus*] 321
   [*cepa*] 321
   [*cidaricola*] 321
   [*ciliatus*] 321
   cirratus 322
   crenatus 322
   [*cristatus*] 321
   [*cymbiformis*] 321
   eburneus 322

Balanus (continued)
  [*hawaiiensis*] 322
  improvisus 322
  [*krugeri*] 321
  littoralis 322
  [*longirostrum*] 321
  [*navicula*] 321
  poecilotheca 322
  pulchellus 322
  reticulatus 322
  [*rosa*] 322
  rostratus rostratus 322
  [*sinensis*] 321
  [*socialis*] 321
  [*tenuis*] 321
  [*tintinnabulum tintinnabulum*] 322
  trigonus 322
  uliginosus 322
  [*variegatus cirratus*] 322
  [*volcano*] 322
  [*xishaensis*] 322
  zhujiangensis 322
Balcis grandis 267
  luchuma 267
  thaanumi 267
[*Balistapus aculeatus*] 459
  [*rectangulus*] 459
  undulatus 459
Balistes capistretus 459
  [*conspicillum*] 459
  [*flavimarginatus*] 459
  [*fuscus*] 459
  vetula 459
  [*vidua*] 459
  [*viridescens*] 459
BALISTIDAE 459
Balistoides conspicillum 459
  viridescens 459
Balladyna parvula 60
Banareia banareias 360
  subglobosa 360
Bangia atropurpurea 133
  breviaticulata 133
  fusco-purpurea 133
  gloiopeitidicola 133
  siliaris 133
  yamadai 133
BANGIACEAE 133
BANGIALES 133
Banjos banjos 434
BANJOSIDAE 434
Bankia campanullata 257
  oryzaformis 257
  saulii 257
  tenuis 257
Baptozius vinosus 360
Barathronus diaphanus 421
Barbatia bistrigata 240
  cometa 240
  decussata 240
  fusca 240
  matsumotoi 240
  parva 240
  signata 240
  stearnsi 240

Barbatia (continued)
  tenella 240
  uwaensis 240
  virescens 240
  yamamotoi 240
Barbourisia rufa 415
BARBOURISIDAE 415
Barentsia discreta 384
  gracilis 384
  laxa 384
BARENTSIIDAE 384
Bargmannia elongata 174
Barnea candida 257
  davidi 257
  [*davidi*] 257
  dilatata 257
  elongata 257
  [*dilatata*] 257
  [*fragilis*] 257
  manilensis 257
Barrimysia siphonosomae 248
Barringtonia asiatica 156
  racemosa 156
Baryspira mammilla 278
Bascanichthys kirkii 417
BASEODISCIDAE 208
Baseodiscus curtus 208
BASIDIOMYCOTINA 132
BASOMMATOPHORA 294
Bassia bassensis 175
Bassobythites macropterus 421
Bassozetus robustus 421
Bathophilus longipinnis 412
Bathropyramis bicornuta 67
Bathyamussium jeffreysi 244
Bathyarca kyurokusimana 240
Bathybembix argenteonitens 263
  crumpii 263
Bathycallicaymus formocanus 447
[*Bathycallionymus kainus*] 447
Bathyconchoecia angeli 303
  crosnieri 303
  galerita 303
  lacunosa 303
  paulula 303
BATHYCRINIDAE 386
Bathydrilus edwardsi 233
Bathygadus antrodes 420
  garretti 420
Bathygobius cocosensis 451
  cotticeps 451
  cyclopterus 451
  fuscus 451
BATHYLAGIDAE 411
Bathylagus argyrogaster 411
  ochotensis 411
Bathylaimus stenolaimus 212
Bathymyrus simus 415
BATHYNOMIDAE 325
Bathynomus doederlein 325
  gigantia 325
Bathypathes affinis 185
  patulis 185
  tenuis 185
BATHYPATHIDAE 185

Bathypectinura heros 397
Bathyplotes moseleyi 387
  natans 387
Bathypontia longicornis 312
  minor 312
  similis 312
  spinifera 312
BATHYPONTIIDAE 312
BATHYPTEROIDAE 415
Bathypterois antennatus 415
  atricolor 415
  guentheri 415
Bathyraja isotrachys 407
Bathyroctes rostratus 412
Bathysiphon albus 70
  arenacea 70
  capillare 70
  echinatus 70
  eocenicus 70
  filiformis 70
  flavidus 70
  folini 70
  giganteus 70
  hystrix 70
  macilentus 70
  major 70
  rufescens 70
  rufus 70
BATHYSIPONIDAE 70
Bathysphyraenops simplex 428
Bathyuroconger vicinas 416
Batillaria bronii 266
  cumingi 266
  multiformis 266
  sordida 266
  zonalis 266
Beania aspinosa 375
  cupulariensis 375
  discordermiae 375
  farreae 375
  hirtssima 375
  intermedia 375
  klugei 375
  mirabilis 375
  petiolata 375
  regularis 375
  spinigera 375
BEANIIDAE 375
Beella digitata 101
Beguina semiorbiculata 248
Belamcanda chinensis 162
Belbolla bohaiensis 212
Beliops batanensis 428
Belizeana tuberculata 130
  cf. *tuberculata* 130
Bellerochea malleus 32
Bellucina civica 247
BELOIDEA 62
Belone platyura 418
Belonepterygion fasciolatum 428
BELONIDAE 418
BELONIFORMES 418
Belonozoum italicum 62
Beltrandia sp. 131
Bembras japonicus 455

Bembras (continued)
   laevis 455
BEMBRIDAE 455
BEMBROPIDAE 444
Bembrops caudimacula 444
   curvatura 444
Benthalbella linguidens 414
Bentharaca asperula 240
   rubrotincta 240
   xenophoricola 240
Bentheogennema intermedia
Benthesicymus altus 335
   investigaloris 335
Bentheuphausia amblyops 334
BENTHEUPHAUSIIDAE 334
Benthobatis moresbyi 409
Benthodeodesmus tonuis 449
Benthogennema intermedia 335
Benthonectes filipes 335
BENTHOPECTINIDAE 389
Benthosema fibulatum 413
   pterotum 413
   simile 413
   suborbitale 413
   [*suborbitale*] 413
Benthotrema pyriformis 194
BERDIDAE
Berenicea ampulliformis 369
   lineata 369
   meanarina 369
   sarmiensis 369
Berghia japonica 292
Bergia serrata 155
Beroe cucumis 190
   ovata 190
BEROIDA 190
BEROIDAE 190
Berthellina delicata 289
BERYCIDAE 422
BERYCIFORMES 421
Beryx splendens 422
Bestiola amoyensis 306
   sinicus 307
Betaeus granulimanus 343
   [*malleodigitus*] 342
   [*microstylus*] 342
   [*yokoyai*] 343
Bhawania brevis 218
   cryptocephata 218
   goodei 218
Bianium dayawanense 195
   hemistoma 195
   lianyungangense 195
   plicitum 195
Bicellariella gracilis 374
   sinica 374
Bicrisia edwardsiana 369
Biddulphia aurita 32
   dubia 32
   granulata 32
   gruendleri 32
   heteroceros 32
   laevis 32
   longicruris 32
   mobiliensis 32

Biddulphia (continued)
   obtusa 32
   plana 32
   pulchella 32
   regia 32
   reticulata 32
   retiformis 32
   rhombus 32
   ———— v. *trigona* 32
   roperiana 32
   schroederiana 32
   seticulosa 32
   sinensis 32
   tridens 32
   [*tuomeyi*] 32
BIDDULPHIACEAE 32
BIDDULPHIALES 32
Biemna fistulosa 165
BIEMNIDAE 165
Bifarina mckinnoni 99
Bifarinella ryukyuensis 109
Bigenerina ammobaculitoidea 78
   curta 78
   irregularis 78
   lytta 78
   nodosaria 78
   shihtiensis 78
   taiwanica 78
Biggaria bermuderse 57
   echinometris 57
   polynucleatum 57
BIGNONIACEAE 158
Biloculinella globula 85
   inflata 85
   irregularis 85
   isabelleana 86
   labiata 86
   subsphaerica 86
   taiwanica 86
   tenuiaperta 86
   toddae 86
Bimeria francisana 166
   vestita 166
Bimonilina sinensis 75
Biovarium cryptocotyle 199
   pomadasydis 199
   schistolecithale 199
   tsingtaoensis 199
[*Biplex aculeata*] 273
   [*microstoma*] 273
   pulchra 273
Birgus latro 349
Birulia kishinouyei 343
Bispira vancouveri 229
Bittium craticulatum 266
Bitubulogenerina convallaria 106
BIVALVIA 239
BIVESCULIDAE 194
Bivesicula auxisae 194
   fistulariae 194
   lutiani 194
   megalopis 194
   ostichthydis 194
   xishaense 194
Blackfordia manhattensis 169

Blackfordia (continued)
   polytentaculata 169
   virginica 169
BLACKFORDIIDAE 169
Blainvillea acmella 158
   [*latifolia*] 158
Blasicrura coffea 271
   hirundo neglecta 271
   interrupta 271
   quadrimaculata 271
   stolida 271
   ursellus 271
BLECHNACEAE 150
Bleekeria anguilliviridis 447
BLENNIIDAE 446
Blennius yatabei 446
Blepharipoda liberata 351
Blepharisma minima 58
Blepharocysta denticulata 55
   splendor-maris 55
Blidingia micrococea 144
   minima 144
Boccardia polychranchia 223
   proboscidea 223
Boccardiella hamata 223
Bodianus anthioides 438
   axillaris 438
   bilunulatus 438
   cylindriatus 438
   diana 438
   hirsutus 438
   izuensis 438
   leucostictus 438
   [*loxozunus*] 438
   luteopunctatus 438
   macrourus 438
   [*macrutus*] 438
   masudai 438
   mesothorax 439
   oxycephalus 439
   pacificus 439
   perditio 439
   prognathus 439
   thoracotaeniatus 439
   [*vulpinus*] 439
Bodotria chinensis 324
   ovalis 324
   scorpioides 324
BODOTRIIDAE 324
Boergesenia forbesii 145
Boerhavia diffusa 151
   erecta 151
Boesemanichthys firmamentum 460
Bohadschia argus 388
   [*bivittata*] 388
   graeffei 388
   [*koellikeri*] 388
   marmorata 388
Bolbolaimus denticulatus 211
Bolbometopon bicolor 441
   muricatus 441
Bolbosoma scomberomuri 213
   vasculosum 213
[*Boleophthalmus chinensis*] 453
   maculatus 452

Boleophthalmus (continued)
   pectinirostris 453
[*Bolinichthys blacki*] 413
   longipes 413
   nanshanensis 413
   pyrsobolus 413
BOLINOPSIDAE 190
Bolinopsis vitrea 190
BOLITAENIDAE 296
BOLIVINACEA 103
Bolivina abbreviata 103
   acerosa 103
   acutula 103
   aenariensis 103
   alata 103
   albatrossi 103
   bassensis 103
   capitata 103
   cochei 103
   compacta 103
   dilatata 103
   göesi 103
   incertum 103
   obscura 103
   ovata 103
   pseudoplicata 103
   pseudopunctata 103
   pusilla 103
   robusta 103
   semicostata 103
   spathulata 103
   spinata 104
   spinea 104
   spissa 104
   striatula 104
   subangularis 104
   subspinescens 104
   subtenuis 104
   vadescens 104
   variabilis 104
Bolivinella basicostata 104
   elegans 105
   folia 105
   leizhouensis 105
BOLIVINELLIDAE 104
BOLIVINIDAE 103
Bolivinita quadrilatera 105
   subangularis 105
   suturornata 105
BOLIVINITACEA 105
BOLIVINITIDAE 105
Bolivinopsis bulbosa 75
   elongata 75
Bolliella adamsi 101
Boloceroides mcmurrichi 180
BOLOCEROIDIDAE 180
BOMOLOCHIDAE 314
Bomolochus decipteri 314
   gassae 314
   managatuwo 314
Bonnemaisonia hamifera 134
   [*nootkana*] 134
BONNEMAISONIACEAE 134
Boodlea coacta 146
   composita 146

Boodlea (continued)
   montagnei 146
   paradoxa 146
   siamensis 146
   struveoides 146
   vanbosseae 146
BOODLEACEAE 146
Boodleopsis aggregata 146
Bopyrella pyrifoma 326
   tanytelson 326
BOPYRIDAE 326
Bopyrinella albida 326
Bopyrione albida 326
   longicapitata 326
   toloensis 326
Bopyrione sp. 326
Bopyrissa pyriforma 326
BORAGINACEAE 157
Borelis melo 90
   pulchra 90
Boreomysis rostrata orientalis 323
Boreotrophon candelabrum 274
Bornella digitata 292
   japonica 292
BORNELLIDAE 292
Bornetella nitida 145
   oligospora 145
   sphaerica 145
Borniopsis ariakensis 247
   nodosa 247
   ochetostomae 247
   tsurumaru 247
Boroecia barealis 303
Borostomias elucens 412
Borzia xishaensis 25
Boscia anglica 320
   oulastreae 320
Bosmina longirostris 301
Bosmina sp. 301
BOSMINIDAE 301
Bossiella cretacea 135
Bostrichthys marmoratus 451
   sinensis 451
Bostrychia binderi 139
   hongkongensis 139
   intricata 139
   kelanensis 139
   radicans 139
   simpliciuscula 139
   tenella 139
Botaurus s. stellaris 465
BOTHIDAE 457
BOTHRIOCEPHALIDAE 206
Bothriocephalus japonicus 206
   scorpi 206
Bothus assimilis 457
   mancus 457
   myriaster 457
   pantherinus 457
BOTRYLLIDAE 402
Botrylloides perspicuum 402
   simodensis 402
   violaceulus 403
Botryllus primigenus 403
   schlosseri 403

Botryllus (continued)
   tsingtaoensis 403
   tuberatus 403
Botryocladia connexa 138
   pyriformis 138
   skottsbergii 138
Botryocyrtis scutum 67
BOTRYODEA 67
Botryopyle setosa 67
Botryostrobus aquilonaris 68
Botrytella micromora 141
Botula silicula 241
Botuloides pauciloculus 90
   perlucida 90
[*Bougainvillia autumnalis*] 166
   bitentaculata 166
   britannica 166
   [*flavida*] 166
   fulva 166
   niobe 166
   paraplatygaster 166
   platygaster 166
   principis 166
   ramosa 166
   superciliaris 166
BOUGAINVILLIDAE 166
Bowerbankia caudata 370
   imbricata 370
Brachiaria subquadripara 159
Brachiella lata 316
   trichiuri 316
   yongxingensis 316
[*Brachidontes curvatus*] 242
   emarginatus 241
   striatulus 241
   variabilis 241
BRACHIONIDAE 214
Brachionus angularis 214
   colyciflorus 214
   leydigi 214
   ureeus 214
BRACHIOPODA 384
BRACHIOTEUTHIDAE 295
Brachioteuthis riisei 295
Brachirus annularis 458
   [*bellus*] 453
   orientalis 458
   pan 458
   swinhonis 458
   [*zebra*] 453
Brachyamblyopus anotus 453
   brachysoma 453
[*Brachydontes aquarius*] 242
   [*hirsutus*] 243
   [*japonica*] 242
   [*senhousei*] 242
Brachypleura novaezeelandiae 456
Brachypodium monshuricum 159
Brachypterois serrulatus 453
[*Brachypteros bellus*] 453
BRACHYSCELIDAE 332
Brachyscelus crusculum 332
   globiceps 332
   [*latipes*] 332
   [*rapacoides*] 332

Index

Brachyscelus (continued)
   rapax 332
Brachysomophis cirrhochilus 417
   crocodilinus 417
Brachystomia kurodai 281
   tawamurai 281
   vexillum 281
[*Brachytoma flavidutus*] 281
   [*vexillium*] 281
Brachytrichia maculans 27
   quoyi 27
BRACHYURA 353
Bractechlamys elegans 244
   quadrilirata 244
   schmeltzii 244
Brada ferruginea 228
   mammillata 228
   talchsapensis 228
   villosa 228
Brama japonica 431
   rayi 431
BRAMIDAE 431
Branchiomma cingulata 229
   nigromaculata 229
   orientalis 229
   pacificum 229
   serratibranchis 229
   [*vesiculosum*] 230
BRANCHIOPODA 300
BRANCHIOSTEGIDAE 429
Branchiostegus albus 429
   argentatus 429
   auratus 429
   japonicus 429
Branchiostoma belcheri 404
   ——— *tsingtauense* 404
[BRANCHIOTREMA] 399
Brania clavata 219
Branta bernicla 466
Bregmaceros arabicus 420
   atlanticus 420
   [*atripinnis*] 420
   bathymaster 420
   japonicus 420
   lanceolatus 420
   macclellandi 420
   nectabanus 420
   pescadorus 420
   rarisquamosus 420
BREGMACEROTIDAE 420
Brentisentis uncinus 213
Bresedium brevipes 365
BRESILIIDAE 339
BRESILIOIDEA 339
Brettia mollis 374
   vectifera 374
Brevibacterium 21
Breviraja tobitukai 407
Breynia officinalis 154
Bridelia tomentosa 154
Brionobutis koilomatodon 451
BRISSIDAE 395
Brissopsis lazonica 395
Brissus latecarinatus 395
Brizalina abbreviata 104

Brizalina (continued)
   acerosa 104
   acutula 104
   alata 104
   anglica 104
   britannica 104
   capitata 104
   doniezi 104
   durrandii 104
   earlandi 104
   hantkeniana 104
   interjuncta 104
   kiiensis 104
   limbata 104
   lowmani 104
   paula 104
   pseudopunctata 104
   pseudoseminuda 104
   punctata 104
   punctatostriata 104
   robusta 104
   rhomboidalis 104
   semicostata 104
   seminuda 104
   spathulata 104
   spinescens 104
   striatula 104
   subcapitata 104
   subreticulata 104
   subspinescens 104
   substriatula 104
   subtenuis 104
   tainanensis 104
   zanzibarica 104
Brooksia rostrata 401
Brotula multibarbata 421
[BROTULIDAE] 421
Brotulina fusca 421
Brucea javanica 153
Bruguiera cylindrica 156
   gymnorhiza 156
   sexangula 156
   ——— v. *rhynchopetala* 156
Bryophyllum pinnatum 152
BRYOPSIDACEAE 146
Bryopsis arbuscula 146
   caespitosa 146
   corticulans 146
   harveyana 146
   hypnoides 146
   muscosa 146
   pennata 146
   plumosa 146
Bryothinusa sp. 300
BRYOZOA 369
Bubulcus ibis coromandus 465
Buccella decora 121
   frigida 121
   inculta 121
   inusitata 121
   modica 121
   radiata 121
   tenerrima 121
   tunicata 121
BUCCINIDAE 276

Buccinium pemphigum 277
   undatumplectrum 277
   yokomaruae 277
Buccinosphaera invaginata 63
Bucephala clangula 466
BUCEPHALIDAE 192
Bucephalopsis ablennus 193
   arcuatus 193
   bennetti 193
   collichthydis 193
   obpyriformis 193
   rhynchobati 193
   scombropsis 193
Bucephalus harpodontis 193
   margaritae 193
   polymorphus 193
   retractilis 193
   trifurcatus 193
   varicus 193
Bufonaria albivaricosa 273
   crumena 273
   echinata 274
   margaritula 274
   nobilis 274
   [*rana*] 274
   [*spinosa*] 274
Bugula aspinosa 374
   californica 374
   dentata 374
   flabellata 374
   intermedia 374
   nanshaensis 374
   neritina 374
   orientalis 374
   pacifica 374
   pseudoaspinosa 374
   robusta 374
   scaphoides 374
   stolonifera 374
   subglobosa 374
   vectifera 374
Bugulella sinica 374
BUGULIDAE 374
BUGULOIDEA 374
Bulbostylis barbata 160
   densa 160
   puberula 160
Bulimina aculeata 106
   barbata 106
   clava 106
   costata 106
   delreyensis 106
   ecuadorana 106
   elegans 106
   exilis 106
   fissura 106
   fossa 106
   gibba 106
   inflata 106
   mapiria 106
   marginata 106
   marginospinata 106
   mauricensis 106
   mexicana 106
   midwayensis 106

Bulimina (continued)
   nipponica 106
   notoensis 106
   notovata 107
   ovata 107
   pagoda 107
   pulchella 107
   pupoides 107
   pyrula 107
   rostrata 107
   spicata 107
   squammigera 107
   submarginata 107
   subornata 107
   subulata 107
   winniana 107
BULIMINACEA 106
Buliminella elegantissima 107
   madagascariensis 107
   milletti 107
   seminuda 107
BULIMINELLIDAE 107
BULIMINIDAE 106
Buliminoides milletti1 112
   parallela 112
   williamsonianus 113
BULIMINOIDIDAE 112
Bulla adamsii 285
   ampulla 285
   ——— *bifasciata* 285
   ——— *trifasciata* 285
   difficilis 285
   orientalis 285
   vernicosa 285
Bullacta exarata 285
[BULLARIIDAE] 285
BULLIDAE 285
Bullina lineata 285
   nobiiis 285
   truncatula 285
[BULLOMORPHA] 285
Bulmeria bulwerii 464
Bunaster ritteri 391
[*Bursa bubo*] 274
   bufo 274
   bufonia 274
   [*bufonia*] 274
   [*corrugata*] 274
   cruentata 274
   [*cruentata*] 274
   [*dunkeri*] 274
   elegans 274
   granularis 274
   [*granularis*] 274
   lamarckii 274
   [*lampas*] 274
   [*leo*] 274
   [*lissostoma*] 274
   [*mammata*] 274
   [*oyamai*] 274
   rana 274
   [*rana*] 274
   rhodostoma 274
   [*rhodostoma*] 274
   rosa 274

Bursa (continued)
   [*rubeta*] 274
   [*siphonata*] 274
   tuberosissima 274
   [*tuberosissima*] 274
Burseolina pacifica 105
BURSIDAE 273
Bushiella argutus 231
Buskia nitens 370
BUSKIIDAE 370
Butis butis 451
   gymnopomus 451
   koilomatodon 451
   melanostigma 451
Butorides striatus connectens 465
Buzasina ringens 73
Byblis calisto 328
   febris 328
   io 328
   japonicus 328
   kallarthra 328
   pilosa 328
BYTHITIDAE 421
Bythocellata cruciformis 167
Bythotiara depressa 167
   murrayi 167

C

Caberea boryi 375
   busifera 375
   climacina 375
   lata 375
   megaceras 375
   symmetrica 375
   transversa 375
CABEREIDAE
Cabira incerta 218
CACTACEAE 155
Cadella delta 250
   ——— *hainanensis* 250
   narutoensis 250
   semitorta 250
   smithi 250
CADIACEAE 146
Cadinella ornatissima 290
Cadlina japonica 291
   sagamiensis 291
CADLINIDAE 291
Cadulus clavatus 261
Caecula longipinnis 417
Caenopedina mirabilis 393
Caelorhynchus anatirostris 420
   asteroides 420
   cingulatus 420
   commutabilis 420
   [*commutabilis*] 420
   formosanus 420
   intermedius 420
   japonicus 420
   jordani 420
   kamoharai 420
   kishinouyei 420
   longissimus 420
   multispinulosus 420

Caelorhynchus (continued)
   parallelus 420
   smithi 420
   tokiensis 420
Caelorinchs dorsalis 420
   gilberti 420
   hubbsi 420
   matsubarai 420
Caesalpinia crista 153
Caesio chrysozona 432
   coerulaureus 432
   diagramma 433
   erythrogaster 433
   lunaris 433
   tile 433
   xanthonotus 433
Cainocreadium epinepheli 197
   gullella 197
Calamagrostis epigejos 159
Calamorhynchus pellucidus 332
   [*rigidus*] 332
CALANIDAE 306
CALANOIDA 306
Calanoides carinatus 306
Calanopia elliptica 311
   minor 311
   thompsoni 311
[*Calanus pacificus*] 306
   sinicus 306
Calappa gallus 355
   hepatica 355
   lophos 355
   philargius 355
   pustulosa 355
   terraercginae 355
CALAPPIDAE 355
CALCAREA 165
Calcarina calcar 124
   hainanensis 124
   hispida 124
   spengleri 124
   viennoti 124
CALCARINIDAE 123
Calcinus elegans 348
   formosus 348
   gaimardii 348
   laevimanus 348
   latens 348
Caleschara laxa 376
   levinseni 376
Calidris acuminata 467
   alpina sakhalina 467
   canutus rogersi 467
   ferruginea 467
   ruficollis 467
   subminuta 467
   temminckii 467
   tenuirostris 467
CALIGIDAE 315
CALIGOIDA 315
Caligus aduncus 315
   bifurcus 315
   biseriodentatus 315
   brevisoris 316
   communis 316

Index

Caligus (continued)
  confusus 316
  coryphaenae 316
  costatus 316
  eleutheronemi 316
  euthynus 316
  hilsae 316
  hirsutus 316
  laticorpus 316
  longicaudus 316
  multispinosus 316
  nibeae 316
  rotundigenitalis 316
  undulatus 316
Callanthias japonicus 425
Callechelys maculatus 417
Calliactis argentacoloratus 179
  conchicola 179
  miriam 179
  polypus 179
  reticulata 179
  rosea 179
  xishaensis 179
[*Callianassa harmandi*] 345
  japonica 345
  joculatrix 345
  modesta 345
  petalura 345
CALLIANASSIDAE 345
Callianidea typa 345
Calliaster cf. *childreni* 390
  quadrispinus 390
Callicanthus hexacanthus 448
  lituratus 448
  lopezi 448
Calliderma emma 390
Callimenellus fumidus 300
Calliodentalium crocinum 261
CALLIONYMIDAE 447
Callionymus altidorsalis 447
  altipinnis 447
  beniteguri 447
  curvicornis 447
  enneactis 447
  flagris 447
  formosanus 447
  hainanensis 447
  hindsi 447
  kaianus 447
  kitaharae 447
  lunaius 447
  marisinensis 447
  meridionalis 447
  monofilispinnus 447
  octostigmatus 447
  olidus 447
  ornatipinnis 447
  planu 447
  richardsoni 447
  schaapi 447
  valenciennei 447
  virgis 447
CALLIOPIIDAE 328
Calliostoma formosense 263
  haliarchus 263

Calliostoma (continued)
  koma 263
  unicum 263
[*Calliscyllium venustum*] 406
Callista chinensis 254
  erycina 254
Callistochiton savigni 238
Callistoctopus arakawai 296
CALLISTOPLACIDAE 238
Callithaca staminea 254
[*Callithamnion callophylloidicola*] 138
  corymbosum 138
Calliurichthys belcheri recurvispinnis 447
  dorysus 447
  filamentosus 447
  japonicus 447
  martinae 447
  recurvispinnis 447
  variegatus 447
Callogobius liolepis 451
  sclateri 451
  snelliusi 451
Callogotrema fistulariae 195
Calloplesiops altivelis 428
Callopora biarmata 373
  bifursca 373
  corniculifera 373
  craticula 373
  granulata 373
  horridoidea 373
  inarmata 373
  inaviculata 373
  inermis 373
  laevigata 373
  lineata 373
  spatulata 373
CALLOPORIDAE 373
Calloporina sculpta 381
CALLOPOROIDEA 372
Callorhinus ursinus 469
Callyspongia diffusa 165
  globosa 165
  orieminens 165
  ramosa 165
Calocalanus contractus 307
  gracilis 307
  monospinus 307
  pavo 307
  pavoninus 307
  plumulosus 307
  styliremis 307
Calocois japonicus 345
Caloglossa chuanshiensis 139
  leprieurii 139
  ogasawaraensis 139
Calogyne pilosa 158
Caloneis aemula 35
  amphisbaena v. *fuscata* 35
  bacillaris 35
  brevis 35
  ——— v. *distoma* 35
  ——— v. *vexans* 35
  castracanei v. *petitiana* 35
  consimilis 35

Caloneis (continued)
  eccentrica 35
  elongata 35
  ——— v. *constricta* 35
  formosa 36
  frater 36
  galapagensis v. *japonica* 36
  janischiana 36
  liber 36
  linearis 36
  ophiocephala 36
  oregonica 36
  permagna 36
  platycephala 36
  robusta 36
  wittii 36
CALOMETRIDAE 387
Calophyllum inophyllum 155
Caloplocamus acutus 289
  ramosus 289
Calothrix aeruginea 26
  confervicola 26
  contarenii 26
  crustacea 26
  fusco-violacea 26
  parasitica 26
  pilosa 26
  robusta 26
  scopulorum 26
  vivipara 26
Calotomus carolinus 441
  japonicus 441
  spinidens 441
Calotropis gigantea 157
Calpurnus lacteus semistriatus 270
  verrucosus 270
Calvactaea tumida 360
Calveriosoma gracile 392
CALWELLIDAE 381
Calycella syringa 169
CALYCOPHORAE 174
CALYCOPSIDAE 167
Calycopsis bigelowi 167
  papillata 167
[CALYPTOBLASTEA] 169
Calyptogena nanshaensis 254
Calyptotheca capitifera 379
  nivea 379
  parviminuta 379
  wasiensis 379
Calyptraea morbida 268
  sakaguchi 268
CALYPTRAEIDAE 268
Calystegia soldanella 157
CAMALLANIDAE 209
Camallanus marinus 209
  mustelus 209
CAMARODONTA 393
Camatopsis rubida 362
Caminus tunghainensis 164
[*Camitia rotellina*] 263
  rotellinus 263
Campanularia calceolifera 172
  denticulata 172
  [*gelatinosa*] 172

Campanularia (continued)
   hincksii 172
   [*integra*] 172
   verticillata 172
   volubjlis 172
CAMPANULARIIDAE 172
CAMPANULINIDAE 169
Campesyllis longicirrus 219
Camposcia retusa 356
Camptandrium elongatum 364
   sexdentatum 364
Camptyolnemopsis major 26
Campylaephora crassa 138
   hypnaeoides 138
Campylaimus gerlachi 211
Campylaspis amblyoda 325
   fusiformis 325
Campylodiscus adriaticus 47
   australis 47
   biangulatus 47
   birostratus 47
   brightwellii 47
   daemelianus 47
   ——— v. *comminuta* 47
   decorus 47
   ecclesianus 47
   echenëis 47
   heufleri 47
   hodgsonii 47
   horologium 47
   incertus 47
   intermedius 47
   kittonianus 47
   latus 47
   ralfsii 47
   triumphans 47
   wallichianus 47
   ——— v. *normanicus* 47
Campyloneis grevillei 44
Campylosira cymbelliformis 43
CANACEIDAE 299
Canavalia cathartica 153
   maritime 153
   [*microcarpa*] 153
Cancellaria mangeloides 279
CANCELLARIIDAE 279
CANCELLOTHYRIDAE 384
Cancer amphioetus 358
   [*anomalus*] 345
   [*chinensis*] 337
   [*custos*] 340
   gibbosulus 358
   [*homarus*] 346
   [*polyphagus*] 346
   [*pygmaeus*] 358
Cancilla filaris 278
   [*isabella*] 279
CANCRIDAE 358
Cancris auriculus 110
   indicus 110
   intermedius 110
   maoricus 110
   oblongus 110
   obesus 110
   ovalis 110
   peroblingus 110

Cancris (continued)
   sagra 110
   segmentalis 110
   torquertus 110
Cancrisocia expansa 180
Canda clypea 375
   pecten 375
   scutata 375
Candacia aethiopica 311
   bipinnata 311
   [*bispinosa*] 311
   bradyi 311
   catula 311
   columbiae 311
   curta 311
   discaudata 311
   guggenheimi 311
   longimana 311
   pachydactyla 311
   [*pacifica*] 311
   [*simplex*] 311
   tenuimana 311
   [*truncata*] 311
   varicans 311
CANDACIIDAE 311
Candeina nitida 101
CANDEINIDAE 101
Candida albicans 129
   amylolenta 129
   azyma 129
   curvata 129
   domerogii 129
   edax 129
   famata 129
   guilliermondii 129
   humicola 129
   hylophilia 129
   insectorum 129
   intermedia 129
   krusei 129
   krusii 129
   lambica 129
   membranaefaciens 129
   multisgemmis 129
   parapsilosis 129
   pelliculosa 129
   sake 129
   silvanorum 129
   tenuis 129
   tepae 129
   terebra 129
   tropicalis 129
   valdiviana 129
   veronae 129
   wiswanathii 129
   zeylancides 129
CANDIDAE 375
Candorbulina suturalis 103
Canepaia brasiliensis 74
CANNOBOTRYIDAE 67
Cantellius albus 320
   gregarea 320
   iwayama 320
   pallidus 320
   secundus 320
   septimus 320

Cantellius (continued)
   sinensis 320
   spinulosa euspinulosa 320
Cantharidus giliberti 263
   infuscatus 263
   [cf. *nitens*] 263
Cantharus cecillei 277
   [cf. *nitens*] 277
   [*fumosus*] 277
   giliberti 277
   melanostomus 277
   undosus 277
   [*undosus*] 277
Cantherhines dumerili 459
   fronticinctus 459
   [*nodestus*] 460
   pardalis 459
Canthidermis maculatus 459
   [*rutundatus*] 459
Canthigaster amboinensis 460
   bennetti 460
   compressus 460
   coronata 460
   jactator 460
   janthinopterus 460
   rivulatus 460
   solandri 460
   valentini 460
Canthocalanus pauper 306
CANTHOCAMPTIDAE 315
Capella g. gallinago 467
   megala 467
   solitaria japonica 467
   stenura 467
Caphyra laevis 358
   rotundifrons 358
Capillaria fudingensis 210
Capillaster multiradiata 386
   sentosa 386
Capitella capitata 225
CAPITELLIDA 225
CAPITELLIDAE 225
Capitulum mitella 318
Capnella imbricata 188
   philippinensis 188
Caprella acanthogaster 334
   [*acutifrons*] 334
   bispinosa 334
   californica 334
   equilibra 334
   iniquilibra 334
   kroyeri 334
   penantis 334
   rhopalochir 334
   scaura 334
   simia 334
   vidua 334
CAPRELLIDAE 334
CAPRELLIDEA 334
Caprodon schlegeli 425
[CAPROIDAE] 423
CAPULIDAE 268
Capulus dilatatus 268
   sagittifer 268
CARANGIDAE 430
[*Carangoides armatus*] 430

Index

Carangoides (continued)
  bucculentus 430
  [*chrysophrys*] 430
  [*coeruleepinnatus*] 430
  delicatissimus 430
  djeddaba 430
  [*ferdau*] 430
  hedlandensis 430
  hippos 430
  ignobilis 430
  ishikqwai 430
  kalla 430
  lugubris 430
  macrurus 430
  malam 430
  mate 430
  melampygus 430
  orthogrammus 430
  oshimai 430
  papuensis 430
  pectoralis 430
  [*plagiotaenia*] 430
  platessa 430
  sansum 430
  stellatus 430
  tille 430
Caranx armatus 430
  chrysophrys 430
  coeruleopinnatus 430
  dinema 430
  equula 430
  ferdau 430
  helvolus 430
  malabaricus 430
  [*melanoptera*] 430
  oblongus 430
  plagiotaenia 430
  plumbeus 430
  praeustus 430
  sexfasciatus 430
  talamparoidas 430
  uraspis 430
  [*vari*] 430
CARAPIDAE 421
Carapus homei 421
  kagoshimanus 421
  lumbricoides 421
  parvipinnis 421
Caratanthus maculatus 455
  unipinna 455
Carbasea carbasea 372
  donghaiensis 372
  [*episcopalis*] 372
  meridionalis 372
  orientalis 372
  sinica 372
CARCHARHINIDAE 406
CARCHARHINIFORMES 405
Carcharhinus albimarginatus 406
  altimus 406
  [*amblyrhy*] 406
  atrodorsus 406
  brachyurus 406
  brevipinna 406
  dussumieri 406
  falcifomis 406

Carcharhinus (continued)
  gangeticus 406
  latistomus 406
  leucas 406
  limbatus 406
  longimanus 406
  macloti 406
  [*macloti*] 406
  melanopterus 406
  menisorrah 406
  microphthalmus 406
  obscunella 406
  obscurus 406
  pleurotaenia 406
  plumbeus 406
  remotoides 406
  sorrah 406
Carcharias arenarias 404
  owstoni 404
  [*taurus*] 404
CARCHARIIDAE 404
Carcharodon carcharias 404
Carcinonemertes mitsukurii 208
CARCINONEMERTIDAE 208
Carcinoplax bispinosa 362
  longimana 362
  longipes 362
  longispinosa 362
  meridionalis 362
  purpurea 363
  sinica 363
  surugensis 363
  vestita 363
Carcinoscopius rotundicauda 297
Carcinosquilla carinata 368
  multicarinata 368
Cardiapoda richardi 269
CARDIIDAE 248
Cardilia semisulcata 250
CARDILIIDAE 250
Cardiomya alcocki 258
  gouldiana 258
  [*nipponica*] 258
  reticulata 258
  [*sagamiana*] 258
  singaporensis 258
  sinica 258
  tosaensis 258
Cardisoma carnifex 367
  hirtipes 367
Cardita crassicosta 248
  leana 248
  nodulosa 248
  variegata 248
Carditella hanzawai 248
CARDITIDAE 248
Caretta c. gigas 463
  [*c. olivacea*] 463
Carex commixta 160
  kobomugi 161
  pumila 161
  scabrifolia 161
Caribbeanella cuspidata 115
  depressa 115
  incerta 115
  irregularis 115

Caribbeanella (continued)
  katasensis 115
  ogiensis 115
CARIDEA 338
Carinaria cristata 269
  galea 269
CARINARIIDAE 269
Carinesta tubulanoides 207
CARNOSA 370
Carpenteria monticularis 116
  proteiformis 116
  utricularis 116
Carpilius convexus 360
  maculatus 360
Carpiscula galearis 271
CARPOCANIIDAE 67
Carpocanium diadema 67
  ensigerum 67
Carpocanopsis obovata 67
Carpocanthum lubricum 63
[*Carpopeltis affinis*] 136
  [*cornea*] 136
  formosana 136
  rigida 136
Carposphaera globosa 62
Carterina spiculotesta 81
CARTERINIDAE 81
Carupa laeviuscula 358
  tenuipes 358
Carybdea rastoni 176
  sivickisi 176
CARYBDEIDAE 176
CARYOPHYLLACEAE 152
Caryophyllia compressa 185
  japonica 185
  pacifica 185
  paucipaliata 185
  scobinosa 185
CARYOPHYLLIIDAE 184, 185
Caryota ochlandra 161
Casella atromarginata 290
  cincia 290
  [*rufomarginata*] 290
Casmaria erinaceus 272
  ponderosa nipponensis 272
  ———— *ponderosa* 272
CASSIDIDAE 272
Cassidula plectorematoides 294
  schmackeriana 294
Cassidulina alternans 105
  bradshawi 105
  carinata 105
  crassa 105
  cuneata 105
  delicata 105
  jonesiana 105
  laevigata 105
  laticamerata 105
  minuta 105
  moluccensis 105
  neocarinata 105
  oblonga 105
  pacifica 105
  pulchella 105
  reniforme 105
  striatostoma 105

Cassidulina (continued)
   subcarinata 105
   subglobosa 105
   teretis 105
CASSIDULINACEA 105
CASSIDULINIDAE 105
Cassidulinoides braziliensis 105
   chapmani 105
   orientale 105
Cassis cornuta 272
   [*rufa*] 272
Cassytha filiformis 152
CASTANELLIDAE 68
Castanidium elegans 68
   moseleyi 68
   variabile 68
Casuarina cunninghamiana 150
   equisetifolia 150
   glauca 150
CASUARINACEAE 150
Catablema vesicarium 167
Catalaphyllia fimbriata 184
Catapaguroides sp. 349
Catapagurus ensifer 349
   japonicus 350
CATAPSYDRACIDAE 101
Catenella impudica 137
   nipae 137
   subumbellata 137
Catenicella elegans 377
   triangulifera 377
   uberrima 377
CATENICELLIDAE
CATENICELLOIDEA 377
Catherinum rossi 318
Catoptometra hartlaubi 386
   magnifica 386
   rubroflava 386
Catoptrus nitidus 358
   truncatifrons 358
CATOSTYLIDAE 176
Catriona ornata 292
   cf. *purpureoanulata* 292
Caudina atacta 389
   intermedia 389
   similis 389
   zhejiangensis 389
CAUDINIDAE 389
[*Caulacanthus okamurai*] 137
   ustulatus 137
Caulastrea furcata 183
Caulerpa brachypus 146
   cupressoides 146
   ———— v. *lycopodium* f. *amicorum* 146
   nummularia 146
   okamurae 146
   peltata 146
   [———— v. *mummularia*] 146
   [———— v. *typica*] 146
   prolifera v. *xishaensis* 146
   racemosa v. *clavifera* 146
   ———— v. *occidentalis* 146
   [———— v. *peltata*] 146
   serrulata 146
   sertularioides f. *longiseta* 146

Caulerpa (continued)
   taxifolia 146
   [———— f. *typica*] 146
   verticillata 146
   webbiana 146
CAULERPACEAE 146
Caulibugula binata 374
   caliculata 374
   ciliata 374
   ciliatoidea 374
   dendrograpta 374
   gracilenta 374
   hainanica 374
   inermis 374
   irregularis 374
   longiconica 374
   longirostris 374
   mortenseni 374
   singulata 374
   sinica 374
   zanzibariensis 374
Caulleriella cf. *typhlops* 224
Caulobacter 22
Cauloraphus cymbaeformis 373
   spiniferus 373
Cavarotalia annectens 123
Cavernularia habereri 190
   obesa 190
Caviplaxus jiaozhouwanensis 330
   longiflagella 330
CAVISOMATIDAE 213
Cavolinia gibbosa 288
   globulosa 288
   inflexa v. *labiata* 288
   longirostris 288
   ———— v. *angulata* 288
   tridentata 288
   uncinata 288
CAVOLINIIDAE 288
CEBIDICHTHYIDAE 447
Celerina heffernani 391
CELASTRACEAE 154
Cellana grata 262
   nigrolineata 262
   radiata 262
   testudinaria 262
   toreuma 262
Cellanthus chapmani 125
   craticulatum 125
   guangdongensis 125
   ibericum 125
   taiwanus 125
   tikutoensis 125
   tungliangensis 125
Cellaria beniensis 377
   [*gracilis*] 377
   punctata 377
   sinica 377
   [*veleronis*] 377
CELLARIIDAE 377
CELLARIOIDEA 377
Celleporaria asperta 378
   columnaris 378
   erectorostris 378
   fusca 378

Celleporaria (continued)
   projecta 378
   sinensis 378
   tridenticulata 378
   umbonatoidea 378
Celleporella expansa 377
   hylina 377
CELLEPORIDAE 381
Celleporina costazii 381
   geminata 381
   pisiformis 381
   porosissima 381
   radiata 381
   spathulata 381
CELLEPOROIDEA 381
CELLULARIOMORPHA 374
Cenchrus calyculata 159
Cenellipsis ellipticum 64
Cenosphaera corota 62
   melifica 62
Centraugaptilus rattrayi 310
CENTRICEAE 27
CENTRISCIDAE 423
Centriscus scutus 423
Centroberyx lineatus 422
   rubricaudus 422
Centrobotrys thermophila 67
Centrobranchus andreae 413
   choerocephalus 413
   nigroocellatus 413
Centroceras apioulatum 138
   clavutatum 138
   japonicum 138
   miniatum 138
Centrocubus cladostylus 64
Centroderes sp. 208
CENTROLEPIDACEAE 162
Centrolepis banksii 162
   [*hainanensis*]
CENTROLOPHIDAE 450
Centropages abdominalis 309
   brevifurcus 309
   calaninus 309
   dorsispinatus 309
   elongatus 309
   furcatus 309
   gracilis 309
   longicornis 309
   [*mcmurrichi*] 309
   orsinii 309
   sinensis 309
   tenuiremis 309
   vioiaceus 309
CENTROPAGIDAE 309
Centrophorus acus 406
   granulosus 406
   lusitanicus 406
   moluccensis 406
   niaukang 406
   robustus 407
   squamosus 407
   squamulosus 407
   tasselatus 407
   uyato 407
Centrophyge bicolor 436

Index

Centrophyge (continued)
  bispinosus 436
  ferrugatus 436
  fisheri 436
  flavicauda 437
  heraldi 437
  interruptus 437
  nox 437
  tibicen 437
  vroliki 437
Centropogon fuscovirens 455
  urostigma 455
Centroscyllium fabricii 407
  kamoharai 407
  nigrum 407
Centroscymnus owstoni 407
CEPHALOPODA 294
CEPHALASPIDEA 285
CEPHALATHRICIDAE 207
Cephalathrix linearis 207
Cephalobothrium longisegmentum 207
CEPHALOCHORDATA 403
Cephalolepidapedon seba 195
Cephalonema brunniceps 208
[*Cephalopholis albomarginatus*] 425
  analis 425
  argus 425
  aurantius 425
  boenack 425
  formosa 425
  igarashiensis 425
  leopardus 425
  miniatus 425
  pachycentron 425
  sexmaculabus 425
  sonnerati 425
  spiloparaeus 425
  urodelus 425
  urodeta 425
Cephaloscyllium fasciatum 405
  formosanum 405
  [*isabellum*] 405
  umbratile 405
CEPHALOSPORIUM 130
Cephalosporium 130
Cephalotrichella alba 207
Cephea conifera 176
CEPHEIDAE
Cepola schlegeli 438
CEPOLIDAE 438
Ceradocus hawaiensis 329
CERAMIACEAE 138
CERAMIALES 138
Ceramium aduncum 138
  boydenii 138
  byssoideum 138
  [*crassum*] 138
  [*fastigiatum*] 138
  fastigiramosum 138
  japonicum 138
  kondoi 138
  masonii 138
  nakamurae 138
  paniculatum 138
  [*rubrum*] 138

Ceramium (continued)
  tenerrimum 138
  tenuissimum 138
CERAMONEMATIDAE 211
Cerapus tubularis 328
Cerataulina bergonii 28
  compacta 28
  pelagica 28
  zhongshaensis 28
Cerataulus smithii 32
  turgidus 32
CERATIACEAE 51
Ceratias holboelli 462
CERATIIDAE 462
[*Ceratium arcuatum*] 52
  arietinurn 51
  ——— v. *gracilentum* 51
  axiale 51
  azoricum 52
  belone 52
  [*biceps*] 52
  bigelowii 52
  breve 52
  ——— v. *paralleum* 52
  [*bucephalum* v. *haterocamptum*] 51
  candelabrum 52
  ——— v. *depressum* 52
  carriense 52
  ——— f. *volans* 52
  cephalatum 52
  contortum 52
  ——— v. *robustum* 52
  ——— v. *saltans* 52
  ——— f. *subcontortum* 52
  contrarium 52
  ——— v. *claviceps* 52
  declinatum 52
  deflexum 52
  dens 52
  digitatum 52
  ——— v. *rotundatum* 52
  ehrenbergii 52
  euarcuatum 52
  extensum 52
  ——— f. *strictum* 52
  falcatiforme 52
  furca 52
  ——— v. *berghii* 52
  ——— v. *eugrammum* 52
  fusus 52
  ——— v. *schuttii* 52
  ——— v. *seta* 52
  genicuiatum 52
  gibberum 52
  ——— v. *dispar* 52
  [*gracile*] 53
  gravidum 52
  ——— v. *angustum* 52
  hexacanthum 52
  ——— v. *contortum* 52
  ——— f. *spirale* 52
  horridum 52
  ——— v. *claviger* 52
  ——— v. *denticulatum* 52
  ——— v. *inclinatum* 52

Ceratium (continued)
  ——— v. *patentissimum* 52
  humile 52
  incisum 52
  inflatum 52
  [*inflexum*] 52
  [*intermedium*] 52
  karstenii 53
  kofoidii 53
  limulus 53
  lineatum 53
  [*longinum*] 52
  longipes 53
  longirostrum 53
  longissimum 53
  lunula 53
  ——— f. *brachyceros* 53
  macroceros 53
  ——— v. *gallicum* 53
  massiliense 53
  ——— v. *armatum* 53
  minutum 53
  molle 53
  paradoxides 53
  [*pennatum*] 52
  pentagonum 53
  ——— v. *longisetum* 53
  platycornia 53
  praelongum 53
  pulchellum 53
  ——— f. *semipulchellum* 53
  ranipes 53
  ——— v. *palmatum* 53
  reflexum 53
  schmidti 53
  [*schrankii*] 53
  schroeteri 53
  setaceum 53
  [*sumatranum*] 53
  symmetricum 53
  ——— v. *coarctatum* 53
  ——— v. *orthoceros* 53
  tenue 53
  teres 53
  trichoceors 53
  tripos 53
  ——— v. *atlanticum* 53
  ——— f. *balticum* 53
  ——— v. *subsalsum* 53
  vultur 53
  ——— f. *angustum* 53
  ——— v. *japonicus* 53
  ——— f. *recuvun* 53
  ——— v. *sumatranum* 53
Ceratobulimina pacifica 98
CERATOBULIMINACEA 98
CERATOBULIMINIDAE 98
Ceratocarcinus trilobatus 357
[*Ceratocephala sibogae*] 221
CERATOCORYACEAE 53
Ceratocorys bipes 53
  gourretii 53
  horrida 53
  magna 53
  reticulata 53

Ceratocottus diceraus 456
Ceratocymba dentata 175
   intermedia 175
   leuckarti 175
   sagitta 175
Ceratodictyon spongiosum 136
Ceratonereis anchylochaeta 220
   burmensis 220
   costae 220
   erythracensis 220
   [*falcaria*] 220
   hircinicola 220
   japonica 220
   marmoraata 220
   mirabilis 220
   pachychaeta 220
   [*tongicanda*] 220
   tripartita 220
Ceratoplax obtusignathus 363
CERATOPOGONIDAE 299
Ceratoscopelus townsendi 413
   warmingi 413
Ceratosoma cornigerum 290
   [*trilobatum*] 290
Ceratostoma fournieri 274
   inornata 274
   rorifluum 274
Ceratothoa oxyrrhynchaena 325
Ceratotrochs brunneus 185
   funicolumna 185
Cerbera manghas 157
Cerberilla albopunctata 292
   asamusiensis 292
Cercaria armata 198
   cloacicola 204
   elegans 195
   gedoelsti 204
   himasthloides 204
   hongkongensis 204
   longicauda 204
   magnicaudata 204
   mesostephanus 204
   minus 198
   mortoni 204
   pernaviridis 193
   spelotremoides 198
Cercarioides hoepplii 204
CEREBRATULIDAE 207
Cerebratulina communis 207
   darvelli 207
   natans 207
   nigra 207
   rubella 207
   sinensis 207
CERIANTHARIA 180
CERIANTHIDAE 180
Cerianthus filiformis 180
   orientalis 180
Ceriodaphnia cornuta 300
   laticaudata 300
   quadrangula 300
Ceriops tagal 156
Ceriosporopsis halima 130
Cerithidea cingulata 266
   djadjariensis 266

Cerithidea (continued)
   largillierti 266
   microptera 266
   [*obtusa*] 266
   ornata 266
   rhizophorarum 266
   sinensis 266
Cerithiella subreticulata 267
CERITHIIDAE 266
CERITHIOPSIDAE 267
Cerithium alutaceum 266
   articulatum 267
   [*asperum*] 267
   bayayi 267
   cedonulli 267
   [*cedonulli*] 267
   citrinum 267
   columna 267
   echinatum 267
   fasciatum 267
   [*fasciatum*] 267
   kobelti 267
   maillaridi 267
   nesioticum 267
   nodulosum 267
   pfefferi 267
   rostratum 267
   rubus 267
   [*sinense*] 267
   [*sinensis*] 266
   vertagus 267
   [*vertagus*] 267
Ceroplastes rubens 300
Cervimunida princeps 350
Cervonema delta 211
CESTIDA 190
CESTIDAE 190
CESTOIDEA 206
Cestopagurus misakiensis 350
Cestum sp. 190
CETACEA 468
Cetoconcha japonica
CETOMIMIFORMES 415
Cetonurus crassiceps 420
CETORHINIDAE 404
Cetorhinus maximus 404
[*Cetoscarus bicolor*] 441
Chaenera vorax 56
Chaenogobius annularis 451
Chaeomia sauliae 273
CHAETANGIACEAE 134
CHAETOCERACEAE 33
Chaetoceros abnormis 33
   aequatoriale 33
   affinis 33
   ——— v. *circinalis* 33
   ——— v. *willei* 33
   anastomosans 33
   atlanticus 33
   ——— v. *neapolitana* 33
   ——— v. *skeleton* 33
   aurivillii 33
   bacteriastroides 33
   borealis 33
   brevis 33

Chaetoceros (continued)
   buceros 33
   castracanei 33
   cinctus 33
   coarctatus 33
   compressus 33
   constrictus 34
   convolutus 34
   costatus 34
   crinitus 34
   curvisetus 34
   dadayi 34
   danicus 34
   debilis 34
   decipiens 34
   ——— f. *singularis* 34
   densus 34
   denticulatus 34
   ——— v. *angusta* 34
   diadema 34
   dichaeta 34
   didymus 34
   ——— v. *anglica* 34
   ——— f. *protubernas* 34
   dipyrenops 34
   distans 34
   diversus 34
   eibenii 34
   filformis 34
   hirundinellus 34
   holsaticus 34
   indicum 34
   knipowitschi 34
   laciniosus 34
   laevis 34
   lauderi 34
   lorenzianus 34
   ——— f. *subsalinus* 34
   messanensis 34
   mitra 34
   muelleri 34
   nipponica 34
   paradox 34
   pelagicus 34
   pendulum 34
   perpusillus 34
   peruvianus 34
   ——— f. *robusta* 34
   pseudoaurivillii 34
   pseudocurvisetus 34
   pseudodichaeta 34
   radians 34
   radicans 34
   rostratus 34
   ——— f. *glandazii* 34
   seiracanthus 34
   setoensis 34
   [*seychellarum*] 33
   siamense 34
   similis 34
   socialis 34
   subsecundus 34
   subtilis 34
   teres 34
   tetrastichon 34

Index

Chaetoceros (continued)
   tortissimus 35
   vanheurcki 35
   weissflogii 35
   xishaensis 35
[*Chaetodermis penicilligerus*] 459
   spinosissimus 459
Chaetodiadema granulatum 392
   japonicum 392
Chaetodon adiergastos 437
   argentatus 437
   auriga 437
   auripes 437
   baronessa 437
   [*bellamaris*] 437
   bennetti 437
   chrysurus 437
   citrinellus 437
   collare 437
   ephippium 437
   falcula 437
   [*fasciatus*] 437
   guencheri 437
   kleini 437
   lineolatus 437
   [*longirostris*] 437
   lunula 437
   melanotus 437
   meyeri 437
   miliaris 437
   modestus 437
   nippon 437
   octofasciatus 437
   ornatissimus 437
   plebeius 437
   punetatofasciatus 437
   quadrimaculatus 437
   rafflesi 437
   reticulatus 437
   selene 437
   semeion 437
   speculum 437
   strigangulus 437
   triangulum 437
   trifascialis 437
   trifasciatus 437
   ulietensis 437
   unimaculatus 437
   vagabundus 437
   wiebeli 437
   xantharus 437
CHAETODONTIDAE 436
Chaetodontoplus chrysocephalus 437
   duboulayi 437
   melanosoma 437
   personifer 437
   septentrionalis 437
CHAETOGNATHA 385
CHAETOMIACEAE 130
Chaetomium globosum 130
Chaetomium sp. 130
Chaetomorpha aerea 145
   ——— f. *versata* 145
   antennia 145
   brachygona 145

Chaetomorpha (continued)
   linum 145
   media 145
   pachynema 145
   spiralis 145
CHAETOPTERIDAE 224
Chaetopterus variopedatus 224
Chaetozone maotienae 224
   setosa 224
   spinosa 224
Chaeturichthys hexanema 451
   stigmatias 451
CHALLENGERIDAE 69
Challengeron dioden 69
   willemoesii 69
Chama brassica 247
   divaricata 247
   dunkeri 247
   fraga 247
   iostoma 247
   jukesi 247
   lazarus 247
   lobata 247
   reflexa 247
   semipurpurata 247
[*Chamaesipho scutelliformis*] 320
CHAMAESIPHONALES 25
CHAMIDAE 247
Champia bififda 138
   japonica 138
   parvula 138
CHAMPIACEAE 138
Champsodon atridorsalis 445
   [*capensis*] 445
   guentheri 445
   snyderi 445
CHAMPSODONTIDAE 445
CHANIDAE 410
Chanos chanos 410
Chaperia acanthina 373
Chaperiopis beniensis 373
   hainanica 373
   pleuroaviculata 373
Characodoma latisinuatum 379
CHARADRIIDAE 467
CHARADRIIFORMES 466
Charadrius alexandrinus dealbatus 467
   asiaticus veredus 467
   dubius curonicus 467
   hiaticula placidus 467
   leschenaultii 467
   m. mongolus 467
Charonia hepatica 273
   tritonis 273
Charybdis acuta 358
   affinis 358
   anisodon 358
   annulata 358
   bimaculata 358
   callianassa 358
   [*cruciata*] 358
   feriatus 358
   hellerii 358
   hongkongensis 358
   japonica 358

Charybdis (continued)
   lucifera 358
   miles 358
   natator 358
   obtusifrons 358
   orientalis 358
   riversandersoni 358
   truncata 358
   vadorum 358
   variegata 358
   ——— *brevispinosa* 358
Chascanopsetta lugubris 457
Chasmagnathus convexus 365
Chasmichthys gulosus 451
Chasmocarcinops gelasimoides 363
CHAULIODONTIDAE 412
Chauliodus sloani 412
CHAUNACIDAE 462
Chaunax abei 462
   fimbriatus 462
Cheilea diaphana 268
   equestris 268
Cheilinus bimaculatus 439
   celebieus 439
   chlorurus 439
   diagrammus 439
   fasciatus 439
   mentalis 439
   oxycephalus 439
   oxyrhynchus 439
   rhodochrous 439
   triobatus 439
   undulatus 439
   unifasciatus 439
Cheilio inermis 439
Cheillonereis cyclurus 220
Cheilodipterus macrodon 429
   quinquelineatus 429
   [*quinquelineatus*] 429
   subulatus 429
Cheilopogon agoo 419
Cheiloporina haddoni 379
CHEILOPORINIDAE 379
Cheiloprion labiatus 443
Cheilosporum jungermannioides 135
CHEILOSTOMATA 371
Cheiraster niasicus 389
Cheiriphotis megacheles 328
Chelidonichthys kumu 454
   spinosus 454
Chelidonura fulvipunctata 287
   hirundinina 287
   inornata 287
   tsurugensis 287
Chelidoperca hirundinacea 425
   margaritifera 425
   pleurospilus 426
Cheliycypraea testudinaria
Chelmon rostratus 437
Chelonia mydas 463
Chelonibia patula 321
   ——— *dentata* 321
   testudinaria 321
CHELONIIDAE 463
Chelonodon patoca 460

Chelophyes appendiculata 174
   contorta 174
Chelycypraea testudinaria 271
Chelyosoma siboja 402
Chemnclzia abseida 284
   acosmia 284
   multigyra 284
Chenia cheni 200
CHENOPODIACEAE 151
Chenopodium acuminatum 151
   album 151
   ambrosioides 151
   glaucum 151
   serotinum 151
Chiasmodon niger 445
CHIASMODONTIDAE 445
[*Chiazacmea pygmaea*] 263
   ——— *lampanicola* 262
Chicoreus aculeatus 274
   [*adustus*] 274
   alabaster 274
   asianus 274
   axicornis 274
   banksi 274
   brevifrons 274
   brunneus 274
   cnissodus 274
   elliscrossi 274
   [*kawamurai*] 274
   laciniatus 274
   microphyllus 274
   orientalis 274
   [*palmarosae*] 274
   pliciferoides 274
   problematics 274
   [*problematics*] 274
   ramosus 274
   rosarius 274
   saulii 274
   superbus 274
   [*superbus*] 274
   torrefactus 274
Childonias hybrida swinhoei 468
   leucoptera 468
CHILODONELLIDAE 57
CHILOGUEMBELINIDAE 99
Chilomycterus affinis 461
   echinatus 461
   orbicularis 461
CHILOPHIURIDA 397
Chiloscyllium colax 405
   griseum 405
   [*indicum*] 405
   [*isabellum*] 405
   plagiosum 405
   punctatum 405
Chilostomella oolina 119
   cushmani 119
   ovoidea 119
CHILOSTOMELLACEA 119
CHILOSTOMELLIDAE 119
Chilostomelloides macrostoma 119
   plummerae 119
Chimaera jordani 409
   phantasma 409

CHIMAERIDAE 409
CHIMAERIFORMES 409
Chinochthamalus scutelliformis 320
Chion dysoni 250
   semigranosus 250
   ticaonicus 250
Chiridiella macrodactyla 307
Chiridius gracilis 307
   poppei 307
Chiridota rigida 389
   stuhlmanni 389
CHIRIDOTIDAE 389
CHIROCENTRIDAE 411
Chirocentrus dorab 411
   nudus 411
Chirona amaryllis 321
   cristatus 321
   krugeri 321
   taiwanensis 321
   tenuis 321
Chironomus sp. 299
CHIROSTYLIDAE 350
CHIROTEUTHIDAE 295
Chiroteuthis imperator 295
Chirundina streetsi 307
Chiton komaianus 238
CHITONIDAE 238
Chlamydalachus anguineus 404
CHLAMYDASELACIDAE 404
Chlamys albolineata 244
   asperulata 244
   farreri 244
   gloriosa 244
   irregularis 244
   jousseaumei 244
   larvatus 244
   lemniscata 244
   madreporarum 244
   nobilis 244
   [*pica*] 245
   [*solaris*] 245
   squamata 244
   [*teilhardi*] 245
   valdecostatus 244
CHLIDONIDAE 376
Chloea castanea 451
   mororana 451
   sarchynnis 451
Chloeia flava 226
   fusca 226
   inermis 226
   parva 226
   rosea 226
   violacea 226
Chlopsis fierasfer 416
   taiwanensis 416
CHLORANGIACEAE 144
Chlorinoides aculeatus 356
   harmandi 356
Chloris barbata 159
   formosana 160
   virgata 160
CHLOROCOCCALES 147
Chlorodesmis hildebrandtii 147
   sinensis 147

Chlorodiella barbata 360
   bidentata 360
   crispipleopa 360
   cytherea 360
   laevissima 360
   nigra 360
   xishaensis 360
CHLOROPHTHALMIDAE 413
Chlorophthalmus acutifrons 413
   agassizi 413
   albatrossis 413
   bicornis 413
   borealis 413
   nigromarginatus 413
   oblongus 413
CHLOROPHYTA 144
Chlorostoma argyrostoma 263
   nigerrima 263
   rustica 263
   [*rusticum*] 263
   xanthostigmata 263
Chlorotocella gracilis 344
Chlorotocoides spinicauda 344
Chlorotocus crassicornis 344
   incertus 344
   [*spinicauda*] 344
CHLOROXYBACTERIA 27
[*Chlorurus bicolor*] 441
   [*muricatus*] 441
Chnoospora implexa 142
   minima 142
CHNOOSPORACEAE
CHOANOLAIMIDAE 212
Choenea limicola 56
Choenicosphaera flammabunda 63
Choerodon anchorago 439
   azurio 439
   gymnogenys 439
   jordani 439
   melanostigma 439
   nectemblema 439
   pescadorensis 439
   quadrifasciatus 439
   robustus 439
   schoenleini 439
   zamboangae 439
Choerodonoides japonicus 439
Choeroichthys sculptus 424
CHONDRACANTHIDAE 316
Chondria armata 139
   crassicaulis 139
   dasyphylla 139
   expansa 139
   [*hapteroclada*] 139
   lancifolia 139
   repens 139
   ryukyuensis 139
   succulenta 139
   tenuissima 139
   xishaensis 139
CHONDRICHTHYES 404
[*Chondrococcus hornemannii*] 137
Chondrus ocellatus 137
Chone filicaudata 229
   infundibuliformis 230

Index

Chone (continued)
    teres 230
Choniognothus reini 356
Chorda filum 143
CHORDACEAE 143
[*Chordaria firma*] 141
CHORDARIACEAE 141
CHORDARIALES 141
CHORDATA 403
Choriaster granulatus 390
Choridactylus multibarbis 454
Chorinemus formosanus 430
    hainanensis 430
    lysan 430
    moadetta 430
    orientalis 430
    sanctipetri 430
    tala 430
    tolooparah 430
Chorististium japonicus 426
    latifasciata 426
    lunulatum 426
Chorizopora bronguiartii 377
CHORIZOPORIDAE 377
Chorlogloea lutea 25
Chrionenma chlorotaenia 444
    chryseres 444
Chromadora quadrilinea 211
CHROMADORIDA 211
CHROMADORIDAE 211
Chromadorina miro 211
Chromis alleni 443
    analis 443
    atripectoralis 443
    atripes 443
    caeruleus 443
    chrysura 443
    cinerascens 443
    dimidiatus 443
    elerae 443
    flavomaculata 443
    fumea 443
    [*isharae*] 443
    lepidolepis 443
    magaritifer 443
    mirationis 443
    notatus 443
    ovatiformis 443
    retrofasciata 443
    ternatensis 443
    vanderbilti 443
    viridis 443
    weberi 443
    xanthochir 443
    xanthura 443
    xanthurus 443
Chromobacterium 21
[CHROMODORIDIDAE] 290
[*Chromodoris alba*] 290
    [*alderi*] 290
    [*atromarginata*] 290
    [*collingwood*] 290
    elizabethina 290
    [*festiva*] 290
    fidelis 290

Chromodoris (continued)
    geometrica 290
    [*inornata*] 290
    [*lineolata*] 290
    [*marginata*] 290
    [*obsoluta*] 290
    odhneri 290
    [*pallescens*] 290
    [*placida*] 290
    [*rubrocornuta*] 290
    sagamiensis 290
    [*shiranae*] 290
    [*sibogae*] 290
    striatella 290
    youngbleuthi 290
CHROOCOCCACEAE 24
CHROOCOCCALES 24
Chroococcus limneticus v. *subsalsus* 24
    membraninus 24
    minor 24
    turgidus 25
    varius 25
Chroothece littorinae 25
Chrysalidinella dimorpha 108
    earlandi 108
    fijiensis 108
    pacifica 108
Chrysanthemodiscus floriatus 28
Chrysaora helovola 176
[*Chrysiptera biocellatus*] 442
    [*cyaneus*] 442
    glauca 443
    [*glaucus*] 442
    leuconoma 443
    rex 443
    starcki 443
    [*uniocellatus*] 442
    [*zonatus*] 443
Chrysochir aureus 431
CHRYSOMONODALES 50
CHRYSOPETALIDAE 218
Chrysopetalum ehlersi 218
    occidentale 218
CHRYSOPHYCEAE 49
CHRYSOPHYTA 49
Chrysopogon orientalis 160
Chrysostoma paradoxum 263
Chrysymenia wrightii 138
CHTHAMALIDAE 320
Chthamalus antennatus 320
    challengeri 320
    [*dalli*] 320
    intertextus 320
    malayensis 320
    moro 320
    sinensis 320
    [*withersi*] 320
Chuniphyes moserae 175
    multidentata 175
CHYDORIDAE 301
Chydorus ventricosus 301
Chymenura aciculata 225
    longicaudata 225
Cibicidella variabilis 115
Cibicides artemi 114

Cibicides (continued)
    beatus 114
    borislavensis 114
    boueanus 114
    bradyi 114
    celebrus 114
    compressus 114
    cookei 114
    cunobelini 114
    cushmani 114
    decoratus 114
    deprimus 114
    deusseni 114
    djaffaensis 114
    dorsitubera 114
    entendus 114
    farafraensis 114
    fletcheri 114
    howelli 114
    laxus 114
    lectus 114
    lobatulus 114
    lobatus 114
    margaritiferus 114
    mayori 114
    mendesi 114
    mollis 114
    multicameratus 114
    nagaoi 114
    praecursorius 114
    praecinctus 114
    pseudoungerianus 114
    punjabensis 115
    refulgens 115
    reinholdi 115
    robertsonianus 115
    robustus 115
    rugosus 115
    sintikuensis 115
    subhaidingeri 115
    succedens 115
    tallahattensis 115
    tani 115
    tenuimargo 115
    umbonatus 115
    ungerianus 115
    wuellerstorfi 115
CIBICIDIDAE 114
Cibicidina expansus 115
    minuta 115
    patinaris 115
    platumbilica 115
Cibicidoides amygdaliformis 113
    anomalos 113
    bellus 113
    bradyi 113
    compressa 113
    haitiansis 113
    hyalinus 113
    midwayensis 113
    mundulus 113
    ovaliformis 113
    phaseoliformis 113
    pseudoungerianus 113
    robertsonianus 113

Cibicidoides (continued)
   subplanospirolus 113
   succedens 113
   temperatus 113
   ungerinus 113
Cicindela anchoralis 300
CICINDELIDAE 300
Ciconia ciconia boyciana 465
   nigra 465
CICONIIDAE 465
CICONIIFORMES 465
CIDARIDAE 392
CIDAROIDEA 392
Cigclisula occlusa 380
   turrids 380
Ciliata pacifica 420
CILIOPHORA 56
Cinachyra porosa 165
   pseudobarbata 165
Cinclopyramis infundibulum 67
Cinereus group 24
Cingulina cingulata 284
Cinnamomum camphora 152
Ciona intestinalis 402
CIONIDAE 402
Circe corrugata 254
   hongkongensis 254
   scripta 254
   stutzeri 254
   sulcata 254
   tumefacia 254
Circeis spirillum 231
Circellobates venustus 298
Circlotoma venusta 264
Circoporus oxyacanthus 69
CIRCOPORIDAE 69
Cirolana albida 325
   cubensis 325
   harfordi japonica 325
   japonensis 325
   minuta 325
   sphaeromiformis 325
CIROLANIDAE 325
Cirratiscala irregularis 267
CIRRATULIDAE 224
Cirratulus annamensis 224
   chrysoderma 224
   *cirratus*224
   filiformis 224
Cirrenalia macrocephala 131
   pseudomacrocephala 131
   pygmea 131
Cirrenalia sp. 131
Cirrhigaleus barbifer 407
Cirrhilabrus cyanopleura 439
   [*diagrammus*] 439
   exquistitus 439
   melanomarginatus 439
   rubimarginatus 439
   solorensis 439
   temminckii 439
Cirrhimuraena chinensis 417
Cirrhitiara simplex 167
Cirrhitichthys aprinus 444
   aureus 444

Cirrhitichthys (continued)
   falco 444
   oxycephalus 444
CIRRHITIDAE 444
Cirrhitus pinnulatus 444
Cirrholovenia polynema 170
   tetranema 170
CIRRHOLOVENIIDAE 170
CIRRHOSCYLLIIDAE 405
Cirrhoscyllium expolitum 405
Cirriformia capensis 224
   chefooensis 224
   dasylophia 224
   filigera 224
   puctata 224
   semicincta 224
   tentaculata 224
Cirripathes anguinis 186
   hainanensis 186
   musculosis 186
   rumphii 186
   sinensis 186
   spiralis 186
Cirripectes castaneus 446
   filamentosus 446
   fuscoguttatus 446
   imitater 446
   perustus 446
   polyzona 446
   quagga 446
   variolosus 446
CIRRIPEDIA 318
Cirrophorus branchiatus 223
   furcatus 223
   neapolitanus pacificus 223
Cirsotrema perplexum 267
   varicosum 267
Cistopus indicus 296
Cistula lorenziana 36
CITHARIDAE 456
Citharoides macrolepidotus 456
   [*macrolepis*] 456
Cladiella humesi 187
   krempfi 187
   madagascarensis 187
   subtilis 187
Cladium chinense 161
Cladococcus indicus 64
   scoparius 64
   sp. cf. *stalactites* 64
CLADOCERA 300
Cladolabes aciculus 387
   crassus 387
   schmeltzi 387
[*Cladonema mayersi*] 167
   radiatum 167
CLADONEMATIDAE 167
Cladophora albida 145
   aokii 145
   boodleoides 145
   catenata 145
   [*composita*] 146
   cymopoliae 145
   delicatula 145
   divergens 145

Cladophora (continued)
   echinus 145
   fascicularis 145
   flexuosa 145
   [*fuliginosa*] 145
   glaucescens 145
   gracilis 145
   [*montagnei*] 146
   montagnei v. *radicans* 145
   ohkuboana 145
   oligoclada 145
   patentiramea 145
   rudolphiana 145
   rugulosa 145
   saviniana 145
   scitula 145
   sibogae 145
   socialis 145
   stimpsonii 145
   utriculosa 145
   wrightiana 145
CLADOPHORACEAE 145
CLADOPHORALES 145
CLADOPYXIDACEAE 53
Cladopyxis brachiolata 53
Cladosporium cladosporiodes 131
Cladosporium sp. 131
Clanculus denticulatus 263
   margaritarius 263
   stigmatarius 263
   [cf. *stigmatarius*] 263
Clangula hyemalis 466
Clathria lipochela 165
CLATHRIIDAE 165
Clathrocanium coarctatum 67
   diadema 67
Clathrocyclas alcmenae 68
   cf. *coscinodiscus* 68
   danaes 68
Clathrodrillia jeffreysii 281
Clathurella subauriformis 281
[*Claudea batanensis*] 139
Claudiconcha japonica 256
CLAUSIDIIDAE 314
Clausinella calophylla 254
   isabellina 254
   tiara 254
Clausocalanus arcuicornis 307
   farrani 307
   furcatus 307
   laticeps 307
   mastigophorus 307
   minor 307
   parapergens 307
   paululus 307
   pergens 307
   [*pergens*] 307
Clausophyes galeata 175
   ovata 175
CLAUSOPHYIDAE 175
Clavadoce nigrimaculata 215
   [*splendida*] 215
CLAVAGELLIDAE
Clavatospora bulbosa 131
[*Clavatula pseudoprinciplis*] 281

Index

Clavellodes macrotrachelus 316
Clavellopsis appendiculata 316
   sargi 316
CLAVIDAE 166
Clavulina crustata 80
   difformis 80
   multicamerata 80
   pacifica 80
Clavulinoides szaboi 80
Cleantiella isopus 326
Cleantis planicauda 326
Cleidochasma bassleri 382
   biavicularium 382
   fallax 382
   porcellanum 382
   protrusum 382
CLEIDOCHASMATIDAE 382
Cleisthenes herzensteini 457
Cleistostoma dilatatum 364
Clementia papyracea 254
   vatheleti 254
Clerodendron inerme 157
Clianella brucei 325
Clibanarius aequabilis 348
   arethusa 348
   bimaculatus 348
   clibanarius 348
   corallinus 349
   cruentatus 349
   eurysternus 349
   formosus 349
   humilis 349
   inaequalis 349
   infraspinatus 349
   japonicus 349
   longitarsus 349
   multipunctatum 349
   padavensis 349
   striolatus 349
   virescens 349
Clidoderma asperrima 457
Climacodium biconcavum 32
   frauenfeldianum 32
Climacosphenia elongata 43
   moniligera 43
CLIMACOSTOMIDAE 58
CLINIDAE 446
Clinocardium californiense 248
Cliona celata 164
   vastifica 164
CLIONIDAE 164, 288
Clistocoeloma merguiensis 365
   sinensis 365
Clithon chlorostoma 264
   corona 264
   ——— *angulosa* 264
   faba 264
   oualaniensis 264
   retropictus 264
   sowerbianus 264
[*Clodophoropsis fasciculatus*] 146
   herpestica 146
   sundanensis 146
   vaucheriaeformis 146
   zollingeri 146

Cloeoascaris trichiuri 209
Cloeosiphon aspergillus 235
Clorida clorida 368
   decorata 368
   japonica 368
   latispina 368
   latreillei 368
   microphthalma 368
   rotundicauda 368
   verrucosa 368
Cloridopsis aquilonaris 368
   scorpio 368
Clupanodon punctatus 410
   thrissaa 410
Clupea harengus pallasi 410
CLUPEIDAE 410
CLUPEIFORMES 410
[*Clymene affinis*] 225
   [*watsoni*] 225
Clymenella cincta 225
   complanata 225
Clypeaster miniaceus 394
   reticulatus 394
   ——— *sundaicus* 394
   virescens 394
CLYPEASTEROIDA 394
Clypemorus bifasciatus 267
   humilis 267
   moniliferum 267
   morus 267
   petrosus 267
   trailli 267
   zonatus 267
Clytemnestra rostrata 314
   scutellata 314
CLYTEMNESTRIDAE 314
Clytia ambiguum 172
   cylindrica 172
   discoida 172
   [*edwardsi*] 172
   folleata 172
   hemispherarica 172
   hexacanalis 172
   languida 172
   malayense 172
   mccradyi 172
   [*minuta*] 172
   ovalis 172
   stechowi 172
   uchidai 172
Cnemidocarpa areolata 403
   chinensis 403
Cnemidocarpa sp. 403
[CNIDARIA] 166
Cnidocodon xiamenensis 168
[*Cnidoscyphyus torresi*] 172
COCCIDAE 300
COCCONEIACEAE 44
Cocconeis britannica 44
   costata 44
   diminuta 44
   dirupta 45
   ——— v. *africana* 45
   ——— v. *flexella* 45
   diruptoides 45

Cocconeis (continued)
   disculoides 45
   distans 45
   ——— v. *bahusiensis* 45
   fasciolata 45
   fluminensis 45
   heteroidea 45
   ——— v. *curvirotunda* 45
   lyra 45
   molesta 45
   notata 45
   pediculus 45
   pellucida 45
   ——— v. *minor* 45
   pelta v. *sinica* 45
   placentula 45
   ——— v. *euglypta* 45
   ——— v. *lineata* 45
   problematica 45
   pseudomarginata 45
   ——— v. *formosa* 45
   ——— v. *intermedia* 45
   scutellum 45
   ——— v. *japonica* 45
   ——— v. *minutissima* 45
   ——— v. *parva* 45
   ——— v. *stauroneiformis* 45
   sublittoralis 45
   subtilis 45
   tenuistriata 45
   versicolor 45
   vitrea 45
Coccorella atrata 415
Cociella cricodilus 455
Cocos nucifera 161
Coculus orbiculatus 152
   [*trilobus*] 152
Codakia punctata 247
   tigerina 247
CODIACEAE 147
Codium adhaerens 147
   arabicum 147
   bartlettii 147
   cylindricum 147
   divaricatum 147
   fragile 147
   [*geppei*] 147
   geppii 147
   intricatum 147
   [*mucronatum*] 147
   ovale 147
   papillatum v. *hainanense* 147
   repens 147
   taitense 147
CODONELLIDAE 58
Codonellina biformis 379
   montferrandii 379
   obtusata 379
CODONELLOPSIDAE 59
Codonellopsis hainanensis 59
   mobilis 59
   ostenfeldi 59
   pehaiensis 59
   rodunda 59
Codonocera elongata 301

Codonocera (continued)
   goniacantha 301
   mortenseni 301
   polygonia 301
   pusilla 301
   stellifera 301
Codonophilus trigoncephalus 325
Coecella chinensis 250
   turgida 250
Coelachne simpliciuscula 160
Coelarthrum boergesenii 138
   [*coactum*] 138
COELENTERATA 166
Coeliodidymocystis abdominalis 199
Coelomactra antiquata 249
Coelopleurus maculatus 392
Coeloseris mayeri 183
Coelothrix irregularis 138
Coenobita brevimanus 349
   cavipes 349
   [*clypeatus*] 349
   perlatus 349
   rugosus 349
COENOBITIDAE 349
COENOBITOIDEA 348
COHNILEMBIDAE 57
Cohnilembus pusillus 57
   reesi 57
Coilia ectenes 410
   grayi 410
   mystus 410
Coitocaecum sigani 197
Coldenia procumbens 157
Coleaspis cronata 61
COLEOPTERA 300
COLLEMBOLA 300
[*Collichthys fragilis*] 431
   lucidus 431
   niveatus 431
Collinsiella cava 144
   tuberculata 144
COLLINSIELLACEAE 144
Collisella cellanica 262
   dorsuosa 262
   formosa 262
   heroldi 262
   kolarovai 262
   luchuana ericae 262
   mortoni 262
   [*schrenckii*] 263
Collocalia fuciphaga 468
COLLOIDEA 62
Collonia donghaiensis 264
   pilula 264
Collosphaera armata 63
   brattstroimi 63
   elliptica 63
   gigantopora 63
   huxleyi 63
   macropora 63
   planca 63
COLLOSPHAERIDAE 62
COLLOZIDAE 62
Collozoum contortum 62
   inerme 62
   ovatum 62

Collybus drachme 431
COLOBOMETRIDAE 386
Colobonema igneum 173
   sericeum 173
Colochirus anceps 387
   quadrangularis 387
Coloconger scholesi 416
Colocyntotrema auxis 199
Cololabis saira 418
Colpomenia bullosa 142
   [*bullosa*] 142
   sinuosa 142
Colsyrnola buxeunnea 284
   ornata 284
Colubraria castanea 277
   compta 277
COLUBRARIIDAE 277
COLUBRIDAE 464
Colubrina asiatica 154
Columbella turturina 276
   varians 276
   versicolor 276
Comantheria delicata 386
   grandicalyx 386
   intermedia 386
   polycnemis 386
Comanthina schlegeli 386
Comanthus bennettic 386
   japonicus 386
   parvicirra 386
   pinguis 386
   solaster 386
Comaster distinctus 386
   gracilis 386
   multibrachiata 386
   multifidus 386
COMASTERIDAE 386
Comatella maculata 386
   stelligera 386
Comatula pectinata 386
   solaris 386
COMBRETACEAE 156
COMESOMATIDAE 211
Comissia peregrina 386
COMMELINACEAE 162
Compages mariae 384
COMPOSITAE 158
CONCHARIIDAE 69
Conchoceras caudatum 69
Conchoderma auritum 319
   hunteri 319
   virgata 319
   [*virgatum hunteri*] 319
Conchodytes monodactylus 340
   nipponensis 340
   tridacnae 340
Conchoecetes artificiosus 353
Conchoecetta acuminata 303
   giesbrechti 303
[*Conchoecia acuminata*] 303
   [*aequiseta*] 304
   [*ametra*] 303
   [*antarotica*] 304
   [*atlantica*] 304
   [*barealis*] 303
   [*bispinosa*] 304

Conchoecia (continued)
   [*brachyaskos*] 304
   [*concentrica*] 305
   [*cophopyga*] 304
   [*curta*] 304
   [*daphnoides*] 303
   [*dasyophthalma*] 304
   [*decipiens*] 304
   [*dentata*] 305
   [*discophora*] 303
   [*dorsotuberculata*] 305
   [*echinata*] 305
   [*elegans*] 303
   [*giesbrechti*] 303
   [*glandulosa*] 304
   [*haddoni*] 304
   hyalophyllum 303
   [*imbricata*] 303
   [*incisa*] 304
   [*inermis*] 305
   [*kampta*] 304
   [*kyrtophora*] 304
   [*leptothrix*] 303
   lophura 303
   ———— *lissoides* 303
   [*loricata*] 304
   macrocheira 303
   [*macromma*] 304
   magna 303
   ———— *rhombica* 303
   [*oblonga*] 305
   [*parthenoda*] 305
   parvidentata 303
   [*porrecta*] 305
   [*procera*] 305
   [*pseudoparthenoda*] 305
   [*pusilla*] 304
   [*reticulata*] 305
   [*rhynchena*] 305
   [*rotundata*] 304
   [*secernenda*] 304
   [*skogsbergi*] 304
   [*spinifera*] 305
   [*spinirostris*] 305
   [*striola*] 304
   subarcuata 303
   [*symmetrica*] 303
   [*tamensis*] 303
   [*valdiviae*] 303
Conchoecilla daphnoides 303
   ———— *minor* 303
Conchoecissa ametra 303
   imbricata 303
   plinthina 303
   symmetrica 303
Condyloderes sp. 208
Condylostoma magnum 58
   patens 58
CONDYLOSTOMATIDAE 58
Conescharellina angusta 382
   breviconica 382
   catella 382
   concava 382
   crassa 382
   elongata 382
   jucunda 382

Index

Conescharellina (continued)
   milleporacea 382
   obliqua 382
   ovalis 382
   papulifera 382
   rectilinea 382
   symmetrica 382
CONESCHARELLINIDEA 382
CONESCHARELLIOIDEA 382
Conger cinereus 416
   japonicus 416
   myriaster 416
CONGIOPODIDAE 455
CONGRIDAE 415
Congrina retrotincta 416
Congriscus megestomus 416
CONGROGADIDAE 447
Congrogadus subducens 447
Conicospirillinoides semidecoratus 81
CONIDAE 280
Conidens laticephalus 462
CONOCHILIDAE 214
Conochilus hippocrepis 214
Conopea calceopus 321
   canaliculatus 321
   cuneiformis 321
   cymbiformis 321
   granulatus 321
   navicula 321
   sinensis 321
Conopeum eriphorum 371
   hirtissimum 371
   [*reticulum*] 371
Conorbella pulvinata 112
[*Conterva utricularis*] 145
Contracaecum amoyensis 209
   clavatum 209
   marinum 209
   paralichthydis 209
   zenopsis 209
Contracaecum sp. 209
Conus achatinus 280
   aculeiformis 280
   arenatus 280
   armadillo 280
   [*aulicus*] 280
   auricomus 280
   australis 280
   betulinus 280
   bullatus 280
   cancellatus 280
   capitaneus 280
   caracteristicus 280
   catus 280
   chaldaeus 280
   comatosa 280
   concolor 280
   coronatus 280
   distant 280
   ebraeus 280
   eburneus 280
   episcopus 280
   eximius 280
   figulinus 280
   flavidus 280
   fulgetrum 280

Conus (continued)
   generalis 280
   geographus 280
   glans 280
   grangeri 280
   hirasei 280
   ichinoseana 280
   imperialis 280
   insculptus 280
   kinoshitai 280
   legatus 280
   leopardus 280
   litoglyphus 280
   litteratus 280
   lividus 280
   lynceus 280
   magnificus 280
   magus 280
   [*marmoreus*] 280
   ———— cf. *vidua* 280
   miles 280
   miliaris 280
   [*minimus*] 280
   mitratus 280
   [*monachus*] 280
   musicus 280
   mustelinus 280
   nussatella 280
   obscurus 280
   [*omaria*] 280
   orbignyi 280
   [*pagodus*] 280
   pennaceus 280
   pertusus 280
   planorbis 281
   praecellens 281
   pseudorbignyi 281
   pulicarius 281
   quercinus 281
   radiatus 281
   rattus 281
   recluzianus 281
   scabriusculus 281
   schepmani 281
   sieboldii 281
   [*sowerbii*] 281
   sponsalis 281
   stercusmuscarum 281
   striatus 281
   stupa 281
   stupella 281
   sugimotonis 281
   sulcatus 281
   teramachii 281
   terebra 281
   tessulatus 281
   textile 281
   tribblei 281
   tulipa 281
   urashimanus 281
   varius 281
   vexillum 281
   virgo 281
   vitulinus 281
   voluminalis 281
CONVOLVULACEAE 157

Conyza canadensis
[*Cookeolus boops*] 428
[COPELATA] 400
COPEPODA 306
COPIDOGNATHINAE 298
Copidognathus cephalocanthus 298
   cerberoideus 298
   consobrinus 298
   gracilunguis 298
   inconspicuus 298
   longiunguis 298
   neptuneus 298
   occultans 298
   paluster 298
   polyporus 298
Copilia lata 313
   longistylis 313
   mirabilis 313
   quadrata 313
   recta 313
   vitrea 313
Coradion chrysozonus 437
[*Coraliophia neritoida*] 276
   [*violacea*] 276
CORALLANIDAE 325
Corallina gracilis 135
   officinalis 135
   pilulifera 135
   ———— f. *filiformis* 135
   ———— f. *sororia* 135
   sessilis 135
CORALLINACEAE 134
Coralliobia deformis 276
   fimbriata 276
   monodonta 276
   violacea 276
   xoaruleia 276
Coralliocaris brevirostris 340
   graminea 340
   [*lucina*] 340
   superba 340
   [———— v. *japonica*] 340
   [*tridentata*] 341
   venusta 340
Coralliophaga coralliophaga 253
[*Coralliophila bulbiformis*] 276
CORALLIOPHILIDAE 276
Corbicula fluminea 254
CORBICULIDAE 254
Corbula fortisulcata 256
   [*tunicata*] 256
CORBULIDAE 256
Corbullella nanshaensis 373
Corculum cardissa 248
   mundum 248
Cordia subcordata 157
CORELLIDAE 402
Corbullela nanshaensis
   tenuirostre
Corethron hystrix 28
   pelagicum 28
Corina granulosa 313
[*Coris awayae*] 440
   aygula 439
   dorsomacula 439
   gaimard 439

Coris (continued)
   musume 439
   [*picta*] 439
   pictoides 439
   [*yamashiroi*] 440
Corispermum puberulum 151
CORIXIDAE 299
Cornuloculina inconstans 82
Cornuspira carinata 81
   crassisepta 81
   involvens 81
   planorbis 81
   selseyensis 81
CORNUSPIRACEA 81
CORNUSPIRIDAE 81
Cornuspiroides foliaceus 81
   lacunosus 81
Cornutella profunda 67
Corocalyptra cervus 68
Corolla ovata 288
Corollospora maritima 130
   pulchella 130
Coronaster volsellatus 392
CORONATAE 176
CORONIDAE 66
CORONIDIDAE 369
Coronidopsis bicuspis 368
   medius 368
Coronopus didymus 152
CORONULIDAE 321
Coronula diadema 321
COROPHIIDAE 328
Corophium acherusicum 328
   baconi 328
   crassicorne 328
   heteroceratum 328
   homoceratum 328
   hongkongensis 328
   insidiosum 328
   kitamorii 328
   lamellatum 328
   major 328
   monospinum 328
   mortonii 328
   sextonae miospinulosum 328
   sinense 328
   triangulapedarum 328
   tridentium 328
   uenoi 328
CORYCAEIDAE 313
Corycaeus affinis 313
   agilis 313
   andrewsi 313
   asiaticus 313
   calaninus 313
   carinata 313
   catus 313
   clausi 313
   concinnus 313
   crassiusculus 313
   dahli 313
   erythraeus 313
   flaccus 313
   furcifer 313
   gibbulus 313

Corycaeus (continued)
   giesbrechti 313
   lautus 313
   limbatus 313
   longicaudis 314
   longistylis 314
   lubbocki 314
   ovalis 314
   pacificus 314
   pumilus 314
   robustus 314
   rostratus 314
   speciosus 314
   subtilis 314
   typicus 314
   viretus 314
CORYMORPHIDAE 168
Coryne crassa 168
   pusilla 168
CORYNEBACTERIACEAE 22
CORYNEBACTERIACES 22
Corynebacterium 22
CORYNIDAE 168
Coryphaena hippurus 431
CORYPHAENIDAE 431
Coryphaenoides marginatus 420
   nasutus 420
[CORYPHAENOIDIDAE] 420
[CORYPHELLINIDAE]
Coryphostoma amygdalaeformis 109
   limbata 109
   lobata 109
CORYSTIDAE 358
Corythoichthys crenulatus 424
   fasciatus 424
   flavorasciatus 424
Coscinaraea exesa 183
Coscinasterias acutispina 392
Cosciniopsis lonchaea 378
   vestila 378
COSCINODISCACEAE 28
COSCINODISCALES 28
Coscinodiscus africanus 28
   agapetos 28
   angstii 28
   anguste-lineatus 28
   apiculatus 28
   ——— v. *ampigus* 28
   argus 28
   asteromphalus 28
   ——— v. *pulchra* 28
   ——— v. *subbuliens* 28
   bathyomphalus v. *hispidus* 28
   bipartitus 28
   blandus 28
   centralis 28
   cinctus 28
   concinnus 28
   confusus 28
   crenulatus 28
   curvatulus 28
   ——— v. *minor* 28
   debilis 28
   decrescens 28
   ——— v. *valida* 28

Coscinodiscus (continued)
   deformatus 28
   denarius 28
   ——— v. *sinensis* 28
   diversus 28
   divisus 28
   excentricus 28
   ——— v. *fasciculata* 28
   fimbriatus 28
   gigas 28
   ——— v. *praetexta* 28
   granii 28
   hainanensis 28
   hexagonus 28
   inclusus 28
   janischii 28
   jonesianus 29
   ——— v. *commutata* 29
   kuetzingii 29
   latimarginatus 29
   lentiginosus 29
   lineatus 29
   marginato-lineatus 29
   marginatus 29
   minor 29
   minutifasciculatus 29
   nitidus 29
   nobilis 29
   nodulifer 29
   obscurus 29
   oculatus 29
   oculus-iridis 29
   ——— v. *borealis* 29
   paradoxus 29
   perforatus 29
   ——— v. *cellulosa* 29
   ——— v. *pavillard* 29
   planithecus 29
   punctulatus 29
   radiatus 29
   reniformis 29
   rothii 29
   ——— v. *grandiuscula* 29
   ——— v. *normani* 29
   scitulus 29
   shantouensis 29
   spiculatus 29
   spinosus 29
   stellaris 29
   ——— v. *symbolophora* 29
   subaulacodiscoidalis 29
   sub-bulliens 29
   subconcavus 29
   ——— v. *tenuior* 29
   subtilis 29
   ——— v. *minorus* 29
   suspectus 29
   temperei 29
   tenuithecus 29
   thorii 29
   turgidus 29
   wailesii 29
   wittianus 29
Cosmonatus genkaiae 354
   grayii 354

Index

[*Cossura coasta*] 223
  longocirrata 223
Cossurella dimorpha 223
COSSURIDA 223
COSSURIDAE 223
Costalynia cf. *costulata* 265
Costasiella pallida 289
Costus speciosus 162
Cothurnia acuta 58
  calix 58
  ceramicola 58
  maritima 58
Cottapistus cottoides 455
COTTIDAE 456
Cottiusculus gonez 456
Courtoisia cyperoides 161
Coxliella annulata 59
COXLIELLIDAE 59
Cranchia scabra 294
CRANCHIIDAE 294
Crangon affinis 344
  [*bidentatus*] 345
  cassiope 344
  [*crangon*] 344
  [*hakodatei*] 344
  [*heterocarpus*] 342
  [*lowisi*] 344
CRANGONIDAE 344
CRANGONOIDEA 344
Craniella spatuliformis 165
[*Cranocephalus goesi*] 332
  scleroticus 332
Cranopsis pelex 262
Cranosina coronata 372
  latimandibutata 373
  philippinensis 373
Craspedochiton laqueatus 238
Craspidaster hesperus 390
Crassatella speciosa 248
CRASSATELLIDAE 248
Crassicutis karwarensis 195
Crassimarginatella crassimarginata 373
  japonica 373
  kumatae 373
  laguncula 373
  lunata 373
  nanshaensis 373
  sinica 373
Crassinarke dormitor 409
[*Crassipenaeus sinensis*] 336
Crassispira pseudoprinciplis 281
Crassostrea ariakensis 246
  gigas 246
  lineata 246
  rivularis 246
  vitrefacta 246
CRASSULACEAE 152
Cratena nigricolor 292
  serrata 292
CREEDIIDAE 445
Crenatula modiolaris 243
  nigrina 244
Crenavolva chinensis 271
  striatula traillii 271
[*Crenella sinica*] 241

Crenimugil crenilabis 424
Crenulilimopsis oblonga 241
Crepidacantha crinispina 380
  posissoni 381
CREPIDACANTHIDAE 380
Crepidula gravispinosa 268
  onyx 268
  [*walshi*] 268
Crepis longipes 376
Creseis acicula 288
  chierchiae 288
  clava 288
  virgula 288
  ———— v. *comica* 288
Creusia indicum 320
  [*spinulosa euspinulosa*] 320
  [*spinulosa pallida*] 320
Cribochalina chinensis 165
Cribraria chinensis 271
  [*chinensis*] 271
  cribraria 271
  [*tere*] 271
  teres 271
Cribrilina annulata 377
  biaviculariata 377
  flabellifera 377
  harmeri 377
  innominata 377
  radiata 377
  setosa 377
CRIBRILINIDAE 377
CRIBRILINOIDEA 377
CRIBRIMORPHA 377
Cribrobaggina reniformis 110
Cribrobigenerina reophaciformis 80
  robustiformis 80
  taiwanica 80
  textularioidea 80
Cribroelphidium articulatum 124
  neominutum 124
  pacificum 124
Cribroeponides cribroconcameratus 111
Cribrolinoides leizhouensis 82
Cribromiliolinella ericsoni 86
  striata 86
Cribrononion albiumbilicatulus 124
  asiaticum 124
  frigidum 124
  fronto 124
  gnythosuturatum 124
  heterocameratus 124
  incertum 124
  limpidus 124
  minutum 124
  pellucens 124
  poeyanum 124
  porisuturalis 124
  rhomboidale 124
  simplex 124
  subincertum 124
  vitreum 124
Cribrostomoides anomalinoides 73
  bradyi 73
  crassimargo 73
  nitidum 73

Cribrostomoides (continued)
  pseudocanariensis 73
  ringens 73
  robusta 73
  scitulum 73
  soldani 73
  spiculotesta 73
  subglobosum 73
  tenuis 73
  turgimentum 73
  weddellensis 73
  wiesneri 73
  yeni 73
CRINOIDEA 386
Crinum asiaticum v. *sinicum* 162
Criocarcinus superciliosus 356
Crisia crisioides 369
  cureata 369
  denticulata 369
  ebuneodenticulata 369
  elongata 369
  geniculata 369
  setosa 369
Crisidia cornuta 369
CRISIIDAE 369
Crisulipora occidentalis 370
Crocethia alba 467
Cromileptes altivelis 426
Cromyechinus antarctica 64
[*Cronia margriticola*] 275
Crossaster papposus 391
Crossorhombus azureus 457
  kanekonis 457
  kobensis 457
  valderostratus 457
Crossota brunnea 173
Crossotonotus gardineri 364
  spinipes 364
Croton crassifolius 154
  laui 154
Crouania attenuata 138
CRUCIFERAE 152
Cruciloculina asanoi 86
  japonica 86
  triangularis 86
CRUSTACEA 300
CRYPHCYSTOMORPHA 376
Cryptarachne okamurai 138
Cryptocentrus albidorsus 451
  cryptocentrus 451
  filifer 452
  gymnocephalus 452
  nigrocellatus 452
  papuanus 452
  pavontnoides 452
  russus 452
  strigilliceps 452
  yatsui 452
CRYPTOCHILIDAE 57
Cryptochilidium echini 57
  sigmoides 57
[*Cryptochirus hongkongensis*] 367
Cryptococcus albidus 129
  ates 129
  hungaricus 129

Cryptococcus (continued)
   infirmo-miniatus 129
   laurentii 129
   luteolus 129
   magnus 129
   uniguttulatus 129
Cryptocoelum opacum 192
Cryptodromia areolata 353
   canaliculata 353
   coronata 353
   dubia 353
   hilgendorfi 353
   planaria 353
   protuberum 353
CRYPTOGONIMIDAE 199
CRYPTOMONADACEAE 50
CRYPTOMONADALES 50
Cryptomonas baltica 50
Cryptomya busoensis 256
Cryptonema producta 254
CRYPTONEMIALES 134
Cryptopecten complanus 244
   nux 244
   tissotii 244
   vesiculosus 244
Cryptopenaeus sinensis 336
Cryptophthalmus smargdinus 285
CRYPTOPHYCEAE 50
CRYPTOPHYTA 50
Cryptoplax eleioti 238
   larvaeformis 238
   oculata 238
   striatus 238
Cryptopodia collifer 357
   fronicata 357
   sinica 357
Cryptopsaras couesi 462
Cryptosula pallasina 379
CRYPTOSULIDAE 379
Cryptocnemus obolus 354
Crystallophyes amygdalina 175
CRYTOCOCCACEAE 129
[*Ctena delicatula*] 247
Ctenamussium sinensis 244
Ctenocalanus vanus 307
Ctenocella pectinata 187
Ctenochatus binotatus 448
   striatus 448
Ctenocheles balssi 345
CTENODRILIDA 231
CTENODRILIDAE 231
Ctenodrilus serratus 231
Ctenogobius aurocingulus 452
   brevirostris 452
   criniger 452
   crocineus 452
   gymnauehen 452
   notophthalmus 452
   pflaumi 452
   puncticeps 452
   tangaroai 452
Ctenoides ales 245
   lischkei 245
CTENOPHORA 190
Ctenophoraster donghaiensis

Ctenopleura astropectinides 390
   ludwigi 390
   sinica 390
CTENOPTERYGIDAE 295
Ctenopteryx siculus 295
CTENOSTOMATA 370
Ctenotrypauchen chinensis 453
   microcephalus 453
Cubiceps squamicepoides 450
   squamiceps 450
[*Cuboides crystallus*] 175
   [*vitreus*] 175
CUBOMEDUSAE 176
CUBOSPHAERIDAE 63
Cubotholonium polystylum 66
Cubotholus regularia 66
Cucullaea labiosa granulosa 241
CUCULLAEIDAE 241
CUCULLANIDAE 209
Cucullanus amadai 209
   anguillae 209
   cirratus 209
   spirocaudus 209
CUCUMARIIDAE 387
Culcita novaeguineae 390
Culex sitiens 299
Culicia rubeola 185
CULICIDAE 299
Culicoides sp. 299
CULTELLIDAE 253
Cultellus attenuatus 253
   cultellus 253
   scalprum 253
CUMACEA 324
Cumella arguta 325
Cuneata arctica 72
Cunina frugifera 173
   octonaria 173
   peregrina 173
CUNINIDAE 173
Cupuladria quinensis 373
CUPULADRIIDAE 373
Curvemysella paula 248
Cushmanella primitiva 98
Cushmanina desmophora 96
[*Cuspidaria abyssicola nipoonica*] 258
   approximata 258
   caduca 258
   corrugata 258
   [*iridella*] 259
   japonica 258
   kawamurai 258
   kyushuensis 258
   macrorhynchus 258
   nobilis 258
   okezoko 258
   prolatissima 258
   steindachneri 258
CUSPIDARIIDAE
[CUTHONIDAE] 292
Cuvlerina columella 288
CYAMIDAE 334
Cyamus erratiaus 334
   ovalis 334
   scammoni 334

Cyanea capillata 176
   ferruginea 176
   nozakii 176
   purpurea 176
CYANEIDAE 176
Cyaneus group 24
CYANOBACTERIA 24
[CYANOPHYTA] 24
CYATHOCOTYLIDAE 204
Cyathodonta granulosa
CYATHOLAIMIDAE 211
CYBIIDAE 449
Cycladicama cumingi 247
   lunaris 247
   oblonga 247
   tsuchii 247
Cyclammina apenninica 74
   asanoi 74
   cancellata 75
   compressa 75
   pusilla 75
   reticulata 75
   samanica 75
   tani 75
   trullisata 75
CYCLAMMINIDAE 74
Cyclaspis linguiloba 324
Cyclasterope bisetosa 302
   fascigera 302
   hilgendorfi 302
Cyclastrum trifastifiatum 64
Cyclax spinicinctus 356
   suborbicularis 356
CYCLIDIDAE 57
Cyclidium amoyensis 57
   citrullus 57
   glaucoma 57
   ozakii 57
Cyclina sinensis 254
CYCLOMYARIA 400
Cycloclypeus carpenteri 126
Cycloes granulosa 355
Cycloforina contorta 83
Cyclograpsus granulatus 365
   [*granulatus*] 365
   incisus 365
   integer 365
   intermedius 365
   longipes 365
   lucidus 365
Cyclogyra carinata 81
   crassisepta 81
   involvens 81
   planorbis 81
   selseyensis 81
Cycloleberis americana 302
   biminiensis 302
   bradyi 302
   brevis 302
   poani 302
   similis 302
CYCLOPIDAE 314
Cyclopina heterospina 314
CYCLOPINIDAE 314
CYCLOPOIDA 312

Index

CYCLOPTERIDAE 456
CYCLORHAGIDA 208
[*Cyclorhynchus planirostris*] 343
Cyclosalpa affinis 401
   bakeri 401
   floridana 401
   pinnata 401
   ―――― *polae* 401
   [―――― *quadriluminis*] 401
   ―――― f. *quadriluminis* 401
   ―――― *sewelli* 401
   [*polae*] 401
   [*virgula*] 401
Cycloseris cyclolites 182
Cyclosorus interruptus 150
CYCLOSTOMATA 404
[CYCLOSTOMATA] 369
[*Cyclostrema fuscopiperata*] 265
CYCLOSTREMATIDAE 264
Cyclosunetta concinna 254
   menstrualis 254
Cyclotella caspia 29
   comta v. *oligactis* 29
   frigida 29
   ladogensis 29
   meneghiniana 29
   stelligera 29
   striata 29
   ―――― v. *ambigua* 29
   ―――― v. *baltica* 29
   ―――― v. *intermedia* 29
   ―――― v. *mucronulata* 29
   stylorum 29
Cyclotellina remies 250
Cyclothone acclinidens 411
   alba 411
   atraria 411
   obscura 411
   pallida 411
   pseudopallida 411
CYDIPPIDA 190
Cygnus c. cygnus 466
   columbianus 466
   olor 466
Cylichna biplicata 286
   [*involuta*] 286
   melampoides 286
   operosa 286
   pyramidata 286
   villica 286
Cylichnatys angustus 285
Cylindrobulla xishaensis 286
CYLINDROBULLIDAE 286
[*Cylindrocarpus rugosus*] 141
Cylindroclavulina bradyi 80
   elongata 80
   ovata 80
Cylindrolepas sinica 321
Cylindrotheca closterium 46
   ―――― v. *californica* 46
   gracilis 46
Cyllometra manca 386
Cymadusa vadosa 328
CYMATIIDAE 273
Cymatium aquatile 273

Cymatium (continued)
   caudatum 273
   [*caudatum*] 273
   cingulatum 273
   clandestinum 273
   comptum 273
   dunkeri 273
   echo 273
   encausticum 273
   exile 273
   flaveolum 273
   gemmatum 273
   grandimaculatum 273
   gutturnium 273
   kiiensis 273
   labiosum 273
   lotarium 273
   lotorium 273
   moniliferum 273
   mundum 273
   muricinum 273
   nicobaricum 273
   [*nicobaricum*] 273
   [*parthenopeum echo*] 273
   parthenopus 273
   [*pferifferia*] 273
   pferifferianus 273
   pileare 273
   [*pileare*] 273
   pyrum 273
   [*pyrum*] 273
   rubercula 273
   sarcostomum 273
   sinense 273
   sinensis 273
   tenuiliratum 273
   testudinarium 273
   tuberosum 273
   vespaceum 273
Cymatodiscus planetophorus 29
Cymatoneis sulcata 36
Cymatosira gibberula 43
   lorenziana 43
Cymatotheca minima 29
   weissflogii 29
   ―――― v. *densestriata* 29
[*Cymba crystallus*] 175
Cymbacephalus nematophthalmus 455
Cymbaloporella tabellaeformis 115
Cymbaloporetta bradyi 115
   solida 115
   squammosa 115
CYMBALOPORIDAE 115
CYMBELLACEAE 42
Cymbiola nobilis 279
Cymbium melo 279
Cymbopogon tortilis 160
Cymbovula deflexa 271
Cymbulia peroni 288
   tricavernosa 288
CYMBULIIDAE 288
[*Cymia mustabilis*] 276
Cymo andreossyi 360
   deplanatus 360
   melanodactylus 360

Cymo (continued)
   [*melanodactylus*] 360
   quadrilobataus 360
Cymodoce acuta 326
   japonica 326
   pubescens 326
Cymodocea rotundata 158
Cymolutes lecluse 439
   torquatus 439
CYMOTHOIDAE 325
Cynanchum insulanum 157
   ―――― v. *lineare* 157
Cynodon dactylon 160
CYNOGLOSSIDAE 458
Cynoglossus abbreviatus 458
   arel 458
   bilineatus 458
   brachycephalus 458
   gracilis 458
   interruptus 458
   itinus 458
   joyneri 458
   kopsi 458
   lida 458
   lighti 458
   lineolatus 458
   macrolepidotus 458
   melampetalus 458
   monopus 458
   nigropinnatus 458
   oligolepis 458
   puncticeps 458
   purpureomaculatus 458
   robustus 458
   roulei 458
   semilaevis 458
   sibogae 458
   sinicus 458
   suyeni 458
   trigrammus 458
   xiphoides 458
Cypassis irregularis 64
CYPERACEAE 160
Cyperus malaccensis 161
   ―――― v. *brevifolius* 161
   rotundus 161
   stoloniferus 161
Cyphastrea serailia 183
   zhongjianensis 183
CYPHINIDAE 64
Cyphocolpus virginis 64
[*Cypraea annulus*] 272
   [*arabica*] 272
   [*argus*] 272
   artuffeli 271
   [*asellus*] 272
   aurantium 271
   boivinii 271
   [*caputserpentis*] 272
   carneola 271
   [*caurica*] 272
   childreni 271
   chinensis 271
   contaminata 271
   [*cribraria*] 271

Cypraea (continued)
  [cylindrica] 272
  depressa 271
  eglantina 271
  [erosa] 272
  [errones] 272
  [felina] 272
  [fimbriata] 272
  [globulus] 272
  [gracilis] 272
  guttata 271
  [helvola] 272
  hirasei 271
  [isabella] 272
  [labrolineata] 272
  longfordi 271
  lulea 271
  lynx 271
  [mappa] 272
  [mauritiana] 272
  microdon 271
  [miliaris] 272
  [moneta] 272
  [onyx] 272
  pallidula 271
  [poraria] 272
  [pulchella] 272
  [punstata] 272
  [scurra] 272
  [staphylaea] 272
  [stolida] 271
  talpa 271
  teramachii 271
  [testudinaria] 271
  tigris 271
  [tigris] 271
  vitellus 272
  [vitellus] 272
  [walkeri] 272
  [ziczac] 272
Cypraecassis rufa 272
CYPRAEIDAE 271
Cypridina acuminata 301
  amphiacantha 301
  [bairdii] 302
  dentata 301
  inermis 301
  japonica 301
  lepidophora 301
  nami 301
  nana 301
  natans 301
  punctata 301
  serrata 301
  sinuosa 301
CYPRIDINIDAE 301
CYPRIDINIFOMES 301
Cypridinodes asymmetrica 302
  avis 302
  bairdi 302
  codonocera 302
  galatheae 302
  minuta 302
Cyprinocirrhites polyactis 444

Cypseluritrematoides microvaei 195
  minor 195
  triangularis 195
[Cypselurus agoo] 419
  [altipennis] 419
  angusticeps 419
  arcticeps 419
  atrisignis 419
  bahiensis 419
  [brevipinnis] 419
  brevis 419
  cyanopterus 419
  katoptron 419
  naresi 419
  nigripennes 419
  oligolepis 419
  opisthopus 419
  oxycephalus 419
  pinnatibarbatus japonicus 419
  poecilopterus 419
  simus 419
  speculiger 419
  spilonotopteus 419
  spilopterus 419
  suttoni 419
  unicolor 419
Cyrenobatissa subsulcata 254
Cyrineum bituberculare 273
Cyrtechinus verruculatus 393
CYRTOCALPIDAE 67
CYRTOIDEA 67
Cyrtomaia hispida 356
  owstoni 356
Cyrtomium falcatum 150
CYRTOPHORIDA 57
Cystammina pauciloculata 74
  spiculifera 74
CYSTONECTAE 173
[Cystophyllum caespitosum] 143
CYSTOSERIACEAE 143
CYTAEIDIDAE 166
Cytaeis tetrastyla 166
Cytophaga 22
CYTOPHAGACEAE 22
CYTOPHAGALES 22
Cytospora spp. 131
CYTTAROCYLIDAE 59
Cyttomimus affinis 423
Cyttopsis roseus 423

D

Dacryopiliumnus rathbunae 360
Dactyliosolen mediterraneus 30
Dactyloctenium aegyptiacum 160
Dactylopodia tisboides 315
Dactyloptena gilberti 456
  orientalis 456
Dactylopus dactylopus 447
DACTYLOPTERIDAE 456
Dactylospora haliotrepha 130
Dactylostomum epinepheli 197
Dactyosaster cylindricus 391
Dadayiella ganymedes 60

[Dakaria typica] 379
Dalatias licha 407
Daicocus peterseni 456
Daldorfia acuta 357
  horrida 357
DALLINIDAE 384
Dalmatella buaensis 25
Daira perlata 360
Dampieria cyclophthalmus 427
  lineata 427
  melanotaenia 427
Danielssenia typica 314
DAPHNIIDAE 300
Daphnia carinata 300
  longispina 300
  obtusa 300
  pulex 300
Daptonema alternum 211
  fissidens 211
  fistulatum 211
  maeoticum 211
DARCOPYLIDAE 65
Dardanus arrosor 349
  aspersus 349
  crassimanus 349
  deformis 349
  guttatus 349
  hessii 349
  imbricatus 349
  impressus 349
  lagopodes 349
  megistos 349
  pedunculatus 349
  scutellatus 349
  tinctor 349
  vulnerans 349
Dascyllus aruanus 443
  marginatus 443
  melanurus 443
  reticulatus 443
  trimaculatus 443
Dasson breviceps 446
  [japonicus] 446
  springeri 446
  trossulus 446
  variabilis 446
Dasya mollis 139
  pedicellata 139
  rigidula 139
  scoparia 139
  sessilis 139
  villosa 139
DASYACEAE 139
DASYATIDAE 408
Dasyatis akajei 408
  atratus 408
  bennetti 408
  gerrardi 408
  imbricatus 408
  kuhli 408
  laevigatus 408
  lata 408
  microphthalmus 408
  navarrae 408

Index

Dasyatis (continued)
  sinensis 408
  uarnak 408
  ushiei 408
  zugei 408
Dasybranchus caducus 225
  lumbricoides 225
[*Dasychone cingulata*] 229
DASYCLADACEAE 145
DASYCLADALES 145
Dasyopsis pilosa 139
Datura metel 157
Daviesina langhami 121
Davila plana 250
Daya jordani 443
Deania aciculata 407
Debaryomyces hansenii 129
  polymorphus 129
Decametra laevininna 386
  mylitta 386
DECAPODA 335
Decapterus fasciatus 430
  kurroides 430
  lajang 430
  macarellus 430
  macrosoma 430
  maruadsi 430
  muroadsi 430
  russelli 430
  tabl 430
Decatopecten striatus 245
Decemtestis asymmetricum 197
  cynoglossi 197
Decorifera globosa 287
  insignis 287
  matusimana 287
Dedignathus inermis 350
DELESSERIACEA 139
Delonovolva formosa 271
Delosina complanata 109
DELOSINACEA 109
DELOSINIDAE 109
Delphineis angustata 43
DELPHINIDAE 469
Delphinus delphis 469
  tropicalis 469
Deltocyathus lens 185
  murrayi 185
Demania baccalipes 360
  rotundata 360
  scaberrima 360
DEMATACEAE 130
Democrinus japonicus 386
Demonax microphthalmus 230
DEMOSPONGIAE 164
Dendostrea crenulifera 246
  folium 246
Dendritina pacifica 90
  striata 90
Dendrobeania longispinosa 375
  sinica 375
DENDROBRANCHIATA 335
DENDROCHIROTA 387
Dendrochirus bellus 453

Dendrochirus (continued)
  biocellatus 453
  zebra 453
Dendrocygna javanica 466
DENDRODORIDIDAE 291
Dendrodoris areolata 291
  denisoni 291
  elongata 291
  [*gemmacea*] 291
  guttata 291
  miniata 291
  nigra 291
  rubra 291
  tuberculosa 291
[*Dendronephya gigantea*] 188
  [*guggenheimi*] 188
  [*rubra*] 188
Dendronereis pinnaricirris 220
Dendronina arborescens 70
DENDRONODIDAE 291
Dendronotus arborescens 291
Dendrophrya ramosa 70
DENDROPHYLLIDAE 183
DENDROPHYLLIIDAE 185
Dendrorhynchus sinensis 207
  zhanjiangensis 208
[*Dendrostoma signifer*] 236
[*Dendrostomum spinulum*] 236
DENORASTERIDAE 394
Dentalina advena 90
  basiplanata 90
  beyrichana 90
  californica 90
  catenulata 90
  communis 90
  consobrina 90
  cuvieri 90
  decepta 90
  emaciata 90
  extensa 90
  farcimen 90
  filiformis 90
  frobisherensis 90
  gracilis 90
  guttifera 91
  insecta 91
  intorta 91
  kreyenhagensis 91
  monroei 91
  mucronata 91
  mutsui 91
  nasuta 91
  roemeri 91
  sidebottomi 91
  subsoluta 91
  tortilis 91
  vertebralis 91
  vistulae 91
DENTALIIDAE 261
Dentalium obtusum 261
  octangulatum 261
  sinuosum 261
Denticula subtilis 46
Dentostomina agglutinans 83

Depressiscala aurita 267
Derbesia marina 146
  [*ryukyuensis*] 146
DERBESIACEAE 146
Dermatobranchus marginlatus 292
  multistriatus 292
  ornatus 292
  tongshanensis 292
Dermatolithon corallinae 135
  tumidulum 135
Dermatomyzon nigripes 315
Dermocarpa fucicola 25
  prasina 25
  solitaria 25
  sphaerica 25
  violacea 25
DERMOCARPACEAE 25
DERMOCHELYIDAD
Dermochelys coriacea
Dermonema frappieri 133
  [*gracilis*] 133
  pulvinata 133
Derogenes bohaiensis 200
  chelidonichthydis 200
  epinepheli 200
  gadi 200
  macrostoma 200
  magnus 200
  minoi 200
  varicus 200
Deropristis paurosoma 195
Derris trifoliata 153
Desmarestia viridis 143
DESMARESTIACEAE 143
DESMARESTIALES 143
Desmechinus anomalus 393
  rufus 393
  versicolor 393
Desmodema polystictum 423
Desmodium gangeticum 153
  podocarpum oxyphyllum 153
  rubrum 153
DESMODORIDA 211
DESMODORIDAE 211
Desmolaimus zealandicus 211
[DESMOMYARIA] 401
DESMOPHYCEAE 50
Desmophyllum alabastrum 185
DESMOPTERIDAE 288
Desmopterus papilio 288
DESMOSCOLECIDA 211
DESMOSCOLECIDAE 211
Desmoscolex americanus 211
Desmospyris stabilis 66
Desmostachya bipinnata 160
Desulfovibrio desulfuricans 21
DESULFOVIBRIONACEAE 21
DESULFOVIBRIONALES 21
DEUTEROMYCOTINA 131
Devonia semperi 248
DEXAMINIDAE 329
Dexiospira alveolatus 231
  foraminosus 231
Diacria quadridentata 288

Diacria (continued)
    ——— v *costata* 288
    trispinosa 288
Diadema savignyi 392
    setosum 392
DIADEMATIDAE 392
Dianella ensifolia 162
Diaperoecia radiata 370
    scalaris 370
DIAPEROECIIDAE 370
Diaphanosoma brachyurun 300
    leuchtenbergianum 300
Diaphus aliciae 413
    brachycephalus 414
    burtoni 414
    chrysorhynchus 414
    coeruleus 414
    diadematus 414
    diademophilus 414
    fragilis 414
    fulgens 414
    garmani 414
    holti 414
    jenseni 414
    lucidus 414
    luetkeni 414
    malayanus 414
    megalops 414
    mollis 414
    parri 414
    perspicillatus 414
    problematicus 414
    regani 414
    richardsoni 414
    sagamiensis 414
    signatus 414
    similis 414
    splendidus 414
    suborbitalis 414
    taaningi 414
    tanakae 414
    termophilus 414
    umbroculus 414
    watasei 414
Diaporthe salsuginosa 130
DIAPTOMIDAE 310
Diaseris fragilis 184
Diasterina odontacantha 391
Diasterope bisetosa 302
DIASTOPORIDAE 369
DIASTYLIDAE 324
Diastylis tricincta 324
Diatoma hyalinum 43
DIATOMACEAE 43
DIATOMALES 43
Diatrype sp. 130
DIAZONIDAE 402
Dibranchus japonicus 462
Diceratias bispinosus 462
DICERATIIDAE 462
DICHELESTHIIDAE 316
Dichelyne longispiculus 209
Dichiloseccus cheloniae 195
Dichothrix bornetiana 26
    fucicola 26

Dichothrix (continued)
    olivacea 26
    penicillata 26
    seriata 26
    zhongjianensis 26
Dichotomia cannoides 170
Dichrometra doederleini 386
    flagellata 386
DICLIDOPHORIDAE 192
Dicodonium jeffersoni 168
DICOTYLEDONEAE 150
DICRANEMACEAE 137
Dicrolene quinquarius 421
    tristis 421
Dicropleon morator 326
Dictyacantha tetragonopa 61
DICTYACANTHIDAE 61
Dictyaspis furcata 61
Dictyoceras prismaticum 68
    virchowii 68
DICTYOCERATIDA 164
Dictyocha fibula 49
    ——— v. *aculeata* 49
    ——— v. *messanensis* 49
    ——— v. *pentagona* 49
    ——— v. *stapedia* 49
    pseudofibula v. *complex* 49
DICTYOCHACEAE 49
Dictyocoryne profunda 65
    tricupiformis 65
    trimaculatum 65
    truncatum 65
Dictyoneis marginata 36
    thumii 36
Dictyophimus arabicus 67
Dictyopteris divaricata 142
    latiuscula 142
    prolifera 142
    repens 142
    undulata 142
Dictyosiphon foeniculaceus 141
DICTYOSIPHONACEAE 141
DICTYOSIPHONALES 141
Dictyosoma burgeri 447
Dictyosphaeria bokotensis 145
    cavernosa 145
    fujianensis 145
    intermedia 145
    spinifera 145
    versluysii 145
Dictyosporium pelagicum 131
Dictyosporium sp. 131
Dictyosquilla foveolata 368
Dictyota cervicornis 142
    dichotoma 142
    divaricata 142
    friabilis 142
    indica 142
    linearis 142
    liqulata 142
    patens 142
    [*radicans*] 142
DICTYOTACEAE 142
DICTYOTAKES 142
DIDEMNIDAE 402

Didemnum areolatum 402
    aspersum 402
    fuscum 402
    granulatum 402
    membranaceum 402
    moseleyi 402
    tonga 402
Didimacar tenebrica 240
DIDINIIDAE 56
Didontoglossa decoratoides 286
    koyasensis 286
Didymocystis bifurcata 199
    crassa 199
    philobranchia 199
    radiatus 199
    soleiformis 199
    wedli 199
Didymocystoides buccalis 199
    dolichorchis 199
    xishaensis 199
DIDYMOZOIDAE 199
Didymozoon biramus 199
    gigas 199
    longicollopsis 199
    platycephali 199
    pneumatophori 199
    priacanthus 199
    scomberomori 199
    sparial 199
Didymozoum simplex 375
Diffalaba picta 267
DIGENEA 192
Digenea simplex 139
Digitaria heterantha 160
    ——— *aequabilis* v. *holosericea* 160
DILLENIACEAE 155
Dilophus okamurae 142
    [*maginatus*] 142
Dimeregramma fulvum 43
    fusiformis 43
    opulens 43
Dimophyes arctica 174
Dimorphina mutsuensis 91
Dimorphocalyx poilanei 154
Dimorphostylis asiatica 324
Dimya japonica 246
DIMYIDAE 246
Dinatys monodonta
Dinematichthys dasyrhynchus 421
    iluocoeteoides 421
    minyomma 421
Dinobothrium septaria 206
DINOPHILIDAE 215
Dinophilus gyrociliatus 215
DINOPHYCEAE 51
DINOPHYSIACEAE 51
DINOPHYSIALES 51
Dinophysis acuminata 51
    caudata 51
    ——— v. *abbreviata* 51
    [*cuneus*] 51
    fortii 51
    hastata 51
    miles 51
    schuttii 51

Index

DINOPONTIIDAE 315
Dinosoma setipinae 200
Dinurus longissimus 200
   megnacetabalum 200
   scombri 200
Diodon bleekeri 461
   eydiuxii 461
   holacanthus 461
   hystrix 461
   iliturosus 461
   novemmaculatus 461
DIODONTIDAE 461
Diodora cruciata 262
   galeata 262
   mus 262
   proxima 262
   quadriradiata 262
   reevei 262
   suprapunicea 262
   ticaonica 262
Diogenes avarus 349
   custas 349
   deflectomanus 349
   edwardsii 349
   gordineri 349
   hainanica 349
   investigatoris 349
   paracristimanus 349
   rectimanus 349
   tormentosus 349
Diogenichthys atlanticus 414
   laternatus 414
   panurgus 414
DIOGENIDAE 348
Diomedea albatrus 464
   nigripes 464
DIOMEDEIDAE 464
DIONCHIDAE 192
Dionchus sp. 192
Diopatra amboinensis 226
   [*bilogata*] 226
   chiliensis 226
   cuprea 226
   dentata 226
   [*neapolitana*] 226
   neotridens 226
   oblique 226
   cf. *ornata* 226
   sugokai 226
   variabilis 226
Diophrys appendiculata 60
   oligothrix 60
   scutum 60
DIOSACCIDAE 315
Diphasia dubia 171
   heurteli 171
   minuta 171
   scalariformis 171
   thornelyi 171
Diphyes bojani 174
   chamissonis 174
   dispar 174
   [*subtiloides*] 175
   [*truncata*] 175
DIPHYIDAE 174

[*Diphyopsis bojani*] 174
   [*chamissonis*] 174
Diplacanthopoma brunnea 421
DIPLANGIDAE 195
Diplangus hainanensis 195
DIPLEUROSOMATIDAE 170
Diploastrea heliopora 183
Diplocirrus erythroporus 228
DIPLOCONIDAE 61
Diploconus fasces 61
   nitidus 61
Diplocyclas bicorona 68
Diplogrammus goramensis 447
   xenicus 448
Diplomeriza duplicata 281
Diploneis adonis v. *ganymedes* 36
   advena 36
   aestiva 36
   beyrichiana 36
   bomboides 36
   ——— v. *media* 36
   bombus 36
   ——— v. *bombiformis* 36
   borealis 36
   campylodiscus 36
   chersonensis 36
   chinensis 36
   coffaeiformis 36
   crabro 36
   ——— v. *excavata* 36
   ——— v. *omeari* 36
   ——— f. *pandura* 36
   ——— v. *subelliptica* 36
   ——— v. *suspecta* 36
   dalmatica 36
   debyi 36
   decipiens 36
   ——— v. *parallela* 36
   diplosticta 36
   eudoxia 36
   eugenia 36
   fusca 36
   ——— v. *pelagi* 36
   gemmata v. *pristiophora* 36
   gorjanovici v. *major* 36
   graeffii 36
   gruendleri 36
   guinardinana 36
   hospes 36
   incurvata 36
   inscripta 36
   interrupta 36
   ——— v. *caffra* 36
   lineata 36
   littoralis 36
   mediterranea 36
   nitescens 36
   notabilis 36
   novaeseelandiae 36
   papula 36
   parca 36
   puella 36
   schmidtii 36
   sejuncta 36
   serratula 36

Diploneis (continued)
   smithii 37
   ——— v. *constricta* 37
   ——— v. *dilatata* 37
   ——— v. *rhombica* 37
   ——— v. *rombica* 37
   splendida 37
   ——— v. *puella* 37
   subcincta 37
   suborbicularis 37
   vacillans 37
   vetula 37
   weissflogii 37
   zanzibarica 37
Diplopeltopsis minor 54
Diplopharyngotrema bipapillosa 199
   latelabraois 199
Diplophos pacifica 411
   taenia 411
Diploprion bifasciatum 426
Diploproctodaeum waki 195
Diplopsalis asymmetrica 54
   excentrica 54
   hainanensis 54
   lenticula 54
   ——— v. *lebourii* 54
   pingi 54
DIPLOSOLENIIDAE 191
Diplosphaera spinosa 64
Dipsacaster pretiosus 390
DIPTERA 299
DIPTEROCARPACEAE 155
Dipterygotus leucogrammicus 433
Dipurena ophiogaster 168
   strangulata 168
DIRETMIDAE 421
Diretmus argenieus 421
Disasterina odontacantha 391
Discammina compressa 73
DISCAMMINIDAE 73
Discanomalina coronata 120
   semipunctata 120
[*Dischistodus melanotus*] 444
   [*perspicilliatus*] 444
   [*prosopotaenia*] 444
   [*taeniurus*] 444
DISCINIDAE 384
Discinisca stella 384
Discoconchoecia discophora 303
   elegans 303
   pseudodiscophora 303
   tamensis 303
Discocyclina pygmaea 127
   sowerbyi 127
DISCOCYCLINIDAE 127
DISCODORIDIDAE 290
Discodoris concinna 290
   fragilis 290
   mauritiana 290
Discoetrema crinophila 462
Discogasteroides hainanensis 194
DISCOIDEA 64
DISCORBACEA 110
DISCORBIDAE 111
Discorbinella advena 113

Discorbinella (continued)
   bertheloti 113
   floridensis 113
   kerimbatica 113
   montereyensis 113
DISCORBINELLACEA 113
DISCORBINELLIDAE 113
Discorbis asterocides 111
   australis 111
   biaperturata 111
   bradyi 111
   bullatus 111
   candeianus 111
   chinensis 111
   distinctus 111
   granulosus 111
   latestoma 111
   mira 111
   minuta 111
   murrayi 111
   obvelatus 111
   orbicularis 111
   placoides 111
   rugosus 111
   subquatus 111
   subvesicularis 111
   taiwanensis 111
   tuberculatus 111
   vilardeboana 111
   williamsoni 111
Discors aurantiacum 248
   biradiatum 248
   multipunctatum 248
Disolenia quadrata 63
Dispinus ruber 422
Disseta palumboi 310
[*Dissodinium gerbautii*] 55
   lunula 55
Distephanus octonarius v. *octonarius* 49
   ———— v. *polyactis* 49
   speculum 49
   ———— v. *octonarium* 49
   ———— v. *septenarius* 49
Distolasterias elegans 392
   nipon 392
Distorsio anus 273
   perdistorta 273
   reticulata 273
Ditrema temmincki 438
Ditrupa arietina 230
Ditylum brightwelli 32
   sol 32
Doclea canalifera 356
   gracilipes 356
   japonica 356
   [*ovis*] 356
   sinensis 356
   tetraptera 356
Dodecaceria concharum 224
   fewkesi 224
Dodonaea viscosa 154
Doederleinia berycoides 426
Dolabella ecaudata 287
   scapula 287
Dolabrifera dolabrifera 287
   fusca 287

Dolichandrone spathacae 158
Dolichopteryx longipes 411
Doliodrilus tener 234
Dolioletta gegenbauri 400
DOLIOLIDAE 400
Doliolina mulleri krohni 400
   obscura 400
   separata 400
Dolioloides rarum 400
Doliolum denticulatum 400
   ———— *ehrenbergi* 400
   nationalis 400
[*Dolium salosum*] 274
   [*zonatum*] 274
Dollfustrema sinica 193
Domecia glabra 360
   hispida 360
DONACIDAE 250
[*Donax cuneatus*] 250
   [*faba*] 250
   [*incarnatus*] 250
   [*kiusiuensis*] 250
   [*nitidus*] 250
   [*semigranosus tropicus*] 250
DORATASPIDAE 61
Dorataspis choanopora 61
   gladiata 61
   loricata 61
   micropora 61
   ———— v. *collosa* 61
Dorcadospyris pentagona 66
DORIDIDAE 289
[DORIDIIDAE] 287
[*Doridium depictum* v. *minor*] 287
   [*gigliolii*] 287
   [*lineolata*] 287
DORIDOPSIDIDAE 290
Doridopsis aurantiaca 290
   granulosa 290
   viridis 290
[*Dorippe astuta*] 354
   callida 354
   [*dorsipes*] 354
   facchino 354
   frascone 354
   granulata 354
   histirio 354
   japonica 354
   polita 354
   [*quadridens*] 354
   sinica 354
   tenuipes 354
DORIPPIDAE 354
[*Doris areolata*] 291
   verrucosa 289
Dorisca nana 254
[*Dorodotes laevicarina*] 344
Dorometra aphrodite 387
Dorothia arenata 77
   exilis 77
   nagaoi 77
   nammalensis 77
   oxyconoides 77
   paupercula 77
   pseudoturris 77
   retusa 77

Dorothia (continued)
   scabra 77
   scrupulosa 77
   tortilis 78
[*Dorvillea japonica*] 228
   [*moniloceras*] 228
   cf. *pseudorubrovittata* 228
   [*rudolphi*] 228
DORVILLEIDAE 228
Dorylaimopsis rabalaisi 211
   turneri 212
[*Doryrhamphus boaja*] 424
   dactyliophorus 424
   excisus 424
   japonicus 424
   melanopleura 424
[*Doryteuthis bleekeri*] 295
Dosinia amphidesmoides 254
   [*angulosa*] 254
   aspera 254
   biscocta 254
   corrugata 254
   cumingii 254
   derupta 254
   exasperata 254
   fibula 254
   [*gibba*] 254
   gruneri 254
   hanleyana 254
   histrio 254
   japonica 254
   [*laminata*] 254
   orbiculata 254
   [*penicillata*] 254
   traillei 254
   troscheli 254
   truncata 255
DOTHIDEACEAE 130
DOTHIDEALES 130
Dotilla wichmanni 364
Dougaloplus echinatus 396
Doxomysis littoralis 323
   longiura 323
   quadrispinosa 323
Drachiella morum 354
Draconetta margarostigma 447
   xenica 447
DRACONETTIDAE 447
Draculo mirabilis 448
DREISSENIDAE 253
Drepane longimana 436
   punctata 436
DREPANIDAE 436
Drepanophora corrugata 382
Drepanopsis frigidus 307
   orbus 307
Drilonereis filum 227
   logani 227
   robustus 227
Dromia dehaani 353
   intermedia 353
Dromidiopsis cranioides 353
   dromid 353
DROMIIDAE 353
Drosera burmannii 152
   indica 152

Index

DROSERACEAE 152
Drupa ambusta 274
  Anaxares 275
  aspera 275
  biconica 275
  borealis 275
  cancellata 275
  cariosa 275
  cavernosa 275
  clathrata 275
  [*concatenata*]
  cornus 275
  [*elata*] 275
  fiscella 275
  fragum 275
  fusca 275
  granulata 275
  grossularia 275
  margariticola 275
  marginatra 275
  morum 275
  musiva 275
  ochrostoma 275
  paucimaculata 275
  ricina 275
  ricina form *albolabris* 275
  [*ricinus*] 275
  rubusidaeus 275
  rugosa 275
  spathulifera 275
  spinosa 275
  uva 275
[*Drupella concatenata*] 275
  [*corpus*] 275
  [*fragum*] 275
Druppatractus irregularia 64
Druppula aspera 64
DRUPPULIDAE 64
Drymosphaera dendrophora 64
DRYOPTERIDACEAE 150
Dryopteris decipiens 150
DRYORHIZOPSIDAE 70
Dugong dugon 469
DUGONGIDAE
Dumontia simplex 134
DUMONTIACEAE 134
Dunckerocampus dactyiiophorus 424
Duosphincter zancli 200
[*Duplicaria dussumieri*] 282
  evoluta 282
Durckheimia caeca 363
Duridnilus piger 234
  tardus 234
Dussumieria acuta 410
  hasselti 410
Duvaucelia exsulans 292
DUVAUCELIIDAE 292
Duymeria flagellifera 439
Dynamena crisioides 171
  nanshaensis 171
  quadridentata 172
Dynamenopsis sp. 326
Dynoides dentisinus 326
Dynomene granulobata 353
  hispida 353
  sinense 353

Dynomene (continued)
  spinosa 353
  tenuilobata 353
DYNOMENIDAE 353
Dyocibicides biserialis 115
  epicharis 115
  perforata 115
Dysidea fragilis 164
DYSIDEIDAE 164
Dysomma anguillaris 417
  dolichosomatum 417
  goslinei 417
  melanurum 417
DYSOMMIDAE 417
Dysteria navicula 57
DYSTERIIDAE 57

E

EACANTHOCEPHALA 213
Eamesiella corrugata 247
Ebalia longimana 354
  malefactrix 354
  scabra 354
  scabriuscula 354
  tosaeusis 354
  tuberculosa 355
Ebisinus cheirophthalmus 456
Ebosia bleekeri 453
ECHELIDAE 417
Echeneibothrium hui 206
  variabile 206
ECHENEIDAE 453
Echeneis naucrates 453
Echidna delicatula 416
  nebulosa 416
  polyzona 416
  zebra 416
Echinaster luzonicus 391
  stereosomus 391
ECHINASTERIDAE 391
Echinimma frugifera 64
  polyacantha 64
Echininus cumingii spinulosus 265
Echinocardium cordatum 394
Echinocephalus pleroplatae 210
Echinochloa crusgalli v. *formosensis* 160
Echinocyamus crispus 394
ECHINODERMATA 386
Echinoderes tchefouensis 208
  sp. 1 208
  sp. 3 209
  sp. 4 209
  sp. 5 209
ECHINODERIDAE 208
Echinodiscus auritus 394
  tenuissimus 394
Echinoecus pentagonus 357
Echinogorgia aurantiaca 186
  coccinea 186
  complexa 186
  flora 186
  lami 187
  mertoni 187
  pseudosassapo 187
  sassapo reticulata 187

ECHINOIDEA 392
[*Echinometra mathaei*] 393
  ——— *mathaei* 393
  ——— *oblonga* 393
ECHINOMETRIDAE 393
Echinomuricea indomalaccensis 187
ECHINONEIDAE 393
Echinoneus abnormalis 393
  cyclostomus 393
Echinophyllia aspera 184
Echinopora gemmacea 183
  horrida 183
  lamellosa 183
ECHINORHINIDAE 407
Echinorhinus cookei 407
ECHINORHYNCHIDAE 213
ECHINORHYNCHIDEA 213
Echinorhynchus gadi 213
  [*manaenae*] 213
  salmonis 213
ECHINOSTOMATIDAE 204
Echinostrephus aciculatus 393
  molaris 393
Echinothrix calamaris 392
  diadema 392
ECHINOTHURIDAE
Echiodon owasianus 421
ECHIURA 237
ECHIURIDA 237
ECHIURIDAE 237
ECHIUROINEA 237
Ecklonia kurome 143
Ecsenius bicolor 446
  frontalis 446
  lineatus 446
  [*namiye*] 446
  oculus 446
  yaeyamacnsis 446
Ectenurus carangis 200
  lepidus 200
  megalaspis 200
  nibeae 200
  pseudosciaenae 200
  theraponae 200
  tianjinensis 201
  trachuri 201
  trichiuri 201
  zonichthyi 201
ECTINOSOMIDAE 314
ECTOCARPACEAE 140
ECTOCARPALES 140
Ectocarpus acutus 140
  arctus 140
  [*breviarticulatus*] 141
  commensalis 140
  [*confervoides*] 140
  corticulatus 140
  [*granulosus*] 141
  [*indicus*] 140
  rallsiae 140
  [*reinboldii*] 141
  rhodochortonoides 140
  siliculosus 140
  simplex 140
  tenellus 140
  terminalis 140

Ectocarpus (continued)
   tomentosus 140
   yezoensis 140
Ectopleura dumortieri 168
   guangdongensis 168
   latitaeniata 168
   minerva 168
   sacculifera 168
   xiamenensis 168
[ECTOPROCTA] 369
Edentostomina cultrata 82
   elongata 82
   milletti 82
   pseudodepressa 82
   rupertiana 82
   vulgaris 82
Edwardsia japonica 179
   sipunculoides 179
EDWARDSIDAE
Eggerella advena 78
   bradyi 78
   conica 78
   decepta 78
   nitens 78
   polita 78
   subconica 78
EGGERELLIDAE 77
Egretta alba modestus 465
   eulophotes 465
   g. garzetta 465
   i. intermedia 465
   s. saora 465
Ehlersia anops 219
   cornuta 219
   cf. *rosea* 219
Ehlersileanira hwanghaiensis 217
   incisa 217
   [*izuensis hwanghaiensis*] 217
   tentaculata 217
Ehrenbergina pacifica 106
   trigona 106
   undulata 106
Eirene brevigona 170
   ceylonensis 170
   chiaochowensis 170
   hexanemalis 170
   kambara 170
   menoni 170
   palkensis 170
   pyramidalis 170
   tenuis 170
   varidula 170
EIRENIDAE 170
Elachista fucicola 141
ELACHISTACEAE 141
ELAEAGNACEAE 155
Elaeagnus oldhami 155
Elagatis bipinnulatus 430
Elamenopsis introversus 356
   sinensis 356
Elaphococcus cervicornis 64
Elasmodes acutus 191
Elasmopus ecuadorensis hawaiensis 329
   japonicus 329
   molokai 329

Elasmopus (continued)
   pectenicrus 329
   spinidactylus 329
   spinimanus 329
Elates ransonneti 455
ELATINACEAE 155
[*Electra anomala*] 372
   axialata 372
   bellula 372
   bengaliensis 372
   deviensis 372
   [*deviensis*] 371
   donghaiensis
   gracilis 372
   inarmata 372
   inermis 372
   monilophora 372
   pilosa 372
   pseudopilosa 372
   pseudospinosa 372
   spiculata 372
   spinigera 372
   tenella 372
   teniuspinosa 372
   xiamenensis 372
ELECTRIDAE 371
Electroma ovata 243
Eleocharis acutangula 161
   geniculata 161
ELEOTRIDAE 451
Eleotriodes immaculatus 451
   longipinnis 451
   muralis 451
   strigatus 451
Eleotris acanthopomus 451
   [*heldsdingenii*] 451
   melanosoma 451
   [*sexguttatus*] 451
   [*strigatus*] 451
ELEUTHEROCARPIDAE 175
Eleutherolaimus stenosoma 211
Eleutheronema tetradactylus
ELLIPSIDAE 64
ELLIPSOLAGENIDAE 96
Ellisella gracilis 187
   laevis 187
ELLISELLIDAE 187
Ellisina canui 373
   costigera 373
   huanghainiana 373
ELLOBIIDAE 294
Ellobium aurismidae 294
   chinensis 294
   poita 294
Ellochelon vaigiensis 424
Elongoparorchis arii
ELOPIDAE 409
ELOPIFORMES 409
[*Elops machnata*] 409
   saurus 409
Elphidiella biseripora 124
   brevicanalis 124
   hannai 124
   kiangsuensis 124
   nutovaensis 124

Elphidiella (continued)
   papilli 124
ELPHIDIIDAE 124
Elphidiononion incertum 124
   neominutum 124
Elphidium advenum 124
   articulatum 124
   asiataicum 124
   chapmani 124
   clavatum 124
   concinnum 124
   crassimargo 124
   craticulatum 124
   crispum 125
   decorum 125
   dellum 125
   excavatum 125
   fissurisuturatum 125
   frgidum 125
   granti 125
   guangdongensis 125
   gunteri 125
   hispidulum 125
   hokkaidoense 125
   hughesi 125
   ibericum 125
   incertum 125
   jenseni 125
   kusiroense 125
   latispatium 125
   limpidum 125
   macellum 125
   magellanicum 125
   margaritaccum 125
   milletti 125
   nakanokawaense 125
   nigarense 125
   oceanicum 125
   owenianum 125
   pacificum 125
   panamense 125
   parvum 125
   poeyanum 125
   pulvereum 125
   pustulatum 125
   reticulosum 125
   rischtanicum 125
   sagrum 125
   shandongensis 125
   simplex 125
   subadvenum 125
   subcrispum 125
   subincertum 125
   taiwanum 125
   texanum 125
   tikutoensis 125
   translucens 125
   tsudai 125
   tungliangensis 125
   verriculata 125
Elysia atroviridis 289
   chilkensis 289
   grandifolia 289
   japonica 289
   nigrocapitata 289

Elysia (continued)
   trisinuata 289
   verrucosa 289
   viridis 289
ELYSIIDAE 289
Elytrophallus coiliae 201
Emarginula bellula 262
   biangulata 262
   bicancellata 262
   capuloidea 262
   clypeus 262
   concinna 262
   eximia 262
   fragilis 262
   genestrella 262
   hosoyai 262
   imaizumi 262
   obovata 262
   planulata 262
   sinica 262
   variegata 262
   xishaensis 262
Emballotheca acutirostris 379
   incisa 379
   pacifica 379
   subsinuata 379
Embelia laeta 156
EMBIOTOCIDAE 438
Embletonia gracile paucipapillata 292
[EMBLETONIIDAE]
Embolichthys mitsukurii 447
EMMELICHTHYIDAE 433
Emmelichthys seintillans 433
   struhsakeri 433
Emplectonema gracile 208
   mitsuii 208
EMPLECTONEMERIIDAE 208
Emydocephalus ijimae 464
Enantiodentalina basitorta 91
ENCHELIDIIDAE 212
Encheliophis gracilis 421
   sagamianus 421
Enchelycore bikiniensis 416
   lichenosa 416
   schismatorhynchus 416
ENCHELYIDAE 56
Enchelyopus elongatus 447
[*Enchelyurus kraussi*] 446
Enchiridium japonicum 191
ENCHYTRAEIDAE 234
Enchytraeus kincaidi 234
[*Encrosicholina chinensis*] 410
   [*commersoni*] 410
   [*heteroloba*] 410
   [*shantungensis*] 410
   [*tri*] 410
   [*zollingeri*] 410
Endarachne binghamiae 142
Endictya oceanica 30
ENDKOCLADIACEAE 136
ENDOMYCETATALES 129
Endopachys grayi 185
Endosiphonia clavigera 139
Enedrias fangi 447
   nebulosus 447

Engina carolinae 277
   [*elegans*] 277
   lineata 277
   mendicaria 277
   pulchra 277
   trifasciata 277
   zonalis 277
ENGRAULIDAE 410
Engraulis japonicus 410
Engyprosopon filipennis 457
   grandisquama 457
   latifrons 457
   longipelvis 457
   [*macroptera*] 457
   mogki 457
   multisquama 457
Enhalus acoroides 159
   [*koenigii*] 159
Enhydris bennetti 464
Enigmonia aenigmatica 246
Enneagonum hyalinum 175
   searsae 175
[*Enneapterygius etheostoma*] 446
   hemimelas 446
   minutus 446
   nanus 446
Enodiotrema hainanensis 198
ENOPLA 208
ENOPLIDA 212
ENOPLIDAE 212
Enoploides delamarei 212
ENOPLOTEUTHIDAE 294
Enoploteuthis chunii 294
Enoplus communis 212
Enosteoides ornata 351
   variabilis 351
Entaliunopsis habutae 261
   intercostatum 261
Entalophoroecia deliculata 370
   proboscidea 370
ENTALOPHOROECIIDAE 370
Entemnotrochus rumphii 261
ENTEROBACTER 21
Enterobacter 21
   coli 21
ENTEROBACTERALES 21
ENTEROGONA 402
Enteromorpha clathrata 144
   compressa 144
   crinita 144
   cruciata 144
   erecta 144
   flexuosa 144
   intestinalis 144
   kylinii 144
   lingulata 144
   linza 144
   [*micrococea*] 144
   plumosa 144
   prolifera 144
   salina v. *polyclados* 144
   spiralis 144
   torta 144
   tubulosa 144
ENTEROPNEUSTA 399

ENTODISCIDAE 57
Entodiscus borealis 57
   fukuii 57
   indomitus 57
   minor 57
Entolingulina inarimensis 98
Entomacrodus caudotasciatus 446
   decussatus 446
   epalzeocheilos 446
   niuafoouensis 446
   stellifer lighti 446
   striatus 446
   thalassinus 446
ENTOMOTAENIATA 284
ENTOPHYSALIDACEAE 25
Entophysalis granulosa 25
ENTOPROCTA 383
ENTORHIPIDIIDAE 57
Entorhipidium triangularis 57
[*Entovalva semperi*] 248
Eocollis harengulae 213
Eocuma lata 324
Eocylichna braunsi 286
   [*cylindrella*] 286
   involuta 286
   musashiensis 286
   sigmolabris 286
   soyoae 286
Eogammarus sinensis 328
Eolis gracilis 292
Eonemertes macrophthalma 208
Eophrixus shojii 326
Eopsetta grigorijewi 457
Eousia verticivarius 207
EOUVIGERINACEA 106
EPHIPPIDAE 436
Ephippodonta oedipus 248
Ephippus orbis 436
Ephydra sp. 299
EPHYDRIDAE 299
Epibrachiella magna 316
Epibulus insidiator 439
Epicodakia delicatula 247
Epidiopatra hupferiana 226
Epinephelus akaara 426
   amblycephalus 426
   areolatus 426
   awoara 426
   bleekeri 426
   bontoides 426
   brunneus 426
   caeruleopunctatus 426
   chlorostigma 426
   coioides 426
   cometae 426
   corallicola 426
   cyanopodus 426
   diacanthus 426
   epistictus 426
   fario 426
   fasciatomaculatus 426
   fasciatus 426
   fuscoguttatus 426
   hexagonatus 426
   hoedii 426

Epinephelus (continued)
   latifasciatus 426
   macrospilos 426
   maculatus 426
   malabaricus 426
   megachir 426
   melanostigma 426
   merra 426
   microdon 426
   moara 426
   morrhua 426
   octofasciatus 426
   poecilonotus 426
   quoyanus 426
   radiatus 426
   retouti 426
   rhyncholepis 426
   [*rivulatus*] 426
   septemfasciatus 426
   sexfasciatus 426
   spilotocep 426
   summana 426
   tauvina 426
   trimaculatus 426
   truncatus 426
   tukula 426
   undulosus 426
Epinula orientalis 449
Epipenaeon ingens 326
Epiplocylis calyx v. *lobiosa* 59
   constricta 59
   undella v. *blanda* 59
Episiphon candelatum 261
   kiaochowwanensis 261
Epistomaria annectens 116
EPISTOMARIIDAE 116
Epistomaroides polystomelloides 116
EPISTOMIDAE 375
Epistominella amakusaensis 113
   exigua 113
   naraensis 113
   nipponica 113
   pacifica 113
   pulchra 113
   takayanagii 113
   tubulifera 113
EPISTOMINIDAE 98
EPISTYLIDAE 58
Epistylis acuminata 58
   aselli 58
   carcini 58
   elongata 58
   harpacticola 58
   uyemurai 58
EPITHEMIACEAE 46
EPITONIIDAE 267
[*Epitonium auritum*] 267
   eboreum 267
   [*japonicus*] 267
   [*latifasciata*] 267
   [*lineolatum*] 267
   [*neglecta*] 267
   [*pretiosa minor*] 267
   scalare 267
   ——— *minor* 267

Epixanthus corrosus 360
   frontalis 360
EPIZOANTHIDAE 179
Epizoanthus fatuus 179
   glacialis 179
   illorieatus 179
   paguriphilus 179
   planus 179
   ramosus 179
   similis 179
   sinensis 179
   tube 179
Eponidella sinensis 116
Eponides berthelotianus 110
   blancoensis 110
   cribroconcameratus 110
   pacifica 110
   parvus 110
   procerus 110
   pusillus 110
   repandus 110
   subornatus 110
   sulcata 110
   umbonatus 110
   varvus 110
   xishaensis 110
EPONIDIDAE 110
Eptatretus burgeri 404
   okinoseanus 404
EPTATRETIDAE 404
Eragalatax contractus 275
Eragrostis cylindrica 160
   nevinii 160
Eratidus foliaceus 74
ERATOIDAE 270
Ercolania coerulea 289
   emarginata 289
   gopalai 289
   tentaculata 289
Eremicaster crassus 391
Eretmochelys imbricata 463
Ereunias grallator 456
EREUNIIDAE 456
Ericthonius brasiliensis 328
   pugnax 328
Eridacnis radcliffei 405
Erignathus barbatus 469
Erilepturus bohaiensis 201
   collichthydis 201
   formosae 201
   pisoodonophis 201
   trichiuri 201
ERIOCAULACEAE 162
Eriocaulon buergerianum 162
Eriocheir japonica 365
   ——— *hepuensis* 365
   leptognathus 365
   rectus 365
   sinensis 365
Erioglossum rubiginosa 154
Eriopisa chilkensis 329
   elongata 329
Eriopisella propagatio 329
   sechellensis 329
[*Eriphia laevimana*] 360

Eriphia (continued)
   [*laevimanus* v. *smithi*] 360
   scabricula 360
   sebana 360
   smithi 360
Erisphex potti 455
   simplex 455
Ernogrammus hexagrammus 447
Erodium cicutarium 153
Erosa erosa 454
Erosaria caputserpentis 272
   erosa 272
   helvola 272
   inocellata 272
   labrolineata 272
   [——— *helenae*] 272
   miliaris 272
   [*miliaris*] 272
   poraria 272
Erpocotyle sp. 192
Erronea caurica 272
   cylindrica 272
   errones 272
   hungerfordi 272
   onyx 272
   pulchella 272
   pyriformis 272
   walkeri 272
ERYONIDAE 346
ERYONIDEA 346
Erythrocladia irregularis 133
   [*subintegra*] 133
Erythrocles schlegeli 434
Erythrocolon podagricum 138
Erythroglossum pinnatum 139
   repens 139
ERYTHROPELTIDACEAE 133
Erythrops minuta 323
Erythrotrichia carnea 133
Escharella immersa 378
   longirostris 378
   ventricosa 378
Escharina grandicella 379
   pesanselis 379
Escharoidea sauroglossa 378
ESCHRICHTIIDAE 468
Eschrichtius robustus 468
Esola longicauda 315
Esosyrinx curta 98
   guttuliniformis 98
ESPERIOPSIDAE 165
Estellarca olivacea 240
Etelis carbunculus 433
   coruscans 433
   radiosus 433
Eteone delta 215
   longa 215
   maculata 215
   ornata 215
   tchangsii 215
Ethmodiscus gazellae 30
Ethusa indica 354
   izuensis 354
   latidactyla 354
   minuta 354

Ethusa (continued)
   quadrata 354
   sexdentata 354
Ethusina desciscens 354
   investigatoris 354
   robusta 354
Etisus anaglyptus 360
   bifrontalis 360
   demani 360
   dentatus 360
   electra 360
   frontalis 360
   laevimanus 360
   splendidus 360
   utilis 360
Etmopterus lucifer 407
   molleri 407
   pusillus 407
   splendidus 407
Etrumeus micropus 410
   [*teres*] 410
Euaetideus acutus 307
   giesbrechti 307
Euaetideus acutus
   giesbrechti
Eualus gracilirostris 343
   leptognathus 343
   sinensis 343
   spathulirostris 343
Euapta godefforyi 389
Euarche maculosa 217
Euaugaptilus affinis 310
   facilis 310
   filigerus 310
   hecticus 310
   mixtus 310
   nodifrons 310
   palumbii 310
Eubalaena glacialis 468
EUBRANCHIDAE 292
Eubranchus misakensis 292
EUCALANIDAE 306
Eucalanus attenuatus 306
   crassus 306
   elongatus 306
   mucronatus 306
   pileatus 306
   pseudattenuatus 306
   subcrassus 306
   subtenuis 306
Eucalyptus citriodora 156
   exserta 156
   robusta 156
Eucampia cornuta 32
   zoodiacus 32
Eucecryphalus cervus 67
   sestrodiscus 67
Euchaeta acutus 308
   concinna 308
   longicornis 308
   marina 308
   media 308
   plana 308
   spinosa 308
   tenuis 308

Euchaeta (continued)
   wolfendeni 308
EUCHAETIDAE 308
Euchaetomera oculata 323
   tenuis 323
Euchaetomeropsis merolepis 323
Eucheilota diademada 171
   duodecimalis 171
   macrogona 171
   menoni 171
   [*menoni*] 171
   multicirrs 171
   paradoxia 171
   taiwanensis 171
   tropica 171
   ventricuiaris 171
EUCHEILOTIDAE 171
Euchelus fussulatus 263
   scaber 263
Eucheuma arnoldii 137
   denticulatum 137
   [*denticulatum*] 137
   gelatinae 137
   [*muricatum*]
   okarnurai 137
   serra 137
[*Euchirella acuta*] 307
   amoena 307
   [*areata*] 307
   bella 307
   bitumida 307
   curticauda 307
   galeata 307
   indica 307
   maxima 307
   messinensis 307
   orientalis 307
   pulchra 307
   rostrata 307
   unispina 307
   venusta 307
Euchitonia aequipondata 64
   elegans 64
   furcata 64
   triangulum 64
   cf. *triangulum* 64
Euchlora rubra 190
Eucidaris metularia 392
Euciroa rostraata 258
   [*teramachii*] 258
Euclia balantium 288
   pyramidata v. *lanceoiata* 288
   ———— v. *microcaudata* 288
Euclymene annandalei 225
   insecta 225
   lombricoides 225
   oerstedii 225
   uncinata 225
Euconchoecia aculeata 303
   bifurata 303
   chierchiae 303
   elongata 303
   maimai 303
   shenghwai 303
Eucopia australis 323

EUCOPIIDAE 323
Eucoronis challengeri 66
   nephrospyris 66
Eucrassatella japonica 248
   nana 248
Eucrate alcocki 363
   costata 363
   crenata 363
   haswelli 363
   [*maculata*] 363
   solaris 363
Eucyrtidium acuminatum 68
   dictyopodium 68
   hexagonatum 68
   urceolatum 68
Eudactylopus latipes 315
EUDENDRIIDAE 166
Eudendrium californicum 166
   capillare 166
   pusillum amoyicum 167
   racemosum 167
   rameum 167
   ramosum 167
Eudesme virescens 141
EUDIOCRINIDAE 386
Eudiocrinus indivisus 386
   venustulus 386
Eudistama laysami 402
Eudistylis vancouveri 230
Eudolium pyriforme 274
Eudorella pacifica 325
Eudoxia macra 174
Eudoxoides mitra 174
   spiralis 174
EUBRANCHIDAE
Eufistulana grandis 256
Eugomphodus taurus 404
EUGONATONOTIDAE 339
Eugonatonotus crassus 339
Eukrohnia bathyantarctica 385
   bathypelagica 385
   fowleri 385
   hamata 385
   sinica 385
[*Eulalia albopicta*] 215
   bilineata 215
   [*macroceros*] 215
   [*nigrimaculata*] 215
   [*notoensis*] 215
   [*sanguinea*] 215
   [*tubiformys*] 215
   viridis 215
EULEPETHIDAE 218
Eulepethus hamifera 218
Euleptorhamphus viridis 419
Eulima bifascialis 267
   [*bilineata*] 267
   maria 268
EULIMIDAE 267
Euloxostomum bradyi 106
Eumegistus illustris 431
Eumetopias jubata 469
Eumida albopicta 215
   sanguinea 215
   tubiformis 215

Eumunida smithii 350
Eunaticina papilla 269
Eunice afra 226
   antennata 226
   aphroditois 227
   australis 227
   coccinea 227
   gracilis 227
   indica 227
   kobiensis 227
   longicirrata 227
   marenzelleri 227
   medicina 227
   multipectinata 227
   northioidea 227
   pennata 227
   [*spinea*] 227
   vittata 227
EUNICIDA 226
EUNICIDAE 226
Euniphysa aculeata 227
   oculata 227
   [*spinea*] 227
   unicusa 227
EUNIPHYSIDAE 227
Eunoe cf. *barbata* 216
   [*oerstedi*] 216
   [cf. *oerstedi*] 216
Eunotia eruca 45
EUNOTIACEAE 45
EUNOTIALES 45
Eunotogramma debile 35
   frauenfeldii 35
   laevis 35
   rostratum 35
[*Eupagurus japonicus*] 350
Eupanthalis edriophthalma 217
   kinbergi 217
   [*maculata*] 217
   [*nigromaculata*] 217
   [*oculata*] 217
Eupasiphae latirostris 338
Eupelte acutispinis 314
Euphausia brevis 334
   diomedeae 334
   gibba 334
   hemigibba 334
   mutica 334
   nana 334
   pacifica 334
   paragibba 334
   pseudogibba 334
   recurva 334
   sanzoi 334
   sibogae 334
   similis 334
   tenera 334
EUPHAUSIACEA 334
EUPHAUSIIDAE 334
Euphilomedes agilis 302
   corrugata 302
   interpuncta 302
   japonica 302
   longiseta 302
   nodosa 302

Euphilomedes (continued)
   sordida 302
Euphlyctis cancrivora 463
Euphorbia atoto 154
   indica 154
   microphylla 154
   tirucalli 154
EUPHORBIACEAE 154
Euphrosine foliosa 226
   myrtosa 226
EUPHROSINIDAE 226
[EUPHURIDAE] 289
Euphyllia turgida 184
Euphysa aurata 168
   [*bigelowi*] 168
Euphysetta elegans 69
EUPHYSIDAE 168
Euphysilla pyramidata 168
Euphysomma brevia 168
Euphysora annulata 168
   bigelowi 168
   furcata 168
Euplectella imperialist 166
   marshalli 166
   oweni 166
EUPLECTELLIDAE 166
Euplerogrammus muticus 449
Euplexaura albida 187
   attenuata 187
   curvata 187
   erecta 187
   robusta 187
Euplica scripta 276
   tururina 276
Euplotes charon 60
   charonopsis 60
   harpa 60
   rariseta 60
   vanus 60
EUPLOTIDAE 60
EUPODIDAE 298
EUPODISCACEAE 31
Eupodiscus radiatus 31
Eupolymnia marenzeller 229
   nebulosa 229
   umbonis 229
Eupolyodontes mitsukurii
Epronoe armata 332
   [*intermedia*] 332
   laticarpa 332
   maculata 332
   minuta 332
Euprymna berryi 296
   morsei 296
Euraphia caudata 320
   withersi 320
Euretaster insignis 391
EURETIDAE 166
EUROTIACEAE 129
EUROTIALES 129
Eurya chinensis 155
   emarginata 155
Euryale aspera 395
   purpura 395
EURYALIDAE 395

EURYALAE 395
Eurynome orientalis 356
Eurynorhynchus pygmeus 467
EURYSQUILLIDAE 368
Eurysquilloides sibogae 368
Eurystomella bilabiata 377
EURYSTOMELLIDAE 377
Eurystomina ophthalmophora 212
Eurytemora pacifica 309
Eurythoe chilensis 226
   complanata 226
   parvecarunculata 226
Eusarsilla longinenna 302
   parvispinosa 302
   spinulosa 302
   tumida 302
Euscalpellum rostratum 318
   stratum 318
Euselenops luniceps 289
[*Eusicyonia cristata*] 338
   [*curvirostris*] 338
   [*lancifer*] 338
   [———— *japonica*] 338
EUSIRIDAE 329
Eustachys tener 160
Eustomias longibarba 412
Eusyllis blomstrandi 219
   inflata 219
   cf. *irregulata* 219
Eutaeniichthys gilli 452
Euterpina acutifrons 314
Euthalenessa digitata 217
   oculata 217
[*Euthemisto bispinosa*] 331
Euthynnus affinis 450
   [*pelamis*] 450
   yaito 450
Euthyonidiella tungshaenesis 387
[*Eutima campanulata*] 170
   curva 170
   gegenbauri 170
   gentiana 170
   gracilis 170
   japonica 170
   levuka 170
   mira 170
   modesta 170
   neucaledonia 170
   orientalis 170
   variabilis 170
Eutintinus tenuis 60
Eutonina scientillans 170
Eutypa sp. 130
Euuvigerina aculeata 107
   peregerina 107
Euxanthus exsculptus 360
   sculptilis 360
Euxiphipops sexstriatus 437
Euzonus arcticus 225
   dillonesis 225
   ezoensis 225
Evadne nordmanni 301
   tergestina 301
Evermannella indica 415
EVERMANNELLIDAE 415

Index

Eviota abax 451
    prasina 451
Evodia meliaefolia 153
Evolutononion shansiensis 117
Evolvocassidulina belfordi 105
Evolvulus alsinoides v. *decumbens* 157
Evoxymetopon poeyi 449
    taeniatus 449
Evynnis cardinalis 434
Exallias brevis 446
Excirolana chiltonii 325
Excoecaria agallocha 154
Exculptina exculpta 95
    pliocenica 95
    semilineata 95
    spinigera 95
Exechonella discoidea 378
    magna 378
    tuberculata 378
EXECHONELLIDAE 378
Exhippolysmata ensirostris 343
Exochella areolata 378
EXOCHELLIDAE 378
EXOCOETIDAE 419
Exocoetus monocirrhus 419
    volitans 419
Exogone dispar 219
    fangopapillata 219
    gemmifera 219
    naidina 219
    verugera 219
Exogonoides antennata 219
Exopalaemon annandalei 340
    carinicauda 340
    modestus 340
    orientalis 340
Exosphaeroma oregonensis 326
Exotica clathrata 250
    obiquistriata 250
Exuviaella marina 50
[*Exyurias puntang*] 451

F

Fabia obtusidentata 363
Fabrea salina 58
Fabulina tsichungyeni 250
FACELENIDAE 292
Facelina zhejiangnensis 292
Facelinella porilophags 292
    quadrilineata 292
FALCONIFORMES 466
[*Falkenbergia hillebrandii*] 134
Falsibugulella sinica 374
Falsoguttulina laevigata 95
Faorina chinensis 394
FARCIMINARIIDAE 375
Farrea occa 166
Fasciolaria filamentosa 278
    trapezium 278
FASCIOLARIIDAE 278
Fasciolites boscii 89
    cylindratus 89
    ellipsoidalis 89
    globosus 89

Fasciolites (continued)
    himalayensis 89
    nuttalli 89
    oblongus 90
    oliviformis 90
    rutimereri 90
    subtilis 90
    tibeticus 90
Faustula qikouensis 194
Favella amoyensis 59
    arcuata 59
    azorica 59
    companula 59
    cylindrica 59
    ehrenbergi 59
    franciscana 59
    hainanensis 59
    ―― v. *parva* 59
    panamensis 59
    shintsuenensis 59
    undulata 59
Favia matthaii 183
    palauensis 183
    rotumana 183
    speciosa 183
    stelligera 183
FAVIIDAE 183
Favites abdita 184
    halicora 184
    pentagona 184
    virens 184
Favocassidulina favus 105
Favulina hexagona 96
    melo 96
    squamosa 96
Felaniella usta 247
Feldmannia indica 140
    Irregularis 141
Fellia bicornis 304
    cornuta 304
    ―― *dispar* 304
FELLODISTOMIDAE 194
Fenestrulina infundibulipora 381
    malusi 381
Feresa attenuata 469
Fibularia acuta 394
    angulipora 394
    ovulum 394
    volva 394
FIBULARIIDAE 394
FICIDAE 274
Ficopomatus macrodon 230
Ficus ficus 274
    filosus 274
    gracilis 274
    subintermedius 274
[FIERASFERIDAE] 421
Figaro melanobranchus 405
Figularia figularis 377
    fissa 377
    fissulata 377
Fijiella simplex 108
Fijinonion fijiensis 119
Filellum serratum 171
    [*serpens*] 171

Filibranchus roseus 229
Filinia maior 214
Filisoma atropi 213
    indicum 213
    oplegnathi 213
Filisparsa candeana 370
    rustica 370
Filocapsularis alexandri 209
[*Fimbria fimbria*] 292
    fimbriata 247
    soverbii 247
FIMBRIIDAE 247, 292
Fimbriosthenelais minor 217
Fimbristylis aestivalis 161
    dichotoma 161
    ferruginea 161
    longispica 161
    longistipitata 161
    ovata 161
    pierotii 161
    polytrichoides 161
    rigidula 161
    sericea 161
    schoenoides 161
    spathacea 161
    subbispicata 161
    tristachya 161
Firoloida desmaresti 269
Fischerinella helix 81
    pellucida 81
    trochoides 81
FISCHERINIDAE 81
Fissidentalium formosum 261
    tenuicostatum 261
    vernedei 261
    yokoyamai 261
FISSURELLIDAE 262
Fissurina annectens 97
    aperta 97
    aradasii 97
    atlantica 97
    barri 97
    biancae 97
    bicarinata 97
    bicaudata 97
    carinata 97
    circularis 97
    clathrata 97
    compressioblonga 97
    contusa 97
    crebra 97
    crenulata 97
    cucullata 97
    cucurbitasema 97
    dendiculifera 97
    diaphana 97
    disjuncta 97
    disjungens 97
    earlandi 97
    eburnea 97
    elliptica 97
    fasciata 97
    fimbriata 97
    formosa 97
    furcicolla 97

Fissurina (continued)
  granulocostulata 97
  jasolini 97
  kerimbatica 97
  lacunata 97
  laevigata 97
  laevis 97
  lucida 97
  marginata 97
  milletti 97
  minipora 97
  occlusa 97
  orbignyana 97
  propinqua 97
  pulchella 97
  pseudoglobosa 97
  pseudolucida 97
  pseudoorbignyana 97
  quadrata 97
  quadricostulata 97
  radiata 97
  radiatomarginata 97
  radiatostoma 97
  robusta 97
  sacculus 97
  semimarginata 97
  sexcostulata 97
  sigmoidella 98
  simplicita 97
  subglobosa 98
  tricaudata 98
  tropicalis 98
  unicospina 98
  walleriana 98
  wrightiana 98
FISTULARIIDAE 423
Fistularia commersonii 423
  petimba 423
  villosa 423
FLABELLIGERIDA 228
FLABELLIGERIDAE 228
FLABELLIIDAE 185
Flabellina ornata 292
  rubrolineata 292
FLABELLINIDAE 292
Flabellopora acutirostris 382
  elegans 382
  irregularis 382
  lenticularis 382
  linguna 382
  pisiformis 382
  pusilla 382
  tuberosa 382
Flabellum deludens 185
  distinctum 185
  pavoninum 185
  rubrum 185
Flacourtia indica 155
FLACOURTIACEAE 155
Flagellaria indica 161
FLAGELLARIACEAE 161
Flammeo argenteus 422
  opercularis 422
  sammara 422
  scythrops 422

Flavobacterium 21
Flavus group 24
Flexibacter 22
Flintina bradyana 86
  crassatina 86
  hainanensis 86
  triquetra 86
Flintinoides labiosa 86
Flongoparorchis arii 203
Florilus atlanticus 117
  belridgensis 117
  boueanus 117
  carinatus 117
  decorus 117
  extensus 117
  florinensis 117
  fuscus 117
  hantkeni 117
  japonicus 117
  labradoricum 117
  limbatostriatus 117
  manpukujiense 117
  scaphum 117
  sloanii 117
  subgrateloupi 117
FLOSCULARIACEA 214
Flustrellidra flabellaris 371
FLUSTRELLIDRIDAE 371
FLUSTRIDAE 372
Foa abocellata 429
Fodiator acutus pacificus 419
[*Foetorepus altivelis*] 448
Folliculina ampulla 58
FOLLICULINIDAE 58
Folliculovarium gymnothoracis 193
  xishaense 193
Fontbotia wuellerstorfi 115
FORAMINIFERA 69
FORAMINIFERIDA 69
Forcipiger falvissimus 437
  longirostris 437
FORCIPULATA 392
Formio niger 431
  [*parastromateus*] 431
FORMIONIDAE 431
Forskalia cuneata 174
  edwardsi 174
FORSKALIDAE 174
Fortuynia sinensis 298
Fortuynia sp. 298
FORTUYNIIDAE 298
Fosliella farinosa 135
  [*lejolisii*] 135
[*Fowleria auritus*] 429
  isostigma 429
  marmorata 429
  ocellata 429
  variegata 429
Fragilaria aurivillii 43
  brevistriata 43
  crotonensis 43
  cylindrus 43
  heidenii 43
  hyalina 43
  [*hyalina*] 43

Fragilaria (continued)
  longissima v. *protenta* 43
  oceanica 43
  striatula 43
Fragum bannoi 248
  carinatum 248
  fragum 248
  peronatum 248
  unedo 248
Fregata andrewsi 465
  a. ariel 465
  m. minor 465
FREGATIDAE 465
Frenulina sanguinolenta 384
Fritillaria bicornis 400
  borealis 400
  ——— *sargassi* 400
  formica 400
  fraudax 400
  haplostoma 400
  megachile 400
  pellucida 400
  tenella 400
  venusta 400
FRITILLARIDAE 400
Fromia milleporella 391
Frondicularia inaequalis 91
  nitida 91
Fronsella fujitaniana 248
Frustulia interposita 37
  ——— v. *chinensis* 37
  ——— v. *dispar* 37
  lewisiana 37
Fryeria ruppelli 291
FUCACEAE 143
FUCALES 143
Fugu alboplumbeus 461
  basilevskianus 461
  bimaculatus 461
  flavidus 461
  niphobles 461
  oblongus 461
  obscurus 461
  ocellatus 461
  pardalis 461
  porphyreus 461
  pseudommus 461
  reticularis 461
  rubripes 461
  stictonotus 461
  vermicularis 461
  xanthopterus 461
Fulgoraria formosana 279
  glabra 279
  hamillei 279
  rupestris 279
Fulica a. atra 466
Fulmarius glacialis rodgersii 465
Fulvia australis 248
  bullata 248
  hungerfordi 248
  mutica 248
  oxygona 248
Funchalia villosa 336
  [*villosa*] 336

Index

FUNGI 129
Fungia danai 182
  echinata 182
  fieldi 182
  fungites 182
  paumotensis 182
  repanda 182
  scutaria 182
Fungiacyathus stephanus 185
  symmetricus 185
FUNGIIDAE 182, 184
Fursenkonia bradyi 109
  carinata 109
  compressa 109
  cornuta 109
  exilis 109
  pauciloculata 109
  rotundata 109
  squammosa 109
  schreibersiana 109
  taiwanica 109
  texturata 109
  tikutoensis 109
FURSENKOINACEA 109
FURSENKOINIDAE 109
Fusarium 130
  graminearum 130
  solani 130
  triciuetum 130
Fusiaphera macrospira 279
Fusinus colus 278
  cf. *forceps* 278
  fragosusw 278
  longicaudus 278
  nicobaricus 278
  perplexus 278
  simplex 278
Fusitriton galea 273
Fusolatirus pilsbryi 278
Fustiaria nipponica 261

G

GADIDAE 420
GADIFORMES 419
Gadilina insoluta 261
Gadomus colletti 420
  melanoptorus 420
  multifilis 420
Gadus macrocephalus 420
Gaetanus brevicornis 307
  miles 307
  minor 307
  pileatus 307
Gaetice depressus 365
Gafrarium dispar 255
  divaricatum 255
  pectinatum 255
Gaidius pungens 308
Galathea balssi 350
  [*elegans*] 350
  formosus 350
  orientalis 350
  pubescens 350
  pusilla 350

Galathea (continued)
  subsquamata 350
GALATHEIDAE 350
GALATHEOIDEA 350
Galaxaura apiculata 134
  [*arborea*] 134
  elongata 134
  fasciculata 134
  filamentosa 134
  [*fruticulosa*] 134
  glabriuscula 134
  oblongata 134
  obtusata 134
  pacifica 134
  robusta 134
  [*rudis*] 134
  stupocaula 134
  subfruticulosa 134
  umbellata 134
  ventricosa 134
  veprecula 134
Galaxea aspera 183
  fascicularis 183
  lamarcki 183
Galene bispinosa 360
Galeoastrea modesta 264
Galeocerdo cuvier 406
Galeodea echinophorella 272
GALEODIDAE 277
Galeomma polita 248
  takii 248
GALEOMMATIDAE 248
Galeoocory leucodoma 273
Galeoocorys leucodoma 274
Galeorhinus hyugaensis 405
  japonicus 406
Galeus eastmani 405
  niponensis 405
  sauteri 405
Gallicrex c. cinerea 466
Gallinula chloropus indica 466
Gallitellia vivans 99
Galwayella trigonomarginata 96
Gambierdiscus toxicus 54
GAMMARIDAE 329
GAMMARIDEA 328
Gammaropsis japonica 328
  laevipalmata 328
  liuruiyui 328
  nilida 328
  sexdentata 328
  togoensis 328
  utinomi 328
Gangliostoma guangdongensis 169
Ganonema farionsa 133
  pinnatiramosa 133
Gardnerula fasciculata 26
  tenuissima 26
  xishaensis 26
Gargariscus prionocephalus 455
Gari anomala 252
  hosoyai 252
  maculosa 252
  [*radiata*] 252
  [*reevei*] 252

Gari (continued)
  schepmani 252
  [*schepmani*] 252
  [*squamosa*] 252
  truncata 252
[*Garveia francisana*] 166
Gascoignella aprica 289
GASCOIGNELLIDAE 289
GASTEROSTEIFORMES 423
Gasteropteron meckeli 287
GASTEROPTRIDAE 287
[*Gastrana contabulata*] 250
[*Gastroblasta chengshanensis*] 170
  [*raffaelei* v. *chengsnanensis*] 170
Gastrochaena cuneiformis 256
  cymbium 256
  [*grandis*] 256
  [*ovata*] 256
GASTROCHAENIDAE 256
Gastroclonium xishaensis 138
Gastrolepidia clavigera 216
Gastrophysus lunaris 461
  [*sceleratus*] 461
  spadiceus 461
  [*suezensis*] 461
GASTROPODA 261
GASTROPTERIDAE
Gastropteron meckeli
Gastrosaccus dunckeri 323
  formosensis 323
  hibbi 323
  indicus 323
  ohshimai 323
  parvus 323
  pelagicus 323
Gattyana deludens 216
  pohaiensis 216
Gaudryina aequa 77
  arenaria 77
  asanoi 77
  chileana 77
  collinsi 77
  contorta 77
  convexa 77
  flintii 77
  haeringensis 77
  hastata 77
  hayasakai 77
  inflata 77
  karihaensis 77
  kokuseiensis 77
  koreaensis 77
  mai 77
  niigataensis 77
  oinomikadoi 77
  pacifica 77
  parva 77
  pliocenica 77
  pseudohayasakai 77
  pyramidata 77
  quadrangularis 77
  siphonifera 77
  subglabrata 77
  taiwanica 77
  transversaria 77

Gaudryina (continued)
   triangularis 77
   trullisata 77
   wrightiana 77
Gaussicia edentata 304
   gaussi 304
   incisa 304
GAVELINELLIDAE 120
Gavelinopsis praegeri 111
Gavia adameii 464
   arctica viridigularis 464
   s. stellata 464
GAVIIDAE 464
GAVIIFORMES 464
Gazelletta hexanema 69
Gazza achlamys 432
   minuta 432
[*Gebia barbata*] 345
   [*carinicauda*] 345
   [*issaeffi*] 346
   [*major*] 346
   [*spinifrons*] 346
[*Gebiopsis darwinii*] 345
GECARCINIDAE 367
Gecarcoidea lalandii 367
Geitodoris granulata 291
GELIDIACEAE 134
[*Gelagna clandestina*] 273
   [*succincta*] 273
Gelidiella acerosa 134
Gelidiopsis intricata 136
Gelidium amansii 134
   ——— f. *elegans* 134
   ——— f. *radicans* 134
   [*clavatum*]
   crinale 134
   divaricatum 134
   [*grubbae*]
   japonicum 134
   johnstonii 134
   kintaroi 134
   latifolium 134
   pacificum 134
   pusillum 134
   tsengii 134
   vagum 134
Gelliodes incrustan 165
   pumila 165
   spinosella 165
Gellius microxes 164
   toxius 164
   varius 164
Gelochelidon n. nilotica 468
Geloina coaxans 254
Geminospira bradyi 98
   mayori 98
   simaensis 99
Gemmula congener cosmoi 281
   cosmoi 281
   deshayesii 281
   diomedea 281
   kieneri 281
   speciosa 281
   unedo 281
GEMPYLIDAE 449

[*Gempylus notha*] 449
   serpens 449
Gena varia 264
Genarchopsis clupeae 201
[*Genetyllis castanea*] 215
Genicanthus lamarck 437
   macclesfieldiensis 437
   melanospilos 437
   semifasciatus 437
   watanabei 438
[*Gennadas alicei*] 335
   incertus 335
   [*incertus*] 335
   [*intermedius*] 335
   parvus 335
   propinquus 335
   scutatus 335
Genolina laticauda 201
Genolopa sparui 195
Geodia japonica
GEODIIDAE 164
Gephyria media 43
Gephyroberyx japonicus 422
Gephyrometra versicolor 387
Geppella echinocaulos 147
   japonica 147
   prolifera 147
GERANIACEAE 153
Gerreomorpha japonica 432
Gerres abbreviatus 432
   acinaces 432
   argyreus 432
   [*equuala*] 432
   filamentosus 432
   lucidus 432
   macrosoma 432
   oblongus 432
   oyena 432
   poeti 432
GERRIDAE 432
Geryon trispinosus 359
Geryonia proboscidalis 173
GERYONIDAE 359
GERYPMOODAE
Gibbula affinis 263
Giffordia acuto-ramuli 141
   fuscata 141
   [*indica*] 140
   [*irregularis*] 141
   [*mitchellae*] 141
   sandriana 141
GIGANTACTINIDAE 462
Gigantactis garganthus 462
   vanhoeffeni 462
Giganthias immaculatus 426
Gigantocypris agassizi 302
   australis 302
   danae 302
   dracontovali 302
Gigantonotum album 289
Gigantopora mutabilis 379
   pupa 379
GIGANTOPORIDAE 379
Gigartacon denticulatus 61
   fragilis 61

GIGARTACONIDAE 61
Gigartina intermedia 137
   pacifica 137
   teedii 137
   tenella 137
GIGARTINACEAE 137
GIGARTINALES 136
Ginglymostoma ferrugineum 405
Girella melanichthys 435
   mezina 435
   punctata 435
GIRELLIDAE 435
Gisekia pharnaceoides 151
Gitanopsis japonica 328
   longus 328
Glabratella crassa 112
   elegans 112
   limbata 112
   makinoi 112
   opercularis 112
   patelliformis 112
   pileolus 112
   pulchella 112
   scabra 112
   tabernacularis 112
GLABRATELLACEA 112
GLABRATELLIDAE 112
Glabrilutjanus nematophorus 433
Glabriscala stigmatica 267
Glabropilumnus dispar 360
   sodalis 361
Glandulina aequalis 98
   apiculata 98
   dimorpha 98
   echinata 98
   elliptica 98
   globosa 98
   granulosa 98
   hainanensis 98
   laevigata 98
   rotundata 98
GLANDULINIDAE 98
Glans hirasei 248
Glaphyrammina americanus 74
Glareola maldivarum 468
GLAREOLIDAE 468
GLAUCIDAE 293
Glaucoammina trilateralis 80
Glaucoma hyalina 57
GLAUCOMIDAE 57
[*Glaucomya chinensis*] 256
Glauconome chinensis 256
   primeana 256
GLAUCONOMIDAE 256
Glaucosoma faurellii 427
   hebraicum 427
GLAUCOSOMIDAE 427
Glaucus group 24
Glaucus marinus 293
Glehnia littoralis 156
Globanomalina wilocoxensis 101
GLOBANOMALINIDAE 101
GLOBICEPHALIDAE 469
Globigerapsis index 101
Globigerina ampliapertura 101

Globigerina (continued)
  angustiumbilicata 101
  apertura 101
  atlantica 101
  borealis 101
  bradyi 101
  bulloides 101
  calida 101
  ciperoensis 102
  decoraperta 102
  digitata 102
  diplostoma 102
  druryi 102
  eggeri 102
  flaconensis 102
  fringa 102
  hexagona 102
  nepenthes 102
  pachyderma 102
  quinqueloba 102
  rubescens 102
  subcreatacea 102
  triloculinoides 102
  varianta 102
  velascoensis 102
  woodi 102
GLOBIGERINACEA 101
Globigerinella adamsi 102
  aequilateralis 102
  calida 102
  siphonifera 102
GLOBIGERINIDAE 101
GLOBIGERININA 99
Globigerinita bradyi 101
  glutinata 101
  incrusta 101
  iota 101
Globigerinoides bolli 102
  conglobatula 102
  conglobatus 102
  cyclostomus 102
  elongatus 102
  extremus 102
  fistulosus 102
  helicina 102
  immaturus 102
  obliquus 102
  parkerae 102
  pyramidalis 102
  quadrilobatus 102
  ruber 102
  sacculifera 102
  sicanus 102
  sinensis 103
  tenellus 103
  trilobus 103
  trilocularis 103
  triloculinoides 103
  varianta 103
Globigerinoidesella fistulosa 103
Globobulimina affinis 107
  arctica 107
  auriculata 107
  glabra 107
  hanzawaia 107

Globobulimina (continued)
  notovata 107
  pacifica 107
  perversa 107
  pyrula 107
  torta 107
Globocassidulina algida 105
  Alternans 105
  complanata 105
  gemma 105
  producta 105
  subglobosa 105
Globophiline huanghenensis 287
Globopilumnus globosus 361
Globoquadrina altispira 101
  conglomerata 101
  dehiscens 101
  dutertrei 101
  hexagona 101
  venezuelana 101
Globorotalia abundocamerata 99
  acostaensis 99
  anfracta 99
  angulata 99
  archeomenardii 99
  bermudezi 99
  clarkei 99
  compressa 99
  conicotruncata 99
  conoidea 99
  continuosa 99
  crasaformis 99
  crassula 99
  crotonensis 99
  cultrata 99
  eastropacia 99
  ehrenbergi 99
  exilis 99
  fimbriata 99
  flexuosa 99
  hirsuta 99
  humerosa 99
  inflata 99
  lata 99
  margaritae 99
  mayeri 99
  menardii 99
  miocenica 99
  multicamerata 99
  obesa 99
  oscitans 99
  padana 99
  parkerae 100
  peripheroacuta 100
  peripheroronda 100
  plesiotumida 100
  praemenardii 100
  pseudobulloides 100
  pseudomenardii 100
  pumilio 100
  puncticulata 100
  redunca 100
  robusta 100
  scitula 100
  siakensis 100

Globorotalia (continued)
  tosaensis 100
  trigonula 100
  truncatulinoides 100
  tumida 100
  ungulata 100
  velascoensis 100
GLOBOROTALIACEA 99
GLOBOROTALIIDAE 99
Globorotaloides hexagona 101
Globotextularia anceps 77
GLOBOTEXTULARIIDAE 77
Globoturborotalita rubescens 103
Globovula margarita 271
  sphaera 271
Globulina gibba 95
  landesi 95
  minuta 95
  trigona 95
Globulotuba entosoleniformis 98
Glochidion obovatum 154
Gloiopeltis complanta 136
  furcata 136
  ——— v. *coliformis* 136
  tenax 136
Gloiophloea chinensis 134
GLOIOSIPHONIACEAE 135
Gloiosiphonia capillaris 135
Glomospira charoides 71
  glomerata 71
  gordialis 71
Glossanodon semifasciata 411
GLOSSIDAE 253
Glossobalanus mortenseni 399
  polybranchioporus 399
Glossocephalus milneedwardsi 332
GLOSSODORIDIDAE 290
[*Glossodoris alba*] 290
  aspersa 290
  aureopurpurea 290
  [*cincta*] 290
  crossei 290
  festiva 290
  [*festive*] 290
  hilaris 290
  lineolata 290
  marginata 290
  mutituberculata 290
  obsoluta 290
  orientalis 290
  pallesens 290
  sibogae 290
  sinensis 290
  tenuis 290
  tincloria 290
  tumlifera 290
  xishaensis 290
Glossogobius biocellatus 452
  brunnoides 452
  circumspectus 452
  [*fasciato-punctotus*] 452
  giuris 452
  obscuripinnis 452
  olivaceus 452
Glossogyne tenuifolia 158

Gluconobacter 21
Glycera alba 221
   armigera 221
   capitata 221
   chirori 221
   clecipiens 221
   convoluta 221
   gigantea 221
   lancadivae 221
   onomichiensis 221
   papillosa nigticans 221
   prashadi 221
   robusta 221
   rouxu 222
   sagittariae 222
   siphonostoma 222
   subaenea 222
   tenuis 222
   tesselata 222
   unicornis 222
GLYCERIDAE 221
Glycinde gurjanovae 222
   kameruniana 222
   oligodon 222
   pacifica 222
GLYCOBOTRYDIDAE 67
GLYCYMERIDIDAE 241
Glycymeris albolineata 241
   pilsbryri 241
   reevei 241
   rotunda 241
   vestita 241
   yessoensis 241
GLYPHAEIDAE 246
[*Glyphis gangeticus*] 406
Glyphocrangon granulosis 345
   hastacauda 345
   pugnax 345
   regalis 345
GLYPHOCRANGONIDAE 345
Glyphus marsupialis 338
Glyptocidaris crenularis 392
Gnathagnus elongatus 445
Gnathanodon speciosus 430
Gnathodentex aurolineatus 434
Gnatholepis cauerensis 452
   otakii 452
Gnathophausia ingens 323
GNATHOPHIURIDA 395
GNATHOPHYLLIDAE 339
Gnathophyllum americanum 339
GNATHOSTOMATIDAE 210
Gnathypops evermanni 445
Gnomoxyala breviseta 211
GNRRIDAE 299
GOBIESOCIDAE 462
GOBIESOCIFORMES 462
GOBIIDAE 451
Gobiodon citrinus 452
   erythrospilus 452
   multilineatus 452
   oculolineatus 452
   okinawae 452
   quinquestrigatus 452
   ruvalabis 452
   verticalis 452

Gobraeus kazusensis 252
Goesella rotundata 80
Goezia gobia 209
   parvus 209
GOEZIIDAE 209
Golfingia elongata 236
   margaritacea margaritacea 236
   vulgaris vulgaris 236
GOLFINGIIDAE 236
GOLFINGIIFORMES 236
Gomeza bicornis 358
Gomophia egyptiaca egeriae 391
Gomphina aequilatera 255
   [*melanaegis*] 255
   [*veneriformis*] 255
Gomphonema salinarum 43
GOMPHONEMACEAE 43
Gomphonitzschia chinensis 46
Gomphosphaeria aponina 25
[*Gomphosus tricolor*] 439
   varius 439
Gomphrena celosioides 151
Gonapodasmius branchialis 199
   hainanensis 199
   pacificus 199
[*Gonatonotus crassus*] 339
GONEPLACIDAE 362
Goneplax renoculis 363
Goniada annulata 222
   emerita 222
   japonica 222
   littorea 222
   maculata 222
   multidentata 222
   paucidens 222
   uncinigera 222
   vorax 222
GONIADIDAE 222
Goniadopsis incerta 222
GONIASTERIDAE 390
Goniastrea aspera 184
   pectinata 184
   retiformis 184
   yamanarii 184
   ———— *profunda* 184
Gonichthys coccoi 414
Goniistius quadricoris 444
   zebra 444
   zonatus 444
Goniocidaris biserialis 392
Goniodiscaster forficulatus 390
   granuliferus 390
   scaber 390
Goniodoma polyedricum 55
   sphaericum 55
GONIODOMACEAE 55
GONIODORIDIDAE 290
Goniodoris castanea 290
   glabra 290
Gonionchus metavillosus 211
[*Gonionemus murbachii* v. *chekiangensis*] 172
   [———— v. *oshoro*] 172
   vertens 172
Goniopneustes pentagonus 393
Goniopora columna 182

Goniopora (continued)
   duofasciata 182
   gracilis 182
   minor 182
   wotouensis 182
GONIOTRICHACEAE 133
[*Goniotrichum alsidii*]
Gonocephalum pseudopubens 300
GONODACTYLIDAE 368
GONODACTYLOIDEA 368
Gonodactylus chiragra 368
   demani 368
   platysoma 368
   smithii 368
   ternatensis 368
GONORHYNCHIDAE 409
GONORHYNCHIFORMES 409
Gonorhynchus abbreviatus 409
Gonostoma atlanticus 411
   elongatus 411
   gracile 411
GONOSTOMATIDAE 411
Gonostomum pediculiforme 60
Gonothyraea clarki 172
   inorata 172
GONYAULACEAE 54
Gonyaulax ceratocoroides 54
   diegensis 54
   digitale 54
   fusiformis 54
   mitra 54
   pacifica 54
   polyedra 54
   polygramma 54
   spinifera 54
GOODENIACEAE 158
Gordoniella yonakuniensis 138
Gorgasia japonica 416
   taiwanensis 416
GORGODERIDAE 195
GORGONOCEPHALIDAE 395
Gorgonocephalus dolichodactylus 395
Gorgorhynchoides epinepheli 213
   satoi 213
Gossleriella tropica 30
Gotocotyle sawara 192
GOTOCOTYLIDAE 192
Gotoea typica 168
   [*Gouldia nana*] 254
   [*sulcata*] 254
Gracila albomarginatus 426
Gracilaria arcuata 136
   articulata 136
   asiatica 136
   ———— v. *zhengii* 136
   bangmeiana 136
   belinae 136
   blodgettii 136
   [*bursa-pastoris*] 136
   [*cacalia*] 137
   [*chorda*] 137
   chouae 136
   [*confervoides*] 136
   constricta 136
   coronopifolia 136
   corticata 136

Gracilaria (continued)
  [*crassa*] 137
  [*denticulata*] 137
  edulis 136
  eucheumoides 136
  firma 136
  [*foliifera*] 137
  [———— f. *textorii*] 136
  gigas 136
  hainanensis 137
  [*heteroclata*] 136
  [*lichenoides*] 136
  lingula 137
  megaspora 137
  mixtra 137
  [*minor*] 137
  punctata 137
  [*purpurascens* f. *spinulosa*] 137
  rubra 137
  salicornia 137
  sjoestedtii 137
  spinulosa 137
  tenuistipitata 137
  ———— v. *liui* 137
  textorii 137
  [*tsengiana*] 137
  [*verrucosa*] 136
  veillardii 137
GRACILARIACEAE 136
Gracilariopsis chorda 137
  lemaneiformis 137
Gradatiscala gradata pygmaea 267
GRAMINEAE 159
Grammatobothus krempfi 457
  polyophthalmus 457
Grammatomya squamosa 252
Grammatophora angulosa 43
  fundata 43
  ———— v. *spinosa* 43
  hamulifera 43
  marina 43
  oceanica 43
  ———— v. *macilenta* 43
  serpentina 43
  undulata 43
  ———— v. *japonica* 43
Grammatorcynus bicarinatus 449
GRAMMICOLEPIDAE
Grammistes sexlineatus 426
Grammoplites scaber 455
Grampus griseus 469
Grandidierella gilesi 329
  japonica 329
  megnae 329
Grania hongkongensis 234
  inermis 234
  stilifera 234
Grantia nipponica 165
Granulifusus nipponica 278
[*Granulilittorina philippiana*] 265
GRANULORETICULOEA 69
Graphium sp. 131
Graphonema amokurae 211
GRAPSIDAE 365
Grapsus albolineatus 365
  crinipes 365

Grapsus (continued)
  grayi 365
  longitarsis 365
  [*strigosus*] 365
  tenuicrustatus 365
Graptacme aciculum 261
  buccinulum 261
Grateloupia divaricata 136
  filicina 136
  ———— v. *lomentaria* 136
  gelatinosa 136
  imbricata 136
  latissima 136
  livida 136
  okamurae 136
  prolongta 136
  [*ramosissima*] 136
Gregariella coralliophaga 242
  splendida 242
  [*striatus*] 241
Grewia piscatorum 154
Griffithsia coacta 138
  japonica 138
  metcalfii 138
  rhizophora 138
  subcylindrica 138
  tenuis 138
  [*tenuis*] 138
Grigelis guttifera 91
  pyrula 91
  semirugosa 91
Grimpoteuthis umbellata 296
Griseobroviolaceus group 24
Griseofuscus group 24
GRUIDAE 466
GRUIFORMES 466
Grus grus lilfordi 466
  japonensis 466
  leucogeranus 466
  monacha 466
  vipio 466
Guembelitria vivans 99
GUEMBELITRIIDAE 99
Guernea longidactyla 329
  mackiei 329
  sombati 329
Guettarda speciosa 158
Guildfordia triumphans 264
  yoca 264
Guinardia blavyana 30
  flaccida 30
GUTTIFERAE 155
Guttulina austriaca 95
  basalis 95
  dawsoni 95
  hantkeni 95
  kisbinouyi 95
  lactea 95
  orientalis 95
  pacifica 95
  praelonga 95
  problema 96
  pulchella 96
  regina 96
  sadoensis 96
  silvestri 96

Gygis alba candida 468
Gyliauchen oligoglandulosus 194
GYLIAUCHENIDAE 194
Gymnangium expansa 171
  gracilicaulis 171
  hians 171
  speciosa 171
  vagae 171
[*Gymnapistes leucogaster*] 455
Gymnapogon annona 429
  japonicus 429
  philippinus 429
  urospilotus 429
GYMNOASCACEAE 130
[GYMNOBLA] 166
Gymnocanthus herzensteini 456
Gymnocranius griseus 434
  japonicus 434
GYMNODINIACEAE 51
GYMNODINIALES 51
Gymnodinium aeruginosum 51
  coeruleum 51
  microadriaticum 51
GYMNODORIDIIDAE 289
Gymnodoris albe 289
  bicolor 289
  [*citrina*] 289
  inornata 289
  japonica 289
  okinawea 289
  striata 289
Gymnogongrus flabelliformis 137
GYMNOGRAMMACEAE 150
GYMNOLAEMATA 370
Gymnomuraena concolor 416
  marmorata 416
  tigrina 416
  zebra 416
Gymnopataqus mirabile 395
Gymnophellia hutchingsae 180
Gymnosarda unicolor 450
GYMNOSOMATA 288
GYMNOSPERMAE 150
[*Gymnosporia diversifolia*] 154
Gymnothorax berndti 416
  boschi 416
  brunneus 416
  buroensis 416
  chilospilus 416
  chlamydatus 416
  eurostus 416
  favagineus 416
  fimbriatus 416
  flavimarginatus 416
  hepaticus 416
  javanicus 416
  kidako 416
  leucostingmus 416
  margaritophorus 416
  melanospilus 416
  melatremus 416
  meleagris 416
  monostigmus 416
  neglectus 416
  nudivomer 416
  pescadoris 416

Gymnothorax (continued)
   petelli 416
   pictus 416
   pindae 417
   polyuranodon 417
   pseudothyrsoideue 417
   punctatofasciatus 417
   reevesi 417
   reticularis 417
   richardsoni 417
   rueppelliae 417
   thyrsoideus 417
   undulatus 417
   zonipectus 417
Gymnura bimaculata 408
   japonica 408
   poecilura 408
GYMNURIDAE 408
GYNODIASTYLIDAE 324
Gynodiastylis anguicephala 324
Gynura divaricata 158
Gypsina globula 116
   vesicularis 116
Gyptis capensis 218
   labatus 218
   pacificus 218
GYRACANTHOCEPHALA 213
Gyractis stimpsoni 180
Gyrineum gyrinum 273
   pusillum 273
   roseum 273
Gyroidina altiformis 120
   broeckhiana 121
   depressa 121
   gemma 121
   lottensis 121
   multilocula 121
   neosoldani 121
   nipponica 121
   orbicularis 121
   quinqueloba 121
   soldanii 121
   tainanensis 121
Gyroidinoides nipponicus 120
   subzelandica 120
Gyrosigma acuminatum 37
   ——— v. *gallica* 37
   balticum 37
   ——— v. *brevius* 37
   ——— v. *sinensis* 37
   ——— v. *sinicum* 37
   compactum 37
   distortum 37
   fasciola v. *arcuata* 37
   ——— v. *sulcata* 37
   ——— v. *tenuirostris* 37
   grovei 37
   macrum 37
   nodiferum 37
   ——— v. *latum* 37
   obliquum 37
   rectum 37
   sciotense 37
   spencerii 37
   strigilis 37

Gyrosigma (continued)
   ——— v. *excentriraphe* 37
   tenuissimum 37
   terryanum 37
   wansbeckii 37
   wormleyi 37

H

Hacelia tuberculata 391
Haddonia torresiensis 74
HADDONIIDAE 74
HADROMERIDA 164
Hadropenaeus lucasii 336
   spinicauda 336
HAEMATOPDIDAE 467
Haematopus ostralegus osculans 467
Haemoxenicon elongatus 198
HALACARIDAE 297
HALACARINAE 297
Halaelurus burgeri 405
Halalaimus alatus 212
   gracilis 212
   inflata 212
HALECIIDAE 171
Halecium cymiforme 171
   flexile 171
   humile 171
   sessile 171
   tenellum 171
Halgerda japonica 291
   orientalis 291
   rubicunda 291
   xishaensis 291
Haliaeetus albicilla 466
   leucogaster 466
[*Halicampus grayi*] 424
   koilomatodon 424
   macrohynchus 424
   mataafae 424
   [*metaatae*] 424
Halicarcinus messor 356
   setirostris 356
Halichoanolaimus duodecimpapillatus 212
   wulaidaoensis 212
Halichoeres amboinensis 439
   argus 439
   bioceatus 439
   centiquadrus 439
   chrysus 439
   cyanopleura 439
   dussumieri 439
   hartzfeldi 439
   hortulanus 439
   hyrili 439
   leparensis 439
   margaritaceus 440
   marginatus 440
   melanochir 440
   melanurus 440
   miniatus 440
   nebulosus 440
   nigrescens 440
   ornatissimus 440

Halichoeres (continued)
   pelicieri 440
   poecilopterus 440
   prosopein 440
   purpurascens 440
   scapularis 440
   tenuispinis 440
   timorensis 440
   trimaculatus 440
Halichondria panicea 164
HALICHONDRIDA 164
HALICHONDRIDAE 164
Haliclona crassiloba 165
   palmata 165
   permolis 165
   subarmigera 165
HALICLONIDAE 165
Haliclystus auricula 175
   steinegeri 175
Halicmetus reticulatus 462
Halicreas conica 173
   mininum 173
   racovitzae 173
HALICREATIDAE 173
Halicyclops sinensis 314
Halieutaea fitzsmonsi 462
   fumosa 462
   indica 462
   nigra 462
   sinica 462
   stellata 462
Haligena viscidula 130
Halimeda cylindraceae 147
   discoidea 147
   gigas 147
   incrassata 147
   macroloba 147
   micronesica 147
   opuntia 147
   ——— f. *triloba* 147
   simulans 147
   taenicola 147
   velasquezii 147
   xishaensis 147
[*Halimede fragifer*] 361
   ochtodes 361
   tyche 361
Halimochirurgus alcocki 459
Haliomma erinaceum 64
   macrodoras 64
Haliommatidium müller 61
HALIOTIDAE 261
Haliotis asinina 261
   clathrata 261
   discus hannai 262
   diversicolor 262
   [——— *aquatilis*] 262
   [*gigantea discus*] 262
   ovina 262
   planata 262
   [*semistriata*] 262
   varia 262
   venusta 262
Haliotrema thysanophridis 192
Haliplanella luciae 180

Haliporoides sibogae 336
[*Haliporus aequalis*] 336
  [*equalis*] 336
  [*propinquus*] 336
  [*sibogae*] 336
  taprobanensis 336
[*Haliseris divaricata*] 142
Halistemma rubrum 174
Halitholus pauper 167
Halitiara formosa 167
Halitiarella ocellata 167
  minutum 167
Halitrephes maasi 173
Haloa binotata 285
  flavescena 285
  margaitoides 285
  ovalis 285
  rotundata 285
  vitrea 285
  yamagutii 285
Halobates germanus 299
  micans 299
  sericeus 299
Halocordyle cavolinii 168
  disticha 168
  grandis 168
  tiarella 168
HALOCORDYLIDAE 168
Halocyphina villosa 132
Halocypria globosa 304
HALOCYPRIDAE 302
HALOCYPRIFORMES 302
Halocypris brevirostris 304
  [*inflata*] 304
Halodule pinifolia 158
  uninervis 158
Halomitra philippinensis 182
Halonectria milfordensis 130
Halophila beccarii 159
  longicauda 374
  ovata 159
Haloplegma duperreyi 138
  polyspora 138
HALOPTERIIDAE 171
Halopteris diaphana 171
Haloptilus acutifrons 310
  austini 310
  fertilis 310
  longicirrus 310
  longicornis 310
  mucronatus 310
  ornatus 310
  oxycephalus 310
  paralongicirrus 310
  [*setuliger*] 310
  spiniceps 310
  tenuis 310
Halosarpheia abonnis 130
  fibrosa 130
  indica 130
  marina 130
  minuta 131
  ratnagiriensis 131
  retorquens 131
  viscosa 131

Halosarpheia sp. 131
HALOSAURIDAE 418
Halosauropsis affinis 418
Halosaurus ovenii 418
  sinensis 418
Halosphaera viridis 50
HALOSPHAERACEAE 50
Halosphaeria hamata 131
  salina 131
Halosydna brevisetosa 216
  [*nebulosa*] 216
Halosydnopsis pilosa 216
Halothrix lumbricalis 141
Halotydeus mollis 298
Halymenia dilatata 136
  durvillaei 136
  ——— v. *ceylanid* 136
  ——— v. *formosa* 136
  sinensis 136
HALYMENIACEAE 136
Halysisis ijimai 377
Hamacreadium hainanensis 197
  lethrini 197
  matabile 197
  synechogobii 197
  xishaene 197
Hamatopeduncularia elegans 192
  simplex 192
[*Haminea binotata*] 285
  [*japonica*] 285
Hamodactylus boschmai 340
Hamopontonia corallicola 340
Hanleya sinica 237
Hansenisca soldani 121
HANTKENINACEA 101
Hantzschia marina 46
  virgata 46
  ——— v. *gracilis* 46
Hanzawaia boueana 121
  convexa 121
  hoppoensis 121
  mantaensis 121
  nipponica 121
  nitidula 121
  rhodiensis 121
  sumitomoi 121
HAPALOCARCINIDAE
Hapalocarcinus marsupialis
Hapalogaster dentata 350
Hapalogenys kishinonyei 435
  mucronatus 435
  [*nigripinnis*] 435
  nitens 435
Hapalotrema flecterotestis 193
HAPLOPHRAGMIACEA 74
Haplophragmoides applanata 73
  australensis 73
  bradyi 73
  breviculus 73
  canariensis 73
  carinatum 73
  chilenum 73
  dibollensis 73
  emaciatum 73
  grandiformis 73

Haplophragmoides (continued)
  hetha 73
  lingfengensis 73
  membranaceum 73
  minimum 73
  neobradyi 73
  praecarinatum 73
  rugosa 73
  shikiyamaensis 73
  shenzhenensis 73
  sphaeriloculum 73
  taiwanensis 73
  trullisatum 73
HAPLOPHRAGMOIDIDAE 73
HAPLOPORIDAE 193
HAPLOSCLERIDA 164
Haploscoloplos elongatus 222
  kerguelensis 222
HAPLOSPANCHNIDAE 193
Haplosplanchnus cuneatus 193
  elongatus 193
  purii 193
Haplosyllis spongicala 219
  ——— *tentaculata* 219
[*Harengula nymphaea*] 410
  [*zunasi*] 410
Harmothoe asiatica 216
  dictyophora 216
  imbricata 216
  indica 216
  minuta 216
  nigricansa 216
Harmotrema linguiforme 204
HARMOTREMATIDAE 204
Harpa amouretta 279
  articularis 279
  conoidalis 279
  harpa 279
  major 279
  nobilis 279
HARPACTICIDAE 315
HARPACTICOIDA 314
Harpacticus uniremis 315
HARPIDAE 279
Harpiliopsis beaupresii 340
  depressa 340
[*Harpilius depressus*] 340
  [*miersi*] 340
Harpiniopsis vadiculus 330
Harpiosquilla annandalei 368
  harpax 368
  japonica 368
  raphidea 368
HARPIOSQUILLIDAE 368
Harpodon microchir 413
  nehereus 413
HARPODONTIDAE 413
HARRIMANIIDAE 399
Harriotta opisthoptera 409
Harrovia bituberculata 357
  elegans 357
  frontodentata 357
  longipes 357
  tuberculata 357
Hartmaniella fujianensis 228

Hartmaniella (continued)
    [cf. *crecta*] 228
HARTMANIELLIDAE 228
Hartmeyeria chinensis 403
HARTORIDA 56
Hastigerina aequilateralis 103
    pelagica 103
    siphonifera 103
Hastigerinella digitata 101, 101
HASTIGERINIDAE 103
Hastula diversa 282
    strigilata 282
    verreauxi 282
Hauerina atlantica 83
    bradyi 83
    diversa 83
    involuta 83
    orientalis 83
    pacifica 83
    speciosa 83
    trilocularis 83
HAUERINIDAE 83
HAUSTORIIDAE 329
Havelockia versicolor 387
[*Hazeus otakii*] 452
Hebella laterocaudata 171
    neglecta 171
    parasitica 171
Hedyotis pinifolia 158
[*Heikea japonica*] 354
Helcogramma habena 446
    obtusirostre 446
    striata 446
Helenina anderseni 111
HELENINIDAE 111
Helgicirrha brevistyla 170
    cornelii 170
    malayensis 170
Heliacus stramineus 266
    variegatus 266
Heliaster hexagonium 64
Helice formosensis 365
    japonica 366
    latimera 366
    leachii 366
    [*pingi*] 366
    tientsinensis 366
    [*tridens sheni*] 366
    [——— *wuane*] 366
    wuana 366
Helicolenus hilgendorfi 453
Helicometra epineli 197
    scorpaenae 197
    selaroidis 197
Heliconema brevispiculum 210
    heliconema 210
    longissima 210
Helicosalpa virgula 401
Helicostomella longa 59
Helicteres angustifolia 155
Heliodiscus asteriscus 64
    echiniscus 64
    phacodiscus 64
Heliofungia actiniformis 182
    ——— *palawensis* 182

Heliolithium aureum 61
Heliopora coerulea 186
HELIOPORACEA 186
HELIOPORIDAE 186
Heliosphaera octacantha 64
Heliotropium marifolium 157
    strigosum 157
Helminthocladia australis 133
    yendoana 133
HELMINTHOCLADIACEAE 133
Helotes cancellatus 435
    sexlineatus 435
HEMEROCOETIDAE 445
HEMEROMETRIDAE 386
Hemiaegina minuta 334
Hemiaulus chinensis 32
    hauckii 32
    indicus 32
    membranaceus 33
    sinensis 33
Hemicentrotus pulcherrimus 393
HEMICHORDATA 399
Hemicristellaria bradyi 93
    gemmata 93
Hemicyclopus dilatatus 314
Hemidiscus cuneiformis 31
    ——— v. *orbicularis* 31
    ——— v. *recta* 31
    ——— v. *ventricosa* 31
    hardmannianus 31
    ovalis 31
Hemifusus colosseus 277
    ternatanus 277
    tuba 277
Hemiglyphidodon plagiometopon 443
Hemigrapsus longitarsis 366
    penicillatus 366
    sanguineus 366
    sinensis 366
Hemigymnus fasciatus 440
    melapterus 440
Hemileucon bidentatus 325
HEMIMYARIA 401
[*Hemipenaeopsis villosus*] 336
[*Hemipenaeus crassipes*] 336
    [*tomentosus*] 335
    [*virilis*] 335
Hemipodus yenourensis 222
HEMIPTERA 299
Hemipteronotus aneitensis 440
    caerulopunctatus 440
    evides 440
    melanopus 440
    pentadactylus 440
    verrens 440
HEMIRAMPHIDAE 419
Hemiramphus dussumieri 419
    far 419
    georgii 419
    gernaeti 419
    intermedius 419
    limbatus 419
    lutkei 419
    marginatu 419
    melanurus 419

Hemiramphus (continued)
    paucirastris 419
    quoyi 419
    sajori 419
    sinensis 419
Hemisiriella parva 323
    pulchra 323
Hemisphaerammina bradyi 71
HEMISPHAERAMMINIDAE 71
Hemistepta lyrata 158
Hemitaurichthys polylepis 438
    zoster 438
Hemitoma panhi 262
[*Hemitriakis japonicus*] 406
Hemitripterus villosus 456
Hemityphis crustatus 333
    tenuimanus 333
HEMIURIDAE 200
Heniochus acuminatus 438
    chrysostomus 438
    monoceros 438
    permutatus 438
    singularius 438
    varius 438
Henricia aspera robusta 391
    leviuscula 391
    spiculifera 391
Hepomedus tener 335
Heptacarpus camtschaticus 343
    futilirostris 343
    geniculatus 343
    pandaloides 343
    rectirostris 343
Heptranchias dakini 404
    perlo 404
Herdmania momus 403
Heritiera littoralis 155
Herklotsichthys punctatus 410
    quadrimaculatus 410
[*Hermadion truncata*] 217
Hermadionella truncata 217
Hermaea dendritica 288
Hermania infantilis 287
Hermonia acantholpis 216
    hystrix 216
Hernandia nymphaeifolia 152
    sonora 152
HERNANDIACEAE 152
Heronallenia laevis 112
    parva 112
HERONALLENIIDAE 112
Heronidrilus bihamis 234
    fastigatus 234
    hutchingsae 234
    keenani 234
    virilis 234
Herpochondria elegans 138
Herpolitha limax 182
    weberi 182
Herposiphonia basilaris 139
    caespitosa 139
    fissidentoides 139
    fujianensis 139
    insidiosa 139
    parca 139

Herposiphonia (continued)
   pecten-veneris 139
   ramosa 139
   secunda 139
   subdisticha 139
   [*tenella*] 139
   [*terminalis*] 139
[*Hervia ceylonica*] 292
   [*serrata*] 292
Herviella yatsui 292
Hesione genetta 218
   intertexta 218
   pantherina 218
   splendida 218
HESIONIDAE 218
Hesionides arenaria 218
Hesionura shandongensis 215
Hespererato nanhaiensis 270
Hesperonoe hwanghaiensis 217
   sinagawaensis 217
Heteracon biformis 61
Heteralepas japonica 319
   quadrata 319
   smilium 319
HETERAULACACEAE 54
Heteraulacus polyedricus 54
Heterobothrium praeorchis 192
Heterobucephalopsis gymnothoracis 193
Heterocardia elliptica 249
Heterocarpoides laevicarina 344
Heterocarpus alphonsi 344
   dorsalis 344
   gibbosus 344
   lavigatus 344
   parvispina 344
   sibogae 344
   tricarinatus 344
   woodmasoni 344
Heterocentrotus mammillatus 393
HETEROCHEILIDAE 209
HETEROCHLORIDALES 50
[*Heterocossura aciculata*] 223
HETEROCOCCALES
Heterocrypta investigatoris 357
   transitans 357
Heterocuma sarsi 324
Heterocyathus acquicostatus 185
   alternatus 185
   japonicus 185
[*Heteroderma zostericola*] 135
Heterodesmus adamsii 302
HETERODINIACEAE 54
Heterodinium blackmanii 54
   gesticulatum 54
   globosum 54
   praetextum 54
   rigdenae 54
   whittingae 54
HETERODONTIDAE 404
HETERODONTIFORMES 404
Heterodontus japonicus 404
   zebra 404
HETEROHELICACEA 99
HETEROHELICIDAE
Heterolebes immaculosus 194

Heterolebes (continued)
   sinensis 194
   spari 194
Heterolepa dutemplei 120
   kezloyensis 120
   loxisutura 120
   praecincta 120
   pseudoungeriana 120
   sigmasutura 120
   subpraecincta 120
   tenera 120
   virosus 120
   xystrota 120
HETEROLEPIDAE 120
Heteromacoma irus 250
Heteromastus filiformis 225
   similis 225
Heteromycteris japonicus 458
HETEROMYOTA 237
HETERONEMERTEA 207
Heteropanope glabra 361
Heteropappus hispidus 158
Heterophyes heterophyes 204
HETEROPHYIDAE 204
Heterophysa tridontesa 227
Heteropilumnus ciliatus 361
   subinteger 361
Heteroplax dentata 363
   nagasakiensis 363
   nitida 363
Heteropodarke heteromorpha africana 218
Heteropora pelluculata 370
HETEROPORIDAE 370
Heteropsammia cochlea 185
   michelini 185
Heteropyramis maculata 175
Heteroralfsia saxicola 141
HETERORHABDIDAE 310
Heterorhabdus abyssalis 310
   clausi 310
   compactus 310
   longicornis 310
   papilliger 310
   spinifrons 310
   vipera 310
Heterosigma akashiwo 50
Heterosiphonia japonica 139
   pulchra 139
Heterosoma heptacanthum 64
[HETEROSOMATA] 456
Heterospio catalinensis 224
   sinica 224
HETEROSPIONIDAE 224
Heterosquilla latifrons 369
Heterostegina depressa 126
   longispina 126
   suborbicularis 126
Heterotiara anonyma 167
   minor 167
HETEROTRICHIDA 58
HEXABOTHRIIDAE 192
HEXABRANCHIDAE 289
Hexabranchus marginatus 289
   sanguineus 289

Hexacontium anaximandrii 63
   aff. *axotrias* 63
   castanetum 63
   enthacanthum 63
   favosum 63
   hostile 63
   mellarium 63
   pachydermum 63
   quadratum 63
   retrospiculum 63
   sarmentum 63
   senticetum 63
   typanum 63
Hexaconus ciliatus 61
Hexacromyum elegans 63
HEXACTINELLIDA 165
Hexadendron bipinnatum 63
HEXAGRAMMIDAE 455
Hexagrammos lagocephalus 455
   octogrammus 455
   otakii 455
HEXANCHIDAE 404
HEXANCHIFORMES 404
Hexanchus griseus 404
   vitulus 404
Hexangium sigani 194
Hexapus anfractus 363
   granuliferus 363
Hexapyle spinulosa 65
HEXASTEROPHORIDA 166
Hexastylus aristarchi 63
   dimensivas 63
   heracliti 63
   phaenaxonius 63
   philosophica 63
   cf. *pythagoraea* 63
   cf. *triaxonius* 63
Hexatrygon bickelli 408
   brevirostra 408
   longirostrum 408
   taiwanensis 408
   yangi 408
HEXATRYGONIDAE 408
Hiantopora intermedia 374
HIANTOPORIDAE 374
[*Hiatella arctica*] 256
   orientalis 256
HIATELLIDAE 256
Hiatula acuta 252
   ambigua 252
   atrata 252
   castanea 252
   chinensis 252
   diphos 252
   inflata 252
   minor 252
   rostrata 252
   togata 252
   violacea 252
   virescens 252
Hibiscus surattensis 154
   tiliaceus 154
Hicksonella guishanensis 187
   princeps 187
Hildenbrandia prototypus 134

Hildenbrandia (continued)
    rivularis 134
HILDENBRANDIACEAE 134
HIMANTOLOPHIDAE 462
Himantolophus groenlandicus 462
Himantopus h. himantopus 467
Hime japonicus 413
Hincksia breviarticulata 141
    granulosus 141
    mitchellae 141
Hincksina curvimandibulata 373
Hindsia sinensis 277
    suturalis 277
Hippa adactyla 351
    truncatifrons 351
Hippaliosina acutirostris 379
    calyciformis 379
    hippocrepis 379
    hongkongensis 379
    ovicellata 379
    triformis 379
Hippellozoum pectinatum 381
[*Hippichthys cyanospilus*] 424
    heptagonus 424
    [*spicifer*] 424
HIPPIDAE 351
HIPPIOIDEA 351
Hippocampus coronatus 424
    [*erinaceus*] 424
    histris 424
    japonicus 424
    kelloggi 424
    kuda 424
    trimaculatus 424
HIPPOCREPINACEA 71
HIPPOCREPINIDAE 71
Hippoflustra bifurcillata 373
    granulata 373
    inarmata 373
Hippoglossoides dubius 457
[*Hippolysmata ensirostris*] 343
    [*vittata*] 343
HIPPOLYTIDAE 343
[*Hippolyte camtschaticus*] 343
    [*cristata*] 338
    [*emboinensis*] 344
    [*ensis*] 344
    [*geniculatus*] 343
    [*gracilirostris*] 343
    [*kuekenthali*] 343
    [*leptognathus*] 343
    [*pandaloides*] 343
    [*paschalis*] 344
    [*pectinifera*] 344
    [*planirostris*] 343
    [*rectirostris*] 343
    ventricosa 343
[HIPPONICIDAE] 268
Hipponix foliacea 268
Hippopetraliella magna 380
HIPPOPODIIDAE
Hippopodina feegeensis 378
    perforata 378
HIPPOPODINIDAE 378
Hippopodius hippopus 174

Hippoporella spinigera 379
Hippoporidra calcarea 381
HIPPOPORIDRIDAE 381
Hippoporina porcellana 379
    [*spinifera*] 379
    verrucosa 379
HIPPOPORINIDAE 378
Hippopus hippopus 249
[*Hipposcarus longiceps*]
Hippothoa distans 377
    divaricata 377
    flagellum 377
HIPPOTHOIDAE 377
HIPPOTHOOIDEA 377
HIPPOTHOOMORPHA 377
Hirudinella beebei 203
    xishaensis 203
HIRUDINEA 235
HIRUDINELLIDAE 203
Hirundapus c. caudcutus 468
    cochinchinensis 468
Hirundichthys oxycephalus 419
    [*rondeleti*] 419
    rondeletii 419
    [*speculiger*] 419
HISPIDOBERYCIDAE 422
Hispidoberyx ambagiosus 422
Histioneis biremis 51
    depressa 51
    hisppoperoides 51
    hyalina 51
    pulchra 51
HISTIOPHORIDAE 450
Histiophorus gladius 450
    orientalis 450
    platypterus 450
Histiophryne cryptacanthus 462
HISTIOPTERIDAE 438
Histiopterus typus 438
HISTIOTEUTHIDAE 294
Histioteuthis celetaria pacific 294
    dofleini 294
    meleagroteuthis 294
Histrio histrio 462
HISTRIOBDELLIDAE 228
Hoglundina bradyi 98
    elegans 98
HOLACANTHA 61
Holacanthus trimaculatus 438
    venustus 438
HOLECTYPOIDA 393
Holenthias borbonius 426
    chrysostictus 426
Holmesiella affinis 323
HOLOCENTRIDAE 422
Holocentrus bleeker 422
    [*caudimaculatus*] 422
    [*cornutus*] 422
    [*furcatus*] 422
    [*lacteoguttatus*] 422
    [*laevis*] 422
    [*microstomus*] 422
    [*opercularis*] 422
    [*ruber*] 422
    [*sammara*] 422

Holocentrus (continued)
    [*spinifer*] 422
    [*tiere*] 422
    [*violaceus*] 422
Hologymnosus annulatus 440
    doliatus 440
    semidiscus 440
Hololapidella commensalis 217
Holosticha diademata 60
    kessleri 60
    lacazei 60
    rubra 60
    simplicis 60
Holotholus histricosa 67
Holothuria albiventer 388
    arenicola 388
    atra 388
    axiologa 388
    cinerascens 388
    dicrepans 388
    difficilis 388
    edulis 388
    flavomaculata 388
    fuscocinerea 388
    gracilis 388
    hilla 388
    impatiens 388
    inhabilis 388
    leucospilota 388
    martensi 388
    moebi 388
    multipilula 388
    nobilis 388
    ocellata 388
    olivacea 388
    pardalis 388
    pervicax 388
    rigida 388
    scabra 388
    sinica 388
    spinifera 388
    verrucosa 388
HOLOTHURIIDAE 388
HOLOTHUROIDEA 387
Homalocantha anatomica 275
    zamboi 275
Homalogastra binucleata 57
HOMALOMETRIDAE 195
Homalopoma amussitatum 264
Homoeothrix rubra 27
HOMOIODORIDIDAE 291
Homoiodoris japonica 291
Homola orientatis 353
HOMOLIDAE 353
Homologenus sinensis 353
    valdiviae 353
HOMOPTERA 300
HOMOSCLEROPHORIDA 164
Homostolus japonicus 421
Homotrema rubrum 116
HOMOTREMATIDAE 116
[*Hopkinsia plana*] 290
Hopkinsiella hiroi 290
Hopkinsina pacifica 106
[*Hoplegnetus fasciatus*] 438

Index

Hoplegnetus (continued)
 [*punctatus*] 438
HOPLICHTHYIDAE 456
Hoplichthys fasciatus 456
 gilberti 456
 langsdorfi 456
 regani 456
Hoplobrotula armata 421
HOPLONEMERTEA 208
Hoplosebastes armatus 453
Hoplostethus crassispinus 422
 mediterraneus 422
Hopperia hexadentata 212
HORMATHIIDAE 179
Hormathonema enilithicum 25
Hormiphora palnata 190
Hormiphora sp. 190
Hormomya mutabilis 242
 sinensis 242
Hormophysa articulata 143
[*Hormosiar articulata*] 143
Hormosina globulifera 73
 monile 73
 mormanni 73
 spiculifera 73
HORMOSINACEA 71
Hormosinella distans 72
HORMOSINIDAE 72
Howella sherborni 429
Hozukius embremarius 453
Huenia brevifrons 356
 proteus 356
Humicola alopallonella 131
 fuscoatra 131
Humicola sp. 132
Hurleytrema hainanensis 195
HYADESIIDAE 298
Hyalarcta sp. 300
Hyale dollfusi 329
 grandicornis 329
 honoluluensis 329
 schmidti 329
HYALIDAE 329
Hyalinea balthica 114
Hyalinoecia tubicola 226
Hyalinonetrion elongata 94
Hyalocyliz striata 288
Hyalodiscus ambiguus 30
 radiatus 30
 [*stelliger*] 30
 subtilis 30
Hyalonema apertum 165
 proximum 165
 toxeres 165
HYALONEMATIDAE 165
Hyalosiphonia caespitosa 134
[*Hyalosynedra laevigata*] 44
Hyaloteuthis pelagica 295
Hyas cocarctatus alutaceus 356
 —— *ursinus* 356
Hyastenus diacanthus 356
 pleione 356
Hybocodon amoyensis 168
 [*forbesii*] 168
 octopleurus 168

Hybocodon (continued)
 prolifer 168
Hyboscolex pacificus 226
Hydatina albocincta 285
 physis 285
 zonata 285
HYDATINIDAE 285
Hydnophora contignatio 184
 exesa 184
 microconos 184
 rigida 184
HYDRACTINIIDAE 167
HYDROBATIDAE 465
HYDROCHARITACEAE 159
Hydroclathrus clathratus 142
 tenuis 142
Hydrocoleus cantharidosmus 25
 glutinossum 25
 lyngbyaceus 25
Hydrocoryne miurensis 168
HYDROCORYNIDAE 168
Hydroides albiceps 230
 centrospina 230
 dirampha 230
 elegans 230
 exaltatus 230
 ezoensis 230
 fusca 230
 fusicola 230
 helmatus 230
 inornata 230
 longispinosa 230
 longistylaris 230
 [*lunulifera*] 230
 minax 230
 multispinosa 230
 nanhaiensis 230
 [*norvegica*] 230
 prisca 230
 protulicola 230
 rhombobulus 230
 tambalalmensis 230
 trilobulus 230
 tuberculata 230
 [*uncinata*] 230
 xishaensis 230
HYDROIDOMEDUSAE 166
HYDROIDSI 166
Hydrolagus isengi 409
[*Hydrolithon boergesenii*] 135
 [*reinboldii*] 135
HYDROMEDUSAE 166
Hydronectria tethys 131
Hydrophasianus chirurgus 466
HYDROPHIIDAE 464
Hydrophis caerulescens 464
 cyanocinctus 464
 fasciatus atriceps 464
 melanocephalus 464
 ornatus 464
[HYDROPOLYPSE] 166
Hydroprogne t. tschegrava 468
Hydrosera triquetra 33
 whampoensis
HYDROZOA 166

Hyella caespitosa 25
 simplex 25
HYELLACEAE 25
Hygophum atratum 414
 macrochir 414
 proximum 414
Hygroscopicus group 24
Hymenachne assamica 160
Hymeniacidon sanguinea 164
HYMENIACIDONIDAE 164
Hymeniastrum euclidis 65
Hymenocephalus gracilis 420
 lethonemus 420
 longiceps 420
 striatissimus 420
Hymenocera elegans 340
 [*nipponensis*] 340
HYMENOCERIDAE 340
Hymenogadus kuronumai 420
Hymenopenaeus aequalis 336
 halli 336
 [*lucasii*] 336
 propinquus 336
 [*sibogae*] 336
HYMENOSOMATIDAE 356
HYMENOSTOMATIDA 57
Hyotissa hyotis 246
Hyperammina bradyi 71
 clavellata 71
 clavigera 71
 cylindrica 71
 distorta 71
 elongata 71
 friabilis 71
 laevigata 71
 maxima 71
 novaezealandiae 71
Hypererythrops spinifera 323
 zimmeri 323
[*Hyperia bengalensis*] 331
 [*dysschistus*] 331
 galba 331
 [*latissimus*] 331
 [*macrophthalmus*] 331
 [*schizogeneios*]
 [*sibaginis*] 331
Hyperiella antarctica 331
Hyperietta luzoni 331
 parviceps 331
 stebbingi 331
 stephenseni 331
 vosseleri 331
HYPERIIDAE 331
HYPERIIDEA 330
Hyperioides longipes 331
 sibaginis 331
Hyperionyx macrodactylus 331
Hyperoche martinezi 331
 mediterranea 331
 medusarum 331
 picta 331
Hyperoglyphe japonicus 450
Hypnea boergesenii 136
 cervicornis 136
 charoides 136

Hypnea (continued)
   chordacea 136
   japonica 136
   musciformis v. *esperi* 136
   [*nidulans*] 136
   pannosa 136
   [*seticulosa*] 136
HYPNEACEAE 136
Hypoatherina tsurugae 418
   valenciennei 418
   [*woodwardi*] 418
Hypocepon globosus 326
Hypocreadium drepanei 195
Hypodytes indicus 455
   longispinis 455
[*Hypogaleus hyugauensis*] 405
Hypoglossum attenuatum 139
   geminatum 139
Hypohepaticola setipinnae 201
Hypophrus longirostris 353
Hypophryxus filiformis 326
Hypoporus phylloides 195
Hypoprion atripinnis 406
   macloti 406
[*Hyporhamphus dussumieri*] 419
   [*gernaeti*] 419
   [*intermedius*] 419
   [*limbatus*] 419
   [*melanurus*] 419
   [*paucirastas*] 419
   [*quoyi*] 419
   [*sajori*] 419
   [*sinensis*] 419
   taiwanensis 419
   yuri 419
HYPOTRICHIDAE 60
Hypoxylon oceanicum 131
Hypselodoris festiva 290
   kanga 290
   maritima 290
   [*placida*] 290
Hypseleotris bipartita 451
Hypsicomus phaeotaenia 230
Hysterolecitha arii 201
   chirocentri 201
   progonimus 201
   rosea 201
Hysterolecithoides epinepheli 201
   multiglandularis 201
Hystrichaspis dorsata 61
   fruticata 61
[*Hyxbocodon forbesii*] 168

I

Iasis zonaria 401
Ibacus ciliatus 346
   novemdentatus 346
Ibla cumingi 318
IBLIDAE 318
Ichthyascaris lophii 209
Ichthyococcus ovatus 411
Ichthyoscopus lebeck 445
[*Icticus pellucidus*] 450
Idanthyrsus pennatus 228

Idasola japonica 242
IDIACANTHIDAE 412
Idiacanthus fasciola 412
Idiellana pristis 172
IDIOSEPIIDAE 296
Idiosepius paradoxa 296
Idmonea irregularis 369
   parasitica 369
   radiata 369
Idotea ochotensis 326
   stenops 326
IDOTHEOIDAE 326
Idunella curvidactyla 329
   janisae 329
   pauli 329
   serra 329
Ihlea punctata 401
Ijimaia dofleini 415
Ikeda taenioides 237
IKEDAIDAE 237
Ikedosoma qingdaoense 237
Ilisha elongata 410
   indica 410
   melastoma 410
ILLIOSENTIDAE 213
Ilyoplax dentimerosa 364
   deschampsi 364
   formosensis 364
   ningpoensis 364
   pingi 364
   serrata 364
   tansuiensis 364
Imperata cylindrica v. *major* 160
Inaequalina concavo-convexa 82
   disparilis 82
INARTICULATA 384
Indigofera hirsuta 153
Indocrassatella oblongata 248
Indocucullanus muraenecis 209
Indopinnixa mortoni 363
Indorhynchus arii 213
Inegocia guttatus 455
   japonicus 455
Inermocheres quadratus 315
Inermonephtys gallardi 222
   cf. *inermis* 222
INGOLFIELLIDAE 334
INGOLFIELLIDEA 334
Iniistius dea 440
   pavo 440
Inimicus cuvieri 454
   didactylus 454
   japonicus 454
   sinensis 454
Inioteuthis japonica 296
   maculosa 296
INOVICELLINA 371
Inquistor flavidula 281
   vulpionis 281
INSECTA 299
Intranstylum asellicola 58
   intermedium 58
Intusatrium crassum 195
Involutina anguillae 80
   intermedia 81

Involutina (continued)
   tenuis 81
INVOLUTINIDAE 80
INVOLUTININA 80
Involvohauerina cribrostoma 86
Iodictyum axillare 381
   gibberosum 381
   polycrenulatum 382
   willeyi 382
IOSPILIDAE 215
Iphiana lischkei 284
   tenuisculpta 284
Iphiculus spongiosus 355
Iphinoe gurjanovae 324
   tenera 324
[*Iphione hirotai*] 217
   muricata 217
IPNOPIDAE 415
Ipnops pristibrachium 415
Ipomoea maxima 157
   pes-caprae 157
   stolonifera 157
Iracundus signifer 453
Iravadia bombayana 265
   ornata 265
   quadrasi 265
Ircinia pinna 164
   variabilis 164
[*Irenopsis hexanemalis*] 170
IRIDACEAE 162
Iris ensata 162
IRONIDAE 212
Irus irus 255
   [*mitis*] 255
Isactis plana 26
[*Ischaemum aristatum* v. *meyenianum*] 160
   barbatum 160
   indicum 160
Ischnochiton alatus 238
   boninensis 238
   comptus 238
   hakodadensis 238
ISCHNOCHITONIDAE 238
ISCHYROCERIDAE 329
[*Ishige foliacea*] 141
   okamurai 141
   sinicola 141
ISHIGEACEAE 141
Isis hippuris 187
   minorbrachyblasta 187
   reticulata 187
Isistius brasiliensis 407
Islandiella californica 105
   islandica 105
   kattoi 105
   seranensis 105
Iso flosmaris 418
   rhothophilus 418
Isobactrus luxtoni 298
   obesus 298
[*Isocardia lamarckii*] 253
   [*vulgaris*] 254
Isochaeta ovalis 310
Isocheles sp. 349

Index

Isocirrus planiceps 225
  cf. *watsoni* 225
Isoconcha sibogai 248
ISOCRINIDAE 386
[*Isognomon acutirostris*] 244
  ephippium 244
  isognomum 244
  legumen 244
  nudeus 244
  pernum 244
ISOGNOMONIDAE 243
Isolda pulchella 229
  whydahaensis 229
Isolima limopsis 245
ISONIDAE 418
ISOPARORCHIIDAE 203
ISOPODA 325
Isthmia japonica 33
  minima 33
  nervosa 33
[*Isthnoles schmidt*] 131
Istiblennius bilitonensis 446
  coronatus 446
  cynostigma 446
  [*dussumieri*] 446
  edentulus 446
  [*edentulus*] 446
  interruptus 446
  [*lineatus*] 446
  mulleri 446
  [*periophthalmus*] 447
[*Istigobius campbelli*] 451
  [*ornatus*] 451
[*Istiophorus platypturus*] 250
ISURIDAE 404
ISURIFORMES 404
Isurus glaucus 404
  [*oxyrinchus*] 404
  paucus 404
Iwatanemertes piperatus 207
Ixa cylindrus 355
  edwardsii 355
Ixeris debilis 158
  japonica salsuginosa 158
  repens 158
Ixobrychus cinnamomeus 465
  eurhythmus 465
  s. sinensis 465
Ixoides cornutus 355

J

JACANIDAE 466
Jaculella acuta 71
  obtusa 71
Jadammina macrescens 76
  planata 76
Jamiesoniella athecata 234
  enigmatica 234
Janczewskia ramiformis 139
Jania adhaerens 135
  capillacea 135
  crassa 135
  [*decussato-dichotoma*] 135
  radiata 135

Jania (continued)
  ungulata 135
  ——— f. *breviov* 135
Janthina janthina 267
  prolongata 267
JANTHINIDAE 267
Japelion latus 277
Japetella diaphana 296
Japonolaeops dentatus 457
Jassa falcata 329
Jatropha curcas 154
[*Joannisiella oblonga*] 247
  [*tsuchii*] 247
Jocaste japonica 340
  lucina 340
Johannesbaptistia pellucida 25
Johinophyllum formiae 201
  qingdaoensis 201
Johnius amblycephalus 431
  belengeri 431
  carutta 431
  dussumieri 431
  fasciatus 431
  macrorhynus 431
  [*sina*] 432
  [*tingi*] 432
Jonas distincta 358
[*Jorunna funebris*] 291
  tomentosa 291
Jouannetia cumingi 257
  globulosa 257
JUNCACEAE 162
JUNCAGINACEAE 159
Junceella fragilis 187
  gemmacea 187
  juncea 187
  racemosa 187
  squamata 187
Juncellus serotinus 161
Juncus effusus 162
  gracillimus 162
  haenkei 162
[*Jousseaumia serratidigicus*] 343
  [*sibogae*] 343
Justitia japonica 346
  longimanus 346

K

Kalinga ornata 289
Kamoharia megastoma 457
Kanaka pelagica 167
Kandelia candel 156
Kareius bicoloratus 457
Karreria fallax 121
Karreriella bradyi 78
  chilostoma 78
  cylindrica 78
  hantkeniana 78
  parkerae 78
  pupiformis 78
  shangtaoensis 78
KARRERIIDAE 121
Karrerulina apicularis 76
  cylindrica 77

KARYORELICTIDA 56
[*Katelysia eugibba*] 255
  [*rimularis*] 255
KATHANIDAE 210
Kathlania leptura 210
Katsuwonus pelamis 450
[*Kefersteima dubia*] 218
Kellia japonica 247
  porculus 247
KELLIELLIDAE 253
Kempina mikado 368
Kentrocapros aculeatus 460
KENTRODORIDIDAE 291
Kentrodoris maculosa 291
  rubescens 291
Keramosphaera tergestina 90
KERAMOSPHAERIDAE 90
Keratella cochlearis 214
  quadrata 214
  valga 214
Kerilia jerdonii siamensis 464
KINETOFRAGMINOPHORA 56
Kinetoskias oblongata 375
KINORHYNCHA 208
Kinorhyncha sp. 209
KIRCHENPAUERIIDAE 171
Kishinouyea nagatensis 176
Kochia scoparia 151
Kochichthys flavofasciata 444
Koellikerina constricta 166
  diforficulata 166
  fasciculata 166
  heteronemalis 166
  multicirrata 166
  octonemalis 166
  taiwanensis 166
KOFOIDINIACEAE 51
Kofoidinium splendens 51
  velelloides 51
Kogia breviceps 469
  simus 469
Koleolepas avis 319
Kollikeria orientalis 199
  sphyraenae 199
[*Konosirus punctatus*] 410
Koremaster evaulus 390
Kormannia leptoderma 144
KORMANNIACEAE 144
Koseiria xishaense 195
Kowala coval 410
Kraussia integra 358
  nitida 358
  obliquefrons 358
  rugulosa 358
Krohnitta pacifica 385
  subtilis 385
Kuhlia marginata 428
  [*mugil*] 428
  rupestris 428
  taeniura 428
KUHLIIDAE 428
Kullinga brevifolia 161
Kumococius detrusus 455
  [*rodericensis*] 455
[*Kurosioia fascialis*] 266

KYPHOSIDAE 435
Kyphosus bigibbus 435
　cinerascens 435
　lembus 435
Kyrtuthrix maculans 26

L

LABIATAE 157
Labidocera acuta 311
　acutifrons 311
　bataviae 311
　bipinnata 311
　detruncata 311
　euchaeta 311
　kroyeri 311
　laevidentata 311
　minuta 311
　pavo 311
　sinilobata 311
Labidodemas semperianum 389
Labidoplax dubia 389
　incerta 389
Labiger sagamiensis
Labioporella cornuta 376
[*Labracinus cyclophthalmus*] 427
Labrichthys cyanotaenia 440
　unilineatus 440
LABRIDAE 438
Labrifer gymnocrini 195
Labroides bicolor 440
　dimidiatus 440
Labropsis manabei 440
Labrorostratus parasiticus 227
Labrospira columbiensis 73
　crassimargo 73
　evolutum 73
　kosterensis 73
Labuanium rotundatum 366
Lachnopodus bidentatus 361
　subacutus 361
Lacrymaria marina 56
LACTARIIDAE 429
Lactarius lactarius 429
LACTOBACILLACEAE 22
Lactobacillus 22
[*Lactophry gibbosus*] 460
Lactoria cornutus 460
　diaphanus 460
　fronasini 460
Lactuca indica 158
Laculatina quadrilatera 95
LACUNIDAE 265
LACYDONIIDAE 216
LAEMOPHIURIDA 395
Laeops kitakarae 457
　lanceolatata 457
　parviceps 457
　tungkongensis 457
Laetmonice brevepinnata 216
　filicornia 216
　japonica 216
Laevicardium attenuatum 248
　lobulatum 248
Laevidentalium eburneum 261

Laevipeneroplis bulloides 90
　crassa 90
　malayensis 90
Laevispinereis fujianensis 220
Lafoea dumosa 171
LAFOEIDAE 171
LAGANIDAE 394
Laganum decagonale 394
　depressum 394
　fudsiyama 394
Lagena acuticosta 94
　alcocki 94
　amphora 94
　apiopleura 94
　aspera 94
　clavata 94
　costata 94
　crenata 94
　dentaliniformis 94
　desmophora 94
　distoma 94
　distorta 94
　doveyensis 94
　elegantissima 94
　elongata 94
　excentrica 94
　exsculpta 94
　feildeniana 94
　flatulenta 94
　flintiana 94
　gibbera 94
　globosa 94
　gracilis 94
　gracillima 94
　gortanii 94
　hispida 94
　hispidula 94
　hyugaensis 95
　interrupta 95
　laevicostata 95
　laevis 95
　longispina 95
　meridionalis 95
　mississippiensis 95
　nebulosa 95
　oxystoma 95
　pellita 95
　perlucida 95
　pliocenica 95
　plumigera 95
　pseudocostata 95
　quadrilatera 95
　semilineata 95
　semistriata 95
　sesquistriata 95
　setigera 95
　spicata 95
　spinigera 95
　stavensis 95
　striata 95
　striatifera 95
　striatopunctata 95
　subamphora 95
　substriata 95
　sulcata 95

Lagena (continued)
　sulcatospicata 95
　tenuis 95
　wiesneri 95
　williamsoni 95
Lagenammina asymmetrica 70
　atlantica 70
　difflugiformis 70
　longicollis 70
　magna 70
　pseudodifflugiformis 70
　rhombiformis 70
　testacea 70
LAGENIDAE 94
LAGENINA 90
Lagenodelphis hosei 469
Lagenonodosaria bilocularis 93
　hirsuta 93
　hispida 93
　leizhouensis 93
　pauciloculata 93
　pyrula 93
　scalaris 93
　semirugosa 93
　variabilis 93
Lagenorhynchus obliquidens 469
Lagenosolenia disjungens 98
Lagis bocki 228
　neapolitana 228
Lagnea radiata 96
Lagocephalus gloveri 461
　inermis 461
　[*lunaris*] 461
　oceanicus 461
　[*pachygaster*] 461
　[*sceleratus*] 461
　[*spadiceus*] 461
　[*suezensis*] 461
　wheeleri 461
Lamarckina fuchsis 98
　scabra 98
　ventricosa 98
Lambis chiragra 268
　[*chiragra*] 268
　crocata 268
　[*crocata*] 268
　lambis 268
　[*lambis*] 268
　millepeda 268
　scorpius 268
　truncata sebae 268
　[*truncata sebae*] 268
Lamellana gymnota 289
Lamellaria kiiensis 270
　[*latens*] 270
LAMELLARIIDAE 270
[LAMELLIBRANCHIA] 239
Laminaria japonica 143
　[*ochotensis*] 143
LAMINARIACEAE 143
LAMINARIALES 143
Laminononion tumidum 119
[LAMNIDAE] 404
[LAMNIFORMES] 404
Lampadena luminosa 414

Index

Lampadena (continued)
   speculingera 414
Lampanyctus alatus 414
   bensoni 414
   hubbsi 414
   macropterus 414
   niger 414
   nobilis 414
   omostigma 414
   punctatissimus 414
   tenuiformis 414
Lampetra japonica 404
LAMPRIDAE 423
LAMPRIDIFORMES 423
Lampris guttatus 423
Lamprocyclas maritalis 68
   ——— *polypora* 68
Lamprocyrtis neoheteroporas danaes 68
Lamprodiscus quadricuspis 67
Lamprohaminoea cymbalum 286
Lamprometra palmata palmata 386
Lampromitra amabilis 67
   cachoni 67
   coronata 67
   quadricuspis 67
   sinuosa 67
LAMPROPIDAE 324
Lamprops hexaspinula 324
Lanassa capensis 229
Lancella bella 284
Lanceola pacifica 330
LANCEOLIDAE 330
Lanceopora formosa 381
LANCEOPORIDAE 381
[*Langerhansia cornuta*] 219
   [*rosea*] 219
Lanice auricula 229
   conchilega 229
   socialis 229
Lanocira gardinieri 325
Laodicea indica 170
   undulata 170
LAODICEIDAE 170
Laomedea calceolifera 172
   flexuosa 172
Laomedia astacina 345
LAOMEDIIDAE 345
Laonice cirrata 223
LAOPHONTIDAE 315
Laophontodes himatus 315
Lapemis hardwickii 464
Laputa kneri 459
LAQUEIDAE 384
[*Laramichthys crocea*] 432
   [*polyactis*] 432
LARCOIDEA 65
Larcopyle butschlii 65
Larcospira minor 66
   quadrangula 66
LARIDAE 468
LARIFORMES 468
Larus a. cachinnans 468
   argentatus 468
   brunnicephalus 468
   canus kamtschatschensis 468

Larus (continued)
   crassirostris 468
   glaucescens 468
   ichthyaetus 468
   ridibundus 468
   saundersi 468
   schistisagus 468
Laryngosigma caudata 98
   hyalascidia 98
   lauta 98
   ovata 98
   williamsoni 98
Lasaea nipponica 247
   rubra 247
   [*undulata*] 247
LASAEIDAE 247
Lasiotocus cryptostoma 195
Latecella reniformis 110
Lateolabrax japonicus 426
Laternula anatina 258
   boschasina 258
   marilina 258
   nanhaiensis 258
   [*pechliensis*] 258
   truncata 258
   [*valenciennesii*] 258
LATERNULIDAE 258
Laterostomella spinosa 99
Lates calcarifer 425
[*Lathophthalmus smargdinus*] 285
Lathyrus maritimus 153
[*Latiaxis gemmatus*] 276
   gyratus 276
   lischkeanus 276
   mawae 276
   pagodus 276
   [*pagodus*] 276
Laticarinina halophora 113
   pauperata 114
Laticauda colubrina 464
   laticaudata 464
   semifasciata 464
LATIDAE 425
Latirus craticulatus 278
   [*filamentosa*] 278
   polygonus 278
   smaragdula 278
   [*smaragdula*] 278
   turritus 278
Latitiara orientalis 167
Latona cuneata 250
   faba 250
Latonopsis sp. 300
Latreillia phalangium 354
   valida 354
LATREILLIDAE 354
Latreillopsis bispinosa 354
   tetraspinosa 354
Latreutes anoplonyx 343
   laminirostris 343
   mucronatus 343
   planirostris 343
Lauderia annulata 30
   borealis 30
Launaea sarmentosa 158

LAURACEAE 152
Laurencia articulata 139
   capitulifomis 139
   cartilaginea 139
   changii 139
   chinensis 139
   composita 139
   decumbens 140
   [*fasciculata*] 140
   flexilis setch v. *tropica* 140
   galtsoffi 140
   hongkongensis 140
   intermedia 140
   [*japonica*] 140
   jejuna 140
   longicaulis 140
   majuscula 140
   mariannensis 140
   microcladia 140
   nipponica 140
   obtusa 140
   ——— v. *divaricata* 140
   okamurai 140
   paniculata 140
   papillosa 140
   parvipapillata 140
   pinnata 140
   silvai 140
   subsimplex 140
   surculigera 140
   tenera 140
   tristicha 140
   undulata 140
   venusta 140
   verticillata 140
Lavendulae group 24
Lavisquilla inermis 368
Laxohyperia vespuliformis 331
Leachia pacifica 294
Leamodonda exatum 294
   punctatostriata 294
   punctigera 294
[*Leander annandalei*] 340
   [——— *stylirostris*] 340
   [*deschampsi*] 340
   [*gravieri*] 340
   [*longirostris* v. *carinatus*] 340
   [*longirostris japonicus*] 340
   [*modestus*] 340
   [*pacifica*] 341
   [*potamaiscus*] 340
   [*serrifer*] 341
   [*sewelli*] 341
   urocaridella 340
Leandrites antonbruunii 340
   [*cyrtorhynchus*] 340
   deschampsi 340
   [*longipes*] 340
[*Leanira izuensis*] 217
   [——— *hwanghaiensis*] 217
   japonica 217
   [*japonica*] 218
Leathesia difformes 141
   nana 141
   saxicola 141

LEATHESIACEAE 141
LECANICEPHALIDAE 207
Lecithaster atropi 201
   confusus 201
   fusiformis 201
   sayori 201
   setipinnae 201
   tylosuri 201
Lecithochirium albulae 201
   apharei 201
   branchialis 201
   caudiparum 201
   dongshanensis 201
   fistulariae 201
   fusiforme 201
   holocentri 201
   kawakawa 201
   kawalea 201
   keokeo 201
   magnaporum 201
   michthydis 201
   microstomum 201
   monticolli 201
   muraenesocis 201
   pacificum 201
   paracanthi 201
   poecilopsettae 201
   pseudosciaenae 201
   rufoviride 201
   savalae 201
   scomberomori 202
   texanum 202
   trichiuri 202
Lecithocladium bulbolabrum 202
   dongshanensis 202
   excisiforme 202
   excisum 202
   glandulum 202
   harpedontis 202
   ilishae 202
   leiognathi 202
   lutiani 202
   megalaspis 202
   miichthydis 202
   pagrosomi 202
   prariovum 202
   putianensis 202
   xishaense 202
Lecithomonoium harpodi 202
Lecithophyllum intermedium 202
Lecithostaphylus ahaaha 195
   fugus 195
LECYTHIDACEAE 156
Lecythorhynchus hilgendorfi 297
LEGUMINOSAE 153
Leiaster spiciosus 391
Leidya sesarmae 326
   ucae 326
Leiochrides australis 225
LEIOGNATHIDAE 432
Leiognathus berbis 432
   bindus 432
   brevirostris 432
   daura 432
   dussumieri 432

Leiognathus (continued)
   elongatus 432
   equulus 432
   fasciatus 432
   insidiator 432
   leuciscus 432
   lineolatus 432
   nuchalis 432
   rivulatus 432
   ruconius 432
   splendens 432
Leipocten sordidulum 364
   trigranulum 364
Leiuranus semicinctus 417
[*Lembeichthys furcocaudalis*] 446
Lemnalia bournei 188
Lenisquilla lata 368
Lensia baryi 174
   campanella 174
   canopusi 174
   challengeri 174
   conoides 174
   cordata 174
   cossack 174
   fowleri 174
   grimaldi 174
   havock 174
   hotspur 175
   leloupi 175
   lelouveteau 175
   meteori 175
   multicristata 175
   multicristatoides 175
   reticulata 175
   subtilis 175
   subtiloides 175
   tottoni 175
Lenticulina angulata 91
   asterizans 91
   calcar 92
   canariensis 92
   changi 92
   chiriguanoi 92
   comptoni 92
   convergens 92
   costata 92
   depressa 92
   dorso-costata 92
   exquisita 92
   gibba 92
   hoppoensis 92
   incavata 92
   iota 92
   kotiensis 92
   limbosa 92
   lobata 92
   lucida 92
   nicobarensis 92
   orbicularis 92
   papillosa 92
   peregrina 92
   platyrhinos 92
   pseudorotulatus 92
   sintikuensis 92
   stachei 92

Lenticulina (continued)
   subangulata 92
   subgibba 92
   submamilligera 92
   suborbicularis 92
   surugaensis 92
   taluensis 92
   tôtômiensis 92
   tumida 92
   xuwenensis 92
Leocrates chinensis 218
   claparedi 218
Leonnates decipiens 220
   jousseaumei 220
   niponicus 220
   persica 220
Lepadichthys frenatus 462
LEPADIDAE 318
Lepas anatifera anatifera 319
   ———— *striata* 319
   anserifera 319
   indica 319
   pectinata 319
   testudinata 319
Lepeophtheirus edwardsi 316
   kareii 316
   lateolabraxi 316
   scutiger 316
Lepidapedon apharei 195
   golphick 195
   longivesculum 195
   sphyraenae 195
[*Lepidasthenia elegans*] 217
   interrupta 217
   izukai 217
   longicirrata 217
   [*longissima*] 217
   maculata 217
   microlepis 217
   ocellata 217
   ohshimai 217
Lepidepecreum nudum 329
Lepidoblepharon ophthalmolepis 456
LEPIDOCENTROIDA 392
Lepidochelys olivacea 463
Lepidochiton rugatus 238
LEPIDOCHITONIDAE 237
Lepidocybium flavobrunneum 449
Lepidocyclina formosa 116
LEPIDOCYCLINIDAE 116
Lepidodeuterammina ochracea 76
Lepidodidymozoon sinicum 200
Lepidonotus carinulatus 217
   [*carinulatus*] 217
   [*chinensis*] 217
   cristatus 217
   [*cristatus*] 217
   dentatus 217
   [*dentatus*] 217
   helotypus 217
   [*helotypus*] 217
   jacksoni 217
   [*jacksoni*] 217
   jukesii 217
   melanogrammus 217

Lepidonotus (continued)
  sagamiana 217
  squamatus 217
  [*squamatus*] 217
  tenuisetosus 217
  [*tenuisetosus*] 217
[*Lepidopleurus assimilis*] 238
LEPIDOPTERA 300
LEPIDORBITOIDIDAE 121
Lepidotrigla abyssalis 454
  alata 454
  guentheri 454
  hime 454
  japonica 454
  kishinouyi 454
  lepidojugulata 454
  longifaciata 454
  longimana 454
  marisinensis 454
  micropterus 454
  oglina 454
  punctipectoralis 454
  spilopterus 454
Lepidozona christiaensis 238
  coreanica 238
  nipponica 238
LEPOCREADIIDAE 195
Lepocreadioides discum 195
  hunghuaensis 195
  indicum 195
  zebrini 195
Lepocreadium dongxiangensis 195
  drepanei 195
  navodoni 195
  trachinoti 195
Leporimetis lacunosus 251
  spectabilis 251
[*Lepralia squadrata*] 379
LEPRALIELLIDAE 378
LEPRALIELLOIDEA 378
LEPRALIOMORPHA 378
Leprotintinnus nordqvisti 58
Leptastrea bottae 184
  immersa 184
  purpurea 184
  transversa 184
[*Lepters huamela*] 449
[*Lepterus muticus*] 449
Leptocarpus disjunctus 162
  potamiscus 340
[*Leptochela aculeocauda*] 339
  gracilis 338
  hainanensis 339
  japonicus 339
  pugnax 339
  robusta 339
Leptochelia dubia 325
LEPTOCHITONIDAE
Leptoclinides madara 402
  reticulatus 402
  rufus 402
Leptoclinum mitsukurii 402
Leptoconchus striatus 276
[*Leptocotis ambobus*] 332
  [*spinifera*] 332

Leptocotis (continued)
  tenuirostris 332
Leptocylindrus adriaticus 30
  danicus 30
Leptoderma retropinnum 412
Leptodius danae 361
  exaratus 361
  gracilis 361
  nigromaculatus 361
  sanguineus 361
Leptodora kindti 300
LEPTODORIDAE 300
Leptojulis lambdastigma 440
LEPTOLAIMIDAE 211
Leptolaimus venustus 211
LEPTOMEDUSAE 169
Leptomithrax compressipes 356
  edwardsi 356
  sinensis 356
Leptomya minuta 252
  nitida 252
[*Leptonereis glauca moniloceras*] 221
Leptopentacta imbricata 387
Leptophytum asperulum 135
  [*zostericolum*] 135
Leptoplana delicatula 191
  fusca 191
  trullaeformis 191
LEPTOPLANIDAE 191
Leptoscarus vaigiensis 441
Loeptoseris gardineri 181
  papyracea 181
LEPTOSOMATIDAE 212
Leptosphaeria australiensis 131
  avicenniae 131
  oraemaris 131
Leptostomias robustus 412
Lepturacanthus savala 449
LERNAEOPODIDAE 316
LERNAEOPODOIDA 316
Lernanthropus abitocephalus 316
  atrox 316
  chirocentrosus 316
  chrysophrys 316
  corniger 316
  cornutus 316
  gazzis 316
  giganteus 316
  gisleri 316
  paenulatus 316
  sanguineus 316
  shishidoi 316
  sillaginis 316
  trifoliatus 316
Lernella inflata 105
Lestidiops mirabilis 415
Lestidium atlanticum 415
  japonica 415
  prolixum 415
Lestrigonus bengalensis 331
  crucipes 331
  latissimus 331
  macrophthalmus 331
  schizogeneios 331
  shoemakeri 331

[*Lestrolepis japonica*] 415
Lethotremus awae 456
LETHRINIDAE 434
Lethrinus haematopterus 434
  harak 434
  kalloperus 434
  [*lentjan*] 434
  leutjanus 434
  mahsena 434
  mahsenoides 434
  miniatus 434
  nebulosus 434
  nematacanthus 434
  ornatus 434
  reticulatus 434
  rhodopterus 434
  rubrioperculatus 434
  semicinctus 434
  variegatus 434
  xanthochilus 434
[*Leucaena glauca*] 153
  leucocephala 153
Leucas chinensis 157
  zeylanica 157
Leuckartiara gardineri 167
  hoepplii 167
  octona 167
  [——— v. *minor*] 167
  orientalis 167
  zacae 167
Leuconia aspera 165
  solida 165
LEUCONIDAE 325
Leucophloeus massalis 164
Leucosia anatum 355
  biminents 355
  compressa 355
  craniolaris 355
  formosensis 355
  haematosticta 355
  latirostrata 355
  longibrachia 355
  [*longifrons*] 355
  minuta 355
  parapulchella 355
  [*pseuolo margaretata*] 355
  puchella 355
  pulchra 355
  rhomboidalis 355
  sinica 355
  unidentata 355
  vittata 355
  whitei 355
LEUCOSIIDAE 354
Leucosolenia botryodes 165
  tenui 165
  variabilis 165
Leucosoma chinensis 411
Leucothoe alata 329
  alcyone 329
  furina 329
Leucothoella bannwarthi 329
LEUCOTHOIDAE 329
Leucotina dinnae 284
[*Leucozonia smaragdula*] 278

Leudugeria janischii 33
Levellea jungermannioides 140
Leydigia ciliata 301
Leymus mollis v. *coreensis* 160
Liachirus melanospilus 458
Liagora boergesenii 133
    [*caenomyce*] 134
    ceranoides 133
    divaricata 133
    [*farinosa*] 133
    filiformis 133
    hainanensis 134
    pinnata 134
    rubra 134
    samaensis 134
    setchellii 134
    sinensis 134
Liagore rubromaculata 361
LIBELLULIDAE 299
Libystes edwardsi 358
    villosus 358
Lichenopora capillata 370
    imperialis 370
    radiata 370
    simplex 370
LICHENOPORIDAE 370
[LICHENS] 132
Licmophora abbreviata 43
    california 43
    ehrenbergii 43
    ———— v. *ovata* 43
    flabellata 43
    gracilis 43
    ———— v. *elongata* 43
    paradoxa 43
    tenuis 43
Licnophora hippocampi 58
LICNOPHORIDAE 58
Lienardella fasciatus 440
Ligia exotica 326
    occidentalis 326
LIGIIDAE 326
Lignincola laevis 131
    longirostris 131
Ligophorus leporinus 192
    vanbenedeni 192
LILIACEAE 162
Lilieborgia serrata 329
    sinica 329
LILIEBORGIIDAE 329
Liloa curta 286
    porcellana 286
[*Lima sowerby*]
    vulgaris 245
Limacina bulimodes 288
    inflata 288
    trochiformis 288
LIMACINIDAE 288
Limaria amakusaensis 245
    basilanica 245
    dentata 245
    fragilis 245
    hakodatensis 245
    perfragilis 245
Limatula bullata 245

Limatula (continued)
    choshiensis 245
    nippona 245
    textilis 245
Limicola falcinellus sibirica 467
LIMIDAE 245
Limnichthys fasciatas 445
Limnodriloides agnes 234
    biforis 234
    fraternus 234
    fuscus 234
    macinnesi 234
    parahastatus 234
    rubicundus 234
    tenuiculus 234
    tenuiductus 234
    toloensis 234
    uniampullatus 234
    victoriensis 234
    virginiae 234
Limnodromus semipalmatus 467
Limnoithona tetraspina 312
LIMNOMEDUSAE 172
Limnoperna fortunei 242
Limnopneustes murrayi 394
[LIMNOPOLYPAE-LIMNOMEDUSAE] 172
Limnoria lignorum 325
    tripunctata 325
LIMNORIIDAE 325
Limnotrachelobdella fujianensis 235
    okae 235
    sinensis 235
Limonium bicolor 157
    sinense 157
LIMOPSIDAE 241
Limopsis tajimae 241
Limosa lapponica novaezealandiae 467
    limosa melanuroides 467
Limulatys angustatus 286
    constrictus 286
    muscarius 286
    okamotoi 286
    ooformis 286
[*Linatella cingulata*] 273
    [*poulseni*] 273
Linchia gracilis 391
    laevigata 391
    multifora 391
LINDSAEACEAE 150
LINEIDAE 207
Lineopselloides albilineus 207
Lineus alborostratus 207
    bilineatus 207
    binigrilinearis 207
    cingulatus 207
    fuscovirids 207
    geniculatus 207
    torquatus 207
    vegetus 207
[*Linguella babai*] 291
    [*punctilucens*] 292
Lingula adamsi 384
    anatina 384
    [*shangtangensis*] 384

Lingula (continued)
    [*unguis*] 384
LINGULIDAE 384
Lingulina costata 91
    kansireiensis 91
    paucicostata 91
    polymorpha 91
    seminuda 91
Lingulinopsis basicostatum 91
    carlofortensis 91
    folium 91
LINHOMOEIDAE 211
Linopherus ambigua 226
    hirsura 226
    microcephala 226
    oligobranchia 226
    pancibranchiata 226
    spiralis 226
Linophryne polypogon 462
LINOPHRYNIDAE 462
Linuche draco 176
LINUCHIDAE 176
Linuparus sordidus
    trigonus
Lioconcha castrensis 255
    fastigiata 255
    lorenziana 255
    ornata 255
Liocranchia globula 294
    reinhardti 295
Liolophura gaimardi 238
    japonica 238
    loochooana 238
Liomera bella 361
    caelata 361
    cinctimana 361
    laevis 361
    margaritata 361
    monticulosa 361
    rugata 361
    stimpsoni 361
    venosa 361
Liopropoma aragai 426
    [*japonicus*] 426
    [*latifasciata*] 426
    susumi 426
Liopsetta obscura 457
Liosaccus cutaneus 461
    pachygaster 461
Liotia peronii 264
LIOSPHAERIDAE 62
LIPARIDAE 456
Liparis chefuensis 456
    choanus 456
    petschiliensis 456
    tanakae 456
Liradiscus reniformis 30
Liriope tetraphylla 173
Liriospyris pulmoformis 66
    rotunda 66
Lischkeia alwinae 263
Lissocarcinus arkati 358
    orbicularis 358
    polybioides 358
Lissodendoryx isodictyalis 165

Index

Lissodendoryx (continued)
   tylostyla 165
Lissoporcellana quadrilobata 351
   spinuligera 351
   streptochiroides 351
Listriolobus brevirostris 237
Lithacrosiphon maldivensis 235
Lithatractus nucleolus 64
LITHELIDAE 66
Lithelius amphistylis 66
   cf. *alveolina* 66
   nerites 66
   solaria 66
   spiralis 66
   xanthiformis 66
LITHOBOTRYIDAE 67
LITHOCAMPIDAE 68
Lithodes longispinus 350
   turritus 350
Lithodesmium undulatus 33
   variabile 33
LITHODIDAE 350
Lithomelissa buetschlii 67
   campanulaeformis 67
   monoceras 67
   spinosissma 67
   thoracites 67
Lithomitra lineata 68
Lithopera bacca 67
Lithophaga antillarum 242
   calyculatus 242
   canaliferus 242
   curta 242
   divarixcalx 242
   hanleyma 242
   lepteces 242
   lessepsiana 242
   lima 242
   [*lima*] 242
   lithura 242
   malaccana 242
   mucronata 242
   nasuta 242
   obesa 242
   [*obesa*] 242
   [*simplex*] 242
   teres 242
   [*teres*] 242
   zitteliana 242
   [*zitteliana*] 242
Lithophyllum kotschyanum 135
   moluccense 135
   okamurae 135
Lithopilium macroceras 68
Lithoporella melobesioides 135
   pacifica 135
Lithothamnium aculeiferum 135
   [*fretense*] 135
   glaciale 135
   intermedium 135
   japonicum 135
   pacificum 135
Lithotrya nicobarica 318
   valentiana 318
LITIOPIDAE 267

Litobopyrus longicaudatus 326
Litonotus fasciola 57
   paracygnus 57
   yinae 57
[*Litorina brevicula* v. *costulata*] 265
   [*cecillei*] 265
   [*chaoi*] 265
   [*columna*] 265
   [*exigua*] 265
   [*malaccana*] 265
   [*rubra*] 265
   [*scabra* v. *suturalis*] 265
   [*sinensis*] 265
   [*strigata*] 265
   [*ventricosa*] 265
   [*vilis*] 265
Litoscalpellum intermedium 318
   sinense 318
Litsea glutinosa 152
   rotundifolia v. *oblongifolia* 152
Littoraria ardouiniana 265
   articulata 265
   melanostoma 265
   pallescens 265
   [*scabra*] 265
   undulata 265
[*Littorina acuminata*] 265
   [*arboricola*] 265
   [*balteata*] 265
   brevicula 265
   coccinea 265
   [*coccinea*] 265
   [*granularis*] 265
   [*heterospiralis*] 265
   [*intermedia*] 265
   [*mandshurica*] 265
   [*melanostoma* v. *articulata*] 265
   [*millerana*] 265
   [*monilifera*] 265
   [*scabra*] 265
   [*sieboldii*] 265
   [*souverbiana*] 265
   [*trochoides*] 265
   [*undulata*] 265
   [—— v. *contracta*] 265
   [—— v. *sulcatula*] 265
   [*vidula*] 265
LITTORINIDAE 265
[*Littorinopsis obesa*] 265
[*Littorivaga brevicula*] 265
Littorophiloscia aldabrana 326
Lituaria amoyensis 190
Lituola hispida 74
LITUOLACEA 73
LITUOLIDAE 74
Lituotuba lituiformis 74
LITUOTUBIDAE 74
Liza alata 424
   carinatus 424
   dussumieri 424
   haematocheila 424
   macrolepis 425
   melinopterus 425
   [*soiuy*] 424
   subviridis 425

Liza (continued)
   tade 425
Lizzia gracilis 166
LOBATE 190
Lobatovitelliovarium fusiforme 195
Lobatula lobatula 115
Lobiger sagamiensis 289
Lobonema smithi 176
LOBONEMATIDAE 176
Lobonemoides gracilis 176
Lobophora variegata 142
Lobophyllia corymbosa 184
   costata 184
Lobophytum angulatum 187
   anomalum 187
   caledonense 187
   caputospiculatum 187
   chevalier 187
   compactum 187
   crassodigitum 187
   crassospiculatum 187
   cristaglli 187
   cristatum 187
   delectum 187
   denticulatum 187
   depressum 187
   gazellae 187
   hirsutum 187
   irregulare 187
   longispiculatum 187
   microspiculatum 187
   oblongum 187
   oligoverrucum 187
   pauciflorum 187
   pulchellum 187
   pygmapedium 187
   ransoni 188
   salvati 188
   schoedei 188
   spicodigitum 188
   spissum 188
   venustum 188
   verrucosum 188
Lobotes surinamensis 434
LOBOTIDAE 434
Lockhartia conditi 122
   haimei 122
   hunti 122
   megapapulata 122
   tipperi 122
LOHMANNELLINAE 298
Loimia arborea 229
   bandera 229
   ingens 229
   medusa 229
   montagui 229
LOLIGINIDAE 295
Loligo beka 295
   bleekeri 295
   chinensis 295
   duvaucelii 295
   edulis 295
   [*formosana*] 295
   gotoi 295
   japonoica 295

Loligo (continued)
   kobiensis 295
   oshimai 295
   sibogae 295
   singhalensis 295
   tagoi 295
   uyii 295
Lomentaria catenate 138
   hakodatensis 138
   pinnata 138
Lonchostaurus rhombicus 61
Longicarpus nodus 229
Longimonorchis pampi 195
Longipedia scotti 314
   weberi 314
LONGIPIDIIDAE 314
LOPADORHYNCHIDAE 215
Lopadorhynchus brevis 215
   krohni 215
   nationaeis 215
   tuberculus 215
   unicinatus 215
Lopha cristagalli 246
LOPHIIDAE 461
LOPHIIFORMES 461
Lophiodes infrabrunneus 461
   naresi 461
Lophiogobius ocellicauda 452
Lophiomus setigerus 461
Lophiotoma acufa 281
   leucotropis 281
Lophius litulon 461
LOPHOGASTRIDAE 323
Lophogaster hawaiensis 323
   pacificus 323
Lophomastrix japonica 351
Lophoplax sculpta 363
   teschi 363
Lophospyris pentagona 66
Lophosquilla costata 368
   tiwarii 368
Lophothrix frontalis 308
   latipes 308
Lophozozymus pictor 361
   pulchellus 361
Loricoecia ctenophora 304
   loricata 304
Lotella phycis 419
Lourinia armata 315
LOURINIIDAE 315
Lovenella assimilis 171
   haichangensis 171
LOVENELLIDAE 171
Lovenia elongata 394
   subcarinata 394
   triforis 394
LOVENIIDAE 394
Loxodon macrorhinus 406
Loxophyllum rostratum 57
   setigerum 57
Loxosoma annulatum 383
   circulare 383
LOXOSOMATIDAE 383
Loxosomella crassicauda 383
   harmeri 383

Loxosomella sp. 384
LOXOSTOMATACEA 104
LOXOSTOMATIDAE 104
Loxostomina mayori 106
Loxostomum amygdaiaeformis 104
   convallarium 104
   formosana 104
   karrerianum 104
   limbatum 104
   mayori 104
   spinosum 104
Lubbockia aculeata 312
   marukawai 312
   squillimana 312
Lucaya sp. 339
Lucibacterium harveyi 21
Lucicutia bicornis 310
   clausi 310
   curta 310
   flavicornis 310
   gemina 310
   longiserrata 310
   magna 310
   maxima 310
   ovalis 310
   tenuicauda 310
   wolfendeni 310
LUCICUTIIDAE 310
Lucifer faxoni 338
   hanseni 338
   intermedius 338
   orientalis 338
   penicillifer 338
   typus 338
Lucilina amanda 238
[*Lucina edentula*] 246
   [*philippiana*] 247
LUCINIDAE 246
Lucinoma annulata 247
   yoshidai 247
Luciobrotula bartschi
Luciogobius guttatus 452
   saikaiensis 452
Lugdunum hantkenianum
Luidia avicularia 390
   hardwicki 390
   longispina 390
   maculata 390
   orientalis 390
   quinaria 390
   yesoensis 390
LUIDIIDAE 390
Lulworthia grandispora 131
Lulwortlhia sp. 131
Lumbricillus sp. 234
[*Lumbriconereis debilis*] 227
   [*impatiens*] 227
LUMBRINERIDAE 227
Lumbrineris acutiformis 227
   amboinensis 227
   bifurcata 227
   caudaensis 227
   cruzensis 227
   hereropoda 227
   inflata 227

Lumbrineris (continued)
   japonica 227
   latreilli 227
   longiforlia 227
   meteorana 227
   mucronata 227
   nagae 227
   ocellata 227
   pterignatha 227
   shiinoi 227
   simplex 227
   sphaerocephala 227
   tetraura 227
Lumnitzera littorea 156
   racemosa 156
Lunatica gilva 269
   [*pallida*] 269
   yokoyamai 269
[*Lunella coronata*] 264
   ——— *coreensis* 264
   ——— *granulata* 264
Lunulicardia retusa 248
Lupocyclus inaequalis 358
   philippinensis 358
   rotundatus 359
   tugelae 359
Luria isabella 272
   [*isabella*] 272
LUTJANIDAE 432
Lutjanus altifrontalis 433
   argentimaculatus 433
   bengalensis 433
   bohar 433
   boutton 433
   carponotatus 433
   chrysotaenia 433
   decussatus 433
   dodecacanthoides 433
   erythopterus 433
   fulviflamma 433
   fulvus 433
   gibbus 433
   johni 433
   kasmira 433
   lineolatus 433
   lutjanus 433
   malabaricus 433
   monostigma 433
   niger 433
   quinquelineatus 433
   rivulatus 433
   russelli 433
   sanguineus 433
   sebae 433
   spilurus 433
   stellatus 433
   vaigiensis 433
   vitta 433
Lutraria arcuata 249
   impar 249
   maxima 249
   philippinarum 249
   sieboldii 249
Lybia caestifer 361
   tessellata 361

Index

Lycaea bajensis 332
   bovalli 332
   bovallioides 332
   nasuta 332
   pachypoda 332
   pulex 332
   vincenti 332
LYCAEIDAE 332
LYCAEOPSIDAE 332
Lycaeopsis themistoides 332
   zamboangae 332
Lycomysis spinicauda 323
Lydia annulipes 361
Lycastopsis augenari 220
Lychnaspis giltschi 61
   longissima 61
   polyancistra 61
   serrata 61
Lychnodictyum challengeri 67
Lycium chinense 158
Lygdamis giardi 228
   indicus 228
   nesiotes 228
   porrectus 228
LYGODIACEAE 150
Lygodium microstachyum 150
Lymnocryptes minimus 467
Lyngbya adherens 25
   aestuarii 25
   baculum 25
   confervoides 25
   epiphytica 25
   infixa 25
   kuetzingii 25
   lulea 25
   majuscula 25
   nordgaardii 25
   semiplena 25
   willei 25
Lyonsia kawamurai 257
   [*praetenuis*] 257
   [*rostrata*] 257
   ventricosa 257
LYONSIIDAE 257
[*Lyreidus integra*] 354
   stenops 354
   tridentatus 354
Lyria kurodai 279
   taiwanica 279
Lyrocardium lyratum 248
Lyrodus singaporeana 257
Lyrula hippocrepis 377
   ortmanni 377
Lysianassa cinghalensis 329
LYSIANASSIDAE 329
Lysidice collaris 227
   ninetta 227
   [——— *collaris*] 227
Lysilla pacifica 229
   sinensis 229
   ubianensis 229
Lysimachia mauritiana 157
Lysiosquilla maculata 369
   sulcirostris 369
LYSIOSQUILLIDAE 369

LYSIOSQUILLOIDEA 369
Lysmata kuekenthali 343
   vittata 343
LYTHRACEAE 156
Lytocarpus niger 171
   nutlingi 171
   philippinus 171
   phoeniceus 171

M

Maackia australis 153
Mabellarca consociata 240
   dautzenbergi 240
Macalia bruguieri 251
Macalifer aprionis 194
   dayawanensis 194
   pacificus 194
   zhoushanensis 194
Macandrewella joanae 308
   tuberculata 308
Maccallozum platycephali 200
[*Macellicephala maculosa*] 217
Macolor niger 433
   [*niger*] 433
Macoma awajiensis 251
   candida 251
   [*candida*] 251
   [*galathaea*] 251
   incongrua 251
   lucerna 251
   murrayi 251
   nobilis 251
   praetexta 251
   praerupta 251
   sectior 251
   tokyoensis 251
   [*truncata*] 251
Macrobrachium equidens 340
   grandimanus 340
   hainanense 340
   latidactylus 340
   mammillodactylus 340
   nipponense 340
   rosenbergii 340
   superbum 340
Macrocheira kaempferi 356
Macrocypridina castanea rotunda 302
MACRODEROIDIDAE 198
Macromedaeus crassimanus 361
   distinguendus 361
Macropharyngodon meleagris 440
   negrosensis 440
Macrophiothrix capillaris 396
   demessa 396
   galateae 396
   longipeda 396
   lorioli 396
   propinqua 396
   robillardi 396
   unicolor 396
   variabilis 396
Macrophthalmus abbreviatus 364
   banzai 364
   boscii 364

Macrophthalmus (continued)
   boteltobagoe 364
   brevis 364
   convexus 364
   crassipes 364
   definitus 364
   dentatus 364
   depressus 364
   dilatum 364
   erato 364
   japonicus 364
   ——— *froequens* 364
   latreillei 364
   pacificus 364
   telescopicus 365
   tomentosus 365
   verreauxi 365
Macropipus corrugatus 359
MACRORHAMPHOSIDAE 423
Macrorhamphosodes uradoi 459
Macrorhamphosus gracilis 423
   japonicus 423
   scolopax 423
Macroschisma sinensis 262
MACROSETELLIDAE 314
Macrospinosa cuja 431
MACROTHRICIDAE 301
Macrothrix brevicornis 301
   laticornis 301
   rosea 301
   spinosa 301
MACROURIDAE 420
MACROUROIDIDAE 421
Macrura kelee 410
   reevesi 410
[*Mactra antiquata*] 249
   aphrodina 249
   chinensis 249
   crossei 249
   cuneata 249
   [*grandis*] 249
   inaequalis 249
   luzonica 249
   maculata 249
   mera 249
   ornata 249
   [*quadriangularis*] 249
   [*spectabilis*] 249
   [*sulcataria*] 249
   veneriformis 249
MACTRIDAE 249
Mactrimula dolabrata 249
Maculotriton serrialis 275
Maeotias inexpectata 172
Maera othonides 329
   pacifica 329
   serratipalma 329
Magelona cineta 224
   crenulifrons 224
   japonica 224
   pacifica 224
   papilliorni 224
MAGELONIDAE 224
MAGILIDAE 276
[MAGILODIDAE]

Magilus antiquatus 276
Mainwaringia rhizophila 265
Maja brevispinosis 356
   gibba 356
   japonica 356
   miersi 356
   sakaii 356
   spinigera 356
MAJIDAE 356
Makaira formosana 450
   indica 450
   marlina 450
   mazara 450
   mitsukurii 450
Malacanthus brevirostris 429
   hoedti 429
   latovittatus 430
Malacocephalus laevis 420
Malacoceros indicus 223
MALACOSTEGINA 371
MALACOSTEIDAE 412
Malacosteus niger 412
MALACOSTRACA 323
Malagazzia carolinae 169
   condensum 169
   curviductum 169
   cyphogonia 169
   taeniogonia 169
MALAGAZZIIDAE 169
Malakichthys elegans 427
   griseus 427
   wakiyai 427
Maldane cristata 225
   globifex 225
   sarsi 225
MALDANIDAE 225
Maldivia palmyrensis 361
   triunguiculata 361
Malea pomum 274
Mallacoota insignis 329
   subcarinata 329
MALLEIDAE 244
Malletia conspicua 239
   sumatrensis 239
MALLETIIDAE 239
Malleus albus 244
   daemoniacus 244
   [*malleus*] 244
   regula 244
   [*regula*] 244
Mallomonas annulata 50
   liturata 50
MALLOMONODACEAE 50
[*Malmgrenia ampulliferoides*] 217
Malthopsis annulifera 462
   jordeni 462
   luteus 462
MALVACEAE 154
Malvastrum coromandelianum 154
MAMMALIA 468
[*Mammillamammata*] 269
[*Mancinella echinulata*] 276
   [*tuberosa*] 276
Mangelia costulata 281
Manta birostris 409

Manteriella chanis 197
   hainanensis 197
Maoricardium mansitlii 249
   setosum 249
Marchia bipinnata 275
   elongata 275
   [*fenestrata*] 275
   martinetana 275
   siro 275
Marcia hiantina 255
   japonica 255
   marmorata 255
[*Marcrua ilisha*] 410
Marcusaedrilus irregularis 234
   tuber 234
   vesiculatus 234
Maretia planulata 394
Margaretta cereoides 381
   gracilis 381
MARGARETTIDAE 381
Marginella dactylus 280
   philippinarum 280
   tricincta 280
MARGINIDAE 280
Marginiosporum aberrans 135
   crassissimum 135
Marginulina costata 93
   glabra 93
   hamuloides 93
   hanzawai 93
   moodysensis 93
   obesa 94
   perobliqua 94
   philippinensis 94
   richteri 94
   robusta 94
   semicostata 94
   sestrona 94
   striatula 94
   tenuis 94
   terquemi 94
   varipapillata 94
Marginulinopsis bradyi 93
   costata 93
   marshalli 93
   waiparaensis 93
Margovula pyriformis 271
   schilderorum 271
   tinctilis 271
Mariametra subcarinata 386
   vicaria 386
MARIAMETRIDAE 386
Marilynia macrodentatus 211
Marinosphaera mangrovei 131
   cf. *mangrovei* 131
Marionia olivacea 292
Marionina coatesae 234
   levitheca 234
   nevisensis 234
   vancouverensis 234
Mariscus dubius 161
   hainanensis 161
   monospermus 161
   radians 161
   ———— v. *floribundus* 161

Marphysa belli 227
   bifurcata 227
   depressa 227
   [*iwamusi*] 227
   macintoshi 227
   mossanbica 227
   sanguinea 227
   sinensis 227
   stragulun 227
   tamurai
Marshallena philippinarum 281
Marsipella cylindrica 70
   dextrospiralis 70
   elongata 70
   gigantea 70
[MARSIPOBRANCHII] 404
Martensia denticulata 139
   fragilis 139
Martesia ova 257
   pygmaea 257
   striata 257
   tubigera 257
   yoshimurai 257
Martinotiella antillarum 78
   bradyana 78
   communis 78
   cylindrica 78
   inculta 78
   milletti 78
   minuta 78
   nodulosa 78
   occidentalis 78
   okinawaensis 78
   primaeva 78
   victoriensis 78
   wrightii 78
Massarina armatispora 131
   thalassiae 131
   velatospora 131
Massarina sp. 131
Massilina corrugata 83
   hachijensis 83
   humblei 83
   inaequalis 83
   intermedia 83
   laevigata 83
   magna 83
   milletti 83
   penglaiensis 83
   pratti 83
   pyristoma 83
   quadrans 83
   secans 83
Mastigias ocellatus 176
   papua 176
MASTIGIIDAE 176
Mastigocoleus testarum 27
[*Mastigophora duterteri*] 379
Mastigopus gracilis 351
MASTIGOTEUTHIDAE 295
Mastigoteuthis cordiformis 295
Mastobranchus indicus 225
Mastogloia achnanthioides 37
   ———— v. *elliptica* 37
   acutiuscula 37

Index

Mastogloia (continued)
———— v. *elliptica* 37
———— v. *vairaensis* 37
adriatica v. *linearis* 37
affirmata 37
amoyensis 37
angulata 37
apiculata 37
aspera 37
———— f. *lanceolata* 37
asperula 37
asperuloides 37
bahamensis 37
baldjikiana 37
bellatula 37
biapiculata 37
binotata 37
———— f. *sparsipunctata* 37
bourrellvana 37
brauni 37
———— v. *constricta* 37
———— f. *elongata* 37
citroides 37
citrus 37
cocconeiformis 37
composita 37
corallum 37
corsicana 38
cribrosa 38
cruciata 38
cruciata v. *elliptica* 38
cucurbita 38
cyclops 38
decipiens 38
decussata 38
densestriata 38
depressa 38
dicephala 38
dissimilis 38
dubitabilis 38
elegantula 38
elliptica v. *dansei* 38
emarginata 38
erythraea 38
———— v. *biocellata* 38
exilis 38
fallax 38
fascistriata 38
fimbriata 38
graciloides 38
grana 38
grevillei 38
grunowi 38
hainanensis 38
horvathiana 38
hustedtii 38
imitatrix 38
inaequalis 38
indonesiana 38
intrita 38
jaoi 38
jelinecki 38
———— v. *extensa* 38
jelineckiana 38
labuensis 38

Mastogloia (continued)
lacrimata 38
lanceolata 38
lata 38
latecostata 38
lemniscata 38
lentiformis 38
levis 38
liaotungensis 38
lineata 38
lunula 38
macdonaldi 38
mammosa 38
manokwariensis 38
mauritiana 38
———— v. *capitata* 38
mediterranea 38
———— v. *elliptica* 38
minutissima 38
muralis 38
nebulosa 38
neorugosa 38
nuiensis 38
obesa 38
occulta 38
omissa 38
ovalis 38
ovata 38
ovulum 38
ovum paschale 38
paracelsiana 38
paradoxa 38
peracuta 38
peragalli 38
pisciculus 38
pseudexilis 38
pseudolatericia 38
pseudomauritiana 38
pulchella 38
pumila 38
———— v. *papuarum* 38
———— v. *rennellensis* 38
punctatissima 38
punctifera 38
qionzhouensis 38
quinquecostata 38
rimosa 38
robusta 38
rostrate 38
savensis 38
schmidtii 38
serians 38
serrata 38
seychellensis 38
similis 38
simplex 38
singaporensis 38
smithii 38
———— v. *excentrica* 39
splendida 39
subaspera 39
sublatericia 39
sulcata 39
tenuis 39
tenuissima 39

Mastogloia (continued)
testudinea 39
umbilicata 39
undulata 39
varians 39
viperina 39
vulnerata 39
woodiana 39
xishaensis 39
Mastogonia crux 32
heptagona 32
ocella 32
Mastophora rosea 135
Masturus lanceolatus 461
[*oxyuropterus*] 461
Matuta banksii 356
granulosa 356
lunaris 356
planipes 356
Maupasia caeca 215
Mauritia arabica 272
[*arabica*] 272
mappa 272
mauritiana 272
[*mauritiana*] 272
scurra 272
Maurolicus muelleri 411
Maytenus diversifolia 154
Meadia roseni 417
Mecynocera clausi 306
Medaeops granulosus 361
Medusetta ansata 69
inflata 69
Megabalanus rosa 322
tintinnabulum crispatus 322
———— *occator* 322
———— *tintinnabulum* 322
———— *zebra* 322
volcano 322
xishaensis 322
Megacardita ferruginosa 248
Megacepon choprai
pleopodata
Megalasma annandalei 319
hamatum 319
striatum 319
Megalaspis cordyla 431
Megalocercus huxleyi 400
Megalomma vesiculosum 230
MEGALOMYCTERIDAE 415
Megalonibes fusca 431
MEGALOPIDAE 409
Megalops cyprinoides 409
Megaluropus agilis 329
Megaptera novaeangliae 469
MEGASCOLECIDAE 234
Mehratrema arii 203
fujianensis 203
Meiacanthus grammistes 446
Meiocardia lamarckii 253
tetragona 253
vulgaris 254
Melamphaes leprus 421
simus 421
MELAMPHAIDAE

Melampus triticeus 294
Melandrium apricum 152
[*Melanella bivittata*] 267
   martinii 268
MELANELLIDAE 267
Melanitta fusca stejnegeri 466
   nigra americana 466
MELANOCETIDAE 462
Melanocetus johnoni 462
Melanostomias melanopogon 412
   valdiviae 412
MELANOSTOMIATIDAE 412
[*Melaraphe blanfordi*] 265
   [*scabra intermedia*] 265
Melaspilea sp. 131
Melavargula japonica 302
MELIACEAE 154
Melibe japonica 292
MELICERTIDAE 170
Melicertoide octolabiatis 170
Melicertum octocostatum 170
Melichthys niger 459
   vidua 459
Melinna aberrans 229
   cristata 229
   elisabethae 229
Melita alluaudi 329
   appendiculata 329
   denticulata 329
   koreana 329
   longidactyla 329
   orgasmos 329
   rylovae 329
   setiflagella 329
   tuberculata 329
Melitodes modesta 186
   ocracea 186
   squarnata 186
MELITODIDAE 186
[*Melo melo*] 279
Melonis affinis 119
   barleeanum 119
   cyrtomatus 119
   formosa 119
   minutus 119
   pompilioides 119
   probliquus 119
   rotulum 119
   suni 119
   umbilicatula 119
   zaandami 119
   zeobesus 119
   zorillus 119
Melophyes melo 174
Melosira architecturalis 30
   arctica 30
   granulata 30
   islandica 30
   juergensi 30
   moniliformis 30
   nummuloides 30
   sol 30
   sulcata 30
   ——— f. *radiata* 30
Membranipora barstchi 371

Membranipora (continued)
   bicornica 371
   conjunctiva 371
   eriophoroidea 371
   falsitenuis 371
   grandicella 371
   hainanica 371
   [*hugliensis*] 372
   irregulata 371
   limosa 371
   limosoidea 371
   lingdingensis 371
   paragrandicella 371
   parasarvatii 371
   pseudoiirregulata 371
   [*quadrata*] 371
   savartii 371
   [*serrilamella*] 371
   serrilamelloides 371
   similus 371
   tachypleusae 371
   tenuis 371
   tuberculata 371
   tuberculatoidea 371
   virgata 371
Membraniporella aragoi 377
   [——— *zhoushanica*] 372
MEMBRANIPORIDAE 371
MEMBRANIPOROIDEA 371
Membraniporopsis bifloris 371
   bisipinosa 371
Menaethius monoceros 356
Membranobalanus longirostrum 321
Mene maculata 431
Menella praelonga 187
   rubescens 187
   spinifera 187
   verrucosa 187
MENIDAE 431
MENISPERMACEAE 152
Mensamaria intercedens 387
Meoncholaimus moles 212
Meretrix castannea 255
   lamarckii 255
   lusoria 255
   meretrix 255
Merga bulbosa 167
   macrobulbosa 167
   tergestina 167
Mergus albellus 466
   serrator 466
   squamatus 466
Merhippolyte calmani 344
Merica asprella 279
   cf. *oblonga* 280
   sinensis 280
Merisca capsoides 251
   diaphana 251
   perplexa 251
Merismopedia glauca 25
   warmingiana 25
Meristotheca papulosa 137
Merocryptus lambriformis 355
Meropesta capillacea 249
   nicobarica 249

Meropesta (continued)
   pellucida 250
   sinojaponica 250
MEROSTOMATA 297
Merremia hirta v. *retusa* 157
Merulina ampliata 184
   laxa 184
   scabricula 184
MERULINIDAE 184
Mesacanthion littoralis 212
Mesochaetopterus japonicus 224
[*Mesodesma striata*] 250
MESODESMATIDAE 250
Mesodinium pujlex 56
MESOGASTROPODA 265
Mesopenaeus mariae 336
Mesophyllum erubescens 135
   mesomorphum 135
   simulans 135
Mesoplodon densirostris 469
   ginkgodens 469
Mesopontonia gorgoniophila 340
Mesorhabdus gracilis 310
Mesosigmoilina minuta 88
Mespilia globulus 393
Mesothuria marginata 388
   media 388
Messerschmidia argentea 157
   sibirica angustiior 157
Metacalanus aurivillii 310
Metacepon choprai 327
   pleopodata 327
Metaconchoecia abyssalis 304
   glandulosa 304
   isocheira 304
   kyrtophora 304
   macromma 304
   pusilla 304
   rotundata 304
   skogsbergi 304
   teretivalvata 304
Metacrangon sinensis 344
Metacrinus interruptus 386
   multisegmentatus 386
   rotundus 386
Metacylis jorgensenii 59
   oviformis 59
   sanyahensis 59
Metadasynemella cassidiniensis 211
Metalia dierana 395
   spatagus 395
Metamicrocotyla gracilis 192
Metamphiascopsis hirsutus bermudae 315
Metanematobothrium serilae 200
Metanephrops andamanicus 345
   armatus 345
   formosanus 345
   japonicus 345
   neptunus 345
   sagamiensis 345
   sinensis 345
   thomsoni 345
Metapenaeopsis acclivis 336
   andamanensis 336

Index

Metapenaeopsis (continued)
  barbata 336
  [*barbeensis*] 337
  [*caliper*] 337
  dalei 336
  dura 336
  hilarula 336
  [*insona*] 337
  lamellata 336
  lata 337
  liui 337
  mogiensis 337
  palmensis 337
  philippi 337
  sinica 337
  stridulans 337
  tenella 337
  toloensis 337
  velutina 337
Metapenaeus affinis 337
  [*burkenroadi*] 337
  ensis 337
  intermedius 337
  joyneri 337
  [*mogiensis*] 337
  [*monoceros*] 337
  moyebi 337
  [*mutatus*] 337
  [*spinulatus*] 337
  [*stridulans*] 337
  tenuipes 337
  [*velutinus*] 337
Metaplax elegans 366
  longipes 366
  sheni 366
  takahashii 366
Metaruncina setoensis 287
Metasepia tullbergi 295
Metasesarma aubryi 366
Metavermilia acanthophora 230
  annobonensis 230
  inflata 230
  cf. *taenia* 230
Metcalfina hexagona 401
METOPIDAE 58
Metopograpsus frontalis 366
  latifrons 366
  quadridentatus 366
  thukuhar 366
Metopus circumlabeus 58
  ellipsoidis 58
  phyllopharius 58
METRIDIIDAE 179, 309
Metridia brevicauda 309
  macrura 309
  princeps 309
  venusta 309
Metridium huanghaiensis 179
  senile fimbriarum 179
  sinensis 179
Metrodira subulata 391
METRODIRIDAE 391
[*Mexichromis multituberculata*] 290
Mexiculepis sineca 218
Micippa philyra 356

Micippa (continued)
  platipes 356
  thalia 356
  xishanensis 356
Micippoides angustifrons 356
MICRABACIIDAE 185
Micracanthorhynchus dakusuensis 213
Microbacterium 22
Microbinia linea 222
Microcalanus pusillus 307
Microcanthus strigatus 436
Microcardium exasperatum 249
  nomurai 249
  torresi 249
Microcephalophis gracilis 464
MICROCHAETACEAE 26
Microchaete aeruginea 27
  grisea 27
  tapahiensis 27
  vitiensis 27
Microchloa indica 160
Microciona prolifera 165
[*Microcladia elegans*] 138
Microclymene caudata 225
MICROCOCCACEAE 22
MICROCOCCALES 22
Micrococcus 22
  candidus 22
  cimnaabareus 22
  cinnabarens Flugge-strain A 22
  ——— B 22
  ——— C 22
  ——— D 22
  ——— E 22
  citreus v. *migula* 22
  ridleyi 22
  corbert-strain A
  corbert v. *marinus* 22
    sulfureus Zimmerman-strain A 22
    ——— B 22
Microcoleus chthonoplastes 25
  confluens 25
  majuscula 25
  tenerrimus 25
  vaginatus 25
Microconchoecia acuticosta 304
  curta 304
  echinulata 304
  stigmatica 304
Microcosmus australis 403
  exasperatus 403
MICROCOTYLIDAE 192
Microcyphus olivaceus 393
Microcystis aeruginosa 25
  ichthyoblabe 25
Microdictyon japonicum 146
  nigrescens 146
  okamurae 146
  pseudohapteron 146
[*Microdonophis erabo*] 418
  [*fasciatus*] 418
  [*intermedius*] 418
  [*polyophthalmus*] 418
Micrognatus brevirotris 424
  metaatae 424

Micromelum falcatum 153
MICROMONOSPORACEAE 24
Micromorospora 24
Micronactra angulifera 250
Micronephtys sphaerocirrata
Microphis boaja 424
  brachyurus 424
  leiaspis 424
  manadensis 424
Microphthalmus biantenna 218
  hartmanae pacificus 218
[*Micropodarke amemiyai*] 218
  dubia 218
Micropolyspora 24
MICROPOLYSPORACEAE 24
Micropora granulata 376
Microporella ciliata 381
  coronata 381
  coscinophora 381
  lunifera 381
  orientalis 381
  vibraculifera 381
MICROPORELLIDAE 381
MICROPORIDAE 376
MICROPOROIDEA 376
Microprosthema validum 338
Microscaphidium japonicum 194
Microsetella norvegica 314
  rosea 314
Microsolenia simplex 30
Microstomus achne 457
Microtetraspora viridis xiamensis 24
Microthalestris forficula 315
Micrura verilli 207
MICTYRIDAE 364
Mictyris longicarpus 364
Migros flintii 80
  spiritensis 80
Miichthys miiuy 431
Mikadotrochus hirasei 261
  salmiana 261
Miliammina earlandi 71
  fusca 71
  obliqua 71
Miliola costata 89
  inflata 89
  natchitochensis 89
  prisca 89
  pseudocarinata 89
  rostrata 89
  saxorum 89
  sublineata 89
MILIOLACEA 82
MILIOLIDAE 89
MILIOLINA 81
Miliolinella australis 86
  californica 86
  chaoyii 86
  chui 86
  chukchiensis 86
  circularis 86
  corrugata 86
  hornibrooki 86
  iongchuanae 86
  labiosa 86

Miliolinella (continued)
   natchirochensis 86
   oblonga 86
   oceanica 86
   pseudoblonga 86
   robusta 86
   striata 86
   subrotunda 86
   temeii 86
   vigilax 86
   webbiana 86
Millepora dichotoma 168
   etaesa 168
   intricata 168
   latifolia 168
   murrayi 168
   platyphylla 169
   tenera 169
   xishaensis 169
MILLEPORIDAE 168
Millettia tessellata 109
Millettiana milletti 115
MILLETTIIDAE 109
Mimoblennius atrocinctus 446
Mimocalanus cultrifera 307
Mimosella verticellata 370
Mimosina orientalis 109
   sidebottomi 109
Miniacina miniacea 116
Minicorbula minutissima 256
Minidiscus chilensis 30
   comicus 30
   ocellatus 30
   spinulosus 30
   subtillis 30
   triocutatus 30
Minolia chinensis 263
Minous coccineus 454
   inermis 454
   monodactylus 454
   pictus 454
   pusillus 454
   quincarinatus 454
   trachycephalus 454
Minuspio cirrifera 223
   polybranchiata 223
Minutocellus polymorphus 43
Miogypsina polymorpha 126
MIOGYPSINIDAE 126
Miracia efferata 314
Mirolabrichthys tuka 427
Miscellanea complanata 126
   minor 126
   miscella 126
   multicolumnata 126
   stampi 126
Mischocarpus sundaicus 154
MISCHOCOCCALES 50
Misophria sinensis 315
MISOPHRIIDAE 315
MISOPHRIOIDA 315
Mississippina concentrica 111
MISSISSIPPINIDAE 111
[*Mitella mitella*] 318

Mitra abyssicola 278
   ambigua 278
   aurantia 278
   cardinalis 278
   chinensis 278
   circula 279
   clathrus 279
   coffea 279
   [*coffea*] 279
   decurtata 279
   edentula 279
   ferruginea 279
   flammea 279
   imperialis 279
   isabella 279
   litterata 279
   lutea 279
   mitra 279
   [*mitra*] 279
   morchii 279
   papalis 279
   papillio 279
   paupercula 279
   pellisserpentis 279
   praestantissima 279
   proscissa 279
   puncticulata 279
   retusa 279
   scutulata 279
   stictica 279
   [*taiwanica*] 279
   virgata 279
   zebra 279
Mitrapus heteropodus 316
Mitrella bella 276
   burchardi 276
MITRIDAE 278
Mithrodia clavigera 391
MITHRODIIDAE 391
Mitropifex obeliscus 279
[*Miyadiella pedunculata*] 337
   podophthalmus 337
Mobula diabolus 409
   formosana 409
   japonica 409
MOBULIDAE 409
Modiolus auriculatus 242
   [*barbatus*] 242
   comptus 242
   elongatus 242
   flavidus 242
   hanleyi 242
   metcalfei 242
   modiolus 242
   nipponicus 242
   philippinarum 242
   plicatus 242
   sirahensis 242
   [*subrugosa*] 242
   vaginus 242
   [*vagina*] 242
   [*watsoni*] 241
MODULIDAE 266
Modulus tectum 266

Moerella culter 251
   iridescens 251
   jedoensis 251
   philippinarum 251
   rutila 251
Moerisia inkermanica 168
   [*lyonsi*] 168
MOERISIIDAE 168
Moina macrocopa 300
   micrura 300
MOINIDAE 300
Moira lachesinella 394
Mola mola 461
Molgula diversa 403
   manhattensis 403
MOLGULIDAE 403
MOLIDAE 461
Mollicia acanthophora 304
   kampta 304
   minki 304
   mollis 304
Mollugo costata 151
   nudicaulia 152
   oppositifolia 152
   pentaphylla 152
MOLLUSCA 237
Molpadia changi 389
   guangdongensis 389
   roretzii 389
MOLPADIIDAE 389
MOLPADONIA 389
Monacanthus chinensis 459
   [*nipponensis*] 460
   [*setifer*] 460
   [*sulcatus*] 459
Monacilla typicus 307
Monalysidium politum 90
MONASCIDAE 195
Monascus filiformis 195
   orientalis 195
MONERA 21
Monetaria annulus 272
   moneta 272
Mongolodiaptomus birulai 310
   [*formosanus*] 310
MONHYSTERIDA 211
MONHYSTERIDAE 211
Monia umbonata 246
Monilea calliferus 263
MONILIALES 130
MONOCENTRIDAE 423
Monocentrus japonicus 423
Monochirus trichodactylus 458
MONOCOTYLEDONEAE 158
Monoculodes limnophilus 330
[MONODACTYLIDAE] 436
Monodactylus argenteus 436
MONODHELMINTHIDAE 203
Monodhelmis philippinensis 203
Monodictys pelagica 132
Monodictys sp. 132
Monodictys-like sp. 132
[*Monodonta australis*] 263
   labio 263

Index

Monodonta (continued)
——— *confusa* 263
neritoides 263
perplexa 263
MONOGENA 192
MONOGONONTA 214
Monolecithotrema lizae 202
Monomitopus kumae 421
longiceps 421
Monopia flaveola 302
tehani 302
Monoporella nodulifera 376
Monoposthia costata 211
MONOPOSTHIIDAE 211
Monopylephorus parvus 234
rubroniveus 234
Monorcheides xishaensis
Monorchicestrahelmins branchiostegi 195
MONORCHIIDAE 195
Monorchis fusiformis
MONORHAPHIIDAE 165
Monorhaphis intermedia 165
Monoserius pennarius 171
Monostaechas quadridens 171
[MONOSTICHOGLOSSA] 288
MONOSTILIFERA 208
Monostroma angicava 144
arcticum 144
crassifolia 144
latissimum 144
nitidum 144
[*undulalum*] 144
MONOSTROMATACEAE 144
Monotaxis grandoculis 434
Monozonium pachystylum 65
Monstrilla grandis 315
MONSTRILLIDAE 315
MONSTRILLOIDA 315
MONTACUTIDAE 248
Montacutona compacta 248
mutsuwanensis 248
Montipora aenigmatica 181
cristagalli 181
danae 181
foliosa 181
foveolata 181
fragilis 181
fruticosa 181
gaimardi 181
hispida 181
ramosa 181
sinensis 181
solanderi 181
striata 181
trabeculata 181
truncata 181
turgescens 181
Mopalia retifera 238
schrencki 238
MOPALIIDAE
Mopsella rubeola 186
MORACEAE 151
Morchellana dollfusi 188

Morchellana (continued)
elongata 188
pulchella 188
rubra 188
MORIDAE 419
Morinda citrifolia 158
parvifolia 158
Moringua abbreviata 417
macrocephalus 417
macrochir 417
MORINGUIDAE 417
Mormula mumia 284
philippiana 284
terebra 284
Morozovella angulata 101
velascoensis 101
Mortensenella forceps 363
[*Morula ambusta*] 274
[*anaxares*] 275
[*aspera*] 275
[*biconica*] 275
[*borealis*] 275
[*cariosa*] 275
[*cavernosa*] 275
[*fiscella*] 275
[*fusca*] 275
[*granulata*] 275
[*margariticola*] 275
[*marginatra*] 275
[*musiva*] 275
[*paucimaculata*] 275
[*spinosa*] 275
[*uva*] 275
Morulaeplecta bulbosa 75
inflata 75
Morum cancellatum 272
[*cancellatum*] 272
grande 272
Mototeuthis lonnbergii 294
Mucropetraliella philippinensis 380
robusta 380
thenardii 380
verrucosa 380
watersi 380
Muggiaea atlantica 175
delsmani 175
[*Mugil affinis*] 425
[*carinatus*] 424
cephalus 425
[*dussumier*] 424
[*engeli*] 425
[*kelaertii*] 425
[*macrolepis*] 425
[*melinopterus*] 425
[*seheli*] 425
[*soiuy*] 424
[*vaigiensis*] 424
MUGILIDAE 424
MUGILIFORMES 424
Mugilogobius abei 452
tagala 452
MULLIDAE 435
Mulloidichthys flavolineatus 435
pflugeri 435

Mulloidichthys (continued)
samoensis 436
suriflamma 436
vanicolensis 436
Multifidiella nodulosa 78
Munida babai 350
compressa 350
exigua 350
incerta 350
japonica 350
perarmata 350
prominula 350
[*scabra*] 350
sinensis 350
Munidopagurus macrocheles 350
Munidopsis longirostris 350
Muraena pardalis 417
MURAENESOCIDAE 416
Muraenesox bagio 416
cinereus 416
talabon 416
talabonoides 416
yamaguchiensis 416
Muraenichthys gymnupterus 417
hattae 417
macropterus 417
malabonensis 417
MURAENIDAE 416
Murdannia kainantensis 162
Murex aduncospenosus 275
haustellum 275
[*haustellum*] 275
hirasei 275
kiiensis 275
[*kiiensis*] 275
[*martinianus*] 275
nigrispinosus 275
pecten 275
rectirostris 275
senkakuensis 275
[*sinensis*] 274
sobrinus 275
[*ternispina*] 275
torrefactsu 275
trapa 275
tribulus 275
[*triremis*] 275
troscheli 275
Murexsul umbilicatus 275
Muricella abnormalis 187
flexuosa 187
rubra 187
sibogae 187
sinensis 187
MURICIDAE 274
[*Muricodrupa granulata*]
Muricoglobigerina soldadoensis 101
Murraya alata v. *hainanensis* 153
microphylla 153
Murrayinella murrayi 112
Murraytrema pricei 192
Mursia armata 356
curtispina 356
trispinosa 356

Musculista japonica 242
   perfragilis 242
   senhausia 242
Musculovesicula trichiuri 202
   zonichthydis 202
Musculus cumingiana 242
   cupreus 242
   [*marmoratus*] 242
   mirandus 242
   nanus 242
   nigrus 242
   [*senhousei*] 242
MUSSIDAE 184
Mustelus griseus 406
   kanekonis 406
   manazo 406
Mya arenaria 256
   [*japonica*] 256
Myagropsis myagroides 143
Mycale macilenta 165
   philippensis 165
   phylohila 165
   vermistyla 165
MYCALIDAE 165
Mycedium elephantotus 184
Mychostomina carinata 81
   peripora 81
   revertens 81
   truncata 81
MYCOPHYCOPHYTA 132
MYCTOPHIDAE 413
MYCTOPHIFORMES 413
Myctophum affine 414
   asperum 414
   aurolaternatum 414
   brachygnathum 414
   lychnobium 414
   nitidulum 414
   obtusirostris 414
   [*pterotum*] 413
   selenoides 414
   spinosum 414
[*Myelophycus caespitosus*] 141
   simplex 141
MYIDAE 256
MYLIOBATIDAE 408
MYLIOBATIFORMES 408
Myliobatis tobijei 408
MYOCHAMIDAE 257
Myodora fluctuosa 257
   [*japonica*] 257
   [*proxima*] 257
MYOIDA 256
MYOPORACEAE 158
Myoporum bontiodes 158
Myoschiston simile 58
Myra affinis 355
   coalita 355
   fugax 355
   kessleri 355
Myrianida cf. *pachycerus* 219
Myrichthys aki 417
   colubrinus 417
   maculosus 418
Myriochele picta 228

[*Myriocladia kuromo*] 141
Myrionema tenue 141
MYRIONEMATACEAE 141
Myripristis adustus 422
   berndti 422
   chryseres 422
   hexagonus 422
   kuntee 422
   melanosticyus 422
   murdjan 422
   parvidens 422
   pralinius 422
   seychellensis 422
   shultzei 422
   violaceus 422
   vittata 422
Myrophis cheni 417
   macrophthalmus 417
MYRSINACEAE 156
MYRTACEAE 156
Mysella triangularis 248
MYSIDACEA 323
MYSIDAE 323
Mysidopsis indica 323
   kempi 323
Mystriophis porphyreus 418
MYTILIDAE 241
MYTILOIDA 241
Mytilopsis sallei 253
Mytilus coruscus 242
   [*edulis*] 242
   galloprovincialis 242
   [*smaragdinus*] 243
Myxicola infundibulum 230
MYXILLIDAE 165
MYXINIDAE 404
Myzostoma antennulata 231
   attenuatum 231
   bocki 231
   dodecaphalcis 231
   lobatum 231
   nasonovi 231
   cf. *pallidum* 231
MYZOSTOMIDA 231
MYZOSTOMIDAE 231

N

Naiades cantrainii 215
NAIDIDAE 234
Naineris dendritica 222
   hainanensis 222
   laevigata 222
Nais glitra 131
   inornata 131
Namalycastis aibiuma 220
   longicirris 220
NANNASTACIDAE 325
Nannocalanus minor 306
Nannosquilla varicosa 369
NANNOSQUILLIDAE 369
Nanocassiops granulipes 361
Nanomia bijuga 174
Nanosesarma batavicum 366
   minutum 366

Nanosesarma (continued)
   pontianacensis 366
Nansenia ardesiaca 411
[*Napalirus japonicus*] 346
Narcetes lloydi 412
Narcine lingula 409
   maculata 409
   timlei 409
NARCOMEDUSAE 173
Nardoa frianti 391
Narke japonica 409
NARKIDAE 409
Naso annulatus 448
   brachycantron 448
   brevirostris 448
   herrei 449
   [*hexacanthus*] 448
   [*lituratus*] 448
   [*lopezi*] 448
   thynnoides 449
   [*thynnoides*] 448
   unicornis 449
   vlamingi 449
Nassa francolinus 275
   serta 275
   [*varicifera*] 278
Nassaria gracilis 277
NASSARIIDAE 277
Nassarius acuminatus 277
   albescens 277
   [*brum*] 277
   caelatus 277
   [*clathratus*] 277
   conoidalis 277
   [*conoidalis*] 277
   conoratus 277
   crematus 277
   cumingi 277
   [*dealbatus*] 277
   dominulus 277
   dorsatus 277
   festivus 277
   fidus 277
   glans 277
   granliferus 277
   gregarius 277
   hepaticus 277
   hiradoensis 277
   horrida 277
   livescens 277
   mangelioider 277
   margaratiferus 277
   nodiferus 278
   olivaceus 278
   papillosus 278
   pullus 278
   reeveanus 278
   [*rutilans*] 277
   semiplicatus 278
   siquijorensis 278
   spiratus 278
   spurcus 278
   succinctus 278
   sufflatus 278
   terefiusculas 278

Index

Nassarius (continued)
   thersites 278
   variciferus 278
   [*variciferus*] 278
   velatus 278
   vitiens 278
NASSELLARIA 66
Natica alapapilionis 269
   albifasciata 269
   ampla 269
   arachnoidea 269
   areolata 269
   asellus 269
   bibalteata 269
   bougei 270
   clausiformis 270
   concavoperculata 270
   concinna 270
   crassoperculata 270
   [*didyma* v. *ampla*] 270
   [*didyma bicolor*] 270
   excellenna 270
   [*fortunei*] 269
   gualteriana 270
   hainanensis 270
   hirasei 270
   [*janthostoma*] 270
   janthostomoides 270
   lineata 270
   lurida 270
   [*maculosa*] 270
   melanoperculata 270
   onca 270
   pygmaea 270
   qizhongyani 270
   sagittata 270
   scopaespira 270
   simplex 270
   sinensis 270
   [*spadicea*] 270
   spadiceoides 270
   superaotanta 270
   tabularis 270
   tenuipicta 270
   tessellata 270
   [*tessellata*] 270
   tigrina 270
   tossansis 270
   trilli 270
   unibalteata 270
   unicolor 270
   vitellus 270
NATICIDAE 269
Naucrates ductor 431
Nausithoe punctata 176
NAUSITHOIDAE 176
[*Nauticaris futilirostris*] 343
NAUTILIDAE 294
NAUTILOIDEA 294
Nautilus pompilius 294
Navicula abrupta 39
   alpha 39
   approximata 39
   ——— v. *niceaensis* 39
   arabica 39

Navicula (continued)
   asymmetrica 39
   australica 39
   biformis 39
   bolleana 39
   brasiliensis 39
   bruchii 39
   caeca 39
   cancellata 39
   ——— v. *apiculata* 39
   ——— v. *retusa* 39
   carinifera 39
   cincta 39
   circumsecta 39
   clavata 39
   ——— v. *indica* 39
   climacospheniae 39
   cluthensis 39
   complanata 39
   consors 39
   corymbosa 39
   crucicula 39
   ——— v. *orientalis* 39
   cruciculoides 39
   cryptocephaloides 39
   cryptotenella 39
   cuspidata 39
   delta 39
   digito-radiata 39
   directa 39
   ——— v. *javanica* 39
   ——— v. *remota* 39
   distans 39
   epsilon 39
   eta 39
   eymei 39
   forcipata 39
   ——— v. *densestriata* 39
   fortis 39
   fujianensis 39
   genuflexa 39
   glacialis 39
   ——— v. *neglecta* 39
   granulata 39
   H-album 39
   hamulifera 39
   hennedyi 39
   ——— f. *california* 39
   ——— v. *nebulosa* 39
   hetero-punctata 39
   hochstetteri 39
   howeana 39
   humerosa 39
   ——— v. *constricta* 39
   ——— v. *minor* 39
   impressa 39
   inhalata 39
   inserata 39
   integra 39
   ——— v. *maculata* 40
   jamalinensis 40
   jejuna 40
   latissima 40
   liaotungiensis 40
   longa 40

Navicula (continued)
   lorenzii 40
   luxuriosa 40
   lyra 40
   ——— v. *dilatata* 40
   ——— v. *elliptica* 40
   ——— v. *insignis* 40
   ——— v. *recta* 40
   ——— v. *signata* 40
   ——— v. *subtypica* 40
   lyroides 40
   maculata 40
   marina 40
   membranacea 40
   mollis 40
   monilifera 40
   mutica 40
   my 40
   nitescens 40
   northumbrica 40
   nummularia 40
   orthoneoides 40
   pantocsekiana 40
   parva 40
   pavillardi 40
   penna 40
   perplexoides 40
   perrhombus 40
   pi 40
   pinna 40
   plicatula 40
   praetexta 40
   punctulata 40
   pupula v. *elliptica* 40
   pygmaea 40
   quincunx 40
   raeana 40
   ramosissima 40
   restituta 40
   rhaphoneis 40
   rho 40
   rhynchocephala 40
   robertisiana 40
   satura 40
   schaarschmidtii 40
   scintilllans 40
   scopulorum 40
   scutelloides 40
   scutiformis 40
   semistauros 40
   sibayiensis 40
   spectabilis 40
   ——— v. *excavata* 40
   stercus muscarum 40
   subcarinata 40
   superimposita 40
   takoradiensis 40
   toulaae 40
   transfuga 40
   tuscula v. *cuneata* 40
   viridula v. *slesvicensis* 40
   yarrensis 40
   zostereti 40
NAVICULACEAE 35
NAVICULALES 35

[*Navodon modestus*] 459
   septentrionalis 459
   tessellatus 460
   xanthopterus 460
Naxioides mammillata 357
   taurus 357
Neallolepidapedon hawaiiense 195
Nealotus tripes 449
Neamia octospina 429
Neanthes donghaiensis 220
   flava 220
   glandicincta 220
   [*ijimai*] 220
   japonica 220
   maculata 220
   nanhaiensis 220
   [*oxypoda*] 220
   succinea 220
   unifasciata 220
NEANURIDAE 300
Nebrius formosanum 405
   macrurus 405 *Nectalia loligo* 174
      NECTALIDAE 174
Nectoneanthes alatopalpis 220
   donghaiensis 220
   fujianensis 220
   ijimai 220
   multignatha 220
   oxypoda 220
NEENCHELYIDAE 417
Neenchelys parvipectoralis 417
Negaprion queenslandicus 406
Negogaleus balfouri 406
   brachygnathus 406
   macrostoma 406
   microstoma 406
Neidium amphirhynchus 40
   iridis v. *amphigomphus* 40
Neilo bisculpta 239
   humilior 239
Neilonella aequatorialis 239
   dubia 239
   guineensis 239
Neisseria 21
NEISSERIACEAE 21
NEISSERIALES 21
Nellia oculata 372
   tenuis 372
Nemacystus decipiens 141
NEMALIACEAE 133
NEMALIALES 133
Nemalion helminthoides v. *vermiculare* 133
   [*pulvinatum*] 133
NEMASTOMACEAE 136
Nematalosa come 410
   japonica 410
   nasus 410
Nemateleotris magnificus 451
Nematobothrium filiforme 200
   schistogonimum 200
Nematobrachian booepis 334
   flexipes 334
   sexspinosus 334
NEMATOCARCINIDAE 339

NEMATOCARCINOIDEA 339
Nematocarcinus cursor 339
   sibogae 339
   undulatipes 339
NEMATODA 209
Nematonereis unicornis 227
Nematopagurus gardimeri 350
   indicus 350
   muricatus 350
   squamichelis 350
   vallatus 350
Nematoscelis atlantica 334
   gracilis 334
   lobata 334
   microps 334
   tenella 335
Nembrotha cristata 289
   kubaryana 289
   lineolata 289
NEMERTINEA 207
Nemertopsis gracilis 208
   quadripunctatus 208
   tetraclitophila 208
NEMICHTHYIDAE 417
Nemichthys scolopaceus 417
NEMIPTERIDAE 434
Nemipterus aurorus 434
   bathybius 434
   furcosus 434
   hexodon 434
   japonicus 434
   oveni 434
   peronii 434
   thosaporni 434
   tolu 434
   virgatus 434
   zysron 434
Nemocardium bechei 249
   samarangae 249
Nemopsis bachei 166
Neoallolepidapedon hawaiiense
Neoamphitrite robusta 229
Neoaphanurus magniprotesticus 202
Neobenthotrema pyriformis 194
Neobythites fasciatus 421
   nigromaculatus 421
   sivicola 421
   stigmosus 421
Neocalanus gracilis 306
   robustior 306
   tenuicornis 306
Neocentropogon japonicus 455
NEOCHEILOSTOMINA 372
Neocirrhilabrus oxyurus 440
Neoconorbina orbicularis 111
   pacifica 111
   terquemi 111
[*Neodexiospira foraminosus*] 231
Neodiaptomus yangtsekiangensis 310
Neodichadena mugilis 202
Neodollfustrema xishaense 193
NEOECHINORHYNCHIDA 213
NEOECHINORHYNCHIDAE 213
Neoechinorhynchus agilis 213
   coiliae 213

Neoechinorhynchus (continued)
   johnii 213
   topseyi 213
   tylosuri 213
Neoepinnula orientalis 449
Neoepisesarma mederi 366
   versicolor 366
Neoeponides berthelinianus 111
   procera 111
   subornatus 111
Neoferdina cumingii 391
Neofibularia chinensis 165
Neogloboquadrina dutertrei 100
   eggeri 100
   pachyderma 100
Neoglyphidodon nigroris 443
Neogoniolithon conicum 135
   fosliei 135
   frutescens 135
   ———— f. *flabelliformis* 135
   megalocystum 135
   pacificum 135
   trichotomum 135
   variabile 135
[*Neohaustator fortilirata*] 266
Neohelicometra dalianensis 197
Neoheteromastus lineus 225
Neolaeops microphthalmus 457
NEOLAGANIDAE 394
Neolenticulina peregrina 93
Neoliomera pubescens 361
Neolithodes nipponensis 350
Neomediomastus glabrus 225
Neomerinthe megalepis 453
   procorva 453
   procurva 453
   rotunda 453
Neomeris annulata 145
   bilimbata 145
   van-bosseae 145
Neometadidymozoon zhejiangensis 200
Neometra alecto 387
Neomultitestis bengalensis 195
Neomysis awatschensis 323
   japonica 323
[*Neoniphon opercularis*] 422
   [*sammara*] 422
Neonotoporus kareii 195
Neoomeia sphyrna 209
Neopanthalis lepidus 217
   muricatus 217
Neopetrolisthes maculatus 351
Neophocaena phocaenoides 469
NEOPHYLLOBIIDAE 298
Neophyllobius sp. 298
[*Neopomacentrus anabatoides*] 442
   azysron 443
   [*bankieri*] 442
   [*brachialis*] 444
   [*mela*] 442
   [*violascens*] 444
Neoprosorhynchus xishaensis 193
Neopygorchis exvitellina 204
Neorhadinorhynchus nudum 213
[*Neorhynchoplax introversus*] 356

Index

Neorhynchoplax (continued)
    [*sinensis*] 356
Neosalanx anderssoni 411
    tangkahkeii 411
Neosarmatium meinerti 366
    smithi 366
Neoscalpellum dicheloplax 318
NEOSCOPELIDAE 413
Neoscopelus macrolepidotus 413
    microchir 413
    porosus 413
Neosebastes entaxis 453
Neosematis distephanus 66
Neostylodactylus amarhynsus 339
NEOTREMATA 384
Neoturris papua 167
Neouvigerina ampullacea 107
Neoxanthias impressus 361
    lineatus 361
    michelae 361
Neoxenophthalmus obscurus 363
NEPENTHACEAE 152
Nepenthes mirabilis 152
Nepanthia belcheri 391
    briareus 391
    variabilis 391
Nephrolepidina rutteni 116
[*Nephrops andamanicus*] 345
    [*japonicus*] 345
NEPHROPSIDAE 345
NEPHROPSIDEA 345
Nephropsis carpenteri 345
    stewarti 345
    sulcata 345
Nephrospyris docris 66
    paradictyum 67
    renilla 67
Nephthea brassica 188
    capnelliformis 188
    erecta 188
    simulata 188
NEPHTHEIDAE 188
NEPHTYIDAE 222
Nephtys caeca 222
    californiensis 222
    ciliata 222
    [*inermis*] 222
    longosetosa 222
    oligobranchia 222
    paradoxa 222
    polybranchia 222
    [*sinensis*] 222
    tulearensis 222
Neptunea arthritica cumingii 277
    [*cumingii*] 277
[*Neptunus argentatus*] 359
NEREIDAE 220
Nereiphysa castanea 215
[*Nereis agassizi*] 221
    [*aibuhitensis*] 221
    [*alatopalpis*] 220
    [*anchylochaeta*] 220
    [*burmensis*] 220
    [*camiguinoides*] 221
    [*cavifrons*] 221

Nereis (continued)
    [*costae*] 220
    coutieri 220
    dayana 220
    denhamensis 220
    donghaiensis 220
    [*dumerilii*] 221
    [*erythraeensis*] 220
    [*ezoensis*] 220
    falcaria 220
    ———— *multignatha* 220
    [*glandicincta*] 220
    grubei 220
    guangdongensis 220
    hainanica 220
    heterocirrata 220
    heteromorpha 220
    huanghaiensis 220
    [*ijimai*] 220
    jacksoni 220
    [*kauderi*] 220
    [*kobiensis*] 221
    [*linea*] 221
    longior 220
    [*mirabilis*] 220
    multignatha 220
    neoneanthes 220
    nichollsi 220
    [*obfascata*] 221
    [*orientalis*] 221
    [*oxypoda*] 220
    [*pachychaeta*] 220
    pelagica 220
    persica 220
    sinensis 220
    [*suluana*] 221
    surugaense nanhaiensis 220
    [*tentaculata*] 220
    trifasciata 220
    [*unifasciata*] 220
    vexillosa 220
    zhongshaensis 220
    zonata 221
    [———— *persica*] 220
    [———— *tigrina*] 221
Nerilla sinica 215
NERILLIDAE 215
[*Nerine cirratulus*] 224
Nerinides globosa
Nerita achatina 264
    albicilla 264
    [*albicilla*] 264
    chamaeleon 264
    costata 264
    [*costata*] 264
    exuvia 264
    incerta 264
    insculpta 264
    japonica 264
    lineata 264
    ocellata 264
    planospira 264
    plicata 264
    [*plicata*] 264
    polita 264

Nerita (continued)
    [*polita*] 264
    reticulata 264
    signata 264
    squamulata 264
    striata 264
    [*striata*] 264
    undata 264
    yoldii 264
NERITIDAE 264
Neritina auriculata 264
    parallela 264
    plumbea 264
    pulligera 264
    turrita 264
    variegata 265
    violacea 265
NERITOPSIDAE 265
Neritopsis radula 265
Nerocila acuminata 325
    depressa 325
    phaeopleura 325
    sundaica 325
Nesiarchus nasutus 449
Netrostoma coerulescens 176
    setouchianum 176
Nettapus coromandelianus 466
Nettastoma parvviceps 416
NETTASTOMIDAE 416
[*Neurocarpus divaricatus*] 142
Neverita didyma 270
Nevillina coronata 86
Nezumia proximus 420
[*Nibea acuta*] 431
    albiflora 431
    chui 431
    diacanthus 431
    japonica 431
    miichthioides 432
    semifasciata 432
Nicolla ditrematis 197
    epinepheli 197
Nicomache inormata 225
    lumbricalis 225
    personata 225
[*Nicon benhami*] 221
    japonica 221
    maculata 221
    moniloceras 221
    sinica 221
NIDALIIDAE 188
Nihonia australis 281
[*Nike aequimana*] 341
    [*japonica*] 341
Nikoides sibogae 341
Ninoe bruuni 227
    gemma 227
    palmata 227
Niobia dendrotentacula 167
NIOBIIDAE 167
Niphon spinosus 427
Nipponatys volvulinus 286
Nipponomysella oblongata 248
Nipponomysis quadrispinosa 323
    sinensis 323

Nipponoscaphander cumingii 286
   japonica 286
   teromachii
Nitidotellina iridella 251
   minuta 251
   nitidula 251
Nitraris sibirica 153
Nitzschia acuminata 46
   aequatorialis 46
   amphibioides v. *chenghaensis* 46
   angularis 46
   ——— v. *affinis* 46
   angustata 46
   antillarum 46
   apiculata v. *liaotungiensis* 46
   bicapitata 46
   brevissima 46
   capuluspalae 46
   coarctata 46
   cocconeiformis 46
   compressa 46
   constricta 46
   corpulenta 46
   cursoria 46
   delicatissima 46
   denticula 46
   didyma 46
   dissipata 46
   distans 46
   distantoides 46
   epithemoides 46
   fasciculata 46
   filiformis 46
   fluminensis 46
   frigida 46
   frustulum 46
   granulata 46
   habirshawii 46
   hungarica 46
   hybrida 46
   insignis 46
   ——— v. *lanceolata* 46
   jelineckii 46
   kolaczeckii 46
   lanceola 46
   lanceolata 46
   ——— v. *chinensis* 46
   ——— v. *incrustoms* 46
   ——— v. *minor* 46
   linkei 46
   littoralis 46
   longissima 46
   ——— v. *chinensis* 46
   ——— v. *costata* 46
   ——— v. *reversa* 46
   lorenziana 46
   ——— v. *densestriata* 46
   macilenta 47
   ——— f. *abbreviata* 47
   majuscula v. *lineata* 47
   marginulata 47
   ——— v. *didyma* 47
   ——— v. *subconstricta* 47
   marina 47
   maxima 47

Nitzschia (continued)
   navicularis 47
   nelsonii 47
   obtusa 47
   ——— v. *scalpelliformis* 47
   ossiformis 47
   palea v. *minuta* 47
   paleacea 47
   panduriformis 47
   ——— v. *elegans* 47
   ——— v. *minor* 47
   [*paradoxa*] 46
   parvula 47
   petitiana 47
   plana 47
   pulcherrima 47
   punctata 47
   ——— v. *coarctata* 47
   ——— v. *elongata* 47
   pungens 47
   ——— v. *atlantica* 47
   scalaris 47
   seriata 47
   sigma 47
   ——— v. *intercedens* 47
   ——— v. *rigida* 47
   ——— v. *sigmatella* 47
   sigmoidea 47
   sinensis 47
   socialis 47
   spathulata 47
   ——— v. *hyalina* 47
   spectabilis 47
   subsalsa 47
   subtilis 47
   tryblionella 47
   ——— v. *subsalina* f. *subconstricta* 47
   ——— v. *victoriae* 47
   ventricosa 47
   vermicularis 47
   vidovichii 47
   vitraea 47
NITZSCHIACEAE 46
Nitzschiella incurva 47
   longissima 47
Nobia conjugatum 320
   grandis 320
   orbicellae 320
   sinica 320
[*Nobilum histirio*] 354
Nocardia 24
NOCARDIACEAE 24
Noctiluca scintillans 51
NOCTILUCACEAE 51
Nodellum membranaceum 69
[*Nodilittorina exigua*] 265
   [*granularis*] 265
   [*hawaiiensis*] 265
   [*miliaris*] 265
   [*millegrana*] 265
   [*picta*] 265
   [*pyramidalis*] 265
   radiata 265
   trochoides 265
   vidua 265

Nodilittorina (continued)
   [*vidua*] 265
Nodobaculariella rustica 81
Nodogenerina antillea 109
   aperturata 109
   lepida 109
   virgula 109
Nodophthalmidium carinata 81
   compressum 82
   simplex 82
   zhongshaensis 82
Nodosaria affinis 91
   albatrossi 91
   calomorpha 91
   cephalota 91
   flintii 91
   inflexa 91
   ittai 91
   koina 91
   nitida 91
   propinqua 91
   pyrula 91
   radicula 91
   raphana 91
   semirugosa 91
   simplex 91
   subsoluta 91
   tainanensis 91
   vertebralis 91
NODOSARIACEA 90
NODOSARIIDAE 90
Nodosinum gaussicum 73
   nodulosus 73
Nodularia harveyana 26
   hawaiiensis 26
Nodulina dentaliniformis 72
Noetiella vivianae 240
Nollela papuensis 370
NOMEIDAE 450
Nomeus gronovii 450
Nonion advenum 117
   affine 117
   akitaense 117
   anomalinoidea 117
   belridgense 117
   bogdanowiczi 117
   boueanum 117
   browni 117
   decorum 117
   delicatum 117
   depressullum 117
   extensum 117
   glabrum 117
   goudkoffi 117
   graniferum 117
   granulosum 117
   grateloupi 117
   ibericum 117
   inexcavatum 117
   japonicum 117
   laevis 117
   ornatissimum 117
   roemeri 117
   rolshauseni 117
   rotulum 117

Index

Nonion (continued)
  scaphum 117
  schwageri 117
  shansiensis 117
  sorachiensis 117
  subdilatatum 117
  sublaeve 118
  tallahattensis 118
  tuberculatum 118
  turgida 118
  usbekistanensis 118
Nonionella africana 118
  alabamensis 118
  ansata 118
  atlantica 118
  auricula 118
  basiloba 118
  coarseperforata 118
  decora 118
  jacksonensis 118
  limbatostriata 118
  magnalingua 118
  modesta 118
  opima 118
  penghuensis 118
  pulchella 118
  reussana 118
  soldadoensis 118
  stella 118
  translucens 118
  tredeca 118
Nonionellina frankei 118
  japonica 118
  labradorica 118
  pizarrense 118
  reniformis 118
NONIONICEA 117
NONIONIDAE 117
Nonionoides grateloupi 118
Nonparahalosydna pleiolepis 217
Norion globosus 316
NOSTOCACEAE 26
NOSTOCALES 26
NOSTOCHOPSIDACEAE 27
Notarchus indicus armatus 287
  leachii cirrosus 287
  longicaudus 287
  risbeci 287
NOTACANTHIDAE 418
NOTACANTHIFORMES 418
Notacanthus abbotti 418
NOTASPIDEA 289
  [*Nothocaris binoculus*] 344
  Notholca sp. 214
Nothria atlantisa 226
  conchylega 226
  [*geophiliformis*] 226
  holobranchiata 226
  pallida 226
Notoacmea concinna 262
  schrenckii 263
Notobranchaea macdonaldi 288
Notobryon wardi 292
Notocaythus conicus 185
Notocirrus japonicus 227

[*Notocochlis lineata*] 270
NOTODORIDIDAE 290
Notolepis rissoi 415
Notomastus cf. *aberans* 225
  fauvel 225
  giganteus 225
  latericeus 225
  polyodon 225
Notophyllum foliosum 215
  imbricatum 215
  splendens 215
Notoplana humilis 191
Notoplax doederleini 238
Notoplax sp. 238
Notopours astrocongeris 197
Notoproctus pacificus 225
Notopygos gigas 226
  sibogae 226
  supragigas 226
[*Notorhynchus cepedianus*] 404
  platycephalus 404
Notoscopelus resplendens 414
Notoseila laqueata 267
Notostomus brevirostris 339
  longirostris 339
  patentissimus 339
Nototodarus hawaiiensis 295
  sloani philippinensis 295
Noumea nivilis 290
Nouria armata 75
  atlantica 75
  carinata 75
  foliacea 75
  gracilenta 75
  harrisii 75
  polymorphinoides 75
  rhombiformis 76
  sinensis 76
  tenuis 76
NOURIIDAE 75
Novaculichthys macrolepidotus 440
  taeniourus 440
Nubecularia lucifaga 82
NUBECULARIIDAE 81
Nubeculina divaricata 82
Nubeculopsis queenslandica 82
Nucia speciosa 355
Nucula convexa 239
  cumingii 239
  cyrenoides 239
  donaciformis 239
  exodonta 239
  izushotoensis 239
  nimbosa 239
  nipponica 239
  pachydonta 239
  parvula 239
  paulula 239
  tenuis 239
  tokyoensis 239
Nuculana forticostata 239
  sadoensis 239
  sinensis 239
  tamseimaruae 239
  yokoyamai 239

NUCULANIDAE 239
NUCULIDAE 239
NUCULOIDA 239
NUDA 190
Nudechinus multicolor 393
NUDIBRANCHIA 289
Numenius arquata orientalis 467
  borealis minutus 467
  madagascariensis 467
  phaeopus variegatus 467
Nummoluculina contraria 88
NUMMULITACEA 126
Nummulites ammonoides 126
  atacicus 126
  baguelensis 126
  complanata 126
  donghaiensis 126
  duimiensis 126
  georgiensis 126
  guayabalensis 126
  laevigatus 126
  mamilla 126
  nuttali 126
  obesus 126
  parvulus 126
  pengaronensis 126
  pernotus 126
NUMMULITIDAE 126
Nummulopyrgo paraglobulus 82
Nursia abbreviata 355
  japonica 355
  minor 355
  rhomboidalis 355
  [*sinica*] 355
Nursilia dentata 355
  sinica 355
  tonsor 355
Nuttallia olivacea 252
Nuttallina alternata 238
NYCTAGINACEAE 151
Nycticorax n. nycticorax 465
Nypa fruticans 161

O

[*Obelia bicuspidata*] 172
  bidentata 172
  [*borealis*] 172
  dichotoma 172
  [*gelatinosa*] 172
  geniculata 172
  [*gracilis*] 172
  [*longissima*] 172
[*Obeliscus buxeus*] 284
Oblimopa japonica 241
Obliquilingulina oblonga 98
Obrussena moeshimaensis 285
Obtusata antarotica 304
Oceanodroma r. monorhis 465
[*Ocenebra japonica*] 274
[*Ochetoclava sinensis*] 267
Ochetostoma erythrogrammon 237
  formolosum 237
[*Ocinebrellus fatcatus aduncus*] 275
Ocosia spinosa 455

Ocosia (continued)
   vespa 455
[*Octocanna polynema*] 169
Octocannoides ocellata 169
Octodendron hamuliferum 64
   nidum 64
   spathillatum 64
Octolasmis angulata 319
   aymonini geryononhila 319
   bathynomi 319
   bullata 319
   cor 319
   geryonophila 319
   neptuni 319
   nierstraszi 319
   orthogonia 319
   scuticosta 319
   tridens 319
   warwickii 319
Octomeris brunnea 320
   sulcata 320
Octophialucium funerarium 169
   indicum 169
   medium 169
   solidum 169
OCTOPODA 296
OCTOPODIDAE 296
OCTOPOTEUTHIDAE 294
Octopoteuthis sicula 294
Octopus aegina 296
   [*aerolatus*] 297
   berenice 296
   bimaculatus 296
   dollfusi 296
   [*faciatus*] 296
   fusiformis 296
   guangdongensis 296
   luteus 296
   maculosa 296
   nanhaiensis 296
   ocellatus 297
   oshimai 297
   ovulum 297
   pallida 297
   rugosus 297
   striolatus 297
   variabilis 297
   vulgaris 297
Octopyle circinata 65
   clypeata 65
   fruticosa 65
   hexagona 65
   octospinosa 65
   polystyle 65
   stenozona 65
[*Octorchis gegenbauri*] 170
Octotiara russelli 167
OCULINIDAE 183
Oculosiphon linearis 70
Ocypode ceratopphthalmus 365
   cordimana 365
   [*macrocera*] 365
   mortoni 365
   sinensis 365
   stimpsoni 365

OCYPODIDAE 364
OCYROPSIDAE 190
Ocyropsis crystallina 190
OCYTHDIDAE 296
Ocythoe tuberculata 296
ODENTODACTYLIDAE 368
Odentodactylus cultrifer 368
   japonicus 368
   scyllarus 368
Odessia microtentaculata 168
ODONATA 299
Odontamblyopus rubicundus 453
Odontanthias rhodopeplus 427
   unimaculatus 427
Odonthosphaera cyrtodon 63
Odontophora axonolaimoides 211
   littoralis 211
   paraxonolaimoides 211
Odontostomps braneri 415
   normalops 415
Odontosyllis enopla 219
   gibba 219
   maculata 219
   rubofasciata 219
Odonus niger 459
Odostomia felix 284
   hilgenoerfi 284
   lecta 284
   lirata 284
   mauritiana 284
   omaensis 284
   physoides 284
   planata 284
   subangulata 284
   subplanata 284
   tenera 284
[*Oedalechilus labiosus*] 425
Oediceroides ornithorhynchus 330
OEDICEROTIDAE 330
Oedignathus inermis
[*Oedipus graminea*] 340
   [*superbus*] 340
Oenothera littoralis 156
Oesophogocystis dissimilis 200
Oestrupia musca 41
OGCOCEPHALIDAE 462
Ogmaster capella 390
Ogyrides orientalis 343
   striaticauda 343
OGYRIDIDAE 343
[*Ogyris orientalis*]
Ohshimella ehrenbergi 388
Oidiodendron sp. 132
Oikopleura albicans 400
   cophocera 400
   cornutogastra 400
   dioica 400
   fusiformis 400
   graciloides 400
   intermedia 400
   longicauda 400
   megastoma 400
   parva 400
   rufescens 400
OIKOPLEURIDAE 400

Oithona attenuata 312
   brevicornis 312
   ——— f. *minor* 312
   ——— f. *typica*
   decipiens 312
   fallax 312
   longispina 312
   nana 312
   plumifera 312
   rigida 312
   robusta 312
   setigera 312
   similis 312
   simplex 312
   spinulosa 312
   tenuis 312
   vivida 312
   wellershausi 312
OITHONIDAE 312
Okadaia elegans 290
OKADAIIDAE 290
[*Okenia distincta*] 290
   japonica 290
   opuntia 290
   plana 290
[OKENIIDAE] 290
OLACACEAE 151
OLIBRINIDAE 326
Olibrinus sp. 326
OLIGOCHAETA 233
OLIGOHYMENOPHORA 57
Oligolecithoides trilobatus 202
Oligolepis acutipinnis 452
   fasciatus 452
Oligometra chinensis 387
   serripinna 387
Oligopus robustus 421
OLIGOTRICHIDA 58
OLINDIASIDAE 172
Oliva annulata 278
   caerulea 278
   emicator 278
   episcopalis 278
   [*erythrostoma*] 278
   hirasei 278
   ispidula 278
   [*lignaria*] 278
   miniacea 278
   multiplicata 278
   mustelina 278
   ——— *concavospira* 278
   [——— *mustelina*] 278
   ornata 278
   reticulata 278
   sericea 278
   textilina 278
   todosina 278
[*Olivella fulgurata*] 278
   lepta 278
   plana 278
OLIVIDAE 278
Ommastrephes bartrami 295
   [*hawaiiensis*] 295
   [*sloani pacificus*] 295
OMMASTREPHIDAE 295

Index

Ommatocarcinus macgillivrayi 363
    pulcher 363
Omniglypta cerina 261
Omobranchus elegans 446
    fasciolaticeps 446
    germaini 446
    japonicus 446
    kallosoma 446
    kraussi 446
    punctatus 446
    ueki 446
OMOSUDIDAE 415
Omosudis lowei 415
[*Omphalius nigerrimus*] 263
OMPHALOMETRIDAE 195
ONAGRACEAE 156
Oncaea clevei 312
    conifera 313
    dentipes 313
    media 313
    mediterranea 313
    minuta 313
    ornata 313
    similis 313
    venusta 313
ONCAEIDAE 312
ONCHIDIIDAE 294
Onchidium verruculatum 294
ONCHIDORIDIDAE 289
Onchnesoma intermedium 236
Onchocalanus cristatus 308
ONCHOLAIMIDAE 212
Oncholaimus oxyuris 212
    qingdaoensis 212
    sinensis 212
Onchoporella selenoides 381
Oncinopus angustefrons 357
ONCOBOTHRIIDAE 206
Oncobyrsa adriatica 27
Oncodiscus sauridae 206
ONCOUSOECIIDAE 370
Oneirodes appendixus 462
ONEIRODIDAE 462
Onigocia macrolepis 455
    spinosus 455
    tuberculatus 455
Onithochiton hirasei 238
Onithochiton sp. 238
Onoba elegantula 265
ONUPHIDAE 226
Onuphis chinensis 226
    cirrobranchiata 226
    eremita 226
    fujianensis 226
    geophiliformis 226
    [*geophiliformis*] 226
    [*holobranchiata*] 226
    irenuta 226
    pseudodibranchiata 226
    willemoesis 226
[*Onustus exuta*] 268
Onuxodon margaritiferae 421
    parvibrachium 421
Onychaspis hexeris 60
    polystyla 60

Onychocamptus mirabilis 315
Onychocella angulosa 376
    subsymmetrica 376
ONYCHOCELLIDAE 376
ONYCHOTEUTHIDAE 294
Onychoteuthis banksii 294
Onycocaris oligodenttata 340
    quadratophthama 340
Onykia carribbaea 294
Oocorys donghaiensis 273
    elongata 273
    lineata 273
OOCORYTHIDAE 273
Oolina ampullodistoma 96
    borealis 96
    brevisolenia 96
    caudigera 96
    costata 96
    favopunctata 96
    globosa 96
    hertwigiana 96
    hexagona 96
    laevigata 96
    lineata 96
    longispina 96
    melo 96
    shunyiense 97
    spiralis 97
    squamosa 97
    taiyanggonense 97
    variata 97
Oothrix bidentata 308
OPECOELIDAE 197
Opecoelina pacifica 197
Opecoeloides fugus 197
Opecoelus arii 197
    bohaiensis 197
    himezi 197
    inimici 197
    lateolabracis 197
    lobatus 197
    nipponicus 197
    pteroisi 197
    sebastisci 197
    spaericus 197
    zhifuensis 197
Opegaster ditrematis 197
    hippocampi 197
    synodi 197
    tamori 197
Opephora gemmata
    martyi
    pacifica
Operculina ammonoides 126
    bartschi 126
    complanata 126
    canalifera 126
    gaimardii 126
    granulosa 127
    madagascariensis 127
    philippinensis 127
    subsalsa 127
    subgranulosa 127
    tuberculata 127
    umbona 127

Operculina (continued)
    venosa 127
Ophelia acuminata 225
    cf. *limacina* 226
OPHELIIDA 225
OPHELIIDAE 225
Ophelina acuminata 226
    anlogaster 226
    grandis 226
Opheodesome australiensis 389
    grisea 389
OPHIACANTHIDAE 395
Ophiacantha composita 395
    pentagona 395
OPHIACTIDAE 396
Ophiactis acosmeta 396
    affinis 396
    hexacantha 396
    modesta 396
    picteci 396
    savignyi 396
Ophiarachna incrassata 397
    ohshimai 397
Ophiarachnella elegans 397
    gorgonia 397
    infernalis 397
    paucispina 397
    septeinspinosa 397
    stablis 397
Ophiarthrum elegans 397
    pictum 397
Ophichthus altipins 418
    apicalis 418
    asakusae 418
    brevicaudatus 418
    celebicus 418
    cephalozona 418
    erabo 418
    evermanni 418
    fasciatus 418
    intermedius 418
    macrochir 418
    polyophthalmus 418
    sternopterus 418
    tsuchidae 418
    urolophus 418
OPHICHTHYIDAE 417
Ophichthys tsuchidai 418
Ophidiaster chinensis 391
    granifer 391
    hemprichi 391
    multispinus 391
OPHIDIASTERIDAE 391
OPHIDIIDAE 421
Ophidion asiro 421
    muraenolenis 421
Ophiocamax rugosa 395
Ophiocara aporos 451
    porocephala 451
Ophiocentrus anomalus 396
    inaequalis 396
    koehleri 396
    putnami 396
Ophiochasma stellatum 397
Ophiochiton fastigatus 397

Ophiochiton (continued)
  megalaspis 397
OPHIOCHITONIDAE 397
Ophiocnemis manmorata 396
Ophiocoma brevipes 397
  dentata 397
  erinaceus 397
  pica 397
  pusilla 397
  schoeleini 397
  scolopendrina 397
Ophiocomella sexradia 397
OPHIOCOMIDAE 397
Ophiodaphna formata 396
  [*materna*]
Ophiodera neglecta 395
OPHIODERMATIDAE 397
Ophiodromus angustifrons 218
  berriofordi 218
  cf. *obscura* 218
  pugettensis 218
Ophioglycera eximia 222
Ophiogymna elegans
Ophiolepis cincta 397
  superba 398
Ophioleuce seminudum 398
OPHIOLEUCIDAE 398
Ophiolipus granulatus 398
Ophiomastix annulosa 397
  caryophyllata 397
  mixta 397
  variabilis 397
Ophiomastus tegulitius 398
Ophiomaza cacaotica 396
Ophiomitra plicata 395
Ophiomusium corticosum 398
  facundun 398
  lymani 398
  morio 398
  scalare 398
  simplex 398
Ophiomymna elegans 396
  funesta 396
  pulchella 396
Ophiomyxa australis 395
OPHIOMYXIDAE 395
Ophionephthy difficilis 396
Ophionereis dubia 397
  ———— *amoyensis* 397
  porrecta 397
  variegata 397
Ophiopallas paradoxa 398
Ophiopeza spinosa 397
Ophiopholis brachyactis 396
  mirabilis 396
Ophioplocus imbricatus 398
  japonicus 398
Ophiopsammus yoldii 397
Ophiopsila abscissa 397
  pantherina 397
  polyacantha 397
Ophiopteron elegans 396
Ophiothela danae 396
Ophiothrix ciliaris 396
  exigua 396
  hybrida 397

Ophiothrix (continued)
  infirma 397
  marenzelleri 397
  melanosticta 397
  nereidina 397
  plana 397
  proteus 397
  purpurea 397
  pusilla 397
  cf. *scotiosa* 397
  signata 397
  striolata 397
  trilineata 397
  virdialba 397
Ophiotreta gratiosa 395
  matura 395
OPHIOTRICHIDAE 396
Ophiotylos brevipes 398
Ophiotypa simplex 398
Ophiozonella subtilis 398
Ophisurus macrorhynchus 418
Ophiura flagellata 398
  kinbergi 398
  lanceolata 398
  micracantha 398
  pteracantha 398
  sarsii 398
OPHIURIDAE 397
OPHIUROIDEA 395
Ophryotrocha puerilins 228
OPHTHALMIDIIDAE 82
Ophthalmidium inconstans 82
  pusillum 82
  tenuimargo 82
  tenuiseptatum 82
Opisthadena fujianensis 202
  marina 202
  setipinnae 202
OPISTHOBRANCHIA 284
Opistholebes microovus 194
OPISTHOLEBETIDAE 194
Opisthomonorcheides decaptei 195
  indicus 195
OPISTHOPROCTIDAE 411
Opisthoproctus soleatus 411
Opisthopterus tardoore 410
Opisthosyllis brunnea 219
  laevis 219
OPISTHOTEUTHIDAE 296
Opisthoteuthis depressa 296
OPISTOGNATHIDAE 445
Opistognathus castelnaui 445
  [*evermanni*] 445
  hongkongiensis 445
OPLEGNATHIDAE 438
Oplegnathus fasciatus 438
  punctatus 438
OPLOPHORIDAE 339
OPLOPHOROIDEA 339
[*Oplophorus brevirostns*] 339
  typus 339
Oplopomus caninoides 452
Opochona glossoides 197
Opostomias mitsuii 412
Opuntia dillenii 155
Oratosquilla anomala 368

Oratosquilla (continued)
  gonypetes 368
  imperialis 368
  interrupta 368
  kempi 368
  nepa 368
  oratoria 368
  ornata 368
  perpensa 368
  quinquedentata 368
  woodmasoni 368
Orbinia curieri 222
  dicrochaeta 222
  exarmata 222
  vietnamensis 222
ORBINIIDA 222
ORBINIIDAE 222
Orbione halipori 327
  penei 327
Orbitacolax uniunguis 314
ORBITOIDACEA 121
Orbitolites complanatus 90
  cotentinensis 90
Orbulina bilobata 103
  circularosuturalis 103
  porosa 103
  suturalis 103
  universa 103
Orbulinelloides agglutinans 71
Orchestia amonala 330
  platensis 330
Orchomene breviceps 329
Orcinus orca 469
OREASTERIDAE 390
ORECTOLOBIDAE 405
ORECTOLOBIFORMES 405
Orectolobus japonicus 405
  maculatus 405
  oregonia gracilis 357
Oreophorus reticulatus 355
ORIDORSALIDAE 120
Oridorsalis stellata 120
  tenera 120
  umbonatus 120
  westi 120
[*Orithyia mammillaris*] 356
  sinica 356
Ornithocercus magnificus 51
  quadratus 51
  splendidus 51
  steinii 51
  thurnii 51
Ornithoteuthis volatilis 295
Orthoconchoecia atlantica 304
  bispinosa 304
  haddoni 304
  secernenda 304
  striola 304
ORTHOPTERA 300
Orthopyxis compressa 172
  integra 172
  platycarpa 172
[*Oryrhynchaxius japonicus*] 345
Osangularia bengalensis 120
  brunswickensis 120
  culter 120

Index

OSANGULARIIDAE 120
Oscanius hilli 289
   hirasei 289
   maricus 289
   xishaensis 289
Oscilla tricordata 284
Oscillatoria amphibia 25
   articulata 25
   bonnemaisonii 25
   brevis 25
   chalybea 25
   corallinae 25
   formosa 25
   laetevirens 26
   limosa 26
   nigro-viridis 26
   sancta 26
   subbrevis 26
   subuliformis 26
OSCILATORIACEAE 25
OSCILLATORIALES 25
OSTEICHTHYES 409
Osteodidymocodium kawakawa 200
Osteomugil ophuyseni 425
   strongylocephalus 425
Ostichthys japonicus 422
   kaianus 422
   sheni 422
[*Ostracion cornutus*]460
   [*cubicus*] 460
   meleagris 460
   solorensis 460
   tuberculatus 460
OSTRACIONTIDAE 460
OSTRACODA 301
[*Ostrea crenulifera*] 246
   [*cristagalli*] 246
   [*cucullata*] 246
   denselamellosa 246
   [*echinata*] 246
   [*folium*] 246
   [*gigas*] 246
   [*hyotis*] 246
   [*imbricata*] 246
   [*paulucciae*] 246
   [*pestigris*] 246
   [*plicatula*] 246
   [*rivularis*] 246
   [*sinensis*] 246
OSTREIDAE 246
OSTREOPSIACEAE 54
Ostreopsis siamensis 54
[*Ostroumovia inkermanica*] 168
OTARIDAE 469
[OTARIIDAE] 469
OTHER FUNGI 129
OTOBOTHRIIDAE 207
Otobothrium linstowi 207
Otolithes argenteus 432
   ruber 432
Otopleura auris cati 284
Otoporpa polystriata 173
Otosphaera auriculata 63
Oudemansia esakii 300
Oulangia stokesiana 185
Ovalipes iridescens 359

Ovalipes (continued)
   punctatus 359
Ovula ovum 271
OVULIDAE 270
Owenia fusiformis 228
OWENIIDA 228
OWENIIDAE 228
Owstonia tosaensis 445
   totomiensis 445
OWSTONIDAE 445
OXALIDACEAE 153
Oxalis corniculata 153
Oxycephalus clausi 332
   latirostris 332
   longipes 332
   [*notabilis*] 332
   piscator 332
   [*porcella*] 333
OXYCEPHALIDAE 332
Oxyconger leptognatus 416
Oxygyrus keraudreni 269
Oxymetopon compressus 451
Oxymonacanthus longirostris 460
Oxynaspis aurivillii 319
   bocki 319
   faroni 319
   pacifica 319
   reducens 319
   sinensis 319
OXYNOEIDAE 289
Oxyperas bernardi 250
OXYPORHAMPHIDAE 419
Oxyporhamphus convexux 419
   micropterus 419
Oxystomina delta 212
   miranda 212
OXYSTOMINIDAE 212
OXYTOXACEAE 54
Oxytoxum challengeroides 54
   sceptrum 54
   scolopax 54
Oxytricha ferruginea 60
   saltans 60
OXYTRICHIDAE 60
Oxyurichthys macrolepis 452
   microlepis 452
   ophthalmonema 452
   papuensis 452
   tentacularis 452
OXYURIDEA 210
Ozakia callyodontis 197
   gerris 197
   latelabracis 198
   sillaginis 198
Ozawaia tongaensis 125
Ozius guttatus 361
   rugulosus 361
   tuberculosus 361

P

Pachychalina brevispiculifera 164
   elongata 164
   punctata 164
   renieroides 165
   similis 165

Pachychalina (continued)
   variabilis 165
Pachycheles garciaensis 351
   johnsoni 351
   pectinicarpus 351
   pisum 351
   sculptus 351
   spinipes 351
Pachydictyon coriaceum 142
Pachygrapsus crassipes 366
   minutus 366
   planifrons 366
   plicatus 366
Pachyseris involuta 182
   rugosa 182
   speciosa 182
Pachysoma dentatum 313
   punctatum 313
Pachystomias microdon 412
Pacinonion novozealandicum 119
Padina arborescens 142
   australis 142
   boryana 142
   [*commersonii*] 142
   crassa 142
   jonesii 142
   minor 142
   tetrastromatica 142
   xishaensis 142
Paecilomyces varioti 130
Pagrosomus major 434
[*Pagrus major*] 434
PAGURIDAE 349
Paguristes acanthomerus 349
   barbatus 349
   ciliatus 349
   hians 349
   kagochimensis 349
   ortmanni 349
   palythophilus 349
   pusillus 349
   ———— *zhejiangensis* 349
   sinensis 349
PAGUROIDEA 349
Pagurus brachiomastus 350
   carpoforaminatus 350
   ———— *nephromma* 350
   dubius 350
   geminus 350
   gracilipes 350
   [*impressus*] 349
   janitor 350
   japonicus 350
   lanuginosus 350
   megalops 350
   ochotensis 350
   pectinatus 350
   pilosipes 350
   rubricatus 350
   sagamiensis 350
   trigonocheirus 350
   triserratus 350
   zebra 350
PALAEACANTHOCEPHALA 213
[*Palaemon beaupresii*] 340
   [*bidens*] 341

Palaemon (continued)
 [*canaliculatus*] 337
 [*carcinus*] 340
 [——— *rosenbergii*] 340
 [*carinicauda*] 340
 concinnus 340
 [*equidens*] 340
 gravieri 340
 guangdongensis 341
 [*hainanse*] 340
 [*hispidus*] 338
 [*idae mammillodactylus*] 340
 [*lampropus*] 340
 [*lancifer*] 338
 macrodactylus 341
 [*modestus*] 340
 [*orientalis*] 340
 ortmanni 341
 pacificus 341
 paucidens 341
 [*potamiscus*]
 [*rosenbergii*] 340
 serrifer 341
 sewelli 341
 [*sinmilis*] 340
 [*sinensis*] 340
 [*sundaicus*] 340
 [*superbus*] 340
 tenuidactylus 341
[*Palaemonella affinis*] 341
 [*laccadivensis*] 341
 [*orientalis*] 341
 pottsi 341
 rotumana 341
PALAEMONIDAE 340
PALAEMONOIDEA 339
[*Palaemonopsis willeyi*] 340
PALAEONEMERTEA 207
PALAEOPNEUSTIDAE 394
Palaeostoma mirabile 395
PALAEOSTOMATIDAE 395
Paleanotus chrysolepis 218
 debilis 218
PALICIDAE 364
[*Palicus trituberculatus*] 364
PALINURIDAE 346
PALINUROIDEA 346
PALINURIDEA 346
[*Palinurus burgeri*] 346
 [*dasypus*] 346
 [*homarus*] 346
 [*japonicus*] 346
 [*longimanus*] 346
 [*longipes*] 346
 [*ornatus*] 346
 [*trigonus*] 346
 [*versicolor*] 346
Palinustus waguensis 346
[*Palio fuitai*] 289
Paliurus ramosissimus 154
[*Pallasia pennata*] 228
Palliolum macrocheiricola 245
Palmadusta asellus 272
 clandestina 272
 felina 272

Palmadusta (continued)
 fimbriata maromorata 272
 gracilis japonica 272
 punctata 272
 ziczac 272
PALMAE 161
PALMARIACEAE 138
PALMARIALES 138
Palmula latifolia 93
Palola siciliensis 227
Palythoa anthoplax 179
 australiae 179
 capensis 179
 haddoni 179
 hainanensis 179
 liscia 179
 natalensis 179
 nelliae 179
 sinensis 179
 singaporensis 179
 stephensoni 179
 titanophila 179
 yongei 179
 xishaensis 179
Pampus argenteus 450
 chinensis 450
 nozawae 450
PANARTIDAE 64
Panartus tetrathalamum 64
Pandaglandulina dinapolii 91
PANDALIDAE 344
PANDALOIDA 344
[*Pandalus alcocki*] 344
 [*crassicornis*] 344
 [*martia*] 344
 meridionalis 344
 [*sindoi*] 344
PANDANACEAE 158
Pandanus forceps 158
 tectorius 158
PANDARIDAE 316
Pandea conica 167
PANDEIDAE 167
Pandeopsis ikarii 167
Pandora elongata 257
 otukai 257
 pseudobilirata 257
 sinica 257
 wardiana 257
PANDORIDAE 257
Panicum repens 160
Panna microdon 432
Panopea japonica 256
Pantala flavescens 299
[*Panthalis edriophthalma*] 217
 [*jogasimae*] 217
 [*melanonotus*] 217
 [*mitsukurii*] 217
 [*nigromaculata*] 217
 oerstedi 217
Pantinonemertes daguilarensis 208
 mortoni 208
 spectaculum 208
PANTOPODA 297
[*Panulirus angulatus*] 346

Panulirus (continued)
 homarus 346
 japonicus 346
 longipes 346
 ornatus 346
 penicillatus 346
 polyphagus 346
 stimpsoni 346
 versicolor 346
Papenfussiella kuromo 141
[*Paphia alapapilionis*] 255
 amabilis 255
 euglypta 255
 exarata 255
 gallus 255
 lirata 255
 paoilionacea 255
 [*sinuosa*] 255
 textile 255
 undulata 255
Papillapogon auritus 429
Papulaspora halima 132
Papyriscala latifasciata 267
 yokoyamai 267
Paraarhynchite hexorenale 237
Parabathymyrus macrophthalmus 416
Parabembras curtus 455
PARABEMBRIDAE 455
[*Parablennius yatabei*] 446
Parabopyrella indica 327
 perplexa 327
Parabothus chlorospilus 457
 coarctatus 457
Parabrachiella sihama 316
Paracaesio caeruleus 433
 kusakarii 433
 xanthurus 433
PARACALANIDAE 306
Paracalanus aculeatus 307
 crassirostris 307
 gracilis 307
 nanus 307
 nudus 307
 parvus 307
 serrulus 307
Paracalliactis sinica 179
Paracalliope karitane 328
Paracandacia bispinosa 311
 simplex 311
 truncata 311
Paracanthomysis hispida 323
Paracanthonchus macrodon 211
Paracanthurus hepatus 449
Paracaprella crassa 334
Paracassidulina minuta 105
 nipponensis 105
Paracaudina chilensis 389
 declicata 389
[*Paracentropogon indicus*] 455
Paracerceis sculpta 326
Parachaetodon ocellatus 438
Parachaeturichthys polynema 452
Paracheilinus carpenteri 440
Paracirrhites arcatus 444
 forsteri 444

Paracleistostoma crassipilum 365
   cristatum 365
   depressum 365
Paraclione longicaudata 288
Paracomesoma heterosetosum 212
Paraconchoecia aequiseta 304
   allotherium 304
   brachyaskos 304
   caudata 304
   cophopyga 304
   dasyophthalma 304
   decipiens 304
   dentata 305
   diacanthus 305
   dorsotuberculata 305
   echinata 305
   gerdhartmanni 305
   inermis 305
   macroprocera 305
   mamillata 305
   microprocera 305
   oblonga 305
   procera 305
   reticulata 305
   spinifera 305
   [*tamensis*] 303
   vitjazi 305
Paracrangon abei 344
Paracrocnida sinensis 396
Paracryptogonimus apharei 199
   elongatus 199
   lutiani 199
   ovatus 199
   ula-ula 199
Paractaea garretti 361
   orientalis 361
   philippinensis 361
   ruppellii 361
   tumulosa 361
Paracyathus pruinosus 185
Paracyclois milneedwardsii 356
Paradella dianae 326
Paradentalina muraii 91
Paradexamine setigera 329
Paradoloria dorsoserrata 302
Paradontophora bohaiensis 211
   cobbi 211
   marina 211
Paradoralaimopsis bohaiensis 212
Paradusa mauritiensis 328
Paraeuniphysa falciseta 227
   spinea 227
   taiwanensis 227
Parafavella elongata 59
Parafissurina agassizi 98
   cuniculifera 98
   himatiostoma 98
   minuta 98
   oblonga 98
   quadrispinata 98
   reflecta 98
   subcarinata 98
   subcircularis 98
   sublata 98
   subovata 98

Paragaleus acutiventralis 406
   tengi 406
Paragobiodon echinocephalus 452
   melanosomus 452
   xanthosomus 452
   zacalles 452
Paragonapodasmius huanghaiensis 200
Paragonaster chinensis 390
   ctenipes 390
Paragyrops edita 434
[*Parahaliporus sibogae*] 336
[*Parahalosydna pleiolepis*] 217
Parahalosydnopsis hartmanae 217
Parahemiurus ambassicola 202
   clupeae 202
   coiliae 202
   collichthydis 202
   harengulae 202
   hemirhamphi 202
   merus 202
   pseudosciaenae 202
   qingdaoensis 202
   sardiniae 202
   seriolae 202
Parahepomadus vaubani 335
Parahyotissa imbricata 246
   sinensis 246
Paraiptasia radiata 180
Paraisocoelium platycephali 198
Paralacydonia paradoxa 216
Paralaophonte congenera 315
Paraleidonotus ampulliferus
Paraleiocapitella mossambica 225
Paralemnalia eburnea 188
   thyrsoides 188
Paraleonnates uschkovi 221
Paralepas nodulosa 319
PARALEPIDIDAE 415
Paralepidonotus ampulliferus 217
Paralepis atlantic indica 415
   elongta 415
Paraleptomysis sinensis 323
   xenops 323
Paraleptus chiloscyllii 210
   syclii 210
[*Paralia sulcata*] 30
PARALICHTHYIDAE 456
Paralichthys olivaceus 456
Paralinhomoeus attenuatus 211
Paraliparis meridionalis 456
Paralomis dofleini 350
   heterotuberculata 350
   hystrix 350
Paralophogaster glaber 323
Paralovenia bitentaculata 170
Paraluteres prionurus 460
Paralycaea gracilis 332
Paramacradenina xiamensis 202
Paramanteriella cantherini 198
Paramedaeus simplex 361
Parameira pendula 315
Parametra orion 387
Paramicrura borborophila 207
Paramina quinquelineatus 429
Paramisophria sinica 310

Paramollicia dichotoma 305
   plactolycos 305
   rhynchena 305
   siboga 305
Paramonacanthus nipponensis 460
   oblongus 460
Paramonhystera pellucida 211
Paramormula aspera 284
Paramphicondrius tetradontus 396
Paramphicteis angustifolia 229
Paramphinome indica 226
PARAMPHISTOMIDAE 194
Paramunida scabra 350
PARAMURICEIDAE 186
Paramya recluzii 256
Paramyxine cheni 404
   fernholmi 404
   nelsoni 404
   sheni 404
   taiwanae 404
   yangi 404
Paranais frici 234
   litoralis 234
   plenus 234
Paranemertes katoi 208
   peregrina 208
Paranibea semiluctuosa 432
Paranophrys carcini spiralis 57
   carnivora 57
   elongata 57
   sarcophaga 57
Paranthura japonica 334
Paranticoma longicaudata 212
Parantipathes cylindrices 186
PARAONIDAE 223
Paraonis gracilis 223
PARAPAGURIDAE 349
Parapagurion calcinicola 327
Parapagurus acutus acutus 349
   ———— *hirsutus* 349
   arcustus monstrosus 349
   [*monstrosus*] 349
   pilosimanus 349
   sibogae 349
Parapalicus nanshaensis 364
   trituberculatus 364
Parapandalus zurstrasseni 344
Parapanope euagora 361
Parapanthalis maculata 217
   muricatus 217
[*Parapasiphae latirostris*] 338
   sulcatifrons 339
Parapenaeon consolidata v. *richardsonae* 327
   japonica 327
[*Parapenaeopsis brevirostris*] 336
   cornuta 337
   cultrirostris 337
   hardwickii 337
   hungertordi 337
   incisa 337
   sinica 337
   [*sculptilis* v. *cultrirostris*] 337
   tenella 337
[*Parapenaeus acclivis*] 336

Parapenaeus (continued)
  [*akayebi*] 336
  [*dalei*] 336
  fissuroides 337
  [*fissurus*] 337
  investigatoris 337
  lanceolatus 337
  longipes 337
  [*mogiensis*] 337
  sextuberculatus 337
PARAPERCIDAE 444
Parapercis alboguttata 444
  aurantiaca 445
  cephalopunctata 445
  clathrata 445
  cylindrica 445
  decemfasciata 445
  hexophthalma 445
  kamoharai 445
  mimaseana 445
  multifasciata 445
  multiplicata 445
  muronis 445
  ommatura 445
  polyophthalma 445
  pulchella 445
  punctata 445
  quadrispinosus 445
  sexfasciata 445
  snyderi 445
  somaliensis 445
  striolata 445
  tetracanthus 445
  xanthozona 445
Parapetalus denticulatus 316
Paraphilomedes tricornuta 302
  unicornuta 302
Parapholas quadrizonata 257
Paraphrixus brevicauda 327
Paraphronima crassipes 331
  gracilis 331
PARAPHRONIMIDAE 331
Paraplagusia bilineata 458
  blochi 458
  guttata 458
  japonica 458
Paraplanocera oligoglena 191
  reticulata 191
Paraploactis kagoshimensis 455
Parapocryptes macrolepis 452
  serperaster 452
Parapriacanthus compressus 436
  ransonneti 436
  vanicolensis 436
Paraprionospio pinnata 223
Parapristipoma trilineatus 435
Parapronoe campbelli 332
  crustulum 332
  elongata 332
  parva 332
Parapterois heterurus 453
Parapterygotrigla macrorhynchus 454
  multiocellata 454
Pararotalia armata 121
  audouni 121

Pararotalia (continued)
  bellatula 121
  calcar 121
  fungiformis 121
  inermis 121
  minuta 121
  murrayi 121
  nipponica 121
  orientalis 121
  ozawai 121
  rosea 121
  taiwanica 121
  tuberculata 121
  umbonata 121
  venusta 121
  xincunensis 121
Parasalenia gratiosa 393
  [——— v. *boninensis*] 393
  pohlii 393
PARASALENIIDAE 393
PARASCELIDAE 333
Parascelus edwardsi 333
  typhoides 333
  [*zebu*] 333
[*Parascolopsis eriomma*] 435
  [*inermis*] 435
  tosensis 434
Parascorpaena mcadamsi 453
  mossambica 453
  picta 453
Parasmittina acumenata 380
  acuta 380
  biformis 380
  californica 380
  deliculata 380
  donghaiensis 380
  glomerata 380
  malleolus 380
  papulata 380
  parsenvalii 380
  projecta 380
  raigii 380
  triformis 380
  trispinosa 380
  tropica 380
  varians 380
[*Paraspirontocaris kishinouyei*] 343
Parasquilla haani 368
Parastenhelia spinosa littoralis 315
PARASTENHELIIDAE 315
Paraster compactus 394
Parastilomysis paradoxa 323
Parastylodactylus bimaxillaris 339
PARATANAIDAE 325
Parathemisto gaudichaudi 331
Parathranites orientalis 359
Paratiara digitalis 167
Paratrachichthys prosthemius 422
Paratrema doederleini 393
Paratriacanthodes retrospinis 459
Paratritonia lutea 292
Paratrochammina charlotensis 76
  simplissima 76
Paratymolus sexspinosus 357
Paratyphis maculatus 333

Paratyphis (continued)
  parvus 333
  promonoiri 333
  spinosus 333
Paraugaptilus buchani 310
Parauronema virginianum 57
Paravargula formosana 302
  hirsuta 302
  taiwantuia 302
Paraxanthias elegans 361
  notatus 361
  pachydactylus 361
Parazen pacificus 423
Pardachirus pavoninus 458
  xenicus 458
Parellisina curvirostris 373
  intercatatoporata 373
  longirostris 373
  silensi 373
Parenchelyurus hepburni 446
Parenchymorpha xishanica 27
PARENCHYMORPHATACEAE 27
Pareuchaeta flava 308
  malayensis 308
  russelli 308
  scaphula 308
  tuberculata 308
  weberi 308
Pareulepis malayana 218
Pareurythoe borealis 226
Parexocoetus brachypterus 419
  mento 419
Parheteromastus tenuis
Parhyale hawaiensis 329
  plumulosa 329
Parhyalella pietschmanni 329
Parilia ovata 355
Parioglossus taeniatus 451
Parione pachychelii 327
Parioninella astridae 327
Pariphiculus mariannae 355
Paromalina coronata 120
  semipunctata 120
Paromola macrochira 354
Paromolopsis boasi 354
Parorchis acanthus 204
  gedoelsti 204
Parrellina hispidula 126
Parrelliodes hyalinus 113
PARRELLOIDIDAE 113
Parribacus antarcticus 346
  japonicus 346
Parrina bradyi 89
Parthenope brevibrachiatus 357
  calappoides 357
  curvispinus 357
  erosa 357
  hayamaensis 357
  [*horrida*] 357
  lamellifrons 357
  longimanus 357
  longispinis 357
  sinensis 357
  tuberculosus 357
  turriger 358

Index

Parthenope (continued)
   validus 358
   ——— *forma intermedius* 358
PARTHENOPIDAE 357
Parupeneus barberinoides 436
   barberinus 436
   bifasciatus 436
   chryserdros 436
   chrysopleuron 436
   ciliatus 436
   cyclostomus 436
   fraterculus 436
   indicus 436
   luteus 436
   megalops 436
   multifasciatus 436
   pleurostigma 436
   spilurus 436
   trifasciatus 436
Parvimalleus irregularis 244
Parvisquilla multituberculata 369
Pasiphaea japonica 339
   pacifica 339
   semispinosa 339
   sinensis 339
PASIPHAEIDAE 338
PASIPHAEOIDEA 338
Paspalum distichum 160
Passeriniella obiones 131
Passeriniella sp. 131
Patella flexuosa 262
   stellaeformis 262
PATELLIDAE 262
Patellina advena 81
   altiformis 81
   corrugata 81
   spinosa 81
Patellinella ambulacrata 110
   hanzawai 110
   inconspicua 110
   jugosa 110
   spinosa 110
PATELLINIDAE 81
Patelloida conoidalis 263
   [*dorsuosa*] 262
   grata 263
   [*kolarovai*] 262
   lampanicola 263
   pygmaea 263
   saccharina lanx 263
   [*schrenckii*] 263
   striata 263
   toloeasis 263
Pateoris hauerinoides 86
Patinapta ooplax 389
Paulonaria beckii 272
Paupidrilus breviductus
Pavona cactus 182
   decussata 182
   frandifera 182
   lata 182
   minikoiensis 182
   minuta 182
   praetorta 182
   seriata 182

Pavona (continued)
   varians 182
Pavonina flabelliformis 109
   ryukyuensis 109
   tasmanensis 109
PAVONINIDAE 109
Peasiella infracostata 265
   lutulenta 265
   roepstorffiana 265
   [*roepstorffiana*] 265
Pecten albicans 245
   excavatus 245
   pyxidatus 245
Pectinaria aegyptia 228
   [*bockis*] 228
   [*capensis*] 228
   conchilega 228
   dimai 228
   [*japonica*] 228
   [*neapolitana*]
   papillosa 228
PECTINARIDAE 228
Pectinia lactuca 184
PECTINIDAE 244
PECTINIIDAE 184
Pectinodrilus disparatus 234
   hoihaensis 234
   molestus 234
Pectinoodonta orientalis 262
[*Pedalion isognomum*] 244
   [*legumen*] 244
   [*padibulum*] 244
   [*quadrangularia*] 244
Pedicellina echinata 384
PEDICELLINIDAE 384
PEDINIDAE 393
Pedinosoma curtum 215
Pedobesia ryukyuensis 146
Pegantha martagon 173
   triloca 173
PEGASIDAE 461
PEGASIFORMES 461
Pegasus draconis 461
   laternarius 461
   volitans 461
Pegea confoederata 401
Pegidia dubia 111
PEGIDIIDAE 111
Pelagia noctiluca 176
   [*panopyra*] 176
   [*perla*] 176
PELAGIIDAE 176
Pelagobis longicirrata 215
Pelagodiscus atlantica 384
Pelamis platurus 464
Pelates quadrilineatus 435
PELECANIDAE 465
PELECANIFORMES 465
Pelecanus p. philippensis 465
Pellamora trochlearis 265
PELLATISPIRIDAE 126
Pellona ditchela 410
Pelosina cylindrica 70
Peltidium falcatum 314
   ovale 314

PELTIDIIDAE 314
[PELTODORIDIIDAE] 290
Peltodoris cf. *aurea* 290
Peltogaster paguri 322
PELTOGASTERIDAE 322
Pelvetia siliquosa 143
   [*minor*] 143
PEMPHERIDAE 436
Pempheris japonicus 436
   [*molucca*]
   nyctereutes 436
   oualensis 436
   xanthopterus 436
Pemphis acidula 156
PENAEIDAE 336
PENAEOIDEA 335
[*Penaeopsis affinis*] 337
   [*challengeri*] 337
   [*coniger* v. *andamanensis*] 336
   eduardoi 337
   [*hilarula*] 336
   rectacutus 337
[*Penaeus acclivis*] 336
   [*affinis barbatus*] 336
   [*ashiaka*] 338
   [*bubulus*] 338
   canaliculatus 337
   [*carinatus*] 338
   chinensis 337
   [*compressipes*] 336
   [*cornutus*] 337
   [*crassicornis*] 336
   [*curvirostris*] 338
   [*edwardsianus*] 336
   [*foliaceus*] 335
   [*hardwickii*] 337
   [*incisipes*] 337
   indicus 337
   [*intermedius*] 337
   japonicus 337
   [*joyneri*] 337
   [*lamellata*] 336
   latisulcatus 337
   [*longipes*] 338
   longistylus 337
   marginatus 337
   [*mastersii*] 337
   [*mastersis*] 337
   merguiensis 338
   [*monoceros ensis*] 337
   monodon 338
   [*monoon*] 338
   [*moyebi*] 337
   [*orientalis*] 337
   [*palmensis*] 337
   penicillatus 338
   [*philippinensis*] 337
   [*podophthalmus*] 337
   [*rectacutus*] 337
   simisulcatus 338
   [*stenodactylus*] 336
   [*tahitensis*] 338
   [*tenellus*] 337
   [*teraei*] 337
   [*velutinus*] 337

PENEROPLIDAE 90
Peneroplis pertusus 90
   planatus 90
Penicillium 129
   chrysogerum 129
Penicillium spp. 132
Penicillus penis
Penilia avirostris 300
Penitella kamakurensis 257
[*Pennahia argentatus*] 431
   [*macrocephalus*] 431
   macrophthalmus 432
   [*pawak*] 431
[*Pennaria cavolinii*] 168
   [*grandis*] 168
   [*tiarella*] 168
PENNATAE 35
Pennatula fimbriata 190
   murrayi 190
   phosphorea 190
PENNATULACEA 190
PENNATULIDAE 190
Pentaceraster alveotatus 390
   chinensis 390
   magnificus 390
   regulus 390
   sibogae 390
Pentaceros japonicus 438
Pentacta nipponensis 387
PENTAPODIDAE 434
Pentapus macrurus 434
   nagasakiensis 434
   setosus 434
Pentarion longimanus 432
Peponocephala electra 469
PEPTOCOCCACEAE 23
PERACLIDAE 288
Peraclis reticulata 288
Peramphithoe orientalis 328
PERCICHTHYIDAE 428
PERCIFORMES 425
Percnon abbreviatum 366
   guinotae 366
   planissimum 366
   sinense 366
PERCOPHIDAE 445
Peregrinamor ohshimai 253
Periclimenaeus arabicus 341
   rastrifer 341
   spinicauda 341
   [*spongicola*] 341
   stylirostris 341
   tridentatus 341
[*Periclimenes aesopius*] 341
   affinis 341
   amymone 341
   [*antonbrnunii*]
   [*arabicus*] 341
   brevicarpalis 341
   commensalis 341
   cristimanus 341
   demani 341
   denticulatus 341
   digitalis 341
   elegans 341

Periclimenes (continued)
   exederons 341
   gorgonicola 341
   holthuisi 341
   hongkongensis 341
   imperator 341
   inornatus 341
   laccadivensis 341
   lanipes 341
   nilandensis 341
   ornatus 341
   paraparvus 341
   pertubans 341
   [*pottsi*] 341
   [*rotumanus*] 341
   [*setoensis*] 341
   sinensis 341
   soror 341
   spiniferus 341
   tennipes 341
   toloensis 341
   tosaensis 341
Periclimenoides odontodactylus 341
Periconia prolifica 132
Periconia sp. 132
Periconiella sp. 132
PERICOSMIDAE 394
Pericosmus melanostomus 394
   porphyrocardius 394
PERIDINIACEAE 54
PERIDINIALES 51
[*Peridiniopsis asymmetrica*] 54
   [*excentrica*] 54
   [*hainanensis*] 54
   [*pingi*] 54
   rotunda 54
Peridinium abei 54
   achromaticum 54
   angustum 54
   asymmetricum 54
   biconicum 54
   brochii 54
   claudicans 54
   compressum 54
   conicoides 54
   conicum 54
   crassipes 54
   deficiens 54
   depressum 54
   diabolus 54
   divergens 54
   elegans 54
   excentricum 54
   fatulipes 54
   gatunense 54
   [*globosum*] 54
   globulus 54
   [——— v. *quarerense*] 55
   grande 54
   granii 54
   humile 54
   latispinum 54
   latissimum 55
   leonis 55
   longipes 55

Peridinium (continued)
   marielebourae 55
   matsenaueri 55
   [*michaelis*] 55
   minutum 55
   murray 55
   nux 55
   oceanicum 55
   ovum 55
   pallidum 55
   parainerme 55
   paralletum 55
   parapentagonum 55
   pellucidum 55
   pentagonum 55
   punctulatum 55
   pyriforme 55
   [*rectum*] 55
   remotum 55
   [*sinaicum*] 55
   solidicorne 55
   [*sphaericum*] 54
   steinii 55
   subinerme 55
   subpyriforme 55
   thorianum 55
   tsingtaoensis 55
   tubum 55
   tumidum 55
   vanustum 55
   [*wiesneri*] 54
Peridium ramosum 67
   spinipes 67
Periglypta chemnitzi 255
   clathrata 255
   compressa 255
   [*crispata*] 255
   [*lacerata*] 255
   puerpera 255
   reticulata 255
Perinea tumida 357
Perinereis aibuhitensis 221
   [*brevicirris*] 221
   camiguinoides 221
   cavifrons 221
   cultrifera 221
   ——— *floridana* 221
   ——— *helleri* 221
   ——— *obfuscata* 221
   ——— *perspicillata* 221
   ——— *striolata* 221
   ——— *typica* 221
   nuntia 221
   ——— *brevicirris* 221
   ——— *majungaensis* 221
   ——— *typica* 221
   ——— *vallata* 221
   [*obfuscata*] 221
   rhombodonta 221
   suluana 221
   [*vallata*] 221
   vancaurica 221
   weizhouensis 221
PERIOPHTHALMIDAE 452
Periophthalmus argentilineatus 453

Periophthalmus (continued)
    cantonensis 453
Periphylla periphylla 176
PERIPHYLLIDAE 176
Periploma otohimeae
PERIPLOMATIDAE
Perissonemertes pyrrhocephalus 207
Perissonoe cruciata 43
PERISTEDIIDAE 455
Peristedion nierstrasi 455
    orientale 455
Peristernia nassatula 278
    [*pilsbryi*] 278
PERITRICHA 57
Perna viridis 243
Peronella lesueuri 394
    pellucida 394
Perotis hordeiformis 160
    indica 160
Perotrochus teramachii 261
Pervagor aspricaudus 460
    janthinosoma 460
    melanocephalus 460
    nitens 460
    tomentosus 460
Petalifera punctulata 287
    qingdaonensis 288
    ramosa 288
Petalomera granulata 353
    [*granulata*] 353
    japonica 353
    lateralis 353
    longipedalis 353
    sheni 353
Petalonia fascia 142
    zosterifolia 142
Petaloproctus cf. *terricolus* 225
Petalotricha aperta 59
PETALOTRICHIDAE 59
PETASIDAE 173
Petasiella asymmetrica 173
Petereaeolidia ianthina 292
    spaperi 292
PETEREAEOLIDIIDAE 292
Petraliella bisinuata 380
    chuakensis 380
PETRALIELLIDAE 380
PETRICOLIDAE 256
Petriella setifera 131
Petriellidium ellipsoideum 131
Petrolisthes asiaticus 351
    boscii 351
    carinipes 351
    coccineus 351
    fimbriatus 351
    hastatus 351
    haswelli 351
    indicus 351
    japonicus 351
    lamarckii 351
    militaris 351
    [*penicillatus*] 351
    scabriculus 351
    tomentosus 351
    unilobatus 351

PETROMYZONIDAE 404
PETROMYZONIFORMES 404
[*Petroscirtes kallosoma*] 446
    mitratus 446
Petrosia testudinaria 165
Petrospongium rugosum 141
Pettia amphophthalma huanghaiensis 219
Peucedanum japonicum 156
Peyssonnelia rubra v. *orientalis* 134
PEYSSONNELIACEAE 134
Phacelotrema claviforme 200
Phacelurus latifolius 160
    —— v. *angustifolia* 160
PHACODISCIDAE 64
Phaenna spinifera 308
PHAENNIDAE 308
PHAENOCALPIDAE 67
PHAEOCALPIA 68
PHAEOCONCHIA 69
PHAEOCYSTINA 68
PHAEODACTYLACEAE 46
PHAEODACTYLALES 46
Phaeodactylum tricornutum 46
PHAEODARIA 68
PHAEOGROMIA 69
PHAEOPHYTA 140
PHAEOSPHAERIA 68
Phaethon aethereus indicus 465
    lepturus dorotheae 465
    rubricauda rothschildi 465
PHAETHONTIDAE 465
Phakellia fusca 164
    pygmaca 164
PHALACROCORACIDAE 465
Phalacrocorax carbo sinensis 465
    filamentosus 465
    p. pelagicus 465
Phalacroma cuneus 51
    doryphorum 51
    favus 51
    mitra 51
    ovum 51
Phalacrophorus pictus 215
    uniformis 215
Phalangipus filiformis 357
    hystrix 357
    longipes 357
PHALAROPODIDAE 468
Phalaropus lobatus 468
Phalium areala 272
    bandatum 272
    [*bandatum*] 272
    bisulcatum 272
    [*bisulcatum*] 272
    [*bulla*] 272
    [*coronadoi wyvillei*] 272
    decussatum 272
    [*decussatum*] 272
    [*flammiferum*] 272
    glabratum bulla 272
    [—— *bulla*] 272
    glaucum 272
    [*glaucum*] 272
    inornatum 272

Phalium (continued)
    kurodai 272
    [*persimilis*] 272
    strigatum breviculum 272
    —— *strigatum* 272
    wyvillei 272
Phallodrilus darvelli 234
    vanus 234
PHANEROZONIA 389
Phanoderma cocksi 212
    daiyawanensis 212
    macrophallum 212
PHANODERMATIDAE 212
Pharaonella perna 251
    rostrata 251
Pharella acutidens 253
PHARELLIDAE 253
Phascolion moskalevi 236
PHASCOLIONIDAE 236
Phascolosoma albolineatum 235
    [*dentigerum*] 235
    [*dunwichi*] 235
    esculenta 235
    formosense 235
    [*hainanicum*] 235
    japonicum 235
    [*japonicum*] 236
    nigrescens 235
    [*onomichianumi*] 235
    pacificum 235
    parvum 235
    pectinatum 235
    perlucens 235
    [*pyriformis*] 236
    rottnesti 235
    [*rueppelii*] 235
    scolops 235
    [*similis*] 235
    sinense 235
    [*uncatum*] 235
    varians 235
    [*vulgare*] 236
PHASCOLOSOMALIFORMES 235
PHASCOLOSOMATIDAE 235
PHASCOLOSOMATIDEA 235
PHASMIDEA 210
Phenacolepas crenulata 265
    senta 265
PHENACOLEPATIDAE 265
Phenacovolva birostris 271
    brevirostris 271
    dancei 271
    gigantea 271
    longirostrata 271
    recuva 271
    rosea 271
    suberflxa 271
Pherallodus indicus 463
Pherecardia striata 226
Pheronema annae 165
    ijimai 165
    raphanus 165
    weberi 165
PHERONEMATIDAE 165
Pherusa cf. *bengalensis* 228

Pherusa (continued)
   eruca 228
   granulosus 228
   parmata 228
   plumosa 228
Phestilla melanobranchia 293
PHESTILLIDAE 293
Phialella macrogona 169
PHIALELLIDAE 169
[*Phialidium ambiguum*] 172
   [*chengshanensis*] 170
   [*discoidum*] 172
   [*folleatum*] 172
   [*hemisphaericum*] 172
   [*malayense*] 172
   [*mccradyi*] 172
   [*ovale*] 172
   [*uchidai*] 172
Phialophora sp. 132
Phialophorophoma litoralis 132
PHIALUCIIDAE 172
[*Phialucium carolinae*] 169
   [*condensum*] 169
   [*curviductum*] 169
   [*cyphogonia*] 169
   mbenga 172
   [*taeniogonia*] 169
   [*virens*] 169
Phidiana indica 292
PHIDIANIDAE 292
Phidolopora pacifica 382
PHIDOLOPORIDAE 381
PHILASTERIDAE 57
[*Philine argentata*] 287
   japonica 287
   kinglipini 287
   kurodii 287
   orientalis 287
   otukai 287
   scalpta 287
   vitrea 287
PHILINIDAE 287
Philinopsis cyaneum 287
   gigliolii 287
   lineolata 287
   minor 287
Philippia radiata 266
Philippidorippe philippinensis 354
Philocheras lowisi 344
Philodicus javanicus 299
Philomachus pugnax 467
Philomedes eugeniae 302
   [*japonica*] 302
   lilljeborg 302
PHILOMETRIDEA 210
[*Philonicus lucasii*] 336
   [*pectinatus*] 336
PHILOPHTHALMIDAE 204
PHILOSCIIDAE 326
PHILYDRACEAE 162
Philydrum lanuginosum 162
Philyra acutidens 355
   biprotubera 355
   carinata 355
   globulosa 355

Philyra (continued)
   heterograna 355
   olivacea 355
   pisum 355
   platychira 355
   tuberculosa 355
[*Phoberus brevirostris*] 345
   [*tenuimanus*] 345
Phoca hispida 469
   largha 469
   [*vitulina*] 469
Phocascaris longispiculum 209
PHOCIDAE 469
PHOCOENIDAE 469
PHOLADIDAE 257
[*Pholadidea acutithyra*] 257
   [*cheveyi*] 257
   [*dolichothyra*] 257
Pholadomya sinica 257
PHOLADOMYIDAE 257
PHOLADOMYOIDA 257
Pholas orientalis 257
PHOLIDAE 447
Pholoe chinensis 218
   minuta 218
Phoma sp. 132
Phomopsis sp. 132
Phormidium corium 26
   crosbyanum 26
   fragile 26
   naveanum v. marina 26
   penicillatum 26
   submenbranaceum 26
   tenue 26
PHORMOCYRTIDAE 68
Phormosoma bursarium 392
PHORMOSPYRIDAE 66
PHORONIDA 384
PHORONIDAE 384
PHORONIDEA 384
Phoronis anchitecta 384
   australis 384
   [*hippocrepia*] 384
   ijimai 384
   muelleri 384
   [*vancouvereusis*] 384
PHORTICIDAE 66
Phorticium polycladum 66
   pylonium 66
[*Phortis ceylonensis*] 170
   [*kambara*] 170
   [*lactea*] 170
   [——— v. *chiaochowensis*] 170
   [*pyramidalis*] 170
Phos roseatum 277
   senticosus 277
Photis digitata 329
   longicaudata 329
Photobacterium 21
   fiseheri 21
   leiognathi 21
   phosphsreum 21
Photonectes albipennis 412
Phoxocampus belcheri 424
PHOXOCEPHALIDAE 330

Phractopelta cruciata 61
PHRACTOPELTIDAE 61
Phragmites communis 160
Phronima atlantica 331
   bucephala 331
   colletti 331
   curvipes 331
   [*megalodous*] 331
   pacifica 331
   sedentaria 331
   solitaria 331
   stebbingi 331
Phronimella elongata 331
PHRONIMIDAE 331
Phronimopsis spinifera 331
Phrosina semilunata 332
PHROSINIDAE 332
Phrynichthys thele 462
Phycodrys fimbriata 139
   radicosa 139
Phycopsis terpnis 164
Phyla nodiflora 157
Phylactella geometrica 379
PHYLACTELLIDAE 379
Phylactellipora collaris 379
Phyllanthus leptoclados 154
Phyllidia elegans 291
   gemmata 291
   honloni 291
   nobilis 291
   ocellata 291
   pustulosa 291
   rotunda 291
   sereni 291
   varicosa 291
   xishaensis 291
PHYLLIDIDAE 291
PHYLLOBOTHRIIDAE 206
Phylloda foliacea 251
Phyllobothrium laciniatum 206
   lactuca 206
   loculatum 206
   ptychocephalum 206
   tumidum 206
Phyllochaetopterus claparedii 224
Phyllodistomum folium 195
   pearsei 195
[*Phyllodoce castanea*] 215
   chinensis 215
   fristedti 215
   gracilis 215
   groenlandica 215
   [*laminosa*] 215
   laminosa laminosa 215
   madeirensis 215
   [*madeirensis*] 215
   malmgreni 215
   papillosa 215
   [*papillosa*] 215
PHYLLODOCIDA 215
PHYLLODOCIDAE 215
PHYLLOPHORACEAE 137
PHYLLOPHORIDAE 387
Phyllophorus hypsipyrgus 388
   konkuliensis 388

Phyllophorus (continued)
   ordinatus 388
   spiculatus 388
Phyllopodopsyllus furciger 315
Phyllopus aequalis 310
   bidentatus 310
   helgae 310
   impar 310
   mutatus 310
Phyllospadix iwatensis 158
   japonica 158
Phyllospongia foliascens 164
PHYLLOSTAURIDAE 61
Phyllostaurus cuspidatus 61
   siculus 61
Phyllotrema bicaudatum 197
   microrchis 197
   quadricaudatum 197
Phylo felix 222
   [—— *asiatieus*] 222
   kupfferi 223
   nudus 223
   ornatus 223
Phymodius drachi 361
   granulosus 361
   laysani 361
   monticulosus 361
   nitidus 361
   ungulatus 361
PHYMOSOMATIDAE 392
Phynchocinetes uritai 339
PHYNCHOCINETIDAE 339
Phyrella fragilis 388
   liuwushanensis 388
Physalia physalis 173
PHYSALIIDAE 173
PHYSALOPTERIDAE 210
Physeter macrocephalus 469
PHYSETERIDAE 469
Physiculus inbarbatus 420
   japonicus 420
   jordani 420
   maximowiczi 420
   nigrescens 420
   roseus 420
Physodon mulleri 406
PHYSONECTAE 173
Physophora hydrostatica 174
PHYSOPHORIDAE 174
Pichia burtonii 129
   carsonii 129
   etchellsii 129
   guilliermondii 129
   heimii 129
   kluyveri 129
   naganishii 129
   ohmeri 129
   philogaea 129
   scolyti 129
Picris japonica v. *koreana* 158
PILARGIIDAE 218
Pilargis berkeleyi 218
   verrucosa pacifica 218
Pillucina pisidia 247
   yamakawai 247

Pilochrota haeckeli 164
Pilodius areolatus 362
   granulatus 362
   melanospinis 362
   nigrocrinitus 362
   pubescens 362
   pugil 362
   scabriculus 362
Pilosabia pilosa 268
   [*trigona*] 268
Pilsbryiscalpellum condensum 318
Pilumnopeus eucratoides 362
   indica 362
   makiana 362
   trispinosus 362
Pilumnus andersoni 362
   cursor 362
   interifrontus 362
   longicornis 362
   minutus 362
   orbitospinis 362
   rotumanus 362
   sinensis 362
   spinulus 362
   tuantaoensis 362
   vespertilio 362
PINACEAE 150
Pinctada anomioides 243
   chemnitzi 243
   maculata 243
   margaritifera 243
   martensi 243
   maxima 243
   nigra 243
   radiata 243
Pinguitellina pinguis 251
Pinjalo pinjalo 433
[*Pinna attenuata*] 243
   [*atropurpurea*] 243
   bicolor 243
   [*bullata*] 243
   incurvata 243
   [*inflata*] 243
   muricata 243
   [*pectinata*] 243
   saccata 243
   [*strangei*] 243
   [*vexillum*] 243
PINNIDAE 243
PINNIPEDIA 469
Pinnixa penultipedalis 363
   tumida 363
[*Pinnotheres affinis*] 363
   boninensis 363
   cyclinus 363
   dilatatus 363
   excussus 363
   gordoni 363
   haiyangensis 363
   latus 363
   luminatus 363
   nigrans 363
   obscurus 363
   paralatissinus 363
   pholadis 363

Pinnotheres (continued)
   pilulus 363
   sinensis 363
   spinidactylus 364
   tsingtaoensis 364
PINNOTHERIDAE 363
Pinnularia bistriata 41
   major 41
   trevelyana 41
Pinquigemmula luzonica 281
Pinus elliottii 150
   massoniana 150
   thunbergii 150
Pionosyllis compacta 219
   magnifica 219
Piromis congoensis 228
Pisania cingulata 277
   ignea 277
   truncata 277
Piscicola magna 235
PISCICOLIDAE 235
Pisidia dispar 351
   gordoni 351
   serratifrons 351
   streptocheles 351
   striata 351
Pisione africana 218
   levisetosa 218
PISIONIDAE 218
Pisonia aculeata 151
   grandis 151
Pisodonophis boro 418
   cancrivorus 418
   rubicandus 418
Pista brevibranchia 229
   cristata 229
   elongata 229
   fasciata 229
   foliigera 229
   macrolobata 229
   pachybranchiata 229
   pacifica 229
   robustiseta 229
   typha 229
   violacea 229
   zachsi 229
Pistacia chinensis 154
Pitar affinis 255
   chordatum 255
   crocea 255
   hebraea 255
   limatula 255
   nipponicum 255
   noguchii 255
   obliquata 255
   pellucida 255
   striatum 255
   sulfurea 255
Pithophorus musculosus 206
Pittacium egrettum 204
   pittacium 204
PITTOSPORACEAE 152
Pittosporum tobira 152
Pityomma drymodes 64
Pityrogramma calomelanos 150

PLACENTULINIDAE 110
Placiphorella japonica 238
Placobranchus ocellatus 289
Placopsilina bradyi 74
PLACOPSILINIDAE 74
PLACUIDAE 56
Placuna ephippium 246
    placenta 246
    [*placenta*] 246
    [*sella*] 246
PLACUNIDAE 246
Placus livida 56
[*Plagiodiscus nervatus*] 48
Plagiogramma antillarum 43
    atomus 44
    pulchellum 44
    reimeri 44
    staurophorum 44
    vanheurckii 44
Plagioporus acanthogobii 198
    apogonichthydis 198
    dorosomatis 198
    epinepheli 198
    hunghuaensis 198
    issaitschikowi 198
    parathlassomatis 198
    rhabdosargi 198
    sillagonis 198
Plagiopsetta fasciatus 457
    glossa 457
PLAGIORCHIIDAE 198
Plagiotata promiscus 191
Plagiotremus rhinorhynchos 446
    spilistius 446
    tapeinosoma 446
Plagusia dentipes 366
    [*depressa*] 366
    ———— *tuberculata* 366
    immaculata 366
    tuberculata 366
PLAKINIDAE 164
[PLAKOBRANCHIDAE] 289
Plakortis simplex 164
PLANAXIDAE 266
Planaxis sulcatus 266
Planes cyaneus 366
    marinus 366
Planispirillina denticulata 81
    paucispira 81
    tuberculatolimbata 81
PLANISPIRILLINIDAE 81
Planispirinella exigua 81
Planispirinoides bucculentus 82
Planktoniella sol 30
PLANOCERIDAE 191
Planococcus 23
Planodiscorbis grossepunctatus 111
    lingi 111
    rarescens 111
Planogypsina squamiformis 116
Planorbulina acervalis 115
    mediterranensis 115
    rubra 115
    variabilis 115

PLANORBULINACEA 114
Planorbulinella larvata 115
PLANORBULINIDAE 115
Planorbulinoides retinaculata 115
Planostrea pestigris 246
PLANTAE 133
Planularia tricarinella 94
Planulina alticamera 114
    ariminensis 114
    bradyi 114
    costata 114
    dohertyi 114
    elegans 114
    foveolata 114
    subdepressus 114
    subtenuissima 114
    wuellerstorfi 114
PLANULINIDAE 114
Planulinoides planoconcavus 113
PLANULINOIDIDAE 113
Platalea leucorodia 465
    minor 465
Platax batavianus 436
    orbicularis 436
    pinnatus 436
    teira 436
Platorchestia platensis 330
Platybelone argalus platyura 418
Platybrissus roemeri 395
PLATYCEPHALIDAE 455
Platycephalus indicus 455
Platyconchoecia prosadene 305
[PLATYDORIDIIDAE] 291
Platygyra astreiformis 184
    crosslandi 184
    gracilis 184
    phrygia 184
    rustica 184
    ryukyuensis 184
PLATYHELMINTHES 191
Platylepas decorada 321
    hexastylos 321
Platymaia alcocki 357
    bartschi 357
    fimbriala 357
Platynereis abnormis 221
    [*agassizi*] 221
    bicanaliculata 221
    dumerilii 221
    pulchella 221
    sinica 221
Platypodia anaglypta 362
    granulosa 362
    semigranosa 362
Platyrhina limboonkenkengi 407
    sinensis 407
PLATYRHINIDAE 407
PLATYSCELIDAE 333
Platyscelus armatus 333
    ovoides 333
    serratulus 333
Platystoma bitentaculata 166
    nanhainensis 166
Platythamnion yezoense 138

Pleagopleura verticalis 400
PLECOGLOSSIDAE 411
Plecoglossus altivelis 411
PLECTANIDAE 66
Plectodiscus circularia 65
Plectofrondicularia advena 91
PLECTOIDEA 66
Plectonemertes sinensis
PLECTONEMERTIDAE 208
Plectorhynchus albovittatus 435
    celebicus 435
    chatodonoides 435
    cinctus 435
    diagrammus 435
    flavomaculatus 435
    foetela 435
    goldmanni 435
    lineatus 435
    nigrus 435
    orientalis 435
    pictus 435
    picus 435
    punctatissimus 435
    reticulatus 435
    schotaf 435
    sinensis 435
Plectranthias anthioides 427
    helenae 427
    [*japonicus*] 427
    kelloggi 427
    longimanus 427
    nanus 427
    wheeleri 427
    whiteheadi 427
    yamakawai 427
Plectrogenium nanum 453
[*Plectroglyphidodon dicki*] 442
    imparipennis 443
    johnstonianus 443
    lacrymatus 443
    leucozona 443
    phoenixensis 443
Plectropomus leopardus 427
    melanoleucus 427
    oligacanthus 427
    truncatus 427
Plectrypops lima 422
Plegmosphaera cf. *entodictyon* 62
    leptoplegma 62
    ovata 62
Pleistacantha japonica 357
    oryx 357
    simplex 357
PLEOCYAMATA 338
Pleonosporium venustissimum 138
PLEORCHIIDAE 198
Pleorchis hainanensis 198
    nibeae 198
    sciaenae 198
    uku 198
Pleospora sp. 131
Plerurus asterocongeri 202
    atulis 202
    carangis 202

Plerurus (continued)
   cynoglossi 203
   hainanensis 203
   longicaudatus 203
   pomadasydis 203
   pseudosciaenae 203
   scomberomori 203
   spatulocirrus 203
   trichiuri 203
Plesidastrea vacua 184
   versipora 184
Plesiomonas 21
Plesionika alcocki 344
   bifurca 344
   binoculus 344
   crosnieri 344
   dentirostris 344
   edwardsii 344
   ensis 344
   indica 344
   izumiae 344
   martia 344
   ortmanni 344
   semilaevis 344
   sindoi 344
   unidens 344
   yui 344
Plesiopenaeus coruscans 336
   edwardsianus 336
PLESIOPIDAE 427
[*Plesiops altivelis*] 428
   coeruleolineatus 428
   [*coeruleolineatus*] 428
   corallicola 428
   melas 428
   nakaharae 428
Pleuranacanthus sceleratus 461
   suezensis 461
Pleuraspis costata 61
   sarmentosa 61
Pleurerythrops inscita 323
   monospinosa 324
Pleurobrachia globosa 190
PLEUROBRACHIDAE 190
Pleurobranchaea brock 289
   novaezealandiae 289
PLEUROBRANCHIDAE 289
[PLEUROBRANCHOMORPHA] 289
Pleurocapsa minuta 25
PLEUROCAPSACEAE 25
PLEUROCAPSALES 25
Pleurocryptosa enosteoidis 327
   megacephalon 327
PLEUROGONA 402
Pleuromamma abdominalis 309
   boraelis 309
   gracilis 309
   robusta 309
   xiphias 309
[PLEURONECTIDA] 456
PLEURONECTIDAE 457
PLEURONECTIFORME 456
Pleuronema coronata v. *marina* 57
PLEURONEMATIDAE 57

Pleuronichthys cornutus 457
[PLEUROPHYLLIDIADAE] 291
Pleurophyllidiopsis amoyensis 292
   orientalis 292
   tsingtaoensis 292
[*Pleuroploca trapezium*]
Pleurosigma acutum 41
   ——— v. *latum* 41
   aestuarii 41
   affine 41
   angulatum 41
   ——— v. *falcatum* 41
   ——— v. *quadratum* 41
   [——— v. *strigosa*] 41
   decorum 41
   delicatulum 41
   diminutum 41
   diverse-striatum 41
   elongatum 41
   ——— v. *sinica* 41
   falx 41
   finmarchicum 41
   formosum 41
   intermedium 41
   ——— v. *dongshanense* 41
   ——— v. *nubecula* 41
   longum v. *inflata* 41
   major 41
   marinum 41
   minutum 41
   naviculaceum 41
   ——— f. *minuta* 41
   normanii 41
   ——— v. *fossilis* 41
   obtusum 41
   pelagicum 41
   rhombeum 41
   rigidium 41
   rostratum 41
   salinarum 41
   speciosum 41
   strigosum 41
   tahitianum 41
PLEUROSTOMATIDA 57
PLEUROTOMARIIDAE 261
Pleuroxus hamulatus 301
   laevis 301
PLEXAURIDAE 187
Plicatula australis 245
   philippinarum 245
   plicata 245
   simplex 245
PLICATULIDAE 245
Plicomugil labiosus 425
PLOCAMIACEAE 136
Plocamium telfairiae 136
   ——— f. *uncinataum* 136
Plocamopherus ceylonicus 289
   tilesii 289
PLOIMA 214
Plotnikovina compressa 80
Plotohelmis alata 215
   capitata 215
PLOTOSIDAE 418

Plotosus anguillaris 418
   lineatus 418
Pluchea indica 158
PLUMBAGINACEAE 157
Plumbago zeylanica 157
Plumularia badia 171
   lagenifera 171
   setacea 171
   setaceoides 171
PLUMULARIDAE 171
Pluvialis dominica fulva 467
   squatarola 467
Plyandrocarpa latericius 403
   monotestis 403
   sagamiensis 403
Pneophyllum lejolisii 135
   zostericolum 135
Pneumatophorus japonicus 449
   tapeinocephalus 449
Pneumoderma atlanticum 288
   heterocotylum 288
   mediterraneum 288
PNEUMODERMALIDAE 288
Pneumodermopsis ciliata 288
   macrocotyla 288
   paucidens 288
   polycatyla 288
Pocillopora brevicornis 181
   damicornis 181
   danae 181
   ligulata 181
   meandrina nobilis 181
   verrucosa 181
POCILLOPORIDAE 181
[*Pocockiella variegata*] 142
Podabacia crustacea 182
   elegans lobata 182
[*Podarke angustifrons*] 218
   [*obscura*] 218
PODICIPEDIDAE 464
PODICIPEDIFORMES 464
PODOCAMPIDAE 68
Podocatactes hamifer 358
PODOCERIDAE 330
Podocerus inconspicuus 330
   tuberculosus 330
Podocoryne apicata 167
   carnea 167
   gracilis 167
   minima 167
   minuta 167
Podocotyle lethrini 198
   lizae 198
   lutiani 198
   nibeae 198
   tetrastyla 198
Podocotyloides plageorchis 198
PODOCYRTIDAE 68
Podocystis adriatica 47
   spathulata 47
PODOLAMPADACEAE 55
Podolampas bipes 55
   ——— v. *reticulata* 55
   elegans 55

Podolampas (continued)
   palmipes 55
   spinifera 55
Podon polyphemoides 301
   schmackeri 301
PODONIDAE 301
Podophthalmus nacreus 359
   vigil 359
Podosira argus 30
   granulata 30
   hormoides 30
   maxima 30
   stelliger 30
Podosphaeraster polyplax 391
Podothecus sturiodes 456
Poecilancistrum sp. *blastocyst* 207
Poecilasma kaempferi 319
   litum 319
   obliqua 319
POECILASMATIDAE 319
POECILOCHAETIDAE 224
Poecilochaetus hystricosus 224
   johnsoni 224
   paratropicus 224
   serpens 224
   spinulosus 224
   tricirratus 224
   tropicus 224
Poecilopsetta colorata 457
   megalepis 457
   natalensis 457
   plinthus 457
   praelonga 457
POECILOSCLERIDA 165
Pogonoculius zebra 451
Pogonoperca ocellatus 427
Polinices albumen 270
   [*didyma*] 269
   effusus 270
   flemingianus 270
   macrostoma 270
   mammata 270
   maurus 270
   melanostomoides 270
   [*melanostomus*] 270
   opacus 270
   peseiphanti 270
   pyriformis 270
   sagamensis 270
   simiae 270
   tumidus 270
   vestitus 270
Pollia fumosus 277
   undosa 277
[*Pollicipes mitella*] 318
Polskiammina asiatica 76
Polyangium linguatula 194
Polyarthra trigla 214
POLYBRACHIORHYNCHIDAE 207
Polycarpa circumarta 403
   irregularis 403
Polycarpaea corymbosa 152
   gaudichaudii 152
Polycarvernosa fastigiata 137
   [*fastigiata*] 136

Polycarvernosa (continued)
   ramulosa 137
   [*remulosa*] 136
Polycaulus uranoscopa 454
   [*yamanokami*] 454
Polycera fuitai 289
   japonica 289
POLYCERIDAE 289
POLYCHAETA 214
Polycheira fusca 389
Polycheles baccatus 346
   enthrix 346
   typhlops 346
Polycirrus dodeka 229
   multus 229
   plumosus 229
   quadratas 229
POLYCITORIDAE 402
POLYCLADIDA 191
POLYCLINIDAE 402
Polyclinum constellatum 402
   festum 402
   sundaicum 402
[*Polydactyeus sextarius*] 425
Polydendrorhynchus papillaris 208
[*Polydora antennata*] 223
   armata 223
   ciliata 223
   flava 223
   giardi 223
   kempi 223
   cf. *pilikia* 223
   socialis 223
   cf. *tentaculata* 223
Polydorella novaegeorgiae 223
POLYGONACEAE 151
Polygonum sibricum 151
POLYHYMENOPHORA 58
POLYIDEACEAE 137
Polyipnus aquavitus 412
   nuttingi 412
   spinosus 412
   sterope 412
   tridentifer 412
   triphanos 412
   unispinus 412
POLYKRIKACEAE 51
Polykrikos schwarzi 51
Polymetme elongata 411
   illustris 411
Polymixia berndti 421
   japonicus 421
   longispina 421
POLYMIXIIDAE 421
Polymorphina elegantissima 96
   ovata 96
Polymorphinella vaginulaeformis 96
POLYMORPHINIDAE 95
POLYMORPHIDA 213
POLYMORPHIDAE 213
Polymorphus minutus 213
POLYNEMIDAE 425
Polynemus indicus 425
   microstoma 425
   plebeius 425

Polynemus (continued)
   sextarius 425
[*Polynices fortuneri*] 269
   [*mammata*] 270
   [*mammatus*] 270
[*Polynoe pleiolepis*] 217
   [*sagamiana*] 217
POLYNOIDAE 216
Polyodontes atromarginatus 217
   [*flagelliformis*] 217
   [*gracilis*] 217
   maxillosus 217
   [*melanonotus*] 217
POLYODONTIDAE 409
[*Polyonyx asiaticus*] 351
   biunguiculata 351
   obesulus 351
   sinensis 351
Pollyopes polyideoides 136
Polyophthalmus pictus 226
Polyphyllia talpina 182
POLYPLACOPHORA 237
Polyplectana kefersteinii 389
Polysegmentina circinata 88
Polysiphonia abscissa 140
   [*akkeshiensis*] 140
   coacta 140
   crassa 140
   decumbens 140
   elongata 140
   ferulacea 140
   fragilis 140
   gracilis 140
   harlandii 140
   japonica 140
   mollis 140
   ———— v. *tongatensis* 140
   morrowii 140
   patens 140
   richardsoni 140
   savatieri 140
   senticulosa 140
   simplex 140
   spiralis 140
   subtilissibma 140
   upolensis 140
   urceolata 140
   yendoi 140
Polystomammina elongata 76
   lobatula 76
Polystomellina discorbinoides 126
Polytretus reinboldii 141
Pomacanthus annularis 438
   imperator 438
   semicirculatus 438
   sexstriatus 438
POMACENTRIDAE 442
Pomacentrus albifasciatus 444
   amboinensis 444
   [*bankanensis*] 444
   chrysurus 444
   [*coelestinus*] 442
   coelestis 444
   dorsalis 444
   grammorhynchus 444

Index

Pomacentrus (continued)
   jenkinsi 444
   lepidogenys 444
   lividus 444
   melanopterus 444
   moluccensis 444
   nagasakiensis 444
   nigricans 444
   nigromarginatus 444
   niomatus 444
   notophthalmus 444
   pavo 444
   perspicilliatus 444
   philippinus 444
   prosopotaenia 444
   stigma 444
   taeniurus 444
   tripunctatus 444
   vaiuli 444
   violascens 444
[*Pomachromis richardsoni*] 442
POMADASYIDAE 435
Pomadasys argenteus 435
   furcatus 435
   grunniens 435
   hasta 435
   kaakan 435
   maculatus 435
   quadrilineatus 435
   umimaculatus 435
[*Pomatoceros triqueter*] 230
[*Pomatoleios crosslandi*] 230
   kraussii 230
Pomatomus saltatris 427
Pomatostegus stellatus 230
Pongamia pinnata 153
Pontella chierchiae 311
   danae 311
   denticauda 311
   fera 311
   kieferi 311
   latifurca 311
   princeps 311
   securifer 311
   sinica 311
   spinicauda 311
   tridactyla 311
PONTELLIDAE 311
Pontellina plumata 311
Pontellopsis armatus 311
   inflatodigitata 311
   krameri 311
   macronyx 311
   regalis 311
   strenua 311
   tenuicauda 311
   villosa 311
   yamadae 311
Pontinus tentacularis 453
Pontobdella loricata 235
   [*moorei*] 235
Pontobdellina macrothela 235
   [*macrothela*] 235
Pontocaris pennata 344
   ruthbunae 344

Pontocaris (continued)
   sibogae 344
Pontocrates altamarimus 330
Pontodora pelagica 215
PONTODORIDAE 215
Pontodrilus litoralis 234
Pontoeciella abyssicola 312
Pontogeneia litorea 329
Pontogenia nuda 216
Pontonia okai 341
   [*quadraphthalma*] 340
Pontophilus angustirostris 345
   bidentatus 345
   carinicauda 345
   incisus 345
   parvirostris 345
Poraster superbus 390
[*Porcellana armata*] 351
   curvifrons 351
   habei 351
   [*ornata*] 351
   pulchra 351
   [*streptocheles*] 351
PORCELLANASTERIDAE 391
Porcellanella triloba 351
PORCELLANIDAE 351
Porcellanopagurus japonoicus 350
PORCELLIDIIDAE 314
Porcellidium acuticaudatum 314
   ovatum 314
Porella compressa 380
   rotundirostris 380
Poricellaria ratoniensis 376
PORIFERA 164
Porites andrewsi 182
   compressa 182
   iwayamaensis 183
   lichen 183
   lutea 183
   matthaii 183
   nigrescens 183
   pukoensis 183
PORITIIDAE 182
Porodiscus ellipticus 65
Poroeponides calida 111
   cribrorepandus 111
   incrassatus 111
   lateralis 111
   repandus 111
   speciosus 111
Porolithon bobergesenii 135
   onkodes 135
   ——— f. *devia* 135
   ——— f. *subramosa* 135
Poromitra crassicepsi 421
   megalops 421
   oscitons 421
Poromya castanea
POROMYIDAE
POROSPATHIDAE 69
Porospathis holostoma 69
Porphyra crispata 133
   dentata 133
   dentimarginata 133
   fujianensis 133

Porphyra (continued)
   guangdongensis 133
   haitanensis 133
   ishigecola 133
   katadai v. *hemiphylla* 133
   marginata 133
   monosporangia 133
   oligospermatangia 133
   pseudocrispata 133
   ramosissima 133
   seriata 133
   suborbiculata 133
   tenera 133
   umbilicalis 133
   vietnamensis 133
   yezoensis 133
PORPHYRIDIALES 133
Porphyropsis coccinea 133
[*Porpita pacifica*] 168
   porpita 168
   [*umbrella*] 168
Porroecia porrecta 305
   spinirostris 305
Portatrochammina eltaninae 76
   wiesneri 76
Portieria hornemannii 137
Portlandia japonica 239
Portulaca pilosa 152
PORTULACACEAE 152
PORTUNIDAE 358
Portunus argentatus 359
   brockii 359
   dayawanensis 359
   [*gladiator*] 359
   gracilimanus 359
   granulatus 359
   haanii 359
   hainanensis 359
   hastatoides 359
   iranjae 359
   orbitosinus 359
   pelagicus 359
   pubescens 359
   pulchricristatus 359
   sanguinolentus 359
   tenuipes 359
   tridentatus 359
   trituberculatus 359
   tuberculosus 359
   tweediei 359
Porzana fusca erythrothorax 466
   p. pusilla 466
POTAMIDIDAE 266
[*Potamilla acuminata*] 230
   cf. *acuminata* 230
   [*chiaochouensis*] 230
   cf. *myriops* 230
   polyophtham 230
   reniformis 230
   torelli 230
Potamocorbula amurensis 256
   laevis 256
   rubromuscula 256
   ustulata 256
POTAMOGETONACEAE 158

Potiarca pilula 240
Praealticus margaritarius 446
   striatus 446
   tanegasimae 446
Praebebalia longidactyla 355
Praeglobobulimina notovata 107
   ovata 107
   spinescens 107
Praescutata viperina 464
Prasinocladus lubricus 144
[*Praxilla praetermissa*] 225
Praxillella cf. *affinis* 225
   gracilis 225
   pacifica 225
   praetermissa 225
Praya dubia 174
   reticulata 174
PRAYIDAE 174
Premna obtusifolia 157
Preptetos cylindricus 197
PRIACANTHIDAE 428
Priacanthus blochi 428
   boops 428
   cruentatus 428
   hamrur 428
   macracanthus 428
   [*multifasciatus*] 428
   [*niphontus*] 428
   tayenus 428
PRIAPULIDA 214
PRIAPULIDAE 214
Priapulus caudatus 214
Priapulopsis cf. *australis* 214
[*Primovula caarctata*] 271
   concinna 271
   dautzenbergi 271
   piriei 271
   [*rhodia*] 271
   singularis 271
Primno brevidens 332
   latreillei 332
   macropa 332
PRIMULACEAE 156
[*Priolepis naraharae*] 452
Prionace glauca 406
Prionalpheus mortoni 343
   triarticulatus 343
Prionechinus forbesianus 393
Prionitis cornea 136
   prolifera 136
   ramosissima 136
Prionocidaris baculosa 392
   ———— v. *annulifera* 392
   bispinosa 392
   verticillata 392
Prionomys aspera 324
Prionospio cirrifera 223
   ehlersi 223
   japonica 223
   krusadensis 223
   malayensis 223
   malmgreni 223
   [*pinnata*] 223
   pygmaea 223
   [*pygmaeus*] 223

Prionospio (cotinued)
   queenslandica 223
   saccifera 223
   xishaensis 223
Prionovolva bulla 271
   pudica 271
   ———— *wilsoniana* 271
Prionurus scalprus 449
PRISTIDAE 407
PRISTIFORMES 407
Pristigenys multifascia 428
PRISTIOPHORIDAE 407
PRISTIOPHORIFORMES 407
Pristiophorus japonicus 407
Pristipomoides auricilla 433
   filamentosus 433
   flavipinnis 433
   microlepis 433
   multidens 433
   sieboldii 433
   typus 433
Pristis cuspidatus 407
   microdon 407
Pristotis jerdoni 444
Probopyria elliptica 327
Probopyriscus novempalensis 327
Proboscidactyla flavicirrata 173
   mutabilis 173
   ornata 173
   [———— v. *gemmifera*] 173
   stellata 173
PROBOSCIDACTYLIDAE 173
Proboscina coapta 370
PROCELLARIIDAE 464
PROCELLARIIFORMES 464
Procephalathrix arenarius 207
   orientalis 207
   spiralis 207
Procerodes graciliceps 191
PROCERODIDAE 191
Procerolagena gracilis 95
Processa aequimana 341
   demani 341
   japonica 341
   sulcata 341
PROCESSOIDAE 341
PROCESSOIDEA 341
PROCHLORACEAE 27
PROCHLORALES 27
Prochloron sp. 27
PROCHLOROPHYCEAE 27
Prochromadora orleji 211
Proctoeces longisaccatus 194
   maculatus 194
   orientalis 195
   parapistipomae 195
Proctotrematoides anguillae 195
   gymnothoracis 195
Progebiophilus sinicus 327
Prognichthys agoo 419
   albimaculatus 419
   brevipinnis 419
   rondeleti 419
   sealei 419
Prohaplosplanchnus diorchis 193

Prolixoplecta exilis 77
PROLIXOPLECTIDAE 76
Promethichthys prometheus 449
Promicrops lanceolatus 427
Promysis orientalis 324
[*Proneomysis quadrispinosa*] 323
   [*sinensis*] 323
Pronoe capito 332
PRONOIDAE 332
Propeamussium caducum 245
   gracilis 245
   sibogai 245
   stellus 245
   watsoni 245
Propeleda soyomaruae 239
Proplectella globosa 60
   ovata 60
PROROCENTRACEAE 50
PROROCENTRALES 50
Prorocentrum cordatum 50
   lenticulatum 51
   magnum 51
   micans 51
   oblongum 51
   triestinum 51
Proscyllium habereri 405
   [*habereri*] 406
Prosimnia semperi 271
PROSOBRANCHIA 261
Prosogonotrema caesionis 203
   plataxum 203
PROSOGONOTREMATIDAE 203
Prosorchiopsis plataxonis 203
Prosorchis provitelletus 203
   rastrelligi 203
   xishaensis 203
PROSORHOCHMIDAE 208
Prosorhynchus clavatum 193
   crucibulum 193
   eleutheronemae 193
   epinepheli 193
   facilis 193
   fujianensis 193
   ozakii 193
   sphraenae 193
   synanceiae 193
   tsengi 193
Prosphorachaeus suluensis 357
PROSTHIOSTOMIDAE 191
Prosthiostomum affine 191
   formosum 191
   obscurum 191
   tenebrosum 192
PROSTOMATIDA 56
Protankyra asymmetrica 389
   bidentata 389
   magnihanula 389
   pseudodigitata 389
   suensoni 389
   verrilli 389
Protatlanta souleyeti 269
Protelphidium anglicum 118
   compressum 118
   fulvofusculus 118
   glabrum 118

Protelphidium (continued)
   granosum 118
   luridus 118
   pauperatum 118
   sublaeve 118
   tersum 118
   tuberculatum 118
Proteonella asymmetrica 70
   atlantica 70
   magna 70
   pseudodifflugiformis 70
   rhombiformis 70
   testacea 70
Proterato callosa 270
   limata 270
   nana 270
   pura 270
   sulcifera 270
Proteus vulgaris 21
PROTIARIDAE 167
PROTISTA 27
Protochondracanthus alatus 316
Protocruzia tuzeti 58
Protocystis nautiloides 69
   xiphodon 69
PROTODRILIDAE 214
Protodrilus huanghaiensis 214
   rubropharzngeus 214
Protoglobobulimina pupoides 107
Protomonostroma undulatum 144
[*Protonibea diacanthus*] 431
Protoreaster nodosus 390
Protorhabdonella curta 59
   simplex 59
Protosalanx hyalocranius 411
Protothaca jadoensis 255
   [*staminea euglypta*] 254
Protula tubularia 230
Provirellotrema crenimugilas 193
Prudhnoeus oligolecithosum 195
PRUNOIDEA 64
Prymnaster rostratus 394
PSALIDOPODIDAE 339
PSALIDOPODOIDEA 339
Psalidopus huxleyi 339
   spiniventris 339
Psammaplysilla purpurea 164
[*Psammobia kazusensis*] 252
   [*maculosa*] 252
   radiata 252
PSAMMOBIIDAE 252
Psammocora contigua 181
   ———— *pulchra* 181
   divaricata 181
   nierstrazi 181
   profundacella 181
Psammofax consociata 70
Psammolyce fijiensis 218
   flava 218
   malayana 218
Psammoperca waigiensis 425
Psammosphaera fusca 70
   parva 70
   rustica 70
PSAMMOSPHAERIDAE 70

Psednobates uncunguis 298
Psenes arafurensis 450
   cyanophrys 450
   [*indicus*] 450
   pellucidus 450
Psenopsis anomala 450
Psephurus gladius 409
PSETTIDAE 436
Psettina filimanus 457
   gigantea 457
   hainanensis 457
   iijimae 457
   tosana 457
Psettodes erumei 456
PSETTODIDAE 456
Pseudalutarius nasicorinis 460
Pseudamiops gracilicauda 429
Pseudanchialina inermis 324
   pusilla 324
Pseudanthessius obscurus 313
   parvus 313
Pseudanthias cichlops 427
   elongatus 426
Pseudapocryptes lanceolatus 452
Pseudarcella semisphaerea 94
Pseudarcopagia minuta 251
Pseudaristeus crassipes 336
Pseudastralium henicus gloriosum 264
Pseudaugaptilus orientalis 310
Pseudemia gelatinosa 429
Pseudeuphausia latifrons 335
   sinica 335
[*Pseudeurythoe ambigua*] 226
   [*hirsura*] 226
   [*oligobranchia*]
   [*pancibranchiata*] 226
   [*spiralis*] 226
Pseudione longicauda
Pseudoantorchis thalassomae 195
Pseudobalistes flavimarginatus 459
   fuscus 459
Pseudoblennius cottoides 456
Pseudoboletia maculata 393
Pseudobolivina antarctica 75
   brevis 75
   nasostoma 75
   torquata 75
PSEUDOBOLIVINIDAE 75
Pseudobranchiomma emersoni 230
Pseudobryopsis hainanensis 146
Pseudobucephalopsis belonnis 193
Pseudobulimina convoluta 99
Pseudobunocotyla clupanodonis 203
Pseudocaenodera alectis 195
   nibeae 195
PSEUDOCALANIDAE 307
Pseudocarchesium aselli 58
Pseudocentrophorus isodon 407
Pseudocepola taeniosoma 438
PSEUDOCERIDAE 191
Pseudoceros exoplatus 191
Pseudochama retroversa 247
Pseudocharopinus markewitschi 316
Pseudocheilinus evanidus 440
   hexataenia 440

Pseudocheilinus (continued)
   octotaenia 440
Pseudochilodonopsis marina 57
Pseudochirella scopularis 308
PSEUDOCHROMIDAE 427
Pseudochromis cynotaenia 427
   fuscus 427
   kikaii 427
   luteus 427
   melanotoenia 427
   porphyreus 427
   sureus 427
   tapeinosoma 427
   xanthochir 427
Pseudoclavulina gracilis 80
   humilis 80
   juncea 80
   mexicana 80
   robusta 80
   scabra 80
   serventyi 80
Pseudocnus echinata 387
Pseudocoeliodidymocystis xishanesis 200
Pseudocollodes demani 357
Pseudocolochirus violaceus
Pseudoconchoecia concentrica 305
   serrulata 305
Pseudocoris awayae 440
   yamashiroi 440
Pseudocreadim hainanensis 197
   monacanthi 197
Pseudocycnus appendiculatus 316
   armatus 316
Pseudocypraea adamsonii 271
   [*adamonii*] 271
PSEUDODIAPTOMIDAE 309
Pseudodiaptomus incisus 309
   marinus 309
   penicillus 309
Pseudodax moluccanus 440
Pseudodichomosiphon constriata 146
Pseudodictomosiphon constricta 50
Pseudoeponides anderseni 116
   angulatus 116
   compressum 116
   heterogeneus 116
   japonicus 116
   nakazatoensis 116
Pseudoetrema fortilirata 281
Pseudoeunotia doliolus 45
Pseudofellodistomum plagiorchis 195
Pseudoflintina bulbosa 83
   triquetra 83
Pseudogaudryina atlantica 80
PSEUDOGAUDRYINIDAE 80
Pseudogaleomma janonica 248
PSEUDOGRAMMIDAE 428
Pseudograpsus albus 366
Pseudogyroidina sinensis 116
Pseudohapalocarcinus ransoni
Pseudohastigerina wilcoxensis 101
Pseudoisocoelium bigonopori 198
Pseudojuloides cerasina 441
Pseudolabrus gracilis 441
   japonicus 441

Pseudoliomera granosimana 362
    helleri 362
    speciosa 362
    variolosa 362
PSEUDOLITHIDAE 61
Pseudolithium compressum 61
Pseudolithophyllum samoense 135
    yendoi 135
Pseudomeira brevifurca 315
[*Pseudolycaea pachypoda*] 332
PSEUDOMALACOSTEGOMORPHA 372
Pseudomaretia alta 394
Pseudomassilina australis 86
    macilenta 86
    reticulata 86
Pseudomma brevicaudum 324
PSEUDOMONADACEAE 21
PSEUDOMONADALES 21
Pseudomonas 21
Pseudomonorcheides ditrematis 195
Pseudomysisdetes cochinensis 324
Pseudoneaera semipellucida 259
Pseudonereis anomala 221
    gallapagensis 221
    variegata 221
Pseudonitzschia delicatissima 47
    [*delicatissima*] 46
    [*pungens*] 47
    [*seriata*] 47
    sicula 47
    ———— v. *bicuneata* 47
    ———— v. *migrans* 47
Pseudonodosaria comatula 91
    glanduliniformis 91
    japonica 91
    laevigata 91
    radicula 91
    torrida 91
    virginiana 91
Pseudonodosinella nodulosus 73
Pseudononion decorum 118
    granuloumbilicatum 118
    japonicus 118
    limbatostriatum 118
    minutum 118
    oinomikadoi 118
Pseudononionella variabilis 98
PSEUDOPARRELLIDAE 113
Pseudopatellina compressa 110
Pseudopecoelina chirocentrosus 198
    platycephali 198
    xishaense 198
Pseudopecoeloides astrocongeris 198
    carcngis 198
    dayawanensis 198
    mugilis 198
Pseudopercoelus alectis 198
    elongatus 198
    epinepheli 198
[*Pseudophilyra olivacea*] 355
PSEUDOPHYLLIDEA 206
Pseudopleuronectes herzensteini 457
    yokohamae 457

Pseudopolydora antennata 223
    kempi 223
    ———— *japonica* 224
    paucibranchiata 224
Pseudopolymorphina indica 96
    netroforms 96
    ovalis 96
    soldani 96
    suboblonga 96
Pseudopriacanthus multifasciatus 428
    niphonius 428
Pseudoprosorhynchus fusiformis 193
Pseudopyrgo milletti 86
    paraglobula 86
    toddae 86
Pseudopythina macrophthalmensis 247
    maipoensis 247
Pseudorathkea macrogastrica 167
Pseudorca crassidens 469
Pseudorertagus aluco 267
Pseudorhombus arsius 456
    cinnamomeus 456
    ctenosquamis 456
    dupliocellatus 456
    elevatus 456
    javanicus 456
    levisquamis 456
    malayanus 456
    neglectus 456
    oligodon 456
    pentophthalmus 456
    quinquocellatus 456
    triocellatus 456
Pseudoringicula sinensis 285
Pseudorotalia compressuiscula 123
    gaimardii 123
    indopacifica 123
    leiqiongensis 123
    papillosa 123
    schroeteriana 123
    tikutoensis 123
    yabei 123
Pseudosciaena crocea 432
    polyactis 432
Pseudoscopelus scriptus 445
Pseudosiderastrea tayamai 183
Pseudosigmoilina minuta 86, 88
Pseudosimnia alabaster 271
    marginata 271
    punctata 271
    sinensis 271
    whitworthi 271
Pseudosiphoderoides lutiani 199
    opakapaka 199
    rotivarijera 199
    xishaensis 199
Pseudosquilla oculata 368
    ornata 368
PSEUDOSQUILLIDAE 368
Pseudostegias dulcilaeuum 327
    setoensis 327
Pseudosteringophorus holognathi 195
Pseudostichopus trachus 388
    unguiculatus 388

Pseudostiotrema rhynchobati 198
Pseudostylochus obscuras 191
Pseudotiara tropica 167
Pseudotrachelocerca trepida 56
PSEUDOTRACHELOCERCIDAE 56
Pseudotriacanthus strigilifer 459
PSEUDOTRIAKIDAE 405
Pseudotriakis acrages 405
    [*microdon*] 405
Pseudotriloculina cyclostoma 86
    lecalvezae 86
    lunata 86
    subglobiformis 86
    subsphaeroides 86
[*Pseudoxymetopon sinensis*] 449
Pseudozius caystrus 362
PSILOCEPHALIDAE 460
Psilocephalus barbatus 460
Psopheticus insignis 363
Psychichthys mitsukurii 409
PSYCHIDAE 300
Psychrolutes inermis 456
PSYCHROLUTIDAE 456
Pteragogus flagellifera 441
PTERASTERIDAE 391
Pterelectroma zebra 243
Pterereotris evides 451
    heteropterus 451
    microlepis 451
[*Pteria alacorvi*] 243
    antelata 243
    brevialata 243
    [*brevialata*] 243
    chinensis 243
    coturnix 243
    [*coturnix*] 243
    cypsellus 243
    dendronephthya 243
    [*lata*] 243
    loveni 243
    [*margaritifera*] 243
    [*martensii*] 243
    penguin 243
    [*tegulata*] 243
PTERIDOPHYTA 150
PTERIIDAE 243
Pterinotrema mirabilis 192
PTERINOTREMATIDAE 192
PTERIOIDA 243
Pterocanium korotnevi 68
    pretextum 68
    trilobum 68
[*Pterocera lambis*]
Pterocirrus macroceros 215
    notoensis 215
Pterocladia capillacea 134
    nana 134
    [*temuis*]
[*Pterocoesio chrysozona*] 432
    [*diagramma*] 433
    [*tile*] 433
Pterocorys campanula 68
    zancleus 68
Pterocypridina alata 302

Index

Pterodroma r. rostrata 465
Pteroeides chinense 190
    sparmannii 190
PTEROEIDIDAE 190
Pterogobius zacalles
Pteroidichthys amboinensis 453
Pterois antennata 453
    lunulata 453
    [*miles*] 453
    radiata 453
    russelli 453
    volitans 453
PTEROMEDUSAE 175
Pterometra pulcherrima 387
    trichopoda 387
[*Pterophryne marmoratus*] 462
Pteropilium clausum 68
Pteropsaron evolans 445
Pteropurpura aduncas 275
    ――― form *expansa* 275
    plorator 275
Pterosagitta draco 385
Pteroscenium pinnatum 67
Pterosiphonia pennata 140
Pterosoma planum 269
[PTEROTA] 288
PTEROTHRISSIDAE 412
Pterothrissus gissus 412
Pterotrachea coronata 269
    hippocampus 269
    minuta 269
    scutata 269
PTEROTRACHEIDAE 269
Pterycombus petersii 431
Pterygia crenulata 279
    dactylus 279
    fenestrate 279
    nucea 279
    sinensis 279
Pterygioteuthis giardi 294
Pterygotrigla hemisticta 454
    ryukyuensis 454
Pterynotus alatus 275
    bipinnatus 275
    elongatus 275
    loebbeckei 275
    [*pinnatus*] 275
    trigonula 275
    tripterus 276
    vespertilio 276
PTYCHOCYLIDAE 59
Ptychodera flava 399
PTYCHODERIDAE 399
Ptychogaster defensa 350
Ptychognathus barbatus 366
Puccinellia chinampoensis 160
    distans 160
Puerulus angulatus 346
Puffinus griseus 465
    leucomelas 465
    pacificus cuneatus 465
Pugettia incisa 357
    minor 357
    nipponensis 357

Pugettia (continued)
    quadridens 357
Pugilina cochlidium 277
Pullenia apertula 119
    bulloides 119
    eocenica 119
    jarvisi 119
    malkinae 119
    marssoni 119
    quaternaria 119
    quinqueloba 119
    subcarinata 119
    subsphaerica 119
Pulleniatina finalis 101
    obliquiloculata 101
    praecursor 101
    primalis 101
PULLENIATINIDAE 101
PULMONATA 294
Pulsiphonina elegans 113
Pulvinus micans 251
Pumiliopes opisthopteri 314
Punctacteon fabreanus 285
    flammeus 285
    kajiyamai 285
    kirai 285
    suturalis 285
    virgatus 285
    yamamurae 285
Punctaria latifolia 141
    plataginea 142
PUNCTARIACEAE 141
Puncturella dorcas 262
    fastigiata 262
    nobilis 262
    pileolus 262
    sinensis 262
Pupa affinis 285
    alveola 285
    coccinata 285
    sinica 285
    solidula 285
    strigosa 285
    sulcata 285
[PUPIDAE] 285
[*Purpura armigera*] 276
    [*bronii* v. *suppressa*] 276
    brunneolabrum 276
    [*hipocastanum*] 276
    [*panama*] 276
    persica 276
    rudolphi 276
Pusia consanguineum 279
[*Pusiostoma mendicaria*]
Pustularia bistrinotata 272
    cicercula 272
    globulus 272
Puteolina bulloides 90
    crassa 90
    malayensis 90
Pycnocraspedum microlepis 421
[*Pycnoderma conogensis*] 228
PYCNOGONIDA 297
Pycnophyes sp. 209

Pycnotheca mirabilis 171
Pycreus polystachyos 161
Pygmaeorota cingulifera 266
Pygmaeoseistron hispidula 95
Pygoplites diacanthus 438
Pygospio elegans 224
PYLOBOTRYDIDAE 67
PYLODISCIDAE 65
Pylodiscus echinatus 65
Pylolena armata 65
PYLONIDAE 65
Pylonium claviflorum 65
    scitulum 65
    scutatulum 65
Pylopagurus serpulophilus 350
Pylospira octopyle 66
Pyramidella nodocincta 284
    ventricosa 284
PYRAMIDELLIDAE 284
Pyramidulina raphana 91
Pyramodon ventralis 421
[*Pyrene bicincta*] 276
    marquesa 276
    [*martensi*] 276
    punctata 276
    testudinaria tylerai 276
PYRENIDAE 276
Pyrenochaeta sp. 132
Pyrgo anomala 86
    bougainvillei 86
    bulloides 86
    calostoma 87
    comata 87
    compressioblonga 87
    denticulata 87
    depressa 87
    elongata 87
    fornasini 87
    grinzingensis 87
    inornata 87
    irregularis 87
    johnsoni 87
    laevis 87
    lucernula 87
    lunula 87
    megastoma 87
    murrhina 87
    oblonga 87
    pacifica 87
    rotalaria 87
    sagittioris 87
    sarsii 87
    serrata 87
    simplex 87
    spinidorsa 87
    striolata 87
    tainanensis 87
    tenuis 87
    ventruosa 87
    vespertilio 87
    williamsoni 87
Pyrgoella sphaera 87
    tenuiaperta 87
Pyrgoma cancellata 320

Pyrgoma (continued)
   stellula 320
PYRGOMATIDAE 320
PYROCYSTACEAE 55
Pyrocystis acuta 55
  fusiformis 55
    —— f. *bicornia* 55
  gerbautii 55
  hamulus 55
    —— v. *inaeaqualis* 55
    —— v. *semicircuralis* 55
  lanceolata 55
  obtusa 55
  pseudonoctiluca 55
  rhomboides 55
  robusta 55
PYROPHACACEAE 55
Pyrophacus horologicum v. *steinii* 55
Pyropiloides elongatus 115
Pyropilus rotundatus 115
Pyrosoma atlanticum 400
PYROSOMATIDAE 400
PYROSOMIDA 400
PYROSTEPHIDAE 174
PYRROPHYTA 50
Pyrulina albatrossi 96
  angusta 96
  cylindroides 96
  extensa 96
  fusiformis 96
  gutta 96
Pyrulinoides rasilis 96
Pyrunculus cylindricus 286
  lagenula 286
  longiformis 286
  phialus 286
  pyriformis obesus 286
  teramachii 286
  tokyoensis 286
Pythia cecillei 294
  fimbriosa 294
Pyura curvigona 403
  elongata 403
  lignosa 403
  mirabilis 403
  vittata 403
Pyura sp. 403
PYURIDAE 403
Pyxidicula mediterranea 30
  weyprechtii 30
Pyxidognathus deianira 366

Q

QUADRICELLARIIDAE 372
QUADRIGYRIDAE 213
Quadrigyrus polyspinosus 213
Quadrimorphina pacifica 119
QUADRIMORPHINIDAE 119
Quasilineus sinicus 207
Quidnipagus palatam 251
QUIMPERIDAE 209
Quinqueloculina agglutinans 83
  agglutinata 83
  akneriana 83
  anguina 83

Quinqueloculina (continued)
  angulostoma 83
  arctica 83
  arenata 83
  argunica 83
  artusoris 83
  aspera 83
  asperula 83
  auberiana 84
  bella 84
  bellatula 84
  berthelotiana 84
  bicarinata 84
  bicornis 84
  bicostata 84
  bicostoides 84
  bidentata 84
  bifossula 84
  bosciana 84
  boueana 84
  bradyana 84
  candeiana 84
  carinatastriata 84
  centrostriata 84
  collumnosa 84
  compressa 84
  complanata 84
  compressiostoma 84
  compta 84
  contorta 84
  costata 84
  crassa 84
  crassicarinata 84
  cultrata 84
  curta 84
  cuvieriana 84
  decora 84
  dimidiata 84
  disparilis 84
  distorqueata 84
  donghaiensis 84
  elongata 84
  ferrussacii 84
  fichteliana 84
  flavescens 84
  frigida 84
  fulgida 84
  funafutiensis 84
  fusiformis 84
  gigas 84
  granuliformis 84
  granulocostata 84
  granulosa 84
  grossa 84
  imperialis 84
  implexa 84
  impolita 84
  inculta 84
  juani 84
  jugosa 84
  kansireiensis 84
  kerimbatica 84
  kuromatunaiensis 84
  laevigata 84
  lamarckiana 84
  lata 84

Quinqueloculina (continued)
  linoreticulata 84
  longidentata 84
  longirostra 84
  mauricensis 85
  microcostata 85
  miles 85
  minuta 85
  multicostata 85
  najaeformis 85
  neosigmoilinoides 85
  neostriatula 85
  notata 85
  oblonga 85
  paravulgaris 85
  parkeri 85
  paucilocula 85
  pauperata 85
  pentagona 85
  philippinensis 85
  poeyana 85
  polygona 85
  praelonga 85
  procera 85
  pseudocandeiana 85
  pseudoproxima 85
  pseudoreticulata 85
  pulchella 85
  reticulata 85
  riveroae 85
  rodolphina 85
  rugosa 85
  sabulosa 85
  sagamiensis 85
  sawanensis 85
  seminula 85
  seminulangulata 85
  septuosa 85
  sigmoilinoides 85
  sinensis 85
  stalkeri 85
  subarenaria 85
  subcurta 85
  subdecorata 85
  suborbicularis 85
  subpolygona 85
  subquadra 85
  subungeriana 85
  sulcata 85
  tikutoensis 85
  totomiensis 85
  tricarinata 85
  tropicalis 85
  tubilocula 85
  ungeriana 85
  venusta 85
  vulgaris 85
  yezoensis 85
Quisquilius eugenius 452

R

RACHYCENTRIDAE 431
Raehycentron canadum 431
Racovitzanus levis 308
RADIOLARIA 61

Index

Raeta pellicula 250
Raetellops pulchella 250
Raja acutispina 407
  boesemani 408
  chinensis 408
  gigas 408
  hollandi 408
  katsukii 408
  kanojei 408
  kwangtungensis 408
  macrocauda 408
  macrophthalma 408
  meerdervoortii 408
  porosa 408
  pulchra 408
  tengu 408
RAJIDAE 407
RAJIFORMES 407
Ralfsia fungiformis 141
  verrucosa 141
RALFSIACEAE 141
RALFSIALES 141
RALLIDAE 466
Ralllus aquaticus indicus 466
  striatus gularis 466
Ramicrusta nanhaiensis 134
Ramphotonotus biformis 373
  bispinigera 373
  circulatis 374
  uniformis 374
  variabilis 374
Ramu lispora 130
Ramulina globulifera 96
[*Ramus xiamenensis*] 168
[*Rana cancrivora*] 463
Randallia eburnea 355
Randia accedens 158
RANIDAE 463
Ranilia horikoshii 354
  orientalis 354
Ranina ranina 354
RANINIDAE 354
Raninoides hendersoni 354
  intermedius 354
  personatus 354
  serratifrons
Rapa bulbiformis 276
  rapa 276
Rapana bezoar 276
  rapiformis 276
  [*thomasiana*] 276
  venosa 276
  [―― *peichiliensis*] 276
Rapanea linearis 156
Raphidascaris acentrogobii 209
  lophii 209
Raphidascaroides halicutaeae 209
  myliobatum 209
Raphidivergens bacilliformis 41
Raphidopus ciliatus 351
Rashgua rubrocincta 225
Rastrelliger faughni 449
  kanagurta 449
Ratabulus megacephalus 455
Ratania flava 312
Rathkea octopunctata 167

RATHKEIDAE 167
RECTANGULATA 370
Rectobolivina aculeata 106
  altilocula 106
  bifrons 106
  digitata 106
  laevis 106
  raphana 106
  subbifrons 106
  subraphanus 106
  virgula 106
  xuwenensis 106
Rectoelphidiella aplata 126
  lepida 126
Recurvirostra avosetta 468
RECURVIROSTRIDAE 467
Recurvoides contortus 74
  crassus 74
  gigas 74
  laevigatum 74
  trochamminiformis 74
  turbinatus 74
REGALECIDAE 423
Regalecus russelli 423
Reginella furcata 377
REMANEICIDAE 76
Remirea maritima 161
Remispora quadriremis 131
Remora albescens 453
  australis 453
  brachyptera 453
  remora 453
[*Remorina albescen*] 453
Reniera implexa v. *baeri*
RENIERIDAE 164
Reophanus gracilis 73
  oviculus 73
Reophax aduncus 72
  advenus 72
  agglutinatus 72
  apiculatus 72
  arayaensis 72
  armatus 72
  atlanticus 72
  barwonensis 72
  bermudezi 72
  bilocularis 72
  bradyi 72
  brevis 72
  capitatus 72
  caribensis 72
  catenulatus 72
  communis 72
  curtus 72
  davepopei 72
  davisi 72
  dentaliniformis 72
  depressus 72
  distans 72
  donghaiensis 72
  enormis 72
  excentricus 72
  eximius 72
  fusiformis 72
  guttifer 72
  hempsteadensis 72

Reophax (continued)
  hispidulus 72
  insectus 72
  irregularis 72
  littoralis 72
  longicollaris 72
  micaceous 72
  miculatus 72
  minimus 72
  moniliforme 72
  nodulosus 72
  obscuratus 72
  orientalis 72
  pauciloculatus 72
  paucus 72
  pilulifer 72
  pisiformis 72
  pseudobacillaris 72
  pseudodistans 72
  pulchrus 72
  pyrifera 72
  regularis 72
  rostratus 72
  scorpiurus 72
  spiculifer 72
  subcapitatus 72
  subdentaliniformis 72
  subfusiformis 72
  tappuensis 72
  tenuis 72
  torquiformis 72
  turbo 72
  tubulus 72
Repmanina charoides 71
[*Repomucenus schaapi*] 447
Reptadeonella joloensis 378
REPTILIA 463
RESTIONACEAE 162
Reteporellina denticulata
Retevirgula kenozooidata 374
  spiniaviculata 374
Reticulophragmium reticulatum 74
  sintikuensis 74
Retiflustra anatina 373
  reticulum 373
  schonauii 373
  sinica 373
Retropluma denticulata 364
  quadrata 364
RETROPLUMIDAE 364
Retusa boenensis 286
  cocillii 286
  concentrica 286
  elegantissima 286
  eumicra 286
  minima 286
  ovulina 286
RETUSIDAE 286
Reussella aculeata 108
  atlantica 108
  costulata 109
  haizumensis 109
  pulchra 109
  simplex 109
  spinosa 109
  spinulosa 109

REUSSELLIDAE 108
Rexea prometheoides 449
    solandri 449
Rhabdamia cypselurus 429
    gracilis 429
Rhabdammina abyssorum 70
    cornuta 70
    discreta 70
    eocenica 70
    fusiformis 70
    neglecta 70
    scabra 70
    triangularis 70
Rhabdamminella cylindrica 70
RHABDAMMINIDAE 70
RHABDITIDA 210
RHABDITIDAE 210
Rhabditis marina 210
Rhabdoblennius ellipes 446
Rhabdochona minjiangensis 210
    synechogobii 210
RHABDOCHONIDAE 210
Rhabdonella amor 59
    conica 59
    elegans 59
    sanyahensis
RHABDONELLIDAE 59
Rhabdonema adriaticum 44
    arcuatum 44
    mirificum 44
    punctatum 44
    sutum 44
Rhabdontus pictus 358
Rhabdosargus sarba 434
Rhabdosoma armatum 332
    brevicaudatum 332
    minor 332
    whitei 332
Rhabdospora sp. 132
Rhabdostyla scyphoides 58
RHADINORHYNCHIDAE 213
RHAMNACEAE 154
Ramphobrachium chuni 226
    diversosetosum 226
Ramphostomella costata 378
    scabra 378
Rhaphidozoum acuferum 62
Rhaphoneis amphiceros 44
    [―― v. *tetragona*] 43
    atlantica 44
    belgica 44
    ―― v. *densestriata* 44
    castracanei 44
    elliptica 44
    lancettula 44
    paralis 44
    rhomoides 44
    surirella 44
    ―― v. *australis* 44
Rhapis excelsa 161
Rhina ancylostoma 407
Rhincalanus cornutus 306
    nasutus 306
Rhincodon typus 405
RHINCODONTIDAE 405

Rhinecanthus aculeatus 459
    echarpe 459
    verrucosus 459
Rhinesomus concatenatus 460
    gibbosus 460
RHINIDAE
RHINOBATIDAE 407
Rhinobatos formosensis 407
    [*granulatus*] 407
    hynnicephalus 407
    schlegeli 407
Rhinobrissus pyramidalis 395
Rhinochimaera pacifica 409
RHINOCHIMAERIDAE 409
Rhinoclavis aspera 267
    fasciata 267
    kochi 267
    sinensis 267
    sordidula 267
    vertagus 267
Rhinomuraena amboinensis 417
    quaesita 417
Rhinopias frondosa 453
RHINOPTERIDAE 409
Rhinopterus hainanica 409
    javanica 409
Rhinothelepus occabus 229
[*Rhipidiphyllin reticulatum*] 146
Rhipidocotyle adbaculum 193
    anguillae 193
    bacullum 193
    clavivesiculum 193
    croceae 193
    pentagonum 193
    xishaensis 193
[*Rhissoplax komaiana*] 238
Rhizammina algaeformis 70
    indivisa 70
RHIZANGIIDAE 185
Rhizocaulus chinensis 172
    verticillatus 172
RHIZOCEPHALA 322
Rhizochilus madreporarum 276
Rhizoclonium hookeri 145
    implexum 145
    riparium 145
Rhizodrilus russus 234
RHIZOLEPADIDAE 322
Rhizolepas gurjanovae 322
Rhizophora apiculata 156
    mucronata 156
    stylosa 156
RHIZOPHORACEAE 156
RHIZOPHYLLIDACEAE 137
Rhizophysa filiformis 173
RHIZOPHYSIDAE 173
Rhizoplecta trithyris 66
Rhizoplegma lychnosphaera 64
RHIZOPODA 69
Rhizopriondon acutus 406
Rhizorus eburnea 286
    kinokuniama 286
    ovulina 286
    radiola 286
    [*rostrata*] 286

Rhizorus (continued)
    tokunagai 286
Rhizosolenia acuminata 35
    alata 35
    ―― f. *curvirostris* 35
    ―― f. *gracillima* 35
    ―― f. *indica* 35
    ―― f. *inermis* 35
    bergonii 35
    calcar-avis 35
    castracanei 35
    ―― v. *rhomboidea* 35
    cleivei 35
    crassispina 35
    cylindrus 35
    delicatula 35
    fragilissima 35
    hebetata 35
    ―― f. *semispina* 35
    hyalina 35
    imbricata 35
    ―― v. *shrubsolei* 35
    robusta 35
    setigera 35
    sinensis 35
    stolterforthii 35
    styliformis 35
    ―― v. *latissima* 35
    ―― v. *longispina* 35
RHIZOSOLENIACEAE 34
RHIZOSOLENIALES 34
Rhizosphaera serrata 64
Rhizospongus arachnoideus 64
RHIZOSTOMATIDAE 176
RHIZOSTOMEAE 176
RHODACARIDAE 298
Rhodacaropsis cheungae 298
[*Rhododermis georgii*] 138
Rhododiscus pulcherrimus 134
Rhodomela confervoides 140
    [*subfusca*] 140
RHODOMELACEAE 139
[*Rhodomonas baltica*] 50
Rhodopeltis borealis 137
Rhodophysema georgii 138
RHODOPHYTA 133
Rhodotorula aurantiaca 129
    glutinis 129
    graminis 129
    minuta 129
    pilimanae 129
    rubra 129
Rhodymenia intricata 138
RHODYMENIACEAE 138
RHODYMENIALES 138
Rhoicosphenia curvata 45
Rhombochirus osteochir 453
RHOMBOGNATHINAE 298
Rhombognathus arenarius 298
    dictyotus 298
    hirtellus 298
    neptunellus 298
    setellus 298
    sinensis 298
    sinensoides 298

Index

Rhombognathus (continued)
   verrucosus 298
Rhopalaea crassa 402
Rhopalastrum cf. *hexaceros* 65
Rhopalione sinensis 327
Rhopalodia gibberula 46
   ―――― v. *vanheurck* 46
   musculus 46
   ―――― v. *constricta* 46
   uncinata 46
Rhopalonema funerarium 173
   velatum 173
RHOPALONEMATIDAE 173
Rhopalophthalmus longipes 324
   orientalis 324
Rhopilema esculentum 176
   hispidum 176
   rhopalophorum 176
Rhus chinensis v. *roxburghii* 154
RHYNCHOBATIDAE 407
Rhynchobatus djiddensis 407
RHYNCHOBDELLIDA 235
Rhynchoconger brevirostris 416
   ectenurus 416
[*Rhynchocymba ectenura*] 416
   nystromi 416
   sivicola 416
Rhynchonerella gracilis 216
   petersii 216
   xishaensis 216
Rhynchopharynx formionis 194
[*Rhynchoplax messor*] 356
   [*setirostris*] 356
[*Rhynchorhamphus georgii*] 419
Rhynchosia volubilis 153
Rhynchospio glutaea 224
Rhynchostracion nasus 460
   rhinorhynchus 460
Rhynchothalestris rufocincta 315
Rhynchozoon globosum 382
   grandicellum 382
   rostratum 382
RHYTIDODIDAE 194
Rhytidodoides cheloniae 194
Richelia intracellularis 26
Rimella canceliata 268
Rimula exquisita 262
[*Ringicula arctata*] 285
   denticulata 285
   doliaris 285
   kurodai 285
   niinoi 285
   orientalis 285
   pilula 285
   propinguan 285
   shenzhenensis 285
   teramachii 285
   yokoyamai 285
RINGICULIDAE 285
[*Risella balteata*] 265
   [*tantillus* v. *subinfracostata*] 265
   [*templiana*] 265
   [―――― v. *nigrofasciata*] 265
   [―――― v. *subimbricata*] 265
RISSOIDAE 265

Rissoina bureri 265
   dunedini 265
Rissolina plicatula 266
Ritteriella amboinensis 401
   picteti 401
Rivularia atra 26
   blasolettiana 26
RIVULARIACEAE 26
Robertsonidra ingens 378
Robertina bradyi 99
   tasmanica 99
   translucens 99
ROBERTINACEA 98
ROBERTINIDAE 98
ROBERTININA 98
Robertinoides bradyi 99
   declivis 99
   normanni 99
Robinia pseudoacacia 153
Robulus asterizans 92
   atlanticus 92
   australis 92
   calcar 92
   convergens 92
   costatus 92
   crassus 92
   cultus 92
   echinatus 92
   expansus 92
   kimituensis 92
   knighti 92
   limbatus 92
   mammilligera 92
   nicobarensis 92
   nitidus 92
   orbicularais 92
   papillosa 92
   pliocenicus 92
   rotulatus 92
   socia 92
   submamilligerus 92
   suborbicularis 92
   tasmanica 93
   thalmanni 93
   tumeyensis 93
   tumidus 93
   vortex 93
Rocella marina 30
Rocinela propodialis 325
Rogadius asper 455
   patriciae 455
Rondeletia loricata 415
RONDELETIIDAE 415
Roperia excentrica 31
   latiovala 31
   tesselata 31
Rosacea plicata 174
ROSACEAE 152
Rosalina australis 112
   bradyi 112
   concinna 112
   crustata 112
   floridana 112
   globularis 112
   neapolitana 112

Rosalina (continued)
   obtusa 112
   opima 112
   orientalis 112
   pacifica 112
   petasiformis 112
   subcomplanata 112
   terquemi 112
   tuberocapitata 112
   vilardeboana 112
   williamsoni 112
ROSALINIDAE 111
Rosaster attenuatus 390
   bipunctus 390
   symbolicus 390
Rosellinia sp. 131
Rosenvingea intricata 142
   Orientalis 142
Roseoporus group 23
Rossia bipapillata 296
   elliptica 41
Rostanga arbutus 290
   [*orientalis*] 290
ROSTANGIDAE 290
Rostratula b. benghalensis 467
ROSTRATULIDAE 467
Rotalia calcarinoides 122
   decipiens 122
   hensoni 122
   microannectens 122
   orientalis 122
   ovata 122
   saxorum 122
   suessonensis 122
   trochidiformis 122
   venusta 122
ROTALIACEA 121
Rotaliammina carinata 76
   chitinosa 76
Rotaliatina buliminoides 120
Rotalidium annectens 123
ROTALIIDAE 121
ROTALIINA 103
Rotalinoides gaimardii 124
ROTIFERA 214
Rouleina guentheri 413
   watasei 413
Roxasia cervicornis 188
RUBIACEAE 158
Rubus parvifolius 152
Rudarius ercodes 460
Rudigaudryina inepta 78
Ruditapes philippinarum 255
   variegata 255
Rugidia corticata 110
Rullierinereis elytrocirra 221
   misakiensis 221
Rumex japonicus 151
[*Runcina setoensis*]
RUNCINIDAE 287
Rupertianella rupertiana 89
Rupertina sinensis 116
[*Ruppia maritima*] 158
   rostellata 158
RUTACEAE 153

RUTILARIACEAE 35
Ruvettus pretiosus 449
    tydemani 449
Rzehakina epigona 71
RZEHAKINACEA 71
RZEHAKINIDAE 71

S

Sabatieria alata 212
    ancudiana 212
    parabyssalis 212
    pulchra 212
    punctata 212
    trivialis 212
[*Sabella pavonina*] 230
    penicillus 230
Sabellaria alcocki 228
    cementarium 228
    ishikawai 228
SABELLARIDAE 228
Sabellastarte indica 230
    sanctijosephi 230
    zebuensis 230
SABELLIDA 229
SABELLIDAE 229
Sabellides octocirrata
Sabia conica 268
Saccammina anglica 70
    atlantica 70
    huanghaiensis 70
    minuta 70
    sphaerica 70
SACCAMMINIDAE 70
Saccaminoides subcarpathicus 74
Saccardoella marinospora 131
Saccella confusa 239
    cuspidata 239
    gordonis 239
    sematensis 239
Saccharomyces cerevisiae 129
    kluyveri 129
SACCHAROMYCETACEAE 129
Saccharum arundinaceum 160
    spontaneum 160
SACCOCIRRIDAE 214
Saccocirrus gabriellae 214
    [*major*] 214
Saccogaster tubercularis 421
Saccoglossus hulangtauensis 399
Saccorhiza ramosa 71
Sacculina confragosa 322
SACCULINIDAE 322
Saccostrea cucullata 246
    echinata 246
SACOGLOSSA 288
Sacura margaritacea 427
Sagartia carcinophilus 180
    rosea 180
SAGARTIIDAE 180
Sagenina divaricans 70
Sageretia thea 154
    [*theezans*] 154
Sagitella kowalevskii 216
Sagitta bedfordii 385
    bedoti 385

Sagitta (continued)
    bipunctata 385
    bruani 385
    crassa 385
    ———— *forma naikaiensis* 385
    decipiens 385
    delicata 385
    enflata 385
    ferox 385
    hexaptera 385
    johorensis 385
    lyra 385
    macrocephala 385
    minima 385
    nagae 385
    neglecta 385
    neodecipiens 385
    oceania 385
    pacifica 385
    planctonis 385
    pseudoserratodentata 385
    pulchra 385
    regularis 385
    robusta 385
    septata 385
    tenuis 385
    tokiokai 385
    zetesios 385
SAGITTIDAE 385
SAGITTOIDEA 385
Saidovina karreriana 106
Sakuraeolis enosimensis 292
SALANGIDAE 411
Salanx acuticeps 411
    ariakensis 411
    cuvieri 411
Salarias dussumieri 446
    edentulus 446
    fasciatus 446
    guttatus 446
    lineatus 446
    margaritatus 446
    periophthalmus 447
Salicornia europaea 151
SALIENTIA 463
Salmaciella dussumieri 393
Salmacina dysteri 230
Salmacis bicolor rarispina 393
    ———— *typica* 393
    sphaeyoides variegata 393
    virgulata 393
Salmoneus serratidigitus 343
    sibogae 343
SALMONIFORMES 411
Saloptia powelli 427
[*Salpa aspera*] 401
    [*cylindrica*] 401
    fusiformis 401
    ———— *aspera* 401
    maxima 401
    ———— *tuberculata*
SALPIDAE 401
Salsola collina 151
    komarovii 151
    pestifer 151
SALVADORACEAE 154

Samacar pacifica 240
Samaris cristatus 457
Samariscus filipectoralis 457
    huysmani 457
    inornatus 457
    latus 458
    longimanus 458
Samytha californiensis 229
    gurjanovae 229
    oculata 229
    sexcirrata 229
    sinica 229
Sandalia rhodia 271
Sandalolitha robusta 182
Sandalops melancholicus 295
Sanderia malayensis 176
[*Sanguinolaria ambigua*] 252
    [*atrata*] 252
    [*castanea*] 252
    [*chinensis*] 252
    [*diphos*] 252
    [*inflata*] 252
    [*minor*] 252
    [*olivacea*] 252
    [*planulata*] 252
    tchangsii 252
    [*violacea*] 252
    [*virescens*] 252
Sanya equulus 316
SAPINDACEAE 154
Sapphirina angusta 313
    auronitens 313
    bicuspidata 313
    darwinii 313
    gastrica 313
    gemma 313
    intestinata 313
    iris 313
    lactens 313
    maculosa 313
    metallina 313
    nigromaculata 313
    opalina 313
    ovatolanceolata 313
    scarlata 313
    sinuicauda 313
    stellata 313
SAPPHIRINIDAE 313
Saracenaria acutauricularis 93
    angularis 93
    hannoverana 93
    italica 93
    latifrons 93
    limbata 93
    midwayensis 93
    perforata 93
    schencki 93
Sarcina 23
SARCODIACEAE 137
SARCODINA
SARCOMASTIGOPHORA 61
Sarconema filiforme 137
    gracilarioides 137
Sarcophyton acutangulum 188
    boletiforme 188
    crassocaule 188

Index

Sarcophyton (continued)
  elegans 188
  furcatum 188
  glaucum 188
  infundibuforme 188
  latum 188
  molle 188
  trocheliophrum 188
Sarcostemma acidum 157
Sarda orientalis 450
[*Sardinella albella*] 410
  aurita 410
  brachysoma 410
  clupeoides 410
  fimbriata 410
  [*gibbosa*] 410
  hualensis 410
  jussieu 410
  melanura 410
  nymphaea 410
  [*nymphaea*] 410
  perforata 410
  richardsoni 410
  sindensis 410
  sirm 410
  zunasi 410
Sardinops sagax melanosticta 410
Sarepta speciosa 239
SARGASSACEAE 143
Sargassum angustifolium 143
  assimile 143
  cinereum 143
  confusum 143
  crassifolium 143
  dazhouensis 143
  dotyi 143
  duplicatum 143
  emarginatum 143
  [*enerve*] 143
  fulvellum 143
  fusiforme 143
  glaucescens 143
  graminifolium 143
  hemiphyllum 143
  henslowianum 143
  herklotsii 143
  horneri 143
  ilicifolium 143
  [*kjellmaniamum*] 143
  kuetzingii 143
  laxifolium 143
  longifructum 143
  [*macrocarpum*] 144
  maclurei 143
  megalocystum 143
  microcanthum 143
  miyabei 143
  naozhouense 143
  nigrifoloides 143
  [*pallidum*] 143
  parvivesiculosum 143
  patens 143
  phyllocystum 143
  ploycystum 143
  [*serratifolia*] 144
  siliquastrum 144

Sargassum (continued)
  subtilissimum 144
  swartzii 144
  tenerrimum 144
  thunbergii 144
  [*tortile*] 144
  turbinatifolium 144
  vachellianum 144
  xishaense 144
[*Sargocentron caudimaculatus*] 422
  [*cornutus*] 422
  [*diadema*] 422
  ittodai 422
  [*lacteoguttatus*] 422
  melanospilos 422
  praslin 422
  rubrum 422
  spiniferum 422
  [*spinosissimus*] 422
  tiere 422
  [*tiere*] 422
Saron mammoratus 344
Sarsia nipponica 168
  [*resplendens*]
SARSIELLIDAE 302
[*Sarsiflastra anatina*]
Sasakiella cruciformis 176
  tsingtaoensis 176
Sassia semitorta 273
Saturnalis circularis 63
Satyrichthys amiscus 455
  fowleri 455
  piercei 455
  rieffeli 455
  welchi 455
[*Saundersella saxicola*] 141
[*Saurenchelys fierasfer*] 416
Saurida elongata 413
  filamentosa 413
  gracilis 413
  tumbil 413
  undosquamis 413
  wanieso 413
Savigniella lafontii 377
SAVIGNIELLIDAE 377
Savignium crenatum 320
  dentatum 320
  elongatum 321
  milleporae 321
  orientale 321
Savoryella lignicola 131
  paucispora 131
Saxidomus purpurata 255
Sayonara satsumae 427
Scaevola hainanensis 158
  sericea 158
[*Scala acuminata*] 267
Scalibregma inflatum 226
SCALIBREGMIDAE 226
Scalisetosus fragilis 226
  longicirrus 226
Scalopidia spinosipes 363
SCALPELLIDAE 318
[*Scalpellum condensum*] 318
  rubrum 318
  stearnsii 318

[SCAPHANDERIDAE] 286
Scapharca anomala 240
  broughtonii 240
  cornea 240
  globosa 240
  gubernaculum 241
  inaequivalvis 241
  indica 241
  japonica 241
  labiosa 241
  satowi 241
  subcrenata 241
  troscheli 241
  vellicata 241
Scaphechinus mirabilis 394
Scaphocalanus brevicornis 308
  echinatus 308
  longifurca 308
  magnus 308
  major 308
Scapholeberis kingi 300
SCAPHOPODA 261
Scapophyllia cylindrica 184
Scaptognathides hawaiiensis 298
Scaptognathus triunguis 298
SCARIDAE 441
Scarops festivus 442
  forsteni 442
  prasiognathos 442
  psittacus 442
  rivulatus 442
  rubroviolaccus 442
  schlegeli 442
Scartela emarginatus 447
Scartelaos gigas 453
  viridis 453
Scarus aeruginosus 441
  atropectoralis 441
  bowersi 441
  chlorodon 441
  dimidiatus 441
  [*dussumieri*] 441
  [*erythrodon*] 442
  fasciatus 441
  forsteri 441
  frenatus 441
  ghobban 441
  gibbus 441
  globiceps 442
  janthochir 442
  lepidus 442
  longiceps 442
  lunula 442
  niger 442
  oedema 442
  oviceps 442
  ovifrons 442
  pyrrhurus 442
  scaber 442
  sordidus 442
  taeniurus 442
  venosus 442
SCATOPHAGIDAE 436
Scatophagus argus 436
SCENEDESMACEAE 147
Scenedesmus brasilieasis 147

Scenedesmus (continued)
   javaensis 147
   quadricauda 147
Schackoinella globosa 112
   lepida 112
Schedophilus maculatus 450
Schenckiella victoriensis 78
Schikhobalotrema acuta 193
Schindleria pictschmanni 447
   praematura 447
SCHINDLERIIDAE 447
Schistocomus hiltoni 229
   sovjeticus 229
Schistomeringos incerta 228
   japonica 228
   rudolphi 228
Schizaea dichotoma 150
SCHIZAEACEAE 150
Schizaster lacunosus 394
SCHIZASTERIDAE 394
Schizocerca diversicornis 214
SCHIZOCHITONIDAE 238
Schizomavella australis 379
   granulosa 379
   pacifica 379
   porata 379
   umbonatoidea 379
Schizophrys aspera 357
Schizoporella biaperta 379
   [*errata*] 379
   erratoidea 379
   tumulosa 380
   unicornis 380
SCHIZOPORELLIDAE 379
SCHIZOPORELLOIDEA 378
Schizoretepora tumescens 382
Schizoscelus ornatus 333
Schizothrix lacustris 26
Schizymenia dubyi 136
Schlumbergerina alveoliniformis 83
   occidentalis 83
Schmackeria dubia 309
   forbesi 309
   inopinus 309
   poplesia 309
Schroederella delicatula 30
   —— f. *schroederi* 30
   [*schroederi*] 30
Schwantzia elegantissima 120
[*Sciaena russelli*] 432
SCIAENIDAE 431
Scina borealis 330
   crassicornis 330
   curvidactyla 330
   incerta 330
   latifrons 330
   nana 330
   similis 330
   submarginata 330
   tullbergi 330
Scinaia boergesenii 134
   [*cottonii*] 134
   japonica 134
   latifroms 134
   tsinglanensis 134

SCINIDAE 330
Scintilla cf. *cuvieri* 248
   nitidella 248
   cf. *opalinus*
Scintillona brissae 248
Scionella japonica 229
Scirpearia erythraea 187
   gracilis 187
Scirpus mariqueter 161
   planiculmis 161
   tabernaemontani 161
   triqueter 161
SCLERACTINIA 181
Sclerodactula meltipes 387
Scleronephthya corymbosa 188
   pustulosa 188
Scobatus granulatus 407
   halavi 407
Scolecithricella abyssalis 308
   altera 308
   arcuata 308
   auropecten 308
   bradyi 308
   ctenopus 308
   dentata 308
   fowleri 308
   gracilis 308
   longispinosa 308
   minor 308
   ovata 308
   tenuiserrata 308
   timida 308
   tropica 308
   valens 308
   vittata 308
SCOLECITHRICIDAE 308
Scolecithrix danae 308
   nicobarica 309
Scolelepis globosa 224
   lefebvrei 224
   squamata 224
Scoliodon dumerili 406
   laticaudus 406
   palasorrah 406
   sorrakowah 406
   walbeehmi 406
Scolionema suvaense 172
Scoliopleura tumida 41
SCOLOPACIDAE 467
Scolopax r. rusticola 467
Scolopia chinensis 155
   hainanensis 155
   henryi 155
Scoloplos acmeceps 223
   armiger 223
   chrysochaeta 223
   dubia 223
   gracilis 223
   johnstonei 223
   marsupialis 223
   rubra 223
   —— *orientalis* 223
   [—— *pacifica*] 223
   spiniferus 223
   tumidus 223

Scoloplos (continued)
   uniramus 223
   [*uschakovi*] 223
SCOLOPSIDAE 434
Scolopsis affinis 434
   bilineatus 435
   bimaculatus 435
   cancellatus 435
   ciliatus 435
   eriomma 435
   inermis 435
   lineatus 435
   margaritifer 435
   monogramma 435
   taeniopterus 435
   temporalis 435
   vosmeri 435
   xenochrous 435
Scomber australasicus 449
   [*japonicus*] 449
   [*Scombermorus australasicus*] 449
   cavalla 449
   commersoni 449
   guttatus 449
   koreana 449
   niphonius 449
   sinensis 450
Scomberoides commersonianus 431
   [*lysan*] 430
   [*orientalis*] 430
   tol 431
SCOMBRESOCIDAE 418
SCOMBRIDAE 449
SCOMBROPIDAE 427
Scombrops boops 427
SCOPELARCHIDAE 414
Scopelarchoides danae 414
   nicholsi 414
Scopelarchus analis 414
   guentheri 414
Scopelengys tristis 413
Scopeloberyx opisthopterus 421
   robustus 421
Scopelogadus mizolepis 421
Scopimera bitympana 365
   curtelsoma 365
   globosa 365
   longidactyla 365
   tuberculata 365
[*Scorpaena bynoensis*] 454
   hatizyoensis 453
   izensis 453
   neglecta 453
   [*onaria*] 453
SCORPAENIDAE 453
SCORPAENIFORMES 453
Scorpaenodes guamensis 453
   hirsutus 453
   [*kelloggi*] 453
   littoralis 453
   minor 454
   parvipinnis 454
   scabra 454
   varipinnis 454
Scorpaenopsis cirrhosa 454

Scorpaenopsis (continued)
    diabolus 454
    gibbosa 454
    neglecta 454
SCORPIDAE 436
Scorzonera mongolica v. *putjatae* 158
Scottiella crispata 302
Scottocalanus australis 309
    farrani 309
    helenae 309
    persecans 309
    rotundatus 309
    securifrons 309
    sedatus 309
    thomasi 309
SCROPHULARIACEAE 158
Scruparia chelata 371
SCRUPARIIDAE 371
SCRUPARIINA 371
SCRUPARIOIDEA 371
Scrupocellaria californica 375
    circulata 375
    curvata 375
    delilii 375
    diadema 375
    [*diegensis*] 375
    ferox 375
    inarmata 375
    maderensis 376
    meijensis 376
    nanshaensis 376
    obtecta 376
    obtectoidea 376
    [*pinigera*] 375
    scabra 376
    [*scrupea*] 375
    seculifera 376
    spatulata 376
    spatulatoidea 376
    unicornis 376
    uniseriata 376
Scutarcopagia scobinata 251
Scutellaria strigillosa 157
SCUTELLIDAE 394
SCUTICOCILIATIDA 57
Scutuloris oblongus 88
    patens 88
    tegminis 88
    translucens 88
Scutus emarginatus 262
    sinensis 262
    unguis 262
SCYLIORHINIDAE 405
Scyliorhinus torazame 405
Scylla serrata 359
Scyllaea pelagica 292
SCYLLAEIDAE 292
SCYLLARIDAE 346
Scyllarides haanii 346
    squamosus 346
[*Scyllarus antarcticus*] 346
    batei 346
    bertholdi 346
    brevicornis 346
    [*ciliatus*] 346

Scyllarus (continued)
    cultrifer 346
    formosanus 346
    [*haahii*] 346
    kitanoviriosus 346
    longidactylus 346
    martensii 346
    rugosus 346
    sordidus 346
    [*squamosus*] 346
    tuberculatus 347
Scymnodon niger 407
    squamulosus 407
Scypha coronatum 165
SCYPHACIDAE 326
Scyphiphora hydrophyllacea 158
SCYPHOMEDUSAE 175
Scyra compressipes 357
Scytonema javanicum 26
    polycystum 26
    rivulare 26
SCYTONEMATACEAE 26
Scytosiphon dotyo 142
    lomentarius 142
SCYTOSIPHONACEAE 142
SCYTOSIPHONALES 142
Sebastapistes albobrunnea 454
    bynoensis 454
    megalepis 454
    nuchalis 454
    vachelli 454
Sebastes hubbsi 454
    intermis 454
    itinus 454
    joyneri 454
    nigricans 454
    nivosus 454
    pachycephalus 454
    schlegeli 454
    thompsoni 454
    trivittatus 454
Sebastiscus albofasciatus 454
    marmoratus 454
    tertius 454
Sebdenia agardhii 137
SEBDENIACEAE 137
[*Secutor insidiator*] 432
    [*ruconius*] 432
Sejunctella laticarinina 81
Selaginopsis trilateralis 172
Selar boops 431
    crumenophthalmus 431
Selaroides leptolepis 431
SELENARIIDAE
Selenanthias analis 427
SELENORIBATIDAE 298
SEMAEOSTOMEAE 176
SEMANTIDIDAE 66
Semele cordiformis 252
    crenulata 252
    scabra 252
    [*sinensis*] 252
    zebuensis 252
SEMELIDAE 252
[*Semicassis japonica*] 273

Semicassis (continued)
    [*pila*] 273
Semipallium fulvicostum 245
    spectabilis 245
    tigris 245
    xishaensis 245
[*Semivertagus nesioticus*] 267
Semnoderes sp. 209
Semperella cucumis 165
SEMPERELLIDAE 165
Semperina brunnea 186
Senecio pseudo-arnica 158
Separatista helicoides 268
Separogermiductus sinicus 203
Sepia aculeata 295
    andreana 295
    brevimana 295
    elliptica 295
    esculenta 295
    [*hercules*] 295
    kobiensis 295
    latimanus 295
    longipes 295
    lycidas 295
    nanshiensis 295
    omani 296
    pharaonis 296
    recurvirostra 296
    robsoni 296
    [*subaculata*] 295
    tenuipes 296
    [*tigris*] 296
    torosa 296
Sepiadarium kochii 296
[*Sepiella japonica*] 296
    maindroni 296
SEPIIDAE 295
SEPIOIDEA 295
Sepiola birostrat 296
SEPIOLIDAE 296
Sepioteuthis lessoniana 295
Septaria janelli 265
    porcellana 265
Septifer bilocularis 243
    excisus 243
    keenae 243
    pulcher 243
    virgatus 243
    xishaensis 243
Septotextularia rugulosa 80
Septotrochammina plicata 76
Ser fukiensis 363
Sergestes gardineri 338
    orientalis 338
    talismani 338
    [*serrulatus*] 338
SERGESTIDAE 338
SERGESTIOIDEA 338
Seriatopora angulata 181
    caliendrum 181
    hystrix 181
    stellata 181
Seriola aureovittata 431
    dumerili 431
    quinqueradiata 431

[*Seriolina nigrofasciata*] 431
SERPENTIFORMES 464
Serpula sinica 230
   vermicularis 230
   watsoni 230
SERPULIDAE 230
Serpulorbis imbricata 266
SERRANIDAE 425
Serranocrrhitus latus 427
Serrasentis longus 213
   sagittifer 213
Serratovola tricarnatus 245
Serrivomer beani 417
SERRIVOMERIDAE 417
Sertella suluensis 382
Sertularella gayi 172
   inabai 172
   indivisa bidentata 172
   mirabilis 172
   miurrensis 172
   philippinensis 172
   sinensis 172
   tenella 172
Sertularia cornicina 172
   distans 172
   fabricii 172
   furcata 172
   similis 172
   [*tubuliformis*] 171
   westindica 172
SERTULARIIDAE 171
Sesarma bidens 366
   dehaani 366
   exquisitum 366
   fasciata 366
   [*gordonae*] 366
   haematocheir 366
   impressum 366
   pictum 366
   plicata 366
   sinensis 366
   tangi 366
   tripectinis 366
Sesbania cannabina 153
Sestronophora arnoldi 111
Sesuvium portulacastrum 152
Setarches fidjiensis 454
   longimanus 454
Setella gracilis 314
Sethoconus myxobrachia 67
Sethocorys odysseus 68
SETHOCYRTIDAE 67
Sethodiscus lenticula 65
   macrococcus 65
Sethophormis pentalactis 67
Sethopyramis quadrata 67
[*Setipinna giberti*] 410
   taty 410
Setosabatieria fibulata 212
   hilarula 212
Setosellina constricta 373
SETOSELLINIDAE 373
[*Sewara koreana*] 449
   [*niphonia*] 449
Shenius anomalus 365

Shinanoeolis emurai 292
Sicyonia cristata 338
   curvirostris 338
   formosa 338
   japonica 338
   lancifer 338
   [——— v. *japonica*] 338
   longicauda 338
   ommanneyi 338
SICYONIDAE 338
Sicyopterus micrurus 452
   [*japonicus*] 452
Sida alnifolia v. *microphylla* 155
   chinensis 155
   cordifolia 155
   corylifolia 155
   crystallina 300
   rhombifolia 155
   szeckuensis 155
SIDERASTREIDAE 183
Siderea picta 417
   thyrsoidea 417
SIDIDAE 300
Sidonops acanthostyles 164
Sigalion asiatica 218
   mathildae 218
   spinosa 218
SIGALIONIDAE 217
Sigambra bassi 218
   hanaokai 218
   tentaculata 218
SIGANIDAE 448
Siganus argenteus 448
   canaliculatus 448
   chrysospilos 448
   corallinus 448
   fuscescens 448
   guttatus 448
   javus 448
   oramin 448
   puellus 448
   rostratus 448
   spinus 448
   virgatus 448
   vulpinus 448
[*Sigaretus papillus*] 269
   [*undulatus*] 269
Sigmadocia symbiotica 165
Sigmamiliolinella australis 86
   denticollaris 86
Sigmavirgulina basistriata 109
   lanceolata 109
   tortuosa 109
Sigmoidella bacomensis 96
   elegantissima 96
   kagaensis 96
   pacifica 96
   subtaiwanica 96
   taiwanensis 96
Sigmoihauerina bradyi 88
   fragilissima 88
Sigmoilina amygdaloides 88
   arenaria 88
   elliptica 88
   foliacea 88

Sigmoilina (continued)
   formosana 88
   inculta 88
   minutissima 88
   obesa 88
   schlumbergeri 88
   sigmoidea 88
   subtenuis 89
   tenuis 89
   tenuissima 89
   syrtica 89
   victoriensis 89
Sigmoilinella tortuosa 89
Sigmoilinita asperula 89
   granulifera 89
   tenuis 89
Sigmoilopsis asperula 89
   carinata 89
   chapmani 89
   finlayi 89
   flintii 89
   herzensteini 89
   moyi 89
   obesa 89
   orientalis 89
   schlumbergeri 89
Sigmomorphina basistriata 96
   gallowayi 96
   ozawai 96
   semitecta 96
   subcircularis 96
   trilocularis 96
   undulosa 96
   williamsoni 96
   yokoyamai 96
Sigmopyrgo vespertilio 89
SILICOFLAGELLATALES 49
Silicosigmoilina calcareoarenacea 71
   californica 71
   elegantissima 71
Siliqua fasciata 253
   grayana 253
   minima 253
   pulchella 253
   radiata 253
Siliquaria anguina 266
   cumingii 266
SILIQUARIIDAE 266
SILLAGINIDAE 429
Sillago asiatica 429
   chondropus 429
   ingenuua 429
   japonica 429
   maculata 429
   microps 429
   parvisquamis 429
   sihama 429
SILURIFORMES 418
SIMARUBACEAE 153
Simenchelys parasiticus 417
Simnialena formicaria 271
Simocephalus vetulus 300
SIMOGNATHINAE 298
Simognathus fovealatus 298
Simorhychotus antennarius 332

Index

Simorhychotus (continued)
    [*lilljeborgi*] 332
Sinaechinocyamus mai 394
    planus 394
Sinocalanus laevidactylus 309
    sinensis 309
    solstitialis 309
    tenellus 309
[*Sinocaligus denticulatus*] 316
Sinoediceros homopalmulus 330
Sinoferdina gigantea 391
Sinoflustra amoyensis 373
    annae 373
Sinolorica scissurata 238
Sinolyonsia sinica
Sinometra acuticirra 387
Sinonereis heteropoda 221
Sinonovacula constricta 253
Sinophiura multispina 398
Sinopontius aesthetascus 315
    punctatus 315
Sinopora sinica 374
Sinosiphonia elegans 140
Sinosquilla hispida 368
    sinica 368
Sinularia capillosa 188
    compressa 188
    corpulenta 188
    fibrillosa 188
    granosa 188
    inexplicita 188
    microclavata 188
    monstrosa 188
    nanolobata 188
    papillosa 188
    partia 188
    polydactyla 188
    prattae 188
    quarciformis 188
    ramulosa 188
    renei 188
    tenella 188
Sinuloculina cyclostoma 86
    lunata 86
    subglobiformis 86
    subsphaeroides 86
Sinum incisum 270
    japonicus 270
    javanicum 270
    cf. *neritoideum* 270
    oblongum 270
    papilla 270
    planulatum 270
Sinupetraliella umbonata 380
Siphamia fuscolineata 429
    majimai 429
    versicolor 429
Siphobigenerina compressa 80
Siphocampe corbula 68
Siphodera vinaledwardsi 199
Siphoderina asiatica 199
Siphoderooides lutiani 199
Siphogaudryina huanghaiensis 77
Siphogenerina bifrons 106
    pacifica 106

Siphogenerina (continued)
    raphana 106
    striata 106
    virgula 106
SIPHOGENERINOIDIDAE 106
Siphonalia fusoides 277
    spadicea 277
    subdilatata 277
Siphonaperta agglutinans 83
    agglutinata 83
    ammophila 83
    crassatina 83
    formosana 83
    inculta 83
    macbeathi 83
    petrophila 83
    prominentis 83
Siphonaria atra 294
    japonica 294
    laciniosa 294
    sirius 294
SIPHONARIIDAE 294
Siphoniferoides siphonifera 80
Siphonina australis 113
    bradyana 113
    pulchra 113
    reticulata 113
    tubulosa 113
SIPHONINACEA 113
SIPHONINIDAE 113
Siphoninoides echinatus 113
    glabra 113
Siphonium maximum 266
Siphonochalina flexa 165
    intermedia 165
SIPHONOCLADACEAE 145
SIPHONOCLADIALES 145
Siphonocladus tropicus 146
    xishaensis 146
SIPHONODENTALIIDAE 261
Siphonodentalium japonica 261
Siphonodosaria insoluta 110
Siphonogorgia cylindrata 188
    gracilis 188
    variabilis 189
SIPHONOLAIMIDAE 211
Siphonolaimus longicaudata 211
SIPHONOPHORAE 173
Siphonosoma australe 236
    cumanense 236
    [*edule*] 236
    [*formosa*] 236
    funatuti 236
    [*nanhaiensis*] 236
    [*pescadolense*] 236
    rotumanum 236
Siphonosphaera ardys 63
    donghaiense 63
    marginata 63
    martensi 63
    pericyclis 63
    polypora 63
    polysiphonia 63
    socialis 63
    tenera 63

SIPHONOSTOMATOIDA 315
Siphopatella walshi 268
Siphoscutula leroyi 79
    pacifica 79
Siphotextularia carinata 79
    concava 79
    cordis 79
    crassaformis 79
    crassisepta 80
    crispata 80
    curta 80
    differens 80
    flintii 80
    foliosa 80
    glabrata 80
    heterostoma 80
    masudai 80
    mestayerae 80
    miniacea 80
    miocenica 80
    obesa 80
    pacifica 80
    philippinensis 80
    pseudoconcava 80
    pulchra 80
    saulcyana 80
    subplana 80
    subplanoides 80
    wairoana 80
Siphouvigerina ampullacea 107
    interrupta 107
    porrecta 107
    proboscidea 107
    pseudoampullacea 107
    succincta 107
SIPUNCULA 235
SIPUNCULIDAE 236
SIPUNCULIDEA 236
SIPUNCULIFORMES 236
[*Sipunculus angasii*] 236
    angasoides 236
    indicus 236
    norvegicus 236
    nudus 236
    robustus 236
Siratus alabaster 276
Sirembo imberbis 421
    marmoratum 421
SIRENIA 469
Siriella aequiremis 324
    dubia 324
    gracilis 324
    japonica 324
    ———— *izuensis* 324
    media 324
    okadai 324
    quadrispinosa 324
    sinensis 324
    thompsoni 324
    trispina 324
    wadai 324
Sirocoleus kurzii 26
Skeletonema costatum 30
    munzelii 30
    potamos 30

Skeletonema (continued)
   tropicum 30
Skogsbergia crenulata 302
   curvata 302
   minuta 302
Skrjabinopsolus sanyaensis 195
Slandenis remiger 461
Smilium acutum 318
   scorpio 318
   sinense 318
Smithea eurygaster 173
Smithsonidrilus minusculus 234
[*Smittina aviculata*] 380
   landsborovii 380
SMITTINIDAE 380
Smittipora abyssicola 376
   cordiformis 376
   harmeriana 376
Smittoidea levis 380
   pacifica 380
   prolifera 380
   reticulata 380
   [*reticulata*] 380
   spinigera 380
Smoutina corpuscula 122
Snyderina yamanokami 454
[*Soa lagenula*] 286
Socarnes dissimulantia 329
Solamen spectabilis 243
SOLANACEAE 157
Solaster dawsoni 391
SOLASTERIDAE 391
Solea ovata 458
SOLECURTIDAE 253
Solecurtus divaricatus 253
   exaratus 253
   wilsoni 253
Solegnathus hardwicki 424
Soleichthys heterorhinos 458
SOLEIDAE 458
SOLEMYIDAE 239
SOLEMYOIDA 239
Solen arcuatus 253
   canaliculatus 253
   dunkerianus 253
   gordonis 253
   [*gouldii*] 253
   gracilis 253
   grandis 253
   linearis 253
   roseomaculatus 253
   sloanii 253
   strictus 253
SOLENIDAE 253
Solenocera alticarinata 336
   [*brevipes*] 336
   [*choprai*] 336
   comata 336
   crassicornis 336
   [*depressa*] 336
   [*distincta*] 336
   faxoni 336
   [*indica*] 336
   koelbeli 336
   [*kuboi*] 336

Solenocera (continued)
   melantho 336
   pectinata 336
   pectinulata 336
   [*prominentis*] 336
   rathbunae 336
   [*rathbuni*] 336
   [*sinensis*] 336
   [*subnuda*] 336
SOLENOCERIDAE 336
[*Solenocurtus divaricatus*] 253
   [*exaratus*] 253
Solenosphaera pandora 63
   zanguebarica 63
SOLENOSTOMIDAE 423
Solenostomus armatus 423
   cyanopterus 423
   paegnius 423
   paradoxus 424
Solidicorbula erythrodon 256
   tunicata 256
Solidobalanus cedaricola 321
   ciliatus 321
   socialis 321
[*Solieria mollis*] 137
   pacifica 137
   [*robusta*] 137
   tenuis 137
SOLIERIACEAE 137
Solmaris flavecens 173
   leucostyla 173
SOLMARISIDAE 173
Solmissus marshalli 173
Solmundella bitentaculata 173
Sonneratia alba 156
   caseolaria 156
   hainanensis 156
   ovata 156
SONNERATIACEAE 156
Soreuma gibbulosum 66
SOREUMIDAE 66
SORITACEA 90
Sorites marginalis 90
   orbiculus 90
SORITIDAE 90
SOROCARPACEAE 141
[*Sorocarpus micromorus*] 141
   pacifica 141
   [*uvaetormis*] 141
[*Sorsogon tuberculatus*] 455
Sousa chinensis 469
   plumbea 469
SPARIDAE 434
Spartina alterniflora 160
   anglica 160
Sparus berda 434
   latus 434
   macrocephalus 434
   [*sarba*] 434
SPATANGIDAE 394
SPATANGOIDA 394
Spatoglossum dichotomum 142
   pacificum 142
   solieri 142
Spergularia salina 152

SPERMATOCHNACEAE 141
[*Spermothammion suychiroi*] 139
   [*yonakuniensis*] 138
Sphacelaria carolinensis 142
   [*furcigera*] 143
   fusca 142
   novae-hollandiae 142
   rigidula 143
   sufrusca 143
   tribuloides 143
   variabilis 143
SPHACELARIACEAE 142
SPHACELARIALES 142
Sphaeramia nematoptera 429
   orbicularis 429
   [*orbicularis*] 428
SPHAERAMMINIDAE 74
SPHAERASTERIDAE 391
Sphaeridia papillata 111
SPHAEROCOCCACEAE 137
Sphaerogypsina globulus 116
SPHAEROIDEA 62
Sphaeroidina bulloides 112
   chilostomata 112
   variabilis 112
Sphaeroidinella dehiscens 103
Sphaeroidinellopsis seminulina 103
   subdehiscens 103
SPHAEROIDINIDAE 112
SPHAEROLAIMIDAE 211
Sphaerolaimus dispar 211
   pacifica 211
Sphaeroma retrolaevis 326
   walkeri 326
SPHAEROMIDAE 325
Sphaeronectes gracilis 175
   [*truncata*] 175
SPHAERONECTIDAE 175
Sphaeropyle mespilus 62
Sphaerosyllis chinensis 219
   erinaceus 219
   glandulata 219
   hirsuta 219
   hystrix 219
   longicauda 219
   periferopsis 219
   pirifera 219
Sphaerothuria bitentaculata
Sphaerotrichia firma 141
Sphaerozius nitidus 362
SPHAEROZOIDAE 62
Sphaerozoum fuscum 62
   punctatum 62
   cf. *strigulosum* 62
   verticillum 62
Spheractis cheungae 180
[*Spheroides alboplumbeus*] 461
   [*obscurus*] 461
   [*ocellatus*] 461
   [*vermicularis*] 461
   [*xanthopterus*]
Sphincetrostoma japonicus 197
Sphoeroides pachygaster 461
Sphyraena acutipinnis 424
   barracuda 424

Sphyraena (continued)
  flavicauda 424
  forsteri 424
  helleri 424
  japonica 424
  jello 424
  nigripinnis 424
  obtusata 424
  pinguis 424
  putnamiae 424
SPHYRAENIDAE 424
Sphyrna blochi 406
  lewini 406
  mokarran 406
  zygaena 406
SPHYRNIDAE 406
Spilophorella paradoxa 211
Spinapsaron barbatus 445
  formosensis 445
Spincterules costatus 93
Spinicapitichthys draconis 448
Spinifex littoreus 160
Spiniplicatula muricata 245
Spiniscala japonica 267
Spinocalanus angusticeps 307
  horridus 307
  magnus 307
  oligospinosus 307
  spinosus 307
Spinoecia crassispina 305
  parthenoda 305
  pseudoparthenoda 305
Spinula tasmanica 239
SPINULOSA 391
Spio filicornis 224
  martinensis 224
Spiochaetopterus costarum 224
SPIONIDA 223
SPIONIDAE 223
Spiophanes bombyx 224
  soederstromi 224
[SPIRALELLIDAE] 288
Spirastrella aurivilli 164
  cuntarix 164
  semilunaris 164
  vagabunda 164
SPIRASTRELLIDAE 164
Spiraulax jollifei 54
Spirema haliomma 66
Spirillina compressa 81
  grosseperforata 81
  limbata 81
  mediospinosa 81
  minima 81
  pectinimarginata 81
  planoconcava 81
  porisuturalis 81
  scalaris 81
  tuberculata 81
  vivipara 81
SPIRILLINIDAE 81
SPIRILLININA 81
Spirobranchus giganteus 230
  jousseaumie 230
  latiscapus 230

Spirobranchus (continued)
  maldivensis 230
  polytrema 230
  semperi 230
  sinensis 231
  tetraceros 231
  tricornis 231
  [*tricornigerus*] 231
Spirocamallanus pereirai 209
Spirocyrtis scalaris 68
  submerospira 68
Spirolina acicularis 90
  arietina 90
  mariei 90
Spirolingulina carlofortensis 94
  polymorpha 94
Spiroloculina affixa 82
  anderseni 82
  angulata 82
  antillarum 82
  arenaria 82
  bicarinata 82
  biformis 82
  bohaiensis 82
  cavernosa 82
  clara 82
  communis 82
  concava 82
  corrugata 82
  dentata 82
  depressa 82
  elongata 82
  excavata 82
  excisa 82
  eximia 82
  foveolata 82
  hadai 82
  henbesti 82
  huanghaiensis 82
  indica 82
  jucunda 82
  laevigata 82
  limbata 82
  lucida 82
  manifesta 82
  norvegica 82
  pauciloculata 82
  planulata 82
  planoconvexa 82
  pulchra 82
  regularis 83
  robusta 83
  rotunda 83
  scita 83
  scrobiculata 83
  soldani 83
  stabilis 83
  terquemiana 83
  valida 83
SPIROLOCULINIDAE 82
Spirontocaris crassirostris 344
  pectinifera 344
  [*sinensis*] 343
  [*spathulirostris*] 343
[*Spirontocens rectirostris*] 343

Spiropagurus spiriger 350
SPIROPHORIDA 165
Spiroplectammina adamsi 75
  atrata 75
  biformis 75
  carinata 75
  compta 75
  cylindroides 75
  esnaensis 75
  fistulosa 75
  floridana 75
  foliosa 75
  howei 75
  mexiaensis 75
  monetalis 75
  nuttalli 75
  phoxa 75
  plummerae 75
  pseudocarinata 75
  sicula 75
  taiwanica 75
  typica 75
  wrightii 75
SPIROPLECTAMMINACEA 75
SPIROPLECTAMMINIDAE 75
Spiroplectinella wrightii 75
Spiroporina longicollis 381
  vertebralis 381
SPIROPORINIDAE 381
SPIRORBIDAE 231
[*Spirorbis foraminosus*] 231
  [*nipponicus*] 231
SPIRORCHIIDAE 198
Spirorutilis fistulosa 75
  kerimbaensis 75
  marielensis 75
  pseudocarinata 75
  wrightii 75
Spirosigmoilina crenata 89
  pulchra 89
  pusilla 89
  tenuis 89
Spirosigmoilinella collaris 71
  compressa 71
  digitata 71
  discofomis 71
  irregularis 71
SPIROSTOMIDAE 58
Spirotextularia floridana 75
  ornatissima 75
Spirulina labyrinthiformis 26
  major 26
  subsalsa 26
  subtillissima 26
  tenerrima 26
  versicolor 26
SPIRURIDAE 210
SPIRURIDEA 209
SPONDYLIDAE 245
Spondylus albibarbatus 245
  anacanthus 245
  barbatus 245
  butleri 245
  candidus 245
  castus 245

Spondylus (continued)
  cruentus 246
  ducalis 246
  flabellum 246
  fragum 246
  imperialis 246
  microlepos 246
  nicobaricus 246
  regius 246
  serraticosta 246
  sinensis 246
  spinosus 246
  squamosus 246
  vesicolor 246
  wrightianus 246
Spongaster pentas 65
  tetras 65
[SPONGIA] 164
Spongia ceylonensis 164
  obique 164
  officinalis 164
Spongicola venusta 338
SPONGIIDAE 164
Spongiomma spinatum 64
Spongionella carteri 164
Spongites reinboldii 135
Spongobrachium froudum 65
  pentagrama 65
  toxon 65
[*Spongocladia vaucheriaeformis*] 146
Spongocore polyacantha 64
  puella 64
Spongodendron macrodoras 64
Spongodes gigantea 188
  guggenheimi 188
  hadzii 188
  spinifera 188
  studeri 188
SPONGODISCIDAE 65
Spongodiscus asiaensis 65
  biconcavus 65
  craticulatus 65
  flos 65
Spongohagiastrum digitatum 65
Spongoliva ellipsoides 64
Spongomorpha arcta 145
Spongophacus flos
Spongopila gracilis 64
  verticillata 64
Spongosorites porites 164
Spongosphaera streptacantha 64
Spongotrochus glacialis
  multispinus
SPONGURIDAE 64
Sporobolus virginicus 160
SPOROCHNACEAE 141
SPOROCHNALES 141
Sporochnus radiciformis 141
Sporotrichum sp. 132
Spriangeratus xanthosome 446
Springeria melanosoma 408
  nanhaiensis 408
SPUMELLARIA 62
SPURILLIDAE 292
Spyridia filamentosa 139

SQUALIDAE 406
SQUALIFORMES 406
Squaliolus laticaudus 407
Squalogadus modificatus 421
Squalus acanthias 407
  blanivillei 407
  brevirostris 407
  japonicus 407
  [*megalops*] 407
  mitsukurii 407
Squamopleura curtisiana 238
Squatina formosa 407
  japonica 407
  nebulosa 407
  tergocellatoides 407
SQUATINIDAE 407
SQUATINIFORMES 407
[*Squilla oratoria*] 368
Squillaconcha subsinuata 247
SQUILLIDAE 368
SQUILLOIDEA 368
Squilloides leptosquilla 368
Stachybotrys-like sp. 132
STACHYPTILIDAE 190
Stachyptilum doflemi 190
Stagnospora sp. 132
Stainforthia complanata 106
  concava 106
  spinosa 106
STAINFORTHIIDAE 106
Stalix immculatus 445
  sheni 445
[*Standella capillacea*] 249
  [*pellucida*] 250
Stanulus seychellensis 447
Staphylaea limacina 272
  nucleus 272
  staphylaea 272
  [—— *staphylaea*] 272
STAPHYLINIDAE 300
STAURACANTHIDAE 61
Stauraspis echinoides 61
Staurodiscus gotoi 170
  vietnanensis 170
STAUROMEDUSAE 175
Stauroneis amphioxys v. *obtum* 41
  constricta 41
  pellucida v. *orientalis* 41
  phoenicenteron 41
STAUROTEUTHIDAE 296
Stauroteuthis albatrossi 296
Stavelia subdistorta 243
Steenstrupia nutans 168
[*Stegastes albifasciatus*] 444
  altus 444
  apicalis 444
  aureus 444
  fasciolatus 444
  insularis 444
  [*lividus*] 444
  [*nigricans*] 444
Steginoporella magniblaris 376
STEGINOPORELLIDAE 376
Stegophryxus minutus 327
Stegophiura hainanensis 398

Stegophiura (continued)
  sladeni 398
  sterilis 398
  vivipara 398
Stegosoma magnum 400
Stegostoma fasciatum 405
Steineria pulchra 211
Steineridora borealis 211
[*Stellaria solaris*] 268
Stellaster equestris 390
Stelletta tenuis 164
  validissina 164
STELLETTIDAE 164
Stemonosudis gracile 415
Stenella attenuata 469
  coeruleoalba 469
  frontalis 469
  longirostris 469
Steno bredanensis 469
Stenochlaena hainanensis 150
  palustris 150
STENOGLOSSA 274
Stenogobius genivittatus 452
  lacrymosus 452
STENOLAEMATA 369
Stenoloma biflora 150
STENOPODIDAE 338
STENOPODIDEA 338
Stenopontius parvus 315
Stenopterobia intermedia 47
Stenopus hispidus 338
Stenoscyphus inabai 176
Stenosemella epunctata 59
  nivalis 59
  pacifica 59
  parvicollis 59
  ventricosa 59
Stenothoe gallensis 330
  qingdaoensis 330
STENOTHOIDAE 330
Stenothyra divalis 265
  formosana 265
  glabar 265
STENOTHYRIDAE 265
Stenotis loui 265
  oxytropis 265
Stephanocyathus nobilis 185
  spiniger 185
Stephanodiscus astraea v. *minutula* 30
Stephanolepis cirrhifer 460
  japonicus 460
Stephanophyllia formosissima 185
  fungulus 185
  japonica 185
Stephanopyxis aculeata 30
  nipponica 30
  palmeriana 30
  turris 30
  —— v. *polaris* 30
Stephanosella bernardii 381
  bolini 381
Stephanostomoides dorabi 195
Stephanostomum argyrosomi 195
  baccatum 195
  bicoronatum 195

Stephanostomum (continued)
   dentatum 195
   ditrematis 195
   fistulariae 195
   lebourae 195
   rachycentronis 195
   seriolae 195
   sphyraenae 195
Stephensoniella marina 234
   sterreri 234
STEPHIDAE 309
STEPHOIDEA 66
STEPHONIDAE 66
Stephos pentacanthos 309
Stercorarius pomarinus 468
STERCULIACEAE 155
Stereocidaris indica philippinensis 392
Stereomastis andamanensis 346
   phospherus 346
   sculpta 346
Stereonephthya inordinata 188
   pumilia 188
   rubiflora 188
Steringophorus congeri 195
Steringotrema sinensis 195
STERNASPIDA 228
STERNASPIDAE 228
Sternaspis scutata 228
Sterna a. anaethetus 468
   albifrons sinensis 468
   dougallii bangsi 468
   fuscata 468
   hirundo longipennis 468
   s. sumatrana 468
STERNOPTYCHIDAE 412
Sternoptyx diaphana 412
   obscura 412
   pseudobscura 412
Sterrhurus chirocentri 203
   gymnothoracis 203
   inimici 203
   longicaudatus 203
   microvatus 203
   pisodonophis 203
Stethojulis axillaris 441
   [*bandanensis*] 441
   interrupta 441
   kalosoma 441
   linearis 441
   phekadopleura 441
   renardi 441
   strigiventer 441
   trilineata 441
Stetsonia altilis 113
   minuta 113
[*Sthenelais boa*] 218
   fusca 218
Sthenocephalus indicus 395
Sthenolepis japonica 218
   [*tentaculata*] 217
Sthenopus mollis 455
STICHAEIDAE 447
Stichopathes abyssicolis 186
   bournei 186
   ceylonensis 186

Stichopathes (continued)
   contortis 186
   desbonni 186
   filiformis 186
   flagellum 186
   gracilis 186
   maldivensis 186
   papillosis 186
   regularis 186
   sacculis 186
   semiglabris 186
   variabilis 186
Stichopilium bicorne 68
   campanulatum 68
   obliqum 68
   rapaeformis 68
   thoracopterum 68
STICHOPODIDAE 389
Stichopus chloronotus 389
   flaccus 389
   horrens 389
   variegatus 389
Stictocardis tiliaefolia 157
Stictodiscus argus 32
   buryanus 32
   ——— f. *subtriangularis* 32
   californicus 32
   ——— v. *nitida* 32
   johnsonianus 32
   [*nitidus*] 32
   varians 32
Stigmatogobius hoeveni 452
   javanicus 452
[*Stigmophora rostrata*] 38
STIGONEMATALES 27
STILIGERIDAE 288
STIRODONTA 392
Stolephorus chinensis 410
   commersoni 410
   heteroloba 410
   indicus 410
   insularis 410
   shantungensis 410
   tri 410
   zollingeri 410
STOLONIFERA 186, 370
STILOSTOMELLACEA 109
STILOSTOMELLIDAE 109
Stolus albescens 387
   buccalis 387
   canescens 387
   molpadioides 387
Stomachetosella collipora 380
STOMACHETOSELLIDAE 380
Stomachicola hainanensis 203
   muraenesocis 203
Stomatella lutea 264
   lyrata 264
   [*varia*] 264
Stomatia angulata 264
   [*decolorata*] 264
   phymotis 264
   planulata 264
   speciosa 264
STOMATIIDAE 264

STOMATOCHESELLIDAE
Stomatolepas elegans 321
   pulchra 321
STOMATOPODA 368
Stomias affinis 412
   nebulosus 412
STOMIATIDAE 412
[STOMOCHORDATA] 399
Stomoloculina lobata 126
   multangula 126
   symmetrica 126
STOMOLOPHIDAE 177
Stomolophus meleagris 177
Stomopneustes variolaris 392
STOMOPNEUSTIDAE 392
[*Stomotoca physophorum*] 167
Stphylococcus 22
Streblacantha circumtexta 66
Streblonema anomalum 141
   bohaiense 141
   codii 141
   corymbiferum 141
   crasscaule 141
   fasciculatum 141
STREBLONIDAE 66
Streblosoma duplicata 229
Streblus compressuicula 124
   tepidus 124
   trispinosus 124
Streetsia challengeri 333
   [*intermedia*] 333
   mindanaonis 333
   porcella 333
   steenstrupi 333
STREPTOCOCCACEAE 22
Streptococcus 22
Streptomyces albidoflavus v. *marine* 24
   diastatochromogenes 24
   lavendulae 24
   rutgersensis gulangyuensis 24
   xiahaiensis 23
STREPTOMYCETACEAE 23
Streptosporangium 24
Streptothece indica 33
   thamesis 33
[*Striarca interplicata*] 240
   thielei 241
Striatella delicatula 44
   interrupta 44
   nanhainica 44
   unipunctata 44
[*Strigilla tomlini*] 250
Striodentalium chinensis 261
   plycostatum 261
   rhabdotum 261
   sedecimcostatum 261
Strioterebrum serotinum 282
STROBILIDIIDAE 58
Strobilidium clavellinae 58
   paraglobosum 58
   raplum 58
   styliferum 58
   typicum 58
STROMATEIDAE 450
[*Stromateus argenteus*] 450

Stromateus (continued)
   [*cinereus*] 450
   [*sinensis*] 450
STROMBIDAE 268
Strombus aratrum 268
   [*aratrum*] 268
  aurisdiane 268
  bulla 268
  [*bulla*] 268
  canarium 268
  [*canarium*] 268
  [―――― *turturellus*] 268
  dentatus 268
  dilatatus 268
  epidromis 268
  erythrinus 268
  fragilis 268
  gibberulus gibbosus 268
  japonicus 269
  labiatus 269
  labiosus 269
  latissimus 269
  lentiginosus 269
  [*lentiginosus*] 269
  luhuanus 269
  [*luhuanus*] 269
  marginatus robustus 269
  microurceus 269
  minimus 269
  mutabilis 269
  [*mutabilis*] 269
  pipus 269
  [*pipus*] 269
  plicatus pulchellus 269
  scalariformis 269
  sinuatus 269
  terebellatus 269
  thersites 269
  urceus 269
  [*urceus*] 269
  [―――― *ustulatus*] 269
  vittatus 269
  [*vittatus*] 269
Strongylidium maritimum 60
STRONGYLOCENTROTIDAE 393
Strongylocentrotus nudus 393
[*Strongylura anastomella*] 419
   [*leiurus*] 419
   [*strongylurus*] 419
Strophidon brummeri 417
  sathete 417
  ui 417
Struvea anastomosans 146
  [*delicatula*] 146
  intermedia 146
Studeriotes debilis 189
  spinosa 189
STURACANTHIDAE
Styela canopus 403
  clava 403
  plicata 403
  qindaoensis 403
  sinensis 403
Styela sp. 403
STYELIDAE 403

Stylacontarium octatigum 63
[*Stylarioides bengalensis*] 228
   [*eruca*] 228
   [*granulosus*] 228
   [*parmata*] 228
   [*plumosa*] 228
Stylater elegans 167
STYLASTERIDAE 167
Stylatractus neptunus 64
Styliola subula 288
Stylocheiron abbreviatum 335
  affine 335
  carinatum 335
  elongatum 335
  indicus 335
  longicorne 335
  maximum 335
  microphthalma 335
  robustum 335
  suhmii 335
STYLOCHIDAE 191
Stylochlamydium aequale 65
  asteriscus 65
  venustum 65
Stylochoplana suoensis 191
  taiwanica 191
  utunomii 191
Stylochus corniculatus 191
  orientalis 191
  pusilla 191
Stylochus sp. 191
Stylocidaris annulosa 392
  bracteata 392
  reini 392
  ―――― v. *cladothrix* 392
  ryukyuensis 392
STYLODACTYLIDAE 339
STYLLODACTYLOIDEA 339
[*Stylodactylus amarhynsus*] 339
   [*bimaxillaris*] 339
  multidentatus 339
Stylodictya arachnia 65
  dujardinii 65
  cf. *gracilis* 65
  lasiacantha 65
  multispina 65
  polygonia 65
  validispina 65
  sp. cf. *validispina* 65
STYLOMMATOPHORA 294
Stylonema alsidii 133
Stylophora danae 181
  pistillata 181
Stylopoma duboisii 380
  parviporosum 380
STYLOSPHAERIDAE 63
Stylotrochus craticulatus
Stypopodium zonale 142
Suaeda australis 151
  glauca 151
  heteroptera 151
  laevissima 151
  liaotungensis 151
  [*salsa*] 151
Subbotina bakeri 101

Subbotina (continued)
  inaequispira 101
  linaperta 101
  triangularis 101
  triloculinoides 101
  velascensis 101
Subergorgia kollikeri 186
  ornata 186
  reticulata 186
  suberosa 186
SUBERGORGIIDAE 186
Suberites carnosa 164
  domuncula 164
  ficus 164
SUBERITIDAE 164
Subreophax aduncus 72
[*Suezichthys gracilis*]
Sufflamen bursa 459
  chrysopterus 459
  fraenatus 459
Suggrundus longirostris 455
  macracanthus 455
  meerdvoorti 455
  rodericensis 455
Sugiura chengshanense 170
SUGIURIDAE 170
Sula leucogaster plotus 465
  sula rubripes 465
Sulcophax palustris 72
Sulculeolaria angusta 175
  bigelowi 175
  biloba 175
  brintoni 175
  chuni 175
  monoica 175
  quadrivalvis 175
  tropica 175
  turgida 175
  xishaensis 175
SULIDAE 465
Sunamphitoe plumosa 328
[*Sunetta concinna*] 254
Sunettina solanderii 255
[*Surcula javana*] 281
Surirella apiae 47
  arabica 47
  armoricana 47
  biseriata 47
  campechiana 47
  collare 47
  elegans v. *norvegica* 47
  eximia 47
  fastuosa 47
  ―――― v. *cuneata* 47
  ―――― v. *plusieura* 47
  ―――― v. *recens* 47
  ―――― v. *spinlifera* 47
  ―――― v. *suborbicularis* 47
  fluminensis 47
  gemma 47
  gravis 47
  japonica 47
  kurzii 47
  lata 47
  ―――― v. *robusta* 47

Surirella (continued)
    liaotungiensis 47
    ────── v. *minuta* 47
    mexicana 47
    nervata 47
    palmeriana 47
    schleinitzii 47
    schmidtii 47
    seychellarum 47
    ────── v. *biseriata* 47
    significans 47
    voigtii 47
SURIRELLACEAE 47
SURIRELLALES 46
Sydaphera spengleriana 280
SYLLIDAE 219
Syllis amica 219
    [*anops*] 219
    [*armillaris*] 219
    [*cornuta*] 219
    [*decorus*] 219
    [*fasciata*] 219
    gracilis 219
    [*inflata*] 219
    longissima 219
    [*sclerolaema*] 219
    spongiphila 219
Symbolophorus boops 414
    evermanni 414
    rufinum 414
[*Sympasiphaea imperialis*] 338
Symphorus nematophorus 433
    spilurus 433
Symphurus novemfasciatus 458
    orientalis 458
    strictus 458
SYMPHYACANTHA 61
Symphyllia agaricia 184
    gigantea 184
    nobilis 184
    radians 184
[*Symphyocladia gracilis*] 140
    latiuscula 140
    marchantioides 140
    pennata 140
Symphysanodon katayamai 427
SYMPHYSANODONTIDAE 427
Symplectoscyphus hozawai 172
    huanghaiensis 172
Symplectoteuthis luminosa 295
    oualaniensis 295
Symplegma oceania 403
    reptans 403
Symploca caespitosa 26
    hydnoides 26
    muscorum 26
[*Sympronoe parva*] 331
Synagrops japonicus 427, 429
    philippinensis 427
    serratospinosus 427
Synagrops japonicus
Synallactes discoidalis 388
SYNALLACTIDAE 388
Synalpheus bispinosus 343
    carinatus 343

Synalpheus (continued)
    charon 343
    coutiere 343
    demani 343
    gravieri 343
    hastilicrassus 343
    iocosta 343
    neomeris 343
    neptunus 343
    nilandensis 343
    odontophorus 343
    paraneomeris 343
    pescadorensis 343
    stimpsoni 343
    stormi 343
    streptodactylus 343
    theano 343
    trispinosus 343
    tumidomanus 343
Synanceia horrida 454
    verrucosa 454
SYNANCEIIDAE 454
Synantheopsis primus 180
SYNAPHOBRANCHIDAE 415
Synaphobranchus affinis 415
    brevidorsalis 415
    kaupii 415
    pinnatus 415
Synapta maculata 389
SYNAPTIDAE
Synaptula reticulata 389
[*Synaptura annularis*] 458
    [*orientalis*] 458
Synasterope knudseni 302
Synchaeta tremula
SYNCHAETIDAE 214
Synchelidium miraculum 330
Synchiropus altivelis 448
    delandi 448
    grinnelli 448
    lateralis 448
    masudai 448
    ocellatus 448
    ornatus 448
    picturatus 448
    splendidus 448
Syncorpozoum hainanensis 200
Synechocystis aquatilis 25
    pevalekii 25
Synechogobius hasta 452
    ommaturus 452
Synedra barbatula 44
    crystallina 44
    formosa 44
    fulgens 44
    gaillonii 44
    hennedyana 44
    investiens 44
    laevigata 44
    pulcherrima 44
    robusta 44
    ────── v. *sinica* 44
    rostrata 44
    tabulata 44
    ────── v. *acuminata* 44

Synedra (continued)
    ────── v. *fasciculata* 44
    ────── v. *parva* 44
    ulna 44
    undulata 44
    ────── v. *curvata* 44
Synedrella nodiflora 158
Synedrosphenia gomphonema 44
Synelmis albini 218
    annamita 218
    sinica 218
Synerythrops intermedia 324
Synestius caliginus 316
SYNGNATHIDAE 424
Syngnathoides biaculeatus 424
Syugnathus acus 424
    argyrostictus 424
    cyanospilus 424
    djarong 424
    pelagicus 424
    spicifer 424
Synidotea laevidorsalis 326
Synnotum aegypticaum 375
    pembaense 375
SYNODIDAE 413
Synodus englemani 413
    fuscus 413
    hoshinonis 413
    indicus 413
    jaculum 413
    kaianus 413
    macrops 413
    rubromarmoratus 413
    ulae 413
    variegatus 413
Synostemon bacciformis 154
SYNTHECIIDAE 172
Synthecium campylocarpum 172
    patulum 172
    tubithecum 172
Synthliboramphus antiquus 468
Synura petersenii 50
SYNURACEAE 50
Syphonota geographica scripta 288
[*Syringidium daemon*] 28
Syringodium isoetifolium 158
Syrnola brunnea 284
    cinctella 284
    cinnamomea 284
Systellaspis debilis 339

T

Taanigichthys bathyphilus 414
Tabellaria flocculosa v. *asterionelloides* 44
    nodosa 44
Tachaea chinensis 325
TACHIDIIDAE 314
Tachybaptus grisegena caspicus 464
    ────── *cristatus* 464
    ────── *holboelii* 464
    ruficollis poggei 464
TACHYPLEIDAE 297
Tachypleu gigas 297

Tachypleu (continued)
   tridentatus 297
Tachypleus sp. 297
Tadorna ferruginea 466
   tadorna 466
TAENIACANTHIDAE 314
Taeniacanthus pteroisi 314
Taenianotus triacanthus 455
Taenioides aguillaris 453
   cirratus 453
TAENIOIDIDAE 453
[*Taenioma macrourum*] 139
   nanum 139
   perpusillum 139
Taeniura melanospilos 408
[*Taius fumifrons*] 434
   tumifrons 434
Takayanagia delicata 105
[*Takifugu niphobles*] 461
   [*oblongus*] 461
   [*poecilonotus*] 461
   [*xanthopterus*] 461
Talismania antillarum 413
   brachycephala 413
TALITRIDAE 330
Talonostrea talonata 246
Talorchestia martensii 330
Talparia argus 272
   talpa 272
TAMARICACEAE 155
Tamarix chinensis 155
   juniperina 155
Tambalagamia fauveli 221
Tambja affinis 289
   morsa 289
[*Tamoya alata*] 176
   gargantua 176
TANAIDACEA 325
TANAIDAE 325
Tanais cavolinii 325
Tanakius kitaharae 458
Tangia carnolabrum 433
Tanystylum sinoabductus 297
Taonius pavo 295
[*Tapes araneosa*] 255
   belcheri 255
   dorsalus 255
   literata 255
   platyptycha 255
   [*quadriradiata*] 255
   [*turgida*] 255
Taphrobothrium japonense 206
Taractes rubescens 431
Taractichthys steindachneri 431
Tarasovium orientale 318
Tarphops oligolepis 457
Taubera gracilis 223
Taxotrophis ilicifolius 151
   [*macrophylla*] 151
Tchthyocampus belcheri 424
Technitella arenaeea 70
   legumen 70
Tectarius coronatus 265
   [*granularis*] 265
   pogodus 265

Tectarius (continued)
   [*vilis*l 265
Tectidrilus achaetus 234
   pictoni 234
Tectillaria fertilis 400
[*Tectus maximus*] 263
   [*pyramis*] 263
   [*triserialis*] 263
Tedania anhelans 165
   strongyla 165
Tegella disincrustata 374
   lamellatoidea 374
   unicornis 374
Tegillarca granosa 241
   nodifera 241
[*Tegula argyrostoma*] 263
   [*nigerrima*] 263
   [*rustica*] 263
   xanthostigma 263
Teissiera australe 169
   macrocystae 169
   medusifera 169
   polypofera 169
TEISSIERIDAE 169
Teixeirichthys formosana 444
   jordani 444
TELAMMINIDAE 72
Telescopium telescopium 266
TELESTACEA 190
TELESTAIDAE 190
Telesto cf. *rubra* 190
   [*Tellina diaphana*] 251
   [*jedoensis*] 251
   [*perna*] 251
   rastella 251
   [*ruglsa*] 251
   spengleri 251
   virgata 251
   [*vulsella*] 251
TELLINIDAE 250
Tellinides chinensis 251
   ovalis 251
   timorensis 252
Tellinimactra maluccensis 252
Telorhynchus astrocongeri 193
   cociellae 193
   hippocampi 193
Teloscalpellum album 318
TELOTREMATA 384
Temanella turrita 265
Temnaspis amygdalum 319
   donghaiense 319
   excavatum 319
   kilepoae 319
   tridens 319
TEMNOPLEURIDAE 393
Temnopleurus apodus 393
   hardwickii 393
   reevesii 393
   toreumaticus 393
Temnotrema muculatum 393
   reticulatum 393
   sculptum 393
   siamense 393
   xishaensis 393

Temora discaudata 309
   stylifera 309
   turbinata 309
TEMORIDAE 309
Temorites brevis 309
Temoropia mayumbaensis 309
Tenagomysis orientalis 324
TENDIPEDIDAE 299
TENEBRIONIDAE 300
TENTACULATA 190
Tentidonax kiusiuensis 250
   nitidus 250
Tentoriceps cristatus 449
Tephrinectes sinensis 457
Teramachia dalli 279
   tibiaeformis 279
[*Terapon cancellatus*] 435
   [*jarbua*] 435
   [*theraps*] 435
Terebella copia 229
   ehrenbergi 229
   [*ingens*] 229
TEREBELLIDA 228
TEREBELLIDAE 229
Terebellides stroemii 229
Terebellum terebellum 269
Terebra affinis 282
   amanda 282
   areolata 282
   bellanodosa 282
   cerithina 282
   chlorata 282
   crenulata 282
   cumingii 282
   dimidiata 282
   [*duplicata*] 281
   dussumieri 282
   felina 282
   fenestrata 282
   funiculata 282
   guttata 282
   koreana 282
   lanceata 282
   maculata 282
   penicillata 282
   pereoa 282
   pretiosa 282
   subulata 282
   torguata 282
   triseriata 282
Terebralia sulcata 266
Terebratalia coreanica 384
TEREBRATALIIDAE 384
Terebratulina hataina 384
TEREBRIDAE 281
TEREDINIDAE 257
Teredo bitubula 257
   clava 257
   diederichseni 257
   furcifera 257
   manni 257
   massa 257
   matacotana 257
   megotara 257
   navalis 257

Index

Teredo (continued)
  parksi 257
  remiformis 257
  samosensis 257
  schizoderma 257
Tergestia atropi 195
  atulis 195
  hainanensis 195
  laticollis 195
  triacanthi 195
TERGIPEDIDAE 292
Terminalia catappa 156
Terminoisocoelium laterolecithale 199
Ternstroemia microphylla 155
  [*pseudoverticillata*] 155
Terschellingia longicaudata 211
  longissimicaudata 211
Tesseropora alba 320
Testudinella mucronata 214
TESTUDINELLIDAE 214
TESTUDIFORMES 463
Testudovolva adaminea 271
  arientis 271
  nipponensis 271
Tethya aurantium 164
  japonica 164
  robusta 164
Tethyaster aulophorus
TETHYIDAE 164
[TETHYIDAE] 292
Tetilla lituiformis 165
  ternatensis 165
TETILLIDAE 165
Tetracera asiatica 155
Tetrachthamalus sinensis 320
Tetraclita coerulescens 320
  japonica 320
  [*squamosa japonica*] 320
  ———— *squamosa* 320
  [*viridis*] 320
TETRACLITIDAE 320
Tetraclitella chinensis 320
  costata 320
  darwini 320
  divisa 320
  multicostata 320
  pilsbryi 320
Tetracyclus javanicus 44
Tetradiella sinica 394
Tetragonia tetragonoides 152
TETRAGONICIPITIDAE 315
Tetralia glaberrima 362
  heterodactyla 362
  ———— *fusca* 362
TETRAODONTIDAE 460
TETRAODONTIFORMES 459
Tetrapentalon elegans 64
TETRAPHYLLIDEA 206
Tetraplatia volitans 175
Tetrapturus angustirostris 450
  audax 450
Tetrapyle circularis 65
  nephropyle 65
  octacantha 65
  pachyderma 65

Tetrapyle (continued)
  quadriloba 65
  rotundospinosa 65
Tetrapylonium pyrum 65
  strobilinum 65
Tetraroge leucogaster 455
Tetrasphaera spongiosa 64
TETRASPORALES 144
Tetrastemma nigrifrons 208
  verinigrum 208
TETRASTEMMIDAE 208
Tetrathyrus arafurae 333
  forcipatus 333
  [*monceuri*] 333
[*Tetreres nesiotes*] 228
Tetrochetus coryphaenae 194
  hainanensis 194
  navodonis 194
  zhoushanensis 194
[*Tetrosomus concatenatus*] 460
  [*gibbosus*] 460
TETTIGONIIDAE 300
TEUTHOIDEA 294
Textilina bocki 79
  crassaformis 79
  lythostrota 79
  semialata 79
Textularia abbreviata 78
  agglutinans 78
  akaminei 78
  alishanensis 78
  articulata 78
  astutia 78
  atrata 78
  aura 78
  australis 78
  awazea 78
  bocki 78
  bradyi 78
  calva 78
  candeiana 78
  concava 78
  conica 78
  contorta 78
  corrugata 78
  crassisepta 78
  crenata 78
  cuneata 79
  curtata 78
  dupla 79
  excavata 79
  farafraensis 79
  foliacea 79
  gramen 79
  hakusikeiensis 79
  halkyardi 79
  hoppoensis 79
  hosonoi 79
  howei 79
  intosiana 79
  kansaiensis 79
  kansireinsis 79
  kapitea 79
  kerimbaensis 79
  lata 79

Textularia (continued)
  lateralis 79
  magallanica 79
  megaloculata 79
  magnifica 79
  midwayensis 79
  mississippiensis 79
  neorugosa 79
  nitens 79
  oceanica 79
  orbica 79
  ornatissima 79
  paragglutinans 79
  parva 79
  parvula 79
  peritubula 79
  pitmani 79
  porrecta 79
  pseudocarinata 79
  pseudogramen 79
  pseudokansaiensis 79
  pseudosolita 79
  pseudotrochus 79
  pseudoturris 79
  rokuzyukeiensis 79
  sagittula 79
  scrupula 79
  secasensis 79
  semialata 79
  sineiensis 79
  sintikuensis 79
  stricta 79
  subantarctica 79
  suttonensis 79
  tainanensis 79
  tenuissima 79
  transversaria 79
  truncata 79
  truncatiformis 79
  tubulosa 79
  valentula 79
  vola 79
  zeaggluta 79
TEXTULARIACEA 77
Textulariella cushmani 77
  simplex 77
TEXTULARIELLIDAE 77
TEXTULARIIDAE 78
TEXTULARIINA 69
Textularioides inflata 80
Textulina obesa 80
Thais aculeata 276
  [*aculeata*] 276
  alouina 276
  armigera 276
  [*armigera*] 276
  bronni 276
  bufo 276
  carinifera 276
  clavigera 276
  [*distinguenda*] 276
  echinata 276
  [*echinata*] 276
  echinulata 276
  [*echinulata*] 276

Thais (continued)
　gradata 276
　hippocastanum 276
　[*hippocastanum*] 276
　luteostoma 276
　[*mancinella*] 276
　mustabilis 276
　pseudodiadema 276
　[*siro*] 275
　squamosa 276
　tissoti 276
　[*trigona*] 276
　tuberosa 276
　[*tuberosa*] 276
Thalamita admete 359
　chaptali 359
　coeruleipes 359
　corrugata 359
　crenata 359
　danae 359
　demani 359
　edwardsi 359
　imparimana 359
　integra 359
　kagosshimaensis 359
　picta 359
　platypedis 359
　procorrugata 359
　prymna 359
　sima 359
　spinifera 359
　spinimana 359
　stephensoni 359
　tenuipes 359
[*Thalamoporella gothica*] 376
　hamata 376
　indica 376
　linearis 376
　lioticha 376
　rozieri 376
　stapiliera 376
　tubifera 377
THALAMOPORELLIDAE 376
THALASOCARIDIDAE 344
Thalasocaris crinita 344
Thalassema fuscum 237
　mortenseni 237
Thalasseus bergii cristatus 468
　zimmermanni 468
Thalassia hemprichii 159
Thaiassina anomalna 345
THALASSINIDAE 345
THALASSINIDEA 345
THALASSINOIDEA 345
Thalassionema baciliaris 44
　nitzschioides 44
　——— v. *parva* 44
Thalassiosira antiqua 30
　baltica 30
　binata 30
　——— v. *bibinata* 30
　——— v. *miner* 30
　condensata 30
　decipiens 30
　eccentrica 30

Thalassiosira (continued)
　gravida 30
　hyalina 30
　hydra 30
　laevissima 30
　lineata 30
　nordenskioldii 30
　oestrupii 30
　pacifica 30
　punctigera 31
　rotula 31
　scrotiformis 31
　simonsenii 31
　subtilis 31
　symmetrica 31
　weissflongii 31
Thalassiothrix delicatula 44
　frauenfeldii 44
　gibberula 44
　longissima 44
　mediterranea 44
　——— v. *pacifica* 44
　sinensis 44
　vanhoeffenii 44
Thalassironus bohaiensis 212
Thalassodrilides briani 234
　gurwitschi 234
Thalassograpsus harpax 367
Thalassoma amblycephatus 441
　cupido 441
　fuscum 441
　hardwicki 441
　jansenii 441
　lunare 441
　lutescens 441
　purpureum 441
　quinquevittatus 441
　trilobatum 441
　umbrostigma 441
THALASSOMETRIDAE 387
[*Thalemessa spinosa*] 218
[——— *asiatica*] 218
[*Thalenessa oculata*] 217
THALESTRIDAE 315
Thalia democratica 401
[——— *democratica*] 401
——— *echinata* 401
——— *orientalis* 401
THALIACEA 400
Thalotia elongatus 263
Thalmannammina parkerae 74
Thalmannita coronata 121
Thalmitoides quadridens 359
　tridens 359
[*Thamnaconus septentrionalis*] 459
THAMNASTERIIDAE 181
Tharacostoma setosum 212
Tharyx acutus 224
　filibranchia 224
　marioni 224
　multifilis 224
　tesselata 224
Thatcheria mirabilis 281
Thaumastocheles japonicus 345
THAUMASTOCHELIDAE 345

Thaumatoplax orientalis 363
THEACEAE 155
[THECATA] 169
Thecacera pennigera 289
[THECATAE-LEPTOMEDUSAE] 169
Thecocarpus brevirostris 171
　myriophyllum orientalis 171
THECOSOMATA 288
Thecosphaera grecoi 62
Thelenota ananas 389
　anax 389
Thelepus opimus 229
　plagiostoma 229
　pulvinus 229
THELYPTERIDACEAE 150
Themiste cymodoceae 236
　dehamata 236
　lageniformis 236
　minor 236
　spinulum 236
Themistella fusca 331
THEMISTIDAE 236
[*Themisto gracilipes*] 331
Thenea grayi 164
THENEIDAE 164
Theocapsa democriti 68
Theoconus hertwigii 68
Theocorythium trachelium trachelium 68
Theophormis callipilium 68
Theopilium cranoides 68
　cucullatum 68
　galeatum 68
　germinis 68
　tricostatum 68
Theora fragilis 252
　lata 252
　[*lubrica*] 252
Therapon jarbua 435
　oxyrhynchus 435
　theraps 435
THERAPONIDAE 435
Theregra chalcogramma 420
Theristus littoralis 211
　macroflevensis 211
　metaflevensis 211
Thespesia populnea 155
Thetys vagina 401
THIOBACILLACAE 21
THIOBACILLALES 21
Thiobacillus denitrificans 21
　thiooxidans 21
　thioparus 21
Thliptodon diaphanus 288
THOLONIIDAE 65
THOLOSPYRIDAE 66
Tholospyris fenestrata 66
　cf. *scaphipes* 66
　tripodiscus 66
Thompsonia sp. 322
Thor amboinensis 344
　paschalis 344
Thoracaster cylindratus 391
THORACICA 318
[*Thoracophrya luciae*] 56
Thordisa maculigera 290

Index

[Thormora jukesii] 217
Thorsonia adversaria 387
Thracia concinna 258
   hainanensis 258
   ovata 258
THRACIIDAE
Threskiornis aethiopicus melanocephalus 465
THRESKIORNITHIDAE 465
Thrissa dussumieri 411
   hamiltonii 411
   kammalensis 411
   mystax 411
   setirostris 411
   vitirostris 411
[Thryssa dussumieri] 411
   [hamiltonii] 411
   [setirostas] 411
Thuaria involuta 160
Thuiaria lonchitis 172
   marktanneri 172
   similis 172
THUNNIDAE 450
Thunnus alalunga 450
   albacora 450
   obesus 450
   thynnus 450
   tonggol 450
Thurammina basispiculata 71
   papillata 71
   papyracea 71
Thyasira tokunagaii 247
THYASIRIDAE 247
THYMELAEACEAE 155
Thyone anomaus 387
   bicornis 387
   fusus v. chinensis 387
   papuensis 387
   pedata 387
   pohaiebsis 387
   spinifera 387
[Thyropus diaphanus] 333
   sphaeroma 333
   [typhoides] 333
Thyroscyphus torresi 172
Thyrsitoides marleyi 449
Thyrsoidea macrurus 417
Thysanactis dentex 412
Thysanichthys crossotus 454
Thysanocardia nigra 236
Thysanophrys arenicda 455
   bataviensis 455
   celebica 455
   chiltonae 455
   otaitensis 455
Thysanopoda acutifrons 335
   aequalis 335
   astylata 335
   cornuta 335
   cristata 335
   egregia 335
   monacantha 335
   obtusifrons 335
   orientalis 335
   pectinata 335

Thysanopoda (continued)
   tricuspidata 335
Thysanostoma flagellatum 176
THYSANOSTOMATIDAE 176
Thysanote appendiculata 316
   fimbriata 317
THYSANOTEUTHIDAE 295
Thysanoteuthis rhombus 295
Tiaricodon coeruleus 168
Tiarinia angusta 357
   depressa 357
Tiaropsis multicirrata 171
Tiberia ebarana 284
   pulchella 285
   pusille 285
   terebolla 285
Tibia cancellata 269
   fusus 269
   [fusus] 269
   martinii 269
   melanocheilus 269
   powisi 269
Tiffaniella suychiroi 139
TILIACEAE 154
Tima formosa 170
Timoclea imbricata 255
   lionota 255
   marica 256
   scabra 256
   subnodulosa 256
Tindaria jinxingae 239
TINDARIIDAE 239
Tinocladia crassa 141
Tinophodella ambitacrena 101
TINTINNIDAE 60
TINTINNIDIDAE 58
Tintinnopsis acuminata 58
   amoyensis 58
   angusta 58
   aperta 58
   beroidea 58
   brasiliensis 59
   brevicollis 59
   chinglanensis 59
   cochleata 59
   compressa 59
   dadayi 59
   digita 59
   directa 59
   gracilis 59
   hemispiralis 59
   inflata 59
   japonica 59
   karajacensis 59
   kiaochowensis 59
   lohmanni 59
   loricata 59
   major 59
   mayeri 59
   minima 59
   nana 59
   nitida 59
   nucula 59
   orientalis 59
   pallida 59

Tintinnopsis (continued)
   parva 59
   radix 59
   rapa 59
   rotundata 59
   schotti 59
   spiralis 59
   tentaculata 59
   tocantinensis 59
   tsingtaoensis 59
   tubulosa 59
   turgida 59
Tintinnus fraknoii 60
   lusus-undae 60
   ——— v. exigua 60
   striatus 60
   tenuis 60
   tubulosa 60
Tiphotrocha convexoconcava 76
   kellettae 76
   minuta 76
Tisbe bermudensis 315
TISBIDAE 315
Tmethypocoelis ceratophora 365
Todarodes pacificus 295
Tolypammina vagans 71
Tolypiocladia calolictyon 140
   glomerulata 140
Tolypothrix subsalsa
TOMOPTERIDAE 216
Tomopteris apsteini 216
   cavallii 216
   duccii 216
   [duccii] 216
   dunckeri 216
   [dunckeri] 216
   elegans 216
   ligulata 216
   mariana 216
   nationalis 216
   pacifica 216
   [pacifica] 216
   planktonis 216
   rolasi 216
   septentrionalis 216
Tonicia sp. 238
Tonna allium 274
   Canaliculata 274
   [cepa] 274
   chinensis 274
   lischkeana 274
   luteostoma 274
   magnifica 274
   [marginata] 274
   olearium 274
   perdix 274
   sulcosa 274
TONNIDAE 274
Tormopsolus fuzhouensis 195
[Tornatina avenaria] 286
   [coarctata] 287
TORNIDAE 266
TORPEDINIDAE 409
TORPIDINIFORMES 409
Torpedo macnilli 409

Torpedo (continued)
    nobilina 409
    tokionis 409
TORTANIDAE 312
Tortanus barbatus 312
    denticulatus 312
    derjugini 312
    dextrilobatus 312
    forcipatus 312
    gracilis 312
    murrayi 312
    scaphus 312
    spinicaudatus 312
    sinicus 312
    vermiculus 312
Tortoplectella crispata 105
TORTOPLECTELLIDAE 105
Torulaspora delbrueckii 129
Tosana niwae 427
[*Tosatrochus attenuata*] 263
[*Tournefortia argentea*] 157
Toxopneustes pileolus 393
TOXOPNEUSTIDAE 393
Toxorchis arcuatus 170
    polynema 170
Tozeuma armatum 344
    lanceolatum 344
    tomentosum 344
Tpedospora radiata 131
Trachelocerca phoenicopterus 56
TRACHELOCERCIDAE 56
TRACHEOPHYTA 150
[*Trachichthodes lineatus*] 422
TRACHICHTHYIDAE 422
Trachidermus fasciatus 456
Trachinocephalus myops 413
Trachinotus bailloni 431
    [*blochii*] 431
    ovatus 431
    russelli 431
TRACHIPTERIDAE 423
Trachonurus villosus 420
Trachurus japonicus 431
Trachycardium angulatum 249
    arenicolum 249
    enode 249
    flavum 249
    foveilatum 249
    impolitum 249
    sewelli 249
    transcendens 249
TRACHYMEDUSAE 173
Trachyneis antillarum 41
    aspera 41
    ——— v. *angusta* 41
    ——— v. *contermina* 41
    ——— v. *oblonga* 41
    ——— v. *orientalis* 41
    ——— v. *perobliqua* 41
    ——— v. *producta* 41
    ——— v. *pulchella* 41
    ——— v. *residua* 41
    ——— v. *unilatera* 41
    ——— v. *vulgaris* 41
    brunii 41
    clepsydra 41

Trachyneis (continued)
    debyi 42
    formosa 42
    johnsiniana 42
    minor 42
    olivaeformis 42
    velata 42
    ——— v. *oblonga* 42
    velatoides 42
Trachypenaeus anchoralis 338
    [*asper*] 338
    curvirostris 338
    [*fulvus*] 338
    [*granulosus*] 338
    longipes 338
    malaiana 338
    pescadoreensis 338
    sedili 338
Trachypterus iris 423
    ishikawae 423
Trachyrhamphus serratus 424
Trapezia areolata 362
    cymodoce 362
    digitalis 362
    ferruginea 362
    flavopunctata 362
    formosa 362
    guttata 362
    [*maculata*] 362
    reticulata 362
    rufopunctata 362
    speciosa 362
TRAPEZIIDAE 253
Trapezium liratum 253
    sublaevigatum 253
Traustedtia multitentaculata 401
Travisia japonica 226
    pupa 226
Travisiopsis dubia 216
    levinseni 216
    lobifera 216
Trematocarpus pygmaeus 137
TREMATODA 192
Trematosphaeria lineolatispora 131
TREMOCTOPODIDAE 296
Tremoctopus violaceus gracilis 296
Tretomphaloides concinnus 112
Tretomphalus bulloides 112
    concinnus 112
    grandis 112
    milletti 112
    planus 112
TRIACANTHIDAE 459
Triacanthinella principale 192
Triacanthodes anomalus 459
TRIACANTHODIDAE 459
Triacanthus biaculeatus 459
    blochi 459
    brevirostris 459
    nieuhofi 459
    [*strigililifer*] 459
Triaenodon obesus 406
Triaenopogon barbatus 452
TRIAKIDAE 405
Triakis scyllium 406
    venustum 406

Trianguloscalpellum balanoides 318
    michelottianun 318
    uniarticulatum 318
Trianthema portulacastrum 152
Triastrum aurivillii 65
Tribonosphaera centripetalis 63
Tribulus cistoides 153
    terrestris 153
Tricellaria gracilis 376
    longispinosa 376
    multispinosa 376
    occidentalis 376
    peachi 376
    porata 376
Triceratium affine 33
    americanum 33
    antedeluvianum 33
    arcticum v. *japonica* 33
    balearicum f. *biquadrata* 33
    broeckii 33
    campechianum 33
    cultum 33
    cuspidatum 33
    dubium 33
    favus 33
    ——— f. *quadrata* 33
    formosum 33
    ——— f. *pentagonale* 33
    ——— f. *quadrangularis* 33
    gallapagense 33
    gibbosum 33
    japonicum 33
    junctum 33
    orbiculatum 33
    pelagicum 33
    pellucida 33
    pentacrinus 33
    ——— f. *quadrata* 33
    perpendiculare 33
    reticulum 33
    rostratum 33
    scitulum 33
    shadboltianum 33
    [*whampoense*] 33
    zonulatum 33
Trichaster acanthifer 395
    flagellifer 395
    palmiferus 395
TRICHIURIDAE 449
Trichiurus brevis 449
    haumela 449
    lepturus 449
    [*muticus*] 449
    nanhaiensis 449
    [*savala*] 449
TRICHOBRACHIDAE 229
Trichobranchus bibranchiatus 229
Trichocanace sinensis 299
Trichocladium achrasporum 132
    linderii 132
Trichocladium sp. 132
Trichocorixa sp. 299
Trichoderma sp. 132
[*Trichodesmium contortum*] 26
    erythraeum 26
    hildebrandtli 26

Index

Trichodesmium (continued)
  maceii 26
  thiebautii 26
Trichogloeopsis hawaiiana 134
TRICHOHYALIDAE 121
Trichomusculus barbatus 243
  semigranata 243
Trichomya hirsuta 243
TRICHONOTIDAE 445
Trichonotus filamentosus 445
  setiger 445
Trichosporon cutaneum 129
  figueirae 129
TRICHOTROPIDAE 268
Trichotropis bicarinata 268
  [*bicarinata*] 268
  unicarinata 268
Trichurus monsoniae 151
TRICLADIDA 191
TRICLIDAE 286
Tridacna crocea 249
  [*cookiana*] 249
  derasa 249
  [*elongata*] 249
  gigas 249
  maxima 249
  squamosa 249
TRIDACNIDAE 249
Tridax procumbens 158
Tridentiger obscurus 452
  trigonocephalus 452
Trifarina angulosa 108
  bradyi 108
  costornata 108
  esnaensis 108
  fluens 108
  granulosa 108
  guangdongensis 108
  lepida 108
  occidentalis 108
  reussi 108
TRIGLIDAE 454
Triglochin maritimum 159
Trigonaphera bocageana 280
  obiquata 280
Trigoniocardia fornicata 249
Trigonothir camelus 357
Trigonothracia jinxingae 258
  pusilla 258
Trigonotrema alatum 197
Trilasmis eburnea 319
Triloculina affinis 87
  alabamensis 87
  basispinata 87
  bertheliniana 87
  bicarinata 87
  bradyana 87
  circularis 87
  complanata 87
  compressa 87
  consobrina 87
  costifera 87
  cuneata 87
  earlandi 87
  elongata 87
  foveata 87

Triloculina (continued)
  gibba 87
  hartingi 87
  inflata 87
  insignis 87
  involuta 87
  irregularis 87
  kerimbatica 88
  labiosa 88
  laeviqata 88
  laevis 88
  lecalvezae 88
  linneiana 88
  marshallana 88
  mindenensis 88
  oblonga 88
  paradoxa 88
  parallela 88
  paratrigonula 88
  pectinata 88
  pentagonalis 88
  peregrina 88
  pinguicula 88
  pinguis 88
  rectilocula 88
  reticulostriata 88
  rupertiana 88
  sidebottomi 88
  subcylindrica 88
  subglobosa 88
  subgranulata 88
  subplanciana 88
  transversestriata 88
  transvoluta 88
  tricarinata 88
  trigonula 88
  trotusa 88
  vespertilio 88
  vicina 88
Triloculinella asymmetrica 88
  laevigata 88
  oblongus 88
  patens 88
  tegminis 88
  translucens 88
Trimosina milletti 109
  orientalis 109
TRIMOSINIDAE 109
Trinacria regina v. *tetragona* 33
Tringa erythropus 467
  glareola 467
  guttifer 467
  hypoleucos 467
  incana bervipes 467
  nebularis 467
  ochropus 467
  stagnatilis 467
  t. totanus 467
Triodon bursarius 461
  macropterus 461
TRIODONTIDAE 461
TRIOPHIDAE 289
Triphora alveolatus 282
  hungerfordi 282
TRIPHORIDAE 282
Triphoturus micropterus 414

Triphoturus (continued)
  nigrescens 414
Triphyllozoon bimunitum 382
  hirsutum 382
  inovicellatum 382
  moniliferum 382
  sinicum 382
Triplophos hemingi 411
Tripneustes gratilla 393
TRIPOCALPIDAE 67
TRIPOCYRTIDAE 67
[*Tripodichthys blochi*] 459
Tripolium vulgare 159
Triposolenia bicornis 51
  intermedia 51
Trippa intecta 291
TRIPTERYGIDAE 446
Tripterygion etheostoma 446
  fuligicauda 446
TRIPYLOIDIDAE 212
Trisidos kiyonoi 241
  semitorta 241
  tortuosa 241
[*Triso dermopterus*] 427
Trisotropis dermopterus 427
Tristichotrochus formosense 263
Tristylospyris palmipes 66
Tritaxia changi 77
  donghaiensis 77
  orientalis 77
TRITAXIIDAE 77
Tritaxilina atlantica 80
  caperata 80
Tritaxis conica 76
  fusca 76
Tritodynamia fujianensis 364
  hainaensis 364
  horvathi 364
  intermedia 364
  longipropodum 364
  rathbunae 364
[*Tritonalia emarginatus*] 274
  [*fatcatus*] 275
[TRITONIIDAE] 292
Triumfetta procumbens 154
TRIVIIDAE 270
Trivirostra exigua 270
  oryza 270
Trizopagurus strigatus 349
Trochammina boltovskoyi 76
  carinata 76
  challengeri 76
  charlottensis 76
  globigenniformis 76
  globorotaliformis 76
  globulosa 76
  hadai 76
  inflata 76
  japonica 76
  macrescens 76
  malovensis 76
  minuta 76
  nobensis 76
  ochracea 76
  ovata 76
  pygmaea 76

Trochammina (continued)
  quadriloba 76
  rotaliformis 76
  squamata 76
  trapeziformis 76
  tricamerata 76
  triloculina 76
  vesicularis 76
  wiesneri 76
  xishaensis 76
TROCHAMMINACEA 76
TROCHAMMINIDAE 76
Trochamminoides coronatum 74
  proteus 74
Trochamminopsis globulosa 76
  pusilla 76
Trochamminula asiatica 76
  elongata 76
  lobatula 76
TROCHIDAE 263
Trochochaeta diverapoda 224
  multisetosum 224
  cf. orissae 224
TROCHOCHAETIDAE 224
Trochocyathus caryophylloides 185
  pileus 185
  pseudowingedus 185
Trochodoson linearis 382
  oplatus 382
[*Trochus calcaratus*] 263
  [calliferus] 263
  chloromphalus 263
  conus 263
  [conus] 263
  creniferus 263
  histrio 263
  [incrassatus] 263
  maculatus 263
  [maculatus] 263
  niloticus 263
  [——— maximus] 263
  [nobulosus] 265
  noduliferus 263
  pyramis 263
  [pyramis] 263
  sacellus rota 263
  stellaris 263
  triserialis 263
Troglocarcinus viridis
Troglopagurus jubatus 349
Tropidinius amoenus 433
  zonatus 433
Tropidocyathus lessonii 185
Tropidoneis chinensis 42
  constricta 42
  gibberula 42
  lepidoptera 42
  longa 42
  maxima 42
  ——— v. sinensis 42
  pusilla 42
Tropiometra afra 387
TROPIOMETRIDAE 387
Tropodiaptomus oryzanus 310

TRUNCOROTALOIDIDAE 100
Tryblioptychus cocconeiformis 47
  hainanensis 47
Trypanedenta taeriaformis
TRYPANORHYNCHA 207
Trypanosyllis taeniaformis 219
  [taeniaformis] 219
  zebra 219
  [zebra] 219
Trypauchen taenlia 453
  vagina 453
Tryphana malmi 332
TRYPHANIDAE 332
Tryptostega venusta
Tsengia nakamurae 136
Tsengiella spinulosa 139
TUBERCULARIACEAE 130
TUBIELLACEAE 25
TUBIFICIDAE 233
Tubificoides imajimai 234
Tubinella funalis 89
Tubipora musica 186
TUBIPORIDAE 186
TUBULANIDAE
Tubulanus punctatus 207
Tubularia larynx 168
  marina 168
  mesembryanthemum 168
TUBULARIIDAE 168
Tubulipora atlantica 369
  cortorta 369
  flabebllaris 369
  flexucosa 369
  lobulata 370
  pulcherrima 370
  pulchra 370
TUBULIPORATA 369
TUBULIPORIDAE 369
TUBULIPORINA 369
Tubulovesicula anguillae 203
  lindbergi 203
  longicorporis 203
  muraenesocis 203
  sauridia 203
  serpentis 203
  spari 203
Tugastraea aurea 183
  coccinea 183
  diaphana 183
  tenuilamellosa 183
  titijimaensis 183
Tucetilla amboinensis 241
  tenuicostata 241
Tugalia gigas 262
Tugonia huanghaiensis 256
  sinensis 256
Tullbergella cuspidata 333
TURBELLARIA 191
Turbinaria agaricia 183
  conoides 144
  contorta 183
  crater 183
  elegans 183
  foliosa 183

Turbinaria (continued)
  globularis 183
  irregularis 183
  mantonae 183
  mesenterina 183
  ornata 144
  parvifolia 144
  peltata 183
  reialata 144
  titzimeensis 183
  undata 183
TURBINIDAE 264
Turbo argyrostomus 264
  [articulatus] 264
  brunneum 264
  chrysostomus 264
  cinereus 264
  cornutus 264
  [coronatus] 264
  [——— granulatus]
  marmoratus 264
  petholatus 264
  reevei 264
  sarmaticus 264
  setosus 264
  sparverius 264
  stenogyrus 264
  [trochiformis] 265
Turbonilla caelata 285
  [mumia] 284
  [ornata] 284
Turborotalia anticompressa 100
  chapmani 100
  crassaformis 100
  dutertrei 100
  inflata 100
  pseudogriffinae 100
  pseudoimitata 100
  mayeri 100
  peripheroronda 100
  pumilio 100
  siakensis 100
  tripartita 100
Turborotalita clarkei 103
  humilis 103
Turcica chinensis 263
  coreensis 263
Turraea pubescens 154
Turricula javana 281
  nelliae spurius 281
TURRIDAE 281
Turridrupa cincta 281
  lamberti 281
Turris crispa 281
  [leucotropis] 281
  [vesicaria] 167
Turrispirillina altispira 81
Turritella bacillum 266
  fascialis 266
  fortilirata 266
  terebra 266
TURRITELLIDAE 266
Turritopsis lata 166
  nutricula 166

Tursiops aduncus 469
   truncatus 469
Tutankhamen pteromerus 358
Tutufa bubo 274
   [*bufo*] 274
   oyamai 274
   ranelloides 274
   rubeta 274
Tydemania expeditionis 147
   japonica 459
   navigatoris 459
TYLIDAE 326
Tylocarcinus sinensis 357
   styx 357
Tylodesma tylostrongyla 165
Tylokepon naxiae 327
Tylonereis bogoyawleskyi 221
Tylorrhynchus heterochaetus 221
Tylos sp. 326
[*Tylosurss anastomella*] 418
[*Tylosurus acus*] 419
   crocodilus 418
   giganteus 418
   leiurus 419
   melanotus 419
   strongylurus 419
Tylotus lichenoides 137
TYMOLIDAE 353
Tymolus japonicus 353
   uncifer 353
[*Tympanotomus cingulatus*] 266
Typha angustifolia 158
TYPHACEAE 158
Typhlocarcinops canaliculata 363
    denticarpes 363
    ocularia 363
Typhlocarcinus nudus 363
   villosus 363
TYPHLOLEPTIDAE 192
TYPHLOSCOLECIDAE 216
Typhloscolex muelleri 216
Typosyllis aciculata orientalis 219
    adamantens kurilensis 219
    alterata 219
    armillaris 219
    benguellana 219
    cirropunctata 219
    fasciata 219
    hyalina 219
    inflata 219
    lutea 219
    maculata 219
    monilata 219
    nigropharynea 219
    prolifera 219
    variegata 219
    vitata 219

U

Uca annulipes 365
   arcuata 365
   borealis 365
   chlorophthalmus crassipes 365

Uca (continued)
   dussumieri 365
   [——— *spinate*] 365
   formosensis 365
   lactea 365
   [*lactea*] 365
    paradussumieri 365
    triangularis 365
    vocans 365
Udotea argentea v. *spumosa* 147
   flabellum 147
   fragilifolia 147
   javensis 147
   reniformis 147
   tenax 147
   tenuifolis 147
   velutina 147
   xishaensis 147
ULMARIDAE 176
ULOTRICHACEAE 144
ULOTRICHALES 144
Ulothrix flacca 144
   implexa 144
   [*pseudoflacca*] 144
Ulua mandibularis 431
   [*mentail*] 431
Ulva conglobata 144
   fasciata 144
   lactuca 144
   lens 144
   [*linza*] 144
   pertusa 144
   reticulata 144
   rigida 144
ULVACEAE 144
ULVALES 144
UMBELLIFERAE 156
Umbellulifera formosa 188
   oreni 188
   striata 188
Umbonium costatum 263
   monoliferum 263
   suturale 263
   thomasi 263
   vestiarium 263
Umbonula alvareziana 378
   arctica 378
   verrucosa 378
UMBONULIDAE 378
UMBONULIOIDEA 378
UMBONULOMORPHA 378
UMBRACULIDAE 289
Umbraculum pulchrum 289
   [*sinicum*] 289
   umbraculum 289
Umbrina russelli 432
Undaria pinnatifida 143
Undeuchaeta intermedia 308
   major 308
   plumosa 308
UNDELLIDAE 60
Undinopsis armatus 308
Undinula darwinii 306
   vulgaris 306

UNGULINIDAE 247
Ungulatella conoides 111
UNGULATELLIDAE
Ungullatelloides imperialis 111
   pagoda 111
Uniporodrilus furcatus 234
Univitellodidymocystis miliaris 200
Upeneus bensasi 436
   luzonius 436
   moluccensis 436
   quadrilineatus 436
   subvittatus 436
   sulphureus 436
   tragula 436
   vittatus 436
Upogebia barbata 345
   carinicauda 345
   darwinii 345
   imperfecta 346
   issaeffi 346
   major 346
   shenjiajuiii 346
   spinifrons 346
   wuhsienweni 346
UPOGEBIIDAE 345
URANOSCOPIDAE 445
Uranoscopus bicinctus 445
   chinensis 445
   japonicus 445
   oligolepis 445
[*Urashimea globosa*] 168
URECHIDAE 237
Urechis unicinctus 237
Urena lobata 155
Urocampus nanus 424
UROCHORDATA 400
Uroconger lepturus 416
Urogymnus africanus 408
UROLOPHIDAE 408
Urolophus aurantiacus 408
   marmoratus 408
Uronema marina 57
URONEMATIDA 57
Uronychia bivalvorum 60
   transfuga 60
   uncinata 60
Uroobovella magna 299
UROPODIDAE 299
Uropterygius macrocephalus 417
   micropterus 417
   tigrinus 417
Uroptychus albatrossae 350
   gracilimanus 350
   granulatus japonicus 350
   naso 351
   scandens 351
Urospora doliifera 145
   penicilliformis 145
Urostyla limboonkengi 60
UROSTYLIDAE 60
Urothoe carda 329
   cuspis 329
   gelasina 329
   orientalis 329

Urothoe (continued)
   spinidigitus 329
Urotrygon daviesis 408
   mundus 408
Uterovesiculurus fujianensis 203
   hamati 203
   lutianius 203
   sinensis 203
   spindis 203
Utica borneensis 367
Utolineus uberis 207
Uvigerina aculeata 107
   adiposa 107
   ampullacea 107
   asperula 107
   attennata 107
   attenuata 107
   auberiana 107
   bassensis 107
   bifurcata 108
   bradyana 108
   brunnensis 108
   bullata 108
   canariensis 108
   chirana 108
   crassa 108
   cushmani 108
   dirupta 108
   finisterrensis 108
   globulosa 108
   graciliformis 108
   hispida 108
   hispido-costata 108
   interrupta 108
   kernensis 108
   mediterranea 108
   miozea 108
   nitidula 108
   peregrina 108
   pigmea 108
   porrecta 108
   proboseidea 108
   rugosa 108
   scheneki 108
   schwageri 108
   segundoensis 108
   senticosa 108
   shukrii 108
   tenuistriata 108
   torquata 108
   vadescens 108
UVIGERINIDAE 107

V

Vaginicola crystalline marina 58
Vaginulina bradyi 94
   legumen 94
   margaritifera 94
   patens 94
   protumida 94
   spinigera 94
   takaoensis 94
VAGINULINIDAE 91
Vaginulinopsis marwicki 94

Vaginulinopsis (continued)
   pacifca 94
   sinuata 94
   sublegumen 94
   tasmanica 94
Valamugil buchanani 425
   cunnesius 425
   formosae 425
   seheli 425
Valenciennea heldsdingenii 451
   sexguttatus 451
   strigatus 451
Valenciennellus tripunctulatus 411
Valonia aegagropila 145
   fastigiata 145
   utricularis 145
   [*ventricosa*] 145
VALONIACEAE 145
Valoniopsis pachynema 146
Valvobifarina mackinnoi 109
Valvulina davidiana 80
Valvulineria laevigata 110
   mexicana 110
   minuta 110
   polita 110
   rugosa 110
   sadonica 110
VALVULINIDAE 80
VAMPYROTEUTHIDAE 296
Vampyroteuthis infernalis 296
Van heurchella admirabilis 47
Vanadis crystallina 216
   fuscapunctata 216
   minuta 216
Vanellus cinereus 467
   vanellus 467
VANIKORIDAE 268
Vanikoro cidaris 268
   delicata 268
   guenniana 268
Vannuccia forbesii 168
Vargula higendorfi 302
   spinulosa 302
Variola albimarginata 427
   louti 427
Varitentaculata yantaiensis 173
Varuna litterata 367
   yui 367
VASIDAE 279
Vasigniella otophora 377
Vasum ceramicum 279
   turbinellum 279
Vatica astrotricha 155
VAUCHRIACEAE 146
VAUCHERIACEAE 50
VAUCHERIALES 50
[*Vayssierea elegans*] 290
[VAYSSIEREIDAE] 290
[*Veiella lata*] 168
   velella 168
VELELLIDAE 168
Velifer hypselopterus 423
VELIFERIDAE 423
Vellitor centropomus 456
Velutina pusio 270

Venatomya truncata 256
[*Venericardita ferruginosa*] 248
VENERIDAE 254
VENEROIDA 246
[*Venerupis philippinarum*] 255
   [*variegata*] 255
Ventricaria ventricosa 145
Ventrifossa bivergens 420
   garmani 420
   nigrodorsalis 420
   rhipidodorsalis 420
[*Venus albina*] 256
   foveolata 256
   toreuma 256
Vepricardium asiaticum 249
   coronatum 249
   multispinosum 249
   sinense 249
Verasper moseri 458
   variegatus 458
VERBENACEAE 157
Verconia verconis 290
VERETILLIDAE 190
VERMETIDAE 266
[*Vermetus imbricatus*] 266
   renisectus 266
[*Vermiliopsis ctenophora*] 231
   [*glandigerus*] 231
   [*infundibulum*] 231
   ——— *glandigera* 231
   pygidialis 231
[*Vermiliopus acanthophora*] 230
Verneuilina advena 77
VERNEUILINACEA 76
VERNEUILINIDAE 77
Verneuilinulla advena 77
   polita 77
Vernonina gravida 111
VERONGIIDAE 164
Veronica undulata 158
Verruca albatrossiana 319
   cristallina 320
   gibbosa 320
   koehleri 320
   nitida 320
   sculpta 320
Verrucariam aura 132
VERRUCIDAE 319
Verruculina enalia 131
Verteberalina insignis 81
   striata 81
VERTEBRATA 404
Verticillium sp. 132
Verticordia deshayesiana 258
VERTICORDIIDAE 258
Vesicocoelium solenphagum 198
VESICOMYIDAE 254
VESICULARIIDAE 370
Vespicula sinensis 455
   trachinoides 455
Vexilla vexillum 276
[*Vexillum coccineum*] 279
   crebriliratum 279
   formosae 279
   cf. *lubens* 279

Index

Vexillum (continued)
  [ornatum] 279
  ———— coccineum 279
  patricarchalis 279
  plicarium 279
  rugosum 279
  vulpeculum 279
Vexitomina chinensis 281
Vibilia armata 330
  australis 330
  chuni 330
  cultripes 330
  gibbosa 331
  longicarpus 331
  propinqua 331
  pyripes 331
  stebbingi 331
  viatrix 331
VIBILIIDAE 330
Vibrio alginolyticus 21
  anguillarum 21
  anguillarum I 21
  campbellii 22
  cholerae 22
  cholerae non-01 22
  costicola 22
  damsela 22
  fischeri 22
  fluvialis 22
  fluvialis I 22
  harveyi 22
  marinus 22
  metschnikovii 22
  migripulchritudo 22
  mimicus 22
  natriegens 22
  nereis 22
  orientalis 22
  parahaemolyticus 22
  pelagius 22
  splendidus biotype I 22
  vulunicus 22
VIBRIONACEAE 21
VICTORIELLIDAE
[Vignadula atrata]
VIGUIERIOTIDAE 189
Villogorgia compressa 187
  intricata 187
Vinciguerra attenuata 411
  nimbaria 411
  poweriae 412
VIPERIDAE 464
Vir orientalis 341
Virgularia gustaviana 190
  reinwardtii 190
VIRGULARIIDAE 190
Virgulina bradyi 109
  pauciloculata 109
  rotundata 109
  schreibersiana 109
  texturata 109
Virgulinella fragilis 109
VIRGULINELLIDAE 109
Virgulopsis orientalis 106
Viridis group 24

[Virroconus ebraeus] 280
Vitex trifolia v. simplicifolia 157
Vitiaziella cubiceps 415
Vogtia glabra 174
  kuruae 174
  microsticella 174
  pentacantha 174
  serrata 174
  spinosa 174
Volachlamys hirasei 245
  singaporinus 245
[Volsella subrugasa] 242
Volutharpa ampullacea perryi 277
VOLUTIDAE 279
[Volva birostris] 271
  [philippinarum] 271
  volva 271
  ———— habei 271
VOLVATELLIDAE 286
Volvatella ayakii 286
Volvula rostrata 286
[Volvulella eburnea] 286
Volvulopsis chinensis 286
Vorticella chydroridicola 57
  cylindrica 57
  fornicata 57
  hamata 57
  marina 57
  nebulifera 57
  patellina 57
  pulchella 57
  striata 58
  utriculus 58
VORTICELLIDAE
Vulsella minor 244
  vulsella 244
VULSELLIDAE 244
Vulvulina arenacea 75
  pennatula 75
  sinensis 75

W

Wak coitor 432
  sina 432
  soldado 432
  tingi 432
Walkeria tuberosa 370
  uva 370
WALKERIIDAE 370
WALKEROIDEA 370
[Waltheria americana] 155
  indica 155
Wangiella dicollaria 59
[Watersipora subovoidea] 379
  subtorguata 379
WATERSIPORIDAE 379
Wedelia biflora 158
  prostrata 158
Weelia cylindrica 401
Wetmorella nigropinnata 441
Wickstroemia indica 155
Wiesnerella auriculata 81
Willetsthenelais diplecirrus 218
  heterochela 218

[Willsia mutabilis] 173
Wrangelia argus 139
  hainanensis 139
  tagoi 139
  tayloriana 139
Wurdemannia miniata 137
WURDEMARNIACEAE 137

X

Xandarovula xanthochila 271
Xanthasia murigera 364
Xanthias canaliculatus 362
  lamarckii 362
  samoensis 362
Xanthichthys auromarginatus 459
  lineopunctatus 459
XANTHIDAE 359
Xanthiopyxis microspinosa v. elliptica 31
Xanthiosphaera lappacea 63
Xanthium strumarium 158
[Xantho reynaudii] 360
Xanthocalanus agilis 308
  dilatus 308
  multispinus 308
  [pulchra] 308
Xanthomonas 21
XANTHOPHYCEAE 50
XANTHOPHYTA 50
Xenia elongata 189
  spicata 189
XENIIDAE 189
Xenocarcinus depressus 357
Xenococcus acervatus 25
  chaetomorphae 25
  cladopharae 25
Xenodermichthys nodulosus 413
Xenolepidichthys dalgleishi 423
[Xenophora calculifera] 268
  exuta 268
  [exuta] 268
  indica 268
  minuta 268
  pallida 268
  sinensis 268
  solarioides 268
  solaris 268
  [solaris] 268
XENOPHORIDAE 268
Xenophthalmodes moebii 363
Xenophthalmus pinnotheroides 364
XENOPNEUSTA 237
Xenostrobus atraus 243
Xenus cinerea 467
Xestospongia exiqua 164
Ximenia americana 151
Xiphacantha alata 61
Xiphasia setifer 447
Xiphatractus trachyphloius 64
  xiphydrion 64
Xiphias gladius 450
XIPHIIDAE 450
[Xiphipops bispinosus] 436
  [ferrugatus] 436
  [flavicauda]

Xiphipops (continued)
  [*nox*] 437
  [*tibicen*] 437
Xiphobotrys clavata 67
  passerina 67
Xiphocheilus quadrimaculatus 441
  typus 441
Xiphosphaera gaea 63
  tesseractis 63
XIPHOSURA 297
XYALIDAE 211
XYLARIALES 130
Xylocarpus granatum 154
Xyrias revulsus 418
[*Xyrichthys aneitensis*] 440
  [*caerulopunctatus*] 440
  [*dea*] 440
  [*des*] 440
  [*evides*] 440
  [*melanopus*] 440
  [*pavo*] 440
  [*pentadactylus*] 440
  [*taeniourus*] 440
  twistii 441
  [*verrens*] 440
  woodi 441
XYRIDACEAE 162
Xyris indica 162
  pauciflora 162

Y

Yamadaella caenomyce 134
[*Yarrella illustris*] 411
YEAST 129
Yokoyamaia argentata 287
  orientalis 287
Yoldia lepidula 239
  notabilis 239
  serotina 239
  similis 239
Yoshithuriz bitentaculata 388
Yozia bicoarctatus 424
YPSILOTHURIIDAE 388

Z

Zachsiella nigromaculata 217
Zafra pumila 276
Zalanthias azumanus 427
Zalerion maritimum 132
  varium 132

Zalerion sp. 132
Zalescopus tosae 445
Zanclea costata 169
  orientalis 169
ZANCLEIDAE 169
ZANCLIDAE 448
Zanclus canescens 448
  cornutus 448
Zannichellia palustris 159
Zanthoxylum nitidum 153
Zebina tridentata 266
Zebrasoma flavescens 449
  scopes 449
  veliferum 449
Zebrias crossolepis 458
  japonicus 458
  quagga 458
  zebra 458
Zebrida adamsi 358
ZEIDAE 423
ZEIFORMES 423
Zellera tawallina 139
Zen cypha 423
Zenarchopterus buffoni 419
Zenion hololepis 423
  japonicus 423
Zenopsis nebulosa 423
Zeuglopora lanceolata 382
Zeugophilomedes polae 302
Zeus japonicus 423
Zeuxapta japonica 192
Zeuxis engylptus 278
  scalaris 278
ZINGIBERACEAE 162
ZIPHIIDAE 469
Ziphius cavirostris 469
Zirfaea crispata 257
  minor 257
ZOANTHIDAE 179
ZOANTHIDEA 179
Zoanthus barnardi 179
  cavernarum 179
  cyanoides 179
  erythrochlores 179
  sinensis 179
  vietnamensis 179
  xishaensis 179
[*Zoarces elongatus*] 447
Zoarchias uchidai 447
ZOARCIDAE 447
Zonaria coriacea 142
  diesingiana 142
  nigrescens 142

Zonichthys nigrofasciata 431
Zonogobius naraharae 452
  semidoliatus 452
Zoobotryon verticellatum 370
ZOOGONIDAE 195
ZOOTHAMNIDAE 58
Zoothamnium affine 58
  alternans 58
  commune 58
  cupiferum 58
  duplicatum 58
  gammari 58
  hadzii 58
  intermedium 58
  maximum 58
  niveum 58
  paraentzii 58
  paragammari 58
  penaei 58
  rigidum 58
  sinensis 58
  thiophilum 58
[*Zooxanthellae*] 51
Zoroaster carinatus philippinensis 392
ZOROASTERIDAE 392
Zostera caespitosa 159
  [*japonica*] 159
  marina 159
  nana 159
  pacifica 159
Zoysia japonica 160
  macrostachys 160
  matrella 160
  sinica 160
  tenuifolia 160
[*Zozia scheepmakeri*] 253
Zozymodes cavipes 362
Zozymus aeneus 362
Zu cristatus 423
[*Zygabikodinium lenticulatum*] 54
Zygocanna vagans 169
Zygocircus longispininus 66
Zygometra comata 386
ZYGOMETRIDAE 386
Zygonemertes grandulosa 208
Zygophylax rufa 171
  tizardensis 171
ZYGOPHYLLACEAE 153
ZYGOSPYRIDAE 66
Zygostaurus amphitectus 61
Zymomonas 21

About the Translator

Dr. Junda Lin grew up in China, received his bachelor of science in marine biology from Xiamen University, China, his masters of science in ecology from Rutgers University, and his doctorate degree in marine sciences from University of North Carolina, Chapel Hill. He currently is an associate professor of biological sciences at Florida Institute of Technology in Melbourne, Florida. Dr. Lin has published over 30 peer-reviewed articles (in English or Chinese) on marine ecology and conservation, fisheries, and aquaculture.